seventh edition

Anatomy Physiology

Rod R. Seeley
Idaho State University

Trent D. Stephens
Idaho State University

Philip Tate
Phoenix College

 Higher Education

Boston   Burr Ridge, IL   Dubuque, IA   Madison, WI   New York   San Francisco   St. Louis
Bangkok   Bogotá   Caracas   Kuala Lumpur   Lisbon   London   Madrid   Mexico City
Milan   Montreal   New Delhi   Santiago   Seoul   Singapore   Sydney   Taipei   Toronto

Mc Graw Hill | Higher Education

ANATOMY & PHYSIOLOGY, SEVENTH EDITION

Published by McGraw-Hill, a business unit of The McGraw-Hill Companies, Inc., 1221 Avenue of the Americas, New York, NY 10020. Copyright © 2006, 2003, 2000, 1998, 1995, 1992, 1989 by The McGraw-Hill Companies, Inc. All rights reserved. No part of this publication may be reproduced or distributed in any form or by any means, or stored in a database or retrieval system, without the prior written consent of The McGraw-Hill Companies, Inc., including, but not limited to, in any network or other electronic storage or transmission, or broadcast for distance learning.

Some ancillaries, including electronic and print components, may not be available to customers outside the United States.

This book is printed on recycled, acid-free paper containing 10% postconsumer waste.

Printed in China

2 3 4 5 6 7 8 9 0 CTP/CTP 0 9 8 7 6

ISBN-13: 978–0–07–111651–0
ISBN-10: 0–07–111651–6

The credits section for this book begins on page C–1 and is considered an extension of the copyright page.

www.mhhe.com

Dedication

This text is dedicated to the students of human anatomy and physiology. Helping students develop a working knowledge of anatomy and physiology is a satisfying challenge, and we have a great appreciation for the effort and enthusiasm of so many who want to know more. It is difficult to imagine anything more exciting, or more important, than being involved in the process of helping people learn about the subject we love.

About the Authors

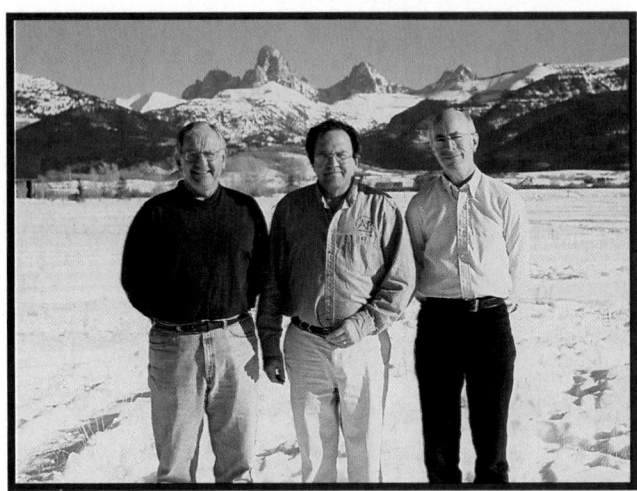

Rod Seeley, Trent Stephens, and Phil Tate in Driggs, Idaho, where they often retreat to collaborate on their textbooks. The Grand Tetons are pictured in the background.

Rod R. Seeley
Professor of Physiology at Idaho State University

With a B.S. in zoology from Idaho State University and an M.S. and Ph.D. in zoology from Utah State University, Rod Seeley has built a solid reputation as a widely published author of journal and feature articles, a popular public lecturer, and an award-winning instructor.

Trent D. Stephens
Professor of Anatomy and Embryology at Idaho State University

An award-winning educator, Trent Stephens teaches human anatomy, neuroanatomy, and embryology. His skill as a biological illustrator has greatly influenced every illustration in his textbooks. He has a B.S. and M.S. in zoology from Brigham Young University and a Ph.D. in anatomy from the University of Pennsylvania.

Philip Tate
Instructor of Anatomy and Physiology at Phoenix College

Phil Tate earned a B.S. in zoology, a B.S. in mathematics, and an M.S. in ecology at San Diego State University; and a Doctor of Arts (D.A.) in biological education from Idaho State University. He is an award-winning instructor who has taught a wide spectrum of students at the four-year and community college levels. Phil has served as the annual conference coordinator, president-elect, president, and past president of the Human Anatomy and Physiology Society (HAPS).

Brief Contents

Contents

Preface

New knowledge continues to accumulate at a rapid pace. Changing technology has not only helped accelerate that process by dramatically improving the ability to uncover previously unknown facts, it has allowed new important questions to be asked. For example, molecular techniques have made it possible to decipher the genetic code for humans. This knowledge increases the ability of investigators to determine the functions of genetic code and the mechanisms that control the expression of all of its components. We are in the process of crossing the threshold of not only knowing the genetic code, but identifying how that code determines the structural and functional characteristics of humans. That knowledge will eventually produce new strategies to treat many diseases for which there is now no cure. The continuing development of electronic instruments and computing capabilities has increased the rate at which data can be collected and analyzed. New imaging systems allow investigators to view small structures, and in some cases view those structures in living animals. Continuing development of modern surgical strategies and instruments is leading to improved procedures that are less invasive than traditional treatments, and newly developed drugs improve the treatment for a variety of conditions.

A major challenge for those who teach anatomy and physiology is to help students learn basic concepts in the subject and, at the same time, help students learn how new knowledge modifies traditional concepts. The accelerating rate at which new knowledge unfolds makes it more important than ever to carefully select critical concepts and to make sure that consistency is maintained between those concepts and current knowledge from the molecular to the systemic level. It is equally important to present these concepts in a way that helps students comprehend the information and teaches them to utilize their knowledge of anatomy and physiology to predict how systems will respond to changing activity, changing environmental conditions, and to trauma and disease.

Despite changing teaching and learning styles and the continuous influx of new information in the field, the principles upon which *Anatomy and Physiology* was founded over twenty years ago still hold true today. Since its inception, this text has been designed to help students develop a solid, basic understanding of anatomy and physiology without an encyclopedic presentation of detail. Great care has been taken to select important concepts and to carefully develop explanations of those concepts in a systematic way. The recipe we have followed for seven editions combines clear and accurate descriptions of anatomy, precise explanations of how structures function, and examples of how they work together to maintain life.

These key ingredients are seasoned with careful examples that illustrate the basic concepts of anatomy and physiology in a familiar context. Explanations of how the systems respond to aging, to changes in physical activity, to injury, and to disease—with a special focus on homeostasis and the regulatory mechanisms that maintain it—clarify difficult topics by making the information accessible to students. Because the vast majority of our readers are pursuing health-related careers, numerous examples demonstrate the application of knowledge in a clinical context. For instance, enough information is presented to allow students to understand the normal structure and function of the heart and how the heart responds to age-related changes. Enough information is presented to allow students to predict the consequences of blood loss and the effects of transfusions on heart function. This information is both relevant and exciting.

To organize the massive amount of information in this textbook into a useful form, we have designed a pedagogical framework that not only helps students study and remember critical concepts, but also challenges them to synthesize the information they gain from their reading and apply it to new and practical uses. This text emphasizes critical thinking exercises, which require a working knowledge of key concepts and stimulate the development of problem-solving skills, as an important route to student success.

Changes to the Seventh Edition

Spurred by the requests of instructors who teach anatomy and physiology, the sixth edition of this text experienced extensive topic reorganization and a dramatic art upgrade. The sixth edition was an intensely laborious revision, but the response to the changes we implemented was resoundingly positive. The seventh edition of *Anatomy and Physiology* builds on the momentum of the sixth edition's success. The major focus of the seventh edition has been polishing our newly remodeled text to insure the accuracy and clarity of the writing and artwork.

Updating the Narrative

Extensive feedback from anatomy and physiology instructors who conscientiously reviewed the sixth edition helped us identify many areas in the text discussions that could be improved. For the seventh edition we have examined the language used to describe structures and functions with the strategy of making those concepts clear and unambiguous. In addition, we have worked diligently to update coverage throughout the book to be in line with the most current information.

We have updated numerous areas of *Anatomy and Physiology* to make the consistency between the molecular, cellular, and systemic levels smooth and clear. For example, more recent molecular information about taste and olfactory receptors will help students more easily understand how these sensory receptors work. We have continued to alter the structural terminology so that it conforms to the most modern views. For example, recent studies show that the traditional classification of the cranial nerves must be modified to be correct. The twelfth cranial nerve receives input from only the spinal cord. Alterations in the descriptions of excitation-contraction coupling in muscle tissue and changes in the descriptions of the mechanism by which sound stimulates hair cells have made these concepts easier for students to understand.

Fine-Tuning the Art Program

In addition to refining the textual discussions, we have made extensive modifications to the artwork in the seventh edition. Based on student and reviewer input, we have increased the number of Process Figures. These figures, which combine illustrations of a physiological process with brief step-by-step explanations, visually summarize complex concepts in an easy-to-follow format.

Also at the request of multiple reviewers, we have enlarged many of the anatomical illustrations in the text to allow higher visibility and distinction of structures. Consistent directional labels have been added to anatomical drawings to help orient students to the exact view that they are studying.

Coordinating the placement of artwork in relation to the text discussion is always difficult when you are dealing with numerous large illustrations. We have been careful to consider page layout and have made every effort to present images so they work in unison with the text that describes them.

Building a Knowledge Base for Solving Problems

The problem-solving pedagogy of *Anatomy and Physiology* has been a defining characteristic since the first edition, and we have continued to improve this aspect of the text in the seventh edition. The infrastructure of pedagogical aids uses a two-pronged approach to learning. Knowledge- and comprehension-level questions are balanced with questions that require more complex reasoning in both the narrative of the text and in the end-of-chapter exercises.

- Review questions at the end of each section of text encourage students to assess their understanding of the material they have read before proceeding to the next section. For the seventh edition, we have refashioned the objectives previously found at the beginning of each section into additional review questions. The result is a greatly expanded number of section review questions that provide a thorough series of probes to test student comprehension after having read the material, rather than a passive list of items to consider before embarking on a new topic. This approach has eliminated redundancy and reduced the number of interruptions to the flow of the text, allowing the development of a more useful student tool without adding undue length to the book.
- Predict questions (many of them new to this edition) are carefully positioned throughout each chapter to prompt students to use newly learned concepts as they solve a problem. These critical thinking activities help students make the connection between basic facts and how those facts translate to broader applications.
- The same hierarchy of knowledge-based and reasoning-based questions is repeated in the end-of-chapter exercises. For more recall practice, Review and Comprehension tests provide a battery of multiple-choice questions that cover all of the key points presented in the chapter.
- The challenging Critical Thinking questions at the end of each chapter help students learn to solve problems by using the information in the text. Tackling questions of this level builds a working knowledge of anatomy and physiology and sharpens reasoning skills.

See the Guided Tour starting on page xii for more details on each of the learning features in *Anatomy and Physiology*.

Guided Tour

A Sound Learning System

Anatomy and Physiology is designed to help you learn in a systematic fashion. Simple facts are the building blocks for developing explanations of more complex concepts. The text discussion is presented within a supporting framework of learning aids that help organize studying, reinforce learning, and promote problem-solving skills.

Chapter Introduction

Each chapter opens with a detailed photomicrograph that ties in with the chapter topic. The opening paragraphs introduce the topic and end with a brief overview of the major section divisions of the chapter.

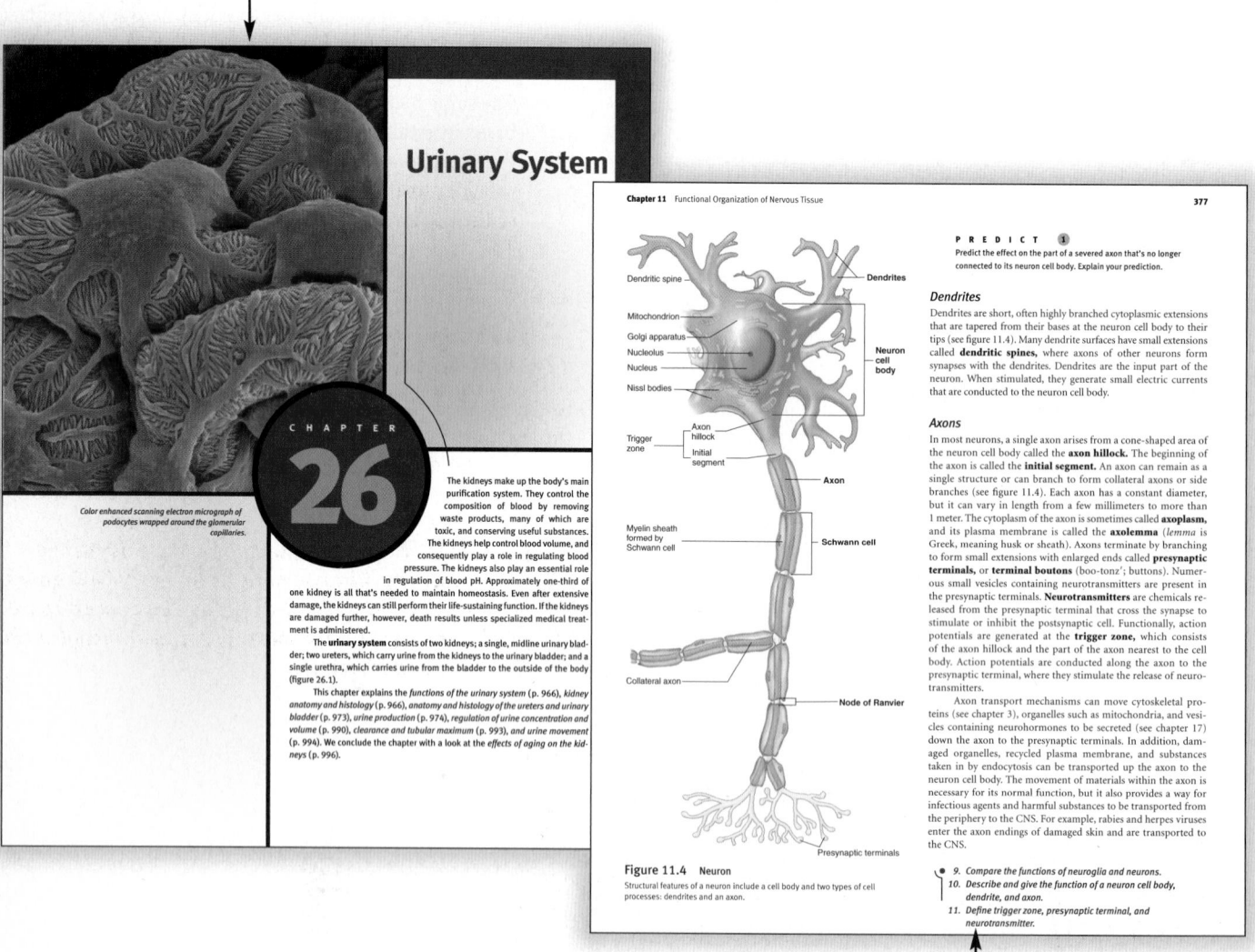

Color enhanced scanning electron micrograph of podocytes wrapped around the glomerular capillaries.

Urinary System

CHAPTER

26

The kidneys make up the body's main purification system. They control the composition of blood by removing waste products, many of which are toxic, and conserving useful substances. The kidneys help control blood volume, and consequently play a role in regulating blood pressure. The kidneys also play an essential role in regulation of blood pH. Approximately one-third of one kidney is all that's needed to maintain homeostasis. Even after extensive damage, the kidneys can still perform their life-sustaining function. If the kidneys are damaged further, however, death results unless specialized medical treatment is administered.

The **urinary system** consists of two kidneys; a single, midline urinary bladder; two ureters, which carry urine from the kidneys to the urinary bladder; and a single urethra, which carries urine from the bladder to the outside of the body (figure 26.1).

This chapter explains the *functions of the urinary system* (p. 966), *kidney anatomy and histology* (p. 966), *anatomy and histology of the ureters and urinary bladder* (p. 973), *urine production* (p. 974), *regulation of urine concentration and volume* (p. 990), *clearance and tubular maximum* (p. 993), *and urine movement* (p. 994). We conclude the chapter with a look at the *effects of aging on the kidneys* (p. 996).

PREDICT 1
Predict the effect on the part of a severed axon that's no longer connected to its neuron cell body. Explain your prediction.

Dendrites

Dendrites are short, often highly branched cytoplasmic extensions that are tapered from their bases at the neuron cell body to their tips (see figure 11.4). Many dendrite surfaces have small extensions called **dendritic spines**, where axons of other neurons form synapses with the dendrites. Dendrites are the input part of the neuron. When stimulated, they generate small electric currents that are conducted to the neuron cell body.

Axons

In most neurons, a single axon arises from a cone-shaped area of the neuron cell body called the **axon hillock.** The beginning of the axon is called the **initial segment.** An axon can remain as a single structure or can branch to form collateral axons or side branches (see figure 11.4). Each axon has a constant diameter, but it can vary in length from a few millimeters to more than 1 meter. The cytoplasm of the axon is sometimes called **axoplasm,** and its plasma membrane is called the **axolemma** (*lemma* is Greek, meaning husk or sheath). Axons terminate by branching to form small extensions with enlarged ends called **presynaptic terminals,** or **terminal boutons** (boo-tonz'; buttons). Numerous small vesicles containing neurotransmitters are present in the presynaptic terminals. **Neurotransmitters** are chemicals released from the presynaptic terminal that cross the synapse to stimulate or inhibit the postsynaptic cell. Functionally, action potentials are generated at the **trigger zone,** which consists of the axon hillock and the part of the axon nearest to the cell body. Action potentials are conducted along the axon to the presynaptic terminal, where they stimulate the release of neurotransmitters.

Axon transport mechanisms can move cytoskeletal proteins (see chapter 3), organelles such as mitochondria, and vesicles containing neurohormones to be secreted (see chapter 17) down the axon to the presynaptic terminals. In addition, damaged organelles, recycled plasma membrane, and substances taken in by endocytosis can be transported up the axon to the neuron cell body. The movement of materials within the axon is necessary for its normal function, but it also provides a way for infectious agents and harmful substances to be transported from the periphery to the CNS. For example, rabies and herpes viruses enter the axon endings of damaged skin and are transported to the CNS.

Figure 11.4 Neuron
Structural features of a neuron include a cell body and two types of cell processes: dendrites and an axon.

9. Compare the functions of neuroglia and neurons.
10. Describe and give the function of a neuron cell body, dendrite, and axon.
11. Define trigger zone, presynaptic terminal, and neurotransmitter.

Section Review

Review questions at the end of each section prompt you to test your understanding of key concepts. Use them as a self-test to determine whether you have a sufficient grasp of the information before proceeding with the next section.

Predict Questions

These innovative critical thinking exercises encourage you to become an active learner as you read. Predict questions challenge you to use your understanding of new concepts to solve a problem. Answers to Predict questions are given at the end of each chapter, allowing you to evaluate your response and discover the logic used to arrive at the correct answer.

P R E D I C T ④

The dermatomal map is important in clinical considerations of nerve damage. Loss of sensation in a dermatomal pattern can provide valuable information about the location of nerve damage. Predict the possible site of nerve damage for a patient who suffered whiplash in an automobile accident and subsequently developed anesthesia (no sensations) in the left arm, forearm, and hand (see figure 12.14 for help).

---

3. *What is the pericardium?*
4. *Describe the parts of the pericardium and their functions.*
5. *Define pericarditis.*
6. *Explain the effects of cardiac tamponade on the heart.*

Heart Wall

The heart wall is composed of three layers of tissue: the epicardium, the myocardium, and the endocardium (figure 20.4). The **epicardium** (ep-i-kar'dē-ŭm) is a thin serous membrane that constitutes the smooth outer surface of the heart. The epicardium and the visceral pericardium are two names for the same structure. The serous pericardium is called the epicardium when considered a part of the heart and the visceral pericardium when considered a part of the pericardium. The thick middle layer of the heart, the **myocardium** (mī-ō-kar'dē-ŭm), is composed of cardiac muscle cells and is responsible for the ability of the heart to contract. The smooth inner surface of the heart chambers is the **endocardium** (en-dō-kar'dē-ŭm), which consists of simple squamous epithelium over a layer of connective tissue. The smooth inner surface allows blood to move easily through the heart. The heart valves result from a fold in the endocardium, thus making a double layer of endocardium with connective tissue in between.

The interior surfaces of the atria are mainly flat, but the interior of both auricles and a part of the right atrial wall contain muscular ridges called **musculi pectinati** (pek'ti-nah'tē; hair comb). The musculi pectinati of the right atrium are separated from the larger, smooth portions of the atrial wall by a ridge called the **crista terminalis** (kris'tă ter'mi-nal'is; terminal crest). The interior walls of the ventricles contain larger muscular ridges and columns called **trabeculae** (tră-bek'ū-lē; beams) **carneae** (kar'nē-ē; flesh).

7. *Describe the three layers of the heart, and state their functions.*
8. *Name the muscular ridges found on the interior of the auricles, and name the ridges and columns found on the interior walls of the ventricles.*

External Anatomy and Coronary Circulation

The heart consists of four chambers: two **atria** (ā'trē-ă; entrance chamber) and two **ventricles** (ven'tri-klz; belly). The thin-walled atria form the superior and posterior parts of the heart, and the thick-walled ventricles form the anterior and inferior portions (figure 20.5). Flaplike **auricles** (aw'ri-klz; ears) are extensions of the atria that can be seen anteriorly between each atrium and ventricle. The entire atrium used to be called the auricle, and some medical personnel still refer to it as such.

Several large veins carry blood to the heart. The **superior vena cava** (vē'nă kā'vă) and the **inferior vena cava** carry blood from the body to the right atrium, and four **pulmonary veins** carry blood from the lungs to the left atrium. In addition, the smaller coronary sinus carries blood from the walls of the heart to the right atrium.

Two arteries, the **aorta** and the **pulmonary trunk**, exit the heart. The aorta carries blood from the left ventricle to the body, and the pulmonary trunk carries blood from the right ventricle to the lungs.

A large **coronary** (kōr'o-nār-ē; circling like a crown) **sulcus** (sool'kŭs; ditch) runs obliquely around the heart, separating the atria from the ventricles. Two more sulci extend inferiorly from the coronary sulcus, indicating the division between the right and left ventricles. The **anterior interventricular sulcus**, or **groove**, is on the anterior surface of the heart, and the **posterior interventricular sulcus**, or **groove**, is on the posterior surface of the heart. In a healthy, intact heart the sulci are covered by fat, and only after this fat is removed can the actual sulci be seen.

The major arteries supplying blood to the tissue of the heart lie within the coronary sulcus and interventricular sulci on the surface of the heart. The **right** and **left coronary arteries** exit the aorta just above the point where the aorta leaves the heart and lie within the coronary sulcus (figure 20.6a). The right coronary artery is usually smaller in diameter than the left one, and it doesn't carry as much blood as the left coronary artery.

A major branch of the left coronary artery, called the **anterior interventricular artery,** or the **left anterior descending artery,** extends inferiorly in the anterior interventricular sulcus and supplies blood to most of the anterior part of the heart. The **left marginal artery** branches from the left coronary artery and extends around to the lateral wall of the left ventricle. The **circumflex** (ser'kŭm-fleks) **artery** branches from the left coronary artery and extends around to the posterior side of the heart in the coronary

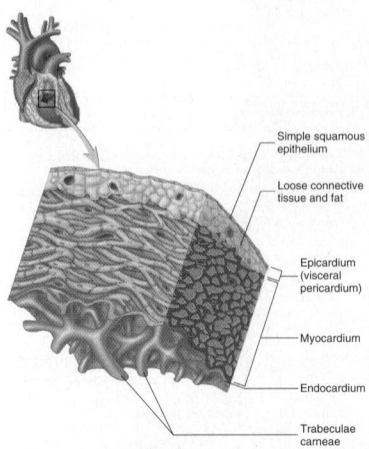

Figure 20.4 Heart Wall

Part of the wall of the heart has been removed to show its structure. The enlarged section illustrates the epicardium, the myocardium, and the endocardium.

Labels (figure 20.4):
- Simple squamous epithelium
- Loose connective tissue and fat
- Epicardium (visceral pericardium)
- Myocardium
- Endocardium
- Trabeculae carneae

Vocabulary Aids

Learning anatomy and physiology is, in many ways, like learning a new language. Mastering the terminology is key to building your knowledge base.

Key terms are set in boldface where they are defined in the chapter, and most terms are included in the glossary at the end of the book. Pronunciation guides are included for difficult words.

Because knowing the original meaning of a term can enhance understanding and retention, derivations of key words are given when they are relevant. Furthermore, a handy list of prefixes, suffixes, and combining forms is printed on the inside back cover as a quick reference to help you identify commonly used word roots. A list of abbreviations used throughout the text is also included.

Instructive Artwork Makes the Difference

A picture is worth a thousand words—especially when you're learning anatomy and physiology. Because words alone cannot convey the nuances of anatomy or the intricacies of physiology, *Anatomy and Physiology* employs a dynamic program of full-color illustrations and photographs that support and further clarify the text explanations. Brilliantly rendered and carefully reviewed for accuracy and consistency, the precisely labeled illustrations and photos provide concrete, visual reinforcement of the topics discussed throughout the text.

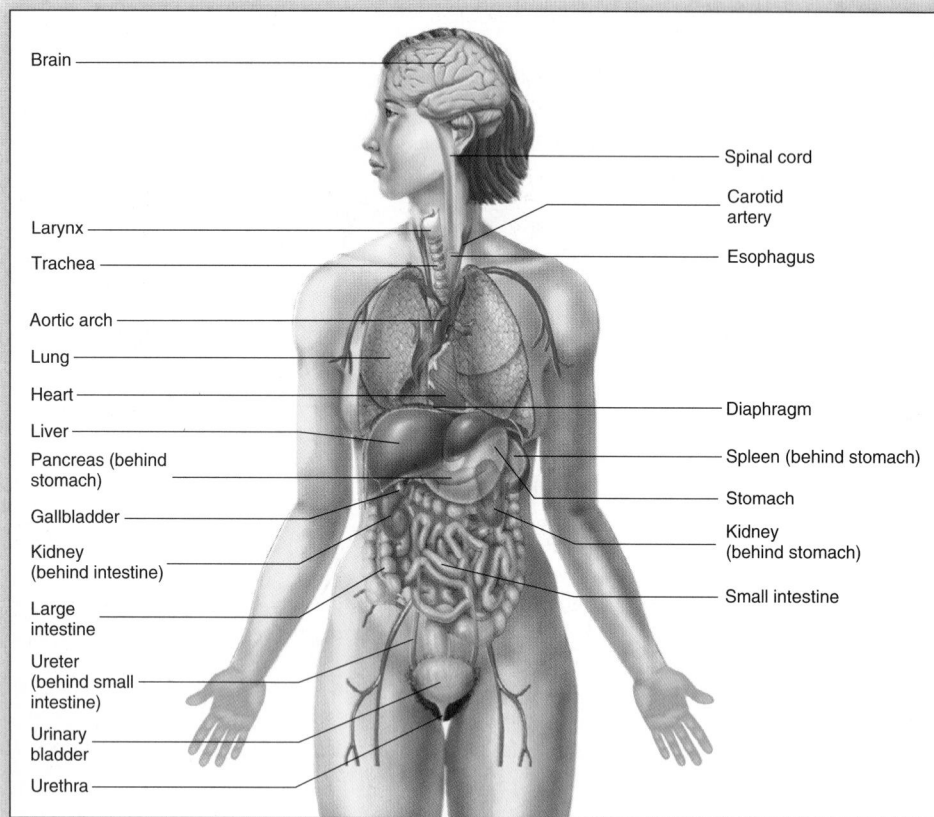

Realistic Anatomical Art

The anatomical figures in Anatomy and Physiology *have been carefully rendered to convey realistic, three-dimensional detail. Richly textured bones and artfully shaded muscles and vessels lend a sense of realism to the figures that helps you envision the appearance of actual structures within in the body.*

The colors used to represent different anatomical structures have been applied consistently throughout the book. This reliable pattern of color consistency helps you easily identify the structures in every figure and promotes continuity between figures.

Atlas-Quality Cadaver Images

Clearly labeled photos of dissected human cadavers provide detailed views of anatomical structures, capturing the intangible characteristics of actual human anatomy that can be appreciated only when viewed in human specimens.

Multi-Level Perspective

Illustrations depicting complex structures or processes combine macroscopic and microscopic views to help you see the relationships between increasingly detailed drawings.

Reference diagrams orient you to the view or plane an illustration represents.

Common bile duct from liver
Duodenum
Minor duodenal papilla
Major duodenal papilla
Jejunum
Pancreatic ducts
Head of pancreas

Circular folds
Epithelium
Submucosa
Circular muscle
Longitudinal muscle
Serosa
(a)

Villi
Blood capillary network
Lacteal
Epithelium
Intestinal gland
Duodenal gland
Top of circular fold
(b)

Microvilli
Epithelial cell
Capillary (blood)
Villus
Lacteal (lymph)
Epithelial cell
(c)

Microvilli of epithelial cell surface
20,000x
(d)

Squamous portion
Zygomatic process
Mandibular fossa
External acoustic meatus
Styloid process
Mastoid process

Combination Art

Drawings are often paired with photographs to enhance visualization of structures.

Thyroid follicle (containing thyroglobulin)
Follicular cells
Parafollicular cell
LM 130x

Histology Micrographs

Light micrographs, as well as scanning and transmission electron micrographs, are used in conjunction with illustrations to present a true picture of anatomy and physiology from the cellular level.

Magnifications are indicated to help you estimate the size of structures shown in the photomicrographs.

Specialized Figures Clarify Tough Concepts

Studying physiology does not have to be an intimidating task mired in memorization. *Anatomy and Physiology* uses two special types of illustrations to help you not only to learn the steps involved in specific processes, but also to apply this knowledge as you predict outcomes in similar situations. Process Figures organize the key occurrences of physiological processes in an easy-to-follow format. Homeostasis Figures summarize the mechanisms of homeostasis by diagramming the means by which a given system regulates a parameter within a narrow range of values.

Process Figures

Process Figures break down physiological processes into a series of smaller steps, allowing you to track the key events and learn them as you go.

Sequence indicators within the artwork correspond to the numbered explanations along the side. These colored circles help you zero in on the site where the action described in each step takes place.

1. The cell membrane of an olfactory hair, unstimulated. The gated ion channel is closed.

2. An odorant binds to a specific odorant receptor.

3. The associated G protein is activated.

4. The α subunit of the G protein binds to and activates adenylate cyclase.

5. Adenylate cyclase catalyzes the conversion of ATP to cyclic AMP (cAMP).

6. cAMP opens ion channels, such as Na^+ and Ca^{2+} channels.

7. Ions entering the olfactory hair cause depolarization of the neuron.

Process Figure 15.2   Action of Odorant Binding to Membrane of Olfactory Hair

Process Figures and Homeostasis Figures are identified next to the figure number. The accompanying caption provides additional explanation.

Homeostasis Figures

*These specialized flowcharts diagram
the mechanisms that body systems
employ to maintain homeostasis.*

*Changes caused by an increase
of a variable outside its normal
range are shown in the green
boxes across the top.*

*The normal range for a
given value is represented
by the graphs in the center
of each figure.*

*Start with the graph on the
left side of the figure and
follow the green arrows to
learn about the chain of
events triggered by an
increase in the variable, or
the red arrows for events
resulting from a decrease
in the variable.*

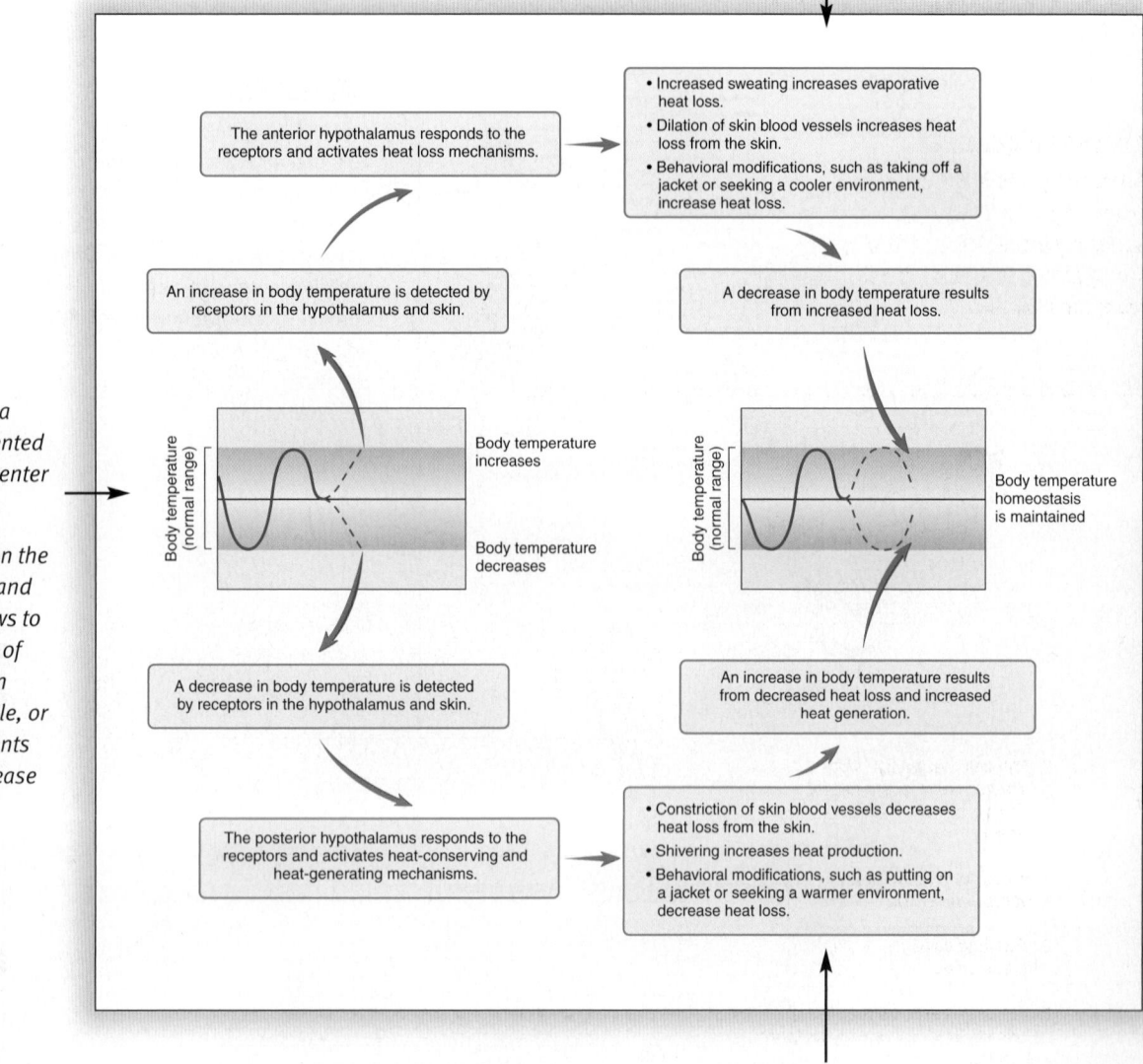

*Changes caused by a decrease of
a variable outside its normal
range are shown in the red boxes
across the bottom.*

Clinical Content Puts Knowledge into Practice

Anatomy and Physiology provides clinical examples to illustrate the application of basic knowledge in an interesting and relevant clinical context. Exposure to clinical information is especially beneficial if you are planning on using your knowledge of anatomy and physiology in a health-related career.

Description of Selected Joints

It's impossible in a limited space to describe all the joints of the body; therefore, we describe only selected joints in this chapter, and they have been chosen because of their representative structure, important function, or clinical significance.

Temporomandibular Joint

The mandible articulates with the temporal bone to form the **temporomandibular joint (TMJ)**. The mandibular condyle fits into the mandibular fossa of the temporal bone. A fibrocartilage articular disk is located between the mandible and the temporal bone, dividing the joint into superior and inferior cavities (figure 8.27). The joint is surrounded by a fibrous capsule to which the articular disk is attached at its margin, and is strengthened by lateral and accessory ligaments.

The temporomandibular joint is a combination plane and ellipsoid joint, with the ellipsoid portion predominating. Depression of the mandible to open the mouth involves an anterior gliding motion of the mandibular condyle and articular disk relative to the temporal bone, which is about the same motion that occurs in protraction of the mandible; it is followed by a hinge motion that occurs between the articular disk and the mandibular head. The mandibular condyle is also capable of slight mediolateral movement, allowing excursion of the mandible.

TMJ Disorders

TMJ disorders are a group of conditions that cause most chronic orofacial pain. The conditions include joint noise; pain in the muscles, joint, or face; headache; and reduction in the range of joint movement. TMJ pain is often felt as referred pain in the ear. Patients may go to a physician complaining of an earache and are then referred to a dentist. As many as 65%–75% of people between ages 20 and 40 experience some of these symptoms. Symptoms appear to affect men and women about equally, but only about 10% of the symptoms are severe enough to cause people to seek medical attention. Women experience severe pain eight times more often than do men.

TMJ disorders are classified as those involving the joint, with or without pain; those involving only muscle pain; or those involving both the joint disorder and muscle pain. TMJ disorders are also classified as acute or chronic. Acute cases are usually self-limiting and have an identifiable cause. Chronic cases are not self-limiting, may be permanent, and often have no apparent cause. Chronic TMJ disorders are not easily treated, and chronic TMJ pain has much in common with other types of chronic pain. Whereas some people learn to live with the pain, others may experience psychologic problems, such as a sense of helplessness and hopelessness, high tension, and loss of sleep and appetite. Drug dependency may occur if strong drugs are used to control the pain; and relationships, lifestyle, vocation, and social interactions may be disrupted. Many of these problems may make the pain worse through positive feedback. Treatment includes teaching the patient to reduce jaw movements that aggravate the problem and to reduce stress and anxiety. Physical therapy may help to relax the muscles and restore function. Analgesic and antiinflammatory drugs may be used, and oral splints may be helpful, especially at night.

Lateral view

Temporal bone
External acoustic meatus
Joint capsule

Zygomatic arch
Lateral ligament
Styloid process
Stylomandibular ligament
Mandible

Temporal bone
Mandibular condyle

Superior joint cavity
Articular disk
Inferior joint cavity
Lateral pterygoid muscle

Sagittal section of temporomandibular joint

Figure 8.27 Right Temporomandibular Joint, Lateral View

Clinical Topics

Interesting clinical sidebars reinforce or expand upon the facts and concepts discussed within the narrative. Once you have learned a concept, applying that information in a clinical context shows you how your new knowledge can be put into practice.

Clinical Focus

These in-depth boxed essays explore relevant topics of clinical interest. Subjects covered include pathologies, current research, sports medicine, exercise physiology, and pharmacology.

Clinical Focus   Kidney Dialysis

The artificial kidney (renal dialysis machine) is a machine used to treat patients suffering from renal failure. The use of this machine often allows people with severe acute renal failure to recover without developing the side effects of renal failure, and the machine can substitute for the kidneys for long periods in people suffering from chronic renal failure.

Renal dialysis allows blood to flow though tubes made of a selectively permeable membrane. On the outside of the dialysis tubes is a fluid that contains the same concentration of solutes as the plasma, except for the metabolic waste products. As a consequence, a diffusion gradient exists for the metabolic waste products from the blood to the dialysis fluid. The dialysis membrane has pores that are too small to allow plasma proteins to pass through them. For smaller solutes, the dialysis fluid contains the same beneficial solutes as the plasma, so the net movement of these substances is zero. In contrast, the dialysis fluid contains no metabolic waste products, so metabolic

waste products diffuse rapidly from the blood into the dialysis fluid.

Blood usually is taken from an artery, passed through tubes of the dialysis machine, and then returned to a vein. The rate of blood flow is normally several hundred milliliters per minute, and the total surface area of exchange in the machine is close to 10,000–20,000 cm^2 (figure A). Kidney dialysis is not convenient for those suffering from kidney failure, and it can be emotionally difficult. Clearly, kidney dialysis is not a good substitute for healthy kidneys.

Figure A   Kidney Dialysis
During kidney dialysis blood flows through a system of tubes composed of a selectively permeable membrane. Dialysis fluid, the composition of which is similar to that of blood, except that the concentration of waste products is very low, flows in the opposite direction on the outside of the dialysis tubes. Consequently, waste products such as urea diffuse from the blood into the dialysis fluid. Other substances such as sodium, potassium, and glucose do not rapidly diffuse from the blood into the dialysis fluid because there is no concentration gradient for these substances between the blood and the dialysis fluid.

Systems Pathology

These boxes explore a specific disorder or condition related to a particular body system. Presented in a simplified case study format, each Systems Pathology box begins with a patient history followed by background information about the featured topic.

Systems Pathology
Stroke

Mr. S, who is approaching middle age, is somewhat overweight and has high blood pressure. He was seated on the edge of his couch, at least most of the time, when he was not jumping to his feet and shouting at the referees for an obviously bad call. He was surrounded by empty pizza boxes, bowls of chips and salsa, empty beer cans, and full ashtrays. As Mr. S cheered on his favorite team in a hotly contested big game, which they would be winning easily if it weren't for the lousy officiating, he noticed that he felt drowsy and that the television screen seemed blurry. He began to feel dizzy. As he tried to stand up, he suddenly vomited and collapsed to the floor, unconscious.

Mr. S was rushed to the local hospital, where the following signs and symptoms were observed. He exhibited weakness in his limbs, especially on the right, and ataxia (inability to walk). He had loss of pain and temperature sensation in his right lower limb and loss of all sensation in the left side of his face. The dizziness persisted and he appeared disoriented and lacked attentiveness. He also exhibited dysphagia (the inability to swallow) and hoarseness. He had nystagmus (rhythmic oscillation of the eyes). His pupils were slightly dilated, his respiration was short and shallow, and his pulse rate and blood pressure were elevated.

Background Information

Mr. S suffered a "stroke," also referred to as a cerebrovascular accident (CVA). The term *stroke* describes a heterogeneous group of conditions involving death of brain tissue resulting from disruption of its vascular supply. Two types of stroke exist: **hemorrhagic stroke,** which results from bleeding of arteries supplying brain tissue, and **ischemic stroke,** which results from blockage of arteries supplying brain tissue (figure B). The blockage in ischemic stroke can result from a thrombus (a clot that develops in place within an artery) or an embolism (a plug,

Anterior

(a) (b)

Posterior

Figure B Images of a Stroke

(*a*) MRI (magnetic resonance imaging) of a massive stroke in the left side (the viewer's right) of the brain. (*b*) Colorized NMR (nuclear magnetic resonance) showing disruption of blood flow to the left side (the viewer's right) of the brain (*yellow*). This disruption could cause a stroke.

composed of a detached t[...]
globule or gas bubble, tha[...]
Mr. S was at high risk for[...]
middle age, was overweig[...]
under stress, and had a po[...]

The combination of m[...]
limbs, and sensory loss, se[...]

A System Interactions table at the end of every box summarizes how the condition impacts each body system.

System Interactions	Effect of Stroke on Other Systems
System	**Interactions**
Integumentary	Decubitus ulcers (bedsores) from immobility; loss of motor function following a stroke leads to immobility.
Skeletal	Loss of bone mass, if muscles are dysfunctional for a prolonged time; in the absence of muscular activity, the bones to which those muscle are attached begin to be resorbed by osteoclasts.
Muscular	Major area of effect; absence of stimulation due to damaged pathways or neurons leads to decreased motor function and may result in muscle atrophy.
Endocrine	Strokes in other parts of the brain could involve the hypothalamus, pineal body, or pituitary gland functions.
Cardiovascular	Risks: Phlebothrombosis (blood clot in a vein) can occur from inactivity. Edema around the brain could apply pressure to the cardioregulatory and vasomotor centers of the brain. This pressure could stimulate these centers, which would result in elevated blood pressure, and congestive heart failure could result. If the cardioregulatory center in the brain is damaged, death may occur rapidly. Bleeding is due to the use of anticoagulants. Hypotension results from use of antihypertensives.
Respiratory	Pneumonia from aspiration of the vomitus or hypoventilation results from decreased function in the respiratory center. If the respiratory center is severely damaged, death may occur rapidly.
Digestive	Vomiting, dysphagia (difficulty swallowing); hypovolemia (decreased blood volume) result from decreased fluid intake; occurs because of dysphagia; may be a loss of bowel control.
Urinary	Control of the micturition reflex may be inhibited. Urinary tract infection results from catheter implantation or from urinary bladder distension.
Reproductive	Loss of libido; innervation of the reproductive organs is often affected.

in his right lower limb and loss of all sensation in the left side of his face; along with the ataxia, dizziness, nystagmus, and hoarseness, suggest that the stroke affected the brainstem and cerebellum. Blockage of the vertebral artery, a major artery supplying the brain, or its branches can result in what is called a lateral medullary infarction (an area of dead tissue resulting from a loss of blood supply to an area). Damage to the descending motor pathways in that area, above the medullary decussation, results in muscle weakness. Damage to ascending pathways can result in loss of pain and temperature sensation (or other sensory modalities depending on the affected tract). Damage to cranial nerve nuclei results in the loss of pain and temperature sensation in the face, dizziness, blurred vision, nystagmus, vomiting, and hoarseness. These signs and symptoms are not observed unless the lesion is in the brainstem, where these nuclei are located. Some damage to the cerebellum, also supplied by branches of the vertebral artery, can account for the ataxia.

Drowsiness, disorientation, inattentiveness, and loss of consciousness are examples of generalized neurologic response to damage.

Seizures may also result from severe neurologic damage. Depression from neurologic damage or from discouragement is also common. Slight dilation of the pupils; short, shallow respiration; and increased pulse rate and blood pressure are all signs of Mr. S's anxiety, not about the outcome of the game but about his current condition and his immediate future. With a loss of consciousness, Mr. S would not remember the last few minutes of what he saw in the game he was watching. People in these circumstances are often worried about how they are going to deal with work tomorrow. They often have no idea that the motor and sensory losses may be permanent, or that they will have a long period of therapy ahead.

P R E D I C T 9

Given that Mr. S exhibited weakness in his right limbs and loss of pain and temperature sensation in his right lower limb and the left side of his face, state which side of the brainstem was most severely affected by the stroke. Explain your answer.

Every Systems Pathology box includes a Predict question specific to the case study.

Study Features Ensure Success

A carefully devised set of learning aids at the end of each chapter helps you review the chapter content, evaluate your grasp of key concepts, and utilize what you have learned. Reading the chapter summary and completing the practice test and critical thinking exercises will greatly improve your understanding of each chapter and is also a great way to study for exams.

Chapter Summary

The summary outline briefly states the important facts and concepts covered in each chapter to provide a convenient "big picture" of the chapter content.

Review and Comprehension

These multiple-choice practice questions cover all of the main points presented in the chapter. Completing this self-test helps you gauge your mastery of the material. Answers are provided in Appendix F.

Critical Thinking

These innovative exercises encourage you to apply chapter concepts to solve a problem. Answering these questions helps build your working knowledge of anatomy and physiology while developing reasoning skills. Answers are provided in Appendix G.

Answers to Predict Questions

The Predict questions that appear throughout the reading are answered at the end of each chapter, allowing you to evaluate your responses and understand the logic used to arrive at the correct answer.

Teaching and Learning Supplements

McGraw-Hill offers various tools and technology products to support *Anatomy and Physiology*, seventh edition. Students can order supplemental study materials by contacting their local bookstore or by calling 800-262-4729. Instructors can obtain teaching aids by calling the Customer Service Department at 800-338-3987, visiting our A&P website at www.mhhe.com/ap, or contacting your local McGraw-Hill sales representative.

For Instructors

Digital Content Manager CD-ROM

This easy-to-use multimedia resource allows instructors to utilize artwork from the text in multiple formats to create customized classroom presentations, visually based tests and quizzes, dynamic course website content, or attractive printed support material. The digital assets on this cross-platform CD-ROM include:

- **Art Library**—Full-color digital files of all illustrations in the book, plus the same art saved in unlabeled and gray scale versions, can be readily incorporated into lecture presentations, exams, or custom-made classroom materials. These images are also pre-inserted into blank PowerPoint slides for ease of lecture preparation.
- **TextEdit Art Library**—Every line art piece is placed in a PowerPoint presentation that allows the user to revise, move, and delete labels as desired to create custom-labeled images for presentations and/or tests.
- **Active Art Library**—Active Art consists of art files that have been converted to a format that allows the artwork to be edited inside of PowerPoint. Each piece can be broken down to its core elements, grouped or ungrouped, and edited to create customized illustrations.
- **Animations Library**—Full-color presentations involving key figures in the book have been brought to life via animation. These animations offer total flexibility for instructors and were designed to be used in lecture. Instructors can pause, rewind, fast forward, and turn audio off and on to create dynamic lectures. Animations are supplied as downloadable files, and are also pre-inserted into PowerPoint slides for easy integration into existing presentations.
- **Photo Library**—Like the Art Library, digital files of all photographs from the book are available.
- **Table Library**—Every table that appears in the book is provided in electronic form.

- **PowerPoint Lecture Outlines**—Ready-made presentations that combine art and lecture notes cover each of the 29 chapters of the text. These lectures can be used as they are, or can be tailored to reflect your preferred lecture topics and sequences.

Instructor's Testing and Resource CD-ROM

This cross-platform CD-ROM provides a wealth of resources for the instructor. Supplements featured on this CD include a computerized test bank that uses testing software to quickly create customized exams. This user-friendly program allows instructors to sort questions by format or difficulty level, edit existing questions or add new ones, and scramble questions for multiple versions of the same test. Word files of the test bank are included for those instructors who prefer to work outside of the test-generator software.

Other assets on the Instructor's Testing and Resource CD-ROM include the Instructor's Manual and Instructor's Manual for Laboratory Manual.

Transparencies

A set of 1,200 transparency overheads includes all illustrations and tables from the textbook, selected photos, and more than 200 unlabeled drawings.

Laboratory Manual

Written by Eric Wise of Santa Barbara City College, this comprehensive manual contains 43 laboratory exercises that are integrated closely with the textbook. Each laboratory exercise aims to demonstrate the anatomical and physiological facts and principles presented in *Anatomy and Physiology* by investigating specific concepts in greater detail. Key features of the lab manual include:

- Step-by-step explanations and a complete materials list for each experiment
- Precisely labeled, full-color drawings and photographs
- Self-contained presentations with the essential background information needed to complete each exercise
- Extensive lab reports at the end of every exercise challenge students to review and apply what they have learned

e-Instruction

The Classroom Performance System (CPS) is an interactive system that allows the instructor to administer in-class questions electronically. Students answer questions via hand-held remote control keypads, and their individual responses are logged into a gradebook. Aggregated responses can be displayed in graphical form. Using this immediate feedback, the instructor can quickly determine if students understand the lecture topic, or if more clarification is needed. CPS promotes student participation, class productivity, and individual student accountability.

Course Delivery Systems

With help from our partners WebCT, Blackboard, Top-Class, eCollege, and other course management systems, professors can take complete control over their course content. Course cartridges containing Online Learning Center content, online testing, and powerful student tracking features are readily available for use within these platforms.

For Students

Anatomy & Physiology Revealed

Anatomy & Physiology Revealed is a unique multimedia tool designed to help students learn and review human anatomy using cadaver specimens. Detailed cadaver photographs blended together with a state of the art layering technique provide a uniquely interactive dissection experience. This easy-to-use program features the following sections:

- **Dissection**—Students can literally peel away layers of the human body to reveal structures beneath the surface. Structures can be pinned and labeled, just like in a real dissection lab. Each labeled structure is accompanied by detailed information and an audio pronunciation. Dissection images can be captured and saved.
- **Animation**—Compelling animations demonstrate muscle actions, clarify anatomical relationships, or explain difficult concepts.
- **Imaging**—Labeled x-ray, MRI, and CT images help students become familiar with the appearance of key anatomical structures as seen through different medical imaging techniques.
- **Self-Test**—Challenging exercises let students test their ability to identify anatomical structures in a timed practical exam format. A results page provides analysis of test scores and links back to all incorrectly identified structures for review.
- **Anatomy Terms**—A visual glossary of general terms helps students learn the language of anatomy. Directional and regional terms are included, as well as planes and terms of movement. Each definition is accompanied by an illustration or animation and an audio pronunciation of the term.

Online Learning Center

The *Anatomy and Physiology* Online Learning Center at www.mhhe.com/seeley7 offers access to a vast array of premium online content to fortify the learning experience.

- **Book-Specific Study Tools**—the OLC features quizzes, interactive learning games, and study tools tailored to coincide with each chapter of the text.
- **Essential Study Partner**—A collection of interactive study modules that contains hundreds of animations, learning activities, and quizzes designed to help students grasp complex concepts.
- **Online Tutoring**—A 24-hour tutorial service moderated by qualified instructors. Help with difficult concepts is only an e-mail away.

Student Study Guide

This comprehensive workbook features multiple question and answer sets designed to test knowledge and understanding of key concepts presented in *Anatomy and Physiology*. Study Guide chapters correspond with the chapters of the textbook and contain a variety of exercises, including matching, completion, ordering, labeling, and short-answer questions.

Student Study Art Notebook

This visual guide contains every illustration from the text to make it easier for students to learn anatomy. This collection of images provides a comprehensive resource for studying anatomical structures and a convenient place to write notes during lecture or lab.

MediaPhys 3.0 CD-ROM

This interactive tool offers detailed explanations, high-quality illustrations, and animations to provide students with a thorough introduction to the world of physiology—giving them a virtual tour of physiological processes. MediaPhys is filled with interactive activities and quizzes to help reinforce physiology concepts that are often difficult to understand.

Physiology Interactive Lab Simulations (Ph.I.L.S.)

The Ph.I.L.S CD-ROM offers dynamic laboratory simulations that explore physiological concepts. Each lab begins with a focused question and review section to help students make connections between lecture concepts and lab activities, followed by a unique wet lab feature that invites students to view each experiment as it would be performed in a traditional lab setting. Each simulation guides the student through measuring, analyzing data, and creating graphs. The easy-to-use software allows students the flexibility to change the parameters of every lab experiment, with no limit to the amount of times a student can repeat experiments or modify variables. Data are dynamically generated so results will vary each time a lab is conducted. Upon completion of each experiment, students can answer and print out a comprehensive lab report. Ph.I.L.S. is the ideal tool for discovering and applying key physiology concepts.

Acknowledgements

No modern textbook is solely the work of the authors. To adequately acknowledge the support of loved ones is not possible. They have had the patience and understanding to tolerate our frustrations and absence, and they have been willing to provide assistance and undying encouragement.

Many hands besides our own have touched this text, guiding it through various stages of development and production. We wish to express our gratitude to the staff of McGraw-Hill for their help and encouragement. We sincerely appreciate Sponsoring Editor Michelle Watnick and Developmental Editor Kris Queck for their many hours of work, suggestions, and tremendous patience and encouragement. We also thank Senior Project Manager Mary Powers, Senior Photo Editor John Leland, and Designer Rick Noel for their hours spent turning manuscript into a book; Media Producer Renee Russian, Media Project Manager Tammy Juran, and Supplement Producer Brenda Ernzen for their assistance in building the various products that support our text; and Marketing Manager Deb Hash for her enthusiasm in promoting this book. The McGraw-Hill employees with whom we have worked are excellent professionals. They have been consistently helpful and their efforts are appreciated. Their commitment to this project has clearly been more than a job to them.

Thanks are gratefully offered to copy editor Cathy Conroy for carefully polishing our words, and to the many illustrators who worked on the development and execution of the illustration program for the seventh edition of *Anatomy and Physiology*.

Special thanks are also extended to Amy Ouchley, University of Louisiana at Monroe, and Jane Johnson-Murray, Houston Community College, for their work in writing the instructor supplements that accompany this book.

Finally, we sincerely thank the many anatomy and physiology instructors who have provided us with excellent constructive criticism through reviews and focus groups. Their helpful suggestions for improvement were significant contributions that we greatly appreciate. We acknowledge them by name in the next section.

Rod Seeley
Trent Stephens
Phil Tate

Reviewers

Pegge Alciatore
University of Louisiana at Lafayette

Dennis Anderson
Oklahoma City Community College

David Bastedo
San Bernardino Valley College

Cynthia Beck
George Mason University

Edmund Burke
University of Nevada–School of Medicine

Jamie Campbell
Truckee Meadows Community College

Ethel Cornforth
San Jacinto College–South

Joseph Crivello
University of Connecticut

Lynette Danzl-Tauer
Rock Valley College

Judith Davis
Florida Community College at Jacksonville

Sandra Denton
Santa Fe Community College

Kathryn Durham
Lorain County Community College

Maria Florez-Duquet
California Polytechnic State University

Margaret Folsom
Methodist College

Cliff Fontenot
Southeastern Louisiana University

Ralph Fregosi
The University of Arizona

Barry Furman
San Jacinto College–Pasadena Campus

Mukul Ganguli
South Texas Community College

Cynthia Hoobler
San Jacinto College–South

Barbara Hunnicutt
Seminole Community College

Jane Johnson-Murray
Houston Community College System

Kamal Kamal
Valencia Community College–West

Ian Kay
Manchester Metropolitan University

Nancy Kincaid
Troy State University Montgomery

Mitchell Lockhart
Valdosta State University

Gayle LoPiccolo
Montgomery College

Carolyn McCracken
Northeast State Community College

Mark Meade
Jacksonville State University

Melissa A. Mills
Anoka-Ramsey Community College

Murad Odeh
South Texas Community College

Tammy Oliver
Eastfield College

Nikki Privacky
Palm Beach Community College

Robert Resau
Community College of Baltimore County

Tim Roye
San Jacinto College–South Campus

Judith Shardo
Middle Tennessee State University

Robert Speed
Wallace Community College

Shawn Thomas
University of Memphis

Janis G. Thompson
Lorain County Community College

Corinne Ulbright
Indiana University-Purdue University Indianapolis

Eva Weinreb
Community College of Philadelphia

Vernon Wiersema
Houston Community College

Victoria Zusman
Miami-Dade College

Focus Group Participants

David Bastedo
San Bernardino Valley College

Bob Broyles
Butler County Community College

Roger D. Choate
Oklahoma City Community College

Susan Conrow-Pate
Daytona Beach Community College

Terry Harrison
Arapahoe Community College

Claudia Hinds
Midland College

Sarah L. Milton
Florida Atlantic University

Ellen Ott-Reeves
Blinn College

Bob Pope
Miami-Dade College–Kendall Campus

James Stittsworth
Florida Community College at Jacksonville

Janis G. Thompson
Lorain County Community College

Anthony Udeogalanya
Medgar Evers College/CUNY

Shirley Whitescarver
Lexington Community College

The Human Organism

The human organism has many membranes that enclose and protect underlying structures. The colorized scanning electron micrograph shows the cells forming the peritoneum, a membrane covering abdominopelvic organs and the inside wall of the abdominopelvic cavity. These cells have many short, hairlike projections called microvilli. The microvilli increase the surface area of the cells, enabling them to secrete a slippery lubricating fluid that protects organs from friction as they rub against one another or the inside of the abdominopelvic wall.

What lies ahead is an astounding adventure—learning about the structure and function of the human body and how they are regulated by intricate systems of checks and balances. For example, tiny collections of cells embedded in the pancreas affect the uptake and use of blood sugar in the body. Eating a candy bar results in an increase in blood sugar, which acts as a stimulus. Pancreatic cells respond to the stimulus by secreting insulin. Insulin moves into blood vessels and is transported to cells throughout the body, where it increases the movement of sugar from the blood into cells, thereby providing the cells with a source of energy and causing blood sugar levels to decrease.

Knowledge of the structure and function of the human body provides the basis for understanding disease. In one type of diabetes mellitus, cells of the pancreas do not secrete adequate amounts of insulin. Not enough sugar moves into cells, which deprives them of a needed source of energy, and they malfunction.

Knowledge of the structure and function of the human body is essential for those planning a career in the health sciences. It is also beneficial to nonprofessionals because it helps with understanding overall health and disease, with evaluating recommended treatments, and with critically reviewing advertisements and articles.

This chapter defines *anatomy* and *physiology* (p. 2). It also explains the body's *structural and functional organization* (p. 5) and provides an overview of the *human organism* (p. 6) and *homeostasis* (p. 7). Finally, the chapter presents *terminology and the body plan* (p. 13).

Anatomy and Physiology

Anatomy is the scientific discipline that investigates the body's structure. For example, anatomy describes the shape and size of bones. In addition, anatomy examines the relationship between the structure of a body part and its function. Just as the structure of a hammer makes it well suited for pounding nails, the structure of a specific body part allows it to perform a particular function effectively. For example, bones can provide strength and support because bone cells surround themselves with a hard, mineralized substance. Understanding the relationship between structure and function makes it easier to understand and appreciate anatomy.

Anatomy can be considered at many different levels. **Developmental anatomy** is the study of the structural changes that occur between conception and adulthood. **Embryology** (em-brē-ol′ō-jē), a subspeciality of developmental anatomy, considers changes from conception to the end of the eighth week of development. Most birth defects occur during embryologic development.

Some structures, such as cells, are so small that they are best studied using a microscope. **Cytology** (sī-tol′ō-jē) examines the structural features of cells, and **histology** (his-tol′ō-jē) examines tissues, which are cells and the materials surrounding them.

Gross anatomy, the study of structures that can be examined without the aid of a microscope, can be approached from either a systemic or regional perspective. In **systemic anatomy** the body is studied system by system, which is the approach taken in this and most other introductory textbooks. A system is a group of structures that have one or more common functions. Examples are the circulatory, nervous, respiratory, skeletal, and muscular systems. In **regional anatomy** the body is studied area by area, which is the approach taken in most graduate programs at medical and dental schools. Within each region, such as the head, abdomen, or arm, all systems are studied simultaneously.

Surface anatomy is the study of the external form of the body and its relation to deeper structures. For example, the sternum (breastbone) and parts of the ribs can be seen and palpated (felt) on the front of the chest. These structures can be used as anatomic landmarks to identify regions of the heart and points on the chest where certain heart sounds can best be heard. **Anatomic imaging** uses radiographs (x-rays), ultrasound, magnetic resonance imaging (MRI), and other technologies to create pictures of internal structures (see Clinical Focus "Anatomic Imaging," p. 3). Both surface anatomy and anatomic imaging provide important information about the body for diagnosing disease.

Anatomic Anomalies

No two humans are structurally identical. For instance, one person may have longer fingers than another person. Despite this variability, most humans have the same basic pattern. Normally, we each have 10 fingers. **Anatomic anomalies** are structures that are unusual and different from the normal pattern. For example, some individuals have 12 fingers.

Anatomic anomalies can vary in severity from the relatively harmless to the life-threatening. For example, each kidney is normally supplied by one blood vessel, but in some individuals a kidney can be supplied by two blood vessels. Either way, the kidney receives adequate blood. On the other hand, in the condition called "blue baby" syndrome certain blood vessels arising from the heart of an infant are not attached in their correct locations; blood is not effectively pumped to the lungs, resulting in tissues not receiving adequate oxygen.

Physiology is the scientific investigation of the processes or functions of living things. Although it may not be obvious at times, living things are dynamic and ever-changing, not static and without motion. The major goals of physiology are to understand and predict the responses of the body to stimuli and to understand how the body maintains conditions within a narrow range of values in a constantly changing environment.

Like anatomy, physiology can be considered at many different levels. **Cell physiology** examines the processes occurring in cells and **systemic physiology** considers the functions of organ systems. **Neurophysiology** focuses on the nervous system and **cardiovascular physiology** deals with the heart and blood vessels. Physiology often examines systems rather than regions because portions of a system in more than one region can be involved in a given function.

The study of the human body must encompass both anatomy and physiology because structures, functions, and processes are interwoven. **Pathology** (pa-thol′ō-jē) is the medical science dealing with all aspects of disease, with an emphasis on the cause and development of abnormal conditions as well as the structural and functional changes resulting from disease. **Exercise physiology** focuses on changes in function, but also structure, caused by exercise.

1. *Define anatomy and physiology. Describe different levels at which each can be considered.*
2. *Define pathology and exercise physiology.*

Clinical Focus Anatomic Imaging

Anatomic imaging has revolutionized medical science. Some estimate that during the past 20 years as much progress has been made in clinical medicine as in all its previous history combined, and anatomic imaging has made a major contribution to that progress. Anatomic imaging allows medical personnel to look inside the body with amazing accuracy and without the trauma and risk of exploratory surgery. Although most of the technology of anatomic imaging is very new, the concept and earliest technology are quite old.

Wilhelm Roentgen (1845–1923) was the first to use **x-rays** in medicine in 1895 to see inside the body. The rays were called x-rays because no one knew what they were. This extremely shortwave electromagnetic radiation (see chapter 2) moves through the body exposing a photographic plate to form a **radiograph** (rā′dē-ō-graf). Bones and radiopaque dyes absorb the rays and create underexposed areas that appear white on the photographic film (figure A). X-rays have been in common use for many years and have numerous applications. Almost everyone has had a radiograph taken, either to visualize a broken bone or to check for a cavity in a tooth. A major limitation of radiographs, however, is that they give only a flat, two-dimensional (2-D) image of the body, which is a three-dimensional (3-D) structure.

Ultrasound is the second oldest imaging technique. It was first developed in the early 1950s as an extension of World War II sonar technology and uses high-frequency sound waves. The sound waves are emitted from a transmitter–receiver placed on the skin over the area to be scanned. The sound waves strike internal organs and bounce back to the receiver on the skin. Even though the basic technology is fairly old, the most important advances in the field occurred only after it became possible to analyze the reflected sound waves by computer. Once the computer analyzes the pattern of sound waves, the information is transferred to a monitor, where the result is visualized as an ultrasound image called a **sonogram** (son′ō-gram) (figure B). One of the more recent advances in ultrasound technology is the ability of more advanced computers to analyze changes in position through time and to display those changes as "real time" movements. Among other medical uses, ultrasound is commonly used to evaluate the condition of the fetus during pregnancy.

Computer analysis is also the basis of another major medical breakthrough in imaging. **Computed tomographic** (tō′mō-graf′ik) **(CT) scans,** developed in 1972 and originally called **computerized axial tomographic (CAT) scans,** are computer-analyzed x-ray images. A low-intensity x-ray tube is rotated through a 360-degree arc around the patient, and the images are fed into a computer. The computer then constructs the image of a "slice" through the body at the point where the x-ray beam was focused and rotated (figure C). It is also possible with some computers to take several scans short distances apart and stack the slices to produce a 3-D image of a part of the body (figure D).

Continued

Figure C Computed Tomography
Transverse section through the skull at the level of the eyes.

Figure A X-ray
Radiograph produced by x-rays shows a lateral view of the head and neck.

Figure B Ultrasound
Sonogram produced with ultrasound shows a lateral view of the head and hand of a fetus within the uterus.

Figure D Computed Tomography (CT)
Stacking of images acquired using CT technology.

Continued

Dynamic spatial reconstruction (DSR) takes CT one step further. Instead of using a single rotating x-ray machine to take single slices and add them together, DSR uses about 30 x-ray tubes. The images from all the tubes are compiled simultaneously to rapidly produce a 3-D image. Because of the speed of the process, multiple images can be compiled to show changes through time, thereby giving the system a dynamic quality. This system allows us to move away from seeing only static structure and toward seeing dynamic structure and function.

Digital subtraction angiography (an-jē-og′ră-fē) **(DSA)** is also one step beyond CT scans. A 3-D radiographic image of an organ such as the brain is made and stored in a computer. A radiopaque dye is injected into the circulation, and a second radiographic computer image is made. The first image is subtracted from the second one, greatly enhancing the differences, with the primary difference being the presence of the injected dye (figure E). These computer images can be dynamic and can be used, for example, to guide a catheter into a carotid artery during angioplasty, which is the insertion of a tiny balloon into a carotid artery to compress material clogging the artery.

Magnetic resonance imaging (MRI) directs radio waves at a person lying inside a large electromagnetic field. The magnetic field causes the protons of various atoms to align (see chapter 2). Because of the large amounts of water in the body, the alignment of hydrogen atom protons is at present most important in this imaging system. Radio waves of certain frequencies, which change the alignment of the hydrogen atoms, then are directed at the patient. When the radio waves are turned off, the hydrogen atoms realign in accordance with the magnetic field. The time it takes the hydrogen atoms to realign is different for various tissues of the body. These differences can be analyzed by computer to produce very clear sections through the body (figure F). The technique is also very sensitive in detecting some forms of cancer and can detect a tumor far more readily than can a CT scan.

Positron emission tomographic (PET) scans can identify the metabolic states of various tissues. This technique is particularly useful in analyzing the brain. When cells are active, they are using energy. The energy they need is supplied by the breakdown of glucose (blood sugar). If radioactively treated, or "labeled," glucose is given to a patient, the active cells take up the labeled glucose. As the radioactivity in the glucose decays, positively charged subatomic particles called positrons are emitted. When the positrons collide with electrons, the two particles annihilate each other, and gamma rays are given off. The gamma rays can be detected, pinpointing the cells that are metabolically active (figure G).

Whenever the human body is exposed to x-rays, ultrasound, electromagnetic fields, or radioactively labeled substances, a potential risk exists. In the medical application of anatomic imaging, the risk must be weighed against the benefit. Numerous studies have been conducted and are still being done to determine the outcomes of diagnostic and therapeutic exposures to x-rays.

The risk of anatomic imaging is minimized by using the lowest possible doses that provide the necessary information. For example, it is well known that x-rays can cause cell damage, particularly to the reproductive cells. As a result of this knowledge, the number of x-rays and the level of exposure are kept to a minimum; the x-ray beam is focused as closely as possible to avoid scattering of the rays; areas of the body not being x-rayed are shielded; and personnel administering x-rays are shielded. No known risks exist from ultrasound or electromagnetic fields at the levels used for diagnosis.

Figure E Digital Subtraction Angiography (DSA)
A lateral view of the head reveals the major blood vessels of the brain.

Figure F Magnetic Resonance Imaging (MRI)
Sagittal section of the head and neck.

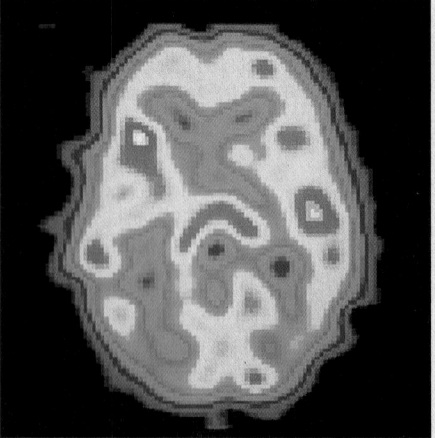

Figure G Positron Emission Tomography (PET)
Shows a transverse section through the skull. The highest level of brain activity is indicated in red, with successively lower levels represented by yellow, green, and blue.

Structural and Functional Organization

The body can be considered to have six levels of organization: the chemical, cell, tissue, organ, organ system, and complete organism (figure 1.1).

1. *Chemical level.* The chemical level involves interactions between atoms, which are tiny building blocks of matter. Atoms can combine to form molecules such as water, sugar, fats, and proteins. The function of a molecule is related intimately to its structure. For example, collagen molecules are ropelike protein fibers that give skin structural strength and flexibility. With old age, the structure of collagen changes, and the skin becomes fragile and is torn more easily. A brief overview of chemistry is presented in chapter 2.

2. *Cell level.* **Cells** are the basic units of all living things. Molecules can combine to form **organelles** (or′gă-nelz), which are the small structures that make up cells. For example, the nucleus contains the cell's hereditary information, and mitochondria manufacture adenosine triphosphate (ATP), which is a molecule used by cells for a source of energy. Although cell types differ in their structure and function, they have many characteristics in common. Knowledge of these characteristics and their variations is essential to a basic understanding of anatomy and physiology. The cell is discussed in chapter 3.

3. *Tissue level.* A **tissue** is a group of similar cells and the materials surrounding them. The characteristics of the cells and surrounding materials determine the functions of the tissue. The numerous different tissues that make up the body are classified into four basic types: epithelial, connective, muscle, and nervous. Tissues are discussed in chapter 4.

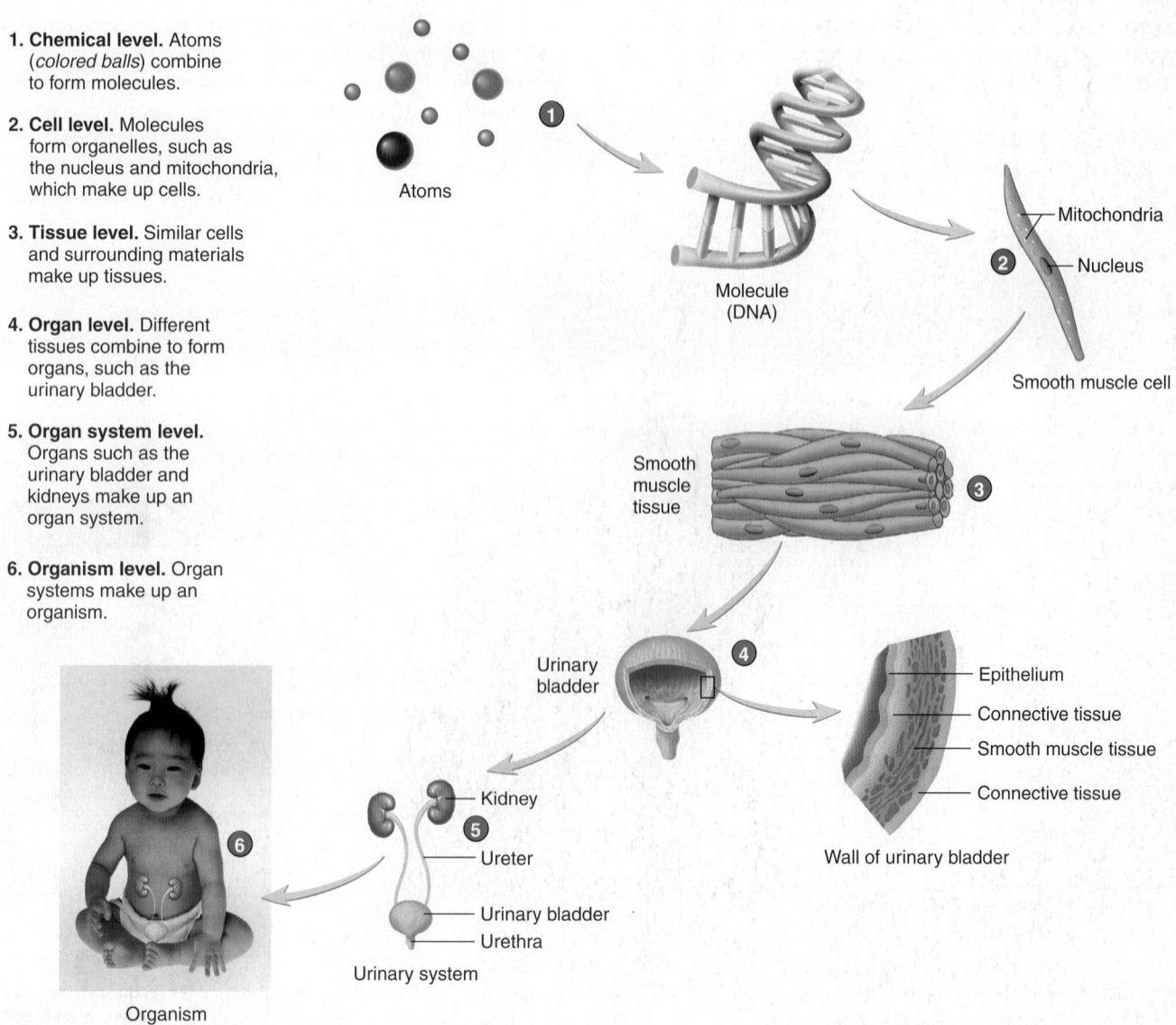

1. **Chemical level.** Atoms (*colored balls*) combine to form molecules.

2. **Cell level.** Molecules form organelles, such as the nucleus and mitochondria, which make up cells.

3. **Tissue level.** Similar cells and surrounding materials make up tissues.

4. **Organ level.** Different tissues combine to form organs, such as the urinary bladder.

5. **Organ system level.** Organs such as the urinary bladder and kidneys make up an organ system.

6. **Organism level.** Organ systems make up an organism.

Atoms

Molecule (DNA)

Mitochondria

Nucleus

Smooth muscle cell

Smooth muscle tissue

Urinary bladder

Epithelium

Connective tissue

Smooth muscle tissue

Connective tissue

Wall of urinary bladder

Kidney

Ureter

Urinary bladder

Urethra

Urinary system

Organism

Figure 1.1 Levels of Organization

Six levels of organization for the human body are the chemical, cell, tissue, organ, organ system, and organism.

4. *Organ level.* An **organ** is composed of two or more tissue types that perform one or more common functions. The urinary bladder, heart, skin, and eye are examples of organs (figure 1.2).

5. *Organ system level.* An **organ system** is a group of organs that have a common function or set of functions and are therefore viewed as a unit. For example, the urinary system consists of the kidneys, ureter, urinary bladder, and urethra. The kidneys produce urine, which is transported by the ureters to the urinary bladder, where it is stored until eliminated from the body by passing through the urethra. In this text the body is considered to have 11 major organ systems: the integumentary, skeletal, muscular, nervous, endocrine, cardiovascular, lymphatic, respiratory, digestive, urinary, and reproductive systems. Figure 1.3 presents a brief summary of the organ systems and their functions.

6. *Organism level.* An **organism** is any living thing considered as a whole, whether composed of one cell such as a bacterium or of trillions of cells such as a human. The human organism is a complex of organ systems, all mutually dependent on one another.

3. From smallest to largest, list and define the body's six levels of organization.
4. What are the four primary tissue types?
5. Which two organ systems are responsible for regulating the other organ systems (see figure 1.3)? Which two are responsible for support and movement?
6. What are the functions of the integumentary, cardiovascular, lymphatic, respiratory, digestive, urinary, and reproductive systems (see figure 1.3)?

P R E D I C T

In one type of diabetes, the pancreas (an organ) fails to produce insulin, which is a chemical normally made by pancreatic cells and released into the circulation. List as many levels of organization as you can in which this disorder could be corrected.

The Human Organism
Characteristics of Life

Humans are organisms and share common characteristics with other organisms. The most important common feature of all organisms is life. Organization, metabolism, responsiveness, growth, development, and reproduction are life's essential characteristics.

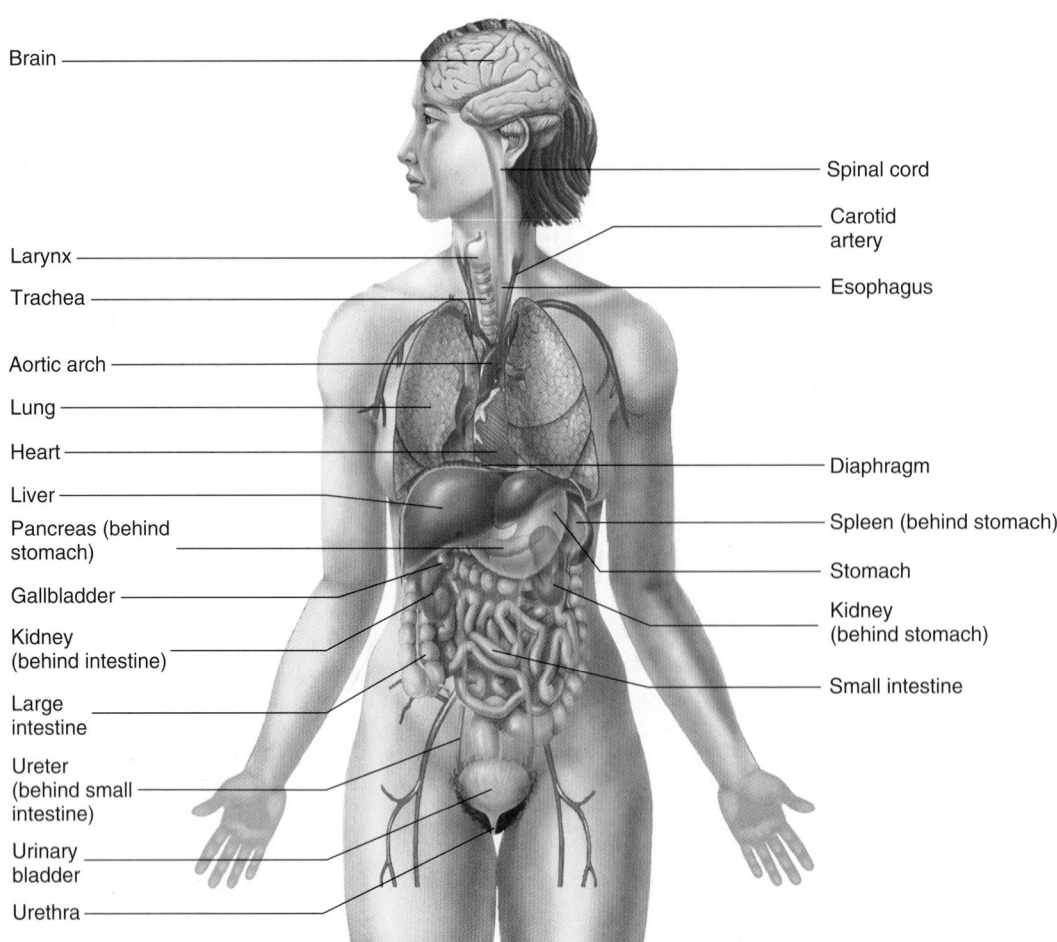

Figure 1.2   **Organs of the Body**

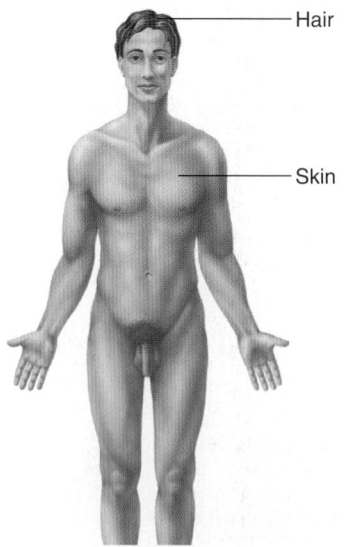

Integumentary System

Provides protection, regulates temperature, prevents water loss, and produces vitamin D precursors. Consists of skin, hair, nails, and sweat glands.

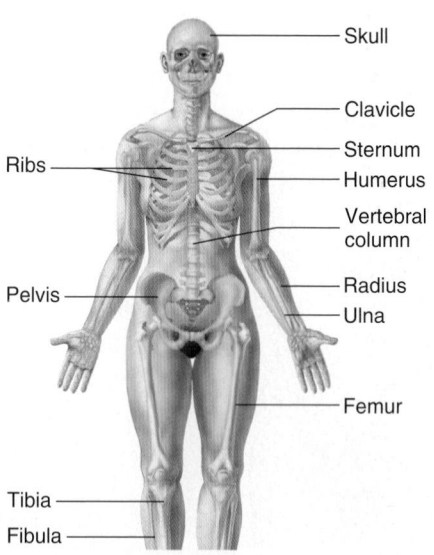

Skeletal System

Provides protection and support, allows body movements, produces blood cells, and stores minerals and fat. Consists of bones, associated cartilages, ligaments, and joints.

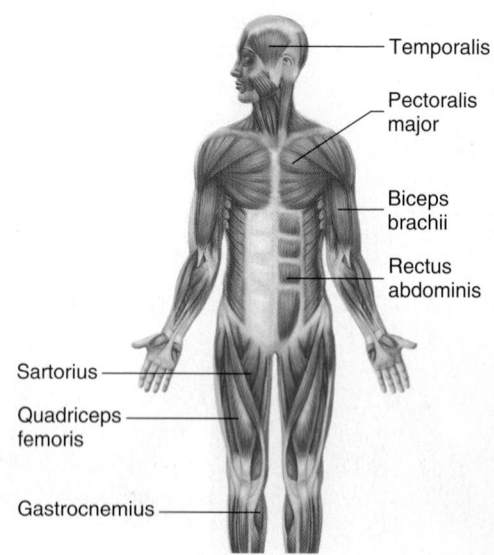

Muscular System

Produces body movements, maintains posture, and produces body heat. Consists of muscles attached to the skeleton by tendons.

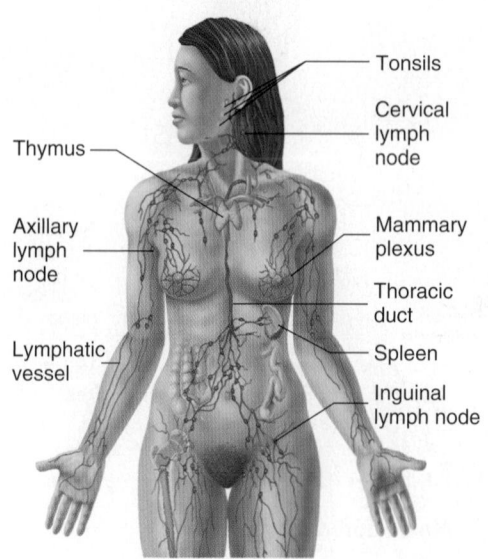

Lymphatic System

Removes foreign substances from the blood and lymph, combats disease, maintains tissue fluid balance, and absorbs fats from the digestive tract. Consists of the lymphatic vessels, lymph nodes, and other lymphatic organs.

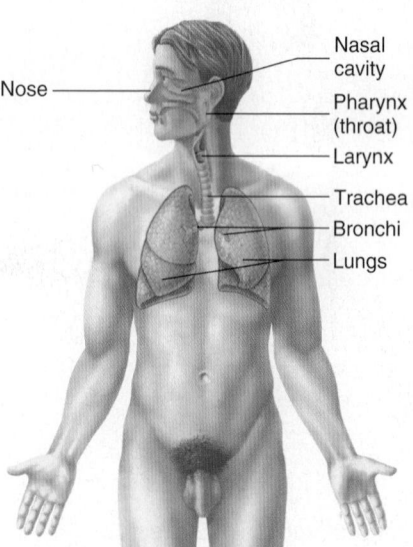

Respiratory System

Exchanges oxygen and carbon dioxide between the blood and air and regulates blood pH. Consists of the lungs and respiratory passages.

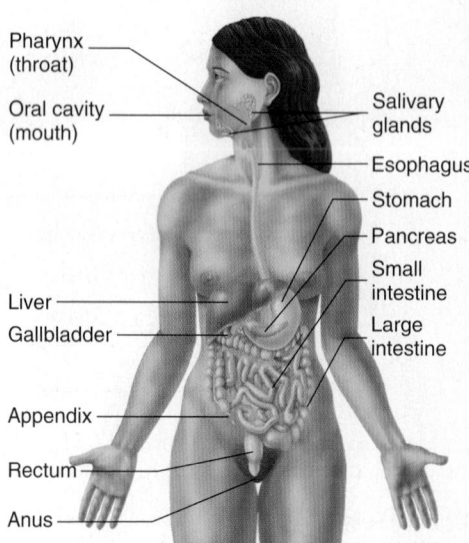

Digestive System

Performs the mechanical and chemical processes of digestion, absorption of nutrients, and elimination of wastes. Consists of the mouth, esophagus, stomach, intestines, and accessory organs.

Figure 1.3 Organ Systems of the Body

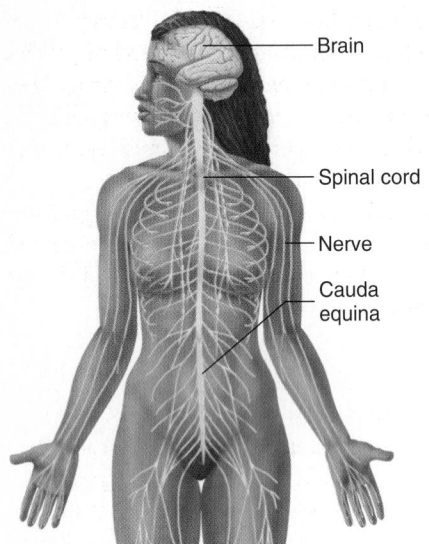

Nervous System

A major regulatory system that detects sensations and controls movements, physiologic processes, and intellectual functions. Consists of the brain, spinal cord, nerves, and sensory receptors.

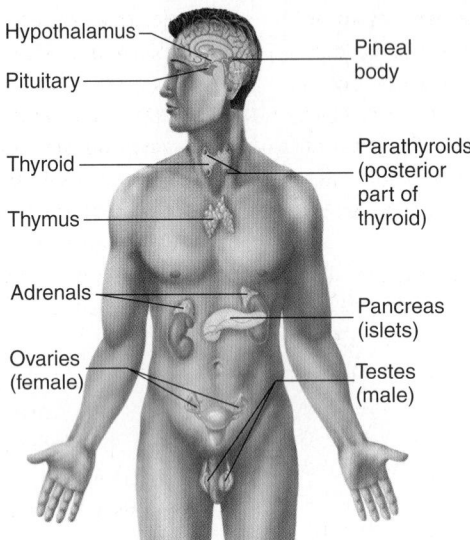

Endocrine System

A major regulatory system that influences metabolism, growth, reproduction, and many other functions. Consists of glands, such as the pituitary, that secrete hormones.

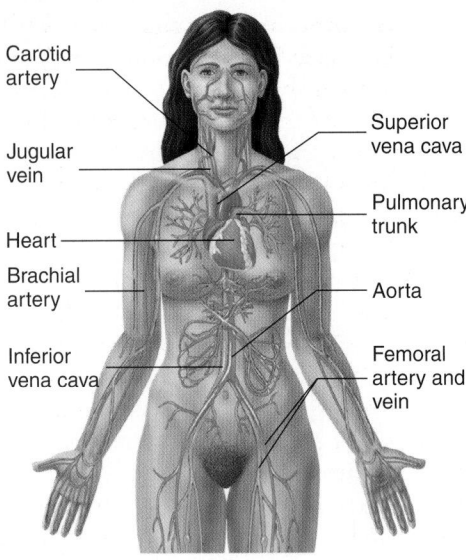

Cardiovascular System

Transports nutrients, waste products, gases, and hormones throughout the body; plays a role in the immune response and the regulation of body temperature. Consists of the heart, blood vessels, and blood.

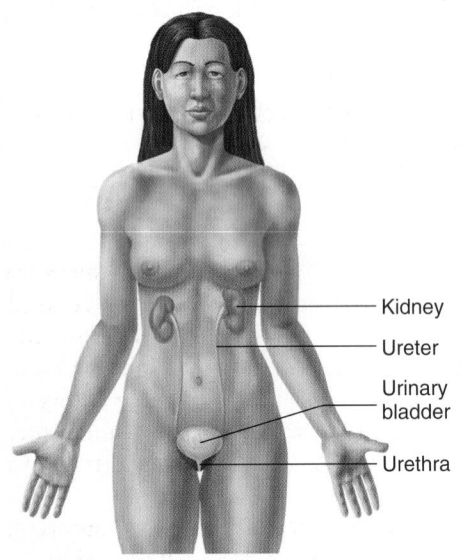

Urinary System

Removes waste products from the blood and regulates blood pH, ion balance, and water balance. Consists of the kidneys, urinary bladder, and ducts that carry urine.

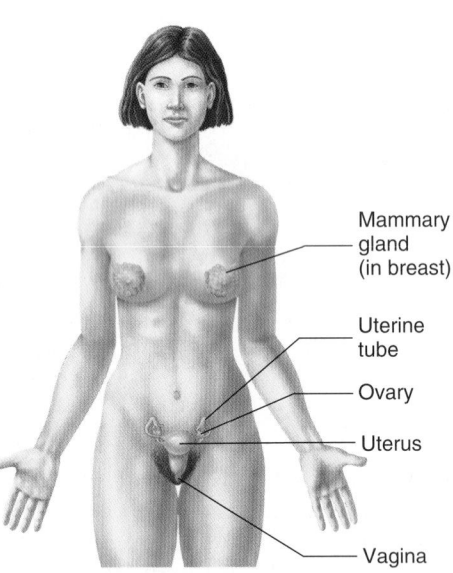

Female Reproductive System

Produces oocytes and is the site of fertilization and fetal development; produces milk for the newborn; produces hormones that influence sexual functions and behaviors. Consists of the ovaries, vagina, uterus, mammary glands, and associated structures.

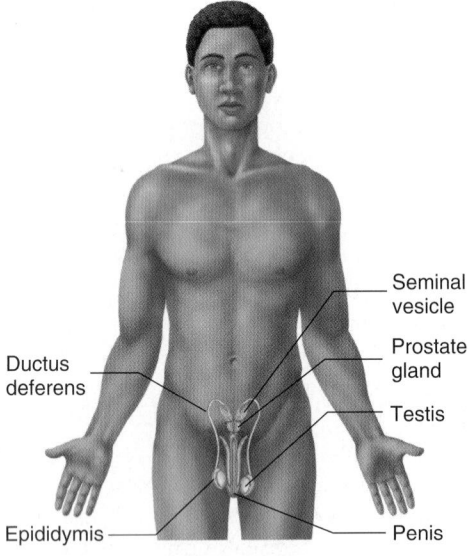

Male Reproductive System

Produces and transfers sperm cells to the female and produces hormones that influence sexual functions and behaviors. Consists of the testes, accessory structures, ducts, and penis.

Figure 1.3   *(continued)*

Organization is the condition in which the parts of an organism have specific relationships to each other and the parts interact to perform specific functions. Living things are highly organized. All organisms are composed of one or more cells. Cells in turn are composed of highly specialized organelles, which depend on the precise organization of large molecules. Disruption of this organized state can result in loss of functions, and even death.

Metabolism (mĕ-tab′ō-lizm) is all of the chemical reactions taking place in an organism. It includes the ability of an organism to break down food molecules, which are used as a source of energy and raw materials to synthesize the organism's own molecules. Energy is also used when one part of a molecule moves relative to another part, resulting in a change in shape of the molecule. Changes in molecular shape, in turn, can change the shape of cells, which can produce movements of the organism. Metabolism is necessary for vital functions, such as responsiveness, growth, development, and reproduction.

Responsiveness is the ability of an organism to sense changes in its external or internal environment and adjust to those changes. Responses include such things as moving toward food or water and away from danger or poor environmental conditions. Organisms can also make adjustments that maintain their internal environment. For example, if body temperature increases in a hot environment, sweat glands produce sweat, which can lower body temperature back toward normal levels.

Growth happens when cells increase in size or number, which produces an overall enlargement of all or part of an organism. For example, a muscle enlarged by exercise has larger muscle cells than an untrained muscle, and the skin of an adult has more cells than the skin of infant. An increase in the materials surrounding cells can also contribute to growth. For instance, the growth of bone results from an increase in cell number and the deposition of mineralized materials around the cells.

Development includes the changes an organism undergoes through time; it begins with fertilization and ends at death. The greatest developmental changes occur before birth, but many changes continue after birth, and some continue throughout life. Development usually involves growth, but it also involves differentiation and morphogenesis. **Differentiation** is change in cell structure and function from generalized to specialized, and **morphogenesis** (mōr-fō-jen′ĕ-sis) is change in the shape of tissues, organs, and the entire organism. For example, following fertilization, generalized cells specialize to become specific cell types, such as skin, bone, muscle, or nerve cells. These differentiated cells form the tissues and organs.

Reproduction is the formation of new cells or new organisms. Without reproduction, growth and development are not possible. Without reproduction of the organism, species become extinct.

Biomedical Research

Studying other organisms has increased our knowledge about humans because humans share many characteristics with other organisms. For example, studying single-celled bacteria provides much information about human cells. Some biomedical research, however, cannot be accomplished using single-celled organisms or isolated cells. Sometimes other mammals must be studied. For example, great progress in open-heart surgery and kidney transplantation was made possible by perfecting surgical techniques on other mammals before attempting them on humans. Strict laws govern the use of animals in biomedical research—laws designed to ensure minimum suffering on the part of the animal and to discourage unnecessary experimentation.

Although much can be learned from studying other organisms, the ultimate answers to questions about humans can be obtained only from humans, because other organisms are often different from humans in significant ways.

Human Versus Animal-Based Knowledge

Failure to appreciate the differences between humans and other animals led to many misconceptions by early scientists. One of the first great anatomists was a Greek physician, Claudius Galen (ca. 130–201). Galen described a large number of anatomic structures supposedly present in humans but observed only in other animals. For example, he described the liver as having five lobes. This is true for rats, but not for humans, who have four-lobed livers. The errors introduced by Galen persisted for more than 1300 years until a Flemish anatomist, Andreas Vesalius (1514–1564), who is considered the first modern anatomist, carefully examined human cadavers and began to correct the textbooks. This example should serve as a word of caution: Some current knowledge in molecular biology and physiology has not been confirmed in humans.

7. *Describe six characteristics of life.*
8. *Why is it important to realize that humans share many, but not all, characteristics with other animals?*

Homeostasis

Homeostasis (hō′mē-ō-stā′sis) is the existence and maintenance of a relatively constant environment within the body. A small amount of fluid surrounds each cell of the body. For cells to function normally, the volume, temperature, and chemical content—conditions known as **variables** because their values can change—of this fluid must remain within a narrow range. Body temperature is a variable that can increase in a hot environment or decrease in a cold one.

Homeostatic mechanisms, such as sweating or shivering, normally maintain body temperature near an ideal normal value, or **set point** (figure 1.4). Note that these mechanisms are not able to maintain body temperature precisely at the set point. Instead, body temperature increases and decreases slightly around the set point to produce a **normal range** of values. As long as body temperature remains within this normal range, homeostasis is maintained.

The organ systems help control the body's internal environment so that it remains relatively constant. For example, the digestive, respiratory, circulatory, and urinary systems function together so that each cell in the body receives adequate oxygen and nutrients and so that waste products do not accumulate to a toxic level. If the fluid surrounding cells deviates from homeostasis, the cells do not function normally and can even die. Disease disrupts homeostasis and sometimes results in death.

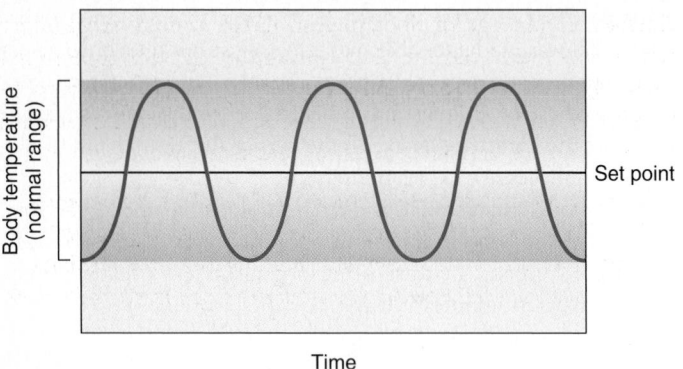

Figure 1.4   Homeostasis

Homeostasis is the maintenance of a variable around an ideal normal value, or set point. The value of the variable fluctuates around the set point to establish a normal range of values.

Negative Feedback

Most systems of the body are regulated by **negative-feedback** mechanisms that maintain homeostasis. *Negative* means that any deviation from the set point is made smaller or is resisted. Many negative-feedback mechanisms have three components: a **receptor,** which monitors the value of some variable such as blood pressure; a **control center,** which establishes the set point around which the variable is maintained; and an **effector,** which can change the value of the variable. A deviation from the set point is called a **stimulus.** The receptor detects the stimulus and informs the control center, which analyzes the input from the receptor. The control center sends output to the effector, and the effector produces a **response,** which tends to return the variable back toward the set point (figure 1.5).

The maintenance of normal blood pressure is an example of a negative-feedback mechanism that maintains homeostasis

4 A control center responds to information from the receptor.

5 The activity of an effector changes.

3 An increase in the variable is detected by a receptor.

6 A decrease in the variable is caused by the response of the effector.

2 Value increases

Value decreases

1 Normal range

7 Normal range

Homeostasis is maintained

A decrease in the variable is detected by a receptor.

An increase in the variable is caused by the response of the effector.

A control center responds to information from the receptor.

The activity of an effector changes.

Homeostasis Figure 1.5   Mechanism of Negative Feedback

Throughout the text, all homeostasis figures have the same format as in this figure. The changes caused by an increase of a variable are shown in the *green boxes,* and the changes caused by a decrease are shown in the *red boxes.* To help you learn how to interpret homeostasis figures, some of the steps in this figure are numbered: (*1*) The variable is within its normal range. (*2*) The value of the variable increases and is outside its normal range. (*3*) The increase in the variable is detected by receptors. (*4*) The control center responds to the change in the variable detected by the receptors. (*5*) The control center causes the activity of the effector to change. (*6*) The change in effector activity causes the value of the variable to decrease. (*7*) The variable returns to its normal range and homeostasis is maintained. See the responses to a decrease of the variable by following the *red boxes.*

(figure 1.6). Normal blood pressure is important because it is responsible for moving blood from the heart to tissues. The blood supplies the tissues with oxygen and nutrients and removes waste products. Thus normal blood pressure is required to ensure that tissue homeostasis is maintained.

Receptors that monitor blood pressure are located within large blood vessels near the heart, the control center for blood pressure is in the brain, and the heart is the effector. Blood pressure depends in part on contraction (beating) of the heart: as heart rate increases, blood pressure increases; as heart rate decreases, blood pressure decreases.

If blood pressure increases slightly, the receptors detect the increased blood pressure and send that information to the control center in the brain. The control center causes heart rate to decrease, resulting in a decrease in blood pressure. If blood pressure decreases slightly, the receptors inform the control center, which increases heart rate, thereby producing an increase in blood pressure. As a result, blood pressure constantly rises and falls within a normal range of values.

Although homeostasis is the maintenance of a normal range of values, this does not mean that all variables are maintained within the same narrow range of values at all times. Sometimes a deviation from the usual range of values can be beneficial. For example, during exercise the normal range for blood pressure differs from the range under resting conditions, and the blood pressure is significantly elevated (figure 1.7). The elevated blood pressure increases blood delivery to muscles so that muscle cells are supplied with the extra nutrients and oxygen they need to maintain their increased rate of activity.

9. *Define homeostasis, variable, and set point. If a deviation from homeostasis occurs, what mechanism restores it?*
10. *What are the three components of many negative-feedback mechanisms? How do they produce a response to a stimulus?*

P R E D I C T
Explain how negative-feedback mechanisms control respiratory rates when a person is at rest and when a person is exercising.

Positive Feedback

Positive-feedback responses are not homeostatic and are rare in healthy individuals. *Positive* implies that, when a deviation from a normal value occurs, the response of the system is to make the deviation even greater (figure 1.8). Positive feedback therefore usually creates a cycle that leads away from homeostasis and, in some cases, results in death.

The control center in the brain that regulates heart rate responds.

The heart rate decreases.

An increase in blood pressure is detected by receptors in blood vessels.

A decrease in blood pressure is caused by a decrease in heart rate.

Blood pressure (normal range)

Blood pressure increases

Blood pressure decreases

Blood pressure (normal range)

Blood pressure homeostasis is maintained

A decrease in blood pressure is detected by receptors in blood vessels.

An increase in blood pressure is caused by an increase in heart rate.

The control center in the brain that regulates heart rate responds.

The heart rate increases.

Homeostasis Figure 1.6   Example of Negative Feedback
Blood pressure is maintained within a normal range by negative-feedback mechanisms.

Figure 1.7   Changes in Blood Pressure During Exercise

During exercise the demand for oxygen by muscle tissue increases. An increase in blood pressure (BP) results in an increase in blood flow to the tissues. The increased blood pressure is not an abnormal or nonhomeostatic condition but is a resetting of the normal homeostatic range to meet the increased demand. The reset range is higher and broader than the resting range. After exercise ceases, the range returns to that of the resting condition.

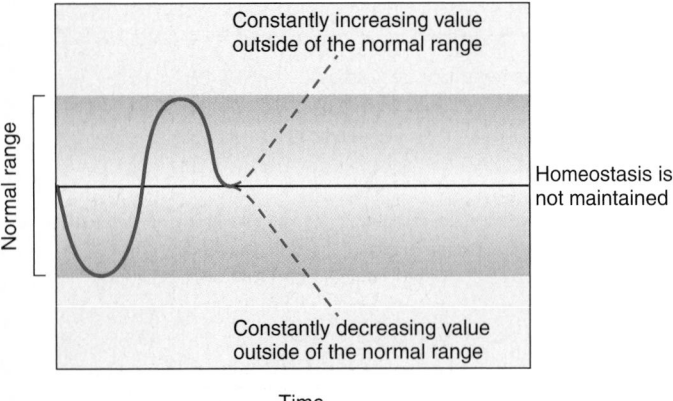

Figure 1.8   Positive Feedback

Deviations from the normal set point value cause an additional deviation away from that value in either a positive or negative direction.

The cardiac (heart) muscle receiving an inadequate amount of blood is an example of positive feedback. Contraction of cardiac muscle generates blood pressure and moves blood through blood vessels to tissues. A system of blood vessels on the outside of the heart provides cardiac muscle with a blood supply sufficient to allow normal contractions to occur. In effect, the heart pumps blood to itself. Just as with other tissues, blood pressure must be maintained to ensure adequate delivery of blood to cardiac muscle. Following extreme blood loss, blood pressure decreases to the point that delivery of blood to cardiac muscle is inadequate. As a result, cardiac muscle homeostasis is disrupted, and cardiac muscle does not function normally. The heart pumps less blood, which causes the blood pressure to drop even further. This additional decrease in blood pressure means that even less blood is delivered to cardiac muscle, and the heart pumps even less blood, which again decreases the blood pressure (figure 1.9). If the process continues

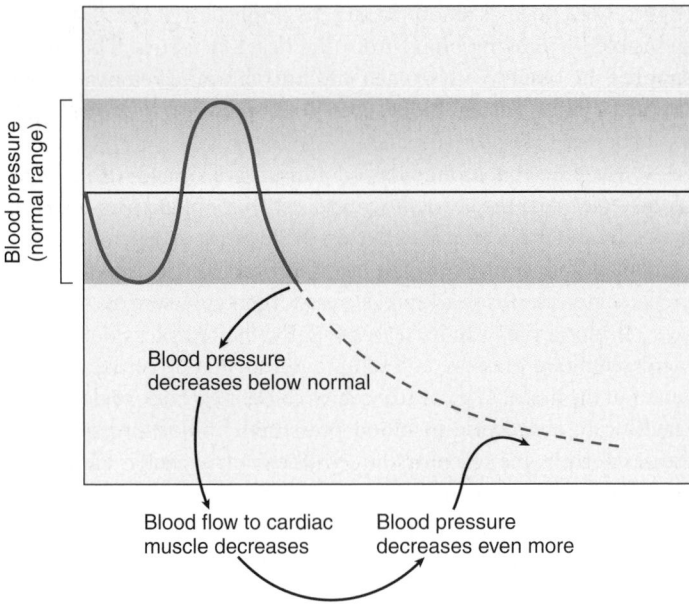

Figure 1.9   Example of Harmful Positive Feedback

A decrease in blood pressure below the normal range causes decreased blood flow to the heart. The heart is unable to pump enough blood to maintain blood pressure, and blood flow to the cardiac muscle decreases. Thus the ability of the heart to pump decreases further, and blood pressure decreases even more.

until the blood pressure is too low to sustain the cardiac muscle, the heart stops beating, and death results.

Following a moderate amount of blood loss (e.g., after a person donates a pint of blood), negative-feedback mechanisms produce an increase in heart rate and other responses that restore blood pressure. If blood loss is severe, however, negative-feedback mechanisms may not be able to maintain homeostasis, and the positive-feedback effect of an ever-decreasing blood pressure can develop. Circumstances in which negative-feedback mechanisms are not adequate to maintain homeostasis illustrate a basic principle. Many disease states result from failure of negative-feedback mechanisms to maintain homeostasis. Medical therapy seeks to overcome illness by aiding negative-feedback mechanisms (e.g., a transfusion reverses a constantly decreasing blood pressure and restores homeostasis).

A few positive-feedback mechanisms do operate in the body under normal conditions, but in all cases they are eventually limited in some way. Birth is an example of a normally occurring positive-feedback mechanism. Near the end of pregnancy, the baby's larger size stretches the uterus. This stretching, especially around the opening of the uterus, stimulates contractions of the uterine muscles. The uterine contractions push the baby against the opening of the uterus and stretch it further. This stimulates additional contractions that result in additional stretching. This positive-feedback sequence ends only when the baby is delivered from the uterus and the stretching stimulus is eliminated.

 11. *Define positive feedback. Why are positive-feedback mechanisms often harmful?*

P R E D I C T   ❸

Is the sensation of thirst associated with a negative- or a positive-feedback mechanism? Explain.

Terminology and the Body Plan

You will be learning many new words as you study anatomy and physiology. Knowing the derivation, or **etymology** (et′uh-mol′ŏ-jē), of these words, can make learning them easy and fun. Most words are derived from Latin or Greek, which are very descriptive languages. For example, *foramen* is a Latin word for hole, and *magnum* means large. The foramen magnum is therefore a large hole in the skull through which the spinal cord attaches to the brain.

Prefixes and suffixes can be added to words to expand their meaning. The suffix -*itis* means an inflammation, so appendicitis is an inflammation of the appendix. As new terms are introduced in this text, their meanings are often explained. The glossary and the list of word roots, prefixes, and suffixes on the inside back cover of the textbook provide additional information about the new terms.

It is very important to learn these new words so that when you speak to colleagues or write reports your message is clear and correct.

Body Positions

The **anatomic position** refers to a person standing erect with the face directed forward, the upper limbs hanging to the sides, and the palms of the hands facing forward (figure 1.10). A person is **supine** when lying face upward and **prone** when lying face downward.

The position of the body can affect the description of body parts relative to each other. In the anatomic position, the elbow is above the hand, but in the supine or prone position, the elbow and hand are at the same level. To avoid confusion, relational descriptions are always based on the anatomic position, no matter the actual position of the body. Thus, the elbow is always described as being above the wrist, whether the person is lying down or is even upside down.

- 12. **What is the anatomic position in humans? Why is it important?**
- 13. **Define supine and prone.**

Directional Terms

Directional terms describe parts of the body relative to each other. Important directional terms are illustrated in figure 1.10 and

Figure 1.10   Directional Terms

All directional terms are in relation to a person in the anatomic position: a person standing erect with the face directed forward, the arms hanging to the sides, and the palms of the hands facing forward.

summarized in table 1.1. It is important to become familiar with these directional terms as soon as possible because you will see them repeatedly throughout the text. Right and left are retained as directional terms in anatomic terminology. Up is replaced by **superior,** down by **inferior,** front by **anterior,** and back by **posterior.**

In humans, *superior* is synonymous with **cephalic** (se-fal′ik), which means toward the head, because, when we are in the anatomic position, the head is the highest point. In humans, the term *inferior* is synonymous with **caudal** (kaw′dăl), which means toward the tail, which would be located at the end of the vertebral column if humans had tails. The terms *cephalic* and *caudal* can be used to describe directional movements on the trunk, but they are not used to describe directional movements on the limbs.

The word *anterior* means that which goes before, and **ventral** means belly. The anterior surface of the human body is therefore the ventral surface, or belly, because the belly "goes first" when we are walking. The word *posterior* means that which follows, and **dorsal** means back. The posterior surface of the body is the dorsal surface, or back, which follows as we are walking.

14. List two terms that in humans indicate toward the head. Name two terms that mean the opposite.
15. List two terms that indicate the back in humans. What two terms mean the front?

P R E D I C T ④

The anatomic position of a cat refers to the animal standing erect on all four limbs and facing forward. On the basis of the etymology of the directional terms, what two terms indicate movement toward the head? What two terms mean movement toward the back? Compare these terms to those referring to a human in the anatomic position.

Proximal means nearest, whereas **distal** means distant. These terms are used to refer to linear structures, such as the limbs, in which one end is near some other structure and the other end is farther away. Each limb is attached at its proximal end to the body, and the distal end, such as the hand, is farther away.

Medial means toward the midline, and **lateral** means away from the midline. The nose is located in a medial position in the face, and the eyes are lateral to the nose. The term **superficial** refers to a structure close to the surface of the body, and **deep** is toward the interior of the body. The skin is superficial to muscle and bone.

16. Define the following terms, and give the word that means the opposite: proximal, lateral, and superficial.

P R E D I C T ⑤
Describe in as many directional terms as you can the relationship between your kneecap and your heel.

Body Parts and Regions

A number of terms are used when referring to different parts or regions of the body (figure 1.11). The upper limb is divided into the arm, forearm, wrist, and hand. The **arm** extends from the shoulder to the elbow, and the **forearm** extends from the elbow to the wrist. The lower limb is divided into the thigh, leg, ankle, and foot. The **thigh** extends from the hip to the knee, and the **leg** extends from the knee to the ankle. Note that, contrary to popular

Table 1.1	Directional Terms for Humans		
Terms	**Etymology***	**Definition**	**Example**
Right		Toward the right side of the body	The right ear.
Left		Toward the left side of the body	The left eye.
Superior	L., higher	A structure above another	The chin is superior to the navel.
Inferior	L., lower	A structure below another	The navel is inferior to the chin.
Cephalic	G. *kephale,* head	Closer to the head than another structure (usually synonymous with superior)	The chin is cephalic to the navel.
Caudal	L. *cauda,* a tail	Closer to the tail than another structure (usually synonymous with inferior)	The navel is caudal to the chin.
Anterior	L., before	The front of the body	The navel is anterior to the spine.
Posterior	L. *posterus,* following	The back of the body	The spine is posterior to the breastbone.
Ventral	L. *ventr-,* belly	Toward the belly (synonymous with anterior)	The navel is ventral to the spine.
Dorsal	L. *dorsum,* back	Toward the back (synonymous with posterior)	The spine is dorsal to the breastbone.
Proximal	L. *proximus,* nearest	Closer to the point of attachment to the body than another structure	The elbow is proximal to the wrist.
Distal	L. *di-* plus *sto,* to stand apart or be distant	Farther from the point of attachment to the body than another structure	The wrist is distal to the elbow.
Lateral	L. *latus,* side	Away from the midline of the body	The nipple is lateral to the breastbone.
Medial	L. *medialis,* middle	Toward the midline of the body	The bridge of the nose is medial to the eye.
Superficial	L. *superficialis,* toward the surface	Toward or on the surface (not shown in figure 1.10)	The skin is superficial to muscle.
Deep	O.E. *deop,* deep	Away from the surface, Internal (not shown in figure 1.10)	The lungs are deep to the ribs.

*Origin and meaning of the word: L., Latin; G., Greek; O.E., Old English.

Cephalic (head)
- Frontal (forehead)
- Orbital (eye)
- Nasal (nose)
- Oral (mouth)
- Cervical (neck)

Trunk
- Thoracic (thorax)
  - Pectoral (chest)
  - Sternal (breastbone)
  - Mammary (breast)
- Abdominal (abdomen)
- Umbilical (navel)
- Pelvic (pelvis)
- Inguinal (groin)
- Pubic (genital)

- Otic (ear)
- Buccal (cheek)
- Mental (chin)
- Clavicular (collar bone)
- Axillary (arm pit)
- Brachial (arm)
- Antecubital (front of elbow)
- Antebrachial (forearm)
- Carpal (wrist)
- Palmar (palm)
- Digital (fingers)
- Manual (hand)
- Coxal (hip)
- Femoral (thigh)
- Patellar (kneecap)
- Crural (leg)
- Talus (ankle)
- Dorsum (top of foot)
- Digital (toes)
- Pedal (foot)

Upper limb

Lower limb

(a)

Figure 1.11   Body Parts and Regions

The anatomic and common (*in parentheses*) names are indicated for some parts and regions of the body. (*a*) Anterior view.

usage, the terms *arm* and *leg* refer to only a part of the respective limb.

The central region of the body consists of the **head, neck,** and **trunk.** The trunk can be divided into the **thorax** (chest), **abdomen** (region between the thorax and pelvis), and **pelvis** (the inferior end of the trunk associated with the hips).

The abdomen is often subdivided superficially into **quadrants** by two imaginary lines—one horizontal and one vertical—that intersect at the navel (figure 1.12*a*). The quadrants formed are the right-upper, left-upper, right-lower, and left-lower quadrants. In addition to these quadrants, the abdomen is sometimes subdivided into nine **regions** by four imaginary lines: two horizontal and two vertical. These four lines create an imaginary tic-tac-toe figure on the abdomen, resulting in nine regions: epigastric, right and left hypochondriac, umbilical, right and left lumbar, hypogastric, and right and left iliac (figure 1.12*b*). Clinicians use the quadrants or regions as reference points for locating underlying organs. For example, the appendix is located in the right-lower quadrant, and the pain of an acute appendicitis is usually felt there.

17. What is the difference between the arm and the upper limb and the difference between the leg and the lower limb?

18. Describe the quadrant and the nine-region methods of subdividing the abdominal region. What is the purpose of these subdivisions?

P R E D I C T
Using figures 1.2 (p. 6) and 1.12 (p. 16), determine in which quadrant each of the following organs is located: spleen, gallbladder, kidneys, most of the stomach, and most of the liver.

Planes

At times it is conceptually useful to describe the body as having imaginary flat surfaces called **planes** passing through it (figure 1.13). A plane divides or sections the body, making it possible to "look inside" and observe the body's structures. A **sagittal** (saj′i-tăl) plane runs vertically through the body and separates it into right and left portions. The word *sagittal* literally means "the flight of an arrow" and refers to the way the body would be split by an arrow passing anteriorly to posteriorly. A **median** plane is a sagittal plane that passes through the midline of the body and divides it into equal right and left halves. A **transverse,** or **horizontal,** plane runs parallel to the ground and divides the body into superior and inferior portions. A **frontal,** or **coronal** (kōr′ŏ-năl,

Cranial (skull)

Occipital (base of skull)

Nuchal (back of neck)

Scapular (shoulder blade)

Acromial (point of shoulder)

Dosal (back)

Vertebral (spinal column)

Lumbar (loin)

Olecranon (point of elbow)

Upper limb

Trunk

Sacral (between hips)

Dorsum (back of hand)

Gluteal (buttock)

Perineal (perineum)

Popliteal (hollow behind knee)

Sural (calf)

Lower limb

(b)

Plantar (sole)

Calcaneal (heel)

Figure 1.11 *(continued)* (*b*) Posterior view.

Right-upper
quadrant

Left-upper
quadrant

Right-lower
quadrant

Left-lower
quadrant

(a)

Right
hypochondriac
region

Epigastric
region

Left
hypochondriac
region

Right
lumbar
region

Umbilical
region

Left
lumbar
region

Right
iliac
region

Hypogastric
region

Left
iliac
region

(b)

Figure 1.12   **Subdivisions of the Abdomen**

Lines are superimposed over internal organs to demonstrate the relationship of the organs to the subdivisions. (*a*) Abdominal quadrants consist of four subdivisions. (*b*) Abdominal regions consist of nine subdivisions.

Figure 1.13 Planes of Section of the Body

Planes of section through the whole body are indicated by "glass" sheets. Actual sections through the head (viewed from the right), abdomen (*inferior view*), and hip (*anterior view*) are also shown.

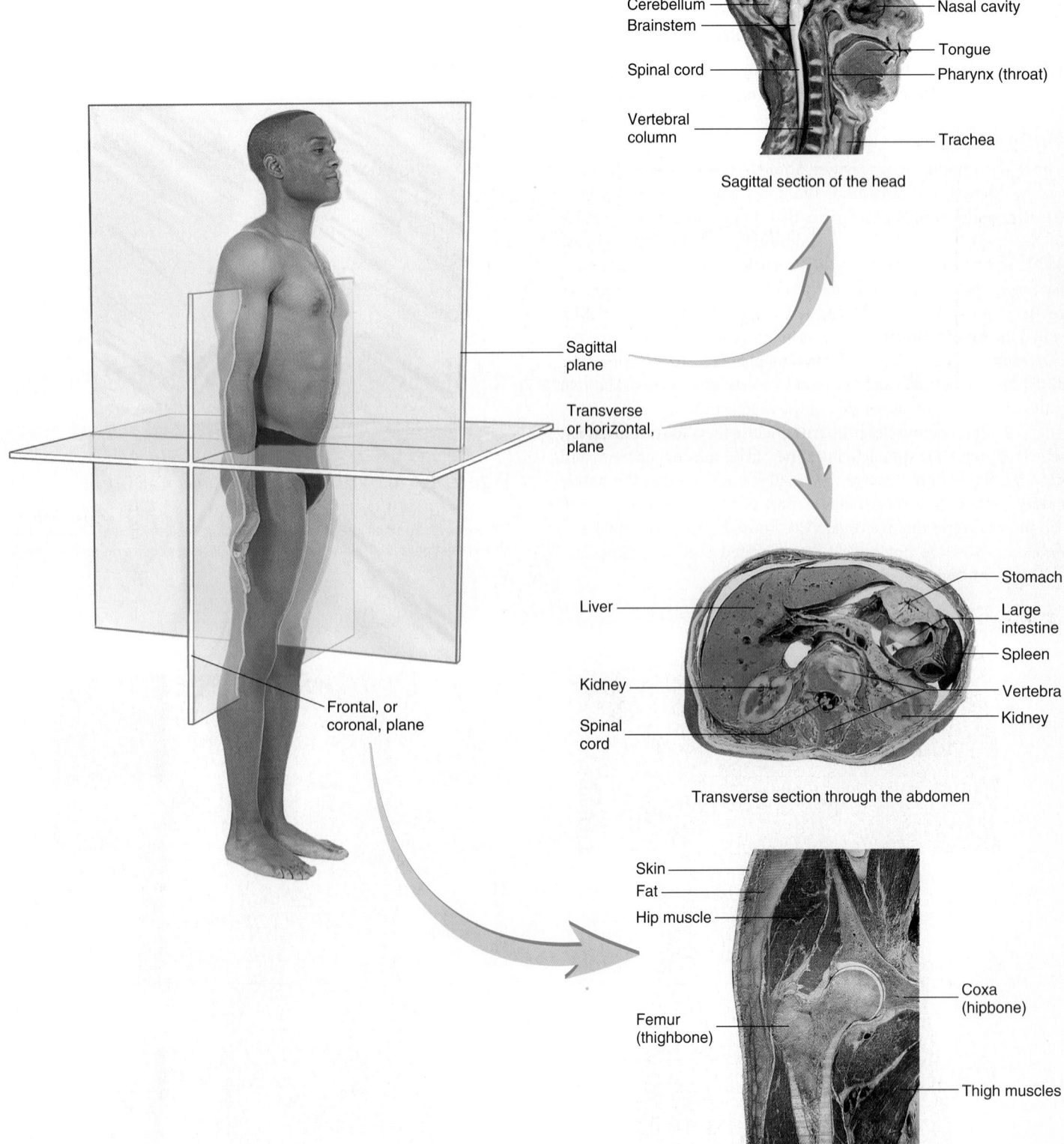

Sagittal plane

Transverse or horizontal, plane

Frontal, or coronal, plane

Cerebrum

Cerebellum
Brainstem

Spinal cord

Vertebral column

Nasal cavity

Tongue

Pharynx (throat)

Trachea

Sagittal section of the head

Liver

Kidney

Spinal cord

Stomach

Large intestine

Spleen

Vertebra

Kidney

Transverse section through the abdomen

Skin
Fat
Hip muscle

Femur (thighbone)

Coxa (hipbone)

Thigh muscles

Frontal section through the right hip

kō-rō′năl), plane runs vertically from right to left and divides the body into anterior and posterior parts.

Organs are often sectioned to reveal their internal structure (figure 1.14). A cut through the long axis of the organ is a **longitudinal** section, and a cut at right angles to the long axis is a **cross,** or **transverse,** section. If a cut is made across the long axis at other than a right angle, it is called an **oblique** section.

● 19. *Define the three planes of the body.*
  20. *In what three ways can an organ be cut?*

Body Cavities

The body contains many cavities. Some of these, such as the nasal cavity, open to the outside of the body, and some do not. The trunk contains three large cavities that do not open to the outside of the body: the thoracic, the abdominal, and the pelvic (figure 1.15). The rib cage surrounds the **thoracic cavity,** and the muscular diaphragm separates it from the abdominal cavity. The thoracic cavity is divided into right and left parts by a median partition called the **mediastinum** (me′dē-as-tī′nŭm; middle wall). The mediastinum contains the heart, thymus gland, trachea, esophagus, and other structures such as blood vessels and nerves. The two lungs are located on either side of the mediastinum.

Abdominal muscles primarily enclose the **abdominal cavity,** which contains the stomach, intestines, liver, spleen, pancreas, and kidneys. Pelvic bones encase the small space known as the **pelvic cavity,** where the urinary bladder, part of the large intestine, and the internal reproductive organs are housed. The abdominal and pelvic cavities are not physically separated and sometimes are called the **abdominopelvic cavity.**

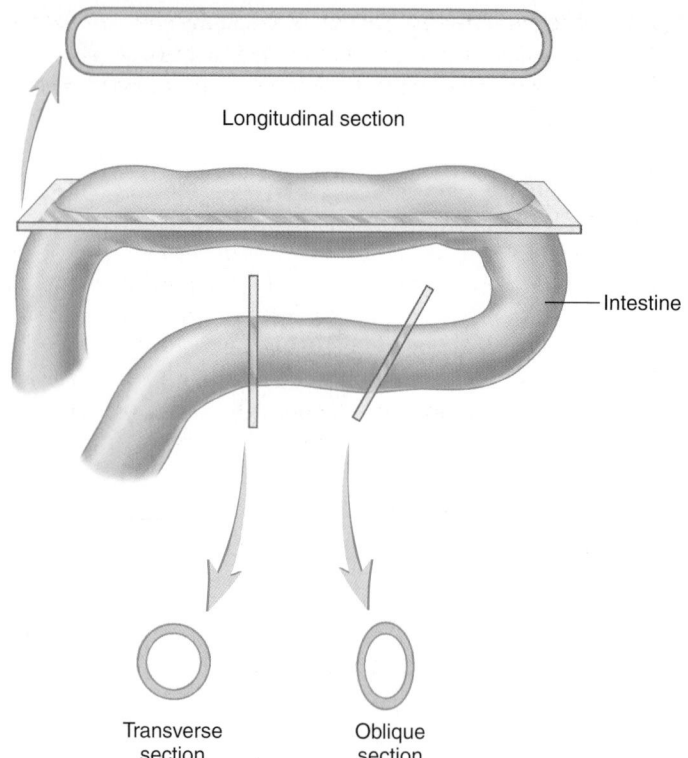

Figure 1.14   **Planes of Section Through an Organ**

Planes of section through the small intestine are indicated by "glass" sheets. The views of the small intestine after sectioning are also shown. Although the small intestine is basically a tube, the sections appear quite different in shape.

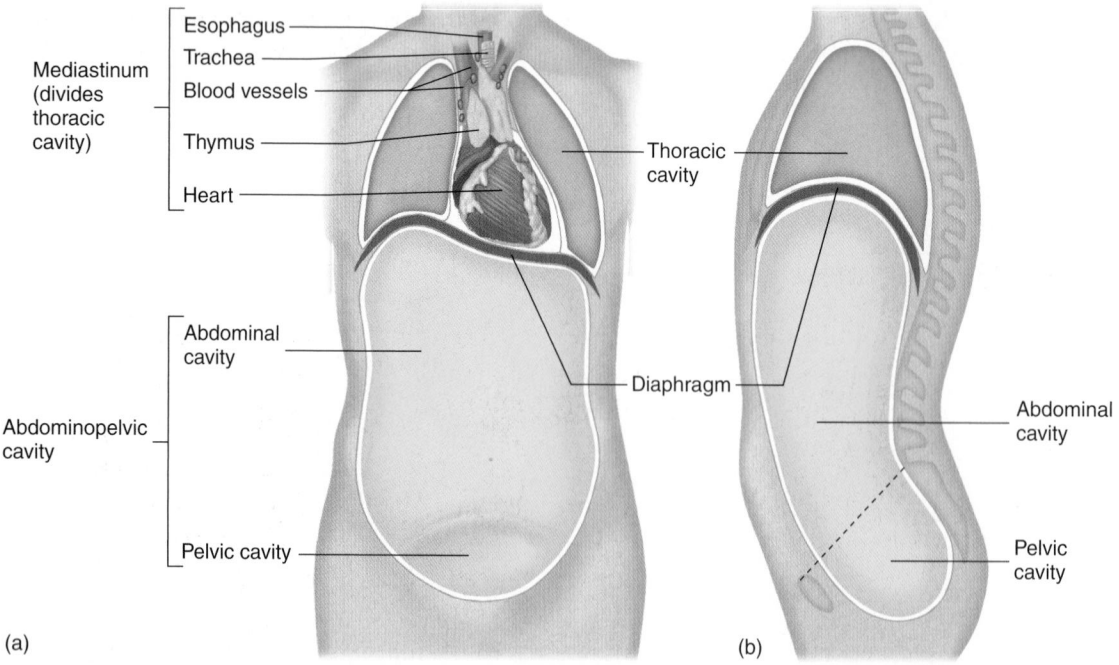

(a)   (b)

Figure 1.15   Trunk Cavities

(*a*) Anterior view showing the major trunk cavities. The diaphragm separates the thoracic cavity from the abdominal cavity. The mediastinum, which includes the heart, is a partition of organs dividing the thoracic cavity. (*b*) Sagittal section of the trunk cavities viewed from the left. The *dashed line* shows the division between the abdominal and pelvic cavities. The mediastinum has been removed to show the thoracic cavity.

21. *What structure separates the thoracic cavity from the abdominal cavity? The abdominal cavity from the pelvic cavity?*
22. *What structure divides the thoracic cavity into right and left parts?*

Serous Membranes

Serous (sēr′ŭs) **membranes** cover the organs of the trunk cavities and line the trunk cavities. Imagine an inflated balloon into which a fist has been pushed (figure 1.16). The fist represents an organ, the inner balloon wall in contact with the fist represents the **visceral** (vis′er-ăl; organ) **serous membrane** covering the organ, and the outer part of the balloon wall represents the **parietal** (pă-rī′ĕ-tăl; wall) **serous membrane.** The cavity or space between the visceral and parietal serous membranes is normally filled with a thin, lubricating film of serous fluid produced by the membranes. As organs rub against the body wall or against another organ, the combination of serous fluid and smooth serous membranes reduces friction.

The thoracic cavity contains three serous membrane-lined cavities: a pericardial cavity and two pleural cavities. The **pericardial** (per-i-kar′dē-ăl; around the heart) **cavity** surrounds the heart (figure 1.17*a*). The visceral pericardium covers the heart, which is contained within a connective tissue sac lined with the parietal pericardium. The pericardial cavity, which contains pericardial fluid, is located between the visceral and parietal pericardia.

A **pleural** (ploor′ăl; associated with the ribs) **cavity** surrounds each lung, which is covered by visceral pleura (figure 1.17*b*). Parietal pleura line the inner surface of the thoracic wall, the lateral surfaces of the mediastinum, and the superior surface of the diaphragm. The pleural cavity lies between the visceral and parietal pleurae and contains pleural fluid.

The abdominopelvic cavity contains a serous membrane-lined cavity called the **peritoneal** (per′i-tō-nē′ăl; to stretch over) **cavity** (figure 1.17*c*). Visceral peritoneum covers many of the organs of the abdominopelvic cavity. Parietal peritoneum lines the wall of the abdominopelvic cavity and the inferior surface of the diaphragm. The peritoneal cavity is located between the visceral and parietal peritonea and contains peritoneal fluid.

Inflammation of Serous Membranes

The serous membranes can become inflamed, usually as a result of an infection. **Pericarditis** (per′i-kar-dī′tis) is inflammation of the pericardium, **pleurisy** (ploor′i-sē) is inflammation of the pleura, and **peritonitis** (per′i-tō-nī′tis) is inflammation of the peritoneum.

Mesenteries (mes′en-ter-ēz), which consist of two layers of peritoneum fused together (see figure 1.17*c*), connect the visceral peritoneum of some abdominopelvic organs to the parietal peritoneum on the body wall or to the visceral peritoneum of other abdominopelvic organs. The mesenteries anchor the organs to the body wall and provide a pathway for nerves and blood vessels to reach the organs. Other abdominopelvic organs are more closely attached to the body wall and do not have mesenteries. Parietal peritoneum covers these other organs which are said to be **retroperitoneal** (re′trō-per′i-tō-nē′ăl; behind the peritoneum). The retroperitoneal organs include the kidneys, the adrenal glands, the pancreas, parts of the intestines, and the urinary bladder (see figure 1.17*c*).

23. *Define serous membranes. Differentiate between parietal and visceral serous membranes. What is the function of the serous membranes?*
24. *Name the serous membrane-lined cavities of the trunk.*
25. *What are mesenteries? Explain their function.*
26. *What are retroperitoneal organs? List five examples.*

P R E D I C T　❼
Explain how an organ can be located within the abdominopelvic cavity but not be within the peritoneal cavity.

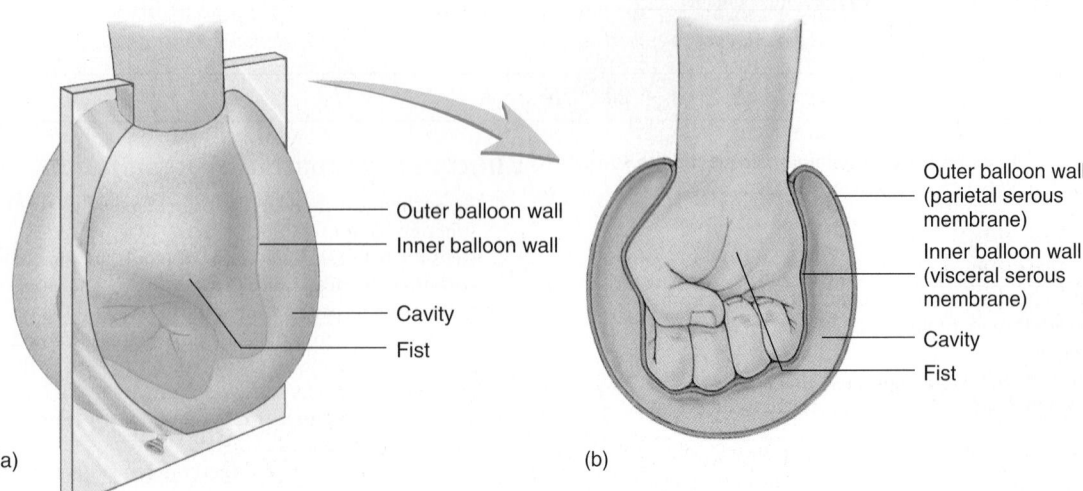

(a)　　　　　　　　　　　　　　　(b)

Figure 1.16　Serous Membranes

(*a*) Fist pushing into a balloon. A "glass" sheet indicates the location of a cross section through the balloon. (*b*) Interior view produced by the section in (*a*). The fist represents an organ, and the walls of the balloon the serous membranes. The inner wall of the balloon represents a visceral serous membrane in contact with the fist (organ). The outer wall of the balloon represents a parietal serous membrane.

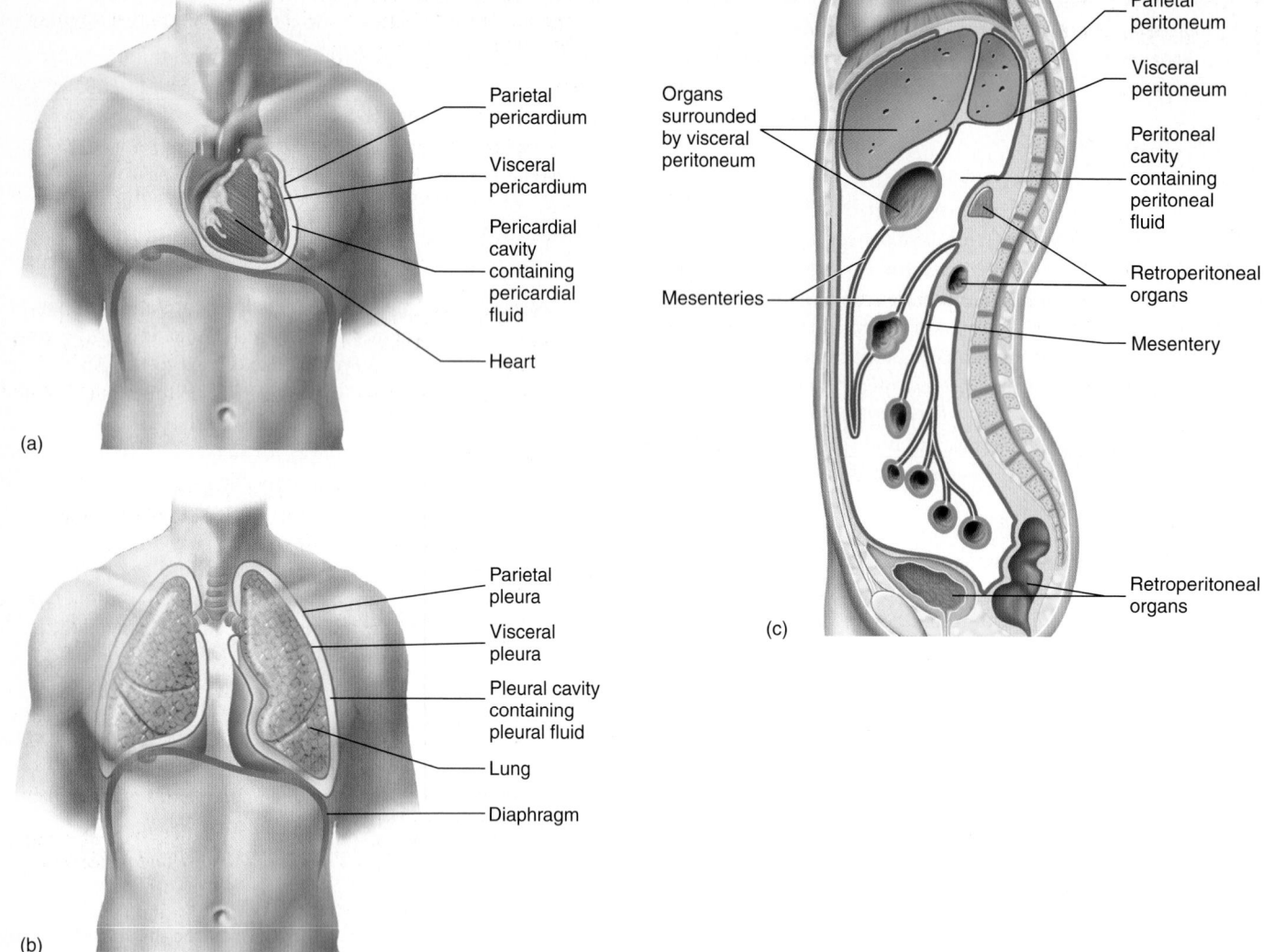

Figure 1.17   Location of Serous Membranes

(a) Frontal section showing the parietal pericardium (*blue*), visceral pericardium (*red*), and pericardial cavity. (b) Frontal section showing the parietal pleural (*blue*), visceral pleural (*red*), and pleural cavities. (c) Sagittal section through the abdominopelvic cavity showing the parietal peritoneum (*blue*), visceral peritoneum (*red*), peritoneal cavity, mesenteries (*purple*), and retroperitoneal organs.

S U M M A R Y

A functional knowledge of anatomy and physiology can be used to solve problems concerning the body when healthy or diseased.

Anatomy and Physiology   (p. 2)

1. Anatomy is the study of the body's structures.
   - Developmental anatomy considers anatomic changes from conception to adulthood. Embryology focuses on the first eight weeks of development.
   - Cytology examines cells, and histology examines tissues.
   - Gross anatomy emphasizes organs from a systemic or regional perspective.
2. Surface anatomy uses superficial structures to locate deeper structures, and anatomic imaging is a noninvasive technique for identifying deep structures.
3. Physiology is the study of the body's functions. It can be approached from a cellular or systems point of view.
4. Pathology deals with all aspects of disease. Exercise physiology examines changes caused by exercise.

Structural and Functional Organization   (p. 5)

1. Basic chemical characteristics are responsible for the structure and functions of life.
2. Cells are the basic living units of plants and animals and have many common characteristics. Organelles are small structures within cells that perform specific functions.
3. Tissues are groups of cells of similar structure and function and the materials surrounding them. The four primary tissue types are epithelial, connective, muscle, and nervous tissues.
4. Organs are structures composed of two or more tissues that perform specific functions.
5. Organs are arranged into the 11 organ systems of the human body (see figure 1.2).
6. Organ systems interact to form a whole, functioning organism.

The Human Organism   (p. 6)
Characteristics of Life

Humans have many characteristics such as organization, metabolism, responsiveness, growth, development, and reproduction in common with other organisms.

Biomedical Research

Much of what is known about humans is derived from research on other organisms.

Homeostasis   (p. 7)

Homeostasis is the condition in which body functions, fluids, and other factors of the internal environment are maintained at levels suitable to support life.

Negative Feedback

1. Negative-feedback mechanisms operate to maintain homeostasis.
2. Many negative-feedback mechanisms consist of a receptor, control center, and effector.

Positive Feedback

1. Positive-feedback mechanisms usually increase deviations from normal.
2. Although a few positive-feedback mechanisms normally exist in the body, most positive-feedback mechanisms are harmful.

Terminology and the Body Plan   (p. 13)
Body Positions

1. A human standing erect with the face directed forward, the arms hanging to the side, and the palms facing forward is in the anatomic position.
2. A person lying face upward is supine and face downward is prone.

Directional Terms

Directional terms always refer to the anatomic position, no matter what the actual position of the body (see table 1.1).

Body Parts and Regions

1. The body can be divided into the limbs, upper and lower, and a central region consisting of the head, neck, and trunk regions.

2. Superficially the abdomen can be divided into quadrants or nine regions. These divisions are useful for locating internal organs or describing the location of a pain or tumor.

Planes

1. Planes of the body
   - A sagittal plane divides the body into right and left parts. A median plane divides the body into equal right and left halves.
   - A transverse (horizontal) plane divides the body into superior and inferior portions.
   - A frontal (coronal) plane divides the body into anterior and posterior parts.
2. Sections of an organ
   - A longitudinal section of an organ divides it along the long axis.
   - A cross (transverse) section cuts at a right angle to the long axis of an organ.
   - An oblique section cuts across the long axis of an organ at an angle other than a right angle.

Body Cavities

1. The mediastinum subdivides the thoracic cavity.
2. The diaphragm separates the thoracic and abdominal cavities.
3. Pelvic bones surround the pelvic cavity.

Serous Membranes

1. Serous membranes line the trunk cavities. The parietal portion of a serous membrane lines the wall of the cavity, and the visceral portion is in contact with the internal organs.
   - The serous membranes secrete fluid that fills the space between the visceral and parietal membranes. The serous membranes protect organs from friction.
   - The pericardial cavity surrounds the heart, the pleural cavities surround the lungs, and the peritoneal cavity surrounds certain abdominal and pelvic organs.
2. Mesenteries are parts of the peritoneum that hold the abdominal organs in place and provide a passageway for blood vessels and nerves to the organs.
3. Retroperitoneal organs are located "behind" the parietal peritoneum.

R E V I E W   A N D   C O M P R E H E N S I O N

1. Physiology
   a. deals with the processes or functions of living things.
   b. is the scientific discipline that investigates the body's structures.
   c. is concerned with organisms and does not deal with different levels of organization, such as cells and systems.
   d. recognizes the static (as opposed to the dynamic) nature of living things.
   e. can be used to study the human body without considering anatomy.

2. Given the following conceptual levels for considering the body:
   1. cell
   2. chemical
   3. organ
   4. organ system
   5. organism
   6. tissue

   Choose the correct order for these conceptual levels, from smallest to largest.
   a. 1,2,3,6,4,5
   b. 2,1,6,3,4,5
   c. 3,1,6,4,5,2
   d. 4,6,1,3,5,2
   e. 1,6,5,3,4,2

For questions 3–8, match each organ system with its correct function.
   a. regulates other organ systems
   b. removes waste products from the blood; maintains water balance
   c. regulates temperature; prevents water loss; provides protection
   d. removes foreign substances from the blood; combats disease; maintains tissue fluid balance
   e. produces movement; maintains posture; produces body heat

3. endocrine system

4. integumentary system

5. lymphatic system

6. muscular system

7. nervous system

8. urinary system

9. The characteristic of life that is defined as "all the chemical reactions taking place in an organism" is
   a. development.
   b. growth.
   c. metabolism.
   d. organization.
   e. responsiveness.

10. Negative-feedback mechanisms
    a. make deviations from the set point smaller.
    b. maintain homeostasis.
    c. are associated with an increased sense of hunger the longer a person goes without eating.
    d. all of the above.

11. The following events are part of a negative-feedback mechanism.
    1. Blood pressure increases.
    2. Control center compares actual blood pressure to the blood pressure set point.
    3. The heart beats faster.
    4. Receptors detect a decrease in blood pressure.

    Choose the arrangement that lists the events in the order they occur.
    a. 1,2,3,4
    b. 1,3,2,4
    c. 3,1,4,2
    d. 4,2,3,1
    e. 4,3,2,1

12. Which of these statements concerning positive feedback is correct?
    a. Positive-feedback responses maintain homeostasis.
    b. Positive-feedback responses occur continuously in healthy individuals.
    c. Birth is an example of a normally occurring positive-feedback mechanism.
    d. When the cardiac muscle receives an inadequate supply of blood, positive-feedback mechanisms increase blood flow to the heart.
    e. Medical therapy seeks to overcome illness by aiding positive-feedback mechanisms.

13. The clavicle (collarbone) is _____ to the nipple of the breast.
    a. anterior
    b. distal
    c. superficial
    d. superior
    e. ventral

14. A term that means nearer the attached end of a limb is
    a. distal.
    b. lateral.
    c. medial.
    d. proximal.
    e. superficial.

15. Which of these directional terms are paired most appropriately as opposites?
    a. superficial and deep
    b. medial and proximal
    c. distal and lateral
    d. superior and posterior
    e. anterior and inferior

16. The part of the upper limb between the elbow and the wrist is called the
    a. arm.
    b. forearm.
    c. hand.
    d. inferior arm.
    e. lower arm.

17. A patient with appendicitis usually has pain in the _____ quadrant of the abdomen.
    a. left-lower
    b. right-lower
    c. left-upper
    d. right-upper

18. A plane that divides the body into anterior and posterior parts is a
    a. frontal (coronal) plane.
    b. sagittal plane.
    c. transverse plane.

19. The pelvic cavity contains the
    a. kidneys.
    b. liver.
    c. spleen.
    d. stomach.
    e. urinary bladder.

20. The lungs are
    a. part of the mediastinum.
    b. surrounded by the pericardial cavity.
    c. found within the thoracic cavity.
    d. separated from each other by the diaphragm.
    e. surrounded by mucous membranes.

21. Given these characteristics:
    1. reduce friction between organs
    2. line fluid-filled cavities
    3. line trunk cavities that open to the exterior of the body

    Which of the characteristics describe serous membranes?
    a. 1,2
    b. 1,3
    c. 2,3
    d. 1,2,3

22. Given these organ and cavity combinations:
    1. heart and pericardial cavity
    2. lungs and pleural cavity
    3. stomach and peritoneal cavity
    4. kidney and peritoneal cavity

    Which of the organs is correctly paired with a space that surrounds that organ?
    a. 1,2
    b. 1,2,3
    c. 1,2,4
    d. 2,3,4
    e. 1,2,3,4

23. Which of these membrane combinations are found on the surface of the diaphragm?
    a. parietal pleura—parietal peritoneum
    b. parietal pleura—visceral peritoneum
    c. visceral pleura—parietal peritoneum
    d. visceral pleura—visceral peritoneum

24. Mesenteries
    a. are found in the pleural, pericardial, and abdominopelvic cavities.
    b. consist of two layers of peritoneum fused together.
    c. anchor organs such as the kidneys and urinary bladder to the body wall.
    d. are found primarily in body cavities that open to the outside.
    e. all of the above.

25. Which of the following organs is *not* retroperitoneal?
    a. adrenal glands
    b. urinary bladder
    c. kidneys
    d. pancreas
    e. stomach

Answers in Appendix F

C R I T I C A L   T H I N K I N G

1. Exposure to a hot environment causes the body to sweat. The hotter the environment, the greater the sweating. Two anatomy and physiology students are arguing about the mechanisms involved: Student A claims that they are positive feedback, and student B claims they are negative feedback. Do you agree with student A or student B and why?

2. A male has lost blood as a result of a gunshot wound. Even though bleeding has been stopped, his blood pressure is low and dropping and his heart rate is elevated. Following a blood transfusion, his blood pressure increases and his heart rate decreases. Which of the following statement(s) is (are) consistent with these observations?
   a. Negative-feedback mechanisms can be inadequate without medical intervention.
   b. The transfusion interrupted a positive-feedback mechanism.
   c. The increased heart rate after the gunshot wound and before the transfusion is a result of a positive-feedback mechanism.
   d. a and b
   e. a, b, and c

3. Provide the correct directional term for the following statement: When a boy is standing on his head, his nose is _____ to his mouth.

4. Complete the following statements, using the correct directional terms for a human being. Note that more than one term can apply.
   a. The navel is _____ to the nose.
   b. The nipple is _____ to the lung.
   c. The arm is _____ to the forearm.
   d. The little finger is _____ to the index finger.

5. In which quadrants and regions are the pancreas and the urinary bladder located?

6. Given the following procedures:
   1. Make an opening into the mediastinum.
   2. Lay the patient supine.
   3. Lay the patient prone.
   4. Make an incision through the pericardial serous membranes.
   5. Make an opening into the abdomen.

   Which of the procedures should be accomplished to expose the anterior surface of a patient's heart?
   a. 2,1,4
   b. 2,5,4
   c. 3,1,4
   d. 3,5,4

7. During pregnancy, which of the mother's body cavities increases most in size?

8. A bullet enters the left side of a man, passes through the left lung, and lodges in the heart. Name in order the serous membranes and their cavities through which the bullet passes.

9. A woman falls while skiing and accidentally is impaled by her ski pole. The pole passes through the abdominal body wall and into and through the stomach, pierces the diaphragm, and finally stops in the left lung. List in order the serous membranes the pole pierces.

Answers in Appendix G

A N S W E R S   T O   P R E D I C T   Q U E S T I O N S

1. The chemical level is the level at which correction is currently being accomplished. Insulin can be purchased and injected into the circulation to replace the insulin normally produced by the pancreas. Therapy at the cellular level includes drugs that stimulate pancreatic cells to increase insulin production or make cells more responsive to insulin. Transplantation of pancreatic cells, pancreatic tissue, and even the entire pancreas are also possibilities.

2. Negative-feedback mechanisms work to control respiratory rates so that body cells have adequate oxygen and are able to eliminate carbon dioxide. The greater the respiratory rate, the greater the exchange of gases between the body and the air. When a person is at rest, there is less of a demand for oxygen, and less carbon dioxide is produced than during exercise. At rest, homeostasis can be maintained with a low respiration rate. During exercise there is a greater demand for oxygen, and more carbon dioxide must be eliminated. Consequently, to maintain homeostasis during exercise, the respiratory rate increases.

3. The sensation of thirst is involved in a negative-feedback mechanism that maintains body fluids. The sensation of thirst increases with a decrease in body fluids. The thirst mechanism causes a person to drink fluids, which returns body fluid levels to normal, thereby maintaining homeostasis.

4. In the cat, cephalic and anterior are toward the head; dorsal and superior are toward the back. In humans, cephalic and superior are toward the head; dorsal and posterior are toward the back.

5. Your kneecap is both proximal and superior to the heel. It is also anterior to the heel because it is on the anterior side of the lower limb, whereas the heel is on the posterior side.

6. The spleen is in the left-upper quadrant, the gallbladder is in the right-upper quadrant, the left kidney is in the left-upper quadrant, the right kidney is in the right-upper quadrant, the stomach is mostly in the left-upper quadrant, and the liver is mostly in the right-upper quadrant.

7. There are two ways in which an organ can be located within the abdominopelvic cavity but not be within the peritoneal cavity. First, the visceral peritoneum wraps around organs. Thus the peritoneal cavity surrounds the organ, but the organ is not inside the peritoneal cavity. The peritoneal cavity contains only peritoneal fluid. Second, retroperitoneal organs are in the abdominopelvic cavity, but they are between the wall of the abdominopelvic cavity and the parietal peritoneal membrane.

Visit the Online Learning Center at www.mhhe.com/seeley7 for chapter quizzes, interactive learning exercises, and other study tools.

The Chemical Basis of Life

The chemical composition of the body's structures determines their function. This colorized scanning electron micrograph shows bundles of collagen fibers (brown) and elastic fibers (blue). The molecules forming collagen fibers are like tiny ropes, and collagen fibers make tissues strong yet flexible. The molecules forming elastic fibers resemble microscopic coiled springs. The elastic fibers allow tissues to be stretched and then recoil back to their original shape.

Chemicals compose the structures of the body, and the interactions of chemicals with one another are responsible for the functions of the body. Nerve impulse generation, digestion, muscle contraction, and metabolism can be described in chemical terms. Many abnormal conditions and illnesses, as well as their treatments, can also be explained in chemical terms. For example, Parkinson's disease, which causes uncontrollable shaking movements, results from a shortage of a chemical called dopamine in certain nerve cells in the brain. It can be treated by giving patients another chemical that is converted to dopamine by brain cells.

To understand anatomy and physiology, it is essential to have a basic knowledge of chemistry—the scientific discipline concerned with the atomic composition and structure of substances and the reactions they undergo. This chapter outlines *basic chemistry* (p. 26), *chemical reactions and energy* (p. 33), *inorganic chemistry* (p. 38), and *organic chemistry* (p. 41). It is not a comprehensive review of chemistry, but it does review some of the basic concepts. When necessary, refer back to this chapter when chemical phenomena are discussed later in the text.

Basic Chemistry

Matter, Mass, and Weight

All living and nonliving things are composed of **matter,** which is anything that occupies space and has mass. **Mass** is the amount of matter in an object, and **weight** is the gravitational force acting on an object of a given mass. For example, the weight of an apple results from the force of gravity "pulling" on the apple's mass.

P R E D I C T

The difference between mass and weight can be illustrated by considering an astronaut. How does an astronaut's mass and weight in outer space compare to the astronaut's mass and weight on the earth's surface?

The **kilogram (kg),** which is the mass of a platinum–iridium cylinder kept at the International Bureau of Weights and Measurements in France, is the international unit for mass. The mass of all other objects is compared to this cylinder. For example, a 2.2-pound lead weight or 1 liter (L) (1.06 qt) of water each has a mass of approximately 1 kg. An object with 1/1000 the mass of a kilogram is defined as having a mass of 1 **gram (g).**

Chemists use a balance to determine the mass of objects. Although we commonly refer to "weighing" an object on a balance, we are actually "massing" the object because the balance compares objects of unknown mass to objects of known mass. When the unknown and known masses are exactly balanced, the gravitational pull of the earth on both of them is the same. Thus, the effect of gravity on the unknown mass is counteracted by the effect of gravity on the known mass. A balance produces the same results on a mountaintop as at sea level because it does not matter if the gravitational pull is strong or weak. It only matters that the effect of gravity on both the unknown and known masses is the same.

Elements and Atoms

An **element** is the simplest type of matter with unique chemical properties. A list of the elements commonly found in the human body is given in table 2.1. About 96% of the weight of the body results from the elements oxygen, carbon, hydrogen, and nitrogen.

An **atom** (*atomos*, indivisible) is the smallest particle of an element that has the chemical characteristics of that element. An element is composed of atoms of only one kind. For example, the element carbon is composed of only carbon atoms, and the element oxygen is composed of only oxygen atoms.

An element, or an atom of that element, often is represented by a symbol. Usually the first letter or letters of the element's name

Table 2.1	Common Elements					
Element	Symbol	Atomic Number	Mass Number	Atomic Mass	Percent in Human Body by Weight (%)	Percent in Human Body by Number of Atoms (%)
Hydrogen	H	1	1	1.008	9.5	63.0
Carbon	C	6	12	12.01	18.5	9.5
Nitrogen	N	7	14	14.01	3.3	1.4
Oxygen	O	8	16	16.00	65.0	25.5
Fluorine	F	9	19	19.00	Trace	Trace
Sodium	Na	11	23	22.99	0.2	0.3
Magnesium	Mg	12	24	24.31	0.1	0.1
Phosphorus	P	15	31	30.97	1.0	0.22
Sulfur	S	16	32	32.07	0.3	0.05
Chlorine	Cl	17	35	35.45	0.2	0.03
Potassium	K	19	39	39.10	0.4	0.06
Calcium	Ca	20	40	40.08	1.5	0.31
Chromium	Cr	24	52	52.00	Trace	Trace
Manganese	Mn	25	55	54.94	Trace	Trace
Iron	Fe	26	56	55.85	Trace	Trace
Cobalt	Co	27	59	58.93	Trace	Trace
Copper	Cu	29	63	63.55	Trace	Trace
Zinc	Zn	30	64	65.39	Trace	Trace
Selenium	Se	34	80	78.96	Trace	Trace
Molybdenum	Mo	42	98	95.94	Trace	Trace
Iodine	I	53	127	126.9	Trace	Trace

are used—for example, C for carbon, H for hydrogen, Ca for calcium, and Cl for chlorine. Occasionally the symbol is taken from the Latin, Greek, or Arabic name for the element—for example, Na from the Latin word *natrium* is the symbol for sodium.

Atomic Structure

The characteristics of living and nonliving matter result from the structure, organization, and behavior of atoms (figure 2.1). Atoms are composed of subatomic particles, some of which have an electric charge. The three major types of subatomic particles are neutrons, protons, and electrons. A **neutron** has no electric charge, a **proton** has one positive charge, and an **electron** has one negative charge. The positive charge of a proton is equal in magnitude to the negative charge of an electron. The number of protons and electrons in each atom is equal, and the individual charges cancel each other. Therefore, each atom is electrically neutral.

Protons and neutrons form the **nucleus,** and electrons are moving around the nucleus (see figure 2.1). The nucleus accounts for 99.97% of an atom's mass but only 1 ten-trillionth of its volume. Most of the volume of an atom is occupied by the electrons. Although it is impossible to know precisely where any given electron is located at any particular moment, the region where it is most likely to be found can be represented by an **electron cloud.**

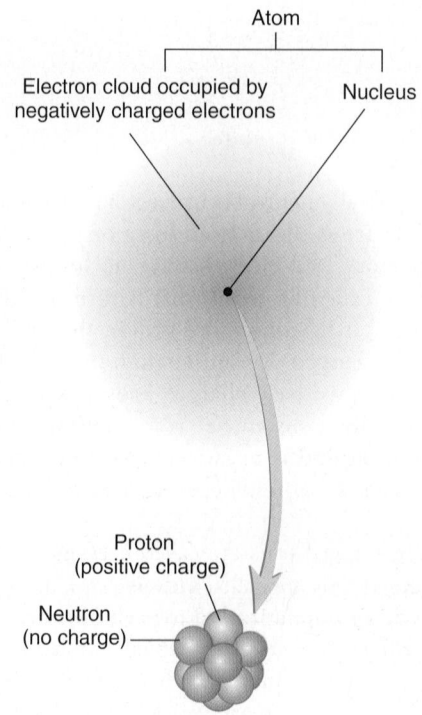

Figure 2.1 **Model of an Atom**

The tiny, dense nucleus consists of positively charged protons and uncharged neutrons. Most of the volume of an atom is occupied by rapidly moving, negatively charged electrons, which can be represented as an electron cloud. The probable location of an electron is indicated by the color of the electron cloud. The darker the color in each small part of the electron cloud, the more likely the electron is located there.

The likelihood of locating an electron at a specific point in a region correlates with the darkness of that region in the diagram. The darker the color, the greater the likelihood of finding the electron there at any given moment.

Atomic Number and Mass Number

The **atomic number** of an element is equal to the number of protons in each atom, and because the number of electrons and protons is equal, the atomic number is also the number of electrons. Each element is uniquely defined by the number of protons in the atoms of that element. For example, only hydrogen atoms have one proton, only carbon atoms have six protons, and only oxygen atoms have eight protons (figure 2.2; see table 2.1).

Scientists have been able to create new elements by changing the number of protons in the nuclei of existing elements. Protons, neutrons, or electrons from one atom are accelerated to very high speeds and then smashed into the nucleus of another atom. The resulting changes in the nucleus produce a new element with a new atomic number. To date, 26 elements with an atomic number greater than 92 have been synthesized in this fashion. These artificially produced elements are usually unstable, and they quickly convert back to more stable elements.

Protons and neutrons have about the same mass, and they are responsible for most of the mass of atoms. Electrons, on the other hand, have very little mass. The **mass number** of an element is the number of protons plus the number of neutrons in each atom. For example, the mass number for carbon is 12 because it has six protons and six neutrons.

P R E D I C T

The atomic number of potassium is 19, and the mass number is 39. What is the number of protons, neutrons, and electrons in an atom of potassium?

Isotopes and Atomic Mass

Isotopes (ī′sō-tōpz) are two or more forms of the same element that have the same number of protons and electrons but a different number of neutrons. Thus isotopes have the *same* atomic number but *different* mass numbers. There are three isotopes of hydrogen: hydrogen, deuterium, and tritium. All three isotopes have one proton and one electron, but hydrogen has no neutrons in its nucleus, deuterium has one neutron, and tritium has two neutrons (figure 2.3). Isotopes can be denoted using the symbol of the element preceded by the mass number (number of protons and neutrons) of the isotope. Thus hydrogen is ^{1}H, deuterium is ^{2}H, and tritium is ^{3}H.

Individual atoms have very little mass. A hydrogen atom has a mass of 1.67×10^{-24} g (see appendix B for an explanation of the scientific notation of numbers). To avoid using such small numbers, a system of relative atomic mass is used. In this system, a **unified atomic mass unit (u),** or **dalton (D),** is 1/12 the mass of ^{12}C, a carbon atom with six protons and six neutrons. Thus ^{12}C has an atomic mass of exactly 12 u. A naturally occurring sample of carbon, however, contains mostly ^{12}C but also a small quantity of other carbon isotopes such as ^{13}C, which has six protons and seven neutrons. The

Figure 2.2 **Hydrogen, Carbon, and Oxygen Atoms**

Within the nucleus, the number of positively charged protons (p^+) and uncharged neutrons (n^0) is indicated. The negatively charged electrons (e^-) are around the nucleus. Atoms are electrically neutral because the number of protons and electrons within an atom are equal.

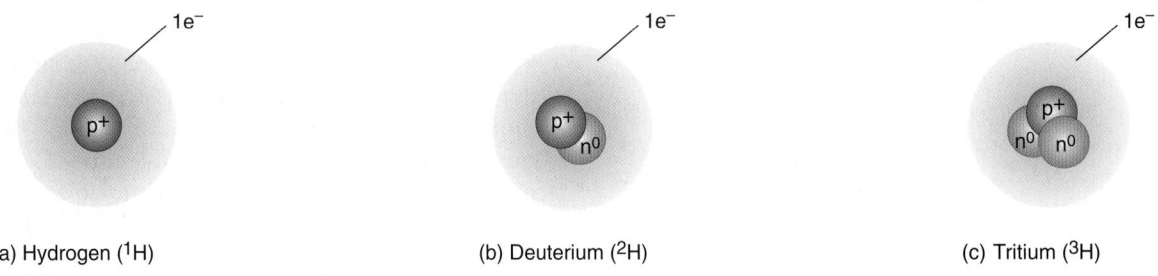

(a) Hydrogen (^1H)   (b) Deuterium (^2H)   (c) Tritium (^3H)

Figure 2.3 **Isotopes of Hydrogen**

(*a*) Hydrogen has one proton and no neutrons in its nucleus. (*b*) Deuterium has one proton and one neutron in its nucleus. (*c*) Tritium has one proton and two neutrons in its nucleus.

atomic mass of an element is the *average* mass of its naturally oc-curring isotopes, taking into account the relative abundance of each isotope. For example, the atomic mass of the element carbon is 12.01 u (see table 2.1), which is slightly more than 12 u because of the additional mass of the small amount of other carbon isotopes. Because the atomic mass is an average, a sample of carbon can be treated as if all the carbon atoms have an atomic mass of 12.01 u.

1. *Define matter. How is the mass and the matter of an object different?*
2. *Define element and atom. What four elements are found in the greatest abundance in humans?*
3. *For each subatomic particle of an atom, state its charge and location. Which subatomic particles are most responsible for the mass and volume of an atom? Which subatomic particles determine atomic number and mass number?*
4. *Define isotopes and give an example. Define atomic mass. Why is the atomic mass of most elements not exactly equal to the mass number?*

Electrons and Chemical Bonding

The outermost electrons of an atom determine its chemical behav-ior. When these outermost electrons are transferred or shared be-tween atoms, **chemical bonding** occurs. Two major types of chemical bonding are ionic and covalent bonding.

Ionic Bonding

An atom is electrically neutral because it has an equal number of protons and electrons. If an atom loses or gains electrons, the number of protons and electrons are no longer equal, and a charged particle called an **ion** (ī'on) is formed. After an atom loses an electron, it has one more proton than it has electrons and is positively charged. A sodium atom (Na) can lose an elec-tron to become a positively charged sodium ion (Na^+) (figure 2.4*a*). After an atom gains an electron, it has one more electron than it has protons and is negatively charged. A chlorine atom (Cl) can accept an electron to become a negatively charged chlo-ride ion (Cl^-).

Positively charged ions are called **cations** (kat'ī-onz), and negatively charged ions are called **anions** (an'ī-onz). Because op-positely charged ions are attracted to each other, cations and an-ions tend to remain close together, which is called **ionic** (ī-on'ik) **bonding.** For example, sodium and chloride ions are held together by ionic bonding to form an array of ions called sodium chloride, or table salt (see figure 2.4*b* and *c*). Some ions commonly found in the body are listed in table 2.2.

Covalent Bonding

Covalent bonding results when atoms share one or more pairs of electrons. The resulting combination of atoms is called a molecule. An example is the covalent bond between two hydrogen atoms to

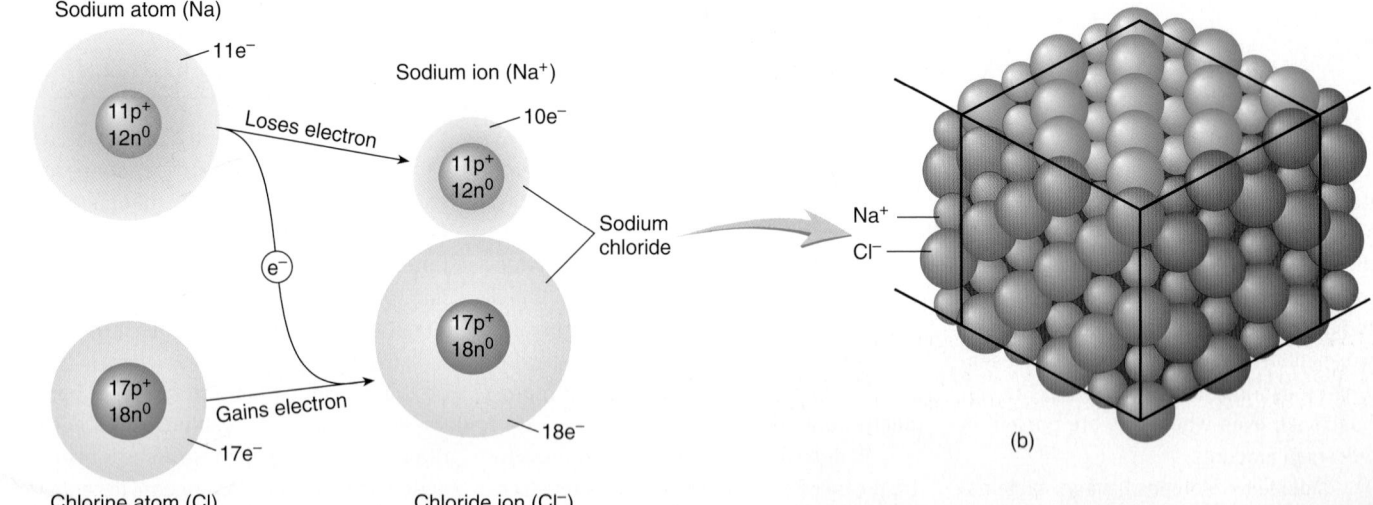

Figure 2.4   Ionic Bonding
(*a*) A sodium atom loses an electron to become a smaller-sized positively charged ion, and a chlorine atom gains an electron to become a larger-sized negatively charged ion. The attraction between the oppositely charged ions results in ionic bonding and the formation of sodium chloride. (*b*) The sodium and chlorine ions are organized to form a cube-shaped array. (*c*) Microphotograph of salt crystals reflects the cubic arrangement of the ions.

Table 2.2	Important Ions	
Common Ions	**Symbols**	**Significance**
Calcium	Ca^{2+}	Part of bones and teeth, blood clotting, muscle contraction, release of neurotransmitters
Sodium	Na^+	Membrane potentials, water balance
Potassium	K^+	Membrane potentials
Hydrogen	H^+	Acid–base balance
Hydroxide	OH^-	Acid–base balance
Chloride	Cl^-	Water balance
Bicarbonate	HCO_3^-	Acid–base balance
Ammonium	NH_4^+	Acid–base balance
Phosphate	PO_4^{3-}	Part of bones and teeth, energy exchange, acid–base balance
Iron	Fe^{2+}	Red blood cell formation
Magnesium	Mg^{2+}	Necessary for enzymes
Iodide	I^-	Present in thyroid hormones

form a hydrogen molecule (figure 2.5). Each hydrogen atom has one electron. As the two hydrogen atoms get closer together, the positively charged nucleus of each atom begins to attract the electron of the other atom. At an optimal distance, the two nuclei mutually attract the two electrons, and each electron is shared by both nuclei. The two hydrogen atoms are now held together by a covalent bond.

When an electron pair is shared between two atoms, a **single covalent bond** results. A single covalent bond is represented by a single line between the symbols of the atoms involved (e.g., H—H). A **double covalent bond** results when two atoms share four electrons, two from each atom. When a carbon atom combines with two oxygen atoms to form carbon dioxide, two double covalent bonds are formed. Double covalent bonds are indicated by a double line between the atoms (O=C=O).

When electrons are shared equally between atoms, as in a hydrogen molecule, the bonds are called **nonpolar covalent bonds.** Atoms bound to one another by a covalent bond do not always share their electrons equally, however, because the nucleus of one atom attracts the electrons more strongly than does the nucleus of the other atom. Bonds of this type are called **polar covalent bonds** and are common in both living and nonliving matter.

Clinical Focus   Radioactive Isotopes and X-Rays

Protons, neutrons, and electrons are responsible for the chemical properties of atoms. They also have other properties that can be useful in a clinical setting. For example, they have been used to develop methods for examining the inside of the body. **Radioactive isotopes** are commonly used by clinicians and researchers because sensitive measuring devices can detect their radioactivity, even when they are present in very small amounts.

Radioactive isotopes have unstable nuclei that spontaneously change to form more stable nuclei. As a result, either new isotopes or new elements are produced. In this process of nuclear change, alpha particles, beta particles, and gamma rays are emitted from the nuclei of radioactive isotopes. Alpha (α) particles are positively charged helium ions (He^{2+}), which consist of two protons and two neutrons. Beta (β) particles are electrons formed as neutrons change into protons. An electron is ejected from the neutron, and the proton that is produced remains in the nucleus. Gamma (γ) rays are a form of electromagnetic radiation (high-energy photons) released from nuclei as they lose energy.

All isotopes of an element have the same atomic number, and their chemical behavior is very similar. For example, ^3H (tritium) can substitute for ^1H (hydrogen), and either ^{125}iodine or ^{131}iodine can substitute for ^{126}iodine in chemical reactions.

Several procedures that are used to determine the concentration of substances such as hormones depend on the incorporation of small amounts of radioactive isotopes, such as ^{125}iodine, into the substances being measured. Clinicians using these procedures can more accurately diagnose disorders of the thyroid gland, the adrenal gland, and the reproductive organs.

Radioactive isotopes are also used to treat cancer. Some of the particles released from isotopes have a very high energy content and can penetrate and destroy tissues. Thus radioactive isotopes can be used to destroy tumors because rapidly growing tissues such as tumors are more sensitive to radiation than healthy cells. Radiation can also be used to sterilize materials that cannot be exposed to high temperatures (e.g., some fabric and plastic items used during surgical procedures). In addition, radioactive emissions can be used to sterilize food and other items.

X-rays are electromagnetic radiations with a much shorter wavelength than visible light. When electric current is used to heat a filament to very high temperatures, energy of the electrons becomes so great that some electrons are emitted from the hot filament. When these electrons strike a positive electrode at high speeds, they release some of their energy in the form of x-rays.

X-rays do not penetrate dense material as readily as they penetrate less dense material, and x-rays can expose photographic film. Consequently, an x-ray beam can pass through a person and onto photographic film. Dense tissues of the body absorb the x-rays, and in these areas the film is underexposed and so appears white or light in color on the developed film. On the other hand, the x-rays readily pass through less dense tissue, and the film in these areas is overexposed and appears black or dark in color. In an x-ray film of the skeletal system the dense bones are white, and the less dense soft tissues are dark, often so dark that no details can be seen. Because the dense bone material is clearly visible, x-rays can be used to determine whether bones are broken or have other abnormalities.

Soft tissues can be photographed by using low-energy x-rays. Mammograms are low-energy x-rays of the breast that can be used to detect tumors, because tumors are slightly denser than normal tissue.

Radiopaque substances are dense materials that absorb x-rays. If a radiopaque liquid is given to a patient, the liquid assumes the shape of the organ into which it is placed. For example, if a barium solution is swallowed, the outline of the upper digestive tract can be photographed using x-rays to detect such abnormalities as ulcers.

Polar covalent bonds can result in polar molecules, which are electrically asymmetric. For example, oxygen atoms attract electrons more strongly than do hydrogen atoms. When covalent bonding between an oxygen atom and two hydrogen atoms forms a water molecule, the electrons are located in the vicinity of the oxygen nucleus more than in the vicinity of the hydrogen nuclei. Because electrons have a negative charge, the oxygen side of the molecule is slightly more negative than the hydrogen side (figure 2.6).

Molecules and Compounds

A **molecule** is formed when two or more atoms chemically combine to form a structure that behaves as an independent unit. The atoms that combine to form a molecule can be of the same type, such as two hydrogen atoms combining to form a hydrogen molecule. More typically, a molecule consists of two or more different types of atoms, such as two hydrogen atoms and an oxygen atom forming water. Thus, a glass of water consists of a collection of individual water molecules positioned next to one another.

A **compound** is a substance composed of two or more *different* types of atoms that are chemically combined. Not all molecules are compounds. For example, a hydrogen molecule is not a compound because it does not consist of different types of atoms. Many molecules are compounds, however. Most covalent substances consist of molecules because their atoms form distinct units as a result of the joining of the atoms to each other by a pair of shared electrons. For example, a water molecule is a covalent compound.

On the other hand, ionic compounds are not molecules because the ions are held together by the force of attraction between

No interaction between the two hydrogen atoms because they are too far apart.

The positively charged nucleus of each hydrogen atom begins to attract the electron of the other.

A covalent bond is formed when the electrons are shared between the nuclei because the electrons are equally attracted to each nucleus.

Figure 2.5 Covalent Bonding

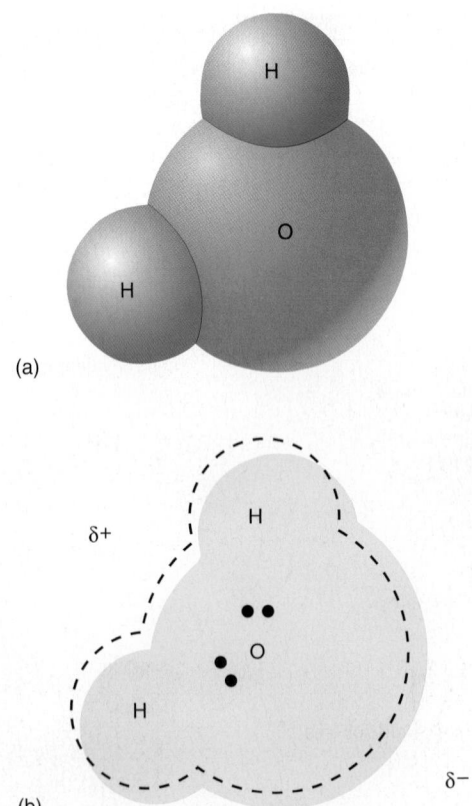

(a)

(b)

Figure 2.6 Polar Covalent Bonds

(*a*) A water molecule forms when two hydrogen atoms form covalent bonds with an oxygen atom. (*b*) Electron pairs (*indicated by dots*) are shared between the hydrogen atoms and oxygen. The dashed outline shows the expected location of the electron cloud if the electrons are shared equally. The electrons are shared unequally, as shown by the electron cloud (*yellow*) not coinciding with the dashed outline. Consequently, the oxygen side of the molecule has a slight negative charge (*indicated by δ^-*) and the hydrogen side of the molecule has a slight positive charge (*indicated by δ^+*).

opposite charges. A piece of sodium chloride does not consist of sodium chloride molecules positioned next to each other. Instead, table salt is an organized array of Na^+ and Cl^- in which each charged ion is surrounded by several ions of the opposite charge (see figure 2.4*b*). Sodium chloride is an example of a substance that is a compound but is not a molecule.

The kinds and numbers of atoms (or ions) in a molecule or compound can be represented by a formula consisting of the symbols of the atoms (or ions) plus subscripts denoting the number of each type of atom (or ion). The formula for glucose (a sugar) is $C_6H_{12}O_6$, indicating that glucose has 6 carbon, 12 hydrogen, and 6 oxygen atoms (table 2.3).

The **molecular mass** of a molecule or compound can be determined by adding up the atomic masses of its atoms (or ions). The term *molecular mass* is used for convenience for ionic compounds, even though they are not molecules. For example, the atomic mass of sodium is 22.99 and chloride is 35.45. The molecular mass of NaCl is therefore 58.44 (22.99 + 35.45).

5. *Describe how ionic bonding occurs. What is a cation and an anion?*
6. *Describe how covalent bonding occurs. What is the difference between polar and nonpolar covalent bonds?*

7. *Distinguish between a molecule and a compound. Are all molecules compounds? Are all compounds molecules?*
8. *Define molecular mass.*

P R E D I C T ❸
What is the molecular mass of a molecule of glucose? (Use table 2.1.)

Intermolecular Forces

Intermolecular forces result from the weak electrostatic attractions between the oppositely charged parts of molecules, or between ions and molecules. Intermolecular forces are much weaker than the forces producing chemical bonding.

Hydrogen Bonds

Molecules with polar covalent bonds have positive and negative "ends." Intermolecular force results from the attraction of the positive end of one polar molecule to the negative end of another polar molecule. When hydrogen forms a covalent bond with

Table 2.3	Picturing Molecules		
Representation	**Hydrogen**	**Carbon Dioxide**	**Glucose**
Chemical Formula Shows the kind and number of atoms present.	H_2	CO_2	$C_6H_{12}O_6$
Electron-Dot Formula The bonding electrons are shown as dots between the symbols of the atoms.	H:H Single covalent bond	O::C::O Double covalent bond	Not used for complex molecules
Bond-Line Formula The bonding electrons are shown as lines between the symbols of the atoms.	H—H Single covalent bond	O=C=O Double covalent bond	
Models Atoms are shown as different-sized and different-colored spheres.	Hydrogen atom	Oxygen atom Carbon atom	

oxygen, nitrogen, or fluorine, the resulting molecule becomes very polarized. If the positively charged hydrogen of one molecule is attracted to the negatively charged oxygen, nitrogen, or fluorine of another molecule, a **hydrogen bond** is formed. For example, the positively charged hydrogen atoms of a water molecule form hydrogen bonds with the negatively charged oxygen atoms of other water molecules (figure 2.7).

Hydrogen bonds play an important role in determining the shape of complex molecules because the hydrogen bonds between different polar parts of the molecule hold the molecule in its normal three-dimensional shape (see the sections "Proteins" p. 46, and "Nucleic Acids: DNA and RNA" p. 48).

Table 2.4 summarizes the important characteristics of chemical bonding (ionic and covalent) and intermolecular forces (hydrogen bonds).

Solubility and Dissociation

Solubility is the ability of one substance to dissolve in another, for example, when sugar dissolves in water. Charged substances such as sodium chloride, and polar substances such as glucose, dissolve

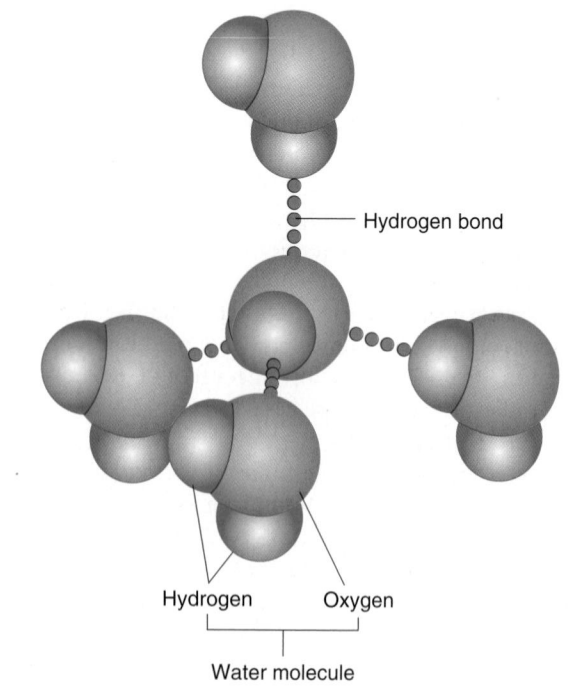

Hydrogen bond

Hydrogen   Oxygen

Water molecule

Figure 2.7 **Hydrogen Bonds**

The positive hydrogen part of one water molecule forms a hydrogen bond (*red dotted line*) with the negative oxygen part of another water molecule. As a result, hydrogen bonds hold the water molecules together.

Table 2.4	Comparison of Bonds	
Definition	**Charge Distribution**	**Example**
Ionic Bond Complete transfer of electrons between two atoms	Separate positively charged and negatively charged ions	Na^+Cl^- Sodium chloride
Polar Covalent Bond Unequal sharing of electrons between two atoms	Slight positive charge (δ^+) on one side of the molecule and slight negative charge (δ^-) on the other side of the molecule	H | δ^+ O δ^- | H Water
Nonpolar Covalent Bond Equal sharing of electrons between two atoms	Charge evenly distributed among the atoms of the molecule	H | H—C—H | H Methane
Hydrogen Bond Attraction of oppositely charged ends of one polar molecule to another polar molecule	Charge distribution within the polar molecules results from polar covalent bonds	H | O---H—O / \ H H Water molecules

in water readily, whereas nonpolar substances such as oils do not. We all have seen how oil floats on water. Substances dissolve in water when they become surrounded by water molecules. If the positive and negative ends of the water molecules are attracted more to the charged ends of other molecules than they are to each other, the hydrogen bonds between the ends of the water molecules are broken, and the water molecules surround the other molecules, which become dissolved in the water.

When ionic compounds dissolve in water, their ions **dissociate,** or separate, from one another because the cations are attracted to the negative ends of the water molecules, and the anions are attracted to the positive ends of the water molecules. When sodium chloride dissociates in water, the sodium and chloride ions separate, and water molecules surround and isolate the ions, thereby keeping them in solution (figure 2.8).

When molecules (covalent compounds) dissolve in water, they usually remain intact even though they are surrounded by water molecules. Thus, in a glucose solution, glucose molecules are surrounded by water molecules.

Cations and anions that dissociate in water are sometimes called **electrolytes** (ē-lek′trō-lītz) because they have the capacity to conduct an electric current, which is the flow of charged particles. An electrocardiogram (ECG) is a recording of electric currents produced by the heart. These currents can be detected by electrodes on the surface of the body because the ions in the body fluids conduct electric currents. Molecules that do not dissociate form solutions that do not conduct electricity and are called **nonelectrolytes.** Pure water is a nonelectrolyte.

9. *Define hydrogen bond, and explain how hydrogen bonds hold polar molecules, such as water, together. How do hydrogen bonds affect the shape of a molecule?*

10. *Define solubility. How do ionic and covalent compounds typically dissolve in water?*

11. *Distinguish between electrolytes and nonelectrolytes.*

Chemical Reactions and Energy

In a **chemical reaction,** atoms, ions, molecules, or compounds interact either to form or to break chemical bonds. The substances that enter into a chemical reaction are called **reactants,** and the substances that result from the chemical reaction are called **products.**

For our purposes, three important points can be made about chemical reactions. First, in some reactions, less complex reactants are combined to form a larger, more complex product. An example is the synthesis of the complex molecules of the human body from basic "building blocks" obtained in food (figure 2.9a). Second, in other reactions, a reactant can be broken down, or decomposed, into simpler, less complex products. An example is the breakdown of food molecules into basic building blocks (figure 2.9b). Third, atoms are generally associated with other atoms through chemical bonding or intermolecular forces; therefore, to synthesize new products or break down reactants it is necessary to change the relationship between atoms.

Synthesis Reactions

When two or more reactants chemically combine to form a new and larger product, the process is called a **synthesis reaction.** An example of a synthesis reaction is the combination of two amino

Figure 2.8   Dissociation

Sodium chloride (table salt) dissociating in water. The positively charged Na$^+$ are attracted to the negatively charged oxygen (*red*) end of the water molecule, and the negatively charged Cl$^-$ are attracted to the positively charged hydrogen (*blue*) end of the water molecule.

Figure 2.9   Synthesis and Decomposition Reactions

(*a*) Synthesis reaction in which amino acids, the basic "building blocks" of proteins, combine to form a protein molecule. (*b*) Decomposition reaction in which a complex carbohydrate breaks down into smaller glucose molecules, which are the "building blocks" of carbohydrates.

acids to form a dipeptide (figure 2.10*a*). In this particular synthesis reaction, water is removed from the amino acids as they are bound together. Synthesis reactions in which water is a product are called **dehydration** (water out) **reactions.** Note that old chemical bonds are broken and new chemical bonds are formed as the atoms rearrange as a result of a synthesis reaction.

Another example of a synthesis reaction in the body is the formation of adenosine triphosphate (ATP). In ATP, *A* stands for adenosine, *T* stands for tri- or three, and *P* stands for phosphate group (PO_4^{3-}). Thus, ATP consists of adenosine and three phosphate groups (see p. 52 for the details of the structure of ATP). ATP is synthesized from adenosine diphosphate (ADP), which has two phosphate groups, and an inorganic phosphate ($H_2PO_4^-$), which is often symbolized as P_i.

$$\begin{array}{ccccc}
\text{A-P-P} & + & P_i & \rightarrow & \text{A-P-P-P} \\
(\text{ADP}) & & (\text{Inorganic} & & (\text{ATP}) \\
& & \text{phosphate}) & &
\end{array}$$

Synthesis reactions produce the molecules characteristic of life, such as ATP, proteins, carbohydrates, lipids, and nucleic acids. All of the synthesis reactions that occur within the body are referred to collectively as **anabolism** (ă-nab′ō-lizm). The growth, maintenance, and repair of the body could not take place without anabolic reactions.

Figure 2.10 Synthesis (Dehydration) and Decomposition (Hydrolysis) Reactions

(*a*) Synthesis reaction in which two amino acids combine to form a dipeptide. This reaction is also a dehydration reaction because it results in the removal of a water molecule from the amino acids. (*b*) Decomposition reaction in which a disaccharide breaks apart to form glucose molecules. This reaction is also a hydrolysis reaction because it involves the splitting of a water molecule.

Decomposition Reactions

The term *decompose* means to break down into smaller parts. A **decomposition reaction** is the reverse of a synthesis reaction—a larger reactant is chemically broken down into two or more smaller products. The breakdown of a disaccharide (a type of carbohydrate) into glucose molecules (figure 2.10*b*) is an example. Note that this particular reaction requires that water be split into two parts and that each part be contributed to one of the new glucose molecules. Reactions that use water in this manner are called **hydrolysis** (hī-drol′i-sis; water dissolution) **reactions.**

The breakdown of ATP to ADP and an inorganic phosphate is another example of a decomposition reaction.

$$\begin{array}{ccccc} \text{A-P-P-P} & \rightarrow & \text{A-P-P} & + & \text{P}_i \\ (\text{ATP}) & & (\text{ADP}) & & (\text{Inorganic} \\ & & & & \text{phosphate}) \end{array}$$

The decomposition reactions that occur in the body are collectively called **catabolism** (kă-tab′-ō-lizm). They include the digestion of food molecules in the intestine and within cells, the breakdown of fat stores, and the breakdown of foreign matter and microorganisms in certain blood cells that function to protect the body. All of the anabolic and catabolic reactions in the body are collectively defined as **metabolism.**

Reversible Reactions

A **reversible reaction** is a chemical reaction in which the reaction can proceed from reactants to products or from products to reactants. When the rate of product formation is equal to the rate of the reverse reaction, the reaction system is said to be at **equilibrium.** At equilibrium the amount of reactants relative to the amount of products remains constant.

The following analogy may help to clarify the concept of reversible reactions and equilibrium. Imagine a trough containing water. The trough is divided into two compartments by a partition, but the partition contains holes that allows water to move freely between the compartments. Because water can move in either direction, this is like a reversible reaction. Let the water in the left compartment be the reactant and the water in the right compartment be the product. At equilibrium, the amount of reactant relative to the amount of product in each compartment is always the same because the partition allows water to pass between the two compartments until the level of water is the same in both compartments. If additional water is added to the reactant compartment, water flows from it through the partition to the product compartment until the level of water is the same in both compartments. Likewise, if additional reactants are added to a reaction system, some will form product until equilibrium is reestablished. Unlike this analogy, however, the amount of the reactants compared to the amount of products of most reversible reactions is not one to one. Depending on the specific reversible reaction, one part reactant to two parts product, two parts reactant to one part product, or many other possibilities can occur.

An important reversible reaction in the human body involves carbon dioxide and hydrogen ions. The reaction between carbon dioxide (CO_2) and water (H_2O) to form carbonic acid (H_2CO_3) is reversible. Carbonic acid then separates by a reversible reaction to form hydrogen ions (H^+) and bicarbonate ions (HCO_3^-).

$$CO_2 + H_2O \rightleftarrows H_2CO_3 \rightleftarrows H^+ + HCO_3^-$$

If CO_2 is added to H_2O, additional H_2CO_3 forms, which, in turn, causes more H^+ and HCO_3^- to form. The amount of H^+ and HCO_3^- relative to CO_2 therefore remains constant. Maintaining a constant level of H^+ is necessary for proper functioning of the nervous system. This can be achieved, in part, by regulating blood CO_2 levels. For example, slowing down the respiration rate causes blood carbon dioxide levels to increase.

P R E D I C T

If the respiration rate increases, CO_2 is eliminated from the blood. What effect does this change have on blood H^+ ion levels?

Oxidation–Reduction Reactions

Chemical reactions that result from the exchange of electrons between the reactants are called oxidation–reduction reactions. When sodium and chlorine react to form sodium chloride, the sodium atom loses an electron, and the chlorine atom gains an electron. The loss of an electron by an atom is called **oxidation,** and the gain of an electron is called **reduction.** The transfer of the electron can be complete, resulting in an ionic bond, or it can be a partial transfer, resulting in a covalent bond. Because the complete or partial loss of an electron by one atom is accompanied by the gain of that electron by another atom, these reactions are called **oxidation–reduction reactions.** Synthesis and decomposition reactions can be oxidation–reduction reactions. Thus, it is possible for a chemical reaction to be described in more than one way.

- 12. *Define a chemical reaction and compare synthesis and decomposition reactions. How do anabolism, catabolism, and metabolism relate to synthesis and decomposition reactions?*
- 13. *Describe a dehydration and a hydrolysis reaction.*
- 14. *Describe reversible reactions. What is meant by the equilibrium condition in reversible reactions?*
- 15. *What is an oxidation–reduction reaction?*

P R E D I C T ⑤

When hydrogen gas combines with oxygen gas to form water, is the hydrogen reduced or oxidized? Explain.

Energy

Energy, unlike matter, does not occupy space, and it has no mass. **Energy** is defined as the capacity to do **work,** that is, to move matter. Energy can be subdivided into potential energy and kinetic energy. **Potential energy** is stored energy that could do work but is not doing so. **Kinetic** (ki-net′ik) **energy** is the form of energy that actually does work and moves matter. A ball held at arm's length above the floor has potential energy. No energy is expended as long as the ball does not move. If the ball is released and falls toward the floor, however, it has kinetic energy.

According to the conservation of energy principle, the total energy of the universe is constant. Therefore, energy is neither created nor destroyed. Potential energy, however, can be converted into kinetic energy, and kinetic energy can be converted into potential energy. For example, the potential energy in the ball is converted into kinetic energy as the ball falls toward the floor. Conversely, the kinetic energy required to raise the ball from the floor is converted into potential energy.

Potential and kinetic energy can be found in many different forms. **Mechanical energy** is energy resulting from the position or movement of objects. Many of the activities of the human body, such as moving a limb, breathing, or circulating blood involve mechanical energy. Other forms of energy are chemical energy, heat energy, electric energy, and electromagnetic (radiant) energy.

Chemical Energy

The **chemical energy** of a substance is a form of stored (potential) energy within its chemical bonds. In any given chemical reaction, the potential energy contained in the chemical bonds of the reactants can be compared to the potential energy in the chemical bonds of the products. If the potential energy in the chemical bonds of the reactants is less than that of the products, then energy must be supplied for the reaction to occur. For example, the synthesis of ATP from ADP.

$$
\begin{array}{cccc}
ADP + H_2PO_4^- & + & Energy & \rightarrow & ATP + H_2O \\
\text{(Less potential} & & & & \text{(More potential} \\
\text{energy in reactants)} & & & & \text{energy in products)}
\end{array}
$$

For simplicity, the H_2O is often not shown in this reaction, and P_i is used to represent inorganic phosphate ($H_2PO_4^-$). For this reaction to occur, bonds in $H_2PO_4^-$ are broken and bonds are formed in ATP and H_2O. As a result of the breaking of existing bonds, the formation of new bonds, and the input of energy, these products have more potential energy than the reactants (figure 2.11*a*).

If the potential energy in the chemical bonds of the reactants is greater than that of the products, energy is released by the reaction. For example, the chemical bonds of food molecules contain more potential energy than the waste products that are produced when food molecules are decomposed. The energy released from the chemical bonds of food molecules is used by living systems to synthesize ATP. Once ATP is produced, the breakdown of ATP to ADP results in the release of energy.

$$
\begin{array}{cccc}
ATP + H_2O & \rightarrow & ADP + H_2PO_4^- & + & Energy \\
\text{(More potential} & & \text{(Less potential} & & \\
\text{energy in reactants)} & & \text{energy in products)} & &
\end{array}
$$

For this reaction to occur, the bonds in ATP and H_2O are broken and bonds in $H_2PO_4^-$ are formed. As a result of breaking the existing bonds and forming new bonds, these products have less potential energy than the reactants, and energy is released (figure 2.11*b*). Note that the energy released does not come from breaking the phosphate bond of ATP, because breaking a chemical

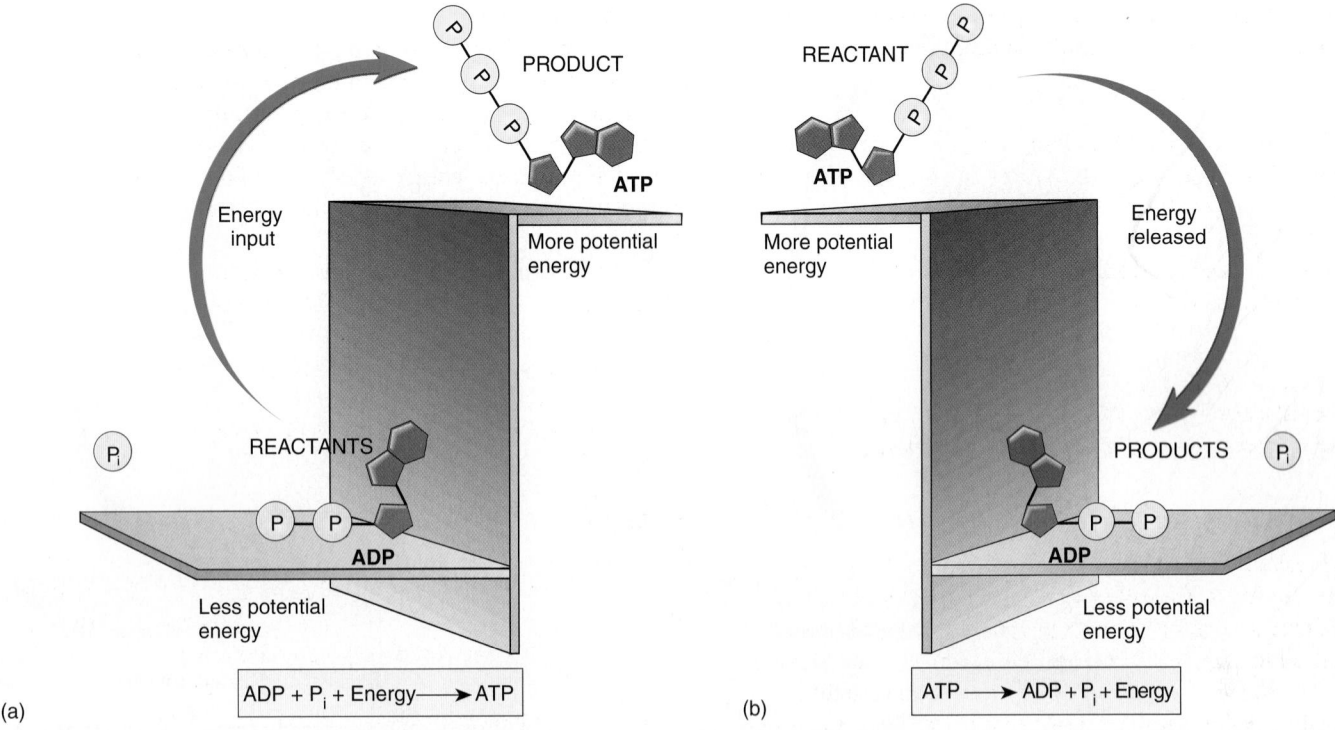

Figure 2.11   Energy and Chemical Reactions

In each figure the upper shelf represents a higher energy level, and the lower shelf represents a lower energy level. (*a*) Reaction in which the input of energy is required for the synthesis of ATP. (*b*) Reaction in which energy is released as a result of the breakdown of ATP.

bond requires the input of energy. It is commonly stated, however, that the breakdown of ATP results in the release of energy, which is true when the overall reaction is considered. The energy released when ATP is broken down can be used in the synthesis of other molecules; to do work, such as muscle contraction; or to produce heat.

Heat Energy

Heat is the energy that flows between objects that are at different temperatures. For example, when you touch someone who has a fever, you can feel the increased heat from the person's body. Temperature is a measure of how hot or cold a substance is relative to another substance. Heat is always transferred from a hotter object to a cooler object, such as from a hot stove top to a finger.

All other forms of energy can be converted into heat energy. For example, when a moving object comes to rest, its kinetic energy is converted into heat energy by friction. Some of the potential energy of chemical bonds is released as heat energy during chemical reactions. The body temperature of humans is maintained by heat produced in this fashion.

> 16. *How is energy different from matter? How are potential and kinetic energy different from each other?*
>
> 17. *Define mechanical energy, chemical energy, and heat energy. How is chemical energy converted to mechanical energy and heat energy in the body?*

18. *Use ATP and ADP to illustrate the release or input of energy in chemical reactions.*

P R E D I C T **6**

Energy from the breakdown of ATP provides the kinetic energy for muscle movement. Why does body temperature increase during exercise?

Speed of Chemical Reactions

Molecules are constantly in motion and therefore have kinetic energy. A chemical reaction occurs only when molecules with sufficient kinetic energy collide with each other. As two molecules move closer together, the negatively charged electron cloud of one molecule repels the negatively charged electron cloud of the other molecule. If the molecules have sufficient kinetic energy, they overcome this repulsion and come together. The nuclei in some atoms attract the electrons of other atoms, resulting in the breaking and formation of new chemical bonds. The **activation energy** is the minimum energy that the reactants must have to start a chemical reaction (figure 2.12*a*). Even reactions that result in a release of energy must overcome the activation energy barrier for the reaction to proceed. For example, heat in the form of a spark is required to start the reaction between oxygen and gasoline vapor. Once some oxygen molecules react with gasoline, the energy released can start additional reactions.

Given any population of molecules, some of them have more kinetic energy and move about faster than others. Even so, at normal

Figure 2.12   **Activation Energy and Enzymes**

(*a*) Activation energy is needed to change ATP to ADP. The upper shelf represents a higher energy level, and the lower shelf represents a lower energy level. The "wall" extending above the upper shelf represents the activation energy. Even though energy is given up moving from the upper to the lower shelf, the activation energy "wall" must be overcome before the reaction can proceed. (*b*) The enzyme lowers the activation energy, making it easier for the reaction to proceed.

body temperatures, most of the chemical reactions necessary for life proceed too slowly to support life because few molecules have enough energy to start a chemical reaction. **Catalysts** (kat′ă-listz) are substances that increase the rate of chemical reactions without being permanently changed or depleted. **Enzymes** (en′zīmz), which are discussed in greater detail on p. 48, are protein catalysts. Enzymes increase the rate of chemical reactions by lowering the activation energy necessary for the reaction to begin (figure 2.12*b*). As a result, more molecules have sufficient energy to undergo chemical reactions. With an enzyme, the rate of a chemical reaction can take place more than a million times faster than without the enzyme.

Temperature can also affect the speed of chemical reactions. As temperature increases, reactants have more kinetic energy, move at faster speeds, and collide with one another more frequently and with greater force, thereby increasing the likelihood of a chemical reaction. When a person has a fever of only a few degrees, reactions occur throughout the body at an accelerated rate, resulting in increased activity in the organ systems such as increased heart and respiratory rates. When body temperature drops, various metabolic processes slow. In cold weather, the fingers are less agile largely because of the reduced rate of chemical reactions in cold muscle tissue.

Within limits, the greater the concentration of the reactants, the greater the rate at which a given chemical reaction proceeds. This occurs because, as the concentration of reactants increases, they are more likely to come into contact with one an-

other. For example, the normal concentration of oxygen inside cells enables oxygen to come into contact with other molecules and produce the chemical reactions necessary for life. If the oxygen concentration decreases, the rate of chemical reactions decreases. A decrease in oxygen in cells can impair cell function and even result in death.

19. *Define activation energy, catalysts, and enzymes. How do enzymes increase the rate of chemical reactions?*
20. *What effect does increasing temperature or increasing concentration of the reactants have on the rate of a chemical reaction?*

Inorganic Chemistry

It was once believed that inorganic substances were those that came from nonliving sources and organic substances were those extracted from living organisms. As the science of chemistry developed, however, it became apparent that organic substances could be manufactured in the laboratory. As defined currently, **inorganic chemistry** generally deals with those substances that do not contain carbon, whereas **organic chemistry** is the study of carbon-containing substances. These definitions have a few exceptions. For example, carbon monoxide (CO), carbon dioxide (CO_2), and bicarbonate ion (HCO_3^-) are classified as inorganic molecules.

Water

A molecule of **water** is composed of one atom of oxygen joined to two atoms of hydrogen by covalent bonds. Water molecules are polar, with a partial positive charge associated with the hydrogen atoms and a partial negative charge associated with the oxygen atom. Hydrogen bonds form between the positively charged hydrogen atoms of one water molecule and the negatively charged oxygen atoms of another water molecule. These hydrogen bonds organize the water molecules into a lattice that holds the water molecules together (see figures 2.6 and 2.7).

Water accounts for approximately 50% of the weight of a young adult female and 60% of a young adult male. Females have a lower percentage of water than males because they typically have more body fat, which is relatively free of water. Plasma, the liquid portion of blood, is 92% water. Water has physical and chemical properties well suited for its many functions in living organisms. These properties are outlined in the following sections.

Stabilizing Body Temperature

Water has a high **specific heat,** meaning that a relatively large amount of heat is required to raise its temperature; therefore, it tends to resist large temperature fluctuations. When water evaporates, it changes from a liquid to a gas, and because heat is required for that process, the evaporation of water from the surface of the body rids the body of excess heat.

Protection

Water is an effective lubricant that provides protection against damage resulting from friction. For example, tears protect the surface of the eye from the rubbing of the eyelids. Water also forms a fluid cushion around organs that helps to protect them from trauma. The cerebrospinal fluid that surrounds the brain is an example.

Chemical Reactions

Many of the chemical reactions necessary for life do not take place unless the reacting molecules are dissolved in water. For example, sodium chloride must dissociate in water into Na^+ and Cl^- before they can react with other ions. Water also directly participates in many chemical reactions. As previously mentioned, a dehydration reaction is a synthesis reaction in which water is produced, and a hydrolysis reaction is a decomposition reaction that requires a water molecule (see figure 2.10).

Mixing Medium

A **mixture** is a combination of two or more substances physically blended together, but not chemically combined. A **solution** is any mixture of liquids, gases, or solids in which the substances are uniformly distributed with no clear boundary between the substances. For example, a salt solution consists of salt dissolved in water, air is a solution containing a variety of gases, and wax is a solid solution of several fatty substances. Solutions are often described in terms of one substance dissolving in another: the **solute** (sol′ūt) dissolves in the **solvent.** In a salt solution, water is the solvent and the dissolved salt is the solute. Sweat is a salt solution in which sodium chloride and other solutes are dissolved in water.

A **suspension** is a mixture containing materials that separate from each other unless they are continually, physically blended together. Blood is a suspension containing red blood cells suspended in a liquid called plasma. As long as the red blood cells and plasma are mixed together as they pass through blood vessels, the red blood cells remain suspended in the plasma. If the blood is allowed to sit in a container, however, the red blood cells and plasma separate from each other.

A **colloid** (kol′oyd) is a mixture in which a dispersed (solutelike) substance is distributed throughout a dispersing (solventlike) substance. The dispersed particles are larger than a simple molecule but small enough that they remain dispersed and do not settle out. Proteins, which are large molecules, and water form colloids. For instance, the plasma portion of blood and the liquid interior of cells are colloids containing many important proteins.

In living organisms the complex fluids inside and outside cells consist of solutions, suspensions, and colloids. Blood is an example of all of these mixtures. It is a solution containing dissolved nutrients such as sugar, a suspension holding red blood cells, and a colloid containing proteins.

The ability of water to mix with other substances enables it to act as a medium for transport, moving substances from one part of the body to another. Body fluids such as plasma transport nutrients, gases, waste products, and a variety of molecules involved with regulating body functions.

Solution Concentrations

The concentration of solute particles dissolved in solvents can be expressed in several ways. One common way is to indicate the percent of solute by weight per volume of solution. A 10% solution of sodium chloride can be made by dissolving 10 g of sodium chloride into enough water to make 100 mL of solution.

Physiologists often determine concentrations in **osmoles** (os′mōlz), which express the number of particles in a solution. A particle can be an atom, ion, or molecule. An osmole (osm) is 6.022×10^{23} particles of a substance in 1 kilogram (kg) of water. Just as a grocer sells eggs in lots of 12 (a dozen), a chemist groups atoms in lots of 6.022×10^{23}. The **osmolality** (os-mō-lal′i-tē) of a solution is a reflection of the number, not the type, of particles in a solution. For example, a 1 osm glucose solution and a 1 osm sodium chloride solution both contain 6.022×10^{23} particles per kg water. The glucose solution, however, has 6.022×10^{23} molecules of glucose, whereas the sodium chloride dissociates into 3.011×10^{23} sodium ions and 3.011×10^{23} chloride ions.

Because the concentration of particles in body fluids is so low, the measurement **milliosmole** (mOsm), 1/1000 of an osmole, is used. Most body fluids have a concentration of about 300 mOsm and contain many different ions and molecules. The concentration of body fluids is important because it influences the movement of water into or out of cells (see chapter 3). Appendix C contains more information on calculating concentrations.

21. *Define inorganic and organic chemistry.*
22. *List four functions that water performs in living organisms and give an example of each.*

23. *Describe solutions, suspensions, and colloids, and give an example of each. Define solvent and solute.*
24. *How is the osmolality of a solution determined? What is a milliosmole?*

Acids and Bases

Many molecules and compounds are classified as acids or bases. For most purposes an **acid** is defined as a proton donor. A hydrogen ion (H^+) is a proton because it results from the loss of an electron from a hydrogen atom, which consists of a proton and an electron. Therefore, a molecule or compound that releases H^+ is an acid. Hydrochloric acid (HCl) forms hydrogen ions (H^+) and chloride ions (Cl^-) in solution and therefore is an acid.

$$HCl \rightarrow H^+ + Cl^-$$

A **base** is defined as a proton acceptor, and any substance that binds to (accepts) H^+ ions is a base. Many bases function as proton acceptors by releasing hydroxide ions (OH^-) when they dissociate. The base sodium hydroxide (NaOH) dissociates to form Na^+ and OH^-.

$$NaOH \rightarrow Na^+ + OH^-$$

The OH^- are proton acceptors that combine with H^+ to form water.

$$OH^- + H^+ \rightarrow H_2O$$

Acids and bases are classified as strong or weak. Strong acids or bases dissociate almost completely when dissolved in water. Consequently, they release almost all of their H^+ or OH^-. The more completely the acid or base dissociates, the stronger it is. For example, HCl is a strong acid because it completely dissociates in water.

$$HCl \rightarrow H^+ + Cl^-$$
Not freely reversible

Weak acids or bases only partially dissociate in water. Consequently, they release only some of their H^+ or OH^-. For example, when acetic acid (CH_3COOH) is dissolved in water, some of it dissociates, but some of it remains in the undissociated form. An equilibrium is established between the ions and the undissociated weak acid.

$$CH_3COOH \rightleftharpoons CH_3COO^- + H^+$$
Freely reversible

For a given weak acid or base, the amount of the dissociated ions relative to the weak acid or base is a constant.

The pH Scale

The **pH scale** is a means of referring to the H^+ concentration in a solution (figure 2.13). Pure water is defined as a **neutral solution** and has a pH of 7. A neutral solution has equal concentrations of H^+ and OH^-. Solutions with a pH less than 7 are **acidic** and have a greater concentration of H^+ than OH^-. **Alkaline** (al′kă-lĭn), or **basic,** solutions have a pH greater than 7 and have fewer H^+ than OH^-.

A change in the pH of a solution by 1 pH unit represents a 10-fold change in the H^+ concentration. For example, a solution of pH 6 has a H^+ concentration 10 times greater than a solution of pH 7 and 100 times greater than a solution of pH 8. As the pH value becomes smaller, the solution has more H^+ and is more acidic, and as the pH value becomes larger, the solution has fewer H^+ and is more basic. Appendix D considers pH in greater detail.

Acidosis and Alkalosis

The normal pH range for human blood is 7.35 to 7.45. **Acidosis** results if blood pH drops below 7.35, in which case the nervous system becomes depressed, and the individual can become disoriented and possibly comatose. **Alkalosis** results if blood pH rises above 7.45. Then the nervous system becomes overexcitable, and the individual can be extremely nervous or have convulsions. Both acidosis and alkalosis can be fatal.

Figure 2.13   **The pH Scale**

A pH of 7 is considered neutral. Values less than 7 are acidic (the lower the number, the more acidic). Values greater than 7 are basic (the higher the number, the more basic). Representative fluids and their approximate pH values are listed.

Salts

A **salt** is a compound consisting of a cation other than a H^+ and an anion other than a OH^-. Salts are formed by the interaction of an acid and a base in which the H^+ of the acid are replaced by the positive ions of the base. For example, in a solution when hydrochloric acid (HCl) reacts with the base sodium hydroxide (NaOH), the salt, sodium chloride (NaCl), is formed.

$$HCl + NaOH \rightarrow NaCl + H_2O$$
(Acid) (Base)   (Salt) (Water)

Typically, when salts such as sodium chloride dissociate in water, they form positively and negatively charged ions (see figure 2.8).

Buffers

The chemical behavior of many molecules changes as the pH of the solution in which they are dissolved changes. For example, many enzymes work best within narrow ranges of pH. The survival of an organism depends on its ability to regulate body fluid pH within a narrow range. Deviations from the normal pH range for human blood are life-threatening.

One way body fluid pH is regulated involves the action of buffers, which resist changes in solution pH when either acids or bases are added. A **buffer** is a solution of a conjugate acid–base pair in which the acid component and the base component occur in similar concentrations. A conjugate base is everything that remains of an acid after the H^+ (proton) is lost. A conjugate acid is formed when a H^+ is transferred to the conjugate base. Two substances related in this way are a **conjugate acid–base pair.** For example, carbonic acid (H_2CO_3) and bicarbonate ion (HCO_3^-), formed by the dissociation of H_2CO_3, are a conjugate acid–base pair.

$$H_2CO_3 \rightleftarrows H^+ + HCO_3^-$$

In the forward reaction, H_2CO_3 loses a H^+ to produce HCO_3^-, which is a conjugate base. In the reverse reaction, a H^+ is transferred to the HCO_3^- (conjugate base) to produce H_2CO_3, which is a conjugate acid.

For a given condition, this reversible reaction results in an equilibrium, in which the amounts of H_2CO_3 relative to the amounts of H^+ and HCO_3^- remains constant. The conjugate acid–base pair can resist changes in pH because of this equilibrium. If an acid is added to a buffer, the H^+ from the added acid can combine with the base component of the conjugate acid–base pair. As a result, the concentration of H^+ does not increase as much as it would without this reaction. If H^+ are added to a H_2CO_3 solution, many of the H^+ combine with HCO_3^- to form H_2CO_3.

On the other hand, if a base is added to a buffered solution, the conjugate acid can release H^+ to counteract the effects of the added base. For example, if H^+ are added to a H_2CO_3 solution, the OH^- combine with H^+ to form water. As the H^+ are incorporated into water, H_2CO_3 dissociates to form H^+ and HCO_3^-, thereby maintaining the H^+ concentration (pH) within a normal range.

The greater the buffer concentration, the more effective it is in resisting a change in pH, but buffers cannot entirely prevent some change in the pH of a solution. For example, when an acid is added to a buffered solution, the pH decreases but not to the extent it would have without the buffer. Important buffers found in living systems have bicarbonate, phosphates, amino acids, and proteins as components.

25. *Define acid and base, and describe the pH scale. What is the difference between a strong acid or base and a weak acid or base?*

26. *Define acidosis and alkalosis, and describe the symptoms of each.*

27. *What is a salt? What is a buffer, and why are buffers important to organisms?*

P R E D I C T 7

Dihydrogen phosphate ion ($H_2PO_4^-$) and monohydrogen phosphate ion (HPO_4^{2-}) form the phosphate buffer system.

$$H_2PO_4^- \rightleftarrows H^+ + HPO_4^{2-}$$

Identify the conjugate acid and conjugate base in the phosphate buffer system. Explain how they function as a buffer when either H^+ or OH^- are added to the solution.

Oxygen

Oxygen (O_2) is an inorganic molecule consisting of two oxygen atoms bound together by a double covalent bond. About 21% of the gas in the atmosphere is oxygen, and it is essential for most animals. Oxygen is required by humans in the final step of a series of reactions in which energy is extracted from food molecules (see chapters 3 and 25).

Carbon Dioxide

Carbon dioxide (CO_2) consists of two oxygen atoms bound to one carbon atom. Each oxygen atom is bound to the carbon atom by a double covalent bond. Carbon dioxide is produced when organic molecules such as glucose are metabolized within the cells of the body (see chapters 3 and 25). Much of the energy stored in the covalent bonds of glucose is transferred to other organic molecules when glucose is broken down, and carbon dioxide is released. Once carbon dioxide is produced, it is eliminated from the cell as a metabolic by-product, transferred to the lungs by blood, and exhaled during respiration. If carbon dioxide is allowed to accumulate within cells, it becomes toxic.

28. *What are the functions of oxygen and carbon dioxide in living systems?*

Organic Chemistry

The ability of carbon to form covalent bonds with other atoms makes possible the formation of the large, diverse, complicated molecules necessary for life. A series of carbon atoms bound together by covalent bonds constitutes the "backbone" of many large molecules. Variation in the length of the carbon chains and the combination of atoms bound to the carbon backbone allows for the formation of a wide variety of molecules. For example, some protein molecules have thousands of carbon atoms bound by covalent bonds to one another or to other atoms, such as nitrogen, sulfur, hydrogen, and oxygen.

The four major groups of organic molecules essential to living organisms are carbohydrates, lipids, proteins, and nucleic acids. Each of these groups has specific structural and functional characteristics.

Carbohydrates

Carbohydrates are composed primarily of carbon, hydrogen, and oxygen atoms and range in size from small to very large. In most carbohydrates, for each carbon atom there are approximately two hydrogen atoms and one oxygen atom. Note that this ratio of hydrogen atoms to oxygen atoms is two to one, the same as in water (H_2O). They are called carbohydrates because carbon (carbo) atoms are combined with the same atoms that form water (hydrated). The large number of oxygen atoms in carbohydrates makes them relatively polar molecules. Consequently, they are soluble in polar solvents such as water.

Carbohydrates are important parts of other organic molecules, and they can be broken down to provide the energy necessary for life. Undigested carbohydrates also provide bulk in feces, which helps to maintain the normal function and health of the digestive tract. Table 2.5 summarizes the roles of carbohydrates in the body.

Monosaccharides

Large carbohydrates are composed of numerous, relatively simple building blocks called **monosaccharides** (mon-ō-sak′ă-rīdz; the prefix *mono-* means one; the term *saccharide* means sugar), or simple sugars. Monosaccharides commonly contain three carbons (trioses), four carbons (tetroses), five carbons (pentoses), or six carbons (hexoses).

The monosaccharides most important to humans include both five- and six-carbon sugars. Common six-carbon sugars, such as glucose, fructose, and galactose, are **isomers** (ī′sō-merz), which are molecules that have the same number and types of atoms but differ in their three-dimensional arrangement (figure 2.14). Glucose, or blood sugar, is the major carbohydrate found in the blood and is a major nutrient for most cells of the body. Fructose and galactose are also important dietary nutrients. Important five-carbon sugars include ribose and deoxyribose (see figure 2.24), which are components of ribonucleic acid (RNA) and deoxyribonucleic acid (DNA), respectively.

Table 2.5	Role of Carbohydrates in the Body
Role	**Example**
Structure	Ribose forms part of RNA and ATP molecules, and deoxyribose forms part of DNA.
Energy	Monosaccharides (glucose, fructose, galactose) can be used as energy sources. Disaccharides (sucrose, lactose, maltose) and polysaccharides (starch, glycogen) must be broken down to monosaccharides before they can be used for energy. Glycogen is an important energy-storage molecule in muscles and in the liver.
Bulk	Cellulose forms bulk in the feces.

Disaccharides

Disaccharides (dī-sak′ă-rīdz; *di-* means two) are composed of two simple sugars bound together through a dehydration reaction. Glucose and fructose, for example, combine to form a disaccharide called **sucrose** (table sugar) plus a molecule of water (figure 2.15a). Several disaccharides are important to humans, including sucrose, lactose, and maltose. Lactose, or milk sugar, is glucose combined with galactose; and maltose, or malt sugar, is two glucose molecules joined together.

Polysaccharides

Polysaccharides (pol-ē-sak′ă-rīdz; *poly-* means many) consist of many monosaccharides bound together to form long chains that are either straight or branched. **Glycogen,** or animal starch, is a polysaccharide composed of many glucose molecules (figure 2.15b). Because glucose can be metabolized rapidly and the resulting energy can be used by cells, glycogen is an important energy-storage molecule. A substantial amount of the glucose that is metabolized to produce energy for muscle contraction during exercise is stored in the form of glycogen in the cells of the liver and skeletal muscles.

Starch and **cellulose** are two important polysaccharides found in plants, and both are composed of long chains of glucose. Plants use starch as an energy storage molecule in the same way that animals use glycogen, and cellulose is an important structural component of plant cell walls. When humans ingest plants, the starch can be broken down and used as an energy source. Humans, however, do not have the digestive enzymes necessary to break down cellulose. The cellulose is eliminated in the feces, where it provides bulk.

29. Name the basic building blocks of carbohydrates.
30. List the five- and six-carbon sugars that are important for humans.
31. What is an isomer?
32. What are disaccharides and polysaccharides and how are they formed?
33. Which carbohydrates are used for energy? What is the function of starch and cellulose in plants? The function of glycogen and cellulose in animals?

Lipids

Lipids are a second major group of organic molecules common to living systems. Like carbohydrates, they are composed principally of carbon, hydrogen, and oxygen; but other elements, such as phosphorus and nitrogen, are minor components of some lipids. Lipids contain a lower ratio of oxygen to carbon than do carbohydrates, which makes them less polar. Consequently, lipids can be dissolved in nonpolar organic solvents, such as alcohol or acetone, but they are relatively insoluble in water.

Lipids have many important functions in the body. They provide protection and insulation, help to regulate many physiologic processes, and form plasma membranes. In addition, lipids are a major energy storage molecule and can be broken down and used as a source of energy. Table 2.6 summarizes the many roles of lipids in the body. Several different kinds of molecules, such as fats, phospholipids, steroids, and prostaglandins, are classified as lipids.

Figure 2.14 Monosaccharides

These monosaccharides almost always form a ring-shaped molecule. They are represented as linear models to more readily illustrate the relationships between the atoms of the molecules. Fructose is a structural isomer of glucose because it has identical chemical groups bonded in a different arrangement in the molecule (*indicated by red shading*). Galactose is a stereoisomer of glucose because it has exactly the same groups bonded to each carbon atom but located in a different three-dimensional orientation (*indicated by yellow shading*).

Table 2.6	Role of Lipids in the Body
Role	**Example**
Protection	Fat surrounds and pads organs.
Insulation	Fat under the skin prevents heat loss. Myelin surrounds nerve cells and electrically insulates the cells from one another.
Regulation	Steroid hormones regulate many physiologic processes. For example, estrogen and testosterone are sex hormones responsible for many of the differences between males and females. Prostaglandins help regulate tissue inflammation and repair.
Vitamins	Fat-soluble vitamins perform a variety of functions. Vitamin A forms retinol, which is necessary for seeing in the dark; active vitamin D promotes calcium uptake by the small intestine; vitamin E promotes wound healing; and vitamin K is necessary for the synthesis of proteins responsible for blood clotting.
Structure	Phospholipids and cholesterol are important components of plasma membranes.
Energy	Lipids can be stored and broken down later for energy; per unit of weight, they yield more energy than carbohydrates or proteins.

Fats are a major type of lipid. Like carbohydrates, fats are ingested and broken down by hydrolysis reactions in cells to release energy for use by those cells. Conversely, if intake exceeds need, excess chemical energy from any source can be stored in the body as fat for later use as energy is needed. Fats also provide protection by surrounding and padding organs, and under-the-skin fats act as an insulator to prevent heat loss.

Triglycerides (trī-glis′er-īdz) constitute 95% of the fats in the human body. Triglycerides, which are sometimes called **triacylglycerols** (tri-as′il-glis′er-olz), consist of two different types of building blocks: one glycerol and three fatty acids. **Glycerol** is a three-carbon molecule with a hydroxyl group attached to each carbon atom, and **fatty acids** consist of a straight chain of carbon atoms with a carboxyl group attached at one end (figure 2.16). A **carboxyl** (kar-bok′sil) **group** (—COOH) consists of both an oxygen atom and a hydroxyl group attached to a carbon atom.

$$\overset{\displaystyle O}{\underset{}{\|}} \quad\quad \overset{\displaystyle O}{\underset{}{\|}}$$
$$—C—OH \quad \text{or} \quad HO—C—$$

The carboxyl group is responsible for the acidic nature of the molecule because it releases hydrogen ions into solution. Glycerides can be described according to the number and kinds of fatty acids that combine with glycerol through dehydration reactions. Monoglycerides have one fatty acid, diglycerides have two fatty acids, and triglycerides have three fatty acids bound to glycerol.

Figure 2.15   **Disaccharide and Polysaccharide**

(*a*) Formation of sucrose, a disaccharide, by a dehydration reaction involving glucose and fructose (monosaccharides). (*b*) Glycogen is a polysaccharide formed by combining many glucose molecules. The photomicrograph shows glycogen granules in a liver cell.

Fatty acids differ from one another according to the length and the degree of saturation of their carbon chains. Most naturally occurring fatty acids contain an even number of carbon atoms, with 14- to 18-carbon chains being the most common. A fatty acid is **saturated** (figure 2.17) if it contains only single covalent bonds between the carbon atoms. Sources of saturated fats include beef, pork, whole milk, cheese, butter, eggs, coconut oil, and palm oil. The carbon chain is **unsaturated** if it has one or more double covalent bonds between carbon atoms. Because the double covalent bonds can occur anywhere along the carbon chain, many types of unsaturated fatty acids with an equal degree of unsaturation are possible. **Monounsaturated** fats, such as olive and peanut oils, have one double covalent bond between carbon atoms. **Polyunsaturated fats,** such as safflower, sunflower, corn, or fish oils, have two or more double covalent bonds between carbon atoms. Unsaturated fats are the best type of fats in the diet because unlike saturated fats they do not contribute to the development of cardiovascular disease.

Phospholipids are similar to triglycerides, except that one of the fatty acids bound to the glycerol is replaced by a molecule containing phosphate and, usually, nitrogen (figure 2.18). They are polar at the end of the molecule to which the phosphate is bound and nonpolar at the other end. The polar end of the molecule is attracted to water, and is said to be **hydrophilic** (water-loving). The nonpolar end is repelled by water and is said to be **hydrophobic** (water-fearing). Phospholipids are important structural components of plasma membranes (see chapter 3).

Figure 2.16   Triglyceride

Production of a triglyceride from one glycerol molecule and three fatty acid molecules. One water molecule (H_2O) is given off for each covalent bond formed between a fatty acid molecule and glycerol.

(a)

(b)

Figure 2.17   Fatty Acids

(a) Palmitic acid (saturated with no double bonds between the carbons). (b) Linolenic acid (unsaturated with three double bonds between the carbons).

The **eicosanoids** (ī′kō-să-noydz) are a group of important chemicals derived from fatty acids. They include **prostaglandins** (pros′tă-glan′dinz), **thromboxanes** (throm′bok-zānz), and **leukotrienes** (loo-kō-trī′ēnz). Eicosanoids are made in most cells and are important regulatory molecules. Among their numerous effects is their role in the response of tissues to injuries. Prostaglandins have been implicated in regulating the secretion of some hormones, blood clotting, some reproductive functions, and many other processes. Many of the therapeutic effects of aspirin and other anti-inflammatory drugs result from their ability to inhibit prostaglandin synthesis.

Steroids differ in chemical structure from other lipid molecules, but their solubility characteristics are similar. All steroid molecules are composed of carbon atoms bound together into four ringlike structures (figure 2.19). Important steroid molecules include cholesterol, bile salts, estrogen, progesterone, and testosterone. Cholesterol is an important steroid because other molecules are synthesized from it. For example, bile salts, which increase fat absorption in the intestines, are derived from cholesterol, as are the reproductive hormones estrogen, progesterone, and testosterone. In addition, cholesterol is an important component of plasma membranes. Although high levels of cholesterol in the blood increase the risk of cardiovascular disease, a certain amount of cholesterol is vital for normal function.

Another class of lipids is the **fat-soluble vitamins.** Their structures are not closely related to one another, but they are

Nitrogen

Phosphorus

Oxygen

Carbon

Hydrogen

Polar (hydrophilic) region
(phosphate-
containing region)

Nonpolar (hydrophobic) region
(fatty acids)

(a)   (b)

Figure 2.18   Phospholipids

(*a*) Molecular model of a phospholipid. (*b*) Simplified way in which phospholipids are often depicted.

Cholesterol

Estrogen (estradiol)

Bile salt (glycocholate)

Testosterone

Figure 2.19   Steroids

Steroids are four-ringed molecules that differ from one another according to the groups attached to the rings. Cholesterol, the most common steroid, can be modified to produce other steroids.

nonpolar molecules essential for many normal functions of the body.

● 34. *Give four functions of fats. Name the most common type of fat in the human body. What are the basic building blocks of this type of fat?*

35. *What is the difference between a saturated and an unsaturated fat? Between monounsaturated and polyunsaturated fats?*

36. *Describe the structure of phospholipids. Which end of the molecule is hydrophilic?*

37. *What are the functions of eicosanoids? Give three examples of eicosanoids.*

38. *Why is cholesterol an important steriod?*

Proteins

All **proteins** contain carbon, hydrogen, oxygen, and nitrogen bound together by covalent bonds, and most proteins contain some sulfur. In addition, some proteins contain small amounts of phosphorus, iron, and iodine. The molecular mass of proteins can be very large. For the purpose of comparison, the molecular mass of water is approximately 18, sodium chloride 58, and glucose 180; but the molecular mass of proteins ranges from approximately 1000 to several million.

Proteins regulate bodily processes, act as a transportation system in the body, provide protection, help muscles contract, and provide structure and energy. Table 2.7 summarizes the functions of proteins in the body.

Protein Structure

The basic building blocks for proteins are the 20 **amino** (ă-mē′nō) **acid** molecules. Each amino acid has an amine (ă-mēn′) group (—NH_2), a carboxyl group (—COOH), a hydrogen atom, and a side chain designated by the symbol *R* attached to the same carbon atom. The side chain can be a variety of chemical structures, and the differences in the side chains make the amino acids different from one another (figure 2.20).

Covalent bonds formed between amino acid molecules during protein synthesis are called **peptide bonds** (figure 2.21). A dipeptide is two amino acids bound together by a peptide bond, a tripeptide is three amino acids bound together by peptide bonds, and a polypeptide is many amino acids bound together by peptide bonds. Proteins are polypeptides composed of hundreds of amino acids.

The **primary structure** of a protein is determined by the sequence of the amino acids bound by peptide bonds (figure 2.22*a*). The potential number of different protein molecules is enormous because 20 different amino acids exist and each amino acid can be located at any position along a polypeptide chain. The characteristics of the amino acids in a protein ultimately determine the three-dimensional shape of the protein, and the shape of the protein

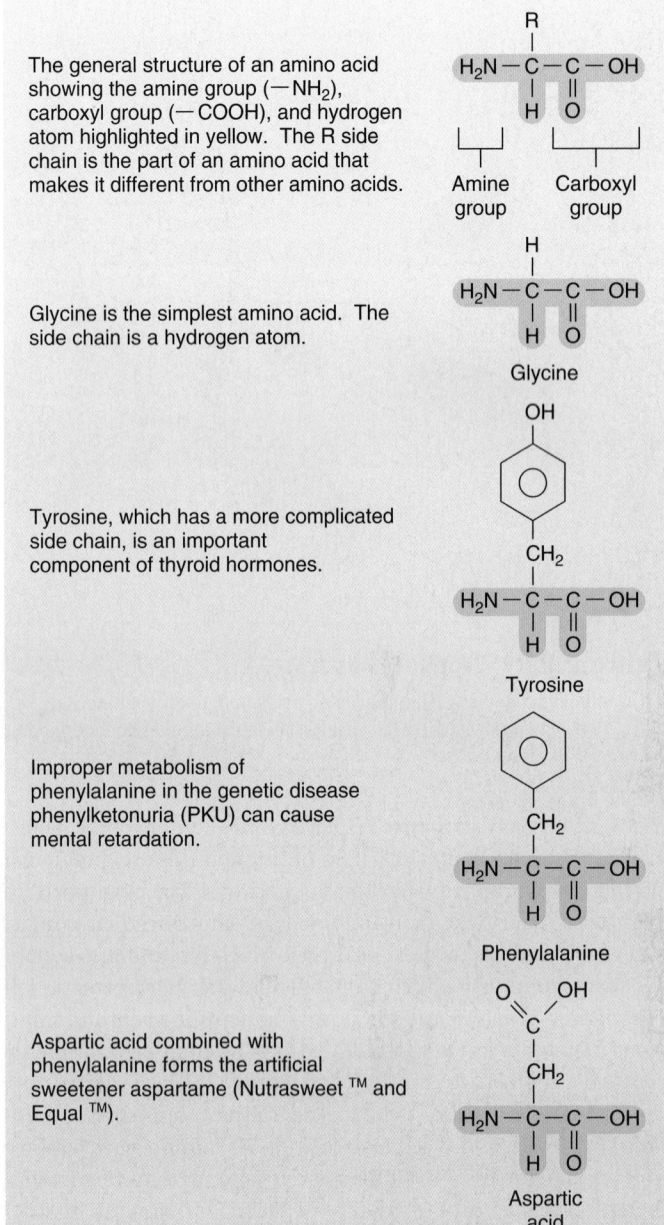

The general structure of an amino acid showing the amine group (—NH_2), carboxyl group (—COOH), and hydrogen atom highlighted in yellow. The R side chain is the part of an amino acid that makes it different from other amino acids.

Amine group   Carboxyl group

Glycine is the simplest amino acid. The side chain is a hydrogen atom.

Glycine

Tyrosine, which has a more complicated side chain, is an important component of thyroid hormones.

Tyrosine

Improper metabolism of phenylalanine in the genetic disease phenylketonuria (PKU) can cause mental retardation.

Phenylalanine

Aspartic acid combined with phenylalanine forms the artificial sweetener aspartame (Nutrasweet ™ and Equal ™).

Aspartic acid

Figure 2.20   Amino Acids

determines its function. A change in one, or a few, amino acids in the primary structure can alter protein function, usually making the protein less or even nonfunctional.

The **secondary structure** results from the folding or bending of the polypeptide chain caused by the hydrogen bonds between amino acids (figure 2.22*b*). Two common shapes that result are helices (coils) or pleated (folded) sheets. If the hydrogen bonds that maintain the shape of the protein are broken, the protein becomes nonfunctional. This change in shape is called **denaturation,** and it can be caused by abnormally high temperatures or changes in the pH of body fluids. An everyday example of denaturation is the change in the proteins of egg whites when they are cooked.

Table 2.7	Role of Proteins in the Body
Role	**Example**
Regulation	Enzymes control chemical reactions. Hormones regulate many physiologic processes; for example, insulin affects glucose transport into cells.
Transport	Hemoglobin transports oxygen and carbon dioxide in the blood. Plasma proteins transport many substances in the blood. Proteins in plasma membranes control the movement of materials into and out of the cell.
Protection	Antibodies and complement protect against microorganisms and other foreign substances.
Contraction	Actin and myosin in muscle are responsible for muscle contraction.
Structure	Collagen fibers form a structural framework in many parts of the body. Keratin adds strength to skin, hair, and nails.
Energy	Proteins can be broken down for energy; per unit of weight, they yield as much energy as carbohydrates.

Figure 2.21   **Peptide Bonds**

Dehydration reaction between three amino acids (*left*) to form a tripeptide (*right*). One water molecule (H_2O) is given off for each peptide bond formed.

The **tertiary structure** results from the folding of the helices or pleated sheets (figure 2.22c). Some amino acids are quite polar and therefore form hydrogen bonds with water. The polar portions of proteins tend to remain unfolded, maximizing their contact with water, whereas the less polar regions tend to fold into a globular shape, minimizing their contact with water. The formation of covalent bonds between sulfur atoms of one amino acid and sulfur atoms in another amino acid located at a different place in the sequence of amino acids can also contribute to the tertiary structure of proteins. The tertiary structure determines the shape of a **domain,** which is a folded sequence of 100–200 amino acids within a protein. The functions of proteins occur at one or more domains. Therefore, changes in the primary or secondary structure that affect the shape of the domain can change protein function.

If two or more proteins associate to form a functional unit, the individual proteins are called subunits. The **quaternary structure** refers to the spatial relationships between the individual subunits (figure 2.22d).

Enzymes

Proteins perform many roles in the body, including acting as enzymes. An **enzyme** is a protein catalyst that increases the rate at which a chemical reaction proceeds without the enzyme being permanently changed. The three-dimensional shape of enzymes is critical for their normal function because it determines the structure of the enzyme's **active site.** According to the **lock-and-key model** of enzyme action, a reaction occurs when the reactants (key) bind to the active site (lock) on the enzyme. This view of enzymes and reactants as rigid structures fitting together has been modified by the **induced fit model,** in which the enzyme is able to slightly change shape and better fit the reactants. The enzyme is

like a glove that does not achieve its functional shape until the hand (reactants) moves into place.

At the active site, reactants are brought into close proximity (figure 2.23). After the reactants combine, they are released from the active site, and the enzyme is capable of catalyzing additional reactions. The activation energy required for a chemical reaction to occur is lowered by enzymes (see figure 2.12) because they orient the reactants toward each other in such a way that it is more likely a chemical reaction will occur.

Slight changes in the structure of an enzyme can destroy the ability of the active site to function. Enzymes are very sensitive to changes in temperature or pH, which can break the hydrogen bonds within them. As a result, the relationship between amino acids changes, thereby producing a change in shape that prevents the enzyme from functioning normally.

To be functional, some enzymes require additional, nonprotein substances called **cofactors.** The cofactor can be an ion, such as magnesium or zinc, or an organic molecule. Cofactors that are organic molecules, such as certain vitamins, may be referred to as **coenzymes.** Cofactors normally form part of the enzyme's active site and are required to make the enzyme functional.

Enzymes are highly specific because their active site can bind only to certain reactants. Each enzyme catalyzes a specific chemical reaction and no others. Many different enzymes are therefore needed to catalyze the many chemical reactions of the body. Enzymes often are named by adding the suffix -*ase* to the name of the molecules on which they act. For example, an enzyme that catalyzes the breakdown of lipids is a **lipase** (lip′ās, lī′pās), and an enzyme that breaks down proteins is called a **protease** (prō′tē-ās).

Enzymes control the rate at which most chemical reactions proceed in living systems. Consequently, they control essentially all cellular activities. At the same time, the activity of enzymes themselves is regulated by several mechanisms that exist within the cells. Some mechanisms control the enzyme concentration by influencing the rate at which the enzymes are synthesized, and others alter the activity of existing enzymes. Much of what is known about the regulation of cellular activity involves knowledge of how enzyme activity is controlled.

> **39.** *What are the building blocks of proteins? Define a peptide bond.*
> **40.** *What determines the primary, secondary, tertiary, and quaternary structures of proteins? Define denaturation and name two things that can cause it to occur.*
> **41.** *Compare the lock-and-key model and the induced fit model of enzyme activity. Define cofactor and coenzyme.*

Nucleic Acids: DNA and RNA

Deoxyribonucleic (dē-oks′ē-rī′bō-noo-klē′ik) **acid (DNA)** is the genetic material of cells, and copies of DNA are transferred from one generation of cells to the next generation. DNA contains the information that determines the structure of proteins. **Ribonucleic** (rī′bō-noo-klē′ik) **acid (RNA)** is structurally related to DNA, and three types of RNA also play important roles in protein synthesis. In chapter 3 the means by which DNA and RNA direct the functions of the cell are described.

(a) Primary structure—the amino acid
    sequence. A protein consists of a chain
    of different amino acids (represented by
    different colored spheres).

Amino acids

Peptide bond

(b) Secondary structure results from
    hydrogen bonding (*dotted red lines*). The
    hydrogen bonds cause the amino acid
    chain to form helices (coils) or pleated
    (folded) sheets.

Pleated sheet

Helix

(c) Tertiary structure with secondary folding
    caused by interactions within the polypeptide
    and its immediate environment

(d) Quaternary structure—the relationships
    between individual subunits

Figure 2.22   Protein Structure

Figure 2.23   **Enzyme Action**

The enzyme brings the two reacting molecules together. After the reaction, the unaltered enzyme can be used again.

Figure 2.24   **Components of Nucleotides**

(*a*) Deoxyribose sugar, which forms nucleotides used in DNA production. (*b*) Ribose sugar, which forms nucleotides used in RNA production. Note that deoxyribose is ribose minus an oxygen atom. (*c*) Deoxyribonucleotide consisting of deoxyribose, a nitrogen base, and a phosphate group.

The **nucleic** (noo-klē′ik, noo-klā′ik) **acids** are large molecules composed of carbon, hydrogen, oxygen, nitrogen, and phosphorus. Both DNA and RNA consist of basic building blocks called **nucleotides** (noo′klē-ō-tīdz). Each nucleotide is composed of a monosaccharide to which a nitrogenous organic base and a phosphate group are attached (figure 2.24). The five-carbon monosaccharide is deoxyribose for DNA, and ribose for RNA. The nitrogenous organic bases consist of carbon and nitrogen atoms organized into rings (figure 2.25). They are bases because the nitrogen atoms tend to take up H^+ from solution. The nitrogenous organic bases are thymine (thī′mēn, thī′min), cytosine (sī′tō-sēn), and uracil (ūr′ă-sil), which are single-ringed pyrimidines (pī-rim′i-dēnz); and adenine (ad′ě-nēn) and guanine (gwahn′ēn), which are double-ringed purines (pūr′ēnz).

DNA has two strands of nucleotides joined together to form a twisted ladderlike structure called a double helix (figure 2.26). The uprights of the ladder are formed by covalent bonds between the deoxyribose molecules and phosphate groups of adjacent nucleotides. The rungs of the ladder are formed by the bases of the nucleotides of one upright connected to the bases of the other upright by hydrogen bonds. Each nucleotide of DNA contains one of the organic bases: adenine, thymine, cytosine, or guanine. Adenine binds only to thymine because the structure of these organic bases allows two hydrogen bonds to form between them. Cytosine binds only to guanine because the structure of these organic bases allows three hydrogen bonds to form between them.

The sequence of organic bases in DNA molecules stores genetic information. Each DNA molecule consists of millions of organic bases, and their sequence ultimately determines the type and sequence of amino acids found in protein molecules. Because enzymes

Figure 2.25   **Nitrogenous Organic Bases**

The organic bases found in nucleic acids are separated into two groups. Purines are double-ringed molecules, and pyrimidines are single-ringed molecules.

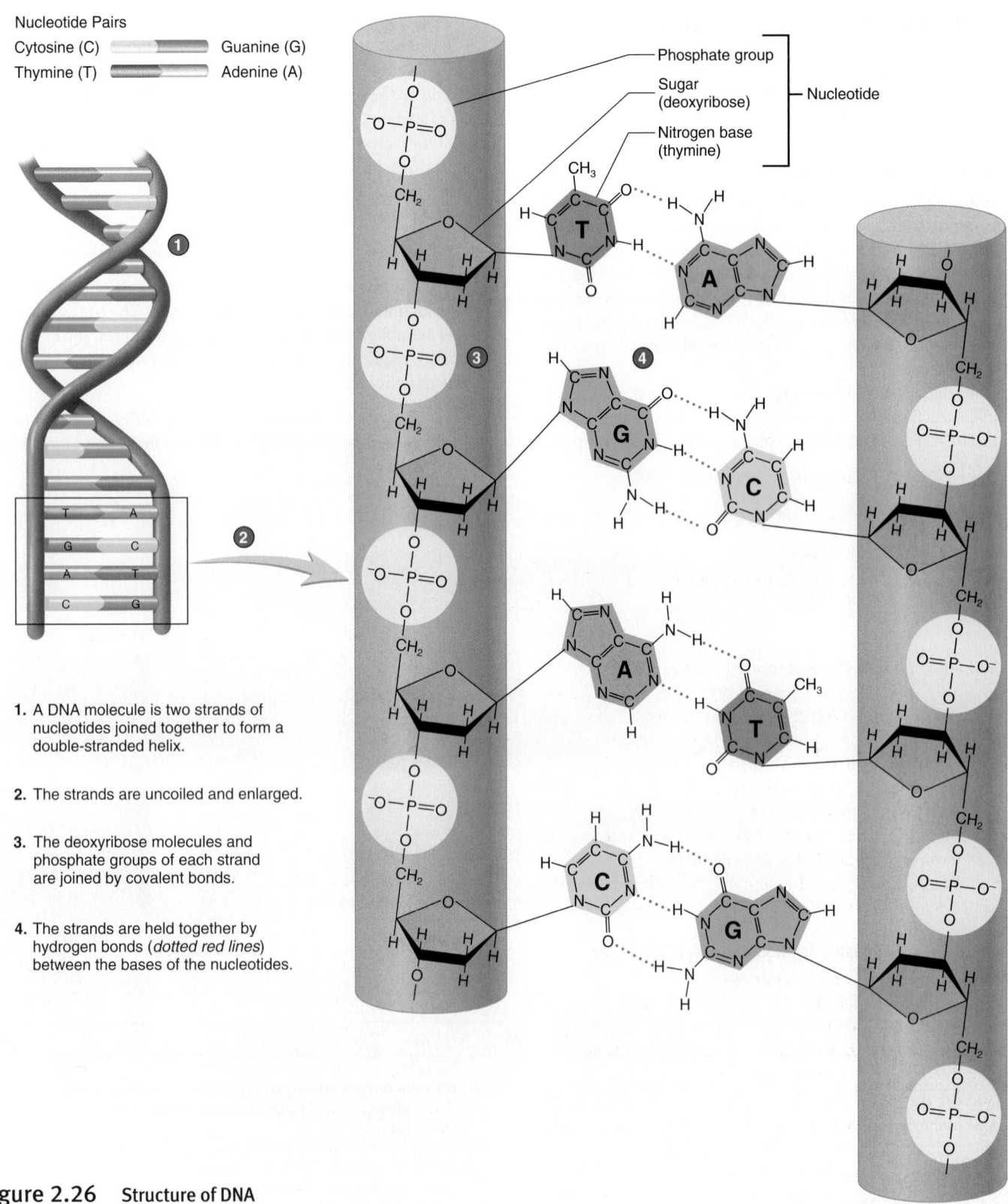

Figure 2.26 Structure of DNA

1. A DNA molecule is two strands of nucleotides joined together to form a double-stranded helix.

2. The strands are uncoiled and enlarged.

3. The deoxyribose molecules and phosphate groups of each strand are joined by covalent bonds.

4. The strands are held together by hydrogen bonds (*dotted red lines*) between the bases of the nucleotides.

are proteins, DNA structure determines the rate and type of chemical reactions that occur in cells by controlling enzyme structure. The information contained in DNA therefore ultimately defines all cellular activities. Other proteins, such as collagen, that are coded by DNA determine many of the structural features of humans.

RNA has a structure similar to a single strand of DNA. Like DNA, four different nucleotides make up the RNA molecule, and the organic bases are the same, except that thymine is replaced with uracil (see figure 2.25). Uracil can bind only to adenine.

42. *Name two types of nucleic acids. What are their functions?*
43. *What are the basic building blocks of nucleic acids? What kinds of sugar and bases are found in DNA? In RNA?*
44. *DNA is like a twisted ladder. What forms the uprights and rungs of the ladder?*

Adenosine Triphosphate

Adenosine triphosphate (ă-den′ō-sēn trī-fos′făt) (**ATP**) is an especially important organic molecule found in all living organisms. It consists of adenosine and three phosphate groups (figure 2.27). Adenosine is the sugar ribose with the organic base adenine. The potential energy stored in the covalent bond between the second and third phosphate groups is important to living organisms because it provides the energy used in nearly all of the chemical reactions within cells.

The catabolism of glucose and other nutrient molecules results in chemical reactions that release energy. Some of that energy is used to synthesize ATP from ADP and an inorganic phosphate group (P_i):

$$ADP + P_i + \text{Energy (from catabolism)} \rightarrow ATP$$

The transfer of energy from nutrient molecules to ATP involves a series of oxidation–reduction reactions in which a high-energy electron is transferred from one molecule to the next molecule in the series. In chapter 25 the oxidation–reduction reactions of metabolism are considered in greater detail.

Once produced, ATP is used to provide energy for other chemical reactions (anabolism) or to drive cell processes such as muscle contraction. In the process, ATP is converted back to ADP and an inorganic phosphate group.

$$ATP \rightarrow ADP + P_i + \text{Energy (for anabolism and other cell processes)}$$

ATP is often called the energy currency of cells because it is capable of both storing and providing energy. The concentration of ATP is maintained within a narrow range of values, and essentially all energy-requiring chemical reactions stop when there is an inadequate quantity of ATP.

45. *Describe the structure of ATP. Where does the energy to synthesize ATP come from? What is the energy stored in ATP used for?*

Figure 2.27   Adenosine Triphosphate (ATP) Molecule

S U M M A R Y

Chemistry is the study of the composition, structure, and properties of substances and the reactions they undergo. Much of the structure and function of healthy or diseased organisms can be understood at the chemical level.

Basic Chemistry   (p. 26)
Matter, Mass, and Weight

1. Matter is anything that occupies space.
2. Mass is the amount of matter in an object.
3. Weight results from the force exerted by earth's gravity on matter.

Elements and Atoms

1. An element is the simplest type of matter with unique chemical and physical properties.

2. An atom is the smallest particle of an element that has the chemical characteristics of that element. An element is composed of only one kind of atom.
3. Atoms consist of protons, neutrons, and electrons.
   • Protons are positively charged, electrons are negatively charged, and neutrons have no charge.
   • Protons and neutrons are found in the nucleus, and electrons, which are located around the nucleus, can be represented by an electron cloud.
4. The atomic number is the unique number of protons in an atom. The mass number is the sum of the protons and the neutrons.
5. Isotopes are atoms that have the same atomic number but different mass numbers.
6. The atomic mass of an element is the average mass of its naturally occurring isotopes weighted according to their abundance.

Electrons and Chemical Bonding

1. The chemical behavior of atoms is determined mainly by their outermost electrons. A chemical bond occurs when atoms share or transfer electrons.
2. Ions are atoms that have gained or lost electrons.
   - An atom that loses one or more electrons becomes positively charged and is called a cation. An anion is an atom that becomes negatively charged after accepting one or more electrons.
   - Ionic bonding is the attraction of the oppositely charged cation and anion to each other.
3. A covalent bond is the sharing of electron pairs between atoms. A polar covalent bond results when the sharing of electrons is unequal and can produce a polar molecule that is electrically asymmetric.

Molecules and Compounds

1. A molecule is two or more atoms chemically combined to form a structure that behaves as an independent unit. A compound is two or more *different* types of atoms chemically combined.
2. The kinds and numbers of atoms (or ions) in a molecule or compound can be represented by a formula consisting of the symbols of the atoms (or ions) plus subscripts denoting the number of each type of atom (or ion).
3. The molecular mass of a molecule or compound can be determined by adding up the atomic masses of its atoms (or ions).

Intermolecular Forces

1. A hydrogen bond is the weak attraction that occurs between the oppositely charged regions of polar molecules. Hydrogen bonds are important in determining the three-dimensional structure of large molecules.
2. Solubility is the ability of one substance to dissolve in another. Ionic substances that dissolve in water by dissociation are electrolytes. Molecules that do not dissociate are nonelectrolytes.

Chemical Reactions and Energy   (p. 33)
Synthesis Reactions

1. Synthesis reactions are the chemical combination of two or more substances to form a new or larger substance.
2. Dehydration reactions are synthesis reactions in which water is produced.
3. Anabolism is the sum of all the synthesis reactions in the body.

Decomposition Reactions

1. Decomposition reactions are the chemical breakdown of a larger substance to two or more different smaller substances.
2. Hydrolysis reactions are decomposition reactions in which water is depleted.
3. All of the decomposition reactions in the body are called catabolism.

Reversible Reactions

Reversible reactions produce an equilibrium condition in which the amount of reactants relative to the amount of products remains constant.

Oxidation–Reduction Reactions

Oxidation–reduction reactions involve the complete or partial transfer of electrons between atoms.

Energy

Energy is the ability to do work. Potential energy is stored energy, and kinetic energy is energy resulting from movement of an object.

Chemical Energy

1. Chemical bonds are a form of potential energy.
2. Chemical reactions in which the products contain more potential energy than the reactants require the input of energy.
3. Chemical reactions in which the products have less potential energy than the reactants release energy.

Heat Energy

1. Heat energy is energy that flows between objects that are at different temperatures.
2. Heat energy is released in chemical reactions and is responsible for body temperature.

Speed of Chemical Reactions

1. Activation energy is the minimum energy that the reactants must have to start a chemical reaction.
2. Enzymes are specialized protein catalysts that lower the activation energy for chemical reactions. Enzymes speed up chemical reactions but are not consumed or altered in the process.
3. Increased temperature and concentration of reactants can increase the rate of chemical reactions.

Inorganic Chemistry   (p. 38)

Inorganic chemistry is mostly concerned with noncarbon-containing substances but does include some carbon-containing substances, such as carbon dioxide and carbon monoxide.

Water

1. Water is a polar molecule composed of one atom of oxygen and two atoms of hydrogen.
2. Water stabilizes body temperature, protects against friction and trauma, makes chemical reactions possible, directly participates in chemical reactions (e.g., dehydration and hydrolysis reactions), and is a mixing medium (e.g., solutions, suspensions, and colloids).
3. A mixture is a combination of two or more substances physically blended together, but not chemically combined.
4. A solution is any liquid, gas, or solid in which the substances are uniformly distributed with no clear boundary between the substances.
5. A solute dissolves in the solvent.
6. A suspension is a mixture containing materials that separate from each other unless they are continually, physically blended together.
7. A colloid is a mixture in which a dispersed (solutelike) substance is distributed throughout a dispersing (solventlike) substance. Particles do not settle out of a colloid.

Solution Concentrations

1. One way to describe solution concentration is an osmole, which contains 6.022×10^{23} of particles (i.e., atoms, ions, or molecules) in 1 kilogram water.
2. A milliosmole is 1/1000 of an osmole.

Acids and Bases

1. Acids are proton (i.e., H^+) donors, and bases (e.g., OH^-) are proton acceptors.
2. A strong acid or base almost completely dissociates in water. A weak acid or base partially dissociates.
3. The pH scale refers to the H^+ concentration in a solution.
   - A neutral solution has an equal number of H^+ and OH^- and is assigned a pH of 7.
   - Acid solutions, in which the number of H^+ is greater than the number of OH^-, have pH values less than 7.
   - Basic, or alkaline, solutions have more OH^- than H^+ and a pH greater than 7.
4. A salt is a molecule consisting of a cation other than H^+ and an anion other than OH^-. Salts are formed when acids react with bases.
5. A buffer is a solution of a conjugate acid–base pair that resists changes in pH when acids or bases are added to the solution.

Oxygen

Oxygen is necessary in the reactions that extract energy from food molecules in living organisms.

Carbon Dioxide

During metabolism when the organic molecules are broken down, carbon dioxide and energy are released.

Organic Chemistry   (p. 41)

Organic molecules contain carbon atoms bound together by covalent bonds.

Carbohydrates

1. Monosaccharides are the basic building blocks of other carbohydrates. They, especially glucose, are important sources of energy. Examples are ribose, deoxyribose, glucose, fructose, and galactose.
2. Disaccharide molecules are formed by dehydration reactions between two monosaccharides. They are broken apart into monosaccharides by hydrolysis reactions. Examples of disaccharides are sucrose, lactose, and maltose.
3. Polysaccharides are many monosaccharides bound together to form long chains. Examples include cellulose, starch, and glycogen.

Lipids

1. Triglycerides are composed of glycerol and fatty acids. One, two, or three fatty acids can attach to the glycerol molecule.
   • Fatty acids are straight chains of carbon molecules with a carboxyl group. Fatty acids can be saturated (only single covalent bonds between carbon atoms) or unsaturated (one or more double covalent bonds between carbon atoms).
   • Energy is stored in fats.
2. Phospholipids are lipids in which a fatty acid is replaced by a phosphate-containing molecule. Phospholipids are a major structural component of plasma membranes.

3. Steroids are lipids composed of four interconnected ring molecules. Examples include cholesterol, bile salts, and sex hormones.
4. Other lipids include fat-soluble vitamins, prostaglandins, thromboxanes, and leukotrienes.

Proteins

1. The building blocks of protein are amino acids, which are joined by peptide bonds.
2. The number, kind, and arrangement of amino acids determine the primary structure of a protein. Hydrogen bonds between amino acids determine secondary structure, and hydrogen bonds between amino acids and water determine tertiary structure. Interactions between different protein subunits determine quaternary structure.
3. Enzymes are protein catalysts that speed up chemical reactions by lowering their activation energy.
4. The active sites of enzymes bind only to specific reactants.
5. Cofactors are ions or organic molecules such as vitamins that are required for some enzymes to function.

Nucleic Acids: DNA and RNA

1. The basic unit of nucleic acids is the nucleotide, which is a monosaccharide with an attached phosphate and organic base.
2. DNA nucleotides contain the monosaccharide deoxyribose and the organic bases adenine, thymine, guanine, or cytosine. DNA occurs as a double strand of joined nucleotides and is the genetic material of cells.
3. RNA nucleotides are composed of the monosaccharide ribose. The organic bases are the same as for DNA, except that thymine is replaced with uracil.

Adenosine Triphosphate

ATP stores energy derived from catabolism. The energy is released from ATP and is used in anabolism and other cell processes.

R E V I E W   A N D   C O M P R E H E N S I O N

1. The smallest particle of an element that still has the chemical characteristics of that element is a(an)
   a. electron.
   b. molecule.
   c. neutron.
   d. proton.
   e. atom.

2. The number of electrons in an atom is equal to the
   a. atomic number.
   b. mass number.
   c. number of neutrons.
   d. isotope number.
   e. molecular mass.

3. ^{12}C and ^{14}C are
   a. atoms of different elements.
   b. isotopes.
   c. atoms with different atomic numbers.
   d. atoms with different numbers of protons.
   e. compounds.

4. A cation is a (an)
   a. uncharged atom.
   b. positively charged atom.
   c. negatively charged atom.
   d. atom that has gained an electron.
   e. both c and d.

5. A polar covalent bond between two atoms occurs when
   a. one atom attracts shared electrons more strongly than another atom.
   b. atoms attract electrons equally.
   c. an electron from one atom is completely transferred to another atom.
   d. the molecule becomes ionized.
   e. a hydrogen atom is shared between two different atoms.

6. Table salt (NaCl) is
   a. an atom.
   b. organic.
   c. a molecule.
   d. a compound.
   e. a cation.

7. The weak attractive force between two water molecules forms a(an)
   a. covalent bond.
   b. hydrogen bond.
   c. ionic bond.
   d. compound.
   e. isotope.

8. Electrolytes are
   a. nonpolar molecules.
   b. covalent compounds.
   c. substances that usually don't dissolve in water.
   d. found in solutions that do not conduct electricity.
   e. cations and anions that dissociate in water.

9. In a decomposition reaction
   a. anabolism occurs.
   b. proteins are formed from amino acids.
   c. large molecules are broken down to form small molecules.
   d. a dehydration reaction may occur.
   e. all of the above.

10. Oxidation–reduction reactions
    a. can be synthesis or decomposition reactions.
    b. have one reactant gaining electrons.
    c. have one reactant losing electrons.
    d. can create ionic or covalent bonds.
    e. all of the above.

11. Potential energy
    a. is energy caused by movement of an object.
    b. is the form of energy that is actually doing work.
    c. includes energy within chemical bonds.
    d. can never be converted to kinetic energy.
    e. all of the above.

12. Which of these descriptions of heat energy is *not* correct?
    a. Heat energy flows between objects that are at different temperatures.
    b. Heat energy can be produced from all other forms of energy.
    c. Heat energy can be released during chemical reactions.
    d. Heat energy must be added to break apart ATP molecules.
    e. Heat energy is always transferred from a hotter object to a cooler object.

13. A *decrease* in the speed of a chemical reaction occurs if
    a. the activation energy requirement is increased.
    b. catalysts are increased.
    c. temperature increases.
    d. the concentration of the reactants increases.
    e. all of the above.

14. Which of these statements concerning enzymes is correct?
    a. Enzymes increase the rate of reactions but are permanently changed as a result.
    b. Enzymes are proteins that function as catalysts.
    c. Enzymes increase the activation energy requirement for a reaction to occur.
    d. Enzymes usually can only double the rate of a chemical reaction.
    e. Enzymes increase the kinetic energy of the reactants.

15. Water
    a. is composed of two oxygen atoms and one hydrogen atom.
    b. has a low specific heat.
    c. is composed of polar molecules into which ionic substances dissociate.
    d. is produced in a hydrolysis reaction.
    e. is a very small organic molecule.

16. When sugar is dissolved in water, the water is called the
    a. solute.
    b. solution.
    c. solvent.

17. Which of these is an example of a suspension?
    a. sweat
    b. water and proteins inside cells
    c. sugar dissolved in water
    d. red blood cells in plasma

18. A solution with a pH of 5 is _____ and contains _____ H^+ than a neutral solution.
    a. a base, more
    b. a base, less
    c. an acid, more
    d. an acid, less
    e. neutral, the same number of

19. A buffer
    a. slows down chemical reactions.
    b. speeds up chemical reactions.
    c. increases the pH of a solution.
    d. maintains a relatively constant pH.
    e. works by forming salts.

20. A conjugate acid–base pair
    a. acts as a buffer.
    b. can combine with H^+ in a solution.
    c. can release H^+ to combine with OH^-.
    d. describes carbonic acid (H_2CO_3) and bicarbonate ion (HCO_3^-)
    e. all of the above.

21. Carbon dioxide
    a. consists of two oxygen atoms ionically bonded to carbon.
    b. becomes toxic if allowed to accumulate within cells.
    c. is mostly eliminated by the kidneys.
    d. is combined with fats to produce glucose during metabolism within cells.
    e. is taken into cells during metabolism.

22. Which of these is an example of a carbohydrate?
    a. glycogen
    b. prostaglandin
    c. steroid
    d. DNA
    e. triglyceride

23. The polysaccharide used for energy storage in the human body is
    a. cellulose.
    b. glycogen.
    c. lactose.
    d. sucrose.
    e. starch.

24. The basic units or building blocks of triglycerides are
    a. simple sugars (monosaccharides).
    b. double sugars (disaccharides).
    c. amino acids.
    d. glycerol and fatty acids.
    e. nucleotides.

25. A _____ fatty acid has one double covalent bond between carbon atoms.
    a. cholesterol
    b. monounsaturated
    c. phospholipid
    d. polyunsaturated
    e. saturated

26. A peptide bond joins together
    a. amino acids.
    b. fatty acids and glycerol.
    c. monosaccharides.
    d. disaccharides.
    e. nucleotides.

27. The _____ structure of a protein results from the folding of the helices or pleated sheets.
    a. primary
    b. secondary
    c. tertiary
    d. quaternary

28. According to the lock-and-key model of enzyme action,
    a. reactants must first be heated.
    b. enzyme shape is not important.
    c. each enzyme can catalyze many types of reactions.
    d. reactants must bind to an active site on the enzyme.
    e. enzymes control only a small number of reactions in the cell.

29. DNA molecules
    a. are the genetic material of cells.
    b. contain a single strand of nucleotides.
    c. contain the nucleotide uracil.
    d. have three different types that have roles in protein synthesis.
    e. contain up to 100 organic bases.

30. ATP
    a. is formed by the addition of a phosphate group to ADP.
    b. is formed with energy released during catabolism reactions.
    c. provides the energy for anabolism reactions.
    d. contains three phosphate groups.
    e. all of the above.

Answers in Appendix F

C R I T I C A L   T H I N K I N G

1. Iron has an atomic number of 26 and a mass number of 56. How many protons, neutrons, and electrons are in an atom of iron? If an atom of iron lost three electrons, what would the charge of the resulting ion be? Write the correct symbol for this ion.

2. Which of the following pairs of terms applies to the reaction that results in the formation of fatty acids and glycerol from a triglyceride molecule?
    a. Decomposition or synthesis reaction
    b. Anabolism or catabolism
    c. Dehydration or hydrolysis reaction

3. A mixture of chemicals is warmed slightly. As a consequence, although no more heat is added, the solution becomes very hot. Explain what occurred to make the solution so hot.

4. Two solutions, when mixed together at room temperature, produce a chemical reaction. When the solutions are boiled and allowed to cool to room temperature before mixing, however, no chemical reaction takes place. Explain.

5. In terms of the potential energy in the food, explain why eating food is necessary for increasing muscle mass.

6. Solution A has a pH of 2, and solution B has a pH of 8. If equal amounts of solutions A and B are mixed, is the resulting solution acidic or basic?

7. Given a buffered solution that is based on the following equilibrium:
$$CO_2 + H_2O \rightleftarrows H_2CO_3 \rightleftarrows H^+ + HCO_3^-$$
what happens to the pH of the solution if $NaHCO_3$ is added?

8. An enzyme E catalyzes the following reaction:
$$A + B \xrightarrow{E} C$$
The product C, however, binds to the active site of the enzyme in a reversible fashion and keeps the enzyme from functioning. What happens if A and B are continually added to a solution that contains a fixed amount of the enzyme?

9. Given the materials commonly found in a kitchen, explain how one could distinguish between a protein and a lipid.

10. A student is given two unlabeled substances: one a typical phospholipid and one a typical protein. She is asked to determine which substance is the protein and which is the phospholipid. The available techniques allow her to determine the elements in each sample. How can she identify each substance?

Answers in Appendix G

A N S W E R S   T O   P R E D I C T   Q U E S T I O N S

1. The mass (amount of matter) of the astronaut on the surface of the earth and in outer space does not change. In outer space, where the force of gravity from the earth is very small, the astronaut is "weightless" compared to his or her weight on the earth's surface.

2. Potassium has 19 protons (the atomic number), 20 neutrons (the mass number minus the atomic number), and 19 electrons (because the number of electrons equals the number of protons).

3. The molecular formula for glucose is $C_6H_{12}O_6$. The atomic mass of carbon is 12.01, hydrogen is 1.008, and oxygen is 16.00. The molecular mass of glucose is therefore $(6 \times 12.01) + (12 \times 1.008) + (6 \times 16.00)$, or 180.2.

4. A decrease in blood CO_2 decreases the amount of H_2CO_3 and therefore the blood H^+ level. Because CO_2 and H_2O are in equilibrium with H^+ and HCO_3^-, with H_2CO_3 as an intermediate, a decrease in CO_2 causes some H^+ and HCO_3^- to join together to form H_2CO_3, which then forms CO_2 and H_2O. Consequently, the H^+ concentration decreases.

5. When two hydrogen atoms combine with an oxygen atom to form water, a polar covalent bond forms between each hydrogen atom and the oxygen atom. Unequal sharing of electrons occurs, and the electrons are associated with the oxygen atom more than with the hydrogen atoms. In this sense, the hydrogen atoms lose their electrons, and the oxygen atom gains electrons. The hydrogen atoms are therefore oxidized, and the oxygen atom is reduced.

6. During exercise, muscle contractions increase, which requires energy. This energy is obtained from the energy in the chemical bonds of ATP. As ATP is broken down, energy is released. Some of the energy is used to drive muscle contractions, and some becomes heat. Because the rate of these reactions increases during exercise, more heat is produced than when at rest, and body temperature increases.

7. Monohydrogen phosphate ion (HPO_4^{2-}) is the conjugate base formed when the conjugate acid, dihydrogen phosphate ion ($H_2PO_4^-$) loses a H^+. If H^+ are added to the solution, they combine with the conjugate base, monohydrogen phosphate ions, to form dihydrogen phosphate ions, which helps to prevent an increase in H^+ concentration. If OH^- are added to the solution, they combine with H^+ to form water. Then the conjugate acid, dihydrogen phosphate ions, dissociate to replace the H^+, which helps to prevent a decrease in H^+ concentration.

Visit the Online Learning Center at www.mhhe.com/seeley7 for chapter quizzes, interactive learning exercises, and other study tools.

Structure and Function of the Cell

Colorized scanning electron micrograph (SEM) of a dividing cell. Cell division is one of the major features of living things and is the basis of growth and repair of tissues and organs.

The cell is the basic structural and functional unit of all living organisms. The characteristic functions of cells include DNA replication, manufacture of macromolecules such as proteins and phospholipids, energy use, and reproduction. Cells are like very complex but minute factories that are always active, carrying out the functions of life.

These microscopic factories are so small that an average-sized cell is only one-fifth the size of the smallest dot you can make on a sheet of paper with a sharp pencil. Each human body is made up of trillions of cells. If each cell was the size of a standard brick, the colossal human statue made from those bricks would be 10 km high!

All the cells of an individual originate from a single fertilized cell. During development, cell division and specialization give rise to a wide variety of cell types, such as nerve, muscle, bone, fat, and blood cells. Each cell type has important characteristics that are critical to the normal function of the body as a whole. One of the important reasons for maintaining homeostasis is to keep the trillions of cells that form the body functioning normally.

This chapter outlines *functions of the cell* (58), *how we see cells* (60), the composition of the *plasma membrane* (60), *membrane lipids* (61), and *membrane proteins* (61). Then it addresses *movement through the plasma membrane* (65) and *endocytosis and exocytosis* (73). The chapter then addresses the *cytoplasm* (75) and *organelles* (78). It then presents an *overview of cell metabolism* (87), *protein synthesis* (88), and *cell life cycle* (92). Finally, some major *cellular aspects of aging* are discussed (96).

Functions of the Cell

Although cells may have quite different structures and functions, they share several common characteristics (figure 3.1; table 3.1). The **plasma** (plaz′mă), or **cell, membrane** forms the outer boundary of the cell, through which the cell interacts with its external environment. The **nucleus** (noo′klē-ŭs) is usually located centrally and functions to direct cell activities, most of which take place in the **cytoplasm** (sī′tō-plazm), located between the plasma membrane and the nucleus. Within cells, specialized structures called **organelles** (or′gă-nelz) perform specific functions.

The main functions of the cell include

1. *Basic unit of life.* The cell is the smallest part of an organism that still retains all the characteristics of life.
2. *Protection and support.* Cells produce and secrete various molecules that provide protection and support of the body. For example, bone cells are surrounded by a mineralized material, making bone a hard tissue that protects the brain and other organs and that supports the weight of the body.

3. *Movement.* All the movements of the body occur because of molecules located within specific cells such as muscle cells.
4. *Communication.* Cells produce and receive chemical and electrical signals that allow them to communicate with one another. For example, nerve cells communicate with one another and with muscle cells by chemical signals. Those signals stimulate muscle cells to contract.
5. *Cell metabolism and energy release.* The chemical reactions that occur within cells are referred to collectively as cell metabolism. Energy released during metabolism is used for cell activities, such as the synthesis of new molecules, muscle contraction, and heat production, which helps maintain body temperature.
6. *Inheritance.* Each cell contains a copy of the genetic information of the individual. Specialized cells are responsible for transmitting that genetic information to the next generation.

🔖 **1.   *What are the major functions of the cell?***

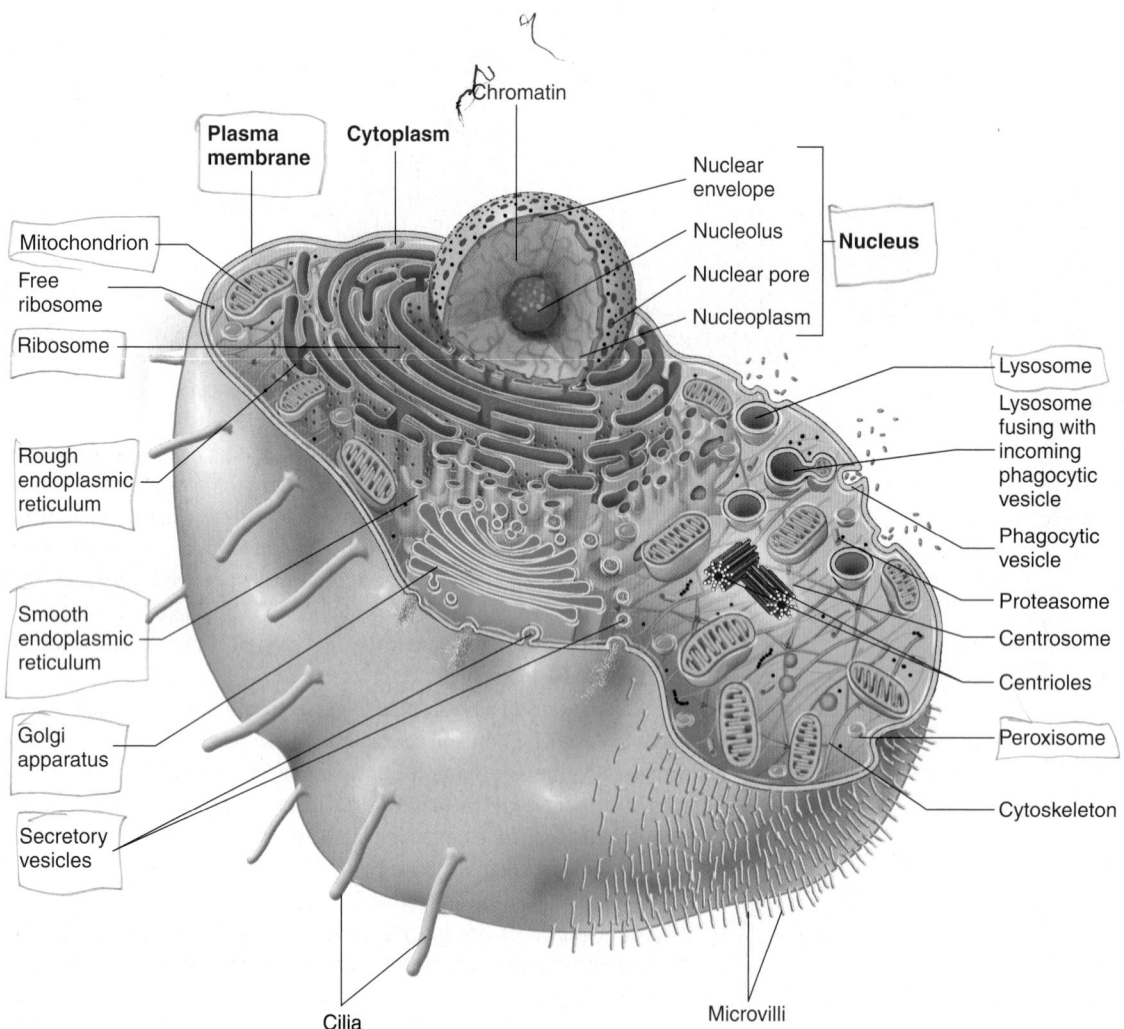

Figure 3.1   The Cell

A generalized human cell showing the plasma membrane, nucleus, and cytoplasm with its organelles. Although no single cell contains all these organelles, many cells contain a large number of them.

Table 3.1   Summary of Cell Parts and Functions

Cell Parts	Structure	Function
Plasma Membrane	Lipid bilayer composed of phospholipids and cholesterol with proteins that extend across or are embedded in either surface of the lipid bilayer	Outer boundary of cells that controls entry and exit of substances; receptor molecules function in intercellular communication; marker molecules enable cells to recognize one another
Cytoplasm: Cytosol		
Fluid part	Water with dissolved ions and molecules; colloid with suspended proteins	Contains enzymes that catalyze decomposition and synthesis reactions; ATP is produced in glycolysis reactions
Cytoskeleton		
Microtubules	Hollow cylinders composed of the protein tubulin; 25 nm in diameter	Support the cytoplasm and form centrioles, spindle fibers, cilia, and flagella; responsible for cell movements
Actin filaments	Small fibrils of the protein actin; 8 nm in diameter	Support the cytoplasm, support microvilli, responsible for cell movements
Intermediate filaments	Protein fibers; 10 nm in diameter	Support the cytoplasm
Cytoplasmic inclusions	Aggregates of molecules manufactured or ingested by the cell; may be membrane-bound	Function depends on the molecules: energy storage (lipids, glycogen), oxygen transport (hemoglobin), skin color (melanin), and others
Cytoplasm: Organelles		
Centrioles	Pair of cylindrical organelles in the centrosome, consisting of triplets of parallel microtubules	Centers for microtubule formation; determine cell polarity during cell division; form the basal bodies of cilia and flagella
Spindle fibers	Microtubules extending from the centrosome to chromosomes and other parts of the cell (i.e., aster fibers)	Assist in the separation of chromosomes during cell division
Cilia	Extensions of the plasma membrane containing doublets of parallel microtubules; 10 μm in length	Move materials over the surface of cells
Flagellum	Extension of the plasma membrane containing doublets of parallel microtubules; 55 μm in length	In humans, responsible for movement of spermatozoa
Microvilli	Extension of the plasma membrane containing microfilaments	Increase surface area of the plasma membrane for absorption and secretion; modified to form sensory receptors
Ribosome	Ribosomal RNA and proteins form large and small subunits; attached to endoplasmic reticulum or free ribosomes are distributed throughout the cytoplasm	Site of protein synthesis
Rough endoplasmic reticulum	Membranous tubules and flattened sacs with attached ribosomes	Protein synthesis and transport to Golgi apparatus
Smooth endoplasmic reticulum	Membranous tubules and flattened sacs with no attached ribosomes	Manufactures lipids and carbohydrates; detoxifies harmful chemicals; stores calcium
Golgi apparatus	Flattened membrane sacs stacked on each other	Modification, packaging, and distribution of proteins and lipids for secretion or internal use
Secretory vesicle	Membrane-bounded sac pinched off Golgi apparatus	Carries proteins and lipids to cell surface for secretion
Lysosome	Membrane-bounded vesicle pinched off Golgi apparatus	Contains digestive enzymes
Peroxisome	Membrane-bound vesicle	One site of lipid and amino acid degradation and breaks down hydrogen peroxide
Proteasomes	Tube-like protein complexes in the cytoplasm	Break down proteins in the cytoplasm
Mitochondria	Spherical, rod-shaped, or threadlike structures; enclosed by double membrane; inner membrane forms projections called cristae	Major site of ATP synthesis when oxygen is available
Nucleus		
Nuclear envelope	Double membrane enclosing the nucleus; the outer membrane is continuous with the endoplasmic reticulum; nuclear pores extend through the nuclear envelope	Separates nucleus from cytoplasm and regulates movement of materials into and out of the nucleus
Chromatin	Dispersed thin strands of DNA, histones, and other proteins; condenses to form chromosomes during cell division	DNA regulates protein (e.g., enzyme) synthesis and therefore the chemical reactions of the cell; DNA is the genetic or hereditary material
Nucleolus	One to four dense bodies consisting of ribosomal RNA and proteins	Assembly site of large and small ribosomal subunits

How We See Cells

Most cells are too small to be seen with the unaided eye. As a result, it is necessary to use microscopes to study them. **Light microscopes** allow us to visualize general features of cells. **Electron microscopes,** however, must be used to study the fine structure of cells. A **scanning electron microscope (SEM)** allows us to see features of the cell surface and the surfaces of internal structures. A **transmission electron microscope (TEM)** allows us to see "through" parts of the cell and thus to discover other aspects of cell structure. If you are not somewhat familiar with these types of microscopes, you should turn to the discussion on microscopic imaging on p. 106.

> ● 2. *What are the differences between light and electron microscopes?*

Plasma Membrane

The plasma membrane is the outermost component of a cell. Substances inside the plasma membrane are **intracellular** and substances outside the cell are **extracellular.** Sometimes extracellular substances are referred to as **intercellular,** meaning between cells. The plasma membrane encloses and supports the cell contents. It attaches cells to the extracellular environment or to other cells. The ability of cells to recognize and communicate with each other takes place through the plasma membrane. In addition, the plasma membrane determines what moves into and out of cells. As a result, the intracellular contents of cells is different from the extracellular environment.

The regulation of ion movement by cells results in a charge difference across the plasma membrane called the **membrane potential.** The outside of the plasma membrane is positively charged compared to the inside because there are more positively charged ions immediately on the outside of the plasma membrane and more negatively charged ions and proteins inside. The membrane potential allows cells to function like tiny batteries with a positive and negative pole. It is an important feature of a living cell's normal function, which will be considered in greater detail in chapters 9 and 11.

The plasma membrane consists of 45%–50% lipids, 45%–50% proteins, and 4%–8% carbohydrates (figure 3.2). The carbohydrates combine with lipids to form glycolipids and with proteins to form glycoproteins. The **glycocalyx** (glī-kō-kā′liks) is

(a)

Integral protein

Receptor protein

Nonpolar (hydrophobic) regions of phospholipid molecules

Polar (hydrophilic) regions of phospholipid molecules

Peripheral protein

Membrane channel

Carbohydrate chains

Glycoprotein

Glycocalyx

Glycolipid

External membrane surface

Phospholipid bilayer

Cholesterol

Internal membrane surface

Cytoskeleton

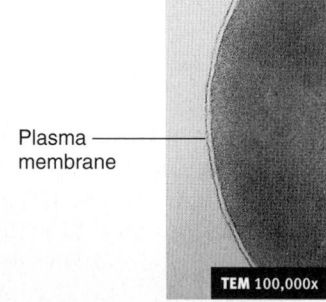

(b)

Plasma membrane

TEM 100,000x

Figure 3.2   Plasma Membrane

(*a*) Fluid-mosaic model of the plasma membrane. The membrane is composed of a bilayer of phospholipids and cholesterol with proteins "floating" in the membrane. The nonpolar hydrophobic region of each phospholipid molecule is directed toward the center of the membrane and the polar hydrophilic region is directed toward the water environment either outside or inside the cell. (*b*) Transmission electron micrograph of a plasma membrane. Proteins at either surface of the lipid bilayer stain more readily than the lipid bilayer does and give the membrane the appearance of consisting of three parts: the two dark outer parts are proteins and the phospholipid heads, and the lighter central part is the phospholipid tails and cholesterol.

the collection of glycolipids, glycoproteins, and carbohydrates on the outer surface of the plasma membrane. The glycocalyx also contains molecules absorbed from the extracellular environment, so there is often no precise boundary where the plasma membrane ends and the extracellular environment begins.

● 3. *Define intracellular, extracellular, and intercellular.*
  4. *What is the membrane potential? Is the outside of the plasma membrane positively or negatively charged compared to the inside.*
  5. *What is the glycocalyx? What types of molecules are found in the glycocalyx?*

Membrane Lipids

The predominant lipids of the plasma membrane are phospholipids and cholesterol. **Phospholipids** readily assemble to form a **lipid bilayer,** a double layer of lipid molecules, because they have a polar (charged) head and a nonpolar (uncharged) tail (see chapter 2). The polar **hydrophilic** (water-loving) heads are exposed to water inside and outside the cell, whereas the nonpolar **hydrophobic** (water-fearing) tails face one another in the interior of the plasma membrane. The other major lipid in the plasma membrane is **cholesterol** (see chapter 2), which is interspersed among the phospholipids and accounts for about a third of the total lipids in the plasma membrane. The hydrophilic OH group of cholesterol extends between the phospholipid heads to the hydrophilic surface of the membrane and the hydrophobic part of the cholesterol molecule lies within the hydrophobic region of the phospholipids. The amount of cholesterol in a given membrane is a major factor in determining the fluid nature of the membrane, which is critical to its function.

● 6. *How do hydrophilic heads and hydrophobic tails of phospholipid molecules result in a plasma membrane?*
  7. *What is the function of cholesterol in plasma membranes?*

Membrane Proteins

The basic structure of the plasma membrane and some of its functions are determined by its lipids, but many functions of the plasma membrane are determined by its proteins. The modern concept of the plasma membrane, the **fluid-mosaic model,** suggests that the plasma membrane is neither rigid nor static in structure but is highly flexible and can change its shape and composition through time. The lipid bilayer functions as a liquid in which other molecules such as proteins are suspended. The fluid nature of the lipid bilayer has several important consequences. It provides an important means of distributing molecules within the plasma membrane. In addition, slight damage to the membrane can be repaired because the phospholipids tend to reassemble around damaged sites and seal them closed. The fluid nature of the lipid bilayer also enables membranes to fuse with one another.

Some protein molecules, called **integral,** or **intrinsic, proteins,** penetrate deeply into the lipid bilayer, in many cases, extending from one surface to the other (figure 3.3), whereas other proteins, called **peripheral,** or **extrinsic, proteins,** are attached to either the inner or outer surfaces of the lipid bilayer. Integral proteins consist of regions made up of amino acids with hydrophobic R groups and other regions of amino acids with hydrophilic R

(a)

(b)

Figure 3.3   Proteins in the Plasma Membrane

(*a*) Proteins are commonly depicted as ribbons (see chapter 2). The space occupied by the protein ribbon can be enclosed by a three-dimensional shaded region. (*b*) The shaded region can be depicted as a three-dimensional integral protein inserted into the plasma membrane.

groups (see chapter 2). The hydrophobic regions are located within the hydrophobic part of the membrane, and the hydrophilic regions are located at the inner or outer surface of the membrane or line channels through the membrane. Peripheral proteins are usually bound to integral proteins. Membrane proteins are markers, attachment sites, channels, receptors, enzymes, or carriers. The ability of membrane proteins to function depends on their three-dimensional shapes and their chemical characteristics.

Marker Molecules

Marker molecules are cell surface molecules that allow cells to identify one another or other molecules. They are mostly **glycoproteins** (proteins with attached carbohydrates) or **glycolipids** (lipids with attached carbohydrates). The protein portions of glycoproteins may be either integral or peripheral proteins (figure 3.4). Examples include recognition of the oocyte by the sperm cell and the ability of the immune system to distinguish between self-cells and foreign cells, such as bacteria or donor cells in an organ transplant. Intercellular communication and recognition are important because cells are not isolated entities and they must work together to ensure normal body functions.

Attachment Sites

Membrane-bound proteins, such as **integrins,** function as **attachment sites,** where cells attach to other cells or to extracellular molecules (figure 3.5). These membrane proteins also attach to intracellular molecules. Integrins function in pairs of two integral proteins, which interact with both intracellular and extracellular molecules.

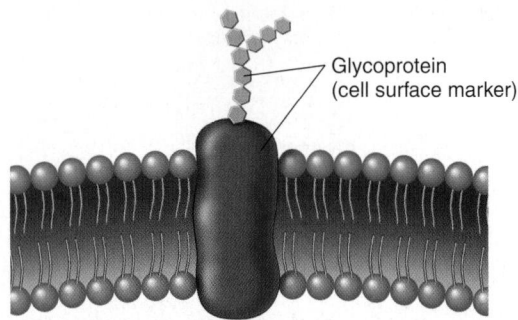

Figure 3.4   **Cell Surface Marker**

Glycoproteins that extend to the cell surface allow cells to identify one another or other molecules.

Figure 3.5   **Attachment Sites**

Proteins (integrins) in the plasma membrane attach to extracellular molecules.

Channel Proteins

Channel proteins are one or more integral proteins arranged so that they form a tiny channel through the plasma membrane (figure 3.6). The hydrophobic regions of the proteins face outward toward the hydrophobic part of the plasma membrane, and the hydrophilic regions of the protein face inward and line the channel. Small molecules or ions of the right shape, size, and charge can pass through the channel. The charges in the hydrophilic part of the channel proteins determine which types of ions can pass through the channel.

     Some channel proteins, called **nongated ion channels,** are always open and are responsible for the permeability of the plasma membrane to ions when the plasma membrane is at rest. Other channels can be open or closed. Some channel proteins open in response to **chemical signals,** or **ligands** (lig′andz, lī′gandz). Ligands are small molecules that bind to proteins or glycoproteins. This is called a **ligand-gated ion channel.** Other channel proteins open the channel when there is a change in charge across the plasma membrane. This is called a **voltage-gated ion channel.**

Receptor Molecules

Receptor molecules (figure 3.7) are proteins in the plasma membrane with an exposed **receptor site** on the outer cell surface, which can attach to specific ligand molecules. Many receptors and the chemical signals they bind are part of an intercellular commu-

a. Some regions of a protein are helical. Each helical region can be depicted as a cylinder.

Protein

b. In some membrane proteins, the helical regions form a circle with a channel in the center.

c. The ring of cylinders can be depicted as a 3-D globular structure with a channel in the center This is called a channel protein.

d. The channel protein can be depicted cut in half to show the channel.

e. The cut channel protein is depicted within the plasma membrane.

Figure 3.6   **Channel Protein**

nication system that facilitates coordination of cell activities. One cell can release a chemical signal that diffuses to another cell and binds to its receptor. The binding acts as a signal that triggers a response. The same chemical signal would have no effect on other cells that lack the specific receptor molecule.

Receptors Linked to Channel Proteins

Some membrane-bound receptors are protein molecules that are part of ligand-gated ion channels in the plasma membrane. When chemical signals, or ligands, bind to these receptors, the

Figure 3.7 **Receptor Protein**

A protein in the plasma membrane with an exposed receptor site, which can attach to specific chemical signals.

combination alters the three-dimensional structure of the proteins of the ion channels, causing the channels either to open or close. The result is a change in the permeability of the plasma membrane to the specific ions passing through the ion channels (figure 3.8). For example, acetylcholine released from nerve cells is a ligand that combines with membrane-bound receptors of skeletal muscle cells. The combination of acetylcholine molecules with the receptor sites of the membrane-bound receptors for acetylcholine opens Na^+ channels in the plasma membrane. Consequently, the ions diffuse into the skeletal muscle cells and trigger events that cause them to contract.

Cystic Fibrosis

Cystic fibrosis is a genetic disorder that affects chloride ion channels. Three types of cystic fibrosis exist. In about 70% of cases, a defective channel protein is produced that fails to reach the plasma membrane from its site of production inside the cell. In the remaining cases, the channel protein is incorporated into the plasma membrane but does not function normally. In some cases, the channel protein fails to bind ATP. In others, ATP binds to the channel protein, but the channel does not open. Failure of these ion channels to function results in the affected cells producing thick, viscous secretions. Although cystic fibrosis affects many cell types, its most profound effects are in the pancreas, causing an inability to digest certain types of food, and in the lungs, where it causes extreme difficulty in breathing.

Receptors Linked to G Proteins

Some membrane-bound receptor molecules function by altering the activity of a **G protein complex** located on the inner surface of the plasma membrane (figure 3.9). The G protein complex consists of three proteins, called alpha, beta, and gamma proteins. A G protein complex attached to a receptor that does not have a chemical signal bound to it is inactive and has guanosine diphosphate (GDP) attached to the alpha subunit (figure 3.9 *1*). When a chemical signal attaches to the receptor, the alpha subunit releases GDP, binds to guanosine triphosphate (GTP), and is activated. The G protein complex separates from the receptor and the alpha subunit separates from the beta and gamma subunits (figure 3.9 *2*). The activated alpha subunit stimulates a cell response, often by means of intracellular chemical signals. Some activated G protein complexes open channels in the plasma

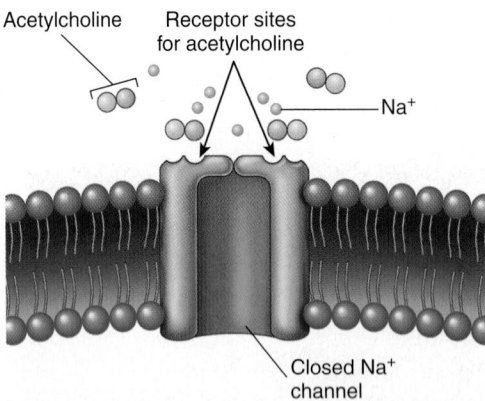

1. The Na^+ channel has receptor sites for the chemical signal, acetylcholine. When the receptor sites are not occupied by acetylcholine, the Na^+ channel remains closed.

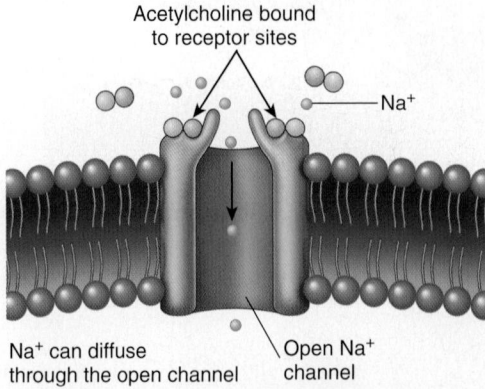

2. When two acetylcholine molecules bind to their receptor sites on the Na^+ channel, the channel opens to allow Na^+ to diffuse through the channel into the cell.

Process Figure 3.8 **Receptors Linked to a Channel Protein**

membrane and others activate enzymes associated with the plasma membrane.

Drugs and Receptors

Drugs with structures similar to specific chemical signals may compete with those chemical signals for their receptor sites. Depending on the exact characteristics of a drug, it may either bind to a receptor site and activate the receptor or bind to a receptor site and inhibit the action of the receptor. For example, drugs exist that compete with the chemical signal epinephrine for its receptor sites. Some of these drugs activate epinephrine receptors and others inhibit them.

Enzymes in the Plasma Membrane

Some membrane proteins function as **enzymes,** which can catalyze chemical reactions on either the inner or outer surface of the plasma membrane. For example, some enzymes on the surface of cells in the small intestine break the peptide bonds of dipeptides to form two single amino acids (figure 3.10). Some membrane-associated enzymes are always active. Others are activated by membrane-bound receptors or G proteins.

1. A G protein complex attached to a receptor without a bound chemical signal has guanosine diphosphate (GDP) bound to it and is inactive.

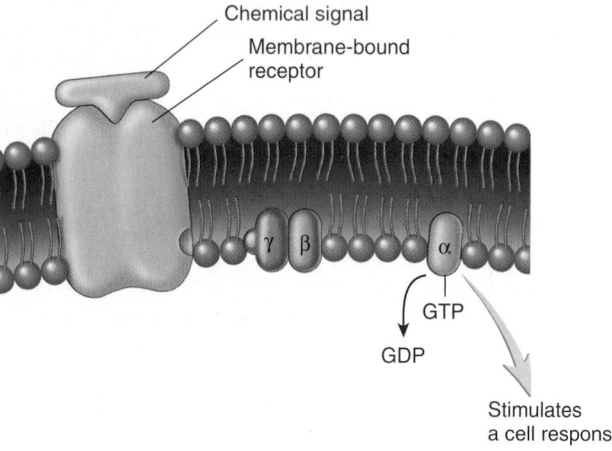

2. When a chemical signal attaches to the receptor, guanosine triphosphate (GTP) replaces GDP on the α-subunit of the G protein complex, which separates from the other subunits. The α-subunit, with GTP attached, stimulates a cell response.

Process Figure 3.9   A Receptor Linked to a G Protein Complex

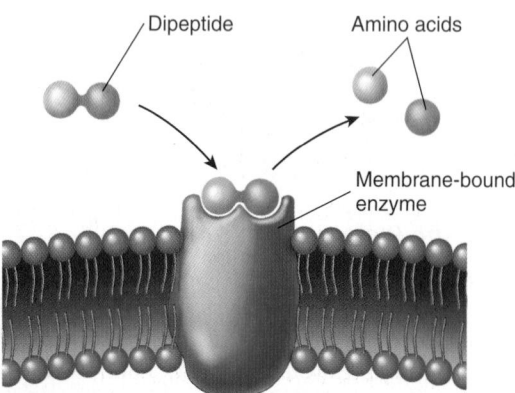

Figure 3.10   Enzyme in the Plasma Membrane

This enzyme in the plasma membrane breaks the peptide bond of a dipeptide to produce two amino acids.

Carrier Proteins

Carrier proteins are integral membrane proteins that move ions or molecules from one side of the plasma membrane to the other. The carrier proteins have specific binding sites to which ions or molecules attach on one side of the plasma membrane. The carrier proteins change shape to move the bound ions or molecules to the other side of the plasma membrane where they are released (figure 3.11)

8. *Describe the difference between integral and peripheral proteins in the plasma membrane. Define glycolipid and glycoprotein.*
9. *List two functions of marker molecules.*
10. *Describe and give the function of integrins.*
11. *Define nongated ion channel, ligand-gated ion channel, and voltage-gated ion channel. What determines the function of a channel protein?*

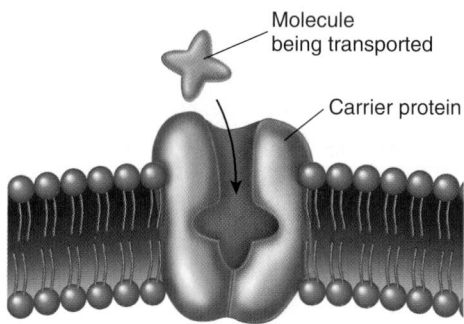

1. A molecule enters the carrier protein from one side of the plasma membrane.

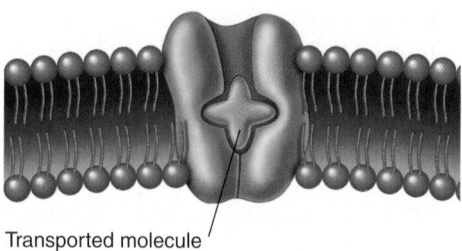

2. The carrier protein briefly binds the transported molecule.

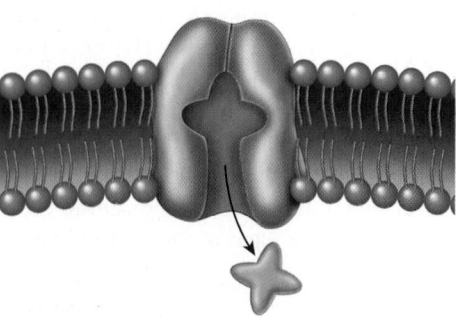

3. The carrier protein changes shape and releases the transported molecule on the other side of the plasma membrane.

Process Figure 3.11   Carrier Protein

12. *To what part of a receptor molecule does a chemical signal attach? Give two examples of how a chemical signal molecule can bind to a receptor in the plasma membrane and cause a response in the cell.*

13. *Give an example of the action of an enzyme in the plasma membrane.*

Movement Through the Plasma Membrane

The plasma membrane separates the extracellular material from the intracellular material and is **selectively permeable,** that is, it allows only certain substances to pass through it. The intracellular material has a different composition from the extracellular material, and the survival of the cell depends on the maintenance of these differences. Enzymes, other proteins, glycogen, and potassium ions are found in higher concentrations intracellularly; sodium, calcium, and chloride ions are found in greater concentrations extracellularly. In addition, nutrients must continually enter the cell, and waste products must exit, but the volume of the cell remains unchanged. Because of the plasma membrane's permeability characteristics and its ability to transport molecules selectively, the cell is able to maintain homeostasis. Rupture of the membrane, alteration of its permeability characteristics, or inhibition of transport processes can disrupt the normal concentration differences across the plasma membrane and lead to cell death.

Molecules and ions can pass through the plasma membrane in four ways:

1. *Directly through the phospholipid membrane.* Molecules that are soluble in lipids, such as oxygen, carbon dioxide, and steroids, pass through the plasma membrane readily by dissolving in the lipid bilayer. The phospholipid bilayer acts as a barrier to most substances that are not lipid-soluble; but some small, nonlipid-soluble molecules, such as carbon dioxide and urea, can diffuse between the phospholipid molecules of the plasma membrane.

2. *Membrane channels.* There are several types of protein channels through the plasma membrane. Each channel type allows only certain molecules to pass through it. The size, shape, and charge of molecules determines whether they can pass through a given channel. For example, sodium ions pass through sodium channels, and potassium and chloride ions pass through potassium and chloride channels, respectively. Rapid movement of water across the cell membrane apparently occurs through membrane channels.

3. *Carrier molecules.* Large polar molecules that are not lipid-soluble, such as glucose and amino acids, cannot pass through the cell membrane in significant amounts unless they are transported by carrier molecules. Substances that are transported across the cell membrane by carrier molecules are said to be transported by **carrier-mediated processes.** Carrier proteins bind to specific molecules and transport them across the cell membrane. Carrier molecules that transport glucose across the cell membrane do not transport amino acids, and carrier molecules that transport amino acids do not transport glucose.

4. *Vesicles.* Large nonlipid-soluble molecules, small pieces of matter, and even whole cells can be transported across the cell membrane in a **vesicle,** which is a small sac surrounded by a membrane. Because of the fluid nature of membranes, the vesicle and the cell membrane can fuse, allowing the contents of the vesicle to cross the cell membrane.

14. *List four ways that substances move across the plasma membrane.*

Diffusion

A solution consists of one or more substances called **solutes** dissolved in the predominant liquid or gas, which is called the **solvent. Diffusion** is the movement of solutes from an area of higher concentration to an area of lower concentration in solution (figure 3.12). Diffusion is a product of the constant random

Concentration gradient for salt — Distilled water

1. When a salt crystal (*green area*) is placed into a beaker of water, there is a concentration gradient for salt from the salt crystal to the water that surrounds it. There is also a concentration gradient for water molecules from the water toward the salt crystal.

2. Salt ions (*green area*) move down their concentration gradient (from an area of high concentration toward an area of low concentration) into the water.

3. Salt ions and water molecules are distributed evenly throughout the solution. Even though the salt ions and water molecules continue to move randomly, an equilibrium exists, and no net movement occurs because no concentration gradient exists.

Process Figure 3.12 **Diffusion**

motion of all atoms, molecules, or ions in a solution. Because more solute particles exist in an area of higher concentration than in an area of lower concentration and because the particles move randomly, the chances are greater that solute particles will move from the higher to the lower concentration than in the opposite direction. Thus the overall, or net, movement is from the area of higher concentration to that of lower concentration. At equilibrium, the net movement of solutes stops, although the random molecular motion continues, and the movement of solutes in any one direction is balanced by an equal movement in the opposite direction. The movement and distribution of smoke or perfume throughout a room in which no air currents exist or of a dye throughout a beaker of still water are examples of diffusion.

A concentration difference exists when the concentration of a solute is greater at one point than at another point in a solvent. The concentration difference between two points, divided by the distance between the two points, is called the **concentration, or density gradient.** Solutes diffuse down their concentration gradients (from a higher to a lower concentration) until an equilibrium is achieved. The greater the concentration gradient, the greater is the rate of diffusion of a solute down that gradient. Increasing the concentration difference or decreasing the distance between the two points increases the concentration gradient, whereas decreasing the concentration difference or increasing the distance between the two points decreases the concentration gradient.

The rate of diffusion is influenced by the magnitude of the concentration gradient, the temperature of the solution, the size of the diffusing molecules, and the viscosity of the solvent. The greater the concentration gradient, the greater is the number of solute particles moving from a higher to a lower concentration. As the temperature of a solution increases, the speed at which all molecules move increases, resulting in a greater diffusion rate. Small molecules diffuse through a solution more readily than do large ones. **Viscosity** is a measure of how easily a liquid flows; thick solutions, such as syrup, are more viscous than water. Diffusion occurs more slowly in viscous solvents than in thin, watery solvents.

Diffusion of molecules is an important means by which substances move between the extracellular and intracellular fluids in the body. Substances that can diffuse through either the lipid bilayer or the membrane channels can pass through the plasma membrane (figure 3.13). Some nutrients enter and some waste products leave the cell by diffusion, and maintenance of the appropriate intracellular concentration of these substances depends to a large degree on diffusion. For example, if the extracellular concentration of oxygen is reduced, inadequate oxygen diffuses into the cell, and normal cell function cannot occur. Some lipid-soluble chemical signals can diffuse through the plasma membrane and attach to receptors inside the cell (figure 3.14).

15. *Define solute, solvent, and concentration gradient. Do solutes diffuse with (down) or against their concentration gradient?*

16. *How is the rate of diffusion affected by an increased concentration gradient? By increased temperature of a solution? By increased viscosity of the solvent?*

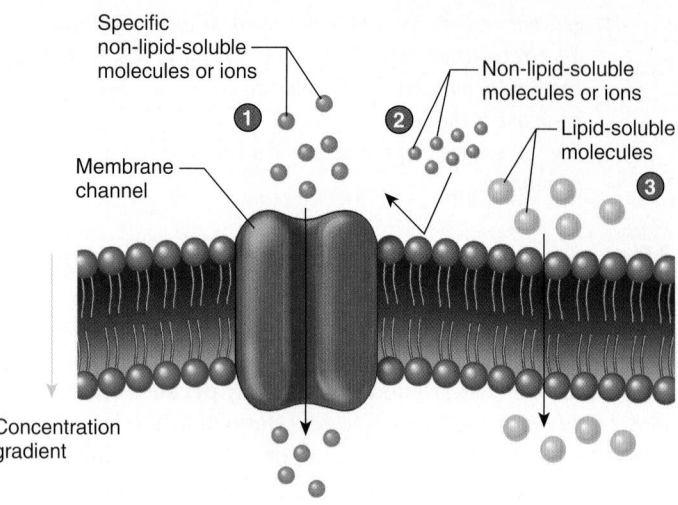

1. Certain, specific non-lipid-soluble molecules or ions diffuse through membrane channels.

2. Other non-lipid-soluble molecules or ions, for which membrane channels are not present in the cell, cannot enter the cell.

3. Lipid-soluble molecules diffuse directly through the plasma membrane.

Process Figure 3.13 **Diffusion Through the Plasma Membrane**

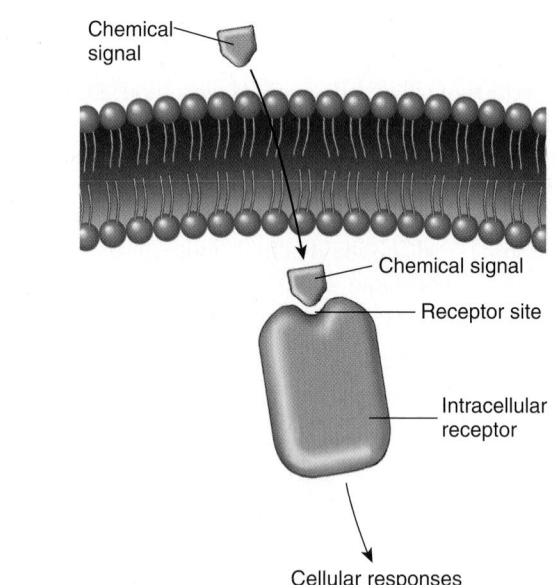

Figure 3.14 **Intracellular Receptor**

This small, lipid-soluble chemical signal diffuses through the plasma membrane and combines with the receptor site of an intracellular receptor.

P R E D I C T **1**

Urea is a toxic waste produced inside cells. It diffuses from the cells into the blood and is eliminated from the body by the kidneys. What would happen to the intracellular and extracellular concentration of urea if the kidneys stopped functioning?

Osmosis

Osmosis (os-mō′sis) is the diffusion of water (solvent) across a selectively permeable membrane, such as a plasma membrane. A selectively permeable membrane is a membrane that allows water but not all the solutes dissolved in the water to diffuse through the membrane. Water diffuses from a solution with proportionately more water, across a selectively permeable membrane, and into a solution with proportionately less water. Because solution concentrations are defined in terms of solute concentrations and not in terms of water content (see chapter 2), water diffuses from the less

concentrated solution (fewer solutes, more water) into the more concentrated solution (more solutes, less water). Osmosis is important to cells because large volume changes caused by water movement disrupt normal cell function.

Osmotic pressure is the force required to prevent the movement of water by osmosis across a selectively permeable membrane. The osmotic pressure of a solution can be determined by placing the solution into a tube that is closed at one end by a selectively permeable membrane (figure 3.15). The tube is then immersed in distilled water. Water molecules move by

* Because the tube contains salt ions (*green and red spheres*) as well as water molecules (*blue spheres*), the tube has proportionately less water than is in the beaker, which contains only water. The water molecules diffuse with their concentration gradient into the tube (*blue arrows*). Because the salt ions cannot leave the tube, the total fluid volume inside the tube increases, and fluid moves up the glass tube (*black arrow*) as a result of osmosis.

1. The end of a tube containing a 3% salt solution (*green*) is closed at one end with a selectively permeable membrane, which allows water molecules to pass through it but retains the salt ions within the tube.

2. The tube is immersed in distilled water. Water moves into the tube by osmosis (see inset above*). The concentration of salt in the tube decreases as water rises in the tube (*lighter green color*).

3. Water moves by osmosis into the tube until the weight of the column of water in the tube (hydrostatic pressure) prevents further movement of water into the tube. The hydrostatic pressure that prevents net movement of water into the tube is equal to the osmotic pressure of the solution in the tube.

Process Figure 3.15 Osmosis

osmosis through the membrane into the tube, forcing the solution to move up the tube. As the solution rises into the tube, its weight produces hydrostatic pressure that moves water out of the tube back into the distilled water surrounding the tube. At equilibrium, net movement of water stops, which means the movement of water into the tube by osmosis is equal to the movement of water out of the tube caused by hydrostatic pressure. The osmotic pressure of the solution in the tube is equal to the hydrostatic pressure that prevents net movement of water into the tube.

The osmotic pressure of a solution provides information about the tendency for water to move by osmosis across a selectively permeable membrane. Because water moves from less concentrated solutions (fewer solutes, more water) into more concentrated solutions (more solutes, less water), the greater the concentration of a solution (the less water it has), the greater the tendency for water to move into the solution, and the greater the osmotic pressure to prevent that movement. Thus, the greater the concentration of a solution, the greater the osmotic pressure of the solution, and the greater the tendency for water to move into the solution.

P R E D I C T   2
Given the demonstration in figure 3.15, what would happen to osmotic pressure if the membrane were not selectively permeable but instead allowed all solutes and water to pass through it?

Three terms describe the osmotic pressure of solutions. Solutions with the same concentration of solute particles (see chapter 2) have the same osmotic pressure and are referred to as **isosmotic**

(ī′sos-mot′ik). The solutions are still isosmotic even if the types of solute particles in the two solutions differ from each other. If one solution has a greater concentration of solute particles and therefore a greater osmotic pressure than another solution, the first solution is said to be **hyperosmotic** (hī′per-oz-mot′ik) compared to the more dilute solution. The more dilute solution, with the lower osmotic pressure, is **hyposmotic** (hī-pos-mot′ik) compared to the more concentrated solution.

Three additional terms describe the tendency of cells to shrink or swell when placed into a solution. If a cell is placed into a solution in which it neither shrinks nor swells, the solution is said to be **isotonic** (ī-sō-ton′ik). In an isotonic solution, the shape of the cell remains constant, maintaining its internal tension or tone, a condition called **tonicity** (tō-nis′i-tē). If a cell is placed into a solution and water moves out of the cell by osmosis, causing the cell to shrink, the solution is called **hypertonic** (hī-per-ton′ik). If a cell is placed into a solution and water moves into the cell by osmosis, causing the cell to swell, the solution is called **hypotonic** (hī-pō-ton′ik) (figure 3.16a).

An isotonic solution may be isosmotic to the cytoplasm. Because isosmotic solutions have the same concentration of solutes and water as the cytoplasm of the cell, no net movement of water occurs, and the cell neither swells nor shrinks (figure 3.16b). Hypertonic solutions can be hyperosmotic and have a greater concentration of solute molecules and a lower concentration of water than the cytoplasm of the cell. Therefore water moves by osmosis from the cell into the hypertonic solution, causing the cell to shrink, a process called **crenation** (krē-nā′shŭn) in red blood cells (figure 3.16c). Hypotonic solutions can be hyposmotic and have a

(a) A hypotonic solution with a low solute concentration results in swelling of the red blood cell placed into the solution. Water enters the cell by osmosis (black arrows), and the red blood cell lyses (bursts; puff of red in the lower part of the cell).

(b) An isotonic solution with a concentration of solutes equal to that inside the cell results in a normally shaped red blood cell. Water moves into and out of the cell at the same rate (black arrows), but there is no net water movement.

(c) A hypertonic solution, with a high solute concentration, causes shrinkage (crenation) of the red blood cell as water moves by osmosis out of the cell and into the hypertonic solution (black arrows).

Figure 3.16   Effects of Hypotonic, Isotonic, and Hypertonic Solutions on Red Blood Cells smaller concentration of solute molecules and a greater concentration of water than the cytoplasm of the cell. Therefore water moves by osmosis into the cell, causing it to swell. If the cell swells enough, it can rupture, a process called **lysis** (lī′sis) (see figure 3.16a). Solutions injected into the circulatory system or the tissues must be isotonic because shrinkage or swelling of cells disrupts their normal function and can lead to cell death.

The -*osmotic* terms refer to the concentration of the solutions, and the -*tonic* terms refer to the tendency of cells to swell or shrink. These terms should not be used interchangeably. Not all isosmotic solutions are isotonic. For example, it is possible to prepare a solution of glycerol and a solution of mannitol that are isosmotic to the cytoplasm of the cell. Because the solutions are isosmotic, they have the same concentration of solutes and water as the cytoplasm. Glycerol, however, can diffuse across the plasma membrane, and mannitol cannot. When glycerol diffuses into the cell, the solute concentration of the cytoplasm increases, and its water concentration decreases. Therefore, water moves by osmosis into the cell, causing it to swell, and the glycerol solution is both isosmotic and hypotonic. In contrast, mannitol cannot enter the cell, and the isosmotic mannitol solution is also isotonic.

- 17. *Define osmosis and osmotic pressure. As the concentration of a solution increases, what happens to its osmotic pressure and to the tendency for water to move into it?*
- 18. *Compare isosmotic, hyperosmotic, and hyposmotic solutions to isotonic, hypertonic, and hypotonic solutions. What type of solution causes crenation of a cell? What type of solution causes lysis of a cell?*

Filtration

Filtration results when a partition containing small holes is placed in a stream of moving liquid. The partition works like a minute sieve. Particles small enough to pass through the holes move through the partition with the liquid, but particles larger than the holes are prevented from moving beyond the partition. In contrast to diffusion, filtration depends on a pressure difference on either side of the partition. The liquid moves from the side of the partition with the greater pressure to the side with the lower pressure.

Filtration occurs in the kidneys as a step in urine formation. Blood pressure moves fluid from the blood through a partition, or filtration membrane. Water, ions, and small molecules pass through the partition, whereas most proteins and blood cells remain in the blood.

- 19. *Define filtration and give an example of where it occurs in the body.*

Mediated Transport Mechanisms

Many essential molecules, such as amino acids and glucose, cannot enter the cell by diffusion, and many products, such as proteins, cannot exit the cell by diffusion. **Mediated transport mechanisms** involve carrier proteins or channels within the plasma membrane that move large, water-soluble molecules or electrically charged molecules or ions across the plasma membrane. Once a molecule to be transported binds to a carrier protein on one side of the membrane, the three-dimensional shape of the carrier protein changes, and the transported molecule is moved to the opposite side of the membrane (see figure 3.11). The carrier protein then resumes its original shape and is available to transport other molecules.

Mediated transport mechanisms have three characteristics: specificity, competition, and saturation. **Specificity** means that each carrier protein or ion channel binds to and transports only a single type of molecule or ion (figure 3.17a). For example, the carrier protein that transports glucose does not transport amino acids or ions. The chemical structure of the binding site determines the specificity of the carrier protein (see figure 3.11). **Competition** is the result of similar molecules binding to the carrier protein. Although the binding sites of carrier proteins exhibit specificity, closely related substances, in which regions of two different molecules have the same shape, may bind to the same binding site. The substance in the greater concentration or the substance that binds to the binding site more readily is transported across the plasma membrane at the greater rate (figure 3.17b). **Saturation** means that the rate of transport of molecules across the membrane is limited by the number of available carrier proteins. As the concentration of a transported substance increases,

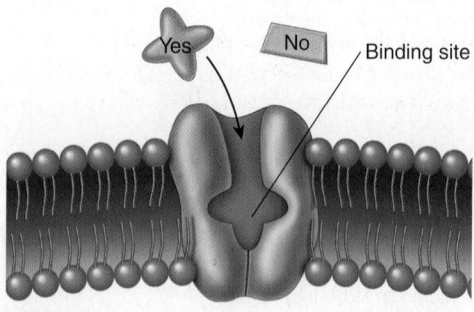

(a) Specificity. Only molecules that are the right shape to bind to the binding site are transported.

(b) Competition. Similarly shaped molecules can compete for the same binding site.

Figure 3.17   Mediated Transport: Specificity and Competition more carrier proteins have their binding sites occupied. The rate at which the substance is transported increases; however, once the concentration of the substance is increased so that all the binding sites are occupied, the rate of transport remains constant, even though the concentration of the substance increases further (figure 3.18).

Ion channels often are thought of as simple tubes, with or without gates, through which ions pass. Many ion channels, however, appear to be more complex than once thought. It now appears that ions briefly bind to specific sites inside channels and there is a change in the shape of those channels as ions are transported through them. The size and charge within a channel determines the channel's specificity. For example, Na^+ channels do not transport K^+, and vice versa. In addition, similar ions moving into and binding within an ion channel are in competition with each other. Furthermore, the number of ions moving into an ion chan-nel can exceed the capacity of the channel, thus saturating the channel. Therefore, ion channels exhibit specificity, competition, and saturation.

Three kinds of mediated transport exist: facilitated diffusion, active transport, and secondary active transport.

Facilitated Diffusion

Facilitated diffusion is a carrier-mediated or channel-mediated process that moves substances into or out of cells from a higher to a lower concentration (figure 3.19). Facilitated diffusion does not require metabolic energy to transport substances across the plasma membrane. The rate at which molecules or ions are transported is directly proportional to their concentration gradient up to the point of saturation, when all the carrier proteins or channels are occupied. Then the rate of transport remains constant at its maxi-mum rate.

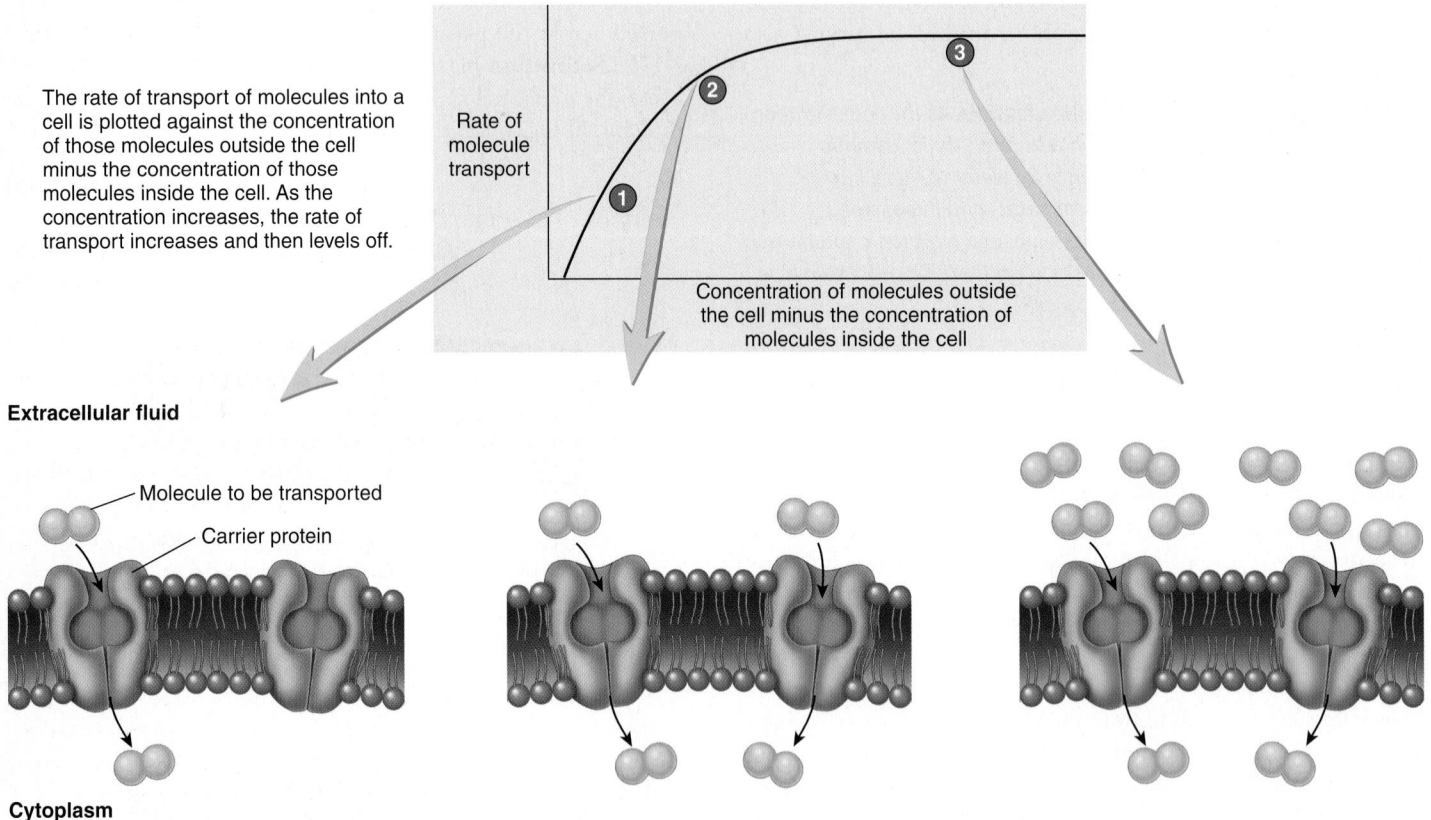

The rate of transport of molecules into a cell is plotted against the concentration of those molecules outside the cell minus the concentration of those molecules inside the cell. As the concentration increases, the rate of transport increases and then levels off.

Rate of molecule transport

Concentration of molecules outside the cell minus the concentration of molecules inside the cell

Extracellular fluid

Molecule to be transported

Carrier protein

Cytoplasm

1. When the concentration of molecules outside the cell is low, the transport rate is low because it is limited by the number of molecules available to be transported.

2. When more molecules are present outside the cell, as long as enough carrier proteins are available, more molecules can be transported, and therefore the transport rate increases.

3. The transport rate is limited by the number of carrier proteins and the rate at which each carrier protein can transport solutes. When the number of molecules outside the cell is so large that the carrier proteins are all occupied, the system is saturated and the transport rate cannot increase.

Process Figure 3.18 Saturation of a Carrier Protein

Extracellular fluid

Carrier protein

Concentration gradient

Cytoplasm

Transported molecule (glucose)

1. The carrier protein binds with a molecule, such as glucose, on the outside of the plasma membrane.

2. The carrier protein changes shape and releases the molecule on the inside of the plasma membrane.

Process Figure 3.19   Facilitated Diffusion

P R E D I C T   ③

The transport of glucose into and out of most cells, such as muscle and fat cells, occurs by facilitated diffusion. Once glucose enters a cell, it is rapidly converted to other molecules, such as glucose-6-phosphate or glycogen. What effect does this conversion have on the ability of the cell to acquire glucose? Explain.

Active Transport

Active transport is a mediated transport process that requires energy provided by ATP (figure 3.20). Movement of the transported substance to the opposite side of the membrane and its subsequent release from the carrier protein are fueled by the breakdown of ATP. The maximum rate at which active transport proceeds depends on the number of carrier proteins in the plasma membrane and the availability of adequate ATP. Active-transport processes are important because they can move substances against their concentration gradients, that is, from lower concentrations to

higher concentrations. Consequently, they have the ability to accumulate substances on one side of the plasma membrane at concentrations many times greater than those on the other side. Active transport can also move substances from higher to lower concentrations.

Some active-transport mechanisms exchange one substance for another. For example, the **sodium–potassium (Na^+–K^+) pump** moves Na^+ out of cells and K^+ into cells (see figure 3.20). The result is a higher concentration of Na^+ outside the cell and a higher concentration of K^+ inside the cell. Because ATP is broken down during the transport of Na^+ and K^+, the pump is also called **sodium/potassium ATP-ase.** The Na^+–K^+ pump is very important to a number of cell functions. These are discussed in chapters 9 and 11.

Secondary Active Transport

Secondary active transport involves the active transport of an ion such as sodium out of a cell, establishing a concentration gradient, with a higher concentration of the ions outside the cell. The tendency for the ions to move back into the cell, down their concentration gradient, provides the energy necessary to transport a different ion or some other molecule into the cell. For example, glucose is transported from the lumen of the intestine into epithelial cells by secondary active transport (figure 3.21). This process requires two carrier proteins: (1) a sodium–potassium exchange pump actively transports Na^+ out of the cell, and (2) the other carrier protein facilitates the movement of Na^+ and glucose into the cell. Both Na^+ and glucose are necessary for the carrier protein to function.

The movement of Na^+ down their concentration gradient provides the energy to move glucose molecules into the cell against their concentration gradient. Thus glucose can accumulate at concentrations higher inside the cell than outside. Because the movement of glucose molecules against their concentration gradient results from the formation of a concentration gradient of Na^+ by an active transport mechanism, the process is called secondary active transport.

The ions or molecules moved by secondary active transport can move in the same direction as or in a different direction across the membrane than the ion that enters the cell by diffusion down its concentration gradient. **Cotransport,** or **symport,** is a type of secondary active transport where movement is in the same direction. For example, glucose, fructose, and amino acids move with Na^+ into cells of the intestine and kidneys. **Countertransport,** or **antiport,** is a type of secondary active transport where ions or molecules move in opposite directions. For example, the internal pH of cells is maintained by countertransport, which moves H^+ out of the cell as Na^+ move into the cell.

● 20. *What is mediated transport? What types of molecules are moved through the plasma membrane by mediated transport?*

1. Three sodium ions (Na⁺) and adenosine
triphosphate (ATP) bind to the Na⁺–K⁺ pump .

2. The ATP breaks down to adenosine diphosphate
(ADP) and a phosphate (P) and releases energy.
That energy is used to power the shape change in
the Na⁺–K⁺ pump. Phosphate remains bound to the
Na⁺–K⁺–ATP binding site.

3. The Na⁺–K⁺ pump changes shape, and the Na⁺ are
transported across the membrane.

4. The Na⁺ diffuse away from the Na⁺–K⁺ pump.

5. Two potassium ions (K⁺) bind to the Na⁺–K⁺ pump.

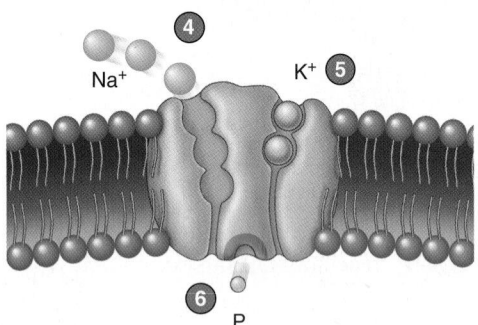

6. The phosphate is released from the Na⁺–K⁺ pump
binding site.

7. The Na⁺–K⁺ pump changes shape, transporting K⁺
across the membrane, and the K⁺ diffuse away from
the pump. The Na⁺–K⁺ pump can again
bind to Na⁺ and ATP.

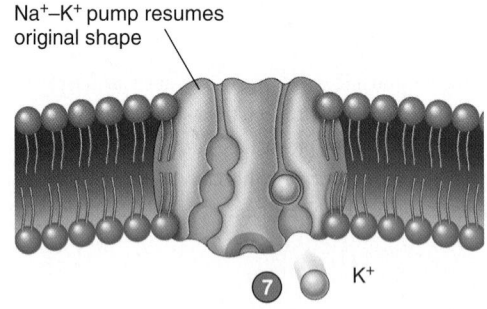

Process Figure 3.20   Sodium–Potassium Pump

21. *Describe specificity, competition, and saturation as*
characteristics of mediated transport mechanisms.

22. *Contrast facilitated diffusion and active transport in relation*
to energy expenditure and movement of substances with or
against their concentration gradients.

23. *What are secondary active transport, cotransport, and*
countertransport?

P R E D I C T   4

In cardiac (heart) muscle cells, the concentration of intracellular
Ca^{2+} affects the force of heart contraction. The higher the intracellular
Ca^{2+} concentration, the greater the force of contraction. Na^+/Ca^{2+}
countertransport helps to regulate intracellular Ca^{2+} levels by
transporting Ca^{2+} out of cardiac muscle cells. Given that digitalis
slows the transport of Na^+, should the heart beat more or less
forcefully when exposed to this drug? Explain.

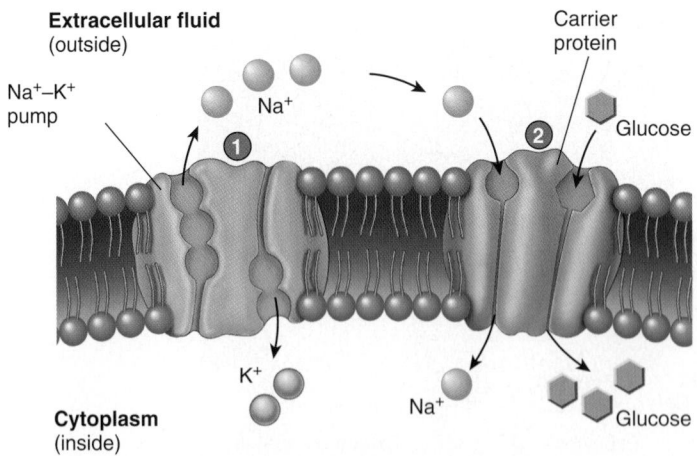

1. A Na⁺–K⁺ pump maintains a concentration of Na⁺ that is higher outside the cell than inside.

2. Na⁺ move back into the cell by a carrier protein that also moves glucose. The concentration gradient for Na⁺ provides energy required to move glucose against its concentration gradient.

Process Figure 3.21 **This example shows contransport of Na⁺ and glucose. Secondary Active Transport (Cotransport)**

Endocytosis and Exocytosis

Endocytosis (en′dō-sī-tō′sis), or the internalization of substances, includes both phagocytosis and pinocytosis and refers to the uptake of material through the plasma membrane by the formation of a vesicle. A **vesicle** is a membrane-bounded sac found within the cytoplasm of a cell. A portion of the plasma membrane wraps around a particle or droplet and fuses so that the particle or droplet is surrounded by a membrane. That portion of the membrane then "pinches off" so that the particle or droplet, surrounded by a membrane, is within the cytoplasm of the cell, and the plasma membrane is left intact.

Phagocytosis (făg-ō-sī-tō′sis) literally means cell-eating (figure 3.22) and applies to endocytosis when solid particles are ingested and phagocytic vesicles are formed. White blood cells and some other cell types phagocytize bacteria, cell debris, and foreign particles. Phagocytosis is therefore important in the elimination of harmful substances from the body.

Pinocytosis (pin′ō-sī-tō′sis) means cell-drinking and is distinguished from phagocytosis in that smaller vesicles are formed and they contain molecules dissolved in liquid rather than particles (figure 3.23). Pinocytosis often forms vesicles near the tips of deep invaginations of the plasma membrane. It is a common transport phenomenon in a variety of cell types and occurs in certain cells of the kidneys, epithelial cells of the intestines, cells of the liver, and cells that line capillaries.

Endocytosis can exhibit specificity. For example, cells that phagocytize bacteria and necrotic tissue do not phagocytize healthy cells. The plasma membrane may contain specific receptor molecules that recognize certain substances and allow them to be transported into the cell by phagocytosis or pinocytosis. This is called **receptor-mediated endocytosis,** and the receptor sites combine only with certain molecules (figure 3.24). This mechanism increases the rate at which specific substances are taken up by the cells. Cholesterol and growth factors are examples of molecules that can be taken into a cell by receptor-mediated endocytosis. Both phagocytosis and pinocytosis require energy in the form of ATP and therefore are active processes. Because they involve the bulk movement of material into the cell, however, phagocytosis and pinocytosis do not exhibit either the degree of specificity or saturation that active transport exhibits.

Hypercholesterolemia

Hypercholesterolemia is a common genetic disorder affecting 1 in every 500 adults in the United States. It consists of a reduction in or absence of low-density lipoprotein (LDL) receptors on cell surfaces. This interferes with receptor-mediated endocytosis of LDL cholesterol. As a result of inadequate cholesterol uptake, cholesterol synthesis within these cells is not regulated, and too much cholesterol is produced. The excess cholesterol accumulates in blood vessels, resulting in atherosclerosis. Atherosclerosis can result in heart attacks or strokes.

(a)

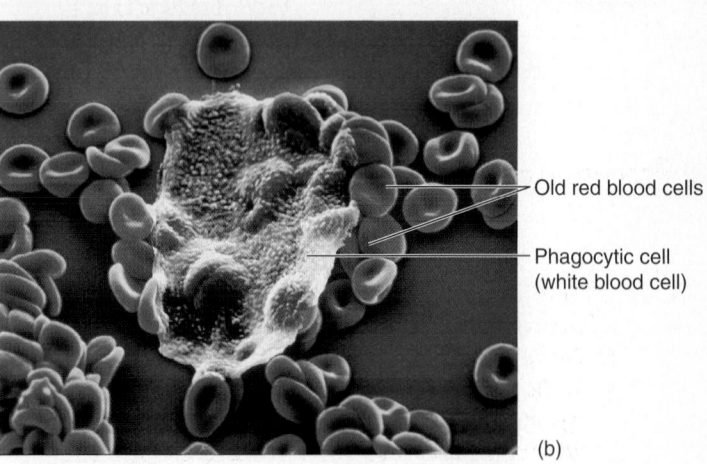

(b)

Figure 3.22 **Endocytosis**

(*a*) Phagocytosis. (*b*) Scanning electron micrograph of phagocytosis of an agarose bead.

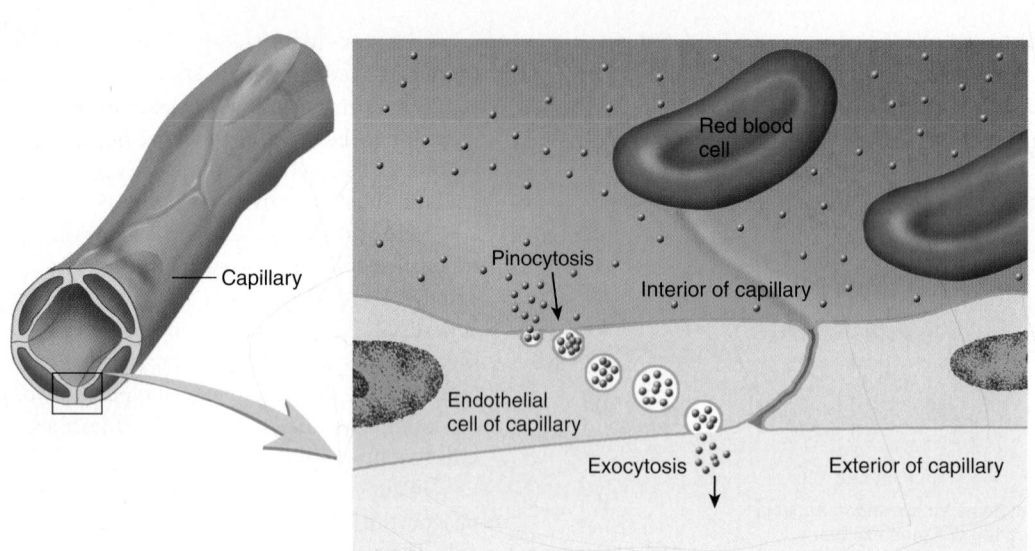

Figure 3.23   Pinocytosis

Pinocytosis is much like phagocytosis, except the cell processes and therefore the vesicles formed are much smaller and the material inside the vesicle is liquid rather than particulate. Pinocytotic vesicles form on the internal side of a capillary, are transported across the cell, and open by exocytosis outside the capillary.

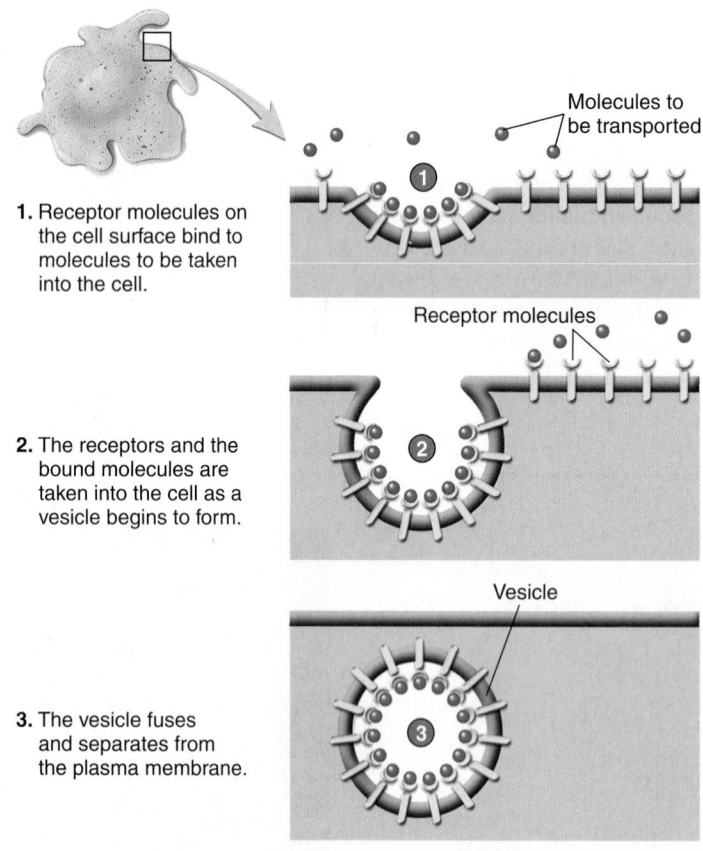

1. Receptor molecules on the cell surface bind to molecules to be taken into the cell.

2. The receptors and the bound molecules are taken into the cell as a vesicle begins to form.

3. The vesicle fuses and separates from the plasma membrane.

Process Figure 3.24   Receptor-Mediated Endocytosis

In some cells, secretions accumulate within vesicles. These secretory vesicles then move to the plasma membrane, where the membrane of the vesicle fuses with the plasma membrane and the content of the vesicle is expelled from the cell. This process is called **exocytosis** (ek′sō-sī-tō′sis) (figure 3.25). Secretion of digestive enzymes by the pancreas, of mucus by the salivary glands, and of milk by the mammary glands are examples of exocytosis. In some respects the process is similar to phagocytosis and pinocytosis but occurs in the opposite direction.

Table 3.2 summarizes and compares the mechanisms by which different kinds of molecules are transported across the plasma membrane.

- **24. Define endocytosis and vesicle. How do phagocytosis and pinocytosis differ from each other?**
- **25. What is receptor-mediated endocytosis?**
- **26. Describe and give examples of exocytosis.**

Cytoplasm

Cytoplasm, the cellular material outside the nucleus but inside the plasma membrane, is about half cytosol and half organelles.

Cytosol

Cytosol (sī′tō-sol) consists of a fluid portion, a cytoskeleton, and cytoplasmic inclusions. The fluid portion of cytosol is a solution with dissolved ions and molecules and a colloid with suspended molecules, especially proteins. Many of these proteins are enzymes that catalyze the breakdown of molecules for energy or the synthesis of sugars, fatty acids, nucleotides, amino acids, and other molecules.

1. A secretory vesicle moves toward the plasma membrane.

Plasma membrane

Secretory vesicle

2. The secretory vesicle fuses with the plasma membrane.

Secretory vesicle fused to plasma membrane

3. The secretory vesicle's contents are released into the extracellular fluid.

Released contents of secretory vesicle

(a)

TEM 30,000x

(b)

Process Figure 3.25   Exocytosis

(*a*) Diagram of exocytosis. (*b*) Transmission electron micrograph of exocytosis.

Table 3.2   Comparison of Membrane Transport Mechanisms

Transport Mechanism	Description	Substances Transported	Example
Diffusion	Random movement of molecules results in net movement from areas of higher to lower concentration.	Lipid-soluble molecules dissolve in the lipid bilayer and diffuse through it; ions and small molecules diffuse through membrane channels.	Oxygen, carbon dioxide, and lipids such as steroid hormones dissolve in the lipid bilayer; Cl^- and urea move through membrane channels.
Osmosis	Water diffuses across a selectively permeable membrane.	Water diffuses through the lipid bilayer.	Water moves from the stomach into the blood.
Filtration	Liquid moves through a partition that allows some, but not all, of the substances in the liquid to pass through it; movement is due to a pressure difference across the partition.	Liquid and substances pass through holes in the partition.	Filtration in the kidneys allows removal of everything from the blood except proteins and blood cells.
Facilitated diffusion	Carrier molecules combine with substances and move them across the plasma membrane; no ATP is used; substances are always moved from areas of higher to lower concentration; it exhibits the characteristics of specificity, saturation, and competition.	Some substances too large to pass through membrane channels and too polar to dissolve in the lipid bilayer and some ions are transported.	Glucose moves by facilitated diffusion into muscle cells and fat cells.
Active transport	Carrier molecules combine with substances and move them across the plasma membrane; ATP is used; substances can be moved from areas of lower to higher concentration; it exhibits the characteristics of specificity, saturation, and competition.	Substances too large to pass through channels and too polar to dissolve in the lipid bilayer are transported; substances that are accumulated in concentrations higher on one side of the membrane than on the other are transported.	Ions such as Na^+, K^+, and Ca^{2+} are actively transported.
Secondary active transport	Ions are moved across the plasma membrane by active transport, which establishes a concentration gradient; ATP is required; ions then move back down their concentration gradient by facilitated diffusion, and another ion or molecule moves with the diffusion ion (cotransport) or in the opposite direction (countertransport).	Some sugars, amino acids, and ions are transported.	A concentration gradient for Na^+ exists in intestinal epithelial cells. This gradient provides the energy for the cotransport of glucose. As Na^+ enter the cell, down their concentration gradient, glucose also enters the cell. In many cells, H^+ is countertransported (in the opposite direction) with Na^+.
Endocytosis	The plasma membrane forms a vesicle around the substances to be transported, and the vesicle is taken into the cell; this requires ATP; in receptor-mediated endocytosis specific substances are ingested.	Phagocytosis takes in cells and solid particles; pinocytosis takes in molecules dissolved in liquid.	Immune system cells called phagocytes ingest bacteria and cellular debris; most cells take in substances through pinocytosis.
Exocytosis	Materials manufactured by the cell are packaged in secretory vesicles that fuse with the plasma membrane and release their contents to the outside of the cell; this requires ATP.	Proteins and other water-soluble molecules are transported out of cells.	Digestive enzymes, hormones, neurotransmitters, and glandular secretions are transported, and cell waste products are eliminated.

Cytoskeleton

The **cytoskeleton** supports the cell and holds the nucleus and other organelles in place. It is also responsible for cell movements, such as changes in cell shape or movement of cell organelles. The cytoskeleton consists of three groups of proteins: microtubules, actin filaments, and intermediate filaments (figure 3.26).

Microtubules are hollow tubules composed primarily of protein units called **tubulin.** The microtubules are about 25 nanometers (nm) in diameter, with walls about 5 nm thick. Microtubules vary in length but are normally several micrometers (μm) long. Microtubules play a variety of roles within cells. They help provide support and structure to the cytoplasm of the cell, much like an internal scaffolding. They are involved in the process of cell division, transport of intracellular materials, and form essential components of certain cell organelles, such as centrioles, spindle fibers, cilia, and flagella.

Actin filaments, or **microfilaments,** are small fibrils about 8 nm in diameter that form bundles, sheets, or networks in the cytoplasm of cells. These filaments have a spiderweb-like appearance within the cell. Actin filaments provide structure to the cytoplasm and mechanical support for microvilli. Actin filaments support the plasma membrane and define the shape of the cell. Changes in cell shape involve the breakdown and reconstruction of actin filaments.

These changes in shape allow some cells to move about. Muscle cells contain a large number of highly organized actin filaments responsible for the muscle's contractile capabilities (see chapter 9).

Intermediate filaments are protein fibers about 10 nm in diameter. They provide mechanical strength to cells. For example, intermediate filaments support the extensions of nerve cells, which have a very small diameter but can be a meter in length.

Cytoplasmic Inclusions

The cytosol also contains **cytoplasmic inclusions,** which are aggregates of chemicals either produced by the cell or taken in by the cell. For example, lipid droplets or glycogen granules store energy-rich molecules; hemoglobin in red blood cells transports oxygen; melanin is a pigment that colors the skin, hair, and eyes; and **lipochromes** (lip′ō-krōmz) are pigments that increase in amount with age. Dust, minerals, and dyes can also accumulate in the cytoplasm.

27. *Define cytoplasm and cytosol.*
28. *What are the two general functions of the cytoskeleton?*
29. *Describe and list the functions of microtubules, actin filaments, and intermediate filaments.*
30. *Define and give examples of cytoplasmic inclusions. What are lipochromes?*

Figure 3.26 Cytoskeleton

(*a*) Diagram of the cytoskeleton. (*b*) Scanning electron micrograph of the cytoskeleton.

Organelles

Organelles are structures within cells that are specialized for particular functions, such as manufacturing proteins or producing ATP. Organelles can be thought of as individual workstations within the cell, each responsible for performing specific tasks. Most, but not all organelles have membranes that are similar to the plasma membrane. The membranes separate the interior of the organelles from the rest of the cytoplasm, creating subcellular compartments with their own enzymes that are capable of carrying out unique chemical reactions. The nucleus is an example of an organelle.

The number and type of cytoplasmic organelles within each cell are related to the specific structure and function of the cell. Cells secreting large amounts of protein contain well-developed organelles that synthesize and secrete protein, whereas cells actively transporting substances such as sodium ions across their plasma membrane contain highly developed organelles that produce ATP. The following sections describe the structure and main functions of the major cytoplasmic organelles found in cells.

Nucleus

The **nucleus,** which contains most of the genetic information of the cell, is a large, membrane-bounded structure usually located near the center of the cell. It may be spherical, elongated, or lobed, depending on the cell type. All cells of the body have a nucleus at some point in their life cycle, although some cells, such as red blood cells (also called red blood corpuscles or erythrocytes), lose their nuclei as they develop. Other cells, such as skeletal muscle cells and certain bone cells, called osteoclasts, contain more than one nucleus. The nucleus consists of **nucleoplasm** surrounded by a **nuclear envelope** (figure 3.27)

Figure 3.27   The Nucleus

(*a*) The nuclear envelope consists of inner and outer membranes that become fused at the nuclear pores. The nucleolus is a condensed region of the nucleus not bounded by a membrane and consisting mostly of RNA and protein. (*b*) Transmission electron micrograph of the nucleus. (*c*) Scanning electron micrograph showing the inner surface of the nuclear envelope and the nuclear pores.

composed of two membranes separated by a space. At many points on the surface of the nuclear envelope, the inner and outer membranes fuse to form porelike structures, the **nuclear pores.** Molecules move between the nucleus and the cytoplasm through these nuclear pores.

Deoxyribonucleic acid (DNA) and associated proteins are dispersed throughout the nucleus as thin strands about 4–5 nm in diameter. The proteins include **histones** (his′tōnz) and other proteins that play a role in the regulation of DNA function. The DNA and protein strands can be stained with dyes and are called **chromatin** (krō′ma-tin; colored material) (figure 3.28). During cell division, the chromatin condenses to form the more densely coiled bodies called **chromosomes** (colored bodies). Each chromosome consists of two **chromatids** (krō′mă-tids), which are attached at a single point called the **centromere** (sen′trō-mēr). The **kinetochore** (ki-nē′tō-kōr, ki-net′ō-kōr) is a specialized granular area within the centromere of each chromosome.

DNA ultimately determines the structure of proteins made in the cell (protein synthesis is described on p. 88). Many structural components of the cell and all the enzymes, which regulate most chemical reactions in the cell, are proteins. By determining

protein structure, DNA therefore ultimately controls the structural and functional characteristics of the cell. DNA does not leave the nucleus but functions by means of an intermediate, **ribonucleic acid (RNA),** which can leave the nucleus. DNA determines the structure of messenger RNA (mRNA), ribosomal RNA (rRNA), and transfer RNA (tRNA) (all described in more detail later). mRNA moves out of the nucleus through the nuclear pores into the cytoplasm, where it determines the sequence of amino acids in proteins.

Human Genome Project

The **Human Genome Project** is an ambitious international project, which began in 1990, with the 15-year goal of mapping and sequencing the entire human genome. The **genome** is the total of all the genes contained within each cell. One goal of the Human Genome Project is to construct a map indicating where each of the approximately 27,000–30,000 genes is located on the human chromosomes. The other major goal of the project is to determine the sequence of the estimated 3 billion base pairs (bp) that make up the human DNA molecules. The sequencing is now complete, and the mapping continues. It is hoped that by knowing for what proteins the genes implicated in genetic disorders are coded, and by determining the functions of those proteins, we will be able to more effectively treat these disorders.

Because mRNA synthesis occurs within the nucleus, cells without nuclei accomplish protein synthesis only as long as the mRNA produced before the nucleus degenerates remains functional. The nuclei of developing red blood cells are expelled from the cells before the red blood cells enter the blood, where they survive without a nucleus for about 120 days. In comparison, many cells with nuclei, such as nerve and skeletal muscle cells, survive as long as the individual person survives.

A **nucleolus** (noo-klē′ō-lŭs) is a somewhat rounded dense region within the nucleus that lacks a surrounding membrane (see figure 3.33). Usually one nucleolus exists per nucleus, but several smaller, accessory nucleoli may also be seen in some nuclei, especially during the latter phases of cell division. The nucleolus incorporates portions of 10 chromosomes (five pairs), called **nucleolar organizer regions.** These regions contain DNA from which rRNA is produced. Within the nucleolus, the subunits of ribosomes are manufactured (see the section on "Ribosomes," p. 82).

- 31. *Define organelles.*
- 32. *Describe the structure of the nucleus and nuclear envelope. What is the function of the nuclear pores?*
- 33. *What molecules are found in chromatin? How does chromatin become a chromosome?*
- 34. *List the types of RNA whose structure is determined by DNA. How can DNA control the structural and functional characteristics of the cell without leaving the nucleus?*
- 35. *Describe the nucleolus. Define and give the function of the nucleolar organizer regions.*

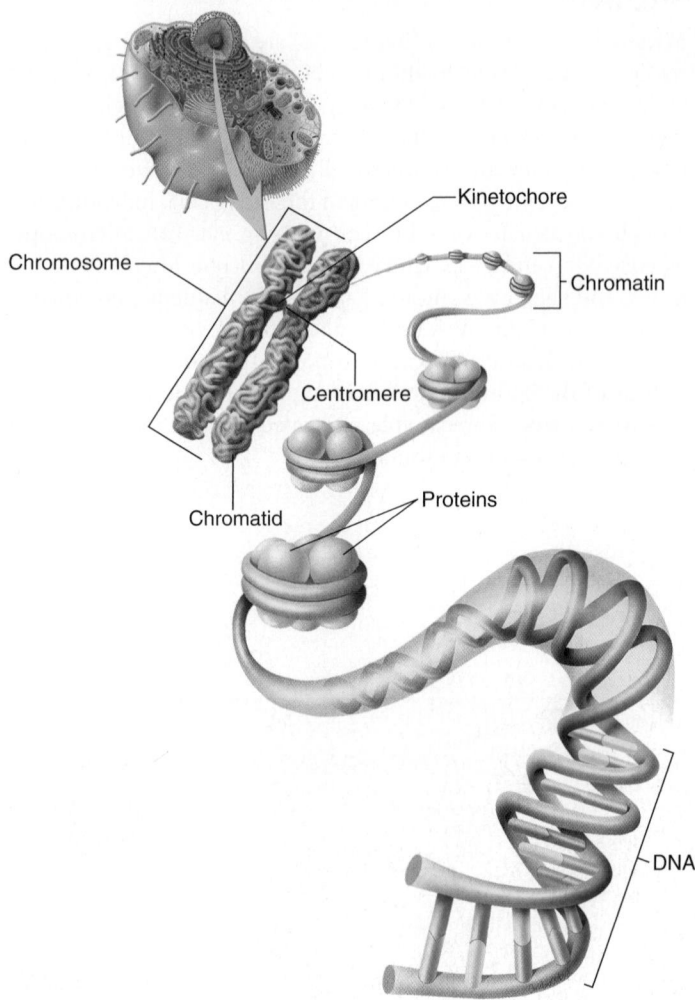

Figure 3.28 Chromosome Structure

DNA is associated with globular histone proteins. Usually the DNA molecule is stretched out, resembling a string of beads, and is called chromatin. During cell division, however, the chromatin condenses to become bodies called chromosomes.

Centrioles and Spindle Fibers

The **centrosome** (sen′trō-sōm) is a specialized zone of cytoplasm close to the nucleus that is the center of microtubule formation. It contains two **centrioles** (sen′trē-ōlz). Each centriole is a small, cylindrical organelle about 0.3–0.5 μm in length and 0.15 μm in diameter, and the two centrioles are normally oriented perpendicular to each other within the centrosome (see figure 3.1). The wall of the centriole is composed of nine evenly spaced, longitudinally oriented, parallel units, or triplets. Each unit consists of three parallel microtubules joined together (figure 3.29).

Microtubules appear to influence the distribution of actin and intermediate filaments. Through its control of microtubule formation, the centrosome is therefore closely involved in determining cell shape and movement. The microtubules extending from the centrosomes are very dynamic—constantly growing and shrinking.

Before cell division, the two centrioles double in number, the centrosome divides into two, and one centrosome, containing two centrioles, moves to each end of the cell. Microtubules called **spindle fibers** extend out in all directions from the centrosome. These microtubules grow and shrink even more rapidly than those of nondividing cells. If the extended end of a spindle fiber comes in contact with a kinetochore, a specialized region in the centromere of each chromosome, the spindle fiber attaches to the kinetochore and stops growing or shrinking. Eventually spindle fibers from each centrosome bind to the kinetochores of all the chromosomes. During cell division, the microtubules facilitate the movement of chromosomes toward the two centrosomes (see the section on "Cell Division," p. 93).

Cilia and Flagella

Cilia (sil′ē-ă) are structures that project from the surface of cells and are capable of movement. They vary in number from one to thousands per cell. Cilia are cylindrical in shape, about 10 μm in length and 0.2 μm in diameter, and the shaft of each cilium is enclosed by the plasma membrane. Two centrally located microtubules and nine peripheral pairs of fused microtubules, the so-called 9+2 arrangement, extend from the base to the tip of each cilium (figure 3.30). Movement of the microtubules past each other, a process that requires energy from ATP, is responsible for movement of the cilia. **Dynein arms,** proteins connecting adjacent pairs of microtubules, push the microtubules past each other. A **basal body** (a modified centriole) is located in the cytoplasm at the base of the cilium. Cilia are numerous on surface cells that line the respiratory tract and the female reproductive tract. In these regions cilia move in a coordinated fashion, with a power stroke in one direction and a recovery stroke in the opposite direction (figure 3.31). Their motion moves materials over the surface of the cells. For example, cilia in the trachea move mucus embedded with dust particles upward and away from the lungs. This action helps keep the lungs clear of debris.

Flagella (flă-jel′ă) have a structure similar to cilia but are longer (45 μm), and usually only one flagellum exists per sperm cell. Sperm cells are the only human cells to possess flagella. Furthermore, whereas cilia move small particles across the cell surface, flagella move the cell. For example, each sperm cell is propelled by a single flagellum. In contrast to cilia, which have a power stroke and a recovery stroke, flagella move in a wavelike fashion.

Microvilli

Microvilli (mī-krō-vil′ī) (figure 3.32) are cylindrically shaped extensions of the plasma membrane about 0.5–1.0 μm in length and 90 nm in diameter. Normally many microvilli are on each cell, and they function to increase the cell surface area. A student looking at photographs may confuse microvilli with cilia. Microvilli, however, are only one-tenth to one-twentieth the size of cilia. Individual microvilli can usually only be seen with an electron microscope, whereas cilia can be seen with a light microscope. Microvilli do not move, and they are supported with actin filaments, not microtubules. Microvilli are found in the intestine, kidney, and other areas in which absorption is an important function. In certain locations of the body, microvilli are highly modified to function as sensory receptors. For example, elongated microvilli in hair cells of the inner ear respond to sound.

(a)   (b)

Microtubule triplet

Centriole (cross-section)

Centriole (longitudinal sectional)

TEM 60,000x

Figure 3.29   Centriole

(*a*) Structure of a centriole, which comprises nine triplets of microtubules. Each triplet contains one complete microtubule fused to two incomplete microtubules.
(*b*) Transmission electron micrograph of a pair of centrioles, which are normally located together near the nucleus. One is shown in cross section and one in longitudinal section.

Microtubules

Plasma
membrane

Microtubule

Dynein arm

Plasma
membrane

Basal body

(a)

Microtubules

TEM 100,000x

(b)

(c)

Figure 3.30   Cilia and Flagella

(*a*) Ciliary or flagellar structures. The shaft is composed of nine microtubule doublets around its periphery and two in the center. Dynein arms are proteins that connect one pair of microtubules to another pair. Dynein arm movement, which requires ATP, causes the microtubules to slide past each other, resulting in bending or movement of the cilium or flagellum. A basal body attaches the cilium or flagellum to the plasma membrane. (*b*) Transmission electron micrograph through cilium. (*c*) Transmission electron micrograph through basal body of cilium.

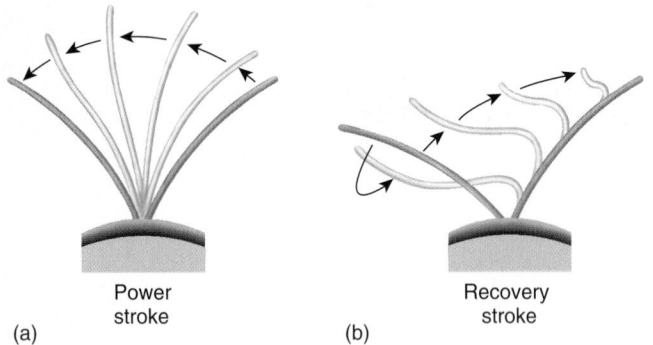

Figure 3.31   **Ciliary Movement**

(*a*) Power and (*b*) recovery strokes.

- 36. *Describe and list the functions of centrosomes. Explain the structure of centrioles.*
- 37. *What are spindle fibers? Explain the relationship between centrosomes, spindle fibers, and the kinetochores of chromosomes during cell division.*
- 38. *Contrast the structure and function of cilia and flagella.*
- 39. *Describe the structure and function of microvilli. How are microvilli different from cilia?*

Ribosomes

Ribosomes (rī′bō-sōms) are the sites of protein synthesis. Each ribosome is composed of a large subunit and a smaller one. The ribosomal subunits, which consist of **ribosomal RNA (rRNA)** and proteins, are produced separately in the nucleolus of the nucleus. The ribosomal subunits then move through the nuclear pores into the cytoplasm, where they assemble to form the functional ribosome during protein synthesis (figure 3.33). Protein synthesis oc-

curs as messenger RNA (mRNA) moves between the two subunits of a ribosome (see figure 3.40). Ribosomes can be found free in the cytoplasm or associated with a membrane called the endoplasmic reticulum. **Free ribosomes** primarily synthesize proteins used inside the cell, whereas ribosomes attached to the endoplasmic reticulum can produce proteins that are secreted from the cell.

Endoplasmic Reticulum

The outer membrane of the nuclear envelope is continuous with a series of membranes distributed throughout the cytoplasm of the cell (see figure 3.1), collectively referred to as the **endoplasmic reticulum** (en′dō-plas′mik re-tik′ū-lŭm; network inside the cytoplasm) (figure 3.34). The endoplasmic reticulum consists of broad, flattened, interconnecting sacs and tubules. The interior spaces of those sacs and tubules are called **cisternae** (sis-ter′nē) and are isolated from the rest of the cytoplasm.

The **rough endoplasmic reticulum** is called "rough" because it has ribosomes attached to it. The ribosomes of the rough endoplasmic reticulum are sites where proteins are produced and modified for secretion and for internal use. The amount and configuration of the endoplasmic reticulum within the cytoplasm depend on the cell type and function. Cells with abundant rough endoplasmic reticulum synthesize large amounts of protein that are secreted for use outside the cell.

Smooth endoplasmic reticulum, which is endoplasmic reticulum without attached ribosomes, manufactures lipids, such as phospholipids, cholesterol, steroid hormones, and carbohydrates like glycogen. Many phospholipids produced in the smooth endoplasmic reticulum help form vesicles within the cell and contribute to the plasma membrane. Cells that synthesize large amounts of lipid contain dense accumulations of smooth endoplasmic reticulum. Enzymes required for lipid synthesis are associated with the membranes of the smooth endoplasmic reticulum. Smooth endoplasmic reticulum also participates in the detoxification processes by

Microvillus — Plasma membrane

Actin filaments

Cytoplasm

(a)   (b)   TEM 60,000x

Figure 3.32   **Microvillus**

(*a*) A microvillus is a tiny tubular extension of the cell and contains cytoplasm and some actin filaments (microfilaments). (*b*) Transmission electron micrograph of microvilli.

1. Ribosomal proteins, produced in the cytoplasm, are transported through nuclear pores into the nucleolus.

2. rRNA, most of which is produced in the nucleolus, is assembled with ribosomal proteins to form small and large ribosomal subunits.

3. The small and large ribosomal subunits leave the nucleolus and the nucleus through nuclear pores.

4. The small and large subunits, now in the cytoplasm, combine with each other and with mRNA.

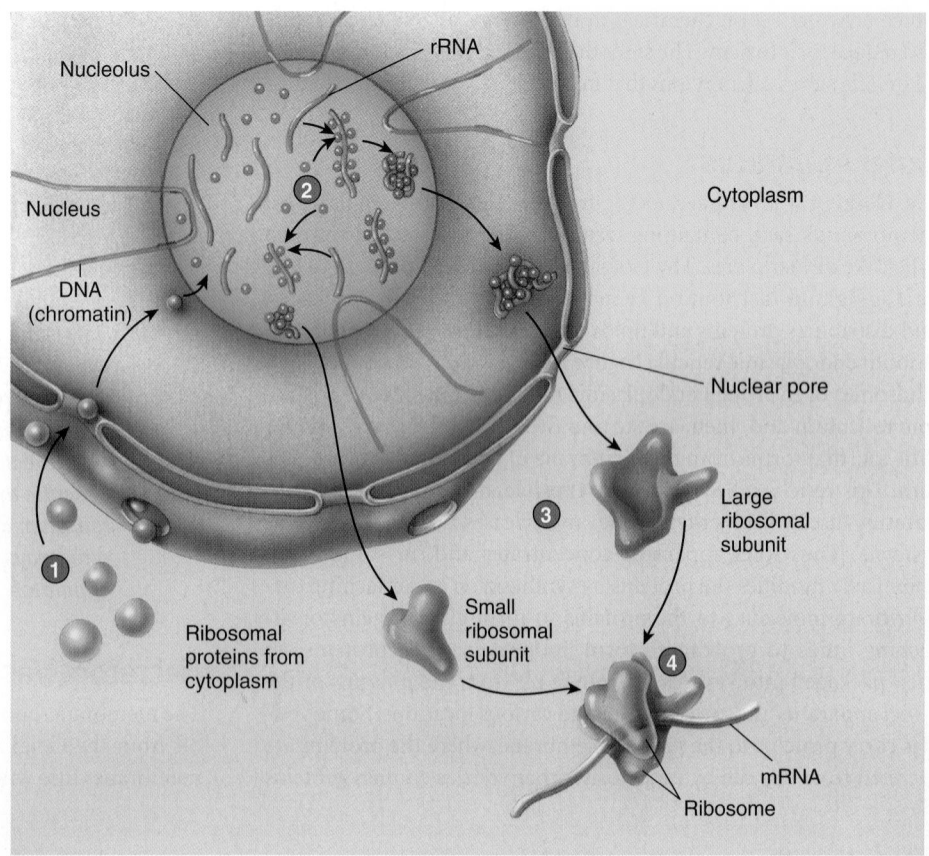

Process Figure 3.33 **Production of Ribosomes**

(a)

(b)

Figure 3.34 **The Endoplasmic Reticulum**

(a) The endoplasmic reticulum is continuous with the nuclear envelope and can exist as either rough endoplasmic reticulum (with ribosomes) or smooth endoplasmic reticulum (without ribosomes). (b) Transmission electron micrograph of the rough endoplasmic reticulum.

which enzymes act on chemicals and drugs to change their structure and reduce their toxicity. The smooth endoplasmic reticulum of skeletal muscle stores calcium ions that function in muscle contraction.

Golgi Apparatus

The **Golgi** (gōl′jē) **apparatus** (figure 3.35) is composed of flattened membranous sacs, containing cisternae, that are stacked on each other like dinner plates. The Golgi apparatus can be thought of as a packaging and distribution center because it modifies, packages, and distributes proteins and lipids manufactured by the rough and smooth endoplasmic reticula (figure 3.36). Proteins produced at the ribosomes of the rough endoplasmic reticulum enter the endoplasmic reticulum and, then, are surrounded by a **vesicle** (ves′i-kl), or little sac, that forms from the membrane of the endoplasmic reticulum. This vesicle, called a **transport vesicle,** moves to the Golgi apparatus, fuses with its membrane, and releases the protein into its cisterna. The Golgi apparatus concentrates and, in some cases, chemically modifies the proteins by synthesizing and attaching carbohydrate molecules to the proteins to form glycoproteins or attaching lipids to proteins to form lipoproteins. The proteins are then packaged into vesicles that pinch off from the margins of the Golgi apparatus and are distributed to various locations. Some vesicles carry proteins to the plasma membrane, where the proteins are secreted from the cell by exocytosis; other vesicles contain proteins

that become part of the plasma membrane; and still other vesicles contain enzymes that are used within the cell.

The Golgi apparatuses are most numerous and most highly developed in cells that secrete large amounts of protein or glycoproteins, such as cells in the salivary glands and the pancreas.

40. *What kinds of molecules are in ribosomes? Where are ribosomal subunits formed and assembled?*
41. *Compare the functions of free ribosomes and endoplasmic reticulum ribosomes.*
42. *How is the endoplasmic reticulum related to the nuclear envelope? How are the cisternae of the endoplasmic reticulum related to the rest of the cytoplasm?*
43. *What are the functions of smooth endoplasmic reticulum?*
44. *Describe the structure and function of the Golgi apparatus.*
45. *Describe the production of a protein at the endoplasmic reticulum and its distribution to the Golgi apparatus. Name three ways in which proteins are distributed from the Golgi apparatus.*

Secretory Vesicles

The membrane-bound **secretory vesicles** (see figure 3.35) that pinch off from the Golgi apparatus move to the surface of the cell, their membranes fuse with the plasma membrane, and the contents of the

(a)

(b) Mitochondrion

Figure 3.35 Golgi Apparatus

(*a*) The Golgi apparatus is composed of flattened membranous sacs, containing cisternae, and resembles a stack of dinner plates or pancakes. (*b*) Transmission electron micrograph of the Golgi apparatus.

1. Some proteins are produced at ribosomes on the surface of the rough endoplasmic reticulum and are transferred into the cisterna as they are produced.

2. The proteins are surrounded by a vesicle that forms from the membrane of the endoplasmic reticulum.

3. The vesicle moves from the endoplasmic reticulum to the Golgi apparatus, fuses with its membrane and releases the proteins into its cisterna.

4. The Golgi apparatus concentrates and, in some cases, modifies the proteins into glycoproteins or lipoproteins.

5. The proteins are packaged into vesicles that form from the membrane of the Golgi apparatus.

6. Some vesicles, such as lysosomes, contain enzymes that are used within the cell.

7. Secretory vesicles carry proteins to the plasma membrane, where the proteins are secreted from the cell by exocytosis.

8. Some vesicles contain proteins that become part of the plasma membrane.

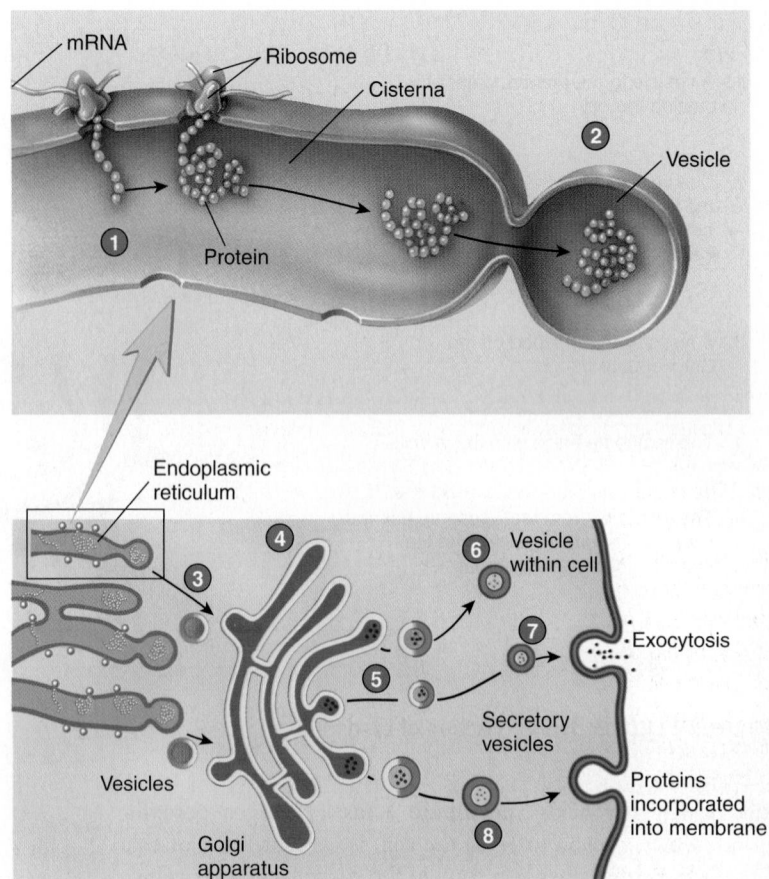

Process Figure 3.36   Function of the Golgi Apparatus

vesicle are released to the exterior by exocytosis. The membranes of the vesicles are then incorporated into the plasma membrane.

Secretory vesicles accumulate in many cells, but their contents frequently are not released to the exterior until a signal is received by the cell. For example, secretory vesicles that contain the hormone insulin do not release it until the concentration of glucose in the blood increases and acts as a signal for the secretion of insulin from the cells.

Lysosomes

Lysosomes (līʹsō-sōmz) are membrane-bound vesicles that pinch off from the Golgi apparatus (see figure 3.35). They contain a variety of hydrolytic enzymes that function in intracellular digestion systems. Vesicles taken into the cell fuse with the lysosomes to form one vesicle and to expose the endocytized materials to hydrolytic enzymes (figure 3.37). Various enzymes within lysosomes digest nucleic acids, proteins, polysaccharides, and lipids. Certain white blood cells have large numbers of lysosomes that contain enzymes to digest phagocytized bacteria. Lysosomes also digest organelles of the cell that are no longer functional in a process called **autophagia** (aw-tō-fāʹjē-ă; self-eating). Furthermore, when tissues are damaged, ruptured lysosomes within the damaged cells release their enzymes, which digest both damaged and healthy cells. In other cells, the lysosomes move to the plasma membrane, and the enzymes are secreted by exocytosis. For example, the normal process of bone remodeling

involves the breakdown of bone tissue by specialized bone cells. Enzymes responsible for that degradation are released into the extracellular fluid from lysosomes produced by those cells.

Diseases of Lysosomal Enzymes

Some diseases result from nonfunctional lysosomal enzymes. For example, **Pompe's disease** results from the inability of lysosomal enzymes to break down glycogen. The glycogen accumulates in large amounts in the heart, liver, and skeletal muscles, an accumulation that often leads to heart failure. **Familial hyperlipoproteinemia** is a group of genetic disorders characterized by the accumulation of large amounts of lipids in phagocytic cells that lack the normal enzymes required to break down the lipid droplets. Symptoms include abdominal pain, enlargement of the spleen and liver, and eruption of yellow nodules in the skin filled with the affected phagocytic cells. **Mucopolysaccharidoses,** such as **Hurler's syndrome,** are diseases in which lysosomal enzymes are unable to break down mucopolysaccharides (glycosaminoglycans), so these molecules accumulate in the lysosomes of connective tissue cells and nerve cells. People affected by these diseases suffer mental retardation and skeletal deformities.

Peroxisomes

Peroxisomes (per-okʹsi-sōmz) are membrane-bounded vesicles that are smaller than lysosomes. Peroxisomes contain enzymes that

1. A vesicle forms around material outside the cell.

2. The vesicle is pinched off from the plasma membrane and becomes a separate vesicle inside the cell.

3. A lysosome is pinched off the Golgi apparatus.

4. The lysosome fuses with the vesicle.

5. The enzymes from the lysosome mix with the material in the vesicle, and the enzymes digest the material.

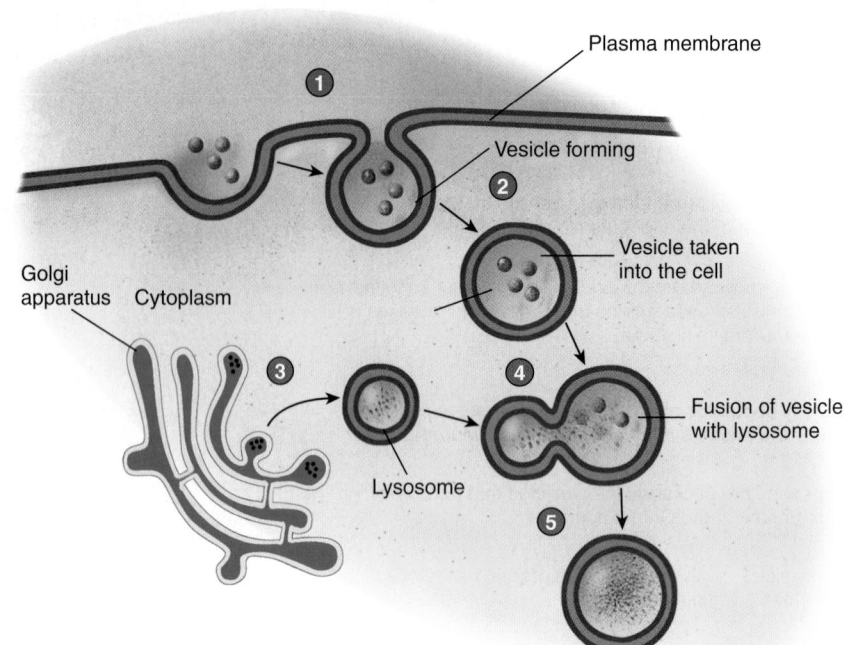

Process Figure 3.37   Action of Lysosomes

break down fatty acids and amino acids. Hydrogen peroxide (H_2O_2), which can be toxic to the cell, is a by-product of that breakdown. Peroxisomes also contain the enzyme **catalase,** which breaks down hydrogen peroxide to water and oxygen. Cells that are active in detoxification, such as liver and kidney cells, have many peroxisomes.

Proteasomes

Proteasomes (prō′tē-ă-sōmz) consist of large protein complexes, including several enzymes that break down and recycle proteins within the cell. Proteasomes are not surrounded by membranes. They are tunnel-like structures, similar to channel protein complexes; the inner surfaces of the tunnel have enzymatic regions that break down proteins. Smaller protein subunits close the ends of the tunnel and regulate which proteins are taken into it for digestion.

Mitochondria

Mitochondria (mī-tō-kon′drē-ă) provide energy for the cell. Consequently, they are often called the cell's power plants. Mitochondria are usually illustrated as small, rod-shaped structures (figure 3.38). In living cells, time-lapse photomicrography reveals that mitochondria constantly change shape from spherical to rod-shaped or even to long, threadlike structures. Mitochondria are the major sites of ATP production, which is the major energy source for most energy-requiring chemical reactions within the cell. Each mitochondrion has an inner and outer membrane separated by an intermembranous space. The outer membrane has a smooth contour, but the inner membrane has numerous infoldings called **cristae** (kris′tē; sing. crista) that project like shelves into the interior of the mitochondria.

A complex series of mitochondrial enzymes forms two major enzyme systems that are responsible for oxidative metabolism and most ATP synthesis (see chapter 25). The enzymes of the citric acid (or Krebs) cycle are found in the **matrix,** which is the substance located in the space formed by the inner membrane. The enzymes of the electron transport chain are embedded within the inner membrane. Cells with a greater energy requirement have more mitochondria with more cristae than cells with lower energy requirements. Within the cytoplasm of a given cell, the mitochondria are more numerous in areas in which ATP is used. For example, mitochondria are numerous in cells that perform active transport and are packed near the membrane where active transport occurs.

Increases in the number of mitochondria result from the division of preexisting mitochondria. When muscles enlarge as a result of exercise, the number of mitochondria within the muscle cells increases to provide the additional ATP required for muscle contraction.

The information for making some mitochondrial proteins is stored in DNA contained within the mitochondria themselves, and those proteins are synthesized on ribosomes within the mitochondria. The structure of many other mitochondrial proteins is determined by nuclear DNA, however, and these proteins are synthesized on ribosomes within the cytoplasm and then transported into the mitochondria. Both the mitochondrial DNA and mitochondrial ribosomes are very different from those within the nucleus and cytoplasm of the cell, respectively. Mitochondrial DNA is a closed circle of about 16,500 base pairs (bp) coding for 37 genes, compared with the open strands of nuclear DNA, which is composed of 3 billion bp coding for 30,000 genes. In addition, unlike nuclear DNA, mitochondrial DNA does not have associated proteins.

Outer membrane
Intermembrane space
Inner membrane
Matrix

DNA

Crista

Enzymes

(a)

Cross section

Longitudinal section

TEM 30,000x

(b)

Figure 3.38   Mitochondrion

(*a*) Typical mitochondrion structure. (*b*) Transmission electron micrograph of mitochondria in longitudinal and cross section.

Mitochondrial DNA

Half of the nuclear DNA of an individual is derived from the mother, and half is derived from the father; but all, or nearly all, mitochondrial DNA comes from the mother. The mitochondria of the sperm cell from the father are not incorporated into the oocyte at the time of fertilization. Because only the mother's mitochondrial DNA is passed down from generation to generation, maternal pedigrees are much easier to trace using mitochondrial DNA than with nuclear DNA. This unique quality of mitochondria has been used in a number of studies, from reuniting mothers or grandmothers with lost children to searching for the origins of the human species. A number of degenerative disorders affecting the nervous system, heart, or kidneys have been linked to mutations in mitochondrial DNA. The study of these disorders is providing valuable clues to the aging process.

46. *Define secretory vesicles.*
47. *Describe the process by which lysosomal enzymes digest phagocytized materials. Define autophagia.*
48. *What is the function of peroxisomes? How does catalase protect cells?*
49. *Describe the structure and function of proteasomes.*
50. *What is the function of mitochondria? What enzymes are found on the cristae and in the matrix? How can the number of mitochondria in a cell increase?*

P R E D I C T   ⑤

Describe the structural characteristics of cells that are highly specialized to do the following: (a) synthesize and secrete proteins, (b) actively transport substances into the cell, (c) synthesize lipids, and (d) phagocytize foreign substances.

Overview of Cell Metabolism

Cell metabolism is the sum of all the catabolic (decomposition) and anabolic (synthesis) reactions in the cell. The breakdown of food molecules such as carbohydrates, lipids, and proteins releases energy that is used to synthesize ATP. Each ATP molecule contains a portion of the energy originally stored in the chemical bonds of the food molecules. The ATP molecules are smaller "packets" of energy that, when released, can be used to drive other chemical reactions or processes such as active transport.

The production of ATP takes place in the cytosol and in mitochondria through a series of chemical reactions (see chapter 25 for details). Energy from food molecules is transferred to ATP in a controlled fashion. If the energy in food molecules were released all at once, the cell literally would burn up.

The breakdown of the sugar glucose, such as from sugar found in a candy bar, is used to illustrate the production of ATP from food molecules. Once glucose is transported into a cell, a series of reactions takes place within the cytosol. These chemical reactions, collectively called **glycolysis** (glī-kol′i-sis), convert the glucose to pyruvic acid. Pyruvic acid can enter different biochemical pathways, depending on oxygen availability (figure 3.39).

Aerobic (ār-ō′bik) **respiration** occurs when oxygen is available. The pyruvic acid molecules enter mitochondria and, through another series of chemical reactions, collectively called the citric acid cycle and the electron-transport chain, are converted to carbon dioxide and water. Aerobic respiration can yield up to 38 ATP molecules from the energy contained in each glucose molecule.

Several important points should be noted about aerobic respiration. First, the quantities of ATP produced through aerobic respiration are absolutely necessary to maintain the energy-requiring chemical reactions of life in human cells. Second, aerobic respiration requires oxygen because the last chemical reaction that takes place in aerobic respiration is the combination of oxygen with hydrogen to form water. If this reaction does not take place, the reactions immediately preceding it do not occur either. This explains why breathing oxygen is necessary for human life: without oxygen, aerobic respiration is inhibited, and the cells do not produce enough ATP to sustain life. Finally, during aerobic respiration the carbon atoms of food molecules are separated from one another to form carbon dioxide. Thus the carbon dioxide humans breathe out comes from the food they eat.

Anaerobic (an-ār-ō′bik) **respiration** occurs without oxygen and includes the conversion of pyruvic acid to lactic acid. A net production of two ATP molecules occurs for each glucose molecule used. Anaerobic respiration does not produce as much ATP as aerobic respiration, but it does allow the cells to function for short periods when oxygen levels are too low for aerobic respiration to provide all the needed ATP. For example, during intense exercise, when aerobic respiration has depleted the oxygen supply, anaerobic respiration can provide additional ATP.

Figure 3.39 Overview of Cell Metabolism

Aerobic respiration requires oxygen and produces more ATP per glucose molecule than does anaerobic respiration.

51. *Define cell metabolism. What molecule is synthesized using the energy released by the breakdown of food molecules?*
52. *Define glycolysis, aerobic respiration, and anaerobic respiration.*
53. *How many ATP molecules are produced from one glucose molecule in aerobic and anaerobic respiration?*
54. *During aerobic respiration, what happens to the oxygen we breathe in? Where does the carbon dioxide we breathe out come from?*
55. *Besides ATP, what molecule is produced as a result of anaerobic respiration? Under what conditions is anaerobic respiration necessary?*

Protein Synthesis

Normal cell structure and function would not be possible without proteins, which form the cytoskeleton and other structural components of cells and function as transport molecules, receptors, and enzymes. In addition, proteins secreted from cells perform vital functions: collagen is a structural protein that gives tissues flexibility and strength, enzymes control the chemical reactions of food digestion in the intestines, and protein hormones regulate the activities of many tissues.

Ultimately, the production of all the proteins in the body is under the control of DNA (figure 3.40). Recall from chapter 2 that the building blocks of DNA are nucleotides containing adenine (A), thymine (T), cytosine (C), and guanine (G). The nucleotides form two antiparallel strands of nucleic acids. The term *antiparallel* means that the strands are parallel but extend in opposite directions. Each strand has a 5′ (phosphate) end and a 3′ (hydroxyl) end. The sequence of the nucleotides in the DNA is a method of storing information. Every three nucleotides, called a **triplet,** or **codon,** code for an amino acid, and amino acids are the building blocks of proteins. All of the triplets required to code for the synthesis of a specific polypeptide or protein are called a **gene.** One DNA strand serves as a template for the production of **messenger RNA (mRNA),** which is a complementary copy of the DNA strand.

The production of proteins from the stored information in DNA involves two steps: transcription and translation, which can be illustrated with an analogy. Suppose a cook wants a recipe that is found only in a reference book in the library. Because the book cannot be checked out, the cook makes a handwritten copy, or **transcription,** of the recipe. Later, in the kitchen the information contained in the copied recipe is used to prepare the meal. The changing of something from one form to another (from recipe to meal) is called **translation.** In this analogy, DNA is the reference book that contains many recipes for making different proteins. DNA, however, is too large a molecule to pass through the nuclear envelope to go to the cytoplasm (the kitchen), where the proteins are prepared. Just as the reference book stays in the library, DNA remains in the nucleus. Therefore, through transcription, the cell makes a copy of the information in DNA (the recipe) necessary to make a particular protein (the meal). The copy, which is called mRNA, travels from the nucleus to ribosomes (the countertop) in the cytoplasm, where the information in the copy is used to construct a protein (i.e., translation). Of course, to turn a recipe into a

1. DNA contains the information necessary to produce proteins.

2. Transcription of one DNA strand results in mRNA, which is a mirror image copy of the information in the DNA strand needed to make a protein.

3. The mRNA leaves the nucleus and goes to a ribosome.

4. Amino acids, the building blocks of proteins, are carried to the ribosome by tRNAs.

5. In the process of translation, the information contained in mRNA is used to determine the number, kinds, and arrangement of amino acids in the protein.

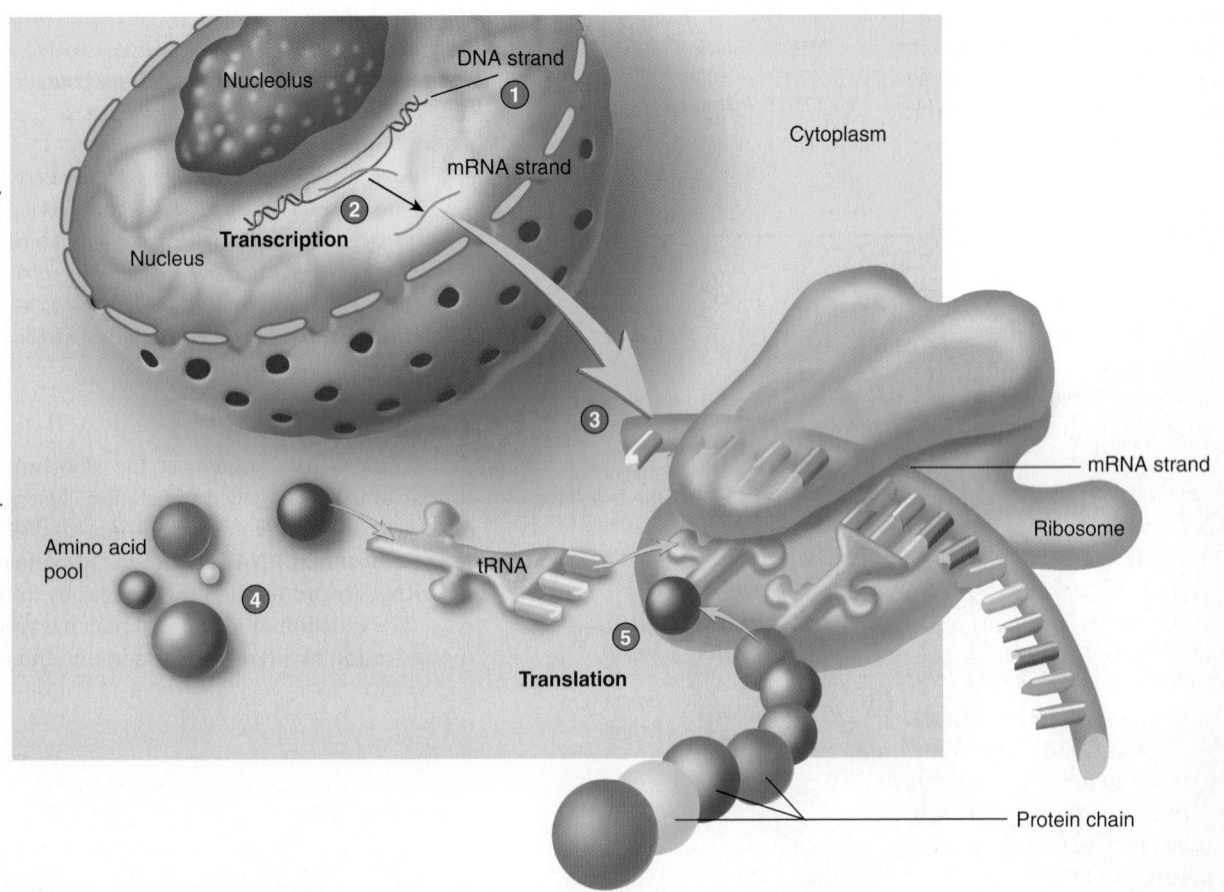

Process Figure 3.40   Overview of Protein Synthesis

meal, the actual ingredients are needed. The ingredients necessary to synthesize a protein are amino acids. Specialized transport molecules, called **transfer RNA (tRNA),** carry the amino acids to the ribosome (see figure 3.40).

In summary, the synthesis of proteins involves transcription, making a copy of part of the stored information in DNA, and translation, converting that copied information into a protein. The details of transcription and translation are considered next.

Transcription

Transcription is the synthesis of mRNA, tRNA, or rRNA on the basis of the sequence of nucleotides in DNA. It occurs when the double strands of a DNA segment separate, one of its strands serves as a template, and RNA nucleotides pair with DNA nucleotides of the template (figure 3.41). Nucleotides pair with each other according to the following rule: adenine pairs with thymine or uracil, and cytosine pairs with guanine. DNA contains thymine, but uracil replaces thymine in RNA. Adenine, thymine, cytosine, and guanine nucleotides of DNA therefore pair with uracil, adenine, guanine, and cytosine nucleotides of mRNA, respectively.

This pairing relationship between nucleotides ensures that the information in DNA is transcribed correctly to mRNA. The RNA nucleotides combine through dehydration reactions catalyzed by RNA polymerase enzymes to form a long mRNA seg-

ment. The elongation of all nucleic acids, both DNA and RNA, occurs in the same chemical direction: from the 5′ to the 3′ end of the molecule. The mRNA molecule contains the information required to determine the sequence of amino acids in a protein. The information, called the **genetic code,** is carried in three-nucleotide codons.

The number and sequence of codons in the mRNA are determined by the number and sequence of sets of three nucleotides, called triplets, in the segments of DNA that were transcribed. For example, the triplet code of CTA in DNA results in the codon GAU in mRNA, which codes for aspartic acid. Each codon codes for a specific amino acid. Sixty-four possible mRNA codons exist, but only 20 amino acids are in proteins. As a result, the genetic code is redundant because more than one codon codes for some amino acids. For example, CGA, CGG, CGT, and CGC all code for the amino acid alanine, and UUU and UAC both code for phenylalanine. Some codons do not code for amino acids but perform other functions. AUG and sometimes GUG act as signals for starting the transcription of a stretch of DNA to RNA. Three codons, UAA, UGA, and UAG, act as signals for stopping the transcription of DNA to RNA.

The region of a DNA molecule between the codon starting transcription and the codon stopping transcription is transcribed into a stretch of RNA and is called a **transcription unit.** A transcription unit codes for a protein or part of a protein. A transcription

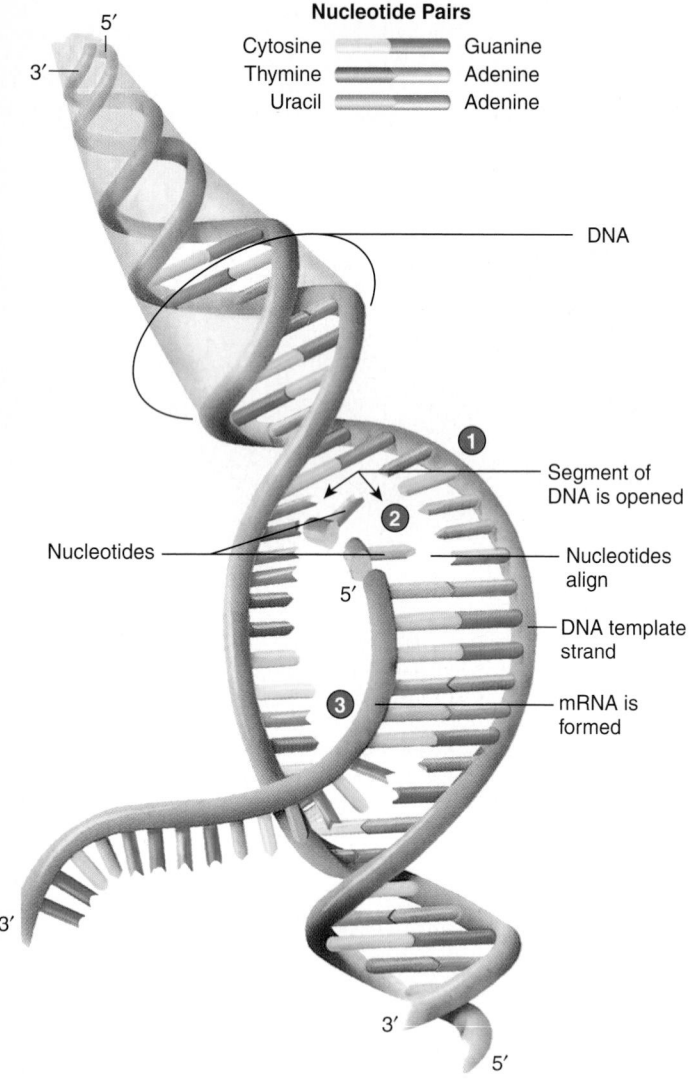

Nucleotide Pairs

Cytosine ▭▭ Guanine
Thymine ▭▭ Adenine
Uracil ▭▭ Adenine

DNA

① Segment of DNA is opened

Nucleotides

②

Nucleotides align

DNA template strand

③ mRNA is formed

1. The strands of the DNA molecule separate from each other. One DNA strand serves as a template for mRNA synthesis.
2. Nucleotides that will form mRNA pair with DNA nucleotides according to the base-pair combinations shown in the key at the top of the figure. Thus the sequence of nucleotides in the template DNA strand (*purple*) determines the sequence of nucleotides in mRNA (*gray*). RNA polymerase (the enzyme is not shown) joins the nucleotides of mRNA together.
3. As nucleotides are added, an mRNA molecule is formed.

Process Figure 3.41 Formation of mRNA by Transcription of DNA

unit is not necessarily a gene. A gene is a functional unit, and some regulatory genes don't code for proteins. A molecular definition of a **gene** is all of the nucleic acid sequences necessary to make a functional RNA or protein.

Not all of a continuous stretch of DNA may code for parts of a protein. Regions of the DNA that code for parts of the protein are called **exons,** whereas those regions of the DNA that do not code for portions of the protein are called **introns.** Both the exon and intron regions of the DNA may be transcribed into mRNA. A mRNA containing introns is called a **pre-mRNA.** After a stretch of pre-mRNA has been transcribed, the introns can be removed and

the exons spliced together by enzyme complexes called **spliceosomes** to produce the functional mRNA (figure 3.42). These changes in the mRNA are called **posttranscriptional processing.**

Thalassemia

Hemoglobin is an oxygen-carrying protein molecule composed of four polypeptides. **Thalassemia** is a group of genetic disorders in which one or more of the polypeptides of hemoglobin is produced in decreased amounts as the result of defective posttranscriptional processing. The decreased amount of hemoglobin in the blood causes anemia, which reduces the oxygen-carrying capacity of the blood.

Translation

The synthesis of a protein at the ribosome in response to the codons of mRNA is called translation (figure 3.43). In addition to mRNA, translation requires ribosomes and tRNA. Ribosomes consist of **ribosomal RNA (rRNA)** and proteins. Like mRNA, tRNA and rRNA are produced in the nucleus by transcription.

The function of tRNA is to match a specific amino acid to a specific codon of mRNA. To do this, one end of each kind of tRNA

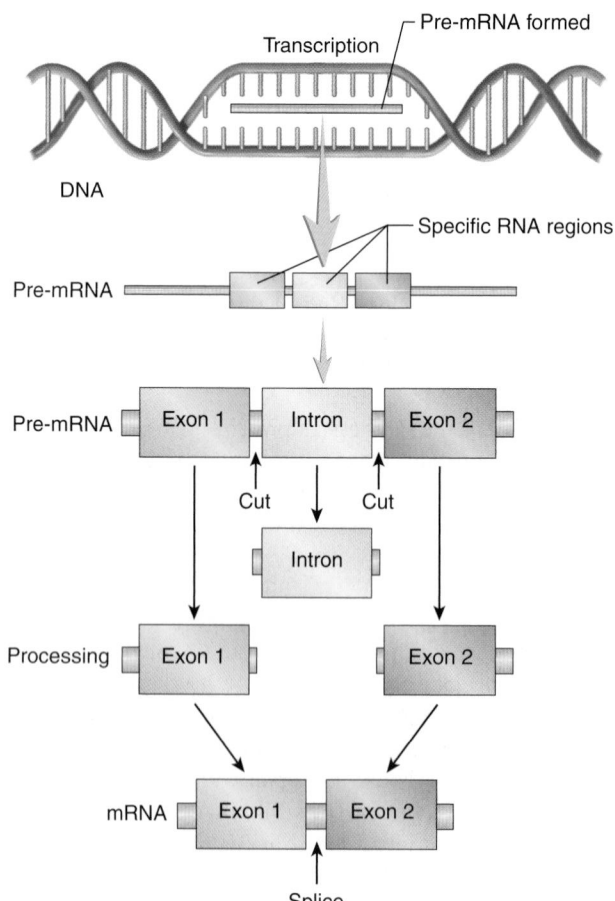

Pre-mRNA formed

Transcription

DNA

Specific RNA regions

Pre-mRNA

Pre-mRNA | Exon 1 | Intron | Exon 2

Cut   Cut

Intron

Processing | Exon 1 | | Exon 2

mRNA | Exon 1 | Exon 2

Splice

Figure 3.42 Posttranscriptional Change in mRNA

The two DNA strands separate and one strand is transcribed to produce a pre-mRNA strand. An intron is cleaved from between two exons and is discarded. The exons are spliced together by spliceosomes (not shown) to make the functional mRNA.

1. To start protein synthesis a ribosome binds to mRNA. The ribosome also has two binding sites for tRNA, one of which is occupied by a tRNA with its amino acid. Note that the codon of mRNA and the anticodon of tRNA are aligned and joined. The other tRNA binding site is open.

2. By occupying the open tRNA binding site the next tRNA is properly aligned with mRNA and with the other tRNA.

3. An enzyme within the ribosome catalyzes a synthesis reaction to form a peptide bond between the amino acids. Note that the amino acids are now associated with only one of the tRNAs.

4. The ribosome shifts position by three nucleotides. The tRNA without the amino acid is released from the ribosome, and the tRNA with the amino acids takes its position. A tRNA binding site is left open by the shift. Additional amino acids can be added by repeating steps *2* through *4*. Eventually a stop codon in the mRNA ends the addition of amino acids to the protein (polypeptide), which is released from the ribosome.

5. Multiple ribosomes attach to a single mRNA. As the ribosomes move down the mRNA, proteins attached to the ribosomes lengthen and eventually detach from the mRNA.

Process Figure 3.43 Translation of mRNA to Produce a Protein combines with a specific amino acid. Another part of the tRNA has an **anticodon,** which consists of three nucleotides. On the basis of the pairing relationships between nucleotides, the anticodon can combine only with its matched codon. For example, the tRNA that binds to aspartic acid has the anticodon CUA, which combines with the codon GAU of mRNA. Therefore the codon GAU codes for aspartic acid.

Ribosomes align the codons of the mRNA with the anti-codons of tRNA and then join the amino acids of adjacent tRNA molecules. As the amino acids are joined together, a chain of amino acids, or a protein, is formed. The step-by-step process of protein synthesis at the ribosome is described in detail in figure 3.43.

Many proteins are longer when first made than they are in their final, functional state. These proteins are called **proproteins,** and the extra piece of the molecule is cleaved off by enzymes to make the proprotein into a functional protein. Many proteins are enzymes, and the proproteins of those enzymes are called **pro-enzymes.** If many proenzymes were made within cells as functional enzymes, they could digest the cell that made them. Instead, they are made as proenzymes and are not converted to active enzymes until they reach a protected region of the body, such as inside the small intestine, where they are functional. Many proteins have side chains, such as polysaccharides, added to them following translation. Some proteins are composed of two or more amino acid chains that are joined after each chain is produced on separate ribosomes. These various modifications to proteins are referred to as **posttranslational processing.**

After the initial part of mRNA is used by a ribosome, another ribosome can attach to the mRNA and begin to make a protein. The resulting cluster of ribosomes attached to the mRNA is called a **polyribosome.** Each ribosome in a polyribosome produces an identical protein, and polyribosomes are an efficient way to use a single mRNA molecule to produce many copies of the same protein.

P R E D I C T   6

Explain how changing one nucleotide within a DNA molecule of a cell could change the structure of a protein produced by that cell. What effect would this change have on the protein's function?

Regulation of Protein Synthesis

All of the cells in the body, except for sex cells, have the same DNA. The transcription of mRNA in cells is regulated, however, so that all portions of all DNA molecules are not continually transcribed. The proteins associated with DNA in the nucleus play a role in regulating the transcription. As cells differentiate and become specialized for specific functions during development, part of the DNA becomes nonfunctional and is not transcribed, whereas other segments of DNA remain very active. For example, in most cells the DNA coding for hemoglobin is nonfunctional, and little if any hemoglobin is synthesized. In developing red blood cells, however, the DNA coding for hemoglobin is functional, and hemoglobin synthesis occurs rapidly.

Protein synthesis in a single cell is not normally constant, but it occurs more rapidly at some times than others. Regulatory molecules that interact with the nuclear proteins can either increase or decrease the transcription rate of specific DNA segments. For example, thyroxine, a hormone released by cells of the thyroid gland, enters cells such as skeletal muscle cells, interacts with specific nuclear proteins, and increases specific types of mRNA transcription. Consequently, the production of certain proteins increases. As a result, an increase in the number of mitochondria and an increase in metabolism occur in these cells.

56. *What type of molecule is produced as a result of transcription? Of translation? Where do these events take place?*
57. *In what molecules are triplets, codons, and anticodons found? What is the genetic code?*
58. *How are triplets, transcription units, and genes related?*
59. *Describe the role of mRNA, rRNA, and tRNA in the production of a protein at a ribosome. What is a polyribosome?*
60. *What are exons and introns? How are they related to pre-mRNA and posttranscriptional processing?*
61. *Define proprotein, proenzyme, and posttranslational processing.*
62. *State two ways the cell controls what DNA is transcribed.*

Cell Life Cycle

The **cell life cycle** includes the changes a cell undergoes from the time it is formed until it divides to produce two new cells. The life cycle of a cell has two stages, an interphase and a cell division stage (figure 3.44).

Interphase

Interphase is the phase between cell divisions. Ninety percent or more of the life cycle of a typical cell is spent in interphase. During this time the cell carries out the metabolic activities necessary for life and performs its specialized functions; for example, secreting digestive enzymes. In addition, the cell prepares to divide. This preparation includes an increase in cell size, because many cell components double in quantity, and a replication of the cell's DNA. The centrioles within the centrosome are also duplicated. Consequently, when the cell divides, each new cell receives the organelles and DNA necessary for continued functioning.

Interphase can be divided into three subphases, called G_1, S, and G_2. During G_1 (the first *gap* phase) and G_2 (the second *gap* phase), the cell carries out routine metabolic activities. During the S phase (the *synthesis* phase), the DNA is replicated (new DNA is synthesized). Many cells in the body do not divide for days, months, or even years. These "resting" cells exit the cell cycle and enter what is called the G_0 phase, in which they remain unless stimulated to divide.

DNA Replication

DNA **replication** is the process by which two new strands of DNA are made, using the two existing strands as templates. During interphase, DNA and its associated proteins appear as dispersed chromatin threads within the nucleus. When DNA replication begins, the two strands of each DNA molecule separate from each other for some distance (figure 3.45). Each strand then functions as a template, or pattern, for the production of a new strand of DNA,

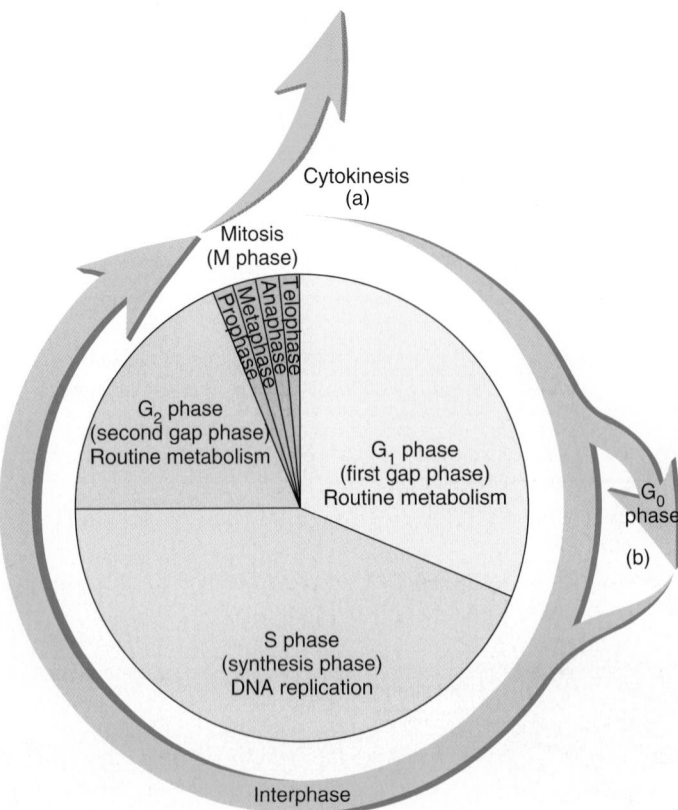

Figure 3.44 Cell Cycle

The cell cycle is divided into interphase and mitosis. Interphase is divided into G_1, S, and G_2 subphases. During G_1 and G_2, the cell carries out routine metabolic activities. During the S phase DNA is replicated. (*a*) Following mitosis, two cells are formed by the process of cytokinesis. Each new cell begins a new cell cycle. (*b*) Many cells exit the cell cycle and enter the G_0 phase, where they remain until stimulated to divide, at which point they reenter the cell cycle.

which is formed as new nucleotides pair with the existing nucleotides of each strand of the separated DNA molecule. The production of the new nucleotide strands is catalyzed by DNA **polymerase,** which adds new nucleotides at the 3′ end of the growing strands. One strand, called the **leading strand,** is formed as a continuous strand, whereas the other strand, called the **lagging strand,** is formed in short segments going in the opposite direction (Okazaki fragments). The short segments are then spliced by **DNA ligase.** As a result of DNA replication, two identical DNA molecules are produced. Each of the two new DNA molecules has one strand of nucleotides derived from the original DNA molecule and one newly synthesized strand.

P R E D I C T ⑦
Suppose a molecule of DNA separates, forming strands 1 and 2. Part of the nucleotide sequence in strand 1 is ATGCTA. From this template, what would be the sequence of nucleotides in the DNA replicated from strand 1 and strand 2?

Cell Division

New cells necessary for growth and tissue repair are produced by cell division. A parent cell divides to form two daughter cells, each

of which has the same amount and type of DNA as the parent cell. Because DNA determines cell structure and function, the daughter cells tend to have the same structure and perform the same functions as the parent cell. During development and cell differentiation, however, the functions of daughter cells may differ from each other and from that of the parent cell.

Cell division involves two major events: the division of the nucleus to form two new nuclei, and the division of the cytoplasm to form two new cells. Each of the new cells contains one of the newly formed nuclei. The division of the nucleus occurs by mitosis, and the division of the cytoplasm is called cytokinesis.

Mitosis

Mitosis (mī-tō′sis) is the division of the nucleus into two nuclei, each of which has the same amount and type of DNA as the original nucleus. The DNA, which was dispersed as chromatin in interphase, condenses in mitosis to form chromosomes. All human **somatic** (sō-mat′ik) cells, which include all cells except the sex cells, contain 46 chromosomes. This number of chromosomes is referred to as a **diploid** (dip′loyd) number. Sex cells have half as many chromosomes as somatic cells (see section on "Meiosis" in chapter 28). The 46 chromosomes in somatic cells are organized into 23 pairs of chromosomes. Twenty-two of these pairs are called **autosomes.** Each member of an autosomal pair of chromosomes looks structurally alike, and together they are called a **homologous** (hŏ-mol′ō-gŭs) pair of chromosomes. One member of each autosomal pair is derived from the person's father, and the other is derived from the mother. The remaining pair of chromosomes are the sex chromosomes. In females, the sex chromosomes look alike, and each is called an **X chromosome.** In males, the sex chromosomes do not look alike. One chromosome is an X chromosome, and the other is smaller and is called a **Y chromosome.** One X chromosome of a female is derived from her mother and the other is derived from her father. The X chromosome of a male is derived from his mother and the Y chromosome is derived from his father.

For convenience of discussion, mitosis is divided into four phases: **prophase, metaphase, anaphase,** and **telophase** (tel′ō-fāz). Although each phase represents major events, mitosis is a continuous process, and no discrete jumps occur from one phase to another. Learning the characteristics associated with each phase is helpful, but a more important concept is how each daughter cell obtains the same number and type of chromosomes as the parent cell. The major events of mitosis are summarized in figure 3.46.

Cytokinesis

Cytokinesis (sī′tō-ki-nē′sis) is the division of the cytoplasm of the cell to produce two new cells. Cytokinesis begins in anaphase, continues through telophase, and ends in the following interphase (see figure 3.46). The first sign of cytokinesis is the formation of a **cleavage furrow,** or indentation of the plasma membrane, which forms midway between the centrioles. A contractile ring composed primarily of actin filaments pulls the plasma membrane inward, dividing the cell into two halves. Cytokinesis is complete when the membranes of the two halves separate at the cleavage furrow to form two separate cells.

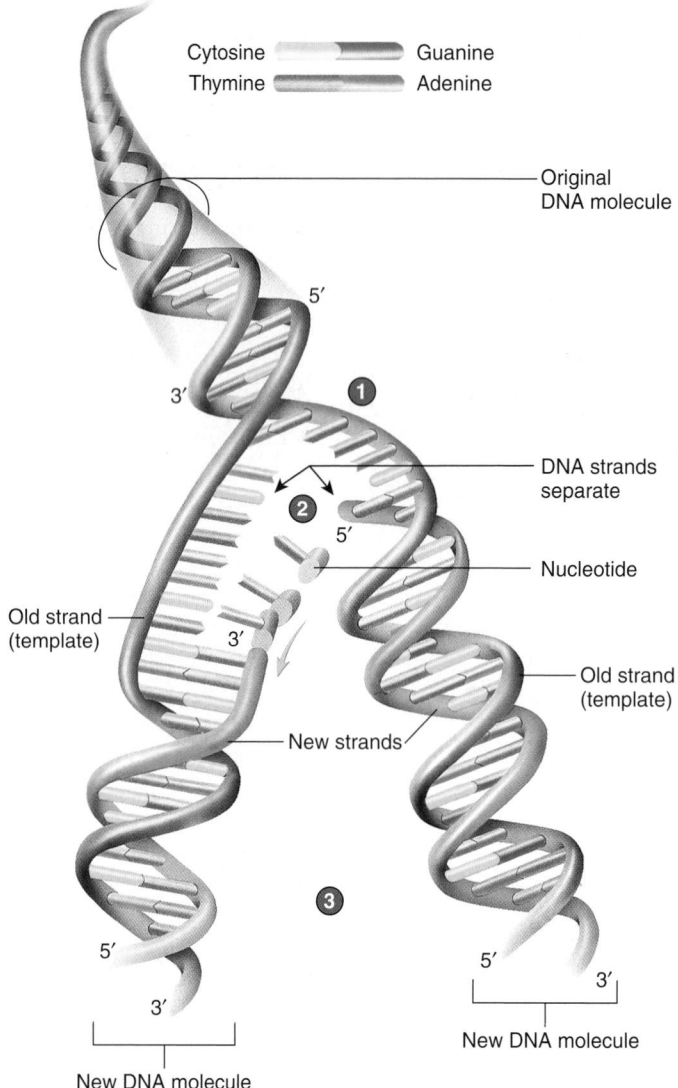

1. The strands of the DNA molecule separate from each other.
2. Each old strand (*dark purple*) functions as a template on which a new strand (*light purple*) is formed. The base-pairing relationship between nucleotides determines the sequence of nucleotides in the newly formed strands.
3. Two identical DNA molecules are produced.

Process Figure 3.45   Replication of DNA
Replication of DNA during interphase produces two identical molecules of DNA.

63. **Define interphase. What percent of the cell life cycle is typically spent in interphase?**
64. **Describe the cell's activities during G_1, S, and G_2 phases of the cell life cycle.**
65. **Describe the process of DNA replication. What are the functions of DNA polymerase and DNA ligase?**
66. **Define mitosis. How do the two nuclei that are produced in mitosis compare to the original nucleus?**
67. **How many chromosomes are contained in a human somatic cell? How are the chromosomes of males and females the same? How are they different?**
68. **List the events that occur during interphase, prophase, metaphase, anaphase, and telophase of mitosis.**
69. **Describe cytokinesis.**

Cloning
Through the process of differentiation, cells become specialized to certain functions and are no longer capable of producing an entire organism if isolated. Over 30 years ago, however, it was demonstrated in frogs that if the nucleus is removed from a differentiated cell and is transferred to an oocyte with the nucleus removed, a complete, normal frog can develop from that oocyte. This process, called **cloning**, demonstrated that during differentiation, genetic information is not irrevocably lost. Because mammalian oocytes are considerably smaller than frog oocytes, cloning of mammalian cells has been technically much more difficult. Dr. Ian Wilmut and his colleagues at the Roslin Institute in Edinburgh, Scotland, overcame those technical difficulties in 1996, when they successfully cloned the first mammal, a sheep named Dolly. Since that time, several other mammalian species have been cloned.

1. **Interphase** is the time between cell divisions. DNA is found as thin threads of chromatin in the nucleus. DNA replication occurs during interphase. Organelles, other than the nucleus, and centrioles duplicate during interphase.

2. In **prophase**, the chromatin condenses into chromosomes. Each chromosome consists of two chromatids joined at the centromere. The centrioles move to the opposite ends of the cell, and the nucleolus and the nuclear envelope disappear. Microtubules form near the centrioles and project in all directions. Some of the microtubules end blindly and are called astral fibers. Others, known as spindle fibers, project toward an invisible line called the equator and overlap with fibers from opposite centrioles.

3. In **metaphase**, the chromosomes align in the center of the cell in association with the spindle fibers. The spindle fibers attach to kinetochores in the centromere of each chromosome.

4. In **anaphase**, the chromatids separate, and each chromatid is then referred to as a chromosome. Thus, when the centromeres divide, the chromosome number is double, and there are two identical sets of chromosomes. The chromosomes, assisted by the spindle fibers, move toward the centrioles at each end of the cell. Separation of the chromatids signals the beginning of anaphase, and by the time anaphase has ended, the chromosomes have reached the poles of the cell. Cytokinesis begins during anaphase.

5. In **telophase**, migration of each set of chromosomes is complete. The chromosomes unravel to become less distinct chromatin threads. The nuclear envelope forms from the endoplasmic reticulum. The nucleoli form, and cytokinesis continues to form two cells.

6. Mitosis is complete, and a new interphase begins. The chromosomes have unraveled to become chromatin. Cell division has produced two daughter cells, each with DNA that is identical to the DNA parent cell.

Process Figure 3.46   Mitosis and Cytokinesis

Clinical Focus   Genetic Engineering

We are living in an exciting era, when the genetic bases of many human illnesses are rapidly being revealed. As we discover the defective genes associated with these diseases and learn the nature and function of the proteins they encode, our ability to understand and therefore to treat many of these diseases is improved. Once we have learned the basis of a given disease, a number of approaches are possible for treating it, such as genetic engineering or other molecular techniques. For example, the gene for insulin has been inserted into a bacterial genome, thereby enabling the bacterium to produce large quantities of human insulin, which increases its availability and functional quality. Antibodies are being developed that will target specific cells or cell surface marker molecules associated with diseases such as arthritis or cancer. Clinical trials are underway to test the efficacy of introducing a functional copy of a gene into the cells of a person who has a defective gene.

A negative side exists to this technology, however. Many people are concerned that the introduction of foreign genes into bacteria and human cells may have unexpected side effects. Many genes have multiple functions, and a danger exists that we may begin using gene therapy before we know all the ramifications. Some people are greatly concerned about how far genetic engineering should be allowed to go. What range of "genetic defects" should humanity be allowed to change, or should no limit be established? For example, when we discover the genes involved in controlling human height, should parents be allowed to use gene therapy to increase a child's height so that he or she can be better at basketball? An even more immediate concern is to what extent a person's genetic code should be made public. For example, should a medical insurance company or employer be allowed to see a person's genetic profile to set insurance premiums or make employment judgments? If a person is shown to have a gene for muscular dystrophy, should the person's insurance company be given that information? Also of concern is whether a person or company should be able to patent and thus to own a human gene.

Apoptosis (Programmed Cell Death)

Apoptosis (ap′op-tō′sis, ap′ō-tō′sis), or **programmed cell death,** is a normal process by which cell numbers within various tissues are adjusted and controlled. During development, extra cells are removed by apoptosis, such as cells between the developing fingers and toes, to fine-tune the contours of the developing fetus. The number of cells in most adult tissues is maintained at a specific level. Apoptosis eliminates excess cells produced by proliferation within some adult tissues to maintain a constant number of cells within the tissue. Damaged or potentially dangerous cells, virus-infected cells, and potential cancer cells are also eliminated by apoptosis.

Apoptosis is regulated by specific genes. The proteins coded for by those genes initiate events within the cell that ultimately lead to the cell's death. As apoptosis begins, the chromatin within the nucleus condenses and fragments. This is followed by fragmentation of the nucleus and finally by death and fragmentation of the cell. The cell fragments are cleaned up by specialized cells called macrophages.

Cellular Aspects of Aging

A number of cellular structures and/or events appear to be involved in the process of aging. The major theories of aging concentrate on molecules within the cell, such as lipids, proteins, and nucleic acids. It is estimated that at least 35% of the factors affecting aging are genetic.

1. *Cellular clock.* One theory of aging suggests that there is a cellular clock, which, after a certain passage of time or a certain number of cell divisions, results in death of the cell line.
2. *Death genes.* Another theory suggests that there are "death genes," which turn on late in life, or sometimes prematurely, causing cells to deteriorate and die.
3. *DNA damage.* Other theories suggest that through time, DNA is damaged, resulting in cell degeneration and death.

It may be that DNA is protected from damage by a specific sequence of nucleotides, TTAGGG, called a **telomere** (tel′ō-mēr), at the end of chromosomes. Apparently, during DNA replication, nucleotides are lost at the extreme distal end of the DNA molecule. Telomeres, at this extreme end, take the brunt of this replicative loss, thereby protecting regions of DNA that code for essential proteins. **Telomerase** is an enzyme that mediates the repair and maintains the integrity of the telomeric region of chromosomes. The enzyme can even add additional nucleotides to the telomeric region. Telomerase appears to be lost from aging populations of somatic cells. Without telomerase to repair the telomeres, they tend to degenerate during replication, and eventually, critical, functional regions of DNA are lost during replication, resulting in cell death.

4. *Free radicals.* The DNA in somatic cells may also be susceptible to more direct damage, resulting in somatic mutations, which may result in cellular dysfunction and, ultimately, cell death. One of the major sources of DNA damage is apparently from **free radicals,** which are atoms or molecules with an unpaired electron.
5. *Mitochondrial damage.* It may be that mitochondrial DNA is more sensitive to free-radical damage than is nuclear DNA. Mitochondrial DNA damage may result in loss of proteins critical to mitochondrial function. Because the mitochondria are the power plants of cells, loss of mitochondrial function could result in the loss of energy critical to cell function and, ultimately, to cell death. One proposal suggests that reduced caloric intake may reduce free radical damage to mitochondria.

70. *How might a cellular clock, death genes, DNA damage, free radicals, or mitochondrial damage contribute to cellular aging?*

S U M M A R Y

1. The plasma membrane forms the outer boundary of the cell.
2. The nucleus directs the activities of the cell.
3. Cytoplasm, between the nucleus and plasma membrane, is where most cell activities take place.

Functions of the Cell (p. 58)

1. Cells are the basic unit of life.
2. Cells provide protection and support.
3. Cells allow for movement.
4. Cells provide a means of communication.
5. Cells metabolize and release energy.
6. Cells provide for inheritance.

How We See Cells (p. 60)

1. Light microscopes allow us to visualize general features of cells.
2. Electron microscopes allow us to visualize the fine structure of cells.

Plasma Membrane (p. 60)

1. The plasma membrane passively or actively regulates what enters or leaves the cell.
2. The plasma membrane is composed of a phospholipid bilayer in which proteins are suspended (fluid-mosaic model).

Membrane Lipids (p. 61)

Lipids give the plasma membrane most of its structure and some of its function.

Membrane Proteins (p. 61)

1. Membrane proteins function as markers, attachment sites, channels, receptors, enzymes, and carriers.
2. Some receptor molecules are linked to and control channel proteins.
3. Some receptor molcules are linked to G proteins, which, in turn, control numerous cellular activities.

Movement Through the Plasma Membrane (p. 65)

1. Lipid-soluble molecules pass through the plasma membrane readily by dissolving in the lipid bilayer.
2. Small molecules pass through membrane channels. Most channels are positively charged, allowing negatively charged ions and neutral molecules to pass through more readily than positively charged ions.
3. Large polar substances (e.g., glucose and amino acids) are transported through the membrane by carrier molecules.
4. Larger pieces of material enter cells in vesicles.

Diffusion

1. Diffusion is the movement of a substance from an area of higher concentration to one of lower concentration (down a concentration gradient).
2. The concentration gradient is the difference in solute concentration between two points divided by the distance separating the points.
3. The rate of diffusion increases with an increase in the concentration gradient, an increase in temperature, a decrease in molecular size, and a decrease in viscosity.
4. The end result of diffusion is a uniform distribution of molecules.
5. Diffusion requires no expenditure of energy.

Osmosis

1. Osmosis is the diffusion of water (solvent) across a selectively permeable membrane.
2. Osmotic pressure is the force required to prevent the movement of water across a selectively permeable membrane.

3. Isosmotic solutions have the same concentration of solute particles, hyperosmotic solutions have a greater concentration, and hyposmotic solutions have a lesser concentration of solute particles than a reference solution.
4. Cells placed in an isotonic solution neither swell nor shrink. In a hypertonic solution they shrink (crenate), and in a hypotonic solution they swell and may burst (lyse).

Filtration

1. Filtration is the movement of a liquid through a partition with holes that allow the liquid, but not everything in the liquid, to pass through them.
2. Liquid movement results from a pressure difference across the partition.

Mediated Transport Mechanisms

1. Mediated transport is the movement of a substance across a membrane by means of a carrier molecule. The substances transported tend to be large, water-soluble molecules.
   - The carrier molecules have binding sites that bind with either a single transport molecule or a group of similar transport molecules. This selectiveness is called specificity.
   - Similar molecules can compete for carrier molecules, with each reducing the rate of transport of the other.
   - Once all the carrier molecules are in use, the rate of transport cannot increase further (saturation).
2. Three kinds of mediated transport can be identified.
   - Facilitated diffusion moves substances down their concentration gradient and does not require energy expenditure (ATP).
   - Active transport can move substances against their concentration gradient and requires ATP. An exchange pump is an active-transport mechanism that simultaneously moves two substances in opposite directions across the plasma membrane.
   - In secondary active transport, an ion is moved across the plasma membrane by active transport, and the energy produced by the ion diffusing back down its concentration gradient can transport another molecule, such as glucose, against its concentration gradient.

Endocytosis and Exocytosis (p. 73)

1. Endocytosis is the bulk movement of materials into cells.
   - Phagocytosis is the bulk movement of solid material into cells by the formation of a vesicle.
   - Pinocytosis is similar to phagocytosis, except that the ingested material is much smaller or is in solution.
2. Exocytosis is the secretion of materials from cells by vesicle formation.
3. Endocytosis and exocytosis use vesicles, can be specific (receptor-mediated endocytosis) for the substance transported, and require energy.

Cytoplasm (p. 75)

The cytoplasm is the material outside the nucleus and inside the plasma membrane.

Cytosol

1. Cytosol consists of a fluid part (the site of chemical reactions), the cytoskeleton, and cytoplasmic inclusions.
2. The cytoskeleton supports the cell and enables cell movements. It consists of protein fibers.
   - Microtubules are hollow tubes composed of the protein tubulin. They form spindle fibers and are components of centrioles, cilia, and flagella.

- Actin filaments are small protein fibrils that provide structure to the cytoplasm or cause cell movements.
- Intermediate filaments are protein fibers that provide structural strength to cells.
3. Cytoplasmic inclusions, such as lipochromes, are not surrounded by membranes.

Organelles   (p. 78)

Organelles are subcellular structures specialized for specific functions.

Nucleus

1. The nuclear envelope consists of two separate membranes with nuclear pores.
2. DNA and associated proteins are found inside the nucleus as chromatin. DNA is the hereditary material of the cell and controls the activities of the cell by producing proteins through RNA.
3. Proteins play a role in the regulation of DNA activity.
4. Nucleoli consist of RNA and proteins and are the sites of ribosomal subunit assembly.

Centrioles and Spindle Fibers

1. Centrioles are cylindrical organelles located in the centrosome, a specialized zone of the cytoplasm. The centrosome is the site of microtubule formation.
2. Spindle fibers are involved in the separation of chromosomes during cell division.

Cilia and Flagella

1. Movement of materials over the surface of the cell is facilitated by cilia.
2. Flagella, much longer than cilia, propel sperm cells.

Microvilli

Microvilli increase the surface area of the plasma membrane for absorption or secretion.

Ribosomes

1. Ribosomes consist of small and large subunits manufactured in the nucleolus and assembled in the cytoplasm.
2. Ribosomes are the sites of protein synthesis.
3. Ribosomes can be free or associated with the endoplasmic reticulum.

Endoplasmic Reticulum

1. The endoplasmic reticulum is an extension of the outer membrane of the nuclear envelope and forms tubules or sacs (cisternae) throughout the cell.
2. The rough endoplasmic reticulum has ribosomes and is a site of protein synthesis and modification.
3. The smooth endoplasmic reticulum lacks ribosomes and is involved in lipid production, detoxification, and calcium storage.

Golgi Apparatus

The Golgi apparatus is a series of closely packed, modified cisternae that function to modify, package, and distribute lipids and proteins produced by the endoplasmic reticulum.

Secretory Vesicles

Secretory vesicles are membrane-bound sacs surrounded by membranes that carry substances from the Golgi apparatus to the plasma membrane, where the contents of the vesicle are released by exocytosis.

Lysosomes

1. Lysosomes are membrane-bounded sacs containing hydrolytic enzymes. Within the cell, the enzymes break down phagocytized material and nonfunctional organelles (autophagia).
2. Enzymes released from the cell by lysis or enzymes secreted from the cell can digest extracellular material.

Peroxisomes

Peroxisomes are membrane-bounded sacs containing enzymes that digest fatty acids and amino acids and enzymes that catalyze the breakdown of hydrogen peroxide.

Proteasomes

Proteasomes are large multienzyme complexes, not bound by membranes, which digest selected proteins within the cell.

Mitochondria

1. Mitochondria are the major sites of the production of ATP, which is used as an energy source by cells.
2. The mitochondria have a smooth outer membrane and an inner membrane that is infolded to produce cristae.
3. Mitochondria contain their own DNA, can produce some of their own proteins, and can replicate independently of the cell.

Overview of Cell Metabolism   (p. 87)

1. Aerobic respiration requires oxygen and produces carbon dioxide, water, and up to 38 ATP molecules from a molecule of glucose.
2. Anaerobic respiration does not require oxygen and produces lactic acid and two ATP molecules from a molecule of glucose.

Protein Synthesis   (p. 88)

1. During transcription, information stored in DNA is copied to form mRNA.
2. During translation, the mRNA goes to ribosomes, where it directs the synthesis of proteins.

Transcription

1. DNA unwinds and, through nucleotide pairing, produces mRNA (transcription).
2. The genetic code, which codes for amino acids, consists of codons, which are sequences of three nucleotides in mRNA.
3. Introns are removed and exons are spliced by spliceosomes during posttranscriptional processing.

Translation

1. The mRNA moves through the nuclear pores to ribosomes.
2. Transfer RNA (tRNA), which carries amino acids, interacts at the ribosome with mRNA. The anticodons of tRNA bind to the codons of mRNA, and the amino acids are joined to form a protein (translation).
3. Proproteins, some of which are proenzymes, are modified into proteins, some of which are enzymes, during posttranslational processing.

Regulation of Protein Synthesis

1. Cells become specialized because of inactivation of certain parts of the DNA molecule and activation of other parts.
2. The level of DNA activity and thus protein production can be controlled internally or can be affected by regulatory substances secreted by other cells.

Cell Life Cycle   (p. 92)

The cell life cycle has two stages: interphase and mitosis.

Interphase

Interphase is the period between cell divisions. This is the time of DNA replication.

DNA Replication

DNA unwinds, and each strand produces a new DNA molecule during replication.

Cell Division

1. Cell division includes nuclear division and cytoplasmic division.
2. Mitosis is the replication of the nucleus of the cell, and cytokinesis is division of the cytoplasm of the cell.
3. Humans have 22 pairs of homologous chromosomes called autosomes. Females also have two X chromosomes, and males also have an X chromosome and a Y chromosome.
4. Mitosis is a continuous process divided into four phases.
   - *Prophase.* Chromatin condenses to become visible as chromosomes. Each chromosome consists of two chromatids joined at the centromere. Centrioles move to opposite poles of the cell, and astral fibers and spindle fibers form. Nucleoli disappear, and the nuclear envelope degenerates.
   - *Metaphase.* Chromosomes align at the equatorial plane.
   - *Anaphase.* The chromatids of each chromosome separate at the centromere. Each chromatid then is called a chromosome. The chromosomes migrate to opposite poles.
   - *Telophase.* Chromosomes unravel to become chromatin. The nuclear envelope and nucleoli reappear.
5. Cytokinesis begins with the formation of the cleavage furrow during anaphase. It is complete when the plasma membrane comes together at the equator, thus producing two new daughter cells.

Cellular Aspects of Aging   (p. 96)

There are five major theories of aging:

1. *Cellular clock.* A cell line may die out after a certain time or a certain number of cell divisions.
2. *Death genes.* There may be "death genes," which turn on late in life, causing cells to die.
3. *DNA damage.* Telomeres normally protect DNA from damage during replication, and telomerase protects these telomeres. Aging cells lack telomerase and, therefore, telomeres. As a result, the DNA in aging cells becomes open to damage.
4. *Free radicals.* Free radicals may also damage DNA.
5. *Mitochondrial damage.* Mitochondrial DNA may be the most sensitive to free-radical damage.

R E V I E W   A N D   C O M P R E H E N S I O N

1. In the plasma membrane, _____ form(s) the lipid bilayer, _____ determine(s) the fluid nature of the membrane, and _____ mainly determine(s) the function of the membrane.
   a. phospholipids, cholesterol, proteins
   b. phospholipids, proteins, cholesterol
   c. proteins, cholesterol, phospholipids
   d. cholesterol, phospholipids, proteins
   e. cholesterol, proteins, phospholipids

2. Which of the following are functions of the proteins found in the plasma membrane?
   a. channel proteins
   b. marker molecules
   c. receptor molecules
   d. enzymes
   e. all of the above

3. Integrins in the plasma membrane function as
   a. channel proteins.
   b. marker molecules.
   c. attachment sites.
   d. enzymes.
   e. receptor molecules.

4. In general, lipid-soluble molecules diffuse through the _____; small, water-soluble molecules diffuse through the _____.
   a. membrane channels, membrane channels
   b. membrane channels, lipid bilayer
   c. lipid bilayer, carrier molecules
   d. lipid bilayer, membrane channels
   e. carrier proteins, membrane channels

5. Small pieces of matter, and even whole cells, can be transported across the plasma membrane in
   a. membrane channels.
   b. carrier molecules.
   c. receptor molecules.
   d. marker molecules.
   e. vesicles.

6. The rate of diffusion increases if the
   a. concentration gradient decreases.
   b. temperature of a solution decreases.
   c. viscosity of a solution decreases.
   d. all of the above.

7. Concerning the process of diffusion, at equilibrium
   a. the net movement of solutes stops.
   b. random molecular motion continues.
   c. there is an equal movement of solute in opposite directions.
   d. concentration of solute is equal throughout the solution.
   e. all of the above.

8. Which of these statements about osmosis is true?
   a. Osmosis always involves a membrane that allows water and all solutes to diffuse through it.
   b. The greater the solute concentration, the smaller the osmotic pressure of a solution.
   c. Osmosis moves water from a greater solute concentration to a lesser solute concentration.
   d. The greater the osmotic pressure of a solution, the greater the tendency for water to move into the solution.
   e. Osmosis occurs because of hydrostatic pressure outside the cell.

9. If a cell is placed in a _____ solution, lysis of the cell may occur.
   a. hypertonic
   b. hypotonic
   c. isotonic
   d. isosmotic

10. Container A contains a 10% salt solution, and container B contains a 20% salt solution. If the two solutions are connected, the net movement of water by diffusion is from _____ to _____, and the net movement of salt by diffusion is from _____ to _____.
    a. A,B; A,B
    b. A,B; B,A
    c. B,A; A,B
    d. B,A; B,A

11. Suppose that a woman ran a long-distance race in the summer. During the race she lost a large amount of hyposmotic sweat. You would expect her cells to
    a. shrink.
    b. swell.
    c. stay the same

12. Suppose that a man is doing heavy exercise in the hot summer sun. He sweats profusely. He then drinks a large amount of distilled water. After he drank the water, you would expect his tissue cells to
    a. shrink.
    b. swell.
    c. remain the same.

13. Unlike diffusion and osmosis, filtration depends on a
_____ on the two sides of the partition.
    a. concentration gradient
    b. pressure difference
    c. difference in electric charge
    d. difference in osmotic pressure
    e. hyposmotic solution

14. Which of these statements about facilitated diffusion is true?
    a. In facilitated diffusion, net movement is down the concentration gradient.
    b. Facilitated diffusion requires the expenditure of energy.
    c. Facilitated diffusion does not require a carrier protein.
    d. Facilitated diffusion moves materials through membrane channels.
    e. Facilitated diffusion moves materials in vesicles.

15. Which of these statements concerning cotransport of glucose into cells is true?
    a. The sodium-potassium exchange pump moves Na^+ into cells.
    b. The concentration of Na^+ outside cells is less than inside cells.
    c. A carrier protein moves Na^+ into cells and glucose out of cells.
    d. The concentration of glucose can be greater inside cells than outside cells.
    e. As Na^+ is actively transported into the cell, glucose is carried along.

16. A white blood cell ingests solid particles by forming vesicles. This describes the process of
    a. exocytosis.
    b. facilitated diffusion.
    c. secondary active transport.
    d. phagocytosis.
    e. pinocytosis.

17. Given these characteristics:
    1. requires energy
    2. requires carrier proteins
    3. requires membrane channels
    4. requires vesicles

    Choose the characteristics that apply to exocytosis.
    a. 1, 2
    b. 1, 4
    c. 1, 3, 4
    d. 1, 2, 3
    e. 1, 2, 3, 4

18. Cytoplasm is found
    a. in the nucleus.
    b. outside the nucleus and inside the plasma membrane.
    c. outside the plasma membrane.
    d. inside mitochondria.
    e. everywhere in the cell.

19. Which of these elements of the cytoskeleton is composed of tubulin and forms essential components of centrioles, spindle fibers, cilia, and flagella?
    a. actin filaments
    b. intermediate filaments
    c. microtubules

20. Cylindrically shaped extensions of the plasma membrane that do not move, and are supported with actin filaments; they may function in absorption or as sensory receptors. This describes
    a. centrioles.
    b. spindle fibers.
    c. cilia.
    d. flagella.
    e. microvilli.

21. A large structure, normally visible in the nucleus of a cell, where ribosomal subunits are produced.
    a. endoplasmic reticulum.
    b. mitochondria.
    c. nucleolus.
    d. lysosome.

22. A cell that synthesizes large amounts of protein for use outside the cell has a large
    a. number of cytoplasmic inclusions.
    b. number of mitochondria.
    c. amount of rough endoplasmic reticulum.
    d. amount of smooth endoplasmic reticulum.
    e. number of lysosomes.

23. Which of these organelles produces large amounts of ATP?
    a. nucleus
    b. mitochondria
    c. ribosomes
    d. endoplasmic reticulum
    e. lysosomes

24. Mature red blood cells cannot
    a. synthesize ATP.
    b. transport oxygen.
    c. synthesize new protein.
    d. use glucose as a nutrient.

25. For each glucose molecule, aerobic respiration may produce up to _____ ATP and 6 CO_2 molecules, whereas anaerobic respiration produces _____ ATP and 2 lactic acid molecules.
    a. 2, 2
    b. 2, 4
    c. 2, 38
    d. 38, 2
    e. 38, 38

26. A portion of an mRNA molecule that determines one amino acid in a polypeptide chain is called a
    a. nucleotide.
    b. gene.
    c. codon.
    d. exon.
    e. intron.

27. In which of these organelles is mRNA synthesized?
    a. nucleus
    b. ribosome
    c. endoplasmic reticulum
    d. nuclear envelope
    e. peroxisome

28. During the cell life cycle, DNA replication occurs during the
    a. G_1 phase.
    b. G_2 phase.
    c. M phase.
    d. S phase.

29. Given the following activities:
    1. repair
    2. growth
    3. gamete production
    4. differentiation

    Which of the activities are the result of mitosis?
    a. 2
    b. 3
    c. 1, 2
    d. 3, 4
    e. 1, 2, 4

Answers in Appendix F

C R I T I C A L   T H I N K I N G

1. Why does a surgeon irrigate a surgical wound from which a tumor has been removed with sterile distilled water rather than with sterile isotonic saline?

2. Solution A is hyperosmotic to solution B. If solution A is separated from solution B by a selectively permeable membrane, does water move from solution A into solution B or vice versa? Explain.

3. A researcher wants to determine the nature of the transport mechanism that moved substance X into a cell. She could measure the concentration of substance X in the extracellular fluid and within the cell, as well as the rate of movement of substance X into the cell. She does a series of experiments and gathers the data shown in the graph. Choose the transport process that is consistent with the data.
   a. diffusion
   b. active transport
   c. facilitated diffusion
   d. not enough information to make a judgment

Graph depicting the rate of movement of substance X from a fluid into a cell (*y* axis) versus the concentration of substance X within the cell minus the concentration outside the cell (*x* axis). At point *A* the extracellular concentration of substance X is equal to the intracellular concentration of substance X (designated *0* on the *x* axis).

4. A dialysis membrane is selectively permeable, and substances smaller than proteins are able to pass through it. If you wanted to use a dialysis machine to remove only urea (a small molecule) from blood, what could you use for the dialysis fluid?
   a. a solution that is isotonic and contains only large molecules, such as protein
   b. a solution that is isotonic and contains the same concentration of all substances except that it has no urea
   c. distilled water, which contains no ions or dissolved molecules
   d. blood, which is isotonic and contains the same concentration of all substances, including urea

5. Predict the consequence of a reduced intracellular K^+ concentration on the resting membrane potential.

6. If you had the ability to inhibit mRNA synthesis with a drug, explain how you could distinguish between proteins released from secretory vesicles in which they had been stored and proteins released from cells in which they have been newly synthesized.

7. Given the following data from electron micrographs of a cell, predict the major function of the cell:
   • moderate number of mitochondria;
   • well-developed rough endoplasmic reticulum;
   • moderate number of lysosomes;
   • well-developed Golgi apparatus;
   • dense nuclear chromatin;
   • numerous vesicles.

Answers in Appendix G

A N S W E R S   T O   P R E D I C T   Q U E S T I O N S

1. Urea is continually produced by metabolizing cells and diffuses from the cells into the interstitial spaces and from the interstitial spaces into the blood. If the kidneys stop eliminating urea, it begins to accumulate in the blood. Because the concentration of urea increases in the blood, urea cannot diffuse from the interstitial spaces. As urea accumulates in the interstitial spaces, the rate of diffusion from cells into the interstitial spaces slows because the urea must pass from a higher to a lower concentration by the process of diffusion. The urea finally reaches concentrations high enough to be toxic to cells, thereby causing cell damage followed by cell death.

2. If the membrane is freely permeable, the solutes in the tube diffuse from the tube (higher concentration of solutes) into the beaker (lower concentration of solutes) until equal amounts of solutes exist inside the tube and beaker (i.e., equilibrium). In a similar fashion, water in the beaker diffuses from the beaker (higher concentration of water) into the tube (lower concentration of water) until equal amounts of water are inside the tube and beaker. Consequently, the solution concentrations inside the tube and

beaker are the same because they both contain the same amounts of solutes and water. Under these conditions, no net movement of water into the tube occurs. This simple experiment demonstrates that osmosis and osmotic pressure require a membrane that is selectively permeable.

3. Glucose transported by facilitated diffusion across the plasma membrane moves from a higher to a lower concentration. If glucose molecules are quickly converted to some other molecule as they enter the cell, a steep concentration gradient is maintained. The rate of glucose transport into the cell is directly proportional to the magnitude of the concentration gradient.

4. Digitalis should increase the force of heart contractions. By interfering with Na^+ transport, digitalis decreases the concentration gradient for Na^+ because fewer Na^+ are pumped out of cells by active transport. Consequently, fewer Na^+ diffuse into cells, and fewer Ca^{2+} move out of the cells by countertransport. The higher intracellular levels of Ca^{2+} promote more forceful contractions.

5. a. Cells highly specialized to synthesize and secrete proteins have large amounts of rough endoplasmic reticulum (ribosomes attached to endoplasmic reticulum) because these organelles are important for protein synthesis. Golgi apparatuses are well developed because they package materials for release in secretory vesicles. Also, numerous secretory vesicles exist in the cytoplasm.

    b. Cells highly specialized to actively transport substances into the cell have a large surface area exposed to the fluid from which substances are actively transported, and numerous mitochondria are present near the membrane across which active transport occurs.

    c. Cells highly specialized to synthesize lipids have large amounts of smooth endoplasmic reticulum. Depending on the kind of lipid produced, lipid droplets may accumulate in the cytoplasm.

    d. Cells highly specialized to phagocytize foreign substances have numerous lysosomes in their cytoplasm and evidence of phagocytic vesicles.

6. By changing a single nucleotide within a DNA molecule, a change in the nucleotide of mRNA produced from that segment of DNA also occurs, and a different amino acid is placed in the amino acid chain for which the mRNA provides direction. Because a change in the amino acid sequence of a protein could change its structure, one substitution of a nucleotide in a DNA chain could result in altered protein structure and function.

7. Because adenine pairs with thymine (no uracil exists in DNA) and cytosine pairs with guanine, the sequence of DNA replicated from strand 1 is TACGAT. This sequence is also the sequence of DNA in the original strand 2. A replicate of strand 2 is therefore ATGCTA, which is the same as the original strand 1.

Visit the Online Learning Center at www.mhhe.com/seeley7 for chapter quizzes, interactive learning exercises, and other study tools.

Colored SEM of simple columnar epithelial cells, with cilia, of the uterine tube.

Histology: The Study of Tissues

CHAPTER

4

In some ways, the human body is like a complex machine such as a car. Both consist of many parts, which are made of materials consistent with their specialized functions. For example, car tires are made of synthetic rubber reinforced with a variety of fibers, the engine is made of a variety of metal parts, and the windows are made of transparent glass. All parts of a car cannot be made of a single type of material. Metal capable of withstanding the heat of the engine cannot be used for windows or tires. Similarly, the many parts of the human body are made of collections of specialized cells and the materials surrounding them. Muscle cells, which contract to produce movements of the body, are structurally different and have different functions than those of epithelial cells, which protect, secrete, or absorb. Also, cells in the retina of the eye, specialized to detect light and allow us to see, do not contract like muscle cells or exhibit the functions of epithelial cells.

The structure and function of tissues are so closely related that you should be able to predict the function of a tissue when given its structure, and vice versa. Knowledge of tissue structure and function is important in understanding the structure and function of organs, organ systems, and the complete organism. This chapter begins with brief discussions of *tissues and histology* (p. 104) and the development of *embryonic tissue* (p. 104) and then describes the structural and functional characteristics of the major tissue types: *epithelial tissue* (p. 104), *connective tissue* (p. 116), *muscle tissue* (p. 128), and *nervous tissue* (p. 130). In addition, the chapter provides an explanation of *membranes* (p. 132), *inflammation* (p. 132), *tissue repair* (p. 134), and *tissue and aging* (p. 136).

Tissues and Histology

Tissues (tish′ūz) are collections of similar cells and the substances surrounding them. Specialized cells and the extracellular matrix surrounding them form all the different tissue types found at the **tissue level of organization.** The classification of tissue types is based on the structure of the cells; the composition of the non-cellular substances surrounding cells, called the **extracellular matrix;** and the functions of the cells. The four primary tissue types, which include all tissues, and from which all organs of the body are formed, are

1. epithelial tissue;
2. connective tissue;
3. muscle tissue;
4. nervous tissue.

Epithelial and connective tissues are the most diverse in form. The different types of epithelial and connective tissues are classified by structure, including cell shapes, relationships of cells to one another, and the materials making up the extracellular matrix. In contrast, muscle and nervous tissues are classified by functional and structural characteristics.

The tissues of the body are interdependent. For example, muscle tissue cannot produce movement unless it receives oxygen carried by red blood cells, and new bone tissue cannot be formed unless epithelial tissue absorbs calcium and other nutrients from the digestive tract. Also, all tissues in the body die if cancer or some other disease destroys the tissues of vital organs such as the liver or kidneys.

Histology (his-tol′ō-jē) is the microscopic study of tissues. Much information about the health of a person can be gained by examining tissues. A **biopsy** (bī′op-sē) is the process of removing tissue samples from patients surgically or with a needle for diagnostic purposes. Examining tissue samples from individuals with various disorders can distinguish the specific disease. For example, some red blood cells have an abnormal shape in people suffering from sickle-cell disease, and red blood cells are smaller than normal in people with iron-deficiency anemia. Cancer is identified and classified based on characteristic changes that occur in tissues. Changes in the structure of epithelial cells are used to identify cancer of the uterine cervix, and changes in white blood cells are used to identify people who have leukemia. Also, the number of white blood cells can be greatly increased in people who have infections. Epithelial cells from respiratory passages have an abnormal structure in people with chronic bronchitis and in people with lung cancer.

Tissue samples can be sent to a laboratory and are prepared for examination. Results are obtained from the examination of the prepared tissues. In some cases tissues can be removed surgically, prepared quickly, and results reported while the patient is still anesthetized. The appropriate therapy is based to a large degree on the results. For example, the amount of tissue removed as part of breast or other types of cancer therapy can be determined by the results.

An **autopsy** (aw′top-sē) is an examination of the organs of a dead body to determine the cause of death or to study the changes caused by a disease. Microscopic examination of tissue is often part of an autopsy.

1. *Name the four primary tissue types, and list three characteristics used to classify them. How does the classification of epithelial and connective tissue differ from the classification of muscle and nervous tissue?*
2. *Define histology. Explain how microscopic examination of tissues taken by biopsy or autopsy can be used to diagnose some diseases.*

Embryonic Tissue

Approximately 13 or 14 days after fertilization, the cells that give rise to a new individual, called embryonic stem cells, form a slightly elongated disk consisting of two layers called ectoderm and endoderm. Cells of the ectoderm then migrate between the two layers to form a third layer called mesoderm. Ectoderm, mesoderm, and endoderm are called germ layers because the beginning of all adult structures can be traced back to one of them (see chapter 29).

The **endoderm** (en′dō-derm), the inner layer, forms the lining of the digestive tract and its derivatives. The **mesoderm** (mez′ō-derm), the middle layer, forms tissues such as muscle, bone, and blood vessels. The **ectoderm** (ek′tō-derm), the outer layer, forms the skin, and a portion of the ectoderm, called **neuro-ectoderm** (noor-ō-ek′tō-derm), becomes the nervous system (see chapter 13). Groups of cells that break away from the neuroectoderm during development, called **neural crest cells,** give rise to parts of the peripheral nerves (see chapter 13), skin pigment (see chapter 5), the medulla of the adrenal gland (chapter 18), and many tissues of the face.

3. *Name the three embryonic germ layers.*
4. *What adult structures are derived from endoderm, mesoderm, ectoderm, neuroectoderm, and neural crest cells?*

Epithelial Tissue

Epithelium (ep-i-thē′lē-ŭm; pl., epithelia, ep-i-thē′lē-ă) or **epithelial tissue** can be thought of as a protective covering of surfaces, both outside and inside the body. Characteristics common to most types of epithelium are (figure 4.1):

1. Epithelium consists almost entirely of cells, with very little extracellular matrix between them.
2. Epithelium covers surfaces of the body and forms glands that are derived developmentally from body surfaces. The body surfaces include the outside surface of the body, the lining of the digestive and respiratory tracts, the heart and blood vessels, and the linings of many body cavities.
3. Most epithelial tissues have one **free,** or **apical** (ap′i-kăl), **surface** not attached to other cells; a **lateral surface,** attached to other epithelial cells; and a **basal surface.** The free surface often lines the lumen of ducts, vessels, or cavities. The basal surface of most epithelial tissues is attached to a **basement membrane.** The basement membrane is a specialized type of extracellular material that is secreted by the epithelial cells and by connective tissue cells. It is like the adhesive on Scotch™ tape. It helps attach

Free surface
Epithelial cells with little extracellular matrix
Nucleus
Basement membrane
Connective tissue
Capillary
Lung
Pleura

Surface view

Cross-sectional view

LM 640x

LM 640x

Figure 4.1 **Characteristics of Epithelium**

Surface and cross-sectional views of epithelium illustrate the following characteristics: little extracellular material between cells, a free surface, a basement membrane attaching epithelial cells to underlying tissues. Capillaries in connective tissue do not penetrate the basement membrane. Nutrients, oxygen, and waste products must diffuse across the basement membrane between the capillaries and the epithelial cells.

the epithelial cells to the underlying tissues, and it plays an important role in supporting and guiding cell migration during tissue repair. A few epithelia, such as in lymphatic capillaries and liver sinusoids, do not have basement membranes, and some epithelial tissues (e.g., in some endocrine glands) do not have a free surface or a basal surface with a basement membrane.

4. Specialized cell contacts, such as tight junctions and desmosomes, bind adjacent epithelial cells together.

5. Blood vessels do not penetrate the basement membrane to reach the epithelium; thus all gases and nutrients carried in the blood must reach the epithelium by diffusing across the basement membrane from blood vessels in the underlying connective tissue. In epithelia with many layers of cells, the most metabolically active cells are close to the basement membrane.

6. Epithelial cells retain the ability to undergo mitosis and therefore are able to replace damaged cells with new epithelial cells. Undifferentiated cells (stem cells) continuously divide and produce new cells. In some types

of epithelia, such as in the skin and in the digestive tract, cells that are lost or die are continuously replaced by new cells.

5. *List six characteristics common to most types of epithelium. Define free (apical), lateral, and basal surfaces of epithelial cells.*

6. *What is the basement membrane and what are its functions? Why must metabolically active epithelial cells be close to the basement membrane?*

Functions of Epithelia

Major functions of epithelia include:

1. *Protecting underlying structures.* Examples include the skin and the epithelium of the oral cavity, which protect the underlying structures from abrasion.

2. *Acting as barriers.* Epithelium prevents the movement of many substances through the epithelial layer. For example, the skin acts as a barrier to water and prevents water loss

Clinical Focus   Microscopic Imaging

We see objects because light either passes through them or is reflected off them and enters our eyes (see chapter 15). We are limited, however, in what we can see with the unaided eye. Without the aid of magnifying lenses, the smallest objects we can resolve, or identify as separate objects, are approximately 100 μm, or 0.1 mm, in diameter, which is approximately the size of a fine pencil dot. The details of cells and tissues, which are much smaller than 100 μm, cannot be examined without the aid of a microscope.

Two basic types of microscopes have been developed: light microscopes and electron microscopes. As their names imply, light microscopes use light to produce an image, and electron microscopes use beams of electrons. **Light microscopes** usually use transmitted light, which is light that passes through the object being examined, but some light microscopes are equipped to use reflected light. Glass lenses are used in light microscopes to magnify images, and images can either be observed directly by looking into the microscope, or the light from the images can be used to expose photographic film to make a photomicrograph of the images. Video cameras are also used to record images. The resolution of light microscopes is limited by the wavelength of light, the lower limit of which is approximately 0.1 μm—about the size of a small bacterium.

A biopsy is the process of removing tissue from living patients for diagnostic examination. For example, changes in tissue structure allow pathologists to identify tumors and to distinguish between noncancerous (benign) and cancerous (malignant) tumors. Light microscopy is used on a regular basis to examine biopsy specimens. Light microscopy is used instead of electron microscopy because less time and effort are required to prepare materials for examination, and the resolution is adequate to diagnose most conditions that cause changes in tissue structure.

Because images are usually produced using transmitted light, tissues to be examined must be cut very thinly to allow the light to pass through them. Sections are routinely cut between 1 and 20 μm thick to make them thin enough for light microscopy. To cut such thin sections, the tissue must be fixed or frozen, which is a process that preserves the tissue and makes it more rigid. Fixed tissues are then embedded in some material, such as wax or plastic, that makes the tissue rigid enough for cutting into sections. Frozen sections, which can be prepared rapidly, are rigid enough for sectioning, but tissue embedded in wax or plastic can be cut much thinner, which makes the image seen through the microscope clearer. Because most tissues are colorless and transparent when thinly sectioned, the tissue must be colored with a stain or dye so that the structural details can be seen. As a result, the colors seen in color photomicrographs are not the true colors of the tissue but instead are the colors of the stains used. The color of the stain can also pro-vide specific information about the tissue, because special stains color only certain structures.

To see objects much smaller than a cell, such as cell organelles, an **electron microscope,** which has a limit of resolution of approximately 0.1 nm, must be used; 0.1 nm is about the size of some molecules. In objects viewed through an electron microscope, a beam of electrons either is passed through objects using a transmission electron microscope (TEM) or is reflected off the surface of objects using a scanning electron microscope (SEM). The electron beam is focused with electromagnets. For both processes, the specimen must be fixed, and for TEM the specimen must be embedded in plastic and thinly sectioned (0.01–0.15 μm thick). Care must be taken when examining specimens in an electron microscope because a focused electron beam can cause most tissues to quickly disintegrate. Furthermore, the electron beam is not visible to the human eye; thus it must be directed onto a fluorescent or photographic plate on which the electron beam is converted into a visible image. Because the electron beam does not transmit color information, electron micrographs are black and white unless color enhancement has been added using computer technology.

The magnification ability of SEM is not as great as that of TEM; however, depth of focus of SEM is much greater and allows for the production of a clearer three-dimensional image of the tissue structure.

---

from the body. The skin is also a barrier that prevents the entry of many toxic molecules and microorganisms into the body.

3. *Permitting the passage of substances.* Epithelium allows the movement of many substances through the epithelial layer. For example, oxygen and carbon dioxide are exchanged between the air and blood by diffusion through the epithelium in the lungs. Epithelium acts as a filter in the kidney allowing many substances to pass from the blood into the urine, but retaining other substances such as blood cells and proteins in the blood.

4. *Secreting substances.* Mucous glands, sweat glands, and the enzyme-secreting portions of the pancreas secrete their products onto epithelial surfaces or into ducts that carry them to other areas of the body.

5. *Absorbing substances.* The cell membranes of certain epithelial tissues contain carrier molecules (see chapter 3) that regulate the absorption of materials.

✐ **7. List the major functions of epithelia.**

Classification of Epithelium

Epithelium is classified primarily according to the number of cell layers and the shape of the superficial cells. There are three major types of epithelium based on the number of cell layers in each type.

1. **Simple epithelium** consists of a single layer of cells, with each cell extending from the basement membrane to the free surface.
2. **Stratified epithelium** consists of more than one layer of cells, but only the basal layer of cells attaches the deepest layer to the basement membrane.
3. **Pseudostratified columnar epithelium** is a special type of simple epithelium. The prefix *pseudo-* means false, so this type of epithelium appears to be stratified but is not. It consists of one layer of cells, with all the cells attached to the basement membrane. There is an appearance of two or more layers of cells because some of the cells are tall and extend to the free surface, whereas others are shorter and do not extend to the free surface.

There are three types of epithelium based on the shape of the epithelial cells.

1. **Squamous** (skwā′mŭs; flat) cells are flat or scalelike.
2. **Cuboidal** (cubelike) cells are cube-shaped; about as wide as they are tall.
3. **Columnar** (tall and thin, similar to a column) cells are taller than they are wide.

In most cases an epithelium is given two names, such as simple squamous, stratified squamous, simple columnar, or pseudostratified columnar. The first name indicates the number of layers, and the second indicates the shape of the cells (table 4.1) at the free surface. Tables 4.2–4.4 provide an overview of the major types of epithelia and their distributions.

Simple squamous epithelium consists of one layer of flat, or scalelike, cells which rest on a basement membrane (see table 4.2a). Stratified squamous epithelium consists of several layers of cells. Near the basement membrane, the cells are more cuboidal, but at the free surface the cells are flat or scalelike (see table 4.3a). Pseudostratified columnar epithelial cells are columnar in shape in that they are taller than they are wide and, although they appear to consist of more than one layer, all of the cells rest on the basement membrane (see table 4.4a).

Stratified squamous epithelium can be classified further as either nonkeratinized (moist) or keratinized, according to the condition of the outermost layer of cells. **Nonkeratinized stratified squamous epithelium** (table 4.3a), found in areas such as the mouth, esophagus, rectum, and vagina, consists of living cells in the deepest and outermost layers. A layer of fluid covers the outermost layers of cells, which makes them moist. In contrast,

keratinized (ker′ă-ti-nizd) **stratified squamous epithelium,** found in the skin (see chapter 5), consists of living cells in the deepest layers, and the outer layers are composed of dead cells containing the protein keratin. The dead, keratinized cells give the tissue a durable, moisture-resistant, dry character.

A unique type of stratified epithelium called **transitional epithelium** (table 4.4b) lines the urinary bladder, ureters, pelvis of the kidney including the major and minor calyces (kal′-i-sēz), and superior part of the urethra. These are structures where considerable expansion can occur. The shape of the cells and the number of cell layers vary, depending on whether the transitional epithelium is stretched or not. The surface cells and the underlying cells are roughly cuboidal or columnar when the epithelium is not stretched, and they become more flattened or squamouslike when the epithelium is stretched. Also, the number of layers of epithelial cells decreases in response to stretch. As the epithelium is stretched, the epithelial cells have the ability to shift on one another so that the number of layers decreases from five or six to two or three.

8. *Describe simple, stratified, and pseudostratified epithelia. Distinguish between squamous, cuboidal, and columnar epithelial cells.*
9. *How do nonkeratinized stratified squamous epithelium and keratinized stratified squamous epithelium differ? Where is each type found?*
10. *Describe the changes in shape and number of layers that occurs in cells of transitional epithelium. Where is transitional epithelium found?*

Table 4.1	Classification of Epithelium
Number of Layers or Category	**Shape of Cells**
Simple (single layer of cells)	Squamous Cuboidal Columnar
Stratified (more than one layer of cells)	Squamous Nonkeratinized (moist) Keratinized Cuboidal (very rare) Columnar (very rare)
Pseudostratified (modification of simple epithelium)	Columnar
Transitional (modification of stratified epithelium)	Roughly cuboidal to columnar when not stretched and squamouslike when stretched

Table 4.2   Simple Epithelium

(a) Simple Squamous Epithelium

Structure: Single layer of flat, often hexagonal cells. The nuclei appear as bumps when viewed as a cross section because the cells are so flat.

Function: Diffusion, filtration, some protection against friction, secretion, and absorption.

Location: Lining of blood and lymphatic vessels (endothelium) and small ducts, alveoli of the lungs, loop of Henle in kidney tubules, lining of serous membranes (mesothelium), and inner surface of the eardrum.

Lung alveoli

Free surface
Nucleus
Basement membrane
Simple squamous epithelial cell

LM 640x

(b) Simple Cuboidal Epithelium

Structure: Single layer of cube-shaped cells; some cells have microvilli (kidney tubules) or cilia (terminal bronchioles of the lungs).

Function: Active transport and facilitated diffusion result in secretion and absorption by cells of the kidney tubules; secretion by cells of glands and choroid plexus; movement of particles embedded in mucus out of the terminal bronchioles by ciliated cells.

Location: Kidney tubules, glands and their ducts, choroid plexus of the brain, lining of terminal bronchioles of the lungs, and surface of the ovaries.

Kidney

Free surface
Nucleus
Simple cuboidal epithelial cell
Basement membrane

LM 640x

Table 4.2 | Simple Epithelium – Continued

(c) Simple Columnar Epithelium

Structure: Single layer of tall, narrow cells. Some cells have cilia (bronchioles of lungs, auditory tubes, uterine tubes, and uterus) or microvilli (intestines).

Function: Movement of particles out of the bronchioles of the lungs by ciliated cells; partially responsible for the movement of the oocyte through the uterine tubes by ciliated cells. Secretion by cells of the glands, the stomach, and the intestine. Absorption by cells of the intestine.

Location: Glands and some ducts, bronchioles of lungs, auditory tubes, uterus, uterine tubes, stomach, intestines, gallbladder, bile ducts, and ventricles of the brain.

Lining of stomach and intestines

Free surface

Goblet cell containing mucus

Nucleus

Simple columnar epithelial cell

Basement membrane

LM 640x

Table 4.3 | Stratified Epithelium

(a) Stratified Squamous Epithelium

Structure: Multiple layers of cells that are cuboidal in the basal layer and progressively flattened toward the surface. The epithelium can be nonkeratinized or keratinized. In nonkeratinized stratified squamous epithelium the surface cells retain a nucleus and cytoplasm. In keratinized stratified epithelium, the cytoplasm of cells at the surface is replaced by a protein called keratin, and the cells are dead.

Function: Protection against abrasion, barrier against infection, and prevents loss of water from the body.

Location: Keratinized—skin. Nonkeratinized—mouth, throat, larynx, esophagus, anus, vagina, inferior urethra, and cornea.

Skin

Cornea

Mouth

Esophagus

Free surface

Nonkeratinized stratified squamous epithelial cell

Nuclei

Basement membrane

LM 286x

Table 4.3   Stratified Epithelium – Continued

(b) Stratified Cuboidal Epithelium

Structure: Multiple layers of somewhat cube-shaped cells.

Function: Secretion, absorption, and protection against infection.

Location: Sweat gland ducts, ovarian follicular cells, and salivary gland ducts.

Parotid gland duct
Sublingual gland duct
Submandibular gland duct

Free surface

Nucleus

Basement membrane

Stratified cuboidal epithelial cell

LM 413x

(c) Stratified Columnar Epithelium

Structure: Multiple layers of cells, with tall, thin cells resting on layers of more cuboidal cells. The cells are ciliated in the larynx.

Function: Protection and secretion.

Location: Mammary gland duct, larynx, and a portion of the male urethra.

Larynx

Free surface

Nucleus

Basement membrane

Stratified columnar epithelial cell

LM 413x

Table 4.4 Pseudostratified Columnar Epithelium and Transitional Epithelium

(a) Pseudostratified Columnar Epithelium

Structure: Single layer of cells; some cells are tall and thin and reach the free surface, and others do not; the nuclei of these cells are at different levels and appear stratified; the cells are almost always ciliated and are associated with goblet cells that secrete mucus onto the free surface.

Function: Synthesize and secrete mucus onto the free surface and move mucus (or fluid) that contains foreign particles over the surface of the free surface and from passages.

Location: Lining of nasal cavity, nasal sinuses, auditory tubes, pharynx, trachea, and bronchi of lungs.

Trachea
Bronchus

Cilia
Free surface
Goblet cell containing mucus
Pseudostratified columnar epithelial cell
Nucleus
Basement membrane

LM 413x

(b) Transitional Epithelium

Structure: Stratified cells that appear cuboidal when the organ or tube is not stretched and squamous when the organ or tube is stretched by fluid.

Function: Accommodates fluctuations in the volume of fluid in an organ or tube; protection against the caustic effects of urine.

Location: Lining of urinary bladder, ureters, and superior urethra.

Ureter
Urinary bladder
Urethra

Free surface
Transitional epithelial cell
Nucleus
Basement membrane

LM 413x

Tissue not stretched

Free surface
Transitional epithelial cell
Nucleus
Basement membrane

LM 413x

Tissue stretched

Functional Characteristics

Epithelial tissues have many functions (table 4.5), including forming a barrier between a free surface and the underlying tissues and secreting, transporting, and absorbing selected molecules. The type and arrangement of organelles within each cell (see chapter 3), the shape of cells, and the organization of cells within each epithelial type reflect these functions. Accordingly, structural specializations of epithelial cells are consistent with the functions they perform.

Cell Layers and Cell Shapes

Simple epithelium, with its single layer of cells, covers surfaces and allows diffusion of gases (lungs), filters blood (kidneys), secretes cellular products (glands), or absorbs nutrients (intestines). The selective movement of materials through epithelium would be hindered by a stratified epithelium, which is found in areas where protection is a major function. The multiple layers of cells in stratified epithelium are well adapted for a protective role because, as the outer cells are damaged, they are replaced by cells from deeper layers and a continuous barrier of epithelial cells is maintained in the tissue. Stratified squamous epithelium is found in areas of the body where abrasion can occur, such as the skin, mouth, throat, esophagus, anus, and vagina.

Differing functions are also reflected in cell shape. Cells that allow substances to diffuse through them and that filter are normally flat and thin. For example, simple squamous epithelium forms blood and lymphatic capillaries, the alveoli (air sacs) of the lungs, and parts of the kidney tubules. Cells that secrete or absorb are usually cuboidal or columnar. They have greater cytoplasmic volume compared to that of squamous epithelial cells; this cytoplasmic volume results from the presence of organelles responsible for the tissues' functions. For example, pseudostratified columnar epithelium, which secretes large amounts of mucus, lines the respiratory tract (see chapter 23) and contains large **goblet cells,** which are specialized columnar epithelial cells. The goblet cells contain abundant organelles responsible for the synthesis and secretion of mucus, such as ribosomes, endoplasmic reticulum, Golgi apparatuses, and secretory vesicles filled with mucus.

11. *What functions would a single layer of epithelial cells be expected to perform? A stratified layer?*
12. *In locations in which diffusion or filtration is occurring, what shape would you expect the epithelial cells to be?*
13. *Why are cuboidal or columnar cells found where secretion or absorption is occurring?*

PREDICT

Explain the consequences of having (a) nonkeratinized stratified epithelium rather than simple columnar epithelium lining the digestive tract, (b) nonkeratinized stratified squamous epithelium rather than keratinized stratified squamous epithelium in the skin, and (c) simple columnar epithelium rather than nonkeratinized stratified squamous epithelium lining the mouth.

Table 4.5	Function and Location of Epithelial Tissue		
Function	**Simple Squamous Epithelium**	**Simple Cuboidal Epithelium**	**Simple Columnar Epithelium**
Diffusion	Blood and lymph capillaries, alveoli of lungs, thin segment of loop of Henle		
Filtration	Bowman's capsule of kidney		
Secretion or absorption	Mesothelium (serous fluid)	Choroid plexus (produces cerebrospinal fluid), part of kidney tubule, many glands and their ducts	Stomach, small intestine, large intestine, uterus, many glands
Protection (against friction and abrasion)	Endothelium (e.g., epithelium of blood vessels) Mesothelium (e.g., epithelium of body cavities)		
Movement of mucus (ciliated)		Terminal bronchioles of lungs	Bronchioles of lungs, auditory tubes, uterine tubes, uterus
Capable of great stretching			
Miscellaneous	Lines the inner part of the eardrum, smallest ducts of glands	Surface of ovary, inside lining of eye (pigmented epithelium of retina), ducts of glands	Bile duct, gallbladder, ependyma (lining of brain ventricles and central canal of spinal cord), ducts of glands

Cell Surfaces

The free surfaces of epithelia can be smooth, contain microvilli, be ciliated, or be folded. Smooth surfaces reduce friction. Simple squamous epithelium with a smooth surface forms the covering of serous membranes. The lining of blood vessels is a simple squamous epithelium that reduces friction as blood flows through the vessels (see chapter 21).

Microvilli and cilia were described in chapter 3. Microvilli are nonmotile and contain microfilaments. They greatly increase surface area and are found in cells that absorb or secrete, such as the lining of the small intestine (see chapter 24). Stereocilia are elongated microvilli. They are found in sensory structures such as the inner ear (chapter 15) and play a role in sound detection. They are found in some places where absorption is important, such as in the epithelium of the epididymis. Cilia contain microtubules and they move materials across the surface of the cell (see chapter 3). Simple ciliated cuboidal, simple ciliated columnar, and pseudostratified ciliated columnar epithelia are in the respiratory tract (see chapter 23), where cilia move mucus that contains foreign particles, like dust, out of the respiratory passages. Cilia are also found on the apical surface of the simple columnar epithelial cells of the uterus and uterine tubes. The cilia help move mucus and oocytes.

Transitional epithelium has a rather unusual plasma membrane specialization: More rigid sections of membrane are separated by very flexible regions in which the plasma membrane is folded. When transitional epithelium is stretched, the folded regions of the plasma membrane can unfold. Transitional epithelium is specialized to expand. It is found in the urinary bladder, ureters, kidney pelvis, calyces of the kidney, and superior part of the urethra.

● 14. *What is the function of an epithelial free surface that is smooth, has cilia, has microvilli, or is folded? Give an example of epithelium in which each surface type is found.*

Cell Connections

Lateral and basilar surfaces have structures that serve to hold cells to one another or to the basement membrane (figure 4.2). These structures do three things: (1) they mechanically bind the cells together, (2) they help form a permeability barrier, and (3) they provide a mechanism for intercellular communication. Epithelial cells secrete glycoproteins that attach the cells to the basement membrane and to one another. This relatively weak binding between cells is reinforced by **desmosomes** (dez′mō-sōmz), disk-shaped structures with especially adhesive glycoproteins that bind cells to one another and intermediate filaments that extend into the cytoplasm of the cells. Many desmosomes are found in epithelia that are subjected to stress, such as the stratified squamous epithelium of the skin. **Hemidesmosomes,** similar to one-half of a desmosome, attach epithelial cells to the basement membrane.

Stratified Squamous Epithelium	Stratified Cuboidal Epithelium	Stratified Columnar Epithelium	Pseudostratified Columnar Epithelium	Transitional Epithelium
Skin (epidermis), cornea, mouth and throat, epiglottis, larynx, esophagus, anus, vagina				
			Larynx, nasal cavity, paranasal sinus, nasopharynx, auditory tube, trachea, bronchi of lungs	
				Urinary bladder, ureter, superior part of urethra
Lower part of urethra, sebaceous gland duct	Sweat gland ducts	Part of male urethra, epididymis, ductus deferens, mammary gland duct	Part of male urethra, salivary gland duct	

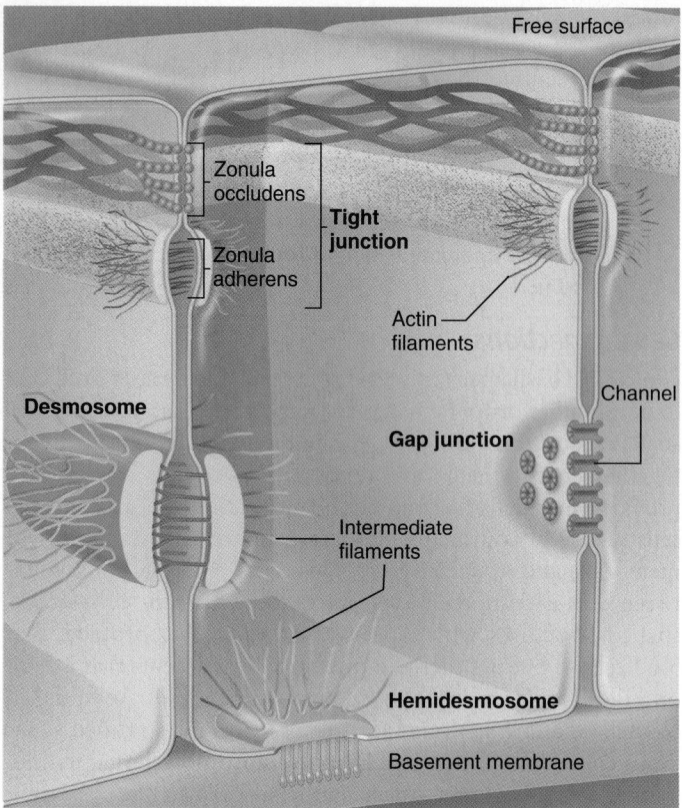

Figure 4.2   **Cell Connections**

Desmosomes anchor cells to one another and hemidesmosomes anchor cells to the basement membrane. Tight junctions consist of a zonula occludens and zonula adherens. Gap junctions allow adjacent cells to communicate with each other. Few cells have all of these different connections.

Tight junctions hold cells together and form a permeability barrier (see figure 4.2). They consist of a zonula adherens and a zonula occludens, which are found in close association with each other. The **zonula adherens** (zōʹnū-lă, zonʹū-lă ad-hērʹenz) is located between the plasma membranes of adjacent cells and acts like a weak glue that holds cells together. The zonulae adherens are best developed in simple epithelial tissues; they form a girdle of adhesive glycoprotein around the lateral surface of each cell and bind adjacent cells together. These connections are not as strong as desmosomes.

The **zonula occludens** (ō-kloodʹenz) forms a permeability barrier. It is formed by plasma membranes of adjacent cells that join one another in a jigsaw fashion to form a tight seal (see figure 4.2). Near the free surface of simple epithelial cells, the zonulae occludens form a ring that completely surrounds each cell and binds adjacent cells together. The zonulae occludens prevent the passage of materials between cells. For example, in the stomach and in the urinary bladder chemicals cannot pass between cells. Thus water and other substances must pass through the epithelial cells, which can actively regulate what is absorbed or secreted. Zonulae occludens are found in areas where a layer of simple epithelium forms a permeability barrier. For example, water can diffuse through epithelial cells, and active transport, cotransport, and facilitated

diffusion move most nutrients through the epithelial cells of the intestine.

A **gap junction** is a small specialized contact region between cells containing protein channels that aid intercellular communication by allowing ions and small molecules to pass from one cell to another (see figure 4.2). The exact function of gap junctions in epithelium is not entirely clear. Gap junctions between ciliated epithelial cells may function to coordinate the movements of the cilia.

Gap junctions are important in coordinating the function of cardiac and smooth muscle tissues. Because ions can pass through the gap junctions from one cell to the next, electric signals can pass from cell to cell to coordinate the contraction of cardiac and smooth muscle cells. Thus electric signals that originate in one cell of the heart can spread from cell to cell and cause the entire heart to contract. The gap junctions between cardiac muscle cells are found in specialized cell-to-cell connections called **intercalated disks** (see chapter 20).

● 15. *Name the ways in which epithelial cells are bound to one another and to the basement membrane.*
16. *In addition to holding cells together, name an additional function of tight junctions. What is the general function of gap junctions?*

P R E D I C T   ②
If a simple epithelial type has well developed zonula occludens, explain how NaCl can be moved from one side of the epithelial layer to the other, what type of epithelium it is likely to be, and how the movement of NaCl causes water to move in the same direction.

Glands

Glands are secretory organs. Many glands are composed primarily of epithelium, with a supporting network of connective tissue. These glands develop from an infolding or outfolding of epithelium in the embryo. If the gland maintains an open contact with the epithelium from which it developed, a duct is present. Glands with ducts are called **exocrine** (ekʹsō-krin) **glands,** and their ducts are lined with epithelium. Alternatively, some glands become separated from the epithelium of their origin. Glands that have no ducts are called **endocrine** (enʹdō-krin) **glands.** Endocrine glands have extensive blood vessels in the connective tissue of the glands. The cellular products of endocrine glands, which are called **hormones** (hōrʹmōnz), are secreted into the bloodstream and are carried throughout the body. Some of the endocrine glands form from non-epithelial tissue. For example, the medulla of the adrenal gland is formed from neural crest cells, and the cortex of the adrenal gland is formed from mesoderm. In addition, cardiac muscle cells can function as endocrine glands.

Most exocrine glands are composed of many cells and are called **multicellular glands,** but some exocrine glands are composed of a single cell and are called **unicellular glands** (figure 4.3a). **Goblet cells** (see table 4.2c) of the respiratory system are unicellular glands that secrete mucus. Multicellular glands can be classified further according to the structure of their ducts (figure 4.3b–i). Glands that have ducts with few branches are called **simple,** and glands with ducts that branch repeatedly are called **compound.**

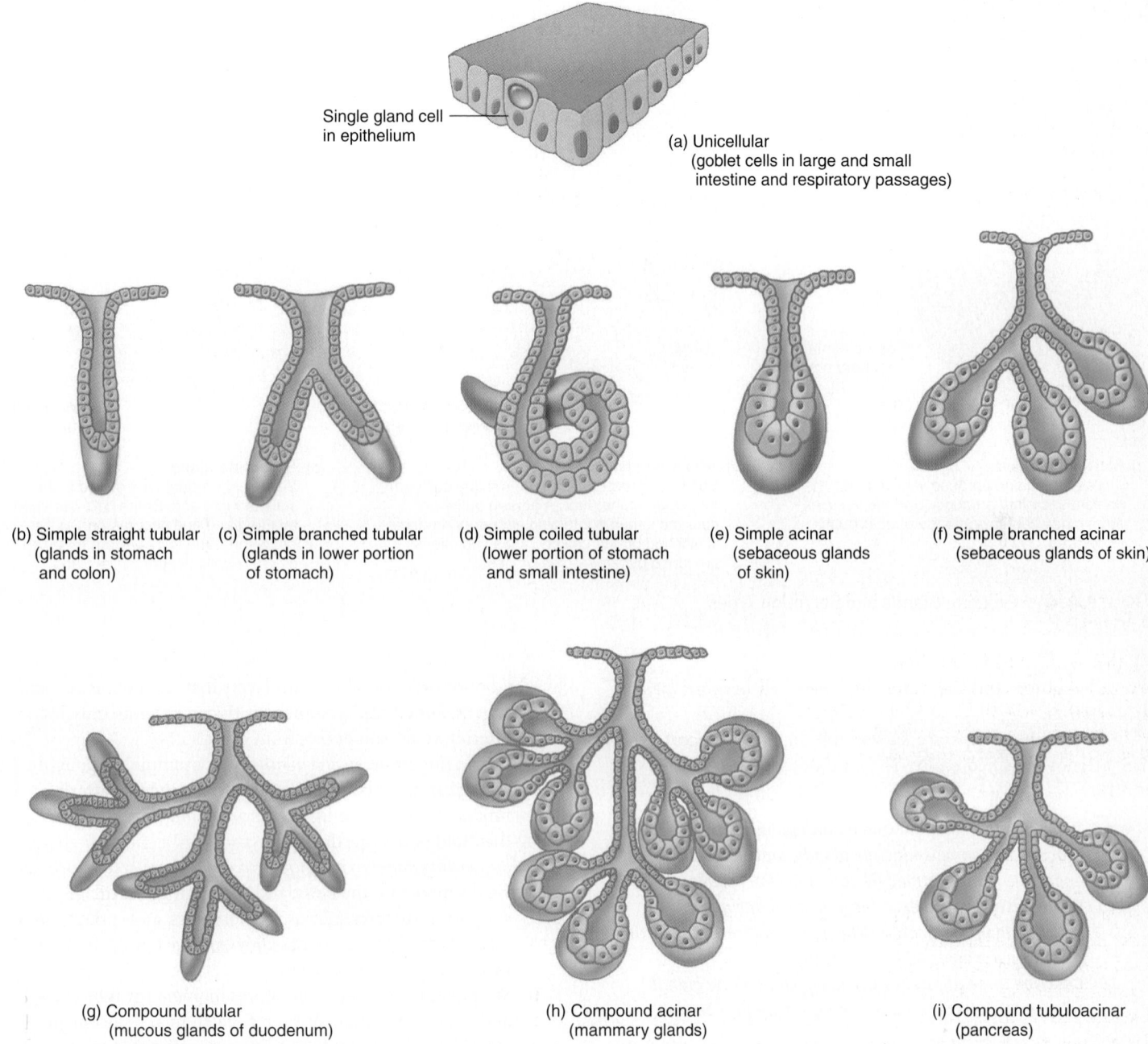

Single gland cell
in epithelium

(a) Unicellular
(goblet cells in large and small
intestine and respiratory passages)

(b) Simple straight tubular
(glands in stomach
and colon)

(c) Simple branched tubular
(glands in lower portion
of stomach)

(d) Simple coiled tubular
(lower portion of stomach
and small intestine)

(e) Simple acinar
(sebaceous glands
of skin)

(f) Simple branched acinar
(sebaceous glands of skin)

(g) Compound tubular
(mucous glands of duodenum)

(h) Compound acinar
(mammary glands)

(i) Compound tubuloacinar
(pancreas)

Figure 4.3 **Structure of Exocrine Glands**

The names of exocrine glands are based on the shapes of their secretory units and their ducts.

Further classification is based on whether the ducts end in **tubules** (small tubes) or saclike structures called **acini** (as′i-nī; grapes, suggesting a cluster of grapes or small sacs) or **alveoli** (al-vē′ō-lī; a hollow sac). Tubular glands can be classified as straight or coiled. Most tubular glands are simple and straight, simple and coiled, or compound and coiled. Acinar glands can be simple or compound.

Exocrine glands can also be classified according to how products leave the cell. **Merocrine** (mer′ō-krin) **glands,** such as water-producing sweat glands and the exocrine portion of the pancreas, secrete products with no loss of actual cellular material

(figure 4.4a). Secretions are either actively transported or packaged in vesicles and then released by the process of exocytosis at the free surface of the cell. **Apocrine** (ap′ō-krin) **glands,** such as the milk-producing mammary glands, discharge fragments of the gland cells in the secretion (figure 4.4b). Products are retained within the cell, and portions of the cell are pinched off to become part of the secretion. Apocrine glands, however, also produce merocrine secretions. Most of the volume secreted by apocrine glands are merocrine in origin. **Holocrine** (hol′ō-krin) **glands,** such as sebaceous (oil) glands of the skin, shed entire cells (figure 4.4c). Products accumulate in the cytoplasm of each epithelial cell,

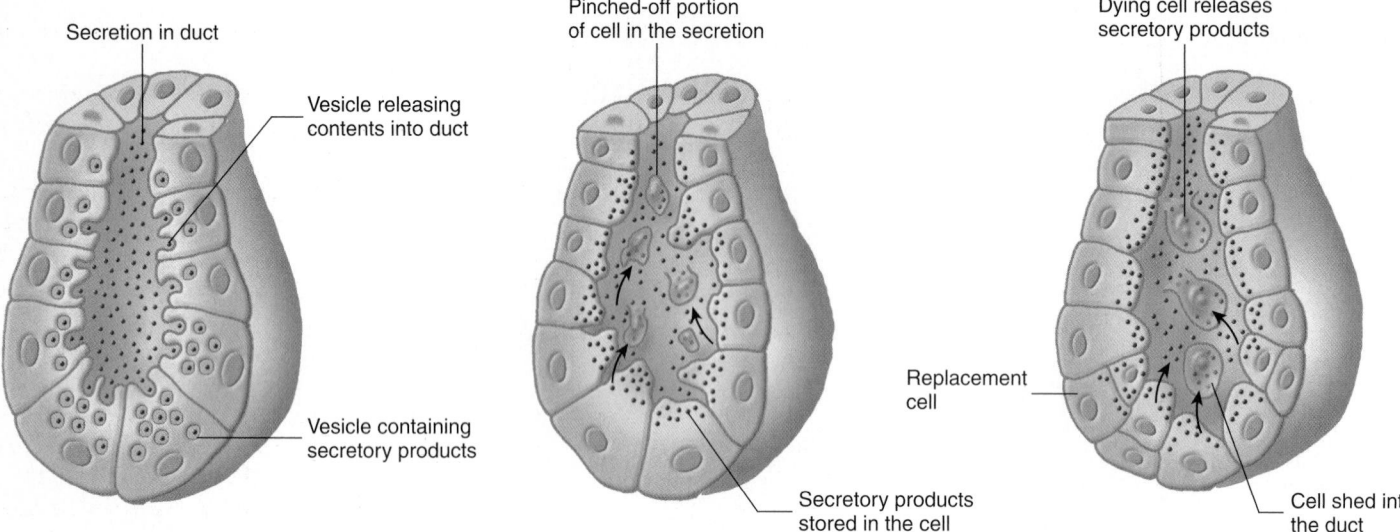

Figure 4.4   Exocrine Glands and Secretion Types
Exocrine glands are classified according to the type of secretion.

the cell ruptures and dies, and the entire cell becomes part of the secretion.

Endocrine glands are so variable in their structure that they are not classified easily. They are described in chapters 17 and 18.

> ● **17. Define the term gland. Explain how epithelia give rise to exocrine and some endocrine glands, and give examples of endocrine glands formed by non-epithelial tissues.**
> **18. Distinguish between exocrine and endocrine glands. Describe the classification scheme for multicellular exocrine glands on the basis of their duct systems.**
> **19. Describe three different ways in which exocrine glands release their secretions. Give an example for each method.**

Connective Tissue

Connective tissue is abundant, and it makes up part of every organ in the body. The major structural characteristic that distinguishes connective tissue from the other three tissue types is that it consists of cells separated from each other by abundant extracellular matrix. Connective tissue structure is diverse, and it performs a variety of important and related functions.

Functions of Connective Tissue

Connective tissues perform the following major categories of functions:

1. *Enclosing and separating.* Sheets of connective tissues form capsules around organs such as the liver and kidneys.

Connective tissue also forms layers that separate tissues and organs. For example, connective tissues separate muscles, arteries, veins, and nerves from one another.
2. *Connecting tissues to one another.* For example, tendons are strong cables, or bands, of connective tissue that attach muscles to bone, and ligaments are connective tissue bands that hold bones together.
3. *Supporting and moving.* Bones of the skeletal system provide rigid support for the body, and the semirigid cartilage supports structures such as the nose, ears, and surfaces of joints. Joints between bones allow one part of the body to move relative to other parts.
4. *Storing.* Adipose tissue (fat) stores high-energy molecules, and bones store minerals such as calcium and phosphate.
5. *Cushioning and insulating.* Adipose tissue cushions and protects the tissue it surrounds and provides an insulating layer beneath the skin that helps conserve heat.
6. *Transporting.* Blood transports substances throughout the body, such as gases, nutrients, enzymes, hormones, and cells of the immune system.
7. *Protecting.* Cells of the immune system and blood provide protection against toxins and tissue injury, as well as from microorganisms. Bones protect underlying structures from injury.

> ● **20. What is the major characteristic that distinguishes connective tissue from other tissues?**
> **21. List the major functions of connective tissues, and give an example of a connectve tissue that performs each function.**

Cells of Connective Tissue

The specialized cells of the various connective tissues produce the extracellular matrix. The names of the cells end with suffixes that identify the cell functions as blasts, cytes, or clasts. **Blasts** create the matrix, **cytes** maintain it, and **clasts** break it down for remodeling. For example, **fibroblasts** are cells that form fibrous connective tissue and **fibrocytes** maintain it. **Chondroblasts** form cartilage (*chondro-* refers to cartilage) and **chondrocytes** maintain it. **Osteoblasts** form bone (*osteo-* means bone), **osteocytes** maintain it, and **osteoclasts** break it down (see chapter 6).

Adipose (ad′i-pōs; fat), or **fat, cells,** also called **adipocytes** (ad′i-pō-sītz), contain large amounts of lipid. The lipid pushes the rest of the cell contents to the periphery, so that each cell appears to contain a large, centrally located lipid droplet with a thin layer of cytoplasm around it. Adipose cells are rare in some connective tissue types such as cartilage, are abundant in others such as loose connective tissue, or are predominant such as in adipose tissue.

Mast cells are commonly found beneath membranes in loose connective tissue and along small blood vessels of organs. They contain chemicals such as heparin, histamine, and proteolytic enzymes. These substances are released in response to injury such as trauma and infection and play important roles in inflammation.

White blood cells continuously move from blood vessels into connective tissues. The rate of movement increases dramatically in response to injury or infection. In addition, accumulations of lymphocytes, a type of white blood cell, are common in some connective tissues, such as in the connective tissue beneath the epithelial lining of certain parts of the digestive system.

Macrophages are found in some connective tissue types. They are derived from monocytes, a white blood cell type. Macrophages are either **fixed** and do not move through the connective tissue in which they are found or are **wandering macrophages** and move by ameboid movement through the connective tissue. Macrophages phagocytize foreign or injured cells, and they play a major role in providing protection against infections.

Undifferentiated mesenchymal cells, sometimes called **stem cells,** are embryonic cells that persist in adult connective tissue. They have the potential to differentiate to form adult cell types such as fibroblasts or smooth muscle cells in response to injury.

- 22. *Explain the difference between connective tissue cells that are termed blast, cyte, or clast cells.*
- 23. *Describe and give the functions of the cells of connective tissue.*

Extracellular Matrix

The extracellular matrix of connective tissue has three major components: (1) protein fibers, (2) ground substance consisting of nonfibrous protein and other molecules, and (3) fluid. The structure of the matrix gives connective tissue types most of their functional characteristics, such as the ability of bones and cartilage to bear weight, of tendons and ligaments to withstand tension, and of dermis of the skin to withstand punctures, abrasions, and other abuses.

Protein Fibers of the Matrix

Three types of protein fibers—collagen, reticular, and elastic fibers—help form connective tissue.

Collagen (kol′lă-jen) **fibers** consist of collagen, which is the most common protein in the body. Collagen accounts for one-fourth to one-third of the total body protein, which is approximately 6% of the total body weight. Each collagen molecule resembles a microscopic rope consisting of three polypeptide chains coiled together. Collagen is very strong and flexible but quite inelastic. There are at least 15 different types of collagen, many of which are specific to certain tissues. Collagen fibers differ in the types of amino acids that make up the polypeptide chains. Of the 15 types of collagen, 6 types are most common. Tendons, ligaments, skin, and bone contain mainly type I collagen, cartilage is mainly type II collagen, and reticular fibers are mainly type III collagen.

Reticular (re-tik′ū-lăr; netlike) **fibers** are very fine collagen fibers and therefore are not a chemically distinct category of fibers. They are very short, thin fibers that branch to form a network and appear different microscopically from other collagen fibers. Reticular fibers are not as strong as most collagen fibers, but networks of reticular fibers fill space between tissues and organs.

Elastic fibers contain a protein called **elastin** (ĕ-las′tin). As the name suggests, this protein is elastic with the ability to return to its original shape after being distended or compressed. Elastin gives the tissue in which it is found an elastic quality. Elastin molecules look like tiny coiled springs, and individual molecules are crosslinked to produce a large, interwoven meshwork of springlike molecules that extend through the entire tissue.

Other Matrix Molecules

Two types of large, nonfibrous molecules called hyaluronic acid and proteoglycans are part of the extracellular matrix. These molecules constitute most of the **ground substance** of the matrix, the "shapeless" background against which the collagen fibers are seen through the microscope. The molecules themselves, however, are not shapeless but are highly structured. **Hyaluronic** (hī′ă-loo-ron′ik; glassy appearance) **acid** is a long, unbranched polysaccharide chain composed of repeating disaccharide units. It gives a very slippery quality to the fluids that contain it; for that reason, it is a good lubricant for joint cavities (see chapter 8). Hyaluronic acid is also found in large quantities in connective tissue and is the major component of the vitreous humor of the eye (see chapter 15). A **proteoglycan** (prō′tē-ō-glī′kan; formed from proteins and polysaccharides) is a large molecule that consists of numerous polysaccharides, called **glycosaminoglycans** (glī′kōs-am-i-nō-glī′kan) each attached at one end to a common protein core. These **proteoglycan monomers** resemble minute pine tree branches. The protein core is the branch of the tree, and the proteoglycans are the needles. The protein cores of proteoglycan monomers can attach to a molecule of hyaluronic acid to form a **proteoglycan aggregate.** The aggregate resembles a complete pine tree, with hyaluronic acid represented by the tree trunk and the proteoglycan monomers forming the limbs. Proteoglycans trap large quantities of water, which gives them the capacity to return to their original shape when compressed or deformed. There are several different types of glycosaminoglycans, and their abundance varies with each connective tissue type.

Several **adhesive molecules** are found in ground substance. These adhesive molecules hold the proteoglycan aggregates together and to structures such as the plasma membranes. A specific adhesive molecule type predominates in certain types of ground substance. For example, **chondronectin** is in the ground substance of cartilage, **osteonectin** is in the ground substance of bone, and **fibronectin** is in the ground substance of fibrous connective tissues.

24. *What three components are found in the extracellular matrix of connective tissue?*
25. *Contrast the structure and characteristics of collagen fibers, reticular fibers, and elastin fibers.*
26. *Describe the structure and function of hyaluronic acid and proteoglycan aggregates. What is the function of adhesive molecules?*

Classification of Connective Tissue

Connective tissue types blend into one another, and the transition points cannot be defined precisely. As a result, the classification scheme for connective tissues is somewhat arbitrary. Classification schemes for connective tissue are influenced by (1) protein fibers and the arrangement of protein fibers in the extracellular matrix, (2) protein fibers and ground substance in the extracellular matrix, and (3) a fluid extracellular matrix. The classification of connective tissues used here is presented in Table 4.6.

The two major categories of connective tissue are embryonic and adult connective tissues.

Embryonic Connective Tissue

Embryonic connective tissue is called **mesenchyme** (mez′en-kīm). It is made up of irregularly shaped fibroblasts surrounded by abundant semifluid extracellular matrix in which delicate collagenous fibers are distributed (table 4.7*a*). It forms in the embryo during the third and fourth weeks of development from mesoderm and neural crest cells (see chapter 29), and all adult connective tissue types develop from it. By 8 weeks of development most of the mesenchyme has become specialized to form types of connective tissue seen in adults as well as muscle, blood vessels, and other tissues. The major source of remaining embryonic connective tissue in the newborn is found in the umbilical cord, where it is called **mucous connective tissue** or **Wharton's jelly** (table 4.7*b*). The structure of mucous connective tissue is similar to mesenchyme.

Adult Connective Tissue

Adult connective tissue consists of six types: loose, dense, connective tissue with special properties, cartilage, bone, and blood.

Loose Connective Tissue

Loose connective tissue (table 4.8) which is sometimes referred to as **areolar** (ă-rē′ō-lăr; area) **tissue,** consists of protein fibers that

Table 4.6	Classification of Connective Tissue

A. Embryonic connective tissue
  1. Mesenchyme
  2. Mucous
B. Adult connective tissue
  1. Loose (areolar)
  2. Dense
    a. Dense regularly arranged collagenous
    b. Dense regularly arranged elastic
    c. Dense irregularly arranged collagenous
    d. Dense irregularly arranged elastic
  3. Special properties
    a. Adipose
    b. Reticular
  4. Cartilage
    a. Hyaline
    b. Fibrocartilage
    c. Elastic
  5. Bone
    a. Cancellous
    b. Compact
  6. Blood and hemopoietic tissue

form a lacy network with numerous fluid-filled spaces. Areolar tissue is the "loose packing" material of most organs and other tissues, and attaches the skin to underlying tissues. It contains collagen, reticular, and elastic fibers and a variety of cells. For example, fibroblasts produce the fibrous matrix, macrophages move through the tissue engulfing bacteria and cell debris, mast cells contain chemicals that help mediate inflammation, and lymphocytes are involved in immunity. The loose packing of areolar tissue is often associated with other connective tissue types such as reticular tissue and fat (adipose tissue).

Dense Connective Tissue

Protein fibers of **dense connective tissue** form thick bundles and fill nearly all of the extracellular space. Most of the cells of developing dense connective tissue are spindle-shaped fibroblasts. Once the fibroblasts become completely surrounded by matrix, they are fibrocytes. Dense connective tissue can be subdivided into two major groups: regular and irregular.

Dense regular connective tissue has protein fibers in the extracellular matrix that are oriented predominantly in one direction. **Dense regular collagenous connective tissue** (table 4.9*a*) has abundant collagen fibers. The collagen fibers give this tissue a white appearance. Dense regular collagenous connective tissue forms structures such as tendons, which connect muscles to bones (see chapter 11), and most ligaments, which connect bones to bones (see chapter 8). The collagen fibers of dense connective tissue resist stretching and give the tissue considerable strength in the direction of the fiber orientation. Tendons and most ligaments consist almost entirely of thick bundles of densely packed parallel collagen fibers with the orientation of the collagen fibers in one

Table 4.7 Embryonic Connective Tissue

(a) Mesenchyme

Structure: The mesenchymal cells are irregularly shaped. The extracellular matrix is abundant and contains scattered reticular fibers.

Location: Mesenchyme is the embryonic tissue from which connective tissues, as well as other tissues, arise.

Intercellular matrix

Nuclei of mesenchyme cells

LM 200x

(b) Mucous Connective Tissue

Structure: Mucous tissue is mesenchymal tissue that remains unspecialized. The cells are irregularly shaped. The extracellular matrix is abundant and contains scattered reticular fibers.

Location: Umbilical cord of newborn.

Umbilical cord

Intercellular matrix

Nuclei of mucous connective tissue cells

LM 200x

Table 4.8 Loose Connective Tissue

Structure: Cells (e.g., fibroblasts, macrophages, and lymphocytes) within a fine network of mostly collagen fibers. Often merges with denser connective tissue.

Function: Loose packing, support, and nourishment for the structures with which it is associated.

Location: Widely distributed throughout the body; substance on which epithelial basement membranes rest; packing between glands, muscles, and nerves. Attaches the skin to underlying tissues.

direction which makes the tendons and ligaments very strong cable-like structures.

The general structure of tendons and ligaments is similar, but major differences between them exist. The differences include the following: (1) collagen fibers of ligaments are often less compact, (2) some fibers of many ligaments are not parallel, and (3) ligaments usually are more flattened than tendons and form sheets or bands of tissues.

Dense regular elastic connective tissue (table 4.9*b*) consists of parallel bundles of collagen fibers and abundant elastic fibers. The elastin in elastic ligaments gives them a slightly yellow color. Dense regular elastic connective tissue forms some elastic ligaments, such as those in the vocal folds and the **nuchal** (noo′kăl; back of the neck) **ligament,** which lies along the posterior of the neck and helps hold the head upright. When elastic ligaments are stretched, they tend to shorten to their original length, much like an elastic band.

P R E D I C T   **3**

Explain the advantages of having elastic ligaments that extend from vertebra to vertebra in the vertebral column and why it would be a disadvantage if tendons, which connect skeletal muscles to bone, were elastic.

Dense irregular connective tissue contains protein fibers arranged as a meshwork of randomly oriented fibers. Alternatively, the fibers within a given layer of dense irregular connective tissue can be oriented in one direction whereas the fibers of adjacent layers are oriented at nearly right angles to that layer. Dense irregular connective tissue forms sheets of connective tissue that have

strength in many directions, but less strength in any single direction than does regular connective tissue.

P R E D I C T   **4**

Scars consist of dense irregular connective tissue made of collagen fibers. Vitamin C is required for collagen synthesis. Predict the effect of scurvy, which is a nutritional disease caused by vitamin C deficiency, on wound healing.

Dense irregular collagenous connective tissue (table 4.9*c*) forms most of the dermis of the skin, which is the tough, inner portion of the skin (see chapter 5) and of the connective tissue capsules that surround organs such as the kidney and spleen.

Dense irregular elastic connective tissue (table 4.9*d*) is found in the wall of elastic arteries. In addition to collagen fibers, oriented in many directions, there are abundant elastic fibers in the layers of this tissue.

27. *List the two types of embryonic connective tissue, and to what does mesenchyme give rise in the adult?*
28. *Describe the fiber arrangement in loose (areolar) connective tissue. What are the functions of this tissue type?*
29. *Structurally and functionally, what is the difference between dense regular connective tissue and dense irregular connective tissue?*
30. *Name the two kinds of dense regular connective tissue, and give an example of each. Do the same for dense irregular connective tissue.*

Table 4.9 | Dense Connective Tissue

(a) Dense Regular Collagenous Connective Tissue

Structure: Matrix composed of collagen fibers running in somewhat the same direction.

Function: Ability to withstand great pulling forces exerted in the direction of fiber orientation, great tensile strength, and stretch resistance.

Location: Tendons (attach muscle to bone) and ligaments (attach bones to each other).

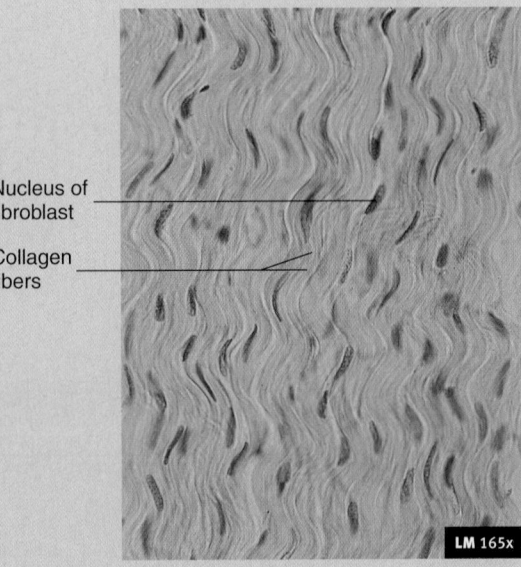

Nucleus of fibroblast

Collagen fibers

LM 165x

Ligament

Tendon

(b) Dense Regular Elastic Connective Tissue

Structure: Matrix composed of regularly arranged collagen fibers and elastin fibers.

Function: Capable of stretching and recoiling like a rubber band with strength in the direction of fiber orientation.

Location: Ligaments between the vertebrae and along the dorsal aspect of the neck (nucha) and in the vocal cords.

Elastin fibers

Nucleus of fibroblast

LM 100x

Base of tongue

Vocal folds (true vocal cords)

Vestibular fold (false vocal cord)

Table 4.9   Dense Connective Tissue – Continued

(c) Dense Irregular Collagenous Connective Tissue

Structure: Matrix composed of collagen fibers that run in all directions or in alternating planes of fibers oriented in a somewhat single direction.

Function: Tensile strength capable of withstanding stretching in all directions.

Location: Sheaths; most of the dermis of the skin; organ capsules and septa; outer covering of body tubes.

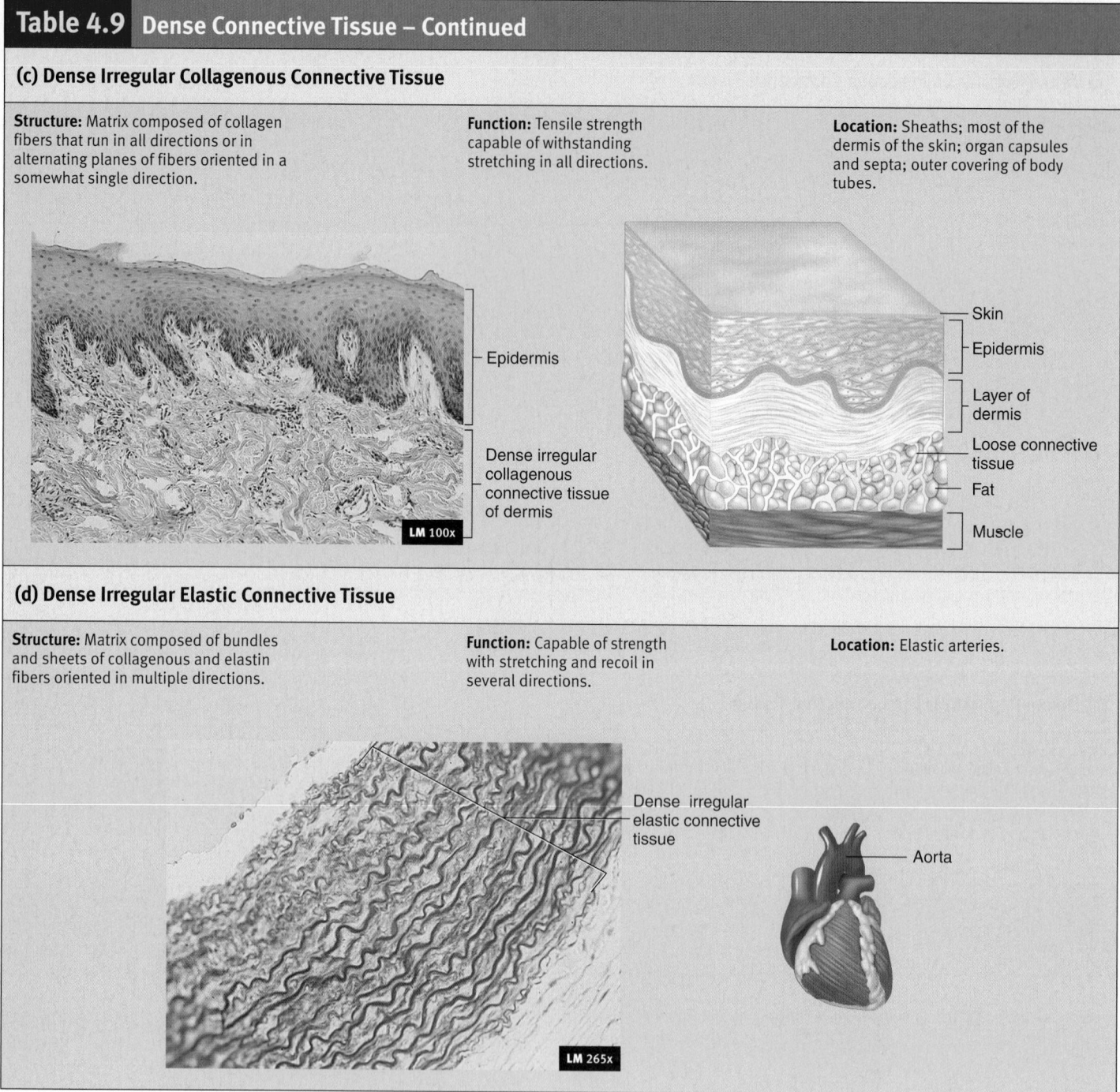

Epidermis

Dense irregular collagenous connective tissue of dermis

LM 100x

Skin

Epidermis

Layer of dermis

Loose connective tissue

Fat

Muscle

(d) Dense Irregular Elastic Connective Tissue

Structure: Matrix composed of bundles and sheets of collagenous and elastin fibers oriented in multiple directions.

Function: Capable of strength with stretching and recoil in several directions.

Location: Elastic arteries.

Dense irregular elastic connective tissue

Aorta

LM 265x

Connective Tissue with Special Properties

Adipose tissue and reticular tissue are connective tissues with special properties. **Adipose tissue** (table 4.10a) consists of adipocytes, or fat cells, which contain large amounts of lipid. Unlike other connective tissue types, adipose tissue is composed of large cells and a small amount of extracellular matrix that consists of loosely arranged collagen and reticular fibers with some scattered elastic fibers. Blood vessels form a network in the extracellular matrix. The fat cells are usually arranged in clusters or lobules separated from one another by loose connective tissue. Adipose tissue functions as an insulator, a protective tissue, and a site of energy storage.

Lipids take up less space per calorie than either carbohydrates or proteins and therefore are well adapted for energy storage.

Adipose tissue exists in both yellow (white) and brown forms. **Yellow adipose** tissue is by far the most abundant. Yellow adipose tissue appears white at birth, but it turns yellow with age because of the accumulation of pigments such as carotene, a plant pigment that humans can metabolize as a source of vitamin A. Storage, insulation, and protection are functions of yellow adipose tissue. Brown adipose tissue is found only in specific areas of the body such as the axillae (armpits), neck, and near the kidneys. The brown color results from the cytochrome pigments in its numerous

Table 4.10   Connective Tissue with Special Properties

(a) Adipose Tissue

Structure: Little extracellular matrix surrounding cells. The adipocytes, or fat cells, are so full of lipid that the cytoplasm is pushed to the periphery of the cell.

Function: Packing material, thermal insulator, energy storage, and protection of organs against injury from being bumped or jarred.

Location: Predominantly in subcutaneous areas, mesenteries, renal pelvis, around kidneys, attached to the surface of the colon, mammary glands, and in loose connective tissue that penetrates into spaces and crevices.

Nucleus

Adipocytes or fat cells

LM 100x

Adipose tissue

Mammary gland

(b) Reticular Tissue

Structure: Fine network of reticular fibers irregularly arranged.

Function: Provides a superstructure for the lymphatic and hemopoietic tissues.

Location: Within the lymph nodes, spleen, and bone marrow.

Nucleus of lymphocyte

Reticular fibers

LM 100x

Lymph node

Spleen

mitochondria and its abundant blood supply. Although brown fat is much more prevalent in babies than in adults, it is difficult to distinguish brown fat from yellow fat in babies because the color difference between them is not great. Brown fat is specialized to generate heat as a result of oxidative metabolism of lipid molecules in mitochondria and can play a significant role in body temperature regulation in newborn babies.

Reticular tissue forms the framework of lymphatic tissue (table 4.10b), such as in the spleen and lymph nodes, as well as in bone marrow and the liver. It is characterized by a network of reticular fibers and reticular cells. **Reticular cells** produce the reticular fibers and remain closely attached to them. The spaces between the reticular fibers can contain a wide variety of other cells, such as dendritic cells, which look very much like reticular cells but are cells of the immune system, macrophages, and blood cells (see chapter 22).

> 31. *What feature of the extracellular matrix distinguishes adipose tissue from other connective tissue types? What is an adipocyte?*
>
> 32. *List the functions of adipose tissue. Name the two types of adipose tissue. Which one is important in generating heat?*
>
> 33. *What is the function of reticular tissue? Where is it found?*

Cartilage

Cartilage (kar′ti-lij) is composed of cartilage cells, or **chondrocytes** (kon′drō-sīts), located in spaces called **lacunae** (lă-koo′nē) within an extensive and relatively rigid matrix. Next to bone, cartilage is the firmest structure in the body. The matrix contains protein fibers, ground substance, and fluid. The protein fibers are collagen fibers or, in some cases, collagen and elastic fibers. The ground substance consists of proteoglycans and other organic molecules. Most of the proteoglycans in the matrix form aggregates with hyaluronic acid. Within the cartilage matrix, proteoglycan aggregates function as minute sponges capable of trapping large quantities of water. This trapped water allows cartilage to spring back after being compressed. The collagen fibers give cartilage considerable strength.

The surface of nearly all cartilage is surrounded by a layer of dense irregular connective tissue called the **perichondrium** (per-i-kon′drē-ŭm). The structure of the perichondrium is described in more detail in chapter 6. Cartilage cells arise from the perichondrium and secrete cartilage matrix. Once completely surrounded by matrix the cartilage cells are called chondrocytes and the spaces in which they are located are called lacunae. Cartilage has no blood vessels or nerves except those of the perichondrium; it therefore heals very slowly after an injury because the cells and nutrients necessary for tissue repair cannot reach the damaged area easily.

There are three types of cartilage.

1. **Hyaline** (hī′ă-lin) **cartilage** has large amounts of both collagen fibers and proteoglycans (table 4.11a). Collagen fibers are evenly dispersed throughout the ground substance, and in joints, hyaline cartilage has a very smooth surface. Specimens appear to have a glassy, translucent matrix when viewed through a microscope. Hyaline cartilage is found in areas in which strong support and some flexibility are needed, such as in the rib cage and the cartilage within the trachea and bronchi (see chapter 23). It also covers the surfaces of bones that move smoothly against each other in joints. Hyaline cartilage forms most of the skeleton before it is replaced by bone in the embryo, and it is involved in growth that increases the length of bones (see chapter 6).

2. **Fibrocartilage** has more collagen fibers than proteoglycans (table 4.11b). Compared to hyaline cartilage, fibrocartilage has much thicker bundles of collagen fibers dispersed through its matrix. Fibrocartilage is slightly compressible and very tough. It is found in areas of the body where a great deal of pressure is applied to joints, such as the knee, the jaw, and between vertebrae.

Table 4.11	Connective Tissue: Cartilage

(a) Hyaline Cartilage

Structure: Collagen fibers are small and evenly dispersed in the matrix, making the matrix appear transparent. The cartilage cells, or chondrocytes, are found in spaces, or lacunae, within the firm but flexible matrix.	**Function:** Allows growth of long bones. Provides rigidity with some flexibility in the trachea, bronchi, ribs, and nose. Forms rugged, smooth, yet somewhat flexible articulating surfaces. Forms the embryonic skeleton.	**Location:** Growing long bones, cartilage rings of the respiratory system, costal cartilage of ribs, nasal cartilage, articulating surface of bones, and the embryonic skeleton.

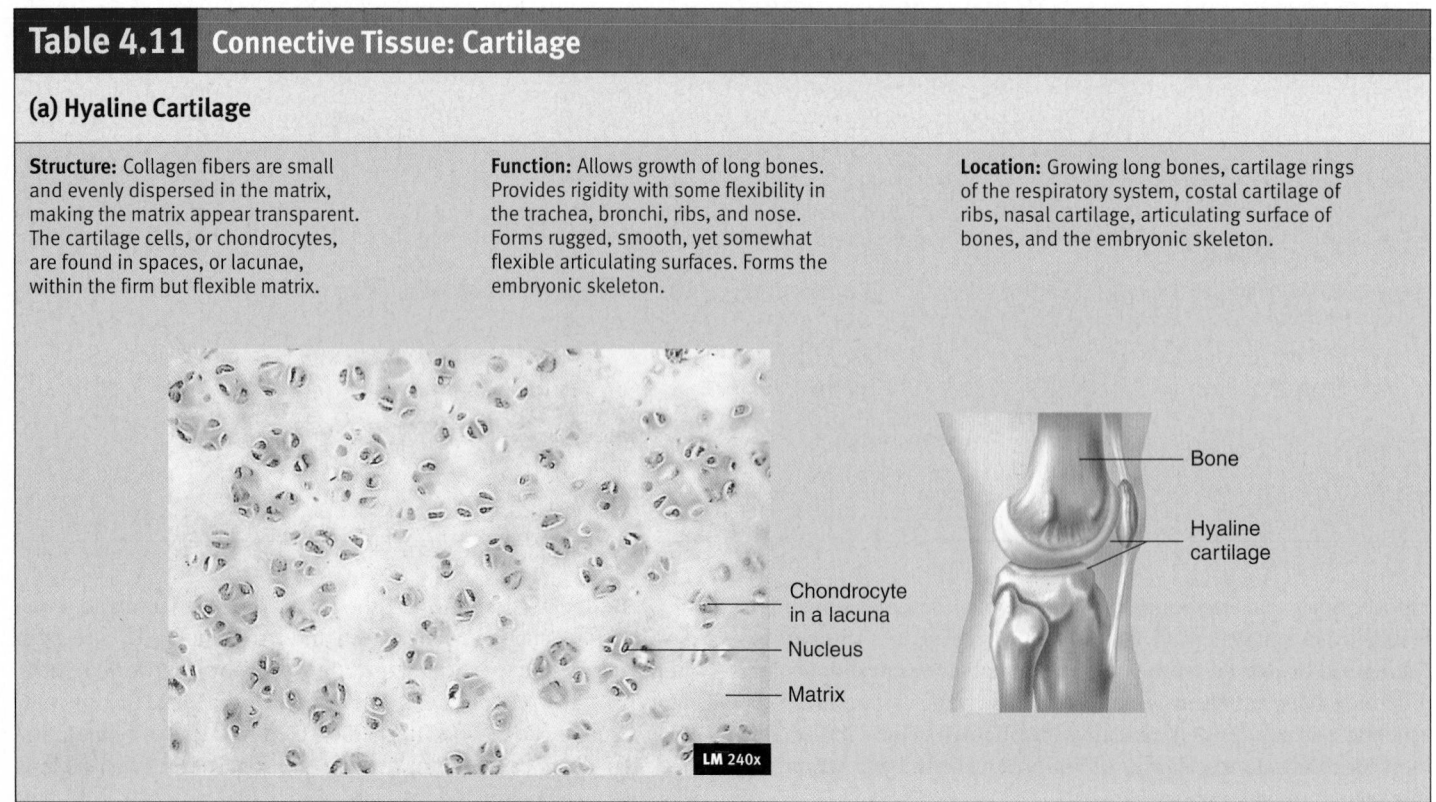

Chondrocyte in a lacuna

Nucleus

Matrix

LM 240x

Bone

Hyaline cartilage

Table 4.11   Connective Tissue: Cartilage – Continued

(b) Fibrocartilage

Structure: Collagenous fibers similar to those in hyaline cartilage. The fibers are more numerous than in other cartilages and are arranged in thick bundles

Function: Somewhat flexible and capable of withstanding considerable pressure. Connects structures subjected to great pressure.

Location: Intervertebral disks, symphysis pubis, articular disks (e.g., knee and temporomandibular [jaw] joints).

Chondrocyte in lacuna

Nucleus

Collagen fibers in matrix

Intervertebral disk

LM 240x

(c) Elastic Cartilage

Structure: Similar to hyaline cartilage, but matrix also contains elastin fibers.

Function: Provides rigidity with even more flexibility than hyaline cartilage because elastic fibers return to their original shape after being stretched.

Location: External ear, epiglottis, and auditory tubes.

Elastic fibers in matrix

Chondrocytes in lacunae

Nucleus

LM 240x

3. **Elastic cartilage** has elastic fibers in addition to collagen and proteoglycans (table 4.11c). The numerous elastic fibers are dispersed throughout the matrix of elastic cartilage. It is found in areas, such as the external ears, that have rigid but elastic properties.

P R E D I C T   5
One of several changes caused by rheumatoid arthritis in joints is the replacement of hyaline cartilage with dense irregular collagenous connective tissue. Predict the effect of replacing hyaline cartilage with fibrous connective tissue.

Bone

Bone is a hard connective tissue that consists of living cells and mineralized matrix. Bone matrix has an organic and an inorganic portion. The organic portion consists of protein fibers, primarily collagen, and other organic molecules. The mineral, or inorganic, portion consists of specialized crystals called **hydroxyapatite** (hī-drok′sē-ap-ă-tīt), which contain calcium and phosphate. The strength and rigidity of the mineralized matrix allow bones to support and protect other tissues and organs of the body. Bone cells, or **osteocytes** (os′tē-ō-sītz), are located within holes in the matrix, which are called lacunae and are similar to the lacunae of cartilage.

Two types of bone exist.

1. **Cancellous** (kan′sĕ-lŭs), or **spongy, bone** has spaces between **trabeculae** (tră-bek′ū-lē; beams), or plates, of bone and therefore resembles a sponge (table 4.12a).
2. **Compact bone** is more solid with almost no space between many thin layers, or **lamellae** (lă-mel′ē; pl., lă-mel′ă; sing.) of bone (table 4.12b).

Bone, unlike cartilage, has a rich blood supply. For this reason, bone can repair itself much more readily than can cartilage. Bone is described more fully in chapter 6.

Blood and Hemopoietic Tissue

Blood is unusual among the connective tissues because the matrix between the cells is liquid (table 4.13a). Like many other connective tissues blood has abundant extracellular matrix. The

---

Table 4.12   Connective Tissue: Bones

(a) Cancellous Bone

Structure: Latticelike network of scaffolding characterized by trabeculae with large spaces between them filled with hemopoietic tissue. The osteocytes, or bone cells, are located within lacunae in the trabeculae.

Function: Acts as a scaffolding to provide strength and support without the greater weight of compact bone.

Location: In the interior of the bones of the skull, vertebrae, sternum, and pelvis; also found in the ends of the long bones.

Osteoblast nuclei
Bone trabecula
Bone marrow
Osteocyte nucleus
Matrix

Cancellous bone

LM 240x

(b) Compact Bone

Structure: Hard, bony matrix predominates. Many osteocytes (not seen in this bone preparation) are located within lacunae that are distributed in a circular fashion around the central canals. Small passageways connect adjacent lacunae.

Function: Provides great strength and support. Forms a solid outer shell on bones that keeps them from being easily broken or punctured.

Location: Outer portions of all bones and the shafts of long bones.

Lacuna
Central canal
Matrix organized into lamellae

Compact bone

LM 240x

Table 4.13 Connective Tissue: Blood and Hemopoietic Tissue

(a) Blood

Structure: Blood cells and a fluid matrix.

Function: Transports oxygen, carbon dioxide, hormones, nutrients, waste products, and other substances. Protects the body from infections and is involved in temperature regulation.

Location: Within the blood vessels. Produced by the hemopoietic tissues. White blood cells frequently leave the blood vessels and enter the interstitial spaces.

White blood cells

Red blood cells

LM 400x

Red blood cell
White blood cell

(b) Bone Marrow

Structure: Reticular framework with numerous blood-forming cells (red marrow).

Function: Production of new blood cells (red marrow); lipid storage (yellow marrow).

Location: Within marrow cavities of bone. Two types: yellow marrow (mostly adipose tissue) in the shafts of long bones; and red marrow (hemopoietic or blood-forming tissue) in the ends of long bones and in short, flat, and irregularly shaped bones.

Cells destined to become red blood cells

Fat

Nuclei

LM 600x

Cancellous bone (red marrow)

Marrow cavity (yellow marrow)

cells of most other connective tissues are more or less stationary within a relatively rigid matrix, but blood cells are free to move within a fluid matrix. Some blood cells leave the bloodstream and wander through other tissues. The liquid matrix of blood allows it to flow rapidly through the body, carrying food, oxygen, waste products, and other materials. The matrix of blood is also unusual in that most of it is produced by cells contained in other tissues rather than by blood cells. Blood is discussed more fully in chapter 19. **Hemopoietic** (hē′mō-poy-et′ik) **tissue** forms blood cells. In adults, hemopoietic tissue is found in **bone marrow** (mar′ō) (table 4.13*b*), which is the soft connective tissue in the cavities of bones. Two types of bone marrow exist: **red marrow** and **yellow marrow** (see chapter 6). Red marrow is hemopoietic tissue surrounded by a framework of reticular fibers. Hemopoietic tissue produces red and white blood cells and is described in detail in chapter 19. In children, the marrow of most bones is red marrow. Yellow marrow consists of yellow adipose tissue and does not produce blood cells. As children grow, yellow marrow replaces much of the red marrow in bones (see chapter 6).

34. *Describe the cells and matrix of cartilage. What are lacunae? What is the perichondrium? Why does cartilage heal slowly?*
35. *How do hyaline cartilage, fibrocartilage, and elastic cartilage differ in structure and function? Give an example of each.*
36. *Describe the cells and matrix of bone. Differentiate between cancellous bone and compact bone.*
37. *What characteristic separates blood from the other connective tissues?*
38. *Describe the function of hemopoietic tissue. Explain the difference between red marrow and yellow marrow.*

Muscle Tissue

The main characteristic of **muscle tissue** is that it contracts or shortens with a force, and therefore is responsible for movement. Muscle contraction is accomplished by the interaction of contractile proteins, which are described in chapter 9. Muscles contract to move the entire body, to pump blood through the heart and blood vessels, and to decrease the size of hollow organs, such as the stomach and urinary bladder. The three types of muscle tissue are skeletal, cardiac, and smooth muscle. The types of muscle tissue are grouped according to both structure and function (table 4.14). Muscle tissue grouped according to structure is either **striated** (strī′āt-ĕd), in which microscopic bands or striations can be seen in muscle cells, or **nonstriated.** When classified according to function, a muscle is **voluntary,** meaning that it can be consciously controlled, although unconscious control can also occur, or **involuntary,** meaning that it is not normally consciously controlled. Thus the three muscle types are striated voluntary, or **skeletal muscle** (table 4.15*a*); striated involuntary, or **cardiac muscle** (table 4.15*b*); and nonstriated involuntary, or **smooth muscle** (table 4.15*c*).

For most people, the term *muscle* means skeletal muscle (see chapter 9), which constitutes the meat of animals and represents a large portion of the total weight of the human body. Most skeletal muscles, as the name implies, attach to the skeleton and, by contracting, cause the major body movements. Some skeletal muscles attach to other connective tissue types, such as the dermis of the skin, and cause movement of nonskeletal structures. Examples include muscles of facial expression, the tongue, and the external muscles of the eye. Cardiac muscle is the muscle of the heart (see chapter 20), and contraction of cardiac muscle is responsible for pumping blood. Smooth muscle is widespread throughout the body and is responsible for a wide range of functions, such as movements in the digestive, urinary, and reproductive systems.

Table 4.14	Comparison of Muscle Types		
Features	**Skeletal Muscle**	**Cardiac Muscle**	**Smooth Muscle**
Location	Attached to bones	Heart	Walls of hollow organs, blood vessels, eyes, glands, and skin
Cell shape	Very long, cylindrical cells (1 mm–4 cm in length and may extend the entire length of the muscle; 10–100 μm in diameter)	Cylindrical cells that branch (100–500 μm in length; 12–20 μm in diameter)	Spindle-shaped cells (15–200 μm in length; 5–8 μm in diameter)
Nucleus	Multinucleated, peripherally located	Single, centrally located	Single, centrally located
Striations	Yes	Yes	No
Control	Voluntary	Involuntary	Involuntary
Ability to contract spontaneously	No	Yes	Yes
Function	Body movement	Contraction provides the major force for moving blood through the blood vessels	Movement of food through the digestive tract, emptying of the urinary bladder, regulation of blood vessel diameter, change in pupil size, contraction of many gland ducts, movement of hair, and many more functions
Special features		Branching fibers, intercalated disks containing gap junctions join the cells to each other.	Gap junctions

Table 4.15 Muscle Tissue

(a) Skeletal Muscle

Structure: Skeletal muscle cells or fibers appear striated (banded). Cells are large, long, and cylindrical, with many nuclei located at the periphery.

Function: Movement of the body; under voluntary control.

Location: Attaches to bone or other connective tissue.

Muscle

Nucleus (near periphery of cell)

Skeletal muscle fiber

Striations

LM 800x

(b) Cardiac Muscle

Structure: Cardiac muscle cells are cylindrical and striated and have a single, centrally located nucleus. They are branched and connected to one another by intercalated disks, which contain gap junctions.

Function: Pumps the blood; under involuntary control.

Location: Cardiac muscle is in the heart.

Nucleus (central)

Cardiac muscle cell

Intercalated disks (special junctions between cells)

Striations

LM 800x

Table 4.15　Muscle Tissue – Continued

(c) Smooth Muscle

Structure: Smooth muscle cells are tapered at each end, are not striated, and have a single nucleus.

Function: Regulates the size of organs, forces fluid through tubes, controls the amount of light entering the eye, and produces "goose flesh" in the skin; under involuntary control.

Location: Smooth muscle is in hollow organs such as the stomach and intestine.

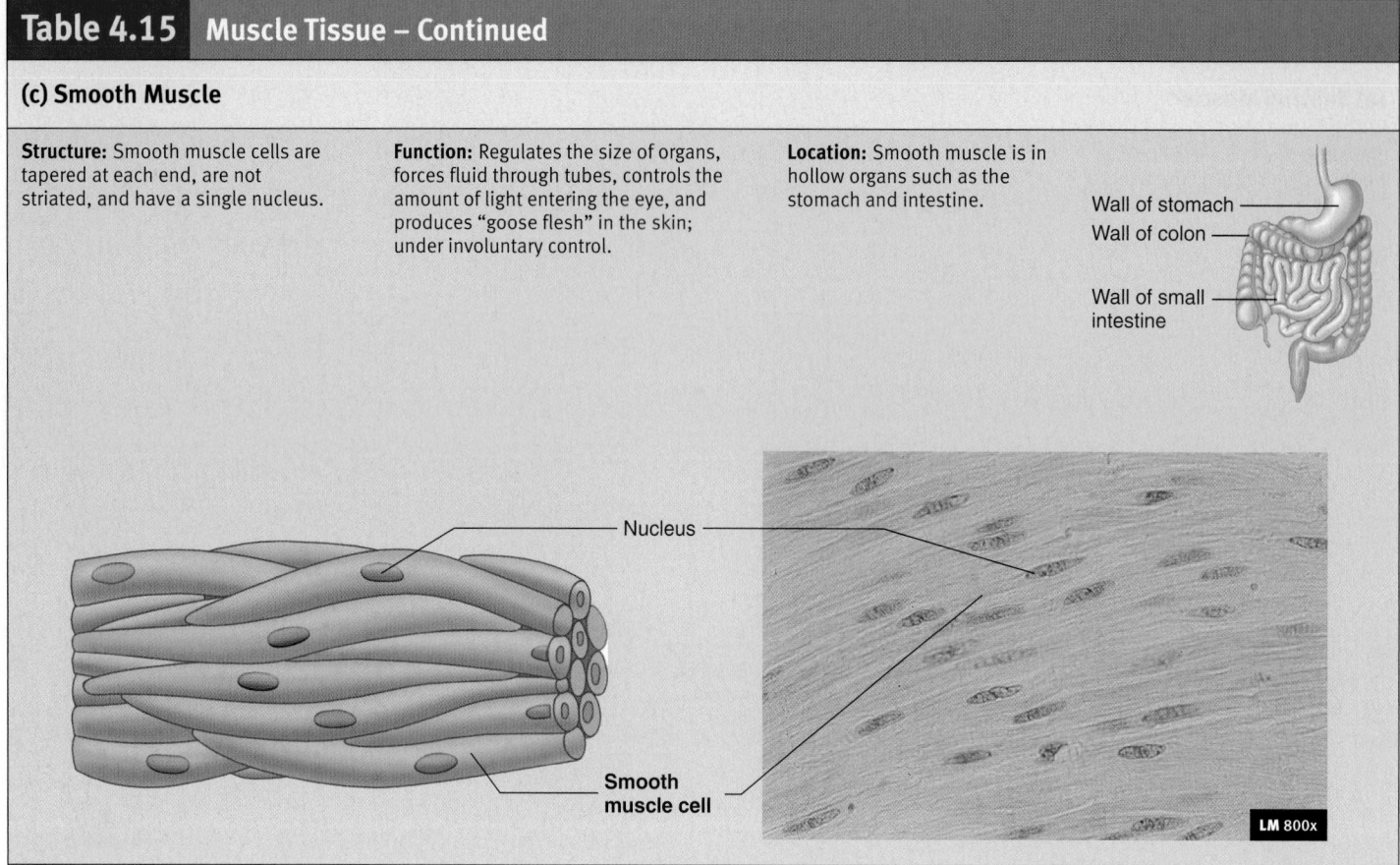

Wall of stomach
Wall of colon
Wall of small intestine

Nucleus

Smooth muscle cell

LM 800x

---

39. *Functionally, what is unique about muscle tissue?*
40. *Compare the structure of skeletal, cardiac, and smooth muscle cells.*
41. *Which of the muscle types is under voluntary control?*
42. *What tasks does each muscle type perform?*

Nervous Tissue

The fourth and final class of tissue is **nervous tissue.** It is found in the brain, spinal cord, and nerves, and is characterized by the ability to conduct electric signals called **action potentials.** It consists of neurons, which are responsible for this conductive ability, and support cells called neuroglia.

Neurons, or **nerve cells** (table 4.16), are the actual conducting cells of nervous tissue. Just as an electrical wiring system transports electricity throughout a house, neurons transport electric signals throughout the body. They are composed of three major parts: cell body, dendrites, and axon. The **cell body** contains the nucleus and is the site of general cell functions. Dendrites and axons are two types of nerve cell processes, both consisting of projections of cytoplasm surrounded by membrane. **Dendrites** (den′drītz) usually receive action potentials. They are much shorter than axons and usually taper to a fine tip.

Axons (ak′sonz) usually conduct action potentials away from the cell body. They can be much longer than dendrites, and they have a constant diameter along their entire length.

Neurons that possess several dendrites and one axon are called **multipolar neurons** (table 4.16a). Neurons that possess a single dendrite and an axon are called **bipolar neurons.** Some very specialized neurons, called **unipolar neurons** (table 4.16b), have only one axon and no dendrites. Within each subgroup are many shapes and sizes of neurons, especially in the brain and the spinal cord.

Neuroglia (noo-rog′lē-ă; nerve glue) are the support cells of the brain, spinal cord, and peripheral nerves (figure 4.5). The term *neuroglia* originally referred only to the support cells of the central nervous system, but it is now applied to cells in the peripheral nervous system as well. Neuroglia nourish, protect, and insulate neurons as well. Neurons and neuroglial cells are described in greater detail in chapter 11.

43. *Functionally, what is unique about nervous tissue?*
44. *Define and list the functions of the cell body, dendrites, and axon of a neuron.*
45. *Differentiate between multipolar, bipolar, and unipolar neurons.*
46. *What is the general function of neuroglia?*

Table 4.16	Types of Neurons

(a) Multipolar Neuron

Structure: The neuron consists of dendrites, a cell body, and a long axon. Neuroglia, or support cells, surround the neurons.

Function: Neurons transmit information in the form of action potentials, store "information," and in some way integrate and evaluate data. Neuroglia support, protect, and form specialized sheaths around axons.

Location: Neurons are located in the brain, spinal cord, and ganglia.

Brain

Spinal cord

Spinal nerves

Dendrite

Cell body of neuron

Nucleus of neuron

Nuclei of neuroglia cells

Neuroglia cells

Axon

LM 240x

(b) Unipolar Neuron

Structure: The neuron consists of a cell body with one axon.

Function: Conducts action potentials from the periphery to the brain or spinal cord.

Location: Cell bodies are located in ganglia outside of the brain and spinal cord.

Nuclei of neuroglia

Cell body of neuron

Nucleus of neuron

Branches of axon (not visible in photomicrograph)

LM 240x

Figure 4.5   Neuroglia

Membranes

A membrane is a thin sheet or layer of tissue that covers a structure or lines a cavity. Most membranes are formed from epithelium and the connective tissue on which it rests. The skin or cutaneous membrane (see chapter 5) is the external membrane. The three major categories of internal membranes are mucous membranes, serous membranes, and synovial membranes.

A **mucous** (mū′kŭs) **membrane** consists of epithelial cells, their basement membrane, a thick layer of loose connective tissue called the **lamina propria** (lam′i-nă prō′prē-ă), and, sometimes, a layer of smooth muscle cells. Mucous membranes line cavities and canals that open to the outside of the body, such as the digestive, respiratory, excretory, and reproductive passages (figure 4.6). Many, but not all, mucous membranes contain goblet cells or multicellular mucous glands, which secrete a viscous substance called **mucus** (mū′kŭs). The functions of the mucous membranes vary, depending on their location, and include protection, absorption, and secretion.

A **serous** (ser′ŭs) **membrane** consists of three components: a layer of simple squamous epithelium called **mesothelium** (mez-ō-thē′lē-ŭm), its basement membrane, and a delicate layer of loose connective tissue. Serous membranes line cavities such as the pericardial, pleural, and peritoneal cavities that do not open to the exterior (see figure 4.6). Serous membranes do not contain glands but are moistened by a small amount of fluid, called **serous fluid,** produced by the serous membranes. The serous fluid lubricates the serous membranes and makes their surfaces slippery. Serous membranes protect the internal organs from friction, help hold them in place, and act as selectively permeable barriers that prevent the accumulation of large amounts of fluid within the serous cavities.

Synovial (si-nō′vē-ăl) **membranes** consist of modified connective tissue cells either intermixed with part of the dense connective tissue of the joint capsule or separated from the capsule by areolar or adipose tissue. Synovial membranes line freely movable joints (see chapter 8) (see figure 4.6). They produce a fluid rich in hyaluronic acid, which makes the joint fluid very slippery, thereby facilitating smooth movement within the joint.

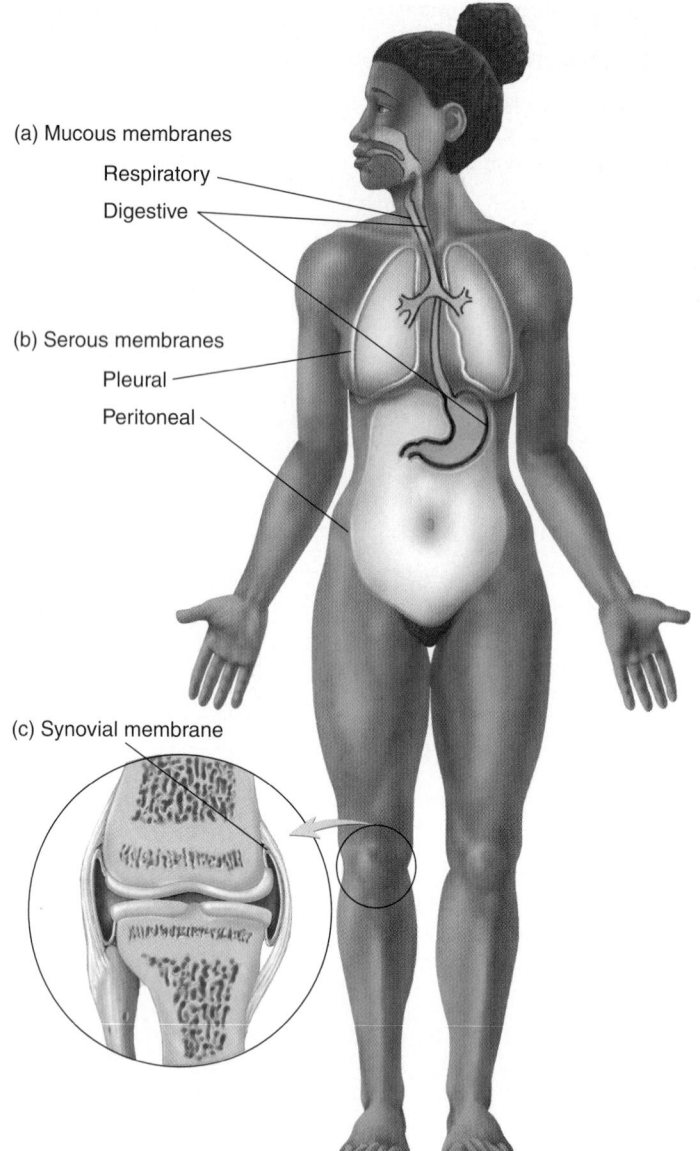

Figure 4.6   Membranes

(*a*) Mucous membranes line cavities that open to the outside and often contain mucous glands, which secrete mucus. (*b*) Serous membranes line cavities that do not open to the exterior, and do not contain glands, but do secrete serous fluid. (*c*) Synovial membranes line cavities that surround synovial joints.

47. *Compare mucous, serous, and synovial membranes according to the type of cavity they line and their secretions.*

Inflammation

The inflammatory response occurs when tissues are damaged (figure 4.7) or in association with an immune response. Although many possible agents cause injury, such as microorganisms, cold, heat, radiant energy, chemicals, electricity, or mechanical trauma, the inflammatory response to all causes is similar. The inflammatory response mobilizes the body's defenses, isolates and destroys microorganisms and other injurious agents, and removes foreign materials and damaged cells so that

1. A splinter in the skin causes damage and introduces bacteria. Mediators of inflammation are released or activated in injured tissues and adjacent blood vessels. Some blood vessels are ruptured causing bleeding.

2. Mediators of inflammation cause capillaries to dilate, causing the skin to become red. Mediators of inflammation also increase capillary permeability, and fluid leaves the capillaries, producing swelling (*arrows*).

Splinter

Bacteria introduced

Epidermis

Dermis

Blood vessel

3. White blood cells (e.g., neutrophils and macrophages) leave the dilated blood vessels and move to the site of bacterial infection, where they begin to phagocytize bacteria and other debris.

Bacteria proliferating

Neutrophil phagocytizing bacteria

Neutrophil migrating through blood vessel wall

Process Figure 4.7   Inflammation

tissue repair can proceed. The details of the inflammatory response are presented in chapter 22.

Inflammation produces five major manifestations: redness, heat, swelling, pain, and disturbance of function. Although unpleasant, these processes usually benefit recovery, and each of the symptoms can be understood in terms of events that occur during the inflammatory response.

After a person is injured, chemical substances called **mediators of inflammation** are released or activated in the tissues and the adjacent blood vessels. The mediators include histamine, kinins, prostaglandins, leukotrienes, and others. Some mediators induce dilation of blood vessels and produce the symptoms of redness and heat. Dilation of blood vessels is beneficial because it increases the speed with which white blood cells and other substances important for fighting infections and repairing the injury arrive at the site of injury.

Mediators of inflammation also stimulate pain receptors and increase the permeability of blood vessels. The increased permeability allows the movement of materials such as clotting proteins and white blood cells out of the blood vessels and into the tissue, where they can deal directly with the injury. As proteins from the blood move into the tissue, they change the osmotic relationship between the blood and the tissue. Water follows the proteins by osmosis, and the tissue swells, producing **edema** (e-dē′mă). Edema increases the pressure in the tissue, which can also stimulate neurons and cause the sensation of pain.

Clotting proteins found in blood diffuse into the interstitial spaces and form a clot. Clotting of blood also occurs in the more severely injured blood vessels. The effect of clotting is to isolate the injurious agent and to separate it from the remainder of the body. Foreign particles and microorganisms present at the site of injury are "walled off" from tissues by the clotting process. Pain, limitation of movement resulting from edema, and tissue destruction all contribute to the disturbance of function. This disturbance can be valuable because it warns the person to protect the injured structure from further damage. Sometimes the inflammatory response lasts longer or is more intense than is desirable, and drugs are used to suppress the symptoms. Antihistamines block the effects of histamine, aspirin prevents the synthesis of prostaglandins, and cortisone reduces the release of several mediators of inflammation. On the other hand, the inflammatory response by itself may not be enough to combat the effects of injury or fight off an infection. Medical intervention such as administering antibiotics may be required.

48. *What is the function of the inflammatory response?*
49. *Name five manifestations of the inflammatory response, and explain how each is produced.*

P R E D I C T 6
In some injuries, tissues are so severely damaged that areas exist where cells are killed and blood vessels are destroyed. For injuries such as these, where do the signs of inflammation such as redness, heat, edema, and pain occur?

Chronic Inflammation

When the agent responsible for an injury is not removed or if some interference occurs with the process of healing, the inflammatory response persists and is called **chronic inflammation.** For example, an infection of the lung can result in a brief period of inflammation followed by repair, but a prolonged infection causes chronic inflammation, which results in tissue destruction and permanent damage to the lung. Also, chronic inflammation of the stomach or small intestine may result in an ulcer. Prolonged infections, prolonged exposure to irritants such as silica in the lung, or abnormal immune responses can result in chronic inflammation. White blood cells invade areas of chronic inflammation, and ultimately healthy tissues are destroyed and replaced by a fibrous connective tissue, which is an important cause of the loss of organ function. Chronic inflammation of the lungs, the liver, the kidney, or other vital organs can lead to death.

Tissue Repair

Tissue repair is the substitution of viable cells for dead cells, and it can occur by regeneration or replacement. In **regeneration** (rē′jen-er-ā′shŭn), the new cells are the same type as those that were destroyed, and normal function is usually restored. In **replacement,** a new type of tissue develops that eventually causes scar production and the loss of some tissue function. Most wounds heal through regeneration and replacement; which process dominates depends on the tissues involved and the nature and extent of the wound.

Cells are classified into three groups called labile, stable, or permanent cells, according to their ability to regenerate. **Labile cells,** including cells of the skin, mucous membranes, and hemopoietic and lymphatic tissues, continue to divide throughout life. Damage to these cells can be repaired completely by regeneration. **Stable cells,** such as connective tissues and glands, including the liver, pancreas, and endocrine glands, do not divide after growth ceases; but they do retain the ability to divide and are capable of regeneration in response to injury. **Permanent cells** have very limited ability to replicate, and, if killed, they are usually replaced by a different type of cell. Neurons fit into this category, although neurons are able to recover from damage. If the cell body of a neuron is not destroyed, most neurons can replace a damaged axon or dendrite; but if the neuron cell body is destroyed, the remainder of the neuron dies. Evidence indicates that some undifferentiated cells of the central nervous system can undergo mitosis and form functional neurons, although the degree to which mitosis occurs and its functional significance is not clear. Undifferentiated cells of skeletal and cardiac muscle also have very limited ability to regenerate in response to injury, although individual skeletal and cardiac muscle cells can repair themselves. In contrast, smooth muscle readily regenerates following injury.

Skin repair is a good example of wound repair (figure 4.8). The basic pattern of the repair is the same for other tissues, especially ones covered by epithelium. If the edges of the wound are close together such as in a surgical incision, the wound heals by a process called **primary union,** or primary intention. If the edges are not close together, or if extensive loss of tissue has occurred, the process is called **secondary union,** or secondary intention.

In primary union, the wound fills with blood, and a clot forms (see chapter 19). The clot contains a threadlike protein, **fibrin** (fī′brin), that binds the edges of the wound together. The surface of the clot dries to form a **scab,** which seals the wound and helps prevent infection. An inflammatory response induces vasodilation and brings increased numbers of blood cells and other substances to the area. Blood vessel permeability increases, resulting in edema. Fibrin and blood cells move into the wounded tissues because of the increased vascular permeability. Fibrin acts to isolate and wall off microorganisms and other foreign matter. Some of the white blood cells that move into the tissue are phagocytic cells called **neutrophils** (noo′trō-filz; figure 4.8). They ingest bacteria, thus helping to fight infection, and they also ingest tissue debris and clear the area for repair. Neutrophils are killed in this process and can accumulate as a mixture of dead cells and fluid called **pus** (pŭs).

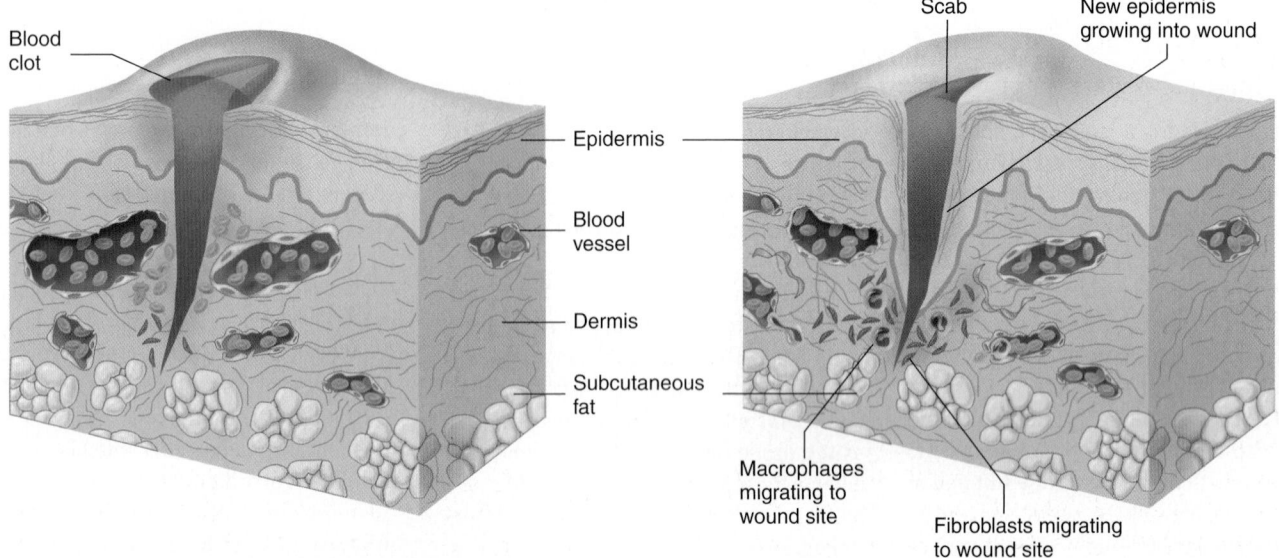

1. Fresh wound cuts through the epithelium (epidermis) and underlying connective tissue (dermis), and a clot forms.

2. Approximately 1 week after the injury, a scab is present, and epithelium (new epidermis) is growing into the wound.

3. Approximately 2 weeks after the injury, the epithelium has grown completely into the wound, and granulation tissue has formed.

4. Approximately 1 month after the injury, the wound has completely closed, the scab has been sloughed, and the granulation tissue is being replaced with dermis.

Process Figure 4.8   Tissue Repair

Fibroblasts from surrounding connective tissue migrate into the clot and produce collagen and other extracellular matrix components. Capillaries grow from blood vessels at the edge of the wound and revascularize the area, and fibrin in the clot is broken down and removed. The result is the replacement of the clot by a delicate connective tissue, called **granulation tissue,** which consists of fibroblasts, collagen, and capillaries. A large amount of granulation tissue sometimes persists as a **scar** (skar), which consists of dense irregular collagenous connective tissue. At first a scar

is bright red because of vascularization of the tissue. Later, the scar blanches and becomes white, as collagen accumulates and the vascular channels are compressed.

Repair by secondary union proceeds in a fashion similar to healing by primary union, but some differences exist. Because the wound edges are far apart, the clot may not close the gap completely, and it takes the epithelial cells much longer to regenerate and cover the wound. With increased tissue damage, the degree of the inflammatory response is greater, more cell debris exists for

Clinical Focus Cancer Tissue

Cancer (kan'ser) is a malignant, spreading tumor and the illness that results from such a tumor. A **tumor** (too'mŏr) is any swelling, although modern usage has limited the term to swellings that involve neoplastic tissue. **Oncology** (ong-kol'ō-jē; the study of tumors) is the study of tumors and their associated problems. **Neoplasm** (nē'ō-plazm) means new growth and refers to abnormal tissue growth resulting in unusually rapid cellular proliferation that continues after normal growth of the tissue has stopped or slowed considerably. A neoplasm can be either **malignant** (mă-lig'nănt; with malice or intent to cause harm), able to spread and become worse, or **benign** (bē-nīn'; kind), not inclined to spread and not likely to become worse. Although benign tumors are usually less dangerous than malignant tumors, they can cause problems. As a benign tumor enlarges, it can compress surrounding tissues and impair their functions. In some cases (e.g., brain tumors), the result can be death.

Malignant tumors can spread by local growth and expansion or by **metastasis** (mĕ-tas'tă-sis, meaning moving to another place), which results from tumor cells separating from the main neoplasm and being carried by the lymphatic or circulatory system to a new site, where a second neo-

plasm forms. A **carcinoma** (kar-si-nō'mă) is a malignant neoplasm derived from epithelial tissue. A **sarcoma** (sar-kō'mă) is a malignant neoplasm derived from connective tissue.

Malignant neoplasms lack the normal growth control that is exhibited by most other adult tissues, and in many ways they resemble embryonic tissue. Rapid growth is one characteristic of embryonic tissue, but as the tissue begins to reach its adult size and function, it slows or stops growing completely. This cessation of growth is controlled at the individual cell level. Cancer results when a cell or group of cells, for some reason, breaks away from that control. This breaking loose involves the genetic machinery and can be induced by viruses, environmental toxins, and other causes. The illness associated with cancer usually occurs as the tumor invades and destroys the healthy surrounding tissues, eliminating their functions.

Cancer therapy concentrates primarily on trying to confine and then kill the malignant cells. This goal is accomplished currently by killing the tissue with x-rays or lasers, by removing the tumor surgically, or by treating the patient with drugs that kill rapidly dividing cells or reduce the blood supply to the tumor. The major

problem with current therapy is that some cancers cannot be removed completely by surgery or killed completely by x-rays and laser therapy. These treatments can also kill normal tissue adjacent to the tumor. Many drugs used in cancer therapy kill not only cancer tissue but also other rapidly growing tissues, such as bone marrow, where new blood cells are produced, and the lining of the intestinal tract. Loss of these tissues can result in anemia, caused by the lack of red blood cells, and nausea, caused by the loss of the intestinal lining. A newer class of drugs eliminates these unwanted side effects. These drugs prevent blood vessel development, thus depriving the cancer tissue of a blood supply, rather than attacking dividing cells. Other normal tissues, in which cells divide rapidly, have well-established blood vessels and are, therefore, not affected by these drugs.

Promising anticancer therapies are being developed in which cells responsible for immune responses can be stimulated to recognize tumor cells and destroy them. A major advantage in such anticancer treatments is that the cells of the immune system can specifically attack the tumor cells and not other, healthy tissues.

---

the phagocytes to remove, and the risk of infection is greater. Much more granulation tissue forms, and **wound contraction** occurs as a result of the contraction of fibroblasts in the granulation tissue. Wound contraction leads to disfiguring and debilitating scars. Thus, it is advisable to suture a large wound so that it can heal by primary rather than secondary union. Healing is faster, the risk of infection is lowered, and the degree of scarring is reduced.

50. *Define tissue repair. Differentiate between tissue repair that occurs by regeneration and by replacement.*
51. *Compare labile cells, stable cells, and permanent cells. Give examples of each type. What is the significance of these cell types to tissue repair?*
52. *Describe the process of wound repair. Contrast healing by primary union and secondary union.*
53. *What is pus? Describe granulation tissue. How does granulation tissue contribute to scars and wound contraction?*

Tissue and Aging

Age-related changes are well documented. For example, reduced visual acuity and reduced smell, taste, and touch sensation have been documented. A clear decline in many types of athletic performance can be measured after approximately age 30–35. Ultimately there is a substantial decrease in the number of neurons and muscle cells. The functional capacity of systems such as the respiratory and cardiovascular systems declines. The rate of healing and scarring are very different in the elderly than in the very young and major changes in the structural characteristics of the skin develop. Characteristic alterations in brain function also develop in the elderly. All of these changes result in the differences between young, middle-age, and older people.

At the tissue level, age-related changes affect cells and the extracellular materials produced by them. In general, cells divide more slowly in older than in younger people. Collagen fibers become more irregular in structure, even though they may increase in number. As a consequence, connective tissues with abundant collagen, such as tendons and ligaments, become less flexible and more fragile. Elastic fibers fragment, bind to calcium ions, and become less elastic. Consequently, elastic connective tissues, such as elastic ligaments, become less elastic.

Changes in the structure of elastic and collagen fibers of arterial walls cause them to become less elastic. Atherosclerosis results as plaques form in the walls of blood vessels, which contain collagen fibers, lipids, and calcium deposits (see chapter 21). These changes result in reduced blood supply to tissues and increased susceptibility to blockage and rupture. The rate of red blood cell synthesis declines in the elderly as well. Reduced flexibility and elasticity of connective tissue is responsible for increased wrinkling of skin as well as the increased tendency for bones to break in older people.

Injuries in the very young heal more rapidly and more completely than in older people. A fracture in the femur of an infant is likely to heal quickly and eventually leave no evidence of the fracture in the bone. A similar fracture in an adult heals more slowly and a scar, seen in x-rays of the bone, is likely to persist throughout life.

- 54. *Describe the age-related changes that occur in cells such as nerve cells, muscle cells, and cells of hemopoietic tissues.*
- 55. *Describe the age-related changes in tissues with abundant collagen and elastic fibers.*

S U M M A R Y

Tissues and Histology (p. 104)

1. Tissues are collections of similar cells and the substances surrounding them.
2. The four primary tissue types are epithelial, connective, muscle, and nervous tissues.
3. Histology is the microscopic study of tissues.

Embryonic Tissue (p. 104)

All four of the primary tissue types are derived from each of the three germ layers (mesoderm, ectoderm, and endoderm).

Epithelial Tissue (p. 104)

1. Epithelium consists of cells with little extracellular matrix, it covers surfaces, it has a basement membrane, and it does not have blood vessels.
2. The basement membrane is secreted by the epithelial cells and attaches the epithelium to the underlying tissues.

Classification of Epithelium

1. Simple epithelium has a single layer of cells, stratified epithelium has two or more layers, and pseudostratified epithelium has a single layer that appears stratified.
2. Cells can be squamous (flat), cuboidal, or columnar.
3. Stratified squamous epithelium can be nonkeratinized or keratinized.
4. Transitional epithelium is stratified, with cells that can change shape from cuboidal to flattened.

Functional Characteristics

1. Simple epithelium is usually involved in diffusion, filtration, secretion, or absorption. Stratified epithelium serves a protective role. Squamous cells function in diffusion and filtration. Cuboidal or columnar cells, with a larger cell volume that contains many organelles, secrete or absorb.
2. A smooth free surface reduces friction (mesothelium and endothelium), microvilli increase absorption (intestines), and cilia move materials across the free surface (respiratory tract and uterine tubes). Transitional epithelium has a folded surface that allows the cell to change shape, and the number of cells making up the epithelial layers changes.
3. Cells are bound together mechanically by glycoproteins, desmosomes, and the zonulae adherens and to the basement membrane by hemidesmosomes. The zonulae occludens and zonulae adherens form a permeability barrier or tight junction, and gap junctions allow intercellular communication.

Glands

1. Glands are organs that secrete. Exocrine glands secrete through ducts, and endocrine glands release hormones that are absorbed directly into the blood.
2. Glands are classified as unicellular or multicellular. Goblet cells are unicellular glands. Multicellular exocrine glands have ducts, which are simple or compound (branched). The ducts can be tubular or end in small sacs (acini or alveoli). Tubular glands can be straight or coiled.
3. Glands are classified according to their mode of secretion. Merocrine glands (pancreas) secrete substances as they are produced, apocrine glands (mammary glands) accumulate secretions that are released when a portion of the cell pinches off, and holocrine glands (sebaceous glands) accumulate secretions that are released when the cell ruptures and dies.

Connective Tissue (p. 116)

Connective tissue is distinguished by its extracellular matrix.

Functions of Connective Tissue

Connective tissues enclose and separate organs and tissue; connect tissues to one another; play a role in support for movement; store high energy molecules; cushion; insulate; transport; and protect.

Cells of Connective Tissue

1. The extracellular matrix results from the activity of specialized connective tissue cells; in general, blast cells form the matrix, cyte cells maintain it, and clast cells break it down. Fibroblasts form protein fibers of many connective tissues, osteoblasts form bone, and chondroblasts form cartilage.
2. Adipose (fat) cells, mast cells, white blood cells, macrophages, and mesenchymal cells (stem cells) are commonly found in connective tissue.

Extracellular Matrix

The extracellular matrix of connective tissue has protein fibers, ground substance, and fluid as major components.

Protein Fibers of the Matrix

1. Collagen fibers structurally resemble ropes. They are strong and flexible but resist stretching.
2. Reticular fibers are fine collagen fibers that form a branching network that supports other cells and tissues.
3. Elastin fibers have a structure similar to a spring. After being stretched they tend to return to their original shape.

Other Matrix Molecules

Ground substance has the following as major components:

1. Hyaluronic acid makes fluids slippery.
2. Proteoglycan aggregates trap water, which gives tissues the capacity to return to their original shape when compressed or deformed.
3. Adhesive molecules hold proteoglycans together and to plasma membranes.

Classification of Connective Tissue (p. 118)

Connective tissue is classified according to the type of protein and the proportions of protein, ground substance, and fluid in the matrix.

Embryonic Connective Tissue

Mesenchyme arises early, consists of irregularly shaped cells and abundant matrix, and gives rise to adult connective tissue.

Adult Connective Tissue

1. Loose Connective Tissue

- Loose (areolar) connective tissue has many different cell types and a random arrangement of protein fibers with space between the fibers. This tissue fills spaces around the organs and attaches the skin to underlying tissues.

2. Dense Connective Tissue

- Dense regular connective tissue is composed of fibers arranged in one direction, which provides strength in a direction parallel to the fiber orientation. Two types of dense regular connective tissue exist: collagenous (tendons and most ligaments) and elastic (ligaments of vertebrae).
- Dense irregular connective tissue has fibers organized in many directions, which produces strength in different directions. Two types of dense irregular connective tissue exist: collagenous (capsules of organs and dermis of skin) and elastic (large arteries).

3. Connective Tissue with Special Properties

- Adipose tissue has fat cells (adipocytes) filled with lipid and very little extracellular matrix (a few reticular fibers).
  Adipose tissue functions as energy storage, insulation, and protection.
  Adipose tissue can be yellow (white) or brown. Brown fat is specialized for generating heat.
- Reticular tissue is a network of reticular fibers and forms the framework of lymphoid tissue, bone marrow, and the liver.
- Hemopoietic tissue, or red bone marrow, is the site of blood cell formation, and yellow bone marrow is a site of fat storage.

4. Cartilage

- Cartilage has a relatively rigid matrix composed of protein fibers and proteoglycan aggregates. The major cell type is the chondrocyte, which is located within lacunae.
  Hyaline cartilage has evenly dispersed collagen fibers that provide rigidity with some flexibility. Examples include the costal cartilage, the covering over the ends of bones in joints, the growing portion of long bones, and the embryonic skeleton.
  Fibrocartilage has collagen fibers arranged in thick bundles; it can withstand great pressure, and it is found between vertebrae, in the jaw, and in the knee.
  Elastic cartilage is similar to hyaline cartilage, but it has elastin fibers. It is more flexible than hyaline cartilage. It is found in the external ear.

5. Bone

  Bone cells, or osteocytes, are located in lacunae that are surrounded by a mineralized matrix (hydroxyapatite) that makes bone very hard.

Cancellous bone has spaces between bony trabeculae, and compact bone is more solid.

6. Blood and Hemopoietic Tissue

- Blood cells are suspended in a fluid matrix.
- Hemopoietic tissue forms blood cells.

Muscle Tissue (p. 128)

1. Muscle tissue has the ability to contract.
2. Skeletal (striated voluntary) muscle attaches to bone and is responsible for body movement. Skeletal muscle cells are long, cylindrically shaped cells with many peripherally located nuclei.
3. Cardiac (striated involuntary) muscle cells are cylindrical, branching cells with a single, central nucleus. Cardiac muscle is found in the heart and is responsible for pumping blood through the circulatory system.
4. Smooth (nonstriated involuntary) muscle forms the walls of hollow organs, the iris of the eye, and other structures. Its cells are spindle-shaped with a single, central nucleus.

Nervous Tissue (p. 130)

1. Nervous tissue has the ability to conduct electric impulses and is composed of neurons (conductive cells) and neuroglia (support cells).
2. Neurons have cell processes called dendrites and axons. The dendrites can receive electric impulses, and the axons can conduct them. Neurons can be multipolar (several dendrites and an axon), bipolar (one dendrite and one axon), or unipolar (one axon).

Membranes (p. 132)

1. Mucous membranes consist of epithelial cells, their basement membrane, the lamina propria, and, sometimes, smooth muscle cells; they line cavities that open to the outside and often contain mucous glands, which secrete mucus.
2. Serous membranes line cavities that do not open to the exterior, do not contain glands, but do secrete serous fluid.
3. Synovial membranes are formed by connective tissue and line joint cavities.

Inflammation (p. 132)

1. The function of the inflammatory response is to isolate injurious agents from the rest of the body and to attack and destroy the injurious agent.
2. The inflammatory response produces five symptoms: redness, heat, swelling, pain, and disturbance of function.

Tissue Repair (p. 134)

1. Tissue repair is the substitution of viable cells for dead ones. Tissue repair occurs by regeneration or replacement.
   - Labile cells divide throughout life and can undergo regeneration.
   - Stable cells do not ordinarily divide after growth is complete but can regenerate if necessary.
   - Permanent cells cannot replicate. If killed, permanent tissue is repaired by replacement.
2. Tissue repair by primary union occurs when the edges of the wound are close together. Secondary union occurs when the edges are far apart.

Tissue and Aging (p. 136)

1. Age-related changes in tissues result from reduced rates of cell division and changes in the extracellular fibers.
2. Collagen fibers become less flexible and have reduced strength.
3. Elastic fibers become fragmented and less elastic.

R E V I E W   A N D   C O M P R E H E N S I O N

1. Given these characteristics:
   1. capable of contraction
   2. covers free body surfaces
   3. lacks blood vessels
   4. composes various glands
   5. anchored to connective tissue by a basement membrane

   Which of these are characteristics of epithelial tissue?
   a. 1,2,3
   b. 2,3,5
   c. 3,4,5
   d. 1,2,3,4
   e. 2,3,4,5

2. Which of these embryonic germ layers gives rise to muscle, bone, and blood vessels?
   a. ectoderm
   b. endoderm
   c. mesoderm

3. A tissue that covers a surface, is one cell layer thick, and is composed of flat cells is
   a. simple squamous epithelium.
   b. simple cuboidal epithelium.
   c. simple columnar epithelium.
   d. stratified squamous epithelium.
   e. transitional epithelium.

4. Epithelium composed of two or more layers of cells with only the deepest layer in contact with the basement membrane is known as
   a. stratified epithelium.
   b. simple epithelium.
   c. pseudostratified epithelium.
   d. columnar epithelium.
   e. cuboidal epithelium.

5. Stratified epithelium is usually found in areas of the body where the principal activity is
   a. filtration.
   b. protection.
   c. absorption.
   d. diffusion.
   e. secretion.

6. Which of these characteristics do *not* describe nonkeratinized stratified squamous epithelium?
   a. many layers of cells
   b. surface cells are flat
   c. surface cells are living
   d. found in the skin
   e. outer layers covered by fluid

7. In parts of the body such as the urinary bladder, where considerable expansion occurs, one can expect to find which type of epithelium?
   a. cuboidal
   b. pseudostratified
   c. transitional
   d. squamous
   e. columnar

8. A tissue that contains cells with these characteristics:
   1. covers a surface
   2. one layer of cells
   3. cells are flat

   Performs which of the following functions?
   a. phagocytosis
   b. active transport
   c. secretion of many complex lipids and proteins
   d. is adapted to allow certain substances to diffuse across it
   e. protection from abrasion

9. Epithelial cells with microvilli are most likely found
   a. lining blood vessels.
   b. lining the lungs.
   c. in serous membranes.
   d. lining the digestive tract.
   e. in the skin.

10. Pseudostratified ciliated columnar epithelium can be found lining the
    a. digestive tract.
    b. trachea.
    c. thyroid gland.
    d. kidney tubules.
    e. urinary bladder.

11. A type of cell connection whose *only* function is to prevent the cells from coming apart is the
    a. desmosome.
    b. gap junction.
    c. tight junction.

12. Those glands that lose their connection with epithelium during embryonic development and secrete their cellular products into the bloodstream are called _____ glands.
    a. apocrine
    b. endocrine
    c. exocrine
    d. holocrine
    e. merocrine

13. Glands that accumulate secretions and release them only when the individual secretory cells rupture and die are called _____ glands.
    a. apocrine
    b. holocrine
    c. merocrine

14. A gland has a duct that branches repeatedly, and the ducts end in saclike structures. This describes a _____ gland.
    a. simple tubular
    b. compound tubular
    c. simple coiled tubular
    d. simple acinar
    e. compound acinar

15. The fibers in dense connective tissue are produced by
    a. fibroblasts.
    b. adipocytes.
    c. osteoblasts.
    d. osteoclasts.
    e. macrophages.

16. Mesenchymal cells
    a. form embryonic connective tissue.
    b. give rise to all adult connective tissues.
    c. in adults produce new connective tissue cells in response to injury.
    d. all of the above

17. A tissue with a large number of collagen fibers organized parallel to each other would most likely be found in
    a. a muscle.
    b. a tendon.
    c. adipose tissue.
    d. a bone.
    e. cartilage.

18. Extremely delicate fibers that make up the framework for organs such as the liver, spleen, and lymph nodes are
    a. elastic fibers.
    b. reticular fibers.
    c. microvilli.
    d. cilia.
    e. collagen fibers.

19. In which of these locations would dense irregular elastic connective tissue be found?
    a. ligaments
    b. nuchal ligament
    c. dermis of skin
    d. large arteries
    e. adipose tissue

20. Which of these is *not* true of adipose tissue?
    a. site of energy storage
    b. a type of connective tissue
    c. acts as a protective cushion
    d. brown adipose is found only in babies
    e. functions as a heat insulator

21. Which of these types of connective tissue has the smallest amount of extracellular matrix?
    a. adipose
    b. bone
    c. cartilage
    d. loose connective tissue
    e. blood

22. Given these characteristics:
    1. cells located in lacunae
    2. proteoglycans in ground substance
    3. no collagen fibers present
    4. perichondrium on surface
    5. heals rapidly after injury

    Which of these characteristics apply to cartilage?
    a. 1,2,3
    b. 1,2,4
    c. 2,4,5
    d. 1,2,4,5
    e. 2,3,4,5

23. Fibrocartilage is found
    a. in the cartilage of the trachea.
    b. in the rib cage.
    c. in the external ear.
    d. on the surface of bones in moveable joints.
    e. between vertebrae.

24. A tissue in which cells are located in lacunae surrounded by a hard matrix of hydroxyapatite is
    a. hyaline cartilage.
    b. bone.
    c. nervous tissue.
    d. dense regular collagenous connective tissue.
    e. fibrocartilage.

25. Which of these characteristics apply to smooth muscle?
    a. striated, involuntary
    b. striated, voluntary
    c. unstriated, involuntary
    d. unstriated, voluntary

26. Which of these statements about nervous tissue is *not* true?
    a. Neurons have cytoplasmic extensions called axons.
    b. Electric signals (action potentials) are conducted along axons.
    c. Bipolar neurons have two axons.
    d. Neurons are nourished and protected by neuroglia.
    e. Dendrites receive electric signals and conduct them toward the cell body.

27. Linings of the digestive, respiratory, excretory, and reproductive passages are composed of
    a. serous membranes.
    b. mucous membranes.
    c. mesothelium.
    d. synovial membranes.
    e. endothelium.

28. Chemical mediators of inflammation
    a. cause blood vessels to constrict.
    b. decrease the permeability of blood vessels.
    c. initiate processes that lead to edema.
    d. help to prevent clotting.
    e. decrease pain.

29. Which of these types of cells are labile?
    a. neurons
    b. skin
    c. liver
    d. pancreas

30. Permanent cells
    a. divide and replace damaged cells in replacement tissue repair.
    b. form granulation tissue.
    c. are responsible for removing scar tissue.
    d. are usually replaced by a different cell type if they are destroyed.
    e. are replaced during regeneration tissue repair.

Answers in Appendix F

C R I T I C A L   T H I N K I N G

1. Given the observation that a tissue has more than one layer of cells lining a free surface, (1) list the possible tissue types that exhibit those characteristics, and (2) explain what additional observations need to be made to identify the tissue as a specific tissue type.

2. A patient suffered from kidney failure a few days after he was exposed to a toxic chemical. A biopsy of his kidney indicated that many of the thousands of epithelium-lined tubules that make up the kidney had lost many of the simple cuboidal epithelial cells that normally line them, although the basement membranes appeared to be mostly intact. Predict how likely this person is to fully recover.

3. Compare the cell shapes and surface specializations of an epithelium that functions to resist abrasion to those of an epithelium that functions to carry out absorption of materials.

4. Tell how to distinguish between a gland that produces a merocrine secretion and a gland that produces a holocrine secretion. Assume that you have the ability to chemically analyze the composition of secretions.

5. Name a tissue that has the following characteristics: abundant extracellular matrix consisting almost entirely of collagen fibers that are parallel to each other. Injuries of which of the following types result from damage to this kind of tissue: dislocated neck vertebrae, a torn tendon, a ruptured intervertebral disk.

6. Indicate whether the following statement is appropriate or not: "If a tissue is capable of contracting, is under involuntary control, and has mononucleated cells, it is smooth muscle." Explain your answer.

7. Antihistamines block the effect of a chemical mediator of inflammation called histamine, which is released during the inflammatory response. What effect does administering antihistamines have on the inflammatory response, and is use of an antihistamine beneficial?

Answers in Appendix G

A N S W E R S   T O   P R E D I C T   Q U E S T I O N S

1. a. Secretion of mucus and digestive enzymes and the absorption of nutrients normally occur in the digestive tract. Simple columnar epithelial cells contain organelles that are specialized to carry out nutrient absorption and secretion of mucus and digestive enzymes. Stratified squamous epithelium is not specialized to either absorb or secrete, and the layers of epithelial cells reduce the ability of nutrient molecules to pass through them and, therefore, to be absorbed. The ability of digestive enzymes to pass through the layers of epithelial cells, and therefore be secreted, is also reduced.

   b. Keratinized stratified epithelium forms a tough layer that is a barrier to the movement of water. Replacing the epithelium of skin with moist stratified squamous epithelium increases the loss of water across the skin because water can diffuse through nonkeratinized stratified squamous epithelium, and it is more delicate and provides less protection than keratinized stratified squamous epithelium.

   c. The stratified squamous epithelium that lines the mouth provides protection. Replacement of it with simple columnar epithelium makes the lining of the mouth much more susceptible to damage because the single layer of epithelial cells is easier to damage.

2. The zonula occludens prevents substances from passing between the epithelial cells. NaCl must pass across the epithelial layer. NaCl could pass by facilitated diffusion across one membrane of the epithelial cells, and by active transport across the other epithelial membrane. If the membranes have a high permeability to water, water will move in the same direction as NaCl because of osmosis.

3. Elastic ligaments attached to the vertebrae help the vertebral column return to its normal upright position after it is flexed. The elastic ligaments act much like elastic bands. Tendons attach muscles to bones. When muscles contract, they pull on the tendons, which in turn pull on bones. Because they are not elastic, when the muscle pulls on the tendon, all of the force is applied to the bone, causing it to move. If tendons were elastic, when the muscle contracted, the tendon would stretch, and not all of the tension would be applied to the bone.

4. Collagen synthesis is required for scar formation. If collagen synthesis does not occur because of a lack of vitamin C or if collagen synthesis is slowed, wound healing does not occur or is slower than normal. One might expect that the density of collagen fibers in a scar is reduced and the scar is not as durable as a normal scar.

5. Hyaline cartilage provides a smooth surface so that bones in joints can move easily. When the smooth surface provided by hyaline cartilage is replaced by dense fibrous connective tissue, the smooth surface is replaced by a less smooth surface, and the movement of bones in joints is much more difficult. The increased friction helps to increase inflammation and pain that occurs in the joints of people who have rheumatoid arthritis.

6. In severely damaged tissue in which cells are killed and blood vessels are destroyed, the usual symptoms of inflammation cannot occur. Surrounding these areas of severe tissue damage, however, where blood vessels are still intact and cells are still living, the classic signs of inflammation do develop. The signs of inflammation therefore appear around the periphery of severely injured tissues.

Visit the Online Learning Center at www.mhhe.com/seeley7 for chapter quizzes, interactive learning exercises, and other study tools.

Integumentary System

Skin, hair, nails, and glands form the integumentary system. This colorized scanning electron micrograph shows the shaft of a hair (yellow) protruding through the surface of the skin. Tightly bound epithelial cells form the hair shaft, and flat, scalelike epithelial cells form the skin's surface. Like the shingles on the roof of a house, the skin's tough cells protect underlying structures. Dandruff is excessive flaking of these epithelial cells from the scalp.

The **integumentary** (in-teg-ū-men′tă-rē) **system** consists of the skin and accessory structures such as hair, nails, and glands. Integument means covering, and the integumentary system is familiar to most people because it covers the outside of the body and is easily observed. In addition, humans are concerned with the appearance of the integumentary system. Skin without blemishes is considered attractive, whereas acne is a source of embarrassment for many people. The development of wrinkles and the graying or loss of hair is a sign of aging that some people find unattractive. Because of these feelings, much time, effort, and money are spent on changing the appearance of the integumentary system. For example, people apply lotion to their skin, color their hair, and trim their nails. They also try to prevent sweating with antiperspirants and body odor with washing, deodorants, and perfumes.

The appearance of the integumentary system can indicate physiological imbalances in the body. Some disorders like acne or warts affect just the integumentary system. Disorders of other parts of the body can be reflected there, and thus the integumentary system is useful for diagnosis. For example, reduced blood flow through the skin during a heart attack can cause a pale appearance, whereas increased blood flow as a result of fever can cause a flushed appearance. Also, the rashes of some diseases are very characteristic, such as the rashes of measles, chicken pox, and allergic reactions.

This chapter provides an *overview of the integumentary system* (p. 144) and an explanation of the *hypodermis* (p. 144), the *skin* (p. 145), and the *accessory skin structures* (p. 150). A *summary of integumentary system functions* (p. 156) and the *effects of aging on the integumentary system* (p. 157) are also presented.

Overview of the Integumentary System

Although we are often concerned with how the integumentary system looks, it has many important functions that go beyond appearance. The integumentary system forms the boundary between the body and the external environment, thereby separating us from the external environment while allowing us to interact with it. Major functions of the integumentary system include:

1. *Protection.* The skin provides protection against abrasion and ultraviolet light. It also prevents the entry of microorganisms and prevents dehydration by reducing water loss from the body.
2. *Sensation.* The integumentary system has sensory receptors that can detect heat, cold, touch, pressure, and pain.
3. *Temperature regulation.* Body temperature is regulated by controlling blood flow through the skin and the activity of sweat glands.
4. *Vitamin D production.* When exposed to ultraviolet light, the skin produces a molecule that can be transformed into vitamin D.
5. *Excretion.* Small amounts of waste products are lost through the skin and in gland secretions.

> 1. *Provide an example for each function of the integumentary system.*

Hypodermis

Just as a house rests on a foundation, the skin rests on the **hypodermis** (hī-pō-der′mis), which attaches it to underlying bone and muscle and supplies it with blood vessels and nerves (figure 5.1). The hypodermis consists of loose connective tissue with collagen and elastin fibers. The main types of cells within the hypodermis are fibroblasts, adipose cells, and macrophages. The hypodermis, which is not part of the skin, is sometimes called **subcutaneous** (sŭb-koo-tā′nē-ŭs) **tissue,** or **superficial fascia** (fash′ē-ă).

Approximately half the body's stored fat is in the hypodermis, where it functions as a source of energy, insulation, and padding. The amount of fat in the hypodermis varies with age, sex, and diet. Most babies have a chubby appearance because they have proportionately more fat than adults. Women have proportionately more fat than men, especially over the thighs, buttocks, and breasts, which accounts for some of the differences in body shape between women and men. The amount of fat in the hypodermis is also responsible for some of the differences in body shape between individuals of the same sex.

> 2. *Name the types of tissue forming the hypodermis.*
> 3. *How is the hypodermis related to the skin?*
> 4. *List the functions of fat contained within the hypodermis.*

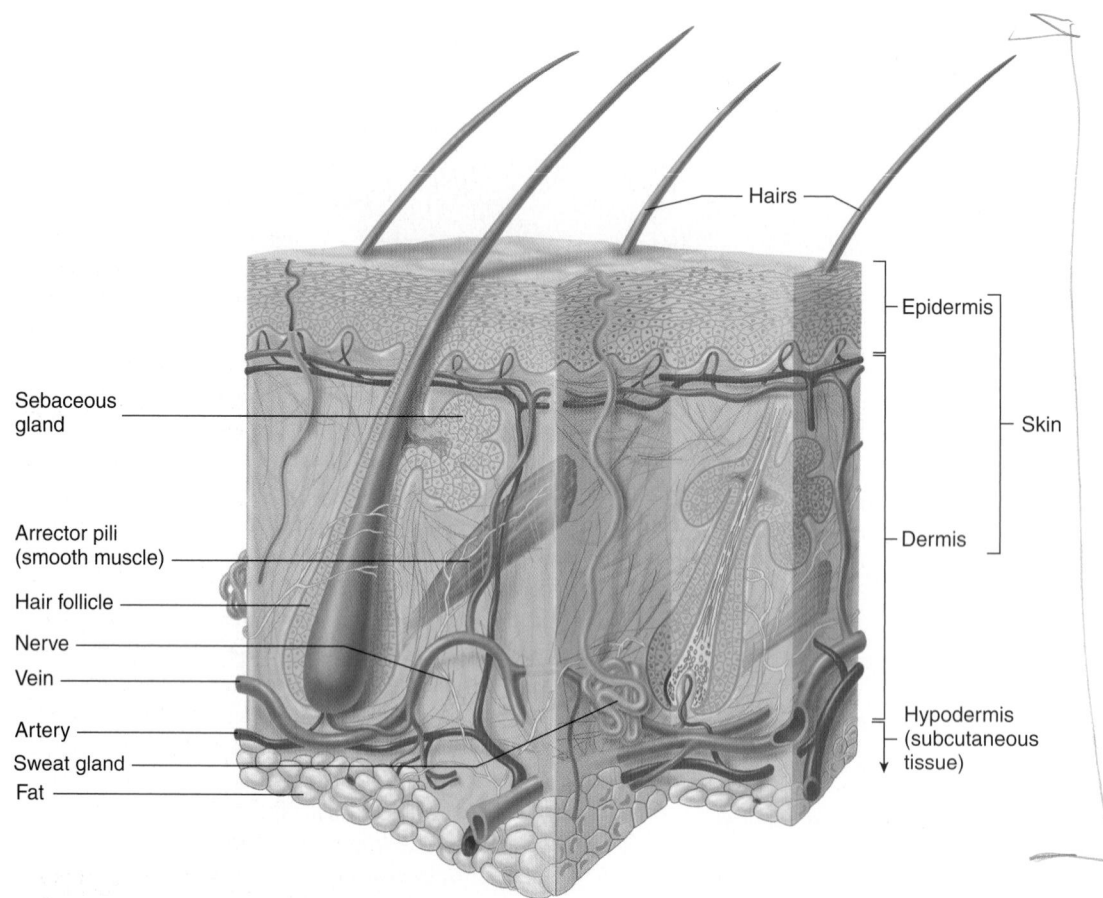

Figure 5.1 **Skin and Hypodermis**

The skin, consisting of the epidermis and the dermis, is connected by the hypodermis to underlying structures. Note the accessory structures (hairs, glands, and arrector pili), some of which project into the hypodermis, and the large amount of fat in the hypodermis.

Uses of the Hypodermis

The hypodermis can be used to estimate total body fat. The skin is pinched at selected locations, and the thickness of the fold of skin and underlying hypodermis is measured. The thicker the fold, the greater is the amount of total body fat. Clinically, the hypodermis is the site of subcutaneous injections.

Skin

The skin is made up of two major tissue layers. The **dermis** (derm′is; skin) is a layer of connective tissue that is connected to the hypodermis. The **epidermis** (ep-i-derm′is; on the dermis) is a layer of epithelial tissue that rests on the dermis (see figure 5.1). If the hypodermis is the foundation on which the house rests, the dermis forms most of the house, and the epidermis is its roof.

Dermis

The dermis is responsible for most of the structural strength of the skin. It is connective tissue with fibroblasts, a few adipose cells, and macrophages. Collagen is the main connective tissue fiber, but elastin

and reticular fibers are also present. Adipose cells and blood vessels are scarce in the dermis compared to the hypodermis. Nerve endings, hair follicles, smooth muscles, glands, and lymphatic vessels are also in the dermis (see figure 5.1). The nerve endings are varied in structure and function: free nerve endings for pain, itch, tickle, and temperature sensations; hair follicle receptors for light touch; Pacinian corpuscles for deep pressure; Meissner's corpuscles for the ability to detect simultaneous stimulation at two points on the skin; and Ruffini's end organs for continuous touch or pressure (see figure 14.1).

Uses of the Dermis

The dermis is that part of an animal hide from which leather is made. The epidermis of the skin is removed, and the dermis is treated with chemicals in a process called tanning. Clinically the dermis in humans is sometimes the site of such injections as the tuberculin skin test.

The dermis is divided into two layers (see figure 5.1, and figure 5.2): the deeper **reticular** (re-tik′ū-lăr) **layer** and the more superficial **papillary** (pap′i-lār-ē) **layer.** The reticular layer, which is composed of dense irregular connective tissue, is the main layer of the dermis. It is continuous with the hypodermis and forms a mat

Figure 5.2   Dermis and Epidermis

(*a*) Photomicrograph of dermis covered by the epidermis. The dermis consists of the papillary and reticular layers. The papillary layer has projections called papillae that extend into the epidermis. (*b*) Higher magnification photomicrograph of the epidermis resting on the papillary layer of the dermis. Note the strata of the epidermis.

of irregularly arranged fibers that are resistant to stretching in many directions. The elastin and collagen fibers are oriented more in some directions than in others and produce **cleavage,** or **tension, lines** in the skin (figure 5.3). Knowledge of cleavage line directions is important because an incision made parallel to the cleavage lines is less likely to gap than is an incision made across them. The closer together the edges of a wound, the less likely is the development of infections and the formation of considerable scar tissue.

If the skin is overstretched, the dermis may rupture and leave lines that are visible through the epidermis. These lines of scar tissue, called **striae** (strī′ē), or **stretch marks,** can develop on the abdomen and breasts of a woman during pregnancy.

The papillary layer derives its name from projections called **papillae** (pă-pil′ē) that extend toward the epidermis (see figure 5.2). The papillary layer is less dense than the reticular layer. Some histologists classify the papillary layer as loose connective tissue because it has thin fibers that are somewhat loosely arranged. The papillary layer also contains blood vessels that supply the overlying epidermis with nutrients, remove waste products, and aid in regulating body temperature.

> 5. *Name and compare the two layers of the dermis. Which layer is responsible for most of the structural strength of the skin?*
> 6. *What are cleavage lines and striae?*

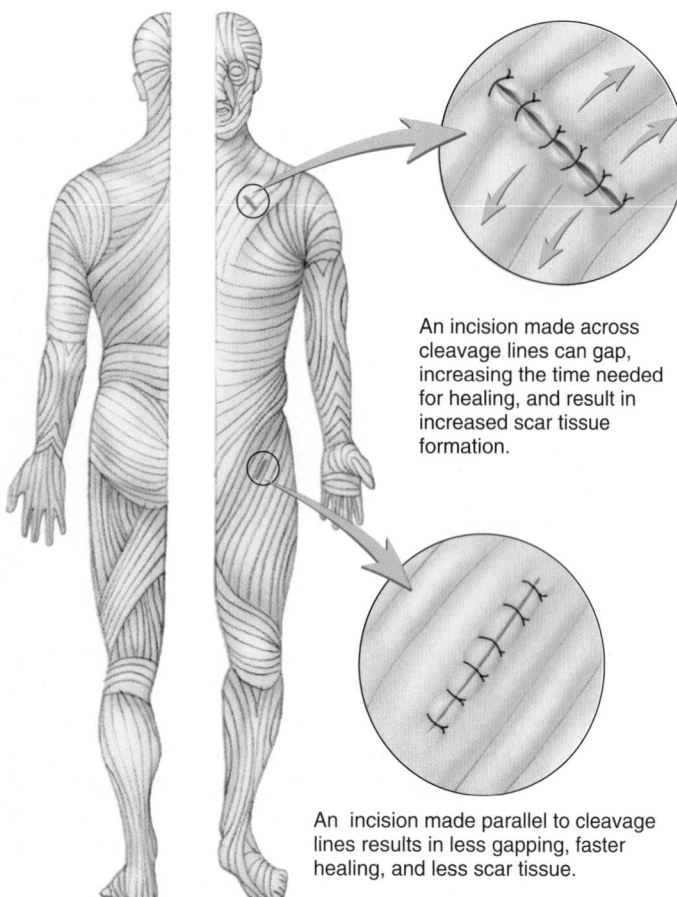

An incision made across cleavage lines can gap, increasing the time needed for healing, and result in increased scar tissue formation.

An incision made parallel to cleavage lines results in less gapping, faster healing, and less scar tissue.

Figure 5.3   Cleavage Lines
The orientation of collagen fibers produces cleavage, or tension, lines in the skin.

Epidermis

The epidermis is stratified squamous epithelium, and it is separated from the papillary layer of the dermis by a basement membrane. The epidermis is not as thick as the dermis, contains no blood vessels, and is nourished by diffusion from capillaries of the papillary layer (see figure 5.2). Most cells of the epidermis are called **keratinocytes** (ke-rat′i-nō-sītz) because they produce a protein mixture called **keratin** (ker′ă-tin). Keratinocytes are responsible for the structural strength and permeability characteristics of the epidermis. Other cells of the epidermis include **melanocytes** (mel′ă-nō-sītz), which contribute to skin color, **Langerhans cells,** which are part of the immune system (see chapter 22), and **Merkel's cells,** which are specialized epidermal cells associated with nerve endings responsible for detecting light touch and superficial pressure (see chapter 14).

Cells are produced by mitosis in the deepest layers of the epidermis. As new cells are formed, they push older cells to the surface where they slough off, or **desquamate** (des′kwă-māt). The outermost cells in this stratified arrangement protect the cells underneath, and the deeper replicating cells replace cells lost from the surface. As they move from the deeper epidermal layers to the surface, the cells change shape and chemical composition. This process is called **keratinization** (ker′ă-tin-i-zā′shŭn) because the cells become filled with keratin. During keratinization, these cells eventually die and produce an outer layer of cells that resists abrasion and forms a permeability barrier.

Keratinization and Disease

The study of keratinization is important because many skin diseases result from malfunctions in this process. For example, large scales of epidermal tissue are sloughed off in **psoriasis** (sō-rī′ă-sis; see "Clinical Focus: Clinical Disorders of the Integumentary System" on p. 158). By comparing normal and abnormal keratinization, scientists may be able to develop effective therapies.

Although keratinization is a continual process, distinct transitional stages can be recognized as the cells change. On the basis of these stages, the many layers of cells in the epidermis are divided into regions, or **strata** (sing., stratum) (see figure 5.2 and figure 5.4). From the deepest to the most superficial, these five strata are observed: stratum basale, stratum spinosum, stratum granulosum, stratum lucidum, and stratum corneum. The number of cell layers in each stratum and even the number of strata in the skin vary, depending on their location in the body.

Stratum Basale

The deepest portion of the epidermis is a single layer of cuboidal or columnar cells called the **stratum basale** (bā′să-lē) or **stratum germinativum** (jer′mi-nă-tīv′um) (see figures 5.2 and 5.4, step 1). Structural strength is provided by hemidesmosomes, which anchor the epidermis to the basement membrane, and by desmosomes, which hold the keratinocytes together (see chapter 4). Keratinocytes are strengthened internally by keratin fibers (intermediate filaments) that insert into the desmosomes. Keratinocytes undergo mitotic divisions approximately every 19 days. One daughter cell becomes a new stratum basale cell and divides

1. In the stratum basale, cells divide by mitosis and some of the newly formed cells become the cells of the more superficial strata.

2. In the stratum spinosum, keratin fibers and lamellar bodies accumulate.

3. In the stratum granulosum, keratohyalin granules accumulate and a hard protein envelope forms beneath the plasma membrane; lamellar bodies release lipids; cells die.

4. In the stratum lucidum, the cells are dead and contain dispersed keratohyalin.

5. In the stratum corneum, the dead cells have a hard protein envelope, contain keratin, and are surrounded by lipids.

Superficial

⑤ **Stratum corneum**

④ **Stratum lucidum**

③ **Stratum granulosum**

② **Stratum spinosum**

① **Stratum basale**

Deep

Intercellular lipids

Keratin

Lamellar body releases lipids

Protein envelope

Keratohyalin granules

Lipid-filled lamellar body

Keratin fiber

Desmosome

Nucleus

Hemidesmosome

Basement membrane

Process Figure 5.4 Epidermal Layers and Keratinization

again, but the other daughter cell is pushed toward the surface and becomes **keratinized** (ker′ă-ti-nīzd). It takes approximately 40–56 days for the cell to reach the epidermal surface and desquamate.

Stratum Spinosum

Superficial to the stratum basale is the **stratum spinosum** (spī-nō′sŭm), consisting of 8–10 layers of many-sided cells (see figures 5.2 and 5.4, step 2). As the cells in this stratum are pushed to the surface, they flatten; desmosomes are broken apart, and new desmosomes are formed. During preparation for microscopic observation, the cells usually shrink from one another, except where they are attached by desmosomes, causing the cells to appear spiny—hence the name stratum spinosum. Additional keratin fibers and lipid-filled, membrane-bounded organelles called **lamellar** (lam′ĕ-lăr, lă-mel′ăr) **bodies** are formed inside the keratinocytes. A limited amount of cell division takes place in the stratum spinosum. Mitosis does not occur in the more superficial strata.

Stratum Granulosum

The **stratum granulosum** (gran-ū-lō′sŭm) consists of two to five layers of somewhat flattened, diamond-shaped cells with long axes that are oriented parallel to the surface of the skin (see figures 5.2 and 5.4, step 3). This stratum derives its name from the nonmembrane-bounded protein granules of **keratohyalin** (ker′ă-tō-hī′ă-lin), which accumulate in the cytoplasm of the cell. The lamellar bodies of these cells move to the plasma membrane and release their lipid contents into the intercellular space. Inside the cell, a

protein envelope forms beneath the plasma membrane. In the most superficial layers of the stratum granulosum, the nucleus and other organelles degenerate, and the cell dies. Unlike the other organelles, however, the keratin fibers and keratohyalin granules do not degenerate.

Stratum Lucidum

The **stratum lucidum** (loo′si-dŭm) appears as a thin, clear zone above the stratum granulosum (see figures 5.2 and 5.4, step 4) and consists of several layers of dead cells with indistinct boundaries. Keratin fibers are present, but the keratohyalin, which was evident as granules in the stratum granulosum, has dispersed around the keratin fibers, and the cells appear somewhat transparent. The stratum lucidum is present in only a few areas of the body (see "Thick and Thin Skin" on p. 148).

Stratum Corneum

The last and most superficial stratum of the epidermis is the **stratum corneum** (kōr′nē-ŭm) (see figures 5.2 and 5.4, step 5). This stratum is composed of approximately 25 or more layers of dead squamous cells joined by desmosomes. Eventually the desmosomes break apart, and the cells are desquamated from the surface of the skin. Excessive desquamation of the stratum corneum of the scalp is called dandruff. Less noticeably, cells are continually shed as clothes rub against the body or as the skin is washed.

The stratum corneum consists of **cornified cells,** which are dead cells with a hard protein envelope that are filled with the protein keratin. **Keratin** is a mixture of keratin fibers and keratohyalin.

The envelope and the keratin are responsible for the structural strength of the stratum corneum. The type of keratin found in the skin is soft keratin. Another type of keratin, hard keratin, is found in nails and the external parts of hair. Cells containing hard keratin are more durable than cells with soft keratin and do not desquamate.

Surrounding the cells are the lipids released from lamellar bodies. The lipids are responsible for many of the permeability characteristics of the skin. Table 5.1 summarizes the structures and functions of the skin and hypodermis.

7. *From deepest to most superficial, name and describe the five strata of the epidermis. In which strata are new cells formed by mitosis? Which strata have live cells, and which have dead cells?*

8. *Describe the structural features resulting from keratinization that make the epidermis structurally strong and resistant to water loss.*

P R E D I C T   1

Some drugs are administered by applying them to the skin (e.g., a nicotine skin patch to help a person stop smoking). The drug diffuses through the epidermis to blood vessels in the dermis. What kind of substances can pass easily through the skin by diffusion? What kind have difficulty?

Thick and Thin Skin

When we say a person has thick or thin skin, we are usually referring metaphorically to the person's ability to take criticism. However, all of us in a literal sense have both thick and thin skin. Skin is classified as thick or thin on the basis of the structure of the epidermis. **Thick skin** has all five epithelial strata, and the stratum corneum has many layers of cells. Thick skin is found in areas subject to pressure or friction, such as the palms of the hands, the soles of the feet, and the fingertips. The papillae of the dermis underlying thick skin are in parallel, curving ridges that shape the overlying epidermis into fingerprints and footprints. The ridges increase friction and improve the grip of the hands and feet.

Fingerprints and Criminal Investigations

Fingerprints were first used in criminal investigation in 1880 by Henry Faulds, a Scottish medical missionary. Faulds used a greasy fingerprint left on a bottle to identify a thief who had been drinking purified alcohol from the dispensary.

Thin skin covers the rest of the body and is more flexible than thick skin. Each stratum contains fewer layers of cells than are found in thick skin; the stratum granulosum frequently consists of

Table 5.1	Comparison of the Skin (Epidermis and Dermis) and Hypodermis	
Part	Structure	Function
Epidermis	Superficial part of skin; stratified squamous epithelium; composed of four or five strata	Barrier that prevents water loss and the entry of chemicals and microorganisms; protects against abrasion and ultraviolet light; produces vitamin D; gives rise to hair, nails, and glands
Stratum corneum	Most superficial strata of the epidermis; 25 or more layers of dead squamous cells	Provision of structural strength by keratin within cells; prevention of water loss by lipids surrounding cells; desquamation of most superficial cells resists abrasion
Stratum lucidum	Three to five layers of dead cells; appears transparent; present in thick skin, absent in most thin skin	Dispersion of keratohyalin around keratin fibers
Stratum granulosum	Two to five layers of flattened, diamond-shaped cells	Production of keratohyalin granules; lamellar bodies release lipids from cells; cells die
Stratum spinosum	A total of 8–10 layers of many-sided cells	Production of keratin fibers; formation of lamellar bodies
Stratum basale	Deepest strata of the epidermis; single layer of cuboidal or columnar cells; basement membrane of the epidermis attaches to the dermis	Production of cells of the most superficial strata; melanocytes produce and contribute melanin, which protects against ultraviolet light
Dermis	Deep part of skin; connective tissue composed of two layers	Responsible for the structural strength and flexibility of the skin; the epidermis exchanges gases, nutrients, and waste products with blood vessels in the dermis
Papillary layer	Papillae projects toward the epidermis; loose connective tissue	Brings blood vessels close to the epidermis; papillae form fingerprints and footprints
Reticular layer	Mat of collagen and elastin fibers; dense, irregular connective tissue	Main fibrous layer of the dermis; strong in many directions; forms cleavage lines
Hypodermis	Not part of the skin; loose connective tissue with abundant fat deposits	Attaches the dermis to underlying structures; fat tissue provides energy storage, insulation, and padding; blood vessels and nerves from the hypodermis supply the dermis

only one or two layers of cells, and the stratum lucidum generally is absent. The dermis under thin skin projects upward as separate papillae and does not produce the ridges seen in thick skin. Hair is found only in thin skin.

The entire skin, including both the epidermis and the dermis, varies in thickness from 0.5 mm in the eyelids to 5.0 mm for the back and shoulders. The terms *thin* and *thick,* which refer to the epidermis only, should not be used when total skin thickness is considered. Most of the difference in total skin thickness results from variation in the thickness of the dermis. For example, the skin of the back is thin skin, whereas that of the palm is thick skin; however, the total skin thickness of the back is greater than that of the palm because more dermis exists in the skin of the back.

In skin subjected to friction or pressure, the number of layers in the stratum corneum greatly increases to produce a thickened area called a **callus** (kal′ŭs). The skin over bony prominences may develop a cone-shaped structure called a **corn.** The base of the cone is at the surface, but the apex extends deep into the epidermis, and pressure on the corn may be quite painful. Calluses and corns can develop in both thin and thick skin.

> **9.** *Compare the structure and location of thick skin and thin skin. Is hair found in thick or thin skin?*

Skin Color

Pigments in the skin, blood circulating through the skin, and the thickness of the stratum corneum together determine skin color. **Melanin** (mel′ă-nin) is the term used to describe a group of pigments responsible for skin, hair, and eye color. Melanin provides protection against ultraviolet light from the sun. Large amounts of melanin are found in certain regions of the skin, such as freckles, moles, nipples, areolae of the breasts, the axillae, and the genitalia. Other areas of the body, such as the lips, the palms of the hands, and the soles of the feet, contain less melanin.

In the production of melanin, the enzyme tyrosinase (tī′rō-si-nās, tir′ō-si-nās) converts the amino acid tyrosine to dopaquinone (dō′pă-kwin′ōn, dō′pă-kwī-nōn). Dopaquinone can be converted to a variety of related molecules, most of which are brown to black pigments, but some of which are yellowish or reddish.

Melanin is produced by **melanocytes** (mel′ă-nō-sītz), irregularly shaped cells with many long processes that extend between the keratinocytes of the stratum basale and the stratum spinosum (figure 5.5). The Golgi apparatuses of the melanocytes package melanin into vesicles called **melanosomes** (mel′ă-nō-sōmz), which move into the cell processes of the melanocytes. Keratinocytes phagocytize (see chapter 3) the tips of the melanocyte cell processes, thereby acquiring melanosomes. Although all keratinocytes can contain melanin, only the melanocytes produce it.

Melanin production is determined by genetic factors, exposure to light, and hormones. Genetic factors are primarily responsible for the variations in skin color between different races and among people of the same race. The amount and types of melanin produced by the melanocytes, and the size, number, and distribution of the melanosomes, is genetically determined. Skin colors are not determined by the number of melanocytes because all races have essentially the same number. Although many genes are responsible for skin color, a single mutation (see chapter 29) can prevent the manufacture of melanin. **Albinism** (al′bi-nizm) usually is a recessive genetic trait causing an inability to produce tyrosinase. The result is a deficiency or absence of pigment in the skin, hair, and eyes.

Exposure to ultraviolet light darkens melanin already present and stimulates melanin production, resulting in tanning of the skin.

During pregnancy, certain hormones, such as estrogen and melanocyte-stimulating hormone, cause an increase in melanin production in the mother, which in turn causes darkening of the

1. Melanosomes are produced by the Golgi apparatus of the melanocyte.
2. Melanosomes move into melanocyte cell processes.
3. Epithelial cells phagocytize the tips of the melanocyte cell processes.
4. These melanosomes are within epithelial cells.

Epithelial cell
Melanocyte
Melanosomes
Nucleus
Golgi apparatus

Process Figure 5.5   Melanin Transfer from Melanocyte to Keratinocytes

Melanocytes make melanin, which is packaged into melanosomes and transferred to many keratinocytes.

nipples, areolae, and genitalia. The cheekbones, forehead, and chest also may darken, resulting in the "mask of pregnancy," and a dark line of pigmentation may appear on the midline of the abdomen. Diseases like Addison's disease that cause an increased secretion of adrenocorticotropic hormone and melanocyte-stimulating hormone also cause increased pigmentation.

Blood flowing through the skin imparts a reddish hue, and, when blood flow increases (e.g., during blushing, anger, and the inflammatory response), the red color intensifies. A decrease in blood flow such as occurs in shock can make the skin appear pale, and a decrease in the blood oxygen content produces **cyanosis** (sī-ă-nō′sis), a bluish skin color.

Carotene (kar′ō-tēn) is a yellow pigment found in plants such as carrots and corn. Humans normally ingest carotene and use it as a source of vitamin A. Carotene is lipid-soluble, and, when large amounts of carotene are consumed, the excess accumulates in the stratum corneum and in the adipose cells of the dermis and hypodermis, causing the skin to develop a yellowish tint that slowly disappears once carotene intake is reduced.

The location of pigments and other substances in the skin affects the color produced. If a dark pigment is located in the dermis or hypodermis, light reflected off the dark pigment can be scattered by collagen fibers of the dermis to produce a blue color. The same effect produces the blue color of the sky as light is reflected from dust particles in the air. The deeper within the dermis or hypodermis any dark pigment is located, the bluer the pigment appears because of the light-scattering effect of the overlying tissue. This effect causes the blue color of tattoos, bruises, and some superficial blood vessels.

10. *Which cells of the epidermis produce melanin? What happens to the melanin once it is produced?*
11. *How do genetic factors, exposure to light, and hormones determine the amount of melanin in the skin?*
12. *How do melanin, carotene, and blood affect skin color?*

P R E D I C T ②
Explain the differences in skin color between (a) the palms of the hands and the lips, (b) the palms of the hands of a person who does heavy manual labor and one who does not, (c) the anterior and posterior surfaces of the forearm, and (d) the genitals and the soles of the feet.

Accessory Skin Structures
Hair

The presence of **hair** is one of the characteristics common to all mammals; if the hair is dense and covers most of the body surface, it is called fur. In humans, hair is found everywhere on the skin except the palms, soles, lips, nipples, parts of the external genitalia, and the distal segments of the fingers and toes.

By the fifth or sixth month of fetal development, delicate unpigmented hair called **lanugo** (lă-noo′gō) develops and covers the fetus. Near the time of birth, **terminal hairs,** which are long, coarse, and pigmented, replace the lanugo of the scalp, eyelids, and eyebrows. **Vellus** (vel′ŭs) **hairs,** which are short, fine,

and usually unpigmented, replace the lanugo on the rest of the body. At puberty, terminal hair, especially in the pubic and axillary regions, replaces much of the vellus hair. The hair of the chest, legs, and arms is approximately 90% terminal hair in males compared with approximately 35% in females. In males, terminal hairs replace the vellus hairs of the face to form the beard. The beard, pubic, and axillary hair are signs of sexual maturity. In addition, pubic and axillary hair may function as wicks for dispersing odors produced by secretions from specialized glands in the pubic and axillary regions. It also has been suggested that pubic hair provides protection against abrasion during intercourse, and axillary hair reduces friction when the arms move.

Hair Structure

A hair is divided into the **shaft** and **root** (figure 5.6*a*). The shaft protrudes above the surface of the skin, and the root is located below the surface. The base of the root is expanded to form the **hair bulb** (figure 5.6*b*). Most of the root and the shaft of the hair are composed of columns of dead keratinized epithelial cells arranged in three concentric layers: the medulla, the cortex, and the cuticle (figure 5.6*c*). The **medulla** (me-dool′ă) is the central axis of the hair and consists of two or three layers of cells containing soft keratin. The **cortex** forms the bulk of the hair and consists of cells containing hard keratin. The **cuticle** (kū′ti-kl) is a single layer of cells that forms the hair surface. The cuticle cells contain hard keratin, and the edges of the cuticle cells overlap like shingles on a roof.

Hard keratin contains more sulfur than does soft keratin. When hair burns, the sulfur combines with hydrogen to form hydrogen sulfide, which produces the unpleasant odor of rotten eggs. In some animals such as sheep, the cuticle edges of the hair are raised and during textile manufacture catch each other and hold together to form threads.

The **hair follicle** consists of a **dermal root sheath** and an **epithelial root sheath.** The dermal root sheath is the portion of the dermis that surrounds the epithelial root sheath. The epithelial root sheath is divided into an external and an internal part (see figure 5.6*b*). At the opening of the follicle, the external epithelial root sheath has all the strata found in thin skin. Deeper in the hair follicle, the number of cells decreases until at the hair bulb only the stratum basale is present. This has important consequences for the repair of the skin. If the epidermis and the superficial part of the dermis are damaged, the undamaged part of the hair follicle that lies deep in the dermis can be a source of new epithelium. The internal epithelial root sheath has raised edges that mesh closely with the raised edges of the hair cuticle and hold the hair in place. When a hair is pulled out, the internal epithelial root sheath usually comes out as well and is plainly visible as whitish tissue around the root of the hair.

The hair bulb is an expanded knob at the base of the hair root (see figure 5.6*a* and *b*). Inside the hair bulb is a mass of undifferentiated epithelial cells, the **matrix,** which produces the hair and the internal epithelial root sheath. The dermis of the skin projects into the hair bulb as a papilla and contains blood vessels that provide nourishment to the cells of the matrix.

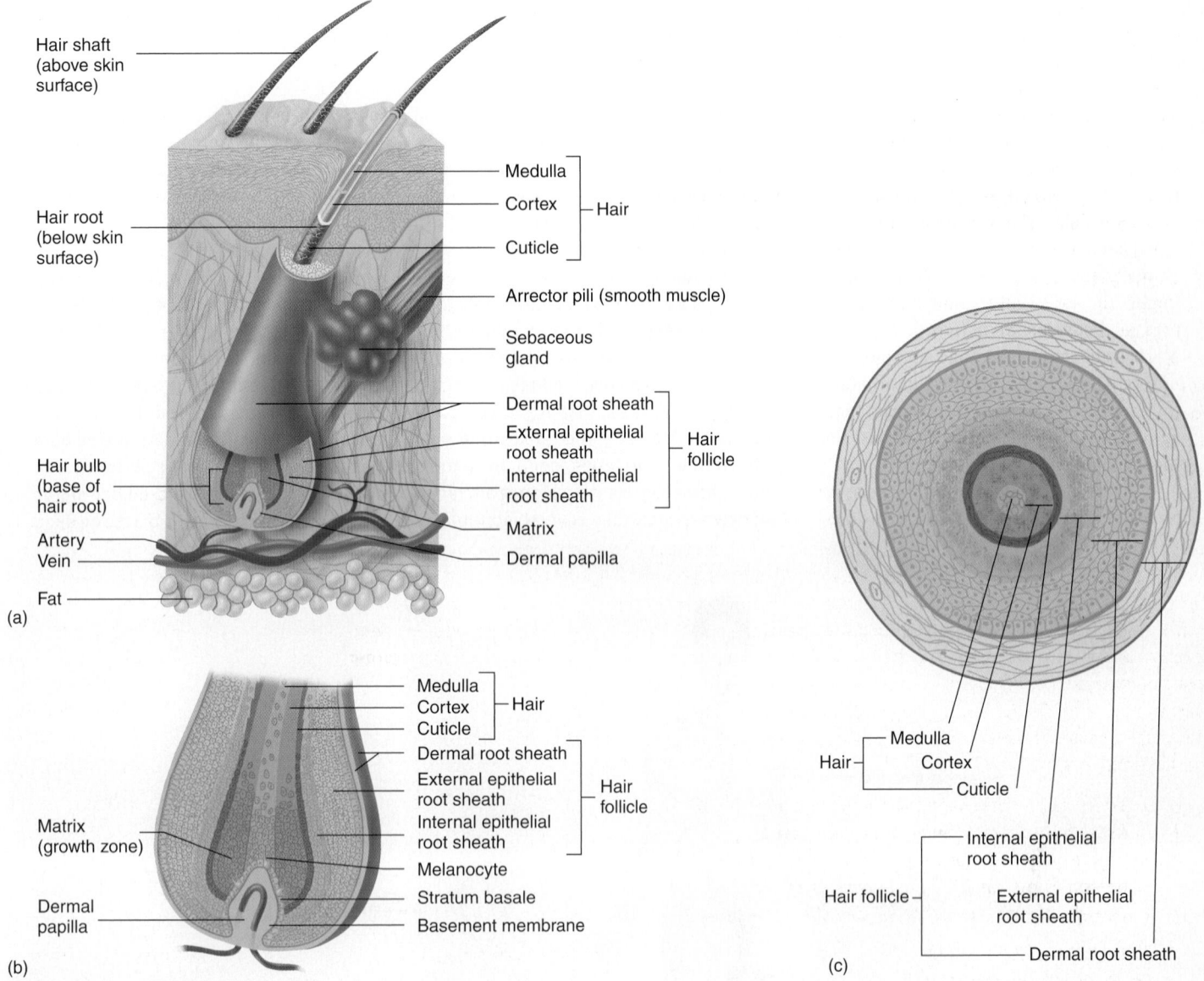

Figure 5.6 Hair Follicle

(*a*) The hair follicle contains the hair and consists of a dermal and epithelial root sheath. (*b*) Enlargement of the hair follicle wall and hair bulb. (*c*) Cross section of a hair within a hair follicle.

Hair Growth

Hair is produced in cycles that involve a **growth stage** and a **resting stage.** During the growth stage, hair is formed by cells of the matrix that differentiate, become keratinized, and die. The hair grows longer as cells are added at the base of the hair root. Eventually hair growth stops; the hair follicle shortens and holds the hair in place. A resting period follows after which a new cycle begins, and a new hair replaces the old hair, which falls out of the hair follicle. Thus loss of hair normally means that the hair is being replaced. The length of each stage depends on the hair—eyelashes grow for approximately 30 days and rest for 105 days, whereas scalp hairs grow for a period of 3 years and rest for 1–2 years. At any given time an estimated 90% of the scalp hairs are in the growing stage, and loss of approximately 100 scalp hairs per day is normal.

The most common kind of permanent hair loss is "pattern baldness." Hair follicles are lost, and the remaining hair follicles revert to producing vellus hair, which is very short, transparent, and for practical purposes invisible. Although more common and more pronounced in certain men, baldness can also occur in women. Genetic factors and the hormone testosterone are involved in causing pattern baldness.

The average rate of hair growth is approximately 0.3 mm per day, although hairs grow at different rates even in the same approximate location. Cutting, shaving, or plucking hair does not alter the growth rate or the character of the hair, but hair can feel coarse and bristly shortly after shaving because the short hairs are less flexible. Maximum hair length is determined by the rate of hair growth and the length of the growing phase. For example, scalp hair can become very long, but eyelashes are short.

Clinical Focus Burns

Burns are classified according to the extent of surface area involved and the depth of the burn. For an adult, the surface area that is burned can be conveniently estimated by "the rule of nines," in which the body is divided into areas that are approximately 9% or multiples of 9% of the total body surface (figure A). For younger patients, surface area relationships are different. For example, in an infant, the head and neck are 21% of total surface area, whereas in an adult they are 9%. For burn victims younger than age 15, a table specifically developed for them should be consulted.

On the basis of depth, burns are classified as either partial-thickness or full-thickness burns (figure B). **Partial-thickness burns** are divided into first- and second-degree burns. **First-degree burns** involve only the epidermis and are red and painful, and slight edema (swelling) may occur. They can be caused by sunburn or brief exposure to hot or cold objects, and they heal in a week or so without scarring.

Second-degree burns damage the epidermis and the dermis. Minimal dermal damage causes redness, pain, edema, and blisters. Healing takes approximately 2 weeks, and no scarring results. If the burn goes deep into the dermis, however, the wound appears red, tan, or white; may take several months to

heal; and might scar. In all second-degree burns the epidermis regenerates from epithelial tissue in hair follicles and sweat glands, as well as from the edges of the wound.

Full-thickness burns are also called **third-degree burns.** The epidermis and dermis are completely destroyed, and deeper tissue may also be involved. Third-degree burns are often surrounded by first- and second-degree burns. Although the areas that have first- and second-degree burns are painful, the region of third-degree burn is usually painless because of destruction of sensory receptors. Third-degree burns appear white, tan, brown, black, or deep

Figure A   The Rule of Nines

(*a*) In an adult, surface areas can be estimated using the rule of nines: each major area of the body is 9%, or a multiple of 9%, of the total body surface area. (*b*) In infants and children the head represents a larger proportion of surface area. The rule of nines is not as accurate for children, as can be seen in this 5-year-old child.

Hair Color

Melanin is produced by melanocytes within the hair bulb matrix and passed to keratinocytes in the hair cortex and medulla. As with the skin, varying amounts and types of melanin cause different shades of hair color. Blonde hair has little black-brown melanin,

whereas jet black hair has the most. Intermediate amounts of melanin account for different shades of brown. Red hair is caused by varying amounts of a red type of melanin. Hair sometimes contains both black-brown and red melanin. With age, the amount of melanin in hair can decrease, causing the color of the hair to fade cherry red in color. Skin can regenerate in a third-degree burn only from the edges, and skin grafts are often necessary.

Deep partial-thickness and full-thickness burns take a long time to heal and form scar tissue with disfiguring and debilitating wound contracture. Skin grafts are performed to prevent these complications and to speed healing. In a split skin graft, the epidermis and part of the dermis are removed from another part of the body

and are placed over the burn. Interstitial fluid from the burned area nourishes the graft until it becomes vascularized. Meanwhile, the donor tissue produces new epidermis from epithelial tissue in the hair follicles and sweat glands such as occurs in superficial second-degree burns.

When it is not possible or practical to move skin from one part of the body to a burn site, artificial skin or grafts from human cadavers or from pigs are used. These

techniques are often unsatisfactory because the body's immune system recognizes the graft as a foreign substance and rejects it. A solution to this problem is laboratory-grown skin. A piece of healthy skin from the burn victim is removed and placed into a flask with nutrients and hormones that stimulate rapid growth. The skin that is produced consists only of epidermis and does not contain glands or hair.

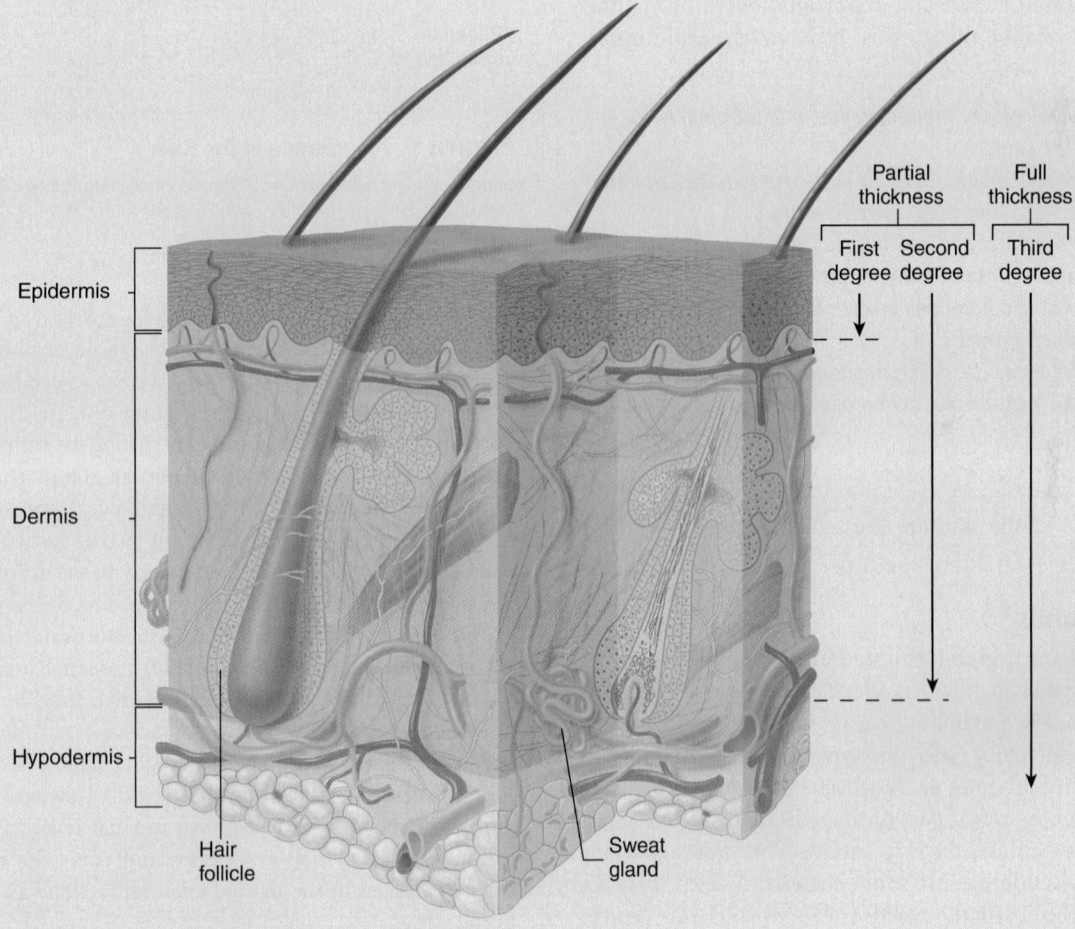

Figure B Burns

Parts of the skin damaged by burns of different degrees.

or become white (i.e., no melanin). Gray hair is usually a mixture of unfaded, faded, and white hairs. Hair color is controlled by several genes, and dark hair color is not necessarily dominant over light.

P R E D I C T

Marie Antoinette's hair supposedly turned white overnight after she heard she would be sent to the guillotine. Explain why you believe or disbelieve this story.

Muscles

Associated with each hair follicle are smooth muscle cells, the **arrector pili** (ă-rek′tōr pī′lī), that extend from the dermal root sheath of the hair follicle to the papillary layer of the dermis (see figure 5.6*a*). Normally, the hair follicle and the hair inside it are at an oblique angle to the surface of the skin. When the arrector pili muscles contract, however, they pull the follicle into a position more perpendicular to the surface of the skin, causing the hair to "stand on end." Movement of the hair follicles produces raised areas called "gooseflesh," or "goose bumps."

Contraction of the arrector pili muscles occurs in response to cold or to frightening situations, and in animals with fur the response increases the thickness of the fur. When the response results from cold temperatures, it is beneficial because the fur traps more air and thus becomes a better insulator. In a frightening situation the animal appears larger and more ferocious, which might deter an attacker. It is unlikely that humans, with their sparse amount of hair, derive any important benefit from either response and probably retain this trait as an evolutionary holdover.

> 13. When and where are lanugo, vellus, and terminal hairs found in the skin?
> 14. Define the root, shaft, and hair bulb of a hair. Describe the three parts of a hair seen in cross section.
> 15. Describe the parts of a hair follicle. How is the epithelial root sheath important in the repair of the skin?
> 16. In what part of a hair does growth take place? What are the stages of hair growth?
> 17. What determines the different shades of hair color?
> 18. Explain the location and action of arrector pili muscles.

Glands

The major glands of the skin are the sebaceous glands and the sweat glands (figure 5.7).

Sebaceous Glands

Sebaceous (sē-bā′shŭs) **glands,** located in the dermis, are simple or compound alveolar glands that produce **sebum** (sē′bŭm), an oily, white substance rich in lipids. Because sebum is released by the lysis and death of secretory cells, sebaceous glands are classified as holocrine glands (see chapter 4). Most sebaceous glands are connected by a duct to the upper part of the hair follicles from which the sebum oils the hair and the skin surface. This prevents drying and provides protection against some bacteria. A few sebaceous glands located in the lips, in the eyelids (meibomian glands), and in the genitalia are not associated with hairs but open directly onto the skin surface.

Sweat Glands

There are two types of **sweat,** or **sudoriferous** (soo-dō-rif′er-ŭs), **glands,** which are called merocrine and apocrine glands. At one time it was believed that secretions were released in a merocrine fashion from merocrine glands and in an apocrine fashion from apocrine glands (see chapter 4). Accordingly, they were called merocrine and apocrine sweat glands. It is now known

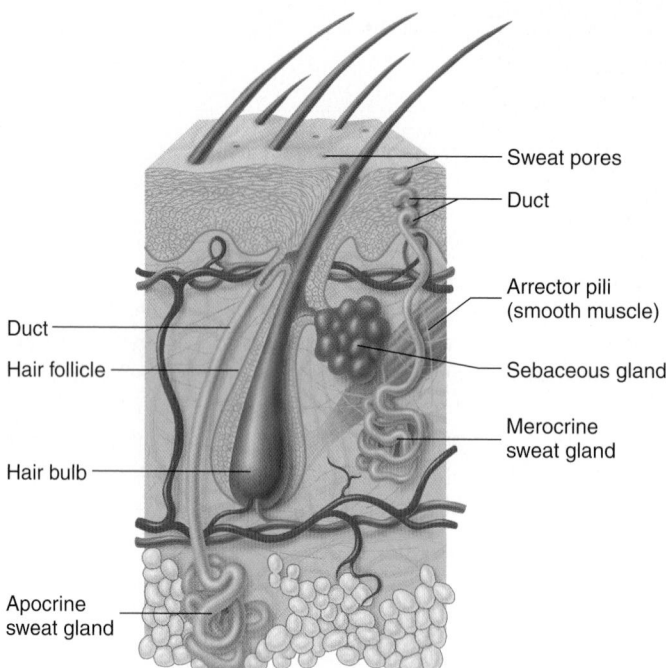

Figure 5.7   Glands of the Skin

Sebaceous and apocrine sweat glands empty into the hair follicle. Merocrine sweat glands empty onto the surface of the skin.

that apocrine sweat glands also release some of their secretions in a merocrine fashion, and possibly some in a holocrine fashion. Traditionally, they are still referred to as apocrine sweat glands.

Merocrine (mer′ō-krin, mer′ō-krīn, mer′ō-krēn), or **eccrine** (ek′rin), **sweat glands** are the most common type of sweat gland. They are simple coiled tubular glands that open directly onto the surface of the skin through sweat pores (see figure 5.7). Merocrine sweat glands can be divided into two parts: the deep coiled portion, which is located mostly in the dermis, and the duct, which passes to the surface of the skin. The coiled part of the gland produces an isotonic fluid that is mostly water but also contains some salts (mainly sodium chloride) and small amounts of ammonia, urea, uric acid, and lactic acid. As this fluid moves through the duct, sodium chloride moves by active transport from the duct back into the body, thereby conserving salts. The resulting hyposmotic fluid that leaves the duct is called **sweat.** When the body temperature starts to rise above normal levels, the sweat glands produce sweat, which evaporates and cools the body. Sweat also can be released in the palms, soles, and axillae as a result of emotional stress.

Detecting Lies

Emotional sweating is used in lie detector (polygraph) tests because sweat gland activity can increase when a person tells a lie. The sweat produced, even in small amounts, can be detected because the salt solution conducts electricity and lowers the electric resistance of the skin.

Merocrine sweat glands are most numerous in the palms of the hands and the soles of the feet but are absent from the margin of the lips, the labia minora, and the tips of the penis and clitoris.

Only a few mammals such as humans and horses have merocrine sweat glands in hairy skin. Dogs, on the other hand, keep cool by water lost through panting instead of sweating.

Apocrine (ap′ō-krin) **sweat glands** are compound coiled tubular glands that usually open into hair follicles superficial to the opening of the sebaceous glands (see figure 5.7). In other mammals, these glands are widely distributed throughout the skin and help to regulate temperature. In humans, apocrine sweat glands are found in the axillae and genitalia (scrotum and labia majora) and around the anus and do not help to regulate temperature. In humans, apocrine sweat glands become active at puberty as a result of the influence of sex hormones. Their secretions contain organic substances, such as 3-methyl-2-hexenoic acid, that are essentially odorless when first released but that are quickly metabolized by bacteria to cause what commonly is known as body odor. Many mammals use scent as a means of communication, and it has been suggested that the activity of apocrine sweat glands may be a sign of sexual maturity.

Other Glands

Other skin glands include the ceruminous glands and the mammary glands. The **ceruminous** (sĕ-roo′mi-nŭs) **glands** are modified merocrine sweat glands located in the ear canal (external auditory meatus). **Cerumen,** or earwax, is the combined secretions of ceruminous glands and sebaceous glands. Cerumen and hairs in the ear canal protect the eardrum by preventing the entry of dirt and small insects. An accumulation of cerumen, however, can block the ear canal and make hearing more difficult.

The **mammary glands** are modified apocrine sweat glands located in the breasts. They function to produce milk. The structure and regulation of mammary glands is discussed in chapter 29.

19. *What secretion is produced by the sebaceous glands? What is the function of the secretion?*
20. *Which glands of the skin are responsible for cooling the body? Which glands are involved with the production of body odor?*
21. *Where are ceruminous and mammary glands located, and what secretions do they produce?*

Nails

The distal ends of primate digits have nails, whereas most other mammals have claws or hooves. Nails protect the ends of the digits, aid in manipulation and grasping of small objects, and are used for scratching.

A **nail** consists of the proximal **nail root** and the distal **nail body** (figure 5.8a). The nail root is covered by skin, and the nail body is the visible portion of the nail. The lateral and proximal edges of the nail are covered by skin called the **nail fold,** and the edges are held in place by the **nail groove** (figure 5.8b). The stratum corneum of the nail fold grows onto the nail body as the **eponychium** (ep-ō-nik′ē-ŭm), or cuticle. Beneath the free edge of the nail body is the **hyponychium** (hī′pō-nik′ē-ŭm), a thickened region of the stratum corneum (figure 5.8c).

The nail root extends distally from the **nail matrix.** The nail also attaches to the **nail bed,** which is located between the nail matrix and the hyponychium. The nail matrix and bed are epithelial tissue with a stratum basale that gives rise to the cells that form the nail.

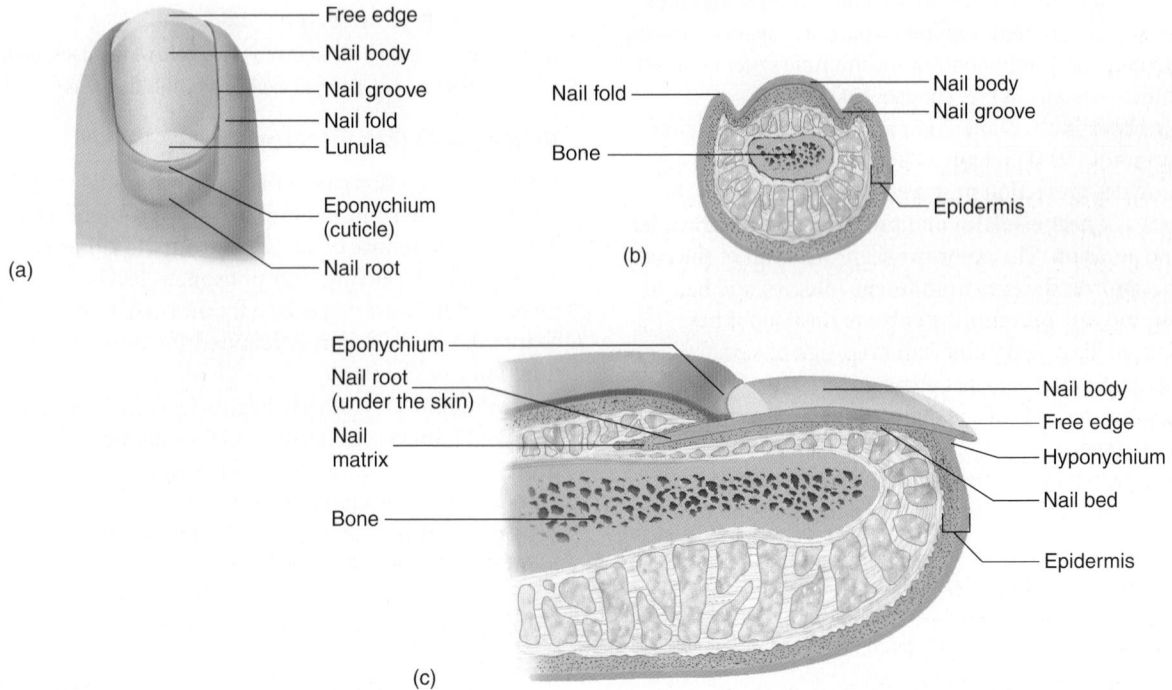

(a)

Free edge
Nail body
Nail groove
Nail fold
Lunula
Eponychium (cuticle)
Nail root

(b)

Nail fold
Bone
Nail body
Nail groove
Epidermis

(c)

Eponychium
Nail root (under the skin)
Nail matrix
Bone
Nail body
Free edge
Hyponychium
Nail bed
Epidermis

Figure 5.8 **Nail**
(*a*) Dorsal view. (*b*) Cross section. (*c*) Longitudinal section.

The nail matrix is thicker than the nail bed and produces most of the nail. The nail bed is visible through the clear nail and appears pink because of blood vessels in the dermis. A small part of the nail matrix, the **lunula** (loo′noo-lă), is seen through the nail body as a whitish, crescent-shaped area at the base of the nail. The lunula, seen best on the thumb, appears white because the blood vessels cannot be seen through the thicker nail matrix.

The nail is stratum corneum. It contains a hard keratin which makes the nail hard. The nail cells produced in the nail matrix and nail bed are pushed distally over the nail bed. Nails grow at an average rate of 0.5–1.2 mm per day, and fingernails grow more rapidly than toenails. Unlike hair, they grow continuously throughout life and do not have a resting phase.

> 22. *Name the parts of a nail. Which part produces most of the nail? What is the lunula?*
> 23. *What makes a nail hard? Do nails have growth stages?*

Summary of Integumentary System Functions

Protection

The integumentary system is the body's fortress, defending it from harm. It performs many protective functions.

1. The stratified squamous epithelium of the skin protects underlying structures against abrasion. As the outer cells of the stratum corneum are desquamated, they are replaced by cells from the stratum basale. Calluses develop in areas subject to heavy friction or pressure.
2. The skin prevents the entry of microorganisms and other foreign substances into the body. Secretions from skin glands produce an environment unsuitable for some microorganisms. The skin contains components of the immune system that act against microorganisms (see chapter 22).
3. Melanin absorbs ultraviolet light and protects underlying structures from its damaging effects.
4. Hair provides protection in several ways. The hair on the head acts as a heat insulator and protects against ultraviolet light and abrasion. The eyebrows keep sweat out of the eyes, eyelashes protect the eyes from foreign objects, and hair in the nose and ears prevents the entry of dust and other materials. Axillary and pubic hair are a sign of sexual maturity and protect against abrasion.
5. Nails protect the ends of the digits from damage and can be used in defense.
6. The intact skin plays an important role in preventing water loss because its lipids act as a barrier to the diffusion of water.

Administering Medications Through the Skin

Some lipid-soluble substances readily pass through the epidermis. Lipid-soluble medications can be administered by applying them to the skin, after which the medication slowly diffuses through the skin into the blood. For example, nicotine patches are used to help reduce withdrawal symptoms in those attempting to quit smoking.

Sensation

The body feels pain, heat, and cold because the integumentary system has sensory receptors in all its layers. For example, the epidermis and dermal papillae are well supplied with touch receptors. The dermis and deeper tissues contain pain, heat, cold, touch, and pressure receptors. Hair follicles (but not the hair) are well innervated, and movement of hair can be detected by sensory receptors surrounding the base of hair follicles. Sensory receptors are discussed in more detail in chapter 14.

Temperature Regulation

Body temperature tends to increase as a result of exercise, fever, or an increase in environmental temperature. Homeostasis is maintained by the loss of excess heat. The blood vessels (arterioles) in the dermis dilate and allow more blood to flow through the skin, thus transferring heat from deeper tissues to the skin (figure 5.9, steps 1 and 2). To counteract environmental heat gain or to get rid of excess heat produced by the body, sweat is produced. The sweat spreads over the surface of the skin, and as it evaporates, heat is lost from the body.

If body temperature begins to drop below normal, heat can be conserved by a decrease in the diameter of dermal blood vessels, thus reducing blood flow to the skin (figure 5.9, steps 3 and 4). With less warm blood flowing through the skin, however, the skin temperature decreases. If the skin temperature drops below approximately 15°C (59°F), blood vessels dilate, which helps to prevent tissue damage from the cold.

Contraction of the arrector pili muscles causes hair to stand on end, but with the sparse amount of hair covering the body, this does not significantly reduce heat loss in humans. Hair on the head, however, is an effective insulator. General temperature regulation is considered in chapter 25.

P R E D I C T

You may have noticed that on very cold winter days, people's ears and noses turn red. Can you explain why this happens?

Vitamin D Production

Vitamin D functions as a hormone to stimulate uptake of calcium and phosphate from the intestines, to promote their release from bones, and to reduce calcium loss from the kidneys, resulting in increased blood calcium and phosphate levels. Adequate levels of these minerals are necessary for normal bone metabolism (see chapter 6), and calcium is required for normal nerve and muscle function (see chapter 9).

Vitamin D synthesis begins in skin exposed to ultraviolet light, and humans can produce all the vitamin D they require by this process if enough ultraviolet light is available. Because humans live indoors and wear clothing, however, their exposure to ultraviolet light may not be adequate for the manufacture of sufficient vitamin D. This is especially likely for people living in cold climates because they remain indoors or are covered by warm clothing when outdoors. Fortunately, vitamin D can also be ingested and absorbed in the intestine. Natural sources of vitamin D are liver (especially fish liver), egg yolks, and dairy products (e.g., butter, cheese, and milk). In addition, the diet can be supplemented with vitamin D in fortified milk or vitamin pills.

1. Blood vessel dilation results in increased blood flow toward the surface of the skin.

2. Increased blood flow beneath the epidermis results in increased heat loss (*gold arrows*).

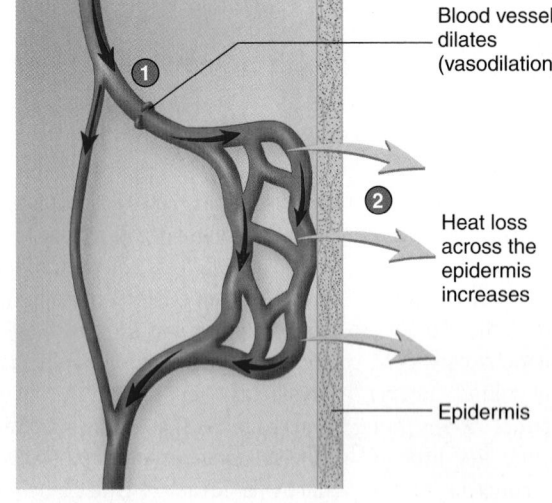

Blood vessel dilates (vasodilation)

Heat loss across the epidermis increases

Epidermis

(a)

3. Blood vessel constriction results in decreased blood flow toward the surface of the skin.

Blood vessel constricts (vasoconstriction)

Heat loss across the epidermis decreases

4. Decreased blood flow beneath the epidermis results in decreased heat loss.

Epidermis

(b)

Process Figure 5.9 **Heat Exchange in the Skin**

Vitamin D synthesis begins when the precursor molecule, 7-dehydrocholesterol (7-dē-hī′drō-kō-les′ter-ol), is exposed to ultraviolet light and is converted into cholecalciferol (kō′lē-kal-sif′er-ol). Cholecalciferol is released into the blood and modified by hydroxylation (hydroxide ions are added) in the liver and kidneys to form active vitamin D (calcitriol; kal-si-trī′ol).

Excretion

Excretion is the removal of waste products from the body. In addition to water and salts, sweat contains a small amount of waste products, such as urea, uric acid, and ammonia. Compared to the kidneys, however, the quantity of waste products eliminated in the sweat is insignificant, even when large amounts of sweat are lost.

- 24. *In what ways does the skin provide protection?*
- 25. *What kind of sensory receptors are found in the skin, and why are they important?*
- 26. *How does the skin assist in the regulation of body temperature?*

27. *Name the locations where cholecalciferol is produced and then modified into vitamin D. What are the functions of vitamin D?*
28. *What substances are excreted in sweat? Is the skin an important site of excretion?*

Effects of Aging on the Integumentary System

As the body ages, the skin is more easily damaged because the epidermis thins and the amount of collagen in the dermis decreases. Skin infections are more likely, and repair of the skin occurs more slowly. A decrease in the number of elastic fibers in the dermis and loss of fat from the hypodermis cause the skin to sag and wrinkle.

The skin becomes drier with age as sebaceous gland activity decreases. A decrease in the activity of sweat glands and a decrease in the blood supply to the dermis result in a poor ability to regulate body temperature. Death from heat prostration can occur in elderly individuals who do not take proper precautions.

Clinical Focus   Clinical Disorders of the Integumentary System

The Integumentary System as a Diagnostic Aid

The integumentary system can be used in diagnosis because it is easily observed and often reflects events occurring in other parts of the body. For example, **cyanosis** (sī-ă-nō'sis), a bluish color to the skin that results from decreased blood oxygen content, is an indication of impaired circulatory or respiratory function. When red blood cells wear out, they are broken down, and part of their contents is excreted by the liver as bile pigments into the intestine. **Jaundice** (jawn'dis), a yellowish skin color, occurs when excess bile pigments accumulate in the blood. If a disease like viral hepatitis damages the liver, bile pigments are not excreted and accumulate in the blood.

Rashes and lesions in the skin can be symptomatic of problems elsewhere in the body. For example, scarlet fever results from a bacterial infection in the throat. The bacteria release a toxin into the blood that causes the pink-red rash for which this disease was named. In allergic reactions (see chapter 22), a release of histamine into the tissues produces swelling and reddening. The development of a rash (hives) in the skin can indicate an allergy to foods or drugs such as penicillin.

The condition of the skin, hair, and nails is affected by nutritional status. In vitamin A deficiency the skin produces excess keratin and assumes a characteristic sandpaper texture, whereas in iron-deficiency anemia the nails lose their normal contour and become flat or concave (spoon-shaped).

Hair concentrates many substances that can be detected by laboratory analysis, and comparison of a patient's hair to "normal" hair can be useful in certain diagnoses. For example, lead poisoning results in high levels of lead in the hair. The use of hair analysis as a screening test to determine the general health or nutritional status of an individual is unreliable, however.

Bacterial Infections

Staphylococcus aureus is commonly found in pimples, boils, and carbuncles and causes **impetigo** (im-pe-tī'gō), a disease of the skin that usually affects children. It is characterized by small blisters containing pus that easily rupture and form a thick, yellowish crust. *Streptococcus pyogenes* causes **erysipelas** (er-i-sip'ĕ-las), swollen red patches in the skin. Burns are often infected by *Pseudomonas aeruginosa,* which produces a characteristic blue-green pus caused by bacterial pigment.

Acne is a disorder of the hair follicles and sebaceous glands that affects almost everyone at some time or another. Although the exact cause of acne is unknown, four factors are believed to be involved: hormones, sebum, abnormal keratinization within hair follicles, and the bacterium *Propionibacterium acnes.* Acne lesions apparently begin with a hyperproliferation of the hair follicle epidermis, and many cells are desquamated. These cells are abnormally sticky and adhere to one another to form a mass of cells mixed with sebum that blocks the hair follicle. During puberty, hormones, especially testosterone, stimulate the sebaceous glands to increase sebum production. Because both the adrenal gland and the testes produce testosterone, the effect is seen in both males and females. An accumulation of sebum behind the blockage produces a whitehead, which may continue to develop into a blackhead or a pimple. A blackhead results if the opening of the hair follicle is pushed open by the accumulating cornified cells and sebum. Although it is generally agreed that dirt is not responsible for the black color of blackheads, the exact cause of the black color is disputed. Once the wall of the follicle ruptures, *P. acnes* and other microorganisms stimulate an inflammatory response that results in the formation of a red pimple filled with pus. If tissue damage is extensive, scarring occurs.

Viral Infections

Some of the well-known viral infections of the skin include **chicken pox** (varicella-zoster), **measles, German measles** (rubella), and **cold sores** (herpes simplex). **Warts,** which are caused by a viral infection of the epidermis, are generally harmless and usually disappear without treatment.

Fungal Infections

Ringworm is a fungal infection that affects the keratinized portion of the skin, hair, and nails and produces patchy scaling and an inflammatory response. The lesions are often circular with a raised edge, and in ancient times they were thought to be caused by worms. Several species of fungus cause ringworm in humans and are usually described by their location on the body; in the scalp the condition is ringworm, in the groin it is jock itch, and in the feet it is athlete's foot.

Decubitus Ulcers

Decubitus (dĕ-kū'bi-tŭs) **ulcers,** also known as bedsores or pressure sores, develop in patients who are immobile (e.g., bedridden or confined to a wheelchair). The weight of the body, especially in areas over bony projections such as the hipbones and heels, compresses tissues and causes **ischemia** (is-kē'mē-ă), or reduced circulation. The consequence is destruction, or **necrosis** (nĕ-krō'sis), of the hypodermis and deeper tissues, which is followed by necrosis of the skin. Once skin necrosis occurs, microorganisms gain entry to produce an infected ulcer.

Bullae

Bullae (bul'ē), or blisters, are fluid-filled areas in the skin that develop when tissues are damaged, and the resultant inflammatory response produces edema. Infections or physical injuries can cause bullae or lesions in different layers of the skin.

Psoriasis

Psoriasis (sō-rī'ă-sis) is characterized by a thicker-than-normal stratum corneum that sloughs to produce large, silvery scales. If the scales are scraped away, bleeding occurs from the blood vessels at the top of the dermal papillae. These changes result from increased cell division in the stratum basale, abnormal keratin production, and elongation of the dermal papillae toward the skin surface. Evidence suggests that the disease has a genetic component and that the immune system stimulates the increased cell divisions. Psoriasis is a chronic disease that can be controlled with drugs and phototherapy (ultraviolet light) but as yet has no cure.

Eczema and Dermatitis

Eczema (ek'zĕ-mă, eg'zĕ-mă, eg-zē'mă) and **dermatitis** (der-mă-tī'tis) are general terms used for inflammatory conditions of the skin. Cause of the inflammation can be allergy, infection, poor circulation, or exposure to physical factors, such as chemicals, heat, cold, or sunlight.

Birthmarks

Birthmarks are congenital (present at birth) disorders of the capillaries in the dermis of the skin. Usually they are only of concern for cosmetic reasons. A **strawberry birthmark** is a mass of soft, elevated tissue that appears bright red to deep purple in color. In 70% of patients, strawberry birthmarks disappear spontaneously by age 7. **Portwine stains** appear as flat, dull red or bluish patches that persist throughout life.

Vitiligo

Vitiligo (vit-i-lī'gō) is the development of patches of white skin because the melanocytes in the affected area are destroyed, apparently by an autoimmune response (see chapter 22).

Moles

A **mole** is an elevation of the skin that is variable in size and is often pigmented and hairy. Histologically, a mole is an aggregation, or "nest," of melanocytes in the epidermis or dermis. They are a normal occurrence, and most people have 10–20 moles, which appear in childhood and enlarge until puberty.

Cancer

Skin cancer is the most common type of cancer (figure C). Although chemicals and radiation (x-rays) are known to induce cancer, the development of skin cancer is most often associated with exposure to ultraviolet (UV) radiation from the sun, and, consequently, most skin cancers develop on the face or neck. The group of people most likely to have skin cancer are fair-skinned (i.e., they have less protection from the sun) or are older than 50 (i.e., they have had long exposure to the sun).

Basal cell carcinoma (kar-si-nō'mă), the most frequent skin cancer, begins in the stratum basale and extends into the dermis to produce an open ulcer. Surgical removal or radiation therapy cures this type of cancer, and fortunately little danger exists that

the cancer will spread, or metastasize (mĕ-tas'tă-sīz), to other areas of the body if treated in time. **Squamous cell carcinoma** develops from stratum spinosum keratinocytes that continue to divide as they produce keratin. Typically, the result is a nodular, keratinized tumor confined to the epidermis, but it can invade the dermis, metastasize, and cause death. **Malignant melanoma** (mel'ă-nō'mă) is a less common form of skin cancer that arises from melanocytes, usually in a preexisting mole. The melanoma can appear as a large, flat, spreading lesion or as a deeply pigmented nodule. Metastasis is common, and, unless diagnosed and treated early in development, this cancer is often fatal. Other types of skin cancer are possible (e.g., metastasis from other parts of the body to the skin).

Limiting exposure to the sun and using sunscreens can reduce the likelihood of developing skin cancer. Some concern over the use of sunscreens, however, has recently arisen because of the different types of UV radiation they can block. Exposure to UVB can cause sunburn and is associated with the development of basal cell and squamous cell carcinomas. The development of malignant melanoma is associated with exposure to UVA. Sunscreens that block primarily UVB allow longer exposure to the sun without sunburning but thereby increase exposure to UVA and the possible development of malignant melanoma. Sunscreens that effectively block UVB and UVA are advisable.

(a) Basal cell carcinoma

(b) Squamous cell carcinoma

(c) Malignant melanoma

Figure C   Cancer of the Skin

Systems Pathology
Burns

Mr. S is a 23-year-old man who had difficulty falling asleep at night. He often stayed up late watching television or reading until he fell asleep. Mr. S was also a chain smoker. One night he took several sleeping pills. Unfortunately, he fell asleep before putting out his cigarette, which started a fire. As a result, Mr. S was severely burned and received full-thickness and partial-thickness burns (figure D*a*). He was rushed to the emergency room and was eventually transferred to a burn unit.

For the first day after his accident, his condition was critical because he went into shock. Administration of large volumes of intravenous fluid stabilized his condition. As part of his treatment, Mr. S was also given a high-protein, high-calorie diet.

A week later, dead tissue was removed from the most serious burns (figure D*b*), and a skin graft was performed. Despite the use of topical antimicrobial drugs and sterile bandages, some of the burns became infected. An additional complication was the development of a venous thrombosis in his leg.

Although the burns were painful and the treatment was prolonged, Mr. S made a full recovery. He no longer smokes.

Background Information

When large areas of skin are severely burned, systemic effects are produced that can be life-threatening. One effect is on capillaries, which are the small blood vessels in which fluid, gases, nutrients, and waste products are normally exchanged between the blood and tissues. Within minutes of a major burn injury, capillaries become more permeable at the burn site and throughout the body. As a result, fluid and electrolytes (see chapter 2) are lost from the burn wound and into tissue spaces. The loss of fluid decreases blood volume, which decreases the ability of the heart to pump blood. The resulting decrease in blood delivery to tissues can cause tissue damage, shock, and even death. Treatment consists of administering intravenous fluid at a faster rate than it leaks out of the capillaries. Although this can reverse the shock and prevent death, fluid continues to leak into tissue spaces causing pronounced **edema,** a swelling of the tissues.

Typically, after 24 hours, capillary permeability returns to normal, and the amount of intravenous fluid administered can be greatly decreased. How burns result in capillary permeability changes is not well understood. It is clear that following a burn, immunologic and metabolic changes occur that affect not only

Full-thickness burn

Partial-thickness burn

(a)

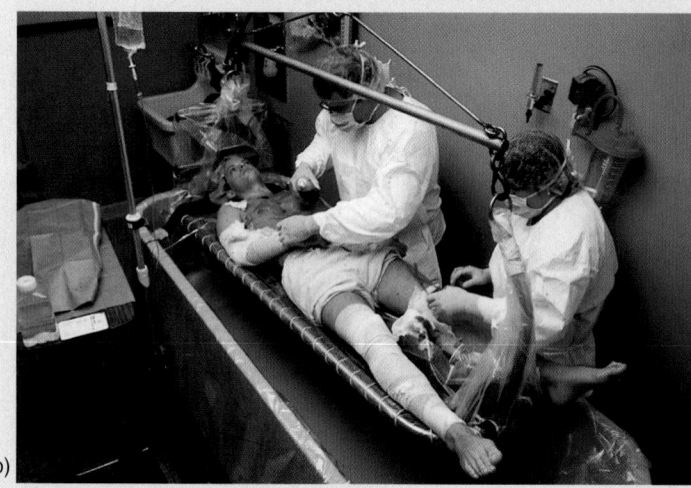

(b)

Figure D   **Burn Victim**
(*a*) Partial and full-thickness burns. (*b*) Patient in a burn unit.

capillaries but the rest of the body as well. For example, mediators of inflammation (see chapter 4), which are released in response to the tissue damage, contribute to changes in capillary permeability throughout the body.

Substances released from the burn may also play a role in causing cells to function abnormally. Burn injuries result in an almost immediate hypermetabolic state that persists until wound closure. Also contributing to the increased metabolism is a resetting of the temperature control center in the brain to a higher temperature and an

System Interactions   Effect of Burns on Other Systems

System	Interactions
Skeletal	Red bone marrow replaces red blood cells destroyed in the burned skin.
Muscular	Loss of muscle mass resulting from the hypermetabolic state caused by the burn.
Nervous	Pain is sensed in the partial-thickness burns. The temperature-regulatory center in the brain is set to a higher temperature, which contributes to increased body temperature. Abnormal K^+ concentrations disturb normal nervous system activity: elevated levels are caused by release of K^+ from damaged tissues; low levels can be caused by rapid loss of K^+ in fluid from the burn.
Endocrine	Increased secretion of epinephrine and norepinephrine from the adrenal gland in response to the injury contributes to increased body temperature by increasing cell metabolism.
Cardiovascular	Increased capillary permeability causes decreased blood volume, resulting in decreased blood delivery to tissues, edema, and shock. The pumping effectiveness of the heart is impaired by electrolyte imbalance and substances released from the burn. Increased blood clotting causes venous thrombosis. Preferential delivery of blood to the injury promotes healing.
Lymphatic and immune	Inflammation increases in response to tissue damage. Later, depression of the immune system can result in infection.
Respiratory	Airway obstruction caused by edema. Increased respiration rate caused by increased metabolism and lactic acid buildup.
Digestive	Decreased blood delivery as a result of the burn causes degeneration of the intestinal lining and liver. Bacteria from the intestine can cause systemic infections. The liver releases blood-clotting factors in response to the injury. Increased nutrients necessary to support increased metabolism and for repair of the integumentary system are absorbed.
Urinary	The kidneys compensate for the increased fluid loss caused by the burn by greatly reducing or even stopping urine production. Decreased blood volume causes decreased blood flow to the kidneys, which reduces urine output but can cause kidney tissue damage. Hemoglobin, released from red blood cells damaged in the burnt skin, can decrease urine production by blocking fluid movement from the blood into the urine.

increase in the hormones released by the endocrine system. For example, epinephrine and norepinephrine from the adrenal glands increase cell metabolism. Compared with a normal body temperature of approximately 37°C (98.6°F), a body temperature of 38.5°C (101.3°F) is typical in burn patients, despite the higher loss of water by evaporation from the burn.

In severe burns, the increased metabolic rate can result in weight loss as great as 30%–40% of the patient's preburn weight. To help compensate, caloric intake may double or even triple. In addition, the need for protein, which is necessary for tissue repair, is greater.

The skin normally maintains homeostasis by preventing the entry of microorganisms. Because burns damage and even completely destroy the skin, microorganisms can cause infections. For this reason, burn patients are maintained in an aseptic environment, which attempts to prevent the entry of microorganisms into the wound. They are also given antimicrobial drugs, which kill microorganisms or suppress their growth. **Debridement,** (dā-brēd-mon′), the removal of dead tissue from the burn, helps to prevent infections by cleaning the wound and removing tissue in which infections could develop. Skin grafts, performed within a week of the injury, also prevent infections by closing the wound and preventing the entry of microorganisms.

Despite these efforts, however, infections still are the major cause of death of burn victims. Depression of the immune system during the first or second week after the injury contributes to the high infection rate. The thermally altered tissue is recognized as a foreign substance that stimulates the immune system. As a result, the immune system is overwhelmed as immune system cells become less effective and production of the chemicals that normally provide resistance to infections decreases (see chapter 22). The greater the magnitude of the burn, the greater the depression of the immune system, and the greater the risk of infection.

Venous thrombosis, the development of a clot in a vein, is also a complication of burns. Blood normally forms a clot when exposed to damaged tissue, such as at a burn site, but clotting can also occur elsewhere, such as in veins. Clots can block blood flow, resulting in tissue destruction. The concentration of chemicals in the blood that cause clotting increases for two reasons: increased release of clotting factors from the liver and loss of fluid from the burn.

P R E D I C T

When Mr. S is first admitted to the burn unit, the nurses carefully monitor his urine output. Why does that make sense in light of his injuries?

The number of functioning melanocytes generally decreases, but in some localized areas, especially on the hands and the face, melanocytes increase in number to produce age spots. (Age spots are different from freckles, which are caused by an increase in melanin production and not an increase in melanocyte numbers.) White or gray hairs also occur because of a decrease in or lack of melanin production.

Skin that is exposed to sunlight appears to age more rapidly than nonexposed skin. This effect is observed on areas of the body, such as the face and hands, that receive sun exposure (figure 5.10). The effects of chronic sun exposure on the skin, however, are different from the effects of normal aging. In skin exposed to sunlight, normal elastic fibers are replaced by an interwoven mat of thick, elasticlike material, the number of collagen fibers decreases, and the ability of keratinocytes to divide is impaired.

29. Compared to young skin, why is aged skin more likely to be damaged, wrinkled, and dry?

30. Why is heat potentially dangerous to the elderly?

31. Explain age spots and white hair.

32. What effect does exposure to sunlight have on skin?

Treatment of Skin Wrinkles

Retin-A (tretinoin; tret′i-nō-in) is a vitamin A derivative that is being used to treat skin wrinkles. It appears to be effective in treating fine wrinkles on the face, such as those caused by long-term exposure to the sun, but is not effective in treating deep lines. One ironic side effect of Retin-A use is increased sensitivity to the sun's ultraviolet rays. Doctors prescribing this cream caution their patients to always use a sunblock when they are going to be outdoors.

(a) A 91-year-old Japanese monk who has spent most of his life indoors.

(b) A 62-year-old Native American woman who has spent most of her life outdoors.

Figure 5.10   Effects of Sunlight on Skin

S U M M A R Y

The integumentary system consists of the skin, hair, nails, and a variety of glands.

Overview of the Integumentary System (p. 144)

The integumentary system separates and protects us from the external environment. Other functions include sensation, temperature regulation, vitamin D production, and excretion of small amounts of waste products.

Hypodermis (p. 144)

1. Located beneath the dermis, the hypodermis is loose connective tissue that contains collagen and elastin fibers.
2. The hypodermis attaches the skin to underlying structures and is a site of fat storage.

Skin (p. 145)
Dermis

1. The dermis is connective tissue divided into two layers.
2. The reticular layer is the main layer. It is dense irregular connective tissue consisting mostly of collagen.
3. The papillary layer has projections called papillae and is loose connective tissue that is well supplied with capillaries.

Epidermis

1. The epidermis is stratified squamous epithelium divided into five strata.
2. The stratum basale consists of keratinocytes, which produce the cells of the more superficial strata.

3. The stratum spinosum consists of several layers of cells held together by many desmosomes.
4. The stratum granulosum consists of cells filled with granules of keratohyalin. Cell death occurs in this stratum.
5. The stratum lucidum consists of a layer of dead transparent cells.
6. The stratum corneum consists of many layers of dead squamous cells. The most superficial cells are desquamated.
7. Keratinization is the transformation of the living cells of the stratum basale into the dead squamous cells of the stratum corneum.
   • Keratinized cells are filled with keratin and have a protein envelope, both of which contribute to structural strength. The cells are also held together by many desmosomes.
   • Intercellular spaces are filled with lipids from the lamellae that contribute to the impermeability of the epidermis to water.
8. Soft keratin is found in skin and the inside of hairs, whereas hard keratin occurs in nails and the outside of hairs. Hard keratin makes cells more durable, and these cells do not desquamate.

Thick and Thin Skin

1. Thick skin has all five epithelial strata. The dermis under thick skin produces fingerprints and footprints.
2. Thin skin contains fewer cell layers per stratum, and the stratum lucidum is usually absent. Hair is found only in thin skin.

Skin Color

1. Melanocytes produce melanin inside melanosomes and then transfer the melanin to keratinocytes. The size and distribution of melanosomes determine skin color. Melanin production is determined genetically but can be influenced by ultraviolet light (tanning) and hormones.
2. Carotene, an ingested plant pigment, can cause the skin to appear yellowish.
3. Increased blood flow produces a red skin color, whereas a decreased blood flow causes a pale skin. Decreased oxygen content in the blood results in a bluish color called cyanosis.

Accessory Skin Structures   (p. 150)
Hair

1. Lanugo (fetal hair) is replaced near the time of birth by terminal hairs (scalp, eyelids, and eyebrows) and vellus hairs. At puberty vellus hairs can be replaced with terminal hairs.
2. Hair is dead keratinized epithelial cells consisting of a central axis of cells with soft keratin, known as the medulla, which is surrounded by a cortex of cells with hard keratin. The cortex is covered by the cuticle, a single layer of cells filled with hard keratin.
3. A hair has three parts: the shaft, the root, and the hair bulb.
4. The hair bulb produces the hair in cycles involving a growth stage and a resting stage.
5. Hair color is determined by the amount and kind of melanin present.
6. Contraction of the arrector pili muscles, which are smooth muscles, causes hair to "stand on end" and produces "gooseflesh."

Glands

1. Sebaceous glands produce sebum, which oils the hair and the surface of the skin.

2. Merocrine sweat glands produce sweat that cools the body. Apocrine sweat glands produce an organic secretion that can be broken down by bacteria to cause body odor.
3. Other skin glands include ceruminous glands, which help to make cerumen (earwax), and the mammary glands, which produce milk.

Nails

1. The nail root is covered by skin, and the nail body is the visible part of the nail.
2. The nail is mostly formed by the nail matrix, but the nail bed does contribute.
3. The lunula is the part of the nail matrix visible through the nail body.
4. The nail is stratum corneum containing hard keratin.

Summary of Integumentary System Functions   (p. 156)
Protection

1. The skin provides protection against abrasion and ultraviolet light, prevents the entry of microorganisms, helps to regulate body temperature, and prevents water loss.
2. Hair protects against abrasion and ultraviolet light and is a heat insulator.
3. Nails protect the ends of the digits.

Sensation

The skin contains sensory receptors for pain, touch, hot, cold, and pressure that allow proper response to the environment.

Temperature Regulation

1. Through dilation and constriction of blood vessels, the skin controls heat loss from the body.
2. Sweat glands produce sweat which evaporates and lowers body temperature.

Vitamin D Production

1. Skin exposed to ultraviolet light produces cholecalciferol, which is modified in the liver and then in the kidneys to form active vitamin D.
2. Vitamin D increases blood calcium levels by promoting calcium uptake from the intestine, release of calcium from bone, and reduction of calcium loss from the kidneys.

Excretion

Skin glands remove small amounts of waste products (e.g., urea, uric acid, and ammonia) but are not important in excretion.

Effects of Aging on the Integumentary System   (p. 157)

1. As the body ages, blood flow to the skin declines, the skin becomes thinner, and elasticity is lost.
2. Sweat and sebaceous glands are less active, and the number of melanocytes decreases.

R E V I E W   A N D   C O M P R E H E N S I O N

1. The hypodermis
   a. is the layer of skin where the hair is produced.
   b. is the layer of skin where nails are produced.
   c. connects the dermis to the epidermis.
   d. is dense irregular connective tissue.
   e. contains approximately half of the body's stored fat.

For questions 2–5, match the layer of the dermis with the correct description or function:
   a. papillary layer
   b. reticular layer

2. The layer of the dermis closest to the epidermis

3. The layer of the dermis responsible for most of the structural strength of the skin

4. The layer of the dermis responsible for fingerprints and footprints

5. The layer of the dermis responsible for cleavage lines and striae

6. A layer of skin (where mitosis occurs) that replaces cells lost from the outer layer of the epidermis is the
   a. stratum corneum.
   b. stratum basale.
   c. stratum lucidum.
   d. reticular layer.
   e. hypodermis.

7. If a splinter penetrates the skin of the palm of the hand to the second epidermal layer from the surface, the last layer damaged is the
   a. stratum granulosum.
   b. stratum basale.
   c. stratum corneum.
   d. stratum lucidum.
   e. stratum spinosum.

For questions 8–12, match the layer of the epidermis with the correct description or function:
   a. stratum basale
   b. stratum corneum
   c. stratum granulosum
   d. stratum lucidum
   e. stratum spinosum

8. Production of keratin fibers; formation of lamellar bodies; limited amount of cell division

9. Desquamation occurs; 25 or more layers of dead squamous cells

10. Production of cells; melanocytes produce and contribute melanin; hemidesmosomes present

11. Production of keratohyalin granules; lamellar bodies release lipids; cells die

12. Dispersion of keratohyalin around keratin fibers; layer appears transparent; cells dead

13. In which of these areas of the body is thick skin found?
    a. back of the hand
    b. abdomen
    c. over the shin
    d. bridge of the nose
    e. heel of the foot

14. The function of melanin in the skin is
    a. lubrication of the skin.
    b. prevention of skin infections.
    c. protection from ultraviolet light.
    d. to reduce water loss.
    e. to help regulate body temperature.

15. Concerning skin color, which of these statements is *not* correctly matched?
    a. skin appears yellow—carotene present
    b. no skin pigmentation (albinism)—genetic disorder
    c. skin tans—increased melanin production
    d. skin appears blue (cyanosis)—oxygenated blood
    e. African-Americans darker than Caucasians—more melanin in African-American skin

16. After birth, the type of hair on the scalp, eyelids, and eyebrows is
    a. lanugo.
    b. terminal hair.
    c. vellus hair.

17. Hair
    a. is produced by the dermal root sheath.
    b. consists of living keratinized epithelial cells.
    c. is colored by melanin.
    d. contains mostly soft keratin.
    e. grows from the tip.

18. Given these parts of a hair and hair follicle:
    1. cortex
    2. cuticle
    3. dermal root sheath
    4. epithelial root sheath
    5. medulla

    Arrange the structures in the correct order from the outside of the hair follicle to the center of the hair.
    a. 1,4,3,5,2
    b. 2,1,5,3,4
    c. 3,4,2,1,5
    d. 4,3,1,2,5
    e. 5,4,3,2,1

19. Concerning hair growth:
    a. Hair falls out of the hair follicle at the end of the growth stage.
    b. Most of the hair on the body grows continuously.
    c. Cutting or plucking the hair increases its growth rate and thickness.
    d. Genetic factors and the hormone testosterone are involved in "pattern baldness."
    e. Eyebrows have a longer growth stage and resting stage than scalp hair.

20. Smooth muscles that produce "goose bumps" when they contract and are attached to hair follicles are called
    a. external root sheaths.
    b. arrector pili.
    c. dermal papillae.
    d. internal root sheaths.
    e. hair bulbs.

For questions 21–23, match the type of gland with the correct description or function.
   a. apocrine sweat gland
   b. merocrine sweat gland
   c. sebaceous gland

21. Alveolar glands that produce a white, oily substance; usually open into hair follicles

22. Coiled tubular glands that secrete a hyposmotic fluid that cools the body; most numerous in the palms of the hands and soles of the feet

23. Secretions from these coiled tubular glands are broken down by bacteria to produce body odor; found in the axillae, genitalia, and around the anus

24. The lunula of the nail appears white because
    a. it lacks melanin.
    b. blood vessels cannot be seen through the thick nail matrix.
    c. the eponychium decreases blood flow to the area.
    d. the nail root is much thicker than the nail body.
    e. the hyponychium is thicker than the eponychium.

25. The stratum corneum of the nail fold grows onto the nail body as the
    a. eponychium.
    b. hyponychium.
    c. lunula.
    d. nail bed.
    e. nail matrix.

26. Most of the nail is produced by the
    a. eponychium.
    b. hyponychium.
    c. nail bed.
    d. nail matrix.
    e. dermis.

27. The skin aids in maintaining the calcium and phosphate levels of the body at optimum levels by participating in the production of
    a. vitamin A.
    b. vitamin B.
    c. vitamin D.
    d. melanin.
    e. keratin.

28. Which of these processes increase(s) heat loss from the body?
    a. dilation of dermal arterioles
    b. constriction of dermal arterioles
    c. increased sweating
    d. both a and c
    e. both b and c

29. In third-degree (full-thickness) burns, both the epidermis and dermis of the skin are destroyed. Which of the following conditions would *not* occur as a result of a third-degree burn?
    a. dehydration (increased water loss)
    b. increased likelihood of infection
    c. increased sweating
    d. loss of sensation in the burned area
    e. poor temperature regulation in the burned area

30. Which of the following factors *increases* with age?
    a. blood flow to the skin
    b. number and diameter of elastic fibers in the skin
    c. number of melanocytes in some localized areas of the skin
    d. melanin production in the hair
    e. activity of sebaceous and sweat glands in the skin

Answers in Appendix F

CRITICAL THINKING

1. A woman has stretch marks on her abdomen, yet she states that she has never been pregnant. Is this possible?

2. The skin of infants is more easily penetrated and injured by abrasion than that of adults. Based on this fact, which stratum of the epidermis is probably much thinner in infants than that in adults?

3. Melanocytes are found primarily in the stratum basale of the epidermis. In reference to their function, why does this location make sense?

4. The rate of water loss from the skin of the hand was measured. Following the measurement, the hand was soaked in alcohol for 15 minutes. After all the alcohol was removed from the hand, the rate of water loss was again measured. Compared with the rate of water loss before soaking the hand in alcohol, what difference, if any, would you expect in the rate of water loss after soaking the hand in alcohol?

5. It has been several weeks since Goodboy Player has competed in a tennis match. After the match he discovers that a blister has formed beneath an old callus on his foot, and the callus has fallen off. When he examines the callus he discovers that it appears yellow. Can you explain why?

6. Harry Fastfeet, a white man, jogs on a cold day. What color would you expect his skin to be (a) just before starting to run, (b) during the run, and (c) 5 minutes after the run?

7. Why are your eyelashes not a foot long? Your fingernails?

8. Pulling on hair can be quite painful, yet cutting hair is not painful. Explain.

9. Given what you know about the cause of acne, propose some ways to prevent or treat the disorder.

10. A patient has an ingrown toenail, a condition in which the nail grows into the nail fold. Would cutting the nail away from the nail fold permanently correct this condition? Why or why not?

11. Consider the following statement: Dark-skinned children are more susceptible to rickets (insufficient calcium in the bones) than fair-skinned children. Defend or refute this statement.

Answers in Appendix G

ANSWERS TO PREDICT QUESTIONS

1. Because the permeability barrier is mainly composed of lipids surrounding the epidermal cells, substances that are lipid-soluble easily pass through, whereas water-soluble substances have difficulty.

2. a. The lips are pinker or redder than the palms of the hand. Several explanations for this are possible: more blood vessels in the lips, increased blood flow could occur in the lips, or the blood vessels could be easier to see through the epidermis of the lips. The last possibility explains most of the difference in color between the lips and the palms. The epidermis of the lips is thinner and not as heavily keratinized as that of the palms. In addition, the papillae containing the blood vessels in the lips are "high" and closer to the surface.

   b. A person who does manual labor has a thicker stratum corneum on the palms (and possibly calluses) than a person who does not perform manual labor. The thicker epidermis masks the underlying blood vessels, and the palms do not appear as pink. In addition, carotene accumulating in the lipids of the stratum corneum might impart a yellowish cast to the palms.

   c. The posterior surface of the forearm appears darker because of the tanning effect of ultraviolet light from the sun.

   d. The genitals normally have more melanin and appear darker than the soles of the feet.

3. The story is not true. Hair color results from the transfer of melanin from melanocytes to keratinocytes in the hair matrix as the hair grows. The hair itself is dead. To turn white, the hair must grow out without the addition of melanin, a process that takes weeks.

4. On cold days, skin blood vessels of the ears and nose can dilate, bringing warm blood to the ears and nose and thus preventing tissue damage from the cold. The increased blood flow makes the ears and nose appear red.

5. Reducing water loss is one of the normal functions of the skin. Loss of skin, or damage to the skin, can greatly increase water loss. In addition, burning large areas of the skin results in increased capillary permeability and additional loss of fluid from the burn and into tissue spaces. The loss of fluid reduces blood volume, which results in reduced blood flow to the kidneys. Consequently, urine output by the kidneys decreases, which reduces fluid loss and thereby helps to compensate for the fluid loss caused by the burn. The reduced blood flow to the kidneys can cause tissue damage, however. To counteract this effect, during the first 24 hours following the injury, part of the treatment for burn victims is the administration of large volumes of fluid. But, how much fluid should be given? The amount of fluid given should be sufficient to match that lost plus enough to prevent kidney damage and allow the kidneys to function. Urine output is therefore monitored. If it is too low, more fluid is administered, and if it is too high, less fluid is given. An adult receiving intravenous fluids should produce 30–50 mL of urine/hour, and children should produce 1 mL/kg of body weight/hour.

Visit the Online Learning Center at www.mhhe.com/seeley7 for chapter quizzes, interactive learning exercises, and other study tools.

Skeletal System:

Bones and Bone Tissue

CHAPTER

6

This is a colorized scanning electron micrograph of a tiny part of a bone called an osteon. The large opening is the space through which blood vessels bring blood to the bone. The surrounding bone matrix is organized into circular layers.

Sitting, standing, walking, picking up a pencil, and taking a breath all involve the skeletal system. It is the structural framework that gives the body its shape and provides protection for internal organs and soft tissues. The term *skeleton* is derived from a Greek word meaning dried, indicating that the skeleton is the dried, hard parts left after the softer parts are removed. Even with the flesh and organs removed, the skeleton is easily recognized as human. Despite its association with death, however, the skeletal system actually consists of dynamic, living tissues that are capable of growth, adapt to stress, and undergo repair after injury.

This chapter describes the *functions of the skeletal system* (p. 168), provides an explanation of *cartilage* (p. 168), and examines *bone anatomy* (p. 169), *bone histology* (p. 171), *bone development* (p. 175), *bone growth* (p. 180), *bone remodeling* (p. 183), *bone repair* (p. 186), *calcium homeostasis* (p. 187), and the *effects of aging on the skeletal system* (p. 189).

Functions of the Skeletal System

The skeletal system has four components: bones, cartilage, tendons, and ligaments. The skeletal system provides support and protection, allows body movements, stores minerals and fats, and is the site of blood cell production.

1. *Support.* Rigid, strong bone is well suited for bearing weight and is the major supporting tissue of the body. Cartilage provides a firm yet flexible support within certain structures, such as the nose, external ear, thoracic cage, and trachea. Ligaments are strong bands of fibrous connective tissue that attach to bones and hold them together.
2. *Protection.* Bone is hard and protects the organs it surrounds. For example, the skull encloses and protects the brain, and the vertebrae surround the spinal cord. The rib cage protects the heart, lungs, and other organs of the thorax.
3. *Movement.* Skeletal muscles attach to bones by tendons, which are strong bands of connective tissue. Contraction of the skeletal muscles moves the bones, producing body movements. Joints, which are formed where two or more bones come together, allow movement between bones. Smooth cartilage covers the ends of bones within some joints, allowing the bones to move freely. Ligaments allow some movement between bones but prevent excessive movements.
4. *Storage.* Some minerals in the blood are taken into bone and stored. Should blood levels of these minerals decrease, the minerals are released from bone into the blood. The principal minerals stored are calcium and phosphorus. Fat (adipose tissue) is also stored within bone cavities. If needed, the fats are released into the blood and used by other tissues as a source of energy.
5. *Blood cell production.* Many bones contain cavities filled with bone marrow that gives rise to blood cells and platelets (see chapter 19).

▼ **1.** *Name the four components of the skeletal system. List the five functions of the skeletal system.*

Cartilage

Cartilage comes in three types: hyaline cartilage, fibrocartilage, and elastic cartilage (see chapter 4). Although each type of cartilage can provide support, hyaline cartilage is most intimately associated with bone. An understanding of hyaline cartilage is important because most of the bones in the body develop from it. In addition, the growth in length of bones and bone repair often involve the production of hyaline cartilage followed by its replacement with bone.

Hyaline cartilage consists of specialized cells that produce a matrix surrounding the cells (figure 6.1). The cells that produce new cartilage matrix are **chondroblasts** (kon′drō-blastz; *chondro* is from the Greek word *chondrion* and means cartilage). When matrix surrounds a chondroblast, it becomes a **chondrocyte** (kon′drō-sīt), which is a rounded cell that occupies a space called a **lacuna** (lă-koo′nă) within the matrix. The matrix contains collagen, which provides strength, and proteoglycans, which make cartilage resilient by trapping water (see chapter 4).

The **perichondrium** (per-i-kon′drē-ŭm) is a double-layered connective tissue sheath covering most cartilage (see figure 6.1). The outer layer of the perichondrium is dense irregular connective tissue containing fibroblasts. The inner, more delicate layer has fewer fibers and contains chondroblasts. Blood vessels and nerves penetrate the outer layer of the perichondrium but do not enter the cartilage matrix, so that nutrients must diffuse through the cartilage matrix to reach the chondrocytes. **Articular** (ar-tik′ū-lăr) **cartilage,** which is the cartilage covering the ends of bones where they come together to form joints, has no perichondrium, blood vessels, or nerves.

Perichondrium

Chondroblast

Lacuna

Chondrocyte

Nucleus

Chondrocytes that have divided

Matrix

LM 400x

Appositional growth
(new cartilage is added to the surface of the cartilage by chondroblasts from the inner layer of the perichondrium)

Interstitial growth
(new cartilage is formed within the cartilage by chondrocytes that divide and produce additional matrix)

Figure 6.1 Hyaline Cartilage
Photomicrograph of hyaline cartilage covered by perichondrium. Chondrocytes within lacunae are surrounded by cartilage matrix.

P R E D I C T
Explain why damaged cartilage takes a long time to heal. What are the advantages of articular cartilage having no perichondrium, blood vessels, or nerves?

Cartilage grows in two different ways. Through **appositional growth,** chondroblasts in the perichondrium lay down new matrix and add new chondrocytes to the outside of the tissue, and through **interstitial growth,** chondrocytes within the tissue divide and add more matrix between the cells (see figure 6.1).

- 2. *Describe the structure of hyaline cartilage. Name two types of cartilage cells. What is a lacuna?*
- 3. *Describe the connective tissue and cells found in both layers of the perichondrium. How do nutrients from blood vessels in the perichondrium reach the chondrocytes?*
- 4. *Describe appositional and interstitial growth of cartilage.*

Bone Anatomy

Bone Shapes

Individual bones are classified according to their shape as long, short, flat, or irregular (figure 6.2). **Long bones** are longer than they are wide. Most of the bones of the upper and lower limbs are long bones. **Short bones** are about as broad as they are long. They are nearly cube-shaped or round and are exemplified by the bones of the wrist (carpals) and ankle (tarsals). **Flat bones** have a

relatively thin, flattened shape and are usually curved. Examples of flat bones are certain skull bones, the ribs, the breastbone (sternum), and the shoulder blades (scapulae). **Irregular bones,** such as the vertebrae and facial bones, have shapes that don't fit readily into the other three categories.

Structure of a Long Bone

Each growing long bone has three major components: a diaphysis, an epiphysis, and an epiphyseal plate (figure 6.3*a* and table 6.1). The **diaphysis** (dī-af′i-sis), or shaft, is composed primarily of **compact bone,** which is mostly bone matrix with a few small spaces. The **epiphysis** (e-pif′i-sis), or end of the bone, consists primarily of **cancellous** (kan′sĕ-lŭs), or **spongy, bone,** which is mostly small spaces or cavities surrounded by bone matrix. The outer surface of the epiphysis is a layer of compact bone, and within joints the epiphyses are covered by articular cartilage. The **epiphyseal** (ep-i-fiz′ē-ăl), or **growth, plate** is hyaline cartilage located between the epiphysis and diaphysis. Growth in bone length occurs at the epiphyseal plate, but, when bone stops growing in length, the epiphyseal plate becomes ossified and is called the **epiphyseal line** (figure 6.3*b*).

In addition to the small spaces within cancellous bone and compact bone, the diaphysis of a long bone can have a large internal space called the **medullary cavity.** The cavities of cancellous bone and the medullary cavity are filled with marrow (mar′ō). **Red marrow** is the site of blood cell formation, and **yellow marrow** is mostly adipose tissue. In children, the spaces within bones are filled with red marrow. As children mature, yellow marrow replaces the red marrow in their skull and limbs. In adults, the bones of the skull and limbs, except for the proximal epiphyses, have yellow marrow (figure 6.4). The rest of the skeleton contains red marrow.

The **periosteum** (per-ē-os′tē-ŭm) is a connective tissue membrane that covers the outer surface of a bone (figure 6.3*c*). The outer fibrous layer is dense, irregular collagenous connective tissue that contains blood vessels and nerves. The inner layer is a single layer of bone cells, which includes osteoblasts, osteoclasts, and osteochondral progenitor cells (see "Bone Cells" on p. 172). Where tendons and ligaments attach to bone, the collagen fibers of the tendon or ligament become continuous with those of the periosteum. In addition, some of the collagen fibers of the tendons or ligaments penetrate the periosteum into the outer part of the bone. These bundles of collagen fibers are called **perforating,** or **Sharpey's, fibers,** and they strengthen the attachment of the tendons or ligaments to the bone.

The **endosteum** (en-dos′tē-ŭm) is a connective tissue membrane that lines the internal surfaces of all cavities within bones, such as the medullary cavity of the diaphysis and the smaller cavities in cancellous and compact bone (see figure 6.3*c*). The endosteum is a single layer of cells, which includes osteoblasts, osteoclasts, and osteochondral progenitor cells.

Structure of Flat, Short, and Irregular Bones

Flat bones usually have no diaphyses or epiphyses, and they contain an interior framework of cancellous bone sandwiched between two layers of compact bone (figure 6.5). Short and irregular bones have a composition similar to the epiphyses of long bones. They have compact bone surfaces that surround a cancellous bone center with small spaces that usually are filled with marrow. Short and irregular bones

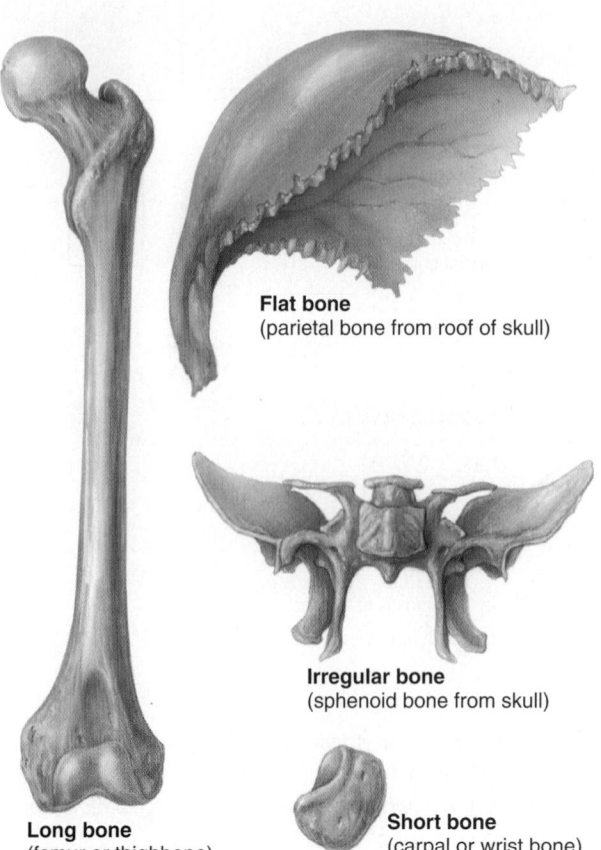

Flat bone
(parietal bone from roof of skull)

Irregular bone
(sphenoid bone from skull)

Long bone
(femur or thighbone)

Short bone
(carpal or wrist bone)

Figure 6.2 Bone Shapes

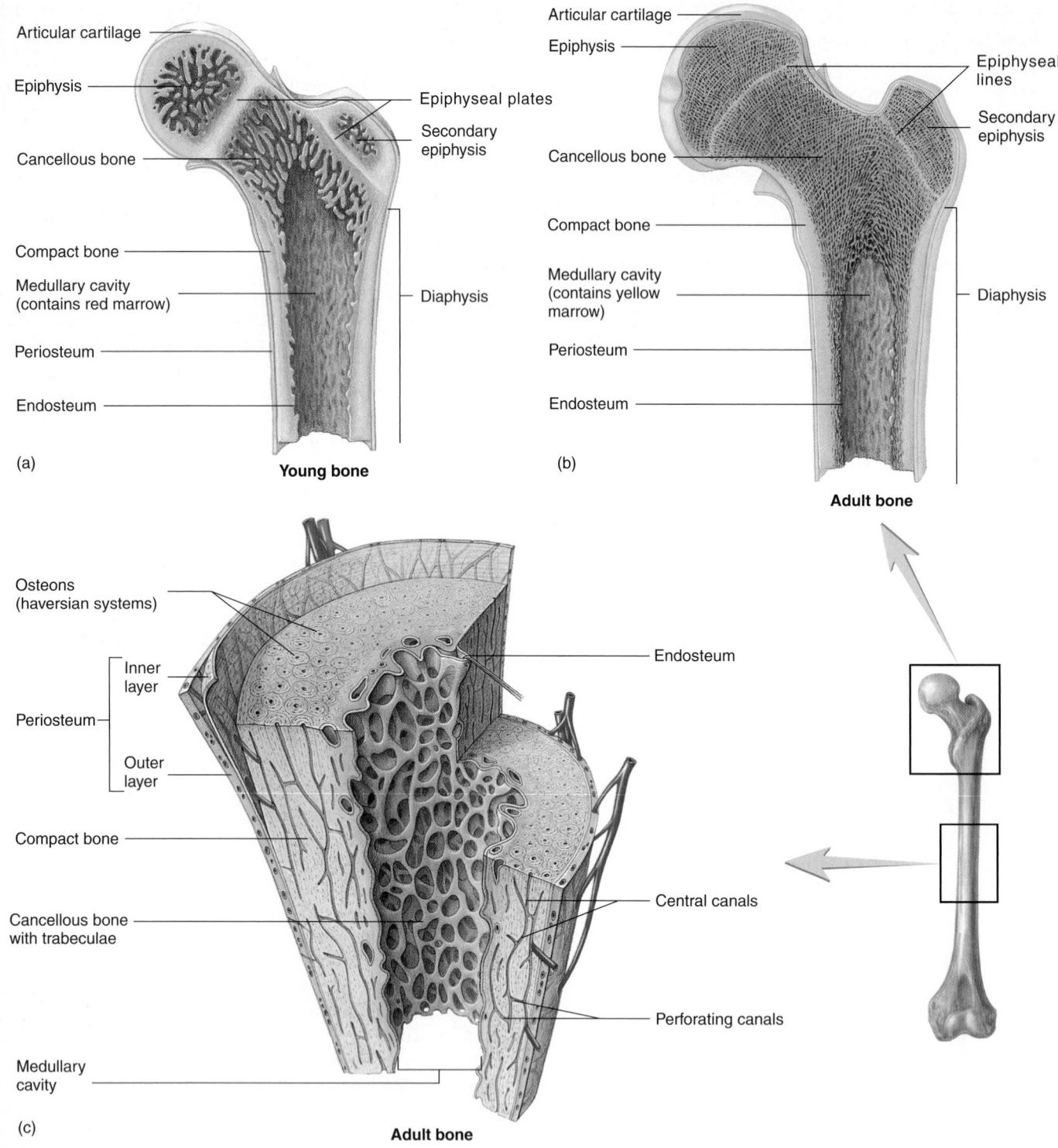

Figure 6.3   Long Bone

(*a*) Young long bone (the femur) showing epiphysis, epiphyseal plates, and diaphysis. (*b*) Adult long bone with epiphyseal lines. (*c*) Internal features of a portion of the diaphysis in (*a*).

are not elongated and have no diaphyses. Certain regions of these bones, however, such as the processes (projections) of irregular bones, possess epiphyseal growth plates and therefore have small epiphyses.

Some of the flat and irregular bones of the skull have air-filled spaces called **sinuses** (sī′nŭs-ĕz) (see chapter 7), which are lined by mucous membranes.

5. *List the four basic shapes of individual bones, and give an example of each.*
6. *Define the diaphysis, epiphysis, epiphyseal plate, and epiphyseal line of a long bone.*
7. *What are red marrow and yellow marrow? Where are they located in a child and in an adult?*

Table 6.1	Gross Anatomy of a Long Bone		
Part	**Description**	**Part**	**Description**
Diaphysis	Shaft of the bone	Epiphyseal plate	Area of hyaline cartilage between the diaphysis and epiphysis; cartilage growth followed by endochondral ossification results in bone growth in length
Epiphyses	Ends of the bone		
Periosteum	Double-layered connective tissue membrane covering the outer surface of bone except where articular cartilage exists; ligaments and tendons attach to bone through the periosteum; blood vessels and nerves from the periosteum supply the bone; the periosteum is the site of bone growth in diameter	Cancellous (spongy) bone	Bone having many small spaces; found mainly in the epiphysis; arranged into trabeculae
		Compact bone	Dense bone with few internal spaces organized into osteons; forms the diaphysis and covers the cancellous bone of the epiphyses
Endosteum	Thin connective tissue membrane lining the inner cavities of bone	Medullary cavity	Large cavity within the diaphysis
Articular cartilage	Thin layer of hyaline cartilage covering a bone where it forms a joint (articulation) with another bone	Red marrow	Connective tissue in the spaces of cancellous bone or in the medullary cavity; the site of blood cell production
		Yellow marrow	Fat stored within the medullary cavity or in the spaces of cancellous bone

Figure 6.4 Bone Marrow

Distribution of red and yellow marrow in an adult.

8. *Where are the periosteum and endosteum located? What types of cells are found in the periosteum and endosteum? What is the function of perforating (Sharpey's) fibers?*

9. *Compare the structure of long bones to the structure of flat, short, and irregular bones. How are compact bone and cancellous bone arranged in each?*

Compact bone

Cancellous bone

Figure 6.5 Structure of a Flat Bone

Outer layers of compact bone surround cancellous bone.

Bone Histology

Bone consists of extracellular bone matrix and bone cells. The composition of the bone matrix is responsible for the characteristics of bone. The bone cells produce the bone matrix, become entrapped within it, and break it down so that new matrix can replace the old matrix.

Bone Matrix

By weight, mature bone matrix normally is approximately 35% organic and 65% inorganic material. The organic material primarily consists of collagen and proteoglycans. The inorganic material primarily consists of a calcium phosphate crystal called **hydroxyapatite** (hī-drok′sē-ap-ă-tīt), which has the molecular formula $Ca_{10}(PO_4)_6(OH)_2$.

The collagen and mineral components are responsible for the major functional characteristics of bone. Bone matrix might be said to resemble reinforced concrete. Collagen, like reinforcing steel bars, lends flexible strength to the matrix, whereas the mineral components, like concrete, give the matrix compression (weight-bearing) strength.

If all the mineral is removed from a long bone, collagen becomes the primary constituent, and the bone becomes overly flexible. On the other hand, if the collagen is removed from the bone, the mineral component becomes the primary constituent, and the bone is very brittle (figure 6.6).

P R E D I C T ②

In general, the bones of elderly people break more easily than the bones of younger people. Give as many possible explanations as you can for this observation.

Bone Cells

Bone cells are categorized as osteoblasts, osteocytes, and osteoclasts, which have different functions and origins.

Osteoblasts

Osteoblasts (os′tē-ō-blastz) have an extensive endoplasmic reticulum and numerous ribosomes. They produce collagen and proteoglycans, which are packaged into vesicles by the Golgi apparatus and released from the cell by exocytosis. Osteoblasts also form vesicles that accumulate calcium ions (Ca^{2+}), phosphate ions (PO_4^{2-}), and various enzymes. The contents of these vesicles are released from the cell by exocytosis and are used to form hydroxyapatite crystals. As a result of these processes, mineralized bone matrix is formed.

Ossification (os′i-fi-kā′shŭn), or **osteogenesis** (os′tē-ō-jen′ĕ-sis), is the formation of bone by osteoblasts. Ossification occurs by appositional growth on the surface of previously existing bone or cartilage. For example, osteoblasts beneath the periosteum cover the surface of preexisting bone (figure 6.7). Elongated cell processes from osteoblasts connect to cell processes of other osteoblasts through gap junctions (see chapter 4). Bone matrix secreted from the osteoblasts covers the older bone surface and surrounds the osteoblast cell bodies and processes. The result is a new layer of bone.

Osteocytes

Once an osteoblast becomes surrounded by bone matrix, it is a mature bone cell called an **osteocyte** (os′tē-ō-sīt). Osteocytes become relatively inactive compared to most osteoblasts, but it's possible for them to produce components needed to maintain the bone matrix.

The spaces occupied by the osteocyte cell bodies are called **lacunae** (lă-koo′nē), and the spaces occupied by the osteocyte cell processes are called **canaliculi** (kan-ă-lik′ū-lī; meaning little canals) (see figure 6.7). In a sense, the cells and their processes form a "mold" around which the matrix is formed. Bone differs from cartilage in that the processes of bone cells are in contact with one another through the canaliculi. Instead of diffusing through the mineralized matrix, nutrients and gases can pass through the small amount of fluid surrounding the cells in the canaliculi and lacunae or pass from cell to cell through the gap junctions connecting the cell processes.

Osteoclasts

Osteoclasts (os′tē-ō-klastz) are large cells with several nuclei and are responsible for the **resorption,** or breakdown, of bone. Where the plasma membrane of osteoclasts contacts bone matrix, it forms many projections called a **ruffled border.** Hydrogen ions are pumped across the ruffled border and produce an acid environment that causes decalcification of the bone matrix. The osteoclasts also release enzymes that digest the protein components of the matrix. Through the process of endocytosis, some of the breakdown products of bone resorption are taken into the osteoclast.

Osteoclasts break down bone best when they are in direct contact with mineralized bone matrix. Osteoblasts assist in the resorption of bone by osteoclasts by producing enzymes that break down the thin layer of unmineralized organic matrix normally covering bone. Removal of this layer by osteoblasts enables the osteoclasts to come into contact with the mineralized bone.

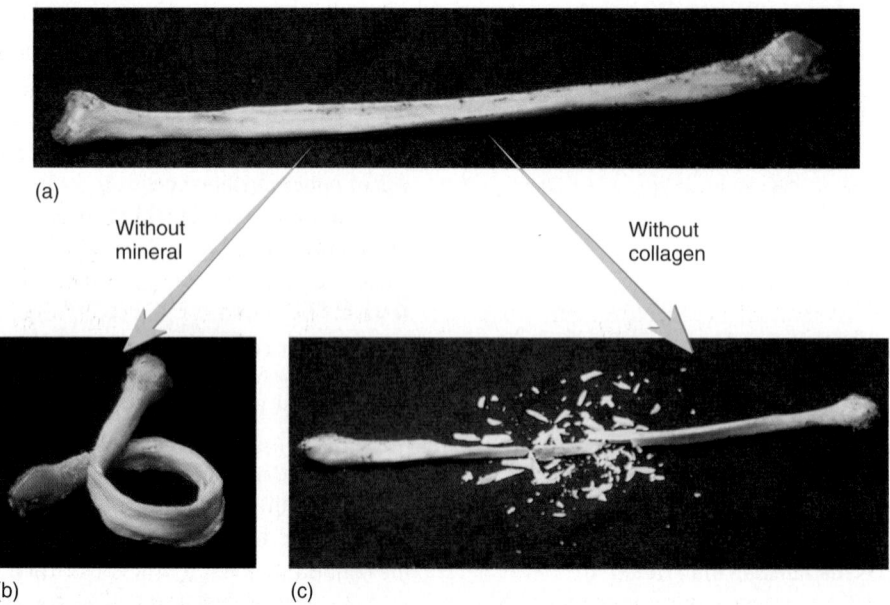

(a)

Without mineral

Without collagen

(b) (c)

Figure 6.6 Effects of Changing the Bone Matrix

(a) Normal bone. (b) Demineralized bone, in which collagen is the primary remaining component, can be bent without breaking. (c) When collagen is removed, mineral is the primary remaining component, thus making the bone so brittle it's easily shattered.

(a)

(b)

(c)

Figure 6.7   Ossification

(a) Osteoblasts on a preexisting surface, such as cartilage or bone. The cell processes of different osteoblasts join together. (b) Osteoblasts have produced bone matrix. The osteoblasts are now osteocytes. (c) Photomicrograph of an osteocyte in a lacuna with cell processes in the canaliculi.

Origin of Bone Cells

Connective tissue develops embryologically from mesenchymal cells (see chapter 4). Some of the mesenchymal cells become **stem cells,** which have the ability to replicate and give rise to more specialized cell types. **Osteochondral progenitor cells** are stem cells that have the ability to become osteoblasts or chondroblasts. Osteochondral progenitor cells are located in the inner layer of the perichondrium, the inner layer of the periosteum, and in the endosteum. From these locations, they can be a potential source of new osteoblasts or chondroblasts.

Osteoblasts are derived from osteochondral progenitor cells, and osteocytes are derived from osteoblasts. Whether or not osteocytes freed from their surrounding bone matrix by resorption can revert to active osteoblasts is a debated issue. Osteoclasts are not derived

from osteochondral progenitor cells but are derived instead from stem cells in red bone marrow (see chapter 19). The bone marrow stem cells that give rise to a type of white blood cell, called a monocyte, also are the source of osteoclasts. The multinucleated osteoclasts probably result from the fusion of many stem cell descendants.

- 10. *Name the components of bone matrix, and explain their contribution to bone flexibility and the ability of bones to bear weight.*
- 11. *What are the functions of osteoblasts, osteocytes, and osteoclasts?*
- 12. *Describe the formation of new bone by appositional growth. Name the spaces that are occupied by osteocyte cell bodies and cell processes.*
- 13. *What cells give rise to osteochondral progenitor cells? What kinds of cells are derived from osteochondral progenitor cells? What types of cells give rise to osteoclasts?*

Woven and Lamellar Bone

Bone tissue is classified as either woven or lamellar bone according to the organization of collagen fibers within the bone matrix. In **woven bone,** the collagen fibers are randomly oriented in many directions. Woven bone is first formed during fetal development or during the repair of a fracture. After its formation, osteoclasts break down the woven bone and osteoblasts build new matrix. This process of removing old bone and adding new bone is called **remodeling.** It is an important process discussed later in this chapter (see p. 183). Woven bone is remodeled to form lamellar bone.

Lamellar bone is mature bone that is organized into thin sheets or layers approximately 3–7 micrometers (μm) thick called **lamellae** (lă-mel′ē). In general, the collagen fibers of one lamella lie parallel to one another but at an angle to the collagen fibers in the adjacent lamellae. Osteocytes, within their lacunae, are arranged in layers sandwiched between lamellae.

Cancellous and Compact Bone

Bone, whether woven or lamellar, can be classified according to the amount of bone matrix relative to the amount of space present within the bone. Cancellous bone has less bone matrix and more space than compact bone, which has more bone matrix and less space than cancellous bone.

Cancellous bone (figure 6.8a) consists of interconnecting rods or plates of bone called **trabeculae** (tră-bek′ū-lē; beam). Between the trabeculae are spaces that in life are filled with bone marrow and blood vessels. Cancellous bone is sometimes called spongy bone because of its porous appearance.

Most trabeculae are thin (50–400 μm) and consist of several lamellae with osteocytes located between the lamellae (figure 6.8b). Each osteocyte is associated with other osteocytes through canaliculi. Usually no blood vessels penetrate the trabeculae, so osteocytes must obtain nutrients through their canaliculi. The surfaces of trabeculae are covered with a single layer of cells consisting mostly of osteoblasts with a few osteoclasts.

Trabeculae are oriented along the lines of stress within a bone (figure 6.9). If the direction of weight-bearing stress is changed slightly (e.g., because of a fracture that heals improperly), the trabecular pattern realigns with the new lines of stress.

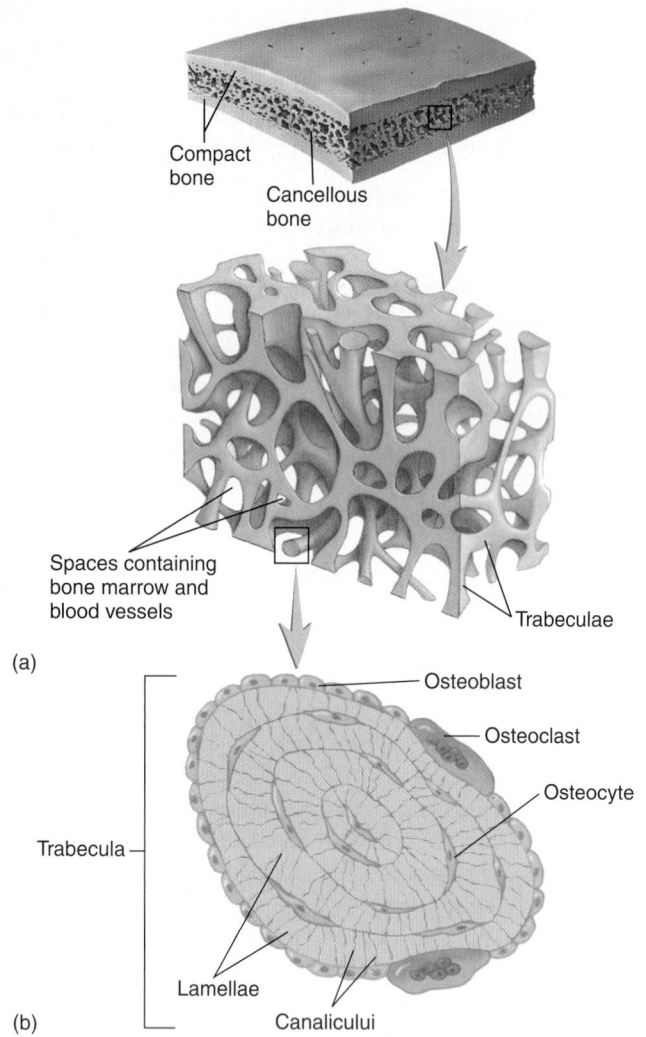

Compact bone

Cancellous bone

Spaces containing bone marrow and blood vessels

Trabeculae

(a)

Osteoblast

Osteoclast

Osteocyte

Trabecula

Lamellae

Canalicului

(b)

Figure 6.8   Cancellous Bone

(*a*) Beams of bone, the trabeculae, surround spaces in the bone. In life, the spaces are filled with red or yellow bone marrow and with blood vessels. (*b*) Transverse section of a trabecula.

Direction of stresses created by the weight of the body

Figure 6.9   Lines of Stress

The proximal end of a long bone (femur) showing trabeculae oriented along lines of stress (*arrows*).

Compact bone (figure 6.10) is denser and has fewer spaces than cancellous bone. Blood vessels enter the substance of the bone itself, and the lamellae of compact bone are primarily oriented around those blood vessels. Vessels that run parallel to the long axis of the bone are contained within **central,** or **haversian** (ha-ver′shan), **canals.** Central canals are lined with endosteum and contain blood vessels, nerves, and loose connective tissue. **Concentric lamellae** are circular layers of bone matrix that surround a common center, the central canal. An **osteon** (os′tē-on), or **haversian system,** consists of a single central canal, its contents, and associated concentric lamellae and osteocytes. In cross section, an osteon resembles a circular target; the "bull's-eye" of the target is the central canal, and 4–20 concentric lamellae form the rings. Osteocytes are located in lacunae between the lamellar rings, and canaliculi radiate between lacunae across the lamellae, producing the appearance of minute cracks across the rings of the target.

The outer surfaces of compact bone are formed by **circumferential lamellae,** which are thin plates that extend around the

bone (see figure 6.10). In some bones, such as certain bones of the face, the layer of compact bone can be so thin that no osteons exist, and the compact bone is composed of only circumferential lamellae. In between the osteons are **interstitial lamellae,** which are remnants of concentric or circumferential lamellae that were partially removed during bone remodeling.

Osteocytes receive nutrients and eliminate waste products through the canal system within compact bone. Blood vessels from the periosteum or medullary cavity enter the bone through **perforating,** or **Volkmann's, canals,** which run perpendicular to the long axis of the bone (see figure 6.10). Perforating canals are not surrounded by concentric lamellae but pass through the concentric lamellae of osteons. The central canals receive blood vessels from perforating canals. Nutrients in the blood vessels enter the central canals, pass into the canaliculi, and move through the cytoplasm of the osteocytes that occupy the canaliculi and lacunae to the most peripheral cells within each osteon. Waste products are removed in the reverse direction.

14. **Distinguish between woven bone and lamellar bone. Where is woven bone found?**
15. **Describe the structure of cancellous bone. What are trabeculae, and what is their function? How do osteocytes within trabeculae obtain nutrients?**
16. **Describe the structure of compact bone. What is an osteon? Name three types of lamellae found in compact bone.**
17. **Trace the pathway nutrients must follow to go from blood vessels in the periosteum to osteocytes within osteons.**

P R E D I C T

Compact bone has perforating and central canals. Why isn't such a canal system necessary in cancellous bone?

Figure 6.10 Compact Bone

(a) Compact bone consists mainly of osteons, which are concentric lamellae surrounding blood vessels within central canals. The outer surface of the bone is formed by circumferential lamellae, and bone between the osteons consists of interstitial lamellae. (b) Photomicrograph of an osteon.

Bone Development

During fetal development, bone formation occurs in two patterns called **intramembranous** and **endochondral ossification.** The terms describe the tissues in which bone formation takes place: intramembranous ossification in connective tissue membranes and endochondral ossification in cartilage. Both methods of ossification initially produce woven bone that is then remodeled. After remodeling, bone formed by intramembranous ossification cannot be distinguished from bone formed by endochondral ossification. Table 6.2 compares intramembranous and endochondral ossification.

Intramembranous Ossification

At approximately the fifth week of development embryonic mesenchyme condenses around the developing brain to form a membrane of connective tissue with randomly oriented, delicate collagen fibers. Intramembranous ossification of the membrane begins at approximately the eighth week of development and is completed by approximately 2 years of age. Many skull bones, part of the mandible (lower jaw), and the diaphyses of the clavicles (collarbones) develop by intramembranous ossification (figure 6.11).

Table 6.2	Comparison of Intramembranous and Endochondral Ossification
Intramembranous Ossification	**Endochondral Ossification**
Embryonic mesenchyme forms a collagen membrane containing osteochondral progenitor cells.	Embryonic mesenchymal cells become chondroblasts that produce a cartilage template surrounded by the perichondrium.
No stage is comparable.	Chondrocytes hypertrophy, the cartilage matrix becomes calcified, and the chondrocytes die.
Embryonic mesenchyme forms the periosteum, which contains osteoblasts.	The perichondrium becomes the periosteum when osteochondral progenitor cells within the periosteum become osteoblasts.
Osteochondral progenitor cells become osteoblasts at centers of ossification; internally the osteoblasts form cancellous bone; externally the periosteal osteoblasts form compact bone.	Blood vessels and osteoblasts from the periosteum invade the calcified cartilage template; internally these osteoblasts form cancellous bone at primary ossification centers (and later at secondary ossification centers); externally the periosteal osteoblasts form compact bone.
Intramembranous bone is remodeled and becomes indistinguishable from endochondral bone.	Endochondral bone is remodeled and becomes indistinguishable from intramembranous bone.

Osteoblast

Osteocyte

Bone matrix of trabecula

LM 500x

LM 250x

1. A cross section of a newly formed trabecula shows the youngest bone in this series of photomicrographs. Osteocytes are surrounded by bone matrix and osteoblasts are forming a ring on the outer surface of the trabecula. As the osteoblasts lay down bone, the trabeculae increase in size.

2. A lower magnification shows older bone than in step 1. Cancellous bone has formed as a result of the enlargement and interconnections of many trabeculae.

Parietal bone

Center of ossification

Superior part of occipital bone

Inferior part of occipital bone

Temporal bone

Vertebrae

Styloid process

Frontal bone

Ethmoid bone

Nasal bone

Maxilla

Zygomatic bone

Mandible

Cartilage of mandible

Sphenoid bone

12 weeks

Compact bone

Periosteum

Cancellous bone

LM 50x

3. A lower magnification than in step 2, with a different stain that makes the bone appear blue, shows the oldest bone in this series. Within the cancellous bone there is trabeculae (*blue*) and developing red bone marrow (*pink*). Beneath the periosteum is an outer layer of developing compact bone.

Process Figure 6.11   Intramembranous Ossification

The inset shows a 12-week-old fetus. Bones formed by intramembranous ossification are *yellow* and bones formed by endochondral ossification are *blue*. Intramembranous ossification starts at a center of ossification and expands outward. Therefore, the youngest bone is at the edge of the expanding bone, and the oldest bone is at the center of ossification.

Centers of ossification are the locations in the membrane where ossification begins. The centers of ossification expand to form a bone by gradually ossifying the membrane. Thus, the centers of ossification have the oldest bone and the expanding edges the youngest bone. The larger membrane-covered spaces between the developing skull bones that have not yet been ossified are called **fontanels,** or soft spots (figure 6.12) (see chapter 8). The bones eventually grow together, and all the fontanels have usually closed by the time an infant is 2 years of age.

1. Intramembranous ossification begins when some of the mesenchymal cells in the membrane become osteochondral progenitor cells, which specialize to become osteoblasts. The osteoblasts produce bone matrix that surrounds the collagen fibers of the connective tissue membrane, and the osteoblasts become osteocytes. As a result of this process, many tiny trabeculae of woven bone develop (figure 6.11, step 1).

2. Additional osteoblasts gather on the surfaces of the trabeculae and produce more bone, thereby causing the trabeculae to become larger and longer (figure 6.11, step 2). Cancellous bone forms as the trabeculae join together, resulting in an interconnected network of trabeculae separated by spaces.

3. Cells within the spaces of the cancellous bone specialize to form red bone marrow, and cells surrounding the developing bone specialize and form the periosteum. Osteoblasts from the periosteum lay down bone matrix to form an outer surface of compact bone (figure 6.11, step 3).

   Thus, the end products of intramembranous bone formation are bones with outer compact bone surfaces and cancellous centers (see figure 6.5). Remodeling converts woven bone to lamellar bone and contributes to the final shape of the bone.

Endochondral Ossification

The formation of cartilage begins at approximately the end of the fourth week of development. Endochondral ossification of some of this cartilage starts at approximately the eighth week of development. Endochondral ossification of some cartilage might not begin until as late as age 18–20 years. Bones of the base of the skull, part of the mandible, the epiphyses of the clavicles, and most of the remaining skeletal system develop through the process of endochondral ossification (see figures 6.11 and 6.12).

1. Endochondral ossification begins as mesenchymal cells aggregate in regions of future bone formation. The mesenchymal cells become chondroblasts, which produce a hyaline **cartilage model** having the approximate shape of the bone that will later be formed (figure 6.13, step 1). As the chondroblasts become surrounded by cartilage matrix, they become chondrocytes. The cartilage model is surrounded by perichondrium, except where a joint will form connecting one bone to another bone. Not shown in figure 6.13, the perichondrium is continuous with tissue that will become the joint capsule (see chapter 8).

Figure 6.12 Bone Formation

Eighteen-week-old fetus showing intramembranous and endochondral ossification. Intramembranous ossification occurs at centers of ossification in the flat bones of the skull. Endochondral ossification has formed bones in the diaphyses of long bones. The epiphyses are still cartilage at this stage of development.

2. When blood vessels invade the perichondrium surrounding the cartilage model (figure 6.13, step 2), osteochondral progenitor cells within the perichondrium become osteoblasts. The perichondrium becomes the periosteum when the osteoblasts begin to produce bone. The osteoblasts produce compact bone on the surface of the cartilage model, forming a **bone collar.** Two other events are occurring at the same time that the bone collar is forming. First, the cartilage model increases in size as a result of interstitial and appositional cartilage growth. Second, the chondrocytes in the center of the cartilage model **hypertrophy** (hī-per′trō-fē), or enlarge, and the matrix between the enlarged cells becomes mineralized with calcium carbonate. At this point, the cartilage is referred to as **calcified cartilage**. The chondrocytes in this calcified area eventually die, leaving enlarged lacunae with thin walls of calcified matrix.

3. Blood vessels grow into the enlarged lacunae of the calcified cartilage (figure 6.13, step 3). The connective tissue surrounding the blood vessels brings in osteoblasts and osteoclasts from the periosteum. The osteoblasts produce bone on the surface of the calcified cartilage, forming bone trabeculae, which changes the calcified cartilage of the diaphysis into cancellous bone. This area of bone formation is called the **primary ossification center.**

4. As bone development proceeds, the cartilage model continues to grow, more perichondrium becomes periosteum, the bone collar thickens and extends further along the diaphysis, and additional cartilage within the diaphysis is calcified and transformed into cancellous bone (figure 6.13, step 4). Remodeling converts woven bone to

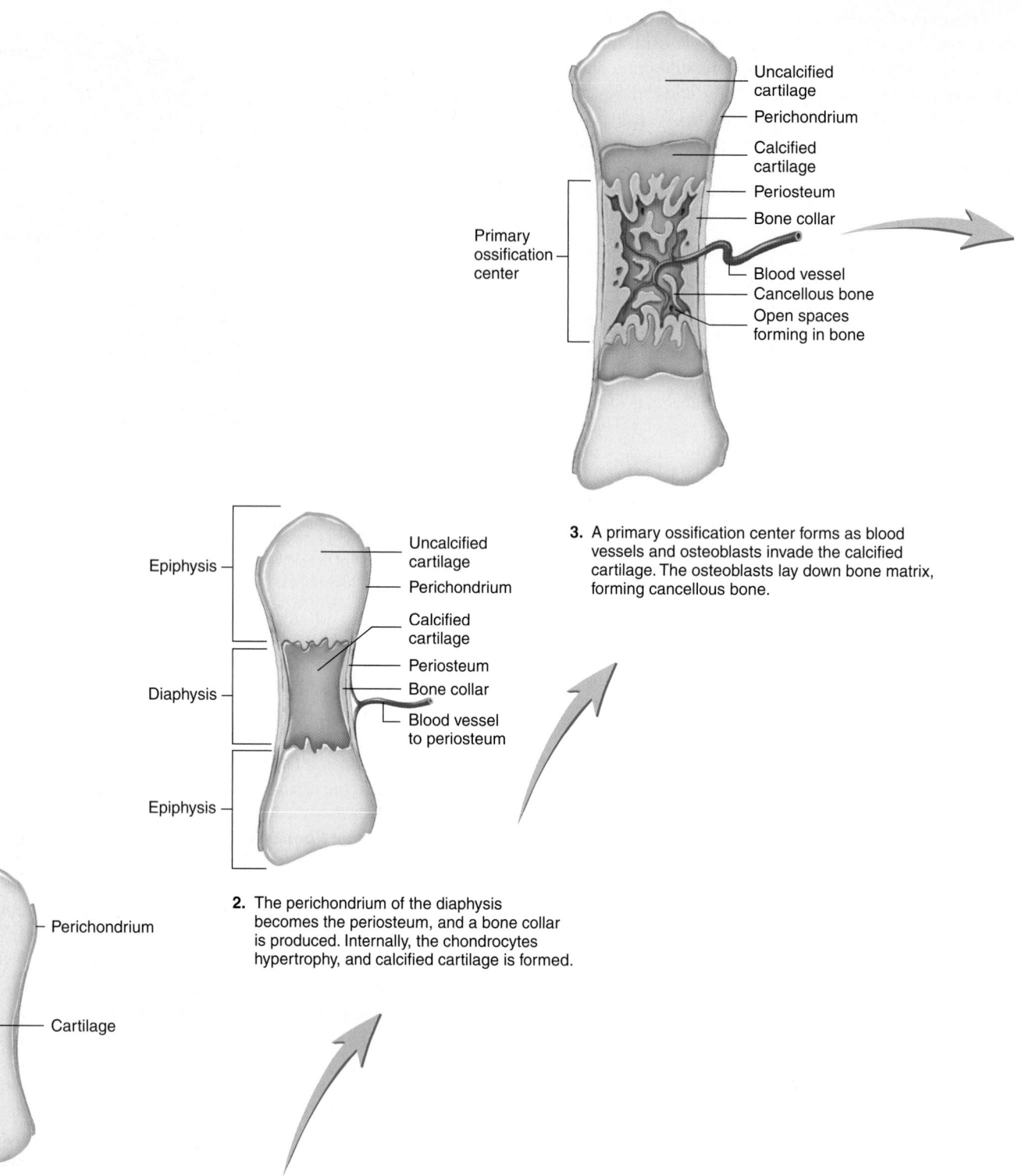

Uncalcified cartilage

Perichondrium

Calcified cartilage

Periosteum

Bone collar

Primary ossification center

Blood vessel

Cancellous bone

Open spaces forming in bone

3. A primary ossification center forms as blood vessels and osteoblasts invade the calcified cartilage. The osteoblasts lay down bone matrix, forming cancellous bone.

Epiphysis

Uncalcified cartilage

Perichondrium

Calcified cartilage

Periosteum

Bone collar

Blood vessel to periosteum

Diaphysis

Epiphysis

2. The perichondrium of the diaphysis becomes the periosteum, and a bone collar is produced. Internally, the chondrocytes hypertrophy, and calcified cartilage is formed.

Perichondrium

Cartilage

1. A cartilage model is produced by chondroblasts. The cartilage model is surrounded by perichondrium, except where joints will form.

Process Figure 6.13 Endochondral Ossification

Endochondral ossification begins with the formation of a cartilage model in the lower left part of the figure. See successive steps as indicated by the *blue arrows*.

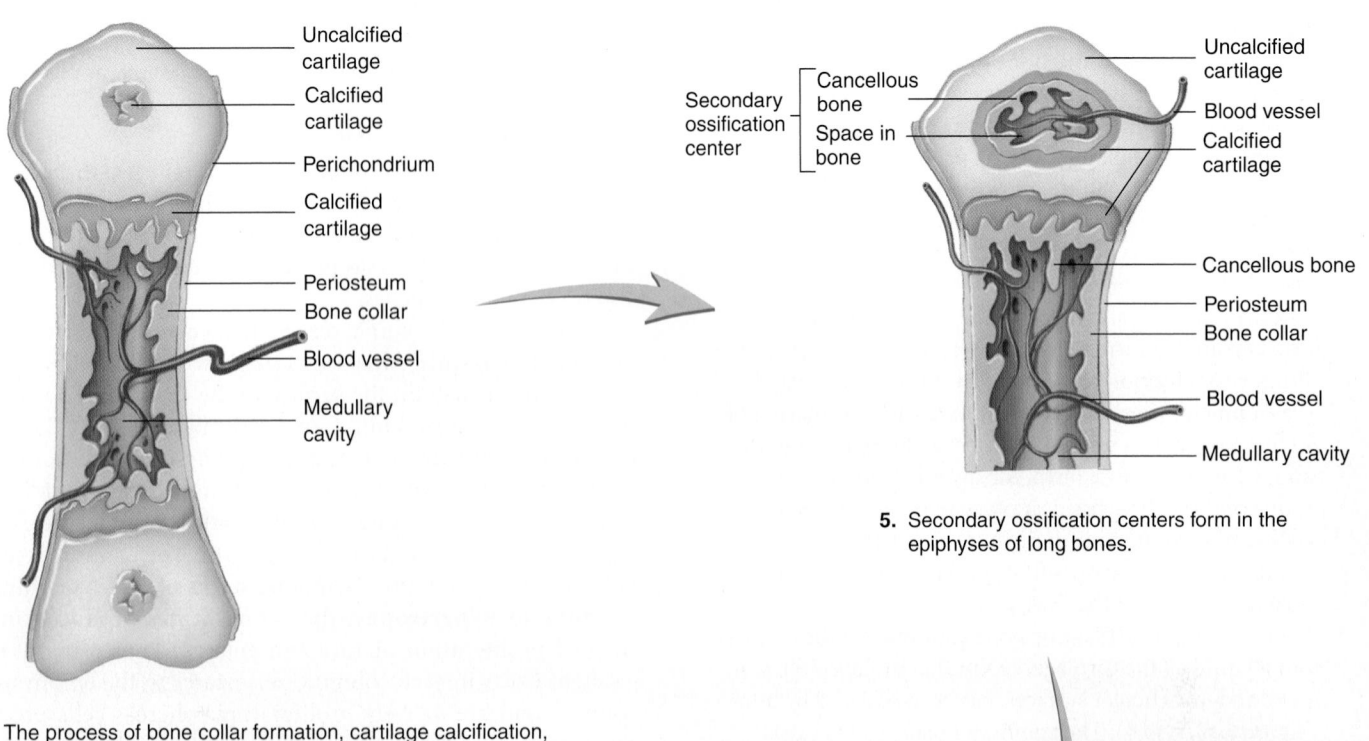

Uncalcified
cartilage

Calcified
cartilage

Perichondrium

Calcified
cartilage

Periosteum

Bone collar

Blood vessel

Medullary
cavity

Secondary
ossification
center

Cancellous
bone

Space in
bone

Uncalcified
cartilage

Blood vessel

Calcified
cartilage

Cancellous bone

Periosteum

Bone collar

Blood vessel

Medullary cavity

5. Secondary ossification centers form in the
epiphyses of long bones.

4. The process of bone collar formation, cartilage calcification,
and cancellous bone production continues. Calcified
cartilage begins to form in the epiphyses. A medullary
cavity begins to form in the center of the diaphysis.

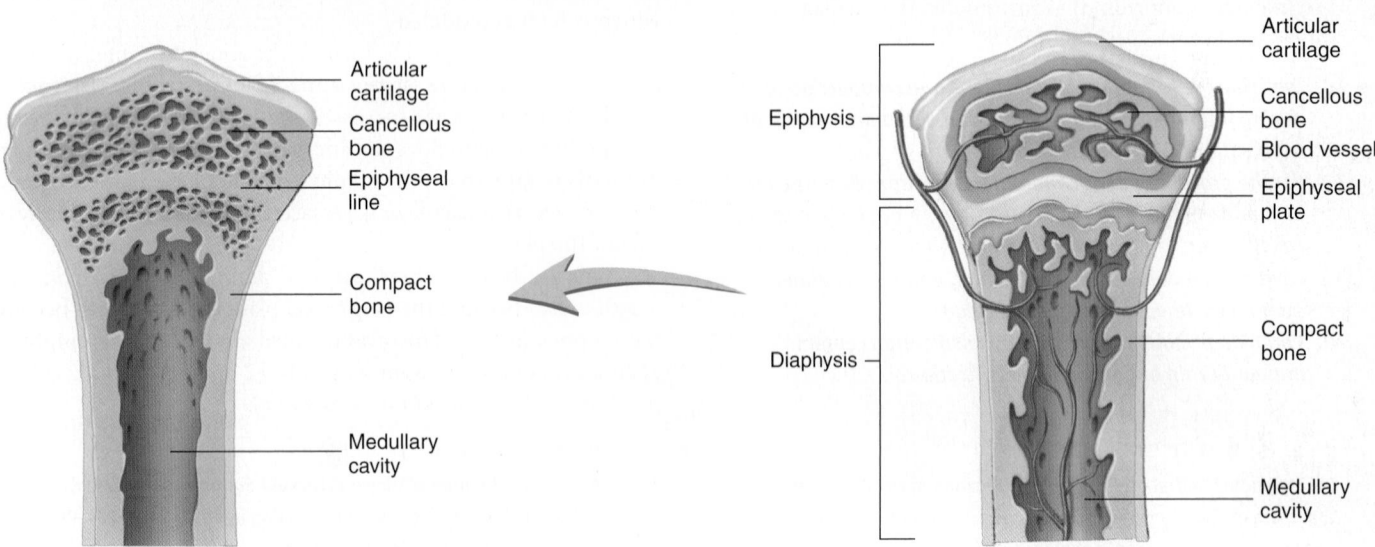

Articular
cartilage

Cancellous
bone

Epiphyseal
line

Compact
bone

Medullary
cavity

Epiphysis

Diaphysis

Articular
cartilage

Cancellous
bone

Blood vessel

Epiphyseal
plate

Compact
bone

Medullary
cavity

7. In a mature bone, the epiphyseal plate has
become the epiphyseal line and all the cartilage in the
epiphysis, except the articular cartilage, has become
bone.

6. The original cartilage model is almost completely
ossified. Unossified cartilage becomes the
epiphyseal plate and the articular cartilage.

Process Figure 6.13   *(continued)*

lamellar bone and contributes to the final shape of the bone. Osteoclasts remove bone from the center of the diaphysis to form the medullary cavity, and cells within the medullary cavity specialize to form red bone marrow.

5. In long bones the diaphysis is the primary ossification center, and additional sites of ossification, called **secondary ossification centers,** appear in the epiphyses (figure 6.13, step 5). The events occurring at the secondary ossification centers are the same as those occurring at the primary ossification centers, except that the spaces in the epiphyses don't enlarge to form a medullary cavity as in the diaphysis. Primary ossification centers appear during early fetal development, whereas secondary ossification centers appear in the proximal epiphysis of the femur, humerus, and tibia about 1 month before birth. A baby is considered full term if one of these three ossification centers can be seen on radiographs at the time of birth. At about 18–20 years of age the last secondary ossification center appears in the medial epiphysis of the clavicle.

6. Replacement of cartilage by bone continues in the cartilage model until all the cartilage, except that in the epiphyseal plate and on articular surfaces, has been replaced by bone (figure 6.13, step 6). The epiphyseal plate, which exists throughout an individual's growth, and the articular cartilage, which is a permanent structure, are derived from the original embryonic cartilage model.

7. In mature bone, cancellous and compact bone are fully developed and the epiphyseal plate has become the epiphyseal line. The only cartilage present is the articular cartilage at the ends of the bone (figure 6.13, step 7). All the original perichondrium that surrounded the cartilage model has become periosteum.

18. *Describe the formation of cancellous and compact bone during intramembranous ossification. What are centers of ossification? What are fontanels?*

19. *For the process of endochondral ossification, describe the formation of these structures: cartilage model, bone collar, calcified cartilage, primary ossification center, medullary cavity, secondary ossification center, epiphyseal plate, epiphyseal line, and articular cartilage.*

20. *When do primary and secondary ossification centers appear during endochondral ossification?*

P R E D I C T   4

During endochondral ossification, calcification of cartilage results in the death of chondrocytes. However, ossification of the bone matrix does not result in the death of osteocytes. Explain.

Bone Growth

Unlike cartilage, bones cannot grow by interstitial growth. Bones increase in size only by appositional growth, the formation of new bone on the surface of older bone or cartilage. For example, trabeculae grow in size by the deposition of new bone matrix by osteoblasts onto the surface of the trabeculae (see figure 6.11).

P R E D I C T   **5**

Explain why bones cannot undergo interstitial growth, as does cartilage.

Growth in Bone Length

Long bones and bony projections increase in length because of growth at the epiphyseal plate. In a long bone, the epiphyseal plate separates the epiphysis from the diaphysis (figure 6.14a). Long projections of bones, such as the processes of vertebrae (see chapter 7), also have epiphyseal plates.

Growth at the epiphyseal plate involves the formation of new cartilage by interstitial cartilage growth followed by appositional bone growth on the surface of the cartilage. The epiphyseal plate is organized into four zones (figure 6.14b). The **zone of resting cartilage** is nearest the epiphysis and contains randomly arranged chondrocytes that do not divide rapidly. The chondrocytes in the **zone of proliferation** produce new cartilage through interstitial cartilage growth. The chondrocytes divide and form columns resembling stacks of plates or coins. In the **zone of hypertrophy,** the chondrocytes produced in the zone of proliferation mature and enlarge. Thus a maturation gradient exists in each column: cells nearer to the epiphysis are younger and are actively proliferating, whereas cells progressively nearer the diaphysis are older and are undergoing hypertrophy. The **zone of calcification** is very thin and consists of cartilage matrix mineralized with calcium carbonate. The hypertrophied chondrocytes die, and blood vessels from the diaphysis grow into the area. The connective tissue surrounding the blood vessels contains osteoblasts from the endosteum. The osteoblasts line up on the surface of the calcified cartilage and through appositional bone growth deposit new bone matrix, which is later remodeled.

As new cartilage cells form in the zone of proliferation, and as these cells enlarge in the zone of hypertrophy, the overall length of the diaphysis increases (figure 6.15). The thickness of the epiphyseal plate does not increase, however, because the rate of cartilage growth on the epiphyseal side of the plate is equal to the rate at which cartilage is replaced by bone on the diaphyseal side of the plate.

As the bones achieve normal adult size, growth in bone length ceases because the epiphyseal plate is ossified and becomes the epiphyseal line. This event, called closure of the epiphyseal plate, occurs between approximately 12 and 25 years of age, depending on the bone and the individual.

P R E D I C T   **6**

A 15-year-old football player is tackled during a game, and the epiphyseal plate of the left femur is damaged (figure 6.16). What are the results of such an injury, and why is recovery difficult?

Growth at Articular Cartilage

Epiphyses increase in size because of growth at the articular cartilage. In addition, growth at the articular cartilage increases the size of bones that do not have an epiphysis, such as short bones. The process of growth in articular cartilage is similar to that occurring in the epiphyseal plate, except that the chondrocyte columns are not as obvious. The chondrocytes near the surface of the articular cartilage are similar to those in the zone of resting

Process Figure 6.14   Epiphyseal Plate

(a) Radiograph of the knee, showing the epiphyseal plate of the tibia (shinbone). Because cartilage does not appear readily on x-ray film, the epiphyseal plate appears as a black area between the white diaphysis and the epiphyses. (b) Zones of the epiphyseal plate.

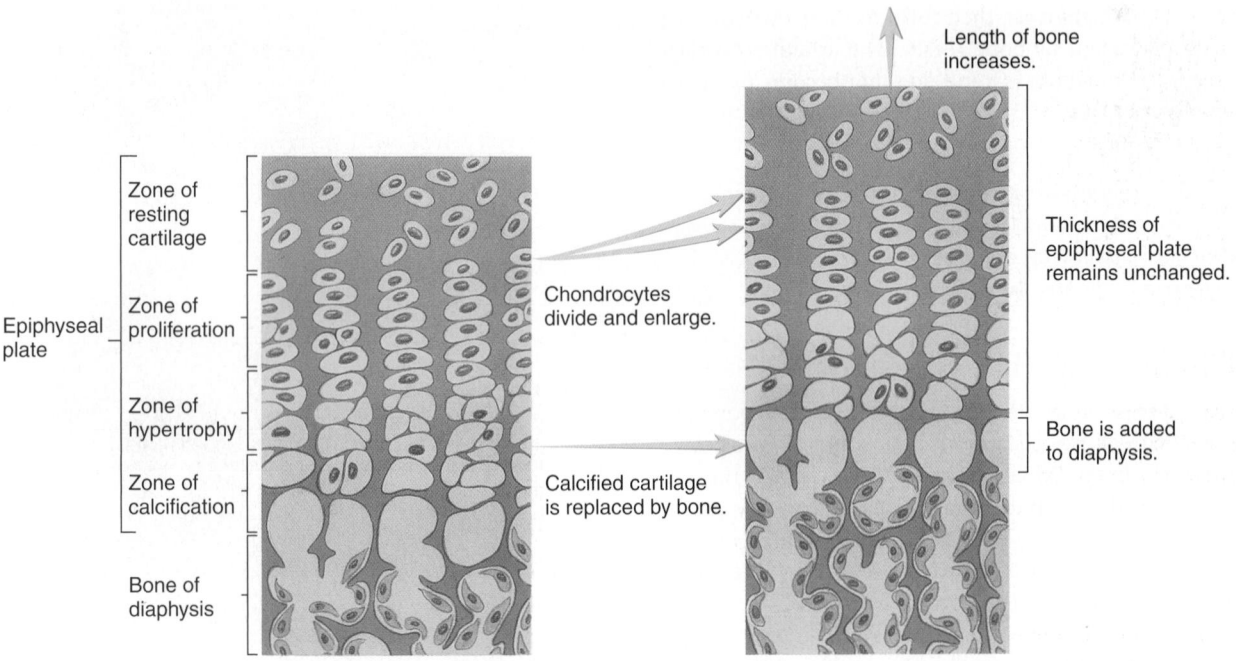

Figure 6.15   Bone Growth in Length at the Epiphyseal Plate

New cartilage is formed on the epiphyseal side of the plate at the same rate that new bone is formed on the diaphyseal side of the plate. Consequently, the epiphyseal plate remains the same thickness, but the length of the diaphysis increases.

Figure 6.16   Fracture of the Epiphyseal Plate

Radiograph of an adolescent's knee. The femur (thighbone) is separated from the tibia (leg bone) by a joint cavity. The epiphyseal plate of the femur is fractured, thereby separating the diaphysis from the epiphysis.

cartilage of the epiphyseal plate. In the deepest part of the articular cartilage, nearer bone tissue, the cartilage is calcified, dies, and is ossified to form new bone.

When the epiphyses reach their full size, the growth of cartilage and its replacement by bone ceases. The articular cartilage, however, persists throughout life and does not become ossified as does the epiphyseal plate.

> **P R E D I C T   7**
>
> Growth at the epiphyseal plate stops when the epiphyseal cartilage becomes ossified. The articular cartilage, however, does not become ossified when growth of the epiphysis ceases. Explain why it is advantageous for the articular cartilage not to be ossified.

Growth in Bone Width

Long bones increase in width (diameter) and other bones increase in size or thickness because of appositional bone growth beneath the periosteum. When bone growth in width is rapid, as in young bones or during puberty, osteoblasts from the periosteum lay down bone to form a series of ridges with grooves between them (figure 6.17, step 1). The periosteum covers the bone ridges and extends down into the bottom of the grooves, and one or more blood vessels of the periosteum lies within each groove. As the osteoblasts continue to produce bone, the ridges increase in size, extend toward each other, and meet to change the groove into a tunnel (figure 6.17, step 2). The name of the periosteum in the tunnel changes to the endosteum because the membrane

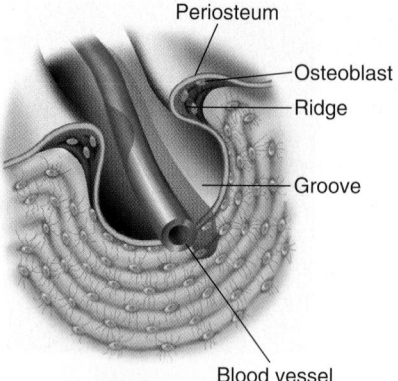

1. Osteoblasts beneath the periosteum lay down bone (*dark brown*) to form ridges separated by grooves. Blood vessels of the periosteum lie in the grooves.

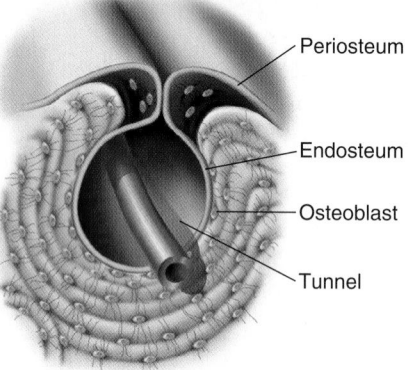

2. The groove is transformed into a tunnel when the bone built on adjacent ridges meets. The periosteum of the groove becomes the endosteum of the tunnel.

3. Appositional growth by osteoblasts from the endosteum results in the formation of a new concentric lamella.

4. The production of additional concentric lamellae fills in the tunnel and completes the formation of the osteon.

Process Figure 6.17   Bone Growth in Width

Bones can increase in width by the formation of new osteons beneath the periosteum.

now lines an internal bone surface. Osteoblasts from the endosteum lay down bone to form a concentric lamella (figure 6.17, step 3). The production of additional lamellae fills in the tunnel, encloses the blood vessel, and produces an osteon (figure 6.17, step 4).

When bone growth in width is slow, the surface of the bone becomes smooth as osteoblasts from the periosteum lay down even layers of bone to form circumferential lamellae. The circumferential lamellae are broken down during remodeling to form osteons (see "Bone Remodeling" on p. 183).

> 21. **Name and describe the events occurring in the four zones of the epiphyseal plate. Explain how the epiphyseal plate remains the same thickness while the bone increases in length.**
> 22. **Describe the process of growth at the articular cartilage. What happens to the epiphyseal plate and the articular cartilage when bone growth ceases?**
> 23. **Describe how new osteons are produced as a bone increases in width.**

Factors Affecting Bone Growth

Bones of an individual's skeleton usually reach a certain length, thickness, and shape through the processes described in the previous sections. The potential shape and size of a bone and an individual's final adult height are determined genetically, but factors such as nutrition and hormones can greatly modify the expression of those genetic factors.

Nutrition

Because bone growth requires chondroblast and osteoblast proliferation, any metabolic disorder that affects the rate of cell proliferation or the production of collagen and other matrix components affects bone growth, as does the availability of calcium or other minerals needed in the mineralization process.

The long bones of a child sometimes exhibit lines of arrested growth, which are transverse regions of greater bone density crossing an otherwise normal bone. These lines are caused by greater calcification below the epiphyseal plate of a bone, where it has grown at a slower rate during an illness or severe nutritional deprivation. They demonstrate that illness or malnutrition during the time of bone growth can cause a person to be shorter than he or she would have been otherwise.

Certain vitamins are important in very specific ways to bone growth. **Vitamin D** is necessary for the normal absorption of calcium from the intestines (see chapters 5 and 24). The body can either synthesize or ingest vitamin D. Its rate of synthesis increases when the skin is exposed to sunlight.

Insufficient vitamin D in children causes **rickets,** a disease resulting from reduced mineralization of the bone matrix. Children with rickets can have bowed bones and inflamed joints. During the winter in northern climates if children are not exposed to sufficient sunlight, vitamin D can be taken as a dietary supplement to prevent rickets. The body's inability to absorb fats in which vitamin D is soluble can also result in vitamin D deficiency. This condition can occur in adults who suffer from digestive disorders and

can be one cause of "adult rickets," or **osteomalacia** (os'tē-ō-mă-lā'shē-ă), which is a softening of the bones as a result of calcium depletion.

Vitamin C is necessary for collagen synthesis by osteoblasts. Normally, as old collagen breaks down, new collagen is synthesized to replace it. Vitamin C deficiency results in bones and cartilage that are deficient in collagen because collagen synthesis is impaired. In children, vitamin C deficiency can cause growth retardation. In children and adults, vitamin C deficiency can result in **scurvy,** which is marked by ulceration and hemorrhage in almost any area of the body because of the lack of normal collagen synthesis in connective tissues. Wound healing, which requires collagen synthesis, is hindered in patients with vitamin C deficiency. In extreme cases the teeth can fall out because the ligaments that hold them in place break down.

Hormones

Hormones are very important in bone growth. **Growth hormone** from the anterior pituitary increases general tissue growth (see chapters 17 and 18), including overall bone growth, by stimulating interstitial cartilage growth and appositional bone growth. **Thyroid hormone** is also required for normal growth of all tissues, including cartilage; therefore, a decrease in this hormone can result in decreased size of the individual. **Sex hormones** also influence bone growth. Estrogen (a class of female sex hormones) and testosterone (a male sex hormone) initially stimulate bone growth, which accounts for the burst of growth at the time of puberty, when production of these hormones increases. Both hormones also stimulate ossification of epiphyseal plates, however, and thus the cessation of growth. Females usually stop growing earlier than males because estrogens cause a quicker closure of the epiphyseal plate than does testosterone. Because their entire growth period is somewhat shorter, females usually don't reach the same height as males. Decreased levels of testosterone or estrogen can prolong the growth phase of the epiphyseal plates, even though the bones grow more slowly. Growth is very complex, however, and is influenced by many factors in addition to sex hormones, such as other hormones, genetics, and nutrition.

> 24. **Explain how illness or malnutrition can affect bone growth. How do vitamins D and C affect bone growth?**
> 25. **What are the effects of growth hormone and thyroid hormone on bone growth?**
> 26. **What effects do estrogen and testosterone have on bone growth? How do these effects account for the average height difference observed in men and women?**

P R E D I C T
A 12-year-old female has an adrenal tumor that produces large amounts of estrogen. If untreated, what effect will this condition have on her growth for the next 6 months? On her height when she is 18?

Bone Remodeling

Just as we renovate or remodel our homes when they become outdated, when bone becomes old, it's replaced with new bone in a process called **bone remodeling.** In this process, osteoclasts remove old bone and osteoblasts deposit new bone. Bone remodeling converts woven bone into lamellar bone, and it is involved in bone growth, changes in bone shape, the adjustment of the bone to stress, bone repair, and calcium ion (Ca^{2+}) regulation in the body. For example, as a long bone increases in length and diameter, the size of the medullary cavity also increases (figure 6.18). Otherwise, the bone would consist of nearly solid bone matrix and would be very heavy. A cylinder with the same height, weight, and composition as a solid rod but with a greater diameter can support much more weight than the rod without bending. Bone therefore has a mechanical advantage as a cylinder rather than as a rod. The relative thickness of compact bone is maintained by the removal of bone on the inside by osteoclasts and the addition of bone to the outside by osteoblasts.

Remodeling is also responsible for the formation of new osteons in compact bone. This process occurs in two ways. First, within already existing osteons, osteoclasts enter a central canal through the blood vessels and begin to remove bone from the center of the osteon, resulting in an enlarged tunnel through the bone. New concentric lamellae are then formed around the vessels until the new osteon fills the area occupied by the old osteon. Second, a few osteoclasts in the periosteum remove bone, resulting in groove formation along the surface of the bone. Periosteal capillaries lie within these grooves and become surrounded to form a tunnel as the osteoblasts of the periosteum form new bone. Additional lamellae then are added to the inside of the tunnel until an osteon results.

Bone is constantly being removed by osteoclasts, and new bone is being formed by osteoblasts. This remodeling process,

however, leaves behind portions of older bone called interstitial lamellae (see figure 6.10).

Mechanical Stress and Bone Strength

Remodeling, the formation of additional bone, alteration in trabecular alignment to reinforce the scaffolding, or other changes can modify the strength of the bone in response to the amount of stress applied to it. Mechanical stress applied to bone increases osteoblast activity in bone tissue, and removal of mechanical stress decreases osteoblast activity. Under conditions of reduced stress, such as when a person is bedridden or paralyzed, osteoclast activity continues at a nearly normal rate, but osteoblast activity is reduced, resulting in a decrease in bone density. In addition, pressure in bone causes an electrical change that increases the activity of osteoblasts. Applying weight (pressure) to a broken bone therefore speeds the healing process. Weak pulses of electric current applied to a broken bone sometimes are used clinically to speed the healing process.

27. *What cells are involved in bone remodeling? Describe how the medullary cavity of a long bone can increase in size as the width of the bone increases.*
28. *Explain two ways that remodeling is responsible for the formation of new osteons in compact bone.*
29. *How does bone adjust to stress? Describe the role of osteoblasts and osteoclasts in this process. What happens to bone that is not subjected to stress?*

Epiphyseal growth
Growth in cartilage surrounding epiphysis
Cartilage replaced by bone
Bone remodeled

Growth in length
Cartilage growth in epiphyseal plate
Cartilage replaced by bone
Bone remodeled
Bone resorption

Growth in diameter
Bone addition
Bone resorption

Growing bone

Articular cartilage

Epiphyseal line

Adult bone

Figure 6.18   Remodeling of a Long Bone

The epiphysis enlarges and the diaphysis increases in length as new cartilage is formed and replaced by bone, which is remodeled. The diameter of the bone increases as a result of bone growth on the outside of the bone, and the size of the medullary cavity increases because of bone resorption.

Clinical Focus   Bone Disorders

Growth and Development Disorders

Giantism is a condition of abnormally increased height that usually results from excessive cartilage and bone formation at the epiphyseal plates of long bones (figure A*a*). The most common type of giantism, **pituitary giantism,** results from excess secretion of pituitary growth hormone. The large stature of some individuals, however, can result from genetic factors rather than from abnormal levels of growth hormone. **Acromegaly** (ak-rō-meg′ă-lē) is also caused by excess pituitary growth hormone secretion; however, acromegaly involves growth of connective tissue, including bones, after the epiphyseal plates have ossified. The effect mainly involves increased diameter of all bones and is most strikingly apparent in the face and hands. Many pituitary giants also develop acromegaly later in life.

Dwarfism, the condition in which a person is abnormally short, is the opposite of giantism (see figure A*a*). **Pituitary dwarfism** results when abnormally low levels of pituitary growth hormone affect the whole body, thus producing a small person who is normally proportioned. **Achondroplastic** (ā-kon-drō-plas′tik) **dwarfism** results in disproportionately short long bones. It's more common than proportionate dwarfing and produces a person with a nearly normal-sized trunk and head but shorter-than-normal limbs. Most cases of achondroplastic dwarfism are the result of genetic defects that cause deficient or improper growth of the cartilage model, especially the epiphyseal plate, and often involve deficient collagen synthesis. Often the cartilage matrix doesn't have its normal integrity, and the chondrocytes of the epiphysis cannot form their normal columns, even though rates of cell proliferation may be normal.

Osteogenesis imperfecta (os′tē-ō-jen′ĕ-sis im-per-fek′tă), a group of genetic disorders producing very brittle bones that are easily fractured, occurs because insufficient

(a) **Giant and dwarf**

Figure A   Bone Disorders

collagen develops to properly strengthen the bones. Intrauterine fractures of the extremities usually heal in poor alignment, thereby causing the limbs to appear bent and short (figure A*b*). Several other hereditary disorders of bone mineralization involve the enzymes responsible for normal phosphate or calcium metabolism. They closely resemble rickets and result in weak bones.

Bacterial Infections

Osteomyelitis (os′tē-ō-mī-ĕ-lī′tis) is bone inflammation that often results from bac-

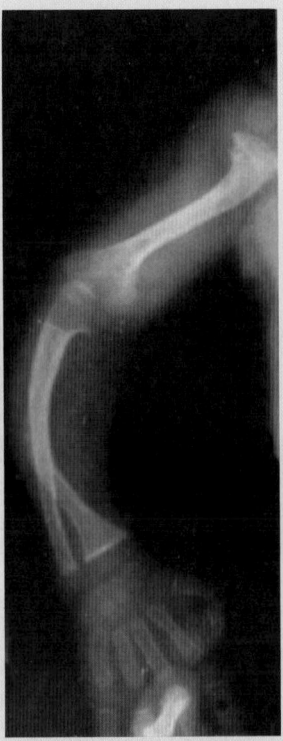

(b) **Osteogenesis imperfecta**

terial infection. It can lead to complete destruction of the bone. *Staphylococcus aureus,* often introduced into the body through wounds, is a common cause of osteomyelitis (figure A*c*). Bone tuberculosis, a specific type of osteomyelitis, results from spread of the tubercular bacterium *(Mycobacterium tuberculosis)* from the initial site of infection such as the lungs to the bones through the circulatory system.

Tumors

Many types of tumors occur that cause a wide range of resultant bone defects with varying prognoses (figure A*d*). Tumors can be benign or malignant. Malignant bone tumors can metastasize to other parts of the body, or they can spread to bone from metastasizing tumors elsewhere in the body.

Continued

Continued

Osteomyelitis

(c) **Osteomyelitis**

Tumor

(d) **Bone tumor**

Decalcification

Osteomalacia (os′-t-ē-ō-mă-lā′shē-ă), or the softening of bones, results from calcium depletion from bones. If the body has an un- usual need for calcium, such as during pregnancy, when growth of the fetus re- quires large amounts of calcium, it can be removed from the mother's bones, which consequently become soft and weakened. **Osteoporosis,** which is a major disorder of decalcification, is discussed in the Systems Pathology section on p. 190.

Bone Repair

Bone is a living tissue that can undergo repair following damage to it. This process has four major steps.

1. *Hematoma formation* (figure 6.19, step 1). When bone is fractured, the blood vessels in the bone and surrounding periosteum are damaged, and a hematoma forms. A **hematoma** (hē-mă-tō′mă, hem-ă-tō′mă) is a localized mass of blood released from blood vessels but confined within an organ or space. Usually the blood in a hematoma forms a clot, which consists of fibrous proteins that stop the bleeding. Disruption of blood vessels in the central canals results in inadequate blood delivery to osteocytes, and bone tissue adjacent to the fracture site dies. Inflammation and swelling of tissues around the bone often occur following the injury.

2. *Callus formation* (figure 6.19, step 2). A **callus** (kal′ŭs) is a mass of tissue that forms at a fracture site and connects the broken ends of the bone. An **internal callus** forms *between* the ends of the broken bone, and in the marrow cavity if the fracture occurs in the diaphysis of a long bone. Several days after the fracture, blood vessels grow into the clot. As the clot dissolves (see chapter 19), macrophages clean up cell debris, osteoclasts break down dead bone tissue, and fibroblasts produce collagen and other extracellular materials to form granulation tissue (see chapter 4). As the fibroblasts continue to produce collagen fibers, a denser fibrous network, which helps to hold the bone together, is produced. Chondroblasts derived from osteochondral progenitor cells of the periosteum and endosteum begin to produce cartilage in the fibrous network. As these events are occurring, osteochondral progenitor cells in the endosteum become osteoblasts and produce new bone that contributes to the internal callus.

   The **external callus** forms a collar *around* the opposing ends of the bone fragments. Osteochondral progenitor cells

Hematoma formation

1. Blood released from damaged blood vessels forms a hematoma.

Callus formation

2. The internal callus forms between the ends of the bones, and the external callus forms a collar around the break.

Callus ossification

3. Woven, cancellous bone replaces the internal and external calluses.

Bone remodeling

4. Compact bone replaces woven bone and part of the internal callus is removed, restoring the medullary cavity.

Process Figure 6.19   Bone Repair

from the periosteum become osteoblasts, which produce bone, and chondroblasts, which produce cartilage. Cartilage production is more rapid than bone production, and cartilage from either side of the break eventually grows together. The external callus is a bone–cartilage collar that stabilizes the ends of the broken bone. In modern medical practice, stabilization of the bone is assisted by using a cast or surgical implantation of metal supports.

3. *Callus ossification* (figure 6.19, step 3). Like the cartilage models formed during fetal development, the cartilage in the external callus is replaced by woven, cancellous bone through endochondral ossification. The result is a stronger external callus. Even as the internal callus is forming and replacing the hematoma, osteoblasts from the periosteum and endosteum enter the internal callus and begin to produce bone. Eventually the fibers and cartilage of the internal callus are replaced by woven, cancellous bone, which further stabilizes the broken bone.

4. *Remodeling of bone* (figure 6.19, step 4). Filling the gap between bone fragments with an internal callus of woven bone is not the end of the repair process because woven bone is not as structurally strong as the original lamellar bone. Repair is not complete until the woven bone of the internal callus and the dead bone adjacent to the fracture site are replaced by compact bone. In this compact bone, osteons from both sides of the break extend across the fracture line to "peg" the bone fragments together. This remodeling process takes time and may not be complete even after a year. As the internal callus is remodeled and becomes stronger, the external callus is reduced in size by osteoclast activity. Eventually, repair may be so complete that no evidence of the break remains, however, the

repaired zone usually remains slightly thicker than the adjacent bone. If the fracture occurred in the diaphysis of a long bone, remodeling also restores the medullary cavity.

30. *How does breaking a bone result in hematoma formation?*
31. *Distinguish between the location and composition of the internal and external callus.*
32. *Why is remodeling of the ossified callus necessary?*

Uniting Broken Bones

Before formation of compact bone between the broken ends of a bone can take place, the appropriate substrate must be present. Normally this is the woven, cancellous bone of the internal callus. If formation of the internal callus is prevented by infections, bone movements, or the nature of the injury, then nonunion of the bone occurs. This condition can be treated by surgically implanting an appropriate substrate such as living bone taken from another site in the body or dead bone from cadavers. Other substrates have also been used. For example, a specific marine coral calcium phosphate is converted into a predominantly hydroxyapatite biomatrix that is very much like cancellous bone.

Calcium Homeostasis

Bones play an important role in regulating blood Ca^{2+} levels, which must be maintained within narrow limits for functions such as muscle contraction and membrane potentials to occur normally (see chapters 9 and 11). Bone is the major storage site for calcium in the body, and movement of Ca^{2+} into and out of bone helps to determine blood Ca^{2+} levels. Calcium ions move into bone as osteoblasts build new bone and out of bone as osteoclasts break down bone (figure 6.20). When osteoblast and osteoclast activity is balanced, the movement of Ca^{2+} into and out of a bone is equal.

When blood Ca^{2+} levels are too low, osteoclast activity increases. More Ca^{2+} are released by osteoclasts from bone into the blood than are removed by osteoblasts from the blood to make new bone. Consequently, a net movement of Ca^{2+} occurs from bone into blood, and blood Ca^{2+} levels increase. Conversely, if blood Ca^{2+} levels are too high, osteoclast activity decreases. Less Ca^{2+} are released by osteoclasts from bone into the blood than are taken from the blood by osteoblasts to produce new bone. As a result, a net movement of Ca^{2+} occurs from the blood to bone, and blood Ca^{2+} levels decrease.

Parathyroid hormone (PTH) from the parathyroid glands (see figure 18.11) is the major regulator of blood Ca^{2+} levels. If the blood Ca^{2+} level decreases, the secretion of PTH increases, resulting in increased numbers of osteoclasts, which causes increased bone breakdown and increased blood Ca^{2+} levels (see figure 6.20). In addition, osteoblasts respond to PTH by releasing enzymes that result in the breakdown of the layer of unmineralized organic bone matrix covering bone, thereby making the mineralized bone matrix available to osteocytes.

The regulation of osteoclast numbers is mediated through osteoblasts and red bone marrow stromal (stem) cells. When PTH levels increase, PTH binds to its receptors on osteoblasts/stromal cells. In response, these cells produce a surface molecule called **receptor for activation of nuclear factor kappa B ligand (RANKL).** When RANKL binds to its receptor on the surface of osteoclast precursor cells, the cells are stimulated to become osteoclasts. Increased PTH

also inhibits the release from osteoblasts/stromal cells of a protein called **osteoprotegerin** (os′tē-ō-prō-teg′er-in) **(OPG).** OPG inhibits the formation of osteoclasts because it binds to RANKL and prevents it from stimulating osteoclast precursor cells. Thus, increased PTH promotes an increase in osteoclast numbers by increasing RANKL, which stimulates osteoclast precursor cells, and by decreasing OPG, which decreases the inhibition of osteoclast precursor cells. Conversely, decreased PTH results in fewer osteoclasts by decreasing RANKL and increasing OPG.

PTH also regulates blood Ca^{2+} levels by increasing Ca^{2+} uptake in the small intestine (see figure 6.20). Increased PTH promotes the formation of vitamin D in the kidneys, and vitamin D increases the absorption of Ca^{2+} from the small intestine. PTH also increases the reabsorption of Ca^{2+} from urine in the kidneys, which reduces Ca^{2+} lost in the urine.

Tumors that secrete large amounts of PTH can cause so much bone breakdown that bones become weakened and fracture easily. On the other hand, an increase in blood Ca^{2+} levels results in less PTH secretion, decreased osteoclast activity, reduced Ca^{2+} release from bone, and decreased blood Ca^{2+} levels.

Calcitonin (kal-si-tō′nin), secreted from the thyroid gland (see figure 18.8), decreases osteoclast activity (see figure 6.20) by binding to receptors on the osteoclasts. An increase in blood Ca^{2+} levels stimulates the thyroid gland to secrete calcitonin, which inhibits osteoclast activity. PTH and calcitonin are described more fully in chapters 18 and 27.

● **33.** *Name the hormone that is the major regulator of Ca^{2+} levels in the body. What stimulates the secretion of this hormone?*

1. Osteoclasts break down bone and release Ca^{2+} into the blood, and osteoblasts remove Ca^{2+} from the blood to make bone. (*Blue arrows represent the movement of Ca^{2+}.*) Parathyroid hormone (PTH) regulates blood Ca^{2+} levels by indirectly stimulating osteoclast activity, resulting in increased Ca^{2+} release into the blood. Calcitonin plays a minor role in Ca^{2+} maintenance by inhibiting osteoclast activity.

2. In the kidneys, PTH increases Ca^{2+} reabsorption from the urine.

3. In the kidneys, PTH also promotes the formation of active vitamin D (*green arrows*), which increases Ca^{2+} absorption from the small intestine.

Bone

PTH promotes and calcitonin inhibits Ca^{2+} release into the blood by osteoclasts

1

Ca^{2+} removed from blood by osteoblasts

Blood

2

PTH promotes Ca^{2+} reabsorption from the urine

Kidney

Ingested Ca^{2+}

Unabsorbed Ca^{2+} lost in the feces

Ca^{2+} lost in the urine

3

PTH promotes active vitamin D formation

Vitamin D promotes Ca^{2+} absorption

Small intestine

Blood

Process Figure 6.20 Calcium Homeostasis

Clinical Focus Classification of Bone Fractures

Bone fractures are classified in several ways. The most commonly used classification involves the severity of injury to the soft tissues surrounding the bone. An **open fracture** (formerly called compound) occurs when an open wound extends to the site of the fracture or when a fragment of bone protrudes through the skin. If the skin is not perforated, the fracture is called a **closed fracture** (formerly called simple). If the soft tissues around a closed fracture are damaged, the fracture is called a **complicated fracture.**

Two other terms to designate fractures are **incomplete,** in which the fracture does not extend completely across the bone, and **complete,** in which the bone is broken into at least two fragments (figure B*a*). An incomplete fracture that occurs on the convex side of the curve of the bone is a **greenstick fracture. Hairline fractures** are incomplete fractures in which the two sections of bone do not separate; they are common in skull fractures.

Comminuted (kom′i-noo-ted) **fractures** are complete fractures in which the bone breaks into more than two pieces—usually two major fragments and a smaller fragment (figure B*b*). **Impacted fractures** are those in which one fragment is driven into the cancellous portion of the other fragment (figure B*c*).

Fractures are also classified according to the direction of the fracture within a bone. **Linear fractures** run parallel to the long axis of the bone, and **transverse fractures** are at right angles to the long axis (figure B*b*). **Spiral fractures** have a helical course around the bone, and **oblique fractures** run obliquely in relation to the long axis (figure B*d*). **Dentate fractures** have rough, toothed, broken ends, and **stellate fractures** have breakage lines radiating from a central point.

Figure B Bone Fractures

(*a*) Complete and incomplete. (*b*) Transverse and comminuted. (*c*) Impacted. (*d*) Spiral and oblique.

34. *Describe how PTH controls the number of osteoclasts. What are the effects of PTH on the formation of vitamin D, Ca^{2+} uptake in the small intestine, and reabsorption of Ca^{2+} from urine?*

35. *What stimulates calcitonin secretion? How does calcitonin affect osteoclast activity?*

Effects of Aging on the Skeletal System

The most significant age-related changes in the skeletal system affect the quality and quantity of bone matrix. Recall that a mineral (hydroxyapatite) in the bone matrix gives bone compression (weight-bearing) strength, but collagen fibers make the bone flexible. The bone matrix in an older bone is more brittle than in a younger bone because decreased collagen production results in a matrix that has relatively more mineral and less collagen fibers. With aging, the amount of matrix also decreases because the rate of matrix formation by osteoblasts becomes slower than the rate of matrix breakdown by osteoclasts.

Bone mass is at its highest around age 30, and men generally have denser bones than women because of the effects of testosterone and greater body weight. Race also affects bone mass. African-Americans and Hispanics have higher bone masses than Caucasians and Asians. After age 35, both men and women have an age-related loss of bone of 0.3%–0.5% a year. This loss can increase 10-fold in women after menopause, and they can have a bone loss of 3%–5% a year for approximately 5–7 years (see "Systems Pathology: Osteoporosis" on p. 190).

Systems Pathology
Osteoporosis

Mrs. B is a 70-year-old grandmother. Since she was a teenager, she has been a heavy smoker. She is typically sedentary, seldom goes outside, has not had the best dietary habits, and is underweight. One of her favorite yearly events is the family picnic on the Fourth of July. During one picnic, misfortune struck when Mrs. B tripped on a lawn sprinkler and fell. She was unable to stand because of severe hip pain, so she was rushed to the hospital, where a radiograph revealed that her femur was broken (figure C*a*) and that she had osteoporosis (figure C*b*).

It was decided that hip replacement surgery was indicated. Before the surgery could be performed, however, a fat embolism from the fracture site lodged in her lungs, making it difficult for her to breathe. The surgery was postponed and the fracture immobilized until she recovered from the fat embolism. Three weeks after the accident, Mrs. B had a successful hip transplant and began physical therapy. She appeared to be on the road to recovery, but 6 weeks after the surgery she developed persistent pain and edema in her hip. A bone biopsy confirmed a postoperative infection that was successfully treated with antibiotics.

Background Information

Osteoporosis (os'tē-ō-pō-rō' sis), or porous bone, results from reduction in the overall quantity of bone tissue. It occurs when the rate of bone resorption exceeds the rate of bone formation. The loss of bone mass makes bones so porous and weakened that they become deformed and prone to fracture. The occurrence of osteoporosis increases with age. In both men and women, bone mass starts to decrease at about age 35 and continually decreases thereafter. Women can eventually lose approximately half, and men a quarter, of their cancellous bone. Osteoporosis is two and a half times more common in women than in men.

In postmenopausal women, the decreased production of the female sex hormone, estrogen, can cause osteoporosis. Estrogen is secreted by the ovaries, and it normally contributes to the maintenance of normal bone mass by inhibiting the stimulatory effects of PTH on osteoclast activity. Following menopause, estrogen production decreases, resulting in degeneration of cancellous bone, especially in the vertebrae of the spine and the bones of the forearm. Collapse of the vertebrae can cause a decrease in height or, in more severe cases, can produce kyphosis, or a "dowager's hump," in the upper back.

Conditions that result in decreased estrogen levels, other than menopause, can also cause osteoporosis. Examples include removal of the ovaries before menopause, extreme exercise to the point of amenorrhea (lack of menstrual flow), anorexia nervosa (self-starvation), and cigarette smoking.

In males, reduction in testosterone levels can cause loss of bone tissue. Decreasing testosterone levels are usually less of a problem for men than decreasing estrogen levels are for women for two reasons. First, because males have denser bones than females, loss of some bone tissue has less of an effect. Second, testosterone levels generally don't decrease significantly until after age 65, and even then the rate of decrease is often slow.

Overproduction of PTH, which results in overstimulation of osteoclast activity, can also cause osteoporosis.

Inadequate dietary intake or absorption of calcium can contribute to osteoporosis. Absorption of calcium from the small intestine decreases with age, and individuals with osteoporosis often have insufficient intake of calcium or vitamin D. Drugs that interfere with calcium uptake or use can also increase the risk of osteoporosis.

Coxa (hipbone)

Break

Femur (thighbone)

(a)

Normal bone

Osteoporotic bone

LM 5x

(b)

Figure C   Osteoporosis

(*a*) Radiograph of a broken hip. A "broken hip" is actually a break of the femur (thighbone) in the hip region. (*b*) Photomicrograph of normal bone and osteoporotic bone.

Finally, osteoporosis can result from inadequate exercise or disuse caused by fractures or paralysis. Significant amounts of bone are lost after 8 weeks of immobilization.

Treatments for osteoporosis are designed to reduce bone loss or increase bone formation, or both. Increased dietary calcium and vitamin D can increase calcium uptake and promote bone formation. Daily doses of 1000–1500 mg of calcium and 800 IU (20 μg) of vitamin D are recommended. Exercise, such as walking or using light weights, also appears to be effective not only in reducing bone loss but in increasing bone mass.

In postmenopausal women, **hormone replacement therapy (HRT)** with estrogen decreases osteoclast numbers by inhibiting the production of RANKL (see p. 188). This reduces bone loss but does not result in an increase in bone mass because osteoclast activity still exceeds osteoblast activity. However, the use of HRT to prevent bone loss is now discouraged because of the results of a study sponsored by the Women's Health Initiative. The study examined HRT in over 16,000 women, and found that HRT increased the risk of breast cancer, uterine cancer, heart attacks, strokes, and blood clots, but decreased the risk of hip fractures and colorectal cancer. **Selective estrogen receptor modulators (SERMs)** are a class of drugs that bind to estrogen receptors. They may be able to protect against bone loss without increasing the risk of breast cancer. For example, raloxifene (ral-ox′ĭ-fēn) stimulates estrogen receptors in bone but inhibits them in the breast and uterus.

Osteoprotegerin, which prevents RANKL from binding to its receptors, is under consideration as a treatment for osteoporosis. Calcitonin (Miacalcin), which inhibits osteoclast activity, is now available as a nasal spray. Calcitonin can be used to treat osteoporosis in men and women and has been shown to produce a slight increase in bone mass. **Statins** (stat′ins) are drugs that inhibit cholesterol synthesis. It has been discovered that statins also stimulate osteoblast activity, and there is some evidence that statins can reduce the risk of fractures. Alendronate (Fosamax) belongs to a class of drugs called **bisphosphonates** (bis-fos′fō-nāts). Bisphosphonates concentrate in bone, and when osteoclasts break down bone, the bisphosphonates are taken up by the osteoclasts. The bisphosphonates interfere with certain enzymes, leading to inactivation and lysis of the osteoclasts. Alendronate increases bone mass and reduces fracture rates even more effectively than calcitonin. Slow-releasing sodium fluoride (Slow Fluoride) in combination with calcium citrate (Citracal) also appears to increase bone mass.

Leptin is a protein hormone produce by adipocytes (fat cells). When released from fat cells into the blood, leptin travels to the brain, where it is a signal involved in the regulation of feeding and energy balance (see chapter 25). There's also evidence that decreased leptin causes the release from the brain of a yet to be identified substance that can increase osteoblast activity. Understanding the leptin pathway may lead to treatments for diseases such as osteoporosis.

Early diagnosis of osteoporosis may lead to the use of more preventative treatments. Instruments that measure the absorption of photons (particles of light) by bone are currently used, of which dual-energy x-ray absorptiometry (DEXA) is considered the best.

P R E D I C T
What advice should Mrs. B give to her granddaughter so that the granddaughter will be less likely to develop osteoporosis when she is Mrs. B's age?

System Interactions — The Effects of Osteoporosis on Other Systems

System	Interactions
Integumentary	Decreased exposure to sunlight because of an indoor lifestyle reduces vitamin D production and decreases Ca^{2+} absorption. Surgical wounds through the skin can allow the entry of bacteria, resulting in postoperative infections.
Muscular	A sedentary lifestyle and decreased body weight reduces stress on bone and contributes to osteoporosis. Muscle atrophy and weakness make it difficult to maintain balance, which increases the likelihood of falling and injury. Following hip replacement surgery, physical therapy places stress on the bones and improves muscular strength.
Nervous	Pain sensations following the injury and during rehabilitation help to prevent further injury.
Endocrine	Although not a factor in this case of osteoporosis, elevated PTH (usually from a benign parathyroid tumor) or elevated thyroid hormone (Graves' disease) can result in excessive osteoclast activity. Calcitonin is being used to treat osteoporosis.
Cardiovascular	Blood clotting following the injury starts the process of tissue repair. Blood cells are carried to the injury site to fight infections and remove cell debris. Blood vessels grow into the recovering tissue, providing nutrients and removing waste products.
Lymphatic and Immune	Immune cells resist infections and release chemicals that promote tissue repair. New immune cells are produced in bone marrow.
Respiratory	Excessive smoking lowers estrogen levels, which increases bone loss. A fat embolism from a fractured bone can impair respiration.
Digestive	Inadequate calcium and vitamin D in the diet or inadequate calcium absorption by the digestive system can contribute to osteoporosis.
Urinary	Calcium ions released from the bones are excreted through the urinary system.
Reproductive	Decreased estrogen levels following menopause contribute to osteoporosis.

Cancellous bone is lost at first as the trabeculae become thinner and weaker. The ability of the trabeculae to provide support also decreases as they become disconnected from each other. Eventually, some of the trabeculae completely disappear. Trabecular bone loss is greatest in the trabeculae that are under the least stress.

A slow loss of compact bone begins about age 40 and increases after age 45. The rate of compact bone loss, however, is approximately half the rate of trabecular bone loss. Bones become thinner, but their outer dimensions change little, because most loss of compact bone occurs under the endosteum on the inner surfaces of bones. In addition, the remaining compact bone becomes weaker as a result of incomplete bone remodeling. In a young bone, when osteons are removed, the resulting spaces are filled with new osteons. With aging, the new osteons fail to completely fill in the spaces produced when the older osteons are removed.

Significant loss of bone increases the likelihood of having bone fractures. For example, loss of trabeculae greatly increases the risk of compression fractures of the vertebrae (backbones) because the weight-bearing body of the vertebrae consists mostly of cancellous bone. In addition, loss of bone can cause deformity, loss of height, pain, and stiffness. For example, compression fractures of the vertebrae can cause an exaggerated curvature of the spine resulting in a bent-forward, stooped posture. Loss of bone from the jaws can also lead to tooth loss.

36. *What effect does aging have on the quality and quantity of bone matrix?*
37. *Describe how cancellous and compact bone change with age. How do these changes affect a person's health?*

S U M M A R Y

Functions of the Skeletal System   (p. 168)

1. The skeletal system consists of bones, cartilage, tendons, and ligaments.
2. The skeletal system supports the body, protects organs it surrounds, allows body movements, stores minerals and fats, and is the site of blood cell production.

Cartilage   (p. 168)

1. Chondroblasts produce cartilage and become chondrocytes. Chondrocytes are located in lacunae surrounded by matrix.
2. The matrix of cartilage contains collagen fibers (for strength) and proteoglycans (trap water).
3. The perichondrium surrounds cartilage.
   • The outer layer contains fibroblasts.
   • The inner layer contains chondroblasts.
4. Cartilage grows by appositional and interstitial growth.

Bone Anatomy   (p. 169)
Bone Shapes
Individual bones can be classified as long, short, flat, or irregular.

Structure of a Long Bone

1. The diaphysis is the shaft of a long bone, and the epiphyses are the ends.
2. The epiphyseal plate is the site of bone growth in length.
3. The medullary cavity is a space within the diaphysis.
4. Red marrow is the site of blood cell production, and yellow marrow consists of fat.
5. The periosteum covers the outer surface of bone.
   • The outer layer contains blood vessels and nerves.
   • The inner layer contains osteoblasts, osteoclasts, and osteochondral progenitor cells.
   • Perforating fibers hold the periosteum, ligaments, and tendons in place.
6. The endosteum lines cavities inside bone and contains osteoblasts, osteoclasts, and osteochondral progenitor cells.

Structure of Flat, Short, and Irregular Bones

Flat, short, and irregular bones have an outer covering of compact bone surrounding cancellous bone.

Bone Histology   (p. 171)
Bone Matrix

1. Collagen provides flexible strength.
2. Hydroxyapatite provides compressional strength.

Bone Cells

1. Osteoblasts produce bone matrix and become osteocytes.
   • Osteoblasts connect to one another through cell processes and surround themselves with bone matrix to become osteocytes.
   • Osteocytes are located in lacunae and are connected to one another through canaliculi.
2. Osteoclasts (with assistance from osteoblasts) break down bone.
3. Osteoblasts originate from osteochondral progenitor cells, whereas osteoclasts originate from stem cells in red bone marrow.
4. Ossification, the formation of bone, occurs through appositional growth.

Woven and Lamellar Bone

1. Woven bone has collagen fibers oriented in many different directions. It's remodeled to form lamellar bone.
2. Lamellar bone is arranged in thin layers, called lamellae, which have collagen fibers oriented parallel to one another.

Cancellous and Compact Bone

1. Cancellous bone has many spaces.
   • Lamellae combine to form trabeculae, beams of bone that interconnect to form a latticelike structure with spaces filled with bone marrow and blood vessels.
   • The trabeculae are oriented along lines of stress and provide structural strength.
2. Compact bone is dense with few spaces.
   • Compact bone consists of organized lamellae: circumferential lamellae form the outer surface of compact bones; concentric lamellae surround central canals, forming osteons; interstitial lamellae are remnants of lamellae left after bone remodeling.
   • Canals within compact bone provide a means for the exchange of gases, nutrients, and waste products. From the periosteum or endosteum perforating canals carry blood vessels to central canals, and canaliculi connect central canals to osteocytes.

Bone Development (p. 175)

Intramembranous Ossification

1. Some skull bones, part of the mandible, and the diaphyses of the clavicles develop from membranes.
2. Within the membrane at centers of ossification, osteoblasts produce bone along the membrane fibers to form cancellous bone.
3. Beneath the periosteum, osteoblasts lay down compact bone to form the outer surface of the bone.
4. Fontanels are areas of membrane that are not ossified at birth.

Endochondral Ossification

1. Most bones develop from a cartilage model.
2. The cartilage matrix is calcified, and chondrocytes die. Osteoblasts form bone on the calcified cartilage matrix, producing cancellous bone.
3. Osteoblasts build an outer surface of compact bone beneath the periosteum.
4. Primary ossification centers form in the diaphysis during fetal development. Secondary ossification centers form in the epiphyses.
5. Articular cartilage on the ends of bones and the epiphyseal plate is cartilage that does not ossify.

Bone Growth (p. 180)

1. Bones increase in size only by appositional growth, the adding of new bone on the surface of older bone or cartilage.
2. Trabeculae grow by appositional growth.

Growth in Bone Length

1. Epiphyseal plate growth involves the interstitial growth of cartilage followed by appositional bone growth on the cartilage.
2. Epiphyseal plate growth results in an increase in the length of the diaphysis and bony processes. Bone growth in length ceases when the epiphyseal plate becomes ossified and forms the epiphyseal line.

Growth at Articular Cartilage

1. Articular cartilage growth involves the interstitial growth of cartilage followed by appositional bone growth on the cartilage.
2. Articular cartilage growth results in larger epiphyses and an increase in the size of bones that don't have epiphyseal plates.

Growth in Bone Width

1. Appositional bone growth beneath the periosteum increases the diameter of long bones and the size of other bones.
2. Osteoblasts from the periosteum form ridges with grooves between them. The ridges grow together, converting the grooves into tunnels that are filled with concentric lamellae to form osteons.

3. Osteoblasts from the periosteum lay down circumferential lamellae, which can be remodeled.

Factors Affecting Bone Growth

1. Genetic factors determine bone shape and size. The expression of genetic factors can be modified.
2. Factors that alter the mineralization process or production of organic matrix, such as deficiencies in vitamins D and C, can affect bone growth.
3. Growth hormone, thyroid hormone, estrogen, and testosterone stimulate bone growth.
4. Estrogen and testosterone cause increased bone growth and closure of the epiphyseal plate.

Bone Remodeling (p. 183)

1. Remodeling converts woven bone to lamellar bone and allows bone to change shape, adjust to stress, repair itself, and regulate body calcium levels.
2. Bone adjusts to stress by adding new bone and by realignment of bone through remodeling.

Bone Repair (p. 186)

1. Fracture repair begins with the formation of a hematoma.
2. The hematoma is replaced by the internal callus consisting of fibers and cartilage.
3. The external callus is a bone–cartilage collar that stabilizes the ends of the broken bone.
4. The internal and external calluses are ossified to become woven bone.
5. Woven bone is remodeled.

Calcium Homeostasis (p. 187)

PTH increases blood Ca^{2+} levels by increasing bone breakdown, Ca^{2+} absorption from the small intestine, and reabsorption of Ca^{2+} from the urine. Calcitonin decreases blood Ca^{2+} by decreasing bone breakdown.

Effects of Aging on the Skeletal System (p. 189)

1. With aging, bone matrix is lost and the matrix becomes more brittle.
2. Cancellous bone loss results from a thinning and a loss of trabeculae. Compact bone loss mainly occurs from the inner surface of bones and involves less osteon formation.
3. Loss of bone increases the risk of fractures and causes deformity, loss of height, pain, stiffness, and loss of teeth.

R E V I E W   A N D   C O M P R E H E N S I O N

1. Which of these is *not* a function of bone?
   a. internal support and protection
   b. provides attachment for the muscles
   c. calcium and phosphate storage
   d. blood cell production
   e. vitamin D storage

2. The extracellular matrix for hyaline cartilage
   a. is produced by chondroblasts.
   b. contains collagen.
   c. contains proteoglycans.
   d. is usually covered by the perichondrium.
   e. all of the above.

3. Chondrocytes are mature cartilage cells found within the _____, and they are derived from _____.
   a. perichondrium, fibroblasts
   b. perichondrium, chondroblasts
   c. lacunae, fibroblasts
   d. lacunae, chondroblasts

4. Which of these statements concerning cartilage is correct?
   a. Cartilage often occurs in thin plates or sheets.
   b. Chondrocytes receive nutrients and oxygen from blood vessels in the matrix.
   c. Articular cartilage has a thick perichondrium layer.
   d. The perichondrium has both chondrocytes and osteocytes.
   e. Appositional growth of cartilage occurs when chondrocytes within the tissue add more matrix from the inside.

5. A fracture in the shaft of a bone is a break in the
   a. epiphysis.
   b. perichondrium.
   c. diaphysis.
   d. articular cartilage.

6. Yellow marrow is
   a. found mostly in children's bones.
   b. associated mostly with flat bones.
   c. found in the epiphyseal plate.
   d. important for blood cell production.
   e. mostly adipose tissue.

7. The periosteum
   a. is an epithelial tissue membrane.
   b. covers the outer and internal surfaces of bone.
   c. contains only osteoblasts.
   d. becomes continuous with collagen fibers of tendons or ligaments.
   e. has a single fibrous layer.

8. Which of these substances makes up the major portion of bone?
   a. collagen
   b. hydroxyapatite
   c. proteoglycan aggregates
   d. osteocytes
   e. osteoblasts

9. The flexible strength of bone occurs because of
   a. osteoclasts.
   b. ligaments.
   c. hydroxyapatite.
   d. collagen fibers.
   e. periosteum.

10. The prime function of osteoclasts is to
    a. prevent osteoblasts from forming.
    b. become osteocytes.
    c. break down bone.
    d. secrete calcium salts and collagen fibers.
    e. form the periosteum.

11. Osteochondral progenitor cells
    a. can become osteoblasts or chondroblasts.
    b. are derived from mesenchymal stem cells.
    c. are located in the perichondrium, periosteum, and endosteum.
    d. do not produce osteoclasts.
    e. all of the above.

12. Lamellar bone
    a. is mature bone.
    b. is remodeled to form woven bone.
    c. is the first type of bone formed during early fetal development.
    d. has collagen fibers randomly oriented in many directions.
    e. all of the above.

13. Central canals
    a. connect perforating canals to canaliculi.
    b. connect cancellous bone to compact bone.
    c. are where blood cells are produced.
    d. are found only in cancellous bone.
    e. are lined with periosteum.

14. The type of lamellae found in osteons is _____ lamellae.
    a. circumferential
    b. concentric
    c. interstitial

15. Cancellous bone consists of interconnecting rods or plates of bone called
    a. osteons.
    b. canaliculi.
    c. circumferential lamellae.
    d. a haversian system.
    e. trabeculae.

16. Given these events:
    1. Osteochondral progenitor cells become osteoblasts.
    2. Connective tissue membrane is formed.
    3. Osteoblasts produce woven bone.

    Which sequence best describes intramembranous bone formation?
    a. 1,2,3
    b. 1,3,2
    c. 2,1,3
    d. 2,3,1
    e. 3,2,1

17. Given these processes:
    1. Chondrocytes die.
    2. Cartilage matrix calcifies.
    3. Chondrocytes hypertrophy.
    4. Osteoblasts deposit bone.
    5. Blood vessels grow into lacunae.

    Which sequence best represents the order in which they occur during endochondral bone formation?
    a. 3,2,1,4,5
    b. 3,2,1,5,4
    c. 5,2,3,4,1
    d. 3,2,5,1,4
    e. 3,5,2,4,1

18. Intramembranous bone formation
    a. occurs at the epiphyseal plate.
    b. is responsible for growth in diameter of a bone.
    c. gives rise to the flat bones of the skull.
    d. occurs within a hyaline cartilage model.
    e. produces articular cartilage in the long bones.

19. The ossification regions formed during early fetal development
    a. are secondary ossification centers.
    b. become articular cartilage.
    c. become medullary cavities.
    d. become the epiphyses.
    e. are primary ossification centers.

20. Growth in the length of a long bone occurs
    a. at the primary ossification center.
    b. beneath the periosteum.
    c. at the center of the diaphysis.
    d. at the epiphyseal plate.
    e. at the epiphyseal line.

21. During growth in length of a long bone, cartilage is formed and then ossified. The location of the ossification is the zone of
    a. calcification.
    b. hypertrophy.
    c. proliferation.
    d. resting cartilage.

22. Given these processes:
    1. An osteon is produced.
    2. Osteoblasts from the periosteum form a series of ridges.
    3. The periosteum becomes the endosteum.
    4. Osteoblasts lay down bone to produce a concentric lamella.
    5. Grooves are changed into tunnels.

    Which sequence best represents the order in which these processes occur during growth in width of a long bone?
    a. 1,4,2,3,5
    b. 2,5,3,4,1
    c. 3,4,2,1,5
    d. 4,2,1,5,3
    e. 5,4,2,1,3

23. Chronic vitamin D deficiency results in which of these consequences?
    a. Bones become brittle.
    b. The percentage of bone composed of hydroxyapatite increases.
    c. Bones become soft and pliable.
    d. Scurvy occurs.
    e. Both a and b.

24. Osteomalacia occurs as a result of a deficiency of
    a. growth hormone.
    b. sex hormones.
    c. thyroid hormone.
    d. vitamin C.
    e. vitamin D.

25. Estrogen
    a. stimulates a burst of growth at puberty.
    b. causes a later closure of the epiphyseal plate than does testosterone.
    c. causes a longer growth period in females than testosterone causes in males.
    d. tends to prolong the growth phase of the epiphyseal plates.
    e. all of the above.

26. Bone remodeling can occur
    a. when woven bone is converted into lamellar bone.
    b. as bones are subjected to varying patterns of stress.
    c. as a long bone increases in diameter.
    d. when new osteons are formed in compact bone.
    e. all of the above.

27. Given these processes:
    1. cartilage ossification
    2. external callus formation
    3. hematoma formation
    4. internal callus formation
    5. remodeling of woven bone into compact bone

    Which sequence best represents the order in which the processes occur during repair of a fracture?
    a. 1,2,3,4,5
    b. 2,4,3,1,5
    c. 3,4,2,1,5
    d. 4,1,5,2,3
    e. 5,3,4,2,1

28. Which of these processes during bone repair requires the longest period of time?
    a. cartilage ossification
    b. external callus formation
    c. hematoma formation
    d. internal callus formation
    e. remodeling of woven bone into compact bone

29. If the secretion of parathyroid hormone (PTH) increases, osteoclast activity _____, and blood Ca^{2+} levels _____.
    a. decreases, decrease
    b. decreases, increase
    c. increases, decrease
    d. increases, increase

30. Osteoclast activity is inhibited by
    a. calcitonin.
    b. growth hormone.
    c. parathyroid hormone.
    d. sex hormones.
    e. thyroid hormone.

Answers in Appendix F

CRITICAL THINKING

1. When a person develops Paget's disease, for unknown reasons the collagen fibers in the bone matrix run randomly in all directions. In addition, the amount of trabecular bone decreases. What symptoms would you expect to observe?

2. When closure of the epiphyseal plate occurs, the cartilage of the plate is replaced by bone. Does this occur from the epiphyseal side of the plate or the diaphyseal side?

3. Assume that two patients have identical breaks in the femur (thighbone). If one is bedridden and the other has a walking cast, which patient's fracture heals faster? Explain.

4. Explain why running helps prevent osteoporosis in the elderly. Does the benefit include all bones or mainly those of the legs and spine?

5. Astronauts can experience a dramatic decrease in bone density while in a weightless environment. Explain how this happens, and suggest a way to slow the loss of bone tissue.

6. Would a patient suffering from kidney failure be more likely to develop osteomalacia or osteoporosis? Explain.

7. In some cultures eunuchs were responsible for guarding harems, which are the collective wives of one male. Eunuchs are males who, as boys, were castrated. Castration removes the testes, the major site of testosterone production in males. Because testosterone is responsible for the sex drive in males, the reason for castration is obvious. As a side effect of this procedure, the eunuchs grew to above-normal heights. Can you explain why?

8. When a long bone is broken, blood vessels at the fracture line are severed. The formation of blood clots stops the bleeding. Within a few days bone tissue on both sides of the fracture site dies. The bone only dies back a certain distance from the fracture line, however. Explain.

9. A patient has hyperparathyroidism because of a tumor in the parathyroid gland that produces excessive amounts of PTH. What effect does this hormone have on bone? Would administration of large doses of vitamin D help the situation? Explain.

Answers in Appendix G

ANSWERS TO PREDICT QUESTIONS

1. In the absence of a good blood supply, nutrients, chemicals, and cells involved in tissue repair enter cartilage tissue very slowly. As a result, the ability of cartilage to undergo repair is poor. Within a joint, the articular cartilage of one bone presses against and moves against the articular cartilage of another bone. If the articular cartilages were covered by perichondrium, or contained blood vessels and nerves, the resulting pressure and friction could damage these structures.

2. In the elderly, the bone matrix contains proportionately less collagen than hydroxyapatite compared to the bones of younger people. Collagen provides bone with flexible strength, and a reduction in collagen results in brittle bones. In addition, the elderly have less dense bones with less matrix. The combination of reduced matrix that is more brittle results in a greater likelihood of bones breaking.

3. Cancellous bone consists of trabeculae with spaces between them. Blood vessels can pass through these spaces. In compact bone, the blood vessels pass through the perforating and central canals. The trabeculae in cancellous bone are thin enough that nutrients and gases can diffuse from blood vessels around the trabeculae to the osteocytes through the canaliculi.

4. Chondroblasts are surrounded by cartilage matrix and receive oxygen and nutrients by diffusion through the matrix. When the matrix becomes calcified, diffusion is reduced to the point the cells die. When osteoblasts form bone matrix, they connect to one another by their cell processes. Thus, when the matrix is laid down, canaliculi are formed. Even though the ossified bone matrix is dense and prevents significant diffusion, it's possible for the osteocytes to receive gases and nutrients through the canaliculi or by movement from one osteocyte to another.

5. Interstitial growth of cartilage results from the division of chondrocytes within the cartilage followed by the addition of new cartilage matrix between the chondrocytes. The resulting expansion of the cartilage matrix is possible because cartilage matrix is not too rigid. Bones cannot undergo interstitial growth because bone matrix is rigid and cannot expand from within. New bone must therefore be added to the surface by apposition.

6. Damage to the epiphyseal plate interferes with bone elongation, and as a result the bone, and therefore the thigh, will be shorter than normal. Recovery is difficult because cartilage repairs very slowly.

7. Growth of articular cartilage results in an increase in the size of epiphyses. This is only one of the functions of articular cartilage; it also forms a smooth, resilient covering over the ends of the epiphyses within joints. Ossified articular cartilage could not perform that function.

8. Her growth for the next few months increases, and she may be taller than a typical 12-year-old female. Because the epiphyseal plates ossify earlier than normal, however, her height at age 18 will be less than otherwise expected.

9. Taking in adequate Ca^{2+} and vitamin D through the digestive system during adulthood increases Ca^{2+} absorption from the small intestine. The increased Ca^{2+} is used to increase bone mass. The greater the bone mass before the onset of osteoporosis, the greater the tolerance for bone loss later in life. For this reason it's important for adults, especially women in their twenties and thirties, to ingest adequate amounts of calcium. Exercising the muscular system places stress on bone, which also increases bone density. The granddaughter shouldn't smoke because this reduces estrogen levels.

Visit the Online Learning Center at **www.mhhe.com/seeley7** for chapter quizzes, interactive learning exercises, and other study tools.

Skeletal System

Gross Anatomy

Colorized SEM of bone trabeculae.

If the body had no skeleton, it may look somewhat like a poorly stuffed rag doll. Without a skeletal system, we would have no framework to help maintain shape and we wouldn't be able to move normally. Most muscles act on bones to produce movement, often pulling on the bones with considerable force. So without the skeleton, muscles wouldn't make the body move. Human bones are very strong and can resist tremendous bending and compression forces without breaking. Nonetheless, each year nearly 2 million Americans manage to break a bone.

The skeletal system includes the bones, cartilage, ligaments, and tendons. To study skeletal gross anatomy, however, dried, prepared bones are used. This allows the major features of individual bones to be seen clearly without being obstructed by associated soft tissues, such as muscles, tendons, ligaments, cartilage, nerves, and blood vessels. As a consequence, however, it's easy to ignore the important relationships among bones and soft tissues and the fact that living bones have soft tissue, such as the periosteum (see chapter 6).

This chapter includes a discussion of *general considerations* (p. 198). It then proceeds to discuss the two catagories of the named bones: the *axial skeleton* (p. 200), which includes the skull, hyoid bone, vertebral column, and thoracic (rib) cage, and the *appendicular skeleton* (p. 229), consisting of the limbs and their girdles.

General Considerations

The average adult skeleton has 206 bones (table 7.1 and figure 7.1). Although this is the traditional number, the actual number of bones varies from person to person and decreases with age as some bones become fused.

Many of the anatomic features of bones are listed in table 7.2. Most of these features are based on the relationship between the bones and associated soft tissues. If a bone possesses a **tubercle** (too′ber-kl; lump) or **process** (projection), such structures usually exist because a ligament or tendon was attached to that tubercle or process during life. If a bone has a smooth, articular surface, that surface was part of a joint and was covered with articular cartilage. If the bone has a **foramen** (fō-rā′men; pl. foramina; fō-ram′i-nă; a hole) in it, that foramen was occupied by something such as a nerve or blood vessel. Some skull bones contain mucous membrane-lined air spaces called **sinuses.** Regions of some skull bones are composed of

Table 7.1 Number of Named Bones Listed by Category

Bones		Number	Bones		Number
Axial Skeleton			**Appendicular Skeleton**		
Skull (cranium)			*Pectoral Girdle*		
Neurocranium (braincase)			Scapula		2
Paired	Parietal	2	Clavicle		2
	Temporal	2	*Upper Limb*		
Unpaired	Frontal	1	Humerus		2
	Sphenoid	1	Ulna		2
	Occipital	1	Radius		2
	Ethmoid	1	Carpals		16
Viscerocranium (face)			Metacarpals		10
Paired	Maxilla	2	Phalanges		28
	Zygomatic	2		Total Upper Limb and Girdle	64
	Palatine	2	*Pelvic Girdle*		
	Lacrimal	2	Coxa		2
	Nasal	2	*Lower Limb*		
	Inferior nasal concha	2	Femur		2
Unpaired	Mandible	1	Tibia		2
	Vomer	1	Fibula		2
	Total Skull	22	Patella		2
Bones Associated with the Skull			Tarsals		14
Auditory ossicles			Metatarsals		10
Malleus		2	Phalanges		28
Incus		2		Total Lower Limb and Girdle	62
Stapes		2		Total Appendicular Skeleton	126
Hyoid		1			
	Total Associated	7		Total Axial Skeleton	80
Vertebral Column				Total Appendicular Skeleton	126
Cervical vertebrae		7		Total Bones	206
Thoracic vertebrae		12			
Lumbar vertebrae		5			
Sacrum		1			
Coccyx		1			
	Total Vertebral Column	26			
Thoracic Cage (rib cage)					
Ribs		24			
Sternum		1			
	Total Thoracic Cage	25			
	Total Axial Skeleton	80			

Axial Skeleton

Appendicular Skeleton

Axial Skeleton

Skull

Mandible

Sternum

Ribs

Vertebral column

Sacrum

Clavicle

Scapula

Humerus

Ulna

Radius

Carpals

Metacarpals

Phalanges

Coxa

Femur

Patella

Tibia

Fibula

Tarsals

Metatarsals

Phalanges

Skull

Mandible

Ribs

Vertebral column

Sacrum

Coccyx

Anterior view

Posterior view

Figure 7.1   The Complete Skeleton

(The skeleton is not shown in the anatomical position.)

Term	Description
Table 7.2	**General Anatomic Terms for Various Features of Bones**
Body	Main part
Head	Enlarged (often rounded) end
Neck	Constriction between head and body
Margin or border	Edge
Angle	Bend
Ramus	Branch off the body (beyond the angle)
Condyle	Smooth, rounded articular surface
Facet	Small, flattened articular surface
Ridges	
Line or linea	Low ridge
Crest or crista	Prominent ridge
Spine	Very high ridge
Projections	
Process	Prominent projection
Tubercle	Small, rounded bump
Tuberosity or tuber	Knob; larger than a tubercle
Trochanter	Tuberosities on the proximal femur
Epicondyle	Upon a condyle
Lingula	Flat, tongue-shaped process
Hamulus	Hook-shaped process
Cornu	Horn-shaped process
Openings	
Foramen	Hole
Canal or meatus	Tunnel
Fissure	Cleft
Sinus or labyrinth	Cavity
Depressions	
Fossa	General term for a depression
Notch	Depression in the margin of a bone
Fovea	Little pit
Groove or sulcus	Deeper, narrow depression

paper-thin, translucent compact bone only and have little or no cancellous center (see chapter 6).

1. How many bones are there in an average adult skeleton?
2. How are lumps, projections, and openings in bones related to soft tissues?

Axial Skeleton

The **axial skeleton** is divided into the skull, auditory ossicles, hyoid bone, vertebral column, and thoracic cage, or rib cage. The axial skeleton forms the upright axis of the body. It also protects the brain, the spinal cord, and the vital organs housed within the thorax.

3. List the parts of the axial skeleton and its functions.

Skull

The **skull,** or **cranium** (krā′nē-ŭm) protects the brain; supports the organs of vision, hearing, smell, and taste; and provides a foundation for the structures that take air, food, and water into the body. When the skull is disassembled, the mandible is easily separated from the rest of the skull, which remains intact. Special effort is needed to separate the other bones. For this reason, it's convenient to think of the skull, except for the mandible, as a single unit. The top of the skull is called the calvaria (kal-vā′rē-ă), or skullcap. It is usually cut off to reveal the skull's interior. The exterior and interior of the skull have ridges, lines, processes, and plates. These structures are important for the attachment of muscles or for articulations between the bones of the skull. Selected features of the intact skull are listed in table 7.3.

Superior View of the Skull

The skull appears quite simple when viewed from above. Only four bones are seen from this view: the frontal bone, two parietal bones, and a small part of the occipital bone. The paired **parietal bones** are joined at the midline by the **sagittal suture,** and the parietal bones are connected to the **frontal bone** by the **coronal suture** (figure 7.2).

Posterior View of the Skull

The parietal and occipital bones are the major structures seen from the posterior view (figure 7.3). The parietal bones are joined to the **occipital bone** by the **lambdoid** (lam′doyd; the shape resembles the Greek letter lambda) **suture.** Occasionally, extra small bones called **sutural** (soo′choor-ăl) **bones** form along the lambdoid suture.

PREDICT **1**
Explain the basis for the names *sagittal* and *coronal sutures.*

Inca Bone
Sutural bones are usually small and bilateral and in many cases are apparently genetically determined. A large midline bone, called an Inca bone, may form at the junction of the lambdoid and sagittal sutures. The bone was common in the skulls of Incas and is still present in their Andean descendants.

An **external occipital protuberance** is present on the posterior surface of the occipital bone (see figure 7.3). It can be felt through the scalp at the base of the head and varies considerably in size from person to person. The external occipital protuberance is the site of attachment of the **ligamentum nuchae** (noo′kē; nape of neck), an elastic ligament that extends down the neck and helps keep the head erect by pulling on the occipital region of the skull. **Nuchal lines** are a set of small ridges that extend laterally from the protuberance and are the points of attachment for several neck muscles.

201

Table 7.3	Processes and Other Features of the Skull	
Feature	**Bone on Which Feature Is Found**	**Description**
External Features		
Alveolar process	Mandible, maxilla	Ridges on the mandible and maxilla containing the teeth
Angle	Mandible	The posterior, inferior corner of the mandible
Coronoid process	Mandible	Attachment point for the temporalis muscle
Genu	Mandible	Chin (resembles a bent knee)
Horizontal plate	Palatine	Posterior third of the hard palate
Mandibular condyle	Mandible	Region where the mandible articulates with the skull
Mandibular fossa	Temporal	Depression where the mandible articulates with the skull
Mastoid process	Temporal	Enlargement posterior to the ear; attachment site for several muscles that move the head
Nuchal lines	Occipital	Attachment points for several posterior neck muscles
Occipital condyle	Occipital	Point of articulation between the skull and the vertebral column
Palatine process	Maxilla	Anterior two-thirds of the hard palate
Pterygoid hamulus	Sphenoid	Hooked process on the inferior end of the medial pterygoid plate, around which the tendon of one palatine muscle passes; an important dental landmark
Pterygoid plates (medial and lateral)	Sphenoid	Bony plates on the inferior aspect of the sphenoid bone; the lateral pterygoid plate is the site of attachment for two muscles of mastication (chewing)
Ramus	Mandible	Portion of the mandible superior to the angle
Styloid process	Temporal	Attachment site for three muscles (to the tongue, pharynx, and hyoid bone) and some ligaments
Temporal lines	Parietal	Where the temporalis muscle, which closes the jaw, attaches
Internal Features		
Crista galli	Ethmoid	Process in the anterior part of the cranial vault to which one of the connective tissue coverings of the brain (dura mater) connects
Petrous portion	Temporal	Thick, interior part of temporal bone containing the middle and inner ears and the auditory ossicles
Sella turcica	Sphenoid	Bony structure resembling a saddle in which the pituitary gland is located

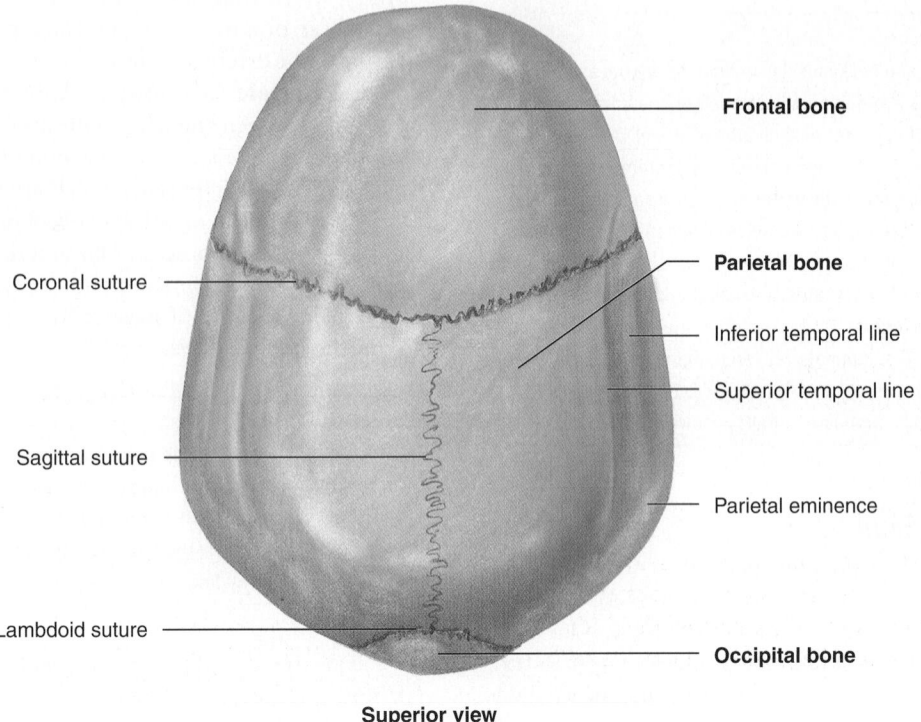

Superior view

Figure 7.2 Skull as Seen from the Superior View

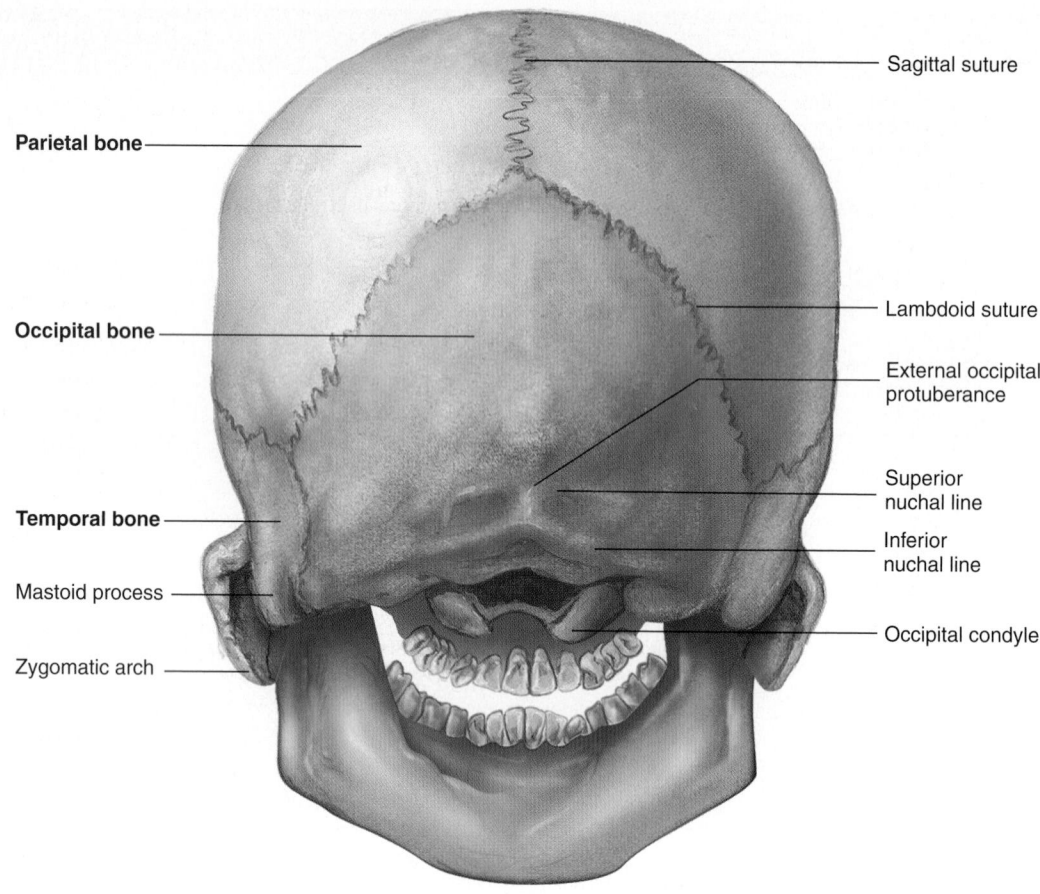

Parietal bone

Occipital bone

Temporal bone

Mastoid process

Zygomatic arch

Sagittal suture

Lambdoid suture

External occipital
protuberance

Superior
nuchal line

Inferior
nuchal line

Occipital condyle

Posterior view

Figure 7.3   Skull as Seen from the Posterior View

Nuchal Lines

The ligamentum nuchae and neck muscles in humans are not as strong as comparable structures in other animals; therefore, the human bony prominence and lines of the posterior skull are not as well developed as they are in those animals. The location of the human foramen magnum allows the skull to balance above the vertebral column and allows for an upright posture. Thus human skulls require less ligamental and muscular effort to balance the head on the spinal column than do the skulls of other animals, including other primates, such as chimpanzees, whose skulls are not balanced over the vertebral column. The presence of small nuchal lines in hominids (i.e., animals with an upright stance like humans) reflects this decreased musculature and is one way used by paleontologists to establish probable upright posture in hominids.

Lateral View of the Skull

The parietal bone and the squamous part of the temporal bone form a large part of the side of the head (figure 7.4). The term *temporal* means related to time, and the temporal bone is so named because the hair of the temples is often the first to turn white, indicating the passage of time. The **squamous suture** joins these bones. A prominent feature of the temporal bone is a large hole, the **external acoustic meatus** (mē-ā′tŭs; passageway or

tunnel), which transmits sound waves toward the eardrum. The external ear, or auricle, surrounds the meatus. Just posterior and inferior to the external auditory meatus is a large inferior projection, the **mastoid** (mas′toyd; resembling a breast) **process.** The process can be seen and felt as a prominent lump just posterior to the ear. The process is not solid bone but is filled with cavities called the **mastoid air cells,** which are connected to the middle ear. Important neck muscles involved in rotation of the head attach to the mastoid process. The superior and inferior **temporal lines,** which are attachment points of the temporalis muscle, one of the major muscles of mastication, arch across the lateral surface of the parietal bone.

Temporal Lines

The temporal lines are important to anthropologists because a heavy temporal line suggests a strong temporalis muscle supporting a heavy jaw. In a male gorilla, the temporalis muscles are so large that the temporal lines meet in the midline of the skull to form a heavy sagittal crest. The temporal lines are much smaller in humans.

The lateral surface of the **greater wing** of the **sphenoid** (sfē′noyd; wedge-shaped) **bone** is immediately anterior to the temporal bone (see figure 7.4). Although appearing to be two bones, one on each side of the skull, the sphenoid bone is actually a single

Coronal suture

Superior temporal line

Inferior temporal line

Parietal bone

Squamous suture

Temporal bone

Occipital bone

Lambdoid suture

Mandibular condyle

External acoustic meatus

Mastoid process

Styloid process

Zygomatic arch

Zygomatic process of temporal bone

Temporal process of zygomatic bone

Angle of mandible

Body of mandible

Frontal bone

Supraorbital foramen

Supraorbital margin

Sphenoid bone (greater wing)

Nasal bone

Lacrimal bone

Nasolacrimal canal

Infraorbital foramen

Zygomatic bone

Coronoid process of mandible

Maxilla

Mandibular ramus

Mental foramen

Mandible

Genu

Lateral view

Figure 7.4   Lateral View of the Skull as Seen from the Right Side

bone that extends completely across the skull. Anterior to the sphenoid bone is the **zygomatic** (zī′gō-mat′ik; a bar or yoke) **bone,** or cheekbone, which can be easily seen and felt on the face (figure 7.5).

The **zygomatic arch,** which consists of joined processes from the temporal and zygomatic bones, forms a bridge across the side of the skull (see figure 7.4). The zygomatic arch is easily felt on the side of the face, and the muscles on either side of the arch can be felt as the jaws are opened and closed (see figure 7.5).

The **maxilla** (mak-sil′ă; upper jaw) is anterior and inferior to the zygomatic bone to which it is joined. The **mandible** (lower jaw) is inferior to the maxilla and articulates posteriorly with the temporal bone (see figure 7.4). The mandible consists of two main portions: the **body,** which extends anteroposteriorly, and the **ramus** (branch), which extends superiorly from the body toward the temporal bone. The superior end of the ramus has a mandibular **condyle,** which articulates with the mandibular fossa of the temporal bone, and the **coronoid** (kōr′ŏ-noyd; shaped like a crow's beak) **process** to which the powerful temporalis muscle, one of the chewing muscles, attaches. The alveolar process of the maxilla contains the superior set of teeth, and the alveolar process of the mandible contains the inferior teeth.

Frontal bone

Supraorbital margin

Zygomatic arch

Nasal bone

Zygomatic bone

Maxilla

Mastoid process

Genu

Mandible

Angle of mandible

Figure 7.5   Lateral View of Bony Landmarks on the Face

Frontal bone

Coronal suture

Glabella

Supraorbital margin

Temporal bone

Nasal bone

Infraorbital margin

Zygomatic bone

Nasal septum

Perpendicular plate of **ethmoid bone**

Vomer

Nasal cavity

Maxilla

Alveolar processes

Body of mandible

Mental foramen

Genu

Parietal bone

Supraorbital foramen

Optic canal

Orbital plate of frontal bone

Sphenoid bone (greater wing)

Superior orbital fissure

Lacrimal bone

Infraorbital foramen

Middle nasal concha

Inferior nasal concha

Anterior nasal spine

Oblique line of mandible

Mandibular symphysis

Anterior view

Figure 7.6 Skull as Seen from the Anterior View

Anterior View of the Skull

The major structures seen from the anterior view are the frontal bone (forehead), the zygomatic bones (cheekbones), the maxillae, and the mandible (figure 7.6). The teeth, which are very prominent in this view, are discussed in chapter 24. Many bones of the face can be easily felt through the skin of the face (figure 7.7).

From this view the most prominent openings into the skull are the orbits and the nasal cavity. The **orbits** are cone-shaped fossae with their apices directed posteriorly (see figures 7.6 and 7.8). They are called orbits because of the rotation of the eyes within the fossae. The bones of the orbits provide both protection for the eyes and attachment points for the muscles that move the eyes. The major portion of each eyeball is within the orbit, and the portion of the eye visible from the outside is relatively small. Each orbit contains blood vessels, nerves, and fat, as well as the eyeball and the muscles that move it. The bones forming the orbit are listed in table 7.4.

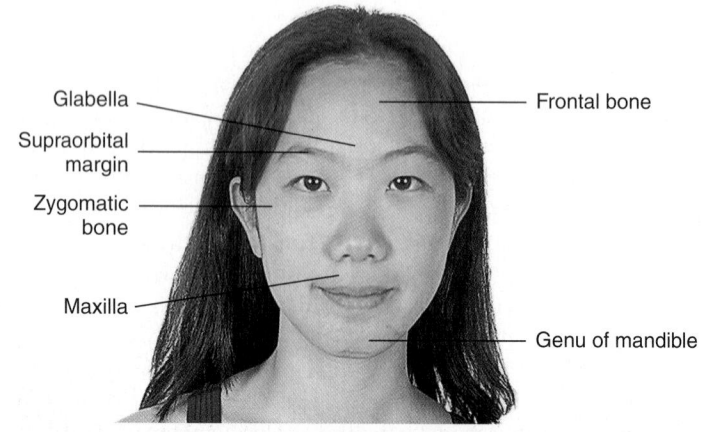

Glabella

Supraorbital margin

Zygomatic bone

Maxilla

Frontal bone

Genu of mandible

Figure 7.7 Anterior View of Bony Landmarks on the Face

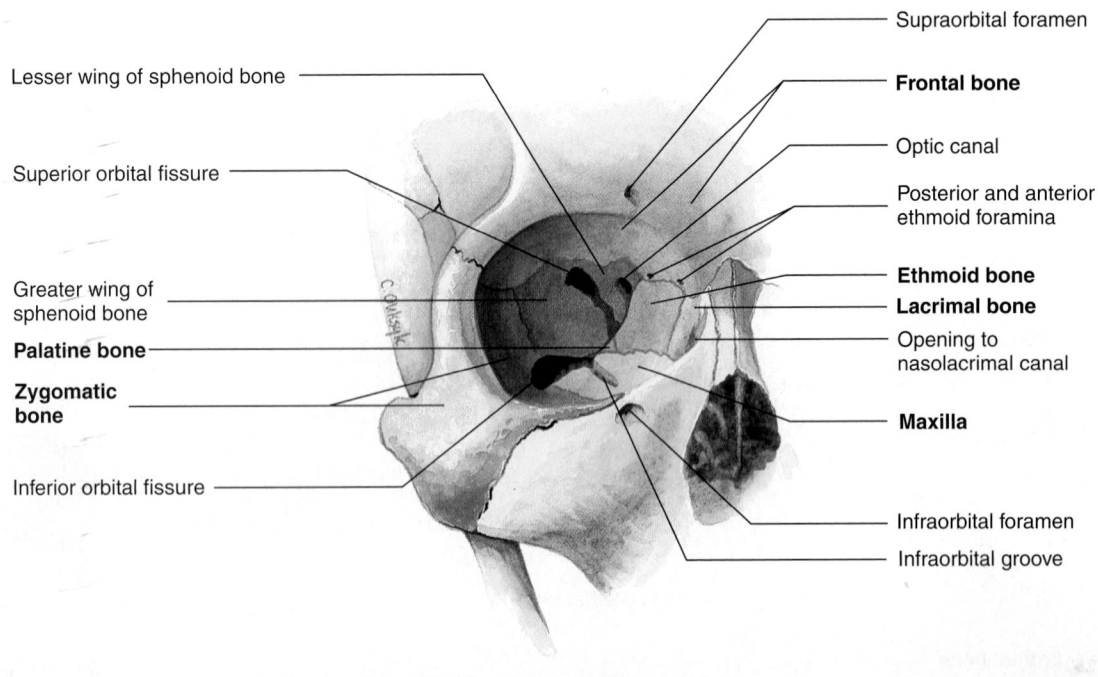

Lesser wing of sphenoid bone

Superior orbital fissure

Greater wing of sphenoid bone

Palatine bone

Zygomatic bone

Inferior orbital fissure

Supraorbital foramen

Frontal bone

Optic canal

Posterior and anterior ethmoid foramina

Ethmoid bone

Lacrimal bone

Opening to nasolacrimal canal

Maxilla

Infraorbital foramen

Infraorbital groove

Anterior view

Figure 7.8 Bones of the Right Orbit

Orbit Weak Point

The superolateral corner of the orbit, where the zygomatic and frontal bones join, is a weak point in the skull that is easily fractured by a blow to that region of the head. The bone tends to collapse into the orbit, resulting in an injury that is difficult to repair.

The orbit has several openings through which structures communicate between it and other cavities. The nasolacrimal duct passes from the orbit into the nasal cavity through the **naso-lacrimal canal** and carries tears from the eyes to the nasal cavity. The optic nerve for the sense of vision passes from the eye through the **optic canal** at the posterior apex of the orbit and enters the cranial cavity. Superior and inferior fissures in the posterior region of the orbit provide openings through which nerves and blood vessels communicate with structures in the orbit or pass to the face.

The **nasal cavity** (table 7.5 and figure 7.9; see figure 7.6) has a pear-shaped opening anteriorly and is divided into right and left halves by a **nasal septum** (sep′tŭm; wall). The bony part of the nasal septum consists primarily of the vomer and the perpendicular plate of the ethmoid bone. Hyaline cartilage forms the anterior part of the nasal septum.

Deviated Nasal Septum

The nasal septum usually is located in the midsagittal plane until a person is 7 years old. Thereafter it tends to deviate, or bulge slightly to one side or the other. The septum can also deviate abnormally at birth or, more commonly, as a result of injury. Deviations can be severe enough to block one side of the nasal passage, and interfere with normal breathing. Repair of severe deviations requires surgery.

Table 7.4	Bones Forming the Orbit (see figures 7.6 and 7.8)
Bone	**Part of Orbit**
Frontal	Roof
Sphenoid	Roof and lateral wall
Zygomatic	Lateral wall
Maxilla	Floor
Lacrimal	Medial wall
Ethmoid	Medial wall
Palatine	Medial wall

Table 7.5	Bones Forming the Nasal Cavity (see figures 7.6 and 7.9)
Bone	**Part of Nasal Cavity**
Frontal	Roof
Nasal	Roof
Sphenoid	Roof
Ethmoid	Roof, septum, and lateral wall
Inferior nasal concha	Lateral wall
Lacrimal	Lateral wall
Maxilla	Floor
Palatine	Floor and lateral wall
Vomer	Septum

Frontal bone

Frontal sinus

Nasal bone

Nasal septum
- Perpendicular plate of ethmoid bone
- Septal cartilage
- **Vomer**

Greater alar cartilage

Anterior nasal spine

Crista galli

Cribriform plate

Olfactory foramina

Sphenoidal sinus

Sphenoid bone

Horizontal plate of palatine bone

Palatine process of maxilla

Incisive canal

Central incisor

(a) **Medial view**

Frontal bone

Frontal sinus

Nasal bone

Maxillary bone

Lateral nasal cartilage

Greater alar cartilage

Lateral incisor

Lacrimal bone

Olfactory recess

Superior nasal concha ⎤ Part of
Middle nasal concha ⎦ ethmoid bone

Sphenoidal sinus

Sphenoid bone

Vertical plate of palatine bone

Inferior nasal concha

Medial pterygoid plate

Horizontal plate of palatine bone

Palatine process of maxilla

(b) **Medial view**

Figure 7.9   Bones of the Nasal Cavity

(*a*) Nasal septum as seen from the left nasal cavity. (*b*) Right lateral nasal wall as seen from inside the nasal cavity (nasal septum removed).

The external part of the nose, formed mostly of hyaline cartilage, is almost entirely absent in the dried skeleton and is represented mainly by the nasal bones and the frontal processes of the maxillary bones, which form the bridge of the nose.

P R E D I C T

A direct blow to the nose may result in a "broken nose." List at least three bones that may be broken.

The lateral wall of the nasal cavity has three bony shelves, the **nasal conchae** (kon′kē; resembling a conch shell), which are directed inferiorly (see figure 7.9). The inferior nasal concha is a separate bone, and the middle and superior nasal conchae are projections

from the ethmoid bone. The conchae function to increase the surface area in the nasal cavity, thereby facilitating moistening, removal of particles, and warming of the air inhaled through the nose.

Several of the bones associated with the nasal cavity have large cavities within them called the **paranasal sinuses,** which open into the nasal cavity (figure 7.10). The sinuses decrease the weight of the skull and act as resonating chambers during voice production. Compare the normal voice to the voice of a person who has a cold and whose sinuses are "stopped up." The sinuses are named for the bones in which they are located and include the frontal, maxillary, and sphenoidal sinuses. The cavities within each ethmoid bone are collectively called the ethmoidal

(a)

(b)

Frontal sinus

(c)

Figure 7.10 Paranasal Sinuses

(a) Lateral view. (b) Anterior view. (c) False-color x-ray of the frontal sinus.

labyrinth because it consists of a maze of interconnected eth-moid air cells. The ethmoid labyrinth is often referred to as the ethmoid sinus.

Interior of the Cranial Cavity

The **cranial cavity** is the cavity in the skull occupied by the brain. The cranial cavity can be exposed by cutting away the **calvaria** (kal-vā′rē-ă), the upper dome-like portion of the skull. With the calvaria removed, the floor of the cranial cavity can be seen (figure 7.11). That floor can be divided roughly into anterior, middle, and posterior fossae, which are formed as the developing neurocranium conforms to the shape of the brain.

A prominent ridge, the **crista galli** (kris′tă găl′ē; rooster's comb), is located in the center of the anterior fossa. The crista galli is a point of attachment for one of the **meninges** (mě-nin′jēz), a thick connective tissue membrane that supports and protects the brain (see chapter 13). On either side of the crista galli is an olfactory fossa. An olfactory bulb rests in each fossa and receives the olfactory nerves for the sense of smell. The **cribriform**

(krib′ri-fōrm; sievelike) **plate** of the ethmoid bone forms the floor of each olfactory fossa. The olfactory nerves extend from the cranial cavity into the roof of the nasal cavity through sievelike perforations in the cribriform plate called **olfactory foramina** (see figure 7.9a and chapter 15).

Fracture of the Cribriform Plate

The cribriform plate may be fractured in an automobile accident involving a car without air bags, if the driver's nose strikes the steering wheel. Cerebrospinal (ser′ě-brō-spī′-năl, sě-rē′brō-spī-năl) fluid from the cranial cavity may leak through the fracture into the nose. This leakage is a dangerous sign and requires immediate medical attention because risk of infection is very high.

The body of the sphenoid bone forms a central promi-nence located within the floor of the cranial cavity. This promi-nence is modified into a structure resembling a saddle, the **sella turcica** (sel′ă tŭr′si-kă; Turkish saddle), which is occupied by the pituitary gland. The petrous (rocky) part of the temporal

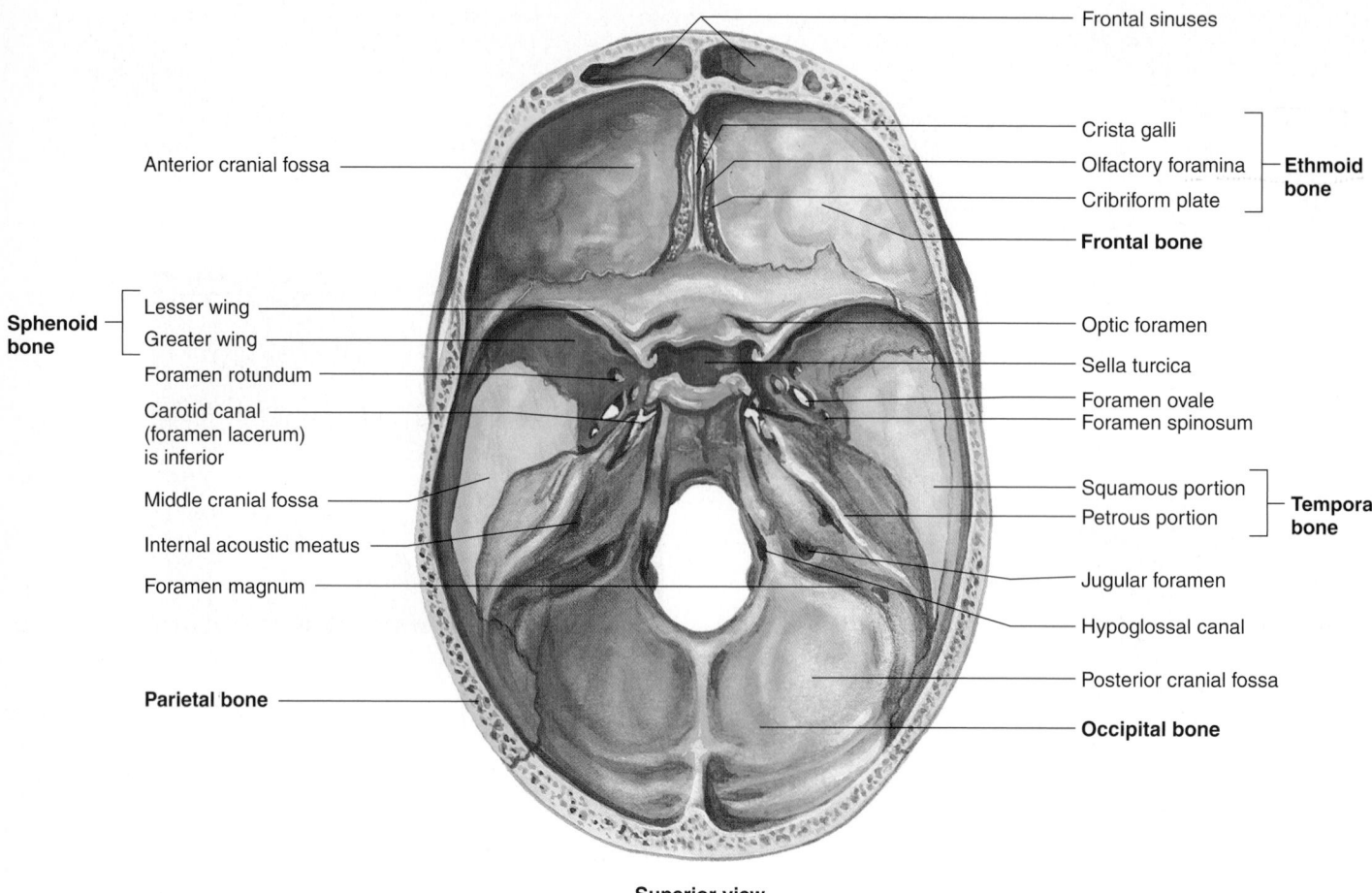

Frontal sinuses

Anterior cranial fossa

Crista galli
Olfactory foramina — **Ethmoid bone**
Cribriform plate

Frontal bone

Sphenoid bone
Lesser wing
Greater wing
Foramen rotundum
Carotid canal (foramen lacerum) is inferior
Middle cranial fossa
Internal acoustic meatus
Foramen magnum

Optic foramen
Sella turcica
Foramen ovale
Foramen spinosum
Squamous portion — **Temporal bone**
Petrous portion
Jugular foramen
Hypoglossal canal
Posterior cranial fossa

Parietal bone

Occipital bone

Superior view

Figure 7.11   Floor of the Cranial Cavity

The roof of the skull has been removed, and the floor is seen from a superior view.

bone is on each side of and slightly posterior to the sella turcica. This thick bony ridge is hollow and contains the middle and inner ears.

The prominent **foramen magnum,** through which the spinal cord and brain are connected, is located in the posterior fossa. The other foramina of the skull and the structures passing through them are listed in table 7.6.

Inferior View of the Skull

Seen from below with the mandible removed, the base of the skull is complex, with a number of foramina and specialized surfaces (figure 7.12). The foramen magnum passes through the occipital bone just slightly posterior to the center of the skull base. **Occipital condyles,** the smooth points of articulation between the skull and the vertebral column, are located on the lateral and anterior margins of the foramen magnum.

The major entry and exit points for blood vessels that supply the brain can be seen from this view. Blood reaches the brain through the internal carotid arteries, which pass through the **carotid** (ka-rot′id; put to sleep) **canals,** and the vertebral arteries,

which pass through the foramen magnum. Immediately after the internal carotid artery enters the carotid canal, it turns medially almost 90 degrees, continues through the carotid canal, again turns almost 90 degrees, and enters the cranial cavity through the superior part of the **foramen lacerum** (lă-ser′um). A thin plate of bone separates the carotid canal from the middle ear, therefore, making it possible for a person to hear his or her own heartbeat, for example, when frightened or after running. Most blood leaves the brain through the internal jugular veins, which exit through the **jugular** (jŭg′ū-lar; throat) **foramina** located lateral to the occipital condyles.

Two long, pointed **styloid** (stī′loyd; stylus- or pen-shaped) **processes** project from the floor of the temporal bone (see figures 7.4 and 7.12). Three muscles involved in movement of the tongue, hyoid bone, and pharynx attach to each process. The **mandibular fossa,** where the mandible articulates with the rest of the skull, is anterior to the mastoid process at the base of the zygomatic arch.

The posterior opening of the nasal cavity is bounded on each side by the vertical bony plates of the sphenoid bone: the **medial**

Table 7.6   Skull Foramina, Fissures, and Canals (see figures 7.11 and 7.12)

Opening	Bone Containing the Opening	Structures Passing Through Openings
Carotid canal	Temporal	Carotid artery and carotid sympathetic nerve plexus
Ethmoid foramina, anterior and posterior	Between frontal and ethmoid	Anterior and posterior ethmoid nerves
External acoustic meatus	Temporal	Sound waves enroute to the eardrum
Foramen lacerum	Between temporal, occipital, and sphenoid	The foramen is filled with cartilage during life; the carotid canal and pterygoid canal cross its superior part but do not actually pass through it
Foramen magnum	Occipital	Spinal cord, accessory nerves, and vertebral arteries
Foramen ovale	Sphenoid	Mandibular division of trigeminal nerve
Foramen rotundum	Sphenoid	Maxillary division of trigeminal nerve
Foramen spinosum	Sphenoid	Middle meningeal artery
Hypoglossal canal	Occipital	Hypoglossal nerve
Incisive foramen (canal)	Between maxillae	Incisive nerve
Inferior orbital fissure	Between sphenoid and maxilla	Infraorbital nerve and blood vessels and zygomatic nerve
Infraorbital foramen	Maxilla	Infraorbital nerve
Internal acoustic meatus	Temporal	Facial nerve and vestibulocochlear nerve
Jugular foramen	Between temporal and occipital	Internal jugular vein, glossopharyngeal nerve, vagus nerve, and accessory nerve
Mandibular foramen	Mandible	Inferior alveolar nerve to the mandibular teeth
Mental foramen	Mandible	Mental nerve
Nasolacrimal canal	Between lacrimal and maxilla	Nasolacrimal (tear) duct
Olfactory foramina	Ethmoid	Olfactory nerves
Optic foramen	Sphenoid	Optic nerve and ophthalmic artery
Palatine foramina, anterior and posterior	Palatine	Palatine nerves
Pterygoid canal	Sphenoid	Sympathetic and parasympathetic nerves to the face
Sphenopalatine foramen	Between palatine and sphenoid	Nasopalatine nerve and sphenopalatine blood vessels
Stylomastoid foramen	Temporal	Facial nerve
Superior orbital fissures	Sphenoid	Oculomotor nerve, trochlear nerve, ophthalmic division of trigeminal nerve, abducens nerve, and ophthalmic veins
Supraorbital foramen or notch	Frontal	Supraorbital nerve and vessels
Zygomaticofacial foramen	Zygomatic	Zygomaticofacial nerve
Zygomaticotemporal foramen	Zygomatic	Zygomaticotemporal nerve

pterygoid (ter′i-goyd; wing-shaped) **plate** and the **lateral pterygoid plate.** The medial and lateral pterygoid muscles, which help move the mandible, attach to the lateral plate (see chapter 10). The **vomer** forms the posterior portion of the nasal septum and can be seen between the medial pterygoid plates in the center of the nasal cavity.

The **hard palate,** or **bony palate,** forms the floor of the nasal cavity. Sutures join four bones to form the hard palate; the palatine processes of the two maxillary bones form the anterior two-thirds of the palate, and the horizontal plates of the two palatine bones form the posterior one-third of the palate. The tissues of the soft palate extend posteriorly from the hard palate. The hard and soft palates separate the nasal cavity from the mouth and enable humans to chew and breathe at the same time.

Cleft Lip or Palate

During development, the facial bones sometimes fail to fuse with one another. A **cleft lip** results if the maxillae don't form normally, and a **cleft palate** occurs when the palatine processes of the maxillae don't fuse with one another. A cleft palate produces an opening between the nasal and oral cavities, making it difficult to eat or drink or to speak distinctly. An artificial palate may be inserted into a newborn's mouth until the palate can be repaired. Approximately one in 700 babies in the U.S. is born with some type of facial cleft. A cleft lip alone, or cleft lip and palate, occurs approximately once in every 1000 births and is more common in males than in females. A cleft palate alone occurs approximately once in every 2000 births and is more common in females than in males. A cleft lip and cleft palate may also occur in the same person.

Inferior view

Figure 7.12 Inferior View of the Skull

4. *List the seven bones that form the orbit of the eye.*

5. *Describe the bones and cartilage found in the nasal septum.*

6. *What is a sinus? What are the functions of sinuses? Give the location of the paranasal sinuses.*

7. *Name the bones that form the hard palate. What is the function of the hard palate?*

8. *Through what foramen does the brainstem connect to the spinal cord? Name the foramina that contain nerves for the senses of vision (optic nerve), smell (olfactory nerves), and hearing (vestibulocochlear nerve)?*

9. *Name the foramina through which the major blood vessels enter and exit the skull.*

10. *List the places where these muscles attach to the skull: neck muscles, throat muscles, muscles of mastication, muscles of facial expression, and muscles that move the eyeballs.*

Bones of the Skull

The skull, or cranium, is composed of 22 separate bones (see tables 7.1 and 7.7). In addition, the skull contains six **auditory ossicles,** which function in hearing (see chapter 15). Each temporal bone holds one set of auditory ossicles, which consists of the malleus, incus, and stapes. These bones cannot be observed unless the temporal bones are cut open.

The 22 bones of the skull are divided into two portions: the neurocranium and the viscerocranium. The **neurocranium,** or **braincase,** consists of eight bones that immediately surround and protect the brain. They include the paired parietal and temporal bones and the unpaired frontal, occipital, sphenoid, and ethmoid bones.

The 14 bones of the **viscerocranium,** or **facial bones,** form the structure of the face in the anterior skull. They are the maxilla (two), zygomatic (two), palatine (two), lacrimal (two), nasal (two), inferior nasal concha (two), mandible (one), and vomer (one) bones. The frontal and ethmoid bones, which are part of the neurocranium, also contribute to the face. The mandible is often listed as a facial bone, even though it is not part of the intact skull.

The facial bones protect the major sensory organs located in the face: the eyes, nose, and tongue. The bones of the face also

Table 7.7 Skull Bones

(a) Parietal bone (right)—lateral view

Parietal eminence

Superior temporal line

Inferior temporal line

Landmark	Description
Parietal eminence	The widest part of the head is from one parietal eminence to the other
Superior and inferior temporal lines	Attachment point for temporalis muscle

Special feature

Forms lateral wall of skull

(b) Temporal bone (right)—lateral and medial views

Squamous portion
Zygomatic process
Mandibular fossa
External acoustic meatus
Styloid process

Mastoid process

Lateral view

Squamous portion
Petrous portion
Internal acoustic canal
Styloid process

Mastoid process

Medial view

Landmark	Description
Carotid canal (shown in figures 7.11 and 7.12)	Canal through which the internal carotid artery enters the cranial cavity
External acoustic meatus	External canal of the ear; carries sound to the ear
Internal acoustic meatus (shown in figure 7.11)	Opening through which the facial (cranial nerve VII) and vestibulocochlear (cranial nerve VIII) nerves enter the petrous portion of the temporal bone
Jugular foramen (shown in figures 7.11 and 7.12)	Foramen through which the internal jugular vein exits the cranial cavity
Mandibular fossa	Articulation point between the mandible and skull
Mastoid process	Attachment point for muscles moving the head and for a hyoid muscle
Middle cranial fossa (shown in figure 7.11)	Depression in the floor of the cranial cavity formed by the temporal lobes of the brain
Petrous portion (shown in figure 7.11)	Thick, "rocky" portion of the temporal bone
Squamous portion	Flat, lateral portion of the temporal bone
Styloid process	Attachment for muscles of the tongue, throat, and hyoid bone
Stylomastoid foramen (shown in figure 7.12)	Foramen through which the facial nerve (cranial nerve VII) exits the skull
Zygomatic process	Helps form the bony bridge from the cheek to just anterior to the ear; attachment for a muscle moving the mandible

Special features

Contains the middle and inner ear, and the mastoid air cells
Place where the mandible articulates with the rest of the skull

Table 7.7 Skull Bones – Continued

(c) Frontal bone—anterior view

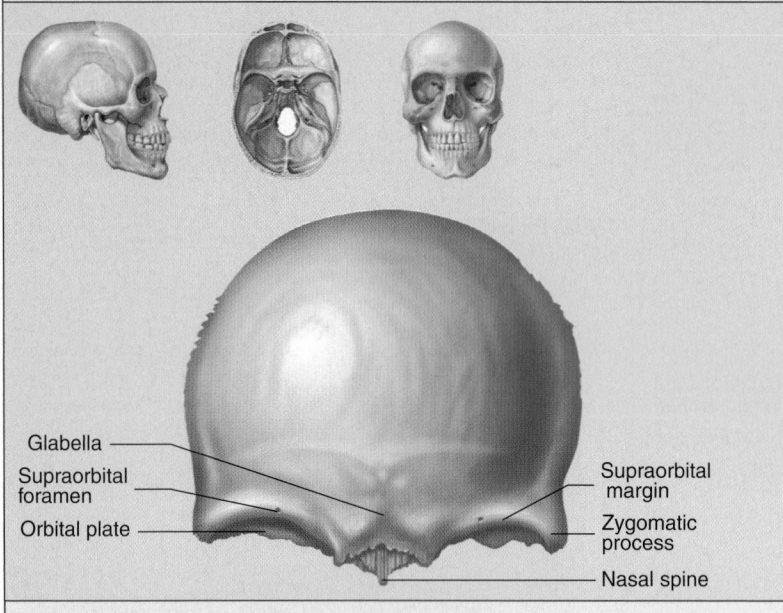

Landmark	Description
Glabella	Area between the supraorbital margins
Nasal spine	Superior part of the nasal bridge
Orbital plate	Roof of the orbit
Supraorbital foramen	Opening through which nerves and vessels exit the skull to the skin of the forehead
Supraorbital margin	Ridge forming the anterior superior border of the orbit
Zygomatic process	Connects to the zygomatic bone; helps form the lateral margin of the orbit

Special features

Forms the forehead and roof of the orbit
Contains the frontal sinus

Glabella
Supraorbital foramen
Orbital plate
Supraorbital margin
Zygomatic process
Nasal spine

(d) Sphenoid bone—superior and posterior views

Landmark	Description
Body	Thickest part of the bone articulates with the occipital bone
Foramen ovale	Opening through which a branch of the trigeminal nerve (cranial nerve V) exits the cranial cavity
Foramen rotundum	Opening through which a branch of the trigeminal nerve (cranial nerve V) exits the cranial cavity
Foramen spinosum	Opening through which a major artery to the meninges (membranes around the brain) enters the cranial cavity
Greater wing	Forms the floor of the middle cranial fossa; several foramina pass through this wing
Lateral pterygoid plate	Attachment point for muscles of mastication (chewing)
Lesser wing	Superior border of the superior orbital fissure
Medial pterygoid plate	Posterolateral walls of the nasal cavity
Optic foramen	Opening through which the optic nerve (cranial nerve II) passes from the orbit to the cranial cavity
Pterygoid canal	Opening through which nerves and vessels exit the cranial cavity
Pterygoid hamulus	Process around which the tendon from a muscle to the soft palate passes
Sella turcica	Fossa containing the pituitary gland
Superior orbital fissure	Opening through which nerves and vessels enter the orbit from the cranial cavity

Special feature

Contains the sphenoidal sinus

Optic foramen
Superior orbital fissure
Lesser wing
Greater wing
Sella turcica
Foramen rotundum
Foramen ovale
Foramen spinosum
Groove of carotid canal

Superior view

Lesser wing
Greater wing
Superior orbital fissure
Body
Foramen rotundum
Pterygoid canal
Pterygoid hamulus
Lateral pterygoid plate
Medial pterygoid plate

Posterior view

Table 7.7 **Skull Bones – Continued**

(e) Occipital bone—inferior view

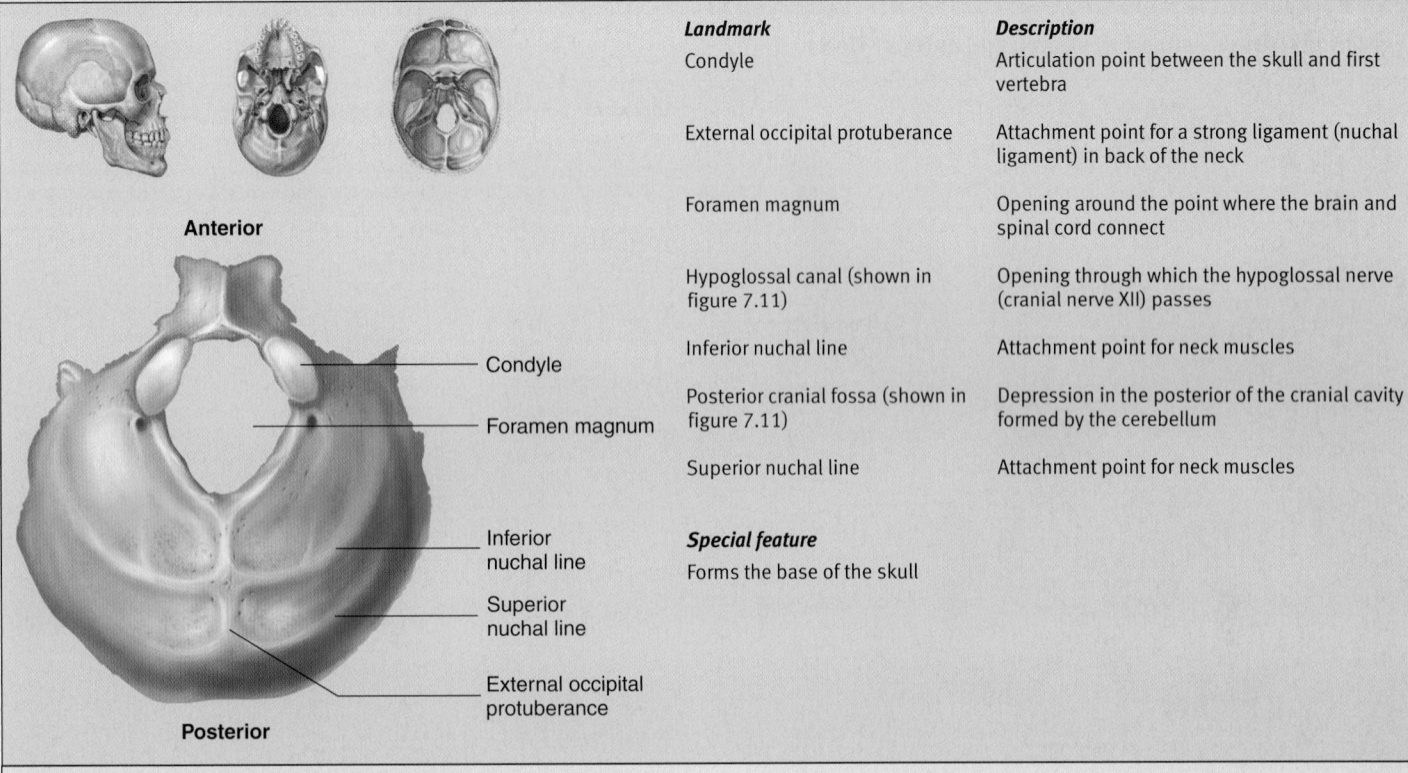

Landmark	Description
Condyle	Articulation point between the skull and first vertebra
External occipital protuberance	Attachment point for a strong ligament (nuchal ligament) in back of the neck
Foramen magnum	Opening around the point where the brain and spinal cord connect
Hypoglossal canal (shown in figure 7.11)	Opening through which the hypoglossal nerve (cranial nerve XII) passes
Inferior nuchal line	Attachment point for neck muscles
Posterior cranial fossa (shown in figure 7.11)	Depression in the posterior of the cranial cavity formed by the cerebellum
Superior nuchal line	Attachment point for neck muscles

Special feature

Forms the base of the skull

(f) Zygomatic bone (right)—lateral view

Landmark	Description
Frontal process	Connection to the frontal bone; helps form the lateral margin of the orbit
Infraorbital margin	Ridge forming the inferior border of the orbit
Temporal process	Helps form the bony bridge from the cheek to just anterior to the ear
Zygomaticofacial foramen	Opening through which a nerve and vessels exit the orbit to the face

Special features

Forms the prominence of the cheek
Forms the anterolateral wall of the orbit

Table 7.7   Skull Bones – Continued

(g) Ethmoid bone—superior, lateral, and anterior views

Landmark	Description
Cribriform plate	Contains numerous openings through which branches of the olfactory nerve (cranial nerve I) enter the cranial cavity from the nasal cavity
Crista galli	Attachment for meninges (membrane around brain)
Ethmoid foramina (shown in figure 7.8)	Openings through which nerves and vessels pass from the orbit to the nasal cavity
Ethmoidal sinus	Spaces in the bone; help lighten the skull
Middle nasal concha	Ridge extending into the nasal cavity; increases surface area, helps warm and moisten air in the cavity
Orbital plate	Forms the medial wall of the orbit
Perpendicular plate	Forms part of the nasal septum
Superior nasal concha	Ridge extending into the nasal cavity; increases surface area, helps warm and moisten air in the cavity

Special features

Forms part of the nasal septum and part of the lateral walls and roof of the nasal cavity

Contains the ethmoidal labyrinth, or ethmoidal sinus. The labyrinth is divided into anterior, middle, and posterior cells.

Table 7.7 Skull Bones – Continued

(h) Maxilla (right)—medial and lateral views

Landmark	Description
Alveolar process	Ridge containing the teeth
Anterior nasal spine	Forms part of the nasal septum
Frontal process	Forms the sides of the nasal bridge
Incisive canal	Opening through which a nerve exits the nasal cavity to the roof of the oral cavity
Infraorbital foramen	Opening through which a nerve and vessels exit the orbit to the face
Maxillary sinus	Cavity in the bone, which helps lighten the skull
Orbital surface	Forms the floor of the orbit
Palatine process	Forms the anterior two-thirds of the hard palate
Tuberosity	Lump posterior to the last maxillary molar tooth
Zygomatic process	Connection to the zygomatic bone; helps form the interior margin of the orbit

Special features

Contains the maxillary sinus and maxillary teeth
Forms part of nasolacrimal canal

Frontal process

Maxillary sinus

Palatine process

Incisive canal
Alveolar process

Molars
Premolars
Canine
Incisors

Medial view

Notch for lacrimal bone

Frontal process

Orbital surface

Zygomatic process

Tuberosity

Infraorbital foramen

Anterior nasal spine

Alveolar process

Incisors
Canine
Premolars
Molars

Lateral view

Table 7.7   Skull Bones – Continued

(i) Palatine bone (right)—medial and anterior views

Landmark	Description
Horizontal plate	Forms the posterior one-third of the hard palate
Vertical plate	Forms part of the lateral nasal wall

Special features

Helps form part of the hard palate and a small part of the wall of the orbit

Vertical plate

Horizontal plate

Vertical plate

Horizontal plate

Medial view

Anterior view

(j) Lacrimal bone (right)—anterolateral view

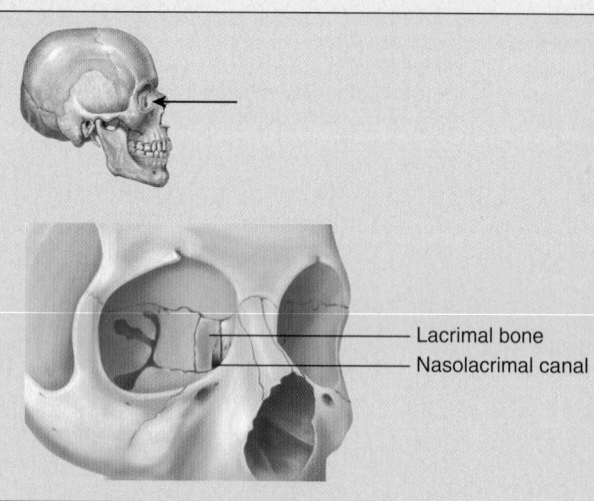

Special feature

Forms a small portion of the orbital wall
Forms part of the nasolacrimal canal

Lacrimal bone
Nasolacrimal canal

(k) Nasal bone (right)—anterolateral view

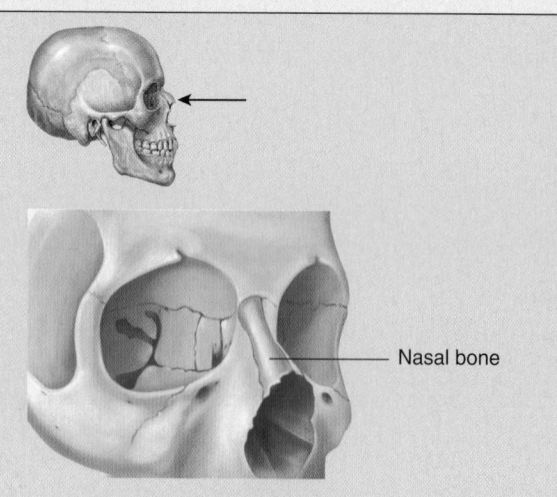

Special feature

Forms the bridge of the nose

Nasal bone

Table 7.7 Skull Bones – Continued

(l) Mandible (right half)—medial and lateral views

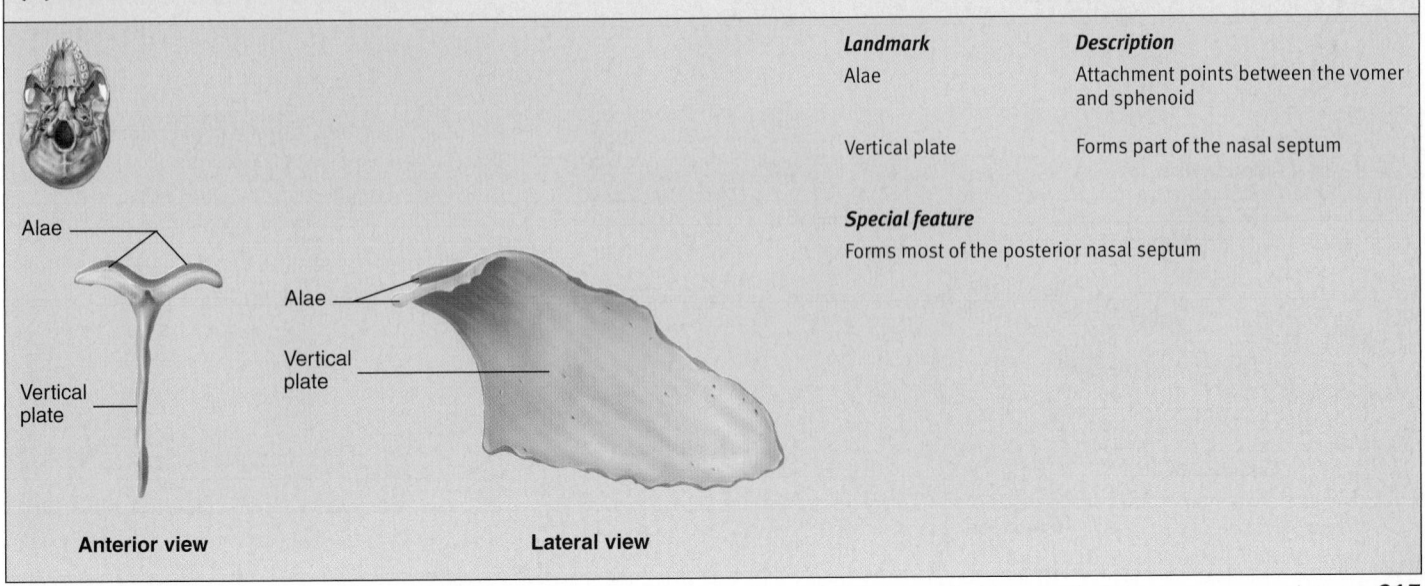

Medial view

Lateral view

Landmark	Description
Alveolar process	Ridge containing the teeth
Angle	Corner between the body and ramus
Body	Major, horizontal portion of the bone
Condylar process	Extension containing the mandibular condyle
Coronoid process	Attachment for a muscle of mastication
Mandibular condyle	Helps form the temporomandibular joint (the point of articulation between the mandible and the rest of the skull)
Mandibular foramen	Opening through which nerves and vessels to the mandibular teeth enter the bone
Mandibular notch	Depression between the condylar process and the coronoid process
Mental foramen	Opening through which a nerve and vessels exit the mandible to the skin of the chin
Mylohyoid line	Attachment point of mylohyoid muscle
Oblique line	A ridge from the anterior edge of the ramus onto the body of the mandible
Ramus	Major, nearly vertical portion of the bone

Special features

The only bone in the skull that is freely movable relative to the rest of the skull bones
Holds the lower teeth

(m) Vomer—anterior and lateral views

Anterior view

Lateral view

Landmark	Description
Alae	Attachment points between the vomer and sphenoid
Vertical plate	Forms part of the nasal septum

Special feature

Forms most of the posterior nasal septum

provide attachment points for muscles involved in **mastication** (mas-ti-kā′shŭn; chewing), facial expression, and eye movement. The jaws (mandible and maxillae) possess **alveolar** (al-vē′ō-lăr) **processes** with sockets for the attachment of the teeth. The bones of the face and their associated soft tissues determine the unique facial features of each individual.

> ● 11. *Name the bones of the neurocranium and viscerocranium. What functions are accomplished by each group?*

Hyoid

The **hyoid bone** (table 7.8), which is unpaired, is often listed as part of the viscerocranium because it has a common developmental origin with the bones of the face. It is not, however, part of the adult skull (see table 7.1). The hyoid bone has no direct bony attachment to the skull but, rather, muscles and ligaments attach it to the skull and the hyoid "floats" in the superior aspect of the neck just below the mandible. The hyoid bone provides an attachment for some tongue muscles, and it's also an attachment point for important neck muscles that elevate the larynx during speech or swallowing.

> ● 12. *Where is the hyoid bone located and what does it do?*

Vertebral Column

The **vertebral column** usually consists of 26 bones, which can be divided into five regions (figure 7.13). Seven **cervical vertebrae** (ver′tĕ-brē), 12 **thoracic vertebrae,** five **lumbar vertebrae,** one **sacral bone,** and one **coccygeal** (kok-sij′ē-ăl) bone make up the vertebral column. The cervical vertebrae are designated "C," thoracic "T," and lumbar "L." A number after the letter indicates the number of the vertebra, from superior to inferior, within each vertebral region. The developing embryo has about 33–34 vertebrae, but the five sacral vertebrae fuse to form one bone, and the four or five coccygeal bones usually fuse to form one bone.

The five regions of the adult vertebral column have four major curvatures (see figure 7.13). Two of the curves appear during embryonic development and reflect the C-shaped curve

Table 7.8	Hyoid Bone

Hyoid—anterior and lateral views

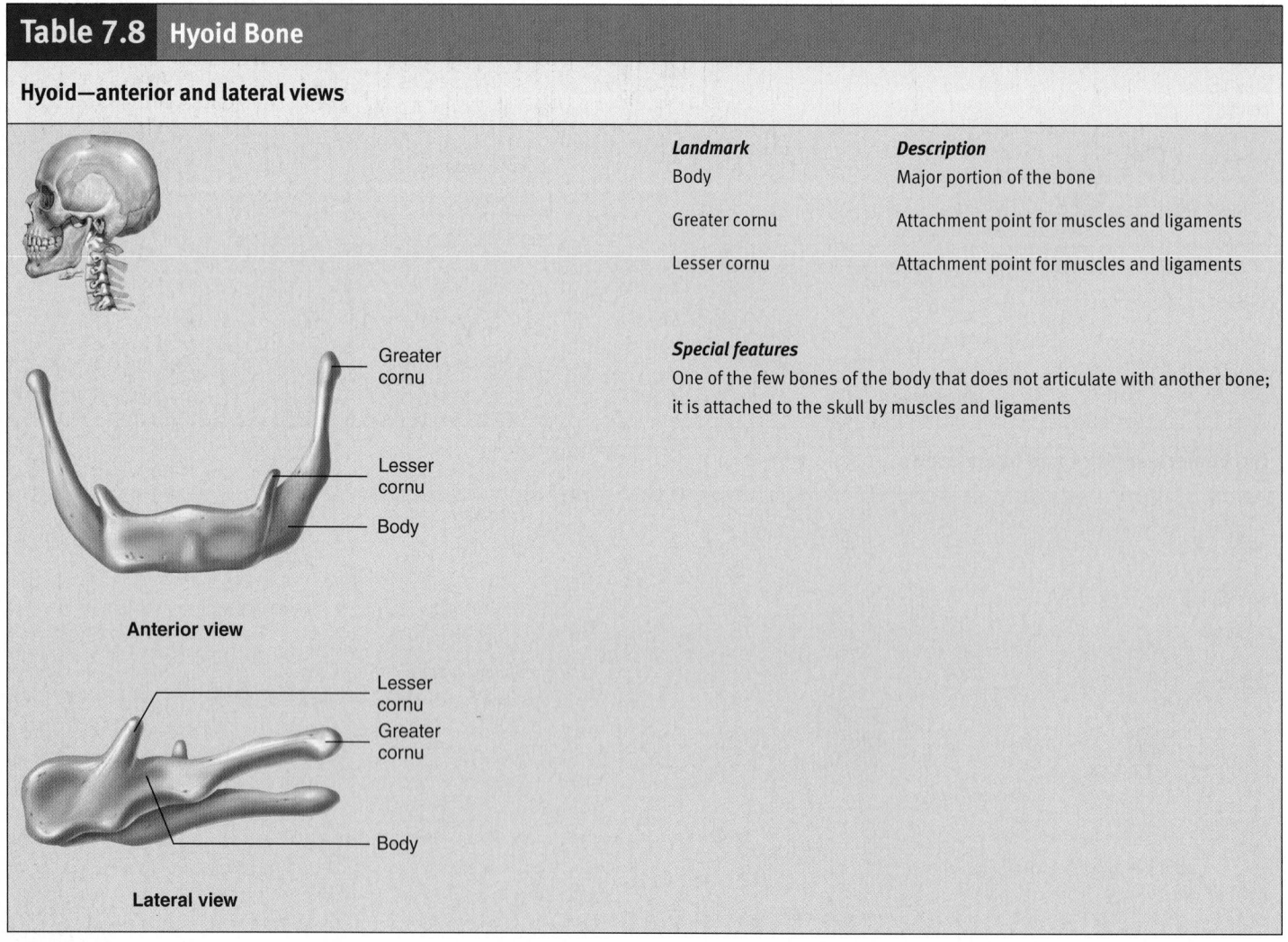

Landmark	Description
Body	Major portion of the bone
Greater cornu	Attachment point for muscles and ligaments
Lesser cornu	Attachment point for muscles and ligaments

Special features

One of the few bones of the body that does not articulate with another bone; it is attached to the skull by muscles and ligaments

Anterior view

Lateral view

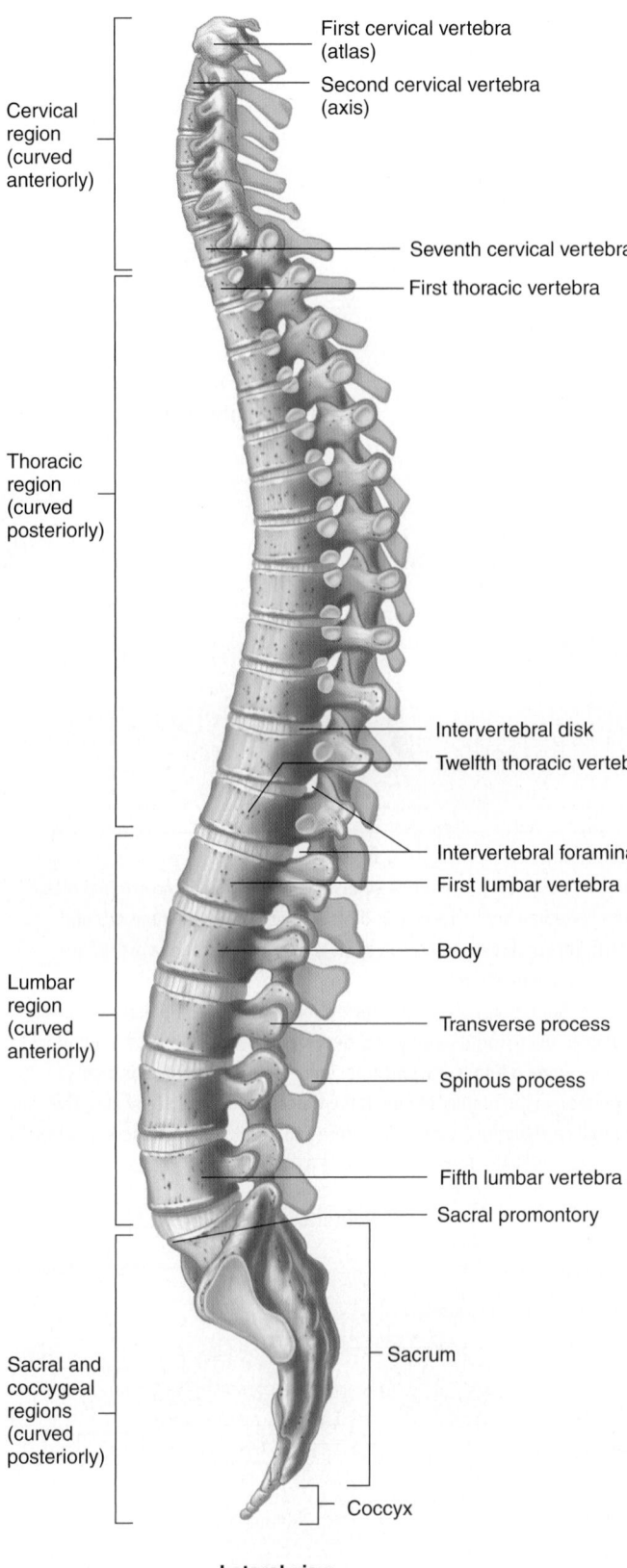

Cervical region (curved anteriorly)

First cervical vertebra (atlas)

Second cervical vertebra (axis)

Seventh cervical vertebra

First thoracic vertebra

Thoracic region (curved posteriorly)

Intervertebral disk

Twelfth thoracic vertebra

Intervertebral foramina

First lumbar vertebra

Body

Lumbar region (curved anteriorly)

Transverse process

Spinous process

Fifth lumbar vertebra

Sacral promontory

Sacral and coccygeal regions (curved posteriorly)

Sacrum

Coccyx

Lateral view

Figure 7.13   Vertebral Column

Complete column viewed from the left side.

of the embryo and fetus within the uterus. When the infant raises its head in the first few months after birth, a secondary curve, which is convex anteriorly, develops in the neck. Later, when the infant learns to sit and then walk, the lumbar portion of the column also becomes convex anteriorly. Thus in the adult vertebral column, the cervical region is convex anteriorly, the thoracic region is concave anteriorly, the lumbar region is convex anteriorly, and the sacral and coccygeal regions are, together, concave anteriorly.

Abnormal Spinal Curvatures

Lordosis (lōr-dō′sis; hollow back) is an exaggeration of the convex curve of the lumbar region. **Kyphosis** (kī-fō′sis; hump back) is an exaggeration of the concave curve of the thoracic region. **Scoliosis** (skō′lē-ō′sis) is an abnormal bending of the spine to the side, which is often accompanied by secondary abnormal curvatures, such as kyphosis (figure A).

Figure A   Scoliosis

Scoliosis is an abnormal lateral curvature of the spine. The abnormality is indicated by the *arrows*.

Intervertebral Disks

During life, **intervertebral disks** of fibrocartilage, which are located between the bodies of adjacent vertebrae (see figures 7.13 and 7.14), provide additional support and prevent the vertebral bodies from rubbing against each other. The intervertebral disks consist of an external **annulus fibrosus** (an′ū-lŭs fī-brō′sŭs; fibrous ring) and an internal gelatinous **nucleus pulposus** (pŭl-pō′sŭs; pulp). The disk becomes more compressed with increasing age so that the distance between vertebrae and therefore the overall height of the individual decreases. The annulus fibrosus also becomes weaker with age and more susceptible to herniation.

(a) **Lateral view**

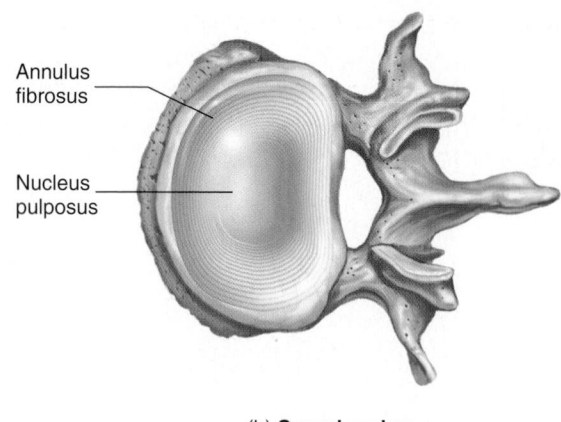

(b) **Superior view**

Figure 7.14 Intervertebral Disk

(*a*) Lateral view. (*b*) Superior view.

Herniated or Ruptured Intervertebral Disk

A **herniated,** or **ruptured, disk** results from the breakage or ballooning of the annulus fibrosus with a partial or complete release of the nucleus pulposus (figure B). The herniated part of the disk may push against the spinal cord or spinal nerves, compromising their normal function and producing pain. Herniation of the inferior lumbar intervertebral disks is most common, but herniation of the inferior cervical disks is almost as common.

Herniated or ruptured disks can be repaired in one of several ways. One procedure uses prolonged bed rest and is based on the tendency for the herniated part of the disk to recede and the annulus fibrosus to repair itself. In many cases, however, surgery is required, and the damaged disk is removed. To enhance the stability of the vertebral column, a piece of hipbone is sometimes inserted into the space previously occupied by the disk, and the adjacent vertebrae become fused by bone across the gap.

Laminectomy and Spina Bifida

In some surgical procedures, such as removal of an intervertebral disk, the vertebrae are in the way and prevent access to the intervertebral disk. This problem can be solved by removing a lamina, a procedure called a **laminectomy.**

Sometimes vertebral laminae may partly or completely fail to fuse (or even fail to form) during fetal development, resulting in a condition called **spina bifida** (spī′nă bif′i-dă; split spine). This defect is most common in the lumbar region. If the defect is severe and involves the spinal cord (figure C), it may interfere with normal nerve function below the point of the defect.

Superior view

Figure B Herniated Disk

Part of the annulus fibrosus has been removed to reveal the nucleus pulposus in the center of the disk.

Superior view

Figure C Spina Bifida

This developmental malformation occurs when two vertebral laminae fail to fuse.

General Plan of the Vertebrae

The vertebral column performs five major functions: (1) it supports the weight of the head and trunk, (2) it protects the spinal cord, (3) it allows spinal nerves to exit the spinal cord, (4) it provides a site for muscle attachment, and (5) it permits movement of the head and trunk. The general structure of a vertebra is outlined in table 7.9. Each vertebra consists of a body, an arch, and various processes (figure 7.15). The weight-bearing portion of the vertebra is a bony disk called the **body.**

The **vertebral arch** projects posteriorly from the body. The arch is divided into left and right halves, and each half has two parts: the **pedicle** (ped'i-kl; foot), which is attached to the body, and the **lamina** (lam'i-na; thin plate), which joins the lamina from the opposite half of the arch. The vertebral arch and the posterior part of the body surround a large opening called the **vertebral foramen.** The vertebral foramina of adjacent vertebrae combine to form the **vertebral canal,** which contains the spinal cord. The vertebral arches and bodies protect the spinal cord.

A **transverse process** extends laterally from each side of the arch between the lamina and pedicle, and a single **spinous process** is present at the point of junction between the two laminae. The spinous processes can be seen and felt as a series of lumps down the midline of the back (figure 7.16). Much vertebral movement is accomplished by the contraction of skeletal muscles that are attached to the transverse and spinous processes (see chapter 10).

Spinal nerves exit the spinal cord through the **intervertebral foramina** (see table 7.9 and figures 7.13 and 7.15*b*). Each intervertebral foramen is formed by notches in the pedicles of adjacent vertebrae.

Movement and additional support of the vertebral column are made possible by the vertebral processes. Each vertebra has two **superior** and two **inferior articular processes,** with the superior processes of one vertebra articulating with the inferior processes of the next superior vertebra. Overlap of these processes increases the rigidity of the vertebral column. The region of overlap and articulation between the superior and inferior articular processes creates a smooth **articular facet** (fas'et, little face), on each articular process.

Table 7.9	General Structure of a Vertebra
Feature	**Description**
Body	Disk-shaped; usually the largest part with flat surfaces directed superiorly and inferiorly; forms the anterior wall of the vertebral foramen; intervertebral disks are located between the bodies
Vertebral foramen	Hole in each vertebra through which the spinal cord passes; adjacent vertebral foramina form the vertebral canal
Vertebral arch	Forms the lateral and posterior walls of the vertebral foramen; possesses several processes and articular surfaces
Pedicle	Foot of the arch with one on each side; forms the lateral walls of the vertebral foramen
Lamina	Posterior part of the arch; forms the posterior wall of the vertebral foramen
Transverse process	Process projecting laterally from the junction of the lamina and pedicle; a site of muscle attachment
Spinous process	Process projecting posteriorly at the point where the two laminae join; a site of muscle attachment; strengthens the vertebral column and allows for movement
Articular processes	Superior and inferior projections containing articular facets where vertebrae articulate with each other; strengthen the vertebral column and allow for movement
Intervertebral notches	Form lateral openings between two adjacent vertebrae through which spinal nerves exit the vertebral canal

Superior view   **Lateral view**   **Lateral view, sagittal section**

Anterior   **Posterior**

Pedicle

Transverse process

Inferior articular process
of superior vertebra

Superior articular
process of inferior vertebra

Body

Spinous processes

(a)

Anterior   **Posterior**

Space for
intervertebral disk

Inferior intervertebral
notch of superior vertebra

Superior intervertebral
notch of inferior vertebra

Intervertebral
foramen

(b)

Figure 7.15   Lateral View (seen from the left) of Two Thoracic Vertebrae

(a) Shows the part of each bone that interacts with the other. (b) Shows the spaces between the vertebrae.

Spinous process of
seventh cervical
vertebra

Superior border
of scapula

Spine of scapula

Scapula

Medial border
of scapula

Inferior angle
of scapula

Lumbar spinous
processes

Figure 7.16   Surface View of the Back Showing the Scapula
and Vertebral Spinous Processes

Regional Differences in Vertebrae

The vertebrae of each region of the vertebral column have specific characteristics that tend to blend at the boundaries between regions (table 7.10). The **cervical vertebrae** (see figures 7.13 and 7.17*a–e*) have very small bodies, most have **bifid** (bī′fid; split) **spinous processes,** and a **transverse foramen** in each transverse process through which the vertebral arteries extend toward the head. Only cervical vertebrae have transverse foramina.

The first cervical vertebra is called the **atlas** (see figure 7.17*a*) because it holds up the head, just as Atlas in classical mythology held up the world. The atlas vertebra has no body and no spinous process, but it has large superior articular facets, where it joins the occipital condyles on the base of the skull. This joint allows the head to move in a "yes" motion or to tilt from side to side. The second cervical vertebra is called the **axis** (figure 7.17*b*) because a considerable amount of rotation occurs at this vertebra to produce a "no" motion of the head. The axis has a highly modified process on the superior side of its small body called the **dens,** or **odontoid** (ō-don′toyd; tooth-shaped) **process** (both dens and odontoid mean tooth-shaped). The dens fits into the enlarged vertebral foramen of the atlas, and the latter rotates around this process. The spinous process of the seventh

Table 7.10	Comparison of Vertebral Regions		
Feature	**Cervical**	**Thoracic**	**Lumbar**
Bodies	Absent in C1, small in others	Medium sized with articular facets for ribs	Large sized
Transverse processes	Transverse foramen	Articular facets for ribs, except T11 and T12	Square
Spinous processes	Absent in C1, bifid in others, except C7	Long, angled inferiorly	Square
Articular facets	Face superior/inferior	Face obliquely	Face medial/lateral

cervical vertebra, which is not bifid, is quite pronounced and often can be seen and felt as a lump between the shoulders (see figure 7.16). The most prominent spinous process in this area is called the **vertebral prominens.** This is usually the spinous process of the seventh cervical vertebra, but may be that of the sixth cervical vertebra or even the first thoracic. The superior articular facets face superiorly and the inferior articular facets face inferiorly.

The **thoracic vertebrae** (see figures 7.13 and 7.18) possess long, thin spinous processes, which are directed inferiorly, and they have relatively long transverse processes. The first 10 thoracic vertebrae have articular facets on their transverse processes, where they articulate with the tubercles of the ribs. Additional articular facets are on the superior and inferior margins of the body where the heads of the ribs articulate. The head of most ribs articulates with the inferior articular facet of one vertebra and the superior articular facet for the rib head on the next vertebra down.

The **lumbar vertebrae** (see figures 7.13 and 7.19) have large, thick bodies and heavy, rectangular transverse and spinous processes. The superior articular facets face medially, and the inferior articular facets face laterally. When the superior articular surface of one lumbar vertebra joins the inferior articulating surface of another lumbar vertebra, the resulting arrangement adds strength to the inferior portion of the vertebral column and limits rotation of the lumbar vertebrae.

P R E D I C T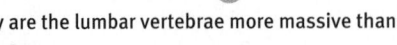
Why are the lumbar vertebrae more massive than the cervical vertebrae?

The **sacral** (sā′krăl) **vertebrae** (see figures 7.13 and 7.20) are highly modified compared to the others. These five vertebrae are fused into a single bone called the **sacrum** (sā′krŭm).

The transverse processes of the sacral vertebrae fuse to form the lateral parts of the sacrum. The superior surfaces of the lateral

parts are wing-shaped areas called the **alae** (ā′lē; wings). Much of the lateral surfaces of the sacrum are ear-shaped **auricular surfaces,** which join the sacrum to the pelvic bones. The spinous processes of the first four sacral vertebrae partially fuse to form projections, called the **median sacral crest,** along the dorsal surface of the sacrum. The spinous process of the fifth vertebra does not form, thereby leaving a **sacral hiatus** (hī-ā′tŭs) at the inferior end of the sacrum, which is often the site of anesthetic injections. The intervertebral foramina are divided into dorsal and ventral foramina, called the **sacral foramina,** which are lateral to the midline. The anterior edge of the body of the first sacral vertebra bulges to form the **sacral promontory** (see figure 7.13), a landmark that separates the abdominal cavity from the pelvic cavity. The sacral promontory can be felt during a vaginal examination, and it's used as a reference point during measurement of the pelvic inlet.

The **coccyx** (kok′siks; shaped like a cuckoo's bill; see figures 7.13 and 7.20), or tailbone, is the most inferior portion of the vertebral column and usually consists of three to five more-or-less fused vertebrae that form a triangle, with the apex directed inferiorly. The coccygeal vertebrae are greatly reduced in size relative to the other vertebrae and have neither vertebral foramina nor well-developed processes.

13. Describe the four major curvatures of the vertebral column, explain what causes them, and when they develop. Define the terms scoliosis, kyphosis, and lordosis.
14. Describe the structures forming the vertebral foramen. Where do spinal nerves exit the vertebral column?
15. Describe how superior and inferior articular processes help support and allow movement of the vertebral column.
16. Name and give the number of each type of vertebra. Describe the characteristics that distinguish the different types of vertebrae.

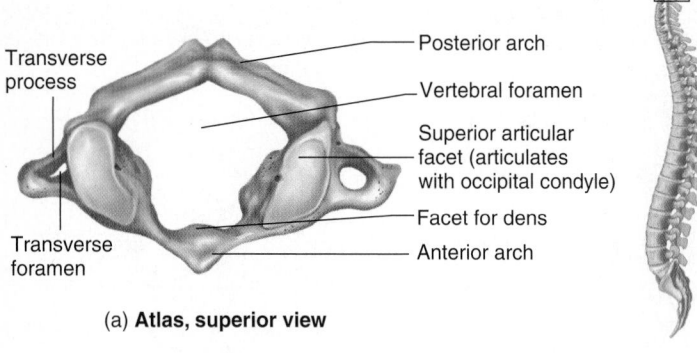

Transverse process — Posterior arch
— Vertebral foramen
— Superior articular facet (articulates with occipital condyle)
— Facet for dens
Transverse foramen — Anterior arch

(a) **Atlas, superior view**

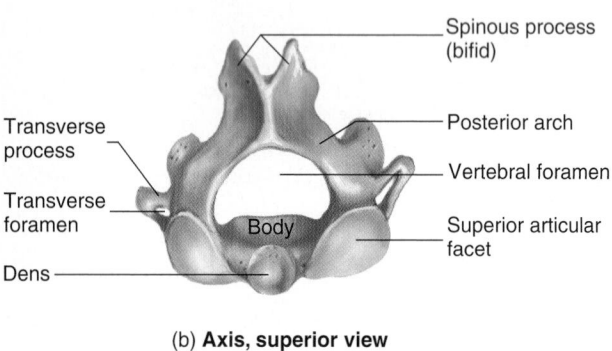

Transverse process — Spinous process (bifid)
— Posterior arch
Transverse foramen — Vertebral foramen
Body
Dens — Superior articular facet

(b) **Axis, superior view**

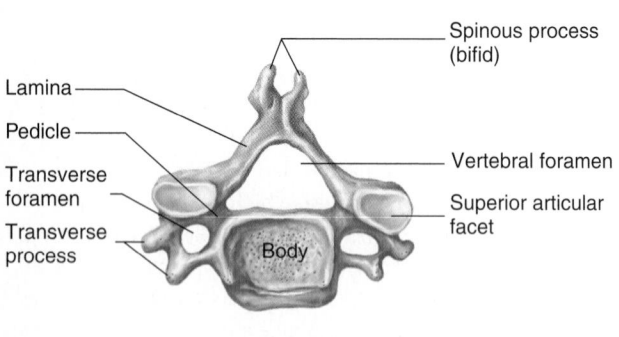

Lamina — Spinous process (bifid)
Pedicle
Transverse foramen — Vertebral foramen
Transverse process — Superior articular facet
Body

(c) **Fifth cervical vertebra, superior view**

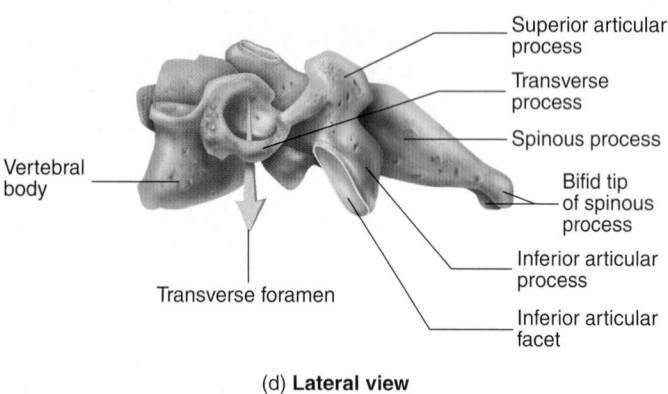

Vertebral body — Superior articular process
— Transverse process
— Spinous process
— Bifid tip of spinous process
— Inferior articular process
Transverse foramen — Inferior articular facet

(d) **Lateral view**

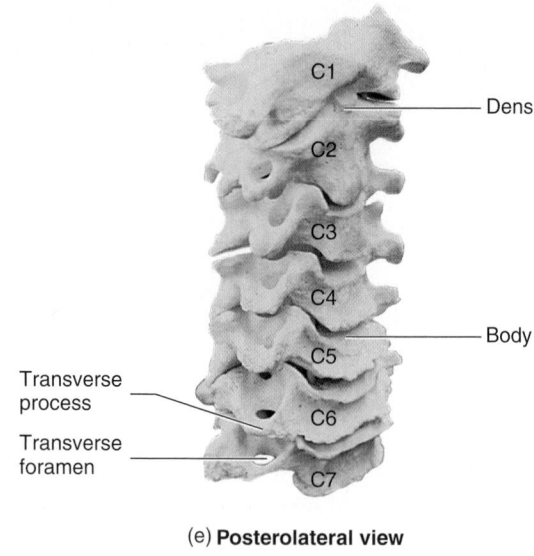

C1 — Dens
C2
C3
C4
Transverse process C5 — Body
Transverse foramen C6
C7

(e) **Posterolateral view**

Figure 7.17 Cervical Vertebrae

(*a*) Atlas (first cervical vertebra), superior view. (*b*) Axis (second cervical vertebra), superior view. (*c*) Fifth cervical vertebra, superior view. (*d*) Fifth cervical vertebra, lateral view. (*e*) Cervical vertebrae together from a lateral view.

Whiplash

Whiplash is a traumatic hyperextension of the cervical vertebrae. The head is a heavy object at the end of a flexible column, and it may become hyperextended when the head "snaps back" as a result of a sudden acceleration of the body. This commonly occurs in "rear-end" automobile accidents, or athletic injuries, in which the body is quickly forced forward while the head remains stationary. Common injuries resulting from whiplash are fracture of the spinous processes of the cervical vertebrae and herniated disks, with an anterior tear of the annulus fibrosus. These injuries can cause posterior pressure on the spinal cord or spinal nerves and strained or torn muscles, tendons, and ligaments.

Lamina

Transverse process

Superior articular process

Pedicle

Body

Spinous process

Vertebral foramen

Articular facet for tubercle of rib

Superior articular facet

Superior articular facet for rib head

(a) **Superior view**

Superior intervertebral notch

Superior articular facet for rib head

Anterior   Body

Inferior articular facet for rib head

Inferior intervertebral notch

Superior articular process

Superior articular facet

Pedicle

Transverse process

Articular facet for tubercle of rib

Lamina

Inferior articular process

Spinous process

Posterior

(b) **Lateral view**

T1

T2

T3

Body

T4

Space for intervertebral disk

T5

T6

T7

Articular facets for rib head

Articular facet for tubercle of rib

Transverse process

Spinous process

Intervertebral foramen

(c) **Posterolateral view**

Vertebral Column Injuries

Because the cervical vertebrae are rather delicate and have small bodies, dislocations and fractures are more common in this area than in other regions of the column. Because the lumbar vertebrae have massive bodies and carry a large amount of weight, fractures are less common, but ruptured intervertebral disks are more common in this area than in other regions of the column. The coccyx is easily broken in a fall in which a person sits down hard on a solid surface.

Figure 7.18 Thoracic Vertebrae

(a) Thoracic vertebra, superior view. (b) Thoracic vertebra, lateral view.
(c) Thoracic vertebrae together from a posterolateral view.

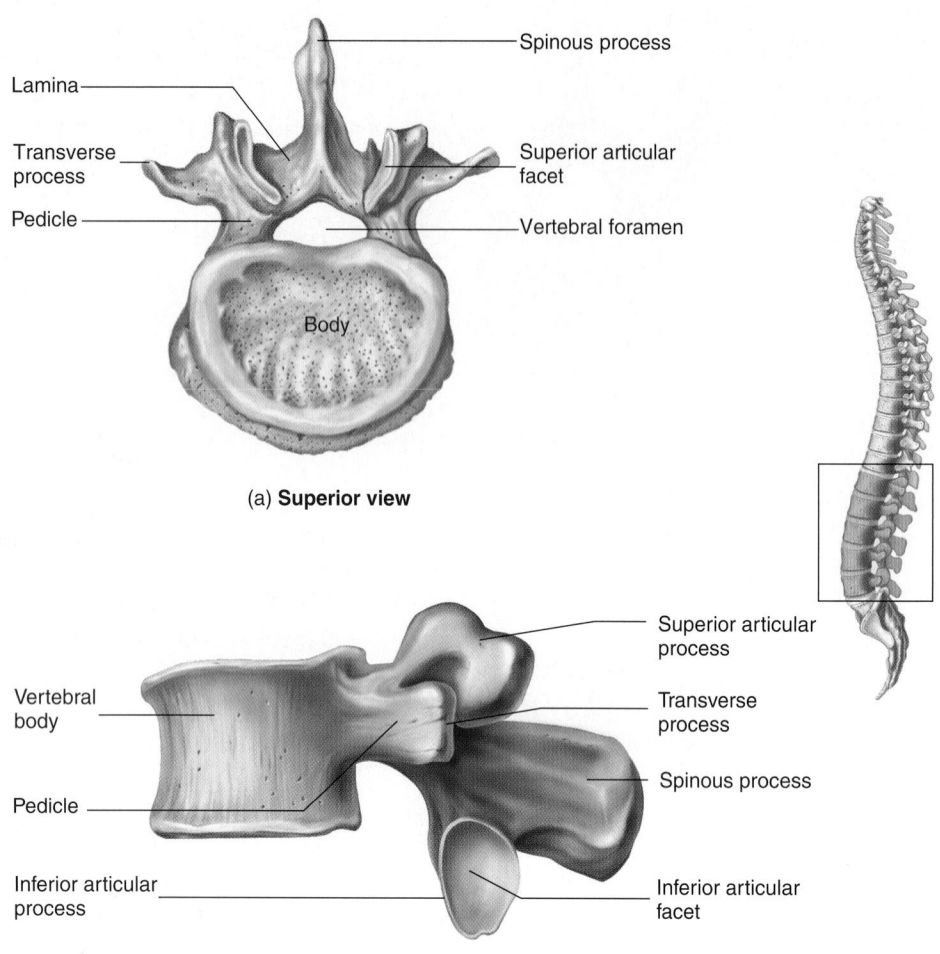

(a) **Superior view**

Spinous process
Lamina
Transverse process
Superior articular facet
Pedicle
Vertebral foramen
Body

(b) **Lateral view**

Vertebral body
Pedicle
Inferior articular process
Superior articular process
Transverse process
Spinous process
Inferior articular facet

(c) **Lateral view**

Space for intervertebral disk
Body
L1
L2
L3
L4
L5
Intervertebral foramen
Spinous process
Transverse process

Figure 7.19   Lumbar Vertebrae

(*a*) Lumbar vertebra, superior view. (*b*) Lumbar vertebra, lateral view.
(*c*) Lumbar vertebrae together from a lateral view.

Variation in Lumbar Vertebrae

The fifth lumbar vertebra or first coccygeal vertebra may become fused into the sacrum. Conversely, the first sacral vertebra may fail to fuse with the rest of the sacrum, resulting in six lumbar vertebrae.

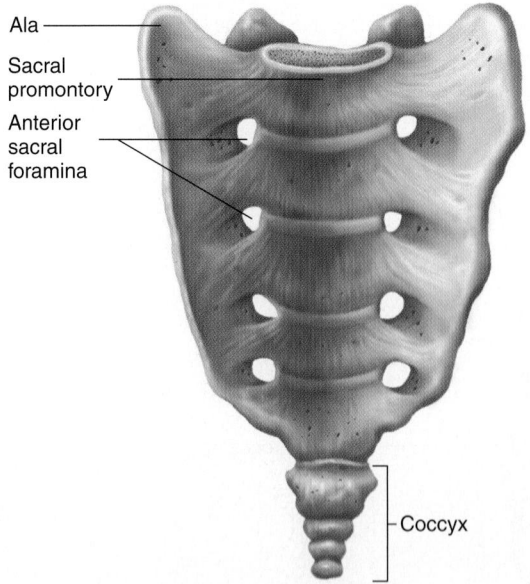

Ala
Sacral promontory
Anterior sacral foramina
Coccyx

(a) **Anterior view**

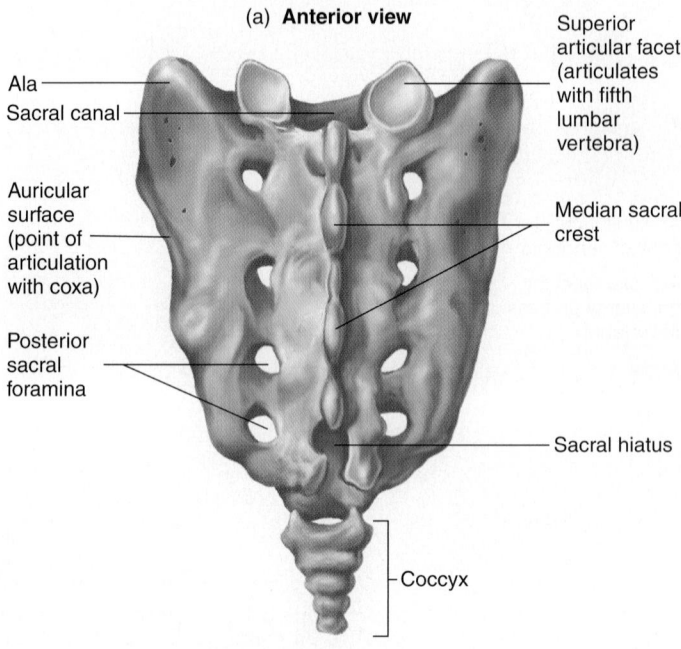

Ala
Sacral canal
Auricular surface (point of articulation with coxa)
Posterior sacral foramina
Superior articular facet (articulates with fifth lumbar vertebra)
Median sacral crest
Sacral hiatus
Coccyx

(b) **Posterior view**

Figure 7.20 Sacrum

(a) Anterior view. (b) Posterior view.

Thoracic Cage

The **thoracic cage,** or **rib cage,** protects the vital organs within the thorax and forms a semi-rigid chamber that can increase and decrease in volume during respiration. It consists of the thoracic vertebrae, the ribs with their associated costal (rib) cartilages, and the sternum (figure 7.21a).

Ribs and Costal Cartilages

The 12 pairs of **ribs** are classified as either true or false ribs. The superior seven pairs are called **true ribs,** or **vertebrosternal** (ver'tě'brō-ster'năl) **ribs,** and articulate with the thoracic verte-

brae and attach directly through their **costal cartilages** to the sternum. The inferior five pairs, or **false ribs,** articulate with the thoracic vertebrae but do not attach directly to the sternum. The false ribs consist of two groups. The eighth, ninth, and tenth ribs, the **vertebrochondral** (ver'tě-brō-kon'drăl) **ribs,** are joined by a common cartilage to the costal cartilage of the seventh rib, which, in turn, is attached to the sternum. Two of the false ribs, the eleventh and twelfth ribs, are also called **floating,** or **vertebral, ribs** because they do not attach to the sternum. The costal cartilages are flexible and permit the thoracic cage to expand during respiration.

Most ribs have two points of articulation with the thoracic vertebrae (figure 7.21b and c). First, the **head** articulates with the bodies of two adjacent vertebrae and the intervertebral disk between them. The head of each rib articulates with the inferior articular facet of the superior vertebra and the superior articular facet of the inferior vertebra. Second, the **tubercle** articulates with the transverse process of one vertebra. The **neck** is between the head and tubercle, and the **body,** or shaft, is the main part of the rib. The **angle** of the rib is located just lateral to the tubercle and is the point of greatest curvature.

Rib Defects

A **separated rib** is a dislocation between a rib and its costal cartilage. As a result of the dislocation, the rib can move, override adjacent ribs, and cause pain. Separation of the tenth rib is the most common.

The angle is the weakest part of the rib and may be fractured in a crushing accident, such as an automobile accident.

The transverse processes of the seventh cervical vertebra may form separate bones called **cervical ribs.** These ribs may be just tiny pieces of bone or may be long enough to reach the sternum. The first lumbar vertebra may develop lumbar ribs.

Sternum

The **sternum,** or breastbone, has been described as being sword-shaped and has three parts (see figure 7.21a). The **manubrium** (mă-noo'brē-ŭm; handle) is the sword handle, the **body** is the blade, and the **xiphoid** (zi'foyd; sword) **process** is the sword tip. The superior margin of the manubrium has a **jugular** (neck) **notch** in the midline, which can be easily felt at the anterior base of the neck (figure 7.22). The first rib and the clavicle articulate with the manubrium. The point at which the manubrium joins the body of the sternum can be felt as a prominence on the anterior thorax called the **sternal angle** (see figure 7.21a). The cartilage of the second rib attaches to the sternum at the sternal angle, the third through seventh ribs attach to the body of the sternum, and no ribs attach to the xiphoid process.

Sternal Angle and Thoracic Landmarks

The sternal angle is important clinically because the second rib is found lateral to it and can be used as a starting point for counting the other ribs. Counting ribs is important because they are landmarks used to locate structures in the thorax, such as areas of the heart. The sternum often is used as a site for taking red bone marrow samples because it is readily accessible. Because the xiphoid process of the sternum is attached only at its superior end, it may be broken during cardiopulmonary resuscitation (CPR) and then may lacerate the liver.

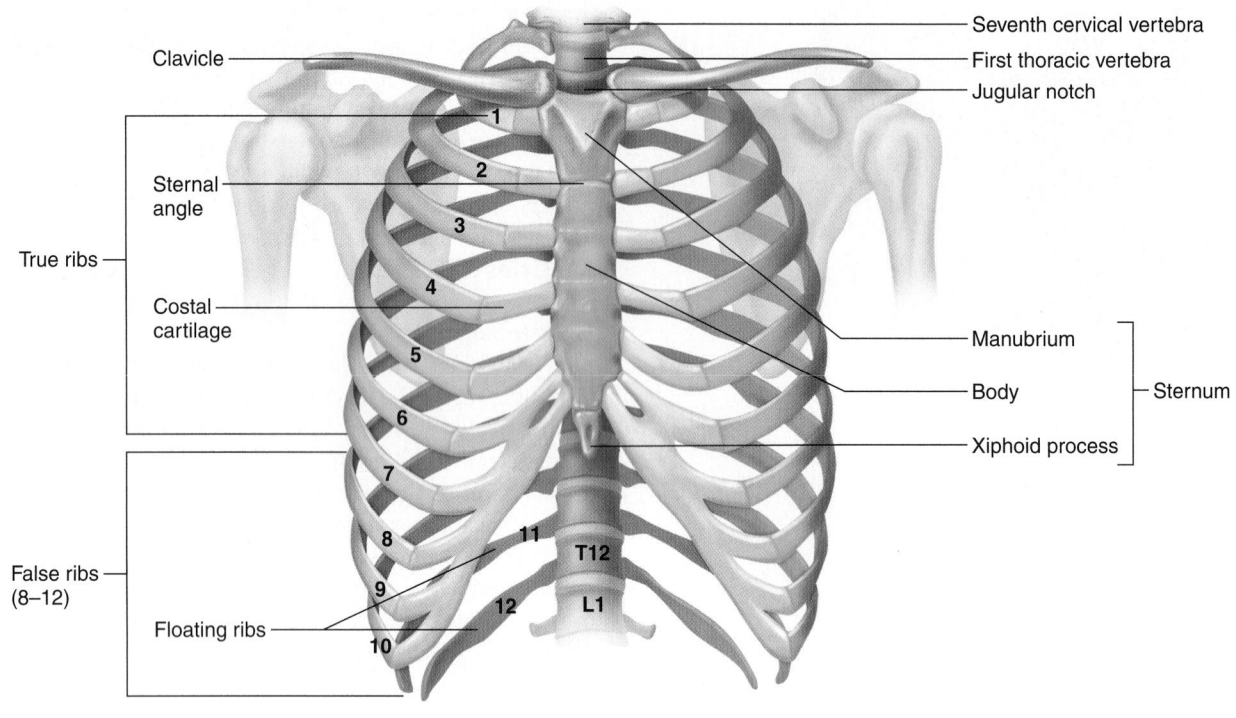

Seventh cervical vertebra

First thoracic vertebra

Jugular notch

Clavicle

1

Sternal angle

2

3

True ribs

4

Costal cartilage

5

6

7

8   11

T12

False ribs (8–12)

9

10   12   L1

Floating ribs

Manubrium

Body

Sternum

Xiphoid process

(a) **Anterior view**

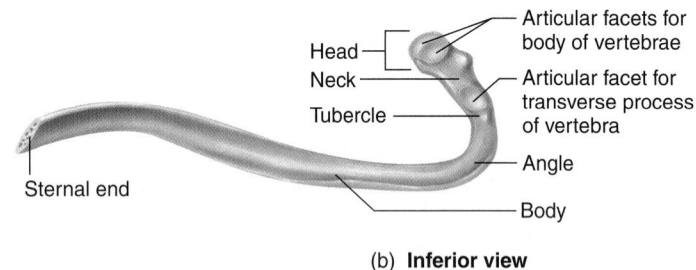

Articular facets for body of vertebrae

Head

Neck

Tubercle

Articular facet for transverse process of vertebra

Angle

Body

Sternal end

(b) **Inferior view**

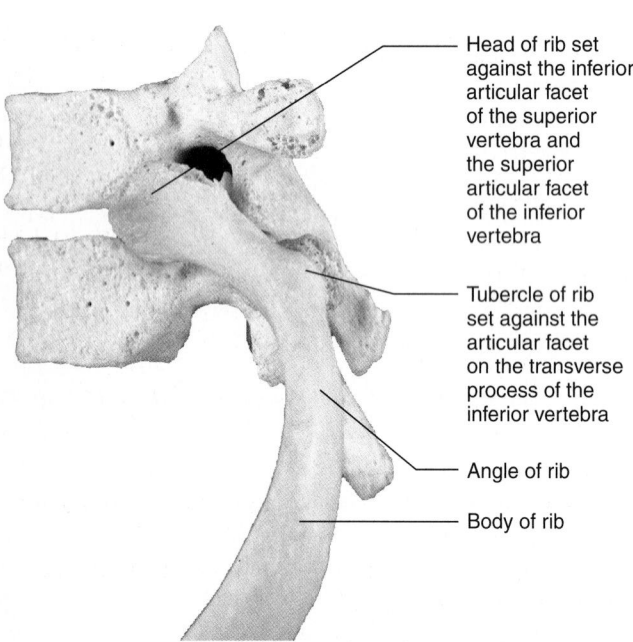

Head of rib set against the inferior articular facet of the superior vertebra and the superior articular facet of the inferior vertebra

Tubercle of rib set against the articular facet on the transverse process of the inferior vertebra

Angle of rib

Body of rib

(c) **Lateral view**

Figure 7.21   Thoracic Cage

(a) Entire thoracic cage as seen from an anterior view. (b) Typical rib, superior view. (c) Photograph of two thoracic vertebrae and the proximal end of a rib, as seen from the left side, showing the relationship between the vertebra and the head and tubercle of the rib.

Figure 7.22 Surface Anatomy Showing Bones of the Upper Thorax

17. *What is the function of the thoracic (rib) cage? Distinguish between true, false, and floating ribs, and give the number of each type.*
18. *Describe the articulation of the ribs with thoracic vertebrae.*
19. *Describe the different parts of the sternum. Name the structures that attach to, or articulate with, the sternum.*

Appendicular Skeleton

The appendicular skeleton (see figure 7.1) consists of the bones of the **upper** and **lower limbs** and the **girdles** by which they are attached to the body. The term *girdle* means a belt or a zone and refers to the two zones, pectoral and pelvic, where the limbs are attached to the body.

Pectoral Girdle and Upper Limb

The human upper limb (figure 7.23) is extremely mobile. It is capable of a wide range of movements, including lifting, grasping, pulling, and touching. Many structural characteristics of the upper limb reflect these functions. The upper limb and its girdle are attached rather loosely by muscles to the rest of the body, an arrangement that allows considerable freedom of movement of this extremity. This freedom of movement allows placement of the hand in a wide range of positions to accomplish its functions.

Pectoral Girdle

The **pectoral** (pek'tŏ-răl), or **shoulder, girdle** consists of two pairs of bones that attach the upper limb to the body: each pair is composed of a **scapula** (skap'ū-lă), or shoulder blade (figure 7.24), and a **clavicle** (klav'i-kl), or collarbone (see figures 7.21, 7.23, and 7.24c). The scapula is a flat, triangular bone that can easily be seen and felt in a living person (see figure 7.16). The base of the triangle, the superior border, faces superiorly; and the apex, the inferior angle, is directed inferiorly.

The large **acromion** (ă-krō'mē-on; shoulder tip) **process** of the scapula, which can be felt at the tip of the shoulder, has three functions: (1) to form a protective cover for the shoulder joint, (2) to form the attachment site for the clavicle, and (3) to provide attachment points for some of the shoulder muscles. The **scapular spine** extends from the acromion process across the posterior surface of the scapula and divides that surface into a small **supraspinous fossa** superior to the spine and a larger **infraspinous fossa** inferior to the

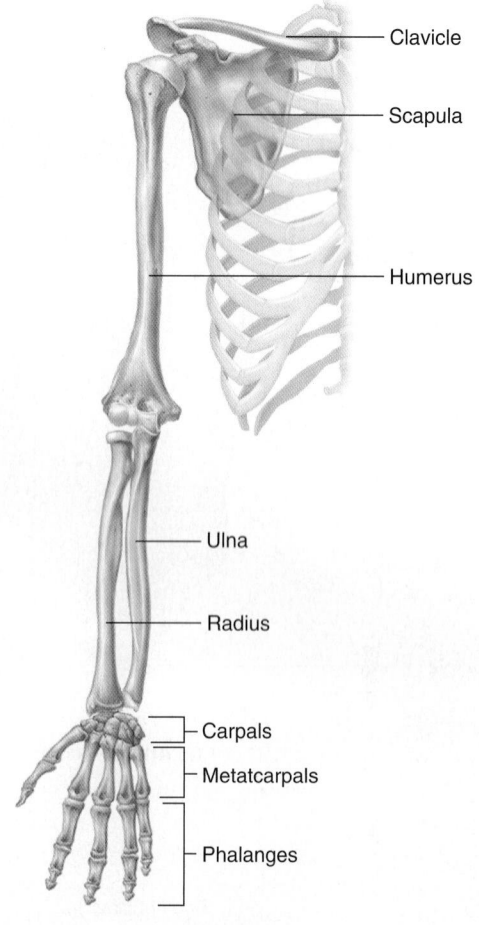

Anterior view

Figure 7.23 Bones of the Pectoral Girdle and Right Upper Limb

spine. The deep, anterior surface of the scapula constitutes the **subscapular fossa.** The smaller **coracoid** (meaning shaped like a crow's beak) **process** provides attachments for some shoulder and arm muscles. A **glenoid** (glē'noyd, glen'oyd) **cavity,** located in the superior lateral portion of the bone, articulates with the head of the humerus.

The clavicle (see figures 7.21, 7.23, and 7.24c) is a long bone with a slight sigmoid (S-shaped) curve and is easily seen and felt in the living human (see figure 7.22). The lateral end of the clavicle articulates with the acromion process, and its medial end articulates with the manubrium of the sternum. These articulations form the only bony connections between the pectoral girdle and the axial skeleton. Because the clavicle holds the upper limb away from the body it facilitates the mobility of the limb.

20. *Name the bones that make up the pectoral girdle. Describe their functions.*
21. *What are the functions of the acromion process and the coracoid process of the scapula?*

P R E D I C T
A broken clavicle changes the position of the upper limb in what way?

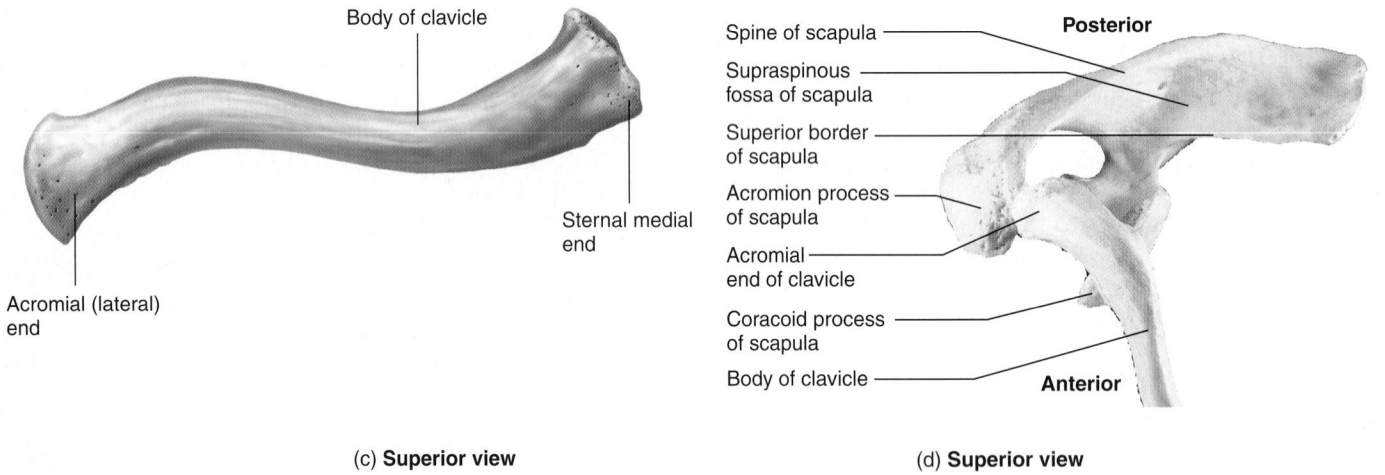

Figure 7.24   Right Scapula and Clavicle

(*a*) Right scapula, anterior view. (*b*) Right scapula, posterior view. (*c*) Right clavicle, superior view. (*d*) Photograph of the right scapula and clavicle from a superior view, showing the relationship between the distal end of the clavicle and the acromion process of the scapula.

Arm

The arm, the part of the upper limb from the shoulder to the elbow, contains only one bone, the **humerus** (figure 7.25). The humeral **head** articulates with the glenoid cavity of the scapula. The **anatomical neck,** immediately distal to the head, is almost non-existent; thus a surgical neck has been designated. The **surgical neck** is so named because it's a common fracture site that often requires surgical repair. If it becomes necessary to remove the humeral head because of disease or injury, it's removed down to the surgical neck. The **greater tubercle** is located on the lateral surface and the **lesser**

tubercle is located on the anterior surface of the proximal end of the humerus, where they function as sites of muscle attachment. The groove between the two tubercles contains one tendon of the biceps brachii muscle and is called the **intertubercular,** or **bicipital** (bī-sip′i-tăl), **groove.** The **deltoid tuberosity** is located on the lateral surface of the humerus a little more than a third of the way along its length and is the attachment site for the deltoid muscle.

The articular surfaces of the distal end of the humerus exhibit unusual features where the humerus articulates with the two forearm bones. The lateral portion of the articular surface is very

Anatomical neck

Head

Greater tubercle

Lesser tubercle

Anatomical neck

Surgical neck

Intertubercular (bicipital) groove

Radial groove

Deltoid tuberosity

Lateral supracondylar ridge

Medial supracondylar ridge

Radial fossa

Coronoid fossa

Olecranon fossa

Lateral epicondyle

Medial epicondyle

Lateral epicondyle

Capitulum

Trochlea

Trochlea

(a) **Anterior view**

(b) **Posterior view**

Figure 7.25   Right Humerus

(*a*) Anterior view. (*b*) Posterior view.

rounded, articulates with the radius, and is called the **capitulum** (kă-pit′ū-lŭm; head-shaped). The medial portion somewhat resembles a spool or pulley, articulates with the ulna, and is called the **trochlea** (trok′lē-ă; spool). Proximal to the capitulum and the trochlea are the **medial** and **lateral epicondyles,** which function as points of muscle attachment for the muscles of the forearm.

Forearm

The forearm has two bones. The **ulna** is on the medial side of the forearm, the side with the little finger. The **radius** is on the lateral, or thumb side, of the forearm (figure 7.26).

The proximal end of the ulna has a C-shaped articular surface, called the **trochlear,** or **semilunar, notch** that fits over

the trochlea of the humerus. The trochlear notch is bounded by two processes. The larger, posterior process is the **olecranon** (ō-lek′ră-non; the point of the elbow) **process.** It can easily be felt and is commonly referred to as "the elbow" (see figure 7.28). Posterior arm muscles attach to the olecranon process. The smaller, anterior process is the **coronoid** (kōr′ŏ-noyd; crow's beak) **process.**

P R E D I C T   ⑤

Explain the function of the olecranon and coronoid fossae on the distal humerus (see figure 7.25).

The distal end of the ulna has a small **head,** which articulates with both the radius and the wrist bones (see figures 7.26 and 7.28). The head can be seen on the posterior, medial (ulnar) side of

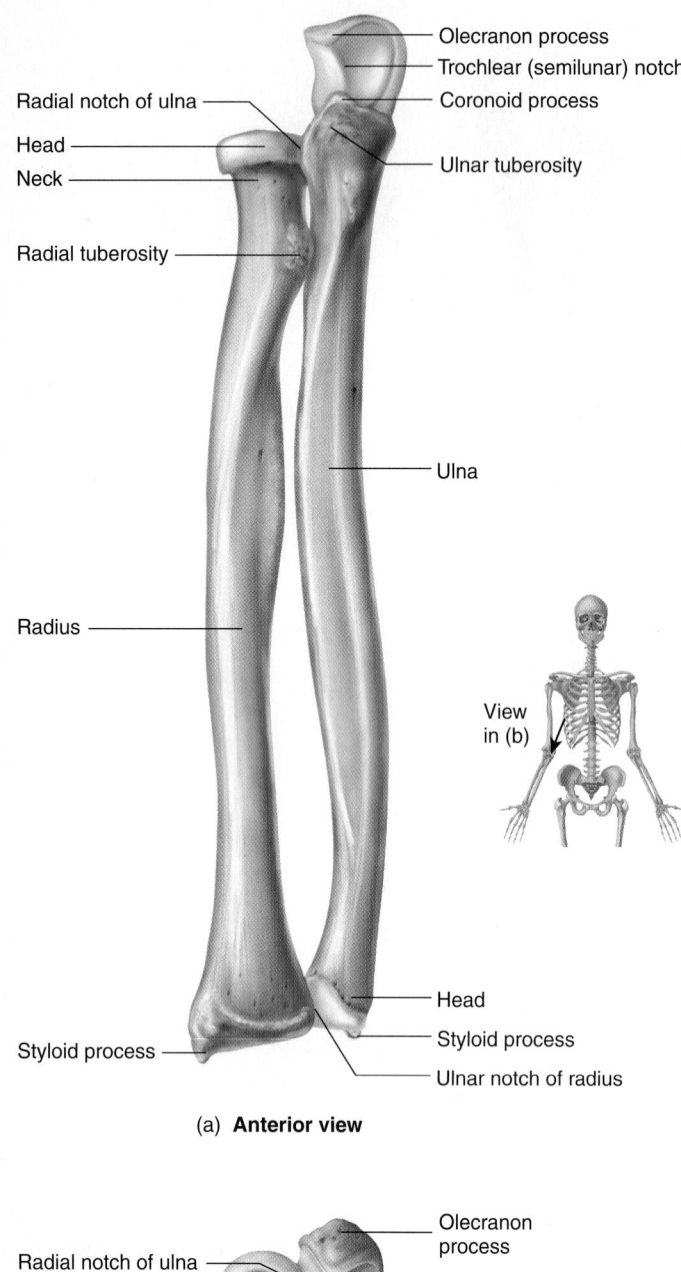

Olecranon process
Trochlear (semilunar) notch
Radial notch of ulna
Coronoid process
Head
Neck
Ulnar tuberosity

Radial tuberosity

Ulna

View in (b)

Radius

Head
Styloid process
Styloid process
Ulnar notch of radius

(a) **Anterior view**

Olecranon process
Radial notch of ulna
Head of radius
Trochlear (semilunar) notch
Coronoid process

(b) **Proximal view**

Figure 7.26   Right Ulna and Radius

(*a*) Anterior view of right ulna and radius. (*b*) Proximal ends of the right ulna and radius.

the distal forearm. The posteromedial side of the head has a small **styloid** (stī′loyd; shaped like a stylus or writing instrument) **process** to which ligaments of the wrist are attached.

The proximal end of the radius is the **head.** It is concave and articulates with the capitulum of the humerus. The lateral surfaces of the head constitute a smooth cylinder, where the radius rotates against the **radial notch** of the ulna. As the forearm rotates (supina-

tion and pronation; see chapter 8), the proximal end of the ulna stays in place, and the radius rotates. The **radial tuberosity** is the point at which a major anterior arm muscle, the biceps brachii, attaches.

The distal end of the radius, which articulates with the ulna and the carpals, is somewhat broadened, and a **styloid process** to which wrist ligaments are attached is located on the lateral side of the distal radius.

Radius Fractures

The radius is the most commonly fractured bone in people over 50 years old. It is often fractured as the result of a fall on an outstretched hand. The most common site of fracture is 2.5 cm proximal to the wrist, and the fracture is often comminuted, or impacted. Such a fracture, which results in posterior displacement of the hand (extension), is called Colles' fracture.

Wrist

The wrist is a relatively short region between the forearm and hand and is composed of eight **carpal** (kar′păl) **bones** arranged into two rows of four each (figure 7.27). The proximal row of carpals, lateral to medial, includes the **scaphoid** (skaf′oyd), which is boat shaped; the **lunate** (loo′nāt), which is moon shaped; the three-cornered **triquetrum** (trī-kwē′trŭm, trī-kwet′rŭm); and the pea-shaped **pisiform** (pis′i-fōrm), which is located on the palmar surface of the triquetrum. The distal row of carpals, from medial to lateral, includes the **hamate** (ha′māt), which has a hooked process on its palmar side, called the hook of the hamate; the head-shaped **capitate** (kap′i-tāt); the **trapezoid** (trap′ĕ-zoyd), which is shaped like a trapezoid (a four-sided geometric form with two sides equal); and the **trapezium** (tra-pē′zē-ŭm), which is shaped like a trapezium (a four-sided geometric form with no two sides equal). A number of mnemonics have been developed to help students remember the carpal bones. This one allows students to remember them in order from lateral to medial for the proximal row (top) and from medial to lateral (by the thumb) for the distal row: **So Long Top Part, Here Comes The Thumb,** that is **S**caphoid, **L**unate, **T**riquetrum, **P**isiform, **H**amate, **C**apitate, **T**rapezoid, and **T**rapezium.

The eight carpals, taken together, are convex posteriorly and concave anteriorly. The anterior concavity of the carpals is accentuated by the tubercle of the trapezium at the base of the thumb and the hook of the hamate at the base of the little finger. A ligament stretches across the wrist from the tubercle of the trapezium to the hook of the hamate to form a tunnel on the anterior surface of the wrist called the **carpal tunnel.** Tendons, nerves, and blood vessels pass through this tunnel to enter the hand.

Carpal Tunnel Syndrome

The bones and ligaments that form the walls of the carpal tunnel do not stretch. Edema (fluid buildup) or connective tissue deposition may occur within the carpal tunnel as the result of trauma or some other problem. The edema or connective tissue may apply pressure against the nerve and vessels passing through the tunnel. This pressure causes carpal tunnel syndrome, which consists of tingling, burning, and numbness in the hand. Carpal tunnel syndrome occurs more frequently in people who use their hands a lot. The number of cases has increased in recent decades among people who perform repetitive tasks such as computer keyboarding.

Radius

Ulna

Carpals
(distal row)

Carpals
(proximal
row)

Scaphoid bone
Lunate bone
Triquetrum bone
Pisiform bone

Hamate bone
Capitate bone
Trapezoid bone
Trapezium bone

Scaphoid bone
Lunate bone
Triquetrum bone
Pisiform bone

Carpals
(proximal
row)

Metacarpals

5   4   3   2

1

1

2   3   4   5

Digits

**Proximal
phalanx**
of thumb

**Distal
phalanx**
of thumb

**Proximal
phalanx**
of finger

Middle phalanx of finger

Distal phalanx of finger

(a) **Posterior view**

(b) **Anterior view**

Figure 7.27 Bones of the Right Wrist and Hand

(*a*) Posterior view. (*b*) Anterior view.

Hand

Five **metacarpals** are attached to the carpal bones and constitute the bony framework of the hand (see figure 7.27). They are numbered one to five starting with the most lateral metacarpal, at the base of the thumb. The metacarpals form a curve so that, in the resting position, the palm of the hand is concave. The distal ends of the metacarpals help form the knuckles of the hand (figure 7.28). The spaces between the metacarpals are occupied by soft tissue.

The five **digits** of each hand include one thumb and four fingers. Each digit consists of small long bones called **phalanges** (fă-lan′jēz; the singular term *phalanx* refers to the Greek word, meaning a line or wedge of soldiers holding their spears, tips outward, in front of them). The thumb has two phalanges, called proximal and distal. Each finger has three phalanges designated proximal, middle, and distal. One or two **sesamoid** (ses′ă-moyd; resembling a sesame seed) **bones** (not illustrated) often form near the junction between the proximal phalanx and the metacarpal of the thumb. Sesamoid bones are small bones located within tendons.

P R E D I C T
Explain why the dried, articulated skeleton appears to have much longer "fingers" than are seen in the hand with the soft tissue intact.

22. *Name the important sites of muscle attachment on the humerus.*
23. *Give the points of articulation between the scapula, humerus, radius, ulna, and wrist bones.*
24. *What is the function of the radial tuberosity? Of the styloid processes? Name the part of the ulna commonly referred to as "the elbow."*
25. *List the eight carpal bones. What is the carpal tunnel?*
26. *What bones form the hand? The knuckles? How many phalanges are in each finger and in the thumb?*

Pelvic Girdle and Lower Limb

The lower limbs support the body and are essential for normal standing, walking, and running. The general pattern of the lower limb (figure 7.29) is very similar to that of the upper limb, except that the pelvic girdle is attached much more firmly to the body than

Figure 7.28   Surface Anatomy Showing Bones of the Pectoral Girdle and Upper Limb

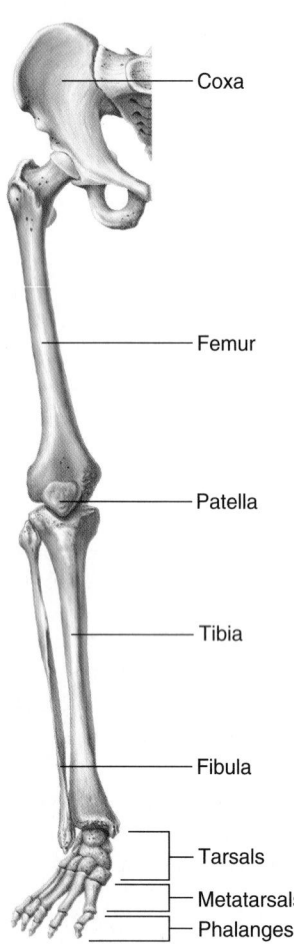

Anterior view

Figure 7.29   Bones of the Pelvic Girdle and Right Lower Limb

is the pectoral girdle, and the bones in general are thicker, heavier, and longer than those of the upper limb. These structures reflect the function of the lower limb in support and movement of the body.

Pelvic Girdle

The right and left **coxae** (kok′sē), or hip bones, join each other anteriorly and the **sacrum** posteriorly to form a ring of bone called the **pelvic girdle.** The **pelvis** (pel′vis, basin) (figure 7.30) includes the pelvic girdle and the coccyx. Each coxa consists of a large, concave bony plate superiorly, a slightly narrower region in the center, and an expanded bony ring inferiorly, which surrounds a large **obturator** (ob′too-rā-tŏr; to occlude or close up, indicating that the foramen is occluded by soft tissue) **foramen.** A fossa called the **acetabulum** (as-ĕ-tab′ū-lŭm; a shallow vinegar cup—a common household item in ancient times) is located on the lateral surface of each coxa and is the point of articulation of the lower limb with the girdle. The articular surface of the acetabulum is crescent-shaped and occupies only the superior and lateral aspects of the fossa. The pelvic girdle is the place of attachment for the lower limbs, supports the weight of the body, and protects internal organs. Because the pelvic girdle is a complete bony ring, it provides more stable support but less mobility than the incomplete ring of the pectoral girdle. In addition, the pelvis in a woman protects the developing fetus and forms a passageway through which the fetus passes during delivery.

The Sacroiliac Joint

Each sacroiliac joint is formed by the junction of the auricular surface of a coxa and one articular surface of the sacrum. The sacroiliac joints receive most of the weight of the upper body and are strongly supported by ligaments. Excessive strain on the joints, however, can cause slight joint movement and can stretch connective tissue and associated nerve endings in the area and cause pain. Thus is derived the expression, "My aching sacroiliac!" This problem sometimes develops in pregnant women because of the forward weight distribution of the fetus.

Sacrum

Sacroiliac joint

Anterior superior
iliac spine

Acetabulum

Symphysis pubis

Obturator
foramen

Subpubic angle

Sacral promontory

Ilium

Coxa

Pubis

Ischium

Anterosuperior view

Figure 7.30   Anterosuperior View of the Pelvis

Each coxa is formed by the fusion of three bones during development: the **ilium** (il′ē-ŭm; groin), the **ischium** (is′kē-ŭm; hip), and the **pubis** (pū′bis; refers to the genital hair). All three bones join near the center of the acetabulum (figure 7.31*a*). The superior portion of the ilium is called the **iliac crest** (figure 7.31*b* and *c*), which extends from the anterior superior iliac spine to the posterior superior iliac spine. The crest ends anteriorly as the **anterior superior iliac spine** and posteriorly as the **posterior superior iliac spine.** The crest and anterior spine can be felt and even seen in thin individuals (figure 7.32). The anterior superior iliac spine is an important anatomic landmark that is used, for example, to find the correct location for giving injections in the hip muscle. A dimple overlies the posterior superior iliac spine just superior to the buttocks. The **greater sciatic** (sī-at′ik) **notch** is on the posterior side of the ilium, just inferior to the inferior posterior iliac spine. The sciatic nerve passes through the greater sciatic notch. The **auricular surface** of the ilium joins the sacrum to form the **sacroiliac joint** (see figure 7.30). The medial side of the ilium consists of a large depression called the **iliac fossa.**

Gluteal Injections

The large gluteal muscles (hip muscles; see chapter 10) are a common site for intermuscular injections. Gluteal injections are made in the superolateral region of the hip (figure D) because a large nerve (the sciatic nerve, see chapter 12) lies deep to the other gluteal regions. The landmarks for such an injection are the anterior superior iliac spine and the tubercle of the iliac crest, which lies about one-third of the way along the iliac crest from anterior to posterior.

Tubercle of iliac crest

Posterior superior
iliac spine

Site of safe area
for injection

Anterior superior
iliac spine

Figure D   Gluteal Injection Site

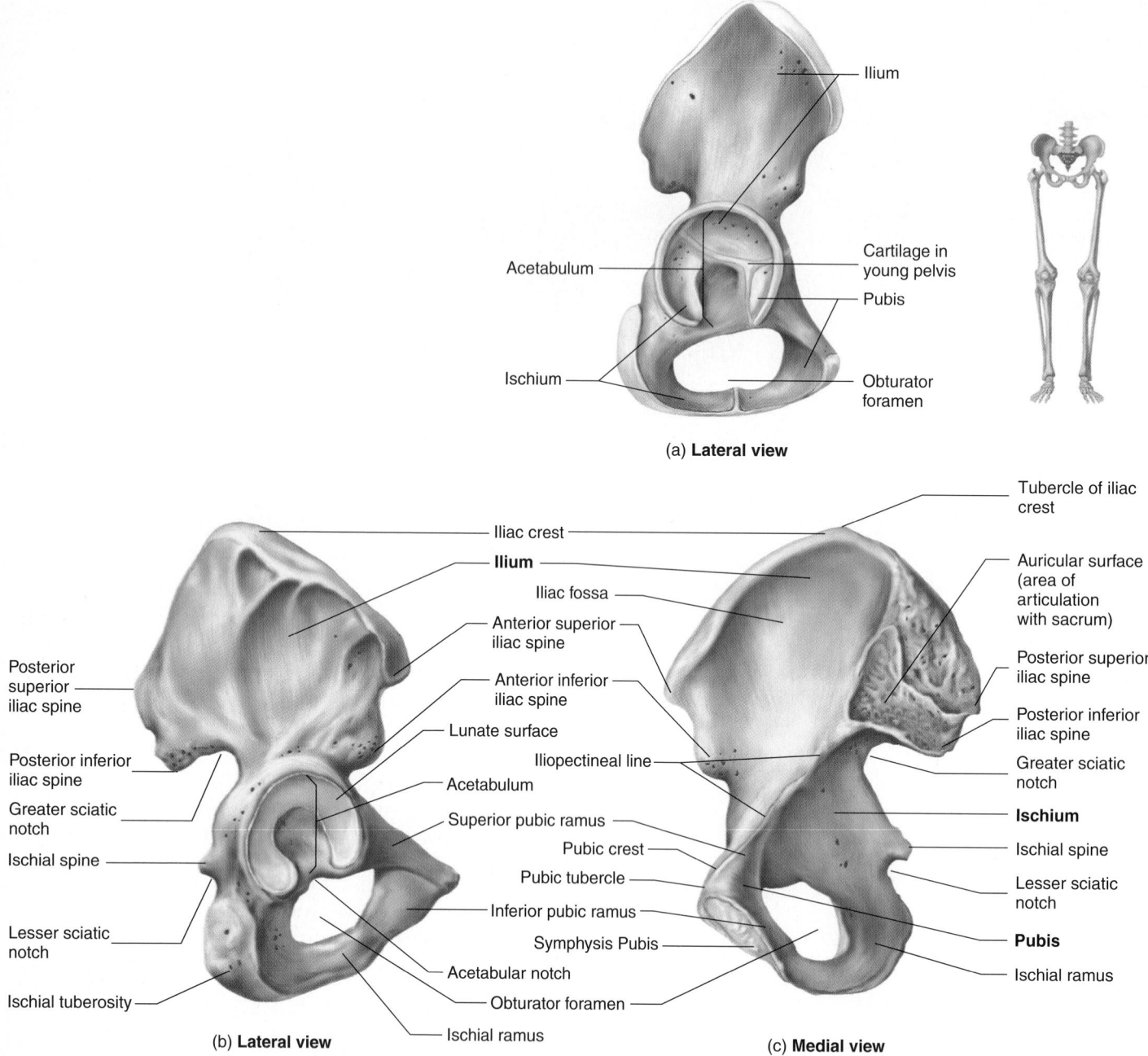

Figure 7.31 Coxa

(*a*) Right coxa of a growing child. Each coxa is formed by fusion of the ilium, ischium, and pubis. The three bones can be seen joining near the center of the acetabulum, separated by lines of cartilage. (*b*) Right coxa, lateral view. (*c*) Right coxa, medial view.

The ischium possesses a heavy **ischial** (is′kē-ăl) **tuberosity,** where posterior thigh muscles attach and on which a person sits (see figure 7.31*b*). The pubis possesses a **pubic crest,** where abdominal muscles attach (see figure 7.31*c*). The pubic crest can be felt anteriorly. Just inferior to the pubic crest is the point of junction, the **symphysis** (sim′fi-sis; a coming together) **pubis,** or **pubic symphysis,** between the two coxae (see figure 7.30).

The pelvis can be thought of as having two parts divided by an imaginary plane passing from the sacral promontory along the **iliopectineal lines** of the ilium to the pubic crest (figure 7.33). The bony boundary of this plane is the **pelvic brim. The false,** or **greater, pelvis** is superior to the pelvic brim and is partially surrounded by bone on the posterior and lateral sides. During life, the abdominal muscles form the anterior wall of the false pelvis. The **true pelvis** is inferior to the pelvic brim and is completely surrounded by bone. The superior opening of the true pelvis, at the level of the pelvic brim, is the **pelvic inlet.** The inferior opening of the true pelvis, bordered by the inferior margin of the pubis, the ischial spines and tuberosities, and the coccyx, is the **pelvic outlet** (see figure 7.33*c*).

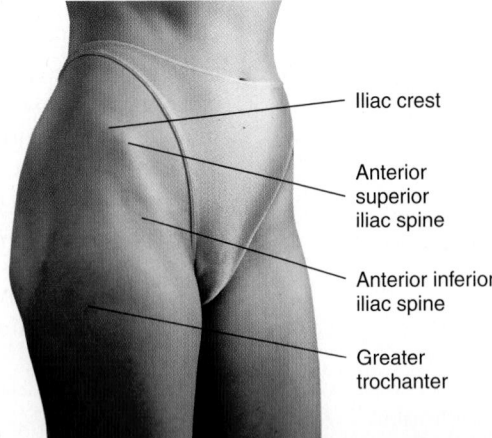

Figure 7.32 Surface Anatomy Showing an Anterior View of the Hipbones

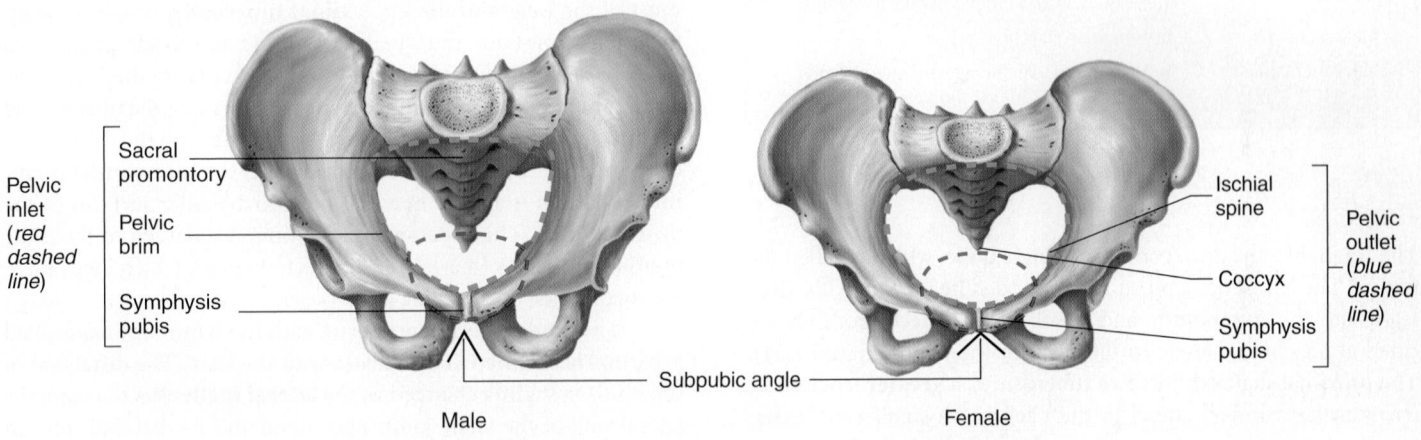

(a) **Anterosuperior view** — Male

(b) **Anterosuperior view** — Female

Figure 7.33 Comparison of the Male and Female Pelvis

(a) Male. The pelvic inlet (*red dashed line*) and outlet (*blue dashed line*) are small, and the subpubic angle is less than 90 degrees. (b) Female. The pelvic inlet (*red dashed line*) and outlet (*blue dashed line*) are larger, and the subpubic angle is 90 degrees or greater. (c) Midsagittal section through the pelvis to show the pelvic inlet (*red arrow* and *red dashed line*) and outlet (*blue arrow* and *blue dashed line*).

Comparison of the Male and Female Pelvis

The male pelvis usually is more massive than the female pelvis as a result of the greater weight and size of the male, and the female pelvis is broader and has a larger, more rounded pelvic inlet and outlet (see figure 7.33a and b), consistent with the need to allow the fetus to pass through these openings in the female pelvis during delivery. Table 7.11 lists additional differences between the male and female pelvis.

- 27. *Define the pelvic girdle. What bones fuse to form each coxa? Where and with what bones do each coxae articulate?*
- 28. *Name the important sites of muscle attachment on the pelvis.*
- 29. *Distinguish between the true pelvis and the false pelvis.*
- 30. *Describe the differences between a male and a female pelvis.*

(c) **Medial view**

Pelvic Outlet and Birth

A wide circular pelvic inlet and a pelvic outlet with widely spaced ischial spines can facilitate delivery. A smaller pelvic outlet can cause problems during delivery; thus the size of the pelvic inlet and outlet is routinely measured during prenatal pelvic examinations of pregnant women. If the pelvic outlet is too small for normal delivery, delivery can be accomplished by **cesarean section,** which is the surgical removal of the fetus through the abdominal wall.

Table 7.11	Differences Between the Male and Female Pelvis (see figure 7.33)
Area	**Description**
General	The female pelvis is somewhat lighter in weight and wider laterally but shorter superiorly to inferiorly and less funnel-shaped; less obvious muscle attachment points exist in females than in males
Sacrum	Broader in females with the inferior part directed more posteriorly; the sacral promontory does not project as far anteriorly in the female
Pelvic inlet	Heart-shaped in males; oval in females
Pelvic outlet	Broader and more shallow in females
Subpubic angle	Less than 90 degrees in males; 90 degrees or more in females
Ilium	More shallow and flared laterally in females
Ischial spines	Farther apart in females
Ischial tuberosities	Turned laterally in females and medially in males

Thigh

The thigh, like the arm, contains a single bone, which is called the **femur.** The femur has a prominent rounded **head,** where it articulates with the acetabulum, and a well-defined **neck;** both are located at an oblique angle to the shaft of the femur (figure 7.34). The proximal shaft exhibits two tuberosities: a **greater trochanter** (trō-kan′ter; runner) lateral to the neck and a smaller, or **lesser, trochanter** inferior and posterior to the neck. Both trochanters are attachment sites for muscles that fasten the hip to the thigh. The greater trochanter and its attached muscles form a bulge that can be seen as the widest part of the hips (see figure 7.32). The distal end of the femur has **medial** and **lateral condyles,** smooth, rounded surfaces that articulate with the tibia. Located proximally to the condyles are the **medial** and **lateral epicondyles,** important sites of ligament attachment.

P R E D I C T

Compare the following in terms of structure and function for the upper and lower limbs: depth of sockets, size of bones, and size of tubercles and trochanters. What is the significance of these differences?

The **patella,** or kneecap, is a large sesamoid bone located within the tendon of the quadriceps femoris muscle group, which is the major muscle group of the anterior thigh (figure 7.35). The patella articulates with the patellar groove of the femur to create a smooth articular surface over the anterior distal end of the femur. The patella holds the tendon away from the distal end of the femur and therefore changes the angle of the tendon between the quadriceps femoris muscle and the tibia, where the tendon attaches. This change in angle increases the force that can be applied from the muscle to the tibia. As the result of this increase in applied force, less muscle contraction force is required to move the tibia.

Leg

The leg is the part of the lower limb between the knee and the ankle. Like the forearm, it consists of two bones: the **tibia** (tib′ē-ă; or shinbone) and the **fibula** (fib′ū-lă; resembling a clasp or buckle; figure 7.36). The tibia is by far the larger of the two and supports most of the weight of the leg. A **tibial tuberosity,** which is the attachment point for the quadriceps femoris muscle group, can easily be seen and felt just inferior to the patella (figure 7.37). The **anterior crest** forms the shin. The proximal end of the tibia has flat **medial** and **lateral condyles,** which articulate with the condyles of the femur. Located between the condyles is the **intercondylar eminence,** which is a ridge between the two articular surfaces of the proximal tibia. The distal end of the tibia is enlarged to form the **medial malleolus** (ma-lē′ō-lŭs; mallet-shaped), which helps form the medial side of the ankle joint.

The fibula does not articulate with the femur but has a small proximal head where it articulates with the tibia. The distal end of the fibula is slightly enlarged as the **lateral malleolus** to create the lateral wall of the ankle joint. The lateral and medial malleoli can be felt and seen as prominent lumps on either side of the ankle (see figure 7.37). The thinnest, weakest portion of the fibula is just proximal to the lateral malleolus.

P R E D I C T

Explain why modern ski boots are designed with high tops that extend partway up the leg.

Head

Greater trochanter

Neck

Intertrochanteric line

Lesser trochanter

Body (shaft) of femur

Medial epicondyle

Lateral epicondyle

Patellar groove

(a) **Anterior view**

Fovea capitis

Head

Greater trochanter

Neck

Intertrochanteric crest

Pectineal line

Gluteal tuberosity

Linea aspera

Lateral epicondyle

Intercondylar fossa

Lateral condyle

Medial condyle

(b) **Posterior view**

Figure 7.34   Right Femur

(*a*) Anterior view. (*b*) Posterior view.

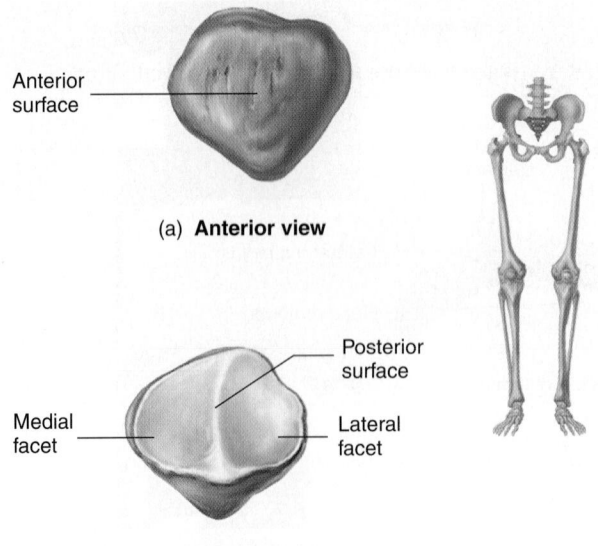

Anterior surface

(a) **Anterior view**

Posterior surface

Medial facet

Lateral facet

(b) **Posterior view**

Figure 7.35   Right Patella

(*a*) Anterior view. (*b*) Posterior view.

Patellar Defects

If the patella is severely fractured, the tendon from the quadriceps femoris muscle group may be torn, resulting in a severe decrease in muscle function. In extreme cases, it may be necessary to remove the patella to repair the tendon. Removal of the patella results in a decrease in the amount of power the quadriceps femoris muscle can generate at the tibia.

The patella normally tracks in the patellar groove on the anterodistal end of the femur. Abnormal tracking of the patella can become a problem in some teenagers, especially females. As the young woman's hips widen during puberty, the angles at the joints between the hips and the tibia may change considerably. As the knee becomes located more medially relative to the hip, the patella may be forced to track more laterally than normal. This lateral tracking may result in pain in the knees of some young athletes.

Anterior view

Figure 7.36   **Right Tibia and Fibula, Anterior View**

Fractures of the Malleoli

Turning the plantar surface of the foot outward so that it faces laterally is called eversion. Forceful eversion of the foot, such as when a person slips and twists the ankle or jumps and lands incorrectly on the foot, may cause the distal ends of the tibia and/or fibula to fracture (figure E*a*). Such fractures are not uncommon in soccer, football, and basketball players. When the foot is forcefully everted, the medial malleolus moves inferiorly toward the ground or floor and the talus slides laterally, forcing the medial and lateral malleoli to separate. The ligament holding the medial malleolus to the tarsal bones is stronger than the bones it connects, and often does not tear as the bones separate because of the eversion. Instead, the medial malleolus breaks. Also, as the talus slides laterally, the force can shear off the lateral malleolus, or more commonly, cause the fibula to break superior to the lateral malleolus. This type of injury to the tibia and fibula is often called a Pott's fracture.

Turning the plantar surface of the foot inward so that it faces medially is called inversion. Forceful inversion of the foot can fracture the fibula just proximal to the lateral malleolus (figure E*b*). More often, because the ligament holding the medial malleolus to the tarsal bones is weaker than the bones it connects, the inversion of the foot causes a sprain in which ligaments are damaged.

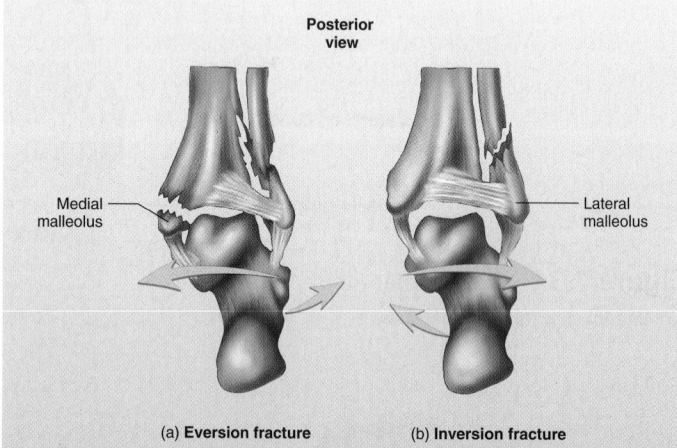

Posterior view

(a) **Eversion fracture**   (b) **Inversion fracture**

Figure E   **Fracture of the Medial or Lateral Malleolus**

Figure 7.37   **Surface Anatomy Showing Bones of the Lower Limb**

Foot

The proximal portion of the foot consists of seven **tarsal** (tar′săl; the sole of the foot) **bones,** which are depicted and named in figure 7.38. The **talus** (tā′lŭs), or "ankle bone," articulates with the tibia and the fibula to form the ankle joint. It also articulates with the calcaneous and navicular. The **calcaneus** (kal-kā′nē-us; heel), the "heel bone," is the largest and strongest bone in the foot. It is located inferior to the talus and supports that bone. The calcaneus protrudes posteriorly where the calf muscles attach to it and where it can be easily felt as

the heel of the foot. The **navicular** (nă-vik′ū-lar), which is boat shaped, lies between the talus posteriorly and the cuneiforms anteriorly. The talus, calcaneus, and navicular are proximal in the foot, but do not form a row. The proximal foot is relatively much larger than the wrist. The distal four bones do form a row. The medial, wedge-shaped three bones of the distal row are called **cuneiforms** (kū′-nē-i-fōrm). They are the **medial cuneiform, intermediate cuneiform,** and **lateral cuneiform.** The **cuboid** (kū′boyd), which is cube shaped, is the most lateral of the distal row. A mnemonic for the distal row

(a) **Superior view**

(b) **Medial view**

Figure 7.38 Bones of the Right Foot

(a) Superior view. (b) Medial view.

is: **MILC** that is, **M**edial, **I**ntermediate, and **L**ateral cuneiforms, and the **C**uboid; That for the proximal three bones is: **No Thanks Cow**, that is **N**avicular, **T**alus, and **C**alcaneus.

The **metatarsals** and **phalanges** of the foot are arranged in a manner very similar to that of the metacarpals and phalanges of the hand, with the great toe analogous to the thumb (see figure 7.38a). Small sesamoid bones often form in the tendons of muscles attached to the great toe. The ball of the foot is the junction between the metatarsals and phalanges. The foot as a unit is convex dorsally and concave ventrally to form the arches of the foot (described more fully in chapter 8).

P R E D I C T   9

A decubitus ulcer is a chronic ulcer that appears in pressure areas of skin overlying a bony prominence in bedridden or otherwise immobilized patients. Where are likely sites for decubitus ulcers to occur?

Arches of the Foot

The foot has three major **arches** that distribute the weight of the body between the heel and the ball of the foot during standing and walking (figure 7.39). As the foot is placed on the ground, weight is transferred from the tibia and the fibula to the talus. From there, the weight is distributed first to the heel (calcaneus) and then through the arch system along the lateral side of the foot to the ball of the foot (head of the metatarsals). This effect can be observed when a person with wet, bare feet walks across a dry surface; the print of the heel, the lateral border of the foot, and the ball of the foot can be seen, but the middle of the plantar surface and the medial border leave no impression. The medial side leaves no mark because the arches on this side of the foot are higher than those on the lateral side. The shape of the arches is maintained by the configuration of the bones, the ligaments connecting them, and the muscles acting on the foot.

31. What is the function of the greater trochanter and the lesser trochanter?
32. Describe the function of the patella.
33. Name the bones of the leg.
34. Give the points of articulation between the pelvis, femur, leg, and ankle.
35. What is the function of the tibial tuberosity?
36. Name the seven tarsal bones. Which bones form the ankle joint? What bone forms the heel?
37. Describe the bones of the foot. How many phalanges are in each toe?

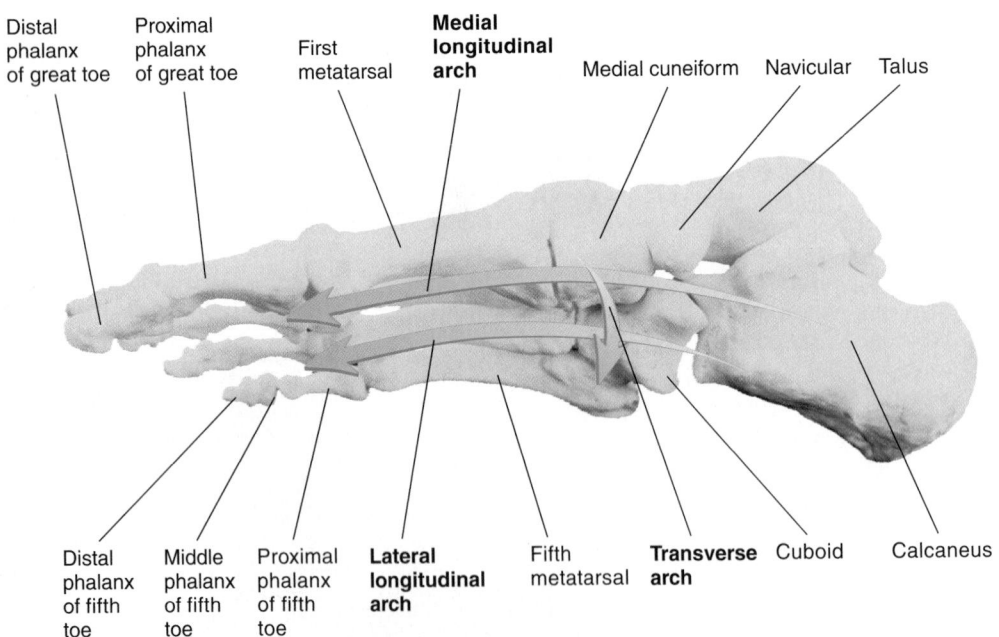

Medioinferior view

Figure 7.39 Arches (*arrows*) of the Right Foot from a Medioinferior View

The medial longitudinal arch is formed by the calcaneus, talus, navicular, cuneiforms, and three medial metatarsals. The lateral longitudinal arch is formed by the calcaneus, cuboid, and two lateral metatarsals. The transverse arch is formed by the cuboid and cuneiforms.

S U M M A R Y

The gross anatomy of the skeletal system considers the features of bone, cartilage, tendons, and ligaments that can be seen without the use of a microscope. Dried, prepared bones display the major features of bone but obscure the relationship between bone and soft tissue.

General Considerations   (p. 198)

Bones have processes, smooth surfaces, and holes that are associated with ligaments, muscles, joints, nerves, and blood vessels.

Axial Skeleton   (p. 200)

The axial skeleton consists of the skull, hyoid bone, vertebral column, and thoracic cage.

Skull

1. The skull, or cranium, can be thought of as a single unit.
2. The parietal bones are joined at the midline by the sagittal suture; they are joined to the frontal bone by the coronal suture, to the occipital bone by the lambdoid suture, and to the temporal bone by the squamous suture.
3. Nuchal lines are the points of attachment for neck muscles.
4. Several skull features are seen from a lateral view.
   • The external auditory meatus transmits sound waves toward the eardrum.
   • Important neck muscles attach to the mastoid process.
   • The temporal lines are attachment points of the temporalis muscle.
   • The zygomatic arch, from the temporal and zygomatic bones, forms a bridge across the side of the skull.
5. Several skull features are seen from a frontal view.
   • The orbits contain the eyes.
   • The nasal cavity is divided by the nasal septum, and the hard palate separates the nasal cavity from the oral cavity.
   • Sinuses within bone are air-filled cavities. The paranasal sinuses, which connect to the nasal cavity, are the frontal, sphenoidal, and maxillary sinuses, and the ethmoidal labyrinth.
   • The mandible articulates with the temporal bone.
6. Several skull features are seen inside the cranial cavity.
   • The crista galli is a point of attachment for one of the meninges.
   • The olfactory nerves extend into the roof of the nasal cavity through the cribriform plate.
   • The sella turcica is occupied by the pituitary gland.
   • The spinal cord and brain are connected through the foramen magnum.
7. Several features are seen on the inferior surface of the skull.
   • Occipital condyles are points of articulation between the skull and the vertebral column.
   • Blood reaches the brain through the internal carotid arteries, which pass through the carotid canals, and the vertebral arteries, which pass through the foramen magnum.
   • Most blood leaves the brain through the internal jugular veins, which exit through the jugular foramina.
   • Styloid processes provide attachment points for three muscles involved in movement of the tongue, hyoid bone, and pharynx.
   • The hard palate forms the floor of the nasal cavity.
8. The skull is composed of 22 bones.
   • The auditory ossicles, which function in hearing, are located inside the temporal bones.
   • The braincase protects the brain.
   • The facial bones protect the sensory organs of the head and function as muscle attachment sites (mastication, facial expression, and eye muscles).
   • The mandible and maxillae possess alveolar processes with sockets for the attachment of the teeth.

Hyoid

The hyoid bone, which "floats" in the neck, is the attachment site for throat and tongue muscles.

Vertebral Column

1. The vertebral column provides flexible support and protects the spinal cord.
2. The vertebral column has four major curvatures: cervical, thoracic, lumbar, and sacral/coccygeal. Abnormal curvatures are lordosis (lumbar), kyphosis (thoracic), and scoliosis (lateral).
3. Adjacent bodies are separated by intervertebral disks. The disk has a fibrous outer covering (annulus fibrosus) surrounding a gelatinous interior (nucleus pulposus).
4. A typical vertebra consists of a body, a vertebral arch, and various processes.
   • Part of the body and the vertebral arch (pedicle and lamina) form the vertebral foramen, which contains and protects the spinal cord.
   • Spinal nerves exit through the intervertebral foramina.
   • The transverse and spinous processes serve as points of muscle and ligament attachment.
   • Vertebrae articulate with one another through the superior and inferior articular processes.
5. Several types of vertebrae can be distinguished.
   • All seven cervical vertebrae have transverse foramina, and most have bifid spinous processes.
   • The 12 thoracic vertebrae are characterized by long, downward-pointing spinous processes and demifacets.
   • The five lumbar vertebrae have thick, heavy bodies and processes.
   • The sacrum consists of five fused vertebrae and attaches to the coxae to form the pelvis.
   • The coccyx consists of four fused vertebrae attached to the sacrum.

Thoracic Cage

1. The thoracic cage (consisting of the ribs, their associated costal cartilages, and the sternum) functions to protect the thoracic organs and changes volume during respiration.
2. Twelve pairs of ribs attach to the thoracic vertebrae. They are divided into seven pairs of true ribs and five pairs of false ribs. Two pairs of false ribs are floating ribs.
3. The sternum is composed of the manubrium, the body, and the xiphoid process.

Appendicular Skeleton   (p. 229)

The appendicular skeleton consists of the upper and lower limbs and the girdles that attach the limbs to the body.

Pectoral Girdle and Upper Limb

1. The upper limb is attached loosely and functions in grasping and manipulation.
2. The pectoral girdle consists of the scapulae and clavicles.
   • The scapula articulates with the humerus and the clavicle. It serves as an attachment site for shoulder, back, and arm muscles.
   • The clavicle holds the shoulder away from the body, permitting free movement of the arm.
3. The arm bone is the humerus.
   • The humerus articulates with the scapula (head), the radius (capitulum), and the ulna (trochlea).
   • Sites of muscle attachment are the greater and lesser tubercles, the deltoid tuberosity, and the epicondyles.
4. The forearm contains the ulna and radius.
   • The ulna and the radius articulate with each other and with the humerus and the wrist bones.
   • The wrist ligaments attach to the styloid processes of the radius and the ulna.
5. Eight carpal, or wrist, bones are arranged in two rows.
6. The hand consists of five metacarpal bones.

7. The phalanges are digital bones. Each finger has three phalanges, and the thumb has two phalanges.

Pelvic Girdle and Lower Limb

1. The lower limb is attached solidly to the coxa and functions in support and movement.
2. The pelvic girdle consists of the right and left coxae and sacrum. Each coxa is formed by the fusion of the ilium, the ischium, and the pubis.
   - The coxae articulate with each other (symphysis pubis) and with the sacrum (sacroiliac joint) and the femur (acetabulum).
   - Important sites of muscle attachment are the iliac crest, the iliac spines, and the ischial tuberosity.
   - The female pelvis has a larger pelvic inlet and outlet than the male pelvis.

3. The thigh bone is the femur.
   - The femur articulates with the coxa (head), the tibia (medial and lateral condyles), and the patella (patellar groove).
   - Sites of muscle attachment are the greater and lesser trochanters.
   - Sites of ligament attachment are the lateral and medial epicondyles.
4. The leg consists of the tibia and the fibula.
   - The tibia articulates with the femur, the fibula, and the talus. The fibula articulates with the tibia and the talus.
   - Tendons from the thigh muscles attach to the tibial tuberosity.
5. Seven tarsal bones form the proximal portion of the foot.
6. The foot consists of five metatarsal bones.
7. The toes have three phalanges each, except for the big toe, which has two.
8. The bony arches transfer weight from the heels to the toes and allow the foot to conform to many different positions.

R E V I E W   A N D   C O M P R E H E N S I O N

1. Which of these is part of the appendicular skeleton?
   a. cranium
   b. ribs
   c. clavicle
   d. sternum
   e. vertebra

2. A knoblike lump on a bone is called a
   a. spine.
   b. facet.
   c. tuberosity.
   d. sulcus.
   e. ramus.

3. The superior and middle nasal conchae are formed by projections of the
   a. sphenoid bone.
   b. vomer bone.
   c. palatine process of maxillae.
   d. palatine bone.
   e. ethmoid bone.

4. The crista galli
   a. separates the nasal cavity into two parts.
   b. attaches the hyoid bone to the skull.
   c. holds the pituitary gland.
   d. is an attachment site for the membranes that surround the brain.
   e. is a passageway for blood vessels.

5. The perpendicular plate of the ethmoid and the _____ form the nasal septum.
   a. palatine process of the maxilla
   b. horizontal plate of the palatine
   c. vomer
   d. nasal bone
   e. lacrimal bone

6. Which of these bones does *not* contain a paranasal sinus?
   a. ethmoid
   b. sphenoid
   c. frontal
   d. temporal
   e. maxilla

7. The mandible articulates with the skull at the
   a. styloid process.
   b. occipital condyle.
   c. mandibular fossa.
   d. zygomatic arch.
   e. medial pterygoid.

8. The nerves for the sense of smell pass through the
   a. cribriform plate.
   b. nasolacrimal canal.
   c. internal auditory meatus.
   d. optic canal.
   e. orbital fissure.

9. The major blood supply to the brain enters through the
   a. foramen magnum.
   b. carotid canals.
   c. jugular foramina.
   d. both a and b.
   e. all of the above.

10. The site of the sella turcica is the
    a. sphenoid bone.
    b. maxillae.
    c. frontal bone.
    d. ethmoid bone.
    e. temporal bone.

11. Which of these bones is *not* in contact with the sphenoid bone?
    a. maxilla
    b. inferior nasal concha
    c. ethmoid
    d. parietal
    e. vomer

12. Which of these statements about vertebral column curvature is *not* true?
    a. The cervical curvature develops before birth.
    b. The thoracic curvature becomes exaggerated in kyphosis.
    c. The lumbar curvature becomes exaggerated in lordosis.
    d. The sacral curvature develops before birth.
    e. The lumbar curvature develops as an infant learns to sit and walk.

13. A herniated disk occurs when
    a. the annulus fibrosus ruptures.
    b. the intervertebral disk slips out of place.
    c. the spinal cord ruptures.
    d. too much fluid builds up in the nucleus pulposus.
    e. all of the above.

14. The weight-bearing portion of a vertebra is the
    a. vertebral arch.
    b. articular process.
    c. body.
    d. transverse process.
    e. spinous process.

15. Transverse foramina are found only in
    a. cervical vertebrae.
    b. thoracic vertebrae.
    c. lumbar vertebrae.
    d. the sacrum.
    e. the coccyx.
16. Articular facets on the bodies and transverse processes are found only on
    a. cervical vertebrae.
    b. thoracic vertebrae.
    c. lumbar vertebrae.
    d. the sacrum.
    e. the coccyx.
17. Medially facing superior articular processes and laterally facing inferior articular processes are found on
    a. cervical vertebrae.
    b. thoracic vertebrae.
    c. lumbar vertebrae.
    d. the sacrum.
    e. the coccyx.
18. Which of these statements concerning ribs is true?
    a. The true ribs attach directly to the sternum with costal cartilage.
    b. There are five pairs of floating ribs.
    c. The head of the rib attaches to the transverse process of the vertebra.
    d. Vertebrochondral ribs are classified as true ribs.
    e. Floating ribs do not attach to vertebrae.
19. The point where the scapula and clavicle connect is the
    a. coracoid process.
    b. styloid process.
    c. glenoid fossa.
    d. acromion process.
    e. capitulum.
20. The distal medial process of the humerus to which the ulna joins is the
    a. epicondyle.
    b. deltoid tuberosity.
    c. malleolus.
    d. capitulum.
    e. trochlea.
21. The depression on the anterior surface of the humerus that receives part of the ulna when the forearm is flexed (bent) is the
    a. glenoid fossa.
    b. capitulum.
    c. coronoid fossa.
    d. olecranon fossa.
    e. radial fossa.
22. Which of these is *not* a point of muscle attachment on the pectoral girdle or upper limb?
    a. epicondyles
    b. mastoid process
    c. radial tuberosity
    d. spine of scapula
    e. greater tubercle
23. Which of these parts of the upper limb are *not* correctly matched with the number of bones in that part?
    a. arm: 1
    b. forearm: 2
    c. wrist: 10
    d. palm of hand: 5
    e. digits: 14
24. The bone of the foot that the tibia rests upon is the
    a. talus.
    b. calcaneus.
    c. metatarsals.
    d. navicular.
    e. phalanges.
25. A place where nerves or blood vessels pass from the trunk to the lower limb is the
    a. pubic crest.
    b. greater sciatic notch.
    c. ischial tuberosity.
    d. iliac crest.
    e. pubis symphysis.
26. A projection on the pelvic girdle that is used as a landmark for finding an injection site is the
    a. ischial tuberosity.
    b. iliac crest.
    c. anterior superior iliac spine.
    d. posterior inferior iliac spine.
    e. ischial spine.
27. When comparing the pectoral girdle to the pelvic girdle, which of these statements is true?
    a. The pectoral girdle has greater mass than the pelvic girdle.
    b. The pelvic girdle is more firmly attached to the body than the pectoral girdle.
    c. The pectoral girdle has the limbs more securely attached than the pelvic girdle.
    d. The pelvic girdle allows greater mobility than the pectoral girdle.
28. When comparing a male pelvis to a female pelvis, which of these statements is true?
    a. The pelvic inlet in males is larger and more circular.
    b. The subpubic angle in females is less than 90 degrees.
    c. The ischial spines in males are closer together.
    d. The sacrum in males is broader and less curved.
29. A site of muscle attachment on the proximal end of the femur is the
    a. greater trochanter.
    b. epicondyle.
    c. greater tubercle.
    d. intercondylar eminence.
    e. condyle.
30. A process that forms the large lateral bump in the ankle is the lateral
    a. malleolus.
    b. condyle.
    c. epicondyle.
    d. tuberosity.
    e. tubercle.

Answers in Appendix F

CRITICAL THINKING

1. A patient has an infection in the nasal cavity. Name seven adjacent structures to which the infection could spread.

2. A patient is unconscious. Radiographic films reveal that the superior articular process of the atlas has been fractured. Which of the following could have produced this condition: falling on the top of the head or being hit in the jaw with an uppercut? Explain.

3. If the vertebral column is forcefully rotated, what part of the vertebra is most likely to be damaged? In what area of the vertebral column is such damage most likely?

4. An asymmetric weakness of the back muscles can produce which of the following: scoliosis, kyphosis, or lordosis? Which could result from pregnancy? Explain.

5. What might be the consequences of a broken forearm involving both the ulna and radius when the ulna and radius fuse to each other during repair of the fracture?

6. Suppose you need to compare the length of one lower limb to the other in an individual. Using bony landmarks, suggest an easy way to accomplish the measurements.

7. A paraplegic individual develops decubitus ulcers (pressure sores) on the buttocks from sitting in a wheelchair for extended periods. Name the bony protuberance responsible.

8. Why are women knock-kneed more often than men?

9. On the basis of bone structure of the lower limb, explain why it's easier to turn the foot medially (sole of the foot facing toward the midline of the body) than laterally. Why is it easier to cock the wrist medially than laterally?

10. Justin Time leaped from his hotel room to avoid burning to death in a fire. If he landed on his heels, what bone was he likely to fracture? Unfortunately for Justin, a 240 lb fire fighter, Hefty Stomper, ran by and stepped heavily on the proximal part of Justin's foot (not the toes). What bones could now be broken?

Answers in Appendix G

ANSWERS TO PREDICT QUESTIONS

1. The sagittal suture is so named because it is in line with the midsagittal plane of the head. The coronal suture is so named because it is in line with the coronal plane (see chapter 1).

2. The bones most often broken in a "broken nose" are the nasals, ethmoid, vomer, and maxillae.

3. The lumbar vertebrae support a greater weight than the other vertebrae. The vertebrae are more massive because of the greater weight they support.

4. The anterior support of the scapula is lost with a broken clavicle, and the shoulder is located more inferiorly and anteriorly than normal. In addition, since the clavicle normally holds the upper limb away from the body, the upper limb moves medially and rests against the side of the body.

5. The olecranon process moves into the olecranon fossa as the elbow is straightened. The coronoid process moves into the coronoid fossa as the elbow is bent.

6. The dried skeleton seems to have longer "fingers" than the hand with soft tissue intact because the soft tissue fills in the space between the metacarpals. With the soft tissue gone, the metacarpals seem to be an extension of the fingers, which appear to extend from the most distal phalanx to the carpals.

7. The depth of the hip socket is deeper, the bone is more massive, and the tubercles are larger than similar structures in the upper limb. All of this correlates with the weight-bearing nature of the lower limb and the more massive muscles necessary for moving the lower limb compared to the upper limb.

8. The top of modern ski boots is placed high up the leg to protect the weakest point of the fibula and make it less susceptible to great strain during a fall. Modern ski boots are also designed to reduce ankle mobility, which increases comfort and performance.

9. Decubitus ulcers form over bony prominences where the bone is close to the overlying skin and where the body contacts the bed when lying down. Such sites are the back and front of the skull and the cheeks (over zygomatic bones), the acromion process, scapula, olecranon process, coccyx, greater trochanter, lateral epicondyle of femur, patella, and lateral malleolus.

Visit the Online Learning Center at www.mhhe.com/seeley7 for chapter quizzes, interactive learning exercises, and other study tools.

Articulations and Movement

This is a colorized scanning electron micrograph of a chondrocyte within a lacuna surrounded by cartilage matrix. Cartilage provides the durable, resilient cushion within joints.

CHAPTER

8

Muscles pull on bones to make them move, but movement would not be possible without joints between the bones. Humans would resemble statues were it not for the joints between bones that allow bones to move once the muscles have provided the pull. Machine parts most likely to wear out are those that rub together and they require the most maintenance. Movable joints are places in the body where the bones rub together, yet we tend to pay little attention to them. Fortunately our joints are self-maintaining, but damage to or disease of a joint can make movement very difficult. We realize then how important the movable joints are for normal function.

An **articulation**, or **joint**, is a place where two bones come together. We usually think of joints as being movable, but that's not always the case. Many joints allow only limited movement, and others allow no apparent movement. The structure of a given joint is directly correlated with its degree of movement. Fibrous joints have much less movement than joints containing fluid and having smooth articulating surfaces.

Joints develop between adjacent bones or areas of ossification, and movement is important in determining the type of joint that develops. If movement is restricted—even in a highly movable joint—at any time during an individual's life, the joint may be transformed into a nonmovable joint.

This chapter presents a scheme for *naming joints* (p. 248) and an explanation of *classes of joints* (p. 248), and *types of movement* (p. 254). It then presents a *description of selected joints* (p. 259) and summarizes the *effects of aging on the joints* (p. 268).

Naming Joints

Joints are commonly named according to the bones or portions of bones that are united at the joint, such as the temporomandibular joint between the temporal bone and the mandible.

P R E D I C T

What is the name of the joint between the metacarpals and the phalanges?

Some joints are given the name of only one of the articulating bones, such as the humeral (shoulder) joint between the humerus and scapula. Still other joints are simply given the Greek or Latin equivalent of the common name, such as **cubital** (kū′bi-tăl) **joint** for the elbow joint.

   1. What criteria are used to name joints?

Classes of Joints

The three major kinds of joints are classified structurally as fibrous, cartilaginous, and synovial. In this classification scheme, joints are categorized according to the major connective tissue type that binds the bones together, and whether or not a fluid-filled joint capsule is present. Joints may also be classified according to their function. This classification is based on the degree of motion at each joint and includes the terms *synarthrosis* (nonmovable joint), *amphiarthrosis* (slightly movable joint), and *diarthrosis* (freely movable joint). This functional classification is somewhat limited and is not used in this text. The structural classification scheme with its various subclasses allows for a more precise classification and is the scheme we use.

Fibrous Joints

Fibrous joints consist of two bones that are united by fibrous connective tissue, have no joint cavity, and exhibit little or no movement. Joints in this group are classified further as sutures, syndesmoses, or gomphoses (table 8.1) based on their structure.

Sutures

Sutures (soo′choorz) are seams between the bones of the skull (figure 8.1). Some sutures may become completely immovable in older adults. Sutures are seldom smooth, and the opposing bones often interdigitate (have interlocking fingerlike processes). This interdigitation adds considerable stability to sutures. The tissue between the two bones is dense, regular collagenous connective tissue, and the periosteum on the inner and outer surfaces of the adjacent bones continues over the joint. The two layers of periosteum plus the dense fibrous connective tissue in between form a **sutural ligament.**

In a newborn, membranous areas called **fontanels** (fon′tă-nelz′) are present within some of the sutures. The fontanels make the skull flexible during the birth process and allow for growth of the head after birth (figure 8.2).

The margins of bones within sutures are sites of continuous intramembranous bone growth, and many sutures eventually become ossified. For example, ossification of the suture between the two frontal bones occurs shortly after birth so that they usually form a single frontal bone in the adult skull. In most normal adults, the coronal, sagittal, and lambdoid sutures are not fused. In some very old adults, however, even these sutures may become ossified. A **synostosis** (sin-os-tō′sis) results when two bones grow together across a joint to form a single bone.

Table 8.1	Fibrous and Cartilaginous Joints	
Class and Example of Joint	**Bones or Structures Joined**	**Movement**
Fibrous Joints		
Sutures		
Coronal	Frontal and parietal	None
Lambdoid	Occipital and parietal	None
Sagittal	The two parietal bones	None
Squamous	Parietal and temporal	Slight
Syndesmoses		
Radioulnar (interosseous membrane)	Ulna and radius	Slight
Stylohyoid	Styloid process and hyoid bone	Slight
Stylomandibular	Styloid process and mandible	Slight
Tibiofibular (interosseous membrane)	Tibia and fibula	Slight
Gomphoses		
Dentoalveolar	Tooth and alveolar process	Slight
Cartilaginous Joints		
Synchrondroses		
Epiphyseal plate	The diaphysis and epiphysis of a long bone	None
Sternocostal	Anterior cartilaginous part of first rib; between rib and sternum	Slight
Sphenooccipital	Sphenoid and occipital	None
Symphyses		
Intervertebral	Bodies of adjacent vertebrae	Slight
Manubriosternal	Manubrium and body of sternum	None
Symphysis pubis	The two coxae	None except during childbirth
Xiphisternal	Xiphoid process and body of sternum	None

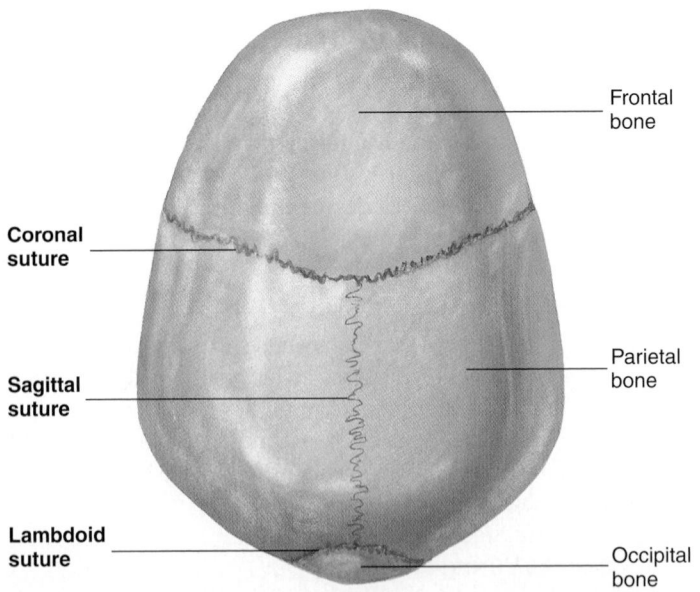

Figure 8.1   Sutures, Superior View

Syndesmoses

A **syndesmosis** (sin′dez-mō′sis; to fasten or bind) is a fibrous joint in which the bones are farther apart than in a suture and are joined by ligaments. Some movement may occur at syndesmoses because of flexibility of the ligaments, such as in the radioulnar syndesmosis, which binds the radius and ulna together (figure 8.3).

Gomphoses

Gomphoses (gom-fō′sēz) are specialized joints consisting of pegs that fit into sockets and that are held in place by fine bundles of regular collagenous connective tissue. The joints between the teeth and the sockets (alveoli) of the mandible and maxillae are gomphoses (figure 8.4). The connective tissue bundles between the teeth and their sockets are called **periodontal** (per′ē-ō-don′tăl) **ligaments** and allow a slight amount of "give" to the teeth during mastication.

Gingivitis

The gingiva, or gums, are the soft tissues covering the alveolar process. Neglect of the teeth can result in **gingivitis,** an inflammation of the gingiva, often resulting from bacterial infection. Left untreated, gingivitis may spread to the tooth socket, resulting in periodontal disease, the leading cause of tooth loss in the United States. Periodontal disease involves an accumulation of plaque and bacteria, and the resulting inflammation, which gradually destroys the periodontal ligaments and the bone. As a result, teeth may become so loose that they come out of their sockets. Proper brushing, flossing, and professional cleaning to remove plaque can usually prevent gingivitis and periodontal disease.

(a) **Lateral view**

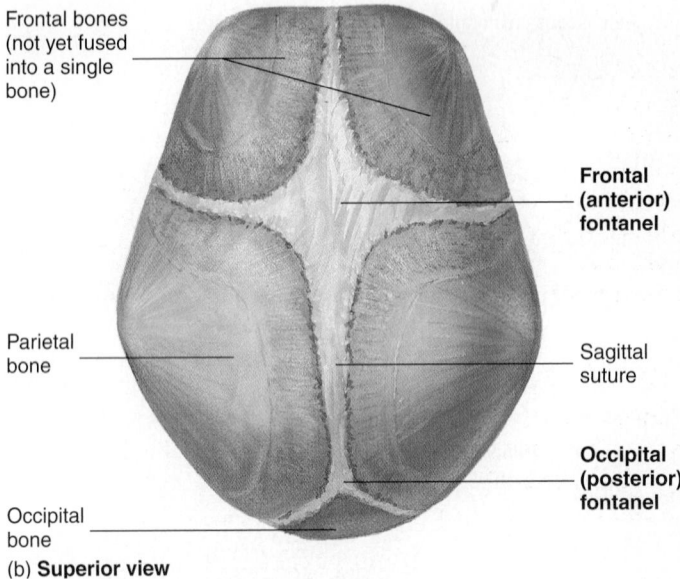

(b) **Superior view**

Figure 8.2   Fetal Skull Showing Fontanels and Sutures
(a) Lateral view. (b) Superior view.

2. *Define the term fibrous joint, describe three different types, and give an example of each. What is a synostosis? Where are periodontal ligaments found?*

Cartilaginous Joints

Cartilaginous joints unite two bones by means of either hyaline cartilage or fibrocartilage (see table 8.1). Joints containing hyaline cartilage are called synchondroses; joints containing fibrocartilage are called symphyses.

Head of radius

Biceps brachii tendon

Radioulnar syndesmosis
(interosseous membrane)

Annular ligament

Radius

Ulna

Anterior view

Figure 8.3   Right Radioulnar Syndesmosis
The interosseous membrane between the ulna and radius.

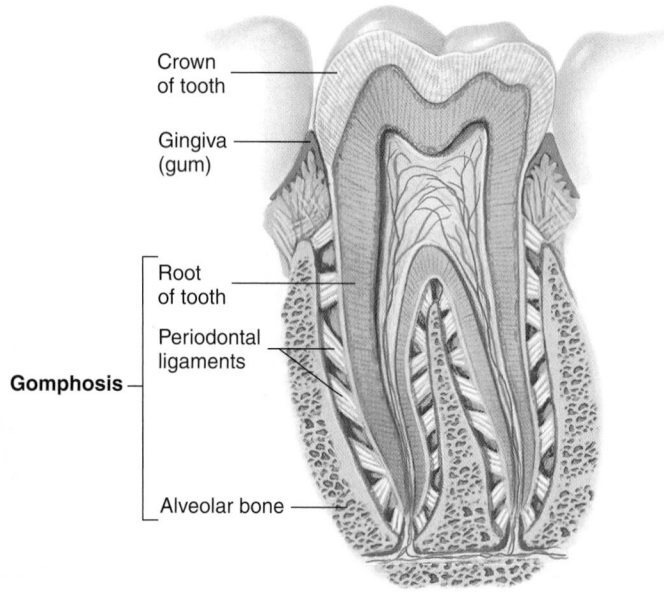

Crown
of tooth

Gingiva
(gum)

Root
of tooth

Periodontal
ligaments

Gomphosis

Alveolar bone

**Figure 8.4   Gomphosis Between a Tooth and Alveolar Bone
of the Mandible**

Synchondroses

A **synchondrosis** (sin′kon-drō′sis; union through cartilage) consists of two bones joined by hyaline cartilage where little or no movement occurs (figure 8.5*a*). The epiphyseal plates of growing bones are synchondroses (figure 8.5*b*). Most synchondroses are temporary, with bone eventually replacing them to form synostoses. On the other hand, some synchondroses persist throughout life. An example is the sternocostal synchondrosis between the first rib

and the sternum by way of the first costal cartilage (figure 8.5*c*). All the costal cartilages begin as synchondroses, but because of the movement that occurs between them and the sternum, all but the first usually develop synovial joints at those junctions. As a result, even though the **costochondral joints** (between the ribs and the costal cartilages) are retained, most costal cartilages no longer qualify as synchondroses because one end of the cartilage attaches to bone (the sternum) by a synovial joint.

Symphyses

A **symphysis** (sim′fi-sis; a growing together) consists of fibrocartilage uniting two bones. Symphyses include the junction between the manubrium and body of the sternum (figure 8.5*c*), the symphysis pubis (figure 8.6), and the intervertebral disks (see figures 7.13 and 7.14). Some of these joints are slightly movable because of the somewhat flexible nature of fibrocartilage.

Joint Changes During Pregnancy

During pregnancy certain hormones, such as estrogen, progesterone, and relaxin, act on the connective tissue of joints, such as the symphysis pubis, causing them to become more stretchable and allowing the joints to loosen. This change allows the pelvic opening to enlarge at the time of delivery. After delivery, the connective tissue of the symphysis pubis returns to its original condition. The enlarged pelvic opening, however, may not return completely to its original size and the woman may have slightly wider hips after the birth of the child.

These same hormones may act on the connective tissue of other joints in the body, such as the arches of the feet, causing them to relax, which may result in fallen arches (see section on "Arch Problems," p. 268). They may also act on some of the baby's joints, such as the hip, causing the joints to become more mobile than normal. Increased mobility of the hip can result in congenital (appearing at birth) subluxation, or congenital dislocation, of the hip. Congenital hip dislocation occurs approximately once in every 670 births.

● *3. Define cartilaginous joints, describe two different types,
    and give an example of each. Why are costochondral joints
    unique?*

Synovial Joints

Synovial (si-nō′vē-ăl; joint fluid; *syn*, coming together, *ovia*, resembling egg albumin) **joints** contain synovial fluid and allow considerable movement between articulating bones (figure 8.7). These joints are anatomically more complex than fibrous and cartilaginous joints. Most joints that unite the bones of the appendicular skeleton are synovial joints, reflecting the far greater mobility of the appendicular skeleton compared to that of the axial skeleton.

The articular surfaces of bones within synovial joints are covered with a thin layer of hyaline cartilage called **articular cartilage,** which provides a smooth surface where the bones meet. In some synovial joints, a flat plate or pad of fibrocartilage, called an **articular disk,** is located between the articular cartilages of bones. The circumference of the disk is attached to the fibrous capsule. Examples of joints with articular disks are the temporomandibular, sternoclavicular, and acromioclavicular joints. A **meniscus** is

Ilium

Epiphysis

Synchondroses (epiphyseal plates)

Secondary epiphysis

Diaphysis

Ischium — Pubis

Synchondroses

(a) **Lateral view**

(b)

First rib

Sternocostal synchondrosis (costal cartilage of first rib)

Manubriosternal symphysis

Manubrium

Body

Xiphoid process

Sternum

Sternal symphyses

Xiphisternal symphysis

Costochondral joint

(c) **Anterior view**

Figure 8.5 Synchondroses

(a) Synchondroses (epiphyseal plates) between the developing bones of the coxa. (b) Epiphyseal plates. (c) Sternocostal synchondroses.

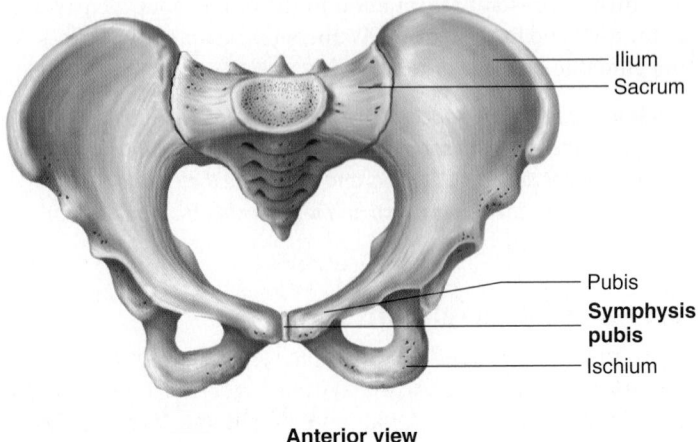

Ilium
Sacrum

Pubis
Symphysis pubis
Ischium

Anterior view

Figure 8.6 Symphysis Pubis

an incomplete, crescent-shaped fibrocartilage pad found in joints such as the knee and wrist.

The articular surfaces of the bones that meet at a synovial joint are enclosed within a synovial **joint cavity,** which is surrounded by a **joint capsule.** This capsule helps to hold the bones together while allowing for movement. The joint capsule consists of two layers: an outer **fibrous capsule** and an inner **synovial membrane** (see figure 8.7). The fibrous capsule consists of dense irregular connective tissue and is continuous with the fibrous layer of the periosteum that covers the bones united at the joint. Portions of the fibrous capsule may thicken to form ligaments. In addition, ligaments and tendons may be present outside the fibrous capsule, thereby contributing to the strength and stability of the joint while limiting movement in some directions.

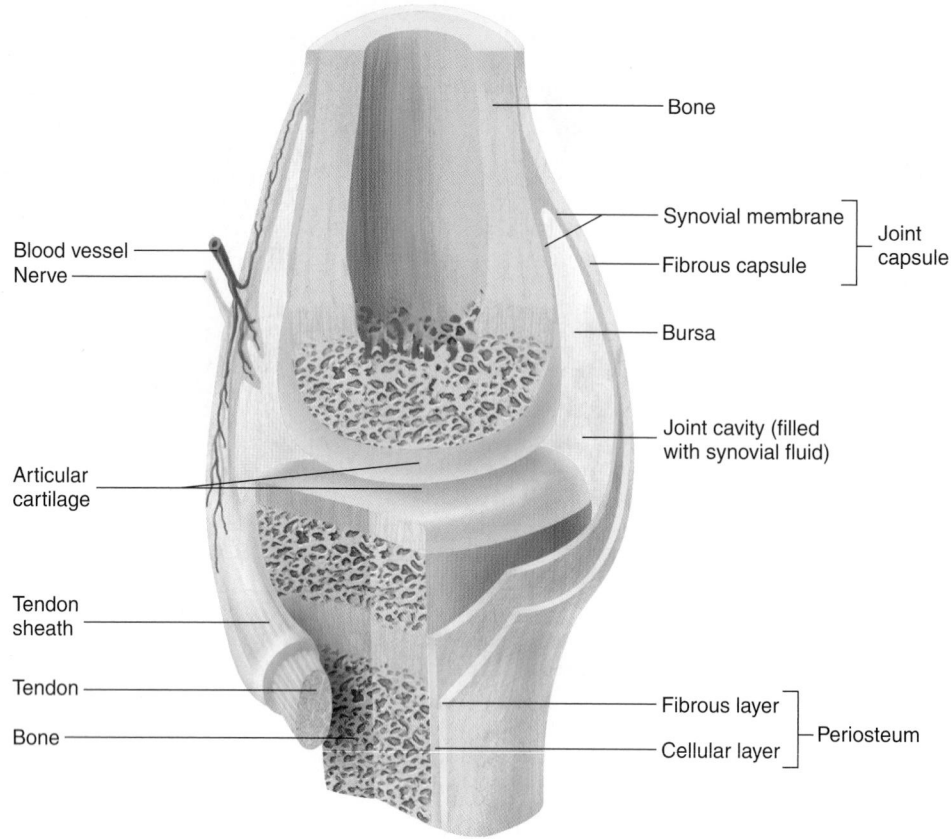

Bone

Synovial membrane ⎤
Fibrous capsule ⎦ Joint capsule

Blood vessel
Nerve

Bursa

Joint cavity (filled with synovial fluid)

Articular cartilage

Tendon sheath

Tendon

Bone

Fibrous layer ⎤
Cellular layer ⎦ Periosteum

Figure 8.7 Structure of a Synovial Joint

The synovial membrane lines the joint cavity, except over the articular cartilage. It is a thin, delicate membrane consisting of a collection of modified connective tissue cells either intermixed with part of the fibrous capsule or separated from it by a layer of areolar tissue or adipose tissue. The membrane produces synovial fluid, which consists of a serum (blood fluid) filtrate and secretions from the synovial cells. Synovial fluid is a complex mixture of polysaccharides, proteins, fat, and cells. The major polysaccharide is hyaluronic acid, which provides much of the slippery consistency and lubricating qualities of synovial fluid. Synovial fluid forms an important thin lubricating film that covers the surfaces of a joint.

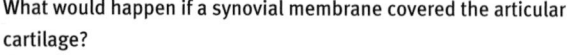

P R E D I C T **3**
What would happen if a synovial membrane covered the articular cartilage?

In certain synovial joints, such as the shoulder and knee, the synovial membrane may extend as a pocket, or sac, called a **bursa** (ber′să; pocket) for a distance away from the rest of the joint cavity (see figure 8.7). Bursae contain synovial fluid and provide a fluid-filled cushion between structures that otherwise would rub against each other, such as tendons rubbing on bones or other tendons. Some bursae, such as the olecranon bursae, are not associated with joints, but provide a cushion between the skin and underlying bony prominences, where friction could damage the tissues. Other bursae extend along tendons for some distance, forming **tendon sheaths. Bursitis** (ber-sī′tis) is an inflammation of a bursa and may cause considerable pain around the joint and restrict movement.

At the peripheral margin of the articular cartilage, blood vessels form a vascular circle that supplies the cartilage with nourishment, but no blood vessels penetrate the cartilage or enter the joint cavity. Additional nourishment to the articular cartilage comes from the underlying cancellous bone and from the synovial fluid covering the articular cartilage. Sensory nerves enter the fibrous capsule and, to a lesser extent, the synovial membrane. They not only supply information to the brain about pain in the joint but also furnish constant information to the brain about the position of the joint and its degree of movement (see chapter 14). Nerves do not enter the cartilage or joint cavity.

4. *Describe the structure of a synovial joint. How do the different parts of the joint function to permit joint movement? What are articular disks and where are they found?*
5. *Define the terms bursa and tendon sheath. What is their function?*

Types of Synovial Joints

Synovial joints are classified according to the shape of the adjoining articular surfaces. The six types of synovial joints are the plane, saddle, hinge, pivot, ball-and-socket, and ellipsoid. These joints are illustrated in figures 8.8 to 8.13 and are listed in table 8.2.

Table 8.2	Types of Joints		
Class and Example of Joint		**Structures Joined**	**Movement**
Plane (multiaxial)	Acromioclavicular	Acromion process of scapula and clavicle	Slight
	Carpometacarpal	Carpals and metacarpals 2–5	Multiple axes as a group
	Costovertebral	Ribs and vertebrae	Slight
	Intercarpal	Between carpals	Slight
	Intertarsal	Between tarsals	Slight
	Intervertebral	Between articular processes of adjacent vertebrae	Slight
	Sacroiliac	Between sacrum and coxa (complex joint with several planes and synchondroses)	Slight
	Tarsometatarsal	Tarsals and metatarsals	Slight
Saddle (biaxial)	Carpometacarpal pollicis	Carpal and metacarpal of thumb	Two axes
	Intercarpal	Between carpals	Slight
	Sternoclavicular	Manubrium of sternum and clavicle	Slight
Hinge (monoaxial)	Cubital (elbow)	Humerus, ulna, and radius	One axis
	Genu (knee)	Femur and tibia	One axis
	Interphalangeal	Between phalanges	One axis
	Talocrural (ankle)	Talus, tibia and fibula	Multiple axes, one predominates
Pivot (monoaxial)	Atlantoaxial	Atlas and axis	Rotation
	Proximal radioulnar	Radius and ulna	Rotation
	Distal radioulnar	Radius and ulna	Rotation
Ball-and-Socket (multiaxial)	Coxal (hip)	Coxa and femur	Multiple axes
	Glenohumeral (shoulder)	Scapula and humerus	Multiple axes
Ellipsoid (biaxial or multiaxial)	Atlantooccipital	Atlas and occipital bone	Two axes
	Metacarpophalangeal (knuckles)	Metacarpals and phalanges	Two axes
	Metatarsophalangeal	Metatarsals and phalanges	Two axes
	Radiocarpal (wrist)	Radius and carpals	Multiple axes
	Temporomandibular	Mandible and temporal bone	Multiple axes, one predominates

Movements at synovial joints are described as **monoaxial** (occurring around one axis), **biaxial** (occurring around two axes situated at right angles to each other), or **multiaxial** (occurring around several axes).

Plane, or **gliding, joints** consist of two opposed flat surfaces of about equal size in which a slight amount of gliding motion can occur between the bones (figure 8.8). These joints are considered monoaxial because some rotation is also possible but is limited by ligaments and adjacent bone. Examples are the articular processes between vertebrae.

Saddle joints consist of two saddle-shaped articulating surfaces oriented at right angles to each other so that complementary surfaces articulate with each other (figure 8.9). Saddle joints are biaxial joints. The carpometacarpal joint of the thumb is an example.

Vertebrae

Figure 8.8 Plane Joint

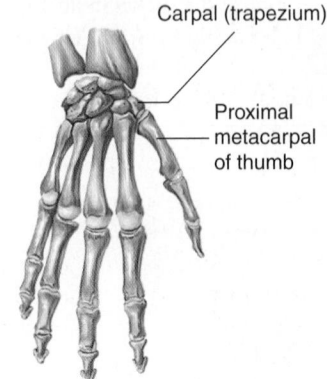

Carpal (trapezium)

Proximal metacarpal of thumb

Figure 8.9 Saddle Joint

Hinge joints are monoaxial joints (figure 8.10). They consist of a convex cylinder in one bone applied to a corresponding concavity in the other bone. Examples include the elbow and knee joints.

Figure 8.10   Hinge Joint

Pivot joints are monoaxial joints that restrict movement to rotation around a single axis (figure 8.11). A pivot joint consists of a relatively cylindrical bony process that rotates within a ring composed partly of bone and partly of ligament. The articulation between the head of the radius and the proximal end of the ulna is an example. The articulation between the dens, a process on the axis (see chapter 7), and the atlas is another example.

Figure 8.11   Pivot Joint

Ball-and-socket joints consist of a ball (head) at the end of one bone and a socket in an adjacent bone into which a portion of the ball fits (figure 8.12). This type of joint is multiaxial, allowing a wide range of movement in almost any direction. Examples are the shoulder and hip joints.

Figure 8.12   Ball-and-Socket Joint

Ellipsoid joints (or condyloid joints) are modified ball-and-socket joints (figure 8.13). The articular surfaces are ellipsoid in shape rather than spherical as in regular ball-and-socket joints. Ellipsoid joints are biaxial, because the shape of the joint limits its range of movement almost to a hinge motion in two axes and restricts rotation. The atlantooccipital joint is an example.

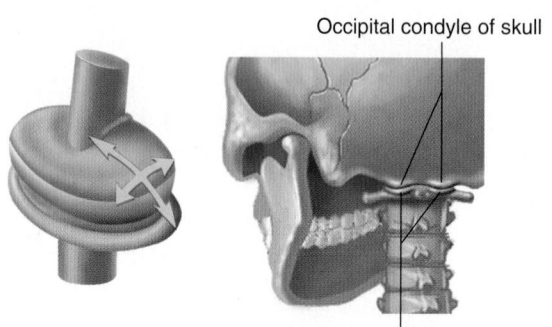

Figure 8.13   Ellipsoid Joint

6. On what basis are synovial joints classified? Describe the different types of synovial joints, and give examples of each. What movements does each type of joint allow?

Types of Movement

A joint's structure relates to the movements that occur at that joint. Some joints are limited to only one type of movement; others can move in several directions. With few exceptions, movement is best described in relation to the anatomic position: (1) movement away from the anatomic position and (2) movement returning a structure toward the anatomic position. Most movements are accompanied by movements in the opposite direction and therefore are listed in pairs.

Gliding Movements

Gliding movements are the simplest of all the types of movement. These movements occur in plane joints between two flat or nearly flat surfaces where the surfaces slide or glide over each other. These joints often give only slight movement, such as between carpal bones.

Angular Movements

Angular movements are those in which one part of a linear structure, such as the body as a whole or a limb, is bent relative to another part of the structure, thereby changing the angle between the two parts. Angular movements also involve the movement of a solid rod, such as a limb, that's attached at one end to the body, so that the angle at which it meets the body is changed. The most common angular movements are flexion and extension and abduction and adduction.

Flexion and Extension

Flexion and extension can be defined in a number of ways, but in each case exceptions to the definition exist. The literal definition is to bend and straighten, respectively. This bending and straightening can easily be seen in the elbow (figure 8.14). We have chosen to use a definition with more utility and fewer exceptions, with reference to the coronal plane. **Flexion** is movement of a body part anterior to the coronal plane, or in the anterior direction. **Extension** is movement of a body part posterior to the coronal plane, or in the posterior direction (figure 8.15). The exception to defining flexion and extension according

to the coronal plane is the knee and foot. At the knee, flexion moves the leg in a posterior direction and extension moves it in an anterior direction (figure 8.16).

Movement of the foot toward the plantar surface, such as when standing on the toes, is commonly called **plantar flexion;** and movement of the foot toward the shin, such as when walking on the heels, is called **dorsiflexion** (figure 8.17).

Hyperextension

Hyperextension is usually defined as an abnormal, forced extension of a joint beyond its normal range of motion. For example, if a person falls and attempts to break the fall by putting out a hand, the force of the fall directed into the hand and wrist may cause hyperextension of the wrist, which may result in sprained joints or broken bones. Some health professionals, however, define hyperextension as the normal movement of structures, except the leg, into the space posterior to the anatomic position.

Abduction and Adduction

Abduction (meaning to take away) is movement away from the midline; **adduction** (meaning to bring together) is movement toward the midline (figure 8.18a). Moving the upper limbs away from the body such as in the outward and then upward portion of doing "jumping jacks" is abduction, and bringing the upper limbs back toward the body is adduction. Abduction of the fingers involves spreading the fingers apart, away from the midline of the hand, and adduction is bringing them back together (figure 8.18b). Abduction of the thumb moves it anteriorly, away from the palm. Abduction of the wrist, which

Figure 8.14 Flexion and Extension of the Elbow

Figure 8.15 Flexion and Extension Defined According to the Coronal Plane

Flexion and extension of (a) the shoulder, (b) the neck, (c) the trunk.

Figure 8.16   Flexion and Extension of the Knee

Figure 8.17   Dorsiflexion and Plantar Flexion of the Foot

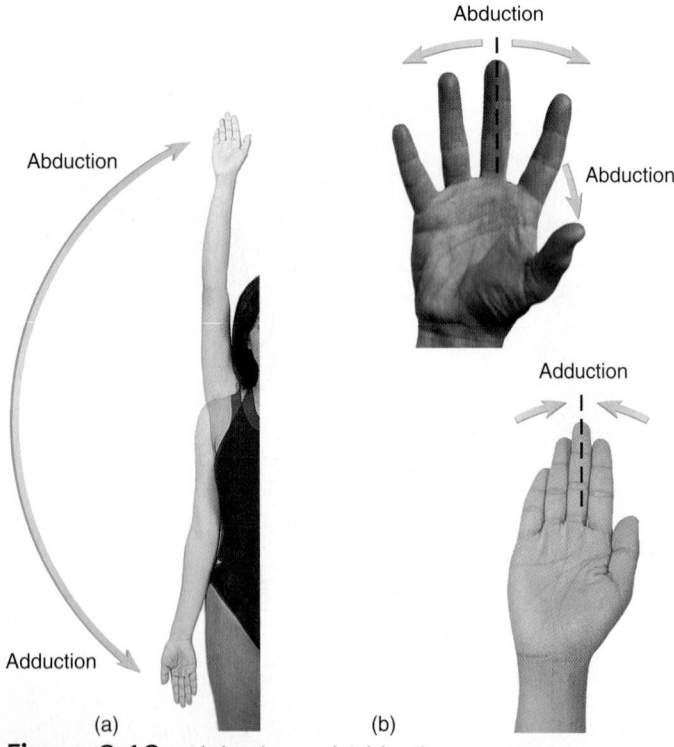

(a)   (b)

Figure 8.18   Abduction and Adduction
Abduction and adduction of (*a*) the upper limb and (*b*) the fingers.

is sometimes called radial deviation, is movement of the hand away from the midline of the body, and adduction of the wrist, which is sometimes called ulnar deviation, results in movement of the hand toward the midline of the body. Abduction of the head is tilting the head to one side or the other and is commonly called lateral flexion of the neck. Bending at the waist to one side or the other is usually called **lateral flexion** of the vertebral column, rather than abduction.

> 7. *Define the terms flexion and extension. How are they different for the upper and lower limbs? What is hyperextension?*
> 8. *Contrast abduction and adduction. Describe these movements for the head, upper limbs, wrist, fingers, lower limbs, and toes. For what part of the body is the term lateral flexion used?*

Circular Movements

Circular movements involve the rotation of a structure around an axis or movement of the structure in an arc.

Rotation

Rotation is the turning of a structure around its long axis, such as rotation of the head, the humerus, or the entire body (figure 8.19). Medial rotation of the humerus with the forearm flexed brings the hand toward the body. Rotation of the humerus so that the hand moves away from the body is lateral rotation.

Pronation and Supination

Pronation (prō-nā′shŭn) and **supination** (soo′pi-nā′shŭn) refer to the unique rotation of the forearm (figure 8.20). The word *prone* means lying facedown; the word *supine* means lying faceup. Pronation is rotation of the forearm so that the palm faces posteriorly in relation to the anatomic position. The palm of the hand faces inferiorly if the elbow is flexed to 90°. Supination is rotation of the forearm so that the palm faces anteriorly in relation to the anatomic position. The palm of the hand faces superiorly if the elbow is flexed to 90°. In pronation the radius and ulna cross; in supination they are in a parallel position. The head of the radius rotates against the radial notch of the ulna during supination and pronation.

Circumduction

Circumduction is a combination of flexion, extension, abduction, and adduction (figure 8.21). It occurs at freely movable joints such

Figure 8.19   Medial and Lateral Rotation of the Arm

Figure 8.20 **Pronation and Supination of the Hand**

Figure 8.22 **Elevation and Depression of the Scapula**

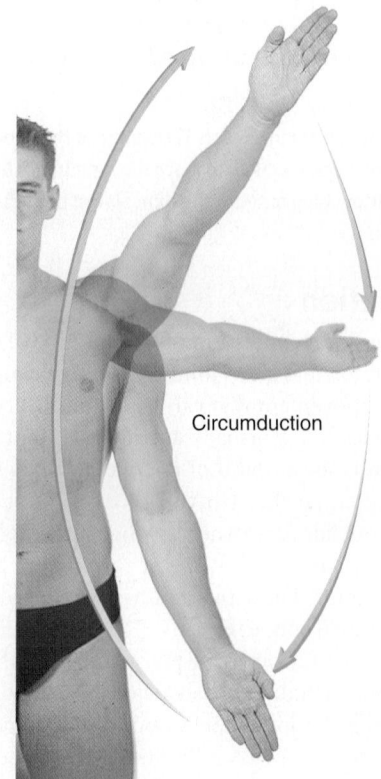

Figure 8.21 **Circumduction**

Shrugging the shoulders is an example of scapular elevation. Depression of the mandible opens the mouth, and elevation closes it.

Protraction and Retraction

Protraction consists of moving a structure in a gliding motion in an anterior direction (figure 8.23). **Retraction** moves the structure back to the anatomic position or even more posteriorly. As with elevation and depression, the mandible and scapulae are primary examples. Pulling the scapulae back toward the vertebral column is retraction.

Figure 8.23 **Protraction and Retraction of the Mandible**

as the shoulder. In circumduction, the arm moves so that it describes a cone with the shoulder joint at the apex.

● *9. Distinguish among rotation, circumduction, pronation, and supination. Give an example of each.*

Special Movements

Special movements are those movements unique to only one or two joints; they don't fit neatly into one of the other categories.

Elevation and Depression

Elevation moves a structure superiorly; **depression** moves it inferiorly (figure 8.22). The scapulae and mandible are primary examples.

Excursion

Lateral excursion refers to moving the mandible to either the right or left of the midline (figure 8.24), such as in grinding the teeth or chewing. **Medial excursion** returns the mandible to the neutral position.

Figure 8.24   Excursion of the Mandible

Opposition and Reposition

Opposition is a unique movement of the thumb and little finger (figure 8.25). It occurs when these two digits are brought toward each other across the palm of the hand. The thumb can also oppose the other digits, but those digits flex to touch the tip of the opposing thumb. **Reposition** is the movement returning the thumb and little finger to the neutral, anatomic position.

Figure 8.25   Opposition and Reposition of the Thumb and Little Finger

Inversion and Eversion

Inversion consists of turning the ankle so that the plantar surface of the foot faces medially, toward the opposite foot. **Eversion** is turning the ankle so that the plantar surface faces laterally (figure 8.26). Inversion of the foot is sometimes called supination, and eversion is called pronation.

Figure 8.26   Inversion and Eversion of the Foot

10. *Define the following jaw movements: protraction, retraction, lateral excursion, medial excursion, elevation and depression.*
11. *Define the terms opposition and reposition.*
12. *What terms are used for flexion and extension of the foot? For turning the side of the foot medially or laterally?*

Combination Movements

Most movements that occur in the course of normal activities are combinations of the movements named previously and are described by naming the individual movements involved in the combined movement. For example, a combination movement occurs when a person raises a hand from the anatomic position out to the side and then brings it toward the median plane, so that the hand is in front of the body at shoulder height. That movement could be considered a combination of abduction and adduction at the shoulder.

P R E D I C T

What combination of movements is required at the shoulder and elbow joints for a person to move the right upper limb from the anatomic position to touch the right side of the head with the fingertips?

Range of Motion

Range of motion is an expression of the amount of mobility that can be demonstrated in a given joint. The **active range of motion** is the amount of movement that can be accomplished by contraction of the muscles that normally act across a joint. The **passive range of motion** is the amount of movement that can be accomplished at a joint when the structures that meet at the joint are moved by some outside force, such as when a therapist holds onto the forearm of a patient and moves it toward the patient's arm, flexing the elbow joint. The active and passive range of motion for normal joints is usually about equal.

The range of motion for a given joint is influenced by a number of factors, including the shape of the articular surfaces of the bones forming the joint, the amount and shape of cartilage covering those articular surfaces, the strength and location of ligaments and tendons surrounding the joint, the strength and location of the muscles associated with the joint, the amount of fluid in and around the joint, the amount of pain in and around the joint, and the amount of use or disuse the joint has received over time. Abnormalities in the range of motion can occur when any of those components changes. For example, damage to a ligament associated with a given joint may increase the range of motion of that joint. A torn piece of cartilage within a joint can limit its range of motion. If the nerve supply to a muscle is damaged so that the muscle is weakened, the active range of motion for the joint acted upon by that muscle may decrease, but the passive range of motion for the joint should remain unchanged. Fluid buildup and/or pain in or around a joint can severely limit both the active and passive range of motion for that joint. With disuse, both the active and passive range of motion for a given joint decrease.

13. *Define range of motion. Contrast active range of motion with passive range of motion. What factors influence range of motion?*

Description of Selected Joints

It's impossible in a limited space to describe all the joints of the body; therefore, we describe only selected joints in this chapter, and they have been chosen because of their representative structure, important function, or clinical significance.

Temporomandibular Joint

The mandible articulates with the temporal bone to form the **temporomandibular joint (TMJ).** The mandibular condyle fits into the mandibular fossa of the temporal bone. A fibrocartilage articular disk is located between the mandible and the temporal bone, dividing the joint into superior and inferior cavities (figure 8.27). The joint is surrounded by a fibrous capsule to which the articular disk is attached at its margin, and is strengthened by lateral and accessory ligaments.

The temporomandibular joint is a combination plane and ellipsoid joint, with the ellipsoid portion predominating. Depression of the mandible to open the mouth involves an anterior gliding motion of the mandibular condyle and articular disk relative to the temporal bone, which is about the same motion that occurs in protraction of the mandible; it is followed by a hinge motion that occurs between the articular disk and the mandibular head. The mandibular condyle is also capable of slight mediolateral movement, allowing excursion of the mandible.

TMJ Disorders

TMJ disorders are a group of conditions that cause most chronic orofacial pain. The conditions include joint noise; pain in the muscles, joint, or face; headache; and reduction in the range of joint movement. TMJ pain is often felt as referred pain in the ear. Patients may go to a physician complaining of an earache and are then referred to a dentist. As many as 65%–75% of people between ages 20 and 40 experience some of these symptoms. Symptoms appear to affect men and women about equally, but only about 10% of the symptoms are severe enough to cause people to seek medical attention. Women experience severe pain eight times more often than do men.

TMJ disorders are classified as those involving the joint, with or without pain; those involving only muscle pain; or those involving both the joint disorder and muscle pain. TMJ disorders are also classified as acute or chronic. Acute cases are usually self-limiting and have an identifiable cause. Chronic cases are not self-limiting, may be permanent, and often have no apparent cause. Chronic TMJ disorders are not easily treated, and chronic TMJ pain has much in common with other types of chronic pain. Whereas some people learn to live with the pain, others may experience psychologic problems, such as a sense of helplessness and hopelessness, high tension, and loss of sleep and appetite. Drug dependency may occur if strong drugs are used to control the pain; and relationships, lifestyle, vocation, and social interactions may be disrupted. Many of these problems may make the pain worse through positive feedback. Treatment includes teaching the patient to reduce jaw movements that aggravate the problem and to reduce stress and anxiety. Physical therapy may help to relax the muscles and restore function. Analgesic and antiinflammatory drugs may be used, and oral splints may be helpful, especially at night.

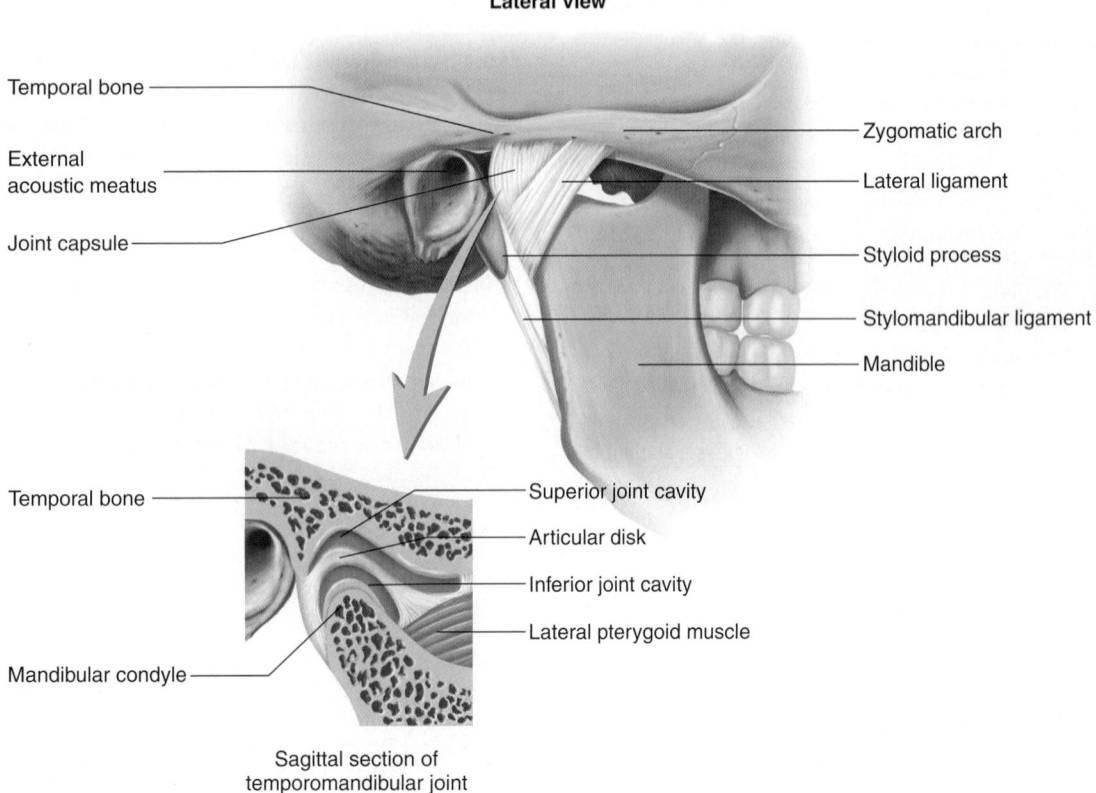

Lateral view

Temporal bone

External acoustic meatus

Joint capsule

Zygomatic arch

Lateral ligament

Styloid process

Stylomandibular ligament

Mandible

Temporal bone

Superior joint cavity

Articular disk

Inferior joint cavity

Lateral pterygoid muscle

Mandibular condyle

Sagittal section of temporomandibular joint

Figure 8.27 Right Temporomandibular Joint, Lateral View

Shoulder Joint

The **shoulder,** or **glenohumeral, joint** is a ball-and-socket joint (figure 8.28) in which stability is reduced and mobility is increased compared to the other ball-and-socket joint, the hip. Flexion, extension, abduction, adduction, rotation, and circumduction can all occur at the shoulder joint. The rounded head of the humerus articulates with the shallow glenoid cavity of the scapula. The rim of the glenoid cavity is built up slightly by a fibrocartilage ring, the **glenoid labrum,** to which the joint capsule is attached. A **subscapular bursa** (not shown in the figure) and a **subacromial bursa** open into the joint cavity.

The stability of the joint is maintained primarily by three sets of ligaments and four muscles. The ligaments of the shoulder are listed in table 8.3. The four muscles, referred to collectively as the **rotator cuff,** pull the humeral head superiorly and medially toward the glenoid cavity. These muscles are discussed in more detail in chapter 10. The head of the humerus is also supported against the glenoid cavity by the tendon from the biceps brachii muscle in the anterior part of the arm. This tendon is unusual in that it passes through the articular capsule of the shoulder joint before crossing the head of the humerus and attaching to the scapula at the supraglenoid tubercle (see figure 7.24*a*).

Shoulder Disorders

The most common traumatic shoulder disorders are **dislocation** and muscle or tendon **tears.** The shoulder is the most commonly dislocated joint in the body. The major ligaments cross the superior part of the shoulder joint, and no major ligaments or muscles are associated with the inferior side. As a result, dislocation of the humerus is most likely to occur inferiorly into the axilla. Because the axilla contains very important nerves and arteries, severe and permanent damage may occur when the humeral head dislocates inferiorily. The axillary nerve is the most commonly damaged (see chapter 12).

Chronic shoulder disorders include tendonitis (inflammation of tendons), bursitis (inflammation of bursae), and arthritis (inflammation of joints). Bursitis of the subacromial bursa can become very painful when the large shoulder muscle, called the deltoid muscle, compresses the bursa during shoulder movement.

PREDICT **5**

Separation of the shoulder consists of stretching or tearing the ligaments of the acromioclavicular joint (acromioclavicular, or AC, separation). Using figure 8.28*a* and your knowledge of the articulated skeleton for assistance, explain the nature of a shoulder separation, and predict the problems that may follow a separation.

Table 8.3	Ligaments of the Shoulder Joint (see figure 8.28)
Ligament	**Description**
Glenohumeral (superior, middle, and inferior)	Three slightly thickened longitudinal sets of fibers on the anterior side of the capsule; extend from the humerus to the margin of the glenoid cavity
Transverse humeral	Lateral, transverse fibrous thickening of the joint capsule; crosses between the greater and lesser tubercles and holds down the tendon from the long head of the biceps muscle
Coracohumeral	Crosses from the root of the coracoid process to the humeral neck
Coracoacromial	Crosses above the joint between the coracoid process and the acromion process; an accessory, protective ligament

Elbow Joint

The **elbow joint** (figure 8.29) is a compound hinge joint consisting of the **humeroulnar joint,** between the humerus and ulna, and the **humeroradial joint,** between the humerus and radius. The **proximal radioulnar joint,** between the proximal radius and ulna, is also closely related. The shape of the trochlear notch and its association with the trochlea of the humerus (figure 8.29*a*) limit movement at the elbow joint to flexion and extension. The rounded radial head, however, rotates in the radial notch of the ulna and against the capitulum of the humerus (figure 8.29*b*), allowing pronation and supination of the hand.

The elbow joint is surrounded by a joint capsule. The humeroulnar joint is reinforced by the **ulnar collateral ligament** (figure 8.29*c*). The humeroradial and proximal radioulnar joints are reinforced by the **radial collateral ligament** and **radial annular ligament** (figure 8.29*d*). A subcutaneous **olecranon bursa** covers the proximal and posterior surfaces of the olecranon process.

Elbow Problems

Olecranon bursitis is an inflammation of the olecranon bursa. This inflammation can be caused by excessive rubbing of the elbow against a hard surface and is sometimes referred to as **student's elbow.** The radial head can become subluxated (partial joint separation) from the annular ligament of the radius. This condition is called **nursemaid's elbow.** If a child is lifted by one hand, the action may subluxate (partially dislocate) the radial head.

Labels for (a) Anterior view:

- Clavicle (cut and elevated)
- Acromioclavicular ligament
- Acromion process
- Coracoacromial ligament
- Subacromial bursa
- Coracohumeral ligament
- Humerus
- Transverse humeral ligament
- Tendon sheath on tendon of long head of biceps brachii
- Biceps brachii (long head) tendon
- Hook retracting subscapularis muscle
- Trapezoid ligament ⎤ Coracoclavicular
- Conoid ligament ⎦ ligament
- Transverse scapular ligament
- Coracoid process
- Superior glenohumeral ligament
- Middle glenohumeral ligament
- Inferior glenohumeral ligament
- Joint capsule
- Triceps brachii tendon (long head)

(a) **Anterior view**

Labels for (b) Frontal section:

- Acromion process (articular surface)
- Subacromial bursa
- Joint cavity
- Tendon sheath on tendon of long head of biceps brachii
- Biceps brachii (long head) tendon
- Humerus
- Biceps brachii (long head) muscle
- Articular cartilage over head of humerus
- Articular cartilage over glenoid cavity
- Scapula (cut surface)
- Glenoid labrum
- Joint capsule

(b) **Frontal section**

Figure 8.28 Right Shoulder Joint

(*a*) Anterior view. (*b*) Frontal section.

(a) **Sagittal section**

(b) **Lateral view**

(c) **Medial view**

(d) **Lateral view**

Figure 8.29   **Right Elbow Joint**

(a) Sagittal section showing the relation between the ulna and humerus.
(b) Lateral side with ligaments cut to show the relation between the radial head, ulna, and humerus. (c) Medial side. (d) Lateral side.

Hip Joint

The femoral head articulates with the relatively deep, concave acetabulum of the coxa to form the **coxal,** or **hip joint** (figure 8.30). The head of the femur is more nearly a complete ball than the articulating surface of any other bone of the body. The acetabulum is deepened and strengthened by a lip of fibrocartilage called the **acetabular labrum,** which is incomplete inferiorly, and by a **transverse acetabular ligament,** which crosses the acetabular notch on the inferior edge of the acetabulum. The hip is capable of a wide range of movement, including flexion, extension, abduction, adduction, rotation, and circumduction.

Hip Dislocation

Dislocation of the hip may occur when the hip is flexed and the femur is driven posteriorly, such as when a person sitting in an automobile is involved in an accident. The head of the femur usually dislocates posterior to the acetabulum, tearing the acetabular labrum, the fibrous capsule, and the ligaments. Fracture of the femur and the coxa often accompany hip dislocation.

An extremely strong joint capsule, reinforced by several ligaments, extends from the rim of the acetabulum to the neck of the femur (table 8.4). The **iliofemoral ligament** is especially strong. When standing, most people tend to thrust the hips anteriorly. This position is relaxing because the iliofemoral ligament supports much of the body's weight. The **ligamentum teres,** which is the ligament of the head of the femur, is located inside the hip joint between the femoral head and the acetabulum. This ligament does not contribute much toward strengthening the hip joint; however, it does carry a small nutrient artery to the head of the femur in about 80% of the population. The acetabular labrum, ligaments of the hip, and the surrounding muscles make the hip joint much more stable but less mobile than the shoulder joint.

Knee Joint

The **knee joint** traditionally is classified as a modified hinge joint located between the femur and the tibia (figure 8.31). Actually, it's a complex ellipsoid joint that allows flexion, extension, and a small amount of rotation of the leg. The distal end of the femur has two large ellipsoid surfaces and a deep fossa between them. The femur articulates with the proximal end of the tibia, which is flattened and smooth laterally, with a crest called the intercondylar eminence in the center (see figure 7.36). The margins of the tibia are built up by thick fibrocartilage articular disks, called **menisci** (mĕ-nis′sī; crescent-shaped; figure 8.31b and d), that deepen the articular surface. The fibula does not articulate with the femur but articulates only with the lateral side of the tibia.

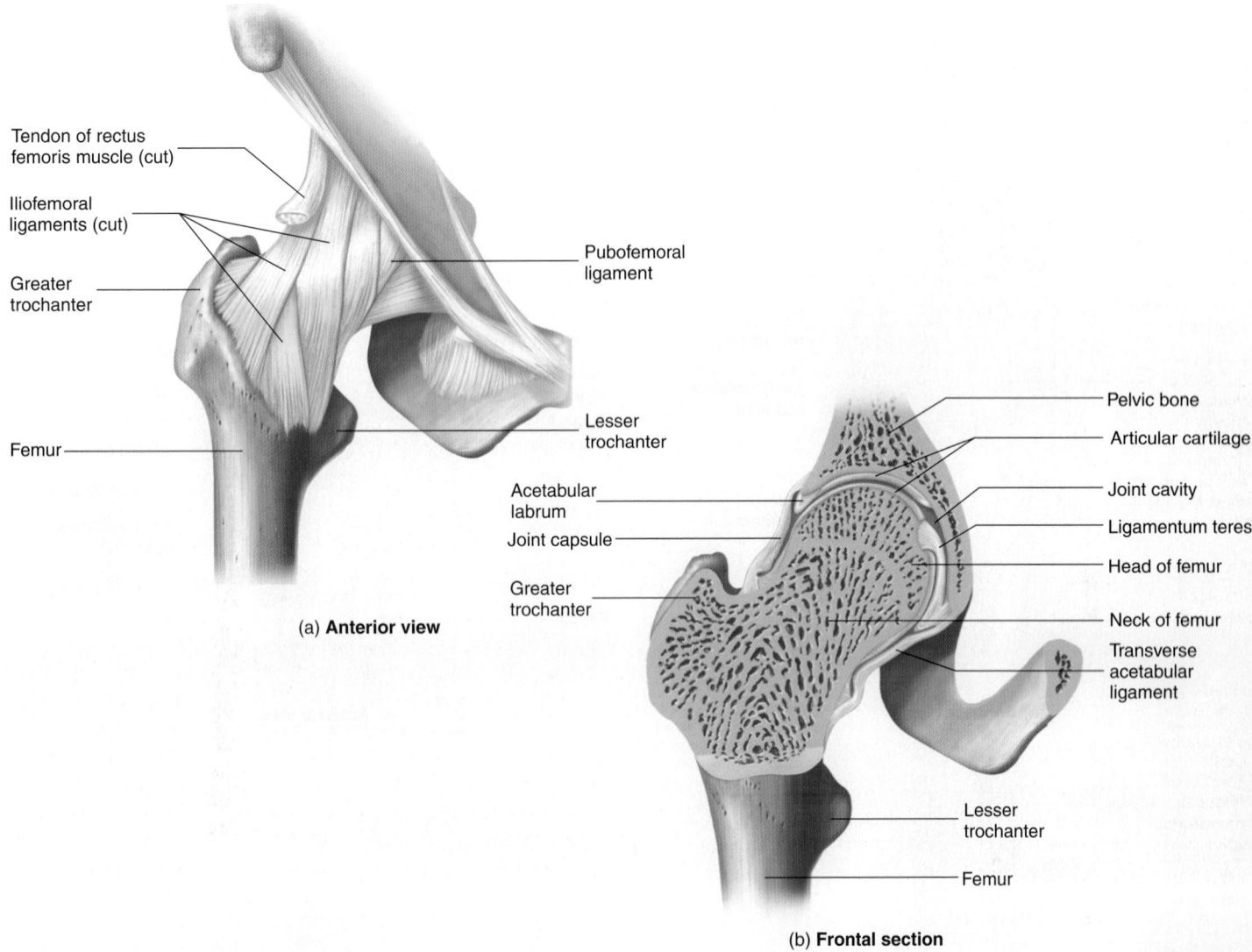

Tendon of rectus
femoris muscle (cut)

Iliofemoral
ligaments (cut)

Greater
trochanter

Femur

Pubofemoral
ligament

Lesser
trochanter

(a) Anterior view

Pelvic bone

Articular cartilage

Joint cavity

Ligamentum teres

Head of femur

Acetabular
labrum

Joint capsule

Greater
trochanter

Neck of femur

Transverse
acetabular
ligament

Lesser
trochanter

Femur

(b) Frontal section

Figure 8.30 Right Hip Joint

(a) Anterior view. (b) Frontal section.

Table 8.4	Ligaments of the Hip Joint (see figure 8.30)
Ligament	**Description**
Transverse acetabular	Bridges gap in the inferior margin of the fibrocartilage acetabular labrum
Iliofemoral	Strong, thick band between the anterior inferior iliac spine and the intertrochanteric line of the femur
Pubofemoral	Extends from the pubic portion of the acetabular rim to the inferior portion of the femoral neck
Ischiofemoral	Bridges the ischial acetabular rim and the superior portion of the femoral neck; less well defined
Ligamentum teres	Weak, flat band from the margin of the acetabular notch and the transverse ligament to a fovea in the center of the femoral head

Two **cruciate** (kroo′shē-āt; crossed) **ligaments** extend between the intercondylar eminence of the tibia and the fossa of the femur (see figure 8.31b, d, and e). The anterior cruciate ligament prevents anterior displacement of the tibia relative to the femur, and the posterior cruciate ligament prevents posterior displacement of the tibia. The joint is also strengthened by **collateral** and **popliteal ligaments** and by the tendons of the thigh muscles, which extend around the knee (table 8.5).

A number of bursae surround the knee (see figure 8.31f). The largest is the **suprapatellar bursa,** which is a superior extension of the joint capsule and allows for movement of the anterior thigh muscles over the distal end of the femur. Other knee bursae include the subcutaneous prepatellar bursa and the deep infrapatellar bursa, as well as the popliteal bursa, the gastrocnemius bursa, and the subcutaneous infrapatellar bursa (not illustrated).

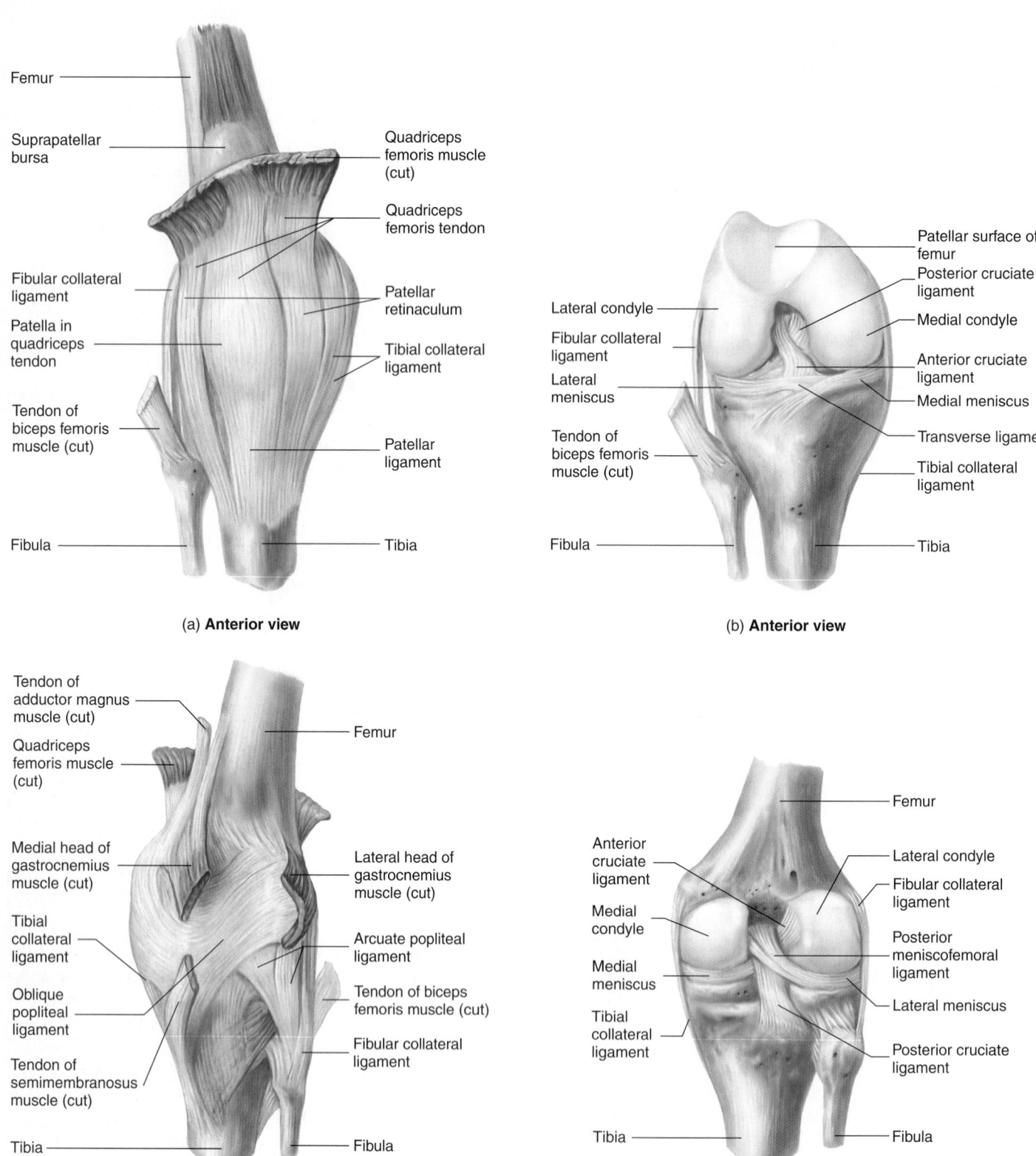

Femur

Suprapatellar bursa

Quadriceps femoris muscle (cut)

Quadriceps femoris tendon

Fibular collateral ligament

Patellar retinaculum

Patella in quadriceps tendon

Tibial collateral ligament

Tendon of biceps femoris muscle (cut)

Patellar ligament

Fibula

Tibia

(a) **Anterior view**

Patellar surface of femur

Posterior cruciate ligament

Lateral condyle

Medial condyle

Fibular collateral ligament

Anterior cruciate ligament

Lateral meniscus

Medial meniscus

Tendon of biceps femoris muscle (cut)

Transverse ligament

Tibial collateral ligament

Fibula

Tibia

(b) **Anterior view**

Tendon of adductor magnus muscle (cut)

Quadriceps femoris muscle (cut)

Femur

Medial head of gastrocnemius muscle (cut)

Lateral head of gastrocnemius muscle (cut)

Tibial collateral ligament

Arcuate popliteal ligament

Oblique popliteal ligament

Tendon of biceps femoris muscle (cut)

Fibular collateral ligament

Tendon of semimembranosus muscle (cut)

Tibia

Fibula

(c) **Posterior view**

Anterior cruciate ligament

Femur

Medial condyle

Lateral condyle

Fibular collateral ligament

Medial meniscus

Posterior meniscofemoral ligament

Tibial collateral ligament

Lateral meniscus

Posterior cruciate ligament

Tibia

Fibula

(d) **Posterior view**

Figure 8.31 Right Knee Joint

(*a*) Anterior superficial view. (*b*) Anterior deep view (knee flexed). (*c*) Posterior superficial view. (*d*) Posterior deep view.

(e) **Anterior view**   (f) **Sagittal section**

Figure 8.31   *(continued)*

(*e*) Photograph of anterior deep view. (*f*) Sagittal section.

Table 8.5	Ligaments of the Knee Joint (see figure 8.31)		
Ligament	**Description**	**Ligament**	**Description**
Patellar	Thick, heavy, fibrous band between the patella and the tibial tuberosity; actually part of the quadriceps femoris tendon	Anterior cruciate	Extends obliquely, superiorly, and posteriorly from the anterior intercondylar eminence of the tibia to the medial side of the lateral femoral condyle
Patellar retinaculum	Thin band from the margins of the patella to the sides of the tibial condyles	Posterior cruciate	Extends superiorly and anteriorly from the posterior intercondylar eminence to the lateral side of the medial condyle
Oblique popliteal	Thickening of the posterior capsule; extension of the semimembranous tendon	Coronary (medial and lateral)	Attaches the menisci to the tibial condyles (not illustrated)
Arcuate popliteal	Extends from the posterior fibular head to the posterior fibrous capsule	Transverse	Connects the anterior portions of the medial and lateral menisci
Tibial collateral	Thickening of the lateral capsule from the medial epicondyle of the femur to the medial surface of the tibia; also called the tibial collateral ligament	Meniscofemoral (anterior and posterior)	Joins the posterior part of the lateral menisci to the medial condyle of the femur, passing anterior and posterior to the posterior cruciate ligament (not illustrated)
Fibular collateral	Round ligament extending from the lateral femoral epicondyle to the head of the fibula; also called the fibular collateral ligament		

Clinical Focus   Knee Injuries and Disorders

Injuries to the medial side of the knee are much more common than injuries to the lateral side. The fibular (lateral) collateral ligament strengthens the joint laterally and is stronger than the tibial (medial) collateral ligament. Damage to the collateral ligaments occurs as a result of blows to the opposite side of the knee. Severe blows to the medial side of the knee, which would damage the **fibular collateral ligament,** are far less common than blows to the lateral side of the knee, which damage the **tibial collateral ligament.**

In addition, the **medial meniscus** is fairly tightly attached to the tibial collateral ligament and is damaged 20 times more often in a knee injury than the lateral meniscus, which is thinner and more loosely attached to the fibular collateral ligament. A **torn meniscus** may result in a "clicking" sound during extension of the leg; or, if the damage is more severe, the torn piece of cartilage may move between the articulating surfaces of the tibia and femur, causing the knee to "lock" in a partially flexed position.

If the knee is driven anteriorly or if it is hyperextended, the **anterior cruciate ligament** may be torn, which causes the knee joint to be very unstable. If the knee is driven posteriorly, the **posterior cruciate ligament** may be torn. Surgical replacement of a cruciate ligament with a transplanted or artificial ligament is a technique used to repair the damage.

A common type of football injury results from a block or tackle to the lateral side of the knee, which can cause the knee to bend inward, opening the medial side of the joint and tearing the tibial collateral ligament. The **medial meniscus** often is torn as well. Because this ligament is strongly attached to the medial meniscus, in severe injuries, the **anterior cruciate ligament,** which is attached to the medial meniscus, is also damaged (figure A).

Bursitis in the **subcutaneous prepatellar bursa** (see figure 8.31*f*), commonly called "housemaid's knee," may result from prolonged work performed while on the hands and knees. Another bursitis, "clergyman's knee," results from excessive kneeling and affects the **subcutaneous infrapatellar bursa** (not illustrated). This type of bursitis is common in carpet layers and roofers.

Other common knee problems include **chondromalacia** (kon'drō-mă-lā'shē-ă), or softening of the cartilage, which results from abnormal movement of the patella within the patellar groove, and the **"fat pad syndrome,"** which consists of an accumulation of fluid in the fat pad posterior to the patella. An acutely swollen knee appearing immediately after an injury is usually a sign of blood accumulation within the joint cavity and is called a **hemarthrosis** (hē'mar-thrō'sis, hem'ar-thrō'sis). A slower accumulation of fluid, "water on the knee," may be caused by bursitis.

Lateral

Medial

Blow

Anterior cruciate ligament

Tibial collateral ligament

Medial meniscus

Anterior view

Figure A   Injury to the Right Knee

Ankle Joint and Arches of the Foot

The distal tibia and fibula form a highly modified hinge joint with the talus called the **ankle,** or **talocrural** (tā′lō-kroo′răl), **joint** (figure 8.32). The medial and lateral malleoli of the tibia and fibula, which form the medial and lateral margins of the ankle, are rather extensive, whereas the anterior and posterior margins are almost nonexistent. As a result, a hinge joint is created from a modified ball-and-socket arrangement. A fibrous capsule surrounds the joint, with the medial and lateral parts thickened to form ligaments. Other ligaments also help stabilize the joint (table 8.6). Dorsiflexion, plantar flexion, and limited inversion and eversion can occur at this joint.

Ankle Injury

The ankle is the most frequently injured major joint in the body. The most common ankle injuries result from forceful inversion of the foot. A **sprained ankle** results when the ligaments of the ankle are torn partially or completely. The calcaneofibular ligament tears most often (figure B), followed in frequency by the anterior talofibular ligament. A fibular fracture can occur with severe inversion because the talus can slide against the lateral malleolus and break it (see chapter 7).

Figure B

(a) **Medial view**

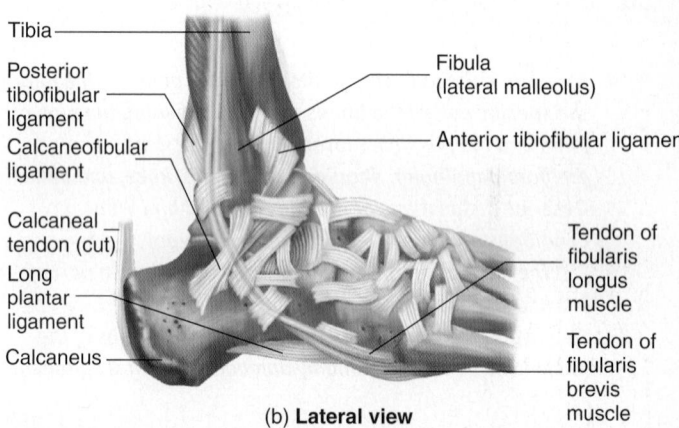

(b) **Lateral view**

Figure 8.32 Ligaments of the Right Ankle Joint

(*a*) Medial view. (*b*) Lateral view.

The ligaments of the arches (see figure 7.39) serve two major functions: to hold the bones in their proper relationship as segments of the arch and to provide ties across the arch somewhat like a bowstring. As weight is transferred through the arch system, some of the ligaments are stretched, giving the foot more flexibility and allowing it to adjust to uneven surfaces. When weight is removed from the foot, the ligaments recoil and restore the arches to their unstressed shape.

Table 8.6	Ligaments of the Ankle and Arch (see figure 8.32)
Ligament	**Description**
Medial	Thickening of the medial fibrous capsule that attaches the medial malleolus to the calcaneus, navicular, and talus; also called the deltoid ligament
Calcaneofibular	Extends from the lateral malleolus to the lateral surface of the calcaneus; separate from the capsule
Anterior talofibular	Extends from the lateral malleolus to the neck of the talus; fused with the joint capsule
Long plantar	Extends from the calcaneus to the cuboid and bases of metatarsals 2–5
Plantar calcaneocuboid	Extends from the calcaneus to the cuboid
Plantar calcaneonavicular (short plantar)	Extends from the calcaneus to the navicular

Arch Problems

The arches of the foot normally form early in fetal life. Failure to form results in congenital flat feet, or fallen arches, a condition in which the arches, primarily the medial longitudinal arch, are depressed or collapsed (see figure 7.39). This condition is not always painful. **Flat feet** may also occur when the muscles and ligaments supporting the arch fatigue and allow the arch, usually the medial longitudinal arch, to collapse. During prolonged standing, the plantar calcaneonavicular ligament may stretch, flattening the medial longitudinal arch. The transverse arch may also become flattened. The strained ligaments can become painful.

The plantar fascia is the deep connective tissue superficial to the ligaments in the central plantar surface of the foot and the thinner fascia on the medial and lateral sides of the plantar surface (see figure 8.32). **Plantar fasciitis,** which is an inflammation of the plantar fascia, can be a problem for distance runners as a result of continuous stretching.

14. *For each of the following joints, name the bones of the joint, the specific part of the bones that form the joint, the type of joint, and the possible movement(s) at the joint: temporomandibular, shoulder, elbow, hip, knee, and ankle.*

15. *Describe dislocations of the shoulder and hip. What conditions are most likely to cause each type?*

16. *List the most common knee injuries, and tell which part of the knee is most often damaged in each type.*

17. *Define the term sprain, and describe which portions of the ankle joint are most commonly damaged when it is sprained.*

Effects of Aging on the Joints

A number of changes occur within many joints as a person ages. Those that occur in synovial joints have the greatest impact and often present major problems for elderly people. In general, as a person ages, the tissues of the body become less flexible and less elastic as protein cross-linking, especially in fibrous connective tissue, increases. The most important proteins related to tissue flexibility are elastin and collagen. Tissue repair slows as cell proliferation rates decline and the rate of new blood vessel development decreases. These general changes can significantly affect synovial joints. With use, the cartilage covering articular surfaces can wear down. When a person is young, production of new, resilient matrix compensates for the wear. As a person ages, the rate of replacement declines and the matrix becomes more rigid, thus adding to its rate of wear. The production rate of lubricating synovial fluid also declines with age, further contributing to the wear of the articular cartilage. In addition, the ligaments and tendons surrounding a joint shorten and become less flexible with age, resulting in a decrease in the range of motion of the joint. With age, muscles, which strengthen the joints, tend to weaken. Older people often experience a general decrease in activity, which causes the joints to become less flexible and their range of motion to decrease.

18. *List the age-related factors that contribute to cartilage wear in synovial joints. List the age-related factors that cause a loss of flexibility and loss of range of motion in synovial joints.*

Clinical Focus   Joint Disorders

Arthritis

Arthritis, an inflammation of any joint, is the most common and best known of the joint disorders, affecting 10% of the world's population and 14% of the U.S. population. There are over 37 million cases of arthritis in the United States alone. More than 100 different types of arthritis exist. Classification is often based on the cause and progress of the arthritis. Causes include infectious agents, metabolic disorders, trauma, and immune disorders. Mild exercise retards joint degeneration and enhances mobility. Swimming and walking are recommended for people with arthritis; but running, tennis, and aerobics are not recommended. Therapy depends on the type of arthritis but usually includes the use of antiinflammatory drugs. Current research is focusing on the possible development of antibodies against the cells that initiate the inflammatory response in the joints or against cell surface markers on those cells.

Osteoarthritis (OA) is the most common type of arthritis, affecting 10% of people in the United States (85% of those over age 70). OA may begin as a molecular abnormality in articular cartilage, with heredity and normal wear-and-tear of the joint important contributing factors. Slowed metabolic rates with increased age also seem to contribute to OA. Inflammation is usually secondary in this disorder. It tends to occur in the weight-bearing joints such as the knees and is more common in overweight individuals.

The first line of treatment for osteoarthritis is to change the lifestyle to reduce stress on affected joints. Synovial joints require movement to remain healthy. Long periods of inactivity may cause joints to stiffen. **Moderate exercise** helps reduce pain and increase flexibility. Exercising also helps people reduce excess weight, which can place stress on joints of the lower limbs. Older people should avoid high-impact sports, such as jogging, tennis, and racquetball, which place stress on the joints. Cycling or walking are recommended, but swimming is the best for

people with osteoarthritis, as it exercises the muscles and joints without stressing the joints. Wearing shock-absorbing shoes can help. Splints or braces worn over an affected joint may sometimes be necessary to properly align the joint and distribute weight around it.

Applying heat, such as with hot soaks, warm paraffin application, heating pads, low-power infrared light, or diathermy (mild electric currents that produce heat), directly over the joint may be helpful. Moving to a warmer climate, however, doesn't seem to make much difference.

The American Geriatrics Society has released guidelines for managing chronic pain in elderly patients with osteoarthritis. They recommend acetaminophen (Tylenol) or other **nonsteroidal antiinflammatory drugs (NSAIDs),** such as aspirin and ibuprofen (Advil), for mild to moderate pain. Capsaicin, a component of hot red peppers, may help relieve pain when applied as a skin cream (Zostrix). Capsaicin seems to reduce levels of a chemical known as substance P that contributes both to inflammation of the joint and to the conduction of pain sensations to the brain. If pain becomes a major problem and over-the-counter pain relievers appear ineffective, physicians may inject corticosteroids directly into the affected joint.

Synvisc and Hyalgan are two drugs derived from **hyaluronic acid,** a natural substance that lubricates joints. They may be administered by injection into the joint when standard medication and exercise programs fail to relieve pain. **Glucosamine** and **chondroitin sulfate** are also natural substances associated with joints. If taken orally or by injection they may help affected joints. However, glucosamine may also raise blood sugar levels, so people with diabetes shouldn't use it without consulting their physician. Injections of genetically treated cells from synovial fluid, which are able to block the immune factors thought to cause the breakdown of joint cartilage, are currently under investigation. An immune system protein called **transforming growth**

factor beta (TGF-β), introduced by gene therapy, is showing some promise in repairing cartilage damaged by osteoarthritis.

If other treatments fail, **surgical procedures** may be employed to relieve pain and increase function in osteoarthritis patients. Using arthroscopy , a surgeon can examine the joint and clean out bone and cartilage fragments that stimulate pain and inflammation. In **osteotomy,** the bones of joint are reshaped to better align the joint. In a procedure called **chondroplasty,** a small amount of healthy cartilage is removed and grown in the laboratory. The newly grown cartilage is then implanted into the joint, where it may stimulate the regeneration of damaged tissue. Research is currently under way to grow new cartilage from stem cells.

Joint replacement is discussed at the end of this Clinical Focus. If the affected joint cannot be replaced, surgeons may perform a procedure called **arthrodesis,** in which the bones meeting at the joint are fused together. This procedure is intended to eliminate the pain, but the joint is eliminated and movement at that point becomes impossible.

Rheumatoid arthritis (RA) is the second most common type of arthritis. It affects about 3% of all women and about 1% of all men in the United States. It is a general connective tissue disorder that affects the skin, vessels, lungs, and other organs, but it is most pronounced in the joints. It is severely disabling and most commonly destroys small joints, such as those in the hands and feet (figure C). The initial cause is unknown but may involve a transient **infection** or an **autoimmune disease** (an immune reaction to one's own tissues; see chapter 22) that develops against collagen. A genetic predisposition may also exist. Whatever the cause, the ultimate course appears to be immunologic. People with classic RA have a protein, **rheumatoid factor,** in their blood. In RA the synovial fluid and associated connective tissue cells

Continued

Continued

proliferate, forming a **pannus** (clothlike layer), which causes the joint capsule to become thickened and which destroys the articular cartilage. In advanced stages, opposing joint surfaces can become fused. **Juvenile rheumatoid arthritis** is similar to the adult type in many ways, but no rheumatoid factor is found in the serum.

Hemophilic arthritis may result from bleeding into the joint cavity caused by hemophilia, a hereditary disease characterized by a deficient clotting mechanism in the blood. Some evidence exists that the iron in the blood is toxic to the chondrocytes, resulting in degeneration of the articular cartilage.

Joint Infections

Lyme disease is the result of a bacterial infection (*Borrelia burgdorferi*) transmitted to humans by a **tick** vector (usually *Ixodes sp.*) that affects the brain, nerves, eyes, heart, and joints. The chronic arthritis and central nervous system dysfunction that are symptoms of the disease are severely disabling but rarely fatal.

The disease is named for an epidemic of childhood arthritis occurring in Lyme, Connecticut, in 1975. Lyme disease has probably existed in Europe for many years and in North America before the first European colonization, but it was not recognized as a distinct disease before 1975. Humans and domestic animals are only incidental hosts to the ticks, which normally infect wild mammals and birds. Deer are of particular concern as carriers of the ticks. The northeastern United States was greatly deforested during the eighteenth and nineteenth centuries, and deer and other wildlife populations declined dramatically. The more recent abandonment and reforestation of farms in New England has lead to an increase in the deer and tick populations, with a resurgence of the associated joint and nervous system disease.

Nearly 150,000 cases of Lyme disease have been reported in the United States since 1982. Although the disease is most common in the northeastern United States, it is relatively rare in Canada. In the U.S., scattered cases have been reported in the north central states, along the West Coast, and scattered throughout the eastern, southeastern, and central states.

Early manifestations of the disease include flulike symptoms, with localized skin rash. If untreated, the bacterium can spread to the nervous system, heart, and joints within a few weeks to months. A human vaccine against Lyme disease is currently being used for high-risk individuals.

Suppurative (pus-forming) **arthritis** may result from a number of infectious agents. These joint infections may be transferred from some other infected site in the body or may be systemic (i.e., throughout the body). Usually only one joint, normally one of the larger joints, is affected, and the course of suppurative arthritis, if treated early, is transitory. With prolonged infection, however, the articular surfaces may degenerate. **Tuberculous arthritis** can occur as a secondary infection from pulmonary tuberculosis and is more damaging than typical suppurative arthritis. It usually affects the spine or large joints and causes ulceration of the articular cartilages and even erosion of the underlying bone. Transient arthritis of multiple joints is a common symptom of rheumatic fever, but permanent damage seldom occurs in joints with this disorder.

Gout

Gout is a group of metabolic disorders involving joints. These disorders are largely idiopathic (of unknown cause), although some cases of gout seem to be familial (occur in families and therefore are probably genetic). Gout is more common in males than in females. The ultimate problem in gout patients is an increase in **uric acid** in the blood because of too much synthesis or decreased removal through the kidneys. The limited solubility of uric

(a)

(b)

Figure C **Rheumatoid Arthritis**
(*a*) Photograph of hands with rheumatoid arthritis. (*b*) Radiographs of the same hands shown in (*a*).

acid salts in the body results in precipitation of **monosodium urate crystals** in various tissues, including the kidneys and joint capsules.

The earliest symptom of gout is transient arthritis resulting from urate crystal accumulation in a joint causing irritation of the synovial membrane. This irritation can ultimately lead to an inflammatory response in the joints, and both the crystal deposition and inflammation can become chronic. This condition is called **gouty arthritis.** Normally only one or two joints are affected. The most commonly affected joints (85% of the cases) are the base of the **great toe** and other foot and leg joints to a lesser extent. Any joint may ultimately be involved, and damage to the kidneys from crystal formation occurs in almost all advanced cases. **Kidney failure** may occur in untreated cases. With modern medications, these complications seldom occur. Weight control and reduced alcohol consumption can help prevent gout.

Pseudogout is a disorder that causes pain and swelling similar to that seen in gout, but it is characterized by calcium hypophosphate crystal deposits in joints.

Hallux Valgus and Bunion

In people who wear pointed shoes, the great toe can be deformed and displaced laterally, a condition called **hallux valgus** (hal′ŭks val′gŭs). Bunions are often associated with hallux valgus. A **bunion** is a bursitis that develops over the first metatarsophalangeal joint because of pressure and rubbing by shoes.

Joint Replacement

As a result of recent advancements in biomedical technology, many joints of the body can now be replaced by artificial joints. Joint replacement, called **arthroplasty,** was first developed in the late 1950s. One of the major reasons for its use is to eliminate unbearable pain in patients near ages 55 to 60 with joint disorders. Osteoarthritis is the leading disease requiring joint replacement and accounts for two-thirds of the patients. Rheumatoid arthritis accounts for more than half of the remaining cases.

The major objectives in the design of joint prostheses (artificial replacements) include the development of stable articulations, low friction, solid fixation to the bone, and normal range of motion. New synthetic replacement materials are being designed by biomedical engineers to accomplish these objectives. Prosthetic joints usually are composed of metal, such as stainless steel, titanium alloys, or cobalt–chrome alloys, in combination with modern plastics, such as high-density polyethylene, silastic, or elastomer.

The bone of the articular area is removed on one side (a procedure called **hemireplacement**) or both sides **(total replacement)** of the joint, and the artificial articular areas are glued to the bone with a synthetic adhesive, such as methylmethacrylate. The smooth metal surface rubbing against the smooth plastic surface provides a low-friction contact with a range of movement that depends on the design.

The success of joint replacement depends on the joint being replaced, the age and condition of the patient, the state of the technology, and the definition of success. Success is usually defined as minimizing pain while maintaining movement. Most reports are based on examination of patients 2–10 years after joint replacement. The technology is improving constantly, so current reports do not adequately reflect the effect of the most recent improvements. Still, reports indicate a success rate of 80%–90% in hip replacements and 60% or more in ankle and elbow replacements. The major reason for failure of prosthetic joints is loosening of the artificial joint from the bone to which it is attached. New prostheses with porous surfaces help to overcome this problem.

S U M M A R Y

An articulation, or joint, is a place where two bones come together.

Naming Joints (p. 248)

Joints are named according to the bones or parts of bones involved.

Classes of Joints (p. 248)

Joints can be classified according to function or according to the type of connective tissue that binds them together and whether fluid is present between the bones.

Fibrous Joints

1. Fibrous joints are those in which bones are connected by fibrous tissue with no joint cavity. They are capable of little or no movement.

2. Sutures involve interdigitating bones held together by dense fibrous connective tissue. They occur between most skull bones.
3. Syndesmoses are joints consisting of fibrous ligaments.
4. Gomphoses are joints in which pegs fit into sockets and are held in place by periodontal ligaments (teeth in the jaws).
5. Some sutures and other joints can become ossified (synostosis).

Cartilaginous Joints

1. Synchondroses are immovable joints in which bones are joined by hyaline cartilage. Epiphyseal plates are examples.
2. Symphyses are slightly movable joints made of fibrocartilage.

Synovial Joints

1. Synovial joints are capable of considerable movement. They consist of the following:
   • Articular cartilage on the ends of bones, which provides a smooth surface for articulation. Articular disks can provide additional support.
   • A joint cavity surrounded by a joint capsule of fibrous connective tissue, which holds the bones together while permitting flexibility, and a synovial membrane, which produces synovial fluid that lubricates the joint.
2. Bursae are extensions of synovial joints that protect skin, tendons, or bone from structures that could rub against them.
3. Synovial joints are classified according to the shape of the adjoining articular surfaces: plane (two flat surfaces), saddle (two saddle-shaped surfaces), hinge (concave and convex surfaces), pivot (cylindrical projection inside a ring), ball-and-socket (rounded surface into a socket), and ellipsoid (ellipsoid concave and convex surfaces).

Types of Movement   (p. 254)

1. Gliding movements occur when two flat surfaces glide over one another.
2. Angular movements include flexion and extension, plantar and dorsiflexion, abduction and adduction.
3. Circular movements include rotation, pronation and supination, and circumduction.
4. Special movements include elevation and depression, protraction and retraction, excursion, opposition and reposition, and inversion and eversion.
5. Combination movements involve two or more of the above-mentioned movements.
6. Range of motion is the amount of movement, active or passive, that can occur at a joint.

Description of Selected Joints   (p. 259)

1. The temporomandibular joint is a complex hinge and gliding joint between the temporal and mandibular bones. It is capable of elevation and depression, protraction and retraction, and lateral and medial excursion movements.
2. The shoulder joint is a ball-and-socket joint between the head of the humerus and the glenoid cavity of the scapula that permits a wide range of movements. It is strengthened by ligaments and the muscles of the rotator cuff. The tendon of the biceps brachii passes through the joint capsule. The shoulder joint is capable of flexion and extension, abduction and adduction, rotation, and circumduction.
3. The elbow joint is a compound hinge joint between the humerus, ulna, and radius. Movement at this joint is limited to flexion and extension.
4. The hip joint is a ball-and-socket joint between the head of the femur and the acetabulum of the coxa. It is greatly strengthened by ligaments and is capable of a wide range of movements, including flexion, extension, abduction, adduction, rotation, and circumduction.
5. The knee joint is a complex ellipsoid joint between the femur and the tibia that is supported by many ligaments. The joint allows flexion and extension and slight rotation of the leg.
6. The ankle joint is a special hinge joint of the tibia, fibula, and talus that allows dorsiflexion and plantar flexion and inversion and eversion.
7. Ligaments of the foot arches hold the bones in an arch and transfer weight in the foot.

Effects of Aging on the Joints   (p. 268)

With age, connective tissue of the joints becomes less flexible and less elastic. The resulting joint rigidity increases the rate of wear in the articulating surfaces. The change in connective tissue also reduces the range of motion.

R E V I E W   A N D   C O M P R E H E N S I O N

1. Which of these is commonly used for classifying joints in the body?
   a. the connective tissue that binds the bones together
   b. the degree of motion at each joint
   c. the number of bones that articulate with each other
   d. the embryonic tissue that formed the joint
   e. both a and b

2. Given these types of joints:
   1. gomphosis
   2. suture
   3. symphysis
   4. synchondrosis
   5. syndesmosis

   Choose the types that are held together by fibrous connective tissue.
   a. 1,2,3
   b. 1,2,5
   c. 2,3,5
   d. 3,4,5
   e. 1,2,3,4,5

3. Given these types of joints:
   1. gomphosis
   2. suture
   3. symphysis
   4. synchondrosis
   5. syndesmosis

   Choose the types that are held together by cartilage.
   a. 1,2
   b. 1,4
   c. 2,3
   d. 3,4
   e. 3,5

4. Which of these joints is *not* matched with the correct joint type?
   a. parietal bone to occipital bone—suture
   b. between the coxae—symphysis
   c. humerus and scapula—synovial
   d. shafts of the radius and ulna—synchondrosis
   e. teeth in alveolar process—gomphosis

5. The epiphyseal plate can be described as a type of joint. Choose the term that describes the joint before growth in the length of the bone has ended.
   a. synchondrosis
   b. synostosis
   c. syndesmosis
   d. symphysis
   e. synovial

6. Which of these types of joints are often temporary, with bone replacing them?
   a. syndesmoses
   b. synovial
   c. symphyses
   d. gomphoses
   e. synchondroses

7. Which of these joints are the most movable?
   a. sutures
   b. syndesmoses
   c. symphyses
   d. synovial
   e. gomphoses

8. In which of these joints are periodontal ligaments found?
   a. sutures
   b. syndesmoses
   c. symphyses
   d. synovial
   e. gomphoses

9. The intervertebral disks are an example of
   a. sutures.
   b. syndesmoses.
   c. symphyses.
   d. synovial joints.
   e. gomphoses.

10. Joints containing hyaline cartilage are called _____ , and joints containing fibrocartilage are called _____ .
    a. sutures; synchondroses
    b. syndesmoses; symphyses
    c. symphyses; syndesmoses
    d. synchondroses; symphyses
    e. gomphoses; synchondroses

11. The inability to produce the fluid that keeps most joints moist would likely be caused by a disorder of the
    a. cruciate ligaments.
    b. synovial membrane.
    c. articular cartilage.
    d. bursae.
    e. tendon sheath.

12. Which of these is *not* associated with synovial joints?
    a. perichondrium on surface of articular cartilage
    b. fibrous capsule
    c. synovial membrane
    d. synovial fluid
    e. bursae

13. All of the costochondral joints, except for the first, usually develop into
    a. bursae.
    b. synovial joints.
    c. syndesmoses.
    d. synostoses.
    e. symphyses.

14. Assume that a sharp object penetrated a synovial joint. From this list of structures:
    1. tendon or muscle
    2. ligament
    3. articular cartilage
    4. fibrous capsule (of joint capsule)
    5. skin
    6. synovial membrane (of joint capsule)

    Choose the order in which they would most likely be penetrated.
    a. 5,1,2,6,4,3
    b. 5,2,1,4,3,6
    c. 5,1,2,6,3,4
    d. 5,1,2,4,3,6
    e. 5,1,2,4,6,3

15. Which of these do hinge joints and saddle joints have in common?
    a. Both are synovial joints.
    b. Both have concave surfaces that articulate with a convex surface.
    c. Both are monoaxial joints.
    d. Both a and b.
    e. All of the above.

16. Which of these joints is correctly matched with the type of joint?
    a. atlas to occipital condyle—pivot
    b. tarsals to metatarsals—saddle
    c. femur to coxa—ellipsoid
    d. tibia to talus—hinge
    e. scapula to humerus—plane

17. Once a doorknob is grasped, what movement of the forearm is necessary to unlatch the door, that is, turn the knob in a clockwise direction? (Assume using the right hand.)
    a. pronation
    b. rotation
    c. supination
    d. flexion
    e. extension

18. After the door is unlatched, what movement of the elbow is necessary to open it? (Assume the door opens in, and you are on the inside.)
    a. pronation
    b. rotation
    c. supination
    d. flexion
    e. extension

19. After the door is unlatched, what movement of the shoulder is necessary to open it? (Assume the door opens in, and you are on the inside.)
    a. pronation
    b. rotation
    c. supination
    d. flexion
    e. extension

20. When grasping a doorknob, the thumb and little finger undergo
    a. opposition.
    b. reposition.
    c. lateral excursion.
    d. medial excursion.
    e. dorsiflexion.

21. Spreading the fingers apart is
    a. rotation.
    b. depression.
    c. abduction.
    d. lateral excursion.
    e. flexion.

22. A runner notices that the lateral side of her right shoe is wearing much more than the lateral side of her left shoe. This could mean that her right foot undergoes more _____ than her left foot.
    a. eversion
    b. inversion
    c. plantar flexion
    d. dorsiflexion
    e. lateral excursion

23. For a ballet dancer to stand on her toes, her feet must
    a. evert.
    b. invert.
    c. plantar flex.
    d. dorsiflex.
    e. abduct.

24. A meniscus is found in the
    a. shoulder joint.
    b. elbow joint.
    c. hip joint.
    d. knee joint.
    e. ankle joint.

25. A lip (labrum) of fibrocartilage deepens the joint cavity of the
    a. temporomandibular joint.
    b. shoulder joint.
    c. elbow joint.
    d. knee joint.
    e. ankle joint.

26. Which of these joints has a tendon inside the joint cavity?
    a. temporomandibular joint
    b. shoulder joint
    c. elbow joint
    d. knee joint
    e. ankle joint
27. Which of these structures help to stabilize the shoulder joint?
    a. rotator cuff muscles
    b. cruciate ligaments
    c. medial and lateral collateral ligaments
    d. articular disk
    e. all of the above

28. Bursitis of the subacromial bursa could result from
    a. flexing the wrist.
    b. kneeling.
    c. overuse of the shoulder joint.
    d. running a long distance.
    e. extending the elbow.
29. Which of these does *not* occur with the aging of joints?
    a. decrease in production of new cartilage matrix
    b. decline in synovial fluid production
    c. ligaments and tendons stretch and increase range of motion
    d. weakening of muscles
    e. increase in protein cross-linking in tissues

Answers in Appendix F

C R I T I C A L   T H I N K I N G

1. What would be the result if the sternal synchondroses and the sternocostal synchondrosis of the first rib were to become synostoses?
2. Using an articulated skeleton, examine the following list of joints. Describe the type of joint and the movement(s) possible.
    a. the joint between the zygomatic bone and the maxilla
    b. the ligamentous connection between the coccyx and the sacrum
    c. the elbow joint
3. For each of the following muscles, describe the motion(s) produced when the muscle contracts. It may be helpful to use an articulated skeleton.
    a. The biceps brachii muscle attaches to the coracoid process of the scapula (one head) and the radial tuberosity of the radius. Name two movements that the muscle accomplishes in the forearm.
    b. The rectus femoris muscle attaches to the anterior superior iliac spine and the tibial tuberosity. How does contraction move the thigh? The leg?

   c. The supraspinatus muscle is located in and attached to the supraspinatus fossa of the scapula. Its tendon runs over the head of the humerus to the greater tubercle. When it contracts, what movement occurs at the glenohumeral (shoulder) joint?
   d. The gastrocnemius muscle attaches to the medial and lateral condyles of the femur and to the calcaneus. What movement of the leg results when this muscle contracts? Of the foot?
4. Crash McBang hurt his knee in an auto accident by ramming it into the dashboard. The doctor tested the knee for ligament damage by having Crash sit on the edge of a table with his knee flexed at a 90-degree angle. The doctor attempted to pull the tibia in an anterior direction (the anterior drawer test) and then tried to push the tibia in a posterior direction (the posterior drawer test). No unusual movement of the tibia occurred in the anterior drawer test but did occur during the posterior drawer test. Explain the purpose of each test, and tell Crash which ligament he has damaged.

Answers in Appendix G

A N S W E R S   T O   P R E D I C T   Q U E S T I O N S

1. The joint between the metacarpals and the phalanges is the metacarpophalangeal joint.
2. Premature sutural synostosis can result in abnormal skull shape, interfere with normal brain growth, and result in brain damage if not corrected. Such an abnormality is usually corrected surgically by removing some of the bone around the suture and creating an artificial fontanel, which then undergoes normal synostosis.
3. The synovial membrane is very thin and delicate. A considerable amount of pressure is exerted on the articular cartilages within a joint, and the articular cartilage is very tough, yet flexible, to withstand the pressure. If the synovial membrane covered the articular cartilage, it would be easily damaged during movement.

4. The movements required are abduction or flexion of the shoulder, flexion of the elbow, and slight pronation of the hand.
5. A shoulder separation involves stretching or tearing of the acromioclavicular ligament and may involve tearing of the coracoclavicular ligament as well. Because the only bony attachment of the upper limb to the body is from the scapula through the clavicle to the sternum, separation of the acromioclavicular joint greatly reduces the stability of the shoulder. The scapula and humerus tend to be displaced inferiorly, and the proximal pivot point for the upper limb is destabilized.

Visit the Online Learning Center at www.mhhe.com/seeley7 for chapter quizzes, interactive learning exercises, and other study tools.

Color-enhanced scanning electron micrograph of skeletal muscle fibers.

Muscular System

Histology and Physiology

Movements of the limbs, heart, and other parts of the body are made possible by muscle cells that function like tiny motors. Muscle cells use energy extracted from nutrient molecules much like motors use energy provided by electric current. The nervous system regulates and coordinates muscle cells so that smooth, coordinated movements are produced much like a computer regulates and coordinates several motors in robotic machines that perform assembly line functions.

This chapter presents the *functions of the muscular system* (p. 276), *general functional characteristics of muscle* (p. 276), and *skeletal muscle structure* (p. 277). The *sliding filament model* (p. 281) of muscle contraction is explained. The *physiology of skeletal muscle fibers* (p. 283), the *physiology of skeletal muscle* (p. 293), the *types of muscle contractions* (p. 296), *fatigue* (p. 299), *energy sources* (p. 300), *slow and fast fibers* (p. 301), and *heat production* (p. 303) are presented. The structure and function of *smooth muscle* (p. 303) and *cardiac muscle* (p. 307) are introduced, but cardiac muscle is discussed in greater detail in chapter 20. Finally the *effects of aging on skeletal muscle* (p. 308) are presented. Because skeletal muscle is more abundant and more is known about it, skeletal muscle is examined in the greatest detail.

Functions of the Muscular System

Movements within the body are accomplished by cilia or flagella on the surface of some cells, by the force of gravity, or by the contraction of muscles. Most of the body's movements, from the beating of the heart to running a marathon, result from muscle contractions. As described in chapter 4, there are three types of muscle tissue: skeletal, smooth, and cardiac. The major functions of muscles are:

1. *Body movement.* Most skeletal muscles are attached to bones, are typically under conscious control, and are responsible for most body movements including walking, running, or manipulating objects with the hands.
2. *Maintenance of posture.* Skeletal muscles constantly maintain tone, which keeps us sitting or standing erect.
3. *Respiration.* Muscles of the thorax are responsible for the movements necessary for respiration.
4. *Production of body heat.* When skeletal muscles contract, heat is given off as a by-product. This released heat is critical to the maintenance of body temperature.
5. *Communication.* Skeletal muscles are involved in all aspects of communication, such as speaking, writing, typing, gesturing, and facial expression.
6. *Constriction of organs and vessels.* The contraction of smooth muscle within the walls of internal organs and vessels causes constriction of those structures. This constriction can help propel and mix food and water in the digestive tract, propel secretions from organs, and regulate blood flow through vessels.
7. *Heart beat.* The contraction of cardiac muscle causes the heart to beat, propelling blood to all parts of the body.

> **1.** List the functions of skeletal, smooth, and cardiac muscles and explain how each is accomplished.

General Functional Characteristics of Muscle

Muscle tissue is highly specialized to contract, or shorten, forcefully. The process of metabolism extracts energy from nutrient molecules. Part of that energy is used for muscle contraction and the remainder is used for other cell processes or is released as heat.

Properties of Muscle

Muscle has four major functional properties: contractility, excitability, extensibility, and elasticity.

1. **Contractility** is the ability of muscle to shorten forcefully. When muscle contracts, it causes movement of the structures to which it is attached, or it may increase pressure inside hollow organs or vessels. Although muscle shortens forcefully during contraction, it lengthens passively; that is, gravity, contraction of an opposing muscle, or the pressure of fluid in a hollow organ or vessel produces a force that acts on the shortened muscle, causing it to lengthen.
2. **Excitability** is the capacity of muscle to respond to a stimulus. Normally skeletal muscle contracts as a result of stimulation by nerves. Smooth muscle and cardiac muscle can contract without outside stimuli, but they also respond to stimulation by nerves and hormones.
3. **Extensibility** means that muscle can be stretched beyond its normal resting length and is still able to contract.
4. **Elasticity** is the ability of muscle to recoil to its original resting length after it has been stretched.

Types of Muscle Tissue

Table 9.1 provides a comparison of the major characteristics of skeletal, smooth, and cardiac muscle. Skeletal muscle with its associated

Table 9.1	Comparison of Muscle Types		
Features	**Skeletal Muscle**	**Smooth Muscle**	**Cardiac Muscle**
Location	Attached to bones	Walls of hollow organs, blood vessels, eyes, glands, and skin	Heart
Cell shape	Very long and cylindrical (1 mm–4 cm in length and may extend the entire length of short muscles; 10–100 μm in diameter)	Spindle-shaped (15–200 μm in length and 5–8 μm in diameter)	Cylindrical and branched (100–500 μm in length; 12–20 μm in diameter)
Nucleus	Multiple, peripherally located	Single, centrally located	Single, centrally located
Special cell–cell attachments	None	Gap junctions join some visceral smooth muscle cells together	Intercalated disks join cells to one another
Striations	Yes	No	Yes
Control	Voluntary and involuntary (reflexes)	Involuntary	Involuntary
Capable of spontaneous contraction	No	Yes (some smooth muscle)	Yes
Function	Body movement	Food movement through the digestive tract, emptying of the urinary bladder, regulation of blood vessel diameter, change in pupil size, contraction of many gland ducts, movement of hair, and many other functions	Pumps blood; contractions provide the major force for propelling blood through blood vessels

connective tissue constitutes about 40% of the body's weight and is responsible for locomotion, facial expressions, posture, respiratory movements, and many other body movements. The nervous system voluntarily, or consciously, controls the functions of the skeletal muscles.

Smooth muscle is the most widely distributed type of muscle in the body, and it has the greatest variety of functions. It's in the walls of hollow organs and tubes, the interior of the eye, the walls of blood vessels, and other areas. Smooth muscle performs a variety of functions, including propelling urine through the urinary tract, mixing food in the stomach and intestine, dilating and constricting the pupils, and regulating the flow of blood through blood vessels.

Cardiac muscle is found only in the heart, and its contractions provide the major force for moving blood through the circulatory system. Unlike skeletal muscle, cardiac muscle and many smooth muscles are autorhythmic, that is, they contract spontaneously at somewhat regular intervals, and nervous or hormonal stimulation is not always required for them to contract. Furthermore, unlike skeletal muscle, smooth muscle and cardiac muscle are not consciously controlled by the nervous system. Rather, they are controlled involuntarily, or unconsciously, by the endocrine and autonomic nervous systems (see chapters 16 and 18).

> 2. *Define contractility, excitability, extensibility, and elasticity of muscle tissue.*
>
> 3. *Describe the structure, function, location, and control of the three major muscle tissue types.*

Figure 9.1 Skeletal Muscle Fibers
Skeletal muscle fibers in longitudinal section.

Skeletal Muscle Structure

Skeletal muscles are composed of skeletal **muscle fibers** associated with smaller amounts of connective tissue, blood vessels, and nerves. Skeletal muscle cells are often called skeletal muscle fibers. Each skeletal muscle fiber is a single, long, cylindrical cell containing several nuclei located around the periphery of the fiber near the plasma membrane (figure 9.1). Muscle fibers develop from less mature multinucleated cells called **myoblasts** (mī′ō-blasts). Their multiple nuclei result from the fusion of myoblast precursor cells and not from the division of nuclei within myoblasts. Myoblasts are converted to muscle fibers as contractile proteins accumulate within their cytoplasm. Shortly after the myoblasts form, nerves grow into the area and innervate the developing muscle fibers.

The number of skeletal muscle fibers remains relatively constant after birth. Enlargement, or hypertrophy, of muscles after birth therefore results not from a substantial increase in the number of muscle fibers, but from an increase in their size. Similarly, hypertrophy of muscles in response to exercise is due mainly to an increase in muscle fiber size, rather than a substantial increase in number.

As seen in longitudinal section, alternating light and dark bands give the muscle fiber a **striated** (strī′āt-ĕd; banded), or striped, appearance (see figure 9.1). A single fiber can extend from one end of a small muscle to the other, but several muscle fibers arranged end to end are required to extend the full length of most longer muscles. Muscle fibers range from approximately 1 mm to about 4 cm in length and from 10–100 μm in diameter. Large muscles contain large-diameter fibers, whereas small, delicate muscles have small-diameter fibers. All the muscle fibers in a given muscle have similar dimensions.

> 4. *Define skeletal muscle fiber. Do the number of muscle fibers increase significantly after birth?*

Connective Tissue

Surrounding each muscle fiber is a delicate **external lamina** (lam′i-nă) composed primarily of reticular fibers. This external lamina is produced by the muscle fiber and, when observed through a light microscope, cannot be distinguished from the plasma membrane of the muscle fiber, the **sarcolemma** (sar′kō-lem′ă). The prefix *sarco-* (G. flesh) refers to muscle or resembling flesh and is used to name some structures found in muscle cells. The **endomysium** (en′dō-miz′ē-ŭm, en′dō-mis′ē-ŭm; G. *mys,* muscle), a delicate network of loose connective tissue with numerous reticular fibers, surrounds each muscle fiber outside the external lamina (figure 9.2). A bundle of muscle fibers with their endomysium is surrounded by another, heavier connective tissue layer called the **perimysium** (per′ĭ-mis′ē-ŭm, per′ĭ-miz′ē-ŭm). Each bundle ensheathed by perimysium is a muscle **fasciculus** (fă-sik′ū-lŭs). A muscle consists of many fasciculi grouped together and surrounded by a third and heavier layer, the **epimysium** (ep-ĭ-mis′ē-ŭm), which is composed of dense, collagenous connective tissue and covers the entire surface of the muscle.

Fascia (fash′ē-ă) is connective tissue that covers the body by forming a sheet of tissue under the skin; it also surrounds individual muscles or groups of muscles. The fascia around an individual muscle is also called epimysium. The connective tissue components of muscles are continuous with one another. At the end of muscles, the connective tissue components are continuous with the

Figure 9.2   **Skeletal Muscle Structure: Connective Tissue, Innervation, and Blood Supply**

Relationship between muscle fibers, fasciculi, and associated connective tissue layers: the epimysium, perimysium, and endomysium. Arteries, veins, and nerves course together through the connective tissue of muscles. They branch frequently as they approach individual muscle fibers. At the level of the perimysium, axons of neurons branch and each branch extends to a muscle fiber.

connective tissue of tendons and the periosteum of bone (see chapter 6). The connective tissue of muscle holds the muscle cells together and attaches muscles to tendons and bones.

> ● 5. *Name the connective tissue structures that surround muscle fibers, muscle fasciculi, and whole muscles. Define sarcolemma and fascia.*

Nerve and Blood Vessels

The nerves and blood vessels that extend to skeletal muscles are abundant. **Motor neurons** are specialized nerve cells. Their cell bodies are located in the brain or spinal cord and their axons extend to skeletal muscle fibers through nerves. The motor neurons stimulate muscles to contract. An artery and either one or two veins extend together with a nerve through the connective tissue layers of

skeletal muscles (see figure 9.2). Numerous branches of the arteries supply the extensive capillary beds surrounding the muscle fibers, and blood is carried away from the capillary beds by branches of the veins. At the level of the perimysium, the axons of motor neurons branch repeatedly, each branch projecting toward the center of one muscle fiber. The contact between the axons and the muscle fibers, called synapses, or neuromuscular junctions, are described later in the chapter (see section on "Neuromuscular Junction" on page 287). Each motor neuron innervates more than one muscle fiber, and every muscle fiber receives a branch of an axon.

> ● 6. *What are motor neurons? How do the axons of motor neurons and blood vessels extend to muscle fibers?*

Muscle Fibers

The many nuclei of each muscle fiber lie just inside the sarcolemma, whereas most of the interior of the fiber is filled with myofibrils (figure 9.3). Other organelles, such as the numerous mitochondria and glycogen granules, are packed between the myofibrils. The cytoplasm without the myofibrils is called **sarcoplasm** (sar'kō-plazm). Each **myofibril** (mī-ō-fī'bril) is a threadlike structure approximately 1–3 μm in diameter that extends from one end of the muscle fiber to the other. Two kinds of protein filaments, called **myofilaments** (mī-ō-fil'ă-ments), are major components of myofibrils. **Actin** (ak'tin) **myofilaments,** or thin myofilaments, are approximately 8 nanometers (nm) in diameter and 1000 nm in length, whereas **myosin** (mī'ō-sin) **myofilaments,** or thick myofilaments, are approximately 12 nm in diameter and 1800 nm in length. The actin and myosin myofilaments form highly ordered units called **sarcomeres** (sar'kō-mērz), which are joined end to end to form the myofibrils (figure 9.4a).

Actin and Myosin Myofilaments

Each actin myofilament is composed of two strands of **fibrous actin (F actin),** a series of **tropomyosin** (trō-pō-mī'ō-sin) **molecules,** and a series of **troponin** (trō'pō-nin) **molecules** (figure 9.4b and c). The two strands of F actin are coiled to form a double helix that extends the length of the actin myofilament. Each F actin strand is a polymer of approximately 200 small globular units called **globular actin (G actin)** monomers. Each G actin monomer has an active site to which myosin molecules can bind during muscle contraction. Tropomyosin is an elongated protein that winds along the groove of the F actin double helix. Each tropomyosin molecule is sufficiently long to cover seven G actin active sites. Troponin is composed of three subunits: one that binds to actin, a second that binds to tropomyosin, and a third that binds to calcium ions. The troponin molecules are spaced between the ends of the tropomyosin molecules in the groove between the F actin strands. The complex of tropomyosin and troponin regulates the interaction between active sites on G actin and myosin.

Myosin myofilaments are composed of many elongated **myosin molecules** shaped like golf clubs (see figure 9.4b and c). Each myosin molecule consists of two **heavy myosin molecules** wound together to form a **rod portion** lying parallel to the myosin myofilament and two **heads** that extend laterally (see figure 9.4b and figure 9.3d). Four light myosin chains are attached to the heads of each myosin molecule. Each myosin myofilament consists of about 300 myosin molecules arranged so that about 150 of them have their

Figure 9.3 **Parts of a Muscle**

(a) Part of a muscle attached by a tendon to a bone. A muscle is composed of muscle fasciculi, each surrounded by perimysium. The fasciculi are composed of bundles of individual muscle fibers (muscle cells), each surrounded by endomysium. (b) Enlargement of one muscle fiber. The muscle fiber contains several myofibrils. (c) A myofibril extended out the end of the muscle fiber. The banding patterns of the sarcomeres are shown in the myofibril. (d) A single sarcomere of a myofibril is composed of actin myofilaments and myosin myofilaments. The Z disk anchors the actin myofilaments, and the myosin myofilaments are held in place by titin molecules and the M line.

heads projecting toward each end. The centers of the myosin myofilaments consist of only the rod portions of the myosin molecules. The myosin heads have three important properties: (1) The heads can bind to active sites on the actin molecules to form **cross-bridges.** (2) The heads are attached to the rod portion by a hinge region that can bend and straighten during contraction. (3) The heads have ATPase activity, the enzymatic activity that breaks down adenosine

triphosphate (ATP), releasing energy. Part of the energy is used to bend the hinge region of the myosin molecule during contraction.

Sarcomeres

Each sarcomere extends from one Z disk to an adjacent Z disk (see figure 9.4 and figure 9.5). A **Z disk** is a filamentous network of protein forming a disklike structure for the attachment of actin myofilaments

Figure 9.4   Structure of Actin and Myosin

(*a*) The sarcomere consists of actin (thin) myofilaments, attached to the Z disks, and myosin (thick) myofilaments, suspended between the actin myofilaments. (*b*) Actin myofilaments are made up of F actin (*chains of purple spheres*), tropomyosin, (*blue strands*), and troponin (*red spheres*). Myosin myofilaments are made up of many golf-club-shaped myosin molecules, with all the heads pointing in one direction at one end and the opposite direction at the other end. (*c*) G actin molecules (*purple spheres*), with their attachment sites (*yellow*), tropomyosin, and troponin make up actin myofilaments. Myosin molecules (*green*) are golf-club-shaped structures composed of two molecules of heavy myosin wound together to form the rod portion, and a double globular head. Four small light myosin molecule are located on the heads of each of the myosin molecules.

(see figure 9.5). The arrangement of the actin myofilaments and myosin myofilaments gives the myofibril a banded, or striated, appearance when viewed longitudinally. Each **isotropic** (ī-sō-trop′ik) (light band), or **I, band** includes a Z disk and extends from either side of the Z disk to the ends of the myosin myofilaments. When seen in longitudinal and cross sections, the I band on either side of the Z disk consists only of actin myofilaments. Each **anisotropic** (an-ī-sō-trop′ik) (dark band), or **A, band** extends the length of the myosin myofilaments within a sarcomere. The actin and myosin myofilaments overlap for some distance at both ends of the A band. In a cross section of the A band in the area where actin and myosin myofilaments overlap, each myosin myofilament is visibly surrounded by six actin myofilaments. In the center of each A band is a smaller band called the **H zone,** where the actin and myosin myofilaments do not overlap and only myosin myofilaments are present. A dark line, called the **M line,** is in the middle of the H zone and consists of delicate filaments that attach to the center of the myosin myofilaments. The M line helps to hold the myosin myofilaments in place similar to the

way the Z disk holds actin myofilaments in place (figure 9.5*b* and *c*). The numerous myofibrils are oriented within each muscle fiber so that A bands and I bands of parallel myofibrils are aligned and thus produce the striated pattern seen through a microscope.

What Makes Muscles Extensible and Elastic?

In addition to actin and myosin, there are other less visible proteins within sarcomeres. These proteins help to hold actin and myosin in place, and one of them accounts for muscle's ability to stretch (extensibility) and recoil (elasticity). **Titin** (tī′tin) (see figure 9.3) is one of the largest known proteins, consisting of a single chain of nearly 27,000 amino acids. It attaches to Z disks and extends along myosin myofilaments to the M line. The myosin myofilaments are attached to the titin molecules, which help to hold them in position. Part of the titin molecule in the I band functions like a spring, allowing the sarcomere to stretch and recoil. Another large protein, called **nebulin** (neb′ū-lin), appears to hold the thin filaments in place. These proteins extend from each side of the Z disk and along the actin myofilaments. Each nebulin molecule is as long as an actin myofilament.

(a) Electron micrograph of longitudinal section of a skeletal muscle fiber showing several sarcomeres, with A bands, I bands, Z disks, H zones, and M lines

(b) The arrangement of I and A bands, H zones, Z disks, and M lines in sarcomeres

(c) Cross sections through regions of the sarcomeres

Actin myofilaments only

Myosin myofilaments surrounded by actin myofilaments

Myosin myofilaments only

Rod portion of myosin myofilaments and M line

Figure 9.5 **Components of Sarcomeres**

7. *Define sarcoplasm, myofibril, and myofilament.*
8. *How do G actin, tropomyosin, and troponin combine to form an actin myofilament? Name the molecule or ion to which each of the three subunits of troponin binds.*
9. *Describe the structure of myosin molecules and how they combine to form a myosin myofilament.*
10. *List three important properties of the myosin head. What is a cross-bridge?*
11. *How are Z disks, actin myofilaments, myosin myofilaments, and M lines arranged to form a sarcomere? Describe how this arrangement produces the I band, A band, and H zone.*

Sliding Filament Model

The **sliding filament model** of muscle contraction includes all the events that result in actin myofilaments sliding over myosin myofilaments to shorten the sarcomeres of muscle fibers. Actin and myosin myofilaments do not change length during contraction of skeletal muscle. Instead, the actin and myosin myofilaments slide past one another in a way that causes the sarcomeres to shorten (figure 9.6). The shortening of sarcomeres is responsible for the contraction of skeletal muscles. When sarcomeres shorten the myofibrils also shorten because the myofibrils consist of sarcomeres joined end to end. The myofibrils extend the length of the muscle fibers, and when they shorten, the muscle fibers shorten. Muscle bundles are made up

1. Actin and myosin myofilaments in a relaxed muscle (*right*) and a contracted muscle (*#4 below*) are the same length. Myofilaments do not change length during muscle contraction.

2. During contraction, actin myofilaments at each end of the sarcomere slide past the myosin myofilaments toward each other. As a result, the Z disks are brought closer together, and the sarcomere shortens.

3. As the actin myofilaments slide over the myosin myofilaments, the H zones (*yellow*) and the I bands (*blue*) narrow. The A bands, which are equal to the length of the myosin myofilaments, do not narrow, because the length of the myosin myofilaments does not change.

4. In a fully contracted muscle, the ends of the actin myofilaments overlap and the H zone disappears.

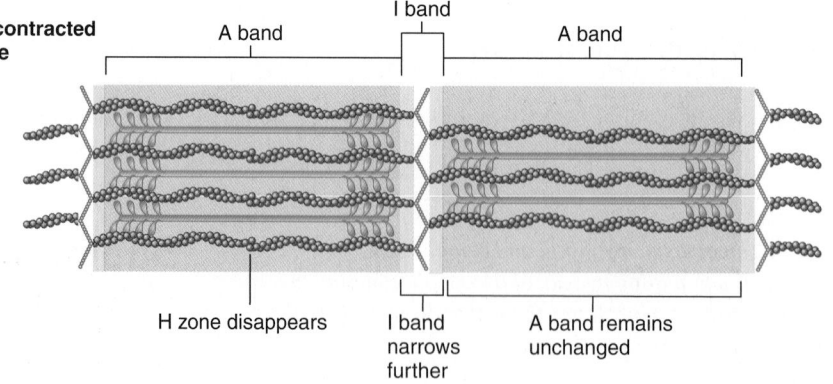

Process Figure 9.6 Sarcomere Shortening

of muscle fibers and muscles are made up of muscle bundles. Therefore, when sarcomeres shorten, myofibrils, muscle fibers, muscle bundles, and muscles shorten to produce muscle contractions.

During relaxation of muscle the sarcomeres lengthen. For this to happen, some external force must be applied to a muscle to cause them to lengthen, such as forces produced by other muscles or by gravity. For example, contraction of a muscle causes a joint such as the elbow or knee to flex. Extension of the joint and lengthening of the muscle results from the contraction of other, antagonistic, muscles (see chapter 10).

> 12. **Why do the I bands and H zones shorten during muscle contraction, but the length of the A band is unchanged?**
>
> 13. **How does shortening of sarcomeres explain muscle contraction?**

P R E D I C T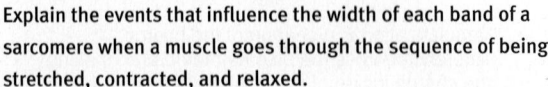

Explain the events that influence the width of each band of a sarcomere when a muscle goes through the sequence of being stretched, contracted, and relaxed.

Physiology of Skeletal Muscle Fibers

Axons of nerve cells extend from the brain and spinal cord to skeletal muscle fibers. The nervous system controls the contraction of skeletal muscles through these axons. Electrical signals, called **action potentials,** travel from the brain or spinal cord along the axons to muscle fibers and cause them to contract.

Membrane Potentials

Plasma membranes are **polarized,** which means there is a voltage difference, or electrical charge difference, across each plasma membrane. This charge difference across the plasma membrane of an unstimulated cell is called the **resting membrane potential** (figure 9.7). Action potentials cannot be produced without the resting membrane potential. The negative charge at the internal surface of the plasma membrane compared to the outer surface results mainly from the concentration differences of ions and charged molecules across the plasma membrane and to its permeability characteristics.

There is a difference in the ion concentrations across the plasma membrane. The Na^+–K^+ pump moves Na^+ to the outside of the cell and K^+ to the inside of the cell. This results in a higher concentration of Na^+ in the extracellular fluid and a lower concentration of Na^+ in the intracellular fluid, and it results in a higher concentration of K^+ in the intracellular fluid and a lower concentration of K^+ in the extracellular fluid.

The plasma membrane is relatively permeable to K^+ and much less permeable to Na^+ and to negatively charged molecules

Figure 9.7 Measuring the Resting Membrane Potential

The recording electrode is inside the membrane, the reference electrode is outside, and a potential difference of about -85 mV is recorded by the recording device (oscilloscope), with the inside of the membrane negative with respect to the outside of the membrane.

found inside of the cell. Consequently, positively charged K^+ tend to diffuse out of the cell through K^+ channels leaving the negatively charged molecules behind, thus polarizing the membrane. The membrane is at equilibrium when the tendency for K^+ to diffuse out of the cell is resisted by the negative charge of the molecules inside of the cell. The details of the resting membrane potential are more fully described in chapter 11.

The resting membrane potential can be measured in units called **millivolts** (mV; mV = 1/1000 Volt). The potential differences across the plasma membranes of nerve cells and muscle fibers are between -70 to -90 mV. The potential difference is reported as a negative number because the inner surface of the plasma membrane is negative compared to the outside.

Ion Channels

The increased permeability of the plasma membrane to K^+ is due to non-gated K^+ channels. These channels are specific for K^+ and they are open continuously. Once the resting membrane potential is established, action potentials can be produced. An action potential is a reversal of the resting membrane potential such that the inside of the plasma membrane becomes positively charged compared to the outside. The permeability characteristics of the plasma membrane change as a result of the opening of certain ion channels, when a cell is stimulated. The diffusion of ions through these channels changes the charge across the plasma membrane and produces an action potential. Two types of gated ion channels play important roles in producing action potentials:

1. *Ligand-gated ion channels.* A **ligand** (lī′gand) is a molecule that binds to a receptor. A **receptor** is a protein or glycoprotein that has a receptor site to which a ligand can bind. **Ligand-gated ion channels** are channels which open in response to a ligand binding to a receptor that is part of the ion channel (see figure 3.8). For example, the axons of

nerve cells supplying skeletal muscle fibers release ligands, called **neurotransmitters** (noor′ō-trans-mit′er), which bind to ligand-gated Na⁺ channels in the membranes of the muscle fibers. As a result, the Na⁺ channels open, allowing Na⁺ to enter the cell.

2. *Voltage-gated ion channels.* These channels are gated membrane channels that open and close in response to small voltage (charge) changes across the plasma membrane, or changes in the membrane potential. When a nerve or muscle cell is stimulated, the charge difference changes and that change causes voltage-gated ion channels to open or close. The major voltage-gated channels that play important roles in an action potential are voltage-gated Na⁺ and K⁺ channels.

Both ligand-gated and voltage-gated ion channels are specific for the type of ion that passes through them. The channels that open determine what ions move across the plasma membrane. For example, opening voltage-gated Na⁺ channels allow Na⁺ to cross the plasma membrane, whereas opening voltage-gated K⁺ channels allow K⁺ to cross.

The concentration gradient for an ion determines whether that ion enters or leaves the cell after the ion channel, specific for that ion, opens (see chapter 3). For example, there is a higher concentration of Na⁺ outside the cell than inside it. Consequently, when voltage-gated Na⁺ channels open, Na⁺ move through them into the cell. In a similar fashion, when gated K⁺ channels open, K⁺ move out of the cell.

- 14. *Define resting membrane potential.*
- 15. *Describe the differences in Na⁺ and K⁺ concentrations across the plasma membrane.*
- 16. *Describe the permeability characteristics of the plasma membrane.*
- 17. *List the two major categories of gated ion channels in the plasma membrane.*
- 18. *List the two major types of voltage-gated ion channels that play important roles in the resting membrane potential.*

P R E D I C T ② 2

When ligand-gated K⁺ channels open, how does this affect the resting membrane potential?

Action Potentials

An action potential takes approximately 1 to a few milliseconds to occur, and it has two phases called depolarization and repolarization. Stimulation of a cell can cause depolarization of its plasma membrane, which is graphed in figure 9.8a. Depolarization occurs when the inside of the plasma membrane becomes less negative, which is indicated by movement of the curve upward toward zero. If the depolarization changes the membrane potential to a value called **threshold** (figure 9.8b), an action potential is triggered. The **depolarization phase** of the action potential is a brief period during which further depolarization occurs and the inside of the cell becomes positively charged. The charge difference across the plasma membrane is said to be

(a) Depolarization is a change of the charge difference across the plasma membrane, making the charge inside of the cell less negative and the outside of the plasma membrane less positive. Once threshold is reached an action potential is produced.

(b) During the depolarization phase of the action potential the membrane potential changes from approximately −85 mV to approximately +20 mV. During the repolarization phase, the inside of the plasma membrane changes from approximately +20 mV to −85 mV.

Figure 9.8 **Depolarization and the Action Potential**

reversed when the membrane potential becomes a positive value. The **repolarization phase** is the return of the membrane potential to its resting value.

The opening and closing of voltage-gated ion channels change the permeability of the plasma membrane to ions, resulting in depolarization and repolarization. Before a nerve or muscle cell is stimulated, these voltage-gated ion channels are closed (figure 9.9, *step 1*). When the cell is stimulated, gated Na⁺ channels open, and Na⁺ diffuse into the cell. The positively charged

1. Resting membrane potential. In addition to some non-gated K⁺ channels (not shown), which are open, voltage-gated Na⁺ channels (*pink*) and voltage-gated K⁺ channels (*purple*) are closed. The outside of the plasma membrane is positively charged compared to the inside.

2. Depolarization. Voltage-gated Na⁺ channels are open. Depolarization results because the inward movement of Na⁺ makes the inside of the membrane more positive.

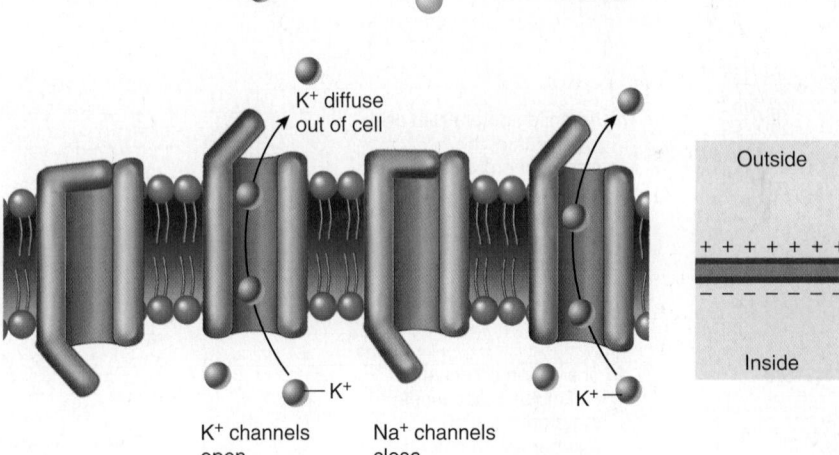

3. Repolarization. Voltage-gated Na⁺ channels are closed and voltage-gated K⁺ channels are open. Na⁺ movement into the cell stops and K⁺ movement out of the cell increases, causing repolarization.

Process Figure 9.9   Gated Ion Channels and the Action Potential

Step 1 illustrates the status of gated Na⁺ and K⁺ channels in a resting cell. Steps 2 and 3 show how the channels open and close to produce an action potential. Next to each step, the charge difference across the plasma membrane is illustrated.

Na⁺ make the inside of the cell membrane less negative. If this depolarization reaches a value called **threshold,** many voltage-gated Na⁺ channels open rapidly, and Na⁺ diffuse into the cell until the inside of the membrane becomes positive for a brief time (figure 9.9, *step 2*). As the inside of the cell becomes positive, this voltage change causes additional permeability changes in the plasma membrane that stop depolarization and start repolarization. Repolarization results from the closing of voltage-gated Na⁺ channels and the opening of voltage-gated K⁺ channels (figure 9.9, *step 3*). Thus, the movement of Na⁺ into the cell stops,

and the movement of K⁺ out of the cell increases. These changes cause the inside of the plasma membrane to become more negative and the outside to become more positive. The action potential ends and the resting membrane potential is reestablished when the voltage-gated K⁺ channels close.

Action potentials occur according to the **all-or-none principle.** If a stimulus is strong enough to produce a depolarization that reaches threshold, or even if it exceeds threshold by a substantial amount, all of the permeability changes responsible for an action potential proceed without stopping. Consequently, all of the action potentials in a given cell are alike (the "all" part). If a stimulus is so weak that the depolarization does not reach threshold, few of the permeability changes occur. The membrane potential returns to its resting level after a brief period without producing an action potential (the "none" part). An action potential can be compared to the flash system of a camera. Once the shutter is triggered (reaches threshold), the camera flashes (an action potential is produced), and each flash is the same brightness (the "all" part) as the previous flashes. If the shutter is depressed, but not triggered (does not reach threshold), no flash results (the "none" part).

An action potential occurs in a very small area of the plasma membrane and does not affect the entire plasma membrane at one time. Action potentials can travel, or **propagate,** across the plasma membrane because an action potential produced at one location in the plasma membrane can stimulate the production of an action potential in an adjacent location (figure 9.10). Note that an action potential does not actually move along the plasma membrane. Rather, an action potential at one location stimulates the production of another action potential in an adjacent location, which, in turn, stimulates the pro-

duction of another and so on. It is much like a long row of toppling dominos in which each domino knocks down the next domino. Each domino falls, but no single domino actually travels the length of the row.

The **action potential frequency** is the number of action potentials produced per unit of time. As the strength of the stimulus applied to a nerve or muscle cell increases (once threshold is reached), the action potential frequency increases as the strength of the stimulus increases. All the action potentials are identical. The action potential frequency can affect the strength of a muscle contraction (see Stimulus Frequency and Muscle Contraction, page 295).

In summary, the resting membrane potential results from a charge difference across the plasma membrane. An action potential, which is a reversal of that charge difference, stimulates cells to respond by contracting. The nervous system controls muscle contractions by sending action potentials along axons to muscle cells and stimulating action potentials in them. An increased frequency of action potentials sent to the muscle cells can result in an increased strength of muscle contraction.

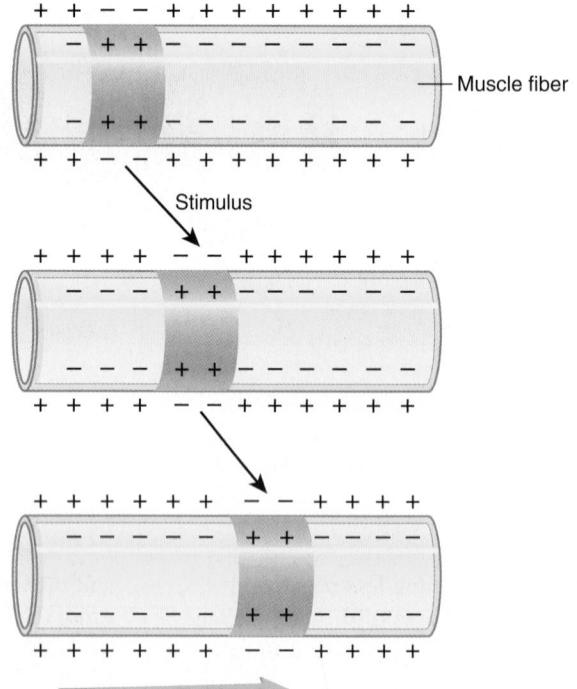

1. An action potential in a local area of the plasma membrane is indicated by the *orange band.* Note the reversal of charge across the membrane.

2. The action potential is a stimulus that causes the production of another action potential in the adjacent plasma membrane.

3. The action potential propagates along the plasma membrane (indicated by *orange arrow*).

Process Figure 9.10   Action Potential Propagation in a Muscle Fiber

19. *What types of gated ion channels are responsible for producing action potentials?*
20. *What value must depolarization reach in a cell to trigger an action potential?*
21. *Describe the changes that occur during the depolarization and repolarization phases of an action potential.*
22. *What is the all-or-none principle of action potentials and what is its significance?*
23. *Describe the propagation of an action potential.*
24. *How does the frequency of action potentials affect muscle contractions?*

Neuromuscular Junction

Action potentials carried by motor neurons cause action potentials to be produced in muscle fibers because of events that occur in the neuromuscular junction. Axons of motor neurons carry action potentials at a high velocity from the brain and spinal cord to skeletal muscle fibers. The axons branch repeatedly, and each branch projects toward one muscle fiber to innervate it. Thus, each muscle fiber receives a branch of an axon, and each axon innervates more than one muscle fiber (see figure 9.2).

Near the muscle fiber it innervates, each axon branch forms a cluster of enlarged axon terminals that rests in an invagination of the sarcolemma to form a **synapse,** or **neuromuscular junction,** which consists of the axon terminals and the area of the muscle fiber sarcolemma they innervate. Each axon terminal is the **presynaptic** (prē′si-nap′tik) **terminal.** The space between the axon terminal and the muscle fiber is the **synaptic** (si-nap′tik) **cleft,** and the muscle plasma membrane in the area of the junction is the **postsynaptic** (pōst-si-nap′tik) **membrane,** or **motor end-plate** (figure 9.11).

Each presynaptic terminal contains numerous mitochondria and many small, spherical sacs approximately 45 μm in diameter, called **synaptic vesicles.** The vesicles contain **acetylcholine** (**ACh;** as-e-til-kō′1ēn), an organic molecule composed of acetic acid and choline, which functions as a neurotransmitter. A **neurotransmitter** (noor′ō-trans-mit′er) is a substance released from a presynaptic membrane that diffuses across the synaptic cleft and stimulates (or inhibits) the production of an action potential in the postsynaptic membrane by binding to ligand-gated ion channels.

(a)

(b)

(c)

Figure 9.11   Neuromuscular Junction

(*a*) Diagram showing the neuromuscular junction. Several branches of an axon form the neuromuscular junction with a single muscle fiber. (*b*) The presynaptic terminal containing synaptic vesicles is separated from the postsynaptic membrane by the synaptic cleft. (*c*) Photomicrograph of neuromuscular junctions.

When an action potential reaches the presynaptic terminal, it causes voltage-gated calcium ion (Ca^{2+}) channels in the plasma membrane of the axon to open, and as a result, Ca^{2+} diffuse into the cell (figure 9.12a). Once inside the cell, the ions cause the contents of a few synaptic vesicles to be secreted by exocytosis from the presynaptic terminal into the synaptic cleft. The acetylcholine molecules released from the synaptic vesicles then diffuse across the cleft and bind to receptor molecules located within the postsynaptic membrane. This causes ligand-gated Na^+ channels to open, increasing the permeability of the membrane to Na^+. Sodium ions then diffuse into the cell causing depolarization. In skeletal muscle, each action potential in the motor neuron causes a depolarization that exceeds threshold, which causes changes in voltage-gated ion channels that result in the production of an action potential in the muscle fiber.

P R E D I C T

Predict the consequence if presynaptic action potentials in an axon could not release sufficient acetylcholine to cause depolarization to threshold in a skeletal muscle fiber.

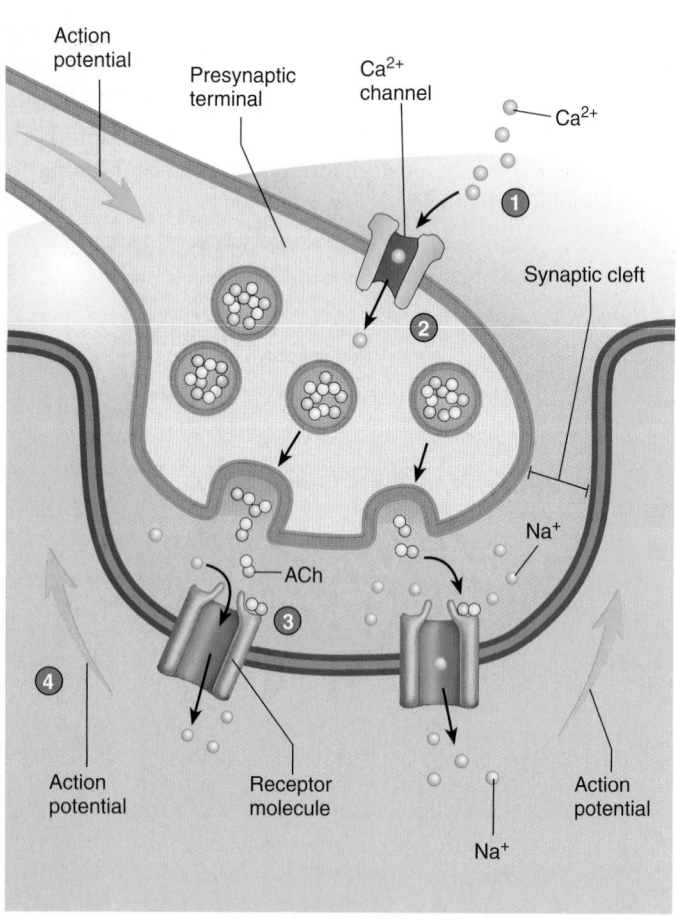

1. An action potential arrives at the presynaptic terminal causing voltage gated Ca^{2+} channels to open, increasing the Ca^{2+} permeability of the presynaptic terminal.

2. Calcium ions enter the presynaptic terminal and initiates the release of a neurotransmitter, acetylcholine (ACh), from synaptic vesicles in the presynaptic cleft.

3. Diffusion of ACh across the synaptic cleft and binding of ACh to ACh receptors on the postsynaptic muscle fiber membrane causes an increase in the permeability of ligand-gated Na^+ channels.

4. The increase in Na^+ permeability results in depolarization of the postsynaptic membrane; once threshold has been reached a postsynaptic action potential results.

(a)

5. Once ACh is released into the synaptic cleft it binds to the receptors for ACh on the postsynaptic membrane and causes Na^+ channels to open.

6. ACh is rapidly broken down in the synaptic cleft by acetylcholinesterase to acetic acid and choline.

7. The choline is reabsorbed by the presynaptic terminal and combined with acetic acid to form more ACh, which enters synaptic vesicles. Acetic acid is taken up by many cell types.

(b)

Process Figure 9.12   Function of the Neuromuscular Junction

(a) Release of ACh in response to an action potential at the neuromuscular junction. (b) Breakdown of ACh in the neuromuscular junction.

Acetylcholine released into the synaptic cleft is rapidly broken down to acetic acid and choline by the enzyme **acetylcholinesterase** (as'e-til-kō-lin-es'ter-ās; figure 9.12*b*). Acetylcholinesterase keeps acetylcholine from accumulating within the synaptic cleft, where it would act as a constant stimulus at the postsynaptic terminal producing many action potentials and continuous contraction in the muscle fiber. The release of acetylcholine and its rapid degradation in the synaptic cleft ensures that one presynaptic action potential yields only one postsynaptic action potential. Choline molecules are actively reabsorbed by the presynaptic terminal and then combined with the acetic acid produced within the cell to form acetylcholine. Recycling choline molecules requires less energy and is more rapid than completely synthesizing new acetylcholine molecules each time they are released from the presynaptic terminal. Acetic acid is an intermediate in the process of glucose metabolism (see chapter 25). A variety of cells can take it up and use it after it diffuses from the area of the neuromuscular junction.

● **25. Describe the neuromuscular junction. How does an action potential in the neuron produce an action potential in the muscle cell?**

26. What is the importance of acetylcholinesterase in the synaptic cleft?

Excitation–Contraction Coupling

Action potentials produced in the sarcolemma of a skeletal muscle fiber can lead to contraction of the fiber. The mechanism by which an action potential causes contraction of a muscle fiber is called **excitation–contraction coupling,** and it involves the sarcolemma, T tubules, sarcoplasmic reticulum, Ca^{2+}, and troponin. The sarcolemma has along its surface many regularly arranged tubelike invaginations called **transverse,** or **T tubules.** T tubules project into the muscle fiber and wrap around sarcomeres in the region where actin and myosin myofilaments overlap (see figure 9.3 and figure 9.13). The lumen of each T tubule is filled with extracellular fluid and is continuous with the exterior of the muscle fiber. Suspended in the sarcoplasm between the T tubules is a highly specialized, smooth endoplasmic reticulum called the **sarcoplasmic reticulum** (sar-kō-plaz'mik re-tik'ū-lŭm). Near the T tubules, the sarcoplasmic reticulum is enlarged to form **terminal cisternae** (sis-ter'nē). A T tubule and the two adjacent terminal cisternae together are called a **triad** (trī'ad) (see figure 9.13). The sarcoplasmic reticulum actively transports Ca^{2+} into its lumen; thus the concentration of Ca^{2+} is approximately 2000 times higher within the sarcoplasmic reticulum than in the sarcoplasm of a resting muscle.

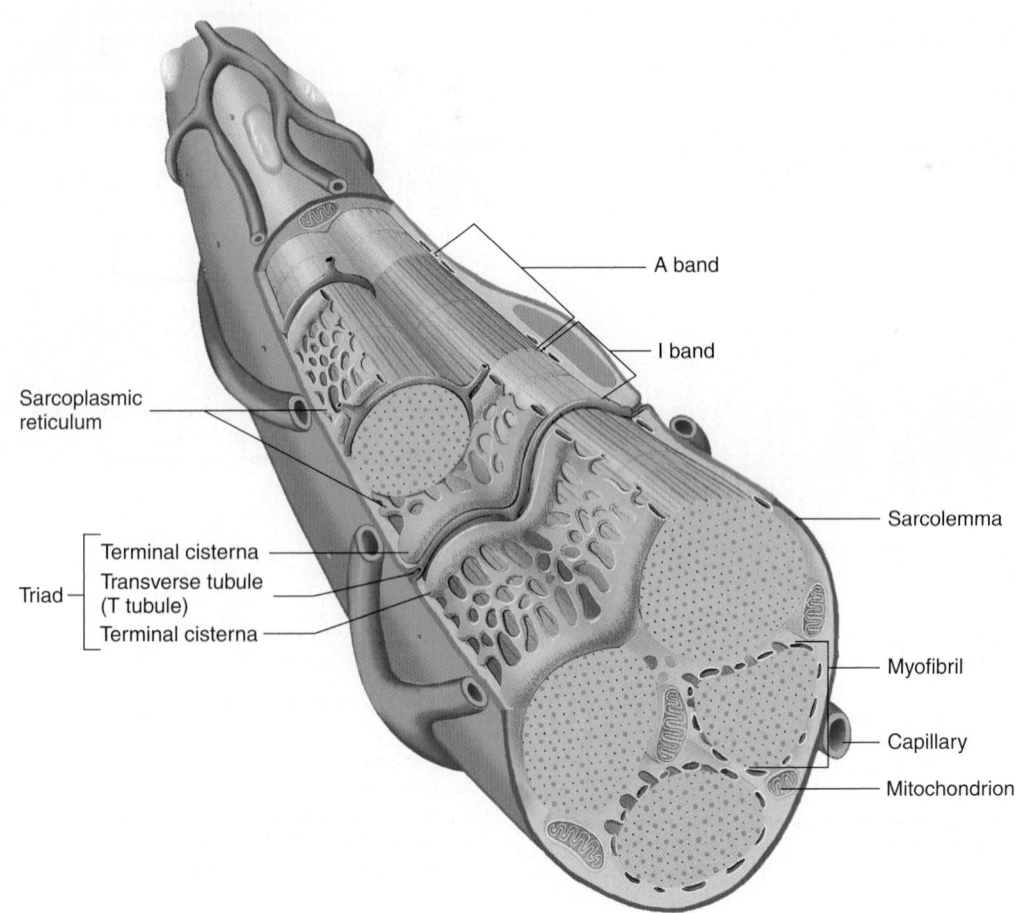

A band

I band

Sarcoplasmic reticulum

Sarcolemma

Terminal cisterna
Transverse tubule (T tubule)
Terminal cisterna

Triad

Myofibril

Capillary

Mitochondrion

Figure 9.13 T Tubules and Sarcoplasmic Reticulum

A T tubule and the sarcoplasmic reticulum on either side of the T tubule (triad).

Excitation–contraction coupling begins at the neuromuscular junction with the production of an action potential in the sarcolemma. The action potential is propagated along the entire sarcolemma of the muscle fiber. When the action potential reaches the T tubules, the membranes of the T tubules undergo depolarization, because the T tubules are invaginations of the sarcolemma. The T tubules carry the depolarizations into the interior of the muscle fiber, and the depolarizations of the T tubules of the triads cause Ca^{2+} channels in the terminal cisternae of the sarcoplasmic reticulum to open. When the Ca^{2+} channels of the sarcoplasmic reticu-

lum open, Ca^{2+} rapidly diffuse the short distance from the sarcoplasmic reticulum into the sarcoplasm surrounding the myofibrils (figure 9.14).

Ca^{2+} bind to troponin of the actin myofilaments. The combination of Ca^{2+} with troponin causes the troponin–tropomyosin complex to move deeper into the groove between the two F actin molecules and thus expose active sites on the actin myofilaments. These exposed active sites bind to the heads of the myosin molecules to form cross-bridges (see figure 9.14). Movement of the cross-bridges results in contraction.

1. An action potential produced at the neuromuscular junction is propagated along the sarcolemma of the skeletal muscle. The depolarization also spreads along the membrane of the T tubules.

2. The depolarization of the T tubule causes gated Ca^{2+} channels in the sarcoplasmic reticulum to open, resulting in an increase in the permeability of the sarcoplasmic reticulum to Ca^{2+}, especially in the terminal cisternae. Calcium ions then diffuse from the sarcoplasmic reticulum into the sarcoplasm.

3. Ca^{2+} released from the sarcoplasmic reticulum bind to troponin molecules in the actin myofilament. Consequently, the troponin molecules bound to G actin molecules are released. This causes tropomyosin molecules to move, thereby exposing active sites on the G actin molecules.

4. Once active sites on G actin molecules are exposed, the heads of the myosin myofilaments bind to them to form cross-bridges.

Process Figure 9.14   Action Potentials and Muscle Contraction

The Effect of Blocking Acetylcholine Receptors and Acetylcholinesterase

Anything that affects the production, release, and degradation of acetylcholine or its ability to bind to its receptor molecule also affects the transmission of action potentials across the neuromuscular junction. For example, some insecticides contain organophosphates that bind to and inhibit the function of acetylcholinesterase. As a result, acetylcholine is not degraded and accumulates in the synaptic cleft, where it acts as a constant stimulus to the muscle fiber. Insects exposed to such insecticides die, partly because their muscles contract and cannot relax—a condition called **spastic paralysis** (spas′tik pă-ral′i-sis), which is followed by fatigue of the muscles.

In humans a similar response to these insecticides occurs. The skeletal muscles responsible for respiration cannot undergo their normal cycle of contraction and relaxation. Instead they remain in a state of spastic paralysis until they become fatigued. Victims die of respiratory failure. Other organic poisons such as curare bind to the acetylcholine receptors, preventing acetylcholine from binding to them. Curare does not allow activation of the receptors, and therefore the muscle is incapable of contracting in response to nervous stimulation—a condition called **flaccid** (flak′sid, flas′id) **paralysis.** Curare is not a poison to which people are commonly exposed, but it has been used to investigate the role of acetylcholine in the neuromuscular synapse and is sometimes used in small doses to relax muscles during certain kinds of surgery.

Myasthenia gravis (mī-as-thē′nē-ă grăv′is) results from the production of antibodies that bind to acetylcholine receptors, eventually causing the destruction of the receptor and thus reducing the number of receptors. As a consequence, muscles exhibit a degree of flaccid paralysis or are extremely weak. A class of drugs that includes neostigmine partially blocks the action of acetylcholinesterase and sometimes is used to treat myasthenia gravis. The drugs cause acetylcholine levels to increase in the synaptic cleft and combine more effectively with the remaining acetylcholine receptor sites.

P R E D I C T
Predict the specific cause of death resulting from a lethal dose of (a) organophosphate poison or (b) curare.

Cross-Bridge Movement

A cycle of events resulting in contraction proceeds very rapidly when the heads of the myosin molecules bind to actin to form cross-bridges (figure 9.15). The heads of myosin molecules move at their hinged region resulting in cross-bridge movement, which forces the actin myofilament, to which the heads of the myosin molecules are attached, to slide over the surface of the myosin myofilament. After cross-bridge movement, each myosin head releases from the actin and returns to its original position. It can then form another cross-bridge at a different site on the actin myofilament, followed by movement, release of the cross-bridge, and return to its original position. During a single contraction, each myosin molecule undergoes the cycle of cross-bridge formation, movement, release, and return to its original position many times.

The energy from one ATP molecule is required for each cycle of cross-bridge formation, movement, and release. Before a cross-bridge binds to the attachment site on an actin myofilament, the head of the myosin molecule is in its resting position, and ADP and phosphate are bound to the head of the myosin molecule (see figure 9.15). Once

Ca^{2+} bind to troponin and the tropomyosin moves, the attachment sites on actin myofilaments are then exposed. The head of a myosin molecule can then bind to an exposed attachment site, and the phosphate is released from the head of the myosin molecule. Energy stored in the head of the myosin molecule causes the head of the myosin molecule to move. Movement of the head causes the actin myofilament to slide past the myosin myofilament and ADP is released from the myosin head. ATP must then bind to the head of the myosin before the cross-bridge can release. As the ATP molecule binds to the head of the myosin molecule, the ATP is broken down to ADP and phosphate, and the myosin head releases from the attachment site on actin. The ADP and phosphate remain attached to the head of the myosin molecule, and the head of the myosin returns to its resting position. Energy released from the breakdown of ATP is stored in the head of the myosin molecule.

Movement of the myosin molecule while the cross-bridge is attached is called the **power stroke,** whereas return of the myosin head to its original position after cross-bridge release is called the **recovery stroke.** Many cycles of power and recovery strokes occur during each muscle contraction. While muscle is relaxed, energy stored in the heads of the myosin molecules is held in reserve until the next contraction. When Ca^{2+} is released from the sarcoplasmic reticulum in response to an action potential, the cycle of cross-bridge formation, movement, and release, which results in contraction, begins (see figures 9.14 and 9.15).

Muscle Relaxation

Relaxation occurs as a result of the active transport of Ca^{2+} back into the sarcoplasmic reticulum. As the Ca^{2+} concentration decreases in the sarcoplasm, the ions diffuse away from the troponin molecules. The troponin–tropomyosin complex then reestablishes its position, which blocks the active sites on the actin molecules. As a consequence, cross-bridges cannot re-form once they have been released, and relaxation occurs.

Energy is needed to make muscles contract, but it is also needed to make muscles relax. The active transport of Ca^{2+} into the sarcoplasmic reticulum requires ATP. The active transport processes that maintain the normal concentrations of Na^+ and K^+ across the sarcolemma also require ATP. The amount of ATP required for cross-bridge formation during contraction is much greater than the other energy requirements in a skeletal muscle.

27. *How does an action potential produced in the postsynaptic membrane of the neuromuscular junction eventually result in contraction of the muscle fiber?*
28. *Where in the contraction and relaxation processes is ATP required? Define power stroke and recovery stroke.*
29. *What conditions are required for relaxation of the muscle fiber?*

P R E D I C T
Predict the consequences of having the following conditions develop in a muscle in response to a stimulus: (a) Na^+ cannot enter the skeletal muscle through voltage-gated Na^+ channels, (b) very little ATP is present in the muscle fiber before a stimulus is applied, and (c) adequate ATP is present within the muscle fiber, but action potentials occur at a frequency so great that calcium is not transported back into the sarcoplasmic reticulum between individual action potentials.

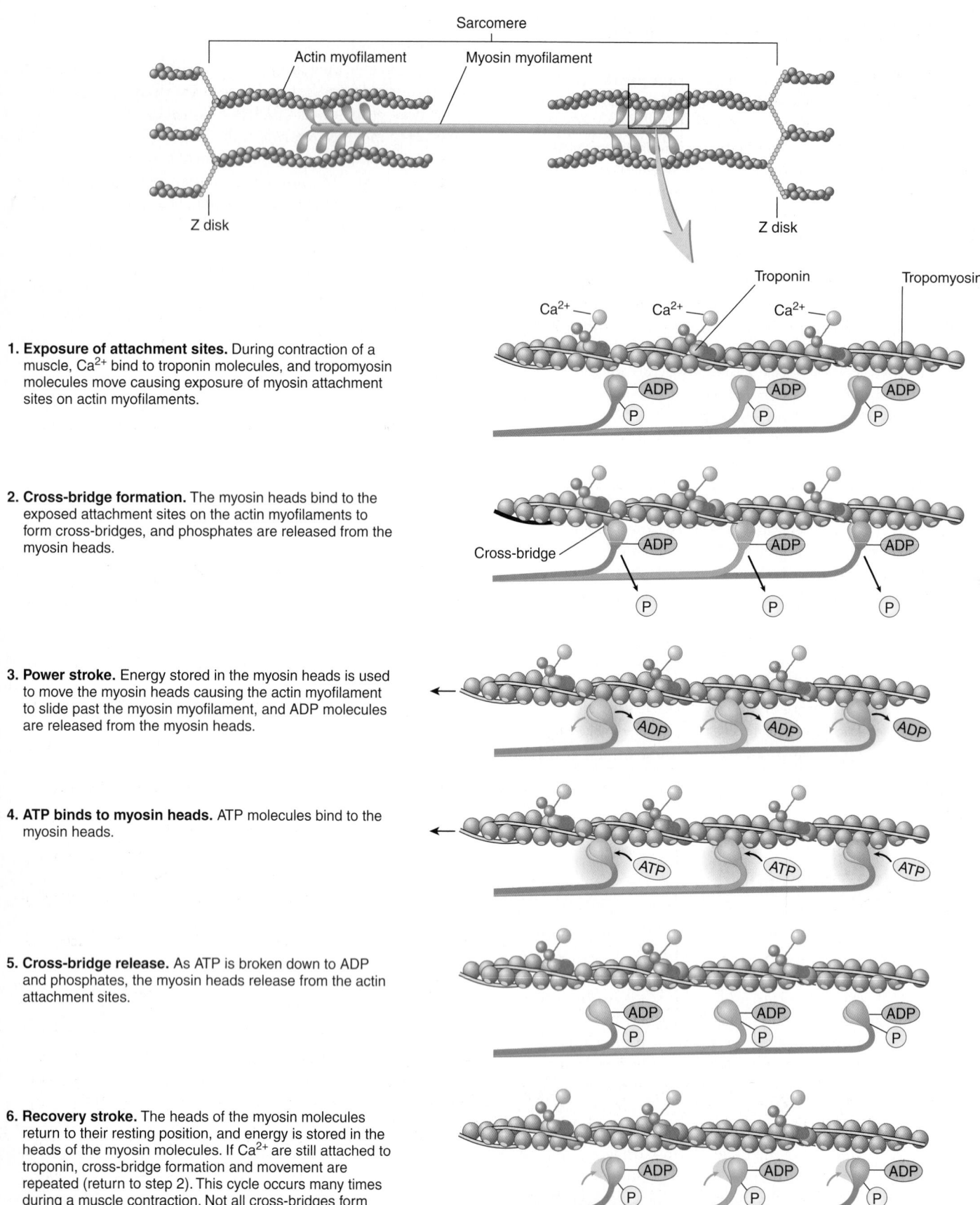

1. **Exposure of attachment sites.** During contraction of a muscle, Ca^{2+} bind to troponin molecules, and tropomyosin molecules move causing exposure of myosin attachment sites on actin myofilaments.

2. **Cross-bridge formation.** The myosin heads bind to the exposed attachment sites on the actin myofilaments to form cross-bridges, and phosphates are released from the myosin heads.

3. **Power stroke.** Energy stored in the myosin heads is used to move the myosin heads causing the actin myofilament to slide past the myosin myofilament, and ADP molecules are released from the myosin heads.

4. **ATP binds to myosin heads.** ATP molecules bind to the myosin heads.

5. **Cross-bridge release.** As ATP is broken down to ADP and phosphates, the myosin heads release from the actin attachment sites.

6. **Recovery stroke.** The heads of the myosin molecules return to their resting position, and energy is stored in the heads of the myosin molecules. If Ca^{2+} are still attached to troponin, cross-bridge formation and movement are repeated (return to step 2). This cycle occurs many times during a muscle contraction. Not all cross-bridges form and release simultaneously.

Process Figure 9.15   Breakdown of ATP and Cross-Bridge Movement During Muscle Contraction

Physiology of Skeletal Muscle

Muscle Twitch

A **muscle twitch** is the contraction of a muscle in response to a stimulus that causes an action potential in one or more muscle fibers. Even though the normal function of muscles is more complex, an understanding of the muscle twitch makes the function of muscles in living organisms easier to comprehend.

A hypothetical contraction of a single muscle fiber in response to a single action potential is illustrated in figure 9.16. The time between application of the stimulus to the motor neuron and the beginning of contraction is the **lag, or latent, phase;** the time during which contraction occurs is the **contraction phase;** and the time during which relaxation occurs is the **relaxation phase** (table 9.2). An action potential is an electrochemical event, but contraction is a mechanical event. An action potential is measured in millivolts and is completed in less than 2 milliseconds. Muscle contraction is measured as a force, also called tension. It is reported as the number of grams lifted, or the distance the muscle shortens, and requires up to 1 second to occur.

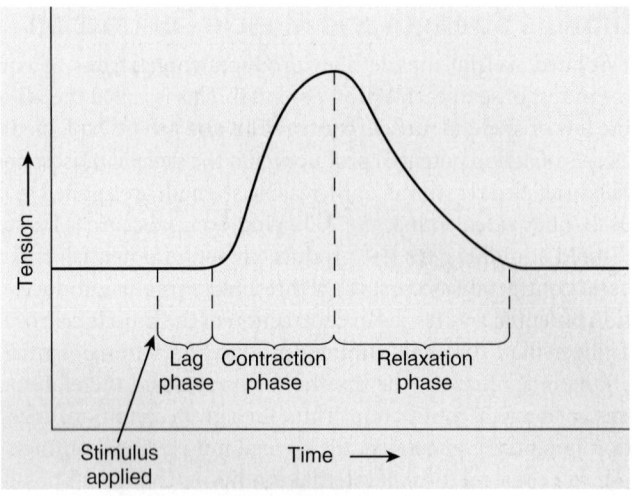

Figure 9.16   **Phases of a Muscle Twitch**

Hypothetical muscle twitch in a single muscle fiber. There is a short lag phase after stimulus application, followed by a contraction phase and a relaxation phase.

   30. *Define the phases of a muscle twitch and describe the events responsible for each phase.*

Table 9.2   Events That Occur During Each Phase of a Muscle Twitch*

Lag Phase

1. An action potential is propagated to the presynaptic terminal of the motor neuron.

2. The action potential causes the permeability of the presynaptic terminal to increase.

3. Calcium ions diffuse into the presynaptic terminal, causing acetylcholine contained within several synaptic vesicles to be released by exocytosis into the synaptic cleft.

4. Acetylcholine released from the presynaptic terminal diffuses across the synaptic cleft and binds to acetylcholine receptor molecules in the postsynaptic membrane of the sarcolemma.

5. The binding of acetylcholine to its receptor site causes ligand-gated Na^+ channels to open, and the postsynaptic membrane becomes more permeable to Na^+.

6. Sodium ions diffuse into the muscle fiber, causing a local depolarization that exceeds threshold and produces an action potential.

7. Acetycholine is rapidly degraded in the synaptic cleft to acetic acid and choline by acetylcholinesterase, thus limiting the length of time acetylcholine is bound to its receptor site. The result is that one presynaptic action potential produces one postsynaptic action potential in each muscle fiber.

8. The action potential produced in a muscle fiber is propagated from the postsynaptic membrane near the middle of the fiber toward both ends and into the T tubules.

9. The depolarization that occurs in the T tubule in response to the action potential causes Ca^{2+} channels of the membrane of the sarcoplasmic reticulum to open, and the membrane of the sarcoplasmic reticulum becomes very permeable to Ca^{2+}.

10. Calcium ions diffuse from the sarcoplasmic reticulum into the sarcoplasm.

11. Calcium ions bind to troponin; the troponin–tropomyosin complex changes its position and exposes the active site on the actin myofilaments.

Contraction Phase

12. Cross-bridges between actin molecules and myosin molecules form, move, release, and re-form many times, causing the sarcomeres to shorten. Energy stored in the head of the myosin molecule allows cross-bridge formation and movement. After cross-bridge movement has occurred, ATP must bind to the myosin head. The ATP is broken down to ADP, and some of the energy is used to release the cross-bridge and cause the head of the myosin molecule to move back to its resting position, where it is ready to form another cross-bridge. Some of the energy from the ATP is stored in the myosin head and is used for the next cross-bridge formation and movement (see figure 9.15). Energy is also released as heat.

Relaxation Phase

13. Calcium ions are actively transported into the sarcoplasmic reticulum.

14. The troponin–tropomyosin complexes inhibit cross-bridge formation.

15. The muscle fibers lengthen passively.

*Assuming that the process begins with a single action potential in the motor neuron.

Stimulus Strength and Muscle Contraction

An isolated skeletal muscle fiber produces contractions of equal force in response to each action potential. This is called the **all-or-none law of skeletal muscle contraction** and can be explained on the basis of action potential production in the skeletal muscle fiber. When brief electric stimuli of increasing strength are applied to the muscle fiber sarcolemma, the following events occur: (1) a subthreshold stimulus does not produce an action potential, and no muscle contraction occurs; (2) a threshold stimulus produces an action potential and results in contraction of the muscle cell; or (3) a stronger-than-threshold stimulus produces an action potential of the same magnitude as the threshold stimulus and therefore produces an identical contraction. Thus, for a given condition, once an action potential is generated, the skeletal muscle fiber contracts to produce a constant force. If internal conditions change, it's possible for the force of contraction to change as well. For example, increasing the amount of calcium available to the muscle cell results in a stronger force of contraction; conversely, muscle fatigue can result in a weaker force of contraction.

Within a given skeletal muscle, such as the biceps brachii (see chapter 10), skeletal muscle fibers form **motor units,** each of which consists of a single motor neuron and all of the muscle fibers it innervates (figure 9.17). Motor units respond as a single unit. All the muscle fibers of a motor unit contract to produce a constant force in response to a threshold stimulus because an action potential in a motor neuron initiates an action potential in each of the muscle fibers it innervates.

Whole muscles exhibit characteristics that are more complex than those of individual muscle fibers or motor units. Instead of responding in an all-or-none fashion, whole muscles respond to stimuli in a **graded fashion,** which means the strength of the contractions can range from weak to strong.

A muscle is composed of many motor units, and the axons of the motor units combine to form a nerve. A whole muscle contracts with either a small force or a large force, depending on the number of motor units stimulated to contract. This relationship is called **multiple motor unit summation** because the force of contraction increases as more and more motor units are stimulated. Multiple motor unit summation resulting in graded responses can be demonstrated by applying brief electric stimuli of increasing strength to the nerve supplying a muscle (figure 9.18). A **subthreshold stimulus** is not strong enough to cause an action potential in any of the axons in a nerve and causes no contraction. As the stimulus strength increases it eventually becomes a **threshold stimulus,** which is strong enough to produce an action potential in a single motor unit axon, causing all the muscle fibers of that motor unit to contract. Progressively stronger stimuli called **submaximal stimuli** produce action potentials in axons of additional motor units. A **maximal stimulus** produces action potentials in the axons of all motor units. Consequently, even greater stimulus strengths, **supramaximal stimuli** have no additional effect. As the stimulus strength increases between threshold and maximum values, motor units are **recruited,** which means that the number of motor units responding to the stimuli increases and the force of contraction produced by the muscle increases in a graded fashion.

Motor units in different muscles do not always contain the same number of muscle fibers. Muscles performing delicate and precise movements have motor units with a small number of muscle fibers, whereas muscles performing more powerful but less precise contractions have motor units with many muscle fibers. For example,

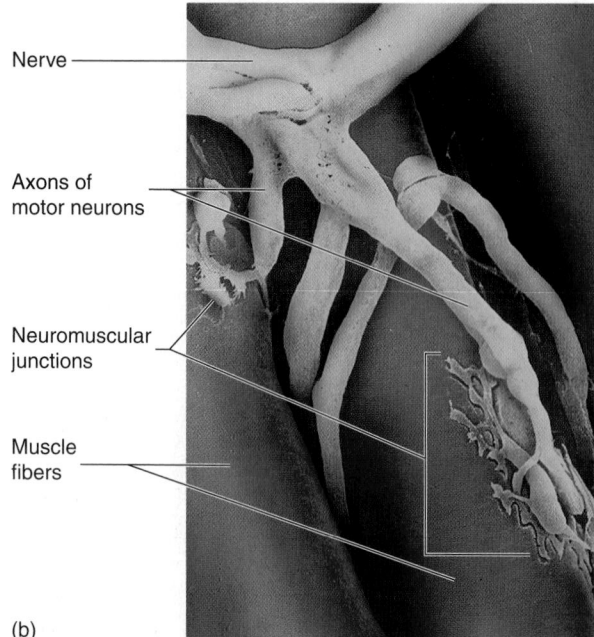

(b)

Figure 9.17   The Motor Unit

(*a*) A motor unit consists of a single motor neuron and all the muscle fibers its branches innervate. (*b*) Photomicrograph of motor units.

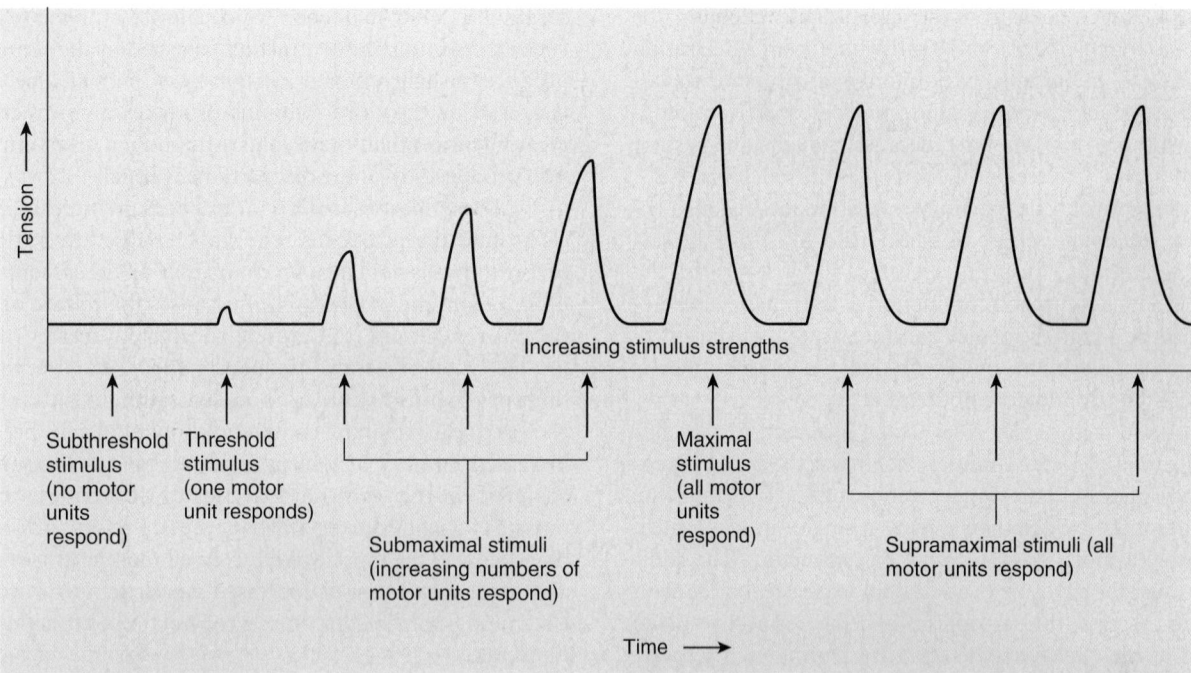

Figure 9.18 Multiple Motor Unit Summation

Multiple motor unit summation occurs as stimuli of increasing strength are applied to a nerve that innervates a muscle. The amount of tension (height of peaks) is influenced by the number of motor units responding.

in very delicate muscles, such as those that move the eye, the number of muscle fibers per motor unit can be fewer than 10, whereas in the heavy muscles of the thigh the number can be several hundred.

31. *Why does a single muscle fiber either not contract or contract with the same force in response to stimuli of different strengths?*
32. *Why does a motor unit either not contract or contract with the same force in response to stimuli of different strengths?*
33. *How does increasing the strength of a stimulus cause a whole muscle to respond in a graded fashion?*
34. *Define multiple motor unit summation.*

P R E D I C T ⑥
In patients with poliomyelitis (pō′lē-ō-mī′ĕ-lī′tis), motor neurons are destroyed, causing loss of muscle function and even flaccid paralysis. Sometimes recovery occurs because of the formation of axon branches from the remaining motor neurons. These branches innervate the paralyzed muscle fibers to produce motor units with many more muscle fibers than usual, resulting in the recovery of muscle function. What effect would this reinnervation of muscle fibers have on the degree of muscle control in a person who has recovered from poliomyelitis?

Stimulus Frequency and Muscle Contraction

An action potential in a single muscle fiber causes it to contract. Although the action potential triggers contraction of the muscle fiber, the action potential is completed long before the contraction phase is completed. In addition, the contractile mechanism in a muscle fiber exhibits no refractory period. That is, relaxation of a muscle fiber is not required before a second action potential can stimulate a second contraction. As the frequency of action potentials in a skeletal

muscle fiber increases, the frequency of contraction also increases (figure 9.19). In **incomplete tetanus** (tet′ă-nŭs), muscle fibers partially relax between the contractions, but in **complete tetanus** action potentials are produced so rapidly in muscle fibers that no muscle relaxation occurs between them. The tension produced by a muscle increases as the frequency of contractions increases. This increased tension is called **multiple-wave summation.**

Tetanus of a muscle caused by stimuli of increasing frequency can be explained by the effect of the action potentials on Ca^{2+} release from the sarcoplasmic reticulum. The first action

Figure 9.19 Multiple-Wave Summation

Stimuli 1–5 are stimuli of increasing frequency. For each stimulus, the up arrow indicates the start of stimulation, and the down arrow indicates the end of stimulation. Stimulus frequency 1 produces successive muscle twitches with complete relaxation between stimuli. Stimuli frequencies 2–4 do not allow complete relaxation between stimuli. Stimulus frequency 5 causes tetanus—no relaxation between stimuli.

potential causes Ca^{2+} release from the sarcoplasmic reticulum, the Ca^{2+} diffuse to the myofibrils, and contraction occurs. Relaxation begins as the Ca^{2+} are pumped back into the sarcoplasmic reticulum. If the next action potential occurs before relaxation is complete, two things happen. First, because enough time has not passed for all the Ca^{2+} to reenter the sarcoplasmic reticulum, Ca^{2+} levels around the myofibrils remain elevated. Second, the next action potential causes the release of additional Ca^{2+} from the sarcoplasmic reticulum. Thus, the elevated Ca^{2+} levels in the sarcoplasm produce continued contraction of the muscle fiber. Action potentials at a high frequency can increase Ca^{2+} concentrations in the sarcoplasm to an extent that the muscle fiber is contracted completely and does not relax at all.

At least two factors play roles in the increased tension observed during multiple-wave summation. First, as the action potential frequency increases, the concentration of Ca^{2+} around the myofibrils becomes greater than during a single muscle twitch, thereby causing a much greater degree of contraction. The additional Ca^{2+} cause the exposure of additional active sites on the actin myofilaments. Second, the sarcoplasm and the connective tissue components of muscle have some elasticity. During each separate muscle twitch, some of the tension produced by the contracting muscle fibers is used to stretch those elastic elements, and the remaining tension is applied to the load to be lifted. In a single muscle twitch, relaxation begins before the elastic components are totally stretched. The maximum tension produced during a single muscle twitch is therefore not applied to the load to be lifted. In a muscle stimulated at a high frequency, the elastic elements are stretched during the very early part of the prolonged contraction. After the elastic components are stretched, all of the tension produced by the muscle is applied to the load to be lifted, and the observed tension produced by the muscle is increased.

Another example of a graded response is **treppe** (trep′eh; staircase), which occurs in muscle that has rested for a prolonged period (figure 9.20). If the muscle is stimulated with a maximal

stimulus at a low frequency, which allows complete relaxation between the stimuli, the contraction triggered by the second stimulus produces a slightly greater tension than the first. The contraction triggered by the third stimulus produces a contraction with a greater tension than the second. After only a few stimuli, the tension produced by all the contractions is equal.

A possible explanation for treppe is an increase in Ca^{2+} levels around the myofibrils. The Ca^{2+} released in response to the first stimulus is not taken up completely by the sarcoplasmic reticulum before the second stimulus causes the release of additional Ca^{2+}, even though the muscle completely relaxes between the muscle twitches. As a consequence, during the first few contractions of the muscle, the Ca^{2+} concentration in the sarcoplasm increases slightly, making contraction more efficient because of the increased number of ions available to bind to troponin. Treppe achieved during warm-up exercises can contribute to improved muscle efficiency during athletic events. Factors such as increased blood flow to the muscle and increased muscle temperature probably are involved as well. Increased muscle temperature causes the enzymes responsible for muscle contraction to function at a more rapid rate.

35. *How does the lack of a refractory period in skeletal muscle fiber contraction explain multiple-wave summation?*
36. *Define incomplete tetanus and complete tetanus.*
37. *Give two reasons why rapid, repeated stimulation of a muscle fiber increases its force of contraction.*
38. *Describe treppe and explain how it occurs.*

Types of Muscle Contractions

Muscle contractions are classified based on the type of contraction that predominates (table 9.3). In **isometric** (ī-sō-met′rik) **contractions,** the length of the muscle doesn't change, but the amount of tension increases during the contraction process. Isometric contractions are responsible for the constant length of the postural muscles of the body such as muscles that hold the spine erect while a person is sitting or standing. In **isotonic** (ī-sō-ton′ik) **contractions,** the amount of tension produced by the muscle is constant during contraction, but the length of the muscle changes. Movements of the arms or fingers are predominantly isotonic contractions. Examples include waving or using a computer keyboard. Most muscle contractions are not strictly isometric or isotonic contractions. For example, both the length and tension of muscles change when a person walks or opens a heavy door. Although some mechanical differences do exist, both types of contractions result from the same contractile process within muscle cells.

Concentric (kon-sen′trik) **contractions** are isotonic contractions in which tension in the muscle is great enough to overcome the opposing resistance, and the muscle shortens. Concentric contractions include contractions that result in an increasing tension as the muscle shortens. A large percentage of the movements performed by muscle contractions are concentric contractions. **Eccentric** (ek-sen′trik) **contractions** are isotonic contractions in which tension is maintained in a muscle, but the opposing resistance is great enough to cause the muscle to increase in length (see table 9.3). Eccentric contractions are performed when a person

Figure 9.20 Treppe

When a rested muscle is stimulated repeatedly with maximal stimuli at a frequency that allows complete relaxation between stimuli, the second contraction produces a slightly greater tension than the first, and the third contraction produces greater tension than the second. After a few contractions, the tension produced by all contractions is equal.

Table 9.3	Types of Muscle Contractions
Contraction Types	**Characteristics**
Multiple motor unit summation	Each motor unit responds in an all-or-none fashion.
	A whole muscle is capable of producing an increasing amount of tension as the number of motor units stimulated increases.
Multiple-wave summation	Summation results when many action potentials are produced in a muscle fiber.
	Contraction occurs in response to the first action potential, but there's not enough time for relaxation to occur between action potentials.
	Because each action potential causes the release of Ca^{2+} from the sarcoplasmic reticulum, the ions remain elevated in the sarcoplasm to produce a tetanic contraction.
	The tension produced as a result of multiple-wave summation is greater than the tension produced by a single muscle twitch. The increased tension results from the greater concentration of Ca^{2+} in the sarcoplasm and the stretch of the elastic components of the muscle early in contraction.
Tetanus of muscles	Tetanus of muscles results from multiple-wave summation.
	Incomplete tetanus occurs when the action potential frequency is low enough to allow partial relaxation of the muscle fibers.
	Complete tetanus occurs when the action potential frequency is high enough that no relaxation of the muscle fibers occurs.
Treppe	Tension produced increases for the first few contractions in response to a maximal stimulus at a low frequency in a muscle that has been at rest for some time.
	Increased tension may result from the accumulation of small amounts of Ca^{2+} in the sarcoplasm for the first few contractions or from an increasing rate of enzyme activity.
Isotonic contraction	A muscle produces a constant tension during contraction.
	A muscle shortens during contraction.
	This type is characteristic of finger and hand movements.
Isometric contraction	A muscle produces an increasing tension during contraction.
	The length of a muscle remains constant during contraction.
	This type is characteristic of postural muscles that maintain a constant tension without changing their length.
Concentric contractions	A muscle produces increasing tension as it shortens.
Eccentric contractions	A muscle produces tension, but the length of the muscle is increasing.

slowly lowers a heavy weight. During eccentric contractions, muscles produce substantial force. Eccentric contractions are of clinical interest because repetitive eccentric contractions, such as seen in the lower limbs of people who run downhill for long distances, tend to injure muscle fibers and the connective tissue of muscles.

Muscle tone refers to the constant tension produced by muscles of the body for long periods of time. Muscle tone is responsible for keeping the back and legs straight, the head upright, and the abdomen flat. Muscle tone depends on a small percentage of all the motor units contracting out of phase with one another at any point in time. The same motor units are not contracting all the time, however. A small percentage of motor units is stimulated with a frequency of nerve impulses that causes incomplete tetanus for short periods. The motor units that are contracting are stimulated in such a way that the tension produced by the whole muscle remains constant.

P R E D I C T   7

Mary Myosin overheard an argument between two students who could not decide if a weight lifter who lifts a weight above the head and then holds it there before lowering it is using isometric, isotonic, concentric, or eccentric muscle contractions. Mary is an expert on muscle contractions, so she settles the debate. What was her explanation?

Movements of the body are usually smooth and occur at widely differing rates—some very slowly and others quite rapidly. Most movements are produced by muscle contractions, but very few of the movements resemble the rapid contractions of individual muscle twitches. Smooth, slow contractions result from an increasing number of motor units contracting out of phase as the muscles shorten, and from a decreasing number of motor units contracting out of phase as muscles lengthen. Each individual motor unit exhibits either incomplete or complete tetanus, but because the contractions are out of phase and because the number of motor units activated varies at each point in time, a smooth contraction results. Consequently, muscles are capable of contracting either slowly or rapidly, depending on the number of motor units stimulated and the rate at which that number increases or decreases.

Length Versus Tension

Active tension is the force applied to an object to be lifted when a muscle contracts. The initial length of a muscle has a strong influence on the amount of active tension it produces. As the length of a muscle increases, its active tension also increases, to a point. If the muscle is stretched farther than that optimum length, the active tension it produces begins to decline. The muscle length plotted against the tension produced by the muscle in response to maximal stimuli is the **active tension curve** (figure 9.21).

If a muscle is stretched so that the actin and myosin myofilaments within the sarcomeres do not overlap or overlap to a very small extent, the muscle produces very little active tension when it is stimulated. Also, if the muscle is not stretched at all,

the myosin myofilaments touch each of the Z disks in each sarcomere, and very little contraction of the sarcomeres can occur. If the muscle is stretched to its optimum length, optimal overlap of the actin and myosin myofilaments takes place. When the muscle is stimulated, cross-bridge formation results in maximal contraction.

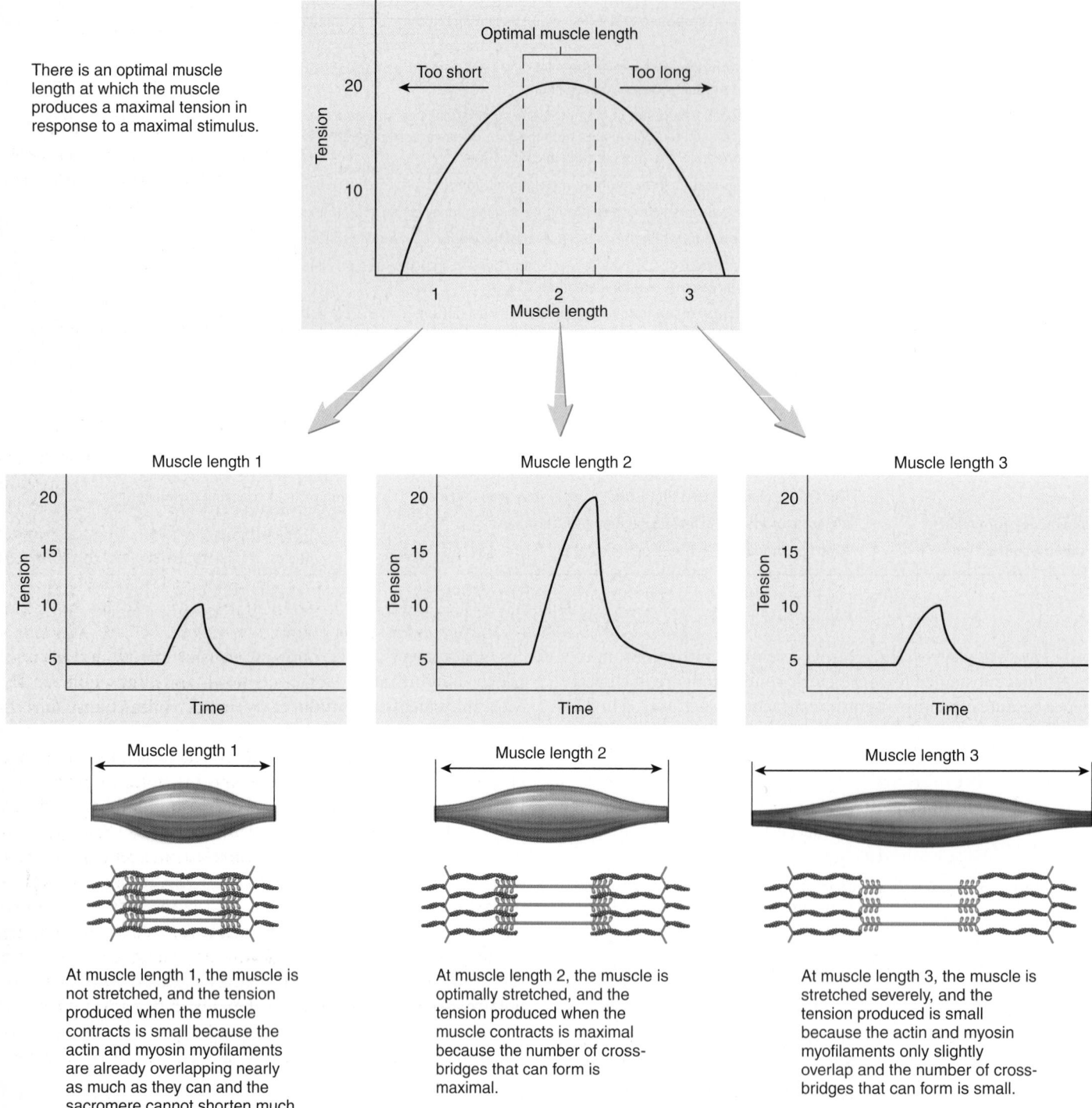

There is an optimal muscle length at which the muscle produces a maximal tension in response to a maximal stimulus.

At muscle length 1, the muscle is not stretched, and the tension produced when the muscle contracts is small because the actin and myosin myofilaments are already overlapping nearly as much as they can and the sarcomere cannot shorten much more.

At muscle length 2, the muscle is optimally stretched, and the tension produced when the muscle contracts is maximal because the number of cross-bridges that can form is maximal.

At muscle length 3, the muscle is stretched severely, and the tension produced is small because the actin and myosin myofilaments only slightly overlap and the number of cross-bridges that can form is small.

Figure 9.21 Muscle Length and Tension

The length of a muscle, before it is stimulated, influences the force of contraction of the muscle.

Passive tension is the tension applied to the load when a muscle is stretched but not stimulated. It is similar to the tension produced if the muscle is replaced with an elastic band. Passive tension exists because the muscle and its connective tissue have some elasticity. The sum of active and passive tension is called **total tension.**

39. *Define isometric, isotonic, concentric, and eccentric contractions.*
40. *What is muscle tone, and how is it maintained?*
41. *How are smooth, slow contractions produced in muscles?*
42. *Draw an active tension curve. How does the overlap of actin and myosin explain the shape of the curve?*

Weight Lifters and Muscle Length

Weight lifters and others who lift heavy objects usually assume positions so that their muscles are stretched close to their optimum length before lifting. For example, the position a weight lifter assumes before power lifting stretches the arm and leg muscles to a near-optimum length for muscle contraction, and the stance a lineman assumes in a football game stretches most muscle groups in the legs so they are near their optimum length for suddenly moving the body forward.

Fatigue

Fatigue (fă-tēg′) is the decreased capacity to do work and the reduced efficiency of performance that normally follows a period of activity. The rate at which individuals develop fatigue is highly variable, but it's a phenomenon that everyone has experienced. Fatigue can develop at three possible sites: the nervous system, the muscles, and the neuromuscular junction.

Psychologic fatigue, the most common type of fatigue, involves the central nervous system. The muscles are capable of functioning, but the individual "perceives" that additional muscular work is not possible. A burst of activity in a tired athlete in response to encouragement from spectators is an illustration of how psychologic fatigue can be overcome. The onset and duration of psychologic fatigue vary greatly and depend on the emotional state of the individual.

The second most common type of fatigue occurs in the muscle fiber. **Muscular fatigue** results from ATP depletion. Without adequate ATP levels in muscle fibers, cross-bridges cannot function normally. As a consequence, the tension that a muscle is capable of producing declines. Fatigue in the lower limbs of marathon runners or in the upper and lower limbs of swimmers are examples.

The least common type of fatigue, called **synaptic fatigue,** occurs in the neuromuscular junction. If the action potential frequency in motor neurons is great enough, the release of acetylcholine from the presynaptic terminals is greater than the rate of acetylcholine synthesis. As a result, the synaptic vesicles become depleted, and insufficient acetylcholine is released to stimulate the muscle fibers. Under normal physiologic conditions, fatigue of neuromuscular junctions is rare; however, it may occur under conditions of extreme exertion.

Muscle Soreness Resulting from Exercise

Pain frequently develops after 1 or 2 days in muscles that are vigorously exercised, and the pain can last for several days. The pain is more common in untrained people who exercise vigorously. In addition, highly repetitive eccentric contractions of muscles produce pain more readily than concentric contractions. The pain is associated with damage to skeletal muscle fibers and with connective tissue surrounding the skeletal muscle fibers. In people with muscle soreness induced by exercise, enzymes that are normally found inside muscle fibers can be detected in the extracellular fluid. Enzymes are able to leave the muscle fibers because injury causes an increase in the permeability or rupture of the plasma membrane. In addition, fragments of collagen molecules can be found in the extracellular fluid of muscles. These observations indicate that injury occurs to both muscle fibers and the connective tissue of muscles. The pain produced appears to be the result of inflammation resulting from damage to muscle fibers and the connective tissue. Exercise schedules that alternate exercise with periods of rest, such as lifting weights every other day, provide time for repair to occur in muscle tissue.

Physiologic Contracture and Rigor Mortis

As a result of extreme muscular fatigue, muscles occasionally become incapable of either contracting or relaxing—a condition called **physiologic contracture** (kon-trak′choor), which is caused by a lack of ATP within the muscle fibers. ATP can decline to very low levels when a muscle is stimulated strongly, such as under conditions of extreme exercise. When ATP levels are very low, active transport of Ca^{2+} into the sarcoplasmic reticulum slows, Ca^{2+} accumulate within the sarcoplasm, and ATP is unavailable to bind to the myosin molecules that have formed cross-bridges with the actin myofilaments. As a consequence, the previously formed cross-bridges cannot release, resulting in physiologic contracture.

Rigor mortis (rig′er mōr′tĭs) is the development of rigid muscles several hours after death and is similar to physiologic contracture. ATP production stops shortly after death, and ATP levels within muscle fibers decline. Because of low ATP levels, active transport of Ca^{2+} into the sarcoplasmic reticulum stops, and Ca^{2+} leak from the sarcoplasmic reticulum into the sarcoplasm. Ca^{2+} can also leak from the sarcoplasmic reticulum as a result of the breakdown of the sarcoplasmic reticulum membrane after cell death. As Ca^{2+} levels increase in the sarcoplasm, cross-bridges form. Too little ATP is available to bind to the myosin molecules, however, so the cross-bridges are unable to release and re-form in a cyclic fashion to produce contractions. As a consequence, the muscles remain stiff until tissue degeneration occurs.

43. *Define the term fatigue, and list three locations in which fatigue can develop.*
44. *Define and explain the cause of physiologic contracture and rigor mortis.*

Energy Sources

ATP provides the immediate source of energy for muscle contractions. As long as adequate amounts of ATP are present, muscles can contract repeatedly for a long time. ATP must be synthesized continuously to sustain muscle contractions, and ATP synthesis must be equal to ATP breakdown because the small amount of ATP stored in muscle fibers is sufficient to support vigorous muscle contractions for only a few seconds. The energy required to produce ATP comes from three sources: (1) creatine phosphate, (2) anaerobic respiration, and (3) aerobic respiration. Only the main points of anaerobic respiration and aerobic respiration are considered here (a more detailed discussion can be found in chapter 25).

Creatine Phosphate

During resting conditions, energy from aerobic respiration is used to synthesize **creatine** (krē′ă-tēn, krē′ă-tin) **phosphate.** Creatine phosphate accumulates in muscle cells and functions to store energy, which can be used to synthesize ATP. As ATP levels begin to fall, ADP reacts with creatine phosphate to produce ATP and creatine.

$$\text{ADP} + \text{Creatine phosphate} \rightarrow \text{Creatine} + \text{ATP}$$

The reaction occurs very rapidly and is able to maintain ATP levels as long as creatine phosphate is available in the cell. During intense muscular contraction, however, creatine phosphate levels are quickly exhausted. ATP and creatine phosphate present in the cell provide enough energy to sustain maximum contractions for about 8–10 seconds.

Anaerobic Respiration

Anaerobic (an-ār-ō′bik) **respiration** occurs in the absence of oxygen and results in the breakdown of glucose to yield ATP and lactic acid. For each molecule of glucose metabolized, a net production of two ATP molecules and two molecules of lactic acid occurs. The first part of anaerobic metabolism and aerobic metabolism are common to each other. In both cases, each glucose molecule is broken down into two molecules of pyruvic acid. Two molecules of ATP are used in this process, but four molecules of ATP are produced, resulting in a net gain of two ATP molecules for each glucose molecule metabolized. In anaerobic metabolism, the pyruvic acid is then converted to lactic acid. Unlike pyruvic acid, much of the lactic acid diffuses out of the muscle fibers into the bloodstream.

Anaerobic respiration is less efficient than aerobic respiration, but it's faster, especially when oxygen availability limits aerobic respiration. By using many glucose molecules, anaerobic respiration can rapidly produce ATP for a short time. During short periods of intense exercise, such as sprinting, anaerobic respiration combined with the breakdown of creatine phosphate provides enough ATP to support intense muscle contraction for up to 3 minutes. ATP formation from creatine phosphate and anaerobic metabolism is limited by depletion of creatine phosphate and glucose and the buildup of lactic acid within muscle fibers.

Aerobic Respiration

Aerobic (ār-ō′bik) **respiration** requires oxygen and breaks down glucose to produce ATP, carbon dioxide, and water. Compared to anaerobic respiration, aerobic respiration is much more efficient. The metabolism of a glucose molecule by anaerobic respiration produces a net gain of two ATP molecules for each glucose molecule. In contrast, aerobic respiration can produce up to 38 ATP molecules for each glucose molecule. In addition, aerobic respiration uses a greater variety of molecules as energy sources, such as fatty acids and amino acids. Some glucose is used as an energy source in skeletal muscles, but fatty acids provide a more important source of energy during sustained exercise and during resting conditions.

In aerobic respiration, pyruvic acid is metabolized by chemical reactions within mitochondria. Two closely coupled sequences of reactions in mitochondria, called the citric acid cycle and the electron-transport chain, produce many ATP molecules. Carbon dioxide molecules are produced and, in the last step, oxygen atoms combine with hydrogen atoms to form water. Thus carbon dioxide, water, and ATP are major end products of aerobic metabolism. The following equation represents aerobic respiration of one molecule of glucose:

$$\text{Glucose} + 6\,O_2 + 38\,\text{ADP} + 38\,\text{P} \rightarrow$$
$$6\,CO_2 + 6\,H_2O + \text{About 38 ATP}$$

Although aerobic metabolism produces many more ATP molecules for each glucose molecule metabolized than does anaerobic metabolism, the rate at which the ATP molecules are produced is slower. Resting muscles or muscles undergoing long-term exercise, such as long-distance running or other endurance exercises, depend primarily on aerobic respiration for ATP synthesis.

Oxygen Debt

After intense exercise, the rate of aerobic metabolism remains elevated for a time. The oxygen taken in by the body, above that required for resting metabolism after exercise, is called the **oxygen debt.** It represents the difference between the amount of oxygen needed for aerobic respiration during muscle activity and the amount that actually was used. ATP produced by anaerobic sources and used during muscle activity contributes to the oxygen debt. The increased aerobic metabolism after exercise reestablishes normal ATP and creatine phosphate levels in muscle fibers. It also converts excess lactic acid to pyruvic acid and then to glucose, primarily in the liver. The glucose is used to help restore glycogen levels in muscle fibers and in liver cells.

Anaerobic Exercise and Oxygen Debt

During brief but intense exercise, such as during a sprint, much of the ATP used by exercising muscles comes from the conversion of creatine phosphate to creatine and from anaerobic respiration. Glycogen is broken down to glucose in the skeletal muscle fibers and in the liver. Glucose is released from the liver into the circulatory system and can be taken up by skeletal muscle fibers. Anaerobic respiration converts the glucose molecules to ATP and lactic acid. Heavy breathing and elevated aerobic respiration after the race results from the oxygen debt. The increased aerobic respiration pays back the oxygen debt by converting creatine to creatine phosphate and converting the excess lactic acid to glucose, which is then stored as glycogen in muscles and in the liver once again. The magnitude of the oxygen debt depends on the intensity of the exercise, the length of time it was sustained, and the physical condition of the individual. Those who are in poor physical condition do not have as great a capacity as well-trained athletes do to perform aerobic metabolism.

45. *List the energy sources used to synthesize ATP for muscle contraction.*
46. *Contrast the efficiency of aerobic and anaerobic respiration. When is each type used by cells?*
47. *What is the function of creatine phosphate and when is it used?*
48. *When does lactic acid increase in a muscle cell?*
49. *Define oxygen debt. What does the body do as it "repays" the oxygen debt?*

P R E D I C T

After a 10 km run with a sprint at the end, a runner continues to breathe heavily for a time. Compare the function of the elevated metabolic processes during the run, near the end, and shortly after the run.

Slow and Fast Fibers

Not all skeletal muscles have identical functional capabilities. They differ in several respects including having muscle fibers that contain slightly different forms of myosin. The myosin of slow-twitch muscle fibers causes these fibers to contract more slowly, and these cells are more resistant to fatigue, whereas the myosin of fast-twitch muscle fibers cause these fibers to contract quickly and these cells fatigue quickly (table 9.4). The proportion of muscle fiber types differs within individual muscles.

Slow-Twitch, or High-Oxidative, Muscle Fibers

Slow-twitch, high-oxidative, or **type I muscle fibers,** contract more slowly, are smaller in diameter, have a better developed blood supply, have more mitochondria, and are more fatigue-resistant than fast-twitch muscle fibers. Slow-twitch muscle fibers respond relatively slowly to nervous stimulation and break down ATP at a limited rate within the heads of their myosin molecules. Aerobic respiration is the primary source for ATP synthesis in slow-twitch muscles, and their capacity to perform aerobic respiration is enhanced by a plentiful blood supply and the presence of numerous mitochondria. They are sometimes called high-oxidative muscle fibers because of their enhanced capacity to carry out aerobic respiration. Slow-twitch fibers also contain large amounts of **myoglobin** (mī-ō-glō′bin), a dark pigment similar to hemoglobin, which binds oxygen and acts as a reservoir for it when the blood does not supply an adequate amount. Myoglobin thus enhances the capacity of the muscle fibers to perform aerobic respiration.

Fast-Twitch, or Low-Oxidative, Muscle Fibers

Fast-twitch, low-oxidative, or **type II muscle fibers,** respond rapidly to nervous stimulation and contain myosin molecules that break down ATP more rapidly than do slow-twitch muscle fibers. This allows their cross-bridges to form, release, and re-form more rapidly than those in slow-twitch muscle fibers. Muscles containing a high percentage of these fibers have a less well-developed blood supply than muscles containing a high percentage of slow-twitch muscle fibers. In addition, fast-twitch muscle fibers have very little myoglobin and fewer and smaller mitochondria. Fast-twitch muscle fibers have large deposits of glycogen and are well adapted to perform anaerobic respiration. The anaerobic respiration of fast-twitch muscle fibers, however, is not adapted for supplying a large amount of ATP for a prolonged period. The muscle fibers tend to contract rapidly for a shorter time and fatigue relatively quickly. Fast-twitch muscle fibers exist in two forms, type IIx and type IIa. Type IIx muscle fibers are classical fast-twitch muscle fibers. Type IIa fibers contain a different form of myosin which breaks ATP down more slowly and contracts more slowly. In addition these muscle fibers are more resistant to fatigue than type IIx fibers.

Distribution of Fast-Twitch and Slow-Twitch Muscle Fibers

The muscles of many animals are composed primarily of either fast-twitch or slow-twitch muscle fibers. The white meat of a chicken or pheasant breast, which is composed mainly of fast-twitch fibers,

Table 9.4	Characteristics of Skeletal Muscle Fiber Types		
Characteristics	**Slow-Twitch High-Oxidative Type I**	**Fast-Twitch Low-Oxidative Type IIa**	**Low-Oxidative Type IIx**
Fiber diameter	Smallest	Intermediate	Largest
Myoglobin content	High	Intermediate	Low
Mitochondria	Many	Intermediate	Few
Capillaries	Many	Intermediate	Few
Metabolism	High aerobic capacity	Intermediate aerobic capacity	Low aerobic capacity
	Low anaerobic capacity	High anaerobic capacity	Highest anaerobic capacity
Fatigue	Resistant	Resistant	Not resistant
Rate of ATP breakdown by ATPase in myosin	Slow	Fast	Fastest
Glycogen concentration	Low	High	High
Location where fibers are numerous (largely genetically determined)	Generally postural muscles and more in lower than upper limbs (e.g., in endurance athletes)	Can predominate in lower limbs (e.g., in athletes who are sprinters)	Upper limbs (more in upper than lower limbs and more in legs of athletes who are sprinters)
Functions	Endurance activities and posture	Endurance activities in endurance-trained muscles	Rapid, intense movements of short duration

appears whitish because of its relatively poor blood supply and lack of myoglobin. The muscles are adapted to contract rapidly for a short time but fatigue quickly. The red, or dark, meat of a chicken leg or of a duck breast is composed of slow-twitch fibers and appears darker because of the relatively well-developed blood supply and a large amount of myoglobin. These muscles are adapted to contract slowly for a longer time and to fatigue slowly. The distribution of slow-twitch and fast-twitch muscle fibers is consistent with the behavior of these animals. For example, pheasants can fly relatively fast for short distances, and ducks fly more slowly for long distances.

Humans exhibit no clear separation of slow-twitch and fast-twitch muscle fibers in individual muscles. Most muscles have both types of fibers, although the number of each varies for each muscle. The large postural muscles contain more slow-twitch fibers, whereas muscles of the upper limbs contain more fast-twitch fibers.

The distribution of slow-twitch and fast-twitch muscle fibers in a given muscle is fairly constant for each individual and apparently is established developmentally. People who are good sprinters have a greater percentage of fast-twitch muscle fibers, whereas good long-distance runners have a higher percentage of slow-twitch muscle fibers in their leg muscles. Athletes who are able to perform a variety of anaerobic and aerobic exercises tend to have a more balanced mixture of fast-twitch and slow-twitch muscle fibers.

> 50. *Contrast the structural and functional differences between slow-twitch and fast-twitch muscle fibers.*
> 51. *Explain the functions for which each type is best adapted and how they are distributed.*

Effects of Exercise

Neither fast-twitch nor slow-twitch muscle fibers can be easily converted to muscle fibers of the other type without specialized training. Training can increase the size and the capacity of both types of muscle fibers to perform more efficiently. Intense exercise resulting in anaerobic metabolism such as weight lifting increases muscular strength and mass and results in an increased enlargement of fast-twitch muscle fibers more than slow-twitch muscle fibers. Aerobic exercise increases the vascularity of muscle and causes enlargement of slow-twitch muscle fibers. Aerobic metabolism can convert some fast-twitch muscle fibers that fatigue readily (type IIx) to fast-twitch muscle fibers that resist fatigue (type IIa). Aerobically trained fast-twitch muscle, with more type IIa fibers, can be called **fatigue-resistant fast-twitch muscles.** In addition to changes in myosin, there is an increase in the number of mitochondria in the muscle cells, and an increase in their blood supply. Weight training followed by periods of rest can convert some muscle fibers from type IIa to type IIx. Through training, a person with more fast-twitch muscle fibers can run long distances, and a person with more slow-twitch muscle fibers can increase the speed at which he or she runs.

P R E D I C T   9

What kind of exercise regimen is appropriate for people who are training to be endurance runners? What effect will the composition of their muscles, in terms of muscle fiber type, have on their ability to perform in an endurance race?

A muscle increases in size, or **hypertrophies** (hī-per′trō-fēz), and increases in strength and endurance in response to exercise. Conversely, a muscle that is not used decreases in size, or **atrophies** (at′rō-fēz). The muscular atrophy that occurs in limbs placed in casts for several weeks is an example. Because muscle cell numbers don't change appreciably during most of a person's life, atrophy and hypertrophy of muscles result from changes in the size of individual muscle fibers. As fibers increase in size, the number of myofibrils and sarcomeres, increases within each muscle fiber. The number of nuclei in each muscle cell increases in response to exercise, but the nuclei of muscle cells cannot divide. New nuclei are added to muscle fibers because small satellite cells near skeletal muscle cells increase in number in response to exercise and then fuse with the skeletal muscle cells. Other elements, such as blood vessels, connective tissue, and mitochondria, also increase. Atrophy of muscles due to lack of exercise results from a decrease in all of these elements without a decrease in muscle cell number. Severe atrophy, such as occurs in elderly people who cannot readily move their limbs, however, does involve an irreversible decrease in the number of muscle cells and can lead to paralysis.

The increased strength of trained muscle is greater than would be expected if that strength were based only on the change in muscle size. Part of the increase in strength results from the ability of the nervous system to recruit a large number of motor units simultaneously in a trained person to perform movements with better neuromuscular coordination. In addition, trained muscles usually are restricted less by excess fat. Metabolic enzymes increase in hypertrophied muscle fibers, resulting in a greater capacity for nutrient uptake and ATP production. Improved endurance in trained muscles is in part a result of improved metabolism, increased circulation to the exercising muscles, increased numbers of capillaries, more efficient respiration, and a greater capacity for the heart to pump blood.

Anabolic Steroids and Growth Hormone

Some people take synthetic hormones called **anabolic steroids** (an-ă-bol′ik stēr′oydz, ster′oydz) to increase the size and strength of their muscles. Anabolic steroids are related to testosterone, a reproductive hormone secreted by the testes, except that they have been altered so that the reproductive effects of these compounds are minimized, but their effect on skeletal muscles is maintained. Testosterone and anabolic steroids cause skeletal muscle tissue to hypertrophy. People who take large doses of an anabolic steroid exhibit an increase in body weight and total skeletal muscle mass, and many athletes believe that anabolic steroids improve performance that depends on strength. Unfortunately, evidence indicates that harmful side effects are associated with taking anabolic steroids, including periods of irritability, testicular atrophy and sterility, cardiovascular diseases such as heart attack or stroke, and abnormal liver function. Most athletic organizations prohibit the use of anabolic steroids, and some even analyze urine samples either randomly or periodically for evidence of their use. Penalties exist for athletes who have evidence of anabolic steroid metabolites in their urine.

Growth hormone is also used inappropriately to increase muscle size by some individuals. Growth hormone increases protein synthesis in muscle tissue although it doesn't produce the same kinds of side effects as those produced by anabolic steroids. The large doses of growth hormone used by athletes, however, can cause harmful side effects if taken over a long period (see chapter 18).

52. *What factors contribute to an increase in muscle size, strength, and endurance?*
53. *How does anaerobic versus aerobic exercise affect muscles?*

Heat Production

The rate of metabolism in skeletal muscle differs before, during, and after exercise. As chemical reactions occur within cells, some energy is released in the form of heat. Normal body temperature results in large part from this heat. Because the rate of chemical reactions increases in muscle fibers during contraction, the rate of heat production also increases, causing an increase in body temperature. After exercise, elevated metabolism resulting from the oxygen debt helps keep the body temperature elevated. If the body temperature increases as a result of increased contraction of skeletal muscle, vasodilation of blood vessels in the skin and sweating function to speed heat loss and keep the body temperature within its normal range (see chapter 25).

When the body temperature declines below a certain level, the nervous system responds by inducing **shivering,** which involves rapid skeletal muscle contractions that produce shaking rather than coordinated movements. The muscle movement increases heat production up to 18 times that of resting levels, and the heat produced during shivering can exceed that produced during moderate exercise. The elevated heat production during shivering helps raise the body temperature to its normal range.

54. *How do muscles contribute to the heat responsible for body temperature before, during, and after exercise? What is accomplished by shivering?*

Smooth Muscle

Smooth muscle is distributed widely throughout the body and is more variable in function than other muscle types. Smooth muscle cells (figure 9.22) are smaller than skeletal muscle cells, ranging from 15 to 500 μm in length and from 5 to 10 μm in diameter. They are spindle-shaped, with a single nucleus located in the middle of the cell. Compared to skeletal muscle, fewer actin and myosin myofilaments are present, and there are more actin than myosin myofilaments. The actin and myosin myofilaments overlap, but they are organized as loose bundles. Consequently, smooth muscle doesn't have a striated appearance. Actin myofilaments are attached to **dense bodies,** which are scattered through the cell cytoplasm, and to **dense areas,** which are in the plasma membrane. Dense bodies and dense areas are considered to be equivalent to the Z disks in skeletal muscle. Noncontractile **intermediate filaments** also attach to the dense bodies. The intermediate filaments and dense bodies form an intracellular cytoskeleton that has a longitudinal or spiral organization. The smooth muscle cells shorten when the actin and myosin slide over one another during contraction (figure 9.23).

Sarcoplasmic reticulum is present in smooth muscle cells, but no T tubule system exists. Some shallow invaginated areas called **caveolae** (kav-ē-ō′lē) are along the surface of the plasma membrane. The function of caveolae is not well known, but it may be

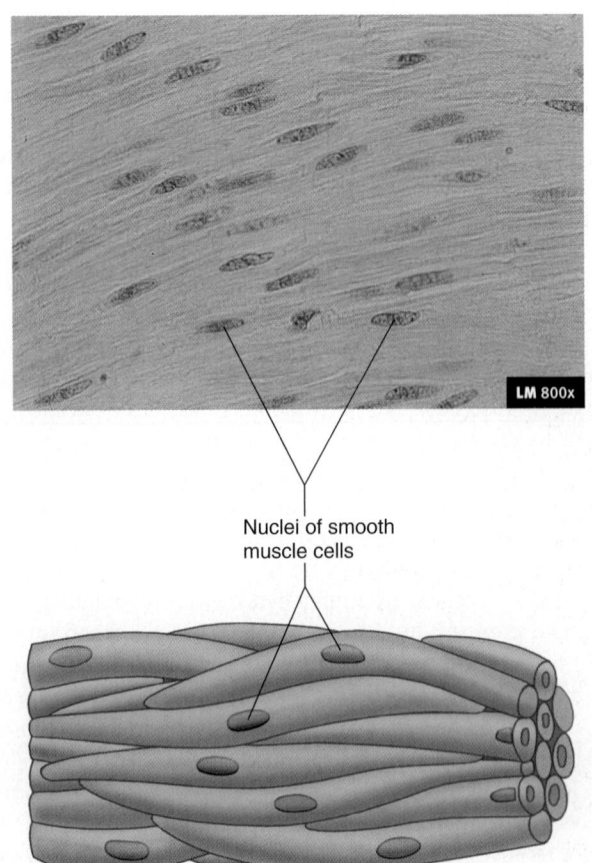

Figure 9.22 **Smooth Muscle Histology**
Smooth muscle tissue is made up of sheets or bundles of spindle-shaped cells, with a single nucleus located in the middle of each cell.

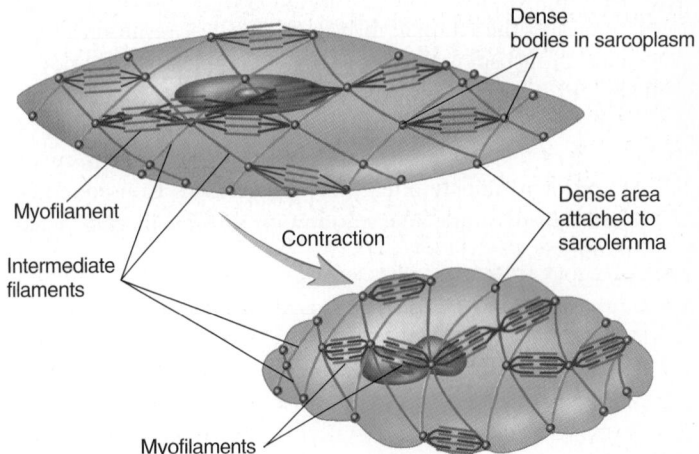

Figure 9.23 **Contractile Proteins in a Smooth Muscle Cell**
Bundles of contractile myofilaments containing actin and myosin are anchored at one end to dense areas in the plasma membrane and at the other end, through dense bodies, to intermediate filament. The contractile myofilaments are oriented with the long axis of the cell, and when actin and myosin slide over one another during contraction, the cell shortens.

similar to that of both the T tubules and the sarcoplasmic reticulum of skeletal muscle.

Some of the Ca^{2+} required to initiate contractions in smooth muscle enters the cell from the extracellular fluid and from the sarcoplasmic reticulum. The greater distance Ca^{2+} must diffuse, the rate at which action potentials are propagated between smooth muscle cells, and the slower rate of cross-bridge formation between actin and myosin myofilaments are responsible for the slower contraction of smooth muscle compared to skeletal muscle.

Ca^{2+} regulate contraction in smooth muscle cells (figure 9.24). The role of Ca^{2+} in smooth muscle differs from that in skeletal muscle cells because there are no troponin molecules associated with actin fibers of smooth muscle cells. Ca^{2+} that enters the cytoplasm binds to a protein called **calmodulin** (kal-mod′ū-lin). Calmodulin molecules with Ca^{2+} bound to them activate an enzyme called **myosin kinase** (kī′nās), which transfers a phosphate group from ATP to light myosin molecules on the heads of myosin molecules. Cross-bridge formation occurs when myosin myofilaments have phosphate groups bound to them. The enzymes responsible for cross-bridge cycling are slower than the enzymes in skeletal muscle resulting in slower cross-bridge formation. Once activated, cross-bridge formation has energy requirements very similar to those for cross-bridge formation in skeletal muscle fibers.

Relaxation of smooth muscle results because of the activity of another enzyme called **myosin phosphatase** (fos′fă-tās). This enzyme removes the phosphate group from the myosin molecules (see figure 9.24). If the phosphate is removed from myosin while the cross-bridges are attached to actin, the cross-bridges release very slowly. This explains how smooth muscle is able to sustain tension for long periods and without extensive energy expenditure. This is often called the **latch state** of smooth muscle contraction. If myosin phosphatase removes the phosphate from myosin molecules while the cross-bridges are not attached, relaxation occurs much more rapidly.

Elevated Ca^{2+} levels in the sarcoplasm of smooth muscle cells result in the activation of myosin molecules and cross-bridge formation. Also, the action of myosin phosphatase results in a high percentage of myosin molecules having their phosphates removed while bound to actin. This process favors sustained contractions, or the latch state, and a low rate of energy consumption because of the slow release of cross-bridges. As long as Ca^{2+} are present, cross-bridges re-form quickly after they are released. Consequently, many cross-bridges are intact at any given time in contracted smooth muscle.

Ca^{2+} levels in the sarcoplasm of smooth muscle are reduced as Ca^{2+} are actively transported across the plasma membrane, including the plasma membrane of caveolae, and into the sarcoplasmic reticulum. Relaxation occurs in response to lower blood levels of Ca^{2+}.

Types of Smooth Muscle

Smooth muscle can be either visceral or multiunit. **Visceral** (vis′er-ăl), or **unitary, smooth muscle** is more common than multiunit smooth muscle. It occurs in sheets and includes smooth muscle of the digestive, reproductive, and urinary tracts. Visceral smooth muscle has numerous gap junctions (see chapter 4), which allow action potentials to pass directly from one cell to another. As a consequence, sheets of smooth muscle cells function as a unit, and a wave of contraction traverses the entire smooth muscle sheet. Visceral smooth muscle is often autorhythmic, but some contracts only when stimulated. For example, visceral smooth muscles of the digestive tract contract spontaneously and at relatively regular intervals, whereas the visceral smooth muscle of the urinary bladder contracts when stimulated by the nervous system.

Multiunit smooth muscle occurs as sheets, such as in the walls of blood vessels; as small bundles, such as in the arrector pili muscles and the iris of the eye; or as single cells, such as in the capsule of the spleen. Multiunit smooth muscle has fewer gap junctions than visceral smooth muscle cells, and cells or groups of cells act as independent units. It normally contracts only when stimulated by nerves or hormones.

Visceral smooth muscle tissue has a different arrangement between neurons and smooth muscle fibers than in skeletal muscle tissue. The synapses are more diffuse than in skeletal muscle. Axons of nerve cells terminate in a series of dilations along the branching axons within the connective tissue among the smooth muscle cells. These dilations have vesicles containing neurotransmitter molecules. Once released, the neurotransmitter molecules diffuse among the smooth muscle cells and bind to receptors on their surfaces. Multiunit smooth muscle has synapses more like those found in skeletal muscle tissue.

55. *Describe a typical smooth muscle cell. How does its structure and its contraction process differ from a skeletal muscle cell?*

56. *Compare visceral smooth muscle and multiunit smooth muscle. Explain why visceral smooth muscle contracts as a single unit.*

1. A hormone combines with a hormone receptor and activates a G protein mechanism, or depolarization of the plasma membrane occurs.

2. An α subunit opens the Ca²⁺ channel in the plasma membrane, or depolarization opens Ca²⁺ channels. Ca²⁺ diffuses through the Ca²⁺ channels and combine with calmodulin.

3. Calmodulin with a Ca²⁺ bound to it binds with myosin kinase and activates it.

4. Activated myosin kinase attaches phosphate from ATP to myosin heads to activate the contractile process.

5. A cycle of cross-bridge formation, movement, detachment, and cross-bridge formation occurs.

6. Relaxation occurs when myosin phosphatase removes phosphate from myosin.

Process Figure 9.24 Smooth Muscle Contraction

Electrical Properties of Smooth Muscle

The resting membrane potential of smooth muscle cells is usually not as negative as that of skeletal muscle fibers. It normally ranges from −55 to −60 mV compared with approximately −85 mV in skeletal muscle fibers. Furthermore, the resting membrane potential of some smooth muscle types fluctuates, with slow depolarization and repolarization phases occurring in many visceral smooth muscle cells. These slow waves of depolarization and repolarization are propagated from cell to cell for short distances (figure 9.25a). More "classic" action potentials can be triggered by the slow waves of depolarization and usually are propagated for longer distances (figure 9.25b). In addition, some smooth muscle types have action potentials with a plateau, or prolonged, depolarization (figure 9.25c). The slow waves in the resting membrane potential may result from a spontaneous and progressive increase in the permeability of the plasma membrane to Na^+ and Ca^{2+}, or they may be controlled by neurons. Sodium ions and Ca^{2+} diffuse into the cell through their respective channels and produce the depolarization.

Smooth muscle does not respond in an all-or-none fashion to action potentials. A series of action potentials in smooth muscle can result in a single, slow contraction followed by slow relaxation instead of individual contractions in response to each action potential, as occurs in skeletal muscle. A slow wave of depolarization that has one to several more classic-appearing action potentials superimposed on it is common in many types of smooth muscle. After the wave of depolarization, the smooth muscle undergoes contraction. Action potentials with plateaus are common in smooth muscle that exhibits periods of sustained contraction.

Spontaneously generated action potentials that lead to contractions are characteristic of visceral smooth muscle in the uterus, the ureter, and the digestive tract. Certain smooth muscle cells in these organs function as **pacemaker cells,** which tend to develop action potentials more rapidly than other cells.

The nervous system can regulate smooth muscle contractions by increasing or decreasing action potentials carried by nerve cell axons to smooth muscle. Responses of smooth muscle cells result in depolarization and increased contraction or hyperpolarization and decreased contraction. The nervous system can also regulate the pacemaker cells.

Hormones and substances produced locally in tissues can bind to hormone receptors on some smooth muscle plasma membranes. The combination of the hormone with the receptor causes ligand-gated Ca^{2+} channels in the plasma membrane to open (see figure 9.24). Ca^{2+} then enter the cell and cause smooth muscle contractions to occur without a major change in the membrane potential. For example, some smooth muscles contract when exposed to the hormone epinephrine because epinephrine combines with epinephrine receptors. Epinephrine combined with its receptors activates G proteins in the plasma membrane (see chapter 3 or 17). The G protein molecules can produce intracellular mediator molecules, which open the ligand-gated Ca^{2+} channels in the plasma membrane or sarcoplasmic reticulum.

(a) **Slow waves of depolarization**

(b) **Action potentials superimposed on a slow wave of depolarization**

(c) **Action potential with prolonged depolarization (plateau)**

Figure 9.25   **Membrane Potential from a Smooth Muscle Preparation**

(a) Slow waves of depolarization. (b) Action potentials in smooth muscle superimposed on a slow wave of depolarization. (c) Action potential with prolonged depolarization (plateau).

Explain how a ligand could bind to a membrane-bound receptor in a smooth muscle cell and cause a sustained contraction of the smooth muscle cell for a prolonged period without a large increase in ATP breakdown.

Functional Properties of Smooth Muscle

Smooth muscle has four functional properties not seen in skeletal muscle: (1) some visceral smooth muscle exhibits autorhythmic contractions; (2) smooth muscle tends to contract in response to being stretched but a slow increase in length produces less of a response than a more rapid increase in length; (3) smooth muscle exhibits a relatively constant tension, called **smooth muscle tone,** over a long period and maintains that same tension in response to a gradual increase in the smooth muscle length; (4) the amplitude of contraction produced by smooth muscle also remains constant, although the muscle length varies. Smooth muscle is therefore well adapted for lining the walls of hollow organs such as the stomach and the urinary bladder. As the volume of the stomach or urinary bladder increases, only a small increase develops in the tension applied to their contents. Also, as the volume of the large and small intestines increases, the contractions that move food through them do not change dramatically in amplitude.

Regulation of Smooth Muscle

Nerves that innervate smooth muscle are part of the autonomic division of the nervous system, whereas skeletal muscle is innervated by the somatic motor nervous system (see chapter 11). The regulation of smooth muscle is therefore involuntary, and the regulation of skeletal muscle is voluntary.

The most important neurotransmitters released from nerves that innervate smooth muscle cells are acetylcholine and norepinephrine. Acetylcholine stimulates some smooth muscle types to contract and inhibits others.

Hormones are also important in regulating smooth muscle. Epinephrine, a hormone from the adrenal medulla, stimulates some smooth muscles, such as those in the blood vessels of the intestine, and inhibits other smooth muscles, such as those in the intestinal wall. Oxytocin stimulates contractions of uterine smooth muscle, especially during delivery of a baby. These and other hormones are discussed more thoroughly in chapters 17 and 18. Other chemical substances produced locally by surrounding tissues, such as histamine, prostaglandins and by-products of metabolism, also influence smooth muscle function. For example, blood flow through capillaries is dramatically influenced by these substances (see chapter 21).

The type of receptors present on the plasma membrane to which the neurotransmitters or hormones bind determines the response of the smooth muscle. Some smooth muscle types have receptors to which acetylcholine binds, and the response of the receptor is to stimulate contractions; other smooth muscle types have receptors to which acetylcholine binds, and the response of

the receptor is to inhibit contractions. A similar relationship exists for smooth muscle receptors for norepinephrine and certain hormones.

The receptor molecules that result in stimulation of smooth muscle contractions often open either Na^+ or Ca^{2+} channels. When these channels open, Na^+ and Ca^{2+} pass through their respective channels into the cell and cause depolarization of the plasma membrane. It's also possible for the receptor to open Ca^{2+} channels in the plasma membrane and sarcoplasmic reticulum. As a result, Ca^{2+} can diffuse into the cytoplasm of the smooth muscle cells without depolarization of the membrane potential to its threshold level, and therefore not produce action potentials.

The receptor molecules that result in inhibition of smooth muscle contractions often close Na^+ and Ca^{2+} channels or open K^+ channels. The result is hyperpolarization of the smooth muscle cells and inhibition. It's also possible for the receptors to increase the activity of the Ca^{2+} pump that transports Ca^{2+} out of the cell or into the sarcoplasmic reticulum. As a result, relaxation may occur without a change in the resting membrane potential.

The response of specific smooth muscle types to either neurotransmitters or hormones is presented in chapters dealing with the smooth muscle types.

57. *How are spontaneous contractions produced in smooth muscle?*

58. *List four functional properties of smooth muscle that are not seen in skeletal muscle. Can smooth muscle develop an oxygen debt?*

59. *How do the nervous system and hormones regulate smooth muscle contraction? How are ion channels affected by receptors that stimulate smooth muscle contractions? How are ion channels affected by receptors that inhibit smooth muscle contractions?*

Cardiac Muscle

Cardiac muscle is found only in the heart and is discussed in detail in chapter 20. Cardiac muscle tissue is striated like skeletal muscle, but each cell usually contains one nucleus located near the center. Adjacent cells join together to form branching fibers by specialized cell-to-cell attachments called **intercalated** (in-ter′kă-lā-ted) **disks,** which have gap junctions that allow action potentials to pass from cell to cell. Cardiac muscle cells are autorhythmic, and one part of the heart normally acts as the pacemaker. The action potentials of cardiac muscle are similar to those in nerve and skeletal muscle but have a much longer duration and refractory period. The depolarization of cardiac muscle results from the influx of both Na^+ and Ca^{2+} across the plasma membrane. Regulation of contraction in cardiac muscle by Ca^{2+} is similar to that of skeletal muscle.

60. *Compare the structural and functional characteristics of cardiac muscle to those of skeletal muscle.*

Clinical Focus   Disorders of Muscle Tissue

Muscle disorders are caused by disruption of normal innervation, degeneration and replacement of muscle cells, injury, lack of use, or disease.

Atrophy

Muscular atrophy is a decrease in the size of muscles. Individual muscle fibers decrease in size, and a progressive loss of myofibrils occurs.

Disuse atrophy is muscular atrophy that results from a lack of muscle use. Bedridden people, people with limbs in casts, or those who are inactive for other reasons experience disuse atrophy in the unused muscles. Disuse atrophy is temporary if a muscle is exercised after it is taken out of a cast, for example. Extreme disuse of a muscle, however, results in muscular atrophy in which skeletal muscle fibers are permanently lost and replaced by connective tissue. Immobility that occurs in bedridden elderly people can lead to permanent and severe muscular atrophy.

Denervation (dē-ner-vā'shŭn) **atrophy** results when nerves that supply skeletal muscles are severed. When motor neurons innervating skeletal muscle fibers are severed, the result is flaccid paralysis. If the muscle is reinnervated, muscle function is restored, and atrophy is stopped. If skeletal muscle is permanently denervated, however, it atrophies and exhibits permanent flaccid paralysis. Eventually muscle fibers are replaced by connective tissue, and the condition cannot be reversed.

Transcutaneous stimulators are used to supply electric stimuli to muscles that have had their nerves temporarily damaged or to muscles that are put in casts for a prolonged period. The electric stimuli keep the muscles functioning and prevent them from permanently atrophying while the nerves resupply the muscles or until the cast is removed.

Muscular Dystrophy

Muscular dystrophy (dis'trō-fē) is one of a group of diseases called **myopathies** (mī-op'ă-thēz) that destroy skeletal muscle tissue. Usually the diseases are inherited and are characterized by degeneration of muscle cells, leading to atrophy and eventual replacement by connective tissue.

Duchenne's muscular dystrophy is an inherited sex-linked (X-linked) recessive disorder that almost exclusively affects males. As muscles atrophy and are replaced by connective tissue, they shorten, causing immobility of joints and postural abnormalities such as scoliosis. By early adolescence, affected individuals are usually confined to wheelchairs (see Systems Pathology, p. 309). Facioscapulohumoral (fa'sĭ-ō-skap-ū-lo-hū'mor-al) muscular dystrophy is generally less severe, and it affects both sexes later in life. The muscles of the face and shoulder girdle are primarily involved. Facioscapulohumoral muscular dystrophy appears to be inherited as an autosomal-dominant condition.

Both types of muscular dystrophy are inherited and progressive, and no drugs can prevent the progression of the disease. Therapy primarily involves exercises. Braces and corrective surgery sometimes help correct abnormal posture caused by the advanced disease. Research is directed at identifying the genes responsible for all types of muscular dystrophy, exploring the mechanism that leads to the disease condition, and finding an effective treatment once the mechanism for the disease is known.

Fibrosis

Fibrosis (fī-brō'sis), or scarring, is the replacement of damaged cardiac muscle or skeletal muscle by connective tissue. Fibrosis is associated with severe trauma to skeletal muscle and with heart attack (myocardial infarction) in cardiac muscle.

Fibrositis

Fibrositis (fī-brō-sī'tis) is an inflammation of fibrous connective tissue, resulting in stiffness, pain, or soreness. It is not progressive, nor does it lead to tissue destruction. Fibrositis can be caused by repeated muscular strain or prolonged muscular tension.

Cramps

Cramps are painful, spastic contractions of muscles that usually result from an irritation within a muscle that causes a reflex contraction (see chapter 12). Local inflammation resulting from a buildup of lactic acid and fibrositis causes reflex contraction of muscle fibers surrounding the irritated region.

Fibromyalgia (fī-brō-mī-al'ja), or chronic muscle pain syndrome, has muscle pain as its main symptom. Fibromyalgia has no known cure, but it is not progressive, crippling, or life-threatening. The pain occurs in muscles or where muscles join their tendons, but not in joints. The pain is chronic and widespread, and is distinguished from other causes of chronic pain by the identification of tender points in muscles, by the length of time the pain persists, and by failure to identify any other cause of the condition.

Effects of Aging on Skeletal Muscle

Several changes occur in aging skeletal muscle that reduce muscle mass, increase the time that muscle takes to contract in response to nervous stimuli, reduce stamina, and increase recovery time. There is a loss of muscle fibers as aging occurs, and the loss begins as early as 25 years of age. By 80 years of age 50% of the muscle mass is gone, and this is due mainly to the loss of muscle fibers. Weight-lifting exercises help slow the loss of muscle mass, but it doesn't prevent the loss of muscle fibers. In addition, fast-twitch muscle fibers decrease in number more rapidly than slow-twitch fibers. Most of the loss of strength and speed is due to the loss of muscle fibers and the loss of fast-twitch muscle fibers. Also, at the synapses the surface area of the synapse decreases. Consequently, action potentials in neurons stimulate action potential production in muscle cells more slowly, and action potentials may not be produced in muscle cells consistently. The number of motor neurons also decreases. Some of the muscle fibers that lose their innervation when a neuron dies are reinnervated by a branch of another motor neuron. This makes motor units in skeletal muscle fewer in number, with a greater number of muscle fibers for each neuron. This may result in less precise control of muscles. Aging is associated with a decrease in the density of capillaries in skeletal muscles, and after exercise a longer period of time is required to recover.

Many of the age-related changes in skeletal muscle can be dramatically slowed if people remain physically active. As people age, they often assume a sedentary life style. Age-related changes develop more rapidly in these people. It has been demonstrated that elderly people who are sedentary can become stronger and more mobile in response to exercise.

● **61.** *Describe the changes in muscle mass and response time that occur in aging skeletal muscle.*

Systems Pathology
Duchenne's Muscular Dystrophy

A couple became concerned about their 3-year-old son when they noticed that he was much weaker than other boys his age and the differences appeared to become more pronounced as time passed. The boy had difficulty sitting, standing, and walking. He seemed clumsy and fell often. He had difficulty climbing stairs, and he often got from a sitting position on the floor to a standing position by using his hands and arms to climb up his legs. His muscles appeared to be poorly developed. The couple took their son to a physician, and after several tests, they were informed that their son had Duchenne's muscular dystrophy.

Background Information

Duchenne's muscular dystrophy (DMD) is usually identified in children at around 3 years of age when the parents notice slow motor development with progressive weakness and muscle wasting (figure A). Typically, muscular weakness begins in the pelvic girdle and causes a waddling gait. Temporary enlargement of the calf muscles is apparent in 80% of cases. Rising from the floor by climbing up the legs is characteristic and is caused by weakness of the lumbar and gluteal muscles. Within 3–5 years, muscles of the shoulder girdle become involved. Wasting of the muscles contributes to muscular atrophy and deformity of the skeleton. People with DMD are usually unable to walk by 10–12 years of age, and few live beyond age 20. No effective treatment exists to prevent the progressive deterioration of muscles in DMD.

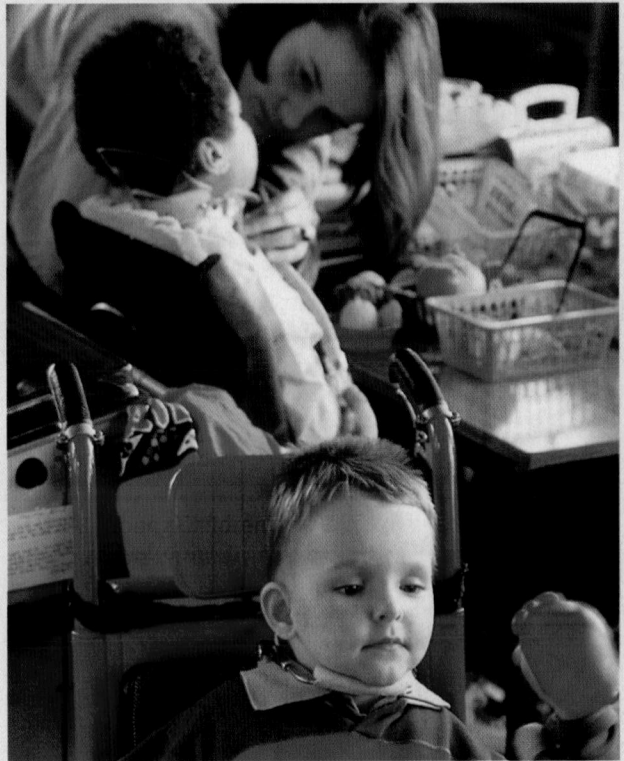

Figure A   Young Children with Duchenne's Muscular Dystrophy

System Interactions   Effect of Duchenne's Muscular Dystrophy on Other Systems

System	Interaction
Skeletal	Replacement of muscles by connective tissue results in shortened inflexible muscles, causing severe deformities of the skeletal system. The shortened muscles are referred to as contractures. Kyphoscoliosis, severe curvature of the spinal column laterally and anteriorly, can be so severe that normal respiratory movements are impaired. Deformities of the limbs result from the contractures. Surgery is sometimes required to prevent contractures from making it impossible for the individual to sit in a wheelchair.
Nervous	Some degree of mental retardation occurs in a large percentage of people with DMD.
Cardiovascular	Cardiac muscle is affected by DMD. Consequently, heart failure occurs in a large number of people with advanced DMD. Heart and respiratory muscles are affected, and death caused by respiratory or cardiac failure usually occurs before age 20. Cardiac involvement becomes serious in as many as 95% of cases.
Lymphatic and Immune	No obvious direct effects occur to the lymphatic system, but phagocytosis of muscle fibers is accomplished mainly by macrophages.
Respiratory	Deformity of the thorax and increasing weakness of the respiratory muscles result in inadequate respiratory movements and an increase in respiratory infections such as pneumonia. Inadequate respiratory movements due to weak respiratory muscles is a major factor in many deaths.
Digestive	Smooth muscle tissue is influenced by muscular dystrophy. The reduced ability of smooth muscle to contract can result in abnormalities of the digestive system, such as an enlarged colon diameter, a twisting of the intestine resulting in increased intestinal obstruction, cramping, and reduced absorption of nutrients.
Urinary	Reduced smooth muscle function and being wheelchair-dependent increase the frequency of urinary tract infections.

DMD results from an abnormal gene located on the X chromosome, at a position called Xp21, and is therefore a sex-linked (X-linked) condition. Although the gene is carried by females, DMD affects males almost exclusively. This chromosome position, or gene locus, is responsible for producing a protein called dystrophin, which plays a role in attaching myofibrils to and regulating the activity of other proteins in the plasma membrane. Dystrophin is thought to protect muscle cells against mechanical stress in the normal individual. In DMD, part of the gene at Xp21 is missing, and the protein it produces malfunctions, resulting in abnormal contractions and progressive muscular weakness.

P R E D I C T

A boy with Duchenne's muscular dystrophy developed pulmonary edema and then pneumonia. His physician diagnosed the condition in the following way: the pulmonary edema was the result of heart failure and the increased fluid in the lungs acted as a site where bacteria invaded and grew. The fact that the boy could not breathe deeply or cough effectively made the condition worse. Explain how a boy with DMD might develop heart failure and ineffective respiratory movements.

S U M M A R Y

Functions of the Muscular System   (p. 276)

Muscle is responsible for movement of the arms, legs, heart, and other parts of the body, maintenance of posture, respiration, production of body heat, communication, contraction of organs and vessels, and the heart beat.

General Functional Characteristics of Muscle   (p. 276)
Properties of Muscle

1. Muscle exhibits contractility (shortens forcefully), excitability (responds to stimuli), extensibility (can be stretched and can still contract), and elasticity (recoils to resting length).
2. Muscle tissue shortens forcefully but lengthens passively.

Types of Muscle Tissue

1. The three types of muscle are skeletal, smooth, and cardiac.

2. Skeletal muscle is responsible for most body movements, smooth muscle is found in the wall of hollow organs and tubes and moves substances through them, and cardiac muscle is found in the heart and pumps blood.

Skeletal Muscle Structure   (p. 277)

Skeletal muscle fibers are multinucleated and appear striated.

Connective Tissue

1. Endomysium surrounds each muscle fiber.
2. Muscle fibers are covered by the external lamina and the endomysium.
3. Muscle fasciculi, bundles of muscle fibers, are covered by the perimysium.
4. Muscle consisting of fasciculi is covered by the epimysium, or fascia.
5. The connective tissue of muscle is bound firmly to the connective tissue of tendons and bone.

Nerve and Blood Vessels

1. Motor neurons extend together with arteries and veins through the connective tissue of skeletal muscles.
2. At the level of the perimysium, axons of motor neurons branch and each branch projects to a muscle fiber to form a synapse.

Muscle Fibers

1. A muscle fiber is a single cell consisting of a plasma membrane (sarcolemma), cytoplasm (sarcoplasm), several nuclei, and myofibrils.
2. Myofibrils are composed of two major protein fibers: actin and myosin.
   - Actin myofilaments consist of a double helix of F actin (composed of G actin monomers), tropomyosin, and troponin.
   - Myosin molecules, consisting of two globular heads and a rodlike portion, constitute myosin myofilaments.
   - A cross-bridge is formed when the myosin binds to the actin.
3. Actin and myosin are organized to form sarcomeres.
   - Sarcomeres are bound by Z disks that hold actin myofilaments.
   - Six actin myofilaments (thin filaments) surround a myosin myofilament (thick filament).
   - Myofibrils appear striated because of A bands and I bands.

Sliding Filament Model   (p. 281)

1. Actin and myosin myofilaments do not change in length during contraction.
2. Actin and myosin myofilaments slide past one another in a way that causes sarcomeres to shorten.
3. The I band and H zones become narrower during contraction, and the A band remains constant in length.

Physiology of Skeletal Muscle Fibers   (p. 283)
Membrane Potentials

Plasma membranes are polarized, which means there is a charge difference called the resting membrane potential, across the plasma membrane. The membrane becomes polarized because the tendency for K^+ to diffuse out of the cell is resisted by the negative charges of ions and molecules inside of the cell.

Ion Channels

1. An action potential is a reversal of the resting membrane potential so the inside of the plasma membrane becomes positive.
2. Ion channels are responsible for producing action potentials.
3. Two types of membrane channels produce action potentials, ligand-gated and voltage-gated channels.

Action Potentials

1. The charge difference across the plasma membrane of cells is the resting membrane potential.
2. Depolarization results from an increase in the permeability of the plasma membrane to Na^+.
3. An all-or-none action potential is produced if depolarization reaches threshold.
4. The depolarization phase of the action potential results from many Na^+ channels opening in an all-or-none fashion.
5. The repolarization phase of the action potential occurs when the Na^+ channels close and K^+ channels open briefly.
6. Propagation of action potentials along the plasma membrane of neurons and skeletal muscle fibers occurs in an all-or-none fashion.

Neuromuscular Junction

1. The presynaptic terminal of the axon is separated from the postsynaptic membrane of the muscle fiber by the synaptic cleft.
2. Acetylcholine released from the presynaptic terminal binds to receptors of the postsynaptic membrane, thereby changing membrane permeability and producing an action potential.
3. After an action potential occurs, acetylcholinesterase splits acetylcholine into acetic acid and choline. Choline is reabsorbed into the presynaptic terminal to re-form acetylcholine.

Excitation–Contraction Coupling

1. Invaginations of the sarcolemma form T tubules that wrap around the sarcomeres.
2. A triad is a T tubule and two terminal cisternae (an enlarged area of sarcoplasmic reticulum).
3. Action potentials move into the T tubule system, causing Ca^{2+} channels to open to release Ca^{2+} from the sarcoplasmic reticulum.
4. Calcium ions diffuse from the sarcoplasmic reticulum to the myofilaments and bind to troponin, causing tropomyosin to move and expose active sites on actin to myosin.
5. Contraction occurs when myosin heads bind to active sites on actin, myosin changes shape, and actin is pulled past the myosin.
6. Relaxation occurs when calcium is taken up by the sarcoplasmic reticulum, ATP binds to myosin, and tropomyosin moves back so active sites on actin are no longer exposed to myosin.

Cross-Bridge Movement

1. ATP is required for the cycle of cross-bridge formation, movement, and release.
2. ATP is also required to transport Ca^{2+} into the sarcoplasmic reticulum and to maintain normal concentration gradients across the plasma membrane.

Muscle Relaxation

1. Calcium ions are transported into the sarcoplasmic reticulum.
2. Calcium ions diffuse away from troponin, preventing further cross-bridge formation.

Physiology of Skeletal Muscle   (p. 293)
Muscle Twitch

1. A muscle twitch is the contraction of a single muscle fiber or a whole muscle in response to a stimulus.
2. A muscle twitch has lag, contraction, and relaxation phases.

Stimulus Strength and Muscle Contraction

1. For a given condition, a muscle fiber or motor unit contracts with a consistent force in response to each action potential, which is called the all-or-none law of skeletal muscle contraction.
2. For a whole muscle, a stimulus of increasing magnitude results in graded contractions of increased force as more motor units are recruited (multiple motor unit summation).

Stimulus Frequency and Muscle Contraction

1. A stimulus of increasing frequency increases the force of contraction (multiple-wave summation).
2. Incomplete tetanus is partial relaxation between contractions, and complete tetanus is no relaxation between contractions.
3. The force of contraction of a whole muscle increases with increased frequency of stimulation because of an increasing concentration of Ca^{2+} around the myofibrils and because of complete stretching of muscle elastic elements.
4. Treppe is an increase in the force of contraction during the first few contractions of a rested muscle.

Types of Muscle Contractions   (p. 296)

1. Isometric contractions cause a change in muscle tension but no change in muscle length.
2. Isotonic contractions cause a change in muscle length but no change in muscle tension.
3. Concentric contractions cause muscles to shorten and tension to increase.

4. Eccentric contractions cause muscles to increase in length and the tension to gradually decrease.
5. Muscle tone is maintenance of a steady tension for long periods.
6. Asynchronous contractions of motor units produce smooth, steady muscle contractions.

Length Versus Tension

Muscle contracts with less-than-maximum force if its initial length is shorter or longer than optimum.

Fatigue   (p. 299)

Fatigue is the decreased ability to do work and can be caused by the central nervous system, depletion of ATP in muscles, or depletion of acetylcholine in the neuromuscular synapse.

Physiologic Contracture and Rigor Mortis

Physiologic contracture (inability of muscles to contract or relax) and rigor mortis (stiff muscles after death) result from inadequate amounts of ATP.

Energy Sources   (p. 300)

Energy for muscle contraction comes from ATP.

Creatine Phosphate

ATP can be synthesized when ADP reacts with creatine phosphate to form creatine and ATP. ATP from this source provides energy for a short time during intense exercise.

Anaerobic Respiration

ATP is synthesized by anaerobic respiration and is used to provide energy for a short time during intense exercise. Anaerobic respiration produces ATP less efficiently but more rapidly than aerobic respiration. Lactic acid levels increase because of anaerobic respiration.

Aerobic Respiration

ATP is synthesized by aerobic respiration. Although ATP is produced more efficiently, it is produced more slowly. Aerobic respiration produces energy for muscle contractions under resting conditions or during exercises such as long-distance running.

Oxygen Debt

After anaerobic respiration, aerobic respiration is higher than normal, thereby restoring creatine phosphate levels and converting lactic acid to glucose.

Slow and Fast Fibers   (p. 301)
Slow-Twitch, or High-Oxidative, Muscle Fibers

Slow-twitch muscle fibers split ATP slowly and have a well-developed blood supply, many mitochondria, and myoglobin.

Fast-Twitch, or Low-Oxidative, Muscle Fibers

Fast-twitch muscle fibers split ATP rapidly.
1. Fatigable fast-twitch fibers have large amounts of glycogen, a poor blood supply, fewer mitochondria, and little myoglobin.
2. Fatigue-resistant fast-twitch fibers have a well-developed blood supply, more mitochondria, and more myoglobin.

Distribution of Fast-Twitch and Slow-Twitch Muscle Fibers

People who are good sprinters have a greater percentage of fast-twitch muscle fibers, and people who are good long-distance runners have a higher percentage of slow-twitch muscle fibers in their leg muscles.

Effects of Exercise

1. Muscles increase (hypertrophy) or decrease (atrophy) in size because of a change in the size of muscle fibers.
2. Anaerobic exercise develops fatigable fast-twitch fibers. Aerobic exercise develops slow-twitch fibers and changes fatigable fast-twitch fibers into fatigue-resistant fast-twitch fibers.

Heat Production   (p. 303)

1. Heat is produced as a by-product of chemical reactions in muscles.
2. Shivering produces heat to maintain body temperature.

Smooth Muscle   (p. 303)

1. Smooth muscle cells are spindle-shaped with a single nucleus. They have actin myofilaments and myosin myofilaments but are not striated.
2. The sarcoplasmic reticulum is poorly developed, and caveolae may function as a T tubule system.
3. Ca^{2+} enter the cell to initiate contraction; calmodulin binds to Ca^{2+} and activates an enzyme that transfers a phosphate group from ATP to myosin. When phosphate groups are attached to myosin, cross-bridges form.
4. Relaxation results when myosin phosphatase removes a phosphate group from the myosin molecule.
   • If phosphate is removed while the cross-bridges are attached, relaxation occurs very slowly, and this is referred to as the catch phase.
   • If phosphate is removed while the cross-bridges are not attached, relaxation occurs rapidly.

Types of Smooth Muscle

1. Visceral smooth muscle fibers contract slowly, have gap junctions (and thus function as a single unit), and can be autorhythmic.
2. Multiunit smooth muscle fibers contract rapidly in response to stimulation by neurons and function independently.

Electrical Properties of Smooth Muscle

1. Spontaneous contractions result from Na^+ and Ca^{2+} leakage into cells. Na^+ and Ca^{2+} movement into the cell is involved in depolarization.
2. The autonomic nervous system, hormones, and chemicals produced locally, can inhibit or stimulate action potentials (and thus contractions). Hormones can also stimulate or inhibit contractions without affecting membrane potentials.

Functional Properties of Smooth Muscle

1. Smooth muscle can contract autorhythmically in response to stretch or when stimulated by the autonomic nervous system or hormones.
2. Smooth muscle maintains a steady tension for long periods.
3. The force of smooth muscle contraction remains nearly constant, despite changes in muscle length.
4. Smooth muscle does not develop an oxygen debt.

Regulation of Smooth Muscle

1. Smooth muscle is innervated by the autonomic nervous system and is involuntary.
2. Hormones are important in regulating smooth muscle. Some hormones can increase the Ca^{2+} permeability of some smooth muscle membranes and, therefore, cause contraction without a change in the resting membrane potential.

Cardiac Muscle   (p. 307)

Cardiac muscle fibers are striated, have a single nucleus, are connected by intercalated disks (thus function as a single unit), and are capable of autorhythmicity.

Effects of Aging on Skeletal Muscle   (p. 308)

1. Aging skeletal muscle is associated with reduced muscle mass, increased response time, and increased time that muscle takes to contract in response to nervous stimuli.

2. Muscle fibers decrease in number, motor units decrease in number, and recovery time increases.

R E V I E W   A N D   C O M P R E H E N S I O N

1. Which of these is true of skeletal muscle?
   a. spindle-shaped cells
   b. under involuntary control
   c. many peripherally located nuclei per muscle cell
   d. forms the walls of hollow internal organs
   e. may be autorhythmic

2. Which of these is *not* a major functional characteristic of muscle?
   a. contractility
   b. elasticity
   c. excitability
   d. extensibility
   e. secretability

3. The connective tissue sheath that surrounds a muscle fasciculus is the
   a. perimysium.
   b. endomysium.
   c. epimysium (fascia).
   d. hypomysium.
   e. external lamina.

4. Given these structures:
   1. whole muscle
   2. muscle fiber (cell)
   3. myofilament
   4. myofibril
   5. muscle fasciculus

   Choose the arrangement that lists the structures in the correct order from the largest to the smallest structure.
   a. 1,2,5,3,4
   b. 1,2,5,4,3
   c. 1,5,2,3,4
   d. 1,5,2,4,3
   e. 1,5,4,2,3

5. Each myofibril
   a. is made up of many muscle fibers.
   b. contains sarcoplasmic reticulum.
   c. is made up of many sarcomeres.
   d. contains T tubules.
   e. is the same thing as a muscle fiber.

6. Myosin myofilaments are
   a. attached to the Z disk.
   b. found primarily in the I band.
   c. thinner than actin myofilaments.
   d. absent from the H zone.
   e. attached to filaments that form the M line.

7. Which of these statements about the molecular structure of myofilaments is true?
   a. Tropomyosin has a binding site for Ca^{2+}.
   b. The head of the myosin molecule binds to an active site on G actin.
   c. ATPase is found on troponin.
   d. Troponin binds to the rodlike portion of myosin.
   e. Actin molecules have a hingelike portion that bends and straightens during contraction.

8. The part of the sarcolemma that invaginates into the interior of the skeletal muscle cells is the
   a. T tubule system.
   b. sarcoplasmic reticulum.
   c. myofibrils.
   d. terminal cisternae.
   e. mitochondria.

9. During the depolarization phase of an action potential, the permeability of the plasma membrane to
   a. Ca^{2+} increases.
   b. Na^+ increases.
   c. K^+ increases.
   d. Ca^{2+} decreases.
   e. Na^+ decreases.

10. During depolarization, the inside of the membrane
    a. becomes more negative than the outside of the membrane.
    b. becomes more positive than the outside of the membrane.
    c. is unchanged.

11. During repolarization of the plasma membrane,
    a. Na^+ move to the inside of the cell.
    b. Na^+ move to the outside of the cell.
    c. K^+ move to the inside of the cell.
    d. K^+ move to the outside of the cell.

12. Given these events:
    1. acetylcholine broken down into acetic acid and choline
    2. acetylcholine moves across the synaptic cleft
    3. action potential reaches the terminal branch of the motor neuron
    4. acetylcholine combines with a receptor molecule
    5. action potential is produced on the muscle fiber's plasma membrane

    Choose the arrangement that lists the events in the order they occur at a neuromuscular junction.
    a. 2,3,4,1,5
    b. 3,2,4,5,1
    c. 3,4,2,1,5
    d. 4,5,2,1,3
    e. 5,1,2,4,3

13. Acetylcholinesterase is an important molecule in the neuromuscular junction because it
    a. stimulates receptors on the presynaptic terminal.
    b. synthesizes acetylcholine from acetic acid and choline.
    c. stimulates receptors within the postsynaptic membrane.
    d. breaks down acetylcholine.
    e. causes the release of Ca^{2+} from the sarcoplasmic reticulum.

14. Given these events:
    1. sarcoplasmic reticulum releases Ca^{2+}
    2. sarcoplasmic reticulum takes up Ca^{2+}
    3. Ca^{2+} diffuse into myofibrils
    4. action potential moves down the T tubule
    5. sarcomere shortens
    6. muscle relaxes

    Choose the arrangement that lists the events in the order they occur following a single stimulation of a skeletal muscle cell.
    a. 1,3,4,5,2,6
    b. 2,3,5,4,6,1
    c. 4,1,3,5,2,6
    d. 4,2,3,5,1,6
    e. 5,1,4,3,2,6
    f. 4,1,3,5,2,6

15. Given these events:
    1. Ca^{2+} combine with tropomyosin
    2. Ca^{2+} combine with troponin
    3. tropomyosin pulls away from actin
    4. troponin pulls away from actin
    5. tropomyosin pulls away from myosin
    6. troponin pulls away from myosin
    7. myosin binds to actin

    Choose the arrangement that lists the events in the order they occur during muscle contraction.
    a. 1,4,7
    b. 2,5,6
    c. 1,3,7
    d. 2,4,7
    e. 2,3,7

16. Which of these regions shortens during skeletal muscle contraction?
    a. A band
    b. I band
    c. H zone
    d. both a and b
    e. both b and c

17. With stimuli of increasing strength, which of these is capable of a graded response?
    a. nerve axon
    b. muscle fiber
    c. motor unit
    d. whole muscle

18. Considering the force of contraction of a skeletal muscle cell, multiple-wave summation occurs because of
    a. increased strength of action potentials on the plasma membrane.
    b. a decreased number of cross-bridges formed.
    c. an increase in Ca^{2+} concentration around the myofibrils.
    d. an increased number of motor units recruited.
    e. increased permeability of the sarcolemma to Ca^{2+}.

19. Which of these events occurs during the lag (latent) phase of muscle contraction?
    a. cross-bridge movement
    b. active transport of Ca^{2+} into the sarcoplasmic reticulum
    c. Ca^{2+} bind to troponin
    d. the sarcomere shortens
    e. ATP is broken down to ADP

20. A weight lifter attempts to lift a weight from the floor, but the weight is so heavy he is unable to move it. The type of muscle contraction the weight lifter used was mostly
    a. isometric.
    b. isotonic.
    c. isokinetic.
    d. concentric.
    e. eccentric.

21. An active tension curve illustrates
    a. how isometric contractions occur.
    b. that the greatest force of contraction occurs if a muscle is not stretched at all.
    c. that passive tension can create active tension.
    d. that optimal overlap of actin and myosin produces the greatest force of contraction.
    e. that the greatest force of contraction occurs with little or no overlap of actin and myosin.

22. Which of these types of fatigue is the most common?
    a. muscular fatigue
    b. psychologic fatigue
    c. synaptic fatigue
    d. army fatigue

23. Given these conditions:
    1. low ATP levels
    2. little or no transport of Ca^{2+} into the sarcoplasmic reticulum
    3. cross-bridges release
    4. Na^+ accumulate in the sarcoplasm
    5. cross-bridges form

    Choose the conditions that occur in both physiologic contracture and rigor mortis.
    a. 1,2,3
    b. 1,2,5
    c. 1,2,3,4
    d. 1,2,4,5
    e. 1,2,3,4,5

24. Jerry Jogger's 3 mile run every morning takes about 30 minutes. Which of these sources provides most of the energy for his run?
    a. aerobic respiration
    b. anaerobic respiration
    c. creatine phosphate
    d. stored ATP

25. Which of these conditions would one expect to find within the leg muscle cells of a world-class marathon runner?
    a. myoglobin-poor
    b. contract very quickly
    c. primarily anaerobic
    d. numerous mitochondria
    e. large deposits of glycogen

26. Which of these increases the least as a result of muscle hypertrophy?
    a. increase in number of sarcomeres
    b. increase in number of myofibrils
    c. increase in number of fibers
    d. increase in blood vessels and mitochondria
    e. increase in connective tissue

27. Relaxation in smooth muscle occurs when
    a. myosin kinase attaches phosphate to the myosin head.
    b. Ca^{2+} bind to calmodulin.
    c. myosin phosphatase removes phosphate from myosin.
    d. Ca^{2+} channels open.
    e. Ca^{2+} are released from the sarcoplasmic reticulum.

28. Compared to skeletal muscle, visceral smooth muscle
    a. has the same ability to be stretched.
    b. when stretched, loses the ability to contract forcefully.
    c. maintains about the same tension, even when stretched.
    d. cannot maintain long, steady contractions.
    e. can accumulate a substantial oxygen debt.

29. Which of these often have spontaneous contractions?
    a. multiunit smooth muscle
    b. visceral smooth muscle
    c. skeletal muscle
    d. both a and b
    e. both b and c

30. Which of these statements concerning aging and skeletal muscle is correct?
    a. There is a loss of muscle fibers with aging.
    b. Slow-twitch fibers decrease in number faster than fast-twitch fibers.
    c. Loss of strength and speed is mainly due to loss of neuromuscular junctions.
    d. There is an increase in density of capillaries in skeletal muscle.
    e. The number of motor neurons remains constant.

Answers in Appendix F

CRITICAL THINKING

1. Bob Canner improperly canned some homegrown vegetables. As a result, he contracted botulism poisoning after eating the vegetables. Symptoms included difficulty in swallowing and breathing. Eventually he died of respiratory failure (his respiratory muscles relaxed and would not contract). Assuming that botulism toxin affects the neuromuscular synapse, propose the ways that the toxin could produce the observed symptoms.

2. A patient is thought to be suffering from either muscular dystrophy or myasthenia gravis. How would you distinguish between the two conditions?

3. Under certain circumstances, the actin and myosin myofilaments can be extracted from muscle cells and placed in a beaker. They subsequently bind together to form long filaments of actin and myosin. Addition of what cell organelle or molecule to the beaker would make the actin and myosin myofilaments unbind?

4. Explain the effect of a lower-than-normal temperature on each of the processes that occur in the lag (latent) phase of muscle contraction.

5. Design an experiment to test the following hypothesis: Muscle A has the same number of motor units as muscle B. Assume you could stimulate the nerves that innervate skeletal muscles with an electronic stimulator and monitor the tension produced by the muscles.

6. Compare the differences that occur when a muscle such as the biceps slowly lifts and lowers a weight and when a whole muscle is stimulated with a maximal stimulus.

7. Predict the shape of an active tension curve for visceral smooth muscle. How does it differ from the active tension curve for skeletal muscle?

8. A researcher is investigating the composition of muscle tissue in the gastrocnemius muscles (in the calf of the leg) of athletes. A needle biopsy is taken from the muscle, and the concentration (or enzyme activity) of several substances is determined. Describe the major differences this researcher sees when comparing the muscles from athletes who perform in the following events: 100 m dash, weight lifting, and 10,000 m run.

9. Harvey Leche milked cows by hand each morning before school. One morning he slept later than usual and had to hurry to get to school on time. As he was milking the cows as fast as he could, his hands became very tired, and for a short time he could neither release his grip nor squeeze harder. Explain what happened.

10. Explain how the force produced by smooth muscle contractions in the stomach change between 5 minutes after eating a meal and 1 hour after eating a meal.

11. Shorty McFleet noticed that his rate of respiration was elevated after running a 100 m race but was not as elevated after running slowly for a much longer distance. Because you studied muscle physiology, he asked you for an explanation. What would you say?

12. It's known that high blood K^+ concentrations cause depolarization of the resting membrane potential. Predict the effect of high blood K^+ levels on smooth muscle function. Explain.

13. Predict and explain the response if the ATP concentration in a muscle that was exhibiting rigor mortis could be instantly increased.

14. A hormone stimulates smooth muscle from a blood vessel to contract. The hormone only causes a small change in the membrane potential, however, even though the smooth muscle tissue contracts substantially. Explain.

15. These experiments were performed in an anatomy and physiology laboratory. The rate and depth of respiration for a resting student were determined. In experiment A, students ran in place for 30 seconds and then immediately sat down and relaxed, and respiration rate and depth were again determined. Experiment B was just like experiment A, except that the students held their breath while running in place. What differences in respiration would you expect for the two different experiments? Explain the basis for your predictions.

Answers in Appendix G

ANSWERS TO PREDICT QUESTIONS

1. When a muscle changes length, the I bands and the H zones change in width, but the A band does not. When a muscle is stretched, the I bands and the H zones increase in width as the length of the sarcomere increases. When a muscle contracts, cross-bridges form and cause the actin myofilaments to slide over the myosin myofilaments. The result is that the I bands and H zones decrease in width as the sarcomeres shorten. When a muscle relaxes, cross-bridges release, and actin myofilaments slide past myosin myofilaments as the sarcomeres lengthen. The I bands and H zones increase in width.

2. When gated K^+ channels open, K^+ diffuse from an area of higher concentration inside of the cell to an area of lower concentration outside of the cell. This leaves more negative charges inside of the cell which causes the resting membrane potential to be more negative.

3. If insufficient acetylcholine is released from the presynaptic terminal of an axon, an action potential is not produced in the muscle fiber and the muscle cannot contract. An action potential must be produced in the muscle fiber for contraction to occur. If inadequate acetylcholine is released from the presynaptic terminal of an axon, several action potentials in the axons would have to occur to cause the presynaptic terminal neurons to release enough acetylcholine to produce an action potential in the muscle fibers. Each action potential would release some acetylcholine. In response, a local potential may be produced in the postsynaptic membrane. If the local potentials were produced over a short period, they could summate (see chapter 11) and reach threshold. If threshold is reached, an action potential is produced.

4. a. Organophosphate poisons inhibit the activity of acetylcholinesterase, which breaks down acetylcholine at the neuromuscular junction and limits the length of time the acetylcholine stimulates the postsynaptic terminal of the muscle fiber. Consequently, acetylcholine accumulates in the synaptic cleft and continuously stimulates the muscle fiber. As a result, the muscle remains contracted until it fatigues. Death is caused by the inability of the victim to breathe. Either the respiratory muscles are in spastic paralysis, or they are so depleted of ATP that they cannot contract at all.
    b. Curare binds to acetylcholine receptors and thus prevents acetylcholine from binding to them. Because curare does not activate the receptors, the muscles don't respond to nervous stimulation. The person suffers from flaccid paralysis and dies from suffocation because the respiratory muscles are not able to contract.

5. a. If Na^+ cannot enter the muscle fiber, no action potentials are produced in the muscle fiber because the influx of Na^+ causes the depolarization phase of the action potential. Without action potentials, the muscle fiber cannot contract at all. The result is flaccid paralysis.
    b. If ATP levels are low in a muscle fiber before stimulation, the following events occur. Energy from the breakdown of ATP already is stored in the heads of the myosin molecules. After stimulation, cross-bridges form. If not enough additional ATP molecules are in the muscle cells to bind to the myosin molecules to allow for cross-bridge release, however, the muscle becomes stiff without contracting or relaxing.
    c. If ATP levels in the muscle fiber are adequate but the action potential frequency is so high that Ca^{2+} accumulates around the myofilaments, the muscle contracts continuously without relaxing. As long as the ions are numerous within the sarcoplasm in the area of the myofilaments, cross-bridge formation is possible. If ATP levels are adequate, cross-bridge formation, release, and formation can proceed again, resulting in a continuously contracting muscle.

6. A decrease occurs in muscle control when reinnervation of muscle fibers occurs after poliomyelitis because the number of motor units in the muscle is decreased. Reinnervation results in a greater number of muscle fibers per motor unit. Control is reduced because the number of motor units that can be recruited is decreased. The greater the number of motor units in a muscle, the greater is the ability to have fine gradations of muscle contraction as motor units are recruited. A smaller number of motor units means that gradations of muscle contraction are not as fine.

7. As a weight is lifted, the muscle contractions are concentric contractions. When a weight lifter lifts a heavy weight above the head, most of the muscle groups contract with a force while the muscle is shortening. Concentric contractions are a category of isotonic contractions in which tension in the muscle increases or remains about the same while the muscle shortens. While the weight is held above the head, the contractions are isometric contractions, because the length of the muscles doesn't change. While the weight is lowered, unless the weight lifter simply drops the weight, the length of the muscles increases as the weight is lowered for most of the muscle groups. Eccentric contractions are contractions in which tension is maintained in a muscle while the muscle increases in length. The major muscle groups are therefore contracting eccentrically while the weight is lowered.

8. During a 10 km run, aerobic metabolism is the primary source of ATP production for muscle contraction. Anaerobic metabolism provides enough ATP for up to 3 minutes during vigorous anaerobic exercise, but running a 10 km race takes much longer. If the runner sprints at the end of the 10 km run, however, anaerobic metabolism accounts for some of the energy production. After the run, aerobic metabolism is elevated for a time to pay back the oxygen debt. Anaerobic metabolism near the end of the run produces lactic acid, which is converted back to glucose after the run, a process that requires ATP. ATP is also required to restore the normal creatine phosphate levels in the muscle fibers and is produced through aerobic metabolism, which uses oxygen. The amount of oxygen used to produce the necessary ATP is the oxygen debt.

9. Long-distance runners should concentrate on running long distances. Slow-twitch muscles function very well for long-distance running. In addition, exercise that causes the muscles to perform aerobic metabolism improves the ability to carry on aerobic exercise and is more effective than exercise done under anaerobic conditions. Aerobic exercise combined with a large percentage of slow-twitch muscle fibers is the best combination for long-distance runners. Aerobic exercise increases the ability of even fast-twitch muscles to resist fatigue.

10. A ligand that binds to its receptor and results in a sustained increase in the permeability of the plasma membrane to Ca^{2+} results in a sustained contraction without a large increase in ATP breakdown. The increased intracellular concentration of Ca^{2+} increases the number of phosphate groups removed from the myosin molecules while cross-bridges are attached. Because these cross-bridges release slowly, the result is a sustained contraction.

11. Muscular dystrophy affects the muscles of respiration and causes deformity of the thoracic cavity. The reduced capacity of muscle tissue to contract is one factor that reduces the ability to breath deeply or cough effectively. In addition, the thoracic cavity can become severely deformed because of the replacement of skeletal muscle with connective tissue. The deformity can result in severe kyphoscoliosis, which also reduces the ability to breathe deeply. In addition, muscular dystrophy can affect the muscle of the heart and cause heart failure. The persistent edema in the lung increases the chance of bacteria multiplying in the lung tissue.

Visit the Online Learning Center at www.mhhe.com/seeley7 for chapter quizzes, interactive learning exercises, and other study tools.

Muscular System

Gross Anatomy

Mannequins are rigid, expression-less, immobile re-creations of the human form. They cannot walk or talk. One of the major characteristics of a living human being is our ability to move about. Without muscles, humans would be little more than mannequins. We wouldn't be able to hold this book. We wouldn't be able to blink, so our eyes would dry out. None of these inconveniences would bother us for long because we wouldn't be able to breathe either.

We use our skeletal muscles all the time—even when we aren't "moving." Postural muscles are constantly contracting to keep us sitting or standing upright. Respiratory muscles are constantly functioning to keep us breathing, even when we sleep. Communication of any kind requires skeletal muscles, whether we are writing, typing, or speaking. Even silent communication with hand signals or facial expression requires skeletal muscle function.

This chapter explains the *general principles* (p. 318) of the muscular system and describes in detail the *head muscles* (p. 324), *trunk muscles* (p. 338), *upper limb muscles* (p. 345), and *lower limb muscles* (p. 358).

Colorized SEM of skeletal muscle.

Part 2   Support and Movement

General Principles

This chapter is devoted to the description of the major named skeletal muscles. The structure and function of cardiac and smooth muscle are considered in other chapters. Most skeletal muscles extend from one bone to another and cross at least one joint. Muscle contraction causes most body movements by pulling one of the bones toward the other across a movable joint. The **action** of a muscle is the movement accomplished when it contracts. Some muscles are not attached to bone at both ends. For example, some facial muscles attach to the skin, which moves as the muscles contract.

The two points of attachment of each muscle are its origin and insertion. At these attachment points, the muscle is connected to the bone by a **tendon.** Some tendons are very short or almost nonexistent; others are long, cable-like structures; and others are broad, sheetlike structures called **aponeuroses** (ap′ō-noo-rō′sēz). The **origin,** also called the **head,** is usually the most stationary end of the muscle. The **insertion** is the end of the muscle attached to the bone undergoing the greatest movement. The part of the muscle between the origin and the insertion is the **belly** (figure 10.1). Some muscles, such as the biceps brachii with two heads and the triceps brachii with three heads, have multiple origins, or heads.

Muscles are typically grouped so that the action of one muscle or group of muscles is opposed by that of another muscle or group of muscles. For example, the biceps brachii flexes the elbow

and the triceps brachii extends the elbow. A muscle that accomplishes a certain movement, such as flexion, is called the **agonist** (ag′ō-nist, a contest). A muscle acting in opposition to an agonist is called an **antagonist** (an-tag′ō-nist). The biceps brachii is the agonist in elbow flexion, whereas the triceps brachii is the antagonist, which extends the elbow.

Muscles also tend to function in groups to accomplish specific movements. For example, the deltoid, biceps brachii, and pectoralis major all help flex the shoulder. Furthermore, many muscles are members of more than one group, depending on the type of movement being considered. For example, the anterior part of the deltoid muscle functions with the flexors of the shoulder, whereas the posterior part functions with the extensors of the shoulder. Members of a group of muscles working together to produce a movement are called **synergists** (sin′er-jistz). For example, the biceps brachii and brachialis are synergists in elbow flexion. Among a group of synergists, if one muscle plays the major role in accomplishing the desired movement, it is the **prime mover.** The brachialis is the prime mover in flexing the elbow. **Fixators** are muscles that hold one bone in place relative to the body while a usually more distal bone is moved. The origin of a prime mover is often stabilized by fixators so that its action occurs at its point of insertion. The muscles of the scapula act as fixators to hold the scapula in place while other muscles contract to move the humerus.

> **1.** Define the terms origin and insertion; agonist and antagonist; and synergist, prime mover, and fixator.

Muscle Shapes

Muscles come in a wide variety of shapes. The shape and size of any given muscle greatly influences the degree to which it can contract and the amount of force it can generate. Muscle shapes are grouped into four classes according to the orientation of the fasciculi: pennate, parallel, convergent, and circular.

Some muscles have their fasciculi (bundles of muscle fibers that can be distinguished by the unaided eye) arranged like the barbs of a feather along a common tendon and therefore are called **pennate** (pen′āt; *pennatus* is Latin, meaning feather) muscles. A muscle with all fasciculi on one side of the tendon is **unipennate,** one with fasciculi on both sides is **bipennate,** and a muscle with fasciculi arranged at many places around the central tendon is **multipennate** (figure 10.2*a*). The long tendons of pennate muscles can extend for some distance between a muscle belly and its insertion. The pennate arrangement allows a large number of fasciculi to attach to a single tendon with the force of contraction concentrated at the tendon. The muscles that extend the knee are examples of multipennate muscles.

In other muscles, called **parallel** muscles, fasciculi are organized parallel to the long axis of the muscle (figure 10.2*b*). As a consequence, the muscles shorten to a greater degree than do pennate muscles because the fasciculi are in a direct line with the tendon; however, they contract with less force because fewer total fascicles are attached to the tendon. The hyoid muscles are an example of parallel muscles.

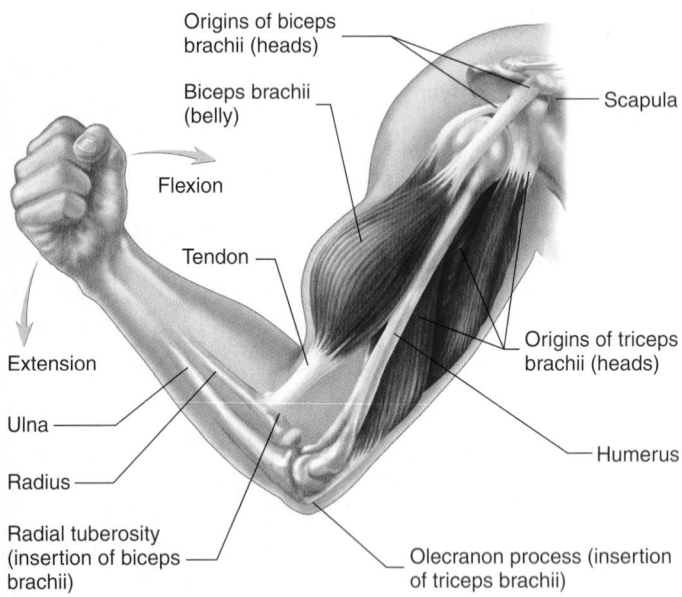

Origins of biceps brachii (heads)

Biceps brachii (belly)

Flexion

Tendon

Extension

Ulna

Radius

Radial tuberosity (insertion of biceps brachii)

Scapula

Origins of triceps brachii (heads)

Humerus

Olecranon process (insertion of triceps brachii)

Figure 10.1 Muscle Attachment

Muscles are attached to bones by tendons. The biceps brachii has two heads that originate on the scapula. The triceps brachii has three heads that originate on the scapula and humerus. The biceps tendon inserts onto the radial tuberosity and onto nearby connective tissue. The triceps brachii inserts onto the olecranon process of the ulna.

Unipennate
muscle

Bipennate
muscle

Multipennate
muscle

(a) **Pennate muscles**

(b) **Parallel
muscle**

(c) **Circular
muscle**

(d) **Convergent
muscle**

Figure 10.2   **Examples of Muscle Types**

Four classes of muscle shapes based on fasciculi arrangement: (*a*) pennate, (*b*) parallel, (*c*) circular, and (*d*) convergent.

Circular muscles, such as the orbicularis oris and orbicularis oculi (figure 10.2*c*) have their fasciculi arranged in a circle around an opening and act as sphincters to close the opening. In **convergent** muscles, such as the deltoid muscle (figure 10.2*d*), the base is much wider than the insertion, giving the muscle a triangular shape and allowing it to contract with more force than could occur in a parallel muscle. Some convergent muscles are also multipennate.

Muscles may also have specific shapes, such as quadrangular, triangular, rhomboidal, or fusiform (figure 10.3*a*). Muscles also may have multiple components, such as two bellies or two heads. A digastric muscle has two bellies separated by a tendon, whereas a bicipital muscle has two origins (heads) and a single insertion (figure 10.3*b*).

> ● 2. *Describe the different shapes of muscles. How are the shapes related to the force of contraction of the muscle and the range of movement the contraction produces?*

Nomenclature

Muscles are named according to their location, size, shape, orientation of fasciculi, origin and insertion, number of heads, or function. Recognizing the descriptive nature of muscle names makes learning those names much easier.

1. *Location.* Some muscles are named according to their location. For example, a pectoralis (chest) muscle is located in the chest, a gluteus (buttock) muscle is located in the buttock, and a brachial (arm) muscle is located in the arm.
2. *Size.* Muscle names may also refer to the relative size of the muscle. For example, the gluteus maximus (large) is the largest muscle of the buttock, and the gluteus minimus (small) is the smallest. A longus (long) muscle is longer than a brevis (short) muscle.
3. *Shape.* Some muscles are named according to their shape. The deltoid (triangular) muscle is triangular, a quadratus (quadrangular) muscle is rectangular, and a teres (round) muscle is round.
4. *Orientation.* Muscles are also named according to their fascicular orientation. A rectus (straight) muscle has muscle fasciculi running straight with the axis of the structure to which the muscle is associated, whereas the fasciculi of an oblique muscle lie oblique to the longitudinal axis of the structure.
5. *Origin and insertion.* Muscles may be named according to their origin and insertion. The sternocleidomastoid originates on the sternum and clavicle and inserts onto the mastoid process of the temporal bone. The brachioradialis originates in the arm (brachium) and inserts onto the radius.
6. *Number of heads.* The number of heads (origins) a muscle has may also be used in naming it. A biceps muscle has two heads, and a triceps muscle has three heads.
7. *Function.* Muscles are also named according to their function. An abductor moves a structure away from the midline, and an adductor moves a structure toward the midline. The masseter (a chewer) is a chewing muscle.

> ● 3. *List the different criteria used to name muscles, and give an example of each.*

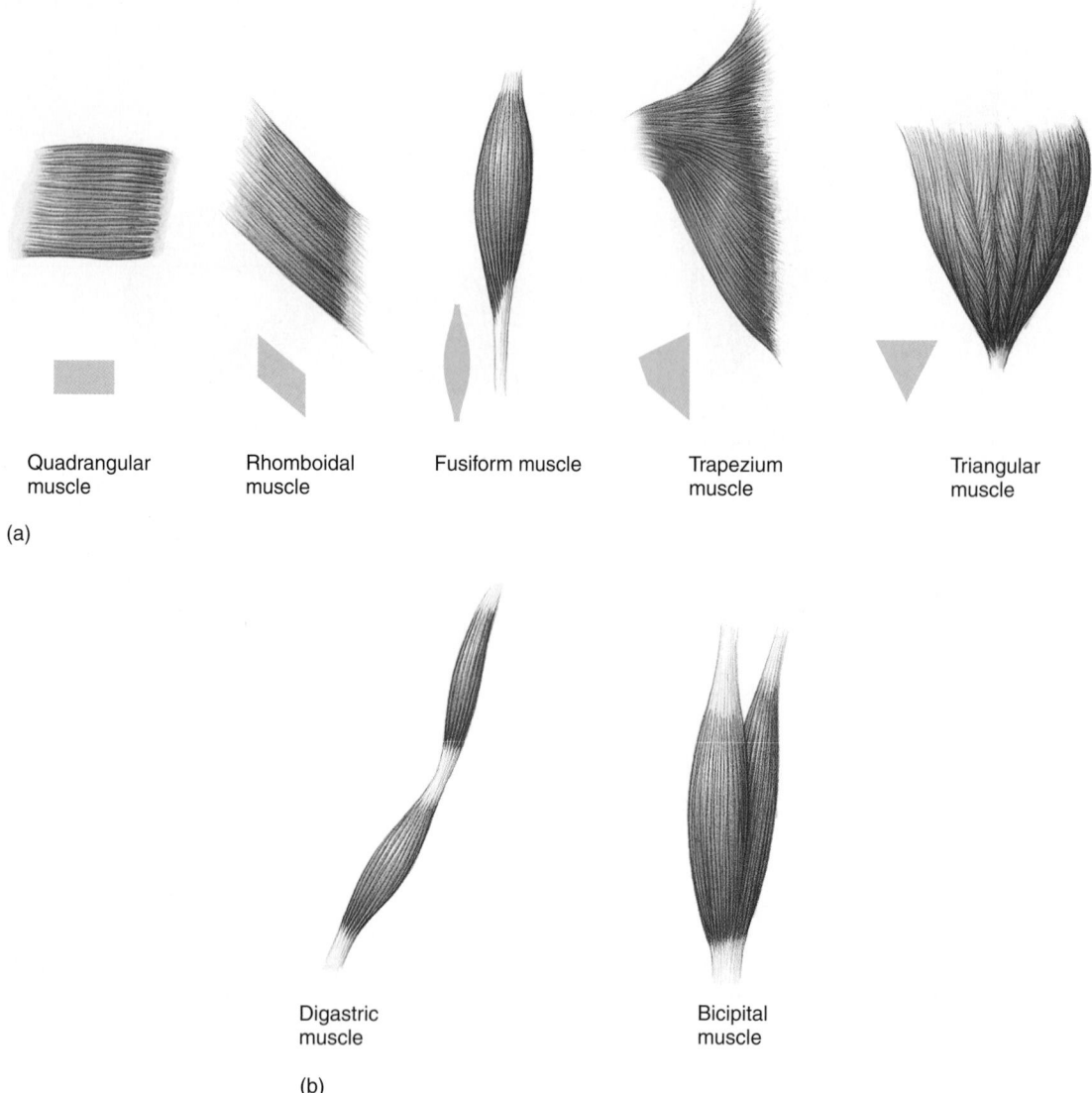

Quadrangular muscle   Rhomboidal muscle   Fusiform muscle   Trapezium muscle   Triangular muscle

(a)

Digastric muscle   Bicipital muscle

(b)

Figure 10.3 Examples of Muscle Shapes

(*a*) Muscles with various shapes: quadrangular (four-sided with right angles), rhomboidal (four-sided with no right angles), trapezium (four-sided with no right angles and no parallel sides), triangular (three sided), and fusiform (spindle-shaped). (*b*) Muscles with various components: digastric (two bellies) and bicipital (two heads).

Movements Accomplished by Muscles

When muscles contract, the **pull** (P), or force, of muscle contraction is applied to levers, such as bones, resulting in movement of the levers (figure 10.4). A **lever** is a rigid shaft capable of turning about a hinge, or pivot point called a **fulcrum** (F) and transferring a force applied at one point along the lever to a **weight** (W), or resistance, placed at some other point along the lever. The joints function as fulcrums, the bones function as levers, and the muscles provide the pull to move the levers. Three classes of levers exist based on the relative positions of the levers, weights, fulcrums, and forces.

Class I Lever

In a **class I lever system,** the fulcrum is located between the force and the weight (figure 10.4*a*). A child's seesaw is an example of this type of lever. The children on the seesaw alternate between being the weight and the pull across a fulcrum in the center of the board. The head is an example of this type of lever in the body. The atlanto-occipital joint is the fulcrum, the posterior neck muscles provide the pull depressing the back of the head, and the face, which is elevated, is the weight. With the weight balanced over the fulcrum, only a small amount of pull is required to lift a weight. For example, only a very small shift in weight is needed

Figure 10.4 Lever Classes

(*a*) **Class I:** The fulcrum (*F*) is located between the weight (*W*) and the force or pull (*P*). The pull is directed downward, and the weight, on the opposite side of the fulcrum, is lifted. In the example in the body, the fulcrum extends through several cervical vertebrae. (*b*) **Class II:** The weight (*W*) is located between the fulcrum (*F*) and the force or pull (*P*). The upward pull lifts the weight. The movement of the mandible is easier to compare to a wheelbarrow if the head is considered upside down. (*c*) **Class III:** The force or pull (*P*) is located between the fulcrum (*F*) and the weight (*W*). The upward pull lifts the weight.

for one child to lift the other on a seesaw. This system is quite limited, however, as to how much weight can be lifted and how high it can be lifted. For example, consider what happens when the child on one end of the seesaw is much larger than the child on the other end.

Class II Lever

In a **class II lever system,** the weight is located between the fulcrum and the pull (figure 10.4*b*). An example is a wheelbarrow, where the wheel is the fulcrum and the person lifting on the handles provides the pull. The weight, or load, carried in the wheelbarrow is placed between the wheel and the operator. In the body, an example of a class II lever is depression of the mandible (as in opening the mouth).

Class III Lever

In a **class III lever system,** the most common type in the body, the pull is located between the fulcrum and the weight (figure

10.4*c*). An example is a person using a shovel. The hand placed on the part of the handle closest to the blade provides the pull to lift the weight, such as a shovel full of dirt, and the hand placed near the end of the handle acts as the fulcrum. In the body, the action of the biceps brachii muscle (force) pulling on the radius (lever) to flex the elbow (fulcrum) and elevate the hand (weight) is an example of a class III lever. This type of lever system does not allow as great a weight to be lifted, but the weight can be lifted a greater distance.

> **4.** *Using the terms fulcrum, lever, and force, explain how contraction of a muscle results in movement. Define the three classes of levers, and give an example of each in the body.*

Muscle Anatomy

An overview of the superficial skeletal muscles is presented in figure 10.5. Muscles of the head, neck, trunk, and limbs are described in the following sections.

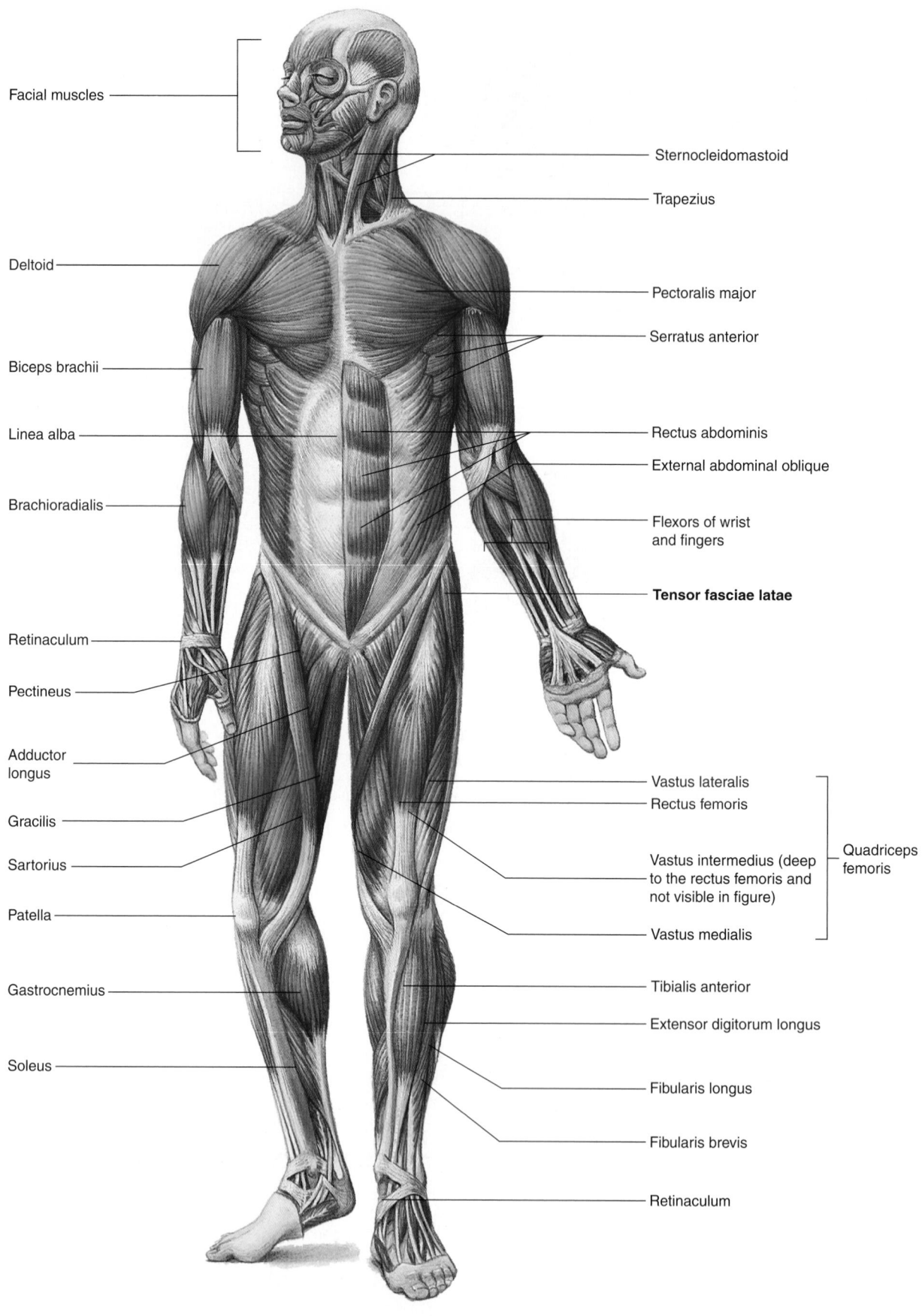

Facial muscles

Deltoid

Biceps brachii

Linea alba

Brachioradialis

Retinaculum

Pectineus

Adductor longus

Gracilis

Sartorius

Patella

Gastrocnemius

Soleus

Sternocleidomastoid

Trapezius

Pectoralis major

Serratus anterior

Rectus abdominis

External abdominal oblique

Flexors of wrist and fingers

Tensor fasciae latae

Vastus lateralis

Rectus femoris

Vastus intermedius (deep to the rectus femoris and not visible in figure)

Vastus medialis

Quadriceps femoris

Tibialis anterior

Extensor digitorum longus

Fibularis longus

Fibularis brevis

Retinaculum

(a) **Anterior view**

Figure 10.5 General Overview of the Superficial Body Musculature

(a) Anterior view.

Sternocleidomastoid

Seventh cervical vertebra

Infraspinatus

Teres minor

Teres major

Triceps brachii

Extensors
of the wrist
and fingers

Hamstring { Semitendinosus
muscles { Biceps femoris
{ Semimembranosus

Fibularis longus

Fibularis brevis

Splenius capitis

Trapezius

Deltoid

Latissimus dorsi

External abdominal
oblique

Gluteus medius

Gluteus maximus

Adductor magnus

Iliotibial tract

Gracilis

Gastrocnemius

Soleus

Calcaneal tendon
(Achilles tendon)

(b) **Posterior view**

Figure 10.5 *(continued)*

(b) Posterior view.

Head Muscles

Head and Neck Muscles

Muscles of the neck are listed in table 10.1 and are illustrated in figures 10.6–10.8. The anterior superficial neck muscles are illustrated in figure 10.6. Most of the flexors of the head and neck lie deep within the neck along the anterior margins of the vertebral bodies (not illustrated). Extension of the neck is accomplished by posterior neck muscles that attach to the occipital bone and mastoid process of the temporal bone (figures 10.7 and 10.8) and function as the force of a class I lever system. They also rotate and laterally flex the neck.

The muscular ridge seen superficially in the posterior part of the neck and lateral to the midline is composed of the trapezius muscle overlying the splenius capitis (see figures 10.6b and 10.7b).

The fasciculi of the trapezius muscles are shorter at the base of the neck and leave a diamond-shaped area over the inferior cervical and superior thoracic vertebral spines (see figure 10.7b).

Rotation and lateral flexion of the neck are accomplished by muscles of both the lateral and posterior groups (see table 10.1). The **sternocleidomastoid** (ster′nō-klī′dō-mas′toyd) muscle is the prime mover of the lateral group. It's easily seen on the anterior and lateral sides of the neck, especially if the head is extended slightly and rotated to one side (figure 10.6b). If the sternocleidomastoid muscle on only one side of the neck contracts, the neck is rotated toward the opposite side. If both contract together, they flex the neck. Lateral flexion of the neck (moving the head back to the midline after it has been tilted to one side or the other) is accomplished by the lateral flexors of the opposite side.

Table 10.1*	Muscles Moving the Head (see figures 10.6–10.8)			
Muscle	**Origin**	**Insertion**	**Nerve**	**Action**
Anterior				
Longus capitis (lon′gŭs ka′pi-tis) (not illustrated)	C3–C6	Occipital bone	C1–C3	Flexes neck
Rectus capitis anterior (rek′tŭs ka′pi-tis) (not illustrated)	Atlas	Occipital bone	C1–C2	Flexes neck
Posterior				
Longissimus capitis (lon-gis′ĭ-mŭs kă′pi-tis)	Upper thoracic and lower cervical vertebrae	Mastoid process	Dorsal rami of cervical nerves	Extends, rotates, and laterally flexes neck
Oblique capitis superior (ka′pi-tis)	Atlas	Occipital bone (inferior nuchal line)	Dorsal ramus of C1	Extends and laterally flexes neck
Rectus capitis posterior (rek′tŭs ka′pi-tis)	Axis, atlas	Occipital bone	Dorsal ramus of C1	Extends and rotates neck
Semispinalis capitis	C4–T6	Occipital bone	Dorsal rami of cervical nerves	Extends and rotates neck
Splenius capitis	C4–T6	Superior nuchal line and mastoid process	Dorsal rami of cervical nerves	Extends, rotates, and laterally flexes neck
Trapezius	Occipital protuberance, nuchal ligament, spinous processes of C7–T12	Clavicle, acromion process, and scapular spine	Accessory (cranial nerve XI)	Extends and laterally flexes neck
Lateral				
Rectus capitis lateralis (not illustrated)	Atlas	Occipital bone	C1	Laterally flexes neck
Sternocleidomastoid	Manubrium and medial clavicle	Mastoid process and superior nuchal line	Accessory (cranical nerve XI)	One contracting alone: rotates and extends neck. Both contracting together: flex neck and elevate face

*The tables in this chapter are to be used as references. As you study the muscular system, first locate the muscle on the figure, and then find its description in the corresponding table.

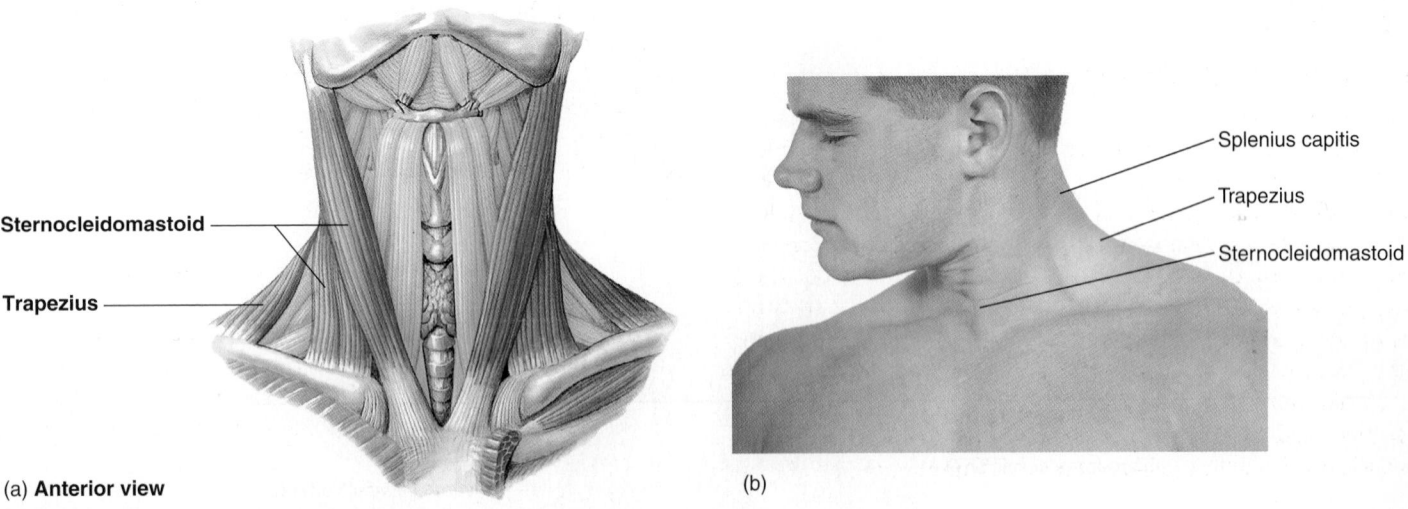

(a) **Anterior view**

(b)

Figure 10.6 Anterior Neck Muscles

(a) Anterior superficial neck muscles. (b) Surface anatomy of anterior neck muscles.

Semispinalis capitis

Splenius capitis

Sternocleidomastoid

Splenius cervicis

Trapezius

Seventh cervical
vertebra

Sternocleidomastoid

Trapezius

Diamond-shaped
bare area

(a) **Posterior view**

(b)

Figure 10.7 Posterior Neck Muscles

(a) Posterior superficial neck muscles. (b) Surface anatomy of anterior neck muscles.

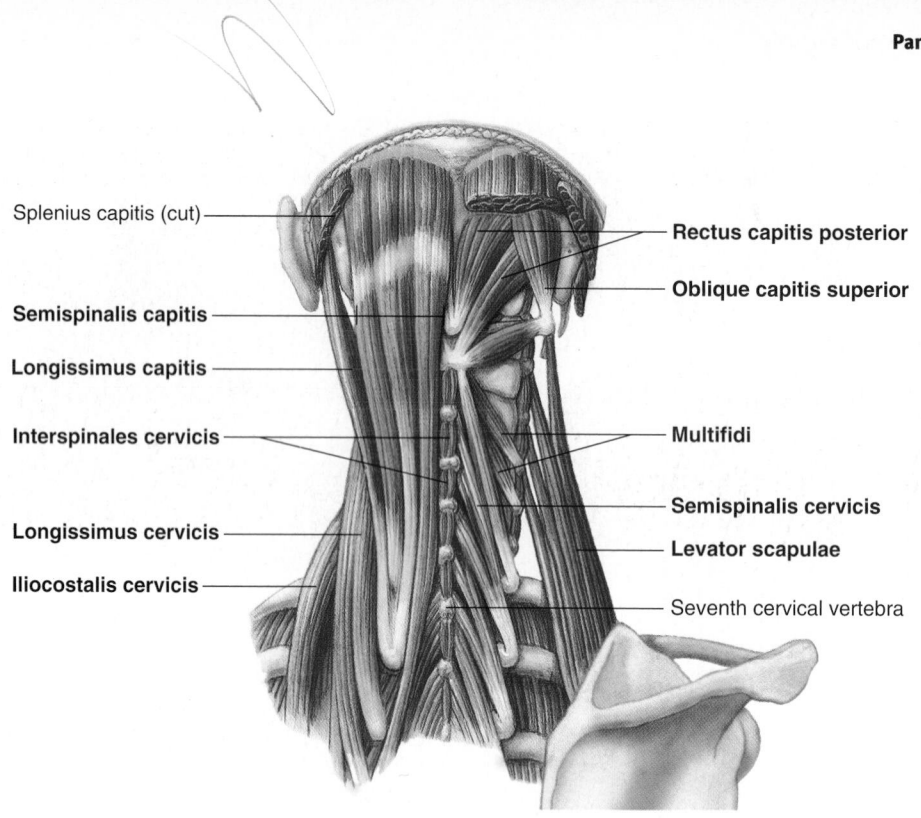

Splenius capitis (cut)

Semispinalis capitis

Longissimus capitis

Interspinales cervicis

Longissimus cervicis

Iliocostalis cervicis

Rectus capitis posterior

Oblique capitis superior

Multifidi

Semispinalis cervicis

Levator scapulae

Seventh cervical vertebra

Posterior view

Figure 10.8 Posterior Deep Neck Muscles

Torticollis

Torticollis (tōr-ti-kol′is; twisted neck, or wry neck), may result from damage to one of the sternocleidomastoid muscles or the accessory nerve (cranial nerve XI) supplying it. Torticollis may also result from a tumor forming in a baby's sternocleidomastoid before birth. Damage to an infant's neck muscles during a difficult birth sometimes causes torticollis and can usually be corrected by exercising the muscle.

Spasmodic torticollis can occur in adults. It is characterized by intermittent contraction of neck muscles, especially the sternocleidomastoid and trapezius, resulting in rotation, flexion, and extention of the neck, and elevation of the shoulder.

P R E D I C T ①

Shortening of the right sternocleidomastoid muscle rotates the head in which direction?

5. *Name the major movements of the head caused by contraction of the anterior, posterior, and lateral neck muscles.*

6. *Name the movements of the head and neck caused by contraction of the sternocleidomastoid muscle. What is torticollis (wry neck)?*

Facial Expression

The skeletal muscles of the face (table 10.2 and figure 10.9) are cutaneous muscles attached to the skin. Many animals have cutaneous muscles over the trunk that allow the skin to twitch to remove irritants such as insects. In humans, facial expressions are important components of nonverbal communication, and the cutaneous muscles are confined primarily to the face and neck.

Several muscles act on the skin around the eyes and eyebrows (figure 10.10 and see figure 10.9). The **occipitofrontalis** (ok-sip′ĭ-tō-frŭn-tă′lis) raises the eyebrows and furrows the skin of the forehead. The **orbicularis oculi** (ōr-bik′ū-lā′ris ok′ū-lī) closes the eyelids and causes "crow's-feet" wrinkles in the skin at the lateral corners of the eyes. The **levator palpebrae** (le-vā′ter, lē-vā′tōr palpē′brē; the palpebral fissure is the opening between the eyelids) **superioris** raises the upper lids (figure 10.10a). A droopy eyelid on one side, called **ptosis** (tō′sis), usually indicates that the nerve to the levator palpebrae superioris has been damaged. The **corrugator supercilii** (kōr′ŭ-gā′ter, kōr′ŭ-gā′tōr soo′per-sil′ē-ī) draws the eyebrows inferiorly and medially, producing vertical corrugations (furrows) in the skin between the eyes (see figures 10.9 and 10.10c).

Several muscles function in moving the lips and the skin surrounding the mouth (see figures 10.9 and 10.10). The **orbicularis oris** (ōr-bik′ū-lā′ris ōr′is) and **buccinator** (buk′si-nā-tōr), the kissing muscles, pucker the mouth. Smiling is accomplished by the **zygomaticus** (zī′gō-mat′i-kŭs) **major** and **minor,** the **levator anguli** (ang′gū-lī) **oris,** and the **risorius** (rī-sōr′ē-ŭs). Sneering is accomplished by the **levator labii** (lā′bē-ī) **superioris** and frowning or pouting by the **depressor anguli oris,** the **depressor labii**

Table 10.2 Muscles of Facial Expression (see figure 10.9)

Muscle	Origin	Insertion	Nerve	Action
Auricularis (aw-rik′ū-lăr′is)				
Anterior	Aponeurosis over head	Cartilage of auricle	Facial	Draws auricle superiorly and anteriorly
Posterior	Mastoid process	Posterior root of auricle	Facial	Draws auricle posteriorly
Superior	Aponeurosis over head	Cartilage of auricle	Facial	Draws auricle superiorly and posteriorly
Buccinator (buk′sĭ-nā′tōr)	Mandible and maxilla	Orbicularis at angle of mouth	Facial	Retracts angle of mouth; flattens cheek
Corrugator supercilii (kōr′ū′gā′ter, soo′per-sil′ē-ī)	Nasal bridge and orbicularis oculi	Skin of eyebrow	Facial	Depresses medial portion of eyebrow and draws eyebrows together as in frowning
Depressor anguli oris (dē-pres′ōr ang′gū-lī ōr′ŭs)	Lower border of mandible	Lip near angle of mouth	Facial	Depresses angle of mouth
Depressor labii inferioris (dē-pres′ōr lā′bē-ī in-fēr′ē-ōr-is)	Lower border of mandible	Skin of lower lip and orbicularis oris	Facial	Depresses lower lip
Levator anguli oris (lē-vā′tor, le-vā′ter ang′gū-lī ōr′ŭs)	Maxilla	Skin at angle of mouth and orbicularis oris	Facial	Elevates angle of mouth
Levator labii superioris (lē-vā′tor, le-vā′ter lā′bē-ī sū-pēr′ē-ōr-is)	Maxilla	Skin and orbicularis oris of upper lip	Facial	Elevates upper lip
Levator labii superioris alaeque nasi (lē-vā′tor, le-vā′ter lā′bē-ī sū-pēr′ē-ōr-is ă-lak′ă nā′zī)	Maxilla	Ala at nose and upper lip	Facial	Elevates ala of nose and upper lip
Levator palpebrae superioris (lē-vā′tor, le-vā′ter pal-pē′brē sū-pēr′ē-ōr-is)	Lesser wing of sphenoid	Skin of eyelid	Oculomotor	Elevates upper eyelid
Mentalis (men-tā′lis)	Mandible	Skin of chin	Facial	Elevates and wrinkles skin over chin; elevates lower lip
Nasalis (nā′ză-lis)	Maxilla	Bridge and ala of nose	Facial	Dilates nostril
Occipitofrontalis (ok-sip′i-tō-frŭn′tā′lis)	Occipital bone	Skin of eyebrow and nose	Facial	Moves scalp; elevates eyebrows
Orbicularis oculi (ōr-bik′ū-lā′ris ok′ū-lī)	Maxilla and frontal bones	Circles orbit and inserts near origin	Facial	Closes eye
Orbicularis oris (ōr-bik′ū-lā′ris ōr′is)	Nasal septum, maxilla, and mandible	Fascia and other muscles of lips	Facial	Closes lip
Platysma (plă-tiz′mă)	Fascia of deltoid and pectoralis major	Skin over inferior border of mandible	Facial	Depresses lower lip; wrinkles skin of neck and upper chest
Procerus (prō-sē′rŭs)	Bridge of nose	Frontalis	Facial	Creates horizontal wrinkle between eyes, as in frowning
Risorius (ri-sōr′ē-ŭs)	Platysma and masseter fascia	Orbicularis oris and skin at corner of mouth	Facial	Abducts angle of mouth
Zygomaticus major (zī′gō-mat′i-kŭs)	Zygomatic bone	Angle of mouth	Facial	Elevates and abducts upper lip
Zygomaticus minor (zī′gō-mat′i-kŭs)	Zygomatic bone	Orbicularis oris of upper lip	Facial	Elevates and abducts upper lip

Epicranial (galea) aponeurosis

Temporalis

Auricularis superior

Auricularis anterior

Occipitofrontalis (occipital portion)

Auricularis posterior

Masseter

Sternocleidomastoid

Trapezius

Occipitofrontalis (frontal portion)

Orbicularis oculi

Corrugator supercilii

Procerus

Levator labii superioris alaeque nasi

Levator labii superioris

Zygomaticus minor

Zygomaticus major

Levator anguli oris

Orbicularis oris

Mentalis

Depressor labii inferioris

Depressor anguli oris

Risorius (cut)

Buccinator

(a) **Lateral view**

Occipitofrontalis (frontal portion)

Orbicularis oculi

Procerus

Orbicularis oculi (palpebral portion)

Levator labii superioris

Zygomaticus minor

Zygomaticus major

Levator anguli oris

Risorius

Depressor anguli oris

Depressor labii inferioris

Corrugator supercilii

Temporalis

Levator labii superioris alaeque nasi

Nasalis

Zygomaticus minor and major (cut)

Levator labii superioris

Levator anguli oris (cut)

Masseter

Buccinator

Orbicularis oris

Mentalis

Platysma

(b) **Anterior view**

Figure 10.9 Muscles of Facial Expression

(*a*) Lateral view. (*b*) Anterior view.

(a)

(b)

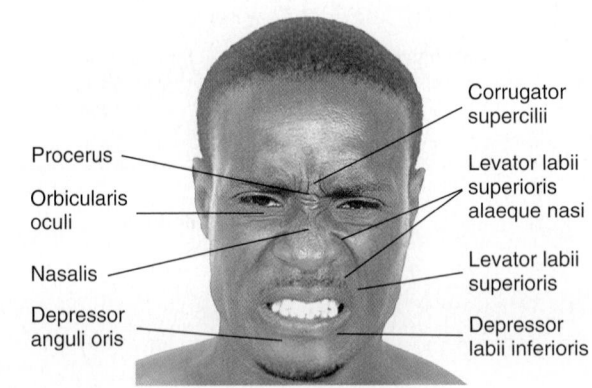

(c)

(d)

Figure 10.10 Surface Anatomy, Muscles of Facial Expression

inferioris, and the **mentalis** (men-tā′lis). If the mentalis muscles are well developed on each side of the chin, a chin dimple, where the skin is tightly attached to the underlying bone or other connective tissue, may appear between the two muscles.

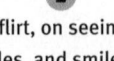

7. *What is unusual about the insertion (and sometimes the origin) of facial muscles?*

8. *Which muscles are responsible for moving the ears, the eyebrows, the eyelids, and the nose? For puckering the lips, smiling, sneering, and frowning? What causes a dimple on the chin? What usually causes ptosis on one side?*

P R E D I C T ❷
Harry Wolf, a notorious flirt, on seeing Sally Gorgeous raises his eyebrows, winks, whistles, and smiles. Name the facial muscles he uses to carry out this communication. Sally, thoroughly displeased with this exhibition, frowns and flares her nostrils in disgust. What muscles does she use?

Mastication

Chewing, or **mastication** (mas-ti-kā′shŭn), involves forcefully closing the mouth (elevating the mandible: temporalis, masseter, and medial pterygoid) and grinding food between the teeth (medial and lateral excursion of the mandible; involving all muscles of mastication). The **muscles of mastication** and

the **hyoid muscles** move the mandible (tables 10.3 and 10.4; figures 10.11 and 10.12). The elevators of the mandible are some of the strongest muscles of the body and bring the mandibular teeth forcefully against the maxillary teeth to crush food. Slight mandibular depression involves relaxation of the mandibular elevators and the pull of gravity. Opening the mouth wide requires the action of the depressors of the mandible (lateral pterygoid, digastric, geniohyoid, mylohyoid); and even though the muscles of the tongue and the buccinator (see tables 10.2 and 10.5) are not involved in the actual process of chewing, they help move the food in the mouth and hold it in place between the teeth.

9. *Name the muscles responsible for opening and closing the jaw and for lateral and medial excursion of the jaw.*

Tongue Movements

The tongue is very important in mastication and speech: (1) it moves food around in the mouth; (2) with the buccinator it holds food in place while the teeth grind it; (3) it pushes food up to the palate and back toward the pharynx to initiate swallowing; and (4) it changes shape to modify sound during speech. The tongue consists of a mass of **intrinsic muscles** (entirely within the tongue) which are involved in changing the shape of the tongue, and **extrinsic muscles** (outside of the tongue but attached to it) which help change the shape and

Table 10.3	Muscles of Mastication (see figures 10.9 and 10.11)			
Muscle	**Origin**	**Insertion**	**Nerve**	**Action**
Temporalis (tem-pŏ-rā′lis)	Temporal fossa	Anterior portion of mandibular ramus and coronoid process	Mandibular division of trigeminal	Elevates and retracts mandible; involved in excursion
Masseter (ma′se-ter)	Zygomatic arch	Lateral side of mandibular ramus	Mandibular division of trigeminal	Elevates and protracts mandible; involved in excursion
Pterygoids (ter′i-goydz)				
Lateral	Lateral side of lateral pterygoid plate and greater wing of sphenoid	Condylar process of mandible and articular disk	Mandibular division of trigeminal	Protracts and depresses mandible; involved in excursion
Medial	Medial side of lateral pterygoid plate and tuberosity of maxilla	Medial surface of mandible	Mandibular division of trigeminal	Protracts and elevates mandible; involved in excursion

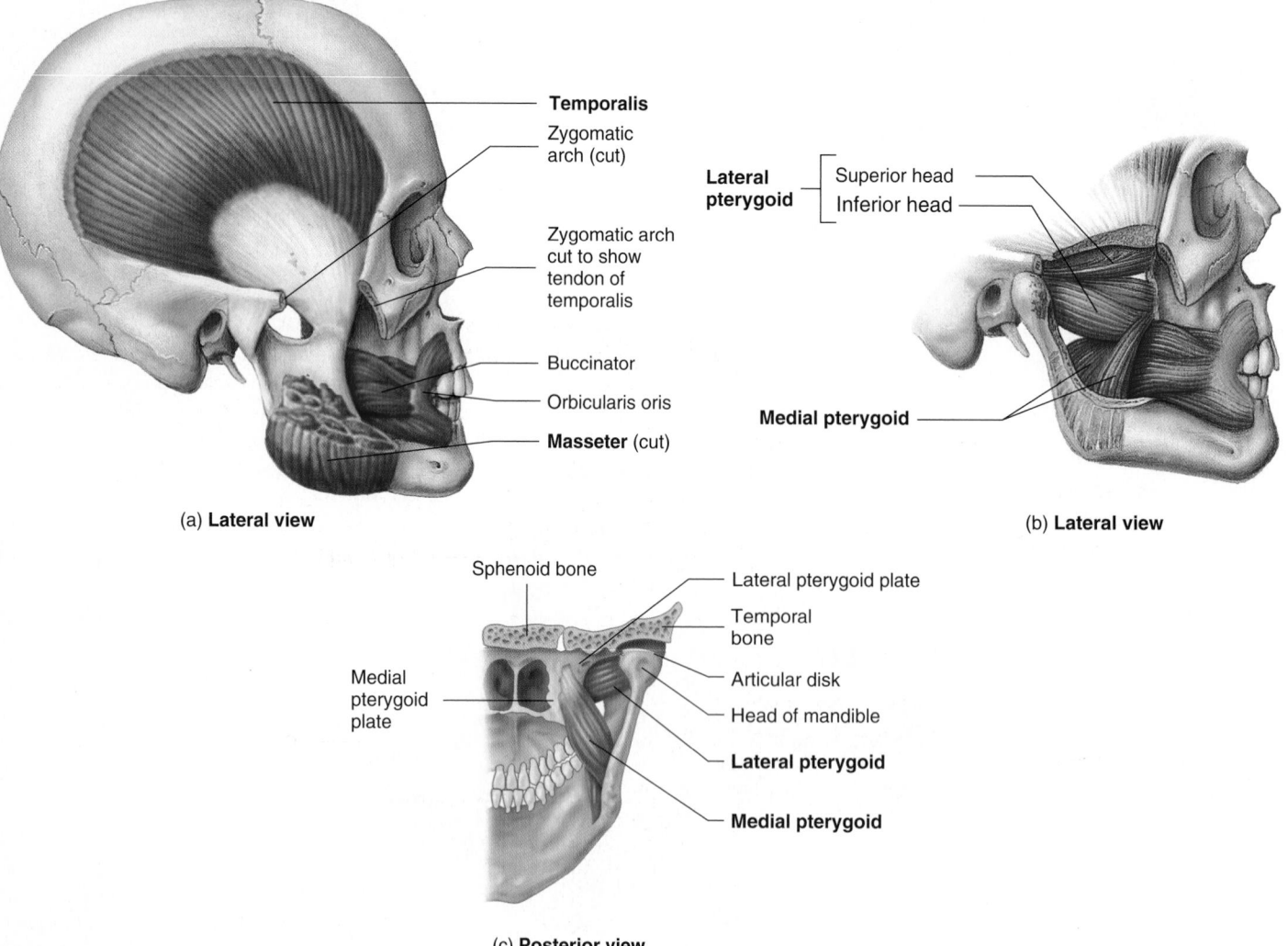

Figure 10.11   Muscles of Mastication

(a) Lateral (superficial) view. Masseter and zygomatic arch are cut away to expose the temporalis. (b) Lateral (deep) view. Masseter and temporalis muscles are removed, and the zygomatic arch and part of the mandible are cut away to reveal the deeper muscles. (c) Frontal section of the head showing the pterygoid muscles from a posterior view.

Table 10.4   Hyoid Muscles (see figures 10.12 and 10.13)

Muscle	Origin	Insertion	Nerve	Action
Suprahyoid Muscles				
Digastric (dī-gas′trik)	Mastoid process (posterior belly)	Mandible near midline (anterior belly)	Posterior belly— facial; anterior belly—mandibular division of trigeminal	Depresses and retracts mandible; elevates hyoid
Geniohyoid (jĕ-nī-ō-hī′oyd)	Genu of mandible	Body of hyoid	Fibers of C1 and C2 with hypoglossal	Protracts hyoid; depresses mandible
Mylohyoid (mī′lō-hī′oyd)	Body of mandible	Hyoid	Mandibular division of trigeminal	Elevates floor of mouth and tongue; depresses mandible when hyoid is fixed
Stylohyoid (stī-lō-hī′oyd)	Styloid process	Hyoid	Facial	Elevates hyoid
Infrahyoid Muscles				
Omohyoid (ō-mō-hī′oyd)	Superior border of scapula	Hyoid	Upper cervical through ansa cervicalis	Depresses hyoid; fixes hyoid in mandibular depression
Sternohyoid (ster′nō-hī′oyd)	Manubrium and first costal cartilage	Hyoid	Upper cervical through ansa cervicalis	Depresses hyoid; fixes hyoid in mandibular depression
Sternothyroid (ster′nō-thī′royd)	Manubrium and first or second costal cartilage	Thyroid cartilage	Upper cervical through ansa cervicalis	Depresses larynx; fixes hyoid in mandibular depression
Thyrohyoid (thī-rō-hī′oyd)	Thyroid cartilage	Hyoid	Upper cervical, passing with hypoglossal	Depresses hyoid and elevates thyroid cartilage of larynx; fixes hyoid in mandibular depression

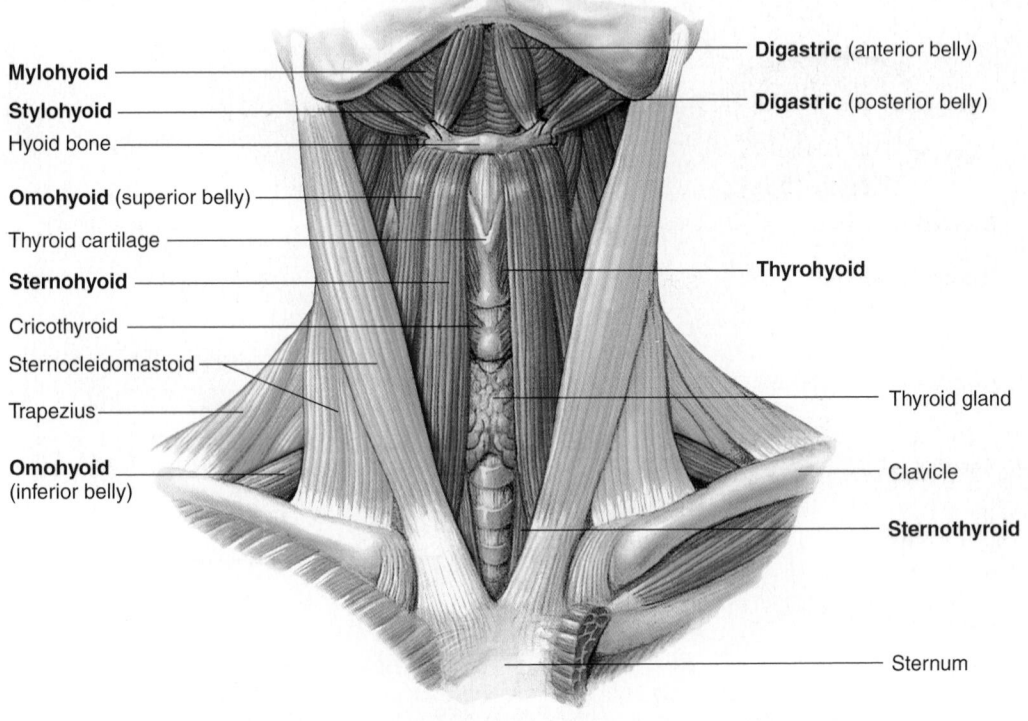

(a) **Anterior view**

Figure 10.12   Hyoid Muscles *(continued)*

(a) Anterior superficial view.

(b) **Anterior view**

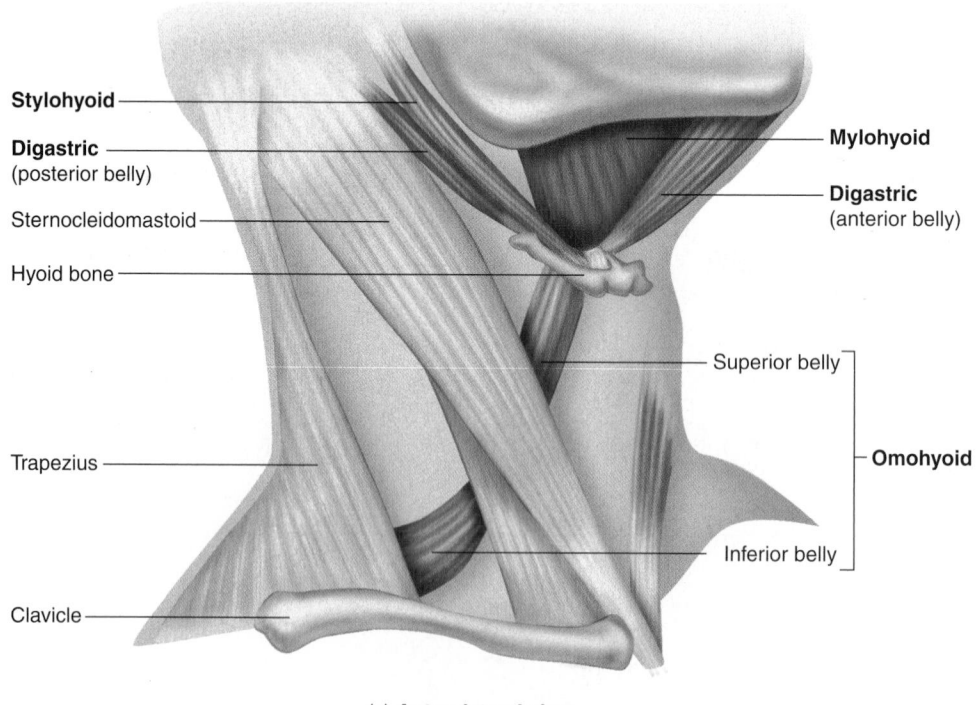

(c) **Anterolateral view**

Figure 10.12   *(continued)*

(*b*) Anterior deep view. (*c*) Anterolateral view.

move the tongue (table 10.5; figure 10.13). The intrinsic muscles are named for their fiber orientation in the tongue. The extrinsic muscles are named for their origin and insertion.

Tongue Rolling

Everyone can change the shape of the tongue, but not everyone can roll the tongue into the shape of a tube. The ability to accomplish such movements apparently is partially controlled genetically, but other factors are involved. In some cases one of a pair of identical twins can roll the tongue but the other twin cannot. It's not known exactly what tongue muscles are involved in **tongue rolling,** and no anatomic differences are reported to exist between tongue rollers and nonrollers.

● 10. *Contrast the movements produced by the extrinsic and intrinsic tongue muscles.*

Swallowing and the Larynx

The hyoid muscles (see table 10.4 and figures 10.12 and 10.13) are divided into a **suprahyoid group** superior to the hyoid bone and an **infrahyoid group** inferior to it. When the hyoid bone is fixed by the infrahyoid muscles so that the bone is stabilized from below, the suprahyoid muscles can help depress the mandible. If the suprahyoid muscles fix the hyoid and thus stabilize it from above, the thyrohyoid muscle (an infrahyoid muscle) can elevate the

Table 10.5	Tongue Muscles (see figure 10.13)			
Muscle	**Origin**	**Insertion**	**Nerve**	**Action**
Intrinsic Muscles				
Longitudinal, transverse, and vertical (not illustrated)	Within tongue	Within tongue	Hypoglossal	Change tongue shape
Extrinsic Muscles				
Genioglossus (jĕ′nī-ō-glos′ŭs)	Genu of mandible	Tongue	Hypoglossal	Depresses and protrudes tongue
Hyoglossus (hī′ō-glos′ŭs)	Hyoid	Side of tongue	Hypoglossal	Retracts and depresses side of tongue
Styloglossus (stī′lō-glos′ŭs)	Styloid process of temporal bone	Tongue (lateral and inferior)	Hypoglossal	Retracts tongue
Palatoglossus (pal-ă-tō-glos′ŭs)	Soft palate	Tongue	Pharyngeal plexus	Elevates posterior tongue

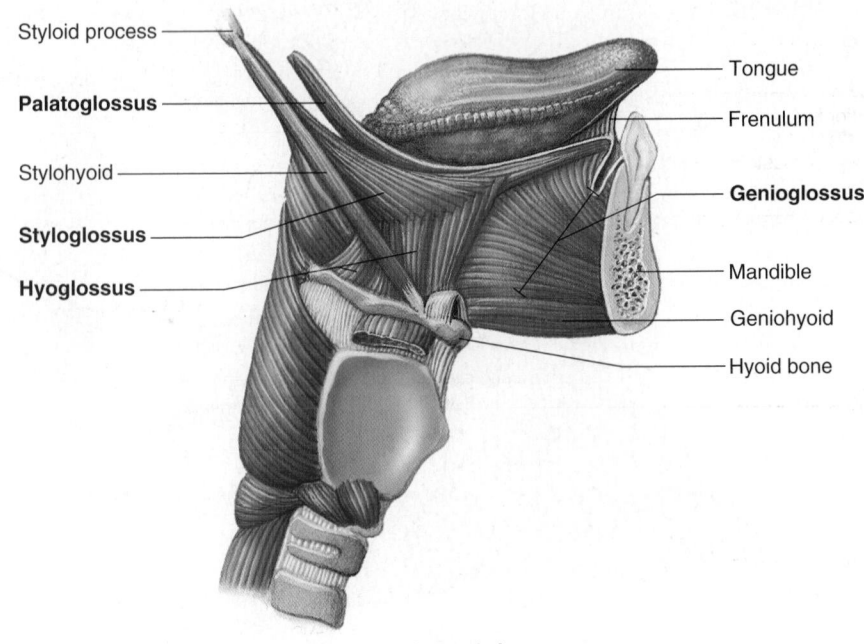

Lateral view

Figure 10.13 Tongue Muscles

As seen from the right side.

larynx. To observe this effect, place your hand on your larynx (Adam's apple) and swallow.

The soft palate, pharynx, and larynx contain several muscles involved in swallowing and speech (table 10.6 and figure 10.14). The muscles of the soft palate close the posterior opening to the nasal cavity during swallowing.

Swallowing (see chapter 24) is accomplished by elevation of the pharynx, which in turn is accomplished by elevation of the larynx, to which the pharynx is attached, and constriction of the **palatopharyngeus** (pal´ă-tō-far-in-jē´ŭs) and **salpingopharyngeus** (sal-pin´gō-far-in-jē´ŭs; *salpingo* means trumpet and refers to the trumpet-shaped opening of the auditory, or eustachian, tube). The pharyngeal constrictor muscles then constrict from superior to inferior, forcing food into the esophagus.

The salpingopharyngeus also opens the auditory tube, which connects the middle ear with the pharynx. Opening the auditory tube equalizes the pressure between the middle ear and the atmosphere; this is why it's sometimes helpful to chew gum or swallow when ascending or descending a mountain in a car or when changing altitudes in an airplane.

The muscles of the larynx are listed in table 10.6 and are illustrated in figure 10.14*b*. Most of the laryngeal muscles help to narrow or close the laryngeal opening so food does not enter the larynx when a person swallows. The remaining muscles shorten the vocal cords to raise the pitch of the voice.

Snoring and Laryngospasm

Snoring is a rough, raspy noise that can occur when a sleeping person inhales through the mouth and nose. The noise usually is made by vibration of the soft palate but also may occur as a result of vocal cord vibration.

Laryngospasm is a tetanic contraction of the muscles that narrows the opening of the larynx (arytenoids, lateral cricoarytenoids). In severe cases, the opening is closed completely, air no longer can pass through the larynx into the lungs, and the victim may die of asphyxiation. Laryngospasm can develop as a result of, for example, severe allergic reactions, tetanus infections, or hypocalcemia.

Table 10.6	Muscles of Swallowing and the Larynx (see figure 10.14)			
Muscle	**Origin**	**Insertion**	**Nerve**	**Action**
Larynx				
Arytenoids (ar-i-tē´noydz)				
Oblique (not illustrated)	Arytenoid cartilage	Opposite arytenoid cartilage	Recurrent laryngeal	Narrows opening to larynx
Transverse (not illustrated)	Arytenoid cartilage	Opposite arytenoid cartilage	Recurrent laryngeal	Narrows opening to larynx
Cricoarytenoids (krī´kō-ar-i-tē´noydz)				
Lateral (not illustrated)	Lateral side of cricoid cartilage	Arytenoid cartilage	Recurrent laryngeal	Narrows opening to larynx
Posterior (not illustrated)	Posterior side of cricoid cartilage	Arytenoid cartilage	Recurrent laryngeal	Widens opening of larynx
Cricothyroid (krī-kō-thī´royd)	Anterior cricoid cartilage	Thyroid cartilage	Superior laryngeal	Tenses vocal cords
Thyroarytenoid (thī´rō-ar´i-tē´noyd) (not illustrated)	Thyroid cartilage	Arytenoid cartilage	Recurrent laryngeal	Shortens vocal cords
Vocalis (vō-kal´ĭs) (not illustrated)	Thyroid cartilage	Arytenoid cartilage	Recurrent laryngeal	Shortens vocal cords
Soft Palate				
Levator veli palatini (lē-vā´tor, le-vā´ter vel´ī pal´ă-tē´nī)	Temporal bone and auditory tube	Soft palate	Pharyngeal plexus	Elevates soft palate
Palatoglossus (pal-ă-tō-glos´ŭs)	Soft palate	Tongue	Pharyngeal plexus	Narrows fauces; elevates posterior tongue

Table 10.6 *continued*

Muscle	Origin	Insertion	Nerve	Action
Soft Palate—cont'd.				
Palatopharyngeus (pal′ă-tō-far-in-jē′ŭs)	Soft palate	Pharynx	Pharyngeal plexus	Narrows fauces; depresses palate; elevates pharynx
Tensor veli palatini (ten′sŏr vel′ī pal′ă-tē′nī)	Sphenoid and auditory tube	Soft palate division of auditory tube	Mandibular, division of trigeminal	Tenses soft palate; opens auditory tube
Uvulae (ū′vū-lē)	Posterior nasal spine	Uvula	Pharyngeal plexus	Elevates uvula
Pharynx				
Pharyngeal constrictors (fă-rin′jē-ăl)				
Inferior	Thyroid and cricoid cartilages	Pharyngeal raphe	Pharyngeal plexus and external laryngeal nerve	Narrows inferior portion of pharynx in swallowing
Middle	Stylohyoid ligament and hyoid	Pharyngeal raphe	Pharyngeal plexus	Narrows pharynx in swallowing
Superior	Medial pterygoid plate, mandible, floor of mouth, and side of tongue	Pharyngeal raphe	Pharyngeal plexus	Narrows superior portion of pharynx in swallowing
Salpingopharyngeus (sal-ping′gō-far-in-jē′ŭs)	Auditory tube	Pharynx	Pharyngeal plexus	Elevates pharynx; opens auditory tube in swallowing
Stylopharyngeus (stī′lō-far-in-jē′ŭs)	Styloid process	Pharynx	Glossopharyngeus	Elevates pharynx

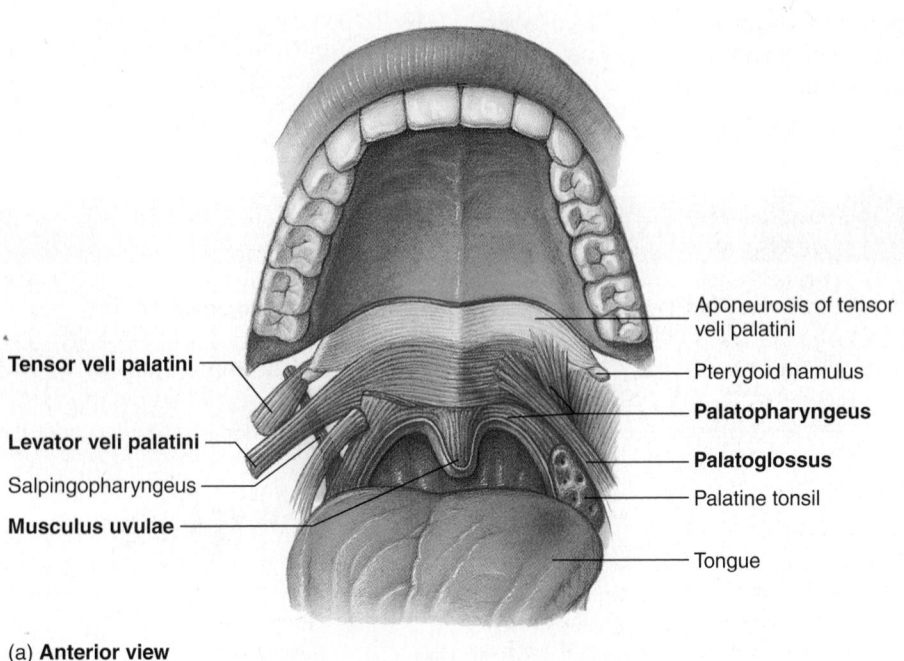

(a) **Anterior view**

Aponeurosis of tensor veli palatini
Pterygoid hamulus
Palatopharyngeus
Palatoglossus
Palatine tonsil
Tongue

Tensor veli palatini
Levator veli palatini
Salpingopharyngeus
Musculus uvulae

Figure 10.14   Muscles of the Palate, Pharynx, and Larynx *(continued)*

(*a*) Anterior-inferior view of the palate. Palatoglossus and part of the palatopharyngeus muscles are cut on one side to reveal the deeper muscles.

Tensor veli palatini
Levator veli palatini
Superior pharyngeal constrictor
Stylopharyngeus
Middle pharyngeal constrictor
Inferior pharyngeal constrictor

Pterygomandibular raphe
Buccinator
Styloglossus
Stylohyoid ligament
Hyoglossus
Mylohyoid
Hyoid bone
Thyroid cartilage
Cricothyroid
Cricoid cartilage

(b) **Lateral view**

Figure 10.14 *(continued)*

(*b*) Lateral view of the palate, pharynx, and larynx. Part of the mandible is removed to reveal the deeper structures.

Movements of the Eyeball

The eyeball rotates within the orbit to allow vision in a wide range of directions. The movements of each eye are accomplished by six muscles named for the orientation of their fasciculi relative to the spherical eye (table 10.7; figure 10.15).

Each rectus muscle (so named because the fibers are nearly straight with the axis of the eye) attaches to the eyeball anterior to the center of the sphere. The superior rectus rotates the anterior portion of the eyeball superiorly so that the pupil, and thus the gaze, are directed superiorly (looking up). The inferior

Table 10.7	Muscles Moving the Eye (see figure 10.15)			
Muscle	**Origin**	**Insertion**	**Nerve**	**Action**
Oblique				
Inferior	Orbital plate of maxilla	Sclera of eye	Oculomotor	Elevates and laterally deviates gaze
Superior	Fibrous ring	Sclera of eye	Trochlear	Depresses and laterally deviates gaze
Rectus				
Inferior	Fibrous ring	Sclera of eye	Oculomotor	Depresses and medially deviates gaze
Lateral	Fibrous ring	Sclera of eye	Abducent	Laterally deviates gaze
Medial	Fibrous ring	Sclera of eye	Oculomotor	Medially deviates gaze
Superior	Fibrous ring	Sclera of eye	Oculomotor	Elevates and medially deviates gaze

Optic nerve

View

Levator palpebrae
superioris (cut)

Lateral rectus

Superior rectus

Inferior oblique

Medial rectus

Superior oblique

Trochlea

(a) **Superior view**

Trochlea

Levator palpebrae
superioris (cut)

Optic nerve

Inferior rectus

Superior oblique

Superior rectus

View

Lateral rectus

Inferior oblique

(b) **Lateral view**

Figure 10.15 Muscles Moving the Right Eyeball

(*a*) Superior view of the right eyeball. (*b*) Lateral view of the right eyeball.

rectus depresses the gaze, the lateral rectus laterally deviates the gaze (looking to the side), and the medial rectus medially deviates the gaze (looking toward the nose). The superior rectus and inferior rectus are not completely straight in their orientation to the eye; thus they also medially deviate the gaze as they contract.

The oblique muscles (so named because their fibers are oriented obliquely to the axis of the eye) insert onto the posterolateral margin of the eyeball so that both muscles laterally deviate the gaze as they contract (see chapter 15, figure 15.10). The superior oblique elevates the posterior part of the eye, thus directing the pupil inferiorly and depressing the gaze. The inferior oblique elevates the gaze.

11. *Explain the interaction of the suprahyoid and infrahyoid muscles to depress the mandible and to elevate the larynx.*
12. *Which muscles open and close the openings to the auditory tube and larynx?*
13. *Describe the muscles of the eye and the movements that they cause.*

P R E D I C T

Strabismus (stra-biz′mŭs) is a condition in which one or both eyes deviate in a medial or lateral direction. In some cases the condition may be caused by a weakness in either the medial or lateral rectus muscle. If the lateral rectus of the right eye is weak, in which direction would the eye deviate?

Trunk Muscles

Muscles Moving the Vertebral Column

The muscles that extend, laterally flex, and rotate the vertebral column are divided into deep and superficial groups (table 10.8). In general, the muscles of the deep group extend from vertebra to vertebra, whereas the muscles of the superficial group extend from the vertebrae to the ribs. In humans, these back muscles are very strong to maintain erect posture. Comparable muscles in cattle are relatively delicate, although quite large. They constitute the area from which tenderloin steaks are cut. The **erector spinae** (spī′nē) group of muscles on each side of the back consists of three subgroups: the **iliocostalis** (il′ē-ō-kos-tā′lis), the **longissimus** (lon-gis′i-mŭs), and the **spinalis** (sp-ī-nā′lis). The longissimus group accounts for most of the muscle mass in the lower back (figure 10.16). The deepest muscles of the back attach

Table 10.8	Muscles Acting on the Vertebral Column (see figures 10.7, 10.8, 10.16, and 10.17)			
Muscle	**Origin**	**Insertion**	**Nerve**	**Action**
Superficial				
Erector spinae (ē-rek′tŏr, ē-rek′tōr spī′nē) (divides into three columns)				
Iliocostalis (il′ē-ō-kos-tā′lis)	Sacrum, ilium, and lumbar spines	Ribs and vertebrae	Dorsal rami of spinal nerves	Extends vertebral column
Cervicis (ser-vī′sis)	Superior six ribs	Middle cervical vertebrae	Dorsal rami of thoracic nerves	Extends, laterally flexes, and rotates vertebral column
Thoracis (thō-ra′sis)	Inferior six ribs	Superior six ribs	Dorsal rami of thoracic nerves	Extends, laterally flexes, and rotates vertebral column
Lumborum (lum-bōr′ŭm)	Sacrum, ilium, and lumbar vertebrae	Inferior six ribs	Dorsal rami of thoracic and lumbar nerves	Extends, laterally flexes, and rotates vertebral column
Longissimus (lon-gis′i-mŭs)				
Capitis (ka′pĭ-tis)	Upper thoracic and lower cervical vertebrae	Mastoid process	Dorsal rami of cervical nerves	Extends head
Cervicis (ser-vī′sis)	Upper thoracic vertebrae	Upper cervical vertebrae	Dorsal rami of cervical nerves	Extends neck
Thoracis (thō-ra′sis)	Ribs and lower thoracic vertebrae	Upper lumbar vertebrae and ribs	Dorsal rami of thoracic and lumbar nerves	Extends vertebral column
Spinalis (spī-nā′lis)				
Cervicis (ser-vī′sis) (not illustrated)	C6–C7	C2–C3	Dorsal rami of cervical nerves	Extends neck
Thoracis (thō-ra′sis)	T11–L2	Middle and upper thoracic vertebrae	Dorsal rami of thoracic nerves	Extends vertebral column
Semispinalis (sem′ē-spī-nā′lis)				
Cervicis (ser-vī′sis)	Transverse processes of T2–T5	Spinous processes of C2–C5	Dorsal rami of cervical nerves	Extends neck
Thoracis (thō-ra′sis)	Transverse processes of T5–T11	Spinous processes of C5–T4	Dorsal rami of thoracic nerves	Extends vertebral column
Splenius cervicis (splē′nē-ŭs ser-vī′sis)	C3–C5	C1–C3	Dorsal rami of cervical nerves	Rotates and extends neck
Longus colli (lon′gŭs kō′lī) (not illustrated)	C3–T3	C1–C6	Ventral rami of cervical nerves	Rotates and flexes neck

Table 10.8 *continued*

Muscle	Origin	Insertion	Nerve	Action
Deep				
Interspinales (in-ter-spī-nā'lēz)	Spinous processes of all vertebrae	Next superior spinous process	Dorsal rami of spinal nerves	Extends back and neck
Intertransversarii (in-ter-trans'ver-săr'ē-ī)	Transverse processes of all vertebrae	Next superior transverse process	Dorsal rami of spinal nerves	Laterally flexes vertebral column
Multifidus (mŭl-tif'i-dŭs)	Transverse processes of vertebrae, posterior surface of sacrum and ilium	Spinous processes of superior vertebrae	Dorsal rami of spinal nerves	Extends and rotates vertebral column
Psoas minor (sō'as mī'ner)	T12–L1	Near pubic crest	L1	Flexes vertebral column
Rotatores (rō-tā'tōrz)	Transverse processes of all vertebrae	Base of spinous process of superior vertebrae	Dorsal rami of spinal nerves	Extends and rotates vertebral column

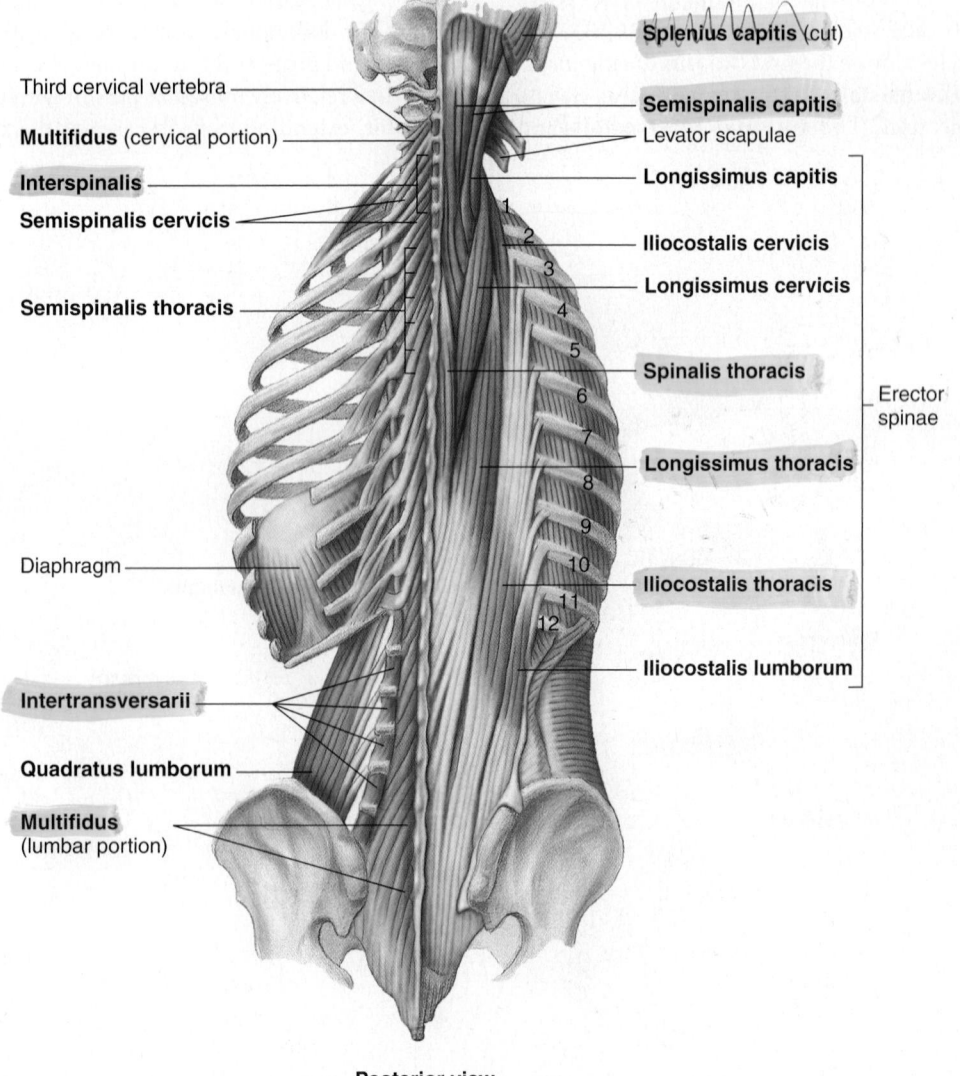

Posterior view

Figure 10.16 Deep Back Muscles

On the right, the erector spinae group of muscles is demonstrated. On the left, these muscles are removed to reveal the deeper back muscles.

between the spinous and transverse processes of individual vertebra (figure 10.17).

● **14.** *List the actions of the group of back muscles that attaches to the vertebrae or ribs (or both). What is the name of the superficial group?*

Back Pain

Low back pain can result from poor posture, from being overweight, or from having a poor fitness level. A few changes may help: sitting and standing up straight; using a low-back support when sitting; losing weight; exercising, especially the back and abdominal muscles; and sleeping on your side on a firm mattress. Sleeping on your side all night, however, may be difficult because most people change position over 40 times during the night.

Thoracic Muscles

The muscles of the thorax are involved mainly in the process of breathing (see chapter 23). Four major groups of muscles are associated with the rib cage (table 10.9 and figure 10.18). The **scalene** (skā′lēn) muscles elevate the first two ribs during inspiration. The **external intercostals** (in-ter-kos′tălz) also elevate the ribs during inspiration. The **internal intercostals** and

transversus thoracis (thō-ra′sis) muscles depress the ribs during forced expiration.

The **diaphragm** (dī′ă-fram; see figure 10.18a) causes the major movement produced during quiet breathing. It is a dome-shaped structure and when it contracts, the dome flattens slightly, causing the volume of the thoracic cavity to increase, resulting in inspiration. If this dome of skeletal muscle or the phrenic nerve supplying it is severely damaged, the amount of air moving into and out of the lungs may be so small that the individual is likely to die without the aid of an artificial respirator.

● **15.** *Name the muscle that is mainly responsible for respiratory movements. How do other muscles aid this movement?*

Abdominal Wall

The muscles of the anterior abdominal wall (table 10.10 and figures 10.19 and 10.20) flex and rotate the vertebral column. Contraction of the abdominal muscles when the vertebral column is fixed decreases the volume of the abdominal cavity and the thoracic cavity and can aid in such functions as forced expiration, vomiting, defecation, urination, and childbirth. The crossing pattern of the abdominal muscles creates a strong anterior wall that holds in and protects the abdominal viscera.

In a relatively muscular person with little fat, a vertical line is visible, extending from the area of the xiphoid process of the

Posterolateral view

Figure 10.17   Deep Muscles Associated with the Vertebrae

Table 10.9   Muscles of the Thorax (see figure 10.18)

Muscle	Origin	Insertion	Nerve	Action
Diaphragm	Interior of ribs, sternum, and lumbar vertebrae	Central tendon of diaphragm	Phrenic	Inspiration; depresses floor of thorax
Intercostalis (in'ter-kos-ta'lis)				
External	Inferior margin of each rib	Superior border of next rib below	Intercostal	Inspiration; elevates ribs
Internal	Superior margin of each rib	Inferior border of next rib above	Intercostal	Expiration; depresses ribs
Scalenus (skā-lē'nŭs)				
Anterior	C3–C6	First rib	Cervical plexus	Elevates first rib
Medial	C2–C6	First rib	Cervical plexus	Elevates first rib
Posterior	C4–C6	Second rib	Cervical and brachial plexuses	Elevates second rib
Serratus posterior (sĕr-ā'tŭs)				
Inferior (not illustrated)	T11–L2	Inferior four ribs	Ninth to twelfth intercostals	Depresses inferior ribs and extends back
Superior (not illustrated)	C6–T2	Second to fifth ribs	First to fourth intercostals	Elevates superior ribs
Transversus thoracis (trans-ver'sus thō-ra'sis) (not illustrated)	Sternum and xiphoid process	Second to sixth costal cartilages	Intercostal	Decreases diameter of thorax

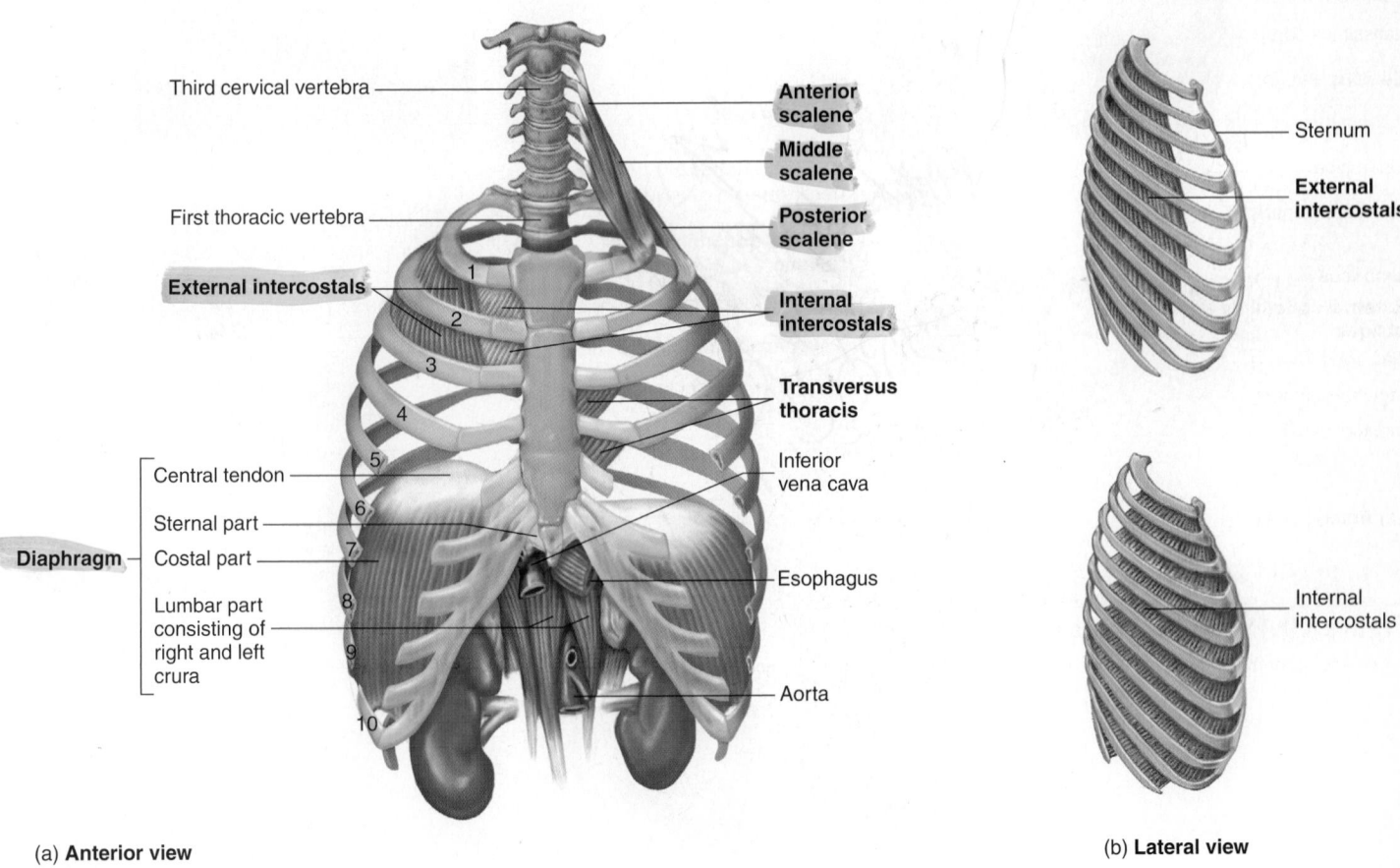

(a) **Anterior view**

(b) **Lateral view**

Figure 10.18   Muscles of the Thorax

(a) Anterior view. A few selected intercostal muscles and the diaphragm are demonstrated. (b) Lateral view.

Table 10.10	Muscles of the Abdominal Wall (see figures 10.5*a*, 10.19, and 10.20)			
Muscle	**Origin**	**Insertion**	**Nerve**	**Action**
Anterior				
Rectus abdominis (rek´tŭs ab-dom´i-nis)	Pubic crest and symphysis pubis	Xiphoid process and inferior ribs	Branches of lower thoracic	Flexes vertebral column; compresses abdomen
External abdominal oblique	Fifth to twelfth ribs	Iliac crest, inguinal ligament, and rectus sheath	Branches of lower thoracic	Flexes and rotates vertebral column; compresses abdomen; depresses thorax
Internal abdominal oblique	Iliac crest, inguinal ligament, and lumbar fascia	Tenth to twelfth ribs and rectus sheath	Lower thoracic	Flexes and rotates vertebral column; compresses abdomen; depresses thorax
Transversus abdominis (trans-ver´sŭs ab-dom´i-nis)	Seventh to twelfth costal cartilages, lumbar fascia, iliac crest, and inguinal ligament	Xiphoid process, linea alba, and pubic tubercle	Lower thoracic	Compresses abdomen
Posterior				
Quadratus lumborum (kwah-drā´tŭs lŭm-bōr´ŭm)	Iliac crest and lower lumbar vertebrae	Twelfth rib and upper lumbar vertebrae	Upper lumbar	Laterally flexes vertebral column and depresses twelfth rib

(a) **Anterior view**

(b)

Figure 10.19 Anterior Abdominal Wall Muscles

(*a*) Anterior abdominal muscles. Windows in the side reveal the various muscle layers. (*b*) Surface anatomy of anterior abdominal muscle.

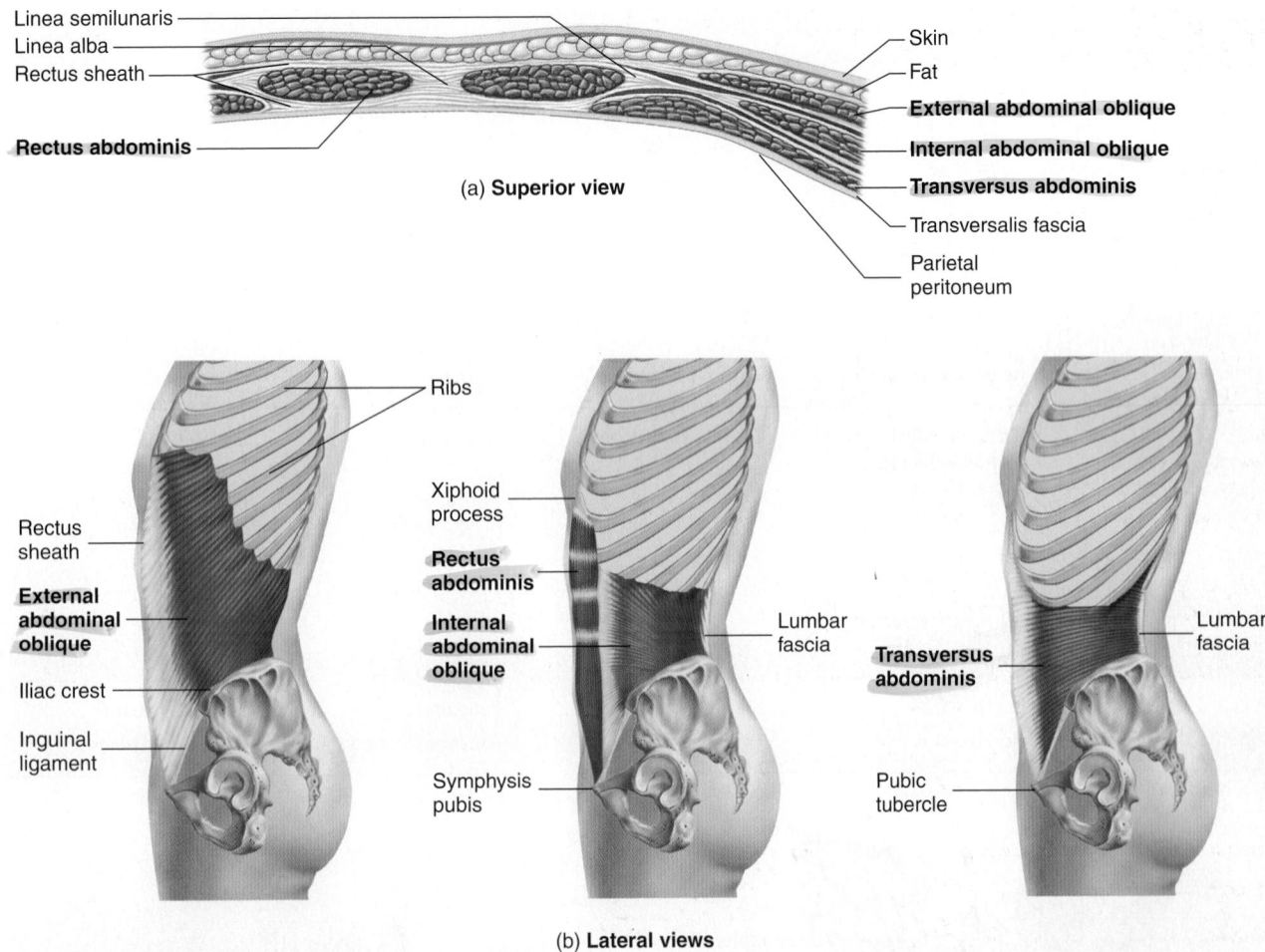

(a) **Superior view**

(b) **Lateral views**

Figure 10.20 Anterior Abdominal Wall Muscles

(a) Cross section superior to the umbilicus. (b) Abdominal muscles shown individually (lateral views).

sternum through the navel to the pubis. This tendinous area of the abdominal wall is devoid of muscle; the **linea alba** (lin′ē-ă al′bă), or white line, is so named because it consists of white connective tissue rather than muscle (see figure 10.19). On each side of the linea alba is the **rectus abdominis** (see figures 10.19 and 10.20), surrounded by a **rectus sheath**. **Tendinous intersections** (tendinous inscriptions) transect the rectus abdominis at three, or sometimes more, locations, causing the abdominal wall of a well-muscled person to appear segmented. Lateral to the rectus abdominis is the **linea semilunaris** (sem-ē-loo-nar′is, meaning a crescent- or half-moon-shaped line); lateral to it are three layers of muscle (see figures 10.19 and 10.20). From superficial to deep, these muscles are the **external abdominal oblique, internal abdominal oblique,** and **transversus abdominis.**

● 16. *Explain the anatomic basis for the segments ("cuts") seen on a well-muscled individual's abdomen. What are the functions of the abdominal muscles? List the muscles of the anterior abdominal wall.*

Pelvic Floor and Perineum

The pelvis is a ring of bone (see chapter 7) with an inferior opening that is closed by a muscular wall through which the anus and the urogenital openings penetrate (table 10.11). Most of the pelvic floor is formed by the **coccygeus** (kok-si′jē-ŭs) muscle and the **levator ani** (a′nī) muscle, referred to jointly as the **pelvic diaphragm.** The area inferior to the pelvic floor is the **perineum** (per′i-nē′ŭm), which is somewhat diamond-shaped (figure 10.21). The anterior half of the diamond is the urogenital triangle, and the posterior half is the anal triangle (see chapter 28). The urogenital triangle contains the **urogenital diaphragm,** which forms a "subfloor" to the pelvis in that area and consists of the **deep transverse perineal** (pĕr′ĭ-nē′ăl) muscle and the **external urethral sphincter** muscle. During pregnancy, the muscles of the pelvic diaphragm and urogenital diaphragm may be stretched by the extra weight of the fetus, and specific exercises are designed to strengthen them.

● 17. *What openings penetrate the pelvic floor muscles? Name the area inferior to the pelvic floor.*

Table 10.11	Muscles of the Pelvic Floor and Perineum (see figure 10.21)			
Muscle	**Origin**	**Insertion**	**Nerve**	**Action**
Bulbospongiosus (bul′bō-spŭn′jē-ō′sŭs)	Male—central tendon of perineum and median raphe of penis	Dorsal surface of penis and bulb of penis	Pudendal	Constricts urethra; erects penis
	Female—central tendon of perineum	Base of clitoris	Pudendal	Erects clitoris
Coccygeus (kok-si′jē-ŭs) (not illustrated)	Ischial spine	Coccyx	S3 and S4	Elevates and supports pelvic floor
Ischiocavernosus (ish′ē-ō-kav′er-nō′sŭs)	Ischial ramus	Corpus cavernosum	Perineal	Compresses base of penis or clitoris
Levator ani (lē-vā′tor, le-vā′ter ā′nī)	Posterior pubis and ischial spine	Sacrum and coccyx	Fourth sacral	Elevates anus; supports pelvic viscera
External anal sphincter (ā′nă l sfingk′ter)	Coccyx	Central tendon of perineum	Fourth sacral and pudenda	Keeps orifice of anal canal closed
External urethral sphincter (ū-rē′thră l sfingk′ter) (not illustrated)	Pubic ramus	Median raphe	Pudendal	Constricts urethra
Transverse perinei (pĕr′i-nē′ī)				
Deep	Ischial ramus	Median raphe	Pudendal	Supports pelvic floor
Superficial	Ischial ramus	Central perineal	Pudendal	Fixes central tendon

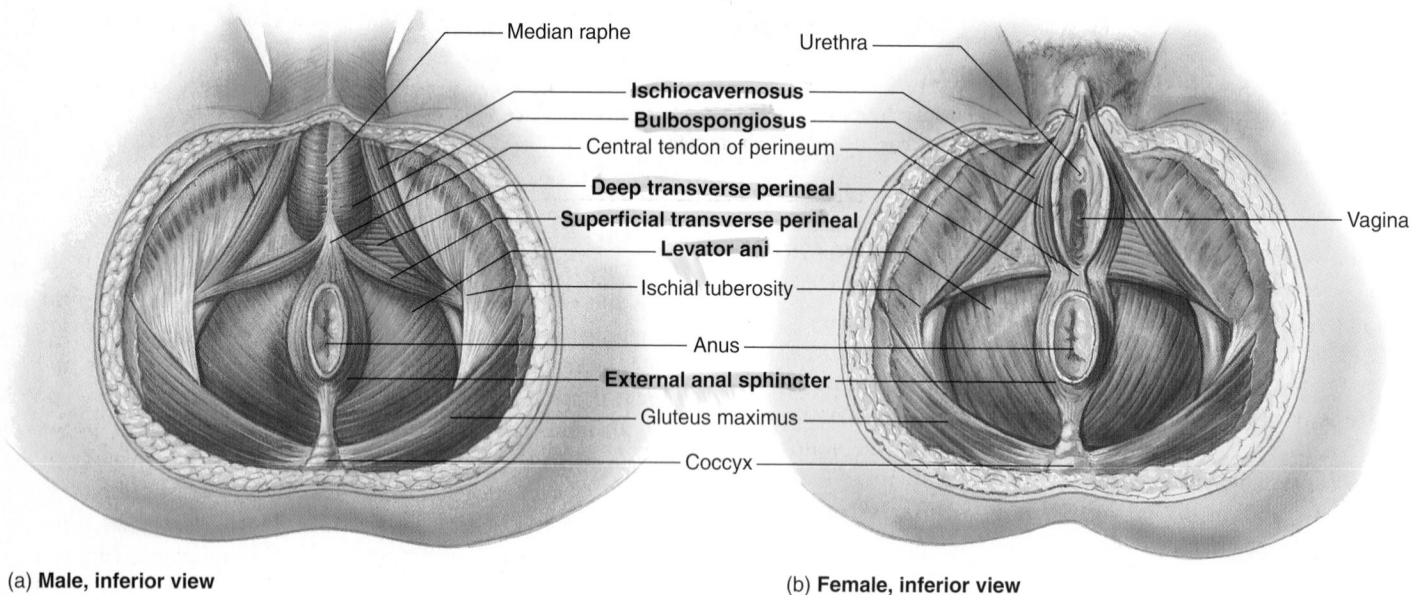

(a) **Male, inferior view**   (b) **Female, inferior view**

Figure 10.21   Muscles of the Pelvic Floor and Perineum

Inferior view. (a) Male. (b) Female.

Upper Limb Muscles

The muscles of the upper limb include those that move the scapula, and those that move the arm, the forearm, and the hand.

Scapular Movements

The major connection of the upper limb to the body is accomplished by muscles (table 10.12 and figure 10.22). The muscles attaching the scapula to the thorax include the **trapezius, levator scapulae** (skap′ū-lē), **rhomboideus** (rom-bō-id′ē-ŭs) **major** and **minor, serratus** (sĕr-ā′tŭs) **anterior,** and **pectoralis** (pek′tō-ra′lis) **minor.** These muscles move the scapula, permitting a wide range of movements of the upper limb, or act as fixators to hold the scapula firmly in position when the muscles of the arm contract. The superficial muscles that act on the scapula can be easily seen on a living person (see figure 10.7b): the trapezius forms the upper line from each shoulder to the neck, and the origin of the serratus anterior from the first eight or nine ribs can be seen along the lateral thorax. The serratus anterior inserts onto the medial border of the scapula (see figures 10.22c and 10.27).

Table 10.12	Muscles Acting on the Scapula (see figure 10.22)			
Muscle	**Origin**	**Insertion**	**Nerve**	**Action**
Levator scapulae (lē-vā′tor, le-vā′ter skap′ū-lē)	C1–C4	Superior angle of scapula	Dorsal scapular	Elevates, retracts, and rotates scapula; laterally flexes neck
Pectoralis minor (pek′tō-ra′lis)	Third to fifth ribs	Coracoid process of scapula	Anterior thoracic	Depresses scapula or elevates ribs
Rhomboideus (rom-bō-id′ē-ŭs)				
Major	T1–T4	Medial border of scapula	Dorsal scapular	Retracts, rotates, and fixes scapula
Minor	C6–C7	Medial border of scapula	Dorsal scapular	Retracts, slightly elevates, rotates, and fixes scapula
Serratus anterior (ser-ā′tŭs)	First to eighth or ninth ribs	Medial border of scapula	Long thoracic	Rotates and protracts scapula; elevates ribs
Subclavius (sŭb-klā′vē-ŭs)	First rib	Clavicle	Subclavian	Fixes clavicle or elevates first rib
Trapezius (tra-pē′zē-ŭs)	External occipital protuberance, ligamentum nuchae, and C7–T12	Clavicle, acromion process, and scapular spine	Accessory and cervical plexus	Elevates, depresses, retracts, rotates, and fixes scapula; extends neck

Trapezius

Seventh cervical vertebra

Levator scapulae

Rhomboideus minor

Rhomboideus major

(a) **Posterior view**

Subclavius

Coracoid process

Pectoralis minor (cut)

Subscapularis

Pectoralis major (cut)

Biceps brachii

Latissimus dorsi

Serratus anterior

Pectoralis major (cut)

Supraspinatus tendon

Subscapularis

Teres minor

Three of four rotator cuff muscles

Teres major (cut)

Pectoralis minor

Latissimus dorsi (cut)

External abdominal oblique

(b) **Anterior view**

Ribs

Scapula

Serratus anterior

Humerus

(c) **Lateral view**

Figure 10.22 Muscles Acting on the Scapula

(a) Posterior view. Trapezius is removed on the right to reveal the deeper muscles.
(b) Anterior view. Pectoralis major is removed on both sides. The pectoralis minor is also removed on the right side. (c) Lateral view of the serratus anterior.

Arm Movements

The arm is attached to the thorax by several muscles, including the **pectoralis major** and the **latissimus dorsi** (lă-tis′i-mŭs dōr′sī) muscles (table 10.13 and figures 10.23 and 10.24; see figure 10.22b). Notice that the pectoralis major muscle is listed in table 10.13 as both a flexor and extensor. The muscle flexes the extended shoulder and extends the flexed shoulder. Try these movements yourself and notice the position and action of the muscle. The **deltoid** (deltoideus) muscle (see figure 10.23) also is listed in table 10.13 as a flexor and extensor. The deltoid muscle is like three

Table 10.13	Muscles Acting on the Arm (see figures 10.22–10.27)			
Muscle	**Origin**	**Insertion**	**Nerve**	**Action**
Coracobrachialis (kōr′ă-kō-brā-kē-ā′lis)	Coracoid process of scapula	Midshaft of humerus	Musculocutaneous	Adducts arm and flexes shoulder
Deltoid (del′toyd)	Clavicle, acromion process, and scapular spine	Deltoid tuberosity	Axillary	Flexes and extends shoulder; abducts and medially and laterally rotates arm
Latissimus dorsi (lă-tis′i-mŭs dōr′sī)	T7–L5, sacrum and iliac crest	Medial crest of intertubercular groove	Thoracodorsal	Adducts and medially rotates arm; extends shoulder
Pectoralis major (pek′tō-rā′lis)	Clavicle, sternum, and abdominal aponeurosis	Lateral crest of intertubercular groove	Anterior thoracic	Flexes shoulder; adducts and medially rotates arm; extends shoulder from flexed position
Teres major (ter′ēz, tēr-ēz)	Lateral border of scapula	Medial crest of intertubercular groove	Subscapular C5 and C6	Extends shoulder; adducts and medially rotates arm
Rotator Cuff				
Infraspinatus (in-fră-spī-nā′tŭs)	Infraspinous fossa of scapula	Greater tubercle of humerus	Suprascapular C5 and C6	Laterally rotates arm; holds head of humerus in place
Subscapularis (sŭb-skap-ū-lā′ris)	Subscapular fossa	Lesser tubercle of humerus	Subscapular C5 and C6	Medially rotates arm; holds head of humerus in place
Supraspinatus (soo-pră-spī-nā′tŭs)	Supraspinous fossa	Greater tubercle of humerus	Suprascapular C5 and C6	Abducts arm; holds head of humerus in place
Teres minor (ter′ēz, tēr-ēz)	Lateral border of scapula	Greater tubercle of humerus	Axillary C5 and C6	Laterally rotates and adducts arm; holds head of humerus in place

(a) **Anterior view**

Figure 10.23 Anterior Muscles Attaching the Upper Limb to the Body

(a) Anterior pectoral muscles.

Deltoid

Sternocleidomastoid

Biceps brachii

Pectoralis major

Serratus anterior

(b) **Anterior view**

Clavicle

Acromion process

Deltoid

Biceps brachii

Trapezius

Sternocleidomastoid

Pectoralis major

Serratus anterior

(c) **Anterior view**

Figure 10.23   *(continued)*

(*b*) Anterior view of the right pectoral region of a cadaver. (*c*) Surface anatomy of the right anterior pectoral region.

Levator scapulae

Rhomboideus minor

Rhomboideus major

Supraspinatus

Infraspinatus

Subscapularis (anterior to scapula and seen in figure 10.25)

Teres minor

Teres major

Latissimus dorsi

Twelfth thoracic vertebra

External abdominal oblique

Rotator cuff

(a) **Posterior view**

Figure 10.24   Posterior Muscles Attaching the Upper Limb to the Body

(*a*) Posterior view of muscles.

(b) **Posterior view**

(c) **Posterior view**

Figure 10.24 *(continued)*

(*b*) Posterior view of the left pectoral region of a cadaver. (*c*) Surface anatomy of the left posterior pectoral region.

muscles in one: the anterior fibers flex the shoulder; the lateral fibers abduct the arm; and the posterior fibers extend the shoulder. The deltoid muscle is part of the group of muscles that binds the humerus to the scapula. The primary muscles holding the head of the humerus in the glenoid cavity, however, are called the **rotator cuff muscles** (listed separately in table 10.13) because they form a cuff or cap over the proximal humerus (figure 10.25). A rotator cuff injury involves damage to one or more of these muscles or their tendons, usually the supraspinatus muscle. The muscles moving the arm are involved in flexion, extension, abduction, adduction, rotation, and circumduction (table 10.14).

Abduction of the arm involves the deltoid, rotator cuff muscles, the trapezius, and the serratus anterior. Abduction from the anatomic position through the first 90 degrees (to the point at which the hand is level to the shoulder) involves movement of the humerus and is accomplished by the deltoid muscle assisted by the rotator cuff muscles, which hold the head of the humerus tightly in place. In the initial phase of abduction (first 15 degrees) the deltoid is assisted by the supraspinatus. Place your hand on your deltoid and feel it contract as you abduct 90 degrees. Abduction from 90 degrees to 180 degrees, so that the hand is held high above the head, primarily involves rotation of the scapula, which is accomplished by the trapezius and serratus anterior muscles. Feel the inferior angle of your scapula as you abduct to 90 degrees and then to 180 degrees. Do you notice a big difference? Abduction from 90 degrees to 180 degrees, however, cannot occur unless the head of the humerus is held tightly in the glenoid cavity by the rotator cuff muscles, especially the supraspinatus. Damage to the supraspinatus muscle can prevent abduction past 90 degrees.

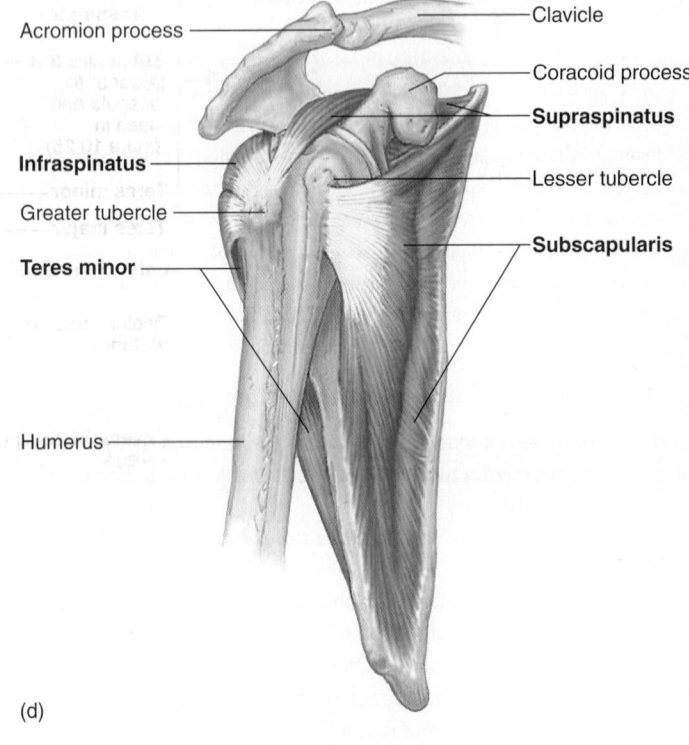

(d)

Anterior view

Figure 10.25 Right Rotator Cuff Muscles

Table 10.14	Summary of Muscle Actions on the Shoulder and Arm				
Flexion	**Extension**	**Abduction**	**Adduction**	**Medial Rotation**	**Lateral Rotation**
Deltoid	Deltoid	Deltoid	Pectoralis major	Pectoralis major	Deltoid
Pectoralis major	Teres major	Supraspinatus	Latissimus dorsi	Teres major	Infraspinatus
Coracobrachialis	Lattissimus dorsi		Teres major	Lattissimus dorsi	Teres minor
Biceps brachii	Pectoralis major		Teres minor	Deltoid	
	Triceps brachii		Triceps brachii	Subscapularis	
			Coracobrachialis		

P R E D I C T   **4**

A tennis player complains of pain in the shoulder when abducting the arm in attempting to serve or when attempting an overhead volley (during extreme abduction). In extreme abduction, the supraspinatus muscle rises superiorly and may be damaged by compression against what bony structure?

Several muscles acting on the arm can be seen very clearly in the living individual (see figures 10.23c and 10.24c). The pectoralis major forms the upper chest, and the deltoids are prominent over the shoulders. The deltoid is a common site for administering injections.

Shoulder Pain

Baseball pitchers, because they throw very hard, may tear their rotator cuffs. Such tears result in pain in the anterosuperior part of the shoulder. Older people may also develop such pain because of **degenerative tendonitis** of the rotator cuff. The supraspinatus tendon is the most commonly affected part of the rotator cuff in either trauma or degeneration, probably because it has a relatively poor blood supply.

Pain in the shoulder can also result from **subacromial bursitis**, which is inflammation of the subacromial bursa. **Biceps tendinitis,** inflammation of the biceps brachii long head tendon, can also cause shoulder pain. This inflammation is also commonly caused by throwing a baseball or football.

18. Name seven muscles that attach the humerus to the scapula. What two muscles attach the humerus directly to the trunk?
19. List the muscles forming the rotator cuff, and describe their function.
20. What muscles cause flexion and extension of the shoulder? Abduction and adduction of the arm? What muscle is involved in abduction of the arm to 90 degrees? Above 90 degrees? What muscles cause rotation of the arm?
21. List the muscles that cause flexion and extension of the elbow. Where are these muscles located?

Forearm Movements
Flexion and Extension of the Elbow

Extension of the elbow is accomplished by the **triceps brachii** (brā′kē-ī) and **anconeus** (ang-kō′nē-ŭs); flexion of the elbow is accomplished by the **brachialis** (brā′-kē-al′is), **biceps brachii,** and **brachioradialis** (brā′kē-ō-rā′dē-al′is; table 10.15; figures 10.26 and 10.27). The triceps constitute the main mass visible on the posterior aspect of the arm (see figures 10.24c and 10.26c). The biceps brachii is readily visible on the anterior aspect of the arm (see figures 10.23c and 10.26c). The brachialis lies deep to the biceps and can be seen only as a mass on the medial and lateral sides of the arm. The brachioradialis forms a bulge on the anterolateral side of the forearm just distal to the elbow (see figures 10.26a, 10.28b, and 10.28d). If the elbow is forcefully flexed in the midprone position (midway between pronation and supination), the brachioradialis stands out clearly on the forearm (see figure 10.28d).

Supination and Pronation

Supination of the forearm and hand is accomplished by the **supinator** and the **biceps brachii** (see figures 10.27 and 10.28c and 10.29b). Pronation is a function of the **pronator quadratus** (kwah-drā′tŭs) and the **pronator teres** (ter′ēz, tēr-ēz) (figure 10.28a and c; and see figure 10.27).

22. Supination and pronation of the forearm and hand are produced by what muscles? Where are these muscles located?

P R E D I C T   **5**

Explain the difference between doing chin-ups with the forearm supinated versus pronated. Which muscle or muscles are used in each type of chin-up? Which type is easier? Why?

Table 10.15	Muscles Acting on the Forearm (see figures 10.26–10.29)			
Muscle	**Origin**	**Insertion**	**Nerve**	**Action**
Arm				
Biceps brachii (bī'seps brā'kē-ī)	Long head—supraglenoid tubercle Short head—coracoid process	Radial tuberosity	Musculocutaneous	Flexes shoulder and elbow; supinates forearm and hand
Brachialis (brā'kē-al'is)	Humerus	Ulnar tuberosity and coronoid process of ulna	Musculocutaneous and radial	Flexes elbow
Triceps brachii (trī'seps brā'kē-ī)	Long head—infraglenoid tubercle on the lateral border of scapula Lateral head—lateral and posterior surface of humerus Medial head—posterior humerus	Olecranon process of ulna	Radial	Extends elbow; extends shoulder and adducts arm
Forearm				
Anconeus (ang-kō'nē-ŭs)	Lateral epicondyle of humerus	Olecranon process and posterior ulna	Radial	Extends elbow
Brachioradialis (brā'kē-ō-rā'dē-al'is)	Lateral supracondylar ridge of humerus	Styloid process of radius	Radial	Flexes elbow
Pronator quadratus (prō-nā-ter, prō-nā-tōr kwah-drā'tŭs)	Distal ulna	Distal radius	Anterior interosseous	Pronates forearm and hand
Pronator teres (prō-nā-tōr ter'ēz, ter-ēz)	Medial epicondyle of humerus and coronoid process of ulna	Radius	Median	Pronates forearm and hand
Supinator (soo'pi-nā-ter, soo'pi-nā-tōr)	Lateral epicondyle of humerus and ulna	Radius	Radial	Supinates forearm and hand

(a) **Lateral view**

Figure 10.26 Lateral Right Arm Muscles *(continued)*

(a) Lateral view of the right shoulder and arm.

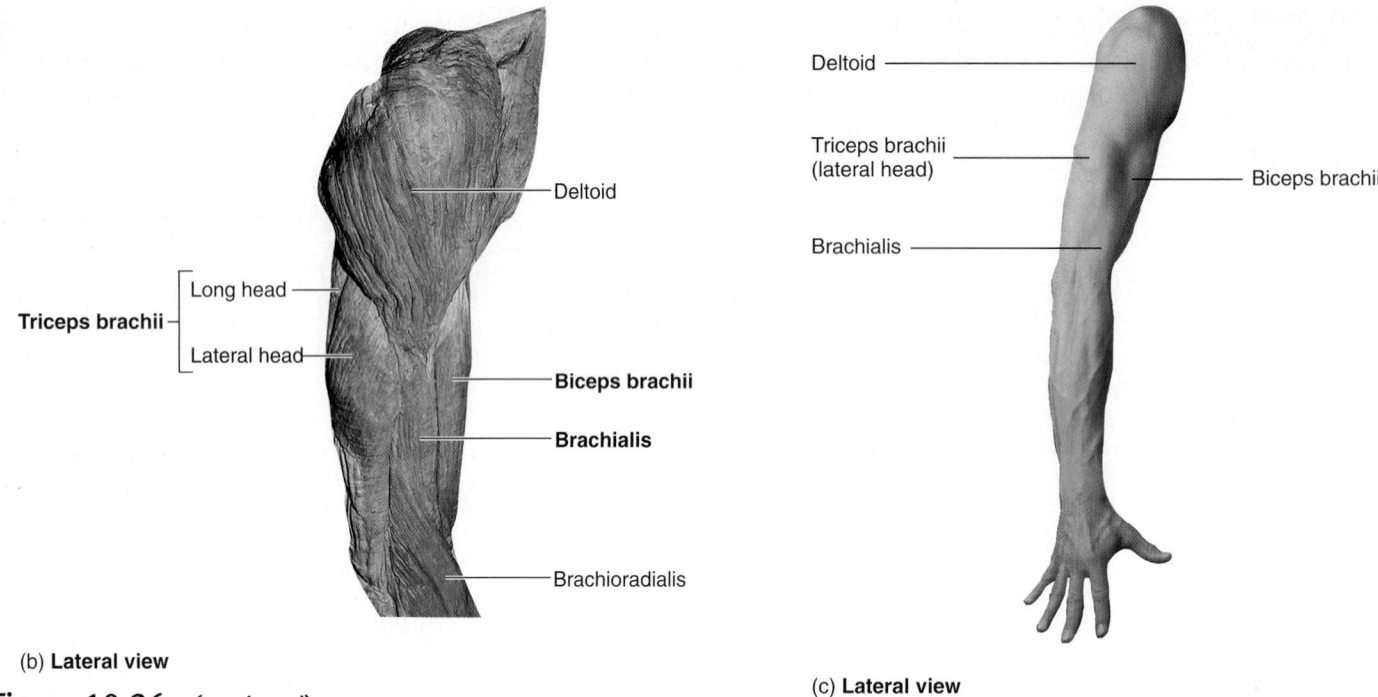

Deltoid

Triceps brachii — Long head

Lateral head

Biceps brachii

Brachialis

Brachioradialis

(b) **Lateral view**

Deltoid

Triceps brachii
(lateral head)

Brachialis

Biceps brachii

(c) **Lateral view**

Figure 10.26   *(continued)*

(*b*) Right shoulder and arm muscles of a cadaver. (*c*) Surface anatomy of the right shoulder and arm.

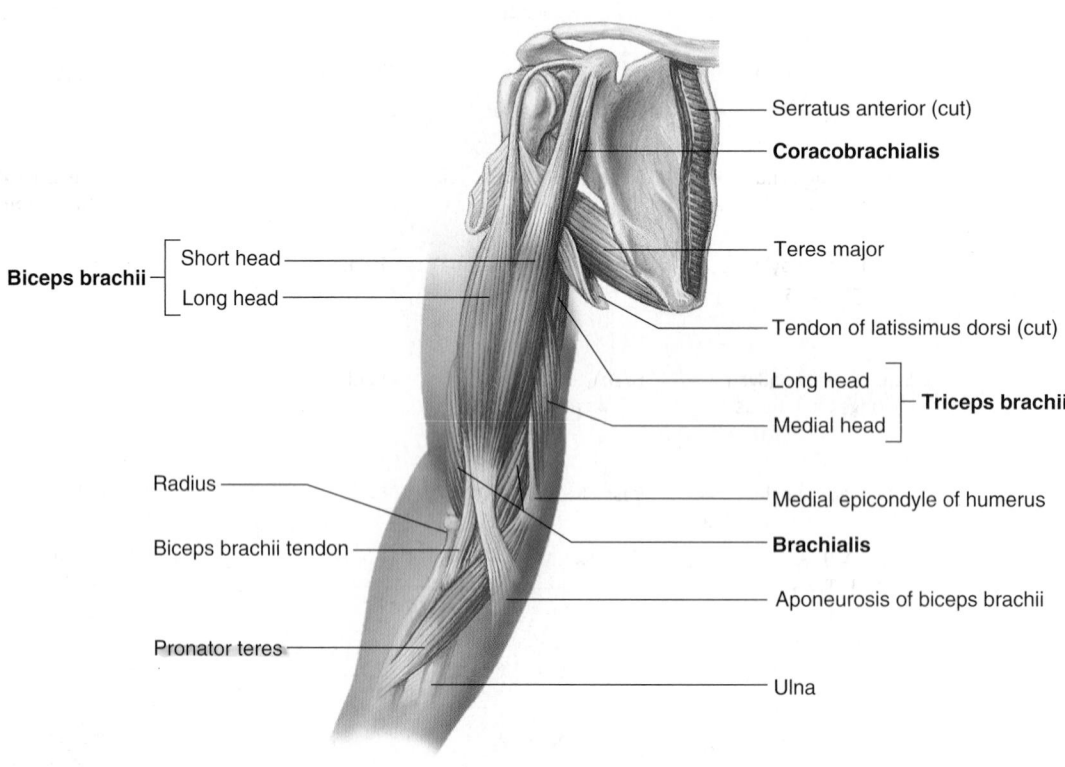

Serratus anterior (cut)

Coracobrachialis

Teres major

Tendon of latissimus dorsi (cut)

Long head

Medial head

Triceps brachii

Medial epicondyle of humerus

Brachialis

Aponeurosis of biceps brachii

Ulna

Biceps brachii — Short head

Long head

Radius

Biceps brachii tendon

Pronator teres

Anteromedial view

Figure 10.27   Anterior Right Arm Muscles

Wrist, Hand, and Finger Movements

The forearm muscles are divided into anterior and posterior groups (table 10.16; see figures 10.28 and 10.29). Most of the anterior forearm muscles are responsible for flexion of the wrist and fingers. Most of the posterior forearm muscles cause extension of the wrist and fingers.

Extrinsic Hand Muscles

The **extrinsic hand muscles** are in the forearm but have tendons that extend into the hand. A strong band of fibrous connective tis-sue, the **retinaculum** (ret-i-nak′ū-lŭm; bracelet), covers the flexor and extensor tendons and holds them in place around the wrist so that they do not "bowstring" during muscle contraction (see figures 10.29a and c).

Two major anterior muscles, the **flexor carpi radialis** (kar′pī rā-dē-ā′lis) and the **flexor carpi ulnaris** (ŭl-nā′ris), flex the wrist; and three posterior muscles, the **extensor carpi radialis longus**, the **extensor carpi radialis brevis**, and the **extensor carpi ulnaris**, extend the wrist. The wrist flexors and extensors are visible on the anterior and posterior surfaces of the forearm (see figures 10.28d

Table 10.16	Muscles of the Forearm Acting on the Wrist, Hand, and Fingers (see figures 10.28 and 10.29)			
Muscle	**Origin**	**Insertion**	**Nerve**	**Action**
Anterior Forearm				
Flexor carpi radialis (kar′pī rā-dē-ā′lis)	Medial epicondyle of humerus	Second and third metacarpals	Median	Flexes and abducts wrist
Flexor carpi ulnaris (kar′pī ŭl-nā′ris)	Medial epicondyle of humerus and ulna	Pisiform	Ulnar	Flexes and adducts wrist
Flexor digitorum profundus (dij′i-tōr′ŭm prō-fŭn′dŭs)	Ulna	Distal phalanges of digits 2–5	Ulnar and median	Flexes fingers and wrist
Flexor digitorum superficialis (dij′i-tōr′ŭm soo′per-fish-ē-ā′lis)	Medial epicondyle of humerus, coronoid process, and radius	Middle phalanges of digits 2–5	Median	Flexes fingers and wrist
Flexor pollicis longus (pol′i-sis lon′gŭs)	Radius	Distal phalanx of thumb	Median	Flexes thumb
Palmaris longus (pawl-mār′is lon′gŭs)	Medial epicondyle of humerus	Palmar fascia	Median	Tenses palmar fascia; flexes wrist
Posterior Forearm				
Abductor pollicis longus (pol′i-sis lon′gŭs)	Posterior ulna and radius and interosseous membrane	Base of first metacarpal	Radial	Abducts and extends thumb; abducts wrist
Extensor carpi radialis brevis (kar′pī rā-dē-ā′lis brev′is)	Lateral epicondyle of humerus	Base of third metacarpal	Radial	Extends and abducts wrist
Extensor carpi radialis longus (kar′pī rā-dē-ā′lis lon′gus)	Lateral supracondylar ridge of humerus	Base of second metacarpal	Radial	Extends and abducts wrist
Extensor carpi ulnaris (kar′pī ŭl-nā′ris)	Lateral epicondyle of humerus and ulna	Base of fifth metacarpal	Radial	Extends and adducts wrist
Extensor digiti minimi (dij′i-tī mi′nī-mī)	Lateral epicondyle of humerus	Phalanges of fifth digit	Radial	Extends little finger and wrist
Extensor digitorum (dij′i-tōr′ŭm)	Lateral epicondyle of humerus	Bases of phalanges of digits 2–5	Radial	Extends fingers and wrist
Extensor indicis (in′di-sis)	Ulna	Second digit	Radial	Extends forefinger and wrist
Extensor pollicis brevis (pol′i-sis brev′is)	Radius	Proximal phalanx of thumb	Radial	Extends and abducts thumb; abducts wrist
Extensor pollicis longus (pol′i-sis lon′gŭs)	Ulna	Distal phalanx of thumb	Radial	Extends thumb

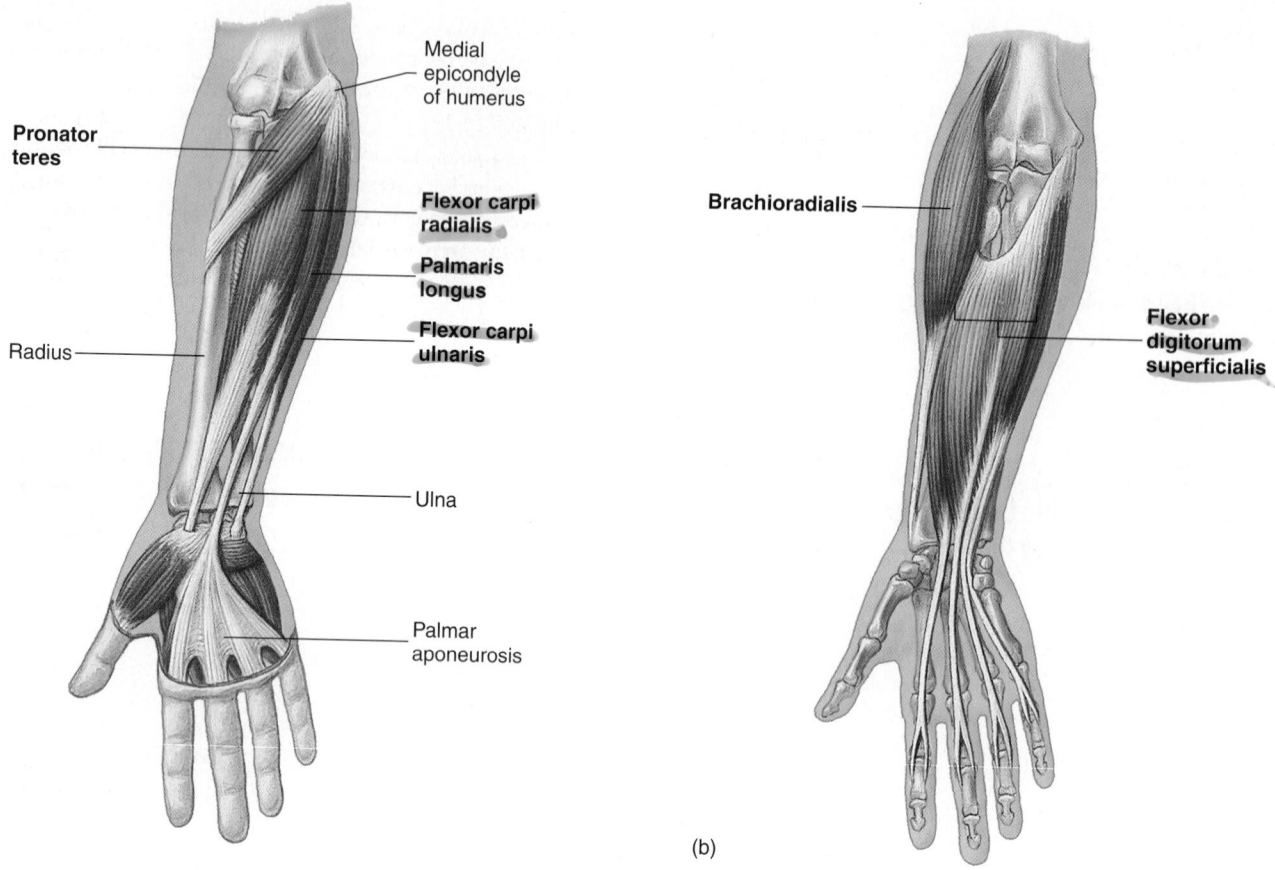

Medial
epicondyle
of humerus

**Pronator
teres**

**Flexor carpi
radialis**

**Palmaris
longus**

**Flexor carpi
ulnaris**

Radius

Ulna

Palmar
aponeurosis

(a)

Brachioradialis

**Flexor
digitorum
superficialis**

(b)

Anterior views

Lateral
epicondyle
of humerus

Radius

Supinator

**Flexor
pollicis
longus**

**Pronator
quadratus**

Medial
epicondyle
of humerus

Ulna

**Flexor
digitorum
profundus**

Lumbricales

Brachioradialis

Forearm
extensors

Forearm flexors

Tendon of
palmaris longus

Tendon of flexor
carpi radialis

(d)

Figure 10.28 Anterior Right Forearm Muscles

(*a*) Anterior view of the right forearm (superficial). Brachioradialis muscle is removed. (*b*) Anterior
view of the right forearm (deeper than *a*). Pronator teres, flexor carpi radialis and ulnaris, and
palmaris longus muscles are removed. (*c*) Anterior view of the right forearm (deeper than *a* or *b*).
Brachioradialis, pronator teres, flexor carpi radialis and ulnaris, palmaris longus, and flexor
digitorum superficialis muscles are removed. (*d*) Surface anatomy of anterior forearm muscles.

(a)

Olecranon of ulna

Brachioradialis

Flexor carpi ulnaris

Anconeus

Extensor carpi ulnaris

Extensor carpi radialis longus

Extensor digitorum

Extensor carpi radialis brevis

Ulna

Abductor pollicis longus

Extensor retinaculum

(b)

Medial epicondyle of humerus

Extensor digitorum (cut and reflected)

Anconeus

Supinator (deep)

Extensor digiti minimi (cut)

Extensor carpi radialis longus

Extensor carpi ulnaris (cut)

Extensor carpi radialis brevis

Abductor pollicis longus

Extensor indicis

Extensor pollicis longus

Cut tendons of extensor digitorum

Extensor pollicis brevis

Posterior views

(c)

Brachioradialis

Extensor carpi radialis longus

Extensor carpi radialis brevis

Extensor digitorum

Extensor carpi ulnaris

Abductor pollicis longus

Extensor pollicis brevis

Extensor pollicis longus

Extensor retinaculum

Extensor indicis tendon

Extensor pollicis longus tendon

First dorsal interosseus

Extensor digitorum tendons

Extensor digiti minimi tendon

(d)

Brachialis

Extensor carpi ulnaris

Extensor carpi radialis

Extensor digitorum

Anatomical snuffbox

Tendons of extensor digitorum

Figure 10.29   Posterior Right Forearm Muscles

(a) Posterior view of the right forearm (superficial). (b) Deep muscles of the right posterior forearm. Extensor digitorum, extensor digiti minimi, and extensor carpi ulnaris muscles are cut to reveal deeper muscles. (c) Photograph showing dissection of the posterior right forearm and hand. (d) Surface anatomy of posterior forearm.

and 10.29d). The tendon of the flexor carpi radialis is an important landmark because the radial pulse can be felt just lateral to the tendon (see figure 10.28d).

Tennis Elbow

Forceful, repetitive use of the forearm extensor muscles can damage them where they attach to the lateral epicondyle. This condition is often called **tennis elbow** because it can result from playing tennis. It is also called **lateral epicondylitis** because it can result from other sports and activities such as shoveling snow.

Flexion of the four medial digits is a function of the **flexor digitorum** (dij′i-tor′ŭm) **superficialis** and **flexor digitorum profundus** (prō-fŭn′dŭs; deep). Extension is accomplished by the **extensor digitorum.** The tendons of this muscle are very visible on the dorsum of the hand (see figure 10.29d). The little finger has an additional extensor, the **extensor digiti minimi** (dij′i-tī min′i-mī). The index finger also has an additional extensor, the **extensor indicis** (in′di-sis).

Movement of the thumb is caused in part by the **abductor pollicis** (pol′i-sis) **longus,** the **extensor pollicis longus,** and the **extensor pollicis brevis.** These tendons form the sides of a depression on the posterolateral side of the wrist called the "anatomical snuffbox" (see figure 10.29). When snuff was in use, a small pinch could be placed into the anatomical snuffbox and inhaled through the nose.

Intrinsic Hand Muscles

The **intrinsic hand muscles** are entirely within the hand (table 10.17 and figure 10.30). Abduction of the fingers is accomplished by the **interossei dorsales** (in′ter-os′e-ī dōr-sa′lēz) and the **abductor digiti minimi,** whereas adduction is a function of the **interossei palmares** (pawl-măr′ēz).

Table 10.17	Intrinsic Hand Muscles (see figure 10.30)			
Muscle	**Origin**	**Insertion**	**Nerve**	**Action**
Midpalmar Muscles				
Interossei (in′ter-os′e-ī)				
Dorsales (dōr-sā′lēz)	Sides of metacarpal bones	Proximal phalanges of second, third, and fourth digits	Ulnar	Abducts second, third, and fourth digits
Palmares (pawl-măr′ez)	Second, fourth, and fifth metacarpals	Second, fourth, and fifth digits	Ulnar	Adducts second, fourth, and fifth digits
Lumbricales (lum-bră-ka′lēz)	Tendons of flexor digitorum profundis	Second through fifth digits	Two on radial side—median; two on ulnar side—ulnar	Flexes proximal and extends middle and distal phalanges
Thenar Muscles				
Abductor pollicis brevis (ab-dŭk-ter, ab-dŭk-tōr pol′i-sis brev′is)	Flexor retinaculum, trapezium, and scaphoid	Proximal phalanx of thumb	Median	Abducts thumb
Adductor pollicis (ab-dŭk-ter, ab-dŭk-tōr pol′i-sis)	Third metacarpal, second metacarpal, trapezoid, and capitate	Proximal phalanx of thumb	Ulnar	Adducts thumb
Flexor pollicis brevis (pol′i-sis brev′is)	Flexor retinaculum and first metacarpal	Proximal phalanx of thumb	Median and ulnar	Flexes thumb
Opponens pollicis (ŏ-pō′nens pol′i-sis)	Trapezium and flexor retinaculum	First metacarpal	Median	Opposes thumb
Hypothenar Muscles				
Abductor digiti minimi (ab-dŭk-ter, ab-dŭk-tōr dij′i-tī min′imī)	Pisiform	Base of fifth digit	Ulnar	Abducts and flexes little finger
Flexor digiti minimi brevis (dij′i-tī min′ī-mī brev′is)	Hamate	Middle and proximal phalanx of fifth digit	Ulnar	Flexes little finger
Opponens digiti minimi (ŏ-pō′nens dij′i-tī min′i-mī)	Hamate and flexor retinaculum	Fifth metacarpal	Ulnar	Opposes little finger

Thenar eminence
- **Abductor pollicis brevis** (cut and reflected)
- **Opponens pollicis**
- **Flexor pollicis brevis**

Adductor pollicis

First dorsal interosseous

Palmar interossei

Flexor retinaculum

- **Abductor digiti minimi**
- **Flexor digiti minimi brevis**
- **Opponens digiti minimi**

Hypothenar eminence

L   **Lumbricales**
FD Flexor digitorum superficialis tendons

(a) **Anterior view**

Thenar eminence

Hypothenar eminence

(c)

Flexor retinaculum

Opponens pollicis

First dorsal interosseous

Dorsal interossei

Palmar interossei

Flexor digitorum tendons (cut)

Opponens digiti minimi

Metacarpals

Phalanges

(b) **Anterior view**

Figure 10.30 Right Hand Muscles

(a) Superficial muscles of the right hand. Abductor pollicis brevis is cut. (b) Deep muscles of the right hand. Flexor digitorum tendons are cut. (c) Surface anatomy of the palmar surface of the hand.

The **flexor pollicis brevis,** the **abductor pollicis brevis,** and the **opponens pollicis** form a fleshy prominence at the base of the thumb called the **thenar** (thē′nar) **eminence** (see figure 10.30). The **abductor digiti minimi, flexor digiti minimi brevis,** and **opponens digiti minimi** constitute the **hypothenar eminence** on the ulnar side of the hand. The thenar and hypothenar muscles are involved in the control of the thumb and little finger.

23. *Describe the muscle groups that cause flexion and extension of the wrist.*
24. *Contrast the location and actions of the extrinsic and intrinsic hand muscles. What is the retinaculum? What is the location and action of the thenar and hypothenar muscles?*
25. *Describe the muscles that move the thumb. The tendons of what muscles form the anatomical snuffbox?*

Lower Limb Muscles
Thigh Movements

Several hip muscles originate on the coxa and insert onto the femur (table 10.18 and figures 10.31–10.35). These muscles are divided into three groups: anterior, posterolateral, and deep.

The anterior muscles (figure 10.31); the **iliacus** (il-ī′ă-kŭs) and the **psoas** (sō′as) **major,** flex the hip. Because these muscles share a common insertion and produce the same movement, they often are referred to as the **iliopsoas** (il′ē-ō-sō′as). When the thigh is fixed, the iliopsoas flexes the trunk on the thigh. For example, the iliopsoas actually does most of the work when a person does sit-ups.

The posterolateral hip muscles consist of the gluteal muscles and the **tensor fasciae latae** (fash′ē-ē lā′tē). The **gluteus** (gloo-tē′ŭs) **maximus** (figure 10.32) contributes most of the

Table 10.18	Muscles Acting on the Thigh (see figures 10.31–10.35)			
Muscle	**Origin**	**Insertion**	**Nerve**	**Action**
Anterior				
Iliopsoas (il′ē-ō-sō′as)				
Iliacus (il-ī′ă-kus)	Iliac fossa	Lesser trochanter of femur and capsule of hip joint	Lumbar plexus	Flexes hip
Psoas major (sō′as)	T12–L5	Lesser trochanter of femur	Lumbar plexus	Flexes hip
Posterior and Lateral				
Gluteus maximus (gloo-tē′ŭs mak′si-mŭs)	Ilium, sacrum, and coccyx	Gluteal tuberosity of femur and the fascia lata	Inferior gluteal	Extends hip; abducts and laterally rotates thigh
Gluteus medius (gloo-tē′ŭs mē′dē-ŭs)	Ilium	Greater trochanter of femur	Superior gluteal	Abducts and medially rotates thigh; tilts pelvis toward supported side
Gluteus minimus (gloo-tē′ŭs min-i-mŭs)	Ilium	Greater trochanter of femur	Superior gluteal	Abducts and medially rotates thigh; tilts pelvis toward supported side
Tensor fasciae latae (ten′sŏr fash′ē-ē lā′tē)	Anterior superior iliac spine	Through iliotibial tract to lateral condyle of tibia	Superior gluteal	Tenses lateral fascia and stabilizes femur on tibia when standing; flexes hip; abducts and medially rotates thigh; tilts pelvis
Deep Thigh Rotators				
Gemellus (jĕ-mel′ŭs)				
Inferior	Ischial tuberosity	Obturator internus tendon	L5 and S1	Laterally rotates and abducts thigh
Superior	Ischial spine	Obturator internus tendon	L5 and S1	Laterally rotates and abducts thigh
Obturator (ob′too-rā-tŏr)				
Externus (eks-ter′nŭs)	Inferior margin of obturator foramen	Greater trochanter of femur	Obturator	Laterally rotates thigh
Internus (in-ter′nŭs)	Margin of obturator foramen	Greater trochanter of femur	Sciatic plexus	Laterally rotates thigh
Piriformis (pir′i-fŏr′mis)	Sacrum and ilium	Greater trochanter of femur	Sciatic plexus	Laterally rotates and abducts thigh
Quadratus femoris (kwah′-drā′tŭs fem′ŏ-ris)	Ischial tuberosity	Intertrochanteric ridge of femur	Sciatic plexus	Laterally rotates thigh

Psoas minor

Anterior superior
iliac spine

Iliacus   ⎤
                 ⎥ Iliopsoas
Psoas major   ⎦

Tensor fasciae latae

Pectineus

Adductor longus

Sartorius

Gracilis

Adductor magnus

Iliotibial tract

Quadriceps
femoris

Anterior view

Figure 10.31   Right Proximal Anterior Thigh Muscles

(a) **Posterior view**

Gluteus medius

Gluteus maximus

Iliac crest

Origin of
gluteus medius

Posterior superior
iliac spine

Gluteus minimus

Origin of
gluteus maximus

Sacrum

Piriformis

Superior gemellus

Obturator internus

Coccyx

Obturator externus

Quadratus femoris

Inferior gemellus

Ischial tuberosity

(b) **Posterior view**

Gluteus medius

Tensor fasciae latae

Gluteus maximus

(c) **Posterior view**

Figure 10.32   Right Posterior Hip Muscles

(a) Posterior view of the right hip, superficial. (b) Posterior view of the right hip, deep. Gluteus maximus and medius are removed to reveal deeper muscles.
(c) Surface anatomy of the right posterior hip muscles.

Table 10.19 Summary of Muscle Actions on the Hip and Thigh

Flexion	Extension	Abduction	Adduction	Medial Rotation	Lateral Rotation
Iliopsoas	Gluteus maximus	Gluteus maximus	Adductor magnus	Tensor fasciae latae	Gluteus maximus
Tensor fasciae latae	Semitendinosus	Gluteus medius	Adductor longus	Gluteus medius	Obturator internus
Rectus femoris	Semimembranosus	Gluteus minimus	Adductor brevis	Gluteus minimus	Obturator externus
Sartorius	Biceps femoris	Tensor fasciae latae	Pectineus		Superior gemellus
Adductor longus	Adductor magnus	Obturator internus	Gracilis		Inferior gemellus
Adductor brevis		Gemellus superior and inferior			Quadratus femoris
Pectineus		Piriformis			Piriformis
					Adductor magnus
					Adductor longus
					Adductor brevis

Table 10.20 Muscles of the Thigh (see figures 10.33 and 10.35)

Muscle	Origin	Insertion	Nerve	Action
Anterior Compartment				
Quadriceps femoris (kwah′dri-seps fem′ŏ-ris)	Rectus femoris—anterior inferior iliac spine; Vastus lateralis—femur; Vastus intermedius—femur; Vastus medialis—linea aspera	Patella and onto tibial tuberosity through patellar ligament	Femoral	Extends knee: rectus femoris also flexes hip
Sartorius (sar-tŏr′ē-ŭs)	Anterior superior iliac spine	Medial side of tibial tuberosity	Femoral	Flexes hip and knee: rotates thigh laterally and leg medially
Medial Compartment				
Adductor brevis (a-dŭk′ter, a-dŭk′tŏr brev′is)	Pubis	Femur	Obturator	Adducts, laterally rotates thigh, and flexes hip
Adductor longus (a-dŭk′ter, a-dŭk′tŏr lon′gŭs)	Pubis	Femur	Obturator	Adducts, laterally rotates thigh, and flexes hip
Adductor magnus (a-dŭk′ter, a-dŭk′tŏr mag′nŭs)	Adductor part: pubis and ischium; Hamstring part: ischial tuberosity	Adductor part: linea aspera of femur; Hamstring part: distal femur	Adductor part: obturator; Hamstring part: tibial	Adductor part: adducts thigh and flexes hip; Hamstring part: extends hip and adducts thigh
Gracilis (gras′i-lis)	Pubis near symphysis	Tibia	Obturator	Adducts thigh; flexes knee
Pectineus (pek′ti-nē′ŭs)	Pubic crest	Pectineal line of femur	Femoral and obturator	Adducts thigh and flexes hip
Posterior Compartment				
Biceps femoris (bī′seps fem′ŏ-ris)	Long head—ischial tuberosity; Short head—femur	Head of fibula	Long head—tibial; Short head—common fibular	Flexes knee and laterally rotates leg; extends hip
Semimembranosus (sem′ē-mem-bră-nō′sŭs)	Ischial tuberosity	Medial condyle of tibia and collateral ligament	Tibial	Flexes knee and medially rotates leg; tenses capsule of knee joint; extends hip
Semitendinosus (sem′ē-ten-di-nō′sŭs)	Ischial tuberosity	Tibia	Tibial	Flexes knee and medially rotates leg; extends hip

mass that can be seen as the buttocks (figure 10.32c); and the **gluteus medius,** a common site for injections, creates a smaller mass just superior and lateral to the maximus. The gluteus maximus functions at its maximum force in extension of the thigh when the hip is flexed at a 45-degree angle so that the muscle is optimally stretched, which accounts for both the sprinter's stance and the bicycle racing posture.

The deep hip muscles, as well as the gluteus maximus, function as lateral thigh rotators. The gluteus medius, gluteus minimus, and tensor fasciae latae are medial hip rotators (see table 10.18 and figure 10.32b). The gluteus medius and minimus muscles help tilt the pelvis and maintain the trunk in an upright posture during walking, as the foot of the opposite limb is raised from the ground. Without the action of these muscles, the pelvis tends to sag downward on the unsupported side.

In addition to the hip muscles, some of the muscles located in the thigh originate on the coxa and can cause movement of the thigh (tables 10.19 and 10.20). Three groups of thigh muscles have been identified based on their location in the thigh: the anterior, which flex the hip (figure 10.33a); the medial, which adduct the thigh (figure 10.34); and the posterior, which extend the hip (figure 10.35).

26. *Name the anterior hip muscle that flexes the hip. What muscles act as synergists to this muscle?*
27. *Describe the movements produced by the gluteus muscles.*

Leg Movements

The anterior thigh muscles are the **quadriceps femoris** (fem′ŏ-ris) and the **sartorius** (sar-tōr′ē-ŭs) (see table 10.20 and figure 10.33a).

The quadriceps femoris is actually four muscles: the **rectus femoris,** the **vastus lateralis,** the **vastus medialis,** and the **vastus intermedius.** The quadriceps group extends the knee. The rectus femoris also flexes the hip because it crosses both the hip and knee joints.

The quadriceps femoris makes up the large mass on the anterior thigh (figure 10.33c). The vastus lateralis sometimes is used as an injection site, especially in infants who may not have well-developed deltoid or gluteal muscles. The muscles of the quadriceps femoris have a common insertion, the patellar tendon, on and around the patella. The patellar ligament is an extension of

(b) **Anterior view**

(a) **Anterior view**

(c) **Anterior view**

Figure 10.33 **Right Anterior Thigh Muscles**

(a) Anterior view of the right thigh. (b) Photograph of the thigh muscles in a cadaver. (c) Surface anatomy of the right anterior thigh.

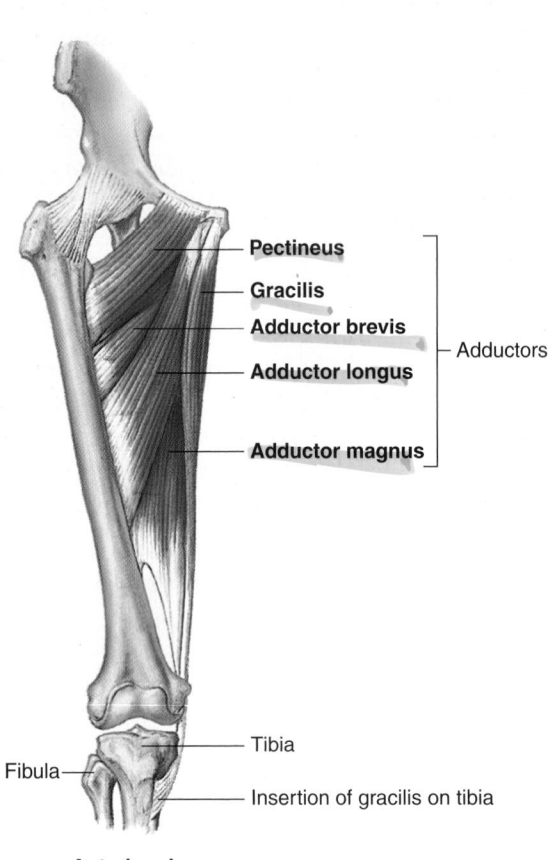

Pectineus

Gracilis

Adductor brevis

Adductor longus } Adductors

Adductor magnus

Tibia

Fibula

Insertion of gracilis on tibia

Anterior view

Figure 10.34   **Right Medial Thigh Muscles**

Ischial tuberosity

Semitendinosus

Hamstrings

Biceps femoris

Semimembranosus

Tibia   Fibula

Posterior view

Figure 10.35   **Right Posterior Thigh Muscles**

Hip muscles are removed.

the patellar tendon onto the tibial tuberosity. The patellar ligament is the point that is tapped with a rubber hammer when testing the knee-jerk reflex in a physical examination.

The sartorius is the longest muscle of the body, crossing from the lateral side of the hip to the medial side of the knee. As the muscle contracts, it flexes the hip and knee and laterally rotates the thigh. This movement is the action required for crossing the legs.

Sartorius—the Tailor's Muscle

The term *sartorius* means tailor. The sartorius muscle is so named because its action is to cross the legs, a common position traditionally preferred by tailors because they can hold their sewing in their lap as they sit and sew by hand.

The medial thigh muscles (see figure 10.34) are involved primarily in adduction of the thigh. Some of these muscles also laterally rotate the thigh and/or flex or extend the hip. The gracilis also flexes the knee.

The posterior thigh muscles (see figure 10.35) are collectively called the hamstring muscles and consist of the **biceps femoris, semimembranosus** (sem′ē-mem-bră-nō′sŭs), and **semitendinosus** (sem′ē-ten-di-nō′sŭs) (see table 10.20). Their tendons are easily seen or felt on the medial and lateral posterior aspect of a slightly bent knee (figure 10.36).

Hamstrings

The **hamstrings** are so named because in pigs these tendons can be used to suspend hams during curing. Some animals such as wolves often bring down their prey by biting through the hamstrings; therefore, "to hamstring" someone is to render the person helpless. A "pulled hamstring" results from tearing one or more of these muscles or their tendons, usually near the origin of the muscle.

28. *Name the muscle compartments of the thigh and the movements produced by the muscles of each compartment. List the muscles of each compartment and the individual action of each muscle.*
29. *How is it possible for thigh muscles to move both the thigh and the leg? Name six muscles that can do this.*

Ankle, Foot, and Toe Movements

Muscles of the leg that move the ankle and the foot are listed in table 10.21 and are illustrated in figures 10.37–10.39. These **extrin-**

sic foot muscles are divided into three groups, each located within a separate compartment of the leg: anterior, posterior, and lateral (figure 10.37). The anterior leg muscles (figure 10.38*a*) are extensor muscles involved in dorsiflexion and eversion or inversion of the foot and extension of the toes.

Shinsplints

Shinsplints is a catchall term involving any one of the following four conditions associated with pain in the anterior portion of the leg:

1. Excessive stress on the tibialis anterior, resulting in pain along the origin of the muscle.
2. Tibial periostitis, an inflammation of the tibial periosteum.
3. Anterior compartment syndrome. During hard exercise, the anterior compartment muscles may swell with blood. The overlying fascia is very tough and does not expand; thus the nerves and vessels are compressed, causing pain.
4. Stress fracture of the tibia 2–5 cm distal to the knee.

Figure 10.36 **Surface Anatomy of the Posterior Lower Limb**

Figure 10.37 **Cross Section Through the Left Leg**

Drawing of the muscular compartments.

Table 10.21	Muscles of the Leg Acting on the Leg, Ankle, and Foot (see figures 10.37–10.39)			
Muscle	**Origin**	**Insertion**	**Nerve**	**Action**
Anterior Compartment				
Extensor digitorum longus (dij'i-tōr-ŭm lon'gŭs)	Lateral condyle of tibia and fibula	Four tendons to phalanges of four lateral toes	Deep fibular*	Extends four lateral toes; dorsiflexes and everts foot
Extensor hallucis longus (hal'i-sis lon'gŭs)	Middle fibula and interosseous membrane	Distal phalanx of great toe	Deep fibular*	Extends great toe; dorsiflexes and inverts foot
Tibialis anterior (tib-ē-a'lis)	Tibia and interosseous membrane	Medial cuneiform and first metatarsal	Deep fibular*	Dorsiflexes and inverts foot
Fibularis tertius (peroneus tertius) (per'ō-nē'ŭs ter'shē-ŭs)	Fibula and interosseous membrane	Fifth metatarsal	Deep fibular*	Dorsiflexes and everts foot
Lateral Compartment				
Fibularis brevis (peroneus brevis) (fib-ū-lā'ris brev'is)	Fibula	Fifth metatarsal	Superficial fibular*	Everts and plantar flexes foot
Fibularis longus (peroneus longus) (fib-ū-lā'ris lon'gŭs)	Fibula	First metatarsal and medial cuneiform	Superficial fibular*	Everts and plantar flexes foot
Posterior Compartment				
Superficial				
Gastrocnemius (gas-trok-nē'mē-ŭs)	Medial and lateral condyles of femur	Through calcaneal (Achilles) tendon to calcaneus	Tibial	Plantar flexes foot; flexes knee
Plantaris (plan-tār'is)	Femur	Through calcaneal tendon to calcaneus	Tibial	Plantar flexes foot; flexes knee
Soleus (sō-lē'ŭs)	Fibula and tibia	Through calcaneal tendon to calcaneus	Tibial	Plantar flexes foot
Deep				
Flexor digitorum longus (dij'i-tōr'ŭm lon'gŭs)	Tibia	Four tendons to distal phalanges of four lateral toes	Tibial	Flexes four lateral toes; plantar flexes and inverts foot
Flexor hallucis longus (hal'i-sis lon'gŭs)	Fibula	Distal phalanx of great toe	Tibial	Flexes great toe; plantar flexes and inverts foot
Popliteus (pop-li-tē'ŭs)	Lateral femoral condyle	Posterior tibia	Tibial	Flexes knee and medially rotates leg
Tibialis posterior (tib-ē-a'lis)	Tibia, interosseous membrane, and fibula	Navicular, cuneiforms, cuboid, and second through fourth metatarsals	Tibial	Plantar flexes and inverts foot

*Formerly referred to as the peroneal nerve.

Soleus

Gastrocnemius

Fibularis longus

Soleus

Tibialis anterior

Anterior compartment muscles

Extensor digitorum longus

Extensor hallucis longus

Fibularis (peroneus) tertius

Lateral compartment muscles

(a) **Anterior view**

Gastrocnemius

Soleus

Fibularis (peroneus) longus (cut)

Fibularis (peroneus) brevis

Tibialis anterior

Extensor digitorum longus

Anterior compartment muscles

Fibularis (peroneus) tertius

Tendon of fibularis longus (cut)

(b) **Lateral view**

Gastrocnemius

Soleus

Tibialis anterior

Fibularis longus

Fibularis brevis

Extensor digitorum longus

Extensor digitorum brevis

Extensor digitorum longus tendons

Fibularis longus tendon

Fibularis brevis tendon

(c) **Lateral view**

Fibularis brevis

(d)

Figure 10.38   Right Lateral Leg Muscles

(*a*) Anterior view of the right leg. (*b*) Lateral view of the right leg. (*c*) Photograph of lateral leg muscles in a cadaver. (*d*) Surface anatomy of the posterolateral leg.

Gastrocnemius

Soleus

Calcaneal tendon
(Achilles tendon)

(a)

Two heads of
gastrocnemius
(cut)

Tibia

Popliteus

Plantaris

Posterior
superficial
compartment
muscles

Soleus

Tendon of
gastrocnemius
(cut)

Medial malleolus

Calcaneal tendon
(Achilles tendon)

Lateral malleolus

(b) **Posterior views**

Tibia

Fibula

**Flexor
digitorum
longus**

**Tibialis
posterior**

Deep
posterior
compartment
muscles

**Flexor
hallucis
longus**

(c)

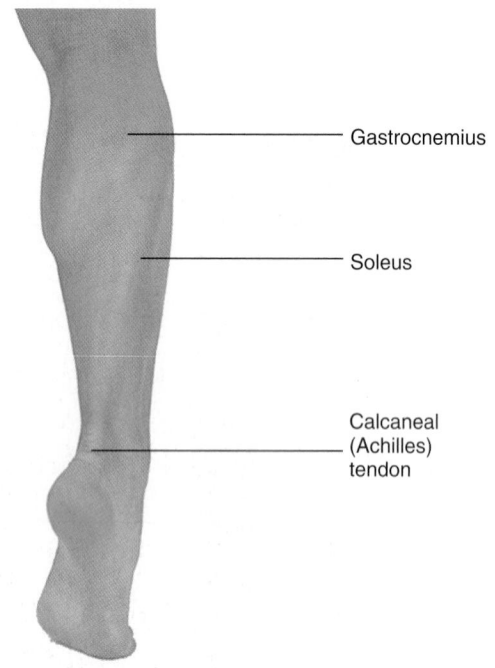

Gastrocnemius

Soleus

Calcaneal
(Achilles)
tendon

(d) **Posterior view**

Figure 10.39 Right Posterior Leg Muscles

(a) Superficial muscles. (b) Posterior view of the right calf, superficial.
Gastrocnemius is removed. (c) Posterior view of the right calf, deep.
Gastrocnemius, plantaris, and soleus muscles are removed. (d) Surface
anatomy of the posterior right leg.

The lateral muscles (figure 10.38*b*) are primarily everters of the foot, but they also aid plantar flexion. The **fibularis brevis** inserts onto the fifth metatarsal and everts and plantar flexes the foot. The **fibularis longus** crosses under the lateral four metatarsals to insert onto the first metatarsal and medial cuneiform. The fibular muscle formerly referred to as *peroneal*, which means "relating to the fibula," has now been simplified by using the term *fibular*. The tendons of the fibularis muscles can be seen on the lateral side of the ankle (figure 10.38*d*).

The superficial muscles of the posterior compartment, the **gastrocnemius** (gas-trok-nē′mē-ŭs) and **soleus,** form the bulge of the calf (posterior leg) (figures 10.39*a, b,* and *d*). They join with the small **plantaris** muscle to form the common **calcaneal** (kal-kā′nē-al), or **Achilles, tendon.** These muscles are involved in plantar flexion of the foot. The deep muscles of the posterior compartment plantar flex and invert the foot and flex the toes.

Achilles Tendon

The **Achilles tendon** derives its name from a hero of Greek mythology. When Achilles was a baby, his mother dipped him into magic water, which made him invulnerable to harm everywhere the water touched his skin. His mother, however, held him by the heel and failed to submerge this part of his body under the water. Consequently, his heel was vulnerable and proved to be his undoing; he was shot in the heel with an arrow at the battle of Troy and died. Thus, saying that someone has an "Achilles' heel" means that the person has a weak spot that can be attacked.

Intrinsic foot muscles, located within the foot itself (table 10.22 and figure 10.40), flex, extend, abduct, and adduct the toes. They are arranged in a manner similar to that of the intrinsic muscles of the hand.

Table 10.22	Intrinsic Muscles of the Foot (see figure 10.40)			
Muscle	**Origin**	**Insertion**	**Nerve**	**Action**
Abductor digiti minimi (ab-dŭk′ter, ab-dŭk′tōr dij′i-tī min′ĭ-mī)	Calcaneus	Proximal phalanx of fifth toe	Lateral plantar	Abducts and flexes little toe
Abductor hallucis (ab-dŭk′ter, ab-dŭk′tōr hal′i-sis)	Calcaneus	Great toe	Medial plantar	Abducts great toe
Adductor hallucis (a-dŭk′ter, a-dŭk′tōr hal′i-sis) (not illustrated)	Lateral four metatarsals	Proximal phalanx of great toe	Lateral plantar	Adducts great toe
Extensor digitorum brevis (dij′i-tōr′ŭm brev′is) (not illustrated)	Calcaneus	Four tendons fused with tendons of extensor digitorum longus	Deep fibular*	Extends toes
Flexor digiti minimi brevis (dij′i-tī min′ĭ-mī brev′is)	Fifth metatarsal	Proximal phalanx of fifth digit	Lateral plantar	Flexes little toe (proximal phalanx)
Flexor digitorum brevis (dij′i-tōr′ŭm brev′is)	Calcaneus and plantar fascia	Four tendons to middle phalanges of four lateral toes	Medial plantar	Flexes lateral four toes
Flexor hallucis brevis (hal′i-sis brev′is)	Cuboid; medial and lateral cuneiforms	Two tendons to proximal phalanx of great toe	Medial and lateral plantar	Flexes great toe
Dorsal interossei (in′ter-os′e-ī) (not illustrated)	Metatarsal bones	Proximal phalanges of second, third, and fourth digits	Lateral plantar	Abduct second, third, and fourth toes; adduct second toe
Plantar interossei (plan′tăr in′ter-os′e-ī)	Third, fourth, and fifth metatarsals	Proximal phalanges of third, fourth, and fifth digits	Lateral plantar	Adduct third, fourth, and fifth toes
Lumbricales (lum′bri-kā-lēz)	Tendons of flexor digitorum longus	Second through fifth digits	Lateral and medial plantar	Flex proximal and extend middle and distal phalanges
Quadratus plantae (kwah′drā′tŭs plan′tē)	Calcaneus	Tendons of flexor digitorum longus	Lateral plantar	Flexes toes

*Formerly referred to as the peroneal nerve.

Flexor digitorum
brevis tendons (cut)

Adductor hallucis

Lumbricales

Flexor hallucis brevis

Flexor digiti minimi brevis

Plantar interossei

Flexor hallucis
longus tendon

Flexor digitorum brevis

Abductor hallucis

Abductor digiti minimi

Plantar aponeurosis
(cut)

Flexor hallucis
longus tendon
(cut)

Flexor digitorum
longus tendons

**Flexor hallucis
brevis**

Quadratus plantae

Flexor digitorum
brevis tendon
(cut)

(a)   **Plantar views**   (b)

Figure 10.40   Right Foot Muscles

(*a*) Superficial muscles of the right foot. Plantar aponeurosis is cut. (*b*) Deep muscles of the right foot. Flexor digitorum brevis and flexor hallucis longus are cut.

Plantar Fasciitis

The muscles in the plantar region of the foot are covered with thick fascia and the plantar aponeurosis. Running on a hard surface wearing poor-fitting or worn-out shoes can result in inflammation of the plantar aponeurosis. The pain resulting from **plantar fasciitis** occurs in the fascia over the heel and along the medial-inferior side of the foot.

30. *What movements are produced by the three muscle compartments of the leg? Name the muscles of each compartment, and describe the movements for which each muscle is responsible.*
31. *What movement do the fibularis (peroneus) muscles have in common? The tibialis muscles?*
32. *Name the leg muscles that flex the knee. Which of them can also plantar flex the foot?*
33. *List the general actions performed by the intrinsic foot muscles.*

Clinical Focus   Bodybuilding

Bodybuilding is a popular sport worldwide. Participants in this sport combine diet and specific weight training to develop maximum muscle mass and minimum body fat, with their major goal being a well-balanced, complete physique. An uninformed, untrained muscle builder can build some muscles and ignore others; the result is a disproportioned body. Skill, training, and concentration are required to build a well-proportioned, muscular body and to know which exercises build a large number of muscles and which are specialized to build certain parts of the body. Is the old adage "no pain, no gain" correct? Not really. Overexercising can cause small tears in muscles and soreness. Torn muscles are weaker, and it may take up to 3 weeks to repair the damage, even though the soreness may only last 5–10 days.

Bodybuilders concentrate on increasing skeletal muscle mass. Endurance tests conducted years ago demonstrated that the cardiovascular and respiratory abilities of bodybuilders were similar to those abilities in normal, healthy persons untrained in a sport. More recent studies, however, indicate that the cardiorespiratory fitness of bodybuilders is similar to that of other well-trained athletes. The difference between the results of the new studies and the older ones is attributed to modern bodybuilding techniques that include aerobic exercise and running, as well as "pumping iron."

Bodybuilding has its own language. Bodybuilders refer to the "lats," "traps," and "delts" rather than the latissimus dorsi, trapezius, and deltoids. The exercises also have special names such as "lat pulldowns," "preacher curls," and "triceps extensions."

Photographs of bodybuilders are very useful in the study of anatomy because they enable easy identification of the surface anatomy of muscles that cannot usually be seen in untrained people (figure A).

Figure A   Bodybuilders

S U M M A R Y

Body movements result from the contraction of skeletal muscles.

General Principles   (p. 318)

1. The less movable end of a muscle attachment is the origin; the more movable end is the insertion.
2. An agonist causes a certain movement and an antagonist acts in opposition to the agonist.
3. Synergists are muscles that function together to produce movement.
4. Prime movers are mainly responsible for a movement. Fixators stabilize the action of prime movers.

Muscle Shapes

Muscle shape is determined primarily by the arrangement of muscle fasciculi.

Nomenclature

Muscles are named according to their location, size, shape, orientation of fasciculi, origin and insertion, number of heads, or function.

Movements Accomplished by Muscles

Contracting muscles generate a force that acts on bones (levers) across joints (fulcrums) to create movement. Three classes of levers have been identified.

Muscle Anatomy

Head Muscles   (p. 324)
Head and Neck Muscles

Origins of these muscles are mainly on the cervical vertebrae (except for the sternocleidomastoid); insertions are on the occipital bone or mastoid process. They cause flexion, extension, rotation, abduction, and adduction of the head.

Facial Expression

Origins of facial muscles are on skull bones or fascia; insertions are into the skin, causing movement of the facial skin, lips, and eyelids.

Mastication

Three pairs of muscles close the jaw; gravity opens the jaw. Forced opening is caused by the lateral pterygoids and the hyoid muscles.

Tongue Movements

Intrinsic tongue muscles change the shape of the tongue; extrinsic tongue muscles move the tongue.

Swallowing and the Larynx
1. Hyoid muscles can depress the jaw and assist in swallowing.
2. Muscles open and close the openings to the nasal cavity, auditory tubes, and larynx.

Movements of the Eyeball
Six muscles with their origins on the orbital bones insert on the eyeball and cause it to move within the orbit.

Trunk Muscles (p. 338)
Muscles Moving the Vertebral Column
1. These muscles extend, abduct, rotate, or flex the vertebral column.
2. A deep group of muscles connects adjacent vertebrae.
3. A more superficial group of muscles runs from the pelvis to the skull, extending from the vertebrae to the ribs.

Thoracic Muscles
1. Most respiratory movement is caused by the diaphragm.
2. Muscles attached to the ribs aid in respiration.

Abdominal Wall
Abdominal wall muscles hold and protect abdominal organs and cause flexion, rotation, and lateral flexion of the vertebral column.

Pelvic Floor and Perineum
These muscles support the abdominal organs inferiorly.

Upper Limb Muscles (p. 345)
Scapular Movements
Six muscles attach the scapula to the trunk and enable the scapula to function as an anchor point for the muscles and bones of the arm.

Arm Movements
Seven muscles attach the humerus to the scapula. Two additional muscles attach the humerus to the trunk. These muscles cause flexion and extension of the shoulder and abduction, adduction, rotation, and circumduction of the arm.

Forearm Movements
1. Flexion and extension of the elbow are accomplished by three muscles located in the arm and two in the forearm.
2. Supination and pronation are accomplished primarily by forearm muscles.

Wrist, Hand, and Finger Movements
1. Forearm muscles that originate on the medial epicondyle are responsible for flexion of the wrist and fingers. Muscles extending the wrist and fingers originate on the lateral epicondyle.
2. Extrinsic hand muscles are in the forearm. Intrinsic hand muscles are in the hand.

Lower Limb Muscles (p. 358)
Thigh Movements
1. Anterior pelvic muscles cause flexion of the hip.
2. Muscles of the buttocks are responsible for extension of the hip and abduction and rotation of the thigh.
3. The thigh can be divided into three compartments.
   - The anterior compartment muscles flex the hip.
   - The medial compartment muscles adduct the thigh.
   - The posterior compartment muscles extend the hip.

Leg Movements
Some muscles of the thigh also act on the leg. The anterior thigh muscles extend the leg, and the posterior thigh muscles flex the leg.

Ankle, Foot, and Toe Movements
1. The leg is divided into three compartments.
   - Muscles in the anterior compartment cause dorsiflexion, inversion, or eversion of the foot and extension of the toes.
   - Muscles of the lateral compartment plantar flex and evert the foot.
   - Muscles of the posterior compartment flex the leg, plantar flex and invert the foot, and flex the toes.
2. Intrinsic foot muscles flex or extend and abduct or adduct the toes.

R E V I E W   A N D   C O M P R E H E N S I O N

1. Muscles that oppose one another are
   a. synergists.
   b. levers.
   c. hateful.
   d. antagonists.
   e. fixators.
2. The most movable attachment of a muscle is its
   a. origin.
   b. insertion.
   c. fascia.
   d. fulcrum.
   e. belly.
3. Which of these muscles is correctly matched with its type of fascicle orientation?
   a. pectoralis major—pennate
   b. transversus abdominis—circular
   c. temporalis—convergent
   d. biceps femoris—parallel
   e. orbicularis oris—parallel
4. The muscle whose name means it is larger and round is the
   a. gluteus maximus.
   b. vastus lateralis.
   c. teres major.
   d. latissimus dorsi.
   e. adductor magnus.
5. In a class III lever system the
   a. fulcrum is located between the pull and the weight.
   b. weight is located between the fulcrum and the pull.
   c. pull is located between the fulcrum and the weight.
6. A prominent lateral muscle of the neck that can cause flexion of the neck or rotate the head is the
   a. digastric.
   b. mylohyoid.
   c. sternocleidomastoid.
   d. buccinator.
   e. platysma.

7. Harry Wolf has just picked up his date for the evening. She's wearing a stunning new outfit. Harry shows his appreciation by moving his eyebrows up and down, winking, smiling, and finally kissing her. Given the muscles listed:
   1. zygomaticus
   2. levator labii superioris
   3. occipitofrontalis
   4. orbicularis oris
   5. orbicularis oculi

   In which order did Harry use these muscles?
   a. 2,3,4,1
   b. 2,5,3,1
   c. 2,5,4,3
   d. 3,5,1,4
   e. 3,5,2,4

8. An aerial circus performer who supports herself only by her teeth while spinning around should have strong
   a. temporalis muscles.
   b. masseter muscles.
   c. buccinator muscles.
   d. both a and b.
   e. all of the above.

9. The tongue curls and folds *primarily* because of the action of the
   a. extrinsic tongue muscles.
   b. intrinsic tongue muscles.

10. The infrahyoid muscles
    a. elevate the mandible.
    b. move the mandible from side to side.
    c. fix (prevent movement of) the hyoid.
    d. both a and b.
    e. all of the above.

11. The soft palate muscles
    a. prevent food from entering the nasal cavity.
    b. close the auditory tube.
    c. force food into the esophagus.
    d. prevent food from entering the larynx.
    e. elevate the mandible.

12. Which of these movements is *not* caused by contraction of the erector spinae muscles?
    a. flexion of the vertebral column
    b. lateral flexion of the vertebral column
    c. extension of the vertebral column
    d. rotation of the vertebral column

13. Which of these muscles is (are) responsible for flexion of the vertebral column (below the neck)?
    a. deep back muscles
    b. superficial back muscles (erector spinae)
    c. rectus abdominis
    d. both a and b
    e. all of the above

14. Which of these muscles is *not* involved with the inspiration of air?
    a. diaphragm
    b. external intercostals
    c. scalene
    d. transversus thoracis

15. Given these muscles:
    1. external abdominal oblique
    2. internal abdominal oblique
    3. transversus abdominis

    Choose the arrangement that lists the muscles from most superficial to deepest.
    a. 1,2,3
    b. 1,3,2
    c. 2,1,3
    d. 2,3,1
    e. 3,1,2

16. Tendinous intersections
    a. attach the rectus abdominis muscles to the xiphoid process.
    b. divide the rectus abdominis muscles into segments.
    c. separate the abdominal wall from the thigh.
    d. are the site of exit of blood vessels from the abdomen into the thigh.
    e. are the central point of attachment for all the abdominal muscles.

17. Which of these muscles can both elevate and depress the scapula?
    a. rhomboideus major and minor
    b. levator scapulae
    c. serratus anterior
    d. trapezius
    e. pectoralis minor

18. Which of these muscles does *not* adduct the arm (humerus)?
    a. latissimus dorsi
    b. deltoid
    c. teres major
    d. pectoralis major
    e. coracobrachialis

19. Which of these muscles is most directly involved in abduction of the arm (humerus)?
    a. supraspinatus
    b. infraspinatus
    c. teres minor
    d. teres major
    e. subscapularis

20. Which of these muscles would you expect to be especially well developed in a boxer known for his powerful jab (punching straight ahead)?
    a. biceps brachii
    b. brachialis
    c. trapezius
    d. triceps brachii
    e. supinator

21. Which of these muscles is an antagonist of the triceps brachii?
    a. biceps brachii
    b. anconeus
    c. latissimus dorsi
    d. brachioradialis
    e. supinator

22. The posterior group of forearm muscles is responsible for
    a. flexion of the wrist.
    b. flexion of the fingers.
    c. extension of the fingers.
    d. both a and b.
    e. all of the above.

23. Which of these muscle(s) is an intrinsic hand muscle that moves the thumb?
    a. flexor pollicis brevis
    b. flexor digiti minimi brevis
    c. flexor pollicis longus
    d. extensor pollicis longus
    e. all of the above

24. Which of these muscles can extend the hip?
    a. gluteus maximus
    b. gluteus medius
    c. gluteus minimus
    d. tensor fasciae latae
    e. sartorius

25. Given these muscles:
    1. iliopsoas
    2. rectus femoris
    3. sartorius

Which of the muscles act to flex the hip?
a. 1
b. 1,2
c. 1,3
d. 2,3
e. 1,2,3

26. Which of these muscles is found in the medial compartment of the thigh?
a. rectus femoris
b. sartorius
c. gracilis
d. vastus medialis
e. semitendinosus

27. Which of these is *not* a muscle that can flex the knee?
a. biceps femoris
b. vastus medialis
c. gastrocnemius
d. gracilis
e. sartorius

28. The _____ muscles evert the foot, whereas the _____ muscles invert the foot.
a. fibularis (peroneus), gastrocnemius
b. fibularis (peroneus), tibialis
c. tibialis, fibularis (peroneus)
d. tibialis, flexor
e. flexor, extensor

29. Which of these muscles causes plantar flexion of the foot?
a. tibialis anterior
b. extensor digitorum longus
c. fibularis (peroneus) tertius
d. soleus
e. sartorius

Answers in Appendix F

C R I T I C A L   T H I N K I N G

1. For each of the following muscles, (1) describe the movement that the muscle produces, and (2) name the muscles that act as synergists and antagonists for them: longus capitis, erector spinae, coracobrachialis.

2. Propose an exercise that would benefit each of the following muscles specifically: biceps brachii, triceps brachii, deltoid, rectus abdominis, quadriceps femoris, and gastrocnemius.

3. Consider only the effect of the brachioradialis muscle for this question. If a weight is held in the hand and the forearm is flexed, what type of lever system is in action? If the weight is placed on the forearm? Which system can lift more weight, and how far?

4. A patient was involved in an automobile accident in which the car was "rear-ended," resulting in whiplash injury of the head (hyperextension). What neck muscles might be injured in this type of accident? What is the easiest way to prevent such injury in an automobile accident?

5. During surgery, a branch of the patient's facial nerve was accidentally cut on one side of the face. As a result, after the operation, the lower eyelid and the corner of the patient's mouth drooped on that side of the face. What muscles were apparently affected?

6. When a person becomes unconscious, the tongue muscles relax, and the tongue tends to retract or fall back and obstruct the airway. Which tongue muscle is responsible? How can this be prevented or reversed?

7. The mechanical support of the head of the humerus in the glenoid fossa is weakest in the inferior direction. What muscles help prevent dislocation of the shoulder when a heavy weight such as a suitcase is carried?

8. How would paralysis of the quadriceps femoris of the left leg affect a person's ability to walk?

9. Speedy Sprinter started a 200 m dash and fell to the ground in pain. Examination of her right leg revealed the following symptoms: inability to plantar flex the foot against resistance, normal ability to evert the foot, dorsiflexion of the foot more than normal, and abnormal bulging of the calf muscles. Explain the nature of her injury.

10. What muscles are required to turn this page?

Answers in Appendix G

A N S W E R S   T O   P R E D I C T   Q U E S T I O N S

1. Shortening the right sternocleidomastoid muscle rotates the head to the left. It also slightly elevates the chin.

2. Raising eyebrows—occipitofrontalis; winking—orbicularis oculi and then levator palpebrae superioris; whistling—orbicularis oris and buccinator; smiling—levator anguli oris, risorius, zygomaticus major, and zygomaticus minor; frowning—corrugator supercilii and procerus; flaring nostrils—levator labii superioris alaeque nasi and nasalis.

3. Weakness of the lateral rectus allows the eye to deviate medially.

4. As the arm is maximally abducted, the supraspinatus tendon rubs against the acromion process of the scapula. This can result in

damage to, and pain in, the supraspinatus muscle, commonly referred to as a torn rotator cuff.

5. Two arm muscles are involved in flexion of the elbow: the brachialis and the biceps brachii. The brachialis only flexes, whereas the biceps brachii both flexes the elbow and supinates the forearm. With the forearm supinated, both muscles can flex the elbow optimally; when pronated, the biceps brachii does less to flex the elbow. Chin-ups with the elbow supinated are therefore easier because both muscles flex the forearm optimally in this position. Bodybuilders who wish to build up the brachialis muscle perform chin-ups with the forearms pronated.

Visit the Online Learning Center at www.mhhe.com/seeley7 for chapter quizzes, interactive learning exercises, and other study tools.

Functional Organization of Nervous Tissue

Light photomicrograph of pyramid-shaped neurons (green) growing on a fibrous network (yellow) in the central nervous system.

The nervous system is made up of the brain, spinal cord, nerves, and sensory receptors. It's responsible for sensory perceptions, mental activities, as well as stimulating muscle movements and the secretions of many glands. For example, as a hungry person prepares to drink a cup of hot soup, he smells the aroma and anticipates the taste of the soup. Feeling the warmth of the cup in his hands, he carefully raises the cup to his lips and takes a sip. The soup is so hot that he burns his tongue. Quickly, he jerks the cup away from his lips and gasps in pain. None of these sensations, thoughts, emotions, and movements would be possible without the nervous system.

This chapter explains the *functions of the nervous system* (p. 374), *divisions of the nervous system* (p. 374), *cells of the nervous system* (p. 376), *organization of nervous tissue* (p. 381), *electric signals* (p. 381), *the synapse* (p. 394), and *neuronal pathways and circuits* (p. 403).

Functions of the Nervous System

The nervous system is involved in some way in most body functions. Some of the major functions of the nervous system are:

1. *Sensory input.* Sensory receptors monitor numerous external and internal stimuli. Some stimuli result in sensations we are aware of, such as sight, vision, hearing, taste, smell, touch, pain, body position, temperature, and so on. Other stimuli, such as blood pH, blood gases, or blood pressure, are processed at an unconscious level.
2. *Integration.* The brain and spinal cord are the major organs for processing sensory input and initiating responses. The input may produce an immediate response, may be stored as memory, or may be ignored.
3. *Homeostasis.* The regulatory and coordinating activities of the nervous system are necessary for maintaining homeostasis. The trillions of cells in the human body do not function independently of each other but must work together to maintain homeostasis. For example, heart cells must contract at a rate that ensures adequate delivery of blood, skeletal muscles of respiration must contract at a rate that ensures oxygenation of blood, and kidney cells must regulate blood volume and remove waste products. The nervous system can stimulate or inhibit the activities of these and other structures to help maintain homeostasis.
4. *Mental activity.* The brain is the center of mental activities, including consciousness, thinking, memory, and emotions.
5. *Control of muscles and glands.* Skeletal muscles normally contract only when stimulated by the nervous system, and the nervous system controls the major movements of the body through the control of skeletal muscle. Some smooth muscle, such as in the walls of blood vessels, contracts only when stimulated by the nervous system or hormones (see chapter 18). Cardiac muscle and some smooth muscle, such as in the wall of the stomach, contract autorhythmically. That is, no external stimulation is necessary for contraction to occur. Although the nervous system does not initiate contraction in these muscles, it can cause the contractions to occur more rapidly or slowly. Finally, the nervous system controls the secretions from many glands, such as sweat glands, salivary glands, and glands of the digestive system.

1. **List and give examples of the general functions of the nervous system.**

Divisions of the Nervous System

Humans have only one nervous system, even though some of its subdivisions are referred to as separate systems (figure 11.1). Thus the central nervous system and the peripheral nervous system are subdivisions of the nervous system, instead of separate organ systems as their names suggest. The **central nervous system (CNS)** consists of the brain and the spinal cord. The brain is located within the skull, and the spinal cord is located within the vertebral canal, formed by the vertebrae (see chapter 7). The brain and spinal cord are continuous with each other at the foramen magnum.

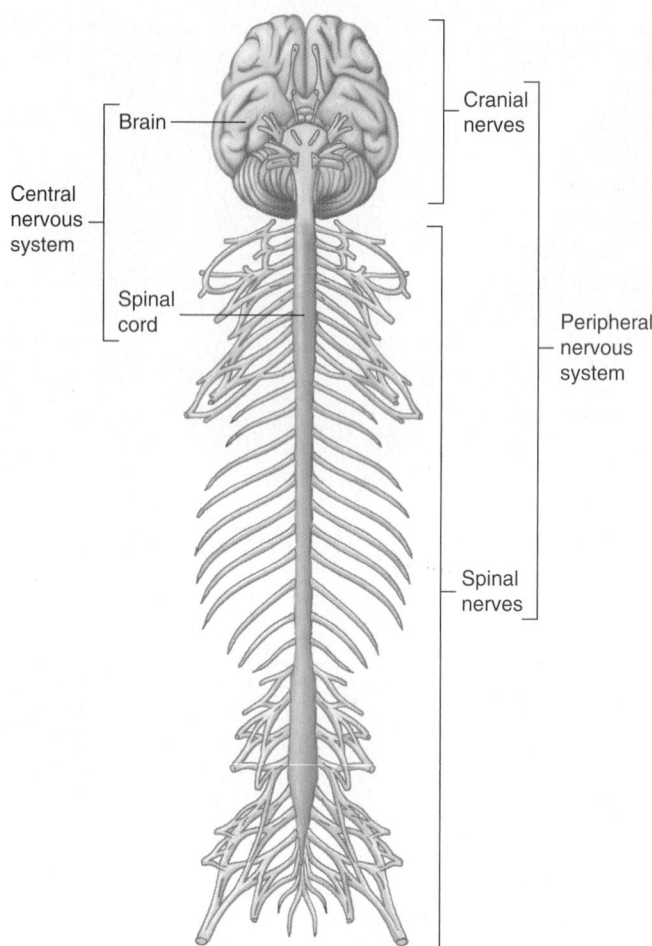

Figure 11.1   The Nervous System
The central nervous system (CNS) consists of the brain and spinal cord. The peripheral nervous system (PNS) consists of cranial nerves, which arise from the brain, and spinal nerves, which arise from the spinal cord. The nerves, which are shown cut in the illustration, actually extend throughout the body.

The **peripheral nervous system (PNS)** is external to the CNS. It consists of sensory receptors, nerves, ganglia, and plexuses. **Sensory receptors** are the endings of nerve cells or separate, specialized cells that detect temperature, pain, touch, pressure, light, sound, odors, and other stimuli. Sensory receptors are located in the skin, muscles, joints, internal organs, and specialized sensory organs such as the eyes and ears. A **nerve** is a bundle of axons and their sheaths that connects the CNS to sensory receptors, muscles, and glands. Twelve pairs of **cranial nerves** originate from the brain, and 31 pairs of **spinal nerves** originate from the spinal cord (see figure 11.1). A **ganglion** (gang′glē-on; pl., ganglia, gang′glē-ă; knot) is a collection of neuron cell bodies located outside the CNS. A **plexus** (plek′sus; braid) is an extensive network of axons and, in some cases, also neuron cell bodies, located outside the CNS.

The PNS can be divided into two subcategories. The **sensory,** or **afferent, division** transmits electric signals, called **action potentials,** from the sensory receptors to the CNS. The cell bodies of sensory neurons are located in ganglia near the spinal cord (figure 11.2*a*) or near the origin of certain cranial nerves. The **motor,** or **efferent, division** transmits action potentials from the CNS to effector organs, such as muscles and glands.

The motor division is divided into the **somatic nervous system** and the **autonomic nervous system (ANS).** The somatic nervous system transmits action potentials from the CNS to skeletal muscles (figure 11.2*b*). Skeletal muscles are voluntarily controlled through the somatic nervous system. The cell bodies of somatic motor neurons are located within the CNS, and their axons extend through nerves to form synapses with skeletal muscle cells. A **synapse** is the junction of a nerve cell with another cell. The neuromuscular junction, the synapse between a neuron and skeletal muscle cell, is discussed in detail in chapter 9. Nerve cells can also form synapses with other nerve cells, smooth muscle cells, cardiac muscle cells, and gland cells.

The ANS transmits action potentials from the CNS to smooth muscle, cardiac muscle, and certain glands. Subconscious, or involuntary, control of smooth muscle, cardiac muscle, and glands occurs through the ANS. The ANS has two sets of neurons that exist in a series between the CNS and the effector organs (figure 11.2*c*). Cell bodies of the first neurons are within the CNS and send their axons to autonomic ganglia, where neuron cell bodies of the second neurons are located. Synapses exist between the first and second neurons within the autonomic ganglia, and the axons of the second neurons extend from the autonomic ganglia to the effector organs.

The ANS is subdivided into the sympathetic and the parasympathetic divisions and the enteric nervous system. In general, the **sympathetic division** is most active during physical activity, whereas the **parasympathetic division** regulates resting or vegetative functions, such as digesting food or emptying the urinary bladder. The **enteric nervous system** consists of plexuses within the wall of the digestive tract (see figure 24.4). Although the enteric nervous system is capable of controlling the digestive tract independently of the CNS, it's considered part of the ANS because of the parasympathetic and sympathetic neurons that contribute to the plexuses. See chapters 16 and 24 for details on the enteric nervous system.

The sensory division of the PNS functions to detect stimuli and transmit information in the form of action potentials to the CNS (figure 11.3). The CNS is the major site for processing information, initiating responses, and integrating mental processes. It's much like a highly sophisticated computer with the ability to receive input, process and store information, and generate responses. The motor division of the PNS conducts action potentials from the CNS to muscles and glands.

2. Define the CNS and the PNS.
3. What is a sensory receptor, nerve, ganglion, and plexus?
4. Based on the direction they transmit action potentials, what are the two subcategories of the PNS?
5. Based on the structures they supply, what are the two subcategories of the motor division?
6. Where are the cell bodies of sensory, somatic motor, and autonomic neurons located? What is a synapse?
7. What are the subcategories of the ANS?
8. Compare the general functions of the CNS and the PNS.

(a)

(b)

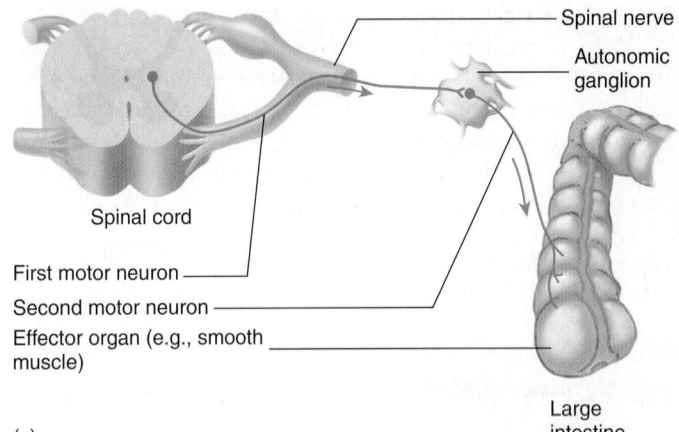

(c)

Figure 11.2   Divisions of the Peripheral Nervous System

(*a*) **Sensory division.** A neuron with its cell body in a dorsal root ganglion. (*b*) **Somatic nervous system.** The neuron extends from the CNS to skeletal muscle. (*c*) **Autonomic nervous system.** Two neurons are in series between the CNS and the effector cells (smooth muscle or glands). The first neuron has its cell body in the CNS, and the second neuron has its cell body in an autonomic ganglion.

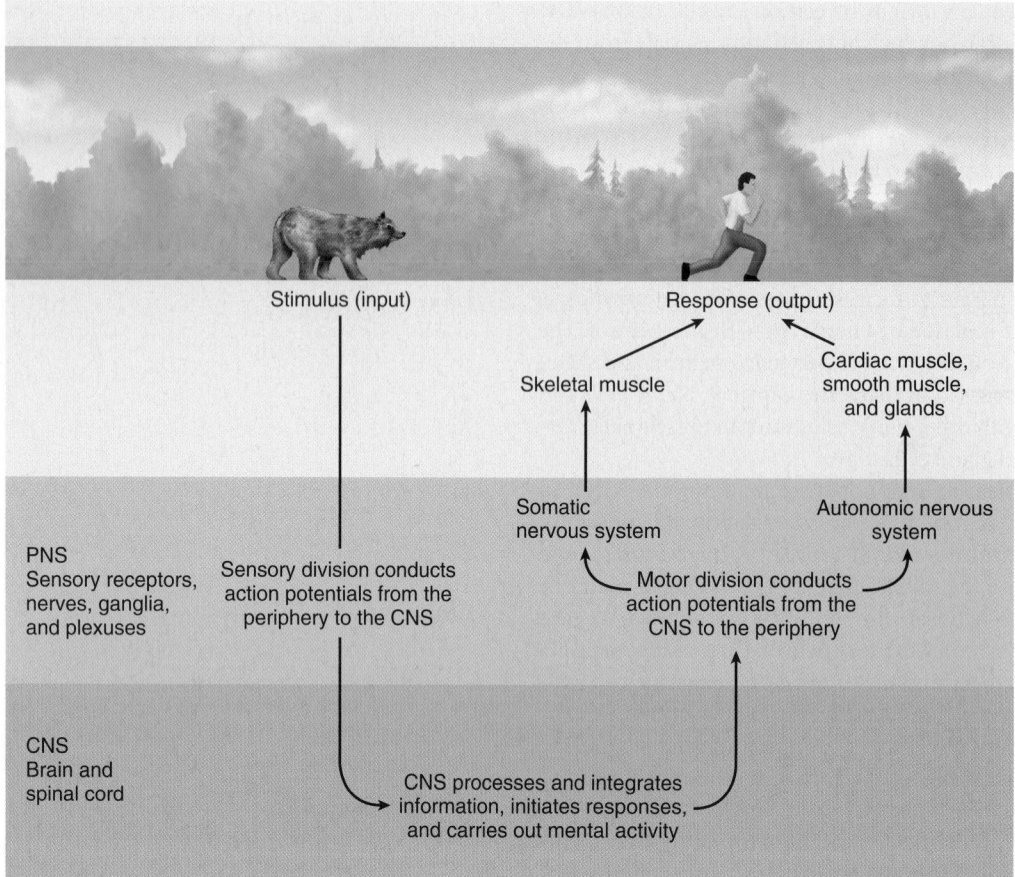

Figure 11.3   Organization of the Nervous System

The sensory division of the peripheral nervous system (PNS) detects stimuli and conducts action potentials to the central nervous system (CNS). The CNS interprets incoming action potentials and initiates action potentials that are conducted through the motor division to produce a response. The motor division is divided into the somatic nervous system and the autonomic nervous system.

Cells of the Nervous System

The nervous system is made up of neurons and nonneural cells. Neurons receive stimuli and conduct action potentials. Nonneural cells are called **neuroglia** (noo-rog′lē-ă; nerve glue), or **glial** (glī′ăl) **cells,** and they support and protect neurons and perform other functions. Neuroglia cells account for over half of the brain's weight, and there can be 10 to 50 times more neuroglia cells than neurons in various parts of the brain.

Neurons

Neurons, or **nerve cells,** receive stimuli and transmit action potentials to other neurons or to effector organs. They are organized to form complex networks that perform the functions of the nervous system. Each neuron consists of a cell body and two types of processes (figure 11.4). The cell body is called the **neuron cell body,** or **soma** (sō′mă; body), and the processes are called **dendrites** (den′drītz) and **axons** (ak′sonz). Dendrite means tree and refers to the branching organization of dendrites. Axon means

axis and refers to the straight alignment and uniform diameter of most axons. Axons are also referred to as **nerve fibers.**

Neuron Cell Body

Each neuron cell body contains a single relatively large and centrally located nucleus with a prominent nucleolus. Extensive rough endoplasmic reticulum and Golgi apparatuses surround the nucleus, and a moderate number of mitochondria and other organelles are also present. Randomly arranged lipid droplets and melanin pigments accumulate in the cytoplasm of some neuron cell bodies. The lipid droplets and melanin pigments increase as humans age, but their functional significance is unknown. Large numbers of intermediate filaments (neurofilaments) and microtubules form bundles that course through the cytoplasm in all directions. The neurofilaments separate areas of rough endoplasmic reticulum called **chromatophilic** (krō-mă-tō-fil′ik) **substance,** or **Nissl** (nis′l) **bodies.** The presence of organelles such as rough endoplasmic reticulum indicates that the neuron cell body is the primary site of protein synthesis within neurons.

Figure 11.4 Neuron

Structural features of a neuron include a cell body and two types of cell processes: dendrites and an axon.

Labels (from top):
- Dendritic spine
- Mitochondrion
- Golgi apparatus
- Nucleolus
- Nucleus
- Nissl bodies
- Trigger zone
  - Axon hillock
  - Initial segment
- Myelin sheath formed by Schwann cell
- Collateral axon
- Dendrites
- Neuron cell body
- Axon
- Schwann cell
- Node of Ranvier
- Presynaptic terminals

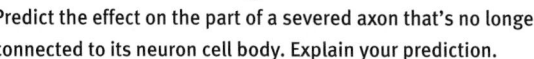

P R E D I C T ①

Predict the effect on the part of a severed axon that's no longer connected to its neuron cell body. Explain your prediction.

Dendrites

Dendrites are short, often highly branched cytoplasmic extensions that are tapered from their bases at the neuron cell body to their tips (see figure 11.4). Many dendrite surfaces have small extensions called **dendritic spines,** where axons of other neurons form synapses with the dendrites. Dendrites are the input part of the neuron. When stimulated, they generate small electric currents that are conducted to the neuron cell body.

Axons

In most neurons, a single axon arises from a cone-shaped area of the neuron cell body called the **axon hillock.** The beginning of the axon is called the **initial segment.** An axon can remain as a single structure or can branch to form collateral axons or side branches (see figure 11.4). Each axon has a constant diameter, but it can vary in length from a few millimeters to more than 1 meter. The cytoplasm of the axon is sometimes called **axoplasm,** and its plasma membrane is called the **axolemma** (*lemma* is Greek, meaning husk or sheath). Axons terminate by branching to form small extensions with enlarged ends called **presynaptic terminals,** or **terminal boutons** (boo-tonz′; buttons). Numerous small vesicles containing neurotransmitters are present in the presynaptic terminals. **Neurotransmitters** are chemicals released from the presynaptic terminal that cross the synapse to stimulate or inhibit the postsynaptic cell. Functionally, action potentials are generated at the **trigger zone,** which consists of the axon hillock and the part of the axon nearest to the cell body. Action potentials are conducted along the axon to the presynaptic terminal, where they stimulate the release of neurotransmitters.

Axon transport mechanisms can move cytoskeletal proteins (see chapter 3), organelles such as mitochondria, and vesicles containing neurohormones to be secreted (see chapter 17) down the axon to the presynaptic terminals. In addition, damaged organelles, recycled plasma membrane, and substances taken in by endocytosis can be transported up the axon to the neuron cell body. The movement of materials within the axon is necessary for its normal function, but it also provides a way for infectious agents and harmful substances to be transported from the periphery to the CNS. For example, rabies and herpes viruses enter the axon endings of damaged skin and are transported to the CNS.

9. *Compare the functions of neuroglia and neurons.*
10. *Describe and give the function of a neuron cell body, dendrite, and axon.*
11. *Define trigger zone, presynaptic terminal, and neurotransmitter.*

Types of Neurons

Neurons are classified according to their function or structure. The functional classification is based on the direction in which action potentials are conducted. **Sensory, or afferent, neurons** conduct action potentials toward the CNS, and **motor, or efferent, neurons** conduct action potentials away from the CNS toward muscles or glands. **Interneurons, or association neurons,** conduct action potentials from one neuron to another within the CNS.

The structural classification scheme is based on the number of processes that extend from the neuron cell body. The three major categories of neurons are multipolar, bipolar, and unipolar.

Multipolar neurons have many dendrites and a single axon. The dendrites vary in number and in their degree of branching (figure 11.5a). Most of the neurons within the CNS and motor neurons are multipolar.

Bipolar neurons have two processes: a dendrite and an axon (figure 11.5b). The dendrite often is specialized to receive the stimulus, and the axon conducts action potentials to the CNS. Bipolar neurons are located in some sensory organs, such as in the retina of the eye and in the nasal cavity.

Unipolar neurons have a single process extending from the cell body (figure 11.5c). This process divides into two branches a short distance from the cell body. One branch extends to the CNS, and the other branch extends to the periphery and has dendritelike sensory receptors. The two branches function as a single axon. The sensory receptors respond to stimuli resulting in the production of action potentials that are transmitted to the CNS. According to a functional definition of a dendrite, the branch of a unipolar neuron that extends from the periphery to the neuron cell body could

be classified as a dendrite because it conducts action potentials toward the neuron cell body. This branch is usually referred to as an axon, however, for two reasons: it cannot be distinguished from an axon on the basis of its structure, and it conducts action potentials in the same fashion as an axon.

12. Describe the three types of neurons based on their function.
13. Describe the three types of neurons based on their structure, and give an example of where each type is found.

Neuroglia of the CNS

Neuroglia are far more numerous than neurons and account for more than half the brain's weight. They are the major supporting cells in the CNS, participate in the formation of a permeability barrier between the blood and the neurons, phagocytize foreign substances, produce cerebrospinal fluid, and form myelin sheaths around axons. There are four types of CNS neuroglial cells and each has unique structural and functional characteristics.

Astrocytes

Astrocytes (as′trō-sītz, *aster* is Greek, meaning star) are neuroglia that are star-shaped because of cytoplasmic processes that extend from the cell body. These extensions widen and spread out to form foot processes which cover the surfaces of blood vessels (figure 11.6), neurons, and the pia mater. (The pia mater is a membrane covering the outside of the brain and spinal cord.) Astrocytes have an extensive cytoskeleton of microfilaments (see chapter 3) that enables them to form a supporting framework for blood vessels and neurons.

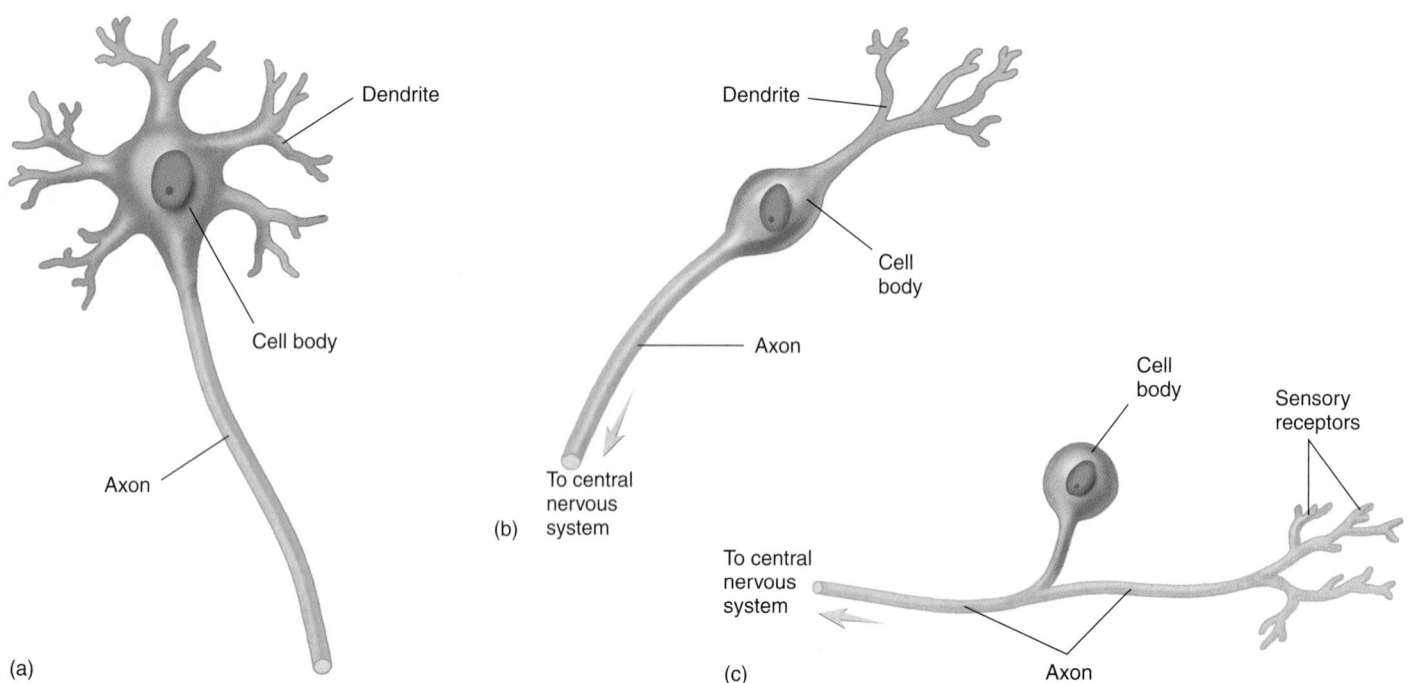

Figure 11.5 Types of Neurons

(*a*) A multipolar neuron has many dendrites and one axon. (*b*) A bipolar neuron has a dendrite and an axon. (*c*) A unipolar neuron has an axon and no dendrites.

Figure 11.6   Astrocytes

Astrocyte processes form feet that cover the surfaces of neurons and blood vessels. The astrocytes provide structural support and play a role in regulating what substances from the blood reach the neurons.

Figure 11.7   Ependymal Cells

(*a*) Ciliated ependymal cells lining the ventricles of the brain and the central canal of the spinal cord help to move cerebrospinal fluid. (*b*) Ependymal cells on the surface of the choroid plexus secrete cerebrospinal fluid.

Astrocytes play a role in regulating the extracellular composition of brain fluid. They release chemicals that promote the formation of tight junctions (see chapter 4) between the endothelial cells of capillaries. The endothelial cells with their tight junctions form the **blood–brain barrier,** which determines what substances can pass from the blood into the nervous tissue of the brain and spinal cord. The blood–brain barrier protects neurons from toxic substances in the blood, allows the exchange of nutrients and waste products between neurons and the blood, and prevents fluctuations in the composition of the blood from affecting the functions of the brain. Astrocytes also help to control the composition of interstitial fluid by regulating the concentration of ions and gases and by synthesizing, absorbing, and recycling neurotransmitters.

Ependymal Cells

Ependymal (ep-en′di-măl) cells line the ventricles (cavities) of the brain and the central canal of the spinal cord (figure 11.7*a*). Specialized ependymal cells and blood vessels form the **choroid plexuses** (ko′royd plek′sŭs-ez) (figure 11.7*b*), which are located within certain regions of the ventricles. The choroid plexuses secrete the cerebrospinal fluid that circulates through the ventricles of the brain (see chapter 13). The free surface of the ependymal cells frequently has patches of cilia that assist in moving cerebrospinal fluid through the cavities of the brain. Ependymal cells also have long processes (not illustrated) at their basal surfaces that extend deep into the brain and the spinal cord and seem, in some cases, to have astrocytelike functions.

Microglia

Microglia (mī-krog′lē-ă) are specialized macrophages in the CNS that become mobile and phagocytic in response to inflammation, and they phagocytize necrotic tissue, microorganisms, and foreign substances that invade the CNS (figure 11.8).

Figure 11.8   Microglia

Microglia within the CNS are similar to macrophages.

Microglia and Brain Damage

Numerous microglia migrate to areas damaged by infection, trauma, or stroke and perform phagocytosis. A pathologist can identify these damaged areas in the CNS during an autopsy because large numbers of microglia are found in them.

Oligodendrocytes

Oligodendrocytes (ol′i-gō-den′drō-sītz) have cytoplasmic extensions that can surround axons. If the cytoplasmic extensions wrap many times around the axons, they form **myelin** (mī′ĕ-lin) **sheaths.** A single oligodendrocyte can form myelin sheaths around portions of several axons (figure 11.9).

Neuroglia of the PNS

Schwann cells, or **neurolemmocytes** (noor-ō-lem′mō-sītz), are neuroglial cells in the PNS that wrap around axons. If the Schwann cell wraps many times around an axon, it forms a myelin sheath. Unlike oligodendrocytes, however, each Schwann cell forms a myelin sheath around a portion of only one axon (figure 11.10).

    Satellite cells surround neuron cell bodies in ganglia, provide support, and can provide nutrients to the neuron cell bodies (figure 11.11).

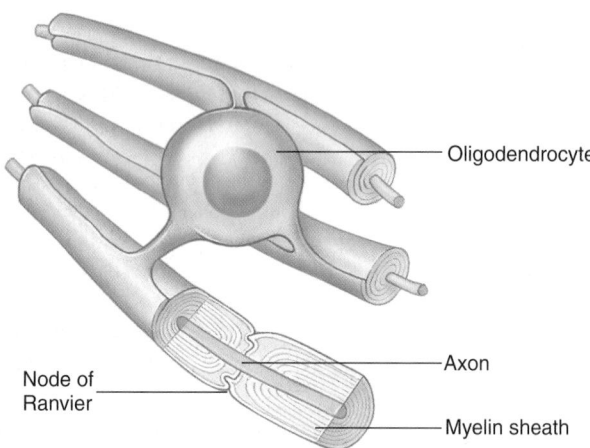

Figure 11.9 Oligodendrocyte

Extensions from the oligodendrocyte form the myelin sheaths of axons within the CNS.

Figure 11.10 Schwann Cell

Extension from the Schwann cell forms the myelin sheath of an axon within the PNS.

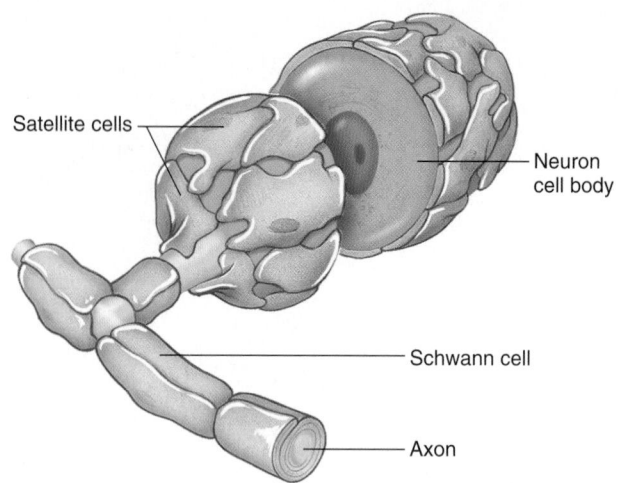

Figure 11.11 Satellite Cells

Neuron cell bodies within ganglia are surrounded by satellite cells.

  ● 14. *Which type of neuroglia supports neurons and blood vessels and promotes the formation of the blood–brain barrier? What is the blood–brain barrier, and what is its function?*

     15. *Name the different kinds of neuroglia responsible for the following functions: produces cerebrospinal fluid, phagocytosis, produces myelin sheaths in the CNS, produces myelin sheaths in the PNS, supports neuron cell bodies in the PNS.*

Myelinated and Unmyelinated Axons

Cytoplasmic extensions of the oligodendrocytes in the CNS and of the Schwann cells in the PNS surround axons to form either myelinated or unmyelinated axons. Myelin protects and electrically insulates axons from one another. In addition, action potentials travel along myelinated axons more rapidly than along unmyelinated axons (see "Propagation of Action Potentials" on p. 392).

    In **myelinated axons,** the extensions from oligodendrocytes or Schwann cells repeatedly wrap around a segment of an axon to form a series of tightly wrapped membranes rich in phospholipids with little cytoplasm sandwiched between the membrane layers (figure 11.12*a*). The tightly wrapped membranes constitute the myelin sheath and give myelinated axons a white appearance because of the high lipid concentration. The myelin sheath is not continuous but is interrupted every 0.3–1.5 mm. At these locations, there are slight constrictions where the myelin sheaths of adjacent cells dip toward the axon but don't cover it, leaving a bare area 2–3 μm in length. These interruptions in the myelin sheath are the **nodes of Ranvier** (ron′vē-ā), and the myelin covered areas between the nodes are called the **internodes.**

    Unmyelinated axons rest in invaginations of the oligodendrocytes or the Schwann cells (figure 11.12*b*). The cell's plasma membrane surrounds each axon, but does not wrap around them many times. Thus each axon is surrounded by a series of cells, and each cell can simultaneously surround more than one unmyelinated axon.

  ● 16. *Define myelin sheath, node of Ranvier, and internode.*

     17. *How are myelinated and unmyelinated axons different from each other?*

(a)

(b)

Figure 11.12   Comparison of Myelinated and Unmyelinated Axons

(a) Myelinated axon with two Schwann cells forming the myelin sheath around a single axon. Each Schwann cell surrounds part of one axon. (b) Unmyelinated axons with two Schwann cells surrounding several axons in parallel formation. Each Schwann cell surrounds part of several axons.

Organization of Nervous Tissue

Nervous tissue is organized so that axons form bundles, and neuron cell bodies and their relatively short dendrites are grouped together. Bundles of parallel axons with their associated myelin sheaths are whitish in color, which accounts for their name, **white matter.** Collections of neuron cell bodies and unmyelinated axons are more gray in color and are called **gray matter.**

The axons that make up the white matter of the CNS form **nerve tracts,** which propagate action potentials from one area in the CNS to another. The gray matter of the CNS performs integrative functions or acts as relay areas in which axons synapse with the cell bodies of neurons. The central area of the spinal cord is gray matter, and the outer surface of much of the brain consists of gray matter called **cortex.** Within the brain are other collections of gray matter called **nuclei.**

In the PNS, bundles of axons form nerves, which conduct action potentials to and from the CNS. Most nerves contain myelinated axons, but some consist of unmyelinated axons. Collections of neuron cell bodies in the PNS are called ganglia.

18. *What is white and gray matter?*
19. *Define and state the location of nerve tracts, nerves, brain cortex, nuclei, and ganglia.*

Electric Signals

Like computers, humans depend on electric signals to communicate and process information. The electric signals produced by cells are called **action potentials.** They are an important means by which cells transfer information from one part of the body to another. For example, stimuli such as light, sound, and pressure act on specialized sensory cells in the eye, ear, and skin to produce action potentials, which are conducted from these cells to the spinal cord and brain. Action potentials originating within the brain and spinal cord are conducted to muscles and certain glands to regulate their activities.

The ability to perceive our environment, perform complex mental activities, and act depends upon action potentials. For example, interpreting the action potentials received from sensory cells results in the sensations of sight, hearing, and touch. Complex mental activities, such as conscious thought, memory, and emotions, result from action potentials. The contraction of muscles and the secretion of certain glands occur in response to action potentials generated in them.

A basic knowledge of the electrical properties of cells is necessary for understanding many of the normal functions and pathologies of the body. These properties result from the ionic concentration differences across the plasma membrane and from the permeability characteristics of the plasma membrane.

Concentration Differences Across the Plasma Membrane

Table 11.1 lists the concentration differences for positively charged ions (cations) and negatively charged ions (anions) between the intracellular and extracellular fluids. The concentration of sodium ions (Na^+) and chloride ions (Cl^-) is much greater outside the cell than inside. The concentration of potassium ions (K^+)

Table 11.1	Representative Concentrations of the Principal Cations and Anions in Extracellular and Intracellular Fluids of Vertebrates	
Ions	**Intracellular Fluid (mEq/L)***	**Extracellular Fluid (mEq/L)**
Cations (Positive)		
Potassium (K^+)	148	5
Sodium (Na^+)	10	142
Calcium (Ca^{2+})	<1	5
Others	41	3
TOTAL	200	155
Anions (Negative)		
Proteins	56	16
Chloride (Cl^-)	4	103
Others	140	36
TOTAL	200	155

*See Appendix C for an explanation of milliequivalents (mEq).

and negatively charged molecules, such as proteins and other molecules containing phosphate, is much greater inside the cell than outside. Note that a steep concentration gradient (see chapter 3) exists for Na$^+$ from outside the cell to the inside. Also, a steep concentration gradient exists for K$^+$ from the inside to the outside of the cell.

Differences in intracellular and extracellular concentrations of ions result primarily from (1) the Na$^+$–K$^+$ pump and (2) the permeability characteristics of the plasma membrane.

The Na$^+$–K$^+$ Pump

The differences in K$^+$ and Na$^+$ concentrations across the plasma membrane are maintained primarily by the action of the **Na$^+$–K$^+$ pump** (figure 11.13). Through active transport, the Na$^+$–K$^+$ pump moves K$^+$ and Na$^+$ through the plasma membrane against their concentration gradients. K$^+$ are transported into the cell, increasing the concentration of K$^+$ inside the cell, and Na$^+$ are transported out of the cell, increasing the concentration of Na$^+$ outside the cell. Approximately three Na$^+$ are transported out of the cell and two K$^+$ are transported into the cell for each ATP molecule used.

Permeability Characteristics of the Plasma Membrane

As noted in chapter 3, the plasma membrane is selectively permeable, thus allowing some, but not all, substances to pass through it. Negatively charged proteins are synthesized inside the cell, and because of their large size and their solubility characteristics, they cannot readily diffuse across the plasma membrane (figure 11.14). Negatively charged Cl$^-$ are repelled by the negatively charged proteins and other negatively charged ions inside the cell. Cl$^-$ diffuse through the plasma membrane and accumulate outside it, resulting in a higher concentration of Cl$^-$ outside of the cell than inside.

Ions pass through the plasma membrane through ion channels. The two major types of ion channels are nongated and gated ion channels.

Nongated Ion Channels

Nongated ion channels, or **leak channels,** are always open and are responsible for the permeability of the plasma membrane to ions when the plasma membrane is unstimulated, or at rest

1. Three Na$^+$ and ATP bind to the carrier molecule.

2. The ATP breaks down to ADP and phosphate and releases energy. The carrier molecule changes shape, and Na$^+$ are transported across the membrane.

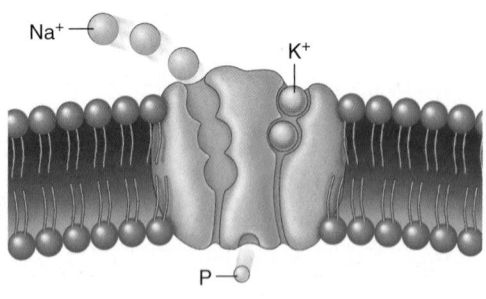

3. Na$^+$ diffuse away from the carrier molecule, two K$^+$ bind to the carrier molecule, and the phosphate is released.

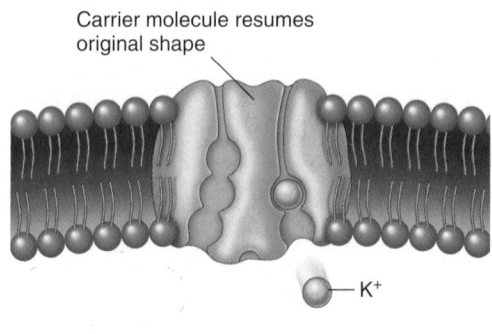

4. The carrier molecule resumes original shape, transporting K$^+$ across the membrane, and K$^+$ diffuse away from the carrier molecule. The carrier molcule can again bind to Na$^+$ and ATP.

Process Figure 11.13   The Na$^+$–K$^+$ Pump

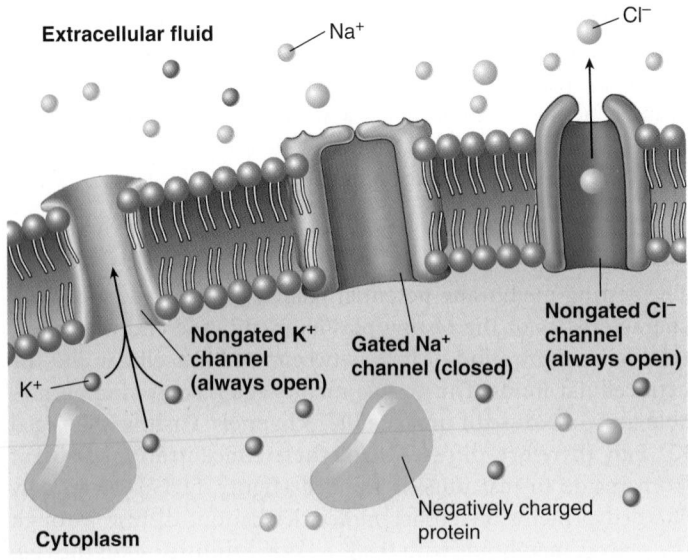

Figure 11.14   Membrane Permeability and Ion Channels

The permeability of the membrane to K⁺ and Cl⁻ is greater than its permeability to Na⁺ because some nongated K⁺ and Cl⁻ channels are open, whereas most gated Na⁺ channels are closed. The membrane is not permeable to the negatively charged proteins inside of the cell because they are too large to pass through membrane channels.

1. The Na⁺ channel has receptor sites for the ligand, acetylcholine. When the receptor sites are not occupied by acetylcholine, the Na⁺ channel remains closed.

Na⁺ diffuse through the open channel

2. When two acetylcholine molecules bind to their receptor sites on the Na⁺ channel, the channel opens to allow Na⁺ to diffuse through the channel into the cell.

Process Figure 11.15   Ligand-Gated Ion Channel

(see figure 11.14). Each ion channel is specific for one type of ion, although the specificity is not absolute. The number of each type of nongated ion channels in the plasma membrane determines the permeability characteristics of the resting plasma membrane to different types of ions. The plasma membrane is more permeable to K⁺ and Cl⁻ and much less permeable to Na⁺ because there are many more K⁺ and Cl⁻ nongated ion channels than Na⁺ nongated ion channels in the plasma membrane.

Gated Ion Channels

Gated ion channels open and close in response to stimuli. By opening and closing, these channels can change the permeability characteristics of the plasma membrane. The major types of gated ion channels are:

1. *Ligand-gated ion channels.* A **ligand** is a molecule that binds to a receptor. A **receptor** is a protein or glycoprotein that has a **receptor site** to which a ligand can bind. Most receptors are located in the plasma membrane. **Ligand-gated ion channels** open or close in response to a ligand binding to a receptor. For example, the neurotransmitter acetylcholine released from the presynaptic terminal of a neuron is a chemical signal that can bind to a ligand-gated Na⁺ channel in the membrane of a muscle cell. As a result, the Na⁺ channel opens, allowing Na⁺ to enter the cell (figure 11.15). Ligand-gated ion channels exist for Na⁺, K⁺, Ca²⁺, and Cl⁻, and these channels are common in tissues such as nervous and muscle tissues, as well as glands.

2. *Voltage-gated ion channels.* These channels open and close in response to small voltage changes across the plasma

membrane. In an unstimulated cell, the inside of the plasma membrane is negatively charged relative to the outside. This charge difference can be measured in units called **millivolts** (mV; 1 mV = 1/1000 V). When a cell is stimulated, the permeability of the plasma membrane changes because gated ion channels open or close. The movement of ions into or out of the cell changes the charge difference across the plasma membrane, which causes voltage-gated ion channels to open or close. Voltage-gated channels specific for Na⁺ and K⁺ are most numerous in electrically excitable tissues, but voltage-gated Ca²⁺ channels are also important, especially in smooth muscle and cardiac muscle cells (see chapters 9 and 20).

3. *Other-gated ion channels.* Gated ion channels that respond to stimuli other than ligands or voltage changes are present in specialized electrically excitable tissues. Examples include touch receptors, which respond to mechanical stimulation of the skin, and temperature receptors, which respond to temperature changes in the skin.

20. Describe the concentration differences for Na⁺ and K⁺ that exist across the plasma membrane.
21. In what direction, into or out of cells, does the Na⁺–K⁺ pump move Na⁺ and K⁺?
22. Define nongated and gated ion channels. How are they responsible for the permeability characteristics of a resting versus a stimulated plasma membrane?
23. Define ligand, receptor, and receptor site.
24. What kinds of stimuli cause gated ion channels to open or close?

The Resting Membrane Potential

Although unequal concentrations of ions exist in the intracellular and extracellular fluids, these fluids are nearly electrically neutral. That is, both intracellular and extracellular fluids have nearly equal numbers of positively and negatively charged ions. However, an unequal distribution of charge exists between the immediate inside and the immediate outside of the plasma membrane. This electric charge difference across the plasma membrane, called a **potential difference,** can be measured between the inside and outside of essentially all cells. By placing the tip of one microelectrode inside a cell and another outside it, and by connecting the electrodes by wires to an appropriate measuring device such as a voltmeter or an oscilloscope, the potential difference can be measured (figure 11.16). The potential

Figure 11.16 Measuring the Resting Membrane Potential

(a) Both recording (needle) and reference (block) electrodes are outside of the cell, and no potential difference (0 mV) is recorded. (b) The recording electrode is inside the cell, the reference electrode is outside, and a potential difference of about −85 mV is recorded, with the inside of the plasma membrane negative with respect to the outside of the membrane.

difference across the plasma membranes of skeletal muscle fibers and nerve cells is −70 to −90 mV. The potential difference is reported as a negative number, because the inside of the plasma membrane is negative compared to the outside. In an unstimulated, or resting, cell, the potential difference across the plasma membrane is called the **resting membrane potential.**

Establishing the Resting Membrane Potential

The resting membrane potential results from the permeability characteristics of the resting plasma membrane and the difference in concentration of ions between the intracellular and the extracellular fluids. The plasma membrane is somewhat permeable to K⁺ because of nongated K⁺ channels. Positively charged K⁺ can therefore diffuse down their concentration gradient from inside to just outside the cell (figure 11.17). Negatively charged proteins and other molecules cannot diffuse through the plasma membrane with the K⁺. As K⁺ diffuse out of the cell, the loss of positive charges makes the inside of the plasma membrane more negative. Because opposite charges attract, the K⁺ are attracted back toward the cell. The accumulation of K⁺ just outside of the plasma membrane makes the outside of the plasma membrane positive relative to the inside. The resting membrane potential is an equilibrium that is established when the tendency for K⁺ to diffuse out of the cell, because of the K⁺ concentration gradient, is equal to the tendency for K⁺ to move into the cell, because of the attraction of the positively charged K⁺ to negatively charged proteins and other molecules.

PREDICT 2
Given that tissue A has significantly more nongated K⁺ channels than tissue B, which tissue has the larger resting membrane potential?

Other ions, such as Na⁺, Cl⁻, and Ca²⁺, do have some small influence on the resting membrane potential, but the major influence is from K⁺. Because the resting plasma membrane is 50–100 times less permeable to Na⁺ than to K⁺, very few Na⁺ can diffuse from the outside to the inside of the resting cell. The resting plasma membrane is not very permeable to Ca²⁺ either. The plasma membrane is relatively permeable to Cl⁻, but these negatively charged ions are repelled by the negative charge inside the cell.

The resting membrane potential is proportional to the tendency for K⁺ to diffuse out of the cell and not to the actual rate of flow for K⁺. At equilibrium, very few of these ions pass through the plasma membrane because their movement out of the cell is opposed by the negative charge inside the cell. Still, some Na⁺ and K⁺ diffuse continuously across the plasma membrane, although at a low rate. The large concentration gradients for Na⁺ and K⁺ would eventually disappear without the continuous activity of the Na⁺–K⁺ pump.

As already noted, the function of the Na⁺–K⁺ pump is to maintain the normal concentration gradients for Na⁺ and K⁺ across the plasma membrane. The pump is also responsible for a small portion of the resting membrane potential, usually less than 15 mV, because it transports approximately three Na⁺ out of the cell and two K⁺ into the cell for each ATP molecule used

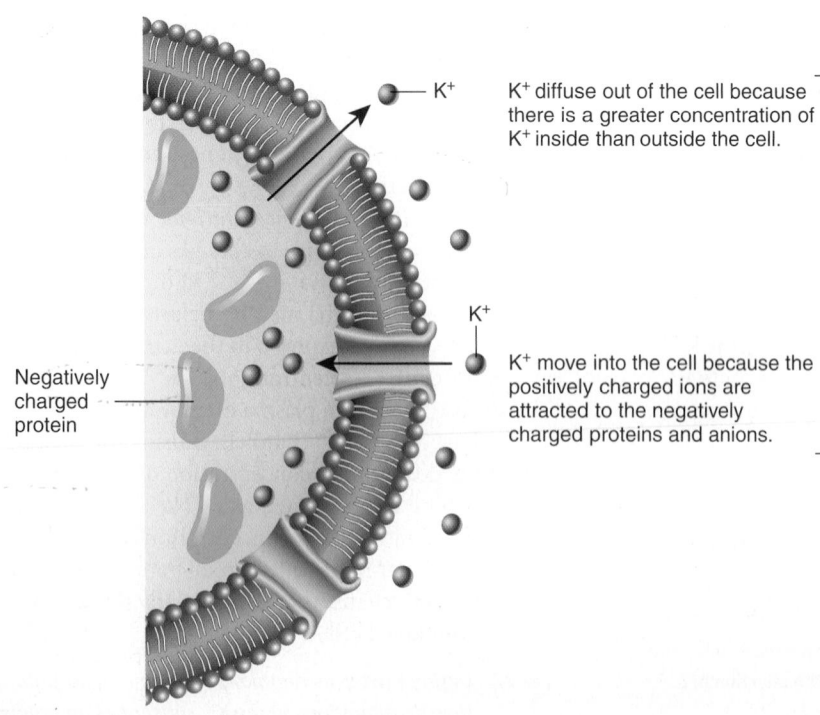

Figure 11.17 Potassium Ions and the Resting Membrane Potential

(see figure 11.13). The outside of the plasma membrane becomes more positively charged than the inside, because more positively charged ions are pumped out of the cell than are pumped into it.

The characteristics responsible for a resting membrane potential are summarized in table 11.2.

25. *Define the resting membrane potential. Is the outside of the plasma membrane positively or negatively charged relative to the inside?*
26. *Explain the role of K⁺ and the Na⁺–K⁺ pump in establishing the resting membrane potential.*

Changing the Resting Membrane Potential

The resting membrane potential can be changed by alterations in the K⁺ concentration gradient, changes in membrane permeability to K⁺ and Na⁺, and changes in extracellular Ca²⁺ concentrations.

1. *K⁺ concentration gradient.* The concentration of K⁺ is higher inside than outside cells. An increase in the extracellular concentration of K⁺ decreases this concentration difference and thus decreases the K⁺ concentration gradient. As a consequence, the tendency for K⁺ to diffuse out of the cell decreases, and a smaller negative charge inside the cell is required to oppose the diffusion of K⁺ out of the cell. At this new equilibrium, the charge difference across the plasma membrane is

Table 11.2	Characteristics Responsible for the Resting Membrane Potential
1.	The number of charged molecules and ions inside and outside the cell is nearly equal.
2.	The concentration of K⁺ is higher inside than outside the cell, and the concentration of Na⁺ is higher outside than inside the cell.
3.	The plasma membrane is 50–100 times more permeable to K⁺ than to other positively charged ions such as Na⁺.
4.	The plasma membrane is impermeable to large intracellular negatively charged molecules such as proteins.
5.	K⁺ tend to diffuse across the plasma membrane from the inside to the outside of the cell.
6.	Because negatively charged molecules cannot follow the positively charged K⁺, a small negative charge develops just inside the plasma membrane.
7.	The negative charge inside the cell attracts positively charged K⁺. When the negative charge inside the cell is great enough to prevent additional K⁺ from diffusing out of the cell through the plasma membrane, an equilibrium is established.
8.	The charge difference across the plasma membrane at equilibrium is reflected as a difference in potential, which is measured in millivolts (mV).
9.	The resting membrane potential is proportional to the potential for K⁺ to diffuse out of the cell but not to the actual rate of flow for K⁺.
10.	At equilibrium there is very little movement of K⁺ or other ions across the plasma membrane.

decreased, which means the resting membrane potential is less negative (figure 11.18*a*). This change is called **depolarization** (dē-pō′lăr-i-zā′shŭn), or **hypopolarization** (hī′pō-pō-lăr-i-zā′shŭn), of the resting membrane potential. That is, the potential difference across the plasma membrane becomes smaller, or less polar.

A decrease in the extracellular concentration of K^+ increases the concentration difference between the inside and the outside of cells and thus increases the K^+ concentration gradient. As a result, the tendency for K^+ to diffuse out of the cell increases and a greater negative charge inside the cell is required to resist that diffusion. Thus the resting membrane potential becomes more negative (figure 11.18*b*), a change called **hyperpolarization** (hī′per-pō′lăr-i-zā′shŭn). That is, the potential difference across the plasma membrane becomes greater, or more polar.

P R E D I C T   **3**

Does the resting membrane potential increase or decrease when the intracellular concentration of K^+ is increased by the injection of a solution of potassium succinate into the cell? Explain.

2. *K^+ membrane permeability.* Although nongated K^+ channels allow some K^+ to diffuse across the plasma membrane, the resting membrane is not freely permeable to K^+. If gated K^+ channels open, membrane permeability to K^+ increases, and more K^+ diffuse out of the cell. The increased tendency for K^+ to diffuse out of the cell is opposed by a greater negative charge that develops inside the plasma membrane (hyperpolarization).

3. *Na^+ membrane permeability.* In an unstimulated cell, the membrane is not very permeable to Na^+ because there are few nongated Na^+ channels. Changes in the concentration of Na^+ on either side of the plasma membrane don't influence the resting membrane potential very much because of this low permeability. If gated Na^+ channels

open, the permeability of the membrane to Na^+ increases (see figure 11.15). Na^+ then diffuse into the cell because the concentration gradient for Na^+ is from the outside to the inside of the cell. As Na^+ diffuse into the cell, the inside of the plasma membrane becomes more positive, resulting in depolarization.

4. *Extracellular Ca^{2+} concentrations.* Voltage-gated Na^+ channels are sensitive to changes in the extracellular concentration of Ca^{2+}. Ca^{2+} in the extracellular fluid are attracted to plasma membrane proteins with negatively charged groups exposed to the extracellular fluid. If the extracellular concentration of Ca^{2+} decreases, these ions diffuse away from plasma membrane proteins, including the voltage-gated Na^+ channels, causing the channels to open. If the extracellular concentration of Ca^{2+} increases, these ions bind to the voltage-gated Na^+ channels, causing them to close. At the Ca^{2+} concentrations normally found in the extracellular fluid, only a small percentage of the voltage-gated Na^+ channels are open at any single moment in an unstimulated cell.

27. *Define the terms depolarization and hyperpolarization. How do alterations in the K^+ concentration gradient, changes in membrane permeability to K^+ or Na^+, and changes in extracellular Ca^{2+} concentration affect depolarization and hyperpolarization?*

P R E D I C T   **4**

Predict the effect of a decrease in the extracellular concentration of Ca^{2+} on the resting membrane potential.

Local Potentials

A stimulus applied at one location on the plasma membrane of a cell normally causes a change in the resting membrane potential called a **local,** or **graded, potential,** which is confined to a small region of the plasma membrane. Local potentials can result from (1) ligands binding to their receptors, (2) changes in the charge across

(a)

(b)

Figure 11.18   **Changes in the Resting Membrane Potential Caused by Changes in Extracellular K^+ Concentration**
(*a*) Elevated extracellular K^+ concentration causes depolarization. (*b*) Decreased extracellular K^+ concentration causes hyperpolarization.

the plasma membrane, (3) mechanical stimulation, (4) temperature changes, or (5) spontaneous changes in membrane permeability.

Changes in membrane permeability to Na^+, K^+, or other ions can produce a local potential, which can be a depolarization or hyperpolarization. For example, if a stimulus causes gated Na^+ channels to open, the diffusion of a few Na^+ into cells results in depolarization. If a stimulus causes gated K^+ channels to open, the diffusion of a few K^+ out of the cell results in hyperpolarization. If the permeability to Cl^- increases, the diffusion of a few Cl^- into the cell results in hyperpolarization.

Local potentials are also called graded potentials because their magnitude can vary from small to large depending on the stimulus strength or frequency. For example, a weak stimulus can cause a few gated Na^+ channels to open. A few Na^+ diffuse into the cell, cause a small depolarization, and produce a small local potential. A stronger stimulus can cause a greater number of gated Na^+ channels to open. A greater number of Na^+ diffusing into the cell causes a larger depolarization and produces a larger local potential (figure 11.19a).

(a)

(b)

Figure 11.19 Local Potentials

(a) Local potentials are proportional to the stimulus strength. A weak stimulus applied briefly causes a small depolarization, which quickly returns to the resting membrane potential (1). Progressively stronger stimuli result in larger depolarizations (2 to 4). (b) A stimulus applied to a cell causes a small depolarization. When a second stimulus is applied before the depolarization disappears, the depolarization caused by the second stimulus is added to the depolarization caused by the first to result in a larger depolarization.

Local potentials can **summate** (sŭm-āt′), or add onto each other (figure 11.19b). For example, if a second stimulus is applied before the local potential produced by the first stimulus has returned to the resting membrane potential, a larger depolarization results than would result from a single stimulus. The first stimulus causes gated Na^+ channels to open, and the second stimulus causes additional Na^+ channels to open. As a result, more Na^+ diffuse into the cell, producing a larger local potential.

Local potentials spread, or are conducted, over the plasma membrane in a decremental fashion. That is, they rapidly decrease in magnitude as they spread over the surface of the plasma membrane. It's much like a teacher talking to a large class. At the front of the class the teacher's voice can be easily heard, but the farther away a student sits, the more difficult it is to hear. Normally a local potential cannot be detected more than a few millimeters from the site of stimulation. As a consequence, a local potential cannot transfer information over long distances from one part of the body to another.

Local potentials are important because of their effect on the generation of action potentials. The characteristics of local potentials are summarized in table 11.3.

 28. *Define a local potential. What does it mean to say a local potential is graded, can summate, and spreads in a decremental fashion?*

P R E D I C T ⑤
Given two cells that are identical in all ways except that the extracellular concentration of Na^+ is less for cell A than for cell B, how would the magnitude of the local potential in cell A differ from that in cell B if stimuli of identical strength were applied to each?

Action Potentials

When a local potential causes depolarization of the plasma membrane to a level called **threshold,** a series of permeability changes

Table 11.3	Characteristics of Local Potentials
1.	A stimulus causes ion channels to open, resulting in increased permeability of the membrane to Na^+, K^+, or Cl^-.
2.	Increased permeability of the membrane to Na^+ results in depolarization. Increased permeability of the membrane to K^+ or Cl^- results in hyperpolarization.
3.	Local potentials are graded; that is, the size of the local potential is proportional to the strength of the stimulus. Local potentials can also summate. Thus, a local potential produced in response to several stimuli is larger than one produced in response to a single stimulus.
4.	Local potentials are conducted in a decremental fashion, meaning that their magnitude decreases as they spread over the plasma membrane. Local potentials cannot be measured a few millimeters from the point of stimulation.
5.	A depolarizing local potential can cause an action potential.

occur that results in an **action potential** (figure 11.20). An action potential is a large change in the membrane potential that propagates, without changing its magnitude, over long distances along the plasma membrane. Thus, action potentials can transfer information from one part of the body to another. It generally takes 1–2 milliseconds (ms) (1 ms = 0.001 s) for an action potential to occur. The characteristics of action potentials are summarized in table 11.4.

The generation of action potentials is dependent on local potentials. Depolarizing local potentials that reach threshold produce an action potential. Hyperpolarizing local potentials can never reach threshold and do not produce action potentials. Thus, depolarizing local potentials can potentially activate a cell by causing the production of an action potential, whereas a hyperpolarizing local potential can prevent the production of an action potential.

In addition, the magnitude of a depolarizing local potential affects the likelihood of generating an action potential. For example, a weak stimulus can produce a small depolarizing local potential that does not reach threshold and therefore does not cause the production of an action potential. A stronger stimulus, however, can produce a larger depolarizing local potential that reaches threshold, resulting in the production of an action potential.

Action potentials occur according to the **all-or-none principle.** If a stimulus produces a depolarizing local potential that is large enough to reach threshold, all the permeability changes responsible for an action potential proceed without stopping and are constant in magnitude (the "all" part). If a stimulus is so weak that the depolarizing local potential does not reach threshold, few of the permeability changes occur. The membrane potential returns to its resting level after a brief period without producing an action potential (the "none" part). An action potential can be compared to the flash system of a camera. When the shutter is triggered (reaches threshold), the camera flashes (an action potential is produced), and each flash is the same brightness (magnitude; the "all" part) as previous flashes. If the shutter is depressed, but not triggered, no flash results (the "none" part).

The action potential has a **depolarization phase,** in which the membrane potential moves away from the resting membrane

Table 11.4	Characteristics of the Action Potential

1. Action potentials are produced when a local potential reaches threshold.
2. Action potentials are all-or-none.
3. Depolarization is a result of increased membrane permeability to Na^+ and movement of Na^+ into the cell. Activation gates of the voltage-gated Na^+ channels open.
4. Repolarization is a result of decreased membrane permeability to Na^+ and increased membrane permeability to K^+, which stops Na^+ movement into the cell and increased K^+ movement out of the cell. The inactivation gates of the voltage-gated Na^+ channels close, and the voltage-gated K^+ channels open.
5. No action potential is produced by a stimulus, no matter how strong, during the absolute refractory period. During the relative refractory period a stronger-than-threshold stimulus can produce an action potential.
6. Action potentials are propagated, and for a given axon or muscle fiber the magnitude of the action potential is constant.
7. Stimulus strength determines the frequency of action potentials.

potential and becomes more positive, and a **repolarization phase,** in which the membrane potential returns toward the resting membrane state and becomes more negative. After the repolarization phase, the plasma membrane may be slightly hyperpolarized for a short period called the **afterpotential** (see figure 11.20).

> 29. *Define an action potential. How do depolarizing and hyperpolarizing local potentials affect the likelihood of generating an action potential?*
>
> 30. *Explain the "all" and the "none" parts of the all-or-none principle of action potentials.*

Depolarization Phase

The change in charge across the plasma membrane caused by a local potential causes increasing numbers of voltage-gated Na^+ channels to open for a brief time. As soon as a threshold depolarization is reached, many voltage-gated Na^+ channels begin to open. Na^+ diffuse into the cell, and the resulting depolarization causes additional voltage-gated Na^+ channels to open. As a consequence, more Na^+ diffuse into the cell, causing a greater depolarization of the membrane, which, in turn, causes still more voltage-gated Na^+ channels to open. This is an example of a positive-feedback cycle, and it continues until most of the voltage-gated Na^+ channels in the plasma membrane are open.

Each voltage-gated Na^+ channel has two voltage-sensitive gates, called **activation** and **inactivation gates.** When the plasma membrane is at rest, the activation gates of the voltage-gated Na^+ channel are closed, and the inactivation gates are open (figure 11.21, *step 1*). Because the activation gates are closed, Na^+ cannot diffuse through the channels. When the local potential reaches threshold, the change in the membrane potential causes many of the activation gates to open, and Na^+ can diffuse through the Na^+ channels into the cell.

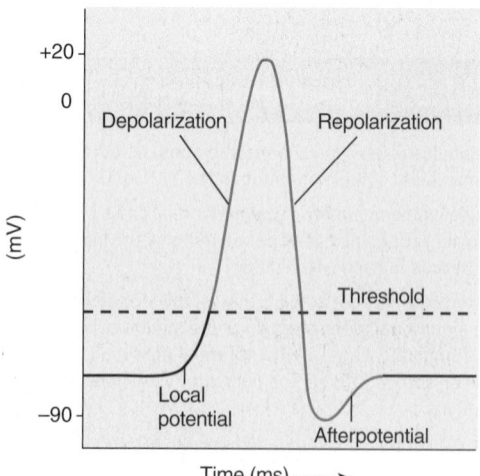

Figure 11.20   The Action Potential

The action potential consists of a depolarization phase and a repolarization phase, often followed by a short period of hyperpolarization called the afterpotential.

1. Resting membrane potential.
Voltage-gated Na$^+$ channels (*pink*) are closed (the activation gates are closed and the inactivation gates are open). Voltage-gated K$^+$ channels (*purple*) are closed.

2. Depolarization. Voltage-gated Na$^+$ channels open because the activation gates open. Voltage-gated K$^+$ channels start to open. Depolarization results because the inward diffusion of Na$^+$ is much greater than the outward diffusion of K$^+$.

3. Repolarization. Voltage-gated Na$^+$ channels are closed because the inactivation gates close. Voltage-gated K$^+$ channels are now open. Na$^+$ diffusion into the cell stops and K$^+$ diffuse out of the cell, causing repolarizaton.

4. End of repolarization and afterpotential. Voltage-gated Na$^+$ channels are closed. Closure of the activation gates and opening of the inactivation gates reestablish the resting condition for Na$^+$ channels (see step 1). Diffusion of K$^+$ through voltage-gated channels produces the afterpotential.

5. Resting membrane potential.
The resting membrane potential is reestablished after the voltage-gated K$^+$ channels close.

Process Figure 11.21   Voltage-Gated Ion Channels and the Action Potential

Step 1 illustrates the status of voltage-gated Na$^+$ and K$^+$ channels in a resting cell. Steps 2–5 show how the channels open and close to produce an action potential. Next to each step, a graph shows in *red* the membrane potential resulting from the condition of the ion channels.

When the plasma membrane is at rest, voltage-gated K⁺ channels, which have one gate, are closed (see figure 11.21, *step 1*). When the local potential reaches threshold, the voltage-gated K⁺ channels begin to open at the same time as the voltage-gated Na⁺ channels, but they open more slowly (figure 11.21, *step 2*). Only a small number of voltage-gated K⁺ channels are open compared to the number of voltage-gated Na⁺ channels because the voltage-gated K⁺ channels open slowly. Depolarization occurs because more Na⁺ diffuse into the cell than K⁺ diffuse out of it.

P R E D I C T
Predict the effect of a reduced extracellular concentration of Na⁺ on the magnitude of the action potential in an electrically excitable cell.

Repolarization Phase

As the membrane potential approaches its maximum depolarization, the change in the potential difference across the plasma membrane causes the inactivation gates in the voltage-gated Na⁺ channels to begin closing, and the permeability of the plasma membrane to Na⁺ decreases. During the repolarization phase, the voltage-gated K⁺ channels, which started to open along with the voltage-gated Na⁺ channels, continue to open (figure 11.21, *step 3*). Consequently, the permeability of the plasma membrane to Na⁺ decreases, and the permeability to K⁺ increases. The decreased diffusion of Na⁺ into the cell and the increased diffusion of K⁺ out of the cell causes repolarization.

At the end of repolarization, the decrease in membrane potential causes the activation gates in the voltage-gated Na⁺ channels to close and the inactivation gates to open. Although this change doesn't affect the diffusion of Na⁺, it does return the voltage-gated Na⁺ channels to their resting state (figure 11.21, *step 4*).

Afterpotential

In many cells, a period of hyperpolarization, or afterpotential, exists following each action potential. The afterpotential exists because the voltage-gated K⁺ channels remain open for a short time (see figure 11.21, *step 4*). The increased K⁺ permeability that develops during the repolarization phase of the action potential lasts slightly longer than the time required to bring the membrane potential back to its resting level. As the voltage-gated K⁺ channels close, the original resting membrane potential is reestablished (figure 11.21, *step 5*).

During an action potential, a small number of Na⁺ diffuse into the cell and a small number of K⁺ diffuse out of the cell. The Na⁺–K⁺ pump functions to restore normal resting ion concentrations by transporting these ions in the opposite direction of their movement during the action potential. That is, Na⁺ are pumped out of the cell and K⁺ are pumped into the cell. The Na⁺–K⁺ pump is too slow to have an effect on either the depolarization or repolarization phase of individual action potentials. As long as the Na⁺ and K⁺ concentrations remain unchanged across the plasma membrane, all the action potentials produced by a cell are identical. They all take the same amount of time, and they all exhibit the same magnitude.

31. *What are the depolarization and repolarization phases of an action potential? Explain how changes in membrane permeability and the movement of Na⁺ and K⁺ cause each phase. What happens when the activation gates in the*

voltage-gated Na⁺ channels open and the inactivation gates close?
32. *Describe the afterpotential and its cause.*

Refractory Period

Once an action potential is produced at a given point on the plasma membrane, the sensitivity of that area to further stimulation decreases for a time called the **refractory** (rē-frak′tōr-ē) **period.** The first part of the refractory period, during which complete insensitivity exists to another stimulus, is called the **absolute refractory period.** In many cells, it occurs from the beginning of the action potential until near the end of repolarization (figure 11.22). At the beginning of the action potential, depolarization occurs when the activation gates in the voltage-gated Na⁺ channel open. At this time, the inactivation gates in the voltage-gated Na⁺ channels are already open (see figure 11.21, *step 2*). Depolarization ends as the inactivation gates close (see figure 11.21, *step 3*). As long as the inactivation gates are closed, further depolarization cannot occur. When the inactivation gates open and the activation gates close near the end of repolarization (see figure 11.21, *step 4*), it once again is possible to stimulate the production of another action potential.

The existence of the absolute refractory period guarantees that once an action potential is begun, both the depolarization and the repolarization phases will be completed, or nearly completed, before another action potential can begin, and that a strong stimulus cannot lead to prolonged depolarization of the plasma membrane. The absolute refractory period has important consequences for the rate at which action potentials can be generated and for the propagation of action potentials (see following sections).

The second part of the refractory period, called the **relative refractory period,** follows the absolute refractory period. A stronger-

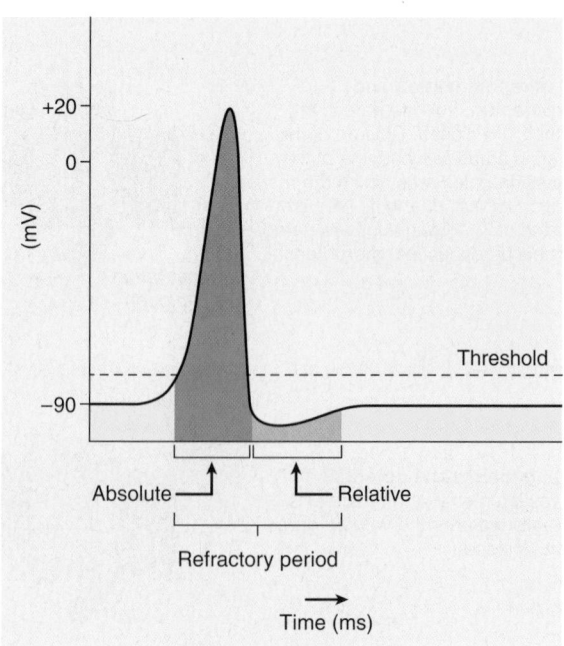

Figure 11.22 Refractory Period

The absolute and relative refractory periods of an action potential. In some cells the absolute refractory period may end during the repolarization phase of the action potential.

than-threshold stimulus can initiate another action potential during the relative refractory period. Thus, after the absolute refractory period, but before the relative refractory period is completed, a sufficiently strong stimulus can produce another action potential. During the relative refractory period, the membrane is more permeable to K$^+$ because many voltage-gated K$^+$ channels are open (see figure 11.21, *step 4*). The relative refractory period ends when the voltage-gated K$^+$ channels close (see figure 11.21, *step 5*).

 33. *What are the absolute and relative refractory periods? Relate them to the depolarization and repolarization phases of the action potential.*

P R E D I C T 7
Does a prolonged threshold stimulus or a prolonged stronger-than-threshold stimulus produce the most action potentials? Explain.

Action Potential Frequency

The **action potential frequency** is the number of action potentials produced per unit of time in response to a stimulus. Action potential frequency is directly proportional to stimulus strength and to the size of the local potential. A **subthreshold stimulus** is any stimulus not strong enough to produce a local potential that reaches threshold. Therefore, no action potential is produced (figure 11.23). A **threshold stimulus** produces a local potential that's just strong enough to reach threshold and cause the production of a single action potential. A **maximal stimulus** is just strong enough to produce a maximum frequency of action potentials. A **submaximal stimulus** includes all stimuli between threshold and the maximal stimulus strength. For submaximal stimuli, the action potential frequency increases in proportion to the strength of the stimulus because the size of the local potential increases with stimulus strength. A **supramaximal stimulus** is any stimulus stronger than a maximal

stimulus. These stimuli cannot produce a greater frequency of action potentials than a maximal stimulus.

The duration of the absolute refractory period determines the maximum frequency of action potentials generated in an excitable cell. During the absolute refractory period, a second stimulus, no matter how strong, cannot stimulate an additional action potential. As soon as the absolute refractory period ends, however, it's possible for a second stimulus to cause the production of an action potential.

P R E D I C T **8**
If the duration of the absolute refractory period of a nerve cell is 1 millisecond (ms), how many action potentials are generated by a maximal stimulus in one second?

The frequency of action potentials provides information about the strength of a stimulus. For example, a weak pain stimulus generates a low frequency of action potentials, whereas a stronger pain stimulus generates a higher frequency of action potentials. The ability to interpret a stimulus as mildly painful versus very painful depends, in part, on the frequency of action potentials generated by individual pain receptors. Communication regarding the strength of stimuli cannot depend on the magnitude of action potentials because, according to the all-or-none principle, the magnitudes are always the same. The magnitude of action potentials produced by weak or strong pain stimuli is the same.

The ability to stimulate muscle or gland cells also depends on action potential frequency. A low frequency of action potentials produces a weaker muscle contraction or less secretion than does a higher frequency. For example, a low frequency of action potentials in a muscle results in incomplete tetanus and a high frequency in complete tetanus (see chapter 9).

In addition to the frequency of action potentials, how long the action potentials are produced provides important information.

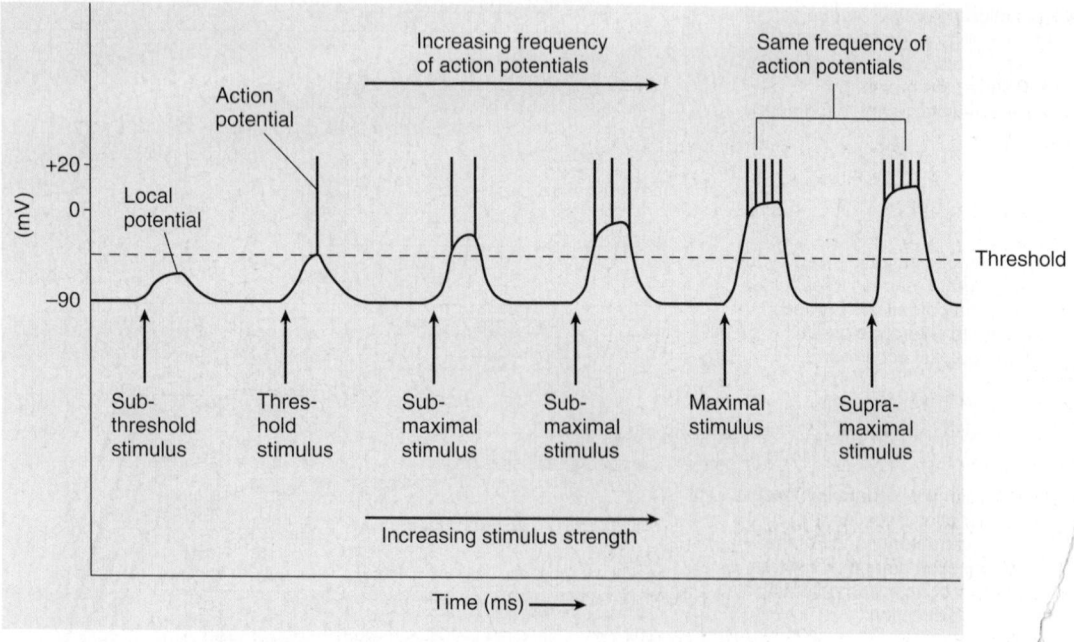

Figure 11.23 Stimulus Strength and Action Potential Frequency
From left to right, each stimulus in the figure is stronger than the previous one. As stimulus strength increases, the frequency of action potentials increases until a maximal rate is produced. Thereafter, increasing stimulus strength does not increase action potential frequency.

For example, a pain stimulus of 1 second is interpreted differently from the same stimulus applied for 30 seconds.

34. *Define action potential frequency. What two factors determine action potential frequency?*

35. *Define a subthreshold, threshold, maximal, submaximal, and supramaximal stimulus. What determines the maximum frequency of action potential generation?*

Propagation of Action Potentials

An action potential occurs in a very small area of the plasma membrane and does not affect the entire membrane at one time. Action potentials can, however, **propagate** or spread across the plasma membrane because an action potential produced at one location in the plasma membrane can stimulate the production of an action potential at an adjacent area of the plasma membrane. Note that an action potential doesn't actually move along an axon. Rather, an action potential at one location stimulates the production of another action potential at an adjacent location which, in turn, stimulates the production of another, and so on. It's like a long row of toppling dominos in which each domino knocks down the next one. Each domino falls, but no one domino actually travels the length of the row.

In a neuron, action potentials are normally produced at the trigger zone and propagate in one direction along the axon (figure 11.24, *step 1*). The location at which the next action potential is generated is different for unmyelinated and myelinated axons. In an unmyelinated axon, the next action potential is generated immediately adjacent to the previous action potential. When an action potential is produced, the inside of the membrane becomes more positive than the outside (figure 11.24, *step 2*). On the outside of the membrane, positively charged ions from the adjacent area are attracted to the negative charges at the site of the action potential. On the inside of the plasma membrane, positively charged ions at the site of the action potential are attracted to the adjacent negatively charged part of the membrane. The movement of positively charged ions is called a **local current.** As a result of the local current, the part of the membrane immediately adjacent to the action potential depolarizes. That is, the outside of the membrane immediately adjacent to the action potential becomes more negative because of the loss of positive charges and the inside becomes more positive because of the gain of positive charges. When the depolarization reaches threshold, an action potential is produced (figure 11.24, *step 3*).

If an action potential is initiated at one end of an axon, it is propagated in one direction down the axon. The absolute refractory

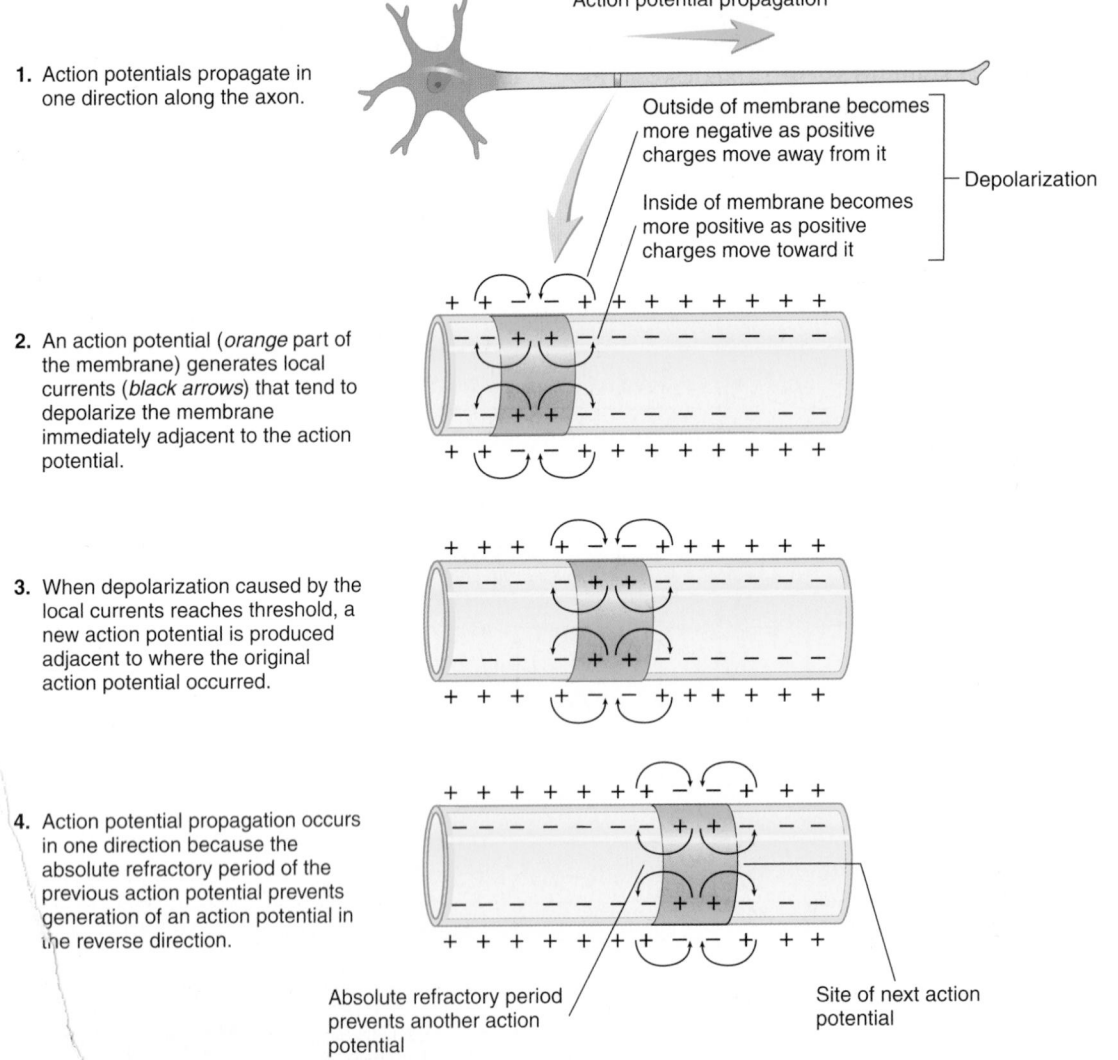

Action potential propagation

1. Action potentials propagate in one direction along the axon.

Outside of membrane becomes more negative as positive charges move away from it

Inside of membrane becomes more positive as positive charges move toward it

Depolarization

2. An action potential (*orange* part of the membrane) generates local currents (*black arrows*) that tend to depolarize the membrane immediately adjacent to the action potential.

3. When depolarization caused by the local currents reaches threshold, a new action potential is produced adjacent to where the original action potential occurred.

4. Action potential propagation occurs in one direction because the absolute refractory period of the previous action potential prevents generation of an action potential in the reverse direction.

Absolute refractory period prevents another action potential

Site of next action potential

Process Figure 11.24   Action Potential Propagation in an Unmyelinated Axon

Clinical Focus   Examples of Abnormal Membrane Potentials

Several important conditions provide examples of the physiology of membrane potentials and the consequence of abnormal ones. **Hypokalemia** (hī-pō-ka-lē′mē-ă) is a lower-than-normal concentration of K^+ in the blood or extracellular fluid. Figure 11.18*b* shows that reduced extracellular K^+ concentrations cause hyperpolarization of the resting membrane potential. Thus, a greater-than-normal stimulus is required to depolarize the membrane to its threshold level and to initiate action potentials in neurons, skeletal muscle, and cardiac muscle. Symptoms of hypokalemia include muscular weakness, an abnormal electrocardiogram, and sluggish reflexes. These symptoms are consistent with the effect of a reduced extracellular K^+ concentration. The symptoms result from the reduced sensitivity of the excitable tissues to stimulation. The several causes of hypokalemia include potassium depletion during starvation, alkalosis, and certain kidney diseases.

Hypocalcemia (hī-pō-kal-sē′mē-ă) is a lower-than-normal concentration of Ca^{2+} in blood or extracellular fluid. Symptoms of hypocalcemia include nervousness and uncontrolled contraction of skeletal muscles, called **tetany** (tet′ă-nē). The symptoms are due to an increased membrane permeability to Na^+ that results because low blood levels of Ca^{2+} cause voltage-gated Na^+ channels in the membrane to open. Na^+ diffuse into the cell, causing depolarization of the plasma membrane to threshold, and initiating action potentials. The tendency for action potentials to occur spontaneously in nervous tissue and muscles accounts for the listed symptoms. A lack of dietary calcium, a lack of vitamin D, or a reduced secretion rate of a parathyroid gland hormone are examples of conditions that cause hypocalcemia.

period ensures one-way propagation of an action potential because it prevents the local current from stimulating the production of an action potential in the reverse direction (figure 11.24, *step 4*).

In a myelinated axon, an action potential is conducted from one node of Ranvier to another in a process called **saltatory conduction** (*saltare* is Latin, meaning to leap). An action potential at one node of Ranvier generates local currents that flow toward the next node of Ranvier (figure 11.25, *step 1*). The lipids within the membranes of the myelin sheath act as a layer of insulation, forcing the local currents to flow from one node of Ranvier to the next. In addition, voltage-gated Na^+ channels are highly concentrated at the nodes of Ranvier. Therefore, the local current quickly flows to a node and stimulates the voltage-gated Na^+ channels to open, resulting in the production of an action potential (figure 11.25, *step 2*).

The speed of action potential conduction along an axon depends on the myelination of the axon. Action potentials are conducted more rapidly in myelinated than unmyelinated axons because they are formed quickly at each successive node of Ranvier (figure 11.25, *step 3*) instead of being propagated more slowly through every part of the axon's membrane as in unmyelinated axons (see figure 11.24). Action potential conduction in a myelinated fiber is like a grasshopper jumping, whereas in an unmyelinated axon it's like a grasshopper walking. The grasshopper (action potential) moves more rapidly by jumping. The generation of action potentials at nodes of Ranvier occurs so rapidly that as many as 30 successive nodes of Ranvier are simultaneously in some phase of an action potential.

The speed of action potential conduction is also affected by the thickness of the myelin sheath, which is determined by how many

1. An action potential (*orange*) at a node of Ranvier generates local currents (*black arrows*). The local currents flow to the next node of Ranvier because the myelin sheath of the Schwann cell insulates the axon of the internode.

2. When the depolarization caused by the local currents reaches threshold at the next node of Ranvier, a new action potential is produced (*orange*).

3. Action potential propagation is rapid in myelinated axons because the action potentials are produced at successive nodes of Ranvier (*1–5*) instead of at every part of the membrane along the axon.

Direction of action potential propagation

Process Figure 11.25   Saltatory Conduction: Action Potential Propagation in a Myelinated Axon times oligodendrocytes or Schwann cells wrap around the axon. Heavily myelinated axons have a thicker myelin sheath and conduct action potentials more rapidly than lightly myelinated axons.

In addition to myelination, the diameter of axons affects the speed of action potential conduction. Large-diameter axons conduct action potentials more rapidly than small-diameter axons because large-diameter axons have a greater surface area. Consequently, at a given site on an axon, more voltage-gated Na^+ channels open during depolarization, resulting in a greater current flow, which more rapidly stimulates adjacent membrane areas.

Nerve fibers (axons) are classified according to their size and myelination. It's not surprising that the structure of nerve fibers reflects their functions. Type A fibers are large-diameter, myelinated axons that conduct action potentials at 15–120 m/s. Motor neurons supplying skeletal muscles and most sensory neurons have type A fibers. Rapid response to the external environment is possible because of the rapid input of sensory information to the CNS and rapid output of action potentials to skeletal muscles.

Type B fibers are medium-diameter, lightly myelinated axons that conduct action potentials at 3–15 m/s, and type C fibers are small-diameter, unmyelinated axons that conduct action potentials at 2 m/s or less. Type B and C fibers are primarily part of the ANS, which stimulates internal organs, such as the stomach, intestines, and heart. The responses necessary to maintain internal homeostasis such as digestion need not be as rapid as responses to the external environment.

36. *What is a local current? How do local currents cause the propagation of action potentials in unmyelinated axons?*
37. *What prevents an action potential from reversing its direction of propagation?*
38. *Describe saltatory conduction of an action potential.*
39. *Compare the speed of action potential conduction in (a) heavily myelinated, lightly myelinated, and unmyelinated axons, and (b) large-diameter and small-diameter axons.*
40. *Compare the functions of type A nerve fibers to type B and C nerve fibers.*

Importance of Myelin Sheaths

Myelin sheaths begin to form late in fetal development. The process continues rapidly until the end of the first year after birth and continues more slowly thereafter. The development of myelin sheaths is associated with the infant's continuing development of more rapid and better coordinated responses.

The importance of myelinated fibers is dramatically illustrated in diseases in which the myelin sheath is gradually destroyed. Action potential transmission is slowed, resulting in impaired control of skeletal and smooth muscles. In severe cases, complete blockage of action potential transmission can occur. Multiple sclerosis and some cases of diabetes mellitus are examples of diseases that result in myelin sheath destruction.

The Synapse

Just as the fire from one lit torch can light another torch, action potentials in one cell can stimulate action potentials in another cell, thereby allowing communication between the cells. For example, if your finger touches a hot pan, the heat is a stimulus that produces action potentials in sensory nerve fibers. The action potentials are prop-

agated along the sensory fibers from the finger toward the CNS. For the CNS to get this information, the action potentials of the sensory neurons must produce action potentials in CNS neurons. After the CNS has received the information, it produces a response. One response is the contraction of the appropriate skeletal muscles causing the finger to move away from the hot pan. CNS action potentials cause motor neurons to produce action potentials that are then transmitted by the motor neurons toward skeletal muscles. The action potentials of the motor neuron produce skeletal muscle action potentials, which are the stimuli that cause muscle fibers to contract (see chapter 9).

The **synapse** (sin′aps), which is a junction between two cells, is the site where action potentials in one cell can cause the production of action potentials in another cell. The cell that carries action potentials toward a synapse is called the **presynaptic cell,** and the cell that carries action potentials away from the synapse is called the **postsynaptic cell.**

The average presynaptic neuron synapses with about 1000 other neurons, but the average postsynaptic neuron has up to 10,000 synapses. Some postsynaptic neurons in a part of the brain called the cerebellum may have up to 100,000 synapses. There are two types of synapses: electrical and chemical.

Electrical Synapses

Electrical synapses are gap junctions (see chapter 4) that allow a local current to flow between adjacent cells (figure 11.26). At these gap

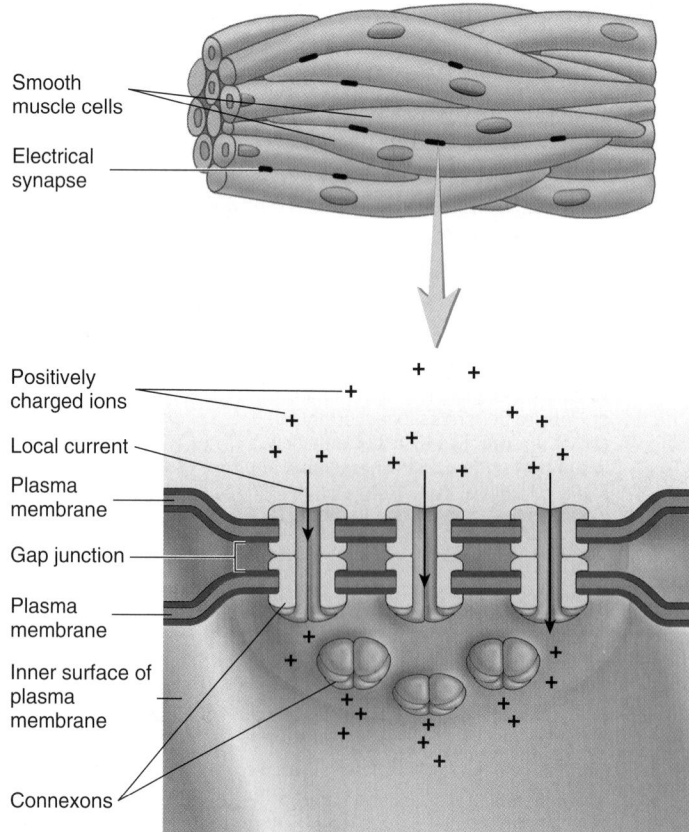

Figure 11.26   **Electrical Synapse**

Electrical synapses are gap junctions in which the plasma membranes of two cells come close together and are joined by connexons. An action potential in one cell can generate local currents (positively charged ions) that flow through the connexons to stimulate an action potential in the other cell.

Clinical Focus   Nervous Tissue Response to Injury

When a nerve is cut, either healing or permanent interruption of the neural pathways occurs. The final outcome depends on the severity of the injury and on its treatment.

Several degenerative changes result when a nerve is cut (figure A). Within about 3–5 days, the axons in the part of the nerve distal to the cut break into irregular segments and degenerate. This occurs because the neuron cell body produces the substances essential to maintain the axon and these substances have no way of reaching parts of the axon distal to the point of damage. Eventually the distal part of the axon completely degenerates. At the same time the axons are degenerating, the myelin part of the Schwann cells around them also degenerates, and macrophages invade the area to phagocytize the myelin. The Schwann cells then enlarge, undergo mitosis, and finally form a column of cells along the re-

gions once occupied by the axons. The columns of Schwann cells are essential for the growth of new axons. If the ends of the regenerating axons encounter a Schwann cell column, their rate of growth increases, and reinnervation of peripheral structures is likely. If the ends of the axons do not encounter the columns, they fail to reinnervate the peripheral structures.

The end of each regenerating axon forms several axonal sprouts. It normally takes about 2 weeks for the axonal sprouts to enter the Schwann cell columns. Only one of the sprouts from each severed neuron forms an axon, however. The other branches degenerate. After the axons grow through the Schwann cell columns, new myelin sheaths are formed, and the neurons reinnervate the structures they previously supplied.

Treatment strategies that increase the probability of reinnervation include bringing the ends of the severed nerve close to-

gether surgically. In some cases in which sections of nerves are destroyed as a result of trauma, nerve transplants are performed to replace damaged segments. The transplanted nerve eventually degenerates, but it does provide Schwann cell columns through which axons can grow.

Regeneration of damaged nerve tracts within the CNS is very limited and is poor in comparison to regeneration of nerves in the PNS. In part, the difference may result from the oligodendrocytes, which exist only in the CNS. Each oligodendrocyte has several processes, each of which forms part of a myelin sheath. The cell bodies of the oligodendrocytes are a short distance from the axons they ensheathe, and fewer oligodendrocytes are present than Schwann cells. Consequently, when the myelin degenerates following damage, no column of cells remains in the CNS to act as a guide for the growing axons.

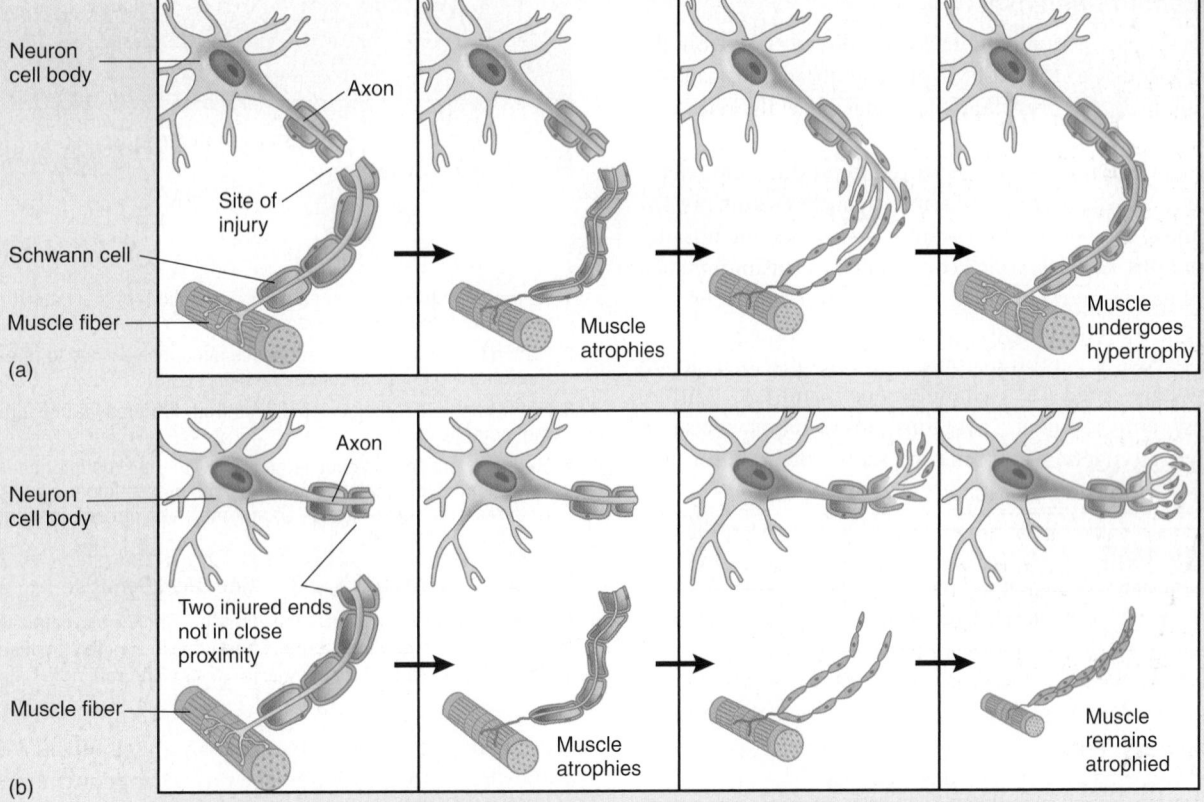

Figure A   Changes That Occur in an Injured Nerve Fiber

(*a*) When the two ends of the injured nerve fiber are aligned in close proximity, healing and regeneration of the axon are likely to occur. Without stimulation from the nerve, the muscle is paralyzed and atrophies (shrinks in size). After reinnervation the muscle can become functional and hypertrophy (increase in size). (*b*) When the two ends of the injured nerve fiber are not aligned in close proximity, regeneration is unlikely to occur. Without innervation from the nerve, muscle function is completely lost, and the muscle remains atrophied.

junctions, the membranes of adjacent cells are separated by a 2 nm gap that is spanned by tubular proteins called **connexons.** The movement of ions through the connexons can generate a local current. Thus, an action potential in one cell produces a local current that generates an action potential in the adjacent cell almost as if the two cells had the same membrane. As a result, action potentials are conducted rapidly between cells, allowing the activity of the cells to be synchronized or occur at about the same time. Electrical synapses are not common in the nervous system of vertebrates, but there are some in humans. Electrical synapses are important, however, in cardiac muscle and in many types of smooth muscle. Coordinated contractions of these muscle cells occur when action potentials in one cell propagate to adjacent cells because of electrical synapses (see chapters 9 and 20).

● **41.** *What is an electrical synapse? Describe its operation. In what kinds of tissue are electrical synapses found?*

Chemical Synapses

The essential components of a **chemical synapse** are the presynaptic terminal, the synaptic cleft, and the postsynaptic membrane (figure 11.27). The **presynaptic terminal** is formed from the end of an axon, and the space separating the axon ending and the cell with which it synapses is the **synaptic cleft.** The membrane of the postsynaptic cell opposed to the presynaptic terminal is the **postsynaptic membrane.** Postsynaptic cells are typically other neurons, muscle cells, or gland cells.

Neurotransmitter Release

In chemical synapses, action potentials don't directly pass from the presynaptic terminal to the postsynaptic membrane. Instead, the action potentials in the presynaptic terminal cause the release of neurotransmitters from the terminal.

Presynaptic terminals are specialized to produce and release neurotransmitters. The major cytoplasmic organelles within presynaptic terminals are mitochondria and numerous membrane-bounded **synaptic vesicles,** which contain neurotransmitters such as acetylcholine (see figure 11.27). Each action potential arriving at the presynaptic terminal initiates a series of specific events that results in the release of neurotransmitters. In response to an action potential, voltage-gated Ca^{2+} channels open, and Ca^{2+} diffuse into the presynaptic terminal. These ions cause synaptic vesicles to fuse with the presynaptic membrane and release their neurotransmitters by exocytosis into the synaptic cleft.

Aging of the Heart

Experiments in rats indicate that the number of voltage-gated Ca^{2+} channels in the presynaptic terminals of neurons that stimulate the heart decreases with age. Consequently, movement of Ca^{2+} into the presynaptic terminals decreases, causing a reduced release of neurotransmitters. The decreased amounts of neurotransmitter result in less stimulation of the heart and may explain the lower ability of the aged heart to pump faster and harder during exercise.

Once neurotransmitters are released from the presynaptic terminal, they diffuse rapidly across the synaptic cleft, which is about 20 nm wide, and bind in a reversible fashion to specific receptors in the postsynaptic membrane (see figure 11.27). De-

1. Action potentials arriving at the presynaptic terminal cause voltage-gated Ca^{2+} channels to open.
2. Ca^{2+} diffuse into the cell and cause synaptic vesicles to release neurotransmitter molecules.
3. Neurotransmitter molecules diffuse from the presynaptic terminal across the synaptic cleft.
4. Neurotransmitter molecules combine with their receptor sites and cause ligand-gated ion channels to open. Ions diffuse into the cell (*shown in illustration*) or out of the cell (*not shown*) and cause a change in membrane potential.

Process Figure 11.27   Chemical Synapse

A chemical synapse consists of the end of a neuron (presynaptic terminal), a small space (synaptic cleft), and the postsynaptic membrane of another neuron or an effector cell such as a muscle or gland cell.

pending on the receptor type, this binding produces a depolarization or hyperpolarization of the postsynaptic membrane. For example, the binding of acetylcholine to ligand-gated Na^+ channels causes them to open, allowing the diffusion of Na^+ into the postsynaptic cell. If the resulting depolarization reaches threshold, an action potential is produced. On the other hand, the opening of K^+ or Cl^- channels results in hyperpolarization.

42. What are the three parts of a chemical synapse?

43. Describe the release of a neurotransmitter in a chemical synapse.

P R E D I C T 9

Is an action potential transmitted fastest between cells connected by electrical or chemical synapses? Explain.

Neurotransmitter Removal

The interaction between a neurotransmitter and a receptor represents an equilibrium.

Neurotransmitter + Receptor $\rightleftharpoons$ Neurotransmitter–receptor complex

When the neurotransmitter concentration in the synaptic cleft is high, many of the receptor molecules have neurotransmitter molecules bound to them, and when the neurotransmitter concentration declines, the neurotransmitter molecules diffuse away from the receptor molecules.

Neurotransmitters have short-term effects on postsynaptic membranes because the neurotransmitter is rapidly destroyed or removed from the synaptic cleft. For example, in the neuromuscular junction (see chapter 9), the neurotransmitter acetylcholine is broken down by the enzyme **acetylcholinesterase** (as'e-til-kō-lin-es'ter-ās) to acetic acid and choline (figure 11.28a). Choline is then transported back into the presynaptic terminal and reacts with acetyl-CoA to form acetylcholine. Acetyl-CoA is synthesized by mitochondria as foods are metabolized to produce ATP (see chapter 25). Acetic acid can be absorbed from the synaptic cleft into the presynaptic terminal, or it can diffuse out of the synaptic cleft and be taken up by a variety of cells. Acetic acid can be used to synthesize acetyl-CoA.

When the neurotransmitter norepinephrine is released into the synaptic cleft, most of it is actively transported back into the presynaptic terminal, where most of it is repackaged into synaptic vesicles for reuse (figure 11.28b). The enzyme **monoamine oxidase (MAO)** (mon-ō-am'īn ok'si-dās) inactivates some of the norepinephrine.

Diffusion of neurotransmitter molecules away from the synapse and into the extracellular fluid also limits the length of time the neurotransmitter molecules remain bound to their receptors. Norepinephrine in the circulation is taken up primarily by liver and kidney cells, where the enzymes monoamine oxidase and **catechol-O-methyltransferase** (kat'ĕ-kol-ō-meth-il-trans'fer-ās) convert it into inactive metabolites.

1. Acetylcholine molecules bind to their receptors.
2. Acetylcholine molecules unbind from their receptors.
3. Acetylcholinesterase splits acetylcholine into choline and acetic acid, which prevents acetylcholine from again binding to its receptors. Choline is taken up by the presynaptic terminal.
4. Choline is used to make new acetylcholine molecules that are packaged into synaptic vesicles.

(a)

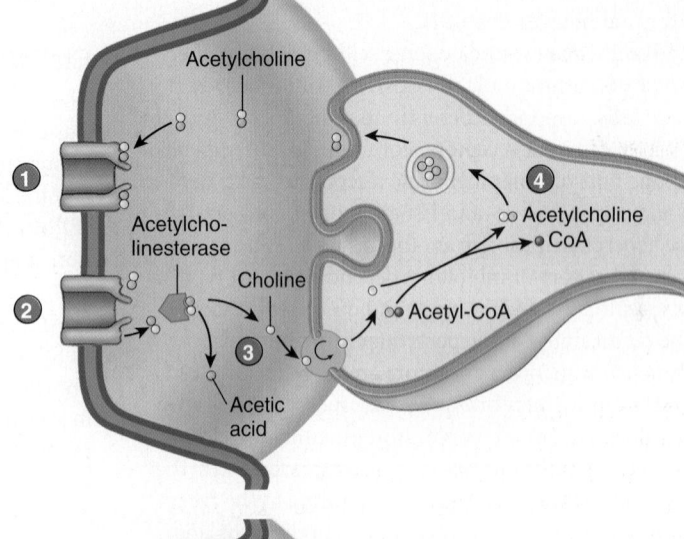

1. Norepinephrine binds to its receptor.
2. Norepinephrine unbinds from its receptor.
3. Norepinephrine is taken up by the presynaptic terminal, which prevents norepinephrine from again binding to its receptor.
4. Norepinephrine is repackaged into synaptic vesicles or is broken down by monoamine oxidase (MAO).

(b)

Process Figure 11.28 Removal of Neurotransmitter from the Synaptic Cleft

(a) In some synapses, neurotransmitters are broken down by enzymes and recycled into the presynaptic terminal. (b) In some synapses, neurotransmitters are taken up whole into the presynaptic terminal.

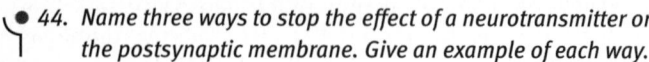

How Amphetamines Work

Amphetamines are known to increase the release of norepinephrine from presynaptic terminals, block the reuptake of norepinephrine from the synaptic cleft by presynaptic terminals, and inhibit the action of monoamine oxidase. The CNS effects of amphetamines include decreased appetite, increased alertness, and increased ability to concentrate and perform physical activities. Overdosage of amphetamines can cause insomnia, tremors, and behavioral changes such as euphoria, anxiety, and panic.

44. *Name three ways to stop the effect of a neurotransmitter on the postsynaptic membrane. Give an example of each way.*

Receptor Molecules in Synapses

Receptor molecules in synapses are membrane-bound, ligand-activated receptors with highly specific receptor sites. Consequently, only neurotransmitter molecules or very closely related substances normally bind to their receptors. For example, acetylcholine binds to acetylcholine receptors, but not norepinephrine receptors, whereas norepinephrine binds to norepinephrine receptors, but not to acetylcholine receptors. Any given cell does not have all possible receptors. Therefore, a neurotransmitter affects only the cells with receptors for that neurotransmitter.

A neurotransmitter can stimulate some cells but inhibit others. More than one type of receptor molecule exists for some neurotransmitters. Different cells respond differently to a neurotransmitter when these cells have different receptors. For example, norepinephrine can bind to one type of norepinephrine receptor to cause depolarization in one synapse and to another type of norepinephrine receptor to cause hyperpolarization in another synapse. Thus, norepinephrine is either stimulatory or inhibitory, depending on the type of norepinephrine receptor to which it binds and on the effect that receptor has on the permeability of the postsynaptic membrane.

Although neurotransmitter receptors are in greater concentrations on postsynaptic membranes, some receptors exist on presynaptic membranes. For example, norepinephrine released from the presynaptic membrane binds to receptors on both the presynaptic and postsynaptic membranes. Its binding to the receptors of the presynaptic membrane decreases the release of additional synaptic vesicles. Norepinephrine can therefore modify its own release by binding to presynaptic receptors. A high frequency of presynaptic action potentials results in the release of fewer synaptic vesicles in response to later action potentials.

Neurotransmitters and Neuromodulators

Several substances have been identified as neurotransmitters, and others are suspected neurotransmitters. It was once thought that each neuron contains only one type of neurotransmitter; however, it's now known that some neurons can secrete more than one type. If a neuron does produce more than one neurotransmitter, it secretes all of them from each of its presynaptic terminals. The physiologic significance of presynaptic terminals that secrete more than one type of neurotransmitter has not been clearly established.

Neuromodulators are substances released from neurons that can presynaptically or postsynaptically influence the likelihood that an action potential in the presynaptic terminal will result in the production of an action potential in the postsynaptic cell. For example, a neuromodulator that decreases the release of an excitatory neurotransmitter from a presynaptic terminal decreases the likelihood of action potential production in the postsynaptic cell. A list of neurotransmitters and neuromodulators is presented in table 11.5.

Preventing Stroke Damage

Glutamate (gloo'tă-māt) is an important excitatory neurotransmitter in the brain and spinal cord. During a stroke, oxygen-deprived presynaptic neurons release large amounts of glutamate. Glutamate binds to postsynaptic neurons and stimulates them to release **nitric oxide (NO),** which in high concentrations can be toxic to cells. The NO diffuses from the postsynaptic neurons and causes damage to surrounding cells. It's possible that stroke damage may be reduced by drugs, not yet developed, that block glutamate receptors or inhibit the production of NO.

45. *Why does a given type of neurotransmitter affect only certain types of cells? How can a neurotransmitter stimulate one type of cell but inhibit another type?*
46. *What is a neuromodulator?*

Excitatory and Inhibitory Postsynaptic Potentials

The combination of neurotransmitters with their specific receptors causes either depolarization or hyperpolarization of the postsynaptic membrane. When depolarization occurs, the response is stimulatory, and the local depolarization is an **excitatory postsynaptic potential (EPSP)** (figure 11.29a). EPSPs are important because the depolarization might reach threshold, thereby producing an action potential and a response from the cell. Neurons releasing neurotransmitter substances that cause EPSPs are **excitatory neurons.** In general, an EPSP occurs because of an increase in the permeability of the membrane to Na^+. For example, glutamate in the brain and acetylcholine in skeletal muscle can bind to their receptors, causing Na^+ channels to open. Because the concentration gradient is large for Na^+ and because the negative charge inside the cell attracts the positively charged Na^+, they diffuse into the cell and cause depolarization. If depolarization reaches threshold, an action potential is produced.

How Local Anesthetics Work

Awareness of pain can only occur if action potentials generated by sensory neurons stimulate the production of action potentials in CNS neurons. Local anesthetics, such as procaine (Novacain), act at their site of application to prevent pain sensations. They do so by blocking voltage-gated Na^+ channels, which prevents the propagation of action potentials along sensory neurons. Consequently, neurotransmitters are not released from the presynaptic terminals of the sensory neurons and EPSPs are not produced in CNS neurons.

When the combination of a neurotransmitter with its receptor results in hyperpolarization of the postsynaptic membrane, the response is inhibitory, and the local hyperpolarization

Table 11.5	Substances That Are Neurotransmitters or Neuromodulators (or both)		
Substance	**Location**	**Effect**	**Clinical Example**
Acetylcholine	Many nuclei scattered throughout the brain and spinal cord. Nerve tracts from the nuclei extend to many areas of the brain and spinal cord. Also found in the neuromuscular junction of skeletal muscle and many ANS synapses.	Excitatory or inhibitory	Alzheimer's disease (a type of senile dementia) is associated with a decrease in acetylcholine-secreting neurons. Myasthenia gravis (weakness of skeletal muscles) results from a reduction in acetylcholine receptors.
Monoamines			
Norepinephrine	A small number of small-sized nuclei in the brainstem. Nerve tracts extend from the nuclei to many areas of the brain and spinal cord. Also in some ANS synapses.	Excitatory or inhibitory	Cocaine and amphetamines increase the release and block the reuptake of norepinephrine, resulting in overstimulation of postsynaptic neurons.
Serotonin	A small number of small-sized nuclei in the brainstem. Nerve tracts extend from the nuclei to many areas of the brain and spinal cord.	Generally inhibitory	Involved with mood, anxiety, and sleep induction. Levels of serotonin are elevated in schizophrenia (delusions, hallucinations, and withdrawal).
Dopamine	Confined to a small number of nuclei and nerve tracts. Distribution is more restricted than that of norepinephrine or serotonin. Also found in some ANS synapses.	Excitatory or inhibitory	Parkinson's disease (depression of voluntary motor control) results from destruction of dopamine-secreting neurons. Drugs used to increase dopamine production induce vomiting and schizophrenia.
Histamine	Hypothalamus, with nerve tracts to many parts of the brain and spinal cord.	Generally inhibitory	No clear indication of histamine-associated pathologies. Histamine apparently is involved with arousal from sleep, pituitary hormone secretion, control of cerebral circulation, and thermoregulation.
Amino Acids			
Gamma-aminobutyric acid (GABA)	GABA-secreting neurons mostly control activities in their own area and are not usually involved with transmission from one part of the CNS to another. Most neurons of the CNS have GABA receptors.	Majority of postsynaptic inhibition in the brain; some presynaptic inhibition in the spinal cord	Drugs that increase GABA function have been used to treat epilepsy (excessive discharge of neurons).
Glycine	Spinal cord and brain. Like GABA, glycine predominantly produces local effects.	Most postsynaptic inhibition in the spinal cord	Glycine receptors are inhibited by the poison strychnine. Strychnine increases the excitability of certain neurons by blocking their inhibition. Strychnine poisoning results in powerful muscle contractions and convulsions. Tetanus of respiratory muscles can cause death.
Glutamate and aspartate	Widespread in the brain and spinal cord, especially in nerve tracts that ascend or descend the spinal cord or in tracts that project from one part of the brain to another.	Excitatory	Drugs that block glutamate or aspartate are under development. These drugs might prevent seizures and neural degeneration from overexcitation.

Table 11.5 *continued*			
Substance	**Location**	**Effect**	**Clinical Example**
Nitric Oxide	Brain, spinal cord, adrenal gland, intramural plexus, nerves to penis.	Excitatory	Blocking nitric oxide production may prevent stroke damage. Stimulating nitric oxide release is used to treat impotence.
Neuropeptides			
Endorphins and enkephalins	Widely distributed in the CNS and PNS.	Generally inhibitory	The opiates morphine and heroin bind to endorphin and enkephalin receptors on presynaptic neurons and reduce pain by blocking the release of a neurotransmitter, such as substance P.
Substance P	Spinal cord, brain, and sensory neurons associated with pain.	Generally excitatory	Substance P is a neurotransmitter in pain transmission pathways. Blocking the release of substance P by morphine reduces pain.

(a)

(b)

Figure 11.29   Postsynaptic Potentials

(*a*) Excitatory postsynaptic potential (EPSP) is closer to threshold.
(*b*) Inhibitory postsynaptic potential (IPSP) is farther from threshold.

is an **inhibitory postsynaptic potential (IPSP)** (figure 11.29*b*). IPSPs are important because they decrease the likelihood of producing action potentials by moving the membrane potential farther from threshold. Neurons releasing neurotransmitter substances that cause IPSPs are **inhibitory neurons.** The IPSP is the result of an increase in the permeability of the plasma membrane to Cl^- or K^+. For example, in the spinal cord, glycine binds to its receptors, directly causing Cl^- channels to open. Because Cl^- are more concentrated outside the cell than inside, when the permeability of the membrane to Cl^- increases, they diffuse into the cell, causing the inside of the cell to become more negative and resulting in hyperpolarization. Acetylcholine can bind to its receptors in the heart, causing G protein–mediated opening of K^+ channels (see chapter 3). The concentration of K^+ is greater inside the cell than outside, and increased permeability of the membrane to K^+ results in diffusion of K^+ out of the cell. Consequently, the outside of the cell becomes more positive than the inside, resulting in hyperpolarization.

Presynaptic Inhibition and Facilitation

Many of the synapses of the CNS are **axoaxonic synapses,** meaning that the axon of one neuron synapses with the presynaptic terminal (axon) of another (figure 11.30). The axoaxonic synapse doesn't initiate an action potential in the presynaptic terminal. When an action potential reaches the presynaptic terminal, however, neuromodulators released in the axoaxonic synapse can alter the amount of neurotransmitter released from the presynaptic terminal.

In **presynaptic inhibition,** there is a reduction in the amount of neurotransmitter released from the presynaptic terminal. For example, sensory neurons for pain can release neurotransmitters from their presynaptic terminals and stimulate the

Figure 11.30 Presynaptic Inhibition at an Axoaxonic Synapse

(a) The inhibitory neuron of the axoaxonic synapse is inactive and has no effect on the release of neurotransmitter from the presynaptic terminal. (b) Release of a neuromodulator from the inhibitory neuron of the axoaxonic synapse reduces the amount of neurotransmitter released from the presynaptic terminal.

postsynaptic membranes of neurons in the brain or spinal cord. Awareness of pain occurs only if action potentials are produced in the postsynaptic membranes of the CNS neurons. Enkephalins and endorphins released from inhibitory neurons of axoaxonic synapses can reduce or eliminate pain sensations by inhibiting the release of neurotransmitter from the presynaptic terminals of sensory neurons (see figure 11.30). Enkephalins and endorphins can block voltage-gated Ca^{2+} channels. Consequently, when action potentials reach the presynaptic terminal, the influx of Ca^{2+} that normally stimulate neurotransmitter release is blocked.

In **presynaptic facilitation,** the amount of neurotransmitter released from the presynaptic terminal increases. For example, in some synapses, the release of neurotransmitter from the presynaptic terminal causes the release of yet more neurotransmitter. Glutamate released from a presynaptic neuron can bind to receptors on the postsynaptic membrane and stimulate the production of nitric oxide by the postsynaptic neuron. The nitric oxide diffuses out of the postsynaptic neuron, crosses the synaptic cleft, diffuses into the presynaptic neuron, and stimulates the release of additional glutamate from the presynaptic neuron.

- 47. *Define and explain the production of EPSPs and IPSPs. Why are they important?*
- 48. *What is presynaptic inhibition and facilitation?*

Spatial and Temporal Summation

Depolarizations produced in postsynaptic membranes are local depolarizations. Within the CNS and in many PNS synapses, a single presynaptic action potential does not cause a local depolarization in the postsynaptic membrane sufficient to reach threshold and produce an action potential. Instead, many presynaptic action potentials cause many local potentials in the postsynaptic neuron. The local potentials combine in a process called **summation** at the

trigger zone of the postsynaptic neuron, which is the normal site of action potential generation for most neurons. If summation results in a local potential that exceeds threshold at the trigger zone, an action potential is produced. Action potentials are readily produced at the trigger zone because the concentration of voltage-gated Na^+ channels is approximately seven times greater there than at the rest of the cell body.

Two types of summation, called spatial summation and temporal summation, are possible. The simplest type of **spatial summation** occurs when two action potentials arrive simultaneously at two different presynaptic terminals that synapse with the same postsynaptic neuron. In the postsynaptic neuron, each action potential causes a local depolarization that undergoes summation at the trigger zone. If the summated depolarization reaches threshold, an action potential is produced (figure 11.31*a*).

Temporal summation results when two or more action potentials arrive in very close succession at a single presynaptic terminal. The first action potential causes a local depolarization in the postsynaptic membrane that remains for a few milliseconds before it disappears, although its magnitude decreases through time. Temporal summation results when another action potential initiates another local depolarization before the local depolarization caused by the previous action potential repolarizes to its resting value (see figure 11.19*b*). Subsequent action potentials cause local depolarizations that summate with previous local depolarizations. If the summated local depolarization reaches threshold at the trigger zone, an action potential is produced in the postsynaptic neuron (figure 11.31*b*).

P R E D I C T

Excitatory neurons A and B both synapse with neuron C. Neuron A releases a neurotransmitter, and neuron B releases the same type and amount of neurotransmitter plus a neuromodulator that produces EPSPs in neuron C. Action potentials produced in neuron A alone can result in action potential production in neuron C. Action potentials produced in neuron B alone also can cause action potential production in neuron C. Which results in more action potentials in neuron C, stimulation by only neuron A or stimulation by only neuron B? Explain.

Excitatory and inhibitory neurons can synapse with the same postsynaptic neuron. Spatial summation of EPSPs and IPSPs occurs in the postsynaptic neuron, and whether a postsynaptic action potential is initiated or not depends on which type of local potential has the greatest influence on the postsynaptic membrane potential (figure 11.31*c*). If the EPSPs (local depolarizations) cancel the IPSPs (local hyperpolarizations) and summate to threshold, an action potential is produced. If the IPSPs prevent the EPSPs from summating to threshold, no action potential is produced.

The synapse is an essential structure for the process of integration carried out by the CNS. For example, action potentials propagated along axons from sensory organs to the CNS can produce a sensation, or they can be ignored. To produce a sensation, action potentials must be transmitted across synapses as they travel through the CNS to the cerebral cortex, where information is interpreted. Stimuli that do not result in action

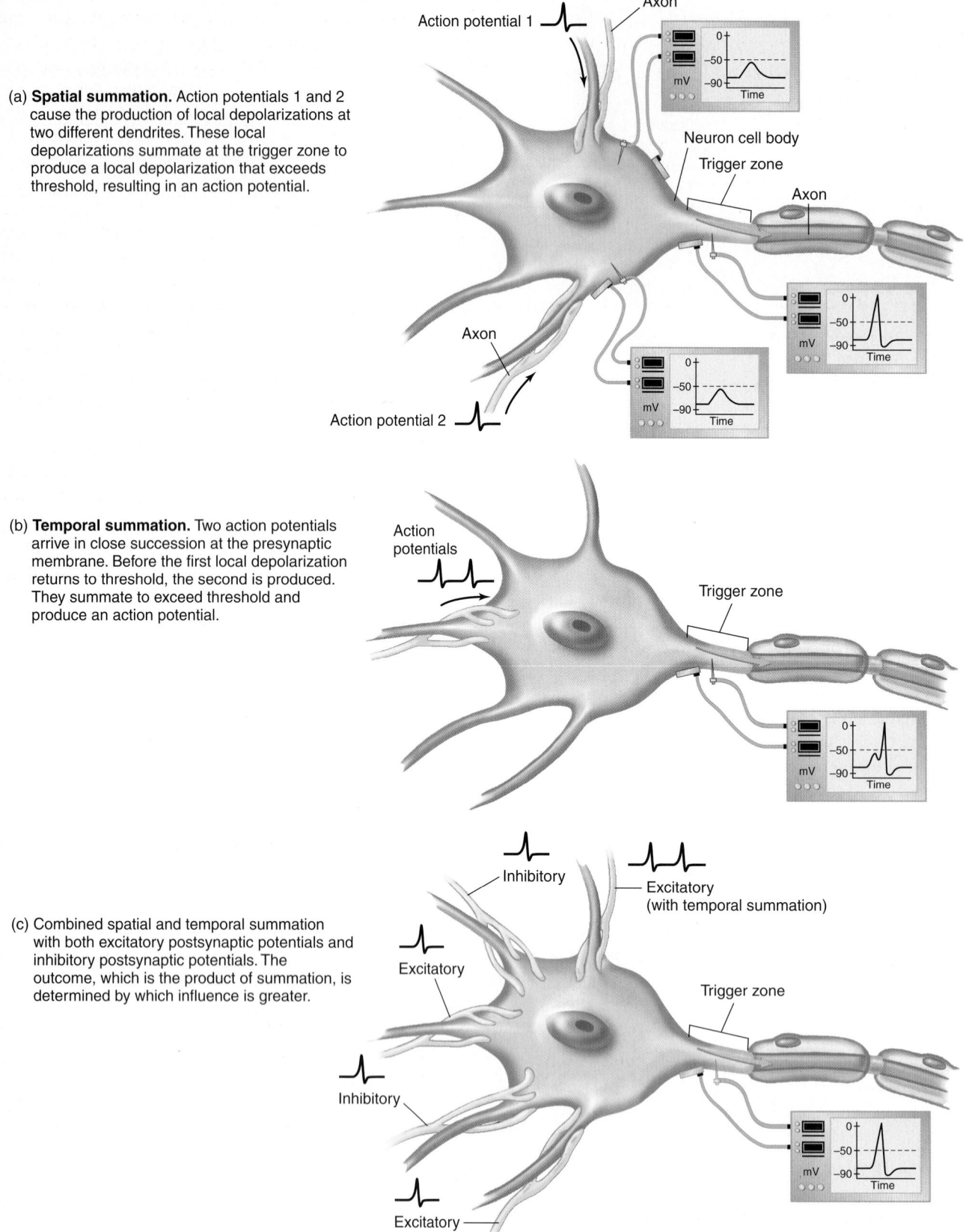

(a) **Spatial summation.** Action potentials 1 and 2 cause the production of local depolarizations at two different dendrites. These local depolarizations summate at the trigger zone to produce a local depolarization that exceeds threshold, resulting in an action potential.

(b) **Temporal summation.** Two action potentials arrive in close succession at the presynaptic membrane. Before the first local depolarization returns to threshold, the second is produced. They summate to exceed threshold and produce an action potential.

(c) Combined spatial and temporal summation with both excitatory postsynaptic potentials and inhibitory postsynaptic potentials. The outcome, which is the product of summation, is determined by which influence is greater.

Figure 11.31   Summation

potential transmission across synapses are ignored because information never reaches the cerebral cortex. The brain can ignore large amounts of sensory information as a result of complex integration.

> **49. Define spatial and temporal summation. In what part of the neuron does summation take place?**
>
> **50. How do EPSPs and IPSPs affect the likelihood that summation will result in an action potential?**

Neuronal Pathways and Circuits

The organization of neurons within the CNS varies from relatively simple to extremely complex patterns. The axon of a neuron can branch repeatedly to form synapses with many other neurons, and hundreds or even thousands of axons can synapse with the cell body and dendrites of a single neuron. Although their complexity varies, three basic patterns can be recognized: convergent pathways, divergent pathways, and oscillating circuits.

In **convergent pathways** many neurons converge and synapse with a smaller number of neurons. The simplest convergent pathway occurs when two presynaptic neurons synapse with a single postsynaptic neuron (figure 11.32a). Convergence makes it

possible for different parts of the nervous system to activate or inhibit the activity of neurons. For example, one part of the nervous system can stimulate the neurons responsible for making a muscle contract, whereas another part can inhibit those neurons. Through summation, action potentials resulting in muscle contraction can be produced if the converging neurons stimulate the production of more EPSPs than IPSPs. Conversely, muscle contraction is inhibited if the converging neurons stimulate the production of more IPSPs than EPSPs.

In **divergent pathways,** a smaller number of presynaptic neurons synapse with a larger number of postsynaptic neurons to allow information transmitted in one neuronal pathway to diverge into two or more pathways. The simplest divergent pathway occurs when a single presynaptic neuron branches to synapse with two postsynaptic neurons (figure 11.32b). Diverging pathways allow one part of the nervous system to affect more than one other part of the nervous system. For example, sensory input to the central nervous system can go to both the spinal cord and the brain.

Oscillating circuits have neurons arranged in a circular fashion, which allows action potentials entering the circuit to cause a neuron farther along in the circuit to produce an action

(a)   (b)

Figure 11.32   Convergent and Divergent Pathways

(a) General model of a convergent pathway, showing two neurons converging on one neuron. (b) General model of a divergent pathway, showing one neuron diverging onto two neurons.

potential more than once (figure 11.33). This response is called **afterdischarge,** and its effect is to prolong the response to a stimulus. Oscillating circuits are similar to positive-feedback systems. Once an oscillating circuit is stimulated, it continues to discharge until the synapses involved become fatigued or until they are inhibited by other neurons. Figure 11.33*a* illustrates a simple circuit in which a collateral axon stimulates its own cell body; figure 11.33*b* shows a more complex circuit. Oscillating circuits play a role in neuronal circuits that are periodically active. Respiration may be controlled by an oscillating circuit that controls inspiration and another that controls expiration.

Neurons that spontaneously produce action potentials are common in the CNS and may activate oscillating circuits, which remain active awhile. The cycle of wakefulness and sleep may involve circuits of this type. Spontaneously active neurons are also capable of influencing the activity of other circuit types. The complex functions carried out by the CNS are affected by the numerous circuits operating together and influencing the activity of one another.

51. *Diagram a convergent pathway, a divergent pathway, and an oscillating circuit, and describe what is accomplished in each.*

(a)

(b)

Figure 11.33   Oscillating Circuits

(*a*) A single neuron stimulates itself. (*b*) A more complex oscillating circuit in which the input neuron is stimulated by two other neurons.

S U M M A R Y

Functions of the Nervous System (p. 374)

The nervous system detects external and internal stimuli (sensory input), processes and responds to sensory input (integration), maintains homeostasis by regulating other systems, is the center for mental activities, and controls body movements through skeletal muscles.

Divisions of the Nervous System (p. 374)

1. The nervous system has two anatomic divisions.
   - The central nervous system (CNS) consists of the brain and spinal cord and is encased in bone.
   - The peripheral nervous system (PNS), the nervous tissue outside of the CNS, consists of sensory receptors, nerves, ganglia, and plexuses.
2. The PNS has two divisions.
   - The sensory division transmits action potentials to the CNS and usually consists of single neurons that have their cell bodies in ganglia.
   - The motor division carries action potentials away from the CNS in cranial or spinal nerves.

3. The motor division has two subdivisions.
   - The somatic nervous system innervates skeletal muscle and is mostly under voluntary control. It consists of single neurons that have their cell bodies located within the CNS.
   - The autonomic nervous system (ANS) innervates cardiac muscle, smooth muscle, and glands. It has two sets of neurons between the CNS and effector organs. The first set has its cell bodies within the CNS, and the second set has its cell bodies within autonomic ganglia.
   - The ANS is subdivided into the sympathetic division, which is most active during physical activity, the parasympathetic division, which regulates resting functions, and the enteric nervous system, which controls the digestive system.
4. The anatomic divisions perform different functions.
   - The PNS detects stimuli and transmits information to and receives information from the CNS.
   - The CNS processes, integrates, stores, and responds to information from the PNS.

Cells of the Nervous System (p. 376)
Neurons

1. Neurons receive stimuli and transmit action potentials.
2. Neurons have three components.
   - The cell body is the primary site of protein synthesis.
   - Dendrites are short, branched cytoplasmic extensions of the cell body that usually conduct electric signals toward the cell body.
   - An axon is a cytoplasmic extension of the cell body that transmits action potentials to other cells.

Types of Neurons

1. Multipolar neurons have several dendrites and a single axon. Interneurons and motor neurons are multipolar.
2. Bipolar neurons have a single axon and dendrite and are found as components of sensory organs.
3. Unipolar neurons have a single axon. Most sensory neurons are unipolar.

Neuroglia of the CNS

1. Neuroglia are nonneural cells that support and aid the neurons of the CNS and PNS.
2. Astrocytes provide structural support for neurons and blood vessels. The endothelium of blood vessels forms the blood–brain barrier, which regulates the movement of substances between the blood and the CNS. Astrocytes influence the functioning of the blood–brain barrier and process substances that pass through it.
3. Ependymal cells line the ventricles and the central canal of the spinal cord. Some are specialized to produce cerebrospinal fluid.
4. Microglia are macrophages that phagocytize microorganisms, foreign substances, or necrotic tissue.
5. An oligodendrocyte forms myelin sheaths around the axons of several CNS neurons.

Neuroglia of the PNS

1. A Schwann cell forms a myelin sheath around part of the axon of a PNS neuron.
2. Satellite cells support and nourish neuron cell bodies within ganglia.

Myelinated and Unmyelinated Axons

1. Myelinated axons are wrapped by several layers of plasma membrane from oligodendrocytes (CNS) or Schwann cells (PNS). Spaces between the wrappings are the nodes of Ranvier. Myelinated axons conduct action potentials rapidly.
2. Unmyelinated axons rest in invaginations of oligodendrocytes (CNS) or Schwann cells (PNS). They conduct action potentials slowly.

Organization of Nervous Tissue (p. 381)

1. Nervous tissue can be grouped into white and gray matter.
   - White matter consists of myelinated axons and functions to propagate action potentials.
   - Gray matter consists of collections of neuron cell bodies or unmyelinated axons. Axons synapse with neuron cell bodies, which is functionally the site of integration in the nervous system.
2. White matter forms nerve tracts in the CNS and nerves in the PNS. Gray matter forms cortex and nuclei in the CNS and ganglia in the PNS.

Electric Signals (p. 381)

Electrical properties of cells result from the ionic concentration differences across the plasma membrane and from the permeability characteristics of the plasma membrane.

Concentration Differences Across the Plasma Membrane

1. The Na^+–K^+ pump moves ions by active transport. K^+ are moved into the cell, and Na^+ are moved out of it.
2. The concentration of K^+ and negatively charged proteins and other molecules is higher inside, and the concentrations of Na^+ and Cl^- are higher outside the cell.
3. Negatively charged proteins and other negatively charged ions are synthesized inside the cell and cannot diffuse out of it, and they repel negatively charged Cl^-.
4. The permeability of the plasma membrane to ions is determined by nongated and gated ion channels.
   - Nongated K^+ channels are more numerous than nongated Na^+ channels, thus the plasma membrane is more permeable to K^+ than to Na^+ when at rest.
   - Gated ion channels in the plasma membrane include ligand-gated ion channels, voltage-gated ion channels, and other-gated ion channels.

The Resting Membrane Potential

1. The resting membrane potential is a charge difference across the plasma membrane when the cell is in an unstimulated condition. The inside of the cell is negatively charged compared to the outside of the cell.
2. The resting membrane potential is due mainly to the tendency of positively charged K^+ to diffuse out of the cell, which is opposed by the negative charge that develops inside the plasma membrane.
3. Depolarization is a decrease in the resting membrane potential and can result from a decrease in the K^+ concentration gradient, a decrease in membrane permeability to K^+, an increase in membrane permeability to Na^+, or a decrease in extracellular Ca^{2+} concentration.
4. Hyperpolarization is an increase in the resting membrane potential that can result from an increase in the K^+ concentration gradient, an increase in membrane permeability to K^+, a decrease in membrane permeability to Na^+, or an increase in extracellular Ca^{2+} concentration.

Local Potentials

1. A local potential is a small change in the resting membrane potential that is confined to a small area of the plasma membrane.
2. An increase in membrane permeability to Na^+ can cause local depolarization, and an increase in membrane permeability to K^+ or Cl^- can result in local hyperpolarization.
3. A local potential is termed *graded* because a stronger stimulus produces a greater potential change than a weaker stimulus.
4. Local potentials can summate, or add together.
5. A local potential decreases in magnitude as the distance from the stimulation increases.

Action Potentials

1. An action potential is a larger change in the resting membrane potential that spreads over the entire surface of the cell.
2. Threshold is the membrane potential at which a local potential depolarizes the plasma membrane sufficiently to produce an action potential.
3. Action potentials occur in an all-or-none fashion. If the action potential occurs at all, it's of the same magnitude, no matter how strong the stimulus.
4. Depolarization occurs as the inside of the membrane becomes more positive because Na^+ diffuse into the cell through voltage-gated ion channels. Repolarization is a return of the membrane potential toward the resting membrane potential because voltage-gated Na^+ channels close and Na^+ diffusion into the cell slows to resting levels and because voltage-gated K^+ channels continue to open and K^+ diffuse out of the cell.

Refractory Period

1. The absolute refractory period is the time during an action potential when a second stimulus, no matter how strong, cannot initiate another action potential.
2. The relative refractory period follows the absolute refractory period and is the time during which a stronger-than-threshold stimulus can evoke another action potential.

Action Potential Frequency

1. The strength of stimuli affects the frequency of action potentials.
   - A subthreshold stimulus produces only a local potential.
   - A threshold stimulus causes a local potential that reaches threshold and results in a single action potential.
   - A submaximal stimulus is greater than a threshold stimulus and weaker than a maximal stimulus. The action potential frequency increases as the strength of the submaximal stimulus increases.
   - A maximal or a supramaximal stimulus produces a maximum frequency of action potentials.
2. A low frequency of action potentials represents a weaker stimulus than a high frequency.

Propagation of Action Potentials

1. An action potential generates local currents, which stimulate voltage-gated Na^+ channels in adjacent regions of the plasma membrane to open, producing a new action potential.
2. In an unmyelinated axon, action potentials are generated immediately adjacent to previous action potentials.
3. In a myelinated axon, action potentials are generated at successive nodes of Ranvier, which are separated from each other by Schwann cells.
4. Reversal of the direction of action potential propagation is prevented by the absolute refractory period.
5. Action potentials propagate most rapidly in myelinated, large-diameter axons.

The Synapse   (p. 394)
Electrical Synapses

1. Electrical synapses are gap junctions in which tubular proteins called connexons allow local currents to move between cells.
2. At an electrical synapse, an action potential in one cell generates a local current that causes an action potential in an adjacent cell.

Chemical Synapses

1. Anatomically, a chemical synapse has three components.
   - The enlarged ends of the axon are the presynaptic terminals containing synaptic vesicles.
   - The postsynaptic membranes contain receptors for the neurotransmitter.
   - The synaptic cleft, a space, separates the presynaptic and postsynaptic membranes.

2. An action potential arriving at the presynaptic terminal causes the release of a neurotransmitter, which diffuses across the synaptic cleft and binds to the receptors of the postsynaptic membrane.
3. The effect of the neurotransmitter on the postsynaptic membrane can be stopped in several ways.
   - The neurotransmitter is broken down by an enzyme.
   - The neurotransmitter is taken up by the presynaptic terminal.
   - The neurotransmitter diffuses out of the synaptic cleft.
4. Neurotransmitters are specific for their receptors. A neurotransmitter can be stimulatory in one synapse and inhibitory in another, depending on the type of receptor present.
5. Neuromodulators influence the likelihood that an action potential in a presynaptic terminal will result in an action potential in a postsynaptic cell.
6. Depolarization of the postsynaptic membrane caused by an increase in membrane permeability to Na^+ is an excitatory postsynaptic potential (EPSP).
7. Hyperpolarization of the postsynaptic membrane caused by an increase in membrane permeability to K^+ is an inhibitory postsynaptic potential (IPSP).
8. Presynaptic inhibition decreases neurotransmitter release. Presynaptic facilitation increases neurotransmitter release.

Spatial and Temporal Summation

1. Presynaptic action potentials through neurotransmitters produce local potentials in postsynaptic neurons. The local potential can summate to produce an action potential at the trigger zone.
2. Spatial summation occurs when two or more presynaptic terminals simultaneously stimulate a postsynaptic neuron.
3. Temporal summation occurs when two or more action potentials arrive in succession at a single presynaptic terminal.
4. Inhibitory and excitatory presynaptic neurons can converge on a postsynaptic neuron. The activity of the postsynaptic neuron is determined by the integration of the EPSPs and IPSPs produced in the postsynaptic neuron.

Neuronal Pathways and Circuits   (p. 403)

1. Convergent pathways have many neurons synapsing with a few neurons.
2. Divergent pathways have a few neurons synapsing with many neurons.
3. Oscillating circuits have collateral branches of postsynaptic neurons synapsing with presynaptic neurons.

R E V I E W   A N D   C O M P R E H E N S I O N

1. The peripheral nervous system includes the
   a. somatic nervous system.
   b. brain.
   c. spinal cord.
   d. nuclei.
   e. all of the above.

2. The part of the nervous system that controls smooth muscle, cardiac muscle, and glands is the
   a. somatic nervous system.
   b. autonomic nervous system.
   c. skeletal division.
   d. sensory division.

3. Neurons have cytoplasmic extensions that connect one neuron to another neuron. Given these structures:
   1. axon
   2. dendrite
   3. dendritic spine
   4. presynaptic terminal

   Choose the arrangement that lists the structures in the order they are found between two neurons.
   a. 1,4,2,3
   b. 1,4,3,2
   c. 4,1,2,3
   d. 4,1,3,2
   e. 4,3,2,1

4. A neuron with many short dendrites and a single long axon is a
   _____ neuron.
   a. multipolar
   b. unipolar
   c. bipolar

5. Motor neurons and interneurons are _____ neurons.
   a. unipolar
   b. bipolar
   c. multipolar
   d. afferent

6. Cells found in the choroid plexuses that secrete cerebrospinal fluid
   are
   a. astrocytes.
   b. microglia.
   c. ependymal cells.
   d. oligodendrocytes.
   e. Schwann cells.

7. Neuroglia that are phagocytic within the central nervous system are
   a. oligodendrocytes.
   b. microglia.
   c. ependymal cells.
   d. astrocytes.
   e. Schwann cells.

8. Unmyelinated axons within nerves may have which of these
   associated with them?
   a. Schwann cells
   b. nodes of Ranvier
   c. oligodendrocytes
   d. all of the above

9. Action potentials are conducted more rapidly
   a. in small-diameter axons than in large-diameter axons.
   b. in unmyelinated axons than in myelinated axons.
   c. along axons that have nodes of Ranvier.
   d. all of the above.

10. Clusters of nerve cell bodies within the peripheral nervous system
    are
    a. ganglia.
    b. fascicles.
    c. nuclei.
    d. laminae.

11. Gray matter contains primarily
    a. myelinated fibers.
    b. neuron cell bodies.
    c. Schwann cells.
    d. oligodendrocytes.

12. Concerning concentration difference across the plasma membrane,
    there are
    a. more K^+ and Na^+ outside the cell than inside.
    b. more K^+ and Na^+ inside the cell than outside.
    c. more K^+ outside the cell than inside and more Na^+ inside the
       cell than outside.
    d. more K^+ inside the cell than outside and more Na^+ outside the
       cell than inside.

13. Compared to the inside of the resting plasma membrane, the
    outside surface of the membrane is
    a. positively charged.
    b. electrically neutral.
    c. negatively charged.
    d. continuously reversing so that it is positive one second and
       negative the next.
    e. negatively charged whenever the Na^+–K^+ pump is operating.

14. Nongated ion channels
    a. open in response to small voltage changes.
    b. open when a ligand binds to its receptor.
    c. are responsible for the ion permeability of the resting plasma
       membrane.
    d. allow substances to move into the cell but not out.
    e. all of the above.

15. The resting membrane potential results when the tendency for
    _____ to diffuse out of the cell is balanced by their
    attraction to opposite charges inside the cell.
    a. Na^+
    b. K^+
    c. Cl^-
    d. negatively charged proteins

16. If the permeability of the plasma membrane to K^+ increases,
    resting membrane potential _____. This is called
    _____.
    a. increases, hyperpolarization
    b. increases, depolarization
    c. decreases, hyperpolarization
    d. decreases, depolarization

17. Decreasing the extracellular concentration of K^+ affects the resting
    membrane potential by causing
    a. hyperpolarization.
    b. depolarization.
    c. no change.

18. Which of these terms are correctly matched with their definition or
    description?
    a. depolarization: membrane potential becomes more negative
    b. hyperpolarization: membrane potential becomes more negative
    c. hypopolarization: membrane potential becomes more negative

19. Which of these statements about ion movement through the plasma
    membrane is true?
    a. Movement of Na^+ out of the cell requires energy (ATP).
    b. When Ca^{2+} binds to proteins in ion channels, the diffusion of
       Na^+ into the cell is inhibited.
    c. There are specific ion channels that regulate the diffusion of Na^+
       through the plasma membrane.
    d. All of the above.

20. The *major* function of the Na^+–K^+ pump is to
    a. pump Na^+ into and K^+ out of the cell.
    b. generate the resting membrane potential.
    c. maintain the concentration gradients of Na^+ and K^+ across the
       plasma membrane.
    d. oppose any tendency of the cell to undergo hyperpolarization.

21. Local potentials
    a. spread over the plasma membrane in decremental fashion.
    b. are not propagated for long distances.
    c. are graded.
    d. can summate.
    e. all of the above.

22. During the depolarization phase of an action potential, the
    permeability of the membrane
    a. to K^+ is greatly increased.
    b. to Na^+ is greatly increased.
    c. to Ca^{2+} is greatly increased.
    d. is unchanged.

23. During repolarization of the plasma membrane,
    a. Na^+ diffuse into the cell.
    b. Na^+ diffuse out of the cell.
    c. K^+ diffuse into the cell.
    d. K^+ diffuse out of the cell.

24. The absolute refractory period
    a. limits how many action potentials can be produced during a given period of time.
    b. prevents an action potential from starting another action potential at the same point on the plasma membrane.
    c. is the period of time when a strong stimulus can initiate a second action potential.
    d. both a and b.
    e. all of the above.

25. A subthreshold stimulus
    a. produces an afterpotential.
    b. produces a local potential.
    c. causes an all-or-none response.
    d. produces more action potentials than a submaximal stimulus.

26. Neurotransmitter substances are stored in vesicles that are located in specialized portions of the
    a. neuron cell body.
    b. axon.
    c. dendrite.
    d. postsynaptic membrane.

27. In a chemical synapse,
    a. action potentials in the presynaptic terminal cause voltage-gated Ca^{2+} channels to open.
    b. neurotransmitters can cause ligand-gated Na^+ channels to open.
    c. neurotransmitters can be broken down by enzymes.
    d. neurotransmitters can be taken up by the presynaptic terminal.
    e. all of the above.

28. An inhibitory presynaptic neuron can affect a postsynaptic neuron by
    a. producing an IPSP in the postsynaptic neuron.
    b. hyperpolarizing the plasma membrane of the postsynaptic neuron.
    c. causing K^+ to diffuse out of the postsynaptic neuron.
    d. causing Cl^- to diffuse into the postsynaptic neuron.
    e. all of the above.

29. Summation
    a. is caused by combining two or more local potentials.
    b. occurs at the trigger zone of the postsynaptic neuron.
    c. results in an action potential if it reaches the threshold potential.
    d. can occur when two action potentials arrive in close succession at a single presynaptic terminal.
    e. all of the above.

30. In convergent pathways,
    a. the response of the postsynaptic neuron depends on the summation of EPSPs and IPSPs.
    b. a smaller number of presynaptic neurons synapse with a larger number of postsynaptic neurons.
    c. information transmitted in one neuronal pathway can go into two or more pathways.
    d. all of the above.

Answers in Appendix F

C R I T I C A L   T H I N K I N G

1. Predict the consequence of a reduced intracellular K^+ concentration on the resting membrane potential.

2. A child eats a whole bottle of salt (NaCl) tablets. What effect would this have on action potentials?

3. Lithium ions reduce the permeability of plasma membranes to Na^+. Predict the effect lithium ions in the extracellular fluid would have on the response of a neuron to stimuli.

4. Some smooth muscle has the ability to contract spontaneously. That is, it contracts without any external stimulation. Propose an explanation for the ability of smooth muscle to contract spontaneously based on what you know about membrane potentials. Assume that an action potential in a smooth muscle cell causes it to contract.

5. Assume that you have two nerve fibers of the same diameter, but one is myelinated and the other is unmyelinated. Along which type of fiber is the conduction of an action potential most energy-efficient? (*Hint:* ATP.)

6. Explain the consequences when an inhibitory neuromodulator is released from a presynaptic terminal and a stimulatory neurotransmitter is released from another presynaptic terminal, both of which synapse with the same neuron.

7. With aging, the speed of action potential propagation and synaptic transmission decreases. List possible explanations.

8. Students in a veterinary school were given the following hypothetical problem. A dog ingests organophosphate poison, and the students are responsible for saving the animal's life. Organophosphate poisons bind to and inhibit acetylcholinesterase. Several substances they could inject include the following: acetylcholine, curare (which blocks acetylcholine receptors), and potassium chloride. If you were a student in the class, what would you do to save the animal?

9. Strychnine blocks receptor sites for inhibitory neurotransmitter substances in the CNS. Explain how strychnine could produce tetanus in skeletal muscles.

Answers in Appendix G

A N S W E R S   T O   P R E D I C T   Q U E S T I O N S

1. When the axon of a neuron is severed, the proximal portion of the axon remains attached to the neuron cell body. The distal portion is detached, however, and has no way to replenish the enzymes and other proteins essential to its survival. Because the DNA in the nucleus provides the information that determines the structure of proteins by directing mRNA synthesis, the distal portion of the axon has no source of new proteins. Consequently, it degenerates and dies. On the other hand, the proximal portion of the axon is still attached to the nucleus and therefore has a source of new proteins. It remains alive and, in many cases, grows to replace the severed distal axon.

2. Tissue A has the larger resting membrane potential. There's a greater tendency for K^+ to diffuse out of the cell because it has significantly more nongated K^+ channels. As a result, a greater negative charge develops on the inside of the plasma membrane, resulting in a larger resting membrane potential.

3. If the intracellular concentration of K^+ is increased, the concentration gradient from the inside to the outside of the plasma membrane increases. This situation is similar to decreasing the extracellular concentration of K^+. The greater concentration gradient for K^+ increases their tendency to diffuse out of the cell

across the plasma membrane. A greater negative charge then develops inside the cell (hyperpolarization).

4. Ca^{2+} binds to gating proteins that regulate the voltage-gated Na^+ channels. Low concentrations of Ca^{2+} cause the voltage-gated Na^+ channels to open, and high concentrations of Ca^{2+} cause the voltage-gated Na^+ channels to close. If the extracellular concentration of Ca^{2+} decreases, the resting membrane potential depolarizes because voltage-gated Na^+ channels open and Na^+ diffuse into the cell.

5. When the cells are stimulated, there is an increase in the permeability of their plasma membranes to Na^+. These ions diffuse into the cells down their concentration gradients and cause depolarization of the plasma membranes. If the concentration gradient for Na^+ is reduced, the tendency for Na^+ to diffuse into the cell decreases. In cell A, with the reduced Na^+ concentration gradient, the local depolarization is of a smaller magnitude than in cell B because fewer ions are able to diffuse into the cell in response to the stimulus.

6. If the extracellular concentration of Na^+ decreases, the magnitude of the action potential is reduced. The smaller extracellular concentration of Na^+ reduces the tendency for Na^+ to diffuse into the cell when the Na^+ channels are open during an action potential. Consequently, the inside of the plasma membrane doesn't become as positive as it does in cells with a high extracellular concentration of Na^+. Even though the magnitude of action potentials is reduced when the extracellular Na^+ concentration is reduced, all of the actions potentials are the same magnitude (all-or-none principle).

7. A prolonged stronger-than-threshold stimulus produces more action potentials than a prolonged threshold stimulus of the same duration. A prolonged stronger-than-threshold stimulus can stimulate more action potentials because the permeability of the membrane to Na^+ is increased. A very strong stimulus can even stimulate action potentials during the relative refractory period, whereas a prolonged threshold stimulus stimulates a low frequency of action potentials. Thus, when a prolonged stronger-than-threshold stimulus is applied, less time elapses between the production of one action potential and the next, resulting in the production of a greater number of action potentials.

8. If the duration of the absolute refractory period is 1 ms, that means action potentials can be generated no faster than every millisecond. The maximal frequency of action potentials is 1000 per second because there is 1000 milliseconds in a second.

9. Action potentials are transmitted fastest by electrical synapses because the local currents can quickly flow through the connexons. In contrast, chemical synapses are slower because the synaptic vesicles must be stimulated to release neurotransmitter, which diffuses across the synaptic cleft. The neurotransmitter must stimulate ligand-gated Na^+ channels to open. The resulting movement of Na^+ into the cell can produce an action potential. All of these events take time.

10. Temporal summation resulting from stimulation by neuron B produces more action potentials in the postsynaptic neuron than temporal summation resulting from stimulation by neuron A. The neuromodulator from neuron B produces EPSPs, which depolarize the membrane potential of neuron C, bringing the membrane potential closer to threshold. A smaller amount of neurotransmitter is therefore required to produce an action potential. Although neuron A and B release the same amount and type of neurotransmitter, the neuromodulator makes the neurotransmitter from neuron B more effective, resulting in more action potentials.

Visit the Online Learning Center at www.mhhe.com/seeley7 for chapter quizzes, interactive learning exercises, and other study tools.

Colorized SEM of nerve fascicles containing bundles of axons.

Spinal Cord and Spinal Nerves

CHAPTER

12

The **central nervous system (CNS)** consists of the brain and spinal cord, with the division between these two parts of the CNS placed at the level of the foramen magnum. The **peripheral nervous system (PNS)** consists of nerves and ganglia outside the CNS (see chapter 11). **Nerves** consist of axons, Schwann cells, and connective tissue sheaths. **Ganglia** are accumulations of cell bodies in the PNS. The PNS includes 12 pairs of cranial nerves and 31 pairs of spinal nerves.

The CNS receives sensory information, integrates and evaluates that information, stores some information, and initiates reactions. The PNS collects information from numerous sources both inside and outside the body and relays it through axons of sensory neurons to the CNS. Axons of motor neurons in the PNS relay information from the CNS to various parts of the body, primarily to muscles and glands, thereby regulating activity in those structures.

The spinal cord and spinal nerves are described in this chapter. The brain and cranial nerves are considered in chapter 13. The specific topics of chapter 12 are the *spinal cord* (p. 412), *reflexes* (p. 415), *interactions with spinal cord reflexes* (p. 420), *structure of peripheral nerves* (p. 420), and *spinal nerves* (p. 422).

Spinal Cord

The **spinal cord** is extremely important to the overall function of the nervous system. It is the major communication link between the brain and the PNS inferior to the head; it integrates incoming information and produces responses through reflex mechanisms.

General Structure

The spinal cord (figure 12.1) extends from the foramen magnum to the level of the second lumbar vertebra. It's considerably shorter than the vertebral column because it doesn't grow as rapidly as the vertebral column during development. The spinal cord

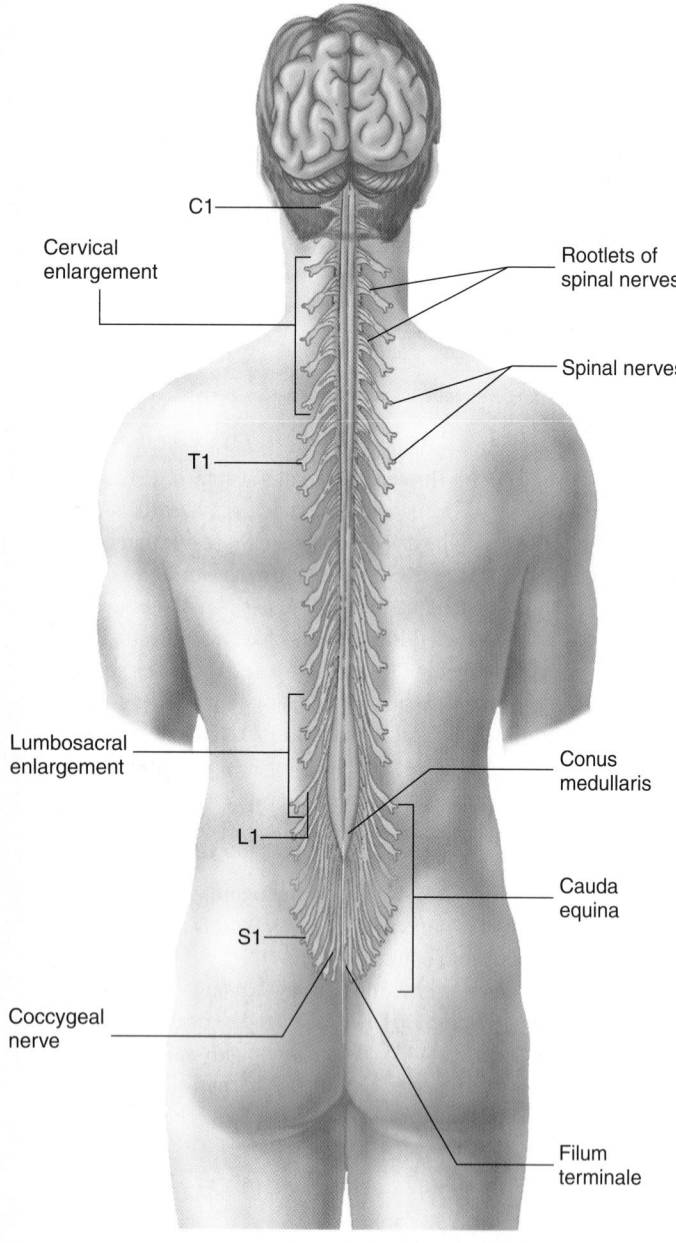

Cervical enlargement

C1

Rootlets of spinal nerves

Spinal nerves

T1

Lumbosacral enlargement

Conus medullaris

L1

Cauda equina

S1

Coccygeal nerve

Filum terminale

Posterior view

Figure 12.1   Spinal Cord and Spinal Nerve Roots

is composed of cervical, thoracic, lumbar, and sacral segments, named according to the portion of the vertebral column from which their nerves enter and exit. The spinal cord gives rise to 31 pairs of spinal nerves, which exit the vertebral column through intervertebral and sacral foramina (see figure 7.13). The nerves from the lower segments descend some distance in the vertebral canal before they exit because the spinal cord is shorter than the vertebral column.

The spinal cord is not uniform in diameter throughout its length. It's larger in diameter at its superior end, and it gradually decreases in diameter toward its inferior end. Two enlargements occur where nerves supplying the upper and lower limbs enter and leave the cord (see figure 12.1). The **cervical enlargement** in the inferior cervical region corresponds to the location where axons that supply the upper limbs enter and leave the cord. The **lumbosacral enlargement** in the inferior thoracic, lumbar, and superior sacral regions is the site where the axons supplying the lower limbs enter or leave the cord.

P R E D I C T
Why is the cord enlarged in the cervical and lumbosacral regions?

Immediately inferior to the lumbosacral enlargement, the spinal cord tapers to form a conelike region called the **conus medullaris.** Its tip is the inferior end of the spinal cord and extends to the level of the second lumbar vertebra. The nerves supplying the lower limbs and other inferior structures of the body arise from the lumbar and sacral nerves. They exit the lumbosacral enlargement and conus medullaris, course inferiorly through the vertebral canal, and exit through the intervertebral foramina from the second lumbar to the fifth sacral vertebrae. The numerous roots (origins) of spinal nerves extending inferiorly from the lumbosacral enlargement and conus medullaris resemble a horse's tail and are therefore called the **cauda** (kaw′dă, tail) **equina** (ē-kwī′nă, horse; see figure 12.1).

1. *Describe the cervical and lumbar enlargements of the spinal cord, the conus medullaris, and the cauda equina. How many pairs of spinal nerves exit the spinal cord?*

Meninges of the Spinal Cord

The spinal cord and brain are surrounded by connective tissue membranes called **meninges** (mĕ-nin′jēz; figure 12.2). The most superficial and thickest membrane is the **dura mater** (doo′ră mā′ter; tough mother). The dura mater forms a sac, often called the **thecal** (the′kal) **sac,** that surrounds the spinal cord. The thecal sac attaches to the rim of the foramen magnum and ends at the level of the second sacral vertebra. The spinal dura mater is continuous with the dura mater surrounding the brain and the connective tissue surrounding the spinal nerves. The dura mater around the spinal cord is separated from the periosteum of the vertebral canal by the **epidural space.** This is a true space between the walls of the vertebral canal and the dura mater of the spinal cord that contains spinal nerve roots, blood vessels, areolar connective tissue, and fat. **Epidural anesthesia** of the spinal nerves is induced by injecting anesthetics into this space. Epidural anesthesia is often given to women during childbirth.

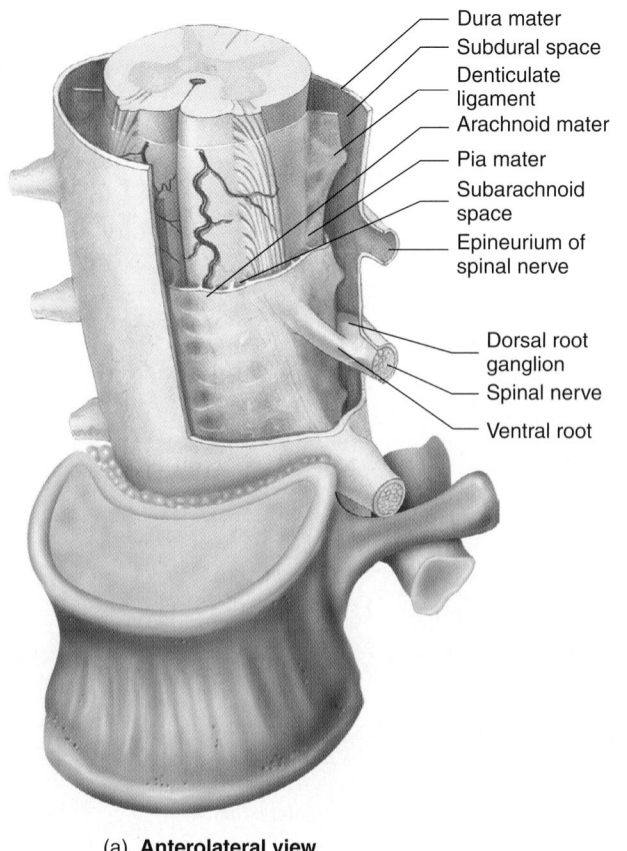

(a) **Anterolateral view**

Labels (top to bottom):
Dura mater
Subdural space
Denticulate ligament
Arachnoid mater
Pia mater
Subarachnoid space
Epineurium of spinal nerve
Dorsal root ganglion
Spinal nerve
Ventral root

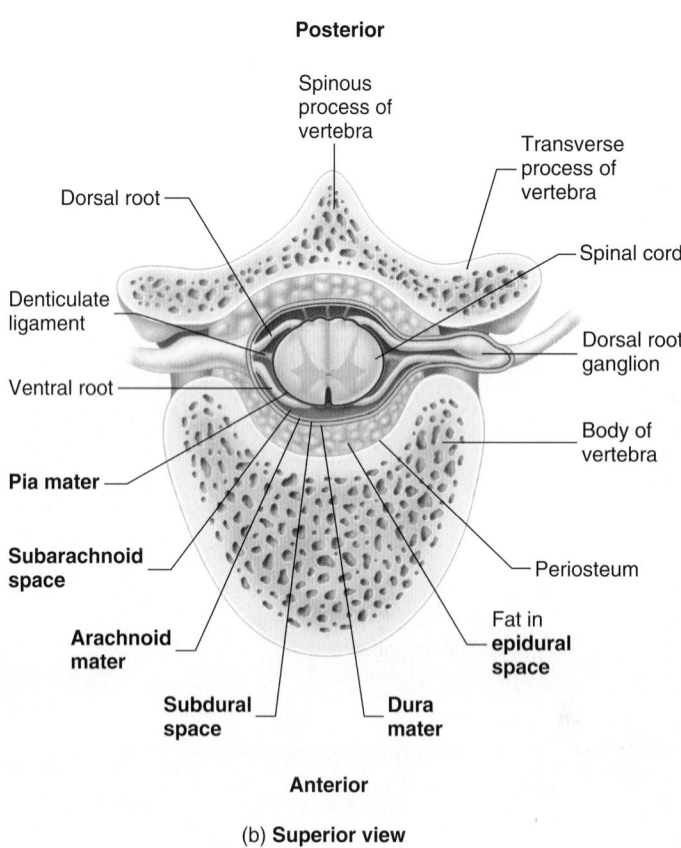

(b) **Superior view**

Labels:
Posterior
Spinous process of vertebra
Transverse process of vertebra
Dorsal root
Spinal cord
Denticulate ligament
Dorsal root ganglion
Ventral root
Body of vertebra
Pia mater
Subarachnoid space
Periosteum
Fat in epidural space
Arachnoid mater
Subdural space
Dura mater
Anterior

Figure 12.2 Meningeal Membranes Surrounding the Spinal Cord

P R E D I C T ②
Epidural anesthesia is usually induced by injection at what level of the vertebral column?

The next deeper meningeal membrane is a very thin, wispy **arachnoid** (ă-rak′noyd; spiderlike; i.e., cobwebs) **mater.** The space between this membrane and the dura mater is the **subdural space** and contains only a very small amount of serous fluid.

The third, deepest, meningeal layer, the **pia** (pī′ă; affectionate) **mater** is bound very tightly to the surface of the spinal cord. The spinal cord is held in place within the thecal sac by the denticulate ligaments and the filum terminale. The paired **denticulate** (den-tik′ū-lāt) **ligaments** are connective tissue septa extending from the lateral sides of the spinal cord to the dura mater (see figure 12.2b). The term *denticulate* refers to having small teeth, and the denticulate ligaments attach to the dura mater by 21 triangular or toothlike processes between the exits of the cervical and thoracic spinal nerves. The denticulate ligaments limit the lateral movement of the spinal cord. The **filum terminale** (fī′lŭm ter′mi-nal′ē) is a connective tissue strand that anchors the conus medullaris and the thecal sac to the first coccygeal vertebra, limiting their superior movement.

Between the arachnoid mater and the pia mater is the **subarachnoid space,** which contains weblike strands of the arachnoid mater, blood vessels, and **cerebrospinal** (ser′ě-brō-spī′năl, sě-rē′brō-spī-nal) **fluid (CSF),** which is described in chapter 13.

Introduction of Needles into the Subarachnoid Space

Several clinical procedures involve the insertion of a needle into the subarachnoid space inferior to the level of the second lumbar vertebra. The needle is introduced into either the L3/L4 or the L4/L5 intervertebral space. The needle doesn't contact the spinal cord because it extends only approximately to the second lumbar vertebra of the vertebral column, but the subarachnoid space extends to level S2 of the vertebral column. Nor does the needle damage the nerve roots of the cauda equina located in the subarachnoid space, because the needle quite easily pushes them aside. In **spinal anesthesia,** or spinal block, drugs that block action potential transmission are introduced into the subarachnoid space to prevent pain sensations in the lower half of the body. A **spinal tap** is the removal of CSF from the subarachnoid space. A spinal tap may be performed to examine the CSF for infectious agents (meningitis), for the presence of blood (hemorrhage), or for the measurement of CSF pressure. A radiopaque substance may also be injected into this area, and a **myelogram** (radiograph of the spinal cord) may be taken to visualize spinal cord defects or damage.

2. Name the meninges surrounding the spinal cord. What is found within the epidural, subdural, and subarachnoid spaces?
3. How is the spinal cord held within the vertebral canal?

Cross Section of the Spinal Cord

A cross section of the spinal cord reveals that the cord consists of a peripheral white portion and a central gray portion (figure 12.3a and b).

(a) **Anterolateral view**

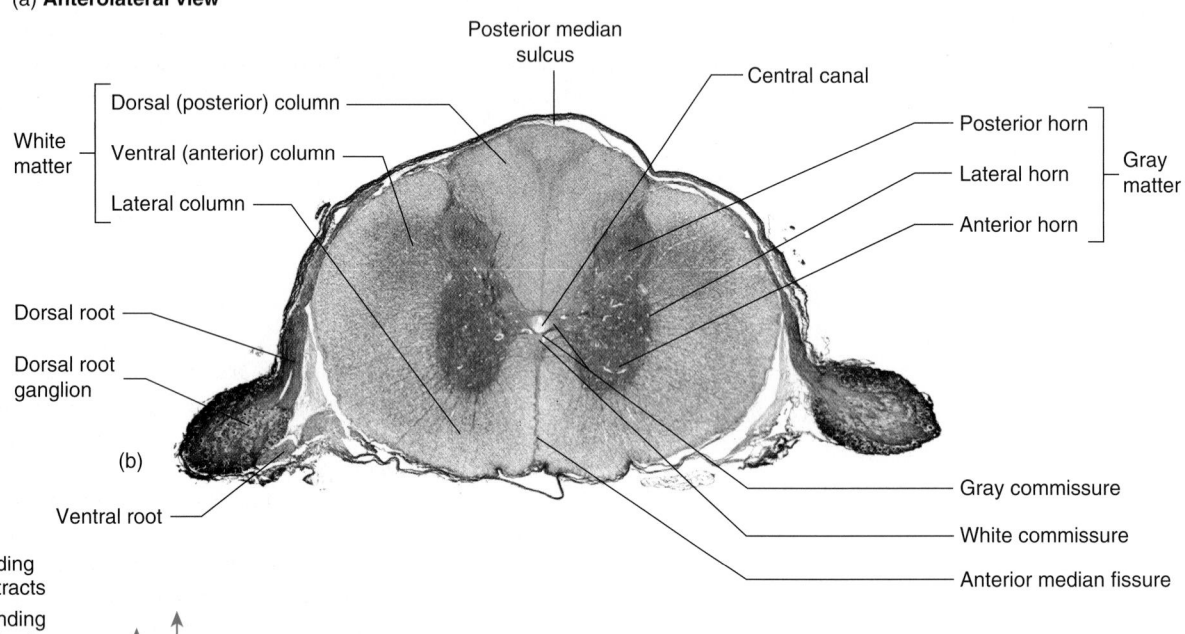

(b)

Ascending
nerve tracts

Descending
nerve tracts

(c) **Anterolateral view**

Figure 12.3   Cross Section of the Spinal Cord

(*a*) A segment of the spinal cord showing one dorsal and one ventral root on each side and the rootlets that form them. (*b*) Photograph of a cross section through the midlumbar region. The *darker-colored areas* are white matter, where tracts are located. The *lighter area* is gray matter, where neuron cell bodies are located. (*c*) Ascending and descending tracts in the spinal cord. Ascending nerve tracts are *blue,* descending nerve tracts are *pink.* The arrows indicate the direction of each pathway.

The white matter consists of myelinated axons forming nerve tracts, and the gray matter consists of neuron cell bodies, dendrites, and axons. An **anterior median fissure** and a **posterior median sulcus** are deep clefts partially separating the two halves of the cord. The white matter in each half of the spinal cord is organized into three **columns,** or **funiculi** (fū-nik′ū-lī), called the **ventral** (anterior), **dorsal** (posterior), and **lateral columns.** Each column of the spinal cord is subdivided into **nerve tracts** or **fasciculi** (fă-sik′ū-lī); also referred to as **pathways.** Individual axons ascending to the brain or descending from the brain are usually grouped together within the nerve tracts (figure 12.3c). Axons within a given nerve tract carry basically the same type of information, although they may overlap to some extent. For example, one ascending nerve tract carries action potentials related to pain and temperature sensations, whereas another functions to carry action potentials related to light touch.

The central gray matter is organized into horns. Each half of the central gray matter of the spinal cord consists of a relatively thin **posterior** (dorsal) **horn** and a larger **anterior** (ventral) **horn.** Small **lateral horns** exist in levels of the cord associated with the autonomic nervous system (see chapter 16). The two halves of the spinal cord are connected by **gray** and **white commissures** (see figure 12.3). The white and gray commissures contain axons that cross from one side of the spinal cord to the other. The **central canal** is in the center of the gray commissure.

Spinal nerves arise from numerous rootlets along the dorsal and ventral surfaces of the spinal cord (see figure 12.3a). About six to eight of these rootlets combine to form each **ventral root** on the ventral (anterior) side of the spinal cord, and another six to eight form each **dorsal root** on the dorsal (posterior) side of the cord at each segment. The ventral and dorsal roots extend laterally from the spinal cord, passing through the subarachnoid space, piercing the arachnoid mater and dura mater, and joining one another to form a spinal nerve. Each dorsal root contains a ganglion, called the **dorsal root,** or **spinal, ganglion** (gang′glē-on; a swelling or knot).

Organization of Neurons in the Spinal Cord and Spinal Nerves

The dorsal root ganglia are collections of cell bodies of the sensory neurons forming the dorsal roots of the spinal nerves (figure 12.4)

(see chapter 11). The axons of these unipolar neurons extend from various parts of the body and pass through spinal nerves to the dorsal root ganglia. The axons do not synapse in the dorsal root ganglion but pass through the dorsal root into the posterior horn of the spinal cord gray matter. The axons either synapse with interneurons in the posterior horn or pass into the white matter and ascend or descend in the spinal cord.

Motor neurons supply muscles and glands. The cell bodies of the motor neurons are located in the anterior and lateral horns of the spinal cord gray matter (see figure 12.4). Multipolar somatic motor neurons are in the anterior horn, also called the motor horn, and autonomic neurons are in the lateral horn. Axons from the motor neurons form the ventral roots and pass into the spinal nerves. Thus, dorsal roots contain sensory axons, ventral roots contain motor axons, and spinal nerves have both sensory and motor axons.

4. *Describe the arrangement of white and gray matter in the spinal cord.*
5. *What are nerve tracts and commissures?*
6. *Where are the cell bodies of sensory, somatic motor, and autonomic neurons located in the gray matter?*
7. *Where do dorsal and ventral roots exit the spinal cord? What kinds of axons are in the dorsal and ventral roots and in the spinal nerves?*

P R E D I C T **3**
Explain why the dorsal root ganglia are larger in diameter than the dorsal roots.

Reflexes

The basic structural unit of the nervous system is the neuron. The **reflex arc** is the basic functional unit of the nervous system and is the smallest, simplest portion capable of receiving a stimulus and producing a response. Because the reflex arc is the basic functional unit of the nervous system, it is possible to learn much about nervous system functions by examining how reflex arcs receive stimuli and produce responses. The reflex arc has five basic components: (1) a sensory receptor, (2) a sensory

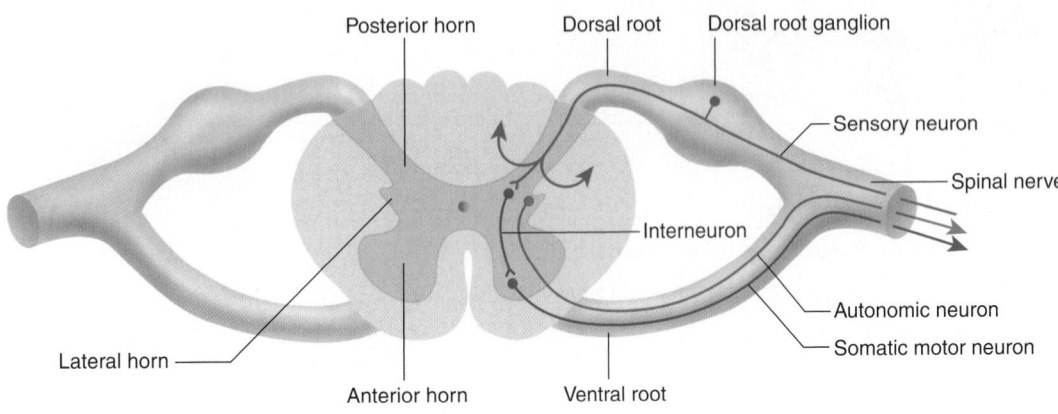

Superior view

Figure 12.4   Relationship of Sensory and Motor Neurons to the Spinal Cord neuron, (3) an interneuron, (4) a motor neuron, and (5) an effector organ (figure 12.5).

A **reflex** is an automatic response to a stimulus produced by a reflex arc. It occurs without conscious thought. Action potentials initiated in sensory receptors are transmitted along the axons of sensory neurons to the CNS, where the axons usually synapse with interneurons. Interneurons synapse with motor neurons, which send axons out of the spinal cord and through the PNS to muscles or glands, where the action potentials of the motor neurons cause these effector organs to respond.

Reflexes are homeostatic. Some reflexes, called somatic reflexes, function to remove the body from painful stimuli that would cause tissue damage or function to keep the body from suddenly falling or moving because of external forces. A number of other reflexes, called autonomic reflexes, are responsible for maintaining relatively constant blood pressure, blood carbon dioxide levels, and water intake.

Individual reflexes vary in their complexity. Some involve simple neuronal pathways and few or even no interneurons so that sensory neurons synapse directly with motor neurons. These are called monosynaptic reflexes. Other reflexes involve complex pathways and integrative centers so that sensory neurons synapse with one or more interneurons. These reflexes involve more than one synapse and are called polysynaptic reflexes. Many are integrated within the spinal cord, and others are integrated within the brain. Some reflexes involve excitatory neurons and result in a response, such as when a muscle contracts (see chapter 11). Other reflexes involve inhibitory neurons and result in inhibition of a response, such as when a muscle relaxes. In addition, higher brain centers influence reflexes by either suppressing or exaggerating them. Major spinal cord reflexes include the stretch reflex, the Golgi tendon reflex, the withdrawal reflex, and the crossed extensor reflex.

8. Name the parts of a reflex arc and explain how they produce a reflex.

Stretch Reflex

The simplest reflex is the **stretch reflex** (figure 12.6), a reflex in which muscles contract in response to a stretching force applied to them. The sensory receptor of this reflex is the **muscle spindle,** which consists of 3–10 small, specialized skeletal muscle cells. The cells are contractile only at their ends and are innervated by specific motor neurons called **gamma motor neurons** (the term *gamma* refers to motor neurons with small diameter axons) originating from the spinal cord and controlling the sensitivity of the muscle spindle cells. Sensory neurons innervate the noncontractile centers of the muscle spindle cells. Axons of these sensory neurons synapse directly with motor neurons in the spinal cord called **alpha motor neurons,** which in turn innervate the muscle in which the muscle spindle is embedded. Neurons can be classified by the diameter of their axons, with alpha motor neurons having the largest diameter. The stretch reflex is a monosynaptic reflex because there is no interneuron between the sensory neuron and the alpha motor neuron.

Stretching a muscle also stretches muscle spindles located among the muscle fibers. The stretch stimulates the sensory neurons that innervate the center of each of the muscle spindles. The increased frequency of action potentials carried to the spinal cord by sensory neurons stimulates the alpha motor neurons in the spinal cord. The alpha motor neurons transmit action potentials to

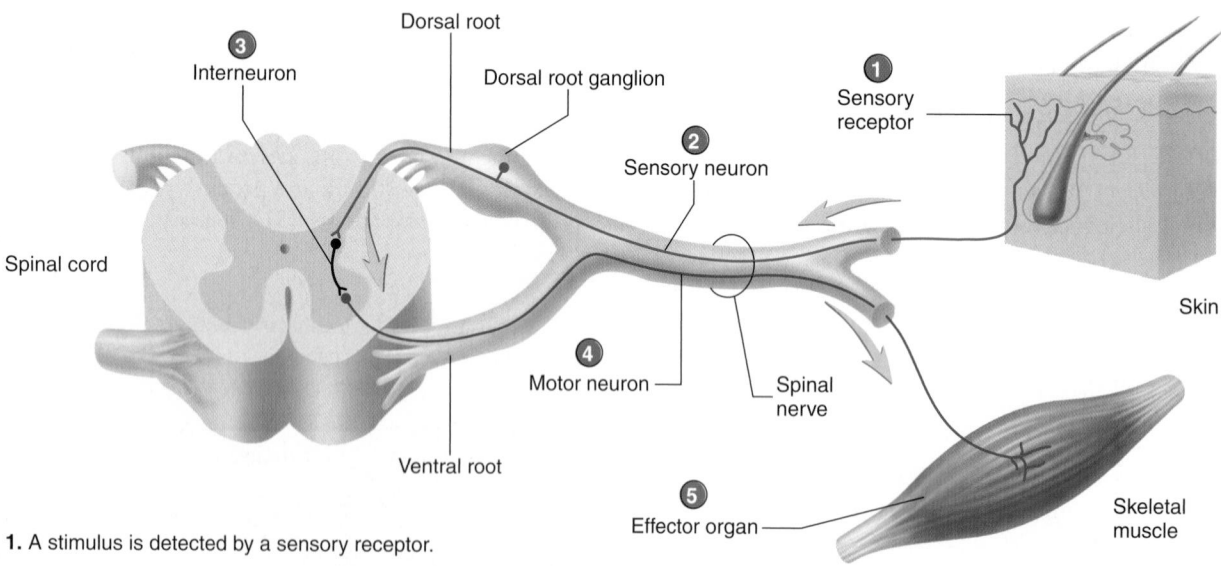

1. A stimulus is detected by a sensory receptor.

2. A sensory neuron conducts an action potential to the spinal cord.

3. In the spinal cord, the sensory neuron synapses with an interneuron.

4. The interneuron synapses with a motor neuron.

5. A motor neuron axon projects to an effector organ.

Process Figure 12.5   Reflex Arc

The parts of a reflex arc are labeled in the order in which action potentials pass through them.

Sudden stretch of a muscle results in:

1. Muscle spindles detect stretch of the muscle.

2. Sensory neurons conduct action potentials to the spinal cord.

3. Sensory neurons synapse directly with alpha motor neurons.

4. Stimulation of the alpha motor neurons causes the muscle to contract and resist being stretched.

Process Figure 12.6 Stretch Reflex

skeletal muscle, causing a rapid contraction of the stretched muscle, which opposes the stretch of the muscle. The postural muscles demonstrate the adaptive nature of this reflex. If a person is standing upright and then begins to tip slightly to one side, the postural muscles associated with the vertebral column on the other side are stretched. As a result, stretch reflexes are initiated in those muscles, which cause them to contract and reestablish normal posture.

Collateral axons from the sensory neurons of the muscle spindles also synapse with neurons whose axons contribute to ascending nerve tracts, which enable the brain to perceive that a muscle has been stretched (see figure 12.3c). Descending neurons within the spinal cord synapse with the neurons of the stretch reflex modifying their activity. This activity is important in maintaining posture and in coordinating muscular activity.

Gamma motor neurons are responsible for regulating the sensitivity of the muscle spindles. As a skeletal muscle contracts, the tension on the centers of muscle spindles within the muscle decreases because the muscle spindles passively shorten as the muscle shortens. The decrease in tension in the centers of the muscle spindles cause them to be less sensitive to stretch. Sensitivity is maintained because at the same time alpha motor neurons are stimulating the muscle to contract, gamma motor neurons stimulate the muscle spindles to contract. The contraction of the muscle fibers at the ends of the muscle spindles pulls on the center part of the muscle spindles and maintains the proper tension. The activity

of the muscle spindles help control and coordinate muscular activity, such as posture, muscle tension, and muscle length.

Knee-Jerk Reflex

The **knee-jerk reflex,** or **patellar reflex,** is a classic example of the stretch reflex. Clinicians use this reflex to determine whether the higher CNS centers that normally influence this reflex are functional. When the patellar ligament is tapped, the tendons and muscles of the quadriceps femoris muscle group are stretched. The muscle spindle fibers within these muscles are also stretched, and the stretch reflex is activated. Consequently, contraction of the muscles extends the leg, thus producing the characteristic knee-jerk response. When the stretch reflex is greatly exaggerated, it indicates that the neurons within the brain that innervate the gamma motor neurons and enhance the stretch reflex are overly active. On the other hand, if the neurons that innervate the gamma motor neurons are depressed, the stretch reflex can be suppressed or absent. Absence of the stretch reflex may indicate that the reflex pathway is not intact.

Golgi Tendon Reflex

The **Golgi tendon reflex** prevents contracting muscles from applying excessive tension to tendons. **Golgi tendon organs** are encapsulated nerve endings that have at their ends numerous terminal branches with small swellings associated with bundles of collagen fibers in tendons. The Golgi tendon organs are located within tendons near the muscle–tendon junction (figure 12.7). As a muscle contracts, the attached tendons are stretched, resulting in increased tension in the tendon. The increased tension stimulates action potentials in the sensory neurons from the Golgi tendon organs. Golgi tendon organs have a high threshold and are sensitive only to intense stretch.

The sensory neurons of the Golgi tendon organs pass through the dorsal root to the spinal cord and enter the posterior gray matter, where they branch and synapse with inhibitory interneurons. The interneurons synapse with alpha motor neurons that innervate the muscle to which the Golgi tendon organ is attached. When a great amount of tension is applied to the tendon, the sensory neurons of the Golgi tendon organs are stimulated. The sensory neurons stimulate the interneurons to release inhibitory neurotransmitters which inhibit the alpha motor neurons of the associated muscle and causes it to relax. This reflex protects muscles and tendons from damage caused by excessive tension. The sudden relaxation of the muscle reduces the tension applied to the muscle and tendons. A weight lifter who suddenly drops a heavy weight after straining to lift it does so, in part, because of the effect of the Golgi tendon reflex.

Tremendous amounts of tension can be applied to muscles and tendons in the legs. Frequently an athlete's Golgi tendon reflex is inadequate to protect muscles and tendons from excessive tension. The large muscles and sudden movements of football players and sprinters can make them vulnerable to relatively frequent hamstring pulls and calcaneal (Achilles) tendon injuries.

Withdrawal Reflex

The function of the **withdrawal,** or **flexor, reflex** is to remove a limb or other body part from a painful stimulus. The sensory receptors are pain receptors (see chapter 15). Action potentials from painful stimuli are conducted by sensory neurons through the dorsal root to the spinal cord, where they synapse with excitatory interneurons, which in turn synapse with alpha motor neurons (figure 12.8). The alpha motor neurons stimulate muscles, usually flexor muscles, that remove the limb from the source of the painful stimulus. Collateral branches of the sensory neurons synapse with ascending fibers to the brain, providing conscious awareness of the painful stimuli.

Reciprocal Innervation

Reciprocal innervation is associated with the withdrawal reflex and reinforces its efficiency (figure 12.9). Collateral axons of sensory neurons that carry action potentials from pain receptors synapse with inhibitory interneurons in the dorsal horn of the spinal cord. The inhibitory interneurons synapse with and inhibit alpha motor neurons of extensor (antagonist) muscles. When the withdrawal reflex is initiated, flexor muscles contract, and reciprocal innervation causes relaxation of the extensor muscles. This reduces the resistance to movement that the extensor muscles would otherwise generate.

Reciprocal innervation is also involved in the stretch reflex. When the stretch reflex causes a muscle to contract, reciprocal innervation causes opposing muscles to relax. In the patellar reflex,

Intense stretch of a skeletal muscle results in:

1. Golgi tendon organs detect tension applied to a tendon.

2. Sensory neurons conduct action potentials to the spinal cord.

3. Sensory neurons synapse with inhibitory interneurons that synapse with alpha motor neurons.

4. Inhibition of the alpha motor neurons causes muscle relaxation, relieving the tension applied to the tendon.

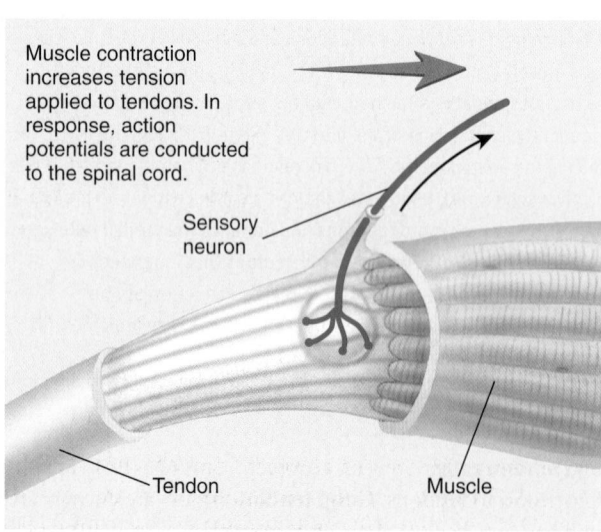

Muscle contraction increases tension applied to tendons. In response, action potentials are conducted to the spinal cord.

Sensory neuron

Tendon

Muscle

Golgi tendon organ

To brain

Sensory neuron

Inhibitory interneuron

Alpha motor neuron

Golgi tendon organ

Golgi tendon reflex

Process Figure 12.7 Golgi Tendon Reflex

Stimulation of pain receptors results in:

1. Pain receptors detect a painful stimulus.

2. Sensory neurons conduct action potentials to the spinal cord.

3. Sensory neurons synapse with excitatory interneurons that synapse with alpha motor neurons.

4. Excitation of the alpha motor neurons results in contraction of the flexor muscles and withdrawal of the limb from the painful stimulus.

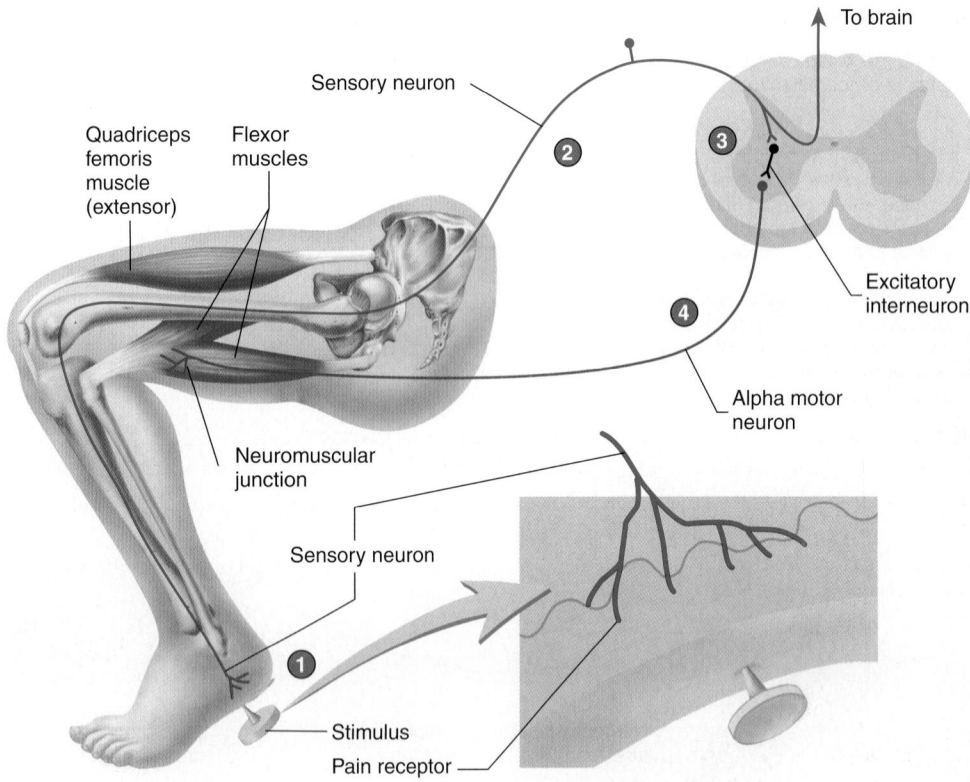

Process Figure 12.8   Withdrawal Reflex

Reciprocal innervation:

1. During the withdrawal reflex, sensory neurons conduct action potentials to the spinal cord.

2. Sensory neurons synapse with excitatory interneurons that are part of the withdrawal reflex.

3. Collateral branches also synapse with inhibitory interneurons that are part of reciprocal innervation.

4. Inhibition of the alpha motor neurons supplying the extensor muscles causes them to relax and not oppose the flexor muscles of the withdrawal reflex.

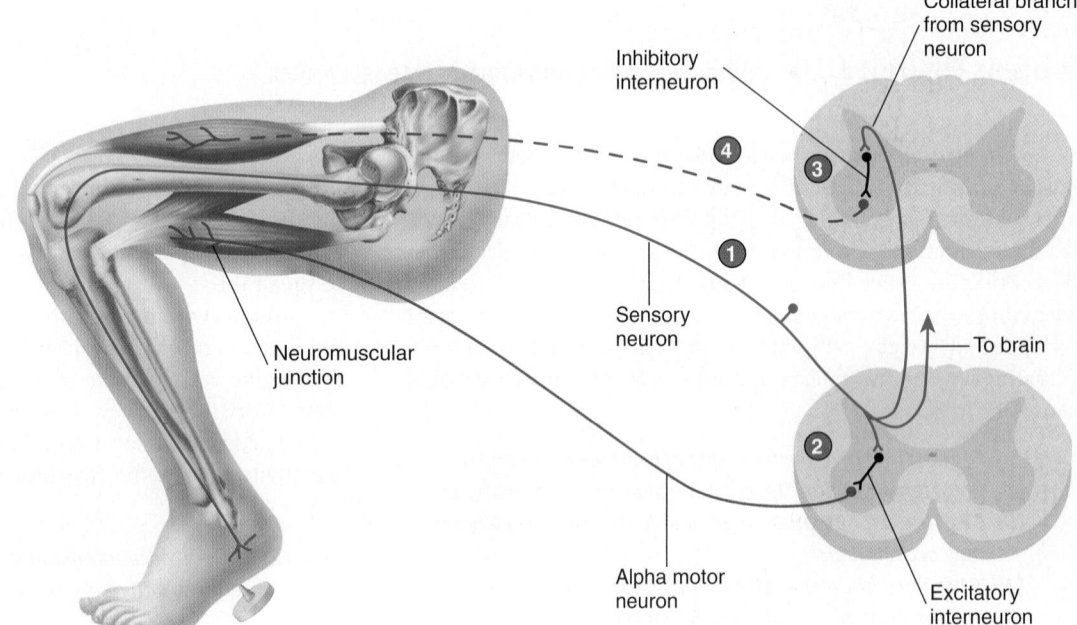

Process Figure 12.9   Withdrawal Reflex with Reciprocal Innervation

for example, the quadriceps femoris muscle contracts and the hamstring muscles relax.

Crossed Extensor Reflex

The **crossed extensor reflex** is another reflex associated with the withdrawal reflex (figure 12.10). Interneurons that stimulate alpha motor neurons, resulting in withdrawal of a limb, have collateral axons that extend through the white commissure to the opposite side of the spinal cord and synapse with alpha motor neurons that innervate extensor muscles in the opposite side of the body. When a withdrawal reflex is initiated in one lower limb, the crossed extensor reflex causes extension of the opposite lower limb.

Crossed extensor reflex:

1. During the withdrawal reflex, sensory neurons conduct action potentials to the spinal cord.

2. Sensory neurons synapse with excitatory interneurons that are part of the withdrawal reflex. Collateral branches also synapse with excitatory interneurons that cross over to the opposite side of the spinal cord as part of the crossed extensor reflex.

3. Stimulation of the alpha motor neurons cause contraction of flexor muscles and stimulation of alpha motor neurons supplying extensor muscles in the opposite limb causes them to contract and support body weight during the withdrawal reflex.

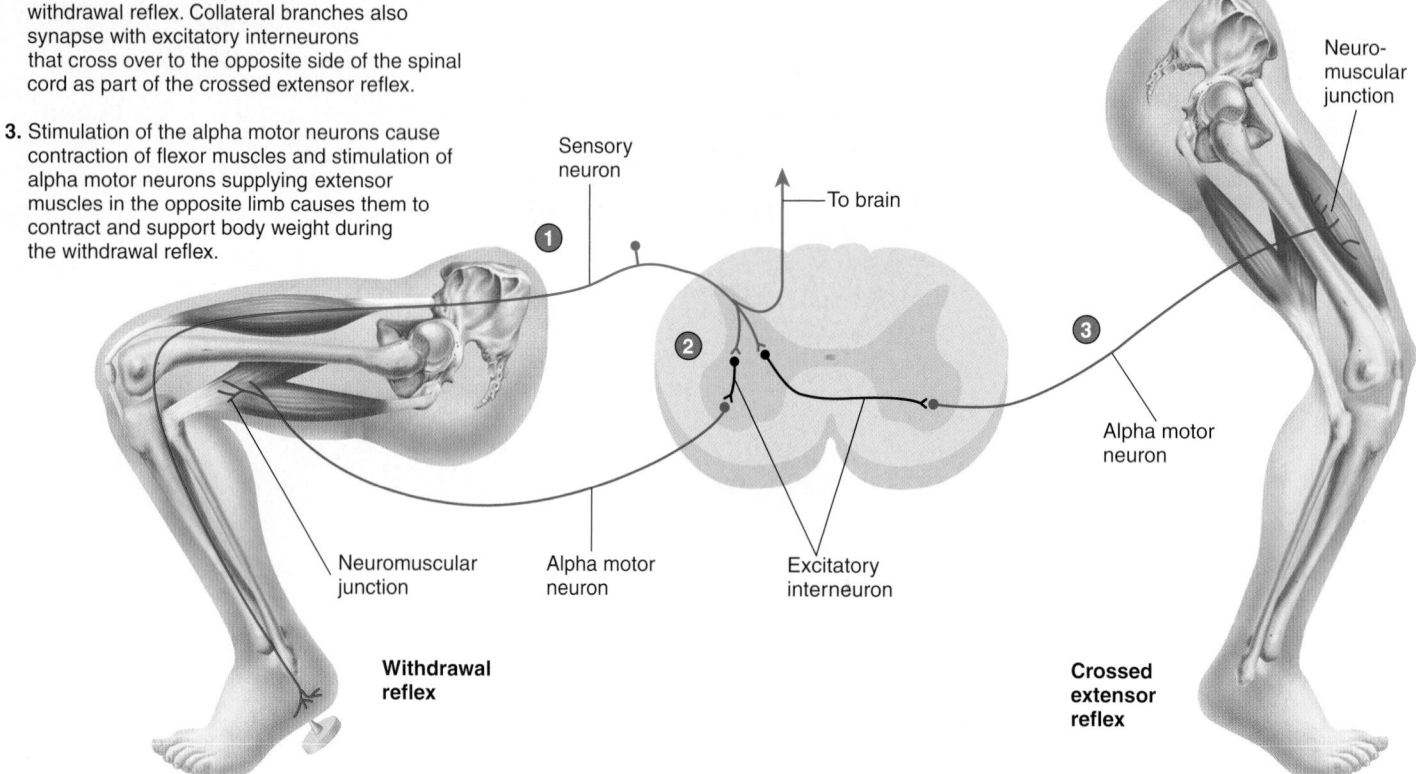

Process Figure 12.10 **Withdrawal Reflex with Crossed Extensor Reflex**

The crossed extensor reflex is adaptive in that it helps prevent falls by shifting the weight of the body from the affected to the unaffected limb. For example, when a person steps on a sharp object, the affected limb is withdrawn from the stimulus (withdrawal reflex) while the other limb is extended (crossed extensor reflex). Therefore, when a person steps on a sharp object with the right foot, the body weight is shifted from the right to the left lower limb. Initiating a withdrawal reflex in both legs at the same time would cause one to fall.

9. *Contrast and give the functions of a stretch reflex and a Golgi tendon reflex. Describe the sensory receptors for each.*
10. *Describe the operation of gamma motor neurons. What do they accomplish?*
11. *What is a withdrawal reflex? How do reciprocal innervation and the crossed extensor reflex assist the withdrawal reflex?*

Interactions with Spinal Cord Reflexes

Reflexes do not operate as isolated entities within the nervous system because of divergent and convergent pathways (see chapter 11). Diverging branches of the sensory neurons or interneurons in a reflex arc send action potentials along ascending nerve tracts to the brain (figure 12.11). A pain stimulus, for example, not only initiates a withdrawal reflex, which removes the affected part of the body from the painful stimulus, but also causes perception of the pain sensation as a result of action potentials sent to the brain.

Axons within descending tracts from the brain carry action potentials to motor neurons in the anterior horn of the spinal cord, converging with neurons of reflex arcs. The neurotransmitters released from the axons of these tracts either stimulate or inhibit the motor neurons in the anterior horn. Neurotransmitters change the sensitivity of the reflex by stimulating (EPSP) or inhibiting (IPSP) the motor neurons.

12. *How do ascending and descending pathways relate to reflexes and other neuron functions?*

Structure of Peripheral Nerves

Peripheral nerves consist of axons, Schwann cells, and connective tissue (figure 12.12). Each axon, or nerve fiber, and its Schwann cell sheath are surrounded by a delicate connective tissue layer, the **endoneurium** (en-dō-noo′rē-ŭm). A heavier connective tissue layer, the **perineurium** (per-i-noo′rē-ŭm), surrounds groups of axons to form nerve **fascicles** (fas′i-klz). A third layer of dense connective

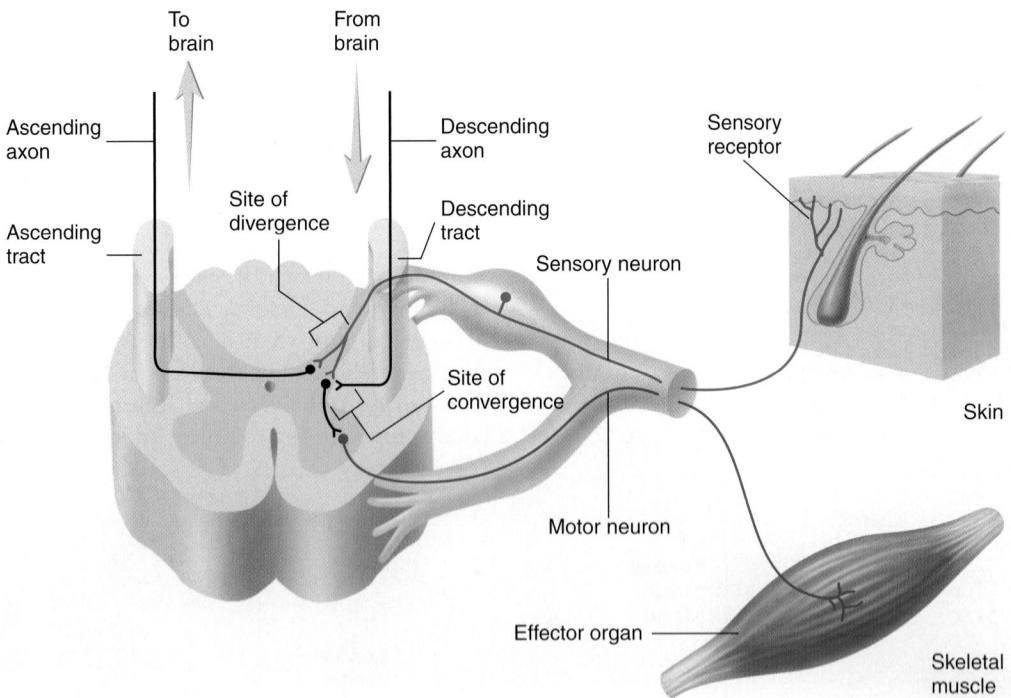

Figure 12.11 **Spinal Reflex, with Ascending and Descending Axons**

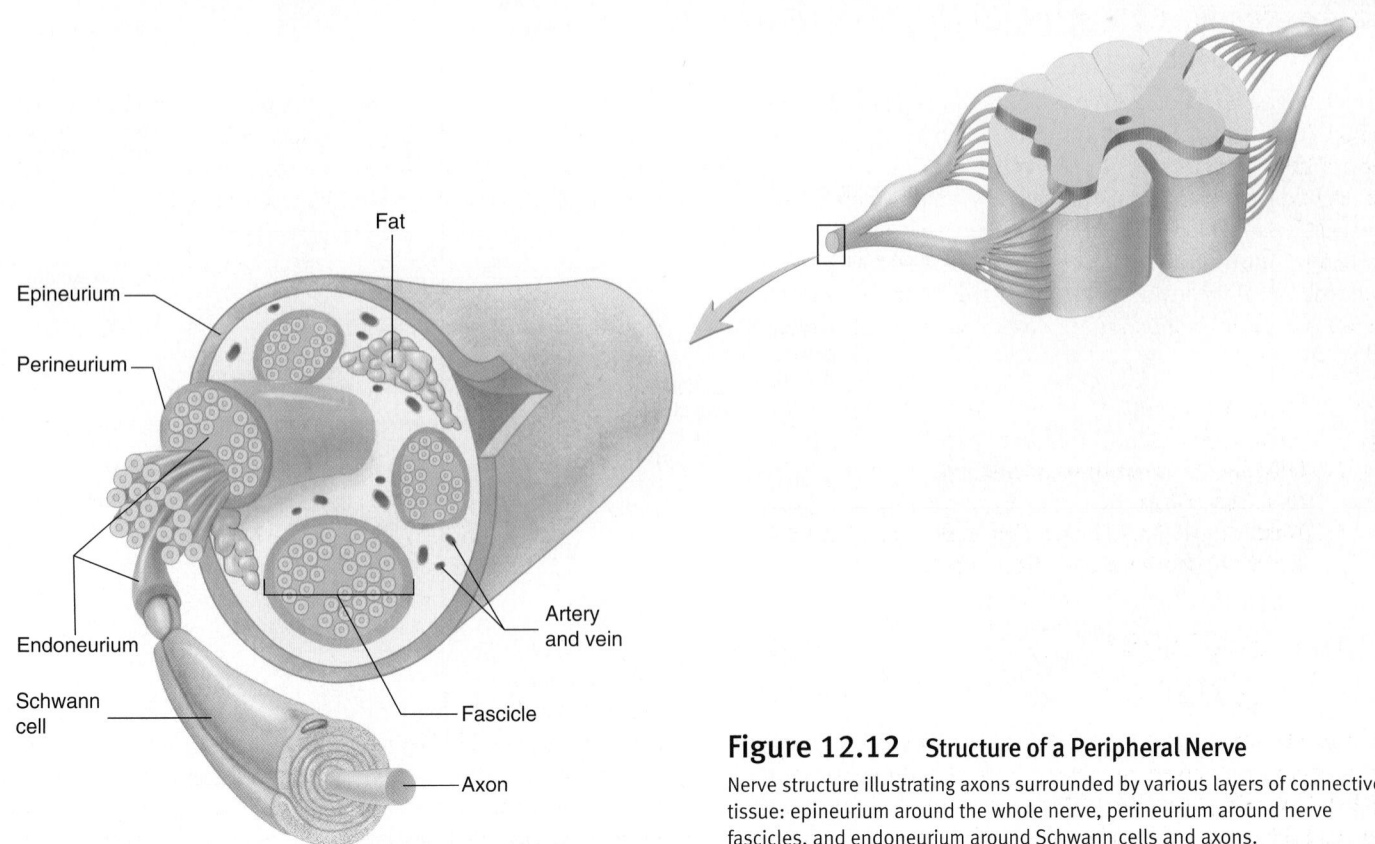

Figure 12.12 **Structure of a Peripheral Nerve**

Nerve structure illustrating axons surrounded by various layers of connective tissue: epineurium around the whole nerve, perineurium around nerve fascicles, and endoneurium around Schwann cells and axons.

tissue, the **epineurium** (ep-i-noo′rē-ŭm), binds the nerve fascicles together to form a nerve. The connective tissue of the epineurium is continuous with the dura mater surrounding the CNS. An analogy of this relationship is a coat (the dura) and its sleeve (epineurium). The connective tissue layers of nerves make them tougher than the nerve tracts in the CNS.

● **13.** *Describe the structure of peripheral nerves.*

Spinal Nerves

All of the 31 pairs of spinal nerves, except the first pair and those in the sacrum, exit the vertebral column through intervertebral foramina located between adjacent vertebrae (see figure 7.13).

The first pair of spinal nerves exits between the skull and the first cervical vertebra. The nerves of the sacrum exit from the single bone of the sacrum through the sacral foramina (see chapter 7). Eight spinal nerve pairs exit the vertebral column in the cervical region, 12 in the thoracic region, 5 in the lumbar region, 5 in the sacral region, and 1 in the coccygeal region (figure 12.13). For convenience, each of the spinal nerves is designated by a letter and number. The letter indicates the region of the vertebral column from which the nerve emerges: C, cervical; T, thoracic; L, lumbar; and S, sacral. The single coccygeal nerve is often not designated, but when it is, the symbol often used is Co. The number indicates the location in each region where the nerve emerges from the vertebral column, with the smallest number always representing the most superior origin. For example, the

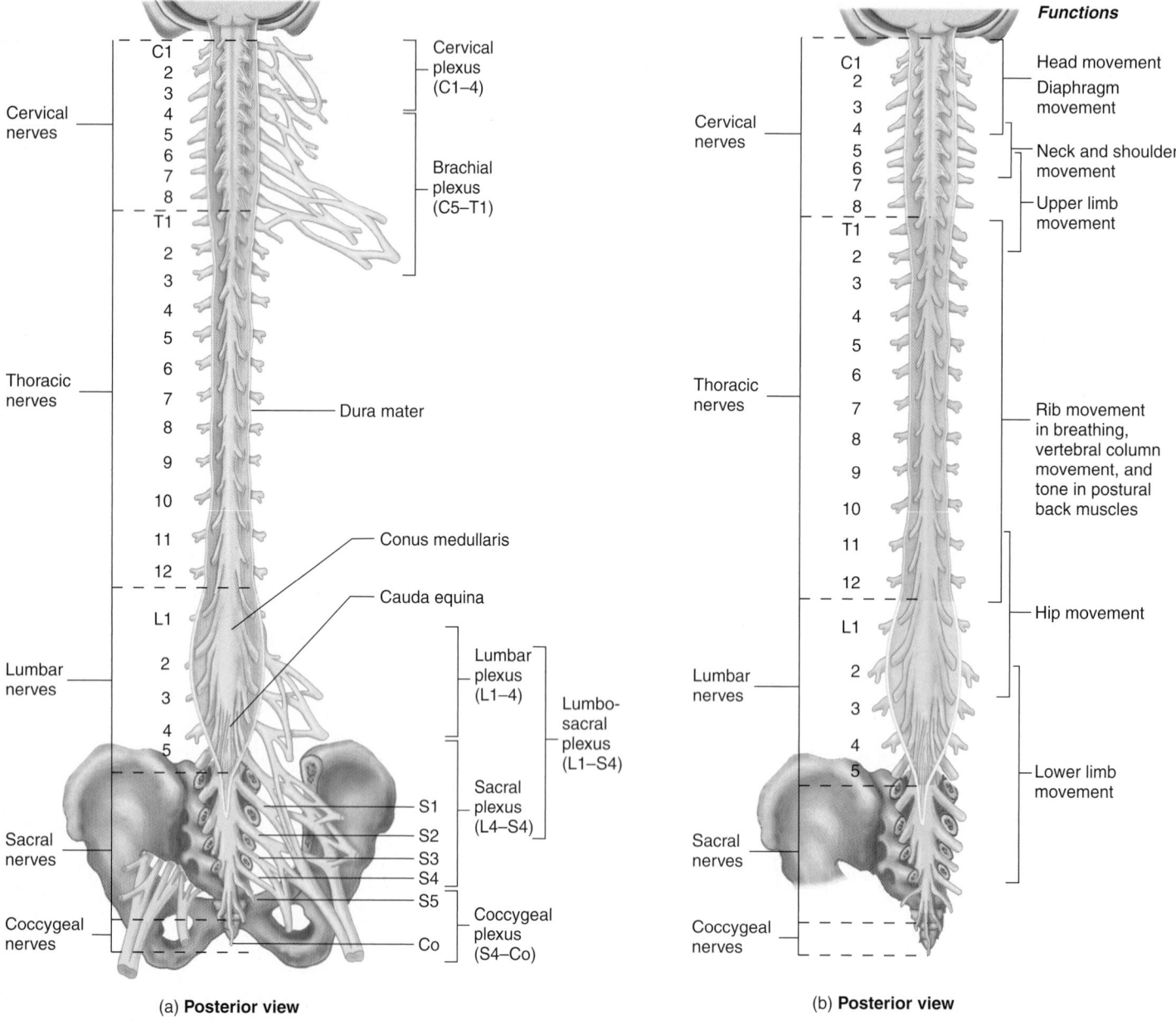

(a) **Posterior view**   (b) **Posterior view**

Figure 12.13   Spinal Nerves

(*a*) Spinal cord, the spinal nerves, their plexuses, and their branches. (*b*) Regions of the spinal cord and their general functions.

most superior nerve exiting from the thoracic region of the vertebral column is designated T1. The cervical nerves are designated C1–C8, the thoracic nerves T1–T12, the lumbar nerves L1–L5, and the sacral nerves S1–S5.

The nerves arising from each region of the spinal cord and vertebral column supply specific regions of the body. Each of the spinal nerves except C1 has a specific cutaneous sensory distribution. Figure 12.14 illustrates the **dermatomal** (der-mă-tō′măl) **map** for the sensory cutaneous distribution of the spinal nerves. A **dermatome** is the area of skin supplied with sensory innervation by a pair of spinal nerves.

P R E D I C T

The dermatomal map is important in clinical considerations of nerve damage. Loss of sensation in a dermatomal pattern can provide valuable information about the location of nerve damage. Predict the possible site of nerve damage for a patient who suffered whiplash in an automobile accident and subsequently developed anesthesia (no sensations) in the left arm, forearm, and hand (see figure 12.14 for help).

Figure 12.15 depicts an idealized section through the trunk. Each spinal nerve has a dorsal and a ventral **ramus** (rā′mŭs; branch). Additional rami (rā′mī), called communicating rami, from the thoracic and upper lumbar spinal cord regions carry axons associated with the sympathetic division of the autonomic nervous system (see chapter 16). The **dorsal rami** (rā′mī) innervate most of the deep muscles of the dorsal trunk responsible for movement of the vertebral column. They also innervate the connective tissue and skin near the midline of the back.

The **ventral rami** are distributed in two ways. In the thoracic region, the ventral rami form **intercostal** (between ribs) **nerves** (see figure 12.15), which extend along the inferior margin of each rib and innervate the intercostal muscles and the skin over the thorax. The ventral rami of the remaining spinal nerves form five major **plexuses** (plek′sŭs-ēz). The term *plexus* means braid and describes the organization produced by the intermingling of the nerves. The ventral rami of different spinal nerves, called the **roots** of the plexus, join with each other to form a plexus. These roots should not be confused with the dorsal and ventral roots from the

Figure 12.14 Dermatomal Map

Letters and *numbers* indicate the spinal nerves innervating a given region of skin.

Dorsal root of
spinal nerve

Ventral root of
spinal nerve

Rootlets

Dorsal root
(spinal) ganglion

Communicating rami

Spinal nerve

Ganglion of
sympathetic chain

Dorsal ramus of
spinal nerve

Ventral ramus of
spinal nerve
(intercostal nerve)

Roots of
splanchnic nerve

(a) **Anterolateral view**

Dorsal rootlets

Intervertebral
foramen

Dorsal
root
ganglion

Spinal
nerve

Transverse
process of
vertebra
(cut)

(b) **Posterior view**

Figure 12.15 Spinal Nerves

(*a*) Typical thoracic spinal nerves. (*b*) Photograph of four dorsal roots
in place along the vertebral column.

Clinical Focus Spinal Cord Injury

Damage to the spinal cord can disrupt ascending tracts from the spinal cord to the brain, resulting in the loss of sensation, and/or descending tracts from the brain to motor neurons in the spinal cord, resulting in the loss of motor functions. About 10,000 new cases of **spinal cord injury** occur each year in the United States. Automobile and motorcycle accidents are leading causes, followed by gunshot wounds, falls, and swimming accidents. Spinal cord injury is classified according to the vertebral level at which the injury occurred, whether the entire cord is damaged at that level or only a portion of the cord, and the mechanism of injury. Most spinal cord injuries occur in the cervical region or at the thoracolumbar junction and are incomplete. The primary mechanisms include concussion (an injury caused by a blow), contusion (an injury resulting in hemorrhage), or laceration (a tear or cut) and involve excessive flexion, extension, rotation, or compression of the vertebral column. The majority of spinal cord injuries are acute contusions of the cord due to bone or disk displacement into the cord and involve a combination of excessive directional movements, such as simultaneous flexion and compression.

At the time of spinal cord injury, two types of tissue damage occur: (1) primary, mechanical damage and (2) secondary, tissue damage. Secondary spinal cord damage, which begins within minutes of the primary damage, is caused by ischemia, edema, ion imbalances, the release of "excitotoxins" such as glutamate, and inflammatory cell invasion. Secondary damage extends into a much larger region of the cord than the primary damage. It is the primary focus of current research in spinal cord injury. The only treatment for primary damage is prevention, such as wearing seat belts when riding in automobiles and not diving in shallow water. Once an accident occurs, however, little can be done at present about the primary damage. On the other hand, it's now known that much of the secondary damage can be prevented or reversed.

Until the 1950s, spinal cord injuries were often ultimately fatal. Now, with quick treatment, directed at the mechanisms of secondary tissue damage, much of the total damage to the spinal cord can be prevented. Treatment of the damaged spinal cord with large doses of methylprednisolone, a synthetic steroid, within 8 hours of the injury, can dramatically reduce the secondary damage to the cord. The objectives of these treatments are to reduce inflammation and edema. Current treatment includes anatomic realignment and stabilization of the vertebral column, decompression of the spinal cord, and administration of methylprednisolone. Rehabilitation is based on retraining the patient to use whatever residual connections exist across the site of damage.

It had long been thought that the spinal cord is incapable of regeneration following severe damage. It's now known that following injury, most neurons of the adult spinal cord survive and begin to regenerate, growing about 1 mm into the site of damage, but then they regress to an inactive, atrophic state. In addition, fetuses and newborns exhibit considerable regenerative ability and functional improvement. The major block to adult spinal cord regeneration is the formation of a scar, consisting mainly of myelin and astrocytes, at the site of injury. Myelin in the scar is apparently the primary inhibitor of regeneration. Implantation of peripheral nerves, Schwann cells, or fetal CNS tissue can bridge the scar and stimulate some regeneration. Certain growth factors can also stimulate some regeneration. Current research continues to look for the right combination of chemicals and other factors to stimulate regeneration of the spinal cord following injury.

spinal cord, which are more medial. Nerves that arise from plexuses usually have axons from more than one spinal nerve and thus more than one level of the spinal cord. The ventral rami of spinal nerves C1–C4 form the cervical plexus, C5–T1 form the brachial plexus, L1–L4 form the lumbar plexus, L4–S4 form the sacral plexus, and S4, S5, and the coccygeal nerve (Co) form the coccygeal plexus.

Several smaller somatic plexuses, such as the pudendal plexus in the pelvis, are derived from more distal branches of the spinal nerves. Some of the somatic plexuses are mentioned where appropriate in this chapter. Autonomic plexuses (described in chapter 16) also exist in the thorax and abdomen.

The five major plexuses derived from the ventral rami of spinal nerves are described more fully in the following sections.

14. *Describe the connective tissue layers within and surrounding spinal nerves.*

15. *Differentiate between rootlet, dorsal root, ventral root, and spinal nerve. Indicate whether each contains motor fibers, sensory fibers, or both.*

16. *List all of the spinal nerves by name and number. Where do they exit the vertebral column?*

17. *What is a dermatome? Why are dermatomes clinically important?*

18. *Contrast dorsal and ventral rami of spinal nerves. What muscles do the dorsal rami innervate?*

19. *Describe the distribution of the ventral rami of the thoracic region.*

20. *What is a plexus? What happens to the axons of spinal nerves as they pass through a plexus?*

21. *Name the main spinal plexuses and the spinal nerves associated with each one.*

Cervical Plexus

The **cervical plexus** is a relatively small plexus originating from spinal nerves C1–C4 (figure 12.16). Branches derived from this plexus innervate superficial neck structures, including several of the muscles attached to the hyoid bone. The cervical plexus innervates the skin of the neck and posterior portion of the head (see figure 12.14). An unusual part of the cervical plexus, the **ansa** (an′sah, bucket handle) **cervicalis,** is a loop between C2 and C3. Nerves to the infrahyoid muscles branch from the ansa cervicalis (see chapter 10).

Hypoglossal nerve (XII)

Accessory nerve (XI)

Lesser occipital nerve

Greater auricular nerve

Nerve to sternocleidomastoid muscle

Superior root of ansa cervicalis

Branch to infrahyoid muscles

Transverse cervical nerve

Ansa cervicalis

Nerve to trapezius muscle

Branches to infrahyoid muscles

Inferior root of ansa cervicalis

Supraclavicular nerves

Phrenic nerve

C1

C2

C3

C4

C5

Anterior view

Ventral rami
Branches
Other nerves (not part of cervical plexus)

C1

C4

Figure 12.16   Cervical Plexus, Anterior View

The roots of the plexus are formed by the ventral rami of the spinal nerves C1–C4.

One of the most important derivatives of the cervical plexus is the **phrenic** (fren′ik) **nerve,** which originates from spinal nerves C3–C5, derived from both the cervical and brachial plexus. The phrenic nerves descend along each side of the neck to enter the thorax. They descend along the sides of the mediastinum to reach the diaphragm, which they innervate. Contraction of the diaphragm is largely responsible for the ability to breathe.

Phrenic Nerve Damage

Damage to the phrenic nerve severely limits a person's ability to breathe. Care must be taken not to damage the phrenic nerve during open thoracic surgery or open heart surgery. Cancer of the bronchus is the most common type of cancer in men, accounting for about 30% of all male cancers, and most often occurs in men who smoke cigarettes. Tumors at the base of the lung can compress the phrenic nerve.

P R E D I C T   5
Explain how damage to or compression of the right phrenic nerve affects the diaphragm. Describe the effect on breathing of a completely severed spinal cord at the level of C2 versus C6.

22. *Name the structures innervated by the cervical plexus. Describe the innervation of the phrenic nerve.*

Brachial Plexus

The **brachial plexus** originates from spinal nerves C5–T1 (figure 12.17). The five ventral rami that constitute the brachial plexus join to form three **trunks,** which separate into six **divisions** and then join again to create three **cords** (posterior, lateral, and medial) from which five **branches,** or nerves of the upper limb, emerge.

The five major nerves emerging from the brachial plexus to supply the upper limb are the axillary, radial, musculocutaneous, ulnar, and median nerves. The axillary nerve innervates part of the shoulder; the radial nerve innervates the posterior arm, forearm, and hand; the musculocutaneous nerve innervates the anterior arm; and the ulnar and median nerves innervate the anterior forearm and hand. Smaller nerves from the brachial plexus innervate the shoulder and pectoral muscles.

Brachial Anesthesia

The entire upper limb can be anesthetized by injecting an anesthetic near the brachial plexus. This is called **brachial anesthesia.** The anesthetic can be injected between the neck and the shoulder posterior to the clavicle.

Roots: C5, C6, C7, C8, T1

Trunks: upper, middle, lower

Anterior divisions

Posterior divisions

Cords: posterior, lateral, medial

Major branches:
Axillary nerve
Radial nerve
Musculocutaneous nerve
Median nerve
Ulnar nerve

C5

T1

C5

Upper trunk

C6

Middle trunk

C7

Long thoracic nerve

C8

Lower trunk

T1

Dorsal scapular nerve

Subclavian nerve

Thoracodorsal nerve

Upper and lower subscapular nerves

Suprascapular nerve

Posterior cord

Axillary nerve

Lateral cord

Radial nerve

Musculo-cutaneous nerve

Medial and lateral pectoral nerves

Median nerve

Ulnar nerve

Medial antebrachial cutaneous nerve

Medial brachial cutaneous nerve

Medial cord

Anterior view

Figure 12.17 Brachial Plexus, Anterior View

The roots of the plexus are formed by the ventral rami of the spinal nerves C5–T1 and join to form an upper, middle, and lower trunk. Each trunk divides into anterior and posterior divisions. The divisions join together to form the posterior, lateral, and medial cords from which the major brachial plexus nerves arise.

Axillary Nerve

The **axillary** (ak′sil-ār-ē) **nerve** innervates the deltoid and teres minor muscles (figure 12.18). It also provides sensory innervation to the shoulder joint and to the skin over part of the shoulder.

Axillary Nerve

Origin
Posterior cord of brachial plexus, C5–C6

Movements/Muscles Innervated
Laterally rotates arm
  • *Teres minor*
Abducts arm
  • *Deltoid*

Cutaneous (sensory) Innervation
Inferior lateral shoulder

Posterior views

Figure 12.18 Axillary Nerve

Route of the axillary nerve and the muscles it innervates (posterior view). The inset depicts the cutaneous (sensory) distribution of the nerve (*blue area*).

Radial Nerve

The **radial nerve** emerges from the posterior cord of the brachial plexus and descends within the deep aspect of the posterior arm (figure 12.19). About midway down the shaft of the humerus, it lies against the bone in the radial groove. The radial nerve innervates all of the extensor muscles of the upper limb, the supinator muscle, and the brachioradialis. Its cutaneous sensory distribution is to the posterior portion of the upper limb, including the posterior surface of the hand.

Radial Nerve Damage

Because the radial nerve lies near the humerus in the axilla, it can be damaged if it's compressed against the humerus. Improper use of crutches (i.e., when the crutch is pushed tightly into the axilla) can result in **"crutch paralysis."** In this disorder, the radial nerve is compressed between the top of the crutch and the humerus. As a result, the radial nerve is damaged, and the muscles it innervates lose their function. The major symptom of crutch paralysis is **"wrist drop"** in which the extensor muscles of the wrist and fingers, which are innervated by the radial nerve, fail to function; as a result, the elbow, wrist, and fingers are constantly flexed.

Radial Nerve

Origin

Posterior cord of brachial plexus, C5–T1

Movements/Muscles Innervated

Extends elbow
- *Triceps brachii*
- *Anconeus*

Flexes elbow
- *Brachialis (part; not shown; sensory only)*
- *Brachioradialis*

Extends and abducts wrist
- *Extensor carpi radialis longus*
- *Extensor carpi radialis brevis*

Supinates forearm and hand
- *Supinator*

Extends fingers
- *Extensor digitorum*
- *Extensor digiti minimi*
- *Extensor indicis*

Extends and adducts wrist
- *Extensor carpi ulnaris*

Abducts thumb
- *Abductor pollicis longus*

Extends thumb
- *Extensor pollicis longus*
- *Extensor pollicis brevis*

Cutaneous (sensory) Innervation

Posterior surface of arm and forearm, lateral two-thirds of dorsum of hand

Figure 12.19 Radial Nerve

Route of the radial nerve and the muscles it innervates (posterior view). The insets depict the cutaneous (sensory) distribution of the nerve (*blue area*).

P R E D I C T **6**

Wrist drop can also result from a compound fracture of the humerus.
Explain how and where damage to the nerve may occur.

Musculocutaneous Nerve

The **musculocutaneous** (mŭs′kū-lō-kū-tā′nē-ŭs) **nerve** provides motor innervation to the anterior muscles of the arm as
well as cutaneous sensory innervation to part of the forearm
(figure 12.20).

Musculocutaneous Nerve

Origin
Lateral cord of brachial plexus, C5–C7

Movements/Muscles Innervated
Flexes shoulder
- *Biceps brachii*
- *Coracobrachialis*

Flexes elbow and supinates forearm
  and hand
- *Biceps brachii*

Flexes elbow
- *Brachialis (also small amount of
  innervation from radial nerve)*

Cutaneous (sensory) Innervation
Lateral surface of forearm

Anterior views

Posterior view

Figure 12.20   Musculocutaneous Nerve
Route of the musculocutaneous nerve and the muscles it innervates (anterior view). The insets depict the cutaneous (sensory) distribution of the nerve (*blue area*).

Ulnar Nerve

The **ulnar nerve** innervates two forearm muscles plus most of the intrinsic hand muscles, except some associated with the thumb. Its sensory distribution is to the ulnar side of the hand (figure 12.21).

Ulnar Nerve Damage

The ulnar nerve is the most easily damaged of all the peripheral nerves, but such damage is almost always temporary. Slight damage to the ulnar nerve may occur where it passes posterior to the medial epicondyle of the humerus. The nerve can be felt just below the skin at this point, and, if this region of the elbow is banged against a hard object, temporary ulnar nerve damage may occur. This damage results in painful tingling sensations radiating down the ulnar side of the forearm and hand. Because of this sensation, this area of the elbow is often called the **"funny bone"** or **"crazy bone."**

Ulnar Nerve

Origin
Medial cord of brachial plexus, C8–T1

Movements/Muscles Innervated

Flexes and adducts wrist
- *Flexor carpi ulnaris*

Flexes fingers
- *Part of the flexor digitorum profundus controlling the distal phalanges of little and ring fingers*

Adducts thumb
- *Adductor pollicis*

Controls hypothenar muscles
- *Flexor digiti minimi brevis*
- *Abductor digiti minimi*
- *Opponens digiti minimi*

Flexes metacarpophalangeal joints and extends interphalangeal joints
- *Two medial (ulnar) lumbricales*

Abducts and adducts fingers
- *Interossei*

Cutaneous (sensory) Innervation
Medial third of hand, little finger, and medial half of ring finger

Figure 12.21 **Ulnar Nerve**
Route of the ulnar nerve and the muscles it innervates (anterior view). The insets depict the cutaneous (sensory) distribution of the nerve (*blue area*).

Median Nerve

The **median nerve** innervates all but one of the flexor muscles of the forearm and most of the hand muscles at the base of the thumb, called the thenar area of the hand. Its cutaneous sensory distribution is to the radial portion of the palm of the hand (figure 12.22).

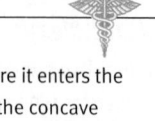

Median Nerve Damage

Damage to the median nerve occurs most commonly where it enters the wrist through the **carpal tunnel.** This tunnel is created by the concave organization of the carpal bones and the flexor retinaculum on the anterior surface of the wrist. None of the connective tissue components of the carpal tunnel expands readily. Inflammation in the wrist or an increase in the size of the tendons in the carpal tunnel can produce pressure within it, thereby compressing the median nerve and resulting in numbness, tingling, and pain in the fingers. This condition is referred to as **carpal tunnel syndrome.** Carpal tunnel syndrome is common among people who perform repetitive movements of the wrists and fingers, such as keyboard operators. Surgery is often required to relieve the pressure.

People attempting suicide by cutting the wrists commonly cut the median nerve proximal to the carpal tunnel.

Median Nerve

Origin
Medial and lateral cords of brachial plexus, C5–T1

Movements/Muscles Innervated
Pronates forearm and hand
- *Pronator teres*
- *Pronator quadratus*

Flexes and abducts wrist
- *Flexor carpi radialis*

Flexes wrist
- *Palmaris longus*

Flexes fingers
- *Part of flexor digitorum profundus controlling the distal phalanx of the middle and index fingers*
- *Flexor digitorum superficialis*

Controls thumb muscle
- *Flexor pollicis longus*

Controls thenar muscles
- *Abductor pollicis brevis*
- *Opponens pollicis*
- *Flexor pollicis brevis*

Flexes metacarpophalangeal joints and extends interphalangeal joints
- *Two lateral (radial) lumbricales*

Cutaneous (sensory) Innervation
Lateral two-thirds of palm of hand, thumb, index and middle fingers, and the lateral half of ring finger and dorsal tips of the same fingers

Figure 12.22   Median Nerve
Route of the median nerve and the muscles it innervates (anterior view). The insets depict the cutaneous (sensory) distribution of the nerve (*blue area*).

Other Nerves of the Brachial Plexus

Several nerves, other than the five just described, arise from the brachial plexus (see figure 12.17). They supply most of the muscles acting on the scapula and arm and include the pectoral, long thoracic, thoracodorsal, subscapular, and suprascapular nerves. In addition, brachial plexus nerves supply the cutaneous innervation of the medial arm and forearm.

● 23. *Name the five major nerves that emerge from the brachial plexus. List the muscles they innervate and the areas of the skin they supply. In addition to these five nerves, name the*

muscles and skin areas supplied by the remaining brachial plexus nerves.

Lumbar and Sacral Plexuses

The **lumbar plexus** originates from the ventral rami of spinal nerves L1–L4 and the **sacral plexus** from L4–S4. Because of their close, overlapping relationship and their similar distribution, however, the two plexuses often are considered together as a single **lumbosacral plexus** (L1–S4; figure 12.23). Four major nerves exit the lumbosacral plexus and enter the

Anterior view

Figure 12.23 Lumbosacral Plexus, Anterior View

The roots of the plexus are formed by the ventral rami of the spinal nerves L1–S4 and form anterior and posterior divisions, which give rise to the lumbrosacral nerves. The lumbosacral trunk joins the lumbar and sacral plexuses.

lower limb: the obturator, femoral, tibial, and common fibular (peroneal). The obturator nerve innervates the medial thigh; the femoral nerve innervates the anterior thigh; the tibial nerve innervates the posterior thigh, the leg, and foot; and the common fibular nerve innervates the posterior thigh, the anterior and lateral leg, and the foot. Other lumbosacral nerves supply the lower back, the hip, and the lower abdomen.

Obturator Nerve

The **obturator** (ob′too-ră-tōr) **nerve** supplies the muscles that adduct the thigh. Its cutaneous sensory distribution is to the medial side of the thigh (figure 12.24).

Obturator Nerve

Origin
Lumbosacral plexus, L2–L4

Movements/Muscles Innervated
Rotates thigh laterally
- *Obturator externus*

Adducts thigh
- *Adductor magnus (adductor part)*
- *Adductor longus*
- *Adductor brevis*

Adducts thigh and flexes knee
- *Gracilis*

Cutaneous (sensory) Innervation
Superior medial side of thigh

Medial view

Anterior view

Figure 12.24 Obturator Nerve
Route of the obturator nerve and the muscles it innervates (anterior view). The inset depicts the cutaneous (sensory) distribution of the nerve (*blue area*).

Femoral Nerve

The **femoral nerve** innervates the iliopsoas and sartorius muscles and the quadriceps femoris group. Its cutaneous sensory distribution is the anterior and lateral thigh and the medial leg and foot (figure 12.25).

Tibial and Common Fibular Nerves

The **tibial** and **common fibular (peroneal)** (per-ō-nē′ăl) **nerves** originate from spinal segments L4–S3 and are bound together within a connective tissue sheath for the length of the thigh

Femoral Nerve

Origin
Lumbosacral plexus, L2–L4

Movements/Muscles Innervated

Flexes hip
- *Psoas major*
- *Iliacus*
- *Pectineus*

Flexes hip and flexes knee
- *Sartorius*

Extends knee
- *Vastus lateralis*
- *Vastus intermedius*
- *Vastus medialis*

Extends knee and flexes hip
- *Rectus femoris*

Cutaneous (sensory) Innervation

Anterior and lateral branches supply the anterior and lateral thigh; saphenous branch supplies the medial leg and foot

Psoas major

Iliacus

Sartorius

Vastus lateralis

Rectus femoris

Vastus intermedius

L2

L3

L4

Femoral nerve

Pectineus

Vastus medialis

Anterior view   **Medial view**

Anterior view

Figure 12.25   Femoral Nerve

Route of the femoral nerve and the muscles it innervates (anterior view). The insets depict the cutaneous (sensory) distribution of the nerve (*blue area*).

(figures 12.26 and 12.27; see figure 12.23). These two nerves, combined within the same sheath, are referred to jointly as the **sciatic** (sī-at′ik, hip), or **ischiadic** (is-kē-ad′ik) **nerve** (see figure 12.23). The sciatic nerve, by far the largest peripheral nerve in the body, passes through the greater sciatic notch in the pelvis and descends in the posterior thigh to the popliteal fossa, where the two portions of the sciatic nerve separate.

The tibial nerve innervates most of the posterior thigh and leg muscle (see figure 12.26). It branches in the foot to form the **medial** and **lateral plantar** (plan′tăr) **nerves,** which innervate the plantar muscles of the foot and the skin over the sole of the foot. Another branch, the **sural** (soo′răl) **nerve,** supplies part of the cutaneous innervation over the calf of the leg and the plantar surface of the foot.

Tibial Nerve

Origin
Lumbosacral plexus, L4–S3

Movements/Muscles Innervated
Extends hip and flexes knee
- *Biceps femoris (long head)*
- *Semitendinosus*
- *Semimembranosus*

Extends hip and adducts thigh
- *Adductor magnus (hamstring part)*

Plantar flexes foot
- *Plantaris*
- *Gastrocnemius*
- *Soleus*
- *Tibialis posterior*

Flexes knee
- *Popliteus*

Flexes toes
- *Flexor digitorum longus*
- *Flexor hallucis longus*

Cutaneous (sensory) Innervation
None

Medial and Lateral Plantar Nerves

Origin
Tibial nerve

Movements/Muscles Innervated
Flex and adduct toes
- *Plantar muscles of foot*

Cutaneous (sensory) Innervation
Sole of foot

Sural Nerve (not shown)

Origin
Tibial nerve

Movements/Muscles Innervated
None

Cutaneous (sensory) Innervation
Lateral and posterior one-third of leg and lateral side of foot

Anterior view   **Posterior view**

Posterior view

Figure 12.26 Tibial Nerve
Route of the tibial nerve and the muscles it innervates (posterior view). The insets depict the cutaneous (sensory) distribution of the nerve (*blue area*).

Chapter 12 Spinal Cord and Spinal Nerves

The common fibular nerve divides into the **deep** and **superficial fibular (peroneal) nerves.** These branches innervate the anterior and lateral muscles of the leg and foot. The cutaneous distribution of the common fibular nerve and its branches is the lateral and anterior leg and the dorsum of the foot (see figure 12.27).

Sciatic Nerve Damage

If a person sits on a hard surface for a considerable time, the sciatic nerve may be compressed against the ischial portion of the coxa. When the person stands up, a tingling sensation described as "pins and needles" can be felt throughout the lower limb, and the limb is said to have "gone to sleep."

The sciatic nerve may be seriously injured in a number of ways. A ruptured intervertebral disk or pressure from the uterus during pregnancy may compress the roots of the sciatic nerve. Other possibilities for causing sciatic nerve damage include hip injury or an improperly administered injection in the hip region.

Common Fibular (Peroneal) Nerve

Origin
Lumbosacral plexus, L4–S2

Movements/Muscles Innervated
Extends hip and flexes knee
- *Biceps femoris (short head)*

Cutaneous (sensory) Innervation
Lateral surface of knee

Deep Fibular (Peroneal) Nerve

Origin
Common fibular (peroneal) nerve

Movements/Muscles Innervated
Dorsiflexes foot
- *Tibialis anterior*
- *Peroneus tertius*

Extends toes
- *Extensor digitorum longus*
- *Extensor hallucis longus*

Cutaneous (sensory) Innervation
Great and second toe

Superficial Fibular (Peroneal) Nerve

Origin
Common fibular (peroneal) nerve

Movements/Muscles Innervated
Plantar flexes and everts foot
- *Peroneus longus*
- *Peroneus brevis*

Extends toes
- *Extensor digitorum brevis*

Cutaneous (sensory) Innervation
Dorsal anterior third of leg and dorsum of foot

Anterior view   Lateral view   Anterior view

Figure 12.27   Fibular Nerve

Route of the common fibular (peroneal) nerve and the muscles it innervates (anterior view). The insets depict the cutaneous (sensory) distribution of the nerve (*blue area*).

Clinical Focus   Nerve Replacement

Patients paralyzed by strokes or spinal cord lesions are now able to regain certain limited functions. Microcomputers are being perfected that stimulate certain programmed activities, such as grasping and walking. The microcomputer initiates electric impulses that are conveyed through fine wire leads to either peripheral nerves or directly to the muscles responsible for the desired movement. The subtle movement of muscles not affected by the paral-ysis initiate the program. Sensors connected to the microcomputer are attached to the skin overlying functional muscles and are able to detect electrical activity associated with movement of the underlying muscles. For example, a person with both legs paralyzed may have such a sensor attached to the abdomen. The abdominal muscles normally involved in stabilizing and moving the pelvis during walking are stimulated by descending tracts when walking is initiated by CNS centers. The resultant abdominal muscle activity is detected by the sensor, which activates the program that stimulates the appropriate sequence of muscles in the lower limbs, and the paralyzed person walks. Similarly, a quadriplegic using subtle movements of the shoulder, neck, or face, where specific sensors can be placed, can initiate certain upper limb and grasping actions.

Other Lumbosacral Plexus Nerves

In addition to the nerves just described, the lumbosacral plexus gives rise to nerves that supply the lower abdominal muscles (iliohypogastric nerve), the hip muscles that act on the femur (gluteal nerves), and the muscles of the abdominal floor (pudendal nerve; see figure 12.23). The **iliohypogastric** (il'ē-ō-hī-pō-gas'trik), **ilioinguinal** (il'ē-ō-ing'gwi-năl), **genitofemoral** (jen'i-tō-fem'ŏ-răl), **cutaneous femoral,** and **pudendal** (pū-den'dăl) **nerves** innervate the skin of the suprapubic area, the external genitalia, the superior medial thigh, and the posterior thigh. The pudendal nerve plays a vital role in sexual stimulation and response.

Pudendal Nerve Anesthesia

Branches of the pudendal nerve are anesthetized before a doctor performs an episiotomy for childbirth. An **episiotomy** (e-piz-ē-ot'ō-mē, e-pis-ē-ot'ō-mē) is a cut in the perineum that makes the opening of the birth canal larger.

Coccygeal Plexus

The **coccygeal** (kok-sij'ē-ăl) **plexus** is a very small plexus formed from the ventral rami of spinal nerves S4, S5, and the coccygeal nerve. This small plexus supplies motor innervation to muscles of the pelvic floor and sensory cutaneous innervation to the skin over the coccyx. The dorsal rami of the coccygeal nerves innervate some skin over the coccyx.

24. Name the four major nerves that arise from the lumbosacral plexus, and describe the muscles and skin area they supply. What is the name applied to the tibial and common fibular nerves bound together?
25. Describe the structures innervated by the remaining lumbosacral nerves.
26. What structures are innervated by the coccygeal plexus?

Clinical Focus Peripheral Nervous System Disorders—Spinal Nerves

General Types of PNS Disorders

Anesthesia is the loss of sensation (the Greek word *esthesis* means sensation). It may be a pathologic condition if it happens spontaneously, or it may be induced to facilitate surgery or some other medical treatment.

Hyperesthesia (hī′per-es-thē′zē-ă) is an abnormal acuteness to sensation, especially an increased sensitivity to pain, pressure, or light.

Paresthesia (par-es-the′zē-ă) is an abnormal spontaneous sensation, such as tingling, prickling, or burning.

Neuralgia (noo-ral′jē-ă) consists of severe spasms of throbbing or stabbing pain resulting from inflammation or damage along the pathway of a nerve.

Sciatica, or **ischiadica** (is′kē-ad′i-kă), is a neuralgia of the sciatic nerve, with pain radiating down the back of the thigh and leg. The most common cause is a herniated lumbar disk, resulting in pressure on the spinal nerves contributing to the lumbar plexus. Sciatica may also be produced by sciatic neuritis arising from a number of causes, including mechanical stretching of the nerve during exertion, vitamin deficiency, or metabolic disorders (such as gout or diabetes).

Neuritis (noo-rī′tis) is a general term referring to inflammation of a nerve that has a wide variety of causes, including mechanical injury or pressure, viral or bacterial infection, poisoning by drugs or other chemicals, and vitamin deficiencies. Neuritis in sensory nerves is characterized by neuralgia or may result in anesthesia and loss of reflexes in the affected area. Neuritis in motor nerves results in loss of motor function.

Infections

Herpes is a family of diseases characterized by skin lesions, which are caused by a group of closely related viruses (the herpes viruses). The term is derived from the Greek word *herpo,* meaning to creep, and indicates a spreading skin eruption. The viruses apparently reside in the ganglia of sensory nerves and cause lesions along the course of the nerves. **Herpes simplex II,** or genital herpes, is usually responsible for a sexually transmitted disease causing lesions on the external genitalia.

The varicella-zoster virus causes the diseases chicken pox in children and **shingles** in older adults, a disease also called **herpes zoster.** Normally, this virus first enters the body in childhood to cause chicken pox. The virus then lies dormant in the spinal ganglia for many years and can become active during a time of reduced resistance to cause **shingles,** a unilateral patch of skin blisters and discoloration along the path of one or more spinal nerves, most commonly around the waist. The symptoms can persist for 3–6 months.

Poliomyelitis (pō′lē-ō-mī′ĕ-lī′tis; "polio" or infantile paralysis; the Greek word *po-lio* means gray matter) is a disease caused by an *Enterovirus.* It's actually a CNS infection, but its major effect is on the peripheral nerves and the muscles they supply. The virus infects the motor neurons in the anterior horn of the spinal cord. The infection causes degeneration of the motor neurons, which results in paralysis and atrophy of the muscles innervated by those nerves.

Anesthetic leprosy is a bacterial infection of the peripheral nerves caused by *Mycobacterium leprae.* The infection results in anesthesia, paralysis, ulceration, and gangrene.

Genetic and Autoimmune Disorders

Myotonic dystrophy is an autosomal dominant hereditary disease characterized by muscle weakness, dysfunction, and atrophy and by visual impairment as a result of nerve degeneration.

Myasthenia (mī-as-thē′nē-ă) **gravis** is an autoimmune disorder resulting in a reduction in the number of functional acetylcholine receptors in the postsynaptic terminals and morphological changes in the neuromuscular synapse. T cells of the immune system break down acetylcholine receptor proteins into two fragments, which trigger antibody production by the immune system. Myasthenia gravis results in fatigue and progressive muscular weakness because of the neuromuscular dysfunction.

S U M M A R Y

Spinal Cord   (p. 412)

General Structure

1. The spinal cord gives rise to 31 pairs of spinal nerves. The spinal cord has cervical and lumbosacral enlargements where nerves of the limbs enter and leave.
2. The spinal cord is shorter than the vertebral column. Nerves from the end of the spinal cord form the cauda equina.

Meninges of the Spinal Cord

Three meningeal layers surround the spinal cord: the dura mater, arachnoid mater, and pia mater.

Cross Section of the Spinal Cord

1. The cord consists of peripheral white matter and central gray matter.
2. White matter is organized into funiculi, which are subdivided into fasciculi, or nerve tracts, which carry action potentials to and from the brain.
3. Gray matter is divided into horns.
   - The dorsal horns contain sensory axons that synapse with interneurons. The ventral horns contain the neuron cell bodies of somatic motor neurons, and the lateral horns contain the neuron cell bodies of autonomic neurons.
   - The gray and white commissures connect each half of the spinal cord.
4. The dorsal root conveys sensory input into the spinal cord, and the ventral root conveys motor output away from the spinal cord.

Reflexes (p. 415)

1. A reflex arc is the functional unit of the nervous system.
   - Sensory receptors respond to stimuli and produce action potentials in sensory neurons.
   - Sensory neurons propagate action potentials to the CNS.
   - Interneurons in the CNS synapse with sensory neurons and with motor neurons.
   - Motor neurons carry action potentials from the CNS to effector organs.
   - Effector organs such as muscles or glands respond to the action potentials.
2. Reflexes don't require conscious thought, and they produce a consistent and predictable result.
3. Reflexes are homeostatic.
4. Reflexes are integrated within the brain and spinal cord. Higher brain centers can suppress or exaggerate reflexes.

Stretch Reflex

Muscle spindles detect stretch of skeletal muscles and cause the muscle to shorten reflexively.

Golgi Tendon Reflex

Golgi tendon organs respond to increased tension within tendons and cause skeletal muscles to relax.

Withdrawal Reflex

1. Activation of pain receptors causes contraction of muscles and the removal of some part of the body from a painful stimulus.
2. Reciprocal innervation causes relaxation of muscles that would oppose the withdrawal movement.
3. In the crossed extensor reflex, during flexion of one limb caused by the withdrawal reflex, the opposite limb is stimulated to extend.

Interactions with Spinal Cord Reflexes (p. 420)

Convergent and divergent pathways interact with reflexes.

Structure of Peripheral Nerves (p. 420)

In the PNS, individual axons are surrounded by the endoneurium. Groups of axons, called fascicles, are bound together by the perineurium. The fascicles form the nerve and are held together by the epineurium.

Spinal Nerves (p. 422)

1. Eight cervical, 12 thoracic, 5 lumbar, 5 sacral pairs, and 1 coccygeal pair make up the spinal nerves.
2. Spinal nerves have specific cutaneous distributions called dermatomes.
3. Spinal nerves branch to form rami.
   - The dorsal rami supply the muscles and skin near the midline of the back.
   - The ventral rami in the thoracic region form intercostal nerves, which supply the thorax and upper abdomen. The remaining ventral rami join to form plexuses (see following summary sections). Communicating rami supply sympathetic nerves (see chapter 16).

Cervical Plexus

Spinal nerves C1–C4 form the cervical plexus, which supplies some muscles and the skin of the neck and shoulder. The phrenic nerves innervate the diaphragm.

Brachial Plexus

1. Spinal nerves C5–T1 form the brachial plexus, which supplies the upper limb.
2. The axillary nerve innervates the deltoid and teres minor muscles and the skin of the shoulder.
3. The radial nerve supplies the extensor muscles of the arm and forearm and the skin of the posterior surface of the arm, forearm, and hand.
4. The musculocutaneous nerve supplies the anterior arm muscles and the skin of the lateral surface of the forearm.
5. The ulnar nerve innervates most of the intrinsic hand muscles and the skin on the ulnar side of the hand.
6. The median nerve innervates the pronator and most of the flexor muscles of the forearm, most of the thenar muscles, and the skin of the radial side of the palm of the hand.
7. Other nerves supply most of the muscles that act on the arm, the scapula, and the skin of the medial arm and forearm.

Lumbar and Sacral Plexuses

1. Spinal nerves L1–S4 form the lumbosacral plexus.
2. The obturator nerve supplies the muscles that adduct the thigh and the skin of the medial thigh.
3. The femoral nerve supplies the muscles that flex the thigh and extend the leg and the skin of the anterior and lateral thigh and the medial leg and foot.
4. The tibial nerve innervates the muscles that extend the thigh and flex the leg and the foot. It also supplies the plantar muscles and the skin of the posterior leg and the sole of the foot.
5. The common fibular nerve supplies the short head of the biceps femoris, the muscles that dorsiflex and plantar flex the foot, and the skin of the lateral and anterior leg and the dorsum of the foot.
6. In the thigh, the tibial nerve and the common fibular nerve are combined as the sciatic nerve.
7. Other lumbosacral nerves supply the lower abdominal muscles, the hip muscles, and the skin of the suprapubic area, external genitalia, and upper medial thigh.

Coccygeal Plexus

Spinal nerves S4, S5, and Co form the coccygeal plexus, which supplies the muscles of the pelvic floor and the skin over the coccyx.

R E V I E W   A N D   C O M P R E H E N S I O N

1. The spinal cord extends from the
   a. medulla oblongata to the coccyx.
   b. level of the third cervical vertebra to the coccyx.
   c. level of the axis to the lowest lumbar vertebra.
   d. medulla oblongata to the level of the second lumbar vertebra.
   e. axis to the sacral hiatus.

2. The structure that anchors the inferior end of the spinal cord to the coccyx is the
   a. conus medullaris.
   b. cauda equina.
   c. filum terminale.
   d. lumbar enlargement.
   e. posterior median sulcus.

3. Axons of sensory neurons synapse with the cell bodies of interneurons in the _____ of spinal cord gray matter.
   a. anterior horn
   b. lateral horn
   c. posterior horn
   d. gray commissure
   e. lateral funiculi

4. Cell bodies for sensory neurons are located in the
   a. anterior horn of spinal cord gray matter.
   b. lateral horn of spinal cord gray matter.
   c. posterior horn of spinal cord gray matter.
   d. dorsal root ganglia.
   e. posterior columns.

5. Given these components of a reflex arc:
   1. effector organ
   2. interneuron
   3. motor neuron
   4. sensory neuron
   5. sensory receptor

   Choose the correct order an action potential follows after a sensory receptor is stimulated.
   a. 5,4,3,2,1
   b. 5,4,2,3,1
   c. 5,3,4,1,2
   d. 5,2,4,3,1
   e. 5,3,2,1,4

6. A reflex response accompanied by the conscious sensation of pain is possible because of
   a. convergent pathways.
   b. divergent pathways.
   c. a reflex arc that contains only one neuron.
   d. sensory perception in the spinal cord.

7. Several of the events that occurred between the time that a physician struck a patient's patellar tendon with a rubber hammer and the time the quadriceps femoris contracted (knee-jerk reflex) are listed below:
   1. increased frequency of action potentials in sensory neurons
   2. stretch of the muscle spindles
   3. increased frequency of action potentials in the alpha motor neurons
   4. stretch of the quadriceps femoris
   5. contraction of the quadriceps femoris

   Which of these lists most closely describes the sequence of events as they normally occur?
   a. 4,1,2,3,5
   b. 4,1,3,2,5
   c. 1,4,3,2,5
   d. 4,2,1,3,5
   e. 4,2,3,1,5

8. _____ are responsible for regulating the sensitivity of the muscle spindle.
   a. Alpha motor neurons
   b. Sensory neurons
   c. Gamma motor neurons
   d. Golgi tendon organs
   e. Inhibitory interneurons

9. Which of these events occurs when a person steps on a tack with the right foot?
   a. The right foot is pulled away from the tack because of the Golgi tendon reflex.
   b. The left leg is extended to support the body because of the stretch reflex.
   c. The flexor muscles of the right thigh contract, and the extensor muscles of the right thigh relax because of reciprocal innervation.
   d. Extensor muscles contract in both thighs because of the crossed extensor reflex.

10. Which of these is a correct count of the spinal nerves?
    a. 9 cervical, 12 thoracic, 5 lumbar, 5 sacral, 1 coccygeal
    b. 8 cervical, 12 thoracic, 5 lumbar, 5 sacral, 1 coccygeal
    c. 7 cervical, 12 thoracic, 5 lumbar, 5 sacral, 1 coccygeal
    d. 8 cervical, 11 thoracic, 4 lumbar, 6 sacral, 1 coccygeal
    e. 7 cervical, 11 thoracic, 5 lumbar, 6 sacral, 1 coccygeal

11. Given these structures:
    1. dorsal ramus
    2. dorsal root
    3. plexus
    4. ventral ramus
    5. ventral root

    Choose the arrangement that lists the structures in the order that an action potential passes through them, given that the action potential originates in the spinal cord and propagates to a peripheral nerve.
    a. 2,1,3
    b. 2,3,1
    c. 3,4,5
    d. 5,3,4
    e. 5,4,3

12. Damage to the dorsal ramus of a spinal nerve results in
    a. loss of sensation.
    b. loss of motor function.
    c. both a and b.

13. A collection of spinal nerves that join together after leaving the spinal cord is called a
    a. ganglion.
    b. nucleus.
    c. projection nerve.
    d. plexus.

14. A dermatome
    a. is the area of skin supplied by a pair of spinal nerves.
    b. exists for each spinal nerve except C1
    c. can be used to locate the site of spinal cord or nerve root damage.
    d. all of the above.

15. Which of these nerves arises from the cervical plexus?
    a. median
    b. musculocutaneous
    c. phrenic
    d. obturator
    e. ulnar

16. The skin on the posterior surface of the hand is supplied by the
    a. median nerve.
    b. musculocutaneous nerve.
    c. ulnar nerve.
    d. axillary nerve.
    e. radial nerve.

17. Most thenar muscles and most of the flexor muscles of the forearm are supplied by the
    a. musculocutaneous nerve.
    b. radial nerve.
    c. median nerve.
    d. ulnar nerve.
    e. axillary nerve.

18. Most intrinsic hand muscles are supplied by the
    a. musculocutaneous nerve.
    b. radial nerve.
    c. median nerve.
    d. ulnar nerve.
    e. axillary nerve.

19. The sciatic nerve is actually two nerves combined within the same sheath. The two nerves are the
    a. femoral and obturator.
    b. femoral and gluteal.
    c. common fibular (peroneal) and tibial.
    d. common fibular (peroneal) and obturator.
    e. tibial and gluteal.

20. The muscles of the anterior compartment of the thigh are supplied by the
    a. obturator nerve.
    b. gluteal nerve.
    c. sciatic nerve.
    d. femoral nerve.
    e. ilioinguinal nerve.

Answers in Appendix F

CRITICAL THINKING

1. Describe how stimulation of a neuron that has its cell body in the cerebrum could inhibit a reflex that is integrated within the spinal cord.

2. A cancer patient has his left lung removed. To reduce the space remaining where the lung is removed, the diaphragm on the left side is paralyzed to allow the abdominal viscera to push the diaphragm upward. What nerve is cut? Where is a good place to cut it, and when would the surgery be done?

3. Based on sensory response to pain in the skin of the hand, how could you distinguish between damage to the ulnar, median, and radial nerves?

4. During a difficult delivery, the baby's arm delivered first. The attending physician grasped the arm and forcefully pulled it. Later a nurse observed that the baby could not abduct or adduct the medial four fingers and flexion of the wrist was impaired. What nerve was damaged?

5. Two patients are admitted to the hospital. According to their charts, both have herniated disks that are placing pressure on the roots of the sciatic nerve. One patient has pain in the buttocks and the posterior aspect of the thigh. The other patient experiences pain in the posterior and lateral aspects of the leg and the lateral part of the ankle and foot. Explain how the same condition, a herniated disk, could produce such different symptoms.

6. In an automobile accident a woman suffers a crushing hip injury. For each of the conditions given here, state what nerve is damaged.
    a. Unable to adduct the thigh
    b. Unable to extend the leg
    c. Unable to flex the leg
    d. Loss of sensation from the skin of the anterior thigh
    e. Loss of sensation from the skin of the medial thigh

7. A skier breaks his ankle. As part of his treatment, the ankle and leg are placed in a plaster cast. Unfortunately, the cast is too tight around the proximal portion of the leg and presses in against the neck of the fibula. Where would you predict the patient would experience tingling or numbness in the leg? Explain.

Answers in Appendix G

ANSWERS TO PREDICT QUESTIONS

1. The cord is enlarged in the cervical and lumbosacral regions because of the large numbers of motor nerve fibers exiting from the cord to the limbs and sensory nerve fibers entering the cord from the limbs. Also, more neuron cell bodies in the spinal cord regions are associated with the increased numbers of sensory and motor fibers.

2. The common site of injections into the vertebral column is between L3 and L4 or between L4 and L5 (see chapter 7).

3. Dorsal root ganglia contain neuron cell bodies, which are larger in diameter than the axons of the dorsal roots.

4. Nerves C5–T1, which innervate the left arm, forearm, and hand, are damaged.

5. Damage to the right phrenic nerve results in the absence of muscular contraction in the right half of the diaphragm. Because the phrenic nerves originate from C3–C5, damage to the upper cervical region of the spinal cord eliminates their functions; damage in the lower cord below the point where the spinal nerves originate doesn't affect the nerves to the diaphragm. Breathing is affected, however, because the intercostal nerves to the intercostal muscles, which move the ribs, are paralyzed.

6. The radial nerve lies along the shaft of the humerus about midway along its length. If the humerus is fractured, the radial nerve can be lacerated by bone fragments or, more commonly, pinched between two fragments of bone, decreasing or eliminating the function of the nerve.

Visit the Online Learning Center at www.mhhe.com/seeley7 for chapter quizzes, interactive learning exercises, and other study tools.

Colorized SEM of a neuron network.

13

Brain and Cranial Nerves

The brain is that part of the CNS contained within the cranial cavity (figure 13.1). It is the control center for many of the body's functions. The brain is much like a complex central computer but with additional functions that no computer can as yet match. Indeed, one goal in computer technology is to make computers that can function more like the human brain. The brain consists of the brainstem, the cerebellum, the diencephalon, and the cerebrum (table 13.1). The brainstem includes the medulla oblongata, pons, midbrain, and reticular formation. The structure of the brain is described in this chapter. Its functions are primarily discussed in chapter 14.

Twelve pairs of cranial nerves, which are part of the PNS, arise directly from the brain. Two pairs arise from the cerebrum, nine pairs arise from the brainstem, and one pair arises from the spinal cord.

This chapter describes the *brainstem* (p. 445), *cerebellum* (p. 447), *diencephalon* (p. 449), *cerebrum* (p. 451), *meninges and cerebrospinal fluid* (p. 454), *blood supply to the brain* (p. 459) *development of the CNS* (p. 459), and the *cranial nerves* (p. 461).

Figure 13.1   Regions of the Right Half of the Brain
(as seen in a midsagittal section)

Table 13.1	Divisions and Functions of the Brain		
Brainstem	Connects the spinal cord to the cerebrum; several important functions (see below); location of cranial nerve nuclei. *damage to small brainstem area causes death*	**Cerebellum**	Control of muscle movement and tone; balance regulates extent of intentional movement; involved in learning motor skills
Medulla oblongata	Pathway for ascending and descending nerve tracts; center for several important reflexes (e.g., heart rate, breathing, swallowing, vomiting)	**Diencephalon**	
Pons	Contains ascending and descending nerve tracts; relay between cerebrum and cerebellum; reflex centers	Thalamus	Major sensory relay center; influences mood and movement
		Subthalamus	Contains nerve tracts and nuclei
		Epithalamus	Contains nuclei responding to olfactory stimulation and contains pineal body
Midbrain	*cerebral peduncles* Contains ascending and descending nerve tracts; visual reflex center; part of auditory pathway	Hypothalamus	Major control center for maintaining homeostasis and regulating endocrine function
		Cerebrum	Conscious perception, thought, and conscious motor activity; can override most other systems
Reticular formation	Scattered throughout brainstem; controls cyclic activities such as the sleep-wake cycle	Basal nuclei	Control of muscle activity and posture; largely inhibit unintentional movement when at rest
		Limbic system	Autonomic response to smell, emotion, mood, and other such functions

Brainstem

The medulla oblongata, pons, and midbrain constitute the **brainstem** (figure 13.2). The brainstem connects the spinal cord to the remainder of the brain and is responsible for many essential functions. Damage to small brainstem areas often causes death because many reflexes essential for survival are integrated in the brainstem, whereas relatively large areas of the cerebrum or cerebellum may be damaged without being life-threatening.

Medulla Oblongata

The **medulla oblongata** (ob-long-gah′tă), often called the medulla, is about 3 cm long, is the most inferior part of the brainstem, and is continuous inferiorly with the spinal cord. It contains sensory and motor nerve tracts; cranial nerve nuclei; other, related nuclei; and part of the reticular formation (see chapter 14). Superficially, the spinal cord blends into the medulla, but internally several differences exist. Discrete **nuclei,** clusters of gray matter composed mostly of neuron cell bodies and having specific functions, are found in the medulla oblongata, whereas the gray matter of the spinal cord extends as a continuous mass in the center of the cord. In addition, the nerve tracts within the medulla don't have the same organization as those of the spinal cord. Several medullary nuclei function as centers for reflexes, such as those involved in the regulation of heart rate, blood vessel diameter, respiration, swallowing, vomiting, hiccuping, coughing, and sneezing.

Two prominent enlargements on the anterior surface of the medulla oblongata are called **pyramids** because they are broader near the pons and taper toward the spinal cord (figure 13.2a). The pyramids are descending nerve tracts involved in the conscious control of skeletal muscles. Near their inferior ends, most of the fibers of the descending nerve tracts cross to the opposite side, or **decussate** (dē′kŭ-sāt, dē-kŭs′āt; the Latin word *decussatus* means to form an X, as in the Roman numeral X). This decussation accounts, in part, for the fact that each half of the brain controls the opposite half of the body. Its role as a conduction pathway is discussed in the description of ascending and descending nerve tracts (see chapter 14).

Two rounded, oval structures, called **olives,** protrude from the anterior surface of the medulla oblongata just lateral to the superior ends of the pyramids (figure 13.2a and b). The olives are nuclei involved in functions such as balance, coordination, and modulation of sound from the inner ear (see chapter 15). The nuclei of cranial nerves V (trigeminal), IX (glossopharyngeal), X (vagus), XI (accessory), and XII (hypoglossal) also are located within the medulla (figure 13.2c).

Pons

The part of the brainstem just superior to the medulla oblongata is the **pons** (see figure 13.2a), which contains ascending and descending nerve tracts and several nuclei. The pontine nuclei, located in the anterior portion of the pons, relay information from the cerebrum to the cerebellum.

The nuclei for cranial nerves V (trigeminal), VI (abducens), VII (facial), VIII (vestibulocochlear), and IX (glossopharyngeal) are contained within the posterior pons. Other important pontine areas include the pontine sleep center and respiratory center, which work with the respiratory centers in the medulla to help control respiratory movements (see chapter 23).

Midbrain

The **midbrain,** or **mesencephalon,** is the smallest region of the brainstem (see figure 13.2b). It's just superior to the pons and contains the nuclei of cranial nerves III (oculomotor), IV (trochlear), and V (trigeminal).

The **tectum** (tek′tŭm; roof) (figure 13.3) of the midbrain consists of four nuclei that form mounds on the dorsal surface, collectively called **corpora** (kōr′pōr-ă; bodies) **quadrigemina** (kwah′dri-jem′i-nă; four twins). Each mound is called a **colliculus** (ko-lik′ū-lŭs; hill); the two superior mounds are called **superior colliculi,** and the two inferior mounds are called **inferior colliculi** (see figure 13.2b). The inferior colliculi are involved in hearing and are an integral part of the auditory pathways in the CNS. Neurons conducting action potentials from the structures of the inner ear (see chapter 15) to the brain synapse in the inferior colliculi. The superior colliculi are involved in visual reflexes, and they receive input from the eyes, the inferior colliculi, the skin, and the cerebrum.

Reflex Movements of the Eyes and Head

The superior colliculi regulate the reflex movements of the eyes and head in response to various stimuli. When a bright object suddenly appears in a person's field of vision, a reflex turns the eyes to focus on it. When a person hears a sudden, loud noise, a reflex turns the head and eyes toward it. When a part of the body, such as the shoulder, is touched, a reflex turns the person's head and eyes toward that part of the body. In each situation, the pathway involves the superior colliculus.

The **tegmentum** (teg-men′tŭm; floor) of the midbrain largely consists of ascending tracts, like the spinal lemniscus and the medial lemniscus, from the spinal cord to the brain. The tegmentum also contains the paired **red nuclei** (see figure 13.3), which are so named because in fresh brain specimens they are

(a) **Anterior view**

- Interthalamic adhesion
- Thalamus
- **Diencephalon**
- Infundibulum
- **Midbrain**
- Cerebral peduncle
- **Pons**
- **Brainstem**
- Pyramid
- Ventral median sulcus
- Olive
- **Medulla oblongata**
- Pyramidal decussation

- Diencephalon
- Brainstem

(b) **Posterolateral view**

- Thalamus
- **Diencephalon**
- Pineal body
- Superior colliculus
- **Midbrain**
- Inferior colliculus
- Cerebral peduncle
- **Pons**
- Superior cerebellar peduncle
- Middle cerebellar peduncle
- Inferior cerebellar peduncle
- Median sulcus
- Nucleus cuneatus
- Nucleus gracilis
- **Medulla oblongata**
- Olive

(c) **Superior-posterior view**

Sensory nuclei (green)

Motor nuclei (purple)

- Edinger-Westphal nucleus (CN III)
- Oculomotor nucleus (CN III)
- Trochlear nucleus (CN IV)
- Sensory trigeminal nuclei (CN V)
- Trigeminal motor nucleus (CN V)
- Abducens nucleus (CN VI)
- Facial motor nucleus (CN VII)
- Cochlear and vestibular nuclei (CN VIII)
- Superior salivatory and lacrimal nuclei (CN VII)
- Inferior salivatory nucleus (CN IX)
- Solitary nucleus — Taste area (CN VII, IX)
- General visceral sensory area (CN IX,X)
- Nucleus ambiguus (CN IX, X, XI)
- Dorsal nucleus of vagus nerve (CN X)
- Hypoglossal nucleus (CN XII)

Figure 13.2   Brainstem and Diencephalon

(a) Anterior view. (b) Posterolateral view. (c) Brainstem nuclei. The sensory nuclei are shown on the left (*green*). The motor nuclei are shown on the right (*purple*). Even though the nuclei are shown on only one side, each half of the brainstem has both sensory and motor nuclei. The inset shows the location of the diencephalon (*red*) and brainstem (*blue*). (*CN* = cranial nerve.)

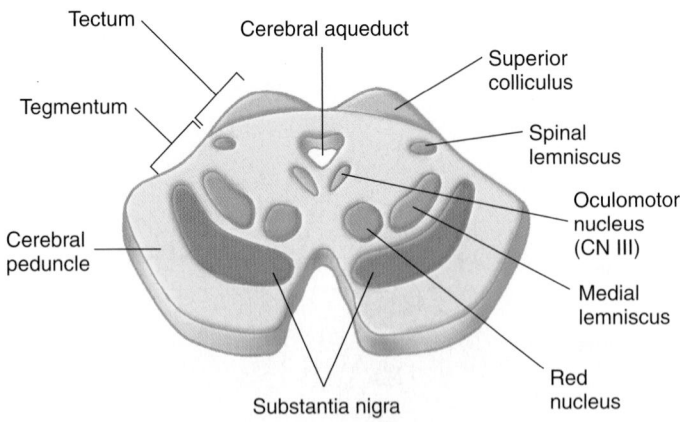

Tectum
Cerebral aqueduct
Superior colliculus
Tegmentum
Spinal lemniscus
Oculomotor nucleus (CN III)
Cerebral peduncle
Medial lemniscus
Red nucleus
Substantia nigra

Anterosuperior view of oblique section

Figure 13.3 **Oblique Section Through the Midbrain**

Inset shows the level of section.

pinkish in color as the result of an abundant blood supply. The red nuclei aid in the unconscious regulation and coordination of motor activities. **Cerebral peduncles** (pe-dŭng′klz, pē′dŭng-klz; the foot of a column) constitute that portion of the midbrain ventral to the tegmentum. They consist primarily of descending tracts from the cerebrum to the brainstem and spinal cord and constitute one of the major CNS motor pathways. The **substantia nigra** (nī′gră; black substance) is a nuclear mass between the tegmentum and cerebral peduncles, containing cytoplasmic melanin granules that give it a dark gray or black color (see figure 13.3). The substantia nigra is interconnected with other basal nuclei of the cerebrum, described later in this chapter, and it's involved in maintaining muscle tone and in coordinating movements.

Reticular Formation

Several loosely packed nuclei, collectively called the **reticular formation** (see table 13.1) is scattered like a cloud throughout the length of the brainstem. The reticular formation receives axons from a large number of sources and especially from nerves that innervate the face. The functions of the reticular formation involve "cycles" of activity, such as the sleep–wake cycle, and are discussed in chapter 14.

> 1. *What are the major components of the medulla oblongata, pons, midbrain, and reticular formation? What are the general functions of each region?*

Cerebellum

The term **cerebellum** (ser-e-bel′ŭm; figure 13.4) means little brain. The cerebellum is attached to the brainstem posterior to the pons. It communicates with other regions of the CNS through three large nerve tracts: the **superior, middle,** and **inferior cerebellar peduncles** (see figure 13.2b), which connect the cerebellum to the midbrain, pons, and medulla oblongata, respectively. The cerebellum has a gray cortex and nuclei, with white medulla in between. The cerebellar cortex has ridges called **folia.** The white matter of the medulla resembles a branching tree and is called the **arbor vitae** (ar′bōr vī′te; tree of life). The **nuclei** of the cerebellum are located in the deep inferior center of the white matter.

The cerebellar cortex contains several cell types, including stellate cells, basket cells, granule cells, Golgi cells, mossy fibers, and Purkinje cells. **Purkinje** (pŭr-kin′jē; Johannes Purkinje, a Bohemian anatomist and physiologist, 1787–1869) **cells** are the largest and probably most interesting cells in the CNS. Purkinje cells receive 200,000 synapses, are inhibitory neurons, and are the only cerebellar cortex neurons sending axons to the cerebellar nuclei. The cerebellar cortex contains more than 10^{12} neurons, more neurons than in the cerebral cortex!

The cerebellum consists of three parts: a small inferior part, the **flocculonodular** (flok′ū-lō-nod′ū-lăr; floccular, meaning a tuft of wool) **lobe;** a narrow central **vermis** (worm-shaped); and two large **lateral hemispheres** (see figure 13.4).

The flocculonodular lobe is the simplest part of the cerebellum and helps control balance and eye movements. The vermis and medial portion of the lateral hemispheres are involved in the control

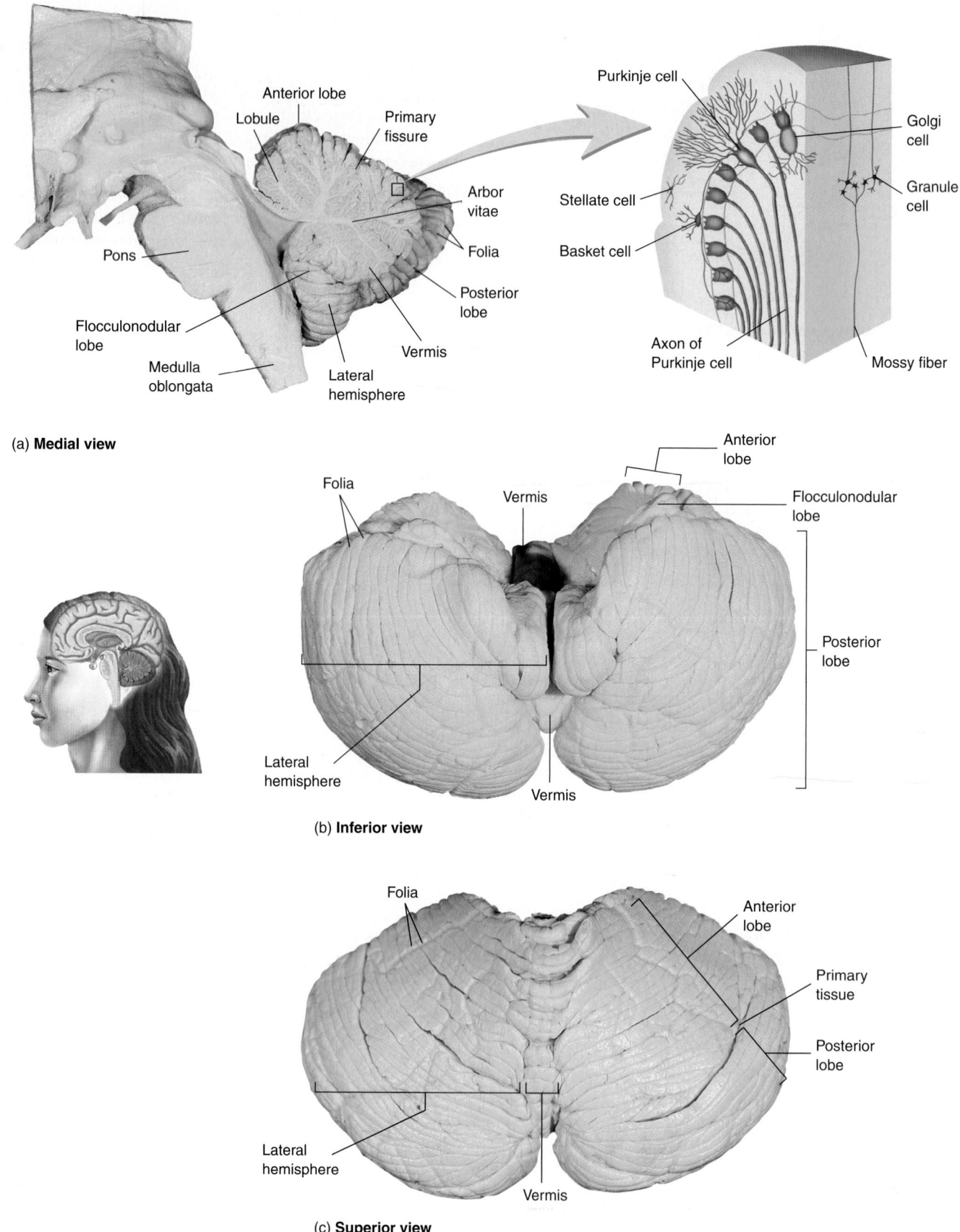

(a) **Medial view**

(b) **Inferior view**

(c) **Superior view**

Figure 13.4 Cerebellum

(*a*) Right half of the cerebellum as seen in a midsagittal section. Inset shows the histology of the cerebellum. (*b*) Inferior view of the cerebellum. (*c*) Superior view of the cerebellum.

of posture, locomotion, and fine motor coordination, thereby producing smooth, flowing movements. The major portion of the lateral hemispheres is involved, with the cerebral cortex of the frontal lobe, in planning, practicing, and learning complex movements.

Each lateral hemisphere is divided by a **primary fissure** into an **anterior lobe** and a **posterior lobe.** The lobes are subdivided into **lobules,** which contain the folia.

● *2. What are the major regions of the cerebellum? Describe the major functions of each.*

Diencephalon

The **diencephalon** (dī-en-sef′ă-lon) is the part of the brain between the brainstem and the cerebrum (see figures 13.1 and 13.5). Its main components are the thalamus, subthalamus, epithalamus, and hypothalamus.

Thalamus

The **thalamus** (thal′ă-mŭs; figure 13.5a and b) is by far the largest part of the diencephalon, constituting about four-fifths of

(a) **Diencephalon, medial view**

Thalamus
Interthalamic adhesion
Hypothalamus
Optic chiasm
Pituitary gland

Corpus callosum
Habenula ⎫
Pineal body ⎬ **Epithalamus**
Subthalamus
Cerebellum

(b) **Thalamus, anterolateral view**

Interthalamic adhesion
Anterior nucleus
Ventral anterior nucleus

Medial nucleus ⎫
Lateral posterior nucleus ⎬ Dorsal tier
Lateral dorsal nucleus ⎭
Pulvinar
Lateral geniculate body
Ventral posterior nucleus
Ventral lateral nucleus

(c) **Hypothalamus, medial view**

Preoptic area
Anterior nucleus
Supraoptic nucleus
Suprachiasmatic nucleus
Optic chiasm

Paraventricular nucleus
Lateral area
Dorsomedial nucleus
Posterior nucleus
Mammillary body
Ventromedial nucleus
Infundibulum
Pituitary gland

Figure 13.5 **Diencephalon**

(*a*) General overview of the right half of the diencephalon as seen in a midsagittal section. (*b*) Thalamus showing the nuclei. (*c*) Hypothalamus showing the nuclei and right half of the pituitary.

its weight. It is a cluster of nuclei shaped somewhat like a yo-yo, with two large, lateral portions connected in the center by a small stalk called the **interthalamic adhesion,** or **intermediate mass.** The space surrounding the interthalamic adhesion and separating the two large portions of the thalamus is the third ventricle of the brain.

a major relay station

A large portion of the sensory input projects to the thalamus, where sensory neurons synapse with thalamic neurons, which send projections from the thalamus to the cerebral cortex. Axons carrying auditory information synapse in the **medial geniculate** (je-nik′ū-lāt; Latin, *genu,* meaning bent like a knee) **nucleus** of the thalamus, axons carrying visual information synapse in the **lateral geniculate nucleus,** and most other sensory impulses synapse in the **ventral posterior nucleus.** The **ventral anterior** and **ventral lateral nuclei** are involved in motor functions, communicating between the basal nuclei, cerebellum, and the motor cortex (these areas are described later in this chapter).

Axons originating in the ventral posterior nucleus project to the **dorsal tier** of nuclei, which register pain (see figure 13.5*b*). Other axons project to the cerebral cortex where sensory input is localized (see chapter 14).

The thalamus also influences mood and actions associated with strong emotions like fear or rage. The **anterior** and **medial nuclei** are connected to the limbic system and to the prefrontal cortex (described later in this chapter and in chapter 14). They are involved in mood modification. The **lateral dorsal nucleus** is connected to other thalamic nuclei and to the cerebral cortex and is involved in regulating emotions. The **lateral posterior nucleus** and the **pulvinar** (pŭl-vī′năr; pillow) also have connections to other thalamic nuclei and are involved in sensory integration.

Subthalamus

The **subthalamus** is a small area immediately inferior to the thalamus (see figure 13.5*a*) that contains several ascending and descending nerve tracts and the **subthalamic nuclei.** A small portion of the red nucleus and substantia nigra of the midbrain extend into this area.

The subthalamic nuclei are associated with the basal nuclei and are involved in controlling motor functions.

Epithalamus

The **epithalamus** is a small area superior and posterior to the thalamus (see figure 13.5*a*). It consists of habenular nuclei and the pineal body. The **habenular** (hă-ben′ū-lăr) **nuclei** are influenced by the sense of smell and are involved in emotional and visceral responses to odors. The **pineal** (pin′ē-ăl) **gland,** or **pineal body,** is shaped somewhat like a pinecone, from which the name *pineal* is derived. It appears to play a role in controlling the onset of puberty, but data are inconclusive, so active research continues in this field. The pineal body also may influence the sleep–wake cycle and other biorhythms.

Brain Sand in the Pineal

In about 75% of adults, the pineal body contains granules of calcium and magnesium salts called "brain sand." These granules can be seen on radiographs and are useful as landmarks in determining whether or not the pineal body has been displaced by a pathologic enlargement of a part of the brain, such as a tumor or a hematoma.

Hypothalamus

The **hypothalamus** is the most inferior portion of the diencephalon (see figure 13.5*a* and *c*) and contains several small nuclei and nerve tracts. The most conspicuous nuclei, called the **mammillary bodies,** appear as bulges on the ventral surface of the diencephalon. They are involved in olfactory reflexes and emotional responses to odors. They may also be involved in memory and sense of direction. A funnel-shaped stalk, the **infundibulum** (in-fŭn-dib′ū-lŭm), extends from the floor of the hypothalamus and connects it to the **posterior pituitary gland,** or **neurohypophysis** (noor′ō-hī-pof′i-sis). The hypothalamus plays an important role in controlling the endocrine system because it regulates the pituitary gland's secretion of hormones, which influence functions as diverse as metabolism, reproduction, responses to stressful stimuli, and urine production (table 13.2; see chapter 18).

Sensory neurons that terminate in the hypothalamus provide input from (1) internal organs; (2) taste receptors of the tongue; (3) the limbic system, which is involved in responses to smell; (4) specific cutaneous areas, such as the nipples and external genitalia; (5) the eyes; and (6) the prefrontal cortex of the cerebrum carrying information relative to "mood" through the thalamus. Efferent fibers from the hypothalamus extend into the brainstem and the spinal cord, where they synapse with neurons of the autonomic nervous system (see chapter 16). These connections regulate functions such as heart rate and digestive functions (see table 13.2). Other fibers extend through the infundibulum to the posterior portion of the pituitary gland (see chapter 18); some extend to trigeminal and facial nerve nuclei to help control the head muscles involved in swallowing; and some extend to motor neurons of the spinal cord to stimulate shivering.

Hypothalamic nuclei directly control temperature by stimulating sweating or shivering. Other hypothalamic nuclei promote or inhibit eating or drinking. The hypothalamus is also very important in a number of functions that are related to mood, motivation, and emotion. Sensations like sexual pleasure, feeling relaxed and "good" after a meal, rage, and fear are related to hypothalamic functions. The hypothalamus interacts with the reticular activating system in the brainstem to coordinate the sleep–wake cycle (see table 13.2).

3. *Name the four main components of the diencephalon.*

4. *What are the functions of the thalamus and hypothalamus? Explain why the hypothalamus is an important link between the nervous system and the endocrine system.*

5. *List the general functions of the subthalamus. Name the parts of the epithalamus and give their functions.*

Table 13.2	Hypothalamic Functions	
Function	**Hypothalamic Nuclei**	**Description**
Autonomic	Preoptic area and anterior nucleus (parasympathetic) Lateral area and posterior nucleus (sympathetic)	Helps control heart rate, urine release from the bladder, movement of food through the digestive tract, and blood vessel diameter
Endocrine	Paraventricular nucleus Supraoptic nucleus	Helps regulate pituitary gland secretions and influences metabolism, ion concentration, sexual development, and sexual functions; site of antidiuretic hormone and oxytocin production (see chapter 18)
Muscle control	Lateral area	Controls muscles involved in swallowing and stimulates shivering
Temperature regulation	Preoptic area Anterior nucleus Posterior nucleus	Promotes heat loss when the hypothalamic temperature increases by increasing sweat production (anterior nucleus) and promotes heat production when the hypothalamic temperature decreases by promoting shivering (posterior nucleus); aspirin affects preoptic area to reduce fever
Regulation of food and water intake	Ventromedial nucleus Lateral area	Hunger center promotes eating and satiety center inhibits eating; thirst center promotes water intake
Emotions	Lateral area Medial area	Large range of emotional influences over body functions; directly involved in stress-related and psychosomatic illnesses and with feelings of fear and rage
Regulation of the sleep–wake cycle	Lateral area Suprachiasmatic nucleus	Coordinates responses to the sleep–wake cycle with the other areas of the brain (e.g., the reticular activating system); suprachiasmatic nucleus receives direct input from the eyes concerning light/dark cycles and is involved in jet lag
Sexual development and behavior	Preoptic area Dorsomedial nucleus Ventromedial nucleus	Stimulates sexual development, sexual arousal, and sexual behavior; the preoptic area is larger in males than in females

Cerebrum

The cerebrum (figure 13.6) is the part of the brain that most people think of when the term *brain* is mentioned. The cerebrum accounts for the largest portion of the total brain weight, which is about 1200 g in females and 1400 g in males. Brain size is related to body size; larger brains are associated with larger bodies, not with greater intelligence.

The cerebrum is divided into left and right hemispheres by a **longitudinal fissure** (figure 13.6*a*). The most conspicuous features on the surface of each hemisphere are numerous folds called **gyri** (jī′rī; sing., gyrus), which greatly increase the surface area of the cortex. The intervening grooves between the gyri are called **sulci** (sŭl′sī; sing., sulcus). A **central sulcus,** which extends across the lateral surface of the cerebrum from superior to inferior, is located about midway along the length of the brain. The central sulcus is located between the **precentral gyrus** anteriorly, which is the **primary motor cortex,** and a **postcentral gyrus** posteriorly, which is the **primary somatic sensory cortex** (see chapter 14). The general pattern of the gyri is similar in all normal human brains, but some variation exists between individuals and even between the two hemispheres of the same cerebrum.

Each cerebral hemisphere is divided into lobes, which are named for the skull bones overlying each one (figure 13.6*b*). The **frontal lobe** is important in voluntary motor function, motivation, aggression, the sense of smell, and mood. The **parietal lobe** is the major center for the reception and evaluation of sensory information, except for smell, hearing, and vision. The frontal and parietal lobes are separated by the central sulcus. The **occipital lobe** functions in the reception and integration of visual input and is not distinctly separate from the other lobes. The **temporal lobe** receives and evaluates input for smell and hearing and plays an important role in memory. Its anterior and inferior portions are referred to as the "psychic cortex," and they are associated with such brain functions as abstract thought and judgment. The temporal lobe is separated from the rest of the cerebrum by a **lateral fissure,** and deep within the fissure is the **insula** (in′soo-lă; island), often referred to as a fifth lobe.

The gray matter on the outer surface of the cerebrum is the **cortex,** and clusters of gray matter deep inside the brain are nuclei. The cerebral cortex contains a number of neuron types, named largely for their shape, such as fusiform cells, stellate cells, and pyramidal cells (figure 13.6*c*). These cells are distributed in layers within the cerebral cortex, ranging from two or three layers in the most "primitive" parts of the cortex to six layers in the more "advanced" regions.

The white matter of the brain between the cortex and nuclei is the **cerebral medulla.** This term should not be confused with the medulla oblongata; medulla is a general term meaning the center of a structure, or marrow. The cerebral medulla consists of nerve tracts that connect the cerebral cortex to other areas of cortex or other parts of the CNS (figure 13.7). These tracts fall into three main categories: (1) **Association fibers** connect areas of the cerebral cortex within the same hemisphere. (2) **Commissural**

(a) **Superior view**

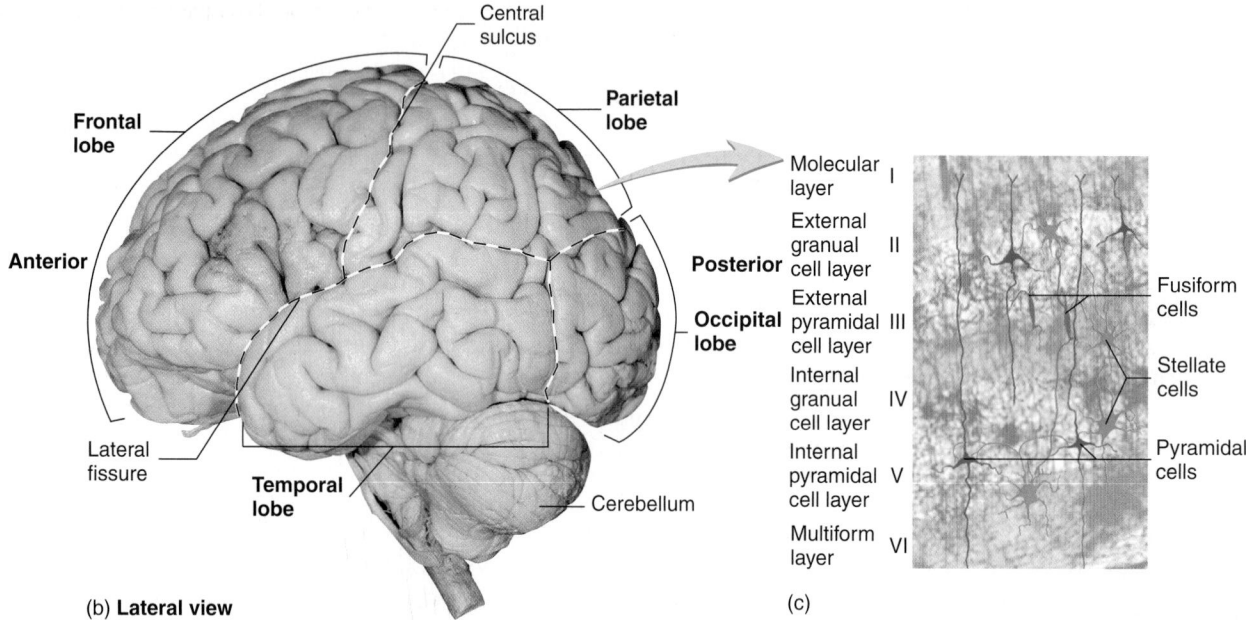

(b) **Lateral view**

(c)

Figure 13.6 The Brain

(a) Superior view. (b) Lateral view of the left cerebral hemisphere. (c) Histology of the cerebral cortex.

fibers connect one cerebral hemisphere to the other. The largest bundle of commissural fibers connecting the two cerebral hemispheres is the **corpus callosum** (kōr′pŭs ka-lō′sŭm; see figures 13.1, 13.7, and 13.8b). (3) **Projection fibers** are between the cerebrum and other parts of the brain and spinal cord (see figure 13.7). The projection fibers form the **internal capsule.**

- 6. *Define the terms gyri and sulci. What structures do the longitudinal fissure, central sulcus, and lateral fissure separate?*
  7. *Define the terms cerebral cortex and cerebral medulla.*

8. *Name the five lobes of the cerebrum, and describe their locations and functions.*
9. *List three categories of nerve tracts in the cerebral medulla.*

Basal Nuclei

The **basal nuclei,** or basal ganglia, are a group of functionally related nuclei located bilaterally in the inferior cerebrum, diencephalon, and midbrain (figure 13.8). These nuclei are involved in the control of motor functions (see chapter 14). The nuclei in the cerebrum are collectively called the **corpus striatum**

Figure 13.7 **Cerebral Medullary Tracts**

(a) Coronal section of the brain showing commissural, association, and projection fibers. (b) Photograph of the left cerebral hemisphere from a lateral view with the cortex and association fibers removed to reveal the projection fibers of the internal capsule deep within the brain.

Figure 13.8 **Basal Nuclei (Ganglia) of the Left Hemisphere**

(a) A "transparent 3-D" drawing of the basal nuclei inside the left hemisphere. (b) Photograph of a frontal section of the brain showing the basal nuclei and other structures. (c) The same photograph as in "b" with nuclei highlighted.

(kōr′pŭs strī-ā′tŭm; striped body) and include the **caudate** (kaw′dāt; having a tail) **nucleus** and **lentiform** (len′ti-fōrm; lens-shaped) **nucleus.** The lentiform nucleus, in turn, is divided into a lateral **putamen** (pū-tā′men; that which falls off in pruning) and the medial **globus pallidus** (glō′bŭs pal′li-dŭs; pale globe). The basal nuclei are the largest nuclei of the brain and occupy a large part of the cerebrum. The subthalamic nucleus is located in the diencephalon, and the substantia nigra is located in the midbrain.

🔖 **10. List the basal nuclei and state their general function.**

Limbic System

Parts of the cerebrum and diencephalon are grouped together under the title **limbic** (lim′bik) **system** (figure 13.9). The limbic system plays a central role in basic survival functions such as memory, reproduction, and nutrition. It is also involved in emotional interpretation of sensory input and emotions in general. *Limbus* means border, and the term *limbic* refers to deep portions of the cerebrum that form a ring around the diencephalon. Structurally the limbic system consists of (1) certain cerebral cortical areas, including the **cingulate** (sin′gū-lāt; to surround) **gyrus,** located along the inner surface of the longitudinal fissure just above the corpus callosum, and the **parahippocampal gyrus,** located on the medial side of the temporal lobe; (2) various nuclei, such as anterior nuclei of the thalamus, the habenula in the epithalamus, and the **dentate nucleus** of the **hippocampus;** (3) parts of the basal nuclei, such as the, **amygdala,** or **amygdaloid nuclear complex;** (4) the hypothalamus, especially the mammillary bodies; (5) the **olfactory cortex;** and (6) tracts connecting the various cortical areas and nuclei, such as the **fornix,** which connects the hippocampus to the thalamus and mammillary bodies.

🔖 **11. List the parts of the limbic system.**

Meninges and Cerebrospinal Fluid
Meninges

Three connective tissue membranes, the **meninges** (mě-nin′jēz), surround and protect the brain and spinal cord (figure 13.10). The most superficial and thickest membrane is the **dura mater** (doo′ră mā′ter; tough mother). Three dural folds, the falx cerebri, the tentorium cerebelli, and the falx cerebelli, extend into the major

Fornix
Anterior nuclei of thalamus
Anterior commissure
Septal area
Olfactory bulb
Olfactory cortex
Mammillary body

Cingulate gyrus
Corpus callosum
Habenula
Dentate nucleus of hippocampus
Parahippocampal gyrus
Amygdaloid body

Medial view

Figure 13.9   Limbic System and Associated Structures of the Right Hemisphere as Seen in a Midsagittal Section

(a) **Anterosuperior view**

(b) **Anterior view**

Figure 13.10 Meninges

(*a*) Meningeal membranes surrounding the brain. (*b*) Frontal section of head to show the meninges.

brain fissures. The **falx cerebri** (falks se-rē′brī; sīckle-shaped) is located between the two cerebral hemispheres in the longitudinal fissure, the **tentorium cerebelli** (ten-tō′rē-ŭm ser′ĕ-bel′ī; tent) is between the cerebrum and cerebellum, and the **falx cerebelli** lies between the two cerebellar hemispheres.

The dura mater surrounding the brain is tightly attached to and continuous with the periosteum of the cranial cavity, forming a single functional layer. The dura mater and dural folds help hold the brain in place within the skull and keep it from moving around too freely. The dura mater around the brain separates in

several places, primarily at the bases of the three dural folds, to form **dural venous sinuses.** The dural venous sinuses collect most of the blood that returns from the brain, as well as cerebrospinal fluid (CSF) from around the brain (see Cerebrospinal Fluid, p. 456). The sinuses then empty into the veins that exit the skull (see chapter 21).

The next meningeal membrane is a very thin, wispy **arachnoid** (ă-rak′noyd; spiderlike; i.e., cobwebs) **mater.** The space between this membrane and the dura mater is the **subdural space** and contains only a very small amount of serous fluid. The third meningeal layer, the **pia** (pī′ă, pē′ă; affectionate) **mater** is bound very tightly to the surface of the brain. Between the arachnoid mater and the pia mater is the **subarachnoid space,** which contains weblike strands of arachnoid mater and the blood vessels supplying the brain, and is filled with CSF. *cerebro spina fluid*

Traumatic Brain Injuries and Hematomas

Head injuries are classified as **open** (when cranial cavity contents are exposed to the outside) or **closed** (when the cranial cavity remains intact). Open head injuries are discussed later on this same page (see CSF and Skull Fractures). Closed injuries are more common and involve the head striking a hard surface or an object striking the head. Such injuries may result in brain trauma. The main trauma may be **coup** (kŭ), occurring at the site of impact, or **contrecoup** (kon′tra-kŭ), occurring on the opposite side of the brain from the impact and resulting from movement of the brain within the skull. Traumatic brain injury is three times more common in males than in females. The most common type of traumatic brain injury (75%–90%) is a **concussion,** which is defined as an immediate, but transient, impairment of neural function, such as loss of consciousness or blurred vision.

Traumatic brain injury may be diffuse or focal. **Diffuse brain injury** usually results from shaking, such as when a person shakes a child, or when a person is thrown about in an automobile accident. As the name suggests, such injury is not localized to one focal point, but involves damage to many small vessels and nerves, especially around the brainstem. **Focal** traumatic brain injury may involve cortical **contusions** (bruisings), involving direct impact injury to the brain, or hemorrhages in or around the brain. Contusions are usually superficial and involve only the gyri.

Hemorrhagic brain injury results from bleeding outside the dura (extradural or epidural), between the dura and the brain (subdural), or within the brain (intracerebral). A hemorrhage, which is bleeding, results in a hematoma, an accumulation of blood. **Extradural hemorrhages,** or **epidural hematomas,** occur in about 1%–2% of major head injuries. They usually occur in the middle cranial fossa and involve a tear of the middle cerebral artery (85%) or in the middle meningeal vein or dural sinus (15%). **Subdural hematomas** are much more common, occurring in 10%–20% of major head injuries. They most commonly involve tears of the cortical veins or dural sinuses, occur in the superior portion of the cranial cavity, and appear within hours of the head injury. **Chronic subdural hematomas,** which involve slow bleeding over an extended period of time (weeks to months), are common in elderly people or in people who abuse alcohol. **Intracerebral hematomas** occur in about 2%–3% of major head injuries and are often associated with contusions. They involve damage of small vessels within the brain itself and are most common in the frontal and temporal lobes. They may occur within 3–10 days of the head trauma.

Ventricles

The CNS is formed as a hollow tube that may be quite reduced in some areas of the adult CNS and expanded in other areas (see discussion of development, p. 459). The interior of the tube is lined with a single layer of epithelial cells called **ependymal** (ep-en′di-măl; see chapter 11) **cells.** Each cerebral hemisphere contains a relatively large cavity, the **lateral ventricle** (figure 13.11).

The lateral ventricles are separated from each other by thin **septa pellucida** (sep′tă pe-loo′sid-ă,; sing., septum pellucidum; translucent walls), which lie in the midline just inferior to the corpus callosum and usually are fused with each other. A smaller midline cavity, the **third ventricle,** is located in the center of the diencephalon between the two halves of the thalamus. The two lateral ventricles communicate with the third ventricle through two **interventricular foramina** (foramina of Monro). The lateral ventricles can be thought of as the first and second ventricles in the numbering scheme, but they are not designated as such. The **fourth ventricle** is in the inferior part of the pontine region and the superior region of the medulla oblongata at the base of the cerebellum. The third ventricle communicates with the fourth ventricle through a narrow canal, the **cerebral aqueduct** (aqueduct of Sylvius), which passes through the midbrain. The fourth ventricle is continuous with the central canal of the spinal cord, which extends nearly the full length of the cord. The fourth ventricle is also continuous with the subarachnoid space through two apertures in its walls and one in the roof.

Cerebrospinal Fluid

Cerebrospinal (ser′ĕ-brō-spī-năl; sĕ-rē′brō-spī-năl) **fluid (CSF)** is a fluid similar to serum with most of the proteins removed. It bathes the brain and the spinal cord and provides a protective cushion around the CNS. It also provides some nutrients to CNS tissues. About 80%–90% of the CSF is produced by specialized ependymal cells within the lateral ventricles, with the remainder produced by similar cells in the third and fourth ventricles. These specialized ependymal cells, their support tissue, and the associated blood vessels together are called **choroid** (kō′royd; lacy) **plexuses** (plek′sŭs-ez; figure 13.12). The choroid plexuses are formed by invaginations of the vascular pia mater into the ventricles, thus producing a vascular connective tissue core covered by ependymal cells.

CSF and Skull Fractures

Open head trauma involves some fracture or hole in the skull, which exposes the contents of the cranial cavity (brain, blood, and/or CSF) to the exterior. Head injuries involving scalp lacerations or damage to the eye, ear, or nose should be carefully evaluated for the possibility of open head trauma involving the cranial cavity. In skull fractures in which the meninges are torn, CSF may leak from the nose if the fracture is in the frontal area or from the ear if the fracture is in the temporal area. Leakage of CSF indicates serious mechanical damage to the head and presents a risk of meningitis, because bacteria may pass from the nose or ear through the tear and into the meninges. A skull fracture involving the base of the skull may result in cranial nerves damage where the nerves exit the skull.

How the choroid plexuses produce CSF is not fully understood. Some portions of the blood plasma simply diffuse across the plexus membranes, whereas other portions require facilitated diffusion or active transport.

Endothelial cells of the blood vessels in the choroid plexuses, which are joined by tight junctions (see chapter 4), form the

Figure 13.11 Ventricles of the Brain Viewed from the Left

Lateral view

Anterior horn of lateral ventricle

Interventricular foramen

Third ventricle

Inferior horn of lateral ventricle

Posterior horn of lateral ventricle

Cerebral aqueduct

Fourth ventricle

Central canal of spinal cord

blood–brain barrier, or, more correctly, the **blood–cerebrospinal fluid barrier.** Consequently, substances do not pass between the cells but must pass through the cells.

CSF fills the ventricles, the subarachnoid space of the brain and spinal cord, and the central canal of the spinal cord. Approximately 23 mL of fluid fills the ventricles, and 117 mL fills the subarachnoid space. The route taken by the CSF from its origin in the choroid plexuses to its return to the circulation is depicted in figure 13.12. The flow rate of CSF from its origin to the point at which it enters the bloodstream is about 0.4 mL/min. CSF passes from the lateral ventricles through the interventricular foramina into the third ventricle and then through the cerebral aqueduct into the fourth ventricle. It can exit the interior of the brain only from the fourth ventricle. One **median aperture** (foramen of Magendie), which opens through the roof of the fourth ventricle, and two **lateral apertures** (foramina of Luschka), which open through the walls, allow the CSF to pass from the fourth ventricle to the subarachnoid space. Masses of arachnoid tissue, **arachnoid granulations,** penetrate into the dural venous sinus along the superior edge of the falx cerebri called the superior sagittal sinus. CSF passes into the blood of the dural venous sinuses through these granulations. The sinuses are blood-filled; thus it is within these dural sinuses that the CSF reenters the bloodstream. From the dural venous sinuses, the blood flows through the internal jugular veins to veins of the general circulation.

Hydrocephalus

If the foramina of the fourth ventricle or the cerebral aqueduct are blocked, CSF can accumulate within the ventricles. This condition is called **internal hydrocephalus** and it results in increased CSF pressure. The production of CSF continues, even when the passages that normally allow it to exit the brain are blocked. Consequently, fluid builds inside the brain, causing pressure that compresses the nervous tissue and dilates the ventricles. Compression of the nervous tissue usually results in irreversible brain damage. If the skull bones are not completely ossified when the hydrocephalus occurs, the pressure may also severely enlarge the head. The cerebral aqueduct may be blocked at the time of birth or may become blocked later in life because of a tumor growing in the brainstem.

Internal hydrocephalus can be successfully treated by placing a drainage tube (shunt) between the brain ventricles and abdominal cavity to eliminate the high internal pressures. There is some risk of infection being introduced into the brain through these shunts, however, and the shunts must be replaced as the person grows. A subarachnoid hemorrhage may block the return of CSF to the circulation. If CSF accumulates in the subarachnoid space, the condition is called **external hydrocephalus.** In this condition, pressure is applied to the brain externally, compressing neural tissues and causing brain damage.

12. *Describe the three meninges that surround the CNS. What are the falx cerebri, tentorium cerebelli, and falx cerebelli?*

13. *Describe and list the contents of the dural sinuses subdural space, and subarachnoid space.*

1. Cerebrospinal fluid (CSF) is produced by the choroid plexuses of each of the four ventricles (*inset*).

2. CSF from the lateral ventricles flows through the interventricular foramina to the third ventricle.

3. CSF flows from the third ventricle through the cerebral aqueduct to the fourth ventricle.

4. CSF exits the fourth ventricle through the lateral and median apertures and enters the subarachnoid space. Some CSF enters the central canal of the spinal cord.

5. CSF flows through the subarachnoid space to the arachnoid granulations in the superior sagittal sinus, where it enters the venous circulation (*inset*).

Skull

Dura mater

Arachnoid mater

Subarachnoid space

Pia mater

Cerebrum

Falx cerebri (dura mater)

Superior sagittal sinus (dural venous sinus)

Arachnoid granulation

Subarachnoid space

Superior sagittal sinus

Arachnoid granulation

Subarachnoid space

Choroid plexus of lateral ventricle

Interventricular foramen

Choroid plexus of third ventricle

Cerebral aqueduct

Lateral aperture

Choroid plexus of fourth ventricle

Median aperture

Subarachnoid space

Central canal of spinal cord

Dura mater

Ependymal cells

Connective tissue

Capillary containing blood

Section of choroid plexus

CSF enters the ventricle

Sagittal section, medial view

Process Figure 13.12   Flow of CSF

CSF flows through the ventricles and subarachnoid space is shown by *white arrows*. Those going through the foramina in the wall and roof of the fourth ventricle depict the CSF entering the subarachnoid space. CSF passes back into the blood through the arachnoid granulations (*white and black arrows*), which penetrate the dural sinus. The *black arrows* show the direction of blood flow in the sinuses.

14. *Name the four ventricles of the brain, and describe their locations and the connections between them. What are the septa pellucida?*
15. *Describe the production and circulation of CSF. Where does the CSF return to the blood?*

Blood Supply to the Brain

The brain requires a tremendous amount of blood to maintain its normal functions. Even though the brain accounts for only about 2% of the total weight of the body, it receives approximately 15%–20% of blood pumped by the heart. Interruption of the brain's blood supply for only seconds can cause unconsciousness, and interruption of the blood supply for minutes can cause irreversible brain damage. This extreme dependence on blood supply results from the brain's very high metabolic rate and, as a result, its extreme dependence on a constant supply of oxygen and glucose. Brain cells are not capable of storing high-energy molecules for any length of time and depend almost exclusively on glucose as their energy source (see chapter 25).

The brain's blood supply is illustrated in chapter 21 (see figures 21.10 and 21.11). Blood reaches the brain through the **internal carotid arteries,** which ascend to the head along the anterior-lateral part of the neck, and the **vertebral arteries,** which ascend along the posterior part of the neck, through the transverse foramina of the cervical vertebrae. The internal carotid arteries enter the cranial cavity through the carotid canals, and the vertebral arteries enter by the foramen magnum. The vertebral arteries join together to form the **basilar artery,** which lies on the ventral surface of the brainstem. The basilar artery and internal carotid arteries contribute to the **cerebral arterial circle** (circle of Willis). Branches from this circle and from the basilar artery supply blood to the brain.

The cerebral cortex on each side of the brain is supplied by three branches from the cerebral arterial circle: the **anterior, middle,** and **posterior cerebral arteries.** The middle cerebral artery supplies most of the lateral surface of each cerebral hemisphere. The anterior cerebral artery supplies the medial portion of the parietal and frontal lobes. The posterior cerebral artery supplies the occipital lobe and the medial surface of the temporal lobe.

The arteries to the brain and their larger branches are located in the subarachnoid space. Small cortical arterial branches leave the subarachnoid space and enter the pia mater, where they branch extensively. Precapillary branches leave the pia mater and enter the substance of the brain. Most of these branches are short and remain in the cortex. Fewer, longer branches extend into the medulla.

The arteries within the substance of the brain quickly divide into capillaries. The epithelial cells of these capillaries are surrounded by the foot processes of astrocytes (see figure 11.6). The astrocytes promote the formation of tight junctions between the epithelial cells. The epithelial cells with their tight junctions form the **blood–brain barrier,** which regulates the movement of materials from the blood into the brain. Materials that would enter many tissues by passing between the epithelial cells of capillaries do not pass through the blood–brain barrier because of the tight junctions. Most materials that enter the brain pass through the epithelial cells. Lipid-soluble substances, such as nicotine, ethanol, and heroin, can diffuse through the plasma membranes of the epithelial cells and enter the brain. Water-soluble molecules such as amino acids and

glucose move across the plasma membranes of the epithelial cells by mediated transport (see chapter 3).

Drugs and the Blood–Brain Barrier

The permeability characteristics of the blood–brain barrier must be considered when developing drugs designed to affect the CNS. For example, Parkinson's disease is caused by a lack of the neurotransmitter dopamine, which normally is produced by certain neurons of the brain. This lack results in decreased muscle control and shaking movements. Administering dopamine is not helpful because dopamine cannot cross the blood–brain barrier. Levodopa (L-dopa), a precursor to dopamine, is administered instead because it can cross the blood–brain barrier. CNS neurons then convert levodopa to dopamine, which helps reduce the symptoms of Parkinson's disease.

16. *Describe the blood supply to the brain. List the arteries supplying each part of the cerebral cortex.*
17. *Describe the blood–brain barrier.*

Development of the CNS

The CNS develops from a flat plate of ectodermal tissue, the **neural plate,** on the upper surface of the embryo, as a result of the influence of the underlying rod-shaped **notochord** (figure 13.13). The lateral sides of the neural plate become elevated as waves, called **neural folds.** The crest of each fold is called a **neural crest,** and the center of the neural plate becomes the **neural groove.** The neural folds move toward each other in the midline, and the crests fuse to create a **neural tube** (see figure 13.13b). The cephalic portion of the neural tube becomes the brain, and the caudal portion becomes the spinal cord. **Neural crest cells** separate from the neural crests and give rise to sensory and autonomic neurons of the peripheral nervous system. They also give rise to all pigment cells of the body, as well as facial bones and dentin of the teeth.

A series of pouches develops in the anterior part of the neural tube (figure 13.14). The pouch walls become the various portions of the adult brain (table 13.3), and the pouch cavities become fluid-filled **ventricles** (ven′tri-klz). The ventricles are continuous with each other and with the **central canal** of the spinal cord. The neural tube develops flexures that cause the brain to be oriented almost 90 degrees to the spinal cord.

Three brain regions can be identified in the early embryo (see table 13.3 and figure 13.14a): a forebrain, or **prosencephalon** (pros-en-sef′ă-lon); a midbrain, or **mesencephalon** (mez-en-sef′ă-lon); and a hindbrain, or **rhombencephalon** (rom-ben-sef′ă-lon). During development, the forebrain divides into the **telencephalon** (tel-en-sef′ă-lon), which becomes the cerebrum, and the **diencephalon** (dī-en-sef′ă-lon). The midbrain remains as a single structure, but the hindbrain divides into the **metencephalon** (met′en-sef′ă-lon), which becomes the pons and cerebellum, and the **myelencephalon** (mī′el-en-sef′ă-lon), which becomes the medulla oblongata (figure 13.14b and c).

18. *Explain how the neural tube forms. Name the five divisions of the neural tube and the parts of the brain that each division becomes.*
19. *What do the cavities of the neural tube become in the adult brain?*

1. The neural plate is formed from ectoderm.

2. Neural folds form as parallel ridges along the embryo.

3. Neural crest cells break away from the crest of the neural folds. Neural crest cells give rise to a number of structures: sensory and autonomic neurons in the PNS, facial pigment cells, facial bones, and dentin of the teeth.

4. The neural folds meet at the midline to form the neural tube. The neural tube becomes the brain and spinal cord.

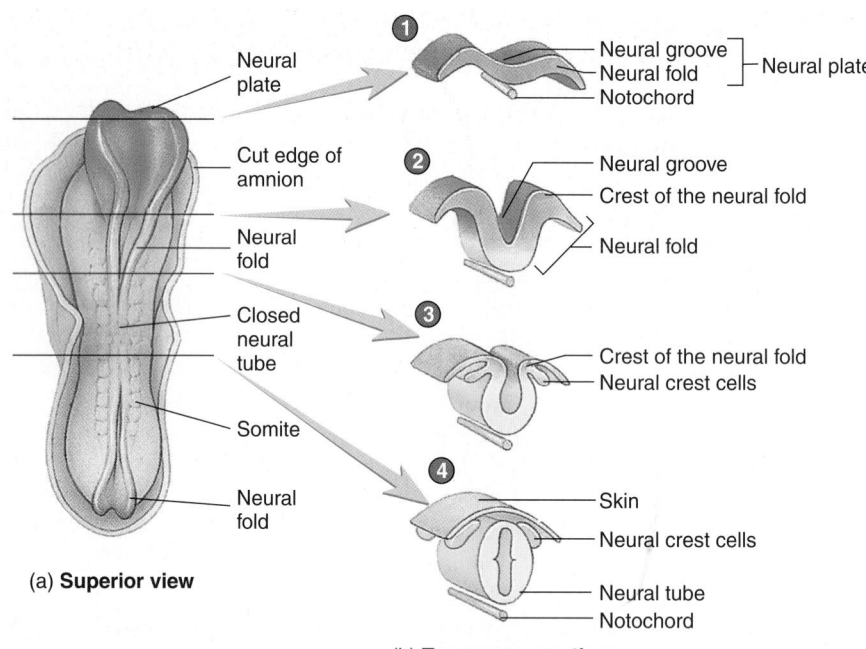

(a) **Superior view**

(b) **Transverse sections**

Process Figure 13.13   Formation of the Neural Tube

(*a*) A 21-day-old human embryo. (*b*) Cross sections through the embryo. The level of each section is indicated by a line in part (*a*).

Prosencephalon

Optic vesicle (eye)

Mesencephalon

Rhombencephalon

Spinal cord

(a) **Early embryo**

Telencephalon

Optic vesicle

Diencephalon

Mesencephalon

Metencephalon

Myelencephalon

Spinal cord

(b) **Late embryo**

Cerebrum (from telencephalon)

Diencephalon

Midbrain (mesencephalon)

Cerebellum (from metencephalon)

Pons (from metencephalon)

Medulla oblongata (from myelencephalon)

Brainstem

Spinal cord

(c) **Adult**

Figure 13.14   Development of the Brain Segments and Ventricles

Table 13.3	Development of the Central Nervous System (see figure 13.14)			
Early Embryo	**Late Embryo**	**Adult**	**Cavity**	**Function**
Prosencephalon (forebrain)	**Telencephalon**	**Cerebrum**	Lateral ventricles	Higher brain functions
	Diencephalon	**Diencephalon** (thalamus, subthalamus, epithalamus, hypothalamus)	Third ventricle	Relay center, autonomic nerve control, endocrine control
Mesencephalon (midbrain)	**Mesencephalon**	**Mesencephalon** (midbrain)	Cerebral aqueduct	Nerve pathways, reflex centers
Rhombencephalon (hindbrain)	**Metencephalon**	**Pons** and **cerebellum**	Fourth ventricle	Nerve pathways, reflex centers, muscle coordination, balance
	Myelencephalon	**Medulla oblongata**	Central canal	Nerve pathways, reflex centers

Cranial Nerves

The 12 pairs of cranial nerves by convention are indicated by Roman numerals (I–XII) from anterior to posterior (figure 13.15). The first two pairs of cranial nerves connect directly to the cerebrum (I) or diencephalon (II). Nine pairs of the cranial nerves connect to the brainstem. One pair of cranial nerves (XI) is connected to the spinal cord and has no direct connection to brain structures. A given cranial nerve may have one or more of three functions: (1) sensory, (2) somatic motor, and (3) parasympathetic (table 13.4). **Sensory** functions include the special senses like vision and the more general senses like touch and pain. **Somatic** (sō-mat′ik) **motor** functions refer to the control of skeletal muscles through motor neurons. **Proprioception** (prō-prē-ō-sep′shun) informs the brain about the position of various body parts, including joints and muscles. The cranial nerves innervating skeletal muscles also contain proprioceptive sensory fibers, which convey action potentials to the CNS from those muscles. Because proprioception is the only sensory function of several otherwise somatic motor cranial nerves, however, that function is usually ignored, and the nerves are designated by convention as motor only. **Parasympathetic** function involves the regulation of glands, smooth muscles, and cardiac muscle. These functions are part of the autonomic nervous system and are discussed in chapter 16. Several of the cranial nerves have associated ganglia, and these ganglia are of two types: parasympathetic and sensory. Table 13.5 lists specific information about each cranial nerve.

The **olfactory (I)** and **optic (II) nerves** are exclusively sensory and are involved in the special senses of smell and vision, respectively. These nerves are discussed in chapter 15.

The **oculomotor nerve (III)** innervates four of the six muscles that move the eyeball (the superior, inferior, and medial rectus muscles and the inferior oblique muscle; see chapter 10), and the levator palpebrae superioris muscle, which raises the superior eyelid. In addition, parasympathetic nerve fibers in the oculomotor nerve innervate smooth muscles in the eye and regulate the size of the pupil and the shape of the lens of the eye.

The **trochlear** (trōk′lē-ar) **nerve (IV)** is a somatic motor nerve that innervates one of the six eye muscles responsible for moving the eyeball (superior oblique).

The **trigeminal** (trī-jem′i-năl) **nerve (V)** has somatic motor, proprioceptive, and cutaneous sensory functions. It supplies motor innervation to the muscles of mastication, one middle ear muscle, one palatine muscle, and two throat muscles. In addition to proprioception associated with its somatic motor functions, the trigeminal nerve also carries proprioception from the temporomandibular joint. Damage to the trigeminal nerve may impede chewing.

The trigeminal nerve has the greatest general sensory function of all the cranial nerves and is the only cranial nerve involved in **sensory cutaneous innervation.** All other cutaneous innervation comes from spinal nerves (see figure 12.14). Trigeminal means three twins, and the sensory distribution of the trigeminal nerve in the face is divided into three regions, each supplied by a branch of the nerve. The three branches—ophthalmic, maxillary, and mandibular—arise directly from the trigeminal ganglion, which serves the same function as the dorsal root ganglia of the spinal nerves. Only the mandibular branch has motor axons, which bypass the trigeminal ganglion, much like the ventral root of a spinal nerve bypasses a dorsal root ganglion.

In addition to these cutaneous functions, the maxillary and mandibular branches are important in dentistry. The maxillary nerve supplies sensory innervation to the maxillary teeth, palate, and gingiva (jin′jī-vă; gum). The mandibular branch supplies sensory innervation to the mandibular teeth, tongue, and gingiva. The various nerves innervating the teeth are referred to as **alveolar** (al-vē′ō-lăr; refers to the sockets in which the teeth are located). The **superior alveolar nerves** to the maxillary teeth are derived from the maxillary branch of the trigeminal nerve, and the **inferior alveolar nerves** to the mandibular teeth are derived from the mandibular branch of the trigeminal nerve.

Anterior

Olfactory bulb (olfactory nerves [I] enter bulb)

Optic nerve (**II**)

Oculomotor nerve (**III**)

Trochlear nerve (**IV**)

Trigeminal nerve (**V**)

Abducens nerve (**VI**)

Facial nerve (**VII**)

Vestibulocochlear nerve (**VIII**)

Glossopharyngeal nerve (**IX**)

Vagus nerve (**X**)

Hypoglossal nerve (**XII**)

Accessory nerve (**XI**)

Olfactory tract

Optic chiasm

Pituitary gland

Mammillary body

Pons

Olive of medulla oblongata

Medulla oblongata

Posterior

Inferior view

Figure 13.15   Inferior Surface of the Brain Showing the Origin of the Cranial Nerves

Table 13.4	Functional Organization of the Cranial Nerves	
Nerve Function	**Cranial Nerve**	
Sensory	I	Olfactory
	II	Optic
	VIII	Vestibulocochlear
Somatic motor	IV	Trochlear
	VI	Abducent
	XI	Accessory
	XII	Hypoglossal
Somatic motor and sensory	V	Trigeminal
Somatic motor and parasympathetic	III	Oculomotor
Somatic motor, sensory, and parasympathetic	VII	Facial
	IX	Glossopharyngeal
	X	Vagus

Dental Anesthesia

Dentists inject anesthetic to block sensory transmission by the alveolar nerves. The superior alveolar nerves are not usually anesthetized directly because they are difficult to approach with a needle. For this reason, the maxillary teeth are usually anesthetized locally by inserting the needle beneath the oral mucosa surrounding the teeth. The inferior alveolar nerve probably is anesthetized more often than any other nerve in the body. To anesthetize this nerve, the dentist inserts the needle somewhat posterior to the patient's last molar.

Several nondental nerves are usually anesthetized during an inferior alveolar block. The mental nerve, which supplies cutaneous innervation to the anterior lip and chin, is a distal branch of the inferior alveolar nerve. When the inferior alveolar nerve is blocked, the mental nerve is blocked also, resulting in a numb lip and chin. Nerves lying near the point where the inferior alveolar nerve enters the mandible often are also anesthetized during inferior alveolar anesthesia. For example, the lingual nerve can be anesthetized to produce a numb tongue. The facial nerve lies some distance from the inferior alveolar nerve, but in rare cases anesthetic can diffuse far enough posteriorly to anesthetize that nerve. The result is a temporary facial palsy (paralysis or paresis), with the injected side of the face drooping because of flaccid muscles, which disappears when the anesthesia wears off. If the facial nerve is cut by an improperly inserted needle, permanent facial palsy may occur.

Table 13.5 Cranial Nerves and Their Functions

Cranial Nerve	Foramen or Fissure*	Function
I: Olfactory	Cribriform plate	Sensory Special sense of smell
II: Optic	Optic foramen	Sensory Special sense of vision
III: Oculomotor	Superior orbital fissure	Motor† and parasympathetic Motor to eye muscles (superior, medial, and inferior rectus; inferior oblique) and upper eyelid (levator palpebrae superioris) Proprioceptive from those muscles Parasympathetic to the sphincter of the pupil (causing constriction) and the ciliary muscle of the lens (causing accommodation)

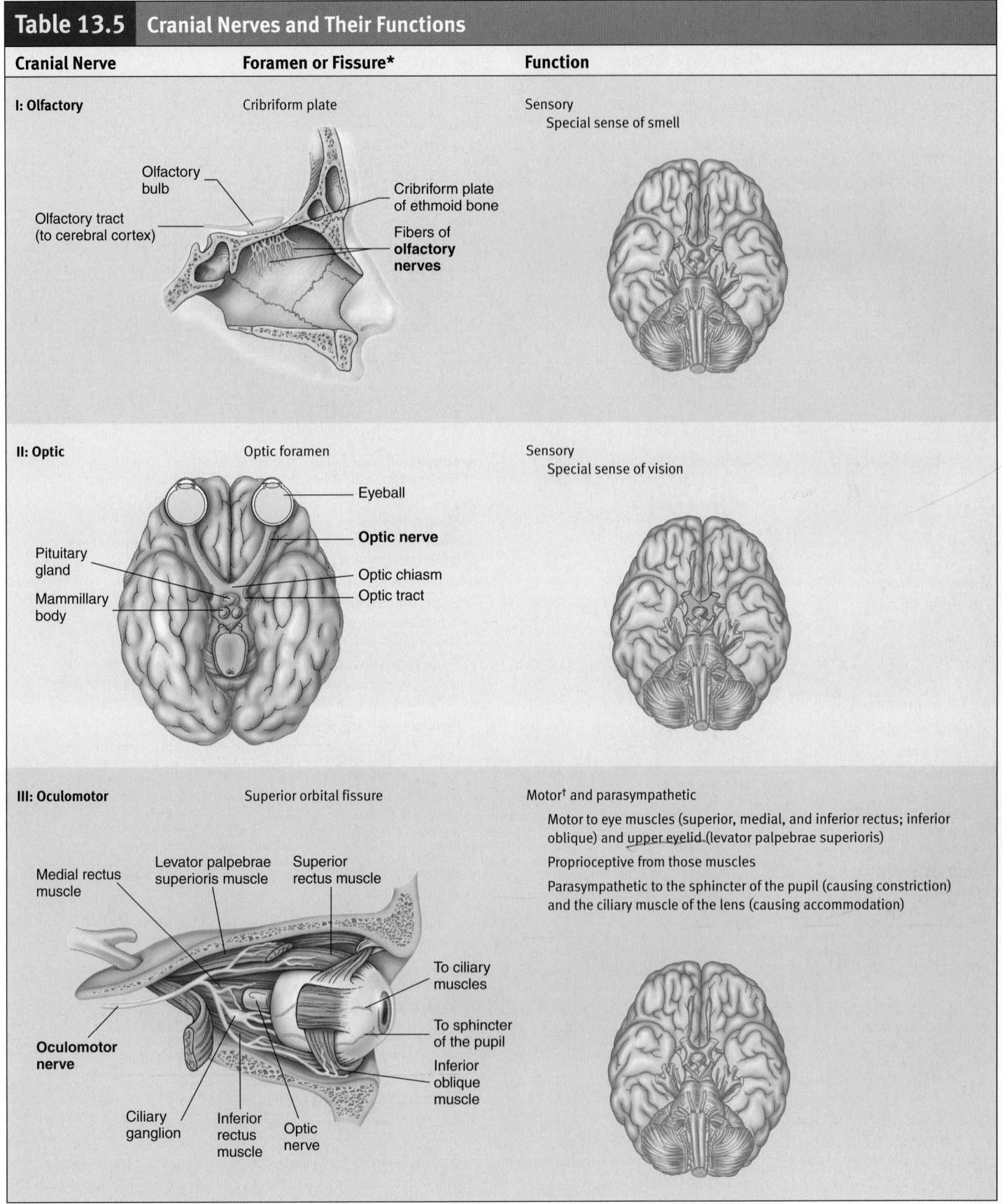

*Route of entry or exit from the skull.

†Proprioception is a sensory function, not a motor function; however, motor nerves to muscles also contain some proprioceptive afferent fibers from those muscles. Because proprioception is the only sensory information carried by some cranial nerves, these nerves still are considered "motor."

continued

Table 13.5 *continued*

Cranial Nerve	Foramen or Fissure*	Function
IV: Trochlear	Superior orbital fissure	Motor[†] Motor to one eye muscle (superior oblique) Proprioceptive from that muscle

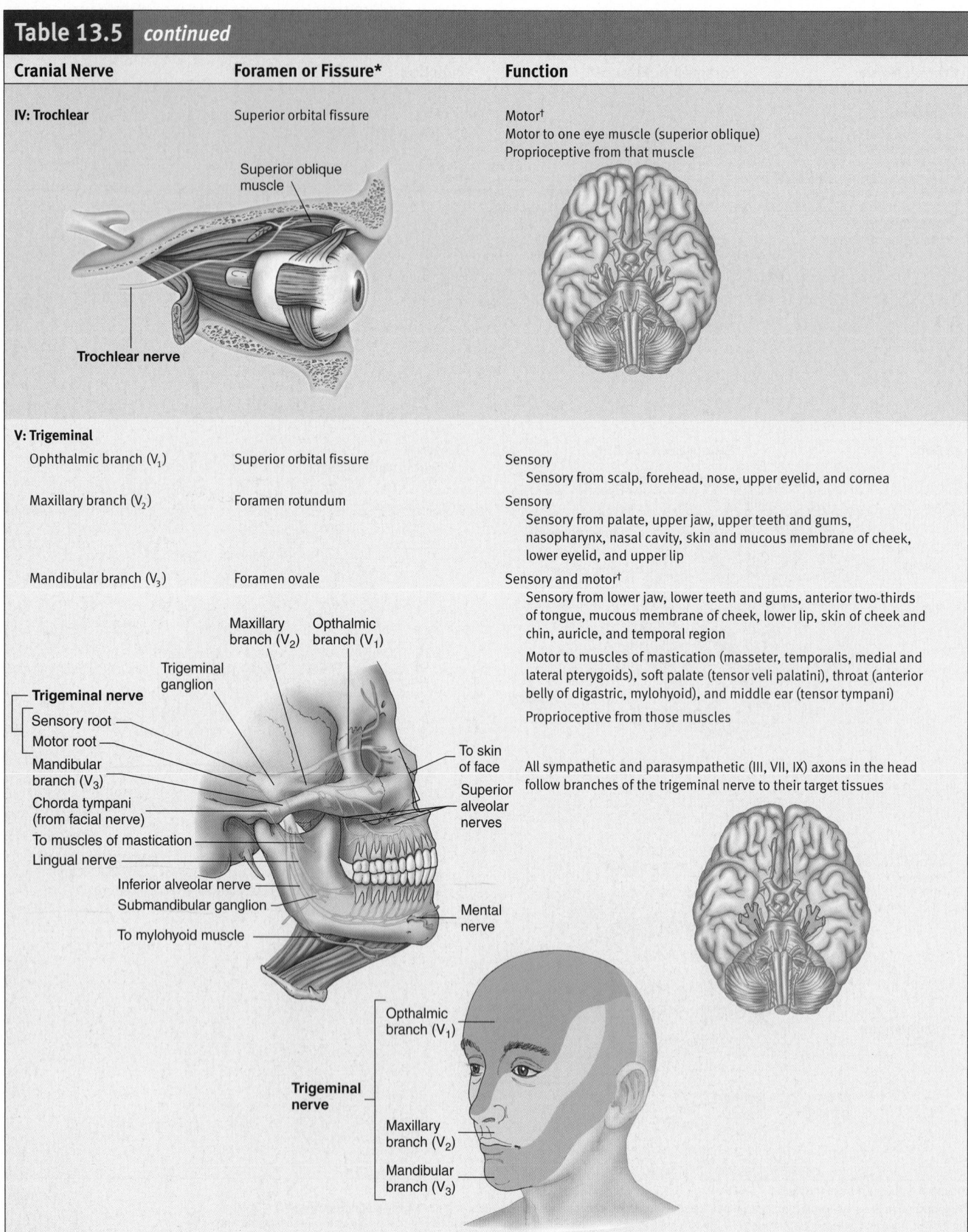

V: Trigeminal

Cranial Nerve	Foramen or Fissure*	Function
Ophthalmic branch (V_1)	Superior orbital fissure	Sensory Sensory from scalp, forehead, nose, upper eyelid, and cornea
Maxillary branch (V_2)	Foramen rotundum	Sensory Sensory from palate, upper jaw, upper teeth and gums, nasopharynx, nasal cavity, skin and mucous membrane of cheek, lower eyelid, and upper lip

Mandibular branch (V_3)	Foramen ovale	Sensory and motor[†]
		  Sensory from lower jaw, lower teeth and gums, anterior two-thirds of tongue, mucous membrane of cheek, lower lip, skin of cheek and chin, auricle, and temporal region
		
		  Motor to muscles of mastication (masseter, temporalis, medial and lateral pterygoids), soft palate (tensor veli palatini), throat (anterior belly of digastric, mylohyoid), and middle ear (tensor tympani)
		
		  Proprioceptive from those muscles
		
		All sympathetic and parasympathetic (III, VII, IX) axons in the head follow branches of the trigeminal nerve to their target tissues

Superior oblique muscle

Trochlear nerve

Maxillary branch (V_2)

Opthalmic branch (V_1)

Trigeminal ganglion

Trigeminal nerve

Sensory root

Motor root

Mandibular branch (V_3)

Chorda tympani (from facial nerve)

To muscles of mastication

Lingual nerve

Inferior alveolar nerve

Submandibular ganglion

To mylohyoid muscle

To skin of face

Superior alveolar nerves

Mental nerve

Opthalmic branch (V_1)

Trigeminal nerve

Maxillary branch (V_2)

Mandibular branch (V_3)

continued

Table 13.5 *continued*

Cranial Nerve	Foramen or Fissure*	Function
VI: Abducent (abducens)	Superior orbital fissure	Motor[†] Motor to one eye muscle (lateral rectus) Proprioceptive from that muscle
VII: Facial	Internal auditory meatus Stylomastoid foramen	Sensory, motor,[†] and parasymathetic Sense of taste from anterior two-thirds of tongue, sensory from some of external ear and palate Motor to muscles of facial expression, throat (posterior belly of digastric, stylohyoid), and middle ear (stapedius) Proprioceptive from those muscles Parasympathetic to submandibular and sublingual salivary glands, lacrimal gland, and glands of the nasal cavity and palate

Figure labels (VI: Abducent):
Abducent nerve
Lateral rectus muscle

Figure labels (VII: Facial):
Trigeminal ganglion
Geniculate ganglion
Pterygopalatine ganglion
To lacrimal gland and nasal mucous membrane
Facial nerve
To forehead muscles
To orbicularis oculi
To orbicularis oris and upper lip muscles
To occipitofrontalis
Chorda tympani (for salivary glands, sense of taste)
To digastric and stylohyoid muscles
To buccinator, lower lip, and chin muscles
To platysma

continued

Table 13.5 *continued*

Cranial Nerve	Foramen or Fissure*	Function
VIII: Vestibulocochlear	Internal auditory meatus	Sensory Special senses of hearing and balance

Cranial Nerve	Foramen or Fissure*	Function

IX: Glossopharyngeal	Jugular foramen	Sensory, motor,[†] and parasympathetic
		  Sense of taste from posterior third of tongue, sensory from pharynx, palatine tonsils, posterior third of tongue, middle ear, carotid sinus and carotid body
		  Motor to pharyngeal muscle (stylopharyngeus)
		  Proprioceptive from that muscle
		  Parasympathetic to parotid salivary gland and the glands of the posterior third of tongue

Vestibular
ganglion

Vestibular nerve

**Vestibulocochlear
nerve**

Cochlear
nerve

Spiral ganglion
of cochlea

Superior and inferior ganglia

To parotid gland

Glossopharyngeal nerve

To stylopharyngeus muscle

To palatine tonsil

To pharynx

To carotid
body and
carotid sinus

To posterior third
of tongue for
taste and
general sensation

continued

Table 13.5 *continued*

Cranial Nerve	Foramen or Fissure*	Function
X: Vagus	Jugular foramen	Sensory, motor,† and parasympathetic Sensory from inferior pharynx, larynx, thoracic and abdominal organs, sense of taste from posterior tongue Motor to soft palate, pharynx, intrinsic laryngeal muscles (voice production), and an extrinsic tongue muscle (palatoglossus) Proprioceptive from those muscles Parasympathetic to thoracic and abdominal viscera
XI: Accessory	Foramen magnum Jugular foramen	Motor† Motor to sternocleidomastoid and trapezius

continued

Table 13.5	*continued*	
Cranial Nerve	**Foramen or Fissure***	**Function**
XII: Hypoglossal	Hypoglossal canal	Motor† Motor to intrinsic and extrinsic tongue muscles (styloglossus, hypoglossus, genioglossus) and throat muscles (thyrohyoid and geniohyoid) Proprioceptive from those muscles

The **abducent** (ab-doo′sent), or **abducens, nerve (VI),** like the trochlear nerve, is a somatic motor nerve that innervates one of the six eye muscles responsible for moving the eyeball (lateral rectus).

P R E D I C T   ①

A drooping upper eyelid on one side of the face is a sign of possible oculomotor nerve damage. Describe how this could possibly be evaluated by examining other oculomotor nerve functions. Describe the movements of the eye that would distinguish among oculomotor, trochlear, and abducens nerve damage.

The **facial nerve (VII)** is somatic motor, sensory, and parasympathetic. It controls all the muscles of facial expression, a small muscle in the middle ear, and two hyoid muscles. It is sensory for the sense of taste in the anterior two-thirds of the tongue (see chapter 15). The facial nerve supplies parasympathetic innervation to the submandibular and sublingual salivary glands and to the lacrimal glands.

The **vestibulocochlear** (ves-tib′ū-lō-kok′lē-ăr) **nerve (VIII),** like the olfactory and optic nerves, is exclusively sensory and transmits action potentials from the inner ear responsible for the special senses of hearing and balance (see chapter 15).

The **glossopharyngeal** (glos′ō-fă-rin′jē-ăl) **nerve (IX),** like the facial nerve, is somatic motor, sensory, and parasympathetic and has both sensory and parasympathetic ganglia. The glossopharyngeal nerve is somatic motor to one muscle of the pharynx and supplies parasympathetic innervation to the parotid salivary glands. The glossopharyngeal nerve is sensory for the sense of taste in the posterior third of the tongue. It also supplies tactile sensory innervation from the posterior tongue, middle ear, and pharynx and transmits sensory stimulation from receptors in the carotid arteries, which monitor blood pressure and blood carbon dioxide, blood oxygen, and blood pH levels (see chapter 21).

The **vagus** (vā′gŭs) **nerve (X),** like the facial and glossopharyngeal nerves, is somatic motor, sensory, and parasympathetic and has both sensory and parasympathetic ganglia. Most muscles of the soft palate, pharynx, and larynx are innervated by the vagus nerve. Damage to the laryngeal branches of the vagus nerve can interfere with normal speech. The vagus nerve is sensory for taste from the root of the tongue (see chapter 15). It's sensory for the inferior pharynx and the larynx and transmits sensory input from receptors in the aortic arch, which monitor blood pressure and carbon dioxide, oxygen, and pH levels in the blood (see chapter 21). In addition, the vagus nerve conveys sensory information from the thoracic and abdominal organs. The parasympathetic part of the vagus nerve is very important in regulating the functions of the thoracic and abdominal organs. It carries parasympathetic fibers to the heart and lungs in the thorax and to the digestive organs and kidneys in the abdomen.

The **accessory (XI) nerve** is a somatic motor nerve. The accessory nerve has been thought for many years to have both a cranial and a spinal component. The cranial component was thought to join the vagus nerve (hence the name *accessory*) and participate in its function. Our understanding of the accessory nerve's structure was based on dissections conducted nearly 150 years ago and, apparently, not repeated until very recently. Extensive dissections of fifteen human brainstems have failed to find any connection between the brainstem and the accessory nerve. These data suggest that the accessory nerve originates entirely from the superior part of the cervical spinal cord, enters the cranial cavity through the foramen magnum, and then exits through the jugular foramen. The accessory nerve provides the major innervation to the sternocleidomastoid and trapezius muscles of the neck and shoulder.

The **hypoglossal (XII) nerve** is a somatic motor nerve that arises from the ventral surface of the medulla oblongata. It supplies the intrinsic tongue muscles, three of the four extrinsic tongue muscles, and the thyrohyoid and the geniohyoid muscles.

20. *What are the three major functions of the cranial nerves?*
21. *Which cranial nerves are sensory only? With what sense is each of these nerves associated?*
22. *Name the cranial nerves that are somatic motor and proprioceptive only. What muscles or muscle groups does each nerve supply?*
23. *The sensory cutaneous innervation of the face is provided by what cranial nerve? How is this nerve important in dentistry? Name the muscles that would not function if this nerve was damaged.*
24. *Which four cranial nerves have a parasympathetic function? Describe the function of each of these nerves.*

25. *Name the cranial nerves that control the movement of the eyeball.*
26. *Which cranial nerves are involved in the sense of taste? What part of the tongue does each supply?*
27. *Speech production involves which cranial nerves? Describe the branches of these nerves.*

P R E D I C T

Injury to the accessory nerve may result in sternocleidomastoid muscle dysfunction, a condition called "wry neck." If the head of a person with wry neck is turned to the left, would this position indicate injury to the left or right spinal component of the accessory nerve?

P R E D I C T

Unilateral damage to the hypoglossal nerve results in loss of tongue movement on one side, which is most obvious when the tongue is protruded. If the tongue is deviated to the right, is the left or right hypoglossal nerve damaged?

Reflexes in the Brainstem Involving Cranial Nerves

Reflexes integrated within the spinal cord were discussed in chapter 12. Many of the body's functions, especially those involved in maintaining homeostasis, involve reflexes that are integrated within the brain. Some of these reflexes, such as those involved in the control of heart rate (see chapter 20), blood pressure (see chapter 21), and respiration (see chapter 23), are integrated in the brainstem and many involve cranial nerve X (vagus nerve).

Many of the brainstem reflexes are associated with cranial nerve function. The circuitry of most of these reflexes is too complex for our discussions, but some general outlines can be presented. These reflexes involve sensory input from the cranial nerves or spinal cord, and the motor output of the motor cranial nerves.

Turning of the eyes toward a flash of light, sudden noise, or a touch on the skin are examples of brainstem reflexes. Moving the eyes to track a moving object is another complex brainstem reflex. Some of the sensory neurons from cranial nerve VIII form a reflex arc with neurons of cranial nerves V and VII, which send axons to muscles of the middle ear and dampen the effects of very loud, sustained noises on delicate inner ear structures (see chapter 15). Reflexes that occur during the process of chewing allow the jaws to react to foods of various hardness and protect the teeth from breakage from very hard food items. Both the sensory and motor components of the reflex arc are carried by cranial nerve V. Reflexes involving input through cranial nerve V and output through cranial nerve XII move the tongue about to position food between the teeth for chewing and then move the tongue out of the way so it isn't bitten!

Clinical Focus   Peripheral Nervous System Disorders—Cranial Nerves

General issues of PNS disorders are described in chapter 12. This chapter addresses only those specific to the cranial nerves.

Trigeminal neuralgia, also called tic douloureux, involves one or more of the trigeminal nerve branches and consists of sharp bursts of pain in the face. This disorder often has a trigger point in or around the mouth, which, when touched, elicits the pain response in some other part of the face. The cause of trigeminal neuralgia is unknown.

Facial palsy (called Bell's palsy) is a unilateral paralysis of the facial muscles. The affected side of the face droops because of the absence of muscle tone. Facial palsy involves the facial nerve and may result from facial nerve neuritis. The

facial nerve passes from deep to superficial through the parotid gland. Although the cause of facial palsy is often unknown, it can result from a stroke or tumor in the cerebral cortex or brainstem (see chapter 14). Temporary facial palsy can result from inflammation of the parotid gland or from anesthesia accidently introduced into the gland during dental anesthesia. Temporary facial palsy can even result from extreme cold in the face, where the superficial branches of the facial nerve are located.

Infections

Herpes simplex I is usually characterized by one or more lesions (sores) on the lips or nose. The virus apparently remain

dormant in the trigeminal ganglion. Eruptions are usually recurrent and often occur in times of reduced resistance, such as during a case of the common cold. For this reason they are called cold sores or fever blisters.

Genetic and Autoimmune Disorders

Neurofibromatosis (noor'ō-fī-brō-mă-tō'sis) is a genetic disorder in which small skin lesions appear in early childhood followed by the development of multiple subcutaneous neurofibromas, which are benign tumors resulting from Schwann cell proliferation. The neurofibromas may slowly increase in size and number over several years and cause extreme disfiguration.

S U M M A R Y

Brainstem   (p. 445)
Medulla Oblongata

1. The medulla oblongata is continuous with the spinal cord and contains ascending and descending nerve tracts.
2. The pyramids are nerve tracts controlling voluntary muscle movement.
3. The olives are nuclei that function in equilibrium, coordination, and modulation of sound from the inner ear.
4. Medullary nuclei regulate the heart, blood vessels, respiration, swallowing, vomiting, coughing, sneezing, and hiccuping. The nuclei of cranial nerves V and IX–XII are in the medulla.

Pons

1. The pons is superior to the medulla.
2. Ascending and descending nerve tracts pass through the pons.
3. Pontine nuclei regulate sleep and respiration. The nuclei of cranial nerves V–IX are in the pons.

Midbrain

1. The midbrain is superior to the pons.
2. The midbrain contains the nuclei for cranial nerves III, IV, and V.
3. The tectum consists of four colliculi. The two inferior colliculi are involved in hearing and the two superior colliculi in visual reflexes.
4. The tegmentum contains ascending tracts and the red nuclei, which are involved in motor activity.
5. The cerebral peduncles are the major descending motor pathway.
6. The substantia nigra connects to other basal nuclei and is involved with muscle tone and movement.

Reticular Formation

The reticular formation consists of nuclei scattered throughout the brainstem. The reticular-activating system extends to the thalamus and cerebrum and maintains consciousness.

Cerebellum   (p. 447)

1. The cerebellar cortex contains more neurons than does the cerebral cortex. The Purkinje cells are the largest cells in the CNS.
2. The cerebellum has three parts that control balance, gross motor coordination, and fine motor coordination.
3. The cerebellum functions to correct discrepancies between intended movements and actual movements.
4. The cerebellum can "learn" highly specific complex motor activities.

Diencephalon   (p. 449)

The diencephalon is located between the brainstem and the cerebrum.

Thalamus

1. The thalamus consists of two lobes connected by the intermediate mass. The thalamus functions as an integration center.
2. Most sensory input synapses in the thalamus. Pain is registered in the thalamus.
3. The thalamus also has some motor functions.

Subthalamus

The subthalamus is inferior to the thalamus and is involved in motor function.

Epithalamus

The epithalamus is superior and posterior to the thalamus and contains the habenular nuclei, which influence emotions through the sense of smell. The pineal body may play a role in the onset of puberty.

Hypothalamus

1. The hypothalamus, the most inferior portion of the diencephalon, contains several nuclei and tracts.
2. The mammillary bodies are reflex centers for olfaction.

3. The hypothalamus regulates many endocrine functions (e.g., metabolism, reproduction, response to stress, and urine production). The pituitary gland attaches to the hypothalamus.
4. The hypothalamus regulates body temperature, hunger, thirst, satiety, swallowing, and emotions.

Cerebrum   (p. 451)

1. The cortex of the cerebrum is folded into ridges called gyri and grooves called sulci, or fissures.
2. The longitudinal fissure divides the cerebrum into left and right hemispheres. Each hemisphere has five lobes.
   • The frontal lobes are involved in smell, voluntary motor function, motivation, aggression, and mood.
   • The parietal lobes contain the major sensory areas receiving general sensory input, taste, and balance.
   • The occipital lobes contain the visual centers.
   • The temporal lobes receive olfactory and auditory input and are involved in memory, abstract thought, and judgment.
3. Nerve tracts connect areas of the cortex within the same hemisphere (association fibers), between different hemispheres (commissural fibers), and with other parts of the brain and the spinal cord (projection fibers).

Basal Nuclei

1. Basal nuclei include the subthalamic nuclei, substantia nigra, and corpus striatum.
2. The basal nuclei are important in controlling motor functions.

Limbic System

1. The limbic system includes parts of the cerebral cortex, basal nuclei, thalamus, hypothalamus, and the olfactory cortex.
2. The limbic system controls visceral functions through the autonomic nervous system and the endocrine system and is also involved in emotions and memory.

Meninges and Cerebrospinal Fluid   (p. 454)
Meninges

1. The brain and spinal cord are covered by the dura, arachnoid, and pia mater.
2. The dura mater attaches to the skull and has two layers that can separate to form dural sinuses.
3. Beneath the arachnoid mater the subarachnoid space contains CSF that helps cushion the brain.
4. The pia mater attaches directly to the brain.

Ventricles

1. The lateral ventricles in the cerebrum are connected to the third ventricle in the diencephalon by the interventricular foramen.
2. The third ventricle is connected to the fourth ventricle in the pons by the cerebral aqueduct. The central canal of the spinal cord is connected to the fourth ventricle.

Cerebrospinal Fluid

1. CSF is produced from the blood in the choroid plexus of each ventricle. CSF moves from the lateral to the third and then to the fourth ventricle.
2. From the fourth ventricle CSF enters the subarachnoid space through three foramina.

3. CSF leaves the subarachnoid space through arachnoid granulations and returns to the blood in the dural sinuses.

Blood Supply to the Brain   (p. 459)

1. The brain receives blood from the internal carotid and vertebral arteries. The latter form the basilar artery. The basilar and internal carotid arteries contribute to the cerebral arterial circle. Branches from the circle and basilar artery supply the brain.
2. The blood–brain barrier is formed from the endothelial cells of the capillaries in the brain, the astrocytes in the brain tissue, and the basement membrane in between.

Development of the CNS   (p. 459)

The brain and spinal cord develop from the neural tube. The ventricles and central canal develop from the lumen of the neural tube.

Cranial Nerves   (p. 461)

1. Cranial nerves perform sensory, somatic motor, proprioceptive, and parasympathetic functions.
2. The olfactory (I) and optic (II) nerves are involved in the sense of smell and vision.
3. The oculomotor nerve (III) innervates four of six extrinsic eye muscles and the upper eyelid. The oculomotor nerve also provides parasympathetic supply to the iris and lens of the eye.
4. The trochlear nerve (IV) controls an extrinsic eye muscle.
5. The trigeminal nerve (V) supplies the muscles of mastication, as well as a middle ear muscle, a palatine muscle, and two throat muscles. The trigeminal nerve has the greatest cutaneous sensory distribution of any cranial nerve. The trigeminal nerve has three branches. Two of the three trigeminal nerve branches innervate the teeth.
6. The abducent nerve (VI) controls an extrinsic eye muscle.
7. The facial nerve (VII) supplies the muscles of facial expression, an inner ear muscle, and two throat muscles. It is involved in the sense of taste. It's parasympathetic to two sets of salivary glands and to the lacrimal glands.
8. The vestibulocochlear nerve (VIII) is involved in the sense of hearing and balance.
9. The glossopharyngeal nerve (IX) is involved in taste and supplies tactile sensory innervation from the posterior tongue, middle ear, and pharynx. It's also sensory for receptors that monitor blood pressure and gas levels in the blood. The glossopharyngeal nerve is parasympathetic to the parotid salivary glands.
10. The vagus nerve (X) innervates the muscles of the pharynx, palate, and larynx. It's also involved in the sense of taste. The vagus nerve is sensory for the pharynx and larynx and for receptors that monitor blood pressure and gas levels in the blood. The vagus nerve is sensory for thoracic and abdominal organs. The vagus nerve provides parasympathetic innervation to the thoracic and abdominal organs.
11. The accessory nerve (XI) has only a spinal component. It supplies the sternocleidomastoid and trapezius muscles.
12. The hypoglossal nerve (XII) supplies the intrinsic tongue muscles, three of four extrinsic tongue muscles, and two throat muscles.

Reflexes in the Brainstem Involving Cranial Nerves

Many reflexes involved in homeostasis involve the cranial nerves and occur in the brainstem.

1. If a section is made that separates the brainstem from the rest of the brain, the cut is between the
   a. medulla oblongata and pons.
   b. pons and midbrain.
   c. midbrain and diencephalon.
   d. thalamus and cerebrum.
   e. medulla oblongata and spinal cord.

2. Important centers for heart rate, blood pressure, respiration, swallowing, coughing, and vomiting are located in the
   a. cerebrum.
   b. medulla oblongata.
   c. midbrain.
   d. pons.
   e. cerebellum.

3. In which of these parts of the brain does decussation of descending nerve tracts involved in the conscious control of skeletal muscles occur?
   a. cerebrum
   b. diencephalon
   c. midbrain
   d. pons
   e. medulla oblongata

4. Important respiratory centers are located in the
   a. cerebrum.
   b. cerebellum.
   c. pons and medulla oblongata.
   d. midbrain
   e. limbic system.

5. The cerebral peduncles are a major descending motor pathway found in the
   a. cerebrum.
   b. cerebellum.
   c. pons.
   d. midbrain.
   e. medulla oblongata.

6. The superior colliculi are involved in _visual reflexes_, whereas the inferior colliculi are involved in _hearing_.
   a. hearing, visual reflexes
   b. visual reflexes, hearing
   c. balance, motor pathways
   d. motor pathways, balance
   e. respiration, sleep

7. The cerebellum communicates with other regions of the CNS through the
   a. flocculonodular lobe.
   b. cerebellar peduncles.
   c. vermis.
   d. lateral hemispheres.
   e. folia.

8. The major relay station for sensory input that projects to the cerebral cortex is the
   a. hypothalamus.
   b. thalamus.
   c. pons.
   d. cerebellum.
   e. midbrain.

9. Which part of the brain is involved with olfactory reflexes and emotional responses to odors?
   a. inferior colliculi
   b. superior colliculi
   c. mammillary bodies
   d. pineal body
   e. pituitary gland

10. The part of the diencephalon directly connected to the pituitary gland is the
    a. hypothalamus.
    b. epithalamus.
    c. subthalamus.
    d. thalamus.

11. Which of the following is a function of the hypothalamus?
    a. regulates autonomic nervous system functions
    b. regulates the release of hormones from the posterior pituitary
    c. regulates body temperature
    d. regulates food intake (hunger) and water intake (thirst)
    e. all of the above

12. The grooves on the surface of the cerebrum are called the
    a. nuclei.
    b. commissures.
    c. tracts.
    d. sulci.
    e. gyri.

13. Which of these areas is located in the postcentral gyrus of the cerebral cortex?
    a. olfactory cortex
    b. visual cortex
    c. primary motor cortex
    d. primary somatic sensory cortex
    e. primary auditory cortex

14. Which of these cerebral lobes is important in voluntary motor function, motivation, aggression, sense of smell, and mood?
    a. frontal
    b. insula
    c. occipital
    d. parietal
    e. temporal

15. Fibers that connect areas of the cerebral cortex within the same hemisphere are
    a. projection fibers.
    b. commissural fibers.
    c. association fibers.
    d. all of the above.

16. The basal nuclei are located in the
    a. inferior cerebrum
    b. diencephalon
    c. midbrain
    d. all of the above

17. The most superficial of the meninges is a thick, tough membrane called the
    a. pia mater.
    b. dura mater.
    c. arachnoid mater.
    d. epidural mater.

18. The ventricles of the brain are interconnected. Which of these ventricles are *not* correctly matched with the structures that connect them?
    a. lateral ventricle to the third ventricle—interventricular foramina
    b. left lateral ventricle to right lateral ventricle—central canal
    c. third ventricle to fourth ventricle—cerebral aqueduct
    d. fourth ventricle to subarachnoid space—median and lateral apertures

19. Cerebrospinal fluid is produced by the _Choroid Plexus_, circulates through the ventricles, and enters the subarachnoid space. The cerebrospinal fluid leaves the subarachnoid space through the _arachnoid granulations_.
    a. choroid plexuses, arachnoid granulations
    b. arachnoid granulations, choroid plexuses
    c. dural sinuses, dura mater
    d. dura mater, dural sinuses

20. Given these spaces:
    1. third ventricle
    2. epidural space
    3. subarachnoid space
    4. subdural space
    5. superior sagittal sinus

    Which of these spaces contains only cerebrospinal fluid (CSF)?
    a. 1, 3
    b. 1,2,3
    c. 1,3,5
    d. 1,2,3,5
    e. 2,3,4,5

21. Some portions of the blood plasma move across the blood–brain barrier by
    a. diffusion.
    b. endocytosis.
    c. exocytosis.
    d. co-transport.
    e. filtration.

22. Which of these parts of the embryonic brain is correctly matched with the structure it becomes in the adult brain?
    a. mesencephalon—midbrain
    b. metencephalon—medulla oblongata
    c. myelencephalon—cerebrum
    d. telencephalon—pons and cerebellum

23. The cranial nerve involved in chewing food is the
    a. trochlear (IV).
    b. trigeminal (V).
    c. abducent (VI).
    d. facial (VII).
    e vestibulocochlear (VIII).

24. The cranial nerve responsible for focusing the eye (innervates the ciliary muscle of the eye) is the
    a. optic (II).
    b. oculomotor (III).
    c. trochlear (IV).
    d. abducent (VI).
    e. facial (VII).

25. The cranial nerve involved in moving the tongue is the
    a. trigeminal (V).
    b. facial (VII).
    c. glossopharyngeal (IX).
    d. accessory (XI).
    e. hypoglossal (XII).

26. The cranial nerve involved in feeling a toothache is the
    a. trochlear (IV).
    b. trigeminal (V).
    c. abducent (VI).
    d. facial (VII).
    e. vestibulocochlear (VIII).

27. From this list of cranial nerves:
    1. olfactory (I)
    2. optic (II)
    3. oculomotor (III)
    4. abducent (VI)
    5. vestibulocochlear (VIII)

    Select the nerves that are sensory only.
    a. 1,2,3
    b. 2,3,4
    c. 1,2,5
    d. 2,3,5
    e. 3,4,5

28. From this list of cranial nerves:
    1. optic (II)
    2. oculomotor (III)
    3. trochlear (IV)
    4. trigeminal (V)
    5. abducent (VI)

    Select the nerves that are involved in moving the eyes.
    a. 1,2,3
    b. 1,2,4,
    c. 2,3,4
    d. 2,4,5
    e. 2,3,5

29. From this list of cranial nerves:
    1. trigeminal (V)
    2. facial (VII)
    3. glossopharyngeal (IX)
    4. vagus (X)
    5. hypoglossal (XII)

    Select the nerves that are involved in the sense of taste.
    a. 1,2,3
    b. 1,4,5
    c. 2,3,4
    d. 2,3,5
    e. 3,4,5

30. From this list of cranial nerves:
    1. trigeminal (V)
    2. facial (VII)
    3. glossopharyngeal (IX)
    4. vagus (X)
    5. hypoglossal (XII)

    Select the nerves that innervate the salivary glands.
    a. 1,2
    b. 2,3
    c. 3,4
    d. 4,5
    e. 3,5

31. From this list of cranial nerves:
    1. oculomotor (III)
    2. trigeminal (V)
    3. facial (VII)
    4. vestibulocochlear (VIII)
    5. glossopharyngeal (IX)
    6. vagus (X)

    Select the nerves that are part of the parasympathetic division of the ANS.
    a. 1,2,4,5
    b. 1,3,5,6
    c. 1,4,5,6
    d. 2,3,4,5
    e. 2,3,5,6

Answers in Appendix F

CRITICAL THINKING

1. A patient looses all sense of feeling in the left side of the back, below the upper limb, and extending in a band around to the chest, also below the upper limb. All sensation on the right is normal. The line between normal and absent sensation is the anterior and posterior midline. Explain this condition.

2. What happens to the developing brain if the CSF is not properly drained, resulting in early hydrocephalus?

3. A patient exhibits enlargement of the lateral and third ventricles, but no enlargement of the fourth ventricle. What would you conclude?

4. During a spinal tap of a patient, blood is discovered in the CSF. What does this finding suggest?

5. Describe a clinical test to evaluate each of the twelve cranial nerves.

Answers in Appendix G

ANSWERS TO PREDICT QUESTIONS

1. The oculomotor nerve innervates four eye muscles and the levator palpebrae superioris muscle. One cause of ptosis, a drooping upper eyelid, can be oculomotor nerve damage and subsequent paralysis of the levator palpebrae superioris muscle. The four eye muscles innervated by the oculomotor nerve move the eyeball so that the gaze is directed superiorly, inferiorly, medially, or superolaterally. Damage to this nerve can be tested by having the patient look in these directions. The abducens nerve directs the gaze laterally, and the trochlear nerve directs the gaze inferolaterally. If the patient can move the eyes in these directions, the associated nerves are intact.

2. The sternocleidomastoid muscle pulls the mastoid process (located behind the ear) toward the sternum, thus turning the face to the opposite side. If the innervation to one sternocleidomastoid muscle is eliminated (accessory nerve injury), the opposite muscle is unopposed and turns the face toward the side of injury. A person with wry neck whose head is turned to the left most likely has an injured left accessory nerve.

3. The tongue is protruded by contraction of the geniohyoid muscle, which pulls the back of the tongue forward, thereby pushing the muscle mass of the tongue forward. With one side pushed forward and unopposed by muscles of the opposite side, the tongue deviates toward the nonfunctional side. In the example, therefore, the right hypoglossal nerve is damaged.

Visit the Online Learning Center at www.mhhe.com/seeley7 for chapter quizzes, interactive learning exercises, and other study tools.

Integration of Nervous System Functions

CHAPTER

14

The nervous system is involved in almost all bodily functions. Although humans have larger, more complex brains than other animals, most human nervous system functions are similar to those of other animals. The sensory input we receive and most of the ways we respond to that input are similar to that in other animals. Yet, the human brain is capable of unique and complex functions, such as recording history, reasoning, and planning, to a degree unparalleled in the animal kingdom. Many of these functions can only be studied in humans. That is a major reason why much of human brain function remains elusive and why it remains one of the most challenging frontiers of anatomy and physiology.

This chapter presents the concept of *sensation* (p. 476) and then discusses the *control of skeletal muscles* (p. 488), the *brainstem functions* (p. 495), *other brain functions* (p. 496), and the *effects of aging on the nervous system* (p. 503).

Sensation

Sensation, or **perception,** is the conscious awareness of the effects of stimuli on sensory receptors. The brain constantly receives action potentials from a wide variety of sensory receptors both inside and outside the body. Sensory receptors respond to stimuli by generating action potentials that are propagated along nerves to the spinal cord and brain. Sensations result when action potentials reach the cerebral cortex. Some other parts of the brain are involved in sensation. For example, the thalamus and amygdala are involved in the sensation of pain.

The **senses** are the means by which the brain perceives information about the environment and the body. Historically, five senses were recognized: smell, taste, sight, hearing, and touch. Today, the senses are divided into two basic groups: general and special senses. The **general senses** are those with receptors distributed over a large part of the body. They are divided into two groups: the somatic and visceral senses (table 14.1). The **somatic senses,** which provide sensory information about the body and the environment, include touch, pressure, itch, vibration, temperature, proprioception, and pain. The **visceral senses,** which provide information about various internal organs, consist primarily of pain and pressure.

Special senses are more specialized in structure, have specialized nerve endings, and are localized to specific parts of the body (see table 14.1). The special senses are smell, taste, sight, hearing, and balance. Chapter 15 considers the special senses in detail.

Awareness

Not all of the sensory information detected by sensory receptors results in sensation. Some action potentials reach areas of the brain where they are not consciously perceived. For example, although we are consciously aware of body position and movements, much of the sensory information from sensory receptors that monitor position and movement is propagated to the cerebellum, where it is processed on an unconscious level. Sensory information from receptors that monitor blood pressure, blood oxygen, and pH levels are processed unconsciously by the medulla oblongata. For example, blood pressure must be regulated to maintain homeostasis, but if we had to consciously regulate blood pressure we might not be able to think of much else.

The cerebral cortex screens much of the sensory information it receives, ignoring many of the action potentials that reach it. In addition, humans exhibit selective awareness. That is, we are more aware of sensations on which we have our attention focused than on other sensations. The CNS cannot be consciously aware of all stimuli. If we were simultaneously aware of all the stimuli that the brain constantly receives, it's unlikely we would be able to function. Being aware of so many stimuli would require us to constantly make conscious decisions about the stimuli to which we should respond. Instead, homeostasis is controlled largely without our conscious involvement. For example, as you read this paragraph, it's unlikely that you are aware of the weight of the book in your hands if you are holding it, or the weight of your arms on the desk or on your lap if you are reading at a desk. It's unlikely that you are aware of the small noises around you or the clothes touching your body until your attention is drawn to them. You certainly aren't aware of changes in your blood pressure, body fluid pH, and blood glucose levels.

Sensation requires the following steps:

1. Stimuli originating either inside or outside of the body are detected by sensory receptors and action potentials are produced, which are propagated to the CNS by nerves (see "Sensory Receptors" on this page).
2. Within the CNS, nerve tracts conduct action potentials to the cerebral cortex and to other areas of the CNS (see "Sensory Nerve Tracts" on p. 480).
3. Many action potentials reaching the cerebral cortex are ignored. Others are translated so the person becomes aware of the stimulus (see "Sensory Areas of the Cerebral Cortex" on p. 484).

Sensory Receptors

Types of Sensory Receptors

The different senses depend upon sensory receptors specialized to respond to specific types of stimuli (see table 14.1). **Mechanoreceptors** respond to mechanical stimuli, such as compression, bending, or stretching of cells. The senses of touch, tickle, itch, vibration, pressure, proprioception, hearing, and balance all depend on a variety of mechanoreceptors. **Chemoreceptors** respond to chemicals that become attached to receptors on their membranes. Smell and taste depend on chemoreceptors. **Thermoreceptors** respond to changes in temperature at the site of the receptor and are necessary for the sense of temperature. **Photoreceptors** respond to light striking the receptor cells and are necessary for vision. **Nociceptors** (nō-si-sep′ters; Latin, *noceo* means hurt), or **pain receptors,** respond to painful mechanical, chemical, or thermal stimuli. Most sensory receptors typically respond to one type of stimulus, but some nociceptors respond to more than one type of stimulus.

At least eight major types of sensory receptors, which differ in their structure and the types of stimuli to which they are most sensitive, are involved in general sensation (table 14.2 and figure 14.1). Many of these sensory receptors are associated with the skin; others are associated with deeper structures, such as tendons, ligaments, and muscles; and some can be found in both the skin and deeper structures. In general, sensory receptors are classified into three groups based on their location: **exteroreceptors** (cutaneous receptors) are associated with the skin, **visceroreceptors** are associated with the viscera or organs, and **proprioceptors** are associated with joints, tendons, and other connective tissue. Exteroreceptors provide information about the external environment, visceroreceptors provide information about the internal environment, and proprioceptors provide information about body position, movement, and the extent of stretch or the force of muscular contractions.

Structurally, the simplest and most common sensory receptors are the **free nerve endings** (see figure 14.1), which are relatively unspecialized neuronal branches similar to dendrites. Free nerve endings are distributed throughout almost all parts of the body. Most visceroreceptors consist of free nerve endings, which are responsible for a number of sensations, including pain, temperature, itch, and movement. The free nerve endings responsible for temperature detection respond to three types of sensations. One type, the **cold receptors,** increases its rate of action potential production as the skin is cooled. The second type, **warm**

Table 14.1 Classification of the Senses

Types of Sense	Nerve Endings	Receptor Type	Initiation of Response
Somatic			
Touch		Mechanoreceptors	Compression of receptors
Stroking	Meissner's corpuscle		
	Hair follicle receptor		
Texture	Merkel's disk		
Vibration	Pacinian corpuscle		
Skin stretch	Ruffini end organ		
Itch	Free nerve endings		
Pressure	Merkel's disk	Mechanoreceptors	Compression of receptors
Proprioception	Pacinian corpuscle	Mechanoreceptors	Compression of receptors
Temperature	Free nerve endings	Thermoreceptors	Temperature around nerve endings
	Cold receptors		
	Warm receptors		
Pain	Free nerve endings	Nociceptors	Irritation of nerve endings (e.g., mechanical, chemical, or thermal)
Visceral			
Pain	Free nerve endings	Nociceptors	Irritation of nerve endings
Pressure	Pacinian corpuscle	Mechanoreceptors	Compression of receptors
Special			
Smell	Specialized	Chemoreceptors	Binding of molecules to membrane receptors
Taste	Specialized	Chemoreceptors	Binding of molecules to membrane receptors
Sight	Specialized	Photoreceptors	Chemical change in receptors initiated by light
Hearing	Specialized	Mechanoreceptors	Bending of microvilli on receptor cells
Balance	Specialized	Mechanoreceptors	Bending of microvilli on receptor cells

Table 14.2 Sensory Receptors

Type of Receptor	Structure	Function
Free nerve ending	Branching, no capsule	Pain, itch, tickle, temperature, joint movement, and proprioception
Merkel's disk	Flattened expansions at the end of axons; each expansion associated with a Merkel's cell	Light touch and superficial pressure
Hair follicle receptor	Wrapped around hair follicles or extending along the hair axis, each axon supplies several hairs, and each hair receives branches from several neurons, resulting in considerable overlap	Light touch; responds to very slight bending of the hair
Pacinian corpuscle	Onion-shaped capsule of several cell layers with a single central nerve process	Deep cutaneous pressure, vibration, and proprioception
Meissner's corpuscle	Several branches of a single axon associated with wedge-shaped epitheloid cells and surrounded by a connective tissue capsule	Two-point discrimination
Ruffini's end organ	Branching axon with numerous small, terminal knobs surrounded by a connective tissue capsule	Continuous touch or pressure; responds to depression or stretch of the skin
Muscle spindle	Three to 10 striated muscle fibers enclosed by a loose connective tissue capsule, striated only at the ends, with sensory nerve endings in the center	Proprioception associated with detection of muscle stretch; important for control of muscle tone
Golgi tendon organ	Surrounds a bundle of tendon fascicles and is enclosed by a delicate connective tissue capsule; nerve terminations are branched with small swellings applied to individual tendon fascicles	Proprioception associated with the stretch of a tendon; important in the control of muscle contraction

Figure 14.1   Sensory Nerve Endings in the Skin

receptors, increases its rate of action potential production as skin temperature increases. Both cold and warm receptors respond most strongly to changes in temperature. Cold receptors are 10–15 times more numerous in any given area of skin than warm receptors. The third type is a pain receptor, which is stimulated by extreme cold or heat. At very cold temperatures (0°–12°C), only pain receptors are stimulated. The pain sensation ends as the temperature increases above 15°C. Between 12° and 35°C, cold fibers are stimulated. Nerve fibers from warm receptors are stimulated between 25° and 47°C. "Comfortable" temperatures, between 25° and 35°C, therefore stimulate both warm and cold receptors. Temperatures above 47°C stimulate cold and pain receptors but don't stimulate warm receptors.

P R E D I C T

How might a very cold object placed in the hand be misperceived as being hot?

Merkel's (mer′kĕlz), or **tactile, disks,** are more complex than free nerve endings (see figure 14.1) and consist of axonal branches that end as flattened expansions, each associated with a specialized epithelial cell. They are distributed throughout the basal layers of the epidermis just superficial to the basement membrane and are associated with dome-shaped mounds of thickened epidermis in hairy skin. Merkel's disks are involved with the sensations of light touch and superficial pressure. These receptors can detect a skin displacement of less than 1 mm (1/25 of an inch).

Hair follicle receptors, or **hair end organs,** respond to very slight bending of the hair and are involved in light touch (see figure 14.1). These receptors are extremely sensitive and require very little stimulation to elicit a response. The sensation, however, is not very well localized. The dendritic tree at the distal end of a sensory axon has several hair follicle receptors. The field of hairs innervated by these receptors overlaps with the fields of hair follicle receptors of adjacent axons. The considerable overlap that exists in the endings of sensory neurons helps explain why light touch is not highly localized, yet because of converging signals within the CNS, it is very sensitive.

Pacinian (pa-sin′ē-an, pa-chin′ē-an), or **lamellated, corpuscles** are complex receptors that resemble an onion (see figure 14.1). A single dendrite extends to the center of each lamellated corpuscle. The corpuscles are located within the deep dermis or hypodermis, where they are responsible for deep cutaneous pressure and vibration. Pacinian corpuscles associated with the joints help relay **proprioceptive** (prō-prē-ō-sep′tiv; perception of position) information about joint positions.

Meissner's (mīs′nerz), or **tactile, corpuscles** are distributed throughout the dermal papillae (see figure 14.1; see chapter 5) and are involved in two-point discrimination touch. **Two-point discrimination** (fine touch) is the ability to detect simultaneous stimulation at two points on the skin. The distance between two points that a person can detect as separate points of stimulation differs for various regions of the body. This sensation is important in evaluating the texture of objects. Meissner's corpuscles are numerous and close together in the tongue and fingertips but are less numerous and more widely separated in other areas such as the back (figure 14.2).

Ruffini's (rū-fē′nēz) **end organs** are located in the dermis of the skin (see figure 14.1), primarily in the fingers. They respond to pressure on the skin directly superficial to the receptor and to stretch of adjacent skin. These receptors are important in responding to continuous touch or pressure.

2 mm

2–3 mm

64 mm

Figure 14.2 Two-Point Discrimination

Two-point discrimination can be demonstrated by touching a person's skin with the two points of a compass. When the two points are closer together than the receptor field, the individual perceives only one point. When the two points of the compass are opened wider, the person becomes aware of two points.

Muscle spindles (figure 14.3) consist of 3–10 specialized skeletal muscle fibers. They are located in skeletal muscles and provide information about the length of the muscle (see "Stretch Reflex" on p. 416). Muscle spindles are important to the control and tone of postural muscles. Brain centers act through descending tracts to either increase or decrease action potentials in gamma motor neurons. Stimulation of the gamma motor system, caused by stretch of the muscle, activates the stretch reflex, which in turn increases the tone of the muscles involved.

Golgi tendon organs are proprioceptive receptors associated with the fibers of a tendon near the junction between the muscle and tendon (figure 14.4). They are activated by an increase in tendon tension, whether it's caused by contraction of the muscle or by passive stretch of the tendon.

1. *In general, into what three groups can sensory receptors be classified?*
2. *List the eight major types of sensory receptors, indicate where they are located, and state the functions they perform.*

Responses of Sensory Receptors

Interaction of a stimulus with a sensory receptor produces a local potential called a **receptor potential.** Some sensory receptor cells, called **primary receptors,** have axons that conduct action potentials in response to the receptor potential. When the ends of these neurons are stimulated, a receptor potential is produced. If it reaches threshold, an action potential is produced and is propagated toward the CNS. Most sensory neurons, including all those in table 14.2, belong to this category. Other receptor cells, called **secondary receptors,** have no axons and the receptor potentials produced in those cells do not result in action potentials. Instead, the receptor potentials cause the release of neurotransmitter molecules from the receptor cell that bind to receptors on the membrane of a neuron. This causes a receptor potential in the neuron, which produces an action potential if threshold is reached. The receptor cells of the special senses of smell, taste, hearing, and balance belong to this category.

Some sensations have the quality of **accommodation,** or **adaptation,** a decreased sensitivity to a continued stimulus. After

Figure 14.3 Muscle Spindle

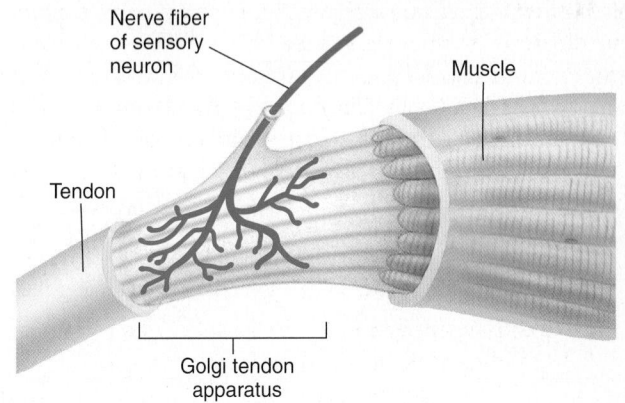

Figure 14.4 Golgi Tendon Organ

exposure to a stimulus for a time, the response of the receptors or the sensory pathways to a certain stimulus strength lessens from that which occurs when the stimulus was first applied. The local depolarization that produces a receptor potential accommodates, or returns to, its resting level even though the stimulus is still applied. For example, when a person first gets dressed, tactile receptors and pathways relay information to the brain that create an awareness that the clothes are touching the skin. After a time, the action potentials from the skin decrease, and the clothes are ignored.

Another way that sensations change through time occurs in proprioception. Proprioception provides information about the precise position and the rate of movement of various body parts, the weight of an object being held in the hand, and the range of movement of a joint. This information is involved in activities like walking, climbing stairs, shooting a basketball, driving a car, eating, or writing. Receptors for this system are located around joints and in muscles. Two types of proprioceptors are involved in providing positional information: tonic receptors and phasic receptors. **Slowly adapting receptors,** or **tonic receptors,** generate action potentials as long as a stimulus is applied and accommodate very slowly. Information from slowly adapting proprioceptors allows a person to know, for example, where the little finger is at all times without having to look for it. **Rapidly adapting receptors,** or **phasic receptors,** by contrast, accommodate rapidly and are most sensitive to changes in stimuli. For example, information from rapidly adapting proprioceptors allows us to know where our little finger is as it moves, thus we can control its movement through space and predict where it will be in the next moment.

We are usually not conscious of tonic or phasic input because most of this peripheral input is ignored by higher brain centers most of the time. Through selective awareness, however, we can call up the information when we wish. For example, where is the thumb of your right hand at this moment? Were you aware of its position a few seconds ago?

3. *What are primary and secondary receptors? What effect does a receptor potential have on them?*
4. *Define adaptation. Describe slowly adapting, or tonic and rapidly adapting, or phasic receptors.*

Sensory Nerve Tracts

The spinal cord and brainstem contain a number of sensory pathways that transmit action potentials from the periphery to various parts of the brain. Each pathway is involved with specific modalities (the type of information transmitted). The neurons that make up each pathway are associated with specific types of sensory receptors. For example, thermoreceptors located in the skin generate action potentials that are propagated along the sensory pathway for pain and temperature, whereas Golgi tendon organs located in tendons generate action potentials that are propagated along the sensory pathway involved with proprioception.

The names of many ascending pathways, or tracts, in the CNS indicate their origin and termination (table 14.3 and figure 14.5). Others indicate their location in the spinal cord. Each pathway usually is given a composite name in which the first half of the word indicates its origin and the second half indicates its termination. Ascending pathways therefore usually begin with the prefix *spino-*, indicating that they originate in the spinal cord. For example, a spinocerebellar (spī′nō-ser-e-bel′ar) tract is one that originates in the spinal cord and terminates in the cerebellum. An example of location nomenclature

Table 14.3	Ascending Spinal Pathways (see figure 14.5)		
Pathway (Tract)	**Modality (Information Transmitted)**	**Origin**	**Termination**
Anterolateral system			
Spinothalamic	Pain and temperature, light touch, pressure, tickle, and itch sensation	Cutaneous receptors	Thalamus and then most to cerebral cortex
Spinoreticular	Pain	Cutaneous receptors	Reticular formation and thalamus
Spinomesencephalic (including spinotectal)	Pain and touch	Cutaneous receptors	Mesencephalon and superior colliculus
Dorsal-column/ medial-lemniscal system	Proprioception, two-point discrimination, pressure, and vibration	Cutaneous receptors, joints	Cerebral cortex and cerebellum
Spinocerebellar	Proprioception to cerebellum	Joints, tendons	Cerebellum
Posterior			
Anterior			
Spinoolivary	Proprioception relating to balance	Joints, tendons	Accessory olivary nucleus, then to cerebellum

Figure 14.5   **Cross Section of the Spinal Cord at the Cervical Level Depicting the Ascending Pathways**

Ascending pathways are labeled on the left side of the figure only (*blue*) although they exist on both sides.

is the dorsal-column/medial-lemniscal system, whose name is a combination of the pathway names in the spinal cord and brainstem. The specific function of each ascending tract, however, is not suggested by its name.

The two major ascending systems of pathways or tracts involved in the conscious perception of external stimuli are the an-

terolateral system and the dorsal-column/medial-lemniscal system (see table 14.3). Those carrying sensory input that we are not consciously aware of include the spinocerebellar, spinoolivary, spinomesencephalic, and spinoreticular tracts, the latter two of which are part of the anterolateral system.

Anterolateral System

The **anterolateral system** is one of the two major systems that convey cutaneous sensory information to the brain. Historically, the anterolateral system was divided into anterior and lateral spinothalamic tracts. It is now known that such division was artificial and not clinically relevant. The anterolateral system includes the spinothalamic, spinoreticular, and spinomesencephalic tracts. The **spinothalamic tract** carries pain and temperature information, as well as light touch and pressure, tickle and itch sensations (figure 14.6).

Three neurons in sequence—the primary, secondary, and tertiary—are involved in the spinothalamic tract from the peripheral receptor to the cerebral cortex. The **primary neuron** cell bodies of the spinothalamic tract are in the dorsal root ganglia. The primary neurons relay sensory input from the periphery to the posterior horn of the spinal cord, where they synapse with interneurons. The interneurons, which are not specifically named in the three-neuron sequence, synapse with secondary neurons. Axons from the **secondary neurons** cross to the contralateral, or opposite side of the spinal cord through the anterior portion of the gray and white commissures and enter the spinothalamic tract, where they ascend to the thalamus. The secondary neurons synapse with cell bodies of tertiary neurons in

Primary Cell Body	Secondary Cell Body	Tertiary Cell Body	Crossover
Dorsal root ganglion	Posterior horn of spinal cord	Thalamus	Level at which primary neuron enters cord for pain and temperature, or eight to 10 segments from where primary neuron entered cord with many collaterals for light touch, etc.
Dorsal root ganglion	Posterior horn of spinal cord	Reticular formation	Some uncrossed; some cross spinal cord at point of entry
Dorsal root ganglion	Posterior horn of spinal cord	Superior colliculus	At point of origin
Dorsal root ganglion	Medulla oblongata	Thalamus	Medulla oblongata
Dorsal root ganglion	Posterior horn of spinal cord	Cerebellum	
			Uncrossed
			Some uncrossed; some cross at point of origin and recross in cerebellum
Dorsal root ganglion	Posterior horn of spinal cord	Accessory olivary nucleus	At point of origin; recross to reach cerebellum

Figure 14.6 The Spinothalamic Portion of the Anterolateral System

Some axons of the spinothalamic tract in the anterolateral system transmit action potentials for pain and temperature. Lines on the inset indicate levels of section.

the thalamus. **Tertiary neurons** from the thalamus project to the somatic sensory cortex.

Some neurons in the spinoreticular tracts do not cross over but ascend on the ipsilateral or same side of the spinal cord on which they enter.

P R E D I C T ❷
Describe the clinical results of a lesion on one side of the spinal cord that interrupts the spinothalamic part of the anterolateral system.

Syringomyelia

Syringomyelia (sĭ-ring′gō-mī-ē′lē-ă) is a degenerative cavitation of the central canal of the spinal cord, often caused by a cord tumor. Symptoms include neuralgia (pain along the course of the nerve), paresthesia (abnormal sensations, burning, tingling, increased sensitivity to pain), specific loss of pain and temperature sensation, and paresis (partial paralysis). This defect is unusual in that it occurs in a distinct band that includes both sides of the body because commissural tracts are destroyed.

The defect causing syringomyelia occurs where the fibers giving rise to the anterolateral system cross over in the center of the spinal cord. As a result of damage to crossing fibers from both sides, there is loss of pain and temperature on both sides of the body in a segment at the level of the cord damage. The loss of function is only at that level. There is not a loss of function below the level of the damage.

Dorsal-Column/Medial-Lemniscal System

The **dorsal-column/medial-lemniscal** (lem-nis′kăl) **system** carries the sensations of two-point discrimination, proprioception, pressure, and vibration (figure 14.7). This system is named for the dorsal column of the spinal cord and the medial lemniscus, which is the continuation of the dorsal column in the brainstem. The term *lemniscus* means ribbon and refers to the thin, ribbonlike appearance of the pathway as it passes through the brainstem.

Primary neurons of the dorsal-column/medial-lemniscal system are located in the dorsal root ganglia. They are the largest cell bodies in the dorsal root ganglia, especially those for two-point discrimination. Many axons of the primary neurons of the dorsal-column/medial-lemniscal system enter the spinal cord and ascend the entire length of the spinal cord, without crossing to its opposite side, and synapse with secondary neurons located in the medulla oblongata. Others synapse in the thoracic portion of the spinal cord.

In the spinal cord, the dorsal-column/medial-lemniscal system is divided into two separate tracts (see figure 14.5) based on the source of the stimulus. The **fasciculus gracilis** (gras′i-lis; thin) conveys sensations from nerve endings below the midthoracic level, and the **fasciculus cuneatus** (kū′nē-ā′tŭs; wedge-shaped) conveys impulses from nerve endings above the midthorax. The fasciculus gracilis terminates by synapsing with secondary neurons in the **nucleus gracilis** or with neurons of the posterior spinocerebellar tracts. The fasciculus cuneatus primarily terminates by synapsing with secondary neurons in the **nucleus cuneatus.** Both the nucleus gracilis and the nucleus cuneatus are in the medulla oblongata. Secondary neurons exit the nucleus gracilis and the nucleus cuneatus, cross to the opposite side of the medulla through the decussations of the medial lemniscus, and ascend through the medial lemniscus to terminate in the thalamus. Tertiary neurons from the thalamus project to the somatic sensory cortex (see page 485).

P R E D I C T ❸
Two people, Bill and Mary, were each involved in an accident and each experienced a loss of proprioception, fine touch, and vibration on the left side of the body below the waist. It was determined that Bill had damage to his spinal cord as a result of the accident and that Mary had damage to her brainstem. Explain which side of the spinal cord was damaged in Bill and which side of the brainstem was damaged in Mary.

neurons in the thalamus. Tertiary neurons from the thalamus project to the somatic sensory cortex.

Spinocerebellar System and Other Tracts

The spinocerebellar tracts (see figure 14.5) carry proprioceptive information to the cerebellum so that information concerning actual movements can be monitored and compared to cerebral information representing intended movements.

Two spinocerebellar tracts extend through the spinal cord: (1) the **posterior spinocerebellar tract** (figure 14.8), which originates in the thoracic and upper lumbar regions and contains uncrossed nerve fibers that enter the cerebellum through the inferior

Figure 14.7 Dorsal-Column/Medial-Lemniscal System

The fasciculus gracilis and fasciculus cuneatus (not illustrated) convey proprioception, pressure, vibration, and two-point discrimination. Only the fasciculus gracilis pathway is shown. Lines on the inset indicate levels of section.

Trigeminothalamic Tract

As the fibers of the spinothalamic tracts pass through the brainstem, they are joined by fibers of the **trigeminothalamic tract** (trigeminal nerve, or cranial nerve V). This tract carries the same sensory information as the spinothalamic tracts and dorsal-column/medial-lemniscal system but from the face, nasal cavity, and oral cavity, including the teeth. The trigeminothalamic tract is similar to the spinothalamic tracts and dorsal-column/medial-lemniscal system in that primary neurons from one side of the face synapse with secondary neurons, which cross to the opposite side of the brainstem. The secondary neurons synapse with tertiary

Figure 14.8 Posterior Spinocerebellar Tract

This tract transmits proprioceptive information from the thorax, upper limbs, and upper lumbar region to the cerebellum. Lines on the inset indicate levels of section.

cerebellar peduncles; and (2) the **anterior spinocerebellar tract,** which carries information from the lower trunk and lower limbs and contains both crossed and uncrossed nerve fibers that enter the cerebellum through the superior cerebellar peduncle. The crossed fibers recross in the cerebellum. Both spinocerebellar tracts transmit proprioceptive information to the cerebellum from the same side of the body as the cerebellar hemisphere to which they project. Why the anterior spinocerebellar tract crosses twice to accomplish this feat is unknown. Much of the proprioceptive information carried from the lower limbs by the fasciculus gracilis of the dorsal-column/medial-lemniscal system is transferred by synapses in the inferior thorax to the spinocerebellar system and enters the cerebellum as unconscious proprioceptive information. The spinocerebellar tracts convey very little information from the upper limbs to the cerebellum. Proprioception from the upper limbs is projected to the thalamus. This input enters the cerebellum through the inferior peduncle from the cuneate nucleus of the dorsal-column/medial-lemniscal system. The dorsal-column/medial-lemniscal system, therefore, is involved not only in conscious awareness of proprioception but also unconscious neuromuscular functions.

P R E D I C T ④

Most of the neurons from the fasciculus gracilis synapse in the inferior thorax and enter the spinocerebellar system, whereas most of the neurons from the fasciculus cuneatus synapse in the nucleus cuneatus and then continue to the thalamus and cerebrum. It can therefore be deduced that most of the proprioception from the lower limbs is unconscious and most of the proprioception from the upper limbs is conscious. Explain why this difference in the two sets of limbs is of value.

The **spinoolivary tracts** project to the accessory olivary nucleus and to the cerebellum, where action potentials carried by these tracts contribute to coordination of movement associated primarily with balance. The **spinotectal** (spī-nō-tek′tăl) **tracts** end in the superior colliculi of the midbrain and transmit action potentials involved in reflexes that turn the head and eyes toward a point of cutaneous stimulation.

Descending Pathways Modifying Sensation

The corticospinal (see p. 490) and other descending pathways send collateral branches to the thalamus, reticular formation, trigeminal nuclei, and spinal cord. Neuromodulators, such as endorphins and enkephalin, released from axons originating in these CNS regions decrease the frequency of action potentials in sensory tracts (see discussion of presynaptic inhibition in chapter 11). Through this route, the cerebral cortex or other brain regions may reduce the conscious perception of sensations.

5. *What are the functions of the anterolateral and dorsal-column/medial-lemniscal systems? Describe where the neurons of these systems cross over and synapse.*
6. *What kind of information is carried in the spinocerebellar tracts? Where do the anterior and posterior spinocerebellar tracts originate? Do these tracts terminate on the same or opposite side of the body from where they originate?*
7. *What are the functions of the spinoolivary and spinomesencephalic tracts?*
8. *How do descending pathways modulate sensation?*

Sensory Areas of the Cerebral Cortex

Figure 14.9 depicts a lateral view of the left cerebral cortex with some of its functional areas labeled. Sensory pathways project to specific regions of the cerebral cortex, called **primary sensory areas,** where these sensations are perceived.

Figure 14.9 Functional Regions of the Lateral Side of the Left Cerebral Cortex

Most of the postcentral gyrus is called the **primary somatic sensory cortex,** or **general sensory area.** The terms *area* and *cortex* are often used interchangeably for the same functional region of the cerebral cortex. Fibers carrying general sensory input, such as pain, pressure, and temperature, synapse in the thalamus, and thalamic neurons relay the information to the primary somatic sensory cortex.

The somatic sensory cortex is organized topographically relative to the general plan of the body (figure 14.10). Sensory impulses conducting input from the feet project to the most superior portion of the somatic sensory cortex, and sensory impulses from the face project to the most inferior portion. The pattern of the somatic sensory cortex in each hemisphere is arranged in the form of an upside-down half homunculus (hō-mŭngk'ū-lŭs; a little

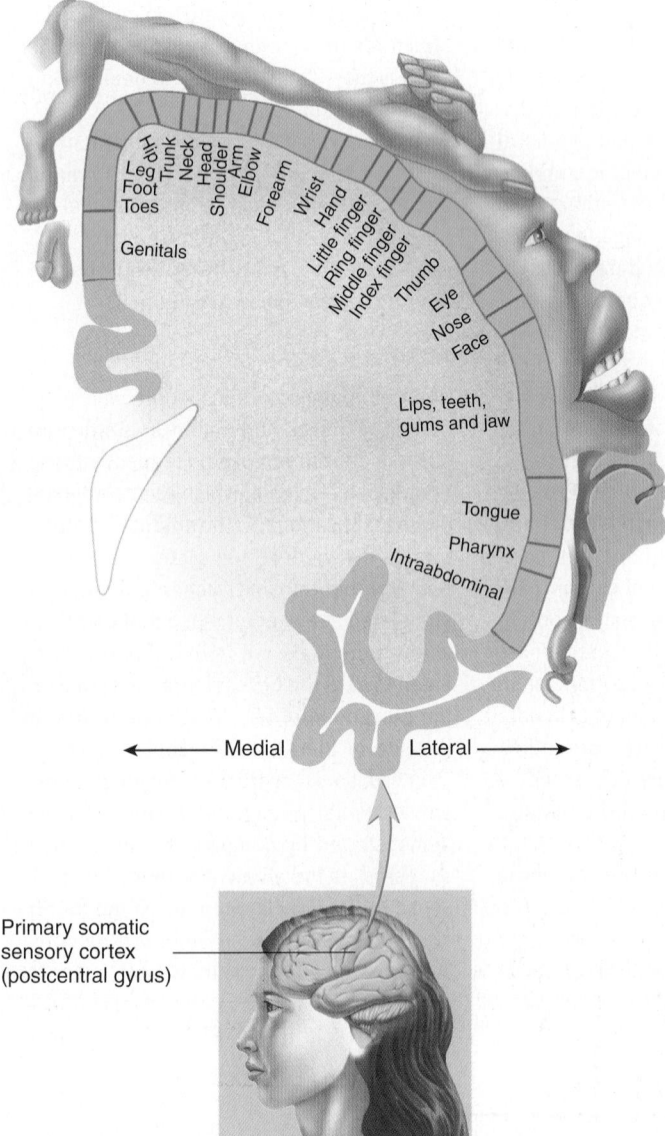

← Medial   Lateral →

Primary somatic sensory cortex (postcentral gyrus)

Figure 14.10 Topography of the Somatic Sensory Cortex

Cerebral cortex seen in coronal section on the left side of the brain. The figure of the body (homunculus) depicts the nerve distributions; the size of each body region shown indicates relative innervation. The cortex occurs on both sides of the brain but appears on only one side in this illustration. The inset shows the somatic sensory region of the left hemisphere (*green*).

human) representing the opposite side of the body, with the feet located superiorly and the head located inferiorly. The size of various regions of the somatic sensory cortex is related to the number of sensory receptors in that area of the body. The density of sensory receptors is much greater in the face than in the legs; therefore, a greater area of the somatic sensory cortex contains sensory neurons associated with the face, and the homunculus has a disproportionately large face.

There are other primary sensory areas of the cerebral cortex (see figure 14.9). The **taste area,** where taste sensations are consciously perceived in the cortex, is located at the inferior end of the postcentral gyrus. The **olfactory cortex** (not shown in figure 14.9) is on the inferior surface of the frontal lobe and is the area in which both conscious and unconscious responses to odor are initiated (see chapter 15). The **primary auditory cortex,** where auditory stimuli are processed by the brain, is located in the superior part of the temporal lobe. The **visual cortex,** where portions of visual images are processed, is located in the occipital lobe. In the visual cortex, color, shape, and movement are processed separately rather than as a complete "color motion picture." These sensory areas are discussed more fully in chapter 15.

The primary sensory areas of the cerebral cortex must be intact for conscious perception, localization, and identification of a stimulus. Cutaneous sensations, although integrated within the cerebrum, are perceived as though they were on the surface of the body. This is called **projection** and indicates that the brain refers a cutaneous sensation to the superficial site at which the stimulus interacts with the sensory receptors.

**P R E D I C T   **

Using the visual association areas as an example, explain the general functions of the association areas around the other primary cortical areas (see figure 14.9).

Sensory Processing

Cortical areas immediately adjacent to the primary sensory centers, called **association areas,** are involved in the process of recognition. The **somatic sensory association area** is posterior to the primary somatic sensory cortex, and the **visual association area** is anterior to the visual cortex (see figure 14.9). Sensory action potentials originating in the retina of the eye reach the visual cortex, where the image is "perceived." Action potentials then pass from the visual cortex to the visual association area, where the present visual information is compared to past visual experience ("Have I seen this before?"). On the basis of this comparison, the visual association area "decides" whether or not the visual input is recognized and passes judgment concerning the significance of the input. For example, we generally pay less attention to people in a crowd we have never seen before than to someone we know.

The visual association area, like other association areas of the cortex, has reciprocal connections with other parts of the cortex that influence decisions. For example, the visual association area has input from the frontal lobe, where emotional value is placed on the visual input. Because of these numerous connections, visual information is judged several times as it passes beyond the visual association area. This may be one of the reasons why two people who witness the same event can present somewhat different versions of what happened.

Clinical Focus Pain

Pain is a sensation characterized by a group of unpleasant and complex perceptual and emotional experiences that trigger autonomic, psychologic, and somatic motor responses. Pain sensation has two components: (1) rapidly conducted action potentials carried by large-diameter, myelinated axons, resulting in sharp, well-localized, pricking, or cutting pain, followed by (2) more slowly propagated action potentials, carried by smaller, less heavily myelinated axons, resulting in diffuse burning or aching pain. Research indicates that pain receptors have very uniform sensitivity that doesn't change dramatically from one instant to another. Variations in pain sensation result from the mechanisms by which pain receptors are stimulated, differences in integration of action potentials from the pain receptors, and complex interactions in the cerebral cortex, cingulate gyrus, and thalamus, where the emotional component of pain is registered. Neurons in the cerebral cortex respond to pain stimuli selectively based on prior experience and context. Stress, for example, can reduce pain perception.

Although the dorsal-column/medial-lemniscal system contains no pain fibers, tactile and mechanoreceptors are often activated by the same stimuli that affect pain receptors. Action potentials from tactile receptors provide information that allows the pain sensation to be localized. Superficial pain is highly localized, in part, because of the simultaneous stimulation of pain receptors and mechanoreceptors in the skin. Deep or visceral pain is not highly localized (diffuse) because of fewer mechanoreceptors in the deeper structures, and it is normally perceived as a diffuse pain.

Dorsal-column/medial-lemniscal system neurons are involved in what is called the **gate-control theory** of pain control. Primary neurons of the dorsal-column/medial-lemniscal system send out collateral branches that synapse with interneurons in the posterior horn of the spinal cord. These interneurons have an inhibitory effect on secondary neurons of the spinothalamic tract. Thus pain action potentials traveling through the spinothalamic tract can be suppressed by action potentials that originate in neurons of the dorsal-column/ medial-lemniscal system. The arrangement may act as a "gate" for pain action potentials transmitted in the spinothalamic tract. Increased activity in the dorsal-column/ medial-lemniscal system tends to close the gate, thereby reducing pain action potentials transmitted in the spinothalamic tract. Descending pathways from the cerebral cortex or other brain regions can also regulate this "gate."

The gate-control theory may explain the physiologic basis for the following methods that have been used to reduce the intensity of chronic pain: electric stimulation of the dorsal-column/medial-lemniscal neurons, transcutaneous electric stimulation (applying a weak electric stimulus to the skin), acupuncture, massage, and exercise. The frequency of action potentials that are transmitted in the dorsal-column/ medial-lemniscal system is increased when the skin is rubbed vigorously and when the limbs are moved and may explain why vigorously rubbing a large area around a source of pricking pain tends to reduce the intensity of the painful sensation. Exercise normally decreases the sensation of pain, and exercise programs are important components in the management of chronic pain not associated with illness. Action potentials initiated by acupuncture procedures may act through a gating mechanism in which inhibition of action potentials in neurons that transmit pain action potentials upward in the spinal cord are influenced by activity sensory cells that send collateral branches to the posterior horn.

Analgesics act in much the same way as gate control. Some analgesics block the transmission of nociception in the spinal cord from primary neurons to neurons of the ascending pathways. Other analgesics function at the level of the cerebral cortex to modulate pain.

Referred Pain

Referred pain is a painful sensation in a region of the body that is not the source of the pain stimulus. Most commonly, referred pain is sensed in the skin or other superficial structures when internal organs are damaged or inflamed. This sensation usually occurs because both the area of skin to which the pain is referred and the visceral area where the actual damage occurs are innervated by neurons that project to the same area of the cerebral cortex. The brain cannot distinguish between the two sources of painful stimuli, and the painful sensation is referred to the most superficial structures innervated by the converging neurons. This referral may occur because the number of receptors is much greater in superficial structures than in deep structures and the brain is more "accustomed" to dealing with superficial stimuli.

Referred pain is clinically useful in diagnosing the actual cause of the painful stimulus. Heart attack victims often feel cutaneous pain radiating from the left shoulder down the arm. Other examples of referred pain are shown in figure A.

Phantom Pain

Phantom pain occurs in people who have had appendages amputated or a structure such as a tooth removed. Frequently these people perceive pain, which can be intense, or other sensations, in the amputated structure as if it were still in place. If a neuron pathway that transmits action potentials is stimulated at any point along that pathway, action potentials are initiated and propagated toward the CNS. Integration results in the perception of pain that is projected to the site of the sensory receptors, even if those sensory receptors are no longer present. A similar phenomenon can be easily demonstrated by bumping the ulnar nerve as it crosses the elbow (the funny bone). A sensation of pain is often felt in the fourth and fifth digits, even though the neurons were stimulated at the elbow.

A factor that may be important in phantom pain results from the lack of touch, pressure, and proprioceptive impulses from the amputated limb. Those action potentials suppress the transmission of pain action potentials in the pain pathways, as explained by the gate-control theory of pain. When a limb is amputated, the inhibitory effect of sensory information is removed. As a consequence, the intensity of phantom pain

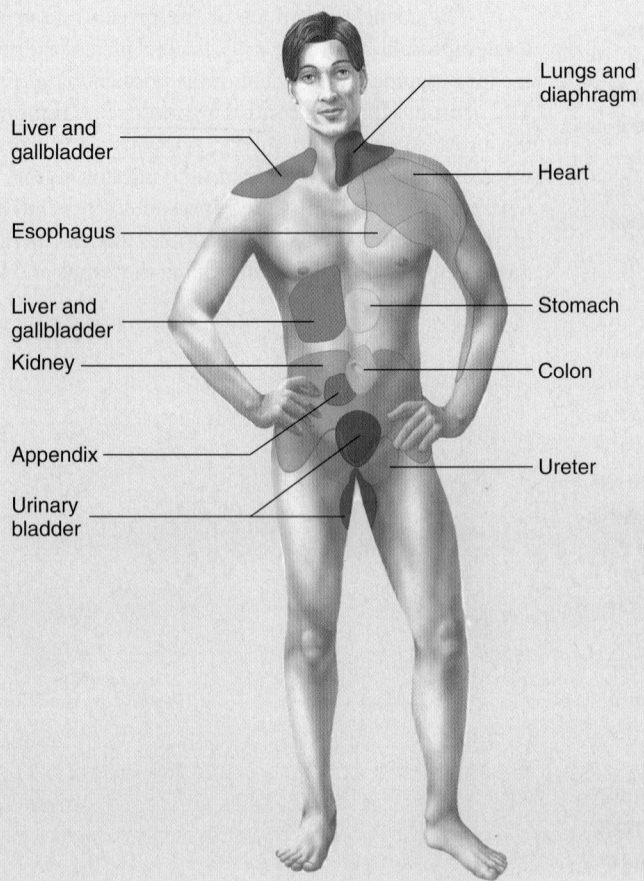

Liver and gallbladder

Esophagus

Liver and gallbladder

Kidney

Appendix

Urinary bladder

Lungs and diaphragm

Heart

Stomach

Colon

Ureter

Figure A   Areas of Referred Pain on the Body Surface
Pain from the indicated internal organs is referred to the surface areas shown.

may be increased. Another factor in phantom pain may be that the cerebral cortex retains an image of the amputated body part.

Chronic Pain

Chronic pain is long lasting. Some chronic pain has a known cause, such as tissue damage, as in the case of arthritis. Other chronic pain cannot be associated with tissue damage and has no known cause.

Pain is important in warning us of potentially injurious conditions because pain receptors are stimulated when tissues are injured. Pain itself, however, can become a problem. Chronic pain, such as migraine headaches, localized facial pain, or back pain, can be very debilitating and pain loses its value of providing information about the condition of the body. People suffering from chronic pain often feel help-

less and hopeless, and they may become dependent on drugs. The pain can interfere with vocational pursuits, and the victims are often unemployed or even housebound and socially isolated. They are easily frustrated or angered, and they suffer symptoms of major depression. These qualities are associated with what is called **chronic pain syndrome.** Over 2 million people in the United States at any given time suffer chronic pain sufficient to impair activity.

Chronic pain may originate with acute pain associated with an injury or may develop for no apparent reason. How sensory signals are processed in the thalamus and cerebrum may determine if the input is evaluated as only a discomfort, a minor pain, or a severe pain and how much distress is associated with the sensation. The brain actively regulates the amount of pain information

that gets through to the level of perception, thereby suppressing much of the input. If this dampening system becomes less functional, pain perception may increase. Other nervous system factors, such as a loss of some sensory modalities from an area, or habituation of pain transmission, which may remain even after the stimulus is removed, may actually intensify otherwise normal pain sensations. Treatment often requires a multidisciplinary approach, including such interventions as surgery or psychotherapy. Some sufferers respond well to drug therapy, but some drugs, such as opiates, have a diminishing effect and may become addictive.

Sensitization in Chronic Pain

Tissue damage within an area of injury, such as the skin, can cause an increase in the sensitivity of nerve endings in the area of damage, a condition called **peripheral sensitization.** One class of pain receptors is not activated by traditional noxious stimuli but is recruited only when tissues become inflamed. These receptors, once activated, add to the total barrage of sensory signals to the brain and intensify the sensation of pain.

The CNS may also respond to tissue damage by decreasing its threshold and increasing its sensitivity to pain. This condition is called **central sensitization.** Central sensitization apparently results from a specific subset of receptors that is only recruited during repetitive neuron firing, such as when intense pain sensations are experienced. These receptors maintain a hyperexcitable state in the CNS cells. This chronic hyperexcitable state can result in persistent, chronic pain states.

This information concerning peripheral and central sensitization, and the knowledge that sensitization involves neuronal and chemical receptors not normally involved in sensation, may lead to the discovery of new drugs for treating chronic pain. Rather than searching for new analgesics, which may decrease a broad range of sensations, an opportunity is now available to develop a new class of drugs, the "antihyperanalgesics," that may block sensitization without diminishing other sensations, including that of normal pain.

P R E D I C T

A man has constipation, which causes distention and painful cramping in his colon. What kind of pain would he experience (local or diffuse) and where would it be perceived? Explain.

9. Describe in the cerebral cortex the locations and functions of the primary sensory areas and their association areas.

10. Describe the topographic organization of the general body plan in the primary somatic sensory cortex. Why are some areas of the body represented as larger than other areas?

11. What are the related functions of the primary motor area, the premotor area, and the prefrontal area of the cerebral cortex?

Control of Skeletal Muscles

The motor system of the brain and spinal cord is responsible for maintaining the body's posture and balance; as well as moving the trunk, head, limbs, and eyes; and communicating through facial expressions and speech. Reflexes mediated through the spinal cord (see chapter 12) and brainstem (see chapter 13) are responsible for some body movements. They occur without conscious thought. **Voluntary movements,** on the other hand, are movements consciously activated to achieve a specific goal, such as walking or typing. Although consciously activated, the details of most voluntary movements occur automatically. After walking begins, it is not necessary to think about the moment-to-moment control of every muscle because neural circuits exist that automatically control the limbs. After learning how to do complex tasks, such as typing, they can be performed relatively automatically.

Voluntary movements depend upon upper and lower motor neurons. **Upper motor neurons** directly or through interneurons connect to lower motor neurons. The cell bodies of upper motor neurons are in the cerebral cortex. **Lower motor neurons** have axons that leave the central nervous system and extend through peripheral nerves to supply skeletal muscles. The cell bodies of lower motor neurons are located in the anterior horns of the spinal cord gray matter and in cranial nerve nuclei of the brainstem.

Voluntary movements depend upon the following:

1. The initiation of most voluntary movement begins in the premotor areas of the cerebral cortex and results in the stimulation of upper motor neurons (see "Motor Areas of the Cerebral Cortex" on this page).

2. The axons of the upper motor neurons form the descending nerve tracts. They stimulate lower motor neurons which stimulate skeletal muscles to contract (see "Motor Nerve Tracts" on p. 489).

3. The cerebral cortex interacts with the basal nuclei and cerebellum in the planning, coordination, and execution of movements (see "Modifying and Refining Motor Activities" on p. 493).

12. Distinguish between upper and lower motor neurons.

Motor Areas of the Cerebral Cortex

The **precentral gyrus** is also called the **primary motor cortex,** or **primary motor area** (see figure 14.9). Action potentials initiated in this region control many voluntary movements, especially the

fine motor movements of the hands. Upper motor neurons are not confined to the precentral gyrus—only about 30% of them are located there. Another 30% are in the premotor area, and the rest are in the somatic sensory cortex.

The cortical functions of the precentral gyrus are arranged topographically according to the general plan of the body—similar to the topographic arrangement of the postcentral gyrus (figure 14.11). The neuron cell bodies controlling motor functions of the feet are in the most superior and medial portions of the precentral gyrus, whereas those for the face are in the inferior region. Muscle groups with many motor units are represented by relatively large areas of the precentral gyrus. For example, muscles performing precise movements, such as those controlling the hands and face, have many

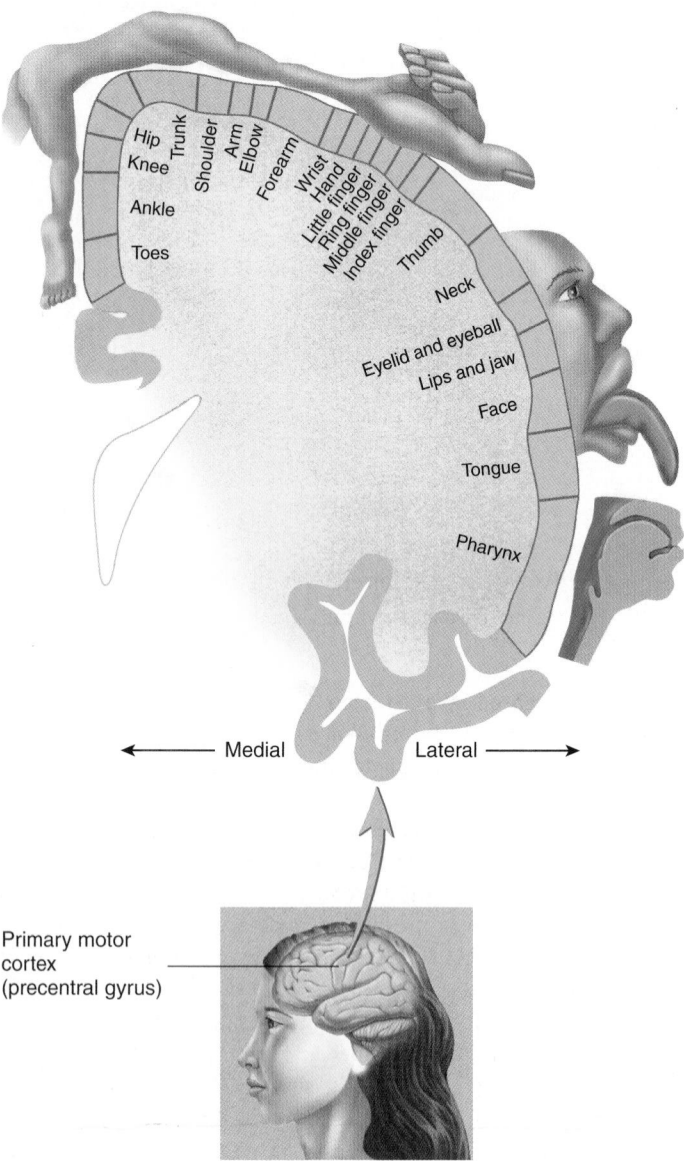

Figure 14.11   **Topography of the Primary Motor Cortex**

Cerebral cortex seen in coronal section on the left side of the brain. The figure of the body (homunculus) depicts the nerve distributions; the size of each body region shown indicates relative innervation. The cortex occurs on both sides of the brain but appears on only one side in this illustration. The inset shows the motor region of the left hemisphere (*pink*).

motor units, each of which has a small number of muscle fibers. Multiple-motor unit summation (see chapter 9) can precisely control the force of contraction of these muscles because only a few muscle fibers at a time are recruited. Muscle groups with few motor units are represented by relatively small areas of the precentral gyrus, even if the muscles innervated are quite large. Muscles, such as those controlling movements of the thigh and leg, have proportionately fewer motor units than hand muscles, but many more and much larger muscle fibers per motor unit. They are less precisely controlled because the activation of a motor unit stimulates the contraction of many large muscle fibers.

The **premotor area,** located anterior to the primary motor cortex (see figure 14.9), is the staging area in which motor functions are organized before they are initiated in the motor cortex. For example, if a person decides to take a step, the neurons of the premotor area are stimulated first. The determination is made in the premotor area as to which muscles must contract, in what order, and to what degree. Action potentials are then passed to the upper motor neurons in the motor cortex, which actually initiate the planned movements.

Apraxia

The premotor area must be intact for a person to carry out complex, skilled, or learned movements, especially ones related to manual dexterity; for example, a surgeon's use of a scalpel or a student's use of a pencil. Impairment in the performance of learned movements, called apraxia (ă-prak′sē-ă), can result from a lesion in the premotor area. Apraxia is characterized by hesitancy and reduced dexterity in performing these movements.

The motivation and foresight to plan and initiate movements occur in the next most anterior portion of the brain, the **prefrontal area,** an association area that is well developed only in primates and especially in humans. It is involved in motivation and regulation of emotional behavior and mood. The large size of this area of the brain in humans may account for their relatively well-developed forethought and motivation and for the emotional complexity of humans.

Prefrontal Lobotomy

In relation to its involvement in motivation, the prefrontal area is also thought to be the functional center for aggression. Beginning in 1935, one method used to eliminate uncontrollable aggression or anxiety in psychiatric hospital patients was to surgically remove or destroy the prefrontal regions of the brain, a procedure called a **prefrontal,** or **frontal, lobotomy.** The operation was sometimes successful in eliminating aggression, but this effect was often only temporary. In addition, some patients developed epilepsy or personality changes, such as lack of inhibition or a lack of initiative and drive. Later studies failed to confirm the usefulness of lobotomies, and the practice was largely discontinued in the late 1950s.

13. *Where are the primary motor, premotor, and prefrontal areas of the cerebral cortex located?*
14. *Why are some areas of the body represented as larger than other areas on the topographic map of the body on the primary motor areas?*

Motor Nerve Tracts

Motor nerve tracts are descending pathways containing axons that carry action potentials from regions of the brain to the brainstem or spinal cord. The names of descending nerve tracts are based on their origin and termination (figure 14.12 and table 14.4). Much like the names of ascending tracts, the prefix indicates its origin and the suffix indicates its destination. For example, the corticospinal tract is a motor tract that originates in the cerebral cortex and terminates in the spinal cord.

Amyotrophic Lateral Sclerosis

Amyotrophic (ă-mī-ō-trō′fik) **lateral sclerosis (ALS),** also called Lou Gehrig's disease, usually affects people between the ages of 40 and 70. About 10% of the cases of ALS are inherited. It begins with weakness and clumsiness and progresses within 2–5 years to loss of muscle control. The disease selectively destroys both upper and lower motor neurons. The inherited form of ALS apparently results from a mutation in DNA coding for the enzyme **superoxide dismutase (SOD)** and is located on chromosome 21. SOD is involved in eliminating a free radical, known as superoxide, from the body. Free radicals are molecules with an odd number of electrons in their outer shells, which makes them highly reactive. They can strip electrons from proteins, lipids, or nucleic acids, thereby destroying their functions and resulting in cell dysfunction or death. Free-radical damage has been implicated in ALS, arteriosclerosis, arthritis, cancer, and aging. Superoxide is one of the most important and toxic free radicals. It forms as the result of oxygen reacting with other free radicals. Although oxygen is critical for aerobic metabolism, it's also dangerous to tissues. SOD catalyzes the conversion of superoxide to hydrogen peroxide, which is then converted by catalase to oxygen and water. Apparently, if SOD is defective, superoxide is not degraded and can destroy cells. Motor neurons appear to be particularly sensitive to superoxide attack.

Figure 14.12 Cross Section of the Spinal Cord at the Cervical Level Depicting the Descending Pathways

Descending pathways are labeled on the left side of the figure only (*pink*), though they exist on both sides.

Table 14.4	Descending Spinal Pathways			
Pathway	**Functions Controlled**	**Origin**	**Termination**	**Crossover**
Direct	Muscle tone and conscious skilled movements, especially of the hands			
Corticospinal	Movements, especially of the hands	Cerebral cortex (upper motor neuron)	Anterior horn of spinal cord (lower motor neuron)	
Lateral				Inferior end of medulla oblongata
Anterior				At level of lower motor neuron
Corticobulbar	Facial and head movements	Cerebral cortex (upper motor neuron)	Cranial nerve nuclei in brainstem (lower motor neuron)	Varies for the various cranial nerves
Indirect	Less precise motor control, overall coordination, posture			
Rubrospinal	Movement coordination	Red nucleus	Anterior horn of spinal cord	Midbrain
Vestibulospinal	Posture, balance	Vestibular nucleus	Anterior horn of spinal cord	Uncrossed
Reticulospinal	Posture adjustment, especially during movement	Reticular formation	Anterior horn of spinal cord	Some uncrossed; some cross at level of termination
Tectospinal	Movement of head and neck in response to visual reflexes	Superior colliculus	Cranial nerve nucleus in medulla oblongata and anterior horn of upper levels of spinal cord (lower motor neurons that turn head and neck)	Midbrain

The descending motor fibers are divided into two groups: direct pathways and indirect pathways (figure 14.13). The **direct pathways,** also called the **pyramidal** (pi-ram′i-dal) **system,** are involved in the maintenance of muscle tone and in controlling the speed and precision of skilled movements, primarily fine movements involved in dexterity. Most of the **indirect pathways,** sometimes called the **extrapyramidal system,** are involved in less precise control of motor functions, especially those associated with overall body coordination and cerebellar function such as posture. Many of the indirect pathways are phylogenetically older and control more "primitive" movements of the trunk and proximal portions of the limbs. The direct pathways, which exist only in mammals, may be thought of as overlying the indirect pathways and are more involved in finely controlled movements of the face and distal portions of the limbs. Some indirect pathways, such as those from the basal nuclei and cerebellum, help in fine control of the direct pathways.

Direct Pathways

Direct pathways are so named because upper motor neurons in the cerebral cortex, whose axons form these pathways, synapse directly with lower motor neurons in the brainstem or spinal cord. They are also called the pyramidal system because the fibers of these pathways primarily pass through the medullary **pyramids.** The direct pathways include groups of nerve fibers arrayed into two tracts: the **corticospinal tract,** which is involved in direct cortical control of

movements below the head, and the **corticobulbar tract,** which is involved in direct cortical control of movements in the head and neck.

The corticospinal tract consists of axons of upper motor neurons located in the primary motor and premotor areas of the frontal lobes and the somatic sensory parts of the parietal lobes. They descend through the **internal capsules** and the cerebral peduncles of the midbrain to the pyramids of the medulla oblongata. At the inferior end of the medulla 75%–85% of the corticospinal fibers cross to the opposite side of the CNS through the **pyramidal decussation,** which is visible on the anterior surface of the inferior medulla. The crossed fibers descend in the **lateral corticospinal tracts** of the spinal cord (figure 14.14). The remaining 15%–25% descend uncrossed in the **anterior corticospinal tracts** and decussate near the level where they synapse with lower motor neurons. The anterior corticospinal tracts supply the neck and upper limbs, and the lateral corticospinal tracts supply all levels of the body.

Most of the corticospinal fibers synapse with interneurons in the lateral portions of the spinal cord central gray matter. The interneurons, in turn, synapse with the lower motor neurons of the anterior horn that innervate primarily distal limb muscles.

Damage to the corticospinal tracts results in reduced muscle tone, clumsiness, and weakness but not in complete paralysis, even if the damage is bilateral. Experiments with monkeys have demonstrated that bilateral sectioning of the medullary pyramids results in (1) loss of contact-related activities such as tactile placing of the foot and grasping, (2) defective fine movements, and (3) hypotonia

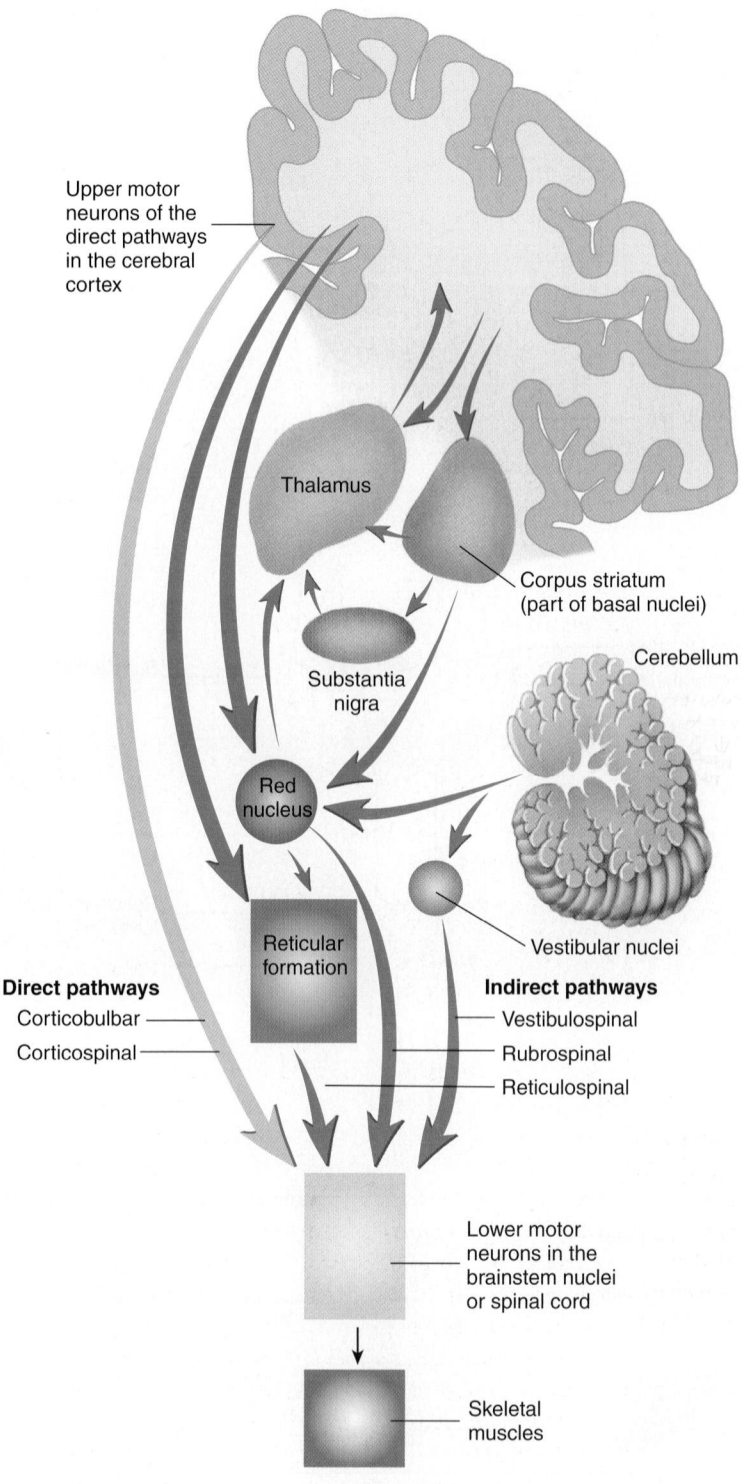

Figure 14.13 Descending Pathways

The direct pathways (corticobulbar and corticospinal) are indicated by the *blue arrow*. The indirect pathways and their interconnections are indicated by the *red arrows*.

Figure 14.14 Direct Pathways

Lateral and anterior corticospinal tract, which are responsible for movement below the head. Lines on the inset indicate levels of section.

(reduced tone). These and other experimental data, support the conclusion that the corticospinal system is superimposed over the older indirect pathways and that it has many parallel functions. It is proposed that the main function of the direct pathways is to add speed and agility to conscious movements, especially of the hands,

and to provide a high degree of fine motor control such as in movements of individual fingers. Spinal cord lesions that affect both the direct and indirect pathways result in complete paralysis.

The corticobulbar tracts are analogous to the corticospinal tracts. The former innervate the head, and the latter innervate the rest of the body. Cells that contribute to the corticobulbar tracts are in regions of the cortex similar to those of the corticospinal tracts, except that they are more laterally and inferiorly located. Corticobulbar tracts follow the same basic route as the corticospinal system down to the level of the brainstem. At that point, most corticobulbar fibers terminate in the reticular formation near the **cranial nerve nuclei.** Interneurons from the reticular formation then enter the cranial nerve nuclei where they synapse with lower motor neurons. These nuclei give rise to nerves that control eye and tongue movements, mastication, facial expression, and palatine, pharyngeal, and laryngeal movements.

Brown-Séquard Syndrome

A lesion of the spinal cord that destroys half the cord at a specific level (hemisection) results in a very specific syndrome, called the **Brown-Séquard syndrome.** The syndrome includes contralateral loss of pain and temperature because fibers entering the anterolateral system cross over where they enter the spinal cord, and ipsilateral loss of proprioception, two-point discrimination, and most upper motor control because the dorsal-column/medial-lemniscal system ascends ipsilaterally and the lateral corticospinal tract descends ipsilaterally.

● 15. **What two tracts form the direct pathways? What area of the body is supplied by each tract? Describe the location of the neurons in each tract, where they cross over, and where they synapse.**

Indirect Pathways

The indirect pathways (figure 14.15) originate in neurons of the cerebrum and cerebellum whose axons synapse in some intermediate nucleus rather than directly with lower motor neurons. Axons from the neurons in those nuclei form the indirect pathways. They do not pass through the pyramids or through the corticobulbar tracts and, therefore, are sometimes called extrapyramidal. The major tracts are the rubrospinal, vestibulospinal, and reticulospinal tracts. Many interconnections and feedback loops are present in this system.

Neurons of the **rubrospinal tract** begin in the red nucleus, which is located at the boundary between the diencephalon and midbrain. The tract decussates in the midbrain, and descends in the lateral column of the spinal cord. The red nucleus receives input from both the motor cortex and the cerebellum. Lesions in the red nucleus result in intention, or action, tremors similar to those seen in cerebellar lesions (see the Clinical Focus on "Dyskinesias"; p. 494). The function of the red nucleus therefore is related closely to cerebellar function. Damage to the rubrospinal tract impairs forearm and hand movements but doesn't greatly affect general body movements. The rubrospinal tract is the one indirect tract that is very closely related to the direct, corticospinal tract. It terminates in the lateral portion of the spinal cord central gray matter with the corticospinal tract, and it transmits action potentials involved in the comparator function of the cerebellum (see p. 493). It plays a major role in regulating fine motor control of muscles in the distal part of the upper limbs.

The **vestibulospinal tracts** (see figure 14.12) originate in the vestibular nuclei, descend in the anterior column, and synapse with interneurons and lower motor neurons in the ventromedial portion

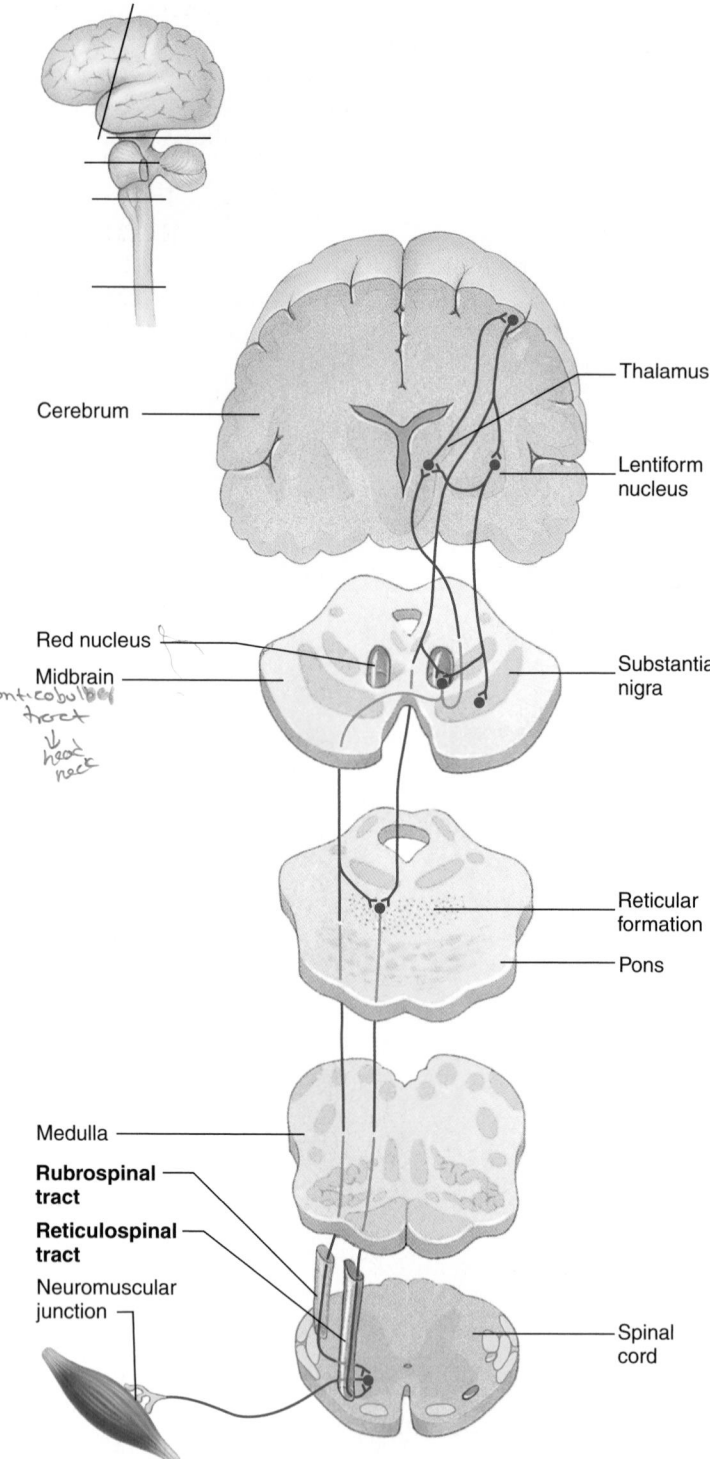

Figure 14.15   **Indirect Pathways**

Examples of indirect pathways: rubrospinal and reticulospinal tracts. Lines on the inset indicate levels of section.

of the spinal cord central gray matter. Their fibers preferentially influence neurons innervating extensor muscles in the trunk and proximal portion of the lower limbs and are involved primarily in the maintenance of upright posture. The vestibular nuclei receive major input from the vestibular nerve (see chapter 15) and the cerebellum.

Neuron cell bodies of the **reticulospinal tract** (see figure 14.12) are in the reticular formation of the pons and medulla oblongata. Their axons descend in the anterior portion of the lateral column and synapse with interneurons and lower motor neurons in the ventromedial portion of the spinal cord central gray matter. The function of this tract involves the maintenance of posture through the action of trunk and proximal upper and lower limb muscles during certain movements. For example, when a person who is standing lifts one foot off the ground, the weight of the body is shifted over to the other limb. The reticulospinal tract apparently enhances the functions of the alpha motor neurons in the crossed extensor reflex during this type of movement so that balance is maintained.

Another major portion of the indirect pathways involves the basal nuclei (see figure 14.13). The basal nuclei have a number of connections with each other, as well as with the thalamus and cerebrum. The basal nuclei also interact with other indirect pathways, like the rubrospinal tract, by which they modulate motor functions.

● 16. *Name the structures and the tracts that form the indirect pathways. What functions do they control? Contrast them with the functions of the direct pathways.*

Modifying and Refining Motor Activities

Basal Nuclei

The **basal nuclei** (see figure 13.8) are important in planning, organizing, and coordinating motor movements and posture. Complex neural circuits link the basal nuclei with each other, with the thalamus, and with the cerebral cortex. These connections form several feedback loops, some of which are stimulatory and others inhibitory. The stimulatory circuits facilitate muscle activity, especially at the beginning of a voluntary movement like rising from a

sitting position or beginning to walk. The inhibitory circuits facilitate the actions of the stimulatory circuits by inhibiting muscle activity in antagonist muscles. Inhibitory circuits also decrease muscle tone when the body, limbs, and head are at rest (eliminating "unwanted" movement). Disorders of the basal nuclei result in difficulty in rising from a sitting position and difficulty in initiating walking. People with basal nuclei disorders exhibit increased muscle tone and exaggerated, uncontrolled movements when they are at rest. A specific feature of some basal nuclei disorders is a "resting tremor," a slight shaking of the hands when a person is not performing a task. Parkinson's disease and cerebral palsy are basal nuclei disorders. They are discussed in the Clinical Focus on "Dyskinesias" (p. 494).

Cerebellum

The **cerebellum** (see figure 13.4) consists of three functional parts: the vestibulocerebellum or flocculonodular lobe, the spinocerebellum, and the cerebrocerebellum. The vestibulocerebellum receives direct input from the vestibular structures, especially the semicircular canals (see chapter 15), and sends axons to the vestibular nuclei of the brainstem. It helps maintain muscle tone in postural muscles. It also helps control balance, especially during movements, and it helps coordinate eye movement.

The **vermis** and medial portion of the **lateral hemisphere,** referred to jointly as the **spinocerebellum,** helps accomplish fine motor coordination of simple movements by means of its **comparator** function. Action potentials from the motor cortex descend into the spinal cord to initiate voluntary movements. At the same time, action potentials are carried from the motor cortex to the cerebellum to give the cerebellar neurons information representing the intended movement (figure 14.16). Simultaneously, action

1. The motor cortex sends action potentials to lower motor neurons in the spinal cord.

2. Action potentials from the motor cortex inform the cerebellum of the intended movement.

3. Lower motor neurons in the spinal cord send action potentials to skeletal muscles, causing them to contract.

4. Proprioceptive signals from the skeletal muscles and joints to the cerebellum convey information concerning the status of the muscles and the structure being moved during contraction.

5. The cerebellum compares the information from the motor cortex to the proprioceptive information from the skeletal muscles and joints.

6. Action potentials from the cerebellum to the spinal cord modify the stimulation from the motor cortex to the lower motor neurons.

7. Action potentials from the cerebellum are sent to the motor cortex, which modify its motor activity.

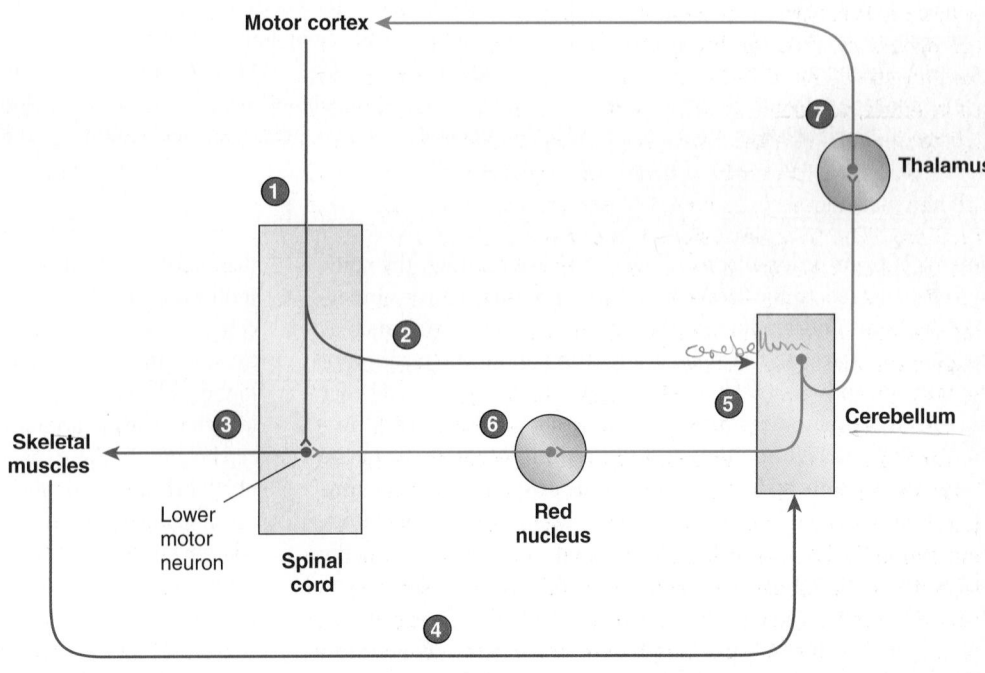

Process Figure 14.16   Cerebellar Comparator Function

Clinical Focus   Dyskinesias

Dyskinesias (dis-ki-nē′zē-ăs) are a group of disorders often involving the basal nuclei in which unwanted, superfluous movements occur. Defects in the basal nuclei may result in brisk, jerky, purposeless movements that resemble fragments of voluntary movements. **Sydenham's chorea** (kōr-ē′ă; also called St. Vitus' dance) is a disease usually associated with a toxic or infectious disorder that apparently causes temporary dysfunction of the corpus striatum. It usually affects children. **Huntington's chorea** is a dominant hereditary disorder that begins in middle life and causes mental deterioration and progressive degeneration of the corpus striatum in affected individuals.

Cerebral palsy (pawl′zē) is a general term referring to defects in motor functions or coordination resulting from several types of brain damage, which may be caused by abnormal brain development or birth-related injury. Some symptoms of cerebral palsy, such as increased muscle tension, are related to basal nuclei dysfunction. **Athetosis** (ath-ĕ-tō′sis), often one of the features of cerebral palsy, is characterized by slow, sinuous, aimless movements. When the face, neck, and tongue muscles are involved, grimacing, protrusion, and writhing of the tongue and difficulty in speaking and swallowing are characteristics.

Damage to the subthalamic nucleus can result in **hemiballismus** (hem-ē-bal-iz′mŭs),

an uncontrolled, purposeless, and forceful throwing or flailing of the arm. Forceful twitching of the face and neck may also result from subthalamic nuclear damage.

Parkinson's disease, characterized by muscular rigidity, loss of facial expression, tremor, a slow, shuffling gait, and general lack of movement, is caused by a dysfunction in the substantia nigra. The disease usually occurs after age 55 and is not contagious or inherited. A resting tremor called "pill-rolling" is characteristic of Parkinson's disease and consists of circular movement of the opposed thumb and index fingertips. The increased muscular rigidity in Parkinson's disease results from defective inhibition of some of the basal nuclei by the substantia nigra. In this disease, dopamine, an inhibitory neurotransmitter produced by the substantia nigra is deficient. The melanin-containing cells of the substantia nigra degenerate, resulting in a loss of pigment.

Parkinson's disease can be treated with levodopa (lē-vō-dō′pă, L-dopa), a precursor to dopamine, or, more effectively, with Sinemet, a combination of L-dopa and carbidopa (kar-bi-dō′pă). Carbidopa is a decarboxylase inhibitor, which prevents the breakdown of L-dopa before it can reach the brain. Because of long-term side effects, including dyskinesias, associated with levodopa, other dopamine agonists, such as ropinirole and pramipexole, are being exam-

ined. A protein called **glial cell line–derived neurotrophic factor (GDNF)** has been discovered that selectively promotes the survival of dopamine-secreting neurons. Chronic stimulation of the globus pallidus (part of the lentiform nucleus) with an electrical pulse generator has shown some success. Experimental treatment of the disorder by transplanting fetal tissues, or stem cells from adult tissues, capable of producing dopamine is also under investigation.

Cerebellar lesions result in a spectrum of characteristic functional disorders. Movements tend to be **ataxic** (jerky) and **dysmetric** (overshooting—for example, pointing past or deviating from a mark that one tries to touch with the finger). Alternating movements such as supination and pronation of the hand are performed in a clumsy manner. **Nystagmus** (nis-tag′mŭs), which is a constant motion of the eyes, may also occur. A cerebellar tremor is an intention tremor (i.e., the more carefully one tries to control a given movement, the greater the tremor becomes). For example, when a person with a cerebellar tremor attempts to drink a glass of water, the closer the glass comes to the mouth, the shakier the movement becomes. This type of tremor is in direct contrast to basal nuclei tremors described previously, in which the resting tremor largely or completely disappears during purposeful movement.

potentials from proprioceptive neurons ascend through the spinocerebellar tracts to the cerebellum. Proprioceptive neurons innervate the joints and tendons of the structure being moved, such as the elbow or knee, and provide information about the position of the body or body parts. These action potentials give the cerebellar neurons information from the periphery about the actual movements. The cerebellum compares the action potentials from the motor cortex to those from the moving structures. That is, it compares the intended movement with the actual movement. If a difference is detected, the cerebellum sends action potentials through the thalamus to the motor cortex and to the spinal cord to correct the discrepancy. The result is smooth and coordinated movements.

The comparator function works to coordinate simple movements like touching your nose. Rapid, complex movements, however, require much greater coordination and training. The **cerebrocerebellum** consists of the lateral two-thirds of the lateral hemispheres. It communicates with the motor, premotor, and prefrontal portions of

the cerebral cortex to help in planning and practicing rapid, complex motor actions. The connections from the cerebrum to the cerebellum constitute a large portion of the axons in the cerebral peduncles. Because of the cerebrocerebellum, with training, a person can learn highly skilled and rapid movements that are accomplished more rapidly than can be accounted for by the comparator function of the cerebellum. In these cases, the cerebellum participates with the cerebrum in learning highly specialized movements like playing the piano or swinging a baseball bat. The cerebrocerebellum is also involved in cognitive functions such as rhythm, conceptualizing time intervals, some word associations, and solving pegboard puzzles—tasks once thought to occur only in the cerebrum.

17. *What are the functions of the basal nuclei?*
18. *Explain the comparator activities of the spinocerebellum.*
19. *Describe the role of the cerebrocerebellum in rapid and skilled motor movements such as playing the piano.*

Cerebellar Dysfunction

Cerebellar dysfunction results in (1) decreased muscle tone, (2) balance impairment, (3) a tendency to overshoot when reaching for or touching an object, and (4) an intention tremor, which is a shaking in the hands that occurs only while attempting to perform a task. Notice that although the cerebellum and basal nuclei both control motor functions, they have opposite effects, and exhibit opposite symptoms when injured. For example, cerebellar dysfunction results in decreased muscle tone and an intention tremor, whereas basal nuclear dysfunction often results in increased muscle tone and a resting tremor.

Brainstem Functions

The major ascending and descending pathways project through the brainstem. In addition, the brainstem contains nuclei, including the nuclei of cranial nerves II–XII nuclei and nuclei of the reticular formation. Only cranial nerve I (olfactory nerve) does not have axons that pass through the brainstem or nuclei in the brainstem.

The brainstem receives sensory input from collateral branches of ascending spinal cord pathways and from the axons of cranial nerves II (vision), V (tactile sensation from the face, nasal cavity, and oral cavity), VII (taste), VIII (hearing and balance), IX (taste; tactile sensation in the throat), and X (taste; tactile sensation in the larynx; visceral sensation in the thorax and abdomen). These cranial nerves all have sensory nuclei in the brainstem. Many of these nuclei are involved in the special senses and are discussed in chapter 15. The brainstem nuclei associated with cranial nerve II are involved in visual reflexes.

As noted earlier, fibers of the spinothalamic tracts passing through the brainstem, are joined by fibers of the trigeminothalamic tract (trigeminal nerve, or cranial nerve V). This tract carries tactile sensations, such as pain and temperature, two-point discrimination, proprioception, and light touch from the face, nasal cavity, and oral cavity, including the teeth. This input is much like that from the spinal nerves in that primary neurons from one side of the face synapse with secondary neurons, which cross to the opposite side of the brainstem. A difference is that the brainstem contains a different nucleus for each of the four tactile sensory modes; pain and temperature, light touch, two point discrimination, and proprioception from the trigeminal nerve. The secondary neurons synapse with tertiary neurons in the thalamus. Tertiary neurons from the thalamus project to the somatic sensory cortex. Collateral branches of trigeminothalamic tract neurons project to the reticular formation where they stimulate wakefulness and consciousness. This part of the reticular formation and its connections constitute the **reticular activating system (RAS),** which is involved in the sleep–wake cycle.

P R E D I C T 7

Describe an effective technique for arousing a sleeping person.

Collateral branches of cranial nerves II (optic) and VIII (vestibulocochlear), ascending tactile sensory pathways, and descending neurons from the cerebrum also project to the RAS. Visual and acoustic stimuli, as well as mental activities, stimulate the RAS to help maintain alertness and attention. Ringing alarm clocks, sudden flashes of bright lights, or cold water being splashed on the face can all arouse consciousness. Removal of visual, auditory, and other stimuli may lead to drowsiness or sleep. For example, consider what happens to many students during a monotonous lecture in a dark lecture hall.

Damage to RAS cells of the reticular formation can result in coma.

Drugs and the Reticular Activating System

Certain drugs can either stimulate or depress the RAS. General anesthetics suppress this system, and many tranquilizers depress it. On the other hand, ammonia (smelling salts) and other irritants stimulate trigeminal nerve endings in the nose. As a result, action potentials are sent to the reticular formation and the cerebral cortex to arouse an unconscious patient.

Several important reflexes are integrated by nuclei in the brainstem. For example, sensory input from cranial nerve IX (glossopharyngeal) conveys tactile information from the back of the tongue, the soft palate, and the throat (pharynx) to the brainstem. Mechanical stimulation of these areas can initiate a gag reflex, whereas other stimulation of the throat can initiate a cough reflex. Sensory input from cranial nerve X conveys tactile information from the larynx (voicebox) and thoracic and abdominal viscera. Tactile input from the larynx can also initiate a cough reflex. In addition, cranial nerve X (vagus nerve) is involved in many complex reflexes associated with vital functions like heart rate, respiration, and digestion. Many of these involve the reticular formation and are discussed in later chapters.

Several critical functions like heart rate, blood pressure, respiration, sleep, swallowing, vomiting, coughing, and sneezing are regulated by nuclei of the brainstem. When a person is involved in a serious accident or is extremely ill, most of the vital functions assessed by medical personnel, such as blood pressure, heart rate, respiration, and dilation of the pupils, are controlled by the brainstem; so many emergency evaluations involve evaluations of brainstem function.

Descending pathways in the brainstem pass to the spinal cord, pass into the cerebellum, or synapse with cranial nerve motor nuclei and other nuclei in the brainstem. Some of the descending pathways originate in the cerebral cortex and pass directly through the brainstem (direct pathways). Others synapse with brainstem nuclei, which, in turn, send descending fibers into the spinal cord (indirect pathways). Descending fibers from the reticular formation constitute one of the body's most important motor pathways. Fibers from the reticular formation are critical in controlling many vital functions, such as respiratory movements and cardiac rhythms. Cranial nerves III, IV, V, VI, VII, IX, X, XI, and XII all have motor nuclei in the brainstem.

Cranial nerves III, IV, and VI control the eye muscles. Collateral branches from the optic nerve tract (II) synapse in the superior colliculi of the midbrain (see figure 13.7). Axons from the superior colliculi project to cranial nerve nuclei II (oculomotor), IV (trochlear), and VI (abducent) and to the cervical part of the spinal cord, spinal nucleus of XI, where they stimulate motor neurons involved in turning the eyes and head toward a visual stimulus. The superior colliculi also receive input from auditory pathways, which can initiate a reflex that turns the eyes and head toward a sudden noise. Action potentials reaching the superior colliculi from the cerebrum are involved in the visual tracking of moving objects. The visual tracking with both eyes to the right involves the lateral rectus muscle and abducens (VI) nerve of the right eye and the medial rectus muscle and oculomotor (III) nerve of the left eye. Coordination of these two nerves and muscles requires nuclei of the reticular formation. Constriction of the pupil involves parasympathetic stimulation through the oculomotor (III) nerve. The visual reflexes resulting in pupil constriction are coordinated through nuclei in the reticular formation. These reflexes are also coordinated by a nuclear region in the diencephalon called the pretectal area (in front of the tectum, the roof of the midbrain).

The Brain's Canary

Function of the pretectal area is critical for normal pupillary constriction in response to light. This area of the brain can be thought of as the "brain's canary." For many years, miners carried caged canaries with them into deep mine shafts to detect poison gas. An unconscious or dead canary would warn the miners of methane gas before enough accumulated to kill them. Like a canary, the pretectal area is more sensitive to brain damage, or symptoms of the damage are more easily observed, than many other parts of the brain. Fixed, dilated pupils in a patient is a sign that the patient may have experienced damage to the brain in addition to the pretectal area.

Motor fibers from the trigeminal nerve (V) innervate the muscles of mastication. Tactile sensory input from the same nerve informs the brainstem and cerebrum of the presence of food or some other object in the mouth. The presence of an object in the mouth, even a nonfood item like a marble, stimulates a reflex between the trigeminal sensory nuclei and the motor nuclei of VII and IX, which innervate the salivary glands to stimulate salivation. A reflex between the trigeminal sensory nuclei and the motor nucleus of V initiates the chewing cycle, which is regulated by the reticular formation. Other reflexes in the trigeminal nerve system detect how hard or soft an item is in the mouth and adjusts the bite accordingly. The motor nucleus of XII innervates the tongue muscles. Reflexes between the trigeminal sensory nuclei and the motor nucleus of XII control the tongue to help place food between the teeth for chewing, while, at the same time, keep the tongue out of harm's way.

20. List the motor nuclei of the brainstem.
21. Describe some of the reflexes that occur in the brainstem.
22. What are some of the vital functions that are regulated in the brainstem?

Other Brain Functions

The human brain is capable of many functions besides awareness of sensory input and the control of skeletal muscles. Speech, mathematical and artistic abilities, sleep, memory, emotions, and judgment are functions of the brain.

Speech

In most people, the speech area is in the left cerebral cortex. Two major cortical areas are involved in speech: **Wernicke's area** (sensory speech area), a portion of the parietal lobe, and **Broca's area** (motor speech area) in the inferior part of the frontal lobe (see figure 14.9). Wernicke's area is necessary for understanding and formulating coherent speech. Broca's area initiates the complex series of movements necessary for speech.

For someone to speak a word that he or she sees, such as when reading out loud, the following sequence of events must take place. Action potentials from the eyes reach the primary visual cortex where the word is seen. The word is then recognized in the visual association area and understood in parts of Wernicke's area. Then action potentials representing the word are conducted through association fibers that connect Wernicke's and Broca's areas. In Broca's area, the word is formulated as it will be spoken. Action potentials are then propagated to the premotor area, where the movements are programmed, and finally to the primary motor cortex, where the proper movements are triggered (figure 14.17).

1. The word is seen in the visual cortex.
2. Information concerning the word is understood in Wernicke's area.
3. Information from Wernicke's area is transferred to Broca's area.
4. Information is transferred from Broca's area to the primary motor cortex.

Figure 14.17 **Demonstration of Cortical Activities During Speech**
The figures show the pathway for reading and naming something that is seen, such as reading aloud. PET scans show the areas of the brain that are most active during various phases of speech. *Red* indicates the most active areas; *blue* indicates the least active areas.

To repeat a word that has been heard is similar. The information passes from the ears to the primary auditory cortex and then passes to the auditory association area, where the word is recognized, and continues to Wernicke's area, where the word is understood and formulated as it will be spoken. From Wernicke's area, it follows the same route as followed for speaking words that are seen.

Aphasia

Aphasia (ă-fā′zē-ă), absent or defective speech or language comprehension, results from a lesion in the language areas of the cortex. The several types of aphasia depend on the site of the lesion. **Receptive aphasia** (Wernicke's aphasia), which includes defective auditory and visual comprehension of language, defective naming of objects, and repetition of spoken sentences, is caused by a lesion in Wernicke's area. Both **jargon aphasia,** in which a person may speak fluently but unintelligibly, and **conduction aphasia,** in which a person has poor repetition but relatively good comprehension, can result from a lesion in the tracts between Wernicke's and Broca's areas. **Anomic** (ă-nō′mik) **aphasia,** caused by the isolation of Wernicke's area from the parietal or temporal association areas, is characterized by fluent but circular speech resulting from poor word-finding ability. **Expressive aphasia** (Broca's aphasia), caused by a lesion in Broca's area, is characterized by hesitant and distorted speech.

23. List the necessary sequence of events that must occur for a person to repeat a word that he or she hears.

P R E D I C T 8
Propose the sequence needed for a blindfolded person to name an object placed in her right hand.

Right and Left Cerebral Cortex

The cortex of the right cerebral hemisphere controls muscular activity in and receives sensory input from the left half of the body. The left cerebral hemisphere controls muscles in and receives sensory input from the right half of the body. Sensory information received by the cortex of one hemisphere is shared with the other through connections between the two hemispheres called **commissures** (kom′i-shŭrz; a joining together). The largest of these commissures is the **corpus callosum** (kōr′pŭs kă-lō′sŭm; callous body), which is a broad band of nerve tracts at the base of the longitudinal fissure (see figure 13.1).

Language and perhaps other functions, such as artistic activities, are not shared equally between the left and right cerebral hemispheres. The left hemisphere is more involved in such skills as mathematics and speech. The right hemisphere is involved in activities like three-dimensional or spatial perception, recognition of faces, and musical ability.

24. Name the largest pathway that connects the right and left cerebral hemispheres.
25. What are the functions localized in the left cerebral hemisphere? In the right cerebral hemisphere?

Hemisphere Dominance and Amorphosynthesis

Dominance of one cerebral hemisphere over the other, for most functions, is probably not very important in most people because the two hemispheres are in constant communication through the corpus callosum, literally allowing the right hand to know what the left hand is doing. Surgical cutting of the corpus callosum has been successful in treating a limited number of epilepsy cases. Under certain conditions, however, interesting functional defects develop in people who have had their corpus callosum severed. For example, if a patient with a severed corpus callosum is asked to reach behind a screen to touch one of several items with one hand without being able to see it and then is asked to point out the same object with the other hand, the person cannot do it. Tactile information from the left hand enters the right somatic sensory cortex but that information is not transferred to the left hemisphere, which controls the right hand. As a result, the left hemisphere cannot direct the right hand to the correct object.

A person suffering a stroke in the right parietal lobe may lose the ability to recognize faces while retaining essentially all other brain functions. A more severe lesion can cause a person to lose the ability to identify simple objects. This defect is called **amorphosynthesis** (ă-mōr′fō-sin′thě-sis). Some people with a similar lesion in the right cerebral hemisphere may tend to ignore the left half of the world, including the left half of their own bodies. These people may completely ignore a person who is to their left but react normally when the person moves to their right. They may also fail to dress the left half of their bodies or eat the food on the left half of their plates.

Brain Waves and Sleep

Electrodes placed on a person's scalp and attached to a recording device can record the electrical activity of the brain, producing an **electroencephalogram** (ē-lek′trō-en-sef′ă-lō-gram; **EEG;** figure 14.18). These electrodes are not sensitive enough to detect individual action potentials, but they can detect the simultaneous action potentials in large numbers of neurons. As a result, the EEG displays wavelike patterns known as **brain waves.** Brain waves are produced continuously, but their intensity and frequency differ from time to time based on the state of brain activity.

Most of the time, EEG patterns from a given individual are irregular with no particular pattern because, although the normal brain is active, most of its electrical activity is not synchronous. At other times, however, specific patterns can be detected. These regular patterns are classified as alpha, beta, theta, or delta waves (see figure 14.18). **Alpha waves** are observed in a normal person who is awake but in a quiet, resting state with the eyes closed. **Beta waves** have a higher frequency than alpha waves and occur during intense mental activity. **Theta waves** usually occur in children, but they can also occur in adults who are experiencing frustration or who have certain brain disorders. **Delta waves** occur in deep sleep, in infancy, and in patients with severe brain disorders.

Brain wave patterns vary during the four stages of sleep (see figure 14.18) A sleeping person arouses several times during a period of sleep. Dreaming occurs during periods when eye movement can be observed in a sleeping person, called rapid eye movement (**REM**) sleep.

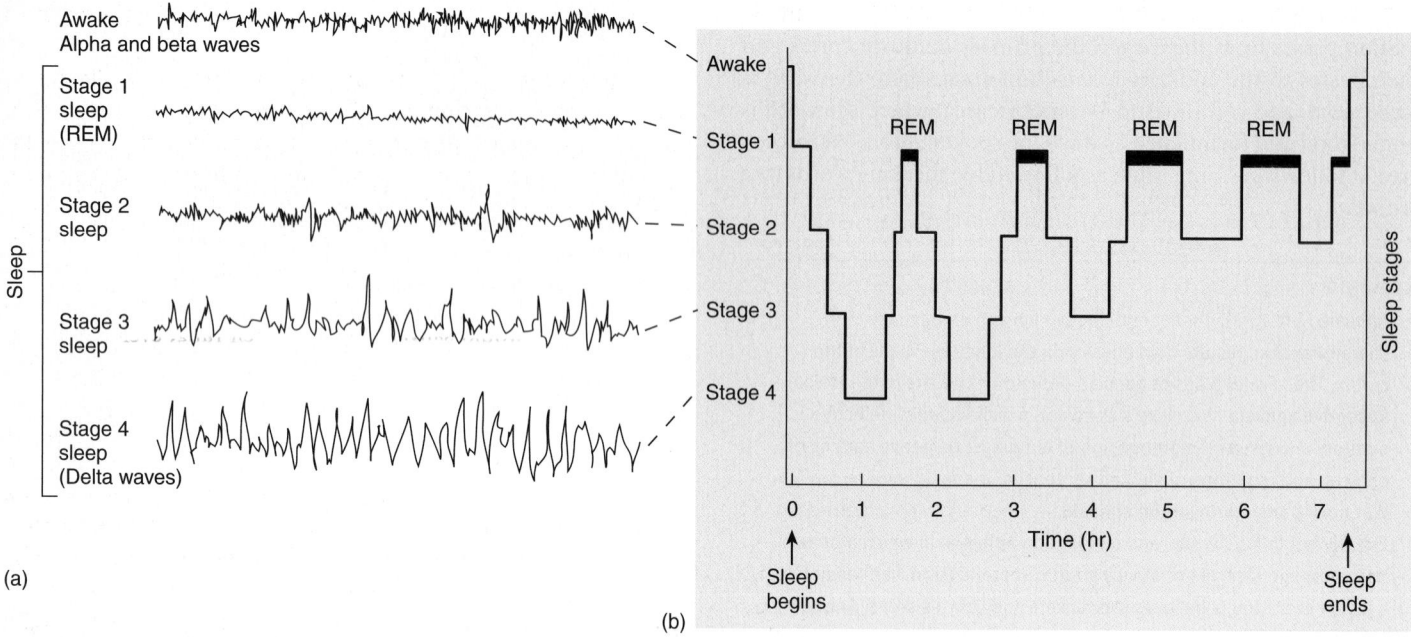

Figure 14.18   Electroencephalograms (EEGs) Showing Brain Waves

(*a*) EEG tracings when a person is awake and during four stages of sleep. (*b*) A typical night's sleep pattern in a young adult. The time spent in REM sleep is labeled and shown by *dark bars.*

Distinct types of EEG patterns can be detected in patients with specific brain disorders, such as epileptic seizures. Neurologists use these patterns to diagnose the disorders and determine the appropriate treatment.

● **26. What is an EEG? What four conditions produce alpha, beta, theta, and delta waves, respectively?**

Memory

Memory is divided into three major types: sensory, short term (or primary), and long term (figure 14.19). **Sensory memory** is the very short-term retention of sensory input received by the brain while something is scanned, evaluated, and acted on. This type of memory lasts less than a second and apparently involves transient changes in membrane potentials.

If a given piece of data held in sensory memory is considered valuable enough, it is moved into **short-term memory,** where information is retained for a few seconds to a few minutes. This memory is limited primarily by the number of bits of information (usually about seven or eight) that can be stored at any one time, although the amount varies from person to person. More bits can be stored when the numbers are grouped into specific segments separated by spaces. When new information is presented, or when the person is distracted, old information previously stored in short-term memory is eliminated; therefore, if a person is given a second telephone number or if the person's attention is drawn to something else, the first number usually is forgotten.

Long-term memory lasts for hours or longer. Some long-term memory may last a lifetime. Two types of long-term memory exist: declarative or explicit and procedural or implicit . **Declarative** or **explicit memory** involves the retention of facts, such as names, dates, and places. Explicit memory is accessed by part of the temporal lobe called the **hippocampus** (hip-ō-kam′pŭs; shaped like a seahorse) and the **amygdala,** or **amygdaloid** (ă-mig′dă-loyd; almond-shaped) **nuclear complex.** The hippocampus is involved in retrieving the actual memory, such as recalling a person's name; and the amygdala is involved in the emotional overtones of that memory, such as feelings of like or dislike, and the recollection of good or bad memories associated with that person. A lesion in the temporal lobe affecting the hippocampus can prevent the brain from moving information from short-term to long-term memory. Emotion and mood apparently serve as gates in the brain and determine what is or is not stored in long-term explicit memory. The amygdala is also a key to the development of fear, which also involves the prefrontal cortex and the hypothalamus.

Figure 14.19   Memory Processing

Fear

Some aspects of fearful responses appear to be "hardwired" in the brain and don't require learning. For example, infant rodents are terrified when exposed to a cat, even though they have never seen a cat. Loud sounds seem to be particularly effective in eliciting fear responses. A direct collateral branch runs from the auditory pathway to the amygdala, which does not involve the cerebral cortex. Fear can be evoked by a loud sound acting directly on the amygdala. Overcoming fear, however, requires the involvement of the cerebral cortex; therefore, the stimulation of fear appears to involve one process, and its suppression another. Flaws in either process could result in fear-related disorders, such as anxiety, depression, panic, phobias, and posttraumatic stress disorder.

Parts of explicit memory appear to be stored separately in various parts of the cerebrum, especially in the parietal lobe, much like storing items in separate "pigeonholes." Memories of people appear to be stored separately from memories of places. People's faces may be stored in yet other pigeonholes. Family members appear to be stored together. Items that are recognized by sight, such as an animal, are stored separately from items that are recognized by feel, such as tools. Damage, such as a stroke, to one part of the brain can remove certain memories without affecting others.

Retrieval of a complete memory requires accessing parts of the memory from different pigeonholes. A complex memory requires accessing and reassembling segments of memory each time the memory is recalled. The memory of an experience, for example, may be stored in at least four different pigeonholes. Where you were is stored in one place, who you were with in another, what happened in another, and how you felt in yet another place. The complexity of this process may be responsible for the changes in what is recalled over time. On occasion, parts of unrelated different memories may be pulled out and put together incorrectly to create a "false memory." Much of what is stored as explicit memory is gradually lost through time.

Procedural or **implicit memory,** also called **reflexive memory,** involves the development of skills like riding a bicycle or playing a piano. Implicit memory is stored primarily in the cerebellum and the premotor area of the cerebrum. Conditioned, or Pavlovian, reflexes are also implicit and can be eliminated in experimental animals by producing cerebellar lesions in the animals. The most famous example of a conditioned reflex is that of Ivan Pavlov's experiments with dogs. Each time he fed the dogs, a bell was rung; soon the dogs would salivate when the bell rang, even if no food was presented. Only a small amount of implicit memory is lost through time.

Several physiologic explanations have been proposed for short-term memory, most of which involve short-term changes in membrane potentials. The changes in membrane potentials are transitory but are longer than those involved in sensory memory, and they can be eliminated by new information reaching the cells.

Certain pieces of information are transferred from short-term to long-term memory. Long-term memory involves changes in neurons, called **long-term potentiation,** which facilitates future transmission of action potentials (figure 14.20). The amount of the neurotransmitter glutamate produced and released by the presynaptic neuron increases. The number of glutamate receptors in the postsynaptic neuron also increases, and the reaction of the postsynaptic neuron to glutamate is potentiated. Long-term memory storage in a single neuron also involves Ca^{2+} influx into the postsynaptic cell. Calcium ions associate with and activate **calmodulin** (kal-mod'ū-lin) inside the cell. Calmodulin, through a cAMP mechanism, stimulates the synthesis of specific proteins. These proteins are involved in changing the shape of the cell. The change in shape is stabilized by the creation of a new cytoskeleton, and the memory becomes more or less permanent.

1. The amount of the neurotransmitter glutamate produced by the presynaptic neuron increases.

2. The amount of glutamate released by the presynaptic neuron also increases.

3. The number of glutamate receptors on the postsynaptic neuron membrane increases.

4. Ca^{2+} channels in the postsynaptic membrane open, allowing Ca^{2+} to enter the cell.

5. The Ca^{2+} that enter the cell associate with the intracellular molecule calmodulin.

6. Activated calmodulin activates a cAMP intracellular mediator, which stimulates synthesis of specific proteins.

7. The cellular effect may involve structural changes in the cell.

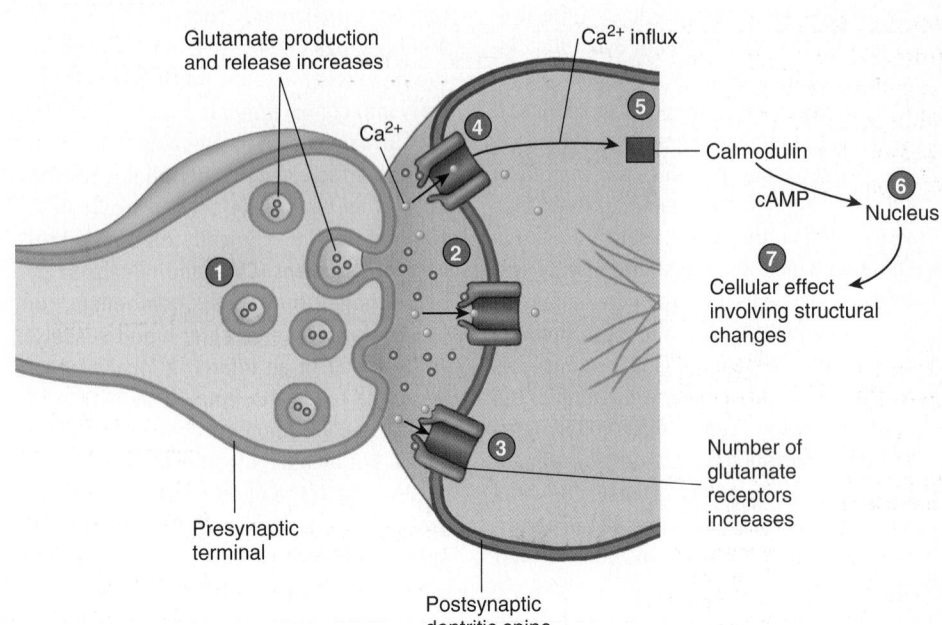

Process Figure 14.20 Cellular Mechanisms of Long-Term Potentiation

Clinical Focus   General CNS Disorders

Infections

Encephalitis (en-sef-ă-lī′tis) is an inflammation of the brain most often caused by a virus and less often by bacteria or other agents. A large variety of symptoms may result, including fever, paralysis, coma, or even death.

Myelitis (mī-ĕ-lī′tis) is an inflammation of the spinal cord caused by trauma, multiple sclerosis, or a number of infectious agents, including viruses, bacteria, or other agents. A large variety of symptoms may result depending on the extent and level of injury or infection.

Meningitis (men-in-jī′tis) is an inflammation of the meninges. It may be virally induced but is more often bacterial. Symptoms usually include stiffness in the neck, headache, and fever. Pus may accumulate in the subarachnoid space, block CSF flow, and result in hydrocephalus. In severe cases, meningitis may also cause paralysis, coma, or death.

Reye's syndrome may develop in children following a viral infection, especially influenza or chickenpox. The use of aspirin in cases of viral infection has been linked to development of the syndrome in the United States. A predisposing disorder in fat metabolism may also be present in some cases. In children affected by the syndrome, the brain cells swell, and the liver and kidneys accumulate fat. Symptoms include vomiting, lethargy, and loss of consciousness and may progress to coma and death or to permanent brain damage.

Rabies is a viral disease transmitted by the bite of an infected mammal. The rabies virus infects the brain, salivary glands (through which it is transmitted), muscles, and connective tissue. When the patient attempts to swallow, the effort can produce pharyngeal muscle spasms; sometimes even the thought of swallowing water or the sight of water can induce pharyngeal spasms. Thus the term *hydrophobia,* fear of water, is applied to the disease. The virus also infects the brain and results in abnormal excitability, aggression, and, in later stages, paralysis and death.

Tabes dorsalis (tā′bēz dōr-sā′lis) is a progressive disorder occurring as a result of untreated syphilis. Tabes means a wasting away, and dorsalis refers to a degeneration of the dorsal roots and dorsal columns of the spinal cord. The symptoms include ataxia, resulting from lack of proprioceptive input; anesthesia, resulting from dorsal root damage; and eventually paralysis as the infection spreads.

Multiple sclerosis (MS), although of unknown cause, possibly involves an autoimmune response to a viral infection. It results in localized brain lesions and demyelination of neurons in the brain and spinal cord, in which the myelin sheaths become sclerotic, or hard—thus the name—causing poor conduction of action potentials. Symptomatic periods of MS are separated by periods of apparent remission. With each recurrence, however, many neurons are permanently damaged so that the progressive symptoms of the disease include exaggerated reflexes, tremor, nystagmus (rhythmic oscillation of the eyes), and speech defects.

Other Disorders

Tumors of the brain develop from neuroglial cells. Symptoms vary widely, depending on the location of the tumor, but may include headaches, neuralgia (pain along the distribution of a peripheral nerve), paralysis, seizures, coma, and death. **Meningiomas** (mĕ-nin′jē-ō′măz), tumors of the meninges, account for 25% of all primary intracranial tumors.

Stroke is a term meaning a blow or sudden attack, suggesting the speed with which this type of defect can occur. It is also referred to clinically as a **cerebrovascular accident (CVA)** and is caused by hemorrhage, thrombosis, embolism, or vasospasm of the cerebral blood vessels, which result in an infarct, a local area of neuronal cell death caused by a lack of blood supply. Symptoms depend on the location but include anesthesia or paralysis on the side of the body opposite the cerebral infarct. Each year 75,000 Americans suffer strokes. Cigarette smokers are 2.5 times more likely to suffer strokes than are nonsmokers. A daily dose of aspirin may reduce a person's risk of stroke by 50%–80% through its ability to interfere with blood clotting.

An **aneurysm** (an′ū-rizm) is a dilation, or ballooning, of an artery. The arteries around the brain are common sites for aneurysms, and hypertension can cause one of these "balloons" to burst or leak, causing a hemorrhage around the brain. With hemorrhaging, blood may enter the epidural space (epidural hematoma), subdural space (subdural hematoma), subarachnoid space, or the brain tissue. Blood in the subdural or subarachnoid space can apply pressure to the brain, causing damage to brain tissue. Blood is toxic to brain tissue, so that blood entering the brain can directly damage brain tissue.

Cerebral compression may occur as a result of hematomas, hydrocephalus, tumors, or edema of the brain, which can occur as the result of a severe blow to the head. The intracranial pressure increases, which may directly damage brain tissue. The cerebellum may compress the fourth ventricle, blocking the foramina and causing internal hydrocephalus, which further increases intracranial pressure. The greatest problem comes from compression of the brainstem. Compression of the midbrain can kink the oculomotor nerves, resulting in dilation of the pupils with no light response. Compression of the medulla oblongata may disrupt cardiovascular and respiratory centers, which can cause death. Compression of any part of the CNS that results in ischemia for as little as 3–5 minutes can result in local neuronal cell death. This is a major problem in spinal cord injuries.

Alzheimer's disease is a severe type of mental deterioration, or dementia, usually affecting older people but occasionally affecting people younger than 60. It accounts for half of all dementias; the other half result from drug and alcohol abuse, infections, or CVAs. Alzheimer's disease is estimated to affect 10% of all people older than 65 and nearly half of those older than 85.

Alzheimer's disease involves a general decrease in brain size resulting from loss of neurons in the cerebral cortex. The gyri become narrower, and the sulci widen. The frontal lobes and specific regions of the temporal lobes are affected most severely. Symptoms include general intellectual deficiency, memory loss, short attention span, moodiness, disorientation, and irritability.

Amyloid plaques and neurofibrillary tangles, which may contain aluminum accumulations, form in the cortex of patients with Alzheimer's disease. **Amyloid** (am'i-loyd) **plaques** are localized axonal enlargements of degenerating nerve fibers, containing large amounts of β-amyloid protein, and **neurofibrillary tangles,** which are filaments inside the cell bodies of the dead or dying neurons.

Some evidence exists that Alzheimer's disease may have characteristics of a chronic inflammatory disease, similar to arthritis, and anti-inflammatory drug therapy has had some affect in slowing its progress. Estrogen treatment may decrease or postpone symptoms in women.

The gene for β-amyloid protein has been mapped to chromosome 21; however, it is thought that only the rare, inherited, early-onset (beginning before age 60) form of Alzheimer's maps to chromosome 21. The more common late-onset form (beginning after age 65), which makes up more than three-fourths of all cases, maps to chromosome 19. It is noteworthy that people with Down's syndrome, or trisomy 21, which means that a person has three copies of chromosome 21, exhibit the cortical and other changes associated with Alzheimer's disease.

Another protein, **apolipoprotein E** (ap'ō-lip-ō-prō'tēn; **apo E**), which binds to β-amyloid protein and is known to transport cholesterol in the blood, has also been associated with Alzheimer's disease. The protein has been found in the plaques and tangles and has been mapped to the same region of chromosome 19 as the late-onset form of Alzheimer's. People with two copies of the apo E-IV gene are eight times more likely to develop the disease than people with no copies of the defective gene. Apo E-IV apparently binds to β-amyloid more rapidly and more tightly than does apo E-III, which is the normal form of the protein.

Apo E may also be involved with regulating phosphorylation of another protein,

called τ (tau), which, in turn, is involved in microtubule formation inside neurons. If τ is overphosphorylated, microtubules are not properly constructed, and the τ proteins intertwine to form neurofibrillary tangles. It has been demonstrated that apo E-III interacts with τ but that apo E-IV does not. It may be that the less stable microtubules, formed with a decreased τ involvement, begin to eventually break down, resulting in neuronal dysfunction. The neurofibrillary tangles of τ proteins may also clog up the cell, further decreasing cell function.

Alzheimer's may be treated with the monoamine oxidase-B inhibitor L-deprenyl. The drug may stimulate nitric oxide production, which could stimulate vasodilation of cerebral blood vessels. There is some evidence that antioxidants; vitamins C, E, and beta carotene; may help prevent Alzheimer's.

Tay-Sachs disease is a hereditary disorder of infants, mostly of Eastern European Jewish descent. Symptoms include paralysis, blindness, and death, usually before age 5. The disease is named for Warren Tay (1843–1927), a British ophthalmologist, and Bernard Sachs (1858–1944) a U.S. neurologist, who first described aspects of the disease. Tay-Sachs disease results from an abnormal gene on chromosome 15, which causes the production of an abnormal enzyme, hexosaminidase A (Hex-A). With Hex-A not functioning properly, the sphingolipid (lipids with long base chains) GM2 ganglioside accumulates in and damages cells, especially neurons of the CNS.

Damage to the CNS begins in the fetus, but clinical manifestations do not appear until several months after birth. The infant appears to develop normally, but then begins to regress, losing motor functions, exhibiting seizures, becoming blind and mentally retarded, and eventually dying by age 4 or 5.

There is no cure or effective treatment at present. Gene replacement therapy is being investigated, but progress is slow.

Chronic mercury poisoning can cause brain disorders, such as intention tremor, exaggerated reflexes, and emotional instability. Mercury causes oxidative damage and

apoptosis in several organs, including the kidneys and CNS.

Lead poisoning is a serious problem, particularly among urban children. Lead is taken into the body from contaminated air, food, and water. Flaking lead paint in older houses and soil contamination can be major sources of lead poisoning in children. Lead usually accumulates slowly in the body until toxic levels are reached.

Brain damage caused by lead poisoning in children includes edema, demyelination, and cortical neuron necrosis with astrocyte proliferation. This damage appears to be permanent and can result in reduced intelligence, learning disabilities, poor psychomotor development, and blindness. In severe cases, psychoses, seizures, coma, or death may occur. Adults exhibit more mild PNS symptoms, including demyelination with decreased neuromuscular function. Other symptoms include abdominal pain and renal disease.

Epilepsy is a group of brain disorders that have seizure episodes in common. The seizure, a sudden massive neuronal discharge, can be either partial or complete, depending on the amount of brain involved and whether or not consciousness is impaired. A seizure involves a change in sensation, consciousness, or behavior due to brief electrical discharge in the brain.

Normally a balance exists between excitation and inhibition in the brain. When this balance is disrupted by increased excitation or decreased inhibition, a seizure may result. The neuronal discharges may stimulate muscles innervated by the neurons involved, resulting in involuntary muscle contractions, or convulsions.

Epilepsy occurs in 1% of all people age 20 and 3% of those over age 75. A number of factors can increase the incidence of epilepsy: 2.4% of children whose fathers have epilepsy develop epilepsy and 8.7% of children whose mothers have epilepsy develop the disorder. Children with cerebral palsy develop epilepsy in 10% of cases, and 10% of those with mental retardation develop epilepsy. Half of children with both

Continued

Continued

cerebral palsy and mental retardation develop epilepsy. Stroke patients develop epilepsy in 22% of cases. Seventy percent of those with epilepsy will eventually experience remission.

Depression may cause more "grief and misery" than any other single disease. Although the illness has been known for over 2000 years, its medical status is still uncertain. Is depression a disease state caused by some chemical excess or deficiency, or is it a psychologic condition that a person can decide to snap out of? The answer is probably that both types of depression exist. Depression is a complex multifaceted group of disorders. Some types of "endogenous" depression can be treated with antidepressants, of which there are five groups: tricyclic antidepressants, nontricyclic compounds, MAO inhibitors, serotonin agonists, and lithium. Many people with depression also have epilepsy. Recent research in which "pacemaker-like" stimulation of the vagus nerve to treat epilepsy has shown some promise in treating depression that does not respond to drugs.

Headaches have a variety of causes that can be grouped into two basic classes: extracranial and intracranial. Extracranial headaches can be caused by inflammation of the sinuses, dental irritations, temporomandibular joint disorders, ophthalmologic disorders, or tension in the muscles moving the head and neck. Intracranial headaches may result from inflammation of the brain or meninges, vascular problems, mechanical damage, or tumors.

Tension headaches are extracranial muscle tension, stress headaches, consisting of a dull, steady pain in the forehead, temples, neck, or throughout the head. Tension headaches are associated with stress, fatigue, and posture.

Migraine headaches (migraine means half a skull) occur in only one side of the head and appear to involve the abnormal dilation and constriction of blood vessels. They often start with distorted vision, shooting spots, and blind spots. Migraines consist of severe throbbing, pulsating pain. About 80% of migraine sufferers have a family history of the disorder, and women are affected four times more often than men. Those suffering migraines are usually women younger than 35. The severity and frequency usually decrease with age.

A **concussion** is a blow to the head producing momentary loss of consciousness without immediate detectable damage to the brain. Often no more problems occur after the person regains consciousness; however, in some cases, **postconcussion syndrome** may occur a short time after the injury. The syndrome includes increased muscle tension or migraine headaches, reduced alcohol tolerance, difficulty in learning new things, reduction in creativity, and motivation, fatigue, and personality changes. The symptoms may be gone in a month or may persist for as much as a year. In some cases, postconcussion syndrome may be the result of a slowly occurring subdural hematoma that may be missed by an early examination. The blood may accumulate from small leaks in the dural sinuses.

Alexia (ă-lek′sē-ă), loss of the ability to read, may result from a lesion in the visual association cortex. **Dyslexia** (dĭs-lek′sē-ă) is a defect in which the reading level is below that expected on the basis of an individual's overall intelligence. Most people with dyslexia have normal or above-normal intelligence quotients. The term means reading deficiency and is also called partial alexia. It is three times more common in males than females. As many as 10% of males in the United States suffer from the disorder. The symptoms vary considerably from person to person and include transposition of letters in a word, confusion between the letters *b* and *d*, and lack of orientation in three-dimensional space. The brains of some dyslexics have abnormal cellular arrangements, including cortical disorganization and the appearance of bits of gray matter in medullary areas. Dyslexia apparently results from abnormal brain development.

Children with **attention deficit disorder (ADD)** are easily distractible, have short attention spans, and may shift from one uncompleted task to another. Children with **attention deficit/hyperactivity disorder (ADHD)** exhibit the characteristics of ADD, but they are also fidgety, have difficulty remaining seated and waiting their turn, engage in excessive talking, and commonly interrupt others. About 3% of all children exhibit ADHD, more so in boys than girls. Symptoms usually occur before age 7. The neurologic basis of both ADD and ADHD is as yet unknown.

A whole series of neurons and their pattern of activity, called a **memory engram,** or memory trace, probably are involved in the long-term retention of information, a thought, or an idea. Repetition of the information and association of the new information with existing memories assist in the transfer of information from short-term to long-term memory.

● 27. *Name the three different types of memory, and describe the processes that result in the transfer of information from short-term to long-term memory.*

28. *Distinguish between implicit and explicit memory.*

Sleep and Memory

Is "pulling an all nighter" a good idea when preparing for an exam, or is it better to quit early and get some sleep? Does sleep actually help in memory processing? The results of studies seem to disagree. A number of studies indicate that REM sleep plays a critical role in learning. Recent research, however, indicates that REM sleep is critical to the consolidation of procedural memory but not of declarative memory. There is no clear evidence that REM sleep helps consolidate declarative memory, the type of memory used in an examination. Some data suggest that deep, slow-wave sleep (SWS), which occurs early in the night, may be important in consolidating declarative memory.

Limbic System

The limbic system (see figure 13.9) influences emotions, the visceral responses to emotions, motivation, mood, and the sensations of pain and pleasure. This system is associated with basic survival instincts: the acquisition of food and water, as well as reproduction. One of the major sources of sensory input into the limbic system is the olfactory nerves. The smell or thought of food stimulates the sense of hunger in the hypothalamus, which motivates us to seek food. Many animals can also smell water, even over great distances. In animals such as dogs and cats, olfactory detection of **pheromones** (fer′ō-mōnz) is important in reproduction. Pheromones are molecules released into the air by one animal that attract another animal of the same species, usually of the opposite sex. Pheromones released by human females can influence the menstrual cycles of other women.

Apparently the cingulate gyrus is a "satisfaction center" for the brain and is associated with the feeling of satisfaction after a meal or after sexual intercourse. The relationship of the hippocampus with the limbic system and with memory is probably important to survival. For example, it's very important for an animal to remember where to obtain food. Once a person has eaten, the satiety center in the hypothalamus is stimulated, the hunger center is inhibited, and the person feels satiated. The hypothalamus interacts with the cingulate gyrus and other parts of the limbic system, causing a sense of satisfaction associated with the satiation.

Lesions in the limbic system can result in a voracious appetite, increased sexual activity, which is often inappropriate, and docility, including the loss of normal fear and anger responses. Because the hippocampus is part of the temporal lobe, damage to that portion can also result in a loss of memory formation.

> 29. *What are the functions of the limbic system? Which of the special senses has a major input into the limbic system?*
> 30. *Define pheromones.*

Addiction, Obsession, and Compulsion

The "satisfaction center" of the brain, the cingulate gyrus, can be directly affected by chemicals—alcohol and drugs. These substances cause a physical dependence. The brain becomes dependent upon these substances for a quick sense of satisfaction. Obsessions and compulsions such as compulsive gambling, kleptomania, fire-setting, and even compulsive overeating exhibit similar neurological patterns to addiction. The same brain areas, such as the cingulate gyrus, show activity during obsessive and compulsive behavior as in addictive behavior.

Effects of Aging on the Nervous System

As a person ages, there's a gradual decline in sensory function because the number of sensory neurons declines, the function of remaining neurons decreases, and CNS processing decreases. In the skin, free nerve endings and hair follicle receptors remain largely unchanged with age. Meissner's corpuscles and pacinian corpuscles,

however, decrease in number. The capsules of those that remain become thicker and structurally distorted and, therefore, exhibit reduced function. As a result of these changes in Meissner's corpuscles and pacinian corpuscles, elderly people are less conscious of something touching or pressing on the skin, have a decreased sense of two-point discrimination, and have a more difficult time identifying objects by touch. These functional changes leave elderly people more prone to skin injuries and with a greater sense of isolation.

Loss of pacinian corpuscles also results in a decreased sense of position of the limbs and in the joints, which can affect balance and coordination. The functions of Golgi tendon organs and muscle spindles also decline with increasing age. As a result, information on the position, tension, and length of tendons and muscles decreases, resulting in additional reduction in the senses of movement, posture, and position, as well as reduced control and coordination of movement.

Other sensory neurons with reduced function include those that monitor blood pressure, thirst, objects in the throat, the amount of urine in the urinary bladder, and the amount of feces in the rectum. As a result, elderly people are more prone to high blood pressure, dehydration, swallowing and choking problems, urinary incontinence, and constipation or bowel incontinence.

There's also a general decline in the number of motor neurons. As many as 50% of the lower motor neurons in the lumbar region of the spinal cord may be lost by age 60. Muscle fibers innervated by the lost motor neurons are also lost, resulting in a general decline in muscle mass. The remaining motor units can compensate for some of the lost function. This, however, often results in a feeling that one must work harder to perform activities that were previously not so difficult. Loss of motor units also leads to more rapid fatigue as the remaining units must perform compensatory work.

Reflexes slow as people age because both the generation and conduction of action potentials and synaptic functions slow. The number of neurotransmitters and receptors declines. Age-related changes in the CNS also slow reflexes. The more complicated the reflex, the more it's affected by age. As reflexes slow, older people are less able to react automatically, quickly, and accurately to changes in internal and external conditions.

The size and weight of the brain decrease as a person ages. At least part of these changes result from the loss of neurons within the cerebrum. The remaining neurons can apparently compensate for much of this loss. In addition to loss of neurons, structural changes occur in the remaining neurons. Neuron plasma membranes become more rigid, the endoplasmic reticulum becomes more irregular in structure, neurofibrillar tangles develop in the cells, and amyloid plaques form in synapses. All of these changes decrease the ability of neurons to function. Age-related changes in brain function include decreased voluntary movement, conscious sensations, reflexes, memory, and sleep. Short-term memory is decreased in most older people. This change varies greatly among individuals, but, in general, such changes are slow until about age 60 and then become more rapid, especially after age 70. However, the total amount of memory loss is normally not great for most people. The most difficult information for older people to assimilate is that which

Systems Pathology
Stroke

Mr. S, who is approaching middle age, is somewhat overweight and has high blood pressure. He was seated on the edge of his couch, at least most of the time, when he was not jumping to his feet and shouting at the referees for an obviously bad call. He was surrounded by empty pizza boxes, bowls of chips and salsa, empty beer cans, and full ashtrays. As Mr. S cheered on his favorite team in a hotly contested big game, which they would be winning easily if it weren't for the lousy officiating, he noticed that he felt drowsy and that the television screen seemed blurry. He began to feel dizzy. As he tried to stand up, he suddenly vomited and collapsed to the floor, unconscious.

Mr. S was rushed to the local hospital, where the following signs and symptoms were observed. He exhibited weakness in his limbs, especially on the right, and ataxia (inability to walk). He had loss of pain and temperature sensation in his right lower limb and loss of all sensation in the left side of his face. The dizziness persisted and he appeared disoriented and lacked attentiveness. He also exhibited dysphagia (the inability to swallow) and hoarseness. He had nystagmus (rhythmic oscillation of the eyes). His pupils were slightly dilated, his respiration was short and shallow, and his pulse rate and blood pressure were elevated.

Background Information

Mr. S suffered a "stroke," also referred to as a cerebrovascular accident (CVA). The term *stroke* describes a heterogeneous group of conditions involving death of brain tissue resulting from disruption of its vascular supply. Two types of stroke exist: **hemorrhagic stroke,** which results from bleeding of arteries supplying brain tissue, and **ischemic stroke,** which results from blockage of arteries supplying brain tissue (figure B). The blockage in ischemic stroke can result from a thrombus (a clot that develops in place within an artery) or an embolism (a plug,

Anterior

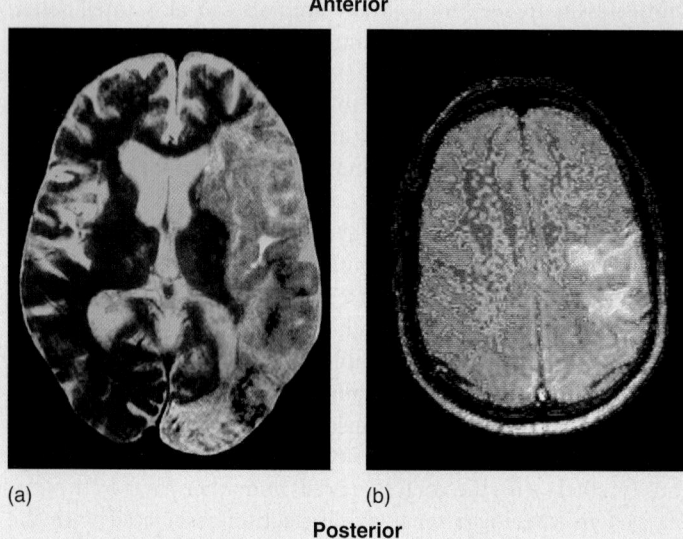

(a)   (b)

Posterior

Figure B   Images of a Stroke

(*a*) MRI (magnetic resonance imaging) of a massive stroke in the left side (the viewer's right) of the brain. (*b*) Colorized NMR (nuclear magnetic resonance) showing disruption of blood flow to the left side (the viewer's right) of the brain (*yellow*). This disruption could cause a stroke.

composed of a detached thrombus or other foreign body, such as a fat globule or gas bubble, that becomes lodged in an artery, blocking it). Mr. S was at high risk for developing a stroke. He was approaching middle age, was overweight, did not exercise enough, smoked, was under stress, and had a poor diet.

The combination of motor loss, which was seen as weakness in his limbs, and sensory loss, seen as loss of pain and temperature sensation

is unfamiliar and presented verbally and rapidly. Some of these problems may occur as older people are required to deal with new information in the face of existing, contradictory memories. Long-term memory appears to be unaffected or even improved in older people.

As with short-term memory, thinking, which includes problem solving, planning, and intelligence, in general, declines slowly to age 60 but more rapidly thereafter. These changes, however, are slight and quite variable. Many older people show no change and about 10% show an increase in thinking ability. Many of these changes are impacted by a person's background, education, health, motivation, and experience.

Among older people, more time is required to fall asleep, there are more periods of waking during the night, and the wakeful periods are of greater duration. Factors that can affect sleep include pain, indigestion, rhythmic leg movements, sleep apnea, decreased urinary bladder capacity, and circulatory problems. There is, on the average, an increase in stage 1 sleep, which is the least restful, and less time spent in stage 4 and REM sleep, which are the most restful.

31. *How does aging affect sensory function? How does loss of motor neurons affect muscle mass?*

32. *Does aging always produce memory loss?*

System Interactions — Effect of Stroke on Other Systems

System	Interactions
Integumentary	Decubitus ulcers (bedsores) from immobility; loss of motor function following a stroke leads to immobility.
Skeletal	Loss of bone mass, if muscles are dysfunctional for a prolonged time; in the absence of muscular activity, the bones to which those muscle are attached begin to be resorbed by osteoclasts.
Muscular	Major area of effect; absence of stimulation due to damaged pathways or neurons leads to decreased motor function and may result in muscle atrophy.
Endocrine	Strokes in other parts of the brain could involve the hypothalamus, pineal body, or pituitary gland functions.
Cardiovascular	Risks: Phlebothrombosis (blood clot in a vein) can occur from inactivity. Edema around the brain could apply pressure to the cardioregulatory and vasomotor centers of the brain. This pressure could stimulate these centers, which would result in elevated blood pressure, and congestive heart failure could result. If the cardioregulatory center in the brain is damaged, death may occur rapidly. Bleeding is due to the use of anticoagulants. Hypotension results from use of antihypertensives.
Respiratory	Pneumonia from aspiration of the vomitus or hypoventilation results from decreased function in the respiratory center. If the respiratory center is severely damaged, death may occur rapidly.
Digestive	Vomiting, dysphagia (difficulty swallowing); hypovolemia (decreased blood volume) result from decreased fluid intake; occurs because of dysphagia; may be a loss of bowel control.
Urinary	Control of the micturition reflex may be inhibited. Urinary tract infection results from catheter implantation or from urinary bladder distension.
Reproductive	Loss of libido; innervation of the reproductive organs is often affected.

in his right lower limb and loss of all sensation in the left side of his face; along with the ataxia, dizziness, nystagmus, and hoarseness, suggest that the stroke affected the brainstem and cerebellum. Blockage of the vertebral artery, a major artery supplying the brain, or its branches can result in what is called a lateral medullary infarction (an area of dead tissue resulting from a loss of blood supply to an area). Damage to the descending motor pathways in that area, above the medullary decussation, results in muscle weakness. Damage to ascending pathways can result in loss of pain and temperature sensation (or other sensory modalities depending on the affected tract). Damage to cranial nerve nuclei results in the loss of pain and temperature sensation in the face, dizziness, blurred vision, nystagmus, vomiting, and hoarseness. These signs and symptoms are not observed unless the lesion is in the brainstem, where these nuclei are located. Some damage to the cerebellum, also supplied by branches of the vertebral artery, can account for the ataxia.

Drowsiness, disorientation, inattentiveness, and loss of consciousness are examples of generalized neurologic response to damage.

Seizures may also result from severe neurologic damage. Depression from neurologic damage or from discouragement is also common. Slight dilation of the pupils; short, shallow respiration; and increased pulse rate and blood pressure are all signs of Mr. S's anxiety, not about the outcome of the game but about his current condition and his immediate future. With a loss of consciousness, Mr. S would not remember the last few minutes of what he saw in the game he was watching. People in these circumstances are often worried about how they are going to deal with work tomorrow. They often have no idea that the motor and sensory losses may be permanent, or that they will have a long period of therapy ahead.

P R E D I C T 9

Given that Mr. S exhibited weakness in his right limbs and loss of pain and temperature sensation in his right lower limb and the left side of his face, state which side of the brainstem was most severely affected by the stroke. Explain your answer.

S U M M A R Y

Sensation (p. 476)

1. The senses include general senses and special senses.
2. Somatic senses include touch, pressure, temperature, proprioception, and pain.
3. Visceral senses are primarily pain and pressure.
4. Special senses are smell, taste, sight, hearing, and balance.
5. Sensation, or perception, is the conscious awareness of stimuli received by sensory receptors.
6. Sensation requires a stimulus, a receptor, conduction of an action potential to the CNS, translation of the action potential, and processing of the action potential in the CNS so that the person is aware of the sensation.

Sensory Receptors

1. Receptors include mechanoreceptors, chemoreceptors, thermoreceptors, photoreceptors, and nociceptors.
2. Free nerve endings detect light touch, pain, itch, tickle, and temperature.
3. Merkel's disks respond to light touch and superficial pressure.
4. Hair follicle receptors wrap around the hair follicle and are involved in the sensation of light touch when the hair is bent.
5. Pacinian corpuscles, located in the dermis and hypodermis, detect pressure. In joints, they serve a proprioceptive function.
6. Meissner's corpuscles, located in the dermis, are responsible for two-point discriminative touch.

7. Ruffini's end organs are involved in continuous touch or pressure.
8. Muscle spindles, located in skeletal muscle, are proprioceptors.
9. Golgi tendon organs, embedded in tendons, respond to changes in tension.
10. A stimulus produces a receptor potential in a sensory receptor. Primary receptors have axons that transmit action potentials toward the CNS. Secondary receptors have no axons but release neurotransmitters.
11. Adaptation is decreased sensitivity to a continued stimulus. Tonic receptors accommodate slowly; phasic receptors accommodate rapidly.

Sensory Nerve Tracts

1. Ascending pathways carry conscious and unconscious sensations. The two major ascending systems are the anterolateral and dorsal-column/medial-lemniscal systems.
2. Anterolateral system:
   • The spinothalamic tract carries pain, temperature, light touch, pressure, tickle, and itch sensations.
   • The spinothalamic tracts are formed by primary neurons that enter the spinal cord and synapse with secondary neurons. The secondary neurons conducting pain and temperature sensations cross the spinal cord and ascend to the thalamus, where they synapse with tertiary neurons that project to the somatic sensory cortex. Axons conducting other sensations, such as light touch or itch, may ascend several spinal cord levels before crossing.
   • The ascending neurons of the spinoreticular tract ascend both contralaterally and ipsilaterally and project to the reticular formation and influence the level of consciousness.
   • The ascending neurons of the spinomesencephalic tract carry action potentials from cutaneous pain receptors. The spinomesencephalic tract also contributes to eye reflexes.
3. The dorsal-column/medial-lemniscal system carries the sensations of two-point discrimination, proprioception, pressure, and vibration. Primary neurons enter the spinal cord and ascend to the medulla, where they synapse with secondary neurons. Secondary neurons cross over and project to the thalamus. Tertiary neurons extend from there to the somatic sensory cortex.
4. The trigeminothalamic tract carries sensory information from the face, nose, and mouth.
5. Spinocerebellar system and other tracts:
   • The spinocerebellar tracts carry unconscious proprioception to the cerebellum from the same side of the body.
   • Neurons of the dorsal-column/medial-lemniscal system synapse with the neurons that carry proprioception information to the cerebellum.
   • The spinoolivary tract contributes to coordination of movement.
6. Descending pathways can reduce conscious perception of sensations.

Sensory Areas of the Cerebral Cortex

1. Sensory pathways project to primary sensory areas in the cerebral cortex.
2. Sensory areas are organized topographically in the somatic sensory cortex.

Sensory Processing

Association areas of the cerebral cortex process sensory input from the primary sensory areas.

Control of Skeletal Muscles   (p. 488)

1. Upper motor neurons are located in the cerebral cortex, cerebellum, and brainstem. Lower motor neurons are found in the cranial nuclei or the anterior horn of the spinal cord gray matter.
2. Upper motor neurons in the cerebral cortex and other brain areas project to lower motor neurons in the brainstem and spinal cord.

Motor Areas of the Cerebral Cortex

1. The primary motor cortex is the precentral gyrus. The premotor and prefrontal areas are staging areas for motor function.
2. The motor cortex is organized topographically.

Motor Nerve Tracts

1. The direct pathways maintain muscle tone and control fine, skilled movements in the face and distal limbs. The indirect pathways control conscious and unconscious muscle movements in the trunk and proximal limbs.
2. The corticospinal tracts control muscle movements below the head.
   • About 75%–85% of the upper motor neurons of the corticospinal tracts cross over in the medulla to form the lateral corticospinal tracts in the spinal cord.
   • The remaining upper motor neurons pass through the medulla to form the anterior corticospinal tracts, which cross over in the spinal cord.
   • The upper motor neurons of both tracts synapse with interneurons that then synapse with lower motor neurons in the spinal cord.
3. The corticobulbar tracts innervate the head muscles. Upper motor neurons synapse with interneurons in the reticular formation that, in turn, synapse with lower motor neurons in the cranial nerve nuclei.
4. The indirect pathways include the rubrospinal, vestibulospinal, and reticulospinal tracts and fibers from the basal nuclei.
5. The indirect pathways are involved in conscious and unconscious trunk and proximal limb muscle movements, posture, and balance.

Modifying and Refining Motor Activities

1. Basal nuclei are important in planning, organizing, and coordinating motor movements and posture.
2. The cerebellum has three parts.
   • The vestibulocerebellum controls balance and eye movement.
   • The spinocerebellum functions to correct discrepancies between intended movements and actual movements.
   • The cerebrocerebellum can "learn" highly specific complex motor activities.

Brainstem Functions   (p. 495)

1. The brainstem contains nuclei of cranial nerves II–XII.
2. Sensory and motor pathways pass through the brainstem.
3. Some sensory pathways synapse in the brainstem.
4. Many important reflexes, some of which are critical to survival, are controlled in the brainstem.

Other Brain Functions   (p. 496)
Speech

1. Speech is located only in the left cortex in most people.
2. Wernicke's area comprehends and formulates speech.
3. Broca's area receives input from Wernicke's area and sends impulses to the premotor and motor areas, which cause the muscle movements required for speech.

Right and Left Cerebral Cortex

1. Each cerebral hemisphere controls and receives input from the opposite side of the body.
2. The right and left hemispheres are connected by commissures. The largest commissure is the corpus callosum, which allows sharing of information between hemispheres.
3. In most people the left hemisphere is dominant, controlling speech and analytic skills. The right hemisphere controls spatial and musical abilities.

Brain Waves and Sleep

1. Electroencephalograms (EEGs) record the electrical activity of the brain as alpha, beta, theta, and delta waves.
2. Some brain disorders can be detected with EEGs.
3. Sleep patterns are characterized by specific EEGs.

Memory

At least three kinds of memory exist: sensory, short term, and long term.

Limbic System

The limbic system controls visceral functions through the autonomic nervous system and the endocrine system and is also involved in emotions and memory.

Effects of Aging on the Nervous System   (p. 503)

1. There is a general decline in sensory and motor functions as a person ages.
2. Short-term memory is decreased in most older people.
3. Thinking ability does not decrease in most older people.

R E V I E W   A N D   C O M P R E H E N S I O N

1. Nociceptors respond to
   a. changes in temperature at the site of the receptor.
   b. compression, bending, or stretching of cells.
   c. painful mechanical, chemical, or thermal stimuli.
   d. light striking a receptor cell.

2. Which of these types of sensory receptors respond to pain, itch, tickle, and temperature?
   a. Merkel's disks
   b. Meissner's corpuscles
   c. Ruffini's end organs
   d. free nerve endings
   e. pacinian corpuscles

3. Which of these types of sensory receptors are involved with proprioception?
   a. free nerve endings
   b. Golgi tendon organs
   c. muscle spindles
   d. pacinian corpuscle
   e. all of the above

4. The sensory receptor in the dermis and hypodermis responsible for sensing continuous touch or pressure are
   a. Merkel's disks.
   b. Meissner's corpuscles.
   c. Ruffini's end organs.
   d. free nerve endings.
   e. pacinian corpuscles.

5. Decreased sensitivity to a continued stimulus is called
   a. adaptation.
   b. projection.
   c. translation.
   d. conduction.
   e. phantom pain.

6. Secondary neurons in the spinothalamic tracts synapse with tertiary neurons in the
   a. medulla oblongata.
   b. gray matter of the spinal cord.
   c. cerebellum.
   d. thalamus.
   e. midbrain.

7. If the spinothalamic tract on the right side of the spinal cord is severed,
   a. pain sensations below the damaged area on the right side are eliminated.
   b. pain sensations below the damaged area on the left side are eliminated.
   c. temperature sensations are unaffected.
   d. neither pain sensations nor temperature sensations are affected.

8. Fibers of the dorsal-column/medial-lemniscal system
   a. carry the sensations of two-point discrimination, proprioception, pressure, and vibration.
   b. cross to the opposite side in the medulla oblongata.
   c. are divided into the fasciculus gracilis and fasciculus cuneatus in the spinal cord.
   d. include secondary neurons that exit the medulla and synapse in the thalamus.
   e. all of the above.

9. Tertiary neurons in both the anterolateral system and dorsal-column/medial-lemniscal systems
   a. project to the somatic sensory cortex.
   b. cross to the opposite side in the medulla oblongata.
   c. are found in the spinal cord.
   d. connect to quaternary neurons in the thalamus.
   e. are part of a descending pathway.

10. Unlike the anterolateral and dorsal-column/medial-lemniscal systems, the spinocerebellar tracts
    a. are descending tracts.
    b. transmit information from the same side of the body as the side of the CNS to which they project.
    c. have four neurons in each pathway.
    d. carry only pain sensations.
    e. have primary neurons that synapse in the thalamus.

11. General sensory inputs (pain, pressure, temperature) to the cerebrum end in the
    a. precentral gyrus.
    b. postcentral gyrus.
    c. central sulcus.
    d. corpus callosum.
    e. arachnoid mater.

12. Neurons from which of these areas of the body occupy the greatest area of the somatic sensory cortex?
    a. foot
    b. leg
    c. torso
    d. arm
    e. face

13. A cutaneous nerve to the hand is severed at the elbow. The distal end of the nerve at the elbow is then stimulated. The subject reports
    a. no sensation because the receptors are gone.
    b. a sensation only in the region of the elbow.
    c. a sensation "projected" to the hand.
    d. a vague sensation on the side of the body containing the cut nerve.

14. Which of these areas of the cerebral cortex is involved in the motivation and foresight to plan and initiate movements?
    a. primary motor cortex
    b. somatic sensory cortex
    c. prefrontal area
    d. premotor area
    e. basal nuclei

15. Which of these pathways is *not* an ascending (sensory) pathway?
    a. spinothalamic tract
    b. corticospinal tract
    c. dorsal column/medial lemniscal system
    d. trigeminothalamic tract
    e. spinocerebellar tract

16. The _____ tracts innervate the head muscles.
    a. corticospinal
    b. rubrospinal
    c. vestibulospinal
    d. corticobulbar
    e. dorsal-column/medial-lemniscal

17. Most fibers of the direct (pyramidal) system
    a. decussate in the medulla oblongata.
    b. synapse in the pons.
    c. descend in the rubrospinal tract.
    d. begin in the cerebellum.

18. A person with a spinal cord injury is suffering from paresis (partial paralysis) in the right lower limb. Which of these pathways is probably involved?
    a. left lateral corticospinal tract
    b. right lateral corticospinal tract
    c. left dorsal column/medial lemniscal system
    d. right dorsal column/medial lemniscal system

19. Which of these pathways is *not* an indirect (extrapyramidal) pathway?
    a. reticulospinal tract
    b. corticobulbar tract
    c. rubrospinal tract
    d. vestibulospinal tract

20. The indirect (extrapyramidal) system is concerned with
    a. posture
    b. trunk movements
    c. proximal limb movements
    d. all of the above

21. The major effect of the basal nuclei is
    a. to act as a comparator for motor coordination.
    b. to decrease muscle tone and inhibit unwanted muscular activity.
    c. affect emotions and emotional responses to odors.
    d. modulate pain sensations.

22. Which of the parts of the cerebellum is correctly matched with its function?
    a. vestibulocerebellum—planning and learning rapid, complex movements
    b. spinocerebellum—comparator function
    c. cerebrocerebellum—balance
    d. none of the above

23. Given the following events:
    1. Action potentials from the cerebellum go to the motor cortex and spinal cord.
    2. Action potentials from the motor cortex go to lower motor neurons and the cerebellum.
    3. Action potentials from proprioceptors go to the cerebellum.

Arrange the events in the order they occur in the cerebellar comparator function.
    a. 1,2,3
    b. 1,3,2
    c. 2,1,3
    d. 2,3,1
    e. 3,2,1

24. The brainstem
    a. consists of ascending and descending pathways.
    b. contains cranial nerve nuclei II–XII.
    c. has nuclei and connections that form the reticular activating system.
    d. has many important reflexes, some of which are necessary for survival.
    e. has all of the above.

25. Given these areas of the cerebral cortex:
    1. Broca's area
    2. premotor area
    3. primary motor cortex
    4. Wernicke's area

    If a person hears and understands a word and then says the word out loud, in what order are the areas used?
    a. 1,4,2,3
    b. 1,4,3,2
    c. 3,1,4,2
    d. 4,1,2,3
    e. 4,1,3,2

26. The main connection between the right and left hemispheres of the cerebrum is the
    a. intermediate mass.
    b. corpus callosum.
    c. vermis.
    d. unmyelinated nuclei.
    e. thalamus.

27. Which of these activities is mostly associated with the left cerebral hemisphere in most people?
    a. sensory input from the left side of the body
    b. mathematics and speech
    c. spatial perception
    d. recognition of faces
    e. musical ability

28. The limbic system is involved in the control of
    a. sleep and wakefulness.
    b. maintaining posture.
    c. higher intellectual processes.
    d. emotion, mood, and sensations of pain or pleasure.
    e. hearing.

29. Long-term memory involves
    a. a change in the cytoskeleton of neurons.
    b. movement of calcium into the neuron.
    c. increase in glutamate release by presynaptic neurons.
    d. specific protein synthesis.
    e. all of the above.

30. Concerning long-term memory,
    a. declarative (explicit) memory involves the development of skills, such as riding a bicycle.
    b. procedural (implicit) memory involves the retention of facts, such as names, dates, or places.
    c. much of declarative (explicit) memory is lost through time.
    d. declarative (explicit) memory is stored primarily in the cerebellum and premotor area of the cerebrum.
    e. all of the above.

Answers in Appendix F

C R I T I C A L   T H I N K I N G

1. Describe all the sensations involved when a woman picks up an apple and bites into it. Explain which of those sensations are special and which are general. What types of receptors are involved? Which aspects of the taste of the apple are actually taste and which are olfaction?

2. Some student nurses are at a party. Because they love anatomy and physiology so much, they are discussing adaptation of the special senses. They make the following observations:
   a. When entering a room, an odor like brewing coffee is easily noticed. A few minutes later, the odor might be barely, if at all, detectable, no matter how hard one tries to smell it.
   b. When entering a room, the sound of a ticking clock can be detected. Later the sound is not noticed until a conscious effort is made to hear it. Then it is easily heard. Explain the basis for each of these observations.

3. A patient suffered a loss of two-point discrimination and proprioception on the right side of the body. Voluntary movement of muscles was not affected, and pain and temperature sensations were normal. Is it possible to conclude that the right side of the spinal cord was damaged?

4. A patient is suffering from the loss of two-point discrimination and proprioceptive sensations on the right side of the body resulting from a lesion in the pons. What tract is affected, and which side of the pons is involved?

5. A patient suffers a lesion in the central core of the spinal cord. It is suspected that the fibers that decussate and that are associated with the lateral spinothalamic tracts are affected in the area of the lesion. What observations would be consistent with that diagnosis?

6. A person in a car accident exhibits the following symptoms: extreme paresis on the right side, including the arm and leg, reduction of pain sensation on the left side, and normal tactile sensation on both sides. Which nerve tracts are damaged? Where did the patient suffer nerve tract damage?

7. If the right side of the spinal cord is completely transected, what symptoms do you expect to observe with regard to motor function, two-point discrimination, light touch, and pain perception?

8. A patient with a cerebral lesion exhibits a loss of fine motor control of the left hand, arm, forearm, and shoulder. All other motor and sensory functions appeared to be intact. Describe the location of the lesion as precisely as possible.

9. A patient suffers brain damage in an automobile accident. It is suspected that the cerebellum is the part of the brain that is affected. On the basis of what you know about cerebellar function, how could you determine that the cerebellum is involved?

10. Woody Knothead was accidentally struck in the head with a baseball bat. He fell to the ground unconscious. Later, when he regains consciousness, he is not able to remember any of the events that happened 10 minutes before the accident. Explain. What complications might be looked for at a later time?

Answers in Appendix G

A N S W E R S   T O   P R E D I C T   Q U E S T I O N S

1. Because hot and cold objects may not be perceived any differently for temperatures of 0°–12°C or above 47°C (both temperature ranges stimulate pain fibers), the nervous system may not be able to discriminate between the two temperatures. At low temperatures, both cold and pain receptors are stimulated; thus, after the object has been in the hand for a very short time, it's possible to discriminate between cold and pain. If, however, the CNS has been preprogrammed to think that the object to be placed in the hand is hot, a cold object can elicit a rapid withdrawal reflex.

2. Lesions on one side of the spinal cord that interrupt the anterolateral system eliminate pain and temperature sensation below that level on the opposite side of the body. This occurs because the pathway is contralateral (ascending on the opposite side of the spinal cord) to the area of skin innervated by the receptors. Light touch, etc., show few if any clinical changes because some fibers cross and ascend contralaterally whereas others ascend ipsilaterally for some distance before crossing.

3. The damage to Bill's spinal cord would be on the left side. The fasciculus gracilis conveys sensations of proprioception, fine touch, and vibration through the spinal cord on the same side of the body as the sensory receptors. The damage to Mary's brainstem would be on the right side if the damage occurred above the medulla oblongata or on the left if it occurred in the inferior part of the medulla oblongata. The secondary neurons in the nucleus gracilis cross over in the medulla through the decussations of the medial lemniscus, and once crossed, are on the opposite side of the body from the nerve endings where the sensations would be initiated.

4. Most proprioception from the lower limbs is unconscious, whereas that from the upper limbs is mostly conscious. This difference is valuable because walking and standing (balance) are not activities on which we want to focus our attention, whereas proprioceptive activities of the arms and hands are essential for gaining information about the environment.

5. In the visual cortex the brain "sees" an object. Without a functional visual cortex, a person is blind. The visual association areas allow us to relate objects seen to previous experiences and to interpret what has been seen. Similarly, other association areas allow us to relate the sensory information integrated in the primary sensory areas with previous experiences and to make judgments about the information.

6. Constipation, with painful distention and cramping of the colon, results in the sensation of diffuse pain. Deep, visceral pain is not highly localized because few mechanoreceptors are present in deeper structures such as the colon. The pain is perceived as occurring in the skin over the lower central portion of the abdomen (in the hypogastric region) because it is referred to that location because of converging CNS pathways.

7. A sleeping person can be aroused by tactile stimulation, especially to the face, so touching the person on the face can arouse the person. Water on the face also works.

8. If a person holds an object in her right hand, tactile sensations of various types travel up the spinal cord to the brain, where they reach the somatic sensory cortex of the left hemisphere and the object is recognized. Action potentials then travel to Wernicke's area (probably on both sides of the cerebrum), where the object is given a name. From there action potentials travel to Broca's area, where the spoken word is initiated. Action potentials from Broca's area travel to the premotor area and primary motor cortex, where action potentials are initiated that stimulate the muscles necessary to form the word.

9. The stroke was on the left side of the brainstem. Both the motor and sensory neurons to the right side of the body are located in the left cerebral cortex. At the level of the upper medulla oblongata, neither the motor nor sensory pathways to the limbs have yet crossed over to the left side of the CNS. Most of the motor fibers cross at the inferior end of the medulla oblongata, whereas sensory pain and temperature fibers cross over at the level where they enter the CNS. Loss of pain and temperature to the left side of the face indicates that the lesion occurred at a level where the nerve fibers from the face had entered the CNS but had not yet crossed.

Visit the Online Learning Center at www.mhhe.com/seeley7 for chapter quizzes, interactive learning exercises, and other study tools.

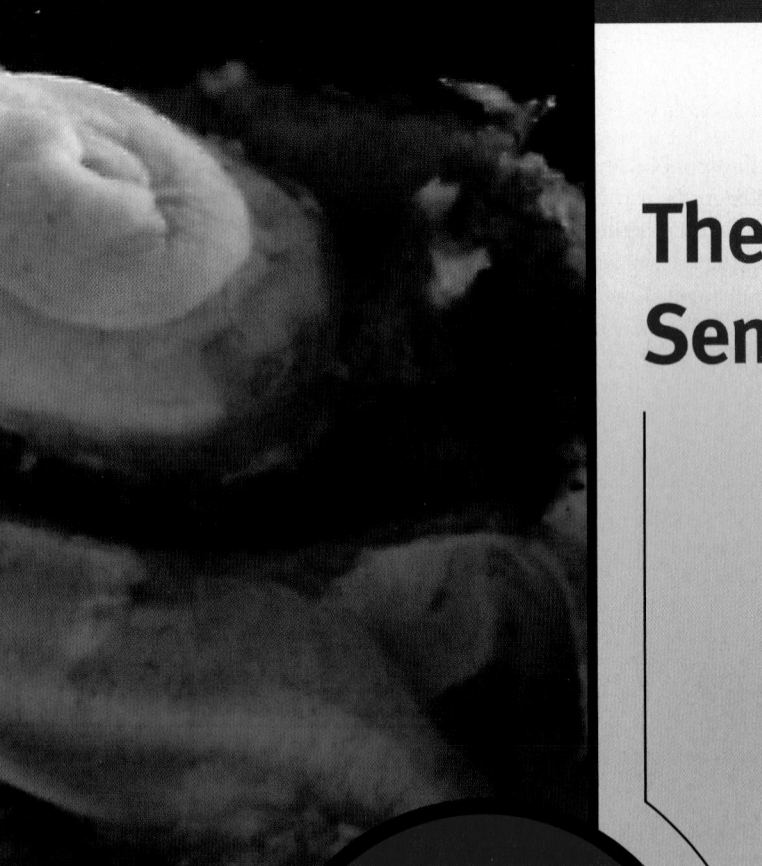

Photograph of an isolated cochlea from the inner ear.

The Special Senses

CHAPTER

15

Historically, it was thought that we had just five senses: smell, taste, sight, hearing, and touch. Today we recognize many more. Some specialists suggest that there are at least 20, or perhaps as many as 40, different senses. Most of these senses are part of what was originally classified as "touch." These "general senses" were discussed in chapter 14. The sense of balance is now recognized as a "special sense," making a total of five special senses: smell, taste, sight, hearing, and balance. **Special senses** are defined as those senses with highly localized receptors that provide specific information about the environment. This chapter describes *olfaction* (p. 512), *taste* (p. 515), the *visual system* (p. 518), and *hearing and balance* (p. 539). We conclude the chapter with a look at the *effects of aging on the special senses* (p. 554).

Olfaction

Olfaction (ol-fak′shŭn), the sense of smell, occurs in response to odors that stimulate sensory receptors located in the extreme superior region of the nasal cavity, called the **olfactory region** (figure 15.1*a*). Most of the nasal cavity is involved in respiration, with only a small superior part devoted to olfaction. The major anatomic features of the nasal cavity are described in chapter 23 in relation to respiration. The specialized nasal epithelium of the olfactory region is called the **olfactory epithelium.**

P R E D I C T

Explain why it sometimes helps to inhale slowly and deeply through the nose when trying to identify an odor.

Olfactory bulb

Olfactory tract

Cribriform plate of ethmoid bone

Cut edge of epithelium

Nasopharynx

Frontal bone

Fibers of olfactory nerve

Olfactory region

Nasal cavity

Hard palate

(a) **Medial view**

Olfactory bulb neurons
- Granule cell
- Tufted cell
- Mitral cell

Olfactory tract

Cribriform plate

Connective tissue

Olfactory epithelium

Mucous layer on epithelial surface

Posterior

Anterior

Olfactory bulb

Foramen

Axon

Basal cell

Supporting cell

Olfactory neuron

Dendrite

Cilia (olfactory hairs)

Olfactory vesicle

(b) **Medial view**

Figure 15.1   Olfactory Region, Epithelium, and Bulb

(*a*) The lateral wall of the nasal cavity (cut in sagittal section), showing the olfactory region and olfactory bulb. (*b*) The olfactory cells within the olfactory epithelium are shown. The olfactory nerve processes passing through the cribriform plate and the fine structure of the olfactory bulb are also shown.

Olfactory Epithelium and Bulb

Ten million **olfactory neurons** are present within the olfactory epithelium (figure 15.1*b*). The axons of these bipolar neurons project through numerous small foramina of the bony cribriform plate (see chapter 7) to the **olfactory bulbs. Olfactory tracts** project from the bulbs to the cerebral cortex.

The dendrites of olfactory neurons extend to the epithelial surface of the nasal cavity, and their ends are modified into bulbous enlargements called **olfactory vesicles** (see figure 15.1*b*). These vesicles possess cilia called **olfactory hairs,** which lie in a thin mucous film on the epithelial surface.

Airborne molecules enter the nasal cavity and are dissolved in the fluid covering the olfactory epithelium. Some of these molecules, referred to as **odorants** (ō′dŏr-ants; a molecule with an odor), bind to transmembrane odorant receptors (chemoreceptors) molecules of the olfactory hair membranes (figure 15.2). A G protein, associated with each odorant receptor, is activated by the binding of the odorant. The α subunit of the activated G protein binds to and activates adenylate cyclase, which, in turn, catalyzes the formation of cyclic AMP (cAMP) from ATP. cAMP in these cells causes

Na^+ and Ca^{2+} channels to open. The influx of ions into the olfactory hairs results in depolarization and the production of action potentials in the olfactory neurons.

Odorant receptors are composed of seven transmembrane polypeptide subunits produced by a large family of genes. Combinations of subunits from approximately 1000 different odorant receptors can be produced. These receptors can react to odorants of different sizes, shapes, and functional groups. These capabilities, together with multiple intracellular pathways involving G proteins, adenylate cyclase, and ion channels allow for a wide variety of detectable smells, which is about 4000 for the average person. With such a wide range of smells, however, only seven primary classes of odors have been proposed: (1) camphoraceous, (2) musky, (3) floral, (4) pepperminty, (5) ethereal, (6) pungent, and (7) putrid. It's very unlikely, however, that this list is an accurate representation of all primary odors, and some studies point to the possibility of as many as 50 primary odors.

The threshold for the detection of odors is very low, so very few odorant molecules are required to trigger the response. Apparently there is rather low specificity in the olfactory epithelium.

1. The cell membrane of an olfactory hair, unstimulated. The gated ion channel is closed.

2. An odorant binds to a specific odorant receptor.

3. The associated G protein is activated.

4. The α subunit of the G protein binds to and activates adenylate cyclase.

5. Adenylate cyclase catalyzes the conversion of ATP to cyclic AMP (cAMP).

6. cAMP opens ion channels, such as Na^+ and Ca^{2+} channels.

7. Ions entering the olfactory hair cause depolarization of the neuron.

Process Figure 15.2   Action of Odorant Binding to Membrane of Olfactory Hair

A given receptor may react to more than one type of odorant. Odorant receptors may become saturated with odorants and no longer respond to odorant molecules. This accommodation makes a person less sensitive to an odorant after being exposed to it for a short time.

The "Odor" of Natural Gas

Methylmercaptan, which has a nauseating odor similar to that of rotten cabbage, is added to natural gas at a concentration of about 1 part per million. A person can detect the odor of about 1/25 billionth of a milligram of the substance and therefore is aware of the presence of the more dangerous but odorless natural gas.

Odor Survey Results

The National Geographic Society conducted a smell survey in 1986, which was the largest sampling of its kind ever conducted. One and a half million people participated. Of six odors studied, 98%–99% of those responding could smell isoamyl acetate (banana), eugenol (cloves), mercaptans, and rose; but 29% could not smell galaxolide (musk), and 35% could not smell androsterone (contained in sweat). Of those responding to the survey, 1.2% could not smell at all, a disorder called **anosmia** (an-oz′mē-ă).

The primary olfactory neurons have the most exposed nerve endings of any neurons, and they are constantly being replaced. The entire olfactory epithelium, including the neurosensory cells, is lost about every 2 months as the olfactory epithelium degenerates and is lost from the surface. Lost olfactory cells are replaced by a proliferation of **basal cells** in the olfactory epithelium. This replacement of olfactory neurons is unique among neurons, most of which are permanent cells that have a very limited ability to replicate (see chapter 4).

Neuronal Pathways for Olfaction

Axons from the olfactory neurons form the olfactory nerves (cranial nerve I), which enter the olfactory bulbs (see figure 15.1*b*), where they synapse with **mitral** (mī′trăl; triangular cells; shaped like a bishop's miter or hat) **cells** or **tufted cells.** The mitral and tufted cells relay olfactory information to the brain through the olfactory tracts and synapse with granule cells in the olfactory bulb. Olfactory bulb neurons also receive input from nerve cell processes entering the olfactory bulb from the brain. As a result of input from both mitral cells and the brain, olfactory bulb neurons can modify olfactory information before it leaves the olfactory bulb. This modification enhances the accommodation occurring in the odorant receptors.

Olfaction is the only major sensation that is relayed directly to the cerebral cortex without first passing through the thalamus. Each olfactory tract terminates in an area of the brain called the **olfactory cortex** (figure 15.3). The olfactory cortex is in the frontal and temporal lobes, within the lateral fissure of the cerebrum, and can be divided structurally and functionally into three areas: lateral, intermediate, and medial. The **lateral olfactory area** is involved in the conscious perception of smell. The **medial olfactory area** is responsible for visceral and emotional reactions to odors and has connections to the limbic system, through which it connects to the hypothalamus. Axons extend from the **intermediate olfactory area** along the olfactory tract to the bulb, synapse with the olfactory bulb neurons, and thus constitute a major

1. Axons of the olfactory neurons in the olfactory epithelium project through foramina in the cribriform plate to the olfactory bulb.

2. Axon of neurons in the olfactory bulb project through the olfactory tract to the olfactory cortex.

3. The lateral olfactory area is involved in the conscious perception of smell.

4. The medial olfactory area is involved in the visceral and emotional reaction to odors.

5. The intermediate olfactory area receives input from the medial and lateral olfactory areas.

6. Axons from the intermediate olfactory area project along the olfactory tract to the olfactory bulb. Action potentials carried by those axons modulate the activity of the neurons in the olfactory bulb.

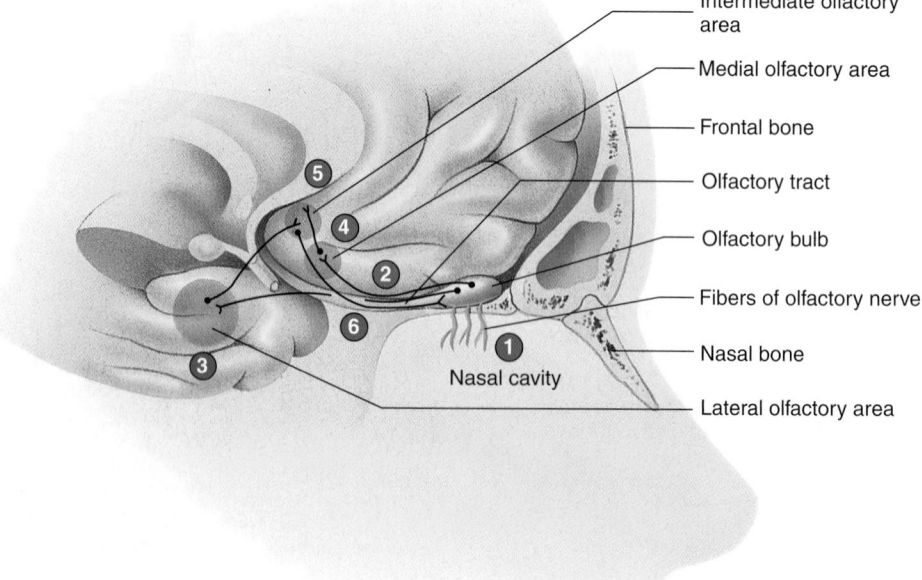

Intermediate olfactory area

Medial olfactory area

Frontal bone

Olfactory tract

Olfactory bulb

Fibers of olfactory nerve

Nasal bone

Lateral olfactory area

Nasal cavity

Medial view

Process Figure 15.3   Olfactory Neuronal Pathways and Cortex

mechanism by which sensory information is modulated within the olfactory bulb. The feedback from the intermediate olfactory area is mostly inhibitory, further enhancing the rapid accommodation of the olfactory system. Accommodation at the receptor level, the olfactory bulb level, and the CNS level make the olfactory system insensitive to an odorant after a brief exposure.

1. Describe the initiation of an action potential in an olfactory neuron. Name all the structures and cells that the action potential would encounter on the way to the olfactory cortex.
2. What is a primary odor? Name seven possible examples. How do the primary odors relate to our ability to smell many different odors?
3. What type of neurons are olfactory neurons? What is unique about olfactory neurons with respect to replacement?
4. How is the sense of smell modified in the olfactory bulb?
5. Name the three areas of the olfactory cortex, and give their functions.
6. Explain how the CNS connections elicit various visceral and conscious responses to smell.

P R E D I C T ②
The olfactory system quickly adapts to continued stimulation, and a particular odor becomes unnoticed before very long, even though the odor molecules are still present in the air. Describe as many sites as you can in the olfactory pathways where such adaptation can occur.

Taste

The sensory structures that detect **gustatory,** or **taste,** stimuli are the **taste buds.** Most taste buds are associated with specialized portions of the tongue called **papillae** (pă-pil´ē). Taste buds, however, are also located on other areas of the tongue, the palate, and even the lips and throat, especially in children. The four major types of papillae are named according to their shape (figure 15.4): **vallate** (val´āt; surrounded by a wall), **fungiform** (fŭn´ji-fōrm; mushroom-shaped), **foliate** (fō´lē-āt; leaf-shaped), and **filiform** (fil´i-fōrm; filament-shaped). Taste buds (figure 15.4c–e) are associated with vallate, fungiform, and foliate papillae. Filiform papillae are the most numerous papillae on the surface of the tongue but have no taste buds.

Vallate papillae are the largest but least numerous of the papillae. Eight to 12 of these papillae form a V-shaped row along the border between the anterior and posterior parts of the tongue (figure 15.4a). Fungiform papillae are scattered irregularly over the entire superior surface of the tongue and appear as small red dots interspersed among the far more numerous filiform papillae. Foliate papillae are distributed in folds on the sides of the tongue and contain the most sensitive of the taste buds. They are most numerous in young children and decrease with age. They are located mostly posteriorly in adults.

Histology of Taste Buds

Taste buds are oval structures embedded in the epithelium of the tongue and mouth (figure 15.4f). Each of the 10,000 taste buds

on a person's tongue consists of three major types of specialized epithelial cells. The sensory cells of each taste bud consist of about 50 **taste** or **gustatory cells.** The remaining two cell types, which are nonsensory cells, are **basal cells** and **supporting cells.** Like olfactory cells, cells of the taste buds are replaced continuously, each having a normal life span of about 10 days. Each taste cell has several microvilli, called **gustatory hairs,** extending from its apex into a tiny opening in the epithelium called the **taste** or **gustatory pore.**

Function of Taste

Substances called **tastants** (tās´tants), dissolved in saliva, enter the taste pore and, by various mechanisms, cause the taste cells to depolarize. These cells have no axons and don't generate their own action potentials. Neurotransmitters are released from the taste cells and stimulate action potentials in the axons of sensory neurons associated with them.

The taste of **salt** results when Na^+ diffuse through Na^+ channels (figure 15.5a) of the gustatory hairs or other cell surfaces of taste cells, resulting in depolarization of the cells. Hydrogen ions (H^+) of **acids** can cause depolarization of taste cells by one of three mechanisms (figure 15.5b): (1) they can enter the cell directly through H^+ channels, (2) they can bind to ligand-gated K^+ channels and block the exit of K^+ from the cell, or (3) they can open ligand-gated channels for other positive ions and allow them to diffuse into the cell. **Sweet** and **bitter** tastants bind to receptors (figure 15.5c and d) on the gustatory hairs of taste cells and cause depolarization through a G protein mechanism. A taste called **umami** (ū-ma´mē; a Japanese term loosely translated as savory) results when amino acids, such as glutamate, bind to receptors (figure 15.5e) on gustatory hairs of taste cells and cause depolarization through a G protein mechanism.

The texture of food in the oral cavity also affects the perception of taste. Hot or cold food temperatures may interfere with the ability of the taste buds to function in tasting food. If a cold fluid is held in the mouth, the fluid becomes warmed by the body, and the taste becomes enhanced. On the other hand, adaptation is very rapid for taste. This adaptation apparently occurs both at the level of the taste bud and within the CNS. Adaptation may begin within 1 or 2 seconds after a taste sensation is perceived, and complete adaptation may occur within 5 minutes.

Even though only five primary tastes have been identified, humans can perceive a fairly large number of different tastes, presumably by combining the five basic taste sensations. As with olfaction, the specificity of the receptor molecules is not perfect. For example, artificial sweeteners have different chemical structures than the sugars they are designed to replace and are often many times more powerful than natural sugars in stimulating taste sensations.

Many of the sensations thought of as being taste are strongly influenced by olfactory sensations. This phenomenon can be demonstrated by pinching one's nose to close the nasal passages, while trying to taste something. With olfaction blocked, it's difficult to distinguish between the taste of a piece of apple and a piece of potato. Much of the "taste" is lost by this action. This is one reason why a person with a cold has a reduced sensation of taste. Although all taste buds are able to detect all five of the basic tastes, each taste cell is usually most sensitive to one.

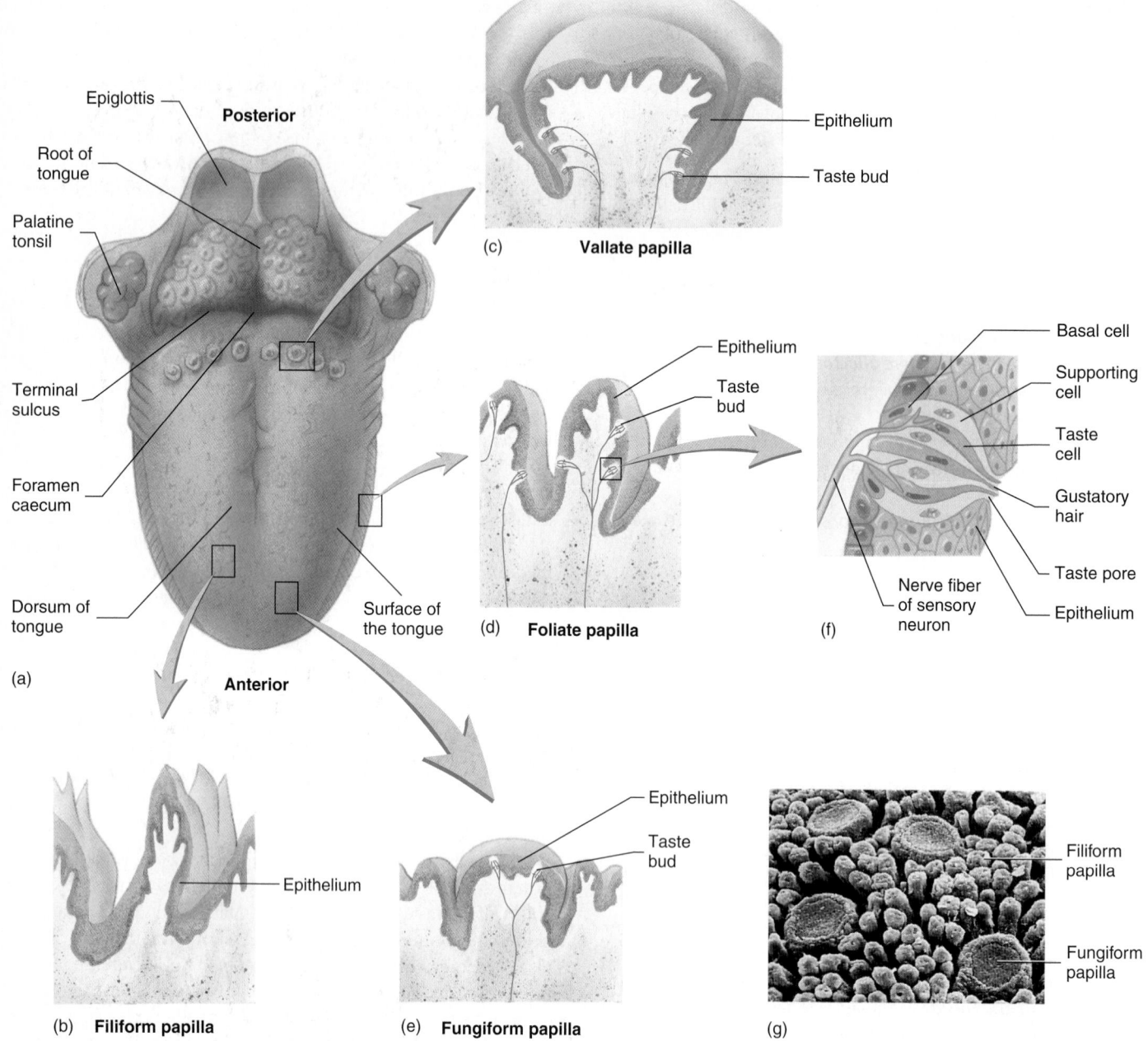

Figure 15.4   Papillae and Taste Buds

(a) Surface of the tongue. (b) Filiform papilla. (c) Vallate papilla. (d) Foliate papilla. (e) Fungiform papilla. (f) A taste bud. (g) Scanning electron micrograph of taste buds (fungiform and filiform papillae) on the surface of the tongue.

Thresholds vary for the five primary tastes. Sensitivity for bitter substances is the highest; sensitivities for sweet and salty tastes are the lowest. Sugars, some other carbohydrates, and some proteins produce sweet tastes; many proteins and amino acids produce umami tastes; acids produce sour tastes; metal ions tend to produce salty tastes; and alkaloids (bases) produce bitter tastes. Many alkaloids are poisonous; thus the high sensitivity for bitter tastes may be protective. On the other hand, humans tend to crave sweet, salty, and umami tastes, perhaps in response to the body's need for sugars, carbohydrates, proteins, and minerals.

Neuronal Pathways for Taste

Taste from the anterior two-thirds of the tongue, except from the circumvallate papillae, is carried by means of a branch of the facial nerve (VII) called the **chorda tympani** (kōr′dă tim′pă-nē; so named because it crosses over the surface of the tympanic membrane of the middle ear). Taste from the posterior one-third of the tongue, the circumvallate papillae, and the superior pharynx is carried by means of the glossopharyngeal nerve (IX). In addition to these two major nerves, the vagus nerve (X) carries a few fibers for taste sensation from the epiglottis.

(a) **Salt:** Na$^+$ diffuse through Na$^+$ channels, resulting in depolarization.

(b) **Sour (acid):** Hydrogen ions (H$^+$) from acids can cause depolarization by one of three mechanisms: (1) they can enter the cell directly through H$^+$ channels, (2) they can bind to gated K$^+$ channels, closing the gate, and preventing K$^+$ from entering the cell, or (3) they can open ligand-gated channels for other positive ions.

(c) **Sweet:** Sugars, such as glucose, or artificial sweeteners bind to receptors and cause the cell to depolarize by means of a G protein mechanism. The α subunit of the G protein activates adenylate cyclase, which produces cAMP. cAMP activates a kinase which phosphorylates K$^+$ channels. The K$^+$ channels close, resulting in depolarization.

(d) **Bitter:** Bitter tastants, such as quinine, bind to receptors and cause depolarization of the cell through a G protein mechanism. The α subunit of the G protein activates phospholipase C, which converts phosphoinositol (PIP$_2$) to inositol triphosphate (IP$_3$). IP$_3$ causes Ca^{2+} release from intracellular stores and depolarization of the cell.

(e) **Glutamate (umami):** Amino acids, such as glutamate, bind to receptors and cause depolarization through a G protein mechanism. The α subunit of the G protein activates adenylate cyclase, which catalyzes the conversion of ATP to cAMP. cAMP opens Ca^{2+} channals. The influx of Ca^{2+} causes depolarization of the cell.

Figure 15.5 Actions of the Major Tastants

These nerves extend from the taste buds to the tractus solitarius of the medulla oblongata (figure 15.6). Fibers from this nucleus decussate and extend to the thalamus. Neurons from the thalamus project to the taste area of the cortex, which is at the extreme inferior end of the postcentral gyrus.

7. *Name and describe the four kinds of papillae found on the tongue. Which ones have taste buds associated with them?*
8. *Starting with the gustatory hair, name the structures and cells that an action potential would encounter on the way to the taste area of the cerebral cortex.*
9. *What is the life span of a normal gustatory cell?*
10. *What are the five primary tastes? Describe how each type of tastant causes depolarization of a taste cell.*
11. *How is the sense of taste related to the sense of smell?*

Visual System

The visual system includes the eyes, the accessory structures, and the optic nerves (II), tracts, and pathways. The eyes respond to light and initiate afferent action potentials, which are transmitted from the eyes to the brain by the optic nerves and tracts. The accessory structures, such as eyebrows, eyelids, eyelashes, and tear glands, help protect the eyes from direct sunlight and damaging particles. Much of the information about the world around us is detected by the visual system. Our education is largely based on visual input and depends on our ability to read words and numbers. Visual input includes information about light and dark, color and hue.

Accessory Structures

Accessory structures protect, lubricate, move, and in other ways aid in the function of the eye. These structures include the eyebrows, eyelids, conjunctiva, lacrimal apparatus, and extrinsic eye muscles.

Eyebrows

The **eyebrows** (figure 15.7) protect the eyes by preventing perspiration, which can irritate the eyes, from running down the forehead and into them, and they help shade the eyes from direct sunlight.

1. Axons of sensory neurons, which synapse with taste receptors, pass through cranial nerves VII, IX, and X and through the ganglion of each nerve (enlarged portion of each nerve).
2. The axons enter the brainstem and synapse in the nucleus of the tractus solitarius.
3. Axons from the nucleus solitarius synapse in the thalamus.
4. Axons from the thalamus terminate in the taste area of the cortex.

Process Figure 15.6   Pathways for the Sense of Taste

The facial nerve (anterior two-thirds of the tongue), glossopharyngeal nerve (posterior one-third of the tongue), and vagus nerve (root of the tongue) all carry taste sensations. The trigeminal nerve is also shown. It carries tactile sensations from the anterior two-thirds of the tongue. The chorda tympani from the facial nerve (carrying taste input) joins the trigeminal nerve.

Superior
palpebra
(eyelid)

Pupil

Caruncle

Medial
angle of
eye

Eyebrow

Iris

Sclera

Lateral
angle of
eye

Inferior
palpebra
(eyelid)

Figure 15.7 **The Left Eye and Its Accessory Structures**

Eyelids

The **eyelids,** also called **palpebrae** (pal-pē′brē), with their associated lashes, protect the eyes from foreign objects. The space between the two eyelids is called the **palpebral fissure,** and the angles where the eyelids join at the medial and lateral margins of the eye are called **canthi** (kan′thī; corners of the eye) (see figure 15.7). The medial canthus contains a small reddish-pink mound called the **caruncle** (kar′ŭng-kl; a mound of tissue). The caruncle contains some modified sebaceous and sweat glands.

The eyelids consist of five layers of tissue (figure 15.8). From the outer to the inner surface, they are (1) a thin layer of integument on the external surface; (2) a thin layer of areolar connective tissue; (3) a layer of skeletal muscle consisting of the

Levator palpebrae
superioris muscle

Smooth muscle to
tarsal plate

Superior rectus
muscle

Inferior rectus
muscle

Inferior oblique
muscle

Eyebrow

Orbicularis oculi muscle

Superior conjunctival
fornix

Bulbar conjunctiva

Palpebral conjunctiva

Tarsal (meibomian)
gland

Tarsal plate

Cornea

Eyelash

Palpebral fissure

Skin

Areolar connective tissue

Orbicularis oculi muscle

Tarsal plate

Palpebral conjunctiva

Inferior conjunctival fornix

Lower eyelid
(inferior palpebra)

Sagittal section

Figure 15.8 **Sagittal Section Through the Eye Showing Its Accessory Structures**

orbicularis oculi and levator palpebrae superioris muscles; (4) a crescent-shaped layer of dense connective tissue called the **tarsal** (tar′săl) **plate,** which helps maintain the shape of the eyelid; and (5) the palpebral conjunctiva (described in the next section), which lines the inner surface of the eyelid and the anterior surface of the eyeball.

If an object suddenly approaches the eye, the eyelids protect the eye by rapidly closing and then opening (blink reflex). Blinking, which normally occurs about 25 times per minute, also helps keep the eye lubricated by spreading tears over the surface of the eye. Movements of the eyelids are a function of skeletal muscles. The orbicularis oculi muscle closes the lids, and the levator palpebrae superioris elevates the upper lid (see chapter 10). The eyelids also help regulate the amount of light entering the eye.

Eyelashes (see figures 15.7 and 15.8) are attached as a double or triple row of hairs to the free edges of the eyelids. **Ciliary glands** are modified sweat glands that open into the follicles of the eyelashes to keep them lubricated. When one of these glands becomes inflamed, it's called a **sty. Meibomian** (mī-bō′mē-an; also called tarsal) **glands** are sebaceous glands near the inner margins of the eyelids and produce **sebum** (sē′bŭm; an oily semifluid substance), which lubricates the lids and restrains tears from flowing over the margin of the eyelids. An infection or blockage of a meibomian gland is called a **chalazion** (ka-lā′zē-on), or **meibomian cyst.**

Conjunctiva

The **conjunctiva** (kon-jŭnk-tī′vă) (see figure 15.8) is a thin, transparent mucous membrane. The **palpebral conjunctiva** covers the inner surface of the eyelids, and the **bulbar conjunctiva** covers the

anterior white surface of the eye. The points at which the palpebral and bulbar conjunctivae meet are the superior and inferior **conjunctival fornices.**

Conjunctivitis

Conjunctivitis is an inflammation of the conjunctiva caused by infection or some other irritation. An example of conjunctivitis caused by a bacterium is **acute contagious conjunctivitis,** also called **pinkeye.**

Lacrimal Apparatus

The **lacrimal** (lak′ri-măl) **apparatus** (figure 15.9) consists of a lacrimal gland situated in the superolateral corner of the orbit and a nasolacrimal duct beginning in the inferomedial corner of the orbit. The **lacrimal gland** is innervated by parasympathetic fibers from the facial nerve (VII). The gland produces tears, which leave the gland through several **lacrimal ducts** and pass over the anterior surface of the eyeball. Tears are produced constantly by the gland at the rate of about 1 mL/day to moisten the surface of the eye, lubricate the eyelids, and wash away foreign objects. Tears are mostly water, with some salts, mucus, and lysozyme, an enzyme that kills certain bacteria. Most of the fluid produced by the lacrimal glands evaporates from the surface of the eye, but excess tears are collected in the medial corner of the eye by the **lacrimal canaliculi.** The opening of each lacrimal canaliculus is called a **punctum** (pŭngk′tŭm). The upper and lower eyelids each have a punctum near the medial canthus. Each punctum is located on a small lump called the **lacrimal papilla.** The lacrimal canaliculi open into a **lacrimal sac,** which in turn continues into the **nasolacrimal duct** (see figure 15.9). The nasolacrimal duct opens into the inferior meatus of the nasal cavity beneath the inferior nasal concha (see chapter 23).

1. Tears are produced in the lacrimal gland and exit the gland through several lacrimal ducts.

2. The tears pass over the surface of the eye.

3. Tears enter the lacrimal canaliculi.

4. Tears are carried through the nasolacrimal duct.

5. Tears enter the nasal cavity from the nasolacrimal duct.

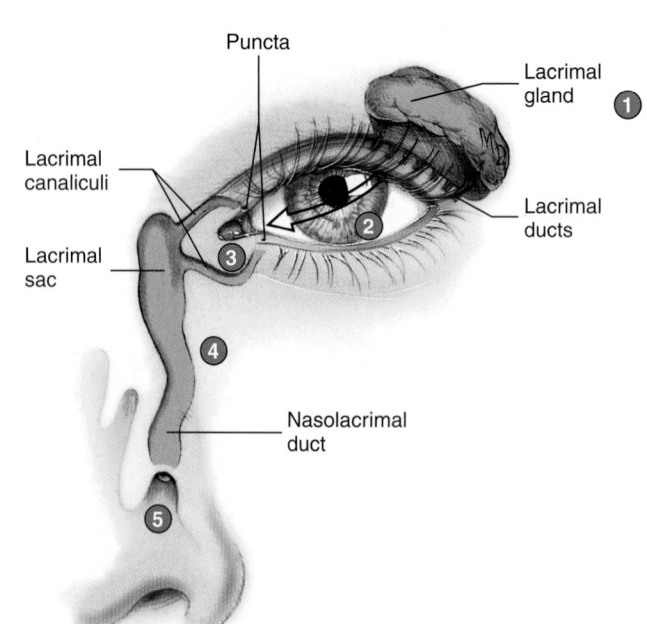

Process Figure 15.9　The Lacrimal Apparatus

Facial Nerve Damage

Facial nerve damage results in the inability to close the eyelid on the affected side. With the ability to blink being lost, tears cannot be washed across the eye, and the conjunctiva and cornea become dry. A dry cornea may become ulcerated, and, if not treated, eyesight may be lost.

P R E D I C T ❸

Explain why it's often possible to "taste" medications, such as eyedrops, that have been placed into the eyes. Why does a person's nose "run" when he or she cries?

Extrinsic Eye Muscles

Six **extrinsic muscles** of the eye (figures 15.10 and 15.11; also see chapter 10) cause the eyeball to move. Four of these muscles run more or less straight anteroposteriorly. They are the superior, inferior, medial, and lateral **rectus muscles.** Two muscles, the superior and inferior **oblique muscles,** are placed at an angle to the globe of the eye.

The movements of the eye can be described graphically by a figure resembling the letter *H*. The clinical test for normal eye movement is therefore called the **H test.** A person's inability to move his eye toward one part of the H may indicate dysfunction of an extrinsic eye muscle or the cranial nerve to the muscle (the actions of the eye muscles are listed in table 10.7).

The superior oblique muscle is innervated by the trochlear nerve (IV). The nerve is so named because the superior oblique muscle goes around a little pulley, or trochlea, in the superomedial corner of the orbit. The lateral rectus muscle is innervated by the abducent nerve (VI), so named because the lateral rectus muscle abducts the eye. The other four extrinsic eye muscles are innervated by the oculomotor nerve (III).

● **12. Describe and state the functions of the eyebrows, eyelids, conjunctiva, lacrimal apparatus, and extrinsic eye muscles.**

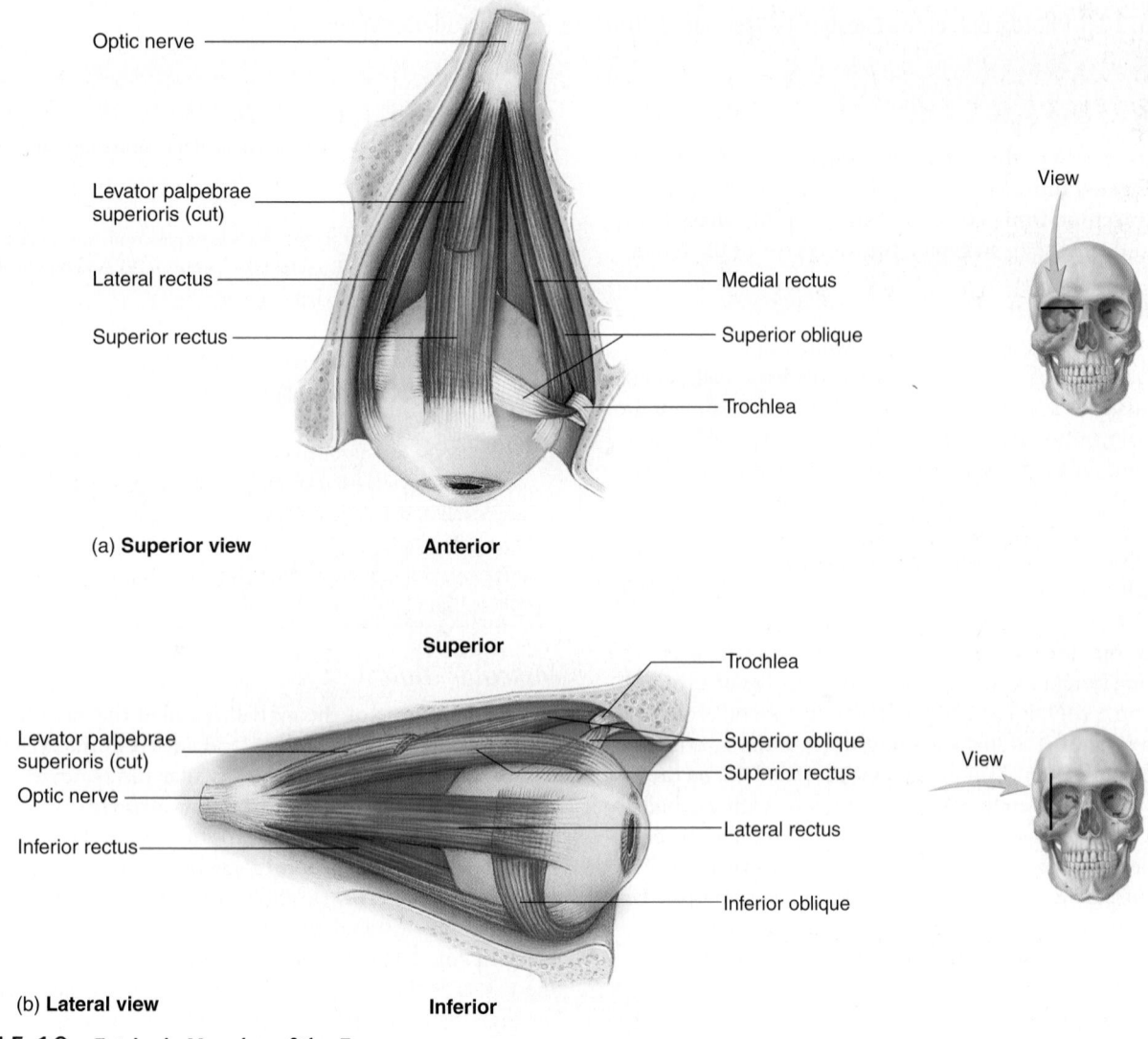

Figure 15.10 Extrinsic Muscles of the Eye

(*a*) Superior view. (*b*) Lateral view.

Superior rectus muscle (cut)

Eyeball

Medial rectus muscle

Lateral rectus muscle

Optic nerve

Optic chiasm

Superior view

Figure 15.11 Photograph of the Eye and Its Associated Structures from a Superior View

Anatomy of the Eye

The eye is composed of three coats, or tunics (figure 15.12). The outer, or **fibrous, tunic** consists of the sclera and cornea; the middle, or **vascular, tunic** consists of the choroid, ciliary body, and iris; and the inner, or **nervous, tunic** consists of the retina.

Fibrous Tunic

The **sclera** (sklēr′ă) is the firm, opaque, white outer layer of the posterior five-sixths of the eye. It consists of dense collagenous connective tissue with elastic fibers. The sclera helps maintain the shape of the eye, protects its internal structures, and provides an attachment point for the muscles that move it. Usually, a small portion of the sclera can be seen as the "white of the eye" when the eye and its surrounding structures are intact (see figure 15.7). The bulbar conjunctiva is loosely attached to the sclera.

The sclera is continuous anteriorly with the cornea. The **cornea** (kōr′nē-ă) is an avascular, transparent structure that permits light to enter the eye and bends, or refracts, that light as part of the focusing system of the eye. The cornea consists of a connective tissue matrix containing collagen, elastic fibers, and proteoglycans, with a layer of stratified squamous epithelium covering the outer surface and a layer of simple squamous epithelium on the inner surface. The outer epithelium is continuous with the bulbar conjunctiva over the sclera. Large collagen fibers are white, whereas smaller collagen fibers and proteoglycans are transparent. The cornea is transparent, rather than white like the sclera, in part because fewer large collagen fibers and more proteoglycans are present in the cornea than in the sclera. The transparency of the cornea also results from its low water content. In the presence of water, proteoglycans trap water and expand, which scatters light. In the absence of water, the proteoglycans decrease in size and do not interfere with the passage of light through the matrix.

PREDICT
Predict the effect of inflammation of the cornea on vision.

The Cornea

The central part of the cornea receives oxygen from the outside air. Soft plastic contact lenses worn for long periods must therefore be permeable to air so that air can reach the cornea.

The most common eye injuries are cuts or tears of the cornea caused by foreign objects like stones or sticks hitting the cornea. Extensive injury to the cornea may cause connective tissue deposition, thereby making the cornea opaque.

The cornea was one of the first organs transplanted. Several characteristics make it relatively easy to transplant: It's easily accessible and relatively easily removed; it's avascular and therefore does not require extensive circulation as do other tissues; and it's less immunologically active and therefore less likely to be rejected than other tissues.

Vascular Tunic

The middle tunic of the eyeball is called the vascular tunic because it contains most of the blood vessels of the eyeball (see figure 15.12). The arteries of the vascular tunic are derived from a number of arteries called **short ciliary arteries,** which pierce the sclera in a circle around the optic nerve. These arteries are branches of the **ophthalmic** (of-thal′mik) **artery,** which is a branch of the internal carotid artery. The vascular tunic contains a large number of melanin-containing pigment cells and appears black in color. The portion of the vascular tunic associated with the sclera of the eye is the **choroid** (ko′royd). The term *choroid* means membrane and suggests that this layer is relatively thin (0.1–0.2 mm thick). Anteriorly, the vascular tunic consists of the ciliary body and iris.

Vitreous chamber
(filled with
vitreous humor)

Conjunctiva

Cornea (fibrous tunic)

Anterior chamber ⎤ Both filled
Posterior chamber ⎦ with aqueous humor

Iris (vascular tunic)

Pupil

Lens

Optic nerve

Vitreous humor

Retina (nervous tunic)

Choroid (vascular tunic)

Sclera (fibrous tunic)

Suspensory ligaments

Ciliary body (vascular tunic)

Figure 15.12 Sagittal Section of the Eye, Demonstrating Its Layers

The **ciliary** (sil'ē-ar-ē) **body** is continuous with the choroid, and the **iris** is attached at its lateral margins to the ciliary body (figure 15.13a and b). The ciliary body consists of an outer **ciliary ring** and an inner group of **ciliary processes,** which are attached to the lens by **suspensory ligaments.** The ciliary body contains smooth muscles called the **ciliary muscles,** which are arranged with the outer muscle fibers oriented radially and the central fibers oriented circularly. The ciliary muscles function as a sphincter, and contraction of these muscles can change the shape of the lens. (This function is described in more detail on p. 527.) The ciliary processes are a complex of capillaries and cuboidal epithelium that produces aqueous humor.

The **iris** is the "colored part" of the eye, and its color differs from person to person. Brown eyes have brown melanin pigment in the iris. Blue eyes are not caused by a blue pigment but result from the scattering of light by the tissue of the iris, overlying a deeper layer of black pigment. The blue color is produced in a fashion similar to the scattering of light as it passes through the atmosphere to form the blue skies from the black background of space.

The iris is a contractile structure, consisting mainly of smooth muscle, surrounding an opening called the **pupil.** Light enters the eye through the pupil, and the iris regulates the amount of light by controlling the size of the pupil. The iris contains two groups of smooth muscles: a circular group called the **sphincter pupillae** (pū-pil'ē), and a radial group called the **dilator pupillae** (figure 15.13c and d). The sphincter pupillae are innervated by parasympathetic fibers from the oculomotor nerve (III). When they contract, the iris decreases or constricts the size of the pupil. The dilator pupillae are innervated by sympathetic fibers. When they contract, the pupil is dilated. The ciliary mus-

cles, sphincter pupillae, and dilator pupillae are sometimes referred to as the intrinsic eye muscles.

Retina

The **retina** is the innermost, nervous tunic of the eye (see figure 15.12). It consists of the outer **pigmented retina,** which is pigmented simple cuboidal epithelium, and the inner **sensory retina,** which responds to light. The sensory retina contains 120 million photoreceptor cells called **rods** and another 6 or 7 million **cones,** as well as numerous relay neurons. The retina covers the inner surface of the eye posterior to the ciliary body. A more detailed description of the histology and function of the retina is presented on page 528 and following.

Eye Pigment

The pupil appears black when you look into a person's eye because of the pigment in the choroid and the pigmented portion of the retina. The eye is a closed chamber, which allows light to enter only through the pupil. Light is absorbed by the pigmented inner lining of the eye; thus looking into it is like looking into a dark room. If a bright light is directed into the pupil, however, the reflected light is red because of the blood vessels on the surface of the retina. This is why the pupils of a person looking directly at a flash camera often appear red in a photograph. People with albinism lack the pigment melanin, and the pupil always appears red because no melanin is present to absorb light and prevent it from being reflected from the back of the eye. The diffusely lighted blood vessels in the interior of the eye contribute to the red color of the pupil.

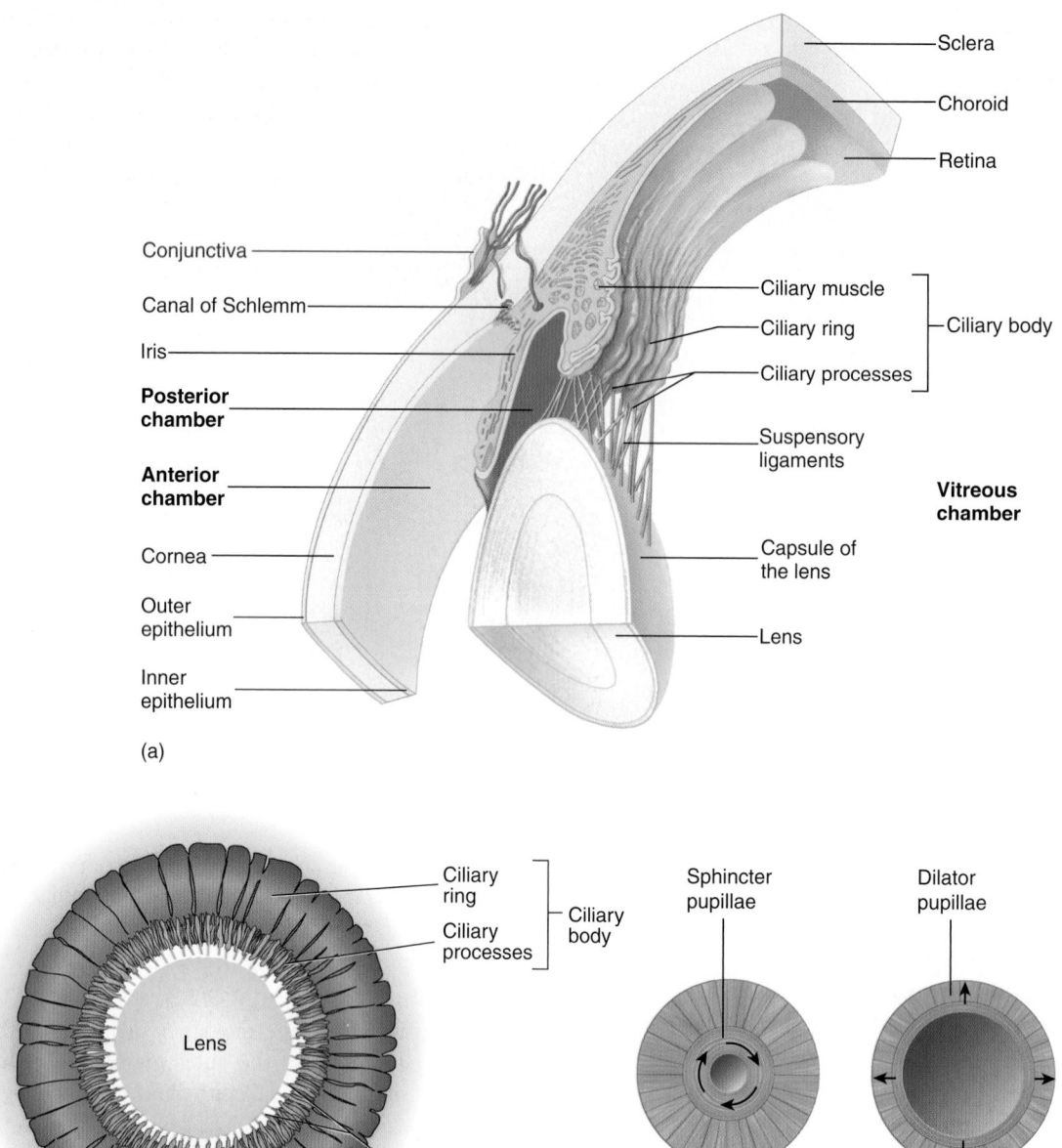

(a)

(b)

(c)

(d)

Figure 15.13   Lens, Cornea, Iris, and Ciliary Body

(*a*) The orientation is the same as in figure 15.12. (*b*) The lens and ciliary body. (*c*) The sphincter pupillae muscles of the iris constrict the pupil. (*d*) The dilator pupillae muscles of the iris dilate the pupil.

When the posterior region of the retina is examined with an **ophthalmoscope** (of-thal′mō-skōp) (figure 15.14), several important features can be observed. Near the center of the posterior retina is a small yellow spot approximately 4 mm in diameter, the **macula lutea** (mak′ū-lă lū′tē-ă). In the center of the macula lutea is a small pit, the **fovea** (fō′vē-ă) **centralis.** The fovea and macula make up the region of the retina where light is focused. The fovea is the portion of the retina with the greatest visual acuity (the ability to see fine im-

ages) because the photoreceptor cells are more tightly packed in that portion of the retina than anywhere else. Just medial to the macula lutea is a white spot, the **optic disc,** through which blood vessels enter the eye and spread over the surface of the retina. This is also the spot where nerve processes from the sensory retina meet, pass through the outer two tunics, and exit the eye as the optic nerve. The optic disc contains no photoreceptor cells and does not respond to light; therefore it's called the **blind spot** of the eye.

through a venous ring at the base of the cornea called the **canal of Schlemm** (shlem), or the **scleral venous sinus** (see figure 15.13). The production and removal of aqueous humor results in "circulation" of aqueous humor and maintenance of a constant intraocular pressure. If circulation of the aqueous humor is inhibited, a defect called **glaucoma** (glaw-kō′mă), which is an abnormal increase in intraocular pressure, can result (see the Clinical Focus on "Eye Disorders," p. 536).

The vitreous chamber of the eye is much larger than the anterior and posterior chambers. It's surrounded almost completely by the retina and is filled with a transparent jellylike substance, the **vitreous** (vit′rē-ŭs) **humor.** The vitreous humor is not produced as rapidly as is the aqueous humor, and its turnover is extremely slow. The vitreous humor helps maintain intraocular pressure and therefore the shape of the eyeball, and it holds the lens and the retina in place. It also functions in the refraction of light in the eye.

Lens

The **lens** is an unusual biologic structure. It is transparent and biconvex, with the greatest convexity on its posterior side. The lens consists of a layer of cuboidal epithelial cells on its anterior surface and a posterior region of very long columnar epithelial cells called **lens fibers.** Cells from the anterior epithelium proliferate and give rise to the lens fibers at the equator of the lens. The lens fibers lose their nuclei and other cellular organelles and accumulate a special set of proteins called **crystallines.** This crystalline lens is covered by a highly elastic transparent **capsule.**

The lens is suspended between the two eye compartments by the suspensory ligaments of the lens, which are connected from the ciliary body to the lens capsule.

> 13. Name the three layers (tunics) of the eye, describe the parts or structures each forms, and explain their functions.
> 14. How does the pupil constrict? How does it dilate? What is the blind spot?
> 15. Name the two compartments of the eye and the substances that fill each compartment.
> 16. What is the function of the canal of Schlemm and the ciliary processes?
> 17. Describe the lens of the eye, and explain how the lens is held in place.

Functions of the Complete Eye

The eye functions much like a camera. The iris allows light into the eye, and the lens, cornea, and humors focus the light onto the retina. The light striking the retina is converted into action potentials that are relayed to the brain.

Light

The electromagnetic spectrum is the entire range of wavelengths or frequencies of electromagnetic radiation from very short gamma waves at one end of the spectrum to the longest radio waves at the other end (figure 15.15). **Visible light** is the portion of the electromagnetic spectrum that can be detected by the human eye. Light has characteristics of both particles (photons) and waves, with a wavelength between 400 and 700 nm. This range sometimes is called the range of visible light or, more correctly, the

(a)

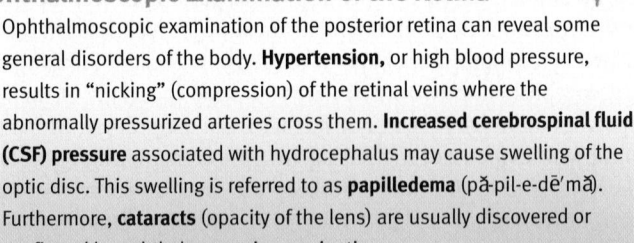

(b)

Figure 15.14 Ophthalmoscopic View of the Left Retina

(a) The posterior wall of the retina as seen when looking through the pupil. Notice the vessels entering the eye through the optic disc (the optic nerve) and the macula lutea with the fovea (the part of the retina with the greatest visual acuity). (b) Demonstration of the blind spot. Close your right eye. Hold the figure in front of your left eye and stare at the +. Move the figure toward your eye. At a certain point, when the image of the spot is over the optic disc, the red spot seems to disappear.

Ophthalmoscopic Examination of the Retina

Ophthalmoscopic examination of the posterior retina can reveal some general disorders of the body. **Hypertension,** or high blood pressure, results in "nicking" (compression) of the retinal veins where the abnormally pressurized arteries cross them. **Increased cerebrospinal fluid (CSF) pressure** associated with hydrocephalus may cause swelling of the optic disc. This swelling is referred to as **papilledema** (pă-pil-e-dē′mă). Furthermore, **cataracts** (opacity of the lens) are usually discovered or confirmed by ophthalmoscopic examination.

Chambers of the Eye

The interior of the eye is divided into three chambers: **anterior chamber, posterior chamber,** and **vitreous** (vit′rē-ŭs), or **postremal, chamber** (see figure 15.12). The anterior chamber lies between the cornea and iris, and a smaller posterior chamber lies between the iris and lens (see figure 15.13). These two chambers are filled with **aqueous humor,** which helps maintain intraocular pressure. The pressure within the eye keeps the eye inflated and is largely responsible for maintaining the shape of the eye. The aqueous humor also refracts light and provides nutrition for the structures of the anterior chamber, such as the cornea, which has no blood vessels. Aqueous humor is produced by the ciliary processes as a blood filtrate and is returned to the circulation

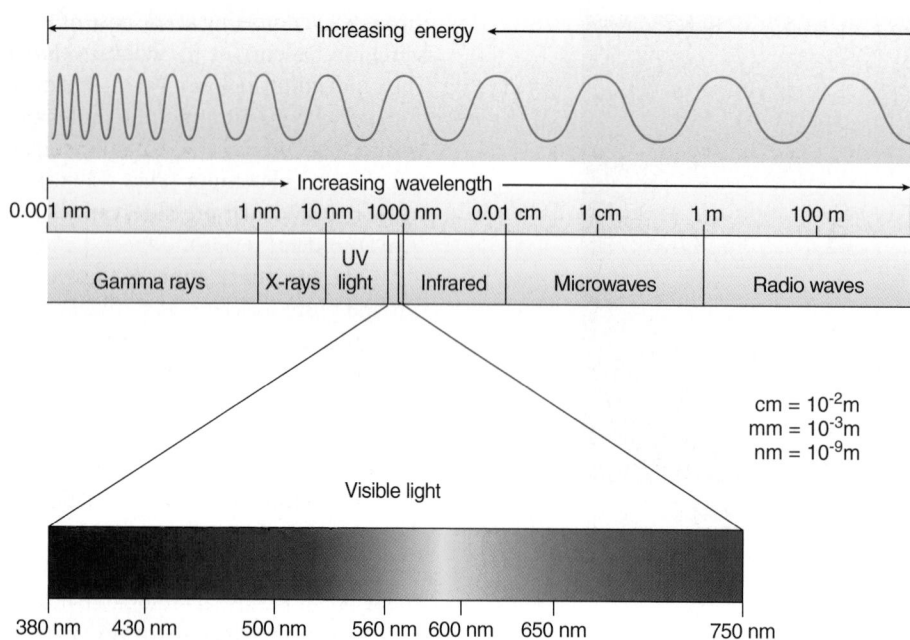

Figure 15.15 The Electromagnetic Spectrum

The spectrum of visible light is pulled out and expanded. The wavelengths of the various colors are also depicted.

visible spectrum. Within the visible spectrum, each color has a different wavelength.

Light Refraction and Reflection

An important characteristic of light is that it can be refracted (bent). As light passes from air to a denser substance like glass or water, its speed is reduced. If the surface of that substance is at an angle other than 90 degrees to the direction the light rays are traveling, the rays are bent as a result of variation in the speed of light as it encounters the new medium. This bending of light is called **refraction.**

If the surface of a lens is concave, with the lens thinnest in the center, the light rays diverge as a result of refraction. If the surface is convex, with the lens thickest in the center, the light rays tend to converge. As light rays converge, they finally reach a point at which they cross. This point is called the **focal point,** and causing light to converge is called **focusing.** No image is formed exactly at the focal point, but an inverted, focused image can form on a surface located some distance past the focal point. How far past the focal point the focused image forms depends on a number of factors. A biconvex lens causes light to focus closer to the lens than does a lens with a single convex surface. Furthermore, the more nearly spherical the lens, the closer to the lens the light is focused; the more flattened the biconcave lens, the more distant is the point where the light is focused.

If light rays strike an object that is not transparent, they bounce off the surface. This phenomenon is called **reflection.** If the surface is very smooth, such as the surface of a mirror, the light rays bounce off in a specific direction. If the surface is rough, the light rays are reflected in several directions and produce a more diffuse reflection. We can see most solid objects because of the light reflected from their surfaces.

Focusing of Images on the Retina

The focusing system of the eye projects a clear image on the retina. Light rays converge as they pass from the air through the convex cornea. Additional convergence occurs as light encounters the aqueous humor, lens, and vitreous humor. The greatest contrast in media density is between the air and the cornea; therefore, the greatest amount of convergence occurs at that point. The shape of the cornea and its distance from the retina are fixed, however, so that no adjustment in the location of the focal point can be made by the cornea. Fine adjustment in focal point location is accomplished by changing the shape of the lens. In general, focusing can be accomplished in two ways. One is to keep the shape of the lens constant and move it nearer or farther from the point at which the image will be focused, such as occurs in a camera, microscope, or telescope. The second way is to keep the distance constant and to change the shape of the lens, which is the technique used in the eye.

As light rays enter the eye and are focused, the image formed just past the focal point is inverted (figure 15.16). Action potentials that represent the inverted image are passed to the visual cortex of the cerebrum, where they are interpreted by the brain as being right side up.

Visual Image Inversion

Because the visual image is inverted when it reaches the retina, the image of the world focused on the retina is upside down. The brain processes information from the retina so that the world is perceived the way "it really is." If, as an experiment, a person wears glasses that invert the image entering the eye, he or she will see the world upside down for a few days, after which time the brain adjusts to the new input to set the world right side up again. If the glasses are then removed, another adjustment period is required before the world is made right by the brain.

Distant vision

Ciliary muscles in the ciliary body relaxed

Suspensory ligaments (tension high)

FP

Lens flattened

(a)

Near vision

Ciliary muscles in the ciliary body contract, moving ciliary body toward lens

Suspensory ligaments (tension low)

FP

Lens thickened

(b)

Figure 15.16 Focus and Accommodation by the Eye

The focal point (FP) is where light rays cross. (a) Distant image. The lens is flattened, and the image is focused on the retina. (b) Accommodation for near vision. The lens is more rounded, and the image is focused on the retina.

its own elastic nature (figure 15.16*b*). The more spherical lens then has a more convex surface, causing greater refraction of light. This process is called accommodation.

As light strikes a solid object, the rays are reflected in every direction from the surface of the object. Only a small portion of the light rays reflected from a solid object, however, pass through the pupil and enter the eye of any given person. An object far away from the eye appears small compared to a nearby object because only nearly parallel light rays enter the eye from a distant object (see figure 15.16*a*). Converging rays leaving an object closer to the eye can also enter the eye (see figure 15.15*b*), and the object appears larger.

When rays from a distant object reach the lens, they don't have to be refracted to any great extent to be focused on the retina, and the lens can remain fairly flat. When an object is closer to the eye, the more obliquely directed rays must be refracted to a greater extent to be focused on the retina.

As an object is brought closer and closer to the eye, accommodation becomes more and more difficult because the lens cannot become any more convex. At some point, the eye no longer can focus the object, and it's seen as a blur. The point at which this blurring occurs is called the **near point of vision,** which is usually about 2–3 inches from the eye for children, 4–6 inches for a young adult, 20 inches for a 45-year-old adult, and 60 inches for an 80-year-old adult. This increase in the near point of vision, called **presbyopia,** occurs because the lens becomes more rigid with increasing age, which is primarily why some older people say they could read with no problem if they only had longer arms.

Vision Charts

When a person's vision is tested, a chart is placed 20 feet from the eye, and the person is asked to read a line of letters that is standardized for normal vision. If the person can read the line, the vision is considered to be 20/20, which means that the person can see at 20 feet what people with normal vision can see at 20 feet. If, on the other hand, the person can see words only at 20 feet that people with normal vision can see at 40 feet, the vision is considered 20/40.

When the ciliary muscles are relaxed, the suspensory ligaments of the ciliary body maintain elastic pressure on the lens, thereby keeping it relatively flat and allowing for distant vision (figure 15.16*a*). The condition in which the lens is flattened so that nearly parallel rays from a distant object are focused on the retina is referred to as **emmetropia** (em-ĕ-trō′pē-ă; measure) and is the normal resting condition of the lens. The point at which the lens does not have to thicken for focusing to occur is called the **far point of vision** and normally is 20 feet or more from the eye.

When an object is brought closer than 20 feet to the eye, three events occur to bring the image into focus on the retina: accommodation by the lens, constriction of the pupil, and convergence of the eyes.

1. *Accommodation.* When focusing on a nearby object, the ciliary muscles contract as a result of parasympathetic stimulation from the oculomotor nerve (III). This sphincterlike contraction pulls the choroid toward the lens to reduce the tension on the suspensory ligaments. This allows the lens to assume a more spherical form because of

2. *Pupil constriction.* Another factor involved in focusing is the **depth of focus,** which is the greatest distance through which an object can be moved and still remain in focus on the retina. The main factor affecting the depth of focus is the size of the pupil. If the pupillary diameter is small, the depth of focus is greater than if the pupillary diameter is large. With a smaller pupillary opening, an object may therefore be moved slightly nearer or farther from the eye without disturbing its focus. This is particularly important when viewing an object at close range because the interest in detail is much greater, and therefore the acceptable margin for error is smaller. When the pupil is constricted, the light entering the eye tends to pass more nearly through the center of the lens and is more accurately focused than light passing through the edges of the lens. Pupillary diameter also regulates the amount of light entering the eye. The dimmer the light, the greater the pupil diameter must be. As the pupil constricts during close vision, therefore, more light is required on the object being observed.

3. *Convergence.* Because the light rays entering the eyes from a distant object are nearly parallel, both pupils can pick up the light rays when the eyes are directed more or less straight ahead. As an object moves closer, however, the eyes must be rotated medially so that the object is kept focused on corresponding areas of each retina. Otherwise the object appears blurry. This medial rotation of the eyes is accomplished by a reflex which stimulates the medial rectus muscle of each eye. This movement of the eyes is called convergence. Convergence can easily be observed. Have someone stand facing you. Have the person reach out one hand and extend an index finger as far in front of his face as possible. While the person keeps his gaze fixed on the finger, have him slowly bring the finger in toward his nose until he finally touches it. Notice the movement of his pupils during this movement. What happens?

P R E D I C T

Explain how several hours of reading can cause eyestrain, or eye fatigue. Describe what structures are involved.

18. *What causes light to refract? What is a focal point? What is emmetropia?*

19. *Describe the changes that occur in the lens, pupil, and extrinsic eye muscles as an object moves from 25 feet away to 6 inches away. What is meant by the terms near point and far point of vision?*

Structure and Function of the Retina

Leonardo da Vinci, in speaking of the eye, said, "Who would believe that so small a space could contain the images of all the universe?" The retina of each eye, which gives us the potential to see the whole world, is about the size and thickness of a postage stamp.

The retina consists of a pigmented retina and a sensory retina. The **sensory retina** contains three layers of neurons: photoreceptor, bipolar, and ganglionic. The cell bodies of these neurons form nuclear layers separated by plexiform layers, where the neurons of adjacent layers synapse with each other (figure 15.17). The outer plexiform (plexuslike) layer is between the photoreceptor and bipolar cell layers. The inner plexiform layer is between the bipolar and ganglionic cell layers.

The **pigmented retina,** or pigmented epithelium, consists of a single layer of cells. This layer of cells is filled with melanin pigment and, together with the pigment in the choroid, provides a black matrix, which enhances visual acuity by isolating individual photoreceptors and reducing light scattering. This pigmentation is not strictly necessary for vision, however. People with albinism (lack of pigment) can see, although their visual acuity is reduced because of some light scattering.

The layer of the sensory retina nearest the pigmented retina is the layer of rods and cones. The rods and cones are photoreceptor cells, which are sensitive to stimulation from "visible" light. The

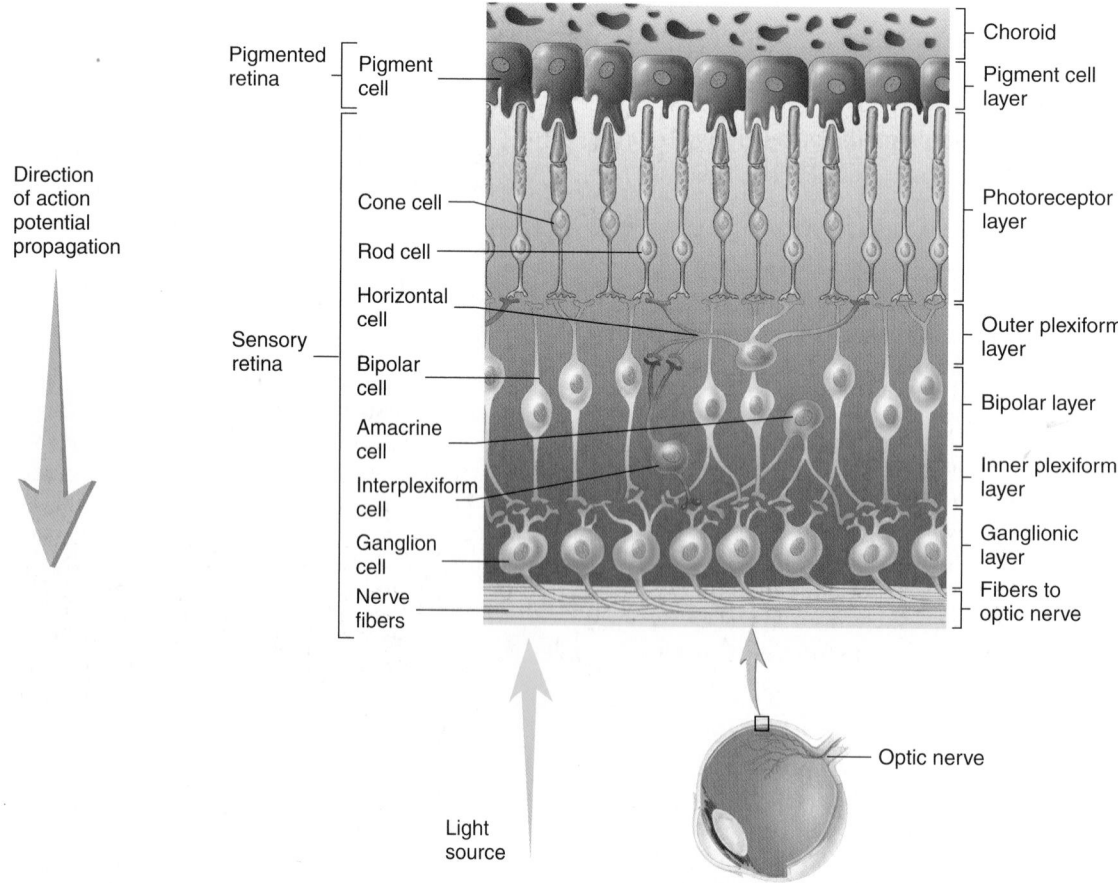

Figure 15.17 Retina

Section through the retina with its major layers labeled.

light-sensitive portion of each photoreceptor cell is adjacent to the pigmented layer.

Rods

Rods are bipolar photoreceptor cells involved in noncolor vision and are responsible for vision under conditions of reduced light (table 15.1). The modified, dendritic, light-sensitive part of rod cells is cylindrical, with no taper from base to apex (figure 15.18*a*). This rod-shaped photoreceptive part of the rod cell contains about 700 double-layered membranous discs. The discs contain **rhodopsin** (rō-dop′sin), a purple pigment, which consists of the protein **opsin** covalently bound to a yellow photosensitive pig-

ment called **retinal** (derived from vitamin A). Opsin is a protein much like a channel protein, consisting of seven transmembrane helical regions. An extracellular plug closes the external opening of the opsin "channel." G proteins and phosphodiesterase enzymes are also associated with the disc membranes. Na^+ channels are located in the outer membrane of the rod cell outer segment (see figure 15.18*d*).

Function of Rhodopsin

Figure 15.19 depicts the changes that rhodopsin undergoes in response to light. In the resting (dark) state, 11-*cis*-retinal is tightly bound to the internal surface of opsin. The extracellular plug helps

Table 15.1	**Rods and Cones**			
Photoreceptive End	**Photoreceptive Molecule**	**Function**		**Location**
Rod Cylindrical	Rhodopsin	Noncolor vision; vision under conditions of low light		Over most of retina; none in fovea
Cone Conical	Iodopsin	Color vision; visual acuity		Numerous in fovea and macula lutea; sparse over rest of retina

Figure 15.18 Sensory Receptor Cells of the Retina

(*a*) Rod cell. (*b*) Cone cell. (*c*) An enlargement of the discs in the outer segment. (*d*) An enlargement of one of the discs, showing the relation of rhodopsin and a gated Na^+ channel to the membrane.

1. Retinal (in an inactive configuration called 11-*cis*-) is attached inside opsin to make rhodopsin.

2. Light causes retinal to change shape from 11-*cis*-retinal to all-*trans*-retinal. This change causes opsin also to change shape (from its dark to light configuration). This activated rhodopsin activates a G protein (called transducin), which activates a phosphodiesterase. The phosphodiesterase catalyzes the conversion of cGMP to GMP, which closes Na$^+$ channels.

3. All-*trans*-retinal detaches from opsin. Transducin is reassociated.

4. All-*trans*-retinal is converted to 11-*cis*-retinal, a process that requires energy.

5. 11-c*is*-retinal attaches to opsin, which returns to its original (dark) configuration. GMP is displaced from the Na$^+$ channel and the channel opens.

Process Figure 15.19 Rhodopsin Cycle

keep retinal from being lost from the cell out of the opsin "channel." As light is absorbed by rod cells, retinal changes shape from 11-*cis*-retinal to all-*trans*-retinal. This change causes opsin to also change shape (dark to light state). The changes in opsin activate the

attached G protein, called **transducin** (trans-doo′sin), which activates a phosphodiesterase. Phosphodiesterase catalyzes the conversion of cGMP to GMP, which closes Na⁺ channels, resulting in hyperpolarization of the cell (figure 15.20).

Process Figure 15.20   Rod Cell Hyperpolarization

(*a*) Changes in the rod cell membrane potential following the opsin and retinal cell shape changes is a hyperpolarization. (*b*) Unstimulated rod cell (dark).
(*c*) Stimulated rod cell (light).

Opsin Mutants

Opsin is a protein composed of 338 amino acids. Mutation at amino acid 23 or 28, in the extracellular plug causes **retinitis pigmentosa.** This is a genetic disorder consisting of progressive retinal degeneration. Because of the mutation, the extracellular plug changes shape and does not adequately close the extracellular opening of opsin thereby allowing retinal to leak out of the cells. This leaking retinal pigment infiltrates the sensory retina, causing it to degenerate, decreasing its function and constricting the visual fields. **Night blindness,** or nyctalopia, (the decreased ability to see in reduced light) may also occur in retinitis pigmentosa. Night blindness also may occur as a result of vitamin A deficiency or as the result of another mutation at amino acid 90 of the opsin molecule. This mutation occurs in the second of the seven helical regions of the protein opsin, changing its shape and affecting the attachment of retinal to opsin.

This hyperpolarization in the photoreceptor cells is somewhat remarkable, because most neurons respond to stimuli by depolarizing. When photoreceptor cells are not exposed to light and are in a resting, nonactivated state, gated Na^+ channels in their membranes are open, and Na^+ flow into the cell. This influx of Na^+, referred to as the dark current, causes the photoreceptor cells to release the neurotransmitter glutamate from their presynaptic terminals (see figure 15.20). Glutamate binds to receptors on the postsynaptic membranes of the bipolar cells of the retina, causing them to hyperpolarize. Thus, glutamate causes an inhibitory postsynaptic potential (IPSP) in the bipolar cells. The influx of Na^+ is offset by the efflux of K^+ through nongated K^+ channels. Equilibrium of Na^+ and K^+ in the cell is maintained by a Na^+/K^+ pump.

When photoreceptor cells are exposed to light, the Na^+ channels close, fewer Na^+ enter the cell, and the amount of glutamate released from the presynaptic terminals decreases. As a result, the hyperpolarization in the bipolar cells decreases, and they depolarize sufficiently to release neurotransmitters, which stimulate ganglionic cells to generate action potentials. The number of Na^+ channels that close and the degree to which they close is proportional to the amount of light exposure.

At the final stage of this light-initiated reaction, retinal is completely released from the opsin. This free retinal may then be converted back to vitamin A, from which it was originally derived. The total vitamin A/retinal pool is in equilibrium so that under normal conditions the amount of free retinal is relatively constant. To create more rhodopsin, the altered retinal must be converted back to its original shape, a reaction that requires energy. Once the retinal resumes its original shape, its recombination with opsin is spontaneous, and the newly formed rhodopsin can again respond to light.

Light and **dark adaptation** is the adjustment of the eyes to changes in light. Adaptation to light or dark conditions, which occurs when a person comes out of a darkened building into the sunlight or vice versa, is accomplished by changes in the amount of available rhodopsin. In bright light excess rhodopsin is broken down so that not as much is available to initiate action potentials, and the eyes become "adapted" to bright light. Conversely, in a dark room more rhodopsin is produced, making the retina more light-sensitive.

P R E D I C T 6

If breakdown of rhodopsin occurs rapidly and production is slow, do eyes adapt more rapidly to light or dark conditions?

Light and dark adaptation also involves pupil reflexes. The pupil enlarges in dim light to allow more light into the eye and contracts in bright light to allow less light into the eye. In addition, rod function decreases and cone function increases in light conditions, and vice versa during dark conditions. This occurs because rod cells are more sensitive to light than cone cells and because rhodopsin is depleted more rapidly in rods than in cones.

Cones

Color vision and visual acuity are functions of cone cells. Color is a function of the wavelength of light, and each color results from a certain wavelength within the visible spectrum. Even though rods are very sensitive to light, they cannot detect color, and sensory input that ultimately reaches the brain from these cells is interpreted by the brain as shades of gray. Cones require relatively bright light to function. As a result, as the light decreases, so does the color of objects that can be seen until, under conditions of very low illumination, the objects appear gray. This occurs because as the light decreases, fewer cone cells respond to the dim light.

Cones are bipolar photoreceptor cells with a conical light-sensitive part that tapers slightly from base to apex (see figure 15.18b). The outer segments of the cone cells, like those of the rods, consist of double-layered discs. The discs are slightly more numerous and more closely stacked in the cones than in the rods. Cone cells contain a visual pigment, **iodopsin** (ī-ō-dop′sin), which consists of retinal combined with a photopigment opsin protein. Three major types of color-sensitive opsin exist: blue, red, and green; each closely resembles the opsin proteins of rod cells but with somewhat different amino acid sequences. These color photopigments function in much the same manner as rhodopsin, but whereas rhodopsin responds to the entire spectrum of visible light, each iodopsin is sensitive to a much narrower spectrum.

Most people have one red pigment gene and one or more green pigment genes located in a tandem array on each X chromosome. An enhancer gene on the X chromosome apparently determines that only one color opsin gene is expressed in each cone cell. Only the first or second gene in the tandem array is expressed in each cone cell, so that some cone cells express only the red pigment gene and others express only one of the green pigment genes.

As can be seen in figure 15.21, although considerable overlap occurs in the wavelength of light to which these pigments are sensitive, each pigment absorbs light of a certain range of wavelengths. As light of a given wavelength, representing a certain color, strikes the retina, all cone cells containing photopigments capable of responding to that wavelength generate action potentials. Because of the overlap among the three types of cones, especially between the green and red pigments, different proportions of cone cells respond to each wavelength, thus allowing color perception over a wide range. Color is interpreted in the visual cortex as combinations of sensory input originating from cone

Figure 15.21 **Wavelengths to Which Each of the Three Visual Pigments are Sensitive: Blue, Green, Red**

There are actually two forms of the red pigment. One, found in 60% of the population, has a serine at position 180; and the other, found in 40% of the population, has an alanine at position 180. Each red pigment has a slightly different wavelength sensitivity.

cells. For example, when orange light strikes the retina, 99% of the red-sensitive cones respond, 42% of the green-sensitive cones respond, and no blue cones respond. When yellow light strikes the retina, the response is shifted so that a greater number of green-sensitive cones respond. The variety of combinations created allows humans to distinguish several million gradations of light and shades of color.

Seeing Red

Not everyone sees the same red. Two forms of the red photopigment are common in humans. Approximately 60% of people have the amino acid serine in position 180 of the red opsin protein, whereas 40% have alanine in that position. That subtle difference in the protein results in slightly different absorption characteristics (see figure 15.21). Even though we were each taught to recognize red when we see a certain color, we apparently don't see that color in quite the same way. This difference may contribute to people having different favorite colors.

Distribution of Rods and Cones in the Retina

Cones are involved in visual acuity, in addition to their role in color vision. The macula lutea, and especially the fovea centralis are used when visual acuity is required, such as for focusing on the

words of this page. The fovea centralis has about 35,000 cones and no rods. The rest of the macula lutea has more cones than rods. The 120 million rods are 10–20 times more plentiful than cones over most of the remaining retina, however. They are more highly concentrated away from the macula and are more important in low-light conditions.

P R E D I C T

Explain why at night a person may notice a movement "out of the corner of her eye," but, when she tries to focus on the area where she noticed the movement, it appears as though nothing is there.

Inner Layers of the Retina

The middle and inner nuclear layers of the retina consist of two major types of neurons: bipolar and ganglion cells. The rod and cone photoreceptor cells synapse with **bipolar cells,** which in turn synapse with **ganglion cells.** Axons from the ganglion cells pass over the inner surface of the retina (see figure 15.17), except in the area of the fovea centralis, converge at the **optic disc,** and exit the eye as the **optic nerve** (II). The fovea centralis is devoid of ganglion cell processes, resulting in a small depression in this area; thus the name *fovea,* meaning small pit. As a result of the absence of ganglion cell processes in addition to the concentration of cone cells mentioned previously, visual acuity is further enhanced in the fovea centralis because light rays don't have to pass through as many tissue layers before reaching the photoreceptor cells.

Rod and cone cells differ in the way they interact with bipolar and ganglion cells. One bipolar cell receives input from numerous rods, and one ganglion cell receives input from several bipolar cells so that spatial summation of the signal occurs and the signal is enhanced, thereby allowing awareness of stimulus from very dim light sources but decreasing visual acuity in these cells. Cones, on the other hand, exhibit little or no convergence on bipolar cells so that one cone cell may synapse with only one bipolar cell. This system reduces light sensitivity but enhances visual acuity.

The **receptive field** of each ganglion cell, the area from which a ganglion cell receives input, is roughly circular in shape, with a smaller portion in the center, called the receptive field center, and a larger surrounding area. The receptive fields in the fovea centralis are very small compared to receptive fields of more peripheral parts of the retina. The small size of these receptive fields is consistent with the visual acuity of the fovea centralis.

Two types of receptive fields exist: on-center ganglion cells and off-center ganglion cells. On-center cells generate more action potentials when light is directed onto the receptive field center. Off-center neurons generate more action potentials when light is turned off in the receptive field center or when light shines on the surrounding area. These receptive fields respond primarily to contrasts in light (edges) rather than the absolute intensity of light.

Within the inner layers of the retina, interneurons are present that modify the signals from the photoreceptor cells before the signal ever leaves the retina (see figure 15.17).

Horizontal cells form the outer plexiform layer and synapse with photoreceptor cells and bipolar cells. **Amacrine** (am′ă-krin) **cells** form the inner plexiform layer and synapse with bipolar and ganglion cells. **Interplexiform cells** form the bipolar layer and synapse with amacrine, bipolar, and horizontal cells to form a feedback loop. The interneurons are either excitatory or inhibitory on the cells with which they synapse. These interneurons enhance borders and contours, thereby increasing the intensity at boundaries, such as the edge of a dark object against a light background.

20. *What is the function of the pigmented retina and of the choroid?*
21. *Describe the changes that occur in a rod cell after light strikes rhodopsin. How does rhodopsin re-form? Why is the response of a rod cell to a stimulus unusual?*
22. *How do dark and light adaptation occur?*
23. *What are the three types of cone cells? How do they function to produce the colors we see?*
24. *Describe the arrangement of rods and cones in the fovea, the macula lutea, and the periphery of the eye.*
25. *Starting with a rod or cone cell, name the cells or structures that an action potential encounters while traveling to the visual cortex.*

Motion Pictures

Action potentials pass from the retina through the optic nerve at the rate of 20–25/s. We "see" a given image for a fraction of a second longer than it actually appears. Motion pictures take advantage of these two facts. When still photographs are flashed on a screen at the rate of 24 frames per second, they appear to flow into each other, and a motion picture results.

Neuronal Pathways for Vision

The optic nerve (II) (figure 15.22) leaves the eye and exits the orbit through the optic foramen to enter the cranial cavity. Just inside the vault and just anterior to the pituitary, the optic nerves are connected to each other at the **optic chiasm** (kī′azm). Ganglion cell axons from the nasal retina (the medial portion of the retina) cross through the optic chiasm and project to the opposite side of the brain. Ganglion cell axons from the temporal retina (the lateral portion of the retina) pass through the optic nerves and project to the brain on the same side of the body without crossing.

Beyond the optic chiasm, the route of the ganglionic axons is called the **optic tract** (see figure 15.22). Most of the optic tract axons terminate in the **lateral geniculate nucleus** of the thalamus. Some axons do not terminate in the thalamus but separate from the optic tract to terminate in the **superior colliculi,** the center for visual reflexes (see chapter 13). Neurons of the lateral geniculate ganglion form the fibers of the **optic radiations,** which project to the **visual cortex** in the **occipital lobe.** Neurons of the visual cortex integrate the messages coming from the retina into a single message, translate that message into a mental image, and then transfer the image to other parts of the brain, where it is evaluated and either ignored or acted on.

The projections of ganglion cells from the retina can be related to the **visual fields** (see figure 15.22). The visual field of one eye can be evaluated by closing the other eye. Everything that can be seen with the one open eye is the visual field of that eye. The visual field of each eye can be divided into temporal (lateral) and nasal (medial) parts. In each eye, the temporal part of the visual field projects onto the nasal retina, whereas the nasal part of the visual field projects to the temporal retina. The projections and nerve pathways are arranged in such a way that images entering the eye from the right part of each visual field project to the left side of the brain. Conversely, the left part of each visual field projects to the right side of the brain.

Tunnel Vision

Because the optic chiasm lies just anterior to the pituitary, a pituitary tumor can put pressure on the optic chiasm and may result in visual defects. Because the nerve fibers crossing in the optic chiasm are carrying information from the temporal halves of the visual fields, a person with optic chiasm damage cannot see objects in the temporal halves of the visual fields, a condition called **tunnel vision.** Tunnel vision is often an early sign of a pituitary tumor.

P R E D I C T   **8**

The lines at *A* and *B* in figure A depict two lesions in the visual pathways. The effect of a lesion at *A* in the optic radiations on the visual fields is depicted (with the right and left fields separated) in the ovals at the bottom of the figure. The black areas indicate what parts of the visual fields are defective. Describe the effect that the lesion at *B* has on the visual fields (see figure 15.22 for help).

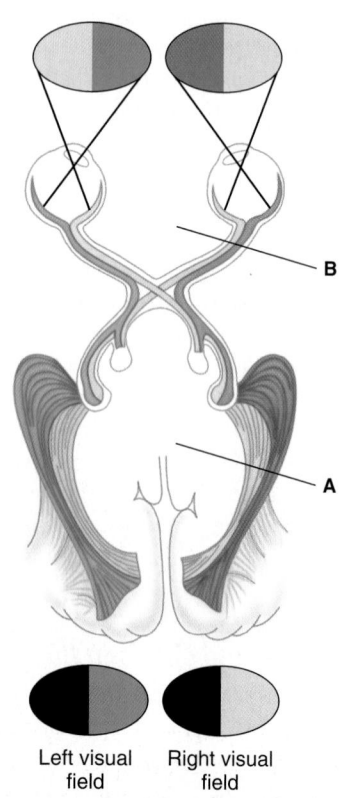

Left visual field   Right visual field

Figure A   Lesions of Visual Pathways

1. Each visual field is divided into a temporal and nasal half.

2. After passing through the lens, light from each half of a visual field projects to the opposite side of the retina.

3. An optic nerve consists of axons extending from the retina to the optic chiasm.

4. In the optic chiasm, axons from the nasal part of the retina cross and project to the opposite side of the brain. Axons from the temporal part of the retina do not cross.

5. An optic tract consists of axons that have passed through the optic chiasm (with or without crossing) to the thalamus.

6. The axons synapse in the lateral geniculate nuclei of the thalamus. Collateral branches of the axons in the optic tracts synapse in the superior colliculi.

7. An optic radiation consists of axons from thalamic neurons that project to the visual cortex.

8. (b) The right part of each visual field (*dark green* and *light blue*) projects to the left side of the brain, and the left part of each visual field projects to the right side of the brain (*light green* and *dark blue*).

Process Figure 15.22   Visual Pathways

(a) Pathways for the left eye (superior view). (b) Pathways for both eyes (superior view). (c) Overlap of the fields of vision (superior view). (d) Photograph of the visual nerves, tracts, and pathways (inferior view).

The visual fields of the eyes partially overlap (see figure 15.22). The region of overlap is the area of **binocular vision,** seen with two eyes at the same time, and it is responsible for **depth perception,** the ability to distinguish between near and far objects and to judge their distance. Because humans see the same object with both eyes, the image of the object reaches the retina of one eye at a slightly different angle from that of the other. With experience, the brain can interpret these differences in angle so that distance can be judged quite accurately.

26. *What is a visual field? How do the visual fields project to the brain?*

27. *Explain how binocular vision allows for depth perception.*

Clinical Focus   Eye Disorders

Myopia

Myopia (mī-ō′pē-ă), or nearsightedness, is the ability to see close objects clearly, but distant objects appear blurry. Myopia is a defect of the eye in which the focusing system, the cornea and lens, is optically too powerful, or the eyeball is too long (axial myopia). As a result, the focal point is too near the lens, and the image is focused in front of the retina (figure B*a*).

Myopia is corrected by a concave lens that counters the refractive power of the eye. Concave lenses cause the light rays coming to the eye to diverge and are therefore called "minus" lenses (figure B*b*).

Another technique for correcting myopia is **radial keratotomy** (ker′ă-tot′ō-mē), which consists of making a series of four to eight radiating cuts in the cornea. The cuts are intended to slightly weaken the dome of the cornea so that it becomes more flattened and eliminates the myopia. One problem with the technique is that it is difficult to predict exactly how much flattening will occur. In one study of 400 patients

5 years after the surgery, 55% had normal vision, 28% were still somewhat myopic, and 17% had become hyperopic. Another problem is that some patients are bothered by glare following radial keratotomy because the slits apparently don't heal evenly.

An alternative procedure being investigated is **laser corneal sculpturing,** in which a thin portion of the cornea is etched away to make the cornea less convex. The advantage of this procedure is that the results can be more accurately predicted than those from radial keratotomy.

Hyperopia

Hyperopia (hī-per-ō′pē-ă), or farsightedness, is the ability to see distant objects clearly, but close objects appear blurry. Hyperopia is a disorder in which the cornea and lens system is optically too weak or the eyeball is too short. The image is focused behind the retina (figure B*c*).

Hyperopia can be corrected by convex lenses that cause light rays to converge as

they approach the eye (figure B*d*). Such lenses are called "plus" lenses.

Presbyopia

Presbyopia (prez-bē-ō′pē-ă) is the normal, presently unavoidable, degeneration of the accommodation power of the eye that occurs as a consequence of aging. It occurs because the lens becomes sclerotic and less flexible. The eye is presbyopic when the near point of vision has increased beyond 9 inches. The average age for onset of presbyopia is the midforties. Avid readers or people engaged in fine, close work may develop the symptoms earlier.

Presbyopia can be corrected by the use of "reading glasses" that are worn only for close work and are removed when the person wants to see at a distance. It's sometimes annoying to keep removing and replacing glasses because reading glasses hamper vision of only a few feet away. This problem may be corrected by the use of half glasses, or by **bifocals,** which have a different lens in the top and the bottom.

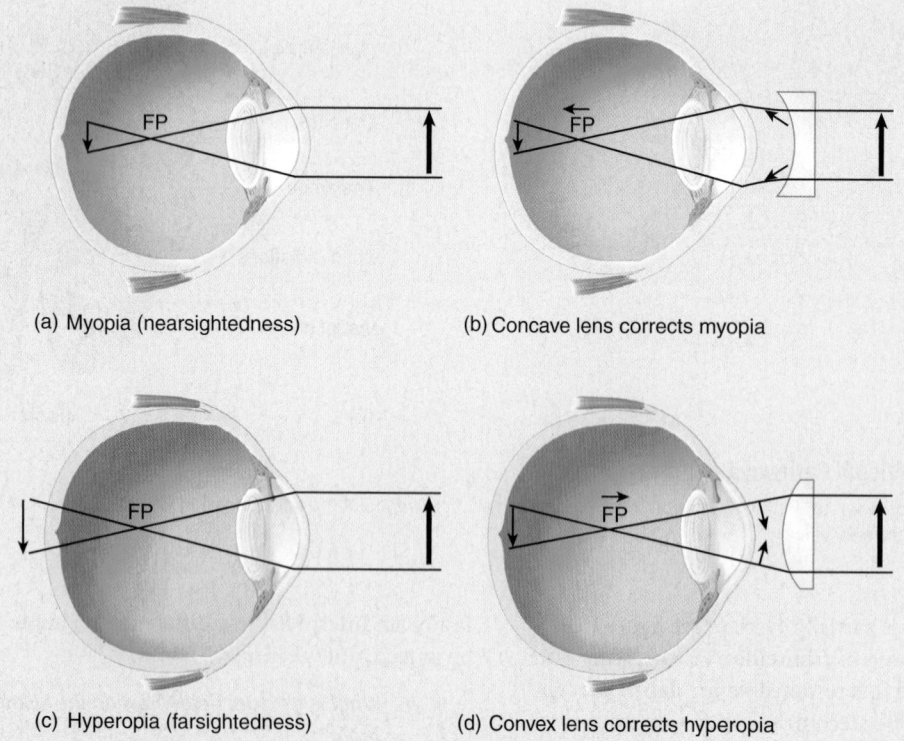

(a) Myopia (nearsightedness)   (b) Concave lens corrects myopia

(c) Hyperopia (farsightedness)   (d) Convex lens corrects hyperopia

Figure B   Visual Disorders and Their Correction by Various Lenses

FP is the focal point. (*a*) Myopia (nearsightedness). (*b*) Correction of myopia with a concave lens. (*c*) Hyperopia (farsightedness). (*d*) Correction of hyperopia with a convex lens.

Astigmatism

Astigmatism (ă-stig′mă-tizm) is a type of refractive error in which the quality of focus is affected. If the cornea or lens is not uniformly curved, the light rays don't focus at a single point but fall as a blurred circle. Regular astigmatism can be corrected by glasses that are formed with the opposite curvature gradation. Irregular astigmatism is a situation in which the abnormal form of the cornea fits no specific pattern and is very difficult to correct with glasses.

Strabismus

Strabismus (stra-biz′mŭs) is a lack of parallelism of light paths through the eyes. Strabismus can involve one eye or both eyes, and the eyes may turn in (convergent) or out (divergent). In **concomitant strabismus,** the most common congenital type, the angle between visual axes remains constant, regardless of the direction of the gaze. In **noncomitant strabismus,** the angle varies, depending on the direction of the gaze, and deviates as the gaze changes.

In some cases, the image that appears on the retina of one eye may be considerably different from that appearing on the other eye. This problem is called **diplopia** (di-plō′pē-ă, -double vision) and is often the result of weak or abnormal eye muscles.

Retinal Detachment

Retinal detachment is a relatively common problem that can result in complete blindness. The integrity of the retina depends on the vitreous humor, which keeps the retina pushed against the other tunics of the eye. If a hole or tear occurs in the retina, fluid may accumulate between the sensory and pigmented retina, thereby separating them. This separation, or detachment, may continue until the sensory retina has become totally detached from the pigmented retina and folded into a funnel-like form around the optic nerve. When the sensory retina becomes separated from its nutrient supply in the choroid, it degenerates, and blindness follows. Causes of retinal detachment include a severe blow to the eye or head; a shrinking of the vitreous humor, which may occur with aging; or diabetes. The space between the sensory and pigmented retina, called the subretinal space, is also important in keeping the retina from detaching, as well as in maintaining the health of the retina. The space contains a gummy substance that glues the sensory retina to the pigmented retina.

Color Blindness

Color blindness results from the dysfunction of one or more of the three photopigments involved in color vision. If one pigment is dysfunctional and the other two are functional, the condition is called **dichromatism.** An example of dichromatism is red-green color blindness (figure C).

The genes for the red and green photopigments are arranged in tandem on the X chromosome, which explains why color blindness is over eight times more common in males than in females (see chapter 29).

Six exons exist for each gene. The red and green genes are 96%–98% identical and, as a result, the exons may be shuffled to form hybrid genes in some people. Some of the hybrid genes produce proteins with nearly normal function, but others do not. Exon 5 is the most critical for determining normal red-green function. If the fifth exon from a green gene replaces a red pigment gene that has the fifth exon, the protein made from the gene responds to wavelengths more toward the green pigment range. The person has a red perception deficiency and is not able to distinguish between red and green. If the fifth exon from a red gene replaces a green pigment gene that has the fifth exon, the protein made from the gene responds to wavelengths more toward the red pigment range. The person has a green perception deficiency and is also not able to distinguish between red and green.

Apparently only about 3 of the over 360 amino acids in the color opsin proteins (those at positions 180 in exon 3 and those at 277 and 285 in exon 5) are key to determining their wavelength absorption characteristics. If those amino acids are altered by hydroxylation, the absorption shifts toward the red end of the spectrum. If they are not hydroxylated, the absorption shifts toward the green end.

Night Blindness

Everyone sees less clearly in the dark than in the light. A person with **night blindness,** however, may not see well enough in a dimly lit environment to function adequately. **Progressive night blindness** results from general retinal degeneration. **Stationary night blindness** results from nonprogressive abnormal rod function. Temporary night blindness can result from a vitamin A deficiency.

Patients with night blindness can now be helped with special electronic optical devices. These include monocular pocket scopes and binocular goggles that electronically amplify light.

Continued

(a) (b)

Figure C Color Blindness Charts

(*a*) A person with normal color vision can see the number 74, whereas a person with red-green color blindness sees the number 21. (*b*) A person with normal color vision can see the number 42. A person with red color blindness sees the number 2, and a person with green color blindness sees the number 4.

Continued

Glaucoma

Glaucoma (figure D*a*) is a disease of the eye involving increased intraocular pressure caused by a buildup of aqueous humor. It usually results from blockage of the aqueous veins or the canal of Schlemm, restricting drainage of the aqueous humor, or from over-production of aqueous humor. If untreated, glaucoma can lead to retinal, optic disc, and optic nerve damage. The damage results from the increased intraocular pressure, which is sufficient to close off the blood vessels, causing starvation and death of the retinal cells.

Glaucoma is one of the leading causes of blindness in the United States, affecting 2% of people over age 35, and accounting for 15% of all blindness. Fifty thousand people in the United States are blind as the result of glaucoma, and it occurs three times more often in black people than in white people. The symptoms include a slow closing in of the field of vision. No pain or redness occurs, nor do light flashes occur.

Glaucoma has a strong hereditary tendency but may develop after surgery or with the use of certain eyedrops containing cortisone. Everyone older than 40 should be checked every 2–3 years for glaucoma; those older than 40 who have relatives with glaucoma should have an annual checkup. During a checkup, the field of vision and the optic nerve are examined. Ocular pressures can also be measured. Glaucoma is usually treated with eyedrops, which do not cure the problem but keep it from advancing. In some cases, laser or conventional surgery may be used.

Cataract

Cataract (figure D*b*) is a clouding of the lens resulting from a buildup of proteins. The lens relies on the aqueous humor for its nutrition. Any loss of this nutrient source leads to degeneration of the lens and, ultimately, opacity of the lens (i.e., a cataract). A cataract may occur with advancing age, infection, or trauma.

A certain amount of lens clouding occurs in 65% of patients older than 50 and 95% of patients older than 65. The decision of whether to remove the cataract depends on the extent to which light passage is blocked. Over 400,000 cataracts are removed in the United States each year. Surgery to remove a cataract is actually the removal of the lens. The posterior portion of the lens capsule is left intact. Although the cornea can still accomplish light convergence, with the lens gone, the rays cannot be focused as well, and an artificial lens must be supplied to help accomplish focusing. In most cases, an artificial lens is implanted into the remaining portion of the lens capsule at the time that the natural lens is removed. The implanted lens helps to restore normal vision, but glasses may be required for near vision.

Macular Degeneration

Macular degeneration (figure D*c*) is very common in older people. It does not cause total blindness but results in the loss of acute vision. This degeneration has a variety of causes, including hereditary disorders, infections, trauma, tumor, or most often, poorly understood degeneration associated with aging. Because no satisfactory medical treatment has been developed, optical aids, such as magnifying glasses, are used to improve visual function.

Diabetes

Loss of visual function is one of the most common consequences of diabetes because a major complication of the disease is dysfunction of the peripheral circulation. Defective circulation to the eye may result in retinal degeneration or detachment. Diabetic retinal degeneration (figure D*d*) is one of the leading causes of blindness in the United States.

Infections

Trachoma (tră-kō′mă) is the leading cause of blindness worldwide. It is caused by an intracellular microbial infection *(Chlamydia trachomatis)* of the conjunctiva (conjunctivitis). As inflammation resulting from the disease progresses and spreads, the eyelids may turn inward, causing the eyelashes to abrade the cornea, resulting in scar tissue formation in the cornea. The bacteria are spread from one eye to another eye by towels, fingers, and other objects. Five hundred million cases of trachoma exist in the world, and 7 million people are blind or visually impaired as a result of it.

Neonatal gonorrheal ophthalmia (of-thal′-mē-ă) is a bacterial infection *(Neisseria gonorrhoeae)* of the eye that causes blindness. If the mother has gonorrhea, which is a sexually transmitted disease of the reproductive tract, the bacteria can infect the newborn during delivery. The disease can be prevented by treating the infant's eyes with silver nitrate, tetracycline, or erythromycin drops.

(a)

(b)

(c)

(d)

Figure D **Defects in Vision**

Visual images as seen with various defects in vision. (*a*) Glaucoma. (*b*) Cataract. (*c*) Macular degeneration. (*d*) Diabetic retinopathy.

Hearing and Balance

The organs of hearing and balance are divided into three parts: external, middle, and inner ears (figure 15.23). The external and middle ears are involved in hearing only, whereas the inner ear functions in both hearing and balance.

The **external ear** includes the **auricle** (aw′ri-kl; ear) and the **external acoustic meatus** (mē-ā′tŭs; the passageway from the outside to the eardrum). The external ear terminates medially at the **eardrum,** or **tympanic** (tim-pan′ik) **membrane.** The **middle ear** is an air-filled space within the petrous portion of the temporal bone, which contains the **auditory ossicles.** The **inner ear** contains the sensory organs for hearing and balance. It consists of interconnecting fluid-filled tunnels and chambers within the petrous portion of the temporal bone.

Auditory Structures and Their Functions

External Ear

The **auricle,** or **pinna** (pin′ă), is the fleshy part of the external ear on the outside of the head and consists primarily of elastic cartilage covered with skin (figure 15.24). Its shape helps to collect sound

waves and direct them toward the external acoustic meatus. The external acoustic meatus is lined with **hairs** and **ceruminous** (sĕ-roo′mi-nŭs) **glands,** which produce **cerumen,** a modified sebum commonly called earwax. The hairs and cerumen help prevent foreign objects from reaching the delicate eardrum. Overproduction of cerumen, however, may block the meatus.

Figure 15.24   **Structures of the Auricle (the Right Ear)**

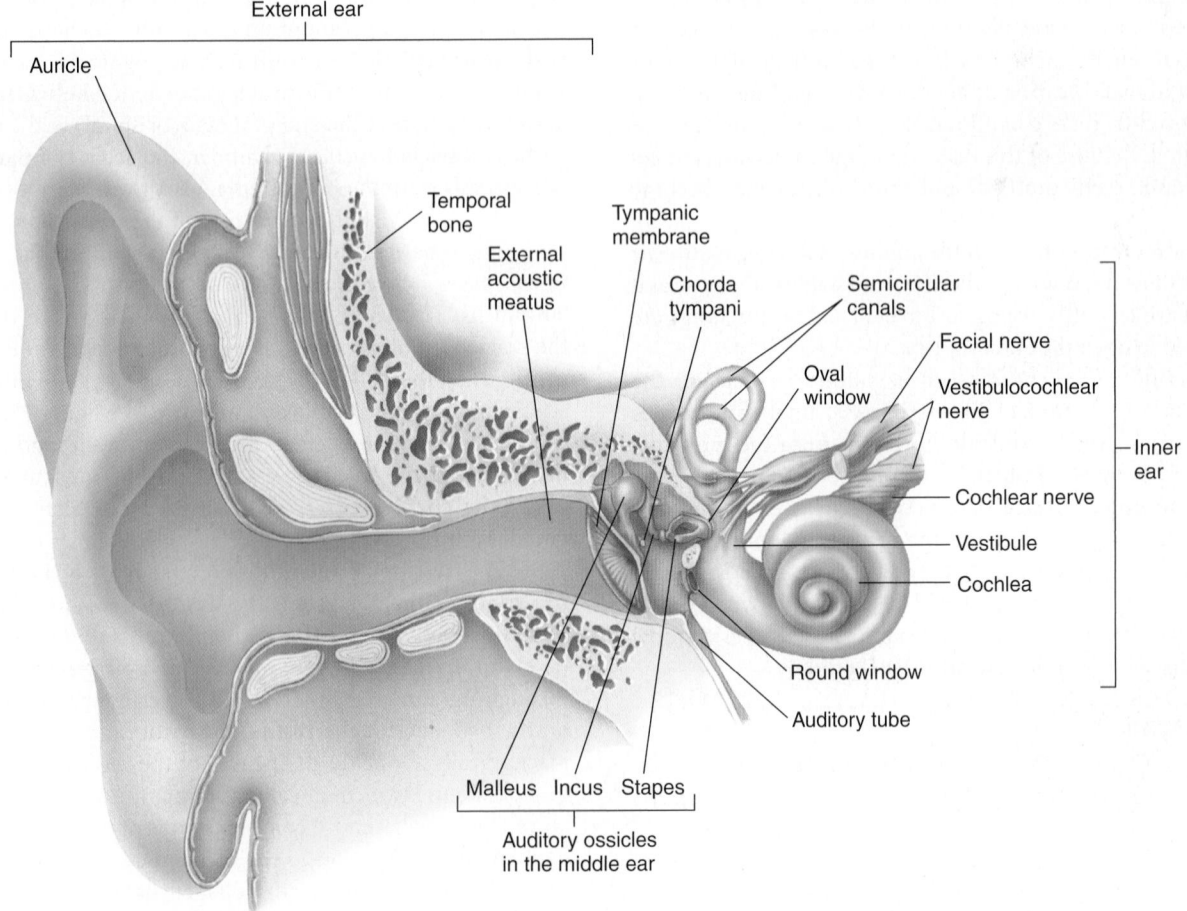

Frontal section

Figure 15.23   **External, Middle, and Inner Ear**

The tympanic membrane, or eardrum, is a thin, semitransparent, nearly oval, three-layered membrane that separates the external ear from the middle ear. It consists of a low, simple cuboidal epithelium on the inner surface and a thin stratified squamous epithelium on the outer surface, with a layer of connective tissue between. Sound waves reaching the tympanic membrane through the external acoustic meatus cause it to vibrate.

Tympanic Membrane Rupture

Rupture of the tympanic membrane results in deafness. Foreign objects thrust into the ear, pressure, or infections of the middle ear can rupture the tympanic membrane. Sufficient differential pressure between the middle ear and the outside air can also cause rupture of the tympanic membrane. This can occur in flyers, divers, or individuals who are hit on the side of the head by an open hand.

Middle Ear

Medial to the tympanic membrane is the air-filled cavity of the middle ear (see figure 15.23). Two covered openings, the round and oval windows, on the medial side of the middle ear separate it from the inner ear. Two openings provide air passages from the middle ear. One passage opens into the **mastoid air cells** in the mastoid process of the temporal bone. The other passageway, the **auditory, pharyngotympanic,** or **eustachian** (ū-stā′shŭn) **tube,** opens into the pharynx and equalizes air pressure between the outside air and the middle ear cavity. Unequal pressure between the middle ear and the outside environment can distort the eardrum, dampen its vibrations, and make hearing difficult. Distortion of the eardrum, which occurs under these conditions, also stimulates pain fibers associated with it. Because of this distortion, when a person changes altitude, sounds seem muffled, and the eardrum may become painful. These symptoms can be relieved by opening the auditory tube to allow air to pass through the auditory tube to equalize air pressure. Swallowing, yawning, chewing, and holding the nose and mouth shut while gently trying to force air out of the lungs are methods used to open the auditory tube.

The middle ear contains three auditory ossicles: the **malleus** (mal′ē-ŭs; hammer), **incus** (ing′kŭs; anvil), and **stapes** (stā′pēz; stirrup), which transmit vibrations from the tympanic membrane to the **oval window.** The handle of the malleus is attached to the inner surface of the tympanic membrane, and vibration of the membrane causes the malleus to vibrate as well. The head of the malleus is attached by a very small synovial joint to the incus, which in turn is attached by a small synovial joint to the stapes. The foot plate of the stapes fits into the oval window and is held in place by a flexible **annular ligament.**

Chorda Tympani

A structure that students might be somewhat surprised to find in the middle ear is the **chorda tympani.** It's a branch of the facial nerve carrying taste impulses from the anterior two-thirds of the tongue. It crosses over the inner surface of the tympanic membrane (see figures 15.23 and 15.32). The chorda tympani has nothing to do with hearing but is just passing through. This nerve can be damaged, however, during ear surgery or by a middle ear infection, resulting in loss of taste sensation from the anterior two-thirds of the tongue on the side innervated by that nerve.

Inner Ear

The tunnels and chambers inside the temporal bone are called the **bony labyrinth** (lab′i-rinth; a maze; figure 15.25). Because the bony labyrinth consists of tunnels within the bone, it cannot easily be removed and examined separately. The bony labyrinth is lined with endosteum, and when the inner ear is shown separately (figure 15.26a), the endosteum is what is depicted. Inside the bony labyrinth is a similarly shaped but smaller set of membranous tunnels and chambers called the **membranous labyrinth.** The inner surface of the endosteum and the outer surface of the membranous labyrinth are covered with a very thin layer of cells called the **perilymphatic cells** (see figure 15.26b). The membranous labyrinth is filled with a clear fluid called **endolymph,** and the space between the membranous and bony labyrinth is filled with a fluid called **perilymph.** Perilymph is very similar to cerebrospinal fluid, but endolymph has a high concentration of potassium and a low concentration of sodium, which is opposite from perilymph and cerebrospinal fluid.

The bony labyrinth is divided into three regions: cochlea, vestibule, and semicircular canals. The **vestibule** (ves′ti-bool) and **semicircular canals** are involved primarily in balance, and the **cochlea** (kok′lē-ă) is involved in hearing. The membranous labyrinth of the cochlea is divided into three parts: the scala vestibuli, the scala tympani, and the cochlear duct.

The oval window communicates with the vestibule of the inner ear, which in turn communicates with a cochlear chamber, the **scala vestibuli** (skā′lă ves-tib′ū-lē; see figure 15.26a). The scala vestibuli extends from the oval window to the **helicotrema** (hel′i-kō-trē′mă; a hole at the end of a helix or spiral) at the apex of the cochlea; a second cochlear chamber, the **scala tympani** (tim′pă-nē), extends from the helicotrema, back from the apex, parallel to the scala vestibuli, to the membrane of the **round window.**

The scala vestibuli and the scala tympani are the perilymph-filled spaces between the walls of the bony and membranous labyrinths. The wall of the membranous labyrinth that bounds the scala vestibuli is called the **vestibular membrane** (Reissner's membrane); the wall of the membranous labyrinth bordering the scala tympani is the **basilar membrane** (see figure 15.26b and c). The space between the vestibular membrane and the basilar membrane is the interior of the membranous labyrinth and is called the **cochlear duct** or **scala media,** which is filled with endolymph.

The vestibular membrane consists of a double layer of squamous epithelium and is the simplest region of the membranous labyrinth. The vestibular membrane is so thin that it has little or no mechanical effect on the transmission of sound waves through the inner ear; therefore, the perilymph and endolymph on the two sides of the vestibular membrane can be thought of mechanically as one fluid. The role of the vestibular membrane is to separate the two chemically different fluids. The basilar membrane is somewhat more complex and is of much greater physiologic interest in relation to the mechanics of hearing. It consists of an acellular portion with collagen fibers, ground substance, and sparsely dispersed elastic fibers and a cellular part with a thin layer of vascular connective tissue that is overlaid with simple squamous epithelium.

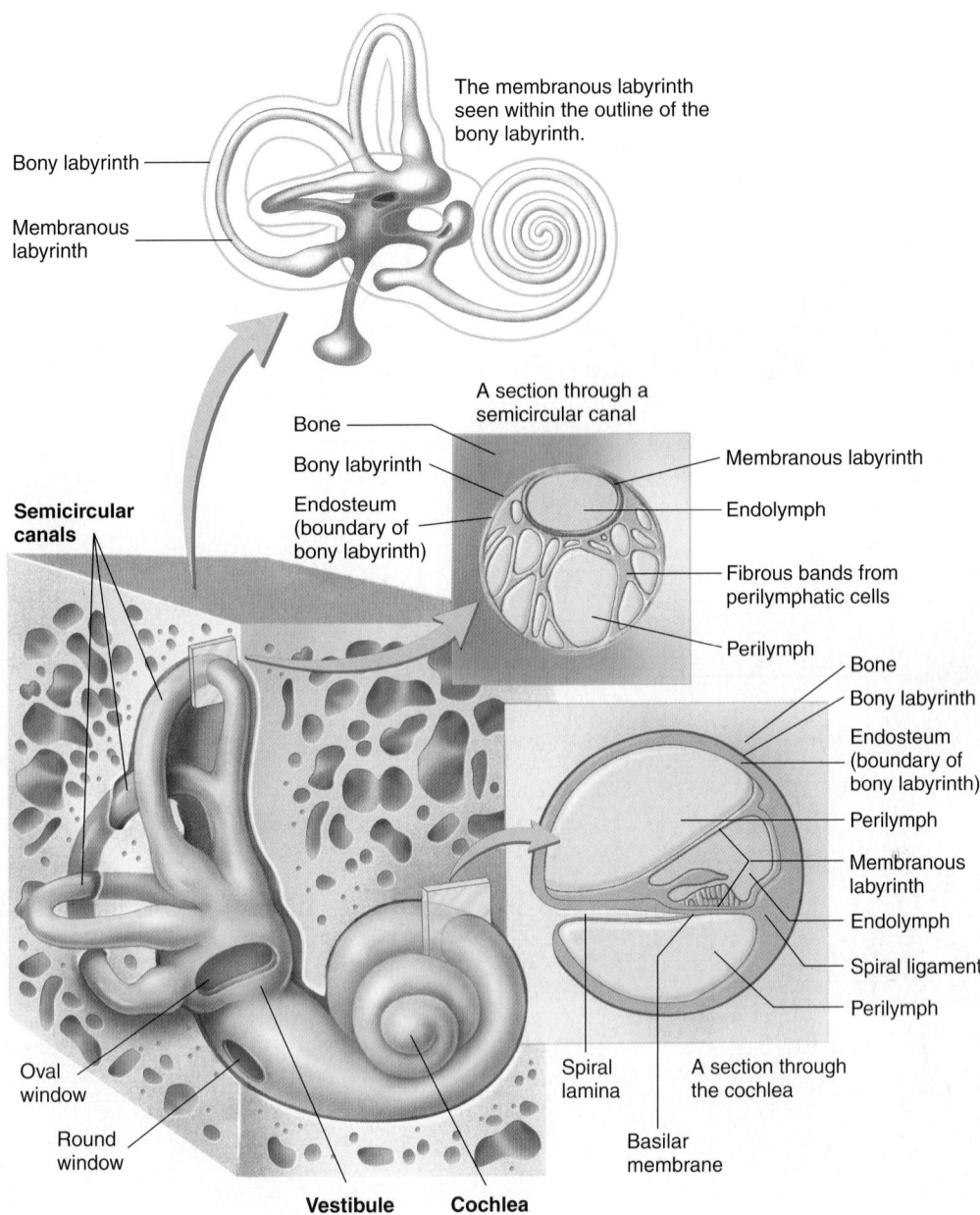

The membranous labyrinth seen within the outline of the bony labyrinth.

Bony labyrinth

Membranous labyrinth

A section through a semicircular canal

Bone

Bony labyrinth

Endosteum (boundary of bony labyrinth)

Membranous labyrinth

Endolymph

Fibrous bands from perilymphatic cells

Perilymph

Semicircular canals

Bone

Bony labyrinth

Endosteum (boundary of bony labyrinth)

Perilymph

Membranous labyrinth

Endolymph

Spiral ligament

Perilymph

Oval window

Round window

Spiral lamina

A section through the cochlea

Basilar membrane

Vestibule   **Cochlea**

Figure 15.25   **The Inner Ear: Bony and Membranous Labyrinths**

The cross sections are taken through a semicircular canal and the cochlea to show the relationship between the bony and membranous labyrinths.

The basilar membrane is attached at one side to the bony **spiral lamina,** which projects from the sides of the **modiolus** (mō′dī′ō-lus), the bony core of the cochlea, like the threads of a screw. At the other side, the basilar membrane is attached to the lateral wall of the bony labyrinth by the **spiral ligament,** a local thickening of the endosteum. The distance between the spiral lamina and the spiral ligament (i.e., the width of the basilar membrane) increases from 0.04 mm near the oval window to 0.5 mm near the helicotrema (see figure 15.33). The collagen fibers of the basilar membrane are oriented across the membrane between the spiral lamina and the spiral ligament, somewhat like the strings of a piano. The collagen fibers near the oval window are both shorter and thicker than those near the

helicotrema. The diameter of the collagen fibers in the membrane decreases as the basilar membrane widens. As a result, the basilar membrane near the oval window is short and stiff, and responds to high-frequency vibrations, whereas that part near the helicotrema is wide and limber and responds to low-frequency vibrations.

The cells inside the cochlear duct are highly modified to form a structure called the **spiral organ,** or the **organ of Corti** (see figure 15.26b and c). The spiral organ contains supporting epithelial cells and specialized sensory cells called **hair cells,** which have hairlike projections at their apical ends. In children, these projections consist of one cilium (kinocilium) and about 80 very long microvilli, often referred to as **stereocilia,** but in adults the cilium is absent from

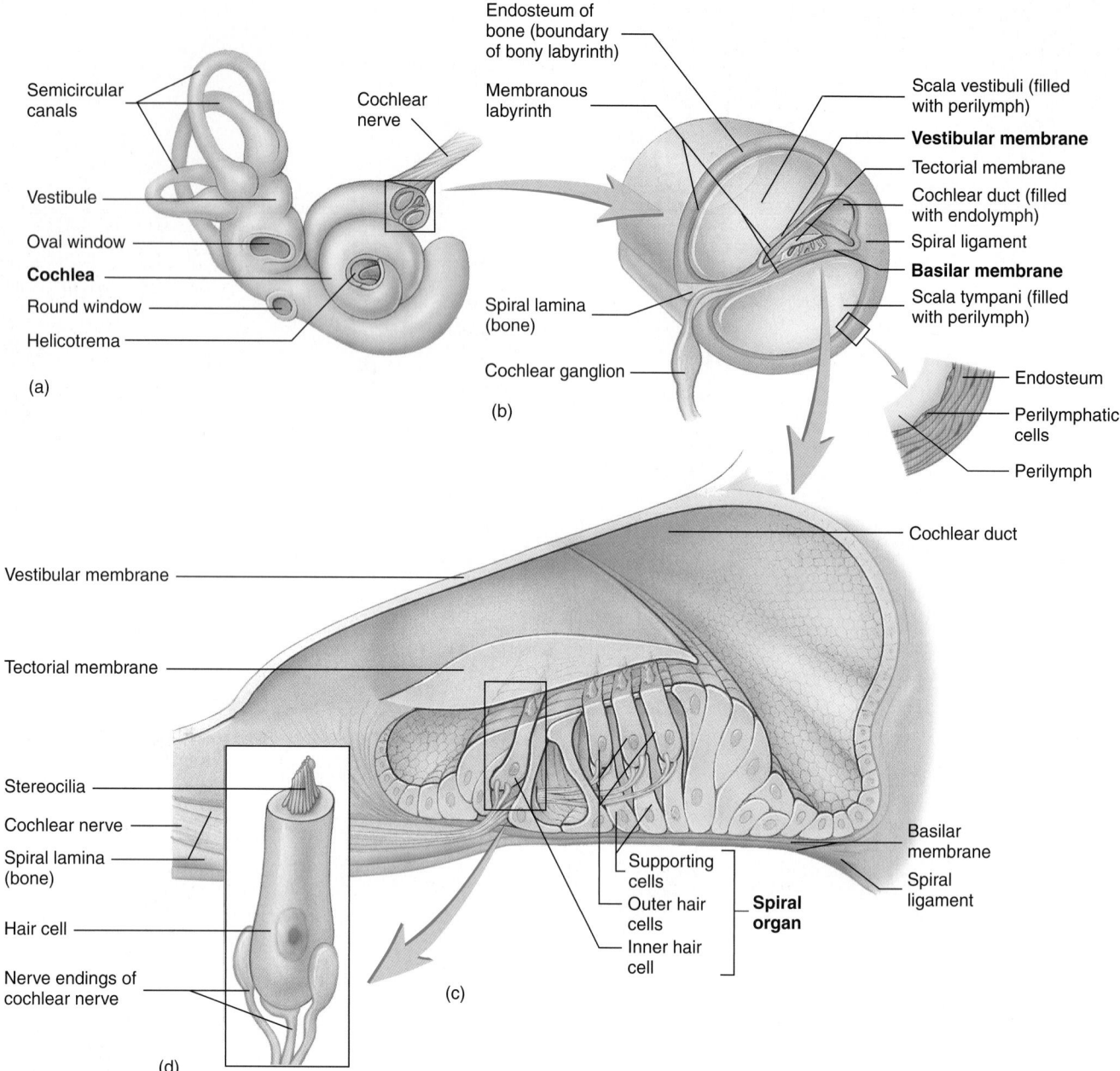

Figure 15.26   Structure of the Cochlea

(*a*) **The inner ear.** The outer surface (*gray*) is the endosteum lining the inner surface of the bony labyrinth. (*b*) **A cross section of the cochlea.** The outer layer is the endosteum lining the inner surface of the bony labyrinth. The membranous labyrinth is very small in the cochlea and consists of the vestibular and basilar membranes. The space between the membranous and bony labyrinth consists of two parallel tunnels: the scala vestibuli and scala tympani. (*c*) An enlarged section of the cochlear duct (membranous labyrinth). (*d*) A greatly enlarged individual sensory hair cell.

most hair cells (see figures 15.26*d* and 15.27). The hair cells are arranged in four long rows extending the length of the cochlear duct. Each row contains 3500–4000 hair cells. The inner row consists of hair cells, called **inner hair cells,** which are the hair cells primarily responsible for hearing. The outer three rows contain **outer hair cells.** The outer hair cells are involved in regulating the tension of the basilar membrane and are separated from the inner hair cells by a gap in the basilar membrane (see figure 15.26). The stereocilia of one inner hair cell form a conical group called a **hair bundle** (figure 15.27*a*). The length of each stereocilium within a hair bundle increases gradu-

ally from one side of the hair cell to the other. The stereocilia of an outer hair cell are arranged in a curved line (figure 15.27*b*). The tips of the longest stereocilia of the outer hair cells are embedded within an acellular gelatinous shelf called the **tectorial** (tek-tōr′ē-ăl) **membrane,** which is attached to the spiral lamina.

A **tip link** connects the tip of each stereocilium in a hair bundle to the side of the next longer stereocilium (figure 15.28). Each tip link is a **gating spring,** a pair of microtubule strands that attaches to the gate of a gated K^+ channel. The gated K^+ channels of hair cells open mechanically. As the stereocilia bend, the gating

(a)

(b)

Figure 15.27   Scanning Electron Micrograph of Cochlear Hair Cell Stereocilia

(*a*) The hair bundle of one inner hair cell. (*b*) The stereocilia of three outer hair cells.

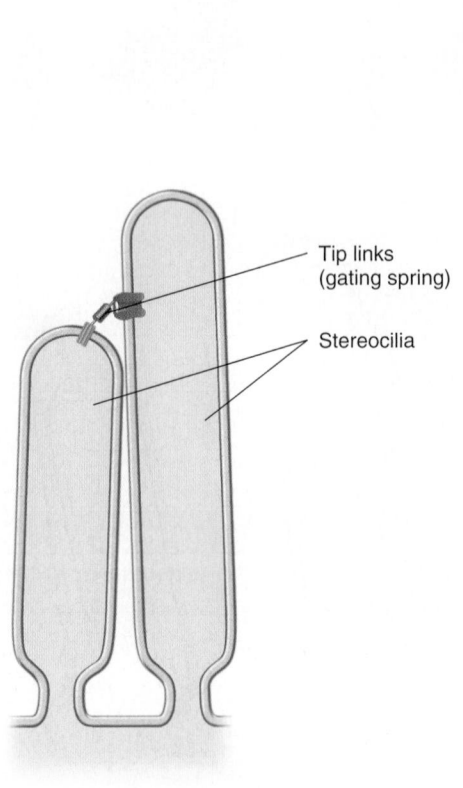

(a)

Figure 15.28   Tip Links or Gating Springs on Inner Hair Cell
Stereocilia

(*a*) Diagram of two stereocilia connected by one tip link. (*b*) Transmission
electron micrograph of three stereocilia and two tip links. (*c*) Scanning
electron micrograph of the tops of stereocilia showing tip links.

(b)

(c)

spring pulls the K$^+$ gate open (process figure 15.29). The response time for such a mechanism is very brief and functions much faster than for a gating mechanism involving the synthesis of intracellular chemical signals such as cAMP.

Hair cells have no axons, but the basilar regions of each hair cell are covered by synaptic terminals of sensory neurons. The cell bodies of afferent neurons are located within the cochlear modiolus and are grouped into a **cochlear,** or **spiral ganglion** (see figures 15.26*b* and 15.34). Afferent fibers of these neurons join to form the **cochlear nerve.** This nerve then joins the vestibular nerve to become the **vestibulocochlear nerve** (VIII), which traverses the internal acoustic meatus and enters the cranial cavity.

28. *Name the three regions of the ear, and list each region's parts.*
29. *Describe the relationship between the tympanic membrane, the ear ossicles, and the oval window of the ear.*
30. *What is the function of the external acoustic meatus and of the auditory tube?*
31. *Explain how the cochlear duct is divided into three compartments. What is found in each compartment?*
32. *Explain the differences between inner and outer hair cells.*
33. *What is the function of tip links or gating springs?*

1. In the unstimulated cell, the gating spring is relaxed and the K$^+$ channel gate is closed.

2. As the stereocilia are bent toward the taller stereocilium (positive stimulus), the gating spring stretches.

3. The force created by stretching of the gating spring opens the gate of the K$^+$ channel and K$^+$ enters the cell.

Process Figure 15.29   Action of the Gating Spring to Open a K$^+$ Channel When Two Stereocilia Bend

Auditory Function

Vibration of matter such as air, water, or a solid material creates sound. No sound occurs in a vacuum. When a person speaks, the vocal cords vibrate, causing the air passing out of the lungs to vibrate. The vibrations consist of bands of compressed air followed by bands of less compressed air (figure 15.30*a*). These vibrations are propagated through the air as sound waves, somewhat like ripples are propagated over the surface of water. **Volume,** or loudness, is a function of wave amplitude, or height, measured in decibels (figure 15.30*b*). The greater the amplitude, the louder is the sound. **Pitch** is a function of the wave frequency (i.e., the number of waves or cycles per second) measured in hertz (Hz) (figure 15.30*c*). The higher the frequency, the higher the pitch. The normal range of human hearing is 20–20,000 Hz and 0 or more decibels (db). Sounds louder than 125 db are painful to the ear.

Timbre (tam′br, tim′br) is the resonance quality or overtones of a sound. A smooth sigmoid curve is the image of a "pure" sound wave, but such a wave almost never exists in nature. The sounds made by musical instruments or the human voice are not smooth sigmoid curves but rather are rough, jagged curves formed by numerous, superimposed curves of various amplitudes and frequencies. The roughness of the curve accounts for the timbre. Timbre allows one to distinguish between, for example, an oboe and a French horn playing a note at the same pitch and volume. The steps involved in the mechanical part of hearing are illustrated in figure 15.31.

Figure 15.30 Sound Waves

(*a*) When something like a tuning fork or vocal cords vibrate, the movements of the object alternate between compressing the air and decompressing the air, or making the air less compressed, thus producing sound. Each sound wave consists of a region of compressed air between two regions of less compressed air (*blue bars*). The sigmoid waves correspond to the regions of more compressed air (peaks) and less compressed air (troughs). The *green shadowed area* represents the width of one cycle (distance between peaks). (*b*) Depicts low- and high-volume sound waves. Compare the relative lengths of the *arrows* indicating the wave height (amplitude). (*c*) Depicts lower and higher pitch sound. Compare the relative number of peaks (frequency) within a given time interval (*between arrows*).

1. Sound waves strike the tympanic membrane and cause it to vibrate.

2. Vibration of the tympanic membrane causes the three bones of the middle ear: the malleus, incus, and stapes to vibrate.

3. The foot plate of the stapes vibrates in the oval window.

4. Vibration of the foot plate causes the perilymph in the scala vestibuli to vibrate.

5. Vibration of the perilymph causes the vestibular membrane to vibrate, which causes vibrations in the endolymph.

6. Vibration of the endolymph causes displacement of the basilar membrane. Short waves (high pitch) cause displacement of the basilar membrane near the oval window, and longer waves (low pitch) cause displacement of the basilar membrane some distance from the oval window. Movement of the basilar membrane is detected in the hair cells of the spiral organ, which are attached to the basilar membrane. Vibrations of the perilymph in the scala vestibuli and of the basement membrane are transferred to the perilymph of the scala tympani.

7. Vibrations in the perilymph of the scala tympani are transferred to the round window, where they are dampened.

Process Figure 15.31   **Effect of Sound Waves on Cochlear Structures**

External Ear

The auricle collects sound waves that are then conducted through the external acoustic meatus toward the tympanic membrane. Sound waves travel relatively slowly in air, 332 m/s, and a significant time interval may elapse between the time a sound wave reaches one ear and the time that it reaches the other. The brain can interpret this interval to determine the direction from which a sound is coming.

Middle Ear

Sound waves strike the tympanic membrane and cause it to vibrate. This vibration causes vibration of the three ossicles of the middle ear, and by this mechanical linkage vibration is transferred to the oval window. More force is required to cause vibration in a liquid like the perilymph of the inner ear than is required in air; thus, the vibrations reaching the perilymph must be amplified as they cross the middle ear. The footplate of the stapes and its annular ligament, which occupy the oval window, are much smaller than the tympanic membrane. Because of this size difference, the mechanical force of vibration is amplified about 20-fold as it passes from the tympanic membrane, through the ossicles, and to the oval window. This amplification occurs because the area of the tympanic membrane is roughly 20 times that of the oval window.

Two small skeletal muscles are attached to the ear ossicles and reflexively dampen excessively loud sounds (figure 15.32). This **sound attenuation reflex** protects the delicate ear structures from damage by loud noises. The **tensor tympani** (ten′sŏr tim′păne̅) muscle is attached to the malleus and is innervated by the trigeminal nerve (V). The **stapedius** (stă-pē′dē-ŭs) muscle is attached to the stapes and is supplied by the facial nerve (VII). The sound attenuation reflex responds most effectively to low-frequency sounds and can reduce by a factor of 100 the energy reaching the oval window. The facial nerve and stapedius are primarily involved in the sound attenuation reflex. The trigeminal nerve and tensor tympani are only stimulated by extremely loud noise. The reflex is too slow to prevent damage from a sudden noise, such as a gunshot, and it cannot function effectively for longer than about 10 minutes, in response to prolonged noise.

P R E D I C T   **9**

What effect does facial nerve damage have on hearing?

Inner Ear

As the stapes vibrates, it produces waves in the perilymph of the scala vestibuli (see figure 15.31). Vibrations of the perilymph are

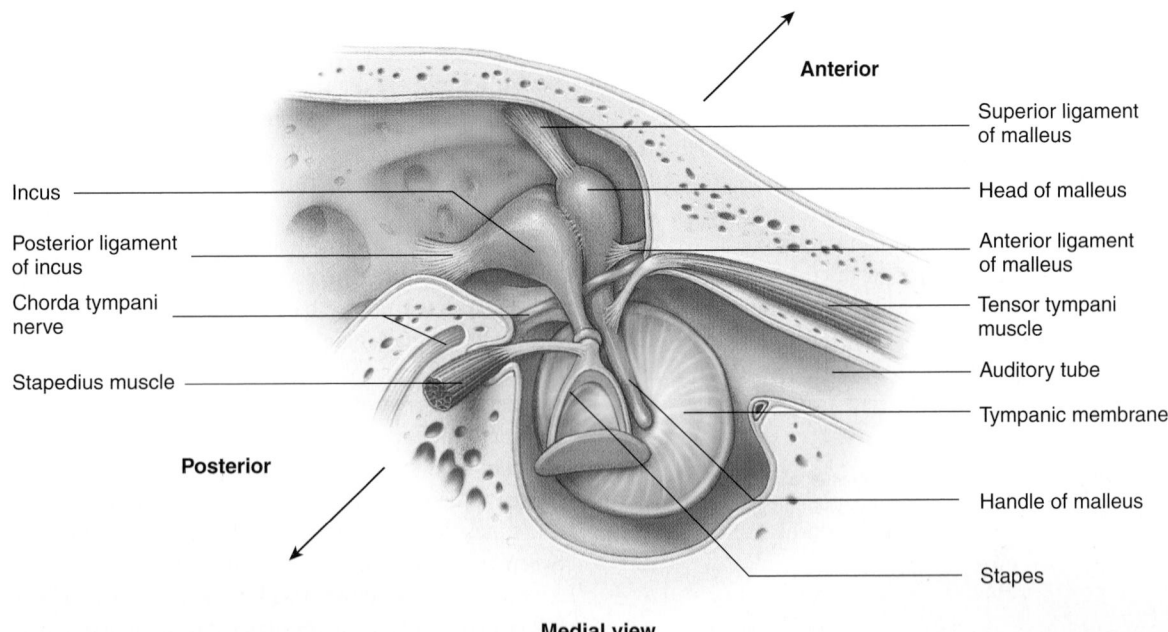

Medial view

Figure 15.32   Muscles of the Middle Ear

Medial view of the middle ear (as though viewed from the inner ear), showing the three ear ossicles with their ligaments and the two muscles of the middle ear: the tensor tympani and the stapedius.

transmitted through the thin vestibular membrane and cause simultaneous vibrations of the endolymph. The mechanical effect is as though the perilymph and endolymph were a single fluid. Vibration of the endolymph causes distortion of the basilar membrane. Waves in the perilymph of the scala vestibuli are transmitted also through the helicotrema and into the scala tympani. Because the helicotrema is very small, however, this transmitted vibration is probably of little consequence. Distortions of the basilar membrane, together with weaker waves coming through the helicotrema, cause waves in the scala tympani perilymph and ultimately result in vibration of the membrane of the round window. Vibration of the round window membrane is important to hearing because it acts as a mechanical release for waves from within the cochlea. If this window were solid, it would reflect the waves, which would interfere with and dampen later waves. The round window also allows relief of pressure in the perilymph because fluid is not compressible, thereby preventing compression damage to the spiral organ.

The distortion of the basilar membrane is most important to hearing. As this membrane distorts, the hair cells resting on the basilar membrane move relative to the tectorial membrane, which remains stationary. The inner hair cell microvilli are bent as they move against the tectorial membrane.

The apical portion of each hair cell is surrounded by endolymph and the basal portion of the cell is surrounded by perilymph. Endolymph has a high K^+ concentration similar to the intracellular K^+ concentration of hair cells. Perilymph has a low concentration of K^+ similar to that of other extracellular fluid. The intracellular charge of hair cells compared to perilymph is −60 mv. The charge of the endolymph is +80 mv compared to the perilymph. This

charge difference is called the **endocochlear potential.** Consequently, the intracellular charge of hair cells compared to endolymph is −140 mv, which is a large charge difference. Consequently, when K^+ channels open, K^+ flow into the hair cells because they are attracted to the negative charge inside of the hair cell even though the intracellular concentration of K^+ is about the same as in the endolymph. The movement of the K^+ cause depolarization of the hair cells. This is a rare instance where an increase in K^+ permeability of the plasma membrane of a cell results in depolarization.

In the unstimulated hair cell, approximately 15% of the gated K^+ channels are open and the resting membrane potential of the cell is approximately −60 mV. If the hair bundle is bent toward the shortest stereocilium (negative stimulus) the gating springs attached to the K^+ channel gates slacken, allowing the open K^+ channels to close and the cell hyperpolarizes. If the hair bundle is bent toward the longest stereocilium (positive stimulus), gating springs pull additional K^+ channel gates open, and K^+ rush into the cell. The influx of K^+ into the hair cell causes a slight depolarization of the cell. This depolarization causes voltage-gated Ca^{2+} channels to open. Ca^{2+} rush into the cell, causing a further depolarization. The cell depolarizes by a total of about 10 mV. Depolarization of hair cells results in the increased release of neurotransmitters which increases the action potential frequency in the afferent neurons. Hyperpolarization decreases neurotransmitter release and decreases action potential frequency in afferent neurons. Depolarization also opens voltage-gated K^+ channels in the basal portion of the hair cell. K^+ tend to leave the cell causing the cell to repolarize.

The neurotransmitter released by the inner hair cells is apparently glutamate, but other neurotransmitters may also be involved.

The release of neurotransmitters from the inner hair cells induces action potentials in the cochlear neurons that synapse on the hair cells. The cell bodies of those neurons are located in the cochlear ganglion.

The part of the basilar membrane that distorts as a result of endolymph vibration depends on the pitch of the sound that created the vibration and, as a result, on the vibration frequency within the endolymph. The width of the basilar membrane and the length and diameter of the collagen fibers stretching across the membrane at each level along the cochlear duct determine the location of the optimum amount of basilar membrane vibration produced by a given pitch (figure 15.33). Higher-pitched tones cause optimal vibration near the base, and lower-pitched tones cause optimal vibration near the apex of the basilar membrane. As the basilar membrane vibrates, hair cells along a large part of the basilar membrane are stimulated. In areas of minimum vibration, the amount of stimulation may not reach threshold. In other areas, a low frequency of afferent action potentials may be transmitted, whereas in the optimally vibrating regions of the basilar membrane, a high frequency of action potentials is initiated.

There are approximately twice as many nerve cells in the cochlear ganglion as there are hair cells. Over 90% of the afferent axons synapse with inner hair cells, ten to thirty axons per hair cell. Only a few, small-diameter afferent axons synapse with the three rows of outer hair cells. The outer cells, however, receive input from efferent axons. Action potentials from those efferent axons stimulate contraction of actin filaments within the hair cells, causing them to shorten. This adjustment in the height of the outer hair cells, attached to both the basilar membrane and the tectorial membrane, fine tunes the tension of the basilar membrane and the distance between the basilar membrane and tectorial membrane. Additional sensitivity can be adjusted within the inner hair cells. The mechanically gated K^+ channels are attached to actin filaments inside the cell. Those filaments can move the K^+ channels along the cell membrane, tightening or loosening the gating spring. By this means, hair cells are tuned to very specific frequencies. Likewise, the response of an inner hair cell to stimulation is graded and can increase up to a point of saturation (when all the K^+ channels are open maximally). Inner hair cells are also very sensitive. A 100 nm (1 degree) deflection of the stereocilia results in a response that is 90% of maximum.

Loud Noises and Hearing Loss

Prolonged or frequent exposure to excessively loud noises can cause degeneration of the spiral organ at the base of the cochlea, resulting in high-frequency deafness. The actual amount of damage can vary greatly from person to person. High-frequency loss can cause a person to miss hearing consonants in a noisy setting. Loud music, amplified to 120 db, can impair hearing. The defects may not be detectable on routine diagnosis, but they include decreased sensitivity to sound in specific narrow frequency ranges and a decreased ability to discriminate between two pitches. Loud music, however, is not as harmful as is the sound of a nearby gunshot, which is a sudden sound occurring at 140 db. The sound is too sudden for the attenuation reflex to protect the inner ear structures, and the intensity is great enough to cause auditory damage. In fact, gunshot noise is the most common recreational cause of serious hearing loss.

Afferent action potentials conducted by cochlear nerve fibers from all along the spiral organ terminate in the **superior olivary nucleus** in the medulla oblongata (figure 15.34; see chapter 13). These action potentials are compared to one another, and the strongest action potential, corresponding to the area of maximum basilar membrane vibration, is taken as standard. Efferent action potentials then are sent from the superior olivary nucleus back to the spiral organ to all regions where the maximum vibration did not occur. These action potentials inhibit the hair cells from initiating additional action potentials in the sensory neurons. Thus, only action potentials from regions of maximum vibration are received by the cortex, where they become consciously perceived.

By this process, tones are localized along the cochlea. As a result of this localization, neurons along a given portion of the

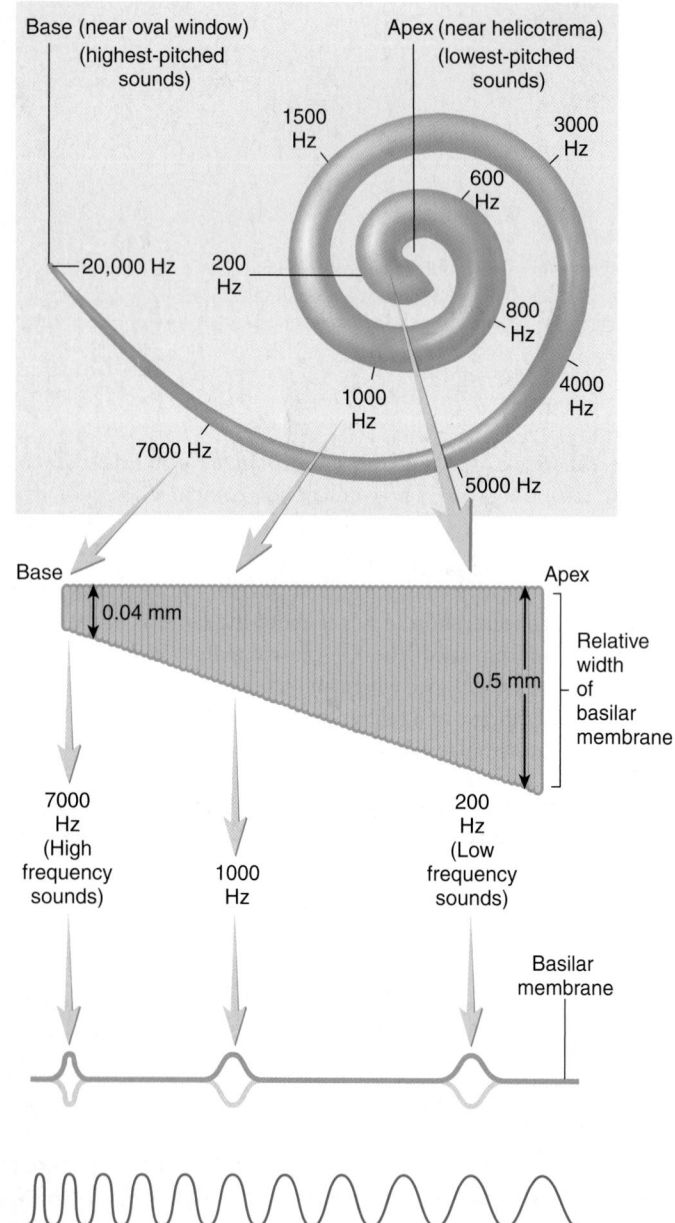

Figure 15.33   **Effect of Sound Waves on Points Along the Basilar Membrane**

Points of maximum vibration along the basilar membrane resulting from stimulation by sounds of various frequencies (in hertz).

1. Sensory axons from the cochlear ganglion terminate in the cochlear nucleus in the brainstem.
2. Axons from the neurons in the cochlear nucleus project to the superior olivary nucleus or to the inferior colliculus.
3. Axons from the inferior colliculus project to the medial geniculate nucleus of the thalamus.
4. Thalamic neurons project to the auditory cortex.
5. Neurons in the superior olivary nucleus send axons to the inferior colliculus, back to the inner ear, or to motor nuclei in the brainstem that send efferent fibers to the middle ear muscles.

Auditory cortex

Thalamus

Auditory cortex

Medial geniculate nucleus

Cochlear ganglion

Nerve to tensor tympani

Inferior colliculus

Vestibulocochlear nerve

Superior olivary nucleus

Cochlear nucleus

Nerve to stapedius

Frontal section

Process Figure 15.34   Central Nervous System Pathways for Hearing

cochlea send action potentials only to the cerebral cortex in response to specific pitches. Action potentials near the base of the basilar membrane stimulate neurons in a certain part of the auditory cortex, which interpret the stimulus as a high-pitched sound, whereas action potentials from the apex stimulate a different part of the cortex, which interprets the stimulus as a low-pitched sound.

P R E D I C T   10
Suggest some possible sites and mechanisms to explain why certain people have "perfect pitch" and other people are "tone deaf."

Sound volume, or loudness, is a function of sound wave amplitude. As high-amplitude sound waves reach the ear, the perilymph, endolymph, and basilar membrane vibrate more intensely, and the hair cells are stimulated more intensely. As a result of the increased stimulation, more hair cells send action potentials at a higher frequency to the cerebral cortex, where this information is perceived as a greater sound volume.

34. *Starting with the auricle, trace a sound wave into the inner ear to the point at which action potentials are generated in the cochlear nerve.*
35. *Explain how pulling on a gating spring results in the generation of an action potential.*

P R E D I C T   11
Explain why it's much easier to perceive subtle musical tones when music is played somewhat softly as opposed to very loudly.

Neuronal Pathways for Hearing

The special senses of hearing and balance are both transmitted by the vestibulocochlear (VIII) nerve. The term *vestibular* refers to the vestibule of the inner ear, which is involved in balance. The term *cochlear* refers to the cochlea and is that portion of the inner ear involved in hearing. The vestibulocochlear nerve functions as two separate nerves carrying information from two separate but closely related structures.

The auditory pathways within the CNS are very complex, with both crossed and uncrossed tracts (see figure 15.34). Unilateral CNS damage therefore usually has little effect on hearing. The neurons from the cochlear ganglion synapse with CNS neurons in the dorsal or ventral **cochlear nucleus** in the superior medulla near the inferior cerebellar peduncle. These neurons in turn either synapse in or pass through the superior olivary nucleus. Neurons terminating in this nucleus may synapse with efferent neurons returning to the cochlea to modulate pitch perception. Nerve fibers from the superior olivary nucleus also project to the trigeminal (V) nucleus, which controls the tensor tympani, and the facial (VII) nucleus, which controls the stapedius muscle. This reflex pathway dampens loud sounds by initiating contractions of these muscles. This is the sound attenuation reflex described previously. Neurons synapsing in the superior olivary nucleus may also join other ascending neurons to the cerebral cortex.

Ascending neurons from the superior olivary nucleus travel in the **lateral lemniscus.** All ascending fibers synapse in the **inferior colliculi,** and neurons from there project to the **medial geniculate nucleus** of the **thalamus,** where they synapse with neurons that project to the cortex. These neurons terminate in the **auditory cortex** in the dorsal portion of the temporal lobe within the lateral fissure and, to a lesser extent, on the superolateral surface of the temporal lobe (see chapter 13). Neurons from the inferior colliculus also project to the **superior colliculus,** where reflexes that turn the head and eyes in response to loud sounds are initiated.

● 36. Describe the neuronal pathways for hearing from the cochlear nerve to the cerebral cortex.

Balance

The organs of balance are divided structurally and functionally into two parts. The first, the **static labyrinth,** consists of the **utricle** (oo′tri-kl) and **saccule** (sak′ūl) of the vestibule and is primarily involved in evaluating the position of the head relative to gravity, although the system also responds to linear acceleration or deceleration, such as when a person is in a car that is increasing or decreasing speed. The second, the **kinetic labyrinth,** is associated with the semicircular canals and is involved in evaluating movements of the head.

Most of the utricular and saccular walls consist of simple cuboidal epithelium. The utricle and saccule, however, each contain a specialized patch of epithelium about 2–3 mm in diameter called the **macula** (mak′ū-lă; figure 15.35*a* and *b*). The macula of the utricle is oriented parallel to the base of the skull, and the macula of the saccule is perpendicular to the base of the skull.

The maculae resemble the spiral organ and consist of columnar supporting cells and hair cells. The "hairs" of these cells, which consist of numerous microvilli, called **stereocilia,** and one cilium, called a **kinocilium** (kī-nō-sil′ē-ŭm), are embedded in a **gelatinous mass** weighted by the presence of **otoliths** (ō′tō-liths) composed of protein and calcium carbonate (figure 15.35*b*). The gelatinous mass moves in response to gravity, bending the hair cells and initiating action potentials in the associated neurons. Deflection of the hairs toward the kinocilium results in depolarization of the hair cell, whereas deflection of the hairs away from the kinocilium results in hyperpolarization of the hair cell. If the head is tipped, otoliths move in response to gravity and stimulate certain hair cells (figure 15.36).

Figure 15.35   Structure of the Macula

(*a*) Vestibule showing the location of the utricular and saccular maculae. (*b*) Enlargement of the utricular macula, showing hair cells and otoliths in the macula. (*c*) An enlarged hair cell, showing the kinocilium and stereocilia.

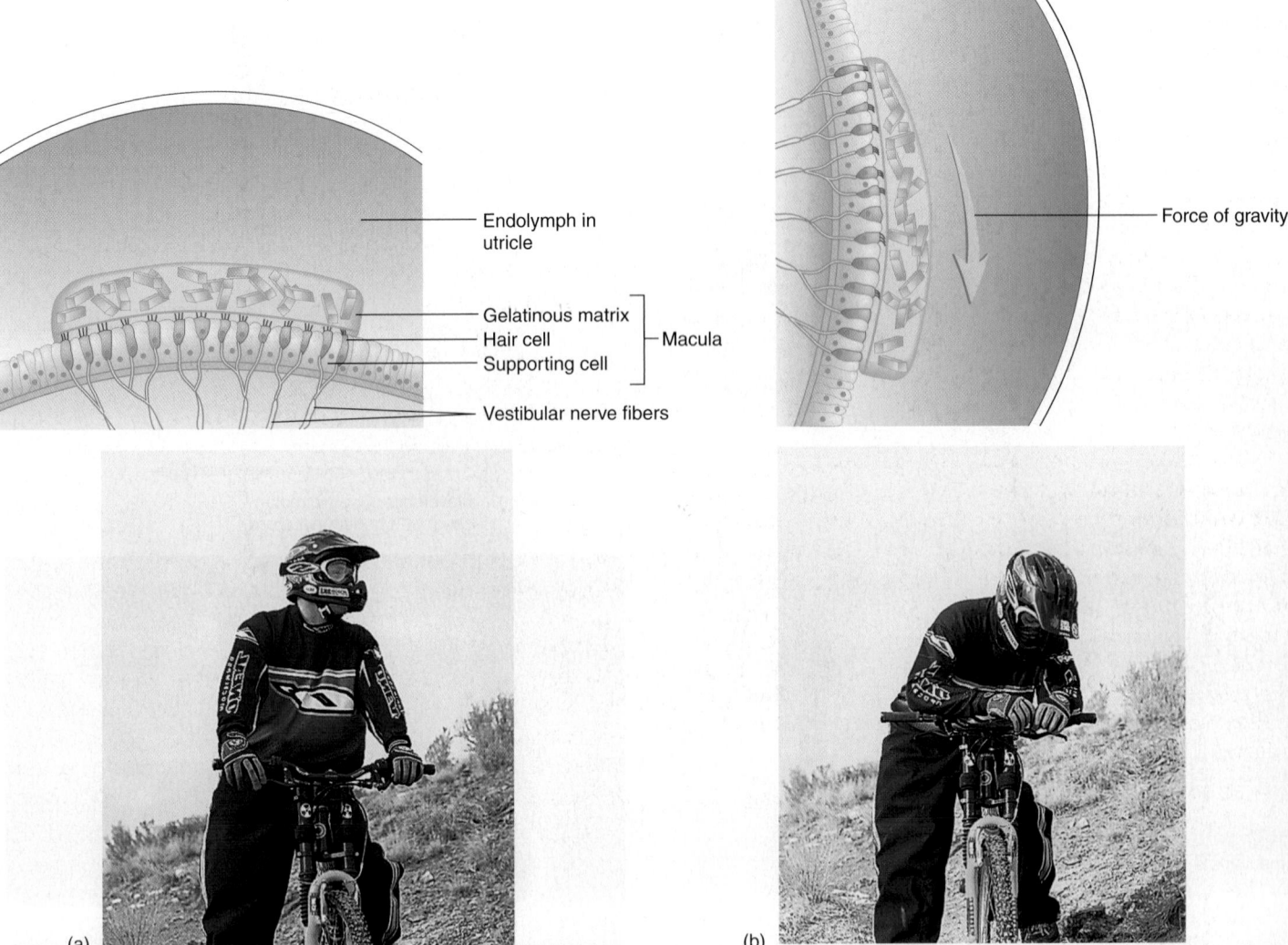

Figure 15.36 **Function of the Vestibule in Maintaining Balance**

(a) As the position of the head changes, such as when a person bends over, the maculae respond to changes in position of the head relative to gravity by moving in the direction of gravity. (b) In an upright position, the maculae don't move.

The hair cells have no axons but synapse directly with neurons of the vestibulocochlear nerve (VIII). Several neurotransmitters are released from the hair cells. One of those neurotransmitters is glutamate. Others have not yet been identified.

The hair cells are constantly being stimulated at a low level by the presence of the otolith-weighted covering of the macula; but as this covering moves in response to gravity, the pattern of intensity of hair cell stimulation changes. This pattern of stimulation and the subsequent pattern of action potentials from the numerous hair cells of the maculae can be translated by the brain into specific information about head position or acceleration. Much of this information is not perceived consciously but is dealt with subconsciously. The body responds by making subtle tone adjustments in muscles of the back and neck, which are intended to restore the head to its proper neutral, balanced position.

The kinetic labyrinth (figure 15.37) consists of three **semicircular canals** placed at nearly right angles to one another, one lying nearly in the transverse plane, one in the coronal plane, and one in the sagittal plane (see chapter 1). The arrangement of the semicircular canals enables a person to detect movement in all directions. The base of each semicircular canal is expanded into an **ampulla** (figure 15.37a). Within each ampulla, the epithelium is specialized to form a **crista ampullaris** (kris′tă am-pū-lar′ŭs). This specialized sensory epithelium is structurally and functionally very similar to that of the maculae. Each crista consists of a ridge or crest of epithelium with a curved gelatinous mass, the **cupula** (koo′poo-lă), suspended over the crest. The hairlike processes of the crista hair cells, similar to those in the maculae, are embedded in the cupula (figure 15.37b). The cupula contains no otoliths and therefore doesn't respond to gravitational pull. Instead, the cupula is a float that is displaced by fluid movements within the semicircular canals. Endolymph movement within each semicircular canal moves the cupula, bends the hairs, and initiates action potentials (figure 15.38).

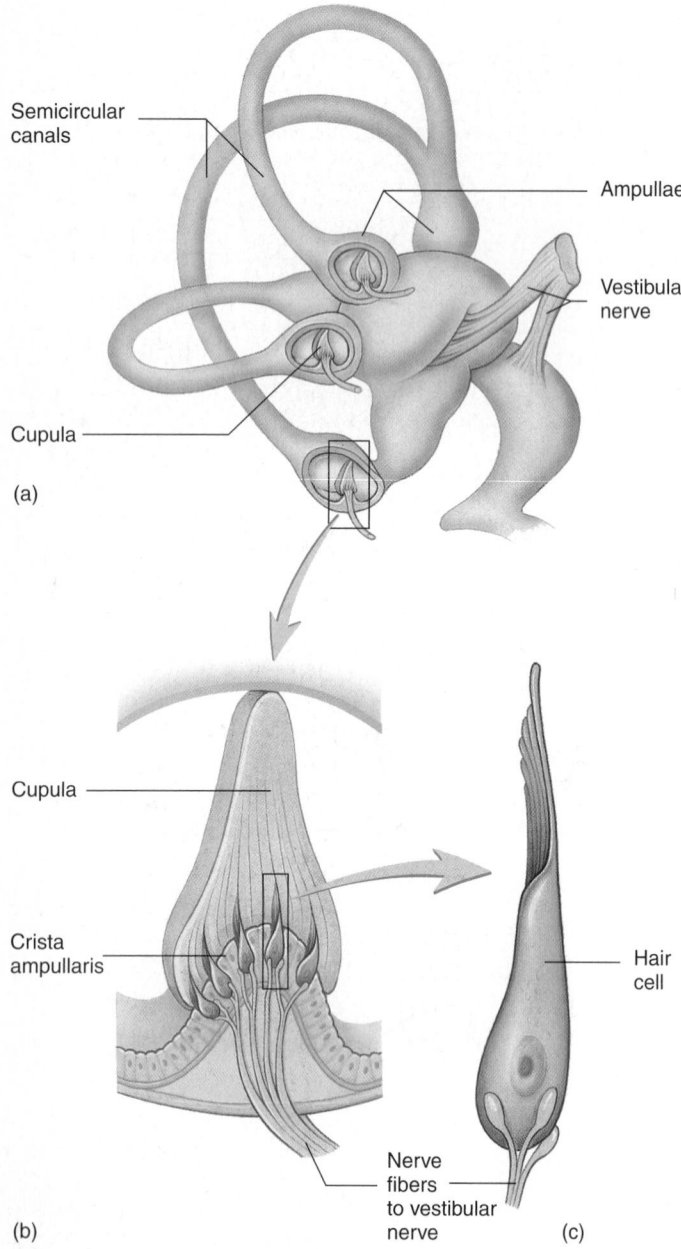

Figure 15.37   **Semicircular Canals**

(*a*) Semicircular canals showing location of the crista ampullaris in the ampullae of the semicircular canals. (*b*) Enlargement of the crista ampullaris, showing the cupula and hair cells. (*c*) Enlargement of a hair cell.

Figure 15.38   **Function of the Semicircular Canals**

The crista ampullaris responds to fluid movements within the semicircular canals. (*a*) When a person is at rest, the crista ampullaris does not move. (*b*) As a person begins to move in a given direction, the semicircular canals begin to move with the body (*blue arrow*), but the endolymph tends to remain stationary relative to the movement (momentum force: *red arrow* pointing in the opposite direction of body and semicircular canal movement), and the crista ampullaris is displaced by the endolymph in a direction opposite to the direction of movement.

As the head begins to move in a given direction, the endolymph does not move at the same rate as the semicircular canals (see figure 15.38). This difference causes displacement of the cupula in a direction opposite to that of the movement of the head, resulting in relative movement between the cupula and the endolymph. As movement continues, the fluid of the semicircular canals begins to move and "catches up" with the cupula, and stimulation is stopped. As movement of the head ceases, the endolymph continues to move because of its momentum, causing displacement of the cupula in the same direction as the head had

been moving. Because displacement of the cupula is most intense when the rate of head movement changes, this system detects changes in the rate of movement rather than movement alone. As with the static labyrinth, the information obtained by the brain from the kinetic labyrinth is largely subconscious.

- **37.** *What are the functions of the saccule and utricle? Describe the macula and its functions.*
- **38.** *What is the function of the semicircular canals? Describe the crista ampullaris and its mode of operation.*

Clinical Focus   Hearing Impairment and Functional Replacement of the Ear

The term **hearing-impaired** refers to any type or degree of hearing loss and may be conductive, sensorineural, or a combination of both. **Conduction deafness** involves a mechanical deficiency in transmission of sound waves from the external ear to the spiral organ. The spiral organ and neuronal pathways for hearing function normally. Conductive hearing loss often can be treated, for example by removing ear wax blocking the external acoustic meatus or by replacing or repairing the auditory ossicles. If the degree of conductive hearing loss does not justify surgical treatment, or if treatment does not resolve the hearing loss, a hearing aid may be beneficial because the amplified (louder) sound waves it produces are transmitted through the conductive blockage and may provide normal stimulation to the spiral organ.

Sensorineural hearing loss involves the spiral organ or neuronal pathways. Sound waves are transmitted normally to the spiral organ, but the ability of the nervous system

to respond to the sound waves is impaired. Hearing aids are commonly used by people with sensorineural hearing loss because the amplified sound waves they produce result in greater than normal stimulation of the spiral organ, helping to overcome the perception of reduced sound volume. Sound clarity also improves with sound amplification, but may never be perceived as normal.

The term **deaf** refers to sensorineural hearing loss so profound that the sense of hearing is nonfunctional, with or without amplification, for ordinary purposes of life. Stimulation of the spiral organ or hearing nerve pathways can help deaf people to hear.

Research is currently being conducted on ways to replace the hearing pathways with electric circuits. One approach involves the direct stimulation of nerves by electric impulses. There has been considerable success in the area of cochlear nerve stimulation. Certain types of sensorineural deafness in which the hair cells of the spiral organ are impaired

can now be partially corrected. Prostheses are available that consist of a microphone for picking up the initial sound waves, a microelectronic processor for converting the sound into electric signals, a transmission system for relaying the signals to the inner ear, and a long, slender electrode that is threaded into the cochlea. This electrode delivers electric signals directly to the endings of the cochlear nerve (figure E). High-frequency sounds are picked up by the microphone and transmitted through specific circuits to terminate near the oval window, whereas low-frequency sounds are transmitted farther up the cochlea to cochlear nerve endings near the helicotrema.

Research is currently underway to develop implants directly into the cochlear nucleus of the brainstem for patients with vestibulocochlear nerve damage. These implants have electrodes of various lengths to stimulate parts of the cochlear nucleus, at various depths from the surface, which respond to sounds of different frequencies.

1. A receiver, transmitter, and antenna are implanted under the skin near the auricle.

2. A small lead from the transmitter is fed through the external acoustic meatus, tympanic membrane, and middle ear into the cochlea.

3. In the cochlea, the cochlear nerve can be directly stimulated by electric impulses from the receiver.

Figure E   Cochlear Implant

Space Sickness

Space sickness is a balance disorder occurring in zero gravity and resulting from unfamiliar sensory input to the brain. The brain must adjust to these unusual signals, or severe symptoms like headaches and dizziness may result. Space sickness is unlike motion sickness in that motion sickness results from an excessive stimulation of the brain, whereas space sickness results from too little stimulation as a result of weightlessness.

Neuronal Pathways for Balance

Neurons synapsing on the hair cells of the maculae and cristae ampullares converge into the **vestibular ganglion,** where their cell bodies are located (figure 15.39). Sensory fibers from these neurons join sensory fibers from the cochlear ganglion to form the vestibulocochlear nerve (VIII) and terminate in the **vestibular nucleus** within the medulla oblongata. Axons run from this nucleus

1. Sensory axons from the vestibular ganglion pass through the vestibular nerve to the vestibular nucleus, which also receives input from several other sources, such as proprioception from the legs.

2. Vestibular neurons send axons to the cerebellum, which influences postural muscles, and to the motor nuclei (oculomotor, trochlear, and abducens), which control extrinsic eye muscles.

3. Vestibular neurons also send axons to the posterior ventral nucleus of the thalamus.

4. Thalamic neurons project to the vestibular area of the cortex.

Vestibular area

Posterior ventral nucleus

Thalamus

Cerebellum

Vestibular nerve

Vestibular ganglion

Vestibular nucleus

Spinovestibular tract

Oculomotor nucleus

Trochlear motor nucleus

Abducens motor nucleus

Vestibulospinal tract

Frontal section

Process Figure 15.39   Central Nervous System Pathways for Balance

to numerous areas of the CNS, such as the spinal cord, cerebellum, cerebral cortex, and the nuclei controlling extrinsic eye muscles.

Balance is a complex process not simply confined to one type of input. In addition to vestibular sensory input, the vestibular nucleus receives input from proprioceptive neurons throughout the body, and from the visual system. People are asked to close their eyes while balance is evaluated in a sobriety test because alcohol affects the proprioceptive and vestibular components of balance (cerebellar function) to a greater extent than it does the visual portion.

Reflex pathways exist between the kinetic part of the vestibular system and the nuclei controlling the extrinsic eye muscles (oculomotor, trochlear, and abducens). A reflex pathway allows maintenance of visual fixation on an object while the head is in motion. This function can be demonstrated by spinning a person around about 10 times in 20 seconds, stopping him or her, and observing eye movements. The reaction is most pronounced if the individual's head is tilted forward about 30 degrees while spinning, thus bringing the lateral semicircular canals into the horizontal plane. A slight oscillatory movement of the eyes occurs. The eyes track in the direction of motion and return with a rapid recovery movement before repeating the tracking motion. This oscillation of the eyes is called

nystagmus (nis-tag′mŭs). If asked to walk in a straight line, the individual deviates in the direction of rotation, and if asked to point to an object, his or her finger deviates in the direction of rotation.

39. Describe the neuronal pathways for balance.

Effects of Aging on the Special Senses

Elderly people experience only a slight loss in the ability to detect odors. However, the ability to correctly identify specific odors is decreased, especially in men over age 70.

In general, the sense of taste decreases as people age. The number of sensory receptors decreases and the ability of the brain to interpret taste sensations declines.

Responses to taste change in some elderly people who are fighting cancer. One side effect of radiation treatment and chemotherapy is the gastrointestinal discomfort resulting from the treatments. The patients experience a loss of appetite because of conditioned taste aversions resulting from treatment.

Clinical Focus   Ear Disorders

Otosclerosis

Otosclerosis (ō′tō-sklē-rō′sis) is an ear disorder in which spongy bone grows over the oval window and immobilizes the stapes, leading to progressive loss of hearing. This disorder can be surgically corrected by breaking away the bony growth and the immobilized stapes. During surgery, the stapes is replaced by a small rod connected by a fat pad or a synthetic membrane to the oval window at one end and to the incus at the other end.

Tinnitus

Tinnitus (ti-nī′tŭs) consists of noises such as ringing, clicking, whistling, or booming in the ears. These noises may occur as a result of disorders in the middle or inner ear or along the central neuronal pathways.

Motion Sickness

Motion sickness consists of nausea, weakness, and other dysfunctions caused by stimulation of the semicircular canals during motion, such as in a boat, automobile, airplane, swing, or amusement park ride. It may progress to vomiting and incapacitation. Antiemetics such as anticholinergic or antihistamine medications can be taken to counter the nausea and vomiting associated with motion sickness. Scopolamine is an anticholinergic drug that blocks acetylcholine-mediated transmission in the parasympathetic nervous system. It depresses parasympathetic activity within areas of the CNS, such as the hypothalamus, in response to vestibular stimulation. It also depresses other CNS functions, which may cause side effects such as restlessness, agitation, psychosis, mania, and Parkinson-like tremors. Cyclizine (Marezine), dimenhydrinate (Dramamine), and diphenhydramine (Benadryl) are antihistamines that affect the neural pathways from the vestibule. Scopolamine can be administered transdermally in the form of a patch placed on the skin behind the ear (Transdermal-Scop). A patch lasts about 3 days.

Otitis Media

Infections of the middle ear, called **otitis media,** are quite common in young children. These infections usually result from the spread of infection from the mucous membrane of the pharynx through the auditory tube to the mucous lining of the middle ear. The symptoms of otitis media, consisting of low-grade fever, lethargy, and irritability, are often not easily recognized by the parent as signs of middle ear infection. The infection can also cause a temporary decrease or loss of hearing because fluid buildup has dampened the tympanic membrane or ossicles.

Earache

Earache can result from otitis media, otitis externa (inflammation of the external acoustic meatus), dental abscesses, or temporomandibular joint pain.

The lenses of the eyes lose flexibility as a person ages because the connective tissue of the lenses becomes more rigid. Consequently there is first a reduction and then an eventual loss in the ability of the lenses to change shape. This condition, called presbyopia, is the most common age-related change in the eyes. It is discussed more fully in the Clinical Focus on "Eye Disorders" earlier in the chapter.

The most common visual problem in older people requiring medical treatment, such as surgery, is the development of cataracts. Macular degeneration is the second most common defect, glaucoma is third, and diabetic retinopathy is fourth. These defects are also described more fully in the Clinical Focus on "Eye Disorders."

The number of cones decreases, especially in the fovea centralis. These changes cause a gradual decline in visual acuity and color perception.

As people age, the number of hair cells in the cochlea decreases. This decline doesn't occur equally in both ears. As a result, because direction is determined by comparing sounds coming into each ear, elderly people may experience a decreased ability to localize the origin of certain sounds. In some people, this may lead to a general sense of disorientation. In addition, CNS defects in the auditory pathways can result in difficulty understanding sounds with echoes or background noise. Such deficit makes it difficult for elderly people to understand rapid or broken speech.

With age, the number of hair cells in the saccule, utricle, and ampullae decreases. The number of otoliths also declines. As a result, elderly people experience a decreased sensitivity to gravity, acceleration, and rotation. Because of these decreases, elderly people experience dizziness (instability) and vertigo (a feeling of spinning). They often feel that they can't maintain posture and are prone to fall.

● 40. *Explain the changes in taste, vision, hearing, and balance that occur with aging.*

S U M M A R Y

Olfaction   (p. 512)

Olfaction is the sense of smell.

Olfactory Epithelium and Bulb

1. Olfactory neurons in the olfactory epithelium are bipolar neurons. Their distal ends are enlarged as olfactory vesicles, which have long cilia.

2. The cilia have receptors that respond to dissolved substances. There are approximately 1000 different odorant receptors.
3. The receptors activate a G protein complex which opens ion channels.
4. At least seven (perhaps 50) primary odors exist. The olfactory neurons have a very low threshold and accommodate rapidly.

Neuronal Pathways for Olfaction

1. Axons from the olfactory neurons extend as olfactory nerves to the olfactory bulb, where they synapse with mitral and tufted cells. Axons from these cells form the olfactory tracts and synapse with granule cells. The olfactory bulb neurons can modulate output to the olfactory tracts.
2. The olfactory tracts terminate in the olfactory and temporal cortex. The lateral olfactory area is involved in the conscious perception of smell, the intermediate area with modulating smell, and the medial area with visceral and emotional responses to smell.

Taste   (p. 515)

Taste buds usually are associated with circumvallate, fungiform, and foliate papillae. Filiform papillae do not have taste buds.

Histology of Taste Buds

1. Taste buds consist of basilar, supporting, and gustatory cells.
2. The gustatory cells have gustatory hairs that extend into taste pores.

Function of Taste

1. Receptors on the hairs detect dissolved substances.
2. Five basic types of taste exist: sour, salty, bitter, sweet, and umami.

Neuronal Pathways for Taste

1. The facial nerve carries taste sensations from the anterior two-thirds of the tongue, the glossopharyngeal nerve from the posterior one-third of the tongue, and the vagus nerve from the epiglottis.
2. The neural pathways for taste extend from the medulla oblongata to the thalamus and to the cerebral cortex.

Visual System   (p. 518)
Accessory Structures

1. The eyebrows prevent perspiration from entering the eyes and help shade the eyes.
2. The eyelids consist of five tissue layers. They protect the eyes from foreign objects and help lubricate the eyes by spreading tears over their surface.
3. The conjunctiva covers the inner eyelid and the anterior part of the eye.
4. Lacrimal glands produce tears that flow across the surface of the eye. Excess tears enter the lacrimal canaliculi and reach the nasal cavity through the nasolacrimal canal. Tears lubricate and protect the eye.
5. The extrinsic eye muscles move the eyeball.

Anatomy of the Eye

1. The fibrous tunic is the outer layer of the eye. It consists of the sclera and cornea.
   - The sclera is the posterior four-fifths of the eye. It is white connective tissue that maintains the shape of the eye and provides a site for muscle attachment.
   - The cornea is the anterior one-fifth of the eye. It is transparent and refracts light that enters the eye.
2. The vascular tunic is the middle layer of the eye.
   - The iris is smooth muscle regulated by the autonomic nervous system. It controls the amount of light entering the pupil.
   - The ciliary muscles control the shape of the lens. They are smooth muscles regulated by the autonomic nervous system. The ciliary process produces aqueous humor.
3. The retina is the inner layer of the eye and contains neurons sensitive to light.
   - The macula lutea (fovea centralis) is the area of greatest visual acuity.
   - The optic disc is the location through which nerves exit and blood vessels enter the eye. It has no photosensory cells and is therefore the blind spot of the eye.
4. The eye has three chambers: anterior, posterior, and vitrus.
   - The anterior and posterior chambers are filled with aqueous humor, which circulates and leaves by way of the canal of Schlemm.
   - The vitrus chamber is filled with vitreous humor.

5. The lens is held in place by the suspensory ligaments, which are attached to the ciliary muscles.

Functions of the Complete Eye

1. Light is that portion of the electromagnetic spectrum that humans can see.
2. When light travels from one medium to another, it can bend or refract. Light striking a concave surface refracts outward (divergence). Light striking a convex surface refracts inward (convergence).
3. Converging light rays meet at the focal point and are said to be focused.
4. The cornea, aqueous humor, lens, and vitreous humor all refract light. The cornea is responsible for most of the convergence, whereas the lens can adjust the focal point by changing shape.
   - Relaxation of the ciliary muscles causes the lens to flatten, producing the emmetropic eye.
   - Contraction of the ciliary muscles causes the lens to become more spherical. This change in lens shape enables the eye to focus on objects that are less than 20 feet away, a process called accommodation.
5. The far point of vision is the distance at which the eye no longer has to change shape to focus on an object. The near point of vision is the closest an object can come to the eye and still be focused.
6. The pupil becomes smaller during accommodation, increasing the depth of focus.

Structure and Function of the Retina

1. The pigmented retina provides a black backdrop for increasing visual acuity.
2. Rods are responsible for vision in low illumination (night vision).
   - A pigment, rhodopsin, is split by light into retinal and opsin, producing hyperpolarization in the rod.
   - Light adaptation is caused by a reduction of rhodopsin; dark adaptation is caused by rhodopsin production.
3. Cones are responsible for color vision and visual acuity.
   - Cones are of three types, each with a different type of iodopsin photopigment. The pigments are most sensitive to blue, red, and green lights.
   - Perception of many colors results from mixing the ratio of the different types of cones that are active at a given moment.
4. Most visual images are focused on the fovea centralis, which has a very high concentration of cones. Moving away from the fovea, fewer cones (the macula lutea) are present; mostly rods are in the periphery of the retina.
5. The rods and the cones synapse with bipolar cells that in turn synapse with ganglion cells, which form the optic nerves.
6. Bipolar and ganglion cells in the retina can modify information sent to the brain.
7. Ganglion cells have receptive fields with on-centers or off-centers. This arrangement enhances contrast.

Neuronal Pathways for Vision

1. Ganglia cell axons extend to the lateral geniculate ganglion of the thalamus, where they synapse. From there neurons form the optic radiations that project to the visual cortex.
2. Neurons from the nasal visual field (temporal retina) of one eye and the temporal visual field (nasal retina) of the opposite eye project to the same cerebral hemisphere. Axons from the nasal retina cross in the optic chiasm, and axons from the temporal retina remain uncrossed.
3. Depth perception is the ability to judge relative distances of an object from the eyes and is a property of binocular vision. Binocular vision results because a slightly different image is seen by each eye.

Hearing and Balance   (p. 539)

The osseous labyrinth is a canal system within the temporal bone that contains perilymph and the membranous labyrinth. Endolymph is inside the membranous labyrinth.

Auditory Structures and Their Functions

1. The external ear consists of the auricle and external acoustic meatus.
2. The middle ear connects the external and inner ears.
   - The tympanic membrane is stretched across the external acoustic meatus.
   - The malleus, incus, and stapes connect the tympanic membrane to the oval window of the inner ear.
   - The auditory tube connects the middle ear to the pharynx and functions to equalize pressure.
   - The middle ear is connected to the mastoid air cells.
3. The inner ear has three parts: the semicircular canals; the vestibule, which contains the utricle and the saccule; and the cochlea.
4. The cochlea is a spiral-shaped canal within the temporal bone.
   - The cochlea is divided into three compartments by the vestibular and basilar membranes. The scala vestibuli and scala tympani contain perilymph. The cochlear duct contains endolymph and the spiral organ (organ of Corti).
   - The spiral organ consists of inner and outer hair cells that attach to the tectorial membrane.

Auditory Function

1. Sound waves are funneled by the auricle down the external acoustic meatus, causing the tympanic membrane to vibrate.
2. The tympanic membrane vibrations are passed along the auditory ossicles to the oval window of the inner ear.
3. Movement of the stapes in the oval window causes the perilymph, vestibular membrane, and endolymph to vibrate, producing movement of the basilar membrane.
4. Movement of the basilar membrane causes bending of the stereocilia of inner hair cells in the spiral organ.
5. Bending of the stereocilia pulls on gating springs which open K^+ channels.
6. K^+ entering the hair cell depolarize the cell, which open Ca^{2+} channels. Ca^{2+} entering the cell cause further depolarization.
7. Depolarization causes the release of glutamate, generating action potentials in the vestibulocochlear nerve.
8. Some vestibulocochlear nerve axons synapse in the superior olivary nucleus. Efferent neurons from this nucleus project back to the cochlea, where they regulate the perception of pitch.
9. The round window protects the inner ear from pressure buildup and dissipates sound waves.

Neuronal Pathways for Hearing

1. Axons from the vestibulocochlear nerve synapse in the medulla. Neurons from the medulla project axons to the inferior colliculi, where they synapse. Neurons from this point project to the thalamus and synapse. Thalamic neurons extend to the auditory cortex.
2. Efferent neurons project to cranial nerve nuclei responsible for controlling muscles that dampen sound in the middle ear.

Balance

1. Static balance evaluates the position of the head relative to gravity and detects linear acceleration and deceleration.
   - The utricle and saccule in the inner ear contain maculae. The maculae consist of hair cells with the hairs embedded in a gelatinous mass that contains otoliths.
   - The gelatinous mass moves in response to gravity.
2. Kinetic balance evaluates movements of the head.
   - Three semicircular canals at right angles to one another are present in the inner ear. The ampulla of each semicircular canal contains the crista ampullaris, which has hair cells with hairs embedded in a gelatinous mass, the cupula.
   - When the head moves, endolymph within the semicircular canal moves the cupula.

Neuronal Pathways for Balance

1. Axons from the maculae and the cristae ampullares extend to the vestibular nucleus of the medulla. Fibers from the medulla run to the spinal cord, cerebellum, cortex, and nuclei that control the extrinsic eye muscles.
2. Balance also depends on proprioception and visual input.

Effects of Aging on the Special Senses   (p. 554)

Elderly people experience a decline in function of all special functions: olfaction, taste, vision, hearing, and balance. These declines can result in loss of appetite, visual impairment, disorientation, and risk of falling.

R E V I E W   A N D   C O M P R E H E N S I O N

1. Olfactory neurons
   a. have projections called cilia.
   b. have axons that combine to form the olfactory nerves.
   c. connect to the olfactory bulb.
   d. have receptors that react with odorants dissolved in fluid.
   e. all of the above.

2. Which of these statements is *not* true with respect to olfaction?
   a. Olfactory sensation is relayed directly to the cerebral cortex without passing through the thalamus.
   b. Olfactory neurons are replaced about every two months.
   c. The lateral olfactory area of the cortex is involved in the conscious perception of smell.
   d. The medial olfactory area of the cortex is responsible for visceral and emotional reactions to odors.
   e. The olfactory cortex is in the occipital lobe of the cerebrum.

3. Gustatory (taste) cells
   a. are found only on the tongue.
   b. extend through tiny openings called taste buds.
   c. have no axons but release neurotransmitter when stimulated.
   d. have axons that extend directly to the taste area of the cerebral cortex.

4. Which of these is *not* one of the basic tastes?
   a. spicy
   b. salt
   c. bitter
   d. umami
   e. sour

5. Which of these types of papillae have no taste buds associated with them?
   a. circumvallate
   b. filiform
   c. foliate
   d. fungiform

6. Tears
   a. are released onto the surface of the eye near the medial corner of the eye.
   b. in excess are removed by the canal of Schlemm.
   c. in excess can cause a sty.
   d. can pass through the nasolacrimal duct into the oral cavity.
   e. contain water, salts, mucus, and lysozyme.

7. The fibrous tunic of the eye includes the
   a. conjunctiva.
   b. sclera.
   c. choroid.
   d. iris.
   e. retina.

8. The ciliary body
   a. contains smooth muscles that attach to the lens by suspensory ligaments.
   b. produces the vitreous humor.
   c. is part of the iris of the eye.
   d. is part of the sclera.
   e. all of the above.

9. The lens normally focuses light onto the
   a. optic disc.
   b. iris.
   c. macula lutea.
   d. cornea.
   e. ciliary body.

10. Given these structures:
    1. lens
    2. aqueous humor
    3. vitreous humor
    4. cornea

    Choose the arrangement that lists the structures in the order that light entering the eye encounters them.
    a. 1,2,3,4
    b. 1,4,2,3
    c. 4,1,2,3
    d. 4,2,1,3
    e. 4,3,2,1

11. Aqueous humor
    a. is the pigment responsible for the black color of the choroid.
    b. exits the eye through the canal of Schlemm.
    c. is produced by the iris.
    d. can cause cataracts if overproduced.
    e. is composed of proteins called crystallines.

12. Contraction of the smooth muscle in the ciliary body causes the
    a. lens to flatten.
    b. lens to become more spherical.
    c. pupil to constrict.
    d. pupil to dilate.

13. Given these events:
    1. medial rectus contracts
    2. lateral rectus contracts
    3. pupils dilate
    4. pupils constrict
    5. lens of the eye flattens
    6. lens of the eye becomes more spherical

    Assume you are looking at an object 30 feet away. If you suddenly look at an object that is 1 foot away, which events occur?
    a. 1,3,6
    b. 1,4,5
    c. 1,4,6
    d. 2,3,6
    e. 2,4,5

14. Given these events:
    1. bipolar cells depolarize
    2. decrease in glutamate released from presynaptic terminals of photoreceptor cells
    3. light strikes photoreceptor cells
    4. photoreceptor cells depolarized
    5. photoreceptor cells hyperpolarized

    Choose the arrangement that lists the correct order of events, starting with the photoreceptor cells in the resting, nonactivated state.
    a. 1,2,3,4,5
    b. 2,4,3,5,1
    c. 3,4,2,5,1
    d. 4,3,5,2,1
    e. 5,3,4,1,2

15. Given these neurons in the retina:
    1. bipolar cells
    2. ganglionic cells
    3. photoreceptor cells

    Choose the arrangement that lists the correct order of the cells encountered by light as it enters the eye and travels toward the pigmented retina.
    a. 1,2,3
    b. 1,3,2
    c. 2,1,3
    d. 2,3,1
    e. 3,1,2

16. Which of these photoreceptor cells is *not* correctly matched with its function?
    a. rods—vision in low light
    b. rods—visual acuity
    c. cones—color vision

17. Concerning dark adaptation,
    a. the amount of rhodopsin increases.
    b. the pupils constrict.
    c. it occurs more rapidly than light adaptation.
    d. all of the above.

18. In the retina there are cones that are most sensitive to a particular color. Given this list of colors:
    1. red
    2. yellow
    3. green
    4. blue

    Indicate which colors correspond to specific types of cones.
    a. 2,3
    b. 3,4
    c. 1,2,3
    d. 1,3,4
    e. 1,2,3,4

19. Given these areas of the retina:
    1. macula lutea
    2. fovea centralis
    3. optic disc
    4. periphery of the retina

    Choose the arrangement that lists the areas according to the density of cones, starting with the area that has the highest density of cones.
    a. 1,2,3,4
    b. 1,3,2,4
    c. 2,1,4,3
    d. 2,4,1,3
    e. 3,4,1,2

20. Concerning axons in the optic nerve from the right eye,
    a. they all go to the right occipital lobe.
    b. they all go to the left occipital lobe.
    c. they all go to the thalamus.
    d. most go to the thalamus, some go to the superior colliculus
    e. some go to the right occipital lobe, and some go to the left occipital lobe.

21. A lesion that destroyed the left optic tract of a boy eliminates vision in his
    a. left nasal visual field.
    b. left temporal visual field.
    c. right temporal visual field
    d. both a and b.
    e. both a and c.

22. A person with an abnormally powerful focusing system
    is _____ and uses a _____ to correct his or her
    vision.
    a. nearsighted, concave lens
    b. nearsighted, convex lens
    c. farsighted, concave lens
    d. farsighted, convex lens

23. Which of these structures is found within or is a part of the external
    ear?
    a. oval window
    b. auditory tube
    c. ossicles
    d. external acoustic meatus
    e. cochlear duct

24. Given these ear bones:
    1. incus
    2. malleus
    3. stapes

    Choose the arrangement that lists the ear bones in order from the
    tympanic membrane to the inner ear.
    a. 1,2,3
    b. 1,3,2
    c. 2,1,3
    d. 2,3,1
    e. 3,2,1

25. Given these structures:
    1. perilymph
    2. endolymph
    3. vestibular membrane
    4. basilar membrane

    Choose the arrangement that lists the structures in the order sound
    waves coming from the outside encounter them in producing sound.
    a. 1,3,2,4
    b. 1,4,2,3
    c. 2,3,1,4
    d. 2,4,1,3
    e. 3,4,2,1

26. The spiral organ is found within the
    a. cochlear duct.
    b. scala vestibuli.
    c. scala tympani.
    d. vestibule.
    e. semicircular canals.

27. An increase in the loudness of sound occurs as a result of an
    increase in the _____ of the sound wave.
    a. frequency
    b. amplitude
    c. resonance
    d. both a and b

28. Interpretation of different sounds is possible because of the ability
    of the _____ to vibrate at different frequencies and
    stimulate the _____.
    a. vestibular membrane, vestibular nerve
    b. vestibular membrane, spiral organ
    c. basilar membrane, vestibular nerve
    d. basilar membrane, spiral organ

29. Which structure is a specialized receptor found within the utricle?
    a. macula
    b. crista ampullaris
    c. spiral organ
    d. cupula

30. Damage to the semicircular canals affects the ability to detect
    a. sound.
    b. the position of the head relative to the ground.
    c. the movement of the head in all directions.
    d. all of the above.

Answers in Appendix F

CRITICAL THINKING

1. An elderly man with normal vision develops cataracts. He is surgically
   treated by removing the lenses of his eyes. What kind of glasses would
   you recommend he wear to compensate for the removal of his lenses?

2. Some animals have a reflective area in the choroid called the
   tapetum lucidum. Light entering the eye is reflected back instead of
   being absorbed by the choroid. What would be the advantage of this
   arrangement? The disadvantage?

3. Perhaps you have heard someone say that eating carrots is good for
   the eyes. What is the basis for this claim?

4. On a camping trip Jean Tights rips her pants. That evening she is
   going to repair the rip. As the sun goes down, the light becomes
   more and more dim. When she tries to thread the needle, it is
   obvious that she is not looking directly at the needle but is looking a
   few inches to the side. Why does she do this?

5. A man stares at a black clock on a white wall for several minutes.
   Then he shifts his view and looks at only the blank white wall.
   Although he is no longer looking at the clock, he sees a light clock
   against a dark background. Explain what happened.

6. Describe the results of a lesion of the optic chiasm.

7. Persistent exposure to loud noise can cause loss of hearing,
   especially for high-frequency sounds. What part of the ear is
   probably damaged? Be as specific as possible.

8. Professional divers are subject to increased pressure as they
   descend to the bottom of the ocean. Sometimes this pressure can
   lead to damage to the ear and loss of hearing. Describe the
   normal mechanisms that adjust for changes in pressure, suggest
   some conditions that might interfere with pressure adjustment,
   and explain how the increased pressure might cause loss of
   hearing.

9. If a vibrating tuning fork is placed against the mastoid process of
   the temporal bone, the vibrations are perceived as sound, even if
   the external acoustic meatus is plugged. Explain how this could
   happen.

Answers in Appendix G

A N S W E R S   T O   P R E D I C T   Q U E S T I O N S

1. Inhaling slowly and deeply allows a large amount of air to be drawn into the olfactory region, whereas not as much air enters during normal breaths. Sniffing (rapid, repeated air intake) is effective for the same reason.

2. Adaptation can occur at several levels in the olfactory system. First, adaptation can occur at the receptor cell membrane, where receptor sites are filled or become less sensitive to a specific odor. Second, olfactory bulb neurons within the olfactory bulb can modify sensitivity to an odor by inhibiting mitral cells or tufted cells. Third, neurons from the intermediate olfactory area of the cerebrum can send action potentials to the olfactory bulb neurons to inhibit further sensory action potentials.

3. Eyedrops placed into the eye tend to drain through the nasolacrimal duct into the nasal cavity. Recall that much of what is considered "taste" is actually smell. The medication is detected by the olfactory neurons and is interpreted by the brain as taste sensation. Crying produces extra tears, which are conducted to the nasal cavity, causing a "runny" nose.

4. Inflammation of the cornea involves edema, the accumulation of fluid. Fluid accumulation in the cornea increases its water content, and because water causes the proteoglycans to expand, the transparency of the lens decreases, interfering with normal vision.

5. Eyestrain, or eye fatigue, occurs primarily in the ciliary muscles. It occurs because close vision requires accommodation. Accommodation occurs as the ciliary muscles contract, releasing the tension of the suspensory ligaments, and allowing the lens to become more rounded. Continued close vision requires maintenance of accommodation, which requires that the ciliary muscles remain contracted for a long time, resulting in their fatigue.

6. Rhodopsin breakdown is associated with adaptation to bright light and occurs rapidly, whereas rhodopsin production occurs slowly and is associated with adaptation to conditions of little light. Eyes adapt rather quickly to bright light but quite slowly to very dim light.

7. Rod cells distributed over most of the retina are involved in both peripheral vision (out of the corner of the eye) and vision under conditions of very dim light. When attempting to focus directly on an object, however, a person relies on the cones within the macula lutea; although the cones are involved in visual acuity, they don't function well in dim light; thus the object may not be seen at all.

8. A lesion in the right optic nerve at *B* results in loss of vision in the right visual field (see following illustration).

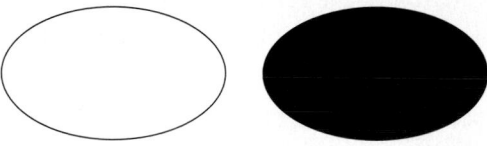

9. The stapedius muscle, attached to the stapes, is innervated by the facial nerve (VII). Loss of facial nerve function eliminates part of the sound attenuation reflex, although not all of it, because the tensor tympani muscle, innervated by the trigeminal nerve, is still functional. A reduction in the sound attenuation reflex results in sounds being excessively loud in the affected ear. A reduced reflex can also leave the ear more susceptible to damage by prolonged loud sounds.

10. "Perfect pitch" is the ability to precisely reproduce a pitch just by being told its name or reading it on a sheet of music, with no other musical support, such as from piano accompaniment. This remarkable talent as well as conditions such as tone deafness (the complete inability to recognize or reproduce musical pitches) or a decreased ability to perceive tone differences could occur at a number of locations. The structure of the basilar membrane may be such that tones are not adequately spaced along the cochlear duct in some people to facilitate clear separation of tones. The reflex from the superior olive to the spiral organ may have a very narrow "window of function" for people with perfect pitch but may not be functioning in some other people. The auditory cortex may not be able to translate as accurately in some people to distinguish differences in tones.

11. It is much easier to perceive subtle musical tones when music is played somewhat softly as opposed to very loudly because loud sounds have sound waves with a greater amplitude, which causes the basilar membrane to vibrate more violently over a wider range. The spreading of the wave in the basilar membrane to some extent counteracts the reflex from the superior olive that is responsible for enabling a person to hear subtle tone differences.

Visit the Online Learning Center at www.mhhe.com/seeley7 for chapter quizzes, interactive learning exercises, and other study tools.

Autonomic Nervous System

CHAPTER

16

Light photomicrograph from a section of the small intestine, showing the nerve cells of the enteric plexus. These nerve cells regulate the contraction of smooth muscle and the secretion of glands within the intestinal wall.

During a picnic on a sunny spring day, it's easy to concentrate on the delicious food and the pleasant surroundings. Maintenance of homeostasis requires no conscious thought. The autonomic nervous system (ANS) helps keep body temperature at a constant level by controlling the activity of sweat glands and the amount of blood flowing through the skin. The ANS helps to regulate the complex activities necessary for the digestion of food. The movement of absorbed nutrients to tissues is possible because the ANS controls heart rate, which helps maintain the blood pressure necessary to deliver blood to tissues. Without the ANS, all of the activities necessary to maintain homeostasis would be overwhelming.

A functional knowledge of the ANS enables you to predict general responses to a variety of stimuli, explain responses to changes in environmental conditions, comprehend symptoms that result from abnormal autonomic functions, and understand how drugs affect the ANS. This chapter examines the autonomic nervous system by *contrasting the somatic and autonomic nervous systems* (p. 562); describing the *anatomy of the autonomic nervous system* (p. 563), the *physiology of the autonomic nervous system* (p. 569), and the *regulation of the autonomic nervous system* (p. 573); and examining *functional generalizations about the autonomic nervous system* (p. 576).

Contrasting the Somatic and Autonomic Nervous Systems

The peripheral nervous system (PNS) is composed of sensory and motor neurons. Sensory neurons carry action potentials from the periphery to the central nervous system (CNS), and motor neurons carry action potentials from the CNS to the periphery. Motor neurons are either somatic motor neurons, which innervate skeletal muscle, or autonomic motor neurons, which innervate smooth muscle, cardiac muscle, and glands.

Although axons of autonomic, somatic, and sensory neurons are in the same nerves, the proportion varies from nerve to nerve. For example, nerves innervating smooth muscle, cardiac muscle, and glands, such as the vagus nerves, consist primarily of autonomic motor neurons and sensory neurons. Nerves innervating skeletal muscles such as the sciatic nerves, consist primarily of somatic motor neurons and sensory neurons. Some cranial nerves such as the olfactory, optic, and vestibulocochlear nerves are composed entirely of sensory neurons.

The cell bodies of somatic motor neurons are in the CNS, and their axons extend from the CNS to skeletal muscle (figure 16.1*a*). The ANS, on the other hand, has two neurons in a series extending between the CNS and the organs innervated (figure 16.1*b*). The first neurons of the series are called **preganglionic neurons.** Their cell bodies are located in the CNS within either the brainstem or the spinal cord, and their axons extend to autonomic ganglia located outside the CNS. The **autonomic ganglia** contain the cell bodies of the second neurons of the series, which are called **postganglionic neurons.** The preganglionic neurons synapse with the postganglionic neurons in the autonomic ganglia. The axons of the postganglionic neurons extend from autonomic ganglia to effector organs, where they synapse with their target tissues.

Many movements controlled by the somatic nervous system are conscious, whereas ANS functions are unconsciously controlled.

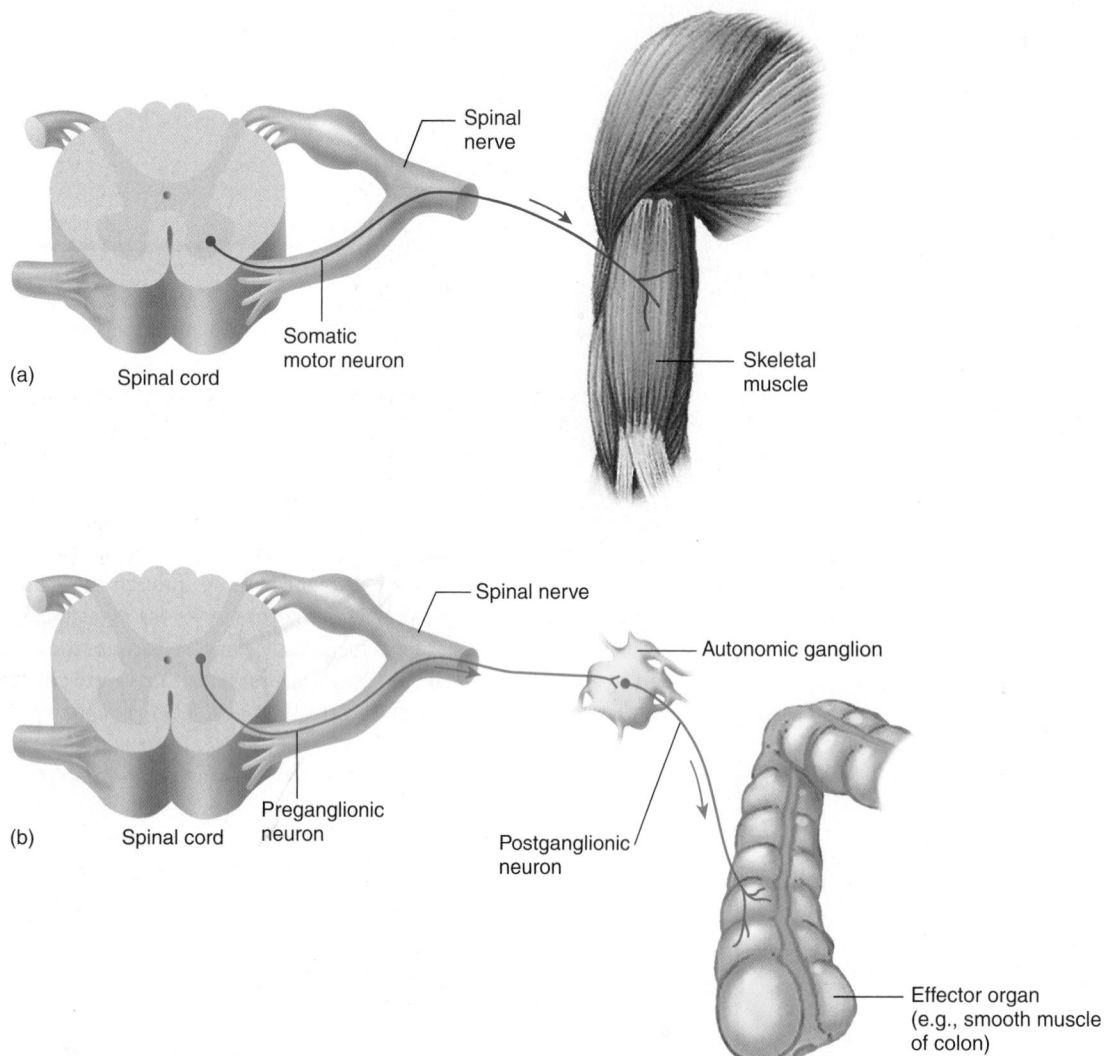

Figure 16.1   Organization of Somatic and Autonomic Nervous System Neurons

(*a*) The cell body of the somatic neuron is in the CNS, and its axon extends to the skeletal muscle. (*b*) The cell body of the preganglionic neuron is in the CNS, and its axon extends to the autonomic ganglion and synapses with the postganglionic neuron. The postganglionic neuron extends to and synapses with its effector organ.

The effect of somatic motor neurons on skeletal muscle is always excitatory, but the effect of the ANS on target tissues can be excitatory or inhibitory. For example, after a meal, the ANS can stimulate stomach activities, but during exercise, the ANS can inhibit those activities. Table 16.1 summarizes the differences between the somatic nervous system and the ANS.

Sensory neurons are not classified as somatic or autonomic. These neurons propagate action potentials from sensory receptors to the CNS and can provide information for reflexes mediated through the somatic nervous system or the ANS. For example, stimulation of pain receptors can initiate somatic reflexes such as the withdrawal reflex and autonomic reflexes such as an increase in heart rate. Although some sensory neurons primarily affect somatic functions and others primarily influence autonomic functions, functional overlap makes attempts to classify sensory neurons as either somatic or autonomic meaningless.

1. Contrast the somatic nervous system with the ANS for each of the following:
   a. the number of neurons between the CNS and effector organ
   b. the location of neuron cell bodies
   c. the structures each innervates
   d. inhibitory or excitatory effects
   e. conscious or unconscious control
2. Why are sensory neurons not classified as somatic or autonomic?
3. Define the terms preganglionic neuron, postganglionic neuron, and autonomic ganglia.

Anatomy of the Autonomic Nervous System

The ANS is subdivided into the **sympathetic** and the **parasympathetic divisions** and the **enteric** (en-ter′ik; bowels) **nervous system (ENS).** The sympathetic and parasympathetic divisions differ structurally in (1) the location of their preganglionic neuron cell bodies within the CNS and (2) the location of their autonomic ganglia.

The enteric nervous system is a complex network of neuron cell bodies and axons within the wall of the digestive tract. An important part of this network is sympathetic and parasympathetic neurons. For this reason, the enteric nervous system is considered to be part of the ANS.

Sympathetic Division

Cell bodies of sympathetic preganglionic neurons are in the lateral horns of the spinal cord gray matter between the first thoracic (T1) and the second lumbar (L2) segments (figure 16.2). Because of the location of the preganglionic cell bodies, the sympathetic division is sometimes called the **thoracolumbar division.** The axons of the preganglionic neurons pass through the ventral roots of spinal nerves T1–L2, course through the spinal nerves for a short distance, leave these nerves, and project to autonomic ganglia on either side of the vertebral column behind the parietal pleura. These ganglia are called **sympathetic chain ganglia,** because they are connected to one another and form a chain, or **paravertebral ganglia,** because they are located along both sides of the vertebral

Table 16.1	Comparison of the Somatic and Autonomic Nervous Systems	
Features	**Somatic Nervous System**	**Autonomic Nervous System**
Target tissues	Skeletal muscle	Smooth muscle, cardiac muscle, and glands
Regulation	Controls all conscious and unconscious movements of skeletal muscle	Unconscious regulation, although influenced by conscious mental functions
Response to stimulation	Skeletal muscle contracts	Target tissues are stimulated or inhibited
Neuron arrangement	One neuron extends from the central nervous system (CNS) to skeletal muscle	Two neurons in series; the preganglionic neuron extends from the CNS to an autonomic ganglion, and the postganglionic neuron extends from the autonomic ganglion to the target tissue
Neuron cell body location	Neuron cell bodies are in motor nuclei of the cranial nerves and in the ventral horn of the spinal cord	Preganglionic neuron cell bodies are in autonomic nuclei of the cranial nerves and in the lateral part of the spinal cord; postganglionic neuron cell bodies are in autonomic ganglia
Number of synapses	One synapse between the somatic motor neuron and the skeletal muscle	Two synapses; first is in the autonomic ganglia; second is at the target tissue
Axon sheaths	Myelinated	Preganglionic axons are myelinated; postganglionic axons are unmyelinated
Neurotransmitter substance	Acetylcholine	Acetylcholine is released by preganglionic neurons; either acetylcholine or norepinephrine is released by postganglionic neurons
Receptor molecules	Receptor molecules for acetylcholine are nicotinic	In autonomic ganglia, receptor molecules for acetylcholine are nicotinic; in target tissues, receptor molecules for acetylcholine are muscarinic, whereas receptor molecules for norepinephrine are either α- or β-adrenergic

— Preganglionic neuron
...... Postganglionic neuron

Preganglionic cell body in lateral horn of gray matter

Preganglionic neuron to sympathetic chain ganglion

Postganglionic neurons

Preganglionic neuron to collateral ganglion

Postganglionic neurons

Collateral ganglia

Sympathetic chain ganglia

T1

L2

Figure 16.2 Sympathetic Division

The location of sympathetic preganglionic (*solid blue*) and postganglionic (*dotted blue*) neurons. The preganglionic cell bodies are in the lateral gray matter of the thoracic and lumbar parts of the spinal cord. The cell bodies of the postganglionic neurons are within the sympathetic chain ganglia or within collateral ganglia.

column. Only the ganglia from T1–L2 receive preganglionic axons from the spinal cord, although the sympathetic chain extends into the cervical and sacral regions so that one pair of ganglia is associated with nearly every pair of spinal nerves. The cervical ganglia usually fuse during fetal development so only two or three pairs exist in the adult.

The axons of preganglionic neurons are small in diameter and myelinated. The short connection between a spinal nerve and a sympathetic chain ganglion through which the preganglionic axons pass is called a **white ramus communicans** (rā′mĭs kŏ-mū′ni-kans; pl., rami communicantes, rā′mī kŏ-mū-ni-kan′tēz) because of the whitish color of the myelinated axons (figure 16.3).

Sympathetic axons exit the sympathetic chain ganglia by the following four routes:

1. *Spinal nerves* (figure 16.3*a*). Preganglionic axons synapse with postganglionic neurons in sympathetic chain ganglia at the same level that the preganglionic axons enter the sympathetic chain. Alternatively, preganglionic axons pass either superiorly or inferiorly through one or more ganglia and synapse with postganglionic neurons in a sympathetic chain ganglion at a different level. Axons of

the postganglionic neurons pass through a **gray ramus communicans** and reenter a spinal nerve. Postganglionic axons are not myelinated, thereby giving the gray ramus communicans its grayish color. The postganglionic axons then project through the spinal nerve to the organs they innervate.

2. *Sympathetic nerves* (figure 16.3*b*). Preganglionic axons enter the sympathetic chain and synapse in a sympathetic chain ganglion at the same or a different level with postganglionic neurons. The postganglionic axons leaving the sympathetic chain ganglion form **sympathetic nerves.**

3. *Splanchnic* (splangk′nik) *nerves* (figure 16.3*c*). Some preganglionic axons enter sympathetic chain ganglia and, without synapsing, exit at the same or a different level to form **splanchnic nerves.** Those preganglionic axons extend to **collateral,** or **prevertebral, ganglia,** where they synapse with postganglionic neurons. Axons of the postganglionic neurons leave the collateral ganglia through small nerves that extend to target organs.

4. *Innervation to the adrenal gland* (figure 16.3*d*). The splanchnic nerve innervation to the adrenal glands is different from other ANS nerves because it consists of only preganglionic neurons. Axons of the preganglionic neurons do not synapse in sympathetic chain ganglia or in collateral ganglia. Instead, the axons pass through those ganglia and synapse with cells in the adrenal medulla. The **adrenal medulla** (me-dool′ă) is the inner portion of the adrenal gland and consists of specialized cells derived during embryonic development from neural crest cells (see figure 13.13), which are the same population of cells that give rise to the postganglionic cells of the ANS. Adrenal medullary cells are round in shape, have no axons or dendrites, and are divided into two groups. About 80% of the cells secrete **epinephrine** (ep′i-nef′rin), also called **adrenaline** (ă-dren′ă-lin), and about 20% secrete **norepinephrine** (nōr′ep-i-nef′rin), also called **noradrenaline** (nōr-ă-dren′ă-lin). Stimulation of these cells by preganglionic axons causes the release of epinephrine and norepinephrine. These substances circulate in the blood and affect all tissues having receptors to which they can bind. The general response to epinephrine and norepinephrine released from the adrenal medulla is to prepare the individual for physical activity. Secretions of the adrenal medulla are considered hormones because they are released into the general circulation and travel some distance to the tissues in which they have their effect (see chapters 17 and 18).

4. *Where are the cell bodies of sympathetic preganglionic neurons located?* lateral horn of gray matter between T1 L2

5. *What types of axon (preganglionic or postganglionic, myelinated or unmyelinated) are found in white and gray rami communicantes?*

6. *Where do preganglionic neurons synapse with postganglionic neurons that are found in spinal and sympathetic nerves?* Sympathetic chain ganglia

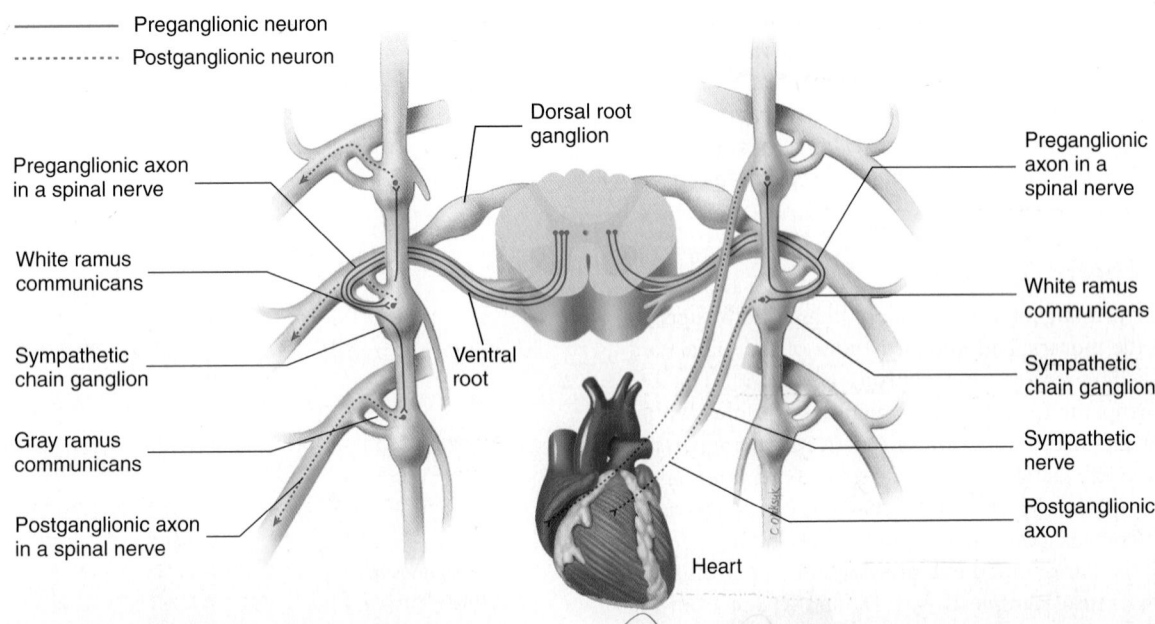

(a) Preganglionic axons from a spinal nerve pass through a white ramus communicans into a sympathetic chain ganglion. Some axons synapse with a postganglionic neuron at the level of entry; others ascend or descend to other levels before synapsing. Each postganglionic axon exits the sympathetic chain through a gray ramus communicans and enters a spinal nerve.

(b) Like part (a), except that each postganglionic neuron exits a sympathetic chain ganglion through a sympathetic nerve.

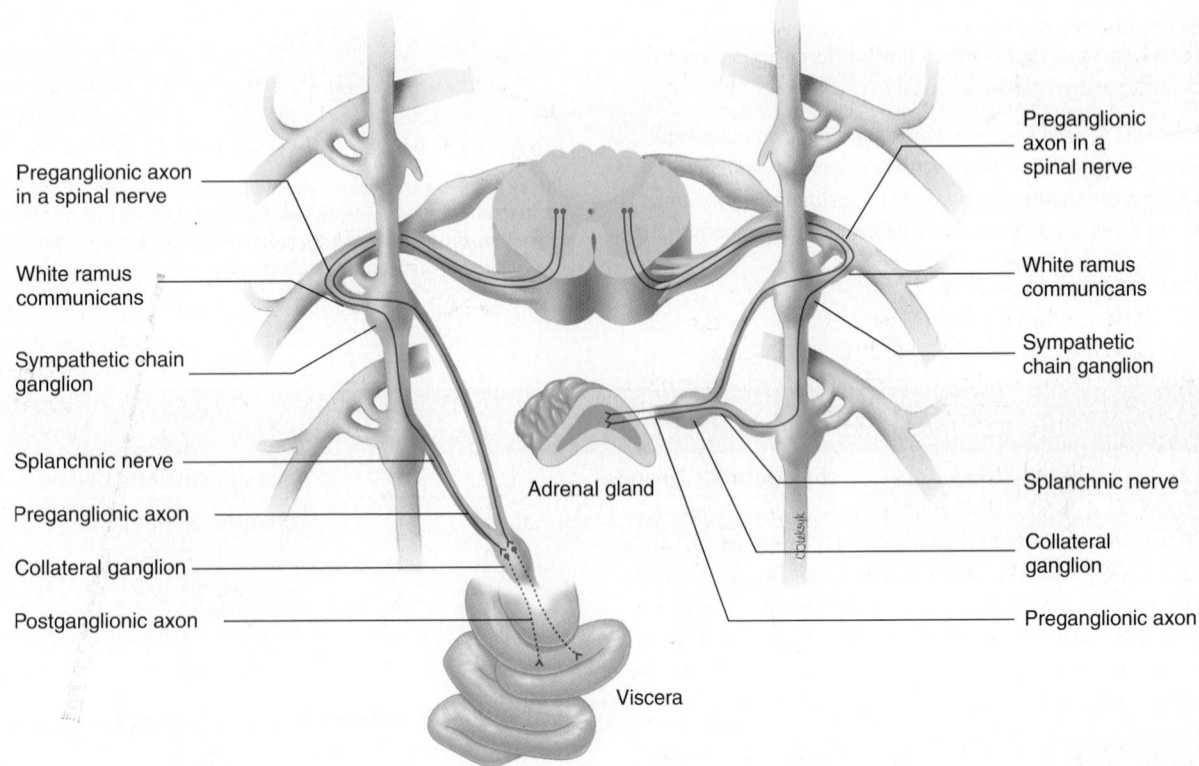

(c) Preganglionic neurons do not synapse in the sympathetic chain ganglia, but exit in splanchnic nerves and extend to a collateral ganglion, where they synapse with postganglionic neurons.

(d) Like part (c), except that the preganglionic axons extend to the adrenal medulla, where they synapse with specialized adrenal medullary cells.

Figure 16.3 Routes Taken by Sympathetic Axons

Preganglionic axons are illustrated as solid lines and postganglionic axons as dashed lines.

7. *Where do preganglionic axons that form splanchnic nerves (except those to the adrenal gland) synapse with postganglionic neurons?* prevertebral ganglia

8. *What is unusual about the splanchnic nerve innervation to the adrenal gland? What do the specialized cells of the adrenal medulla secrete, and what is the effect of these substances?* epinephrine physical activity

Parasympathetic Division

Parasympathetic preganglionic neurons are located both superior and inferior to the thoracic and lumbar regions of the spinal cord where sympathetic preganglionic neurons are found. The cell bodies of parasympathetic preganglionic neurons are either within cranial nerve nuclei in the brainstem or within the lateral parts of the gray matter in the sacral region of the spinal cord from S2–S4 (figure 16.4). For that reason, the parasympathetic division is sometimes called the **craniosacral** (krā′nē-ō-sā′krăl) **division.**

Axons of the parasympathetic preganglionic neurons from the brain are in **cranial nerves** III, VII, IX, and X; and from the spinal cord in **pelvic nerves.** The preganglionic axons course through these nerves to **terminal ganglia** where they synapse with postganglionic neurons. The axons of the postganglionic neurons extend relatively short distances from the terminal ganglia to the target organs. The terminal ganglia are either near or embedded within the walls of the organs innervated by the parasympathetic neurons. Many of the parasympathetic ganglia are small in size, but some, such as those in the wall of the digestive tract, are large.

Table 16.2 summarizes the structural differences between the sympathetic and parasympathetic divisions.

9. *Where are the cell bodies of parasympathetic preganglionic neurons located? In what structure do parasympathetic preganglionic neurons synapse with postganglionic neurons?* brainstem an inferior spinal cord (sacral) region

10. *What nerves are formed by the axons of parasympathetic preganglionic neurons?*

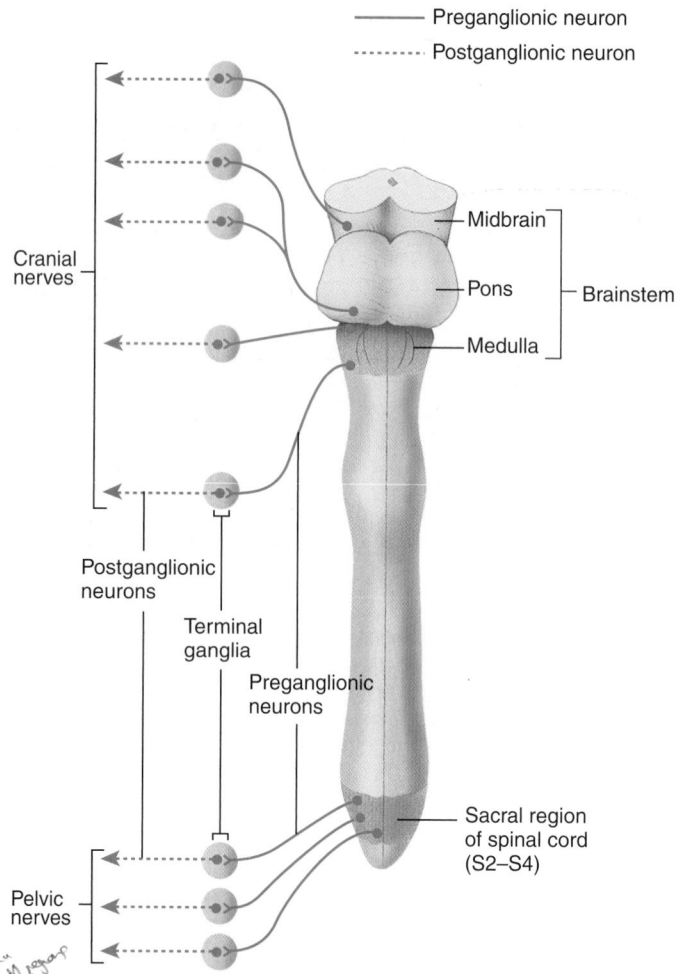

Figure Legend:
— Preganglionic neuron
----- Postganglionic neuron

Cranial nerves
Midbrain
Pons — Brainstem
Medulla
Postganglionic neurons
Terminal ganglia
Preganglionic neurons
Sacral region of spinal cord (S2–S4)
Pelvic nerves

Figure 16.4   Parasympathetic Division

The location of parasympathetic preganglionic (*solid red*) and postganglionic (*dotted red*) neurons. The preganglionic neuron cell bodies are in the brainstem and the lateral gray matter of the sacral part of the spinal cord, and the postganglionic neuron cell bodies are within terminal ganglia.

Table 16.2	Comparison of the Sympathetic and Parasympathetic Divisions	
Features	**Sympathetic Division**	**Parasympathetic Division**
Location of preganglionic cell body	Lateral horns of spinal cord gray matter (T1–L2)	Brainstem and lateral parts of spinal gray matter (S2–S4)
Outflow from the CNS	Spinal nerves Sympathetic nerves Splanchnic nerves	Cranial nerves Pelvic nerves
Ganglia	Sympathetic chain ganglia along spinal cord for spinal and sympathetic nerves; collateral ganglia for splanchnic nerves	Terminal ganglia near or on effector organ
Number of postganglionic neurons for each preganglionic neuron	Many (much divergence)	Few (less divergence)
Relative length of neurons	Short preganglionic Long postganglionic	Long preganglionic Short postganglionic

Enteric Nervous System

The enteric nervous system consists of nerve plexuses within the wall of the digestive tract (see figure 24.4). The plexuses have contributions from three sources: (1) sensory neurons that connect the digestive tract to the CNS, (2) ANS motor neurons that connect the CNS to the digestive tract, and (3) enteric neurons, which are confined to the enteric plexuses. The CNS is capable of monitoring the digestive tract and controlling its smooth muscle and glands through autonomic reflexes (see "Regulation of the Autonomic Nervous System" on p. 573). For example, stretch of the digestive tract is detected by sensory neurons and action potentials are transmitted to the CNS. In response, the CNS sends action potentials to glands in the digestive tract, causing them to secrete.

There are several major types of enteric neurons: (1) Enteric sensory neurons can detect changes in the chemical composition of the contents of the digestive tract or detect stretch of the digestive tract wall. (2) Enteric motor neurons can stimulate or inhibit smooth muscle contraction and gland secretion. (3) Enteric interneurons connect enteric sensory and motor neurons to each other. Enteric neurons are capable of monitoring and controlling the digestive tract independently of the CNS through local reflexes (see "Regulation of the Autonomic Nervous System" on p. 575). For example, stretch of the digestive tract is detected by enteric sensory neurons, which stimulate enteric interneurons. The enteric interneurons stimulate enteric motor neurons, which stimulate glands to secrete. Although the enteric nervous system is capable of controlling the activities of the digestive tract completely independently of the CNS, normally the two systems work together.

> **11. What is the enteric nervous system and where is it located?**
> **12. How does the CNS monitor and control the digestive tract?**
> **13. Name three major types of enteric neurons. How do the enteric neurons monitor and control the digestive tract?**

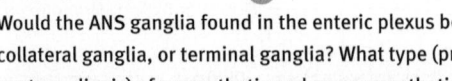

P R E D I C T   1

Would the ANS ganglia found in the enteric plexus be chain ganglia, collateral ganglia, or terminal ganglia? What type (preganglionic or postganglionic) of sympathetic and parasympathetic axons contribute to the enteric plexus?

The Distribution of Autonomic Nerve Fibers

Sympathetic Division

Sympathetic axons pass from the sympathetic chain ganglia to their target tissues through spinal, sympathetic, and splanchnic nerves. The sympathetic and splanchnic nerves can join **autonomic nerve plexuses,** which are complex, interconnected neural networks formed by neurons of the sympathetic and parasympathetic divisions. In addition, the axons of sensory neurons contribute to these plexuses.

The autonomic nerve plexuses typically are named according to organs they supply or to blood vessels along which they are found. For example, the cardiac plexus supplies the heart and the thoracic aortic plexus is found along the thoracic aorta.

Plexuses following the route of blood vessels is a major means by which autonomic axons are distributed throughout the body.

The major means by which sympathetic axons reach organs include the following:

1. *Spinal nerves.* From all levels of the sympathetic chain, some postganglionic axons project through gray rami communicates to spinal nerves. The axons extend to the same structures innervated by the spinal nerves and supply sweat glands in the skin, smooth muscle in skeletal and skin blood vessels, and the smooth muscle of the arrector pili. See figure 12.14 for the distribution of spinal nerves to the skin.

2. *Head and neck nerve plexuses.* Most of the sympathetic nerve supply to the head and neck is derived from the superior cervical sympathetic chain ganglion (figure 16.5). Postganglionic axons of sympathetic nerves form plexuses that extend superiorly to the head and inferiorly to the neck. The plexuses give off branches to supply sweat glands in the skin, smooth muscle in skeletal and skin blood vessels, and the smooth muscle of the arrector pili. Axons from the plexuses also join branches of the trigeminal nerves (cranial nerve V) to supply the skin of the face, the salivary glands, the iris, and the ciliary muscles of the eye.

3. *Thoracic nerve plexuses.* The sympathetic supply for organs of the thorax is mainly derived from the cervical and upper five thoracic sympathetic chain ganglia. Postganglionic axons in sympathetic nerves contribute to the **cardiac plexus,** supplying the heart, the **pulmonary plexus,** supplying the lungs, and other thoracic plexuses (see figure 16.5).

4. *Abdominopelvic nerve plexuses.* Sympathetic chain ganglia from T5 and below mainly supply the abdominopelvic organs. The preganglionic axons of splanchnic nerves synapse with postganglionic neurons in the collateral ganglia of abdominopelvic nerve plexuses. Postganglionic axons from the collateral ganglia innervate smooth muscle and glands in the abdominopelvic organs. There are several abdominopelvic nerve plexuses (see figure 16.5). The **celiac** (sē′lē-ak) **plexus** has two large celiac ganglia and other smaller ganglia. It supplies the diaphragm, stomach, spleen, liver, gallbladder, adrenal glands, kidneys, testes, and ovaries. The **superior mesenteric** (mez-en-ter′ik) **plexus** includes the superior mesenteric ganglion and supplies the pancreas, small intestine, ascending colon, and the transverse colon. The **inferior mesenteric plexus** includes the inferior mesenteric ganglion and supplies the transverse colon to the rectum. The **superior** and **inferior hypogastric plexuses** supply the descending colon to the rectum, the urinary bladder, and reproductive organs in the pelvis.

Parasympathetic Division

Parasympathetic outflow is through cranial and pelvic nerves. Branches of these nerves either supply organs or join nerve plexuses

Facial nerve
Glossopharyngeal nerve
Internal carotid plexus
Superior cervical sympathetic chain ganglion
Sympathetic nerves
Cervicothoracic ganglion
Sympathetic nerves
Fifth thoracic sympathetic chain ganglion
Greater splanchnic nerve
Spinal nerve
White ramus communicans
Gray ramus communicans
Lesser splanchnic nerve
Kidney
Second lumbar sympathetic chain ganglion
Lumbar splanchnic nerves
Sacral splanchnic nerves
Pelvic nerves
Sacral plexus
Rectum

Oculomotor nerve
Ciliary ganglion
Pterygopalatine ganglion
Otic ganglion
Submandibular ganglion
Vagus nerve
Pulmonary plexus
Cardiac plexus
Esophagus and esophageal plexus
Heart
Aorta and thoracic aortic plexus
Stomach
Celiac ganglion and plexus
Superior mesenteric ganglion and plexus
Aorta and abdominal aortic plexus
Small intestine
Inferior mesenteric ganglion and plexus
Superior hypogastric plexus
Colon
Inferior hypogastric plexus
Urinary bladder
Prostate gland

Sympathetic
Parasympathetic

Figure 16.5 Distribution of Autonomic Nerve Fibers

Sympathetic supply: (1) spinal nerves to limbs and body, (2) head and neck by sympathetic nerves from the superior cervical chain ganglia, (3) thoracic organs by sympathetic nerves from the cervical and thoracic chain ganglia (to T5) supplying thoracic nerve plexuses, and (4) abdominopelvic nerves by splanchnic nerves from chain ganglia below T5 supplying abdominopelvic nerve plexuses. *Parasympathetic supply:* (1) head and neck by cranial nerves and their ganglia, (2) thoracic organs by vagus nerves supplying thoracic plexuses, (3) abdominal organs by vagus nerves supplying abdominal nerve plexuses, and (4) pelvic nerves from S2–S4.

to be distributed to organs. The major means by which parasympathetic axons reach organs include the following:

1. *Cranial nerves supplying the head and neck.* Three pairs of cranial nerves have parasympathetic preganglionic axons that extend to terminal ganglia in the head. Postganglionic neurons from the terminal ganglia supply nearby structures. The parasympathetic cranial nerves, their terminal ganglia, and the structures innervated are (see figure 16.5 and table 13.5):
   a. The **oculomotor (III) nerve,** through the **ciliary** (sil′ē-ar-ē) **ganglion,** supplies the ciliary muscles and the iris of the eye.
   b. The **facial (VII) nerve,** through the **pterygopalatine** (ter′i-gō-pal′ă-tīn) **ganglion,** supplies the lacrimal gland and mucosal glands of the nasal cavity and palate. The facial nerve, through the **submandibular ganglion,** also supplies the submandibular and sublingual salivary glands.
   c. The **glossopharyngeal (IX) nerve,** through the **otic** (ō′tik) **ganglion,** supplies the parotid salivary gland.
2. *The vagus nerve and thoracic nerve plexuses.* Although cranial nerve X, the **vagus nerve,** has somatic motor and sensory functions in the head and neck, its parasympathetic distribution is to the thorax and abdomen. Preganglionic axons extend through the vagus nerves to the thorax, where they pass through branches of the vagus nerves to contribute to the cardiac plexus, which supplies the heart, and the pulmonary plexus, which supplies the lungs. The vagus nerves continue down the esophagus, and give off branches to form the **esophageal plexus.**
3. *Abdominal nerve plexuses.* After the esophageal plexus passes through the diaphragm, some of the vagal preganglionic axons supply terminal ganglia in the wall of the stomach, while others contribute to the celiac and superior mesenteric plexuses. Through these plexuses, the preganglionic axons supply terminal ganglia in the walls of the gallbladder, biliary ducts, pancreas, small intestine, ascending colon, and the transverse colon.
4. *Pelvic nerves and pelvic nerve plexuses.* Parasympathetic preganglionic axons whose cell bodies are in the S2–S4 region of the spinal cord pass to the ventral rami of spinal nerves and enter the pelvic nerves. The pelvic nerves supply the transverse colon to the rectum, and they also contribute to the hypogastric plexus. The hypogastric plexus and its derivatives supply the lower colon, rectum, urinary bladder, and organs of the reproductive system in the pelvis.

Sensory Neurons in Autonomic Nerve Plexuses

Although not strictly part of the ANS, the axons of sensory neurons run alongside ANS axons within ANS nerves and plexuses. Some of these sensory neurons are part of reflex arcs regulating organ activities. Sensory neurons also transmit pain and pressure sensations from organs to the CNS. The cell bodies of these sensory neurons are found in the dorsal root ganglia and in the sensory ganglia of certain cranial nerves, which are swellings on the nerves close to their attachment to the brain.

Effects of Spinal Cord Injury on ANS Functions

Spinal cord injury can damage nerve tracts and interrupt control of autonomic neurons by ANS centers in the brain. For the parasympathetic division, effector organs innervated through the sacral region of the spinal cord are affected, but most effector organs still have normal parasympathetic function because they are innervated by the vagus nerve. For the sympathetic division, brain control of sympathetic neurons is lost below the site of the injury. The higher the level of injury, the greater the number of body parts affected.

14. *Define autonomic nerve plexuses. How are they typically named?*
15. *Describe the four major ways by which sympathetic axons pass from sympathetic chain ganglia to reach organs. Name four thoracic and four abdominopelvic autonomic nerve plexuses.*
16. *List the four major means by which parasympathetic axons reach organs. List the cranial nerves and ganglia that supply the head and neck. What cranial nerve supplies the thoracic and abdominal nerve plexuses? To what plexus do pelvic nerves contribute?*

P R E D I C T   2
Starting in the small intestine and ending with the ganglia where their cell bodies are located, trace the route for sensory axons passing alongside sympathetic axons. Name all of the plexuses, nerves, ganglia, etc., that the sensory axon passes through. Also trace the route for sensory neurons passing alongside parasympathetic axons.

Physiology of the Autonomic Nervous System

Neurotransmitters

Sympathetic and parasympathetic nerve endings secrete one of two neurotransmitters. If the neuron secretes acetylcholine, it is a **cholinergic** (kol-in-er′jik) **neuron,** and if it secretes norepinephrine, it is an **adrenergic** (ad-rĕ-ner′jik) **neuron.** All preganglionic neurons of the sympathetic and parasympathetic divisions and all postganglionic neurons of the parasympathetic division are cholinergic. Almost all postganglionic neurons of the sympathetic division are adrenergic, but a few postganglionic neurons that innervate thermoregulatory sweat glands are cholinergic (figure 16.6).

In recent years, substances in addition to the regular neurotransmitters have been extracted from ANS neurons. These substances include nitric oxide; fatty acids, such as eicosanoids; peptides, such as gastrin, somatostatin, cholecystokinin, vasoactive intestinal peptide, enkephalins, and substance P; and monoamines, such as dopamine, serotonin, and histamine. The specific role that many of these compounds play in the regulation of the ANS is unclear, but they appear to function as either neurotransmitters or neuromodulator substances (see chapter 11).

Sympathetic division

Most target tissues innervated by the sympathetic division have adrenergic receptors. When norepinephrine (NE) binds to adrenergic receptors, some target tissues are stimulated, and others are inhibited. For example, smooth muscle cells in blood vessels are stimulated to constrict, and stomach glands are inhibited.

Sympathetic division

Some sympathetic target tissues, such as sweat glands, have muscarinic receptors, which respond to acetylcholine (ACh). Stimulation of sweat glands results in increased sweat production.

Parasympathetic division

All parasympathetic target tissues have muscarinic receptors. The general response to ACh is excitatory, but some target tissues, such as the heart, are inhibited.

Figure 16.6 Location of ANS Receptors

Nicotinic receptors are on the cell bodies of both sympathetic and parasympathetic postganglionic cells in the autonomic ganglia. *Abbreviations: NE, norepinephrine; ACh, acetylcholine.*

Receptors

Receptors for acetylcholine and norepinephrine are located in the plasma membrane of certain cells (table 16.3). The combination of neurotransmitter and receptor functions as a signal to cells, causing them to respond. Depending on the type of cell, the response can be excitatory or inhibitory.

Cholinergic Receptors

Receptors to which acetylcholine binds are called **cholinergic receptors.** They have two major structurally different forms. **Nicotinic** (nik-ō-tin′ik) **receptors** also bind to nicotine, an alkaloid substance found in tobacco; and **muscarinic** (mŭs-kă-rin′ik) **receptors** also bind to muscarine, an alkaloid extracted from some poisonous mushrooms. Although nicotine and muscarine are not naturally in

the human body, they demonstrate differences in the two classes of cholinergic receptors. Nicotine binds to nicotinic receptors but not to muscarinic receptors, whereas muscarine binds to muscarinic receptors but not to nicotinic receptors. On the other hand, nicotinic and muscarinic receptors are very similar because acetylcholine binds to and activates both types of receptors.

The membranes of all postganglionic neurons in autonomic ganglia and the membranes of skeletal muscle cells have nicotinic receptors. The membranes of effector cells that respond to acetylcholine released from postganglionic neurons have muscarinic receptors.

P R E D I C T

Would structures innervated by the sympathetic division or the parasympathetic division be affected after the consumption of nicotine? After the consumption of muscarine? Explain.

Table 16.3	Effects of the Sympathetic and Parasympathetic Divisions on Various Tissues	
Organ	**Sympathetic Effects and Receptor Type***	**Parasympathetic Effects and Receptor Type***
Adipose tissue	Fat breakdown and release of fatty acids (α_2, β_1)	None
Arrector pili muscle	Contraction (α_1)	None
Blood (platelets)	Increases coagulation (α_2)	None
Blood vessels		
Arterioles (carry blood to tissues)		
Digestive organs	Constriction (α_1)	None
Heart	Constriction (α_1), dilation (β_2)[†]	None
Kidneys	Constriction (α_1, α_2); dilation (β_1, β_2)	None
Lungs	Constriction (α_1), dilation (β_2)	None
Skeletal muscle	Constriction (α_1), dilation (β_2)	None
Skin	Constriction (α_1, α_2)	None
Veins (carry blood away from tissues)	Constriction (α_1, α_2), dilation (β_2)	
Eye		
Ciliary muscle	Relaxation for far vision (β_2)	Contraction for near vision (m)
Pupil	Dilated (α_1)[‡]	Constricted (m)[‡]
Gallbladder	Relaxation (β_2)	Contraction (m)
Glands		
Adrenal	Release of epinephrine and norepinephrine (n)	None
Gastric	Decreases gastric secretion (α_2)	Increases gastric secretion (m)
Lacrimal	Slight tear production (α)	Increases tear secretion (m)
Pancreas	Decreases insulin secretion (α_2)	Increases insulin secretion (m)
	Decreases exocrine secretion (α)	Increases exocrine secretion (m)
Salivary	Constriction of blood vessels and slight production of a thick, viscous saliva (α_1)	Dilation of blood vessels and thin, copious saliva (m)
Sweat		
Apocrine	Thick, organic secretion (m)	None
Merocrine	Watery sweat from most of the skin (m); sweat from the palms and soles (α_1)	None
Heart	Increases rate and force of contraction (β_1, β_2)	Decreases rate of contraction (m)
Liver	Glucose released into blood (α_1, β_2)	None
Lungs	Dilates air passageways (β_2)	Constricts air passageways (m)
Metabolism	Increases up to 100% (α, β)	None
Sex organs	Ejaculation (α_1), erection[§]	Erection (m)
Skeletal muscles	Breakdown of glycogen to glucose (β_2)	None
Stomach and intestines		
Wall	Decreases tone (α_1, α_2, β_2)	Increases motility (m)
Sphincter	Increases tone (α_1)	Decreases tone (m)
Urinary bladder		
Wall (detrusor)	None	Contraction (m)
Neck of bladder	Contraction (α_1)	Relaxation (m)
Internal urinary sphincter	Contraction (α_1)	Relaxation (m)

*When known, receptor subtypes are indicated. The receptors are α_1- and α_2-adrenergic, β_1- and β_2-adrenergic, nicotinic cholinergic (n), and muscarinic cholinergic (m).

[†]Normally blood flow increases through coronary arteries because of increased demand by cardiac tissue for oxygen (local control of blood flow is discussed in chapter 21). In experiments that isolate the coronary arteries, sympathetic nerve stimulation, acting through α-adrenergic receptors, causes vasoconstriction. The β-adrenergic receptors are relatively insensitive to sympathetic nerve stimulation but can be activated by epinephrine released from the adrenal gland and by drugs. As a result, coronary arteries vasodilate.

[‡]Contraction of the radial muscles of the iris causes the pupil to dilate. Contraction of the circular muscles causes the pupil to constrict (see chapter 15).

[§]Decreased stimulation of alpha receptors by the sympathetic division can cause vasodilation of penile blood vessels, resulting in an erection.

Clinical Focus The Influence of Drugs on the Autonomic Nervous System

Some drugs that affect the ANS have important therapeutic value in treating certain diseases because they can increase or decrease activities normally controlled by the ANS. Chemicals that affect the ANS are also found in medically hazardous substances such as tobacco and insecticides.

Direct-acting and indirect-acting drugs influence the ANS. Direct-acting drugs bind to ANS receptors to produce their effects. For example, **agonists** or **stimulating agents** bind to specific receptors and activate them, and **antagonists** or **blocking agents** bind to specific receptors and prevent them from being activated. The main topic of this Clinical Focus is direct-acting drugs. It should be noted, however, that some indirect-acting drugs produce a stimulatory effect by causing the release of neurotransmitters or by preventing the metabolic breakdown of neurotransmitters. Other indirect-acting drugs produce an inhibitory effect by preventing the biosynthesis or release of neurotransmitters.

Drugs That Bind to Nicotinic Receptors

Drugs that bind to nicotinic receptors and activate them are **nicotinic agents.** Although these agents have little therapeutic value and are mainly of interest to researchers, nicotine is medically important because of its presence in tobacco. Nicotinic agents bind to the nicotinic receptors on all postganglionic neurons within autonomic ganglia and produce stimulation. Responses to nicotine are variable and depend on the amount taken into the body. Because nicotine stimulates the postganglionic neurons of both the sympathetic and parasympathetic divisions, much of the variability of its effects results from the opposing actions of these divisions. For example, in response to the nicotine contained in a cigarette, the heart rate may either increase or decrease. Heart rate rhythm tends to become less regular as a result of the simultaneous actions on the sympathetic division, which increases the heart rate, and the parasympathetic division, which decreases the heart rate. Blood pressure tends to increase because of the constriction of blood vessels, which are almost exclusively innervated by sympathetic neurons. In addition to its influence on the ANS, nicotine also affects the CNS; therefore, not all of its effects can be explained on the basis of action on the ANS. Nicotine is extremely toxic, and small amounts can be lethal.

Drugs that bind to and block nicotinic receptors are called **ganglionic blocking agents** because they block the effect of acetylcholine on both parasympathetic and sympathetic postganglionic neurons. The effect of these substances on the sympathetic division, however, overshadows the effect on the parasympathetic division. For example, trimethaphan camsylate (trĭ-meth′ă-fan kam′sil-āt), used to treat high blood pressure, blocks sympathetic stimulation of blood vessels, causing the blood vessels to dilate, which decreases blood pressure. Ganglionic blocking agents have limited uses because they affect both sympathetic and parasympathetic ganglia. Whenever possible, more selective drugs, which affect receptors of target tissues, are now used.

Drugs That Bind to Muscarinic Receptors

Drugs that bind to and activate muscarinic receptors are **muscarinic,** or **parasympathomimetic** (par-ă-sim′pă-thō-mi-met′ik), **agents.** These drugs activate the muscarinic receptors of target tissues of the parasympa-

---

Acetylcholine binding to nicotinic receptors has an excitatory effect because it results in the direct opening of Na^+ channels and the production of action potentials. When acetylcholine binds to muscarinic receptors, the cell's response is mediated through G proteins (see chapters 3 and 17). The response is either excitatory or inhibitory, depending on the target tissue in which the receptors are found. For example, acetylcholine binds to muscarinic receptors in cardiac muscle, thereby reducing heart rate; and acetylcholine binds to muscarinic receptors in smooth muscle cells of the stomach, thus increasing its rate of contraction.

- 17. *Define cholinergic and adrenergic neurons. Which neurons of the ANS are cholinergic and adrenergic?*
- 18. *Name the two major subtypes of cholinergic receptors. Where are they located? When acetylcholine binds to each subtype, does it result in an excitatory or inhibitory cell response?*

Adrenergic Receptors

Norepinephrine or epinephrine can bind to **adrenergic receptors.** Norepinephrine that is released from adrenergic postganglionic neurons of the sympathetic division (see figure 16.6) diffuses across the synapse and binds to receptor molecules within the plasma membranes of effector organs. Epinephrine and norepinephrine released from the adrenal glands and carried to effector organs by the blood can also bind to adrenergic receptors. The response of cells to norepinephrine or epinephrine binding to adrenergic receptors is mediated through G proteins (see chapters 3 and 17).

Adrenergic receptors are subdivided into two major categories: **alpha (α) receptors** and **beta (β) receptors,** each of which has subtypes. The main subtypes for alpha receptors are α_1- and α_2-adrenergic receptors and for beta receptors are β_1- and β_2-adrenergic receptors. Activation of α_1 and β_1 receptors generally produces a stimulatory response. For example, stimulation of α_1 receptors in most smooth muscle and β_1 receptors in cardiac muscle results in contraction. The response to the activation of α_2 and β_2 receptors varies so much with different target cells that no simple generalization about their effects is appropriate. Activation of α_2 receptors on platelets promotes blood clotting but decreases insulin secretion by the pancreas; activation of β_2 receptors stimulates the liver to release glucose but causes smooth muscle relaxation.

The α_1 and β_1 receptors are typically found in the membranes of target cells in the vicinity of sympathetic nerve terminals. Thus, the sympathetic division controls target cells with α_1 and β_1 receptors through sympathetic nerves. For example, at rest, stimulation of α_1 receptors at sympathetic nerve terminals in smooth muscle cells of blood vessels results in partial constriction of the vessels.

thetic division and the muscarinic receptors of sweat glands, which are innervated by the sympathetic division. Muscarine causes increased sweating, increased secretion of glands in the digestive system, decreased heart rate, constriction of the pupils, and contraction of respiratory, digestive, and urinary system smooth muscles. Bethanechol (be-than′ĕ-kol) chloride is a parasympathomimetic agent used to stimulate the urinary bladder following surgery, because the general anesthetics used for surgery can temporarily inhibit a person's ability to urinate.

Drugs such as atropine that bind to and block the action of muscarinic receptors are **muscarinic, or parasympathetic, blocking agents.** These drugs dilate the pupil of the eye and are used during eye examinations to help the examiner see the retina through the pupil. They also decrease salivary secretion and are used during surgery to prevent patients from choking on excess saliva while they are anesthetized.

Drugs That Bind to Alpha and Beta Receptors

Drugs that activate adrenergic receptors are **adrenergic,** or **sympathomimetic** (sim′pă-thō-mi-met′ik) **agents.** Drugs such as phenylephrine (fen-il-ef′rin) stimulate alpha receptors, which are numerous in the smooth muscle cells of certain blood vessels, especially in the digestive tract and the skin. These drugs increase blood pressure by causing vasoconstriction. On the other hand, albuterol (al-bū′ter-ol) is a drug that selectively activates beta receptors in cardiac muscle and bronchiolar smooth muscle. β-adrenergic-stimulating agents are sometimes used to dilate bronchioles in respiratory disorders such as asthma and are occasionally used as cardiac stimulants.

Drugs that bind to and block the action of alpha receptors are **α-adrenergic-blocking agents.** For example, prazosin (prā′zō-sin) hydrochloride is used to treat hypertension. By binding to alpha receptors in the smooth muscle of blood vessel walls, prazosin hydrochloride blocks the normal effects of norepinephrine released from sympathetic postganglionic neurons. Thus, the blood vessels relax, and blood pressure decreases.

Propranolol (prō-pran′ō-lōl) is an example of a **β-adrenergic-blocking agent.** These drugs are sometimes used to treat high blood pressure, some types of cardiac arrhythmias, and patients recovering from heart attacks. Blockage of the beta receptors within the heart prevents sudden increases in heart rate and thus decreases the probability of arrhythmic contractions.

Future Research

Our present knowledge of the ANS is more complicated than the broad outline presented here. In fact, each of the major receptor types has subtype receptors. For example, α-adrenergic receptors are subdivided into the following subgroups: α_{1A}-, α_{1B}-, α_{2A}-, and α_{2B}-adrenergic receptors. The exact number of subtypes in humans is not yet known; however, their existence suggests the possibility of designing drugs that affect only one subtype. For example, a drug that affects the blood vessels of the heart but not other blood vessels might be developed. Such drugs could produce specific effects yet would not produce undesirable side effects because they would act only on specific target tissues.

The sympathetic division regulates blood flow by slightly increasing or decreasing stimulation of the blood vessels. Increased stimulation causes further constriction and reduces blood flow, whereas decreased stimulation results in dilation and increases blood flow. Control of blood vessel diameter plays an important role in the regulation of blood flow and blood pressure (see chapter 20).

The α_2 and β_2 receptors are typically found in parts of the membrane that are not near nerve terminals releasing norepinephrine. These receptors respond to epinephrine and norepinephrine released from the adrenal glands into the blood. During exercise, epinephrine and norepinephrine bind to β_2 receptors and causes blood vessel dilation in skeletal muscles.

Dopamine and the Treatment of Shock

Norepinephrine is produced from a precursor molecule called dopamine. Certain sympathetic neurons release dopamine, which binds to dopamine receptors. Dopamine is structurally similar to norepinephrine and also binds to beta receptors. Dopamine hydrochloride has been used successfully to treat circulatory shock because it can bind to dopamine receptors in kidney blood vessels. The resulting vasodilation increases blood flow to the kidneys and prevents kidney damage. At the same time, dopamine can bind to beta receptors in the heart, causing stronger contractions.

19. *Name the two major subtypes of adrenergic receptors. Where are they located?*
20. *On what part of a cell are α_1- and β_1-adrenergic receptors typically found? How are they typically stimulated? What type of response is generally produced when they are stimulated?*
21. *On what part of a cell are α_2- and β_2-adrenergic receptors typically found? How are they typically stimulated? What type of responses are produced when they are stimulated?*

P R E D I C T   4

Injection of a small dose of epinephrine causes vasodilation of skeletal muscle blood vessels. An injection of a large dose, however, causes vasoconstriction. Explain.

Regulation of the Autonomic Nervous System

Much of the regulation of structures by the ANS occurs through autonomic reflexes, but input from the cerebrum, hypothalamus, and other areas of the brain allows conscious thoughts and actions, emotions, and other CNS activities to influence autonomic functions. Without the regulatory activity of the ANS, an individual has limited ability to maintain homeostasis.

Autonomic reflexes, like other reflexes, involve sensory receptors, sensory neurons, interneurons, motor neurons, and effector cells (figure 16.7; see chapter 12). For example, **baroreceptors** (stretch receptors) in the walls of large arteries near the heart detect changes in blood pressure, and sensory neurons transmit information from the baroreceptors through the glossopharyngeal and vagus nerves to the medulla oblongata. Interneurons in the medulla oblongata integrate the information, and action potentials are produced in auto-nomic neurons that extend to the heart. If baroreceptors detect a change in blood pressure, autonomic reflexes change heart rate, which returns blood pressure to normal. A sudden increase in blood pressure initiates a parasympathetic reflex that inhibits cardiac muscle cells and reduces heart rate, thus bringing blood pressure down toward its normal value. Conversely, a sudden decrease in blood pressure initiates a sympathetic reflex, which stimulates the heart to increase its rate and force of contraction, thus increasing blood pressure.

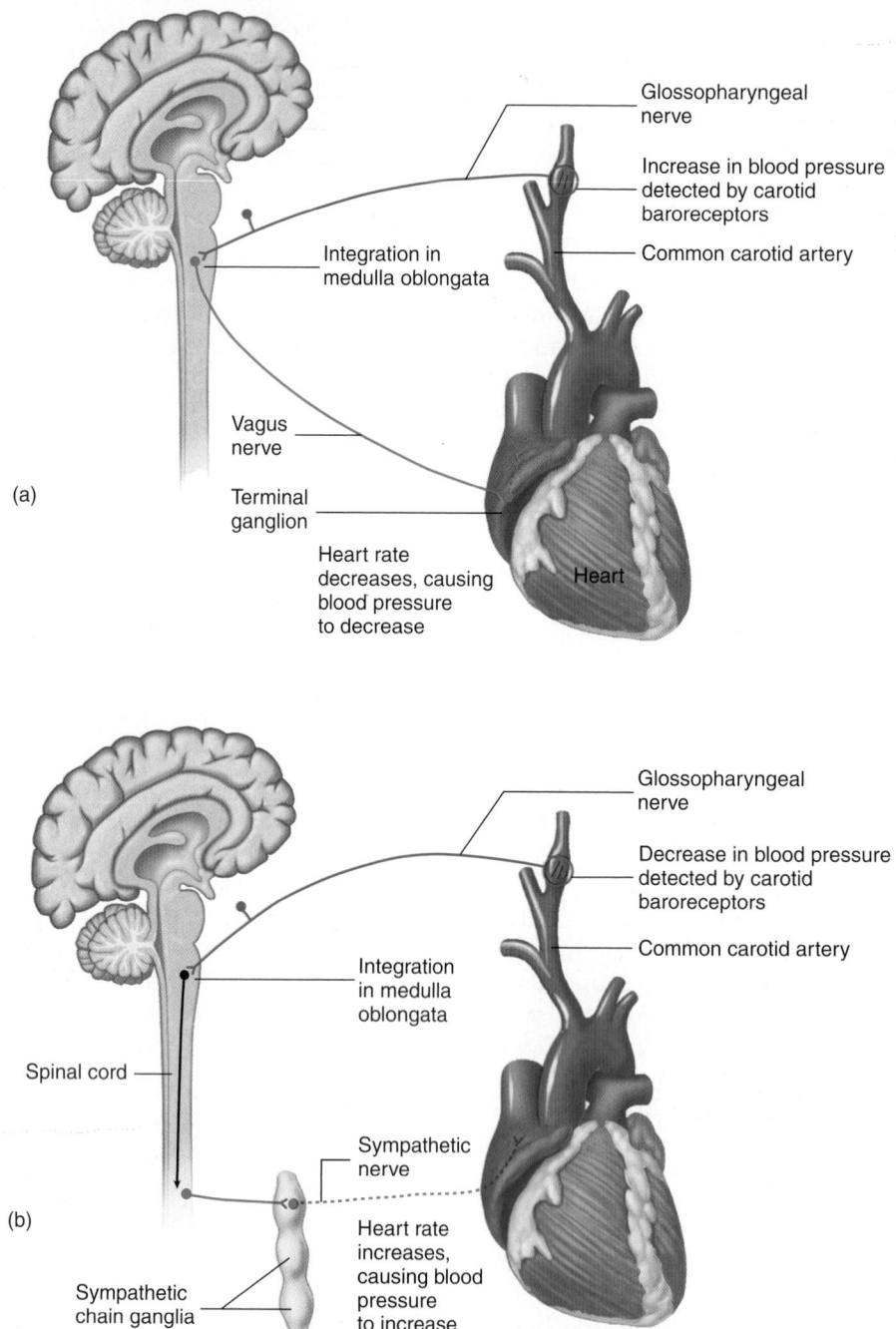

Figure 16.7   Autonomic Reflexes

Sensory input from the carotid baroreceptors is sent along the glossopharyngeal nerves to the medulla oblongata. The input is integrated in the medulla, and motor output is sent to the heart. (*a*) **Parasympathetic reflex.** Increased blood pressure results in increased stimulation of the heart by the vagus nerves, which increases inhibition of the heart and lowers heart rate. (*b*) **Sympathetic reflex.** Decreased blood pressure results in increased stimulation of the heart by sympathetic nerves, which, in turn, increases stimulation of the heart and increases heart rate and the force of contraction.

P R E D I C T

Sympathetic neurons stimulate sweat glands in the skin. Predict how they function to control body temperature during exercise and during exposure to cold temperatures.

Other autonomic reflexes participate in the regulation of blood pressure (see chapter 21). For example, numerous sympathetic neurons transmit a low but relatively constant frequency of action potentials that stimulate blood vessels throughout the body, keeping them partially constricted. If the vessels constrict further, blood pressure increases; and if they dilate, blood pressure decreases. Thus, altering the frequency of action potentials delivered to blood vessels along sympathetic neurons can either raise or lower blood pressure.

P R E D I C T

How would sympathetic reflexes that control blood vessels respond to a sudden decrease and a sudden increase in blood pressure?

The brainstem and the spinal cord contain important autonomic reflex centers responsible for maintaining homeostasis (figure 16.8). The hypothalamus, however, is in overall control of the ANS. Almost any type of autonomic response can be evoked by stimulating some part of the hypothalamus, which, in turn, stimulates ANS centers in the brainstem or spinal cord. Although there is overlap, stimulation of the posterior hypothalamus produces sympathetic responses, whereas stimulation of the anterior hypothalamus produces parasympathetic responses. In addition, the hypothalamus monitors and controls body temperature.

The hypothalamus has connections with the cerebrum and is an important part of the limbic system, which plays an important role in emotions. The hypothalamus integrates thoughts and emotions to produce ANS responses. Pleasant thoughts of a delicious banquet initiate increased secretion by salivary glands and by glands within the stomach and increased smooth muscle contractions within the digestive system. These responses are controlled by parasympathetic neurons. Emotions like anger increase blood pressure by increasing heart rate and constricting blood vessels through sympathetic stimulation.

The enteric nervous system is involved with autonomic and local reflexes that regulate the activity of the digestive tract. Autonomic reflexes help control the digestive tract because sensory neurons of the enteric plexuses supply the CNS with information about intestinal contents and ANS neurons to the enteric plexuses affect the responses of smooth muscle and glands within the digestive tract wall. For example, sensory neurons detecting stretch of the digestive tract wall send action potentials to the CNS. In response, the CNS sends action potentials out the ANS, causing smooth muscle in the digestive tract wall to contract.

The neurons of the enteric nervous system also operate independently of the CNS to produce local reflexes. A **local reflex** does not involve the CNS, but it does produce an involuntary, unconscious, stereotypic response to a stimulus. For example, sensory neurons not connected to the CNS detect stretch of the digestive tract

Cerebrum and limbic system — Thoughts and emotions can influence ANS functions through the hypothalamus

Hypothalamus — ANS integrating center that interacts with the cerebrum, limbic system, brainstem, and spinal cord; also regulates body temperature

Brainstem — ANS reflex centers for controlling pupil size, accommodation, tear production, salivation, coughing, swallowing, digestive activities, heart rate and force of contraction, blood vessel diameter, and respiration

Spinal cord — ANS reflex centers for regulating defecation, urination, penile and clitoral erection, and ejaculation

Figure 16.8 **Influence of Higher Parts of the Brain on Autonomic Functions**

The hypothalamus and the cerebrum influence the ANS. Neural pathways extend from the cerebrum to the hypothalamus and from the hypothalamus to neurons of the ANS.

wall. These sensory neurons send action potentials through the enteric plexuses to motor neurons, causing smooth muscle contraction or relaxation. See chapter 24 for more information on local reflexes.

> 22. *Name the components of an autonomic reflex. Describe the autonomic reflex that maintains blood pressure by altering heart rate or the diameter of blood vessels.*
> 23. *What part of the CNS stimulates ANS reflexes and integrates thoughts and emotions to produce ANS responses?*
> 24. *Define a local reflex. Explain how the enteric nervous system operates to produce local reflexes.*

Functional Generalizations About the Autonomic Nervous System

Generalizations can be made about the function of the ANS on effector organs, but most of the generalizations have exceptions.

Stimulatory Versus Inhibitory Effects

Both divisions of the ANS produce stimulatory and inhibitory effects. For example, the parasympathetic division stimulates contraction of the urinary bladder and inhibits the heart, causing a decrease in heart rate. The sympathetic division causes vasoconstriction by stimulating smooth muscle contraction in blood vessel walls and produces dilation of lung air passageways by inhibiting smooth muscle contraction in the walls of the passageways. Thus, it is *not* true that one division of the ANS is always stimulatory and the other is always inhibitory.

Dual Innervation

Most organs that receive autonomic neurons are innervated by both the parasympathetic and the sympathetic divisions (figure 16.9). The gastrointestinal tract, heart, urinary bladder, and reproductive tract are examples (see table 16.3). Dual innervation of organs by both divisions of the ANS is not universal, however. Sweat glands and blood vessels, for example, are innervated by sympathetic neurons almost exclusively. In addition, most structures receiving dual innervation are not regulated equally by both divisions. For example, parasympathetic innervation of the gastrointestinal tract is more extensive and exhibits a greater influence than does sympathetic innervation.

Opposite Effects

When a *single* structure is innervated by both autonomic divisions, the two divisions usually produce opposite effects on the structure. As a consequence, the ANS is capable of both increasing and decreasing the activity of the structure. In the gastrointestinal tract, for example, parasympathetic stimulation increases secretion from glands, whereas sympathetic stimulation decreases secretion. In a few instances, however, the effect of the two divisions is not clearly opposite. For example, both divisions of the ANS increase salivary secretion: the parasympathetic division initiates the production of a large volume of thin, watery saliva, and the sympathetic division causes the secretion of a small volume of viscous saliva.

Cooperative Effects

One autonomic division can coordinate the activities of *different* structures. For example, the parasympathetic division stimulates the pancreas to release digestive enzymes into the small intestine and stimulates contractions to mix the digestive enzymes with food within the small intestine. These responses enhance the digestion and absorption of the food.

Both divisions of the ANS can act together to coordinate the activity of *different* structures. In the male, the parasympathetic division initiates erection of the penis, and the sympathetic division stimulates the release of secretions from male reproductive glands and helps initiate ejaculation in the male reproductive tract.

General Versus Localized Effects

The sympathetic division has a more general effect than the parasympathetic division because activation of the sympathetic division often causes secretion of both epinephrine and norepinephrine from the adrenal medulla. These hormones circulate in the blood and stimulate effector organs throughout the body. Because circulating epinephrine and norepinephrine can persist for a few minutes before being broken down, they can also produce an effect for a longer time than the direct stimulation of effector organs by postganglionic sympathetic axons.

The sympathetic division diverges more than the parasympathetic division. Each sympathetic preganglionic neuron synapses with many postganglionic neurons, whereas each parasympathetic preganglionic neuron synapses with about two postganglionic neurons. Consequently, stimulation of sympathetic preganglionic neurons can result in greater stimulation of an effector organ.

Sympathetic stimulation often activates many different kinds of effector organs at the same time as a result of CNS stimulation or epinephrine and norepinephrine release from the adrenal medulla. It's possible, however, for the CNS to selectively activate effector organs. For example, vasoconstriction of cutaneous blood vessels in a cold hand is not always associated with an increased heart rate or other responses controlled by the sympathetic division.

Functions at Rest Versus Activity

In cases in which both sympathetic and parasympathetic neurons innervate a single organ, the sympathetic division has a major influence under conditions of physical activity or stress, whereas the parasympathetic division tends to have a greater influence under resting conditions. The sympathetic division does play a major role during resting conditions, however, by maintaining blood pressure and body temperature.

In general, the sympathetic division decreases the activity of organs not essential for the maintenance of physical activity and shunts blood and nutrients to structures that are active during physical exercise. This is sometimes referred to as the **fight-or-flight response** (see Clinical Focus on "Biofeedback, Meditation, and the Fight-or-Flight Response" on p. 578). Typical responses produced by the sympathetic division during exercise include:

1. Increased heart rate and force of contraction increase blood pressure and the movement of blood.
2. As skeletal or cardiac muscle contracts, oxygen and nutrients are used and waste products are produced. During exercise, a decrease in oxygen and nutrients and an accumulation of waste products are stimuli that cause vasodilation of muscle blood vessels (see local control of

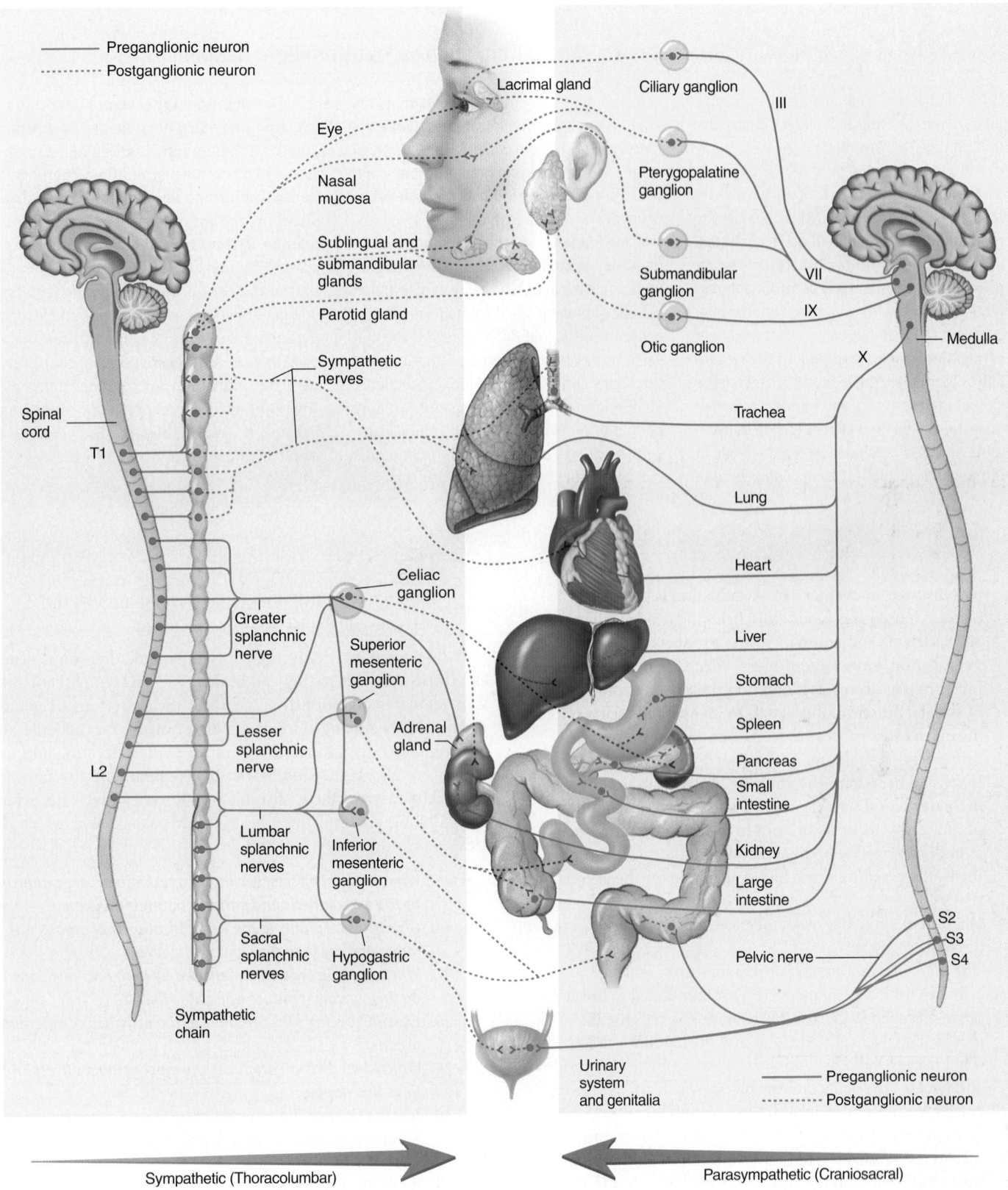

Figure 16.9 Innervation of Organs by the ANS

Preganglionic fibers are indicated by *solid lines,* and postganglionic fibers are indicated by *dashed lines.*

Clinical Focus Biofeedback, Meditation, and the Fight-or-Flight Response

Biofeedback takes advantage of electronic instruments or other techniques to monitor and change subconscious activities, many of which are regulated by the ANS. Skin temperature, heart rate, and brain waves are monitored electronically. By watching the monitor and using biofeedback techniques, a person can learn how consciously to reduce heart rate and blood pressure and regulate blood flow in the limbs. For example, people claim that they can prevent the onset of migraine headaches or reduce their intensity by learning to dilate blood vessels in the skin of their forearms and hands. Increased blood vessel dilation increases skin temperature, which is correlated with a decrease in the severity of the migraine. Some people

use biofeedback methods to relax by learning to reduce their heart rate or change the pattern of their brain waves. The severity of some stomach ulcers, high blood pressure, anxiety, and depression may be reduced by using biofeedback techniques.

Meditation is another technique that influences autonomic functions. Although numerous claims about the value of meditation include improving one's spiritual well-being, consciousness, and holistic view of the universe, it has been established that meditation does influence autonomic functions. Meditation techniques are useful in some people in reducing heart rate, blood pressure, severity of ulcers, and other symptoms that are frequently associated with stress.

The fight-or-flight response occurs when an individual is subjected to stress, such as a threatening, frightening, embarrassing, or exciting situation. Whether a person confronts or avoids a stressful situation, the nervous system and the endocrine system are involved either consciously or unconsciously. The autonomic part of the fight-or-flight response results in a general increase in sympathetic activity, including heart rate, blood pressure, sweating, and other responses, that prepare the individual for physical activity. The fight-or-flight response is adaptive because it also enables the individual to resist or move away from a threatening situation.

---

blood vessels in chapter 21). Vasodilation is beneficial because it increases blood flow, bringing needed oxygen and nutrients and removing waste products. Too much vasodilation, however, can cause a decrease in blood pressure that decreases blood flow. Increased stimulation of skeletal muscle blood vessels by sympathetic nerves during exercise causes vasoconstriction that prevents a drop in blood pressure (see chapter 21).

3. Increased heart rate and force of contraction potentially increases blood flow through tissues. Vasoconstriction of blood vessels in tissues not involved in exercise, such as abdominopelvic organs, reduces blood flow through them, thus making more blood available for the exercising tissues.

4. Dilation of air passageways increases air flow into and out of the lungs.

5. The availability of energy sources increases. Skeletal muscle cells and liver cells (hepatocytes) are stimulated to break down glycogen to glucose. Skeletal muscle cells use the glucose and liver cells release it into the blood for use by other tissues. Fat cells (adipocytes) break down triglycerides and release fatty acids into the blood, which are used as an energy source by skeletal and cardiac muscle.

6. As exercising muscles generate heat, body temperature increases. Vasodilation of blood vessels in the skin brings warm blood close to the surface, where heat is lost to the environment. Sweat gland activity increases, resulting in increased sweat production, and evaporation of the sweat removes additional heat.

7. The activities of organs not essential for exercise decrease. For example, the process of digesting food slows as digestive glands decrease their secretions and the contractions of

smooth muscle that mix and move food through the gastrointestinal tract decrease.

Increased activity of the parasympathetic division is generally consistent with resting conditions. The parasympathetic division regulates digestion by increasing the secretions of glands, promoting the mixing of food with digestive enzymes and bile, and moving materials through the digestive tract. Defecation and urination are also controlled by the parasympathetic division. Increased parasympathetic stimulation lowers heart rate, which lowers blood pressure, and constricts air passageways, which decreases air movement through them.

25. *What kinds of effects, excitatory or inhibitory, are produced by the sympathetic and parasympathetic divisions?*

26. *Give two exceptions to the generalization that organs are innervated by both divisions of the ANS.*

27. *When a single organ is innervated by both ANS divisions, do they usually produce opposite effects?*

28. *Explain how the ANS coordinates the activities of different organs.*

29. *Which ANS division produces the most general effects? How does this happen?*

30. *Use the fight-or-flight response to describe the responses produced by the sympathetic division.*

P R E D I C T

Bethanechol (be-than′ĕ-kol) chloride is a drug that binds to muscarinic receptors. Explain why this drug can be used to promote emptying of the urinary bladder. Which of the following side effects would you predict: abdominal cramps, asthmatic attack, decreased tear production, decreased salivation, dilation of the pupils, or sweating.

Clinical Focus Disorders of the Autonomic Nervous System

Normal function of all components of the ANS is not required to maintain life, as long as environmental conditions are constant and optimal. Abnormal autonomic functions, however, markedly affect an individual's ability to respond to changing conditions. **Sympathectomy,** the removal of sympathetic ganglia, demonstrates this. The normal regulation of body temperature is lost following sympathectomy. In a hot environment, the ability to lose heat by increasing blood flow to the skin and by sweating is decreased. When exposed to the cold, the ability to reduce blood flow to the skin and conserve heat is decreased. Sympathectomy also results in low blood pressure caused by dilation of peripheral blood vessels and in the inability to increase blood pressure during periods of physical activity.

Orthostatic hypotension is a drop in blood pressure that occurs when a person who was sitting or lying down suddenly stands up. It is sometimes caused by disorders, such as diabetes mellitus, that decrease the frequency of action potentials in sympathetic nerves innervating blood vessels. Consequently, on standing, blood pools in dilated blood vessels in the lower extremities, less blood returns to the heart, and the amount of blood the heart pumps decreases. Blood pressure decreases, resulting in reduced blood flow to the brain, which causes fainting because of a lack of oxygen.

Raynaud's disease involves the spasmodic contraction of blood vessels in the periphery of the body, especially in the digits, and results in pale, cold hands that are prone to ulcerations and gangrene because of poor circulation. This condition can be caused by exaggerated sensitivity of blood vessels to sympathetic innervation. Preganglionic denervation (cutting the preganglionic neurons) is occasionally performed to alleviate the condition.

Hyperhidrosis (hī′per-hī-drō′sis), or excessive sweating, is caused by exaggerated sympathetic innervation of the sweat glands.

Achalasia (ak-ă-lā′zē-ă) is characterized by difficulty in swallowing and in controlling contraction of the esophagus where it enters the stomach, therefore interrupting normal peristaltic contractions of the esophagus. The swallowing reflex is controlled partly by somatic reflexes and partly by parasympathetic reflexes. The cause of achalasia can be abnormal parasympathetic regulation of the swallowing reflex. The condition is aggravated by emotions.

Dysautonomia (dis′aw-tō-nō′mē-ă), an inherited condition involving an autosomal-recessive gene, causes reduced tear gland secretion, poor vasomotor control, trouble in swallowing, and other symptoms. It is the result of poorly controlled autonomic reflexes.

Hirschsprung's disease, or **megacolon,** is caused by a functional obstruction in the lower colon and rectum. Ineffective parasympathetic stimulation and a predominance of sympathetic stimulation of the colon inhibit peristaltic contractions, causing feces to accumulate above the inhibited area. The resulting dilation of the colon can be so great that surgery is required to alleviate the condition.

S U M M A R Y

Contrasting the Somatic and Autonomic Nervous Systems (p. 562)

1. The cell bodies of somatic neurons are located in the CNS, and their axons extend to skeletal muscles, where they have an excitatory effect that usually is controlled consciously.
2. The cell bodies of the preganglionic neurons of the ANS are located in the CNS and extend to ganglia, where they synapse with postganglionic neurons. The postganglionic axons extend to smooth muscle, cardiac muscle, or glands and have an excitatory or inhibitory effect that usually is controlled unconsciously.

Anatomy of the Autonomic Nervous System (p. 563)

Sympathetic Division

1. Preganglionic cell bodies are in the lateral horns of the spinal cord gray matter from T1–L2.
2. Preganglionic axons pass through the ventral roots to the white rami communicantes to the sympathetic chain ganglia. From there, four courses are possible.
   - Preganglionic axons synapse (at the same or a different level) with postganglionic neurons, which exit the ganglia through the gray rami communicantes and enter spinal nerves.
   - Preganglionic axons synapse (at the same or a different level) with postganglionic neurons, which exit the ganglia through sympathetic nerves.
   - Preganglionic axons pass through the chain ganglia without synapsing to form splanchnic nerves. Preganglionic axons then synapse with postganglionic neurons in collateral ganglia.
   - Preganglionic axons synapse with the cells of the adrenal medulla.

Parasympathetic Division

1. Preganglionic cell bodies are in nuclei in the brainstem or the lateral parts of the spinal cord gray matter from S2–S4.
   - Preganglionic axons from the brain pass to ganglia through cranial nerves.
   - Preganglionic axons from the sacral region pass through the pelvic nerves to the ganglia.
2. Preganglionic axons pass to terminal ganglia within the wall of or near the organ that is innervated.

Enteric Nervous System

1. The enteric nerve plexus is within the wall of the digestive tract.
2. The enteric plexus consists of sensory neurons, ANS motor neurons, and enteric neurons.

The Distribution of Autonomic Nerve Fibers

1. Sympathetic axons reach organs through spinal nerves, head and neck nerve plexuses, thoracic nerve plexuses, and abdominopelvic nerve plexuses.

2. Parasympathetic axons reach organs through cranial nerves, thoracic nerve plexuses, abdominopelvic nerve plexuses, and pelvic nerves.
3. Sensory neurons run alongside sympathetic and parasympathetic neurons within nerves and nerve plexuses.

Physiology of the Autonomic Nervous System   (p. 569)
Neurotransmitters

1. Acetylcholine is released by cholinergic neurons (all preganglionic neurons, all parasympathetic postganglionic neurons, and some sympathetic postganglionic neurons).
2. Norepinephrine is released by adrenergic neurons (most sympathetic postganglionic neurons).

Receptors

1. Acetylcholine binds to nicotinic receptors (found in all postganglionic neurons) and muscarinic receptors (found in all parasympathetic and some sympathetic effector organs).
2. Norepinephrine and epinephrine bind to alpha and beta receptors (found in most sympathetic effector organs).
3. Activation of nicotinic receptors is excitatory, whereas activation of the alpha and beta receptors are either excitatory or inhibitory.
4. The main subtypes for alpha receptors are α_1- and α_2-adrenergic receptors, and for beta receptors are β_1- and β_2-adrenergic receptors.

Regulation of the Autonomic Nervous System   (p. 573)

1. Autonomic reflexes control most of the activity of visceral organs, glands, and blood vessels.
2. Autonomic reflex activity can be influenced by the hypothalamus and higher brain centers.
3. The sympathetic and parasympathetic divisions can influence the activities of the enteric nervous system through autonomic reflexes. The enteric nervous system can function independently of the CNS through local reflexes.

Functional Generalizations About the Autonomic Nervous System   (p. 576)

1. Both divisions of the ANS produce stimulatory and inhibitory effects.
2. Most organs are innervated by both divisions. Usually each division produces an opposite effect on a given organ.
3. Either division alone or both working together can coordinate the activities of different structures.
4. The sympathetic division produces more generalized effects than the parasympathetic division.
5. Sympathetic activity generally prepares the body for physical activity, whereas parasympathetic activity is more important for vegetative functions.

R E V I E W   A N D   C O M P R E H E N S I O N

1. Given these statements:
   1. neuron cell bodies in the nuclei of cranial nerves
   2. neuron cell bodies in the lateral gray matter of the spinal cord (S2–S4)
   3. two synapses between the CNS and effector organs
   4. regulates smooth muscle

   Which of the statements are true for the autonomic nervous system?
   a. 1,3
   b. 2,4
   c. 1,2,3
   d. 2,3,4
   e. 1,2,3,4

2. Given these structures:
   1. gray ramus communicans
   2. white ramus communicans
   3. sympathetic chain ganglion

   Choose the arrangement that lists the structures in the order an action potential passes through them from a spinal nerve to an effector organ.
   a. 1,2,3
   b. 1,3,2
   c. 2,1,3
   d. 2,3,1
   e. 3,2,1

3. Given these structures:
   1. collateral ganglion
   2. sympathetic chain ganglion
   3. white ramus communicans
   4. splanchnic nerve

   Choose the arrangement that lists the structures in the order an action potential travels through them on the way from a spinal nerve to an effector organ.
   a. 1,3,2,4
   b. 1,4,2,3
   c. 3,1,4,2
   d. 3,2,4,1
   e. 4,3,1,2

4. The white ramus communicans contains
   a. preganglionic sympathetic fibers.
   b. postganglionic sympathetic fibers.
   c. preganglionic parasympathetic fibers.
   d. postganglionic parasympathetic fibers.

5. The cell bodies of the postganglionic neurons of the sympathetic division are located in the
   a. sympathetic chain ganglia.
   b. collateral ganglia.
   c. terminal ganglia.
   d. dorsal root ganglia.
   e. both a and b.

6. Splanchnic nerves
   a. are part of the parasympathetic division.
   b. have preganglionic neurons that synapse in the collateral ganglia.
   c. exit from the cervical region of the spinal cord.
   d. travel from the spinal cord to the sympathetic chain ganglia.
   e. all of the above.

7. Which of the following statements regarding the adrenal gland is true?
   a. The parasympathetic division stimulates the adrenal gland to release acetylcholine.
   b. The parasympathetic division stimulates the adrenal gland to release epinephrine.
   c. The sympathetic division stimulates the adrenal gland to release acetylcholine.
   d. The sympathetic division stimulates the adrenal gland to release epinephrine.

8. The parasympathetic division
   a. is also called the craniosacral division.
   b. has preganglionic axons in cranial nerves.
   c. has preganglionic axons in pelvic nerves.
   d. has ganglia near or in the wall of effector organs.
   e. all of the above.

9. Which of these is *not* a part of the enteric nervous system?
   a. ANS motor neurons
   b. neurons located only in the digestive tract
   c. sensory neurons
   d. somatic neurons

10. Sympathetic axons reach organs through all of the following *except*
    a. abdominopelvic nerve plexuses.
    b. head and neck nerve plexuses.
    c. thoracic nerve plexuses.
    d. pelvic nerves.
    e. spinal nerves.

11. Which of these cranial nerves does *not* contain parasympathetic fibers?
    a. oculomotor (III)
    b. facial (VII)
    c. glossopharyngeal (IX)
    d. trigeminal (V)
    e. vagus (X)

12. Which of the following statements concerning the preganglionic neurons of the ANS is true?
    a. All parasympathetic preganglionic neurons secrete acetylcholine.
    b. Only parasympathetic preganglionic neurons secrete acetylcholine.
    c. All sympathetic preganglionic neurons secrete norepinephrine.
    d. Only sympathetic preganglionic neurons secrete norepinephrine.

13. A cholinergic neuron
    a. secretes acetylcholine.
    b. has receptors for acetylcholine.
    c. secretes norepinephrine.
    d. has receptors for norepinephrine.
    e. secretes both acetylcholine and norepinephrine.

14. When acetylcholine binds to nicotinic receptors,
    a. the cell's response is mediated by G proteins.
    b. the response can be excitatory or inhibitory.
    c. Na^+ channels open.
    d. it occurs at the effector organ.
    e. all of the above.

15. Nicotinic receptors are located in
    a. postganglionic neurons of the parasympathetic division.
    b. postganglionic neurons of the sympathetic division.
    c. membranes of skeletal muscle cells.
    d. both a and b.
    e. all of the above.

16. The activation of α_1- and β_1-adrenergic receptors
    a. generally produces a stimulatory response.
    b. generally produces an inhibitory response.
    c. most commonly occurs when epinephrine from the adrenal glands binds to them.
    d. occurs when acetylcholine binds to them.

17. The sympathetic division
    a. is always stimulatory.
    b. is always inhibitory.
    c. is usually under conscious control.
    d. generally opposes the actions of the parasympathetic division.
    e. both a and c.

18. A sudden increase in blood pressure
    a. initiates a sympathetic reflex that decreases heart rate.
    b. initiates a local reflex that decreases heart rate.
    c. initiates a parasympathetic reflex that decreases heart rate.
    d. both a and b.
    e. both b and c.

19. Which of these structures is innervated almost exclusively by the sympathetic division?
    a. gastrointestinal tract
    b. heart
    c. urinary bladder
    d. reproductive tract
    e. blood vessels

20. Which of these is expected if the sympathetic division is activated?
    a. Secretion of watery saliva increases.
    b. Tear production increases.
    c. Air passageways dilate.
    d. Glucose release from the liver decreases.
    e. All of the above.

Answers in Appendix F

C R I T I C A L   T H I N K I N G

1. When a person is startled or sees a "pleasurable" object, the pupils of the eyes may dilate. What division of the ANS is involved in this reaction? Describe the nerve pathway involved.

2. Reduced secretion from salivary and lacrimal glands could indicate damage to what nerve?

3. In a patient with Raynaud's disease, blood vessels in the skin of the hand may become chronically constricted, thereby reducing blood flow and producing gangrene. These vessels are supplied by nerves that originate at levels T2 and T3 of the spinal cord and eventually exit through the first thoracic and inferior cervical sympathetic ganglia. Surgical treatment for Raynaud's disease severs this nerve supply. At which of the following locations would you recommend that the cut be made: white rami of T2–T3, gray rami of T2–T3, spinal nerves T2–T3, or spinal nerves C1–T1? Explain.

4. Patients with diabetes mellitus can develop autonomic neuropathy, which is damage to parts of the autonomic nerves. Given the following parts of the ANS—vagus nerve, splanchnic nerve, pelvic nerve, cranial nerve, outflow of gray ramus—match the part with the symptom it would produce if the part were damaged:
   a. impotence
   b. subnormal sweat production
   c. gastric atony and delayed emptying of the stomach
   d. diminished pupil reaction (constriction) to light
   e. bladder paralysis with urinary retention

5. Explain why methacholine, a drug that acts like acetylcholine, is effective for treating tachycardia (heart rate faster than normal). Which of the following side effects would you predict: increased salivation, dilation of the pupils, sweating, and difficulty in taking a deep breath?

6. A patient has been exposed to the organophosphate pesticide malathion, which inactivates acetylcholinesterase. Which of the following symptoms would you predict: blurring of vision, excess tear formation, frequent or involuntary urination, pallor (pale skin), muscle twitching, or cramps? Would atropine be an effective drug to treat the symptoms (see p. 573 for the action of atropine)? Explain.

7. Epinephrine is routinely mixed with local anesthetic solutions. Why?

8. A drug blocks the effect of the sympathetic division on the heart. Careful investigation reveals that after administration of the drug, normal action potentials are produced in the sympathetic preganglionic and postganglionic neurons. Also, injection of norepinephrine produces a normal response in the heart. Explain, in as many ways as you can, the mode of action of the unknown drug.

9. A drug is known to decrease heart rate. After cutting the white rami of T1–T4, the drug still causes heart rate to decline. After cutting the vagus nerves, the drug no longer affects heart rate. Which division of the ANS does the drug affect? Does the drug have its effect at the synapse between preganglionic and postganglionic neurons, at the synapse between postganglionic neurons and effector organs, or in the CNS? Is the effect of the drug excitatory or inhibitory?

10. Make a list of the responses controlled by the ANS in (a) a person who is extremely angry and (b) a person who has just finished eating and is relaxing.

Answers in Appendix G

ANSWERS TO PREDICT QUESTIONS

1. Terminal ganglia are found near or embedded within the wall of organs supplied by the parasympathetic division and contribute to the enteric nervous system. Postganglionic parasympathetic axons from the terminal ganglia also contribute to the enteric nervous system. Chain ganglia and collateral ganglia contain the cell bodies of sympathetic neurons. They are not embedded within the walls of organs supplied by the sympathetic division. Instead, postganglionic neurons extend from them to organs. Thus, postganglionic sympathetic axons are found in the enteric nervous system.

2. For a sensory axon running alongside sympathetic axons, the sensory axon leaves the wall of the small intestine, joins the superior mesenteric plexus, and passes through the superior mesenteric ganglion and from there through a splanchnic nerve to a sympathetic chain ganglion. From the sympathetic chain ganglion the sensory axon passes through a white ramus communicans, the ventral rami of a spinal nerve, a spinal nerve, the dorsal root of a spinal nerve, to a dorsal root ganglion. For a sensory axon running alongside parasympathetic axons, the sensory axon leaves the wall of the small intestine, joins the superior mesenteric plexus, and passes to the esophageal plexus. From there, the sensory axon passes through a vagus nerve to its sensory ganglion.

3. Nicotinic receptors are located within the autonomic ganglia as components of the membranes of the postganglionic neurons of the sympathetic and parasympathetic divisions. Nicotine binds to the nicotinic receptors of the postganglionic neurons, resulting in action potentials. Consequently, the postganglionic neurons stimulate their effector organs. After consumption of nicotine, structures innervated by both the sympathetic and parasympathetic divisions are affected.

   After the consumption of muscarine, only the effector organs that respond to acetylcholine are affected. This includes all the effector organs innervated by the parasympathetic division, and the sweat glands, which are innervated by the sympathetic division.

4. The low dose of epinephrine stimulates β_2 receptors and causes vasodilation. Although the large dose also stimulates β_2 receptors, it stimulates so many α_1 receptors that the vasoconstriction effect dominates the vasodilation effect.

5. The frequency of action potentials in sympathetic neurons to the sweat glands increases as the body temperature increases. The increasing body temperature is detected by the hypothalamus, which activates the sympathetic neurons. Sweating cools the body by evaporation. As the body temperature declines, the frequency of action potentials in sympathetic neurons to the sweat glands decreases. A lack of sweating helps prevent heat loss from the body.

6. In response to an increase in blood pressure, information is transmitted in the form of action potentials along sensory neurons to the medulla oblongata. From the medulla oblongata, the frequency of action potentials delivered along sympathetic nerve fibers to blood vessels decreases. As a result, blood vessels dilate, causing the blood pressure to decrease.

   In response to a decrease in blood pressure, fewer action potentials are transmitted along sensory neurons to the medulla oblongata, which responds by increasing the frequency of action potentials delivered along sympathetic nerves to blood vessels. As a result, blood vessels constrict, causing blood pressure to increase.

7. The parasympathetic division releases acetylcholine, which binds to muscarinic receptors on organs. Bethanechol chloride produces effects similar to stimulation of organs by the parasympathetic division. Thus, this drug should stimulate the urinary bladder to contract. Side effects can be produced by stimulation of muscarinic receptors elsewhere in the body. Stimulation of smooth muscle in the digestive tract can produce abdominal cramps. Stimulation of air passageways can cause an asthmatic attack. Decreased tear production, salivation, and dilation of the pupils are not expected side effects because parasympathetic stimulation causes increased tear production, salivation, and constriction of the pupils. Sweat glands are innervated by the sympathetic division but have muscarinic receptors. Bethanechol chloride can increase sweating.

Visit the Online Learning Center at www.mhhe.com/seeley7 for chapter quizzes, interactive learning exercises, and other study tools.

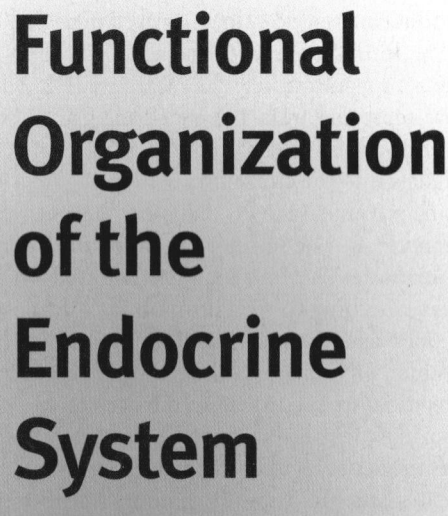

Functional Organization of the Endocrine System

Colorized transmission electron micrograph of a growth hormone-secreting cell from the anterior pituitary gland. The secretory vesicles (brown) contain growth hormone.

The nervous and endocrine systems are the two major regulatory systems of the body. Together, they regulate and coordinate the activities of essentially all other body structures. The nervous system functions something like telephone messages sent along telephone wires to their destination. It transmits information in the form of action potentials along the axons of nerve cells. Chemical signals in the form of neurotransmitters are released at synapses between neurons and the cells they control. The endocrine system is more like radio signals broadcast widely that everyone with radios tuned to the proper channel can receive. It sends information to the cells it controls in the form of chemical signals released from endocrine glands. The chemical signals are carried to all parts of the body by the circulatory system. Cells that are able to recognize the chemical signals respond to them whereas other cells do not.

This chapter introduces the general characteristics of the endocrine system. It compares some of the functions of the nervous and endocrine systems, emphasizes the role of the endocrine system in the maintenance of homeostasis, and illustrates the means by which the endocrine system regulates the functions of cells. This chapter explains the *general characteristics of the endocrine system* (p. 584), the *chemical structure of hormones* (p. 585), the *control of secretion rate* (p. 585), *transport and distribution in the body* (p. 591), *metabolism and excretion* (p. 593), *interaction of hormones, and other chemical signals with their target tissues* (p. 593), and *classes of receptors* (p. 595). The structure and function of the endocrine glands, the chemicals they secrete, and the means by which activities are regulated are described in chapter 18.

General Characteristics of the Endocrine System

The term **endocrine** (en′dō-krin) is derived from the Greek words *endo,* meaning within, and *crino,* to separate. The term implies that cells of endocrine glands secrete chemical signals that influence tissues that are separated from the endocrine glands by some distance. The **endocrine system** is composed of glands that secrete chemical signals into the circulatory system (figure 17.1). In contrast, exocrine glands have ducts that carry their secretions to surfaces (see chapter 4). The secretory products of endocrine glands are called **hormones** (hōr′mōnz), a term derived from the Greek word *hormon,* meaning to set into motion. Traditionally, a hormone is defined as a chemical signal, or **ligand,** that (1) is produced in minute amounts by a collection of cells; (2) is secreted into the interstitial spaces; (3) enters the circulatory system, where it is transported some distance; and (4) acts on specific tissues called **target tissues** at another site in the body to influence the activity of those tissues in a specific fashion. All hormones exhibit most components of this definition, but some components don't apply to every hormone.

Both the endocrine system and the nervous system regulate the activities of structures in the body, but they do so in different ways. For example, hormones secreted by most endocrine glands can be described as **amplitude-modulated signals** (am′pli-tood mod-ū-lāt′ed), which consist mainly of increases or decreases in the concentration of hormones in the body fluids (figure 17.2),

over periods ranging usually from minutes to hours. The responses to hormones either increase or decrease as a function of the hormone concentration. For example, antidiuretic hormone released from the posterior pituitary gland acts on the kidney to decrease the volume of urine produced. It produces a response within minutes, and the urine volume produced is a function of the amount of hormone released. On the other hand, the all-or-none action potentials carried along axons can be described as **frequency-modulated signals** (see figure 17.2), which vary in frequency but not in amplitude. A low frequency of action potentials is a weak stimulus, whereas a high frequency of action potentials is a strong stimulus (see chapter 11). Each action potential usually lasts from

Amplitude-modulated system. The concentration of the hormone determines the strength of the signal and the magnitude of the response. For most hormones, a small concentration of a hormone is a weak signal and produces a small response, whereas a larger concentration is a stronger signal and results in a greater response.

Frequency-modulated system. The strength of the signal depends on the frequency, not the size, of the action potentials. All action potentials are the same size in a given tissue. A low frequency of action potentials is a weak stimulus, and a higher frequency is a stronger stimulus.

Figure 17.2   Regulatory Systems

Hypothalamus

Pituitary glands

Thyroid gland

Thymus

Adrenals glands

Ovaries (female)

Pineal body

Parathyroid glands (posterior part of thyroid)

Pancreas (islets)

Testes (male)

Figure 17.1   Endocrine Glands

The location of major endocrine glands in the human body.

one to a few milliseconds (msec). For example, action potentials carried by axons to a muscle cause a response within msec. The muscle response is a product of the action potential frequency and the length of time the action potential frequency is maintained. The responses of the endocrine system are usually slower and of longer duration, and its effects are usually more generally distributed than those of the nervous system.

Although the stated differences between the endocrine and nervous systems are generally true, exceptions exist. Some endocrine responses are more rapid than some neural responses, and some endocrine responses have a shorter duration than some neural responses. In addition, a few hormones act as both amplitude- and frequency-modulated signals, in which the concentrations of the hormones and the frequencies at which the increases in hormone concentrations occur are important.

At one time, the endocrine system was believed to be relatively independent and different from the nervous system. An intimate relationship between these systems is now recognized, and the two systems cannot be completely separated either anatomically or functionally. Some neurons secrete chemical signals called **neurohormones** (noor-ō-hōr′mōnz) into the circulatory system, which function like hormones. Also, some neurons directly innervate endocrine glands and influence their secretory activity. Neurons release chemical signals at synapses in the form of neurotransmitters and neuromodulators, and the membrane potentials of some endocrine glands undergo depolarization or hyperpolarization, which results in either an increase or a decrease in the rate of hormone secretion. Conversely, some hormones secreted by endocrine glands affect the nervous system and markedly influence its activity.

Intercellular chemical signals allow one cell to communicate with other cells. These signals coordinate and regulate the activities of most cells. Several categories of intercellular chemical signals exist. Neurotransmitters and neuromodulators play essential roles in the function of the nervous system (see chapter 11). Hormones are secreted by endocrine glands.

Autocrine (aw′tō-krin) **chemical signals** are released by cells and have a local effect on the same cell type from which the chemical signals are released. Examples include prostaglandin-like chemicals released from smooth muscle cells and platelets in response to inflammation. These chemicals cause the relaxation of blood vessel smooth muscle cells and the aggregation of platelets. As a result, the blood vessels dilate and blood clots.

Paracrine (par′ă-krin) **chemical signals** are released by cells and affect other cell types locally without being transported in the blood. For example, a peptide called somatostatin is released by cells in the pancreas and functions locally to inhibit the secretion of insulin from other cells of the pancreas (see chapter 18).

Pheromones (fer′ō-mōnz) are special chemical signals because they are secreted into the environment and modify the behavior and the physiology of other individuals. For example, pheromones released in the urine of cats and dogs at certain times are olfactory signals that indicate fertility. Evidence supports the existence of pheromones produced by women that influence the length of menstrual cycles in other women (table 17.1).

Many intercellular chemical signals consistently fit one specific definition, but others do not. For example, norepinephrine functions both as a neurotransmitter and a neurohormone; prostaglandins function as neurotransmitters, neuromodulators, parahormones, and autocrine chemical signals. The schemes used to classify chemicals on the basis of their functions are useful, but they don't indicate that a specific molecule always performs as the same type of chemical signal in every place it's found. For that reason, the study of endocrinology often includes the study of autocrine and paracrine chemical signals in addition to hormones.

1. *Define the terms endocrine gland, endocrine system, and hormone. Explain why a simple definition for hormone is difficult to create.*
2. *Contrast the endocrine system and the nervous system for the following: amplitude versus frequency modulation; speed and duration of target cell response.*
3. *Explain why, despite their differences, the nervous and endocrine systems cannot be completely separated.*
4. *Name and describe five categories of intercellular chemical signals, other than hormones.*

Chemical Structure of Hormones

Hormones, including neurohormones, can be either proteins or short sequences of amino acids called polypeptides, derivatives of amino acids, or lipids. Some protein hormones, called glycoprotein hormones, are composed of one or more polypeptide chains combined with carbohydrate molecules. Lipid hormones are either steroids or derivatives of fatty acids. Table 17.2 presents major categories and subcategories of hormones based on their chemical structures, and figure 17.3 provides examples.

5. *List three categories of hormones and their subcategories, based on chemical structure, and give an example of each.*

Control of Secretion Rate

Most hormones are not secreted at a constant rate. Instead, the secretion of most hormones increases and decreases dramatically over time. Specific mechanisms regulating the secretion rates of hormones are presented in chapter 18, but the general patterns of regulation are introduced in this chapter. Hormones function to regulate the rates of many activities in the body. Generally, the secretion rate of each hormone is controlled by a negative-feedback mechanism (see chapter 1), so that the body activity it regulates is maintained within a normal range and homeostasis is maintained. In a few instances, positive-feedback systems play a role in the regulation of hormone secretion. In these instances the positive-feedback system is limited.

Hormones have three major patterns of regulation. One pattern involves changes in the extracellular concentration of a substance

Table 17.1 Functional Classification of Intercellular Chemical Signals

Intercellular Chemical Signal	Description	Example	
Autocrine	Secreted by cells in a local area and influences the activity of the same cell type from which it was secreted	Prostaglandins	Autocrine chemical signal
Paracrine	Produced by a wide variety of tissues and secreted into tissue spaces; usually has a localized effect on other tissues	Histamine, prostaglandins	Paracrine chemical signal
Hormone	Secreted into the blood by specialized cells; travels some distance to target tissues; influences specific activities	Thyroxine, insulin	Hormone
Neurohormone	Produced by neurons and functions like hormones	Oxytocin, antidiuretic hormone	Neuron, Neurohormone
Neurotransmitter or neuromodulator	Produced by neurons and secreted into extracellular spaces by presynaptic nerve terminals; travels short distances; influences postsynaptic cells	Acetylcholine, epinephrine	Neurotransmitter, Neuron
Pheromone	Secreted into the environment; modifies physiology and behavior of other individuals	Sex pheromones are released by humans and many other animals. They are released in the urine of animals, such as dogs and cats. Pheromones produced by women influence the length of the menstrual cycle of other women.	Pheromone

Table 17.2 Structural Categories of Hormones

Structural Category	Examples	Structural Category	Examples
Proteins	Growth hormone Prolactin Insulin	Amino acid derivatives	Epinephrine Norepinephrine Thyroid hormones (both T_4 and T_3) Melatonin
Glycoproteins (protein and carbohydrate)	Follicle-stimulating hormone Luteinizing hormone Thyroid-stimulating hormone Parathyroid hormone	Lipids Steroids (cholesterol is a precursor for all steroids)	Estrogens Progestins (progesterone) Testosterone Mineralocorticoids (aldosterone) Glucocorticoids (cortisol)
Polypeptides	Thyrotropin-releasing hormone Oxytocin Antidiuretic hormone Calcitonin Glucagon Adrenocorticotropic hormone Endorphins Thymosin Melanocyte-stimulating hormones Hypothalamic hormones Lipotropins Somatostatin	Fatty acid derivatives	Prostaglandins Thromboxanes Prostacyclins Leukotrienes

Abbreviations: T_4 = tetraiodothyronine or thyroxine; T_3 = triiodothyronine.

Figure 17.3
Examples of Hormone Chemical Structure

(a) Insulin is an example of a protein hormone.
(b) Oxytocin is an example of a peptide hormone.
(c) The thyroid hormones, triiodothyronine (T_3) and tetraiodothyronine (T_4), are examples of amino acid derivatives. (d) Testosterone, a steroid, and prostaglandin $F_{2\alpha}$ are examples of lipid hormones.

Protein hormone
Insulin

Peptide hormone
Oxytocin

Hormones derived from amino acids
Triiodothyronine (T_3)
Tetraiodothyronine or thyroxine (T_4)

Lipid hormones
Steroids — Testosterone
Fatty acid derivative (formed from a fatty acid) — Prostaglandin $F_{2\alpha}$ ($PGF_{2\alpha}$)

other than a hormone and the effect of those changes on the endocrine gland. Figure 17.4 describes the influence of blood glucose on insulin secretion from the pancreas. An increasing blood glucose level causes an increase in insulin secretion from the pancreas. Insulin increases glucose movement into cells, resulting in a decrease in blood glucose levels, which in turn causes a decrease in insulin secretion. Thus insulin levels increase and decrease in response to changes in blood glucose levels.

A second pattern of hormone regulation involves neural control of the endocrine gland. Neurons synapse with the cells that produce the hormone, and, when action potentials result, the neurons release a neurotransmitter. In some cases, the neurotransmitter is stimulatory and causes the cells to increase hormone secretion. In other cases the neurotransmitter is inhibitory and decreases hormone secretion. Thus sensory input and emotions acting through the nervous system can influence hormone secretion. Figure 17.5 illustrates the neural control of epinephrine and norepinephrine secretion from the adrenal gland. In response to stimuli such as stress or exercise, the nervous system stimulates the adrenal gland to secrete epinephrine and norepinephrine, which help the body respond to the stimuli. When the stimuli are no longer present, secretion of epinephrine and norepinephrine decreases.

A third pattern of hormone regulation involves the control of the secretory activity of one endocrine gland by a hormone or a neurohormone secreted by another endocrine gland. Figure 17.6 illustrates how thyroid-releasing hormone (TRH) from the hypothalamus of the brain stimulates the secretion of thyroid-stimulating hormone (TSH) from the anterior pituitary gland, which, in turn, stimulates the secretion of the thyroid hormones, T_3 and T_4 (see chapter 18), from the thyroid gland. A negative-feedback mecha-

nism for regulating thyroid hormone secretion exists because T_3 and T_4 inhibit the secretion of TRH and TSH. Thus, the concentrations of TRH, TSH, and T_3 and T_4 increase and decrease within a normal range (see chapter 18).

Neural Control of Insulin Secretion

Blood glucose levels regulate insulin secretion, but insulin secretion is also regulated by the nervous system. When action potentials in parasympathetic neurons that innervate the pancreas increase, the neurotransmitter acetylcholine is released. Acetylcholine causes depolarization of pancreatic cells, and insulin is secreted. When action potentials in sympathetic neurons that innervate the pancreas increase, the neurotransmitter norepinephrine is released. Norepinephrine causes hyperpolarization of pancreatic cells, and insulin secretion decreases. Thus, nervous stimulation of the pancreas can either increase or decrease insulin secretion.

P R E D I C T   **1**

For a person having normal thyroid function, the rate at which TSH and thyroid hormones are secreted remains within a normal range of concentrations. In some people, however, the immune system begins to produce large amounts of an abnormal substance that functions like TSH. Predict what that substance will do to the rate of TSH secretion and the rate of thyroid hormone secretion.

One of these three major patterns by which hormone secretion is regulated applies to each hormone, but the complete picture isn't quite so simple. The regulation of hormone secretion often involves more than one mechanism. For example, both the concentration of blood glucose and the autonomic nervous system influence insulin secretion from the pancreas.

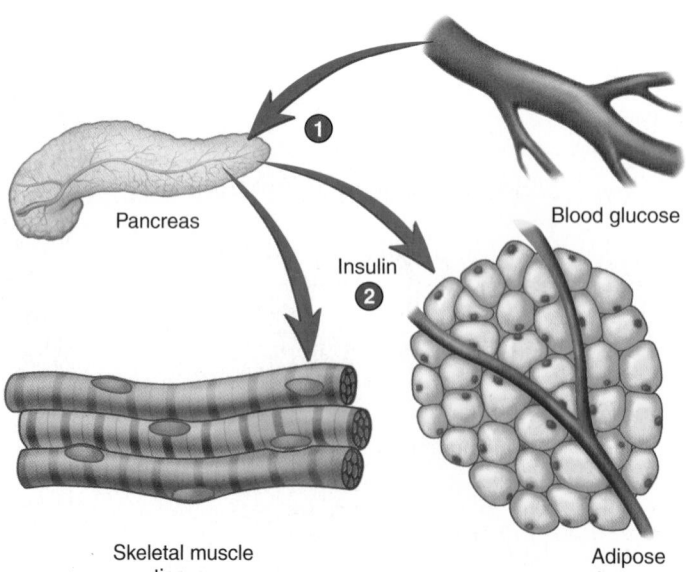

1. Increased blood glucose stimulates increased insulin secretion from the pancreas. Decreased blood glucose inhibits insulin secretion.

2. Increasing insulin levels in the blood stimulate glucose uptake by tissues, which decreases blood glucose levels. Lower blood insulin levels slow glucose uptake by tissues.

Process Figure 17.4   Nonhormonal Regulation of Hormone Secretion

Glucose, which is not a hormone, regulates the secretion of insulin from the pancreas.

1. Stimuli such as stress or exercise activate the sympathetic division of the autonomic nervous system.

2. Sympathetic neurons stimulate the release of epinephrine and smaller amounts of norepinephrine from the adrenal medulla. Epinephrine and norepinephrine prepare the body to respond to stressful conditions.

Once the stressful stimuli are removed, less epinephrine is released as a result of decreased stimulation from the autonomic nervous system.

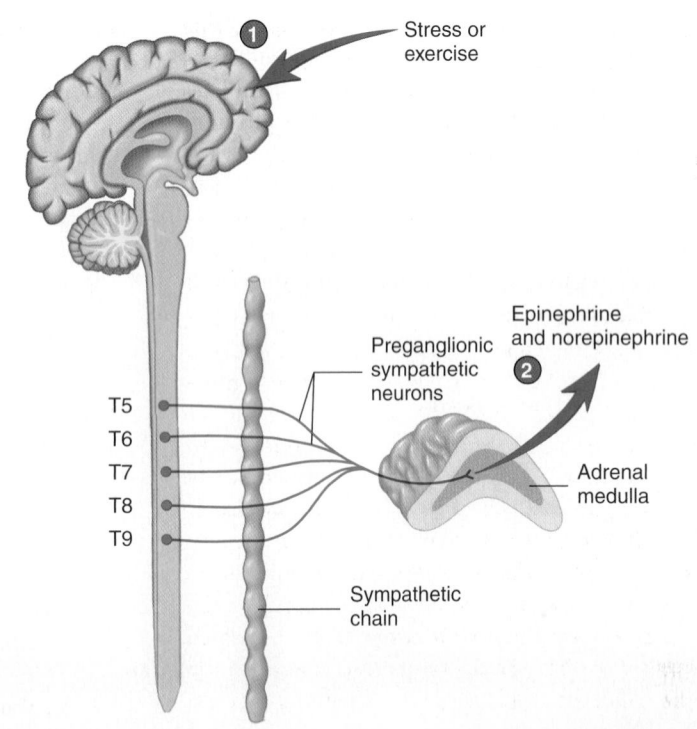

Process Figure 17.5 Nervous System Regulation of Hormone Secretion

The sympathetic division of the autonomic nervous system stimulates the adrenal gland to secrete epinephrine and norepinephrine.

1. Thyroid-releasing hormone (TRH) is released from neurons in the hypothalamus and travels in the blood to the anterior pituitary gland.

2. TRH stimulates the release of thyroid-stimulating hormone (TSH) from the anterior pituitary gland. TSH travels in the blood to the thyroid gland.

3. TSH stimulates the secretion of thyroid hormones (T_3 and T_4) from the thyroid gland into the blood.

4. T_3 and T_4 act on tissues to produce responses.

5. T_3 and T_4 also have a negative-feedback effect on the hypothalamus and the anterior pituitary to inhibit both TRH secretion and TSH secretion. The negative feedback helps keep blood thyroid hormone levels within a narrow range.

Process Figure 17.6 Hormonal Regulation of Hormone Secretion

Hormones can stimulate or inhibit the secretion of other hormones.

Examples of positive-feedback regulation in the endocrine system are represented by components of the female reproductive system. Prior to ovulation, estrogen from the ovary stimulates luteinizing hormone (LH) secretion from the anterior pituitary gland. LH, in turn, stimulates estrogen secretion from the ovary. Consequently, blood levels of estrogen and LH increase prior to ovulation (figure 17.7*a*). The release of oxytocin during delivery of an infant is another example (see chapters 28 and 29). In cases of positive feedback, negative feedback limits the degree to

which positive feedback proceeds (figure 17.7*b*). For example, after ovulation the ovary secretes progesterone, which inhibits LH secretion.

Some hormones are in the circulatory system at relatively constant levels, some change suddenly in response to certain stimuli, and others change in relatively constant cycles (figure 17.8). For example, thyroid hormones in the blood vary within a small range of concentrations that remain relatively constant. Epinephrine is released in large amounts in response to stress or physical exercise;

1. During the menstrual cycle, before ovulation, small amounts of estrogen are secreted from the ovary.

2. Estrogen stimulates the release of gonadotropin-releasing hormone (GnRH) from the hypothalamus and luteinizing hormone (LH) from the anterior pituitary.

3. GnRH also stimulates the release of LH from the anterior pituitary.

4. LH causes the release of additional estrogen from the ovary. The GnRH and LH levels in the blood increase because of this positive-feedback effect.

(a)

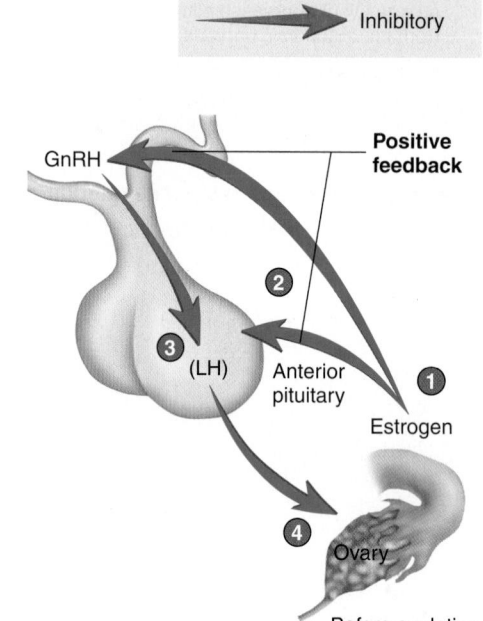

1. During the menstrual cycle, after ovulation, the ovary begins to secrete progesterone in response to LH.

2. Progesterone inhibits the release of GnRH from the hypothalamus and LH from the anterior pituitary.

3. Decreased GnRH release from the hypothalamus reduces LH secretion from the anterior pituitary. GnRH and LH levels in the blood decrease because of this negative-feedback effect.

(b)

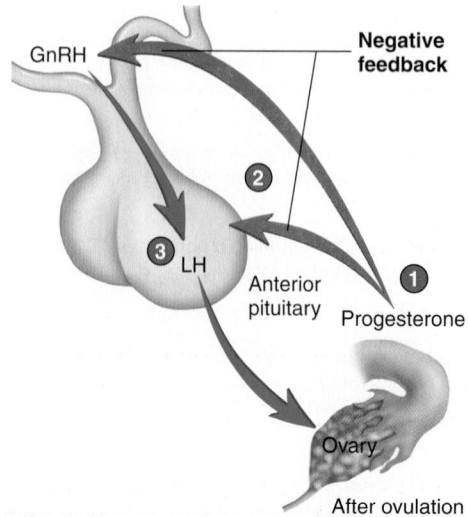

Process Figure 17.7 Positive and Negative Feedback

(a) **Chronic hormone regulation.** The maintenance of a relatively constant concentration of hormone in the circulating blood over a relatively long period.

(b) **Acute hormone regulation.** A hormone rapidly increases in the blood for a short time in response to a stimulus.

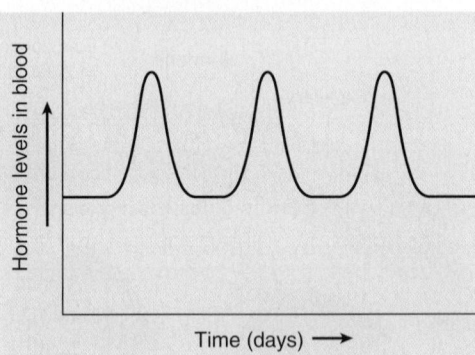

(c) **Cyclic hormone regulation.** A hormone is regulated so that it increases and decreases in the blood at a relatively constant time and to roughly the same amount.

Figure 17.8 **Changes in Hormone Secretion Through Time**

At least three basic patterns of hormone secretion exist.

thus its concentration can change suddenly. Reproductive hormones increase and decrease in a cyclic fashion in women during their reproductive years.

6. *Describe and give examples of the three major patterns by which hormone secretion is regulated. Give an example of a hormone that is controlled by more than one mechanism.*

7. *Is hormone secretion generally regulated by negative-feedback or positive-feedback mechanisms?*

8. *Describe chronic, acute, and cyclic patterns of hormone secretion.*

Transport and Distribution in the Body

Hormones are dissolved in blood plasma and transported either in a free form or bound to plasma proteins. Hormones that are free in the plasma readily diffuse from capillaries into interstitial spaces. As the concentration of free hormone molecules increases in the blood, more hormone molecules diffuse from the capillaries into the interstitial spaces and bind to target cells. As the concentration of the free hormone molecules decreases in the blood, fewer hormone molecules diffuse from the capillaries into the interstitial spaces and bind to target cells (figure 17.9).

Hormones that bind to plasma proteins do so in a reversible fashion. An equilibrium is established between the free plasma hormones and hormones bound to plasma proteins called binding proteins.

$$
\begin{array}{ccccc}
\text{H} & + & \text{BP} & \longleftrightarrow & \text{HBP} \\
\text{Hormone} & & \text{Binding} & & \text{Hormone bound} \\
& & \text{protein} & & \text{to binding protein}
\end{array}
$$

Many hormones bind only to certain types of plasma proteins. For example, a specific type of plasma protein binds to thyroid hormones, and a different type of plasma protein binds to sex hormones, such as testosterone. The equilibrium between the unbound hormone and the hormone bound to the plasma proteins is important because only the free hormone is able to diffuse through capillary walls and bind to target tissues. The hormone molecules bound to plasma proteins act as a reservoir for the hormone. Hormones bound to plasma proteins tend to remain at a relatively constant level in the blood for longer periods of time (see next section). A large decrease in the plasma protein concentration can result in the loss of a hormone from the blood because free hormones are more rapidly eliminated from the circulation through either the kidney or the liver (figure 17.10).

Because hormones circulate in the blood, they are distributed quickly throughout the body. They diffuse through the capillary endothelium and enter the interstitial spaces, although the rate at which this movement occurs varies from one hormone to the next. Lipid-soluble hormones readily diffuse through the walls of all capillaries. In contrast, water-soluble hormones, such as proteins, must pass through pores called fenestrae (see chapter 21) in the capillary endothelium. The capillary endothelia of organs that are regulated by protein hormones have large pores. Large pores exist in the capillaries of the endocrine glands that secrete these hormones as well (see chapter 21).

9. *What effect does a hormone binding to a plasma protein have on the amount of free hormone in the blood? On the amount of time the hormone remains in the blood?*

10. *Why do the capillary endothelia of organs regulated by protein hormones have large pores?*

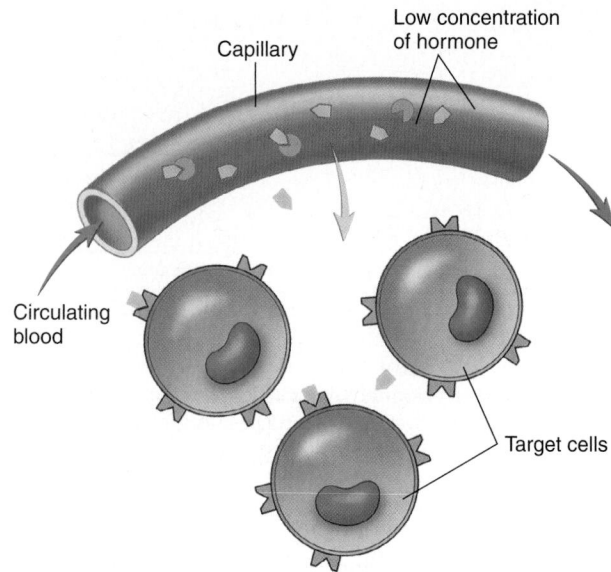

(a) As the concentration of free hormone molecules increases in the blood, more molecules diffuse from the capillary to the target cells.

(b) As the concentration of hormone molecules decreases in the blood, fewer diffuse from the capillary to the target cells.

Figure 17.9   Hormone Concentrations at the Target Cell

Hormone molecules diffuse from the blood through the walls of the capillaries into the interstitial spaces. Once within the interstitial spaces, they diffuse to the target cells.

(a) An equilibrium exists between free hormone molecules and hormone molecules bound to plasma proteins. The free hormone molecules can diffuse from the capillaries to the interstitial spaces. Free hormones are also more readily eliminated from the body by the kidneys and liver. Hormones bound to plasma proteins remain in the blood for a longer period of time.

(b) A decrease in plasma protein concentration reduces the number of hormone molecules bound to plasma proteins. This increases the rate at which free hormone molecules diffuse from the capillaries, and increases the rate at which hormones are eliminated from the blood by the kidneys and liver. The rapid loss of hormone from the circulatory system reduces the hormone concentration in the blood.

Figure 17.10   Effect of Changes in Plasma Protein Concentration on the Concentration of Free Hormone

Metabolism and Excretion

The destruction and elimination of hormones limit the length of time during which they are active, and body activities can increase and decrease quickly when hormones are secreted and remain active for only short periods. The length of time it takes for half a dose of a substance to be eliminated from the circulatory system is called its **half-life.** The half-life of a hormone is a standard measurement used by endocrinologists because it allows them to predict the rate at which hormones are eliminated from the body. The length of time required for total removal of a hormone from the body is not as useful because that measurement is influenced dramatically by the starting concentration. Water-soluble hormones, such as proteins, glycoproteins, epinephrine, and norepinephrine, have relatively short half-lives because they are degraded rapidly by enzymes within the circulatory system or organs, such as the kidneys, liver, or lungs. Hormones with short half-lives normally have concentrations that increase and decrease rapidly within the blood. They generally regulate activities that have a rapid onset and a short duration.

Hormones that are lipid-soluble, such as the steroids and T_3 and T_4 secreted by the thyroid gland, commonly circulate in the blood in combination with plasma proteins. The rate at which hormones are eliminated from the circulation is greatly reduced when the hormones bind to plasma proteins. The combination reduces the rate at which they diffuse through the wall of blood vessels and increases their half-life. Hormones with a long half-life have blood levels that are maintained at a relatively constant level through time. Table 17.3 outlines the ways hormone half-life is shortened or lengthened.

Hormones are removed from the blood in four major ways: excretion, metabolism, active transport, and conjugation. The kidney excretes hormones into the urine, or the liver excretes them into the bile. Enzymes in the blood or in tissues like the liver, kidney, lungs, or other target cells metabolize or chemically modify hormones. The end products can be excreted in the urine or bile, or they can be taken up by cells and used in metabolic processes. For example, epinephrine is modified enzymatically and then excreted by the kidney. Protein hormones are broken down to their amino acid building blocks. The amino acids can then be taken up by cells and used to synthesize new proteins. Some hormones can be actively transported into cells and recycled. For example, both epinephrine and norepinephrine can be actively transported into cells and secreted again.

The liver conjugates some hormones. **Conjugation** (kon-jŭ-gā′shŭn) is accomplished when cells in the liver attach water-soluble molecules to the hormone. These molecules are usually sulfate or glucuronic acid. Once they are conjugated, hormones are excreted by the kidney and liver at a greater rate.

> 11. *Define the half-life of a hormone. What happens to this half-life when a hormone binds to a plasma protein? What kinds of hormones bind to plasma proteins?*
>
> 12. *What kinds of activities do hormones with a short half-life regulate? With a long half-life?*
>
> 13. *What are the ways by which the half-life of a hormone is shortened or lengthened?*

Table 17.3	Factors That Influence the Half-Life of Hormones

A. Means by which the half-life of hormones is shortened:

1. **Excretion**
   Hormones are excreted by the kidney into the urine or excreted by the liver into the bile.

2. **Metabolism**
   Hormones are enzymatically degraded in the blood, liver, kidney, lungs, or target tissues. End products of metabolism are either excreted in urine or bile or used in other metabolic processes by cells in the body.

3. **Active Transport**
   Some hormones are actively transported into cells and are used again as either hormones or neurotransmitter substances.

4. **Conjugation**
   Substances such as sulfate or glucuronic acid groups are attached to hormones primarily in the liver, normally making them less active as hormones and increasing the rate at which they are excreted in the urine or bile.

B. Means by which the half-life of hormones is lengthened:

1. Some hormones are protected from rapid excretion or metabolism by binding reversibly with plasma proteins.

2. Some hormones are protected by their structure. The carbohydrate components of the glycoprotein hormones protect them from proteolytic enzymes in the circulatory system.

P R E D I C T **2**

How is the half-life of a hormone affected by a decrease in the concentration of the specific plasma protein to which that hormone binds?

Interaction of Hormones and Other Chemical Signals with Their Target Tissues

Intercellular chemical signals, commonly called **ligands** (lig′and, lī′gand), are molecules that bind to proteins or glycoproteins and change their functions. Hormones make up a major category of intercellular chemical signals. Other intercellular chemical signals include substances such as neurotransmitters and chemical mediators of inflammation. The portion of each protein or glycoprotein molecule where a chemical signal binds is called a **binding site.** If the protein or glycoprotein molecule is a receptor, the binding site is called a **receptor site.** The shape and chemical characteristics of each receptor site allow only a specific type of chemical signal to bind to it (figure 17.11). The tendency for each type of chemical signal to bind to a specific type of receptor site, and not to others, is called **specificity.** Insulin therefore binds to insulin receptors but not to receptors for growth hormone. However, chemical signals,

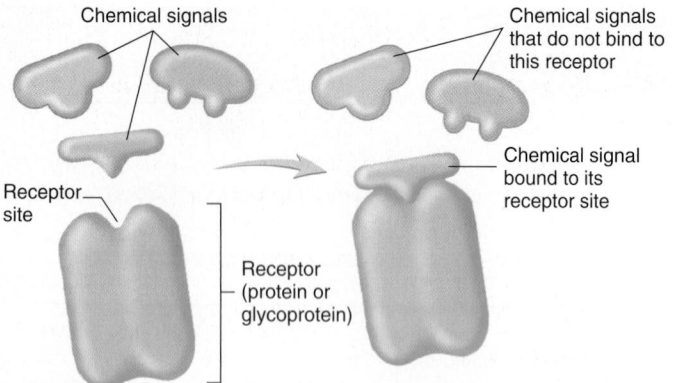

Figure 17.11 Receptors and Specificity of Receptor Sites

Chemical signals bind to receptor molecules. The shape and chemical characteristics of each receptor site allow certain chemical signals that have a compatible shape and compatible chemical characteristics to bind to it, not others. This relationship is called specificity.

such as some hormones, can bind to a number of different receptors that are closely related. For example, epinephrine can bind to more than one type of epinephrine receptor.

Hormones are secreted and distributed throughout the body by the circulatory system, but the presence or absence of specific receptor molecules in cells determines which cells will or will not respond to each hormone (figure 17.12). For example, there are receptors for TSH in cells of the thyroid gland, but there are no such receptors in most other cells of the body. Consequently, cells

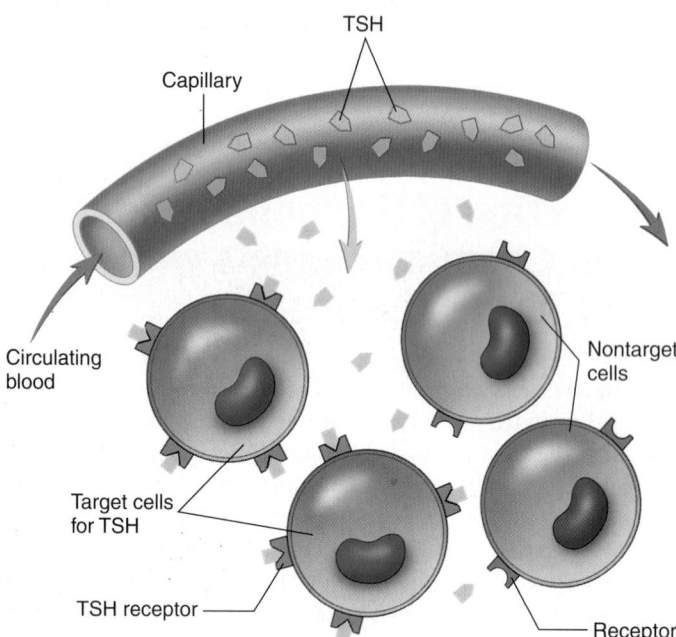

Figure 17.12 Response of Target Cells to Hormones

TSH is secreted into the blood and distributed throughout the body, where TSH diffuses from the blood into the interstitial fluid. Only target cells, however, have receptors for TSH; therefore, although TSH is distributed throughout the body, only target cells for that hormone can respond to it.

of the thyroid gland produce a response when exposed to TSH, but cells without receptor molecules do not respond to it.

Drugs with structures similar to specific chemical signals may compete with those chemical signals for their receptor sites (see chapter 3). Depending on the exact characteristics of a drug, it may either bind to a receptor site and activate the receptor or it may bind to a receptor site and inhibit the action of the receptor. For example, drugs exist that compete with epinephrine for its receptor sites. Some of these drugs activate epinephrine receptors and others inhibit them.

The response to a given concentration of a chemical signal is constant in some cases but variable in others. In some cells the response rapidly decreases through time. Fatigue of the target cells after prolonged stimulation explains some decreases in responsiveness. Also, the number of receptors can rapidly decrease after exposure to certain chemical signals—a phenomenon called **down-regulation** (figure 17.13a). Two known mechanisms are responsible for down-regulation. First, the rate at which receptors are synthesized decreases in some cells after the cells are exposed to a chemical signal. Because most receptor molecules are degraded after a time, a decrease in the synthesis rate reduces the total number of receptor molecules in a cell. Second, the combination of chemical signals and receptors can increase the rate at which receptor molecules are degraded. In some cases, when a chemical signal binds to a receptor, both the chemical signal and the receptor are taken into the cell by phagocytosis. Once the chemical signal and receptor are inside the cell, the cell can break them down.

Gonadotropin-releasing hormone (GnRH), which is released from neurons of the hypothalamus, causes the secretion of LH and follicle-stimulating hormone (FSH) from the anterior pituitary cells. In addition, exposure of the anterior pituitary cells to GnRH causes the number of receptor molecules for GnRH in the pituitary gland cells to dramatically decrease several hours after exposure to the hormone. The down-regulation of GnRH receptors causes the pituitary gland to become less sensitive to additional GnRH. The normal response of the pituitary gland cells to GnRH, therefore, depends on periodic rather than constant exposure of the gland to the hormone.

In general, tissues that exhibit down-regulation of receptor molecules are adapted to respond to short-term increases in hormone concentrations, and tissues that respond to hormones maintained at constant levels normally do not exhibit down-regulation.

In addition to down-regulation, periodic increases in the sensitivity of some cells to certain hormones also occur. This is called **up-regulation,** and it results from an increase in the rate of receptor molecule synthesis (figure 17.13b). An example of up-regulation is the increased number of receptor molecules for LH in cells of the ovary during each menstrual cycle. FSH molecules secreted by the pituitary gland increase the rate of LH receptor molecule synthesis in cells of the ovary. Thus, exposure of a tissue to one hormone can increase its sensitivity to a second by causing up-regulation of hormone receptors.

● **14.** *Define chemical signal (ligand) and receptor site. What characteristics of the receptor site make it specific for one type of chemical signal?*

Down-regulation

Number of receptors decreases

Time ⟶

(a) Down-regulation occurs when the number of receptors for a hormone decreases within target cells. For example, gonadotropin-releasing hormone (GnRH) released from the hypothalamus binds to GnRH receptors in the anterior pituitary. GnRH bound to its receptors causes down-regulation of the GnRH receptors so that eventually the target cells become less sensitive to the GnRH.

Up-regulation

Number of receptors increases

Time ⟶

(b) Up-regulation occurs when some stimulus causes the number of receptors for a hormone to increase within a target cell. For example, FSH acts on cells of the ovary to up-regulate the number of receptors for LH. Thus the ovary becomes more sensitive to the effect of LH.

Figure 17.13 Down-Regulation and Up-Regulation

15. *What is down-regulation? What two mechanisms are responsible for down-regulation? Give an example of down-regulation in the body.*
16. *What is up-regulation? Give an example of up-regulation in the body.*

P R E D I C T **③**

Estrogen is a hormone secreted by the ovary. It is secreted in greater amounts after menstruation and a few days before ovulation. Among its many effects is causing up-regulation of receptors in the uterus for another hormone secreted by the ovary called progesterone. Progesterone is secreted after ovulation. A major effect of progesterone is to cause the uterus to become ready for the embryo to attach to its wall following ovulation. Pregnancy cannot occur unless the embryo attaches to the wall of the uterus. Predict the consequence if the ovary secretes too little estrogen.

Classes of Receptors

Chemical signals, including hormones can be placed into two major categories, and bind to two major categories of receptors.

1. *Chemical signals that bind to membrane-bound receptors.* These chemical signals include large molecules and water-soluble molecules that cannot pass through the plasma membrane. They interact with **membrane-bound receptors,** which are receptors that extend across the plasma membrane and have their receptor sites exposed to the outer surface of the plasma membrane (figure 17.14*a*). When a chemical signal binds to the receptor site on the outside of the plasma membrane, the receptor initiates a response inside the cell. These chemical signals include many large hormones that are proteins, glycoproteins,

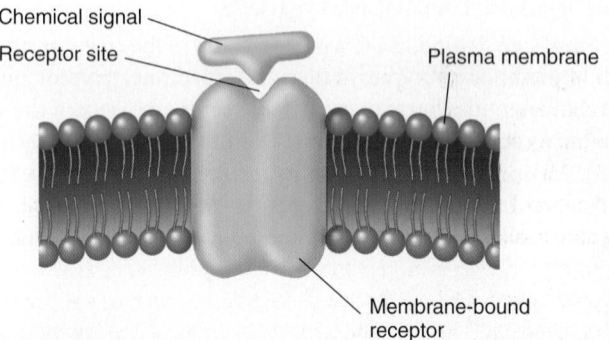

(a) Chemical signals that are large and water-soluble interact with membrane-bound receptors that extend across the plasma membrane and have receptor sites exposed to the outer surface of the plasma membrane. The portion of the receptor inside the cell produces the response.

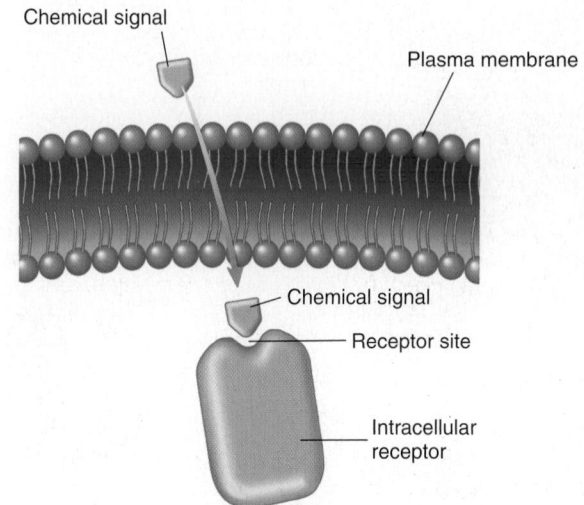

(b) Small lipid-soluble chemical signals diffuse through the plasma membrane and combine with the receptor sites of intracellular receptors. The combination of chemical signals and intracellular receptors produces a response.

Figure 17.14 Membrane-Bound and Intracellular Receptors

polypeptides, and some smaller molecules such as epinephrine and norepinephrine.

2. *Chemical signals that bind to intracellular receptors.* These chemical signals are lipid-soluble and relatively small. They diffuse through the plasma membrane and bind to **intracellular receptors,** which are receptors in the cytoplasm or in the nucleus of the cell (figure 17.14*b*). Subsequently, the receptors, with the chemical signals bound to their receptor sites, interact with DNA in the nucleus of the cell or interact with existing enzymes to produce a response. Thyroid hormones and steroid hormones, such as testosterone, estrogen, progesterone, aldosterone, and cortisol are examples.

● 17. *Define membrane-bound receptor and intracellular receptor. Describe the types of molecules that bind to each type of receptor.*

Membrane-Bound Receptors

Chemical signals bind in a reversible fashion to the receptor sites of membrane-bound receptor molecules. Hormone receptor molecules have peptide chains that cross the membrane once in the case of some receptors and several times for other receptors (see chapter 3). After a hormone binds to its receptor site, the intracellular part of the receptor initiates events that lead to a response. The mechanisms by which all membrane-bound receptors produce an intracellular response is not known, but at least three major mechanisms exist. Chemical signals bind to membrane-bound receptors that (1) directly change the permeability of the plasma membrane by opening or closing ion channels, (2) alter the activity of G proteins at the inner surface of the plasma membrane, or (3) alter the activity of intracellular enzymes (table 17.4). These changes produce specific responses in cells.

Receptors That Directly Alter Membrane Permeability

Some membrane-bound receptors are protein molecules that make up part of ion channels in the plasma membrane (see chapter 3). When chemical signals bind to the receptor sites of this type of receptor, the combination alters the three-dimensional structure of the proteins of the ion channels, causing the channels either to open or close. These channels are called **ligand-gated ion channels.** The result is a change in the permeability of the plasma membrane to the specific ions passing through the ion channels (figure 17.15). For example, serotonin molecules bind to serotonin receptor sites that are part of ligand-gated Na^+ channels and cause them to open. Na^+ diffuse into the cell and cause depolarization of the plasma membrane. Depolarization of target cells may lead to action potential initiation in those cells. Similarly, the neurotransmitter acetylcholine, released from nerve cells, is a chemical signal that combines with membrane-bound receptors of skeletal muscle cells. The combination of acetylcholine molecules with the receptor sites of the membrane-bound receptors for acetylcholine opens

Table 17.4   Overview of Responses to Hormones Binding to Their Receptors

Figure 17.15 Membrane-Bound Receptors That Directly Control Membrane Channels

Membrane-bound receptors for serotonin are part of the Na^+ channel. When a serotonin molecule binds to its receptor site on the serotonin receptor, the Na^+ channel opens and the permeability of the membrane to Na^+ increases. Na^+ then diffuse through the channels into the cell.

Na^+ channels in the plasma membrane. Consequently, Na^+ diffuse into the skeletal muscle cells causing depolarization and action potential initiation, and contraction (see chapter 9). Table 17.5 lists examples of ligand-gated ion channels. Many of these channels respond to neurotransmitters and not hormones, but some play important roles in regulating hormone secretion or mediating responses to paracrine chemical signals.

Receptors That Activate G Proteins

Many membrane-bound receptors produce responses through the action of a complex of proteins of the plasma membrane called **G proteins** (table 17.6 and figure 17.16). G proteins consist of three subunits; from the largest to smallest, they are called alpha (α), beta (β), and gamma (γ). The G proteins are so named because one of the subunits binds to guanine nucleotides. In the inactive state, a guanine diphosphate (GDP) molecule is bound to the α subunit of each G protein.

Table 17.5	Chemical Signals That Bind to Receptors and Directly Control Ion Channels	
Ligand	**Channel Type**	**Response**
Acetylcholine	Cation channel (primarily Na^+ channels)	Excitatory
Serotonin	Cation channel (primarily Na^+ channels)	Excitatory
Glutamate	Cation channel (primarily Na^+ channels)	Excitatory
Glycine	Cl^- channels	Inhibitory
GABA	Cl^- channels	Inhibitory

Abbreviations: GABA = gamma(γ)-aminobutyric acid.

Table 17.6	Examples of Hormones That Bind to Membrane-Bound Receptors and Activate G Proteins	
Hormone	**Source**	**Target Tissue**
Luteinizing hormone	Anterior pituitary	Ovary or testis
Follicle-stimulating hormone	Anterior pituitary	Ovary or testis
Prolactin	Anterior pituitary	Ovary or testis
Thyroid-stimulating hormone	Anterior pituitary	Thyroid gland
Adrenocorticotropic hormone	Anterior pituitary	Adrenal cortex
Oxytocin	Posterior pituitary	Uterus
Vasopressin	Posterior pituitary	Kidney
Calcitonin	Thyroid gland (parafollicular cells)	Osteoclasts and osteocytes
Parathyroid hormone	Parathyroid gland	Osteoclasts
Glucagon	Pancreas	Liver
Epinephrine	Medulla of adrenal gland	Cardiac muscle

1. The membrane-bound receptor has a receptor site exposed to the outside of the cell. The portion of the receptor inside of the cell can bind to the G protein.

2. The chemical signal binds to the receptor site of the membrane-bound receptor. The combination alters the G protein. GDP is released and GTP replaces the GDP on the α subunit. The α subunit separates from the γ and β subunits, and the active α subunit can influence ion channels in the plasma membrane or the synthesis of intracellular mediators.

3. When the chemical signal separates from the receptor site, additional G proteins are no longer activated. Inactivation of the α subunit occurs when phosphate (P$_i$) is removed from the GTP, leaving GDP bound to the α subunit.

4. The subunits of the G protein recombine.

Process Figure 17.16   **Membrane-Bound Receptors That Activate G Proteins**

G proteins can bind with receptors at the inner surface of the plasma membrane. After a ligand binds to the receptor on the outside of a cell, the receptor changes shape. As a result, the receptor combines with a G protein complex on the inner surface of the plasma membrane, and GDP is released from the α subunit. Guanine triphosphate (GTP), which is more abundant than GDP, binds to the α subunit, thereby activating it. The G proteins separate from the receptor and the activated α subunit separates from the β and γ subunits (see figure 17.16 steps *1* and *2*). The activated α subunit can alter the activity of molecules within the plasma membrane or inside the cell, thus producing cellular responses. After a short time, the activated α subunit is turned off because a phosphate group is removed from GTP and converts it to GDP. The α subunit then recombines with the β and γ subunits (see figure 17.16 steps *3* and *4*).

Some activated α subunits of G proteins can combine with ion channels, causing them to open or close (figure 17.17). For example, activated α subunits can open Ca^{2+} channels in smooth muscle cells resulting in the movement of Ca^{2+} into those cells. The Ca^{2+} function as **intracellular mediators.** Intracellular chemical mediators are ions or molecules that either enter the cell, or are synthesized in the cell, and regulate enzyme activities inside of the cell. The Ca^{2+} combine with calmodulin (kal-mod′ū-lin) molecules, and the calcium-calmodulin complexes activate enzymes that cause contraction of the smooth muscle cells (figure 17.17 steps *1* and *2* and see figure 9.24). After a short time, the activated α subunit is inactivated because GTP is converted to GDP. The α subunit then recombines with the β and γ subunits (see 17.17 steps *3* and *4*).

Activated α subunits of G proteins can also alter the activity of enzymes inside of the cell. For example, activated α subunits can

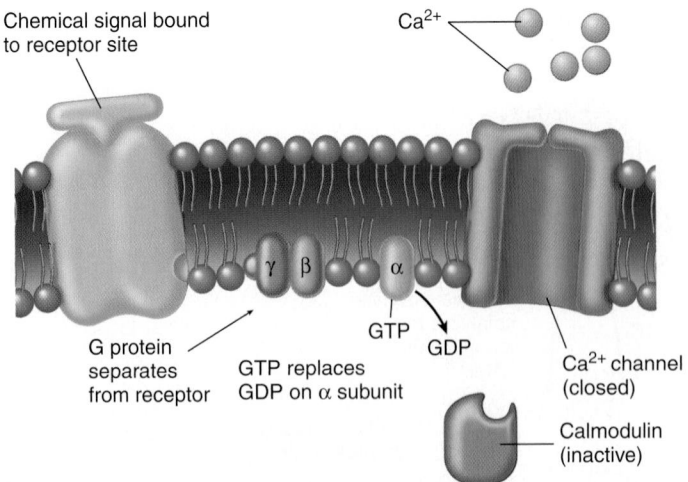

1. A chemical signal binds to the receptor site of the membrane-bound receptor. The combination alters the G protein. GTP replaces GDP on the α subunit, and the α subunit separates from the γ and β subunits.

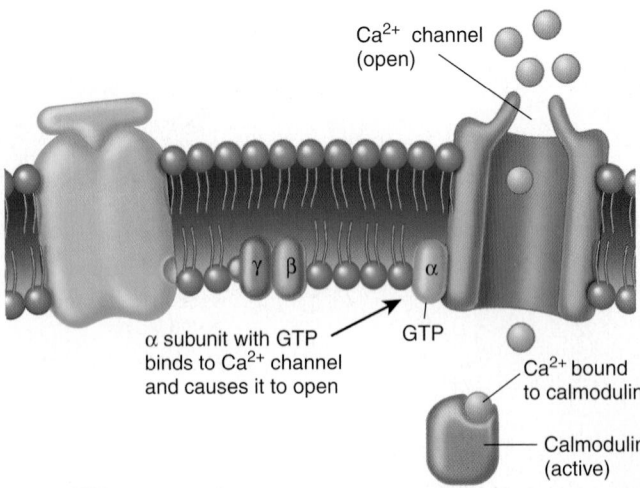

2. The α subunit, with GTP bound to it, combines with the Ca²⁺ channel, and the combination causes the Ca²⁺ channel to open. The ions diffuse into the cell and combine with calmodulin. The combination of Ca²⁺ with calmodulin produces the response of the cell to the ligand.

3. Phosphate is removed from the GTP bound to the α subunit, leaving GDP bound to the α subunit. The α subunit can no longer stimulate a cellular response, it separates from the Ca²⁺ channel, and the channel closes.

4. The α subunit recombines with γ and β subunits.

Process Figure 17.17 **Membrane-Bound Receptors, G Proteins, and Ca²⁺ Channels**

influence the rate of cyclic adenosine monophosphate (cAMP) formation (figure 17.18). **Adenylate cyclase** (a-den′i-lāt sī′klās), an enzyme activated by G proteins, converts ATP to cAMP. The cAMP molecules act as intracellular mediator molecules. Cyclic AMP binds to protein kinases and activate them. **Protein kinases** are enzymes that regulate the activity of other enzymes by attaching phosphates to them, a process called **phosphorylation.** Depending on the enzyme, phosphorylation increases or decreases the activity of the enzyme. The amount of time cAMP is present to produce a response in a cell is limited. An enzyme in the cytoplasm, called **phosphodiesterase** (fos′fō-dī-es′ter-ās), breaks down cAMP to AMP. The response of the cell is terminated after cAMP levels are reduced below a certain level.

Cyclic AMP acts as an intracellular mediator in many cell types. The response in each cell type is different because the enzymes activated by cAMP in each cell type are different. For example, glucagon combines with receptors on the surface of liver cells, activating G proteins and causing an increase in cAMP synthesis, which increases the release of glucose from liver cells (see figure 17.18). In contrast, LH combines with receptors on the surface of cells of the ovary, activating G proteins, and increasing cAMP synthesis. The major response to the increased cAMP is ovulation.

The combination of chemical signals with their receptors doesn't always result in increased cAMP synthesis. There are other common intracellular mediators (table 17.7). In some cell types, the

Figure 17.18   Membrane-Bound Receptors That Activate G Proteins and Increase the Synthesis of cAMP

Membrane-bound receptors for glucagon are associated with G proteins in liver cells. When glucagon binds to glucagon receptors, the α subunit of the G proteins dissociates from the other subunits. GTP replaces GDP on the α subunit. The active α subunit then binds to adenylate cyclase and activates it. The resulting increase in cAMP activates protein kinase enzymes, which phosphorylate other specific enzymes that break down glycogen and release glucose from the liver cells.

Table 17.7	Common Intracellular Mediators	
Intracellular Mediator	**Example of Cell Type**	**Example of Response**
Cyclic guanine monophosphate (cGMP)	Kidney cells	Increases Na^+ and water excretion by the kidney
Cyclic adenosine monophosphate (cAMP)	Liver cells	Increases the breakdown of glycogen and the release of glucose into the circulatory system
Calcium ions (Ca^{2+})	Smooth muscle cells	Contraction of smooth muscle cells
Inositol triphosphate (IP_3)	Smooth muscle cells	Contraction of certain smooth muscle cells in response to epinephrine
Diacylglycerol (DAG)	Smooth muscle cells	Contraction of certain smooth muscle cells in response to epinephrine
Nitric oxide (NO)	Smooth muscle cells	Relaxation of smooth muscle cells of blood vessels resulting in vasodilation

combination of chemical signals with their receptors causes the G proteins to inhibit the synthesis of cAMP, producing a response.

G proteins can also alter the concentration of intracellular mediators other than Ca^{2+} or cAMP (see table 17.7). For example, **diacylglycerol** (dī′as-il-glis′er-ol) **(DAG)** and **inositol** (in-ō′si-tōl, in-ō′si-tol) **triphosphate (IP₃)** are intracellular mediator molecules that are influenced by G proteins (figure 17.19). Epinephrine binds to certain membrane-bound receptors in some types of smooth muscle. The combination activates a G protein mechanism, which, in turn, increases the activity of phospholipase C. Phospholipase C converts phosphoinositol diphosphate (PIP_2) to DAG and IP₃. DAG activates enzymes that synthesize prostaglandins.

Prostaglandins increase smooth muscle contraction. IP₃ releases Ca^{2+} from the endoplasmic reticulum or opens Ca^{2+} channels in the plasma membrane. The ions enter the cytoplasm and increase contraction of the smooth muscle cells.

Receptors That Directly Alter the Activity of Intracellular Enzymes

Some chemical signals bind to membrane-bound receptors and directly change the activity of an intracellular enzyme. The altered enzyme activity either increases or decreases the synthesis of intracellular mediator molecules, or it results in the phosphorylation of intracellular proteins. The intracellular mediators or

1. Epinephrine binds to its receptor in the smooth muscle plasma membrane.

2. G proteins dissociate and GTP replaces GDP on the α subunit.

3. The active α subunit then binds with phospholipase C, which acts on phosphoinositol (PIP_2) and produces inositol triphosphate (IP_3) and diacylglycerol (DAG).

4. IP_3 releases Ca^{2+} from the endoplasmic reticulum or opens Ca^{2+} channels in the plasma membrane. Ca^{2+} then regulates enzyme activity.

5. DAG regulates enzymes such as those that synthesize prostaglandin.

These responses increase smooth muscle contraction.

Process Figure 17.19 Membrane-Bound Receptors That Activate G Proteins and Increase the Synthesis of IP_3 and DAG

Epinephrine receptors in some smooth muscle cells are associated with G proteins.

phosphorylated proteins activate processes that produce the response of cells to the chemical signals.

Intracellular enzymes that are controlled by membrane-bound receptors can be part of the membrane-bound receptor, or they may be separate molecules. The intracellular mediator molecules act as chemical signals that move from the enzymes that produce them into the cytoplasm of the cell, where they activate processes that produce the response of the cell.

Cyclic guanine (gwahn′ēn) **monophosphate (cGMP)** is an intracellular mediator molecule that is synthesized in response to a chemical signal binding with a membrane-bound receptor (figure 17.20). The chemical signal binds to its receptor, and the combination activates an enzyme called **guanylate cyclase** (gwahn′i-lāt sī′klās) located at the inner surface of the plasma membrane. The guanylate cyclase enzyme converts guanine triphosphate (GTP) to cGMP and two inorganic phosphate groups. The cGMP molecules then combine with specific enzymes in the cytoplasm of the cell and activate them. The activated enzymes, in turn, produce the response of the cell to the chemical signal. For example, atrial natriuretic hormone is a chemical signal that combines with its receptor in the plasma membrane of kidney cells. The result is an increase in the rate of cGMP synthesis at the inner surface of the plasma membranes (see figure 17.20). Cyclic GMP influences the action of enzymes in the kidney cells, which increase the rate of Na^+ and water excretion by the kidney (see chapter 26). The amount of time the

Figure 17.20 Membrane-Bound Receptor That Directly Activates an Intracellular Mediator

Atrial natriuretic hormone binds with its receptor site. At the inner surface of the plasma membrane, guanylate cyclase is activated to produce cGMP from GTP. Cyclic GMP is an intracellular mediator that mediates the response of the cell. Phosphodiesterase is an enzyme that breaks down cGMP to inactive GMP.

cGMP is present to produce a response in the cell is limited. Phosphodiesterase breaks down cGMP to GMP. Consequently, the length of time a chemical signal increases cGMP synthesis and has an effect on a cell is brief after the ligand is no longer present.

Some chemical signals bind to membrane-bound receptors, and the portion of the receptor on the inner surface of the plasma membrane acts as a phosphorylase enzyme that phosphorylates several specific proteins. Some of the phosphorylated proteins are part of the membrane-bound receptor, and others are in the cytoplasm of the cell (figure 17.21). The phosphorylated proteins influence the activity of other enzymes in the cytoplasm of the cell. For example, insulin binds to its membrane-bound receptor, resulting in the phosphorylation of parts of the receptor on the inner surface of the plasma membrane and the phosphorylation of certain other intracellular proteins. The phosphorylated proteins produce the responses of the cells to insulin. Some receptors for hormones that phosphorylate intracellular proteins are listed in table 17.8.

Hormones that stimulate the synthesis of an intracellular mediator molecule often produce rapid responses. This is possible because the mediator influences already-existing enzymes and causes a **cascade effect,** which results when a few mediator molecules

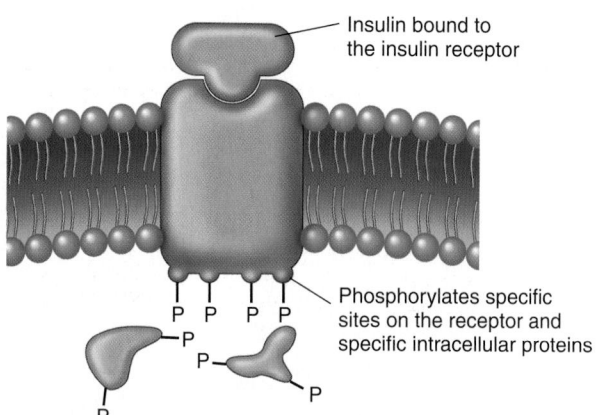

Figure 17.21   **Membrane-Bound Receptors That Phosphorylate Intracellular Proteins**

Insulin receptors are membrane-bound receptors. When insulin binds to the insulin receptor, the receptor acts as an enzyme that phosphorylates specific sites on the receptor and on intracellular proteins. The phosphorylated proteins produce the normal response to insulin.

Labels in figure:
- Insulin bound to the insulin receptor
- Phosphorylates specific sites on the receptor and specific intracellular proteins

activate several enzymes and each of the activated enzymes in turn activates several other enzymes that produce the final response. Thus, an amplification system exists in which a few molecules, such as cAMP, cGMP, or phosphorylated proteins, can control the activity of many enzymes within a cell (figure 17.22)

18. *Describe how membrane permeability can be changed when a hormone binds to a membrane-bound receptor. Give an example.*

19. *Explain how the combination of a chemical signal and its receptor can alter the G proteins on the inner surface of the plasma membrane. Which activated subunit of the G protein alters the activity of molecules inside the plasma membrane or inside the cell?*

20. *Describe how G proteins can alter the permeability of the plasma membrane and how they can alter the synthesis of an intracellular mediator molecule such as cAMP. Give examples.*

21. *Other than cAMP and Ca^{2+}, list two additional intracellular mediators affected by G proteins.*

22. *Describe how a chemical signal can combine with a membrane-bound receptor and change enzyme activity inside the cell and increase phosphorylation of intracellular proteins. Give examples.*

23. *What limits the activity of intracellular mediator molecules, such as cAMP, and phosphorylated proteins?*

24. *Explain what is meant by the cascade effect for the intracellular mediator model of hormone action. Does the intracellular mediator mechanism produce a slow or rapid response?*

P R E D I C T

When smooth muscle cells in the airways of the lungs contract, as in asthma, breathing becomes very difficult, whereas breathing is easy if the smooth muscle cells are relaxed. During asthma attacks, the smooth muscle cells in the airways of the lungs contract. Some of the drugs used to treat asthma increase cAMP in smooth muscle cells. Explain as many ways as possible how these drugs might work.

Intracellular Receptors

Intracellular receptors are either in the cytoplasm or in the nucleus of cells. Lipid-soluble chemical signals cross the plasma membrane into the cytoplasm or into the nucleus and bind to intracellular receptors by the process of diffusion (figure 17.23). After a chemical

Table 17.8	Hormones That Bind to Receptors Linked to Intracellular Enzymes That Phosphorylate Intracellular Proteins	
Hormone	**Source**	**Target Tissue and Effect**
Insulin	Pancreatic islets	Most cells; increases glucose and amino acid uptake
Growth hormone	Anterior pituitary gland	Most cells; increases protein synthesis and resists protein breakdown
Prolactin	Anterior pituitary gland	Mammary glands and ovary; initiates milk production following pregnancy
Growth factors	Various tissues	Stimulate growth in certain cell types
Some intercellular immune signal molecules	Cells of the immune system	Immune-competent cells; help mediate responses of the immune system
Atrial nutriuretic hormone	Cells of kidney tubules	Increases Na^+ excretion by kidney tubule cells and increases urine volume

Figure 17.22 The Cascade Effect

The combination of a hormone with a membrane-bound receptor activates several G proteins. The G proteins, in turn, activate many inactive adenylate cyclase enzymes, which cause the synthesis of a large number of cAMP molecules. The large number of cAMP molecules, in turn, activate many inactive protein kinase enzymes, which produce a rapid and amplified response.

1. Aldosterone is a lipid-soluble hormone and can easily diffuse through the plasma membrane.

2. Once aldosterone is inside of the cell, it binds with an aldosterone receptor molecule in the cytoplasm.

3. The aldosterone–receptor complex moves into the nucleus and binds to DNA.

4. The binding of the aldosterone–receptor complex to DNA stimulates the synthesis of messenger RNA (mRNA), which codes for specific proteins.

5. The mRNA leaves the nucleus, passes into the cytoplasm of the cell, and binds to ribosomes, where it directs the synthesis of the specific proteins.

6. The proteins synthesized on the ribosomes produce the response of the cell to aldosterone. For example, aldosterone increases the synthesis of proteins that transport Na^+ and K^+ across epithelial plasma membranes in the kidney.

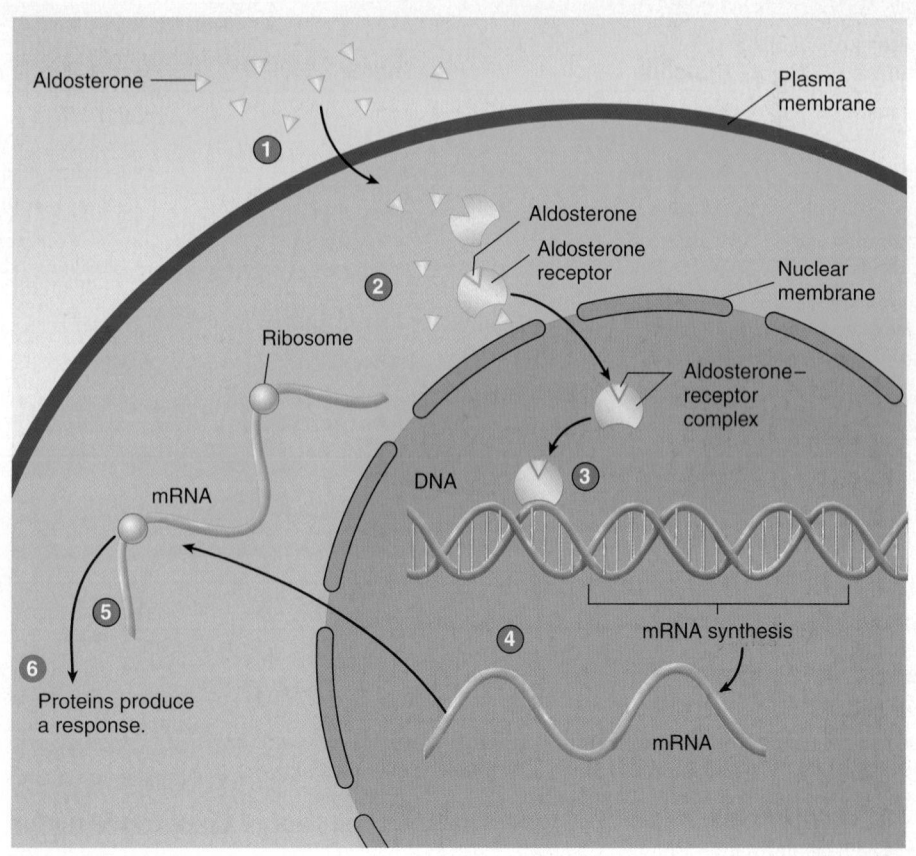

Process Figure 17.23 Intracellular Receptor Model

signal binds with an intracellular receptor, the receptor can alter the activity of enzymes in the cell, or it can bind to DNA to produce a response (see table 17.4). Some intracellular receptors that influence the expression of DNA are located in the cytoplasm. Once a chemical signal binds to its receptor, the receptor and chemical signal diffuse into the nucleus and bind to DNA. Other intracellular receptors are located in the nucleus. A chemical signal diffuses into the nucleus and binds to its receptor, and the receptor then binds to DNA.

Receptors that interact with DNA have specific "fingerlike" projections that interact with specific parts of a DNA molecule. The combination of the chemical signal and its receptor with DNA increases the synthesis of specific **messenger ribonucleic acid (mRNA)** molecules. The mRNA molecules then move to the cytoplasm and increase the synthesis of specific proteins at the ribosomes. The newly synthesized proteins produce the cell response to the chemical signal. For example, testosterone from the testes and estrogen from the ovaries stimulate the synthesis of proteins that are responsible for the secondary sex characteristics of males and females. The effect of the steroid aldosterone on its target cells in the kidney is to stimulate the synthesis of proteins that increase the rate of Na^+ and K^+ transport. The result is an increase in the reabsorption of Na^+ from, and increased secretion of K^+ into, the filtrate in the kidney and a reduction in the amount of Na^+ and an increase in K^+ lost in the urine. Other hormones that produce responses through intracellular receptor mechanisms include T_3 and T_4 from the thyroid follicles, and vitamin D which has some hormonelike characteristics (table 17.9).

Cells that synthesize new protein molecules in response to hormonal stimuli normally have a latent period of several hours between the time the hormones bind to their receptors and the time responses are observed. During this latent period, mRNA and new proteins are synthesized. Receptor–hormone complexes normally are degraded within the cell, limiting the length of time hormones influence the activities of cells, and the cells slowly return to their previous functional states.

25. *Describe how a chemical signal that crosses the plasma membrane interacts with its receptor and how it alters the rate of protein synthesis. Why is there normally a latent period between the time hormones bind to their receptors and the time responses are observed?*

26. *What finally limits the processes activated by the intracellular receptor mechanism?*

P R E D I C T **5**

Of membrane-bound receptors and intracellular receptors, which is better adapted for mediating a response that lasts a considerable length of time and which is better for mediating a response with a rapid onset and a short duration? Explain why.

Table 17.9	Major Hormones That Combine with Intracellular Receptors		
Category of Hormone	**Hormone**	**Source**	**Target Tissue and Effect**
Sex steroids	Testosterone	Testis	Responsible for development of the reproductive structures and development of male secondary sex characteristics
	Progesterone	Ovary	Causes increased size of cells lining the uterus
	Estrogen	Ovary	Causes increased cell division in the lining of the uterus
Mineralocorticoid steroids	Aldosterone	Adrenal cortex	Increased reabsorption of Na^+ and increased secretion of K^+ in the kidney
Glucocorticoid steroid hormones	Cortisol	Adrenal cortex	Increased breakdown of proteins and fats and increased blood levels of glucose
Thyroid hormones	Triiodothyronine (T_3)	Thyroid gland	Regulate development and metabolism
Vitamin D*	1,25-dihydroxycholecalciferol	Combination of the skin, liver, and kidney	Increased reabsorption of Ca^{2+} in the kidney and absorption of Ca^{2+} in the gastrointestinal tract

*Although vitamin D is not a hormone, it has hormone-like characteristics.

S U M M A R Y

General Characteristics of the Endocrine System (p. 584)

1. Endocrine glands produce hormones that are released into the interstitial fluid and diffuse into the blood. Hormones act on target tissues, which produce specific responses.
2. Endocrine glands produce other chemical messengers, including neurohormones, neurotransmitters, neuromodulators, parahormones, and pheromones.

3. Generalizations about the differences between the endocrine and nervous systems include the following: (a) the endocrine system is amplitude-modulated, whereas the nervous system is frequency-modulated; and (b) the response of target tissues to hormones is usually slower and of longer duration than that to neurons.

Chemical Structure of Hormones (p. 585)

Hormones are proteins, glycoproteins, peptides, derivatives of amino acids, or lipids (steroids or derivatives of fatty acids).

Control of Secretion Rate (p. 585)

1. Most hormones are not secreted at a constant rate.
2. Negative-feedback mechanisms, which function to maintain homeostasis, control the secretion of most hormones.
3. Hormone secretion from an endocrine tissue is regulated by one or more of three mechanisms: changes in the extracellular concentration of a nonhormone substance, stimulation by the nervous system, or stimulation by a hormone from another endocrine tissue.

Transport and Distribution in the Body (p. 591)

Hormones are dissolved in plasma or bind to plasma proteins. The blood quickly distributes hormones throughout the body.

Metabolism and Excretion (p. 593)

1. Nonpolar, readily diffusible hormones bind to plasma proteins and have an increased half-life.
2. Water-soluble hormones, such as proteins, epinephrine, and norepinephrine, do not bind to plasma proteins or readily diffuse out of the blood. Instead, they are broken down by enzymes or are taken up by tissues. They have a short half-life.
3. Hormones with a short half-life regulate activities that have a rapid onset and a short duration.
4. Hormones with a long half-life regulate activities that remain at a constant rate through time.
5. Hormones are eliminated from the blood by excretion from the kidneys and liver, enzymatic degradation, conjugation, or active transport.

Interaction of Hormones and Other Chemical Signals with Their Target Tissues (p. 593)

1. Target tissues have receptor molecules that are specific for a particular hormone.
2. Hormones bound with receptors affect the rate at which already existing processes occur.

3. Down-regulation is a decrease in the number of receptor molecules in a target tissue, and up-regulation is an increase in the number of receptor molecules.

Classes of Receptors (p. 595)

1. Membrane-bound receptors bind to water-soluble or large-molecular-weight hormones.
2. Intracellular receptors bind to lipid-soluble hormones.

Membrane-Bound Receptors

1. Membrane-bound receptors are proteins or glycoproteins that have polypeptide chains that are folded to cross the cell several times.
2. When a hormone binds to a membrane-bound receptor:
   - A change in the structure of membrane channels can result in a change in permeability of the plasma membrane to ions.
   - G proteins are activated. The α subunit of the G protein can bind to ion channels and cause them to open or change the rate of synthesis of intracellular mediator molecules, such as cAMP, cGMP, IP_3, and DAG.
   - Intracellular enzymes can be directly activated, which in turn synthesizes intracellular mediators, such as cGMP, or adds a phosphate group to intracellular enzymes, which alters their activity.
3. Intracellular mediator mechanisms are rapid-acting because they act on already-existing enzymes and produce a cascade effect.

Intracellular Receptors

1. Intracellular receptors are proteins in the cytoplasm or nucleus.
2. Hormones bind with the intracellular receptor, and the receptor–hormone complex activates genes. Consequently, DNA is activated to produce mRNA. The mRNA initiates the production of certain proteins (enzymes) that produce the response of the target cell to the hormone.
3. Intracellular receptor mechanisms are slow-acting because time is required to produce the mRNA and the protein.
4. Intracellular receptor–activated processes are limited by the breakdown of the receptor–hormone complex.

R E V I E W   A N D   C O M P R E H E N S I O N

1. When comparing the endocrine system and the nervous system, generally speaking, the endocrine system
   a. is faster-acting than the nervous system.
   b. produces effects that are of shorter duration.
   c. uses amplitude-modulated signals.
   d. produces more localized effects.
   e. relies less on chemical signals.

2. A chemical signal released from a cell that has a local effect on the same cell type from which the chemical signal is released is a(n)
   a. paracrine chemical signal.
   b. pheromone.
   c. autocrine chemical signal.
   d. hormone.
   e. intracellular mediator.

3. Given this list of molecule types:
   1. nucleic acid derivatives
   2. fatty acid derivatives
   3. polypeptides
   4. proteins
   5. phospholipids

   Which could be hormone molecules?
   a. 1,2,3
   b. 2,3,4
   c. 1,2,3,4
   d. 2,3,4,5
   e. 1,2,3,4,5

4. Which of these regulates secretion of a hormone from an endocrine tissue?
   a. other hormones
   b. negative-feedback mechanisms
   c. nonhormone substance in the blood
   d. the nervous system
   e. all of the above

5. Hormones are released into the blood
   a. at relatively constant levels.
   b. in large amounts in response to a stimulus.
   c. increasing and decreasing in a cyclic fashion.
   d. all of the above.

6. Lipid-soluble hormones readily diffuse through capillary walls, whereas water-soluble hormones, such as proteins, must
   a. pass through capillary cells.
   b. pass through pores in the capillary endothelium.
   c. be moved out of the capillary by active transport.
   d. remain in the blood.
   e. be broken down to amino acids before leaving the blood.

7. Concerning the half-life of hormones,
   a. lipid-soluble hormones generally have a longer half-life.
   b. hormones with shorter half-lives regulate activities with a slow onset and long duration.
   c. hormones with a shorter half-life are maintained at more constant levels in the blood.
   d. lipid-soluble hormones are degraded rapidly by enzymes in the circulatory system.
   e. water-soluble hormones usually combine with plasma proteins.

8. Given these observations:
   1. A hormone will affect only a specific tissue (not all tissues).
   2. A tissue can respond to more than one hormone.
   3. Some tissues respond rapidly to a hormone, whereas others take many hours to respond.

   Which of these observations can be explained by the characteristics of hormone receptors?
   a. 1
   b. 1,2
   c. 2,3
   d. 1,3
   e. 1,2,3

9. Which of these is *not* a means by which hormones are eliminated from the circulatory system?
   a. excreted into urine or bile
   b. bound to plasma proteins
   c. metabolism (enzymatically degraded in the blood)
   d. actively transported into cells
   e. conjugated with sulfate or glucuronic acid

10. Down-regulation
    a. produces a decrease in the number of receptors in the target cells.
    b. produces an increase in the sensitivity of the target cells to a hormone.
    c. is found in target cells that respond to hormones that are maintained at constant levels.
    d. occurs partly because of an increase in receptor synthesis by the target cell.
    e. all of the above.

11. A chemical signal
    a. can function as an enzyme.
    b. is also a G protein.
    c. can bind to a receptor site.
    d. is an intracellular mediator.
    e. all of the above.

12. Activated G proteins can
    a. cause ion channels to open or close.
    b. activate adenylate cyclase.
    c. inhibit the synthesis of cAMP.
    d. alter the activity of IP_3.
    e. all of the above.

13. Given these events:
    1. GTP is converted to GDP.
    2. The α subunit separates from the β and γ units.
    3. GDP is released from the α subunit.

    List the order in which the events occur after a chemical signal binds to a membrane-bound receptor.
    a. 1,2,3
    b. 1,3,2
    c. 2,3,1
    d. 3,2,1
    e. 3,1,2

14. Which of these can limit the response of a cell to a chemical signal?
    a. phosphodiesterase
    b. converting GTP to GDP
    c. decreasing the number of receptors
    d. blocking binding sites
    e. all of the above

15. Given these events:
    1. Na^+ channels open.
    2. Na^+ channels close.
    3. The plasma membrane depolarizes.
    4. The plasma membrane hyperpolarizes.

    Choose the arrangement that lists the events in the order they occur after serotonin binds to its receptor.
    a. 1,3
    b. 1,4
    c. 2,3
    d. 2,4

16. Given these events:
    1. The α subunit of a G protein interacts with Ca^{2+} channels.
    2. Ca^{2+} diffuse into the cell.
    3. The α subunit of a G protein is activated.

    Choose the arrangement that lists the events in the order they occur after a chemical signal combines with a receptor on a smooth muscle cell.
    a. 1,2,3
    b. 1,3,2
    c. 2,1,3
    d. 3,1,2
    e. 3,2,1

17. Given these events:
    1. cAMP is synthesized.
    2. The α subunit of G protein is activated.
    3. Phosphodiesterase breaks down cAMP.

    Choose the arrangement that lists the events in the order they occur after a chemical signal binds to a receptor.
    a. 1,2,3
    b. 1,3,2
    c. 2,1,3
    d. 2,3,1
    e. 3,2,1

18. Which of these events can occur after a G protein activates phospholipase C?
    a. DAG and IP_3 are synthesized from PIP_2.
    b. IP_3 causes Ca^{2+} channels to open.
    c. DAG activates enzymes that synthesize prostaglandins.
    d. All of the above.

19. When a chemical signal binds to an intracellular receptor
    a. DNA produces mRNA.
    b. G proteins are activated.
    c. the receptor–hormone complex causes ion channels to open or close.
    d. the cell's response is faster than when a chemical signal binds to a membrane-bound receptor.
    e. the chemical signal is usually a large, water-soluble molecule.

20. Given these events:
    1. activation of cAMP
    2. activation of genes
    3. enzyme activity altered

    Which of these events can occur when a hormone binds to an intracellular hormone receptor?
    a. 1
    b. 1,2
    c. 2,3
    d. 1,2,3

Answers in Appendix F

C R I T I C A L   T H I N K I N G

1. Consider a hormone that is secreted in large amounts at a given interval, modified chemically by the liver, and excreted by the kidney at a rapid rate, thus making the half-life of the hormone in the circulatory system very short. The hormone therefore rapidly increases in the blood and then decreases rapidly. Predict the consequences of liver and kidney disease on the blood levels of that hormone.

2. Consider a hormone that controls the concentration of some substance in the circulatory system. If a tumor begins to produce that substance in large amounts in an uncontrolled fashion, predict the effect on the secretion rate for the hormone.

3. How could you determine whether or not a hormone-mediated response resulted from the intracellular mediator mechanism or the intracellular receptor mechanism?

4. If the effect of a hormone on a target tissue is through a membrane-bound receptor that has a G protein associated with it, predict the consequences if a genetic disease causes the α subunit of the G protein to have a structure that prevents it from binding to GTP.

5. Prostaglandins are a group of hormones produced by many cells of the body. Unlike other hormones, they don't circulate but usually have their effect at or very near their site of production. Prostaglandins apparently affect many body functions, including blood pressure, inflammation, induction of labor, vomiting, fever, and inhibition of the clotting process. Prostaglandins also influence the formation of cAMP. Explain how an inhibitor of prostaglandin synthesis could be used as a therapeutic agent. Inhibitors of prostaglandin synthesis can produce side effects. Why?

6. For a hormone that binds to a membrane-bound receptor and has cAMP as the intracellular mediator, predict and explain the consequences if a drug is taken that strongly inhibits phosphodiesterase.

7. When an individual is confronted with a potentially harmful or dangerous situation, epinephrine (adrenaline) is released from the adrenal gland. Epinephrine prepares the body for action by increasing the heart rate and blood glucose levels. Explain the advantages or disadvantages associated with a short half-life for epinephrine and those associated with a long half-life.

8. Thyroid hormones are important in regulating the basal metabolic rate of the body. What are the advantages or disadvantages of
   a. a long half-life for thyroid hormones?
   b. a short half-life?

9. An increase in thyroid hormones causes an increase in metabolic rate. If liver disease results in reduced production of the plasma proteins to which thyroid hormones normally bind, what is the effect on metabolic rate? Explain.

10. Predict the effect on LH and FSH secretion if a small tumor in the hypothalamus of the brain secretes large concentrations of GnRH continuously. Given that LH and FSH regulate the function of the male and female reproductive systems, predict whether the condition increases or decreases the activity of these systems.

11. Insulin levels normally change in order to maintain normal blood sugar levels, despite periodic fluctuations in sugar intake. A constant supply of insulin from a skin patch might result in insulin levels that are too low when blood sugar levels are high (after a meal) and might be too high when blood sugar levels are low (between meals). In addition, insulin is a protein hormone that would not readily diffuse through the lipid barrier of the skin (see chapter 5). Estrogen is a lipid soluble steroid hormone.

Answers in Appendix G

A N S W E R S   T O   P R E D I C T   Q U E S T I O N S

1. Because the abnormal substance acts like TSH, it acts on the thyroid gland to increase the rate of secretion of T_3 and T_4, which increase in concentration in the circulatory system. The thyroid hormones have a negative-feedback effect on the secretion of TSH from the anterior pituitary gland, thereby decreasing the concentration of TSH in the circulatory system to low levels. Because the abnormal substance is not regulated, it can cause T_3 and T_4 levels to become very elevated.

2. A major function of plasma proteins, to which hormones bind, is to increase the half-life of the hormone. If the concentration of the plasma protein decreases, the half-life and, consequently, the concentration of the hormone in the circulatory system decrease. The half-life of the hormone decreases because the rate the hormone leaves the circulatory system increases. If the secretion rate for the hormone does not increase, its concentration in the blood declines.

3. If too little estrogen is secreted, the up-regulation of receptors in the uterus for progesterone cannot occur. As a result, progesterone cannot prepare the uterus for the embryo to attach to its wall following ovulation, and pregnancy cannot occur. Because of the lack of up-regulation, the uterus cannot respond adequately to progesterone, regardless of how much is secreted. If some

progesterone receptors are present, the uterus will require a much larger amount of progesterone to produce its normal response.

4. A drug could increase the cAMP concentration in a cell by stimulating its synthesis or by inhibiting its breakdown. Drugs that bind to a receptor that increases adenylate cyclase activity will increase cAMP synthesis. Because phosphodiesterase normally causes the breakdown of cAMP, an inhibitor of phosphodiesterase decreases the rate of cAMP breakdown and causes cAMP to increase in the smooth muscle cells of the airway and produces relaxation.

5. Intracellular receptor mechanisms result in the synthesis of new proteins that exist within the cell for a considerable amount of time. Intracellular receptors are therefore better adapted for mediating responses that last a relatively long time (i.e., for many minutes, hours, or longer). On the other hand, membrane-bound receptors that increase the synthesis of intracellular mediators such as cAMP normally activate enzymes already existing in the cytoplasm of the cell for shorter periods. The synthesis of cAMP occurs quickly, but the duration is short because cAMP is broken down quickly, and the activated enzymes are then deactivated. Membrane-bound receptor mechanisms are therefore better adapted to short-term and rapid responses.

Visit the Online Learning Center at www.mhhe.com/seeley7 for chapter quizzes, interactive learning exercises, and other study tools.

Endocrine Glands

Light micrograph of a pancreatic islet showing insulin-secreting beta cells (green) and the glucagon-secreting cells (red).

Homeostasis depends on the precise regulation of the organs and organ systems of the body. Together the nervous and endocrine systems regulate and coordinate the activity of nearly all other body structures. When either the nervous or endocrine system fails to function properly, conditions can rapidly deviate from homeostasis.

Disorders of the endocrine system can result in diseases like insulin-dependent diabetes and Addison's disease. Early in the 1900s, people who developed those diseases died. No effective treatments were available for those and other diseases of the endocrine system, such as diabetes insipidus, Cushing's syndrome, and many reproductive abnormalities. Advances have been made in understanding the endocrine system, so the outlook for people with these and other endocrine diseases has improved.

The endocrine system is small in size compared to its importance to healthy body functions. It consists of several small glands distributed throughout the body that could escape notice if not for the importance of the small amounts of hormones they secrete.

This chapter first explains the *functions of the endocrine system* (p. 610) and then profiles the *pituitary gland and hypothalamus* (p. 610), *hormones of the pituitary gland* (p. 613), *thyroid gland* (p. 619), *parathyroid glands* (p. 624), *adrenal glands* (p. 627), and *pancreas* (p. 632). It then moves to discussions about *hormonal regulation of nutrients* (p. 637), *hormones of the reproductive system* (p. 639), *hormones of the pineal body* (p. 640), *hormones of the thymus* (p. 640), and *hormones of the gastrointestinal tract* (p. 640), and *hormonelike substances* (p. 640). The chapter concludes with a look at the *effects of aging on the endocrine system* (p. 642).

Functions of the Endocrine System

Several pieces of information are needed to understand how the endocrine system regulates body functions.

1. the anatomy of each gland and its location;
2. the hormone secreted by each gland;
3. the target tissues and the response of target tissues to each hormone;
4. the means by which the secretion of each hormone is regulated;
5. the consequences and causes, if known, of hypersecretion and hyposecretion of the hormone.

The main regulatory functions of the endocrine system include:

1. *Metabolism and tissue maturation.* The endocrine system regulates the rate of metabolism and influences the maturation of tissues such as those of the nervous system.
2. *Ion regulation.* The endocrine system helps regulate blood pH as well as Na^+, K^+, and Ca^{2+} concentrations in the blood.
3. *Water balance.* The endocrine system regulates water balance by controlling the solute concentration of the blood.
4. *Immune system regulation.* The endocrine system helps control the production of immune cells.
5. *Heart rate and blood pressure regulation.* The endocrine system helps regulate the heart rate and blood pressure and helps prepare the body for physical activity.
6. *Control of blood glucose and other nutrients.* The endocrine system regulates blood glucose levels and other nutrient levels in the blood.
7. *Control of reproductive functions.* The endocrine system controls the development and functions of the reproductive systems in males and females.
8. *Uterine contractions and milk release.* The endocrine system regulates uterine contractions during delivery and stimulates milk release from the breasts in lactating females.

1. What pieces of information are needed to understand how the endocrine system regulates body functions?
2. List eight regulatory functions of the endocrine system.

Pituitary Gland and Hypothalamus

The **pituitary** (pi-too'i-tār-rē) **gland,** or **hypophysis** (hī-pof'i-sis; an undergrowth), secretes nine major hormones that regulate numerous body functions and the secretory activity of several other endocrine glands.

The **hypothalamus** (hī'pō-thal'ă-mŭs) of the brain and the pituitary gland are major sites where the nervous and endocrine systems interact (figure 18.1). The hypothalamus regulates the secre-

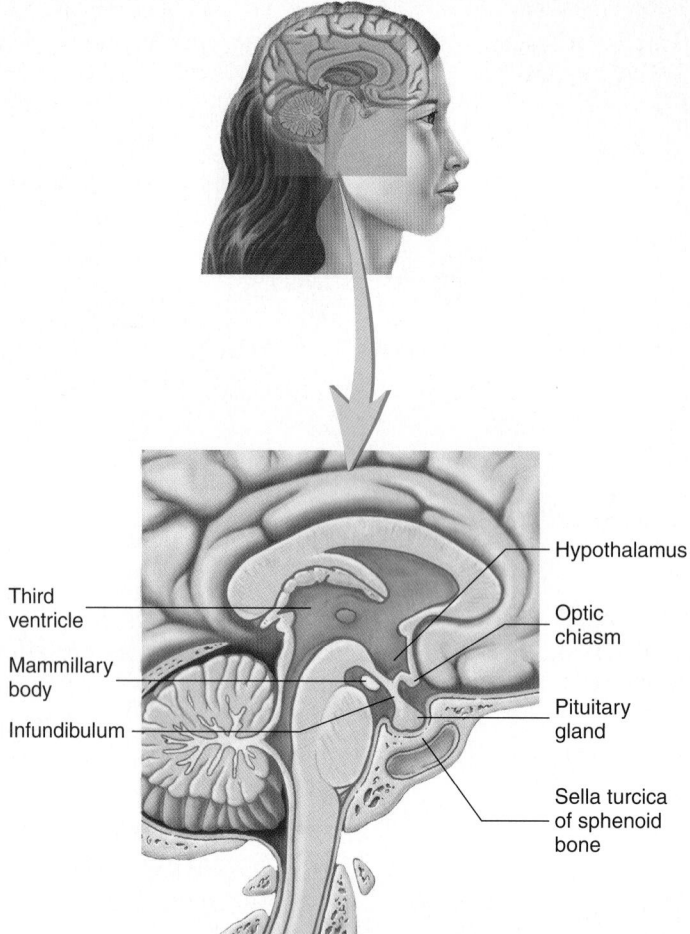

Figure 18.1   The Hypothalamus and Pituitary Gland
A midsagittal section of the head through the pituitary gland showing the location of the hypothalamus and the pituitary. The pituitary gland is in a depression called the sella turcica in the floor of the skull. It's connected to the hypothalamus of the brain by the infundibulum.

tory activity of the pituitary gland. Indeed, the posterior pituitary is an extension of the hypothalamus. Hormones, sensory information that enters the central nervous system, and emotions, in turn, influence the activity of the hypothalamus.

Structure of the Pituitary Gland

The pituitary gland is roughly 1 cm in diameter, weighs 0.5–1.0 g, and rests in the sella turcica of the sphenoid bone (see figure 18.1). It is located inferior to the hypothalamus and is connected to it by a stalk of tissue called the **infundibulum** (in-fŭn-dib'ū-lŭm).

The pituitary gland is divided functionally into two parts: the **posterior pituitary,** or **neurohypophysis** (noor'ō-hī-pof'i-sis), and the **anterior pituitary,** or **adenohypophysis** (ad'ĕ-nō-hī-pof'i-sis).

Posterior Pituitary, or Neurohypophysis

The posterior pituitary is called the neurohypophysis because it is continuous with the brain (*neuro-* refers to the nervous system). It is formed during embryonic development from an outgrowth of the inferior part of the brain in the area of the hypothalamus (see chapter 29). The outgrowth of the brain forms the infundibulum, and the distal end of the infundibulum enlarges to form the posterior pituitary (figure 18.2). Secretions of the posterior pituitary are **neurohormones** (noor-ō-hōr′mōnz) because the posterior pituitary is an extension of the nervous system.

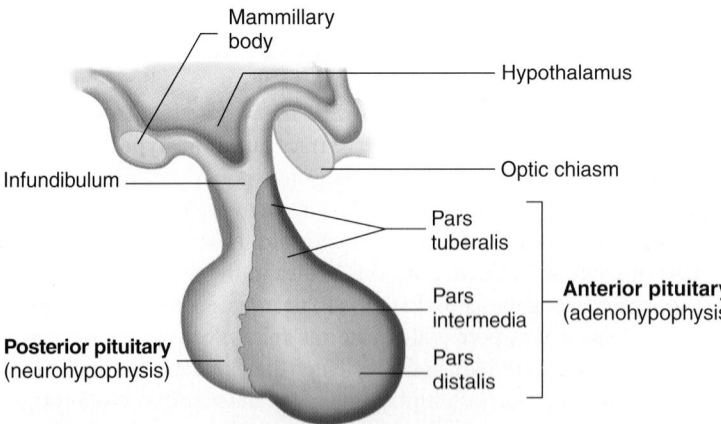

Figure 18.2   Subdivisions of the Pituitary Gland

The pituitary gland is divided into the anterior pituitary, or adenohypophysis, and the posterior pituitary, or neurohypophysis. The anterior pituitary is subdivided further into the pars distalis, pars intermedia, and pars tuberalis. The posterior pituitary consists of the enlarged distal end of the infundibulum, which connects the posterior pituitary to the hypothalamus.

Anterior Pituitary, or Adenohypophysis

The anterior pituitary, or adenohypophysis (*adeno-* means gland), arises as an outpocketing of the roof of the embryonic oral cavity called the pituitary diverticulum or Rathke's pouch, which grows toward the posterior pituitary. As it nears the posterior pituitary, the pituitary diverticulum loses its connection with the oral cavity and becomes the anterior pituitary. The anterior pituitary is subdivided into three areas with indistinct boundaries: the pars tuberalis, which extends superiorly and partially wraps around the infundibulum; the pars distalis, which is the enlarged distal portion of the anterior pituitary; and the pars intermedia, which is adjacent to the posterior pituitary (see figure 18.2). The hormones secreted from the anterior pituitary, in contrast to those from the posterior pituitary, are not neurohormones because the anterior pituitary is derived from epithelial tissue of the embryonic oral cavity and not from neural tissue.

Relationship of the Pituitary to the Brain

Portal vessels are blood vessels that begin in a primary capillary network, extend some distance and end in a secondary capillary network. The **hypothalamohypophysial** (hī′pō-thal′ă-mō-hī′pō-fiz′ē-ăl) **portal system** is one of two major portal systems. The other is the hepatic portal system (see chapter 21). The hypothalamohypophysial portal system extends from a part of the hypothalamus to the anterior pituitary (figure 18.3). The primary capillary

1. Stimuli within the nervous system cause releasing hormones (*blue balls*) to be secreted from neurons of the hypothalamus.

2. Releasing hormones pass through the hypothalamohypophysial portal system to the anterior pituitary.

3. Releasing hormones leave capillaries, bind to membrane-bound receptors, and stimulate the release of hormones from anterior pituitary cells.

4. Anterior pituitary hormones (*yellow squares*) are carried in the blood to their target tissues (*green arrow*), which, in some cases, are other endocrine glands.

Process Figure 18.3   Relationship Among the Hypothalamus, Anterior Pituitary, and Target Tissues

network in the hypothalamus is supplied with blood from arteries that deliver blood to the hypothalamus. From the primary capillary network, the hypothalamohypophysial portal vessels carry blood to a secondary capillary network in the anterior pituitary. Veins from the secondary capillary network eventually merge with the general circulation.

Neurohormones, produced and secreted by neurons of the hypothalamus, enter the primary capillary network and are carried to the secondary capillary network. There the neurohormones leave the blood and act on cells of the anterior pituitary. They act either as **releasing hormones,** increasing the secretion of anterior pituitary hormones, or as **inhibiting hormones,** decreasing the secretion of anterior pituitary hormones. Each releasing hormone stimulates and each inhibiting hormone inhibits the production and secretion of a specific hormone by the anterior pituitary. In response to the releasing hormones, anterior pituitary cells secrete hormones that enter the secondary capillary network and are carried by the general circulation to their target tissues. Thus, the hypothalamohypophysial portal system provides a means by which the hypothalamus, using neurohormones as chemical signals, regulates the secretory activity of the anterior pituitary (see figure 18.3).

Several major releasing and inhibiting hormones are released from hypothalamic neurons. **Growth hormone-releasing hormone (GHRH)** is a small peptide that stimulates the secretion of growth hormone from the anterior pituitary gland, and **growth hormone-inhibiting hormone (GHIH),** also called **somatostatin,** is a small peptide that inhibits growth hormone secretion. **Thyroid-releasing hormone (TRH)** is a small peptide that stimulates the secretion of thyroid-stimulating hormone from the anterior pituitary gland. **Corticotropin-releasing hormone (CRH)** is a peptide that stimulates adrenocorticotropic hormone from the anterior pituitary gland. **Gonadotropin-releasing hormone (GnRH)** is a small peptide that stimulates luteinizing hormone and follicle-stimulating hormone from the anterior pituitary gland.

Prolactin-releasing hormone (PRH) and **prolactin-inhibiting hormone (PIH)** regulate the secretion of prolactin from the anterior pituitary gland (table 18.1). Secretions of the anterior pituitary gland are described in a following section (p. 616).

There is no portal system to carry hypothalamic neurohormones to the posterior pituitary. Neurohormones released from the posterior pituitary are produced by neurosecretory cells with their cell bodies located in the hypothalamus. The axons of these cells extend from the hypothalamus through the infundibulum into the posterior pituitary and form a nerve tract called the **hypothalamohypophysial tract** (figure 18.4). Neurohormones produced in the hypothalamus pass down these axons in tiny vesicles and are stored in secretory vesicles in the enlarged ends of the axons. Action potentials originating in the neuron cell bodies in the hypothalamus are propagated along the axons to the axon terminals in the posterior pituitary. The action potentials cause the release of neurohormones from the axon terminals, and they enter the circulatory system. Secretions of the posterior pituitary gland are described in a following section (p. 613).

3. Where is the pituitary gland located? Contrast the embryonic origin of the anterior pituitary and the posterior pituitary.
4. Name the parts of the pituitary gland and the function of each part.
5. Define portal system. Describe the hypothalamohypophysial portal system. How does the hypothalamus regulate the secretion of the anterior pituitary hormones?
6. List the releasing and inhibiting hormones that are released from hypothalamic neurons.
7. Describe the hypothalamohypophysial tract, including the production of neurohormones in the hypothalamus and their release from the posterior pituitary.

Table 18.1	Hormones of the Hypothalamus		
Hormones	**Structure**	**Target Tissue**	**Response**
Growth hormone-releasing hormone (GHRH)	Small peptide	Anterior pituitary cells that secrete growth hormone	Increased growth hormone secretion
Growth hormone-inhibiting hormone (GHIH), or somatostatin	Small peptide	Anterior pituitary cells that secrete growth hormone	Decreased growth hormone secretion
Thyroid-releasing hormone (TRH)	Small peptide	Anterior pituitary cells that secrete thyroid-stimulating hormone	Increased thyroid-stimulating hormone secretion
Corticotropin-releasing hormone (CRH)	Peptide	Anterior pituitary cells that secrete adrenocorticotropic hormone	Increased adrenocorticotropic hormone secretion
Gonadotropin-releasing hormone (GnRH)	Small peptide	Anterior pituitary cells that secrete luteinizing hormone and follicle-stimulating hormone	Increased secretion of luteinizing hormone and follicle-stimulating hormone
Prolactin-inhibiting hormone (PIH)	Unknown (possibly dopamine)	Anterior pituitary cells that secrete prolactin	Decreased prolactin secretion
Prolactin-releasing hormone (PRH)	Unknown	Anterior pituitary cells that secrete prolactin	Increased prolactin secretion

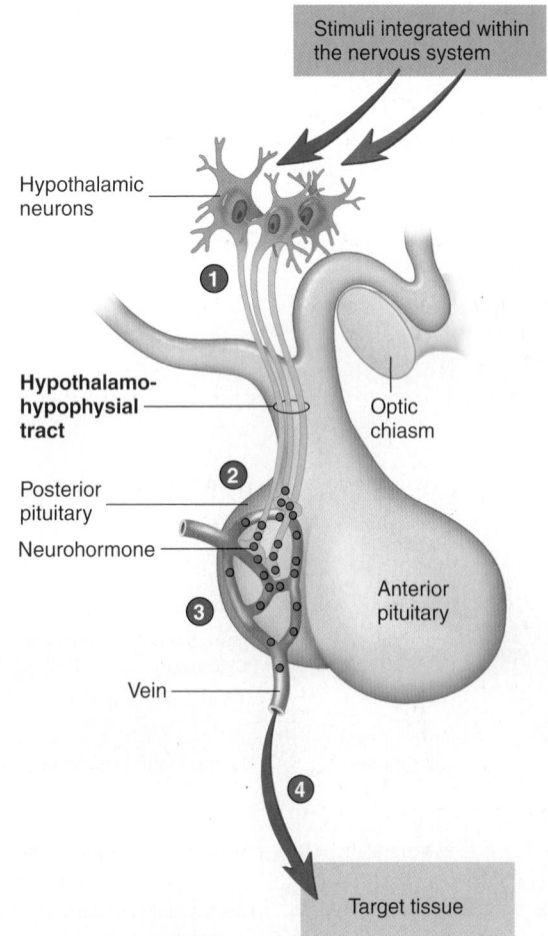

1. Stimuli within the nervous system stimulate hypothalamic neurons to either increase or decrease their action potential frequency.

2. Action potentials are carried by axons through the hypothalamohypophysial tract to the posterior pituitary. The axons of neurons store hormones in the posterior pituitary.

3. In the posterior pituitary gland, action potentials cause the release of neurohormones (*red balls*) from axon terminals into the circulatory system.

4. The neurohormones pass through the circulatory system and influence the activity of their target tissues (*green arrow*).

Stimulatory

Inhibitory

Process Figure 18.4 Relationship Among the Hypothalamus, Posterior Pituitary, and Target Tissues

P R E D I C T **1**
Surgical removal of the posterior pituitary in experimental animals results in marked symptoms, but these symptoms associated with hormone shortage are temporary. Explain these results.

Hormones of the Pituitary Gland

This section describes the hormones secreted from the pituitary gland (table 18.2), their effects on the body, and the mechanisms that regulate their secretion rate. In addition, some major consequences of abnormal hormone secretion are stressed.

Posterior Pituitary Hormones

The posterior pituitary stores and secretes two polypeptide neurohormones called antidiuretic hormone and oxytocin. A separate population of cells secretes each neurohormone.

Antidiuretic Hormone

Antidiuretic (an′tē-d-ī-ū-ret′ik) **hormone (ADH)** is so named because it prevents (*anti-*) the output of large amounts of urine (*diuresis*). ADH is sometimes called **vasopressin** (vā-sō-pres′in, vas-ō-pres′in) because it constricts blood vessels and raises blood pressure when large amounts are released. ADH molecules are

synthesized predominantly by neuron cell bodies in the supraoptic nuclei of the hypothalamus, and are transported within the axons of the hypothalamohypophysial tract to the posterior pituitary, where they are stored in axon terminals. ADH molecules are released from these axon terminals into the blood and carried to its primary target tissue, the kidney tubules. Kidney tubules are the sites of urine production in the kidneys. ADH molecules promote the retention of water in kidney tubules and reduce urine volume (see chapter 26).

The secretion rate for ADH changes in response to alterations in blood osmolality and blood volume (figure 18.5). The osmolality of a solution increases as the concentration of solutes in the solution increases. Specialized neurons, called **osmoreceptors** (os′mō-rē-sep′terz, os′mō-rē-sep′tōrz), synapse with the ADH neurosecretory cells in the hypothalamus. When blood osmolality increases, the frequency of action potentials in the osmoreceptors increases, resulting in a greater frequency of action potentials in the ADH neurosecretory cells. As a consequence, ADH secretion increases. Alternatively, an increase in blood osmolality can directly stimulate the ADH neurosecretory cells. Because ADH stimulates the kidney tubules to retain water, it functions to reduce blood osmolality and resists any further increase in the osmolality of body fluids.

As the osmolality of the blood decreases, the action potential frequency in the osmoreceptors and the neurosecretory cells decreases.

Table 18.2 Hormones of the Pituitary Gland

Hormones	Structure	Target Tissue	Response
Posterior Pituitary (Neurohypophysis)			
Antidiuretic hormone (ADH)	Small peptide	Kidney	Increased water reabsorption (less water is lost in the form of urine)
Oxytocin	Small peptide	Uterus; mammary glands	Increased uterine contractions; increased milk expulsion from mammary glands; unclear function in males
Anterior Pituitary (Adenohypophysis)			
Growth hormone (GH), or somatotropin	Protein	Most tissues	Increased growth in tissues; increased amino acid uptake and protein synthesis; increased breakdown of lipids and release of fatty acids from cells; increased glycogen synthesis and increased blood glucose levels; increased somatomedin production
Thyroid-stimulating hormone (TSH)	Glycoprotein	Thyroid gland	Increased thyroid hormone secretion
Adrenocorticotropic hormone (ACTH)	Peptide	Adrenal cortex	Increased glucocorticoid hormone secretion
Lipotropins	Peptides	Fat tissues	Increased fat breakdown
β endorphins	Peptides	Brain, but not all target tissues are known	Analgesia in the brain; inhibition of gonadotropin-releasing hormone secretion
Melanocyte-stimulating hormone (MSH)	Peptide	Melanocytes in the skin	Increased melanin production in melanocytes to make the skin darker in color
Luteinizing hormone (LH)	Glycoprotein	Ovaries in females; testes in males	Ovulation and progesterone production in ovaries; testosterone synthesis and support for sperm cell production in testes
Follicle-stimulating hormone (FSH)	Glycoprotein	Follicles in ovaries in females; seminiferous tubes in males	Follicle maturation and estrogen secretion in ovaries; sperm cell production in testes
Prolactin	Protein	Ovaries and mammary glands in females	Milk production in lactating women; increased response of follicle to LH and FSH; unclear function in males

Thus, less ADH is secreted from the posterior pituitary gland, and the volume of water eliminated in the form of urine increases.

Urine volume increases within minutes to a few hours in response to the consumption of a large volume of water. In contrast, urine volume decreases and urine concentration increases within hours if little water is consumed. ADH plays a major role in these changes in urine formation. The effect is to maintain the osmolality and the volume of the extracellular fluid within a normal range of values.

Sensory receptors that detect changes in blood pressure send action potentials through sensory nerve fibers of the vagus nerve that eventually synapse with the ADH neurosecretory cells. A decrease in blood pressure, which normally accompanies a decrease in blood volume, causes an increased action potential frequency in the neurosecretory cells and increased ADH secretion, which stimulates the kidneys to retain water. Because the water in urine is derived from blood as it passes through the kidneys, ADH slows any further reduction in blood volume.

An increase in blood pressure decreases the action potential frequency in neurosecretory cells. This leads to the secretion of less ADH from the posterior pituitary. As a result, the volume of urine produced by the kidneys increases (see figure 18.5). The effect of ADH on the kidney and its role in the regulation of extracellular osmolality and volume are described in greater detail in chapters 26 and 27.

Diabetes Insipidus

A lack of ADH secretion is one cause of **diabetes insipidus** and leads to the production of a large amount of dilute urine, which can approach 20 L/day. The loss of many liters of water in the form of urine causes an increase in the osmolality of the body fluids, and a decrease in extracellular fluid volume, but negative-feedback mechanisms fail to stimulate ADH release. The volume of urine produced each day increases rapidly as the rate of ADH secretion becomes less than 50% of normal. Diabetes insipidus can also result from either damage to the kidney tubules or a genetic disorder that makes the kidney tubules incapable of responding to ADH. Damage to the kidney tubules can result from infection or other diseases that make them insensitive to ADH. In genetic disorders either receptors for ADH are abnormal or the intracellular signal molecules fail to produce a normal response. The consequences of diabetes insipidus are not obvious until the condition becomes severe. When the condition is severe, dehydration and death can result unless the intake of water is adequate to accommodate its loss.

An excess of ADH secretion is called the **syndrome of inappropriate ADH secretion.** The excess ADH molecules act on the kidney tubules. As a consequence, a small volume of concentrated urine is produced and water is retained in the blood, which increases blood volume and blood pressure.

1. Osmoreceptors in the hypothalamus detect changes in blood osmolality, and sensory neurons that send action potentials through the vagus nerves to the hypothalamus detect changes in blood presure.

2. An increase in osmolality and a decrease in blood pressure increase action potentials in ADH-secreting neurons.

3. Action potentials are carried by axons of ADH-secreting neurons through the hypothalamohypophysial tract to the posterior pituitary.

4. In the posterior pituitary, increased action potentials increase the release of ADH from the axon terminals into the circulatory system.

5. Increasing ADH acts on the kidney tubules to increase water reabsorption, resulting in a reduced urine volume, increased urine osmolality, and decreased blood osmolality. This helps to maintain blood volume.

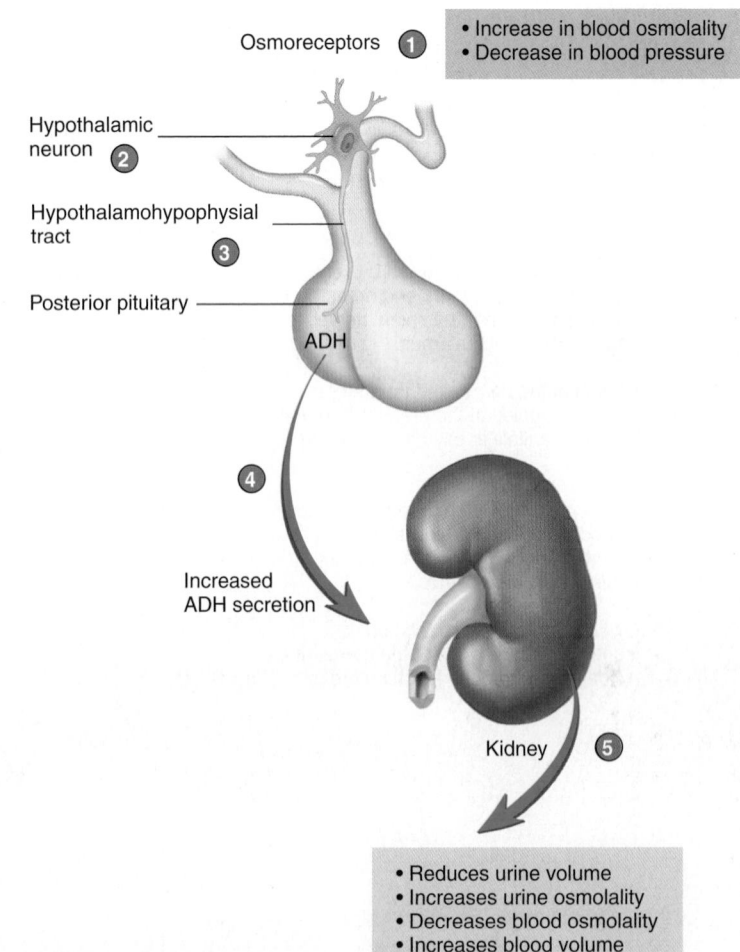

Osmoreceptors ①
- Increase in blood osmolality
- Decrease in blood pressure

Hypothalamic neuron ②

Hypothalamohypophysial tract ③

Posterior pituitary

ADH

④

Increased ADH secretion

Kidney ⑤

- Reduces urine volume
- Increases urine osmolality
- Decreases blood osmolality
- Increases blood volume

Process Figure 18.5 Control of Antidiuretic Hormone (ADH) Secretion

Small changes in blood osmolality are important in regulating ADH secretion. Larger changes in blood pressure are required to influence ADH secretion.

---

8. *Where is ADH produced, from where is it secreted, and what is its target tissue? When ADH levels increase, how are urine volume, blood osmolality, and blood volume affected?*

9. *The secretion rate for ADH changes in response to alterations in what two factors? Name the types of sensory cells that respond to alterations in those factors.*

Oxytocin

Oxytocin (ok-sē-tō′sin) is synthesized by neuron cell bodies in the paraventricular nuclei of the hypothalamus and then is transported through axons to the posterior pituitary, where it is stored in the axon terminals.

Oxytocin stimulates smooth muscle cells of the uterus. This neurohormone plays an important role in the expulsion of the fetus from the uterus during delivery by stimulating uterine smooth muscle contraction. It also causes contraction of uterine smooth muscle in nonpregnant women, primarily during menses and sexual intercourse. The uterine contractions help in expelling of the uterine epithelium and small amounts of blood during menses and

can participate in the movement of sperm cells through the uterus after sexual intercourse. Oxytocin is also responsible for milk ejection in lactating females by promoting contraction of smooth musclelike cells surrounding the alveoli of the mammary glands (see chapter 29). Little is known about the effect of oxytocin in males.

Stretch of the uterus, mechanical stimulation of the cervix, or stimulation of the nipples of the breast when a baby nurses activate nervous reflexes that stimulate oxytocin release. Action potentials are carried by sensory neurons from the uterus and from the nipples to the spinal cord. Action potentials are then carried up the spinal cord to the hypothalamus, where they stimulate an increase in the action potential frequency in oxytocin-secreting neurons. Action potentials in the oxytocin-secreting neurons pass along the axons in the hypothalamohypophysial tract to the posterior pituitary, where they cause the axon terminals to release oxytocin (figure 18.6). The role of oxytocin in the reproductive system is described in greater detail in chapter 29.

10. *Where is oxytocin produced and secreted, and what effects does it have on its target tissues? What factors stimulate the secretion of oxytocin?*

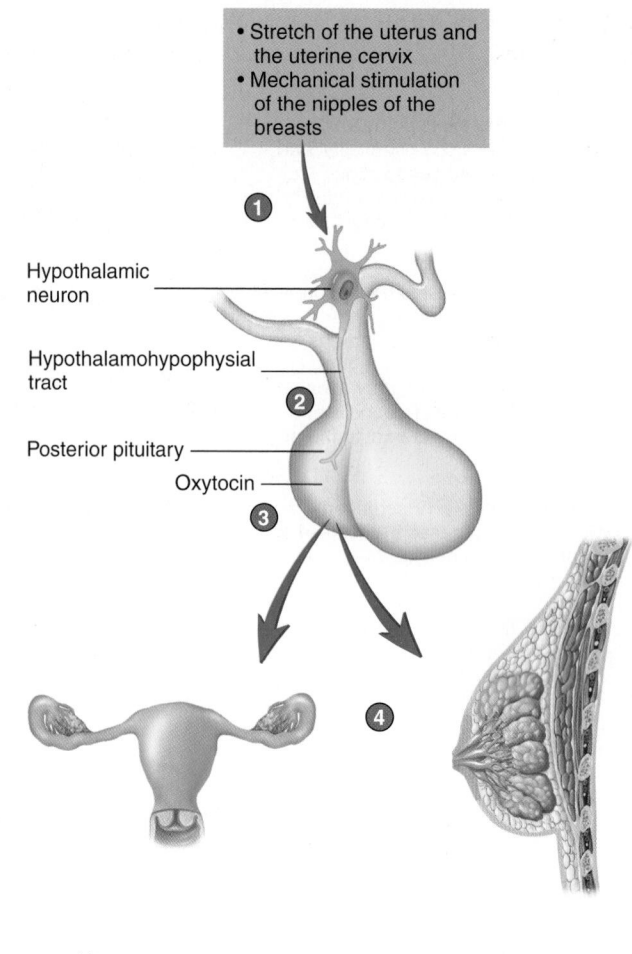

1. Action potentials are carried by sensory neurons from the uterus and breast to the spinal cord, and up spinal nerve tracts to the hypothalamus.

2. Stretch of the uterus and the uterine cervix or nipples of the breasts increase action potentials in oxytocin-secreting neurons.

3. Action potentials are carried by oxytocin-secreting neurons in the hypothalamohypophysial tract to the posterior pituitary gland.

4. Oxytocin enters the circulation. It increases contractions of the uterus and milk ejection from the lactating breast.

Process Figure 18.6 **Control of Oxytocin Secretion**

Anterior Pituitary Hormones

Releasing and inhibiting hormones that pass from the hypothalamus through the hypothalamohypophysial portal system to the anterior pituitary influence anterior pituitary secretions. For some anterior pituitary hormones, the hypothalamus produces both releasing hormones and inhibiting hormones. For others regulation is primarily by releasing hormones (see table 18.1).

The hormones released from the anterior pituitary are proteins, glycoproteins, or polypeptides. They are transported in the circulatory system, have a half-life measured in minutes, and bind to membrane-bound receptor molecules on their target cells. For the most part, each hormone is secreted by a separate cell type. Adrenocorticotropic hormone and lipotropin are exceptions because these hormones are derived from the same precursor protein.

Anterior pituitary hormones are called **tropic** (trop′ik, trō′ pik) **hormones.** They are released from the anterior pituitary gland and regulate target tissues. This regulation includes controlling the secretion of hormones as well as the growth of target tissues. The tropic hormones include growth hormone, adrenocorticotropic hormone and related substances, luteinizing hormone, follicle-stimulating hormone, prolactin, and thyroid-stimulating hormone.

11. *Structurally, what kind of hormones are released from the posterior pituitary and the anterior pituitary? Do these hormones bind to plasma proteins, how long is their half-life, and how do they activate their target tissues?*

Growth Hormone

Growth hormone (GH), sometimes called **somatotropin** (sō′ mă-tō-trō′ pin), stimulates growth in most tissues, plays a major role in regulating growth, and therefore, plays an important role in determining how tall a person becomes. It is also a regulator of metabolism. GH increases the number of amino acids entering cells, favors their incorporation into proteins, and it slows protein breakdown. It increases lipolysis, or the breakdown of lipids and the release of fatty acids from fat cells into the blood. Fatty acids then can be used as energy sources to drive chemical reactions, including anabolic reactions, by other cells. GH increases glucose synthesis by the liver, which releases glucose into the blood. This and the increased use of fats as an energy source decreases glucose usage. GH plays an important role in regulating blood nutrient levels after a meal and during periods of fasting.

GH binds directly to membrane-bound receptors on target cells (see chapter 17), such as fat cells, to produce responses. These responses are called the direct effects of GH and include the increased breakdown of lipids and decreased use of glucose as an energy source.

GH also has indirect effects on some tissues. It increases the production of a number of polypeptides, primarily by the liver but also by skeletal muscle and other tissues. These polypeptides, called **somatomedins** (sō′mă-tō-mē′dinz), circulate in the blood and bind to receptors on target tissues. The best understood effects of the somatomedins are the stimulation of growth in cartilage and bone and the increased synthesis of protein in skeletal muscles. The best known somatomedins are two polypeptide hormones produced by the liver called **insulin-like growth factor I** and **II** because of the similarity of their structure to insulin and because the receptor molecules function through a mechanism similar to the receptors for insulin. Growth hormone and growth factors, like somatomedins, bind to membrane-bound receptors that phosphorylate intracellular proteins (see chapter 17).

Two neurohormones released from the hypothalamus regulate the secretion of GH (figure 18.7). One factor, growth hormone-releasing hormone (GHRH), stimulates the secretion of GH, and the other, growth hormone-inhibiting hormone (GHIH), or **somatostatin** (sō′mă-tō-stat′in), inhibits the secretion of GH. Stimuli that influence GH secretion act on the hypothalamus to increase or decrease the secretion of the releasing and inhibiting hormones. Low blood glucose levels and stress stimulate secretion of GH, and high blood glucose levels inhibit secretion of GH. Rising blood levels of certain amino acids also increases GH secretion.

In most people, a rhythm of GH secretion occurs. Daily peak levels of GH are correlated with deep sleep. A chronically elevated blood GH level during periods of rapid growth does not occur, although children tend to have somewhat higher blood levels of GH than adults. In addition to GH, factors like genetics, nutrition, and sex hormones influence growth.

Several pathologic conditions are associated with abnormal GH secretion. In general, the causes for hypersecretion or hyposecretion of GH are the result of tumors in the hypothalamus or pituitary, the synthesis of structurally abnormal GH, the inability of the liver to produce somatomedins, or the lack of functional receptors in target tissues. The consequences of hypersecretion and hyposecretion of growth hormone are described in the Clinical Focus on "Growth Hormone and Growth Disorders" (p. 618); also see chapter 6.

1. Stress and decreased blood glucose levels increase the release of growth hormone-releasing hormone (GHRH) and decrease the release of growth hormone-inhibiting hormone (GHIH).

2. GHRH and GHIH travel through the hypothalamohypophysial portal system to the anterior pituitary.

3. Increased GHRH and reduced GHIH act on the anterior pituitary and result in increased GH secretion.

4. GH acts on target tissues.

5. Increasing GH levels have a negative-feedback effect on the hypothalamus, resulting in decreased GHRH and decreased GHIH.

Stimulatory

Inhibitory

① Stress / Low blood glucose

Increased growth hormone-releasing hormone (GHRH)

Decreased growth hormone-inhibiting hormone (GHIH)

② Hypothalamohypophysial portal system

⑤

Anterior pituitary

③ GH

④ Target tissue
- Increases protein synthesis and decreases protein breakdown
- Increases tissue growth
- Increases fat breakdown
- Increases glucose synthesis and reduces glucose usage

Process Figure 18.7 Control of Growth Hormone (GH) Secretion

Secretion of GH is controlled by two neurohormones released from the hypothalamus: growth hormone-releasing hormone (GHRH), which stimulates GH secretion, and growth hormone-inhibiting hormone (GHIH), which inhibits GH secretion. Stress increases GHRH secretion and inhibits GHIH secretion. High levels of GH have a negative-feedback effect on the production of GHRH by the hypothalamus.

Clinical Focus   Growth Hormone and Growth Disorders

Chronic hyposecretion of GH in infants and children leads to a type of **dwarfism** (dwŏrf'izm) sometimes called pituitary dwarfism. It results in a short stature due to delayed bone growth. The bones usually have a normal shape, and people with this condition exhibit normal intelligence, in contrast to dwarfism caused by hyposecretion of thyroid hormones. Other symptoms resulting from the lack of GH include mild obesity and retarded development of adult reproductive functions. Two types of dwarfism result from a lack of GH secretion: (1) In approximately two-thirds of the cases, GH and other anterior pituitary hormones are secreted in reduced amounts. The decrease in other anterior pituitary hormones can result in additional disorders caused by reduced secretion of thyroid hormones, adrenal cortex hormones, and reproductive hormones. (2) In the remaining approximately one-third of cases, a reduced amount of GH is observed, and the secretion of other anterior pituitary hormones is closer to normal. No disorders caused by a lack of thyroid hormones or adrenal cortex hormones are observed, and normal reproduction is possible for these individuals. No obvious pathology is associated with hyposecretion of GH in adults, although some evidence suggests that lack of GH can lead to reduced bone mineral content in adults.

The gene responsible for determining the structure of GH has been transferred successfully from human cells to bacterial cells, which produce GH that is identical to human GH. The GH produced in this fashion is available to treat patients who suffer from a lack of GH secretion.

Chronic hypersecretion of GH leads to **giantism** (jī'an-tizm) or **acromegaly** (ak-rō-meg'ă-lē), depending on whether the hypersecretion occurs before or after complete ossification of the epiphysial plates in the skeletal system. Chronic hypersecretion of GH before the epiphysial plates have ossified causes exaggerated and prolonged growth in long bones, resulting in giantism. Some individuals thus affected have grown to be 8 feet tall or more.

In adults, chronically elevated GH levels result in acromegaly. No increase in height occurs because of the ossified epiphysial plates. The condition does result in an increased diameter of fingers, toes, hands, and feet; the deposition of heavy bony ridges above the eyes; and a prominent jaw. The influence of GH on soft tissues results in a bulbous or broad nose, an enlarged tongue, thickened skin, and sparse subcutaneous adipose tissue. Nerves frequently are compressed as a result of the proliferation of connective tissue. Because GH spares glucose usage, chronic hyperglycemia results, frequently leading to diabetes mellitus and the development of severe atherosclerosis. Treatment for chronic hypersecretion of GH often involves surgical removal or irradiation of a GH-producing tumor.

---

12. *What effects do stress, amino acid levels in the blood, and glucose levels in the blood have on GH secretion?*

13. *What stimulates somatomedin production, where is it produced, and what are its effects?*

P R E D I C T   ❷

Mr. Hoops has a son who wants to be a basketball player almost as much as Mr. Hoops wants him to be one. Mr. Hoops knows a little bit about growth hormone and asks his son's doctor if he would prescribe some for his son, so he can grow tall. What do you think the doctor tells Mr. Hoops?

Thyroid-Stimulating Hormone

Thyroid-stimulating hormone (TSH), also called **thyrotropin** (thī-rot'rō-pin, thī-rō-trō'pin), stimulates the synthesis and secretion of thyroid hormones from the thyroid gland. TSH is a glycoprotein consisting of α and β subunits, which bind to membrane-bound receptors of the thyroid gland. The receptors respond through a G protein mechanism that increases the intracellular chemical signal, cAMP. In higher concentrations, TSH also increases the activity of phospholipase. Phospholipase activates mechanisms that open Ca^{2+} channels and increases the Ca^{2+} concentration in cells of the thyroid gland (see chapter 17).

TSH secretion is controlled by TRH from the hypothalamus and thyroid hormones from the thyroid gland. TRH binds to membrane-bound receptors in cells of the anterior pituitary gland and activates G proteins, which results in increased TSH secretion. In contrast, thyroid hormones inhibit both TRH and TSH secretion. TSH is secreted in a pulsatile fashion and its blood levels are highest at night, but the blood levels of thyroid hormones are maintained within a narrow range of values (see "Thyroid Hormones" p. 619).

Adrenocorticotropic Hormone and Related Substances

Adrenocorticotropic (ă-drē'nō-kōr'ti-kō-trō'pik) **hormone (ACTH)** is one of several peptide hormones derived from a precursor protein called **proopiomelanocortin** (prō-ō'pē-ō-mel'ă-nō-kōr'tin) and secreted from the anterior pituitary. This precursor protein gives rise to ACTH, lipotropins, β endorphin, and melanocyte-stimulating hormone.

ACTH binds to membrane-bound receptors and activates a G protein mechanism that increases cAMP, which produces a response. ACTH increases the secretion of hormones, primarily cortisol, from the adrenal cortex. ACTH and melanocyte-stimulating hormone also bind to melanocytes in the skin and increase skin pigmentation (see chapter 5). In pathologic conditions like Addison's disease, blood levels of ACTH and related hormones are chronically elevated, and the skin becomes markedly darker. Regulation of ACTH secretion and the effect of hypersecretion and hyposecretion of ACTH are described in the section on "Adrenal Glands" on page 627.

The **lipotropins** (li-pō-trō′pinz) secreted from the anterior pituitary bind to membrane-bound receptor molecules on adipose tissue cells. They cause fat breakdown and the release of fatty acids into the circulatory system.

The **β endorphins** (en′dōr-finz) have the same effects as opiate drugs like morphine, and they can play a role in analgesia in response to stress and exercise. Other functions have been proposed for the β endorphins, including regulation of body temperature, food intake, and water balance. Both ACTH and β-endorphin secretions increase in response to stress and exercise.

Melanocyte-stimulating hormone (MSH) binds to membrane-bound receptors on skin melanocytes and stimulates increased melanin deposition in the skin. The regulation of MSH secretion and its function in humans is not well understood, although it's an important regulator of skin pigmentation in some other vertebrates.

Luteinizing Hormone, Follicle-Stimulating Hormone, and Prolactin

Gonadotropins (gō′nad-ō-trō′pinz) are glycoprotein hormones capable of promoting growth and function of the **gonads,** which include the ovaries and testes. The two major gonadotropins secreted from the anterior pituitary are **luteinizing** (loo′tē-ĭ-nīz-ing) **hormone (LH)** and **follicle-stimulating hormone (FSH).** LH, FSH, and another anterior pituitary hormone called **prolactin** (prō-lak′tin) play important roles in regulating reproduction.

LH and FSH secreted into the blood bind to membrane-bound receptors, increase the intracellular synthesis of cAMP through G protein mechanisms, and stimulate the production of **gametes** (gam′ēts)—sperm cells in the testes and oocytes in ovaries. LH and FSH also control the production of reproductive hormones—estrogens and progesterone in the ovaries and testosterone in the testes.

LH and FSH are released from anterior pituitary cells under the influence of the hypothalamic-releasing hormone, **gonadotropin-releasing hormone (GnRH).** GnRH is also called **luteinizing hormone-releasing hormone (LHRH).**

Prolactin plays an important role in milk production in the mammary glands of lactating females. It binds to a membrane-bound receptor that phosphorylates intracellular proteins. The phosphorylated proteins produce the response in the cell. Prolactin also can enhance progesterone secretion of the ovary after ovulation. No role for this hormone has been clearly established in males. Several hypothalamic neurohormones can be involved in the complex regulation of prolactin secretion. One neurohormone is prolactin-releasing hormone (PRH), and another is prolactin-inhibiting hormone (PIH). The regulation of gonadotropin and prolactin secretion and their specific effects are explained more fully in chapter 28.

14. *For each of the following hormones secreted by the anterior pituitary—GH, TSH, ACTH, LH, FSH, and prolactin—name its target tissue and the effect of the hormone on its target tissue.*

15. *How are ACTH, MSH, lipotropins, and β endorphins related? What are the functions of these hormones?*

16. *Define gonadotropins, and name two gonadotropins produced by the anterior pituitary.*

Thyroid Gland

The **thyroid gland** is composed of two lobes connected by a narrow band of thyroid tissue called the **isthmus.** The lobes are lateral to the upper portion of the trachea just inferior to the larynx, and the isthmus extends across the anterior aspect of the trachea (figure 18.8a). The thyroid gland is one of the largest endocrine glands, with a weight of approximately 20 g. It is highly vascular and appears more red than its surrounding tissues.

The thyroid gland contains numerous **follicles,** which are small spheres whose walls are composed of a single layer of cuboidal epithelial cells (figure 18.8b and c). The center, or lumen, of each thyroid follicle is filled with proteins called **thyroglobulin** (thī-rō-glob′ū-lin) synthesized and secreted by cells of the thyroid follicles. Large amounts of the thyroid hormones are stored in the thyroid follicles as part of the thyroglobulin molecules.

Between the follicles, a delicate network of loose connective tissue contains numerous capillaries. Scattered **parafollicular** (par-ă-fo-lik′ū-lăr) **cells** are found between the follicles and among the cells that make up the walls of the follicle. **Calcitonin** (kal-si-tō′nin) is secreted from the parafollicular cells and plays a role in reducing the concentration of calcium in the body fluids when calcium levels become elevated.

Thyroid Hormones

The thyroid hormones include both **triiodothyronine** (trī-ī′ō-dō-thī′rō-nēn; T_3) and **tetraiodothyronine** (tet′ră-ī′ō-dō-thī′rō-nēn; T_4). T_4 is also called **thyroxine** (thī-rok′sēn, thī-rok′sin). T_3 and T_4 constitute major secretory products of the thyroid gland, consisting of 10% T_3 and 90% T_4 (table 18.3). Although calcitonin is secreted by the parafollicular cells of the thyroid gland, T_3 and T_4 are considered to be the thyroid hormones because they are more clinically important and because they are secreted from the thyroid follicles.

T_3 and T_4 Synthesis

Thyroid-stimulating hormone (TSH) from the anterior pituitary must be present to maintain thyroid hormone synthesis and secretion. TSH causes an increase in synthesis of T_3 and T_4, which are then stored inside of the thyroid follicles as part of thyroglobulin. TSH also causes T_3 and T_4 to be released from thyroglobulin and enter the circulatory system. Because iodine is a component of T_3 and T_4, an adequate amount of iodine in the diet is required for thyroid hormone synthesis. The following

(a) **Anterior view**

Figure 18.8 Anatomy and Histology of the Thyroid Gland

(*a*) Anterior view of the thyroid gland. (*b*) Histology of the thyroid gland. The gland is made up of many spheric thyroid follicles containing thyroglobulin. Parafollicular cells are in the tissue between the thyroid follicles. (*c*) Low-power photomicrograph of thyroid follicles.

Table 18.3	Hormones of the Thyroid and Parathyroid Glands		
Hormones	**Structure**	**Target Tissue**	**Response**
Thyroid Gland			
Thyroid Follicles			
Thyroid hormones (triiodothyronine and tetraiodothyronine)	Amino acid derivative	Most cells of the body	Increased metabolic rate; essential for normal process of growth and maturation
Parafollicular Cells			
Calcitonin	Polypeptide	Bone	Decreased rate of breakdown of bone by osteoclasts; prevention of a large increase in blood Ca^{2+} levels
Parathyroid			
Parathyroid hormone	Peptide	Bone; kidney; small intestine	Increased rate of breakdown of bone by osteoclasts; increased reabsorption of Ca^{2+} in kidneys; increased absorption of Ca^{2+} from the small intestine; increased vitamin D synthesis; increased blood Ca^{2+} levels

events in the thyroid follicles result in T_3 and T_4 synthesis and secretion (figure 18.9):

1. Iodide ions (I^-) are taken up by thyroid follicle cells by active transport. The active transport of the I^- is against a concentration gradient of approximately 30-fold in healthy individuals.

2. Thyroglobulins, which contain numerous tyrosine amino acid molecules, are synthesized within the cells of the follicle.

3. Nearly simultaneously, the I^- are oxidized to form iodine (I) and either one or two iodine atoms are bound to each of the tyrosine molecules of thyroglobulin. This occurs close to the time the thyroglobulin molecules are secreted by the process of exocytosis into the lumen of the follicle. As a result, the secreted thyroglobulin contains many iodinated tyrosines either mono- or diiodotyrosine amino acids.

4. In the lumen of the follicle, two diiodotyrosine molecules of thyroglobulin combine to form tetraiodothyronine (T_4), or one monoiodotyrosine and one diiodotyrosine molecule combine to form triiodothyronine (T_3). Large amounts of T_3 and T_4 are stored within the thyroid follicles as part of thyroglobulin. A reserve sufficient to supply thyroid hormones for approximately 2 weeks is stored in this form.

5. Thyroglobulin is taken into the thyroid follicle cells by endocytosis where lysosomes fuse with the endocytotic vesicles.

6. Proteolytic enzymes break down thyroglobulin to release T_3 and T_4. T_3 and T_4 are lipid soluble so they diffuse through the plasma membranes of the follicular cells into the interstitial spaces and finally into the capillaries of the thyroid gland. The remaining amino acids of thyroglobulin are used again to synthesize more thyroglobulin.

Transport in the Blood

T_3 and T_4 are transported in combination with plasma proteins in the circulatory system. Approximately 70%–75% of the circulating T_3 and T_4 are bound to **thyroxine-binding globulin (TBG),** which is synthesized by the liver and 20% to 30% are bound to other plasma proteins, including albumen. T_3 and T_4, bound to these plasma proteins, form a large reservoir of circulating thyroid hormones, and the half-life of these hormones is increased greatly because of this binding. After thyroid gland removal in experimental animals, it takes approximately 1 week for T_3 and T_4 levels in the blood to decrease by 50%. As free T_3 and T_4 levels decrease in the interstitial spaces, additional T_3 and T_4 dissociate from the plasma proteins to maintain the levels in the tissue spaces. When sudden secretion of T_3 and T_4 occurs, the excess binds to the plasma proteins. As a consequence, the concentration of thyroid hormones in the tissue spaces fluctuates very little.

Approximately 33%–40% of the T_4 is converted to T_3 in the body tissues. This conversion can be important in the action of thyroid hormones on their target tissues because T_3 is the major hormone that interacts with target cells. In addition, T_3 is several times more potent than T_4.

Much of the circulating T_4 is eliminated from the body by being converted to tetraiodothyroacetic acid and then excreted in

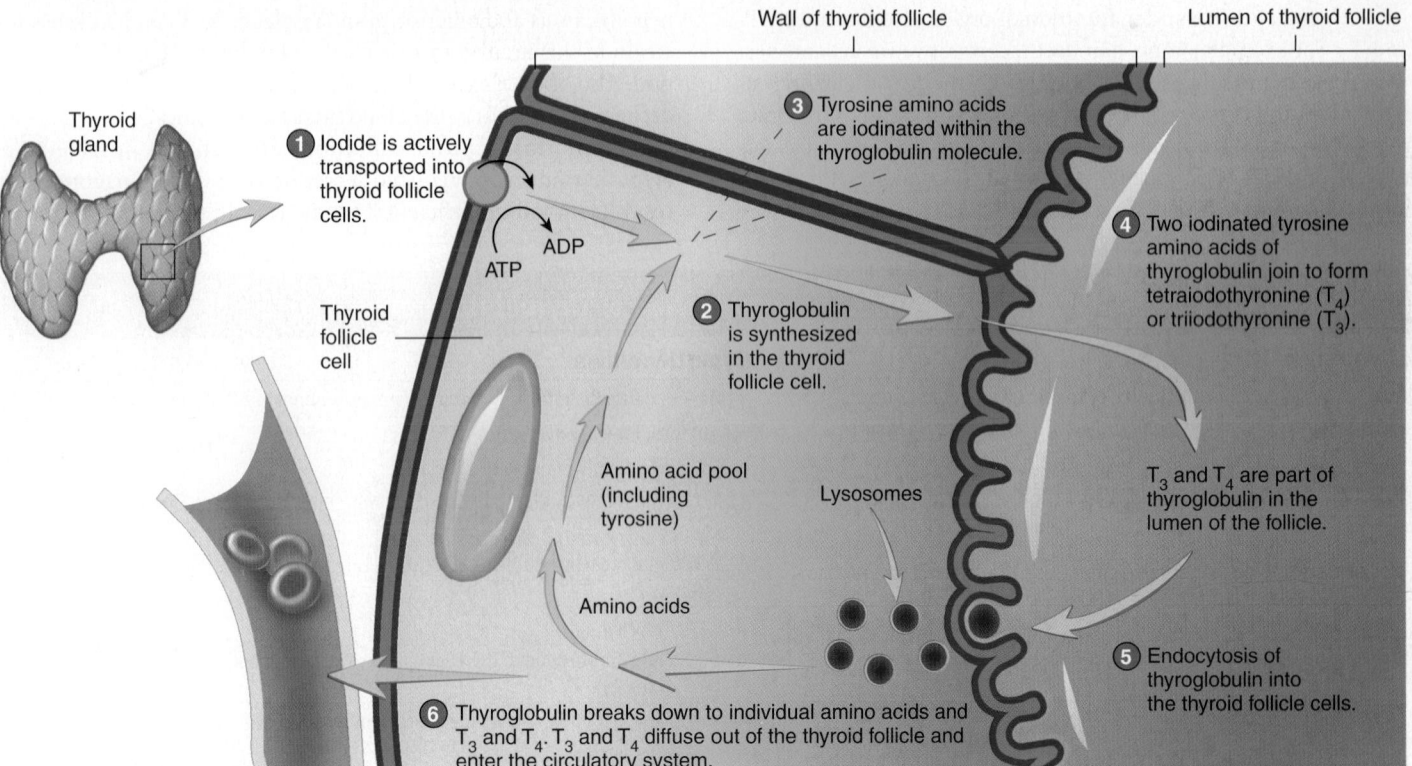

Process Figure 18.9 Biosynthesis of Thyroid Hormones

The numbered steps describe the synthesis and the secretion of T_3 and T_4 from the thyroid gland. See text for details of each numbered step.

the urine or bile. In addition, a large amount is converted to an inactive form of T_3 and rapidly metabolized and excreted.

Mechanism of Action of T_3 and T_4

T_3 and T_4 interact with their target tissues in a fashion similar to that of the steroid hormones. They readily diffuse through plasma membranes into the cytoplasm of cells. Within cells, they bind to receptor molecules in the nuclei. Thyroid hormones combined with their receptor molecules interact with DNA in the nucleus to influence regulatory genes and initiate new protein synthesis. The newly synthesized proteins within the target cells mediate the response of the cells to thyroid hormones. It takes up to a week after the administration of thyroid hormones for a maximal response to develop, and new protein synthesis occupies much of that time.

Effects of T_3 and T_4

T_3 and T_4 affect nearly every tissue in the body, but not all tissues respond identically. Metabolism is primarily affected in some tissues, and growth and maturation are influenced in others.

The normal rate of metabolism for an individual depends on an adequate supply of thyroid hormone, which increases the rate at which glucose, fat, and protein are metabolized. The increased rate of metabolism produces heat. Blood levels of cholesterol decline. Thyroid hormones increase the activity of Na^+–K^+ pump, which helps increase the body temperature. T_3 and T_4 can alter the number and activity of mitochondria, resulting in greater ATP synthesis and heat production. The metabolic rate can increase from 60%–100% when blood T_3 and T_4 are elevated. Low levels of T_3 and T_4 lead to the opposite effect. Normal body temperature depends on an adequate amount of T_3 and T_4.

Normal growth and maturation of organs also depend on T_3 and T_4. For example, bone, hair, teeth, connective tissue, and nervous tissue require thyroid hormone for normal growth and development. Both normal growth and normal maturation of the brain require thyroid hormones. Also, T_3 and T_4 play a permissive role for GH, and GH does not have its normal effect on target tissues if T_3 and T_4 are not present.

The specific effects of hyposecretion and hypersecretion of thyroid hormones are outlined in table 18.4. Hypersecretion of T_3 and T_4 increases the rate of metabolism. High body temperature, weight loss, increased appetite, rapid heart rate, and an enlarged thyroid gland are major symptoms.

Hyposecretion of T_3 and T_4 decreases the rate of metabolism. Low body temperature, weight gain, reduced appetite, reduced heart rate, reduced blood pressure, weak skeletal muscles, and apathy are major symptoms. If hyposecretion of T_3 and T_4 occurs during development there is a decreased rate of metabolism, abnormal nervous system development, abnormal growth, and abnormal maturation of tissues. The consequence is a mentally retarded person of short stature and distinctive form called a **cretin** (krē′tin).

Regulation of Thyroid Hormone Secretion

Thyroid-releasing hormone (TRH) from the hypothalamus and TSH from the anterior pituitary function together to increase T_3 and T_4 secretion from the thyroid gland. Exposure to cold and stress cause increased TRH secretion and prolonged fasting decreases TRH secretion. TRH stimulates the secretion of TSH from the anterior pituitary. When TRH release increases, TSH secretion from the anterior pituitary gland also increases. When TRH release decreases, TSH secretion decreases. Small fluctuations in blood levels of TSH occur on a daily basis, with a small nocturnal increase. TSH stimulates T_3 and T_4 secretion from the thyroid gland. TSH also increases the synthesis of T_3 and T_4 as well as causing hypertrophy (increased cell size) and hyperplasia (increased cell number) of the thyroid gland. Decreased blood levels of TSH lead to decreased T_3 and T_4 secretion and thyroid gland atrophy. Figure 18.10 illustrates the regulation of T_3 and T_4 secretion. T_3 and T_4 have a negative-feedback effect on the hypothalamus and anterior pituitary gland. As T_3 and T_4 levels increase in the circulatory system, they inhibit TRH and TSH secretion. Also, if the thyroid gland is removed or if T_3 and T_4 secretion declines, TSH levels in the blood increase dramatically.

Abnormal thyroid conditions are outlined in table 18.5. Hypothyroidism, or reduced secretion of thyroid hormones, can result from iodine deficiency, taking certain drugs, and exposure

Table 18.4   Effects of Hyposecretion and Hypersecretion of Thyroid Hormones

Hypothyroidism	Hyperthyroidism
Decreased metabolic rate, low body temperature, cold intolerance	Increased metabolic rate, high body temperature, heat intolerance
Weight gain, reduced appetite	Weight loss, increased appetite
Reduced activity of sweat and sebaceous glands, dry and cold skin	Copious sweating, warm and flushed skin
Reduced heart rate, reduced blood pressure, dilated and enlarged heart	Rapid heart rate, elevated blood pressure, abnormal electrocardiogram
Weak, flabby skeletal muscles, sluggish movements	Weak skeletal muscles that exhibit tremors, quick movements with exaggerated reflexes
Constipation	Bouts of diarrhea
Myxedema (swelling of the face and body) as a result of mucoprotein deposits	Exophthalmos (protruding of the eyes) as a result of mucoprotein and other deposits behind the eye
Apathetic, somnolent	Hyperactivity, insomnia, restlessness, irritability, short attention span
Coarse hair, rough and dry skin	Soft, smooth hair and skin
Decreased iodide uptake	Increased iodide uptake
Possible goiter (enlargement of the thyroid gland)	Almost always develops goiter

1. TRH is released from neurons within the hypothalamus. It passes through the hypothalamohypophysial portal system to the anterior pituitary.

2. TRH causes cells of the anterior pituitary to secrete TSH, which passes through the general circulation to the thyroid gland.

3. TSH causes increased release of T_3 and T_4 into the general circulation.

4. T_3 and T_4 act on target tissues to produce a response.

5. T_3 and T_4 also have an inhibitory effect on the secretion of TRH from the hypothalamus and TSH from the anterior pituitary.

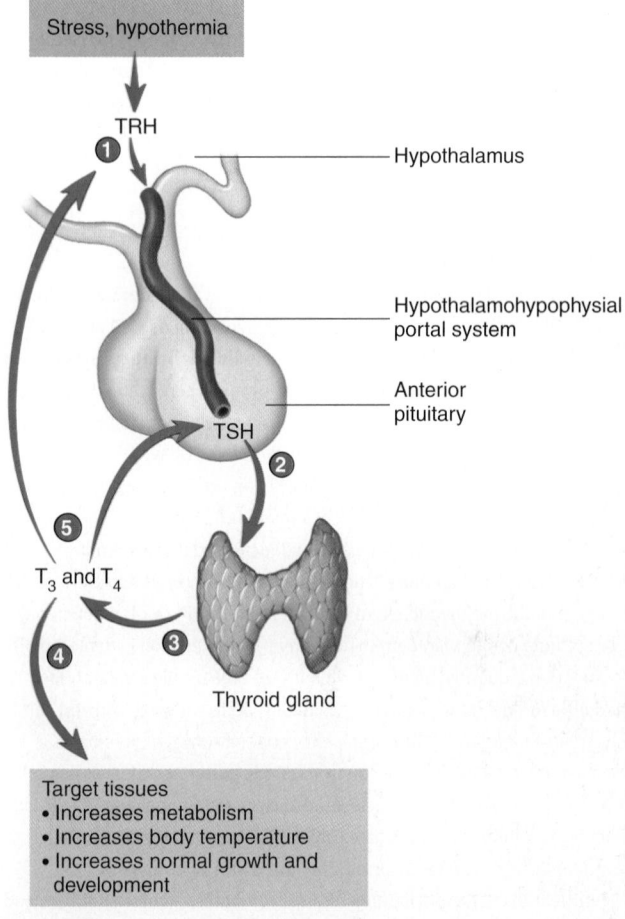

Stress, hypothermia

TRH

Hypothalamus

Hypothalamohypophysial portal system

Anterior pituitary

TSH

T_3 and T_4

Thyroid gland

Target tissues
• Increases metabolism
• Increases body temperature
• Increases normal growth and development

Stimulatory

Inhibitory

Process Figure 18.10 Regulation of Thyroid Hormone (T_3 and T_4) Secretion

Table 18.5	Abnormal Thyroid Conditions
Cause	**Description**
Hypothyroidism	
Iodine deficiency	Causes inadequate T_3 and T_4 synthesis, which results in elevated thyroid-stimulating hormone (TSH) secretion; thyroid gland enlarges (goiter) as a result of TSH stimulation; T_3 and T_4 frequently remain in the low to normal range
Goiterogenic substances	Found in certain drugs and in small amounts in certain plants such as cabbage; inhibit T_3 and T_4 synthesis
Cretinism	Caused by maternal iodine deficiency or congenital errors in thyroid hormone synthesis; results in mental retardation and a short, grotesque appearance
Lack of thyroid gland	Partial or complete surgical removal or drug-induced destruction of the thyroid gland as a treatment for Graves' disease (hyperthyroidism)
Pituitary insufficiency	Results from lack of TSH secretion; often associated with inadequate secretion of other anterior pituitary hormones
Hashimoto's disease	Autoimmune disease in which thyroid hormone secretion can be normal or depressed
Hyperthyroidism	
Graves' disease	Characterized by goiter and exophthalmos; apparently an autoimmune disease; most patients have long-acting thyroid stimulator, a TSH-like immune globulin, in their plasma
Tumors—benign adenoma or cancer	Result in either normal secretion or hypersecretion of thyroid hormones (rarely hyposecretion)
Thyroiditis—a viral infection	Produces painful swelling of the thyroid gland with normal or slightly increased T_3 and T_4 production
Elevated TSH levels	Result from a pituitary tumor
Thyroid storm	Sudden release of large amounts of T_3 and T_4; caused by surgery, stress, infections, and unknown reasons

to other chemicals that inhibit T_3 and T_4 synthesis. It can also be due to inadequate secretion of TSH, an autoimmune disease that depresses thyroid hormone function, or surgical removal of the thyroid gland. Hypersecretion of T_3 and T_4 can result from the synthesis of an immune globulin that can bind to TSH receptors and acts like TSH, and from TSH-secreting tumors of the pituitary gland.

Goiter and Exophthalmos

An abnormal enlargement of the thyroid gland is called a **goiter.** Goiters can result from conditions that cause hypothyroidism as well as conditions that cause hyperthyroidism. An **iodine deficiency goiter** results when dietary iodine intake is very low and there is too little iodine to synthesize T_3 and T_4 (see table 18.5). As a result, blood levels of T_3 and T_4 decrease and the person may exhibit symptoms of hypothyroidism. The reduced negative feedback of T_3 and T_4 on the anterior pituitary and hypothalamus result in elevated TSH secretion. TSH causes hypertrophy and hyperplasia of the thyroid gland and increased thyroglobulin synthesis even though there is not enough iodine to synthesize T_3 and T_4. Consequently, the thyroid gland enlarges. Historically, iodine deficiency goiters were common in people from areas where the soil is depleted of iodine. Consequently, plants grown in these areas, called goiter belts, had little iodine in them and caused iodine-deficient diets. Iodized salt has nearly eliminated iodine deficiency goiters. However, it remains a problem in some developing countries. **Toxic goiter** secretes excess T_3 and T_4, and it can result from elevated TSH secretion or elevated TSH-like immune globulin molecules (see Graves' disease in table 18.5). Toxic goiter results in elevated T_3 and T_4 secretion and symptoms of hyperthyroidism. **Exophthalmos** often accompanies hyperthyroidism and is caused by the deposition of excess connective tissue proteins behind the eyes. The excess tissue makes the eyes move anteriorly, and consequently they appear to be larger than normal.

Graves, disease is the most common cause of hyperthyroidism. Elevated T_3 and T_4 resulting from this condition suppresses TSH and TRH, but the T_3 and T_4 levels remain elevated. Exophthalmos is common. Treatment often involves use of antithyroid drugs that reduce T_3 and T_4 secretion, partial destruction of the thyroid gland using radioactive iodine, or partial or complete surgical removal of the thyroid gland. These treatments are followed by the oral administration of the appropriate amount of T_3 and T_4. Unfortunately removal of the thyroid gland normally does not reverse exophthalmos.

---

17. *Where is the thyroid gland located? Describe the follicles and the parafollicular cells within the thyroid. What hormones do they produce?*
18. *Starting with the uptake of iodide by the follicles, describe the production and secretion of thyroid hormones.*
19. *How are the thyroid hormones transported in the blood? What effect does this transportation have on their half-life?*
20. *What are the target tissues of thyroid hormone? By what mechanism do thyroid hormones alter the activities of their target tissues? What effects are produced?*

21. *Starting in the hypothalamus, explain how chronic exposure to cold, food deprivation, or stress can affect thyroid hormone production.*
22. *Diagram two negative-feedback mechanisms involving hormones that function to regulate production of thyroid hormones.*

P R E D I C T   **3**

Predict the effect of surgical removal of the thyroid gland on blood levels of TRH, TSH, T_3 and T_4. Predict the effect of oral administration of T_3 and T_4 on TRH and TSH.

Calcitonin

The parafollicular cells of the thyroid gland, which secrete calcitonin, are dispersed between the thyroid follicles throughout the thyroid gland. The major stimulus for increased calcitonin secretion is an increase in calcium levels in the body fluids.

The primary target tissue for calcitonin is bone (see chapter 6). Calcitonin binds to membrane-bound receptors, decreases osteoclast activity, and lengthens the life span of osteoblasts. The result is a decrease in blood calcium and phosphate levels caused by increased bone deposition.

The importance of calcitonin in the regulation of blood Ca^{2+} levels is unclear. Its rate of secretion increases in response to elevated blood Ca^{2+} levels, and it may function to prevent large increases in blood Ca^{2+} levels following a meal. Blood levels of calcitonin decrease with age to a greater extent in females than males. Osteoporosis increases with age and occurs to a greater degree in females than males. Complete thyroidectomy does not result in high blood Ca^{2+} levels, however. It's possible that the regulation of blood Ca^{2+} levels by other hormones, such as parathyroid hormone, and vitamin D compensates for the loss of calcitonin in individuals who have undergone a thyroidectomy. No pathologic condition is associated directly with a lack of calcitonin secretion.

---

23. *What effect does calcitonin have on osteoclasts, osteoblasts, and blood calcium levels? What stimulus can cause an increase in calcitonin secretion?*

Parathyroid Glands

The **parathyroid** (par-ă-thī′royd) **glands** are usually embedded in the posterior part of each lobe of the thyroid gland, and are made up of two cell types, the chief cells and oxyphils. The chief cells secrete parathyroid hormone, but the function of the oxyphils is unknown. Usually four parathyroid glands are present, with their cells organized in densely packed masses or cords rather than in follicles (figure 18.11). In some cases one or more of the parathyroid glands does not become embedded in the thyroid gland and remains in the nearby connective tissue.

Parathyroid hormone (PTH), also called parathormone, is a polypeptide hormone that is important in the regulation of calcium

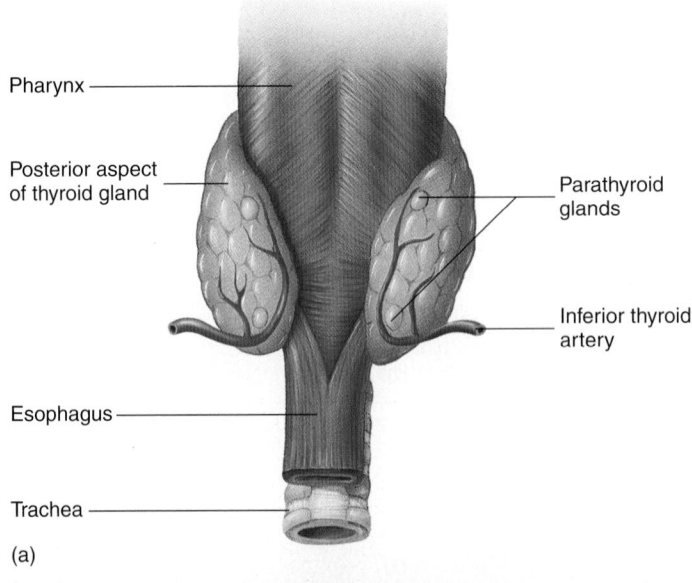

- Pharynx
- Posterior aspect of thyroid gland
- Parathyroid glands
- Inferior thyroid artery
- Esophagus
- Trachea

(a)

Posterior view

- Thyroid follicles
- Parathyroid gland

LM 100x

(b)

Figure 18.11 Anatomy and Histology of the Parathyroid Glands

(a) The parathyroid glands are embedded in the posterior part of the thyroid gland. (b) The parathyroid glands are composed of densely packed cords of cells.

levels in body fluids (see table 18.3). Bone, the kidneys, and the intestine are its major target tissues. PTH binds to membrane-bound receptors and activates a G protein mechanism that increases intracellular cAMP levels in target tissues. Without functional parathyroid glands, the ability to adequately regulate blood calcium levels is lost.

PTH stimulates osteoclast activity in bone and can cause the number of osteoclasts to increase. The increased osteoclast activity results in bone resorption and the release of calcium and phosphate, causing an increase in blood calcium levels. PTH receptors are not present on osteoclasts but are present on osteoblasts and on red bone marrow stromal (stem) cells. PTH binds to receptors on osteoblasts which then promote an increase in osteoclast activity (see chapter 6).

PTH induces calcium reabsorption within the kidneys so that less calcium leaves the body in urine. It also increases the enzymatic formation of active vitamin D in the kidneys. Calcium is actively absorbed by the epithelial cells of the small intestine, and the synthesis of transport proteins in the intestinal cells requires active vitamin D. PTH increases the rate of active vitamin D synthesis, which in turn increases the rate of calcium and phosphate absorption in the intestine, thereby elevating blood levels of calcium.

Although PTH increases the release of phosphate ions (PO_4^{3-}) from bone and increases PO_4^{3-} absorption in the gut, it increases PO_4^{3-} excretion in the kidney. The overall effect of PTH is to decrease blood phosphate levels. A simultaneous increase in both Ca^{2+} and PO_4^{3-} results in the precipitation of calcium phosphate in soft tissues of the body, where they cause irritation and inflammation.

The regulation of PTH secretion and its role in regulating blood Ca^{2+} levels are outlined in figure 18.12. The primary stimulus for the secretion of PTH is a decrease in blood Ca^{2+} levels, whereas elevated blood Ca^{2+} levels inhibit PTH secretion. This regulation keeps blood Ca^{2-} levels fluctuating within a normal range of values. Both hypersecretion and hyposecretion of PTH cause serious symptoms (table 18.6). The regulation of blood Ca^{2+} levels is discussed more thoroughly in chapter 27.

PREDICT

Predict the effect of an inadequate dietary intake of calcium on PTH secretion and on target tissues for PTH.

Inactive parathyroid glands result in hypocalcemia. Reduced extracellular calcium levels cause voltage-gated Na^+ channels in plasma membranes to open, which increases the permeability of plasma membranes to Na^+. As a consequence, Na^+ diffuse into cells and cause depolarization (see chapter 11). Symptoms of hypocalcemia are nervousness, muscle spasms, cardiac arrhythmias, and convulsions. In extreme cases, tetany of skeletal muscles results and tetany of the respiratory muscles can cause death.

Decreased secretion of PTH
from the parathyroid glands results.

- Decreased breakdown of bone by osteoclasts
  results in decreased release of Ca^{2+} from bone.
- Decreased reabsorption of Ca^{2+} by the kidneys
  results in increased Ca^{2+} loss in the urine.
- Decreased synthesis of active vitamin D by the
  kidneys results in decreased Ca^{2+} absorption
  from the small intestine.

An increase in blood Ca^{2+} levels is detected
by the cells of the parathyroid glands.

A decrease in blood Ca^{2+} levels results because
fewer Ca^{2+} enter the blood than leave the blood.

Blood Ca^{2+}
(normal range)

Blood Ca^{2+}
levels increase

Blood Ca^{2+}
levels decrease

Blood Ca^{2+}
(normal range)

Blood Ca^{2+}
homeostasis
is maintained

A decrease in blood Ca^{2+} levels is detected
by the cells of the parathyroid glands.

An increase in blood Ca^{2+} levels results because
more Ca^{2+} enter the blood than leave the blood.

An increased secretion of PTH from
the parathyroid glands results.

- Increased breakdown of bone by osteoclasts
  results in increased release of Ca^{2+} from bone.
- Increased reabsorption of Ca^{2+} by the kidneys
  results in decreased Ca^{2+} loss in the urine.
- Increased synthesis of active vitamin D by the
  kidneys results in increased Ca^{2+} absorption
  from the small intestine.

Homeostasis Figure 18.12 Parathyroid Hormone (PTH) Regulation of Blood Ca^{2+} Levels

24. *Where are the parathyroid glands located, and what*
    hormone do they produce?
25. *What effect does PTH have on osteoclasts, osteoblasts, the*
    kidneys, the small intestine, and blood calcium and blood
    phosphate levels? What stimulus can cause an increase in
    PTH secretion?

P R E D I C T

A patient with a malignant tumor had his thyroid gland removed. What
effect would this removal have on blood levels of Ca^{2+}? If the
parathyroid glands are inadvertently removed along with the thyroid
gland during surgery, death can result because muscles of respiration
undergo sustained contractions. Explain.

Table 18.6	Causes and Symptoms of Hypersecretion and Hyposecretion of Parathyroid Hormone
Hypoparathyroidism	**Hyperparathyroidism**
Causes	
Accidental removal during thyroidectomy	Primary hyperparathyroidism: a result of abnormal parathyroid function—adenomas of the parathyroid gland (90%), hyperplasia of parathyroid idiopathic (unknown cause) cells (9%), and carcinomas (1%) Secondary hyperparathyroidism: caused by conditions that reduce blood Ca^{2+} levels, such as inadequate Ca^{2+} in the diet, inadequate levels of vitamin D, pregnancy, or lactation
Symptoms	
Hypocalcemia	Hypercalcemia or normal blood Ca^{2+} levels; calcium carbonate salts may be deposited throughout the body, especially in the renal tubules (kidney stones), lungs, blood vessels, and gastric mucosa
Normal bone structure	Bones weaken and are eaten away as a result of resorption; some cases are first diagnosed when a radiograph is taken of a broken bone
Increased neuromuscular excitability; tetany, laryngospasm, and death from asphyxiation can result	Neuromuscular system less excitable; muscular weakness may be present
Flaccid heart muscle; cardiac arrhythmia may develop	Increased force of contraction of cardiac muscle; at very high blood Ca^{2+} levels, cardiac arrest during contraction is possible
Diarrhea	Constipation

Adrenal Glands

The **adrenal** (ă-drē′năl) **glands,** also called the **suprarenal** (soo′pră-rē′năl) **glands,** are near the superior poles of the kidneys. Like the kidneys, they are retroperitoneal, and they are surrounded by abundant adipose tissue. The adrenal glands are enclosed by a connective tissue capsule and have a well-developed blood supply (figure 18.13a).

The adrenal glands are composed of an inner **medulla** and an outer **cortex,** which are derived from two separate embryonic tissues. The adrenal medulla arises from neural crest cells, which also give rise to postganglionic neurons of the sympathetic division of the autonomic nervous system (see chapters 16 and 29). Unlike most glands of the body, which develop from invaginations of epithelial tissue, the adrenal cortex is derived from mesoderm.

Trabeculae of the connective tissue capsule penetrate into the adrenal gland in several locations, and numerous small blood vessels course with them to supply the gland. The medulla consists of closely packed polyhedral cells centrally located in the gland (figure 18.13b). The cortex is composed of smaller cells and forms three indistinct layers: the **zona glomerulosa** (glō-mār′ū-lōs-ă), the **zona fasciculata** (fa-sik′ū-lă-tă), and the **zona reticularis** (re-tik′ū-lăr′is). These three layers are functionally and structurally specialized. The zona glomerulosa is immediately beneath the capsule and is composed of small clusters of cells. Beneath the zona glomerulosa is the thickest part of the adrenal cortex, the zona fasciculata. In this layer, the cells form long columns, or fascicles, of cells that extend from the surface toward the medulla of the gland. The deepest layer of the adrenal cortex is the zona reticularis, which is a thin layer of irregularly arranged cords of cells.

Hormones of the Adrenal Medulla

The adrenal medulla secretes two major hormones: **epinephrine** (**adrenaline;** ă-dren′ă-lin), 80%, and **norepinephrine** (**noradrenaline;** nor-ă-dren′ă-lin), 20% (table 18.7). Epinephrine and norepinephrine are closely related to each other. In fact, norepinephrine is a precursor to the formation of epinephrine. Because the adrenal medulla consists of cells derived from the same cells that give rise to postganglionic sympathetic neurons, its secretory products are neurohormones.

Epinephrine and norepinephrine combine with adrenergic receptors, which are membrane-bound receptors in target cells. They are classified as either α-adrenergic or β-adrenergic receptors, and each of these categories has subcategories. All of the adrenergic receptors function through G protein mechanisms. The α-adrenergic receptors cause Ca^{2+} channels to open, cause the release of Ca^{2+} from endoplasmic reticulum by activating phospholipase enzymes, open K^+ channels, decrease cAMP synthesis, or stimulate the synthesis of eicosanoid molecules such as prostaglandins. The β-adrenergic receptors all increase cAMP synthesis. The effects of epinephrine and norepinephrine released from the adrenal medulla are described when the systems these hormones affect are discussed (see chapters 16, 20, 21, 24, and 26).

Epinephrine increases blood levels of glucose. It combines with membrane-bound receptors in liver cells. Cyclic AMP, in turn, activates enzymes that catalyze the breakdown of glycogen to glucose, and the release of glucose from the liver cells into the blood. Epinephrine also increases the breakdown of glycogen in muscle cells, but muscle cells do not release glucose into the blood. The glucose in muscle cells is metabolized in the muscle cells. Epinephrine also increases fat breakdown in adipose tissue, and fatty acids are released into the blood. The fatty acids can be taken up and metabolized by tissues such as skeletal and cardiac muscle.

(a)

(b)

Figure 18.13   **Anatomy and Histology of the Adrenal Gland**

(*a*) An adrenal gland is at the superior pole of each kidney. (*b*) The adrenal glands have an outer cortex and an inner medulla. The cortex is surrounded by a connective tissue capsule and consists of three layers: the zona glomerulosa, the zona fasciculata, and the zona reticularis.

Table 18.7	Hormones of the Adrenal Gland		
Hormones	**Structure**	**Target Tissue**	**Response**
Adrenal Medulla			
Epinephrine primarily; norepinephrine	Amino acid derivatives	Heart, blood vessels, liver, fat cells	Increased cardiac output; increased blood flow to skeletal muscles and increased blood flow to the heart (see chapter 20); vasoconstriction of blood vessels, especially visceral and skin blood vessels; increased release of glucose and fatty acids into blood; in general, preparation for physical activity
Adrenal Cortex			
Glucocorticoids (Cortisol)	Steroids	Most tissues	Increased protein and fat breakdown; increased glucose production; inhibition of immune response and decreased inflammation
Mineralocorticoids (Aldosterone)	Steroids	Kidney	Increased Na^+ reabsorption and K^+ and H^+ excretion; enhances water reabsorption
Androgens	Steroids	Many tissues	Minor importance in males; in females, development of some secondary sexual characteristics, such as axillary and pubic hair

Epinephrine and norepinephrine increase the heart's rate and force of contraction and cause blood vessels to constrict in the skin, kidneys, gastrointestinal tract, and other viscera. Also, epinephrine causes dilation of blood vessels in skeletal muscles and cardiac muscle.

Secretion of adrenal medullary hormones prepares the individual for physical activity and is a major component of the fight-or-flight response (see chapter 16). The response results in reduced activity in organs not essential for physical activity and in increased blood flow and metabolic activity in organs that participate in physical activity. In addition, it mobilizes nutrients that can be used to sustain physical exercise.

The effects of epinephrine and norepinephrine are short-lived because they are rapidly metabolized, excreted, or taken up by tissues. Their half-life in the circulatory system is measured in minutes.

The release of adrenal medullary hormones primarily occurs in response to stimulation by sympathetic neurons because the adrenal medulla is a specialized part of the autonomic nervous system. Several conditions, including emotional excitement, injury, stress, exercise, and low blood glucose levels, lead to the release of adrenal medullary neurohormones (figure 18.14).

Pheochromocytoma and Neuroblastoma

The two major disorders of the adrenal medulla are both tumors: pheochromocytoma (fē′ō-krō′mō-sī-tō′mă), a benign tumor, and neuroblastoma (noor′ō-blas-tō′mă), a malignant tumor. Symptoms result from the release of large amounts of epinephrine and norepinephrine and include hypertension (high blood pressure), sweating, nervousness, pallor, and tachycardia (rapid heart rate). The high blood pressure results from the effect of these hormones on the heart and blood vessels and is correlated with an increased chance of heart disease and stroke.

26. *Where are the adrenal glands located? Describe the embryonic origin of the adrenal medulla and adrenal cortex.*
27. *Name two hormones secreted by the adrenal medulla, and list the effects of these hormones.*
28. *List several conditions that can stimulate the production of adrenal medullary hormones. What role does the nervous system play in the release of adrenal medullary hormones? How does this role relate to the embryonic origin of the adrenal medulla?*

Hormones of the Adrenal Cortex

The adrenal cortex secretes three hormone types: **mineralocorticoids** (min′er-al-ō-kōr′ti-koydz), **glucocorticoids** (gloo-kō-kōr′ti-koydz), and **androgens** (an′drō-jenz) (see table 18.7). All are similar in structure in that they are steroids, highly specialized lipids that are derived from cholesterol. Because they are lipid-soluble, they are not stored in the adrenal gland cells but diffuse from the cells as they are synthesized. Adrenal cortical hormones are transported in the blood in combination with specific plasma proteins; they are metabolized in the liver and excreted in the bile and urine. The hormones of the adrenal cortex bind to intracellular receptors and stimulate the synthesis of specific proteins that are responsible for producing the cell's responses.

Mineralocorticoids

The major secretory products of the zona glomerulosa are the mineralocorticoids. **Aldosterone** (al-dos′ter-ōn) is produced in the greatest amounts, although other closely related mineralocorticoids are also secreted. Aldosterone increases the rate of sodium reabsorption by the kidneys, thereby increasing blood levels of sodium. Sodium reabsorption can result in increased water reabsorption by the kidneys and an increase in blood volume providing ADH is also

1. Stress, physical activity, and low blood glucose levels act as stimuli to the hypothalamus, resulting in increased sympathetic nervous system activity.

2. An increased frequency of action potentials conducted through the sympathetic division of the autonomic nervous system stimulates the adrenal medulla to secrete epinephrine and some norepinephrine into the circulatory system.

3. Epinephrine and norepinephrine act on their target tissues to produce responses.

- Stress
- Physical activity
- Low blood glucose levels

Hypothalamus

Spinal cord

Sympathetic nerve fiber

Sympathetic chain

Adrenal medulla

- Increases the release of glucose from the liver into the blood
- Increases the release of fatty acids from adipose tissue into the blood
- Increases heart rate
- Decreases blood flow through blood vessels of internal organs
- Increases blood flow through blood vessels of skeletal muscle and the heart
- Increases blood pressure
- Decreases the function of visceral organs
- Increases the metabolic rate of skeletal muscles

Process Figure 18.14 Regulation of Adrenal Medullary Secretions

Stress, physical exercise, and low blood glucose levels cause increased activity of the sympathetic nervous system, which increases epinephrine and norepinephrine secretion from the adrenal medulla.

secreted. Aldosterone increases K$^+$ excretion into the urine by the kidneys, thereby decreasing blood levels of K$^+$. It also increases the rate of H$^+$ excretion into the urine. When aldosterone is secreted in high concentrations, it can result in reduced blood levels of K$^+$ and alkalosis (elevated pH of body fluids). The details of the effects of aldosterone and the mechanisms controlling aldosterone secretion are discussed along with kidney functions in chapters 26 and 27 and with the cardiovascular system in chapter 21.

P R E D I C T   **6**

Alterations in blood levels of sodium and potassium have profound effects on the electrical properties of cells. Because high blood levels of aldosterone cause retention of sodium and excretion of potassium, predict and explain the effects of high aldosterone levels on nerve and muscle function. Conversely, because low blood levels of aldosterone cause low blood levels of sodium and elevated blood levels of potassium, predict the effects of low aldosterone levels on nerve and muscle function.

Glucocorticoids

The zona fasciculata of the adrenal cortex primarily secretes **glucocorticoid hormones,** the major one of which is **cortisol** (kōr′ti-sol). The target tissues and responses to the glucocorticoids are numerous (table 18.8). The responses are classified as metabolic, developmental, or anti-inflammatory. Glucocorticoids increase fat catabolism, decrease glucose and amino acid uptake in skeletal muscle, increase **gluconeogenesis** (gloo′kō-nē-ō-jen′ĕ-sis; the synthesis of glucose from precursor molecules like amino acids in the liver), and increase protein degradation. Thus, some major effects of glucocorticoids are an increased use of fats and proteins to provide energy for cells. There is an increase in blood glucose levels and glycogen deposits in cells. The glucose and the glycogen are a reservoir of molecules that can be metabolized rapidly. Glucocorticoids are also required for the maturation of tissues like fetal lungs and for the development of receptor molecules in target tissues for epinephrine and norepinephrine. Glucocorticoids decrease the intensity of the inflammatory response by decreasing both the number of white blood cells and the secretion of inflammatory chemicals from tissues. This anti-inflammatory effect is most important under conditions of stress, when the rate of glucocorticoid secretion is relatively high.

ACTH is necessary to maintain the secretory activity of the adrenal cortex, which rapidly atrophies without this hormone. Corti-

cotropin-releasing hormone (CRH) is released from the hypothalamus and stimulates the anterior pituitary to secrete ACTH. The zona fasciculata is very sensitive to ACTH, and responds by increasing cortisol secretion. At high concentrations, ACTH acts on the zona glomerulosa to enhance aldosterone secretion. The regulation of ACTH and cortisol secretion is outlined in figure 18.15. Both ACTH and cortisol inhibit CRH secretion from the hypothalamus and thus constitute a negative-feedback influence on CRH secretion. In addition, high concentrations of cortisol in the blood inhibit ACTH secretion from the anterior pituitary, and low concentrations stimulate it. This negative-feedback loop is important in maintaining blood cortisol levels within a narrow range of concentrations. In response to stress or hypoglycemia, blood levels of cortisol increase rapidly because these stimuli trigger a large increase in CRH release from the hypothalamus. Table 18.9 outlines several abnormalities associated with hypersecretion and hyposecretion of adrenal hormones.

P R E D I C T   **7**

Cortisone, a drug similar to cortisol, is sometimes given to people who have severe allergies or extensive inflammation or who suffer from autoimmune diseases. Taking this substance chronically can damage the adrenal cortex. Explain how this damage can occur.

Adrenal Androgens

Some adrenal steroids, including **androstenedione** (an-drŏ-stēn′dī-ōn), are weak androgens. They are secreted by the zona reticularis and converted by peripheral tissues to the more potent androgen, testosterone. Adrenal androgens stimulate pubic and axillary hair growth and sexual drive in females. Their effects in males are negligible in comparison to testosterone secreted by the testes. Chapter 28 presents additional information about androgens.

29. Describe the three layers of the adrenal cortex, and name the hormones produced by each layer.
30. Name the target tissue of aldosterone, and list the effects of an increase in aldosterone secretion on the concentration of ions in the blood.
31. Describe the effects produced by an increase in cortisol secretion. Starting in the hypothalamus, describe how stress or low blood sugar levels can stimulate cortisol release.
32. What effects do adrenal androgens have on males and females?

Table 18.8   Target Tissues and Their Responses to Glucocorticoid Hormones

Target Tissues	Responses
Peripheral tissues, such as skeletal muscle, liver, and adipose tissue	Inhibits glucose use; stimulates formation of glucose from amino acids and, to some degree, from fats (gluconeogenesis) in the liver, which results in elevated blood glucose levels; stimulates glycogen synthesis in cells; mobilizes fats by increasing lipolysis, which results in the release of fatty acids into the blood and an increased rate of fatty acid metabolism; increases protein breakdown and decreases protein synthesis
Immune tissues	Anti-inflammatory—depresses antibody production, white blood cell production, and the release of inflammatory components in response to injury. The immune system is suppressed.
Target cells for epinephrine	Receptor molecules for epinephrine and norepinephrine decrease without adequate amounts of glucocorticoid hormone

1. Cortiocotropin-releasing hormone (CRH) is released from hypothalamic neurons in response to stress or low blood glucose and passes, by way to the hypothalamohypophysial portal system, to the anterior pituitary.

2. In the anterior pituitary CRH binds to and stimulates cells that secrete adrenocorticotropic hormone (ACTH).

3. ACTH binds to membrane-bound receptors on cells of the adrenal cortex and stimulates the secretion of glucocorticoids, primarily cortisol.

4. Cortisol acts on target tissues, resulting in increased fat and protein breakdown, increased glucose levels, and anti-inflammatory effects.

5. Cortisol has a negative-feedback effect because it inhibits CRH from the hypothalamus and ACTH secretion from the anterior pituitary.

Process Figure 18.15 Regulation of Cortisol Secretion

Table 18.9	Symptoms of Hyposecretion and Hypersecretion of Adrenal Cortex Hormones	
Hyposecretion	**Hypersecretion**	
Mineralocorticoids (Aldosterone)		
Hyponatremia (low blood levels of sodium)	Slight hypernatremia (high blood levels of sodium)	
Hyperkalemia (high blood levels of potassium)	Hypokalemia (low blood levels of potassium)	
Acidosis	Alkalosis	
Low blood pressure	High blood pressure	
Tremors and tetany of skeletal muscles	Weakness of skeletal muscles	
Polyuria	Acidic urine	
Glucocorticoids (Cortisol)		
Hypoglycemia (low blood glucose levels)	Hyperglycemia (high blood glucose levels; adrenal diabetes)—leads to diabetes mellitus	
Depressed immune system	Depressed immune system	
Protein and fats from diet are unused, resulting in weight loss	Destruction of tissue proteins, causing muscle atrophy and weakness, osteoporosis, weak capillaries (easy bruising), thin skin, and impaired wound healing; mobilization and redistribution of fats, causing depletion of fat from limbs and deposition in face (moon face), neck (buffalo hump), and abdomen	
Loss of appetite, nausea, and vomiting	Emotional effects, including euphoria and depression	
Increased skin pigmentation (caused by elevated ACTH)		
Androgens		
In women reduction of pubic and axillary hair	In women hirsutism (excessive facial and body hair), acne, increased sex drive, regression of breast tissue, and loss of regular menses	

Clinical Focus   Hormone Pathologies of the Adrenal Cortex

Several pathologies are associated with abnormal secretion of adrenal cortex hormones.

Addison's disease results from abnormally low levels of aldosterone and cortisol. The cause of many cases of Addison's disease is unknown, but it is a suspected autoimmune disease in which the body's defense mechanisms inappropriately destroy the adrenal cortex. Bacteria like tuberculosis bacteria, acquired immunodeficiency syndrome (AIDS), fungal infections, adrenal hemorrhage, and cancer can also damage the adrenal cortex, thus causing some cases of Addison's disease. Prolonged treatment with glucocorticoids, which suppresses pituitary gland function, can cause Addison's disease, as can tumors that damage the hypothalamus. Symptoms of Addison's disease include weakness, fatigue, weight loss, anorexia, and in many cases increased pigmentation of the skin. Reduced blood pressure results from the loss of Na^+ and water through the kidney. Reduced blood pressure is the most critical manifestation and requires immediate treatment. Low blood levels of Na^+, high blood levels of K^+, and reduced blood pH are consistent with the condition.

Aldosteronism (al-dos'ter-on-izm) is caused by excess production of aldosterone. Primary aldosteronism results from an adrenal cortex tumor, and secondary aldosteronism occurs when some extraneous factor like overproduction of renin, a substance produced by the kidney, increases aldosterone secretion. Major symptoms of aldosteronism include reduced blood levels of K^+, increased blood pH, and elevated blood pressure. Elevated blood pressure is a result of the retention of water and Na^+ by the kidneys.

Cushing's syndrome (figure A) is a disorder characterized by hypersecretion of cortisol and androgens and possibly by excess aldosterone production. The majority of cases are caused by excess ACTH production by nonpituitary tumors, which usually result from a type of lung cancer. Some cases of increased ACTH secretion do result from pituitary tumors. Sometimes adrenal tumors or unidentified causes can be responsible for hypersecretion of the adrenal cortex without increases in ACTH secretion. Elevated secretion of glucocorticoids results in muscle wasting, the accumulation of adipose tissue in the face and trunk of the body, and increased blood glucose levels.

Hypersecretion of androgens from the adrenal cortex causes a condition called **adrenogenital** (ă-drē'nō-jen'i-tăl) **syndrome,** in which secondary sexual characteristics develop early in male children, and female children are masculinized. If the condition develops before birth in females, the external genitalia can be masculinized to the extent that the infant's reproductive structures are neither clearly female nor

Figure A   Male Patient with Cushing's Syndrome

male. Hypersecretion of adrenal androgens in male children before puberty results in rapid and early development of the reproductive system. If not treated, early sexual development and short stature result. The short stature results from the effect of androgens on skeletal growth (see chapter 6). In adult females partial development of male secondary sexual characteristics, such as facial hair and a masculine voice, occurs.

Pancreas

The **pancreas** (pan'krē-us) lies behind the peritoneum between the greater curvature of the stomach and the duodenum. It is an elongated structure approximately 15 cm long weighing approximately 85–100 g. The head of the pancreas lies near the duodenum, and its body and tail extend toward the spleen.

The pancreas is both an exocrine gland and an endocrine gland. The exocrine portion consists of **acini** (as'i-nī), which produce pancreatic juice, and a duct system, which carries the pancreatic juice to the small intestine (see chapter 24). The endocrine part, consisting of **pancreatic islets** (islets of Langerhans), which (figure 18.16) produce hormones that enter the circulatory system.

Between 500,000 and 1 million pancreatic islets are dispersed among the ducts and acini of the pancreas. Each islet is composed of **alpha (α) cells** (20%), which secrete **glucagon,** a small polypeptide hormone; **beta (β) cells** (75%), which secrete **insulin,** a small protein hormone consisting of two polypeptide chains bound together; and other cell types (5%). The remaining cells are either immature cells of questionable function or **delta (δ) cells,** which secrete somatostatin, a small polypeptide hormone. Nerves from both divisions of the autonomic nervous system innervate the pancreatic islets, and a well-developed capillary network surrounds each islet.

Effect of Insulin and Glucagon on Their Target Tissues

The pancreatic hormones play an important role in regulating the concentration of critical nutrients in the circulatory system, especially

Clinical Focus Stress

The adrenal cortex and the adrenal medulla play major roles in response to stress.

In general, stress activates nervous and endocrine responses that prepare the body for physical activity, even when physical activity is not the most appropriate response to the stressful conditions, such as during an examination or other mentally stressful situations. The endocrine response to stress involves increased CRH release from the hypothalamus and increased sympathetic stimulation of the adrenal medulla. CRH stimulates ACTH secretion from the anterior pituitary, which in turn stimulates cortisol from the adrenal cortex. Increased sympathetic stimulation of the adrenal medulla increases epinephrine and norepinephrine secretion.

Together, epinephrine and cortisol increase blood glucose levels and the release of fatty acids from adipose tissue and the liver. Sympathetic innervation of the pancreas decreases insulin secretion. Consequently, most tissues do not readily take up and use glucose. Thus, glucose is available primarily to the nervous system, and fatty acids are used by skeletal muscle, cardiac muscle, and other tissues.

Epinephrine and sympathetic stimulation also increase cardiac output, increase blood pressure, and act on the central nervous system to increase alertness and aggressiveness. Cortisol also decreases the initial inflammatory response.

Responses to stress illustrate the close relationship between the nervous and endocrine systems and provide an example of their integrated functions. Our ability to respond to stressful conditions depends on the nervous and endocrine responses to stress.

Although responses to stress are adaptive under many circumstances, they can become harmful. For example, if stress is chronic, the elevated secretion of cortisol and epinephrine produces harmful effects.

Figure 18.16 Histology of the Pancreatic Islets

A pancreatic islet consists of clusters of specialized cells among the acini of the exocrine portion of the pancreas. The stain used for this slide does not distinguish between alpha and beta cells.

glucose, or blood sugar, and amino acids (table 18.10). The major target tissues of insulin are the liver, adipose tissue, muscles, and the satiety center within the hypothalamus of the brain. The **satiety** (sa′-tī-ĕ-tē) **center** is a collection of neurons in the hypothalamus that controls appetite, but insulin doesn't directly affect most areas of the nervous system. The specific effects of insulin on these target tissues are listed in table 18.11.

Insulin molecules bind to membrane-bound receptors on target cells. Once insulin molecules bind their receptors, the receptors cause specific proteins in the membrane to become phosphorylated. Part of the cells' response to insulin is to increase the number of active-transport proteins in the membrane of cells for glucose and amino acids. Finally, insulin and receptor molecules are taken by endocytosis into the cell. The insulin molecules are released from the insulin receptors and broken down within the cell, and the insulin receptor can once again become associated with the plasma membrane.

In general, the target tissue response to insulin is an increase in its ability to take up and use glucose and amino acids. Glucose molecules that are not needed immediately as an energy source to maintain cell metabolism are stored as glycogen in skeletal muscle, the liver, and other tissues and are converted to fat in adipose tissue. Amino acids can be broken down and used as an energy source

or to synthesize glucose, or they can be converted to protein. Without insulin, the ability of these tissues to take up glucose and amino acids and use them is minimal.

The normal regulation of blood glucose levels requires insulin. Blood glucose levels can increase dramatically when too little insulin is secreted or when insulin receptors do not respond to it (see Clinical Focus on "Diabetes Mellitus," p. 635). In the absence of insulin, the movement of glucose and amino acids into cells declines dramatically, even though blood levels of these molecules may increase to very high levels. The satiety center requires insulin to take up glucose. In the absence of insulin, the satiety center cannot detect the presence of glucose in the extracellular fluid even when high levels are present. The result is an intense sensation of hunger in spite of high blood glucose levels, a condition called polyphagia. High blood glucose levels cause increased urine volume (polyuria), and loss of water in the urine (see chapter 26). Increased blood glucose levels also cause an increased sensation of thirst (polydipsia) (see chapter 27).

Blood glucose levels can fall to very low levels when too much insulin is secreted. When too much insulin is present, target tissues rapidly take up glucose from the blood, causing blood levels of glucose to decline to very low levels. Although the nervous system, except for cells of the satiety center, is not a target tissue for

Table 18.10	Pancreatic Hormones			
Cells In Islets	Hormone	Structure	Target Tissue	Response
Beta (β)	Insulin	Protein	Especially liver, skeletal muscle, fat tissue	Increased uptake and use of glucose and amino acids
Alpha (α)	Glucagon	Polypeptide	Liver primarily	Increased breakdown of glycogen; release of glucose into the circulatory system
Delta (δ)	Somatostatin	Peptide	Alpha and beta cells (some somatostatin is produced in the hypothalamus)	Inhibition of insulin and glucagon secretion

Table 18.11	Effect of Insulin and Glucagon on Target Tissues	
Target Tissue	Response to Insulin	Response to Glucagon
Skeletal muscle, cardiac muscle, cartilage, bone, fibroblasts, leukocytes, and mammary glands	Increased glucose uptake and glycogen synthesis; increased uptake of certain amino acids	Little effect
Liver	Increased glycogen synthesis; increased use of glucose for energy (glycolysis)	Causes rapid increase in the breakdown of glycogen to glucose (glycogenolysis) and release of glucose into the blood Increased formation of glucose (gluconeogenesis) from amino acids and, to some degree, from fats Increased metabolism of fatty acids, resulting in increased ketones in the blood
Adipose cells	Increased glucose uptake, glycogen synthesis, fat synthesis, and fatty acid uptake; increased glycolysis	High concentrations cause breakdown of fats (lipolysis); probably unimportant under most conditions
Nervous system	Little effect except to increase glucose uptake in the satiety center	No effect

Clinical Focus Diabetes Mellitus

Diabetes mellitus results primarily from inadequate secretion of insulin or the inability of tissues to respond to insulin. **Insulin-dependent diabetes mellitus (IDDM),** also called **type I diabetes mellitus,** affects approximately 3% of people with diabetes mellitus and results from diminished insulin secretion. It develops as a result of autoimmune destruction of the pancreatic islets, and symptoms appear after approximately 90% of the islets are destroyed. IDDM most commonly develops in young people. Heredity may play some role in the condition, although initiation of pancreatic islet destruction may involve a viral infection of the pancreas (see the Systems Pathology essay, p. 643).

Noninsulin-dependent diabetes mellitus (NIDDM), also called **type II diabetes mellitus,** results from the inability of tissues to respond to insulin. NIDDM usually develops in people older than 40–45 years of age, although the age of onset varies considerably. A strong genetic component exists in the disease, but its actual cause is unknown. A peptide hormone called leptin (see chapter 25) produced by fat cells has been shown to decrease the response of target tissues to insulin. It is possible that overproduction of substances like this could be responsible for NIDDM. In some cases, abnormal receptors for insulin or antibodies may bind to and damage insulin receptors, or, in other cases, abnormalities may occur in the mechanisms that the insulin receptors activate.

NIDDM is more common than IDDM. Approximately 97% of people who have diabetes mellitus have NIDDM. The reduced number of functional receptors for insulin make the uptake of glucose by cells very slow, which results in elevated blood glucose levels after a meal. Obesity is common, although not universal, in patients with NIDDM. Elevated blood glucose levels cause fat cells to convert glucose to fat, even though the rate at which adipose cells take up glucose is impaired. Increased blood glucose and increased urine production lead to hyperosmolality of blood and dehydration of cells. The poor use of nutrients and dehydration of cells leads to lethargy, fatigue, and periods of irritability. The elevated blood glucose levels lead to recurrent infections and prolonged wound healing.

Patients with NIDDM don't suffer sudden, large increases in blood glucose and severe tissue wasting because a slow rate of glucose uptake does occur, even though the insulin receptors are defective. In some people with NIDDM, insulin production eventually decreases because pancreatic islet cells atrophy and IDDM develops. Approximately 25%–30% of patients with NIDDM take insulin, 50% take oral medication to increase insulin secretion and increase the efficiency of glucose utilization, and the remainder control blood glucose levels with exercise and diet.

Glucose tolerance tests are used to diagnose diabetes mellitus. In general, the test involves feeding the patient a large amount of glucose after a period of fasting. Blood samples are collected for a few hours, and a sustained increase in blood glucose levels strongly indicates that the person is suffering from diabetes mellitus.

Too much insulin relative to the amount of glucose ingested leads to insulin shock. The high levels of insulin cause target tissues to take up glucose at a very high rate. As a result, blood glucose levels rapidly fall to a low level. Because the nervous system depends on glucose as its major source of energy, neurons malfunction because of a lack of metabolic energy. The result is a series of nervous system responses that include disorientation, confusion, and convulsions. Taking too much insulin, too little food intake after an injection of insulin, or increased metabolism of glucose due to excess exercise by a diabetic patient can cause insulin shock.

It appears that damage to blood vessels and reduced nerve function can be reduced in diabetic patients suffering from either IDDM or NIDDM by keeping blood glucose well within normal levels at all times. Doing so, however, requires increased attention to diet, frequent blood glucose testing, and increased chance of suffering from low blood glucose levels, which leads to symptoms of insulin shock. A strict diet and routine exercise are often effective components of a treatment strategy for diabetes mellitus, and in many cases diet and exercise are adequate to control NIDDM.

---

insulin, the nervous system depends primarily on blood glucose for a nutrient source. Consequently, low blood glucose levels cause the central nervous system to malfunction.

Glucagon primarily influences the liver, although it has some effect on skeletal muscle and adipose tissue (see table 18.11). Glucagon binds to membrane-bound receptors, activates G proteins, and increases cAMP synthesis. In general, glucagon causes the breakdown of glycogen and increased glucose synthesis in the liver. It also increases the breakdown of fats. The amount of glucose released from the liver into the blood increases dramatically after glucagon secretion increases. Because glucagon is secreted into the hepatic portal circulation, which carries blood from the intestine and pancreas to the liver, it is delivered in a relatively high concentration to the liver, where it has its major effect. The liver also rapidly metabolizes it. Thus, glucagon has less of an effect on skeletal muscles and adipose tissue than on the liver.

Regulation of Pancreatic Hormone Secretion

Blood levels of nutrients, neural stimulation, and hormones control the secretion of insulin. Hyperglycemia, or elevated blood levels of glucose, directly affects the beta cells and stimulates insulin secretion. Hypoglycemia, or low blood levels of glucose, directly inhibits insulin secretion. Thus, blood glucose levels play a major role in the regulation of insulin secretion. Certain amino acids also stimulate insulin secretion by acting directly on the beta cells. After a meal, when glucose and amino acid levels increase in the circulatory system, insulin secretion increases. During periods of fasting, when blood glucose levels are low, the rate of insulin secretion declines (figure 18.17).

The autonomic nervous system also controls insulin secretion. Parasympathetic stimulation is associated with food intake, and its stimulation acts with the elevated blood glucose levels to increase insulin secretion. Sympathetic innervation inhibits insulin secretion and helps prevent a rapid fall in blood glucose levels. Because most tissues, except nervous tissue, require insulin to take up glucose, sympathetic stimulation maintains blood glucose levels in a normal range during periods of physical activity or excitement. This response is important for maintaining normal functioning of the nervous system.

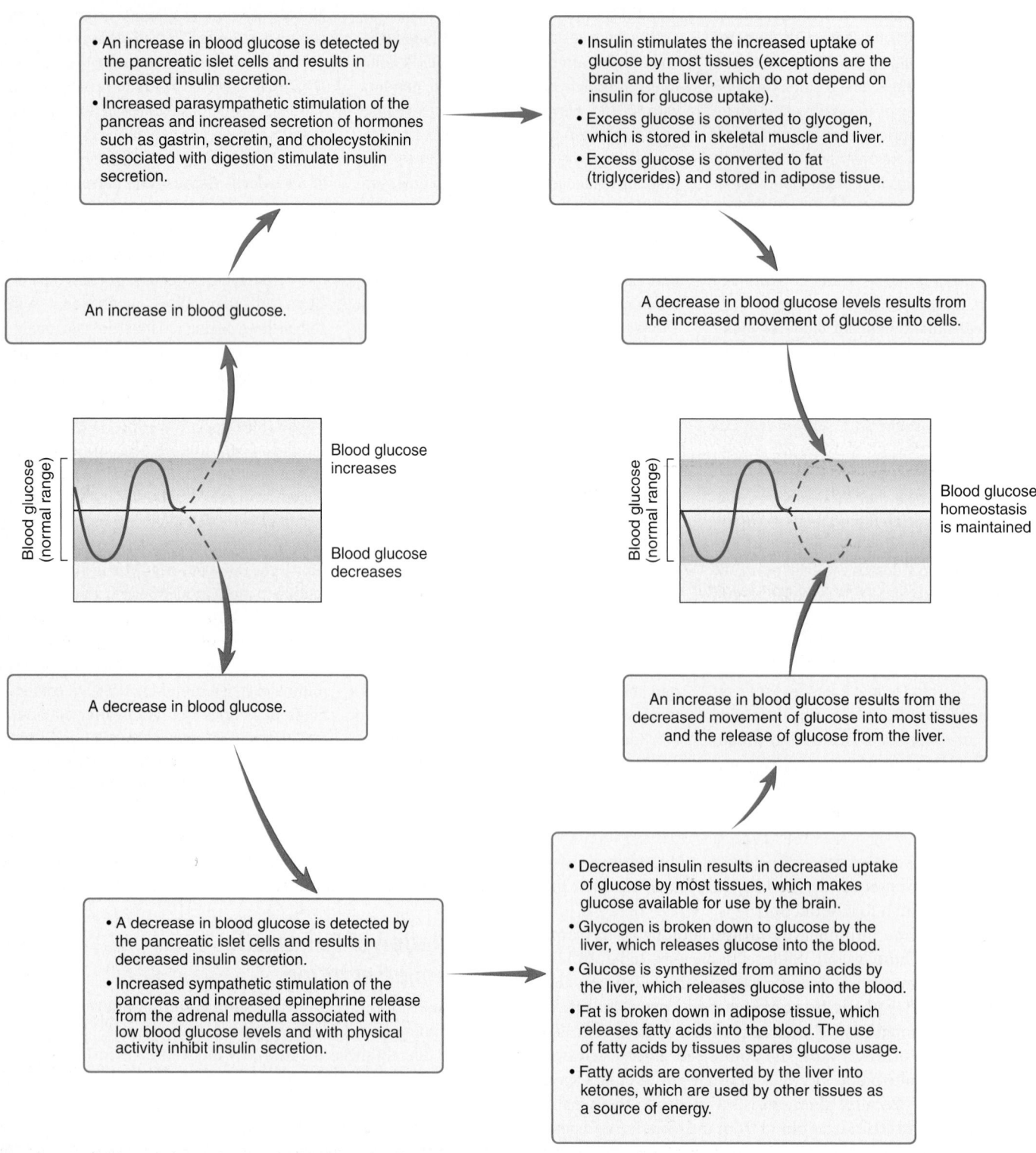

Homeostasis Figure 18.17   Regulation of Insulin Secretion

Gastrointestinal hormones involved with the regulation of digestion, such as gastrin, secretin, and cholecystokinin (see chapter 24), increase insulin secretion. Somatostatin inhibits insulin and glucagon secretion, but the factors that regulate somatostatin secretion are not clear. It can be released in response to food intake, in which case somatostatin may prevent oversecretion of insulin.

P R E D I C T

Explain why the increase in insulin secretion in response to parasympathetic stimulation and gastrointestinal hormones is consistent with the maintenance of blood glucose levels in the circulatory system.

Low blood glucose levels stimulate glucagon secretion, and high blood glucose levels inhibit it. Certain amino acids and sympathetic stimulation also increase glucagon secretion. After a high-protein meal, amino acids increase both insulin and glucagon secretion. Insulin causes target tissues to accept the amino acids for protein synthesis, and glucagon increases the process of glucose synthesis from amino acids in the liver (gluconeogenesis). Both protein synthesis and the use of amino acids to maintain blood glucose levels result from the low, but simultaneous, secretion of insulin and glucagon induced by meals high in protein content.

33. *Where is the pancreas located? Describe the exocrine and endocrine parts of this gland and the secretions produced by each portion.*

34. *Name the target tissues for insulin and glucagon, and list the effects they have on their target tissues.*

35. *How does insulin affect the nervous system in general and the satiety center in the hypothalamus in particular?*

36. *What effect do blood glucose levels, blood amino acid levels, the autonomic nervous system, and somatostatin have on insulin and glucagon secretion?*

P R E D I C T

Compare the regulation of glucagon and insulin secretion after a meal high in carbohydrates, after a meal low in carbohydrates but high in proteins, and during physical exercise.

Hormonal Regulation of Nutrients

Two different situations—after a meal and during exercise—can illustrate how several hormones function together to regulate blood nutrient levels.

After a meal and under resting conditions, secretion of glucagon, cortisol, GH, and epinephrine is reduced (figure 18.18). Both increasing blood glucose levels and parasympathetic stimulation elevate insulin secretion to increase the uptake of glucose, amino acids, and fats by target tissues. Substances not immediately used for cell metabolism are stored. Glucose is converted to glycogen in skeletal muscle and the liver, and is used for fat synthesis in adipose tissue and the liver. The rapid uptake and storage of glucose prevent too large an increase in blood glucose levels. Amino acids are incorporated into proteins, and fats that were ingested as part of the meal are stored in adipose tissue and the liver. If the meal is high in protein, a small amount of glucagon is secreted, thereby increasing the rate at which the liver uses amino acids to form glucose.

Within 1–2 hours after the meal, absorption of digested materials from the gastrointestinal tract declines, and blood glucose levels decline. As a result, secretion of glucagon, cortisol, GH, and epinephrine increases, thereby stimulating the release of glucose from tissues. As blood glucose decreases, insulin secretion decreases, and the rate of glucose entry into the target tissues for insulin decreases. Glycogen is converted back to glucose and is used as an energy source. Glucose is released into the blood by the liver. The decreased uptake of glucose by most tissues, combined with its release from the liver, helps maintain blood glucose at levels necessary for normal brain function. Cells that use less glucose start using more fats and proteins. Adipose tissue releases fatty acids, and the liver releases triglycerides (in lipoproteins) and ketones into the blood. Tissues take up these substances from the blood and use them for energy. Fat molecules are a major source of energy for most tissues when blood glucose levels are low.

The interactions of insulin, GH, glucagon, epinephrine, and cortisol are excellent examples of negative-feedback mechanisms. When blood glucose levels are high, these hormones cause rapid uptake and storage of glucose, amino acids, and fats. When blood glucose levels are low, they cause release of glucose and a switch to fat and protein metabolism as a source of energy for most tissues.

During exercise, skeletal muscles require energy to support the contraction process (see chapter 9). Although metabolism of intracellular nutrients can sustain muscle contraction for a short time, additional energy sources are required during prolonged activity. Sympathetic nervous system activity, which increases during exercise, stimulates the release of epinephrine from the adrenal medulla and of glucagon from the pancreas (figure 18.19). These hormones induce the conversion of glycogen to glucose in the liver and the release of glucose into the blood, thus providing skeletal muscles with a source of energy. Because epinephrine and glucagon have short half-lives, they can rapidly adjust blood glucose levels for varying conditions of activity.

During sustained activity, glucose released from the liver and other tissues is not adequate to support muscle activity, and a danger exists that blood glucose levels will become too low to support brain function. A decrease in insulin prevents uptake of glucose by most tissues, thus conserving glucose for the brain. Epinephrine, glucagon, cortisol, and GH cause an increase of fatty acids, triglycerides, and ketones in the blood. Because GH increases protein synthesis and it slows the breakdown of proteins, thereby decreasing the rate muscles use themselves as an energy source. Consequently, glucose metabolism decreases, and fat metabolism in skeletal muscles increases. At the end of a long race, for example, muscles rely to a large extent on fat metabolism for energy.

37. *Describe the hormonal effects after a meal that result in the movement of nutrients into cells and their storage. Describe the hormonal effects that later cause the release of stored materials for use as energy.*

38. *During exercise, how does sympathetic nervous system activity regulate blood glucose levels? Name five hormones that interact to ensure that both the brain and muscles have adequate energy sources.*

P R E D I C T

Explain why long-distance runners may not have much of a "kick" left when they try to sprint to the finish line.

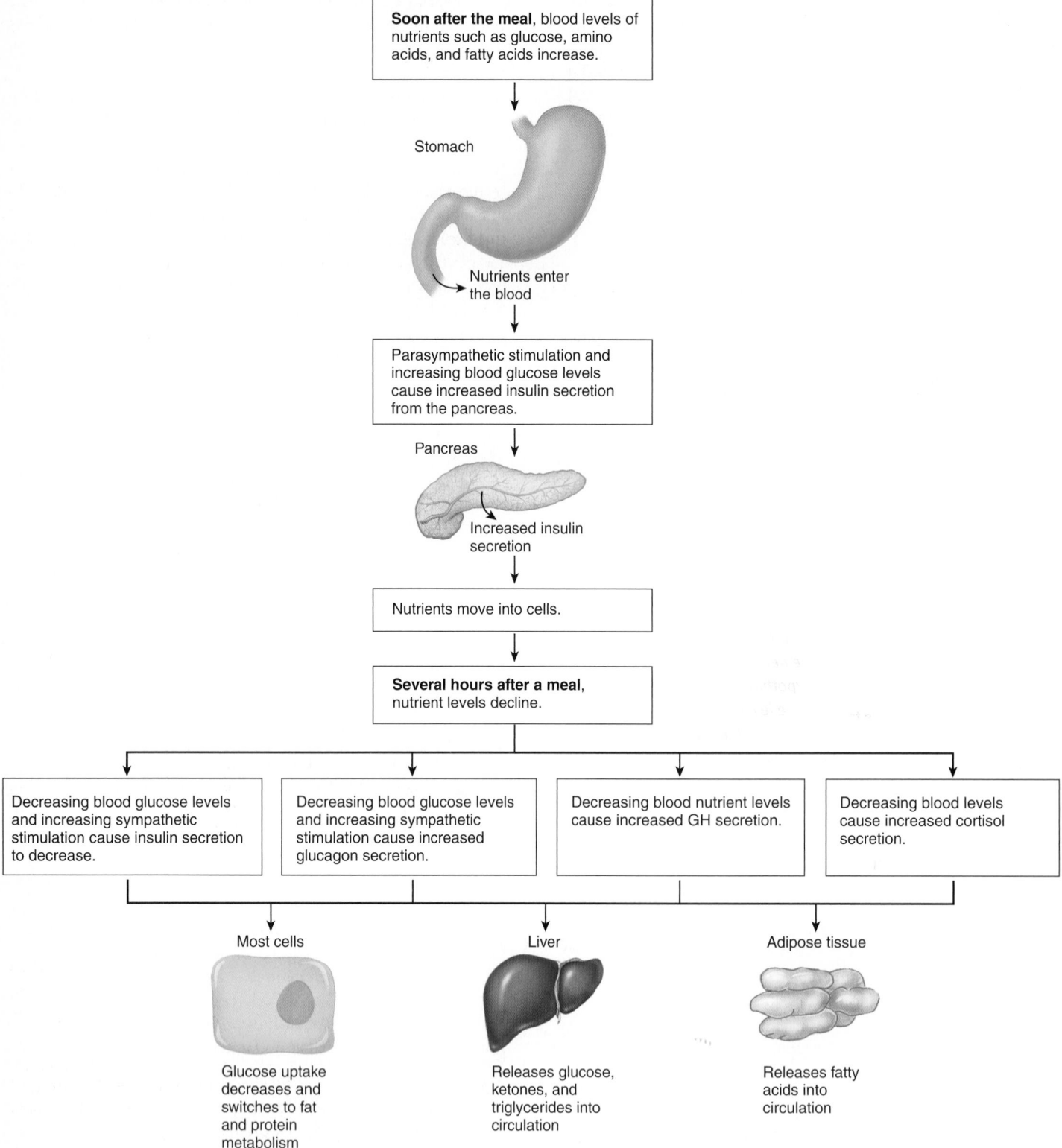

Soon after the meal, blood levels of nutrients such as glucose, amino acids, and fatty acids increase.

Stomach

Nutrients enter the blood

Parasympathetic stimulation and increasing blood glucose levels cause increased insulin secretion from the pancreas.

Pancreas

Increased insulin secretion

Nutrients move into cells.

Several hours after a meal, nutrient levels decline.

Decreasing blood glucose levels and increasing sympathetic stimulation cause insulin secretion to decrease.

Decreasing blood glucose levels and increasing sympathetic stimulation cause increased glucagon secretion.

Decreasing blood nutrient levels cause increased GH secretion.

Decreasing blood levels cause increased cortisol secretion.

Most cells

Glucose uptake decreases and switches to fat and protein metabolism

Liver

Releases glucose, ketones, and triglycerides into circulation

Adipose tissue

Releases fatty acids into circulation

Figure 18.18   **Regulation of Blood Nutrient Levels**

Blood nutrient levels are maintained immediately after a meal and several hours after a meal.

Exercise

Muscle

Short-term exercise

Sympathetic stimulation increases epinephrine secretion from the adrenal medulla and glucagon secretion from the pancreas. It also inhibits insulin secretion from the pancreas.

Epinephrine increases the rate at which glycogen in **muscle cells** is broken down to glucose. The glucose is used as an energy source in muscle cells. The muscle cells take up less glucose from the blood and increase their rate of fatty acid metabolism.

Liver

Epinephrine and glucagon increase glycogen breakdown to glucose molecules in the **liver**. The glucose molecules are released into the circulatory system.

Fat

Epinephrine and sympathetic stimulation also increase the breakdown of **fat** and the release of fatty acids from adipose tissue.

Blood nutrients

Blood glucose levels are maintained for normal nervous system function.

Prolonged exercise

ACTH and GH release from the anterior pituitary increase. ACTH stimulates increased cortisol secretion from the adrenal cortex.

Cortisol increases protein breakdown in **muscle** and the **liver** to amino acids, and increases glucose synthesis from amino acids and from some components of fat, such as glycerol.

Cortisol increases the breakdown of **fats** and the use of fatty acids by **muscle cells** as an energy source.

GH slows the breakdown of proteins and helps conserve them, thus increasing the dependence on fats as an energy source.

Blood glucose levels are maintained for normal nervous system function and there is an increasing dependence on fats and proteins for energy sources.

Figure 18.19 Regulation of Blood Nutrient Levels During Exercise

Hormones of the Reproductive System

Reproductive hormones are secreted primarily from the ovaries, testes, placenta, and pituitary gland (table 18.12). These hormones are discussed in chapter 28. The main endocrine glands of the male reproductive system are the testes. The functions of the testes depend on the secretion of FSH and LH from the anterior pituitary gland. The main hormone secreted by the testes is testosterone, an androgen. **Testosterone** regulates the production of sperm cells by the testes and the development and maintenance of male reproductive organs and secondary sex characteristics. The testes secrete another hormone called **inhibin,** which inhibits the secretion of FSH from the anterior pituitary.

The main endocrine glands of the female reproductive system are the ovaries. Like the testes, the functions of the ovaries depend on the secretion of FSH and LH from the anterior pituitary gland. The main hormones secreted by the ovaries are **estrogen** and **progesterone.** These hormones, along with FSH and LH, control the female reproductive cycle, prepare the mammary glands for lactation, and maintain pregnancy. Estrogen and progesterone are also responsible for the development of the female reproductive organs and female secondary sex characteristics. The ovaries also secrete inhibin, which inhibits FSH secretion.

During the first one-third of pregnancy the placenta secretes an LH-like substance that is necessary to maintain pregnancy (see chapter 28). Throughout most of pregnancy, the ovaries and placenta secrete increasing amounts of estrogen and progesterone, which are also necessary to maintain pregnancy. In addition, the ovaries secrete **relaxin,** which increases the flexibility of connective tissue of the symphysis pubis and helps dilate the cervix of the uterus. This facilitates delivery by making the birth canal larger.

Table 18.12	Hormones of the Reproductive Organs		
Hormones	**Structure**	**Target Tissue**	**Response**
Testis			
Testosterone	Steroid	Most cells	Aids in spermatogenesis; maintenance of functional reproductive organs; secondary sex characteristics; sexual behavior
Inhibin	Polypeptide	Anterior pituitary gland	Inhibits FSH secretion
Ovary			
Estrogens	Steroids	Most cells	Uterine and mammary gland development and function; external genitalia structure; secondary sex characteristics; sexual behavior and menstrual cycle
Progesterone	Steroid	Most cells	Uterine and mammary gland development and function; external genitalia structure; secondary sex characteristics; menstrual cycle
Inhibin	Polypeptide	Anterior pituitary gland	Inhibits FSH secretion
Relaxin	Polypeptide	Connective tissue cells	Increases flexibility of connective tissue in the pelvic area, especially the symphysis pubis

39. *List the hormones secreted by the testes, and give their functions. What hormones regulate the testes?*
40. *List the hormones secreted by the ovaries, and give their functions. During pregnancy, what other organ, in addition to the ovaries, secretes hormones? Upon what hormones does ovarian function depend?*

Hormones of the Pineal Body

The **pineal** (pin′ē-ăl) **body** in the epithalamus of the brain secretes hormones that act on the hypothalamus or the gonads to inhibit reproductive functions. Two substances have been proposed as secretory products: **melatonin** (mel-ă-tōn′in) and **arginine vasotocin** (ar′ji-nēn vā-sō-tō′sin, vas-ō-tos′in) (table 18.13). Melatonin can decrease GnRH secretion from the hypothalamus and may inhibit reproductive functions through this mechanism. It may also help regulate sleep cycles by increasing the tendency to sleep.

The **photoperiod** is the amount of daylight and darkness that occurs each day and changes with the seasons of the year. In some animals, the photoperiod regulates pineal secretions (figure 18.20). For example, increased daylight initiates action potentials in the retina of the eye that are propagated to the brain and cause a decrease in the action potentials sent first to the spinal cord and then through sympathetic neurons to the pineal body. Decreased pineal secretion results. In the dark, action potentials delivered by sympathetic neurons to the pineal body increase, thereby stimulating the secretion of pineal hormones. Humans secrete larger amounts of melatonin at night than in the daylight. In animals that breed in the spring, the increased length of a day decreases pineal secretions. Because pineal secretions inhibit reproductive functions in these species, the increased length of a day results in hypertrophy of the reproductive structures.

The function of melatonin in the regulation of reproductive functions in humans is not clear, but it is recommended by some to enhance sleep. Because melatonin causes atrophy of reproductive structures in some species there's a possibility of undesirable side effects on the reproductive system.

The function of the pineal body in humans is not clear, but tumors that destroy the pineal body correlate with early sexual

development, and tumors that result in pineal hormone secretion correlate with retarded development of the reproductive system. It's not clear, however, if the pineal body controls the onset of puberty.

Arginine vasotocin works with melatonin to regulate the function of the reproductive system in some animals. Evidence for the role of melatonin is more extensive, however.

41. *Where is the pineal body located? Name the hormones it produces and their possible effects.*

Hormones of the Thymus

The **thymus** (thī′mŭs) is in the neck and superior to the heart in the thorax, and it secretes a hormone called **thymosin** (thī′mō-sin) (see table 18.13). Both the thymus and thymosin play an important role in the development and maturation of the immune system, and are discussed in chapter 22.

Hormones of the Gastrointestinal Tract

Several hormones are released from the gastrointestinal tract. They regulate digestive functions by influencing the activity of the stomach, intestines, liver, and pancreas. They are discussed in chapter 24.

Hormonelike Substances

Autocrine chemical signals are released from cells that influence the same cell type from which they are released. **Paracrine chemical signals** are released from one cell type, diffuse short distances, and influence the activity of another cell type, which is its target tissue. Autocrine and paracrine chemical signals differ from hormones in that they are not secreted from discrete endocrine glands, they have local effects rather than systemic effects, or they have functions that are not understood adequately to explain their role in the body. Examples of autocrine chemical signals include chemical mediators of inflammation derived from the fatty acid **arachidonic**

Table 18.13 Other Hormones and Hormonelike Substances

Chemical Signal	Structure	Target Tissue	Response
Pineal Body			
Melatonin	Amino acid derivative	At least the hypothalamus	Inhibition of gonadotropin-releasing hormone secretion, thereby inhibiting reproduction; significance is not clear in humans; may help regulate sleep–wake cycles
Arginine vasotocin	Amino acid derivative	Possibly the hypothalamus	Possible inhibition of gonadotropin-releasing hormone secretion
Thymus Gland			
Thymosin	Peptide	Immune tissues	Development and function of the immune system
Several Tissues (autocrine and paracrine regulatory substances)			
Eicosanoids			
Prostaglandins	Modified fatty acid	Most tissues	Mediation of the inflammatory response increased uterine contractions; ovulation, possible inhibition of progesterone synthesis; blood coagulation; and other functions
Prostacyclins	Modified fatty acid	Most tissues	Mediation of the inflammatory response and other functions
Thromboxanes	Modified fatty acid	Most tissues	Mediation of the inflammatory response and other functions
Leukotrienes	Modified fatty acid	Most tissues	Mediation of the inflammatory response and other functions
Enkephalins and endorphins	Peptides	Nervous system	Reduction of pain sensation and other functions
Epidermal growth factor	Protein	Many tissues	Stimulates division in many cell types and plays a role in embryonic development
Fibroblast growth factor	Protein	Many tissues	Stimulates cell division in many cell types and plays a role in embryonic development
Interleukin-2	Protein	Certain immune competent cells	Stimulates cell division of T lymphocytes

1. Light entering the eye stimulates neurons in the retina of the eye to produce action potentials.

2. Action potentials are transmitted to the hypothalamus in the brain.

3. Action potentials from the hypothalamus are transmitted through the sympathetic division to the pineal body.

4. A decrease in light (dark) results in increased sympathetic stimulation of the pineal gland and increased melatonin secretion. An increase in light results in decreased sympathetic stimulation of the pineal gland and decreased melatonin secretion.

5. Melatonin inhibits GnRH secretion from the hypothalamus and may help regulate sleep cycles.

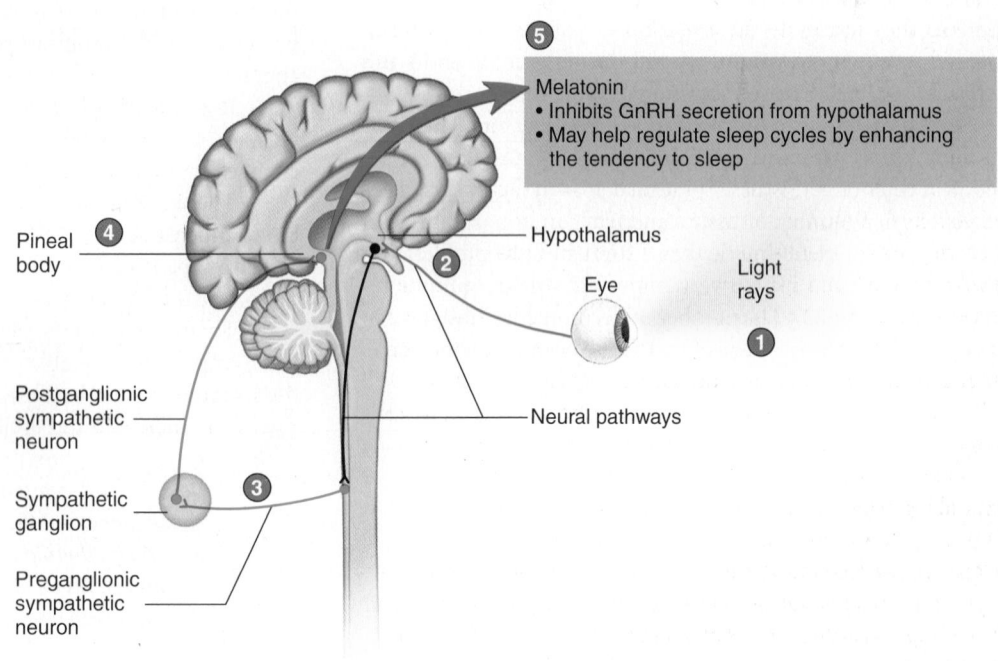

Process Figure 18.20 Regulation of Melatonin Secretion from the Pineal Body

Light entering the eye inhibits and dark stimulates the release of melatonin from the pineal body.

(ă-rak-i-don'ik) **acid,** such as eicosanoids and modified phospholipids. The **eicosanoids** include **prostaglandins** (pros'-stă-glandinz), **thromboxanes** (throm'bok-zānz), **prostacyclins** (pros-tă-sī'klinz), and **leukotrienes** (loo'kō-trī'ēnz). Modified phospholipids include platelet activating factor (see chapter 19). Paracrine chemical signals include substances that play a role in modulating the sensation of pain, such as **endorphins** (en'dōr-finz) and **enkephalins** (en-kef'ă-linz), and several peptide growth factors, such as **epidermal growth factor, fibroblast growth factor,** and **interleukin-2** (in-ter-loo'kin) (see table 18.13).

Prostaglandins, thromboxanes, prostacyclins, and leukotrienes are released from injured cells and are responsible for initiating some of the symptoms of inflammation (see chapter 22), in addition to being released from certain healthy cells. For example, prostaglandins are involved in the regulation of uterine contractions during menstruation and childbirth, the process of ovulation, the inhibition of progesterone synthesis by the corpus luteum, the regulation of coagulation, kidney function, and modification of the effect of other hormones on their target tissues. Pain receptors are stimulated directly by prostaglandins and other inflammatory compounds, or prostaglandins cause vasodilation of blood vessels, which is associated with headaches. Anti-inflammatory drugs like aspirin inhibit prostaglandin synthesis and, as a result, reduce inflammation and pain. These examples are paracrine regulatory substances because they are synthesized and secreted by the cells near their target cells. Once prostaglandins enter the circulatory system, they are metabolized rapidly.

Three classes of peptide molecules, which are endogenously produced analgesics, bind to the same receptor molecules as morphine. They include enkephalins, endorphins, and **dynorphins** (dī'nōr-finz). They are produced in several sites in the body, such as parts of the brain, pituitary, spinal cord, and gut. They act as neurotransmitters in some neurons of both the central and peripheral nervous systems and as hormones or paracrine regulatory substances. In general, they moderate the sensation of pain (see chapter 14). Decreased sensitivity to painful stimuli during exercise and stress may result from the increased secretion of these substances.

Several proteins can be classified as growth factors. They generally function as paracrine chemical signals because they are secreted near their target tissues. Epidermal growth factor stimulates cell divisions in a number of tissues and plays an important role in embryonic development. Interleukin-2 stimulates the proliferation of T lymphocytes and plays a very important role in immune responses (see chapter 22). The number of hormonelike substances in the body is large, and only a few of them have been mentioned here. Chemical communication among cells in the body is complex, well developed, and necessary for maintenance of homeostasis. Investigations into chemical regulation increase our knowledge of body functions—knowledge that can be used in the development of techniques for the treatment of pathologic conditions.

42. *Define autocrine chemical signals. List eicosanoids and modified phospholipids that function as autocrine chemical signals, and explain their function.*

43. *Define paracrine chemical signals. List examples of substances that play a role in modulating pain or are peptide growth factors. How can prostaglandins function as both autocrine and paracrine chemical signals?*

Effects of Aging on the Endocrine System

Age-related changes in the endocrine system are not the same for all of the endocrine glands. There's a gradual decrease in the secretory activity of some endocrine glands, but not in all of them. In addition, some decreases in secretory activity of endocrine glands appear to be secondary to a decrease in physical activity as people age.

There is a decrease in the secretion of GH as people age. The decrease is greater in people who do not exercise, and it may not occur in people who exercise regularly. Decreasing GH secretion may explain the gradual decrease in lean body mass. For example, bone mass and muscle mass decrease as GH levels decline. At the same time adipose tissue increases.

Melatonin secretion decreases in aging people. The decrease may influence age-related changes in sleep patterns and the secretory patterns of other hormones such as GH and testosterone.

The secretion of thyroid hormones decreases slightly with increasing age, and there's a decrease in the T_3/T_4 ratio. This may be less of a decrease in the secretory activity of the thyroid gland than it is compensating for the decrease in the lean body mass in aging people. Age-related damage to the thyroid gland by the immune system can occur. This change occurs in women more than in men. The result is that approximately 10% of elderly women have thyroid glands that don't produce enough T_3 and T_4.

Parathyroid hormone secretion doesn't appear to decrease with age. Blood levels of Ca^{2+} may decrease slightly because of reduced dietary calcium intake and vitamin D levels. The greatest risk is a loss of bone matrix as parathyroid hormone increases to maintain blood levels of Ca^{2+} within their normal range.

The kidneys of the elderly secrete less renin. Consequently, there's a reduced ability to respond to decreases in blood pressure by activating the renin-angiotensin-aldosterone mechanism (see chapter 26).

Reproductive hormone secretion gradually declines in elderly men, and women experience menopause. These age-related changes are described in chapter 28.

There are no age-related decreases in the ability to regulate blood glucose levels. However, there's an age-related tendency to develop type II diabetes for those who have a familial tendency to do so, and it is correlated with age-related increases in body weight.

Thymosin from the thymus decreases with age. Fewer immature lymphocytes are able to mature and become functional, and the immune system becomes less effective in protecting the body. There's an increased susceptibility to infection and to cancer.

44. *Describe age-related changes in the secretion and the consequences of these changes in the following: GH, melatonin, thyroid hormones, renin, and reproductive hormones. Name one hormone that doesn't appear to decrease with age.*

Systems Pathology
Insulin-Dependent Diabetes Mellitus

Billy, a 10-year-old boy, was diagnosed as having insulin-dependent diabetes mellitus (IDDM). Billy's mother took him to a physician after noticing that he was constantly hungry and was losing weight rapidly in spite of his unusually large food intake. More careful observation made it clear that Billy was constantly thirsty and that he urinated frequently. In addition, he felt weak and lethargic, and his breath occasionally had a distinctive sweet, or acetone, odor. Diagnostic tests confirmed that he had IDDM.

Background Information

IDDM is caused by diminished insulin secretion. In patients with IDDM, nutrients are absorbed from the intestine after a meal, but skeletal muscle, adipose tissue, and other target tissues don't readily take glucose into their cells, and liver cells cannot convert glucose to glycogen. Consequently, blood levels of glucose increase dramatically. Glucagon and glucocorticoid secretion increase because the glucose in the blood cannot enter the cells that produce these hormones, so their rate of secretion is similar to when blood glucose levels are low. Epinephrine secretion also increases. In response to these hormones, glycogen, fats, and proteins are broken down and metabolized to produce the ATP required by cells.

When blood glucose levels are very high, glucose is excreted in the urine, which results in an increase in urine volume. The rapid loss of water in the urine increases the osmotic concentration of blood, which increases the sensation of thirst. The increased osmolality of blood and the ionic imbalances caused by the loss of Ca^{2+} and K^+ in the large amount of urine produced cause neurons to malfunction and result in diabetic coma in severe cases. When insulin levels in the blood are low and cells of the nervous system that control appetite appear to be unable to take up glucose even when blood glucose levels are high, the result is an increased appetite. **Polyuria** (pol-ē-ū′rē-ă; increased urine volume), **polydipsia** (pol-ē-dip′sē-ă; increased thirst), and **polyphagia** (pol-ē-fā′jē-ă; increased appetite) are major symptoms of IDDM. Acidosis is caused by rapid fat catabolism, which results in increased levels of **acetoacetic** (as′e-tō-a-sē′tik) **acid,** which is converted to **acetone** (as′e-tōn) and **β-hydroxybutyric** (bā′tă hī-drok′sē-bū-tir′ik) **acid.** These three substances collectively are referred to as **ketone** (kē′tōn) **bodies.** The presence of excreted ketone bodies in the urine and in expired air ("acetone breath") suggests that the person has diabetes mellitus.

Billy's physician explained that prior to the late 1920s people with his condition always died in a relatively short time. They suffered from massive weight loss and appeared to starve to death in spite of eating a large amount of food. The physician explained that because of

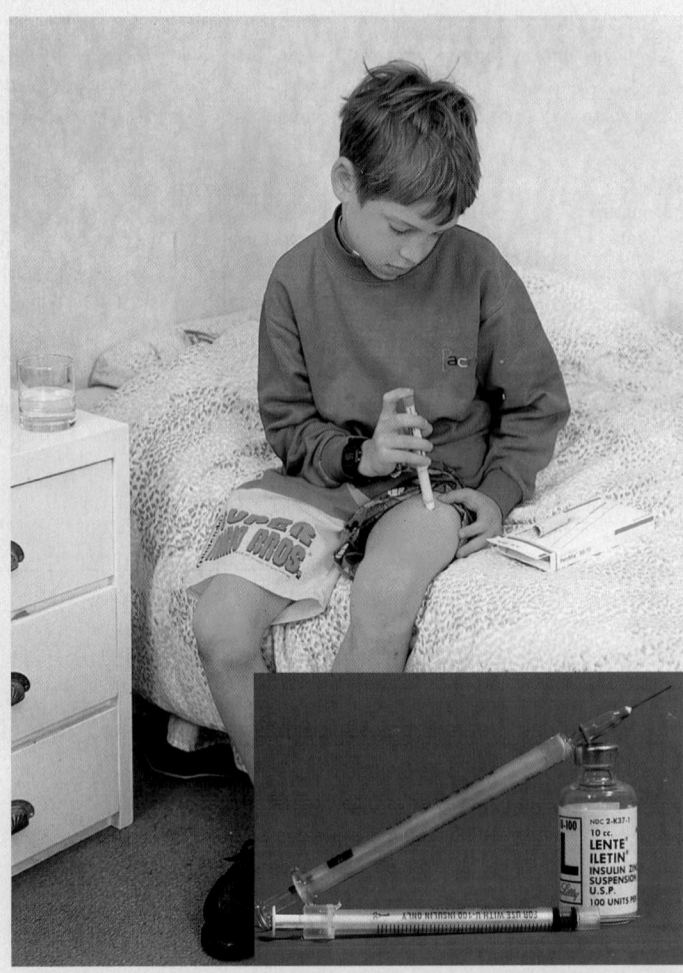

Figure B   A 10-Year-Old Boy Giving Himself Insulin

the discovery of insulin, many people with his type of diabetes mellitus are able to live nearly normal lives. Taking insulin injections (figure B), monitoring blood glucose levels, and following a strict diet to keep blood glucose levels within a normal range of values are the major treatments for IDDM.

P R E D I C T   11

After Billy was diagnosed with diabetes mellitus, he followed a strict diet and took insulin for a few months. He began to feel much better than before. In fact, he felt so well that he began to sneak candy and soft drinks when his parents were not around. Predict the consequences of his actions on his health.

System Interactions — Effect of IDDM on Other Systems

System	Interaction
Muscular	Untreated diabetes mellitus, especially IDDM, results in severe muscle atrophy because glycogen, stored fat, and proteins of muscles are broken down and used as energy sources. Ionic imbalances can also lead to muscular weakness.
Nervous	Untreated IDDM can have dramatic effects on the nervous system. When the blood glucose reaches very high levels, the osmolality of the extracellular fluid is increased. Thus, water diffuses from the neurons of the brain. In addition, acidosis develops because of the rapid metabolism of fats. As a result, the nervous system cannot function normally, and diabetic coma can result. A long-term effect is the degeneration of the myelin sheaths of neurons, resulting in abnormal nerve functions.
Cardiovascular	Atherosclerosis develops more rapidly in diabetics than in the healthy population. Changes in the capillary structure and high blood glucose levels increase the probability of reduced circulation and gangrene.
Lymphatic and Immune	The tendency to develop infections increases, and the rate of healing is slower. In some cases, an allergic reaction to the injected insulin occurs.
Respiratory	Acidosis causes hyperventilation, which increases blood pH back toward normal levels by decreasing blood CO_2 levels.
Urinary	High blood glucose levels cause polyuria, the urine contains glucose and has a high osmolality, and people with diabetes are more likely to develop urinary tract infections.
Reproductive	Pregnant women with diabetes mellitus may have babies with a larger-than-normal birth weight because the blood glucose levels may be high in the mother and fetus, and the fetus's pancreas produces insulin. Glucose is therefore taken up by cells of the fetus, where it is converted to fat.

S U M M A R Y

Functions of the Endocrine System (p. 610)

Main regulatory functions include water balance, uterine contractions and milk release, metabolism and tissue maturation, ion regulation, heart rate and blood pressure regulation, control of blood glucose and other nutrients, immune system regulation, and control of reproductive functions.

Pituitary Gland and Hypothalamus (p. 610)

1. The pituitary gland secretes at least nine hormones that regulate numerous body functions and other endocrine glands.
2. The hypothalamus regulates pituitary gland activity through neurohormones and action potentials.

Structure of the Pituitary Gland

1. The posterior pituitary develops from the floor of the brain and consists of the infundibulum and pars nervosa.
2. The anterior pituitary develops from the roof of the mouth and consists of the pars distalis, pars intermedia, and pars tuberalis.

Relationship of the Pituitary to the Brain

1. The hypothalamohypophysial portal system connects the hypothalamus and the anterior pituitary.
   • Neurohormones are produced in hypothalamic neurons.
   • Through the portal system, the neurohormones inhibit or stimulate hormone production in the anterior pituitary.
2. The hypothalamohypophysial tract connects the hypothalamus and the posterior pituitary.
   • Neurohormones are produced in hypothalamic neurons.
   • The neurohormones move down the axons of the nerve tract and are secreted from the posterior pituitary.

Hormones of the Pituitary Gland (p. 613)
Posterior Pituitary Hormones

1. ADH promotes water retention by the kidneys.
2. Oxytocin promotes uterine contractions during delivery and causes milk ejection in lactating women.

Anterior Pituitary Hormones

1. GH, or somatotropin
   • GH stimulates growth in most tissues and is a regulator of metabolism.
   • GH stimulates the uptake of amino acids and their conversion into proteins and stimulates the breakdown of fats and the synthesis of glucose.
   • GH stimulates the production of somatomedins; together they promote bone and cartilage growth.
   • GH secretion increases in response to an increase in blood amino acids, low blood glucose, or stress.
   • GH is regulated by GHRH and GHIH, or somatostatin.
2. TSH, or thyrotropin, causes the release of thyroid hormones.
3. ACTH is derived from proopiomelanocortin; it stimulates cortisol secretion from the adrenal cortex and increases skin pigmentation.
4. Several hormones in addition to ACTH are derived from proopiomelanocortin.
   • Lipotropins cause fat breakdown.
   • β endorphins play a role in analgesia.
   • MSH increases skin pigmentation.
5. LH and FSH
   • Both hormones regulate the production of gametes and reproductive hormones (testosterone in males; estrogen and progesterone in females).
   • GnRH from the hypothalamus stimulates LH and FSH secretion.
6. Prolactin stimulates milk production in lactating females. Prolactin-releasing hormone and prolactin-inhibiting hormone from the hypothalamus affect prolactin secretion.

Thyroid Gland (p. 619)

1. The thyroid gland is just inferior to the larynx.
2. The thyroid gland is composed of small, hollow balls of cells called follicles, which contain thyroglobulin.
3. Parafollicular cells are scattered throughout the thyroid gland.

Thyroid Hormones

1. T_3 and T_4 synthesis
   - Iodide ions are taken into the follicles by active transport, are oxidized, and are bound to tyrosine molecules in thyroglobulin.
   - Thyroglobulin is secreted into the follicle lumen. Tyrosine molecules with iodine combine to form T_3 and T_4, thyroid hormones.
   - Thyroglobulin is taken into the follicular cells and is broken down; T_3 and T_4 diffuse from the follicles to the blood.
2. T_3 and T_4 transport in the blood
   - T_3 and T_4 bind to thyroxine-binding globulin and other plasma proteins.
   - The plasma proteins prolong the half-life of T_3 and T_4 and regulate the levels of T_3 and T_4 in the blood.
   - Approximately one-third of the T_4 is converted into functional T_3.
3. Mechanism of action of T_3 and T_4
   - T_3 and T_4 bind with intracellular receptor molecules and initiate new protein synthesis.
4. Effects of T_3 and T_4
   - T_3 and T_4 increase the rate of glucose, fat, and protein metabolism in many tissues, thus increasing body temperature.
   - Normal growth of many tissues is dependent on T_3 and T_4.
5. Regulation of T_3 and T_4 secretion
   - Increased TSH from the anterior pituitary increases T_3 and T_4 secretion.
   - TRH from the hypothalamus increases TSH secretion. TRH increases as a result of chronic exposure to cold, food deprivation, and stress.
   - T_3 and T_4 inhibit TSH and TRH secretion.

Calcitonin

1. The parafollicular cells secrete calcitonin.
2. An increase in blood calcium levels stimulates calcitonin secretion.
3. Calcitonin decreases blood calcium and phosphate levels by inhibiting osteoclasts.

Parathyroid Glands   (p. 624)

1. The parathyroid glands are embedded in the thyroid glands.
2. PTH increases blood calcium levels.
   - PTH stimulates osteoclasts.
   - PTH promotes calcium reabsorption by the kidneys and the formation of active vitamin D by the kidneys.
   - Active vitamin D increases calcium absorption by the intestine.
3. A decrease in blood calcium levels stimulates PTH secretion.

Adrenal Glands   (p. 627)

1. The adrenal glands are near the superior poles of the kidneys.
2. The adrenal medulla arises from neural crest cells and functions as part of the sympathetic nervous system. The adrenal cortex is derived from mesoderm.
3. The adrenal medulla is composed of closely packed cells.
4. The adrenal cortex is divided into three layers: the zona glomerulosa, the zona fasciculata, and the zona reticularis.

Hormones of the Adrenal Medulla

1. Epinephrine accounts for 80% and norepinephrine for 20% of the adrenal medulla hormones.
   - Epinephrine increases blood glucose levels, use of glycogen and glucose by skeletal muscle, and heart rate and force of contraction, and it causes vasoconstriction in the skin and viscera and vasodilation in skeletal and cardiac muscle.
   - Norepinephrine stimulates cardiac muscle and causes constriction of most peripheral blood vessels.
2. The adrenal medulla hormones prepare the body for physical activity.
3. Release of adrenal medulla hormones is mediated by the sympathetic nervous system in response to emotions, injury, stress, exercise, and low blood glucose levels.

Hormones of the Adrenal Cortex

1. The zona glomerulosa secretes the mineralocorticoids, especially aldosterone. Aldosterone acts on the kidneys to increase sodium and to decrease potassium and hydrogen levels in the blood.
2. The zona fasciculata secretes glucocorticoids, especially cortisol.
   - Cortisol increases fat and protein breakdown, increases glucose synthesis from amino acids, decreases the inflammatory response, and is necessary for the development of some tissues.
   - ACTH from the anterior pituitary stimulates cortisol secretion. CRH from the hypothalamus stimulates ACTH release. Low blood glucose levels or stress stimulate CRH secretion.
3. The zona reticularis secretes androgens. In females, androgens stimulate axillary and pubic hair growth and sexual drive.

Pancreas   (p. 632)

1. The pancreas is located along the small intestine and the stomach. It is both an exocrine and an endocrine gland.
2. The exocrine portion of the pancreas consists of a complex duct system that ends in small sacs called acini that produce pancreatic digestive juices.
3. The endocrine portion consists of the pancreatic islets. Each islet is composed of alpha cells, which secrete glucagon, beta cells, which secrete insulin, and delta cells, which secrete somatostatin.

Effect of Insulin and Glucagon on Their Target Tissues

1. Insulin's target tissues are the liver, adipose tissue, muscle, and the satiety center in the hypothalamus. The nervous system is not a target tissue, but it does rely on blood glucose levels maintained by insulin.
2. Insulin increases the uptake of glucose and amino acids by cells. Glucose is used for energy or is stored as glycogen. Amino acids are used for energy or are converted to glucose or proteins.
3. Glucagon's target tissue is mainly the liver.
4. Glucagon causes the breakdown of glycogen and fats for use as an energy source.

Regulation of Pancreatic Hormone Secretion

1. Insulin secretion increases because of elevated blood glucose levels, an increase in some amino acids, parasympathetic stimulation, and gastrointestinal hormones. Sympathetic stimulation decreases insulin secretion.
2. Glucagon secretion is stimulated by low blood glucose levels, certain amino acids, and sympathetic stimulation.
3. Somatostatin inhibits insulin and glucagon secretion.

Hormonal Regulation of Nutrients   (p. 637)

1. After a meal, the following events take place:
   - High glucose levels inhibit glucagon, cortisol, GH, and epinephrine, which reduces the release of glucose from tissues.
   - Insulin secretion increases as a result of the high blood glucose levels, thereby increasing the uptake of glucose, amino acids, and fats, which are used for energy or are stored.
   - Sometime after the meal, blood glucose levels drop. Glucagon, cortisol, GH, and epinephrine levels increase, insulin levels decrease, and glucose is released from tissues.
   - Adipose tissue releases fatty acids, triacylglycerols, and ketones, which most tissues use for energy.
2. During exercise the following events occur:
   - Sympathetic activity increases epinephrine and glucagon secretion, causing a release of glucose into the blood.
   - Low blood sugar levels, caused by uptake of glucose by skeletal muscles, stimulate epinephrine, glucagon, GH, and cortisol secretion, causing an increase in fatty acids, triacylglycerols, and ketones in the blood, all of which are used for energy.

Hormones of the Reproductive System   (p. 639)

The ovaries, testes, placenta, and pituitary gland secrete reproductive hormones.

Hormones of the Pineal Body   (p. 640)

The pineal body produces melatonin and arginine vasotocin, which can inhibit reproductive maturation and may regulate sleep–wake cycles.

Hormones of the Thymus   (p. 640)

The thymus gland produces thymosin, which is involved in the development of the immune system.

Hormones of the Gastrointestinal Tract   (p. 640)

The gastrointestinal tract produces several hormones that regulate digestive functions.

Hormonelike Substances   (p. 640)

1. Autocrine and paracrine chemical signals are produced by many cells of the body and usually have a local effect. They affect many body functions.

2. Eicosanoids such as prostaglandins, prostacyclins, thromboxanes, and leukotrienes are derived from fatty acids and mediate inflammation and other functions. Endorphins, enkephalins, and dynorphins are analgesic substances. Growth factors influence cell division and growth in many tissues, and interleukin-2 influences cell division in T cells of the immune system.

Effects of Aging on the Endocrine System   (p. 642)

There is a gradual decrease in the secretion rate of most, but not all, hormones. Some decreases are secondary to gradual decreases in physical activity.

R E V I E W   A N D   C O M P R E H E N S I O N

1. The pituitary gland
   a. develops from the floor of the brain.
   b. develops from the roof of the mouth.
   c. is stimulated by neurohormones produced in the midbrain.
   d. secretes only three major hormones.
   e. both a and b.

2. The hypothalamohypophysial portal system
   a. contains one capillary bed.
   b. carries hormones from the anterior pituitary to the body.
   c. carries hormones from the posterior pituitary to the body.
   d. carries hormones from the hypothalamus to the anterior pituitary.
   e. carries hormones from the hypothalamus to the posterior pituitary.

3. Which of these hormones is *not* a hormone that is secreted into the hypothalamohypophysial portal system?
   a. GHRH
   b. TRH
   c. PIH
   d. GnRH
   e. ACTH

4. Hormones secreted from the posterior pituitary
   a. are produced in the anterior pituitary.
   b. are transported to the posterior pituitary within axons.
   c. include GH and TSH.
   d. are steroids.
   e. all of the above.

5. Which of these stimulates the secretion of ADH?
   a. elevated blood osmolality
   b. decreased blood osmolality
   c. releasing hormones from the hypothalamus
   d. ACTH
   e. increased blood pressure

6. Oxytocin is responsible for
   a. preventing milk release from the mammary glands.
   b. preventing goiter.
   c. causing contraction of the uterus.
   d. maintaining normal calcium levels.
   e. increasing metabolic rate.

7. Growth hormone
   a. increases the usage of glucose.
   b. increases the breakdown of lipids.
   c. decreases the synthesis of proteins.
   d. decreases the synthesis of glycogen.
   e. all of the above.

8. Which of these hormones stimulates somatomedin secretion?
   a. FSH
   b. GH
   c. LH
   d. Prolactin
   e. TSH

9. Hypersecretion of growth hormone
   a. results in giantism if it occurs in children.
   b. causes acromegaly in adults.
   c. increases the probability that one will develop diabetes.
   d. can lead to severe atherosclerosis.
   e. all of the above.

10. LH and FSH
    a. are produced in the hypothalamus.
    b. production is increased by TSH.
    c. promote the production of gametes and reproductive hormones.
    d. inhibit the production of prolactin.
    e. all of the above.

11. T_3 and T_4
    a. require iodine for their production.
    b. are made from the amino acid tyrosine.
    c. are transported in the blood bound to thyroxine-binding globulin.
    d. all of the above.

12. Which of these symptoms is associated with hyposecretion of the thyroid gland?
    a. hypertension
    b. nervousness
    c. diarrhea
    d. weight loss with a normal or increased food intake
    e. decreased metabolic rate

13. Which of these conditions most likely occurs if a healthy person receives an injection of T_3 and T_4?
    a. The secretion rate of TSH declines.
    b. The person develops symptoms of hypothyroidism.
    c. The person develops hypercalcemia.
    d. The person secretes more TRH.

14. Which of these occurs as a response to a thyroidectomy (removal of the thyroid gland)?
    a. increased calcitonin secretion
    b. increased T_3 and T_4 secretion
    c. decreased TRH secretion
    d. increased TSH secretion

15. Choose the statement that most accurately predicts the long-term effect of a substance that prevents active transport of iodide by the thyroid gland.
    a. Large amounts of T_3 and T_4 accumulate within the thyroid follicles, but little is released.
    b. The person exhibits hypothyroidism.
    c. The anterior pituitary secretes smaller amounts of TSH.
    d. The circulating levels of T_3 and T_4 increase.

16. Calcitonin
    a. is secreted by the parathyroid glands.
    b. levels increase when blood calcium levels decrease.
    c. causes blood calcium levels to decrease.
    d. insufficiency results in weak bones and tetany.

17. Parathyroid hormone secretion increases in response to
    a. a decrease in blood calcium levels.
    b. increased production of parathyroid-stimulating hormone from the anterior pituitary.
    c. increased secretion of parathyroid-releasing hormone from the hypothalamus.
    d. increased secretion of calcitonin.
    e. a decrease in secretion of ACTH.

18. If parathyroid hormone levels increase, which of these conditions is expected?
    a. Osteoclast activity is increased.
    b. Calcium absorption from the small intestine is inhibited.
    c. Calcium reabsorption from the urine is inhibited.
    d. Less active vitamin D is formed in the kidneys.
    e. All of the above.

19. The adrenal medulla
    a. produces steroids.
    b. has cortisol as its major secretory product.
    c. decreases its secretions during exercise.
    d. is formed from a modified portion of the sympathetic division of the ANS.
    e. all of the above.

20. Pheochromocytoma is a condition in which a benign tumor results in hypersecretion of the adrenal medulla. The symptoms that one would expect include
    a. hypotension.
    b. bradycardia.
    c. pallor (decreased blood flow to the skin).
    d. lethargy.
    e. hypoglycemia.

21. Which of these is *not* a hormone secreted by the adrenal cortex?
    a. aldosterone
    b. androgens
    c. cortisol
    d. epinephrine

22. If aldosterone secretions increase
    a. blood potassium levels increase.
    b. blood hydrogen levels increase.
    c. acidosis results.
    d. blood sodium levels decrease.
    e. blood volume increases.

23. Glucocorticoids (cortisol)
    a. increase the breakdown of fats.
    b. increase the breakdown of proteins.
    c. increase blood glucose levels.
    d. decrease inflammation.
    e. all of the above.

24. The release of cortisol from the adrenal cortex is regulated by other hormones. Which of these hormones is correctly matched with its origin and function?
    a. CRH—secreted by the hypothalamus; stimulates the adrenal cortex to secrete cortisol
    b. CRH—secreted by the anterior pituitary; stimulates the adrenal cortex to secrete cortisol
    c. ACTH—secreted by the hypothalamus; stimulates the adrenal cortex to secrete cortisol
    d. ACTH—secreted by the anterior pituitary; stimulates the adrenal cortex to produce cortisol

25. Which of these would be expected in Cushing's syndrome?
    a. loss of hair in women
    b. deposition of fat in the face, neck, and abdomen
    c. low blood glucose
    d. low blood pressure
    e. all of the above

26. Within the pancreas, the pancreatic islets produce
    a. insulin.
    b. glucagon.
    c. digestive enzymes.
    d. both a and b.
    e. all of the above.

27. Insulin increases
    a. the uptake of glucose by its target tissues.
    b. the breakdown of protein.
    c. the breakdown of fats.
    d. glycogen breakdown in the liver.
    e. all of the above.

28. Which of these tissues is least affected by insulin?
    a. adipose tissue
    b. heart
    c. skeletal muscle
    d. brain
    e. liver

29. Glucagon
    a. primarily affects the liver.
    b. causes glycogen to be stored.
    c. causes blood glucose levels to decrease.
    d. decreases fat metabolism.
    e. all of the above.

30. When blood glucose levels increase, the secretion of which of these hormones increases?
    a. glucagon
    b. insulin
    c. GH
    d. cortisol
    e. epinephrine

31. If a person who has diabetes mellitus forgot to take an insulin injection, symptoms that may soon appear include
    a. acidosis.
    b. hyperglycemia.
    c. increased urine production.
    d. lethargy and fatigue.
    e. all of the above.

32. Which of these is *not* a hormone produced by the ovaries?
    a. estrogen
    b. progesterone
    c. prolactin
    d. inhibin
    e. relaxin

33. Melatonin
    a. is produced by the posterior pituitary.
    b. production increases as day length increases.
    c. inhibits the development of the reproductive system.
    d. increases GnRH secretion from the hypothalamus.
    e. decreases the tendency to sleep.
34. Which of these substances, produced by many tissues of the body, can promote inflammation, pain, and vasodilation of blood vessels?
    a. endorphin
    b. enkephalin
    c. thymosin
    d. epidermal growth factor
    e. prostaglandin

35. Which of the changes listed does *not* decrease with aging of the endocrine system?
    a. GH secretion
    b. melatonin secretion
    c. thyroid hormone secretion
    d. parathyroid hormone secretion
    e. renin secretion by the kidneys

Answers in Appendix F

C R I T I C A L   T H I N K I N G

1. The hypothalamohypophysial portal system connects the hypothalamus with the anterior pituitary. Why is such a special circulatory system advantageous?

2. The secretion of ADH can be affected by exposure to hot or cold environmental temperatures. Predict the effect of a hot environment on ADH secretion, and explain why it is advantageous. Propose a mechanism by which temperature produces a change in ADH secretion.

3. A patient exhibits polydipsia (thirst), polyuria (excess urine production), and urine with a low specific gravity (contains few ions and no glucose). If you want to reverse the symptoms, would you administer insulin, glucagon, ADH, or aldosterone? Explain.

4. A patient complains of headaches and visual disturbances. A casual glance reveals that the patient's finger bones are enlarged in diameter, a heavy deposition of bone exists over the eyes, and the patient has a prominent jaw. The doctor tells you that the headaches and visual disturbances result from increased pressure within the skull and that the patient is suffering from a pituitary tumor that is affecting hormone secretion. Name the hormone that is causing the problem, and explain why an increase in pressure exists within the skull.

5. Most laboratories have the ability to determine blood levels of TSH, T_3, and T_4. Given that ability, design a method of determining whether hyperthyroidism in a patient results from a pituitary abnormality or from the production of a nonpituitary thyroid stimulatory substance.

6. An anatomy and physiology instructor asks two students to predict a patient's response to chronic vitamin D deficiency. One student claims that the person would suffer from hypocalcemia and the symptoms associated with that condition. The other student claims that calcium levels would remain within their normal range, although at the low end of the range, and that bone resorption would occur to the point that advanced osteomalacia might be seen. With whom do you agree, and why?

7. Given the ability to measure blood glucose levels, design an experiment that distinguishes between a person with diabetes, a healthy person, and a person who has a pancreatic tumor that secretes large amounts of insulin.

8. A patient arrives in an unconscious condition. A medical emergency bracelet reveals that he has diabetes. The patient can be in diabetic coma or insulin shock. How could you tell which, and what treatment would you recommend for each condition?

9. Diabetes mellitus can result from a lack of insulin, which results in hyperglycemia. Adrenal diabetes and pituitary diabetes also produce hyperglycemia. What hormones produce the last two conditions?

10. Predict some of the consequences of exposure to intense and prolonged stress.

Answers in Appendix G

A N S W E R S   T O   P R E D I C T   Q U E S T I O N S

1. The cell bodies of the neurosecretory cells that produce ADH are in the hypothalamus, and their axons extend into the posterior pituitary, where ADH is stored and secreted. Removing the posterior pituitary severs the axons, resulting in a temporary reduction in secretion. The cell bodies still produce ADH, however, and as the ADH accumulates at the ends of severed axons, ADH secretion resumes.

2. If GH is administered to young people before growth of their long bones is complete, it causes their long bones to grow and they will grow taller. To accomplish this, however, GH would have to be administered over a considerable length of time. It's likely that some symptoms of acromegaly would develop. In addition to undesirable changes in the skeleton, nerves frequently are compressed as a result of the proliferation of connective tissue. Because GH spares glucose usage, chronic hyperglycemia results, frequently leading to diabetes mellitus and the development of severe atherosclerosis. Mr. Hoops's doctor would therefore not prescribe GH.

3. Surgical removal of the thyroid gland cause T_3 and T_4 levels to decline in the blood. TRH and TSH levels in the blood increase because, as T_3 and T_4 levels in the blood decrease, the negative-feedback effect of T_3 and T_4 on TRH and TSH are removed. Oral administration of T_3 and T_4 cause blood levels of T_3 and T_4 to increase and, because of negative feedback, TRH and TSH levels decline.

4. In response to a reduced dietary intake of calcium, the blood levels of calcium begin to decline. In response to the decline in blood levels of calcium, an increase of PTH secretion from the parathyroid glands occurs. The PTH functions to increase calcium resorption from bone. Consequently, blood levels of calcium are maintained within the normal range but, at the same time, bones are being decalcified. Severe dietary calcium deficiency results in bones that become soft and eaten away because of the decrease in calcium content.

5. Removal of the thyroid gland means that the tissue responsible for thyroid hormone (T_3 and T_4) secretion from thyroid follicles, and calcitonin from parafollicular cells, would no longer occur. However, blood Ca^{2+} would remain within its normal range. Calcitonin is not essential for the maintenance of normal blood Ca^{2+} levels. Removal of the parathyroid gland would eliminate PTH secretion. Without PTH, blood levels of calcium fall. When the blood levels of calcium fall below normal, the permeability of nerve and muscle cells to Na^+ increases. As a consequence, spontaneous action potentials are produced that cause tetanus of muscles. Death can result from tetany of respiratory muscles.

6. High aldosterone levels in the blood lead to elevated Na^+ levels in the circulatory system and low blood levels of K^+. The effect of low blood levels of K^+ is hyperpolarization of muscle and neurons. The hyperpolarization results from the lower levels of K^+ in the extracellular fluid and a greater tendency for K^+ to diffuse from the cell. As a result, a greater-than-normal stimulus is required to cause the cells to depolarize to threshold and generate an action potential. Symptoms of low serum K^+ levels therefore include lethargy and muscle weakness. Elevated Na^+ concentrations result in a greater-than-normal amount of water retention in the circulatory system, which can result in elevated blood pressure. The major effect of a low rate of aldosterone secretion is elevated blood K^+ levels. As a result, nerve and muscle cells partially depolarize. Because of their partial depolarization, they produce action potentials spontaneously or in response to very small stimuli. The result is muscle spasms, or tetanus.

7. Large doses of cortisone can damage the adrenal cortex because cortisone inhibits ACTH secretion from the anterior pituitary. ACTH is required to keep the adrenal cortex from undergoing atrophy. Prolonged use of large doses of cortisone can cause the adrenal gland to atrophy to the point at which it cannot recover if ACTH secretion does increase again.

8. An increase in insulin secretion in response to parasympathetic stimulation and gastrointestinal hormones is consistent with the maintenance of homeostasis because parasympathetic stimulation and increased gastrointestinal hormones result from conditions such as eating a meal. Insulin levels therefore increase just before large amounts of glucose and amino acids enter the circulatory system. The elevated insulin levels prevent a large increase in blood glucose and the loss of glucose in the urine.

9. In response to a meal high in carbohydrates, insulin secretion is increased, and glucagon secretion is reduced. The stimulus for the insulin secretion comes from parasympathetic stimulation and, more importantly, from elevated blood levels of glucose. Target tissues take up glucose and blood glucose levels remain within a normal range. In response to a meal high in protein but low in carbohydrates, insulin secretion is increased slightly, and glucagon secretion is also increased. Insulin secretion is stimulated by the parasympathetic system and an increase in blood amino acid levels. Glucagon is stimulated by low blood glucose levels and by some amino acids. The lower insulin secretion causes some increase in the rate of glucose uptake and amino acid uptake, but the rate of uptake is not great enough to cause blood glucose levels to fall below normal values. Glucagon also causes glucose to be released from the liver. During periods of exercise, sympathetic stimulation inhibits insulin secretion. As blood glucose levels decline, an increase of glucagon secretion occurs. The lower rate of insulin secretion decreases the rate at which tissues such as skeletal muscle take up glucose. Muscle depends on intracellular glycogen and fatty acids for energy. Blood glucose levels are maintained within its normal range of values. Glucagon prevents glucose levels from decreasing too much.

10. Sympathetic stimulation during exercise inhibits insulin secretion. Blood glucose levels are not high because skeletal muscle tissue continues to take up some glucose and metabolizes it. Muscle contraction depends on glucose stored in the form of glycogen in muscles and fatty acid metabolism. During a long run, glycogen levels are depleted. The "kick" at the end of the race results from increased energy production through anaerobic respiration, which uses glucose or glycogen as an energy source. Because blood glucose levels and glycogen levels are low, the source of energy is insufficient for greatly increased muscle activity.

11. Increased sugar intake will result in elevated blood glucose levels. The elevated blood glucose levels can lead to polyuria and to increased osmolality of the body fluids. That results in dehydration of neurons. As a result some of the neural symptoms of untreated diabetes, such as irritability and a general sensation of not feeling well, occur. Billy may also experience a sudden increase in weight gain because of increased sugar intake and insulin administration. In addition, he may have an increased chance of infections, such as urinary tract infections. Many of the long-term consequences of diabetes, such as nephropathies, neuropathies, atherosclerosis, and others, develop much more rapidly.

Visit the Online Learning Center at www.mhhe.com/seeley7 for chapter quizzes, interactive learning exercises, and other study tools.

Cardiovascular System

Blood

Colorized scanning electron micrograph of a blood clot. The red discs are red blood cells, the blue particles are platelets, and the yellow strands are fibrin.

Many cultures around the world, both ancient and modern, share beliefs in the magical qualities of blood. Blood was considered the "essence of life" because the uncontrolled loss of it can result in death. Blood was also thought to define our character and emotions. People of a noble bloodline were described as "blue bloods," whereas criminals were considered to have "bad" blood. It was said that anger caused the blood to "boil," and fear resulted in blood "curdling." The scientific study of blood reveals characteristics as fascinating as any of these fantasies. Blood performs many functions essential to life and often can reveal much about our health.

Blood is a type of connective tissue, consisting of cells and cell fragments surrounded by a liquid matrix. The cells and cell fragments are the formed elements, and the liquid is the plasma. The formed elements make up about 45%, and plasma makes up about 55% of the total blood volume (figure 19.1). The total blood volume in the average adult is about 4–5 L in females and 5–6 L in males. Blood makes up about 8% of the total weight of the body.

Cells require constant nutrition and waste removal because they are metabolically active. The cardiovascular system, which consists of the heart, blood vessels, and blood, connects the various tissues of the body. The heart pumps blood through blood vessels, and the blood delivers nutrients and picks up waste products.

This chapter explains the *functions of blood* (p. 652), *plasma* (p. 653), and the *formed elements* (p. 654) of blood. *Hemostasis* (p. 662), *blood grouping* (p. 667), and *diagnostic blood tests* (p. 670) are also described.

Functions of Blood

Blood is pumped by the heart through blood vessels, which extend throughout the body. Blood helps to maintain homeostasis in several ways.

1. *Transport of gases, nutrients, and waste products.* Oxygen enters blood in the lungs and is carried to cells. Carbon dioxide, produced by cells, is carried in the blood to the lungs, from which it is expelled. Ingested nutrients, ions, and water are transported by the blood from the digestive tract to cells, and waste products of cells are transported by the blood to the kidneys for elimination.

2. *Transport of processed molecules.* Many substances are produced in one part of the body and transported in the blood to another part where they are modified. For example, the precursor to vitamin D is produced in the skin (see

chapter 5) and transported by the blood to the liver and then to the kidneys for processing into active vitamin D. Active vitamin D is transported in the blood to the small intestines, where it promotes the uptake of calcium. Another example is lactic acid produced by skeletal muscles during anaerobic respiration (see chapter 9). Lactic acid is carried by the blood to the liver, where it is converted into glucose.

3. *Transport of regulatory molecules.* Many of the hormones and enzymes that regulate body processes are carried from one part of the body to another by the blood.

4. *Regulation of pH and osmosis.* Buffers (see chapter 2), which help keep the blood's pH within its normal limits of 7.35–7.45, are in the blood. The osmotic composition of blood is also critical for maintaining normal fluid and ion balance.

5. *Maintenance of body temperature.* Blood is involved with body temperature regulation because warm blood is

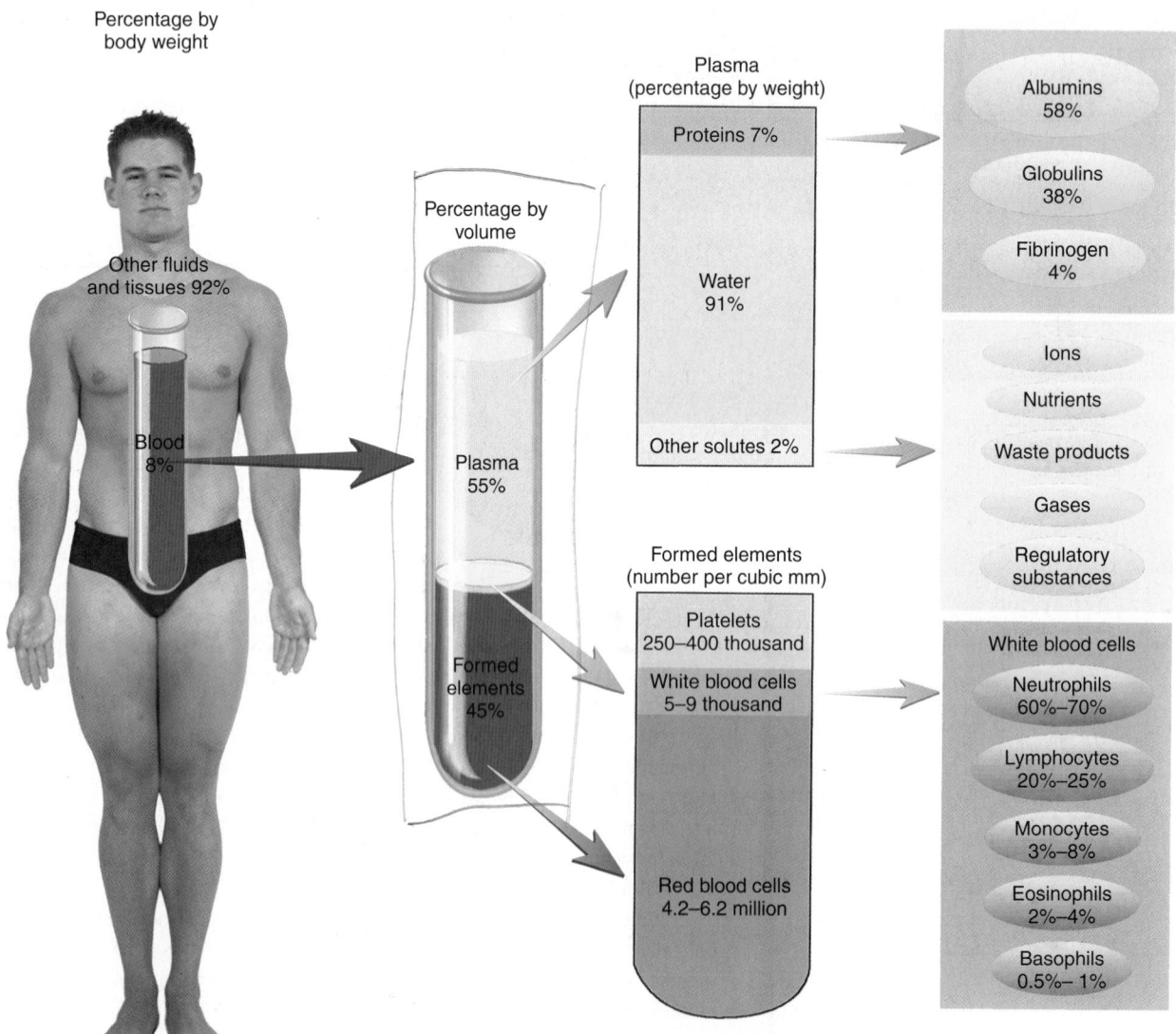

Figure 19.1   **Composition of Blood**

Approximate values for the components of blood in a normal adult.

transported from the interior to the surface of the body, where heat is released from the blood.

6. *Protection against foreign substances.* Cells and chemicals of the blood make up an important part of the immune system, protecting against foreign substances such as microorganisms and toxins.

7. *Clot formation.* Blood clotting provides protection against excessive blood loss when blood vessels are damaged. When tissues are damaged, the blood clot that forms is also the first step in tissue repair and the restoration of normal function (see chapter 4).

1. **List the ways that blood helps to maintain homeostasis in the body.**

Plasma

Plasma (plaz′mă) is the liquid part of blood. It's a pale yellow fluid that consists of about 91% water and 9% other substances, such as proteins, ions, nutrients, gases, and waste products (table 19.1). Plasma is a **colloid** (kol′oyd), which is a liquid containing suspended substances that don't settle out of solution. Most of the suspended substances are plasma proteins, which include albumin, globulins,

and fibrinogen. **Albumin** (al-bū′min) makes up 58% of the plasma proteins and is important in the regulation of water movement between tissues and blood. Because albumin doesn't easily pass from the blood into tissues, it plays an important role in maintaining blood colloid osmotic pressure (see chapters 21 and 26). Other molecules, such as fatty acids, bilirubin, and thyroid hormones, are transported in the blood attached to albumin. **Globulins** (glob′ū-linz) account for 38% of the plasma proteins. Some globulins, such as antibodies and complement, are part of the immune system (see chapter 22), whereas others function as transport molecules (see chapter 17). **Fibrinogen** (fī-brin′ō-jen) constitutes 4% of the plasma proteins and is responsible for the formation of blood clots (see "Coagulation" on p. 663).

The water, proteins, and other substances in the blood, such as ions, nutrients, waste products, gases, and regulatory substances, are maintained within narrow limits. Normally, water intake through the digestive tract closely matches water loss through the kidneys, lungs, digestive tract, and skin. Therefore, plasma volume remains relatively constant. Suspended or dissolved substances in the blood come from the liver, kidneys, intestines, endocrine glands, and immune tissues like the lymph nodes and spleen. Oxygen enters blood in the lungs and leaves the blood as it flows through tissues. Carbon dioxide enters blood from the tissues and leaves the blood as it flows through the lungs.

Table 19.1	Composition of Plasma
Plasma Components	**Function**
Water	Acts as a solvent and suspending medium for blood components
Plasma Proteins	
Albumin	Partly responsible for blood viscosity and osmotic pressure; acts as a buffer; transports fatty acids, free bilirubin, and thyroid hormones
Globulins	Transports lipids, carbohydrates, hormones, iron, and copper; antibodies and complement are involved in immunity
Fibrinogen	Functions in blood clotting
Ions	
Sodium, potassium, calcium, magnesium, chloride, iron, phosphate, hydrogen, hydroxide, bicarbonate	Involved in osmosis, membrane potentials, and acid–base balance
Nutrients	
Glucose, amino acids, triacylglycerol, cholesterol	Source of energy and basic "building blocks" of more complex molecules
Vitamins	Promote enzyme activity
Waste Products	
Urea, uric acid, creatinine, ammonia salts	Breakdown products of protein metabolism; excreted by the kidneys
Bilirubin	Breakdown product of red blood cells; excreted as part of the bile from the liver into the intestine
Lactic acid	End product of anaerobic respiration; converted to glucose by the liver
Gases	
Oxygen	Necessary for aerobic respiration; terminal electron acceptor in electron-transport chain
Carbon dioxide	Waste product of aerobic respiration; as bicarbonate, helps buffer blood
Nitrogen	Inert
Regulatory Substances	Enzymes catalyze chemical reactions; hormones stimulate or inhibit many body functions

2. *Define the term plasma. What are the functions of albumin, globulins, and fibrinogen in plasma? What other substances are found in plasma?*

Formed Elements

About 95% of the volume of the **formed elements** consists of red blood cells, or erythrocytes (ĕ-rith′rō-sītz). The remaining 5% consists of white blood cells, or leukocytes (loo′kō-sītz), and cell fragments called platelets, or thrombocytes (throm′bō-sītz). The formed elements of the blood are outlined and illustrated in table 19.2. In healthy adults, white blood cells are the only formed elements possessing nuclei, whereas red blood cells and platelets lack nuclei.

White blood cells are named according to their appearance in stained preparations. **Granulocytes** (gran′yū-lō-sītz) are white blood cells with large cytoplasmic granules and lobed nuclei (see table 19.2). Their granules stain with dyes that make the cells more visible when viewed through a light microscope. The three types of granulocytes are named according to the staining characteristics of their granules: **neutrophils** (nu′trō-filz) stain with acidic and basic dyes, **eosinophils** (ē-ō-sin′ō-filz) stain red with acidic dyes, and **basophils** (bā′sō-filz) stain dark purple with basic dyes. **Agranulocytes**

(ă-gran′yū-lō-sītz) are white blood cells that appear to have no granules when viewed in the light microscope. Agranulocytes actually have granules, but they are so small that they cannot be seen easily with the light microscope. The two types of agranulocytes are **lymphocytes** (lim′fō-sītz) and **monocytes** (mon′ō-sītz). They have nuclei that are not lobed.

Production of Formed Elements

The process of blood cell production, called **hematopoiesis** (hē′mă-tō-poy-ē′sis, hem′ă-to-poy-ē′sis) or **hemopoiesis** (hē′mō-poy-ē′sis), occurs in the embryo and fetus in tissues like the yolk sac, liver, thymus, spleen, lymph nodes, and red bone marrow. After birth, hematopoiesis is confined primarily to red bone marrow, with some lymphoid tissue helping in the production of lymphocytes (see chapter 22). In young children, nearly all the marrow is red bone marrow. In adults, however, red marrow is confined to the ribs, sternum, vertebrae, pelvis, proximal femur, and proximal humerus. Yellow marrow replaces red marrow in other locations in the body (see chapter 6).

All the formed elements of the blood are derived from a single population of **stem cells** called **hemocytoblasts** located in the red bone marrow. Hemopoietic stem cells are precursor cells capable of dividing to produce daughter cells that can differentiate into various types of blood cells (figure 19.2): **proerythroblasts**

Table 19.2	Formed Elements of the Blood			
Cell Type	**Illustration**	**Description**		**Function**
Red blood cell		Biconcave disk; no nucleus; contains hemoglobin, which colors the cell red; 7.5 μm in diameter		Transports oxygen and carbon dioxide
White blood cells		Spherical cells with a nucleus; white in color because they lack hemoglobin		Five types of white blood cells, each with specific functions
Granulocytes				
Neutrophil		Nucleus with two to four lobes connected by thin filaments; cytoplasmic granules stain a light pink or reddish purple; 10–12 μm in diameter		Phagocytizes microorganisms and other substances
Basophil		Nucleus with two indistinct lobes; cytoplasmic granules stain blue-purple; 10–12 μm in diameter		Releases histamine, which promotes inflammation, and heparin, which prevents clot formation
Eosinophil		Nucleus often bilobed; cytoplasmic granules stain orange-red or bright red; 11–14 μm in diameter		Releases chemicals that reduce inflammation; attacks certain worm parasites
Agranulocytes				
Lymphocyte		Round nucleus; cytoplasm forms a thin ring around the nucleus; 6–14 μm in diameter		Produces antibodies and other chemicals responsible for destroying microorganisms; contributes to allergic reactions, graft rejection, tumor control, and regulation of the immune system
Monocyte		Nucleus round, kidney-shaped, or horseshoe-shaped; contains more cytoplasm than does lymphocyte; 12–20 μm in diameter		Phagocytic cell in the blood; leaves the blood and becomes a macrophage, which phagocytizes bacteria, dead cells, cell fragments, and other debris within tissues
Platelet		Cell fragment surrounded by a plasma membrane and containing granules; 2–4 μm in diameter		Forms platelet plugs; releases chemicals necessary for blood clotting

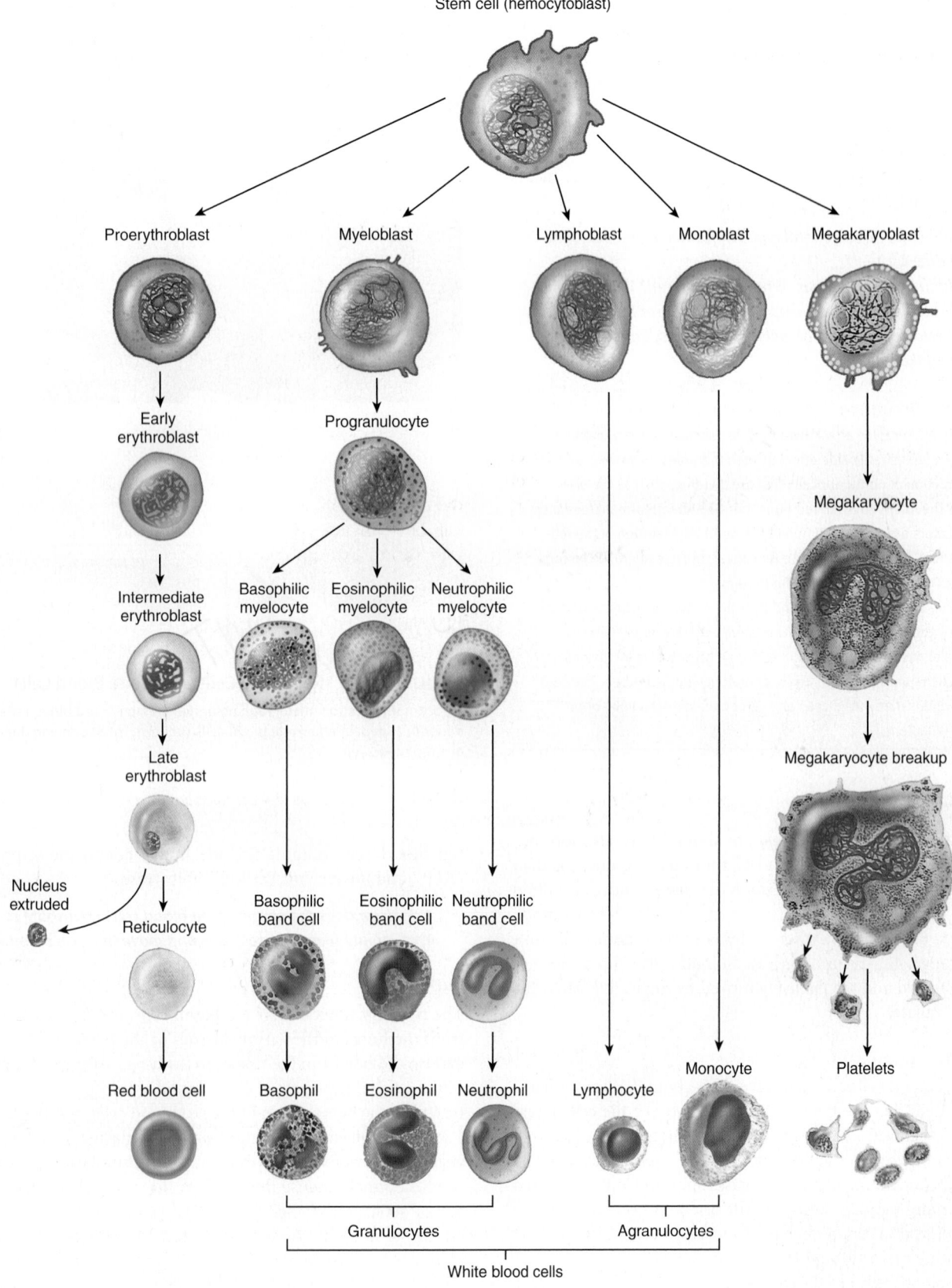

Figure 19.2 Hematopoiesis

Stem cells give rise to the cell lines that produce the formed elements.

(prō-ĕ-rith′rō-blastz), from which red blood cells develop; **myeloblasts** (mī′ĕ-lō-blastz), from which basophils, eosinophils, and neutrophils develop; **lymphoblasts** (lim′fō-blastz), from which lymphocytes develop; **monoblasts** (mon′ō-blastz), from which monocytes develop; and **megakaryoblasts** (meg-ă-kar′ē-ō-blastz), from which platelets develop. The development of the cell lines is regulated by growth factors. That is, the type of formed element derived from the stem cells and how many formed elements are produced are determined by different growth factors.

> 3. *Name the three general types of formed elements in the blood.*
>
> 4. *Define hematopoiesis. What is a stem cell? What types of formed elements develop from proerythroblasts, myeloblasts, lymphoblasts, monoblasts, and megakaryoblasts?*

Stem Cells and Cancer Therapy

Many cancer therapies affect dividing cells, such as those found in tumors. An undesirable side effect of such therapies, however, can be the destruction of nontumor cells that are dividing, such as the stem cells and their derivatives in red bone marrow. After treatment for cancer, growth factors are used to stimulate the rapid regeneration of the red bone marrow. Although not a cure for cancer, the use of growth factors can speed recovery from the cancer therapy.

Some types of leukemia and genetic immune deficiency diseases can be treated with a bone marrow or stem cell transplant. To avoid problems of tissue rejection, families with a history of these disorders can freeze the umbilical cord blood of their newborn children. The cord blood contains many stem cells and can be used instead of a bone marrow transplant.

Red Blood Cells

Red blood cells, or **erythrocytes,** are about 700 times more numerous than white blood cells and 17 times more numerous than platelets in the blood (figure 19.3*a*). Males have about 5.4 million red blood cells per microliter (μL; 1 mm^3 or 10^{-6} L) of blood (range: 4.6–6.2 million), whereas females have about 4.8 million/μL (range: 4.2–5.4 million). Red blood cells cannot move of their own accord and are passively moved by forces that cause the blood to circulate.

Structure

Normal red blood cells are biconcave discs about 7.5 μm in diameter with edges that are thicker than the center of the cell (figure 19.3*b*). The biconcave shape increases the surface area of the red blood cell compared to a flat disc of the same size. The greater surface area makes the movement of gases into and out of the red blood cell more rapid. In addition, the red blood cell can bend or fold around its thin center, thereby decreasing its size and enabling it to pass more easily through small blood vessels.

Red blood cells are derived from specialized cells that lose their nuclei and nearly all their cellular organelles during maturation. The main component of the red blood cell is the pigmented protein **hemoglobin** (hē-mō-glō′bin), which occupies about one-third of the total cell volume and accounts for its red color. Other

(a)

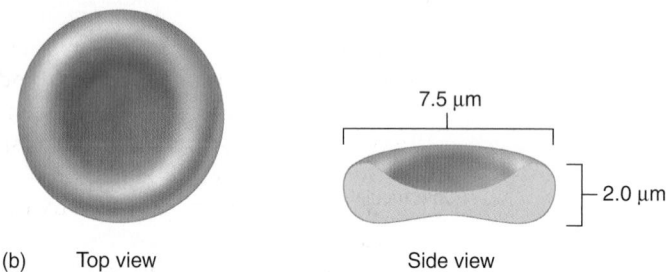

(b)   Top view   Side view

Figure 19.3   **Red Blood Cells and White Blood Cells**

(*a*) Scanning electron micrograph of formed elements: red blood cells (*red doughnut shapes*) and white blood cells (*yellow*). (*b*) Shape and dimensions of a red blood cell.

red blood cell contents include lipids, adenosine triphosphate (ATP), and the enzyme carbonic anhydrase.

> 5. *How does the shape of red blood cells contribute to their ability to exchange gases and move through blood vessels?*

Function

The primary functions of red blood cells are to transport oxygen from the lungs to the various tissues of the body and to transport carbon dioxide from the tissues to the lungs. Approximately 98.5% of the oxygen transported in the blood is transported in combination with the hemoglobin in the red blood cells, and the remaining 1.5% is dissolved in the water part of the plasma. If red blood cells rupture, the hemoglobin leaks out into the plasma and becomes nonfunctional because the shape of the molecule changes as a result of denaturation (see chapter 2). Red blood cell rupture followed by hemoglobin release is called **hemolysis** (hē-mol′i-sis). Hemolysis occurs in hemolytic anemia (see Clinical Focus "Disorders of the Blood," p. 672), transfusion reactions, hemolytic disease of the newborn, and malaria.

Carbon dioxide is transported in the blood in three major ways: approximately 7% is transported as carbon dioxide dissolved in the plasma, approximately 23% is transported in combination with blood proteins (mostly the globin part of hemoglobin), and 70% is transported in the form of bicarbonate ions. The bicarbonate ions (HCO_3^-) are produced when carbon dioxide (CO_2) and water (H_2O) combine to form carbonic acid (H_2CO_3), which dissociates to form hydrogen (H^+) and bicarbonate ions. The combination of carbon dioxide and water is catalyzed by an enzyme, **carbonic anhydrase,** which is located primarily within red blood cells.

$$CO_2 + H_2O \overset{\text{Carbonic}}{\underset{\text{anhydrase}}{\rightleftarrows}} H_2CO_3 \rightleftarrows H^+ + HCO_3^-$$

Carbon dioxide   Water   Carbonic acid   Hydrogen ion   Bicarbonate ion

● 6. *Give the percentage for each of the ways that oxygen and carbon dioxide are transported in the blood. What is the function of carbonic anhydrase?*

Hemoglobin

Hemoglobin consists of four polypeptide chains and four heme groups. Each polypeptide chain, called a **globin** (glō'bin), is bound to one **heme** (hēm). Each heme is a red-pigment molecule containing one **iron atom** (figure 19.4). Several types of globin exist, each having a slightly different amino acid composition. The four globins in normal adult hemoglobin consist of two alpha (α) chains and two beta (β) chains.

Embryonic hemoglobin is the first type of hemoglobin produced. It is replaced with fetal hemoglobin by about the third month of development. After birth, fetal hemoglobin is rapidly replaced with adult hemoglobin, but small amounts of fetal hemoglobin may be found in children (less than 2%) or adults (less than 0.5%). Embryonic and fetal hemoglobins are more effective at binding oxygen than is adult hemoglobin. Abnormal hemoglobins are less effective at attracting oxygen than is normal hemoglobin and can result in anemia (see Clinical Focus "Disorders of the Blood" on p. 672).

P R E D I C T **1**

What would happen to a fetus if maternal blood had an equal or greater affinity for oxygen than does fetal blood?

Iron is necessary for the normal function of hemoglobin because each oxygen molecule that is transported is associated with an iron atom. The adult human body normally contains about 4 g of iron, two-thirds of which is associated with hemoglobin. Small amounts of iron are regularly lost from the body in waste products like urine and feces. Females lose additional iron as a result of menstrual bleeding and, therefore, require more dietary iron than do males. Dietary iron is absorbed into the circulation from the upper part of the intestinal tract. Stomach acid and vitamin C in food increase the absorption of iron by converting ferric iron (Fe^{3+}) to ferrous iron (Fe^{2+}), which is more readily absorbed.

Effect of Carbon Monoxide on Oxygen Transport

Various types of poisons affect the hemoglobin molecule. Carbon monoxide (CO), which is produced by the incomplete combustion of gasoline, binds almost irreversibly to the iron of hemoglobin to form the relatively stable compound **carboxyhemoglobin** (kar-bok'sē-hē-mō-glō'bin). As a result of the stable binding of carbon monoxide, hemoglobin cannot transport oxygen, and death may occur. Carbon monoxide is found in cigarette smoke, and the blood of smokers can contain 5%–15% carboxyhemoglobin.

When hemoglobin is exposed to oxygen, one oxygen molecule can become associated with each heme group. This oxygenated form of hemoglobin is called **oxyhemoglobin** (ok'sē-hē-mō-glō'bin). The oxyhemoglobin in one red blood cell transports about one billion molecules of oxygen. Each heme molecule binds to one oxygen molecule; there are four heme molecules per hemoglobin and there are 250,000 hemoglobin molecules per red blood cell. Hemoglobin containing no oxygen is called **deoxyhemoglobin.** Oxyhemoglobin is bright red, whereas deoxyhemoglobin has a darker red color.

Figure 19.4 Hemoglobin

(*a*) Four polypeptide chains, each with a heme, form a hemoglobin molecule. (*b*) Each heme contains one iron atom.

Hemoglobin transports carbon dioxide, which doesn't combine with the iron atoms but is attached to amino groups of the globin molecule. This hemoglobin form is **carbaminohemoglobin** (kar-bam′i-nō-hē-mō-glō′bin). The transport of oxygen and carbon dioxide by the blood is discussed more fully in chapter 23.

Hemoglobin also transports nitric oxide, which is produced by the endothelial cells lining blood vessels. In the lungs, at the same time that heme picks up oxygen, in each β-globin a sulfur-containing amino acid, cysteine, binds with a nitric oxide molecule to form *S*-nitrosothiol (nī-trōs′ō-thī-ol; SNO). When oxygen is released in tissues so is the nitric oxide, which functions as a chemical signal that induces the smooth muscle of blood vessels to relax. By affecting the amount of nitric oxide in tissues, hemoglobin may play a role in regulating blood pressure, because relaxation of blood vessels results in a decrease in blood pressure (see chapter 21).

Blood Substitutes

Research is being conducted to develop blood substitutes that will deliver oxygen to tissues. One such substitute is Hemopure. It is an ultrapurified, chemically cross-linked cow hemoglobin in a balanced salt solution. Thus, Hemopure is a stabilized hemoglobin that is no longer within red blood cells. The use of Hemopure for blood transfusions has several benefits compared to using blood. Hemopure has a longer shelf life than blood and can be used when blood is not available. The free oxygen-carrying hemoglobin molecule of Hemopure is 1000 times smaller than red blood cells, thus allowing it to flow past partially blocked arteries. There are no transfusion reactions because there are no red blood cell surface antigens (see "Blood Grouping" on p. 667). The possibility of transferring human diseases such as hepatitis or AIDS is eliminated. Stringent manufacturing techniques are necessary, however, to ensure the removal of disease-causing agents from cows, such as Creutzfeldt-Jakob disease and bovine spongiform encephalopathy (mad cow disease).

7. *Describe the two basic parts of a hemoglobin molecule. Which part is associated with iron? What gases are transported by each part?*

Life History of Red Blood Cells

Under normal conditions about 2.5 million red blood cells are destroyed every second. This loss seems staggering until you realize that it represents only 0.00001% of the total 25 trillion red blood cells contained in the normal adult circulation. Homeostasis is maintained by replacing the 2.5 million cells lost every second with an equal number of new red blood cells. Thus, approximately 1% of the total number of red blood cells is replaced each day.

The process by which new red blood cells are produced is called **erythropoiesis** (ĕ-rith′rō-poy-ē′sis; see figure 19.2), and the time required for the production of a single red blood cell is about 4 days. Stem cells, from which all blood cells originate, give rise to **proerythroblasts.** After several mitotic divisions, proerythroblasts become **early (basophilic) erythroblasts** (ĕ-rith′rō-blastz), which stain with a basic dye. The dye stains the cytoplasm a purplish color because it binds to the large numbers

of ribosomes, which are sites of synthesis for the protein hemoglobin. Early erythroblasts give rise to **intermediate (polychromatic) erythroblasts,** which stain different colors with basic and acidic dyes. As hemoglobin is synthesized and accumulates in the cytoplasm, it's stained a reddish color by an acidic dye. Intermediate erythroblasts continue to produce hemoglobin, and then most of their ribosomes and other organelles degenerate. The resulting **late erythroblasts** have a reddish color because about one-third of the cytoplasm is now hemoglobin.

The late erythroblasts lose their nuclei by a process of extrusion to become immature red blood cells, which are called **reticulocytes** (re-tik′ū-lō-sītz), because a reticulum, or network, can be observed in the cytoplasm when a special staining technique is used. The reticulum is artificially produced by the reaction of the dye with the few remaining ribosomes in the reticulocyte. Reticulocytes are released from the bone marrow into the circulating blood, which normally consists of mature red blood cells and 1%–3% reticulocytes. Within 1 to 2 days, reticulocytes become mature red blood cells when the ribosomes degenerate.

PREDICT   2

What does an elevated reticulocyte count indicate? Would the reticulocyte count change during the week after a person had donated a unit (about 500 mL) of blood?

Cell division requires the B vitamins folate and B_{12}, which are necessary for the synthesis of DNA (see chapter 3). Hemoglobin production requires iron. Consequently, adequate amounts of folate, vitamin B_{12}, and iron are necessary for normal red blood cell production.

Red blood cell production is stimulated by low blood oxygen levels, typical causes of which are decreased numbers of red blood cells, decreased or defective hemoglobin, diseases of the lungs, high altitude, inability of the cardiovascular system to deliver blood to tissues, and increased tissue demands for oxygen, for example, during endurance exercises.

Low blood oxygen levels stimulate red blood cell production by increasing the formation of the glycoprotein **erythropoietin** (ĕ-rith-rō-poy′ĕ-tin), which is a hormone produced mostly by the kidneys (figure 19.5). Erythropoietin stimulates red bone marrow to produce more red blood cells by increasing the number of proerythroblasts formed and by decreasing the time required for red blood cells to mature. Thus, when oxygen levels in the blood decrease, erythropoietin production increases, which increases red blood cell production. The increased number of red blood cells increases the ability of the blood to transport oxygen. This mechanism returns blood oxygen levels to normal and maintains homeostasis by increasing the delivery of oxygen to tissues. Conversely, if blood oxygen levels increase, less erythropoietin is released, and red blood cell production decreases.

PREDICT   3

Cigarette smoke produces carbon monoxide. If a nonsmoker smoked a pack of cigarettes a day for a few weeks, what would happen to the number of red blood cells in the person's blood? Explain.

Red blood cells normally stay in the circulation for about 120 days in males and 110 days in females. These cells have no nuclei and, therefore, cannot produce new proteins or divide. As

Figure 19.5 Red Blood Cell Production

In response to decreased blood oxygen, the kidneys release erythropoietin into the general circulation. The increased erythropoietin stimulates red blood cell production in the red bone marrow. This process increases blood oxygen levels.

their existing proteins, enzymes, plasma membrane components, and other structures degenerate, the red blood cells are less able to transport oxygen and their plasma membranes become more fragile. Eventually the red blood cells rupture as they squeeze through some tight spot in the circulation.

Macrophages located in the spleen, liver, and other lymphatic tissue (figure 19.6) take up the hemoglobin released from ruptured red blood cells. Within the macrophage, lysosomal enzymes digest the hemoglobin to yield amino acids, iron, and bilirubin. The globin part of hemoglobin is broken down into its component amino acids, most of which are reused in the production of other proteins. Iron atoms released from heme can be carried by the blood to red bone marrow, where they are incorporated into new hemoglobin molecules. The non-iron part of the heme groups are converted to **biliverdin** (bil-i-ver′din) and then to **bilirubin** (bil-i-roo′bin), which is released into the plasma. Bilirubin binds to albumin and is transported to liver cells. This bilirubin is called **free bilirubin** because it is not yet conjugated. Free bilirubin is taken up by the liver cells and is conjugated, or joined, to glucuronic acid to form **conjugated bilirubin,** which is more water-soluble than free bilirubin. The conjugated bilirubin becomes part of the **bile,** which is the fluid secreted from the liver into the small intestine. In the intestine, bacteria convert bilirubin into the pigments that give feces its characteristic brownish color. Some of these pigments are absorbed from the intestine, modified in the kidneys, and excreted in the urine, thus contributing to the characteristic yellowish color of urine. **Jaundice** (jawn′dis) is a yellowish staining of the skin and sclerae caused by a buildup of bile pigments in the circulation and interstitial spaces.

8. *Define erythropoiesis. Describe the formation of red blood cells, starting with the stem cells in the red bone marrow.*
9. *What is erythropoietin, where is it produced, what causes it to be produced, and what effect does it have on red blood cell production?*
10. *Where are red blood cells removed from the blood? List the three breakdown products of hemoglobin and explain what happens to them.*

White Blood Cells

White blood cells, or **leukocytes,** are clear or whitish-colored cells that lack hemoglobin but have a nucleus. In stained preparations, white blood cells attract stain, whereas red blood cells remain relatively unstained (figure 19.7; see table 19.2).

White blood cells protect the body against invading microorganisms and remove dead cells and debris from the body. Most white blood cells are motile, exhibiting **ameboid movement,** which is the ability to move like an ameba by putting out irregular cytoplasmic projections. White blood cells leave the circulation and enter tissues by **diapedesis** (dī′ă-pĕ-dē′sis), a process in which they become thin and elongated and slip between or, in some cases, through the cells of blood vessel walls. The white blood cells can then be attracted to foreign materials or dead cells within the tissue by **chemotaxis** (kē-mō-tak′sis). At the site of an infection, white blood cells accumulate and phagocytize bacteria, dirt, and dead cells; then they die. The accumulation of dead white blood cells and bacteria, along with fluid and cell debris, is called **pus.**

The five types of white blood cells are neutrophils, eosinophils, basophils, lymphocytes, and monocytes.

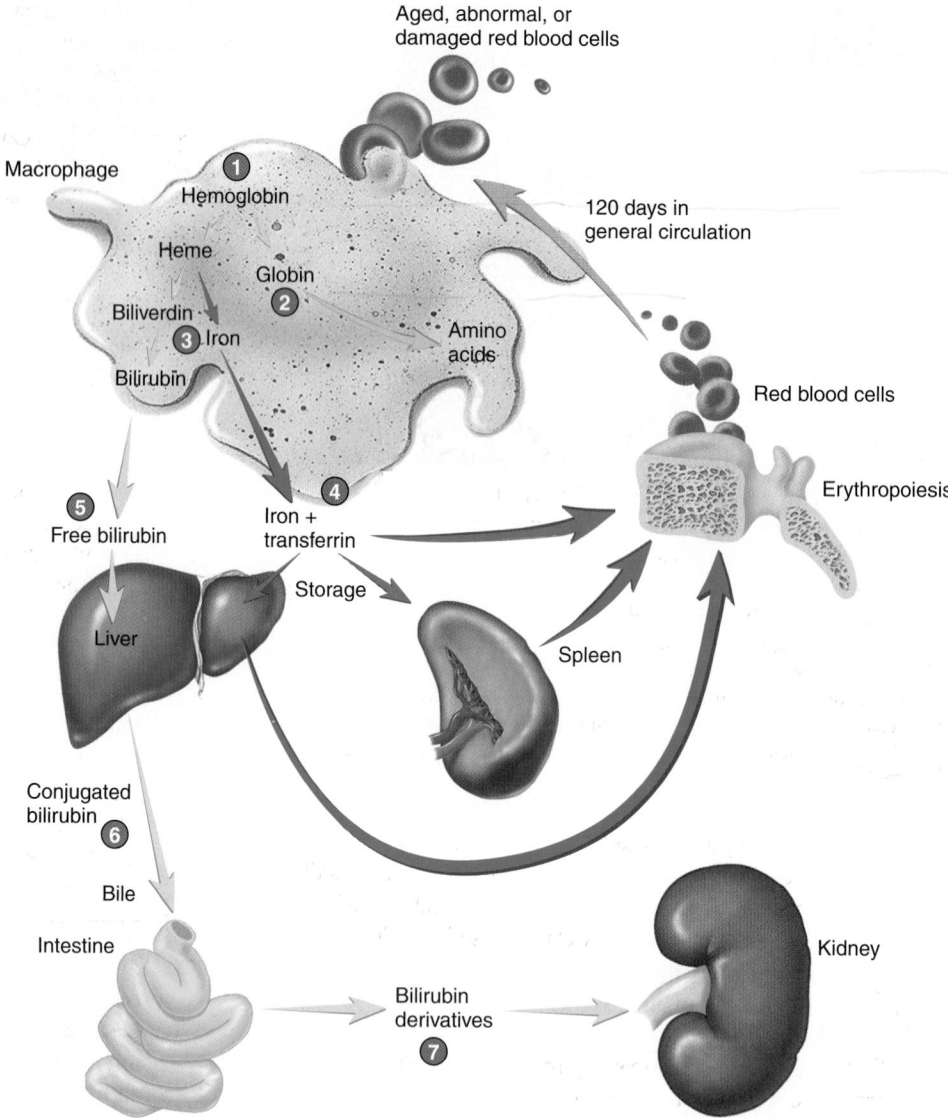

1. Hemoglobin is broken down into heme and globin chains.

2. The globin chains of hemoglobin are broken down to individual amino acids (*pink arrow*) and are metabolized or used to build new proteins.

3. Iron is released from the heme of hemoglobin. The heme is converted into biliverdin, which is converted into bilirubin.

4. Iron is transported in combination with transferrin in the blood to various tissues for storage or transported to the red bone marrow and used in the production of new hemoglobin (*green arrows*).

5. Free bilirubin (*blue arrow*) is transported in the blood to the liver.

6. Conjugated bilirubin is excreted as part of the bile into the small intestine.

7. Bilirubin derivatives contribute to the color of feces or are reabsorbed from the intestine into the blood and excreted from the kidneys in the urine.

Process Figure 19.6 Hemoglobin Breakdown

Hemoglobin is broken down in macrophages, and the breakdown products are used or excreted.

Neutrophils

Neutrophils (see table 19.2), the most common type of white blood cells in the blood, have small cytoplasmic granules that stain with both acidic and basic dyes. Their nuclei are commonly lobed, with the number of lobes varying from two to five. Neutrophils are often called **polymorphonuclear** (pol′ē-mōr-fō-noo′klē-ăr) **neutrophils,** or PMNs, to indicate that their nuclei can occur in more than one (*poly*) form (*morph*). Neutrophils usually remain in the circulation for about 10–12 hours and then move into other tissues, where they become motile and seek out and phagocytize bacteria, antigen–antibody complexes (antigens and antibodies bound together), and other foreign matter. Neutrophils also secrete a class of enzymes called **lysozymes** (lī′sō-zīmz), which are capable of destroying certain bacteria. Neutrophils usually survive for 1–2 days after leaving the circulation.

Eosinophils

Eosinophils (see table 19.2) contain cytoplasmic granules that stain bright red with eosin, an acidic stain. They are motile cells that leave the circulation to enter the tissues during an inflammatory reaction. They are most common in tissues undergoing an allergic response, and their numbers are elevated in the blood of people with allergies. Eosinophils apparently reduce the inflammatory response by producing enzymes that destroy inflammatory chemicals like histamine. Eosinophils also release toxic chemicals that attack certain worm parasites, such as tapeworms, flukes, pinworms, and hookworms.

Basophils

Basophils (see table 19.2), the least common of all white blood cells, contain large cytoplasmic granules that stain blue or purple with basic dyes. Basophils, like eosinophils and neutrophils, leave

Platelet   Red blood cells

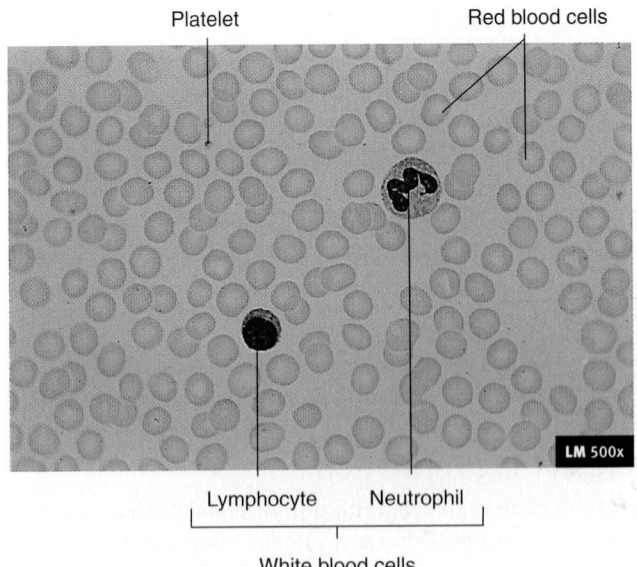

Lymphocyte   Neutrophil

White blood cells

Figure 19.7 Standard Blood Smear

A thin film of blood is spread on a microscope slide and stained. The white blood cells have pink-colored cytoplasm and purple-colored nuclei. The red blood cells do not have nuclei. The center of a red blood cell appears whitish because light more readily shines through the thin center of the disc than through the thicker edges. The platelets are purple-colored cell fragments.

the circulation and migrate through the tissues. They increase in number in both allergic and inflammatory reactions. Basophils contain large amounts of **histamine,** which they release within tissues to increase inflammation. They also release **heparin,** which inhibits blood clotting.

Lymphocytes

The smallest white blood cells are **lymphocytes,** most of which are slightly larger in diameter than red blood cells (see table 19.2). The lymphocytic cytoplasm consists of only a thin, sometimes imperceptible ring around the nucleus. Although lymphocytes originate in red bone marrow, they migrate through the blood to lymphatic tissues, where they can proliferate and produce more lymphocytes. The majority of the body's total lymphocyte population is in the lymphatic tissues: the lymph nodes, spleen, tonsils, lymphatic nodules, and thymus.

Although they cannot be identified by standard microscopic examination, a number of different kinds of lymphocytes play important roles in immunity (see chapter 22 for details). For example,

B cells can be stimulated by bacteria or toxins to divide and form cells that produce proteins called **antibodies.** Antibodies can attach to bacteria and activate mechanisms that result in destruction of the bacteria. **T cells** protect against viruses and other intracellular microorganisms by attacking and destroying the cells in which they are found. In addition, T cells are involved in the destruction of tumor cells and tissue graft rejections.

Monocytes

Monocytes are typically the largest of the white blood cells (see table 19.2). They normally remain in the circulation for about 3 days, leave the circulation, become transformed into macrophages, and migrate through various tissues. They phagocytize bacteria, dead cells, cell fragments, and other debris within the tissues. An increase in the number of monocytes is often associated with chronic infections. In addition, macrophages can break down phagocytized foreign substances and present the processed substances to lymphocytes, which results in activation of the lymphocytes (see chapter 22).

11. *What are the two major functions of white blood cells? Define ameboid movement, diapedesis, and chemotaxis.*
12. *Describe the morphology of the five types of white blood cells.*
13. *Name the two white blood cells that function primarily as phagocytic cells. Define lysozymes.*
14. *Which white blood cell reduces the inflammatory response? Which white blood cell releases histamine and promotes inflammation?*
15. *B and T cells are examples of what type of white blood cell? How do these cells protect us against bacteria and viruses?*

P R E D I C T   4

Based on their morphology, identify each of the white blood cells shown in figure 19.8.

Platelets

Platelets, or **thrombocytes** (see figure 19.7 and table 19.2), are minute fragments of cells consisting of a small amount of cytoplasm surrounded by a plasma membrane. Platelets are roughly disk-shaped and average about 3 μm in diameter. The surface of platelets has glycoproteins and proteins that allow platelets to attach to other molecules, such as collagen in connective tissue. Some of these surface molecules, as well as

(a)

(b)

(c)

(d)

(e)

Figure 19.8 Identification of White Blood Cells

See Predict question 4.

molecules released from granules in the platelet cytoplasm, play important roles in controlling blood loss. The platelet cytoplasm also contains actin and myosin, which can cause contraction of the platelet (see section on "Clot Retraction and Dissolution" on p. 666).

The life expectancy of platelets is about 5–9 days. They are produced within the red marrow and are derived from **megakaryocytes** (meg-ă-kar′ē-ō-sītz), which are extremely large cells with diameters up to 100 μm. Small fragments of these cells break off and enter the circulation as platelets.

Platelets play an important role in preventing blood loss by (1) forming platelet plugs, which seal holes in small vessels, and (2) by promoting the formation and contraction of clots, which help seal off larger wounds in the vessels.

> 16. *What is a platelet? How are platelets formed?*
> 17. *What are the two major roles of platelets in preventing blood loss?*

Hemostasis

Hemostasis (hē′mō-stā-sis, hē-mos′tă-sis), the stoppage of bleeding, is very important to the maintenance of homeostasis. If not stopped, excessive bleeding from a cut or torn blood vessel can result in a positive-feedback cycle, consisting of ever-decreasing blood volume and blood pressure, which disrupts homeostasis, and results in death. Fortunately, when a blood vessel is damaged, a number of events occur that help prevent excessive blood loss. Vascular spasm, platelet plug formation, and coagulation can cause hemostasis.

Vascular Spasm

Vascular spasm is an immediate but temporary constriction of a blood vessel resulting from contraction of smooth muscle within the wall of the vessel. This constriction can close small vessels completely and stop the flow of blood through them. Nervous system reflexes and chemicals produce vascular spasms. For example, during the formation of a platelet plug, platelets release **thromboxanes** (throm′bok-zānz), which are derived from certain prostaglandins, and endothelial cells release the peptide **endothelin** (en-dō′thē-lin).

> 18. *What is a vascular spasm? Name two factors that produce it. What is the source of thromboxanes and endothelin?*

Platelet Plug Formation

A **platelet plug** is an accumulation of platelets that can seal up small breaks in blood vessels. Platelet plug formation is very important in maintaining the integrity of the circulatory system because small tears occur in the smaller vessels and capillaries many times each day, and platelet plug formation quickly closes them. People who lack the normal number of platelets tend to develop numerous small hemorrhages in their skin and internal organs.

The formation of a platelet plug can be described as a series of steps, but in actuality many of the steps take place simultaneously (figure 19.9).

1. **Platelet adhesion** occurs when platelets bind to collagen exposed by blood vessel damage. Most platelet adhesion is mediated through **von Willebrand factor (vWF),** which is

1. Platelet adhesion occurs when von Willebrand factor connects collagen and platelets.

2. The platelet release reaction is the release of ADP, thromboxanes, and other chemicals that activate other platelets.

3. Platelet aggregation occurs when fibrinogen receptors on activated platelets bind to fibrinogen, connecting the platelets to one another. A platelet plug is formed by the accumulating mass of platelets.

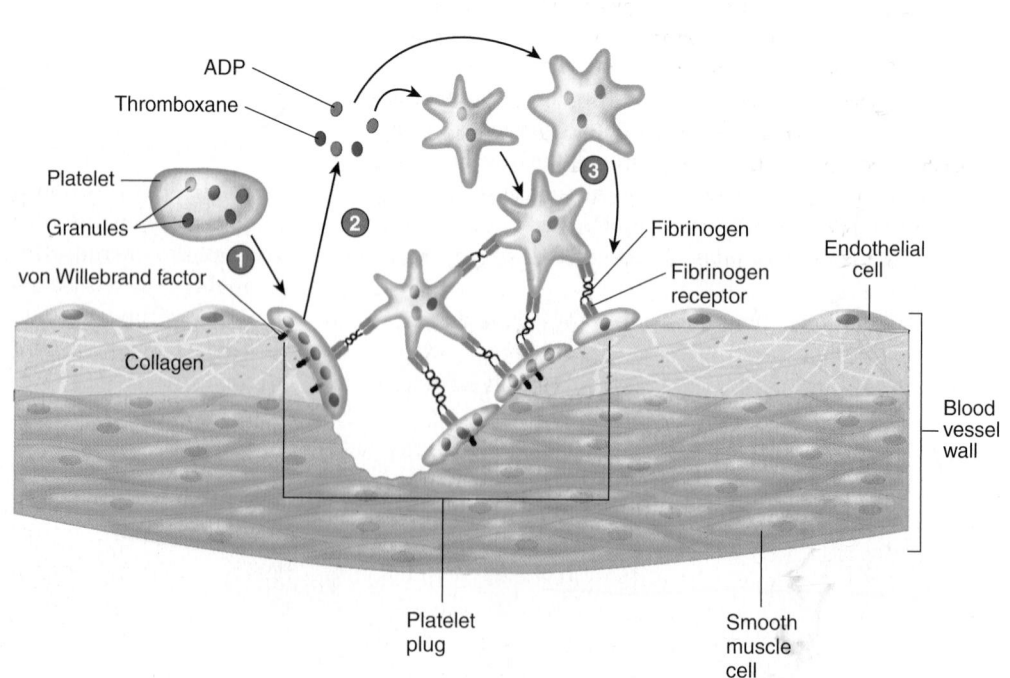

Process Figure 19.9 Platelet Plug Formation

a protein produced and secreted by blood vessel endothelial cells. Von Willebrand factor forms a bridge between collagen and platelets by binding to platelet surface receptors and collagen. In addition, other platelet surface receptors can bind directly to collagen.

2. After platelets adhere to collagen, they become activated, and in the **platelet release reaction,** adenosine diphosphate (ADP), thromboxanes, and other chemicals are extruded from the platelets by exocytosis. The ADP and thromboxanes stimulate other platelets to become activated and release additional chemicals, thereby producing a cascade of chemical release by the platelets. Thus, more and more platelets become activated.

3. As platelets become activated, they change shape and express fibrinogen receptors that can bind to fibrinogen, a plasma protein. In **platelet aggregation,** fibrinogen forms a bridge between the fibrinogen receptors of different platelets, resulting in the formation of a platelet plug.

4. Activated platelets express phospholipids (platelet factor III) and coagulation factor V, which are important in clot formation (see following section on "Coagulation").

How Aspirin Increases the Risk of Bleeding

Eicosanoids (ī-kō′să-noydz, *eicosa-*, twenty + *eidos,* form) are a group of related compounds derived from 20-carbon essential fatty acids (see chapter 25). Examples of eicosanoids are prostaglandins, thromboxanes, and leukotrienes. In humans, arachidonic acid is the most common precursor molecule for the eicosanoids. The enzyme cyclooxygenase-1 (COX-1) converts arachidonic acid into a prostaglandin that can be converted into thromboxane. Aspirin inhibits COX-1, which inhibits prostaglandin synthesis and therefore thromboxane synthesis. As a result, aspirin reduces platelet activation.

If an expectant mother ingests aspirin near the end of pregnancy, thromboxane synthesis is inhibited and several effects are possible. Two of these effects are (1) the mother can experience excessive bleeding after delivery because of decreased platelet function, and (2) the baby can exhibit numerous localized hemorrhages called **petechiae** (pe-tē′kē-ē) over the surface of its body as a result of decreased platelet function. If the quantity of ingested aspirin is large, the infant, mother, or both may die as a result of hemorrhage.

In a stroke or heart attack, platelet plugs and clots can form in vessels and be life-threatening. The United States Preventive Services Task Force (USPSTF) and the American Heart Association (AHA) recommend low-dose aspirin therapy (75–160 mg/day) for the prevention of cardiovascular disease for all men and women at high risk for cardiovascular disease. Determining high risk involves analyzing many risk factors and should be done in consultation with a physician. Decreased risk of cardiovascular disease from aspirin therapy must be weighed against the increased risk of hemorrhagic stroke and gastrointestinal bleeding. Risk factors for cardiovascular disease include age (men over 40 and postmenopausal women), high cholesterol, high blood pressure, a history of smoking, diabetes, a family history of cardiovascular disease, or a previous clotting event, such as a heart attack, or ischemic stroke (see Systems Pathology, "Stroke" in chapter 14).

---

19. *What is the function of a platelet plug? Describe the process of platelet plug formation. How are platelets an important part of clot formation?*

Coagulation

Vascular spasms and platelet plugs alone are not sufficient to close large tears or cuts. When a blood vessel is severely damaged, **coagulation** (kō-ag-ū-lā′shŭn), or **blood clotting,** results in the formation of a clot. A **blood clot** is a network of threadlike protein fibers, called **fibrin,** that traps blood cells, platelets, and fluid (figure 19.10).

The formation of a blood clot depends on a number of proteins, called **coagulation factors,** found within plasma (table 19.3). Normally the coagulation factors are in an inactive state and don't cause clotting. After injury, the clotting factors are activated to produce a clot. This activation is a complex process involving many chemical reactions, some of which require calcium ions (Ca^{2+}) and molecules on the surface of activated platelets, such as phospholipids and coagulation factor V.

P R E D I C T   5

Why is it advantageous for clot formation to involve molecules on the surface of activated platelets?

The activation of clotting proteins occurs in three main stages (figure 19.11). **Stage 1** consists of the formation of **prothrombinase, stage 2** is the conversion of **prothrombin** to **thrombin** by prothrombinase, and **stage 3** consists of the conversion of soluble **fibrinogen** to insoluble **fibrin** by thrombin. Depending on how prothrombinase is formed in stage 1, two separate pathways for coagulation can occur: the **extrinsic clotting pathway** and the **intrinsic clotting pathway.**

Platelet

Fibrin network

Trapped red blood cells

SEM 1400x

Figure 19.10 Blood Clot

A blood clot consists of fibrin fibers that trap red blood cells, platelets, and fluid.

Table 19.3   Coagulation Factors

Factor Number	Name (synonym)	Description and Function
I	Fibrinogen	Plasma protein synthesized in liver; converted to fibrin in stage 3
II	Prothrombin	Plasma protein synthesized in liver (requires vitamin K); converted to thrombin in stage 2
III	Thromboplastin (tissue factor)	Mixture of lipoproteins released from damaged tissue; required in extrinsic stage 1
IV	Calcium ion	Required throughout entire clotting sequence
V	Proaccelerin (labile factor)	Plasma protein synthesized in liver; activated form functions in stages 1 and 2 of both intrinsic and extrinsic clotting pathways
VI		Once thought to be involved but no longer accepted as playing a role in coagulation; apparently the same as activated factor V
VII	Serum prothrombin conversion accelerator (stable factor, proconvertin)	Plasma protein synthesized in liver (requires vitamin K); functions in extrinsic stage 1
VIII	Antihemophilic factor (antihemophilic globulin)	Plasma protein synthesized in megakaryocytes and endothelial cells; required for intrinsic stage 1
IX	Plasma thromboplastin component (Christmas factor)	Plasma protein synthesized in liver (requires vitamin K); required for intrinsic stage 1
X	Stuart factor (Stuart-Prower factor)	Plasma protein synthesized in liver (requires vitamin K); required in stages 1 and 2 of both intrinsic and extrinsic clotting pathways
XI	Plasma thromboplastin antecedent	Plasma protein synthesized in liver; required for intrinsic stage 1
XII	Hageman factor	Plasma protein required for intrinsic stage 1
XIII	Fibrin-stabilizing factor	Protein found in plasma and platelets; required for stage 3
Platelet Factors		
I	Platelet accelerator	Same as plasma factor V
II	Thrombin accelerator	Accelerates thrombin (intrinsic clotting pathway) and fibrin production
III		Phospholipids necessary for the intrinsic and extrinsic clotting pathways
IV		Binds heparin, which prevents clot formation

Extrinsic Clotting Pathway

The extrinsic clotting pathway is so named because it begins with chemicals that are outside of, or extrinsic to, the blood (see figure 19.11). In stage 1, damaged tissues release a mixture of lipoproteins and phospholipids called **thromboplastin** (throm-bō-plas′tin), also known as **tissue factor (TF),** or factor III. Thromboplastin, in the presence of Ca^{2+}, forms a complex with factor VII, which activates factor X. On the surface of platelets, activated factor X, factor V, platelet phospholipids, and Ca^{2+} complex to form prothrombinase. In stage 2, prothrombinase converts the soluble plasma protein prothrombin into the enzyme thrombin. During stage 3, thrombin converts the soluble plasma protein fibrinogen into the insoluble protein fibrin. Fibrin forms the fibrous network of the clot. Thrombin also stimulates factor XIII activation, which is necessary to stabilize the clot.

Intrinsic Clotting Pathway

The intrinsic clotting pathway is so named because it begins with chemicals that are inside, or intrinsic to, the blood (see figure 19.11). In stage 1, damage to blood vessels can expose collagen in the connective tissue beneath the epithelium lining the blood vessel. When plasma factor XII comes into contact with collagen, factor XII is activated and it stimulates factor XI, which in turn activates factor IX. Activated factor IX joins with factor VIII, platelet phospholipids, and Ca^{2+} to activate factor X. On the surface of platelets, activated factor X, factor V, platelet phospholipids, and Ca^{2+} complex form prothrombinase. Stages 2 and 3 then are activated, and a clot results.

Although once considered distinct pathways, it's now known that the extrinsic pathway can activate the clotting proteins in the intrinsic pathway. The thromboplastin/factor VII complex from the extrinsic pathway can stimulate the formation of activated factors IX in the intrinsic pathway. When tissues are damaged, thromboplastin also rapidly leads to the production of thrombin, which can activate many of the clotting proteins such as factor XI and prothrombinase. Thus, thrombin is part of a positive-feedback system in which thrombin production stimulates the production of additional thrombin. Thrombin also has a positive-feedback effect on coagulation by stimulating platelet activation.

1. The extrinsic clotting pathway of stage 1 starts with thromboplastin, which is released outside of the plasma in damaged tissue.

2. The intrinsic clotting pathway of stage 1 starts when inactive factor XII, which is in the plasma, is activated by coming into contact with a damaged blood vessel.

3. Activation of the extrinsic or intrinsic clotting pathways results in the production of prothrombinase.

4. In stage 2, prothrombin is converted to thrombin by prothrombinase.

5. In stage 3, fibrinogen is converted to fibrin (the clot) by thrombin.

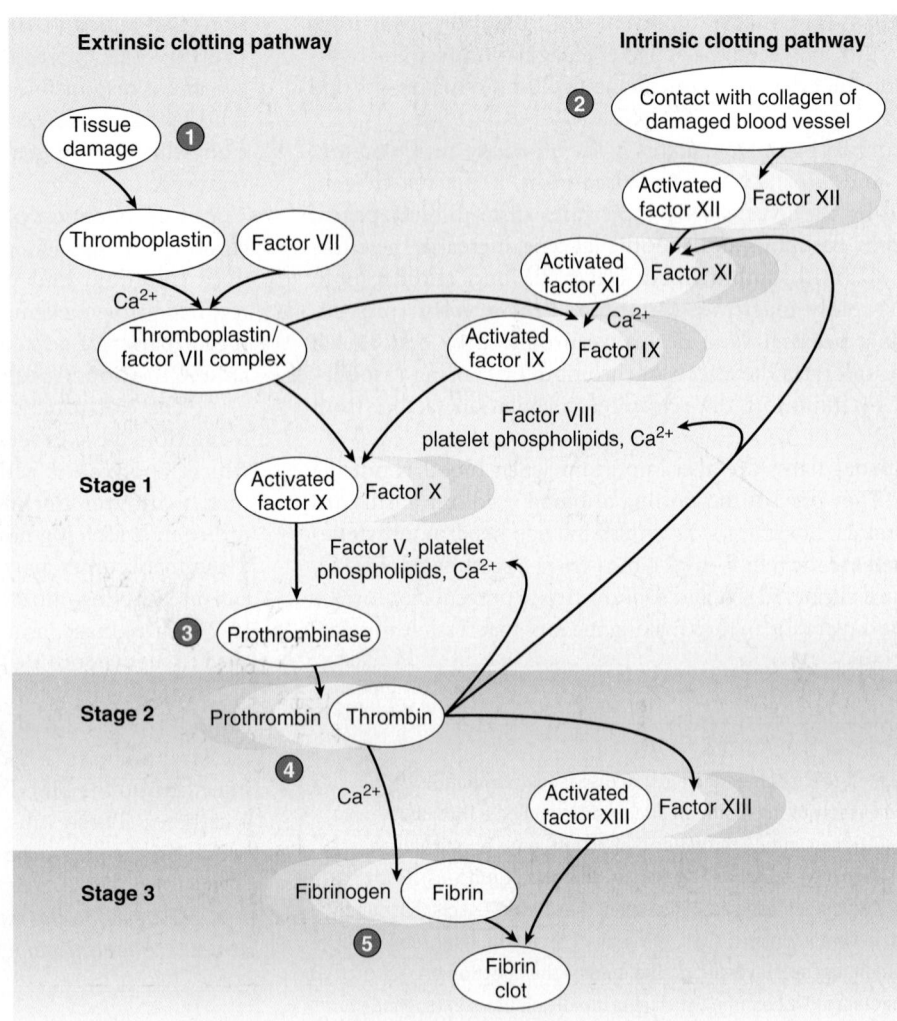

Process Figure 19.11   Clot Formation

Clot formation is divided into three stages. Stage 1 can begin through the extrinsic or intrinsic clotting pathways. A sequence of chemical reactions occurs as activated coagulation factors (*white ovals*) activate inactive coagulation factors (*yellow ovals*).

How Vitamin K Helps to Prevent Bleeding

Many of the factors involved in clot formation require vitamin K for their production (see table 19.3). Humans rely on two sources for vitamin K. About half comes from the diet, and half comes from bacteria within the large intestine. Antibiotics taken to fight bacterial infections sometimes kill these intestinal bacteria, thereby reducing vitamin K levels and resulting in bleeding problems. Vitamin K supplements may be necessary for patients on prolonged antibiotic therapy. Newborns lack these intestinal bacteria, and a vitamin K injection is routinely given to infants at birth. Infants can also obtain vitamin K from food such as milk. Because cow's milk contains more vitamin K than does human milk, breast-fed infants are more susceptible to hemorrhage than bottle-fed infants.

The absorption of vitamin K, which is a fat-soluble vitamin, from the intestine requires the presence of bile. Disorders like obstruction of bile flow to the intestine can interfere with vitamin K absorption and lead to insufficient clotting. Liver diseases that result in the decreased synthesis of clotting factors can also lead to insufficient clot formation.

20. *What is a clot and what is its function?*
21. *What are coagulation factors?*
22. *Clotting is divided into three stages. Describe the final event that occurs in each stage.*
23. *What is the difference between extrinsic and intrinsic activation of clotting?*

Control of Clot Formation

Without control, coagulation would spread from the point of initiation to the entire circulatory system. Furthermore, vessels in a healthy person contain rough areas that can stimulate clot formation, and small amounts of prothrombin are constantly being converted into thrombin. To prevent unwanted clotting, the blood contains several **anticoagulants** (an′tē-kō-ag′ū-lantz), which prevent coagulation factors from initiating clot formation. Only when coagulation factor concentrations exceed a given threshold in a local area does coagulation occur. At the site of injury, so many coagulation factors are activated that the anticoagulants are unable to prevent clot formation. Away from the injury site, however, the activated coagulation factors are diluted in the blood, anticoagulants neutralize them, and clotting is prevented.

Examples of anticoagulants in the blood are antithrombin, heparin, and prostacyclin. **Antithrombin,** a plasma protein produced by the liver, slowly inactivates thrombin. **Heparin,** produced by basophils and endothelial cells, increases the effectiveness of antithrombin because heparin and antithrombin together rapidly inactivate thrombin. **Prostacyclin** (pros-tă-sī′klin) is a prostaglandin derivative produced by endothelial cells. It counteracts the effects of thrombin by causing vasodilation and by inhibiting the release of coagulation factors from platelets.

Anticoagulants are also important when blood is outside the body. They prevent the clotting of blood used in transfusions and laboratory blood tests. Examples include heparin, **ethylenediaminetetraacetic** (eth′il-ēn-dī′ă-mēn-tet-ră-ă-sē′tik) **acid (EDTA),** and sodium citrate. EDTA and sodium citrate prevent clot formation by binding to Ca^{2+}, thus making the ions inaccessible for clotting reactions.

The Danger of Unwanted Clots

When platelets encounter damaged or diseased areas on the walls of blood vessels or the heart, an attached clot called a **thrombus** (throm′bŭs) may form. A thrombus that breaks loose and begins to float through the circulation is called an **embolus** (em′bō-lŭs). Both thrombi and emboli can result in death if they block vessels that supply blood to essential organs, such as the heart, brain, or lungs. Abnormal coagulation can be prevented or hindered by the injection of anticoagulants like heparin, which acts rapidly. Coumadin (koo′mă-din), or warfarin (war′fă-rin), acts more slowly than heparin. Coumadin prevents clot formation by suppressing the production of vitamin K–dependent coagulation factors (II, VII, IX, and X) by the liver. Interestingly, coumadin was first used as a rat poison by causing rats to bleed to death. In small doses, warfarin is a proven, effective anticoagulant in humans. Caution is necessary with anticoagulant treatment, however, because the patient can hemorrhage internally or bleed excessively when cut.

 ● 24. *What is the function of anticoagulants in blood? Name three anticoagulants in blood, and explain how they prevent clot formation.*
   25. *Define the terms thrombus and embolus, and explain why they are dangerous.*

Clot Retraction and Dissolution

The fibrin meshwork constituting the clot adheres to the walls of the blood vessel. Once a clot has formed, it begins to condense into a denser, compact structure through a process known as **clot retraction.** Platelets contain the contractile proteins actin and myosin, which operate in a similar fashion to that of actin and myosin in smooth muscle (see chapter 9). Platelets form exten-

sions that attach to fibrinogen through fibrinogen receptors (see figure 19.9). Contraction of the extensions pulls on the fibrinogen and is responsible for clot retraction. As the clot condenses, a fluid called **serum** (sēr′ŭm) is squeezed out of it. Serum is plasma from which fibrinogen and some of the clotting factors have been removed.

Consolidation of the clot pulls the edges of the damaged blood vessel together, which can help to stop the flow of blood, reduce infection, and enhance healing. The damaged vessel is repaired by the movement of fibroblasts into the damaged area and the formation of new connective tissue. In addition, epithelial cells around the wound proliferate and fill in the torn area.

The clot usually is dissolved within a few days after clot formation by a process called **fibrinolysis** (fī-bri-nol′i-sis), which involves the activity of **plasmin** (plaz′min), an enzyme that hydrolyzes fibrin. Plasmin is formed from inactive plasminogen, which is a normal blood protein produced by the liver. Plasmin becomes part of the clot as it is formed. Plasmin is activated by thrombin, factor XII, tissue plasminogen activator (t-PA), urokinase, and lysosomal enzymes released from damaged tissues (figure 19.12). In disorders that are caused by blockage of a vessel by a clot, such as a heart attack, dissolving the clot can restore blood flow and reduce damage to tissues. For example, streptokinase (a bacterial enzyme), t-PA, or urokinase can be injected into the blood or introduced at the clot site by means of a catheter. These substances convert plasminogen to plasmin, which breaks down the clot.

 ● 26. *Describe clot retraction. What is serum?*
   27. *What is fibrinolysis? How does it occur?*

1. Inactive plasminogen is converted to the active enzyme plasmin.

2. Plasmin breaks the fibrin molecules, and therefore the clot, into smaller pieces, which are washed away in the blood or phagocytized.

Process Figure 19.12   Fibrinolysis

Blood Grouping

If large quantities of blood are lost during surgery or in an accident, the patient can go into shock and die unless a transfusion or infusion is performed. A **transfusion** is the transfer of blood or blood components from one individual to another. When large quantities of blood are lost, red blood cells must be replaced so that the oxygen-carrying capacity of the blood is restored. An **infusion** is the introduction of a fluid other than blood, such as a saline or glucose solution, into the blood. In many cases, the return of blood volume to normal levels is all that is necessary to prevent shock. Eventually, the body produces red blood cells to replace those that were lost.

Early attempts to transfuse blood from one person to another were often unsuccessful because they resulted in transfusion reactions, which included clotting within blood vessels, kidney damage, and death. It's now known that transfusion reactions are caused by interactions between antigens and antibodies (see chapter 22). In brief, the surfaces of red blood cells have molecules called **antigens** (an′ti-jenz), and, in the plasma, molecules called **antibodies** are present. Antibodies are very specific, meaning that each antibody can combine only with a certain antigen. When the antibodies in the plasma bind to the antigens on the surfaces of the red blood cells, they form molecular bridges that connect the red blood cells. As a result, **agglutination** (ă-gloo-ti-nă′shŭn), or clumping, of the cells occurs. The combination of the antibodies with the antigens can also initiate reactions that cause **hemolysis,** or rupture of the red blood cells. Because the antigen–antibody combinations can cause agglutination, the antigens are often called **agglutinogens** (ă-gloo-tin′ō-jenz), and the antibodies are called **agglutinins** (ă-gloo′ti-ninz).

The antigens on the surface of red blood cells have been categorized into **blood groups,** and more than 35 blood groups, most of which are rare, have been identified. For transfusions, the ABO and Rh blood groups are among the most important. Other well-known groups include the Lewis, Duffy, MNSs, Kidd, Kell, and Lutheran groups.

> ● 28. *What are blood groups, and how do they cause transfusion reactions? Define the terms agglutination and hemolysis.*

ABO Blood Group

In the **ABO blood group,** type A blood has type A antigens, type B blood has type B antigens, type AB blood has both A and B antigens, and type O blood has neither A nor B antigens on the surface of red blood cells (figure 19.13). In addition, plasma from type A blood contains anti-B antibodies, which act against type B antigens, whereas plasma from type B blood contains anti-A antibodies, which act against type A antigens. Type AB blood has neither type of antibody, and type O blood has both anti-A and anti-B antibodies.

The ABO blood types are not found in equal numbers. In Caucasians in the United States, the distribution is type O, 47%; type A, 41%; type B, 9%; and type AB, 3%. Among African-Americans, the distribution is type O, 46%; type A, 27%; type B, 20%; and type AB, 7%.

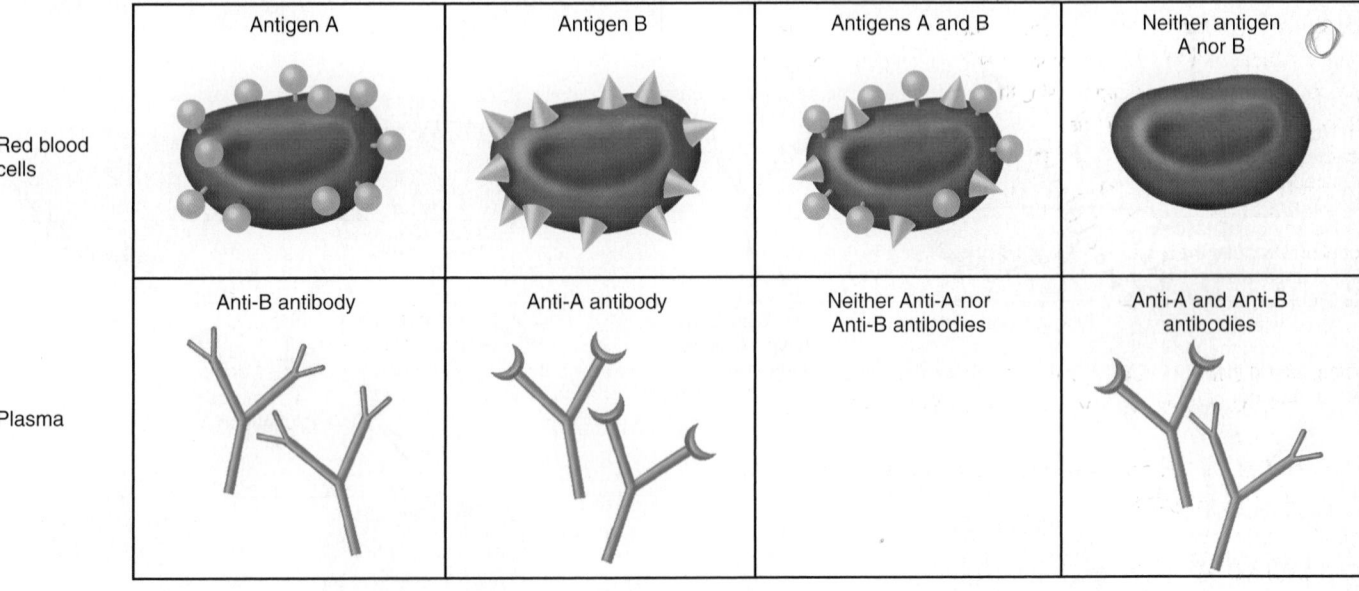

Type A
Red blood cells with type A surface antigens and plasma with anti-B antibodies

Type B
Red blood cells with type B surface antigens and plasma with anti-A antibodies

Type AB
Red blood cells with both type A and type B surface antigens, and neither anti-A nor anti-B plasma antibodies

Type O
Red blood cells with neither type A nor type B surface antigens, but both anti-A and anti-B plasma antibodies

Figure 19.13 ABO Blood Groups

Antibodies normally don't develop against an antigen unless the body is exposed to that antigen. This means, for example, that a person with type A blood should not have anti-B antibodies unless he or she has received a transfusion of type B blood, which contains type B antigens. People with type A blood do have anti-B antibodies, however, even though they have never received a transfusion of type B blood. One possible explanation is that type A or B antigens on bacteria or food in the digestive tract stimulate the formation of antibodies against antigens that are different from one's own antigens. Thus a person with type A blood would produce anti-B antibodies against the B antigens on the bacteria or food. In support of this hypothesis is the observation that anti-A and anti-B antibodies are not found in the blood until about 2 months after birth.

A blood **donor** gives blood, and a **recipient** receives blood. Usually a recipient can receive blood from a donor if they both have the same blood type. For example, a person with type A blood can receive blood from another person with type A blood. No ABO transfusion reaction would occur because the recipient has no anti-A antibodies against the type A antigen. On the other hand, if type A blood were donated to a person with type B blood, a transfusion reaction would occur because the person with type B blood has anti-A antibodies against the type A antigen, and agglutination would result (figure 19.14).

Historically, people with type O blood have been called universal donors because they usually can give blood to the other ABO blood types without causing an ABO transfusion reaction. Their red blood cells have no ABO surface antigens and, therefore, do not react with the recipient's anti-A or anti-B antibodies. For example, if type O blood is given to a person with type A blood, the type O red blood cells do not react with the anti-B antibodies

in the recipient's blood. In a similar fashion, if type O blood is given to a person with type B blood, no reaction occurs to the recipient's anti-A antibodies.

The term *universal donor* is misleading, however. Transfusion of type O blood, in some cases, produces a transfusion reaction for two reasons. First, other blood groups can cause a transfusion reaction. Second, antibodies in the blood of the donor can react with antigens in the blood of the recipient. For example, type O blood has anti-A and anti-B antibodies. If type O blood is transfused into a person with type A blood, the anti-A antibodies (in the type O blood) react against the A antigens (in the type A blood). Usually such reactions are not serious because the antibodies in the donor's blood are diluted in the larger volume of the recipient's blood, and few reactions take place. Because type O blood sometimes causes transfusion reactions, it's given to a person with another blood type only in life-or-death emergency situations.

29. What kinds of antigens and antibodies are found in each of the four ABO blood types?

30. Why is a person with type O blood considered to be a universal donor?

P R E D I C T

Historically, people with type AB blood were called universal recipients. What is the rationale for this term? Explain why the term is misleading.

Rh Blood Group

Another important blood group is the **Rh blood group,** so named because it was first studied in rhesus monkeys. People are Rh-positive if they have certain Rh antigens (the D antigens) on the

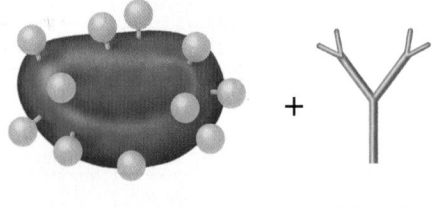

(a) **No agglutination reaction.** Type A blood donated to a type A recipient does not cause an agglutination reaction because the anti-B antibodies in the recipient do not combine with the type A antigens on the red blood cells in the donated blood.

Type A blood of donor

Anti-B antibody in type A blood of recipient

Antigen and antibody do not match

No agglutination

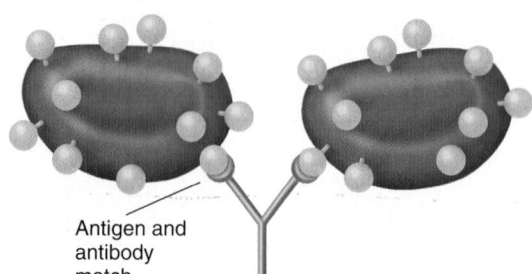

(b) **Agglutination reaction.** Type A blood donated to a type B recipient causes an agglutination reaction because the anti-A antibodies in the recipient combine with the type A antigens on the red blood cells in the donated blood.

Type A blood of donor

Anti-A antibody in type B blood of recipient

Antigen and antibody match

Agglutination

Figure 19.14 Agglutination Reaction

surface of their red blood cells, and people are Rh-negative if they do not have these Rh antigens. About 85% of Caucasians in the United States and 88% of African-Americans are Rh-positive. The ABO blood type and the Rh blood type usually are designated together. For example, a person designated as A positive is type A in the ABO blood group and Rh-positive. The rarest combination in the United States is AB negative, which occurs in less than 1% of all Americans.

Antibodies against the Rh antigen do not develop unless an Rh-negative person is exposed to Rh-positive blood. This can occur through a transfusion or by transfer of blood between a mother and her fetus across the placenta. When an Rh-negative person receives a transfusion of Rh-positive blood, the recipient becomes sensitized to the Rh antigen and produces anti-Rh antibodies. If the Rh-negative person is unfortunate enough to receive a second transfusion of Rh-positive blood after becoming sensitized, a transfusion reaction results.

Rh incompatibility can pose a major problem in some pregnancies when the mother is Rh-negative and the fetus is Rh-positive (figure 19.15). If fetal blood leaks through the placenta and mixes with the mother's blood, the mother becomes sensitized to the Rh antigen. The mother produces anti-Rh antibodies that cross the placenta and cause agglutination and hemolysis of fetal red blood cells. This disorder is called **hemolytic disease of the newborn (HDN),** or **erythroblastosis fetalis** (ĕ-rith′rō-blas-tō′sis fē-ta′lis), and it may be fatal to the fetus. In the mother's first

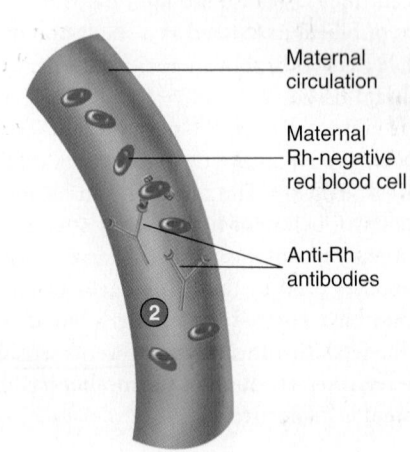

1. Before or during delivery, Rh-positive red blood cells from the fetus enter the blood of an Rh-negative woman through a tear in the placenta.

2. The mother is sensitized to the Rh antigen and produces anti-Rh antibodies. Because this usually happens after delivery, there is no effect on the fetus in the first pregnancy.

3. During a subsequent pregnancy with an Rh-positive fetus, Rh-positive red blood cells cross the placenta, enter the maternal circulation, and stimulate the mother to produce antibodies against the Rh antigen. Antibody production is rapid because the mother has been sensitized to the Rh antigen.

4. The anti-Rh antibodies from the mother cross the placenta, causing agglutination and hemolysis of fetal red blood cells, and hemolytic disease of the newborn (HDN) develops.

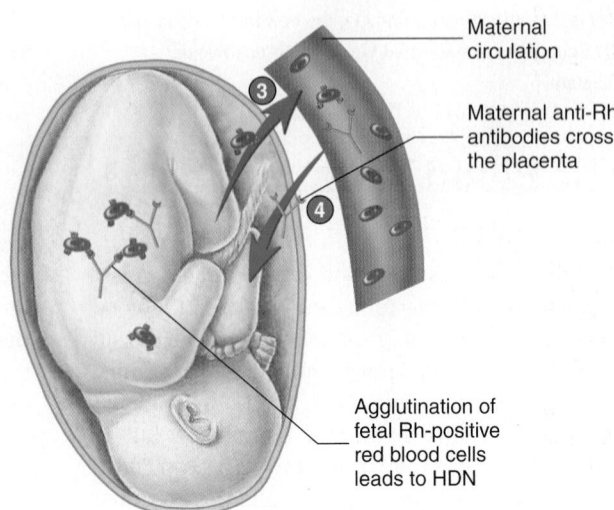

Process Figure 19.15   Hemolytic Disease of the Newborn (HDN)

pregnancy, however, usually no problem occurs. The leakage of fetal blood is usually the result of a tear in the placenta that takes place either late in the pregnancy or during delivery. Thus, not enough time exists for the mother to produce enough anti-Rh antibodies to harm the fetus. In later pregnancies, however, a problem can arise because the mother has already been sensitized to the Rh antigen. Consequently, if the fetus is Rh-positive and if any leakage of fetal blood into the mother's blood occurs, she rapidly produces large amounts of anti-Rh antibodies, and HDN develops. Therefore, the levels of anti-Rh antibodies in the mother should be monitored. If they increase to unacceptable levels, the fetus should be tested to determine the severity of the HDN. In severe cases, a transfusion to replace lost red blood cells can be performed through the umbilical cord, or the baby can be delivered if mature enough.

After birth, treatment of HDN consists of slowly removing the newborn's blood and replacing it. The newborn can also be exposed to fluorescent light, because the light helps to break down the large amounts of bilirubin formed as a result of red blood cell destruction. High levels of bilirubin are toxic to the nervous system and can damage brain tissue.

HDN can be prevented if the Rh-negative mother is given an injection of a specific type of antibody preparation, called $Rh_o(D)$ immune globulin (RhoGAM). The injection can be administered during the pregnancy or before or immediately after each delivery or abortion. The injection contains antibodies against Rh antigens. The injected antibodies bind to the Rh antigens of any fetal red blood cells that may have entered the mother's blood. This treatment inactivates the fetal Rh antigens and prevents sensitization of the mother. However, if sensitization of the mother has already occurred, the treatment is ineffective.

31. What is meant by the term Rh-positive?
32. What Rh blood types must the mother and fetus have before HDN can occur?
33. How is HDN harmful to the fetus?
34. Why doesn't HDN usually develop in the first pregnancy?
35. How can HDN be prevented? How is HDN treated?

P R E D I C T   7
In HDN, what is the rationale for removing the newborn's blood and replacing it? Should the replacement blood be Rh-positive or negative? Explain.

Diagnostic Blood Tests

Type and Crossmatch

To prevent transfusion reactions the blood is typed, and a **crossmatch** is made. **Blood typing** determines the ABO and Rh blood groups of the blood sample. Typically, the cells are separated from the serum. The cells are tested with known antibodies to determine the type of antigen on the cell surface. For example, if a patient's blood cells agglutinate when mixed with anti-A antibodies but do not agglutinate when mixed with anti-B antibodies, it's concluded that the cells have type A antigen. In a similar fashion, the serum is mixed with known cell types (antigens) to determine the type of antibodies in the serum.

Normally, donor blood must match the ABO and Rh type of the recipient. Because other blood groups can also cause a transfusion reaction, however, a crossmatch is performed. In a crossmatch, the donor's blood cells are mixed with the recipient's serum, and the donor's serum is mixed with the recipient's cells. The donor's blood is considered safe for transfusion only if no agglutination occurs in either match.

Complete Blood Count

The **complete blood count (CBC)** is an analysis of the blood that provides much information. It consists of a red blood count, hemoglobin and hematocrit measurements, a white blood count, and a differential white blood count.

Red Blood Count

Blood cell counts usually are done automatically with an electronic instrument, but they can also be done manually with a microscope. The normal range for a **red blood count (RBC)** is the number (expressed in millions) of red blood cells per microliter of blood. It is 4.6–6.2 million/μL of blood for a male, and 4.2–5.4 million/μL of blood for a female. **Erythrocytosis** (ĕ-rith′rō-sī-tō′sis) is an overabundance of red blood cells (see Clinical Focus "Disorders of the Blood" on p. 672).

Hemoglobin Measurement

The **hemoglobin measurement** determines the amount of hemoglobin in a given volume of blood, usually expressed as grams of hemoglobin per 100 mL of blood. The normal hemoglobin count for a male is 14–18 g/100 mL of blood, and for a female it is 12–16 g/100 mL of blood. Abnormally low hemoglobin is an indication of **anemia** (ă-nē′mē-ă), which is a reduced number of red blood cells per 100 mL of blood or a reduced amount of hemoglobin in each red blood cell (see Clinical Focus "Disorders of the Blood" on p. 672).

Hematocrit Measurement

The percentage of total blood volume composed of red blood cells is the **hematocrit** (hē′mă-tō-krit, hem′ă-tō-krit). One way to determine hematocrit is to place blood in a tube and spin the tube in a centrifuge. The formed elements are heavier than the plasma and are forced to one end of the tube (figure 19.16). White blood cells and platelets form a thin, whitish layer, called the **buffy coat,** between the plasma and the red blood cells. The red blood cells account for 40%–54% of the total blood volume in males and 38%–47% in females.

The number and size of red blood cells affect the hematocrit measurement. **Normocytes** (nōr′mō-sītz) are normal sized red blood cells with a diameter of 7.5 μm. **Microcytes** (mī′krō-sītz) are smaller than normal with a diameter of 6 μm or less, and **macrocytes** (mak′krō-sītz) are larger than normal with a diameter 9 μm or greater. Blood disorders can result in abnormal hematocrit measurement because they cause red blood cells numbers to be abnormally high or low, or cause red blood cells to be abnormally small or large (see "Disorders of the Blood" on p. 672). A decreased hematocrit indicates that the volume of red blood cells is less than normal. It can result from a decreased number of normocytes or a normal number of microcytes. For example, inadequate

Centrifuge blood in the hematocrit tube

Hematocrit tube

Plasma

White blood cells and platelets form the buffy coat

Red blood cells

Withdraw blood into hematocrit tube

Male Female

Figure 19.16 Hematocrit

Blood is withdrawn into a capillary tube and placed in a centrifuge. The blood is separated into plasma, red blood cells, and a small amount of white blood cells and platelets, which rest on the red blood cells. The hematocrit measurement is the percent of the blood volume that is red blood cells. It doesn't measure the white blood cells and platelets. Normal hematocrits for a male and a female are shown.

iron in the diet can impair hemoglobin production. Consequently, during their formation red blood cells do not fill with hemoglobin, and they remain smaller than normal.

White Blood Count

A **white blood count (WBC)** measures the total number of white blood cells in the blood. Normally 5000–10,000 white blood cells are present in each microliter of blood. **Leukopenia** (loo-kō-pē′nē-ă) is a lower-than-normal WBC and can indicate depression or destruction of the red marrow by radiation, drugs, tumor, or a deficiency of vitamin B_{12} or folate. **Leukocytosis** (loo′kō-sī-tō′sis) is an abnormally high WBC. **Leukemia** (loo-kē′mē-ă) (a cancer of the red marrow) often results in leukocytosis, but the white blood cells have an abnormal structure and function. Bacterial infections also can cause leukocytosis by stimulating neutrophils to increase in number.

Differential White Blood Count

A **differential white blood count** determines the percentage of each of the five kinds of white blood cells in the WBC. Normally neutrophils account for 60%–70%; lymphocytes, 20%–30%; monocytes, 2%–8%; eosinophils, 1%–4%; and basophils, 0.5%–1%. A differential WBC can provide much insight about a patient's condition. For example, in patients with bacterial infections the neutrophil

count is often greatly increased, whereas in patients with allergic reactions the eosinophil and basophil counts are elevated.

Clotting

Two measurements that test the ability of the blood to clot are the platelet count and prothrombin time.

Platelet Count

A normal **platelet count** is 250,000–400,000 platelets per microliter of blood. **Thrombocytopenia** (throm′bō-sī-tō-pē′nē-ă) is a condition in which the platelet count is greatly reduced, resulting in chronic bleeding through small vessels and capillaries. It can be caused by decreased platelet production as a result of hereditary disorders, lack of vitamin B_{12}, drug therapy, or radiation therapy.

Prothrombin Time Measurement

Prothrombin time measurement is a measure of how long it takes for the blood to start clotting, which normally is 9–12 seconds. Prothrombin time is determined by adding thromboplastin to whole plasma. Thromboplastin is a chemical released from injured tissues that starts the process of clotting (see figure 19.11). Prothrombin time is officially reported as the International Normalized Ratio (INR), which standardizes the time it takes to clot based on the slightly different thromboplastins used by different labs. Because many clotting factors must be activated to form fibrin, a deficiency of any one of them can cause an abnormal prothrombin time. Vitamin K deficiency, certain liver diseases, and drug therapy can cause an increased prothrombin time.

Blood Chemistry

The composition of materials dissolved or suspended in the plasma can be used to assess the functioning of many of the body's systems (Appendix E). For example, high blood glucose levels can indicate that the pancreas is not producing enough insulin; high blood urea nitrogen (BUN) can be a sign of reduced kidney function; increased bilirubin can indicate liver dysfunction or hemolysis; and high cholesterol levels can indicate an increased risk of developing cardiovascular disease. A number of blood chemistry tests are routinely done when a blood sample is taken, and additional tests are available.

> **36.** For each of the following tests, define the test and give an example of a disorder that would cause an abnormal test result:
> a. red blood count
> b. hemoglobin measurement
> c. hematocrit measurement
> d. white blood count
> e. differential white blood count
> f. platelet count
> g. prothrombin time measurement
> h. blood chemistry tests

P R E D I C T **8**

When a patient complains of acute pain in the abdomen, the physician suspects appendicitis, which is often caused by a bacterial infection of the appendix. What blood test should be done to support the diagnosis?

Clinical Focus   Disorders of the Blood

Erythrocytosis

Erythrocytosis (ĕ-rith′rō-sī-tō′sis) is an over-abundance of red blood cells, resulting in increased blood viscosity, reduced flow rates, and, if severe, plugging of the capillaries. **Relative erythrocytosis** results from decreased plasma volume, such as that caused by dehydration, diuretics, and burns. **Primary erythrocytosis**, often called **polycythemia vera** (pol′ē-sī-thē′mē-ă ve′ra), is a stem cell defect of unknown cause that results in the overproduction of red blood cells, granulocytes, and platelets. Erythropoietin levels are low and the spleen can be enlarged. **Secondary erythrocytosis (polycythemia)** results from a decreased oxygen supply, such as that which occurs at high altitudes, in chronic obstructive pulmonary disease, or in congestive heart failure. The resulting decrease in oxygen delivery to the kidneys stimulates erythropoietin secretion and causes an increase in red blood cell production. In primary and secondary erythrocytosis, the increased number of red blood cells increases blood viscosity and blood volume. There can be clogging of capillaries and the development of hypertension.

Anemia

Anemia (ă-nē′mē-ă) is a deficiency of hemoglobin in the blood. It can result from a decrease in the number of red blood cells, a decrease in the amount of hemoglobin in each red blood cell, or both. The decreased hemoglobin reduces the ability of the blood to transport oxygen. Anemic patients suffer from a lack of energy and feel excessively tired and listless. They can appear pale and quickly become short of breath with only slight exertion.

One general cause of anemia is nutritional deficiencies. **Iron-deficiency anemia** results from a deficient intake or absorption of iron or from excessive iron loss. Consequently, not enough hemoglobin is produced, and the red blood cells are smaller than normal (microcytic). **Folate deficiency** can also cause anemia. An inadequate amount of folate in the diet is the usual cause of folate deficiency, with the disorder developing most often in the poor, in preg-

nant women, and in chronic alcoholics. Because folate helps in the synthesis of DNA, folate deficiency results in fewer cell divisions. There is decreased red blood cell production, but the cells grow larger than normal (macrocytic). Another type of nutritional anemia is **pernicious** (per-nish′ŭs) **anemia,** which is caused by inadequate amounts of vitamin B_{12}. Because vitamin B_{12} is important for folate synthesis, inadequate amounts of vitamin B_{12} causes a secondary folate deficiency. Thus, a deficiency of vitamin B_{12} also causes a decreased production of red blood cells that are larger than normal. Although inadequate levels of vitamin B_{12} in the diet can cause pernicious anemia, the usual cause is insufficient absorption of the vitamin. Normally the stomach produces intrinsic factor, a protein that binds to vitamin B_{12}. The combined molecules pass into the small intestine, where intrinsic factor facilitates the absorption of the vitamin. Without adequate levels of intrinsic factor, insufficient vitamin B_{12} is absorbed, and pernicious anemia develops. It may take 2–3 years for pernicious anemia to develop because a 2- to 3-year supply of vitamin B_{12} can be stored in the liver. Present evidence suggests that the inability to produce intrinsic factor is due to an autoimmune disease in which the body's immune system damages the cells in the stomach that produce intrinsic factor.

Another general cause of anemia is loss or destruction of red blood cells. **Hemorrhagic** (hem-ŏ-raj′ik) **anemia** results from a loss of blood, such as can result from trauma, ulcers, or excessive menstrual bleeding. Chronic blood loss, in which small amounts of blood are lost over time, can result in iron-deficiency anemia. **Hemolytic** (hē-mō-lit′ik) **anemia** is a disorder in which red blood cells rupture or are destroyed at an excessive rate. It can be caused by inherited defects within the red blood cells. For example, one kind of inherited hemolytic anemia results from a defect in the plasma membrane that causes red blood cells to rupture easily. Many kinds of hemolytic anemia result from unusual damage to the red blood cells by drugs, snake venom, artificial

heart valves, autoimmune disease, or hemolytic disease of the newborn.

Aplastic anemia is caused by an inability of the red bone marrow to produce red blood cells. It's usually acquired as a result of damage to the red marrow by chemicals (e.g., benzene), drugs (e.g., certain antibiotics and sedatives), or radiation. Aplastic anemia is usually associated with an inability to produce granulocytes and platelets.

Some anemias result from inadequate or defective hemoglobin production. **Thalassemia** (thal-ă-sē′mē-ă) is an autosomal recessive hereditary disease found predominantly in people of Mediterranean, Asian, and African ancestry. It's caused by insufficient production of the globin part of the hemoglobin molecule. The major form of the disease results in death by age 20, the minor form in a mild anemia. **Sickle-cell disease** is an autosomal recessive hereditary disease found mostly in people of African ancestry but also occasionally among people of Mediterranean heritage. It results in the formation of an abnormal hemoglobin, in which the red blood cells assume a rigid sickle shape and plug up small blood vessels (figure A). They are also more fragile than normal red blood cells. In its severe form, sickle-cell disease is usually fatal before the person is 30 years of age, whereas in its minor form, sickle-cell trait, symptoms usually do not occur.

Von Willebrand's Disease

Von Willebrand's disease is an autosomal dominant hereditary disease. It is the most common inherited bleeding disorder, occurring as frequently as 1 in 1000 individuals. Von Willebrand factor (vWF) helps platelets to stick to collagen (platelet adhesion) and is the plasma carrier for factor VIII (see discussion on "Coagulation" on p. 663 and table 19.3). One treatment for von Willebrand's disease involves injections of vWF or concentrates of factor VIII to which vWF is attached. Another therapeutic approach is to administer a drug that increases vWF levels in the blood.

Hemophilia

Hemophilia (hē-mō-fil′ē-ă) is a genetic disorder in which clotting is abnormal or

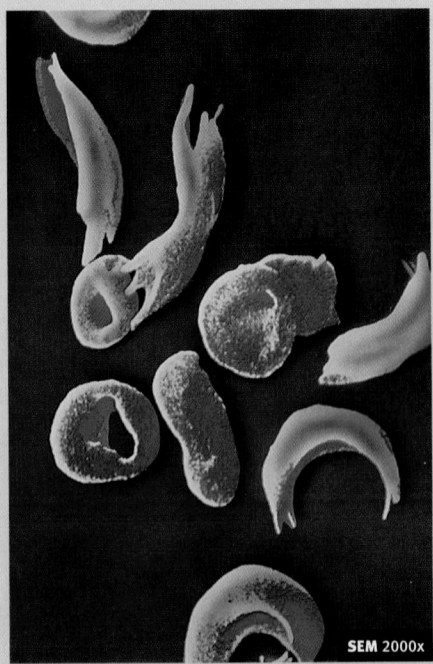

Figure A Sickle-Cell Disease

Red blood cells in a person with sickle-cell disease appear normal in oxygenated blood. In deoxygenated blood, hemoglobin changes shape and causes the cells to become sickle-shaped and rigid.

absent. It's most often found in people from northern Europe and their descendants. Because hemophilia is an X-linked trait (see chapter 29), it occurs almost exclusively in males. **Hemophilia A** (classic hemophilia) results from a deficiency of plasma coagulation factor VIII, and **hemophilia B** is caused by a deficiency in plasma factor IX. Hemophilia A occurs in approximately 1 in 10,000 male births, and hemophilia B occurs in approximately 1 in 100,000 male births. Treatment of hemophilia involves injection of the missing clotting factor taken from donated blood.

Disseminated Intravascular Coagulation

Disseminated intravascular coagulation (DIC) is a complex disorder involving clotting throughout the vascular system followed by bleeding. Normally excessive clotting is prevented by anticoagulants. DIC can develop when these control mechanisms are over-whelmed. Many conditions can cause DIC by overstimulating blood clotting. Examples include massive tissue damage, such as burns, or alteration of the lining of blood vessels caused by infections or snake bites. If DIC occurs slowly, the predominant effect is thrombosis and blockage of blood vessels. If DIC occurs rapidly, massive clot formation occurs, quickly using up available blood clotting factors and platelets. The result is continual bleeding around wounds, intravenous lines, catheters, and internal bleeding. The best therapy for DIC is to treat the condition that is causing it.

Thrombocytopenia

Thrombocytopenia (throm′bō-sī-tō-pē′nē-ă) is a condition in which the number of platelets is greatly reduced, resulting in chronic bleeding through small vessels and capillaries. Thrombocytopenia has several causes, including increased platelet destruction, caused by autoimmune disease (see chapter 22) or infections, and decreased platelet production, resulting from hereditary disorders, pernicious anemia, drug therapy, radiation therapy, or leukemias.

Leukemia

The leukemias are cancers of the red bone marrow in which abnormal production of one or more of the white blood cell types occur. Because these cells are usually immature or abnormal and lack their normal immunologic functions, patients are very susceptible to infections. The excess production of white blood cells in the red marrow can also interfere with red blood cell and platelet formation and thus lead to anemia and bleeding.

Infectious Diseases of the Blood

Microorganisms don't normally survive in the blood. Blood can transport microorganisms, however, and they can multiply in the blood. Microorganisms can enter the body and be transported by the blood to the tissues they infect. For example, the poliomyelitis virus enters through the gastrointestinal tract and is carried to nervous tissue. After microorganisms are established at a site of infection, some can enter the blood. They can then be transported to other locations in the body, multiply within the blood, or be eliminated by the body's immune system.

Septicemia (sep-ti-sē′mē-ă), or blood poisoning, is the spread of microorganisms and their toxins by the blood. Often septicemia results from the introduction of microorganisms by a medical procedure, such as the insertion of an intravenous tube into a blood vessel. The release of toxins by microorganisms can cause septic shock, which is a decrease in blood pressure that can result in death.

In a few diseases, microorganisms actually multiply within blood cells. **Malaria** (mă-lār′ē-ă) is caused by a protozoan (*Plasmodium*) that is introduced into the blood by the bite of the *Anopheles* mosquito. Part of the development of the protozoan occurs inside red blood cells. The symptoms of chills and fever in malaria are produced by toxins released when the protozoan causes the red blood cells to rupture. **Infectious mononucleosis** (mon′ō-noo-klē-ō′sis) is caused by a virus (Epstein-Barr virus) that infects lymphocytes (B cells). The lymphocytes are altered by the virus, and the immune system attacks and destroys the lymphocytes. The immune system response is believed to produce the symptoms of fever, sore throat, and swollen lymph nodes. **Acquired immunodeficiency syndrome (AIDS)** is caused by the human immunodeficiency virus (HIV), which infects lymphocytes and suppresses the immune system (see chapter 22).

The presence of microorganisms in the blood is a concern with blood transfusions, because it's possible to infect the blood recipient. Blood is routinely tested, especially for AIDS and hepatitis, in an effort to eliminate this risk. One cause of **hepatitis** (hep-ă-tī′tis) is an infection of the liver caused by viruses (see chapter 24). After recovering, hepatitis victims can become virus carriers. Although they show no signs of the disease, they release the virus into their blood or bile. To prevent infection of others, anyone who has had hepatitis is asked not to donate blood products.

S U M M A R Y

Functions of Blood   (p. 652)

1. Blood transports gases, nutrients, waste products, and hormones.
2. Blood is involved in the regulation of homeostasis and the maintenance of pH, body temperature, fluid balance, and electrolyte levels.
3. Blood protects against disease and blood loss.

Plasma   (p. 653)

1. Plasma is mostly water (91%) and contains proteins, such as albumin (maintains osmotic pressure), globulins (function in transport and immunity), fibrinogen (involved in clot formation), and hormones and enzymes (involved in regulation).
2. Plasma also contains ions, nutrients, waste products, and gases.

Formed Elements   (p. 654)

The formed elements include red blood cells (erythrocytes), white blood cells (leukocytes), and platelets (cell fragments).

Production of Formed Elements

1. In the embryo and fetus, the formed elements are produced in a number of locations.
2. After birth, red bone marrow becomes the source of the formed elements.
3. All formed elements are derived from stem cells.

Red Blood Cells

1. Red blood cells are biconcave discs containing hemoglobin and carbonic anhydrase.
   - A hemoglobin molecule consists of four heme and four globin molecules. The heme molecules transport oxygen, and the globin molecules transport carbon dioxide and nitric oxide. Iron is required for oxygen transport.
   - Carbonic anhydrase is involved with the transport of carbon dioxide.
2. Erythropoiesis is the production of red blood cells.
   - Stem cells in red bone marrow eventually give rise to late erythroblasts, which lose their nuclei and are released into the blood as reticulocytes. Loss of the endoplasmic reticulum by a reticulocyte produces a red blood cell.
   - In response to low blood oxygen, the kidneys produce erythropoietin, which stimulates erythropoiesis.
3. Hemoglobin from ruptured red blood cells is phagocytized by macrophages. The hemoglobin is broken down, and heme becomes bilirubin, which is secreted in bile.

White Blood Cells

1. White blood cells protect the body against microorganisms and remove dead cells and debris.
2. Five types of white blood cells exist.
   - Neutrophils are small phagocytic cells.
   - Eosinophils function to reduce inflammation.
   - Basophils release histamine and are involved with increasing the inflammatory response.
   - Lymphocytes are important in immunity, including the production of antibodies.
   - Monocytes leave the blood, enter tissues, and become large phagocytic cells called macrophages.

Platelets

Platelets, or thrombocytes, are cell fragments pinched off from megakaryocytes in the red bone marrow.

Hemostasis   (p. 662)

Hemostasis, the stoppage of bleeding, is very important to the maintenance of homeostasis.

Vascular Spasm

Vasoconstriction of damaged blood vessels reduces blood loss.

Platelet Plug Formation

1. Platelets repair minor damage to blood vessels by forming platelet plugs.
   - In platelet adhesion, platelets bind to collagen in damaged tissues.
   - In the platelet release reaction, platelets release chemicals that activate additional platelets.
   - In platelet aggregation, platelets bind to one another to form a platelet plug.
2. Platelets also release chemicals involved with coagulation.

Coagulation

1. Coagulation is the formation of a blood clot.
2. Coagulation consists of three stages.
   - Activation of prothrombinase.
   - Conversion of prothrombin to thrombin by prothrombinase.
   - Conversion of fibrinogen to fibrin by thrombin. The insoluble fibrin forms the clot.
3. The first stage of coagulation occurs through the extrinsic or intrinsic clotting pathway. Both pathways end with the production of prothrombinase.
   - The extrinsic clotting pathway begins with the release of thromboplastin from damaged tissues.
   - The intrinsic clotting pathway begins with the activation of factor XII.

Control of Clot Formation

1. Heparin and antithrombin inhibit thrombin activity. Fibrinogen is, therefore, not converted to fibrin, and clot formation is inhibited.
2. Prostacyclin counteracts the effects of thrombin.

Clot Retraction and Dissolution

1. Clot retraction results from the contraction of platelets, which pull the edges of damaged tissue closer together.
2. Serum, which is plasma minus fibrinogen and some clotting factors, is squeezed out of the clot.
3. Factor XII, thrombin, tissue plasminogen activator, and urokinase activate plasmin, which dissolves fibrin (the clot).

Blood Grouping   (p. 667)

1. Blood groups are determined by antigens on the surface of red blood cells.
2. Antibodies can bind to red blood cell antigens, resulting in agglutination or hemolysis of red blood cells.

ABO Blood Group

1. Type A blood has A antigens, type B blood has B antigens, type AB blood has A and B antigens, and type O blood has neither A nor B antigens.
2. Type A blood has anti-B antibodies, type B blood has anti-A antibodies, type AB blood has neither anti-A nor anti-B antibodies, and type O blood has both anti-A and anti-B antibodies.
3. Mismatching the ABO blood group results in transfusion reactions.

Rh Blood Group

1. Rh-positive blood has certain Rh antigens (the D antigens), whereas Rh-negative blood does not.

2. Antibodies against the Rh antigen are produced by an Rh-negative person when the person is exposed to Rh-positive blood.
3. The Rh blood group is responsible for hemolytic disease of the newborn.

Diagnostic Blood Tests   (p. 670)
Type and Crossmatch
Blood typing determines the ABO and Rh blood groups of a blood sample. A crossmatch tests for agglutination reactions between donor and recipient blood.

Complete Blood Count
The complete blood count consists of the following: red blood count, hemoglobin measurement (grams of hemoglobin per 100 mL of blood), hematocrit measurement (percent volume of red blood cells), and white blood count.

Differential White Blood Count
The differential white blood count determines the percentage of each type of white blood cell.

Clotting
Platelet count and prothrombin time measure the ability of the blood to clot.

Blood Chemistry
The composition of materials dissolved or suspended in plasma (e.g., glucose, urea nitrogen, bilirubin, and cholesterol) can be used to assess the functioning and status of the body's systems.

REVIEW AND COMPREHENSION

1. Which of these is a function of blood?
   a. clot formation
   b. protection against foreign substances
   c. maintenance of body temperature
   d. regulation of pH and osmosis
   e. all of the above

2. Which of these is *not* a component of plasma?
   a. nitrogen
   b. sodium ions
   c. platelets
   d. water
   e. urea

3. Which of these plasma proteins plays an important role in maintaining the osmotic concentration of the blood?
   a. albumin
   b. fibrinogen
   c. platelets
   d. hemoglobin
   e. globulins

4. The cells that give rise to the red blood cells are
   a. lymphoblasts.
   b. megakaryoblasts.
   c. monoblasts.
   d. myeloblasts.
   e. proerythroblasts.

5. Red blood cells
   a. are the least numerous formed element in the blood.
   b. are phagocytic cells.
   c. are produced in the yellow marrow.
   d. do not have a nucleus.
   e. all of the above.

6. Given these ways of transporting carbon dioxide in the blood:
   1. bicarbonate ions
   2. combined with blood proteins
   3. dissolved in plasma

   Choose the arrangement that lists them in the correct order from largest to smallest percentage of carbon dioxide transported.
   a. 1, 2, 3
   b. 1, 3, 2
   c. 2, 3, 1
   d. 2, 1, 3
   e. 3, 1, 2

7. Which of these components of a red blood cell is correctly matched with its function?
   a. heme group of hemoglobin—oxygen transport
   b. globin portion of hemoglobin—carbon dioxide transport
   c. carbonic anhydrase—carbon dioxide transport
   d. cysteine on β-globin—nitric oxide transport
   e. all of the above

8. Each hemoglobin molecule can become associated with _____ oxygen molecules.
   a. one
   b. two
   c. three
   d. four
   e. unlimited

9. Which of these substances is *not* required for normal red blood cell production?
   a. folate
   b. vitamin K
   c. iron
   d. vitamin B_{12}

10. Erythropoietin
    a. is produced mainly by the heart.
    b. inhibits the production of red blood cells.
    c. production increases when blood oxygen decreases.
    d. production is inhibited by testosterone.
    e. all of the above.

11. Which of these changes occurs in the blood in response to the initiation of a vigorous exercise program?
    a. increased erythropoietin production
    b. increased concentration of reticulocytes
    c. decreased bilirubin formation
    d. both a and b
    e. all of the above

12. Which of the components of hemoglobin is correctly matched with its fate following the destruction of a red blood cell?
    a. heme: reused to form a new hemoglobin molecule
    b. globin: broken down into amino acids
    c. iron: mostly secreted in bile
    d. all of the above

13. If you live near sea level and are training for a track meet in Denver (5280 ft elevation), you would want to spend a few weeks before the meet training at
    a. sea level.
    b. an altitude similar to Denver's.
    c. a facility with a hyperbaric chamber.
    d. any location—it doesn't matter.
14. The blood cells that function to inhibit inflammation are
    a. eosinophils.
    b. basophils.
    c. neutrophils.
    d. monocytes.
    e. lymphocytes.
15. The most numerous type of white blood cell, whose primary function is phagocytosis, is
    a. eosinophils.
    b. basophils.
    c. neutrophils.
    d. monocytes.
    e. lymphocytes.
16. Monocytes
    a. are the smallest white blood cells.
    b. increase in number during chronic infections.
    c. give rise to neutrophils.
    d. produce antibodies.
17. The white blood cells that release large amounts of histamine and heparin are
    a. eosinophils.
    b. basophils.
    c. neutrophils.
    d. monocytes.
    e. lymphocytes.
18. The smallest white blood cells, which include B cells and T cells, are
    a. eosinophils.
    b. basophils.
    c. neutrophils.
    d. monocytes.
    e. lymphocytes.
19. Platelets
    a. are derived from megakaryocytes.
    b. are cell fragments.
    c. have surface molecules that attach to collagen.
    d. play an important role in clot formation.
    e. all of the above.
20. Given these processes in platelet plug formation:
    1. platelet adhesion
    2. platelet aggregation
    3. platelet release reaction

    Choose the arrangement that lists the processes in the correct order after a blood vessel is damaged.
    a. 1, 2, 3
    b. 1, 3, 2
    c. 3, 1, 2
    d. 3, 2, 1
    e. 2, 3, 1
21. A constituent of blood plasma that forms the network of fibers in a clot is
    a. fibrinogen.
    b. tissue factor.
    c. platelets.
    d. thrombin.
    e. prothrombinase.

22. Given these chemicals:
    1. activated factor XII
    2. fibrinogen
    3. prothrombinase
    4. thrombin
    Choose the arrangement that lists the chemicals in the order they are used during clot formation.
    a. 1, 3, 4, 2
    b. 2, 3, 4, 1
    c. 3, 2, 1, 4
    d. 3, 1, 2, 4
    e. 3, 4, 2, 1
23. The extrinsic clotting pathway
    a. begins with the release of thromboplastin (tissue factor).
    b. leads to the production of prothrombinase.
    c. requires Ca^{2+}.
    d. all of the above.
24. Which of these is *not* an anticoagulant found in the blood?
    a. ethylenediamenetetraacetic acid (EDTA)
    b. antithrombin
    c. heparin
    d. prostacyclin
25. The chemical that is involved in the breakdown of a clot (fibrinolysis) is
    a. antithrombin.
    b. fibrinogen.
    c. heparin.
    d. plasmin.
    e. sodium citrate.
26. A person with type A blood
    a. has anti-A antibodies.
    b. has type B antigens.
    c. will have a transfusion reaction if given type B blood.
    d. all of the above.
27. In the United States, the most common blood type is
    a. A positive.
    b. B positive.
    c. O positive.
    d. O negative.
    e. AB negative.
28. Rh-negative mothers who receive a RhoGAM injection are given that injection to
    a. initiate the synthesis of anti-Rh antibodies in the mother.
    b. initiate anti-Rh antibody production in the baby.
    c. prevent the mother from producing anti-Rh antibodies.
    d. prevent the baby from producing anti-Rh antibodies.
29. The blood test that distinguishes between leukocytosis and leukopenia is
    a. type and crossmatch.
    b. hematocrit.
    c. platelet count.
    d. complete blood count.
    e. prothrombin time measurement.
30. An elevated neutrophil count is usually indicative of
    a. an allergic reaction.
    b. a bacterial infection.
    c. a viral infection.
    d. a parasitic infection.
    e. increased antibody production.

Answers in Appendix F

CRITICAL THINKING

1. In hereditary hemolytic anemia, massive destruction of red blood cells occurs. Would you expect the reticulocyte count to be above or below normal? Explain why one of the symptoms of the disease is jaundice. In 1910, it was discovered that hereditary hemolytic anemia could be successfully treated by removing the spleen. Explain why this treatment is effective.

2. Red Packer, a physical education major, wanted to improve his performance in an upcoming marathon race. About 6 weeks before the race, 500 mL of blood was removed from his body, and the formed elements were separated from the plasma. The formed elements were frozen, and the plasma was reinfused into his body. Just before the competition, the formed elements were thawed and injected into his body. Explain why this procedure, called blood doping or blood boosting, would help Red's performance. Can you suggest any possible bad effects?

3. Chemicals like benzene and chloramphenicol can destroy red bone marrow and cause aplastic anemia. What symptoms develop as a result of the lack of (a) red blood cells, (b) platelets, and (c) white blood cells?

4. Some people habitually use barbiturates to depress feelings of anxiety. Barbiturates cause hypoventilation, which is a slower-than-normal rate of breathing, because they suppress the respiratory centers in the brain. What happens to the red blood count of a habitual user of barbiturates? Explain.

5. What blood problems would you expect to observe in a patient after total gastrectomy (removal of the stomach)? Explain.

6. According to the old saying, "Good food makes good blood." Name three substances in the diet that are essential for "good blood." What blood disorders develop if these substances are absent from the diet?

Answers in Appendix G

ANSWERS TO PREDICT QUESTIONS

1. The reason fetal hemoglobin must be more effective at binding oxygen than adult hemoglobin is so that the fetal circulation can draw the needed oxygen away from the maternal circulation. If maternal blood had an equal or greater oxygen affinity, the fetal blood would not be able to draw away the required oxygen, and the fetus would die.

2. An elevated reticulocyte count indicates that erythropoiesis and the demand for red blood cells are increased and that immature red blood cells (reticulocytes) are entering the circulation in large numbers. An elevated reticulocyte count can occur for a number of reasons, including loss of blood; therefore, after a person donates a unit of blood, the reticulocyte count increases.

3. Carbon monoxide binds to the iron of hemoglobin and prevents the transport of oxygen. The decreased oxygen stimulates the release of erythropoietin from the kidneys. Erythropoietin increases red blood cell production in red bone marrow, thereby causing the number of red blood cells in the blood to increase.

4. The white blood cells shown in figure 19.8 are (a) lymphocyte, (b) basophil, (c) monocyte, (d) neutrophil, and (e) eosinophil.

5. Platelets become activated at sites of tissue damage, which is the location where it's advantageous to form a clot to stop bleeding.

6. People with type AB blood were called universal recipients because they could receive type A, B, AB, or O blood with little likelihood of a transfusion reaction. Type AB blood does not have antibodies against type A or B; therefore, transfusion of these antigens in type A, B, or AB blood does not cause a transfusion reaction in a person with type AB blood. The term is misleading, however, for

two reasons. First, other blood groups can cause a transfusion reaction. Second, antibodies in the donor's blood can cause a transfusion reaction. For example, type O blood contains anti-A and anti-B antibodies that can react against the A and B antigens in type AB blood.

7. In HDN, red blood cells have been destroyed, releasing bilirubin into the blood. The replacement blood increases the number of red blood cells in the newborn's blood. Removing the newborn's blood removes bilirubin, which is toxic to the nervous system. It also removes anti-Rh antibodies from the plasma, which decreases the likelihood of agglutination and lysis of red blood cells. The replacement blood should be Rh-negative, even though the newborn is Rh-positive. Rh-negative red blood cells do not have Rh antigens. Therefore, any anti-Rh antibodies in the newborn's blood do not react with them. Giving Rh-negative blood to a Rh-positive newborn does not change the blood type of the newborn, because blood type is genetically determined. Eventually, all of the Rh-negative red blood cells die, and only Rh-positive red blood cells are produced by the newborn.

8. A white blood count (WBC) should be done. An elevated WBC, leukocytosis, can be an indication of bacterial infections. A differential WBC should also be done. An increase in the number of neutrophils supports the diagnosis of a bacterial infection. Coupled with other symptoms, this could mean appendicitis. If these tests are normal, appendicitis is still a possibility and the physician must rely on other clinical signs. Diagnostic accuracy for appendicitis is approximately 75%–85% for experienced physicians.

Visit the Online Learning Center at www.mhhe.com/seeley7 for chapter quizzes, interactive learning exercises, and other study tools.

Cardiovascular System

The Heart

Colorized scanning electron micrograph of Purkinje fibers of the heart.

Approximately 370 years ago, it was established that the heart's pumping action is essential to maintain the continuous circulation of blood throughout the body. Our current understanding of the detailed function of this amazing pump, its regulation, and modern treatments for heart disease is, in comparison, very recent.

The heart is actually two pumps in one. The right side of the heart receives blood from the body and pumps blood through the **pulmonary** (pŭl′mō-nār-ē) **circulation,** which carries blood to the lungs and returns it to the left side of the heart. In the lungs, carbon dioxide diffuses from the blood into the lungs, and oxygen diffuses from the lungs into the blood. The left side of the heart pumps blood through the **systemic circulation,** which delivers oxygen and nutrients to all remaining tissues of the body. From those tissues carbon dioxide and other waste products are carried back to the right side of the heart (figure 20.1).

The heart of a healthy 70 kg person pumps approximately 7200 L (approximately 1900 gallons) of blood each day at a rate of 5 L/min. For most people, the heart continues to pump for more than 75 years. During periods of vigorous exercise, the amount of blood pumped per minute increases severalfold. The life of the individual is in danger if the heart loses its ability to pump blood for even a few minutes. **Cardiology** (kar-dē-ol′ō-jē) is a medical specialty concerned with the diagnosis and treatment of heart disease.

This chapter describes the *functions of the heart* (p. 680), *size, shape, and location of the heart* (p. 680), the *anatomy of the heart* (p. 682), the *route of blood flow through the heart* (p. 688), and its *histology* (p. 690) and *electrical properties* (p. 693). The *cardiac cycle* (p. 698), *mean arterial blood pressure* (p. 704), *regulation of the heart* (p. 706), and the *heart and homeostasis* (p. 708) are described. The chapter ends with the *effects of aging on the heart* (p. 711).

Figure 20.1   Systemic and Pulmonary Circulation

The circulatory system consists of the pulmonary and systemic circulations. The right side of the heart pumps blood through vessels to the lungs and back to the left side of the heart through the pulmonary circulation. The left side of the heart pumps blood through vessels to the tissues of the body and back to the right side of the heart through the systemic circulation.

Functions of the Heart

The functions of the heart include:

1. *Generating blood pressure.* Contractions of the heart generate blood pressure, which is responsible for blood movement through the blood vessels.
2. *Routing blood.* The heart separates the pulmonary and systemic circulations and ensures better oxygenation of blood flowing to tissues.
3. *Ensuring one-way blood flow.* The valves of the heart ensure a one-way flow of blood through the heart and blood vessels.
4. *Regulating blood supply.* Changes in the rate and force of contraction match blood delivery to the changing metabolic needs of the tissues, such as during rest, exercise, and changes in body position.

● **1. List four major functions of the heart.**

Size, Shape, and Location of the Heart

The adult heart is shaped like a blunt cone and is approximately the size of a closed fist. It is larger in physically active adults than in other healthy adults, and it generally decreases in size after approximately age 65, especially in those who are not physically active.

The blunt, rounded point of the cone is the **apex;** and the larger, flat part at the opposite end of the cone is the **base.**

The heart is located in the thoracic cavity between the lungs. The heart, trachea, esophagus, and associated structures form a midline partition, the **mediastinum** (me′dē-as-tī′nŭm; see figure 1.14).

It's important for clinical reasons to know the location of the heart in the thoracic cavity. Positioning a stethoscope to hear the heart sounds and positioning electrodes to record an **electrocardiogram** (ē-lek-trō-kar′dē-ō-gram; **ECG** or **EKG**) from chest leads depend on this knowledge. Effective cardiopulmonary resuscitation (kar′dē-ō-pŭl′mo-nār-ē rē-sŭs′i-tā-shŭn; **CPR**) also depends on a reasonable knowledge of the position and shape of the heart. The heart lies obliquely in the mediastinum, with its base directed posteriorly and slightly superiorly and the apex directed anteriorly and slightly inferiorly. The apex is also directed to the left so that approximately two-thirds of the heart's mass lies to the left of the midline of the sternum (figure 20.2). The base of the heart is located deep to the sternum and extends to the second intercostal space. The apex is located deep to the fifth intercostal space, approximately 7–9 centimeters (cm) to the left of the sternum and medial to the midclavicular line, which is a perpendicular line that extends down from the middle of the clavicle.

● **2. Give the approximate size and shape of the heart. Where is it located?**

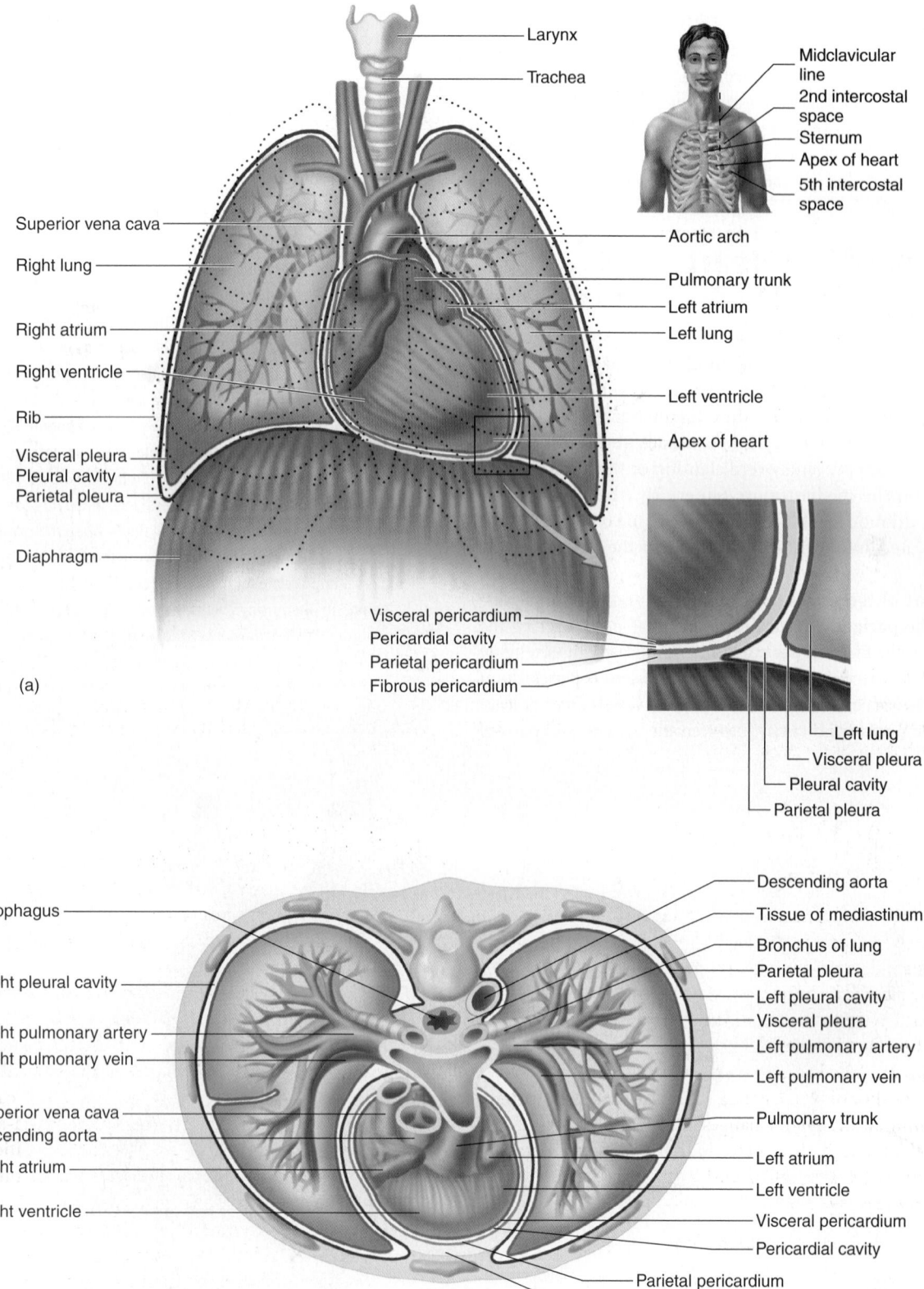

Figure 20.2 Location of the Heart in the Thorax

(*a*) The heart is located in the thoracic cavity between the lungs, deep and slightly to the left of the sternum. The base of the heart, located deep to the sternum, extends to the second intercostal space, and the apex of the heart is located deep to the fifth intercostal space, approximately 7–9 cm to the left of the sternum, or where the midclavicular line intersects with the fifth intercostal space (see inset). (*b*) Cross section of the thorax showing the position of the heart in the mediastinum and its relationship to other structures.

Cardiopulmonary Resuscitation (CPR)

In cases in which the heart suddenly stops beating, CPR can save lives. CPR involves rhythmic compression of the chest combined with artificial ventilation of the lungs. Applying pressure to the sternum compresses the chest wall, which also compresses the heart and causes it to pump blood. In many cases, CPR can provide an adequate blood supply to the heart wall and brain until emergency medical assistance arrives.

Anatomy of the Heart

Pericardium

The **pericardium** (per-i-kar′dē-ŭm), or **pericardial sac,** is a double-layered closed sac that surrounds the heart (figure 20.3). It consists of a tough, fibrous connective tissue outer layer called the **fibrous pericardium** and a thin, transparent inner layer of simple squamous epithelium called the **serous pericardium.** The fibrous pericardium prevents overdistention of the heart and anchors it within the mediastinum. Superiorly, the fibrous pericardium is continuous with the connective tissue coverings of the great vessels, and inferiorly it is attached to the surface of the diaphragm.

The part of the serous pericardium lining the fibrous pericardium is the **parietal pericardium,** and that part covering the heart surface is the **visceral pericardium,** or epicardium (see figure 20.3). The parietal and visceral portions of the serous pericardium are continuous with each other where the great vessels enter or leave the heart. The **pericardial cavity,** between the visceral and parietal pericardia, is filled with a thin layer of serous **pericardial fluid,** which helps reduce friction as the heart moves within the pericardial sac.

Even though the pericardium contains fibrous connective tissue, it can accommodate changes in heart size by gradually enlarging. The pericardial cavity can also increase in volume to hold a significant volume of pericardial fluid.

Pericarditis and Cardiac Tamponade

Pericarditis (per′i-kar-dī′tis) is an inflammation of the serous pericardium. The cause is frequently unknown, but it can result from infection, diseases of connective tissue, or damage due to radiation treatment for cancer. It can be extremely painful, with sensations of pain referred to the back and chest, which can be confused with the pain of a myocardial infarction (heart attack). Pericarditis can result in fluid accumulation within the pericardial sac.

Cardiac tamponade (tam-pŏ-nād′) is a potentially fatal condition in which a large volume of fluid or blood accumulates in the pericardial sac. The fluid compresses the heart from the outside. Although the heart is a powerful muscle, it relaxes passively. When it is compressed by fluid within the pericardial sac, it cannot dilate when the cardiac muscle relaxes. Consequently, it cannot fill with blood during relaxation, which makes it impossible for it to pump blood. Cardiac tamponade can cause a person to die quickly unless the fluid is removed. Causes of cardiac tamponade include rupture of the heart wall following a myocardial infarction, rupture of blood vessels in the pericardium after a malignant tumor invades the area, damage to the pericardium resulting from radiation therapy, and trauma (e.g., a traffic accident).

Pericardium {
    Fibrous pericardium
    Serous pericardium {
        Parietal pericardium
        Visceral pericardium (or epicardium)
        Pericardial cavity filled with pericardial fluid

Figure 20.3   Heart in the Pericardium

The heart is located in the pericardium, which consists of an outer fibrous pericardium and an inner serous pericardium. The serous pericardium has two parts: the parietal pericardium lines the fibrous pericardium, and the visceral pericardium (epicardium) covers the surface of the heart. The pericardial cavity, between the parietal and visceral pericardium, is filled with a small amount of pericardial fluid.

3. *What is the pericardium?*
4. *Describe the parts of the pericardium and their functions.*
5. *Define pericarditis.*
6. *Explain the effects of cardiac tamponade on the heart.*

Heart Wall

The heart wall is composed of three layers of tissue: the epicardium, the myocardium, and the endocardium (figure 20.4). The **epicardium** (ep-i-kar′dē-ŭm) is a thin serous membrane that constitutes the smooth outer surface of the heart. The epicardium and the visceral pericardium are two names for the same structure. The serous pericardium is called the epicardium when considered a part of the heart and the visceral pericardium when considered a part of the pericardium. The thick middle layer of the heart, the **myocardium** (mī-ō-kar′dē-ŭm), is composed of cardiac muscle cells and is responsible for the ability of the heart to contract. The smooth inner surface of the heart chambers is the **endocardium** (en-dō-kar′dē-ŭm), which consists of simple squamous epithelium over a layer of connective tissue. The smooth inner surface allows blood to move easily through the heart. The heart valves result from a fold in the endocardium, thus making a double layer of endocardium with connective tissue in between.

The interior surfaces of the atria are mainly flat, but the interior of both auricles and a part of the right atrial wall contain muscular ridges called **musculi pectinati** (pek′ti-nah′tĕ; hair

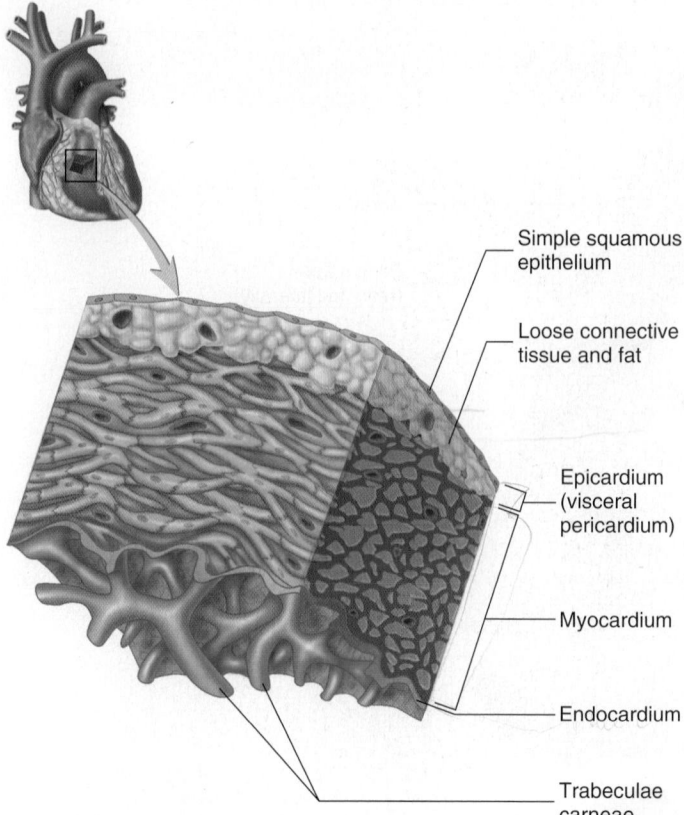

Simple squamous epithelium

Loose connective tissue and fat

Epicardium (visceral pericardium)

Myocardium

Endocardium

Trabeculae carneae

Figure 20.4 Heart Wall

Part of the wall of the heart has been removed to show its structure. The enlarged section illustrates the epicardium, the myocardium, and the endocardium.

comb). The musculi pectinati of the right atrium are separated from the larger, smooth portions of the atrial wall by a ridge called the **crista terminalis** (kris′tă ter′mi-nal′is; terminal crest). The interior walls of the ventricles contain larger muscular ridges and columns called **trabeculae** (tră-bek′ū-lē; beams) **carneae** (kar′nē-ē; flesh).

7. *Describe the three layers of the heart, and state their functions.*
8. *Name the muscular ridges found on the interior of the auricles, and name the ridges and columns found on the interior walls of the ventricles.*

External Anatomy and Coronary Circulation

The heart consists of four chambers: two **atria** (ā′trē-ă; entrance chamber) and two **ventricles** (ven′tri-klz; belly). The thin-walled atria form the superior and posterior parts of the heart, and the thick-walled ventricles form the anterior and inferior portions (figure 20.5). Flaplike **auricles** (aw′ri-klz; ears) are extensions of the atria that can be seen anteriorly between each atrium and ventricle. The entire atrium used to be called the auricle, and some medical personnel still refer to it as such.

Several large veins carry blood to the heart. The **superior vena cava** (vē′nă kā′vă) and the **inferior vena cava** carry blood from the body to the right atrium, and four **pulmonary veins** carry blood from the lungs to the left atrium. In addition, the smaller coronary sinus carries blood from the walls of the heart to the right atrium.

Two arteries, the **aorta** and the **pulmonary trunk,** exit the heart. The aorta carries blood from the left ventricle to the body, and the pulmonary trunk carries blood from the right ventricle to the lungs.

A large **coronary** (kōr′o-nār-ē; circling like a crown) **sulcus** (sool′kŭs; ditch) runs obliquely around the heart, separating the atria from the ventricles. Two more sulci extend inferiorly from the coronary sulcus, indicating the division between the right and left ventricles. The **anterior interventricular sulcus,** or **groove,** is on the anterior surface of the heart, and the **posterior interventricular sulcus,** or **groove,** is on the posterior surface of the heart. In a healthy, intact heart the sulci are covered by fat, and only after this fat is removed can the actual sulci be seen.

The major arteries supplying blood to the tissue of the heart lie within the coronary sulcus and interventricular sulci on the surface of the heart. The **right** and **left coronary arteries** exit the aorta just above the point where the aorta leaves the heart and lie within the coronary sulcus (figure 20.6a). The right coronary artery is usually smaller in diameter than the left one, and it doesn't carry as much blood as the left coronary artery.

A major branch of the left coronary artery, called the **anterior interventricular artery,** or the **left anterior descending artery,** extends inferiorly in the anterior interventricular sulcus and supplies blood to most of the anterior part of the heart. The **left marginal artery** branches from the left coronary artery to supply blood to the lateral wall of the left ventricle. The **circumflex** (ser′kŭm-fleks) **artery** branches from the left coronary artery and extends around to the posterior side of the heart in the coronary

Aortic arch

Left pulmonary artery

Superior vena cava

Aorta

Branches of left
pulmonary artery

Branches of right
pulmonary artery

Pulmonary trunk

Left pulmonary veins

Right pulmonary veins

Left atrium

Right atrium

Great cardiac vein

Coronary sulcus

Right coronary artery

Anterior interventricular artery

Right ventricle

Left ventricle

Inferior vena cava

(a) Anterior view

Superior
vena cava

Aorta

Pericardium
(reflected laterally)

Right auricle

Right atrium

Pulmonary trunk

Right coronary
artery (in coronary
sulcus)

Great cardiac vein

Right ventricle

Anterior interventricular artery
(in anterior interventricular sulcus)

Small
cardiac vein

Left ventricle

Right
marginal
artery

(b) Anterior view

Figure 20.5   Anterior Surface View of the Heart

(a) The anterior view of the heart. The two atria (right and left) are located superiorly, and the two ventricles (right and left) are located inferiorly. The superior and inferior venae cava enter the right atrium. The pulmonary veins enter the left atrium. The pulmonary trunk exits the right ventricle, and the aorta exits the left ventricle. (b) Photograph of the anterior surface.

Aorta

Left pulmonary artery

Left pulmonary veins

Left atrium

Great cardiac vein
(in coronary sulcus)

Coronary sinus

Left ventricle

Middle cardiac vein
(in posterior inter-
ventricular sulcus)

Apex

(c)

Superior vena cava

Right pulmonary artery

Right pulmonary veins

Right atrium

Inferior vena cava

Right coronary artery

Small cardiac vein

Posterior interventricular artery

Right ventricle

Posterior view

Figure 20.5 *(continued)*

(c) The posterior surface view of the heart. The two atria (right and left) are located superiorly, and the two ventricles (right and left) are located inferiorly. The superior and inferior venae cava enter the right atrium, and the four pulmonary veins enter the left atrium.

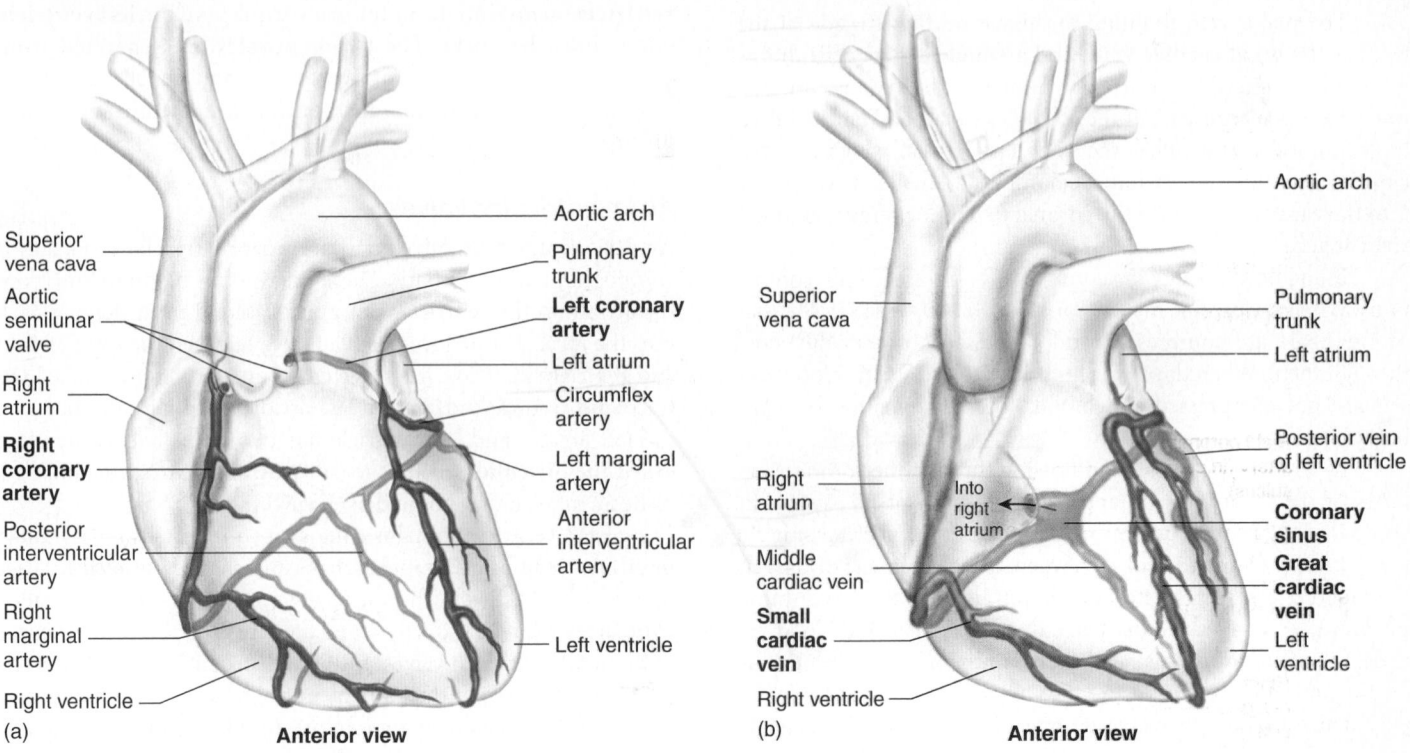

Superior
vena cava

Aortic
semilunar
valve

Right
atrium

**Right
coronary
artery**

Posterior
interventricular
artery

Right
marginal
artery

Right ventricle

(a)

Aortic arch

Pulmonary
trunk

**Left coronary
artery**

Left atrium

Circumflex
artery

Left marginal
artery

Anterior
interventricular
artery

Left ventricle

Anterior view

Superior
vena cava

Right
atrium

Middle
cardiac vein

**Small
cardiac
vein**

Right ventricle

(b)

Into
right
atrium

Aortic arch

Pulmonary
trunk

Left atrium

Posterior vein
of left ventricle

**Coronary
sinus**

**Great
cardiac
vein**

Left
ventricle

Anterior view

Figure 20.6 Coronary Circulation

(a) Arteries supplying blood to the heart. The arteries of the anterior surface are seen directly and are darker in color; the arteries of the posterior surface are seen through the heart and are lighter in color. (b) Veins draining blood from the heart. The veins of the anterior surface are seen directly and are darker in color; the veins of the posterior surface are seen through the heart and are lighter in color.

sulcus. Its branches supply blood to much of the posterior wall of the heart.

The right coronary artery lies within the coronary sulcus and extends from the aorta around to the posterior part of the heart. A larger branch of the right coronary artery, called the **right marginal artery,** and other branches supply blood to the lateral wall of the right ventricle. A branch of the right coronary artery called the **posterior interventricular artery** lies in the posterior interventricular sulcus and supplies blood to the posterior and inferior part of the heart.

P R E D I C T

Predict the effect on the heart if blood flow through a coronary artery, such as the anterior interventricular artery, is restricted or completely blocked.

Most of the myocardium receives blood from more than one arterial branch. Furthermore, there are many anastamoses, or direct connections, between the arterial branches. The anastamoses are either between branches of a given artery or between branches of different arteries. In the event that one artery is blocked, the areas primarily supplied by that artery may still receive some blood through other arterial branches and through anastamoses with other branches. Aerobic exercise tends to increase the density of blood vessels supplying blood to the myocardium and the number and extent of the anastamoses increase. Consequently, aerobic exercise increases the chance that a person will survive the blockage of a small coronary artery. Blockage of larger coronary blood vessels still have the potential to permanently damage large areas of the heart wall.

The major vein draining the tissue on the left side of the heart is the **great cardiac vein,** and a **small cardiac vein** drains the right margin of the heart (figure 20.6*b*). These veins converge toward the posterior part of the coronary sulcus and empty into a large venous cavity called the **coronary sinus,** which in turn empties into the right atrium. A number of smaller veins empty into the cardiac veins, into the coronary sinus, or directly into the right atrium.

Blood flow through the coronary blood vessels is not continuous. When the cardiac muscle contracts, blood vessels in the wall of the heart are compressed and blood does not readily flow through them. When the cardiac muscle is relaxing, the blood vessels are not compressed and blood flow through the coronary blood vessels resumes.

In a resting person, blood flowing through the coronary arteries of the heart gives up approximately 70% of its oxygen. In comparison, blood flowing through arteries to skeletal muscle gives up only about 25% of its oxygen. The percentage of oxygen the blood releases to skeletal muscle can increase to 70% or more during exercise. Because the percentage of oxygen the blood releases to cardiac muscle is near its maximum at rest, it cannot increase substantially during exercise. Cardiac muscle is therefore very dependent on an increased rate of blood flow through the coronary arteries above its resting level to provide an adequate oxygen supply during exercise.

9. *Name the chambers of the heart and describe their structure.*
10. *Name the major blood vessels that enter and leave the heart. Which chambers of the heart do they enter or exit?*
11. *Describe the flow of blood through the coronary arteries and their branches.*
12. *Is blood flow through the coronary vessels continuous?*
13. *Describe the flow of blood through the cardiac veins.*

Heart Chambers and Valves

Right and Left Atria

The **right atrium** has three major openings: the openings from the superior vena cava and the inferior vena cava receive blood from the body, and the opening of the coronary sinus receives blood from the heart itself (figure 20.7). The **left atrium** has four relatively uniform openings that receive the four pulmonary veins from the lungs. The two atria are separated from each other by the **interatrial septum.** A slight oval depression, the **fossa ovalis** (fos′ă ō-va′lis), on the right side of the septum marks the former location of the **foramen ovale** (ō-va′lē), an opening between the right and left atria in the embryo and the fetus. This opening allows blood to flow from the right to the left atrium in the fetus to bypass the pulmonary circulation (see chapter 29).

Right and Left Ventricles

The atria open into the ventricles through **atrioventricular canals** (see figure 20.7). Each ventricle has one large, superiorly placed outflow route near the midline of the heart. The **right ventricle** opens into the pulmonary trunk, and the **left ventricle** opens into the aorta. The two ventricles are separated from each other by the **interventricular septum,** which has a thick muscular part toward the apex and a thin membranous part toward the atria.

Atrioventricular Valves

An **atrioventricular valve** is in each atrioventricular canal and is composed of cusps, or flaps. These valves allow blood to flow from the atria into the ventricles but prevent blood from flowing back into the atria. The atrioventricular valve between the right atrium and the right ventricle has three cusps and is therefore called the **tricuspid** (trī-kŭs′pid) **valve.** The atrioventricular valve between the left atrium and left ventricle has two cusps and is therefore called the **bicuspid** (bī-kŭs′pid), or **mitral** (mī′trăl; resembling a bishop's miter, a two-pointed hat), **valve.**

Each ventricle contains cone-shaped muscular pillars called **papillary** (pap′i-lār-ē; nipple, or pimple-shaped) **muscles.** These muscles are attached by thin, strong connective tissue strings called **chordae tendineae** (kōr′dē ten′di-nē-ē; heart strings) to the cusps of the atrioventricular valves (see figure 20.7 and figure 20.8*a*). The papillary muscles contract when the ventricles contract and prevent the valves from opening into the atria by pulling on the chordae tendineae attached to the valve cusps. Blood flowing from the atrium into the ventricle pushes the valve open into the ventricle,

Figure 20.7 Internal Anatomy of the Heart

The heart is cut in a frontal plane to show the internal anatomy.

(a) **Anterior view**

(b) **Superior view**

Figure 20.8 Heart Valves

(a) View of the tricuspid valve, the chordae tendineae, and the papillary muscles. (b) A superior view of the heart valves. Note the three cusps of each semilunar valve meeting to prevent the backflow of blood.

but, when the ventricle contracts, blood pushes the valve back toward the atrium. The atrioventricular canal is closed as the valve cusps meet (figure 20.9).

Semilunar Valves

Within the aorta and pulmonary trunk are **aortic** and **pulmonary semilunar** (sem-ē-loo′năr; half-moon-shaped) **valves,** respectively. Each valve consists of three pocketlike semilunar cusps, the free inner borders of which meet in the center of the artery to block blood flow (see figures 20.7 and 20.8b). Blood flowing out of the ventricles pushes against each valve, forcing it open, but when blood flows back from the aorta or pulmonary trunk toward the ventricles, it enters the pockets of the cusps, causing them to meet in the center of the aorta or pulmonary trunk, thus closing them and keeping blood from flowing back into the ventricles (see figure 20.9).

● 14. Describe the openings into the right and left atria. What structure separates the atria from each other?
15. Describe the openings of the right and left ventricles. What structure separates the ventricles from each other?
16. Name the valves that separate the right atrium from the right ventricle and the left atrium from the left ventricle.

What are the functions of the papillary muscles and the chordae tendineae?
17. *Name the valves found in the aorta and pulmonary trunk.*

Route of Blood Flow Through the Heart

Blood flow through the heart is depicted in figure 20.10. Even though it's more convenient to discuss blood flow through the heart one side at a time, it's important to understand that both atria contract at about the same time and both ventricles contract at about the same time. This concept is particularly important when considering electrical activity, pressure changes, and heart sounds.

Blood enters the right atrium from the systemic circulation, which returns blood from all the tissues of the body. Blood flows from an area of higher pressure in the systemic circulation to the right atrium, which has a lower pressure. Most of the blood in the right atrium then passes into the right ventricle as the ventricle relaxes following the previous contraction. The right atrium then contracts, and most of the blood remaining in the atrium is pushed into the ventricle to complete right ventricular filling.

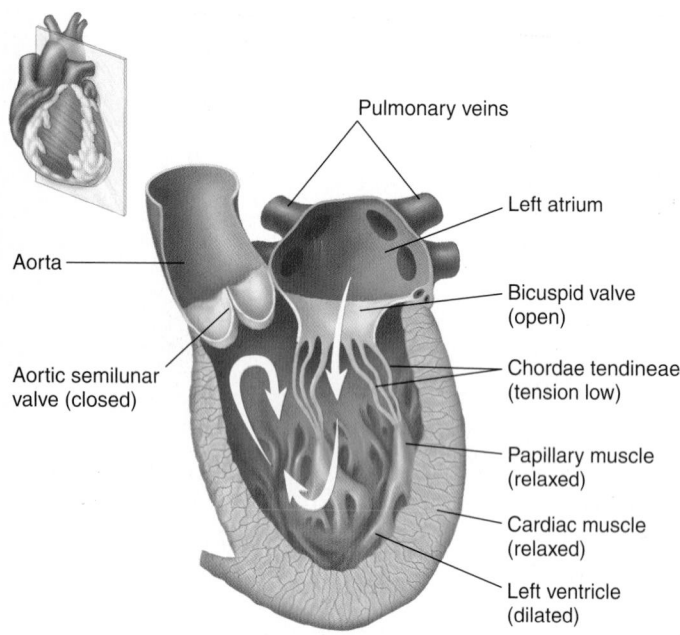

(a) When the bicuspid valve is open, the cusps of the valve are pushed by blood into the ventricle. Papillary muscles are relaxed and tension on the chordae tendineae is low. Blood flows from the left atrium into the left ventricle. When the aortic semilunar valve is closed, the cusps of the valve overlap as they are pushed by the blood in the aorta toward the ventricle. There is no blood flow from the aorta into the ventricle.

(b) When the bicuspid valve is closed, the cusps of the valves overlap as they are pushed by the blood toward the left atrium. There is no blood flow from the ventricle into the atrium. Papillary muscles are contracted and tension on the chordae tendineae is increased. When the aortic semilunar valve is open, the cusps of the valve are pushed by the blood toward the aorta. Blood then flows from the left ventricle into the aorta.

Figure 20.9 Function of the Heart Valves

(a) Valve positions when blood is flowing into the left ventricle. (b) Valve positions when blood is flowing out of the left ventricle. Illustrations of the bicuspid and aortic semilunar valves are presented to illustrate the function of those valves as well as the tricuspid and pulmonary semilunar valves.

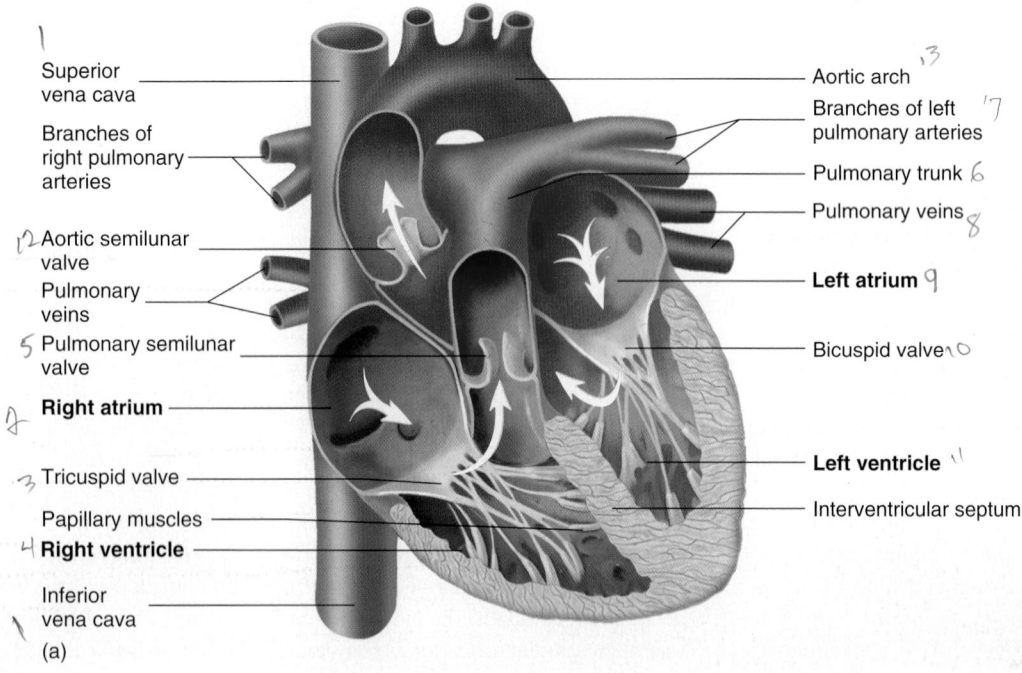

Superior vena cava
Branches of right pulmonary arteries
Aortic semilunar valve
Pulmonary veins
Pulmonary semilunar valve
Right atrium
Tricuspid valve
Papillary muscles
Right ventricle
Inferior vena cava
(a)

Aortic arch
Branches of left pulmonary arteries
Pulmonary trunk
Pulmonary veins
Left atrium
Bicuspid valve
Left ventricle
Interventricular septum

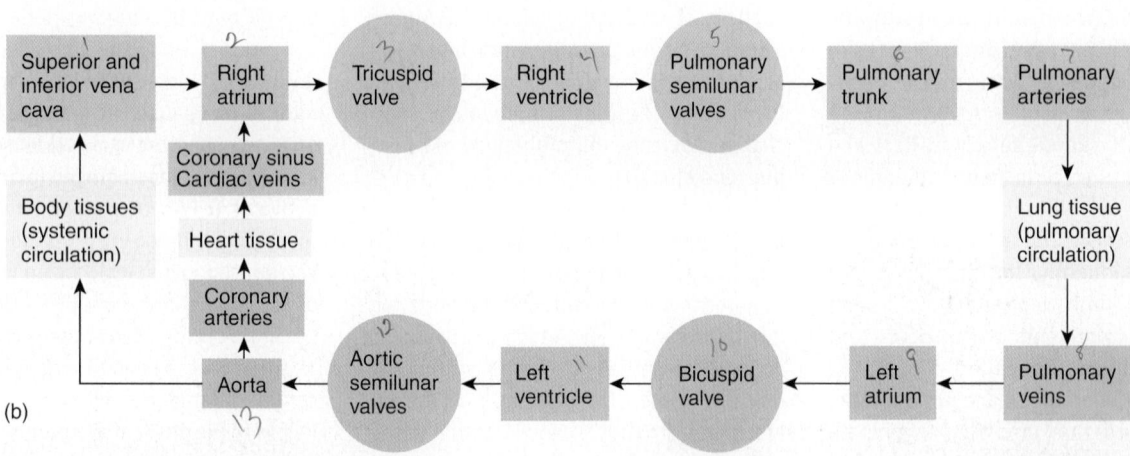

Superior and inferior vena cava → Right atrium → Tricuspid valve → Right ventricle → Pulmonary semilunar valves → Pulmonary trunk → Pulmonary arteries

Coronary sinus Cardiac veins

Body tissues (systemic circulation)

Heart tissue

Coronary arteries

Aorta ← Aortic semilunar valves ← Left ventricle ← Bicuspid valve ← Left atrium ← Pulmonary veins

Lung tissue (pulmonary circulation)

(b)

Figure 20.10 Blood Flow Through the Heart

(a) Frontal section of the heart revealing the four chambers and the direction of blood flow through the heart. (b) Diagram listing in order the structures through which blood flows in the systemic and pulmonary circulations. The heart valves are indicated by *circles:* deoxygenated blood (*blue*); oxygenated blood (*red*).

Contraction of the right ventricle pushes blood against the tricuspid valve, forcing it closed, and against the pulmonary semilunar valve, forcing it open, thus allowing blood to enter the pulmonary trunk.

The pulmonary trunk branches to form the **pulmonary arteries** (see figure 20.5), which carry blood to the lungs, where carbon dioxide is released and oxygen is picked up (see chapters 21 and 23). Blood returning from the lungs enters the left atrium through the four pulmonary veins. The blood passing from the left atrium to the left ventricle opens the bicuspid

valve, and contraction of the left atrium completes left ventricular filling.

Contraction of the left ventricle pushes blood against the bicuspid valve, closing it, and against the aortic semilunar valve, opening it and allowing blood to enter the aorta. Blood flowing through the aorta is distributed to all parts of the body except to the parts of the lungs supplied by the pulmonary blood vessels (see chapter 23).

● **18. Starting at the venae cavae and ending at the aorta, describe the flow of blood through the heart.**

Clinical Focus   Angina, Infarctions, and Treatment of Blocked Coronary Arteries

Angina pectoris (an'ji-nă, an-jī'nă pek'tō-ris) is pain that results from a reduction in blood supply to cardiac muscle. The pain is temporary and, if blood flow is restored, little permanent change or damage results. Angina pectoris is characterized by chest discomfort deep to the sternum, often described as heaviness, pressure, or moderately severe pain. It is often mistaken for indigestion. The pain can also be referred to the neck, lower jaw, left arm, and left shoulder (see chapter 14, p. 487).

Most often, angina pectoris results from narrowed and hardened coronary arterial walls. The reduced blood flow results in a reduced supply of oxygen to cardiac muscle cells. As a consequence, the limited anaerobic metabolism of cardiac muscle results in a buildup of lactic acid and reduced pH in affected areas of the heart. Pain receptors are stimulated by the lactic acid. The pain is predictably associated with exercise because the increased pumping activity of the heart requires more oxygen, and the narrowed blood vessels cannot supply it. Rest and drugs like nitroglycerin frequently relieve angina pectoris. Nitroglycerin dilates the blood vessels, including the coronary arteries. Consequently, the drug increases the oxygen supply to cardiac muscle and reduces the workload of the heart. Because peripheral arteries are dilated, the heart has to pump blood against a smaller pressure, and the need for oxygen decreases. The heart also pumps less blood because blood tends to remain in the dilated blood vessels and less blood is returned to the heart.

Myocardial infarction (mī-ō-kar'dē-ăl in-fark'shŭn) results from a prolonged lack of blood flow to a part of the cardiac muscle, resulting in a lack of oxygen and ultimately cellular death. Myocardial infarctions vary with the amount of cardiac muscle and the part of the heart that is affected. If blood supply to cardiac muscle is reestablished within 20 minutes, no permanent damage occurs. If the lack of oxygen lasts longer, cell death results. Within 30–60 seconds after blockage of a coronary blood vessel, however, functional changes are obvious. The electrical properties of the cardiac muscle are altered, and the ability of the heart to function properly is lost. The most common cause of myocardial infarction is thrombus formation that blocks a coronary artery. Coronary arteries narrowed by **atherosclerotic** (ath'er-ō-skler-ot'ik) **lesions** provide one of the conditions that increase the chances of myocardial infarction. Atherosclerotic lesions partially block blood vessels, resulting in turbulent blood flow, and the surfaces of the lesions are rough. These changes increase the probability of thrombus formation.

Angioplasty (an'jē-ō-plas-tē) is a process whereby a small balloon is inserted, usually into the femoral artery (see chapter 21), and is threaded through the aorta and into a coronary artery. After the balloon has entered the partially occluded coronary artery, it is inflated, thereby flattening the atherosclerotic deposits against the vessel walls and opening the occluded blood vessel. This technique improves the function of cardiac muscle in patients suffering from an inadequate blood flow to the cardiac mus-

cle through the coronary arteries. Some controversy exists about its effectiveness. At least in some patients, dilation of the coronary arteries can be reversed within a few weeks or months and blood clots can form in coronary arteries following angioplasty. To help prevent future blockage, a metal-mesh tube called a **stent** is inserted into the vessel. Although the stent is better able to hold the vessel open, it too can eventually become blocked. Small rotating blades and lasers are also used to remove lesions from coronary vessels.

A **coronary bypass** is a surgical procedure that relieves the effects of obstructions in the coronary arteries. The technique involves taking healthy segments of blood vessels from other parts of the patient's body and using them to bypass obstructions in the coronary arteries. The technique is common for those who suffer from severe occlusion in specific parts of coronary arteries.

Special enzymes are used to break down blood clots that form in the coronary arteries and cause heart attacks. The major enzymes used are **streptokinase** (strep-tō-kī'nās), **tissue plasminogen** (plaz-min'ō-jen) **activator (t-PA),** or, sometimes, **urokinase** (ūr-ō-kī'nās). These enzymes function to activate plasminogen, which is an inactive form of an enzyme in the body that breaks down the fibrin of clots. The strategy is to administer these drugs to people suffering from myocardial infarctions as soon as possible following the onset of symptoms. Removal of the occlusions produced by clots reestablishes blood flow to the cardiac muscle and reduces the amount of cardiac muscle permanently damaged by the occlusion.

Histology

Heart Skeleton

The **heart skeleton** consists of a plate of fibrous connective tissue between the atria and ventricles. This connective tissue plate forms **fibrous rings** around the atrioventricular and semilunar valves and provides a solid support for them (figure 20.11). The fibrous connective tissue plate also serves as electrical insulation between the atria and the ventricles and provides a rigid site for attachment of the cardiac muscles.

19. Describe and list the functions of the skeleton of the heart.

Cardiac Muscle

Cardiac muscle cells are elongated, branching cells that contain one or occasionally two centrally located nuclei. Cardiac muscle cells contain actin and myosin myofilaments organized to form sarcomeres, which join end to end to form myofibrils (see chapter 9). The actin and myosin myofilaments are responsible for muscle contraction, and their organization gives cardiac muscle a striated (banded) appearance. The striations are less

Skeleton of the heart including fibrous rings around valves

Pulmonary semilunar valve

Aortic semilunar valve

Bicuspid valve

Tricuspid valve

Cardiac muscle of the right ventricle

Cardiac muscle of the left ventricle

Figure 20.11   Skeleton of the Heart

The skeleton of the heart consists of fibrous connective tissue rings that surround the heart valves and separate the atria from the ventricles. Cardiac muscle attaches to the fibrous connective tissue. The muscle fibers are arranged so that when the ventricles contract a wringing motion is produced and the distance between the apex and base of the heart shortens.

regularly arranged and less numerous than in skeletal muscle (figure 20.12).

Cardiac muscle has a **smooth sarcoplasmic reticulum,** but it is not as regularly arranged as it is in skeletal muscle fibers, and there are no dilated cisternae, as in skeletal muscle. The sarcoplasmic reticulum comes into close association at various points with membranes of **transverse tubules (T tubules).** The T tubules of

cardiac muscle are found near the Z disks of the sarcomeres instead of where the actin and myosin overlap as in skeletal muscle. The T tubules in cardiac muscle are larger in diameter than in skeletal muscle, and there are extensions of T tubules that are parallel with the sarcoplasmic reticulum. The loose association between the sarcoplasmic reticulum and the T tubules is partly responsible for the slow onset of contraction and the prolonged contraction phase in cardiac muscle. Depolarizations of the cardiac muscle plasma membrane are not carried from the surface of the cell to the sarcoplasmic reticulum as efficiently as they are in skeletal muscles, and calcium must diffuse a greater distance from the sarcoplasmic reticulum to the actin myofilaments. In addition, some Ca^{2+} enter the cardiac muscle cells from the extracellular fluid and from the T tubules.

Adenosine triphosphate (ATP) provides the energy for cardiac muscle contraction, and, as in other tissues, ATP production depends on oxygen availability. Cardiac muscle, however, cannot develop a large oxygen debt, a characteristic that is consistent with the function of the heart. Development of a large oxygen debt would result in muscular fatigue and cessation of cardiac muscle contraction. Cardiac muscle cells are rich in mitochondria, which perform oxidative metabolism at a rate rapid enough to sustain normal myocardial energy requirements. The extensive capillary network provides an adequate oxygen supply to the cardiac muscle cells.

PREDICT 2
Under resting conditions, most ATP produced in cardiac muscle is derived from the metabolism of fatty acids. During periods of heavy exercise, however, cardiac muscle cells use lactic acid as an energy source. Explain why this arrangement is an advantage.

Cardiac muscle cells are organized in spiral bundles or sheets. The cells are bound end to end and laterally to adjacent cells

Branching muscle fibers

Intercalated disks

Nucleus of cardiac muscle cell

Striations

T tubule

Sarcoplasmic reticulum

Sarcomere

Myofibrils

Mitochondrion

Sarcolemma (cell membrane)

Connective tissue

LM 400x

(a)

(b)

Figure 20.12   Histology of the Heart

(*a*) Cardiac muscle cells are branching cells with centrally located nuclei. As in skeletal muscle, sarcomeres join end-to-end to form myofibrils, and mitochondria provide ATP for contraction. The cells are joined to one another by intercalated disks, which allow action potentials to pass from one cardiac muscle cell to the next. Sarcoplasmic reticulum and T tubules are visible but are not as numerous as they are in skeletal muscle. (*b*) A light micrograph of cardiac muscle tissue. The cardiac muscle fibers appear to be striated because of the arrangement of the individual myofilaments.

by specialized cell–cell contacts called **intercalated** (in-ter′kă-lā-ted) **disks** (see figure 20.12). The membranes of the intercalated disks have folds, and the adjacent cells fit together, thus greatly increasing contact between them. Specialized plasma membrane structures called **desmosomes** (dez′mō-sōmz) hold the cells together, and **gap junctions** function as areas of low electric resistance between the cells, allowing action potentials to pass from one cell to adjacent cells (see figure 4.2). Electrically, the cardiac muscle cells behave as a single unit, and the highly coordinated contractions of the heart depend on this functional characteristic.

20. *Describe the similarities and differences between cardiac muscle and skeletal muscle.*
21. *Why does cardiac muscle have a slow onset of contraction and a prolonged contraction?*
22. *What substances do cardiac muscle cells use as an energy source? Do cardiac muscle cells develop an oxygen debt?*
23. *What anatomic features are responsible for the ability of cardiac muscle cells to contract as a unit?*

Conducting System

The conducting system of the heart, which relays electric action potentials through the heart, consists of modified cardiac muscle cells that form two nodes (meaning a knot or lump) and a conducting bundle (figure 20.13). The two nodes are contained within the walls of the right atrium and are named according to their position in the atrium. The **sinoatrial (SA) node** is medial to the opening of the superior vena cava, and the **atrioventricular (AV) node** is medial to the right atrioventricular valve. The AV node gives rise to a conducting bundle of the heart, the **atrioventricular (AV) bundle.**

This bundle passes through a small opening in the fibrous skeleton to reach the interventricular septum, where it divides to form the **right** and **left bundle branches,** which extend beneath the endocardium on either side of the interventricular septum to the apices of the right and left ventricles, respectively.

The inferior terminal branches of the bundle branches are called **Purkinje** (per-kin′jē) **fibers,** which are large-diameter cardiac muscle fibers. They have fewer myofibrils than most cardiac muscle cells and don't contract as forcefully. Intercalated disks are well developed between the Purkinje fibers and contain numerous gap junctions. As a result of these structural modifications, action potentials travel along the Purkinje fibers much more rapidly than through other cardiac muscle tissue.

Cardiac muscle cells have the capacity to generate spontaneous action potentials, but cells of the SA node do so at a greater frequency. As a result, the SA node is called the **pacemaker** of the heart. The SA node is made up of specialized, small-diameter, cardiac muscle cells that merge with the other cardiac muscle cells of the right atrium. Thus, the heart contracts spontaneously and rhythmically. Once action potentials are produced, they spread from the SA node to adjacent cardiac muscle fibers of the atrium. Preferential pathways conduct action potentials from the SA node to the AV node at a greater velocity than they are transmitted in the remainder of the atrial muscle fibers, although such pathways cannot be distinguished structurally from the remainder of the atrium.

When the heart beats under resting conditions, approximately 0.04 of a second is required for action potentials to travel from the SA node to the AV node. Within the AV node, action potentials are propagated slowly compared to the remainder of the conducting system. The slow rate of action potential conduction in the AV

1. Action potentials originate in the sinoatrial (SA) node (the pacemaker) and travel across the wall of the atrium (*arrows*) from the SA node to the atrioventricular (AV) node.

2. Action potentials pass through the AV node and along the atrioventricular (AV) bundle, which extends from the AV node, through the fibrous skeleton, into the interventricular septum.

3. The AV bundle divides into right and left bundle branches, and action potentials descend to the apex of each ventricle along the bundle branches.

4. Action potentials are carried by the Purkinje fibers from the bundle branches to the ventricular walls.

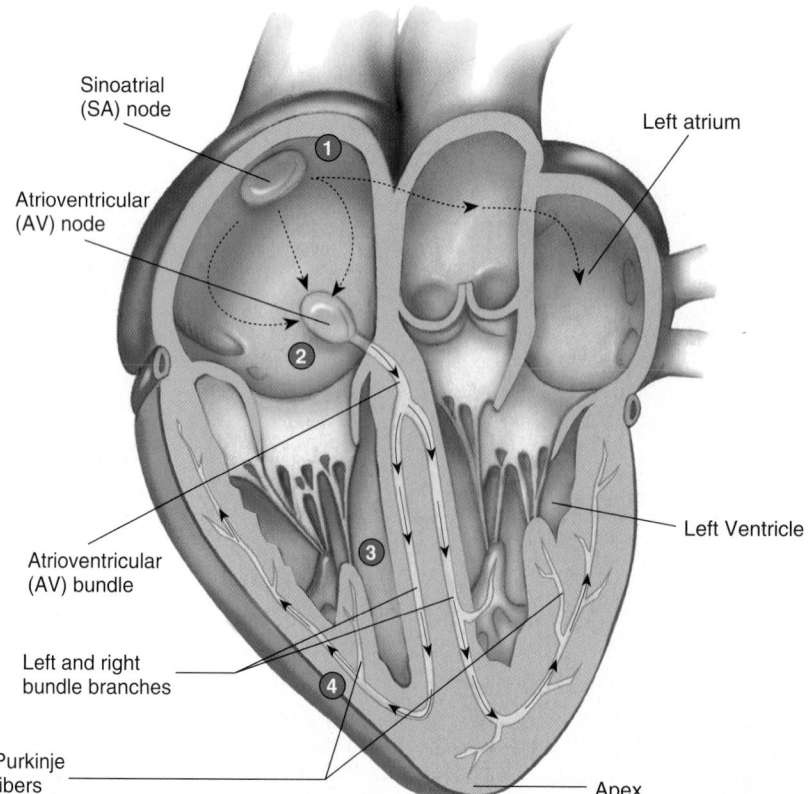

Process Figure 20.13   Conducting System of the Heart

node is due, in part, to the smaller diameter muscle fibers and fewer gap junctions in their intercalated disks. Like the other specialized conducting fibers in the heart, they have fewer myofibrils than most cardiac muscle cells. As a consequence, a delay occurs of 0.11 of a second from the time action potentials reach the AV node until they pass to the AV bundle. The total delay of 0.15 of a second allows completion of the atrial contraction before ventricular contraction begins.

After action potentials pass from the AV node to the highly specialized conducting bundles, the velocity of conduction increases dramatically. The action potentials pass through the left and right bundle branches and through the individual Purkinje fibers that penetrate into the myocardium of the ventricles (see figure 20.13).

Because of the arrangement of the conducting system, the first part of the myocardium that is stimulated is the inner wall of the ventricles near the apex. Thus ventricular contraction begins at the apex and progresses throughout the ventricles. Once stimulated, the spiral arrangement of muscle layers in the wall of the heart results in a wringing action that proceeds from the apex toward the base of the heart. During the process, the distance between the apex and the base of the heart decreases.

24. *List the parts of the conducting system of the heart. Explain how the conducting system coordinates contraction of the atria and ventricles. Explain why Purkinje fibers conduct action potentials more rapidly than other cardiac muscle cells.*
25. *Explain why the SA node is the pacemaker of the heart.*

P R E D I C T   3

Explain why it's more efficient for contraction of the ventricles to begin at the apex of the heart than at the base.

Electrical Properties

Cardiac muscle cells, like other electrically excitable cells such as neurons and skeletal muscle fibers, have a **resting membrane potential (RMP).** The RMP depends on a low permeability of the plasma membrane to Na^+ and Ca^{2+} and a higher permeability to K^+. When neurons, skeletal muscle cells, and cardiac muscle cells are depolarized to their threshold level, action potentials result (see chapter 11).

Action Potentials

Like action potentials in skeletal muscle, those in cardiac muscle exhibit depolarization followed by repolarization of the RMP. Alterations in membrane channels are responsible for the changes in the permeability of the plasma membrane that produce the action potentials. Action potentials in cardiac muscle last longer than those in skeletal muscle, and the membrane channels differ from those in skeletal muscle. In contrast to action potentials in skeletal muscle, which take less than 2 milliseconds (ms) to complete, action potentials in cardiac muscle take approximately 200–500 ms to complete.

In cardiac muscle, the action potential consists of a rapid **depolarization phase,** followed by a rapid, but partial, **early repolarization** phase. Then a prolonged period of slow repolarization occurs, called the **plateau phase.** At the end of the plateau, a more rapid **final repolarization phase** takes place, during which the membrane potential returns to its resting level (figure 20.14).

Membrane channels, called **voltage-gated Na^+ channels,** open bringing about the depolarization phase of the action potential. As the voltage-gated Na^+ channels open, Na^+ diffuses into the

cell, causing rapid depolarization until the cell is depolarized to approximately +20 millivolts (mV).

The voltage change occurring during depolarization affects other ion channels in the plasma membrane. Several different types of **voltage-gated K^+ channels** exist, each of which opens and closes at different membrane potentials, causing changes in membrane permeability to K^+. For example, at rest, the movement of K^+ through open voltage-gated K^+ channels is primarily responsible for establishing the resting membrane potential in cardiac muscle cells. Depolarization causes these voltage-gated K^+ channels to close, thereby decreasing membrane permeability to K^+. Depolarization also causes **voltage-gated Ca^{2+}** channels to begin to open. Compared to sodium channels, the calcium channels open and close slowly.

Repolarization is the result of changes in membrane permeability to Na^+, K^+, and Ca^{2+}. Early repolarization occurs when the voltage-gated Na^+ channels and some voltage-gated Ca^{2+} channels close, and a small number of voltage-gated K^+ channels open. Na^+ movement into the cell slows, and some K^+ move out of the cell. The plateau phase occurs as voltage-gated Ca^{2+} channels remain open, and the movement of Ca^{2+} and some Na^+ through the voltage-gated Ca^{2+} channels into the cell counteracts the potential change produced by the movement of K^+ out of the cell. The plateau phase ends and final repolarization begins as the voltage-gated Ca^{2+} channels close and many more voltage-gated K^+ channels open. Thus Ca^{2+} and Na^+ stop diffusing into the cell, and the tendency for K^+ to diffuse out of the cell increases. These permeability changes cause the membrane potential to return to its resting level.

Action potentials in cardiac muscle are conducted from cell to cell, whereas action potentials in skeletal muscle fibers are conducted along the length of a single muscle fiber, but not from fiber to fiber. Also, the rate of action potential propagation is slower in cardiac muscle than in skeletal muscle because cardiac muscle cells are smaller in diameter and much shorter than skeletal muscle fibers. Although the gap junctions of intercalated disks allow transfer of action potentials between cardiac muscle cells, they do slow the rate of action potential conduction between the cardiac muscle cells.

The movement of Ca^{2+} through the plasma membrane, including the membranes of the T tubules, into cardiac muscle cells stimulates the release of Ca^{2+} from the sarcoplasmic reticulum, a process called **calcium-induced calcium release (CICR).** When an action potential occurs in a cardiac muscle cell, Ca^{2+} enter the cell and bind to receptors in the membranes of sarcoplasmic reticulum resulting in the opening of Ca^{2+} channels. Ca^{2+} then move out of the sarcoplasmic reticulum and activate the interaction between actin and myosin to produce contraction of the cardiac muscle cells.

Autorhythmicity of Cardiac Muscle

The heart is said to be **autorhythmic** (aw′tō-rith′mik) because it stimulates itself (*auto*) to contract at regular intervals (*rhythmic*). If the heart is removed from the body and maintained under physiologic conditions with the proper nutrients and temperature, it will continue to beat autorhythmically for a long time.

In the SA node, pacemaker cells generate action potentials spontaneously and at regular intervals. These action potentials spread through the conducting system of the heart to other cardiac muscle cells, causing voltage-gated Na^+ channels to open. As a result, action potentials are produced and the cardiac muscle cells contract.

(a)

Permeability changes during an action potential in skeletal muscle:

1. **Depolarization phase**
   - Voltage-gated Na⁺ channels open.
   - Voltage-gated K⁺ channels begin to open.

2. **Repolarization phase**
   - Voltage-gated Na⁺ channels close.
   - Voltage-gated K⁺ channels continue to open.
   - Voltage-gated K⁺ channels close at the end of repolarization and return the membrane potential to its resting value.

(b)

Permeability changes during an action potential in cardiac muscle:

1. **Depolarization phase**
   - Voltage-gated Na^+ channels open.
   - Voltage-gated K^+ channels close.
   - Voltage-gated Ca^{2+} channels begin to open.

2. **Early repolarization and plateau phases**
   - Voltage-gated Na^+ channels close.
   - Some voltage-gated K^+ channels open, causing early repolarization.
   - Voltage-gated Ca^{2+} channels are open, producing the plateau by slowing further repolarization.

3. **Final repolarization phase**
   - Voltage-gated Ca^{2+} channels close.
   - Many voltage-gated K^+ channels open.

Process Figure 20.14   **Comparison of Action Potentials in Skeletal and Cardiac Muscle**

(*a*) An action potential in skeletal muscle consists of depolarization and repolarization phases. (*b*) An action potential in cardiac muscle consists of depolarization, early repolarization, plateau, and final repolarization phases. Cardiac muscle does not repolarize as rapidly as skeletal muscle (*indicated by the break in the curve*) because of the plateau phase.

     The generation of action potentials in the SA node results when a spontaneously developing local potential, called the **prepotential,** reaches threshold (figure 20.15). Changes in ion movement into and out of the pacemaker cells cause the prepotential. Na⁺ cause depolarization by moving into the cells through specialized nongated Na⁺ channels. A decreasing permeability to K⁺ also causes depolarization as fewer K⁺ move out of the cells. The decreasing K⁺ permeability occurs due to the voltage changes at the end of the previous action potential. As a result of the depolarization, voltage-gated Ca²⁺ channels open, and the movement of Ca²⁺ into the pacemaker cells causes further depolarization. When the prepotential reaches threshold, many voltage-gated Ca²⁺ channels open. Unlike other cardiac muscle cells, the movement of Ca²⁺ into the pacemaker cells is primarily responsible for the depolarization phase of the action potential. Repolarization occurs, as in other cardiac muscle cells, when the voltage-gated Ca²⁺ channels close and the voltage-gated K⁺ channels open. After the RMP is reestablished, production of another prepotential starts the generation of the next action potential.

Drugs that Block Calcium Channels

Various chemical agents like manganese ions (Mn^{2+}) and verapamil (ver-ap′ă-mil) block voltage-gated Ca^{2+} channels. Voltage-gated Ca^{2+} channel-blocking agents prevent the movement of Ca^{2+} through voltage-gated Ca^{2+} channels into the cell and, for that reason, are called **calcium channel blockers.** Some calcium channel blockers are widely used clinically in the treatment of various cardiac disorders, including tachycardia and certain arrhythmias. Calcium channel blockers slow the development of the prepotential and thus reduce the heart rate. If action potentials arise prematurely within the SA node or other areas of the heart, calcium channel blockers reduce that tendency. Calcium channel blockers also reduce the amount of work performed by the heart because less calcium enters cardiac muscle cells to activate the contractile mechanism. On the other hand, epinephrine and norepinephrine increase the heart rate and its force of contraction by opening voltage-gated Ca^{2+} channels.

Permeability changes in pacemaker cells:
1. **Prepotential**
   - A small number of Na^+ channels are open.
   - Voltage-gated K^+ channels that opened in the repolarization phase of the previous action potential are closing.
   - Voltage-gated Ca^{2+} channels begin to open.

2. **Depolarization phase**
   - Voltage-gated Ca^{2+} channels are open.
   - Voltage-gated K^+ channels are closed.

3. **Repolarization phase**
   - Voltage-gated Ca^{2+} channels close.
   - Voltage-gated K^+ channels open.

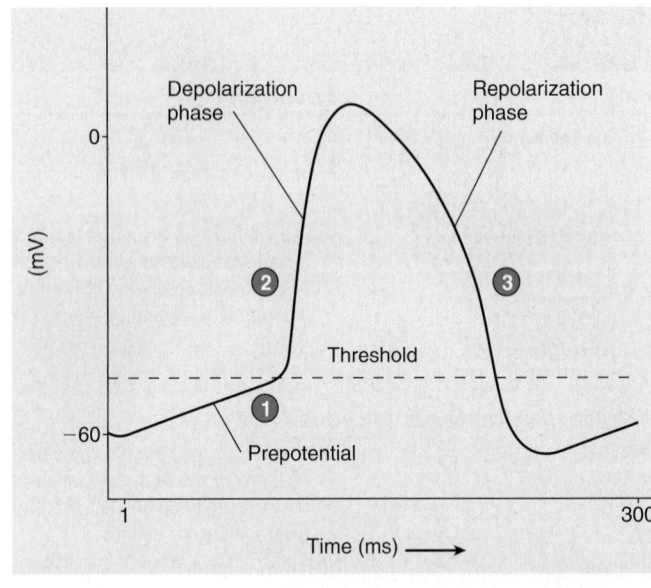

Process Figure 20.15   SA Node Action Potential

The production of action potentials by the SA node is responsible for the autorhythmicity of the heart.

Although most cardiac muscle cells respond to action potentials produced by the SA node, some cardiac muscle cells in the conducting system can generate spontaneous action potentials. Normally, the SA node controls the rhythm of the heart because its pacemaker cells generate action potentials at a faster rate than other potential pacemaker cells to produce a heart rate of 70–80 beats per minute (bpm). An **ectopic focus** (ek-top'ik fō'kŭs; pl., foci, fō'sī) is any part of the heart other than the SA node that generates a heartbeat. For example, if the SA node does not function properly, the part of the heart to produce action potentials at the next highest frequency is the AV node, which produces a heart rate of 40–60 bpm. Another cause of an ectopic focus is blockage of the conducting pathways between the SA node and other parts of the heart. For example, if action potentials do not pass through the AV node, an ectopic focus can develop in an AV bundle, resulting in a heart rate of 30 bpm.

Ectopic foci can also appear when the rate of action potential generation in cardiac muscle cells outside of the SA node becomes enhanced. For example, when cells are injured their plasma membranes become more permeable, resulting in depolarization. Inflammation or lack of adequate blood flow to cardiac muscle tissue can injure cardiac muscle cells. These injured cells can be the source of ectopic action potentials. Also, alterations in blood levels of K^+ and Ca^{2+} can change the cardiac muscle membrane potential and certain drugs, such as those that mimic the effect of epinephrine on the heart, can alter cardiac muscle membrane permeability. Changes in cardiac muscle cells' membrane potentials or permeability can produce ectopic foci.

P R E D I C T   4
Predict the consequences for the pumping effectiveness of the heart if numerous ectopic foci in the ventricles produce action potentials at the same time.

Refractory Period of Cardiac Muscle

Cardiac muscle, like skeletal muscle, has **refractory** (rē-frak'tōr-ē) **periods** associated with its action potentials. During the **absolute refractory period,** the cardiac muscle cell is completely insensitive to further stimulation, and during the **relative refractory period** the cell exhibits reduced sensitivity to additional stimulation. Because the plateau phase of the action potential in cardiac muscle delays repolarization to the RMP, the refractory period is prolonged. The long refractory period ensures that, after contraction, relaxation is nearly complete before another action potential can be initiated, thus preventing tetanic contractions in cardiac muscle.

P R E D I C T   5
Predict the consequences if cardiac muscle could undergo tetanic contraction.

Electrocardiogram

The conduction of action potentials through the myocardium during the cardiac cycle produces electric currents that can be measured at the surface of the body. Electrodes placed on the surface of the body and attached to an appropriate recording device can detect small voltage changes resulting from action potentials in the cardiac muscle. The electrodes detect a summation of all the action potentials that are transmitted through the heart at a given time. Electrodes do not detect individual action potentials. The summated record of the cardiac action potentials is an electrocardiogram (ECG or EKG).

The ECG is not a direct measurement of mechanical events in the heart, and neither the force of contraction nor blood pressure can be determined from it. Each deflection in the ECG record, however, indicates an electrical event within the heart and correlates with a subsequent mechanical event. Consequently, it's an extremely valuable diagnostic tool in identifying a number of abnormal cardiac rhythms (arrythmias) (table 20.1) and other

Table 20.1 Major Cardiac Arrhythmias

Conditions	Symptoms	Possible Causes
Abnormal Heart Rhythms		
Tachycardia	Heart rate in excess of 100 bpm	Elevated body temperature; excessive sympathetic stimulation; toxic conditions
Paroxysmal atrial tachycardia	Sudden increase in heart rate to 95–150 bpm for a few seconds or even for several hours; P wave precedes every QRS complex; P wave inverted and superimposed on T wave	Excessive sympathetic stimulation; abnormally elevated permeability of slow channels
Ventricular tachycardia	Frequently causes fibrillation	Often associated with damage to AV node or ventricular muscle
Abnormal Rhythms Resulting from Ectopic Action Potentials		
Atrial flutter	300 P waves/min; 125 QRS complexes/min resulting in two or three P waves (atrial contraction) for every QRS complex (ventricular contraction)	Ectopic action potentials in the atria
Atrial fibrillation	No P waves; normal QRS complexes; irregular timing; ventricles constantly stimulated by atria; reduced pumping effectiveness and filling time	Ectopic action potentials in the atria
Ventricular fibrillation	No QRS complexes; no rhythmic contraction of the myocardium; many patches of asynchronously contracting ventricular muscle	Ectopic action potentials in the ventricles
Bradycardia	Heart rate less than 60 bpm	Elevated stroke volume in athletes; excessive vagal stimulation; carotid sinus syndrome
Sinus Arrhythmia	Heart rate varies 5% during respiratory cycle and up to 30% during deep respiration	Cause not always known; occasionally caused by ischemia or inflammation or associated with cardiac failure
SA Node Block	Cessation of P wave; new low heart rate due to AV node acting as pacemaker; normal QRS complex and T wave	Ischemia; tissue damage due to infarction; causes unknown
AV Node Block		
First-degree	PR interval greater than 0.2 second	Inflammation of AV bundle
Second-degree	PR interval 0.25–0.45 second; some P waves trigger QRS complexes and others do not; 2:1, 3:1, and 3:2 P wave/QRS complex ratios may occur	Excessive vagal stimulation
Complete heart block	P wave dissociated from QRS complex; atrial rhythm approximately 100 bpm; ventricular rhythm less than 40 bpm	Ischemia of AV nodal fibers or compression of AV bundle
Premature Atrial Contractions	Occasional shortened intervals between one contraction and the succeeding contraction; frequently occurs in healthy people P wave superimposed on QRS complex	Excessive smoking; lack of sleep; too much caffeine; alcoholism
Premature Ventricular Contractions (PVCs)	Prolonged QRS complex; exaggerated voltage because only one ventricle may depolarize; inverted T wave; increased probability of fibrillation	Ectopic foci in ventricles; lack of sleep; too much caffeine, irritability; occasionally occurs with coronary thrombosis

Abbreviations: SA = sinoatrial; AV = atrioventricular.

abnormalities, particularly because it is painless, easy to record, and noninvasive (meaning that it doesn't require surgical procedures). Abnormal heart rates or rhythms, abnormal conduction pathways, hypertrophy or atrophy of portions of the heart, and the approximate location of damaged cardiac muscle can be determined from analysis of an ECG (table 20.1).

The normal ECG consists of a P wave, a QRS complex, and a T wave (figure 20.16). The **P wave,** which is the result of action potentials that cause depolarization of the atrial myocardium, signals the onset of atrial contraction. The **QRS complex** is composed of three individual waves: the Q, R, and S waves. The QRS complex results from ventricular depolarization and signals the onset of ventricular contraction. The **T wave** represents repolarization of the ventricles and precedes ventricular relaxation. A wave representing repolarization of the atria cannot be seen because it occurs during the QRS complex.

The time between the beginning of the P wave and the beginning of the QRS complex is the PQ interval, commonly called the PR interval because the Q wave is often very small. During the PR interval, which lasts approximately 0.16 second, the atria contract and begin to relax. The ventricles begin to depolarize at the end of the PR interval. The QT interval extends from the beginning of the QRS complex to the end of the T wave, lasts approximately 0.36 second, and represents the approximate length of time required for the ventricles to contract and begin to relax.

Alterations in the Electrocardiogram

Elongation of the PR interval can result from (1) a delay in action potential conduction through the atrial muscle because of damage, such as that caused by **ischemia** (is-kē′mē-ă), which is the obstruction of the blood supply to the walls of the heart, (2) a delay of action potential conduction through atrial muscle because of a dilated atrium, or (3) a delay of action potential conduction through the AV node and bundle because of ischemia, compression, or necrosis of the AV node or bundle. These conditions result in slow conduction of action potentials through the bundle branches. An unusually long QT interval reflects the abnormal conduction of action potentials through the ventricles, which can result from myocardial infarctions or from an abnormally enlarged left or right ventricle.

Examples of alteration in the form of the electrocardiogram due to cardiac abnormalities are illustrated in figure 20.17. Examples include complete heart block, premature ventricular contraction, bundle branch block, atrial fibrillation, and ventricular fibrillation.

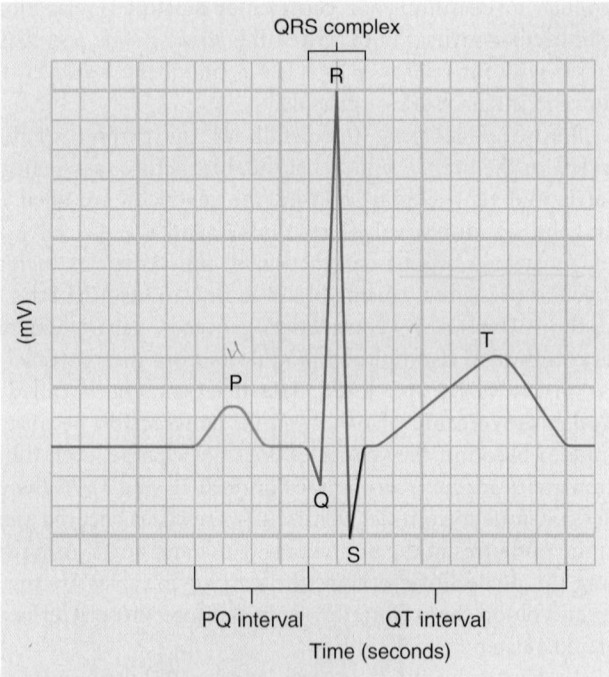

Figure 20.16 Electrocardiogram

The major waves and intervals of an electrocardiogram are labeled. Each thin horizontal line on the ECG recording represents 1 mV, and each thin vertical line represents 0.04 second.

26. *For cardiac muscle action potentials, describe ion movement during the depolarization, early repolarization, plateau, and final repolarization phases. What ions are associated with fast channels and slow channels?*

27. *Why is cardiac muscle referred to as autorhythmic? What are ectopic foci?*

28. *How does the depolarization of pacemaker cells differ from the depolarization of other cardiac muscle cells? What is the prepotential?*

29. *Why does cardiac muscle have a prolonged refractory period? What is the advantage of a prolonged refractory period?*

30. *What does an ECG measure? Name the waves produced by an ECG, and state what events occur during each wave.*

Complete heart block (P waves and QRS complexes are not coordinated)

Premature ventricular contraction (PVC) (no P waves precede PVCs)

Bundle branch block

Atrial fibrillation (no clear P waves and rapid QRS complexes)

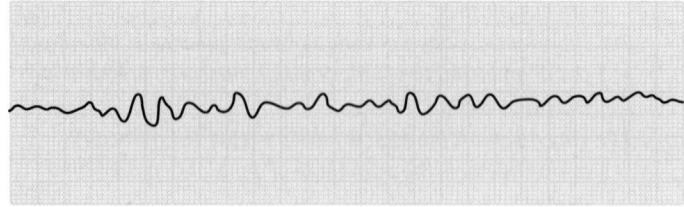

Ventricular fibrillation (no P, QRS, or T waves)

Figure 20.17 **Examples of Alterations in the Electrocardiogram**

Cardiac Cycle

The heart is actually two separate pumps that work together, one in the right half and the other in the left half of the heart. Each pump consists of a primer pump—the atrium—and a power pump—the ventricle. Both atrial primer pumps complete the filling of the ventricles with blood, and both ventricular power pumps produce

the major force that causes blood to flow through the pulmonary and systemic arteries. The term **cardiac cycle** refers to the repetitive pumping process that begins with the onset of cardiac muscle contraction and ends with the beginning of the next contraction (figures 20.18 and 20.19; table 20.2). Pressure changes produced within the heart chambers as a result of cardiac muscle contraction are responsible for blood movement because blood moves from areas of higher pressure to areas of lower pressure.

The duration of the cardiac cycle varies considerably among humans and also during an individual's lifetime. It can be as short as 0.25–0.3 second in a newborn infant or as long as 1 or more seconds in a well-trained athlete. The normal cardiac cycle of 0.7–0.8 second depends on the capability of cardiac muscle to contract and on the functional integrity of the conducting system.

The term **systole** (sis′tō-lē) means to contract, and **diastole** (dī-as′tō-lē) means to dilate. **Atrial systole** is contraction of the atrial myocardium, and **atrial diastole** is relaxation of the atrial myocardium. Similarly, **ventricular systole** is contraction of the ventricular myocardium, and **ventricular diastole** is relaxation of the ventricular myocardium. When the terms *systole* and *diastole* are used without reference to specific chambers, however, they mean ventricular systole or diastole.

Before considering the details of the cardiac cycle, an overview of the main events is helpful. Just before systole begins, the atria and ventricles are relaxed, the ventricles are filled with blood, the semilunar valves are closed, and the AV valves are open. As systole begins, contraction of the ventricles increases ventricular pressures, causing blood to flow toward the atria and close the AV valves. As contraction proceeds, ventricular pressures continue to rise, but no blood flows from the ventricles because all the valves are closed. This brief interval is called the **period of isovolumic** (ī′sō-vol-ū′mik) **contraction** because the volume of blood in the ventricles does not change, even though the ventricles are contracting (see figure 20.18 step 1). As the ventricles continue to contract, ventricular pressures become greater than the pressures in the pulmonary trunk and aorta. As a result, during the **period of ejection,** the semilunar valves are pushed open and blood flows from the ventricles into those arteries (see figure 20.18 step 2).

As diastole begins, the ventricles relax and ventricular pressures decrease below the pressures in the pulmonary trunk and aorta. Consequently, blood begins to flow back toward the ventricles, causing the semilunar valves to close (see figure 20.18 step 3). With closure of the semilunar valves, all the heart valves are closed and no blood flows into the relaxing ventricles during the **period of isovolumic relaxation.**

Throughout ventricular systole and the period of isovolumic relaxation, the atria relax and blood flows into them from the veins. As the ventricles continue to relax, ventricular pressures become lower than atrial pressures, the AV valves open, and blood flows from the atria into the relaxed ventricles (see figure 20.18 step 4). At rest, most ventricular filling is a passive process resulting from the greater pressure of blood in the veins and atria than in the completely relaxed ventricles. Completion of ventricular filling is an active process resulting from increased atrial pressure produced by contraction of the atria (see figure 20.18 step 5).

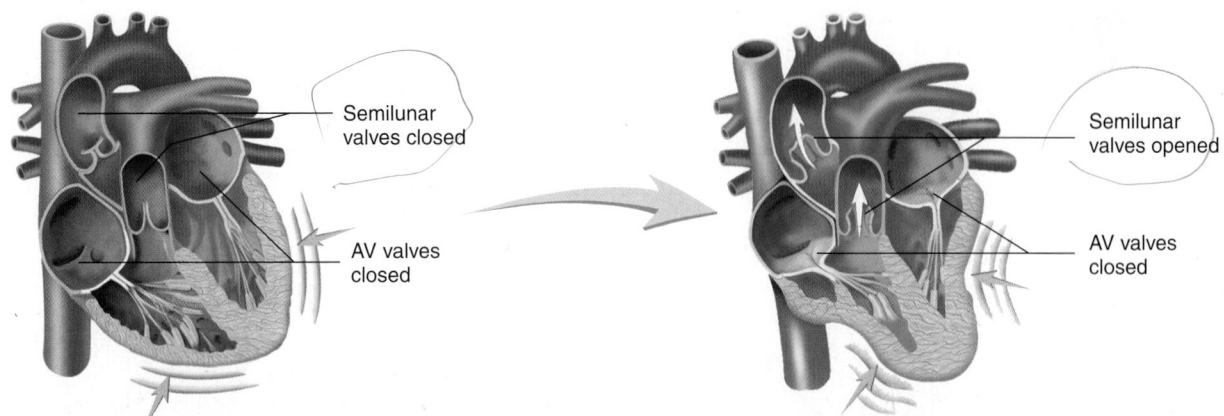

1. *Systole: Period of isovolumic contraction.* Ventricular contraction causes the AV valves to close, which is the beginning of ventricular systole. The semilunar valves were closed in the previous diastole and remain closed during this period.

2. *Systole: Period of ejection.* Continued ventricular contraction pushes blood out of the ventricles, causing the semilunar valves to open.

5. *Diastole: Active ventricular filling.* The atria contract and complete ventricular filling.

3. *Diastole: Period of isovolumic relaxation.* Blood flowing back toward the relaxed ventricles causes the semilunar valves to close, which is the beginning of ventricular diastole. Note that the AV valves are closed, also.

4. *Diastole: Passive ventricular filling.* The AV valves open and blood flows into the relaxed ventricles, accounting for most of the ventricular filling.

Process Figure 20.18 The Cardiac Cycle

The cardiac cycle is a repeating series of contraction and relaxation that moves blood through the heart. See figure 20.19 and table 20.2 for additional details and explanations. (*AV* = atrioventricular)

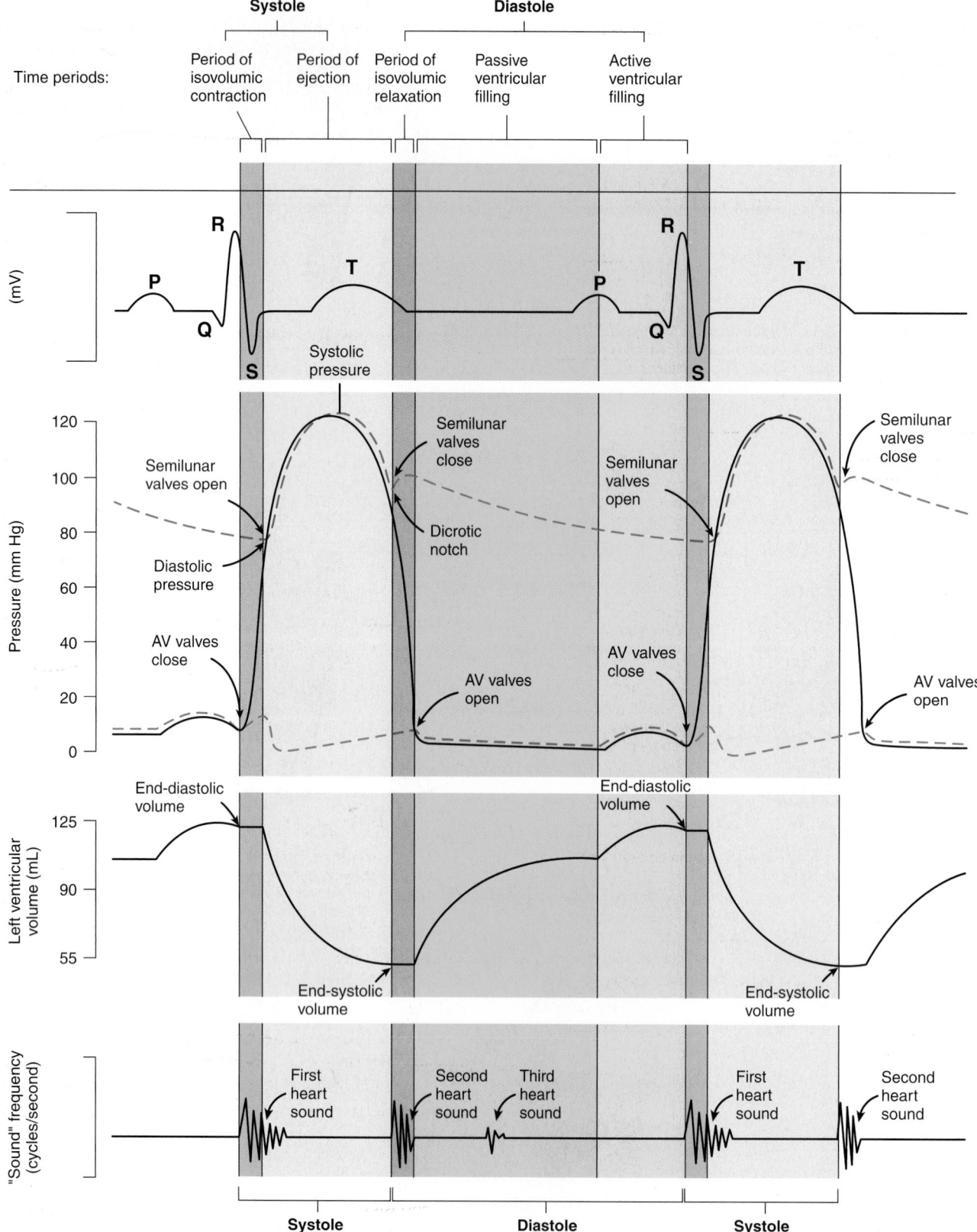

Figure 20.19   Events Occurring During the Cardiac Cycle

The cardiac cycle is divided into five periods (*see top of figure*). Within these periods, four graphs are presented. From *top* to *bottom,* the electrocardiogram; pressure changes for the left atrium (*blue line*), left ventricle (*black line*), and aorta (*red line*); left ventricular volume curve; and heart sounds are illustrated. See table 20.2 for explanations of events during each period and figure 20.18 for illustrations of the valves and blood flow movement.

During exercise, atrial contraction is more important for ventricular filling because, as heart rate increases, less time is available for passive ventricular filling.

Events Occurring During Ventricular Systole

Figure 20.19 displays the main events of the cardiac cycle in graphic form and should be examined from top to bottom for each period of the cardiac cycle. The ECG indicates the electrical events that cause contraction and relaxation of the atria and ventricles. The pressure graph shows the pressure changes within the left atrium, left ventricle, and aorta resulting from atrial and ventricular contraction and relaxation. Although pressure changes in the right side of the heart are not shown, they are similar to those in the left side, only lower. The volume graph presents the changes in left ventricular volume as blood flows into and out of the left ventricle as a result of the pressure changes. The sound graph records the closing of valves caused by blood flow. See also figure 20.18 for illustrations of the valves and blood flow and table 20.2 for a summary of the events occurring during each period.

Period of Isovolumic Contraction

Completion of the QRS complex initiates contraction of the ventricles. Ventricular pressure rapidly increases, resulting in closure of the AV valves. During the previous ventricular diastole, the ventricles were filled with 120–130 mL of blood, which is called the **end-diastolic volume.** Ventricular volume doesn't change during the period of isovolumic contraction because all the heart valves are closed at this time.

> **P R E D I C T** **6**
>
> Is the cardiac muscle contracting isotonically or isometrically during the period of isovolumic contraction?

Period of Ejection

As soon as ventricular pressures exceed the pressures in the aorta and pulmonary trunk, the semilunar valves open. The aortic semilunar valve opens at approximately 80 millimeters of mercury (mm Hg) ventricular pressure, whereas the pulmonary semilunar valve opens at approximately 8 mm Hg. Although the pressures are different, both valves open at nearly the same time.

As blood flows from the ventricles during the period of ejection, the left ventricular pressure continues to climb to approximately 120 mm Hg, and the right ventricular pressure increases to approximately 25 mm Hg. The larger left ventricular pressure causes blood to flow throughout the body (systemic circulation), whereas the lower right ventricle pressure causes blood to flow through the lungs (pulmonary circuit). Even though the pressure generated by the left ventricle is much higher than that of the right ventricle, the amount of blood pumped by each is almost the same.

> **P R E D I C T** **7**
>
> Which ventricle has the thickest wall? Why is it important for each ventricle to pump approximately the same volume of blood?

During the first part of ejection, blood flows rapidly out of the ventricles. Toward the end of ejection, very little blood flow occurs, which causes the ventricular pressure to decrease despite continued ventricular contraction. At the end of ejection, the volume has decreased to 50–60 mL, which is called the **end-systolic volume.**

Events Occurring During Ventricular Diastole

Period of Isovolumic Relaxation

Completion of the T wave results in ventricular repolarization and relaxation. The already decreasing ventricular pressure falls very rapidly as the ventricles suddenly relax. When the ventricular pressures fall below the pressures in the aorta and pulmonary trunk, the recoil of the elastic arterial walls, which were stretched during the period of ejection, forces the blood to flow back toward the ventricles, thereby closing the semilunar valves. Ventricular volume doesn't change during the period of isovolumic relaxation because all the heart valves are closed at this time.

Passive Ventricular Filling

During ventricular systole and the period of isovolumic relaxation, the relaxed atria fills with blood. As ventricular pressure drops below atrial pressure, the atrioventricular valves open and allow blood to flow from the atria into the ventricles. Blood flows from the area of higher pressure in the veins and atria toward the area of lower pressure in the relaxed ventricles. Most ventricular filling occurs during the first one-third of ventricular diastole. At the end of passive ventricular filling, the ventricles are approximately 70% filled.

> **P R E D I C T** **8**
>
> Fibrillation is abnormal, rapid contractions of different parts of the heart that prevent the heart muscle from contracting as a single unit. Explain why atrial fibrillation does not immediately cause death, but ventricular fibrillation does.

Active Ventricular Filling

Depolarization of the SA node generates action potentials that spread over the atria, producing the P wave and stimulating both atria to contract (atrial systole). The atria contract during the last one-third of ventricular diastole and complete ventricular filling.

Under most conditions, the atria function primarily as reservoirs, and the ventricles can pump sufficient blood to maintain homeostasis even if the atria do not contract at all. During exercise, however, the heart pumps 300%–400% more blood than during resting conditions. It is under these conditions that the pumping action of the atria becomes important in maintaining the pumping efficiency of the heart.

Table 20.2   Summary of the Events of the Cardiac Cycle

	Ventricular Systole	
	Period of Isovolumic Contraction	**Period of Ejection**
Time Period	The ventricles begin to contract, but ventricular volume doesn't change.	The ventricles continue to contract and blood is pumped out of the ventricles.
Condition of Valves	Semilunar valves closed; AV valves closed (see figure 20.18a).	Semilunar valves opened; AV valves closed (see figure 20.18b).
ECG	The QRS complex is completed and the ventricles are depolarized. As a result, the ventricles begin to contract. Atrial repolarization is masked by the QRS complex. The atria are relaxed (atrial diastole).	The **T wave** results from ventricular repolarization.
Atrial Pressure Graph	Atrial pressure decreases in the relaxed atria. When atrial pressure is less than venous pressure, blood flows into the atria. Atrial pressure increases briefly as the contracting ventricles push blood back toward the atria.	Atrial pressure increases gradually as blood flows from the veins into the relaxed atria.
Ventricular Pressure Graph	Ventricular contraction causes an increase in ventricular pressure, which causes blood to flow toward the atria, closing the AV valves. Ventricular pressure increases rapidly.	Ventricular pressure becomes greater than pressure in the aorta as the ventricles continue to contract. The semilunar valves are pushed open as blood flows out of the ventricles. Ventricular pressure peaks as the ventricles contract maximally; then pressure decreases as blood flow out of the ventricles decreases.
Aortic Pressure Graph	Just before the semilunar valves open, pressure in the aorta decreases to its lowest value, called the **diastolic pressure** (approximately 80 mm Hg).	As ventricular contraction forces blood into the aorta, pressure in the aorta increases to its highest value, called the **systolic pressure** (approximately 120 mm Hg).
Volume Graph	During the **period of isovolumic contraction,** ventricular volume doesn't change because the semilunar and AV valves are closed.	After the semilunar valves open, blood volume decreases as blood flows out of the ventricles during the **period of ejection.** The amount of blood left in a ventricle at the end of the period of ejection is called the **end-systolic volume.**
Sound Graph	Blood flowing from the ventricles toward the atria closes the AV valves. Vibrations of the valves and the turbulent flow of blood produce the **first heart sound,** which marks the beginning of ventricular systole.	

Heart Sounds

Distinct sounds are heard when a stethoscope is used to listen to the heart (figure 20.20 and see figure 20.19). The **first heart sound** is a low-pitched sound, often described as a "lubb" sound. It's caused by vibration of the atrioventricular valves and surrounding fluid as the valves close at the beginning of ventricular systole. The **second heart sound** is a higher-pitched sound often described as a "dupp" sound. It results from closure of the aortic and pulmonary semilunar valves, at the beginning of ventricular diastole. Systole is, therefore, approximately the time between the first and second heart sounds. Diastole, which lasts somewhat longer, is approximately the time between the second heart sound and the next first heart sound.

Occasionally, a faint **third heart sound,** can be heard in some normal people, particularly in those who are thin and young. It is caused by blood flowing in a turbulent fashion into

Ventricular Diastole		
Period of Isovolumic Relaxation	**Passive Ventricular Filling**	**Active Ventricular Filling**
The ventricles relax, but ventricular volume doesn't change.	Blood flows into the ventricles because blood pressure is higher in the veins and atria than in the relaxed ventricles.	Contraction of the atria pumps blood into the ventricles.
Semilunar valves closed; AV valves closed (see figure 20.18c).	Semilunar valves closed; AV valves opened (see figure 20.18d).	Semilunar valves closed; AV valves opened (see figure 20.18e).
The T wave is completed and the ventricles are repolarized. The ventricles relax.	The **P wave** is produced when the SA node generates action potentials and a wave of depolarization begins to propagate across the atria.	The P wave is completed and the atria are stimulated to contract. Action potentials are delayed in the AV node for 0.11 second, allowing time for the atria to contract. The **QRS complex** begins as action potentials are propagated from the AV node to the ventricles.
Atrial pressure continues to increase gradually as blood flows from the veins into the relaxed atria.	After the AV valves open, atrial pressure decreases as blood flows out of the atria into the relaxed ventricles.	Atrial contraction (systole) causes an increase in atrial pressure, and blood is forced to flow from the atria into the ventricles.
Elastic recoil of the aorta pushes blood back toward the heart, causing the semilunar valves to close. After closure of the semilunar valves, the pressure in the relaxing ventricles rapidly decreases.	No significant change occurs in ventricular pressure during this time period.	Atrial contraction (systole) and the movement of blood into the ventricles cause a slight increase in ventricular pressure.
After the semilunar valves close, elastic recoil of the aorta causes a slight increase in aortic pressure, producing the **dicrotic notch,** or **incisura.**	Aortic pressure gradually decreases as blood runs out of the aorta into other systemic blood vessels.	Aortic pressure gradually decreases as blood runs out of the aorta into other systemic blood vessels.
During the **period of isovolumic relaxation,** ventricular volume doesn't change because the semilunar and AV valves are closed.	After the AV valves open, blood flows from the atria and veins into the ventricles because of pressure differences. Most ventricular filling occurs during the first one-third of diastole. Little ventricular filling occurs during the middle one-third of diastole.	Atrial contraction (systole) completes ventricular filling during the last one-third of diastole.
Blood flowing from the ventricles toward the aorta and pulmonary trunk closes the semilunar valves. Vibrations of the valves and the turbulent flow of blood produce the **second heart sound,** which marks the beginning of ventricular diastole.	Sometimes the turbulent flow of blood into the ventricles produces a **third heart sound.**	The amount of blood in a ventricle at the end of ventricular diastole is called the **end-diastolic volume.**

Abbreviation: AV = atrioventricular.

the ventricles, and it can be detected near the end of the first one-third of diastole.

Aortic Pressure Curve

The elastic walls of the aorta are stretched as blood is ejected into the aorta from the left ventricle. Aortic pressure remains slightly below ventricular pressure during this period of ejection. As ventricular pressure drops below that in the aorta, blood flows back toward the ventricle because of the elastic recoil of the aorta. Consequently, the aortic semilunar valve closes, and pressure within the aorta increases slightly, producing a **dicrotic** (dī-krot'ik) **notch** in the aortic pressure curve (see figure 20.19). The term *dicrotic* means double-beating; when increased pressure caused by recoil is large, a double pulse can be felt. The dicrotic notch is also called an **incisura** (in'sī-soo'ră; a cutting into). Aortic pressure then gradually falls throughout the

Clinical Focus   Abnormal Heart Sounds

Heart sounds provide important information to clinicians about the normal function of the heart and assist in diagnosing cardiac abnormalities. Abnormal heart sounds are called **murmurs** (mer′merz), and certain murmurs are important indicators of specific cardiac abnormalities. For example, an **incompetent valve** leaks significantly. After an incompetent valve closes, blood flows through it but in a reverse direction. The movement of blood in a direction opposite to normal results in turbulence, which causes a gurgling or swishing sound immediately after the valve closes. An incompetent tricuspid valve

or bicuspid valve makes a swish sound immediately after the first heart sound, and the first heart sound may be muffled. An incompetent aortic or pulmonary semilunar valve results in a swish sound immediately after the second heart sound.

Stenosed (sten′ōzd) valves have an abnormally narrow opening and also produce abnormal heart sounds. Blood flows through stenosed valves in a very turbulent fashion and produces a rushing sound before the valve closes. A stenosed atrioventricular valve, therefore, results in a rushing sound immediately before the first heart

sound, and a stenosed semilunar valve results in a rushing sound immediately before the second heart sound.

Inflammation of the heart valves, resulting from conditions like rheumatic fever, can cause valves to become either incompetent or stenosed. In addition, myocardial infarctions that make papillary muscles nonfunctional can cause bicuspid or tricuspid valves to be incompetent.

Either incompetent or stenosed valves increase the amount of work the cardiac muscle must perform. Consequently, these conditions can lead to heart failure.

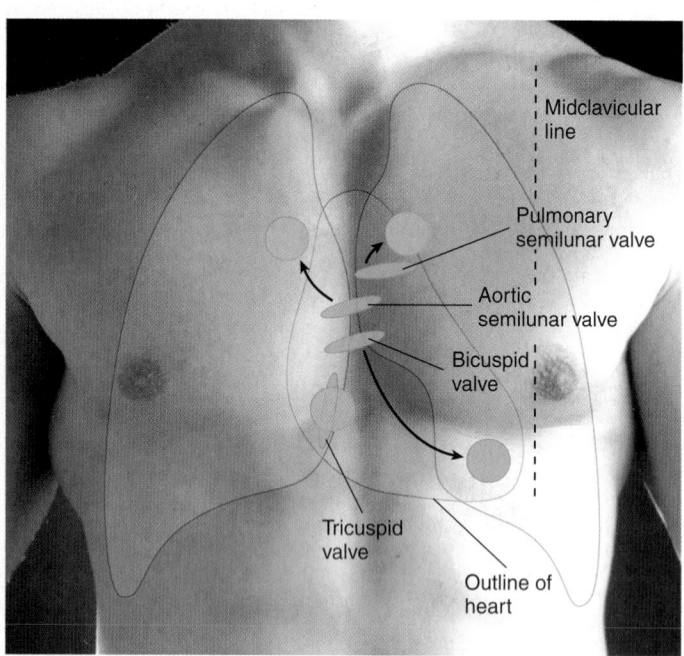

Figure 20.20   Location of the Heart Valves in the Thorax

Surface markings of the heart in the male. The positions of the four heart valves are indicated by *blue ellipses,* and the sites where the sounds of the valves are best heard with the stethoscope are indicated by *pink circles.*

rest of ventricular diastole as blood flows through the peripheral vessels. By the time aortic pressure has fallen to approximately 80 mm Hg, the ventricles again contract, forcing blood once more into the aorta.

Blood pressure measurements performed for clinical purposes reflect the pressure changes that occur in the aorta rather

than in the left ventricle (see chapter 21). The blood pressure in the aorta fluctuates between systolic pressure, which is about 120 mm Hg, and diastolic pressure, which is about 80 mm Hg for the average young adult at rest.

31. *Define systole and diastole.*
32. *List the five periods of the cardiac cycle (see figure 20.19 and table 20.2), and state whether the AV and semilunar valves are open or closed during each period.*
33. *Define isovolumic. When does most ventricular filling occur?*
34. *Define end-diastolic volume and end-systolic volume.*
35. *What produces the first heart sound, the second heart sound, and the third heart sound?*
36. *Explain the production in the aorta of systolic pressure, diastolic pressure, and the dicrotic notch, or incisura.*

Mean Arterial Blood Pressure

Blood pressure is necessary for blood movement and, therefore, is critical to the maintenance of homeostasis. Blood flows from areas of higher to areas of lower pressure. For example, during one cardiac cycle, blood flows from the higher pressure in the aorta, resulting from contraction of the left ventricle, toward the lower pressure in the relaxed right atrium.

Mean arterial pressure (MAP) is slightly less than the average of the systolic and diastolic pressure in the aorta. It's proportional to **cardiac output (CO)** times **peripheral resistance (PR).** Cardiac output, or **minute volume,** is the amount of blood pumped by the heart per minute, and peripheral resistance is the total resistance against which blood must be pumped.

$$MAP = CO \times PR$$

Changes in cardiac output and peripheral resistance (figure 20.21) can alter mean arterial pressure. Cardiac output is discussed in this chapter, and peripheral resistance is explained in chapter 21.

Cardiac output is equal to heart rate times stroke volume. **Heart rate (HR)** is the number of times the heart beats (contracts) per minute. **Stroke volume (SV),** which is the volume of blood pumped during each heartbeat (cardiac cycle), is equal to end-diastolic volume minus end-systolic volume. During diastole, blood flows from the atria into the ventricles, and end-diastolic volume normally increases to approximately 125 mL. After the ventricles partially empty during systole, end-systolic volume decreases to approximately 55 mL. The **stroke volume** is therefore equal to 70 mL (125−55).

To better understand stroke volume, imagine that you're rinsing out a sponge under a running water faucet. As you relax your hand, the sponge fills with water; as your fingers contract, water is squeezed out of the sponge; and, after you have squeezed it, some water is left in the sponge. In this analogy, the amount of water you squeeze out of the sponge (stroke volume) is the difference between the amount of water in the sponge when your hand is relaxed (end-diastolic volume) and the amount that is left in the sponge after you squeeze it (end-systolic volume).

Stroke volume can be increased by increasing end-diastolic volume or by decreasing end-systolic volume (see figure 20.21).

During exercise, end-diastolic volume increases because of an increase in **venous return,** which is the amount of blood returning to the heart from the peripheral circulation. End-systolic volume decreases because the heart contracts more forcefully. For example, stroke volume could increase from a resting value of 70 mL to an exercising value of 115 mL by increasing end-diastolic volume to 145 mL and decreasing end-systolic volume to 30 mL.

Under resting conditions, the heart rate is approximately 72 bpm, and the stroke volume is approximately 70 mL/beat, although these values can vary considerably from person to person. The cardiac output is therefore

$$CO = HR \times SV$$
$$= 72 \text{ bpm} \times 70 \text{ mL/beat}$$
$$= 5040 \text{ mL/min (approximately 5 L/min)}$$

During exercise, heart rate can increase to 190 bpm, and the stroke volume can increase to 115 mL. Consequently, cardiac output is

$$CO = 190 \text{ bpm} \times 115 \text{ mL/beat}$$
$$= 21,850 \text{ mL/min (approximately 22 L/min)}$$

The difference between cardiac output when a person is at rest and maximum cardiac output is called **cardiac reserve.** The greater a person's cardiac reserve, the greater his or her capacity for doing

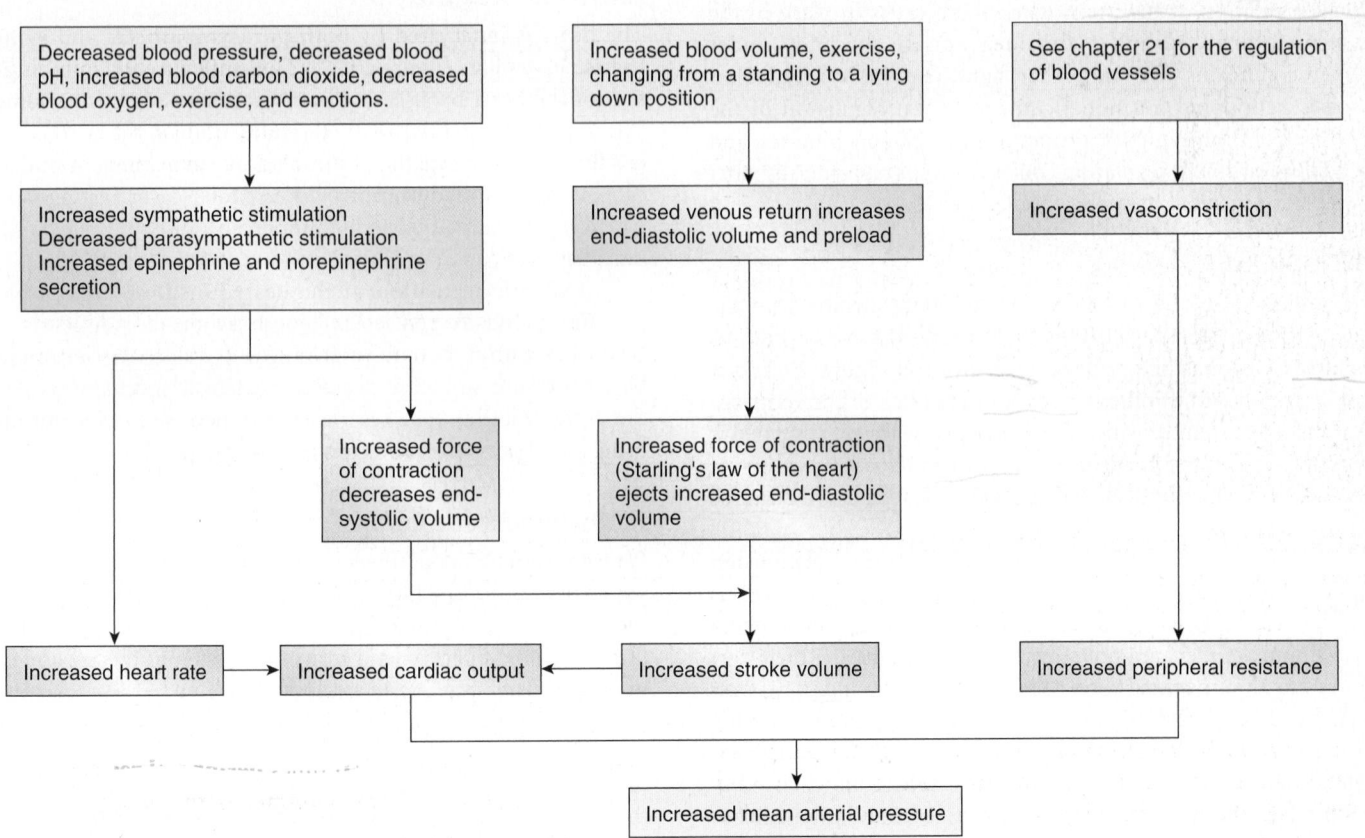

Figure 20.21 Factors Affecting Mean Arterial Pressure

Mean arterial pressure is regulated by controlling cardiac output and peripheral resistance.

exercise. Lack of exercise and cardiovascular diseases can reduce cardiac reserve and affect a person's quality of life. Exercise training can greatly increase cardiac reserve by increasing cardiac output. In well-trained athletes, stroke volume during exercise can increase to over 200 mL/beat, resulting in cardiac outputs of 40 L/min or more.

P R E D I C T **9**

Predict the effect of an incompetent bicuspid valve on the stroke volume, the volume of the left atrium at the end of ventricular systole, and the volume of the left ventricle at the end of ventricular diastole.

37. *Define mean arterial pressure, cardiac output, and peripheral resistance. Explain the role of mean arterial pressure in causing blood flow.*

38. *Define stroke volume, and state two ways to increase stroke volume.*

39. *What is cardiac reserve? How can exercise training influence cardiac reserve?*

Regulation of the Heart

To maintain homeostasis, the amount of blood pumped by the heart must vary dramatically. For example, during exercise cardiac output can increase several times over resting values. Either intrinsic or extrinsic regulatory mechanisms control cardiac output. **Intrinsic regulation** results from the normal functional characteristics of the heart and does not depend on either neural or hormonal regulation. It functions when the heart is in place in the body or is removed and maintained outside the body under proper conditions. On the other hand, **extrinsic regulation** involves neural and hormonal control. Neural regulation of the heart results from sympathetic and parasympathetic reflexes, and the major hormonal regulation comes from epinephrine and norepinephrine secreted from the adrenal medulla.

Intrinsic Regulation

The amount of blood that flows into the right atrium from the veins during diastole is called the venous return. As venous return increases, end-diastolic volume increases (see figure 20.21). A greater end-diastolic volume increases the stretch of the ventricular walls. The extent to which the ventricular walls are stretched is sometimes called the **preload.** An increased preload causes an increase in cardiac output, and a decreased preload causes a decrease in cardiac output.

Cardiac muscle exhibits a length-versus-tension relationship similar to that of skeletal muscle. Skeletal muscle, however, is stretched to nearly its optimal length before contraction, whereas cardiac muscle fibers are not stretched to the point at which they contract with a maximal force (see chapter 9). An increased preload, therefore, causes the cardiac muscle fibers to contract with a greater force and produce a greater stroke volume. This relationship between preload and stroke volume is commonly referred to as **Starling's law of the heart,** which describes the relationship between changes in the pumping effectiveness of the heart and changes in preload (see figure 20.21). Venous return can decrease

to a value as low as 2 L/min or increase to as much as 24 L/min, which has a major effect on the preload.

Afterload is the pressure the contracting ventricles must produce to overcome the pressure in the aorta and move blood into the aorta. Although the pumping effectiveness of the heart is greatly influenced by relatively small changes in the preload, it is very insensitive to large changes in afterload. Aortic blood pressure must increase to more than 170 mm Hg before it hampers the ability of the ventricles to pump blood.

During physical exercise, blood vessels in exercising skeletal muscles dilate and allow increased flow of blood through the vessels. The increased blood flow increases oxygen and nutrient delivery to the exercising muscles. In addition, skeletal muscle contractions repeatedly compress veins and cause an increased rate of blood flow from the skeletal muscles toward the heart. As blood rapidly flows through skeletal muscles and back to the heart, venous return to the heart increases, resulting in an increased preload. The increased preload causes an increased force of cardiac muscle contraction, which increases stroke volume. The increase in stroke volume results in increased cardiac output, and the volume of blood flowing to the exercising muscles increases. When a person rests, venous return to the heart decreases because arteries in the skeletal muscles constrict and because muscular contractions no longer repeatedly compress the veins. As a result blood flow through skeletal muscles decreases, and there is a decrease in preload and cardiac output.

Extrinsic Regulation

The heart is innervated by both **parasympathetic** and **sympathetic** nerve fibers (figure 20.22). They influence the pumping action of the heart by affecting both heart rate and stroke volume. The influence of parasympathetic stimulation on the heart is much less than that of sympathetic stimulation. Sympathetic stimulation can increase cardiac output by 50%–100% over resting values, whereas parasympathetic stimulation can cause only a 10%–20% decrease.

Extrinsic regulation of the heart functions to keep blood pressure, blood oxygen levels, blood carbon dioxide levels, and blood pH within their normal ranges of values. For example, if blood pressure suddenly decreases, extrinsic mechanisms detect the decrease and initiate responses that increase cardiac output to bring blood pressure back to its normal range.

Parasympathetic Control

Parasympathetic nerve fibers are carried to the heart through the **vagus nerves.** Preganglionic fibers of the vagus nerve extend from the brainstem to terminal ganglia within the wall of the heart, and postganglionic fibers extend from the ganglia to the SA node, AV node, coronary vessels, and atrial myocardium.

Parasympathetic stimulation has an inhibitory influence on the heart, primarily by decreasing the heart rate. When a person is at rest, continuous parasympathetic stimulation inhibits the heart to some degree. An increase in heart rate during exercise results, in part, from decreased parasympathetic stimulation. Strong parasympathetic stimulation can decrease the heart rate

1. Sensory (*green*) neurons carry action potentials from baroreceptors to the cardioregulatory center. Chemoreceptors in the medulla oblongata influence the cardioregulatory center.

2. The cardioregulatory center controls the frequency of action potentials in the parasympathetic (*red*) neurons extending to the heart through the vagus nerves. The parasympathetic neurons decrease the heart rate.

3. The cardioregulatory center controls the frequency of action potential in the sympathetic (*blue*) neurons. The sympathetic neurons extend through the cardiac nerves and increase the heart rate and the stroke volume.

4. The cardioregulatory center influences the frequency of action potentials in the sympathetic (*blue*) neurons extending to the adrenal medulla. The sympathetic neurons increase the secretion of epinephrine and some norepinephrine into the general circulation. Epinephrine and norepinephrine increase the heart rate and stroke volume.

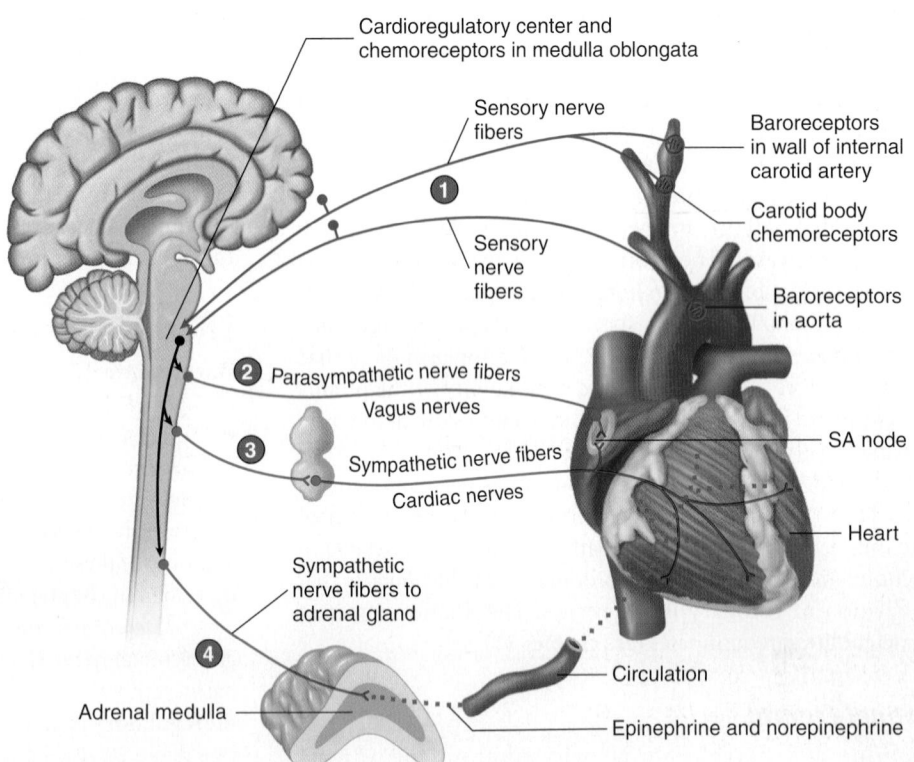

Process Figure 20.22   Baroreceptor and Chemoreceptor Reflexes

Sensory (*green*) nerves carry action potentials from sensory receptors to the medulla oblongata. Sympathetic (*blue*) and parasympathetic (*red*) nerves exit the spinal cord or medulla oblongata and extend to the heart to regulate its function. Epinephrine and norepinephrine from the adrenal gland also help regulate the heart's action. (*SA* = sinoatrial)

below resting levels by at least 20–30 bpm, but it has little effect on stroke volume. In fact, if venous return remains constant while the heart is inhibited by parasympathetic stimulation, stroke volume actually can increase. The longer time between heartbeats allows the heart to fill to a greater capacity, resulting in an increased preload, which increases stroke volume because of Starling's law of the heart.

Acetylcholine, the neurotransmitter produced by postganglionic parasympathetic neurons, binds to ligand-gated channels that cause cardiac plasma membranes to become more permeable to K^+. As a consequence, the membrane hyperpolarizes. Heart rate decreases because the hyperpolarized membrane takes longer to depolarize and cause an action potential.

Sympathetic Control

Sympathetic nerve fibers originate in the thoracic region of the spinal cord as preganglionic neurons. These neurons synapse with postganglionic neurons of the inferior **cervical** and upper **thoracic sympathetic chain ganglia,** which project to the heart as **cardiac nerves** (see figure 20.22 and chapter 16). The postganglionic sympathetic nerve fibers innervate the SA and AV nodes, the coronary vessels, and the atrial and ventricular myocardium.

Sympathetic stimulation increases both the heart rate and the force of muscular contraction. In response to strong sympathetic stimulation, the heart rate can increase to 250 or, occasionally, 300 bpm. Stronger contractions also can increase stroke volume. The increased force of contraction resulting from sympathetic stimulation causes a lower end-systolic volume in the heart; therefore, the heart empties to a greater extent (see figure 20.21).

P R E D I C T

What effect does sympathetic stimulation have on stroke volume if the venous return remains constant? Dilation of the coronary blood vessels occurs in response to an increased heart rate and stroke volume. Explain the functional advantage of that effect.

Limitations exist, however, to the relationship between increased heart rate and cardiac output. If the heart rate becomes too fast, diastole is not long enough to allow complete ventricular filling, end-diastolic volume decreases, and stroke volume actually decreases. In addition, if heart rate increases beyond a critical level, the strength of contraction decreases, probably as a result of the accumulation of metabolites in cardiac muscle cells. The limit of the heart's ability to increase the volume of blood pumped is 170–250 bpm in response to intense sympathetic stimulation.

Sympathetic stimulation of the ventricular myocardium plays a significant role in regulation of its contraction force when a person is at rest. Sympathetic stimulation maintains the strength of ventricular contraction at a level approximately 20% greater than it would be with no sympathetic stimulation.

Norepinephrine, the postganglionic sympathetic neurotransmitter, increases the rate and degree of cardiac muscle depolarization so that both the frequency and amplitude of the action potentials are increased. The effect of norepinephrine on the heart involves the association between norepinephrine and cell surface β-adrenergic receptors. This combination causes a G protein–mediated synthesis and accumulation of cAMP in the cytoplasm of cardiac muscle cells. Cyclic AMP increases the permeability of the plasma membrane to Ca^{2+}, primarily by opening calcium slow channels in the plasma membrane.

Increased sympathetic stimulation causes coronary arteries to constrict, to some degree. However, increased metabolism of cardiac muscle, in response to sympathetic stimulation, results in the accumulation of metabolic by-products in cardiac muscle that cause dilation of coronary blood vessels. The dilation effect of these metabolites predominates (see chapter 21).

Hormonal Control

Epinephrine and norepinephrine released from the adrenal medulla can markedly influence the pumping effectiveness of the heart. Epinephrine has essentially the same effect on cardiac muscle as norepinephrine and, therefore, increases the rate and force of heart contractions (see figure 20.21). The secretion of epinephrine and norepinephrine from the adrenal medulla is controlled by sympathetic stimulation of the medulla and occurs in response to increased physical activity, emotional excitement, or stressful conditions. Many stimuli that increase sympathetic stimulation of the heart also increase release of epinephrine and norepinephrine from the adrenal gland (see chapter 18). Epinephrine and norepinephrine are transported in the blood through the vessels of the heart to the cardiac muscle cells, where they bind to β-adrenergic receptors and stimulate cAMP synthesis. Epinephrine takes a longer time to act on the heart than sympathetic stimulation does, but the effect lasts longer.

40. *Define the term venous return, and explain how it affects preload. How does preload affect cardiac output? State Starling's law of the heart.*
41. *Define the term afterload, and describe its effect on the pumping effectiveness of the heart.*
42. *What part of the brain regulates the heart? Describe the autonomic nerve supply to the heart.*
43. *What effect do parasympathetic stimulation and sympathetic stimulation have on heart rate, force of contraction, and stroke volume?*
44. *What neurotransmitters are released by the parasympathetic and sympathetic postganglionic neurons of the heart? What effects do they have on membrane permeability and excitablity?*
45. *Name the two main hormones that affect the heart. Where are they produced, what causes their release, and what effects do they have on the heart?*

Heart and Homeostasis

The pumping efficiency of the heart plays an important role in the maintenance of homeostasis. Blood pressure in the systemic vessels must be maintained at a level which is high enough to achieve nutrient and waste product exchange across the walls of the capillaries that meets metabolic demands. The activity of the heart must be regulated because the metabolic activities of the tissues change under such conditions as exercise and rest.

Effect of Blood Pressure

Baroreceptor (bar′ō-rē-sep′ter, bar′ō-rē-sep′tōr) **reflexes** detect changes in blood pressure and result in changes in heart rate and in the force of contraction. The sensory receptors of the baroreceptor reflexes are stretch receptors. They are in the walls of certain large arteries, such as the internal carotid arteries and the aorta, and they function to measure blood pressure (see figure 20.22). The anatomy of these sensory structures and their afferent pathways are described in chapter 21.

Afferent neurons project primarily through the glossopharyngeal (cranial nerve IX) and vagus (cranial nerve X) nerves from the baroreceptors to an area in the medulla oblongata called the **cardioregulatory center,** where sensory action potentials are integrated (see figure 20.22). The part of the cardioregulatory center that functions to increase heart rate is called the **cardioacceleratory center,** and the part that functions to decrease heart rate is called the **cardioinhibitory center.** Efferent action potentials then are sent from the cardioregulatory center to the heart through both the sympathetic and parasympathetic divisions of the autonomic nervous system.

Increased blood pressure within the internal carotid arteries and aorta causes their walls to stretch, thereby stimulating an increase in action potential frequency in the baroreceptors (figure 20.23). At normal blood pressures (80–120 mm Hg), afferent action potentials are sent from the baroreceptors to the medulla oblongata at a relatively constant frequency. When blood pressure increases, the arterial walls are stretched further, and the afferent action potential frequency increases. When blood pressure decreases, the arterial walls are stretched to a lesser extent, and the afferent action potential frequency decreases. In response to increased blood pressure, the baroreceptor reflexes decrease sympathetic stimulation and increase parasympathetic stimulation of the heart, causing the heart rate to decrease. Decreased blood pressure causes decreased parasympathetic and increased sympathetic stimulation of the heart, resulting in an increased heart rate and force of contraction. Withdrawal of parasympathetic stimulation is primarily responsible for increases in heart rate up to approximately 100 bpm. Larger increases in heart rate, especially during exercise, result from sympathetic stimulation. The baroreceptor reflexes are homeostatic because they keep the blood pressure within a narrow range of values, which is adequate to maintain blood flow to the tissues.

Effect of pH, Carbon Dioxide, and Oxygen

Chemoreceptor (kē′mō-rē-sep′tor) **reflexes** help regulate the activity of the heart. Chemoreceptors sensitive to changes in pH and carbon dioxide levels exist within the medulla oblongata. A drop in pH and a rise in carbon dioxide decrease parasympathetic and

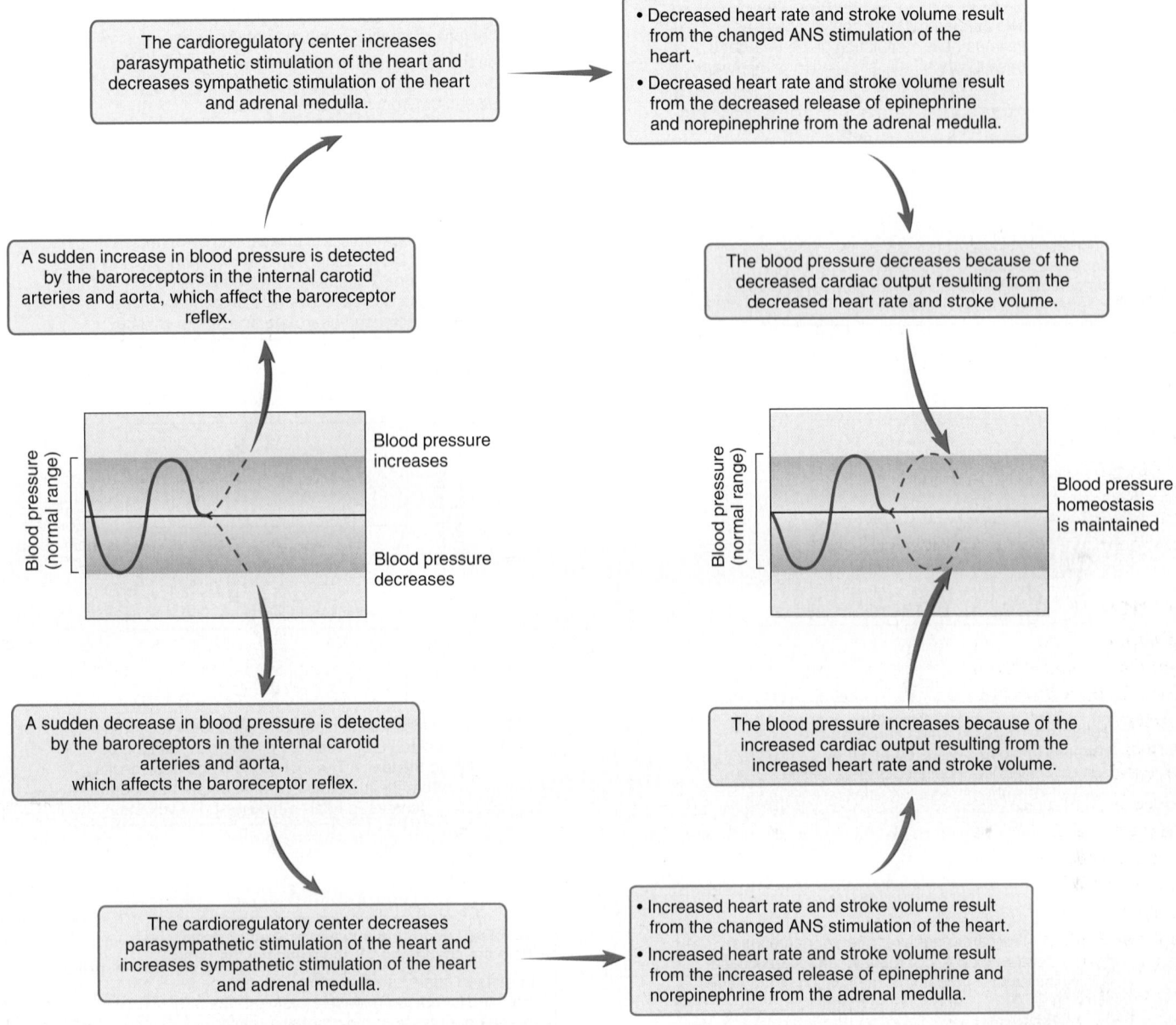

Homeostasis Figure 20.23 **Baroreceptor Reflex**

The baroreceptor reflex maintains homeostasis in response to changes in blood pressure. (*ANS* = autonomic nervous system)

increase sympathetic stimulation of the heart, resulting in an increased heart rate and force of contraction (figure 20.24).

The increased cardiac output causes greater blood flow through the lungs, where carbon dioxide is eliminated from the body. This helps bring the blood carbon dioxide level down to its normal range of values and helps to increase the blood pH.

Chemoreceptors primarily sensitive to blood oxygen levels are found in the carotid and aortic bodies. These small structures are located near large arteries close to the brain and heart, and they monitor blood flowing to the brain and to the rest of the body. A dramatic decrease in blood oxygen levels, such as during asphyxiation, activates the carotid and aortic body chemoreceptor reflexes. In carefully controlled experiments, it's possible to isolate the effects of the carotid and aortic body chemoreceptor reflexes from other

reflexes, such as the medullary chemoreceptor reflexes. These experiments indicate that a decrease in blood oxygen results in a decrease in heart rate and an increase in vasoconstriction. The increased vasoconstriction causes blood pressure to rise, which promotes blood delivery despite the decrease in heart rate. The carotid and aortic body chemoreceptor reflexes may protect the heart for a short time by slowing the heart and thereby reducing its need for oxygen. The carotid and aortic body chemoreceptor reflexes normally don't function independently of other regulatory mechanisms. When all regulatory mechanisms function together, the effect of large, prolonged decreases in blood oxygen levels is to increase the heart rate. Low blood oxygen levels result in increased stimulation of respiratory movements (see chapter 23). Increased inflation of the lungs stimulates stretch receptors in the lungs. Afferent action potentials

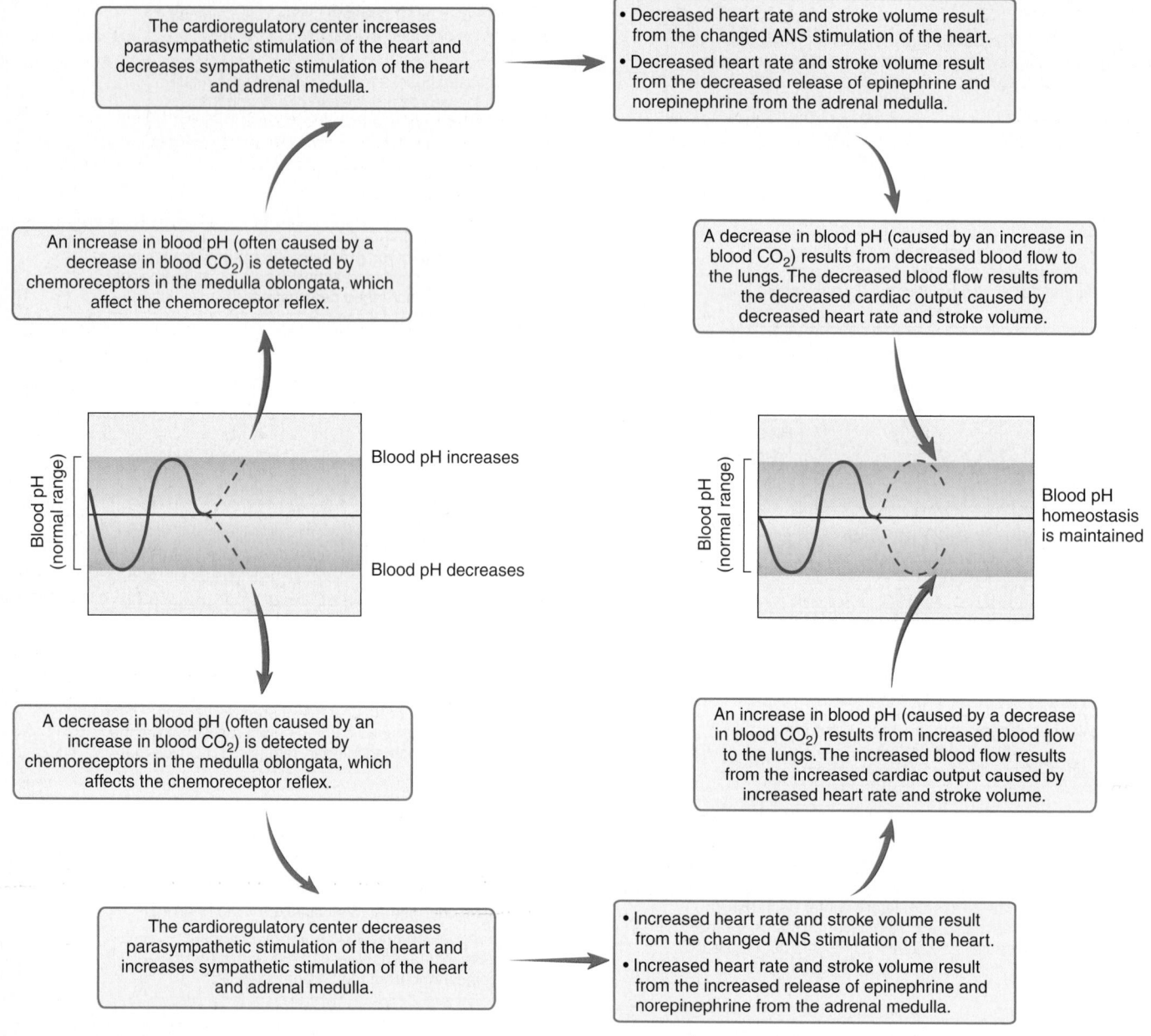

Homeostasis Figure 20.24   **Chemoreceptor Reflex-pH**

The chemoreceptor reflex maintains homeostasis in response to changes in blood concentrations of CO_2 and H^+. (*ANS* = autonomic nervous system)

from these stretch receptors influence the cardioregulatory center, which causes an increase in heart rate. The reduced oxygen levels that exist at high altitudes can cause an increase in heart rate even when blood carbon dioxide levels remain low. The carotid and aortic body chemoreceptor reflexes are more important in the regulation of respiration (see chapter 23) and blood vessel constriction (see chapter 21) than in the regulation of heart rate.

Effect of Extracellular Ion Concentration

Ions that affect cardiac muscle function are the same ions (potassium, calcium, and sodium) that influence membrane potentials in other electrically excitable tissues. Some differences do exist, however, between the response of cardiac muscle and that of nerve or muscle tissue to these ions. For example, the extracellular levels of

Na^+ rarely deviate enough from the normal value to affect the function of cardiac muscle significantly.

Excess K^+ in cardiac tissue causes the heart rate and stroke volume to decrease. A twofold increase in extracellular K^+ results in **heart block,** which is loss of the functional conduction of action potentials through the conducting system of the heart. The excess K^+ in the extracellular fluid causes partial depolarization of the resting membrane potential, resulting in a decreased amplitude of action potentials and, because of the decreased amplitude, a decreased rate at which action potentials are conducted along cardiac muscle fibers. As the conduction rates decrease, ectopic action potentials can occur. In many cases, partially depolarized cardiac muscle cells spontaneously produce action potentials because the membrane potential reaches threshold. Increased blood levels of

K^+ can produce enough ectopic action potentials to cause fibrillation. The reduced action potential amplitude also results in less calcium entering the sarcoplasm of the cell; thus, the strength of cardiac muscle contraction decreases.

Although the extracellular concentration of K^+ normally is small, a decrease in extracellular K^+ results in a decrease in the heart rate because the resting membrane potential is hyperpolarized; as a consequence, it takes longer for the membrane to depolarize to threshold. The force of contraction is not affected, however.

An increase in the extracellular concentration of Ca^{2+} produces an increase in the force of cardiac contraction because of a greater influx of Ca^{2+} into the sarcoplasm during action potential generation. Elevated plasma Ca^{2+} levels have an indirect effect on heart rate because they reduce the frequency of action potentials in nerve fibers, thus reducing sympathetic and parasympathetic stimulation of the heart (see chapter 11). Generally, elevated blood Ca^{2+} levels reduce the heart rate.

A low blood Ca^{2+} level increases the heart rate, although the effect is imperceptible until blood Ca^{2+} levels are reduced to approximately one-tenth of their normal value. The reduced extracellular Ca^{2+} levels cause Na^+ channels to open, which allows Na^+ to diffuse more readily into the cell, resulting in depolarization and action potential generation. Reduced Ca^{2+} levels, however, usually cause death as a result of tetany of skeletal muscles before they decrease enough to markedly influence the heart's function.

Effect of Body Temperature

Under resting conditions, the temperature of cardiac muscle normally doesn't change dramatically in humans, although alterations in temperature influence the heart rate. Small increases in cardiac muscle temperature cause the heart rate to increase, and decreases in temperature cause the heart rate to decrease. For example, during exercise or fever, increased heart rate and force of contraction accompany temperature increases, but the heart rate decreases under conditions of hypothermia. During heart surgery, the body temperature sometimes is reduced dramatically to slow the heart rate and other metabolic functions in the body.

46. *How does the nervous system detect and respond to (a) a decrease in blood pressure, (b) an increase in carbon dioxide levels, (c) a decrease in blood pH, and (d) a decrease in blood oxygen levels?*

47. *Describe the baroreceptor reflex and the response of the heart to an increase in venous return.*

48. *What effect does an increase or decrease in extracellular potassium, calcium, and sodium ions have on heart rate and the force of contraction of the heart?*

49. *What effect does temperature have on heart rate?*

Effects of Aging on the Heart

Aging results in gradual changes in the function of the heart, which are minor under resting conditions, but become more significant in response to exercise and when age-related diseases develop. The mechanisms that regulate the heart effectively compensate for most of the changes under resting conditions.

Hypertrophy of the left ventricle is a common age-related change. This appears to result from a gradual increase in the pressure in the aorta against which the left ventricle must pump blood and a gradual increase in the stiffness of cardiac muscle tissue. The increased pressure in the aorta results from a gradual decrease in arterial elasticity resulting in an increased stiffness of the aorta and other large arteries. Myocardial cells accumulate lipid, and collagen fibers increase in cardiac tissue. These changes make the cardiac muscle tissue stiffer and less compliant. The increased volume of the left ventricle can sometimes result in an increase in left atrial pressure and increased pulmonary capillary pressure. This can cause pulmonary edema and a tendency for people to feel out of breath when they exercise strenuously.

There is a gradual decrease in the maximum heart rate. This can be roughly predicted by the following formula: Maximum heart rate = 220 − age of the individual. There is an increase in the rate at which ATP is broken down by cardiac muscle and a decrease in the rate of Ca^{2+} transport. There is a decrease in the maximum rate at which cardiac muscle can carry out aerobic metabolism. In addition, there is a decrease in the degree to which epinephrine and norepinephrine can increase the heart rate. These changes are consistent with longer contraction and relaxation times for cardiac muscle and a decrease in the maximum heart rate. Both the resting and maximum cardiac output slowly decrease as people age and, by 85 years of age, the cardiac output may be decreased by 30%–60%.

Age-related changes in the connective tissue of the heart valves occur. The connective tissue becomes less flexible and Ca^{2+} deposits increase. The result is an increased tendency for heart valves to function abnormally. There is especially an increased tendency for the aortic semilunar valve to become stenosed, but other heart valves, such as the bicuspid valve, may become either stenosed or incompetent.

Atrophy and replacement of cells of the left bundle branch and a decrease in the number of SA node cells alter the electrical conduction system of the heart and lead to a higher rate of cardiac arrhythmias in elderly people.

The enlarged and thickened cardiac muscle, especially in the left ventricle, consumes more oxygen to pump the same amount of blood pumped by a younger heart. This change is not significant except if the coronary circulation is decreased by coronary artery disease. However, the development of coronary artery disease is age-related. Congestive heart disease is also age-related. Approximately 10% of elderly people over 80 have congestive heart failure, and a major contributing factor is coronary artery disease. Because of the age-related changes in the heart, many elderly people are limited in their ability to respond to emergencies, infections, blood loss, or stress.

Exercise has many beneficial effects on the heart. Regular aerobic exercise improves the functional capacity of the heart at all ages, providing no conditions develop which cause the increased workload of the heart to be harmful.

50. *Explain how age-related changes affect the function of the left ventricle.*

51. *Describe the age-related changes in the heart rate.*

52. *Describe how increasing age affects the function of the conduction system and the heart valves.*

53. *Describe the effect of two age-related heart diseases on functions of the aging heart.*

Clinical Focus   Conditions and Diseases Affecting the Heart

Inflammation of Heart Tissues

Endocarditis (en′dō-kar-dī′tis) is inflammation of the endocardium. It affects the valves more severely than other areas of the heart and can lead to deposition of scar tissue, causing valves to become stenosed or incompetent.

Myocarditis (mī′ō-kar-dī′tis) is inflammation of the myocardium and can lead to heart failure.

Pericarditis is inflammation of the pericardium. Pericarditis can result from bacterial or viral infections and can be extremely painful.

Rheumatic (roo-mat′ik) **heart disease** can result from a streptococcal infection in young people. Toxin produced by the bacteria can cause an immune reaction called rheumatic fever about 2–4 weeks after the infection. The immune reaction can cause inflammation of the endocardium, called **rheumatic endocarditis.** The inflamed valves, especially the bicuspid valve, can become stenosed or incompetent. The effective treatment of streptococcal infections with antibiotics has reduced the frequency of rheumatic heart disease.

Reduced Blood Flow to Cardiac Muscle

Coronary heart disease reduces the amount of blood that the coronary arteries are able to deliver to the myocardium. The reduction in blood flow damages the myocardium. The degree of damage depends on the size of the arteries involved, whether occlusion (blockage) is partial or complete, and whether occlusion is gradual or sudden. As the walls of the arteries thicken and harden with age, the volume of blood they can supply to the heart muscle declines, and the ability of the heart to pump blood decreases. Inadequate blood flow to the heart muscle can result in angina pectoris, which is a poorly localized sensation of pain in the region of the chest, left arm, and left shoulder.

Degenerative changes in the artery wall can cause the inside surface of the artery to become roughened. The chance of platelet aggregation increases at the rough surface, which increases the chance of **coronary thrombosis** (throm-bō′sis; formation of a blood clot in a coronary vessel). Inadequate blood flow can cause an **infarct** (in′farkt), an

area of damaged cardiac tissue. A heart attack is often referred to as a coronary thrombosis or a **myocardial infarct.** The outcome of coronary thrombosis depends on the extent of the damage to heart muscle caused by inadequate blood flow and whether other blood vessels can supply enough blood to maintain the heart's function. Death can occur swiftly if the infarct is large; if the infarct is small, the heart can continue to function. In most cases, scar tissue replaces damaged cardiac muscle in the area of the infarct.

People who survive infarctions often lead fairly normal lives if they take precautions. Most cases call for moderate exercise, adequate rest, a disciplined diet, and reduced stress.

Congenital Conditions Affecting the Heart

Congenital heart disease is the result of abnormal development of the heart. The following conditions are common congenital defects.

Septal defect is a hole in a septum between the left and right sides of the heart. The hole may be in the interatrial or interventricular septum. These defects allow blood to flow from one side of the heart to the other and, as a consequence, greatly reduce the pumping effectiveness of the heart (see chapter 29).

Patent ductus arteriosus (dŭk′tŭs ar-tēr′ē-ō-sŭs) results when a blood vessel called the **ductus arteriosus,** which is present in the fetus, fails to close after birth. The ductus arteriosus extends between the pulmonary trunk and the aorta. It allows blood to pass from the pulmonary trunk to the aorta, thus bypassing the lungs. This is normal before birth because the lungs are not functioning (see chapter 29). If the ductus arteriosus fails to close after birth, blood flows in the opposite direction, from the aorta to the pulmonary trunk. As a consequence, blood flows through the lungs under higher pressure, causing damage to the lungs. In addition, the amount of work required of the left ventricle to maintain adequate systemic blood pressure increases.

Stenosis (ste-nō′sis) **of a heart valve** is a narrowed opening through one of the heart valves. In aortic or pulmonary valve stenosis, the workload of the heart is increased because the ventricles must contract with a much greater force to pump blood from the

ventricles. Stenosis of the bicuspid valve prevents the flow of blood into the left ventricle, causing blood to back up in the left atrium and in the lungs, resulting in congestion of the lungs. Stenosis of the tricuspid valve causes blood to back up in the right atrium and systemic veins, causing swelling in the periphery.

An **incompetent heart valve** is one that leaks. Blood, therefore, flows through the valve when it's closed. The workload of the heart is increased because incompetent valves reduce the pumping efficiency of the heart. For example, an incompetent aortic semilunar valve allows blood to flow from the aorta into the left ventricle during diastole. Thus, the left ventricle fills with blood to a greater degree than normal. The increased filling of the left ventricle results in a greater stroke volume because of Starling's law of the heart. The pressure produced by the contracting ventricle and the pressure in the aorta is greater than normal during ventricular systole. The pressure in the aorta, however, decreases very rapidly as blood leaks into the left ventricle during diastole.

An incompetent bicuspid valve allows blood to flow back into the left atrium from the left ventricle during ventricular systole. This increases the pressure in the left atrium and pulmonary veins, which results in pulmonary edema. Also, the stroke volume of the left ventricle is reduced, which causes a decrease in systemic blood pressure. Similarly, an incompetent tricuspid valve allows blood to flow back into the right atrium and systemic veins, causing edema in the periphery.

Cyanosis (sī-ă-nō′sis) is a symptom of inadequate heart function in babies suffering from congenital heart disease. The term *blue baby* is sometimes used to refer to infants with cyanosis. Low blood oxygen levels in the peripheral blood vessels cause the skin to look blue.

Heart Failure

Heart failure is the result of progressive weakening of the heart muscle and the failure of the heart to pump blood effectively. Hypertension (high blood pressure) increases the afterload on the heart, can produce significant enlargement of the heart, and can finally result in heart failure. Advanced age, malnutrition, chronic infections, toxins, severe anemias, or hyperthyroidism can cause degeneration of the heart muscle, resulting in heart failure. Hereditary factors can also be responsible for increased susceptibility to heart failure.

Heart Medications

Digitalis (dij-i-tal'is, dij-i-ta'lis) slows and strengthens contractions of the heart muscle. This drug is frequently given to people who suffer from heart failure, although it also can be used to treat atrial tachycardia.

Nitroglycerin (nī-trō-glis'er-in) causes dilation of all of the veins and arteries, including coronary arteries, without an increase in heart rate or stroke volume. When all blood vessels dilate, a greater volume of blood pools in the dilated blood vessels, causing a decrease in the venous return to the heart. The flow of blood through coronary arteries also increases. The reduced preload causes cardiac output to decrease, resulting in a decreased amount of work performed by the heart. Nitroglycerin is frequently given to people who suffer from coronary artery disease, which restricts coronary blood flow. The decreased work performed by the heart reduces the amount of oxygen required by the cardiac muscle. Consequently, the heart doesn't suffer from a lack of oxygen, and angina pectoris doesn't develop.

Beta-adrenergic-blocking agents reduce the rate and strength of cardiac muscle contractions, thus reducing the heart's demand for oxygen. These blocking agents bind to receptors for norepinephrine and epinephrine and prevent these substances from having their normal effects. They are often used to treat people who suffer from rapid heart rates, certain types of arrhythmias, and hypertension.

Calcium channel blockers reduce the rate at which Ca^{2+} diffuse into cardiac muscle cells and smooth muscle cells. Because the action potentials that produce cardiac muscle contractions depend in part on the flow of Ca^{2+} into cardiac muscle cells, calcium channel blockers can be used to control the force of heart contractions and reduce arrhythmia, tachycardia, and hypertension. Because entry of Ca^{2+} into smooth muscle cells causes contraction, calcium channel blockers cause dilation of coronary blood vessels and can be used to treat angina pectoris.

Antihypertensive (an'tē-hī-per-ten'siv) **agents** comprise several drugs used specifically to treat hypertension. These drugs reduce blood pressure and, therefore, reduce the work required by the heart to pump blood. In addition, the reduction of blood pressure reduces the risk of heart attacks and strokes. Drugs used to treat hypertension include those that reduce the ac-

tivity of the sympathetic nervous system, that dilate arteries and veins, that increase urine production (diuretics), and that block the conversion of angiotensinogen to angiotensin I.

Anticoagulants (an'tē-kō-ag'ū-lantz) prevent clot formation in persons with damage to heart valves or blood vessels or in persons who have had a myocardial infarction. Aspirin functions as a weak anticoagulant.

Instruments and Selected Procedures

An **artificial pacemaker** is an instrument placed beneath the skin, equipped with an electrode that extends to the heart. The instrument provides an electric stimulus to the heart at a set frequency. Artificial pacemakers are used in patients in whom the natural pacemaker of the heart doesn't produce a heart rate high enough to sustain normal physical activity. Modern electronics has made it possible to design artificial pacemakers that can increase the heart rate as increases in physical activity occur. Pacemakers can also detect cardiac arrest, extreme arrythmias, or fibrillation. In response, strong stimulation of the heart by the pacemaker may restore heart function.

A **heart lung machine** serves as a temporary substitute for a patient's heart and lungs. It oxygenates the blood, removes carbon dioxide, and pumps blood throughout the body. It has made possible many surgeries on the heart and lungs.

Heart valve replacement or repair is a surgical procedure performed on those who have diseased valves that are so deformed and scarred from conditions like endocarditis that the valves are severely incompetent or stenosed. Substitute valves made of synthetic materials like plastic or Dacron are effective; valves transplanted from pigs are also used.

A **heart transplant** is a surgical procedure made possible when the immune characteristics of a donor and the recipient are closely matched (see chapter 22). The heart of a recently deceased donor is transplanted to the recipient, and the diseased heart of the recipient is removed. People who have received heart transplants must continue to take drugs that suppress their immune responses for the rest of their lives. If they don't, their immune system will reject the transplanted heart.

An **artificial heart** is a mechanical pump that replaces the heart. It is still experimental

and cannot be viewed as a permanent substitute for the heart. It has been used to keep a patient alive until a donor heart can be found.

Cardiac assistance involves temporarily implanting a mechanical device that assists the heart in pumping blood. In some cases, the decreased workload on the heart provided by the device appears to promote recovery of failing hearts, and the device has been successfully removed. In **cardiomyoplasty,** a piece of a back muscle (latissimus dorsi) is wrapped around the heart and stimulated to contract in synchrony with the heart.

Prevention of Heart Disease

Heart disease is a major cause of death. Several precautions can be taken to help prevent heart disease. Proper nutrition is important in reducing the risk of heart disease (see chapter 25). A recommended diet is low in fats, especially saturated fats and cholesterol, and low in refined sugar. Diets should be high in fiber, whole grains, fruits, and vegetables. Total food intake should be limited to avoid obesity, and sodium chloride intake should be reduced.

Tobacco and excessive use of alcohol should be avoided. Smoking increases the risk of heart disease at least 10-fold, and excessive use of alcohol also substantially increases the risk of heart disease.

Chronic stress, frequent emotional upsets, and a lack of physical exercise can increase the risk of cardiovascular disease. Remedies include relaxation techniques and aerobic exercise programs involving gradual increases in duration and difficulty in activities, such as walking, swimming, jogging, or aerobic dancing.

Hypertension (hī'per-ten'shŭn) is abnormally high systemic blood pressure. It affects about one-fifth of the U.S. population. Regular blood pressure measurements are important because hypertension does not produce obvious symptoms. If hypertension cannot be controlled by diet and exercise, it's important to treat the condition with prescribed drugs. The cause of hypertension in the majority of cases is unknown.

Some data suggest that taking an aspirin daily reduces the chance of a heart attack. Aspirin inhibits the synthesis of prostaglandins in platelets, thereby helping to prevent clot formation (see chapter 19).

Systems Pathology
Myocardial Infarction

Mr. P was an overweight, out-of-shape executive who regularly smoked and consumed food with a high fat content. He viewed his job as frustrating because he was frequently confronted with stressful deadlines. He had not had a physical examination for several years, so he was not aware that his blood pressure was high. One evening, Mr. P was walking to his car after work when he began to feel chest pain that radiated down his left arm. Shortly after the onset of pain, he felt out of breath, developed marked pallor, became dizzy, and had to lie down on the sidewalk. The pain in his chest and arm was poorly localized, but intense, and he became anxious and then disoriented. Mr. P lost consciousness, although he did not stop breathing. After a short delay, one of his coworkers noticed him and called for help. When paramedics arrived they determined that Mr. P's blood pressure was low and he exhibited arrhythmia and tachycardia. The paramedics transmitted the electrocardiogram they took to a physician by way of their electronic communications system, and they discussed Mr. P's symptoms with the physician who was at the hospital. The paramedics were directed to administer oxygen and medication to control arrhythmias and transport him to the hospital. At the hospital, tissue plasminogen activator (t-PA) was administered, which improved blood flow to the damaged area of the heart by activating plasminogen, which dissolves blood clots. Enzymes, like creatine phosphokinase, increased in Mr. P's blood over the next few days, which confirmed that damage to cardiac muscle resulted from an infarction.

In the hospital, Mr. P began to experience shortness of breath because of pulmonary edema, and after a few days in the hospital, he developed pneumonia. He was treated for pneumonia and gradually improved over the next few weeks. An angiogram performed several days after Mr. P's infarction indicated that he had suffered damage to a significant part of the lateral wall of his left ventricle and that neither angioplasty nor bypass surgery were necessary, although Mr. P has some serious restrictions to blood flow in his coronary arteries.

Background Information

Mr. P experienced a myocardial infarction. A thrombosis in one of the branches of the left coronary artery reduces the blood supply to the lateral wall of the left ventricle, resulting in ischemia of the left ventricle wall. That t-PA is effective in treating a heart attack is consistent with the conclusion that the infarction was caused by a thrombosis. An ischemic area of the heart wall is not able to contract normally and, therefore, the pumping effectiveness of the heart is dramatically reduced. The reduced pumping capacity of the heart is responsible for the low blood pressure, which causes the blood flow to the brain to decrease resulting in confusion, disorientation, and unconsciousness.

Low blood pressure, increasing blood carbon dioxide levels, pain, and anxiousness increase sympathetic stimulation of the heart and adrenal glands. Increased sympathetic stimulation of the adrenal medulla results in release of epinephrine. Increased parasympathetic stimulation of the heart results from pain sensations. In such cases, the heart is periodically arrhythmic due to the combined effects of parasympathetic stimulation, epinephrine and norepinephrine from the adrenal gland, and sympathetic stimulation. In addition, ectopic beats are produced by the ischemic areas of the left ventricle.

Pulmonary edema results from the increased pressure in the pulmonary veins because of the inability of the left ventricle to pump blood. The edema allows bacteria to infect the lungs and cause pneumonia.

Mr. P's heart began to beat rhythmically in response to medication because the infarction did not damage the conducting system of the heart, which is an indication that no permanent arrhythmias developed. Permanent arrhythmias are indications of damage done to cardiac muscle specialized to conduct action potentials in the heart.

Analysis of the electrocardiogram, blood pressure measurements, and the angiogram (figure A) indicate that the infarction, in this case, was located on the left side of Mr. P's heart. Mr. P exhibited several characteristics that are correlated with an increased probability of myocardial infarction: lack of physical exercise, being overweight, smoking, and stress.

Mr. P's physician made it very clear to him that he was lucky to have survived a myocardial infarction, and the physician recommended a weight-loss program, a low-sodium and low-fat diet, and that Mr. P should stop smoking. He explained that Mr. P would have to take medication for high blood pressure if his blood pressure did not decrease in response to the recommended changes. After a period of recovery, Mr. P's physician recommended an aerobic exercise program to him. He advised Mr. P to seek ways to reduce the stress associated with his job. His physician also recommended that Mr. P regularly take a small amount of aspirin. The aspirin was prescribed to reduce the probability of thrombosis. Because aspirin inhibits prostaglandin synthesis, it reduces the tendency for blood to clot. Mr. P followed the doctor's recommendations, and after several months, he began to feel better than he had in years, and his blood pressure was normal.

Occluded coronary artery

Figure A Angiogram

An angiogram (an′jē-ō-gram) is a picture of a blood vessel. It is usually obtained by placing a catheter into a blood vessel and injecting a dye that can be detected with x-rays. Note the occluded (blocked) coronary blood vessel in this angiogram, which has been computer-enhanced to show colors.

P R E D I C T **11**

Severe ischemia in the wall of a ventricle can result in the death of cardiac muscle cells. Inflammation around the necrotic tissue results, and macrophages invade the necrotic tissue and phagocytize dead cells. At the same time, blood vessels and connective tissue grow into the necrotic area and begin to deposit connective tissue to replace the necrotic tissue. Assume that Mr. P had a myocardial infarction and was recovering. After about a week, however, his blood pressure suddenly decreased to very low levels, and he died within a very short time. At autopsy, a large amount of blood was found in the pericardial sac, and the wall of the left ventricle was ruptured. Explain.

System Interactions   Effect of Myocardial Infarction on Other Systems

System	Interactions
Integumentary	Pallor of the skin resulted from intense constriction of peripheral blood vessels, including those in the skin.
Muscular	Reduced skeletal muscle activity required for activities such as walking results from lack of blood flow to the brain and because blood is shunted from blood vessels that supply skeletal muscles to those that supply the heart and brain.
Nervous	Decreased blood flow to the brain, decreased blood pressure, and pain due to ischemia of heart muscle result in increased sympathetic and decreased parasympathetic stimulation of the heart. Loss of consciousness occurs when the blood flow to the brain decreases enough to result in too little oxygen to maintain normal brain function, especially in the reticular activating system.
Endocrine	When blood pressure decreases to low values, antidiuretic hormone (ADH) is released from the posterior pituitary gland and renin, released from the kidney, activates the renin-angiotensinogen-aldosterone mechanism. ADH, secreted in large amounts, and angiotensin II cause vasoconstriction of peripheral blood vessels. ADH and aldosterone act on the kidneys to retain water and electrolytes. An increased blood volume increases venous return, which results in an increased stroke volume of the heart and an increase in blood pressure unless damage to the heart is very severe.
Lymphatic and Immune	White blood cells, including macrophages, move to the area of cardiac muscle damaged and phagocytize any dead cardiac muscle cells.
Respiratory	Decreased blood pressure results in a decreased blood flow to the lungs. The decrease in gas exchange results in increased blood CO_2 levels, acidosis, and decreased blood O_2 levels. Initially, respiration becomes deep and labored because of the elevated CO_2 levels, decreased blood pH, and depressed O_2 levels. If the blood O_2 levels decrease too much, the person loses consciousness. Pulmonary edema can result when the pumping effectiveness of the left ventricle is substantially reduced.
Digestive	Decreased blood flow to the digestive system to very low levels often results in increased nausea and vomiting.
Urinary	Blood flow to the kidney decreases dramatically in response to sympathetic stimulation. If the kidney becomes ischemic, damage to the kidney tubules can occur, resulting in acute renal failure. Acute renal failure reduces urine production. Increased blood urea nitrogen, increased blood levels of K^+, and edema are indications that the kidneys cannot eliminate waste products and excess water. If damage is not too great, the period of reduced urine production may last up to 3 weeks and then the rate of urine production slowly returns to normal as the kidney tubules heal.

S U M M A R Y

Functions of the Heart   (p. 680)

The heart produces the force that causes blood circulation.

Size, Shape, and Location of the Heart   (p. 680)

The heart is approximately the size of a closed fist and is shaped like a blunt cone. The heart lies obliquely in the mediastinum, with its base directed posteriorly and slightly superiorly and the apex directed anteriorly, inferiorly, and to the left. The base is deep to the second intercostal space and the apex extends to the fifth intercostal space. It is in the mediastinum.

Anatomy of the Heart   (p. 682)

The heart consists of two atria and two ventricles.

Pericardium

1. The pericardium is a sac that surrounds the heart and consists of the fibrous pericardium and the serous pericardium.
2. The fibrous pericardium helps hold the heart in place.
3. The serous pericardium reduces friction as the heart beats. It consists of the following parts:
   • The parietal pericardium lines the fibrous pericardium.
   • The visceral pericardium lines the exterior surface of the heart.
   • The pericardial cavity lies between the parietal and visceral pericardium and is filled with pericardial fluid.

Heart Wall

1. The heart wall has three layers:
   • The outer epicardium (visceral pericardium) provides protection against the friction of rubbing organs.
   • The middle myocardium is responsible for contraction.
   • The inner endocardium reduces the friction resulting from the blood's passing through the heart.
2. The inner surfaces of the atria are mainly smooth. The auricles have raised areas called musculi pectinati.
3. The ventricles have ridges called trabeculae carneae.

External Anatomy and Coronary Circulation

1. Each atrium has a flap called the auricle.
2. The coronary sulcus separates the atria from the ventricles. The interventricular grooves separate the right and left ventricles.
3. The inferior and superior venae cavae and the coronary sinus enter the right atrium. The four pulmonary veins enter the left atrium.
4. The pulmonary trunk exits the right ventricle, and the aorta exits the left ventricle.
5. Coronary arteries branch off the aorta to supply the heart. Blood returns from the heart tissues to the right atrium through the coronary sinus and cardiac veins.

Heart Chambers and Valves

1. The interatrial septum separates the atria from each other, and the interventricular septum separates the ventricles.
2. The tricuspid valve separates the right atrium and ventricle. The bicuspid valve separates the left atrium and ventricle. The chordae tendineae attach the papillary muscles to the atrioventricular valves.
3. The semilunar valves separate the aorta and pulmonary trunk from the ventricles.

Route of Blood Flow Through the Heart   (p. 688)

1. Blood from the body flows through the right atrium into the right ventricle and then to the lungs.
2. Blood returns from the lungs to the left atrium, enters the left ventricle, and is pumped back to the body.

Histology   (p. 690)
Heart Skeleton

The fibrous heart skeleton supports the openings of the heart, electrically insulates the atria from the ventricles, and provides a point of attachment for heart muscle.

Cardiac Muscle

1. Cardiac muscle cells are branched and have a centrally located nucleus. Actin and myosin are organized to form sarcomeres. The sarcoplasmic reticulum and T tubules are not as organized as in skeletal muscle.
2. Cardiac muscle cells are joined by intercalated disks, which allow action potentials to move from one cell to the next. Thus, cardiac muscle cells function as a unit.
3. Cardiac muscle cells have a slow onset of contraction and a prolonged contraction time caused by the length of time required for calcium to move to and from the myofibrils.
4. Cardiac muscle is well supplied with blood vessels that support aerobic respiration.
5. Cardiac muscle aerobically uses glucose, fatty acids, and lactic acid to produce ATP for energy. Cardiac muscle does not develop a significant oxygen debt.

Conducting System

1. The SA node and the AV node are in the right atrium.
2. The AV node is connected to the bundle branches in the interventricular septum by the AV bundle.
3. The bundle branches give rise to Purkinje fibers, which supply the ventricles.
4. The SA node is made up of small-diameter cardiac muscle cells that initiate action potentials, which spread across the atria and cause them to contract.
5. Action potentials are slowed in the AV node, allowing the atria to contract and blood to move into the ventricles. Then, the action potentials travel through the AV bundles and bundle branches to the Purkinje fibers, causing the ventricles to contract, starting at the apex. The AV node is also made up of small-diameter cardiac muscle fibers.

Electrical Properties   (p. 693)
Action Potentials

1. After depolarization and partial repolarization, a plateau is reached, during which the membrane potential only slowly repolarizes.
2. The movement of Na^+ through the voltage-gated Na^+ channels causes depolarization.
3. During depolarization, voltage-gated K^+ channels close and voltage-gated Ca^{2+} channels begin to open.
4. Early repolarization results from closure of the voltage-gated Na^+ channels and the opening of some voltage-gated K^+ channels.
5. The plateau exists because voltage-gated Ca^{2+} channels remain open.
6. The rapid phase of repolarization results from closure of the voltage-gated Ca^+ channels and the opening of many voltage-gated K^+ channels.
7. The entry of Ca^{2+} into cardiac muscle cells causes Ca^{2+} to be released from the sarcoplasmic reticulum to trigger contractions.

Autorhythmicity of Cardiac Muscle

1. Cardiac pacemaker muscle cells are autorhythmic because of the spontaneous development of a prepotential.
2. The prepotential results from the movement of Na^+ and Ca^{2+} into the pacemaker cells.

3. Ectopic foci are areas of the heart that regulate heart rate under abnormal conditions.

Refractory Period of Cardiac Muscle

Cardiac muscle has a prolonged depolarization and thus a prolonged refractory period, which allows time for the cardiac muscle to relax before the next action potential causes a contraction.

Electrocardiogram

1. The ECG records only the electrical activities of the heart.
   - Depolarization of the atria produces the P wave.
   - Depolarization of the ventricles produces the QRS complex. Repolarization of the atria occurs during the QRS complex.
   - Repolarization of the ventricles produces the T wave.
2. Based on the magnitude of the ECG waves and the time between waves, ECGs can be used to diagnose heart abnormalities.

Cardiac Cycle   (p. 698)

1. The cardiac cycle is repetitive contraction and relaxation of the heart chambers.
2. Blood moves through the circulatory system from areas of higher pressure to areas of lower pressure. Contraction of the heart produces the pressure.
3. The cardiac cycle is divided into five periods.
   - Although the heart is contracting, during the period of isovolumic contraction ventricular volume doesn't change because all the heart valves are closed.
   - During the period of ejection, the semilunar valves open, and blood is ejected from the heart.
   - Although the heart is relaxing, during the period of isovolumic relaxation, ventricular volume doesn't change because all the heart valves are closed.
   - Passive ventricular filling results when blood flows from the higher pressure in the veins and atria to the lower pressure in the relaxed ventricles.
   - Active ventricular filling results when the atria contract and pump blood into the ventricles.

Events Occurring During Ventricular Systole

1. Contraction of the ventricles closes the AV valves, opens the semilunar valves, and ejects blood from the heart.
2. The volume of blood in a ventricle just before it contracts is the end-diastolic volume. The volume of blood after contraction is the end-systolic volume.

Events Occurring During Ventricular Diastole

1. Relaxation of the ventricles results in closing of the semilunar valves, opening of the AV valves, and the movement of blood into the ventricles.
2. Most ventricular filling occurs when blood flows from the higher pressure in the veins and atria to the lower pressure in the relaxed ventricles.
3. Contraction of the atria completes ventricular filling.

Heart Sounds

1. Closure of the atrioventricular valves produces the first heart sound.
2. Closure of the semilunar valves produces the second heart sound.

Aortic Pressure Curve

1. Contraction of the ventricles forces blood into the aorta, thus producing the peak systolic pressure.
2. Blood pressure in the aorta falls to the diastolic level as blood flows out of the aorta.
3. Elastic recoil of the aorta maintains pressure in the aorta and produces the dicrotic notch.

Mean Arterial Blood Pressure   (p. 704)

1. Mean arterial pressure is the average blood pressure in the aorta. Adequate blood pressure is necessary to ensure delivery of blood to the tissues.
2. Mean arterial pressure is proportional to cardiac output (amount of blood pumped by the heart per minute) times peripheral resistance (total resistance to blood flow through blood vessels).
3. Cardiac output is equal to stroke volume times heart rate.
4. Stroke volume, the amount of blood pumped by the heart per beat, is equal to end-diastolic volume minus end-systolic volume.
   - Venous return is the amount of blood returning to the heart. Increased venous return increases stroke volume by increasing end-diastolic volume.
   - Increased force of contraction increases stroke volume by decreasing end-systolic volume.
5. Cardiac reserve is the difference between resting and exercising cardiac output.

Regulation of the Heart   (p. 706)
Intrinsic Regulation

1. Venous return is the amount of blood that returns to the heart during each cardiac cycle.
2. Starling's law of the heart describes the relationship between preload and the stroke volume of the heart. An increased preload causes the cardiac muscle fibers to contract with a greater force and produce a greater stroke volume.

Extrinsic Regulation

1. The cardioregulatory center in the medulla oblongata regulates the parasympathetic and sympathetic nervous control of the heart.
2. Parasympathetic control
   - Parasympathetic stimulation is supplied by the vagus nerve.
   - Parasympathetic stimulation decreases heart rate.
   - Postganglionic neurons secrete acetylcholine, which increases membrane permeability to K^+, producing hyperpolarization of the membrane.
3. Sympathetic control
   - Sympathetic stimulation is supplied by the cardiac nerves.
   - Sympathetic stimulation increases heart rate and the force of contraction (stroke volume).
   - Postganglionic neurons secrete norepinephrine, which increases membrane permeability to Na^+ and Ca^{2+} and produces depolarization of the membrane.
4. Epinephrine and norepinephrine are released into the blood from the adrenal medulla as a result of sympathetic stimulation.
   - The effects of epinephrine and norepinephrine on the heart are long lasting compared to those of neural stimulation.
   - Epinephrine and norepinephrine increase the rate and force of heart contraction.

Heart and Homeostasis   (p. 708)
Effect of Blood Pressure

1. Baroreceptors monitor blood pressure.
2. In response to a decrease in blood pressure, the baroreceptor reflexes increase sympathetic stimulation and decrease parasympathetic stimulation of the heart, resulting in an increase in heart rate and force of contraction.

Effect of pH, Carbon Dioxide, and Oxygen

1. Chemoreceptors monitor blood carbon dioxide, pH, and oxygen levels.
2. In response to increased carbon dioxide and decreased pH, medullary chemoreceptor reflexes increase sympathetic stimulation and decrease parasympathetic stimulation of the heart.
3. Carotid body chemoreceptor receptors stimulated by low oxygen levels result in a decreased heart rate and vasoconstriction.

4. All regulatory mechanisms functioning together in response to low blood pH, high blood carbon dioxide, and low blood oxygen levels usually produce an increase in heart rate and vasoconstriction. Decreased oxygen levels stimulate an increase in heart rate indirectly by stimulating respiration, and the stretch of the lungs activates a reflex that increases sympathetic stimulation of the heart.

Effect of Extracellular Ion Concentration

1. An increase or decrease in extracellular K^+ decreases heart rate.
2. Increased extracellular Ca^{2+} increase the force of contraction of the heart and decrease the heart rate. Decreased Ca^{2+} levels produce the opposite effect.

Effect of Body Temperature

Heart rate increases when body temperature increases, and it decreases when body temperature decreases.

Effects of Aging on the Heart (p. 711)

1. Aging results in gradual changes in the function of the heart which are minor under resting conditions but are more significant during exercise.
2. Hypertrophy of the left ventricle is a common age-related condition.
3. The maximum heart rate decreases and by age 85 the cardiac output may be decreased by 30%–60%.
4. There is an increased tendency for valves to function abnormally and for arrhythmias to occur.
5. An increased oxygen consumption, required to pump the same amount of blood, makes age-related coronary artery disease more severe.
6. Exercise improves the functional capacity of the heart at all ages.

R E V I E W   A N D   C O M P R E H E N S I O N

1. The fibrous pericardium
   a. is in contact with the heart.
   b. is a serous membrane.
   c. is also known as the epicardium.
   d. forms the outer layer of the pericardial sac.
   e. all of the above.

2. Which of these structures returns blood to the right atrium?
   a. coronary sinus
   b. inferior vena cava
   c. superior vena cava
   d. both b and c
   e. all of the above

3. The valve located between the right atrium and the right ventricle is the
   a. aortic semilunar valve.
   b. pulmonary semilunar valve.
   c. tricuspid valve.
   d. bicuspid (mitral) valve.

4. The papillary muscles
   a. are attached to chordae tendineae.
   b. are found in the atria.
   c. contract to close the foramen ovale.
   d. are attached to the semilunar valves.
   e. surround the openings of the coronary arteries.

5. Given these blood vessels:
   1. aorta
   2. inferior vena cava
   3. pulmonary trunk
   4. pulmonary vein

   Choose the arrangement that lists the vessels in the order a red blood cell would encounter them in going from the systemic veins back to the systemic arteries.
   a. 1,3,4,2
   b. 2,3,4,1
   c. 2,4,3,1
   d. 3,2,1,4
   e. 3,4,2,1

6. Which of these does *not* correctly describe the skeleton of the heart?
   a. electrically insulates the atria from the ventricles
   b. provides a rigid source of attachment for the cardiac muscle
   c. functions to reinforce or support the valve openings
   d. is composed mainly of cartilage

7. The bulk of the heart wall is
   a. epicardium.
   b. pericardium.
   c. myocardium.
   d. endocardium.
   e. exocardium.

8. Muscular ridges on the interior surface of the auricles are called
   a. trabeculae carneae.
   b. crista terminalis.
   c. musculi pectinati.
   d. endocardium.
   e. papillary muscles.

9. Cardiac muscle has
   a. sarcomeres.
   b. a sarcoplasmic reticulum.
   c. transverse tubules.
   d. many mitochondria.
   e. all of the above.

10. Action potentials pass from one cardiac muscle cell to another
    a. through gap junctions.
    b. by a special cardiac nervous system.
    c. because of the large voltage of the action potentials.
    d. because of the plateau phase of the action potentials.
    e. by neurotransmitters.

11. During the transmission of action potentials through the conducting system of the heart, there is a temporary delay at the
    a. bundle branches.
    b. Purkinje fibers.
    c. AV node.
    d. SA node.
    e. AV bundle.

12. Given these structures of the conduction system of the heart:
    1. atrioventricular bundle
    2. AV node
    3. bundle branches
    4. Purkinje fibers
    5. SA node

    Choose the arrangement that lists the structures in the order an action potential passes through them.
    a. 2,5,1,3,4
    b. 2,5,3,1,4
    c. 2,5,4,1,3
    d. 5,2,1,3,4
    e. 5,2,4,3,1

13. Purkinje fibers
    a. are specialized cardiac muscle cells.
    b. conduct impulses much more slowly than ordinary cardiac muscle.
    c. conduct action potentials through the atria.
    d. connect between the SA node and the AV node.
    e. ensure that ventricular contraction starts at the base of the heart.

14. T waves on an ECG represent
    a. depolarization of the ventricles.
    b. repolarization of the ventricles.
    c. depolarization of the atria.
    d. repolarization of the atria.

15. Which of these conditions observed in an electrocardiogram suggests that the AV node is not conducting action potentials?
    a. complete lack of the P wave
    b. complete lack of the QRS complex
    c. more QRS complexes than P waves
    d. a prolonged PR interval
    e. P waves and QRS complexes are not synchronized

16. The greatest amount of ventricular filling occurs during
    a. the first one-third of diastole.
    b. the middle one-third of diastole.
    c. the last one-third of diastole.
    d. ventricular systole.

17. While the semilunar valves are open during a normal cardiac cycle, the pressure in the left ventricle is
    a. greater than the pressure in the aorta.
    b. less than the pressure in the aorta.
    c. the same as the pressure in the left atrium.
    d. less than the pressure in the left atrium.

18. The pressure within the left ventricle fluctuates between
    a. 120 and 80 mm Hg.
    b. 120 and 0 mm Hg.
    c. 80 and 0 mm Hg.
    d. 20 and 0 mm Hg.

19. Blood flows neither into nor out of the ventricles during
    a. the period of isovolumic contraction.
    b. the period of isovolumic relaxation.
    c. diastole.
    d. systole.
    e. both a and b.

20. Stroke volume is the
    a. amount of blood pumped by the heart per minute.
    b. difference between end-diastolic and end-systolic volume.
    c. difference between the amount of blood pumped at rest and that pumped at maximum output.
    d. amount of blood pumped from the atria into the ventricles.

21. Cardiac output is defined as
    a. blood pressure times peripheral resistance.
    b. peripheral resistance times heart rate.
    c. heart rate times stroke volume.
    d. stroke volume times blood pressure.
    e. blood pressure minus peripheral resistance.

22. Pressure in the aorta is at its lowest
    a. at the time of the first heart sound.
    b. at the time of the second heart sound.
    c. just before the AV valves open.
    d. just before the semilunar valves open.

23. Just after the dicrotic notch on the aortic pressure curve,
    a. the pressure in the aorta is greater than the pressure in the left ventricle.
    b. the pressure in the left ventricle is greater than the pressure in the aorta.
    c. the pressure in the left atrium is greater than the pressure in the left ventricle.
    d. the pressure in the left atrium is greater than the pressure in the aorta.
    e. blood pressure in the aorta is 0 mm Hg.

24. The "lubb" sound (first heart sound) of the heart is caused by the
    a. closing of the AV valves.
    b. closing of the semilunar valves.
    c. blood rushing out of the ventricles.
    d. filling of the ventricles.
    e. ventricular contraction.

25. Increased venous return results in
    a. increased stroke volume.
    b. increased cardiac output.
    c. decreased heart rate.
    d. both a and b.

26. Parasympathetic nerve fibers are found in the _____ nerves and release _____ at the heart.
    a. cardiac, acetylcholine
    b. cardiac, norepinephrine
    c. vagus, acetylcholine
    d. vagus, norepinephrine

27. Increased parasympathetic stimulation of the heart
    a. increases the force of ventricular contraction.
    b. increases the rate of depolarization in the SA node.
    c. decreases the heart rate.
    d. increases cardiac output.

28. Because of the baroreceptor reflex, when normal arterial blood pressure decreases
    a. heart rate decreases.
    b. stroke volume decreases.
    c. the frequency of afferent action potentials from baroreceptors decreases.
    d. the cardioregulatory center stimulates parasympathetic neurons.
    e. all of the above.

29. A decrease in blood pH and an increase in blood carbon dioxide levels result in
    a. increased heart rate.
    b. increased stroke volume.
    c. increased sympathetic stimulation of the heart.
    d. increased cardiac output.
    e. all of the above.

30. An increase in extracellular potassium levels could cause
    a. an increase in stroke volume.
    b. an increase in the force of contraction.
    c. a decrease in heart rate.
    d. both a and b.

Answers in Appendix F

C R I T I C A L   T H I N K I N G

1. Explain why the walls of the ventricles are thicker than the walls of the atria.

2. In most tissues, peak blood flow occurs during systole and decreases during diastole. In heart tissue, however, the opposite is true, and peak blood flow occurs during diastole. Explain why this difference occurs.

3. A patient has tachycardia. Would you recommend a drug that prolongs or shortens the plateau of cardiac muscle cell action potentials?

4. Endurance-trained athletes often have a decreased heart rate compared to that of a nonathlete when both are resting. Explain why an endurance-trained athlete's heart rate decreases rather than increases.

5. A doctor lets you listen to a patient's heart with a stethoscope at the same time that you feel the patient's pulse. Once in a while you hear two heartbeats very close together, but you feel only one pulse beat. Later, the doctor tells you that the patient has an ectopic focus in the right atrium. Explain why you hear two heartbeats very close together. The doctor also tells you that the patient exhibits a pulse deficit (i.e., the number of pulse beats felt is fewer than the number of heartbeats heard). Explain why a pulse deficit occurs.

6. Heart rate and cardiac output were measured in a group of nonathletic students. After 2 months of aerobic exercise training, their measurements were repeated. It was found that heart rate had decreased, but cardiac output remained the same for many activities. Explain these findings.

7. Explain why it's sufficient to replace the ventricles, but not the atria, in artificial heart transplantation.

8. During an experiment in a physiology laboratory, a student named Cee Saw was placed on a table that could be tilted. The instructor asked the students to predict what would happen to Cee Saw's heart rate if the table were tilted so that her head was lower than her feet. Some students predicted an increase in heart rate, and others claimed it would decrease. Can you explain why both predictions might be true?

9. After Cee Saw is tilted so that her head is lower than her feet for a few minutes, the regulatory mechanisms that control blood pressure adjust so that the heart pumps sufficient blood to supply the needs of her tissues. If she is then tilted so that her head is higher than her feet, gravity causes blood to flow toward her feet, and the blood pressure in the carotid sinus and aortic arch decreases. The decrease in blood pressure is detected by the baroreceptors in these vessels and activates baroreceptor reflexes. The result is increased sympathetic and decreased parasympathetic stimulation of the heart and an increase in the heart rate. The increased heart rate functions to increase the blood pressure to its normal value.

10. A friend tells you that her son had an ECG and it revealed that he has a slight heart murmur. Should you be convinced that he has a heart murmur? Explain.

11. Predict the effect of an incompetent aortic semilunar valve on stroke volume, the volume of blood in the left ventricle at the end of diastole, and on heart sounds?

12. An experiment on a dog was performed in which the mean arterial blood pressure was monitored before and after the common carotid arteries were partially clamped (at time A). The results are graphed below:

Explain the change in mean arterial blood pressure (*Hint:* baroreceptors are located in the internal carotid arteries, which are superior to the site of clamping of the common carotid arteries).

13. During hemorrhagic shock (caused by loss of blood) the blood pressure may fall dramatically, although the heart rate is elevated. Explain why the blood pressure falls despite the increase in heart rate.

Answers in Appendix G

A N S W E R S   T O   P R E D I C T   Q U E S T I O N S

1. The heart tissues supplied by the artery lose their oxygen and nutrient supply and die. This part of the heart (and possibly the entire heart) stops functioning. Restricted blood flow through the anterior interventricular artery reduces blood flow to the anterior portion of the ventricles. If this condition develops rapidly, it's called a heart attack, or myocardial infarction.

2. The heart must continue to function under all conditions and requires energy in the form of ATP. During heavy exercise, lactic acid is produced in skeletal muscle as a by-product of anaerobic metabolism. The ability to use lactic acid provides the heart with an additional energy source.

3. Contraction of the ventricles, beginning at the apex and moving toward the base of the heart, forces blood out of the ventricles and toward their outflow vessels—the aorta and pulmonary trunk. The aorta and pulmonary trunks are located at the base of the heart.

4. Ectopic foci cause various regions of the heart to contract at different times. As a result, pumping effectiveness is reduced. Cardiac muscle contraction is not coordinated, which interrupts the cyclic filling and emptying of the ventricles.

5. If cardiac muscle could undergo tetanic contraction, it would contract for a long time without relaxing. Its pumping action then would stop because that action requires alternating contraction and relaxation.

6. During isovolumic contraction, the volume of the ventricles does not change because no blood leaves the ventricle. Therefore, the pressure increases but the length of the cardiac muscle doesn't change significantly. Therefore, the contraction is isometric (see chapter 9).

7. The left ventricle has the thickest wall. The pressure produced by the left ventricle is much higher than the pressure produced by the right ventricle, when the ventricles contract. It's important for each ventricle to pump the same amount of blood because, with two connected circulation loops, the blood flowing into one must equal the blood flowing into the other so that one doesn't become overfilled with blood at the expense of the other. For example, if the right ventricle pumps less blood than the left ventricle, blood must accumulate in the systemic blood vessels. If the left ventricle pumps less blood than the right ventricle, blood accumulates in the pulmonary blood vessels.

8. Fibrillation makes cardiac muscle an ineffective pump. The pumping action of the heart depends on coordinated contraction of cardiac muscle. Fibrillation destroys the coordinated contractions and results in the loss of the ability for cardiac muscle to function as a pump. The ventricles are the primary pumping chambers of the heart. Ventricular fibrillation results in death because of the inability of the heart to pump blood. The atria function primarily as reservoirs. Their pumping action is most important during exercise. Therefore atrial fibrillation does not destroy the ability of the ventricles to pump blood.

9. An incompetent bicuspid valve allows blood to flow from the left ventricle to the left atrium when the left ventricle contracts. As a result, the stroke volume (the volume of blood ejected into the aorta) is reduced. At the end of ventricular systole, the volume of the left atrium is increased because blood flowed back into the left atrium during ventricular systole. At the end of ventricular diastole, the volume of the left ventricle is greater than normal because the blood that flowed back into the left atrium during the previous ventricular systole now flows into the ventricle, along with the blood that normally returns from the lungs through the pulmonary veins.

10. Sympathetic stimulation increases heart rate. If venous return remains constant, stroke volume decreases as the number of beats per minute increases. Dilation of the coronary arteries is important because, as the heart does more work, the cardiac tissue requires more energy and, therefore, a greater blood supply to carry more oxygen.

11. Rupture of the left ventricle, as experienced by Mr. P, is more likely several days after a myocardial infarction. As the necrotic tissues are removed by macrophages, the wall of the ventricle becomes thinner and may bulge during systole. If the wall of the ventricle becomes very thin before new connective tissue is deposited, it may rupture. If the left ventricle ruptures, blood flows from the left ventricle into the pericardial sac. As blood fills the pericardial sac, it compresses the ventricle from the outside. This is called cardiac tamponade (tam-pŏ-nād′). Thus the ventricle is not able to fill with blood and its pumping ability is eliminated. Death occurs quickly in response to a ruptured wall of the left ventricle.

Visit the Online Learning Center at www.mhhe.com/seeley7 for chapter quizzes, interactive learning exercises, and other study tools.

Cardiovascular System

Peripheral Circulation and Regulation

Part 4 Regulations and Maintenance

Color enhanced scanning electron micrograph of an artery. The inner surface of the artery appears to be scalloped, or folded, because the circular smooth muscle cells in the wall of the artery partially contract when tissues are prepared for microscopic examination.

Complex urban water systems seem rather simple when compared to the intricacy and coordinated functions of blood vessels. The heart is the pump that provides the major force causing blood to circulate, and the blood vessels are the pipes that carry blood to tissues of the body and back to the heart. In addition, the blood vessels participate in the regulation of blood pressure and help direct blood flow to tissues that are most active.

The peripheral circulatory system comprises two sets of blood vessels: systemic and pulmonary vessels. **Systemic vessels** transport blood through essentially all parts of the body from the left ventricle and back to the right atrium. **Pulmonary vessels** transport blood from the right ventricle through the lungs and back to the left atrium (see figure 20.1). Both the blood vessels and the heart are regulated to ensure that the blood pressure is high enough to cause blood flow in sufficient quantities to meet the metabolic needs of the tissues. The cardiovascular system ensures the survival of the tissues in the body by supplying nutrients to and removing waste products from them.

This chapter explains the functions of the peripheral circulation (p. 724) *general features of blood vessel structure* (p. 724), *pulmonary circulation* (p. 730), *systemic circulation: arteries* (p. 730), *systemic circulation: veins* (p. 741), *dynamics of blood circulation* (p. 753), *physiology of systemic circulation* (p. 757), *control of blood flow in tissues* (p. 762), and *regulation of mean arterial pressure* (p. 766).

Functions of the Peripheral Circulation

The heart provides the major force that causes blood to circulate, and the peripheral circulation functions to

1. *Carry blood.* Blood vessels carry blood from the heart to all tissues of the body and back to the heart.
2. *Exchange nutrients, waste products, and gases.* Nutrients and oxygen diffuse from blood vessels to cells in essentially all areas of the body. Waste products and carbon dioxide diffuse from the cells, where they are produced, to blood vessels.
3. *Transport.* Hormones, components of the immune system, molecules required for coagulation, enzymes, nutrients, gases, waste products, and other substances are transported in the blood to all areas of the body.
4. *Regulate blood pressure.* The peripheral circulatory system and the heart work together to regulate blood pressure within a normal range of values.
5. *Direct blood flow.* The peripheral circulatory system directs blood to tissues when increased blood flow is required to maintain homeostasis.

General Features of Blood Vessel Structure

The ventricles pump blood from the heart into large elastic arteries that branch repeatedly to form many progressively smaller arteries. As they become smaller, the arteries undergo a gradual transition from having walls that contain a large amount of elastic tissue and a smaller amount of smooth muscle to having walls with a smaller amount of elastic tissue and a relatively large amount of smooth muscle. Although the arteries form a continuum from the largest to the smallest branches, they normally are classified as (1) elastic arteries, (2) muscular arteries, or (3) arterioles.

Blood flows from arterioles into capillaries. Most of the exchange that occurs between the blood and interstitial spaces occurs across the walls of capillaries. Their walls are the thinnest of all the blood vessels, blood flows through them slowly, and a greater number of them exist than any other blood vessel type.

From the capillaries, blood flows into the venous system. When compared to arteries, the walls of the veins are thinner and contain less elastic tissue and fewer smooth muscle cells. The veins increase in diameter and decrease in number, and their walls increase in thickness as they project toward the heart. They are classified as (1) venules, (2) small veins, or (3) medium or large veins.

> ● 1. *Name, in order, all the types of blood vessels, starting at the heart, going into the tissues, and returning to the heart.*

Capillaries

All blood vessels have an internal lining of simple squamous epithelial cells called the **endothelium** (en-dō-thē′lē-ŭm), which is continuous with the endocardium of the heart.

The capillary wall consists primarily of endothelial cells (figure 21.1), which rest on a basement membrane. Outside the

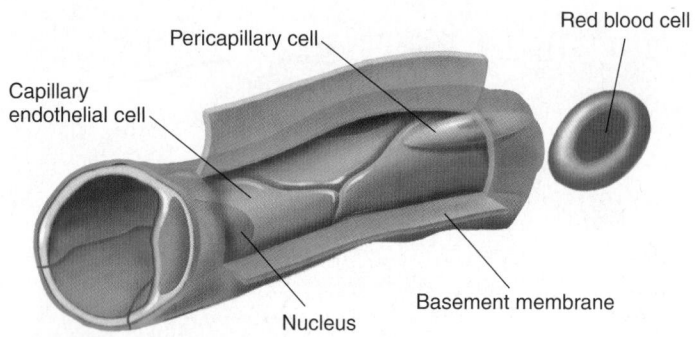

Figure 21.1   Capillary

Section of a capillary showing that it is composed of flattened endothelial cells.

basement membrane is a delicate layer of loose connective tissue that merges with the connective tissue surrounding the capillary.

Along the length of the capillary are some scattered cells that are closely associated with the endothelial cells. These scattered cells lie between the basement membrane and the endothelial cells and are called **pericapillary cells.** They are apparently fibroblasts, macrophages, or undifferentiated smooth muscle cells.

Most capillaries range from 7–9 μm in diameter, and they branch without a change in their diameter. Capillaries are variable in length, but in general, they are approximately 1 mm long. Red blood cells flow through most capillaries in a single file and frequently are folded as they pass through the smaller-diameter capillaries.

Types of Capillaries

Capillaries are classified as continuous, fenestrated, or sinusoidal, depending on their diameter and permeability characteristics.

Continuous capillaries are approximately 7–9 μm in diameter, and their walls exhibit no gaps between the endothelial cells (figure 21.2a). Continuous capillaries are less permeable to large molecules than are other capillary types and occur in muscle, nervous tissue, and many other locations.

In **fenestrated** (fen′es-trā′ted) **capillaries,** endothelial cells have numerous fenestrae (figure 21.2b). The **fenestrae** (fe-nes′trē; windows) are areas approximately 70–100 nm in diameter in which the cytoplasm is absent and the plasma membrane consists of a porous diaphragm that's thinner than the normal plasma membrane. In some capillaries the diaphragm is not present. Fenestrated capillaries are in tissues where capillaries are highly permeable, such as in the intestinal villi, ciliary process of the eye, choroid plexuses of the central nervous system, and glomeruli of the kidney.

Sinusoidal (sī-nŭ-soy′dăl) **capillaries** are larger in diameter than either continuous or fenestrated capillaries, and their basement membrane is less prominent (see figure 21.2c) or absent. Their fenestrae are larger than those in fenestrated capillaries and gaps may exist between endothelial cells. The sinusoidal capillaries occur in such places as endocrine glands, where large molecules cross their walls.

Sinusoids are large-diameter sinusoidal capillaries. Their basement membrane is sparse and often missing, and their structure suggests that large molecules and sometimes cells can move readily across their walls between the endothelial cells (see figure 21.2c).

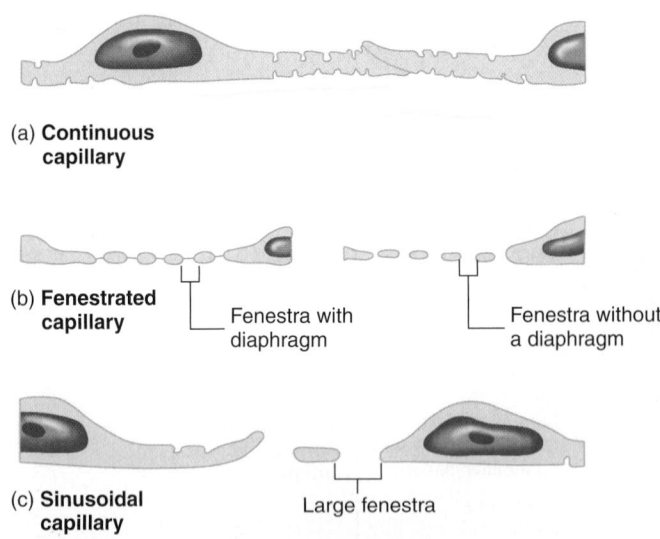

(a) Continuous capillary

(b) Fenestrated capillary — Fenestra with diaphragm — Fenestra without a diaphragm

(c) Sinusoidal capillary — Large fenestra

Figure 21.2 The Structure of Capillary Walls

(*a*) Continuous capillaries have no gaps between endothelial cells and no fenestrae. They are common in muscle, nervous, and connective tissue. (*b*) Fenestrated capillaries have fenestrae 7–100 nm in diameter covered by thin porous diaphragms, which are not present in some capillaries. They are found in intestinal villi, ciliary process of the eye, choroids plexus of the central nervous system, and glomerulus of the eye. (*c*) Sinusoidal capillaries have larger fenestrae without diaphragms and can have gaps between endothelial cells. They are found in bone marrow, liver, spleen, and lymphoid organs.

Arteriole — Metarteriole — Thoroughfare channel

Arterial capillaries — Precapillary sphincter

Venous capillaries

Venule

Figure 21.3 Capillary Network

The metarteriole giving rise to the network feeds directly from an arteriole into the thoroughfare channel, which feeds into the venule. The network forms numerous branches that transport blood from the thoroughfare channel and can return to the channel. Smooth muscle cells, called precapillary sphincters, regulate blood flow through the capillaries. Blood flow decreases when the precapillary sphincters constrict and increases when they dilate.

Sinusoids are common in the liver and the bone marrow. Macrophages are closely associated with the endothelial cells of the liver sinusoids. **Venous sinuses** are similar in structure to the sinusoidal capillaries but are even larger in diameter. They occur primarily in the spleen, and they have large gaps between the endothelial cells that make up their walls.

Substances cross capillary walls by diffusing through the endothelial cells, through fenestrae, or between the endothelial cells. Lipid-soluble substances, such as oxygen and carbon dioxide, and small water-soluble molecules readily diffuse through the plasma membrane. Larger water-soluble substances must pass through the fenestrae or gaps between the endothelial cells. In addition, transport by pinocytosis occurs, but little is known about its role in the capillaries. The walls of the capillaries are effective permeability barriers because red blood cells and large water-soluble molecules like proteins cannot readily pass through them.

Capillary Network

Arterioles supply blood to each capillary network (figure 21.3). Blood then flows through the capillary network and into the venules. The ends of capillaries closest to the arterioles are **arterial capillaries,** and the ends closest to venules are **venous capillaries.**

Blood flows from arterioles through **metarterioles** (met′ar-tēr′ē-ōlz), which have isolated smooth muscle cells along their walls. Blood flows from a metarteriole into a **thoroughfare channel,** which extends in a relatively direct fashion from a metarteriole to a venule. Blood flow through thoroughfare channels is relatively continuous. Several capillaries branch from the thoroughfare channels, and in these branches blood flow is intermittent. Smooth muscle

cells called **precapillary sphincters,** which are located at the origin of the branches (see figure 21.3), regulate flow in these capillaries.

Capillary networks are more numerous and more extensive in highly metabolic tissues, such as the lung, liver, kidney, skeletal muscle, and cardiac muscle. Capillary networks in the skin have many more thoroughfare channels than capillary networks in cardiac or skeletal muscle. Capillaries in the skin function in thermoregulation, and heat loss results from the flow of a large volume of blood through them. In muscle, however, nutrient and waste product exchange is the major function of the capillaries.

2. *Describe the structure of a capillary.*
3. *Compare the structure of the three types of capillaries. Explain the ways that materials pass through capillary walls.*
4. *Describe a capillary network. Where is the smooth muscle that regulates blood flow into and through the capillary network located? What is the function of a thoroughfare channel?*
5. *Contrast the function of capillaries in the skin with the function of capillaries in muscle tissue.*

Structure of Arteries and Veins

General Features

Except for the capillaries and the venules, the blood vessel walls consist of three relatively distinct layers, which are most apparent in the muscular arteries and least apparent in the veins. From the lumen to the outer wall of the blood vessels, the layers, or **tunics** (too′niks), are (1) the tunica intima, (2) the tunica media, (3) and the tunica adventitia, or tunica externa (figure 21.4 and figure 21.5).

Figure 21.4   Histology of a Blood Vessel

The layers, or tunics, of the blood vessel wall include the intima, media, and adventitia. A vasa vasorum is a blood vessel that supplies blood to the wall of the blood vessel.

Figure 21.5   Photomicrograph of an Artery and a Vein

The typical structure of a medium-sized artery (A) and vein (V). Note that the artery has a thicker wall than the vein. The tunica intima, tunica media, and tunica adventitia make up the walls of the blood vessels. The predominant layer in the wall of the artery is the tunica media with its circular layers of smooth muscle. The predominant layer in the wall of the vein is the tunica adventitia, and the tunica media is thinner.

The **tunica intima** consists of endothelium, a delicate connective tissue basement membrane, a thin layer of connective tissue called the lamina propria, and a fenestrated layer of elastic fibers called the **internal elastic membrane.** The internal elastic membrane separates the tunica intima from the next layer, the tunica media.

The **tunica media,** or middle layer, consists of smooth muscle cells arranged circularly around the blood vessel. The amount of blood flowing through a blood vessel can be regulated by contraction or relaxation of the smooth muscle in the tunica media. A decrease in blood flow results from **vasoconstriction** (vā′sō-kon-strik′shŭn, vas′ō-kon-strik′shŭn), a decrease in blood vessel diameter caused by smooth muscle contraction, whereas an increase in blood flow is produced by **vasodilation** (vā′sō-dī-lā′shŭn, vas-ō-dī-lā′shŭn), an increase in blood vessel diameter because of smooth muscle relaxation. The tunica media also contains variable amounts of elastic and collagen fibers, depending on the size of the vessel. An **external elastic membrane,** which separates the tunica media from the tunica adventitia, can be identified at the outer border of the tunica media in some arteries. A few longitudinally oriented smooth muscle cells occur in some arteries near the tunica intima.

The **tunica adventitia** (too′ni-kă ad-ven-tish′ă) is composed of connective tissue, which varies from dense connective tissue near the tunica media to loose connective tissue that merges with the connective tissue surrounding the blood vessels.

The relative thickness and composition of each layer varies with the diameter of the blood vessel and its type. The transition from one artery type or from one vein type to another is gradual, as are the structural changes.

Large Elastic Arteries

Elastic arteries have the largest diameters (figure 21.6a) and often are called **conducting arteries.** Pressure is relatively high in these vessels, and it fluctuates between systolic and diastolic values. A greater amount of elastic tissue and a smaller amount of smooth muscle occur in their walls compared to the elastic tissue and smooth muscle of other arteries. The elastic fibers are responsible for the elastic characteristics of the blood vessel wall, but collagenous connective tissue determines the degree to which the arterial wall can be stretched.

The tunica intima is relatively thick. The elastic fibers of the internal and external elastic membranes merge and are not recognizable as distinct layers. The tunica media consists of a meshwork of elastic fibers with interspersed circular smooth muscle cells and some collagen fibers. The tunica adventitia is relatively thin.

Muscular Arteries

The larger **muscular arteries,** often called **medium arteries,** can be observed in a gross dissection. They include most of the smaller *unnamed* arteries. Their walls are relatively thick compared to their diameter, mainly because the tunica media contains 25–40 layers of smooth muscle (figure 21.6b). The tunica intima of the muscular arteries has a well-developed internal elastic membrane. The tunica adventitia is composed of a relatively thick layer of collagenous connective tissue that blends with the surrounding connective tissue. Muscular arteries frequently are called **distributing arteries** because the smooth muscle cells allow these vessels to partially regulate blood supply to different regions of the body by either constricting or dilating.

Smaller muscular arteries range from 40–300 μm in diameter, and those that are 40 μm in diameter have approximately three or four layers of smooth muscle in their tunica media, whereas arteries that are 300 μm across have essentially the same structure as the larger muscular arteries. The small muscular arteries are adapted for vasodilation and vasoconstriction.

Arterioles

The **arterioles** (ar-tĕr′ē-ōlz) transport blood from small arteries to capillaries and are the smallest arteries in which the three tunics can be identified. They range from approximately 40 μm to as small as 9 μm in diameter. The tunica intima has no observable internal elastic membrane, and the tunica media consists of one or two layers of circular smooth muscle cells. The arterioles, like the small arteries, are capable of vasodilation and vasoconstriction.

Venules and Small Veins

Venules (ven′-oolz, vē′noolz), with a diameter of up to 40–50 μm, are tubes composed of endothelium resting on a delicate basement membrane. Their structure, except for their diameter, is very similar to that of capillaries. A few isolated smooth muscle cells exist outside the endothelial cells, especially in the larger venules. As the vessels increase to 0.2–0.3 mm in diameter, the smooth muscle cells form a continuous layer; the vessels then are called **small veins.** The small veins also have a tunica adventitia composed of collagenous connective tissue.

The venules collect blood from the capillaries and transport it to small veins, which in turn transport it to medium-sized veins. Nutrient exchange occurs across the venule walls, but, as the walls of the small veins increase in thickness, the degree of nutrient exchange decreases.

Medium and Large Veins

Most of the veins observed in gross anatomic dissections, except for the large veins, are **medium veins.** They collect blood from small veins and deliver it to large veins. The **large veins** transport blood from the medium veins to the heart. Their tunica intima is thin and consists of endothelial cells, a relatively thin layer of collagenous connective tissue, and a few scattered elastic fibers. The tunica media is also thin and is composed of a thin layer of circularly arranged smooth muscle cells containing some collagen fibers and a few sparsely distributed elastic fibers. The tunica adventitia, which is composed of collagenous connective tissue, is the predominant layer (figure 21.6c).

Valves

Veins having diameters greater than 2 mm contain **valves** that allow blood to flow toward the heart but not in the opposite direction (figure 21.6d). The valves consist of folds in the tunica intima that form two flaps that are shaped and function like the semilunar valves of the heart. The two folds overlap in the middle of the vein so that, when blood attempts to flow in a reverse direction, the valves occlude the vessel. Many valves are present in the medium veins, and the number is greater in veins of the lower extremities than in veins of the upper extremities.

Varicose Veins, Phlebitis, and Gangrene

Stretching of the vein walls in the lower limbs causes valves to become incompetent, which results in **varicose veins.** The veins become so dilated that the flaps of the venous valves no longer overlap and prevent the backflow of blood. As a consequence, the venous pressure is greater than normal in the veins of the lower limbs, resulting in edema. Blood flow in the veins can become sufficiently stagnant that the blood clots. The condition can result in **phlebitis** (fle-bī′tis), which is inflammation of the veins. If the inflammation is severe and blood flow becomes stagnant in a large area, it can lead to **gangrene** (gang′grēn), which is tissue death caused by a reduction or loss of blood supply. Some people have a genetic propensity for the development of varicose veins. For women with that genetic propensity, some medical conditions increase the pressure in veins, causing them to stretch, and varicose veins can develop. One such condition is pregnancy, in which the venous pressure in the veins that drain the lower limbs increases because of compression of the veins by the expanded uterus.

(a) **Elastic arteries.** The tunica media is mostly elastic connective tissue. Elastic arteries recoil when stretched, which prevents blood pressure from falling rapidly.

Tunica adventitia

Tunica media (elastic tissue and smooth muscle)

Tunica intima (endothelium and basement membrane)

(b) **Muscular arteries.** The tunica media is a thick layer of smooth muscle. Muscular arteries regulate blood flow to different regions of the body.

Tunica adventitia

External elastic membrane

Smooth muscle

Internal elastic membrane

Lamina propria

Endothelium and basement membrane

Tunica media

Tunica intima

(c) **Medium and large veins.** All three tunics are present. The tunica media is thin, but can regulate vessel diameter because blood pressure in the venous system is low. The predominant layer is the tunica adventitia.

Tunica adventitia

Tunica media

Internal elastic membrane

Endothelium and basement membrane

Tunica intima

(d) Folds in the endothelium form the valves of veins, which allow blood to flow toward the heart but not in the opposite direction.

Valve closed

Vein

Valve open

Direction of blood flow

Figure 21.6 Structural Comparison of Blood Vessel Types

(a) Elastic arteries are large-diameter arteries with thick walls that contain a large amount of elastic connective tissue in the tunica media. (b) Muscular arteries have a distinctive layer of smooth muscle cells in the tunica media, and they are capable of constriction and dilation. (c) Medium veins have thinner walls. The tunica media is thinner than the tunica media in arteries and contains fewer smooth muscle cells. The dominant layer in the veins is the tunica adventitia. (d) The valves in veins are folds in the endothelium that allow blood to flow toward the heart but not in the opposite direction.

6. *Name the three layers of a blood vessel. What kinds of tissue are in each layer?*

7. *Compare the amount of elastic fibers and smooth muscle found in each different type of artery and vein.*

8. *What is the function of valves in blood vessels? In which blood vessels are they found?*

Vasa Vasorum

For arteries and veins greater than 1 mm in diameter, nutrients cannot diffuse from the lumen of the vessel to all of the layers of the wall. Nutrients are, therefore, supplied to their walls by way of small blood vessels called **vasa vasorum** (vā′să vā′sor-ŭm), which penetrate from the exterior of the vessel to form a capillary network in the tunica adventitia and the tunica media (see figure 21.4).

Arteriovenous Anastomoses

Arteriovenous anastomoses (ă-nas′tō-mō′sez) allow blood to flow from arterioles to small veins without passing through capillaries. A **glomus** (glō′mŭs) is an arteriovenous anastomosis that consists of arterioles arranged in a convoluted fashion surrounded by collagenous connective tissue.

Naturally occurring arteriovenous anastomoses are present in large numbers in the sole of the foot, the palm of the hand, the terminal phalanges, and the nail beds. They function in temperature regulation. **Pathologic arteriovenous anastomoses** can result from injury or tumors. They cause the direct flow of blood from arteries to veins and can, if they are sufficiently large, lead to heart failure because of the tremendous increase in venous return to the heart.

Portal Veins

Portal (pōr′tăl, door) **veins** are veins that begin in a primary capillary network, extend some distance and end in a secondary capillary network. There is no pumping mechanism like the heart between the two capillary networks. There are two systems of portal veins in humans. The hepatic portal veins carry blood from the capillaries in the gastrointestinal tract and spleen to dilated capillaries, called sinusoids, in the liver (see figure 21.26). The hypothalamohypophysial portal veins carry blood from the hypothalamus of the brain to the anterior pituitary gland (see figure 18.3).

9. *Define the terms vasa vasorum and arteriovenous anastamosis, and give their function. Define portal veins and name two examples.*

Nerves

Unmyelinated sympathetic nerve fibers (see figure 21.4) richly innervate the walls of most blood vessels. Some blood vessels, such as those in the penis or clitoris, are innervated by parasympathetic fibers. The nerve fibers form plexuses in the tunica adventitia, and nerve terminals project among the smooth muscle cells of the tunica media. Synapses consist of enlargements of the nerve fibers. Small arteries and arterioles are innervated to a greater extent than other blood vessel types. The response of the blood vessels to sympathetic stimulation is vasoconstriction.

Parasympathetic stimulation of blood vessels in the penis or clitoris results in vasodilation.

The smooth muscle cells of blood vessels act to some extent as a functional unit. Frequent gap junctions occur between adjacent smooth muscle cells, and as a consequence, stimulation of a few smooth muscle cells in the vessel wall results in constriction of a relatively large segment of the blood vessel.

A few myelinated sensory neurons innervate some blood vessels and function as baroreceptors. They monitor stretch in the blood vessel wall and detect changes in blood pressure.

10. *Describe the innervation of the walls of blood vessels. Which types of vessels have the greatest innervation?*

Aging of the Arteries

The walls of all arteries undergo changes as they age, although some arteries change more rapidly than others and some individuals are more susceptible to change than others. The most significant changes occur in the large elastic arteries like the aorta, large arteries that carry blood to the brain, and the coronary arteries. The age-related changes described here refer to these blood vessel types. Changes in muscular arteries do occur, but they are less dramatic and often do not result in disruption of normal blood vessel function.

Degenerative changes in arteries that make them less elastic are referred to collectively as **arteriosclerosis** (ar-tēr′ē-ō-skler-ō′sis; hardening of the arteries). These changes occur in many individuals and become more severe with advancing age. A related term, **atherosclerosis** (ath′er-ō-skler-ō′sis), refers to the deposition of material in the walls of arteries to form plaques. The material is a fatlike substance containing cholesterol (figure 21.7). The fatty material can be replaced later with dense connective tissue and calcium deposits. The initial signs of arteriosclerosis have been identified in the arteries of people in their teens, and it develops earlier and progresses more rapidly in some individuals than in others.

In arteriosclerosis, the tunica intima thickens and the tunica media becomes less elastic, because of a chemical change that takes place in the elastic fibers. Fat gradually accumulates between the elastic and collagen fibers to produce a lesion that protrudes into the lumen of the vessel, which can eventually

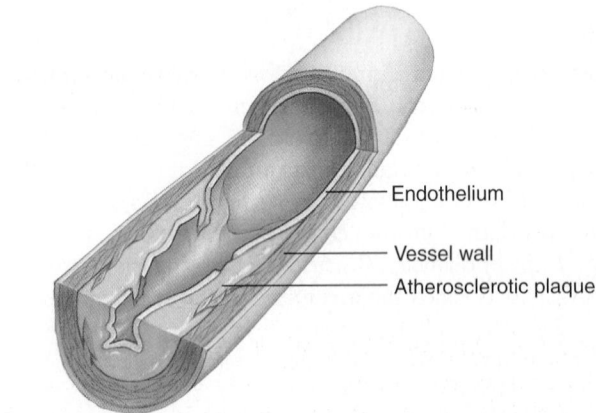

Figure 21.7 **Atherosclerotic Plaque in an Artery**
Atherosclerotic plaques develop within the tissue of the artery wall.

hamper normal blood flow. In advanced forms of arteriosclerosis, calcium deposits, primarily in the form of calcium carbonate, accumulate in the walls of the blood vessels.

Arteriosclerosis greatly increases resistance to blood flow. Advanced arteriosclerosis, as a consequence, adversely affects the normal circulation of blood and greatly increases the work performed by the heart.

Some investigators think that arteriosclerosis may not be a pathologic process. Instead, it may be simply an aging or wearing-out process. Evidence also suggests that arteriosclerosis may result from inflammation, which, in some cases, may be the result of an autoimmune disease. In either case, several factors increase the rate at which it develops. Obesity, high dietary cholesterol and other fat consumption, and smoking are some of the factors correlated with the premature development of arteriosclerosis.

> 11. *Describe the changes that occur in arteries due to aging. In which vessels do the most significant changes occur?*
>
> 12. *Name the factors associated with premature arteriosclerosis.*

Pulmonary Circulation

The heart pumps blood from the right ventricle into the **pulmonary** (pŭl′mō-nār-ē; relating to the lungs) **trunk** (figure 21.8). This short vessel, 5 cm long, branches into the right and left **pulmonary arteries,** one transporting blood to each lung. Within the lungs, gas exchange occurs between air in the lungs and the blood. Two **pulmonary veins** exit each lung and enter the left atrium (see figure 20.10).

> 13. *For the vessels of the pulmonary circulation, give their starting point, ending point, and function.*

Systemic Circulation: Arteries

Oxygenated blood entering the heart from the pulmonary veins passes through the left atrium into the left ventricle and from the left ventricle into the aorta. Blood flows from the aorta to all parts of the body (see figure 21.8).

Aorta

All **arteries** of the systemic circulation are derived either directly or indirectly from the **aorta** (ā-ōr′tă), which usually is divided into three general parts: the ascending aorta, the aortic arch, and the descending aorta. The descending aorta is divided further into a thoracic aorta and an abdominal aorta (see figure 21.14).

At its origin from the left ventricle, the aorta is approximately 2.8 cm in diameter. Because it passes superiorly from the heart, this part is called the **ascending aorta.** It's approximately 5 cm long and has only two arteries branching from it: the right and left **coronary arteries,** which supply blood to the cardiac muscle (see figure 20.6a).

The aorta then arches posteriorly and to the left as the **aortic arch.** Three major branches, which carry blood to the head and upper limbs, originate from the aortic arch: the brachiocephalic artery, the left common carotid artery, and the left subclavian artery.

Trauma and the Aorta

Trauma that ruptures the aorta is almost immediately fatal. Trauma can also lead to an **aneurysm** (an′ū-rizm), however, which is a bulge caused by a weakened spot in the aortic wall. If the weakened aortic wall leaks blood slowly into the thorax, the aneurysm must be corrected surgically. Also, once the aneurysm is formed it is likely to enlarge and becomes more likely to rupture. The majority of traumatic aortic arch ruptures occur during automobile accidents and result from the great force with which the body is thrown into the steering wheel, dashboard, or other objects. Waist-type safety belts alone do not prevent this type of injury as effectively as shoulder-type safety belts and air bags.

The next part of the aorta is the longest part, the **descending aorta.** It extends through the thorax in the left side of the mediastinum and through the abdomen to the superior margin of the pelvis. The **thoracic aorta** is that portion of the descending aorta located in the thorax. It has several branches that supply various structures between the aortic arch and the diaphragm. The **abdominal aorta** is that part of the descending aorta between the diaphragm and the point at which the aorta ends by dividing into the two **common iliac** (il′ē-ak; relating to the flank area) **arteries.** The abdominal aorta has several branches that supply the abdominal wall and organs. Its terminal branches, the common iliac arteries, supply blood to the pelvis and lower limbs.

> 14. *Name the different parts of the aorta.*

Coronary Arteries

The **coronary** (kōr′o-năr-ē; encircling the heart like a crown) **arteries,** which are the only branches of the ascending aorta, are described in chapter 20.

> 15. *Name the arteries that branch from the aorta to supply the heart.*

Arteries to the Head and the Neck

The first vessel to branch from the aortic arch is the **brachiocephalic** (brā′kē-ō-se-fal′ik; arm and head) **artery** (figure 21.9). It is a very short artery, and it branches at the level of the clavicle to form the **right common carotid** (ka-rot′id) **artery,** which transports blood to the right side of the head and neck, and the **right subclavian** (sŭb-klā′vē-an; below the clavicle) **artery,** which transports blood to the right upper limb (see figures 21.8, 21.9, 21.11, 21.12, and 21.13).

The second and third branches of the aortic arch are the **left common carotid artery,** which transports blood to the left side of the head and neck, and the **left subclavian artery,** which transports blood to the left upper limb.

The common carotid arteries extend superiorly, without branching, along either side of the neck from their base to the inferior angle of the mandible, where each common carotid artery branches into **internal** and **external carotid arteries** (see figures 21.9 and 21.11). At the point of bifurcation on each side of the neck, the common carotid artery and the base of the internal carotid artery are dilated slightly to form the **carotid sinus,** which is important in monitoring blood pressure (baroreceptor reflex). The external carotid arteries have several branches that supply the structures of the neck and face (table 21.1; see figures 21.9 and 21.11).

Figure 21.8 **The Major Arteries**

The arteries carry blood from the heart to the tissues of the body.

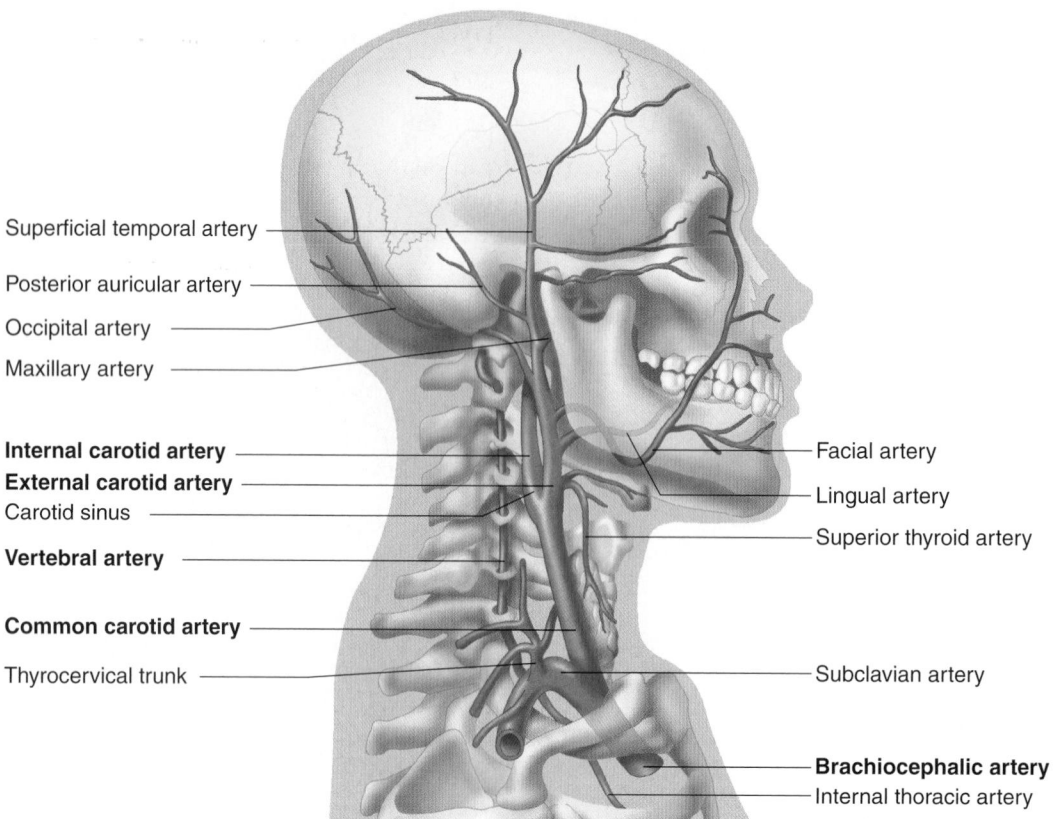

Figure 21.9 Arteries of the Head and Neck

The brachiocephalic artery, the right common carotid artery, and the right vertebral artery are major arteries that supply the head and neck. The right common carotid artery branches from the brachiocephalic artery, and the vertebral artery branches from the subclavian artery.

Table 21.1	**Arteries of the Head and Neck (see figures 21.9, 21.10, and 21.11)**
Arteries	**Tissues Supplied**
Common Carotid Arteries	Head and neck by branches listed below
External Carotid	
Superior thyroid	Neck, larynx, and thyroid gland
Lingual	Tongue, mouth, and submandibular and sublingual glands
Facial	Mouth, pharynx, and face
Occipital	Posterior head and neck and meninges around posterior brain
Posterior auricular	Middle and inner ear, head, and neck
Ascending pharyngeal	Deep neck muscles, middle ear, pharynx, soft palate, and meninges around posterior brain
Superficial temporal	Temple, face, and anterior ear
Maxillary	Middle and inner ears, meninges, lower jaw and teeth, upper jaw and teeth, temple, external eye structures, face, palate, and nose
Internal Carotid	
Posterior communicating	Joins the posterior cerebral artery
Anterior cerebral	Anterior portions of the cerebrum and forms the anterior communicating arteries
Middle cerebral	Most of the lateral surface of the cerebrum
Vertebral Arteries (branches of the subclavian arteries)	
Anterior spinal	Anterior spinal cord
Posterior inferior cerebellar	Cerebellum and fourth ventricle
Basilar Artery (formed by junction of vertebral arteries)	
Anterior inferior cerebellar	Cerebellum
Superior cerebellar	Cerebellum and midbrain
Posterior cerebral	Posterior portions of the cerebrum

The internal carotid arteries, together with the vertebral arteries, which are branches of the subclavian arteries, supply the brain (see table 2.1 and figures 21.9, 21.10, and 21.11).

P R E D I C T

The term *carotid* means to put to sleep, implying that if the carotid arteries are occluded for even a short time, the patient could lose consciousness (go to sleep). The blood supply to the brain is extremely important to its function. Elimination of this supply for even a relatively short time can result in permanent brain damage because the brain is dependent on oxidative metabolism and quickly malfunctions in the absence of oxygen. What is the physiologic significance of arteriosclerosis, which slowly reduces blood flow through the carotid arteries?

Branches of the subclavian arteries, the **left** and **right verte- bral arteries,** enter the cranial cavity through the foramen mag-

num, give off arteries to the cerebellum, and then unite to form a single, midline **basilar** (bas'i-lăr) **artery** (see figures 21.10 and 21.11; see table 21.1). The basilar artery gives off branches to the pons and the cerebellum and then branches to form the **posterior cerebral arteries,** which supply the posterior part of the cerebrum (see figure 21.10).

The internal carotid arteries enter the cranial vault through the carotid canals and terminate by forming the **middle cerebral arter- ies,** which supply large parts of the lateral cerebral cortex (see figure 21.10). Posterior branches of these arteries, the **posterior communi- cating arteries,** unite with the posterior cerebral arteries; and ante- rior branches, the **anterior cerebral arteries,** supply blood to the frontal lobes of the brain. The anterior cerebral arteries are in turn connected by an **anterior communicating artery,** which completes a circle around the pituitary gland and the base of the brain called the **cerebral arterial circle** (circle of Willis) (see figures 21.10 and 21.11).

Middle cerebral artery

Part of temporal lobe removed to reveal branches of the middle cerebral artery

Pituitary gland (attachment site)

Posterior cerebral artery

Basilar artery

Vertebral artery

Part of cerebellum removed to reveal branches of the posterior cerebral artery

Anterior communicating artery

Anterior cerebral artery

Internal carotid artery

Posterior communicating artery

Posterior cerebral artery

Cerebral arterial circle (circle of Willis)

Superior cerebellar artery

Anterior inferior cerebellar artery

Posterior inferior cerebellar artery

Inferior view

Figure 21.10 The Cerebral Arterial Circle (Circle of Willis)

The blood supply to the brain is carried by the internal carotid and vertebral arteries. The vertebral arteries join to form the basilar artery. Branches of the internal carotid arteries and basilar artery supply blood to the brain and complete a circle of arteries around the pituitary gland and the base of the brain called the cerebral arterial circle (circle of Willis).

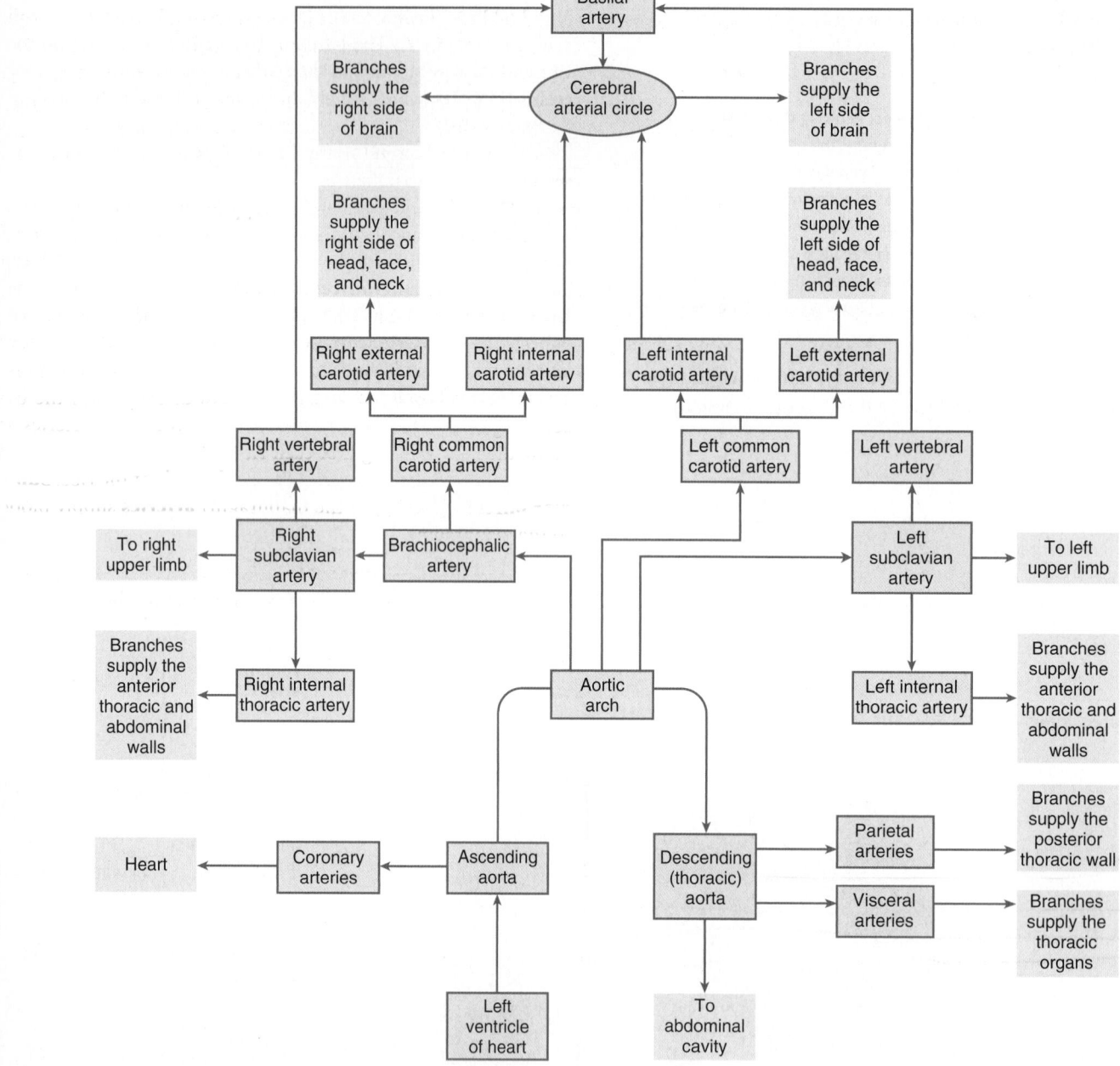

Figure 21.11   Major Arteries of the Head and Thorax

16. *Name the arteries that branch from the aorta to supply the head and neck.*
17. *List the arteries that are part of, and branch from, the cerebral arterial circle.*

Arteries of the Upper Limb

The three major arteries of the upper limb, called the **subclavian, axillary,** and **brachial arteries,** are a continuum rather than a branching system. The axillary artery is the continuation of the subclavian artery, and the brachial artery is the continuation of the axillary artery. The subclavian artery is located deep to the clavicle, the axillary artery is within the axilla, and the brachial artery lies within the arm itself (table 21.2 and figures 21.12 and 21.13).

The brachial artery divides at the elbow into **ulnar** and **radial arteries,** which form two arches within the palm of the hand, referred to as the superficial and deep palmar arches. The **superficial palmar arch** is formed by the ulnar artery and is completed by anastomosing with the radial artery. The **deep palmar arch** is formed by the radial artery and is completed by anastomosing with the ulnar artery. This arch is not only deep to the superficial arch but is proximal as well.

Digital (dij'i-tăl; relating to the digits—the fingers and the thumb) **arteries** branch from each of the two palmar arches and unite to form single arteries on the medial and lateral sides of each digit.

18. *Name the arteries that branch from the aorta to supply the upper limbs.*
19. *List, in order, the arteries that travel through the upper limb to the digits.*

Thoracic Aorta and Its Branches

The branches of the thoracic aorta are divided into two groups: the **visceral branches** supplying the thoracic organs, and the **parietal branches** supplying the thoracic wall (figure 21.14*a* and *b* and table 21.3). The visceral branches supply a portion of the lungs, including the bronchi and bronchioles (see chapter 23); the esophagus; and the pericardium. Even though a large quantity of blood flows to the lungs through the pulmonary arteries, the bronchi and bronchioles require a separate oxygenated blood supply through small bronchial branches from the thoracic aorta.

The walls of the thorax are supplied with blood by the **intercostal** (in-ter-kos'tăl; between the ribs) **arteries,** which consist of two sets: the anterior intercostals and the posterior intercostals. The **anterior intercostals** are derived from the **internal thoracic arteries,** which are branches of the subclavian arteries and lie on the inner surface of the anterior thoracic wall (see table 21.3 and figure 21.14*a* and *b*). The **posterior intercostals** are derived as bilateral branches directly from the descending aorta. The anterior and posterior intercostal arteries lie along the inferior margin of each rib and anastomose with each other approximately midway between the ends of the ribs. **Superior phrenic** (fren'ik; to the diaphragm) **arteries** supply blood to the diaphragm.

20. *Name the two types of branches arising from the thoracic aorta. What structures are supplied by each group?*

Table 21.2	Arteries of the Upper Limbs (see figure 21.12)
Arteries	**Tissues Supplied**
Subclavian Arteries	
(right subclavian originates from the brachiocephalic artery, and left subclavian originates directly from the aorta)	
Vertebral	Spinal cord and cerebellum form the basilar artery (see table 21.1)
Internal thoracic	Diaphragm, mediastinum, pericardium, anterior thoracic wall, and anterior abdominal wall
Thyrocervical trunk	Inferior neck and shoulder
Axillary Arteries	
(continuation of subclavian)	
Thoracoacromial	Pectoral region and shoulder
Lateral thoracic	Pectoral muscles, mammary gland, and axilla
Subscapular	Scapular muscles
Brachial Arteries	
(continuation of axillary arteries)	
Deep brachial	Arm and humerus
Radial	Forearm
Deep palmar arch	Hand and fingers
Digital arteries	Fingers
Ulnar	Forearm
Superficial palmar arch	Hand and fingers
Digital arteries	Fingers

Vertebral artery

Thoracoacromial artery

Humeral circumflex arteries

Deep brachial artery

Brachial artery

Radial artery

Thyrocervical trunk

Right subclavian artery

Right common carotid artery

Brachiocephalic artery

Internal thoracic artery

Lateral thoracic artery

Axillary artery

Subscapular artery

Ulnar artery

Deep palmar arch

Superficial palmar arch

Digital arteries

Anterior view

Figure 21.12   Arteries of the Upper Limb

Anterior view of the arteries of the right upper limb and their branches. The right brachiocephalic, subclavian, axillary, radial, and ulnar arteries and their branches are shown.

Subclavian artery

Shoulder, chest, and back

Axillary artery

Arm

Brachial artery

Lateral forearm ← Radial artery

Ulnar artery → Medial forearm

Superficial and deep palmar arches → Palm of hand

Digital arteries → Thumb and fingers

Figure 21.13   Major Arteries of the Shoulder and Upper Limb

(a)

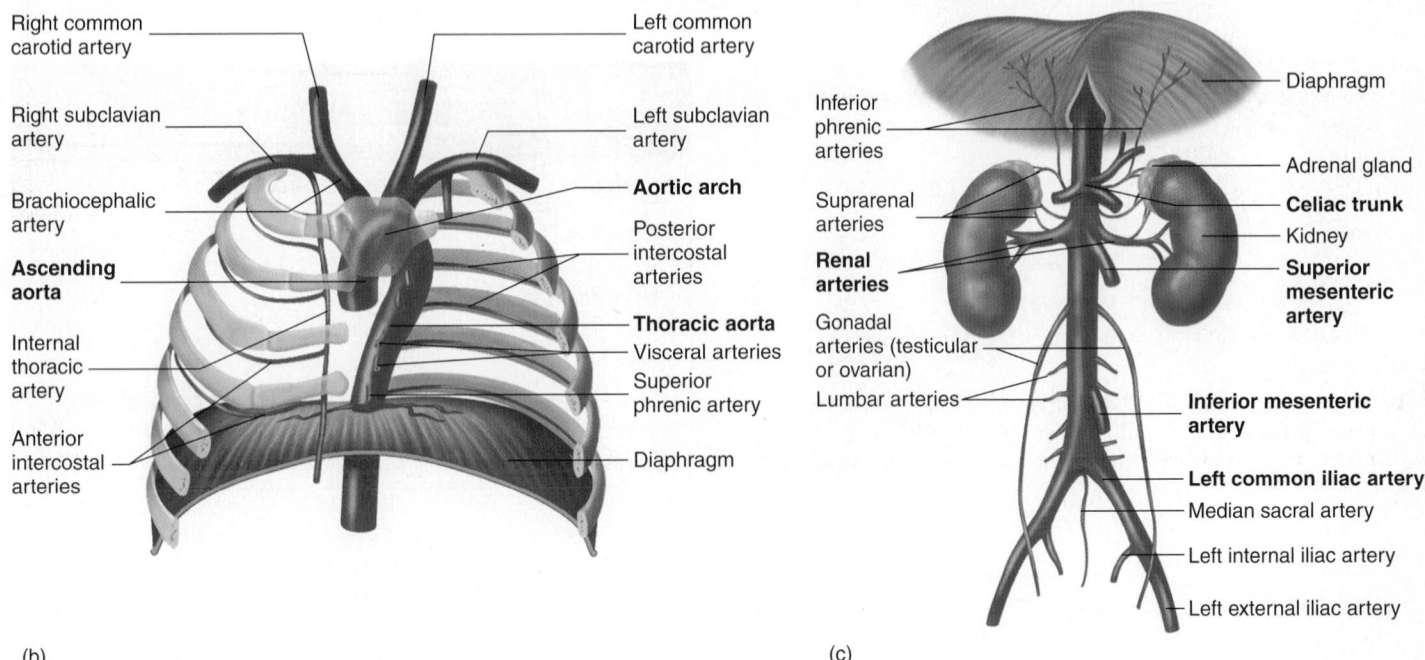

(b)

(c)

Figure 21.14   Branches of the Aorta

(a) The aorta is considered in three portions: the ascending aorta, the aortic arch, and the descending aorta. The descending aorta consists of the thoracic and abdominal aorta. *(b)* The thoracic aorta. *(c)* The abdominal aorta.

Abdominal Aorta and Its Branches

The branches of the abdominal aorta, like those of the thoracic aorta, are divided into visceral and parietal parts (table 21.3 and figures 21.14*a* and *c* and 21.15). The visceral arteries are in turn divided into paired and unpaired branches. Three major un-

paired branches exist: the **celiac** (sē′lē-ak; belly) **trunk,** the **superior mesenteric artery** (mez-en-ter′ik; relating to the mesenteries), and the **inferior mesenteric artery** (see figure 21.14*a* and *c*). Each has several major branches supplying the abdominal organs.

The paired visceral branches of the abdominal aorta supply the kidneys, adrenal glands, and gonads (testes or ovaries). The parietal arteries of the abdominal aorta supply the diaphragm and abdominal wall. The arteries of the abdomen and the areas they supply are shown schematically in figure 21.15.

● **21.** *What areas of the body are supplied by the paired arteries that branch from the abdominal aorta? The unpaired arteries? Name the three major unpaired branches.*

Arteries of the Pelvis

The abdominal aorta divides at the level of the fifth lumbar vertebra into two **common iliac arteries.** They divide to form the **external iliac arteries,** which enter the lower limbs, and the **internal iliac arteries,** which supply the pelvic area. Visceral branches supply the pelvic organs, such as the urinary bladder, rectum, uterus, and vagina; and parietal branches supply blood to the walls and floor of the pelvis; the lumbar, gluteal, and proximal thigh muscles; and the external genitalia (table 21.4 and figures 21.15 and 21.16).

● **22.** *Name the arteries that branch from the aorta to supply the pelvic area. List the organs of the pelvis that are supplied by branches of these arteries.*

Table 21.3	Thoracic and Abdominal Aorta (see figures 21.14 and 21.15)
Arteries	**Tissues Supplied**
Thoracic Aorta	
Visceral Branches	
Bronchial	Lung tissue
Esophageal	Esophagus
Parietal Branches	
Intercostal	Thoracic wall
Superior phrenic	Superior surface of diaphragm
Abdominal Aorta	
Visceral Branches	
Unpaired	
Celiac trunk	
Left gastric	Stomach and esophagus
Common hepatic	
Gastroduodenal	Stomach and duodenum
Right gastric	Stomach
Hepatic	Liver
Splenic	Spleen and pancreas
Left gastroepiploic	Stomach
Superior mesenteric	Pancreas, small intestine, and colon
Inferior mesenteric	Descending colon and rectum
Paired	
Suprarenal	Adrenal gland
Renal	Kidney
Gonadal	
Testicular (male)	Testis and ureter
Ovarian (female)	Ovary, ureter, and uterine tube
Parietal Branches	
Inferior phrenic	Adrenal gland and inferior surface of diaphragm
Lumbar	Lumbar vertebrae and back muscles
Median sacral	Inferior vertebrae
Common iliac	
External iliac	Lower limb (see table 21.5)
Internal iliac	Lower back, hip, pelvis, urinary bladder, vagina, uterus, rectum, and external genitalia (see table 21.4)

Table 21.4	Arteries of the Pelvis (see figures 21.15 and 21.16)
Arteries	**Tissues Supplied**
Internal Iliac	Pelvis through the branches listed below
Visceral Branches	
Middle rectal	Rectum
Vaginal	Vagina and uterus
Uterine	Uterus, vagina, uterine tube, and ovary
Parietal Branches	
Lateral sacral	Sacrum
Superior gluteal	Muscles of the gluteal region
Obturator	Pubic region, deep groin muscles, and hip joint
Internal pudendal	Rectum, external genitalia, and floor of pelvis
Inferior gluteal	Inferior gluteal region, coccyx, and proximal thigh

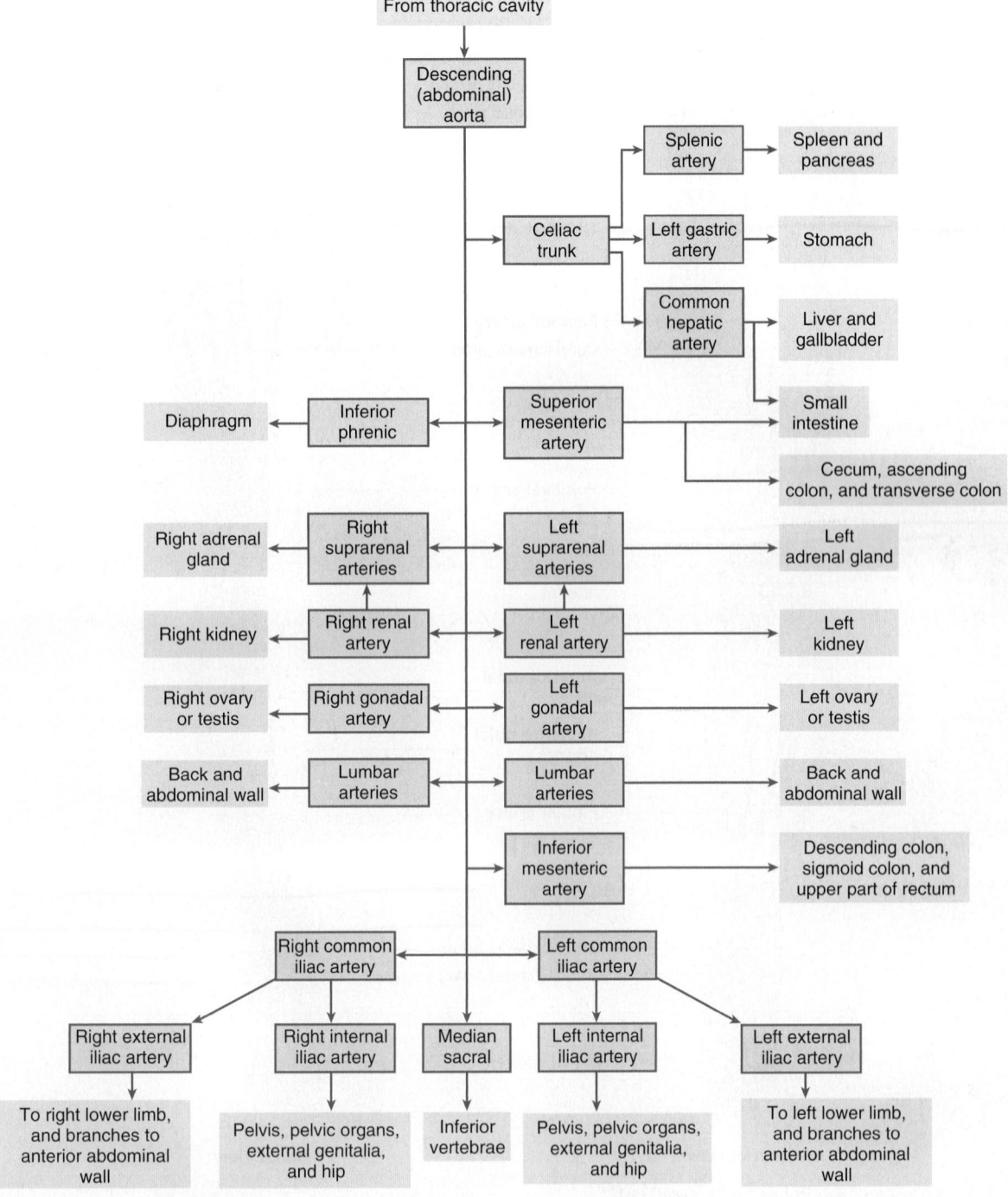

Figure 21.15 Major Arteries of the Abdomen and Pelvis

Visceral branches include those that are unpaired (celiac trunk, superior mesenteric, inferior mesenteric and those that are paired (renal, suprarenal, testicular, or ovarian). Parietal branches include inferior phrenic, lumbar, and median sacral.

Inferior vena cava

Common iliac artery

External iliac artery

Lateral circumflex
artery

Descending branch of
lateral circumflex artery

Abdominal aorta

Median sacral artery

Internal iliac artery

Lateral sacral artery

Internal pudendal artery

Obturator artery

Femoral artery

Deep femoral artery

Popliteal artery

Genicular
arteries

**Anterior tibial
artery**

**Posterior tibial
artery**

Fibular artery

Dorsalis pedis artery

Medial plantar artery

Digital arteries

Superior gluteal artery

Lateral plantar artery

Anterior view

Posterior view

Figure 21.16   Arteries of the Pelvis and Lower Limb

The internal and external iliac arteries and their branches are shown. The internal iliac artery supplies the pelvis and hip, and the external iliac artery supplies the lower limb through the femoral artery.

Arteries of the Lower Limb

The arteries of the lower limb form a continuum similar to that of the arteries of the upper limb. The **external iliac artery** becomes the **femoral** (fem′ŏ-răl; relating to the thigh) **artery** in the thigh, which becomes the **popliteal** (pop-lit′ē-ăl, pop-li-tē′ăl; ham, the hamstring area posterior to the knee) **artery** in the popliteal space. The popliteal artery gives off the **anterior tibial artery** just inferior to the knee and then continues as the **posterior tibial artery.** The anterior tibial artery becomes the **dorsalis**

pedis artery at the foot. The posterior tibial artery gives off the **fibular,** or **peroneal, artery** and then gives rise to **medial** and **lateral plantar** (plan′tăr; the sole of the foot) **arteries,** which in turn give off **digital branches** to the toes. The arteries of the lower limb are illustrated in figures 21.16 and 21.17 and are listed in table 21.5.

● *23. List, in order, the arteries that travel from the aorta to the digits of the lower limbs.*

Table 21.5	Arteries of the Lower Limb (see figures 21.16 and 21.17)	
Arteries	**Tissues Supplied**
Femoral	Thigh, external genitalia, and anterior abdominal wall
Deep femoral	Thigh, knee, and femur
Popliteal	
(continuation of the femoral artery)	
Posterior tibial	Knee and leg
Fibular (peroneal)	Calf and peroneal muscles and ankle
Medial plantar	Plantar region of foot
Digital arteries	Digits of foot
Lateral plantar	Plantar region of foot
Digital arteries	Digits of foot
Anterior tibial	Knee and leg
Dorsalis pedis	Dorsum of foot
Digital arteries	Digits of foot

Systemic Circulation: Veins

Three major veins return blood from the body to the right atrium: the **coronary sinus,** returning blood from the walls of the heart (see figures 20.6*b* and 20.7); the **superior vena cava** (vē′nă kă′vă, kā′vă; venous cave), returning blood from the head, neck, thorax, and upper limbs; and the **inferior vena cava,** returning blood from the abdomen, pelvis, and lower limbs (figure 21.18).

In a very general way, the smaller veins follow the same course as the arteries and often are given the same names. The veins, however, are more numerous and more variable. The larger veins often follow a very different course and have names different from the arteries.

● **24. List, the veins that carry blood from each of the major body areas.**

Three major types of veins exist: superficial veins, deep veins, and sinuses. The superficial veins of the limbs are, in general, larger than the deep veins, whereas in the head and trunk the opposite is the case. Venous sinuses occur primarily in the cranial cavity and the heart.

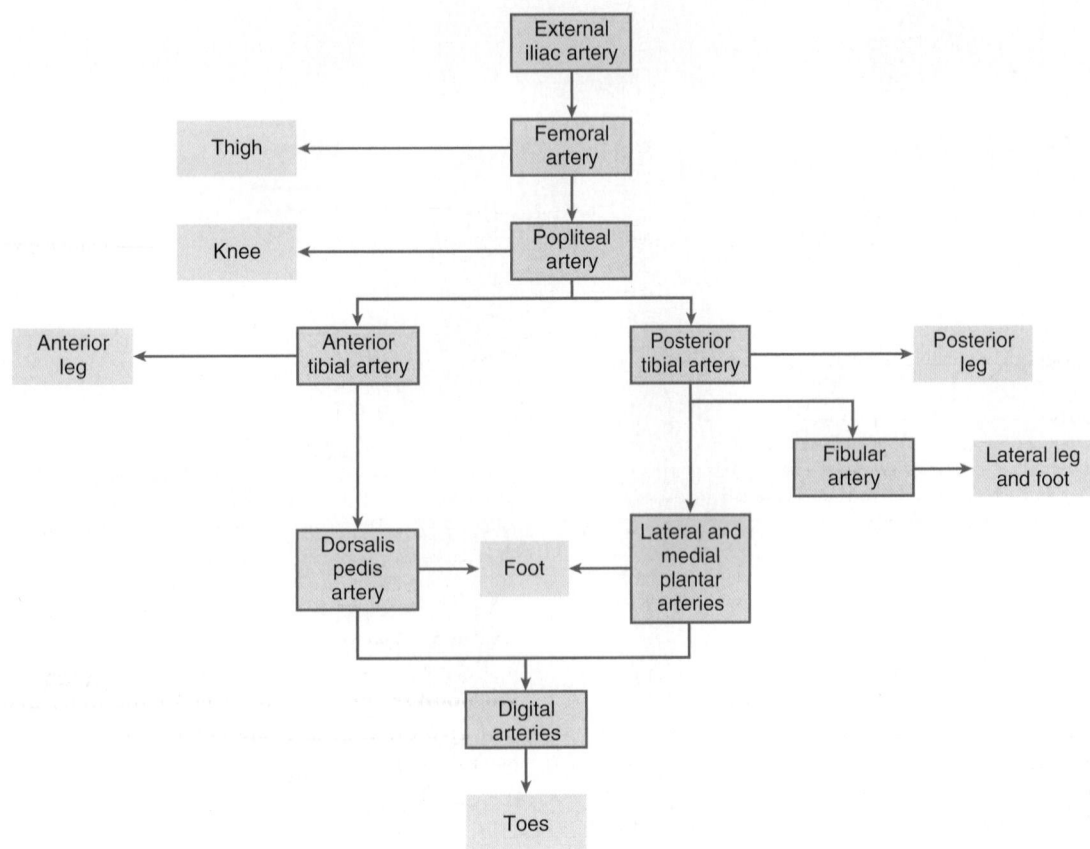

Figure 21.17 Major Arteries of the Lower Limb

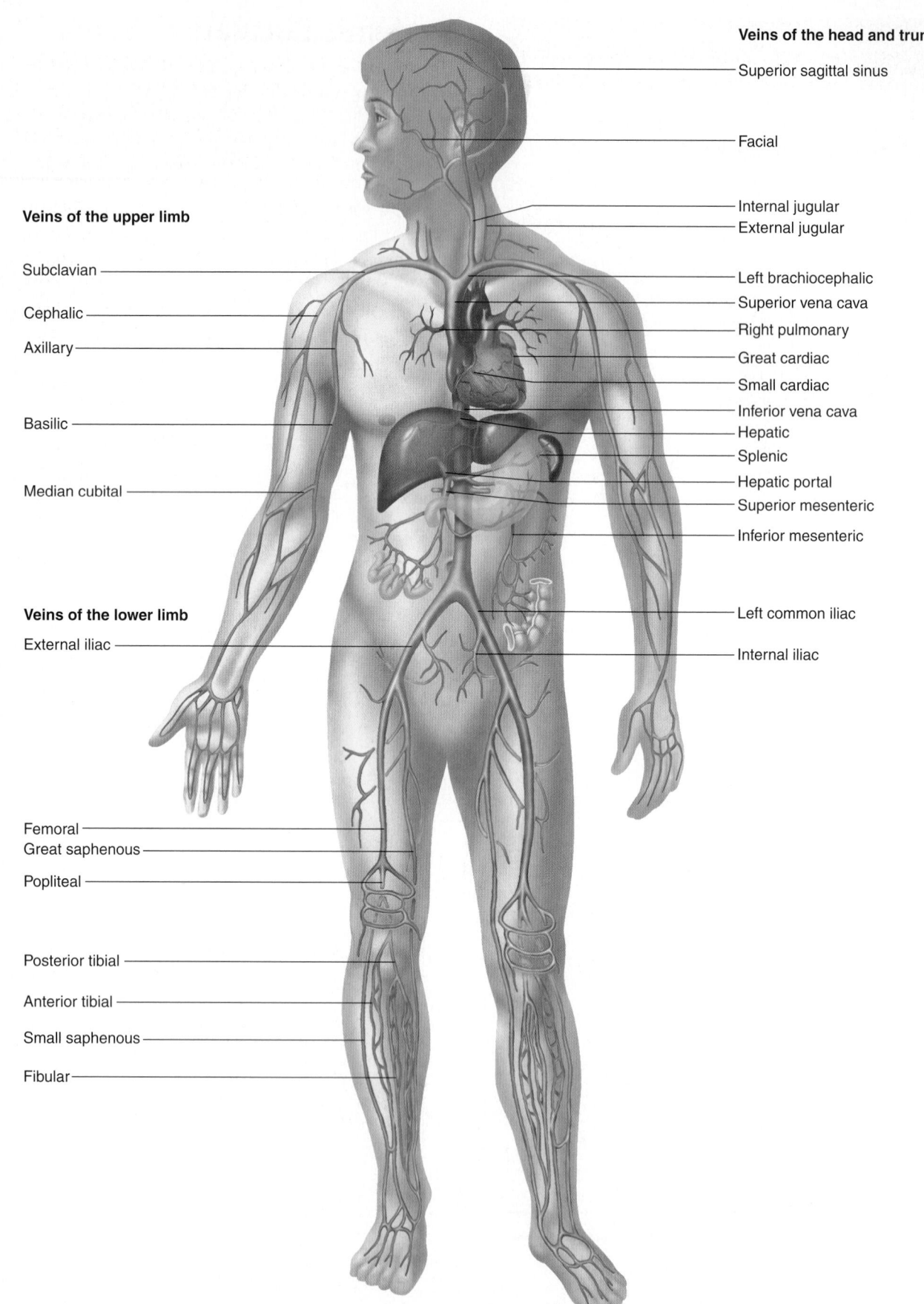

Veins of the head and trunk

— Superior sagittal sinus

— Facial

— Internal jugular
— External jugular

— Left brachiocephalic
— Superior vena cava
— Right pulmonary
— Great cardiac
— Small cardiac
— Inferior vena cava
— Hepatic
— Splenic
— Hepatic portal
— Superior mesenteric
— Inferior mesenteric

— Left common iliac

— Internal iliac

Veins of the upper limb

Subclavian —

Cephalic —

Axillary —

Basilic —

Median cubital —

Veins of the lower limb

External iliac —

Femoral —
Great saphenous —

Popliteal —

Posterior tibial —

Anterior tibial —

Small saphenous —

Fibular —

Figure 21.18   The Major Veins

The veins carry blood to the heart from the tissues of the body.

Veins Draining the Heart

The **cardiac veins,** which transport blood from the walls of the heart and return it through the coronary sinus to the right atrium, are described in chapter 20.

Veins of the Head and Neck

The two pairs of major veins that drain blood from the head and neck are the **external** and **internal jugular** (jŭg′ū-lar; neck) **veins.** The external jugular veins are the more superficial of the two sets, and they drain blood primarily from the posterior head and neck. The external jugular vein usually drains into the subclavian vein. The internal jugular veins are much larger and deeper than the external jugular veins. They drain blood from the cranial cavity and the anterior head, face, and neck.

The internal jugular vein is formed primarily as the continuation of the **venous sinuses** of the cranial cavity. The venous sinuses are actually spaces within the dura mater surrounding the brain (see chapter 13). They are depicted in figure 21.19 and are listed in table 21.6.

Facial Pimples and Meningitis

Because venous communication exists between the facial veins and venous sinuses through the ophthalmic veins, infections can potentially be introduced into the cranial cavity through this route. A superficial infection of the face on either side of the nose can enter the facial vein. The infection can then pass through the ophthalmic veins to the venous sinuses and result in meningitis. For this reason people are warned not to aggravate pimples or boils on the face on either side of the nose.

Table 21.6	Venous Sinuses of the Cranial Cavity (see figure 21.19)
Veins	**Tissues Drained**
Internal Jugular Vein	
Sigmoid sinus	
Superior and inferior petrosal sinuses	Anterior portion of cranial cavity
Cavernous sinus	
Ophthalmic veins	Orbit
Transverse sinus	
Occipital sinus	Central floor of posterior fossa of skull
Superior sagittal sinus	Superior portion of cranial cavity and brain
Straight sinus	
Inferior sagittal sinus	Deep portion of longitudinal fissure

Figure 21.19   Venous Sinuses Associated with the Brain

Once the internal jugular veins exit the cranial cavity, they receive several venous tributaries that drain the external head and face (table 21.7 and figures 21.20 and 21.21). The **internal jugular veins** join the **subclavian veins** on each side of the body to form the **brachiocephalic veins.**

> 25. *List the two pairs of major veins that drain blood from the head and neck. Describe the venous sinuses. To what large vein do the venous sinuses connect?*

Table 21.7	Veins Draining the Head and Neck (see figures 21.20 and 21.21)
Veins	**Tissues Drained**
Brachiocephalic	
Internal jugular	Brain
Lingual	Tongue and mouth
Superior thyroid	Thyroid and deep posterior facial structures (also empties into external jugular)
Facial	Superficial and anterior facial structures
External jugular	Superficial surface of posterior head and neck

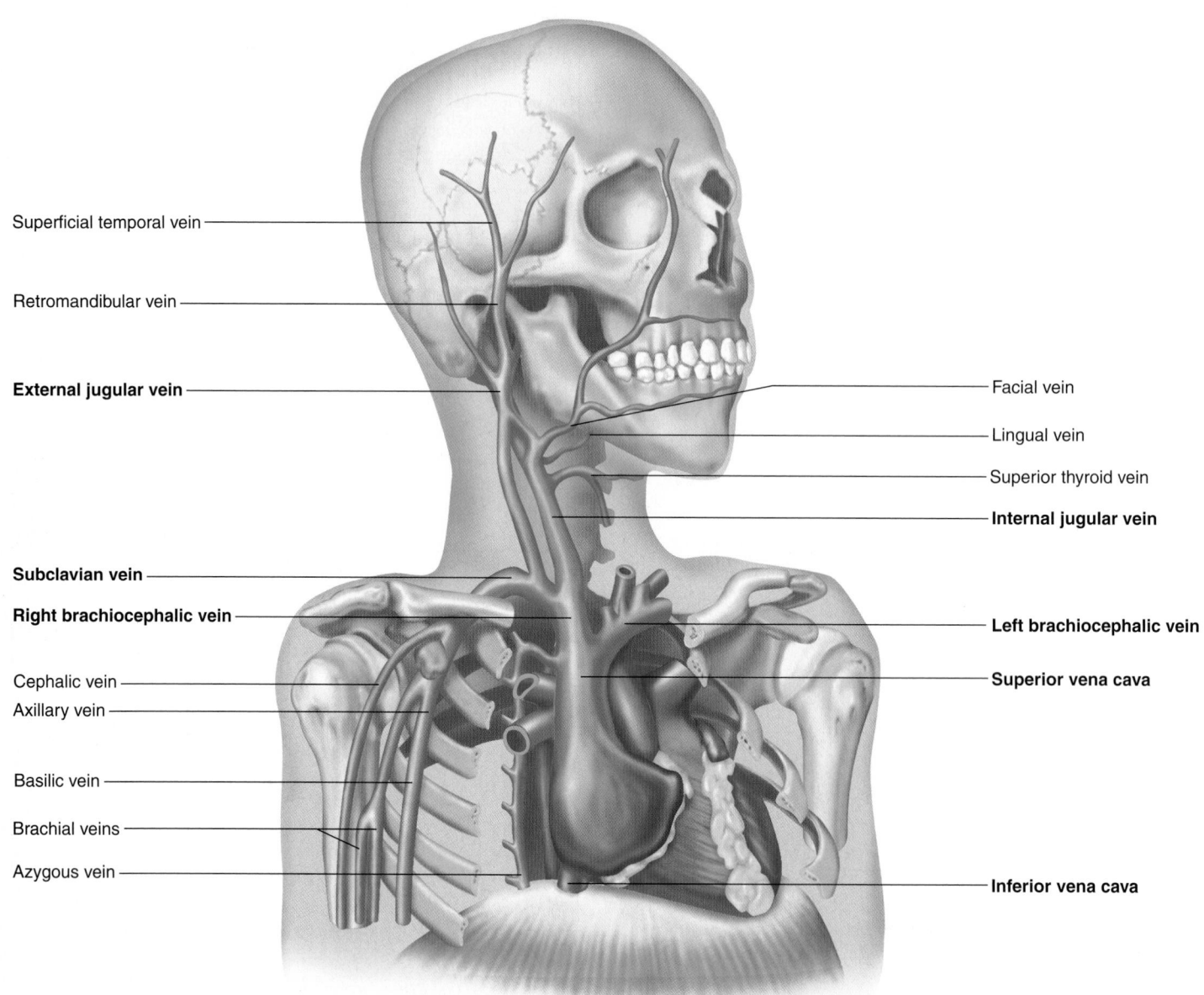

Figure 21.20   Veins of the Head and Neck

The right brachiocephalic vein and its tributaries. The major veins draining the head and neck are the internal and external jugular veins.

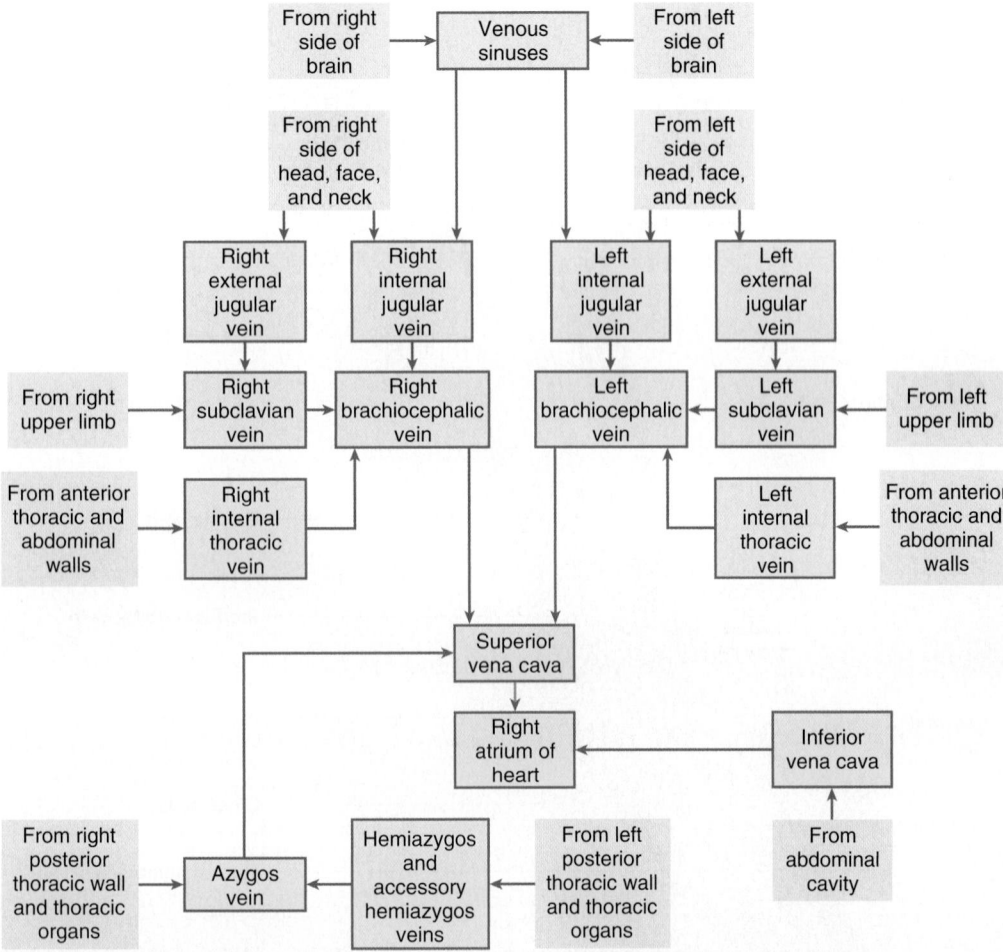

Figure 21.21 Major Veins of the Head and Thorax

Veins of the Upper Limb

The **cephalic** (se-fal′ik; toward the head), **basilic** (ba-sil′ik), and **brachial veins** are responsible for draining most of the blood from the upper limbs (table 21.8 and figures 21.22 and 21.23). Many of the tributaries of the cephalic and basilic veins in the forearm and hand can be seen through the skin. Because of the considerable variation in the tributary veins of the forearm and hand, they often are left unnamed. The basilic vein of the arm becomes the **axillary vein** as it courses through the axillary region. The axillary vein then becomes the **subclavian vein** at the margin of the first rib. The cephalic vein enters the axillary vein.

The **median cubital** (kū′bi-tăl; pertaining to the elbow) **vein** is a variable vein that usually connects the cephalic vein or its tributaries with the basilic vein. In many people, this vein is quite prominent on the anterior surface of the upper limb at the level of the elbow (cubital fossa) and is, therefore, often used as a site for drawing blood from a patient.

The deep veins draining the upper limb follow the same course as the arteries. The **radial** and **ulnar veins,** therefore, are named for the arteries they attend. They usually are paired, with one small vein lying on each side of the artery, and they have

Table 21.8	Veins of the Upper Limb (see figures 21.22 and 21.23)
Veins	**Tissues Drained**
Subclavian (continuation of the axillary vein)	
Axillary (continuation of the basilic vein)	
Cephalic	Lateral arm, forearm, and hand (superficial veins of the forearm and hand are variable)
Brachial (paired, deep veins)	Deep structures of the arm
Radial vein	Deep forearm
Ulnar vein	Deep forearm
Basilic	Medial arm, forearm, and hand (superficial veins of the forearm and hand are variable)
Median cubital	Connects basilic and cephalic veins
Deep and superficial palmar venous arches	Drain into superficial and deep veins of the forearm
Digital veins	Fingers

Internal jugular vein

Subclavian vein

Brachiocephalic vein

Clavicle

Cephalic vein

Axillary vein

Brachial veins

Basilic vein

Median cubital vein

Cephalic
vein

Basilic vein

Ulnar veins

Median antebrachial vein

Radial veins

Deep palmar arch

Superficial palmar arch

Digital veins

Anterior view

Figure 21.22   Veins of the Upper Limb

The subclavian vein and its tributaries. The major veins draining the superficial structures of the limb are the cephalic and basilic veins. The brachial veins drain the deep structures.

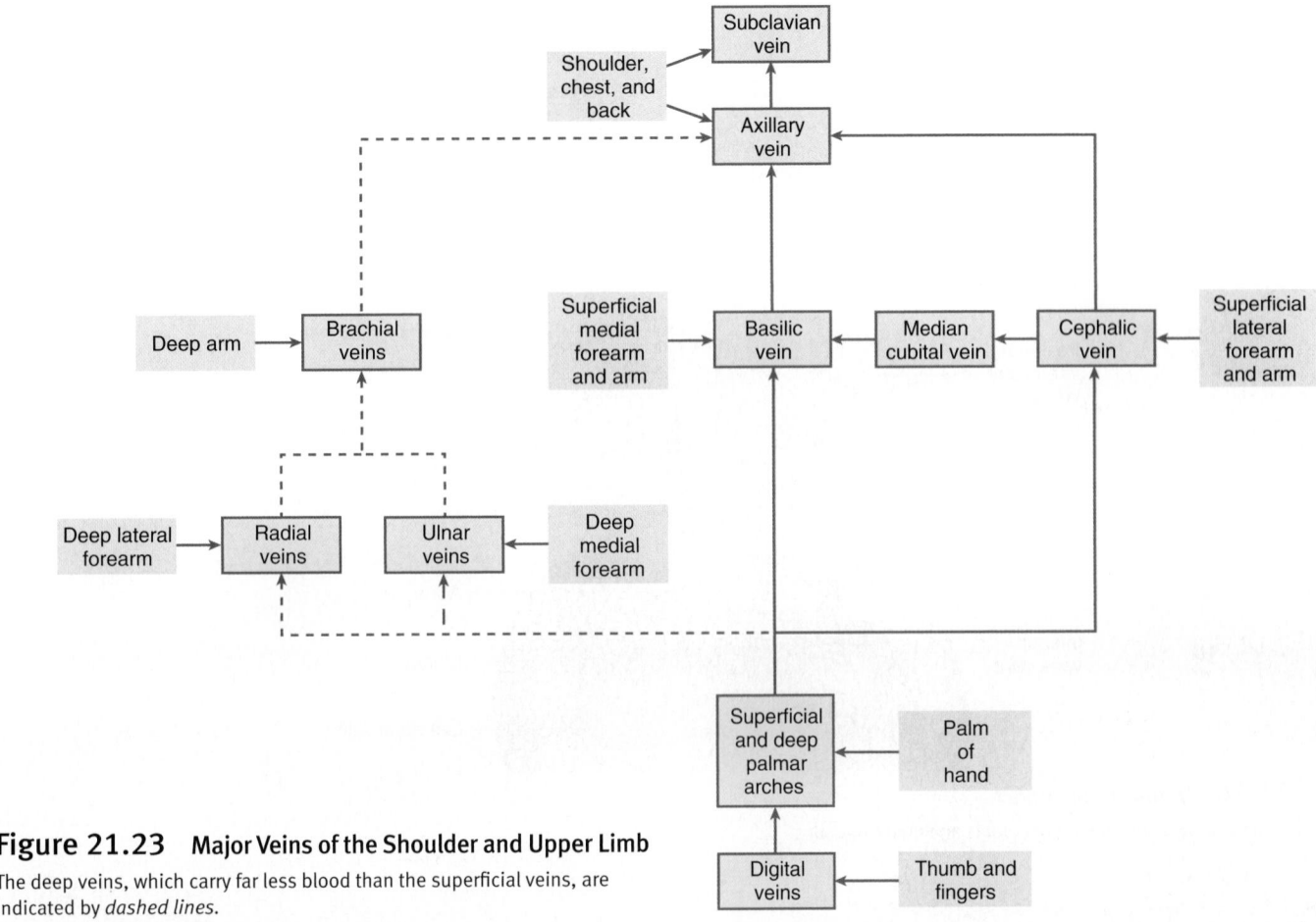

Figure 21.23   Major Veins of the Shoulder and Upper Limb

The deep veins, which carry far less blood than the superficial veins, are indicated by *dashed lines*.

numerous connections with one another and with the superficial veins. The radial and ulnar veins empty into the **brachial veins,** which accompany the brachial artery and empty into the axillary vein (see figures 21.22 and 21.23).

● *26. List the major deep and superficial veins of the upper limb.*

Veins of the Thorax

Three major veins return blood from the thorax to the superior vena cava: the right and left brachiocephalic veins and the **azygos** (az′ī-gos; unpaired) **vein.** The thoracic drainage to the brachiocephalic veins is through the anterior thoracic wall by way of the **internal thoracic veins.** They receive blood from the **anterior intercostal veins.** Blood from the posterior thoracic wall is collected by **posterior intercostal veins** that drain into the azygos vein on the right and the **hemiazygos** (hem′ē-az′ī-gos) or **accessory hemiazygos vein** on the left. The hemiazygos and accessory hemiazygos veins empty into the azygos vein, which drains into the superior vena cava. The thoracic veins are listed in table 21.9 and illustrated in figure 21.24 (see also figure 21.21).

Table 21.9	Veins of the Thorax (see figure 21.24)
Veins	**Tissues Drained**
Superior Vena Cava	
Brachiocephalic	
Azygos vein	Right side, posterior thoracic wall and posterior abdominal wall; esophagus, bronchi, pericardium, and mediastinum
Hemiazygos	Left side, inferior posterior thoracic wall and posterior abdominal wall; esophagus and mediastinum
Accessory hemiazygos	Left side, superior posterior thoracic wall

Figure 21.24 Veins of the Thorax

Anterior view of the azygos and hemiazygos veins and their tributaries.

27. *List the three major veins that return blood from the thorax to the superior vena cava.*

Veins of the Abdomen and Pelvis

Blood from the posterior abdominal wall drains into the **ascending lumbar veins.** These veins are continuous superiorly with the hemiazygos on the left and the azygos on the right. Blood from the rest of the abdomen, pelvis, and lower limbs returns to the heart through the inferior vena cava. The gonads (testes or ovaries), kidneys, and adrenal glands are the only abdominal organs outside the pelvis that drain directly into the inferior vena cava. The **internal iliac veins** drain the pelvis and join the **external iliac veins** from the lower limbs to form the **common iliac veins,** which unite to form the inferior vena cava. The major abdominal and pelvic veins are listed in table 21.10 and illustrated in figures 21.25 and 21.27.

Table 21.10	Veins Draining the Abdomen and Pelvis (see figures 21.25 and 21.27)
Veins	**Tissues Drained**
Inferior Vena Cava	
Hepatic veins	Liver (see hepatic portal system)
Common iliac	
External iliac	Lower limb (see table 21.12)
Internal iliac	Pelvis and its viscera
Ascending lumbar	Posterior abdominal wall (empties into common iliac, azygos, and hemiazygos veins)
Renal	Kidney
Suprarenal	Adrenal gland
Gonadal	
Testicular (male)	Testis
Ovarian (female)	Ovary
Phrenic	Diaphragm

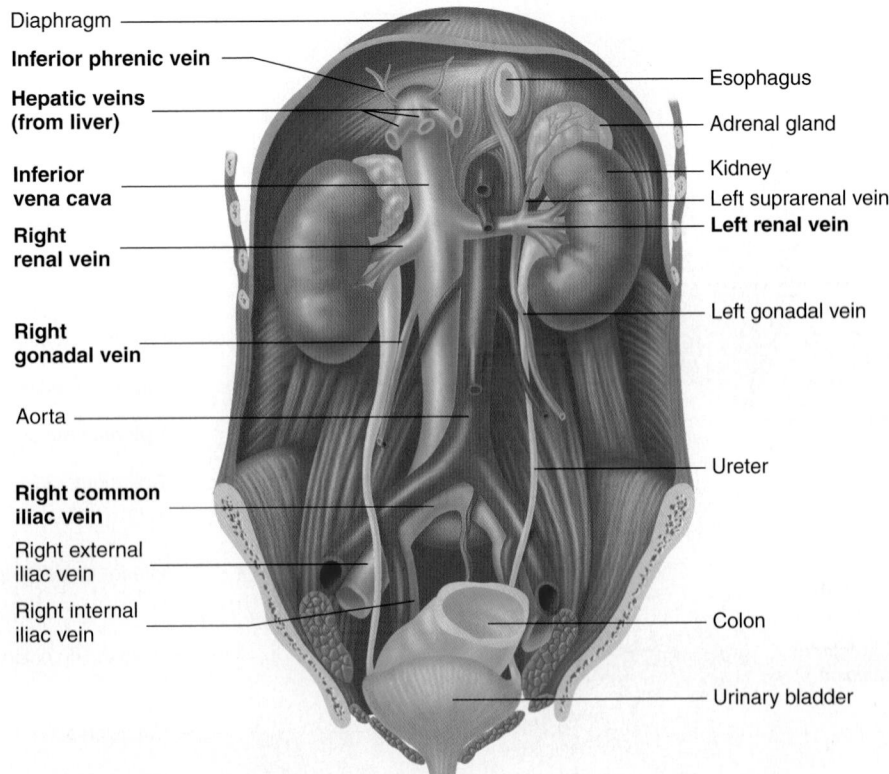

Diaphragm

Inferior phrenic vein

Hepatic veins (from liver)

Inferior vena cava

Right renal vein

Right gonadal vein

Aorta

Right common iliac vein

Right external iliac vein

Right internal iliac vein

Esophagus

Adrenal gland

Kidney

Left suprarenal vein

Left renal vein

Left gonadal vein

Ureter

Colon

Urinary bladder

Figure 21.25 Inferior Vena Cava and Its Tributaries

The hepatic veins transport blood to the inferior vena cava from the hepatic portal system, which ends as a series of blood sinusoids in the liver (see figure 21.26).

Hepatic Portal System

The **hepatic** (he-pat′ik; relating to the liver) **portal system** (table 21.11 and figures 21.26 and 21.27) carries blood through veins from capillaries within most of the abdominal viscera, such as the stomach, intestines, and spleen, to a series of dilated capillaries called **sinusoids** in the liver. Nutrients and other substances absorbed from the stomach or intestine are delivered to the liver by the hepatic portal system (see chapter 24).

The **hepatic portal vein,** the largest vein of the system, is formed by the union of the **superior mesenteric vein,** which drains the small intestine, and the **splenic vein,** which drains the spleen. The splenic vein receives the **inferior mesenteric vein,** which drains part of the large intestine, and the **pancreatic veins,** which drain the pancreas. The hepatic portal vein also receives gastric veins before entering the liver.

Table 21.11	Hepatic Portal System (see figures 21.26 and 21.27)
Veins	**Tissues Drained**
Hepatic Portal	
Superior mesenteric	Small intestine and most of the colon
Splenic	Spleen
Inferior mesenteric	Descending colon and rectum
Pancreatic	Pancreas
Gastroomental	Stomach
Gastric	Stomach
Cystic	Gallbladder

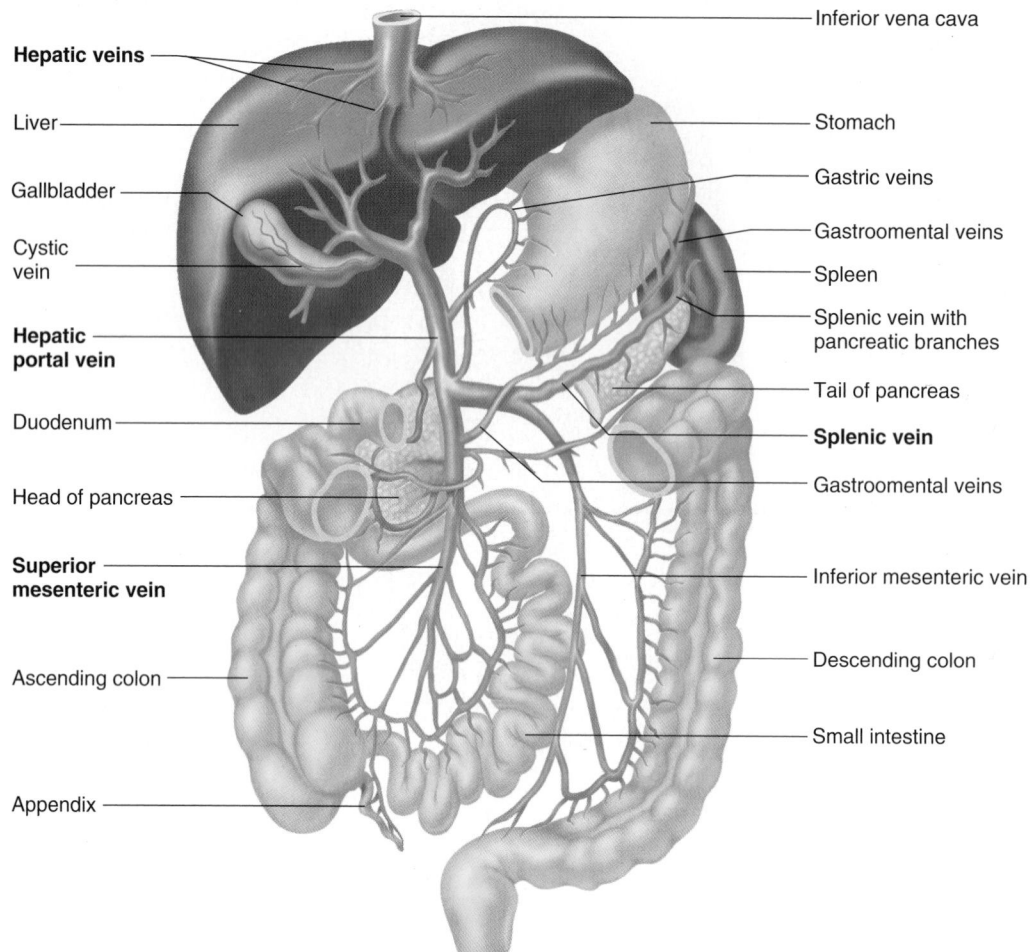

Figure 21.26   Veins of the Hepatic Portal System

The hepatic portal system begins as capillary beds in the stomach, pancreas, spleen, small intestine, and large intestine. The veins of the hepatic portal system converge on the hepatic portal vein, which carries blood to a series of capillaries (sinusoids) in the liver. Hepatic veins carry blood from capillaries in the liver to the hepatic veins, which then enter the inferior vena cava (see figure 21.25).

Blood from the liver sinusoids is collected into **central veins,** which empty into **hepatic veins.** Blood from the cystic veins, which drain the gallbladder, also enters the hepatic veins. The hepatic veins join the inferior vena cava. Blood entering the liver through the hepatic portal vein is rich with nutrients collected from the intestines, but it also can contain a number of toxic substances harmful to the tissues of the body. Within the liver, the nutrients are either taken up and stored or are modified chemically and used by other cells of the body (see chapter 24). The cells of the liver also help remove toxic substances by altering their structure or making them water-soluble, a process called **biotransformation.** The water-soluble substances can then be transported in the blood to the kidneys, from which they are excreted in the urine (see chapter 26).

> ● *28. Explain the three ways that blood from the abdomen returns to the heart.*

> *29. List the vessels that carry blood from abdominal organs to the hepatic portal vein.*

Veins of the Lower Limb

The veins of the lower limb, like those of the upper limb, consist of superficial and deep groups. The distal deep veins of each limb are paired and follow the same path as the arteries, whereas the proximal deep veins are unpaired. The **anterior** and **posterior tibial veins** are paired and accompany the anterior and posterior tibial arteries. They unite just inferior to the knee to form the single **popliteal vein,** which ascends through the thigh and becomes the **femoral vein.** The femoral vein becomes the external iliac vein. **Fibular,** or **peroneal** (per-ō-nē′ăl), **veins** also are paired in each leg and accompany the fibular arteries. They empty into the posterior tibial veins just before those veins contribute to the popliteal vein.

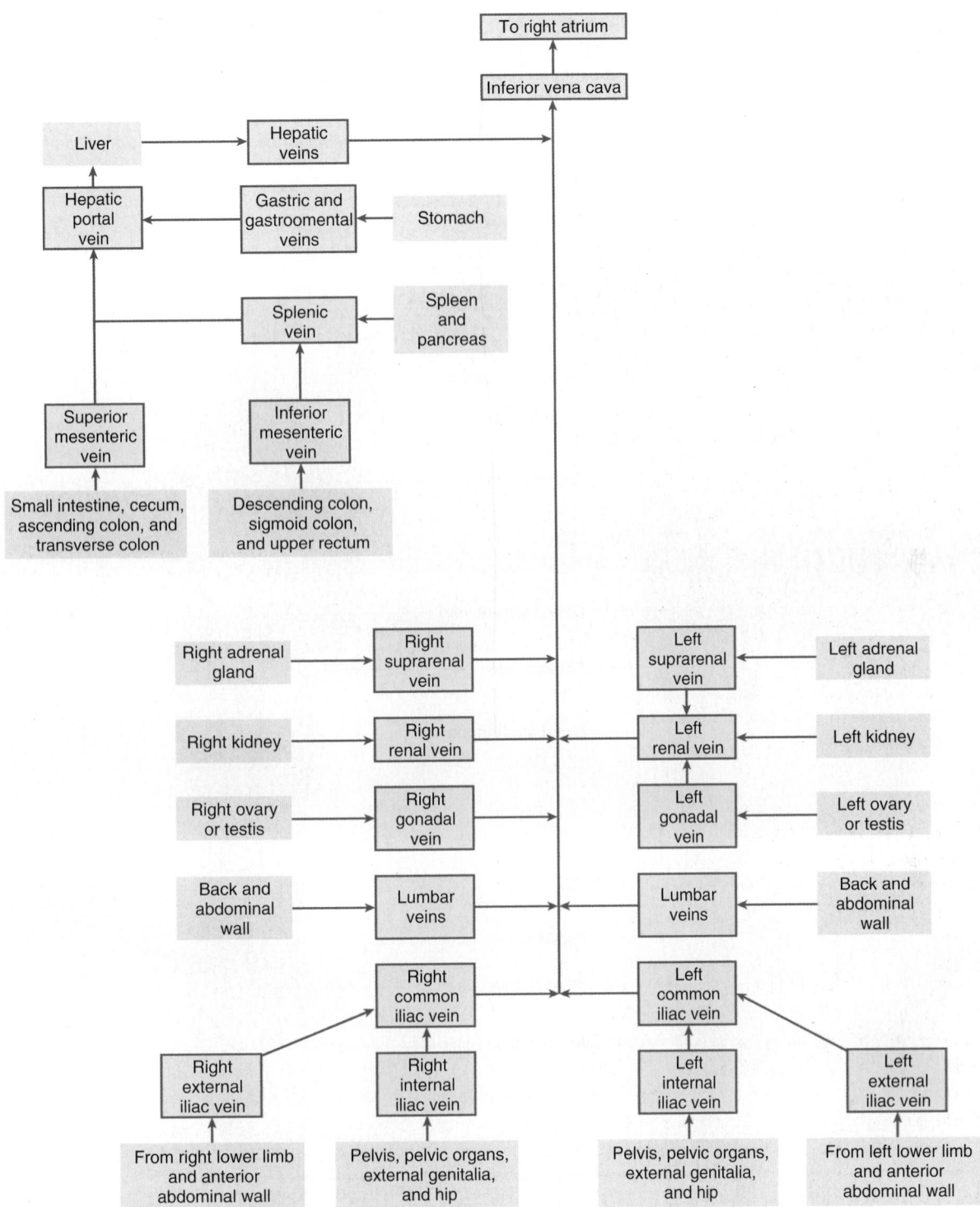

Figure 21.27 Major Veins of the Abdomen and Pelvis

The superficial veins consist of the great and small saphenous veins. The **great saphenous** (să-fē′nŭs; visible) **vein,** the longest vein of the body, originates over the dorsal and medial side of the foot and ascends along the medial side of the leg and thigh to empty into the femoral vein. The **small saphenous vein** begins over the lateral side of the foot and ascends along the posterior leg to the popliteal space, where it empties into the popliteal vein. The veins of the lower limb are illustrated in figures 21.28 and 21.29 and are listed in table 21.12. The saphenous veins can be removed and used as a source of blood vessels for coronary bypass surgery (see chapter 20).

● 30.   List the major deep and superficial veins of the lower limbs.

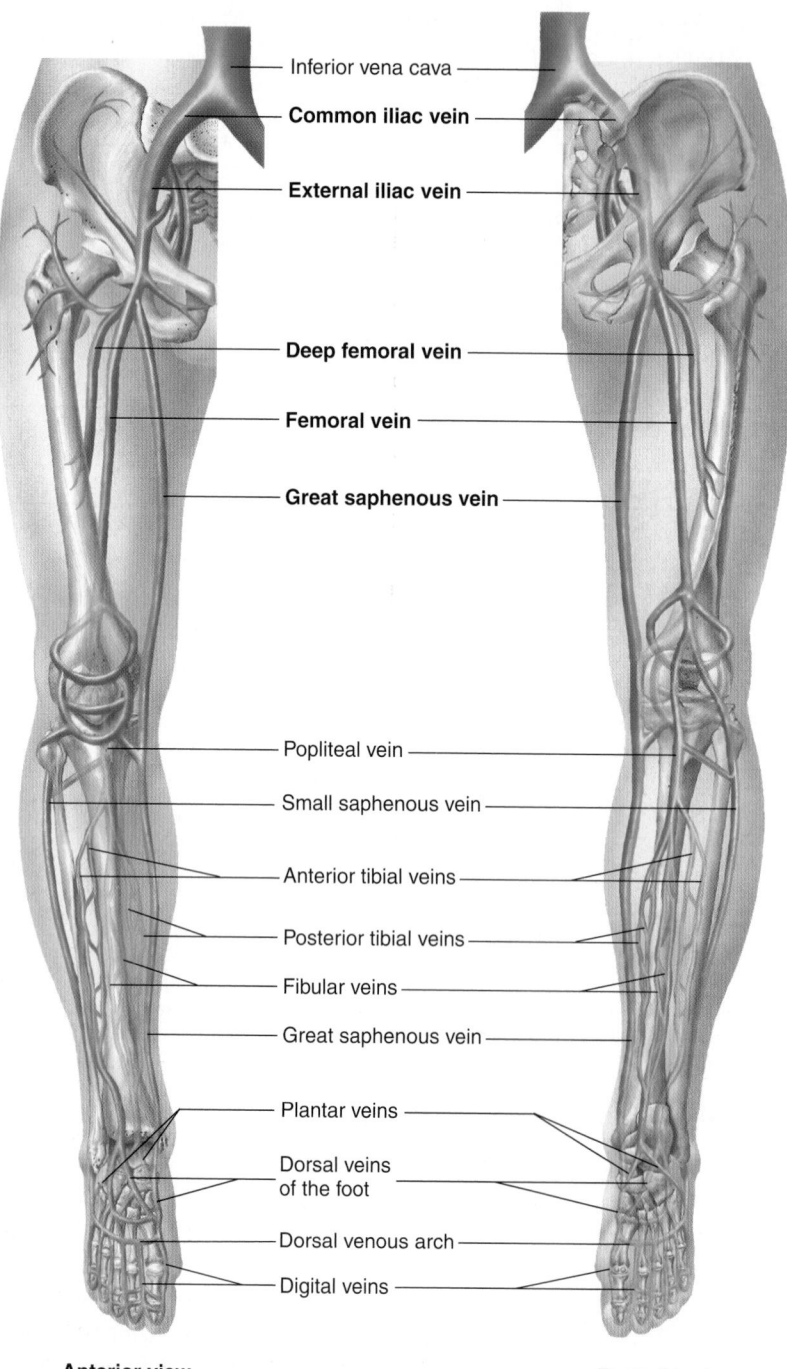

Inferior vena cava
Common iliac vein
External iliac vein
Deep femoral vein
Femoral vein
Great saphenous vein
Popliteal vein
Small saphenous vein
Anterior tibial veins
Posterior tibial veins
Fibular veins
Great saphenous vein
Plantar veins
Dorsal veins of the foot
Dorsal venous arch
Digital veins

Anterior view   **Posterior view**

Figure 21.28   Veins of the Pelvis and Lower Limb

The right common iliac vein and its tributaries.

Figure 21.29 Major Veins of the Lower Limb

Table 21.12	Veins of the Lower Limb (see figures 21.28 and 21.29)
Veins	**Tissues Drained**
External Iliac Vein	
(continuation of the femoral vein)	
Femoral	Thigh
(continuation of the popliteal vein)	
Popliteal	
Anterior tibial	Deep anterior leg
Dorsal vein of foot	Dorsum of foot
Posterior tibial	Deep posterior leg
Plantar veins	Plantar region of foot
Fibular (peroneal)	Deep lateral leg and foot
Small saphenous	Superficial posterior leg and lateral side of foot
Great saphenous	Superficial anterior and medial leg, thigh, and dorsum of foot
Dorsal vein of foot	Dorsum of foot
Dorsal venous arch	Foot
Digital veins	Toes

Dynamics of Blood Circulation

Dynamics of blood circulation through blood vessels are the same as those affecting the flow of water or liquids through pipes. The interrelationships between pressure, flow, resistance, and the control mechanisms that regulate blood pressure and blood flow through vessels play a critical role in the function of the circulatory system. Many of these interrelationships are clinically significant.

Laminar and Turbulent Flow in Vessels

Fluid, including blood, tends to flow through long, smooth-walled tubes in a streamlined fashion called **laminar flow** (figure 21.30a). Fluid behaves as if it were composed of a large number of concentric layers. The layer nearest the wall of the tube experiences the greatest resistance to flow because it moves against the stationary wall. The innermost layers slip over the surface of the outermost layers and experience less resistance to movement. Thus, flow in a vessel consists of movement of concentric layers, with the outermost layer moving slowest and the layer at the center moving fastest.

(a)

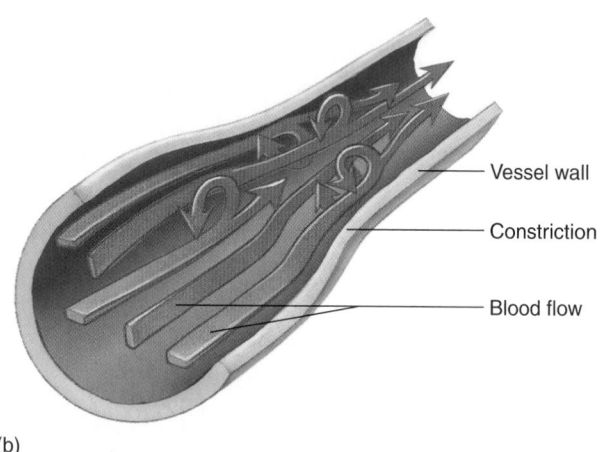

(b)

Figure 21.30 Laminar and Turbulent Flow

(a) In laminar flow, fluid flows in long smooth-walled tubes as if it is composed of a large number of concentric layers. *(b)* Turbulent flow is caused by numerous small currents flowing crosswise or obliquely to the long axis of the vessel, resulting in flowing whorls and eddy currents.

Laminar flow is interrupted and becomes **turbulent flow** when the rate of flow exceeds a critical velocity or when the fluid passes a constriction, a sharp turn, or a rough surface (figure 21.30*b*). Vibrations of the liquid and blood vessel walls during turbulent flow cause the sounds produced when blood pressure is measured using a blood pressure cuff. Turbulent flow is also common as blood flows past the valves in the heart and is partially responsible for the heart sounds (see chapter 20).

Turbulent flow of blood through vessels occurs primarily in the heart and to a lesser extent where arteries branch. Sounds caused by turbulent blood flow in arteries are not normal and usually indicate that the blood vessel is constricted abnormally. In addition, turbulent flow in abnormally constricted arteries increases the probability that thromboses will develop in the area of turbulent flow.

● 31. Describe laminar flow and turbulent flow through a tube. What conditions cause turbulent flow of blood?

Blood Pressure

Blood pressure is a measure of the force blood exerts against blood vessel walls. An instrument for measuring blood pressure is the **mercury (Hg) manometer,** which measures pressure in millimeters of mercury (mm Hg). If the blood pressure is 100 mm Hg, the pressure is great enough to lift a column of mercury 100 mm.

Blood pressure can be measured directly by inserting a **cannula** (or tube) into a blood vessel and connecting a manometer or an electronic pressure transducer to it. Electronic transducers are very sensitive to changes in pressure and can precisely detect rapid fluctuation in pressure.

Placing catheters into blood vessels or into chambers of the heart to monitor pressure changes is possible, but not appropriate for routine clinical determinations of systemic blood pressure. The **auscultatory** (aws-kŭl′tă-tō′rē) **method** can be used to measure blood pressure without surgical procedures or causing discomfort, so it's used under most clinical conditions. A blood pressure cuff connected to a **sphygmomanometer** (sfig′mō-mă-nom′ĕ-ter) is placed around the patient's arm just above the elbow, and a stethoscope is placed over the brachial artery (figure 21.31). Some sphygmomanometers have mercury manometers, and others have digital manometers, but they all measure pressure in terms of millimeters of mercury. The blood pressure cuff is inflated until the brachial artery is completely collapsed. Because no blood flows through the constricted area, no sounds can be heard. The pressure in the cuff is gradually lowered. As soon as it declines below the systolic pressure, blood flows through the constricted area during systole. The blood flow is turbulent and produces vibrations in the blood and surrounding tissues that can be heard through the stethoscope. These sounds are called **Korotkoff** (kō-rot′kof) **sounds,** and the pressure at which a Korotkoff sound is first heard represents the **systolic pressure.**

As the pressure in the blood pressure cuff is lowered still more, the Korotkoff sounds change tone and loudness. When the pressure has dropped until continuous laminar blood flow is reestablished, the sound disappears completely. The pressure at which continuous laminar flow is reestablished is the **diastolic pressure.** This method for determining systolic and diastolic pressures is not entirely accurate, but its results are within 10% of methods that are more direct. In healthy people blood pressure values are maintained within a normal range of values.

● 32. Define the term blood pressure. Describe how blood pressure can be measured.

Blood Flow and Poiseuille's Law

The **rate** at which blood or any other liquid flows through a tube can be expressed as the volume that passes a specific point per unit of time. Blood flow usually is reported in either milliliters (mL) per minute or liters (L) per minute. For example, when a person is resting, the **cardiac output** of the heart is approximately 5 L/min; thus blood flow through the aorta is approximately 5 L/min.

1. No sound is heard because there is no blood flow when the cuff pressure is high enough to keep the brachial artery closed.

2. **Systolic pressure** is the pressure at which a Korotkoff sound is first heard. When cuff pressure decreases and is no longer able to keep the brachial artery closed during systole, blood is pushed through the partially opened brachial artery to produce turbulent blood flow and a sound. The brachial artery remains closed during diastole.

3. As cuff pressure continues to decrease, the brachial artery opens even more during systole. At first, the artery is closed during diastole, but as cuff pressure continues to decrease, the brachial artery partially opens during diastole. Turbulent blood flow during systole produces Korotkoff sounds, although the pitch of the sounds change as the artery becomes more open.

4. **Diastolic pressure** is the pressure at which the sound disappears. Eventually cuff pressure decreases below the pressure in the brachial artery and it remains open during systole and diastole. Nonturbulent flow is reestablished and no sounds are heard.

Process Figure 21.31 **Blood Pressure Measurement**

The rate of blood flow in a vessel can be described by the following:

$$\text{Flow} = \frac{P_1 - P_2}{R}$$

Where P_1 and P_2 are the pressures in the vessel at points one and two, respectively, and R is the resistance to flow. Blood always flows from an area of higher pressure to an area of lower pressure, and the greater the pressure difference, the greater the rate of flow. For example, the average blood pressure in the aorta (P_1) is greater than the blood pressure in the relaxed right atrium (P_2). Therefore, blood flows from the aorta to tissues and from tissues to the right atrium. If the heart should stop contracting, the pressure in the aorta would become equal to that in the right atrium and blood would no longer flow.

The flow of blood, resulting from a pressure difference between the two ends of a blood vessel, is opposed by a resistance to flow. As the resistance increases, blood flow decreases, and as the

resistance decreases, blood flow increases. Factors that affect resistance can be represented as follows:

$$\text{Resistance} = \frac{8vl}{\pi r^4}$$

Where v is the viscosity of blood, l is the length of the vessel, and r is the radius of the vessel. Both 8 and π are constants and, for practical purposes, the length of the blood vessel is constant. Thus, the radius of the blood vessel and the viscosity (see below) of the blood determine resistance. The viscosity of blood changes slowly.

When the equation for resistance is combined with the equation for flow, the following relationship, called **Poiseuille's** (pwah-zuh'yes) **law** results.

$$\text{Flow} = \frac{P_1 - P_2}{R} = \frac{\pi(P_1 - P_2)r^4}{8vl}$$

According to Poiseuille's law, a small change in the radius of a vessel dramatically changes the resistance to flow and therefore the amount of blood that moves through the vessel because the radius is raised to the fourth power. Vasoconstriction decreases the radius of a vessel, increases resistance to flow, and decreases blood flow through the vessel. For example, decreasing the radius of a vessel by half increases the resistance to flow sixteenfold and decreases flow sixteenfold. Vasodilation increases the radius of a vessel, decreases resistance to flow, and increases blood flow through the vessel.

Changes in blood pressure and blood vessel radius produce the major changes in blood flow through blood vessels. During exercise, heart rate and stroke volume increase, causing blood pressure in the aorta to increase. In addition, blood vessels in skeletal muscles vasodilate and resistance to flow decreases. As a consequence, blood flow through blood vessels in exercising skeletal muscles increases dramatically.

Viscosity (vis-kos′i-tē) is a measure of the resistance of a liquid to flow. As the viscosity of a liquid increases, the pressure required to force it to flow increases. A common means for reporting the viscosity of liquids is to consider the viscosity of distilled water as 1 and to compare the viscosity of other liquids to it. Using this procedure, whole blood has a viscosity of 3.0–4.5, which means that about three times as much pressure is required to force whole blood to flow through a given tube at the same rate as water.

The viscosity of blood is influenced largely by **hematocrit** (hē′mă-tō-krit, hem′ă-tō-krit), which is the percentage of the total blood volume composed of red blood cells (see chapter 19). As the hematocrit increases, the viscosity of blood increases logarithmically. Blood with a hematocrit of 45% has a viscosity about three times that of water, whereas blood with a very high hematocrit of 65% has a viscosity about seven to eight times that of water. The plasma proteins have only a minor effect on the viscosity of blood. Dehydration or uncontrolled production of erythrocytes can increase hematocrit and the viscosity of blood substantially. Viscosity above its normal range of values increases the workload on the heart, and, if this workload is great enough, heart failure can result.

33. *Define the terms blood flow and resistance.*
34. *Describe the relationship between blood flow, blood pressure, and resistance.*
35. *According to Poiseuille's law, what effect do viscosity, blood vessel diameter, and blood vessel length have on resistance? On blood flow?*
36. *Define the term viscosity, and state the effect of hematocrit on viscosity.*

P R E D I C T

Predict the effect of each of the following conditions on blood flow: (a) vasoconstriction of blood vessels in the skin in response to cold exposure, (b) vasodilation of the blood vessels in the skin in response to an elevated body temperature, (c) erythrocytosis, which results in a greatly increased hematocrit.

Critical Closing Pressure and Laplace's Law

Each blood vessel exhibits a **critical closing pressure,** the pressure below which the vessel collapses and blood flow through the vessel stops. Under conditions of shock, blood pressure can decrease below the critical closing pressure in vessels (see the Clinical Focus on "Shock," p. 772). As a consequence, the blood vessels collapse, and flow ceases. Tissues supplied by those vessels can become necrotic because of the lack of blood supply.

Laplace's (la-plas′ez) **law** states that the force that stretches the vascular wall is proportional to the diameter of the vessel times the blood pressure. Laplace's law helps explain the critical closing pressure. As the pressure in a vessel decreases, the force that stretches the vessel wall also decreases. Some minimum force is required to keep the vessel open. If the pressure decreases so that the force is below that minimum requirement, the vessel will close. As the pressure in a vessel increases, the force that stretches the vessel wall also increases.

Laplace's law is expressed by the following formula:

$$F = D \times P$$

where F is force, D is vessel diameter, and P is pressure.

According to Laplace's law, as the diameter of a vessel increases, the force applied to the vessel wall increases, even if the pressure remains constant. If a part of an arterial wall becomes weakened so that a bulge forms in it, the force applied to the weakened part is greater than at other points along the blood vessel because its diameter is greater. The greater force causes the weakened vessel wall to bulge even more, further increasing the force applied to it. This series of events can proceed until the vessel finally ruptures. As the bulges in weakened blood vessel walls, called **aneurysms,** enlarge, the danger of their rupturing increases. Ruptured aneurysms in the blood vessels of the brain or in the aorta often result in death.

37. *State Laplace's law. How does it explain critical closing pressure and aneurysms?*

Vascular Compliance

Compliance (kom-plī′ans) is the tendency for blood vessel volume to increase as the blood pressure increases. The more easily the vessel wall stretches, the greater is its compliance. The less easily the vessel wall stretches, the smaller is its compliance.

Compliance is expressed by the following formula:

$$\text{Compliance} = \frac{\text{Increase in volume (mL)}}{\text{Increase in pressure (mm Hg)}}$$

Vessels with a large compliance exhibit a large increase in volume when the pressure increases a small amount. Vessels with a small compliance do not show a large increase in volume when the pressure increases.

Venous compliance is approximately 24 times greater than the compliance of arteries. As venous pressure increases, the volume of the veins increases greatly. Consequently, veins act as storage areas, or reservoirs, for blood because their large compliance allows them to hold much more blood than other areas of the vascular system (table 21.13).

Table 21.13	Distribution of Blood Volume in Blood Vessels
Vessels	**Total Blood Volume (%)**
Systemic	
Veins	64
Large veins (39%)	
Small veins (25%)	
Arteries	15
Large arteries (8%)	
Small arteries (5%)	
Arterioles (2%)	
Capillaries	5
TOTAL IN SYSTEMIC VESSELS	84
Pulmonary vessels	9
Heart	7
TOTAL BLOOD VOLUME	100

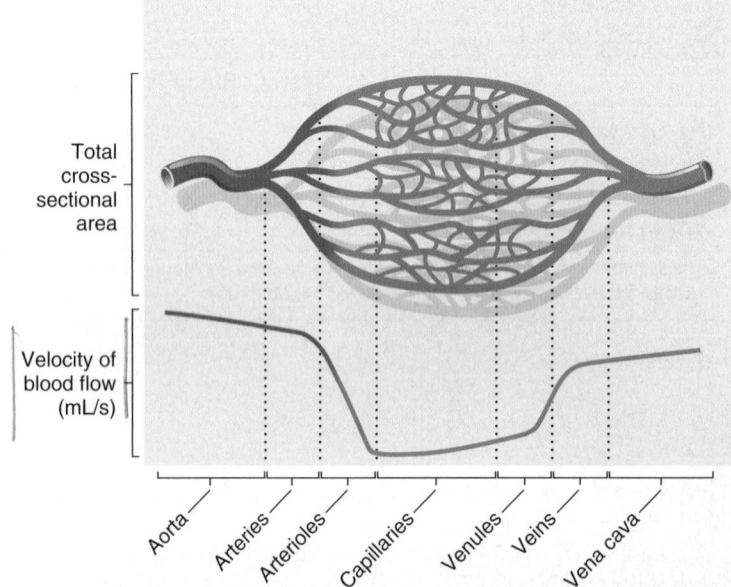

Figure 21.32 Blood Vessel Types and Velocity of Blood Flow

Total cross-sectional area for each of the major blood vessel types is illustrated. The cross-sectional area of each blood vessel is the space through which blood flows, measured in square centimeters. The cross-sectional area of the aorta is about 5 cm². The cross-sectional area of each capillary is much smaller, but there are so many that the total cross-sectional area of all capillaries is much greater (2500 cm²) than the cross-sectional area of the aorta. The graph shows that blood velocity drops dramatically in arterioles, capillaries, and venules. As the total cross-sectional area increases, the velocity of blood flow decreases.

● 38. *Define the term vascular compliance. Do veins or arteries have greater compliance?*

Physiology of Systemic Circulation

The anatomy of the circulatory system, the dynamics of blood flow, and the regulatory mechanisms that control the heart and blood vessels determine the physiologic characteristics of the circulatory system. The entire circulatory system functions to maintain adequate blood flow to all tissues.

Approximately 84% of the total blood volume is contained in the systemic blood vessels. Most of the blood volume is in the veins (64%), which are the vessels with the greatest compliance. Smaller volumes of blood are in the arteries (15%) and capillaries (5%) (see table 21.13).

● 39. *List the percent distribution of blood in the large arteries, small arteries, arterioles, capillaries, small veins, and large veins.*

Cross-Sectional Area of Blood Vessels

If the cross-sectional area of each blood vessel type is determined and multiplied by the number of each type of blood vessel, the result is the total cross-sectional area for each blood vessel type. For example, only one aorta exists, and it has a cross-sectional area of 5 square centimeters (cm²). On the other hand, millions of capillaries exist, and each has a very small cross-sectional area. The total cross-sectional area of all capillaries, however, is 2500 cm², which is much greater than the cross-sectional area of the aorta (figure 21.32).

The velocity of blood flow is greatest in the aorta, but the total cross-sectional area is small. In contrast, the total cross-sectional area for the capillaries is large, but the velocity of blood flow is low. As the veins become larger in diameter, their total cross-sectional area decreases, and the velocity of blood flow increases. The relationship between blood vessel diameter and velocity of blood flow is much like a stream that flows rapidly through a narrow gorge but flows slowly through a broad plane (see figure 21.32).

● 40. *Describe the total cross-sectional areas of the aorta, arteries, arterioles, capillaries, venules, veins, and vena cava.*
41. *Describe how the velocity of blood flow changes as blood flows through the aorta to the superior vena cava and inferior vena cava.*

Pressure and Resistance

The left ventricle of the heart forcefully ejects blood from the heart into the aorta. Because the pumping action of the heart is pulsatile, the aortic pressure fluctuates between a systolic pressure of 120 mm Hg and a diastolic pressure of 80 mm Hg (table 21.14

Table 21.14　Blood Pressure Classification (Adults)

	Systolic Blood Pressure (mm Hg)	Diastolic Blood Pressure (mm Hg)
Normal Blood Pressure	< 120	< 80
Prehypertension	120–139	80–89
Stage 1 Hypertension	140–159	90–99
Stage 2 Hypertension	≥ 160	≥ 100

The seventh Report of the Joint National Committee on Prevention, Detection, and Evaluation, and Treatment of High Blood Pressure, U.S. Dept. of Health and Human Services, NIH Publication No. 03–5233, May 2003.

and figure 21.33). As blood flows from arteries through the capillaries and the veins, the pressure falls progressively to a minimum of approximately 0 mm Hg or even slightly lower by the time it returns to the right atrium.

The decrease in arterial pressure in each part of the systemic circulation is directly proportional to the resistance to blood flow. Resistance is small in the aorta, so the average pressure at the end

of the aorta is nearly the same as at the beginning of the aorta; about 100 mm Hg. The resistance in medium arteries, which are as small as 3 mm in diameter, is also small, so that their average pressure is only decreased to 95 mm Hg. In the smaller arteries, however, the resistance to blood flow is greater; by the time blood reaches the arterioles, the average pressure is approximately 85 mm Hg. Within the arterioles, the resistance to flow is higher than in any other part of the systemic circulation, and at their ends, the average pressure is only approximately 30 mm Hg. The resistance is also fairly high in the capillaries. The blood pressure at the arterial end of the capillaries is approximately 30 mm Hg, and it decreases to approximately 10 mm Hg at the venous end. Resistance to blood flow in the veins is low because of their relatively large diameter; by the time the blood reaches the right atrium in the venous system, the average pressure has decreased from 10 mm Hg to approximately 0 mm Hg.

The muscular arteries and arterioles are capable of constricting or dilating in response to autonomic and hormonal stimulation. If constriction occurs, the resistance to blood flow increases, less blood flows through the constricted blood vessels, and blood is shunted to other, nonconstricted areas of the body. Muscular arteries help control the amount of blood flowing to each region of the body, and arterioles regulate blood flow through specific tissues. Constriction of an arteriole decreases blood flow through the local area it supplies, and vasodilation increases the blood flow.

42. Describe the changes in resistance and blood pressure as blood flows through the aorta to the superior vena cava and/ or inferior vena cava.

43. Explain how constriction and dilation of muscular arteries shunts blood from one area of the body to another and constriction and dilation of arterioles changes blood flow through local areas.

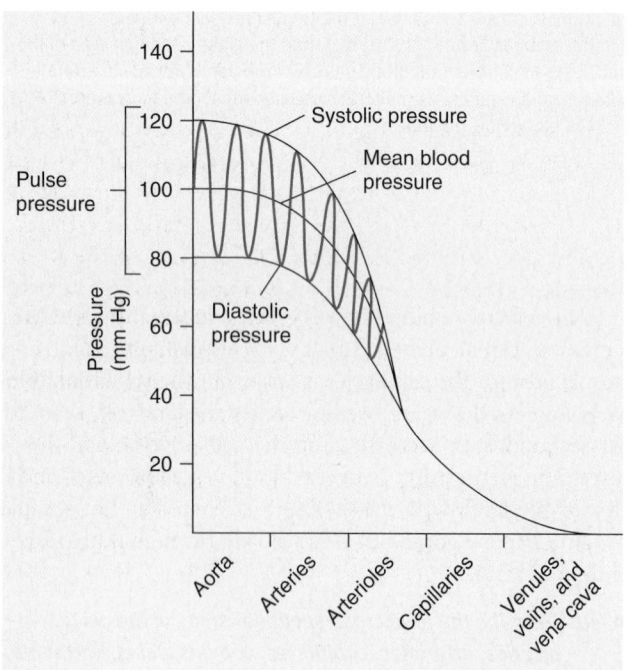

Figure 21.33　Blood Pressure in Major Blood Vessel Types
Blood pressure fluctuations between systole and diastole are damped in small arteries and arterioles. There are no large fluctuations in blood pressure in capillaries and veins.

Pulse Pressure

The difference between systolic and diastolic pressures is called **pulse pressure** (see figure 21.33). In a healthy young adult at rest, systolic pressure is approximately 120 mm Hg, and diastolic pressure is approximately 80 mm Hg; thus, the pulse pressure is approximately 40 mm Hg. Two major factors influence pulse pressure: stroke volume of the heart and vascular compliance. When stroke volume decreases, pulse pressure also decreases; and when stroke volume increases, pulse pressure increases. For example, during exercise such as running, the stroke volume increases and, as a consequence, the pulse pressure also increases. After running, the pulse pressure gradually returns to its resting value as the stroke volume of the heart decreases. The compliance of blood vessels decreases as arteries age. Arteries in older people become less elastic, or arteriosclerotic, and the resulting decrease in compliance causes the pressure in the aorta to rise more rapidly and to a greater degree during systole and to fall more rapidly to its diastolic value. Thus, for a given stroke volume, systolic pressure and pulse pressure are higher as vascular compliance decreases.

> **P R E D I C T** 3
>
> Explain the consequences of arteriosclerosis that is getting progressively more severe on a large aortic aneurysm.

The pulse pressure caused by the ejection of blood from the left ventricle into the aorta produces a pressure wave, or pulse, that travels rapidly along the arteries. Its rate of transmission is approximately 15 times greater in the aorta (7–10 m/s) and 100 times greater (15–35 m/s) in the distal arteries than the velocity of blood flow. The pulse is monitored frequently, especially in the radial artery, where it's called the **radial pulse,** to determine heart rate and rhythm. Also, weak pulses usually indicate a decreased stroke volume or increased constriction of the arteries as a result of intense sympathetic stimulation of the arteries.

As the pulse passes through the smallest arteries and arterioles, it's gradually damped so that there is a smaller fluctuation between the systolic and diastolic pressure. This difference is almost absent at the end of the arterioles (see figure 21.33). At the beginning of the capillary there is a steady pressure of close to 30 mm Hg. As a consequence, there is adequate pressure to cause blood to flow through capillaries if the precapillary sphincters dilate.

 44. *What is pulse pressure? How do stroke volume and vascular compliance affect pulse pressure?*

> **P R E D I C T** 4
>
> Explain each of the following: weak pulses in response to ectopic and premature beats of the heart, strong bounding pulses in a person who received too much saline solution intravenously, weak pulses in a person who is suffering from hemorrhagic shock.

Capillary Exchange and Regulation of Interstitial Fluid Volume

Approximately 10 billion capillaries exist in the body. The heart and blood vessels all function to maintain blood flow through those capillaries and to support **capillary exchange,** which is the movement of substances into and out of capillaries. Capillary exchange is the process by which cells receive everything they need to survive and to eliminate metabolic waste products. If blood flow through capillaries is not maintained, cells cannot survive.

By far, the most important means by which capillary exchange occurs is **diffusion.** Nutrients, such as glucose and amino acids, O_2, and hormones diffuse from a higher concentration in capillaries to a lower concentration in the interstitial fluid. Waste products, including CO_2, diffuse from a higher concentration in the interstitial fluid to a lower concentration in the capillaries. Lipid-soluble molecules cross capillary walls by diffusing through the plasma membranes of the endothelial cells of the capillaries. Examples include O_2, CO_2, steroid hormones, and fatty acids. Water-soluble substances, such as glucose and amino acids, diffuse through intercellular spaces or through fenestrations of capillaries. In a few areas of the body, such as the spleen and liver, the spaces between the endothelial cells are large enough to allow proteins to pass through them. In other areas, the connections between endothelial cells are extensive and few molecules pass between the endothelial cells, such as in the capillaries of the brain that form the blood–brain barrier. In these capillaries mediated transport processes move water-soluble substances across the capillary walls (see chapter 13 for a description of the blood–brain barrier).

Endothelial cells of capillaries appear to take up small pinocytotic vesicles and transport them across the capillary wall. The pinocytotic vesicles, however, don't appear to be a major means by which molecules move across the wall of the capillary.

Clinical Focus   Pulse

The pulse is important clinically because one can determine heart rate, rhythmicity, and other characteristics by feeling it. A pulse can be felt at 10 major locations on each side of the body where large arteries are close to the surface.

On the head and neck, a pulse can be felt in three arteries: the common carotid artery in the neck, the superficial temporal artery immediately anterior to the ear, and the facial artery at the point where it crosses the inferior border of the mandible approximately midway between the angle and the genu (figure A).

On the upper limb, a pulse can also be felt in three arteries: the axillary artery in the axilla, the brachial artery on the medial side of the arm slightly proximal to the elbow, and the radial artery on the lateral side of the anterior forearm just proximal to the wrist. The radial artery is traditionally the most common site for taking a pulse, because it is the most easily accessible artery in the body.

In the lower part of the body, a pulse can be felt in four locations: the femoral artery in the groin, the popliteal artery just proximal to the knee, and the dorsalis pedis artery and posterior tibial artery at the ankle.

Figure A   **Location of Major Points at Which the Pulse Can Be Monitored**
Each pulse point is named after the artery on which it occurs.

A small amount of fluid moves out of capillaries at their arterial ends, and most, but not all, of that fluid reenters capillaries at their venous ends (figure 21.34). The remaining fluid enters lymphatic vessels, which eventually return it to the venous circulation (see chapter 22). Alterations in the forces affecting fluid movement across capillary walls are responsible for edema.

Net filtration pressure (NFP) is the force responsible for moving fluid across capillary walls. It is the difference between net hydrostatic pressure and net osmotic pressure.

$$\text{NFP} = \text{Net hydrostatic pressure} - \text{Net osmotic pressure}$$

Net hydrostatic pressure is the difference in pressure between the blood and interstitial fluid. **Blood pressure (BP)** at the arterial end of a capillary is about 30 mm Hg. It results mainly from the force of contraction of the heart, but it can be modified by the effect of gravity on fluids within the body (see Blood Pressure and the Effect of Gravity on p. 762).

Interstitial fluid pressure (IFP) is the pressure of interstitial fluid within the tissue spaces. It is −3 mm Hg. IFP is a negative number because of the suction effect produced by the lymphatic vessels as they pump excess fluid from the tissue spaces. The lym-

phatic system is described in chapter 22. Here, it is only necessary to understand that excess interstitial fluid enters lymphatic capillaries and is eventually returned to the blood.

At the arterial end of capillaries, the net hydrostatic pressure that moves fluid across capillary walls into the tissue spaces is the difference between BP and IFP.

$$\begin{aligned}
\textbf{Net hydrostatic pressure} &= \textbf{BP} - \textbf{IFP} \\
&= 30 - (-3) \\
&= 33 \text{ mm Hg}
\end{aligned}$$

Net osmotic pressure is the difference in osmotic pressure between the blood and the interstitial fluid. The osmotic pressure caused by the plasma proteins is called the **blood colloid osmotic pressure (BCOP),** and the osmotic pressure caused by proteins in the interstitial fluid is called the **interstitial colloid osmotic pressure (ICOP).** Large proteins do not freely pass through the capillary walls, and the difference in protein concentrations between the blood and interstitial fluid is responsible for osmosis. Ions and small molecules do not make a significant contribution to osmosis across the capillary wall because they freely pass through it and their concentrations are approximately the same in the blood as in the interstitial fluid.

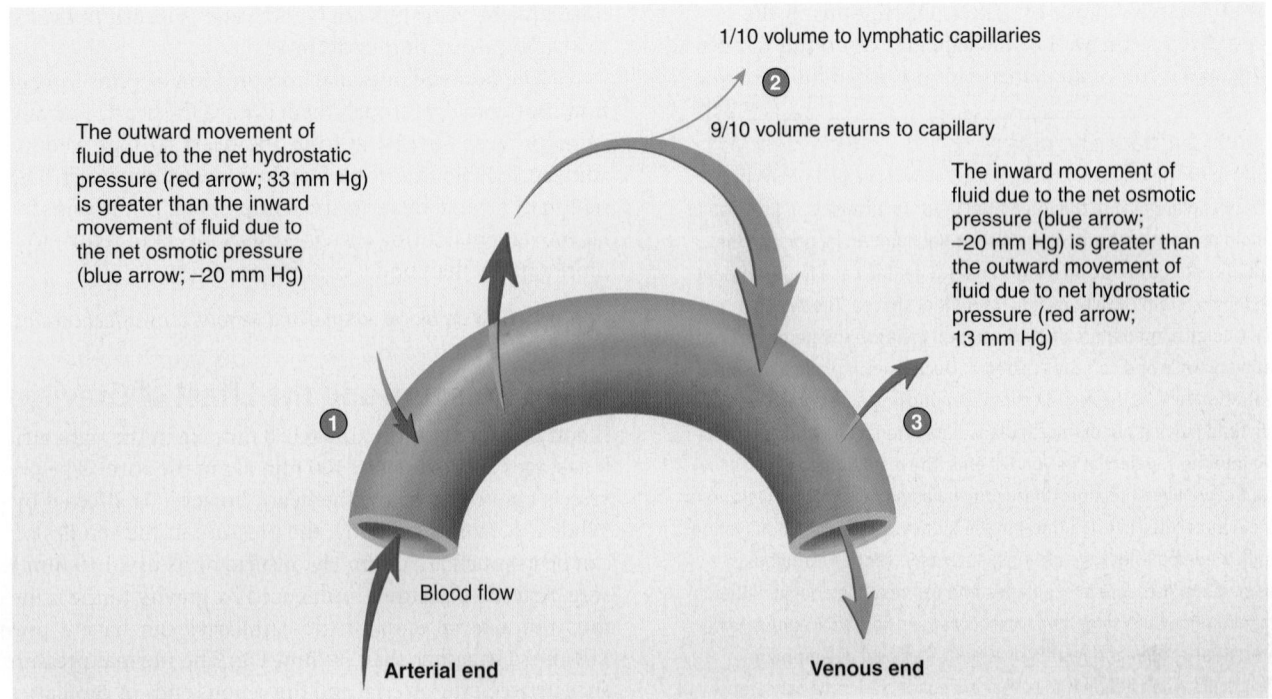

The outward movement of fluid due to the net hydrostatic pressure (red arrow; 33 mm Hg) is greater than the inward movement of fluid due to the net osmotic pressure (blue arrow; –20 mm Hg)

1/10 volume to lymphatic capillaries

②

9/10 volume returns to capillary

The inward movement of fluid due to the net osmotic pressure (blue arrow; –20 mm Hg) is greater than the outward movement of fluid due to net hydrostatic pressure (red arrow; 13 mm Hg)

①

③

Blood flow

Arterial end

Venous end

1. At the arterial end of the capillary, the **net hydrostatic pressure** is greater than the **net osmotic pressure,** resulting in a positive **net filtration pressure** that causes fluid to move out of the capillary.

   Net hydrostatic pressure = 33 mm Hg
   Net osmotic pressure = –20 mm Hg
   Net filtration pressure = 13 mm Hg

2. Approximately nine-tenths of the fluid that leaves the capillary at its arterial end reenters the capillary at its venous end. About one-tenth of the fluid passes into the lymphatic capillaries.

3. At the venous end of the capillary, the **net hydrostatic pressure** is less than the **net osmotic pressure,** resulting in a negative **net filtration pressure** that causes fluid to move into the capillary.

   Net hydrostatic pressure = 13 mm Hg
   Net osmotic pressure = –20 mm Hg
   Net filtration pressure = –7 mm Hg

Process Figure 21.34   Fluid Exchange Across the Walls of Capillaries

The total pressure differences between the inside and the outside of the capillary at its arterial and venous ends are illustrated.

The BCOP (28 mm Hg) is several times larger than the ICOP (8 mm Hg) because of the presence of albumin and other proteins in the plasma (see chapter 19). Therefore, the net osmotic pressure is equal to BCOP–ICOP.

$$\textbf{Net osmotic pressure} = \textbf{BCOP} - \textbf{ICOP}$$
$$= 28 - 8$$
$$= 20 \text{ mm Hg}$$

The greater the osmotic pressure of a fluid, the greater is the tendency for water to move into that fluid (see chapter 3). The net osmotic pressure results in the osmosis of water into the capillary because there is a greater tendency for water to move into the blood than into the interstitial fluid.

The net filtration pressure at the arterial end of the capillary is equal to the net hydrostatic pressure, which moves fluid out of the capillary, minus the net osmotic pressure, which moves fluid into the capillary.

$$\textbf{NFP} = \textbf{Net hydrostatic pressure} - \textbf{Net osmotic pressure}$$
$$= 33 - 20$$
$$= 13 \text{ mm Hg}$$

Between the arterial ends of capillaries and their venous ends, the blood pressure decreases from about 30 mm Hg to 10 mm Hg, which reduces the net hydrostatic pressure moving fluid out of the venous end of the capillary.

$$\textbf{Net hydrostatic pressure} = \textbf{BP} - \textbf{IFP}$$
$$= 10 - (-3)$$
$$= 13 \text{ mm Hg}$$

The concentration of proteins within capillaries and the concentration of proteins within interstitial fluid do not change significantly because only a small amount of fluid passes from the capillaries into the tissue spaces. Therefore, the net osmotic pressure moving fluid into capillaries by osmosis is still approximately 20 mm Hg. At the venous end of capillaries the NFP now causes fluid to reenter the capillary:

$$\textbf{NFP} = \textbf{Net hydrostatic pressure} - \textbf{Net osmotic pressure}$$
$$= 13 - 20$$
$$= -7 \text{ mm Hg}$$

Exchange of fluid across the capillary wall and movement of fluid into lymphatic capillaries keep the volume of the interstitial fluid within a narrow range of values. Disruptions in the movement of fluid across the wall of the capillary can result in edema, or swelling, as a result of an increase in interstitial fluid volume.

Edema and Capillary Exchange

Increases in the permeability of capillaries allow plasma proteins to move from capillaries into the interstitial fluid. This causes an increase in the colloid osmotic pressure of the interstitial fluid. An increase in the interstitial colloid osmotic pressure causes a net increase in the amount of fluid moving from capillaries into interstitial spaces. The result is edema. Chemical mediators of inflammation increase the permeability of the capillary walls and can cause edema. Decreases in plasma protein concentration reduce the blood colloid osmotic pressure, which results in more fluid moving out of the capillary at its arterial end and less fluid moving into the capillary at its venous end. The result once again is edema. Severe liver infections that reduce plasma protein synthesis, loss of protein molecules in urine through the kidneys, and protein starvation all result in edema. Blockage of veins, such as in venous thrombosis, increases blood pressure in capillaries and can result in edema. Either blockage or removal of lymphatic vessels causes fluid to accumulate in the interstitial spaces and results in edema. Removal of lymphatic vessels occurs when lymph nodes that are suspected to be cancerous are removed.

45. What is the most important means by which capillary exchange occurs?
46. Describe the factors that influence the movement of fluid from capillaries into the tissues.
47. What happens to the fluid in the tissues? What is edema?

PREDICT 5
Edema often results from a disruption in the normal inwardly and outwardly directed pressures across the capillary wall. On the basis of what you know about fluid movement across the wall of the capillary and the regulation of capillary blood pressure, explain why large fluctuations in arterial blood pressure occur without causing significant edema and why small increases in venous pressure can lead to edema.

Functional Characteristics of Veins

Cardiac output depends on the preload, which is determined by the volume of blood that enters the heart from the veins (see chapter 20). The factors that affect flow in the veins are, therefore, of great importance to the overall function of the cardiovascular system. If the volume of blood is increased because of a rapid transfusion, the amount of blood flow to the heart through the veins increases. This increases the preload, which causes the cardiac output to increase because of Starling's law of the heart. On the other hand, a rapid loss of a large volume of blood decreases venous return to the heart, which decreases the preload and cardiac output.

Venous tone is a continual state of partial contraction of the veins as a result of sympathetic stimulation. Increased sympathetic stimulation increases venous tone by causing constriction of the veins, which forces the large venous volume to flow toward the heart. Consequently, venous return and preload increase, causing an increase in cardiac output. Conversely, decreased sympathetic stimulation decreases venous tone, allowing veins to relax and

dilate. As the veins fill with blood, venous return to the heart, preload, and cardiac output decrease.

The periodic muscular compression of veins forces blood to flow more rapidly through them toward the heart. The valves in the veins prevent flow away from the heart so that when veins are compressed, blood is forced to flow toward the heart. The combination of arterial dilation and compression of the veins by muscular movements during exercise causes blood to return to the heart more rapidly than under conditions of rest.

48. How do blood volume and venous tone affect cardiac output?

Blood Pressure and the Effect of Gravity

Blood pressure is approximately 0 mm Hg in the right atrium, and it averages approximately 100 mm Hg in the aorta. The pressure in vessels above and below the heart, however, is affected by gravity. While a person is standing, the pressure in the venules of the feet can be as much as 90 mm Hg, instead of its usual 10 mm Hg pressure. Arterial pressure is influenced by gravity to the same degree; thus the arterial ends of the capillaries can have a pressure of 110 mm Hg rather than 30 mm Hg. The normal pressure difference between the arterial and the venous ends of capillaries still remains the same, so that flow continues through the capillaries. The major effect of the high pressure in the feet and legs when a person stands for a prolonged time without moving is edema. Without muscular movement, the pressure at the venous end of the capillaries increases. Up to 15%–20% of the total blood volume can pass through the walls of the capillaries into the interstitial spaces of the lower limbs during 15 minutes of standing still.

49. What effect does standing have on blood pressure in the feet and the head? Explain why this effect occurs.

PREDICT 6
Explain why people who are suffering from edema in the legs are told to keep them elevated.

Control of Blood Flow in Tissues

Blood flow provided to the tissues by the cardiovascular system is highly controlled and matched closely to the metabolic needs of tissues. Mechanisms that control blood flow through tissues are classified as (1) local control and (2) nervous and hormonal control.

Local Control of Blood Flow by the Tissues

Blood flow is much greater in some organs than in others. For example, blood flow through the brain, kidneys, and liver is relatively high. The muscle mass of the body is large so that flow through resting skeletal muscles, although not high, is greater than that through other tissue types because skeletal muscle constitutes 35%–40% of the total body mass. Flow through exercising skeletal muscles can increase up to 20-fold, however, and the blood flow through the viscera, including the kidneys and liver, either remains the same or decreases.

In most tissues, blood flow is proportional to the metabolic needs of the tissue; therefore, as the activity of skeletal muscle increases, blood flow increases to supply the increased need for oxygen and other nutrients. Blood flow also increases in response to a buildup of metabolic end products.

In some tissues, however, blood flow serves purposes other than the delivery of nutrients and the removal of waste products. In the skin, blood flow also dissipates heat from the body. In the kidneys, it eliminates metabolic waste products, regulates water balance, and controls the pH of body fluids. Among other functions, blood flow delivers nutrients that enter the blood from the small intestine to the liver for processing.

Functional Characteristics of the Capillary Bed

The innervation of the metarterioles and the precapillary sphincters in capillary beds is sparse (table 21.15). Local factors regulate these structures primarily. As the rate of metabolism increases in a tissue, blood flow through its capillaries increases. The precapillary sphincters relax, allowing blood to flow into the local capillary bed. Blood flow can increase sevenfold to eightfold as a result of vasodilation of the metarterioles and the precapillary sphincters in response to an increased rate of metabolism.

Vasodilator substances are produced as the rate of metabolism increases. The vasodilator substances then diffuse from the tissues supplied by the capillary to the area of the precapillary sphincter, the metarterioles, and the arterioles, to cause vasodilation (figure 21.35a). Several chemicals, including carbon dioxide, lactic acid, adenosine, adenosine monophosphate, adenosine diphosphate, endothelium-derived relaxation factor (EDRF), potassium ions, and hydrogen ions, cause vasodilation, and they increase in concentration in the extracellular fluid as the rate of metabolism in tissues increases.

Lack of nutrients can also be important in regulating local blood flow. For example, oxygen and other nutrients are required to maintain vascular smooth muscle contraction. An increased rate of metabolism decreases the amount of oxygen and other nutrients in the tissues. Smooth muscle cells of the precapillary sphincter relax in response to a lack of oxygen and other nutrients, resulting in vasodilation (see figure 21.35a).

Blood flow through capillaries is not continuous but cyclic. The cyclic fluctuation is the result of periodic contraction and relaxation of the precapillary sphincters called **vasomotion** (vā-sō-mō′shŭn, vas-ō-mō′shŭn). Blood flows through the capillaries until

Table 21.15	Homeostasis: Local Control of Blood Flow
Stimulus	**Response**
Regulation by Metabolic Need of Tissues	
Increased vasodilator substances (e.g., CO_2, lactic acid, adenosine, adenosine monophosphate, adenosine diphosphate, endothelium-derived relaxation factor, K^+, decreased pH) or decreased nutrients (e.g., O_2, glucose, amino acids, fatty acids, and other nutrients) as a result of increased metabolism	Relaxation of precapillary sphincters and subsequent increase in blood flow through capillaries
Decreased vasodilator substances and a reduced need for O_2 and other nutrients	Contraction of precapillary sphincters and subsequent decrease in blood flow through capillaries
Regulation by Nervous Mechanisms	
Increased physical activity or increased sympathetic activity	Constriction of blood vessels in skin and viscera
Increased body temperature detected by neurons of the hypothalamus	Dilation of blood vessels in skin (see chapter 5)
Decreased body temperature detected by neurons of the hypothalamus	Constriction of blood vessels in skin (see chapter 5)
Decrease in skin temperature below a critical value	Dilation of blood vessels in skin (protects skin from extreme cold)
Anger or embarrassment	Dilation of blood vessels in skin of face and upper thorax
Regulation by Hormonal Mechanisms	
(reinforces increased activity of the sympathetic nervous system)	
Increased physical activity and increased sympathetic activity causing release of epinephrine and small amounts of norepinephrine from the adrenal medulla	Constriction of blood vessels in skin and viscera; dilation of blood vessels in skeletal and cardiac muscle
Autoregulation	
Increased blood pressure	Contraction of precapillary sphincters to maintain constant capillary blood flow
Decreased blood pressure	Relaxation of precapillary sphincters to maintain constant capillary blood flow
Long-Term Local Blood Flow	
Increased metabolic activity of tissues over a long period	Increased diameter and number of capillaries
Decreased metabolic activity of tissues over a long period	Decreased diameter and number of capillaries

Clinical Focus　Hypertension

Hypertension, or high blood pressure, affects approximately 20% of the human population at sometime in their lives. Generally, a person is considered hypertensive if the systolic blood pressure is greater than 140 mm Hg and the diastolic pressure is greater than 90 mm Hg. Current methods of evaluation, however, take into consideration diastolic and systolic blood pressures in determining whether a person is suffering from hypertension (see table 21.14). In addition, normal blood pressure is age-dependent, so classification of an individual as hypertensive depends on the person's age.

Chronic hypertension has an adverse effect on the function of both the heart and blood vessels. Hypertension requires the heart to work harder than normal. This extra work leads to hypertrophy of the cardiac muscle, especially in the left ventricle, and can lead to heart failure. Hypertension also increases the rate at which arteriosclerosis

develops. Arteriosclerosis, in turn, increases the probability that blood clots, or thromboemboli (throm'bō-em'bō-lī), may form and that blood vessels will rupture. Common medical problems associated with hypertension are cerebral hemorrhage, coronary infarction, hemorrhage of renal blood vessels, and poor vision caused by burst blood vessels in the retina.

Some conditions leading to hypertension include a decrease in functional kidney mass, excess aldosterone or angiotensin production, and increased resistance to blood flow in the renal arteries. All of these conditions cause an increase in total blood volume, which causes cardiac output to increase. Increased cardiac output forces blood to flow through tissue capillaries, causing the precapillary sphincters to constrict. Thus increased blood volume increases cardiac output and peripheral resistance, both of which result in greater blood pressure.

Although these conditions result in hypertension, roughly 90% of the diagnosed cases of hypertension are called **idiopathic,** or **essential, hypertension,** which means the cause of the condition is unknown. Drugs that dilate blood vessels (called vasodilators), drugs that increase the rate of urine production (called diuretics), or drugs that decrease cardiac output normally are used to treat essential hypertension. The vasodilator drugs increase the rate of blood flow through the kidneys and thus increase urine production, and the diuretics also increase urine production. Increased urine production reduces blood volume, which reduces blood pressure. Substances that decrease cardiac output, such as β-adrenergic-blocking agents, decrease the heart rate and force of contraction. In addition to these treatments, low-salt diets normally are recommended to reduce the amount of sodium chloride and water absorbed from the intestine into the bloodstream.

(a) Precapillary sphincters relax as the tissue concentration of nutrients such as O_2, glucose, amino acids, fatty acids, and other nutrients decreases.

Precapillary sphincters relax as the concentration of tissue metabolic by-products, such as CO_2, lactic acid, adenosine, adenosine monophosphate, adenosine diphosphate, nitric oxide, and K^+ increase, and as pH decreases.

(b) Precapillary sphincters contract as the tissue concentration of nutrients such as O_2, glucose, amino acids, fatty acids, and other nutrients increases.

Precapillary sphincters contract as the tissue concentration of metabolic by-products, such as CO_2, lactic acid, adenosine, adenosine monophosphate, adenosine diphosphate, nitric oxide decrease, and as K^+ and pH increase.

Figure 21.35　Control of Local Blood Flow Through Capillary Beds

(a) Dilation of precapillary sphincters. *(b)* Constriction of precapillary sphincters.

the by-products of metabolism are reduced in concentration and until nutrient supplies to precapillary smooth muscles are replenished. Then the precapillary sphincters constrict and remain constricted until the by-products of metabolism increase and nutrients decrease (figure 21.35b).

Autoregulation of Blood Flow

Arterial pressure can change over a wide range, whereas blood flow through tissues remains relatively constant. The maintenance of blood flow by tissues is called **autoregulation** (aw'tō-reg-ū-lā'shŭn). Between arterial pressures of approximately 75 mm Hg

Smooth muscle of precapillary sphincter relaxes

Blood flow increases

Blood flow

Smooth muscle of precapillary sphincter contracts

Blood flow decreases

Blood flow

and 175 mm Hg, blood flow through tissues remains within 10%–15% of its normal value. The mechanisms responsible for autoregulation are the same as those for vasomotion. The need for nutrients and the buildup of metabolic by-products cause precapillary sphincters to dilate, and blood flow through tissues increases if a minimum blood pressure exists. On the other hand, once the supply of nutrients and oxygen to tissues is adequate, the precapillary sphincters constrict, and blood flow through the tissues decreases, even if blood pressure is very high.

P R E D I C T **7**
When blood flow to a tissue has been blocked for a short time, the blood flow through that tissue increases to as much as five times its normal value after the removal of the blockade. The response is called reactive hyperemia. Develop an explanation for this phenomenon based on the local control of blood flow.

Long-Term Local Blood Flow

The long-term regulation of blood flow through tissues is matched closely to the metabolic requirements of the tissue. If the metabolic activity of a tissue increases and remains elevated for an extended period, the diameter and the number of capillaries in the tissue increase, and local blood flow increases. The increased density of capillaries in the well-trained skeletal muscles of athletes compared to that in poorly trained skeletal muscles is an example.

The availability of oxygen to a tissue can be a major factor in determining the adjustment of the vascularity of a tissue to its long-term metabolic needs. If oxygen is scarce, capillaries increase in diameter and in number, and if the oxygen levels remain elevated in a tissue, the vascularity decreases.

Occlusion of Blood Vessels and Collateral Circulation

Blockage, or occlusion, of a blood vessel leads to an increase in the diameter of smaller blood vessels that bypass the occluded vessel. In many cases, the development of these collateral vessels is marked. For example, if a vessel such as the femoral artery becomes occluded, the small vessels that bypass the occluded vessel become greatly enlarged. An adequate blood supply to the lower limb is often reestablished over a period of weeks. If the occlusion is sudden and so complete that tissues supplied by a blood vessel suffer from ischemia (lack of blood flow), cell death (necrosis) can occur. In this instance, collateral circulation doesn't have a chance to develop before necrosis occurs.

Nervous and Hormonal Regulation of Local Circulation

Nervous control of arterial blood pressure is important in the minute-to-minute regulation of local circulation. The blood pressure must be adequate to cause blood flow through capillaries while at rest, during exercise, or in response to circulatory shock, during which blood pressure decreases to a very low value. For example, during exercise, increased arterial blood pressure is needed to sustain increased blood flow through the capillaries of skeletal muscles, which have dilated precapillary sphincters. The increased blood flow is needed to supply oxygen and nutrients to the exercising skeletal muscles.

Nervous regulation also provides a means by which blood can be shunted from one large area of the peripheral circulatory system to another. For example, in response to blood loss, blood flow to the viscera and the skin is reduced dramatically. This helps maintain the arterial blood pressure within a range sufficient to allow adequate blood flow through the capillaries of the brain and cardiac muscle.

Nervous regulation, by the autonomic nervous system, can function rapidly (within 1–30 seconds). The most important part of the autonomic nervous system for this regulation is the sympathetic division (figure 21.36). Sympathetic vasomotor fibers innervate all blood vessels of the body except the capillaries, precapillary sphincters, and most metarterioles. The innervation of the small arteries and arterioles allows the sympathetic nervous system to increase or decrease resistance to blood flow.

P R E D I C T **8**
A strong athlete just finished a 1-mile run and sat down to have a drink with her friends. Her blood pressure was not dramatically elevated during the run, but her cardiac output was greatly increased. After the run, her cardiac output decreased dramatically, but her blood pressure only decreased to its resting level. Predict how sympathetic stimulation of her large veins, arteries in her digestive system, and arteries in her skeletal muscles change while she is relaxing. Explain why this is consistent with the decrease in her cardiac output.

Sympathetic vasoconstrictor fibers extend to most parts of the circulatory system, but they are less prominent in skeletal

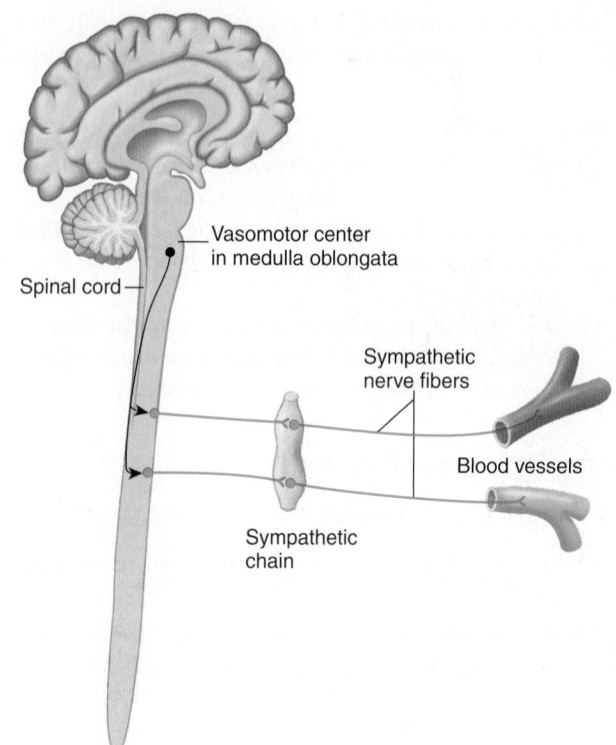

Figure 21.36 **Nervous Regulation of Blood Vessels**
Most blood vessels are innervated by sympathetic nerve fibers. The vasomotor center within the medulla oblongata plays a major role in regulating the frequency of action potentials in nerve fibers that innervate blood vessels.

muscle, cardiac muscle, and the brain and more prominent in the kidneys, gut, spleen, and skin.

An area of the lower pons and upper medulla oblongata, called the **vasomotor** (vā-sō-mō′ter, vas-ō-mō′ter) **center** (see figure 21.36), is tonically active. A low frequency of action potentials is transmitted continually through the sympathetic vasoconstrictor fibers. As a consequence, the peripheral blood vessels are partially constricted, a condition called **vasomotor tone.**

Part of the vasomotor center inhibits vasomotor tone. Thus, the vasomotor center consists of an excitatory part, which is tonically active, and an inhibitory part, which can induce vasodilation. Vasoconstriction results from an increase and vasodilation from a decrease in vasomotor tone.

Areas throughout the pons, midbrain, and diencephalon can either stimulate or inhibit the vasomotor center. For example, the hypothalamus can exert either strong excitatory or inhibitory effects on the vasomotor center. Increased body temperature detected by temperature receptors in the hypothalamus causes vasodilation of blood vessels in the skin (see chapter 5). The cerebral cortex also can either excite or inhibit the vasomotor center. For example, action potentials that originate in the cerebral cortex during periods of emotional excitement activate hypothalamic centers, which in turn increase vasomotor tone (see table 21.15).

The neurotransmitter for the vasoconstrictor fibers is norepinephrine, which binds to α-adrenergic receptors on vascular smooth muscle cells to cause vasoconstriction. Sympathetic action potentials also cause the release of epinephrine and norepinephrine into the blood from the adrenal medulla. These hormones are transported in the blood to all parts of the body. In most vessels, they cause vasoconstriction, but in some vessels, especially those in skeletal muscle, epinephrine binds to β-adrenergic receptors, which are present in larger numbers, and can cause the skeletal muscle blood vessels to dilate.

50. *Explain how vasodilator substances and nutrients are involved with local control of blood flow. What is vasomotion? What is autoregulation of local blood flow?*
51. *How is long-term regulation of blood flow through tissues accomplished?*
52. *Describe nervous and hormonal control of blood flow. Under what conditions is nervous control of blood flow important? Define the term vasomotor tone.*

Regulation of Mean Arterial Pressure

Blood flow to all areas of the body depends on the maintenance of an adequate pressure in the arteries. As long as arterial blood pressure is adequate, local control of blood flow through tissues is appropriately matched to their metabolic needs. Blood flow through tissues cannot be adequate if arterial blood pressure is too low, and damage, including heart and blood vessel damage, can result if arterial blood pressure is too high. This section describes the mechanisms that operate to maintain arterial blood pressure within a normal range of values.

Mean arterial pressure (MAP) is slightly less than the average of systolic and diastolic pressures because diastole lasts longer than systole. MAP is approximately 70 mm Hg at birth, is approxi-

mately 100 mm Hg from adolescence to middle age, and reaches 110 mm Hg in the healthy older person, but it can be as high as 130 mm Hg. The range of normal systolic and diastolic blood pressures for adults is presented in table 21.14.

Cardiac output (CO) is the volume of blood pumped by the heart each minute. It is equal to the **heart rate (HR)** times the **stroke volume (SV). Peripheral resistance (PR)** is the resistance to blood flow in all the blood vessels. MAP in the body is proportional to the cardiac output times the peripheral resistance: Blood flow through the entire circulatory system is determined by the cardiac output *(CO)*, which is equal to the heart rate *(HR)* times the stroke volume *(SV)* and peripheral resistance *(PR)*, which is the resistance to blood flow in all the blood vessels.

$$MAP = CO \times PR \quad or \quad MAP = HR \times SV \times PR$$

This equation expresses the effect of heart rate, stroke volume, and peripheral resistance on blood pressure. An increase in any one of them results in an increase in blood pressure. Conversely, a decrease in any one of them produces a decrease in blood pressure. The mechanisms that control blood pressure do so by changing peripheral resistance, heart rate, or stroke volume. Because stroke volume depends on the amount of blood entering the heart, regulatory mechanisms that control blood volume also affect blood pressure. For example, an increase in blood volume increases venous return, which increases preload, and the increased preload increases stroke volume.

When blood pressure suddenly drops because of hemorrhage or some other cause, the control systems respond by increasing blood pressure to a value consistent with life and by increasing blood volume to its normal value. Two major types of control systems operate to achieve these responses: (1) those that respond in the short term and (2) those that respond in the long term.

The regulatory mechanisms that control pressure on a short-term basis respond quickly but begin to lose their capacity to regulate blood pressure a few hours to a few days after blood pressure is maintained at higher or lower values. This occurs because sensory receptors adapt to the altered pressures. Long-term regulation of blood pressure is controlled primarily by mechanisms that influence kidney function, and those mechanisms don't adapt rapidly to altered blood pressures.

53. *Define mean arterial pressure, cardiac output, and peripheral resistance.*
54. *Describe the factors that determine mean arterial pressure.*

Short-Term Regulation of Blood Pressure

The short-term, rapidly acting mechanisms controlling blood pressure include the baroreceptor reflexes, the adrenal medullary mechanism, chemoreceptor reflexes, and the central nervous system ischemic response. Some of these reflex mechanisms operate on a minute-to-minute basis and help regulate blood pressure within a narrow range of values. Other mechanisms respond primarily to emergency situations. The mechanisms responsible for the short-term regulation of the blood pressure are summarized in figures 21.39 and 21.41.

Baroreceptor Reflexes

Baroreceptor reflexes are very important in regulating blood pressure on a minute-to-minute basis. They detect even small changes

Clinical Focus Blood Flow Through Tissues During Exercise

During exercise, blood flow through tissues is changed dramatically. Its rate of flow through exercising skeletal muscles can be 15–20 times greater than through resting muscles. Increased blood flow is the product of local, nervous, and hormonal regulatory mechanisms. When skeletal muscle is resting, only 20%–25% of the capillaries in the skeletal muscle are open, whereas during exercise 100% of the capillaries are open.

Low oxygen tensions resulting from greatly increased muscular activity or the release of vasodilator substances, such as lactic acid, carbon dioxide, and potassium ions, cause dilation of precapillary sphincters. Increased sympathetic stimulation and epinephrine released from the adrenal medulla cause vasoconstriction in the blood vessels of the skin and viscera and some vasoconstriction in the blood vessels of

skeletal muscles. The resistance to blood flow in skeletal muscle decreases even though some vasoconstriction in skeletal muscle blood vessels occurs because the capillaries are all open. Resistance to blood flow in the skin and viscera increases. Blood is therefore shunted from the viscera and the skin to the vessels in skeletal muscles.

The movement of skeletal muscles that compresses veins in a cyclic fashion and the constriction of veins greatly increase the venous return to the heart. The resulting increase in the preload and increased sympathetic stimulation of the heart result in elevated heart rate and stroke volume, which increases cardiac output. As a consequence, blood pressure usually increases by 20–60 mm Hg, which helps sustain the increased blood flow through skeletal muscle blood vessels.

In response to sympathetic stimulation, some decrease in blood flow through the skin can occur at the beginning of exercise. As body temperature increases in response to the increased muscular activity, however, temperature receptors in the hypothalamus are stimulated. As a result, action potentials in sympathetic nerve fibers causing vasoconstriction decrease, resulting in vasodilation of blood vessels in the skin. As a consequence, the skin turns a red or pinkish color, and a great deal of excess heat is lost as blood flows through the dilated blood vessels.

The overall effect of exercise on circulation is to greatly increase blood flow through exercising muscles and to keep blood flow through other organs at a value just adequate to supply their metabolic needs.

---

in blood pressure and respond quickly. However, they are not as important as other mechanisms in regulating blood pressure over long periods of time.

Baroreceptors, or **pressoreceptors,** are sensory receptors sensitive to stretch. They are scattered along the walls of most of the large arteries of the neck and thorax, and are most numerous in the area of the carotid sinus at the base of the internal carotid artery and in the walls of the aortic arch. Action potentials are transmitted from the carotid sinus baroreceptors through the glossopharyngeal nerves to the cardioregulatory and vasomotor centers in the medulla oblongata and from the aortic arch through the vagus nerves to the medulla oblongata (figure 21.37). Stimulation of baroreceptors in the carotid sinus activates the **carotid sinus reflex,** and stimulation of baroreceptors in the aortic arch activates the **aortic arch reflex.** Both of these reflexes, are baroreceptor reflexes, and they both function to control blood pressure within a narrow range of values.

In the carotid sinus and the aortic arch, normal blood pressure partially stretches the arterial wall so that the baroreceptors produce a constant, but low, frequency of action potentials. Increased pressure in the blood vessels stretches the vessel walls and causes the baroreceptors to increase the frequency of action potentials. Conversely, a decrease in blood pressure reduces the stretch of the arterial wall and causes the baroreceptors to decrease the frequency of action potentials.

A sudden increase in blood pressure increases the frequency of action potentials produced in the baroreceptors. The increase in action potentials influences the vasomotor and cardioregulatory centers of the medulla oblongata. The vasomotor center responds by decreasing sympathetic stimulation of blood vessels, and the cardioregulatory center responds by increasing parasympathetic stimu-

lation of the heart. As a result, peripheral blood vessels dilate, the heart rate decreases, and the blood pressure decreases (see figure 21.40).

A sudden decrease in blood pressure results in a decreased frequency of action potentials produced by the baroreceptors. The decreased action potentials influence the vasomotor center and cardioregulatory centers of the medulla oblongata. The vasomotor center responds by increasing sympathetic stimulation of blood vessels, and the cardioregulatory center responds by increasing sympathetic stimulation of the heart. As a result peripheral blood vessels constrict, the heart rate and stroke volume increase, and the blood pressure increases. This increase is accompanied by a decrease in parasympathetic stimulation of the heart (see figures 21.37 and 21.39).

The carotid sinus and aortic arch baroreceptor reflexes are important in regulating blood pressure moment to moment. When a person rises rapidly from a sitting or lying position to a standing position, a dramatic drop in blood pressure in the neck and thoracic regions occurs because of the pull of gravity on the blood. This reduction can be so great that blood flow to the brain becomes sufficiently sluggish to cause dizziness or loss of consciousness. The falling blood pressure activates the baroreceptor reflexes, which reestablish normal blood pressure within a few seconds. A healthy person may experience only a temporary sensation of dizziness.

P R E D I C T

Explain how the baroreceptor reflex responds when a person does a headstand.

The baroreceptor reflexes are short term and rapid-acting. They don't change the average blood pressure in the long run. The baroreceptors adapt within 1–3 days to any new sustained blood

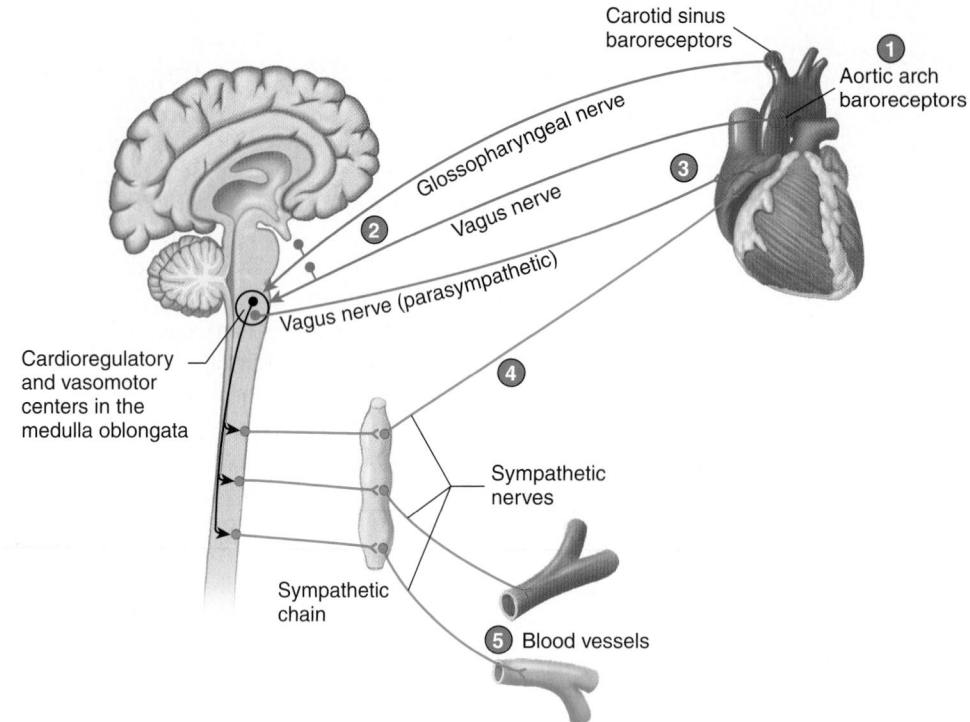

1. Baroreceptors in the carotid sinus and aortic arch monitor blood pressure.

2. Action potentials are conducted by the glossopharyngeal and vagus nerves to the cardioregulatory and vasomotor centers in the medulla oblongata.

3. Increased parasympathetic stimulation of the heart decreases the heart rate.

4. Increased sympathetic stimulation of the heart increases the heart rate and stroke volume.

5. Increased sympathetic stimulation of blood vessels increases vasoconstriction.

Process Figure 21.37   Baroreceptor Reflex Control of Blood Pressure

An increase in blood pressure increases parasympathetic stimulation of the heart and decreases sympathetic stimulation of the heart and blood vessels, resulting in a decrease in blood pressure. A decrease in blood pressure decreases parasympathetic stimulation of the heart and increases sympathetic stimulation of the heart and blood vessels, resulting in an increase in blood pressure.

pressure to which they are exposed. If the blood pressure is elevated for more than a few days, the baroreceptors adapt to the elevated pressure and the baroreceptor reflex does not reduce blood pressure to its original value. This adaptation is common in people who have hypertension.

The Carotid Sinus Syndrome

Occasionally, the application of pressure to the carotid arteries in the upper neck results in a dramatic decrease in blood pressure. This condition, called the **carotid sinus syndrome,** is most common in patients in whom arteriosclerosis of the carotid artery is advanced. In such patients a tight collar can apply enough pressure to the region of the carotid sinuses to stimulate the baroreceptors. The increased action potentials from the baroreceptors initiate reflexes that result in a decrease in vasomotor tone and an increase in parasympathetic action potentials to the heart. As a result of the decreased peripheral resistance and heart rate, blood pressure decreases dramatically. As a consequence, blood flow to the brain decreases to such a low level that the person becomes dizzy or may even faint. People suffering from this condition must avoid applying external pressure to the neck region. If the carotid sinus becomes too sensitive, a treatment for this condition is surgical destruction of the innervation to the carotid sinuses.

Adrenal Medullary Mechanism

The adrenal medullary mechanism is activated when stimuli result in a substantial increase in sympathetic stimulation of the heart and blood vessels (figure 21.38 and figure 21.39). Large de-

creases in blood pressure, sudden and substantial increases in physical activity, and other stressful conditions are examples. The adrenal medullary mechanism results from stimulation of the ad-

Figure 21.38   **The Adrenal Medullary Mechanism**

Stimuli that increase sympathetic stimulation of the heart and blood vessels also result in increased sympathetic stimulation of the adrenal medulla and result in epinephrine and some norepinephrine secretion.

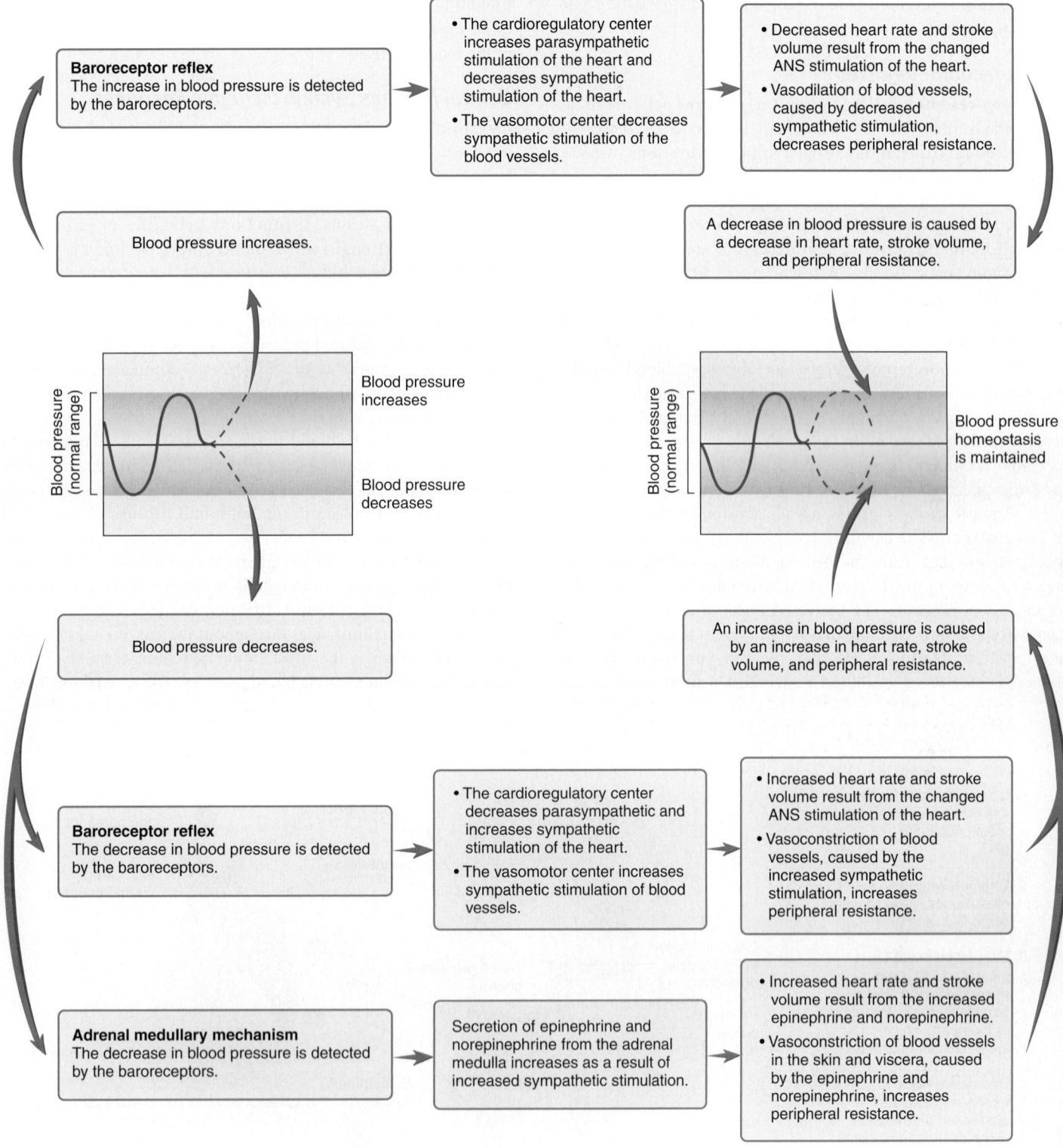

Homeostasis Figure 21.39   Baroreceptor Effects on Blood Pressure

renal medulla by the sympathetic nerve fibers. The adrenal medulla releases epinephrine and smaller amounts of norepinephrine into the circulatory system (see figure 21.38 and figure 26.39). These hormones affect the cardiovascular system in a fashion similar to direct sympathetic stimulation, causing increased heart rate, increased stroke volume, and vasoconstriction in blood vessels to the skin and viscera. Epinephrine can indirectly cause vasodilation in blood vessels to the heart because of the increased rate of cardiac muscle metabolism (see chapter 20). The adrenal medullary mechanism is short term and rapid-acting, whereas other hormonal mechanisms are long term and slow-acting (see following sections).

Chemoreceptor Reflexes

The **chemoreceptor** (kē′mō-rē-sep′tor) **reflexes** help maintain homeostasis when oxygen tension in the blood decreases or when carbon dioxide and hydrogen ion concentrations increase (figure 21.40 and figure 21.41).

Carotid bodies, small organs approximately 1–2 mm in diameter, lie near the carotid sinuses, and several **aortic bodies** lie adjacent to the aorta. Chemoreceptors are located in the carotid and aortic bodies. Afferent nerve fibers pass to the medulla oblongata through the glossopharyngeal nerve (IX) from the carotid bodies and through the vagus nerve (X) from the aortic bodies.

The chemoreceptors receive an abundant blood supply. When oxygen availability decreases in the chemoreceptor cells, the frequency of action potentials increases and stimulates the vasomotor center, resulting in increased vasomotor tone. The chemoreceptors act under emergency conditions and don't regulate the cardiovascular system under resting conditions. They normally don't respond strongly unless oxygen tension in the blood decreases markedly. The chemoreceptor cells are also stimulated by increased carbon dioxide and hydrogen ion concentrations to increase vasomotor tone. The increased vasomotor tone increases the mean arterial pressure. The increased mean arterial pressure increases blood flow through tissues in which blood vessels do not constrict. Blood vessels that are not constricted by the chemoreceptor reflex are blood vessels that deliver blood to the brain and cardiac muscle. Thus, the reflex helps provide an adequate oxygen supply to the brain and the heart when oxygen levels in the blood decrease.

Central Nervous System Ischemic Response

Elevation in blood pressure in response to a lack of blood flow to the medulla oblongata of the brain is called the **central nervous system (CNS) ischemic response.** The CNS ischemic response doesn't play an important role in regulating blood pressure under normal conditions. It functions primarily in response to emergency situations in which blood flow to the brain is severely restricted or when blood pressure falls below approximately 50 mm Hg.

Reduced blood flow results in reduced oxygen, increased carbon dioxide, and reduced pH within the medulla oblongata. Neurons of the vasomotor center are strongly stimulated. As a result, vasoconstriction is stimulated by the vasomotor center, and the systemic blood pressure rises dramatically.

The increase in blood pressure that occurs in response to CNS ischemia increases blood flow to the CNS, provided the blood vessels are intact. However, if severe ischemia lasts longer than a few minutes, metabolism in the brain fails because of the lack of oxygen. The vasomotor center becomes inactive, and extensive vasodilation occurs in the periphery as vasomotor tone decreases. Prolonged ischemia of the medulla oblongata leads to a massive decline in blood pressure and ultimately death.

In most circumstances throughout the day, the baroreceptor reflex mechanism is the most important short-term regulatory mechanism for maintaining blood pressure. The adrenal medullary

1. Chemoreceptors in the carotid and aortic bodies monitor blood O_2, CO_2, and pH.

2. Chemoreceptors in the medulla oblongata monitor blood CO_2 and pH.

3. Decreased blood O_2, increased CO_2, and decreased pH decrease parasympathetic stimulation of the heart, which increases the heart rate.

4. Decreased blood O_2, increased CO_2, and decreased pH increase sympathetic stimulation of the heart, which increases the heart rate and stroke volume.

5. Increased sympathetic stimulation of blood vessels increases vasoconstriction.

Process Figure 21.40 Chemoreceptor Reflex Control of Blood Pressure

An increase in blood CO_2 and a decrease in pH and blood O_2 result in an increased heart rate and vasoconstriction. A decrease in blood CO_2 and an increase blood pH result in a decreased heart rate and vasodilation.

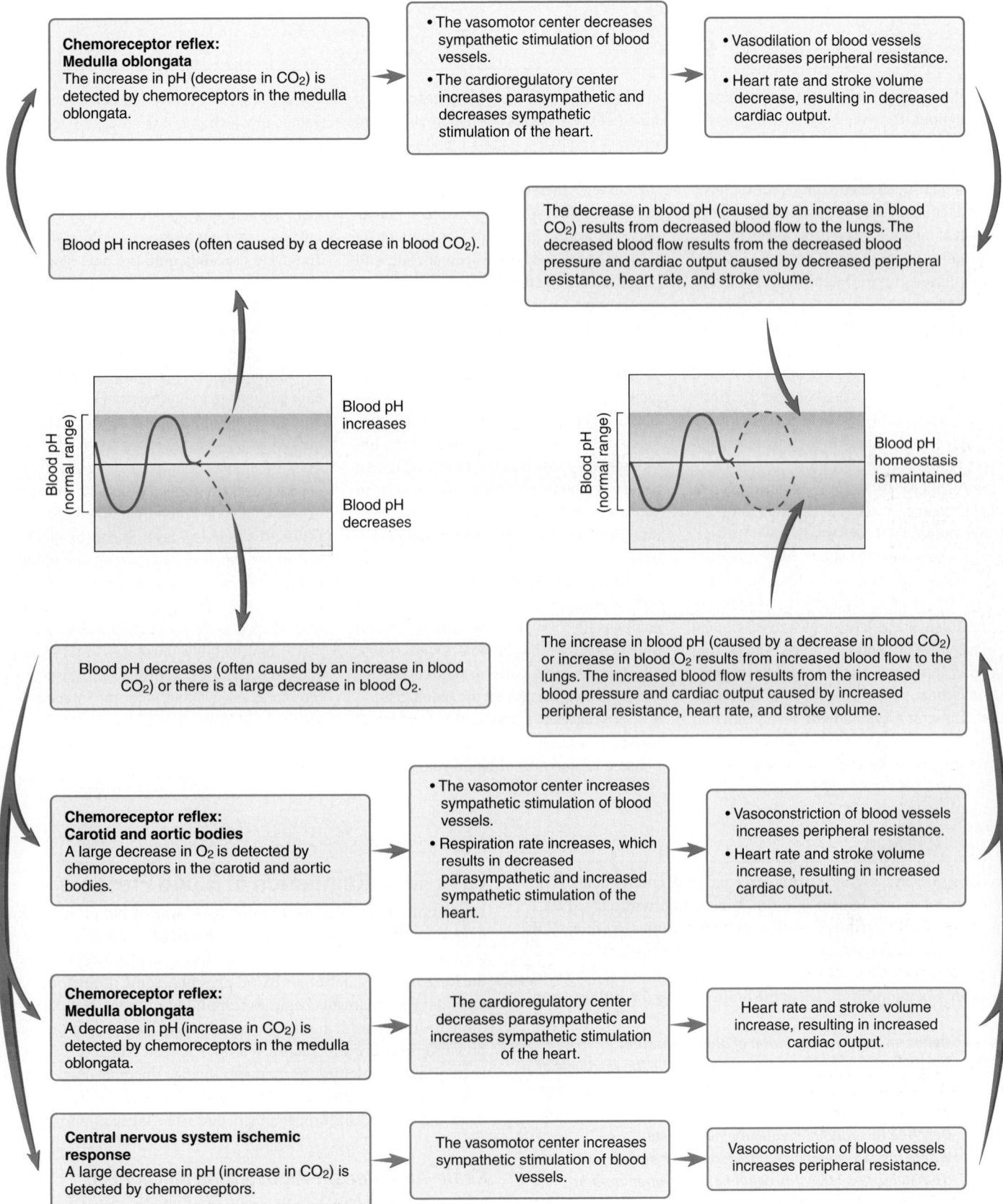

Homeostasis Figure 21.41 Effects of pH and Gases on Blood Pressure

Clinical Focus   Shock

Circulatory shock is inadequate blood flow throughout the body. Failure of mechanisms that function to maintain blood pressure within a normal range of values results in dramatic decreases in blood pressure. As a consequence, tissues suffer damage as a result of too little delivery of oxygen to cells. Severe circulatory shock can damage vital body tissues to the extent that the individual dies.

Depending on its severity, circulatory shock can be divided into three separate stages: (1) nonprogressive or compensated shock, (2) progressive shock, and (3) irreversible shock. All types of circulatory shock exhibit one or more of these stages, regardless of their cause. Several causes of circulatory shock exist, but hemorrhagic, or hypovolemic, shock is used to illustrate the characteristics of each stage.

In **compensated shock,** blood pressure decreases only a moderate amount, and the mechanisms that regulate blood pressure function successfully to reestablish normal blood pressure and blood flow. The baroreceptor reflexes, chemoreceptor reflexes, and ischemia within the medulla oblongata initiate strong sympathetic responses that result in intense vasoconstriction and increased heart rate. As blood volume decreases, the stress–relaxation response of blood vessels causes the blood vessels to contract and helps sustain blood pressure. In response to reduced blood flow through the kidneys, thereby increased amounts of renin are released. The elevated renin release results in a greater rate of angiotensin II formation, causing vasoconstriction and increased aldosterone release from the adrenal cortex. Aldosterone, in turn, promotes water and salt retention by the kidneys, thereby conserving water. In addition, ADH is released from the posterior pituitary gland and enhances the retention of water by the kidneys. Because of the fluid shift mechanism, water also moves from the interstitial spaces and the intestinal lumen to restore normal blood volume. An intense sensation of thirst increases water intake, also helping to elevate normal blood volume.

In mild cases of compensated shock, baroreceptor reflexes can be adequate to compensate for blood loss until blood volume is restored, but in more severe cases, all of the mechanisms described are required to compensate for the blood loss.

In **progressive shock,** the compensatory mechanisms are inadequate to compensate for the loss of blood volume. As a consequence, a positive-feedback cycle develops in which the blood pressure regulatory mechanisms are unable to compensate for circulatory shock. As circulatory shock worsens, regulatory mechanisms become even less able to compensate for the increasing severity of the circulatory shock. The cycle proceeds until the next stage of shock is reached or until medical treatment is applied that assists the regulatory mechanisms in reestablishing adequate blood flow to tissues.

During progressive shock, blood pressure declines to a very low level that is inadequate to maintain blood flow to cardiac muscle; thus, the heart begins to deteriorate. Substances that are toxic to the heart are released from tissues that suffer from severe ischemia. When blood pressure declines to a very low level, blood begins to clot in the small vessels. Eventually blood vessel dilation begins as a result of decreased sympathetic activity and because of the lack of oxygen in capillary beds. Capillary permeability increases under ischemic conditions, allowing fluid to leave the blood vessels and enter the interstitial spaces, and finally intense tissue deterioration begins in response to inadequate blood flow.

mechanism plays a role during exercise and emergencies. The chemoreceptor mechanism is more important when oxygen tension in the blood is reduced, such as at high altitudes or when CO_2 is elevated or pH is reduced. Thus, it is more important in emergency situations. The CNS ischemic response is only activated in rare emergency conditions when the brain receives too little oxygen.

55. *Define short-term regulation of blood pressure.*
56. *Where are baroreceptors located? Describe the response of the baroreceptor reflex when blood pressure increases and decreases.*
57. *Describe the adrenal medullary mechanism.*
58. *Where are the chemoreceptors for carbon dioxide, pH, and oxygen located? Describe what happens when oxygen levels in the blood decrease.*
59. *Describe the CNS ischemic response. Under what conditions does this mechanism operate?*
60. *What mechanism is most important for short-term regulation of blood pressure under resting conditions?*

Long-Term Regulation of Blood Pressure

Regulation of the concentration and volume of blood by the kidneys, the movement of fluid across the wall of blood vessels, and alterations in the volume of the blood vessels all play a central role in the long-term regulation of blood pressure. Some of the long-term regulatory mechanisms begin to respond in minutes, but they continue to function for hours, days, or longer. They adjust the blood pressure precisely and keep it within a narrow range of values for years. Major regulatory mechanisms include the renin-angiotensin-aldosterone mechanism, vasopressin mechanism, atrial natriuretic mechanism, fluid shift mechanism, and stress–relaxation response.

Renin-Angiotensin-Aldosterone Mechanism

The **renin-angiotensin-aldosterone mechanism** helps regulate kidney functions. This mechanism can also influence peripheral resistance by causing vasoconstriction. The kidneys increase urine output as the blood volume and arterial pressure increase, and they decrease urine output as the blood volume and arterial pressure

Without medical intervention, progressive shock leads to irreversible shock. **Irreversible shock** leads to death, regardless of the amount or type of medical treatment applied. In this stage of shock, the damage to tissues, including cardiac muscle, is so extensive that the patient is destined to die, even if adequate blood volume is reestablished and blood pressure is elevated to its normal value. Irreversible shock is characterized by decreasing heart function and progressive dilation of and increased permeability of peripheral blood vessels.

Patients suffering from shock are normally placed in a horizontal plane, usually with the head slightly lower than the feet, and oxygen is often supplied. **Replacement therapy** consists of transfusions of whole blood, plasma, artificial solutions called plasma substitutes, and physiologic saline solutions administered to increase blood volume. In some circumstances, drugs that enhance vasoconstriction are also administered. Occasionally, such as in patients in anaphylactic (an'ă-fi-lak'tik; allergic) shock, anti-inflammatory substances like glucocorticoids and antihistamines are administered. The basic objective in treating shock is to reverse the condition so that progressive shock is arrested, to prevent it from progressing to the irreversible stage, and to reverse the condition so that normal blood flow through tissues is reestablished.

Several types of shock are classified here by cause:

- **Hemorrhagic shock** is external or internal bleeding that causes a reduction in blood volume.
- **Plasma loss shock** is reduced blood volume that results from a loss of plasma into the interstitial spaces and greatly increased blood viscosity. Plasma loss shock includes **intestinal obstruction,** resulting in the movement of a large amount of plasma from the blood into the intestine, and **severe burns,** resulting in the loss of large amounts of plasma from the burned surface.
- **Dehydration** results from a severe and prolonged shortage of fluid intake.
- **Severe diarrhea** or **vomiting** cause a loss of plasma through the intestinal wall.
- **Neurogenic shock** is a rapid loss of vasomotor tone leading to vasodilation so extensive that a severe decrease in blood pressure results.
- **Anesthesia** includes deep general anesthesia or spinal anesthesia that decreases the activity of the medullary vasomotor center or the sympathetic nerve fibers.
- **Brain damage** leads to an ineffective medullary vasomotor function.
- **Emotional shock (vasovagal syncope)** results from emotions that cause strong parasympathetic stimulation of the heart and results in vasodilation in skeletal muscles and in the viscera.
- **Anaphylactic shock** results from an allergic response that causes the release of inflammatory substances that increase vasodilation and capillary permeability.
- **Septic shock,** or **"blood poisoning,"** results from peritoneal, systemic, and gangrenous infections that cause the release of toxic substances into the circulatory system, depressing the activity of the heart, leading to vasodilation, and increasing capillary permeability.
- **Cardiogenic shock** occurs when the heart stops pumping in response to conditions such as heart attack or electrocution.

decrease. Increased urine output reduces the blood volume and blood pressure, and decreased urine output resists a further decrease in the blood volume and blood pressure. The control of urine output is not only an important means by which blood pressure is regulated, it continues to operate until the blood pressure is precisely within its normal range of values.

The kidneys release an enzyme called **renin** (rē'nin) into the circulatory system (see chapter 26) from specialized structures called the **juxtaglomerular** (jŭks'tă-glŏ-mer'ū-lăr) **apparatuses** (figure 21.42). Renin acts on plasma proteins, synthesized by the liver, called **angiotensinogen** (an'jē-ō-ten-sin'ō-jen) to split a fragment off one end. The fragment, called **angiotensin** (an-jē-ō-ten'sin) **I,** contains 10 amino acids. Another enzyme, called **angiotensin-converting enzyme,** found primarily in small blood vessels of the lung, cleaves two additional amino acids from angiotensin I to produce a fragment consisting of eight amino acids called **angiotensin II,** or **active angiotensin.**

Angiotensin II causes vasoconstriction in arterioles and to some degree in veins. As a result, it increases peripheral resistance and venous return to the heart, both of which function to raise blood pressure. Angiotensin II also stimulates aldosterone secretion from the adrenal cortex. **Aldosterone** (al-dos'ter-ōn) acts on the kidneys to increase the reabsorption of sodium and chloride ions from the filtrate into the extracellular fluid. If antidiuretic hormone (ADH; see chapter 18) is present, water moves by osmosis with the sodium and chloride ions. Consequently, aldosterone causes the kidney to retain solutes such as sodium and chloride ions and water. The results are to decrease the production of urine and to conserve water to prevent further reduction in blood volume caused by the formation of urine. If water intake is adequate, the effect of aldosterone is to increase blood volume (see chapter 26). Angiotensin II also increases the salt appetite, thirst, and ADH secretion.

Decreased blood pressure stimulates renin secretion and increased blood pressure decreases renin secretion. The renin-angiotensin-aldosterone mechanism is important in maintaining blood pressure on a daily basis. It also reacts strongly under conditions of circulatory shock, but it requires many hours to become

1. Decreased blood pressure is detected by the kidneys, resulting in increased renin secretion.

2. Renin converts angiotensinogen, a protein secreted from the liver, to angiotensin I.

3. Angiotensin-converting enzyme in the lungs converts angiotensin I to angiotensin II.

4. Angiotensin II is a potent vasoconstrictor, resulting in increased blood pressure.

5. Angiotensin II stimulates the adrenal cortex to secrete aldosterone.

6. Aldosterone acts on the kidneys to increase Na+ reabsorption. As a result, urine volume decreases and blood volume increases, resulting in increased blood pressure.

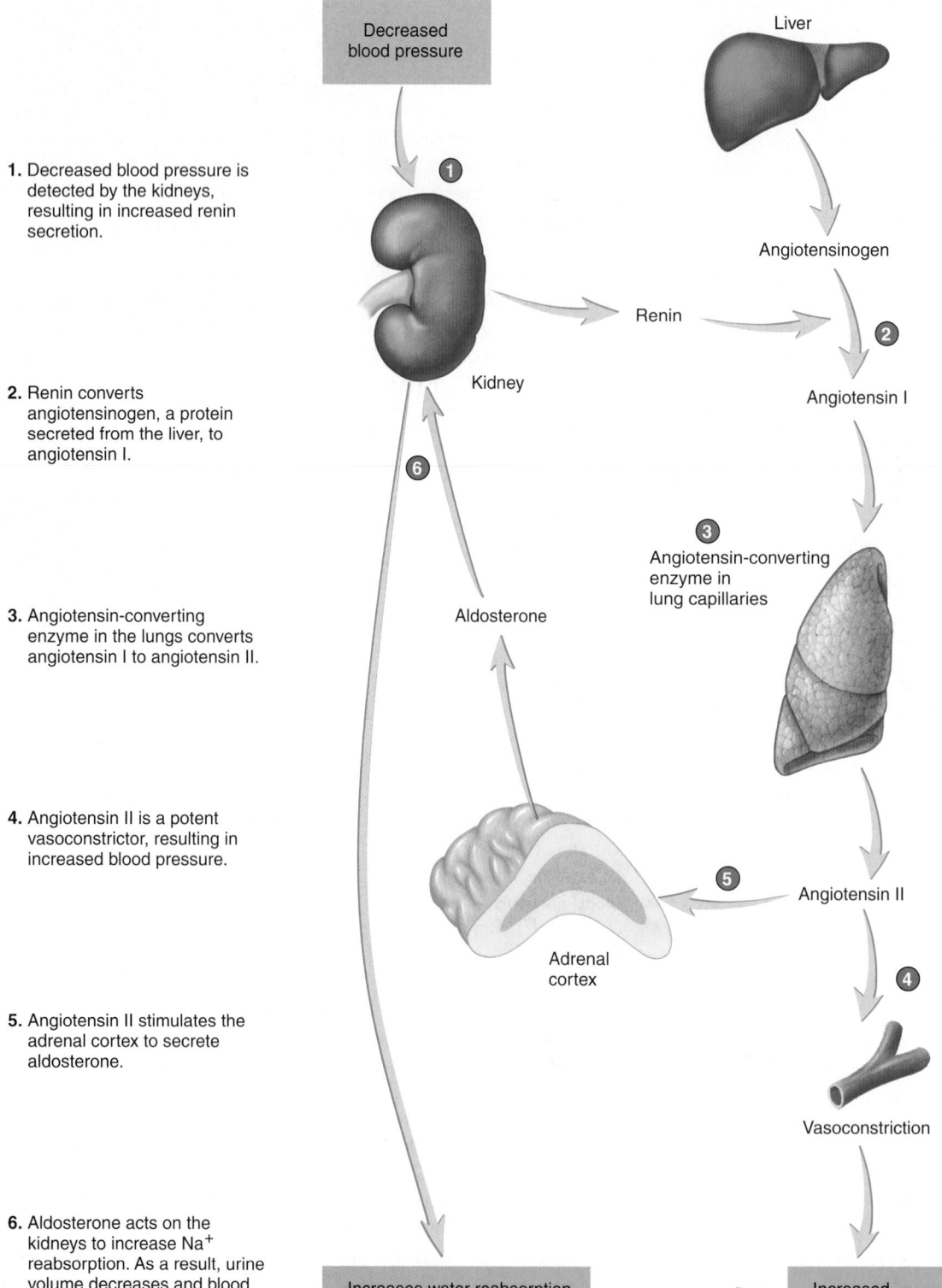

Process Figure 21.42   The Renin-Angiotensin-Aldosterone Mechanism

Decreased blood pressure is detected by the kidney, resulting in increased renin secretion. The result is vasoconstriction, increased water reabsorption, and decreased urine volume. These changes function to maintain blood pressure.

maximally effective. Its onset of action is not as fast as nervous reflexes or the adrenal medullary response, but its duration of action is longer. Once renin is secreted, it remains active for approximately 1 hour and the effect of aldosterone is much longer (many hours).

Some stimuli can directly stimulate aldosterone secretion. For example, an increased plasma ion concentration of K^+ and a reduced plasma concentration of Na^+ directly stimulate aldosterone secretion from the adrenal cortex (see chapters 18 and 27). The action of aldosterone is to regulate the concentration of these ions in the plasma. A decreased blood pressure and elevated K^+ concentration occur during plasma loss, dehydration, and in response to tissue damage, such as burns and crushing injuries.

ACE Inhibitors and Hypertension

Angiotensin-converting enzyme (ACE) inhibitors are a class of drugs that inhibit angiotensin-converting enzyme, which converts angiotensin I to angiotensin II. These drugs were first identified as components of the venoms of pit vipers. Subsequently, several ACE inhibitors were synthesized. ACE inhibitors are effective in lowering blood pressure in many people who suffer from hypertension and have become one of the drugs commonly administered to people to combat hypertension.

Vasopressin (ADH) Mechanism

The **vasopressin mechanism** works in harmony with the renin-angiotensin-aldosterone mechanism in response to changes in blood pressure (figure 21.43). Baroreceptors are sensitive to changes in blood pressure, and decreases in blood pressure detected by the baroreceptors result in the release of vasopressin (vā-sō-pres′in, vas-ō-pres′in), or ADH, from the posterior pituitary, although the blood pressure must decrease substantially before the mechanism is activated.

ADH acts directly on blood vessels to cause vasoconstriction, although it's not as potent as other vasoconstrictor agents. Within minutes after a rapid and substantial decline in blood pressure, ADH is released in sufficient quantities to help reestablish normal blood pressure. ADH also decreases the rate of urine production by the kidneys, thereby helping to maintain blood volume and blood pressure.

Neurons of the hypothalamus are sensitive to changes in the solute concentration of the plasma. Even small increases in the plasma concentration of solutes directly stimulate hypothalamic neurons that increase ADH secretion (see figure 21.43 and chapter 26). Increases in the concentration of the plasma and decreases in blood pressure stimulate ADH secretion. Dehydration and injuries involving plasma loss, such as extensive burns or crushing injuries, are examples.

Atrial Natriuretic Mechanism

A polypeptide called **atrial natriuretic** (ā′trē-ăl nā′trē-ū-ret′ik) **hormone** is released from cells in the atria of the heart. A major stimulus for its release is increased venous return, which stretches atrial cardiac muscle cells. Atrial natriuretic hormone acts on the kidneys to increase the rate of urine production and Na^+ loss in the urine. It also dilates arteries and veins. Loss of water and Na^+ in the urine causes the blood volume to decrease, which decreases venous return, and vasodilation results in a decrease in peripheral resistance. These effects function to cause a decrease in blood pressure.

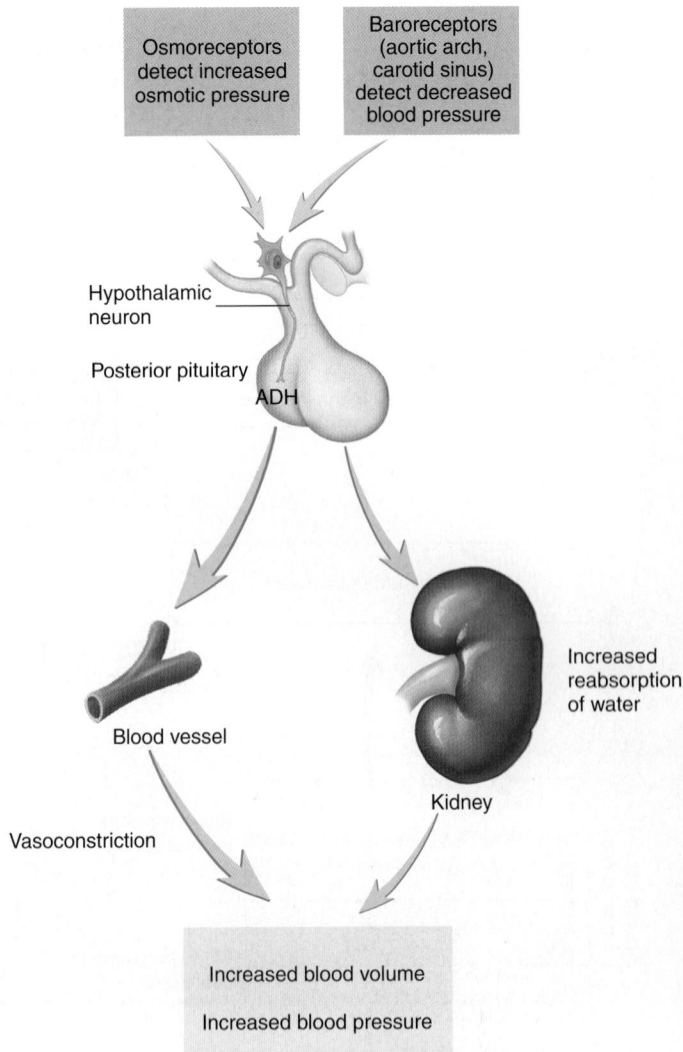

Figure 21.43 The Vasopressin (ADH) Mechanism

Increases in osmolality of blood or decreases in blood pressure result in ADH secretion. ADH increases water reabsorption by the kidney, and large amounts of ADH result in vasoconstriction. These changes function to maintain blood pressure.

The renin-angiotensin-aldosterone, vasopressin (ADH), and atrial natriuretic mechanisms work simultaneously to help regulate blood pressure by controlling urine production by the kidneys. If blood pressure drops below 50 mm Hg, the volume of urine produced by the kidneys is reduced to nearly zero. If blood pressure is increased to 200 mm Hg, the urine volume produced is approximately six to eight times greater than normal. The mechanisms that regulate blood pressure in the long term are summarized in figure 21.44.

Fluid Shift Mechanism

The **fluid shift mechanism** begins to act within a few minutes but requires hours to achieve its full functional capacity. It plays a very important role when dehydration develops over several hours, or when a large volume of saline is administered over several hours.

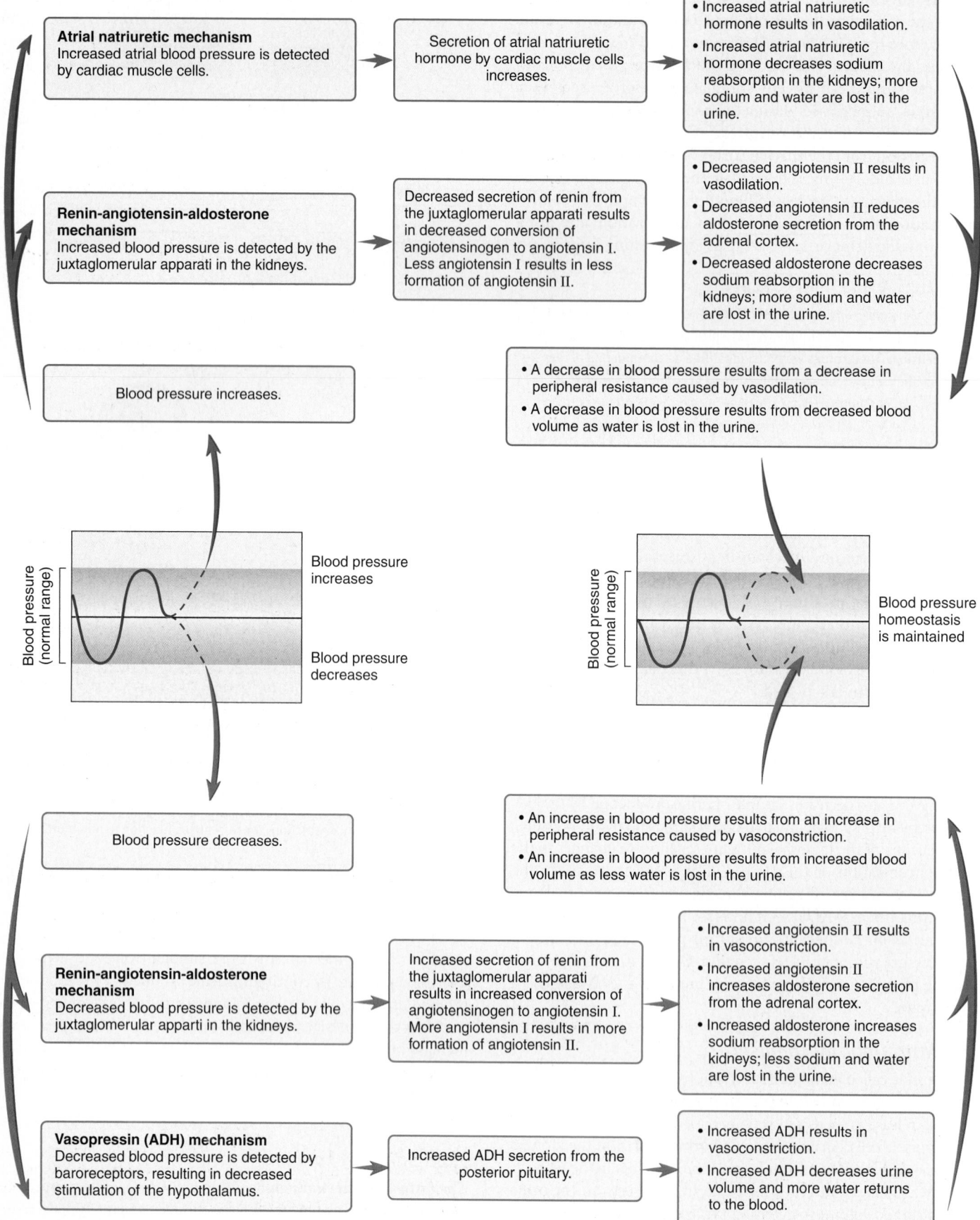

Atrial natriuretic mechanism
Increased atrial blood pressure is detected by cardiac muscle cells.

Secretion of atrial natriuretic hormone by cardiac muscle cells increases.

• Increased atrial natriuretic hormone results in vasodilation.
• Increased atrial natriuretic hormone decreases sodium reabsorption in the kidneys; more sodium and water are lost in the urine.

Renin-angiotensin-aldosterone mechanism
Increased blood pressure is detected by the juxtaglomerular apparati in the kidneys.

Decreased secretion of renin from the juxtaglomerular apparati results in decreased conversion of angiotensinogen to angiotensin I. Less angiotensin I results in less formation of angiotensin II.

• Decreased angiotensin II results in vasodilation.
• Decreased angiotensin II reduces aldosterone secretion from the adrenal cortex.
• Decreased aldosterone decreases sodium reabsorption in the kidneys; more sodium and water are lost in the urine.

Blood pressure increases.

• A decrease in blood pressure results from a decrease in peripheral resistance caused by vasodilation.
• A decrease in blood pressure results from decreased blood volume as water is lost in the urine.

Blood pressure (normal range)

Blood pressure increases

Blood pressure decreases

Blood pressure (normal range)

Blood pressure homeostasis is maintained

Blood pressure decreases.

• An increase in blood pressure results from an increase in peripheral resistance caused by vasoconstriction.
• An increase in blood pressure results from increased blood volume as less water is lost in the urine.

Renin-angiotensin-aldosterone mechanism
Decreased blood pressure is detected by the juxtaglomerular apparti in the kidneys.

Increased secretion of renin from the juxtaglomerular apparati results in increased conversion of angiotensinogen to angiotensin I. More angiotensin I results in more formation of angiotensin II.

• Increased angiotensin II results in vasoconstriction.
• Increased angiotensin II increases aldosterone secretion from the adrenal cortex.
• Increased aldosterone increases sodium reabsorption in the kidneys; less sodium and water are lost in the urine.

Vasopressin (ADH) mechanism
Decreased blood pressure is detected by baroreceptors, resulting in decreased stimulation of the hypothalamus.

Increased ADH secretion from the posterior pituitary.

• Increased ADH results in vasoconstriction.
• Increased ADH decreases urine volume and more water returns to the blood.

Homeostasis Figure 21.44 Control of Blood Pressure Long-Term (Slow-Acting) Mechanisms

The fluid shift mechanism occurs in response to small changes in pressures across capillary walls. As blood pressure increases, some fluid is forced from the capillaries into the interstitial spaces. The movement of fluid into the interstitial spaces helps prevent the development of high blood pressure. As blood pressure falls, interstitial fluid moves into capillaries, which resists a further decline in blood pressure.

The fluid shift mechanism is a powerful method through which blood pressure is maintained because the interstitial volume acts as a reservoir, and it is in equilibrium with the large volume of intercellular fluid.

Stress–Relaxation Response

A **stress–relaxation response** is characteristic of smooth muscle cells (see chapter 9). When blood volume suddenly declines, blood pressure also decreases, causing a reduction in the force applied to smooth muscle cells in blood vessel walls. As a result, during the next few minutes to an hour, the smooth muscle cells contract, reducing the volume of the blood vessels and thus resisting a further

decline in blood pressure. Conversely, when blood volume increases rapidly, such as during a transfusion, blood pressure increases and smooth muscle cells of the blood vessel walls relax, resulting in a more gradual increase in blood pressure. The stress–relaxation mechanism is most effective when changes in blood pressure occur over a period of many minutes.

61. *Define long-term regulation of blood pressure.*
62. *For each of these hormones—renin, angiotensin, aldosterone, antidiuretic hormone, and atrial natriuretic hormone—state where the hormone is produced and what effects it has on the circulatory system.*
63. *What is fluid shift, and what does it accomplish? Describe the stress–relaxation response of a blood vessel.*

P R E D I C T

Explain the differences in mechanisms that regulate blood pressure in response to hemorrhage that results in the rapid loss of a large volume of blood compared to hemorrhage that results in the loss of the same volume of blood but over a period of several hours.

S U M M A R Y

Functions of the Peripheral Circulation   (p. 724)

General Features of Blood Vessel Structure   (p. 724)

1. Blood flows from the heart through elastic arteries, muscular arteries, and arterioles to the capillaries.
2. Blood returns to the heart from the capillaries through venules, small veins, and large veins.

Capillaries

1. The entire circulatory system is lined with simple squamous epithelium called endothelium. Capillaries consist only of endothelium.
2. Capillaries are surrounded by loose connective tissue, the adventitia, that contains pericapillary cells.
3. Three types of capillaries exist.
   - Fenestrated capillaries have pores called fenestrae that extend completely through the cell.
   - Sinusoidal capillaries are large-diameter capillaries with large fenestrae.
   - Continuous capillaries do not have fenestrae.
4. Materials pass through the capillaries in several ways: between the endothelial cells, through the fenestrae, and through the plasma membrane.
5. Blood flows from arterioles through metarterioles and then through the capillary network. Venules drain the capillary network.
   - Smooth muscle in the arterioles, metarterioles, and precapillary sphincters regulates blood flow into the capillaries.
   - Blood can pass rapidly through the thoroughfare channel.

Structure of Arteries and Veins

1. Except for capillaries and venules, blood vessels have three layers. The inner tunica intima consists of endothelium, basement membrane, and internal elastic lamina.
   - The tunica media, the middle layer, contains circular smooth muscle and elastic fibers.
   - The outer tunica adventitia is connective tissue.

2. The thickness and the composition of the layers vary with blood vessel type and diameter.
   - Large elastic arteries are thin-walled with large diameters. The tunica media has many elastic fibers and little smooth muscle.
   - Muscular arteries are thick-walled with small diameters. The tunica media has abundant smooth muscle and some elastic fibers.
   - Arterioles are the smallest arteries. The tunica media consists of smooth muscle cells and a few elastic fibers.
   - Venules are composed of endothelium surrounded by a few smooth muscle cells.
   - Small veins are venules covered with a layer of smooth muscle.
   - Medium-sized veins and large veins contain less smooth muscle and fewer elastic fibers than arteries of the same size.
3. Valves prevent the backflow of blood in the veins.

Vasa Vasorum

1. Vasa vasorum are blood vessels that supply the tunica adventitia and tunica media.
2. Arteriovenous anastomoses allow blood to flow from arteries to veins without passing through the capillaries. They function in temperature regulation.

Nerves

Sympathetic nerve fibers supply the smooth muscle of the tunica media.

Aging of the Arteries

Arteriosclerosis results from a loss of elasticity in the aorta, large arteries, and coronary arteries.

Pulmonary Circulation   (p. 730)

The pulmonary circulation moves blood to and from the lungs. The pulmonary trunk arises from the right ventricle and divides to form the pulmonary arteries, which project to the lungs. From the lungs, the pulmonary veins return to the left atrium.

Systemic Circulation: Arteries   (p. 730)

Aorta

The aorta leaves the left ventricle to form the ascending aorta, aortic arch, and descending aorta (consisting of the thoracic and abdominal aortae).

Coronary Arteries

Coronary arteries supply the heart.

Arteries to the Head and the Neck

1. The brachiocephalic, left common carotid, and left subclavian arteries branch from the aortic arch to supply the head and the upper limbs. The brachiocephalic artery divides to form the right common carotid and the right subclavian arteries. The vertebral arteries branch from the subclavian arteries.
2. The common carotid arteries and the vertebral arteries supply the head.
   - The common carotid arteries divide to form the external carotids, which supply the face and mouth, and the internal carotids, which supply the brain.
   - The vertebral arteries join within the cranial cavity to form the basilar artery, which supplies the brain.

Arteries of the Upper Limb

1. The subclavian artery continues (without branching) as the axillary artery and then as the brachial artery. The brachial artery divides into the radial and ulnar arteries.
2. The radial artery supplies the deep palmar arch, and the ulnar artery supplies the superficial palmar arch. Both arches give rise to the digital arteries.

Thoracic Aorta and Its Branches

The thoracic aorta has visceral branches that supply the thoracic organs and parietal branches that supply the thoracic wall.

Abdominal Aorta and Its Branches

1. The abdominal aorta has visceral branches that supply the abdominal organs and parietal branches that supply the abdominal wall.
2. The visceral branches are paired and unpaired. The paired arteries supply the kidneys, adrenal glands, and gonads. The unpaired arteries supply the stomach, spleen, and liver (celiac trunk); the small intestine and upper part of the large intestine (superior mesenteric); and the lower part of the large intestine (inferior mesenteric).

Arteries of the Pelvis

1. The common iliac arteries arise from the abdominal aorta, and the internal iliac arteries branch from the common iliac arteries.
2. The visceral branches of the internal iliac arteries supply pelvic organs, and the parietal branches supply the pelvic wall and floor and the external genitalia.

Arteries of the Lower Limb

1. The external iliac arteries branch from the common iliac arteries.
2. The external iliac artery continues (without branching) as the femoral artery and then as the popliteal artery. The popliteal artery divides to form the anterior and posterior tibial arteries.
3. The posterior tibial artery gives rise to the fibular (peroneal) and plantar arteries. The plantar arteries form the plantar arch from which the digital arteries arise.

Systemic Circulation: Veins   (p. 741)

1. The three major veins returning blood to the heart are the superior vena cava (head, neck, thorax, and upper limbs), the inferior vena cava (abdomen, pelvis, and lower limbs), and the coronary sinus (heart).
2. Veins are of three types: superficial, deep, and sinuses.

Veins Draining the Heart

Coronary veins enter the coronary sinus or the right atrium.

Veins of the Head and Neck

1. The internal jugular veins drain the venous sinuses of the anterior head and neck.
2. The external jugular veins and the vertebral veins drain the posterior head and neck.

Veins of the Upper Limb

1. The deep veins are the small ulnar and radial veins of the forearm, which join the brachial veins of the arm. The brachial veins drain into the axillary vein.
2. The superficial veins are the basilic, cephalic, and median cubital. The basilic vein becomes the axillary vein, which then becomes the subclavian vein. The cephalic vein drains into the axillary vein.

Veins of the Thorax

The left and right brachiocephalic veins and the azygos veins return blood to the superior vena cava.

Veins of the Abdomen and Pelvis

1. Ascending lumbar veins from the abdomen join the azygos and hemiazygos veins.
2. Vessels from the kidneys, adrenal gland, and gonads directly enter the inferior vena cava.
3. Vessels from the stomach, intestines, spleen, and pancreas connect with the hepatic portal vein. The hepatic portal vein transports blood to the liver for processing. Hepatic veins from the liver join the inferior vena cava.

Veins of the Lower Limb

1. The deep veins are the fibular (peroneal), anterior and posterior tibials, popliteal, femoral, and external iliac.
2. The superficial veins are the small and great saphenous veins.

Dynamics of Blood Circulation   (p. 753)

Laminar and Turbulent Flow in Vessels

Blood flow through vessels normally is streamlined, or laminar. Turbulent flow is disruption of laminar flow.

Blood Pressure

1. Blood pressure is a measure of the force exerted by blood against the blood vessel wall. Blood moves through vessels because of blood pressure.
2. Blood pressure can be measured by listening for Korotkoff sounds produced by turbulent flow in arteries as pressure is released from a blood pressure cuff.

Blood Flow and Poiseuille's Law

1. Blood flow is the amount of blood that moves through a vessel in a given period. Blood flow is directly proportional to pressure differences and is inversely proportional to resistance.
2. Resistance is the sum of all the factors that inhibit blood flow. Resistance increases when viscosity increases and when blood vessels become smaller in diameter or increase in length.
3. Viscosity is the resistance of a liquid to flow. Most of the viscosity of blood results from red blood cells. The viscosity of blood increases when the hematocrit increases.

Critical Closing Pressure and Laplace's Law

1. As pressure in a vessel decreases, the force holding it open decreases, and the vessel tends to collapse. The critical closing pressure is the pressure at which a blood vessel closes.

2. Laplace's law states that the force acting on the wall of a blood vessel is proportional to the diameter of the vessel times blood pressure.

Vascular Compliance

Vascular compliance is a measure of the change in volume of blood vessels produced by a change in pressure. The venous system has a large compliance and acts as a blood reservoir.

Physiology of Systemic Circulation (p. 757)

The greatest volume of blood is contained in the veins. The smallest volume is in the arterioles.

Cross-Sectional Area of Blood Vessels

As the diameter of vessels decreases, their total cross-sectional area increases, and the velocity of blood flow through them decreases.

Pressure and Resistance

Blood pressure averages 100 mm Hg in the aorta and drops to 0 mm Hg in the right atrium. The greatest drop occurs in the arterioles, which regulate blood flow through tissues.

Pulse Pressure

1. Pulse pressure is the difference between systolic and diastolic pressures. Pulse pressure increases when stroke volume increases or vascular compliance decreases.
2. Pulse pressure waves travel through the vascular system faster than the blood flows. Pulse pressure can be used to take the pulse.

Capillary Exchange and Regulation of Interstitial Fluid Volume

1. Blood pressure, capillary permeability, and osmosis affect movement of fluid from the capillaries.
2. A net movement of fluid occurs from the blood into the tissues. The fluid gained by the tissues is removed by the lymphatic system.

Functional Characteristics of Veins

Venous return to the heart increases because of an increase in blood volume, venous tone, and arteriole dilation.

Blood Pressure and the Effect of Gravity

In a standing person, hydrostatic pressure caused by gravity increases blood pressure below the heart and decreases pressure above the heart.

Control of Blood Flow in Tissues (p. 762)
Local Control of Blood Flow by the Tissues

1. Blood flow through a tissue is usually proportional to the metabolic needs of the tissue. Exceptions are tissues that perform functions that require additional blood.
2. Control of blood flow by the metarterioles and precapillary sphincters can be regulated by vasodilator substances or by lack of nutrients.

3. Only large changes in blood pressure have an effect on blood flow through tissues.
4. If the metabolic activity of a tissue increases, the number and the diameter of capillaries in the tissue increases over time.

Nervous and Hormonal Regulation of Local Circulation

1. The sympathetic nervous system (vasomotor center in the medulla) controls blood vessel diameter. Other brain areas can excite or inhibit the vasomotor center.
2. Vasomotor tone is a state of partial contraction of blood vessels.
3. The nervous system is responsible for routing the flow of blood and maintaining blood pressure.
4. Sympathetic action potentials stimulate epinephrine and norepinephrine release from the adrenal medulla, and these hormones cause vasoconstriction in most blood vessels.

Regulation of Mean Arterial Pressure (p. 766)

Mean arterial pressure (MAP) is proportional to cardiac output times peripheral resistance.

Short-Term Regulation of Blood Pressure

1. Baroreceptors are sensory receptors sensitive to stretch.
   • Baroreceptors are located in the carotid sinuses and the aortic arch.
   • The baroreceptor reflex changes peripheral resistance, heart rate, and stroke volume in response to changes in blood pressure.
2. Chemoreceptors are sensory receptors sensitive to oxygen, carbon dioxide, and pH levels in the blood.
3. Epinephrine and norepinephrine are released from the adrenal medulla as a result of sympathetic stimulation. They increase heart rate, stroke volume, and vasoconstriction.
4. The CNS ischemic response results from high carbon dioxide or low pH levels in the medulla and increases peripheral resistance.

Long-Term Regulation of Blood Pressure

1. Renin-angiotensin-aldosterone mechanism. Renin is released by the kidneys in response to low blood pressure. Renin promotes the production of angiotensin II, which causes vasoconstriction and an increase in aldosterone secretion.
2. Vasopressin (ADH) mechanism. ADH released from the posterior pituitary in response to a substantial decrease in blood pressure causes vasoconstriction.
3. Atrial natriuretic mechanism. Atrial natriuretic hormone is released from the cardiac muscle cells when atrial blood pressure increases. It stimulates an increase in urinary production, causing a decrease in blood volume and blood pressure.
4. Fluid shift mechanism. Fluid shift is a movement of fluid from the interstitial spaces into capillaries in response to a decrease in blood pressure to maintain blood volume.
5. Stress–relaxation response. The stress–relaxation response is an adjustment of the smooth muscles of blood vessels in response to a change in blood volume.

R E V I E W   A N D   C O M P R E H E N S I O N

1. Given these blood vessels:
   1. arteriole
   2. capillary
   3. elastic artery
   4. muscular artery
   5. vein
   6. venule

Choose the arrangement that lists the blood vessels in the order a red blood cell passes through them as it leaves the heart, travels to a tissue, and returns to the heart.
   a. 3,4,2,1,5,6
   b. 3,4,1,2,6,5
   c. 4,3,1,2,5,6
   d. 4,3,2,1,6,5
   e. 4,2,3,5,1,6

2. Given these structures:
   1. metarteriole
   2. precapillary sphincter
   3. thoroughfare channel

   Choose the arrangement that lists the structures in the order a red blood cell encounters them as it passes through a tissue.
   a. 1,3,2
   b. 2,1,3
   c. 2,3,1
   d. 3,1,2
   e. 3,2,1

3. In which of these blood vessels are elastic fibers present in the largest amounts?
   a. large arteries
   b. medium arteries
   c. arterioles
   d. venules
   e. large veins

4. Comparing and contrasting arteries and veins, veins have
   a. thicker walls.
   b. a greater amount of smooth muscle than arteries.
   c. a tunica media and arteries do not.
   d. valves and arteries do not.
   e. all of the above.

5. The structure that supplies the walls of blood vessels with blood is the
   a. venous shunt.
   b. tunic channel.
   c. arteriovenous anastomosis.
   d. vasa vasorum.
   e. coronary artery.

6. Given these blood vessels:
   1. aorta
   2. inferior vena cava
   3. pulmonary arteries
   4. pulmonary veins

   Which vessels carry oxygen-rich blood?
   a. 1,3
   b. 1,4
   c. 2,3
   d. 2,4
   e. 3,4

7. Given these arteries:
   1. basilar
   2. common carotid
   3. internal carotid
   4. vertebral

   Which of these arteries have *direct* connections with the cerebral arterial circle (circle of Willis)?
   a. 1,2
   b. 2,4
   c. 1,3
   d. 3,4
   e. 2,3

8. Given these blood vessels:
   1. axillary artery
   2. brachial artery
   3. brachiocephalic artery
   4. radial artery
   5. subclavian artery

Choose the arrangement that lists the vessels in order, going from the aorta to the right hand.
   a. 2,5,4,1
   b. 5,2,1,4
   c. 5,3,1,4,2
   d. 3,5,1,2,4
   e. 4,5,1,2,3

9. A branch of the aorta that supplies the liver, stomach, and spleen is the
   a. celiac trunk.
   b. common iliac.
   c. inferior mesenteric.
   d. superior mesenteric.
   e. renal.

10. Given these arteries:
    1. common iliac
    2. external iliac
    3. femoral
    4. popliteal

    Choose the arrangement that lists the arteries in order going from the aorta to the knee.
    a. 1,2,3,4
    b. 1,2,4,3
    c. 2,1,3,4
    d. 2,1,4,3
    e. 3,1,2,4

11. Given these veins:
    1. brachiocephalic
    2. internal jugular
    3. superior vena cava
    4. venous sinus

    Choose the arrangement that lists the veins in order going from the brain to the heart.
    a. 1,2,4,3
    b. 2,4,1,3
    c. 2,4,3,1
    d. 4,2,1,3
    e. 4,2,3,1

12. Blood returning from the arm to the subclavian vein passes through which of these veins?
    a. cephalic
    b. basilic
    c. brachial
    d. both a and b
    e. all of the above

13. Given these blood vessels:
    1. inferior mesenteric vein
    2. superior mesenteric vein
    3. hepatic portal vein
    4. hepatic vein

    Choose the arrangement that lists the vessels in order going from the small intestine to the inferior vena cava.
    a. 1,3,4
    b. 1,4,3
    c. 2,3,4
    d. 2,4,3
    e. 3,1,4

14. Given this list of veins:
    1. small saphenous
    2. great saphenous
    3. fibular (peroneal)
    4. posterior tibial

Which are superficial veins?
a. 1,2
b. 1,3
c. 2,3
d. 2,4
e. 3,4

15. If you could increase any of these factors that affect blood flow by twofold, which one would cause the greatest increase in blood flow?
    a. blood viscosity
    b. the pressure gradient
    c. vessel radius
    d. vessel length

16. Vascular compliance is
    a. greater in arteries than in veins.
    b. the increase in vessel volume divided by the increase in vessel pressure.
    c. the pressure at which blood vessels collapse.
    d. proportional to the diameter of the blood vessel times pressure.
    e. all of the above.

17. The resistance to blood flow is greatest in the
    a. aorta.
    b. arterioles.
    c. capillaries.
    d. venules.
    e. veins.

18. Pulse pressure
    a. is the difference between systolic and diastolic pressure.
    b. increases when stroke volume increases.
    c. increases as vascular compliance decreases.
    d. all of the above.

19. Veins
    a. increase their volume because of their large compliance.
    b. increase venous return to the heart when they vasodilate.
    c. vasodilate because of increased sympathetic stimulation.
    d. all of the above.

20. Local direct control of blood flow through a tissue
    a. maintains an adequate rate of flow despite large changes in arterial blood pressure.
    b. results from relaxation and contraction of precapillary sphincters.
    c. occurs in response to a buildup in carbon dioxide in the tissues.
    d. occurs in response to a decrease in oxygen in the tissues.
    e. all of the above.

21. An increase in mean arterial pressure can result from
    a. an increase in peripheral resistance.
    b. an increase in heart rate.
    c. an increase in stroke volume.
    d. all of the above.

22. In response to an increase in mean arterial pressure, the baroreceptor reflex causes
    a. an increase in sympathetic nervous system activity.
    b. a decrease in peripheral resistance.
    c. stimulation of the vasomotor center.
    d. vasoconstriction.
    e. an increase in cardiac output.

23. When blood oxygen levels markedly decrease, the chemoreceptor reflex causes
    a. peripheral resistance to decrease.
    b. mean arterial blood pressure to increase.
    c. vasomotor tone to decrease.
    d. vasodilation.
    e. all of the above.

24. When blood pressure is suddenly decreased a small amount (10 mm Hg), which of these mechanisms are activated to restore blood pressure to normal levels?
    a. chemoreceptor reflexes
    b. baroreceptor reflexes
    c. CNS ischemic response
    d. all of the above

25. A sudden release of epinephrine from the adrenal medulla
    a. increases heart rate.
    b. increases stroke volume.
    c. causes vasoconstriction in visceral blood vessels.
    d. all of the above.

26. When blood pressure decreases,
    a. renin secretion increases.
    b. angiotensin II formation decreases.
    c. aldosterone secretion decreases.
    d. all of the above.

27. In response to a decrease in blood pressure,
    a. ADH secretion increases.
    b. the kidneys decrease urine production.
    c. blood volume increases.
    d. all of the above.

28. In response to a decrease in blood pressure,
    a. more fluid than normal enters the tissues (fluid shift mechanism).
    b. smooth muscles in blood vessels relax (stress–relaxation response).
    c. the kidneys retain more salts and water than normal.
    d. all of the above.

29. A patient is found to have severe arteriosclerosis of his renal arteries, which reduced renal blood pressure. Which of these is consistent with that condition?
    a. hypotension
    b. hypertension
    c. decreased vasomotor tone
    d. exaggerated sympathetic stimulation of the heart
    e. both a and c

30. During exercise the blood flow through skeletal muscle may increase up to 20-fold. However, the cardiac output does not increase that much. This occurs because of
    a. vasoconstriction in the viscera.
    b. vasoconstriction in the skin (at least temporarily).
    c. vasodilation of skeletal muscle blood vessels.
    d. both a and b.
    e. all of the above.

Answers in Appendix F

C R I T I C A L   T H I N K I N G

1. For each of the following destinations, name all the arteries that a red blood cell would encounter if it started its journey in the left ventricle.
   a. Posterior interventricular groove of the heart
   b. Anterior neck to the brain (give two ways)
   c. Posterior neck to the brain (give two ways)
   d. External skull
   e. Tip of the fingers of the left hand (What other blood vessel would be encountered if the trip were through the right upper limb?)
   f. Anterior compartment of the leg
   g. Liver
   h. Small intestine
   i. Urinary bladder

2. For each of the following starting places, name all the veins that a red blood cell would encounter on its way back to the right atrium.
   a. Anterior interventricular groove of the heart (give two ways)
   b. Venous sinus near the brain
   c. External posterior of skull
   d. Hand (return deep and superficial)
   e. Foot (return deep and superficial)
   f. Stomach
   g. Kidney
   h. Left inferior wall of the thorax

3. In a study of heart valve functions, it's necessary to inject a dye into the right atrium of the heart by inserting a catheter into a blood vessel and moving the catheter into the right atrium. What route would you suggest? If you wanted to do this procedure into the left atrium, what would you do differently?

4. In endurance-trained athletes, the hematocrit can be lower than normal because plasma volume increases more than red blood cell numbers increase. Explain why this condition would be beneficial.

5. All the blood that passes through the aorta, except the blood that flows into the coronary vessels, returns to the heart through the venae cavae. (*Hint:* The diameter of the aorta is 26 mm, and the diameter of a vena cava is 32 mm.) Explain why the resistance to blood flow in the aorta is greater than the resistance to blood flow in the venae cavae. Because the resistances are different, explain why blood flow can be the same.

6. As blood vessels increase in diameter, the amount of smooth muscle decreases and the amount of connective tissue increases. Explain why. (*Hint:* Remember Laplace's law.)

7. A patient is suffering from edema in the lower right limb. Explain why massage helps to remove the excess fluid.

8. A very short nursing student is asked to measure the blood pressure of a very tall person. She decides to measure the blood pressure at the level of the tall person's foot while he is standing. What artery does she use? After taking the blood pressure, she decides that the tall person is suffering from hypertension because the systolic pressure is 200 mm Hg. Is her diagnosis correct? Why or why not?

9. Mr. D. was suffering from severe cirrhosis of the liver and hepatitis. He developed severe edema over a period of time. Explain how decreased liver function can result in edema.

10. During hyperventilation, carbon dioxide is "blown off," and carbon dioxide levels in the blood decrease. What effect does this decrease have on blood pressure? Explain. What symptoms do you expect to see as a result?

11. Epinephrine causes vasodilation of blood vessels in cardiac muscle but vasoconstriction of blood vessels in the skin. Explain why this is a beneficial arrangement.

12. One cool evening, Skinny Dip jumps into a hot Jacuzzi. Predict what happens to Skinny's heart rate.

Answers in Appendix G

A N S W E R S   T O   P R E D I C T   Q U E S T I O N S

1. Arteriosclerosis slowly reduces blood flow through the carotid arteries and therefore the amount of blood that flows to the brain. As the resistance to flow increases in the carotid arteries during the late stages of arteriosclerosis, the blood flow to the brain is compromised, resulting in reduced oxygen delivery. Confusion, loss of memory, and loss of the ability to perform other normal brain functions occur.

2. a. Vasoconstriction of blood vessels in the skin in response to exposure to cold results in a decreased flow of blood through the skin and in a dramatic increase in resistance (see Poiseuille's law). Vasoconstriction makes the skin appear pale.
   b. Vasodilation of blood vessels in the skin results in increased blood flow through the skin. Vasodilation makes the skin appear flushed or red in color.
   c. In a patient with erythrocytosis, the hematocrit increases dramatically. As a result, the viscosity of the blood increases, which increases resistance to flow. Consequently, flow decreases or a greater pressure is needed to maintain the same flow.

3. An aneurysm in the aorta is a major problem because the tension applied to the aneurysm becomes greater as its size increases (see Laplace's law). The aneurysm usually develops because of a weakness in the wall of the aorta. Arteriosclerosis complicates the matter by making the wall of the artery less elastic and by increasing the systolic blood pressure. The decreased elasticity and the increased blood pressure increase the probability that the aneurysm will rupture.

4. Premature beats of the heart and ectopic beats result in contraction of the heart muscle before the heart has had time to fill to its normal capacity. Consequently, a reduced stroke volume occurs, which results in a weak pulse in response to that premature contraction. Other contractions and the resulting pulses are normal. Strong bounding pulses in a person who received too much saline solution in an intravenous transfusion result from an increase in venous return to the heart because of the increased volume of fluid in the circulatory system. Because of the increased venous return (increased preload), the heart contracts with greater force and produces a larger stroke volume. The strong bounding pulse results from the increased stroke volume. Weak pulses occur in response to hemorrhagic shock because of a decreased venous return. The heart does not fill with blood between contractions (decreased preload); the stroke volume is therefore reduced, and the pulse is weak.

5. Arterial blood pressure can increase substantially without resulting in edema. As arterial blood pressure increases the precapillary sphincters constrict to match capillary blood flow with the metabolic needs of tissues. Thus, the capillary blood pressure doesn't change substantially even though the blood pressure may increase to high levels. The blood pressure must increase above approximately 175 mm Hg before edema results. In contrast, a small increase in venous pressure leads to edema because there is no sphincter muscle that protects the capillary from an increase in pressure. Thus, small increases in venous pressure can lead to edema. Blockage of veins by blood clots or increases in venous pressure due to heart failure result in edema.

6. Keeping the legs elevated reduces the blood pressure in the capillaries of the legs because of the effect of gravity on blood flow. A major effect is that the force that moves fluid out of the capillary is decreased. As a result, the net movement of fluid out of the arterial ends decreases and the net movement into the venous ends of the capillaries increases. Therefore excess interstitial fluid is carried away from the legs. In addition, the effect of gravity increases lymph flow into the lymphatic capillaries, which also increases the rate that interstitial fluid is drained from the legs.

7. Reactive hyperemia can be explained on the basis of any of the theories for the local control of blood pressure. When a blood vessel is occluded, nutrients are depleted, and waste products accumulate in tissue that is suffering from a lack of adequate blood supply. Both of these effects cause vasodilation and a greatly increased blood flow through the area after the occlusion has been removed.

8. While this athlete is relaxing, the sympathetic stimulation of arteries in her skeletal muscles, arteries in her digestive system, and large veins decrease. As a result, vasoconstriction increases in the arteries of her muscles, and vasodilation occurs in blood vessels of her digestive system and in the large veins. Blood flow decreases to her skeletal muscles, and blood flow increases to her digestive system. In addition, more blood accumulates in the large veins. Consequently, venous return to the heart decreases, which is consistent with the reduced cardiac output.

9. During a headstand, gravity acting on the blood causes the blood pressure in the area of the aortic arch and carotid sinus baroreceptors to increase. The increased pressure activates the baroreceptor reflexes, increasing parasympathetic stimulation of the heart and decreasing sympathetic stimulation. Thus the heart rate decreases. Because standing on one's head also causes blood from the periphery to run downhill to the heart, the venous return increases, causing the stroke volume to increase because of Starling's law of the heart. Some peripheral vasodilation also can occur because the elevated baroreceptor pressure causes a decrease in vasomotor tone.

10. The baroreceptor reflex, ADH, and renin-angiotensin-aldosterone mechanisms function similarly in both cases. The fluid shift mechanism, however, is important when the loss of blood occurs over several hours, but it doesn't operate within a short period. The fluid shift mechanism plays a very important role in the maintenance of blood volume when blood loss or dehydration develops over several hours. When the blood pressure decreases, interstitial fluids pass into the capillaries, which prevents a further decline in blood pressure. The fluid shift mechanism is a powerful method through which blood pressure is maintained because the interstitial fluid acts as a fluid reservoir. The stress–relaxation mechanism responds to changes in blood pressure, but it is most responsive to sudden changes in blood volume and it responds within minutes to hours.

Visit the Online Learning Center at www.mhhe.com/seeley7 for chapter quizzes, interactive learning exercises, and other study tools.

Lymphatic System and Immunity

A macrophage (larger cell) is about to phagocytize a bacterial cell (E. coli).

One of the basic themes of life is that many organisms consume or use other organisms to survive. For example, deer graze on grasses and wolves feed on deer. A parasite lives on or in another organism called the host. The host provides the parasite with the conditions and food necessary for survival. For example, hookworms can live in the sheltered environment of the human intestine, where they feed on blood. Humans are host to many different kinds of organisms, including microorganisms, such as bacteria, viruses, fungi, and protozoans; insects; and worms. It's often the case that parasites harm humans, causing disease and sometimes death. However, our bodies have ways to resist or destroy harmful microorganisms. The lymphatic system and immunity are the body's defense systems. This chapter considers the *lymphatic system* (p. 786), *immunity* (p. 796), *innate immunity* (p. 796), *adaptive immunity* (p. 801), *immune interactions* (p. 815), *immunotherapy* (p. 817), *acquired immunity* (p. 817), and the *effects of aging on the lymphatic system and immunity* (p. 820).

Lymphatic System

The **lymphatic** (lim-fat′ik) **system** includes lymph, lymphatic vessels, lymphatic tissue, lymphatic nodules, lymph nodes, tonsils, the spleen, and the thymus (figure 22.1).

Functions of the Lymphatic System

The lymphatic system helps to maintain fluid balance in tissues and absorb fats from the digestive tract. It's also part of the body's defense system against microorganisms and other harmful substances.

1. *Fluid balance.* Approximately 30 L of fluid pass from the blood capillaries into the interstitial fluid each day, whereas only 27 L pass from the interstitial fluid back into the blood capillaries. If the extra 3 L of fluid were to remain in the interstitial fluid, edema would result, causing tissue damage and eventual death. Instead, the 3 L of fluid enters the lymphatic capillaries, where the fluid is called **lymph** (limf; clear spring water), and passes through the lymphatic vessels back to the blood (see chapter 21). In addition to water, lymph contains solutes derived from two sources: (1) substances in plasma, such as ions, nutrients, gases, and some proteins, pass from blood capillaries into the interstitial fluid and become part of the lymph; and (2) substances derived from cells, such as hormones, enzymes, and waste products, are also found in the lymph.

2. *Fat absorption.* The lymphatic system absorbs fats and other substances from the digestive tract (see chapter 24). Special lymphatic vessels called **lacteals** (lak′tē-ălz) are located in the lining of the small intestine. Fats enter the lacteals and pass through the lymphatic vessels to the venous circulation. The lymph passing through these lymphatic vessels has a milky appearance because of its fat content and is called **chyle** (kīl).

3. *Defense.* Microorganisms and other foreign substances are filtered from lymph by lymph nodes and from blood by the spleen. In addition, lymphocytes and other cells are capable of destroying microorganisms and other foreign substances.

● **1. List the parts of the lymphatic system, and describe the three main functions of the lymphatic system.**

Lymphatic Vessels

The lymphatic vessels are essential for the maintenance of fluid balance. They begin as small, dead-end tubes called **lymphatic capillaries** (figure 22.2a). Fluids tend to move out of blood capillaries into tissue spaces (see "Capillary Exchange and Regulation of Interstitial Fluid Volume" in chapter 21). Excess fluid passes through the tissue spaces and enters lymphatic capillaries to become lymph. Lymphatic capillaries are in almost all tissues of the body, with the exception of the central nervous system, the bone marrow, and tissues without blood vessels, such as cartilage,

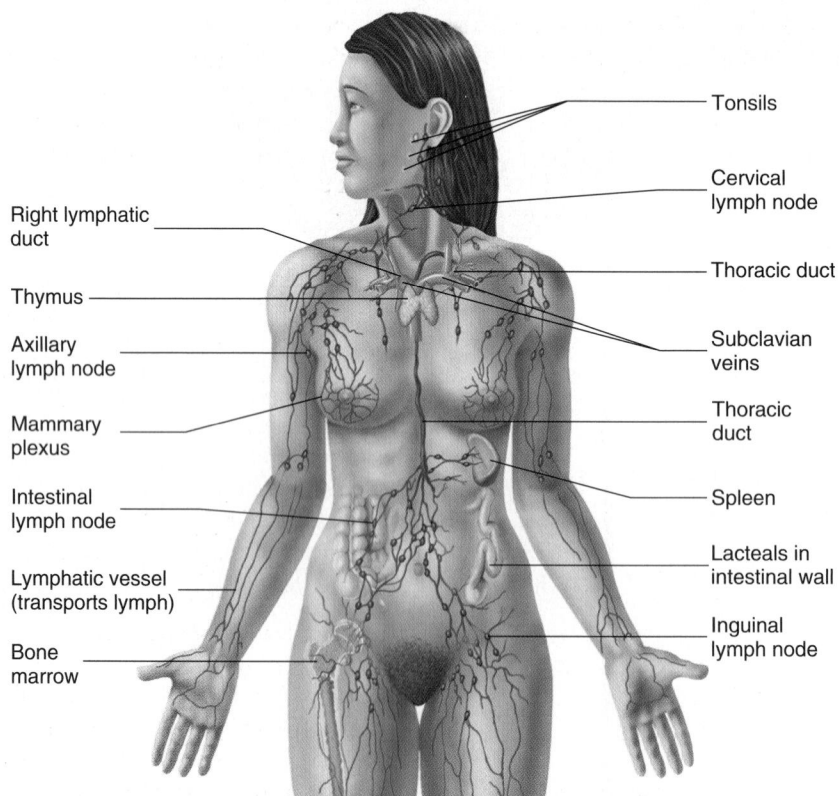

Figure 22.1　Lymphatic System
The major lymphatic organs and vessels are shown.

Tonsils

Cervical lymph node

Thoracic duct

Subclavian veins

Thoracic duct

Spleen

Lacteals in intestinal wall

Inguinal lymph node

Right lymphatic duct

Thymus

Axillary lymph node

Mammary plexus

Intestinal lymph node

Lymphatic vessel (transports lymph)

Bone marrow

Figure 22.2 **Lymph Formation and Movement**

(*a*) Movement of fluid from blood capillaries into tissues and from tissues into lymphatic capillaries to form lymph. (*b*) The overlap of epithelial cells of the lymphatic capillary allows easy entry of fluid but prevents movement back into the tissue. Valves, located farther along in lymphatic vessels, also ensure one-way flow of lymph.

epidermis, and the cornea. A superficial group of lymphatic capillaries is in the dermis of the skin and the hypodermis. A deep group of lymphatic capillaries drains the muscles, joints, viscera, and other deep structures.

Lymphatic capillaries differ from blood capillaries in that they lack a basement membrane and the cells of the simple squamous epithelium slightly overlap and are attached loosely to one another (figure 22.2*b*). Two things occur as a result of this structure. First, the lymphatic capillaries are far more permeable than blood capillaries, and nothing in the interstitial fluid is excluded from the lymphatic capillaries. Second, the lymphatic capillary epithelium functions as a series of one-way valves that allow fluid to enter the capillary but prevent it from passing back into the interstitial spaces.

The lymphatic capillaries join to form larger **lymphatic vessels,** which resemble small veins. The inner layer of the lymphatic vessel consists of endothelium surrounded by an elastic membrane, the middle layer consists of smooth muscle cells and elastic fibers, and the outer layer is a thin layer of fibrous connective tissue.

Small lymphatic vessels have a beaded appearance because of the presence of one-way valves along their lengths that are similar to the valves of veins (see figure 22.2*b*). When a lymphatic vessel is compressed, backward movement of lymph is prevented by the valves; as a consequence, the lymph moves forward through the lymphatic vessel. Three factors are responsible for the compression of lymphatic vessels: (1) contraction of surrounding skeletal muscles during activity, (2) periodic contraction of the smooth muscles in the lymphatic vessel walls, and (3) pressure changes in the thorax during respiration.

Lymph nodes are round, oval, or bean-shaped bodies distributed along the various lymphatic vessels (see "Lymph Nodes" on p. 790). They function to filter lymph, which enters and exits the lymph nodes through the lymphatic vessels. The lymph nodes are

connected together in a series, so that lymph leaving one lymph node is carried to another lymph node, and so on.

After passing through the lymph nodes, the lymphatic vessels converge to form larger vessels called **lymphatic trunks,** each of which drains a major portion of the body (figure 22.3*a* and *b*). The **jugular trunks** drain the head and neck; the **subclavian trunks** drain the upper limbs, superficial thoracic wall, and mammary glands; the **bronchomediastinal** (brong′kō-mē′dē-as-tī′năl) **trunks** drain thoracic organs and the deep thoracic wall; the **intestinal trunks** drain abdominal organs such as the intestines, stomach, pancreas, spleen, and liver; and the **lumbar trunks** drain the lower limbs, pelvic and abdominal walls, pelvic organs, ovaries or testes, kidneys, and adrenal glands.

The lymphatic trunks connect to large veins in the thorax or join to yet larger vessels called **lymphatic ducts,** which then connect to the large veins. The connections of the lymphatic trunks and ducts to veins are quite variable. Many connect at the junction of the internal jugular and subclavian veins, but connections on the subclavian, jugular, and even brachiocephalic vein exist.

On the right side, the jugular, subclavian, and bronchomediastinal trunks typically join a right thoracic vein separately (see figure 22.3*b*). About 20% of the time, the three trunks join together to form a short duct 1 cm in length called the **right lymphatic duct** (see figure 22.1; not shown in figure 22.3*b*), which joins a right thoracic vein. These trunks drain the right side of the head, right-upper limb, and right thorax (figure 22.3*c*).

The right side of the body inferior to the thorax and the entire left side of the body (see figure 22.3*c*) mostly drain through the **thoracic duct** (see figure 22.3*a* and *b*). The thoracic duct is the largest lymphatic vessel. It is approximately 38–45 cm in length, extending from the twelfth thoracic vertebra to the base of the neck (see figure 22.3*c*). The jugular and subclavian trunks join the thoracic duct. The

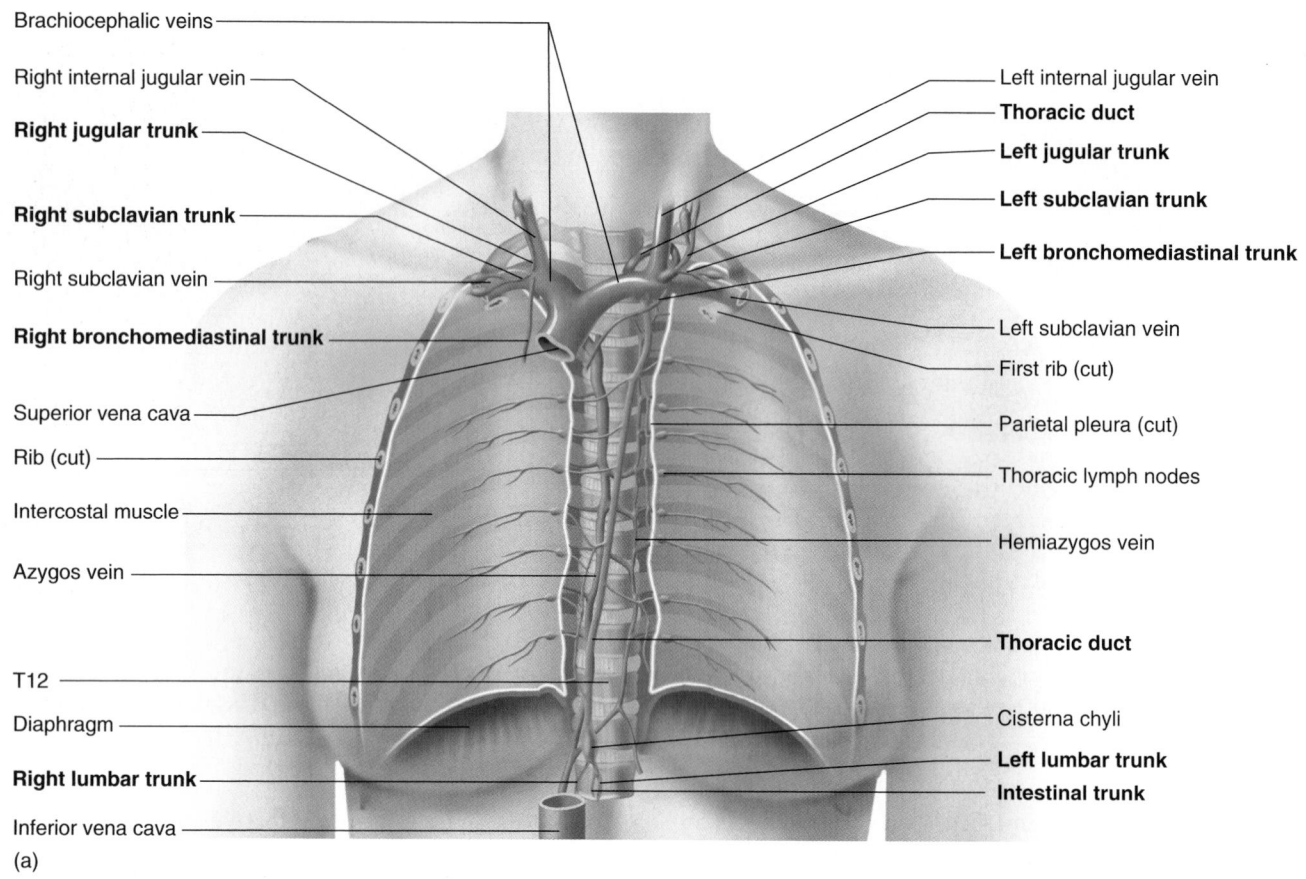

Brachiocephalic veins

Right internal jugular vein

Right jugular trunk

Right subclavian trunk

Right subclavian vein

Right bronchomediastinal trunk

Superior vena cava

Rib (cut)

Intercostal muscle

Azygos vein

T12

Diaphragm

Right lumbar trunk

Inferior vena cava

(a)

Left internal jugular vein

Thoracic duct

Left jugular trunk

Left subclavian trunk

Left bronchomediastinal trunk

Left subclavian vein

First rib (cut)

Parietal pleura (cut)

Thoracic lymph nodes

Hemiazygos vein

Thoracic duct

Cisterna chyli

Left lumbar trunk

Intestinal trunk

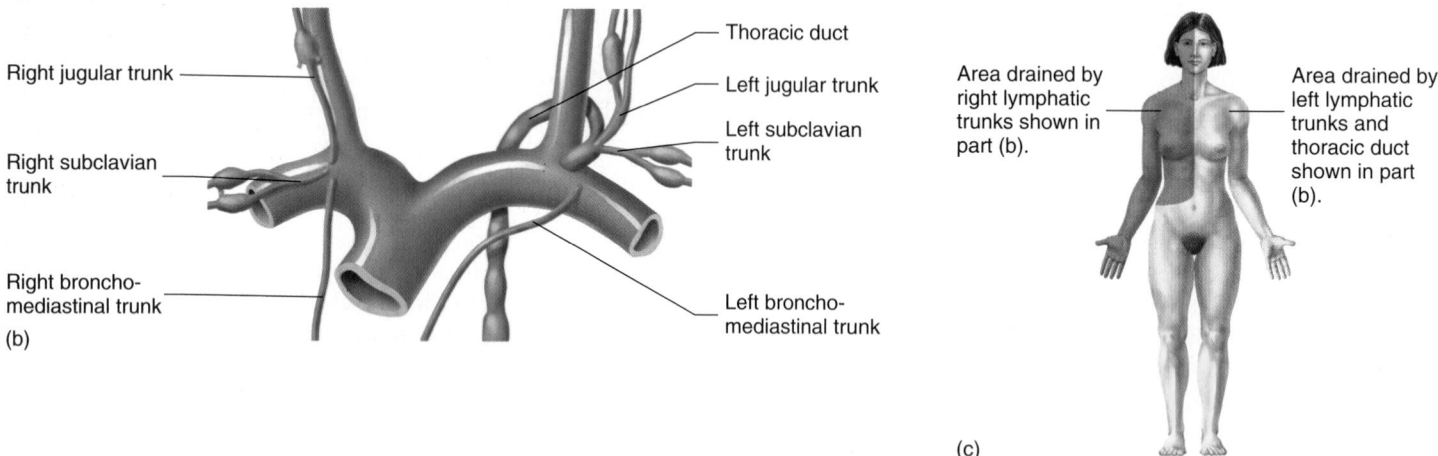

Right jugular trunk

Right subclavian trunk

Right broncho-mediastinal trunk

(b)

Thoracic duct

Left jugular trunk

Left subclavian trunk

Left broncho-mediastinal trunk

Area drained by right lymphatic trunks shown in part (b).

Area drained by left lymphatic trunks and thoracic duct shown in part (b).

(c)

Figure 22.3   Lymph Drainage Into Veins

(a) Anterior view of the major lymphatic vessels in the thorax and abdomen. (b) Close-up view of the lymphatic vessels from which lymph enters the blood.
(c) Regions of the body drained by the right and left lymphatic vessels.

bronchomediastinal trunk sometimes connects to the thoracic duct, but typically joins a vein. The intestinal and lumbar trunks, which drain lymph inferior to the diaphragm, supply the inferior end of the thoracic duct. They can directly join the thoracic duct or merge to form a network that connects to the thoracic duct. In a small proportion of cases, the lymphatic trunks form a sac called the **cisterna chyli** (sis-ter′nă kī′lī; a cistern or tank that contains juice).

In summary, lymph enters lymphatic capillaries, which converge to form larger lymphatic vessels. Lymph passes through the lymphatic vessels and through associated lymph nodes, where it is filtered. Lymphatic vessels converge to form larger lymphatic trunks, which drain lymph from major regions of the body. Lymphatic trunks empty directly into thoracic veins or combine to form larger lymphatic ducts, which empty into thoracic veins.

2. How is lymph formed?

3. Describe the structure of a lymphatic capillary. Why is it easy for fluid and other substances to enter a lymphatic capillary?

4. What is the function of the valves in lymphatic vessels? Name three things that cause lymph to move through the lymphatic vessels.

5. What are lymphatic trunks and ducts? Name the largest lymphatic vessel. What is the cisterna chyli?

6. What areas of the body are drained by the right lymphatic trunks, left lymphatic trunks, and thoracic duct?

P R E D I C T

During radical cancer surgery, malignant lymph nodes are often removed, and the lymphatic vessels to them are tied off to prevent metastasis, or spread, of the cancer. Predict the consequences of tying off the lymphatic vessels.

Lymphatic Tissue and Organs

Lymphatic organs contain **lymphatic tissue,** which consists primarily of lymphocytes; but it also includes macrophages, dendritic cells, reticular cells, and other cell types. **Lymphocytes** are a type of white blood cell (see chapter 19). They originate from red bone marrow and are carried by the blood to lymphatic organs and other tissues. When the body is exposed to microorganisms or foreign substances, the lymphocytes divide, increase in number, and are part of the immune response that destroys microorganisms and foreign substances. Lymphatic tissue also has very fine collagen fibers, called **reticular fibers,** which are produced by **reticular cells.** Lymphocytes and other cells attach to these fibers. When lymph or blood filters through lymphatic organs, the fiber network traps microorganisms and other particles in the fluid.

Lymphatic tissue surrounded by a connective tissue capsule is said to be encapsulated, whereas lymphatic tissue without a capsule is called nonencapsulated. Lymphatic organs with a capsule include lymph nodes, the spleen, and the thymus. **Mucosa-associated lymphoid tissue (MALT)** is aggregates of nonencapsulated lymphatic tissue found in and beneath the mucous membranes lining the digestive, respiratory, urinary, and reproductive tracts. In these locations, the lymphatic tissue is well located to intercept microorganisms as they enter the body. Examples of MALT include diffuse lymphatic tissue, lymphatic nodules, and the tonsils.

Diffuse Lymphatic Tissue and Lymphatic Nodules

Diffuse lymphatic tissue contains dispersed lymphocytes, macrophages, and other cells; has no clear boundary; and blends with surrounding tissues (figure 22.4). It is located deep to mucous membranes, around lymphatic nodules, and within the lymph nodes and spleen.

Lymphatic nodules are denser arrangements of lymphoid tissue organized into compact, somewhat spherical structures, ranging in size from a few hundred microns to a few millimeters or more in diameter (see figure 22.4). Lymphatic nodules are numerous in the loose connective tissue of the digestive, respiratory, urinary, and reproductive systems. **Peyer's patches** are aggregations of lymphatic nodules found in the distal half of the small intestine

Figure 22.4 Diffuse Lymphatic Tissue and Lymphatic Nodule

Diffuse lymphatic tissue surrounding a lymphatic nodule in the small intestine (Peyer's patch).

and the appendix. In addition to MALT, lymphatic nodules are found within lymph nodes and the spleen, where they are usually referred to as **lymphatic follicles.**

Tonsils

Tonsils are large groups of lymphatic nodules and diffuse lymphatic tissue located deep to the mucous membranes within the pharynx (throat) (figure 22.5). The tonsils provide protection against bacteria and other potentially harmful material entering the pharynx from the nasal or oral cavities. In adults, the tonsils decrease in size and eventually may disappear.

There are three groups of tonsils, but the **palatine tonsils** usually are referred to as "the tonsils." They are relatively large, oval lymphoid masses on each side of the junction between the oral cavity and the pharynx. The **pharyngeal** (fă-rin′jē-ăl) **tonsil,** or adenoid

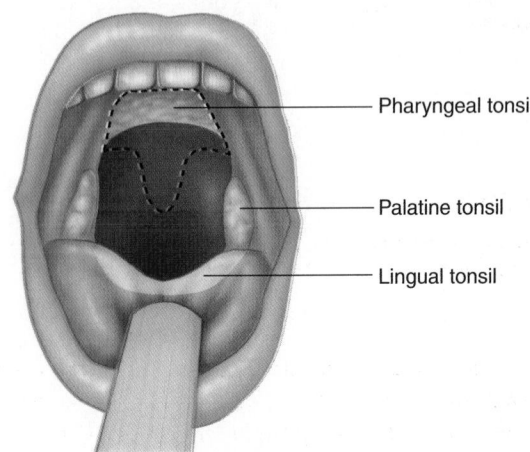

Figure 22.5 Location of the Tonsils

Anterior view of the oral cavity showing the tonsils. Part of the palate is removed (*dotted line*) to show the pharyngeal tonsil.

(ad′ĕ-noyd), is a collection of somewhat closely aggregated lymphatic nodules near the junction between the nasal cavity and the pharynx. An enlarged pharyngeal tonsil can interfere with normal breathing. The **lingual tonsil** is a loosely associated collection of lymphatic nodules on the posterior surface of the tongue.

Sometimes the palatine or pharyngeal tonsils become chronically infected and must be removed. The lingual tonsil becomes infected less often than the other tonsils and is more difficult to remove.

⦿ 7. **What are the functions of lymphocytes and reticular fibers in lymphatic tissue?**

8. **What is mucosa-associated lymphoid tissue (MALT)? In what way is the location of MALT beneficial?**

9. **Define diffuse lymphatic tissue, lymphatic nodule, Peyer's patches, and lymphatic follicle.**
10. **Describe the structure, function, and location of the tonsils.**

Lymph Nodes

Lymph nodes are small, round, or bean-shaped structures, ranging in size from 1–25 mm long, and are distributed along the course of the lymphatic vessels (see figures 22.1 and 22.6). They filter the lymph, removing bacteria and other materials. In addition, lymphocytes congregate, function, and proliferate within lymph nodes.

Lymph nodes are categorized as superficial or deep. **Superficial lymph nodes** are in the hypodermis beneath the skin and

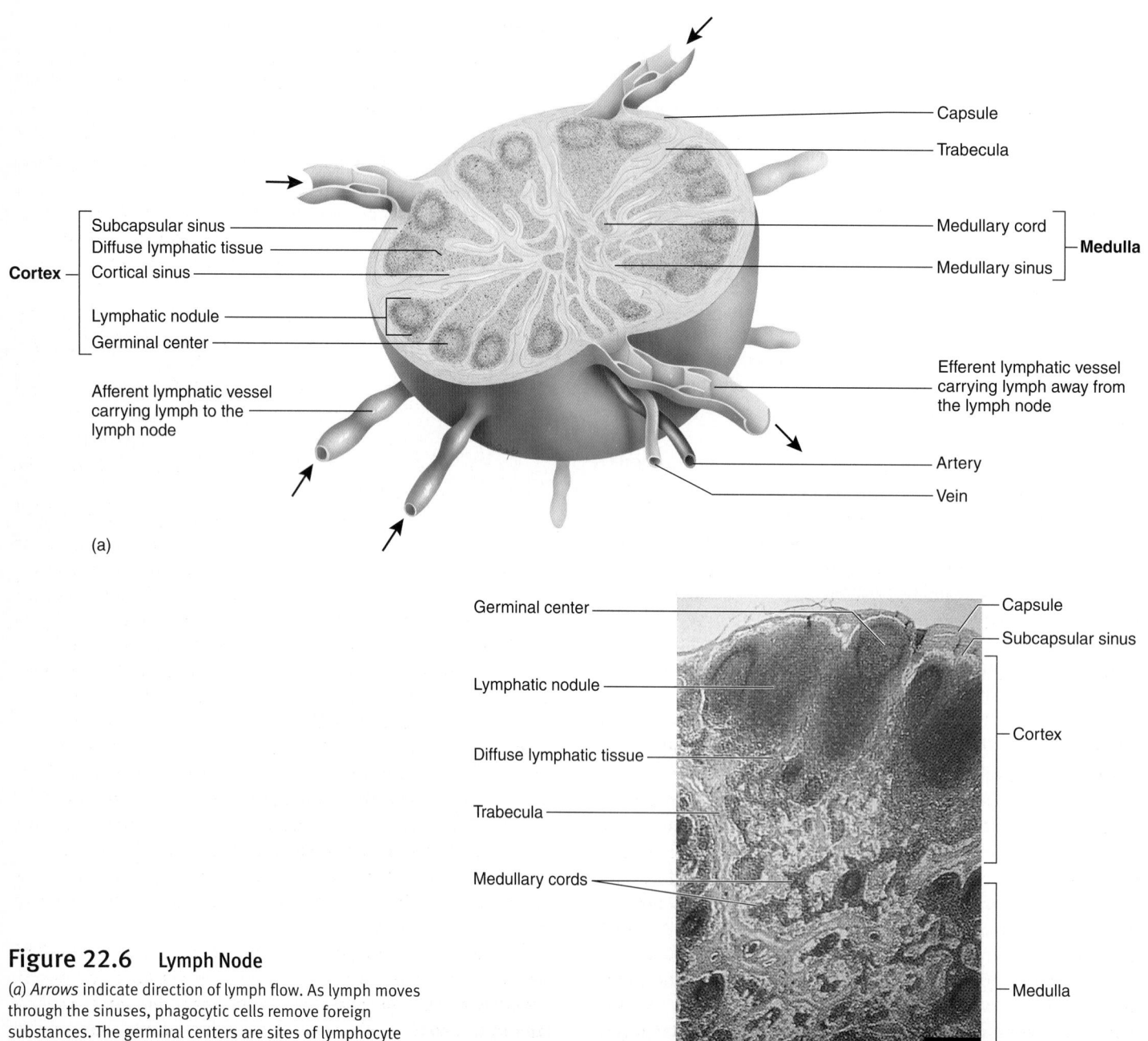

Figure 22.6   Lymph Node

(a) Arrows indicate direction of lymph flow. As lymph moves through the sinuses, phagocytic cells remove foreign substances. The germinal centers are sites of lymphocyte production. (b) Histology of a lymph node.

deep lymph nodes are everywhere else. Both superficial and deep lymph nodes typically are located in adipose tissue near or on blood vessels. Approximately 450 lymph nodes are found throughout the body. Cervical and head nodes (about 70) filter lymph from the head and neck, axillary nodes (about 30) filter lymph from the upper limbs and superficial thorax, thoracic nodes (about 100) filter lymph from the thoracic wall and organs, abdominopelvic nodes (about 230) filter lymph from the abdomen and pelvis, and inguinal and popliteal nodes (about 20) filter lymph from the lower limbs and superficial pelvis.

A dense connective tissue **capsule** surrounds each lymph node. Extensions of the capsule, called **trabeculae** (tră-bek′ū-lē), form a delicate internal skeleton in the lymph node. Reticular fibers extend from the capsule and trabeculae to form a fibrous network throughout the entire node. In some areas of the lymph node, lymphocytes and macrophages are packed around the reticular fibers to form lymphatic tissue, and in other areas the reticular fibers extend across open spaces to form **lymphatic sinuses.** The lymphatic tissue and sinuses within the node are arranged into two somewhat indistinct layers, an outer cortex and an inner medulla. The **cortex** consists of a subcapsular sinus, beneath the capsule, and cortical sinuses, which are separated by diffuse lymphatic tissue, trabeculae, and lymphatic nodules. The inner **medulla** is organized into branching, irregular strands of diffuse lymphatic tissue, the **medullary cords,** separated by medullary sinuses.

Lymph nodes are the only structures to filter lymph. They have **afferent lymphatic vessels,** which carry lymph to the lymph nodes, where it is filtered, and **efferent lymphatic vessels,** which carry lymph away from the nodes. Lymph from afferent lymphatic vessels enters the subcapsular sinus and filters through the cortex to the medulla, passing through the cortical sinuses and lymphatic tissue of the cortex. Lymph then passes through the sinuses and lymphatic tissue of the medulla and exits the lymph node through efferent lymphatic vessels. The efferent vessels of one lymph node may become the afferent vessels of another node or may converge to form lymphatic trunks, which carry lymph to the blood at thoracic blood vessels.

Macrophages line the lymphatic sinuses, and they remove bacteria and other foreign substances from the lymph as it slowly filters through the sinuses. Microorganisms or other foreign substances in the lymph can stimulate lymphocytes throughout the lymph node to undergo cell division, with proliferation especially evident in the lymphatic nodules of the cortex. These areas of rapid lymphocyte division are called **germinal centers.** The newly produced lymphocytes are released into the lymph and eventually reach the bloodstream, where they circulate. Subsequently, the lymphocytes can leave the blood and enter other lymphatic tissues.

- ● 11. *Where are lymph nodes found? Describe the parts of a lymph node and explain how lymph flows through a lymph node.*
- 12. *What are the functions of lymph nodes? How is this accomplished? What is a germinal center?*

Spleen

The **spleen,** which is roughly the size of a clenched fist, is located on the left side in the extreme, superior part of the abdominal cavity (figure 22.7). The average weight of the adult spleen is 180 g in males and 140 g in females. The size and weight of the spleen tends to decrease in older people, but in certain diseases the spleen can achieve weights of 2000 g or more.

The spleen has an outer **capsule** of dense irregular connective tissue and a small amount of smooth muscle. Bundles of connective tissue fibers from the capsule form **trabeculae,** which extend into the organ, subdividing it into small, interconnected compartments. Arteries, veins, and lymphatic vessels extend through the trabeculae to supply the compartments, which are filled with white and red pulp. **White pulp** is lymphatic tissue surrounding the arteries within the spleen. **Red pulp** is associated with the veins. It consists of a fibrous network, filled with macrophages and red blood cells, and enlarged capillaries that connect to the veins. Approximately one-fourth of the volume of the spleen is white pulp and three-fourths is red pulp.

Branches of the **splenic** (splen′ik) **artery** enter the spleen at the **hilum,** and their branches follow the various trabeculae into the spleen (see figure 22.7*a* and *b*). From the trabeculae, arterial branches extend into the white pulp, which consists of the periarterial lymphatic sheath and lymphatic nodules (see figure 22.7*c*). The **periarterial lymphatic sheath** is diffuse lymphatic tissue surrounding arteries and arterioles extending to lymphatic nodules. Arterioles enter lymphatic nodules and give rise to capillaries supplying the red pulp, which consists of the splenic cords and venous sinuses. The **splenic cords** are a network of reticular cells which produce reticular fibers (see chapter 4). The spaces between the reticular cells are occupied by splenic macrophages and blood cells that have come from the capillaries. The **venous sinuses** are enlarged capillaries between the splenic cords. The venous sinuses typically connect to trabecular veins, which unite to form vessels that leave the spleen to form the **splenic vein.**

Blood flows through the spleen at three different rates. The fast flow takes a few seconds, intermediate flow a few minutes, and slow flow an hour or more. Most blood flows through the spleen rapidly, but about 10% moves at the intermediate rate, and 2% flows at the slow rate.

1. Branches from the trabecular arteries are surrounded by periarterial lymphatic sheaths.

2. An arteriole enters a lymphatic nodule and divides.

3. A few capillaries directly connect to a venous sinus (closed, fast circulation).

4. The ends of most capillaries are separated from the beginning of the venous sinuses by a small gap (open, fast circulation).

5. Some capillaries empty into the splenic cords (open, intermediate and slow circulations). Blood percolates through the splenic cords and passes through the walls of the venous sinuses.

6. The venous sinuses connect to the trabecular vein.

Process Figure 22.7   Spleen

(*a*) Inferior view of the spleen. (*b*) Section showing the arrangement of arteries, veins, white pulp, and red pulp. White pulp is associated with arteries, and red pulp is associated with veins. (*c*) Blood flow through white and red pulp. (*d*) Histology of spleen.

The fast flow is typical of flow through organs with a **closed circulation,** in which there is a direct capillary connection between the arterial and venous vessels (see figure 22.7*c*). In the spleen, however, direct connections only occur rarely. Most circulation in the spleen is an **open circulation,** in which there is no direct capillary connection between the arterial and venous vessels. Instead, blood empties into the boundary between the white and red pulp or into the splenic cords. In most cases, blood quickly enters into the open ends of venous sinuses, which originate near the boundary. Other-wise, the blood percolates through the splenic cords and passes through the walls of the venous sinuses, which have intercellular slits. The fast flow through the spleen results from blood moving through the closed circulation or quickly into the open ends of the venous sinuses in the open circulation. The intermediate flow is the passage of blood through the splenic cords and through the walls of the venous sinuses. The slow flow follows the same pathway as the intermediate flow, but it takes longer because of the temporary adhesion of blood cells to splenic cord cells.

The spleen destroys defective red blood cells, detects and responds to foreign substances in the blood, and acts as a blood reservoir. As red blood cells age, they lose their ability to bend and fold. Consequently, the cells can rupture as they pass through the meshwork of the splenic cords or the intercellular slits of the venous sinus walls. Splenic macrophages then phagocytize the cellular debris.

Foreign substances in the blood passing through the spleen can stimulate an immune response because of the presence in the white pulp of specialized lymphocytes, described later in this chapter. There are high concentrations of T cells in the periarterial lymphatic sheath and B cells in the lymphatic nodules.

The human spleen is a limited reservoir for blood. For example, during exercise splenic volume can be reduced by approximately 40%–50%. The resulting small increase in circulating red blood cells can promote better oxygen delivery to muscles during exercise or emergency situations. It's not presently known if in humans this reduction results from contraction of smooth muscle within the capsule, from contraction of smooth muscle (myofibroblast) within the trabeculae, or from reduced blood flow through the spleen caused by constriction of blood vessels.

The Ruptured Spleen

Although the spleen is protected by the ribs, it is often ruptured in traumatic abdominal injuries. A ruptured spleen can cause severe bleeding, shock, and possibly death. Surgical intervention may stop the bleeding. Cracks in the spleen are repaired using sutures and blood-clotting agents. Mesh wrapped around the spleen can hold it together. A **splenectomy** (splē-nek′tō-mē), removal of the spleen, may be necessary if these techniques do not stop the bleeding. Other lymphatic organs and the liver compensate for the loss of the spleen's functions.

● **13. Where is the spleen located? Name the two components of white pulp. Of red pulp.**
  14. Explain the fast, intermediate, and slow flow of blood through the spleen.
  15. What are three functions of the spleen?

Thymus

The **thymus** (thī′mŭs) is a bilobed gland (figure 22.8) located in the superior mediastinum, the partition dividing the thoracic cavity into left and right parts. It was once thought that the thymus

Figure 22.8 Thymus

(a) Location and shape of the thymus.
(b) Section showing a thymic lobule.
(c) Histology of the thymus, showing outer cortex and inner medulla.

Clinical Focus   Disorders of the Lymphatic System

It's not surprising that many infectious diseases produce symptoms associated with the lymphatic system. The lymphatic system is involved with the production of lymphocytes, which fight infectious diseases, and the lymphatic system filters blood and lymph to remove microorganisms. **Lymphadenitis** (lim-fad´ĕ-nī´tis) is an inflammation of the lymph nodes, which causes them to become enlarged and tender. This inflammation is an indication that microorganisms are being trapped and destroyed within the lymph nodes. Sometimes the lymphatic vessels become inflamed to produce **lymphangitis** (lim-fan-jī´tis). This often results in visible red streaks in the skin that extend away from the site of infection. If microorganisms pass through the lymphatic vessels and nodes to reach the blood, **septicemia** (sep-ti-sē´mē-ă), or blood poisoning, can result (see chapter 19).

Bubonic plague and elephantiasis are diseases of the lymphatic system. **Bubonic** (boo-bon´ik) **plague** is caused by bacteria *(Yersinia pestis)*, which are transferred to humans from rats by the bite of the rat flea *(Xenopsylla)*. The bacteria localize in the lymph nodes, causing them to enlarge. The term *bubonic* is derived from a Greek word referring to the groin because the disease often causes the inguinal lymph nodes of the groin to swell. Without treatment, the bacteria enter the blood, multiply, and infect tissues throughout the body, rapidly causing death in 70%–90% of those infected. In the sixth, fourteenth, and nineteenth centuries, the bubonic plague killed large numbers of people in Europe. Because of improved sanitation and the advent of antibiotics, fortunately relatively few cases occur today. **Elephantiasis** (el-ĕ-fan-tī´ă-sis) is caused by long, slender roundworms *(Wuchereria bancrofti)*. The adult worms lodge in the lymphatic vessels and block lymph flow. The resulting accumulation of fluid in the interstitial spaces and lymphatic vessels can cause permanent swelling and enlargement of a limb. The affected limb supposedly resembles an elephant's leg, providing the basis for the name of the disease. The offspring of the adult worms pass through the lymphatic system into the blood, from which they can be transferred to another human by mosquitoes.

A **lymphoma** (lim-fō´mă) is a neoplasm (tumor) of lymphatic tissue. Lymphomas are usually divided into two groups: (1) Hodgkin's disease and (2) all other lymphomas, which are called non-Hodgkin's lymphomas. Typically, lymphomas begin as an enlarged, painless mass of lymph nodes. The immune system is depressed, and the patient has an increased susceptibility to infections. Enlargement of the lymph nodes can also compress surrounding structures and produce complications. Fortunately, treatment with drugs and radiation is effective for many people who suffer from lymphoma.

---

increases in size until puberty, after which it dramatically decreases in size. It's now believed that the thymus increases in size until the first year of life, after which it remains approximately the same size, even though the size of the individual increases. After 60 years of age, it decreases in size, and in older adults, the thymus may be so small that it is difficult to find during dissection. Although the size of the thymus is fairly constant throughout much of life, by 40 years of age much of the thymic lymphatic tissue has been replaced with adipose tissue.

Each lobe of the thymus is surrounded by a thin connective tissue **capsule. Trabeculae** extend from the capsule into the substance of the gland, dividing it into **lobules.** Unlike other lymphatic tissue, which has a fibrous network of reticular fibers, the framework of thymic tissue consists of epithelial cells. The processes of the epithelial cells are joined by desmosomes, and the cells form small, irregularly shaped compartments filled with lymphocytes. Near the capsule and trabeculae, the lymphocytes are numerous and form dark staining areas of the lobules called the **cortex.** A lighter staining central portion of the lobules, called the **medulla,** has fewer lymphocytes. The medulla also contains rounded epithelial structures, called **thymic corpuscles** (Hassall's corpuscles), whose function is unknown.

The thymus is the site of maturation of certain lymphocytes called T cells. Large numbers of lymphocytes are produced in the thymus, but most degenerate. The lymphocytes that survive the maturation process are capable of reacting to foreign substances, but they normally do not react to and destroy healthy body cells (see the "Origin and Development of Lymphocytes" on p. 802). These surviving thymic lymphocytes migrate to the medulla, enter the blood, and travel to other lymphatic tissues.

> **16.** *Where is the thymus located? Describe its structure. What is the blood–thymic barrier? How is it related to the function of the thymus?*

Overview of the Lymphatic System

Figure 22.9 summarizes the parts of the lymphatic system and their functions. Lymphatic capillaries and vessels remove fluid from tissues and absorb fats from the small intestines. Lymph nodes filter lymph and the spleen filters blood.

Figure 22.9 also illustrates two types of lymphocytes called B and T cells. B cells originate and mature in red bone marrow. Pre-T cells are produced in red bone marrow and migrate to the thymus, where they mature to become T cells. B cells from red bone marrow and T cells from the thymus circulate to, and populate, other lymphatic tissues.

B and T cells are responsible for much of immunity. In response to infections, B and T cells increase in number and circulate to lymphatic and infected tissues. How B and T cells protect us is discussed in the section on "Adaptive Immunity" on p. 801.

(a) Lymphatic capillaries remove fluid from tissues. The fluid becomes lymph (see figure 22.2a).

(b) Lymph flows through lymphatic vessels, which have valves that prevent the backflow of lymph (see figure 22.2b).

(c) Lymph nodes filter lymph (see figure 22.6) and are sites where lymphocytes respond to infections, etc.

(d) Lymph enters lymphatic trunks and ducts (see figure 22.3b).

(e) Lymph enters the blood.

(f) Lacteals in the small intestine (see figure 24.16c) absorb fats, which enter the thoracic duct (see figure 22.3a).

(g) Chyle, which is lymph containing fats, enters the blood.

(h) The spleen (see figure 22.7) filters blood and is a site where lymphocytes respond to infections, etc.

(i) Lymphocytes (pre-B and pre-T cells) originate from stem cells in the red bone marrow (see figure 22.12). The pre-B cells become mature B cells in the red bone marrow and are released into the blood. The pre-T cells enter the blood and migrate to the thymus.

(j) The thymus (see figure 22.8) is where pre-T cells derived from red bone marrow increase in number and become mature T cells that are released into the blood (see figure 22.12).

(k) B and T cells from the blood enter and populate all lymphatic tissues. These lymphocytes can remain in the lymphatic tissues or pass through them and return to the blood. B and T cells can also respond to infections, etc., by dividing and increasing in number (see figures 22.18 and 22.22).

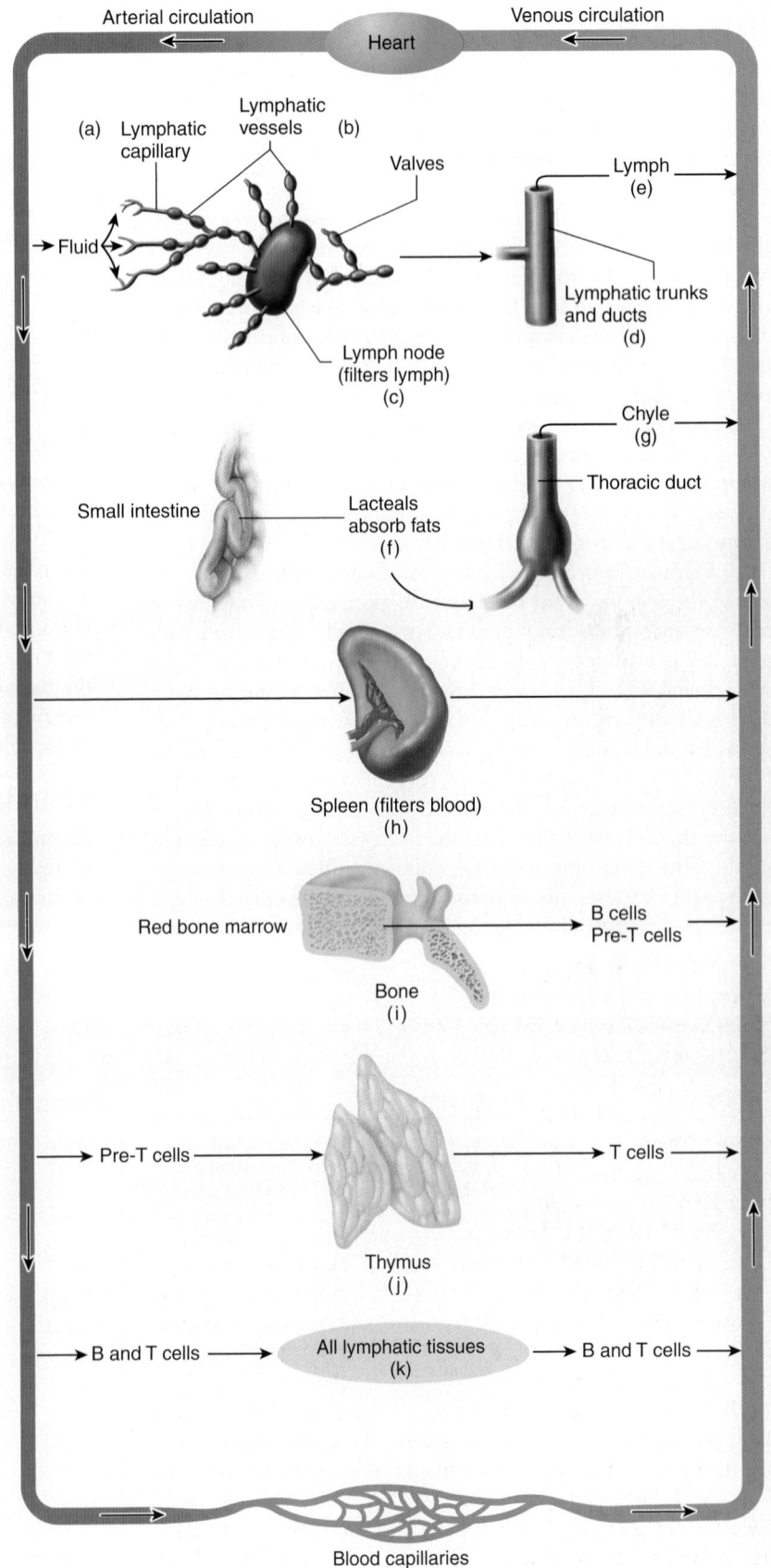

Figure 22.9 Overview of the Lymphatic System

Immunity

Immunity is the ability to resist damage from foreign substances such as microorganisms and harmful chemicals. Immunity is categorized as **innate immunity** (also called nonspecific resistance) or **adaptive immunity** (also called specific immunity). In innate immunity, the body recognizes and destroys certain foreign substances, but the response to them is the same each time the body is exposed to them. In adaptive immunity, the body recognizes and destroys foreign substances, but the response to them improves each time the foreign substance is encountered.

Specificity and memory are characteristics of adaptive immunity but not innate immunity. **Specificity** is the ability of adaptive immunity to recognize a particular substance. For example, innate immunity can act against bacteria in general, whereas adaptive immunity can distinguish among different kinds of bacteria. **Memory** is the ability of adaptive immunity to "remember" previous encounters with a particular substance. As a result, the response is faster, stronger, and lasts longer.

In innate immunity, each time the body is exposed to a substance, the response is the same because specificity and memory of previous encounters is not present. For example, each time a bacterial cell is introduced into the body, it is phagocytized with the same speed and efficiency. In adaptive immunity, the response during the second exposure is faster and stronger than the response to the first exposure because the immune system remembers the bacteria from the first exposure. For example, following the first exposure to the bacteria, the body can take many days to destroy them. During this time, the bacteria damage tissues and produce the symptoms of disease. Following the second exposure to the same bacteria, the response is rapid and effective. Bacteria are destroyed before any symptoms develop, and the person is said to be **immune.**

17. *Define the terms immunity, specificity, and memory.*
18. *What are the differences between innate and adaptive immunity?*

Innate Immunity

The main components of innate immunity include (1) mechanical mechanisms that prevent the entry of microbes into the body or that physically remove them from body surfaces; (2) chemical mediators that act directly against microorganisms or that activate other mechanisms, leading to the destruction of the microorganisms; and (3) cells involved in phagocytosis and the production of chemicals that participate in the response of the immune system.

Mechanical Mechanisms

Mechanical mechanisms, such as the skin and mucous membranes, form barriers that prevent the entry of microorganisms and chemicals into the tissues of the body. They also remove microorganisms and other substances from the surface of the body in several ways. The substances are washed from the eyes by tears, from the mouth by saliva, and from the urinary tract by urine. In the respiratory tract, ciliated mucous membranes sweep microbes trapped in the mucus to the back of the throat, where they are swallowed. Coughing and sneezing also remove microorganisms from the respiratory tract. Microorganisms cannot cause disease if they cannot get into the body.

Chemical Mediators

Chemical mediators are molecules responsible for many aspects of innate immunity (table 22.1). Some chemical mediators found on the surface of cells, such as lysozyme, sebum, and mucus, kill microorganisms or prevent their entry into the cells. Other chemical mediators, such as histamine, complement, and eicosanoids

Table 22.1	Chemicals of Innate Immunity and Their Functions		
Chemical	**Description**	**Chemical**	**Description**
Surface chemicals	Lysozymes (in tears, saliva, nasal secretions, and sweat) lyse cells; acid secretions (sebum in the skin and hydrochloric acid in the stomach) prevent microbial growth or kill microorganisms; mucus on the mucous membranes traps microorganisms until they can be destroyed	Complement	A group of plasma proteins that increase vascular permeability, stimulate the release of histamine, activate kinins, lyse cells, promote phagocytosis, and attract neutrophils, monocytes, macrophages, and eosinophils
Histamine	An amine released from mast cells, basophils, and platelets; histamine causes vasodilation, increases vascular permeability, stimulates gland secretions (especially mucus and tear production), causes smooth muscle contraction of airway passages (bronchioles) in the lungs, and attracts eosinophils	Prostaglandins	A group of lipids (PGEs, PGFs, thromboxanes, and prostacyclins), produced by mast cells, that cause smooth muscle relaxation and vasodilation, increase vascular permeability, and stimulate pain receptors
Kinins	Polypeptides derived from plasma proteins; kinins cause vasodilation, increase vascular permeability, stimulate pain receptors, and attract neutrophils	Leukotrienes	A group of lipids, produced primarily by mast cells and basophils, that cause prolonged smooth muscle contraction (especially in the lung bronchioles), increase vascular permeability, and attract neutrophils and eosinophils
Interferon	A protein, produced by most cells, that interferes with virus production and infection	Pyrogens	Chemicals, released by neutrophils, monocytes, and other cells, that stimulate fever production

Abbreviations: PGE = prostaglandin E; PGF = prostaglandin F.

(e.g., prostaglandins and leukotrienes), promote inflammation by causing vasodilation, increasing vascular permeability, attracting white blood cells, and stimulating phagocytosis. In addition, interferons protect cells against viral infections.

Complement

Complement is a group of about 20 proteins that make up approximately 10% of the globulin part of serum. They include proteins

named C1–C9 and factors B, D, and P (properdin). Normally, complement proteins circulate in the blood in an inactive, nonfunctional form. They become activated in the **complement cascade,** a series of reactions in which each component of the series activates the next component (figure 22.10). The complement cascade begins through either the alternative pathway or the classical pathway. The **alternative pathway** is part of innate immunity and is initiated when the complement protein C3 becomes spontaneously active.

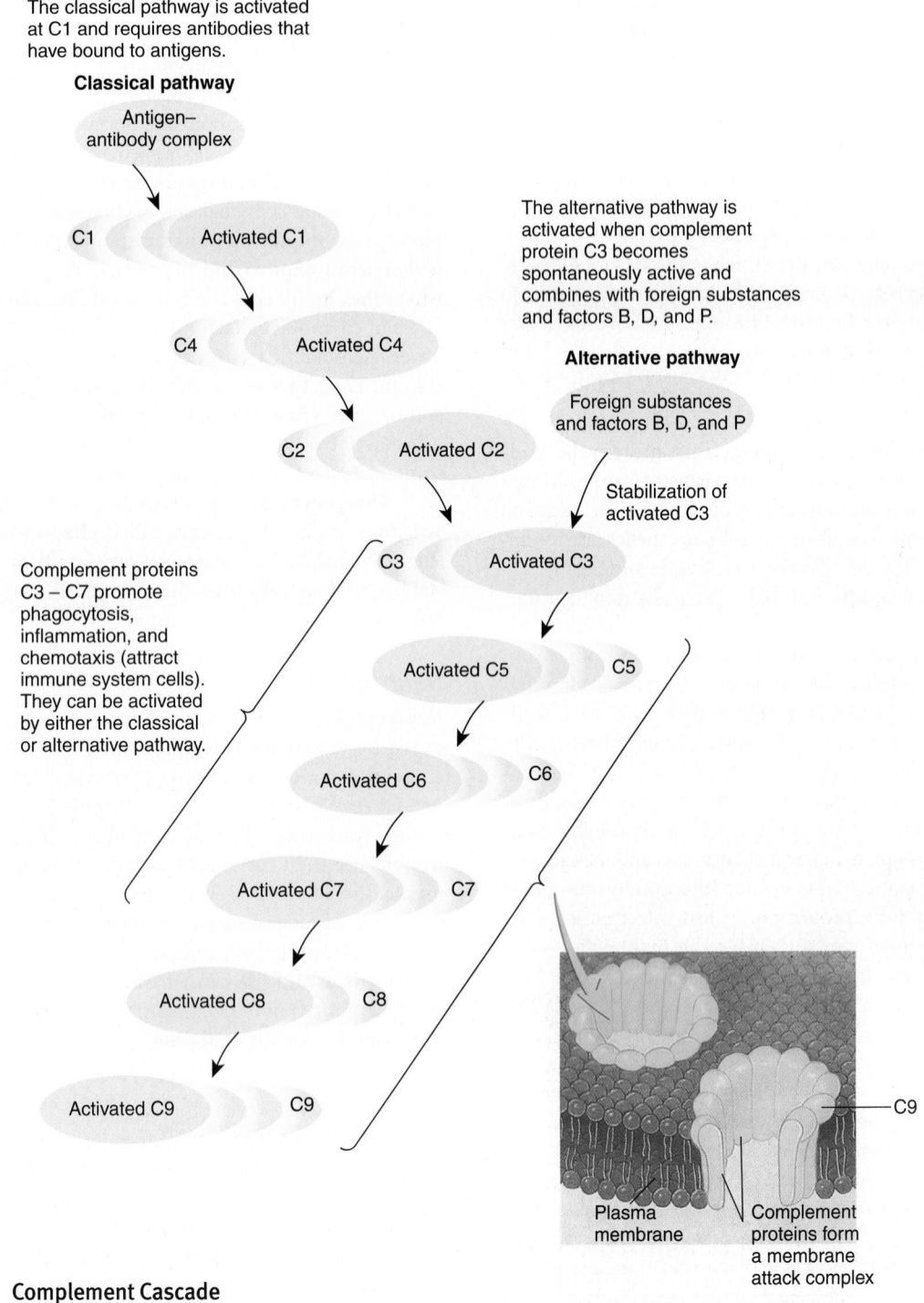

The classical pathway is activated at C1 and requires antibodies that have bound to antigens.

Classical pathway

Antigen–antibody complex

C1 Activated C1

C4 Activated C4

C2 Activated C2

The alternative pathway is activated when complement protein C3 becomes spontaneously active and combines with foreign substances and factors B, D, and P.

Alternative pathway

Foreign substances and factors B, D, and P

Stabilization of activated C3

Complement proteins C3 – C7 promote phagocytosis, inflammation, and chemotaxis (attract immune system cells). They can be activated by either the classical or alternative pathway.

C3 Activated C3

Activated C5 C5

Activated C6 C6

Activated C7 C7

Activated C8 C8

Activated C9 C9

C9

Plasma membrane

Complement proteins form a membrane attack complex

Figure 22.10 Complement Cascade

Inactive complement proteins become active complement proteins (*blue ovals*) in a cascade reaction: each activated complement protein activates the next protein in the sequence.

The membrane attack complex (MAC; only C9 is shown) forms a channel in the plasma membrane of a target cell.

Activated C3 normally is quickly inactivated by proteins on the surface of the body's cells. Activated C3 can combine with some foreign substances, such as part of a bacterial cell or virus. It can become stabilized and cause activation of the complement cascade. The **classical pathway** is part of the adaptive immune system, discussed on p. 809.

Activated complement proteins provide protection in several ways (see figure 22.10). They can form a **membrane attack complex (MAC),** which produces a channel through the plasma membrane. MAC formation begins when activated C3 attaches to a plasma membrane, stimulating a series of reactions (C5–C9). The main component of a MAC is activated C9 molecules, which change shape, attach to each other, and form a channel through the membrane. When MACs form in the plasma membrane of a nucleated cell, Na^+ and water enter the cell through the channel and cause the cell to lysis. When MACs form in the outer membrane of certain bacteria (Gram negative), an enzyme called lysozyme passes through the channel and digests the bacterial cell wall. When the wall breaks apart, the bacterial cell undergoes lysis.

Complement proteins can also attach to the surface of bacterial cells and stimulate macrophages to phagocytize the bacteria. In addition, complement proteins attract immune system cells to sites of infection and promote inflammation.

Interferons

Interferons (in-ter-fēr′onz) are proteins that protect the body against viral infection and perhaps some forms of cancer. After a virus infects a cell, viral replication can occur. Viral nucleic acids and proteins, which are produced using the cell's organelles, are assembled into new viruses. The new viruses are released from the infected cell to infect other cells. Because infected cells usually stop their normal functions or die during viral replication, viral infections are clearly harmful to the body. Fortunately, viruses and other substances can stimulate infected cells to produce interferons. Interferons neither protect the cell that produces them nor act directly against viruses. Instead, they bind to the surface of neighboring cells, where they stimulate them to produce antiviral proteins. These antiviral proteins stop viral reproduction in the neighboring cells by preventing the production of viral nucleic acids and proteins. Interferon viral resistance is innate rather than adaptive, and the same interferons act against many different viruses. Infection by one kind of virus actually can produce protection against infection by other kinds of viruses. Some interferons also play a role in the activation of immune cells such as macrophages and natural killer cells.

Treating Viral Infections and Cancer with Interferons

Because some cancers are induced by viruses, interferons may play a role in controlling cancers. Interferons activate macrophages and natural killer cells (a type of lymphocyte), which attack tumor cells. Through genetic engineering, interferons currently are produced in sufficient quantities for clinical use and, along with other therapies, have been effective in treating certain viral infections and cancers. For example, interferons are used to treat hepatitis C, a viral disorder that can cause cirrhosis and cancer of the liver, and to treat genital warts, caused by the herpes virus. Interferons are also approved for the treatment of Kaposi's sarcoma, a cancer that can develop in AIDS patients.

19. List the three main components of innate immunity.
20. Name two mechanical mechanisms that form barriers preventing the entry of microorganisms. In what ways are microorganisms removed from the surfaces of the body?
21. What is complement? In what two ways is it activated? How does complement provide protection?
22. What are interferons? How do they provide protection against viruses?

Cells

White blood cells and the cells derived from them (see table 19.2) are the most important cellular components of the immune system (table 22.2). White blood cells are produced in red bone marrow and lymphatic tissue and are released into the blood, where they are transported throughout the body. To be effective, white blood cells must move into the tissues where they are needed. **Chemotactic** (kē-mō-tak′tik) **factors** are parts of microbes or chemicals released by tissue cells that act as chemical signals to attract white blood cells. Important chemotactic factors include complement, leukotrienes, kinins, and histamine. They diffuse from the area where they are released. White blood cells can detect small differences in chemotactic factor concentration and move from areas of lower chemotactic factor concentration to areas of higher concentration. Thus, they move toward the source of these substances, an ability called **chemotaxis.** White blood cells can move by ameboid movement over the surface of cells, can squeeze between cells, and sometimes pass directly through other cells.

Phagocytosis (fag-ō-sī-tō′sis) is the endocytosis and destruction of particles by cells called **phagocytes** (see figure 3.22). The particles can be microorganisms or their parts, foreign substances, or dead cells from the individual's body. The most important phagocytic cells are neutrophils and macrophages.

Neutrophils

Neutrophils are small phagocytic cells produced in large numbers in red bone marrow that are released into the blood, where they circulate for a few hours. Approximately 126 billion neutrophils per day leave the blood and pass through the wall of the gastrointestinal tract, where they provide phagocytic protection. The neutrophils are then eliminated as part of the feces. Neutrophils are usually the first cells to enter infected tissues in large numbers, and they often die after a single phagocytic event.

Neutrophils also release lysosomal enzymes that kill microorganisms and also cause tissue damage and inflammation. **Pus** is an accumulation of dead neutrophils, dead microorganisms, debris from dead tissue, and fluid.

Macrophages

Macrophages are monocytes that leave the blood, enter tissues, enlarge about fivefold, and increase their number of lysosomes and mitochondria. They are large phagocytic cells that outlive neutrophils, and they can ingest more and larger phagocytic particles than neutrophils. Macrophages usually accumulate in tissues after neutrophils and are responsible for most of the phagocytic activity in the late stages of an infection, including the cleanup of dead neutrophils and other cellular debris. In addition to their

Table 22.2   Immune System Cells and Their Primary Functions

Cell	Primary Function	Cell	Primary Function
Innate Immunity		**Adaptive Immunity**	
Neutrophil	Phagocytosis and inflammation; usually the first cell to leave the blood and enter infected tissues	B cell	After activation, differentiates to become plasma cell or memory B cell
Monocyte	Leaves the blood and enters tissues to become a macrophage	Plasma cell	Produces antibodies that are directly or indirectly responsible for the destruction of the antigen
Macrophage	Most effective phagocyte; important in later stages of infection and in tissue repair; located throughout the body to "intercept" foreign substances; processes antigens; involved in the activation of B and T cells	Memory B cell	Quick and effective response to an antigen against which the immune system has previously reacted; responsible for immunity
		Cytotoxic T cell	Responsible for the destruction of cells by lysis or by the production of cytokines
Basophil	Motile cell that leaves the blood, enters tissues, and releases chemicals that promote inflammation	Delayed hypersensitivity T cell	Produces cytokines that promote inflammation
		Helper T cell	Activates B and effector T cells
Mast cell	Nonmotile cell in connective tissues that promotes inflammation through the release of chemicals	Suppressor T cell	Inhibits B and effector T cells
Eosinophil	Enters tissues from the blood and releases chemicals that inhibit inflammation	Memory T cell	Quick and effective response to an antigen against which the immune system has previously reacted; responsible for adaptive immunity
Natural killer cell	Lyses tumor and virus-infected cells	Dendritic cell	Processes antigen and is involved in the activation of B and T cells

phagocytic role, macrophages produce a variety of chemicals, such as interferons, prostaglandins, and complement, that enhance the immune response.

Macrophages are beneath the free surfaces of the body, such as the skin (dermis), hypodermis, mucous membranes, and serous membranes, and around blood and lymphatic vessels. In these locations, macrophages provide protection by trapping and destroying microorganisms entering the tissues.

If microbes do gain entry to the blood or lymphatic system, macrophages are waiting within enlarged spaces, called sinuses, to phagocytize them. Blood vessels in the spleen, bone marrow, and liver have sinuses, as do lymph nodes. Within the sinuses, reticular cells produce a fine network of reticular fibers that slows the flow of blood or lymph and provides a large surface area for the attachment of macrophages. In addition, macrophages are on the endothelial lining of the sinuses.

Because macrophages on the reticular fibers and endothelial lining of the sinuses were among the first macrophages studied, these cells were referred to as the **reticuloendothelial system.** It's now recognized that macrophages are derived from monocytes and are in locations other than the sinuses. Because monocytes and macrophages have a single, unlobed nucleus, they are now called the **mononuclear phagocytic system.** Sometimes macrophages are given specific names, for instance dust cells in the lungs, Kupffer cells in the liver, and microglia in the central nervous system.

Basophils, Mast Cells, and Eosinophils

Basophils, which are derived from red bone marrow, are motile white blood cells that can leave the blood and enter infected tissues.

Mast cells, which are also derived from red bone marrow, are nonmotile cells in connective tissue, especially near capillaries. Like macrophages, mast cells are located at potential points of entry of microorganisms into the body, such as the skin, lungs, gastrointestinal tract, and urogenital tract.

Basophils and mast cells can be activated through innate immunity (e.g., by complement) or through adaptive immunity (see "Antibodies" on p. 808). When activated, they release chemicals, for example, histamine and leukotrienes, that produce an inflammatory response or activate other mechanisms, for example, smooth muscle contraction in the lungs.

Eosinophils are produced in red bone marrow, enter the blood, and within a few minutes enter tissues. Enzymes released by eosinophils break down chemicals released by basophils and mast cells. Thus, at the same time that inflammation is initiated, mechanisms are activated that contain and reduce the inflammatory response. This process is similar to the blood clotting system in which clot prevention and removal mechanisms are activated while the clot is being formed (see chapter 19). In patients with parasitic infections or allergic reactions with much inflammation, eosinophil numbers greatly increase. Eosinophils also secrete enzymes that effectively kill some parasites.

Natural Killer Cells

Natural killer (NK) cells are a type of lymphocyte produced in red bone marrow, and they account for up to 15% of lymphocytes. NK cells recognize classes of cells, such as tumor cells or virus-infected cells in general, rather than specific tumor cells or cells infected by a specific virus. For this reason and because NK cells don't exhibit a memory response, they are classified as part of innate immunity. NK cells use a variety of methods to kill their target cells, including the release of chemicals that damage plasma membranes, causing the cells to lyse.

> 23. Define the terms chemotactic factor, chemotaxis, and phagocytosis.
> 24. What are the functions of neutrophils and macrophages? What is pus?
> 25. What effects are produced by the chemicals released from basophils, mast cells, and eosinophils?
> 26. Describe the function of NK cells.

Inflammatory Response

The **inflammatory response** is a complex sequence of events involving many of the chemical mediators and cells of innate immunity. Tissue injury, regardless of the type, can cause inflammation. Trauma, burns, chemicals, or infections can damage tissues, resulting in inflammation. A bacterial infection is used here to illustrate an inflammatory response (figure 22.11). The bacteria, or damage to tissues, cause the release or activation of chemical mediators, such as histamine, complement, kinins, and eicosanoids (e.g., prostaglandins and leukotrienes). The chemical mediators produce several effects: (1) vasodilation, which increases blood flow and brings phagocytes and other white blood cells to the area; (2) chemotactic attraction of phagocytes, which leave the blood and enter the tissue; and (3) increased vascular permeability, which allows fibrinogen and complement to enter the tissue from the blood. Fibrinogen is converted to fibrin, which prevents the spread of infection by walling off the infected area. Complement further enhances the inflammatory response and attracts additional phagocytes. The process of releasing chemical mediators and attracting phagocytes and other white blood cells continues until the bacteria are destroyed. Phagocytes, such as neutrophils and macrophages, remove microorganisms and dead tissue, and the damaged tissues are repaired.

Inflammation can be localized or systemic. **Local inflammation** is an inflammatory response confined to a specific area of the body (see chapter 4). Symptoms of local inflammation include redness, heat, swelling, pain, and loss of function. Redness, heat, and swelling result from increased blood flow and increased vascular permeability. Pain is caused by swelling and by chemical mediators acting on pain receptors. Loss of function results from tissue destruction, swelling, and pain.

Systemic inflammation is an inflammatory response that occurs in many parts of the body. In addition to the local symptoms at the sites of inflammation, three additional features can be present. First, red bone marrow produces and releases large numbers of neutrophils, which promote phagocytosis. Second, **pyrogens** (pī′rō-jenz, fire producing), chemicals released by microorganisms, macrophages, neutrophils, and other cells, stimulate fever production. Pyrogens affect the body's temperature-regulating mechanism in the hypothalamus, heat is conserved, and body temperature increases. Fever promotes the activities of the immune system, such as phagocytosis, and inhibits the growth of some microorganisms. Third, in severe cases of systemic inflammation, increased vascular permeability is so widespread that large amounts

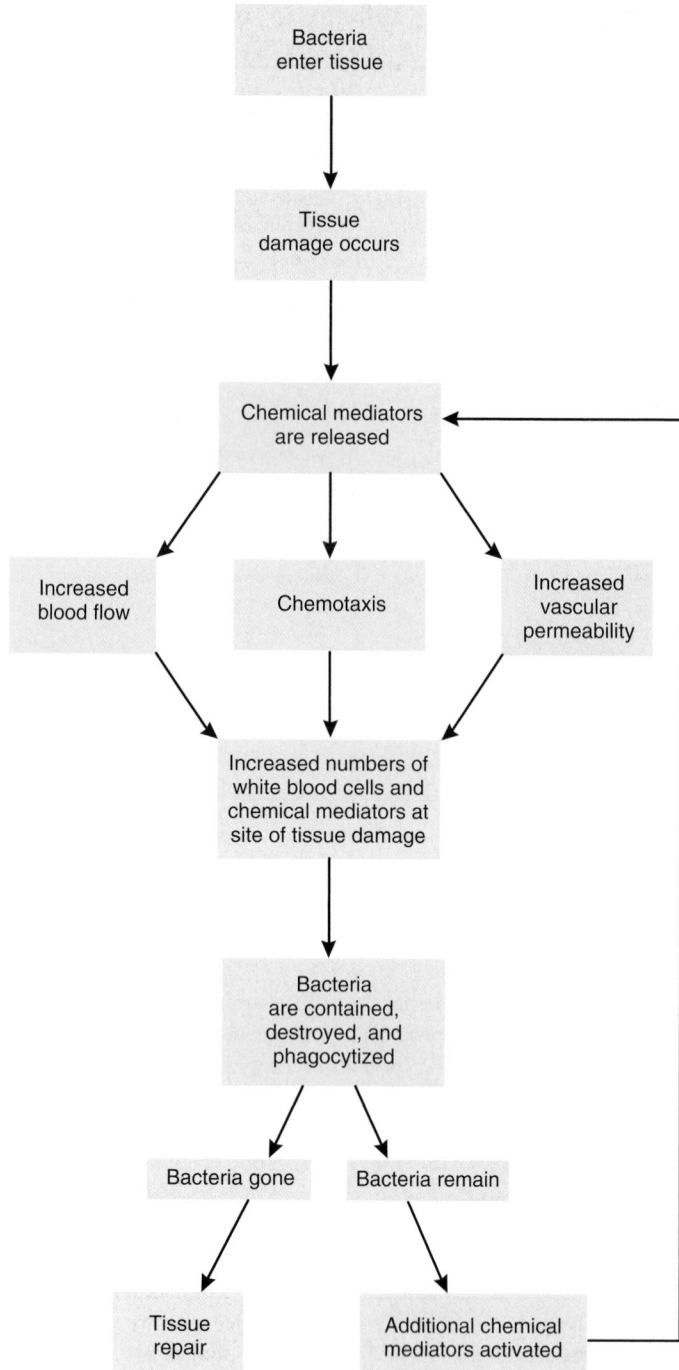

Figure 22.11 **Inflammatory Response**

Bacteria cause tissue damage and the release of chemical mediators that initiate inflammation and phagocytosis, resulting in the destruction of the bacteria. If any bacteria remain, additional chemical mediators are activated. After all the bacteria are destroyed, the tissue is repaired.

of fluid are lost from the blood into the tissues. The decreased blood volume can cause shock and death.

> 27. Describe the events that take place during an inflammatory response.
> 28. What are the symptoms of local and systemic inflammations?

Adaptive Immunity

Adaptive immunity exhibits the ability to recognize, respond to, and remember a particular substance. Substances that stimulate adaptive immunity are called **antigens** (an′ti-jenz). They usually are large molecules with a molecular weight of 10,000 or more. **Haptens** (hap′tenz) are small molecules (low molecular weight) capable of combining with larger molecules like blood proteins to stimulate an adaptive immune system response.

Allergic Reactions to Penicillin

Penicillin is an example of a hapten of clinical importance. It's a small molecule that doesn't evoke an immune system response. Penicillin can, however, break down and bind to serum proteins to form a combined molecule that can produce an allergic reaction. Most commonly, the reaction produces a rash and fever, but rarely a severe reaction can cause death.

Antigens are divided into two groups: foreign antigens and self-antigens. **Foreign antigens** are not produced by the body but are introduced from outside it. Components of bacteria, viruses, and other microorganisms are examples of foreign antigens that cause disease. Pollen, animal dander (scaly, dried skin), feces of house dust mites, foods, and drugs are also foreign antigens and can trigger an overreaction of the immune system in some people, called an **allergic reaction.** Transplanted tissues and organs that contain foreign antigens result in the rejection of the transplant. **Self-antigens** are molecules produced by the body that stimulate an adaptive immune system response. The response to self-antigens can be beneficial or harmful. For example, the recognition of tumor antigens can result in tumor destruction, whereas **autoimmune disease** can result when self-antigens stimulate unwanted tissue destruction.

Adaptive immunity historically has been divided into two types: **humoral** (hū′mōr-ăl) and **cell-mediated immunity.** Early investigators of the immune system found that, when plasma from an immune animal was injected into the blood of a nonimmune animal, the nonimmune animal became immune. Because this process involved body fluids (humors), it was called humoral immunity. It was also discovered that blood cells transferred from an immune animal could be responsible for immunity, and this process was called cell-mediated immunity.

It's now known that immunity results from the activities of lymphocytes called B and T cells (see table 22.2). **B cells** give rise to cells that produce proteins called **antibodies,** which are found in the plasma. Humoral immunity is now called **antibody-mediated immunity** because antibodies are responsible.

T cells are responsible for cell-mediated immunity. Several subpopulations of T cells exist, each of which is responsible for a particular aspect of cell-mediated immunity. For example, **effector T cells,** such as **cytotoxic T cells** and **delayed hypersensitivity T cells,** are responsible for producing the effects of cell-mediated immunity; whereas **regulatory T cells,** such as **helper T cells** and **suppressor T cells,** can promote or inhibit the activities of both antibody-mediated immunity and cell-mediated immunity.

Table 22.3 summarizes and contrasts the main features of innate and adaptive immunity.

Table 22.3	Comparison of Innate and Adaptive Immunity		
		Adaptive Immunity	
Characteristics	**Innate Immunity**	**Antibody-Mediated Immunity**	**Cell-Mediated Immunity**
Primary cells	Neutrophils, eosinophils, basophils, mast cells, monocytes, and macrophages	B cells	T cells
Origin of cells	Red bone marrow	Red bone marrow	Red bone marrow
Site of maturation	Red bone marrow (neutrophils, eosinophils, basophils, monocytes) and tissues (mast cells and macrophages)	Red bone marrow	Thymus
Location of mature cells	Blood, connective tissue, and lymphatic tissue	Blood and lymphatic tissue	Blood and lymphatic tissue
Primary secretory products	Histamine, kinins, complement, prostaglandins, leukotrienes, and interferon	Antibodies	Cytokines
Primary actions	Inflammatory response and phagocytosis	Protection against extracellular antigens (bacteria, toxins, parasites, and viruses outside of cells)	Protection against intracellular antigens (viruses, intracellular bacteria, and intracellular fungi) and tumors: regulates antibody-mediated immunity and cell-mediated immunity responses (helper T and suppressor T cells)
Hypersensitivity reactions	None	Immediate hypersensitivity (atopy, anaphylaxis, cytotoxic reactions, and immune complex disease)	Delayed hypersensitivity (allergy of infection and contact hypersensitivity)

29. *Define the terms antigen and hapten. Distinguish between a foreign antigen and a self-antigen.*
30. *What are allergic reactions and autoimmune diseases?*

Origin and Development of Lymphocytes

All blood cells, including lymphocytes, are derived from stem cells in the red bone marrow (see chapter 19). The process of blood cell formation begins during embryonic development and continues throughout life. Some stem cells give rise to pre-T cells that migrate through the blood to the thymus, where they divide and are processed into T cells (see figures 22.9 and 22.12). The thymus produces hormones such as thymosin, which stimulates T-cell maturation. Other stem cells produce pre-B cells, which are processed in the red bone marrow into B cells. A **positive selection** process results in the survival of pre-B and pre-T cells that are capable of an immune response. Cells that are incapable of an immune response die.

The B and T cells that can respond to antigens are composed of small groups of identical lymphocytes called **clones.** Although each clone can respond only to a particular antigen, such a large number of clones exist that the immune system can react to most molecules. Some of the clones can also respond to self-antigens. A **negative selection** process eliminates or suppresses clones acting against self-antigens, thereby preventing the destruction of self-cells. Although the negative selection process mostly occurs during prenatal development, it continues throughout life (see section on "Inhibition of Lymphocytes" on p. 807).

B cells are released from red bone marrow, T cells are released from the thymus, and both types of cells move through the blood to lymphatic tissue. There are approximately five T cells for every B cell in the blood. These lymphocytes live for a few months to many years and continually circulate between the blood and the lymphatic tissues. Antigens can come into contact with and activate lymphocytes, resulting in cell divisions that increase the number of lymphocytes that can recognize the antigen. These lymphocytes can circulate in blood and lymph to reach antigens in tissues throughout the body.

The **primary lymphatic organs** are the sites where lymphocytes mature into functional cells. These organs are the red bone marrow and thymus. The **secondary lymphatic organs and tissues** are the sites where lymphocytes interact with each other, antigen-presenting cells, and antigens to produce an immune response. The secondary lymphatic organs and tissues include diffuse lymphatic tissue, lymphatic nodules, tonsils, lymph nodes, and the spleen.

31. *Describe the origin and development of B and T cells.*
32. *What are lymphocyte clones? Distinguish between positive and negative lymphocyte selection.*
33. *What are primary and secondary lymphatic organs and tissues?*

Activation of Lymphocytes

Antigens activate lymphocytes in different ways, depending on the type of lymphocyte and the type of antigen involved. Despite these differences, however, two general principles of lymphocyte activation exist: (1) lymphocytes must be able to recognize the antigen, and (2) after recognition, the lymphocytes must increase in number to effectively destroy the antigen.

Antigenic Determinants and Antigen Receptors

If an adaptive immune system response is to occur, lymphocytes must recognize an antigen. Lymphocytes don't interact with an entire antigen, however. Instead, **antigenic determinants,** or **epitopes** (ep'i-tōpz), are specific regions of a given antigen recognized by a lymphocyte, and each antigen has many different antigenic determinants (figure 22.13). All the lymphocytes of a given clone have on their surfaces identical proteins called **antigen receptors,** which combine with a specific antigenic determinant. The immune system response to an antigen with a particular antigenic determinant is similar to the lock-and-key model for enzymes (see chapter 2), and

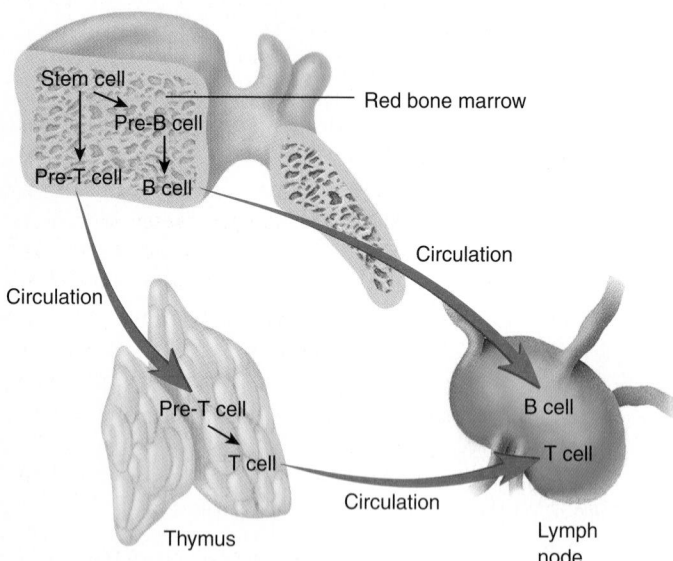

Figure 22.12   **Origin and Processing of B and T Cells**

Pre-B and pre-T cells originate from stem cells in red bone marrow. The pre-B cells remain in the red bone marrow and become B cells. The pre-T cells circulate to the thymus where they become T cells. Both B and T cells circulate to other lymphatic tissues, such as lymph nodes, where they can divide and increase in number in response to antigens.

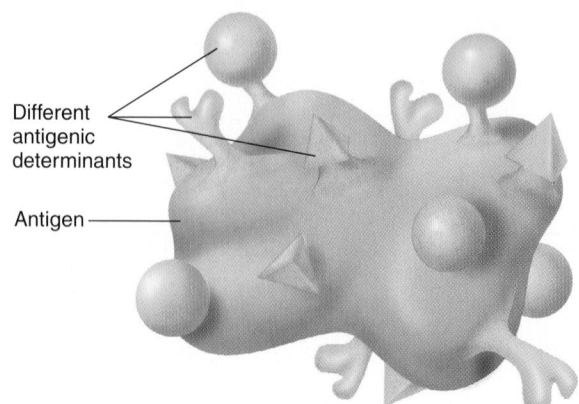

Figure 22.13   **Antigenic Determinants**

An antigen has many antigenic determinants to which lymphocytes can respond.

any given antigenic determinant can combine only with a specific antigen receptor. The **T-cell receptor** consists of two polypeptide chains, which are subdivided into a variable and a constant region (figure 22.14). The variable region can bind to an antigen. The many different types of T-cell receptors respond to different antigens because they have different variable regions. The **B-cell receptor** consists of four polypeptide chains with two identical variable regions. It is a type of antibody and is considered in greater detail on p. 808.

Major Histocompatibility Complex Molecules

Although some antigens bind to their receptors and directly activate B cells and some T cells, most lymphocyte activation involves glycoproteins on the surfaces of cells called **major histocompatibility complex (MHC) molecules.** MHC molecules are attached to plasma membranes, and they have a variable region that can bind to foreign and self-antigens.

MHC class I molecules are found on nucleated cells and function to display antigens produced inside the cells on their surfaces (figure 22.15a). This is necessary because the immune system cannot directly respond to an antigen inside a cell. For example, viruses reproduce inside cells, forming viral proteins that are foreign antigens. Some of these viral proteins are broken down in the cytoplasm. The protein fragments enter the rough endoplasmic reticulum and combine with MHC class I molecules to form complexes that move through the Golgi apparatus to be distributed on the surface of the cell (see chapter 3).

MHC class I/antigen complexes on the surface of cells can bind to T-cell receptors on the surface of T cells. This combination is a signal that activates T cells. As described later in this chapter, activated T cells can destroy infected cells, which effectively stops viral replication. Thus, the MHC class I/antigen complex functions as a signal, or "red flag," that prompts the immune system to destroy the displaying

cell. In essence, the cell is displaying a sign that says, "Kill me!" This process is said to be **MHC-restricted,** because both the antigen and the individual organism's own MHC molecule are required.

P R E D I C T 2

In mouse A, T cells can respond to virus X. If these T cells are transferred to mouse B, which is infected with virus X, will the T cells respond to the virus? Explain.

The same process that moves foreign protein fragments to the surface of cells can also inadvertently transport self-protein fragments (see figure 22.15a). As part of normal protein metabolism, cells continually break down old proteins and synthesize new ones. Some self-protein fragments that result from protein breakdown can combine with MHC class I molecules and be displayed on the surface of the cell, thus becoming self-antigens. Normally, the immune system doesn't respond to self-antigens in combination with MHC molecules because the lymphocytes that could respond have been eliminated or inactivated (see section on "Inhibition of Lymphocytes" on p. 807).

MHC class II molecules are found on **antigen-presenting cells,** which include B cells, macrophages, monocytes, and dendritic cells. **Dendritic** (den-drit′ik) **cells** are large, motile cells with long cytoplasmic extensions, and they are scattered throughout most tissues (except the brain), with their highest concentrations in lymphatic tissues and the skin. Dendritic cells in the skin are often called **Langerhans' cells.**

Antigen-presenting cells can take in foreign antigens by endocytosis (figure 22.15b). Within the endocytotic vesicle, the antigen is broken down into fragments to form processed antigens. Vesicles from the Golgi apparatus containing MHC class II molecules combine with the endocytotic vesicles. The MHC class II molecules and processed antigens combine, and the MHC class II/antigen complexes are transported to the surface of the cell, where they are displayed to other immune cells.

MHC class II/antigen complexes on the surface of cells can bind to T-cell receptors on the surface of T cells. The presentation of antigen using MHC class II molecules is MHC-restricted because both the antigen and the individual's own MHC class II molecule are required. Unlike MHC class I molecules, however, this display does not result in the destruction of the antigen-presenting cell. Instead the MHC class II/antigen complex is a "rally around the flag" signal that stimulates other immune system cells to respond to the antigen. The displaying cell is like Paul Revere, who spread the alarm for the militia to arm and organize. The militia then went out and killed the enemy. For example, when the lymphocytes of the B-cell clone that can recognize the antigen come into contact with the MHC class II/antigen complex, they are stimulated to divide. The activities of these lymphocytes, such as the production of antibodies, then result in the destruction of the antigen.

34. *Define the terms antigenic determinant and antigen receptor. How are they related to each other?*
35. *What type of antigens are displayed by MHC class I and II molecules?*
36. *What type of cells display MHC class I and II antigen complexes, and what happens as a result?*
37. *Define MHC-restricted.*

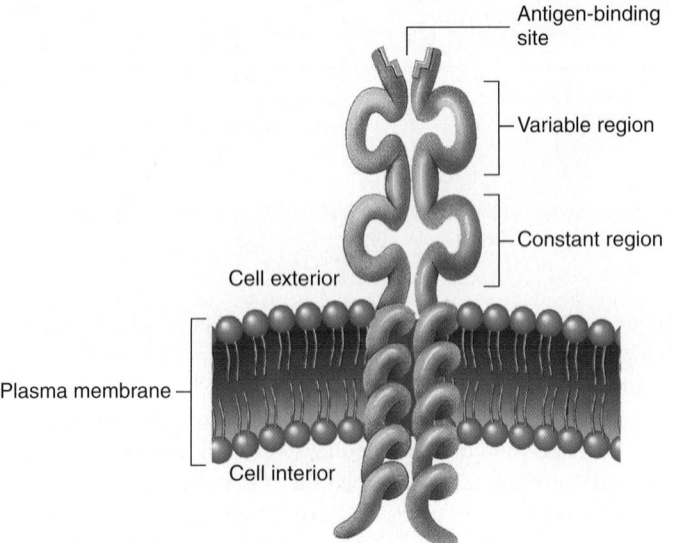

Figure 22.14 The T-Cell Receptor

The T-cell receptor consists of two polypeptide chains. The variable region of each type of T-cell receptor is specific for a given antigen. The constant region attaches the T-cell receptor to the plasma membrane.

1. Foreign proteins or self-proteins within the cytosol are broken down into fragments that are antigens.
2. Antigens are transported into the rough endoplasmic reticulum.
3. Antigens combine with MHC class I molecules.
4. The MHC class I/antigen complex is transported to the Golgi apparatus, packaged into a vesicle, and transported to the plasma membrane.
5. Foreign antigens combined with MHC class I molecules stimulate cell destruction.
6. Self-antigens combined with MHC class I molecules do not stimulate cell destruction.

(a)

1. A foreign antigen is ingested by endocytosis and is within a vesicle.
2. The antigen is broken down into fragments to form processed foreign antigens.
3. The vesicle containing the processed foreign antigens fuses with vesicles produced by the Golgi apparatus that contain MHC class II molecules. Processed foreign antigens and MHC class II molecules combine.
4. The MHC class II/antigen complex is transported to the plasma membrane.
5. The displayed MHC class II/antigen complex can stimulate immune cells.

(b)

Process Figure 22.15 **Antigen Processing**

(a) Foreign proteins, such as viral proteins, or self-proteins in the cytosol, are processed and presented at the cell surface by MHC class I molecules. (b) Foreign antigens are taken into an antigen-presenting cell, processed, and presented at the cell surface by MHC class II molecules.

P R E D I C T   3

How does elimination of the antigen stop the production of antibodies?

Costimulation

The combination of an MHC class II/antigen complex with an antigen receptor is usually only the first signal necessary to produce a response from a B or T cell. In many cases, **costimulation** by additional signals is also required. Costimulation is accomplished by molecules released from cells and by molecules attached to the surface of cells. **Cytokines** (sī′tō-kīnz), which are proteins or peptides secreted by one cell as a regulator of neighboring cells, promote costimulation

(figure 22.16a). Cytokines produced by lymphocytes are often called **lymphokines** (lim′fō-kīnz). Cytokines are involved in the regulation of immunity, inflammation, tissue repair, cell growth, and other processes. Table 22.4 lists important cytokines and their functions.

Certain pairs of surface molecules can also be involved in costimulation (figure 22.16b). When the surface molecule on one cell combines with the surface molecule on another, the combination can act as a signal that stimulates a response from one of the cells, or the combination can hold the cells together. Typically, several different kinds of surface molecules are necessary to produce a response. For example, a molecule called B7 on macrophages must bind with a molecule called CD28 on helper T cells before the helper T cells can

(a) Helper T cells are activated by a first signal and by costimulation. The first signal is the binding of the MHC class II/antigen complex to the T-cell receptor. Costimulation is an additional signal, such as molecules released from another cell. For example, the release by the macrophage of cytokines that bind to receptors on the helper T cell.

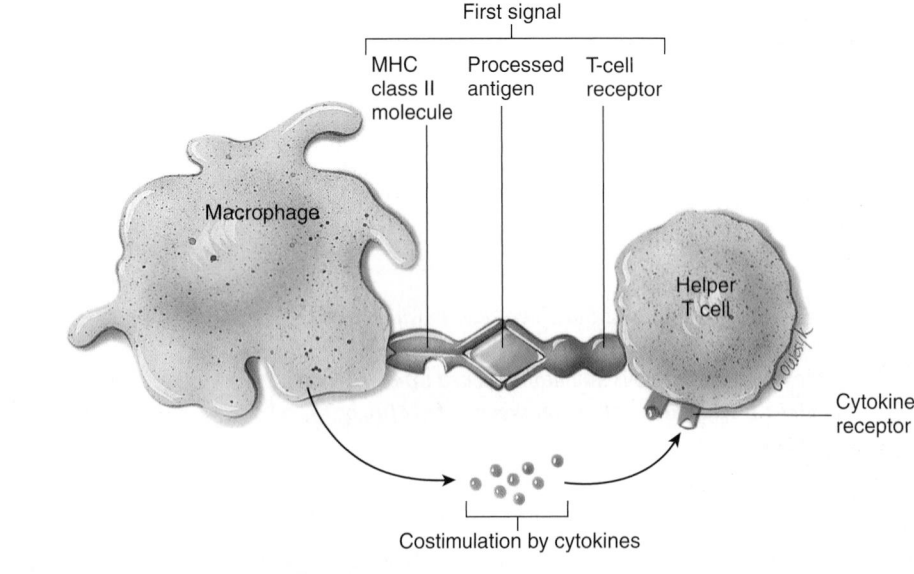

(b) Other costimulatory signals are the combining of surface molecules between cells, such as the binding of a B7 molecule of the macrophage with a CD28 molecule of the helper T cell. The CD4 molecule of the helper T cell binds to the macrophage's MHC class II molecule and helps to hold the cells together.

Figure 22.16 Costimulation

Table 22.4 Cytokines and Their Functions

Cytokine*	Description
Interferon alpha (IFNα)	Prevents viral replication and inhibits cell growth; secreted by virus-infected cells
Interferon beta (IFNβ)	Prevents viral replication, inhibits cell growth, and decreases the expression of major histocompatibility complex (MHC) class I and II molecules; secreted by virus-infected fibroblasts
Interferon gamma (IFNγ)	About 20 different proteins that activate macrophages and natural killer (NK) cells, stimulate adaptive immunity by increasing the expression of MHC class I and II molecules, and prevent viral replication; secreted by helper T, cytotoxic T, and NK cells
Interleukin-1 (IL-1)	Costimulation of B and T cells, promotes inflammation through prostaglandin production, and induces fever acting through the hypothalamus (pyrogen); secreted by macrophages, B cells, and fibroblasts
Interleukin-2 (IL-2)	Costimulation of B and T cells, activation of macrophages and NK cells; secreted by helper T cells
Interleukin-4 (IL-4)	Plays a role in allergic reactions by activation of B cells, resulting in the production of immunoglobulin E (IgE); secreted by helper T cells
Interleukin-5 (IL-5)	Part of the response against parasites by stimulating eosinophil production; secreted by helper T cells
Interleukin-8 (IL-8)	Chemotactic factor that promotes inflammation by attracting neutrophils and basophils; secreted by macrophages
Interleukin-10 (IL-10)	Inhibits the secretion of interferon gamma and interleukins; secreted by suppressor T cells
Lymphotoxin	Kills target cells; secreted by cytotoxic T cells
Perforin	Makes a hole in the membrane of target cells, resulting in lysis of the cell; secreted by cytotoxic T cells
Tumor necrosis factor (TNF)	Activates macrophages and promotes fever (pyrogen); secreted by macrophages

*Some cytokines were named according to the laboratory test first used to identify them; however, these names rarely are a good description of the actual function of the cytokine.

respond to the antigen presented by the macrophage. In addition, helper T cells have a glycoprotein called CD4, which helps to connect helper T cells to the macrophage by binding to MHC class II molecules. For this reason, helper T cells are sometimes referred to as **CD4, or T4, cells.** In a similar fashion, cytotoxic T cells are sometimes called **CD8, or T8, cells** because they have a glycoprotein called CD8, which helps to connect cytotoxic T cells to cells displaying MHC class I molecules. The CD designation stands for "cluster of differentiation," which is a system used to classify many surface molecules.

● 38. *What is costimulation? State two ways in which it can happen.*
  39. *Why are helper T cells sometimes called CD4, or T4, cells? Why are cytotoxic T cells sometimes called CD8, or T8, cells?*

Lymphocyte Proliferation

Before exposure to an antigen, the number of lymphocytes in a clone is too small to produce an effective response against the antigen. Exposure to an antigen results in an increase in lymphocyte number. First, there is an increase in the number of helper T cells. This is important because the increased number of helper T cells responding to the antigen can find and stimulate B or effector T cells. Second, the number of B or effector T cells increases. This is important because it is these cells that are responsible for the immune response that destroys the antigen.

1. *Proliferation of helper T cells* (figure 22.17). Antigen-presenting cells use MHC class II molecules to present processed antigens

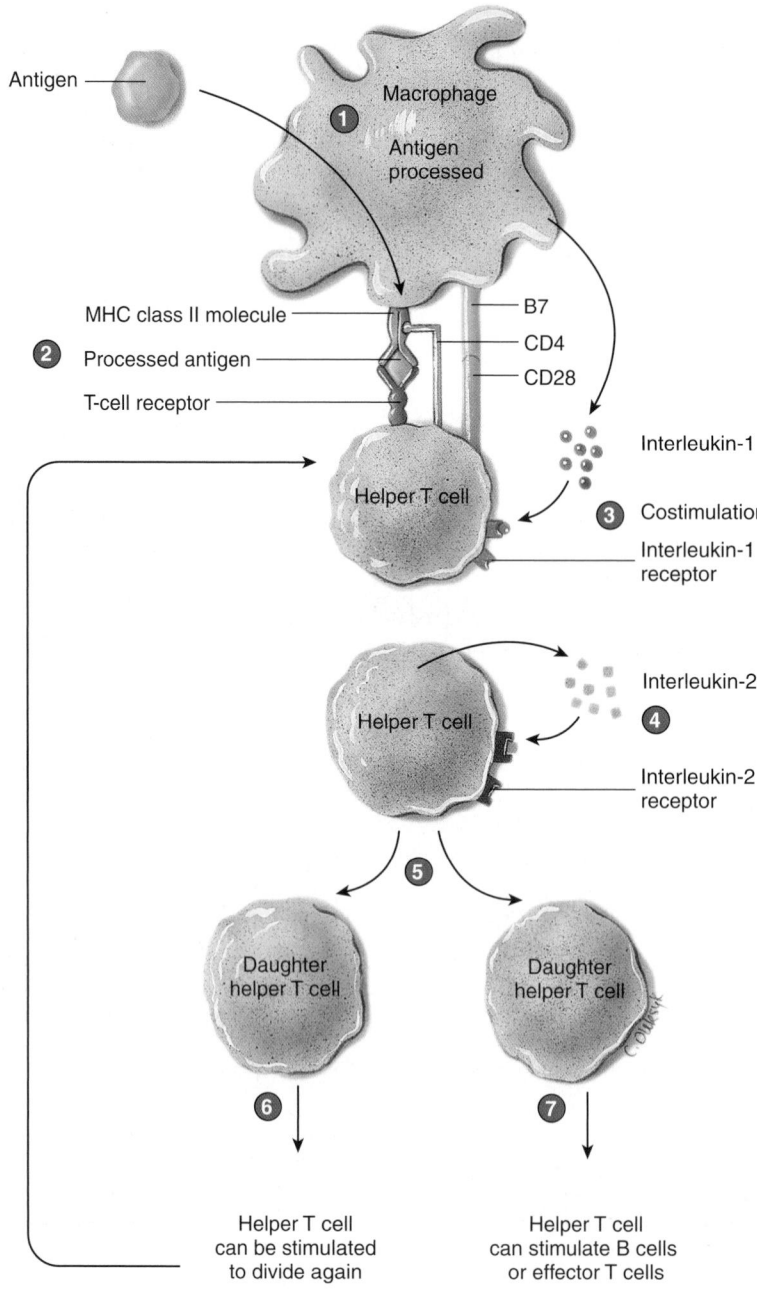

1. Antigen-presenting cells such as macrophages take in, process, and display antigens on the cell's surface.

2. The antigen is bound to a MHC class II molecule, which functions to present the processed antigen to a T-cell receptor of a helper T cell for recognition.

3. Costimulation occurs by a CD4 glycoprotein of the helper T cell or by cytokines. The macrophage secretes a cytokine called interleukin-1.

4. Interleukin-1 attaches to interleukin receptors and stimulates the helper T cell to secrete the cytokine interleukin-2 and to produce interleukin-2 receptors.

5. The helper T cell stimulates itself to divide when interleukin-2 binds to interleukin-2 receptors.

6. The "daughter" helper T cells resulting from this division can be stimulated to divide again if they are exposed to the same antigen that stimulated the "parent" helper T cell. This greatly increases the number of helper T cells.

7. The increased number of helper T cells can facilitate the activation of B cells or effector T cells.

Process Figure 22.17   Proliferation of Helper T Cells
An antigen-presenting cell (macrophage) stimulates helper T cells to divide.

to helper T cells. Only the helper T cells with the T-cell receptors that can bind to the antigen respond. These helper T cells respond to the MHC class II/antigen complex and costimulation by dividing. As a result, the number of helper T cells that recognize the antigen increases.

2. *Proliferation and activation of B or effector T cells.* Typically, the proliferation and activation of B or effector T cells involves helper T cells. This process is illustrated in figure 22.18 for B cells (see "Cell-Mediated Immunity" on p. 814 for activation of effector T cells). The clone of B cells that can recognize a particular antigen have B-cell receptors that can bind to that antigen. The antigens and receptors are taken into the B cell by receptor-mediated endocytosis. The antigens are broken down into fragments to form processed antigens that combine with MHC class II molecules. B cells use MHC class II/antigen complexes to present antigens to the helper T cells that increased in number in response to the antigen. These helper T cells stimulate the B cells to divide and produce antibodies. The increased number of cells, each producing antibodies, can produce an immune response that destroys the antigens (see "Effects of Antibodies" on p. 809).

- 40. *Describe how antigen-presenting cells stimulate an increase in the number of helper T cells. Why is this important?*
- 41. *Describe how helper T cells stimulate an increase in the number of B or T cells. Why is this important?*

Inhibition of Lymphocytes

Tolerance is a state of unresponsiveness of lymphocytes to a specific antigen. Although foreign antigens can induce tolerance, the most important function of tolerance is to prevent the immune system from responding to self-antigens. The need to maintain tolerance and to avoid the development of autoimmune disease is obvious. Tolerance can be induced in many ways.

1. *Deletion of self-reactive lymphocytes.* During prenatal development and after birth, stem cells in red bone marrow and the thymus give rise to immature lymphocytes that develop into mature lymphocytes capable of an immune response. When immature lymphocytes are exposed to antigens, instead of responding in ways that result in the elimination of the antigen, they respond by dying. Because

1. Before a B cell can be activated by a helper T cell, the B cell must phagocytize and process the same antigen that activated the helper T cell. The antigen binds to a B-cell receptor, and both the receptor and antigen are taken into the cell by endocytosis.

2. The B cell uses a MHC class II molecule to present the processed antigen to the helper T cell.

3. The T-cell receptor binds to the MHC class II/antigen complex.

4. There is costimulation of the B cell by CD4 and other surface molecules.

5. There is costimulation by interleukins (cytokines) released from the helper T cell.

6. The B cell divides, and the resulting daughter cells divide, and so on, eventually producing many cells (only two are shown here).

7. The daughter cells produce antibodies, which are part of the antibody-mediated immune system response that eliminates the antigen.

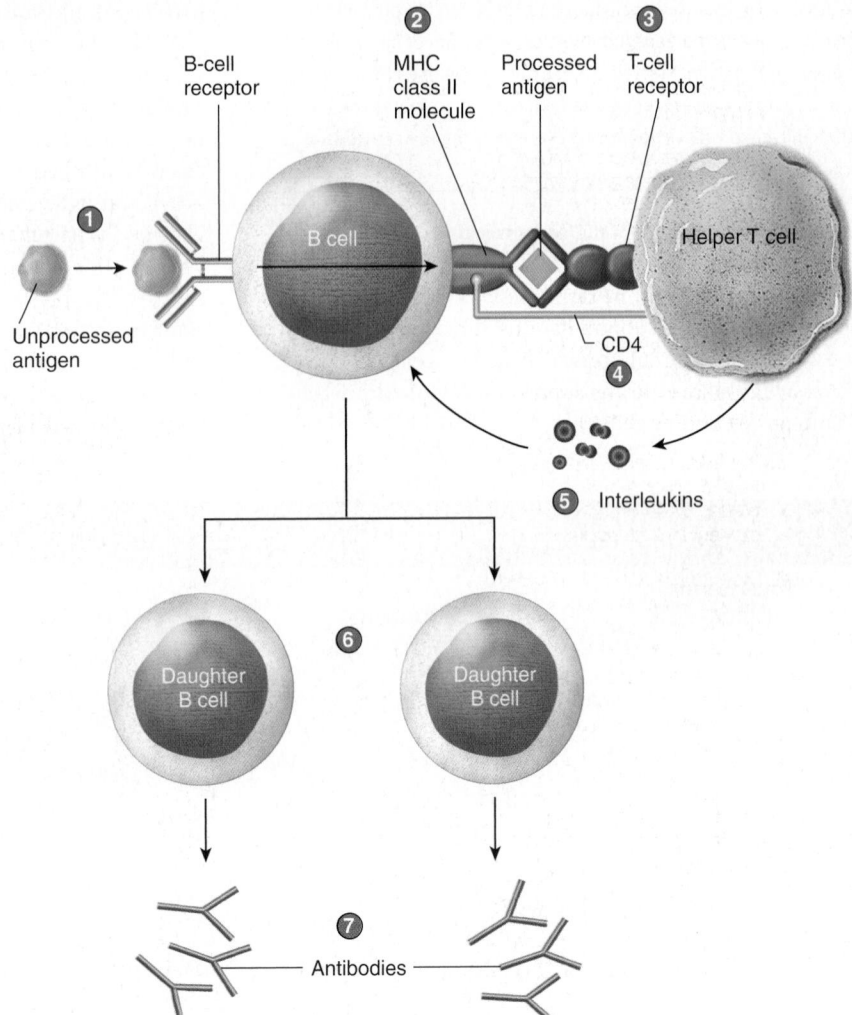

Process Figure 22.18   Proliferation of B Cells

A helper T cell stimulates a B cell to divide.

immature lymphocytes are exposed to self-antigens, this process eliminates self-reactive lymphocytes. In addition, immature lymphocytes that escape deletion during their development and become mature, self-reacting lymphocytes can still be deleted in ways that are not clearly understood.

2. *Preventing activation of lymphocytes.* For activation of lymphocytes to take place, two signals are usually required: (1) the MHC/antigen complex binding with an antigen receptor and (2) costimulation. Preventing either of these events stops lymphocyte activation. For example, blocking, altering, or deleting an antigen receptor prevents activation. **Anergy** (an′er-jē), which means "without working," is a condition of inactivity in which a B or T cell does not respond to an antigen. Anergy develops when an MHC/antigen complex binds to an antigen receptor, and no costimulation occurs. For example, if a helper T cell encounters a self-antigen on a cell that cannot provide costimulation, the T cell is turned off. It's likely that only antigen-presenting cells can provide costimulation.

Inhibiting and Stimulating Immunity

Decreasing the production or activity of cytokines can suppress the immune system. For example, cyclosporine, a drug used to prevent the rejection of transplanted organs, inhibits the production of interleukin-2. Conversely, genetically engineered interleukins can be used to stimulate the immune system. Administering interleukin-2 has promoted the destruction of cancer cells in some cases by increasing the activities of effector T cells.

3. *Activation of suppressor T cells.* **Suppressor T cells** are a poorly understood group of T cells that are defined by their ability to suppress immune responses. It's likely that suppressor T cells are subpopulations of helper T cells and cytotoxic T cells. The suppressor (helper) T cells release suppressive cytokines, or the suppressor (cytotoxic) T cells kill antigen-presenting cells.

● 42. *What is tolerance? List three ways it is accomplished.*

Antibody-Mediated Immunity

Exposure of the body to an antigen can lead to activation of B cells and to production of antibodies, which are responsible for destruction of the antigen. Because antibodies occur in body fluids, **antibody-mediated immunity** is effective against extracellular antigens. These include bacteria, viruses, protozoans, fungi, parasites, and toxins when they are outside cells. Antibody-mediated immunity can also cause immediate hypersensitivity reactions (see "Clinical Focus: Immune System Problems of Clinical Significance" on p. 812).

Antibodies

Antibodies are proteins produced in response to an antigen. Large amounts of antibodies occur in plasma, although plasma also contains other proteins. On the basis of protein type and associated lipids, plasma proteins are separated into albumin and alpha-(α), beta-(β), and gamma-(γ)globulin parts. As a group, antibodies are sometimes called **gamma globulins** because they are mostly found in the γ-globulin part of plasma. They are also called **immunoglobulins (Ig)** because they are globulin proteins involved in immunity.

The five general classes of immunoglobulins are denoted IgG, IgM, IgA, IgE, and IgD (table 22.5). All classes of antibodies have a similar structure, consisting of four polypeptide chains (figure 22.19): two identical heavy chains and two identical light chains. Each light chain is attached to a heavy chain, and the ends of the combined heavy and light chains form the **variable region** of the antibody, which is the part that combines with the antigenic determinant of the antigen. Different antibodies have different variable regions, and they are specific for different antigens. The rest of the antibody is the **constant region,** which is responsible for activities of antibodies like the ability to activate complement or to attach the antibody to such cells as macrophages, basophils, mast cells, and eosinophils. All the antibodies of a particular class have nearly the same constant regions.

Table 22.5	**Classes of Antibodies and Their Functions**		
Antibody	**Total Serum Antibody (%)**	**Structure**	**Description**
IgG	80–85	IgG	Activates complement and functions as an opsonin to increase phagocytosis; can cross the placenta and provide immune protection to the fetus and newborn; responsible for Rh reactions, such as hemolytic disease of the newborn
IgM	5–10	IgM / IgA	Activates complement and acts as an antigen-binding receptor on the surface of B cells; responsible for transfusion reactions in the ABO blood system; often the first antibody produced in response to an antigen
IgA	15		Secreted into saliva, tears, and onto mucous membranes to provide protection on body surfaces; found in colostrum and milk to provide immune protection to the newborn
IgE	0.002	IgE / IgD (Heavy chain)	Binds to mast cells and basophils and stimulates the inflammatory response
IgD	0.2	(Light chain)	Functions as antigen-binding receptors on B cells

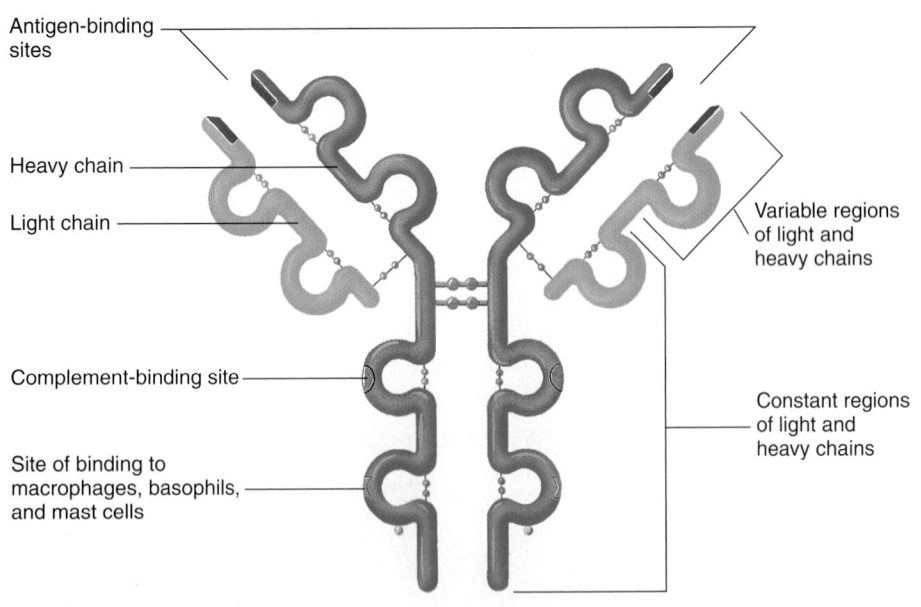

Antigen-binding sites

Heavy chain

Light chain

Variable regions of light and heavy chains

Complement-binding site

Constant regions of light and heavy chains

Site of binding to macrophages, basophils, and mast cells

Figure 22.19   Structure of an Antibody

Antibodies consist of two heavy and two light polypeptide chains. The variable region of the antibody binds to the antigen. The constant region of the antibody can activate the classical pathway of the complement cascade. The constant region can also attach the antibody to the plasma membrane of cells such as macrophages, basophils, or mast cells.

43. *What type of lymphocyte is responsible for antibody-mediated immunity? What are the functions of antibody-mediated immunity?*

44. *What are the functions of the constant and variable regions of an antibody? List the five classes of antibodies, and state their functions.*

Uses of Monoclonal Antibodies

Each type of **monoclonal antibody** is a pure antibody preparation that is specific for only one antigen. When the antigen is injected into a laboratory animal, it activates a B-cell clone against the antigen. The B cells are removed from the animal and fused with tumor cells. The resulting hybridoma cells have two ideal characteristics: they divide to form large numbers of cells, and the cells of a given clone produce only one kind of antibody.

Monoclonal antibodies are used for determining pregnancy and for diagnosing diseases like gonorrhea, syphilis, hepatitis, rabies, and cancer. These tests are specific and rapid because the monoclonal antibodies bind only to the antigen being tested. Monoclonal antibodies may someday be used to effectively treat cancer by delivering drugs to cancer cells (see "Immunotherapy" on p. 817).

Effects of Antibodies

Antibodies can directly affect antigens in two ways. The antibody can bind to the antigenic determinant and interfere with the ability of the antigen to function (figure 22.20a). Alternatively, the antibody can combine with an antigenic determinant on two different antigens, rendering the antigens ineffective (figure 22.20b). The ability of antibodies to join antigens together is the basis for many clinical tests, such as blood typing, because, when enough antigens are bound together, they become visible as a clump or a precipitate.

Although antibodies can directly affect antigens, most of their effectiveness results from other mechanisms. When an antibody (IgG or IgM) combines with an antigen through the variable region, the constant region can activate the complement cascade through the **classical pathway** (see figure 22.10 and 22.20c). Activated complement stimulates inflammation; attracts neutrophils, monocytes, macrophages, and eosinophils to sites of infection; and kills bacteria by lysis.

Antibodies (IgE) can initiate an inflammatory response (figure 22.20d). The antibodies attach to mast cells or basophils through their constant region. When antigens combine with the variable region of the antibodies, the mast cells or basophils release chemicals through exocytosis, and inflammation results.

Opsonins (op′sŏ-ninz) are substances that make an antigen more susceptible to phagocytosis. An antibody (IgG) acts as an opsonin by connecting to an antigen through the variable region of the antibody and to a macrophage through the constant region of the antibody. The macrophage then phagocytizes the antigen and the antibody (figure 22.20e).

Antibody Production

The production of antibodies after the first exposure to an antigen is different from that after a second or subsequent exposure. The **primary response** results from the first exposure of a B cell to an antigen for which it is specific and includes a series of cell divisions, cell differentiation, and antibody production. The B-cell receptors on the surface of B cells are antibodies, usually IgM and IgD. The receptors have the same variable region as the antibodies that are eventually produced by the B cell. Before stimulation by an antigen, B cells are small lymphocytes. After activation, the B cells undergo a series of divisions to produce large lymphocytes. Some of these enlarged cells become **plasma cells,** which produce antibodies, and others revert back to small lymphocytes and become **memory B cells** (figure 22.21). Usually, IgM is the first antibody produced in response to an antigen, but later other classes of antibodies are produced as well. The primary response normally takes 3–14 days to produce enough antibodies to be effective against the antigen. In the meantime, the individual usually develops disease symptoms because the antigen has had time to cause tissue damage.

(a) **Inactivates the antigen.** An antibody binds to an antigen and inactivates it.

(b) **Binds antigens together.** Antibodies bind several antigens together.

(c) **Activates the complement cascade.** An antigen binds to an antibody. As a result, the antibody can activate complement proteins, which can produce inflammation, chemotaxis, and lysis.

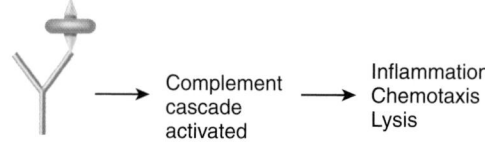

(d) **Initiates the release of inflammatory chemicals.** An antibody binds to a mast cell or basophil. When an antigen binds to the antibody, it triggers a release of chemicals that cause inflammation.

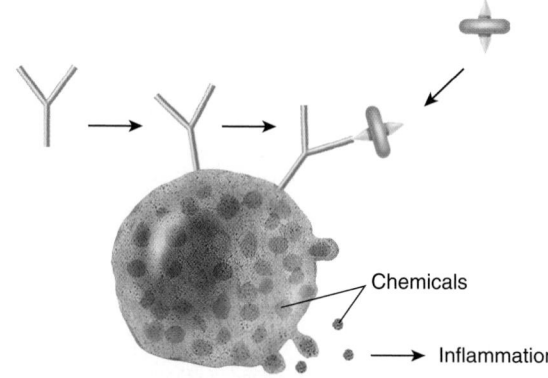

(e) **Facilitates phagocytosis.** An antibody binds to an antigen and then to a macrophage, which phagocytizes the antibody and antigen.

Figure 22.20   Effects of Antibodies

Antibodies directly affect antigens by inactivating the antigens or binding the antigens together. Antibodies indirectly affect antigens by activating other mechanisms through the constant region of the antibody. Indirect mechanisms include activation of complement, increased inflammation resulting from the release of inflammatory chemicals from mast cells or basophils, and increased phagocytosis resulting from antibody attachment to macrophages.

Process Figure 22.21   Antibody Production

Memory
B cells

B cell
1

Fewer
plasma
cells

More
memory
B cells

Memory
B cells
2

More
plasma
cells

More
antibodies

Fewer
antibodies

Magnitude
of response

First
exposure

Primary
response

Longer
response
time (3–14 days)

Secondary
response

Second
exposure

Shorter
response
time (hours to a few days)

1. **Primary response.** The primary response occurs
   when a B cell is first activated by an antigen. The
   B cell proliferates to form plasma cells and
   memory cells. The plasma cells produce
   antibodies.

2. **Secondary response.** The secondary response
   occurs when another exposure to the same antigen
   causes the memory cells to rapidly form plasma
   cells and additional memory cells. The secondary
   response is faster and produces more antibodies
   than the primary response.

The **secondary, or memory, response** occurs when the immune system is exposed to an antigen against which it has already produced a primary response. The secondary response results from memory B cells, which rapidly divide to produce plasma cells and large amounts of antibody when exposed to the antigen. The secondary response provides better protection than the primary response for two reasons. First, the time required to start producing antibodies is less (hours to a few days); and second, the amount of antibody produced is much larger. As a consequence, the antigen is quickly destroyed, no disease symptoms develop, and the person is immune.

The memory response also includes the formation of new memory B cells, which provide protection against additional exposures to the antigen. Memory B cells are the basis for adaptive immunity. After destruction of the antigen, plasma cells die, the antibodies they released are degraded, and antibody levels decline to the point at which they can no longer provide adequate protection. Memory B cells may persist for many years and probably for

life in some cases. If memory cell production is not stimulated, however, or if the memory B cells produced are short-lived, repeated infections of the same disease are possible. For example, the same cold virus can cause the common cold more than once in the same person.

45. *Describe the different ways that antibodies participate in the destruction of antigens.*

46. *What are plasma cells and memory cells, and what are their functions?*

47. *What are the primary and secondary antibody responses? Why doesn't the primary response prevent illness but the secondary response does?*

P R E D I C T   4

One theory for long-lasting immunity assumes that humans are continually exposed to the disease-causing agent. Explain how this exposure could produce lifelong immunity.

Clinical Focus   Immune System Problems of Clinical Significance

Hypersensitivity Reactions

Immune and hypersensitivity (allergy) reactions involve the same mechanisms, but the differences between them are unclear. Both require exposure to an antigen and subsequent stimulation of antibody-mediated immunity or cell-mediated immunity (or both). If immunity to an antigen is established, later exposure to the antigen results in an immune system response that eliminates the antigen, and no symptoms appear. In **hypersensitivity reactions,** the antigen is called an **allergen,** and later exposure to the allergen stimulates much the same process that occurs during the normal immune system response. The processes that eliminate the allergen, however, also produce undesirable side effects, such as a very strong inflammatory reaction. This immune system response can be more harmful than beneficial and can produce many unpleasant symptoms. Hypersensitivity reactions are categorized as immediate or delayed.

Immediate Hypersensitivities

An **immediate hypersensitivity reaction** occurs when antibodies interact with allergens and cause symptoms to appear within a few minutes of exposure to the allergens. Immediate hypersensitivity reactions include atopy, anaphylaxis, cytotoxic reactions, and immune complex disease.

Atopy (at'ō-pē) is a localized IgE-mediated hypersensitivity reaction. For example, plant pollens can be allergens that cause hay fever when they are inhaled and absorbed through the respiratory mucosa. The resulting localized inflammatory response produces swelling of the mucosa and excess mucus production. In asthma patients, allergens can stimulate the release of leukotrienes and histamine in the bronchioles of the lung, causing constriction of the smooth muscles of the bronchioles and difficulty in breathing. Hives (urticaria) is an allergic reaction that results in a skin rash or localized swellings and is usually caused by an ingested allergen.

Anaphylaxis (an'ă-fi-lak'sis) is a systemic IgE-mediated reaction and can be life-threatening. Introduction of allergens, such as drugs (e.g., penicillin) and insect stings, is the most common cause. The chemicals released from mast cells and basophils cause systemic vasodilation, a drop in blood pressure, and cardiac failure. Symptoms of hay fever, asthma, and hives may also be observed.

In **cytotoxic reactions,** IgG or IgM combines with the antigen on the surface of a cell, resulting in the activation of complement and subsequent lysis of the cell. A cytotoxic reaction against a bacterial cell can be protective, but against a human cell it can be harmful. Transfusion reactions caused by incompatible blood types, hemolytic disease of the newborn (see chapter 19), and some types of autoimmune disease are examples of harmful cytotoxic reactions.

Immune complex disease occurs when too many immune complexes are formed. Immune complexes are combinations of soluble antigens and IgG or IgM. When too many immune complexes are present, too much complement is activated, and an acute inflammatory response develops. Complement attracts neutrophils to the area of inflammation and stimulates the release of lysosomal enzymes. This release causes tissue damage, especially in small blood vessels, where the immune complexes tend to lodge; and lack of blood supply causes tissue necrosis. Arthus reactions, serum sickness, some autoimmune diseases, and chronic graft rejection are examples of immune complex diseases.

An **Arthus reaction** is a localized immune complex reaction. For example, suppose an individual has been sensitized to antigens in the tetanus toxoid vaccine because of repeated vaccinations. If that individual were vaccinated again, large amounts of antigen in the vaccine would be present at the injection site. Antibodies could complex with the antigens, causing a localized inflammatory response, neutrophil infiltration, and tissue necrosis.

Serum sickness is a systemic Arthus reaction in which the antibody–antigen complexes circulate and lodge in many different tissues. Serum sickness can develop from prolonged exposure to an antigen, which provides enough time for an antibody response and the formation of many immune complexes. Examples of antigens include long-lasting drugs and proteins found in the serum used for achieving passive artificial immunity. Symptoms include fever, swollen lymph nodes and spleen, and arthritis. Symptoms of anaphylaxis, such as hives, may also be present because IgE involvement is a part of serum sickness. If large numbers of the circulating antibody–antigen complexes are removed from the blood by the kidney, immune complex glomerulonephritis can develop, in which kidney blood vessels are destroyed and the kidneys fail to function.

Delayed Hypersensitivity

Delayed hypersensitivity is mediated by T cells, and symptoms usually take several hours or days to develop. Like immediate hypersensitivity, delayed hypersensitivity is an acute extension of the normal operation of the immune system. Exposure to the allergen causes activation of T cells and the production of cytokines. The cytokines attract basophils and monocytes, which differentiate into macrophages. The activities of these cells result in progressive tissue destruction, loss of function, and scarring.

Delayed hypersensitivity can develop as allergy of infection and contact hypersensitivity. **Allergy of infection** is a side effect of cell-mediated efforts to eliminate intracellular microorganisms, and the amount of tissue destroyed is determined by the persistence and distribution of the antigen. The minor rash of measles results from tissue damage as cell-mediated immunity destroys virus-infected cells.

In patients with chronic infections with long-term antigenic stimulation, the allergy-of-infection response can cause extensive tissue damage. The destruction of lung tissue in tuberculosis is an example.

Contact hypersensitivity is a delayed hypersensitivity reaction to allergens that contact the skin or mucous membranes. Poison ivy, poison oak, soaps, cosmetics, drugs, and a variety of chemicals can induce contact hypersensitivity, usually after prolonged exposure. The allergen is absorbed by epithelial cells, and T cells invade the affected area, causing inflammation and tissue destruction. Although itching can be intense, scratching is harmful because it damages tissues and causes additional inflammation.

Autoimmune Diseases

In **autoimmune disease,** the immune system fails to differentiate between self-antigens and foreign antigens. Consequently, an immune system response is produced against some self-antigens, resulting in tissue destruction. In many instances, autoimmunity probably results from a breakdown of tolerance, which normally prevents an immune system response to self-antigens. In a situation called molecular mimicry, a foreign antigen that is very similar to a self-antigen stimulates an immune system response. After the foreign antigen is eliminated, the immune system continues to act against the self-antigen. It's hypothesized that type I diabetes (see chapter 18) develops in this fashion. In susceptible people, a foreign antigen can stimulate adaptive immunity, especially cell-mediated immunity, which destroys the insulin-producing beta cells of the pancreas. Other autoimmune diseases that involve antibodies are rheumatoid arthritis, rheumatic fever, Graves' disease, systemic lupus erythematosus, and myasthenia gravis.

Immunodeficiencies

Immunodeficiency is a failure of some part of the immune system to function properly. A deficient immune system is not uncommon because it can have many causes. Inadequate protein in the diet inhibits protein synthesis, thereby allowing antibody levels to decrease. Stress can depress the immune system, and fighting an infection can deplete lymphocyte and granulocyte reserves and make a person more susceptible to further infection. Diseases that cause proliferation of lymphocytes, such as mononucleosis, leukemias, and myelomas,

can result in an abundance of lymphocytes that don't function properly. Finally, the immune system can purposefully be suppressed by drugs to prevent graft rejection.

Congenital (present at birth) immunodeficiencies can involve inadequate B-cell formation, inadequate T-cell formation, or both. **Severe combined immunodeficiency disease (SCID)** in which both B and T cells fail to differentiate, although rare, is probably the best known. Unless the person suffering from SCID is kept in a sterile environment or is provided with a compatible bone marrow transplant, death from infection results.

Tumor Control

Tumor cells have tumor antigens that distinguish them from normal cells. According to the concept of **immune surveillance,** the immune system detects tumor cells and destroys them before a tumor can form. T cells, natural killer cells, and macrophages are involved in the destruction of tumor cells. Immune surveillance may exist for some forms of cancer caused by viruses. The immune response appears to be directed more against the viruses, however, than against tumors in general. Only a few cancers are known to be caused by viruses in humans. For most tumors, the response of the immune system may be ineffective and too late.

Transplantation

Genes that code for the production of MHC molecules are generally called major histocompatibility complex genes. Histocompatibility refers to the ability of tissues (Greek, *histo*) to get along (compatibility) when tissues are transplanted from one individual to another. In humans, the major histocompatibility complex genes are often referred to as **human leukocyte antigen (HLA) genes** because they were first identified in leukocytes. The HLA genes control the production of HLAs, also called MHC antigens, which are inserted onto the surface of cells. The immune system can distinguish between self-cells and foreign cells because they are both marked with HLAs. Rejection

of a transplanted tissue is caused by a normal immune system response to the foreign HLAs. Millions of possible combinations of the HLA genes exist, and it's very rare for two individuals (except identical twins) to have the same set of HLA genes. Because they are genetically determined, however, the closer the relationship between two individuals, the greater the likelihood of sharing the same HLA genes.

Acute rejection of a graft occurs several weeks after transplantation and results from a delayed hypersensitivity reaction and cell lysis. Lymphocytes and macrophages infiltrate the area, a strong inflammatory response occurs, and the foreign tissue is destroyed. If acute rejection doesn't develop, **chronic rejection** may occur at a later time. In chronic rejection, immune complexes form in the arteries supplying the graft, blood supply fails, and the graft is rejected.

Graft rejection can occur in two different directions. In **host-versus-graft rejection,** the recipient's immune system recognizes the donor's tissue as foreign and rejects the transplant. In a **graft-versus-host rejection,** the donor tissue recognizes the recipient's tissue as foreign, and the transplant rejects the recipient, causing destruction of the recipient's tissues, and death.

To reduce graft rejection, a tissue match is performed. Only tissues with HLAs similar to the recipient's have a chance of acceptance. Even when the match is close, immunosuppressive drugs must be administered throughout the person's life to prevent rejection. Unfortunately, the person then has a drug-produced immunodeficiency and is more susceptible to infections. An exact match is possible only for a graft from one part to another part of the same person's body or between identical twins.

HLAs are important in ways in addition to organ transplants. Because they are genetically determined, characterization of HLAs can help resolve paternity suits. In forensic medicine, the HLAs in blood, semen, and other tissues help identify the person from whom the tissue came.

Cell-Mediated Immunity

Cell-mediated immunity is a function of T cells and is most effective against intracellular microorganisms, such as viruses, fungi, intracellular bacteria, and parasites. Delayed hypersensitivity reactions and control of tumors also involve cell-mediated immunity (see "Clinical Focus: Immune System Problems of Clinical Significance" on p. 812).

Antibody-mediated immunity is not effective against intracellular microorganisms while they are inside cells because antibodies cannot cross the plasma membrane. Cell-mediated immunity is effective against intracellular microorganisms because it destroys the cells in which they are sequestered. For example, when viruses infect cells, they enter the cells and direct the cells to make new viruses, which are then released to infect other cells. Thus, cells are turned into virus manufacturing plants. Cell-mediated immunity fights viral infections by destroying virally infected cells. When viruses infect cells, some viral proteins are broken down and become processed foreign antigens that are combined with MHC class I molecules and displayed on the surface of infected cell (figure 22.22). T cells can distinguish between virally infected cells and noninfected cells because T-cell receptors can bind to MHC class I/antigen complexes on the surface of infected cells. Binding of the T-cell receptor to the MHC class I/antigen complex is a signal for activating cytotoxic T cells. Costimulation by other surface molecules such as CD8 also occurs. Helper T cells provide costimulation by releasing cytokines, such as interleukin-2, which stimulates activation and cell division of T cells. Unlike their interactions with macrophages and B cells, however, helper T cells do not connect to T cells through MHC class II/antigen complexes or other surface molecules.

When the number of helper T cells increases, it results in greater stimulation of cytotoxic T cells. In cell-mediated responses, helper T cells are activated and stimulated to divide in the same fashion as in antibody-mediated responses (see figure 22.17).

After T cells are activated by antigen on the surface of a target cell, they undergo a series of divisions to produce cytotoxic T cells and memory T cells (figure 22.23). The cytotoxic T cells are responsible for the cell-mediated immunity response. Memory T cells can provide a secondary response and long-lasting immunity in the same fashion as memory B cells.

1. A MHC class I molecule displays an antigen, such as a viral protein, on the surface of a target cell.

2. Activation of a cytotoxic T cell begins when a T-cell receptor binds to the MHC class I/antigen complex.

3. There is costimulation of the cytotoxic T cell by CD8 and other surface molecules.

4. There is costimulation by cytokines, such as interleukin-2, released from helper T cells.

5. The activated cytotoxic T cell divides, and the resulting daughter cells divide, and so on, eventually producing many cytotoxic T cells (only two are shown here).

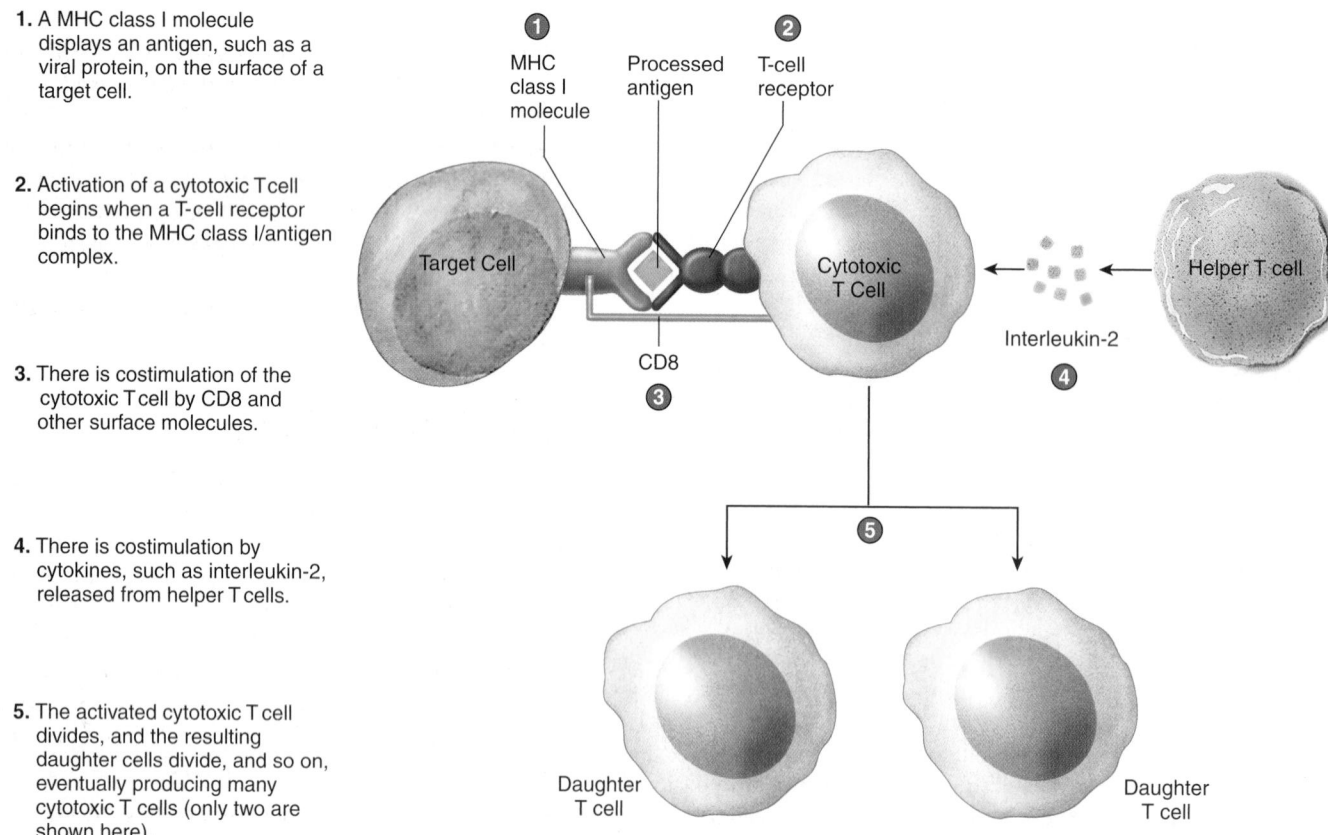

Process Figure 22.22   Proliferation of Cytotoxic T Cells

Figure 22.23   Stimulation and Effects of T Cells

When activated, cytotoxic T cells form many cytotoxic T cells and memory T cells. The cytotoxic T cells release cytokines that promote the destruction of the antigen or cause the lysis of target cells, such as virus-infected cells, tumor cells, or transplanted cells. The memory T cells are responsible for the secondary response.

Cytotoxic T Cells

Cytotoxic T cells have two main effects: they lyse cells and they produce cytokines (see figure 22.23). Cytotoxic T cells can come into contact with other cells and cause them to lyse. Virus-infected cells have viral antigens, tumor cells have tumor antigens, and tissue transplants have foreign antigens on their surfaces that can stimulate cytotoxic T-cell activity. A cytotoxic T cell binds to a target cell and releases chemicals that cause the target cell to lyse. The major method of lysis involves a protein called **perforin**, which is similar to the complement protein C9 (see figure 22.10). Perforin forms a channel in the plasma membrane of the target cell through which water enters the cell, causing lysis. The cytotoxic T cell then moves on to destroy additional target cells.

In addition to lysing cells, cytotoxic T cells release cytokines that activate additional components of the immune system. For example, one important function of cytokines is the recruitment of cells like macrophages. These cells are then responsible for phagocytosis and inflammation.

PREDICT ⑤

In patients with acquired immunodeficiency syndrome (AIDS), helper T cells are destroyed by a viral infection. The patients can die of pneumonia caused by an intracellular fungus *(Pneumocystis carinii)* or from Kaposi's sarcoma, which consists of tumorous growths in the skin and lymph nodes. Explain what is happening.

Delayed Hypersensitivity T Cells

Delayed hypersensitivity T cells respond to antigens by releasing cytokines. Consequently, they promote phagocytosis and inflammation, especially in allergic reactions (see "Clinical Focus: Immune System Problems of Clinical Significance" on p. 812).

For example, poison ivy antigens can be processed by Langerhans' cells in the skin, which present the antigen to delayed hypersensitivity T cells, resulting in an intense inflammatory response.

48. *What type of lymphocyte is responsible for cell-mediated immunity? What are the functions of cell-mediated immunity?*

49. *How do intracellular microorganisms stimulate cytotoxic T cells? What role do helper T cells play in this process?*

50. *State the two main responses of cytotoxic T cells.*

51. *What kind of immune response is produce by delayed hypersensitivity T cells?*

52. *How is long-lasting immunity achieved in cell-mediated immunity?*

Immune Interactions

Although the immune system can be described in terms of innate, antibody-mediated, and cell-mediated immunity, only one immune system really exists. These categories are an artificial division that is used to emphasize particular aspects of immunity. Actually, immune system responses often involve components of more than one type of immunity (figure 22.24). For example, although adaptive immunity can recognize and remember specific antigens, once recognition has occurred, many of the events that lead to the destruction of the antigen are innate immunity activities, such as inflammation and phagocytosis.

53. *Describe how interactions between innate, antibody-mediated, and cell-mediated immunity can eliminate an antigen.*

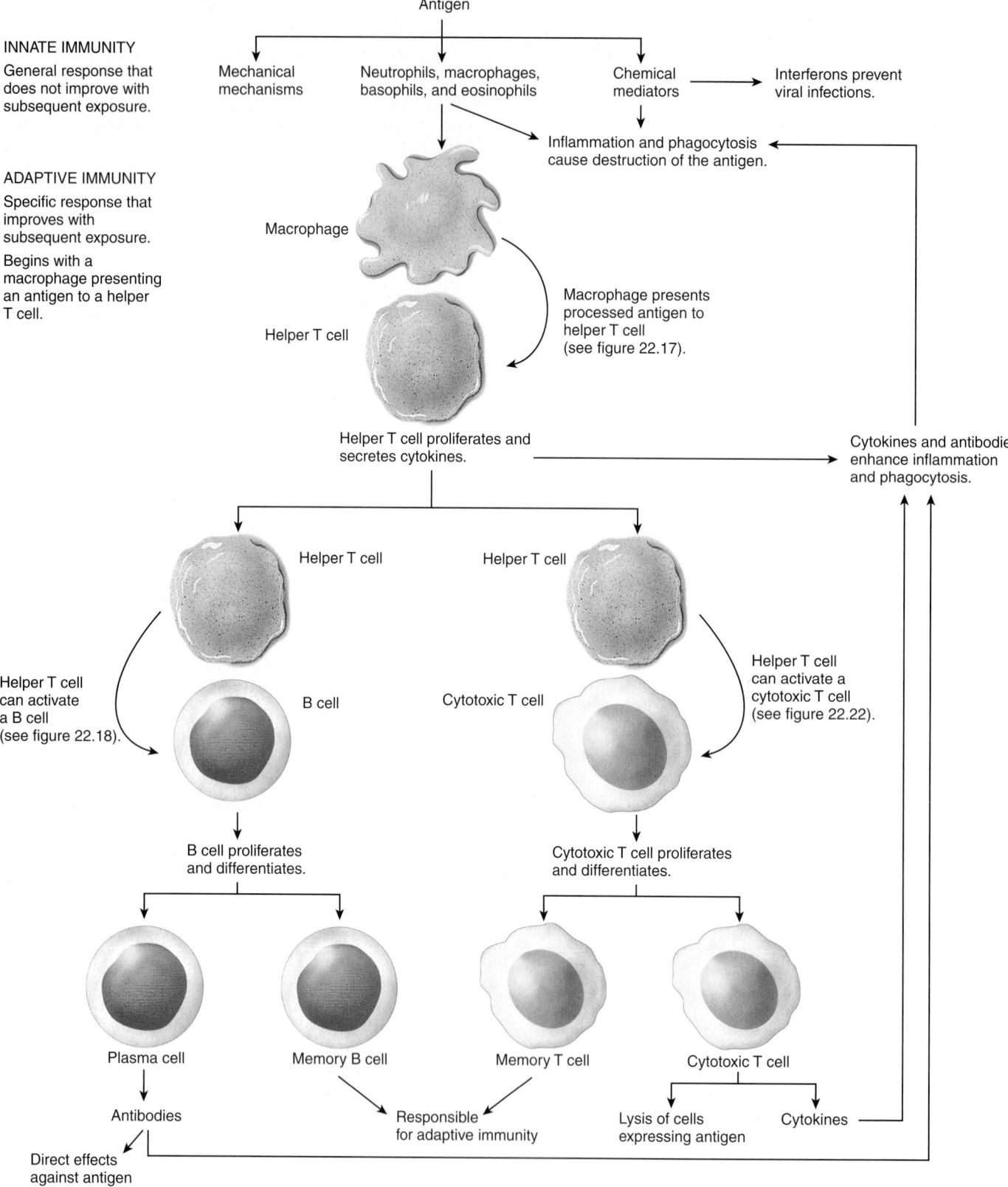

Antigen

INNATE IMMUNITY

General response that does not improve with subsequent exposure.

Mechanical mechanisms

Neutrophils, macrophages, basophils, and eosinophils

Chemical mediators

Interferons prevent viral infections.

Inflammation and phagocytosis cause destruction of the antigen.

ADAPTIVE IMMUNITY

Specific response that improves with subsequent exposure.

Begins with a macrophage presenting an antigen to a helper T cell.

Macrophage

Helper T cell

Macrophage presents processed antigen to helper T cell (see figure 22.17).

Helper T cell proliferates and secretes cytokines.

Cytokines and antibodies enhance inflammation and phagocytosis.

Helper T cell

Helper T cell

Helper T cell can activate a B cell (see figure 22.18).

B cell

Cytotoxic T cell

Helper T cell can activate a cytotoxic T cell (see figure 22.22).

B cell proliferates and differentiates.

Cytotoxic T cell proliferates and differentiates.

Plasma cell

Memory B cell

Memory T cell

Cytotoxic T cell

Antibodies

Responsible for adaptive immunity

Lysis of cells expressing antigen

Cytokines

Direct effects against antigen

Antibody-mediated immunity

Antibodies act against antigens in solution or on the surfaces of extracellular microorganisms.

Cell-mediated immunity

Cytotoxic T cells act against antigens bound to MHC molecules on the surface of cells; effective against intracellular microorganisms, tumors, and transplanted cells.

Figure 22.24   Immune Interactions

The major interactions and responses of innate and adaptive immunity to an antigen.

Immunotherapy

Knowledge of the basic ways that the immune system operates has produced two fundamental benefits: (1) an understanding of the cause and progression of many diseases, and (2) the development or proposed development of effective methods to prevent, stop, or even reverse diseases.

Immunotherapy treats disease by altering immune system function or by directly attacking harmful cells. Some approaches attempt to boost immune system function in general. For example, administering cytokines or other agents can promote inflammation and the activation of immune cells, which can help in the destruction of tumor cells. On the other hand, sometimes inhibiting the immune system is helpful. For example, multiple sclerosis is an autoimmune disease in which the immune system treats self-antigens as foreign antigens, thereby destroying the myelin that covers axons. Interferon beta (IFNβ) blocks the expression of MHC molecules that display self-antigens and is now being used to treat multiple sclerosis.

Some immunotherapy takes a more specific approach. For example, vaccination can prevent many diseases (see section on "Acquired Immunity" on p. 817). The ability to produce monoclonal antibodies may result in therapies that are effective for treating tumors. If an antigen unique to tumor cells can be found, then monoclonal antibodies could be used to deliver radioactive isotopes, drugs, toxins, enzymes, or cytokines that can kill the tumor cell or can activate the immune system to kill the cell. Unfortunately, no antigen on tumor cells has been found that is not also found on normal cells. Nonetheless, this approach may be useful if damage to normal cells is minimal. For example, tumor cells may have more surface antigens of a particular type than normal cells, resulting in greater treatment delivery. Tumor cells may also be more susceptible to damage, or normal cells may be better able to recover from the treatment.

One problem with monoclonal antibody delivery systems is that the immune system recognizes the monoclonal antibody as a foreign antigen. After the first exposure, a memory response quickly destroys the monoclonal antibodies, rendering the treatment ineffective. In a process called **humanization,** the monoclonal antibodies are modified to resemble human antibodies. This approach has allowed monoclonal antibodies to sneak past the immune system.

The use of monoclonal antibodies to treat tumors is mostly in the research stage of development, but a few clinical trials are now yielding promising results. For example, monoclonal antibodies with radioactive iodine (^{131}I) have caused regression of B-cell lymphomas and produced few side effects. Herceptin is a monclonal antibody that binds to a growth factor that is overexpressed in 25%–30% of primary breast cancers. The antibodies serve to "tag" cancer cells, which are then lysed by natural killer cells. Herceptin slows disease progression and increases survival time, but it's not a cure for breast cancer.

Many other approaches for immunotherapy are being studied, and the development of treatments that use the immune system are certain to increase in the future. Your knowledge of the immune system will enable you to understand and appreciate these therapies.

● **54. What is immunotherapy? Give examples.**

Neuroendocrine Regulation of Immunity

An intriguing possibility for reducing the severity of diseases or even curing them is to use neuroendocrine regulation of immunity. The nervous system regulates the secretion of hormones, such as cortisol, epinephrine, endorphins, and enkephalins, for which lymphocytes have receptors. For example, cortisol released during times of stress inhibits the immune system. In addition, most lymphatic tissues, including some individual lymphocytes, receive sympathetic innervation. That a neuroendocrine connection exists with the immune system is clear. The question we need to answer is: Can we use this connection to control our own immunotherapy?

Acquired Immunity

It's possible to acquire adaptive immunity in four ways: active natural, active artificial, passive natural, and passive artificial immunity (figure 22.25). The terms *natural* and *artificial* refer to the method of exposure. Natural exposure implies that contact

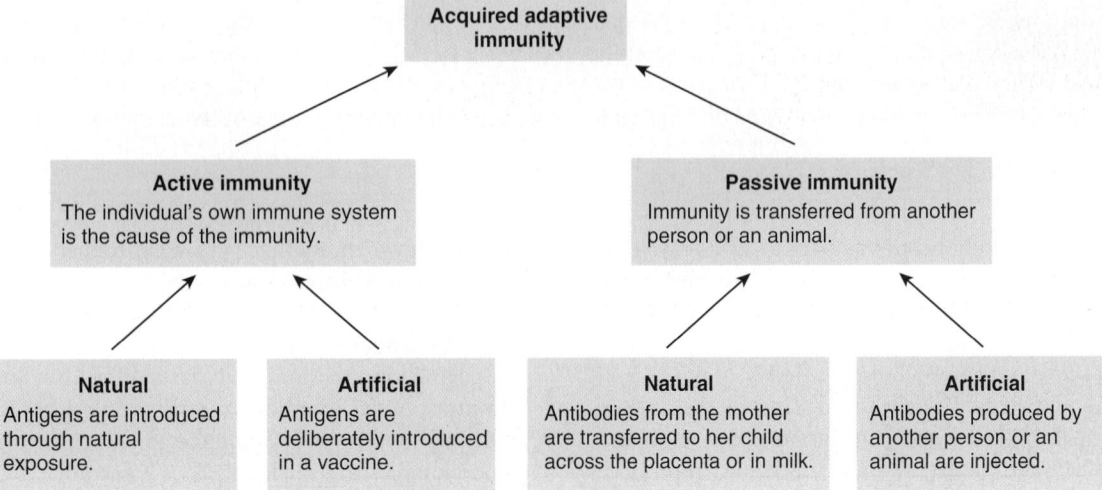

Figure 22.25 Ways to Acquire Adaptive Immunity

Clinical Focus Acquired Immunodeficiency Syndrome

Acquired immunodeficiency syndrome (AIDS) is a life-threatening disease caused by the **human immunodeficiency virus (HIV)**. Two strains of HIV are recognized: HIV-1 is responsible for most cases of AIDS, whereas HIV-2 is increasingly being found in West Africa. AIDS was first reported in 1981 in the United States. Since then, over 800,000 cases have been reported in the United States to the Centers for Disease Control and Prevention (CDC). The United Nations Program on AIDS (UNAIDS) estimates that 60 million people have been infected by HIV worldwide, and 18 million have died. The course of HIV infection varies. After contracting an HIV infection, some people die within a year; most, however, survive for 10 or 11 years, and some have survived beyond 20 years.

HIV is transmitted from an infected to a noninfected person in body fluids, such as blood, semen, or vaginal secretions. The major methods of transmission are unprotected intimate sexual contact, contaminated needles used by intravenous drug users, tainted blood products, and from a pregnant woman to her fetus. Present evidence indicates that household, school, or work contacts do not result in transmission.

In the United States, most cases of AIDS during the 1980s occurred in homosexual or bisexual men and in intravenous drug users. A small percentage of cases have resulted from transfusions or contaminated clotting factors used by hemophiliacs. Children can be infected before birth, during delivery, or after birth from breast-feeding. A few cases of AIDS have occurred in health-care workers accidentally exposed to HIV-infected blood or body fluids, and even fewer cases of health-care workers infecting patients have been documented. The most rapidly increasing group of AIDS patients in the United States is heterosexual women or men who have had sexual contact with an infected person. In other countries the pattern of AIDS cases is different from that in the United States. UNAIDS estimates that over 90% of all HIV infections globally are transmitted heterosexually.

Preventing transmission of HIV is presently the only way to prevent AIDS. The risk of transmission can be reduced by educating the public about relatively safe sexual practices, such as reducing the number of one's sexual partners, avoiding anal intercourse, and using condoms. Public education also includes warnings to intravenous drug users of the dangers of using contaminated needles. Ensuring the safety of the blood supply is another important preventive measure. In 1985, a test for HIV antibodies in blood became available. Heat treatment of clotting factors taken from blood has also been effective in preventing transmission of HIV to hemophiliacs.

HIV infection begins when a protein on the surface of the virus, called gp120, binds to a CD4 molecule on the surface of a cell. The CD4 molecule is found primarily on helper T cells, and it normally enables helper T cells to adhere to other lymphocytes, for example, during the process of antigen presentation. Certain monocytes, macrophages, neurons, and neuroglial cells also have CD4 molecules. Once attached to the CD4 molecules, the virus injects its genetic material (RNA) and enzymes into the cell and begins to replicate. Copies of the virus are manufactured using the organelles and materials within the cell. Replicated viruses escape from the cell and infect other cells.

Following infection by HIV, within 3 weeks to 3 months, many patients develop mononucleosis-like symptoms, such as fever, sweats, fatigue, muscle and joint aches, headache, sore throat, diarrhea, rash, and swollen lymph nodes. Within 1–3 weeks, these symptoms disappear as the immune system responds to the virus by producing antibodies and activating cytotoxic T cells that kill HIV-infected cells. The immune system is not able to completely eliminate HIV, however, and by about 6 months a kind of "set point" is achieved in which the virus continues to replicate at a low, but steady, rate. This chronic stage of infection lasts, on the average, for 8–10 years, and the infected person feels good and exhibits few, if any, symptoms.

Although helper T cells are infected and destroyed during the chronic stage of HIV infection, the body responds by producing large numbers of helper T cells. Nonetheless, over a period of years the HIV numbers gradually increase and helper T cell numbers decrease. Normally approximately 1200 helper T cells are present per cubic millimeter of blood. An HIV-infected person is considered to have AIDS when one or more of the following conditions appear: the helper T cell count falls below 200 cells/mm^3, an opportunistic infection occurs, or Kaposi's sarcoma develops.

Opportunistic infections involve organisms that normally don't cause disease but can do so when the immune system is depressed. Without helper T cells, cytotoxic T-

---

with an antigen or antibody occurs as part of everyday living and is not deliberate. Artificial exposure, also called **immunization,** is a deliberate introduction of an antigen or antibody into the body.

The terms *active* and *passive* indicate whether or not an individual's immune system is directly responding to the antigen. When an individual is naturally or artificially exposed to an antigen, an adaptive immune system response can occur that produces antibodies. This is called **active immunity** because the individual's own immune system is the cause of the immunity. **Passive immu-nity** occurs when another person or animal develops antibodies and the antibodies are transferred to a nonimmune individual. This is called passive immunity because the nonimmune individual didn't produce the antibodies.

How long the immunity lasts differs for active and passive immunity. Active immunity can persist for a few weeks (common cold) to a lifetime (whooping cough and chickenpox). Immunity can be long lasting if enough B or T memory cells are produced and persist to respond to later antigen exposure. Passive immunity is not long lasting because the individual doesn't produce his or her and B-cell activation is impaired, and adaptive resistance is suppressed. Examples of opportunistic infections include pneumocystis (noo-mō-sis′tis) pneumonia (caused by an intracellular fungus, *Pneumocystis carinii*), tuberculosis (caused by an intracellular bacterium, *Myocobacterium tuberculosis*), syphilis (caused by a sexually transmitted bacterium, *Treponema pallidum*), candidiasis (kan-di-dī′ă-sis; a yeast infection of the mouth or vagina caused by *Candida albicans*), and protozoans that cause severe, persistent diarrhea. Kaposi's sarcoma is a type of cancer that produces lesions in the skin, lymph nodes, and visceral organs. Also associated with AIDS are symptoms resulting from the effects of HIV on the nervous system, including motor retardation, behavioral changes, progressive dementia, and possibly psychosis.

No cure for AIDS has yet been discovered. Management of AIDS can be divided into two categories: (1) management of secondary infections or malignancies associated with AIDS and (2) treatment of HIV. In order for HIV to replicate, the viral RNA is used to make viral DNA, which is inserted into the host cell's DNA. The inserted viral DNA directs the production of new viral RNA and proteins, which are assembled to form new HIV. Key steps in the replication of HIV require viral enzymes. **Reverse transcriptase** promotes the formation of viral DNA from viral RNA, and **integrase** (in′te-grās) inserts the viral DNA into the host cell's DNA. A viral **protease** (prō′tē-ās) breaks large viral proteins into smaller proteins, which are incorporated into the new HIV.

Blocking the activity of HIV enzymes can inhibit replication of HIV. The first effective treatment of AIDS was the drug azidothymidine (AZT) (az′i-dō-thī′mi-dēn), also called zidovudine (zī-dō′voo-dēn). AZT is a **reverse transcriptase inhibitor,** which prevents HIV RNA from producing viral DNA. AZT can delay the onset of AIDS but doesn't appear to increase the survival time of AIDS patients. However, the number of babies who contract AIDS from their HIV-infected mothers can be dramatically reduced by giving AZT to the mothers during pregnancy and to the babies following birth. AZT can produce serious side effects such as anemia or even total bone marrow failure.

Often after 6–18 months of treatment with AZT, viral mutations result in HIV that are resistant to AZT. Other drugs that inhibit viral nucleic acid replication, such as dideoxyinosine (DDI) (dī′dē-oks-ē-ī′nō-sēn), have been developed. These drugs have been used for patients who are resistant to, or do not respond to, AZT.

Protease inhibitors are drugs that interfere with viral proteases. Examples of protease inhibitors are ritonavir and indinavir. The current treatment for suppressing HIV replication is a combination of three drugs, such as two reverse transcriptase inhibitors and one protease inhibitor. It's less likely that HIV will develop resistance to all three drugs. This strategy has proven very effective in reducing the death rate from AIDS and partially restoring health in some individuals.

Still in the research stage are **integrase inhibitors,** which prevent the insertion of viral DNA into the host cell's DNA. Perhaps someday integrase inhibitors will be part of a combination drug therapy for AIDS.

Another advance in AIDS treatment is a test for measuring **viral load,** which measures the number of viral RNA molecules in a milliliter of blood. The actual level of HIV is one-half the RNA count because each HIV has two RNA strands. Viral load is a good predictor of how soon a person will develop AIDS. If viral load is high, the onset of AIDS is much sooner than if it is low. It's also possible to detect developing viral resistance by an increase in viral load. In response, a change in drug dose or type may slow viral replication. Current treatment guidelines are to keep viral load below 500 RNA molecules per milliliter of blood.

An effective treatment for AIDS is not a cure. Even if viral load decreases to the point that the virus is undetected in the blood, the virus still remains in cells throughout the body. It's possible that the virus will eventually mutate and escape drug suppression. In addition, the long-term effects of these drug therapies are unknown.

The long-term goal for dealing with AIDS is to develop a vaccine that prevents HIV infection. Although research to develop vaccines, as well as clinical trials to test them, is occurring around the world, an effective vaccine has not yet been demonstrated.

Because of improved treatment, people with HIV/AIDS can now live for many years. HIV/AIDS is, therefore, being viewed increasingly as a chronic disease, not as a death sentence. A multidisciplinary team that includes occupational therapists, physical therapists, nutritionists/dieticians, psychologists, infectious disease physicians, and others can work together to manage patients with HIV/AIDS to help them have a better quality of life.

own memory cells. Because active immunity can last longer than passive immunity, it's the preferred method. Passive immunity is preferred, however, in some situations when immediate protection is needed.

Active Natural Immunity

Natural exposure to an antigen, such as a disease-causing microorganism, can cause an individual's immune system to mount an adaptive immune system response against the antigen and achieve **active natural immunity.** Because the individual is not immune during the first exposure, he or she usually develops the symptoms of the disease. Interestingly, exposure to an antigen doesn't always produce symptoms. Many people, if exposed to the poliomyelitis virus at an early age, have an immune system response and produce poliomyelitis antibodies, yet they don't exhibit any disease symptoms.

Active Artificial Immunity

In **active artificial immunity,** an antigen is deliberately introduced into an individual to stimulate the immune system. This process is **vaccination,** and the introduced antigen is a **vaccine.** Injection of the vaccine is the usual mode of administration. Examples of injected vaccinations are the DTP injection against diphtheria, tetanus, and pertussis (whooping cough); and the MMR injection against mumps, measles, and rubella (German measles). Sometimes the vaccine is ingested, as in the oral poliomyelitis vaccine (OPV).

The vaccine usually consists of some part of a microorganism, a dead microorganism, or a live, altered microorganism. The antigen has been changed so that it will stimulate an immune response but will not cause the symptoms of disease. Because active artificial immunity produces long-lasting immunity without disease symptoms, it's the preferred method of acquiring adaptive immunity.

PREDICT **6**

In some cases, a booster shot is used as part of a vaccination procedure. A booster shot is another dose of the original vaccine given some time after the original dose was administered. Why are booster shots given?

Passive Natural Immunity

Passive natural immunity results from the transfer of antibodies from a mother to her child across the placenta before birth. During her life, the mother has been exposed to many antigens, either naturally or artificially, and she has antibodies against many of these antigens. These antibodies protect the mother and the developing fetus against disease. Some of the antibodies (IgG) can cross the placenta and enter the fetal blood. Following birth, the antibodies provide protection for the first few months of the baby's life. Eventually the antibodies are broken down, and the baby must rely on his or her own immune system. If the mother nurses her baby, antibodies (IgA) in the mother's milk may also provide some protection for the baby.

Passive Artificial Immunity

Achieving **passive artificial immunity** usually begins with vaccinating an animal, such as a horse. After the animal's immune system responds to the antigen, antibodies (sometimes T cells) are removed from the animal and injected into the individual requiring immunity. In some cases, a human who has developed immunity through natural exposure or vaccination is used as a source of antibodies. Passive artificial immunity provides immediate protection for the individual receiving the antibodies and is therefore preferred when time might not be available for the individual to develop his or her own immunity. This technique provides only temporary immunity, however, because the antibodies are used or eliminated by the recipient.

Antiserum is the general term used for serum, which is plasma minus the clotting factors, that contains antibodies responsible for passive artificial immunity. Antisera are available against microorganisms that cause diseases such as rabies, hepatitis, and measles; bacterial toxins such as tetanus, diphtheria, and botulism; and venoms from poisonous snakes and black widow spiders.

55. *Distinguish between active and passive immunity.*
56. *State four general ways of acquiring adaptive immunity. Which two provide the longest lasting immunity?*

Effects of Aging on the Lymphatic System and Immunity

Aging appears to have little effect on the ability of the lymphatic system to remove fluid from tissues, absorb fats from the digestive tract, or remove defective red blood cells from the blood.

Aging also seems to have little direct effect on the ability of B cells to respond to antigens, and the number of circulating B cells remains stable in most individuals. With age, thymic tissue is replaced with adipose tissue, and the ability to produce new, mature T cells in the thymus is eventually lost. Nonetheless, the number of T cells remains stable in most individuals due to the replication (not maturation) of T cells in secondary lymphatic tissues. In many individuals, however, there is a decreased ability of helper T cells to proliferate in response to antigens. Thus, antigen exposure produces fewer helper T cells, which results in less stimulation of B cells and effector T cells. Consequently, both antibody-mediated immunity and cell-mediated immunity responses to antigens decrease.

Primary and secondary antibody responses decrease with age. More antigen is required to produce a response, the response is slower, less antibody is produced, and fewer memory cells result. Thus, the ability to resist infections and develop immunity decreases. It's recommended that vaccinations should be given well before age 60 because these declines are most evident after age 60. Vaccinations, however, can be beneficial at any age, especially if the individual has reduced resistance to infection.

The ability of cell-mediated immunity to resist intracellular pathogens decreases with age. For example, the elderly are more susceptible to influenza (flu) and should be vaccinated every year. Some pathogens cause disease but are not eliminated from the body. With age, a decrease in immunity can result in reactivation of the pathogen. For example, the virus that causes chickenpox in children can remain latent within nerve cells even though the disease seems to have disappeared. Later in life, the virus can leave the nerve cells and infect skin cells, causing painful lesions known as herpes zoster or shingles.

Autoimmune disease occurs when immune responses destroy otherwise healthy tissue (see "Autoimmune Diseases" on p. 813). There is very little increase in the number of new-onset autoimmune diseases in the elderly. However, the chronic inflammation and immune responses that began earlier in life have a cumulative, damaging effect. The increased incidence of cancer in the elderly is assumed to be related to a decrease in the immune response.

57. *What effect does aging have on the major functions of the lymphatic system?*
58. *Describe the effects of aging on B cells and T cells. Give examples of how this affects antibody-mediated immunity and cell-mediated immunity responses.*

Systems Pathologies
Systemic Lupus Erythematosus

Mrs. L is a 30-year-old divorced woman with two children. Despite the fact that she has to work to support herself and the children, she entered college, determined to become a nurse and provide a better life for her family. Mrs. L was an excellent student, but her class attendance and her performance on tests were somewhat erratic. Sometimes she seemed very energetic and earned high grades, but other times she seemed depressed and didn't do as well. Toward the end of the course, she developed a rash on her face (figure A), a large red lesion on her arm, and was obviously not feeling well. Mrs. L went to the instructor to ask if she could take an incomplete grade and take the last exam at a later time. She explained that she has had lupus since she was 25 years old. Normally, medication helps to control her symptoms, but the stress of being a single parent combined with the challenges of school seemed to be making her condition worse. She further explained that the symptoms of lupus come and go, and bed rest was often helpful. Mrs. L finished the course requirements later that summer. She went on to complete her education and now has a full-time job as a nurse at a local hospital.

Background Information

Systemic lupus erythematosus (SLE) is a disease of unknown cause in which tissues and cells are damaged by the immune system. The name describes some of the characteristics of the disease. The term *lupus* literally means wolf and was originally used to refer to eroded (as if gnawed by a wolf) lesions of the skin. Erythematosus refers to a redness of the skin resulting from inflammation. Unfortunately, as the term *systemic* implies, the disorder is not confined to the skin but can affect tissues and cells throughout the body. Another systemic effect is the presence of low-grade fever in most cases of active SLE.

SLE is an autoimmune disorder in which a large variety of antibodies are produced that recognize self-antigens, such as nucleic acids, phospholipids, coagulation factors, red blood cells, and platelets. The combination of the antibodies with self-antigens forms immune complexes that circulate throughout the body to be deposited in various tissues, in which they stimulate inflammation and tissue destruction. Thus, SLE is a disease that can affect many different systems of the body. For example, the most common antibodies act against DNA that is released from damaged cells. Normally the liver removes the DNA, but when DNA and antibodies form immune complexes, they tend to be deposited in the kidneys and other tissues.

Figure A   Systemic Lupus Erythematosus

The butterfly rash resulting from inflammation in the skin caused by systemic lupus erythematosus.

Approximately 40%–50% of individuals with SLE develop renal disease. In some cases, the antibodies can bind to antigens on cells, resulting in lysis of the cells. For example, the binding of antibodies to red blood cells results in hemolysis and the development of anemia.

The cause of SLE is unknown. The most popular hypothesis is that a viral infection disrupts the function of suppressor T cells, resulting in loss of tolerance to self-antigens. The picture is probably more complicated, however, because not all SLE patients have reduced numbers of suppressor T cells. In addition, some patients have decreased numbers of the helper T cells that normally stimulate suppressor T-cell activity.

Genetic factors probably contribute to the development of the disease. The likelihood of developing SLE is much higher if a family member also has it. In addition, family members of SLE patients who don't have SLE are much more likely to have DNA antibodies than does the general population.

Approximately 1 out of 2000 individuals in the United States has SLE. The first symptoms usually appear between 15 and 25 years of age and affect women approximately nine times as often as men. The progress of the disease is unpredictable, with flare-ups of symptoms followed by periods of remission. The survival after diagnosis is greater than 90% after 10 years. The most frequent causes of death involve kidney failure, central nervous system dysfunction, infections, and cardiovascular disease.

System Interactions   The Effect of Systemic Lupus Erythematosus on Other Systems

System	Interactions
Integumentary	Skin lesions frequently occur and are made worse by exposure to the sun. There are three forms: (1) an inflammatory redness that can take the form of the butterfly rash, which extends from the bridge of the nose to the cheeks; (2) small, localized pimplelike eruptions accompanied by scaling of the skin; (3) areas of atrophied, depigmented skin with borders of increased pigmentation. Diffuse thinning of the hair results from hair loss.
Skeletal	Arthritis, tendonitis, and death of bone tissue can occur.
Muscular	Destruction of muscle tissue and muscular weakness can occur.
Nervous	Memory loss, intellectual deterioration, disorientation, psychosis, reactive depression, headache, seizures, nausea, and loss of appetite can occur. Stroke is a major cause of dysfunction and death. Cranial nerve involvement results in facial muscle weakness, drooping of the eyelid, and double vision. Central nervous system lesion can cause paralysis.
Endocrine	Sex hormones may play a role in SLE because 90% of the cases occur in females, and females with SLE have reduced levels of androgens.
Cardiovascular	Inflammation of the pericardium (pericarditis) with chest pain can develop. Damage to heart valves, inflammation of cardiac tissue, tachycardia, arrhythmias, angina, and myocardial infarction can also occur. Hemolytic anemia, and leukopenia can be present (see chapter 19). Antiphospholipid antibody syndrome, through an unknown mechanism, increases coagulation and thrombus formation, which increases the risk of stroke and heart attack.
Respiratory	Chest pain caused by inflammation of the pleural membranes; fever, shortness of breath, and hypoxemia caused by inflammation of the lungs; and alveolar hemorrhage can develop.
Digestive	Ulcers develop in the oral cavity and pharynx. Abdominal pain and vomiting are common, but no cause can be found. Inflammation of the pancreas and occasionally enlargement of the liver and minor abnormalities in liver function tests occur.
Urinary	Renal lesions and glomerulonephritis can result in progressive failure of kidney function. Excess proteins are lost in the urine, resulting in lower-than-normal blood proteins, which can produce edema.

No cure for SLE exists, nor does one standard of treatment, because the course of the disease is highly variable and many differences can be found among patients. Treatment usually begins with mild medications and proceeds to more and more potent therapies as conditions warrant. Aspirin and nonsteroidal anti-inflammatory drugs are used to suppress inflammation. Antimalarial drugs are used to treat skin rash and arthritis in SLE, but the mechanism of action is unknown. Patients who don't respond to these drugs or those with severe SLE are helped by steroids. Although steroids effectively suppress inflammation, they can produce undesirable side effects, including suppression of normal adrenal gland functions. In patients with life-threatening SLE, very high doses of steroids are used.

P R E D I C T

The red lesion Mrs. L developed on her arm is called purpura (pŭr′poo-rä), which is caused by bleeding into the skin. The lesions gradually change color and disappear in 2–3 weeks. Explain how SLE produces purpura.

S U M M A R Y

Lymphatic System   (p. 786)

The lymphatic system consists of lymph, lymphatic vessels, lymphatic tissue, lymphatic nodules, lymph nodes, tonsils, the spleen, and the thymus.

Functions of the Lymphatic System

The lymphatic system maintains fluid balance in tissues, absorbs fats from the small intestine, and defends against microorganisms and foreign substances.

Lymphatic Vessels

1. Lymphatic vessels carry lymph away from tissues.
2. Lymphatic capillaries lack a basement membrane and have loosely overlapping epithelial cells. Fluids and other substances easily enter the lymphatic capillary.

3. Lymphatic capillaries join to form lymphatic vessels.
   - Lymphatic vessels have valves that ensure one-way flow of lymph.
   - Skeletal muscle action, contraction of lymphatic vessel smooth muscle, and thoracic pressure changes move the lymph.
4. Lymph nodes are along the lymphatic vessels. After passing through lymph nodes, lymphatic vessels form lymphatic trunks and lymphatic ducts.
5. Lymphatic trunks and ducts empty into the blood at thoracic veins (junctions of the internal jugular and subclavian veins).
   - Lymph from the right thorax, the upper-right limb, and the right side of the head and the neck enters right thoracic veins.
   - Lymph from the lower limbs, pelvis, and abdomen; the left thorax; the upper-left limb; and the left side of the head and the neck enters left thoracic veins.

6. The jugular, subclavian, and brochomediastinal trunks may unite to form the right lymphatic duct.
7. The thoracic duct is the largest lymphatic vessel.
8. The intestinal and lumbar trunks may converge on the cisterna chyli, a sac that joins the inferior end of the thoracic duct.

Lymphatic Tissue and Organs

1. Lymphatic tissue is reticular connective tissue that contains lymphocytes and other cells.
2. Lymphatic tissue can be surrounded by a capsule (lymph nodes, spleen, thymus).
3. Lymphatic tissue can be nonencapsulated (diffuse lymphatic tissue, lymphatic nodules, tonsils). Mucosa-associated lymphoid tissue is nonencapsulated lymphatic tissue located in and below the mucous membranes of the digestive, respiratory, urinary, and reproductive tracts.
4. Diffuse lymphatic tissue consists of dispersed lymphocytes and has no clear boundaries.
5. Lymphatic nodules are small aggregates of lymphatic tissue (e.g., Peyer's patches in the small intestines).
6. The tonsils
   • The tonsils are large groups of lymphatic nodules in the oral cavity and nasopharynx.
   • The three groups of tonsils are the palatine, pharyngeal, and lingual tonsils.
7. Lymph nodes
   • Lymphatic tissue in the node is organized into the cortex and the medulla. Lymphatic sinuses extend through the lymphatic tissue.
   • Substances in lymph are removed by phagocytosis, or they stimulate lymphocytes (or both).
   • Lymphocytes leave the lymph node and circulate to other tissues.
8. The spleen
   • The spleen is in the left superior side of the abdomen.
   • Foreign substances stimulate lymphocytes in the white pulp (periarterial lymphatic sheath and lymphatic nodules).
   • Foreign substances and defective red blood cells are removed from the blood by phagocytes in the red pulp (splenic cords and venous sinuses).
   • The spleen is a limited reservoir for blood.
   • Most blood flows through the spleen in a few seconds. About 20% of the blood takes a few minutes to flow through the spleen, and about 2% takes an hour or more.
9. The thymus
   • The thymus is a gland in the superior mediastinum and is divided into a cortex and a medulla.
   • Lymphocytes in the cortex are separated from the blood by reticular cells.
   • Lymphocytes produced in the cortex migrate through the medulla, enter the blood, and travel to other lymphatic tissues, where they can proliferate.

Overview of the Lymphatic System
See figure 22.9.

Immunity (p. 796)
Immunity is the ability to resist the harmful effects of microorganisms and other foreign substances.

Innate Immunity (p. 796)
Mechanical Mechanisms
Mechanical mechanisms prevent the entry of microbes (skin and mucous membranes) or remove them (tears, saliva, and mucus).

Chemical Mediators
1. Chemical mediators promote phagocytosis and inflammation.
2. Complement can be activated by either the alternative or the classical pathway. Complement lyses cells, increases phagocytosis, attracts immune system cells, and promotes inflammation.

3. Interferons prevent viral replication. Interferons are produced by virally infected cells and move to other cells, which are then protected.

Cells
1. Chemotactic factors are parts of microorganisms or chemicals that are released by damaged tissues. Chemotaxis is the ability of white blood cells to move to tissues that release chemotactic factors.
2. Phagocytosis is the ingestion and destruction of materials.
3. Neutrophils are small phagocytic cells.
4. Macrophages are large phagocytic cells.
   • Macrophages can engulf more than neutrophils can.
   • Macrophages in connective tissue protect the body at locations where microbes are likely to enter, and macrophages clean blood and lymph.
5. Basophils and mast cells release chemicals that promote inflammation.
6. Eosinophils release enzymes that reduce inflammation.
7. Natural killer cells lyse tumor cells and virus-infected cells.

Inflammatory Response
1. The inflammatory response can be initiated in many ways.
   • Chemical mediators cause vasodilation and increase vascular permeability, which allows the entry of other chemical mediators.
   • Chemical mediators attract phagocytes.
   • The amount of chemical mediators and phagocytes increases until the cause of the inflammation is destroyed. Then the tissue undergoes repair.
2. Local inflammation produces the symptoms of redness, heat, swelling, pain, and loss of function. Symptoms of systemic inflammation include an increase in neutrophil numbers, fever, and shock.

Adaptive Immunity (p. 801)
1. Antigens are large molecules that stimulate an adaptive immune system response. Haptens are small molecules that combine with large molecules to stimulate an adaptive immune system response.
2. B cells are responsible for humoral, or antibody-mediated, immunity. T cells are involved with cell-mediated immunity.

Origin and Development of Lymphocytes
1. B cells and T cells originate in red bone marrow. T cells are processed in the thymus, and B cells are processed in bone marrow.
2. Positive selection ensures the survival of lymphocytes that can react against antigens, and negative selection eliminates lymphocytes that react against self-antigens.
3. A clone is a group of identical lymphocytes that can respond to a specific antigen.
4. B cells and T cells move to lymphatic tissue from their processing sites. They continually circulate from one lymphatic tissue to another.
5. Primary lymphatic organs (red bone marrow and thymus) are where lymphocytes mature into functional cells. Secondary lymphatic organs and tissues are where lymphocytes produce an immune response.

Activation of Lymphocytes
1. The antigenic determinant is the specific part of the antigen to which the lymphocyte responds. The antigen receptor (T-cell receptor or B-cell receptor) on the surface of lymphocytes combines with the antigenic determinant.
2. MHC class I molecules display antigens on the surface of nucleated cells, resulting in the destruction of the cells.
3. MHC class II molecules display antigens on the surface of antigen-presenting cells, resulting in the activation of immune cells.
4. MHC/antigen complex and costimulation are usually necessary to activate lymphocytes. Costimulation involves cytokines and certain surface molecules.
5. Antigen-presenting cells stimulate the proliferation of helper T cells, which stimulate the proliferation of B or T effector cells.

Inhibition of Lymphocytes

1. Tolerance is suppression of the immune system's response to an antigen.
2. Tolerance is produced by deletion of self-reactive cells, by preventing lymphocyte activation, and by suppressor T cells.

Antibody-Mediated Immunity

1. Antibodies are proteins.
   • The variable region of an antibody combines with the antigen. The constant region activates complement or binds to cells.
   • Five classes of antibodies exist: IgG, IgM, IgA, IgE, and IgD.
2. Antibodies affect the antigen in many ways.
   • Antibodies bind to the antigen and interfere with antigen activity or bind the antigens together.
   • Antibodies act as opsonins (a substance that increases phagocytosis) by binding to the antigen and to macrophages.
   • Antibodies can activate complement through the classical pathway.
   • Antibodies attach to mast cells or basophils and cause the release of inflammatory chemicals when the antibody combines with the antigen.
3. The primary response results from the first exposure to an antigen. B cells form plasma cells, which produce antibodies and memory B cells.
4. The secondary response results from exposure to an antigen after a primary response, and memory B cells quickly form plasma cells and additional memory B cells.

Cell-Mediated Immunity

1. Cells infected with intracellular microorganisms process antigens that combine with MHC class I molecules.
2. Cytotoxic T cells are stimulated to divide, producing more cytotoxic T cells and memory T cells, when MHC class I/antigen complexes are presented to T-cell receptors. Cytokines released from helper T cells also stimulate cytotoxic T cells.
3. Cytotoxic T cells lyse virus-infected cells, tumor cells, and tissue transplants.
4. Cytotoxic T cells produce cytokines, which promote phagocytosis and inflammation.

Immune Interactions   (p. 815)

Innate immunity, antibody-mediated immunity, and cell-mediated immunity can function together to eliminate an antigen.

Immunotherapy   (p. 817)

Immunotherapy stimulates or inhibits the immune system to treat diseases.

Acquired Immunity   (p. 817)
Active Natural Immunity

Active natural immunity results from natural exposure to an antigen.

Active Artificial Immunity

Active artificial immunity results from deliberate exposure to an antigen.

Passive Natural Immunity

Passive natural immunity results from the transfer of antibodies from a mother to her fetus or baby.

Passive Artificial Immunity

Passive artificial immunity results from transfer of antibodies (or cells) from an immune animal to a nonimmune animal.

Effects of Aging on the Lymphatic System and Immunity   (p. 820)

1. Aging has little effect on the ability of the lymphatic system to remove fluid from tissues, absorb fats from the digestive tract, or remove defective red blood cells from the blood.
2. Decreased helper T cell proliferation results in decreased antibody-mediated immunity and cell-mediated immunity responses to antigens.
3. The primary and secondary antibody responses decrease with age.
4. The ability to resist intracellular pathogens decreases with age.

R E V I E W   A N D   C O M P R E H E N S I O N

1. The lymphatic system
   a. removes excess fluid from tissues.
   b. absorbs fats from the digestive tract.
   c. defends the body against microorganisms and other foreign substances.
   d. all of the above.

2. Lymphatic capillaries
   a. have a basement membrane.
   b. are less permeable than blood capillaries.
   c. prevent backflow of lymph into the tissues.
   d. all of the above.

3. Lymph is moved through lymphatic vessels because of
   a. contraction of surrounding skeletal muscles.
   b. contraction of the heart.
   c. pressure changes in the blood vessels.
   d. flapping of the lymphatic valves.
   e. pumping by lymph nodes.

4. Which of the following statements is true?
   a. Lymphatic vessels do not have valves.
   b. Lymphatic vessels empty into lymph nodes.
   c. Lymph from the right-lower limb passes into the right jugular or subclavian vein.
   d. Lymph from the jugular and subclavian trunks empties into the cisterna chyli.
   e. All of the above.

5. The tonsils
   a. consist of three groups of lymphatic nodules.
   b. are located in the nasal cavity.
   c. are located in the oral cavity.
   d. increase in size in adults.
   e. all of the above.

6. Lymph nodes
   a. filter lymph.
   b. are where lymphocytes divide and increase in number.
   c. contain a network of reticular fibers.
   d. contain lymphatic sinuses.
   e. all of the above.

7. Which of these statements about the spleen is *not* correct?
   a. The spleen has white pulp associated with the arteries.
   b. The spleen has red pulp associated with the veins.
   c. The spleen destroys defective red blood cells.
   d. The spleen is surrounded by trabeculae located outside the capsule.
   e. The spleen is a limited reservoir for blood.

8. The thymus
   a. increases in size in adults.
   b. produces macrophages that move to other lymphatic tissue.
   c. responds to foreign substances in the blood.
   d. has a blood–thymic barrier.
   e. all of the above.

9. Which of these is an example of innate immunity?
   a. Tears and saliva wash away microorganisms.
   b. Basophils release histamine and leukotrienes.
   c. Neutrophils phagocytize a microorganism.
   d. The complement cascade is activated.
   e. All of the above.

10. Neutrophils
    a. enlarge to become macrophages.
    b. account for most of the dead cells in pus.
    c. are usually the last cell type to enter infected tissues.
    d. are usually located in lymphatic and blood sinuses.

11. Macrophages
    a. are large phagocytic cells that outlive neutrophils.
    b. develop from mast cells.
    c. often die after a single phagocytic event.
    d. have the same function as eosinophils.
    e. all of the above.

12. Which of these cells is the most important in the release of histamine, which promotes inflammation?
    a. monocyte
    b. macrophage
    c. eosinophil
    d. mast cell
    e. natural killer cell

13. Which of these conditions does *not* occur during the inflammatory response?
    a. histamine and other chemical mediators are released
    b. chemotaxis of phagocytes
    c. fibrinogen enters tissues from the blood
    d. vasoconstriction of blood vessels
    e. increased permeability of blood vessels

14. Which of these is a symptom of systemic inflammation?
    a. large numbers of neutrophils are produced and released
    b. pyrogens stimulate fever production
    c. greatly increased vascular permeability
    d. shock
    e. all of the above

15. Antigens
    a. are foreign substances introduced into the body.
    b. are molecules produced by the body.
    c. stimulate an adaptive immune system response.
    d. all of the above.

16. B cells
    a. are processed in the thymus.
    b. originate in red bone marrow.
    c. once released into the blood, remain in the blood.
    d. are responsible for cell-mediated immunity.
    e. all of the above.

17. MHC molecules
    a. are glycoproteins.
    b. attach to the plasma membrane.
    c. have a variable region that can bind to foreign and self-antigens.
    d. may form an MHC/antigen complex that activates T cells.
    e. all of the above.

18. Antigen-presenting cells can
    a. take in foreign antigens.
    b. process antigens.
    c. use MHC class II molecules to display the antigens.
    d. stimulate other immune system cells.
    e. all of the above.

19. Which of these participates in costimulation?
    a. cytokines
    b. complement
    c. antibodies
    d. histamine
    e. natural killer cells

20. Helper T cells
    a. respond to antigens from macrophages.
    b. respond to cytokines from macrophages.
    c. stimulate B cells with cytokines.
    d. all of the above.

21. The most important function of tolerance is to
    a. increase lymphocyte activity.
    b. increase complement activation.
    c. prevent the immune system from responding to self-antigens.
    d. prevent excessive immune system response to foreign antigens.
    e. process antigens.

22. Variable amino acid sequences on the arms of the antibody molecule
    a. make the antibody specific for a given antigen.
    b. enable the antibody to activate complement.
    c. enable the antibody to attach to basophils and mast cells.
    d. are part of the constant region.
    e. all of the above.

23. Antibodies
    a. prevent antigens from binding together.
    b. promote phagocytosis.
    c. inhibit inflammation.
    d. block complement activation.
    e. block the function of opsonins.

24. The secondary antibody response
    a. is slower than the primary response.
    b. produces fewer antibodies than the primary response.
    c. prevents disease symptoms from occurring.
    d. occurs because of cytotoxic T cells.

25. The type of lymphocyte that is responsible for the secondary antibody response is the
    a. memory B cell.
    b. B cell.
    c. T cell.
    d. helper T cell.

26. The largest percentage of antibodies in the blood are
    a. IgA.
    b. IgD.
    c. IgE.
    d. IgG.
    e. IgM.

27. Antibody-mediated immunity
    a. works best against intracellular antigens.
    b. is involved in tumor control.
    c. cannot be transferred from one person to another person.
    d. is responsible for immediate hypersensitivity reactions.

28. The activation of cytotoxic T cells can result in the
    a. lysis of virus-infected cells.
    b. production of cytokines.
    c. production of memory T cells.
    d. all of the above.

29. Cytokines
    a. promote inflammation.
    b. activate macrophages.
    c. kill target cells by causing them to lyse.
    d. all of the above.

30. Delayed hypersensitivity is
    a. caused by activation of B cells.
    b. a result of antibodies reacting with an allergen.
    c. mediated by T cells.
    d. caused by natural killer cells.
    e. caused by interferon.

Answers in Appendix F

C R I T I C A L   T H I N K I N G

1. A patient is suffering from edema in the lower-right limb. Explain why elevation of the limb and massage helps to remove the excess fluid.

2. If the thymus of an experimental animal is removed immediately after its birth, the animal exhibits the following characteristics: (a) it is more susceptible to infections, (b) it has decreased numbers of lymphocytes in lymphatic tissue, and (c) its ability to reject grafts is greatly decreased. Explain these observations.

3. If the thymus of an adult experimental animal is removed, the following observations can be made: (a) no immediate effect occurs and (b) after 1 year, the number of lymphocytes in the blood decreases, the ability to reject grafts decreases, and the ability to produce antibodies decreases. Explain these observations.

4. Adjuvants are substances that slow but do not stop the release of an antigen from an injection site into the blood. Suppose injection A of a given amount of antigen is given without an adjuvant and injection B of the same amount of antigen is given with an adjuvant that causes the release of antigen over a period of 2–3 weeks. Does injection A or B result in the greater amount of antibody production? Explain.

5. Tetanus is caused by bacteria that enter the body through wounds in the skin. The bacteria produce a toxin that causes spastic muscle contractions. Death often results from failure of the respiration muscles. A patient comes to the emergency room after stepping on a nail. If the patient has been vaccinated against tetanus, the patient is given a tetanus booster shot, which consists of the toxin altered so that it is harmless. If the patient has never been vaccinated against tetanus, the patient is given an antiserum shot against tetanus. Explain the rationale for this treatment strategy. Sometimes both a booster and an antiserum shot are given, but at different locations of the body. Explain why this is done, and why the shots are given in different locations.

6. An infant appears to be healthy until about 9 months of age. Then he develops severe bacterial infections, one after another. Fortunately, the infections are successfully treated with antibiotics.

When infected with the measles and other viral diseases, the infant recovers without unusual difficulty. Explain the different immune responses to these infections. Why did it take so long for this disorder to become apparent? (*Hint:* IgG.)

7. A baby is born with severe combined immunodeficiency disease (SCID). In an attempt to save her life, a bone marrow transplant is performed. Explain how this procedure might help the baby. Unfortunately, there is a graft rejection, and the baby dies. Explain what happened.

8. A patient has many allergic reactions. As part of the treatment scheme, doctors decide to try to identify the allergen that stimulates the immune system's response. A series of solutions, each containing an allergen that commonly causes a reaction, is composed. Each solution is injected into the skin at different locations on the patient's back. The following results are obtained: (a) at one location, the injection site becomes red and swollen within a few minutes; (b) at another injection site, swelling and redness appear 2 days later; and (c) no redness or swelling develops at the other sites. Explain what happened for each observation by describing what part of the immune system was involved and what caused the redness and swelling.

9. Ivy Hurtt developed a poison ivy rash after a camping trip. Her doctor prescribed a cortisone ointment to relieve the inflammation. A few weeks later Ivy scraped her elbow, which became inflamed. Because she had some of the cortisone ointment left over, she applied it to the scrape. Explain why the ointment was or was not a good idea for the poison ivy and for the scrape.

10. Suzy Withitt has just had her ears pierced. To her dismay, she finds that when she wears inexpensive (but tasteful) jewelry, by the end of the day there is an inflammatory (allergic) reaction to the metal in the jewelry. Is this because of antibodies or cytokines?

Answers in Appendix G

A N S W E R S   T O   P R E D I C T   Q U E S T I O N S

1. Cutting and tying off the lymphatic vessels prevents the movement of interstitial fluid from the interstitial spaces. The small amount of fluid that fails to reenter the venous end of the capillaries after it leaves the arteriolar end of the capillaries is normally carried by the lymphatic vessels away from the tissue spaces and back to the general circulation. If the lymphatic vessels are tied off, the fluid accumulates in the interstitial spaces and results in edema.

2. The T cells transferred to mouse B don't respond to the antigen. The T cells are MHC-restricted and must have the MHC proteins of mouse A as well as antigen X to respond.

3. When the antigen is eliminated, it's no longer available for processing and combining with MHC class II molecules. Consequently, no signal takes place to cause lymphocytes to proliferate and produce antibodies.

4. The first exposure to the disease-causing agent (antigen) evokes a primary response. Gradually, however, antibodies degrade, and memory cells die. If, before all the memory cells are eliminated, a second exposure to the antigen occurs, a secondary response results. The memory cells produced then could provide immunity until the next exposure to the antigen.

5. With depression of helper T-cell activity, the ability of antigens to activate effector T cells is greatly decreased. Depression of cell-mediated immunity results in an inability to resist intracellular microorganisms and cancer.

6. The booster shot stimulates a secondary (memory) response, resulting in the formation of large amounts of antibodies and memory cells. Consequently there is better, longer-lasting immunity.

7. SLE is an autoimmune disorder in which self-antigens activate immune responses. Often, this results in the formation of immune complexes and inflammation. But sometimes antibodies bind to antigens on cells, resulting in the lysis of the cells. Purpura results from bleeding into the skin, which means that platelet plug formation, the normal mechanism for repairing small breaks in blood vessels, is not working. In this case of SLE, antibodies are causing the destruction of platelets, and the decreased number of platelets results in decreased platelet plug formation and coagulation (see chapter 19). The condition is called thrombocytopenia.

Visit the Online Learning Center at www.mhhe.com/seeley7 for chapter quizzes, interactive learning exercises, and other study tools.

Respiratory System

This chapter explains the *functions of the respiratory system* (p. 830), the *anatomy and histology of the respiratory system* (p. 830), *ventilation* (p. 844), *measuring lung function* (p. 849), *physical principles of gas exchange* (p. 851), *oxygen and carbon dioxide transport in the blood* (p. 854), *rhythmic ventilation* (p. 860), *modification of ventilation* (p. 861), and *respiratory adaptations to exercise* (p. 865). We conclude the chapter by looking at the *effects of aging on the respiratory system* (p. 865).

CHAPTER

23

Colorized scanning electron micrograph of the lung, showing alveoli, which are small chambers where gas exchange takes place between the air and the blood.

From our first breath at birth, the rate and depth of our respiration is unconsciously matched to our activities, whether studying, sleeping, talking, eating, or exercising. We can voluntarily stop breathing, but within a few seconds we must breathe again. Breathing is so characteristic of life that, along with the pulse, it's one of the first things we check for to determine if an unconscious person is alive.

Breathing is necessary because all living cells of the body require oxygen and produce carbon dioxide. The respiratory system allows exchange of these gases between the air and the blood, and the cardiovascular system transports them between the lungs and the cells of the body. The capacity to carry out normal activity is reduced without healthy respiratory and cardiovascular systems.

Respiration includes: (1) ventilation, the movement of air into and out of the lungs; (2) gas exchange between the air in the lungs and the blood, sometimes called external respiration; (3) transport of oxygen and carbon dioxide in the blood; and (4) gas exchange between the blood and the tissues, sometimes called internal respiration. The term *respiration* is also used in reference to cell metabolism, which is considered in chapter 25.

Functions of the Respiratory System

Respiration is necessary because all living cells of the body require oxygen and produce carbon dioxide. The respiratory system assists in gas exchange and performs other functions as well.

1. *Gas exchange.* The respiratory system allows oxygen from the air to enter the blood and carbon dioxide to leave the blood and enter the air. The cardiovascular system transports oxygen from the lungs to the cells of the body and carbon dioxide from the cells of the body to the lungs. Thus, the respiratory and cardiovascular systems work together to supply oxygen to all cells and to remove carbon dioxide.
2. *Regulation of blood pH.* The respiratory system can alter blood pH by changing blood carbon dioxide levels.
3. *Voice production.* Air movement past the vocal folds makes sound and speech possible.
4. *Olfaction.* The sensation of smell occurs when airborne molecules are drawn into the nasal cavity.
5. *Protection.* The respiratory system provides protection against some microorganisms by preventing their entry into the body and by removing them from respiratory surfaces.

1. Explain the functions of the respiratory system.

Anatomy and Histology of the Respiratory System

The respiratory system consists of the external nose, the nasal cavity, the pharynx, the larynx, the trachea, the bronchi, and the lungs (figure 23.1). Although air frequently passes through the oral cavity, it is considered to be part of the digestive system instead of the respiratory system. The term **upper respiratory tract** refers to the nose, the pharynx, and associated structures; and the term **lower respiratory tract** refers to the larynx, trachea, bronchi, and lungs. These terms are not official anatomical terms, however, and there are several alternate definitions. For example, one alternate definition places the larynx in the upper respiratory tract. The diaphragm and the muscles of the thoracic and abdominal walls are responsible for respiratory movements.

2. Define upper and lower respiratory tract.

Nose

The **nasus** (nā′sŭs), or **nose,** consists of the external nose and the nasal cavity. The **external nose** is the visible structure that forms a prominent feature of the face. The largest part of the external nose is composed of cartilage plates (see figure 7.9b). The bridge of the nose consists of the nasal bones plus extensions of the frontal and maxillary bones.

The **nasal cavity** extends from the nares to the choanae (figure 23.2). The **nares** (nā′res; sing., nā′ris), or **nostrils,** are the external openings of the nasal cavity and the **choanae** (kō′an-ē) are the openings into the pharynx. The anterior part of the nasal cavity, just inside each naris, is the **vestibule** (ves′ti-bool; entry room). The vestibule is lined with stratified squamous epithelium that is continuous with the stratified squamous epithelium of the skin. The **hard palate** (pal′ăt) is a bony plate covered by a mucous membrane that forms the floor of the nasal cavity. It separates the nasal cavity from the oral cavity. The **nasal septum** is a partition dividing the nasal cavity into right and left parts (see figure 7.9a). The anterior part of the nasal septum is cartilage, and the posterior part consists of the vomer bone and the perpendicular plate of the ethmoid bone. A deviated nasal septum occurs when the septum bulges to one side or the other.

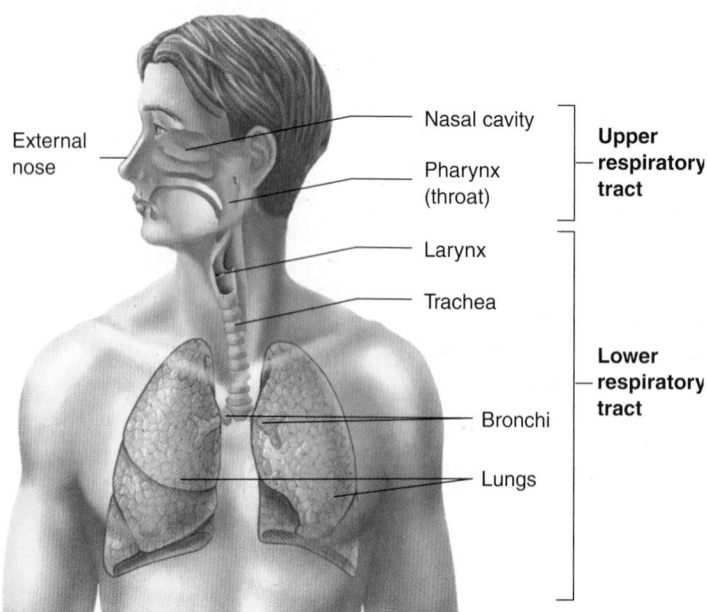

Figure 23.1   The Respiratory System

The upper respiratory tract consists of the nasal cavity and pharynx (throat). The lower respiratory tract consists of the larynx, trachea, bronchi, and lungs.

Cribriform plate

Frontal sinus — **Paranasal sinuses**
Sphenoidal sinus

Nasal cavity
Superior concha
Middle concha
Inferior concha
Vestibule
Naris

Superior meatus
Middle meatus — **Nasal cavity**
Inferior meatus
Choana

Hard palate
Oral cavity
Tongue
Palatine tonsil
Lingual tonsil

Pharyngeal tonsil
Opening of auditory tube
Nasopharynx
Soft palate
Uvula
Oropharynx — **Pharynx**

Hyoid bone

Laryngopharynx

Larynx
Epiglottis
Vestibular fold
Vocal fold
Thyroid cartilage
Cricoid cartilage

Trachea

Esophagus

(a)   **Medial view**

Superior nasal concha
Middle nasal concha
Inferior nasal concha
Hard palate

Sphenoidal sinus

Superior meatus
Middle meatus
Inferior meatus
Soft palate

(b)   **Medial view**

Figure 23.2   Nasal Cavity and Pharynx

(a) Sagittal section through the nasal cavity and pharynx viewed from the medial side. (b) Photograph of sagittal section of the head.

Three bony ridges called **conchae** (kon′kē; resembling a conch shell) modify the lateral walls of the nasal cavity. Beneath each concha is a passageway called a **meatus** (mē-ā′tŭs; a tunnel or passageway). Within the superior and middle meatus are openings from the various **paranasal sinuses** (see figure 7.10), and the opening of a **nasolacrimal** (nā-zō-lak′ri-măl) **duct** is within each inferior meatus (see figure 15.9).

Sinusitis

Sinusitis (sī-nŭ-sī′tis) is an inflammation of the mucous membrane of any sinus, especially of one or more paranasal sinuses. Viral infections, such as the common cold, can cause mucous membranes to become inflamed, swell, and produce excess mucus. As a result, the sinus opening into the nasal cavity can be partially or completely blocked. Mucus accumulates within the sinus, which can promote the development of a bacterial infection. Treatments consist of antibiotics against the bacteria and promoting sinus drainage with decongestants, hydration, and steam inhalation. Sinusitis can also result from swelling caused by allergies or by polyps that obstruct a sinus opening into the nasal cavity.

The nasal cavity has several functions:

1. The nasal cavity is a passageway for air that's open even when the mouth is full of food.
2. The nasal cavity cleans the air. The vestibule is lined with hairs that trap some of the large particles of dust in the air. The nasal septum and nasal conchae increase the surface area of the nasal cavity and make airflow within the cavity more turbulent, thereby increasing the likelihood that air comes into contact with the mucous membrane lining the nasal cavity. This mucous membrane consists of pseudostratified ciliated columnar epithelium with goblet cells, which secrete a layer of mucus. The mucus traps debris in the air, and the cilia on the surface of the mucous membrane sweep the mucus posteriorly to the pharynx, where it is swallowed and eliminated by the digestive system.
3. The nasal cavity humidifies and warms the air. Moisture from the mucous epithelium and from excess tears that drain into the nasal cavity through the nasolacrimal duct is added to the air as it passes through the nasal cavity. Warm blood flowing through the mucous membrane warms the air within the nasal cavity before it passes into the pharynx, thus preventing damage from cold air to the rest of the respiratory passages.

P R E D I C T   ❶
Explain what happens to your throat when you sleep with your mouth open, especially when your nasal passages are plugged as a result of having a cold. Explain what may happen to your lungs when you run a long way in very cold weather while breathing rapidly through your mouth.

4. The olfactory epithelium, the sensory organ for smell, is located in the most superior part of the nasal cavity (see figure 15.1).
5. The nasal cavity and paranasal sinuses are resonating chambers for speech.

▶ **3.** *How are the structures of the nasal cavity responsible for its functions?*

Pharynx

The **pharynx** (far′ingks; throat) is the common opening of both the digestive and respiratory systems. It receives air from the nasal cavity and air, food, and drink from the oral cavity. Inferiorly, the pharynx is connected to the respiratory system at the larynx and to the digestive system at the esophagus. The pharynx is divided into three regions: the nasopharynx, the oropharynx, and the laryngopharynx (see figure 23.2).

The **nasopharynx** (nā′zō-far′ingks) is located posterior to the choanae and superior to the **soft palate,** which is an incomplete muscle and connective tissue partition separating the nasopharynx from the oropharynx. The **uvula** (ū′vū-lă; a grape) is the posterior extension of the soft palate. The soft palate prevents swallowed materials from entering the nasopharynx and nasal cavity. The nasopharynx is lined with a mucous membrane containing pseudostratified ciliated columnar epithelium with goblet cells. Debris-laden mucus from the nasal cavity is moved through the nasopharynx and swallowed. Two auditory tubes from the middle ears open into the nasopharynx (see figures 15.23 and 23.2*a*). Air passes through them to equalize air pressure between the atmosphere and the middle ears. The posterior surface of the nasopharynx contains the pharyngeal tonsil, or adenoid (ad′ĕ-noyd), which aids in defending the body against infection (see chapter 22). An enlarged pharyngeal tonsil can interfere with normal breathing and the passage of air through the auditory tubes.

The **oropharynx** (ōr′ō-far′ingks) extends from the soft palate to the epiglottis. The oral cavity opens into the oropharynx through the **fauces** (faw′sēz). Thus, air, food, and drink all pass through the oropharynx. Moist stratified squamous epithelium lines the oropharynx and protects it against abrasion. Two sets of tonsils called the palatine tonsils and the lingual tonsils are located near the fauces.

The **laryngopharynx** (lă-ring′gō-far-ingks) extends from the tip of the epiglottis to the esophagus and passes posterior to the larynx. Food and drink pass through the laryngopharynx to the esophagus. A small amount of air is usually swallowed with the food and drink. Swallowing too much air can cause excess gas in the stomach and belching. The laryngopharynx is lined with moist stratified squamous epithelium.

▶ **4.** *Name the three parts of the pharynx. With what structures does each part communicate?*

Larynx

The **larynx** (lar′ingks) is located in the anterior part of the throat. It is connected by membranes and/or muscles superiorly to the hyoid bone and inferiorly to the trachea (figure 23.3). The larynx consists of an outer casing of nine cartilages that are connected to one another by muscles and ligaments. Six of the nine cartilages are paired, and three are unpaired. The largest of the cartilages is the unpaired **thyroid** (shield; refers to the shape of the cartilage) **cartilage,** or Adam's apple.

The most inferior cartilage of the larynx is the unpaired **cricoid** (krī′koyd; ring-shaped) **cartilage,** which forms the base of the larynx on which the other cartilages rest.

Figure 23.3 Anatomy of the Larynx

The third unpaired cartilage is the **epiglottis** (ep-i-glot′is; on the glottis). It's attached to the thyroid cartilage and projects as a free flap toward the tongue. The epiglottis differs from the other cartilages in that it consists of elastic rather than hyaline cartilage. During swallowing, the epiglottis covers the opening of the larynx and prevents materials from entering it.

The paired **arytenoid** (ar-i-tē′noyd; ladle-shaped) **cartilages** articulate with the posterior, superior border of the cricoid cartilage, and the paired **corniculate** (kōr-nik′ū-lāt; horn-shaped) **cartilages** are attached to the superior tips of the arytenoid cartilages. The paired **cuneiform** (kū′nē-i-fōrm; wedge-shaped) **cartilages** are contained in a mucous membrane anterior to the corniculate cartilages (see figure 23.3b).

Two pairs of ligaments extend from the anterior surface of the arytenoid cartilages to the posterior surface of the thyroid cartilage. The superior ligaments are covered by a mucous membrane called the **vestibular folds,** or **false vocal cords** (see figures 23.3c and 23.4a and b). When the vestibular folds come together, they prevent food and liquids from entering the larynx during swallowing and prevent air from leaving the lungs, as when a person holds his or her breath.

The inferior ligaments are covered by a mucous membrane called the **vocal folds,** or **true vocal cords** (see figure 23.4). The

vocal folds and the opening between them are called the **glottis** (glot′is). The vestibular folds and the vocal folds are lined with stratified squamous epithelium. The remainder of the larynx is lined with pseudostratified ciliated columnar epithelium. An inflammation of the mucosal epithelium of the vocal folds is called **laryngitis** (lar-in-jī′tis).

The larynx performs three important functions.

1. The thyroid and cricoid cartilages maintain an open passageway for air movement.
2. The epiglottis and vestibular folds prevent swallowed material from moving into the larynx.
3. The vocal folds are the primary source of sound production. Air moving past the vocal folds causes them to vibrate and produce sound. The greater the amplitude of the vibration, the louder is the sound. The force of air moving past the vocal folds determines the amplitude of vibration and the loudness of the sound. The frequency of vibrations determines pitch, with higher frequency vibrations producing higher pitched sounds and lower frequency fibrations producing lower pitched sounds. Variations in the length of the vibrating segments of the vocal folds affect the

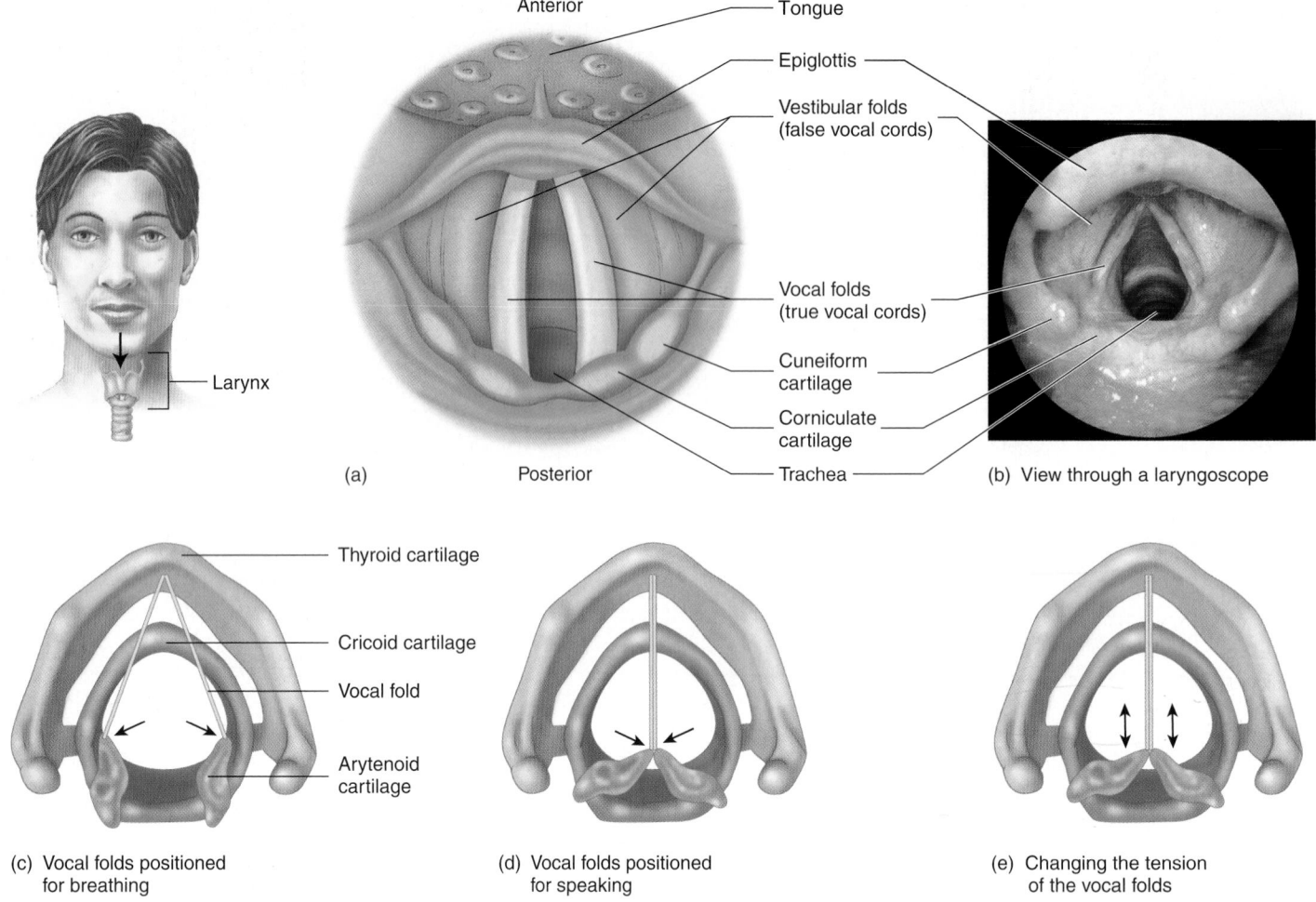

Larynx

(a)

Anterior

Tongue

Epiglottis

Vestibular folds
(false vocal cords)

Vocal folds
(true vocal cords)

Cuneiform
cartilage

Corniculate
cartilage

Posterior   Trachea

(b) View through a laryngoscope

Thyroid cartilage

Cricoid cartilage

Vocal fold

Arytenoid
cartilage

(c) Vocal folds positioned
for breathing

(d) Vocal folds positioned
for speaking

(e) Changing the tension
of the vocal folds

Figure 23.4 **Vestibular Vocal Folds**

Arrow shows the direction of viewing the vestibular and vocal folds. (*a*) The relationship of the vestibular folds to the vocal folds and the laryngeal cartilages. (*b*) Laryngoscopic view of the vestibular and vocal folds. (*c*) Lateral rotation of the arytenoid cartilages positions the vocal folds for breathing. (*d*) Medial rotation of the arytenoid cartilages positions the vocal folds for speaking. (*e*) Anterior/posterior movement of the arytenoid cartilages changes the length and tension of the vocal folds.

frequency of the vibrations. Higher-pitched tones are produced when only the anterior parts of the folds vibrate, and progressively lower tones result when longer sections of the folds vibrate. Because males usually have longer vocal folds than females, they usually have lower-pitched voices. The sound produced by the vibrating vocal folds is modified by the tongue, lips, teeth, and other structures to form words. A person whose larynx has been removed because of carcinoma of the larynx can produce sound by swallowing air and causing the esophagus to vibrate.

Movement of the arytenoid and other cartilages is controlled by skeletal muscles, thereby changing the position and length of the vocal folds. When only breathing, lateral rotation of the arytenoid cartilages abducts the vocal folds, which allows greater movement of air (figure 23.4*c*). Medial rotation of the arytenoid cartilages adducts the vocal folds, places them in position for producing sounds, and changes the tension on them. (figure 23.4*d*). Anterior/posterior movement of the

arytenoid cartilages also changes the length and tension of the vocal folds (figure 23.4*e*).

 5. *Name and describe the three unpaired cartilages of the larynx. What are their functions?*

 6. *Distinguish between the vestibular and vocal folds. How are sounds of different loudness and pitch produced by the vocal folds?*

 7. *How does the position of the arytenoid cartilages change when just breathing versus making low-pitched and high-pitched sounds?*

Trachea

The **trachea** (trā′kē-ă), or windpipe, is a membranous tube that consists of dense regular connective tissue and smooth muscle reinforced with 15–20 C-shaped pieces of cartilage. The cartilages support the anterior and lateral sides of the trachea (figure 23.5*a*). They protect the trachea and maintain an open passageway for air. The posterior wall of the trachea is devoid of cartilage and contains an elastic

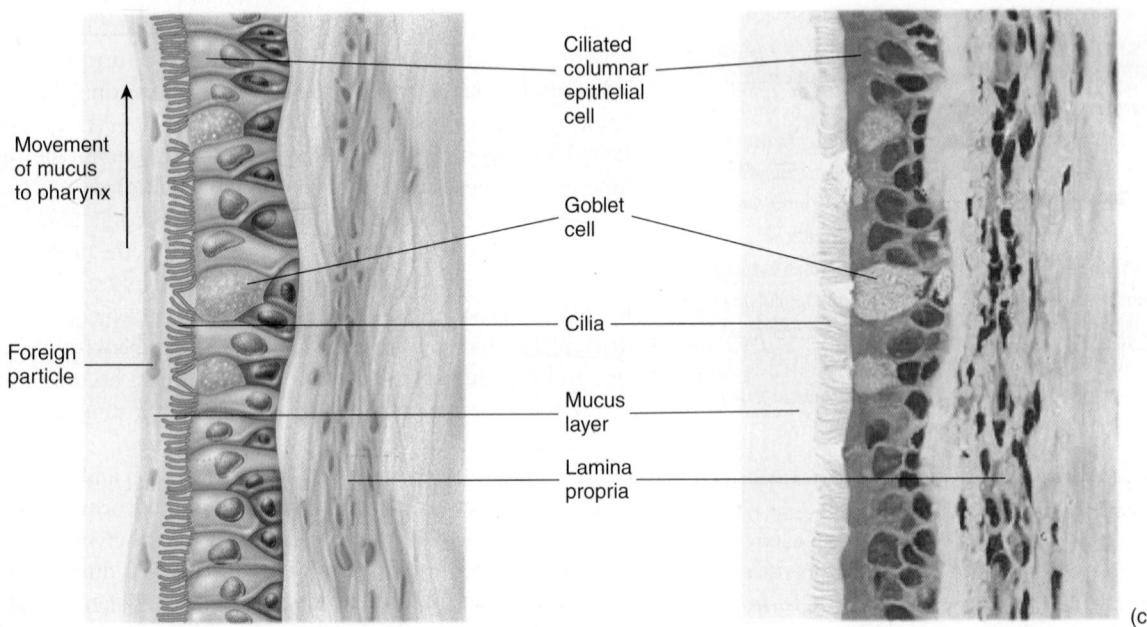

Figure 23.5 Trachea

(*a*) Light micrograph of a transverse section of the trachea. The esophagus is located posterior to the trachea, next to the smooth muscle connecting the ends of the C-shaped cartilages of the trachea. (*b*) Mucus, produced by the goblet cells, traps particles in the air. Movement of the cilia moves the mucus and particles to the laryngopharynx. (*c*) Light micrograph of the surface of the mucous membrane lining the trachea. Goblet cells are interspersed between ciliated cells.

ligamentous membrane and bundles of smooth muscle called the **trachealis** (trā′kē-ā-lis) **muscle.** Contraction of the smooth muscle can narrow the diameter of the trachea. During coughing, this action causes air to move more rapidly through the trachea, which helps to expel mucus and foreign objects. The esophagus lies immediately posterior to the cartilage-free posterior wall of the trachea.

P R E D I C T

Explain what happens to the shape of the trachea when a person swallows a large mouthful of food. Why is this change of shape advantageous?

The mucous membrane lining the trachea consists of pseudostratified ciliated columnar epithelium with numerous goblet cells (figure 23.5*b*). The goblet cells produce mucus that traps inhaled foreign particles. The cilia move the mucus and foreign particles into the larynx, from which they enter the pharynx and are swallowed (figure 23.5*c*). Constant irritation to the trachea, such as occurs in smokers, can cause the tracheal epithelium to become moist stratified squamous epithelium that lacks cilia and goblet cells. Consequently, the normal function of the tracheal epithelium is lost.

Establishing Airflow

In cases of extreme emergency when the upper air passageway is blocked by a foreign object to the extent that the victim cannot breathe, quick reaction is required to save the person's life. The **Heimlich maneuver** is designed to force such an object out of the air passage by the sudden application of pressure to the abdomen. The person who performs the maneuver stands behind the victim with arms under the victim's arms and hands over the victim's abdomen between the navel and the rib cage. With one hand formed into a fist and the other hand over it, both hands are suddenly pulled toward the abdomen with an accompanying upward motion. This maneuver, if done properly, forces air up the trachea and dislodges most foreign objects.

There are other ways to establish air flow, but they should be performed only by trained medical personnel. **Intubation** is the insertion of a tube into an opening, canal, or hollow organ. A tube can be passed through the mouth or nose into the pharynx and then through the larynx to the trachea.

Sometimes, it is necessary to make an opening through which to pass the tube. The preferred point of entry in emergency cases is through the membrane between the cricoid and thyroid cartilages, a procedure referred to as a **cricothyrotomy** (krī'kō-thī-rot'ō-mē).

A **tracheostomy** (trā'-kē-os'tō-mē, tracheo- + *stoma*, mouth) is an operation to make an opening into the trachea, usually between the second and third cartilage rings. Usually the opening is intended to be permanent, and a tube is inserted into the trachea to allow airflow and provide a way to remove secretions. The term **tracheotomy** (trā-kē-ot'ō-mē, tracheo- + *tome*, incision) refers to the actual cutting into the trachea. Sometimes the terms *tracheostomy* and *tracheotomy* are used interchangeably. It is not advisable to enter the air passageway through the trachea in emergency cases because arteries, nerves, and the thyroid gland overlie the anterior surface of the trachea.

The trachea has an inside diameter of 12 mm and a length of 10–12 cm, descending from the larynx to the level of the fifth thoracic vertebra (figure 23.6). The trachea divides to form two smaller tubes called **main,** or **primary bronchi** (brong'kī; sing., bronchus, brong'kŭs; windpipe), each of which extends to a lung. The most inferior tracheal cartilage forms a ridge called the **carina** (kă-rī'nă), which separates the openings into the main bronchi. The carina is an important radiologic landmark. In addition, the mucous membrane of the carina is very sensitive to mechanical stimulation, and materials reaching the carina stimulate a powerful cough reflex. Materials in the air passageways inferior to the carina do not usually stimulate a cough reflex.

Tracheobronchial Tree

The trachea divides to form main bronchi, which, in turn, divide to form smaller and smaller bronchi, until, eventually, many microscopically small tubes and sacs are formed. Beginning with the trachea, all the respiratory passageways are called the **tracheobronchial** (trā'kē-ō-brong'kē-ăl) **tree** (see figure 23.6). Based on function, the tracheo-bronchial tree can be subdivided into the conducting zone and the respiratory zone.

Conducting Zone

The **conducting zone** extends from the trachea to small tubes called terminal bronchioles (see figure 23.6). Approximately 16 generations of branching occur from the trachea to the terminal bronchioles. The conducting zone functions as a passageway for air movement and contains epithelial tissue that helps to remove debris from the air and move it out of the tracheo-bronchial tree.

The trachea divides into the left and right main bronchi, which extend to the lungs (see figure 23.6). The right main bronchus is shorter, has a wider diameter, and is more vertical than the left main bronchus.

P R E D I C T ❸
Into which lung would a foreign object that's small enough to pass into a main bronchus most likely become lodged and block air movement?

The main bronchi divide into **lobar,** or **secondary bronchi** within each lung. Two lobar bronchi exist in the left lung, and three exist in the right lung. The lobar bronchi, in turn, give rise to **segmental,** or **tertiary bronchi.** The bronchi continue to branch, finally giving rise to **bronchioles** (brong'kē-ōlz), which are less than 1 mm in diameter. The bronchioles also subdivide several times to become even smaller **terminal bronchioles.**

As the air passageways of the lungs become smaller, the structure of their walls changes. Like the trachea, the main bronchi are supported by C-shaped cartilage connected by smooth muscle. In the lobar bronchi, the C-shaped cartilages are replaced with cartilage plates, and smooth muscle forms a layer between the cartilage and the mucous membrane. As the bronchi become smaller, the cartilage becomes more sparse and smooth muscle becomes more abundant. The terminal bronchioles have no cartilage, and the smooth muscle layer is prominent. Relaxation and contraction of the smooth muscle within the bronchi and bronchioles can change the diameter of the air passageways and thereby change the volume of air moving through them. For example, during exercise, the diameter can increase, which reduces the resistance to airflow and thereby increases the volume of air moved. During an **asthma attack,** however, contraction of the smooth muscle in the terminal bronchioles, which have no cartilage in their walls, can result in decreased diameter, increased resistance to airflow, and greatly reduced airflow. In severe cases, air movement can be so restricted that death results.

The bronchi are lined with a pseudostratified ciliated columnar epithelium. The larger bronchioles are lined with ciliated simple columnar epithelium, which changes to ciliated simple cuboidal epithelium in the terminal bronchioles. The epithelium in the conducting part of the air passageways functions as a mucus–cilia escalator, which traps debris in the air and removes it from the respiratory system.

Respiratory Zone

The **respiratory zone** extends from the terminal bronchioles to small air-filled chambers called **alveoli** (al-vē'ō-lī; hollow cavity), which are the sites of gas exchange between the air and blood. Approximately seven generations of branching are present in the

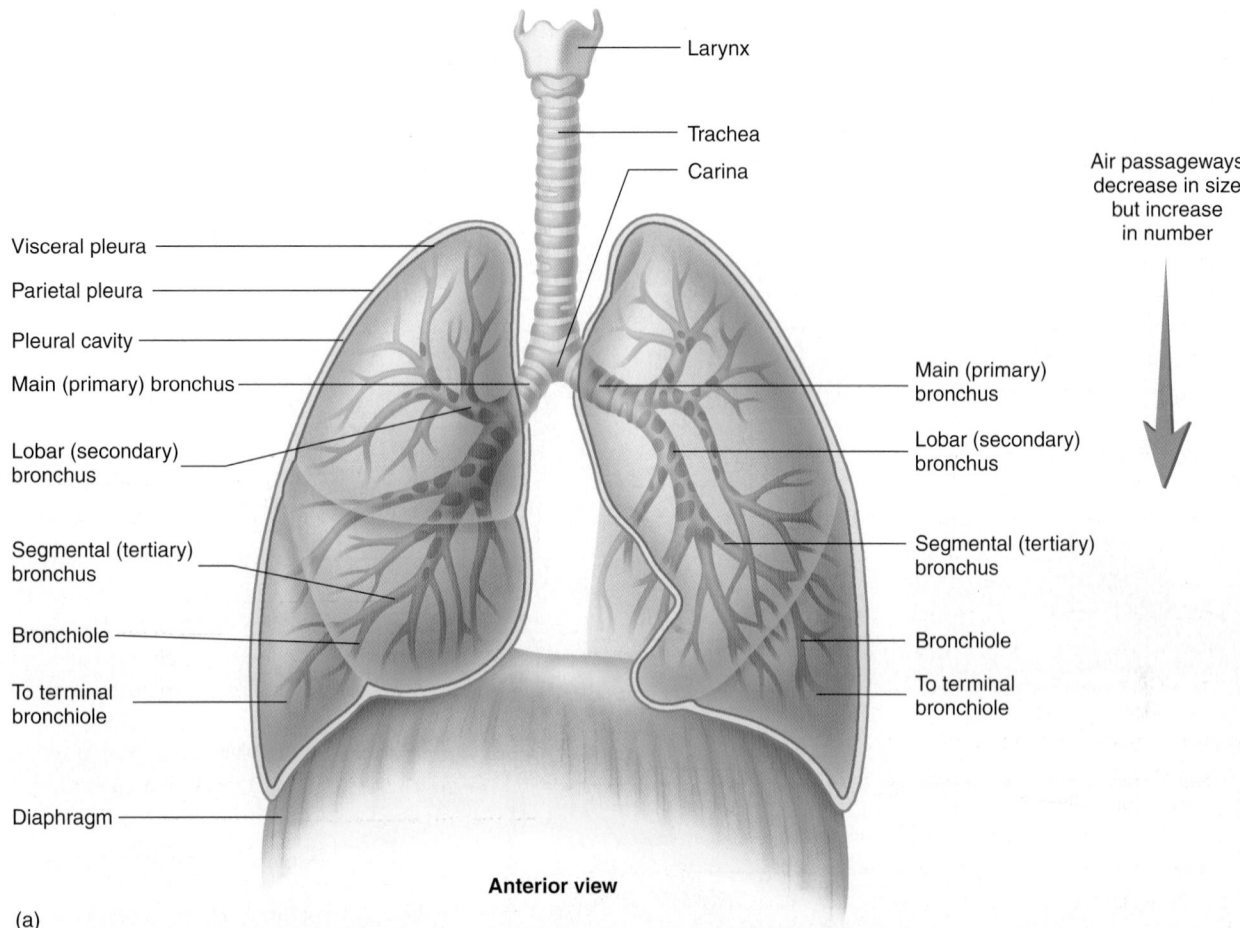

Larynx

Trachea

Carina

Air passageways
decrease in size
but increase
in number

Visceral pleura

Parietal pleura

Pleural cavity

Main (primary) bronchus

Lobar (secondary)
bronchus

Segmental (tertiary)
bronchus

Bronchiole

To terminal
bronchiole

Diaphragm

Main (primary)
bronchus

Lobar (secondary)
bronchus

Segmental (tertiary)
bronchus

Bronchiole

To terminal
bronchiole

Anterior view

(a)

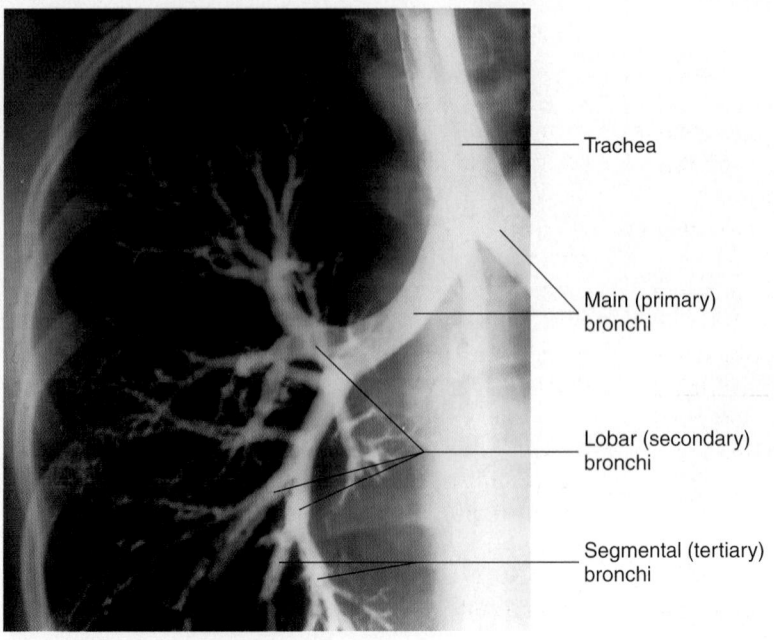

Trachea

Main (primary)
bronchi

Lobar (secondary)
bronchi

Segmental (tertiary)
bronchi

Anterior view

(b)

Figure 23.6 Tracheobronchial Tree

(a) The conducting zone of the tracheobronchial tree
begins at the trachea and ends at the terminal
bronchioles. (b) A bronchogram is a radiograph of the
tracheobronchial tree. A contrast medium, which makes
the passageways visible, is injected through a catheter
after a topical anesthetic is applied to the mucous
membranes of the nose, pharynx, larynx, and trachea.

respiratory zone. The terminal bronchioles divide to form **respiratory bronchioles** (figure 23.7), which have a limited ability for gas exchange because of a few attached alveoli. As the respiratory bronchioles divide to form smaller respiratory bronchioles, the number of attached alveoli increases. The respiratory bronchioles give rise to **alveolar** (al-vē′ō-lăr) **ducts,** which are like long branching hallways

with many open doorways. The doorways open into alveoli, which become so numerous that the alveolar duct wall is little more than a succession of alveoli. The alveolar ducts end as two or three **alveolar sacs,** which are chambers connected to two or more alveoli.

The tissue surrounding the alveoli contains elastic fibers that allow the alveoli to expand during inspiration and recoil during

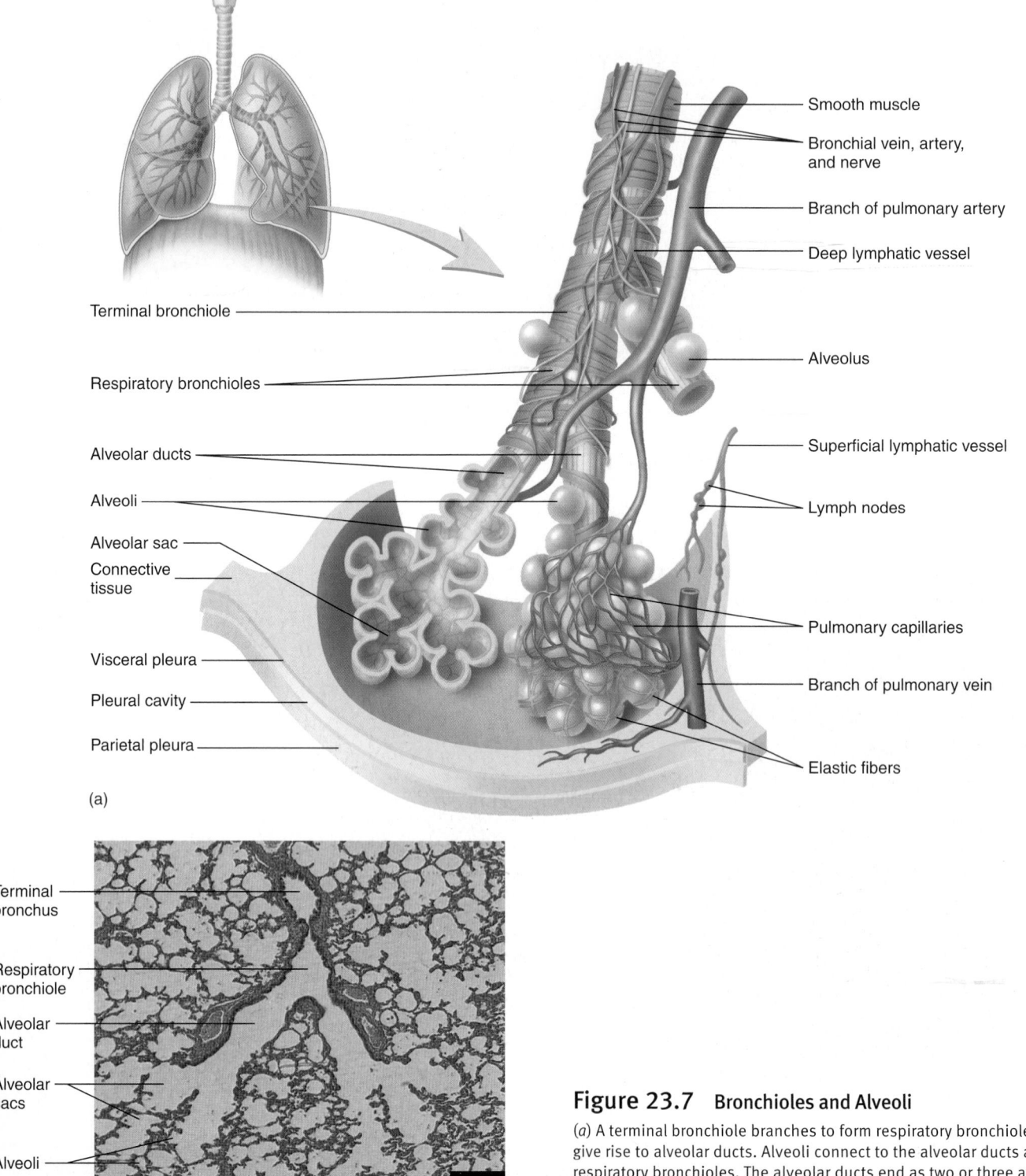

Figure 23.7   **Bronchioles and Alveoli**

(*a*) A terminal bronchiole branches to form respiratory bronchioles, which give rise to alveolar ducts. Alveoli connect to the alveolar ducts and respiratory bronchioles. The alveolar ducts end as two or three alveolar sacs. (*b*) Photomicrograph of lung tissue.

expiration. The lungs are very elastic, and when inflated, they are capable of expelling the air and returning to their original, uninflated state. Even when not inflated, however, the lungs retain some air, which gives them a spongy quality.

The walls of respiratory bronchioles consists of collagenous and elastic connective tissue with bundles of smooth muscle. The epithelium in the respiratory bronchioles is a simple cuboidal epithelium. The alveolar ducts and alveoli consist of simple squamous epithelium. Although the epithelium of the respiratory zone

is not ciliated, debris from the air can be removed by macrophages that move over the surfaces of the cells. The macrophages don't accumulate in the respiratory zone because they either move into nearby lymphatic vessels or enter terminal bronchioles, thereby becoming entrapped in mucus that is swept to the pharynx.

Approximately 300 million alveoli are in the two lungs. The average diameter of the alveoli is approximately 250 μm, and their walls are extremely thin. Two types of cells form the alveolar wall (figure 23.8a). **Type I pneumocytes** are thin, squamous epithelial

(a)

Macrophage

Air space within alveolus

Nucleus

Mitochondrion

Type II pneumocyte (surfactant-secreting cell)

Alveolar epithelium (wall)

Type I pneumocyte

Capillary endothelium (wall)

Red blood cell

(b)

Alveolus

Capillary

Alveolar fluid (with surfactant)

Alveolar epithelium

Basement membrane of alveolar epithelium

Interstitial space

Basement membrane of capillary endothelium

Capillary endothelium

Respiratory membrane

Diffusion of O$_2$

Diffusion of CO$_2$

Red blood cell

Figure 23.8   Alveolus and the Respiratory Membrane

(a) Section of an alveolus showing the air-filled interior and thin walls composed of simple squamous epithelium. The alveolus is surrounded by elastic connective tissue and blood capillaries. (b) Diffusion of oxygen and carbon dioxide across the six thin layers of the respiratory membrane.

cells that form 90% of the alveolar surface. Most gas exchange between alveolar air and the blood takes place through these cells. **Type II pneumocytes** are round or cube-shaped secretory cells that produce surfactant, which makes it easier for the alveoli to expand during inspiration (see "Lung Recoil" on p. 847).

The **respiratory membrane** of the lungs is where gas exchange between the air and blood takes place. It is mainly formed by the alveolar walls and surrounding pulmonary capillaries (figure 23.8*b*), but there's some contribution by the respiratory bronchioles and alveolar ducts. The respiratory membrane is very thin to facilitate the diffusion of gases. It consists of

1. a thin layer of fluid lining the alveolus;
2. the alveolar epithelium composed of simple squamous epithelium;
3. the basement membrane of the alveolar epithelium;
4. a thin interstitial space;
5. the basement membrane of the capillary endothelium;
6. the capillary endothelium composed of simple squamous epithelium.

    8. *What are the parts of the conducting and respiratory zones of the tracheobronchial tree?*

    9. *Describe the arrangement of cartilage, smooth muscle, and epithelium in the tracheobronchial tree. Explain why breathing becomes more difficult during an asthma attack.*

   10. *How is debris removed from the conducting and respiratory zones?*

   11. *Name the two types of cells in the alveolar wall, and state their functions.*

   12. *List the parts of the respiratory membrane.*

Lungs

The **lungs** are the principal organs of respiration, and on a volume basis they are among the largest organs of the body. Each lung is conical in shape, with its base resting on the diaphragm and its apex extending superiorly to a point approximately 2.5 cm superior to the clavicle. The right lung is larger than the left and weighs an average of 620 g, whereas the left lung weighs an average of 560 g.

The **hilum** (hī′lŭm) is a region on the medial surface of the lung where structures, such as the main bronchus, blood vessels, nerves, and lymphatic vessels, enter or exit the lung. All the structures passing through the hilum are referred to as the **root of the lung.**

The right lung has three **lobes,** and the left lung has two (figure 23.9). The lobes are separated by deep, prominent **fissures** on the surface of the lung, and each lobe is supplied by a lobar bronchus. The lobes are subdivided into **bronchopulmonary segments,** which are supplied by the segmental bronchi. Nine bronchopulmonary segments are present in the left lung, and 10 are present in the right lung. The bronchopulmonary segments are separated from each other by connective tissue partitions, which are not visible as surface fissures. Individual diseased bronchopulmonary segments can be surgically removed, leaving the rest of the lung relatively intact, because major blood vessels and bronchi don't cross the connective tissue partitions. The bronchopulmonary segments are subdivided into **lobules** by incomplete connective tissue walls. The lobules are supplied by the bronchioles.

   13. *Distinguish among a lung, a lung lobe, a bronchopulmonary segment, and a lobule. How are they related to the tracheobronchial tree?*

Thoracic Wall and Muscles of Respiration

The **thoracic wall** consists of the thoracic vertebrae, ribs, costal cartilages, the sternum, and associated muscles (see chapters 7 and 10). The **thoracic cavity** is the space enclosed by the thoracic wall and the **diaphragm** (dī′ă-fram, meaning partition), which separates the thoracic cavity from the abdominal cavity. The diaphragm and other skeletal muscles associated with the thoracic wall are responsible for respiration (figure 23.10). The **muscles of inspiration** include the diaphragm, external intercostals, pectoralis minor, and scalenes. Contraction of the diaphragm is responsible for approximately two-thirds of the increase in thoracic volume during inspiration. The external intercostals, pectoralis minor and scalene muscles also increase thoracic volume by elevating the ribs. The **muscles of expiration** consist of muscles that depress the ribs and sternum, such as the abdominal muscles and the internal intercostals. Although the internal intercostals are most active during expiration, and the external intercostals are most active during inspiration, the primary function of these muscles is to stiffen the thoracic wall by contracting at the same time. By so doing, they prevent inward collapse of the thoracic cage during inspiration.

Effect of Spinal Cord Injury on Ventilation

The diaphragm is supplied by the phrenic nerves, which arise from spinal nerves C3–C5 (see figure 12.16), descend along each side of the neck to enter the thorax, and pass to the diaphragm. The intercostal muscles are supplied by the intercostal nerves (see figure 12.15), which arise from spinal nerves T1–T11 and extend along the spaces between the ribs. Spinal cord injury superior to the origin of the phrenic nerves causes paralysis of the diaphragm and intercostal muscles and results in death unless artificial respiration is provided. A high spinal cord injury below the origin of the phrenic nerves causes paralysis of the intercostal muscles. Even though the diaphragm can function maximally, ventilation is drastically reduced because the intercostal muscles no longer prevent the thoracic wall from collapsing inward. Vital capacity is reduced to about 300 mL. With low spinal cord injury, below the origin of the intercostal nerves, both the diaphragm and intercostal muscles function normally.

The diaphragm is dome-shaped, and the base of the dome attaches to the inner circumference of the inferior thoracic cage (see figure 10.18). The top of the dome is a flat sheet of connective tissue called the **central tendon.** In normal, quiet inspiration, contraction of the diaphragm results in inferior movement of the central tendon with very little change in the overall shape of the dome. Inferior movement of the central tendon can occur because of relaxation of the abdominal muscles, which allows the abdominal organs to move out of the way of the diaphragm. As the depth of inspiration increases, inferior movement of the central tendon is prevented by the abdominal organs. Continued contraction of the diaphragm causes it to flatten as the lower ribs are elevated. In addition, other muscles of inspiration can elevate the ribs. As the ribs

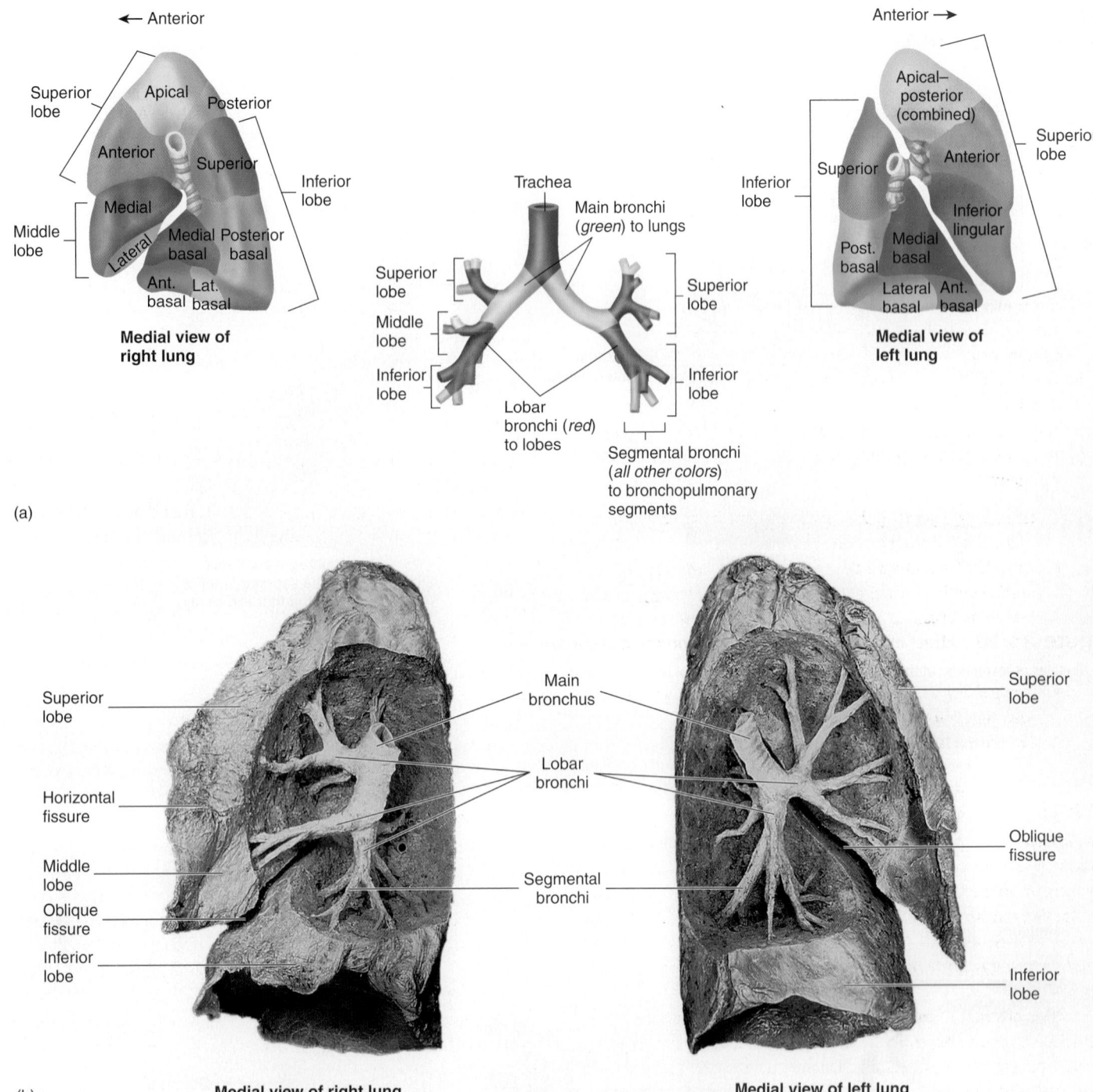

Figure 23.9 **Lobes and Bronchopulmonary Segments of the Lungs**

(a) The trachea (*blue*), main bronchi (*green*), lobar bronchi (*red*), and segmental bronchi (*all other colors*) are in the center of the figure, surrounded by two views of each lung, showing the bronchopulmonary segments. In general, each bronchopulmonary segment is supplied by a segmental bronchus (*color-coded to match the bronchopulmonary segment it supplies*). (b) Photograph of the lungs showing the lung lobes and the bronchi. The right lung is divided into three lobes by the horizontal and oblique fissures. The left lung is divided into two lobes by the oblique fissure. A main bronchus supplies each lung, a lobar bronchus supplies each lung lobe, and segmental bronchi supply the bronchopulmonary segments (*not visible*).

are elevated, the costal cartilages allow lateral rib movement and lateral expansion of the thoracic cavity (figure 23.11). The ribs slope inferiorly from the vertebrae to the sternum, and elevation of the ribs also increases the anterior–posterior dimension of the thoracic cavity.

Expiration during quiet breathing occurs when the diaphragm and external intercostals relax and the elastic properties of the thorax and lungs cause a passive decrease in thoracic volume. In addition, contraction of the abdominal muscles helps to push abdominal organs and the diaphragm in a superior direction.

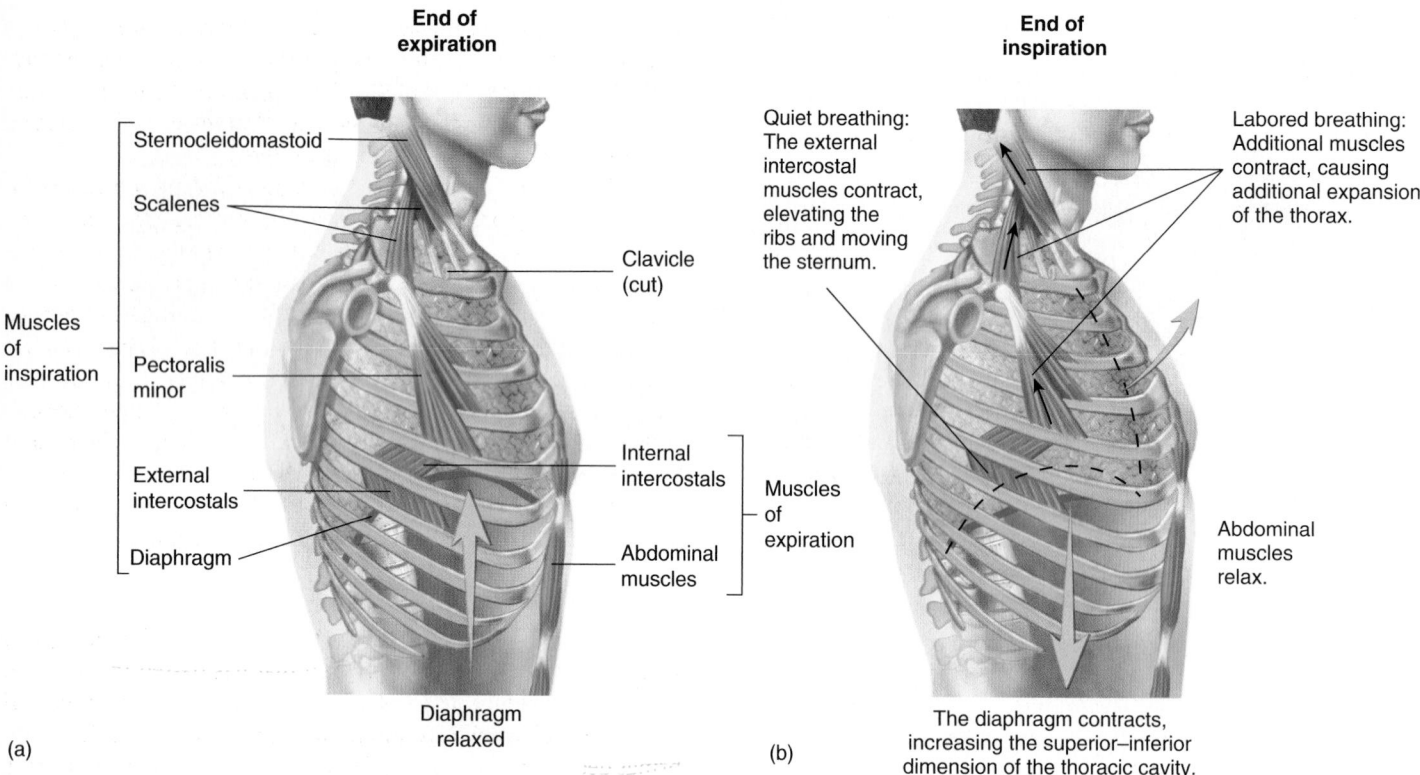

End of expiration

Muscles of inspiration
- Sternocleidomastoid
- Scalenes
- Pectoralis minor
- External intercostals
- Diaphragm

Clavicle (cut)

Internal intercostals

Muscles of expiration
- Abdominal muscles

Diaphragm relaxed

(a)

End of inspiration

Quiet breathing: The external intercostal muscles contract, elevating the ribs and moving the sternum.

Labored breathing: Additional muscles contract, causing additional expansion of the thorax.

Abdominal muscles relax.

(b)

The diaphragm contracts, increasing the superior–inferior dimension of the thoracic cavity.

Figure 23.10 Effect of the Muscles of Respiration on Thoracic Volume

(a) Muscles of respiration at the end of expiration. (b) Muscles of respiration at the end of inspiration.

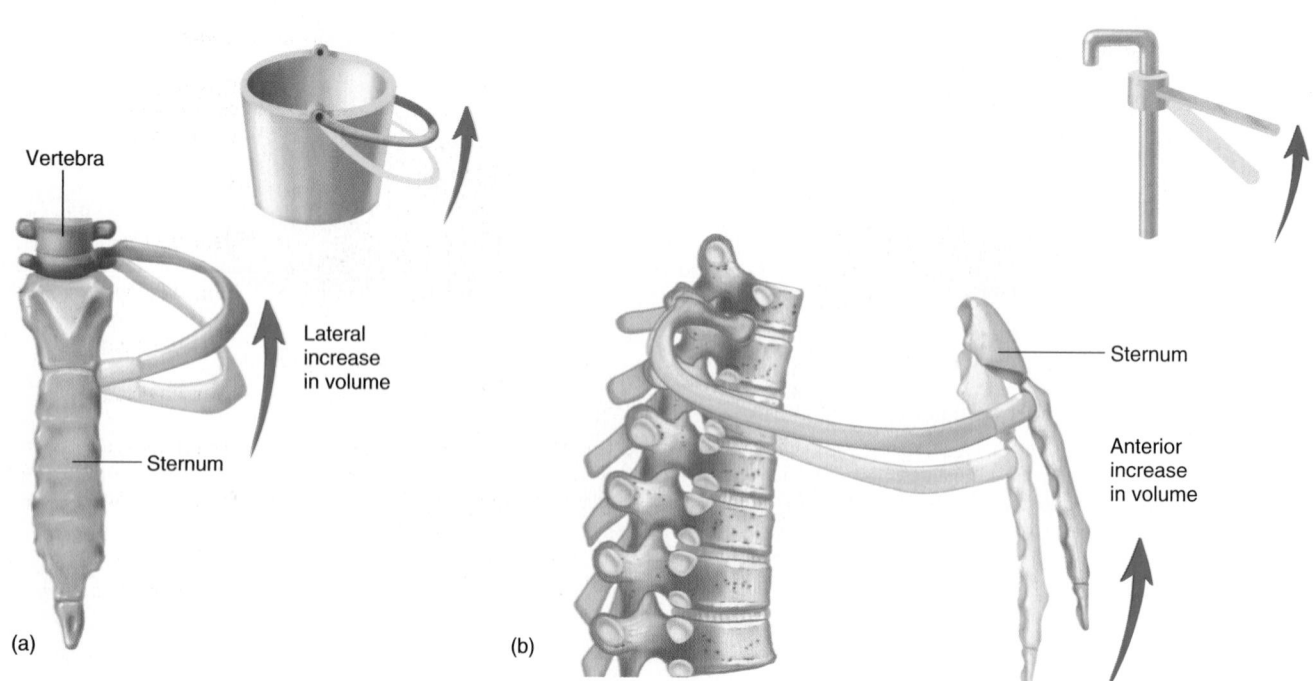

Vertebra

Sternum

Lateral increase in volume

(a)

(b)

Sternum

Anterior increase in volume

Figure 23.11 Effect of Rib and Sternum Movement on Thoracic Volume

(a) Elevation of the rib in the "bucket-handle" movement laterally increases thoracic volume. (b) As the rib is elevated, rotation of the rib in the "pump-handle" movement increases thoracic volume anteriorly.

Several differences can be recognized between normal, quiet breathing and labored breathing. During labored breathing, all of the inspiratory muscles are active, and they contract more forcefully than during quiet breathing, causing a greater increase in thoracic volume (see figure 23.10*b*). During labored breathing, forceful contraction of the internal intercostals and the abdominal muscles produces a more rapid and greater decrease in thoracic volume than would be produced by the passive recoil of the thorax and lungs.

Pleura

The lungs are contained within the thoracic cavity, but each lung is surrounded by a separate **pleural** (ploor′ăl; relating to the ribs) **cavity** formed by the pleural serous membranes (figure 23.12). The **mediastinum** (mē′dē-as-tī′nŭm), a midline partition formed by the heart, trachea, esophagus, and associated structures, separates the pleural cavities. The **parietal pleura** covers the inner thoracic wall, the superior surface of the diaphragm, and the mediastinum. At the hilum, the parietal pleura is continuous with the **visceral pleura,** which covers the surface of the lung.

The pleural cavity is filled with pleural fluid, which is produced by the pleural membranes. The pleural fluid does two things: (1) it acts as a lubricant, allowing the parietal and visceral pleural membranes to slide past each other as the lungs and the thorax change shape during respiration, and (2) it helps hold the parietal and visceral pleural membranes together. When thoracic volume changes during respiration, lung volume changes because the parietal pleura is attached to the diaphragm and inner thoracic wall, and the visceral pleura is attached to the lungs. The pleural fluid is analogous to a thin film of water between two sheets of glass (the visceral and parietal pleurae); the glass sheets can easily slide over each other, but it's difficult to separate them.

Blood Supply

Blood that has passed through the lungs and picked up oxygen is called **oxygenated blood,** and blood that has passed through the tissues and released some of its oxygen is called **deoxygenated blood.** Two blood flow routes to the lungs exist. The major route brings deoxygenated blood to the lungs, where it is oxygenated (see chapter 21 and figure 23.12). The deoxygenated blood flows

Superior view

Figure 23.12 Pleural Cavities and Membranes

Transverse section of the thorax showing the relationship of the pleural cavities to the thoracic organs. Each lung is surrounded by a pleural cavity. The parietal pleura lines the wall of each pleural cavity, and the visceral pleura covers the surface of the lungs. The space between the parietal and visceral pleurae is small and filled with pleural fluid.

Clinical Focus   Cough and Sneeze Reflexes

The function of both the cough reflex and the sneeze reflex is to dislodge foreign matter or irritating material from respiratory passages. The bronchi and trachea contain sensory receptors that are sensitive to foreign particles and irritating substances. The cough reflex is initiated when the sensory receptors detect such substances and initiate action potentials that pass along the vagus nerves to the medulla oblongata, where the cough reflex is triggered.

The movements resulting in a cough occur as follows: approximately 2.5 L of air is in-spired; the vestibular and vocal folds close tightly to trap the inspired air in the lungs; the abdominal muscles contract to force the abdominal contents up against the diaphragm; and the muscles of expiration contract forcefully. As a consequence, the pressure in the lungs increases to 100 mm Hg or more. Then the vestibular and vocal folds open suddenly, the soft palate is elevated, and the air rushes from the lungs and out the oral cavity at a high velocity, carrying foreign particles with it.

The sneeze reflex is similar to the cough reflex, but it differs in several ways. The source of irritation that initiates the sneeze reflex is in the nasal passages instead of in the trachea and bronchi, and the action potentials are conducted along the trigeminal nerves to the medulla oblongata, where the reflex is triggered. During the sneeze reflex the soft palate is depressed so that air is directed primarily through the nasal passages, although a considerable amount passes through the oral cavity. The rapidly flowing air dislodges particulate matter from the nasal passages and can propel it a considerable distance from the nose.

---

through pulmonary arteries to pulmonary capillaries, becomes oxygenated, and returns to the heart through pulmonary veins. The other route brings oxygenated blood to the tissues of the bronchi down to the respiratory bronchioles. The oxygenated blood flows from the thoracic aorta through bronchial arteries to capillaries, where oxygen is released. Deoxygenated blood from the proximal part of the bronchi returns to the heart through the bronchial veins and the azygos venous system (see chapter 21). More distally, the venous drainage from the bronchi enters the pulmonary veins. Thus, the oxygenated blood returning from the alveoli in the pulmonary veins is mixed with a small amount of deoxygenated blood returning from the bronchi.

Lymphatic Supply

The lungs have two lymphatic supplies. The **superficial lymphatic vessels** are deep to the visceral pleura and function to drain lymph from the superficial lung tissue and the visceral pleura. The **deep lymphatic vessels** follow the bronchi and function to drain lymph from the bronchi and associated connective tissues. No lymphatic vessels are located in the walls of the alveoli. Both the superficial and deep lymphatic vessels exit the lung at the hilum.

Phagocytic cells within the lungs phagocytize carbon particles and other debris from inspired air and move them to the lymphatic vessels. In older people, the surface of the lungs can appear gray to black because of the accumulation of these particles, especially if the person smokes or has lived most of his or her life in a city with air pollution. Cancer cells from the lungs can spread to other parts of the body through the lymphatic vessels.

14. *List the muscles of respiration and describe their role in quiet inspiration and expiration. How does this change during labored breathing?*

15. *Name the pleurae of the lungs. What is their function?*

16. *What are the two major routes of blood flow to and from the lungs? What is the function of each route?*

17. *Describe the lymphatic supply of the lungs.*

Ventilation

Pressure Differences and Airflow

Ventilation is the process of moving air into and out of the lungs. The flow of air into the lungs requires a pressure gradient from the outside of the body to the alveoli, and airflow from the lungs requires a pressure gradient in the opposite direction. The physics of airflow in tubes, such as the ones that make up the respiratory passages, is the same as the flow of blood in blood vessels (see chapter 21). Thus, the following relationships hold:

$$F = \frac{P_1 - P_2}{R}$$

where F is airflow (milliliters per minute) in a tube, P_1 is pressure at point one, P_2 is pressure at point two, and R is resistance to airflow.

Air moves through tubes because of a pressure difference. When P_1 is greater than P_2, gas flows from P_1 to P_2 at a rate that's proportional to the pressure difference. For example, during inspiration, air pressure outside the body is greater than air pressure in the alveoli, and air flows through the trachea and bronchi to the alveoli.

Disorders That Decrease the Radius of Air Passageways

The flow of air decreases when the resistance to airflow is increased by conditions that reduce the radius of the respiratory passageways. According to Poiseuille's law (see chapter 21), the resistance to airflow is proportional to the radius (r) of a tube raised to the fourth power (r^4). Thus, a small change in radius results in a large change in resistance, which greatly decreases airflow. For example, asthma results in the release of inflammatory chemicals such as leukotrienes that cause severe constriction of the bronchioles. Emphysema produces increased airway resistance because the bronchioles are obstructed as a result of inflammation and because damaged bronchioles collapse during expiration, thus trapping air within the alveoli. Cancer can also occlude respiratory passages as the tumor replaces lung tissue. Increasing the pressure difference between alveoli and the atmosphere can help to maintain airflow despite increased resistance. Within limits, this can be accomplished by increased contraction of the muscles of respiration.

Pressure and Volume

The pressure in a container, such as the thoracic cavity or an alveolus, is described according to the **general gas law.**

$$P = \frac{nRT}{V}$$

where P is pressure, n is the number of gram moles of gas (a measure of the number of molecules present), R is the gas constant, T is absolute temperature, and V is volume.

The value of R is a constant, and the values of n and T (body temperature) are considered constants in humans. Thus, the general gas law reveals that air pressure is inversely proportional to volume. As volume increases, pressure decreases; and as volume decreases, pressure increases (table 23.1).

- **18.** *Define the term ventilation.*
- **19.** *How do pressure differences and resistance affect airflow through a tube?*
- **20.** *What happens to the pressure within a container when the volume of the container increases?*

Airflow into and out of Alveoli

Respiratory physiologists use three conventions to help simplify the numbers used to express pressures. First, **barometric air pressure**

Table 23.1	Gas Law
Description	**Importance**
General Gas Law	
The pressure of a gas is inversely proportional to its volume (at a constant temperature, this is referred to as Boyle's law).	Air flows from areas higher to lower pressure. When alveolar volume increases, causing pleural pressure to decrease below atmosphereic pressure, air moves into the lungs. When alveolar volume decreases, causing pleural pressure to increase above atmospheric pressure, air moves out of the lungs.
Dalton's Law	
The partial pressure of a gas in a mixture of gases is the percentage of the gas in the mixture times the total pressure of the mixture of gases.	Gases move from areas of higher to areas of lower partial pressures. The greater the difference in partial pressure between two points, the greater the rate of gas movement. Maintaining partial pressure differences ensures gas movements.
Henry's Law	
The concentration of a gas dissolved in a liquid is equal to the partial pressure of the gas over the liquid times the solubility coefficient of the gas.	Only a small amount of the gases in air dissolves in the fluid lining the alveoli. Carbon dioxide, however, is 24 times more soluble than oxygen; therefore, carbon dioxide passes out through the respiratory membrane more readily than oxygen enters.

(P_B), which is atmospheric air pressure outside the body, is assigned a value of zero. Thus, whether at sea level with a pressure of 760 mm Hg or at 10,000 feet above sea level on a mountaintop with a pressure of 523 mm Hg, P_B is always zero. Second, the small pressures in respiratory physiology are usually expressed in centimeters of water (cm H_2O). A pressure of 1 cm H_2O is equal to 0.74 mm Hg. Third, other pressures are measured in reference to barometric air pressure. For example, **alveolar pressure (P_{alv})** is the pressure inside an alveolus. An alveolar pressure of 1 cm H_2O is 1 cm H_2O greater pressure than barometric air pressure, and an alveolar pressure of -1 cm H_2O is 1 cm H_2O less pressure than barometric air pressure.

Movement of air into and out of the lungs results from changes in thoracic volume, which cause changes in alveolar volume. The changes in alveolar volume produce changes in alveolar pressure. The pressure difference between barometric air pressure and alveolar pressure ($P_B - P_{alv}$) results in air movement. The details of this process during quiet breathing are described as follows:

1. *End of expiration* (figure 23.13, *step 1*). At the end of expiration, barometric air pressure and alveolar pressure are equal. Therefore, no movement of air into or out of the lungs takes place.
2. *During inspiration* (figure 23.13, *step 2*). As inspiration begins, contraction of inspiratory muscles increases thoracic volume, which results in expansion of the lungs and an increase in alveolar volume (see following section on "Changing Alveolar Volume"). The increased alveolar volume causes a decrease in alveolar pressure below barometric air pressure to approximately -1cm H_2O. Air flows into the lungs because barometric air pressure is greater than alveolar pressure.
3. *End of inspiration* (figure 23.13, *step 3*). At the end of inspiration, the thorax stops expanding, the alveoli stop expanding, and alveolar pressure becomes equal to barometric air pressure because of airflow into the lungs. No movement of air occurs after alveolar pressure becomes equal to barometric pressure, but the volume of the lungs is larger than at the end of expiration.
4. *During expiration* (figure 23.13, *step 4*). During expiration, the volume of the thorax decreases as the diaphragm relaxes, and the thorax and lungs recoil. The decreased thoracic volume results in a decrease in alveolar volume and an increase in alveolar pressure over barometric air pressure to approximately 1cm H_2O. Air flows out of the lungs because alveolar pressure is greater than barometric air pressure. As expiration ends, the decrease in thoracic volume stops and the alveoli stop changing size. The process repeats beginning at step *1*.

- **21.** *Define barometric and alveolar pressures.*
- **22.** *Explain how changes in alveolar volume cause air to move into and out of the lungs.*

Changing Alveolar Volume

It's important to understand how alveolar volume is changed because these changes cause the pressure differences resulting in ventilation. In addition, many respiratory disorders affect how alveolar volume changes. Lung recoil and changes in pleural pressure cause changes in alveolar volume.

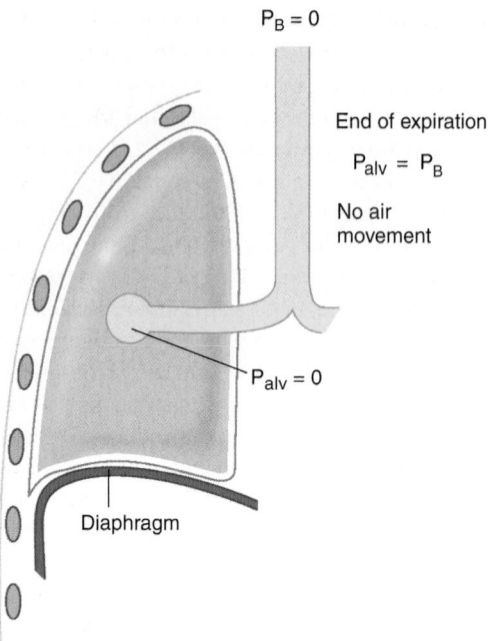

$P_B = 0$

End of expiration

$P_{alv} = P_B$

No air movement

$P_{alv} = 0$

Diaphragm

1. At the end of expiration, alveolar pressure (P_{alv}) is equal to barometric air pressure (P_B) and there is no air movement.

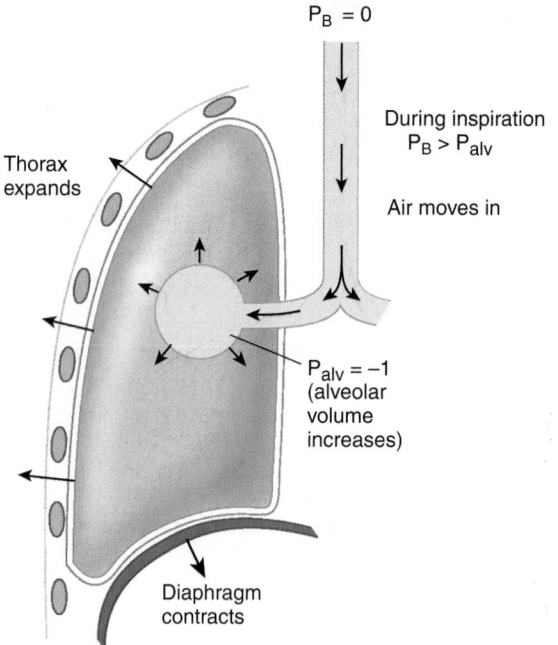

$P_B = 0$

Thorax expands

During inspiration $P_B > P_{alv}$

Air moves in

$P_{alv} = -1$ (alveolar volume increases)

Diaphragm contracts

2. During inspiration, increased thoracic volume results in increased alveolar volume and decreased alveolar pressure. Barometric air presure is greater than alveolar pressure, and air moves into the lungs.

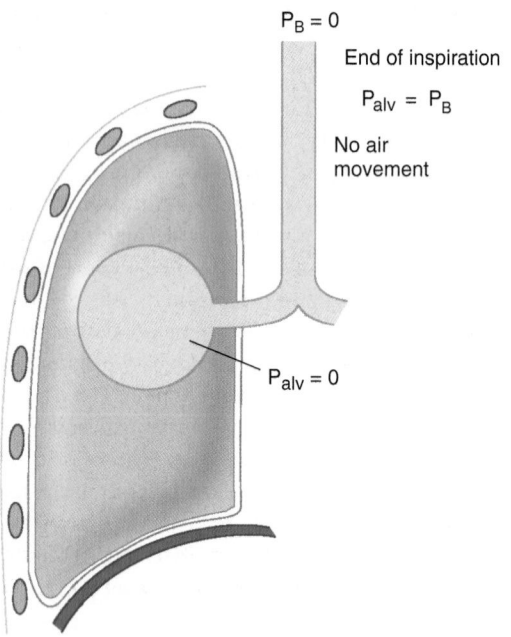

$P_B = 0$

End of inspiration

$P_{alv} = P_B$

No air movement

$P_{alv} = 0$

3. At the end of inspiration, alveolar pressure is equal to barometric air pressure and there is no air movement.

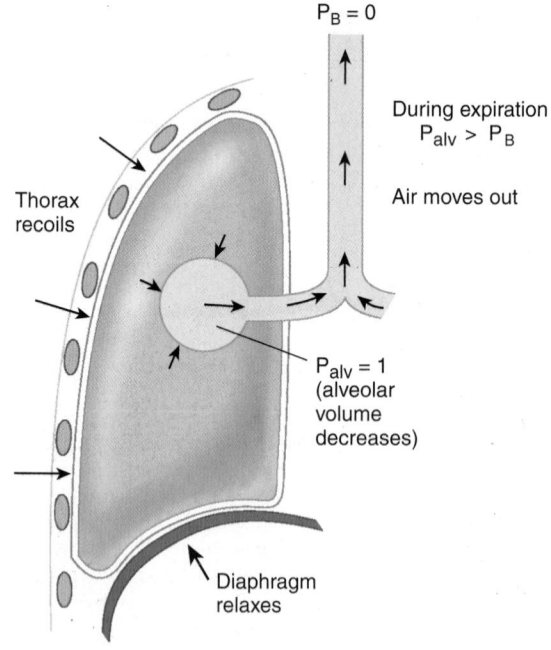

$P_B = 0$

During expiration $P_{alv} > P_B$

Air moves out

Thorax recoils

$P_{alv} = 1$ (alveolar volume decreases)

Diaphragm relaxes

4. During expiration, decreased thoracic volume results in decreased alveolar volume and increased alveolar pressure. Alveolar pressure is greater than barometric air pressure, and air moves out of the lungs.

Process Figure 23.13   **Alveolar Pressure Changes During Inspiration and Expiration**

The combined space of all the alveoli is represented by a large "bubble." The alveoli are actually microscopic in size and cannot be seen in the illustration.

Lung Recoil

Lung recoil is a decrease in size of an expanded lung. A lung decreases in size as a result of a decrease in size (volume) of its alveoli. Expanded alveoli decrease in size for two reasons: (1) elastic recoil caused by the elastic fibers in the alveolar walls and (2) surface tension of the film of fluid that lines the alveoli. Surface tension occurs at the boundary between water and air because the polar water molecules are attracted to one another more than they are attracted to the air molecules. Consequently, the water molecules are drawn together, tending to form a droplet. Because the water molecules of the alveolar fluid are also attracted to the surface of the alveoli, formation of a droplet causes the alveoli to collapse, thus producing fluid-filled alveoli with smaller volumes than air-filled alveoli.

Surfactant (ser-fak′tănt, surface acting agent) is a mixture of lipoprotein molecules produced by the type II pneumocytes of the alveolar epithelium. The surfactant molecules form a monomolecular layer over the surface of the fluid within the alveoli to reduce the surface tension. With surfactant, the force produced by surface tension is approximately 4 cm H_2O; without surfactant, the force can be as high as 40 cm H_2O. Thus, surfactant greatly reduces the tendency of the lungs to collapse.

Respiratory Distress Syndrome

In premature infants, **respiratory distress syndrome**, or **hyaline** (hī′ă-lin) **membrane disease,** is common, especially for infants with a gestation age of less than 7 months. This occurs because surfactant is not produced in adequate quantities until approximately 7 months of development. Thereafter, the amount produced increases as the fetus matures. Cortisol can be given to pregnant women who are likely to deliver prematurely, because it crosses the placenta into the fetus and stimulates surfactant synthesis.

If insufficient surfactant is produced by a newborn, the lungs tend to collapse. Thus, a great deal of energy must be exerted by the muscles of respiration to keep the lungs inflated, and even then inadequate ventilation occurs. Without specialized treatment, most babies with this disease die soon after birth as a result of inadequate ventilation of the lungs and fatigue of the respiratory muscles. Positive end-expiratory pressure delivers oxygen-rich, pressurized air to the lungs through a tube passed through the respiratory passages. The pressure helps to keep the alveoli inflated. In addition, surfactant administered with the pressurized air can reduce surface tension in the alveoli. Surfactant can be obtained from cow, pig, and human lungs; human amniotic fluid; and genetically-modified bacteria. Synthetic surfactant is also available.

> ● **23.** Name two things that cause the lungs to recoil. How does surfactant reduce lung recoil? What happens if there are inadequate amounts of surfactant in the alveoli?

Pleural Pressure

Pleural pressure (P_{pl}) is the pressure in the pleural cavity. When pleural pressure is less than alveolar pressure, the alveoli tend to expand. This principle can be understood by considering a balloon. The balloon expands when the pressure outside the balloon is less than the pressure inside. This pressure difference is normally achieved by increasing the pressure inside the balloon when a person forcefully blows into it. This pressure difference, however, can also be achieved by decreasing the pressure outside the balloon. For example, if the balloon is placed in a chamber from which air is removed, the pressure around the balloon becomes lower than atmospheric pressure, and the balloon expands. The lower the pressure outside the balloon, the greater the tendency for the higher pressure inside the balloon to cause it to expand. In a similar fashion, decreasing pleural pressure can result in expansion of the alveoli.

Normally the alveoli are expanded because of a negative pleural pressure that is lower than alveolar pressure. At the end of a normal expiration, pleural pressure is −5 cm H_2O, and alveolar pressure is 0 cm H_2O. Pleural pressure is lower than alveolar pressure because of a "suction effect" caused by fluid removal by the lymphatic system and by lung recoil. As the lungs recoil, the visceral and parietal pleurae tend to be pulled apart. Normally the lungs don't pull away from the thoracic wall because pleural fluid holds the visceral and parietal pleurae together. Nonetheless, this pull decreases pressure in the pleural cavity, an effect that can be appreciated by putting water on the palms of the hands and putting them together. A sensation of negative pressure is felt as the hands are gently pulled apart.

When pleural pressure is lower than alveolar pressure, the alveoli tend to expand. This expansion is opposed by the tendency of the lungs to recoil. Therefore, the alveoli expand when the pleural pressure is low enough that lung recoil is overcome. If the pleural pressure is not low enough to overcome lung recoil, then the alveoli collapse.

Pneumothorax

A **pneumothorax** is the introduction of air into the pleural cavity through an opening in the thoracic wall or lung. Pneumothorax can result from penetrating trauma by a knife, bullet, broken rib, or other object; nonpenetrating trauma such as a blow to the chest; medical procedures such as inserting a catheter to withdraw pleural fluid; disease, such as infections or emphysema; or can be of unknown cause.

Pleural pressure increases and becomes equal to barometric air pressure when the pleural cavity is connected to the outside through an opening in the thoracic wall or the surface of the lung. The alveoli, therefore, don't tend to expand, lung recoil is unopposed, and the lung collapses and pulls away from the thoracic wall. Pneumothorax can occur in one lung while the lung on the opposite side remains inflated because the two pleural cavities are separated by the mediastinum.

The most common symptoms of pneumothorax are chest pain and shortness of breath. Treatment of pneumothorax depends upon its cause and severity. In patients with mild symptoms, the pneumothorax may resolve on its own. In other cases, a chest tube that aspirates the pleural cavity and restores a negative pressure can cause re-expansion of the lung. Surgery may also be necessary to close the opening into the pleural cavity.

In a **tension pneumothorax,** the pressure within the thoracic cavity is always higher than barometric air pressure. A tissue flap or air passageway forms a flutter valve that allows air to enter the pleural cavity during inspiration but not exit during expiration. The result is an increase in air and pressure within the pleural cavity that can compress blood vessels returning blood to the heart, causing decreased venous return, low blood pressure, and inadequate delivery of oxygen to tissues. Insertion of a large bore needle into the pleural cavity allows air to escape and releases the pressure.

24. *Define pleural pressure. What happens to alveolar volume when pleural pressure decreases? What causes pleural pressure to be lower than alveolar pressure?*

25. *How does an opening in the chest wall cause the lung to collapse?*

26. *During inspiration, what causes pleural pressure to decrease? What effect does this have on alveolar pressure and air movement?*

27. *During expiration, what causes pleural pressure to increase? What effect does this have on alveolar pressure and air movement?*

Pressure Changes During Inspiration and Expiration

At the end of a normal expiration, pleural pressure is -5 cm H_2O, and alveolar pressure is equal to barometric pressure (0 cm H_2O) (figure 23.14). During normal, quiet inspiration, pleural pressure decreases to -8 cm H_2O. Consequently, the alveolar volume increases, alveolar pressure decreases below barometric air pressure,

and air flows into the lungs. As air flows into the lungs, alveolar pressure increases and becomes equal to barometric pressure at the end of inspiration.

The decrease in pleural pressure during inspiration occurs for two reasons. First, because of the effect of changing volume on pressure (general gas law), when the volume of the thoracic cavity increases, pleural pressure decreases. Second, as the thoracic cavity expands, the lungs expand because they adhere to the inner thoracic wall through the pleurae. As the lungs expand, the tendency for the lungs to recoil increases, resulting in an increased suction effect and a lowering of pleural pressure. The tendency for the lungs to recoil increases as the lungs are stretched, similar to the increased force generated in a stretched rubber band.

During expiration, pleural pressure increases because of decreased thoracic volume and decreased lung recoil (see figure 23.14). As pleural pressure increases, alveolar volume decreases, alveolar pressure increases above barometric air pressure, and air flows out of the lungs. As air flows out of the lungs, alveolar pressure decreases and becomes equal to barometric pressure at the end of expiration.

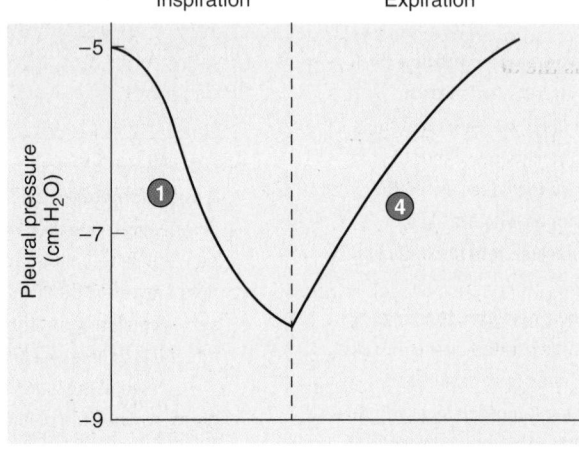

Changes during inspiration

1. Pleural pressure decreases because thoracic volume increases.

2. As inspiration begins, alveolar pressure decreases below barometric air pressure (0 on the graph) because the decreased pleural pressure causes alveolar volume to increase. By the end of inspiration, alveolar and barometric air pressure are equal.

3. During inspiration, air flows into the lungs because alveolar pressure is lower than barometric air pressure.

Changes during expiration

4. Pleural pressure increases because thoracic volume decreases.

5. As expiration begins, alveolar pressure increases above barometric air pressure (0 on the graph) because the increased pleural pressure causes alveolar volume to decrease. By the end of expiration, alveolar and barometric air pressure are equal.

6. During expiration, air flows out of the lungs because alveolar pressure is greater than barometric air pressure.

Process Figure 23.14   Dynamics of a Normal Breathing Cycle

P R E D I C T
How does the pleural pressure at the end of expiration in a newborn with respiratory distress syndrome compare to that of a healthy newborn? How does the pleural pressure compare during inspiration? Explain.

Measuring Lung Function

A variety of measurements can be used to assess lung function. Each of these tests compares a subject's measurements to a normal range. These measurements can be used to diagnose diseases, track the progress of diseases, or track recovery from diseases.

Compliance of the Lungs and the Thorax

Compliance is a measure of the ease with which the lungs and the thorax expand. The compliance of the lungs and thorax is the volume by which they increase for each unit of pressure change in alveolar pressure. It is usually expressed in liters (volume of air) per centimeter of water (pressure), and for the normal person the compliance of the lungs and thorax is 0.13 L/cm H_2O. That is, for every 1 cm H_2O change in alveolar pressure, the volume changes by 0.13 L.

The greater the compliance, the easier it is for a change in pressure to cause expansion of the lungs and thorax. For example, one possible result of emphysema is the destruction of elastic lung tissue. This reduces the elastic recoil force of the lungs, thereby making expansion of the lungs easier and resulting in a higher-than-normal compliance. A lower-than-normal compliance means that it's harder to expand the lungs and thorax. Conditions that decrease compliance include deposition of inelastic fibers in lung tissue (pulmonary fibrosis), collapse of the alveoli (respiratory distress syndrome and pulmonary edema), increased resistance to airflow caused by airway obstruction (asthma, bronchitis, and lung cancer), and deformities of the thoracic wall that reduce the ability of the thoracic volume to increase (kyphosis and scoliosis).

Effects of Decreased Compliance

Pulmonary diseases can markedly affect the total amount of energy required for ventilation, as well as the percentage of the total amount of energy expended by the body. Diseases that decrease compliance can increase the energy required for breathing up to 30% of the total energy expended by the body.

> 28. *Define the term compliance. What is the effect on lung expansion when compliance increases or decreases?*

Pulmonary Volumes and Capacities

Spirometry (spī-rom′ĕ-trē) is the process of measuring volumes of air that move into and out of the respiratory system, and a **spirometer** (spī-rom′ĕ-ter) is a device used to measure these pulmonary volumes (figure 23.15a). The four pulmonary volumes and representative values (figure 23.15b) for a young adult male follow:

1. **Tidal volume** is the volume of air inspired or expired with each breath. At rest, quiet breathing results in a tidal volume of approximately 500 mL.

2. **Inspiratory reserve volume** is the amount of air that can be inspired forcefully after inspiration of the tidal volume (approximately 3000 mL at rest).

3. **Expiratory reserve volume** is the amount of air that can be forcefully expired after expiration of the tidal volume (approximately 1100 mL at rest).

4. **Residual volume** is the volume of air still remaining in the respiratory passages and lungs after the most forceful expiration (approximately 1200 mL).

The tidal volume increases when a person is more active. Because the maximum volume of the respiratory system does not change from moment to moment, an increase in tidal volume causes a decrease in the inspiratory and expiratory reserve volumes.

Pulmonary capacities are the sum of two or more pulmonary volumes (see figure 23.15b). Some pulmonary capacities follow:

1. **Inspiratory capacity** is the tidal volume plus the inspiratory reserve volume, which is the amount of air that a person can inspire maximally after a normal expiration (approximately 3500 mL).

2. **Functional residual capacity** is the expiratory reserve volume plus the residual volume, which is the amount of air remaining in the lungs at the end of a normal expiration (approximately 2300 mL).

3. **Vital capacity** is the sum of the inspiratory reserve volume, the tidal volume, and the expiratory reserve volume, which is the maximum volume of air that a person can expel from the respiratory tract after a maximum inspiration (approximately 4600 mL).

4. **Total lung capacity** is the sum of the inspiratory and expiratory reserve volumes plus the tidal volume and the residual volume (approximately 5800 mL).

Factors like sex, age, body size, and physical conditioning cause variations in respiratory volumes and capacities from one individual to another. For example, the vital capacity of adult females is usually 20%–25% less than that of adult males. The vital capacity reaches its maximum amount in the young adult and gradually decreases in the elderly. Tall people usually have a greater vital capacity than short people, and thin people have a greater vital capacity than obese people. Well-trained athletes can have a vital capacity 30%–40% above that of untrained people. In patients whose respiratory muscles are paralyzed by spinal cord injury or diseases like poliomyelitis or muscular dystrophy, vital capacity can be reduced to values not consistent with survival (less than 500–1000 mL). Factors that reduce compliance also reduce vital capacity.

The **forced expiratory vital capacity** is a simple and clinically important pulmonary test. The individual inspires maximally and then exhales maximally into a spirometer as rapidly as possible. The volume of air expired at the end of the test is the person's vital capacity. The spirometer also records the volume of air that enters it per second. The **forced expiratory volume in one second (FEV$_1$)** is the amount of air expired during the first second of the test. In some conditions, the vital capacity may not be dramatically affected, but how rapidly air is expired can be greatly decreased.

(a)

Figure 23.15　Spirometer, Lung Volumes, and Lung Capacities

(*a*) A spirometer used to measure lung volumes and capacities. (*b*) Lung volumes and capacities. The tidal volume in the figure is the tidal volume during resting conditions.

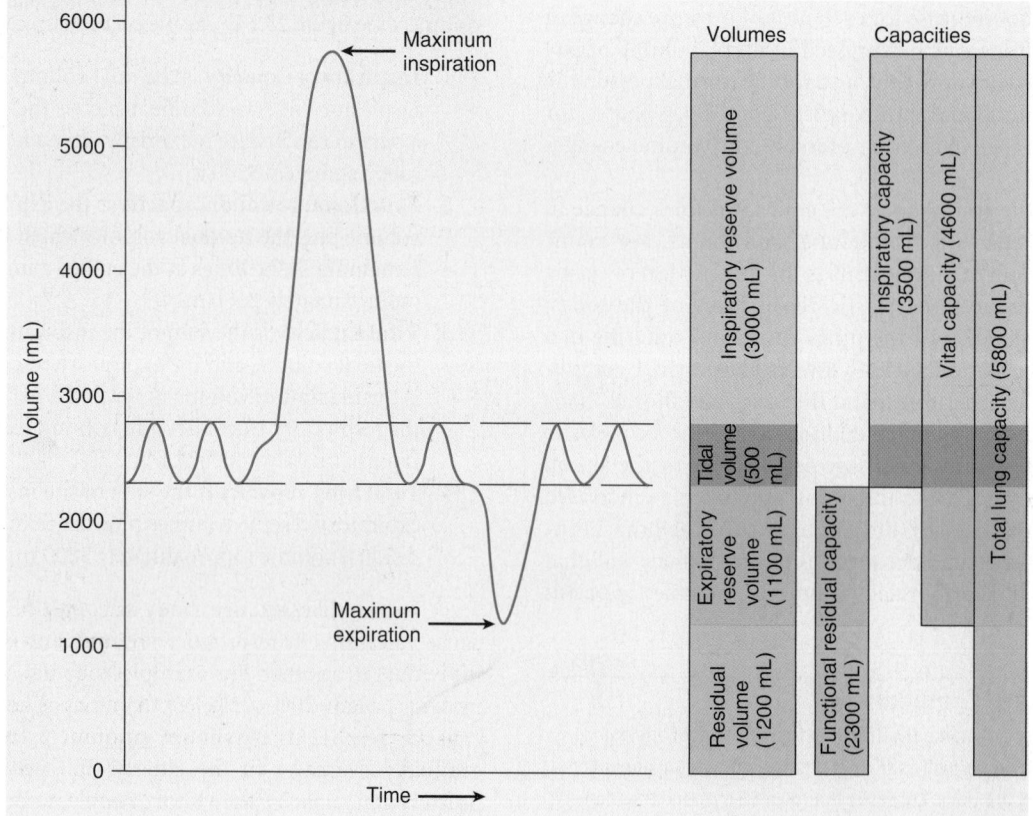

(b)

Airway obstruction, caused by asthma, collapse of bronchi in emphysema, or a tumor, and disorders that reduce the ability of the lungs or chest wall to deflate, such as pulmonary fibrosis, silicosis, kyphosis, and scoliosis, can cause a decreased FEV$_1$.

29. *Define the terms tidal volume, inspiratory reserve volume, expiratory reserve volume, and residual volume.*

30. *Define the terms inspiratory capacity, functional residual capacity, vital capacity, and total lung capacity.*

31. *What is forced expiratory volume in one second, and why is it clinically important?*

Minute Ventilation and Alveolar Ventilation

Minute ventilation is the total amount of air moved into and out of the respiratory system each minute, and it is equal to tidal volume times the respiratory rate. **Respiratory rate,** or **respiratory frequency,** is the number of breaths taken per minute. Because resting tidal volume is approximately 500 mL and respiratory rate is approximately 12 breaths per minute, minute ventilation averages approximately 6 L/min.

Although minute ventilation measures the amount of air moving into and out of the lungs per minute, it's not a measure of the amount of air available for gas exchange because gas exchange takes place mainly in the alveoli and to a lesser extent in the alveolar ducts and the respiratory bronchioles. The part of the respiratory system where gas exchange does not take place is called the dead space. A distinction can be made between anatomic and physiologic dead space. **Anatomic dead space,** which measures 150 mL, is formed by the nasal cavity, pharynx, larynx, trachea, bronchi, bronchioles, and terminal bronchioles. **Physiologic dead space** is anatomic dead space plus the volume of any alveoli in which gas exchange is less than normal. In a healthy person, anatomic and physiologic dead spaces are nearly the same, meaning that few nonfunctional alveoli exist.

Emphysema and Dead Space

In patients with **emphysema,** alveolar walls degenerate, and small alveoli combine to form larger alveoli. The result is fewer alveoli, but alveoli with an increased volume and decreased surface area. Although the enlarged alveoli are still ventilated, surface area is inadequate for complete gas exchange, and the physiologic dead space increases.

During inspiration, much of the inspired air fills the dead space first before reaching the alveoli and, thus, is unavailable for gas exchange. The volume of air available for gas exchange per minute is called **alveolar ventilation ($\dot{V}_A$),** and it is calculated as follows:

$$\dot{V}_A = f(V_T - V_D)$$

where $\dot{V}_A$ is alveolar ventilation (milliliters per minute), f is respiratory rate (frequency; breaths per minute), V_T is tidal volume (milliliters per respiration), and V_D is dead space (milliliters per respiration).

32. **Define the terms minute ventilation and alveolar ventilation.**
33. **What is dead space? What is the difference between anatomic and physiologic dead space?**

P R E D I C T

What is the alveolar ventilation of a resting person with a tidal volume of 500 mL, a dead space of 150 mL, and a respiratory rate of 12 breaths per minute? If the person exercises and tidal volume increases to 4000 mL, dead space increases to 300 mL as a result of dilation of the respiratory passageways, and respiratory rate increases to 24 breaths per minute, what is the alveolar ventilation? How is the change in alveolar ventilation beneficial for doing exercise?

Physical Principles of Gas Exchange

Ventilation supplies atmospheric air to the alveoli. The next step in the process of respiration is the diffusion of gases between alveoli and blood in the pulmonary capillaries. The molecules of gas move randomly, and if a gas is in a higher concentration at one point than at another, random motion ensures that the net movement of gas is from the higher concentration toward the lower concentration until a homogeneous mixture of gases is achieved. One measurement of the concentration of gases is partial pressure.

Partial Pressure

At sea level, atmospheric pressure is approximately 760 mm Hg, which means that the mixture of gases that constitute atmospheric air exerts a total pressure of 760 mm Hg. The major components of dry air are nitrogen (approximately 79%) and oxygen (approximately 21%). According to **Dalton's law,** in a mixture of gases, the part of the total pressure resulting from each type of gas is determined by the percentage of the total volume represented by each gas type (see table 23.1). The pressure exerted by each type of gas in a mixture is referred to as the **partial pressure** of that gas. Because nitrogen makes up 78.62% of the volume of atmospheric air, the partial pressure resulting from nitrogen is 0.7862 times 760 mm Hg, or 597.5 mm Hg. Because oxygen is 20.84% of the volume of atmospheric air, the partial pressure resulting from oxygen is 0.2084 times 760 mm Hg, or 158.4 mm Hg. It's traditional to designate the partial pressure of individual gases in a mixture with a capital P followed by the symbol for the gas. Thus, the partial pressure of nitrogen is denoted P_{N_2}, oxygen is P_{O_2}, and carbon dioxide is P_{CO_2}.

When air comes into contact with water, some of the water turns into a gas and evaporates into the air. Water molecules in gaseous form also exert a partial pressure. This partial pressure (P_{H_2O}) is sometimes referred to as **water vapor pressure.** The composition of dry, humidified, alveolar, and expired air is presented in table 23.2. The composition of alveolar air and of expired air is not identical to the composition of dry atmospheric air for three reasons. First, air entering the respiratory system during inspiration is humidified; second, oxygen diffuses from the alveoli into the blood, and carbon dioxide diffuses from the pulmonary capillaries into the alveoli; and third, the air within the alveoli is only partially replaced with atmospheric air during each inspiration.

Table 23.2	**Partial Pressures of Gases at Sea Level**							
	Dry Air		**Humidified Air**		**Alveolar Air**		**Expired Air**	
Gases	**mm Hg**	**%**	**mm Hg**	**%**	**mm Hg**	**%**	**mm Hg**	**%**
Nitrogen	597.5	78.62	563.4	74.09	569.0	74.9	566.0	74.5
Oxygen	158.4	20.84	149.3	19.67	104.0	13.6	120.0	15.7
Carbon dioxide	0.3	0.04	0.3	0.04	40.0	5.3	27.0	3.6
Water vapor	0.0	0.0	47.0	6.20	47.0	6.2	47.0	6.2

34. *According to Dalton's law, what is the partial pressure of a gas? What is water vapor pressure?*

35. *Why is the composition of inspired, alveolar, and expired air different?*

Diffusion of Gases Through Liquids

When a gas comes into contact with a liquid such as water, it tends to dissolve in the liquid. At equilibrium, the concentration of a gas in the liquid is determined by its partial pressure in the gas and by its solubility in the liquid. This relationship is described by **Henry's law** (see table 23.1).

$$\text{Concentration of dissolved gas} = \text{Partial pressure of gas} \times \text{Solubility coefficient}$$

The solubility coefficient is a measure of how easily the gas dissolves in the liquid. In water, the solubility coefficient for oxygen is 0.024, and for carbon dioxide it is 0.57. Thus, carbon dioxide is approximately 24 times more soluble in water than is oxygen.

Gases don't actually produce a partial pressure in a liquid as they do when in the gaseous state. Using the general gas law equation and the concentration of a gas in a liquid, however, the partial pressure of the gas if it were in a gaseous state can be calculated. Because the calculated partial pressure of a gas in a liquid is a measure of concentration, it can be used to determine the direction of diffusion of the gas through the liquid: the gas moves from areas of higher to areas of lower partial pressure.

36. *According to Henry's law, how does the partial pressure and solubility of a gas affect its concentration in a liquid?*

P R E D I C T   **6**

As a SCUBA diver descends, the pressure of the water on the body prevents normal expansion of the lungs. To compensate, the diver breathes pressurized air, which has a greater pressure than air at sea level. What effect does the increased pressure have on the amount of gas dissolved in the diver's body fluids? A SCUBA diver who suddenly ascends to the surface from a great depth can develop decompression sickness (the bends) in which bubbles of nitrogen gas form. The expanding bubbles damage tissues or block blood flow through small blood vessels. Explain the development of the bubbles.

Diffusion of Gases Through the Respiratory Membrane

The factors that influence the rate of gas diffusion across the respiratory membrane include (1) the thickness of the membrane; (2) the diffusion coefficient of the gas in the substance of the membrane, which is approximately the same as the diffusion coefficient for the gas through water; (3) the surface area of the membrane; and (4) the difference of the partial pressures of the gas between the two sides of the membrane.

Respiratory Membrane Thickness

Increasing the thickness of the respiratory membrane decreases the rate of diffusion. The thickness of the respiratory membrane normally averages 0.6 μm, but diseases can cause an increase in the thickness. If the thickness of the respiratory membrane increases two or three times, the rate of gas exchange markedly decreases. Pulmonary edema caused by failure of the left side of the heart is the most common cause of an increase in the thickness of the respiratory membrane. Left side heart failure increases venous pressure in the pulmonary capillaries and results in the accumulation of fluid in the alveoli. Conditions such as tuberculosis, pneumonia, or advanced silicosis that result in inflammation of the lung tissues can also cause fluid accumulation within the alveoli.

Diffusion Coefficient

The **diffusion coefficient** is a measure of how easily a gas diffuses through a liquid or tissue, taking into account the solubility of the gas in the liquid and the size of the gas molecule (molecular weight). If the diffusion coefficient of oxygen is assigned a value of 1, then the relative diffusion coefficient of carbon dioxide is 20, which means carbon dioxide diffuses through the respiratory membrane about 20 times more readily than oxygen does.

When the respiratory membrane becomes progressively damaged as a result of disease, its capacity for allowing the movement of oxygen into the blood is often impaired enough to cause death from oxygen deprivation before the diffusion of carbon dioxide is dramatically reduced. If life is being maintained by extensive oxygen therapy, which increases the concentration of oxygen in the lung alveoli, the reduced capacity for the diffusion of carbon dioxide across the respiratory membrane can result in substantial increases in carbon dioxide in the blood.

Surface Area

In a healthy adult, the total surface area of the respiratory membrane is approximately 70 m^2 (approximately the floor area of a 25- by 30-foot room). Several respiratory diseases, including emphysema and lung cancer, cause a decrease in the surface area of the respiratory membrane. Even small decreases in this surface area adversely affect the respiratory exchange of gases during strenuous exercise. When the total surface area of the respiratory membrane is decreased to one-third or one-fourth of normal, the exchange of gases is significantly restricted even under resting conditions.

A decreased surface area for gas exchange can also result from the surgical removal of lung tissue, the destruction of lung tissue by cancer, the degeneration of the alveolar walls by emphysema, or the replacement of lung tissue by connective tissue caused by tuberculosis. More acute conditions that cause the alveoli to fill with fluid also reduce the surface area for gas exchange, because the increased thickness of the respiratory membrane caused by the fluid makes the alveoli nonfunctional. Examples include pneumonia and pulmonary edema resulting from failure of the left ventricle.

Partial Pressure Difference

The partial pressure difference of a gas across the respiratory membrane is the difference between the partial pressure of the gas in the alveoli and the partial pressure of the gas in the blood of the pulmonary capillaries. When the partial pressure of a gas is greater on one side of the respiratory membrane than on the other side, net diffusion occurs from the higher to the lower partial pressure (see figure 23.8b). Normally, the partial pressure of oxygen (PO_2) is greater in the alveoli than in the blood of the pulmonary capillaries, and the partial pressure of carbon dioxide (PCO_2) is greater in the blood than in the alveolar air.

By increasing alveolar ventilation, the partial pressure difference for oxygen and carbon dioxide can be raised. The greater volume of atmospheric air exchanged with the residual volume raises alveolar PO_2, lowers alveolar PCO_2, and thus promotes gas exchange. Conversely, inadequate ventilation causes a lower-than-normal partial pressure difference for oxygen and carbon dioxide, resulting in inadequate gas exchange.

- 37. *Describe four factors that affect the diffusion of gases across the respiratory membrane. Give examples of diseases that decrease diffusion by altering these factors.*
- 38. *Does oxygen or carbon dioxide diffuse most easily through the respiratory membrane?*

Relationship Between Ventilation and Pulmonary Capillary Blood Flow

Under normal conditions, ventilation of the alveoli and blood flow through pulmonary capillaries is such that effective gas exchange occurs between the air and the blood. During exercise, effective gas exchange is maintained because both ventilation and cardiac output increase.

The normal relationship between ventilation and pulmonary capillary blood flow can be disrupted in two different ways. One way occurs when ventilation exceeds the ability of the blood to pick up oxygen, which can happen because of inadequate cardiac output after a heart attack. Another way occurs when ventilation is not great enough to provide the oxygen needed to oxygenate the blood flowing through the pulmonary capillaries. For example, constriction of the bronchioles in asthma can decrease air delivery to the alveoli.

Blood that isn't completely oxygenated is called shunted blood. Two sources of shunted blood exist in the lungs. An **anatomic shunt** results when deoxygenated blood from the bronchi and bronchioles mixes with blood in the pulmonary veins (see section on "Blood Supply" on p. 843). The other source of shunted blood is blood that passes through pulmonary capillaries but doesn't become fully oxygenated. The **physiologic shunt** is the combination of deoxygenated blood from the anatomic shunt and deoxygenated blood that passes through the pulmonary capillaries. Normally, 1%–2% of cardiac output makes up the physiologic shunt.

Disorders That Increase Shunted Blood

Any condition that decreases gas exchange between the alveoli and the blood can increase the amount of shunted blood. For example, obstruction of the bronchioles in conditions such as asthma can decrease ventilation beyond the obstructed areas. The result is a large increase in shunted blood because the blood flowing through the pulmonary capillaries in the obstructed area remains unoxygenated. In pneumonia or pulmonary edema, a buildup of fluid in the alveoli results in poor gas diffusion and less oxygenated blood.

When a person is standing, greater blood flow and ventilation occur in the base of the lung than in the top of the lung because of the effects of gravity. Arterial pressure at the base of the lung is 22 mm Hg greater than at the top of the lung because of hydrostatic pressure caused by gravity (see chapter 21). This greater pressure increases blood flow and distends blood vessels. The decreased pressure at the top of the lung results in less blood flow and vessels that are less distended, some of which are even collapsed during diastole.

During exercise, cardiac output and ventilation increase. The increased cardiac output increases pulmonary blood pressure throughout the lung, which increases blood flow. Blood flow increases most at the top of the lung, however, because the increased pressure expands the less distended vessels and opens the collapsed vessels. Thus, the effectiveness of gas exchange at the top of the lung increases because of greater blood flow.

Although gravity is the major factor affecting regional blood flow in the lung, under certain circumstances alveolar PO_2 can have an effect also. In most tissues, low PO_2 results in increased blood flow through the tissues (see chapter 21). In the lung, low PO_2 has the opposite effect, causing arterioles to constrict and reducing blood flow. This response reroutes blood away from areas of low oxygen toward parts of the lung that are better oxygenated. For example, if a bronchus becomes partially blocked, ventilation of alveoli past the blockage site decreases, which decreases gas exchange between the air and blood. The effect of this decreased gas exchange on overall gas exchange in the lungs is reduced by rerouting the blood to better-ventilated alveoli.

- 39. *What effect do ventilation and pulmonary capillary blood flow have on gas exchange? What is the physiologic shunt?*
- 40. *What are the effects of gravity and alveolar PO_2 on blood flow in the lung?*

P R E D I C T

Even people in "good shape" can have trouble breathing at high altitudes. Explain how this can happen, even when ventilation of the lungs increases.

Oxygen and Carbon Dioxide Transport in the Blood

Once oxygen diffuses across the respiratory membrane into the blood, most of it combines reversibly with hemoglobin, and a smaller amount dissolves in the plasma. Hemoglobin transports oxygen from the pulmonary capillaries through the blood vessels to the tissue capillaries, where some of the oxygen is released. The oxygen diffuses from the blood to tissue cells, where it is used in aerobic respiration.

Cells produce carbon dioxide during aerobic metabolism, and it diffuses from the cells into the tissue capillaries. Once carbon dioxide enters the blood, it is transported dissolved in the plasma, in combination with hemoglobin, and in the form of bicarbonate ions.

Oxygen Diffusion Gradients

The P_{O_2} within the alveoli averages approximately 104 mm Hg, and as blood flows into the pulmonary capillaries, it has a P_{O_2} of approximately 40 mm Hg (figure 23.16). Consequently, oxygen diffuses from the alveoli into the pulmonary capillary blood because the P_{O_2} is greater in the alveoli than in the capillary blood. By the time blood flows through the first third of the pulmonary capillary beds, an equilibrium is achieved, and the P_{O_2} in the blood is 104 mm Hg, which is equivalent to the P_{O_2} in the alveoli. Even with the greater velocity of blood flow associated with exercise, by the time blood reaches the venous ends of the pulmonary capillaries, the P_{O_2} in the capillaries has achieved the same value as that in the alveoli.

Blood leaving the pulmonary capillaries has a P_{O_2} of 104 mm Hg, but blood leaving the lungs in the pulmonary veins has a P_{O_2} of approximately 95 mm Hg. The decrease in the P_{O_2} occurs because the blood from the pulmonary capillaries mixes with deoxygenated (shunted) blood from the bronchial veins.

The blood that enters the arterial end of the tissue capillaries has a P_{O_2} of approximately 95 mm Hg. The P_{O_2} of the interstitial spaces, in contrast, is close to 40 mm Hg and is probably near 20 mm Hg in the individual cells. Oxygen diffuses from the tissue capillaries to the interstitial fluid and from the interstitial fluid into the cells of the body, where it's used in aerobic metabolism. Because oxygen is continuously used by cells, a constant diffusion gradient exists for oxygen from the tissue capillaries to the cells.

Carbon Dioxide Diffusion Gradients

Carbon dioxide is continually produced as a by-product of cellular respiration, and a diffusion gradient is established from tissue cells to the blood within the tissue capillaries. The intracellular P_{CO_2} is approximately 46 mm Hg, and the interstitial fluid P_{CO_2} is approximately 45 mm Hg. At the arterial end of the tissue capillaries, the P_{CO_2} is close to 40 mm Hg. As blood flows through the tissue capillaries, carbon dioxide diffuses from a higher P_{CO_2} to a lower P_{CO_2} until an equilibrium in P_{CO_2} is established. At the venous end of the capillaries, blood has a P_{CO_2} of 45 mm Hg (see figure 23.16).

After blood leaves the venous end of the capillaries, it's transported through the cardiovascular system to the lungs. At the arterial end of the pulmonary capillaries, the P_{CO_2} is 45 mm Hg. Because the P_{CO_2} is approximately 40 mm Hg in the alveoli, carbon dioxide diffuses from the pulmonary capillaries into the alveoli. At the venous end of the pulmonary capillaries, the P_{CO_2} has again decreased to 40 mm Hg.

41. Describe the partial pressures of oxygen and carbon dioxide in the alveoli, lung capillaries, tissue capillaries, and tissues. How do these partial pressures account for the movement of oxygen and carbon dioxide between air and blood and between blood and tissues?

Hemoglobin and Oxygen Transport

Approximately 98.5% of the oxygen transported in the blood from the lungs to the tissues is transported in combination with hemoglobin in red blood cells, and the remaining 1.5% is dissolved in the water part of the plasma. The combination of oxygen with hemoglobin is reversible. In the pulmonary capillaries, oxygen binds to hemoglobin, and in the tissue spaces oxygen diffuses away from hemoglobin and enters the tissues.

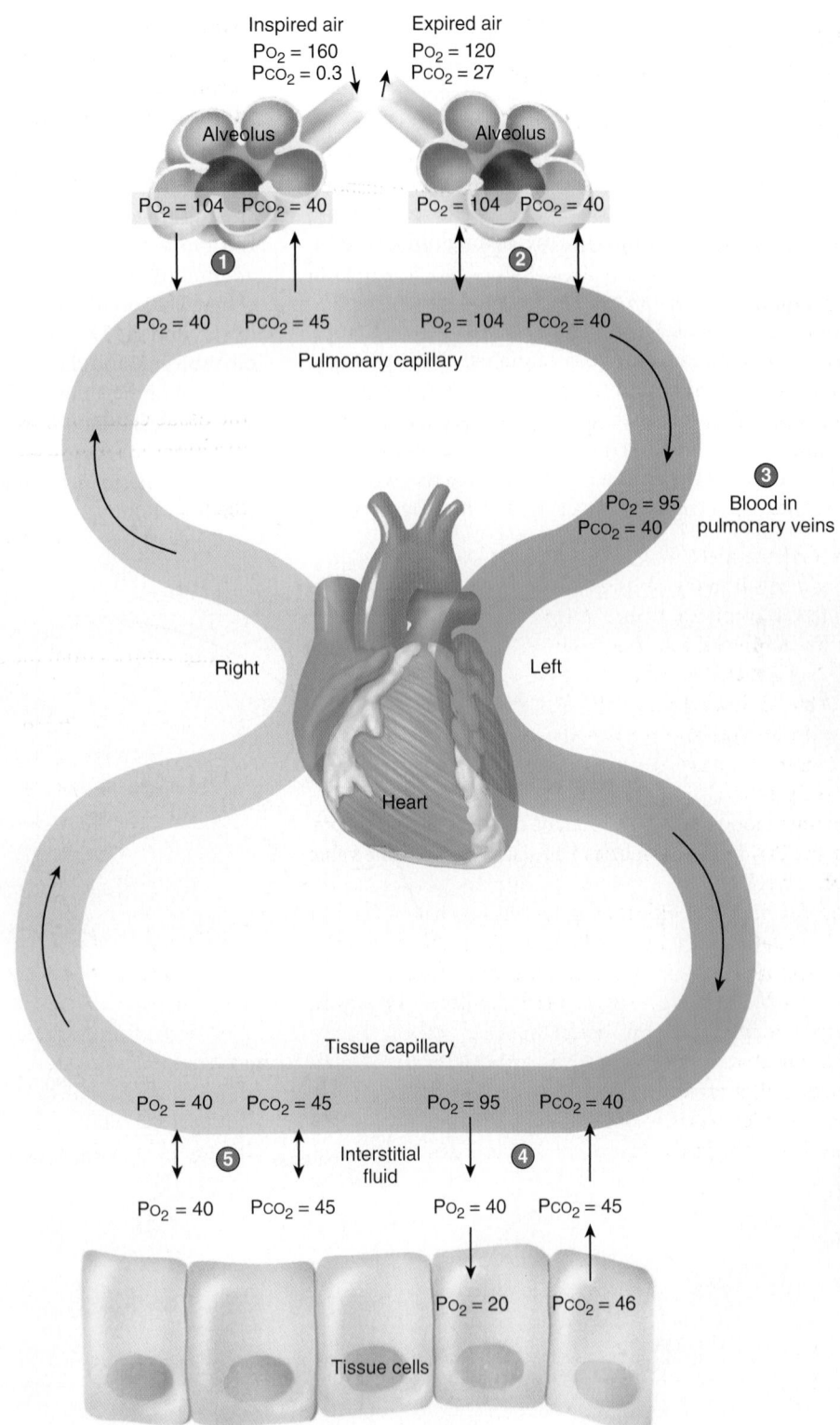

Inspired air
$Po_2 = 160$
$Pco_2 = 0.3$

Expired air
$Po_2 = 120$
$Pco_2 = 27$

Alveolus

Alveolus

$Po_2 = 104$ $Pco_2 = 40$

$Po_2 = 104$ $Pco_2 = 40$

① ②

$Po_2 = 40$ $Pco_2 = 45$

$Po_2 = 104$ $Pco_2 = 40$

Pulmonary capillary

③

$Po_2 = 95$
$Pco_2 = 40$

Blood in
pulmonary veins

Right

Left

Heart

Tissue capillary

$Po_2 = 40$ $Pco_2 = 45$

$Po_2 = 95$ $Pco_2 = 40$

⑤

④

Interstitial
fluid

$Po_2 = 40$ $Pco_2 = 45$

$Po_2 = 40$ $Pco_2 = 45$

$Po_2 = 20$ $Pco_2 = 46$

Tissue cells

1. Oxygen diffuses into the arterial ends of pulmonary capillaries and CO_2 diffuses into the alveoli because of differences in partial pressures.

2. As a result of diffusion at the venous ends of pulmonary capillaries, the Po_2 in the blood is equal to the Po_2 in the alveoli and the Pco_2 in the blood is equal to the Pco_2 in the alveoli.

3. The Po_2 of blood in the pulmonary veins is less than in the pulmonary capillaries because of mixing with deoxygenated blood from veins draining the bronchi and bronchioles.

4. Oxygen diffuses out of the arterial ends of tissue capillaries and CO_2 diffuses out of the tissue because of differences in partial pressures.

5. As a result of diffusion at the venous ends of tissue capillaries, the Po_2 in the blood is equal to the Po_2 in the tissue and the Pco_2 in the blood is equal to the Pco_2 in the tissue.
Go back to step 1.

Process Figure 23.16 Gas Exchange

Oxygen and carbon dioxide partial pressure diffusion gradients between the alveoli and the pulmonary capillaries and between the tissues and the tissue capillaries are responsible for gas exchange.

Effect of PO₂

The **oxygen–hemoglobin dissociation curve** describes the percentage of hemoglobin saturated with oxygen at any given PO_2. Hemoglobin is saturated when an oxygen molecule is bound to each of its four heme groups (see chapter 19). At any PO_2 above 80 mm Hg, approximately 95% of the hemoglobin is saturated with oxygen (figure 23.17). Because the PO_2 in the pulmonary capillaries is normally 104 mm Hg, the hemoglobin is 98% saturated.

In a resting person, the normal PO_2 of blood leaving the tissue capillaries of skeletal muscle is 40 mm Hg. At a PO_2 of 40 mm Hg, hemoglobin is approximately 75% saturated. Thus, approximately 23% of the oxygen bound to hemoglobin is released into the blood and can diffuse into the tissue spaces. During conditions of vigorous exercise, the blood PO_2 can decline to levels as low as 15 mm Hg because the skeletal muscle cells are using the oxygen in aerobic respiration (see chapter 9). At a PO_2 of 15 mm Hg, approx-

imately 25% of the hemoglobin is saturated with oxygen, and it releases 73% of the bound oxygen (figure 23.18). Thus, when the oxygen needs of tissues increase, blood PO_2 decreases, and more oxygen is released for use by the tissues.

Effect of pH, PCO₂, and Temperature

In addition to PO_2, other factors influence the degree to which oxygen binds to hemoglobin. As the pH of the blood declines, the amount of oxygen bound to hemoglobin at any given PO_2 also declines. This occurs because decreased pH results from an increase in H^+, and the H^+ combine with the protein part of the hemoglobin molecule and change its three-dimensional structure, causing a decrease in the ability of hemoglobin to bind oxygen. Conversely, an increase in blood pH results in an increased ability of hemoglobin to bind oxygen. The effect of pH on the oxygen–hemoglobin dissociation curve is called the **Bohr effect** after its discoverer, Christian Bohr.

Figure 23.17 Oxygen–Hemoglobin Dissociation Curve at Rest

(*a*) At the PO_2 in the lungs, hemoglobin is 98% saturated. At the PO_2 of resting tissues, hemoglobin is 75% saturated. Consequently, 23% of the oxygen picked up in the lungs is released to the tissues. (*b*) The ability of hemoglobin to pick up and release oxygen is like a glass filling and emptying.

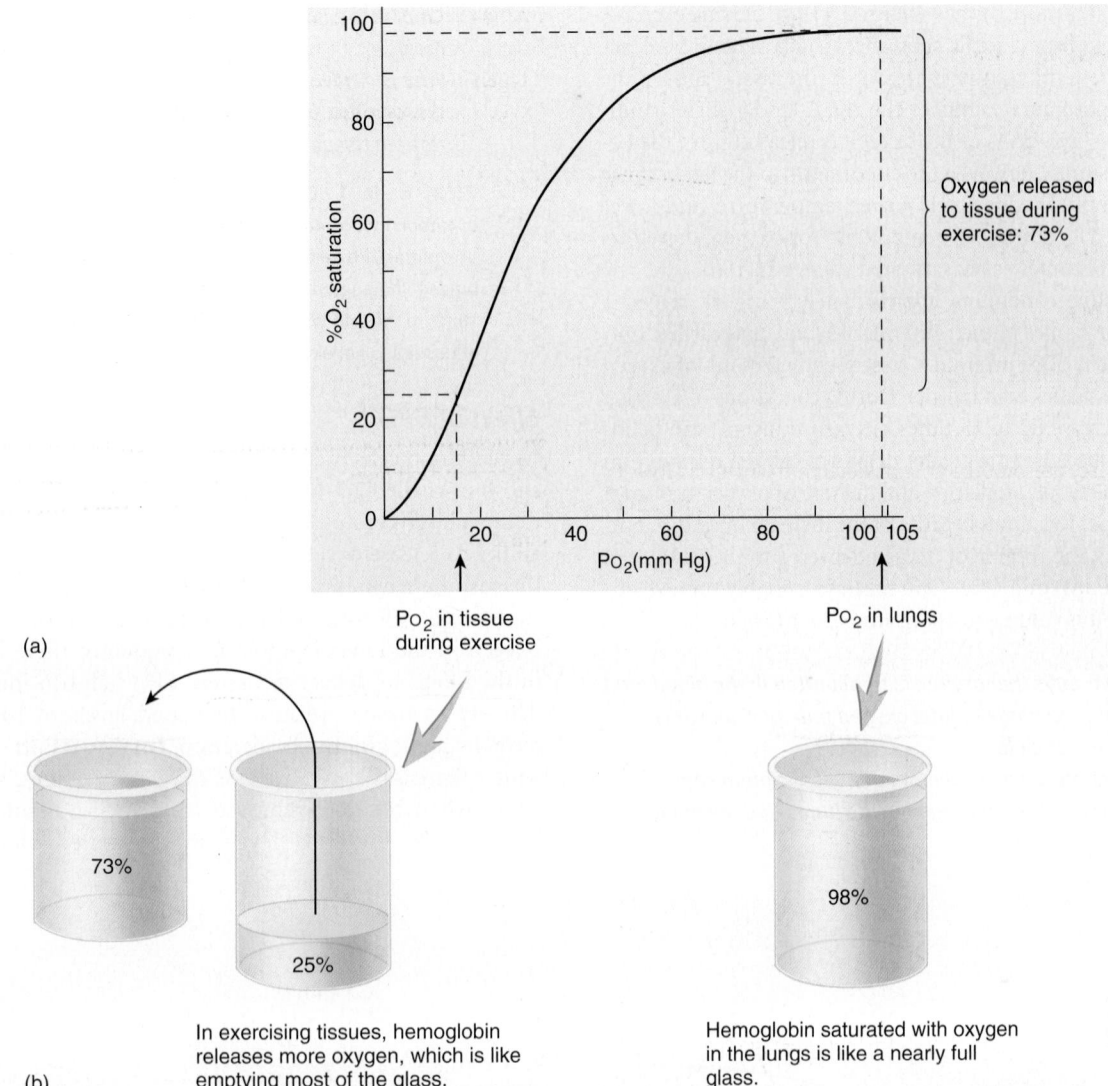

Figure 23.18 **Oxygen–Hemoglobin Dissociation Curve During Exercise**

(*a*) At the Po$_2$ in the lungs, hemoglobin is 98% saturated. At the Po$_2$ of exercising tissues, hemoglobin is 25% saturated. Consequently, 73% of the oxygen picked up in the lungs is released to the tissues. (*b*) The ability of hemoglobin to pick up and release oxygen is like a glass filling and emptying.

An increase in PCO$_2$ also decreases the ability of hemoglobin to bind oxygen because of the effect of carbon dioxide on pH. Within red blood cells, an enzyme called **carbonic anhydrase** catalyzes this reversible reaction.

$$\underset{\substack{\text{Carbon} \\ \text{dioxide}}}{CO_2} + \underset{\text{Water}}{H_2O} \rightleftarrows \underset{\substack{\text{Carbonic} \\ \text{acid}}}{H_2CO_3} \rightleftarrows \underset{\substack{\text{Hydrogen} \\ \text{ion}}}{H^+} + \underset{\substack{\text{Bicarbonate} \\ \text{ion}}}{HCO_3^-}$$

As carbon dioxide levels increase, more H$^+$ are produced, and the pH declines. As carbon dioxide levels decline, the reaction proceeds in the opposite direction, resulting in a decrease in H$^+$ concentration and an increase in pH.

As blood passes through tissue capillaries, carbon dioxide enters the blood from the tissues. As a consequence, blood carbon dioxide levels increase, hemoglobin has less affinity for oxygen in the tissue capillaries, and a greater amount of oxygen is released in the tissue capillaries than would be released if carbon dioxide were not present. When blood is returned to the lungs and passes through the pulmonary capillaries, carbon dioxide leaves the capillaries and enters the alveoli. As a consequence, carbon dioxide levels in the pulmonary capillaries are reduced, and the affinity of hemoglobin for oxygen increases.

An increase in temperature also decreases the tendency for oxygen to remain bound to hemoglobin. Elevated temperatures resulting from increased metabolism, therefore, increase the amount of oxygen released into the tissues by hemoglobin. In less metabolically active tissues in which the temperature is lower, less oxygen is released from hemoglobin.

When the affinity of hemoglobin for oxygen decreases, the oxygen–hemoglobin dissociation curve is shifted to the right, and hemoglobin releases more oxygen (figure 23.19*a*). During exercise, when carbon dioxide and acidic substances, such as lactic acid, accumulate and the temperature increases in the tissue spaces, the oxygen–hemoglobin curve shifts to the right. Under these conditions, as much as 75%–85% of the oxygen is released from the hemoglobin. In the lungs, however, the curve shifts to the left because of the lower carbon dioxide levels, lower temperature, and lower lactic acid levels. The affinity of hemoglobin for oxygen, therefore, increases, and it becomes easily saturated (figure 23.19*b*).

During resting conditions, approximately 5 mL of oxygen is transported to the tissues in each 100 mL of blood, and cardiac output is approximately 5000 mL/min. Consequently, 250 mL of oxygen is delivered to the tissues each minute. During conditions of exercise, this value can increase up to 15 times. Oxygen transport can be increased threefold because of a greater degree of oxygen release from hemoglobin in the tissue capillaries, and the rate of oxygen transport is increased another five times because of the increase in cardiac output. Consequently, the volume of oxygen delivered to the tissues can be as high as 3750 mL/min (15 × 250 mL/min). Highly trained athletes can increase this volume to as high as 5000 mL/min.

42. *Name two ways that oxygen is transported in the blood, and state the percentage of total oxygen transport for which each is responsible.*

43. *How does the oxygen–hemoglobin dissociation curve explain the uptake of oxygen in the lungs and the release of oxygen in tissues?*

44. *What is the Bohr effect? How is it related to blood carbon dioxide?*

45. *Why is it advantageous for the oxygen–hemoglobin dissociation curve to shift to the left in the lungs and to the right in tissues?*

P R E D I C T **8**

In carbon monoxide (CO) poisoning, CO binds to hemoglobin, thereby preventing the uptake of oxygen by hemoglobin. In addition, when CO binds to hemoglobin, the oxygen–hemoglobin dissociation curve shifts to the left. What are the consequences of this shift on the ability of tissues to get oxygen? Explain.

Effect of BPG

As red blood cells break down glucose for energy, they produce a substance called **2,3-bisphosphoglycerate** (**BPG;** formerly called diphosphoglycerate). BPG binds to hemoglobin and increases its ability to release oxygen. Compared to sea level, at high altitudes the barometric pressure is lower, the partial pressure of oxygen in the alveoli is lower, and the percent saturation of blood with oxygen in the lungs is lower. Consequently, there is less oxygen in the blood to deliver to tissues. BPG helps to increase oxygen delivery to tissues because increased levels of BPG cause the oxygen–hemoglobin dissociation curve to shift to the right, which increases the release of oxygen in tissues. On the other hand, when blood is removed from the body and stored in a blood bank, the BPG levels in the stored blood gradually

(a) **In the tissues, the oxygen–hemoglobin dissociation curve shifts to the right.** As pH decreases, P_{CO_2} increases, or temperature increases, the curve (*black*) shifts to the right (*red*), resulting in an increased release of oxygen.

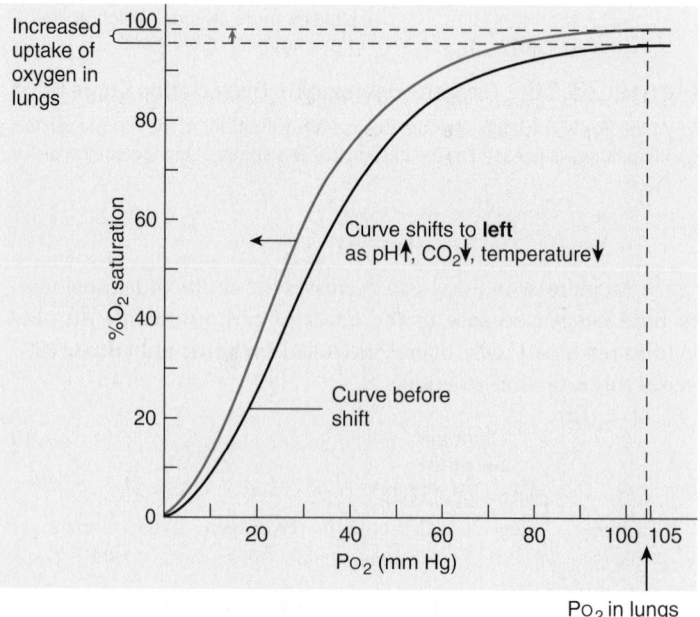

(b) **In the lungs, the oxygen–hemoglobin dissociation curve shifts to the left.** As pH increases, P_{CO_2} decreases, or temperature decreases, the curve (*black*) shifts to the left (*red*), resulting in an increased ability of hemoglobin to pick up oxygen.

Figure 23.19   Effects of Shifting the Oxygen–Hemoglobin Dissociation Curve decrease. As BPG levels decrease, the blood becomes unsuitable for transfusion purposes because the hemoglobin releases less oxygen to the tissues.

 46. *How does BPG affect the release of oxygen from hemoglobin?*

P R E D I C T 9

If a person lacks the enzyme necessary for BPG synthesis, would she exhibit anemia (lower-than-normal number of red blood cells) or erythrocytosis (higher-than-normal number of red blood cells)? Explain.

Fetal Hemoglobin

As fetal blood circulates through the placenta, oxygen is released from the mother's blood into the fetal blood and carbon dioxide is released from fetal blood into the mother's blood. Fetal blood is very efficient at picking up oxygen for several reasons.

1. The concentration of fetal hemoglobin is approximately 50% greater than the concentration of maternal hemoglobin.
2. Fetal hemoglobin is different from maternal hemoglobin. It has an oxygen–hemoglobin dissociation curve that's to the left of the maternal oxygen–hemoglobin dissociation curve. Thus, for a given P_{O_2} fetal hemoglobin can hold more oxygen than maternal hemoglobin.
3. BPG has little effect on fetal hemoglobin. That is, BPG does not cause fetal hemoglobin to release oxygen.
4. The movement of carbon dioxide out of the fetal blood causes the fetal oxygen–hemoglobin dissociation curve to shift to the left. At the same time, the movement of carbon dioxide into the mother's blood causes the maternal oxygen–hemoglobin dissociation curve to shift to the right. Thus, the mother's blood releases more oxygen and the fetal blood picks up more oxygen. This is called the **double Bohr effect.**

47. *How does the affinity for oxygen of fetal hemoglobin compare to maternal hemoglobin?*
48. *What is the double Bohr effect?*

Transport of Carbon Dioxide

Carbon dioxide is transported in the blood in three major ways: approximately 7% is transported as carbon dioxide dissolved in the plasma, approximately 23% is transported in combination with blood proteins (mostly hemoglobin), and 70% is transported in the form of HCO_3^-.

The most abundant protein to which carbon dioxide binds in the blood is hemoglobin. Carbon dioxide binds in a reversible fashion to the globin part of the hemoglobin molecule, and many carbon dioxide molecules can combine to a single hemoglobin molecule.

Hemoglobin that has released its oxygen binds more readily to carbon dioxide than hemoglobin that has oxygen bound to it. This is called the **Haldane effect.** In tissues, after hemoglobin has released oxygen, the hemoglobin has an increased ability to pick up carbon dioxide. In the lungs, as hemoglobin binds to oxygen, the hemoglobin more readily releases carbon dioxide.

Chloride Shift

Carbon dioxide from tissues diffuses into red blood cells within the capillaries (figure 23.20*a*). Some of the carbon dioxide binds to hemoglobin, but most of it reacts with water inside the red blood cells to form carbonic acid, a reaction catalyzed by carbonic anhydrase. The carbonic acid then dissociates to form HCO_3^- and H^+. Thus, most of the carbon dioxide becomes part of a HCO_3^-.

Lowering the amount of HCO_3^- and H^+ inside red blood cells promotes carbon dioxide transport, because as these reaction products are removed and their ion concentrations decrease, more carbon dioxide combines with water to form additional HCO_3^- and H^+ (see section on "Reversible Reactions" on p. 34). In a process called the **chloride shift** (see figure 23.20*a*), HCO_3^- concentrations inside red blood cells are lowered by exchanging them for chloride ions (Cl^-). As HCO_3^- are produced, carrier molecules in red blood cell membranes move HCO_3^- out of the red blood cells and Cl^- into the red blood cells. The exchange of negatively charged ions maintains electrical balance in the red blood cells and the plasma. Hemoglobin, which binds H^+, decreases the concentration of H^+ inside the red blood cells. Thus, hemoglobin functions as a buffer and resists an increase in pH within the red blood cells.

P R E D I C T 10

How is the ability of hemoglobin to release oxygen and pick up carbon dioxide in tissues affected by the change in the concentration of H^+ inside red blood cells? Explain.

The reverse of the previous events occurs in the lungs (figure 23.20*b*). Carbon dioxide diffuses from the red blood cells into the alveoli. As carbon dioxide levels in the red blood cells decrease, carbonic acid is converted to carbon dioxide and water. In response, HCO_3^- join with H^+ to form carbonic acid. As the HCO_3^- and H^+ decrease because of this reaction, they are replaced. Bicarbonate ions enter the red blood cell in exchange for Cl^-, and H^+ are released from hemoglobin.

Carbon Dioxide and Blood pH

Blood pH refers to the pH in plasma, not inside red blood cells. In plasma, carbon dioxide can combine with water to form carbonic acid, a reaction that is catalyzed by carbonic anhydrase on the surface of capillary endothelial cells. The carbonic acid then dissociates to form HCO_3^- and H^+. Thus, as plasma carbon dioxide levels increase, H^+ levels increase, and blood pH decreases. An important function of the respiratory system is to regulate blood pH by changing plasma carbon dioxide levels (see chapter 27). Hyperventilation decreases plasma carbon dioxide, and hypoventilation increases it.

49. *List three ways that carbon dioxide is transported in the blood, and state the percentage of total carbon dioxide transport for which each is responsible.*
50. *What is the Haldane effect?*
51. *Where and why does the chloride shift take place?*

P R E D I C T 11

What effect does hyperventilation and holding one's breath have on blood pH? Explain.

(a) In the tissue capillaries, CO_2 enters red blood cells and reacts with H_2O to form H_2CO_3, which dissociates to form HCO_3^- and H^+. HCO_3^- are exchanged for Cl^- in the chloride shift. H^+ combine with hemoglobin. Lowering the concentration of HCO_3^- and H^+ inside red blood cells promotes the conversion of CO_2 to HCO_3^-.

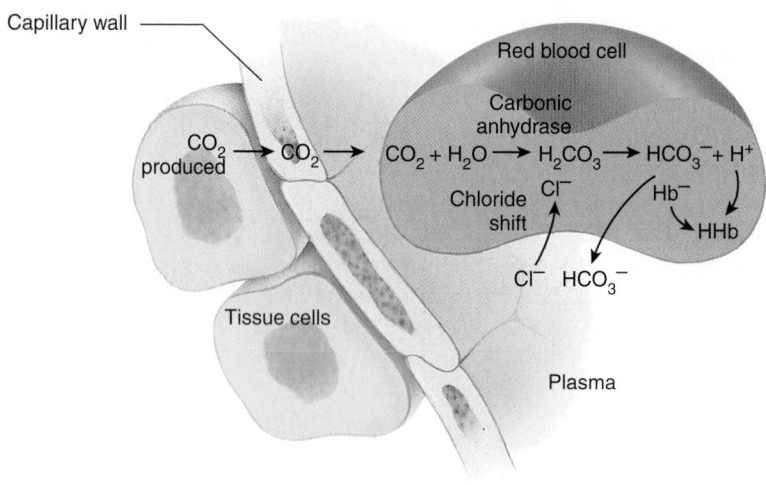

(b) In the pulmonary capillaries, CO_2 leaves red blood cells, resulting in the formation of additional CO_2 from H_2CO_3. HCO_3^- and H^+ combine to replace the H_2CO_3. The HCO_3^- are exchanged for Cl^-, and the H^+ are released from hemoglobin.

Figure 23.20   Carbon Dioxide Transport and Chloride Movement

Rhythmic Ventilation

The generation of the basic rhythm of ventilation is controlled by neurons within the medulla oblongata that stimulate the muscles of respiration. Recruitment of muscle fibers and the increased frequency of stimulation of muscle fibers result in stronger contractions of the muscles and an increased depth of respiration. The rate of respiration is determined by how frequently the respiratory muscles are stimulated.

Respiratory Areas in the Brainstem

The classic view of respiratory areas held that distinct inspiratory and expiratory centers were located in the brainstem. This view is now known to be too simplistic. Although neurons involved with respiration are aggregated in certain parts of the brainstem, neurons that are active during inspiration are intermingled with neurons that are active during expiration.

The **medullary respiratory center** consists of two **dorsal respiratory groups,** each forming a longitudinal column of cells lo-

cated bilaterally in the dorsal part of the medulla oblongata, and two **ventral respiratory groups,** each forming a longitudinal column of cells located bilaterally in the ventral part of the medulla oblongata (figure 23.21). Although the dorsal and ventral respiratory groups are bilaterally paired, cross communication exists between the pairs so that respiratory movements are symmetric. In addition, communication exists between the dorsal and ventral respiratory groups.

Each dorsal respiratory group is a collection of neurons that are most active during inspiration, but some are active during expiration. The dorsal respiratory groups are primarily responsible for stimulating contraction of the diaphragm. They receive input from other parts of the brain and peripheral receptors that allows modification of respiration.

Each ventral respiratory group is a collection of neurons that are active during inspiration and expiration. These neurons primarily stimulate the external intercostal, internal intercostal, and abdominal muscles.

The **pontine respiratory group,** formerly called the pneumotaxic center, is a collection of neurons in the pons (see figure 23.21).

Pons

Pontine respiratory group

Dorsal respiratory group

Ventral respiratory group

Medullary respiratory center

Medulla oblongata

Lateral view

Spinal cord

Phrenic nerve

Intercostal nerves

Internal intercostal muscles (involved in expiration)

External intercostal muscles (involved in inspiration)

Diaphragm (involved in inspiration)

Anterior view

Figure 23.21   Respiratory Structures in the Brainstem

The relationship of respiratory structures to each other and to the nerves innervating the muscles of respiration.

Some of the neurons are only active during inspiration, some only during expiration, and some during both inspiration and expiration. The precise function of the pontine respiratory group is unknown, but it has connections with the medullary respiratory center and appears to play a role in switching between inspiration

and expiration. It's not considered to be essential for the generation of the respiratory rhythm.

Generation of Rhythmic Ventilation

One explanation for the generation of rhythmic ventilation involves the integration of stimuli that start and stop inspiration.

1. *Starting inspiration.* Certain neurons in the medullary respiratory center spontaneously establish the basic rhythm of respiration. The medullary respiratory center constantly receives stimulation from receptors that monitor blood gas levels, blood temperature, and movements of muscles and joints. In addition, stimulation from parts of the brain concerned with voluntary respiratory movements and emotions can occur. Inspiration starts when the combined input from all these sources causes the production of action potentials in the neurons that stimulate respiratory muscles.

2. *Increasing inspiration.* Once inspiration begins, more and more neurons are gradually activated. The result is progressively stronger stimulation of the respiratory muscles that lasts for approximately 2 seconds.

3. *Stopping inspiration.* The neurons stimulating the muscles of respiration also stimulate other neurons in the medullary respiratory center that are responsible for stopping inspiration. The neurons responsible for stopping inspiration also receive input from the pontine respiratory group, stretch receptors in the lungs, and probably other sources. When these inhibitory neurons are activated, they cause the neurons stimulating respiratory muscles to be inhibited. Relaxation of respiratory muscles results in expiration, which lasts approximately 3 seconds. For the next inspiration, go back to step 1.

52. *Name the three respiratory groups and describe their main functions.*
53. *How is rhythmic ventilation generated?*

Modification of Ventilation

Although the medullary neurons establish the basic rate and depth of breathing, their activities can be influenced by input from other parts of the brain and by input from peripherally located receptors.

Cerebral and Limbic System Control

Through the cerebral cortex, it's possible to consciously or unconsciously increase or decrease the rate and depth of the respiratory movements (figure 23.22). For example, during talking or singing, air movement is controlled to produce sounds as well as to facilitate gas exchange.

Apnea (ap′nē-ă) is the absence of breathing. A person may stop breathing voluntarily. As the period of voluntary apnea increases, a greater and greater urge to breathe develops. That urge is primarily associated with increasing P_{CO_2} levels in the arterial blood. Finally, the P_{CO_2} reaches levels that cause the respiratory decrease back to normal levels. The role of the respiratory system in maintaining pH is considered in greater detail in chapter 27.

Effect of Carbon Dioxide

Blood carbon dioxide levels are a major regulator of respiration during resting conditions and conditions when the carbon dioxide levels are elevated, for example, during intense exercise. Even a small increase in carbon dioxide in the circulatory system triggers a large increase in the rate and depth of respiration. An increase in P_{CO_2} of 5 mm Hg, for example, causes an increase in ventilation of 100%. A greater-than-normal amount of carbon dioxide in the blood is called **hypercapnia** (hī-per-kap′nē-ă). Conversely, lower-than-normal carbon dioxide levels, a condition called **hypocapnia** (hī-pō-kap′nē-ă), result in periods in which respiratory movements are reduced or do not occur.

> **P R E D I C T**   **12**
>
> Explain why a person who breathes rapidly and deeply (hyperventilates) for several seconds experiences a short period during which respiration does not occur (apnea) before normal breathing resumes.

The chemoreceptors in the chemosensitive area of the medulla oblongata and in the carotid and aortic bodies respond to changes in carbon dioxide because of the effects of carbon dioxide on blood pH (see figure 23.23). The carotid and aortic bodies, however, are responsible for, at most, 15%–20% of the total response to changes in P_{CO_2} or pH. The chemosensitive area in the medulla oblongata is far more important for the regulation of P_{CO_2} and pH than are the carotid and aortic bodies. During intense exercise, however, the carotid bodies respond more rapidly to changes in blood pH than does the chemosensitive area of the medulla.

Effect of Oxygen

Changes in P_{O_2} can affect respiration (see figure 23.23), although P_{CO_2} levels detected by the chemosensitive area are responsible for most changes in respiration. A decrease in oxygen levels below normal values is called **hypoxia** (hī-pok′sē-ă). If P_{O_2} levels in the arterial blood are markedly reduced while the pH and P_{CO_2} are held constant, an increase in ventilation occurs. Within a normal range of P_{O_2} levels, however, the effect of oxygen on the regulation of respiration is small. Only after arterial P_{O_2} decreases to approximately 50% of its normal value does it begin to have a large stimulatory effect on respiratory movements.

At first, it's somewhat surprising that small changes in P_{O_2} don't cause changes in respiratory rate. Consideration of the oxygen–hemoglobin dissociation curve, however, provides an explanation. Because of the S shape of the curve, at any P_{O_2} above 80 mm Hg nearly all of the hemoglobin is saturated with oxygen. Consequently, until P_{O_2} levels change significantly, the oxygen-carrying capacity of the blood is unaffected.

The carotid and aortic body chemoreceptors respond to decreased P_{O_2} by increased stimulation of the respiratory center, which can keep it active, despite decreasing oxygen levels. If P_{O_2} decreases sufficiently, however, the respiratory center can fail to function, resulting in death.

Importance of Reduced P_{O_2}

Carbon dioxide is much more important than oxygen as a regulator of normal alveolar ventilation, but under certain circumstances a reduced P_{O_2} in the arterial blood does play an important stimulatory role. During conditions of shock in which blood pressure is very low, the P_{O_2} in arterial blood can drop to levels sufficiently low to strongly stimulate carotid and aortic body sensory receptors. At high altitudes where barometric air pressure is low, the P_{O_2} in arterial blood can also drop to levels sufficiently low to stimulate carotid and aortic bodies. Although P_{O_2} levels in the blood are reduced, the ability of the respiratory system to eliminate carbon dioxide is not greatly affected by low barometric air pressure. Thus, blood carbon dioxide levels become lower than normal because of the increased alveolar ventilation initiated in response to low P_{O_2}.

A similar situation exists in people who have emphysema. Because carbon dioxide diffuses across the respiratory membrane more readily than oxygen, the decreased surface area of the respiratory membrane caused by the disease results in low arterial P_{O_2} without elevated arterial P_{CO_2}. The elevated rate and depth of respiration are due, to a large degree, to the stimulatory effect of low arterial P_{O_2} levels on carotid and aortic bodies. More severe emphysema, in which the surface area of the respiratory membrane is reduced to a minimum, can also result in elevated P_{CO_2} levels in arterial blood.

Hering-Breuer Reflex

The **Hering-Breuer** (her′ing-broy′er) **reflex** limits the degree to which inspiration proceeds and prevents overinflation of the lungs (see figure 23.22). This reflex depends on stretch receptors in the walls of the bronchi and bronchioles of the lung. Action potentials are initiated in these stretch receptors when the lungs are inflated and are passed along sensory neurons within the vagus nerves to the medulla oblongata. The action potentials have an inhibitory influence on the respiratory center and result in expiration. As expiration proceeds, the stretch receptors are no longer stimulated, and the decreased inhibitory effect on the respiratory center allows inspiration to begin again.

In infants, the Hering-Breuer reflex plays a role in regulating the basic rhythm of breathing and in preventing overinflation of the lungs. In adults, however, the reflex is important only when the tidal volume is large, such as during exercise.

Effect of Exercise on Ventilation

The mechanisms by which ventilation is regulated during exercise are controversial, and no one factor can account for all of the observed responses. Ventilation during exercise is divided into two phases.

1. *Ventilation increases abruptly.* At the onset of exercise, ventilation immediately increases. This initial increase can be as much as 50% of the total increase that occurs during exercise. The immediate increase in ventilation occurs too quickly to be explained by changes in metabolism or blood gases. As axons pass from the motor cortex of the cerebrum through the motor pathways, numerous collateral fibers project into the reticular formation of the brain. During exercise, action potentials in the motor

pathways stimulate skeletal muscle contractions, and action potentials in the collateral fibers stimulate the respiratory center (see figure 23.22).

Furthermore, during exercise, body movements stimulate proprioceptors in the joints of the limbs. Action potentials from the proprioceptors pass along sensory nerve fibers to the spinal cord and along ascending nerve tracts (the dorsal-column/medial-lemniscal system) of the spinal cord to the brain. Collateral fibers project from these ascending pathways to the respiratory center in the medulla oblongata. Movement of the limbs has a strong stimulatory influence on the respiratory center (see figure 23.22).

A learned component may also exist to the ventilation response during exercise. After a period of training, the brain "learns" to match ventilation with the intensity of the exercise. Well-trained athletes match their respiratory movements more efficiently with their level of physical activity than do untrained individuals. Thus, centers of the brain involved in learning have an indirect influence on the respiratory center, but the exact mechanism for this kind of regulation is unclear.

2. *Ventilation increases gradually.* After the immediate increase in ventilation, a gradual increase occurs that levels off within 4–6 minutes after the onset of exercise. Factors responsible for the immediate increase in ventilation may play a role in the gradual increase as well.

Despite large changes in oxygen consumption and carbon dioxide production during exercise, the *average* arterial P_{O_2}, P_{CO_2}, and pH remain constant and close to resting levels as long as the exercise is aerobic (see chapter 9). This suggests that changes in blood gases and pH do not play an important role in regulating ventilation during aerobic exercise. During exercise, however, the values of arterial P_{O_2}, P_{CO_2}, and pH rise and fall more than at rest. Thus, even though their average values don't change, their oscillations may be a signal for helping to control ventilation.

The highest level of exercise that can be performed without causing a significant change in blood pH is called the **anaerobic threshold.** If the exercise intensity is high enough to exceed the anaerobic threshold, then skeletal muscles produce and release lactic acid into the blood. The resulting change in blood pH stimulates the carotid bodies, resulting in increased ventilation. In fact, ventilation can increase so much that arterial P_{CO_2} decreases below resting levels and arterial P_{O_2} increases above resting levels.

Other Modifications of Ventilation

The activation of touch, thermal, and pain receptors can also affect the respiratory center (see figure 23.22). For example, irritants in the nasal cavity can initiate a sneeze reflex, and irritants in the lungs can stimulate a cough reflex. An increase in body temperature can stimulate increased ventilation.

54. *Describe cerebral and limbic system control of ventilation.*
55. *Define central and peripheral chemoreceptors. Which are most important for the regulation of blood pH and carbon dioxide?*

56. *Define hypercapnia and hypocapnia.*
57. *What effect does a decrease in blood pH or carbon dioxide have on respiratory rate?*
58. *Describe the Hering-Breuer reflex and its function.*
59. *Define hypoxia. Why must arterial P_{O_2} change significantly before it affects respiratory rate?*
60. *What mechanisms regulate ventilation at the onset of exercise and during exercise? What is the anaerobic threshold?*

P R E D I C T **13**

Describe the respiratory response when cold water is splashed onto a person. In the past, newborn babies were sometimes swatted on the buttocks. Explain the rationale for this procedure.

Respiratory Adaptations to Exercise

In response to training, athletic performance increases because the cardiovascular and respiratory systems become more efficient at delivering oxygen and picking up carbon dioxide. Ventilation in most individuals does not limit performance because ventilation can increase to a greater extent than does cardiovascular function.

After training, vital capacity increases slightly and residual volume decreases slightly. Tidal volume at rest and during submaximal exercise does not change. At maximal exercise, however, tidal volume increases. After training, the respiratory rate at rest or during submaximal exercise is slightly lower than in an untrained person, but at maximal exercise respiratory rate is generally increased.

Minute ventilation is affected by the changes in tidal volume and respiratory rate. After training, minute ventilation is essentially unchanged or slightly reduced at rest and is slightly reduced during submaximal exercise. Minute ventilation is greatly increased at maximal exercise. For example, an untrained person with a minute ventilation of 120 L/min can increase to 150 L/min after training. Increases to 180 L/min are typical of highly trained athletes.

Gas exchange between the alveoli and blood increases at maximal exercise following training. The increased minute ventilation results in increased alveolar ventilation. In addition, increased cardiovascular efficiency results in greater blood flow through the lungs, especially in the superior parts of the lungs.

61. *What effect does training have on resting, submaximal, and maximal tidal volumes and on minute ventilation?*

Effects of Aging on the Respiratory System

Almost all aspects of the respiratory system are affected by aging. Even though vital capacity, maximum ventilation rates, and gas exchange decrease with age, the elderly can engage in light to moderate exercise because the respiratory system has a large reserve capacity.

Clinical Focus   Disorders of the Respiratory System

Bronchi and Lungs

Bronchitis (brong-kī′tis) is an inflammation of the bronchi caused by irritants, such as cigarette smoke, air pollution, or infections. The inflammation results in swelling of the mucous membrane lining the bronchi, increased mucus production, and decreased movement of mucus by cilia. Consequently, the diameter of the bronchi is decreased, and ventilation is impaired. Bronchitis can progress to emphysema.

Emphysema (em-fi-zē′mă) results in the destruction of the alveolar walls. Many smokers have both bronchitis and emphysema, which are often referred to as **chronic obstructive pulmonary disease (COPD)**. Chronic inflammation of the bronchioles, usually caused by cigarette smoke or air pollution, probably initiates emphysema. Narrowing of the bronchioles restricts air movement, and air tends to be retained in the lungs. Coughing to remove accumulated mucus increases pressure in the alveoli, resulting in rupture and destruction of alveolar walls. Loss of alveolar walls has two important consequences. The respiratory membrane has a decreased surface area, which decreases gas exchange, and loss of elastic fibers decreases the ability of the lungs to recoil and expel air. Symptoms of emphysema include shortness of breath and enlargement of the thoracic cavity. Treatment involves removing sources of irritants (e.g., stopping smoking), promoting the removal of bronchial secretions, using bronchiodilators, retraining people to breathe so that expiration of air is maximized, and using antibiotics to prevent infections. The progress of emphysema can be slowed, but no cure exists.

Cystic fibrosis is an inherited disease that affects the secretory cells lining the lungs, pancreas, sweat glands, and salivary glands. The defect produces an abnormal chloride transport protein that doesn't reach the cell surface or doesn't function normally if it does reach the cell surface. The result is decreased chloride ion secretion out of cells and increased sodium ion movement into cells. Normally, the presence of chloride and sodium ions outside of the cells causes water to move to the outside by osmosis. In the lungs, the water forms a thin fluid layer over which mucus is moved by ciliated cells. In cystic fibrosis, the decreased chloride and sodium ions outside the cells results in dehydrated respiratory secretions. The mucus is more viscous, resisting movement by cilia, and it accumulates in the lungs. For reasons not completely understood, the mucus accumulation increases the likelihood of infections. Chronic airflow obstruction causes difficulty in breathing, and coughing in an attempt to remove the mucus can result in pneumothorax and bleeding within the lungs. Many victims of cystic fibrosis are now surviving into young adulthood because of improved treatments. Future treatments could include the development of drugs that correct or assist the normal ion transport mechanism. Alternatively, cystic fibrosis may someday be cured through genetic engineering by inserting a functional copy of the defective gene into a person with the disease. Research on this exciting possibility is currently underway.

Pulmonary fibrosis is the replacement of lung tissue with fibrous connective tissue, thereby making the lungs less elastic and breathing more difficult. Exposure to asbestos, silica (silicosis), or coal dust is the most common cause.

Lung, or **bronchiogenic, cancer** arises from the epithelium of the respiratory tract. Cancers arising from tissues other than respiratory epithelium are not called lung cancer, even though they occur in the lungs. Lung cancer is the most common cause of cancer death in males and females in the United States, and almost all cases occur in smokers. Because of the rich lymph and blood supply in the lungs, cancer in the lung can readily spread to other parts of the lung or body. In addition, the disease is often advanced before symptoms become severe enough for the victim to seek medical aid. Typical symptoms include coughing, sputum production, and blockage of the airways. Treatments include removal of part or all of the lung, chemotherapy, and radiation.

Nervous System

Sudden infant death syndrome (SIDS), or crib death, is the most frequent cause of death of infants between 2 weeks and 1 year of age. Death results when the infant stops breathing during sleep. Although the cause of SIDS remains controversial, evidence exists that damage to the respiratory center during development is a factor. No treatment has yet been found, but at-risk babies can be placed on a monitor that sounds an alarm if the baby stops breathing.

Paralysis of the respiratory muscles can result from damage of the spinal cord in the cervical or thoracic regions. The damage interrupts nerve tracts that transmit action potentials to the muscles of respiration. Transection of the spinal cord can result

---

Vital capacity decreases with age because of a decreased ability to fill the lungs (decreased inspiratory reserve volume) and a decreased ability to empty the lungs (decreased expiratory reserve volume). As a result, maximum minute ventilation rates decrease, which in turn decreases the ability to perform intense exercise. These changes are related to weakening of respiratory muscles and to decreased compliance of the thoracic cage caused by stiffening of cartilage and ribs. Lung compliance actually increases with age,

but this effect is offset by the decreased thoracic cage compliance. Lung compliance increases because parts of the alveolar walls are lost, which reduces lung recoil. There are no significant age-related changes in lung elastic fibers or surfactant.

Residual volume increases with age as the alveolar ducts and many of the larger bronchioles increase in diameter. This increases the dead space, which decreases the amount of air available for gas exchange (alveolar ventilation). In addition, gas exchange across from trauma, such as automobile accidents or diving into water that is too shallow. Another cause of paralysis is poliomyelitis, a viral infection that damages neurons of the respiratory center or motor neurons that stimulate the muscles of respiration. Anesthetics or central nervous system depressants can also depress the function of the respiratory center if they are taken or administered in large enough doses.

Diseases of the Upper Respiratory Tract

Strep throat is caused by a streptococcal bacteria *(Streptococcus pyogenes)* and is characterized by inflammation of the pharynx and by fever. Frequently, inflammation of the tonsils and middle ear is involved. Without a throat analysis, the infection cannot be distinguished from viral causes of pharyngeal inflammation. Current techniques allow rapid diagnosis within minutes to hours, and antibiotics are an effective treatment.

Diphtheria (dif-thē'rē-ă) was once a major cause of death among children. It is caused by a bacterium *(Corynebacterium diphtheriae)*. A grayish membrane forms in the throat and can block the respiratory passages totally. A vaccine against diphtheria is part of the normal immunization program for children in the United States.

The **common cold** is the result of a viral infection. Symptoms include sneezing, excessive nasal secretions, and congestion. The infection can easily spread to sinus cavities, lower respiratory passages, and the middle ear. Laryngitis and middle ear infections are common complications. The common cold usually runs its course to recovery in about 1 week.

Diseases of the Lower Respiratory Tract

Laryngitis (lar-in-jī'tis) is an inflammation of the larynx, especially the vocal folds, and **bronchitis** is an inflammation of the bronchi. Bacterial or viral infections can move from the upper respiratory tract to cause laryngitis or bronchitis. Bronchitis is also often caused by continually breathing air containing harmful chemicals, such as those found in cigarette smoke.

Whooping cough (pertussis; per-tŭs'is) is a bacterial infection *(Bordetella pertussis)* that causes a loss of cilia of the respiratory epithelium. Mucus accumulates, and the infected person attempts to cough up the mucous accumulations. The coughing can be severe. A vaccine for whooping cough is part of the normal vaccination procedure for children in the United States.

Tuberculosis (tū-ber'kyū-lō'sis) is caused by a tuberculosis bacterium *(Mycobacterium tuberculosis)*. In the lung, the bacteria form lesions called tubercles. The small lumps contain degenerating macrophages and tuberculosis bacteria. An immune reaction is directed against the tubercles, which causes the formation of larger lesions and inflammation. The tubercles can rupture and release bacteria that infect other parts of the lung or body. Recently, a strain of the tuberculosis bacteria has developed that is resistant to treatment, and this strain is increasing concern that tuberculosis will again become a widespread infectious disease.

Pneumonia (noo-mō'nē-ă) is a general term that refers to many infections of the lung. Most pneumonias are caused by bacteria, but some result from viral, fungal, or protozoan infections. Symptoms include fever, difficulty in breathing, and chest pain. Inflammation of the lungs results in the accumulation of fluid within alveoli (pulmonary edema) and poor inflation of the lungs with air. A fungal infection *(Pneumocystis carinii)* that results in pneumocystosis pneumonia is rare, except in persons who have a compromised immune system. This type of pneumonia has become one of the infections commonly suffered by persons who have AIDS.

Flu (influenza) is a viral infection of the respiratory system and does not affect the digestive system as is commonly assumed. Flu is characterized by chills, fever, headache, and muscular aches, in addition to coldlike symptoms. Several strains of flu viruses have been identified. The mortality rate from flu is approximately 1%, and most of those deaths occur among the very old and very young. During a flu epidemic, the infection rate is so rapid and the disease so widespread that the total number of deaths is substantial, even though the percentage of deaths is relatively low. Flu vaccines can provide some protection against the flu.

A number of fungal diseases, such as **histoplasmosis** (his'tō-plaz-mō'sis) and **coccidioidomycosis** (kok-sid-ē-oy'dō-mī-kō'sis), affect the respiratory system. The fungal spores *(Histoplasma capsulatum; Coccidioides immitis)* usually enter the respiratory system through dust particles. Spores in soil and feces of certain animals make the rate of infection higher in farm workers and in gardeners. The infections usually result in minor respiratory infections, but in some cases they can cause infections throughout the body.

the respiratory membrane is reduced because parts of the alveolar walls are lost, which decreases the surface area available for gas exchange, and the remaining walls thicken, which decreases diffusion of gases. A gradual increase in resting tidal volume with age compensates for these changes.

With age, mucus accumulates within the respiratory passageways. The mucus-cilia escalator is less able to move the mucus because it becomes more viscous and because the number of cilia

and their rate of movement decrease. As a consequence, the elderly are more susceptible to respiratory infections and bronchitis.

62. *Why do vital capacity, alveolar ventilation, and diffusion of gases across the respiratory membrane decrease with age?*

63. *Why are the elderly more likely to develop respiratory infections and bronchitis?*

Systems Pathology
Asthma

Mr. W is an 18-year-old track athlete in seemingly good health. One day he came down with a common cold, resulting in the typical symptoms of nasal congestion and discomfort. After several days, he began to cough and wheeze, and he thought that his cold had progressed to his lungs. Determined not to get "out of shape" because of his cold, Mr. W took a few aspirins to relieve his discomfort and went to the track to jog. After a few minutes of exercise, he began to wheeze very forcefully and rapidly, and he felt that he could hardly get enough air. Even though he stopped jogging, his condition did not improve (figure A). Fortunately, a concerned friend who was also at the track took him to the emergency room.

Although Mr. W had no previous history of asthma, careful evaluation by the emergency room doctor convinced her that he probably was having an asthma attack. Mr. W inhaled a bronchiodilator drug, which resulted in rapid improvement in his condition. He was released from the emergency room and referred to his personal physician for further treatment and education about asthma.

Background Information

Asthma (az′mă) is a disease characterized by increased constriction of the trachea and bronchi in response to various stimuli, resulting in a narrowing of the air passageways and decreased ventilation efficiency. Symptoms include wheezing, coughing, and shortness of breath. In contrast to many other respiratory disorders, however, the symptoms of asthma typically reverse either spontaneously or with therapy.

It's estimated that the prevalence of asthma in the United States is from 3%–6% of the general population. Approximately half the cases first appear before age 10, and twice as many boys as girls develop asthma. Anywhere from 25%–50% of childhood asthmatics are symptom-free from adolescence onward.

The exact cause or causes of asthma are unknown, but asthma and allergies run strongly in some families. No definitive pathologic feature or diagnostic test for asthma has been discovered, but three important features of the disease are chronic airway inflammation, airway hyperreactivity, and airflow obstruction. The inflammatory response results in tissue damage, edema, and mucus buildup, which can block airflow through the bronchi. Airway hyperreactivity is greatly

Figure A   **Jogger with Asthma**

increased contraction of the smooth muscle in the trachea and bronchi in response to a stimulus. As a result of airway hyperreactivity, the diameter of the airway decreases, and resistance to airflow increases. The effects of inflammation and airway hyperreactivity combine to cause airflow obstruction.

Many cases of asthma appear to be associated with a chronic inflammatory response by the immune system. The number of immune cells in the bronchi increases, including mast cells, eosinophils, neutrophils, macrophages, and lymphocytes. These cells release chemical mediators, such as interleukins, leukotrienes, prostaglandins, platelet-activating factor, thromboxanes, and chemotactic factors. These chemical mediators promote inflammation, increase mucus secretion, and attract additional immune cells to the bronchi, resulting in chronic airway inflammation. Airway hyperreactivity and inflammation appear to be linked by some of the chemical mediators, which increase the sensitivity of the airway to stimulation and cause smooth muscle contraction.

System Interactions — Effect of Asthma on Other Systems

System	Interactions
Integumentary	Cyanosis, a bluish skin color, results from a decreased blood oxygen content.
Muscular	Skeletal muscles are necessary for respiratory movements and the cough reflex. Increased muscular work during a severe asthma attack can cause metabolic acidosis because of anaerobic respiration and excessive lactic acid production.
Skeletal	Red bone marrow is the site of production of many of the immune cells responsible for the inflammatory response of asthma. The thoracic cage is necessary for respiration.
Nervous	Emotional upset or stress can evoke an asthma attack. Peripheral and central chemoreceptor reflexes affect ventilation. The cough reflex helps to remove mucus from respiratory passages. Pain, anxiety, and death from asphyxiation can result from the altered gas exchange caused by asthma. One theory of the cause of asthma is an imbalance in the autonomic nervous system (ANS) control of bronchiolar smooth muscle, and drugs that enhance sympathetic effects or block parasympathetic effects are used in asthma treatment.
Endocrine	Steroids from the adrenal gland play a role in regulating inflammation, and they are used in asthma therapy.
Cardiovascular	Increased vascular permeability of lung blood vessels results in edema. Blood carries ingested substances that provoke an asthma attack to the lungs. Blood carries immune cells from the red bone marrow to the lungs. Tachycardia commonly occurs, and the normal effects of respiration on venous return of blood to the heart are exaggerated, resulting in large fluctuations in blood pressure.
Lymphatic and Immune	Immune cells release chemical mediators that promote inflammation, increase mucus production, and cause bronchiolar constriction (believed to be a major factor in asthma). Ingested allergens, such as aspirin or sulfites in food, can evoke an asthma attack.
Digestive	Reflux of stomach acid into the esophagus can evoke an asthma attack.
Urinary	Modifying hydrogen ion secretion into the urine helps to compensate for acid–base imbalances caused by asthma.

The stimuli that prompt airflow obstruction vary from one individual to another. Some asthmatics have reactions to particular allergens, which are foreign substances that evoke an inappropriate immune system response (see chapter 22). Examples include inhaled pollen, animal dander, and dust mites. Many cases of asthma may be caused by an allergic reaction to substances in the droppings and carcasses of cockroaches, which may explain the higher rate of asthma in poor, urban areas.

On the other hand, inhaled substances, such as chemicals in the workplace or cigarette smoke, can provoke an asthma attack without stimulating an allergic reaction. Over 200 substances have been associated with occupational asthma. An asthma attack can also be stimulated by ingested substances like aspirin, nonsteroidal anti-inflammatory compounds like ibuprofen (i-boo′prō-fen), sulfites in food preservatives, and tartrazine (tar′tră-zēn) in food colorings. Asthmatics can substitute acetaminophen (as-et-ă-mē′nō-fen; Tylenol) for aspirin.

Other stimuli, such as strenuous exercise, especially in cold weather, can precipitate an asthma attack. Such episodes can often be avoided by using a bronchiodilator drug prior to exercise. Viral infections, emotional upset, stress, and even reflux of stomach acid into the esophagus are known to elicit an asthma attack.

Treatment of asthma involves avoiding the causative stimulus and administering drug therapy. Steroids and mast cell–stabilizing agents, which prevent the release of chemical mediators from mast cells, are used to reduce airway inflammation. Theophylline (thē-of′i-lēn, thē-of′i-lin) and β-adrenergic agents (see chapter 16) are commonly used to cause bronchiolar dilation. Although treatment is generally effective in controlling asthma, in rare cases death by asphyxiation may occur. Earlier and more intensive therapy will in most cases prevent death by asphyxiation.

P R E D I C T

It is not usually necessary to assess arterial blood gases in the diagnosis and treatment of asthma. This information, however, can sometimes be useful in cases of severe asthma attacks. Suppose that Mr. W had a P_{O_2} of 60 mm Hg and a P_{CO_2} of 30 mm Hg when he first came to the emergency room. Explain how that could happen.

Respiration includes the movement of air into and out of the lungs, the exchange of gases between the air and the blood, the transport of gases in the blood, and the exchange of gases between the blood and tissues.

Functions of the Respiratory System   (p. 830)

Major functions associated with the respiratory system include gas exchange, regulation of blood pH, voice production, olfaction, and protection against some microorganisms.

Anatomy and Histology of the Respiratory System   (p. 830)

Nose

1. The nose consists of the external nose and the nasal cavity.
2. The bridge of the nose is bone, and most of the external nose is cartilage.
3. Openings of the nasal cavity
   - The nares open to the outside, and the choane lead to the pharynx.
   - The paranasal sinuses and the nasolacrimal duct open into the nasal cavity.
4. Parts of the nasal cavity
   - The nasal cavity is divided by the nasal septum.
   - The anterior vestibule contains hairs that trap debris.
   - The nasal cavity is lined with pseudostratified ciliated columnar epithelium that traps debris and moves it to the pharynx.
   - The superior part of the nasal cavity contains the olfactory epithelium.

Pharynx

1. The nasopharynx joins the nasal cavity through the internal nares and contains the openings to the auditory tube and the pharyngeal tonsils.
2. The oropharynx joins the oral cavity and contains the palatine and lingual tonsils.
3. The laryngopharynx opens into the larynx and the esophagus.

Larynx

1. Cartilage
   - Three unpaired cartilages exist. The thyroid cartilage and cricoid cartilage form most of the larynx. The epiglottis covers the opening of the larynx during swallowing.
   - Six paired cartilages exist. The vocal folds attach to the arytenoid cartilages.
2. Sounds are produced as the vocal folds vibrate when air passes through the larynx. Tightening the folds produces sounds of different pitches by controlling the length of the fold, which is allowed to vibrate.

Trachea

The trachea connects the larynx to the main bronchi.

Tracheobronchial Tree

1. The trachea divides to form two main bronchi, which go to the lungs. The main bronchi divide to form lobar bronchi, which divide to form segmental bronchi, which divide to form bronchioles, which divide to form terminal bronchioles.
2. The conducting zone, from the trachea to the terminal bronchioles, is a passageway for air movement.
   - The area from the trachea to the terminal bronchioles is ciliated to facilitate removal of debris.
   - Cartilage helps to hold the tube system open (from the trachea to the bronchioles).
   - Smooth muscle controls the diameter of the tubes (terminal bronchioles).
3. Terminal bronchioles divide to form respiratory bronchioles, which give rise to alveolar ducts. Air-filled chambers called alveoli open

into the respiratory bronchioles and alveolar ducts. The alveolar ducts end as alveolar sacs, which are chambers that connect to two or more alveoli.
4. The respiratory zone, from the respiratory bronchioles to the alveoli, is a site of gas exchange.
5. The components of the respiratory membrane include a film of water, the walls of the alveolus and the capillary, and an interstitial space.

Lungs

1. The body contains two lungs.
2. The lungs are divided into lobes, bronchopulmonary segments, and lobules.

Thoracic Wall and Muscles of Respiration

1. The thoracic wall consists of vertebrae, ribs, sternum, and muscles that allow expansion of the thoracic cavity.
2. Contraction of the diaphragm increases thoracic volume.
3. Muscles can elevate the ribs and increase thoracic volume or can depress the ribs and decrease thoracic volume.

Pleura

The pleural membranes surround the lungs and provide protection against friction.

Blood Supply

1. Deoxygenated blood is transported to the lungs through the pulmonary arteries, and oxygenated blood leaves through the pulmonary veins.
2. Oxygenated blood is mixed with a small amount of deoxygenated blood from the bronchi.

Lymphatic Supply

The superficial and deep lymphatic vessels drain lymph from the lungs.

Ventilation   (p. 844)

Pressure Differences and Airflow

1. Ventilation is the movement of air into and out of the lungs.
2. Air moves from an area of higher pressure to an area of lower pressure.

Pressure and Volume

Pressure is inversely related to volume.

Airflow into and out of Alveoli

1. Inspiration results when barometric air pressure is greater than alveolar pressure.
2. Expiration results when barometric air pressure is less than alveolar pressure.

Changing Alveolar Volume

1. Lung recoil causes alveoli to collapse.
   - Lung recoil results from elastic fibers and water surface tension.
   - Surfactant reduces water surface tension.
2. Pleural pressure is the pressure in the pleural cavity.
   - A negative pleural pressure can cause the alveoli to expand.
   - Pneumothorax is an opening between the pleural cavity and the air that causes a loss of pleural pressure.
3. Changes in thoracic volume cause changes in pleural pressure, resulting in changes in alveolar volume, alveolar pressure, and airflow.

Measuring Lung Function   (p. 849)

Compliance of the Lungs and the Thorax

1. Compliance is a measure of lung expansion caused by alveolar pressure.

2. Reduced compliance means that it's more difficult than normal to expand the lungs.

Pulmonary Volumes and Capacities

1. Four pulmonary volumes exist: tidal volume, inspiratory reserve volume, expiratory reserve volume, and residual volume.
2. Pulmonary capacities are the sum of two or more pulmonary volumes and include inspiratory capacity, functional residual capacity, vital capacity, and total lung capacity.
3. The forced expiratory vital capacity measures vital capacity as the individual exhales as rapidly as possible.

Minute Ventilation and Alveolar Ventilation

1. The minute ventilation is the total amount of air moved in and out of the respiratory system per minute.
2. Dead space is the part of the respiratory system in which gas exchange does not take place.
3. Alveolar ventilation is how much air per minute enters the parts of the respiratory system in which gas exchange takes place.

Physical Principles of Gas Exchange   (p. 851)
Partial Pressure

1. Partial pressure is the contribution of a gas to the total pressure of a mixture of gases (Dalton's law).
2. Water vapor pressure is the partial pressure produced by water.
3. Atmospheric air, alveolar air, and expired air have different compositions.

Diffusion of Gases Through Liquids

The concentration of a gas in a liquid is determined by its partial pressure and by its solubility coefficient (Henry's law).

Diffusion of Gases Through the Respiratory Membrane

1. The respiratory membrane is thin and has a large surface area that facilitates gas exchange.
2. The rate of diffusion of gases through the respiratory membrane depends on its thickness, the diffusion coefficient of the gas, the surface area of the membrane, and the partial pressure of the gases in the alveoli and the blood.

Relationship Between Ventilation and Pulmonary Capillary Blood Flow

1. Increased ventilation or increased pulmonary capillary blood flow increases gas exchange.
2. The physiologic shunt is the deoxygenated blood returning from the lungs.

Oxygen and Carbon Dioxide Transport in the Blood   (p. 853)
Oxygen Diffusion Gradients

1. Oxygen moves from the alveoli (P_{O_2} = 104 mm Hg) into the blood (P_{O_2} = 40 mm Hg). Blood is almost completely saturated with oxygen when it leaves the capillary.
2. The P_{O_2} in the blood decreases (P_{O_2} = 95 mm Hg) because of mixing with deoxygenated blood.
3. Oxygen moves from the tissue capillaries (P_{O_2} = 95 mm Hg) into the tissues (P_{O_2} = 40 mm Hg).

Carbon Dioxide Diffusion Gradients

1. Carbon dioxide moves from the tissues (P_{CO_2} = 45 mm Hg) into tissue capillaries (P_{CO_2} = 40 mm Hg).
2. Carbon dioxide moves from the pulmonary capillaries (P_{CO_2} = 45 mm Hg) into the alveoli (P_{CO_2} = 40 mm Hg).

Hemoglobin and Oxygen Transport

1. Oxygen is transported by hemoglobin (98.5%) and is dissolved in plasma (1.5%).
2. The oxygen–hemoglobin dissociation curve shows that hemoglobin is almost completely saturated when P_{O_2} is 80 mm Hg or above. At lower partial pressures, the hemoglobin releases oxygen.
3. A shift of the oxygen–hemoglobin dissociation curve to the right because of a decrease in pH (Bohr effect), an increase in carbon dioxide, or an increase in temperature results in a decrease in the ability of hemoglobin to hold oxygen.
4. A shift of the oxygen–hemoglobin dissociation curve to the left because of an increase in pH (Bohr effect), a decrease in carbon dioxide, or a decrease in temperature results in an increase in the ability of hemoglobin to hold oxygen.
5. The substance 2,3-bisphosphoglycerate increases the ability of hemoglobin to release oxygen.
6. Fetal hemoglobin has a higher affinity for oxygen than does maternal hemoglobin.

Transport of Carbon Dioxide

1. Carbon dioxide is transported as HCO_3^- (70%), in combination with blood proteins (23%), and in solution in plasma (7%).
2. Hemoglobin that has released oxygen binds more readily to carbon dioxide than hemoglobin that has oxygen bound to it (Haldane effect).
3. In tissue capillaries, carbon dioxide combines with water inside the red blood cells to form carbonic acid, which dissociates to form HCO_3^- and H^+.
4. The chloride shift is the movement of Cl^- into red blood cells as HCO_3^- move out.
5. In lung capillaries, HCO_3^- and H^+ move into red blood cells, and Cl^- move out. Bicarbonate ions combine with H^+ to form carbonic acid. The carbonic acid is converted to carbon dioxide and water. The carbon dioxide diffuses out of the red blood cells.
6. Increased plasma carbon dioxide lowers blood pH. The respiratory system regulates blood pH by regulating plasma carbon dioxide levels.

Rhythmic Ventilation   (p. 860)
Respiratory Areas in the Brainstem

1. The medullary respiratory center consists of the dorsal and ventral respiratory groups.
   - The dorsal respiratory groups stimulate the diaphragm.
   - The ventral respiratory groups stimulate the intercostal and abdominal muscles.
2. The pontine respiratory group is involved with switching between inspiration and expiration.

Generation of Rhythmic Ventilation

1. When stimuli from receptors or other parts of the brain exceed a threshold level, inspiration begins.
2. At the same time that respiratory muscles are stimulated, neurons that stop inspiration are stimulated. When the stimulation of these neurons exceeds a threshold level, inspiration is inhibited.

Modification of Ventilation   (p. 861)
Cerebral and Limbic System Control

Respiration can be voluntarily controlled and can be modified by emotions.

Chemical Control of Ventilation

1. Carbon dioxide is the major regulator of respiration. An increase in carbon dioxide or a decrease in pH can stimulate the chemosensitive area, causing a greater rate and depth of respiration.
2. Oxygen levels in the blood affect respiration when a 50% or greater decrease from normal levels exists. Decreased oxygen is detected by receptors in the carotid and aortic bodies, which then stimulate the respiratory center.

Hering-Breuer Reflex

Stretch of the lungs during inspiration can inhibit the respiratory center and contribute to a cessation of inspiration.

Effect of Exercise on Ventilation

1. Collateral fibers from motor neurons and from proprioceptors stimulate the respiratory centers.
2. Chemosensitive mechanisms and learning fine-tune the effects produced through the motor neurons and proprioceptors.

Other Modifications of Ventilation

Touch, thermal, and pain sensations can modify ventilation.

Respiratory Adaptations to Exercise   (p. 865)

Tidal volume, respiratory rate, minute ventilation, and gas exchange between the alveoli and blood remain unchanged or slightly lower at rest or during submaximal exercise but increase at maximal exercise.

Effects of Aging on the Respiratory System   (p. 865)

1. Vital capacity and maximum minute ventilation decrease with age because of weakening of respiratory muscles and decreased thoracic cage compliance.
2. Residual volume and dead space increase because of increased diameter of respiratory passageways. As a result, alveolar ventilation decreases.
3. An increase in resting tidal volume compensates for decreased alveolar ventilation, loss of alveolar walls (surface area), and thickening of alveolar walls.
4. The ability to remove mucus from the respiratory passageways decreases with age.

R E V I E W   A N D   C O M P R E H E N S I O N

1. The nasal cavity
   a. has openings for the paranasal sinuses.
   b. has a vestibule, which contains the olfactory epithelium.
   c. is connected to the pharynx by the nares.
   d. has passageways called conchae.
   e. is lined with squamous epithelium, except for the vestibule.

2. The nasopharynx
   a. is lined with moist stratified squamous epithelium.
   b. contains the pharyngeal tonsil.
   c. opens into the oral cavity through the fauces.
   d. extends to the tip of the epiglottis.
   e. is an area that food, drink, and air pass through.

3. The larynx
   a. connects the oropharynx to the trachea.
   b. has three unpaired and six paired cartilages.
   c. contains the vocal folds.
   d. contains the vestibular folds.
   e. all of the above.

4. The trachea contains
   a. skeletal muscle.
   b. pleural fluid glands.
   c. C-shaped pieces of cartilage.
   d. all of the above.

5. The conducting zone of the tracheobronchial tree ends at the
   a. alveolar duct.
   b. alveoli.
   c. bronchioles.
   d. respiratory bronchioles.
   e. terminal bronchioles.

6. During an asthma attack, the patient has difficulty breathing because of constriction of the
   a. trachea.
   b. bronchi.
   c. terminal bronchioles.
   d. alveoli.
   e. respiratory membrane.

7. During quiet expiration, the
   a. abdominal muscles relax.
   b. diaphragm moves inferiorly.
   c. external intercostal muscles contract.
   d. thorax and lungs passively recoil.
   e. all of the above.

8. The parietal pleura
   a. covers the surface of the lung.
   b. covers the inner surface of the thoracic cavity.
   c. is the connective tissue partition that divides the thoracic cavity into right and left pleural cavities.
   d. covers the inner surface of the alveoli.
   e. is the membrane across which gas exchange occurs.

9. Contraction of the bronchiolar smooth muscle has which of these effects?
   a. a smaller pressure gradient is required to get the same rate of airflow when compared to normal bronchioles
   b. increases airflow through the bronchioles
   c. increases resistance to airflow
   d. increases alveolar ventilation

10. During the process of expiration, the alveolar pressure is
    a. greater than the pleural pressure.
    b. greater than the barometric pressure.
    c. less than the barometric pressure.
    d. unchanged.

11. The lungs do not normally collapse because of
    a. surfactant.
    b. pleural pressure.
    c. elastic recoil.
    d. both a and b.

12. Immediately after the creation of an opening through the thorax into the pleural cavity,
    a. air flows through the hole and into the pleural cavity.
    b. air flows through the hole and out of the pleural cavity.
    c. air flows neither out nor in.
    d. the lung protrudes through the hole.

13. Compliance of the lungs and thorax
    a. is the volume by which the lungs and thorax change for each unit change of alveolar pressure.
    b. increases in emphysema.
    c. decreases because of lack of surfactant.
    d. all of the above.

14. Given these lung volumes:
    1. tidal volume = 500 mL
    2. residual volume = 1000 mL
    3. inspiratory reserve volume = 2500 mL
    4. expiratory reserve volume = 1000 mL
    5. dead space = 1000 mL

The vital capacity is
a. 3000 mL.
b. 3500 mL.
c. 4000 mL.
d. 5000 mL.
e. 6000 mL.

15. The alveolar ventilation is the
    a. tidal volume times respiratory rate.
    b. minute ventilation plus the dead space.
    c. amount of air available for gas exchange in the lungs.
    d. vital capacity divided by respiratory rate.
    e. inspiratory reserve volume times minute ventilation.

16. If the total pressure of a gas is 760 mm Hg and its composition is 20% oxygen, 0.04% carbon dioxide, 75% nitrogen, and 5% water vapor, the partial pressure of oxygen is
    a. 15.2 mm Hg.
    b. 20 mm Hg.
    c. 118 mm Hg.
    d. 152 mm Hg.
    e. 740 mm Hg.

17. The rate of diffusion of a gas across the respiratory membrane increases as the
    a. respiratory membrane becomes thicker.
    b. surface area of the respiratory membrane decreases.
    c. partial pressure difference of the gas across the respiratory membrane increases.
    d. diffusion coefficient of the gas decreases.
    e. all of the above.

18. In which of these sequences does P_{O_2} progressively decrease?
    a. arterial blood, alveolar air, body tissues
    b. body tissues, arterial blood, alveolar air
    c. body tissues, alveolar air, arterial blood
    d. alveolar air, arterial blood, body tissues
    e. arterial blood, body tissues, alveolar air

19. The partial pressure of carbon dioxide in the venous blood is
    a. greater than in the tissue spaces.
    b. less than in the tissue spaces.
    c. less than in the alveoli.
    d. less than in arterial blood.

20. Oxygen is mostly transported in the blood
    a. dissolved in plasma.
    b. bound to blood proteins.
    c. within HCO_3^-.
    d. bound to the heme portion of hemoglobin.

21. The oxygen–hemoglobin dissociation curve is adaptive because it
    a. shifts to the right in the pulmonary capillaries and to the left in the tissue capillaries.
    b. shifts to the left in the pulmonary capillaries and to the right in the tissue capillaries.
    c. doesn't shift.

22. Carbon dioxide is mostly transported in the blood
    a. dissolved in plasma.
    b. bound to blood proteins.
    c. within HCO_3^-.
    d. bound to the heme portion of hemoglobin.
    e. bound to the globin portion of hemoglobin.

23. When blood passes through the tissues, the hemoglobin in blood is better able to combine with carbon dioxide because of the
    a. Bohr effect.
    b. Haldane effect.
    c. chloride shift.
    d. Boyle effect.
    e. Dalton effect.

24. The chloride shift
    a. occurs primarily in pulmonary capillaries.
    b. occurs when Cl^- replace HCO_3^- within red blood cells.
    c. decreases the formation of bicarbonate ions.
    d. decreases the number of hydrogen ions.

25. Which of these parts of the brainstem is correctly matched with its main function?
    a. ventral respiratory groups—stimulate the diaphragm
    b. dorsal respiratory groups—limit inflation of the lungs
    c. pontine respiratory group—switching between inspiration and expiration
    d. all of the above

26. The chemosensitive area
    a. stimulates the respiratory center when blood carbon dioxide levels increase.
    b. stimulates the respiratory center when blood pH increases.
    c. is located in the pons.
    d. stimulates the respiratory center when blood oxygen levels increase.
    e. all of the above.

27. Blood oxygen levels
    a. are more important than carbon dioxide in the regulation of respiration.
    b. need to change only slightly to cause a change in respiration.
    c. are detected by sensory receptors in the carotid and aortic bodies.
    d. all of the above.

28. The Hering-Breuer reflex
    a. limits inspiration.
    b. limits expiration.
    c. occurs in response to changes in carbon dioxide levels in the blood.
    d. is stimulated when oxygen decreases in the blood.
    e. does not occur in infants.

29. At the onset of exercise, respiration rate and depth increases primarily because of
    a. increased blood carbon dioxide levels.
    b. decreased blood oxygen levels.
    c. decreased blood pH.
    d. input to the respiratory center from the cerebral motor cortex and proprioceptors.

30. In response to exercise training,
    a. the tidal volume at rest does not change.
    b. minute ventilation during maximal exercise increases.
    c. the brain learns to match ventilation to exercise intensity.
    d. all of the above.

Answers in Appendix F

C R I T I C A L T H I N K I N G

1. What effect does rapid (respiratory rate equals 24 breaths per minute), shallow (tidal volume equals 250 mL per breath) breathing have on minute ventilation, alveolar ventilation, and alveolar P_{O_2} and P_{CO_2}?

2. A person's vital capacity is measured while standing and while lying down. What difference, if any, in the measurement do you predict and why?

3. Ima Diver wanted to do some underwater exploration. She didn't want to buy expensive SCUBA equipment, however. Instead, she bought a long hose and an inner tube. She attached one end of the hose to the inner tube so that the end was always out of the water, and she inserted the other end of the hose in her mouth and went diving. What happened to her alveolar ventilation and why? How would she compensate for this change? How would diving affect lung compliance and the work of ventilation?

4. The bacteria that cause gangrene *(Clostridium perfringens)* are anaerobic microorganisms that don't thrive in the presence of oxygen. Hyperbaric oxygenation (HBO) treatment places a person in a chamber that contains oxygen at three to four times normal atmospheric pressure. Explain how HBO helps in the treatment of gangrene.

5. Cardiopulmonary resuscitation (CPR) has replaced older, less efficient methods of sustaining respiration. The back-pressure/arm-lift method is one such technique that's no longer used. This procedure is performed with the victim lying face down. The rescuer presses firmly on the base of the scapulae for several seconds and then grasps the arms and lifts them. The sequence is then repeated. Explain why this procedure results in ventilation of the lungs.

6. Another technique for artificial respiration is mouth-to-mouth resuscitation. The rescuer takes a deep breath, blows air into the victim's mouth, and then lets air flow out of the victim. The process is repeated. Explain the following: (1) Why do the victim's lungs expand? (2) Why does air move out of the victim's lungs? and (3) What effect do the P_{O_2} and the P_{CO_2} of the rescuer's air have on the victim?

7. The left phrenic nerve supplies the left side of the diaphragm and the right phrenic nerve supplies the right side. Damage to the left phrenic nerve results in paralysis of the left side of the diaphragm. During inspiration, does the left side of the diaphragm move superiorly, inferiorly, or stay in place?

8. Suppose that the thoracic wall is punctured at the end of a normal expiration, producing a pneumothorax. Would the thoracic wall move inward, outward, or not move?

9. During normal quiet respiration, when does the maximum rate of diffusion of oxygen in the pulmonary capillaries occur? The maximum rate of diffusion of carbon dioxide?

10. Is the oxygen–hemoglobin dissociation curve in humans who live at high altitudes to the left or to the right of a person who lives at low altitudes?

11. Predict what would happen to tidal volume if the vagus nerves were cut. The phrenic nerves? The intercostal nerves?

12. You and your physiology instructor are trapped in an overturned ship. To escape, you must swim underwater a long distance. You tell your instructor it would be a good idea to hyperventilate before making the escape attempt. Your instructor calmly replies, "What good would that do, since your pulmonary capillaries are already 100% saturated with oxygen?" What would you do and why?

Answers in Appendix G

A N S W E R S T O P R E D I C T Q U E S T I O N S

1. When you sleep with your mouth open, less air passes through the nasal passages. This is especially true when the nasal passages are plugged because you have a cold. As a consequence, inspired air is not humidified and warmed. The dry air dries and thus irritates the throat. Breathing through the mouth while running in very cold weather results in inspired air that is not warmed and humidified, which can irritate the throat, larynx, and trachea.

2. When food moves down the esophagus, the normally collapsed esophagus expands. If the cartilage rings were solid, expansion of the esophagus, and, therefore, swallowing, would be more difficult.

3. A foreign object is more likely to become lodged in the right main bronchus because it has a larger diameter and is more directly in line with the trachea.

4. Respiratory distress syndrome results from inadequate surfactant, which results in increased water surface tension. Consequently, lung recoil is increased. At the end of expiration, pleural pressure is lower than normal because of the increased lung recoil. Although the decreased pleural pressure increases the tendency for the alveoli to expand, the alveoli don't expand because the increased force of expansion is only counteracting the increased lung recoil. The alveoli collapse if the lung recoil becomes larger than the force of expansion caused by the difference between alveolar and pleural pressure. During inspiration, pleural pressure has to be lower than normal to overcome the effect of the larger-than-normal lung recoil. A larger-than-normal increase in thoracic volume can cause a greater-than-normal decrease in pleural pressure. The effort of overcoming the increased lung recoil, however, can cause muscular fatigue and death.

5. The alveolar ventilation is 4200 mL/min ($12 \times [500 - 150]$). During exercise, the alveolar ventilation is 88,800 mL/min ($24 \times [4000 - 300]$), a 21-fold increase. The increased air exchange increases P_{O_2} and decreases P_{CO_2} in the alveoli, thus increasing gas exchange between the alveoli and the blood.

6. The air the diver is breathing has a greater total pressure than atmospheric pressure at sea level. Consequently, the partial pressure of each gas in the air increases. According to Henry's law, as the partial pressure of a gas increases, the amount (concentration) of gas dissolved in the liquid (e.g., body fluids) with which the gas is in contact increases. When the diver suddenly ascends, the partial pressure of gases in the body returns toward sea level barometric pressure. As a result, the amount (concentration) of gas that can be dissolved in body fluids suddenly decreases. When the fluids can no longer hold all the gas, gas bubbles form.

7. At high altitudes, the atmospheric P_{O_2} decreases because of a decrease in atmospheric pressure. The decreased atmospheric P_{O_2} results in a decrease in alveolar P_{O_2} and less oxygen diffusion into lung tissue. If the person's arterioles are especially sensitive to the decreased oxygen levels, constriction of the arterioles reduces blood flow through the lungs, and the ability to oxygenate blood decreases. Such generalized hypoxemia can also be caused by certain respiratory diseases, such as emphysema and cystic fibrosis.

8. Remember that the oxygen–hemoglobin dissociation curve normally shifts to the right in tissues. The shift of the curve to the left caused by CO reduces the ability of hemoglobin to release oxygen to tissues, which contributes to the detrimental effects of CO poisoning. In the lungs, the shift to the left could slightly increase

the ability of hemoglobin to pick up oxygen, but this effect is offset by the decreased ability of hemoglobin to release oxygen to tissues.

9. A person who cannot synthesize BPG has mild erythrocytosis. Her hemoglobin releases less oxygen to tissues. Consequently, one would expect increased erythropoietin release from the kidneys and increased red blood cell production in red bone marrow.

10. In tissues, carbon dioxide moves into red blood cells, resulting in an increase in H^+. According to the Bohr effect, as H^+ bind to hemoglobin the oxygen–hemoglobin dissociation curve shifts to the right and there is increased release of oxygen. According to the Haldane effect, hemoglobin that has released oxygen picks up more carbon dioxide.

11. Hyperventilation decreases blood carbon dioxide levels, causing an increase in blood pH. Holding one's breath increases blood carbon dioxide levels and decreases blood pH.

12. When a person hyperventilates, P_{CO_2} in the blood decreases. Consequently, carbon dioxide moves out of cerebrospinal fluid into the blood. As carbon dioxide levels in cerebrospinal fluid decrease, H^+ and HCO_3^- combine to form carbonic acid, which forms carbon dioxide. The result is a decrease in H^+ concentration in cerebrospinal fluid and decreased stimulation of the respiratory center by the chemosensitive area. Until blood P_{CO_2} levels increase, the chemosensitive area is not stimulated, and apnea results.

13. Through touch, thermal, or pain receptors, the respiratory center can be stimulated to cause a sudden inspiration of air.

14. A P_{O_2} of 60 mm Hg and a P_{CO_2} of 30 mm Hg are both below normal. The movement of air into and out of the lungs is restricted because of the asthma and a mismatch occurs between ventilation of the alveoli and blood flow to the alveoli. Consequently, because of the ineffective ventilation, blood oxygen levels decrease. Mr. W hyperventilates, which helps to maintain blood oxygen levels but also results in lower-than-normal blood carbon dioxide levels. (If no hyperventilation occurred, one would expect decreased blood oxygen but increased blood carbon dioxide levels.)

Visit the Online Learning Center at www.mhhe.com/seeley7 for chapter quizzes, interactive learning exercises, and other study tools.

Digestive System

Colorized scanning electron micrograph of the interior surface of the small intestine showing villi.

Every cell of the body needs nourishment, yet most cells cannot leave their position in the body and travel to a food source, so the food must be converted to a usable form and delivered. The digestive system, with the help of the circulatory system, acts like a gigantic "meals on wheels," providing nourishment to over a hundred trillion "customer" cells in the body. It also has its own quality control and waste disposal system.

The digestive system provides the body with water, electrolytes, and other nutrients. To do this, the digestive system is specialized to ingest food, propel it through the digestive tract, digest it, and absorb water, electrolytes, and other nutrients from the lumen of the gastrointestinal tract. Once these useful substances are absorbed, they are transported through the circulatory system to cells, where they are used. The undigested portion of the food is moved through the digestive tract and eliminated through the anus.

This chapter presents the general *anatomy of the digestive system* (p. 878), followed by descriptions of the *functions of the digestive system* (p. 878), the *histology of the digestive tract* (p. 880), the *regulation of the digestive system* (p. 881) and the *peritoneum* (p. 882). The anatomy and physiology of each section of the digestive tract and its accessory structures are then presented: the *oral cavity* (p. 884), *pharynx* (p. 890), *esophagus* (p. 890), along with a section on *swallowing* (p. 890), *stomach* (p. 892), *small intestine* (p. 900), *liver* (p. 903), *gallbladder* (p. 908), *pancreas* (p. 909), and *large intestine* (p. 912). *Digestion, absorption, and transport* (p. 915) of nutrients are then discussed, along with the *effects of aging on the digestive system* (p. 921).

Anatomy of the Digestive System

The **digestive system** (figure 24.1) consists of the **digestive tract,** a tube extending from the mouth to the anus, and its associated **accessory organs,** primarily glands located outside the digestive tract, which secrete fluids into the digestive tract. The digestive tract is also called the **alimentary tract,** or **alimentary canal.** The term **gastrointestinal** (gas′trō-in-tes′tin-ăl; **GI**) **tract** technically only refers to the stomach and intestines but is often used as a synonym for the digestive tract.

The regions of the digestive tract include

1. the *oral cavity,* or *mouth,* which has salivary glands and tonsils as accessory organs;
2. the *pharynx,* or throat, with tubular mucous glands;
3. the *esophagus,* with tubular mucous glands;
4. the *stomach,* which contains many tubelike glands, as well as mucous cells;
5. the *small intestine,* consisting of the duodenum, jejunum, and ileum, with the liver, gallbladder, and pancreas as major accessory organs;
6. the *large intestine,* including the cecum, colon, rectum, and anal canal, with mucous glands;
7. the *anus.*

● **1.** **List the major regions of the digestive tract.**

Functions of the Digestive System

The major functions of the digestive system are outlined as follows (table 24.1):

1. *Ingestion* is the introduction of solid or liquid food into the stomach. The normal route of ingestion is through the oral cavity, but food can be introduced directly into the stomach by a nasogastric, or stomach, tube.
2. *Mastication* is the process by which food taken into the mouth is chewed by the teeth. Digestive enzymes cannot easily penetrate solid food particles and can only work effectively on the surfaces of the particles. It's vital, therefore, to normal digestive function that solid foods be mechanically broken down into small particles. Mastication breaks large food particles into many smaller particles, which have a much larger total surface area than do a few large particles.
3. *Propulsion* in the digestive tract is the movement of food from one end of the digestive tract to the other. The total time that it takes food to travel the length of the digestive tract is usually about 24–36 hours. Each segment of the digestive tract is specialized to assist in moving its contents from the oral end to the anal end. **Deglutition** (dē′gloo-tish′ŭn), or swallowing, moves food and liquids, called a **bolus,** from the oral cavity into the esophagus. **Peristalsis** (per-i-stal′sis; figure 24.2) is responsible for moving material through most of the digestive tract. Muscular contractions occur in **peristaltic** (per-i-stal′tik) **waves,** consisting of a wave of relaxation of the circular

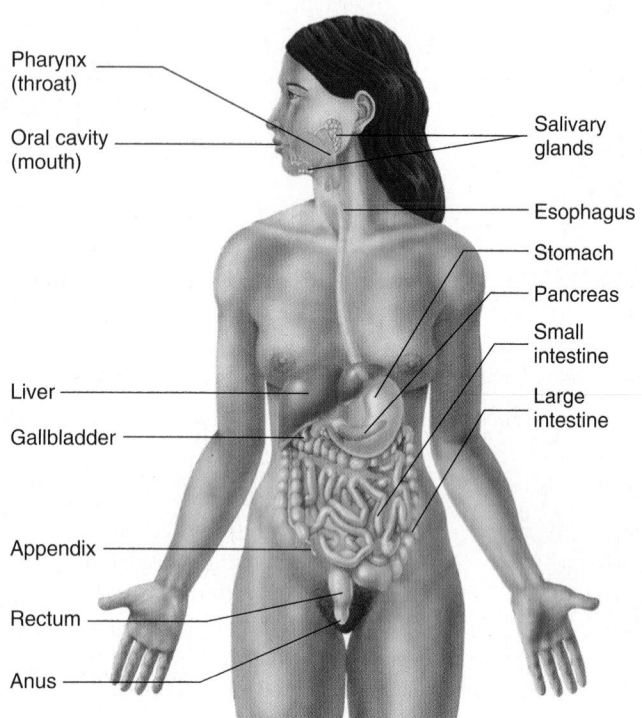

Pharynx (throat)

Oral cavity (mouth)

Salivary glands

Esophagus

Stomach

Pancreas

Small intestine

Liver

Large intestine

Gallbladder

Appendix

Rectum

Anus

Figure 24.1   **The Digestive System**

1. A wave of circular smooth muscle relaxation moves ahead of the bolus of food allowing the digestive tract to expand.

2. A wave of contraction of the circular smooth muscles behind the bolus of food propels it through the digestive tract.

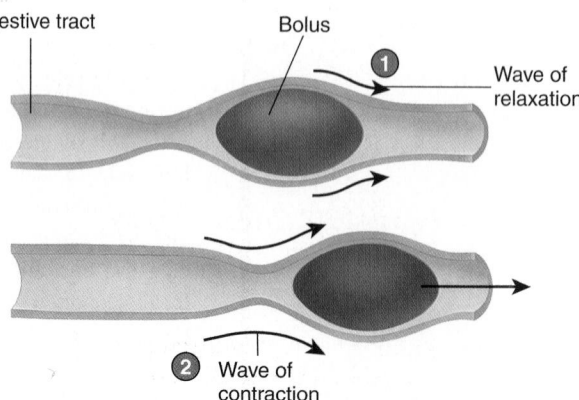

Digestive tract

Bolus

① Wave of relaxation

② Wave of contraction

Process Figure 24.2   **Peristalsis**

Table 24.1   Functions of the Digestive Tract

Organ	Functions
Oral cavity	*Ingestion*. Solid food and fluids are taken into the digestive tract through the oral cavity. *Taste*. Tastants dissolved in saliva stimulate taste buds in the tongue. *Mastication*. Movement of the mandible by the muscles of mastication cause the teeth to break food down into smaller pieces. The tongue and cheeks help to place the food between the teeth. *Digestion*. Amylase in saliva begins carbohydrate (starch) digestion. *Swallowing*. The tongue forms food into a bolus and pushes the bolus into the pharynx. *Communication*. The lips, cheeks, teeth, and tongue are involved in speech. The lips change shape as part of facial expressions. *Protection*. Mucin and water in saliva provide lubrication, and lysozyme (an enzyme that lyses cells) kills microorganisms. Nonkeratinized stratified squamous epithelium prevents abrasion.
Pharynx	*Swallowing*. The involuntary phase of swallowing moves the bolus from the oral cavity to the esophagus. Materials are prevented from entering the nasal cavity by the soft palate and from entering the lower respiratory tract by the epiglottis and vestibular folds. *Breathing*. Air passes from the nasal or oral cavity through the pharynx to the lower respiratory tract. *Protection*. Mucus provides lubrication. Nonkeratinized stratified squamous epithelium prevents abrasion.
Esophagus	*Propulsion*. Peristaltic contractions move the bolus from the pharynx to the stomach. The lower esophageal sphincter limits reflux of the stomach contents into the esophagus. *Protection*. Glands produce mucus that provides lubrication and protects the inferior esophagus from stomach acid.
Stomach	*Storage*. Rugae allow the stomach to expand and hold food until it can be digested. *Digestion*. Protein digestion begins as a result of the actions of hydrochloric acid and pepsin. Intrinsic factor prevents the breakdown of vitamin B_{12} by stomach acid. *Absorption*. Except for a few substances (e.g., water, alcohol, aspirin) little absorption takes place in the stomach. *Mixing and propulsion*. Mixing waves churn ingested materials and stomach secretions into chyme. Peristaltic waves move the chyme into the small intestine. *Protection*. Mucus provides lubrication and prevents digestion of the stomach wall. Stomach acid kills most microorganisms.
Small intestine	*Neutralization*. Bicarbonate ions from the pancreas and bile from the liver neutralize stomach acid to form a pH environment suitable for pancreatic and intestinal enzymes. *Digestion*. Enzymes from the pancreas and the lining of the small intestine complete the breakdown of food molecules. Bile salts from the liver emulsify fats. *Absorption*. The circular folds, villi, and microvilli increase surface area. Most nutrients are actively or passively absorbed. Most of the ingested water or the water in digestive tract secretions is absorbed. *Mixing and propulsion*. Segmental contractions mix the chyme, and peristaltic contractions move the chyme into the large intestine. *Excretion*. Bile from the liver contains bilirubin, cholestrol, fats, and fat-soluble hormones. *Protection*. Mucus provides lubrication, prevents the digestion of the intestinal wall, and protects the small intestine from stomach acid. Peyer's patches protect against microorganisms.
Large intestine	*Absorption*. The proximal half of the colon absorbs salts (e.g., sodium chloride), water, and vitamins (e.g., K) produced by bacteria. *Storage*. The distal half of the colon holds feces until it is eliminated. *Mixing and propulsion*. Slight segmental mixing occurs. Mass movements propel feces toward the anus and defecation eliminates the feces. *Protection*. Mucus provides lubrication; mucus and bicarbonate ions protect against acids produced by bacteria.

muscles, which forms a leading wave of distention in front of the bolus, followed by a wave of strong contraction of the circular muscles behind the bolus, which forces the bolus along the digestive tube. Each peristaltic wave travels the length of the esophagus in about 10 seconds. Peristaltic waves in the small intestine usually only travel for short distances. In some parts of the large intestine, material is moved by **mass movements,** which are contractions that

extend over much larger parts of the digestive tract than peristaltic movements.

4. *Mixing*. Some contractions don't propel food from one end of the digestive tract to the other but rather move it back and forth within the digestive tract to **mix** it with digestive secretions and to help break it into smaller pieces. **Segmental contractions** (figure 24.3) are mixing contractions that occur in the small intestine.

1. A secretion introduced into the digestive tract or food within the tract begins in one location.

2. Segments of the digestive tract alternate between contraction and relaxation.

3. Material (*brown*) in the intestine is spread out in both directions from the site of introduction.

4. The secretion or food is spread out in the digestive tract and becomes more diffuse (*lighter color*) through time.

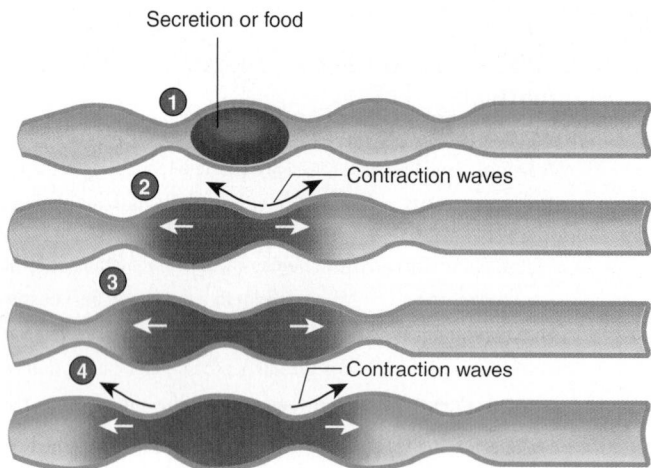

Process Figure 24.3   **Segmental Contractions**

5. *Secretion.* As food moves through the digestive tract, **secretions** are added to lubricate, liquefy, buffer, and digest the food. **Mucus,** secreted along the entire digestive tract, lubricates the food and the lining of the tract. The mucus coats and protects the epithelial cells of the digestive tract from mechanical abrasion, from the damaging effect of acid in the stomach, and from the digestive enzymes of the digestive tract. The secretions also contain large amounts of **water,** which liquefies the food, thereby making it easier to digest and absorb. Water also moves into the intestine by osmosis. Liver secretions break large fat droplets into much smaller droplets, which makes possible the digestion and absorption of fats. **Enzymes** secreted by the oral cavity, stomach, intestine, and pancreas break large food molecules down into smaller molecules that can be absorbed by the intestinal wall.

6. *Digestion* is the breakdown of large organic molecules into their component parts: carbohydrates into monosaccharides, proteins into amino acids, and triglycerides into fatty acids and glycerol. Digestion consists of **mechanical digestion,** which involves mastication and mixing of food, and **chemical digestion,** which is accomplished by digestive enzymes that are secreted along the digestive tract. Digestion of large molecules into their component parts must be accomplished before they can be absorbed by the digestive tract. Minerals and water are not broken down before being absorbed. Vitamins are also absorbed without digestion and lose their function if their structure is altered by digestion.

7. *Absorption* is the movement of molecules out of the digestive tract and into the circulation or into the lymphatic system. The mechanism by which absorption occurs depends on the type of molecule involved. Molecules pass out of the digestive tract by diffusion, facilitated diffusion, active transport, cotransport, or endocytosis (see chapter 3).

8. *Elimination* is the process by which the waste products of digestion are removed from the body. During this process, occurring primarily in the large intestine, water and salts are absorbed and change the material in the digestive tract from a liquefied state to a semisolid state. These semisolid waste products, called **feces,** are then eliminated from the digestive tract by the process of **defecation.**

● *2. Describe each of the processes involved in the normal functions of the digestive system.*

Histology of the Digestive Tract

Figure 24.4 depicts a generalized view of the digestive tract histology. The digestive tube consists of four major layers, or tunics: an internal mucosa and an external serosa with a submucosa and muscularis in between. These four tunics are present in all areas of the digestive tract from the esophagus to the anus. Three major types of glands are associated with the intestinal tract: (1) unicellular mucous glands in the mucosa, (2) multicellular glands in the mucosa and submucosa, and (3) multicellular glands (accessory glands) outside the digestive tract.

Mucosa

The innermost tunic, the **mucosa** (mū-kō′să), or mucous membrane, consists of three layers: (1) the inner **mucous epithelium,** which is moist stratified squamous epithelium in the mouth, oropharynx, esophagus, and anal canal and simple columnar epithelium in the remainder of the digestive tract; (2) a loose connective tissue called the **lamina propria** (lam′i-nă prō′prē-ă); and (3) an outer thin smooth muscle layer, the **muscularis mucosae.** The epithelium extends deep into the lamina propria in many places to form **intestinal glands** or **crypts.** Specialized cells in the mucosa are mechanoreceptors involved in peristaltic reflexes.

Submucosa

The **submucosa** is a thick connective tissue layer containing nerves, blood vessels, lymphatic vessels, and small glands that lies beneath the mucosa. The plexus of nerve cells in the submucosa form the **submucosal plexus** (plek′sŭs; Meissner's plexus), an enteric nerve plexus consisting of axons, many scattered neuron cell bodies, and

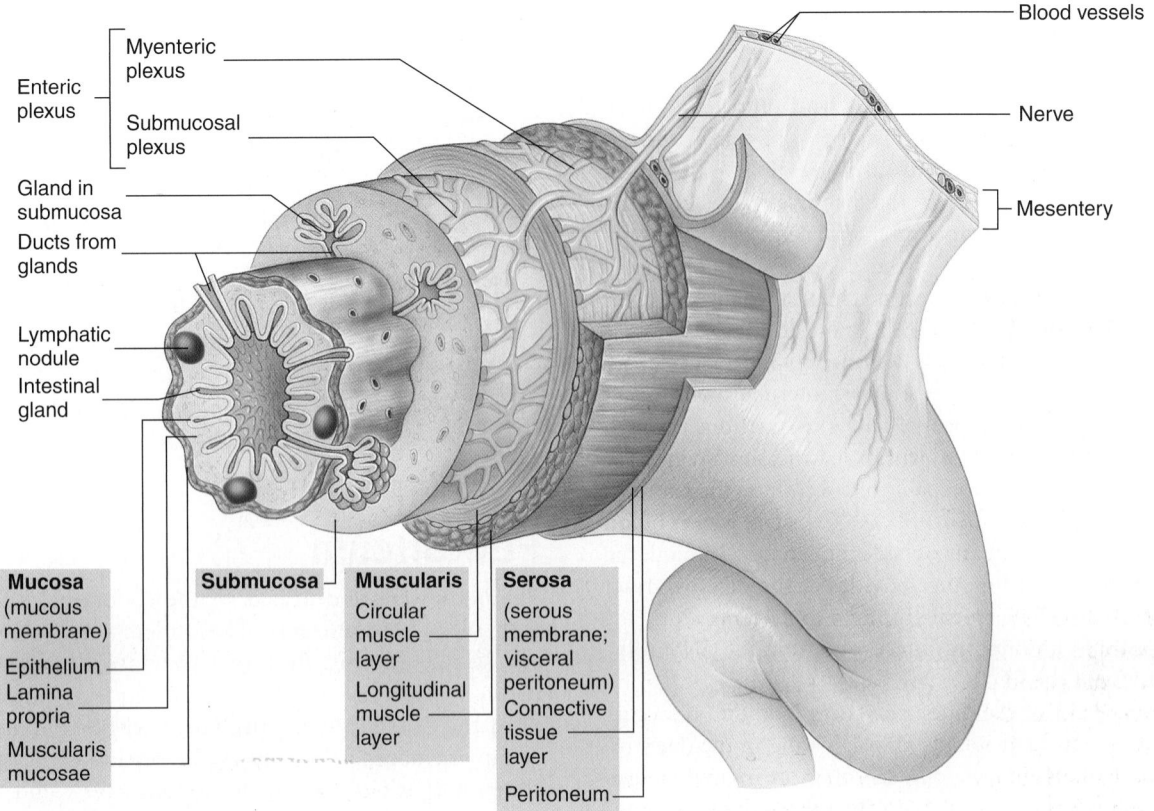

Figure 24.4 Digestive Tract Histology

The four tunics are the mucosa, submucosa, muscularis, and serosa or adventitia. Glands may exist along the digestive tract as part of the epithelium, within the submucosa, or as large glands that are outside the digestive tract.

neuroglial cells. Axons from the submucosal plexus extend to cells in epithelial intestinal glands, stimulating their secretion. The esophagus and stomach lack a submucosal plexus, but the plexus is extensive throughout the rest of the digestive tract.

Muscularis

The next tunic is the **muscularis,** which consists of an inner layer of circular smooth muscle and an outer layer of longitudinal smooth muscle. Two exceptions are the upper esophagus, where the muscles are striated, and the stomach, which has three layers of smooth muscle. Another nerve plexus, the **myenteric plexus** (mī-en-ter′ik; Auerbach's plexus), which also consists of axons, many scattered neuron cell bodies, and neuroglial cells, is between these two muscle layers (see figure 24.4). The myenteric plexus is much more extensive than the submucosal plexus and controls the motility of the intestinal tract.

Together, the submucosal and myenteric plexuses constitute the **enteric plexus** (en-tĕr′ik; relating to the intestine) or **intramural** (in′tră-mŭ′răl; within the walls) **plexus.** The enteric plexus is extremely important in the control of secretion and movement (see also chapter 16).

Serosa or Adventitia

The fourth layer of the digestive tract is either the **serosa** or the **adventitia** (ad-ven-tish′ă; foreign or coming from outside), depending on the structure of the layer. Parts of the digestive tract that protrude into the peritoneal cavity have a serosa as the outermost layer. This serosa, or serous membrane, is called the visceral peritoneum. It consists of a thin layer of connective tissue and a simple squamous epithelium. When the outer layer of the digestive tract is derived from adjacent connective tissue, the tunic is called the adventitia and consists of a connective tissue covering that blends with the surrounding connective tissue. These areas include the esophagus and the retroperitoneal organs (discussed later in relation to the peritoneum, p. 882).

3. *What are the major layers of the digestive tract? How do the serosa and adventitia differ?*

4. *Describe the enteric plexus. In what layers of the digestive tract are the submucosal and myenteric plexuses found?*

Regulation of the Digestive System

Elaborate nervous and chemical mechanisms regulate the movement, secretion, absorption, and elimination processes.

Nervous Regulation of the Digestive System

The enteric plexus is extensive, it actually contains more neurons than the spinal cord! Most of the nervous control of the GI tract is local, occurring as the result of local reflexes within the enteric plexus, and some is more general, mediated largely by the parasympathetic division of the ANS through the vagus nerves and by sympathetic nerves.

Local neuronal control of the digestive tract occurs within the **enteric nervous system (ENS).** The ENS consists of the enteric plexus, made up of enteric neurons, and extensive associated neuroglial cells, within the wall of the digestive tract (see figure 24.4). There are three major types of enteric neurons: (1) Enteric sensory neurons detect changes in the chemical composition of the digestive tract contents or detect mechanical changes such as stretch of the digestive tract wall. (2) Enteric motor neurons stimulate or inhibit smooth muscle contraction and glandular secretion in the digestive system. (3) Enteric interneurons connect enteric sensory and motor neurons. The ENS functions through **local reflexes** to control activities within specific, short regions of the digestive tract. The ENS is capable of controlling the complex peristaltic and mixing movements of the GI tract, as well as blood flow to the GI tract, without any outside influences. Although the ENS is capable of controlling the activities of the digestive tract independent of the CNS, normally the two systems work together. For example, autonomic innervation from the CNS influences the activity of the ENS.

General control of the digestive system by the CNS occurs when reflexes are activated by stimuli originating in the digestive tract. Action potentials are carried by sensory neurons in the vagus and sympathetic nerves to the CNS, where the reflexes are integrated. In addition, reflexes within the CNS may be activated by the sight, smell, or taste of food, which stimulate the sensation of hunger. All of these reflexes influence activity in parasympathetic neurons of the CNS. Parasympathetic neurons extend to the digestive tract through the vagus nerves to control responses or alter the activity of the ENS and local reflexes. Some sympathetic neurons inhibit muscle contraction and secretion in the digestive system and decrease blood flow to the digestive system.

Chemical Regulation of the Digestive System

Over 30 neurotransmitters are associated with the ENS. One of the major ENS neurotransmitters is serotonin. Specialized interstitial cells mediate neurotransmission from ENS neurons to GI smooth muscle cells. Groups of these interstitial cells form "pacemakers" along the GI tract which regulate GI movement. Dysfunction of these pacemakers causes hypomobility in the GI tract.

The digestive tract produces a number of hormones, such as gastrin, secretin, and others, which are secreted by endocrine cells of the digestive system and carried through the circulation to target organs of the digestive system or to target tissues in other systems. These hormones help regulate many gastrointestinal tract functions as well as the secretions of associated glands such as the liver and pancreas.

In addition to the hormones produced by the digestive system, which enter the circulation, other paracrine chemicals, such as histamine, are released locally within the digestive tract and influence the activity of nearby cells. These localized chemical regulators help local reflexes within the ENS control local digestive tract environments, such as pH levels.

● **5. What are the nervous and chemical mechanisms that regulate the digestive system?**

Peritoneum

The body walls and organs of the abdominal cavity are lined with **serous membranes.** These membranes are very smooth and secrete a serous fluid that provides a lubricating film between the layers of membranes. These membranes and fluid reduce the friction as organs move within the abdomen. The serous membrane that covers the organs is the **visceral peritoneum** (per′i-tō-nē′ŭm; to stretch over), and the one that covers the interior surface of the body wall is the **parietal peritoneum** (figure 24.5).

Connective tissue sheets called **mesenteries** (mes′en-ter′ēz; middle intestine) hold many of the organs in place within the abdominal cavity. The mesenteries consist of two layers of serous membranes with a thin layer of loose connective tissue between them. They provide a route by which vessels and nerves can pass from the body wall to the organs. Other abdominal organs lie against the abdominal wall, have no mesenteries, and are referred to as **retroperitoneal** (re′trō-per′i-tō-nē′ăl; behind the peritoneum; see chapter 1). The retroperitoneal organs include the duodenum, the pancreas, the ascending colon, the descending colon, the rectum, the kidneys, the adrenal glands, and the urinary bladder.

Some mesenteries are given specific names. The mesentery connecting the lesser curvature of the stomach and the proximal end of the duodenum to the liver and diaphragm is called the **lesser omentum** (ō-men′tŭm; membrane of the bowels), and the mesentery extending as a fold from the greater curvature and then to the transverse colon is called the **greater omentum** (see figure 24.5). The greater omentum forms a long, double fold of mesentery that

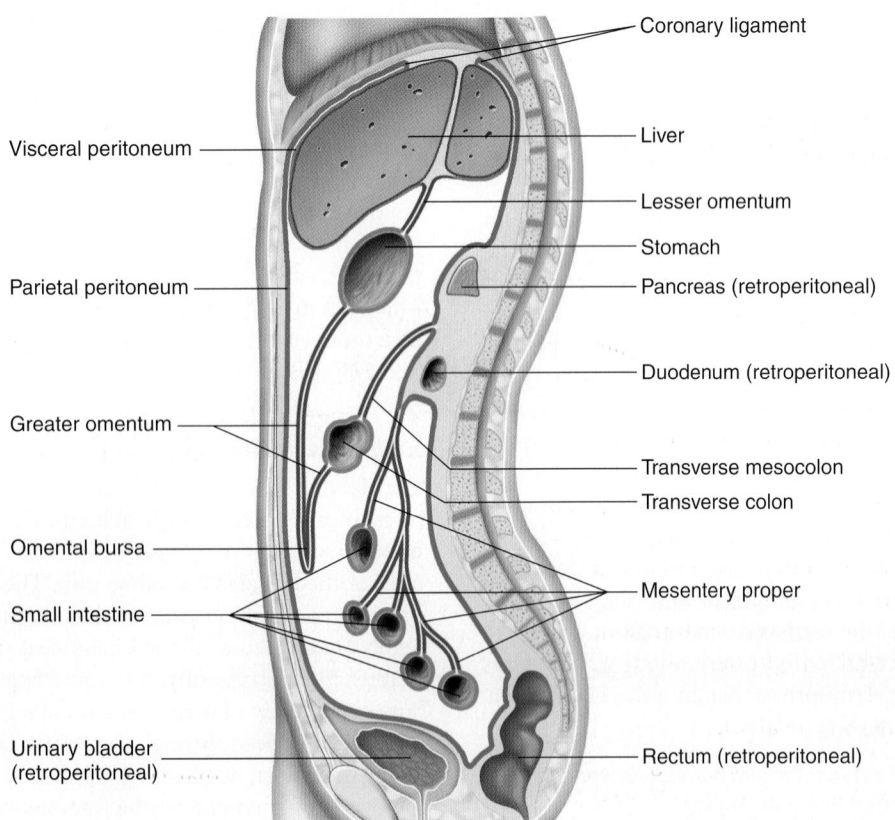

Coronary ligament

Visceral peritoneum

Liver

Lesser omentum

Stomach

Parietal peritoneum

Pancreas (retroperitoneal)

Duodenum (retroperitoneal)

Greater omentum

Transverse mesocolon

Transverse colon

Omental bursa

Mesentery proper

Small intestine

Urinary bladder (retroperitoneal)

Rectum (retroperitoneal)

(a) **Medial view**

Liver

Falciform ligament

Greater omentum

(b)   **Anterior view**

Liver

Gallbladder

Stomach

Transverse mesocolon

Transverse colon

Mesentery proper

Small intestine

(c)   **Anterior view**

Figure 24.5   Peritoneum and Mesenteries

(*a*) Sagittal section through the trunk showing the peritoneum and mesenteries associated with some abdominal organs. (*b*) Photograph of the abdomen of a cadaver with the greater omentum in place. (*c*) Photograph of the abdomen of a cadaver with the greater omentum removed to reveal the underlying viscera.

extends inferiorly from the stomach over the surface of the small intestine. Because of this folding, a cavity, or pocket, called the **omental bursa** (ber′să; pocket) is formed between the two layers of mesentery. A large amount of fat accumulates in the greater omentum, and it is sometimes referred to as the "fatty apron." The greater omentum has considerable mobility in the abdomen.

PREDICT

If you placed a pin through the greater omentum, through how many layers of simple squamous epithelium would the pin pass?

The **coronary ligament** attaches the liver to the diaphragm. Unlike other mesenteries, the coronary ligament has a wide space in the center, the bare area of the liver, where no peritoneum exists. The **falciform ligament** attaches the liver to the anterior abdominal wall (see figure 24.5).

Although the term *mesentery* is a general term referring to the serous membranes attached to the abdominal organs, it is also used specifically to refer to the mesentery associated with the small intestine, sometimes called the **mesentery proper.** The mesenteries of parts of the colon are the **transverse mesocolon,** which extends from the transverse colon to the posterior body wall, and the **sigmoid mesocolon.** The vermiform appendix even has its own little mesentery called the **mesoappendix.**

6. *Where are visceral peritoneum and parietal peritoneum found? What is a retroperitoneal organ?*
7. *Define the term mesentery. Name and describe the location of the mesenteries found in the abdominal cavity.*

Oral Cavity

The **oral cavity** (figure 24.6), or mouth, is that part of the digestive tract bounded by the lips anteriorly, the **fauces** (faw′sēz; throat; opening into the pharynx) posteriorly, the cheeks laterally, the palate superiorly, and a muscular floor inferiorly. The oral cavity is divided into two regions: (1) the **vestibule** (ves′ti-bool; entry), which is the space between the lips or cheeks and the alveolar processes, which contain the teeth; and (2) the **oral cavity proper,** which lies medial to the alveolar processes. The oral cavity is lined with moist stratified squamous epithelium, which provides protection against abrasion.

Lips and Cheeks

The **lips,** or **labia** (lā′bē-ă) (see figure 24.6), are muscular structures formed mostly by the **orbicularis oris** (ōr-bik′ū-lā′ris ōr′is) **muscle** (see figure 10.9), as well as connective tissue. The outer surfaces of the lips are covered by skin. The keratinized stratified epithelium of the skin is thin at the margin of the lips and is not as highly keratinized as the epithelium of the surrounding skin (see chapter 5); consequently, it is more transparent than the epithelium over the rest of the body. The color from the underlying blood vessels can be seen through the relatively transparent epithelium, giving the lips a reddish pink to dark red appearance, depending on the overlying pigment. At the internal margin of the lips, the epithelium is continuous with the moist stratified squamous epithelium of the mucosa in the oral cavity.

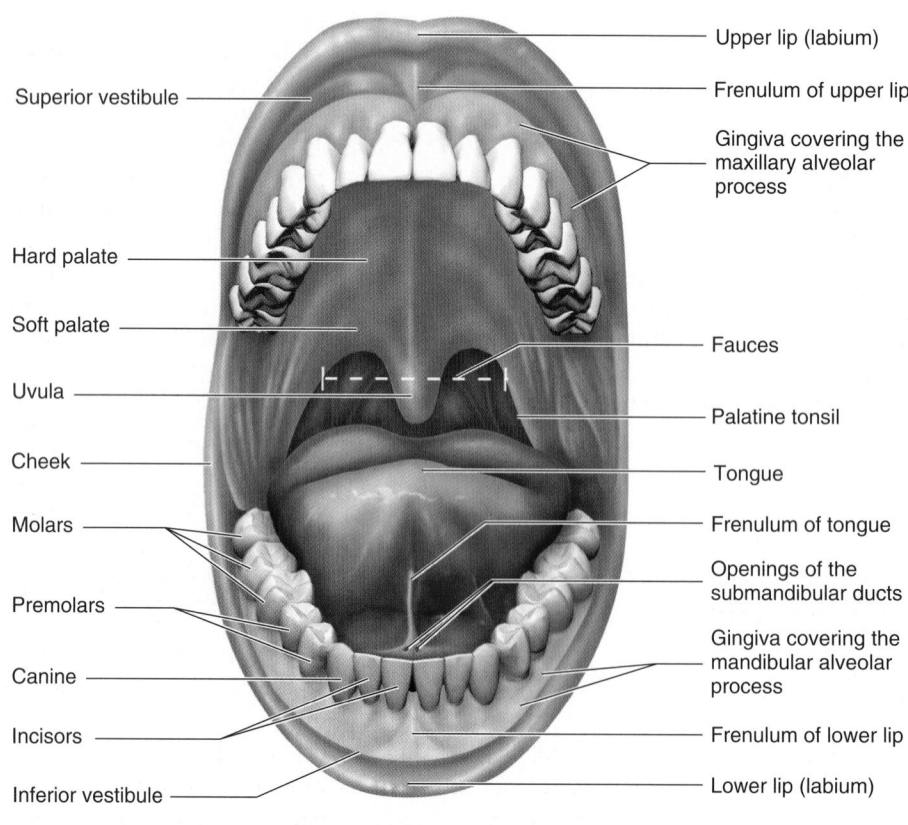

Anterior view

Figure 24.6   Oral Cavity

One or more **frenula** (fren′ū-lă; bridle), which are mucosal folds, extend from the alveolar processes of the maxilla to the upper lip and from the alveolar process of the mandible to the lower lip.

The **cheeks** form the lateral walls of the oral cavity. They consist of an interior lining of moist stratified squamous epithelium and an exterior covering of skin. The substance of the cheek includes the **buccinator muscle** (see chapter 10), which flattens the cheek against the teeth, and the **buccal fat pad,** which rounds out the profile on the side of the face.

The lips and cheeks are important in the processes of mastication and speech. They help manipulate food within the mouth and hold it in place while the teeth crush or tear it. They also help form words during the speech process. A large number of the muscles of facial expression are involved in movement of the lips. They are listed in chapter 10.

Palate and Palatine Tonsils

The **palate** (see figure 24.6) consists of two parts, an anterior bony part, the **hard palate** (see chapter 7), and a posterior, non-bony part, the **soft palate,** which consists of skeletal muscle and connective tissue. The **uvula** (ū′vū-lă; a grape) is the projection from the posterior edge of the soft palate. The palate is important in the swallowing process; it prevents food from passing into the nasal cavity.

Palatine tonsils are located in the lateral wall of the fauces (see chapter 22).

Tongue

The **tongue** is a large, muscular organ that occupies most of the oral cavity proper when the mouth is closed. Its major attachment in the oral cavity is through its posterior part. The anterior part of the tongue is relatively free and is attached to the floor of the mouth by a thin fold of tissue called the **frenulum.** The muscles associated with the tongue are divided into two categories: **intrinsic muscles,** which are within the tongue itself; and **extrinsic muscles,** which are outside the tongue but attached to it. The intrinsic muscles are largely responsible for changing the shape of the tongue, such as flattening and elevating the tongue during drinking and swallowing. The extrinsic tongue muscles protrude and retract the tongue, move it from side to side, and change its shape (see chapter 10).

Tongue-Tied

A person is "tongue-tied" in a more literal sense if the frenulum extends too far toward the tip of the tongue, thereby inhibiting normal movement of the tongue and interfering with normal speech. Surgically cutting the frenulum can correct the condition.

A groove called the **terminal sulcus** divides the tongue into two parts. The part anterior to the terminal sulcus accounts for about two-thirds of the surface area and is covered by papillae, some of which contain taste buds (see chapter 15). The posterior one-third of the tongue is devoid of papillae and has only a few scattered taste buds. It has, instead, a few small glands and a large amount of lymphoid tissue, the **lingual tonsil** (see chapter 22). Moist stratified squamous epithelium covers the tongue.

Lipid-Soluble Drugs

Certain drugs that are lipid-soluble and can diffuse through the plasma membranes of the oral cavity can be quickly absorbed into the circulation. An example is nitroglycerin, which is a vasodilator used to treat cases of angina pectoris. The drug is placed under the tongue, where, in less than 1 minute, it dissolves and passes through the very thin oral mucosa into the lingual veins.

The tongue moves food in the mouth and, in cooperation with the lips and gums, holds the food in place during mastication. It also plays a major role in the mechanism of swallowing (discussed on p. 890). It is a major sensory organ for taste (see chapter 15) and one of the primary organs of speech.

Glossectomy and Speech

Patients who have undergone glossectomies (tongue removal) as a result of glossal carcinoma can compensate for loss of the tongue's function in speech, and they can learn to speak fairly well. These patients, however, have substantial difficulty chewing and swallowing food.

Teeth

Normal adults have 32 **teeth,** which are distributed in two **dental arches.** One is called the maxillary arch and the other is called the mandibular arch. The teeth in the right and left halves of each dental arch are roughly mirror images of each other. As a result, the teeth are divided into four quadrants: right upper, left upper, right lower, and left lower. The teeth in each quadrant include one central and one lateral **incisor,** one **canine,** first and second **premolars,** and first, second, and third **molars** (figure 24.7a). The third molars are called **wisdom teeth** because they usually appear in a person's late teens or early twenties, when the person is old enough to have acquired some wisdom.

Impacted Wisdom Teeth

In some people with small dental arches, the third molars may not have room to erupt into the oral cavity and remain embedded within the jaw. Embedded wisdom teeth are referred to as impacted. They may cause pain or irritation and their surgical removal is often necessary.

The teeth of the adult mouth are **permanent,** or **secondary, teeth.** Most of them are replacements for **primary,** or **deciduous** (dē-sid′ū-ŭs; those that fall out; also called milk teeth), **teeth** that are lost during childhood (figure 24.7b). The deciduous teeth erupt (the crowns appear within the oral cavity) between about 6 months and 24 months of age (see figure 24.7b). The permanent teeth begin replacing the deciduous teeth by about 5 years and the process is completed by about 11 years.

(a)

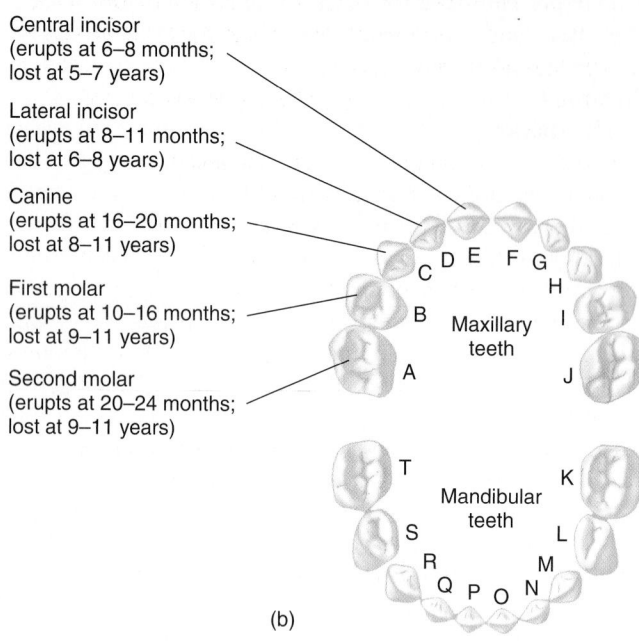

(b)

Figure 24.7   Teeth

(*a*) Permanent teeth. (*b*) Deciduous teeth.
A "universal" numbering and lettering system has been developed by dental professionals for convenience in identifying individual teeth.

Each tooth consists of a **crown** with one or more **cusps** (points), a **neck,** and a **root** (figure 24.8). The **clinical crown** is that part of the tooth exposed in the oral cavity. The **anatomical crown** is the entire enamel-covered part of the tooth. The center of the tooth is a **pulp cavity,** which is filled with blood vessels, nerves, and connective tissue called **pulp.** The pulp cavity within the root is called the **root canal.** The nerves and blood vessels of the tooth enter and exit the pulp through a hole at the point of each root called the **apical foramen.** The pulp cavity is surrounded by a living, cellular, and calcified tissue called **dentin.** The dentin of the tooth crown is covered by an extremely hard, nonliving, acellular substance called **enamel,** which protects the tooth against abrasion and acids produced by bacteria in the mouth. The surface of the dentin in the root is covered with a bonelike substance, called **cementum,** which helps anchor the tooth in the jaw.

The teeth are set in **alveoli** (al-vē′ō-lī; sockets) along the alveolar processes of the mandible and maxilla. Dense fibrous connective tissue and stratified squamous epithelium, referred to as the **gingiva** (jin′ji-vă; gums) cover the alveolar processes (see figure 24.6). **Periodontal** (per′ē-ō-don′tăl; around a tooth) **ligaments** secure the teeth in the alveoli.

The teeth play an important role in mastication and a role in speech.

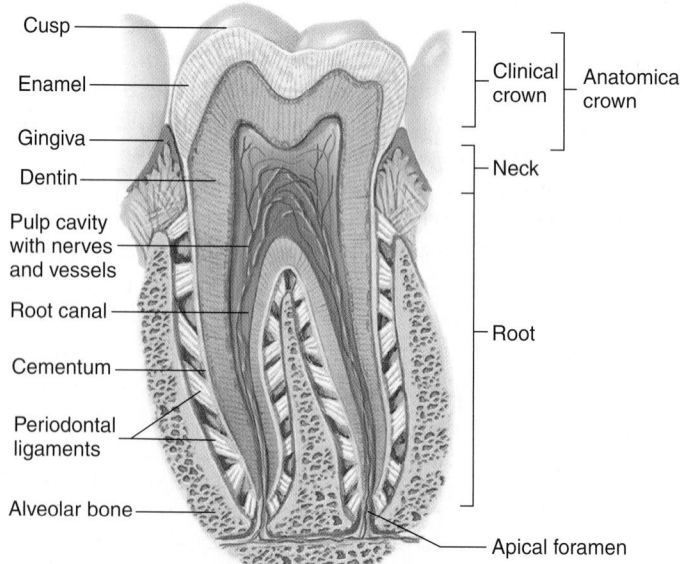

Figure 24.8   Molar Tooth in Place in the Alveolar Bone

The tooth consists of a crown and root. The root is covered with cementum, and the tooth is held in the socket by periodontal ligaments. Nerves and vessels enter and exit the tooth through the apical foramen.

8. *Distinguish between the vestibule and the oral cavity proper.*
9. *What are the functions of the lips and cheeks? What muscle forms the substance of the cheek?*
10. *What are the hard and soft palate? Where is the uvula found?*
11. *List the functions of the tongue. Distinguish between intrinsic and extrinsic tongue muscles.*
12. *What are deciduous and permanent teeth? Name the different kinds of teeth.*
13. *Describe the parts of a tooth. What are dentin, enamel, cementum, and pulp?*

Dental Diseases

Dental caries, or tooth decay, is caused by a breakdown of enamel by acids produced by bacteria on the tooth surface. Because the enamel is nonliving and cannot repair itself, a dental filling is necessary to prevent further damage. If the decay reaches the pulp cavity with its rich supply of nerves, toothache pain may result. In some cases in which decay has reached the pulp cavity, it may be necessary to perform a dental procedure called a "root canal," which consists of removing the pulp from the tooth.

Periodontal disease is the inflammation and degradation of the periodontal ligaments, gingiva, and alveolar bone. This disease is the most common cause of tooth loss in adults. **Gingivitis** (jin-ji-vī′tis) is an inflammation of the gingiva, often caused by food deposited in gingival crevices and not promptly removed by brushing and flossing. Gingivitis may eventually lead to periodontal disease. **Pyorrhea** (pī-ō-rē′ă) is a condition in which pus occurs with periodontal disease. **Halitosis** (hal-i-tō′sis), or bad breath, often occurs with periodontal disease and pyorrhea.

Mastication

Food taken into the mouth is **chewed,** or **masticated,** by the teeth. The anterior teeth, the incisors, and the canines primarily cut and tear food, whereas the premolars and molars primarily crush and grind it. Mastication breaks large food particles into smaller ones, which have a much larger total surface area. Because digestive enzymes digest food molecules only at the surface of the particles, mastication increases the efficiency of digestion.

Four pairs of muscles move the mandible during mastication: the **temporalis, masseter, medial pterygoid,** and **lateral pterygoid** (see chapter 10 and figure 10.11). The temporalis, masseter, and medial pterygoid muscles close the jaw; and the lateral pterygoid muscle opens it. The medial and lateral pterygoids and the masseter muscles accomplish protraction and lateral and medial excursion of the jaw. The temporalis retracts the jaw. All these movements are involved in tearing, crushing, and grinding food.

The **chewing,** or **mastication, reflex,** which is integrated in the medulla oblongata, controls the basic movements involved in chewing. The presence of food in the mouth stimulates sensory receptors, which activate a reflex that causes the muscles of mastication to relax. The muscles are stretched as the mandible is lowered, and stretch of the muscles activates a reflex that causes contraction of the muscles of mastication. Once the mouth is closed, the food again stimulates the muscles of mastication to relax, and the cycle is repeated. Descending pathways from the cerebrum strongly influence the activity of the mastication reflex so that chewing can be initiated or stopped consciously. The rate and intensity of chewing movements can also be influenced by the cerebrum.

Salivary Glands

A considerable number of **salivary glands** are scattered throughout the oral cavity. Three pairs of large multicellular glands exist: the parotid, the submandibular, and the sublingual glands (figure 24.9). In addition to these large consolidations of glandular tissue, numerous small, coiled tubular glands are located deep to the epithelium of the tongue (lingual glands), palate (palatine glands), cheeks (buccal glands), and lips (labial glands). The secretions from these glands help keep the oral cavity moist and begin the process of digestion. They also suspend molecules in solution so they can be tasted.

All of the major large salivary glands are compound **alveolar glands,** which are branching glands with clusters of alveoli resembling grapes (see chapter 4). They produce thin serous secretions or thicker mucous secretions. Thus, saliva is a combination of serous and mucous secretions from the various salivary glands.

The largest salivary glands, the **parotid** (pă-rot′id; beside the ear) **glands,** are serous glands, which produce mostly watery saliva, and are located just anterior to the ear on each side of the head. Each **parotid duct** exits the gland on its anterior margin, crosses the lateral surface of the masseter muscle, pierces the buccinator muscle, and enters the oral cavity adjacent to the second upper molar (see figure 24.9).

Saliva and the Second Molar

Because the parotid secretions are released directly onto the surface of the second upper molar, it tends to have a considerable accumulation of mineral, secreted from the gland, on its surface.

Mumps

Inflammation of the parotid gland is called parotiditis. **Mumps,** which is caused by a virus, is the most common type of parotiditis.

The **submandibular** (below the mandible) **glands** are mixed glands with more serous than mucous alveoli. Each gland can be felt as a soft lump along the inferior border of the posterior half of the mandible. A submandibular duct exits each gland, passes anteriorly deep to the mucous membrane on the floor of the oral cavity, and opens into the oral cavity beside the frenulum of the tongue (see figure 24.6). In certain people, if the mouth is opened and the tip of the tongue is elevated, the submandibular ducts are compressed and saliva may squirt out of the mouth from the openings of these ducts.

The **sublingual** (below the tongue) **glands,** the smallest of the three large, paired salivary glands, are mixed glands containing some serous alveoli but consisting primarily of mucous alveoli. They lie immediately below the mucous membrane in the floor of the mouth. These glands do not have single, well-defined ducts like those of the submandibular and parotid glands. Instead, each sublingual gland opens into the floor of the oral cavity through 10–12 small ducts.

(a)

(b)

Figure 24.9   Salivary Glands

(*a*) The large salivary glands are the parotid glands, the submandibular glands, and the sublingual glands. The parotid duct extends anteriorly from the parotid gland. (*b*) An idealized schematic drawing of the histology of the large salivary glands. The figure is representative of all the glands and does not depict any one salivary gland.
(*c*) Photomicrograph of the parotid gland.   (c)

Saliva is secreted at the rate of about 1–1.5 L/day. The serous part of saliva, produced mainly by the parotid and submandibular glands, contains a digestive enzyme called **salivary amylase** (am′il-ās; starch-splitting enzyme), which breaks the covalent bonds between glucose molecules in starch and other polysaccharides to produce the disaccharides maltose and isomaltose (tables 24.2 and 24.4). The release of maltose and isomaltose gives starches a sweet taste in the mouth. Food spends very little time in the mouth, however; therefore, only about 3%–5% of the total carbohydrates are digested in the mouth. Most of the starches are covered by cellulose in plant tissues and are inaccessible to salivary amylase. Cooking and thorough chewing of food destroy the cellulose covering and increase the efficiency of the digestive process.

Saliva prevents bacterial infection in the mouth by washing the oral cavity. Saliva also contains substances, such as **lysozyme,** which has a weak antibacterial action, and immunoglobulin A, which helps prevent bacterial infection. Any lack of salivary gland secretion increases the chance of ulceration and infection of the oral mucosa and of caries in the teeth.

The mucous secretions of the submandibular and sublingual glands contain a large amount of **mucin** (mū′sin), a proteoglycan that gives a lubricating quality to the secretions of the salivary glands.

Salivary gland secretion is stimulated by the parasympathetic and sympathetic nervous systems, with the parasympathetic system being more important. Salivary nuclei in the brainstem increase salivary secretions by sending action potentials through parasympathetic fibers of the facial (VII) and glossopharyngeal (IX) cranial nerves in response to a variety of stimuli, such as tactile stimulation in the oral cavity or certain tastes, especially sour. Higher centers of the brain also affect the activity of the salivary glands. Odors that trigger thoughts of food or the sensation of hunger can increase salivary secretions.

Table 24.2 Functions of Major Digestive Secretions

Fluid or Enzyme	Function
Oral Cavity Secretions	
Serous (watery) saliva	Moistens food and mucous membrane; lysozyme kills bacteria
Salivary amylase	Starch digestion (conversion to maltose and isomaltose)
Mucus	Lubricates food; protects gastrointestinal tract from digestion by enzymes
Lingual lipase	Digests a small amount (<10%) of lipids
Esophagus Secretions	
Mucus	Lubricates the esophagus; protects the esophagus lining from abrasion and allows food to move more smoothly through the esophagus
Gastric Secretions	
Hydrochloric acid	Decreases stomach pH to activate pepsinogen
Pepsinogen	Pepsin, the active form of pepsinogen, digests protein into smaller peptide chains
Mucus	Protects stomach lining from digestion
Gastric amylase and gelatinase	Digest a minor amount of carbohydrates
Gastric lipase	Digests a minor amount of lipid
Liver Secretions	

Bile
  Sodium glycocholate (bile salt)
  Sodium taurocholate (bile salt)
  Cholesterol
  Biliverdin
  Bilirubin
  Mucus
  Fat
  Lecithin
  Cells and cell debris
	Bile salts emulsify fats, making them available to intestinal lipases; help make end products soluble and available for absorption by the intestinal mucosa; aid peristalsis. Many of the other bile contents are waste products transported to the intestine for disposal.

Pancreas Secretions	
Trypsin	Digests proteins (breaks polypeptide chains at arginine or lysine residues)
Chymotrypsin	Digests proteins (cleaves carboxyl links of hydrophobic amino acids)
Carboxypeptidase	Digests proteins (removes amino acids from the carboxyl end of peptide chains)
Pancreatic amylase	Digests carbohydrates (hydrolyzes starches and glycogen to form maltose and isomaltose)
Pancreatic lipase	Digests fat (breaks down lipids into monoglycerides and free fatty acids)
Ribonuclease	Digests ribonucleic acid
Deoxyribonuclease	Digests deoxyribonucleic acid (hydrolyzes phosphodiester bonds)
Cholesterol esterase	Hydrolyzes cholesterol esters to form cholesterol and free fatty acids
Bicarbonate ions	Provides appropriate pH for pancreatic enzymes
Small Intestine Secretions	
Mucus	Protects duodenum from stomach acid, gastric enzymes, and intestinal enzymes; provides adhesion for fecal matter; protects intestinal wall from bacterial action and acid produced in the feces
Aminopeptidase	Splits polypeptides into amino acids (from amino end of chain)
Peptidase	Splits amino acids from polypeptides
Enterokinase	Activates trypsin from trypsinogen
Amylase	Digests carbohydrates
Sucrase	Splits sucrose into glucose and fructose
Maltase	Splits maltose into two glucose molecules
Isomaltase	Splits isomaltose into two glucose molecules
Lactase	Splits lactose into glucose and galactose
Lipase	Splits fats into monoglycerides and fatty acids

14. *List the muscles of mastication and the actions they produce. Describe the mastication reflex.*
15. *Name and give the location of the three largest salivary glands. Name the other kinds of salivary glands.*
16. *What substances are contained in saliva?*
17. *What is the difference between serous and mucous saliva?*

Pharynx

The **pharynx** was described in detail in chapter 23; thus, only a brief description is provided here. The pharynx consists of three parts: the nasopharynx, the oropharynx, and the laryngopharynx. Normally, only the oropharynx and laryngopharynx transmit food. The **oropharynx** communicates with the nasopharynx superiorly, the larynx and **laryngopharynx** inferiorly, and the mouth anteriorly. The laryngopharynx extends from the oropharynx to the esophagus and is posterior to the larynx. The epiglottis covers the opening of the larynx and keeps food and drink from entering the larynx. The posterior walls of the oropharynx and laryngopharynx consist of three muscles: the superior, middle, and inferior **pharyngeal constrictors,** which are arranged like three stacked flowerpots, one inside the other. The oropharynx and the laryngopharynx are lined with moist stratified squamous epithelium, and the nasopharynx is lined with ciliated pseudostratified columnar epithelium.

18. *Name the three parts of the pharynx. What are the pharyngeal constrictors?*

PREDICT ②

Explain the functional significance of the differences in epithelial types among the three pharyngeal regions.

Esophagus

The **esophagus** is that part of the digestive tube that extends between the pharynx and the stomach. It is about 25 cm long and lies in the mediastinum, anterior to the vertebrae and posterior to the trachea. It passes through the esophageal hiatus (opening) of the diaphragm and ends at the stomach. The esophagus transports food from the pharynx to the stomach.

Hiatal Hernia

A **hiatal hernia** is a widening of the esophageal hiatus. Hiatal hernias occur most commonly in adults and allow part of the stomach to extend through the opening into the thorax. The hernia can decrease the resting pressure in the lower esophageal sphincter, allowing gastroesophageal reflux and subsequent esophagitis to occur. Hiatal herniation can also compress the blood vessels in the stomach mucosa, which can lead to gastritis or ulcer formation. Esophagitis, gastritis, and ulceration can be very painful.

The esophagus has thick walls consisting of the four tunics common to the digestive tract: mucosa, submucosa, muscularis,

and adventitia. The muscular tunic has an outer longitudinal layer and an inner circular layer, as is true of most parts of the digestive tract, but it's different because it consists of skeletal muscle in the superior part of the esophagus and smooth muscle in the inferior part. An **upper esophageal sphincter** and a **lower esophageal sphincter,** at the upper and lower ends of the esophagus, respectively, regulate the movement of materials into and out of the esophagus. The mucosal lining of the esophagus is moist stratified squamous epithelium. Numerous mucous glands in the submucosal layer produce a thick, lubricating mucus that passes through ducts to the surface of the esophageal mucosa.

19. *Where is the esophagus located? Describe the layers of the esophageal wall and the esophageal sphincters.*

Swallowing

Swallowing, or **deglutition,** is divided into three separate phases: voluntary, pharyngeal, and esophageal. During the **voluntary phase** (figure 24.10, *step 1*), a bolus of food is formed in the mouth and pushed by the tongue against the hard palate, forcing the bolus toward the posterior part of the mouth and into the oropharynx.

The **pharyngeal phase** (see figure 24.10, *steps 2–4*) of swallowing is a reflex that is initiated by stimulation of tactile receptors in the area of the oropharynx. Afferent action potentials travel through the trigeminal (V) and glossopharyngeal (IX) nerves to the **swallowing center** in the medulla oblongata. There, they initiate action potentials in motor neurons, which pass through the trigeminal (V), glossopharyngeal (IX), vagus (X), and accessory (XI) nerves to the soft palate and pharynx. This phase of swallowing begins with the elevation of the soft palate, which closes the passage between the nasopharynx and oropharynx. The pharynx elevates to receive the bolus of food from the mouth and moves the bolus down the pharynx into the esophagus. The superior, middle, and inferior **pharyngeal constrictor muscles** contract in succession, forcing the food through the pharynx. At the same time, the upper esophageal sphincter relaxes, the elevated pharynx opens the esophagus, and food is pushed into the esophagus. This phase of swallowing is unconscious and is controlled automatically, even though the muscles involved are skeletal. The pharyngeal phase of swallowing lasts about 1–2 seconds.

PREDICT

Why is it important to close the opening between the nasopharynx and oropharynx during swallowing? What may happen if a person has an explosive burst of laughter while trying to swallow a liquid?

During the pharyngeal phase, the vestibular folds are moved medially, the **epiglottis** (ep-i-glot′is; on the glottis) is tipped posteriorly so that the epiglottic cartilage covers the opening into the larynx, and the larynx is elevated. These movements of the larynx prevent food from passing through the opening into the larynx.

PREDICT

What happens if you try to swallow and speak at the same time?

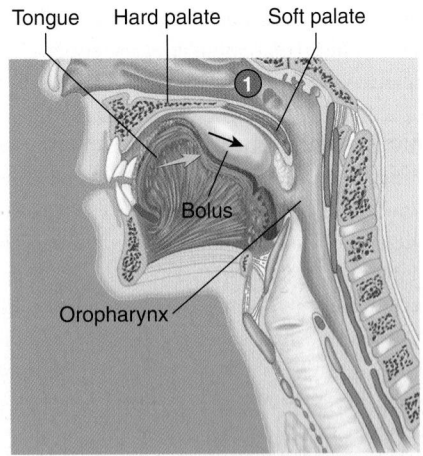

Tongue Hard palate Soft palate

Bolus

Oropharynx

1. During the voluntary phase, a bolus of food (*yellow*) is pushed by the tongue against the hard and soft palates and posteriorly toward the oropharynx (*blue arrow* indicates tongue movement; *black arrow* indicates movement of the bolus). *Tan*: bone, *purple*: cartilage, *red*: muscle.

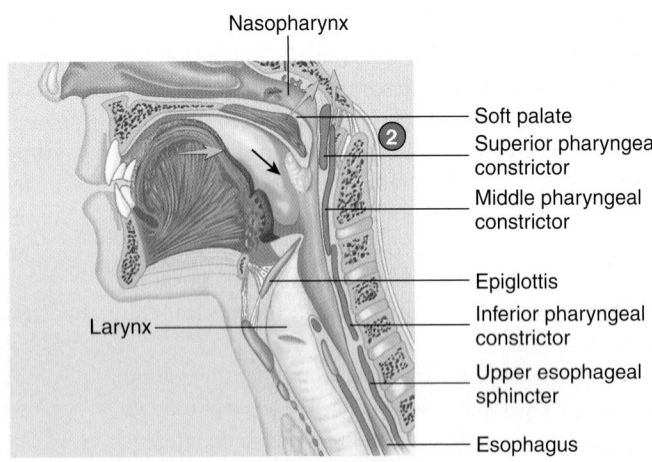

Nasopharynx

Soft palate
Superior pharyngeal constrictor
Middle pharyngeal constrictor
Epiglottis
Inferior pharyngeal constrictor
Upper esophageal sphincter
Esophagus

Larynx

2. During the pharyngeal phase, the soft palate is elevated, closing off the nasopharynx. The pharynx and larynx are elevated (*blue arrows* indicate muscle movement).

Superior pharyngeal constrictor

Middle pharyngeal constrictor

Epiglottis
Opening of larynx

3. Successive constriction of the pharyngeal constrictors from superior to inferior (*blue arrows*) forces the bolus through the pharynx and into the esophagus. As this occurs, the epiglottis is bent down over the opening of the larynx largely by the force of the bolus pressing against it.

Inferior pharyngeal constrictor
Upper esophageal sphincter
Esophagus

4. As the inferior pharyngeal constrictor contracts, the upper esophageal sphincter relaxes (outwardly directed *blue arrows*), allowing the bolus to enter the esophagus.

Esophagus

5. During the esophageal phase, the bolus is moved by peristaltic contractions of the esophagus toward the stomach (inwardly directed *blue arrows*).

Process Figure 24.10   **Three Phases of Swallowing (Deglutition)**

The **esophageal phase** (figure 24.10, *step 5*) of swallowing takes about 5–8 seconds and is responsible for moving food from the pharynx to the stomach. Muscular contractions in the wall of the esophagus occur in peristaltic waves.

The peristaltic waves associated with swallowing cause relaxation of the lower esophageal sphincter in the esophagus as the peristaltic waves, and bolus of food, approach the stomach. This sphincter is not anatomically distinct from the rest of the esophagus, but it can be identified physiologically because it remains tonically constricted to prevent the reflux of stomach contents into the lower part of the esophagus.

The presence of food in the esophagus stimulates the enteric plexus, which controls the peristaltic waves. The presence of food in the esophagus also stimulates tactile receptors, which send afferent impulses to the medulla oblongata through the vagus nerves. Motor impulses, in turn, pass along the vagal efferent fibers to the striated and smooth muscles within the esophagus, thereby stimulating their contractions and reinforcing the peristaltic contractions.

Swallowing and Gravity

Gravity assists the movement of material through the esophagus, especially when liquids are swallowed. The peristaltic contractions that move material through the esophagus are sufficiently forceful, however, to allow a person to swallow, even while doing a headstand or floating in the zero-gravity environment of space.

> **20.** *What are the three phases of deglutition? List sequentially the processes involved in the last two phases, and describe how they are regulated.*

Stomach

The **stomach** is an enlarged segment of the digestive tract in the left superior part of the abdomen (see figure 24.1). Its shape and size vary from person to person; even within the same individual its size and shape change from time to time, depending on its food content and the posture of the body. Nonetheless, several general anatomic features can be described.

Anatomy of the Stomach

The opening from the esophagus into the stomach is the **gastro-esophageal,** or **cardiac** (located near the heart), **opening,** and the region of the stomach around the cardiac opening is the **cardiac region** (figure 24.11). The lower esophageal sphincter, also called the **cardiac sphincter,** surrounds the cardiac opening. Recall that although this is an important structure in the normal function of the stomach, it is a physiologic constrictor only and cannot be seen anatomically. A part of the stomach to the left of the cardiac region, the **fundus** (fŭn′dŭs; the bottom of a round-bottomed leather bottle), is actually superior to the cardiac opening. The largest part of the stomach is the **body,** which turns to the right, thus creating a **greater curvature** and a **lesser curvature.** The body narrows to form the funnel-shaped **pyloric part** of the stomach. The wider part of the funnel, toward the body of the stomach, is the **pyloric antrum.** The narrow part of the funnel is the **pyloric canal.** The pyloric canal opens through the pyloric

orifice into the small intestine. The pyloric orifice is surrounded by the **pylorus** (pī-lōr′us; gatekeeper) a relatively thick ring of smooth muscle, also called the **pyloric sphincter.**

Hypertrophic Pyloric Stenosis

Hypertrophic pyloric stenosis is a common defect of the stomach in infants, occurring in 1 in 150 males and 1 in 750 females, in which the pylorus is greatly thickened, resulting in interference with normal stomach emptying. Infants with this defect develop symptoms 3 to 10 weeks after birth and often exhibit projectile vomiting. Because the pylorus is blocked, little food enters the intestine, and the infant fails to gain weight. Constipation is also a frequent complication. The defect can be corrected with surgery in which the pyloric orifice is enlarged by cutting through the mucosa and sphincter.

Histology of the Stomach

The **serosa,** or visceral peritoneum, is the outermost layer of the stomach. It consists of an inner layer of connective tissue and an outer layer of simple squamous epithelium. The **muscularis** of the stomach consists of three layers: an outer longitudinal layer, a middle circular layer, and an inner oblique layer (figure 24.11*a*). In some areas of the stomach, such as in the fundus, the three layers blend with one another and cannot be separated. Deep to the muscular layer are the submucosa and the mucosa, which are thrown into large folds called **rugae** (roo′gē; wrinkles) when the stomach is empty. These folds allow the mucosa and submucosa to stretch, and the folds disappear as the stomach volume increases as it is filled.

The stomach is lined with simple columnar epithelium. The epithelium forms numerous tubelike **gastric pits,** which are the openings for the **gastric glands** (figure 24.11*b*). The epithelial cells of the stomach are of five types. The first type, **surface mucous cells,** which produce mucus, is on the surface and lines the gastric pit. The remaining four cell types are in the gastric glands. They are **mucous neck cells,** which produce mucus; **parietal** (oxyntic) **cells,** which produce hydrochloric acid and intrinsic factor; **chief** (zymogenic) **cells,** which produce pepsinogen; and **endocrine cells,** which produce regulatory hormones. The mucous neck cells are located near the openings of the glands; whereas the parietal, chief, and endocrine cells are interspersed in the deeper parts of the glands. There are several types of endocrine cells. Enterochromaffin-like cells produce histamine, which stimulates acid secretion by parietal cells. Gastrin-containing cells secrete gastrin, and somatostatin-containing cells secrete somatostatin, which inhibits gastrin and insulin secretion.

Surface mucous cells of the stomach protect the wall of the stomach from being damaged by the acid and digestive enzymes in the stomach. These cells produce an alkaline mucus on their surface that neutralizes the acid and is a barrier to the digestive enzymes. The surface mucous cells are connected by tight junctions, which provide an additional barrier that prevents acids and enzymes from reaching deeper tissues. In addition, when surface mucous cells are damaged, they are rapidly replaced.

> **21.** *Describe the parts of the stomach. List the layers of the stomach wall. How is the stomach different from the esophagus?*
>
> **22.** *What are gastric pits and gastric glands? Name the different cell types in the stomach and the secretions they produce.*

Figure 24.11 Anatomy and Histology of the Stomach

(*a*) Cutaway section reveals muscular layers and internal anatomy. (*b*) A section of the stomach wall that illustrates its histology, including several gastric pits and glands. (*c*) Photomicrograph of gastric glands.

Secretions of the Stomach

Ingested food and stomach secretions, mixed together, form a semifluid material called **chyme** (kīm; juice). The stomach functions primarily as a storage and mixing chamber for the chyme. Although some digestion and absorption occur in the stomach, they are not its major functions.

Stomach secretions include mucus, hydrochloric acid, gastrin, histamine, intrinsic factor, and pepsinogen. Pepsinogen is the inactive form of the protein-digesting enzyme pepsin.

The surface mucous cells and mucous neck cells secrete a viscous and alkaline **mucus** that covers the surface of the epithelial cells and forms a layer 1–1.5 mm thick. The thick layer of mucus lubricates and protects the epithelial cells of the stomach wall from the damaging effect of the acidic chyme and pepsin. Irritation of the stomach mucosa results in stimulation of the secretion of a greater volume of mucus.

Parietal cells in the gastric glands of the pyloric region secrete intrinsic factor and a concentrated solution of hydrochloric acid. **Intrinsic factor** is a glycoprotein that binds with vitamin B_{12} and makes the vitamin more readily absorbed in the ileum. Vitamin B_{12} is important in deoxyribonucleic acid (DNA) synthesis.

Hydrochloric acid produces the low pH of the stomach, which is normally between 1 and 3. Although the hydrochloric acid secreted into the stomach has a minor digestive effect on ingested food, one of its main functions is to kill bacteria that are ingested with essentially everything humans put into their mouths. Some pathogenic bacteria may avoid digestion in the stomach, however, because they have an outer coat that resists stomach acids.

The low pH of the stomach also stops carbohydrate digestion by inactivating salivary amylase. Stomach acid also denatures many proteins so that proteolytic enzymes can reach internal peptide bonds, and it provides the proper pH environment for the function of pepsin.

Hydrogen ions are derived from carbon dioxide and water, which enter the parietal cell from its serosal surface, which is the side opposite the lumen of the gastric pit (figure 24.12). Once inside

1. Carbon dioxide (CO_2) diffuses into the cell.

2. Carbon dioxide combines with water (H_2O) in an enzymatic reaction that is catalyzed by carbonic anhydrase (CA) to form carbonic acid (H_2CO_3).

3. Carbonic acid dissociates into a bicarbonate ion (HCO_3^-) and a hydrogen ion (H^+).

4. Bicarbonate ions are transported back into the bloodstream. An anion exchange molecule in the plasma membrane exchanges HCO_3^- for a chloride ion (Cl^-) (counter transport).

5. Hydrogen ions are actively transported into the duct of the gastric gland.

6. Chloride ions diffuse with the charged H^+.

7. Some potassium ions (K^+) are counter transported into the cell in exchange for H^+.

Process Figure 24.12   Hydrochloric Acid Production by Parietal Cells in the Gastric Glands of the Stomach the cell, carbonic anhydrase catalyzes the reaction between carbon dioxide and water to form carbonic acid. Some of the carbonic acid molecules then dissociate to form hydrogen ions and bicarbonate ions. The hydrogen ions are actively transported across the mucosal surface of the parietal cell into the lumen of the stomach; some potassium ions are moved into the cell in exchange for the hydrogen ions. Although hydrogen ions are actively transported against a steep concentration gradient, chloride ions diffuse with the hydrogen ions from the cell through the plasma membrane. Diffusion of chloride ions with the positively charged hydrogen ions reduces the amount of energy needed to transport the hydrogen ions against both a concentration gradient and an electrical gradient. Bicarbonate ions move down their concentration gradient from the parietal cell into the extracellular fluid. During this process, bicarbonate ions are exchanged for chloride ions through an anion exchange molecule, which is located in the plasma membrane, and the chloride ions subsequently move into the cell.

P R E D I C T

Explain why a slight increase in the blood pH may occur following a heavy meal. The elevated pH of blood, especially in the veins that carry blood away from the stomach, is called "the postenteric alkaline tide."

Chief cells within the gastric glands secrete **pepsinogen** (pep-sin′ō-jen). Pepsinogen is packaged in **zymogen** (zī-mō-jen; related to enzymes) **granules,** which are released by exocytosis when pepsinogen secretion is stimulated. Once pepsinogen enters the lumen of the stomach, hydrochloric acid and previously formed pepsin molecules convert it to **pepsin.** Pepsin exhibits optimum enzymatic activity at a pH of 3 or less. Pepsin catalyzes the cleavage of some covalent bonds in proteins, thus breaking them into smaller peptide chains.

Heartburn

Heartburn, or **pyrosis** (pī-rō′sis), is a painful or burning sensation in the chest usually associated with reflux of acidic chyme into the esophagus. The pain is usually short-lived but may be confused with the pain of an ulcer or a heart attack. Overeating, eating fatty foods, lying down immediately after a meal, consuming too much alcohol or caffeine, smoking, or wearing extremely tight clothing can all cause heartburn. A hiatal hernia can also cause heartburn, especially in older people.

Regulation of Stomach Secretion

Approximately 2–3 L of gastric secretions (gastric juice) are produced each day. The amount and type of food entering the stomach and intestine dramatically affects the secretion amount, but up to 700 mL is secreted as a result of a typical meal. Both nervous and hormonal mechanisms regulate gastric secretions. The neural mechanisms involve reflexes integrated within the medulla oblongata and local reflexes integrated within the enteric plexus of the GI tract. In addition, higher brain centers influence the reflexes. Chemical signals that regulate stomach secretions include the hormones gastrin, secretin, gastric inhibitory polypeptide, and cholecystokinin (table 24.3), as well as the paracrine chemical signal histamine.

Regulation of stomach secretion is divided into three phases: cephalic, gastric, and gastrointestinal.

Table 24.3	Functions of the Gastrointestinal Hormones		
Site of Production	**Method of Stimulation**	**Secretory Effects**	**Motility Effects**
Gastrin			
Stomach and duodenum	Distention; partially digested proteins, autonomic stimulation, ingestion of alcohol or caffeine	Increases gastric secretion	Increases gastric emptying by increasing stomach motility and relaxing the pyloric sphincter
Secretin			
Duodenum	Acidity of chyme	Inhibits gastric secretion; stimulates pancreatic secretions high in bicarbonate ions; increases the rate of bile and increases intestinal secretion; mucus secretion	Decreases gastric motility
Cholecystokinin			
Intestine	Fatty acids and other lipids	Slightly inhibits gastric secretion; stimulates pancreatic secretions high in digestive enzymes; and causes contraction of the gallbladder and relaxation of the hepatopancreatic ampullar sphincter	Decreases gastric motility
Gastric Inhibitory Polypeptide			
Duodenum and proximal jejunum	Fatty acids and other lipids	Inhibits gastric secretions	Decreases gastric motility

1. **Cephalic phase.** In the cephalic phase of gastric regulation, the sensations of the taste and smell of food, stimulation of tactile receptors during the process of chewing and swallowing, and pleasant thoughts of food stimulate centers within the medulla oblongata that influence gastric secretions (figure 24.13a). Action potentials are sent from the medulla along parasympathetic neurons within the vagus (X) nerves to the stomach. Within the stomach wall, the preganglionic neurons stimulate postganglionic neurons in the enteric plexus. The postganglionic neurons, which are primarily cholinergic, stimulate secretory activity in the cells of the stomach mucosa.

   Parasympathetic stimulation of the stomach mucosa results in the release of the neurotransmitter acetylcholine, which increases the secretory activity of both the parietal and chief cells and stimulates the secretion of **gastrin** (gas′trin) and histamine from endocrine cells. Gastrin is released into the circulation and travels to the parietal cells, where it stimulates additional hydrochloric acid and pepsinogen secretion. In addition, gastrin stimulates enterochromaffin-like cells to release histamine, which stimulates parietal cells to secrete hydrochloric acid. Acetylcholine, histamine, and gastrin working together cause a greater secretion of hydrochloric acid than any of them does separately. Of the three, histamine has the greatest stimulatory effect.

Inhibitors of Gastric Acid Secretion

Histamine receptors on parietal cells are called H_2 receptors, and are different from the H_1 receptors involved in allergic reactions. Drugs that block allergic reactions do not affect histamine-mediated stomach acid secretion and vice versa. Cimetidine (Tagamet) and ranitidine (Zantac) are histamine receptor antagonists on parietal cells, and prevent histamine binding. These chemicals are extremely effective inhibitors of gastric acid secretion. Cimetidine, one of the most commonly prescribed drugs, is used to treat cases of gastric acid hypersecretion associated with gastritis and gastric ulcers.

---

2. **Gastric phase.** The greatest volume of gastric secretions is produced during the gastric phase of gastric regulation. The presence of food in the stomach initiates the gastric phase (figure 24.13b). The primary stimuli are distention of the stomach and the presence of amino acids and peptides in the stomach.

   Distention of the stomach wall, especially in the body or fundus, results in the stimulation of mechanoreceptors. Action potentials generated by these receptors initiate reflexes that involve both the CNS and enteric reflexes, resulting in secretion of mucus, hydrochloric acid, pepsinogen, intrinsic factor, and gastrin. The presence of partially digested proteins or moderate amounts of alcohol or caffeine in the stomach also stimulates gastrin secretion.

   When the pH of the stomach contents falls below 2, increased gastric secretion produced by distention of the stomach is blocked. This negative-feedback mechanism limits the secretion of gastric juice.

Amino acids and peptides released by the digestive action of pepsin on proteins directly stimulate parietal cells of the stomach to secrete hydrochloric acid. The mechanism by which this response is mediated is not clearly understood. It doesn't involve known neurotransmitters, and, when the pH drops below 2, the response is inhibited.

3. **Gastrointestinal phase.** The entrance of acidic stomach contents into the duodenum of the small intestine controls the gastrointestinal phase of gastric regulation (figure 24.13c). The presence of chyme in the duodenum activates both neural and hormonal mechanisms. When the pH of the chyme entering the duodenum drops to 2 or below, or if the chyme contains fat digestion products, gastric secretions are inhibited.

   Acidic solutions in the duodenum cause the release of the hormone **secretin** (se-krē′tin) into the circulatory system. Secretin inhibits gastric secretion by inhibiting both parietal and chief cells.

   Fatty acids and certain other lipids in the duodenum and the proximal jejunum initiate the release of two hormones: **gastric inhibitory polypeptide** and **cholecystokinin** (kō′lē-sis-tō-kī′nin). Gastric inhibitory polypeptide strongly inhibits gastric secretion, and cholecystokinin inhibits gastric secretions to a lesser degree. Hypertonic solutions in the duodenum and jejunum also inhibit gastric secretions. The mechanism appears to involve the secretion of a hormone referred to as **enterogastrone** (en′ter-ō-gas′trōn), but the actual existence of this hormone has never been established.

   Inhibition of gastric secretions is also under nervous control. Distention of the duodenal wall, the presence of irritating substances in the duodenum, reduced pH, and hypertonic or hypotonic solutions in the duodenum activate the enterogastric reflex. The **enterogastric reflex** consists of a local reflex and a reflex integrated within the medulla oblongata. It reduces gastric secretions.

Movements of the Stomach

Stomach Filling

As food enters the stomach, the rugae flatten, and the stomach volume increases. Despite the increase in volume, the pressure within the stomach doesn't increase until the volume nears maximum capacity because smooth muscle can stretch without an increase in tension (see chapter 9) and because of a reflex integrated within the medulla oblongata. This reflex inhibits muscle tone in the body of the stomach.

Mixing of Stomach Contents

Ingested food is thoroughly mixed with the secretions of the stomach glands to form chyme. This mixing is accomplished by gentle **mixing waves,** which are peristaltic-like contractions that occur about every 20 seconds and proceed from the body toward the pyloric sphincter to mix the ingested material with the secretions of the stomach. **Peristaltic waves** occur less frequently, are significantly more powerful than mixing waves, and force the chyme near the periphery of the stomach toward the pyloric

Cephalic Phase

1. The taste, smell, or thought of food or tactile sensations of food in the mouth stimulate the medulla oblongata (*green arrow*).

2. Parasympathetic action potentials are carried by the vagus nerves to the stomach (*pink arrow*).

3. Preganglionic parasympathetic vagus nerve fibers stimulate postganglionic neurons in the enteric plexus of the stomach.

4. Postganglionic neurons stimulate secretion by parietal and chief cells and stimulate gastrin secretion by endocrine cells.

5. Gastrin is carried through the circulation back to the stomach (*purple arrow*), where it stimulates secretion by parietal and chief cells.

(a)

Gastric Phase

1. Distention of the stomach stimulates mechanoreceptors (stretch receptors) and activates a parasympathetic reflex. Action potentials generated by the mechanoreceptors are carried by the vagus nerves to the medulla oblongata (*green arrow*).

2. The medulla oblongata increases action potentials in the vagus nerves that stimulate stomach secretions (*pink arrow*).

3. Distention of the stomach also activates local reflexes that increase stomach secretions (*purple arrow*).

(b)

Gastrointestinal Phase

1. Chyme in the duodenum with a pH less than 2 or containing fat digestion products (lipids) inhibits gastric secretions by three mechanisms (2–4).

2. Chemoreceptors in the duodenum are stimulated by H+ (low pH) or lipids. Action potentials generated by the chemoreceptors are carried by the vagus nerves to the medulla oblongata (*green arrow*), where they inhibit parasympathetic action potentials (*pink arrow*).

3. Local reflexes activated by H+ or lipids also inhibit gastric secretion (*orange arrows*).

4. Secretin, gastric inhibitory polypeptide, and cholecystokinin produced by the duodenum (*brown arrows*) inhibit gastric secretions in the stomach.

(c)

Process Figure 24.13   The Three Phases of Stomach Secretion

(*a*) Cephalic phase. (*b*) Gastric phase. (*c*) Gastrointestinal phase.

Clinical Focus Peptic Ulcer

Approximately 10% of the U.S. population will develop peptic ulcers during their lifetime. Most cases of peptic ulcer are apparently due to the infection of a specific bacterium, *Helicobacter pylori*. It's also thought that the bacterium is involved in many cases of gastritis and gastric cancer. Conventional wisdom has focused for years on the notion that stress, diet, smoking, or alcohol cause excess acid secretion in the stomach, resulting in ulcers.

Antacids remain very popular in treating ulcers, as well as for the relief of temporary stomach problems. Close to $1 billion is spent on antacids in the United States annually. Antacid therapy does relieve the ulcer in most cases. A 50% incidence of relapse occurs within 6 months with antacid treatment, and a 95% incidence of relapse occurs after 2 years. On the other hand, studies using antibiotic therapy in addition to bismuth and ranitidine have demonstrated a 95% eradication of gastric ulcers and 74% healing of duodenal ulcers within 2 months. Dramatically reduced relapse rates have also been obtained. One such study reported a recurrence rate of 8% following antibiotic therapy, compared with a recurrence rate of 86% in controls.

Other treatments include H₂ receptor antagonists, which bind histamine receptors and prevent histamine-stimulated HCl secretion. Proton pump inhibitors directly inhibit HCl secretion. Prostaglandins are naturally produced by the mucosa of the GI tract and help the mucosa resist injury. Synthetic prostaglandins can supplement this resistance as well as inhibit HCl secretion.

The infection rate from *H. pylori* in the United States population is about 1% per year of age: 30% of people that are 30 years old have the bacterium, and 80% of those age 80 are infected. In developing countries, as many as 100% of people age 25 or older are infected. This may relate to the high rates of stomach cancer in some of those countries. We still have much to learn before we can understand this bacterium. Very little is known concerning how people become infected. Also, with such high rates of infection, it's not known why only a small fraction of those infected actually develop ulcers. It may be that factors such as diet and stress predispose a person who is infected by the bacterium to actually develop an ulcer.

Peptic ulcer is classically viewed as a condition in which the stomach acids and pepsin digest the mucosal lining of the GI tract itself. The most common site of a peptic ulcer is near the pylorus, usually on the duodenal side (i.e., a duodenal ulcer; 80% of peptic ulcers are duodenal). Ulcers occur less frequently along the lesser curvature of the stomach or at the point at which the esophagus enters the stomach. The most common presumed cause of peptic ulcers is the oversecretion of gastric juice relative to the degree of mucous and alkaline protection of the small intestine. One reason that bacterial involvement in ulcers was dismissed for such a long time is that it was assumed that the extreme acid environment killed all bacteria. Apparently not only can *H. pylori* survive in such an environment, but it may even thrive there.

People experiencing severe anxiety for a long time are the most prone to develop duodenal ulcers. They often have a high rate of gastric secretion (as much as 15 times the normal amount) between meals.

This secretion results in highly acidic chyme entering the duodenum. The duodenum is usually protected by sodium bicarbonate (secreted mainly by the pancreas), which neutralizes the chyme. When large amounts of acid enter the duodenum, however, the sodium bicarbonate is not adequate to neutralize it. The acid tends to reduce the mucous protection of the duodenum, perhaps leaving that part of the digestive tract open to the action of *H. pylori*, which may further destroy the mucous lining.

In one study, it was determined that ulcer patients prefer their hot drinks extra hot, 62°C compared with 56°C for a control group without ulcers. The high temperatures of the drinks may cause thinning of the mucous lining of the stomach, thus making these people more susceptible to ulcers, again perhaps by increasing their sensitivity to *H. pylori* invasion.

In some patients with gastric ulcers, often normal or even low levels of gastric hydrochloric acid secretion exist. The stomach has a reduced resistance to its own acid, however. Such inhibited resistance can result from excessive ingestion of alcohol or aspirin.

Reflux of duodenal contents into the pylorus can also cause gastric ulcers. In this case, bile, which is present in the reflux, has a detergent effect that reduces gastric mucosal resistance to acid and bacteria.

An ulcer may become perforated (a hole in the stomach or duodenum), causing peritonitis. The perforation must be corrected surgically. Selective vagotomy, cutting branches of the vagus (X) nerve going to the stomach, is sometimes performed at the time of surgery to reduce acid production in the stomach.

sphincter. The more solid material near the center of the stomach is pushed superiorly toward the cardiac region for further digestion (figure 24.14). Roughly 80% of the contractions are mixing waves, and 20% are peristaltic waves.

Stomach Emptying

The amount of time food remains in the stomach depends on a number of factors, including the type and volume of food. Liquids exit the stomach within 1½–2½ hours after ingestion. After a typical meal, the stomach is usually empty within 3–4 hours. The pyloric sphincter usually remains partially closed because of mild tonic contraction. Each peristaltic contraction is sufficiently strong to force a small amount of chyme through the pyloric opening and into the duodenum. The peristaltic contractions responsible for movement of chyme through the partially closed pyloric opening are called the **pyloric pump.**

1. A mixing wave initiated in the body of the stomach progresses toward the pylorus (*pink arrows directed inward*).

2. The more fluid part of the chyme is pushed toward the pylorus (*blue arrows*) as the mixing wave passes over the more solid center of the chyme. As a result, the more solid center of the chyme is pushed past the constriction back toward the body of the stomach (*orange arrow*).

3. Peristaltic waves (*purple arrows*) move in the same direction and in the same way as the mixing waves but are stronger. As one wave reaches the pyloric part of the stomach, a second wave begins in the body.

4. As with the mixing waves, the more fluid part of the chyme is pushed toward the pylorus (*blue arrows*), whereas the more solid center of the chyme is pushed back toward the body of the stomach (*orange arrow*).

5. Peristaltic contractions force a few milliliters of some of the most fluid chyme through the pyloric orifice into the duodenum (*small blue arrows*). Most of the chyme, including the more solid portion, is forced back toward the body of the stomach for further mixing (*orange arrows*).

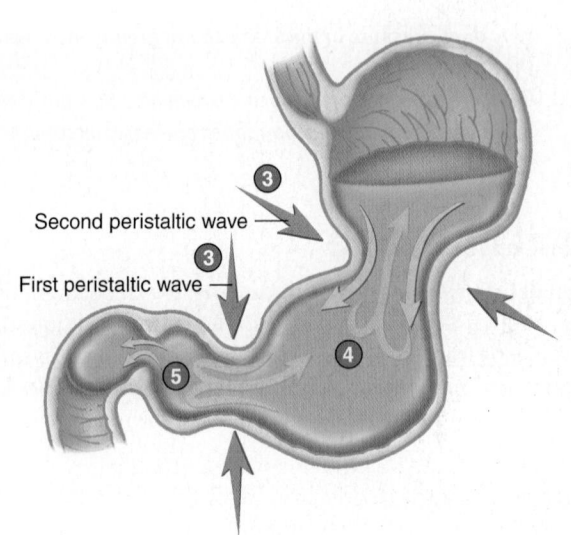

Process Figure 24.14   **Movements in the Stomach**

Hunger Contractions

Hunger contractions are peristaltic contractions that approach tetanic contractions for periods of about 2–3 minutes. Low blood glucose levels cause the contractions to increase and become sufficiently strong to create uncomfortable sensations called "hunger pangs." Hunger pangs usually begin 12–24 hours after the previous meal or in less time for some people. If nothing is ingested, they reach their maximum intensity within 3–4 days and then become progressively weaker.

Regulation of Stomach Emptying

If the stomach empties too fast, the efficiency of digestion and absorption is reduced, and acidic gastric contents dumped into the duodenum may damage its lining. If the rate of emptying is too slow, the highly acidic contents of the stomach may damage the stomach wall and reduce the rate at which nutrients are digested and absorbed. Stomach emptying is regulated to prevent these two extremes. Some of the hormonal and neural mechanisms that stimulate stomach secretions also are involved with increasing stomach motility. For example, during the gastric phase of stomach secretion, distention of the stomach stimulates local reflexes, CNS reflexes, and the release of gastrin, all of which increase stomach motility and cause relaxation of the pyloric sphincter. The result is an increase in stomach emptying. Conversely, some of the hormonal and neural mechanisms that decrease gastric secretions also inhibit gastric motility, increase constriction of the pyloric sphincter, and decrease the rate of stomach emptying.

Vomiting

Vomiting can result from irritation (e.g., overdistention or overexcitation) anywhere along the GI tract. Action potentials travel through the vagus nerve and spinal visceral afferent nerves to the vomiting center in the medulla oblongata. Once the vomiting center is stimulated and the reflex is initiated, the following events occur: (1) a deep breath is taken; (2) the hyoid bone and larynx are elevated, opening the upper esophageal sphincter; (3) the opening of the larynx is closed; (4) the soft palate is elevated, closing the posterior nares; (5) the diaphragm and abdominal muscles are forcefully contracted, strongly compressing the stomach and increasing the intragastric pressure; (6) the lower esophageal sphincter is relaxed; and (7) the gastric contents are forced out of the stomach, through the esophagus and oral cavity, to the outside.

23. *Describe the three phases of regulation of stomach secretion, and discuss the cause and result of each phase.*
24. *How are gastric secretions inhibited? Why is this inhibition necessary?*
25. *Why does pressure in the stomach not greatly increase as the stomach fills?*
26. *What are two kinds of stomach movements? How are stomach movements regulated by hormones and nervous control?*

Small Intestine

The **small intestine** consists of three parts: the duodenum, the jejunum, and the ileum (figure 24.15). The entire small intestine is about 6 m long (range: 4.6–9 m). The duodenum is about 25 cm long (the term *duodenum* means 12, suggesting that it is 12 inches long). The jejunum, constituting about two-fifths of the total length of the

Stomach

Duodenum

Ascending colon

Jejunum

Mesentery

Ileocecal junction

Ileum

Cecum

Appendix

Anterior view

Figure 24.15   **The Small Intestine**

small intestine, is about 2.5 m long; and the ileum, constituting three-fifths of the small intestine, is about 3.5 m long. Two major accessory glands, the liver and the pancreas, are associated with the duodenum.

The small intestine is the site at which the greatest amount of digestion and absorption occur. Each day, about 9 L of water enters the digestive system. It comes from water that is ingested and from fluid secretions produced by glands along the length of the digestive tract. Most of the water, 8–8.5 L, moves by osmosis, with the absorbed solutes, out of the small intestine. A small part, 0.5–1 L, enters the colon.

Anatomy and Histology of the Small Intestine

Duodenum

The **duodenum** nearly completes a 180-degree arc as it curves within the abdominal cavity (figure 24.16), and the head of the pancreas lies within this arc. The duodenum begins with a short superior part, which is where it exits the pylorus of the stomach, and ends in a sharp bend, which is where it joins the jejunum. Two small mounds are within the duodenum about two-thirds of the way down the descending part: the **major duodenal papilla** and the **minor duodenal papilla.** Ducts from the liver and/or pancreas open at these papillae (see p. 904).

The surface of the duodenum has several modifications that increase its surface area about 600-fold to allow for more efficient digestion and absorption of food. The mucosa and submucosa form a series of folds called the **circular folds,** or **plicae** (plī´sē; folds) **circulares** (figure 24.16*a*), which run perpendicular to the long axis of the digestive tract. Tiny fingerlike projections of the mucosa form numerous **villi** (vil´ī; shaggy hair), which are 0.5–1.5 mm in length (figure 24.16*b*). Each villus is covered by simple columnar epithelium and contains a blood capillary network and a lymphatic capillary called a **lacteal** (lak´tē-ăl) (figure 24.16*c*). Most of the cells that make up the surface of the villi have numerous cytoplasmic extensions (about 1 μm long) called **microvilli,** which further increase the surface area (figure 24.16*d*). The combined microvilli on the entire epithelial surface form the **brush border.** These various modifications greatly increase the surface area of the small intestine and, as a result, greatly enhance absorption.

The mucosa of the duodenum is simple columnar epithelium with four major cell types: (1) **absorptive cells** are cells with microvilli, which produce digestive enzymes and absorb digested food; (2) **goblet cells,** which produce a protective mucus; (3) **granular cells** (Paneth's cells), which may help protect the intestinal epithelium from bacteria; and (4) **endocrine cells,** which produce regulatory hormones. The epithelial cells are produced within tubular invaginations of the mucosa, called **intestinal glands** (crypts of Lieberkühn), at the base of the villi. The absorptive and goblet cells migrate from the intestinal glands to cover the surface of the villi and eventually are shed from its tip. The granular and endocrine cells remain in the bottom of the glands. The submucosa of the duodenum contains coiled tubular mucous glands called **duodenal glands** (Brunner's glands), which open into the base of the intestinal glands.

Jejunum and Ileum

The **jejunum** and **ileum** are similar in structure to the duodenum (see figure 24.15), except that a gradual decrease occurs in the

Figure 24.16 Anatomy and Histology of the Duodenum

(*a*) Wall of the duodenum, showing the circular folds. (*b*) The villi on a circular fold. (*c*) A single villus, showing the lacteal and capillary network. (*d*) Transmission electron micrograph of microvilli on the surface of a villus.

diameter of the small intestine, the thickness of the intestinal wall, the number of circular folds, and the number of villi as one progresses through the small intestine. The duodenum and jejunum are the major sites of nutrient absorption, although some absorption occurs in the ileum. Lymphatic nodules called **Peyer's patches** are numerous in the mucosa and submucosa of the ileum.

The junction between the ileum and the large intestine is the **ileocecal junction.** It has a ring of smooth muscle, the **ileocecal sphincter,** and a one-way **ileocecal valve** (see figure 24.25).

27. *Name and describe the three parts of the small intestine. What are the major and lesser duodenal papilla?*
28. *What are the circular folds, villi, and microvilli in the small intestine? What are their functions?*
29. *Name the four types of cells found in the duodenal mucosa, and state their functions.*

Secretions of the Small Intestine

The mucosa of the small intestine produces secretions that primarily contain mucus, electrolytes, and water. Intestinal secretions lubricate and protect the intestinal wall from the acidic chyme and the action of digestive enzymes. They also keep the chyme in the small intestine in a liquid form to facilitate the digestive process (see table 24.2). The intestinal mucosa produces most of the secretions that enter the small intestine, but the secretions of the liver and the pancreas also enter the small intestine and play essential roles in the process of digestion. Most of the digestive enzymes entering the small intestine are secreted by the pancreas. The intestinal mucosa also produces enzymes, but these remain associated with the intestinal epithelial surface.

The duodenal glands, intestinal glands, and goblet cells secrete large amounts of mucus. This mucus provides the wall of the intestine with protection against the irritating effects of acidic chyme and against the digestive enzymes that enter the duodenum from the pancreas. Secretin and cholecystokinin are released from the intestinal mucosa and stimulate hepatic and pancreatic secretions (see figures 24.21 and 24.24).

The vagus nerve, secretin, and chemical or tactile irritation of the duodenal mucosa stimulate secretion from the duodenal glands. Goblet cells produce mucus in response to the tactile and chemical stimulation of the mucosa.

Duodenal Ulcer

Sympathetic nerve stimulation inhibits duodenal gland secretion, thus reducing the coating of mucus on the duodenal wall, which protects it against acid and gastric enzymes. If a person is highly stressed, elevated sympathetic activity may therefore inhibit duodenal gland secretion and increase the person's susceptibility to **duodenal ulcers.**

Enzymes of the intestinal mucosa are bound to the membranes of the absorptive cell microvilli. These surface-bound enzymes include **disaccharidases,** which break disaccharides down to monosaccharides; **peptidases,** which hydrolyze the peptide bonds between small amino acid chains; and **nucleases,** which break down nucleic acids (see table 24.2). Although these enzymes are not

secreted into the intestine, they influence the digestive process significantly, and the large surface area of the intestinal epithelium brings these enzymes into contact with the intestinal contents. Small molecules, which are breakdown products of digestion, are absorbed through the microvilli and enter the circulatory or lymphatic systems.

Movement in the Small Intestine

Mixing and propulsion of chyme are the primary mechanical events that occur in the small intestine. These functions are the result of segmental or peristaltic contractions, which are accomplished by the smooth muscle in the wall of the small intestine and which are only propagated for short distances. **Segmental contractions** (see figure 24.3) mix the intestinal contents, and **peristaltic contractions** propel the intestinal contents along the digestive tract. A few peristaltic contractions may proceed the entire length of the intestine. Frequently, intestinal peristaltic contractions are continuations of peristaltic contractions that begin in the stomach. These contractions both mix and propel substances through the small intestine as the wave of contraction proceeds. The contractions move at a rate of about 1 cm/min. The movements are slightly faster at the proximal end of the small intestine and slightly slower at the distal end. It usually takes 3–5 hours for chyme to move from the pyloric region to the ileocecal junction.

Local mechanical and chemical stimuli are especially important in regulating the motility of the small intestine. Smooth muscle contraction increases in response to distention of the intestinal wall. Solutions that are either hypertonic or hypotonic, solutions with a low pH, and certain products of digestion like amino acids and peptides also stimulate contractions of the small intestine. Local reflexes, which are integrated within the enteric plexus of the small intestine, mediate the response of the small intestine to these mechanical and chemical stimuli. Stimulation through parasympathetic nerve fibers may also increase the motility of the small intestine, but the parasympathetic influences in the small intestine are not as important as those in the stomach.

The ileocecal sphincter at the juncture between the ileum and the large intestine remains mildly contracted most of the time, but peristaltic waves reaching it from the small intestine cause it to relax and allow movement of chyme from the small intestine into the cecum. Cecal distention, however, initiates a local reflex that causes more intense constriction of the ileocecal sphincter. Closure of the sphincter facilitates digestion and absorption in the small intestine by slowing the rate of chyme movement from the small intestine into the large intestine and prevents material from returning to the ileum from the cecum.

30. *What are the functions of the intestinal glands and duodenal glands? State the factors that stimulate secretion from the duodenal glands and from goblet cells.*
31. *List the enzymes of the small intestine wall and give their functions.*
32. *What are two kinds of movement of the small intestine? How are they regulated?*
33. *What is the function of the ileocecal sphincter?*

Liver
Anatomy of the Liver

The **liver** is the largest internal organ of the body, weighing about 1.36 kg (3 pounds), and it is in the right-upper quadrant of the abdomen, tucked against the inferior surface of the diaphragm (see figures 24.1 and 24.17). The liver consists of two major **lobes, left** and **right,** and two minor lobes, **caudate** and **quadrate.**

A **porta** (gate) is on the inferior surface of the liver, where the various vessels, ducts, and nerves enter and exit the liver. The **hepatic** (he-pat'ik; associated with the liver) **portal vein,** the **hepatic artery,** and a small hepatic nerve plexus enter the liver through the porta (see figure 24.17b). Lymphatic vessels and two hepatic ducts, one each from the right and left lobes, exit the liver at the porta. The hepatic ducts transport bile out of the liver. The right and left hepatic ducts unite to form a single **common hepatic duct** (figure 24.18). The **cystic duct** from the gallbladder

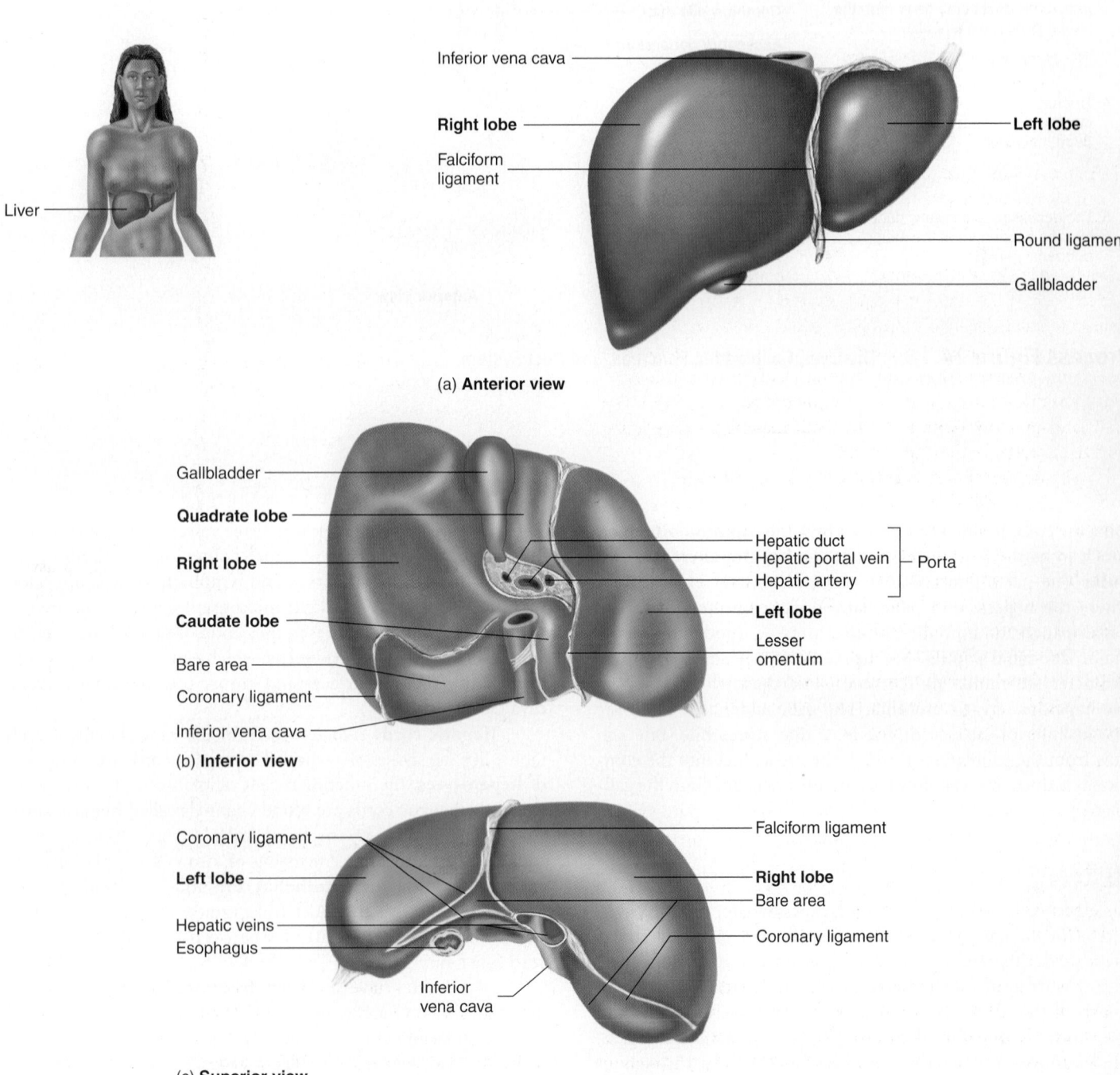

(a) **Anterior view**

(b) **Inferior view**

(c) **Superior view**

Figure 24.17 The Liver

1. The hepatic ducts from the liver lobes combine to form the common hepatic duct.

2. The common hepatic duct combines with the cystic duct from the gallbladder to form the common bile duct.

3. The common bile duct and the pancreatic duct combine to form the hepatopancreatic ampulla.

4. The hepatopancreatic ampulla empties into the duodenum at the major duodenal papilla.

5. The accessory pancreatic duct empties into the duodenum at the minor duodenal papilla.

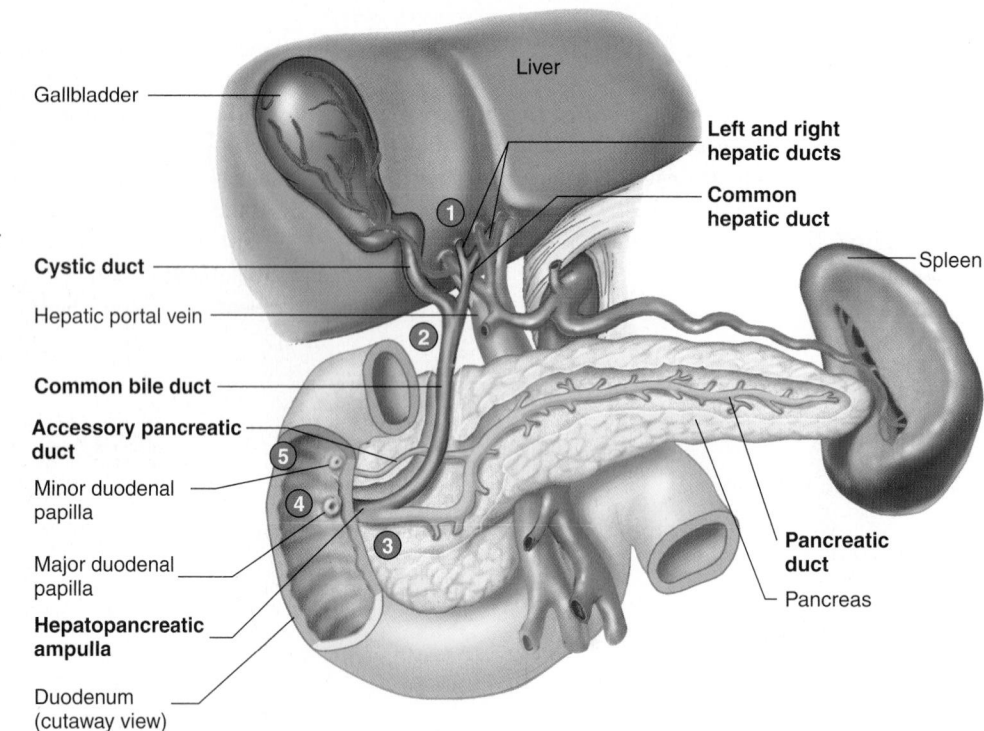

Process Figure 24.18   The Liver, Gallbladder, Pancreas, and Duct System

joins the common hepatic duct to form the **common bile duct,** which joins the pancreatic duct at the **hepatopancreatic ampulla** (hĕ-pat′ō-pan-crē-at′ik am-pul′lă), an enlargement where the hepatic and pancreatic ducts come together. The hepatopancreatic ampulla empties into the duodenum at the major duodenal papilla (see figure 24.18). A smooth muscle sphincter surrounds the common bile duct where it enters the hepatopancreatic ampulla. The gallbladder is a small sac on the inferior surface of the liver that stores bile. Bile can flow from the gallbladder through the cystic duct into the common bile duct, or it can flow back up the cystic duct into the gallbladder.

Histology of the Liver

A connective tissue capsule and visceral peritoneum cover the liver, except for the **bare area,** which is a small area on the diaphragmatic surface that lacks a visceral peritoneum and is surrounded by the coronary ligament (see figure 24.17c). At the porta, the connective tissue capsule sends a branching network of septa (walls) into the substance of the liver to provide its main support. Vessels, nerves, and ducts follow the connective tissue branches throughout the liver.

The connective tissue septa divide the liver into hexagon-shaped **lobules** with a **portal triad** at each corner (figure 24.19).

The triads are so named because three vessels—the hepatic portal vein, hepatic artery, and hepatic duct—are commonly located in them. Hepatic nerves and lymphatic vessels, often too small to be easily seen in light micrographs, are also located in these areas. A **central vein** is in the center of each lobule. Central veins unite to form **hepatic veins,** which exit the liver on its posterior and superior surfaces and empty into the inferior vena cava (see figure 24.19).

Hepatic cords radiate out from the central vein of each lobule like the spokes of a wheel. The hepatic cords are composed of **hepatocytes,** the functional cells of the liver. The spaces between the hepatic cords are blood channels called **hepatic sinusoids.** The sinusoids are lined with a very thin, irregular squamous endothelium consisting of two cell populations: (1) extremely thin, sparse **endothelial cells** and (2) **hepatic phagocytic cells** (Kupffer cells). A cleftlike lumen, the **bile canaliculus** (kan-ă-lik′ū-lŭs; little canal), lies between the cells within each cord (see figure 24.19).

Hepatocytes have six major functions (described in more detail starting on the next page): (1) bile production, (2) storage, (3) interconversion of nutrients, (4) detoxification, (5) phagocytosis, and (6) synthesis of blood components. Nutrient-rich, oxygen-poor blood from the viscera enters the hepatic sinusoids from branches of the hepatic portal vein and mixes with oxygen-rich, nutrient-depleted blood from the hepatic arteries

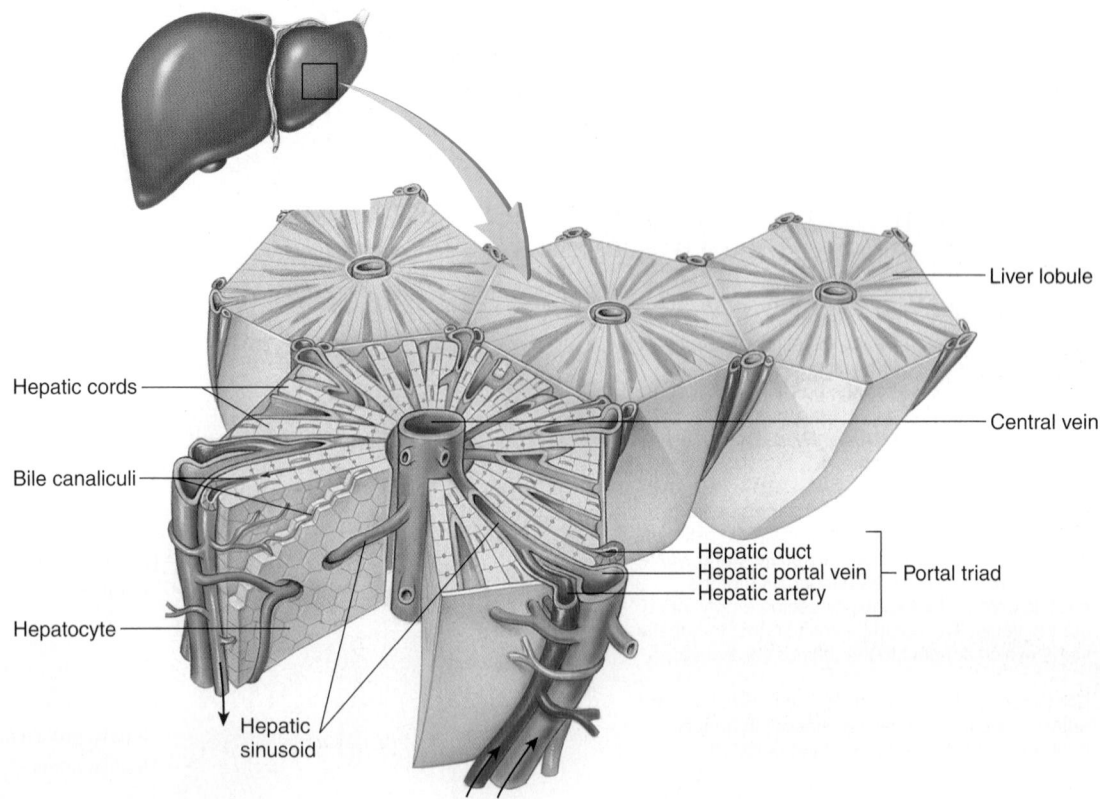

Figure 24.19 Histology of the Liver

(figure 24.20). From the blood, the hepatocytes can take up the oxygen and nutrients, which are stored, detoxified, used for energy, or used to synthesize new molecules. Molecules produced by or modified in the hepatocytes are released into the hepatic sinusoids or into the bile canaliculi.

Mixed blood in the hepatic sinusoids flows to the central vein, where it exits the lobule and then exits the liver through the hepatic veins. **Bile,** produced by the hepatocytes and consisting primarily of metabolic by-products, flows through the bile canaliculi toward the hepatic triad and exits the liver through the hepatic ducts. Blood, therefore, flows from the triad toward the center of each lobule, whereas bile flows away from the center of the lobule toward the triad.

In the fetus, special blood vessels bypass the liver sinusoids. The remnants of fetal blood vessels can be seen in the adult as the round ligament (ligamentum teres) and the ligamentum venosum (see chapter 29).

Liver Rupture or Enlargement

The liver is easily ruptured because it is large, fixed in position, and fragile, or it can be lacerated by a broken rib. Liver rupture or laceration results in severe internal bleeding.

The liver may become enlarged as a result of heart failure, hepatic cancer, cirrhosis, or Hodgkin's disease (a lymphatic cancer). However, as long as part of the liver contains normal tissue, diseased portions of the liver can be removed and the normal tissue may restore the complete liver by regeneration. Likewise, an entire liver or a portion of the liver may be transplanted from one person to another.

34. *Describe the lobes of the liver. What is the porta?*
35. *Diagram the duct system from the liver, gallbladder, and pancreas that empties into the major duodenal papilla.*
36. *What are the hepatic cords and the sinusoids?*
37. *Describe the flow of blood to and through the liver. Describe the flow of bile away from the liver.*

Functions of the Liver

The liver performs important digestive and excretory functions, stores and processes nutrients, synthesizes new molecules, and detoxifies harmful chemicals.

Bile Production

The liver produces and secretes about 600–1000 mL of bile each day (see table 24.2). Bile contains no digestive enzymes, but it plays a role in digestion because it neutralizes and dilutes stomach acid and emulsifies fats. The pH of chyme as it leaves the stomach is too low for the normal function of pancreatic enzymes. Bile helps to neutralize the acidic chyme and to bring the pH up to a level at which pancreatic enzymes can function. **Bile salts** emulsify fats (described in more detail on p. 907). Bile also contains excretory products like bile pigments. Bilirubin is a bile pigment that results from the breakdown of hemoglobin. Bile also contains cholesterol, fats, fat-soluble hormones, and lecithin.

1. The hepatic artery carries oxygen-rich blood from the aorta through the porta of the liver. Hepatic artery branches become part of the portal triads. Blood from the hepatic artery branches enters the hepatic sinusoids and supplies hepatocytes in the hepatic cords with oxygen.

2. The hepatic portal vein carries nutrient-rich, oxygen-poor blood from the intestines through the porta of the liver. Hepatic portal vein branches become part of the portal triads. Blood from the hepatic portal vein branches enters the hepatic sinusoids and supplies hepatocytes in the hepatic cords with nutrients.

3. Blood in the hepatic sinusoids that comes from the hepatic artery and hepatic portal vein picks up plasma proteins, processed molecules, and waste products produced by the hepatocytes of the hepatic cords. The hepatic sinusoids empty into central veins. The central veins connect to hepatic veins, which connect to the inferior vena cava.

4. Bile produced by hepatocytes in the hepatic cords enters bile canaliculi, which connect to hepatic duct branches that are part of the portal triads.

5. The hepatic duct branches converge to form the left and right hepatic ducts, which carry bile out the porta of the liver.

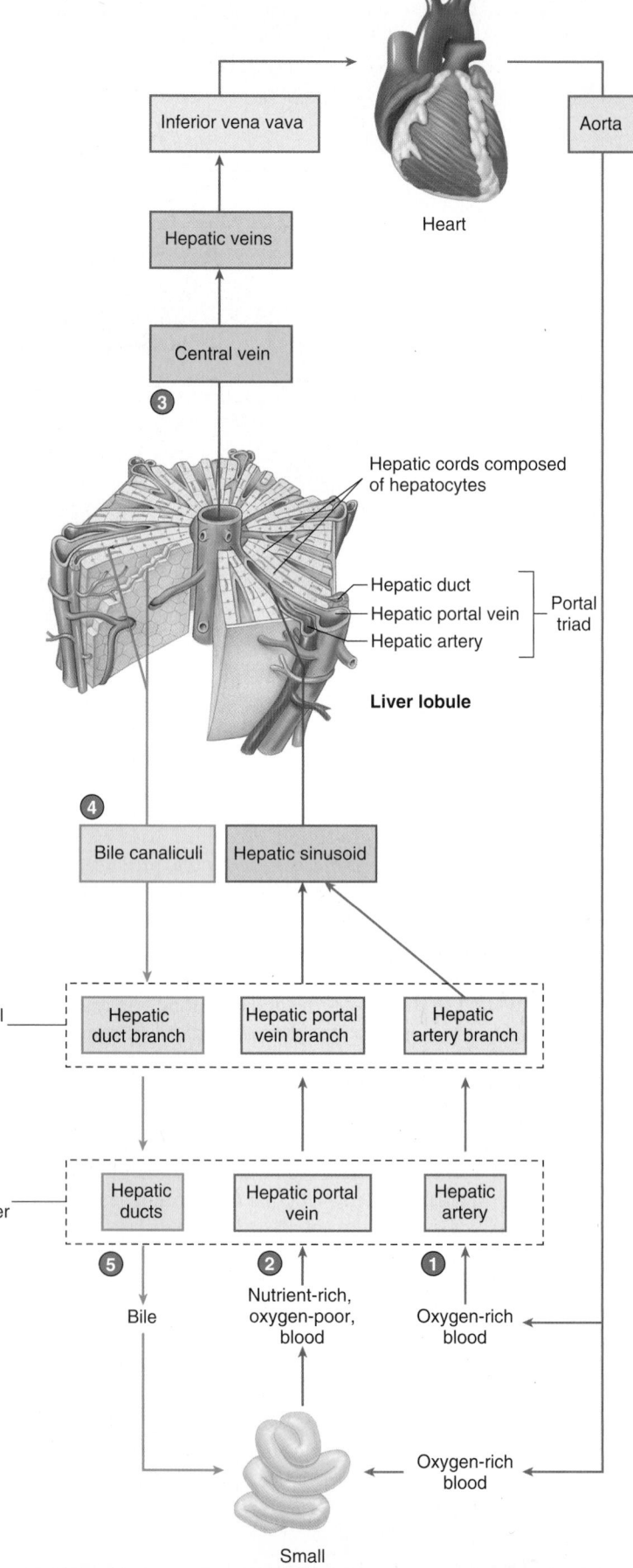

Process Figure 24.20   **Blood and Bile Flow Through the Liver**

Secretin stimulates bile secretion, primarily by increasing the water and bicarbonate ion content of bile (figure 24.21). Bile salts also increase bile secretion through a positive-feedback system. Over 90% of bile salts are reabsorbed in the ileum and carried in the blood back to the liver, where they contribute to further bile secretion. The loss of bile salts in the feces is reduced by this recycling process. Bile secretion into the duodenum continues until the duodenum empties.

Storage

Hepatocytes can remove sugar from the blood and store it in the form of **glycogen.** They can also store fat, vitamins (A, B_{12}, D, E, and K), copper, and iron. This storage function is usually short term, and the amount of stored material in the hepatocytes and, thus, the cell size fluctuate during a given day.

Hepatocytes help control blood sugar levels within very narrow limits. If a large amount of sugar enters the general circulation after a meal, it will increase the osmolality of the blood and produce hyperglycemia. This is prevented because the blood from the intestine passes through the hepatic portal vein to the liver, where glucose and other substances are removed from the blood by hepatocytes, stored, and secreted back into the circulation when needed.

Nutrient Interconversion

Interconversion of nutrients is another important function of the liver. Ingested nutrients are not always in the proportion needed by the tissues. If this is the case, the liver can convert some nutrients into others. If, for example, a person is on a diet that is excessively high in protein, an oversupply of amino acids and an undersupply of lipids and carbohydrates may be delivered to the liver. The hepatocytes break down the amino acids and cycle many of them through metabolic pathways so they can be used to produce adenosine triphosphate, lipids, and glucose (see chapter 25).

Hepatocytes also transform substances that cannot be used by most cells into more readily usable substances. For example, ingested fats are combined with choline and phosphorus in the liver to produce phospholipids, which are essential components of plasma membranes. Vitamin D is hydroxylated in the liver. The hydroxylated form of vitamin D is the major circulating form of vitamin D, which is transported through the circulation to the kidney, where it's again hydroxylated. The double-hydroxylated vitamin D is the active form of the vitamin, which functions in calcium maintenance.

Detoxification

Many ingested substances are harmful to the cells of the body. In addition, the body itself produces many by-products of metabolism

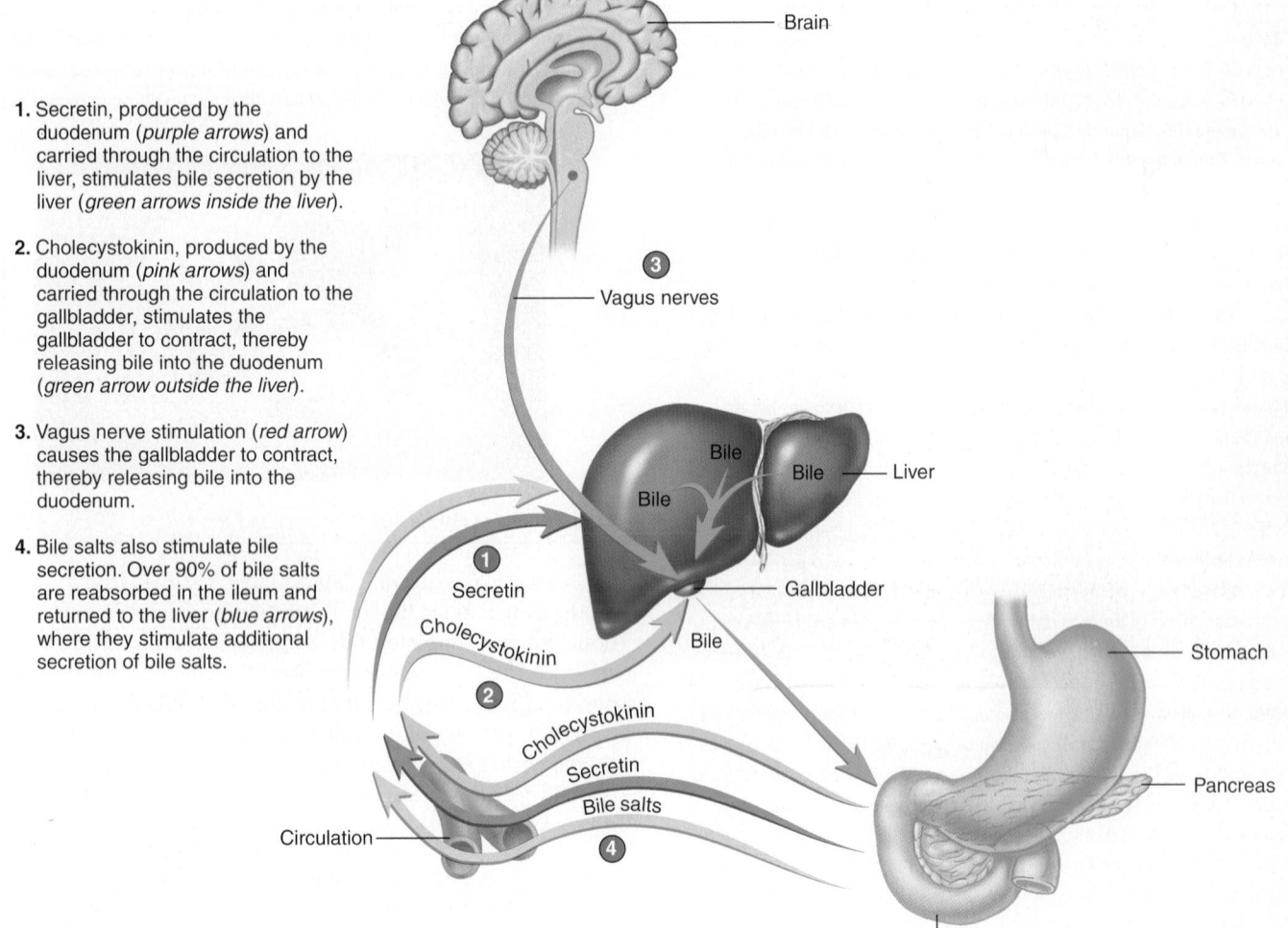

1. Secretin, produced by the duodenum (*purple arrows*) and carried through the circulation to the liver, stimulates bile secretion by the liver (*green arrows inside the liver*).

2. Cholecystokinin, produced by the duodenum (*pink arrows*) and carried through the circulation to the gallbladder, stimulates the gallbladder to contract, thereby releasing bile into the duodenum (*green arrow outside the liver*).

3. Vagus nerve stimulation (*red arrow*) causes the gallbladder to contract, thereby releasing bile into the duodenum.

4. Bile salts also stimulate bile secretion. Over 90% of bile salts are reabsorbed in the ileum and returned to the liver (*blue arrows*), where they stimulate additional secretion of bile salts.

Process Figure 24.21 Control of Bile Secretion and Release that, if accumulated, are toxic. The liver forms a major line of defense against many of these harmful substances. It detoxifies many substances by altering their structure to make them less toxic or make their elimination easier. Ammonia, for example, a by-product of amino acid metabolism, is toxic and is not readily removed from the circulation by the kidneys. Hepatocytes remove ammonia from the circulation and convert it to urea, which is less toxic than ammonia and is secreted into the circulation and then eliminated by the kidneys in the urine. Other substances are removed from the circulation and excreted by the hepatocytes into the bile.

Phagocytosis

Hepatic phagocytic cells (Kupffer cells), which lie along the sinusoid walls of the liver, phagocytize "worn-out" and dying red and white blood cells, some bacteria, and other debris that enters the liver through the circulation.

Synthesis

The liver can also produce its own unique new compounds. It produces many blood proteins, such as albumins, fibrinogen, globulins, heparin, and clotting factors, which are released into the circulation (see chapter 19).

Hepatitis, Cirrhosis, and Liver Damage

Strictly defined, **hepatitis** is an inflammation of the liver and does not necessarily result from an infection. Hepatitis can be caused by alcohol consumption or infection. Infectious hepatitis is caused by viral infections. Hepatitis A, also called infectious hepatitis, is responsible for about 30% of hepatitis cases in the U.S. Hepatitis B, also called serum hepatitis, is a more chronic infection responsible for half the hepatitis cases in the U.S. Hepatitis C, also called non-A and non-B hepatitis, causes 20% of the U.S. hepatitis cases. It's caused by one or more virus types that cannot be identified in blood tests. It's spread by blood transfusions or sexual intercourse. If hepatitis is not treated, liver cells die and are replaced by scar tissue, resulting in loss of liver function. Death caused by liver failure can occur.

Cirrhosis (sir-rō′sis) of the liver involves the death of hepatocytes and their replacement by fibrous connective tissue. The liver becomes pale in color (the term *cirrhosis* means a tawny or orange condition) because of the presence of excess white connective tissue. It also becomes firmer, and the surface becomes nodular. The loss of hepatocytes eliminates the function of the liver, often resulting in jaundice, and the buildup of connective tissue can impede blood flow through the liver. Cirrhosis frequently develops in alcoholics and may develop as a result of biliary obstruction, hepatitis, or nutritional deficiencies.

Under most conditions, mature hepatocytes can proliferate and replace lost parts of the liver. If the liver is severely damaged, however, the hepatocytes may not have enough regenerative power to replace the lost parts. In this case, a liver transplant may be necessary. Recent evidence suggests that the liver also maintains an undifferentiated stem cell population, called "oval" cells, which gives rise to two cell lines, one forming bile duct epithelium and the other producing hepatocytes. It is hoped that these stem cells can be used to reconstitute a severely damaged liver. It may even be possible at some time in the future to remove stem cells from a person with hemophilia, genetically engineer the cells to produce the missing clotting factors, and then reintroduce the altered stem cells into the person's liver.

---

38. **Explain and give examples of the major functions of the liver.**
39. **What stimulates bile secretion from the liver?**

Gallbladder

The **gallbladder** is a saclike structure on the inferior surface of the liver that is about 8 cm long and 4 cm wide (see figure 24.18). Three tunics form the gallbladder wall: (1) an inner mucosa folded into rugae that allow the gallbladder to expand; (2) a muscularis, which is a layer of smooth muscle that allows the gallbladder to contract; and (3) an outer covering of serosa. The cystic duct connects the gallbladder to the common bile duct.

Bile is continually secreted by the liver and flows to the gallbladder, where 40–70 mL of bile can be stored. While the bile is in the gallbladder, water and electrolytes are absorbed, and bile salts and pigments become as much as 5–10 times more concentrated than they were when secreted by the liver. Shortly after a meal, the gallbladder contracts in response to stimulation by cholecystokinin and, to a lesser degree, in response to vagal stimulation, thereby dumping large amounts of concentrated bile into the small intestine (see figure 24.21).

Gallstones

Cholesterol, secreted by the liver, may precipitate in the gallbladder to produce **gallstones** (figure A). Occasionally, a gallstone can pass out of the gallbladder and enter the cystic duct, blocking release of bile. Such a condition interferes with normal digestion, and the gallstone often must be removed surgically. If the gallstone moves far enough down the duct, it can also block the pancreatic duct, resulting in pancreatitis.

Figure A   Gallstones

Drastic dieting with rapid weight loss may lead to gallstone production. Food in the small intestine stimulates secretin production, which stimulates bile secretion. During drastic dieting episodes, this stimulation decreases. A decrease in bile salt production can result in bile retention within the gallbladder and the precipitation of bile salt as gallstones. In one study, 25% of obese people participating in an 8-week, quick-weight-loss program developed gallstones. Six percent required surgical removal of the stones. No gallstones developed in nondieting obese controls.

---

40. **Describe the three tunics of the gallbladder wall.**
41. **What is the function of the gallbladder? What stimulates the release of bile from the gallbladder?**

Pancreas

Anatomy of the Pancreas

The **pancreas** is a complex organ composed of both endocrine and exocrine tissues that perform several functions. The pancreas consists of a **head,** located within the curvature of the duodenum (figure 24.22a), a **body,** and a **tail,** which extends to the spleen.

The endocrine part of the pancreas consists of **pancreatic islets** (islets of Langerhans; figure 24.22b). The islet cells produce insulin and glucagon, which are very important in controlling blood levels of nutrients, such as glucose and amino acids, and somatostatin, which regulates insulin and glucagon secretion and may inhibit growth hormone secretion (see chapter 18).

The exocrine part of the pancreas is a compound acinar gland (see discussion of glands in chapter 4). The **acini** (as′i-nī; grapes; see figure 24.22b) produce digestive enzymes. Clusters of acini form lobules that are separated by thin septa. Lobules are connected by small **intercalated ducts** to **intralobular ducts,** which leave the lobules to join **interlobular ducts** between the lobules. The interlobular ducts attach to the main **pancreatic duct,** which joins the common bile duct at the hepatopancreatic ampulla (Vater's ampulla; see figures 24.18 and 24.22). The hepatopancreatic ampulla empties into the duodenum at the major duodenal papilla. A smooth muscle sphincter, the **hepatopancreatic ampullar sphincter** (sphincter of Oddi) regulates the opening of the ampulla. An accessory pancreatic duct, present in most people, opens at the minor duodenal papilla. The ducts are lined with simple cuboidal epithelium, and the epithelial cells of the acini are pyramid-shaped. A smooth muscle sphincter surrounds the pancreatic duct where it enters the hepatopancreatic ampulla.

● **42. Describe the parts of the pancreas responsible for endocrine and exocrine secretions. Diagram the duct system of the pancreas.**

Pancreatic Secretions

The exocrine secretions of the pancreas are called **pancreatic juice** and have two major components: an aqueous component and an enzymatic component. Pancreatic juice is produced in the pancreas

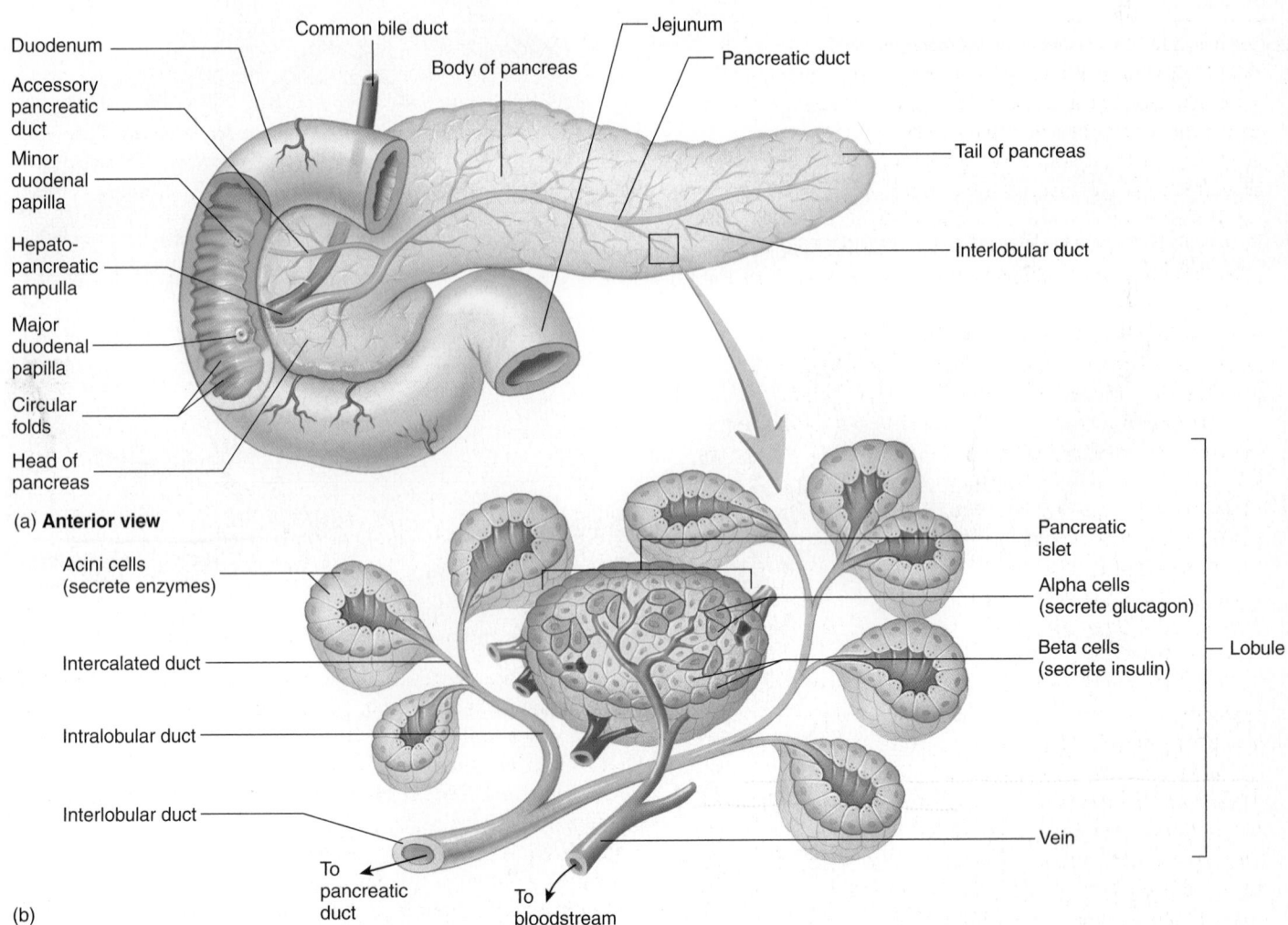

(a) Anterior view

Duodenum
Accessory pancreatic duct
Minor duodenal papilla
Hepato-pancreatic ampulla
Major duodenal papilla
Circular folds
Head of pancreas
Common bile duct
Body of pancreas
Jejunum
Pancreatic duct
Tail of pancreas
Interlobular duct

(b)

Acini cells (secrete enzymes)
Intercalated duct
Intralobular duct
Interlobular duct
To pancreatic duct
To bloodstream
Pancreatic islet
Alpha cells (secrete glucagon)
Beta cells (secrete insulin)
Lobule
Vein

Figure 24.22   Anatomy and Histology of the Duodenum and Pancreas

(a) The head of the pancreas lies within the duodenal curvature, with the pancreatic duct emptying into the duodenum. (b) Histology of the pancreas showing both the acini and the pancreatic duct system.

and is then delivered through the pancreatic ducts to the small intestine, where it functions in digestion. The **aqueous component** is produced principally by columnar epithelial cells that line the smaller ducts of the pancreas. It contains Na^+ and K^+ ions in about the same concentration found in extracellular fluid. Bicarbonate ions are a major part of the aqueous component, and they neutralize the acidic chyme that enters the small intestine from the stomach. The increased pH caused by pancreatic secretions in the duodenum stops pepsin digestion but provides the proper environment for the function of pancreatic enzymes. Bicarbonate ions are actively secreted by the duct epithelium, and water follows passively to make the pancreatic juice isotonic. The cellular mecha-

nism that is responsible for the secretion of bicarbonate ions is diagrammed in figure 24.23.

The enzymes of the pancreatic juice are produced by the acinar cells of the pancreas and are important for the digestion of all major classes of food. Without the enzymes produced by the pancreas, lipids, proteins, and carbohydrates are not adequately digested (see tables 24.1 and 24.2).

The proteolytic pancreatic enzymes, which digest proteins, are secreted in inactive forms, whereas many of the other enzymes are secreted in active form. The major proteolytic enzymes are **trypsin, chymotrypsin,** and **carboxypeptidase.** They are secreted in their inactive forms as trypsinogen, chymotrypsinogen,

1. Water (H_2O) and carbon dioxide (CO_2) combine under the influence of carbonic anhydrase (CA) to form carbonic acid (H_2CO_3).

2. Carbonic acid dissociates to form hydrogen ions (H^+) and bicarbonate ions (HCO_3^-).

3. The H^+ are exchanged for sodium ions (Na^+) so H^+ enter and Na^+ are removed from the bloodstream.

4. The HCO_3^- are actively transported into the intercalated ducts. Sodium ions and H_2O follow the HCO_3^- into the ducts.

5. The ions move through the intercalated duct toward the interlobular duct.

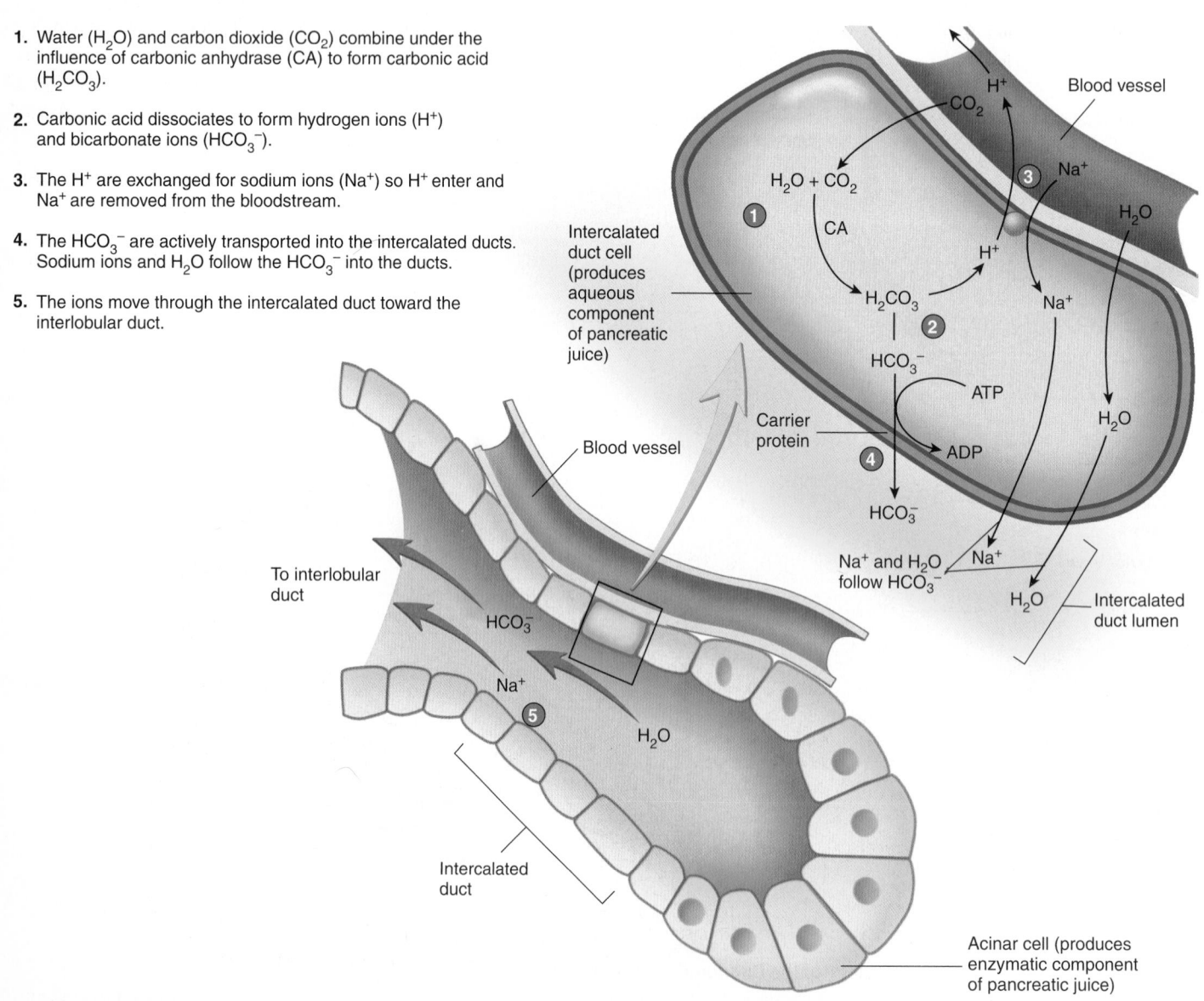

Process Figure 24.23   Bicarbonate Ion Production in the Pancreas and procarboxypeptidase and are activated by the removal of certain peptides from the larger precursor proteins. If these were produced in their active forms, they would digest the tissues producing them. The proteolytic enzyme **enterokinase** (en´tĕr-ō-kī´nās; intestinal enzyme), which is an enzyme attached to the brush border of the small intestine, activates trypsinogen. Trypsin then activates more trypsinogen, as well as chymotrypsinogen and procarboxypeptidase.

Pancreatic juice also contains **pancreatic amylase,** which continues the polysaccharide digestion that was initiated in the oral cavity. In addition, pancreatic juice contains a group of lipid-digesting enzymes called **pancreatic lipases,** which break down lipids into monoglycerides, free fatty acids, cholesterol, and other components.

Enzymes that reduce DNA and ribonucleic acid to their component nucleotides, **deoxyribonucleases** and **ribonucleases,** respectively, are also present in pancreatic juice.

Pancreatitis and Pancreatic Cancer

Pancreatitis is an inflammation of the pancreas and occurs quite commonly. Pancreatitis involves the release of pancreatic enzymes within the pancreas itself, which digest pancreatic tissue. It may result from alcoholism, use of certain drugs, pancreatic duct blockage, cystic fibrosis, viral infection, or pancreatic cancer. Symptoms can range from mild abdominal pain to systemic shock and coma.

Cancer of the pancreas can obstruct the pancreatic and the common hepatic ducts, resulting in jaundice. Pancreatic cancer may not be detected until the mass has become fairly large and may become so large as to block off the pyloric region of the stomach.

Regulation of Pancreatic Secretion

Both hormonal and neural mechanisms (figure 24.24) control the exocrine secretions of the pancreas. An acidic chyme in the duodenum stimulates the pancreas to release secretin. Secretin stimulates the secretion of a watery solution that contains a large amount of bicarbonate ions from the pancreas into the duodenum. Bicarbonate ions increase the pH of chyme in the duodenum so that the duodenum is not damaged by the low pH. In addition, pancreatic and brush border enzymes do not function at a low pH.

P R E D I C T   6
Explain why secretin production in response to acidic chyme and its stimulation of bicarbonate ion secretion constitute a negative-feedback mechanism.

Cholecystokinin stimulates the release of bile from the gallbladder and the secretion of pancreatic juice rich in digestive enzymes. The major stimulus for the release of cholecystokinin is the presence of fatty acids and other lipids in the duodenum.

Parasympathetic stimulation through the vagus (X) nerves also stimulates the secretion of pancreatic juices rich in pancreatic enzymes, and sympathetic impulses inhibit secretion. The effect of vagal stimulation on pancreatic juice secretion is greatest during the cephalic and gastric phases of stomach secretion.

43. Name the two kinds of exocrine secretions produced by the pancreas. What stimulates their production and what is their function?
44. What are the enzymes present in pancreatic juice? Explain the function of each.

1. Secretin (*purple arrows*) released from the duodenum, stimulates the pancreas to release a watery secretion, rich in bicarbonate ions.

2. Cholecystokinin (*pink arrows*) released from the duodenum, causes the pancreas to release a secretion rich in digestive enzymes.

3. Parasympathetic stimulation from the vagus nerve (*red arrow*) causes the pancreas to release a secretion rich in digestive enzymes.

Process Figure 24.24   Control of Pancreatic Secretion

Large Intestine

The **large intestine** is the portion of the digestive tract extending from the ileocecal junction to the anus. It consists of the cecum, colon, rectum, and anal canal. Normally 18–24 hours are required for material to pass through the large intestine, in contrast to the 3–5 hours required for movement of chyme through the small intestine. Thus, the movements of the colon are more sluggish than those of the small intestine. While in the colon, chyme is converted to feces. Absorption of water and salts, the secretion of mucus, and extensive action of microorganisms are involved in the formation of feces, which the colon stores until the feces are eliminated by the process of defecation. About 1500 mL of chyme enters the cecum each day, but more than 90% of the volume is reabsorbed so that only 80–150 mL of feces is normally eliminated by defecation.

Anatomy of the Large Intestine

Cecum

The **cecum** (sē′kŭm; blind) is the proximal end of the **large intestine.** It's where the large and small intestines meet at the ileocecal junction. The cecum extends inferiorly about 6 cm past the ileocecal junction in the form of a blind sac (figure 24.25). Attached to the cecum is a small blind tube about 9 cm long called the **vermiform** (ver′mi-fōrm; worm-shaped) **appendix.** The walls of the appendix contain many lymphatic nodules.

Appendicitis

Appendicitis is an inflammation of the vermiform appendix and usually occurs because of obstruction of the appendix. Secretions from the appendix cannot pass the obstruction and accumulate, causing enlargement and pain. Bacteria in the area can cause infection of the appendix. Symptoms include sudden abdominal pain, particularly in the right lower portion of the abdomen, along with a slight fever, loss of appetite, constipation or diarrhea, nausea, and vomiting. In the right-lower quadrant of the abdomen, about midway along a line between the umbilicus and the right anterior superior iliac spine, is an area on the body's surface called McBurney's point. This area becomes very tender in patients with acute appendicitis because of pain referred from the inflamed appendix to the body's surface. Each year, 500,000 people in the United States suffer an appendicitis. An appendectomy is removal of the appendix. If the appendix bursts, the infection can spread throughout the peritoneal cavity, causing peritonitis, with life-threatening results.

Colon

The **colon** (kō′lon) is about 1.5–1.8 m long and consists of four parts: the ascending colon, transverse colon, descending colon, and sigmoid colon (see figure 24.25). The **ascending colon** extends superiorly from the cecum and ends at the right colic flexure (hepatic flexure) near the right inferior margin of the liver. The **transverse colon** extends from the right colic flexure to the left colic flexure (splenic flexure), and the **descending colon** extends from the left

(a) **Anterior view**   (b) **Anterior view**

Figure 24.25   Large Intestine

(*a*) Large intestine (i.e., cecum, colon, and rectum) and anal canal. The teniae coli and epiploic appendages are along the length of the colon. (*b*) A radiograph of the large intestine following a barium enema.

colic flexure to the superior opening of the true pelvis, where it becomes the sigmoid colon. The **sigmoid colon** forms an S-shaped tube that extends into the pelvis and ends at the rectum.

The circular muscle layer of the colon is complete, but the longitudinal muscle layer is incomplete. The longitudinal layer doesn't completely envelop the intestinal wall but forms three bands, called the **teniae coli** (tē′nē-ē kō′lī; a band or tape along the colon), that run the length of the colon (see figures 24.25 and 24.26). Contractions of the teniae coli cause pouches called **haustra** (haw′stră; to draw up) to form along the length of the colon, giving it a puckered appearance. Small, fat-filled connective tissue pouches called **epiploic** (ep′i-plō′ik; related to the omentum) **appendages** are attached to the outer surface of the colon along its length.

The mucosal lining of the large intestine consists of simple columnar epithelium. This epithelium is not formed into folds or villi like that of the small intestine but has numerous straight tubular glands called **crypts** (see figure 24.26). The crypts are somewhat similar to the intestinal glands of the small intestine, with three cell types that include absorptive, goblet, and granular cells. The major difference is that in the large intestine goblet cells predominate and the other two cell types are greatly reduced in number.

Figure 24.26   Histology of the Large Intestine

(a) Section of the transverse colon cut open to show the inner surface.
(b) Enlargement of the inner surface, showing openings of the crypts.
(c) Higher magnification of a single crypt.

Rectum

The **rectum** is a straight, muscular tube that begins at the termination of the sigmoid colon and ends at the anal canal (see figure 24.25). The mucosal lining of the rectum is simple columnar epithelium, and the muscular tunic is relatively thick compared to the rest of the digestive tract.

Anal Canal

The last 2–3 cm of the digestive tract is the **anal canal** (see figure 24.25). It begins at the inferior end of the rectum and ends at the **anus** (external GI tract opening). The smooth muscle layer of the anal canal is even thicker than that of the rectum and forms the **internal anal sphincter** at the superior end of the anal canal. Skeletal muscle forms the **external anal sphincter** at the inferior end of the canal. The epithelium of the superior part of the anal canal is simple columnar and that of the inferior part is stratified squamous.

Hemorrhoids

Hemorrhoids are the enlargement, or inflammation, of the rectal veins, which supply the anal canal. The condition is also called varicose rectal veins. Hemorrhoids cause pain, itching, and bleeding around the anus. Hemorrhoids can be internal, involving the internal rectal veins, or external, involving the external rectal veins. Treatments include increasing the bulk (indigestible fiber) in the diet, taking sitz baths, and using hydrocortizone suppositories. Surgery may be necessary if the condition is extreme and doesn't respond to other treatments.

45. *Describe the parts of the large intestine. What are teniae coli, haustra, and crypts?*
46. *Explain the difference in structure between the internal anal sphincter and the external anal sphincter.*

Secretions of the Large Intestine

The mucosa of the colon has numerous goblet cells that are scattered along its length and numerous crypts that are lined almost entirely with goblet cells. Little enzymatic activity is associated with secretions of the colon when mucus is the major secretory product (see tables 24.1 and 24.2). Mucus lubricates the wall of the colon and helps the fecal matter stick together. Tactile stimuli and irritation of the wall of the colon trigger local enteric reflexes that increase mucous secretion. Parasympathetic stimulation also increases the secretory rate of the goblet cells.

Diarrhea

When the large intestine is irritated and inflamed, such as in patients with bacterial enteritis (inflamed intestine resulting from bacterial infection of the bowel), the intestinal mucosa secretes large amounts of mucus and electrolytes, and water moves by osmosis into the colon. An abnormally frequent discharge of fluid feces is called **diarrhea.** Although such discharge increases fluid and electrolyte loss, it also moves the infected feces out of the intestine more rapidly and speeds recovery from the disease.

A molecular pump exchanges bicarbonate ions for chloride ions in epithelial cells of the colon in response to acid produced by colic bacteria. Another pump exchanges sodium ions for hydrogen ions. Water crosses the wall of the colon through osmosis with the sodium chloride gradient.

The feces that leave the digestive tract consist of water, solid substances (e.g., undigested food), microorganisms, and sloughed-off epithelial cells.

Numerous microorganisms inhabit the colon. They reproduce rapidly and ultimately constitute about 30% of the dry weight of the feces. Some bacteria in the intestine synthesize vitamin K, which is passively absorbed in the colon, and break down a small amount of cellulose to glucose.

Bacterial actions in the colon produce gases called **flatus** (flă′tŭs; blowing). The amount of flatus depends partly on the bacterial population present in the colon and partly on the type of food consumed. For example, beans, which contain certain complex carbohydrates, are well known for their flatus-producing effect.

Movement in the Large Intestine

Segmental mixing movements occur in the colon much less often than in the small intestine. Peristaltic waves are largely responsible for moving chyme along the ascending colon. At widely spaced intervals (normally three or four times each day), large parts of the transverse and descending colon undergo several strong contractions, called **mass movements.** Each mass movement contraction extends over a much longer part of the digestive tract (≥ 20 cm) than does a peristaltic contraction and propels the colon contents a considerable distance toward the anus (figure 24.27). Mass movements are very common after meals because the presence of food in the stomach or duodenum initiates them. Mass movements are most common about 15 minutes after breakfast. They usually persist for 10–30 minutes and then stop for perhaps half a day. Local reflexes in the enteric plexus, which are called **gastrocolic reflexes** if initiated by the stomach or **duodenocolic reflexes** if initiated by the duodenum, integrate mass movements.

Distention of the rectal wall by feces acts as a stimulus that initiates the **defecation reflex.** Local reflexes cause weak contractions of the rectum and relaxation of the internal anal sphincter. Parasympathetic reflexes cause strong contractions of the rectum and are normally responsible for most of the defecation reflex. Action potentials produced in response to the distention travel along afferent nerve fibers to the sacral region of the spinal cord, where efferent action potentials are initiated that reinforce peristaltic contractions in the lower colon and rectum. The defecation reflex reduces action potentials to the internal anal sphincter, causing it to relax. The external anal sphincter, which is composed of skeletal muscle and is under conscious cerebral control, prevents the movement of feces out of the rectum and through the anal opening. If this sphincter is relaxed voluntarily, feces are expelled. The defecation reflex persists for only a few minutes and quickly declines. Generally, the reflex is reinitiated after a period that may be as long as several hours. Mass movements in the colon are usually the reason for the reinitiation of the defecation reflex.

Defecation is usually accompanied by voluntary movements that support the expulsion of feces. These voluntary movements include a large inspiration of air followed by closure of the larynx and forceful contraction of the abdominal muscles. As a

1. The presence of food in the stomach and chyme in the duodenum stimulate mass movement in the colon.

2. Mass movements are integrated by the enteric plexus, located throughout the colon.

3. They propel the contents of the colon toward the rectum.

4. The presence of feces in the rectum stimulates parasympathetic and local reflexes that result in defecation.

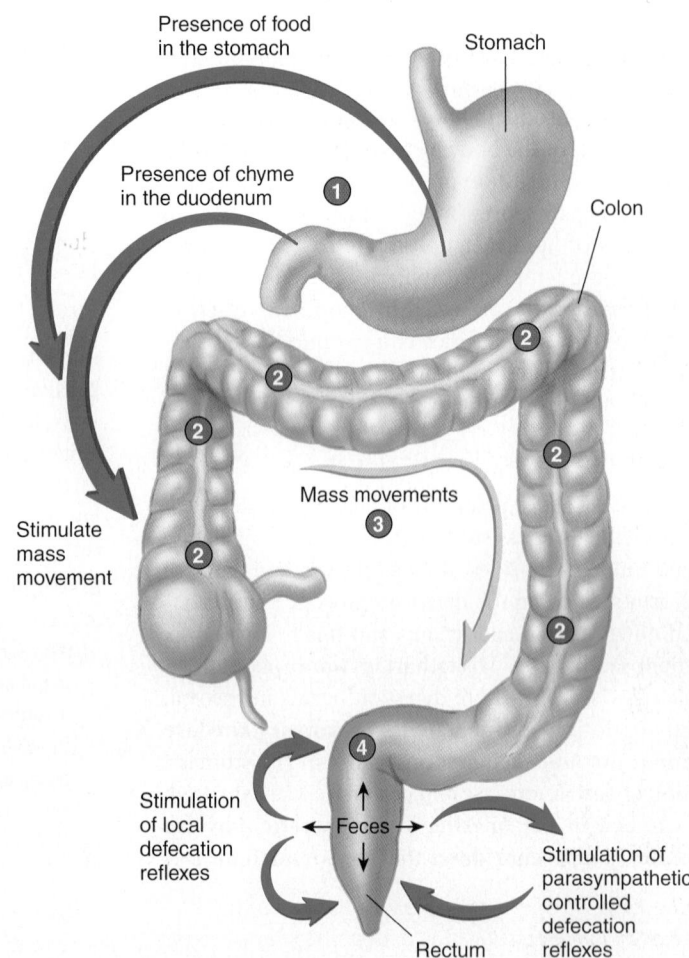

Process Figure 24.27   Reflexes in the Colon and Rectum

consequence, the pressure in the abdominal cavity increases, thereby helping force some of the distal colon contents through the anal canal and out of the anus.

47. *Name the substances secreted and absorbed by the large intestine. What is the role of microorganisms in the colon?*
48. *What kind of movements occur in the colon? Describe the defecation reflex.*

On Being "Regular"

The importance of regularity of defecation has been greatly overestimated. Many people have the misleading notion that a daily bowel movement is critical for good health. As with many other body functions, what is "normal" differs from person to person. Whereas many people defecate one or more times per day, some normal, healthy adults defecate on the average only every other day. A defecation rate of only twice per week, however, is usually described as constipation. Habitually postponing defecation when the defecation reflex occurs can lead to constipation and may eventually result in desensitization of the rectum so that the defecation reflex is greatly diminished.

Digestion, Absorption, and Transport

Digestion is the breakdown of food to molecules that are small enough to be absorbed into the circulation. **Mechanical digestion** breaks large food particles down into smaller ones. **Chemical digestion** involves the breaking of covalent chemical bonds in organic molecules by digestive enzymes. Carbohydrates are broken down into monosaccharides, proteins are broken down into amino acids, and fats are broken down into fatty acids and glycerol. **Absorption** and **transport** are the means by which molecules are moved out of the digestive tract and into the circulation for distribution throughout the body. Not all molecules (e.g., vitamins, minerals, and water) are broken down before being absorbed. Digestion begins in the oral cavity and continues in the stomach, but most digestion occurs in the proximal end of the small intestine, especially in the duodenum.

Absorption of certain molecules can occur all along the digestive tract. A few chemicals, such as nitroglycerin, can be absorbed through the thin mucosa of the oral cavity below the tongue. Some small molecules (e.g., alcohol and aspirin) can diffuse through the stomach epithelium into the circulation. Most absorption, however, occurs in the duodenum and jejunum, although some absorption occurs in the ileum.

Once the digestive products have been absorbed, they are transported to other parts of the body by two different routes. Water, ions, and water-soluble digestion products, such as glucose and amino acids, enter the hepatic portal system and are transported to the liver. The products of lipid metabolism are coated with proteins and transported into lymphatic capillaries called lacteals (see figure 24.16c). The lacteals are connected by lymphatic vessels to the thoracic duct (see chapter 21), which empties into the left subclavian vein. The protein-coated lipid products then travel in the circulation to adipose tissue or to the liver.

Carbohydrates

Ingested **carbohydrates** consist primarily of polysaccharides, such as starches; disaccharides, such as sucrose (table sugar) and lactose (milk sugar); and monosaccharides, such as glucose and fructose (found in many fruits). During the digestion process, polysaccharides are broken down into smaller chains and finally into disaccharides and monosaccharides. Disaccharides are broken down into monosaccharides. Carbohydrate digestion begins in the oral cavity with the partial digestion of starches by **salivary amylase** (am′il-ās). A minor amount of digestion occurs in the stomach through the action of gastric amylase and gelatinase. Carbohydrate digestion is continued in the intestine by **pancreatic amylase** (table 24.4). A series of **disaccharidases** that are bound to the mi-

crovilli of the intestinal epithelium digest disaccharides into monosaccharides.

Lactose Intolerance

Lactase deficiency results in **lactose intolerance,** which is an inability to digest milk products. This disorder is primarily hereditary, affecting 5%–15% of Europeans and 80%–90% of Africans and Asians. Symptoms include cramps, bloating, and diarrhea.

Monosaccharides such as glucose and galactose are taken up into intestinal epithelial cells by cotransport, powered by a sodium ion gradient (figure 24.28). Monosaccharides such as fructose are taken up by facilitated diffusion. The monosaccharides are transferred by facilitated diffusion to the capillaries of the intestinal villi and are carried by the hepatic portal system to the liver, where the nonglucose sugars are converted to glucose. Glucose enters the cells through facilitated diffusion. The rate of glucose transport into most types of cells is greatly influenced by **insulin** and may increase 10-fold in its presence.

Type I Diabetes Mellitus

In patients with type I diabetes mellitus, insulin is lacking, and insufficient glucose is transported into the cells of the body. As a result, the cells do not have enough energy for normal function, blood glucose levels become significantly elevated, and abnormal amounts of glucose are released into the urine. This condition is discussed more fully in chapter 18.

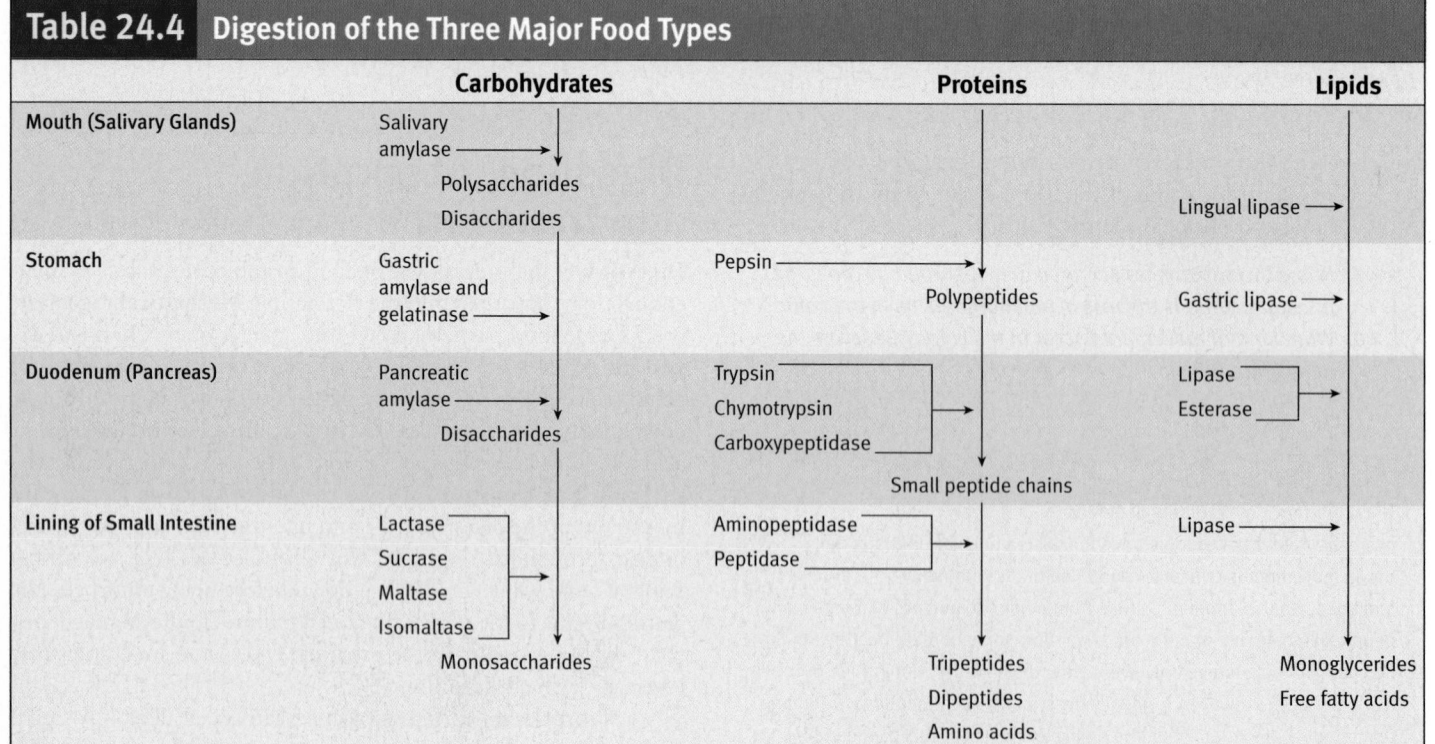

Table 24.4	**Digestion of the Three Major Food Types**		
	Carbohydrates	**Proteins**	**Lipids**
Mouth (Salivary Glands)	Salivary amylase ⟶ Polysaccharides Disaccharides		Lingual lipase ⟶
Stomach	Gastric amylase and gelatinase ⟶	Pepsin ⟶ Polypeptides	Gastric lipase ⟶
Duodenum (Pancreas)	Pancreatic amylase ⟶ Disaccharides	Trypsin Chymotrypsin Carboxypeptidase ⟶ Small peptide chains	Lipase Esterase ⟶
Lining of Small Intestine	Lactase Sucrase Maltase Isomaltase ⟶ Monosaccharides	Aminopeptidase Peptidase ⟶ Tripeptides Dipeptides Amino acids	Lipase ⟶ Monoglycerides Free fatty acids

Villus

Capillary   Lacteal

Intestinal
epithelial cell

Monosaccharide transport

1. Monosaccharides are absorbed by cotransport into intestinal epithelial cells.

2. Monosaccharides move out of intestinal epithelial cells by facilitated diffusion.

3. They enter the capillaries of the intestinal villi and are carried through the hepatic portal vein to the liver.

Na⁺

Na⁺

②

③

Monosaccharide

①

Cotransport

Facilitated
diffusion

To liver

Process Figure 24.28   Transport of Monosaccharides Across the Intestinal Epithelium

Lipids

Lipids are molecules that are insoluble or only slightly soluble in water. They include triglycerides, phospholipids, cholesterol, steroids, and fat-soluble vitamins. **Triglycerides** (trī-glis′er-īdz), also called **triacylglycerol** (trī-as′il-glis′er-ol), consist of three fatty acids and one glycerol molecule covalently bound together. The first step in lipid digestion is **emulsification** (ē-mŭl′si-fi-kā′shŭn), which is the transformation of large lipid droplets into much smaller droplets. The enzymes that digest lipids (pancreatic lipase) are water-soluble and can digest the lipids only by acting at the surface of the droplets. The emulsification process increases the surface area of the lipid exposed to the digestive enzymes by decreasing the droplet size. Emulsification is accomplished by **bile salts** secreted by the liver and stored in the gallbladder.

Lipase (lip′ās) digests lipid molecules (see table 24.4). The vast majority of lipase is secreted by the pancreas. A minor amount of lingual lipase is secreted in the oral cavity, is swallowed with the food, and digests a small amount (<10%) of lipid in the stomach. The stomach also produces very small amounts of gastric lipase. The primary products of lipase digestion are free fatty acids and monoglycerides. Cholesterol and phospholipids also constitute part of the lipid digestion products.

Once lipids are digested in the intestine, bile salts aggregate around the small droplets to form **micelles** (mi-selz′, mī-selz′; a small morsel; figure 24.29). The hydrophobic ends of the bile salts are directed toward the free fatty acids, cholesterol, and monoglycerides at the center of the micelle; and the hydrophilic ends are directed outward toward the water environment. When a micelle comes into contact with the epithelial cells of the small intestine, the contents of the micelle pass by means of simple diffusion through the plasma membrane of the epithelial cells.

Cystic Fibrosis

Cystic fibrosis is a hereditary disorder that occurs in 1 of every 2000 births and affects 33,000 people in the United States; it's the most common lethal genetic disorder among Caucasians. The most critical effects of the disease, accounting for 90% of the deaths, are on the respiratory system. Several other problems occur, however, in affected people. Because the disease is a disorder in chloride ion transport channel proteins, which affects chloride transport and, as a result, movement of water, all exocrine glands are affected. The buildup of thick mucus in the pancreatic and hepatic ducts causes blockage of the ducts so that bile salts and pancreatic digestive enzymes are prevented from reaching the duodenum. As a result, fats and fat-soluble vitamins, which require bile salts to form micelles and which cannot be adequately digested without pancreatic enzymes, are not well digested and absorbed. The patient suffers from vitamin A, D, E, and K deficiencies, which result in conditions like night blindness, skin disorders, rickets, and excessive bleeding. Therapy includes administering the missing vitamins to the patient and reducing dietary fat intake.

Lipid transport

1. Bile salts surround fatty acids and monoglycerides to form micelles.

2. Micelles attach to the cell membranes of intestinal epithelial cells, and the fatty acids and monoglycerides pass by simple diffusion into the intestinal epithelial cells.

3. Within the intestinal epithelial cell, the fatty acids and monoglycerides are converted to triglycerides; proteins coat the triglycerides to form chylomicrons, which move out of the intestinal epithelial cells by exocytosis.

4. The chylomicrons enter the lacteals of the intestinal villi and are carried through the lymphatic system to the general circulation.

Process Figure 24.29   Transport of Lipids Across the Intestinal Epithelium

Lipid Transport

Within the smooth endoplasmic reticulum of the intestinal epithelial cells, free fatty acids are combined with monoglyceride molecules to form triglycerides. Proteins synthesized in the epithelial cells attach to droplets of triglycerides, phospholipids, and cholesterol to form **chylomicrons** (kī-lō-mi′kronz; small particles in the chyle, or fat-filled lymph). The chylomicrons leave the epithelial cells and enter the lacteals of the lymphatic system within the villi. Chylomicrons enter the lymphatic capillaries rather than the blood capillaries because the lymphatic capillaries lack a basement membrane and are more permeable to large particles like chylomicrons (about 0.3 mm in diameter). Chylomicrons are about 90% triglyceride, 5% cholesterol, 4% phospholipid, and 1% protein (figure 24.30). They are carried through the lymphatic system to the bloodstream and then by the blood to adipose tissue. Before entering the adipose cells, triglyceride is broken back down into fatty acids and glycerol, which enter the fat cells and are once more converted to triglyceride. Triglycerides are stored in adipose tissue until an energy source is needed elsewhere in the body. In the liver, the chylomicron lipids are stored, converted into other molecules, or used as energy. The chylomicron remnant, minus the triglyceride, is conveyed through the circulation to the liver, where it is broken up.

Because lipids are either insoluble or only slightly soluble in water, they are transported through the blood in combination with proteins, which are water-soluble. Lipids combined with proteins are called **lipoproteins.** Chylomicrons are one type of lipoprotein. Other lipoproteins are referred to as high- or low-density lipoproteins. Density describes the compactness of a substance and is the ratio of mass to volume. Lipids are less dense than water and tend to float in water. Proteins, which are denser than water, tend to sink in water. A lipoprotein with a high lipid content has a very low density, whereas a lipoprotein with a high protein content has a relatively high density. Chylomicrons, which are made up of 99% lipid and only 1% protein, have an extremely low density. The other major transport lipoproteins are **very low-density lipoprotein (VLDL),** which is 92% lipid and 8% protein, **low-density lipoprotein (LDL),** which is 75% lipid and 25% protein, and **high-density lipoprotein (HDL),** which is 55% lipid and 45% protein (see figure 24.30).

About 15% of the cholesterol in the body is ingested in the food we eat, and the remaining 85% is manufactured in the cells of the body, mostly in the liver and intestinal mucosa. Most of the lipid taken into or manufactured in the liver leaves the liver in the form of VLDL. Most of the triglycerides are removed from the VLDL to be stored in adipose tissue and, as a result, VLDL becomes LDL.

The cholesterol in LDL is critical for the production of steroid hormones in the adrenal cortex and the production of bile acids in the liver. It's also an important component of plasma membranes. LDL is delivered to cells of various tissues through the circulation. Cells have **LDL receptors** in "pits" on their surfaces, which bind the LDL. Once LDL is bound to the receptors, the pits on the cell surface become endocytotic vesicles,

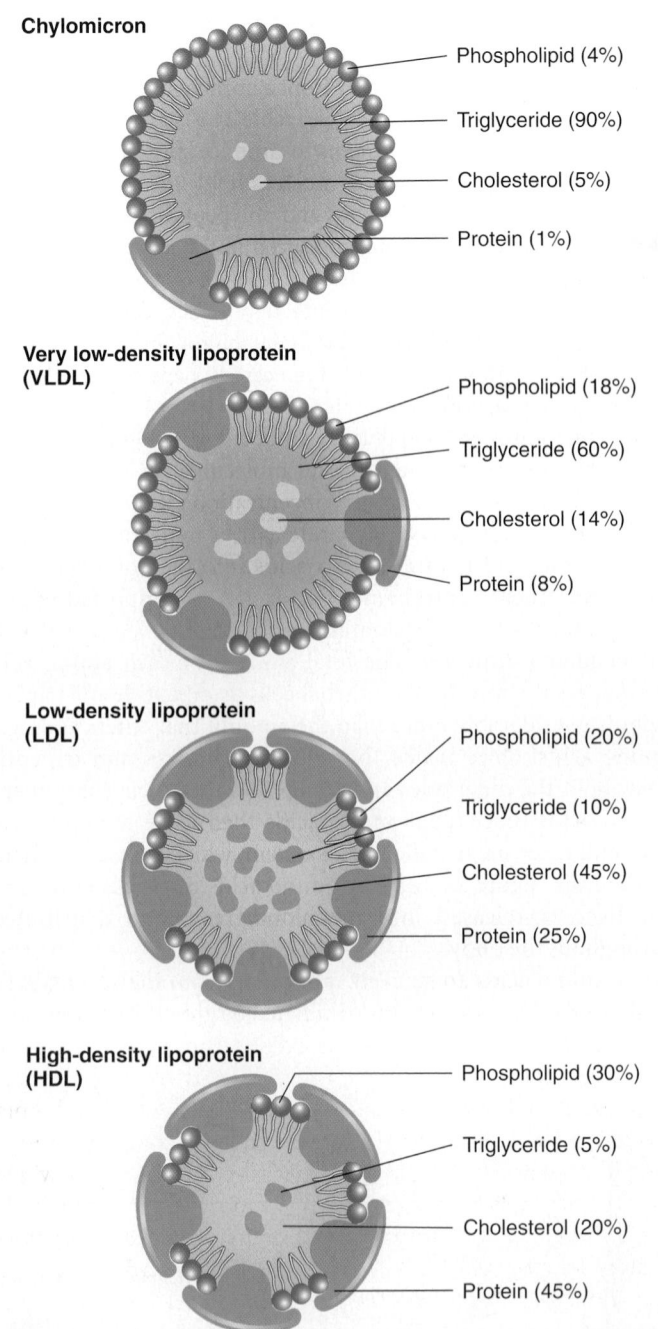

Chylomicron

Phospholipid (4%)

Triglyceride (90%)

Cholesterol (5%)

Protein (1%)

Very low-density lipoprotein (VLDL)

Phospholipid (18%)

Triglyceride (60%)

Cholesterol (14%)

Protein (8%)

Low-density lipoprotein (LDL)

Phospholipid (20%)

Triglyceride (10%)

Cholesterol (45%)

Protein (25%)

High-density lipoprotein (HDL)

Phospholipid (30%)

Triglyceride (5%)

Cholesterol (20%)

Protein (45%)

Figure 24.30   Lipoproteins

LDL

LDL receptor

"Pit" on cell surface

1. Cells have pits on the surface, which contain LDL receptors.

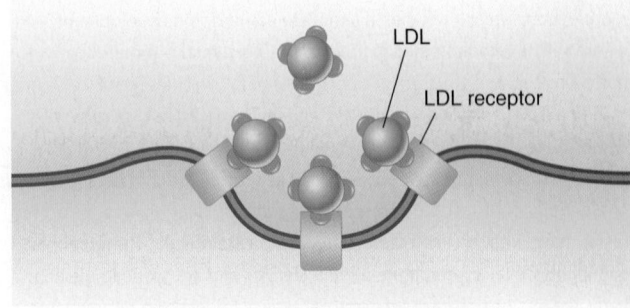

LDL

LDL receptor

2. LDL binds to the LDL receptors in the pits.

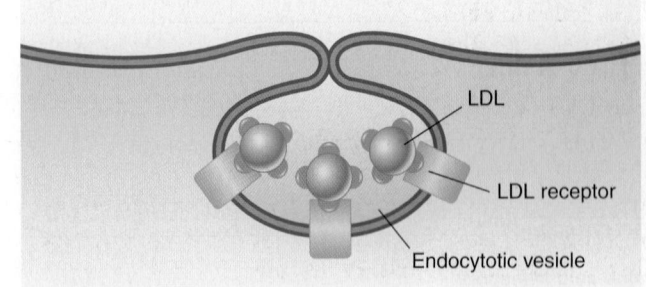

LDL

LDL receptor

Endocytotic vesicle

3. The LDL, bound to LDL receptors, is taken into the cell by endocytosis.

Process Figure 24.31   Transport of LDL into Cells

and the LDL is taken into the cell by receptor-mediated endocytosis (figure 24.31). Each fibroblast, as an example of a tissue cell, has 20,000–50,000 LDL receptors on the surface. Those receptors are confined to cell surface pits, however, which occupy only 2% of the cell surface. Once inside the cell, the endocytotic vesicle combines with a lysosome, and the LDL components are separated for use in the cell.

Cells not only take in cholesterol and other lipids from LDLs, but they also make their own cholesterol. When the combined intake and manufacture of cholesterol exceeds a cell's needs, a negative-feedback system functions, which reduces the amount of LDL receptors and cholesterol manufactured by the cell. Excess lipids are also packaged into HDLs by the cells. These are transported back to the liver for recycling or disposal.

Cholesterol and Coronary Heart Disease

Cholesterol is a major component of atherosclerotic plaques. The level of plasma cholesterol is positively linked to coronary heart disease (CHD). Cholesterol levels of over 200 mg/100 mL increase the risk of CHD. Other risk factors, which are additive to high cholesterol levels, are hypertension, diabetes mellitus, cigarette smoking, and low plasma HDL levels. Low HDL levels are linked to obesity, and weight reduction increases HDL levels. Aerobic exercise can decrease LDL levels and increase HDL levels. Ingestion of saturated fatty acids raises plasma cholesterol levels by stimulating LDL production and inhibiting LDL receptor production. Ingestion of unsaturated fatty acids lowers plasma cholesterol levels. Replacing fats by carbohydrates in the diet can also reduce blood cholesterol levels. The American Heart Association recommends that no more than 30% of an adult's total caloric intake should be from fats and that only 10% be from saturated fats. Our total cholesterol intake should be no more than 300 mg/day. People should eat no more than 7 ounces of meat per day, and that should be chicken, fish, or lean meat. We should eat only two eggs per week and drink milk with 1% or less butter fat. Young children, however, require more fats in their diet to stimulate normal brain development, and whole milk is recommended in their diets. Cholesterol is required for normal membrane structure in cells. Abnormally low cholesterol levels, may lead to weakened blood vessel walls and an increased risk for cerebral hemorrhage.

A small number of people have a genetic disorder in the production or function of LDL receptors, resulting in poor LDL clearing, and, as a result, have what is called **familial hypercholesterolemia.** These people commonly develop premature atherosclerosis and are prone to die at an early age of a heart attack. In some of these disorders, the LDL receptor is not produced. In other cases, the receptor is produced, but it has a lower-than-normal affinity for LDL. In yet other cases, the receptor binds to LDL but the receptor–LDL complex is not taken into the cell by endocytosis.

Proteins

Proteins are taken into the body from a number of dietary sources. **Pepsin** secreted by the stomach (see table 24.3) catalyzes the cleavage of covalent bonds in proteins to produce smaller polypeptide chains. Gastric pepsin digests as much as 10%–20% of the total ingested protein. Once the proteins and polypeptide chains leave the stomach, proteolytic enzymes produced in the pancreas continue the digestive process and produce small peptide chains (see table 24.4). These are broken down into tripeptides, dipeptides, and amino acids by **peptidases** bound to the microvilli of the small intestine. Each peptidase is specific for a certain peptide chain length or for a certain peptide bond.

Dipeptides and tripeptides enter intestinal epithelial cells through a group of related carrier molecules, by a cotransport mechanism, powered by a Na^+ concentration gradient similar to that described for glucose. Separate carrier molecules transport basic, acidic, and neutral amino acids into the epithelial cells. Acidic and most neutral amino acids are cotransported with a Na^+ gradient, whereas basic amino acids enter the epithelial cells by facilitated diffusion. The total amount of each amino acid that enters the intestinal epithelial cells as dipeptides or tripeptides is considerably more than the amount that enters as single amino acids. Once inside the cells, dipeptidases and tripeptidases split the dipeptides and tripeptides into their component amino acids. Individual amino acids then leave the epithelial cells and enter the hepatic portal system, which transports them to the liver (figure 24.32). The amino acids may be modified in the liver or released into the bloodstream and distributed throughout the body.

Amino acids are actively transported into the various cells of the body. This transport is stimulated by growth hormone and

Amino acid transport

1. Tripeptides, dipeptides, as well as acidic and most neutral amino acids, are absorbed by cotransport into intestinal epithelial cells.

2. Amino acids move out of intestinal epithelial cells by active transport.

3. They enter the capillaries of the intestinal villi and are carried through the hepatic portal vein to the liver.

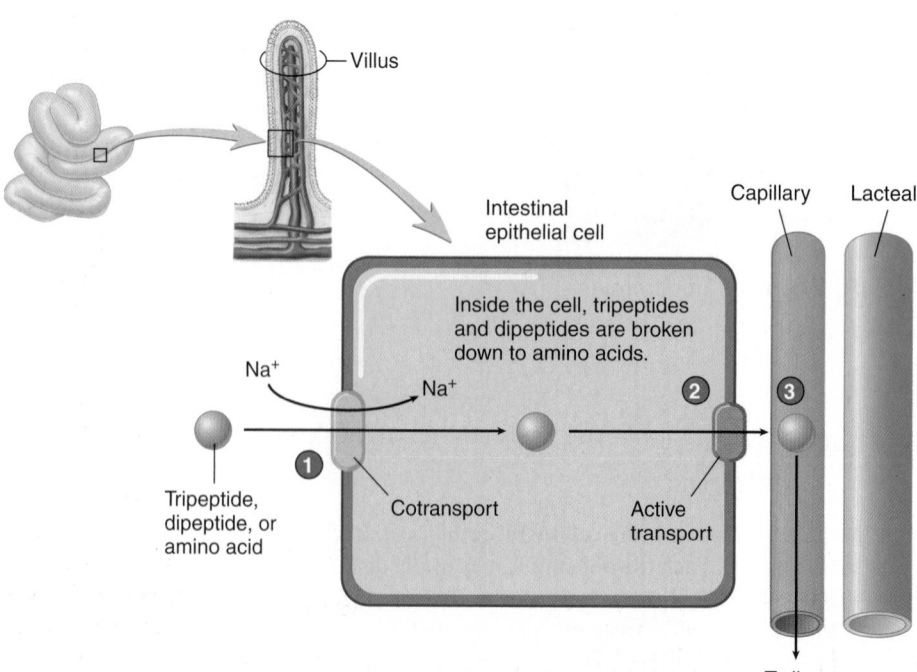

Process Figure 24.32   Amino Acid Transport Across the Intestinal Epithelium insulin. Most amino acids are used as building blocks to form new proteins (see chapter 2), but some amino acids may be used for energy.

Water

About 9 L of **water** enters the digestive tract each day, of which about 92% is absorbed in the small intestine, and another 6%–7% is absorbed in the large intestine (figure 24.33). Water can move in either direction across the wall of the small intestine. Osmotic gradients across the epithelium determine the direction of its diffusion. When the chyme is dilute, water is absorbed by osmosis across the intestinal wall into the blood. When the chyme is very concentrated and contains very little water, water moves by osmosis into the lumen of the small intestine. As nutrients are absorbed in the small intestine, its osmotic pressure decreases; as a consequence, water moves from the intestine into the surrounding extracellular fluid. Water in the extracellular fluid can then enter the circulation. Because of the osmotic gradient produced as nutrients are absorbed in the small intestine, 92% of the water that enters the small intestine by way of the oral cavity, stomach, or intestinal secretions is reabsorbed.

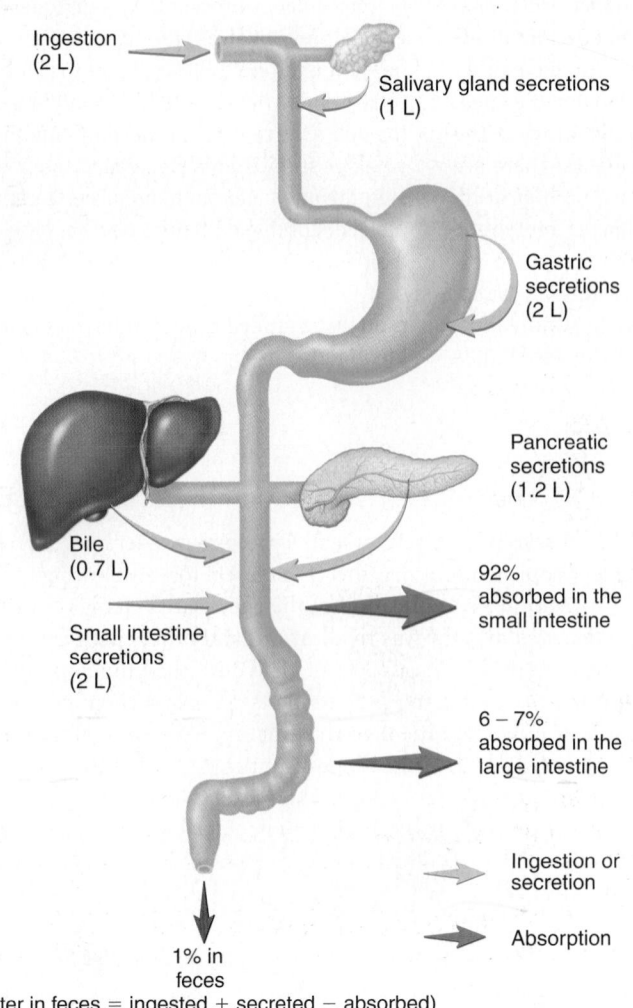

Figure 24.33 **Fluid Volumes in the Digestive Tract**

Ions

Active transport mechanisms for **sodium** ions are present within the epithelial cells of the small intestine. **Potassium, calcium, magnesium,** and **phosphate** are also actively transported. **Chloride** ions move passively through the intestinal wall of the duodenum and the jejunum following the positively charged sodium ions, but chloride ions are actively transported from the ileum. Although calcium ions are actively transported along the entire length of the small intestine, vitamin D is required for that transport process. The absorption of calcium is under hormonal control, as is its excretion and storage. Parathyroid hormones, calcitonin, and vitamin D all play a role in regulating blood levels of calcium in the circulatory system (see chapters 6, 18, and 27).

49. *Describe the mechanism of absorption and the route of transport for water-soluble and lipid-soluble molecules.*
50. *Describe the enzymatic digestion of carbohydrates, lipids, and proteins, and list the breakdown products of each.*
51. *Explain how fats are emulsified. Describe the role of micelles, chylomicrons, VLDLs, LDLs, and HDLs in the absorption and transport of lipids in the body.*
52. *Explain how tripeptides, dipeptides, and amino acids enter intestinal epithelial cells.*
53. *Describe the movement of water through the intestinal wall.*
54. *When and where are various ions absorbed?*

Effects of Aging on the Digestive System

As a person ages, gradual changes occur throughout the entire digestive tract. The connective tissue layers of the digestive tract, the submucosa and serosa, tend to thin. The blood supply to the digestive tract decreases. There is also a decrease in the number of smooth muscle cells in the muscularis, resulting in decreased motility in the digestive tract. In addition, goblet cells within the mucosa secrete less mucus. Glands along the digestive tract, such as the gastric glands, the liver, and the pancreas, also tend to secrete less with age. These changes by themselves don't appreciably decrease the function of the digestive system.

Through the years the digestive tract, like the skin and lungs, is directly exposed to materials from the outside environment. Some of those substances can cause mechanical damage to the digestive tract and others may be toxic to the tissues. Because the connective tissue of the digestive tract becomes thin with age and because the protective mucus covering is reduced, the digestive tract of elderly people becomes less and less protected from these outside influences. In addition, the mucosa of elderly people tends to heal more slowly following injury. The liver's ability to detoxify certain chemicals tends to decline, the ability of the hepatic phagocytic cells to remove particulate contaminants decreases, and the liver's ability to store glycogen decreases. These problems are increased in people who smoke.

This overall decline in the defenses of the digestive tract with advancing age leaves elderly people more susceptible to infections and to the effects of toxic agents. Elderly people are more likely to

Clinical Focus   Intestinal Disorders

Inflammatory Bowel Disease

Inflammatory bowel disease (IBD) is the general name given to either Crohn's disease or ulcerative colitis. IBD occurs at a rate in Europe and North America of approximately 4 to 8 new cases per 100,000 people per year, which is much higher than in Asia and Africa. Males and females are affected about equally. IBD is of unknown cause, but infectious, autoimmune, and hereditary factors have been implicated. **Crohn's disease** involves localized inflammatory degeneration that may occur anywhere along the digestive tract but most commonly involves the distal ileum and proximal colon. The degeneration involves the entire thickness of the digestive tract wall. The intestinal wall often becomes thickened, constricting the lumen, with ulcerations and fissures in the damaged areas. The disease causes diarrhea, abdominal pain, fever, and weight loss. Treatment centers around anti-inflammatory drugs, but other treatments, including avoiding foods that increase symptoms and even surgery, are employed. **Ulcerative colitis** is limited to the mucosa of the large intestine. The involved mucosa exhibits inflammation, including edema, vascular congestion, hemorrhage, and the accumulation of plasma cells, lymphocytes, neutrophils, and eosinophils. Patients may experience abdominal pain, fever, malaise, fatigue, and weight loss, as well as diarrhea and hemorrhage. In rare cases, severe diarrhea and hemorrhage may require transfusions. Treatment includes the use of anti-inflammatory drugs and, in some cases, avoiding foods that increase symptoms.

Irritable Bowel Syndrome

Irritable bowel syndrome (IBS) is a disorder of unknown cause in which intestinal mobility is abnormal. The disorder accounts for over half of all referrals to gastroenterologists. Male and female children are affected equally, but adult females are affected twice as often as males. IBS patients experience abdominal pain mainly in the left lower quadrant, especially after eating. They also have alternating bouts of constipation and diarrhea. There is no specific histopathology in the digestive tracts of IBS patients. There are no anatomic abnormalities, no indication of infection, and no sign of metabolic causes. Patients with IBS appear to exhibit greater-than-normal levels of psychological stress or depression and show increased contractions of the esophagus and small intestine during times of stress. There is a high familial incidence. Some patients might present with a history of traumatic events such as physical or sexual abuse. Treatments include psychiatric counseling and stress management, diets with increased fiber and limited gas-producing foods, and loose clothing. In some patients, drugs that reduce parasympathetic stimulation of the digestive system may be useful.

Malabsorption Syndrome

Malabsorption syndrome (sprue) is a spectrum of disorders of the small intestine that results in abnormal nutrient absorption. One type of malabsorption results from an immune response to gluten, which is present in certain types of grains and involves the destruction of newly formed epithelial cells in the intestinal glands. These cells fail to migrate to the villi surface, the villi become blunted, and the surface area decreases. As a result, the intestinal epithelium is less capable of absorbing nutrients. Another type of malabsorption (called tropical malabsorption) is apparently caused by bacteria, although no specific bacterium has been identified.

---

develop ulcerations and cancers of the digestive tract. Colorectal cancers, for example, are the second leading cause of cancer deaths in the United States, with an estimated 135,000 new cases and 57,000 deaths each year.

    Gastroesophageal reflux disorder (GERD) increases with advancing age. It is probably the main reason that elderly people take antacids, H_2 antagonists, and proton pump inhibitors. Disorders that are not necessarily age-induced, such as hiatal hernia and irregular or inadequate esophageal motility, may be worsened by the effects of aging, because of a general decreased motility in the digestive tract.

    The enamel on the surface of elderly people's teeth becomes thinner with age and may expose the underlying dentin. In addition, the gingiva covering the tooth root recedes, exposing additional dentin. Exposed dentin may become painful and change the person's eating habits. Many elderly people also lose teeth, which can have a marked effect on eating habits unless artificial teeth are provided.

The muscles of mastication tend to become weaker and, as a result, older people tend to chew their food less before swallowing.

    Another complication of the age-related changes in the digestive system is the way medications and other chemicals are absorbed from the digestive tract. The decreased mucous covering and the thinned connective tissue layers allow chemicals to pass more readily from the digestive tract into the circulatory system. However, a decline in the blood supply to the digestive tract hinders the absorption of such chemicals. Drugs administered to treat cancer, which occurs in many elderly people, may irritate the mucosa of the digestive tract, resulting in nausea and loss of appetite.

● *55. What is the general effect of aging on digestive tract secretions?*

56. What are the effects of the overall decline in the defenses of the digestive tract with advancing age?

Enteritis

Enteritis is any inflammation of the intestines that can result in diarrhea, dehydration, fatigue, and weight loss. It may result from an infection, chemical irritation, or from some unknown cause. **Regional enteritis,** or Crohn's disease, is a local enteritis of unknown cause characterized by patchy, deep ulcers developing in the intestinal wall, usually in the distal end of the ileum. The disease results in overproliferation of connective tissue and invasion of lymphatic tissue into the involved area, with a subsequent thickening of the intestinal wall and narrowing of the lumen.

Colitis is an inflammation of the colon.

Colon Cancer

Colon cancer is the second leading cause of cancer-related deaths in the United States and accounts for 55,000 deaths a year. Susceptibility to colon cancer can be familial; however, a correlation exists between colon cancer and diets low in fiber and high in fat. People who eat beef, pork, or lamb daily have 2.5 times the risk of developing colon cancer compared to people who eat these meats less than once per month. Eating processed meats increases the risk by an additional 50%–100%. Ingesting calcium in the form of calcium carbonate antacid tablets at twice the recommended daily allowances may prevent 75% of colon cancers. Greatly increased calcium levels may also cause constipation.

A gene for colon cancer may be present in as many as 1 in 200 people, making colon cancer one of the most common inherited diseases. Nine different genes have been found to be associated with colon cancer. Most of those genes are involved in cell regulation, that is, keeping cell growth in check, but one gene mutation results in a high degree of genetic instability. As a result of this mutation, the DNA is not copied accurately during cell division of the colon cancer cells, causing wholesale errors and mutations throughout the genome (all the genes). Such genetic instability has been identified in 13% of sporadic (not occurring in families) colon cancer. Screening for colon cancer includes testing the stool for blood content and performing a colonoscopy, which allows the physician to see into the colon.

Constipation

Constipation is the slow movement of feces through the large intestine. The feces often become dry and hard because of increased fluid absorption during the extended time they are retained in the large intestine. In the United States, 2.5 million doctor visits occur each year from people complaining of constipation, and $400 million dollars is spent each year on laxatives.

Constipation often results after a prolonged time of inhibiting normal defecation reflexes. A change in habits, such as travel, dehydration, depression, disease, metabolic disturbances, certain medications, pregnancy, or dependency on laxatives can all cause constipation. Irritable bowel syndrome, also called spastic colon, which is of unknown cause but is stress-related, can also cause constipation. Constipation can also occur with diabetes, kidney failure, colon nerve damage, or spinal cord injuries or as the result of an obstructed bowel; of greatest concern, the obstruction could be caused by colon cancer. Chronic constipation can result from the slow movement of feces through the entire colon, in just the distal part (descending colon and rectum), or in just the rectum. Interestingly, in one large study of people who claimed to be suffering from chronic constipation, one-third were found to have normal movement of feces through the large intestine. Defecation frequency was often normal. Many of those people were suffering from psychologic distress, anxiety, or depression and just thought they had abnormal defecation frequencies.

S U M M A R Y

Anatomy of the Digestive System   (p. 878)

1. The digestive system consists of a digestive tube and its associated accessory organs.
2. The digestive system consists of the oral cavity, pharynx, esophagus, stomach, small intestine, large intestine, and anus.
3. Accessory organs such as the salivary glands, liver, gallbladder, and pancreas are located along the digestive tract.

Functions of the Digestive System   (p. 878)

The functions of the digestive system are ingestion, mastication, propulsion, mixing, secretion, digestion, absorption, and elimination.

Histology of the Digestive Tract   (p. 880)

The digestive tract is composed of four tunics: mucosa, submucosa, muscularis, and serosa or adventitia.

Mucosa

1. The mucosa consists of a mucous epithelium, a lamina propria, and a muscularis mucosae.
2. The epithelium extends into the lamina propria to form intestinal glands.

Submucosa

The submucosa is a connective tissue layer containing the submucosal plexus (part of the enteric plexus), blood vessels, and small glands.

Muscularis

1. The muscularis consists of an inner layer of circular smooth muscle and an outer layer of longitudinal smooth muscle.
2. The myenteric plexus is between the two muscle layers.
3. Interstitial cells of Cajal are located throughout the myenteric plexus.

Systems Pathology
Diarrhea

While on vacation in Mexico, Mr. T was shopping with his wife when he started to experience sharp pains in his abdominal region (figure B). He also began to feel hot and sweaty and felt an extreme urge to defecate. His wife quickly looked up the word *toilet* in their handy Spanish–English pocket travel dictionary, and Mr. T anxiously inquired of a local resident where the nearest facility could be found. Once the immediate need was taken care of, Mr. and Mrs. T went back to their hotel room, where they remained while Mr. T recovered. Over the next 2 days, his stools were frequent and watery. He also vomited a couple of times. Because they were in a foreign country, Mr. T didn't consult a physician. He rested, took plenty of fluids, and was feeling much better, although a little weak, in a couple of days.

Background Information

Diarrhea is one of the most common complaints in clinical medicine and affects more than half of the tourists in developing countries. **Diarrhea** is defined as any change in bowel habits in which stool frequency or volume is increased or in which stool fluidity is increased. Diarrhea is not itself a disease but is a symptom of a wide variety of disorders. Normally, about 600 mL of fluid enters the colon each day and all but 150 mL is reabsorbed. The loss of more than 200 mL of stool per day is considered abnormal.

Mucus secretion by the colon increases dramatically in response to diarrhea. This mucus contains large quantities of bicarbonate ions, which comes from the dissociation of carbonic acid into bicarbonate ions (HCO_3^-) and hydrogen (H^+) ions within the blood supply to the colon. The HCO_3^- enter the mucus secreted by the colon, whereas the H^+ remain in the circulation and, as a result, the blood pH decreases. Thus, a condition called metabolic acidosis can develop (see chapter 27).

Diarrhea in tourists usually results from the ingestion of food or water contaminated with bacteria or bacterial toxins. Acute diarrhea is

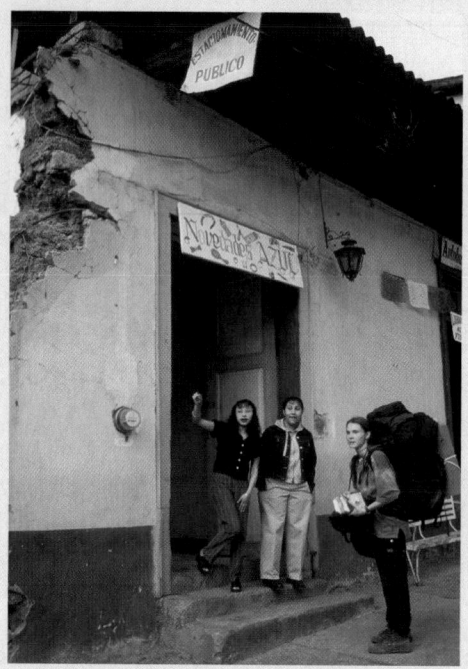

Figure B   **Many Tourists Develop Diarrhea**

defined as lasting less than 2–3 weeks, and diarrhea lasting longer than that is considered chronic. Acute diarrhea is usually self-limiting, but some forms of diarrhea can be fatal if not treated. Diarrhea results from either a decrease in fluid absorption in the gut or an increase in fluid secretion. Some bacterial toxins and other chemicals can also cause an increase in bowel motor activity. As a result, chyme is moved more rapidly through the digestive tract, fewer nutrients and water are

Serosa or Adventitia

The serosa or adventitia forms the outermost layer of the digestive tract.

Regulation of the Digestive System   (p. 881)

1. Nervous, hormonal, and local chemical mechanisms regulate digestion.
2. Nervous regulation involves the enteric nervous system and CNS reflexes.
3. Over 30 neurotransmitters are associated with the ENS.
4. The digestive tract produces hormones that regulate digestion.
5. Other chemicals are produced by the digestive tract that exercise local control of digestion.

Peritoneum   (p. 882)

1. The peritoneum is a serous membrane that lines the abdominal cavity and organs.
2. Mesenteries are peritoneum that extends from the body wall to many of the abdominal organs.
3. Retroperitoneal organs are located behind the peritoneum.

Oral Cavity   (p. 884)

The oral cavity includes the vestibule and oral cavity proper.

System Interactions | The Effects of Diarrhea on Other Systems

System	Interactions
Integumentary	Pallor occurs due to vasoconstriction of blood vessels in the skin, resulting from a decrease in blood fluid levels. Pallor and sweating increase in response to abdominal pain and anxiety.
Muscular	Muscular weakness may result due to electrolyte loss, metabolic acidosis, fever, and general malaise. The involuntary stimulus to defecate may become so strong as to overcome the voluntary control mechanisms.
Nervous	Local reflexes in the colon respond to increased colon fluid volume by stimulating mass movements and the defecation reflex. Abdominal pain, much of which is felt as referred pain, can occur as the result of inflammation and distention of the colon. Decreased function is due to electrolyte loss. Reduced blood fluid levels stimulate a sensation of thirst in the CNS.
Endocrine	A decrease in extracellular fluid volume, due to the loss of fluid in the feces, stimulates the release of hormones (antidiuretic hormone from the posterior pituitary and aldosterone from the adrenal cortex) that increase water retention and electrolyte reabsorption in the kidney. In addition, decreased extracellular fluid volume and anxiety result in increased release of epinephrine and norepinephrine from the adrenal medulla.
Cardiovascular	Movement of extracellular fluid into the colon results in a decreased blood volume. The reduced blood volume activates the baroreceptor reflex, antidiuretic hormone release, the renin-angiotensin-aldosterone mechanism, and atrial naturitic hormone release, which all function to elevate blood volume or increase blood pressure.
Lymphatic and Immune	White blood cells migrate to the colon in response to infection and inflammation. In the case of bacterial diarrhea, the immune response is initiated to begin production of antibodies against bacteria and bacterial toxins.
Respiratory	As the result of reduced blood pH, the rate of respiration increases to eliminate carbon dioxide, which helps eliminate excess H^+.
Urinary	A decrease in urine volume and an increase in urine concentration results from activation of the baroreceptor reflex, which decreases blood flow to the kidney; antidiuretic hormone secretion, which increases water reabsorption in the kidney; and aldosterone secretion, which increases electrolyte and water reabsorption in the kidney. After a period of approximately 24 hours, the kidney is activated to compensate for metabolic acidosis by increasing hydrogen ion secretion and bicarbonate ion reabsorption.

absorbed out of the small intestine, and more water enters the colon. Symptoms can occur in as little as 1–2 hours after bacterial toxins are ingested to as long as 24 hours or more for some strains of bacteria.

In cases of short-term acute diarrhea, the infectious agent is seldom identified. Nearly any bacterial species is capable of causing diarrhea. Some types of bacterial diarrhea include severe vomiting, whereas others do not. Some bacterial toxins also induce fever. Some viruses and amebic parasites can also cause diarrhea. In most cases, laboratory analysis of food or stool is necessary to identify the causal organism. In cases of mild diarrhea away from home, laboratory evaluation is not practical, and empiric therapy is usually applied. Fluids and electrolytes must be replaced, and consumption of fluids with electrolytes is important. The diet should be limited to clear fluids during at least the first day or so. Bismuth subsalicylate (Pepto-Bismol) or loperamide (Imodium; except in cases of fever) may also be used to help combat secretory diarrhea. Milk and milk products should be avoided. Breads, toast, rice, and baked fish or chicken can be added to the diet with improvement. A normal diet can be resumed after 2–3 days.

PREDICT
Predict the effects of prolonged diarrhea.

Lips and Cheeks

The lips and cheeks are involved in facial expression, mastication, and speech.

Palate and Palatine Tonsils

1. The roof of the oral cavity is divided into the hard and soft palates.
2. The palatine tonsils are located in the lateral wall of the fauces.

Tongue

1. The tongue is involved in speech, taste, mastication, and swallowing.
2. The intrinsic tongue muscles change the shape of the tongue, and the extrinsic tongue muscles move the tongue.
3. The anterior two-thirds of the tongue is covered with papillae, the posterior one-third is devoid of papillae.

Teeth

1. Twenty deciduous teeth are replaced by 32 permanent teeth.
2. The types of teeth are incisors, canines, premolars, and molars.
3. A tooth consists of a crown, a neck, and a root.
4. The root is composed of dentin. Within the dentin of the root is the pulp cavity, which is filled with pulp, blood vessels, and nerves. The crown is dentin covered by enamel.
5. Periodontal ligaments hold the teeth in the alveoli.

Mastication

The muscles of mastication are the masseter, the temporalis, the medial pterygoid, and the lateral pterygoid.

Salivary Glands

1. Salivary glands produce serous and mucous secretions.
2. The three pairs of large salivary glands are the parotid, submandibular, and sublingual.

Pharynx   (p. 890)

The pharynx consists of the nasopharynx, oropharynx, and laryngopharynx.

Esophagus   (p. 890)

1. The esophagus connects the pharynx to the stomach. The upper and lower esophageal sphincters regulate movement.
2. The esophagus consists of an outer adventitia, a muscular layer (longitudinal and circular), a submucosal layer (with mucous glands), and a stratified squamous epithelium.

Swallowing   (p. 890)

1. During the voluntary phase of deglutition, a bolus of food is moved by the tongue from the oral cavity to the pharynx.
2. The pharyngeal phase is a reflex caused by stimulation of stretch receptors in the pharynx.
   - The soft palate closes the nasopharynx, and the epiglottis and vestibular folds close the opening into the larynx.
   - Pharyngeal muscles move the bolus to the esophagus.
3. The esophageal phase is a reflex initiated by the stimulation of stretch receptors in the esophagus. A wave of contraction (peristalsis) moves the food to the stomach.

Stomach   (p. 892)

Anatomy of the Stomach

The openings of the stomach are the gastroesophageal (to the esophagus) and the pyloric (to the duodenum).

Histology of the Stomach

1. The wall of the stomach consists of an external serosa, a muscle layer (longitudinal, circular, and oblique), a submucosa, and simple columnar epithelium (surface mucous cells).
2. Rugae are the folds in the stomach when it is empty.
3. Gastric pits are the openings to the gastric glands which contain mucous neck cells, parietal cells, chief cells, and endocrine cells.

Secretions of the Stomach

1. Mucus protects the stomach lining.
2. Pepsinogen is converted to pepsin, which digests proteins.
3. Hydrochloric acid promotes pepsin activity and kills microorganisms.
4. Intrinsic factor is necessary for vitamin B_{12} absorption.
5. The sight, smell, taste, or thought of food initiates the cephalic phase. Nerve impulses from the medulla stimulate hydrochloric acid, pepsinogen, gastrin, and histamine secretion.
6. Distention of the stomach, which stimulates gastrin secretion and activates CNS and local reflexes that promote secretion, initiates the gastric phase.
7. Acidic chyme, which enters the duodenum and stimulates neuronal reflexes and the secretion of hormones that inhibit gastric secretions, initiates the gastrointestinal phase.

Movements of the Stomach

1. The stomach stretches and relaxes to increase volume.
2. Mixing waves mix the stomach contents with stomach secretions to form chyme.
3. Peristaltic waves move the chyme into the duodenum.

4. Gastrin and stretching of the stomach stimulate stomach emptying.
5. Chyme entering the duodenum inhibits movement through neuronal reflexes and the release of hormones.

Small Intestine   (p. 900)

The small intestine is divided into the duodenum, jejunum, and ileum.

Anatomy and Histology of the Small Intestine

1. The wall of the small intestine consists of an external serosa, muscles (longitudinal and circular), submucosa, and simple columnar epithelium.
2. Circular folds, villi, and microvilli greatly increase the surface area of the intestinal lining.
3. Absorptive, goblet, and endocrine cells are in intestinal glands. Duodenal glands produce mucus.

Secretions of the Small Intestine

1. Mucus protects against digestive enzymes and stomach acids.
2. Digestive enzymes (disaccharidases and peptidases) are bound to the intestinal wall.
3. Chemical or tactile irritation, vagal stimulation, and secretin stimulate intestinal secretion.

Movement in the Small Intestine

1. Segmental contractions mix intestinal contents. Peristaltic contractions move materials distally.
2. Stretch of smooth muscles, local reflexes, and the parasympathetic nervous system stimulate contractions. Distention of the cecum initiates a reflex that inhibits peristalsis.

Liver   (p. 903)
Anatomy of the Liver

1. The liver has four lobes: right, left, caudate, and quadrate.
2. The liver is divided into lobules.
   - The hepatic cords are composed of columns of hepatocytes that are separated by the bile canaliculi.
   - The sinusoids are enlarged spaces filled with blood and lined with endothelium and hepatic phagocytic cells.

Histology of the Liver

1. The portal triads supply the lobules.
   - The hepatic arteries and the hepatic portal veins bring blood to the lobules and empty into the sinusoids.
   - The sinusoids empty into central veins, which join to form the hepatic veins, which leave the liver.
   - Bile canaliculi converge to form hepatic ducts, which leave the liver.
2. Bile leaves the liver through the hepatic duct system.
   - The hepatic ducts receive bile from the lobules.
   - The cystic duct from the gallbladder joins the hepatic duct to form the common bile duct.
   - The common bile duct joins the pancreatic duct at the point at which it empties into the duodenum.

Functions of the Liver

1. The liver produces bile, which contains bile salts that emulsify fats. Secretin increases bile production.
2. The liver stores and processes nutrients, produces new molecules, and detoxifies molecules.
3. Hepatic phagocytic cells phagocytize red blood cells, bacteria, and other debris.
4. The liver produces blood components.

Gallbladder   (p. 908)

1. The gallbladder is a small sac on the inferior surface of the liver.
2. The gallbladder stores and concentrates bile.
3. Cholecystokinin stimulates gallbladder contraction.

Pancreas   (p. 909)
Anatomy of the Pancreas

1. The pancreas is an endocrine and an exocrine gland. Its exocrine function is the production of digestive enzymes.
2. The pancreas is divided into lobules that contain acini. The acini connect to a duct system that eventually forms the pancreatic duct, which empties into the duodenum.

Pancreatic Secretions

Secretin stimulates the release of a watery bicarbonate solution that neutralizes acidic chyme.

Regulation of Pancreatic Secretion

Cholecystokinin and the vagus nerve stimulate the release of digestive enzymes.

Large Intestine   (p. 912)
Anatomy of the Large Intestine

1. The cecum forms a blind sac at the junction of the small and large intestines. The vermiform appendix is a blind tube off the cecum.
2. The ascending colon extends from the cecum superiorly to the right colic flexure. The transverse colon extends from the right to the left colic flexure. The descending colon extends inferiorly to join the sigmoid colon.
3. The sigmoid colon is an S-shaped tube that ends at the rectum.
4. Longitudinal smooth muscles of the large intestine wall are arranged into bands called teniae coli that contract to produce pouches called haustra.
5. The mucosal lining of the large intestine is simple columnar epithelium with mucus-producing crypts.
6. The rectum is a straight tube that ends at the anus.
7. An internal anal sphincter (smooth muscle) and an external anal sphincter (skeletal muscle) surround the anal canal.

Secretions of the Large Intestine

1. Mucus provides protection to the intestinal lining.
2. Epithelial cells secrete bicarbonate ions. Sodium is absorbed by active transport, and water is absorbed by osmosis.
3. Microorganisms are responsible for vitamin K production, gas production, and much of the bulk of feces.

Movement in the Large Intestine

1. Segmental movements mix the colon's contents.
2. Mass movements are strong peristaltic contractions that occur three to four times a day.
3. Defecation is the elimination of feces. Reflex activity moves feces through the internal anal sphincter. Voluntary activity regulates movement through the external anal sphincter.

Digestion, Absorption, and Transport   (p. 915)

1. Digestion is the breakdown of organic molecules into their component parts.
2. Absorption and transport are the means by which molecules are moved out of the digestive tract and are distributed throughout the body.
3. Transportation occurs by two different routes.
   - Water, ions, and water-soluble products of digestion are transported to the liver through the hepatic portal system.
   - The products of lipid digestion are transported through the lymphatic system to the circulatory system.

Carbohydrates

1. Carbohydrates consist of starches, glycogen, sucrose, lactose, glucose, and fructose.
2. Polysaccharides are broken down into monosaccharides by a number of different enzymes.
3. Monosaccharides are taken up by intestinal epithelial cells by active transport or by facilitated diffusion.
4. The monosaccharides are carried to the liver where the nonglucose sugars are converted to glucose.
5. Glucose is transported to the cells that require energy.
6. Glucose enters the cells through facilitated diffusion.
7. Insulin influences the rate of glucose transport.

Lipids

1. Lipids include triglycerides, phospholipids, steroids, and fat-soluble vitamins.
2. Emulsification is the transformation of large lipid droplets into smaller droplets and is accomplished by bile salts.
3. Lipase digests lipid molecules to form free fatty acids and glycerol.
4. Micelles form around lipid digestion products and move to epithelial cells of the small intestine, where the products pass into the cells by simple diffusion.
5. Within the epithelial cells, free fatty acids are combined with glycerol to form triglyceride.
6. Proteins coat triglycerides, phospholipids, and cholesterol to form chylomicrons.
7. Chylomicrons enter lacteals within intestinal villi and are carried through the lymphatic system to the bloodstream.
8. Triglyceride is stored in adipose tissue, converted into other molecules, or used as energy.
9. Lipoproteins include chylomicrons, VLDL, LDL, and HDL.
10. LDL transports cholesterol to cells, and HDL transports it from cells to the liver.
11. LDLs are taken into cells by receptor-mediated endocytosis, which is controlled by a negative-feedback mechanism.

Proteins

1. Pepsin in the stomach breaks proteins into smaller polypeptide chains.
2. Proteolytic enzymes from the pancreas produce small peptide chains.
3. Peptidases, bound to the microvilli of the small intestine, break down peptides.
4. Amino acids are absorbed by cotransport, which requires transport of sodium.
5. Amino acids are transported to the liver, where the amino acids can be modified or released into the bloodstream.
6. Amino acids are actively transported into cells under the stimulation of growth hormone and insulin.
7. Amino acids are used as building blocks or for energy.

Water

Water can move in either direction across the wall of the small intestine, depending on the osmotic gradients across the epithelium.

Ions

1. Sodium, potassium, calcium, magnesium, and phosphate are actively transported.
2. Chloride ions move passively through the wall of the duodenum and jejunum but are actively transported from the ileum.
3. Calcium ions are actively transported, but vitamin D is required for transport, and the transport is under hormonal control.

Effects of Aging on the Digestive System   (p. 921)

The mucous layer, the connective tissue, the muscles, and the secretions all tend to decrease as a person ages. These changes make an older person more open to infections and toxic agents.

R E V I E W   A N D   C O M P R E H E N S I O N

1. Which layer of the digestive tract is in direct contact with the food that is consumed?
   a. mucosa
   b. muscularis
   c. serosa
   d. submucosa

2. The enteric plexus is found in the
   a. submucosa layer.
   b. muscularis layer.
   c. serosa layer.
   d. both a and b.
   e. all of the above.

3. The tongue
   a. holds food in place during mastication.
   b. plays a major role in swallowing.
   c. helps to form words during speech.
   d. is a major sense organ for taste.
   e. all of the above.

4. Dentin
   a. forms the surface of the crown of the teeth.
   b. holds the teeth to the periodontal ligaments.
   c. is found in the pulp cavity.
   d. makes up most of the structure of the teeth.
   e. is harder than enamel.

5. The number of premolar deciduous teeth is
   a. 0.
   b. 2.
   c. 4.
   d. 8.
   e. 12.

6. Which of these glands does *not* secrete saliva into the oral cavity?
   a. submandibular glands
   b. pancreas
   c. sublingual glands
   d. parotid glands

7. The portion of the digestive tract in which digestion begins is the
   a. oral cavity.
   b. esophagus.
   c. stomach.
   d. duodenum.
   e. jejunum.

8. During deglutition (swallowing),
   a. movement of food results primarily from gravity.
   b. the swallowing center in the medulla oblongata is activated.
   c. food is pushed into the oropharynx during the pharyngeal phase.
   d. the soft palate closes off the opening into the larynx.

9. The stomach
   a. has large folds in the submucosa and mucosa called rugae.
   b. has two layers of smooth muscle in the muscularis layer.
   c. opening from the esophagus is the pyloric opening.
   d. has an area closest to the duodenum called the fundus.
   e. all of the above.

10. Which of these stomach cell types is *not* correctly matched with its function?
    a. surface mucous cells: produce mucus
    b. parietal cells: produce hydrochloric acid
    c. chief cells: produce intrinsic factor
    d. endocrine cells: produce regulatory hormones

11. HCl
    a. is an enzyme.
    b. creates the acid condition necessary for pepsin to work.
    c. is secreted by the small intestine.
    d. activates salivary amylase.
    e. all of the above.

12. Why doesn't the stomach digest itself?
    a. The stomach wall is not composed of protein, so it's not affected by proteolytic enzymes.
    b. The digestive enzymes of the stomach are not strong enough to digest the stomach wall.
    c. The lining of the stomach wall has a protective layer of epithelial cells.
    d. The stomach wall is protected by large amounts of mucus.

13. Which of these hormones stimulates stomach secretions?
    a. cholecystokinin
    b. gastric inhibitory peptide
    c. gastrin
    d. secretin

14. Which of these phases of stomach secretion is correctly matched?
    a. Cephalic phase: the largest volume of secretion is produced.
    b. Gastric phase: gastrin secretion is inhibited by distention of the stomach.
    c. Gastric phase: initiated by chewing, swallowing, or thinking of food.
    d. Gastrointestinal phase: stomach secretions are inhibited.

15. Which of these structures function to increase the mucosal surface of the small intestine?
    a. circular folds
    b. villi
    c. microvilli
    d. length of the small intestine
    e. all of the above

16. Given these parts of the small intestine:
    1. duodenum
    2. ileum
    3. jejunum

    Choose the arrangement that lists the parts in the order food encounters them as it passes from the stomach through the small intestine.
    a. 1,2,3
    b. 1,3,2
    c. 2,1,3
    d. 2,3,1
    e. 3,1,2

17. Which cells in the small intestine have digestive enzymes attached to their surfaces?
    a. mucuous cells
    b. goblet cells
    c. endocrine cells
    d. absorptive cells

18. The hepatic sinusoids
    a. receive blood from the hepatic artery.
    b. receive blood from the hepatic portal vein.
    c. empty into the central veins.
    d. all of the above.

19. Given these ducts:
    1. common bile duct
    2. common hepatic duct
    3. cystic duct
    4. hepatic ducts

    Choose the arrangement that lists the ducts in the order bile passes through them when moving from the bile canaliculi of the liver to the small intestine.
    a. 3,4,2
    b. 3,2,1
    c. 3,4,1
    d. 4,1,2
    e. 4,2,1

20. Which of these might occur if a person suffers from a severe case of hepatitis that impairs liver function?
    a. Fat digestion is difficult.
    b. By-products of hemoglobin breakdown accumulate in the blood.
    c. Plasma proteins decrease in concentration.
    d. Toxins in the blood increase.
    e. All of the above.

21. The gallbladder
    a. produces bile.
    b. stores bile.
    c. contracts and releases bile in response to secretin.
    d. contracts and releases bile in response to sympathetic stimulation.
    e. both b and c.

22. The aqueous component of pancreatic secretions
    a. is secreted by the pancreatic islets.
    b. contains bicarbonate ions.
    c. is released primarily in response to cholecystokinin.
    d. passes directly into the blood.
    e. all of the above.

23. Given these structures:
    1. ascending colon
    2. descending colon
    3. sigmoid colon
    4. transverse colon

    Choose the arrangement that lists the structures in the order that food encounters them as it passes between the small intestine and the rectum.
    a. 1,2,3,4
    b. 1,4,2,3
    c. 2,3,1,4
    d. 2,4,1,3
    e. 3,4,1,2

24. Which of these is *not* a function of the large intestine?
    a. absorption of fats
    b. absorption of certain vitamins
    c. absorption of water and salts
    d. production of mucus
    e. all of the above

25. Defecation
    a. can be initiated by stretch of the rectum.
    b. can occur as a result of mass movements.
    c. involves local reflexes.
    d. involves parasympathetic reflexes mediated by the spinal cord.
    e. all of the above.

26. Which of these structures produces enzymes that digest carbohydrates?
    a. salivary glands
    b. pancreas
    c. lining of the small intestine
    d. both a and b
    e. all of the above

27. Bile
    a. is an important enzyme for the digestion of fats.
    b. is made by the gallbladder.
    c. contains breakdown products from hemoglobin.
    d. emulsifies fats.
    e. both c and d.

28. Micelles are
    a. lipids surrounded by bile salts.
    b. produced by the pancreas.
    c. released into lacteals.
    d. stored in the gallbladder.
    e. reabsorbed in the colon.

29. If the thoracic duct were tied off, which of these classes of nutrients would *not* enter the circulatory system at their normal rate?
    a. amino acids
    b. glucose
    c. lipids
    d. fructose
    e. nucleotides

30. Which of these lipoprotein molecules transports excess lipids from cells back to the liver?
    a. high-density lipoprotein (HDL)
    b. low-density lipoprotein (LDL)
    c. very low-density lipoprotein (VLDL)

    Answers in Appendix F

C R I T I C A L   T H I N K I N G

1. While anesthetized, patients sometimes vomit. Given that the anesthetic eliminates the swallowing reflex, explain why it's dangerous for an anesthetized patient to vomit.

2. Achlorhydria is a condition in which the stomach stops producing hydrochloric acid and other secretions. What effect would achlorhydria have on the digestive process? On red blood cell count?

3. Victor Worrystudent experienced the pain of a duodenal ulcer during final examination week. Describe the possible reasons. Explain what habits could have caused the ulcer, and recommend a reasonable remedy.

4. Gallstones sometimes obstruct the common bile duct. What are the consequences of such a blockage?

5. A patient has a spinal cord injury at level L2 of the spinal cord. How will this injury affect the patient's ability to defecate? What components of the defecation response are still present, and which are lost?

6. The bowel (colon) occasionally can become impacted. Given what you know about the functions of the colon and the factors that determine the movement of substances across the colon wall, predict the effect of the impaction on the contents of the colon above the point of impaction.

Answers in Appendix G

A N S W E R S   T O   P R E D I C T   Q U E S T I O N S

1. A pin placed through the greater omentum passes through four layers of simple squamous epithelium. The greater omentum is actually a folded mesentery, with each part consisting of two layers of serous squamous epithelium.

2. The moist stratified squamous epithelium of the oropharynx and the laryngopharynx protects these regions from abrasive food when it is first swallowed. The ciliated pseudostratified epithelium of the nasopharynx helps move mucus produced in the nasal cavity and the nasopharynx into the oropharynx and esophagus. It's not as necessary to protect the nasopharynx from abrasion because food does not normally pass through this cavity.

3. It's important for the nasopharynx to be closed during swallowing so that food doesn't reflux into it or the nasal cavity. An explosive burst of laughter can relax the soft palate, open the nasopharynx, and cause the liquid to enter the nasal cavity.

4. Usually if a person tries to swallow and speak at the same time, the epiglottis is elevated, the laryngeal muscles closing the opening to the larynx are mostly relaxed, and food or liquid could enter the larynx, causing the person to choke.

5. After a heavy meal, blood pH may increase because, as bicarbonate ions pass from the cells of the stomach into the extracellular fluid, the pH of the extracellular fluid increases. As the extracellular fluid exchanges ions with the blood, the blood pH also increases.

6. Secretin production and its stimulation of bicarbonate ion secretion constitute a negative-feedback mechanism because, as the pH of the chyme in the duodenum decreases as a result of the presence of acid, secretin causes an increase in bicarbonate ion secretion, which increases the pH and restores the proper pH balance in the duodenum.

7. The major effect of prolonged diarrhea is on the cardiovascular system and is much like massive blood loss. Hypovolemia continues to increase. Blood pressure declines in a positive-feedback cycle and without intervention can lead to heart failure.

Visit the Online Learning Center at **www.mhhe.com/seeley7** for chapter quizzes, interactive learning exercises, and other study tools.

Nutrition, Metabolism, and Temperature Regulation

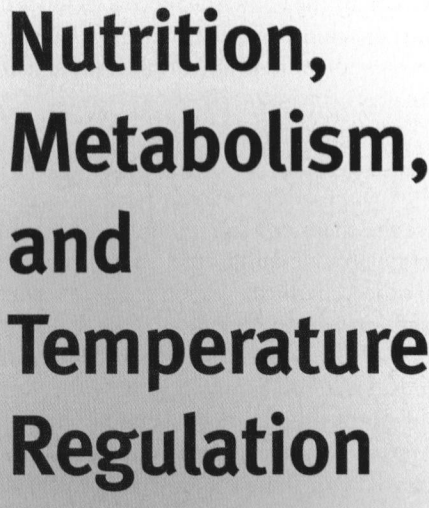

This colorized scanning electron micrograph shows a mitochondrion (pink) in the cytoplasm of an intestinal epithelial cell. The mitochondrion has an outer and inner membrane. The inner membrane has numerous folds that project into the interior of the mitochondrion. Enzymes, necessary for producing ATP, are located in these folds.

We are often more concerned with the taste of food than with its nutritional value when choosing from a menu or when selecting food to prepare. Knowing about nutrition is important, however, because food provides us with energy and the building blocks necessary to synthesize new molecules. What happens if we don't obtain enough vitamins, or if we eat too much sugar and fats? Health claims about foods and food supplements bombard us every day. Which ones are ridiculous, and which ones have merit? A basic understanding of nutrition can help us to answer these and other questions so that we can develop a healthy diet.

This chapter explains *nutrition* (p. 932), *metabolism* (p. 940), *carbohydrate metabolism* (p. 942), *lipid metabolism* (p. 949), *protein metabolism* (p. 950), *interconversion of nutrient molecules* (p. 952), *metabolic states* (p. 953), *metabolic rate* (p. 955), and *body temperature regulation* (p. 956).

Nutrition

Nutrition is the process by which certain components of food are obtained and used by the body. The process includes digestion, absorption, transportation, and cell metabolism. Nutrition is also defined as the evaluation of food and drink requirements for normal body function.

Nutrients

Nutrients are the chemicals taken into the body that are used to produce energy, provide building blocks for new molecules, or function in other chemical reactions. Some important substances in food, such as nondigestible plant fibers, are not nutrients. Nutrients are divided into six major classes: carbohydrates, proteins, lipids, vitamins, minerals, and water. Carbohydrates, proteins, and lipids are the major organic nutrients and are broken down by enzymes into their individual components during digestion. Many of these subunits are broken down further to supply energy, whereas others are used as building blocks for other macromolecules. Carbohydrates, proteins, lipids, and water are required in fairly substantial quantities, whereas vitamins and minerals are required in only small amounts. Vitamins, minerals, and water are taken into the body without being digested.

    Essential nutrients are nutrients that must be ingested because the body cannot manufacture them or is unable to manufacture adequate amounts of them. The essential nutrients include certain amino acids, certain fatty acids, most vitamins, minerals, water, and a minimum amount of carbohydrates. The term *essential* doesn't mean, however, that only the essential nutrients are required by the body. Other nutrients are necessary, but, if they are not part of the diet, they can be synthesized from other ingested nutrients. Most of this synthesis takes place in the liver, which has a remarkable ability to transform and manufacture molecules.

    The U.S. Department of Agriculture (USDA) provides recommendations for obtaining the proper amounts of carbohydrates, lipids, proteins, vitamins, minerals, and fiber in the form of a "food guide pyramid" (figure 25.1). The six food groups shown in the pyramid are (1) bread, cereal, rice, and pasta; (2) vegetables; (3) fruits; (4) milk, yogurt, and cheese; (5) meat, poultry, fish, dry beans, eggs, and nuts; and (6) fats, oils, and sweets. The shape of the pyramid suggests that grains, vegetables, and fruits should be the main part of the diet. Fats, oils, and sweets can be used in moderation to improve the flavor of foods. A balanced diet includes a variety of foods from each of the major food groups. Variety is necessary because no one food contains all the nutrients necessary for good health.

    The USDA presented the food pyramid in 1992. This food pyramid may be too simplistic and misleading, and it is now being reassessed. In particular, not all carbohydrates and fats have the same effects on health. Not all complex carbohydrates are good for you, and not all fats are bad for you. Furthermore, the proportions of carbohydrates, fats, and proteins in the diet is being questioned.

Benefits of a Healthy Diet

Two studies completed in 2000 compared the eating habits of 67,272 women and 51,529 men to the government's Healthy Eating Index, a measure of how well a diet conforms to dietary guidelines and the food pyramid. Those who ate best, according to the index, were compared to those who ate the worst. Women who ate best had a 14% reduction in heart disease but no significant decrease in chronic diseases compared to women who ate worst. Men who ate best had a 28% reduction in heart disease and a 11% decrease in chronic diseases compared to men who ate worst. There was no significant difference in cancer between the women and men who ate best compared to those who ate worst.

1. *Define the terms nutrient and essential nutrient, and list the six major classes of nutrients.*
2. *List the six food groups shown in a food guide pyramid. What food groups are at the bottom and top of the pyramid?*

Kilocalories

The energy stored within the chemical bonds of certain nutrients is used by the body. A **calorie** (kal′ō-rē; **cal**) is the amount of energy (heat) necessary to raise the temperature of 1 g of water 1°C. A **kilocalorie** (kil′ō-kal-ō-rē; **kcal**) is 1000 calories and is used to express the larger amounts of energy supplied by foods and released through metabolism.

What Is a Calorie?

A kilocalorie is often called a Calorie (with a capital *C*). Unfortunately, this usage has resulted in confusion between the terms *calorie* (with a lowercase *c*) and *Calorie* (with a capital *C*). It is common practice on food labels and in nutrition books to use calorie when Calorie (kilocalorie) is the proper term.

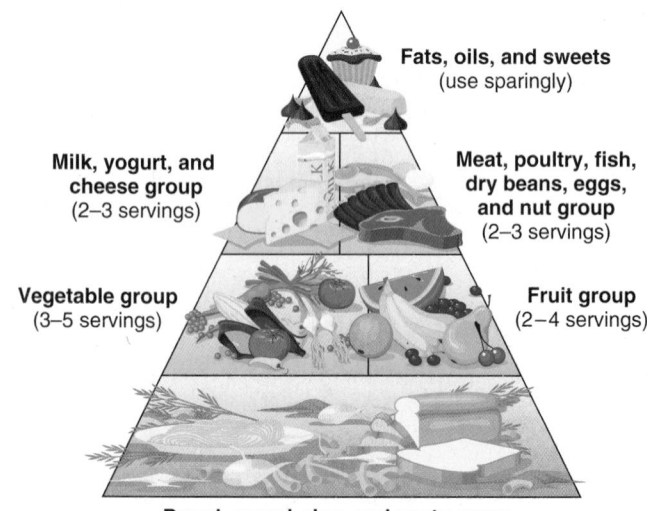

Figure 25.1   Food Guide Pyramid

The pyramid suggests three approaches to a healthy diet: eat different amounts of foods from each basic food group, use fats and sugars sparingly, and choose variety by eating the indicated number of servings per day of the different foods from each major food group.

Almost all of the kilocalories supplied by food come from carbohydrates, proteins, or fats. For each gram of carbohydrate or protein that the body metabolizes, about 4 kcal of energy is released. Fats contain more energy per unit of weight than carbohydrates and proteins and yield about 9 kcal/g. Table 25.1 lists the kilocalories supplied by some typical foods. A typical diet in the United States consists of 50%–60% carbohydrates, 35%–45% fats, and 10%–15% protein. Table 25.1 also lists the carbohydrate, fat, and protein composition of some foods.

> 3. *Define the term kilocalorie, and state the number of kilocalories supplied by a gram of carbohydrate, lipid, and protein.*

Carbohydrates

Sources in the Diet

Carbohydrates include monosaccharides, disaccharides, and polysaccharides (see chapter 2). Most of the carbohydrates humans ingest come from plants. An exception is lactose (milk sugar), which is found in animal and human milk.

The most common monosaccharides in the diet are glucose and fructose. Plants capture the energy in sunlight and use the energy to produce glucose which can be found in vegetables. Fructose (fruit sugar) and galactose are isomers of glucose (see figure 2.14). Glucose is found in vegetables and fructose is found in fruits, berries, honey, and high-fructose corn syrup, which is used to sweeten soft drinks and desserts. Galactose is usually found in milk.

The disaccharide sucrose (table sugar) is what most people think of when they use the term *sugar*. Sucrose is a glucose and a fructose molecule joined together, and its principal sources are sugarcane, sugar beets, maple sugar, and honey. Maltose (malt sugar), derived from germinating cereals, is a combination of two glucose molecules, and lactose (in milk) consists of a glucose and a galactose molecule (see figure 2.14).

The **complex carbohydrates** are the polysaccharides: starch, glycogen, and cellulose. These polysaccharides consist of many glucose molecules bound together to form long chains. Starch is an energy storage molecule in plants and is found primarily in vegetables, fruits, and grains. Glycogen is an energy storage molecule in animals and is located primarily in muscle and in the liver. By the time meats are processed, they contain little, if any, glycogen because it is degraded during slaughter and later storage. Cellulose forms cell walls, which surround plant cells.

Uses in the Body

During digestion, polysaccharides and disaccharides are split into monosaccharides, which are absorbed into the blood (see chapter 24). Humans can digest starch and glycogen because they can break the bonds between the glucose molecules of starch and glycogen. Humans are unable to digest cellulose because they can't break the bonds between its glucose molecules. Instead, cellulose provides fiber, or "roughage," thereby increasing the bulk of feces, making it easier to defecate.

The liver converts fructose, galactose, and other monosaccharides absorbed by the blood into glucose. Glucose, whether absorbed from the digestive tract or synthesized in the liver, is an energy source used to produce **adenosine triphosphate (ATP)** molecules (see "Anaerobic Respiration" on p. 945 and "Aerobic Respiration" on p. 946). Because the brain relies almost entirely on glucose for its energy, blood glucose levels are carefully regulated (see chapter 18).

If excess amounts of glucose are present, the glucose is converted into glycogen, which is stored in muscle and in the liver. Because cells can store only a limited amount of glycogen, any additional glucose is converted into fat, which is stored in adipose tissue. Glycogen can be rapidly converted back to glucose when energy is needed. For example, during exercise, muscles convert glycogen to glucose, and in-between meals, the liver helps to maintain blood sugar levels by converting glycogen to glucose, which is released into the blood.

In addition to being used as a source of energy, sugars have other functions. They form part of deoxyribonucleic acid (DNA), ribonucleic acid (RNA), and ATP molecules (see chapter 2); and they combine with proteins to form glycoproteins, such as the glycoprotein receptor molecules on the outer surface of the plasma membrane (see chapter 3).

Recommended Amounts

No minimum recommended daily intake for carbohydrates has been established. Carbohydrates account for approximately 50% of the daily kilocalorie intake for adults in the United States and about 70% worldwide. The USDA food pyramid (see figure 25.1) recommendation is that 55% to 60% of the daily kilocalorie intake be from carbohydrates.

Complex carbohydrates are recommended because starchy foods often contain other valuable nutrients like vitamins and minerals. Although foods like soft drinks and candy are rich in carbohydrates, they are mostly sugar and they may have little other nutritive value. For example, a typical soft drink contains 9 teaspoons of sugar. In excess, the consumption of foods high in sugar content can result in obesity and tooth decay.

> 4. *What are the most common monosaccharides in the diet? What are sucrose, maltose, and lactose?*
> 5. *Give three examples of complex carbohydrates. How are they used by the body?*
> 6. *How does the body use glucose and other monosaccharides?*
> 7. *What quantities of carbohydrate should be ingested daily?*

Lipids

Sources in the Diet

About 95% of the lipids in the human diet are **triglycerides** (trī-glis′er-īdz). Triglycerides, which are sometimes called **triacylglycerols** (trī-as′il-glis′er-olz), consist of three fatty acids attached to a glycerol molecule (see chapter 2). Triglycerides are often referred to as fats or oils. Fats are solid at room temperature, whereas oils are a liquid. Fats and oils can be categorized as saturated or unsaturated. **Saturated fats** or **oils** have only single covalent bonds between the

Table 25.1   Food Consumption

Food	Quantity	Food Energy (kcal)	Carbohydrate (g)	Fat (g)	Protein (g)
Dairy Products					
Whole milk (3.3% fat)	1 cup	150	11	8	8
Low fat milk (2% fat)	1 cup	120	12	5	8
Butter	1 tablespoon	100	—	12	—
Grain					
Bread, white enriched	1 slice	75	24	1	2
Bread, whole wheat	1 slice	65	14	1	3
Fruit					
Apple	1	80	20	1	—
Banana	1	100	26	—	1
Orange	1	65	16	—	1
Vegetables					
Corn, canned	1 cup	140	33	1	4
Peas, canned	1 cup	150	29	1	8
Lettuce	1 cup	5	2	—	—
Celery	1 cup	20	5	—	1
Potato, baked	1 large	145	33	—	4
Meat, Fish, and Poultry					
Lean ground beef (10% fat)	3 ounces	185	—	10	23
Shrimp, french fried	3 ounces	190	9	9	17
Tuna, canned	3 ounces	170	—	7	24
Chicken breast, fried	3 ounces	160	1	5	26
Bacon	2 slices	85	—	8	4
Hot dog	1	170	1	15	7
Fast Foods					
McDonald's Egg McMuffin	1	327	31	15	19
McDonald's Big Mac	1	563	41	33	26
Taco Bell's beef burrito	1	466	37	21	30
Arby's roast beef	1	350	32	15	22
Pizza Hut Super Supreme	1 slice	260	23	13	15
Long John Silver's fish	2 pieces	366	21	22	22
Dairy Queen malt, large	1	840	125	28	22
Desserts					
Chocolate chip cookie	4	200	29	9	2
Apple pie	1 piece	345	51	15	3
Dairy Queen cone, large	1	340	52	10	10
Beverage					
Cola soft drink	12 ounces	145	37	—	—
Beer	12 ounces	144	13	—	1
Wine	31/2 ounces	73	2	—	—
Hard liquor (86 proof)	11/2 ounces	105	—	—	—
Miscellaneous					
Egg	1	80	1	6	6
Mayonnaise	1 tablespoon	100	—	11	—
Sugar	1 tablespoon	45	12	—	—

carbon atoms of their fatty acids (see figure 2.17). They are found in the fats of meats (e.g., beef, pork), dairy products (e.g., whole milk, cheese, butter), eggs, coconut oil, and palm oil. **Unsaturated fats** or **oils** have one (monounsaturated) or more (polyunsaturated) double covalent bonds between the carbon atoms of their fatty acids (see figure 2.17). Monounsaturated fats include olive and peanut oils; and polyunsaturated fats occur in fish, safflower, sunflower, and corn oil.

Saturating Fats

Solid fats, such as shortening and margarine, work better than liquid oils do for preparing some foods, such as pastries. Polyunsaturated vegetable oils can be changed from a liquid to a solid by making them more saturated, that is, by decreasing the number of double covalent bonds in their polyunsaturated fatty acids. Hydrogen gas is bubbled through the oil. As hydrogen binds to the fatty acids, double covalent bonds are converted to single covalent bonds to produce a change in molecular shape that solidifies the oil. The more saturated the product, the harder it becomes at room temperature.

The remaining 5% of lipids include cholesterol and phospholipids like **lecithin** (les′i-thin). Cholesterol is a steroid (see chapter 2) found in high concentrations in liver and egg yolks, but it's also present in whole milk, cheese, butter, and meats. Cholesterol is not found in plants. Phospholipids are major components of plasma membranes, and they are found in a variety of foods. A good source of lecithin is egg yolks.

Uses in the Body

Triglycerides are important sources of energy that are used to produce ATP molecules. A gram of triglyceride delivers more than twice as many kilocalories as a gram of carbohydrate. Some cells, such as skeletal muscle cells, derive most of their energy from triglycerides.

After a meal, excess triglycerides that are not immediately used are stored in adipose tissue or the liver. Later, when energy is required, the triglycerides are broken down, and their fatty acids are released into the blood, where they are taken up and used by various tissues. In addition to storing energy, adipose tissue surrounds and pads organs, and under the skin adipose tissue is an insulator, which prevents heat loss.

Cholesterol is an important molecule with many functions in the body. It can either be obtained in food or manufactured by the liver and most other tissues. Cholesterol is a component of the plasma membrane, and it can be modified to form other useful molecules, such as bile salts and steroid hormones. Bile salts are necessary for fat digestion and absorption. Steroid hormones include the sex hormones estrogen, progesterone, and testosterone, which regulate the reproductive system.

The eicosanoids (ī′kō-să-noydz), which include prostaglandins and leukotrienes, are derived from fatty acids. The molecules are involved in activities like inflammation, blood clotting, tissue repair, and smooth muscle contraction. Phospholipids, such as lecithin, are part of the plasma membrane and are used to construct the myelin sheath around the axons of nerve cells. Lecithin is also found in bile and helps to emulsify fats.

Recommended Amounts

The American Heart Association recommends that fats account for 30% or less of the total daily kilocaloric intake, with 8%–10% coming from saturated fats, up to 10% from polyunsaturated fats, and up to 15% from monounsaturated fats. Furthermore, saturated fats should contribute no more than 10% of total fat intake, and cholesterol should be limited to 300 mg (the amount in an egg yolk) or less per day. These guidelines reflect the belief that excess amounts of fats, especially saturated fats and cholesterol, contribute to cardiovascular disease. Evidence also suggests that high-fat intake is associated with colon cancer. The typical U.S. diet derives 35%–45% of its kilocalories from fats, indicating that most Americans need to reduce fat consumption. On the other hand, total fat intake may account for as little as 10% of the kilocalories in a healthy person's diet.

Most of the lecithin consumed in the diet is broken down in the digestive tract. The liver has the ability to manufacture all of the lecithin necessary to meet the body's needs, and it's not necessary to consume lecithin supplements.

Linoleic (lin-ō-lē′ik) **acid** and **α-linolenic** (lin-ō-len′ik) **acid** are **essential fatty acids** because the body cannot synthesize them and they must be ingested. Seeds, nuts, and legumes are good sources of linoleic acid and α-linolenic acids. Linoleic acid is also found in grains, and α-linolenic acid is in the green leaves of plants.

Fatty Acids and Blood Clotting

The essential fatty acids are used to synthesize prostaglandins that affect blood clotting. Linoleic acid can be converted to **arachidonic** (ă-rak-i-don′ik) **acid,** which is used to produce prostaglandins that increase blood clotting. Alpha-linolenic acid can be converted to **eicosapentaenoic** (ī′kō-să-pen-tă-nō′ik) **acid (EPA),** which is used to produce prostaglandins that decrease blood clotting. Normally, most prostaglandins are synthesized from linoleic acid because it's more plentiful in the body. Individuals who consume foods rich in EPA, however, such as herring, salmon, tuna, and sardines, increase the synthesis of prostaglandins from EPA. Individuals who eat these fish twice or more times per week have a lower risk of heart attack than those who don't, possibly because of reduced blood clotting. Although EPA can be obtained using fish oil supplements, this is not currently recommended because fish oil supplements contain high amounts of cholesterol, vitamins A and D, and uncommon fatty acids, all of which can cause health problems when taken in large amounts.

8. What is the major source of lipids in the diet? What are other sources?

9. Define saturated and unsaturated fats.

10. How are triglycerides, cholesterol, prostaglandins, and lecithin used by the body?

11. Describe the recommended dietary intake of lipids. List the essential fatty acids.

Proteins

Sources in the Diet

Proteins are chains of amino acids (see chapter 2). Proteins in the human body are constructed of 20 different kinds of amino acids, which are divided into two groups: essential and nonessential amino acids. The body cannot synthesize **essential amino acids,** so they must be obtained in the diet. The nine essential amino acids are histidine, isoleucine, leucine, lysine, methionine, phenylalanine, threonine, tryptophan, and valine. The body can synthesize **nonessential amino acids** from other molecules. If adequate amounts of the essential amino acids are ingested, they can be converted to the nonessential amino acids. Like the essential amino acids, the nonessential amino acids are necessary for good health.

A **complete protein** food contains adequate amounts of all nine essential amino acids, whereas an **incomplete protein** food does not. Examples of complete proteins are meat, fish, poultry, milk, cheese, and eggs, and examples of incomplete proteins are leafy green vegetables, grains, and legumes (peas and beans).

Uses in the Body

Essential and nonessential amino acids are used to synthesize proteins. Proteins perform numerous functions in the human body as the following examples illustrate. Collagen provides structural strength in connective tissue as does keratin in the skin, and the combination of actin and myosin makes muscle contraction possible. Enzymes are responsible for regulating the rate of chemical reactions, and protein hormones regulate many physiologic processes (see chapter 18). Proteins in the blood act as buffers to prevent changes in pH, and hemoglobin transports oxygen and carbon dioxide in the blood. Proteins also function as carrier molecules to move materials across plasma membranes, and other proteins in the plasma membrane function as receptor molecules and ion channels. Antibodies, lymphokines, and complement are part of the immune system response that protects against microorganisms and other foreign substances.

Proteins are also used as a source of energy, yielding the same amount of energy as carbohydrates. If excess proteins are ingested, the energy in the proteins can be stored by converting their amino acids into glycogen or fats.

Recommended Amounts

The recommended daily consumption of protein for a healthy adult is 0.8 g/kg of body weight, or about 10% of total kilocalories (55 g protein/day for a 2000 kcal/day intake). A cup of skim milk contains 8 g protein, 1 ounce of meat contains 7 g protein, and a slice of bread provides 2 g protein. If two incomplete proteins, such as rice and beans are ingested, each can provide amino acids lacking in the other. Thus, a correctly balanced vegetarian diet can provide all of the essential amino acids.

When protein intake is adequate, the synthesis and breakdown of proteins in a healthy adult occurs at the same rate. The amino acids of proteins contain nitrogen; so saying that a person is in **nitrogen balance** means that the nitrogen content of ingested protein is equal to the nitrogen excreted in urine and feces. A starving person is in negative nitrogen balance because the nitrogen gained in the diet is less than that lost by excretion. In other words, when proteins are broken down for energy, more nitrogen is lost than is replaced in the diet. A growing child or a healthy pregnant woman, on the other hand, is in positive nitrogen balance because more nitrogen is going into the body to produce new tissues than is lost by excretion.

12. *Distinguish between essential and nonessential amino acids. Between complete and incomplete protein foods.*
13. *Describe some of the functions performed by proteins in the body.*
14. *What is the recommended daily consumption of proteins? Define the term nitrogen balance.*

Vitamins

Vitamins (vīt′ă-minz; life-giving chemicals) are organic molecules that exist in minute quantities in food and are essential to normal metabolism (table 25.2). **Essential vitamins** cannot be produced by the body and must be obtained through the diet. Because no single food item or nutrient class provides all the essential vitamins, it's necessary to maintain a balanced diet by eating a variety of foods. The absence of an essential vitamin in the diet can result in a specific deficiency disease. A few vitamins, such as vitamin K, are produced by intestinal bacteria, and a few can be formed by the body from substances called provitamins. A **provitamin** is a part of a vitamin that can be assembled or modified by the body into a functional vitamin. Beta carotene is an example of a provitamin that can be modified by the body to form vitamin A. The other provitamins are **7-dehydrocholesterol** (dē-hī′dro-kō-les′ter-ol), which can be converted to vitamin D, and **tryptophan** (trip′tō-fan), which can be converted to niacin.

Vitamins are not broken down by catabolism but are used by the body in their original or slightly modified forms. After the chemical structure of a vitamin is destroyed, its function is usually lost. The chemical structure of many vitamins is destroyed by heat, such as when food is overcooked.

Many vitamins function as **coenzymes,** which combine with enzymes to make the enzymes functional (see chapter 2). Without coenzymes and their enzymes, many chemical reactions would occur too slowly to support good health and even life. For example, vitamins B_2 and B_3, biotin (bī′ō-tin), and pantothenic (pan-tō-then′ik) acid are critical for some of the chemical reactions involved in the production of ATP. Folate (fō′lāt) and vitamin B_{12} are involved in nucleic acid synthesis. Vitamins A, B_1, B_6, B_{12}, C, and D are necessary for growth. Vitamin K is necessary for the synthesis of proteins involved in blood clotting (see table 25.2).

Vitamins are either fat-soluble or water-soluble. **Fat-soluble vitamins,** such as vitamins A, D, E, and K, dissolve in lipids. They are absorbed from the intestine along with lipids. Some of them can be stored in the body for a long time. Because they can be stored, it's possible to accumulate these vitamins in the body to the point of toxicity. **Water-soluble vitamins,** such as the B vitamins and vitamin C, dissolve in water. They are absorbed from the water in the intestinal tract and remain in the body only a short time before being excreted in the urine.

Vitamins were discovered at the beginning of the twentieth century. They were found to be associated with certain foods that were known to protect people from diseases like rickets and beriberi. In 1941, the first Food and Nutrition Board established the **Recommended Dietary Allowances (RDAs),** which are the nutrient intakes sufficient to meet the needs of nearly all people in certain age and gender groups. RDAs have been established for different-aged males and females, starting with infants and

Table 25.2	The Principal Vitamins				
Vitamin	**Fat (F)– or Water (W)– Soluble**	**Source**	**Function**	**Symptoms of Deficiency**	**Reference Daily Intake (RDI)s***
A (retinol)	F	From provitamin carotene found in yellow and green vegetables: preformed in liver, egg yolk, butter, and milk	Necessary for rhodopsin synthesis, normal health of epithelial cells, and bone and tooth growth	Rhodopsin defiency, night blindness, retarded growth, skin disorders and increase in infection risk	1000 RE[†]
B_1 (thiamine)	W	Yeast, grains, and milk	Involved in carbohydrate and amino acid metabolism, necessary for growth	Beriberi—muscle weakness (including cardiac muscle), neuritis, and paralysis	1.5 mg
B_2 (riboflavin)	W	Green vegetables, liver, wheat germ, milk, and eggs	Component of flavin adenine dinucleotide; involved in citric acid cycle	Eye disorders and skin cracking, especially at corners of the mouth	1.7 mg
B_3 (niacin)	W	Fish, liver, red meat, yeast, grains, peas, beans, and nuts	Component of nicotinamide adenine dinucleotide; involved in glycolysis and citric acid cycle	Pellagra—diarrhea, dermatitis, and nervous system disorder	20 mg
Pantothenic acid	W	Liver, yeast, green vegetables, grains, and intestinal bacteria	Constituent of coenzyme-A, glucose production from lipids and amino acids, and steroid hormone synthesis	Neuromuscular dysfunction and fatigue	10 mg
Biotin	W	Liver, yeast, eggs, and intestinal bacteria	Fatty acid and nucleic acid synthesis; movement of pyruvic acid into citric acid cycle	Mental and muscle dysfunction, fatigue, and nausea	0.3 mg
B_6 (pyridoxine)	W	Fish, liver, yeast, tomatoes, and intestinal bacteria	Involved in amino acid metabolism	Dermatitis, retarded growth, and nausea	2.0 mg
Folate	W	Liver, green leafy vegetables, and intestinal bacteria	Nucleic acid synthesis, hematopoiesis; prevents birth defects	Macrocytic anemia (enlarged red blood cells) and spina bifida	0.4 mg
B_{12} (cobalamins)	W	Liver, red meat, milk, and eggs	Necessary for red blood cell production, some nucleic acid and amino acid metabolism	Pernicious anemia and nervous system disorders	6 μg
C (ascorbic acid)	W	Citrus fruit, tomatoes, and green vegetables	Collagen synthesis; general protein metabolism	Scurvy—defective bone formation and poor wound healing	60 mg
D (cholecalciferol, ergosterol)	F	Fish liver oil, enriched milk, and eggs; provitamin D converted by sunlight to cholecalciferol in the skin	Promotes calcium and phosphorus use; normal growth and bone and teeth formation	Rickets—poorly developed, weak bones, osteomalacia; bone reabsorption	400 IU[‡]
E (tocopherol, tocotrienols)	F	Wheat germ, cotton seed, palm, and rice oils; grain, liver, and lettuce	Prevents the oxidation of plasma membranes and DNA	Hemolysis of red blood cells	30 IU
K (phylloquinone)	F	Alfalfa, liver, spinach, vegetable oils, cabbage, and intestinal bacteria	Required for synthesis of a number of clotting factors	Excessive bleeding due to retarded blood clotting	80μg

* RDIs for people over 4 years of age; IU = international units.
† Retinol equivalents (RE). 1 retinol equivalent = 1 μg retinol or 6 μg beta carotene.
‡ As cholecalciferol. 1 μg cholecalciferol = 40 IU (international units) vitamin D.

continuing on to adults. RDAs are also set for pregnant and lactating women. The RDAs have been reevaluated every 4–5 years and updated, when necessary, on the basis of new information.

The RDAs establish a minimum intake of vitamins and minerals that should protect almost everyone (97%) in a given group from diseases caused by vitamin or mineral deficiencies. Although personal requirements can vary, the RDAs are a good benchmark. The further dietary intake is below the RDAs, the more likely a nutritional deficiency can occur. On the other hand, the consumption of too large a quantity of some vitamins and minerals can have harmful effects. For example, the long-term ingestion of 3–10 times the RDA for vitamin A can cause bone and muscle pain, skin disorders, hair loss, and increased liver size. The long-term consumption of 5–10 times the RDA of vitamin D can result in the deposition of calcium in the kidneys, heart, and blood vessels, and the regular consumption of more than 2 g of vitamin C daily can cause stomach inflammation and diarrhea.

Free Radicals and Antioxidants

Damage from free radicals may contribute to aging and certain diseases, such as atherosclerosis and cancer. **Free radicals** are molecules, produced as part of normal metabolism, that are missing an electron. Free radicals can replace the missing electron by taking an electron from cell molecules, such as fats, proteins, or DNA, resulting in damage to the cell. The loss of an electron from a molecule is called oxidation. **Antioxidants** are substances that prevent oxidation of cell components by donating an electron to free radicals. Examples of antioxidants include beta carotene (provitamin A), vitamin C, and vitamin E.

Many studies have been done to determine whether or not taking large doses of antioxidants is beneficial. Although future research may suggest otherwise, the consensus among scientists establishing the RDAs is that the best evidence presently available doesn't support the claims that taking large doses of antioxidants prevents chronic disease or otherwise improves health. On the other hand, the amount of antioxidants normally found in a balanced diet that includes fruits and vegetables rich in antioxidants, combined with the complex mix of other chemicals found in food, can be beneficial.

15. *What are vitamins, essential vitamins, and provitamins? Name the water-soluble vitamins and the fat-soluble vitamins.*

16. *List some of the functions of vitamins.*

17. *What are Recommended Dietary Allowances (RDAs)? Why are they useful?*

PREDICT

What would happen if vitamins were broken down during the process of digestion rather than being absorbed intact into the circulation?

Minerals

Minerals (min′er-ălz) are inorganic nutrients that are necessary for normal metabolic functions. They constitute about 4%–5% of the total body weight and are components of coenzymes, a few vitamins, hemoglobin, and other organic molecules. Minerals are involved in a number of important functions, such as establishing resting membrane potentials and generating action potentials, adding mechanical strength to bones and teeth, combining with organic molecules, or acting as coenzymes, buffers, or regulators of osmotic pressure. Table 25.3 lists important minerals and their functions.

Minerals are ingested by themselves or in combination with organic molecules. Minerals are obtained from animal and plant sources. Mineral absorption from plants, however, can be limited because the minerals tend to bind to plant fibers. Refined breads and cereals have hardly any minerals or vitamins because they are lost in the processing of the seeds used to make them. The seeds are crushed and the outer parts of the seeds, which contains most of their minerals and vitamins, are removed. The inner part of the seeds, which has few minerals and vitamins, is used to make the refined breads and cereals. Minerals and vitamins are often added to refined breads and cereals to compensate for their loss during the refinement process. A balanced diet can provide all the necessary minerals, with a few possible exceptions. For example, women who suffer from excessive menstrual bleeding may need an iron supplement.

18. *What are minerals? List some of the important functions of minerals.*

Caloric Intake and Life Span

Studies in mice, rats, and other animals indicate that life span can be increased by approximately one-third by decreasing normal caloric intake 30%–50%, provided the diet includes enough protein, fat, vitamins, and minerals. Why life span increases is not understood, but one proposed explanation for this phenomenon is that decreased caloric intake in some way reduces free radical damage to mitochondria. It has been suggested that humans might derive a similar benefit by reducing caloric input, starting at age 20. Unlike laboratory animals, however, humans would have to voluntarily restrict their caloric intake by 30%–50%, which is an unlikely behavioral change for most humans. Much more needs to be learned before it will be known if the restriction of caloric intake to increase longevity is beneficial to humans.

Daily Values

Daily Values are dietary reference values now appearing on food labels to help consumers plan a healthy diet. Daily Values are based on two other sets of reference values: Reference Daily Intakes and Daily Reference Values. The **Reference Daily Intakes (RDIs)** are based on the 1968 RDAs for certain vitamins and minerals. RDIs have been set for four categories of people: infants, toddlers, people over 4 years of age, and pregnant or lactating women. Generally, the RDIs are set to the highest 1968 RDA value of an age category. For example, the highest RDA for iron in males over 4 years of age is 10 mg/day and for females over 4 years of age is 18 mg/day. Thus, the RDI for iron is set at 18 mg/day. The **Daily Reference Values (DRVs)** are set for total fat, saturated fat, cholesterol, total carbohydrate, dietary fiber, sodium, potassium, and protein.

Having two standards on food labels, RDIs for vitamins and minerals and DRVs for other nutrients, was thought to be more

Table 25.3 Important Minerals

Mineral	Function	Symptoms of Deficiency	Reference Daily Intake (RDIs)*
Calcium	Bone and teeth formation, blood clotting, muscle activity, and nerve function	Spontaneous action potential generation in neurons and tetany	1 g
Chloride	Blood acid–base balance; hydrochloric acid production in stomach	Acid–base imbalance	3.4 g
Chromium	Associated with enzymes in glucose metabolism	Unknown	120 µg
Cobalt	Component of vitamin B_{12}; red blood cell production	Anemia	Unknown
Copper	Hemoglobin and melanin production, electron-transport system	Anemia and loss of energy	2.0 mg
Fluorine	Provides extra strength in teeth; prevents dental caries	No real pathology	2.5 mg
Iodine	Thyroid hormone production, maintenance of normal metabolic rate	Goiter and decrease in normal metabolism	150 µg
Iron	Component of hemoglobin; ATP production in electron-transport system	Anemia, decreased oxygen transport, and energy loss	18 mg
Magnesium	Coenzyme constituent; bone formation; muscle and nerve function	Increased nervous system irritability, vasodilation, and arrhythmias	400 mg
Manganese	Hemoglobin synthesis; growth; activation of several enzymes	Tremors and convulsions	3.5 mg
Molybdenum	Enzyme component	Unknown	75µg
Phosphorus	Bone and teeth formation; important in energy transfer (ATP); component of nucleic acids	Loss of energy and cellular function	1 g
Potassium	Muscle and nerve function	Muscle weakness, abnormal electrocardiogram, and alkaline urine	2 g
Selenium	Component of many enzymes	Unknown	55 µg
Sodium	Osmotic pressure regulation; nerve and muscle function	Nausea, vomiting, exhaustion, and dizziness	500 mg†
Sulfur	Component of hormones; several vitamins, and proteins	Unknown	Unknown
Zinc	Component of several enzymes; carbon dioxide transport and metabolism; necessary for protein metabolism	Deficient carbon dioxide transport and deficient protein metabolism	15 mg

* RDIs for people over 4 years of age, except for sodium.
† The estimated minimum for people over 10 years of age. The maximum Daily Value for sodium is 2400 mg.

confusing for consumers than having one standard. Therefore, the RDIs and DRVs are combined to form the Daily Values. In addition, not all possible Daily Values are required to be listed.

The Daily Values appearing on food labels are based on a 2000 kcal reference diet, which approximates the weight maintenance requirements of postmenopausal women, women who exercise moderately, teenage girls, and sedentary men (figure 25.2). On large food labels, additional information is listed based on a daily intake of 2500 kcal, which is adequate for young men.

The Daily Values for energy-producing nutrients are determined as a percentage of daily kilocaloric intake: 60% for carbohydrates, 30% for total fats, 10% for saturated fats, and 10% for proteins. The Daily Value for fiber is 11.5 g for each 1000 kcal of intake. The Daily Values for a nutrient in a 2000 kcal/day diet can be calculated on the basis of the recommended daily percentage of the nutrient and the kilocalories in a gram of the nutrient. For example, carbohydrates should be 60% of a 2000 kcal/day diet, or 1200 kcal/day (0.60 × 2000). Since there are 4 kcal in a gram of carbohydrate, the Daily Value for carbohydrate is 300 g/day (1200/4).

The Daily Value for some nutrients is the uppermost limit considered desirable because of the link between these nutrients and certain diseases. Thus, the Daily Values for total fats is less than 65 g, saturated fats is less than 20 g, and cholesterol is less than 300 mg because of their association with increased risk of heart disease. The Daily Value for sodium is less than 2400 mg because of its association with high blood pressure in some people.

For a particular food, the Daily Value is used to calculate the **Percent Daily Value (% Daily Value)** for some of the nutrients in one serving of the food (see figure 25.2). For example, if a serving of food has 3 g of fat and the Daily Value for total fat is 65 g, then the % Daily Value is 5% (3/65 = 0.05, or 5%). The Food and Drug Administration (FDA) requires % Daily Values to be on food labels so that the public has useful and accurate dietary information.

P R E D I C T
One serving of a food has 30 g of carbohydrate. What % Daily Value for carbohydrate is on the food label for this food?

Nutrition Facts

Serving Size 1 oz. (28g/About 32 chips)
Servings Per Container 2.5

Amount Per Serving

Calories 160	Calories from Fat 90

	% Daily Value*
Total Fat 10g	**16%**
Saturated Fat 1.5g	**7%**
Cholesterol 0mg	**0%**
Sodium 170mg	**7%**
Total Carbohydrate 15g	**5%**
Dietary Fiber 1g	**4%**
Sugars less than 1g	
Protein 2g	

Vitamin A 0%	•	Vitamin C 0%
Calcium 2%	•	Iron 0%

* Percent Daily Values are based on a 2,000 calorie diet. Your daily values may be higher or lower depending on your calorie needs:

	Calories:	2,000	2,500
Total Fat	Less than	65g	80g
Sat Fat	Less than	20g	25g
Cholesterol	Less than	300mg	300mg
Sodium	Less than	2,400mg	2,400mg
Total Carbohydrate		300g	375g
Dietary Fiber		25g	30g

Calories per gram:
Fat 9   •   Carbohydrate 4   •   Protein 4

Figure 25.2   Food Label

The % Daily Values for nutrients related to energy consumption are based on a 2000 kcal/day diet. For people who maintain their weight on a 2000 kcal/day diet, the total of the % Daily Values for each of these nutrients should add up to no more than 100%. For individuals consuming more or fewer kilocalories per day than 2000 kcal, however, the total of the % Daily Values can be more or fewer than 100%. For example, for a person consuming 2200 kcal/day, the total of the % Daily Values for each of these nutrients should add up to no more than 110% because 2200/2000 = 1.10, or 110%.

P R E D I C T   ③

Suppose a person consumes 1800 kcal/day. What total % Daily Values for energy-producing nutrients is recommended?

When using the % Daily Values of a food to determine how the amounts of certain nutrients in the food fit into the overall diet,

the number of servings in a container or package needs to be considered. For example, suppose a small (2.25-ounce) bag of corn chips has a % Daily Value of 16% for total fat. One might suppose that eating the bag of chips accounts for 16% of total fat for the day. The bag, however, contains 2.5 servings. Therefore, if all the chips in the bag are consumed, they account for 40% (16% × 2.5) of the maximum recommended total fat.

● 19. *What are the Reference Daily Intakes and the Daily Reference Values? When combined, what reference set of values is established?*

20. *Define % Daily Values. The % Daily Values appearing on food labels are based on how many kilocalories per day?*

Metabolism

Metabolism (mĕ-tab′ō-lizm; change) is the total of all the chemical reactions that occur in the body. It consists of **anabolism** (ă-nab′ō-lizm), the energy-requiring process by which small molecules are joined to form larger molecules, and **catabolism** (kă-tab′ō-lizm), the energy-releasing process by which large molecules are broken down into smaller molecules.

The cellular metabolic processes are often referred to as cellular metabolism or cellular respiration. Anabolism occurs in all cells of the body as they divide to form new cells, maintain their own intracellular structure, and produce molecules like hormones, neurotransmitters, or extracellular matrix molecules for export. Catabolism begins during the process of digestion and is concluded within individual cells. The energy derived from catabolism is used to drive anabolic reactions and processes such as active transport and muscle contraction.

Large nutrient molecules, such as carbohydrates, lipids, and proteins, are broken down by digestion into smaller molecules, such as glucose, amino acids, and fatty acids, which are absorbed from the digestive tract into the blood (see chapter 24). These smaller molecules are taken into cells, catabolized, and the energy from them is used to combine adenosine diphosphate (ADP) and an inorganic phosphate group (P_i) to form ATP (figure 25.3).

$$ADP + P_i + Energy \rightarrow ATP$$

The energy in small nutrient molecules is used to produce many ATP molecules, each of which stores a small amount of energy. The smaller amount of energy in each of the many ATP molecules is more readily available for use in cells than the larger

Process Figure 25.3 ATP Coupling of Catabolic and Anabolic Reactions

Energy released by catabolic reactions is used to form ATP, which releases the energy for use in anabolic reactions.

amount of energy stored in nutrient molecules. ATP is often called the energy currency of the cell because when it is spent, or broken down to ADP, energy becomes available for use by the cell. If a quarter represents an ATP molecule, then a twenty dollar bill would be analogous to a small nutrient molecule. The quarter (ATP) can be used in various vending machines (chemical reactions), but the twenty dollar bill (nutrient molecule) cannot.

The chemical reactions responsible for the transfer of energy from the chemical bonds of nutrient molecules to ATP molecules involve oxidation–reduction reactions (see chapter 2). A molecule is reduced when it gains electrons and is oxidized when it loses electrons. A nutrient molecule has many hydrogen atoms covalently bonded to the carbon atoms that form the "backbone" of the molecule. Because a hydrogen atom is a hydrogen ion (proton) and an electron, the nutrient molecule has many electrons and is, therefore, highly reduced. When a hydrogen ion and an associated electron are lost from the nutrient molecule, the molecule loses energy and becomes oxidized. The energy in the electron is used to synthesize ATP. The major events of cellular metabolism are summarized in figure 25.4.

21. Define metabolism, anabolism, and catabolism. How is the energy derived from catabolism used to drive anabolic reactions?

22. How does the removal of hydrogen atoms from nutrient molecules result in a loss of energy from the nutrient molecule?

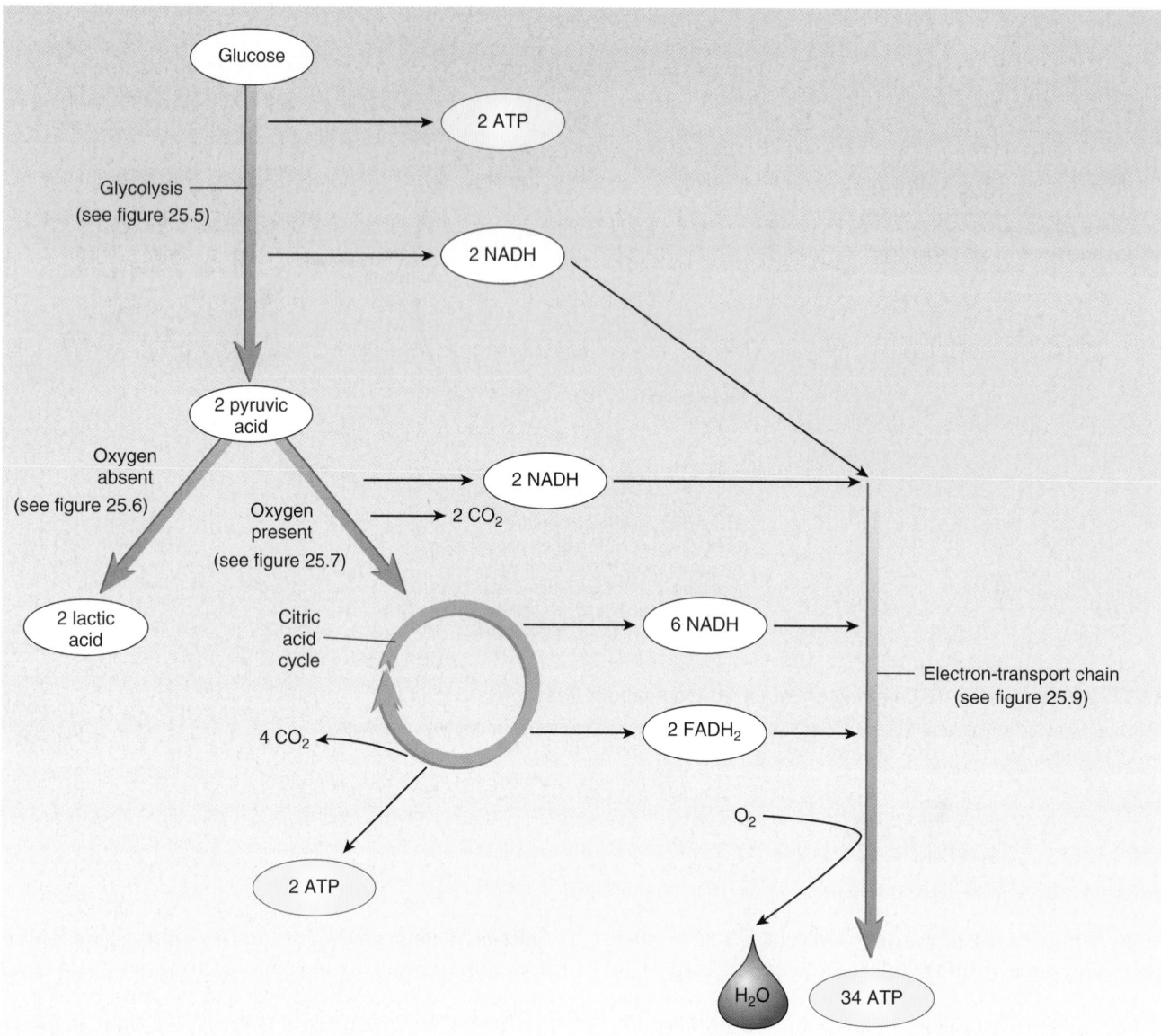

Figure 25.4   Cellular Metabolism

Overview of cellular metabolism, including glycolysis, citric acid cycle, and electron-transport system.

Carbohydrate Metabolism

Glycolysis

Carbohydrate metabolism begins with **glycolysis** (glī-kol′i-sis), which is a series of chemical reactions in the cytosol that results in the breakdown of glucose into two **pyruvic** (pī-roo′vik) **acid** molecules (figure 25.5).

Glycolysis is divided into four phases.

1. *Input of ATP.* The first steps in glycolysis require the input of energy in the form of two ATP molecules. A phosphate group is transferred from ATP to the glucose molecule, a process called **phosphorylation** (fos′fōr-i-lā′shŭn), to form glucose-6-phosphate. The glucose-6-phosphate atoms are rearranged to form fructose-6-phosphate, which is then converted to fructose-1,6-bisphosphate by the addition of another phosphate group from another ATP.

2. *Sugar cleavage.* Fructose-1,6-bisphosphate is cleaved into two three-carbon molecules, glyceraldehyde (glis-er-al′dĕ-hīd)-3-phosphate and dihydroxyacetone (dī′hī-drok-sē-as′e-tōn) phosphate. Dihydroxyacetone phosphate is rearranged to form glyceraldehyde-3-phosphate; consequently, two molecules of glyceraldehyde-3-phosphate result.

3. *NADH production.* Each glyceraldehyde-3-phosphate molecule is oxidized (loses two electrons) to form 1,3-bisphosphoglyceric (biz′phos-fo-gli′sēr′ik) acid, and **nicotinamide adenine** (nik-ō-tin′ă-mīd ad′ĕ-nēn) **dinucleotide (NAD$^+$)** is reduced (gains two electrons) to **NADH.** Glyceraldehyde-3-phosphate also loses two H$^+$, one of which binds to NAD$^+$.

$$NAD^+ + 2\,e^- + 2\,H^+ \rightarrow NADH + H^+$$

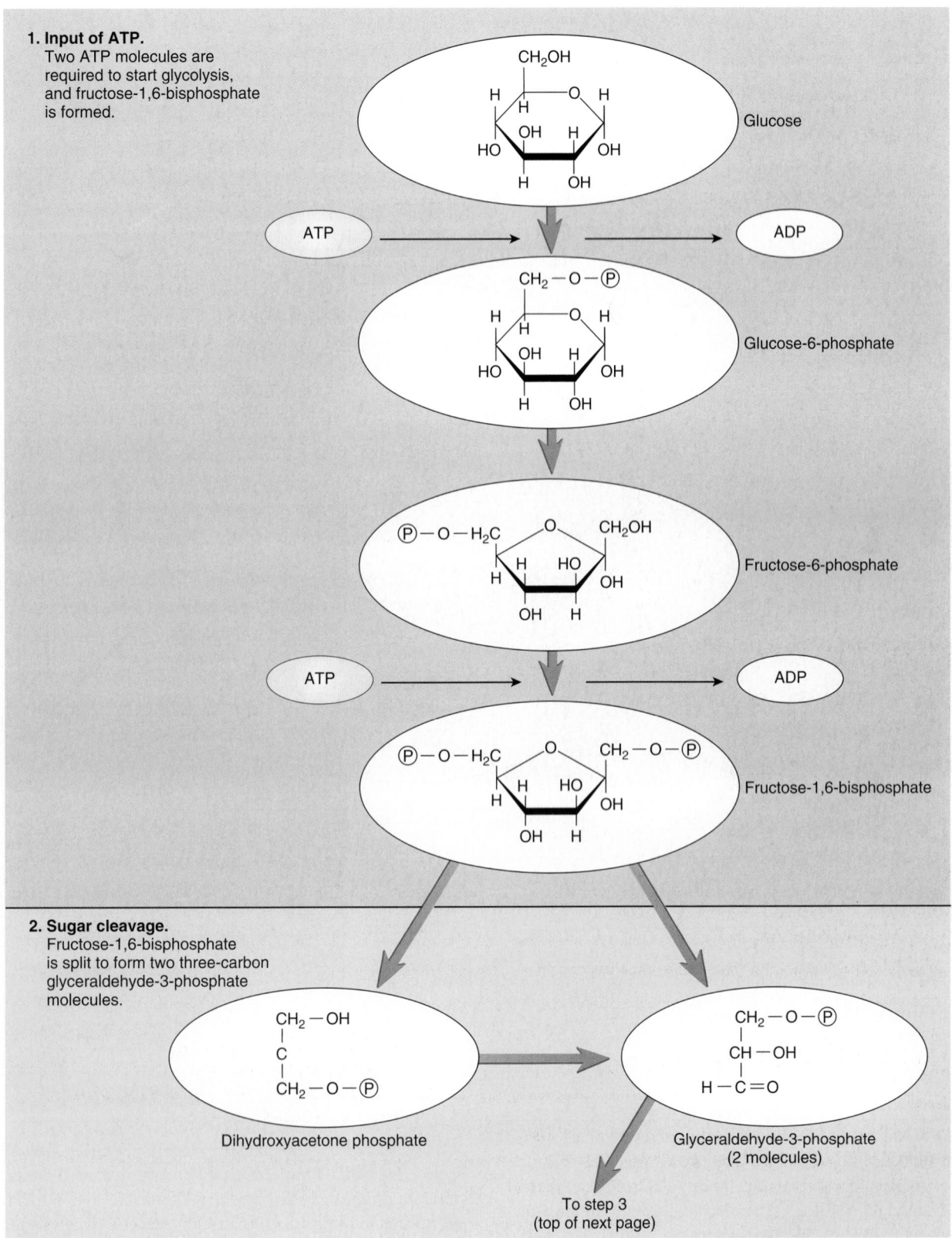

1. Input of ATP.
Two ATP molecules are required to start glycolysis, and fructose-1,6-bisphosphate is formed.

Glucose

ATP → ADP

Glucose-6-phosphate

Fructose-6-phosphate

ATP → ADP

Fructose-1,6-bisphosphate

2. Sugar cleavage.
Fructose-1,6-bisphosphate is split to form two three-carbon glyceraldehyde-3-phosphate molecules.

Dihydroxyacetone phosphate

Glyceraldehyde-3-phosphate
(2 molecules)

To step 3
(top of next page)

Process Figure 25.5   Glycolysis

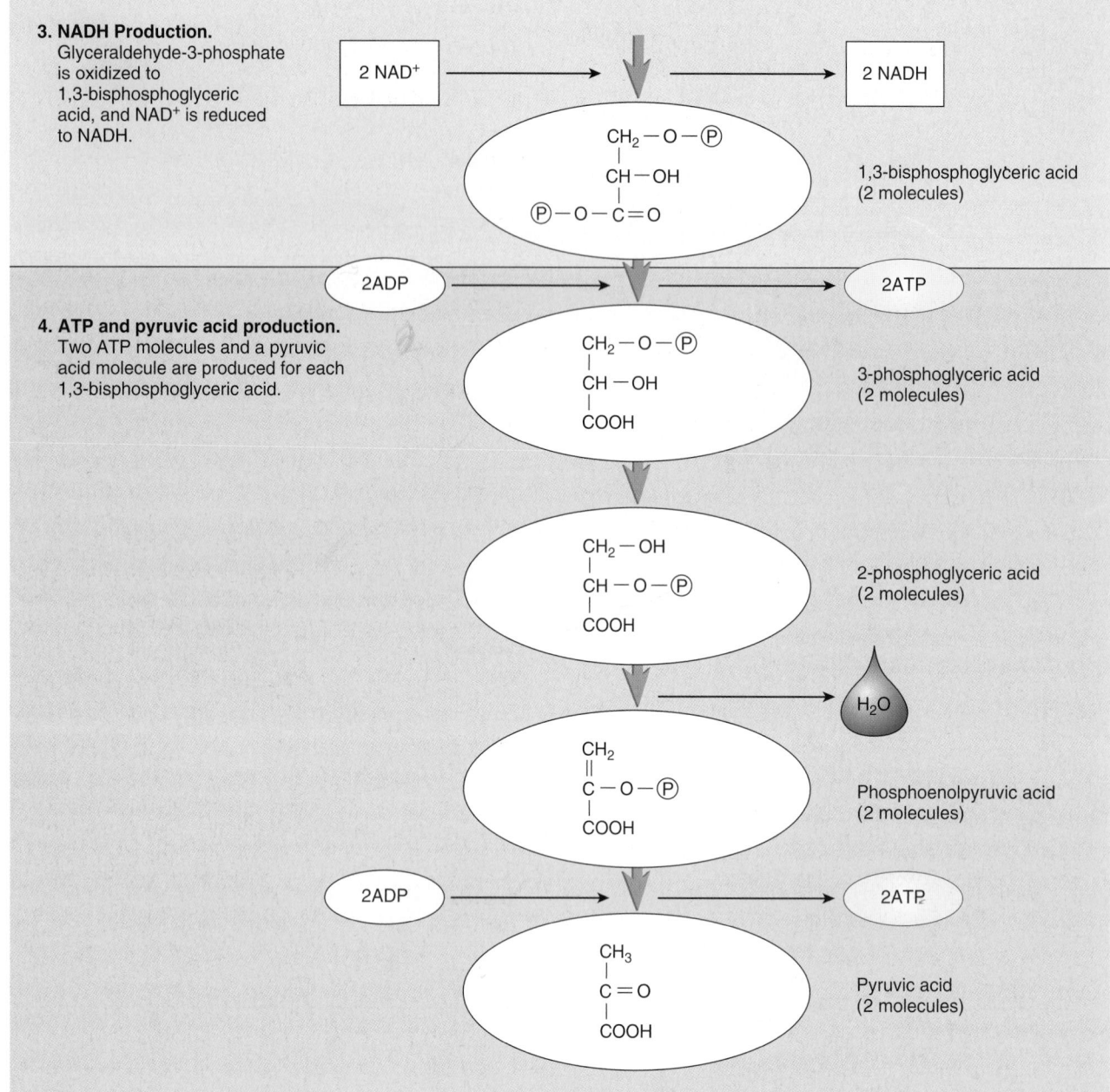

3. NADH Production.
Glyceraldehyde-3-phosphate is oxidized to 1,3-bisphosphoglyceric acid, and NAD⁺ is reduced to NADH.

4. ATP and pyruvic acid production.
Two ATP molecules and a pyruvic acid molecule are produced for each 1,3-bisphosphoglyceric acid.

Process Figure 25.5   *(continued)*

NAD^+ is the oxidized form of nicotinamide adenine dinucleotide, and NADH is the reduced form. NADH is a carrier molecule with two high-energy electrons (e^-) that can be used to produce ATP molecules through the electron-transport chain (see "Electron-Transport Chain" on p. 948).

4. *ATP and pyruvic acid production.* The last four steps of glycolysis produce two ATP molecules and one pyruvic acid molecule from each 1,3-bisphosphoglyceric acid molecule.

The events of glycolysis are summarized in table 25.4. Each glucose molecule that enters glycolysis forms two glyceraldehyde-3-phosphate molecules at the sugar cleavage phase. Each

glyceraldehyde-3-phosphate molecule produces two ATP molecules, one NADH molecule, and one pyruvic acid molecule. Each glucose molecule, therefore, forms four ATP, two NADH, and two pyruvic acid molecules. Because the start of glycolysis requires the input of two ATP molecules, however, the final yield of each glucose molecule is two ATP, two NADH, and two pyruvic acid molecules (see figure 25.4).

If the cell has adequate amounts of oxygen, the NADH and pyruvic acid molecules are used in aerobic respiration to produce ATP. In the absence of sufficient oxygen, they are used in anaerobic respiration.

Table 25.4	ATP Production from One Glucose Molecule	
Process	Product	Total ATP Produced*
Glycolysis	4 ATP	2 ATP (4 ATP produced minus 2 ATP to start)
	2 NADH	6 ATP (or 4 ATP; see text)
Acetyl-CoA production	2 NADH	6 ATP
Citric acid cycle	2 ATP	2 ATP
	6 NADH	18 ATP
	2 FADH$_2$	4 ATP
Total		38 ATP (or 36 ATP; see text)

*NADH and FADH$_2$ are used in the production of ATP in the electron-transport chain. *Abbreviations:* ATP = adenosine triphosphate, NADH = reduced nicotinamide adenine dinucleotide, FADH$_2$ = reduced flavin adenine diphosphate, acetyl-CoA = acetyl coenzyme A.

• 23. **Describe the four phases of glycolysis. What are the products of glycolysis?**

24. **What determines whether the pyruvic acid produced in glycolysis is used in aerobic or anaerobic respiration?**

Anaerobic Respiration

Anaerobic (an-ār-ō′bik) **respiration** is the breakdown of glucose in the absence of oxygen to produce two molecules of **lactic** (lak′tik) **acid** and two molecules of ATP (figure 25.6). The ATP thus produced is a source of energy during activities such as intense exercise, when insufficient oxygen is delivered to tissues. Anaerobic respiration can be divided into two phases.

1. *Glycolysis.* The first phase of anaerobic respiration is glycolysis, in which glucose undergoes several reactions to produce two pyruvic acid molecules and two NADH. There's also a net gain of two ATP molecules.
2. *Lactic acid formation.* The second phase is the conversion of pyruvic acid to lactic acid, a reaction that requires the input of energy from the NADH produced in phase 1 of anaerobic respiration.

Lactic acid is released from the cells that produce it and is transported by the blood to the liver. When oxygen becomes available, the lactic acid in the liver can be converted through a series of chemical reactions into glucose. The glucose then can be released from the liver and transported in the blood to cells that use glucose as an energy source. This process of converting lactic acid to glucose is called the **Cori cycle.** Some of the reactions involved in converting lactic acid into glucose require the input of energy derived from ATP that is produced by aerobic respiration. The oxygen necessary for the synthesis of the ATP is part of the **oxygen debt** (see chapter 9).

1. Glycolysis converts glucose to two pyruvic acid molecules. The many reactions within the pathway are not shown. There is a net gain of two ATP and two NADH from glycolysis.

2. Anaerobic respiration, which does not require oxygen, includes glycolysis and converts the two pyruvic acid molecules produced by glycolysis to two lactic acid molecules. This conversion requires energy, which is derived from the NADH generated in glycolysis.

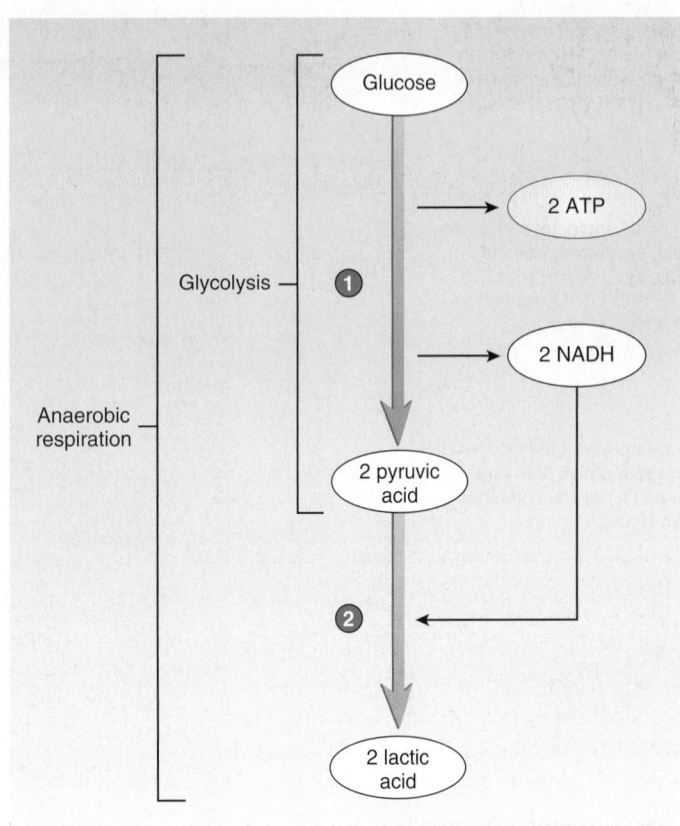

Process Figure 25.6 Glycolysis and Anaerobic Respiration

● 25. *Describe the two phases of anaerobic respiration. How many ATP molecules are produced by anaerobic respiration?*

26. *What happens to the lactic acid produced in anaerobic respiration when oxygen becomes available?*

Aerobic Respiration

Aerobic (ār-ō′bik) **respiration** is the breakdown of glucose in the presence of oxygen to produce carbon dioxide, water, and 38 ATP molecules. Most of the ATP molecules required to sustain life are produced through aerobic respiration, which can be considered in four phases: glycolysis, acetyl-CoA formation, the citric acid cycle, and the electron-transport chain. The first phase of aerobic respiration, as in anaerobic respiration, is glycolysis. The remaining phases are acetyl-CoA formation, the citric acid cycle, and the electron-transport chain (figure 25.7).

Acetyl-CoA Formation

In the second phase of aerobic respiration, pyruvic acid moves from the cytosol into a mitochondrion, which is separated into an inner and outer compartment by the inner mitochondrial membrane. Within the inner compartment, enzymes remove a carbon and two oxygen atoms from the three-carbon pyruvic acid molecule to form carbon dioxide and a two-carbon acetyl (as′e-til) group (figure 25.8). Energy is released in the reaction and is used to reduce NAD^+ to NADH. The acetyl group combines with coenzyme A (CoA) to form acetyl-CoA. For each two pyruvic acid molecules from glycolysis, two acetyl-CoA molecules, two carbon dioxide molecules, and two NADH are formed (see figure 25.4).

Citric Acid Cycle

The third phase of aerobic respiration is the **citric acid cycle,** which is named after the six-carbon citric acid molecule formed in the first

1. Glycolysis converts glucose to two pyruvic acid molecules and produces two ATP and two NADH. The NADH can go to the electron-transport chain.

2. The two pyruvic acid molecules produced in glycolysis are converted to two acetyl-CoA molecules, producing two CO_2 and two NADH. The NADH can go to the electron-transport chain.

3. The two acetyl-CoA molecules enter the citric acid cycle, which produces four CO_2, six NADH, two $FADH_2$, and two ATP. The NADH and $FADH_2$ can go to the electron-transport chain.

4. The electron-transport chain uses NADH and $FADH_2$ to produce 34 ATP. This process requires O_2, which combines with H^+ to form H_2O.

Process Figure 25.7   Aerobic Respiration

Aerobic respiration involves four steps: (1) glycolysis, (2) acetyl-CoA formation, (3) citric acid cycle, and (4) electron-transport chain. The number of carbon atoms in a molecule is indicated after the molecule's name. As glucose is broken down, the carbon atoms from glucose are incorporated into carbon dioxide.

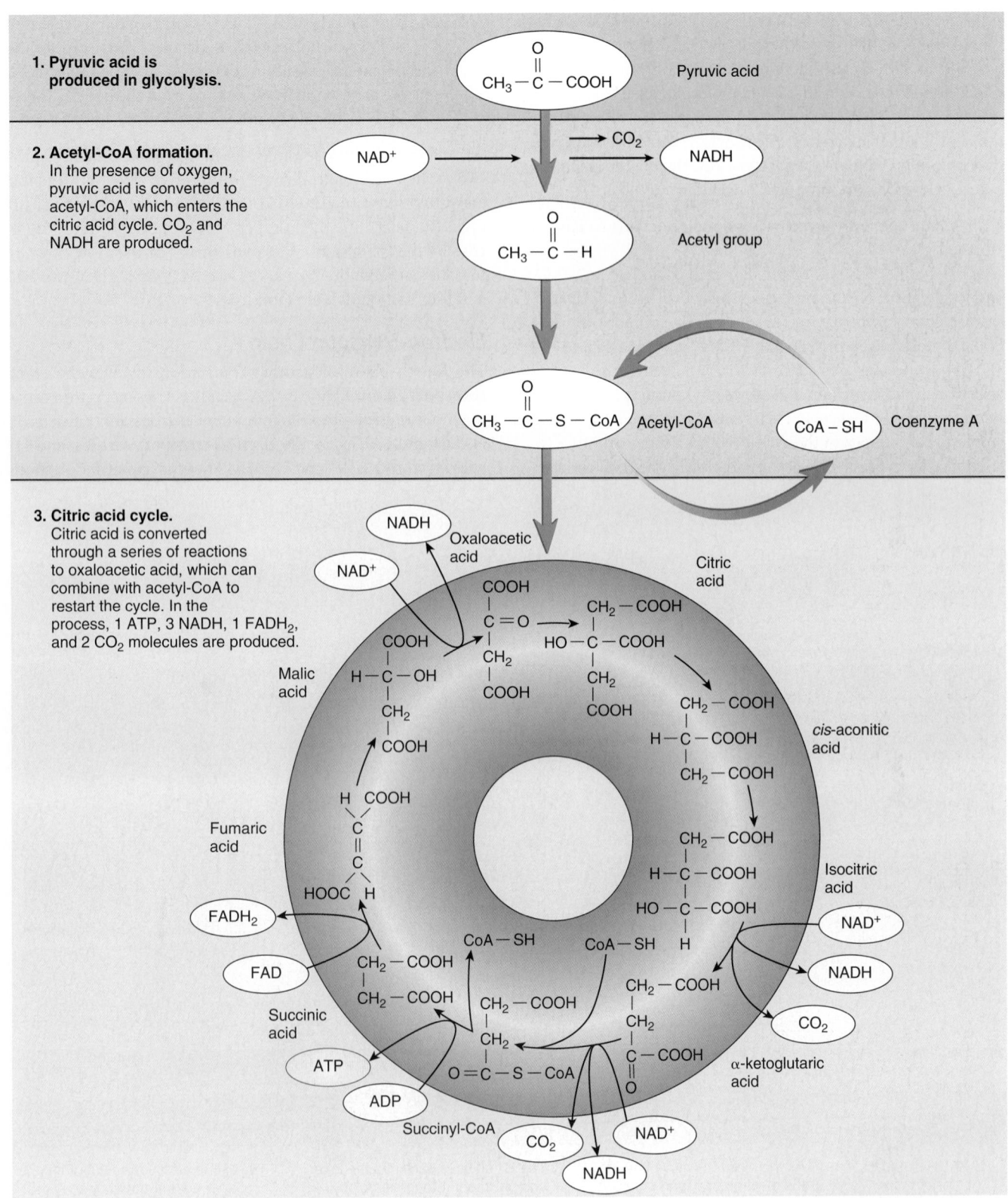

1. Pyruvic acid is produced in glycolysis.

Pyruvic acid

2. Acetyl-CoA formation.
In the presence of oxygen, pyruvic acid is converted to acetyl-CoA, which enters the citric acid cycle. CO_2 and NADH are produced.

Acetyl group

Acetyl-CoA

Coenzyme A

3. Citric acid cycle.
Citric acid is converted through a series of reactions to oxaloacetic acid, which can combine with acetyl-CoA to restart the cycle. In the process, 1 ATP, 3 NADH, 1 $FADH_2$, and 2 CO_2 molecules are produced.

Oxaloacetic acid

Citric acid

Malic acid

cis-aconitic acid

Fumaric acid

Isocitric acid

Succinic acid

α-ketoglutaric acid

Succinyl-CoA

Process Figure 25.8 Aerobic Respiration

step of the cycle (see figure 25.8). It is also called the Krebs' cycle after its discoverer, the British biochemist Sir Hans Krebs. The citric acid cycle begins with the production of citric acid from the combination of acetyl-CoA and a four-carbon molecule called oxaloacetic (ok′să-lō-ă-sē′tik) acid. A series of reactions occurs, resulting in the formation of another oxaloacetic acid, which can start the cycle again by combining with another acetyl-CoA. During the reactions of the citric acid cycle, three important events occur.

1. *ATP production.* For each citric acid molecule, one ATP is formed.
2. *NADH and FADH$_2$ production.* For each citric acid molecule, three NAD$^+$ molecules are converted to NADH molecules, and one flavin (flā′vin) adenine dinucleotide (FAD) molecule is converted to FADH$_2$. The NADH and FADH$_2$ molecules are electron carriers that enter the electron-transport chain and are used to produce ATP.
3. *Carbon dioxide production.* Each six-carbon citric acid molecule at the start of the cycle becomes a four-carbon oxaloacetic acid molecule at the end of the cycle. Two carbon

and four oxygen atoms from the citric acid molecule are used to form two carbon dioxide molecules. Thus, some of the carbon and oxygen atoms that make up food molecules like glucose are eventually eliminated from the body as carbon dioxide. We literally breathe out part of the food we eat!

For each glucose molecule that begins aerobic respiration, two pyruvic acid molecules are produced in glycolysis, and they are converted into two acetyl-CoA molecules that enter the citric acid cycle. To determine the number of molecules produced from glucose by the citric acid cycle, two "turns" of the cycle must, therefore, be counted; the results are two ATP, six NADH, two FADH$_2$, and four carbon dioxide molecules (see figure 25.4).

Electron-Transport Chain

The fourth phase of aerobic respiration involves the **electron-transport chain** (figure 25.9), which is a series of electron carriers in the inner mitochondrial membrane. Electrons are transferred from NADH and FADH$_2$ to the electron-transport carriers, and H$^+$ are released from NADH and FADH$_2$. After the loss of the electrons and

1. NADH or FADH$_2$ transfer their electrons to the electron-transport chain.

2. As the electrons move through the electron-transport chain, some of their energy is used to pump H$^+$ into the outer compartment, resulting in a higher concentration of H$^+$ in the outer than in the inner compartment.

3. The H$^+$ diffuses back into the inner compartment through special channels (ATP synthase) that couple the H$^+$ movement with the production of ATP. The electrons, H$^+$, and O$_2$ combine to form H$_2$O.

4. ATP is transported out of the inner compartment by a carrier molecule that exchanges ATP for ADP. A different carrier molecule moves phosphate into the inner compartment.

Process Figure 25.9   The Electron-Transport Chain

The electron-transport chain in the inner membrane consists of four protein complexes (*purple;* numbered I to IV) with carrier molecules. As electrons are transferred from one carrier molecule to another, they lose energy that is used to move H$^+$ out of the inner compartment. Hydrogen ions move back into the inner compartment through special channels (ATP synthase; *green*), which produce ATP. Carrier molecules (*brown*) move ATP out of and ADP and P$_i$ into the inner compartment.

the H^+, the oxidized NAD^+ and FAD are reused to transport additional electrons from the citric acid cycle to the electron-transport chain.

The electrons released from NADH and $FADH_2$ pass from one electron carrier to the next through a series of oxidation–reduction reactions. Three of the electron carriers also function as proton pumps that move the H^+ from the inner mitochondrial compartment into the outer mitochondrial compartment. Each proton pump accepts an electron, uses some of the electron's energy to export a H^+, and passes the electron to the next electron carrier. The last electron carrier in the series collects the electrons and combines them with oxygen and H^+ to form water.

$$1/2\ O_2 + 2\ H^+ + 2\ e^- \rightarrow H_2O$$

Without oxygen to accept the electrons, the reactions of the electron-transport chain cease, effectively stopping aerobic respiration.

The H^+ released from NADH and $FADH_2$ are moved from the inner mitochondrial compartment to the outer mitochondrial compartment by active transport. As a result, the concentration of H^+ in the outer compartment exceeds that of the inner compartment, and H^+ diffuse back into the inner compartment. The H^+ pass through certain channels formed by an enzyme called **ATP synthase.** As the H^+ diffuse down their concentration gradient they lose energy that is used to produce ATP. This process is called the **chemiosmotic** (kem-ē-os-mot′ik) **model** because the chemical formation of ATP is coupled to a diffusion force similar to osmosis.

27. *Define the term aerobic respiration, and list the products produced by it. Describe the four phases of aerobic respiration.*
28. *Why is the citric acid cycle a cycle? What molecules are produced as a result of the citric acid cycle?*
29. *What is the function of the electron-transport chain? Describe the chemiosmotic model of ATP production.*

P R E D I C T 4

Many poisons function by blocking certain steps in the metabolic pathways. For example, cyanide blocks the last step in the electron-transport chain. Explain why this blockage would cause death.

Summary of ATP Production

For each glucose molecule, aerobic respiration produces a net gain of 38 ATP molecules: 2 from glycolysis, 2 from the citric acid cycle, and 34 from the NADH molecules and $FADH_2$ molecules that pass through the electron-transport chain (see table 25.4). For each NADH molecule formed, three ATP molecules are produced by the electron-transport chain, and for each $FADH_2$ molecule, two ATP molecules are produced.

The number of ATP molecules produced is also reported as 36 ATP molecules. The two NADH molecules produced by glycolysis in the cytosol cannot cross the inner mitochondrial membrane; thus, their electrons are donated to a shuttle molecule that carries the electrons to the electron-transport chain. Depending on the shuttle molecule, each glycolytic NADH molecule can produce 2 or 3 ATP molecules. In skeletal muscle and the brain, 2 ATP molecules are produced for each NADH molecule, resulting in a total number of 36 ATP molecules; but in the liver, kidneys, and heart, 3 ATP molecules are produced for each NADH molecule, and the total number of ATP molecules formed is 38.

Six carbon dioxide molecules are produced in aerobic respiration. Water molecules are reactants in some of the chemical reactions of aerobic respiration and products in others. Six water molecules are used, but 12 are formed, for a net gain of 6 water molecules. Thus, aerobic respiration can be summarized as follows:

$$C_6H_{12}O_6 + 6\ O_2 + 6\ H_2O + 38\ ADP + 38\ P_i \rightarrow$$
$$6\ CO_2 + 12\ H_2O + 38\ ATP$$

30. *In aerobic respiration, how many ATP molecules are produced from one molecule of glucose through glycolysis, the citric acid cycle, and the electron-transport chain?*
31. *Why is the total number of ATP produced in aerobic respiration listed as 38 or 36?*

The Quantity of ATP Produced from Glucose

The number of ATP molecules produced per glucose molecule is a theoretical number that assumes two hydrogen ions are necessary for the formation of each ATP. If the number required is more than two, the efficiency of aerobic respiration decreases. In addition, it's now understood that it costs energy to get ADP and phosphates into the mitochondria and to get ATP out. Considering all these factors, it's estimated that each glucose molecule yields about 25 ATP molecules instead of 38 ATP molecules.

Lipid Metabolism

Lipids are the body's main energy-storage molecules. In a healthy person, lipids are responsible for about 99% of the body's energy storage, and glycogen accounts for about 1%. Although proteins are used as an energy source, they are not considered storage molecules because the breakdown of proteins normally involves the loss of molecules that perform other functions.

Lipids are stored primarily as triglycerides in adipose tissue. There is constant synthesis and breakdown of triglycerides; thus, the fat present in adipose tissue today isn't the same fat that was there a few weeks ago. Between meals when triglycerides are broken down in adipose tissue, some of the fatty acids produced are released into the blood, where they are called **free fatty acids.** Other tissues, especially skeletal muscle and the liver, use the free fatty acids as a source of energy.

The metabolism of fatty acids occurs by **beta-oxidation,** a series of reactions in which two carbon atoms at a time are removed from the end of a fatty acid chain to form acetyl-CoA. The process of beta-oxidation continues to remove two carbon atoms at a time until the entire fatty acid chain is converted into acetyl-CoA molecules. Acetyl-CoA can enter the citric acid cycle and be used to generate ATP (figure 25.10).

Acetyl-CoA is also used in **ketogenesis** (kē-tō-jen′ĕ-sis), the formation of ketone bodies. In the liver when large amounts of acetyl-CoA are produced, not all of the acetyl-CoA enters the citric acid cycle. Instead, two acetyl-CoA molecules combine to form a molecule of acetoacetic (as′ e-tō-a-sē′tik) acid, which is converted mainly into β-hydroxybutyric (hī-drōk′sē-bū-tir′ik) acid and a smaller amount of acetone (as′e-tōn). Acetoacetic acid, β-hydroxybutyric acid, and acetone are called **ketone** (kē′tōn) **bodies** and are released into the blood, where they travel to other tissues, especially skeletal muscle. In these tissues, the ketone bodies are converted back into acetyl-CoA that enters the citric acid cycle to produce ATP.

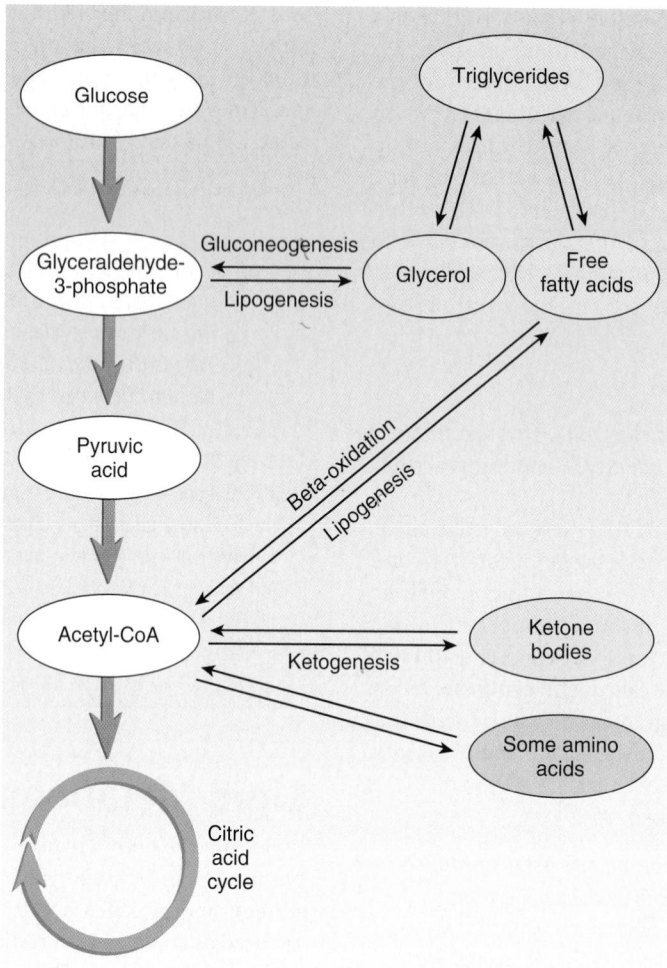

Figure 25.10 Lipid Metabolism

Triglyceride is broken down into glycerol and fatty acids. Glycerol enters glycolysis to produce ATP. The fatty acids are broken down by beta-oxidation into acetyl-CoA, which enters the citric acid cycle to produce ATP. Acetyl-CoA can also be used to produce ketone bodies (ketogenesis). Lipogenesis is the production of lipids. Glucose is converted to glycerol, and amino acids are converted to acetyl-CoA molecules. Acetyl-CoA molecules can combine to form fatty acids. Glycerol and fatty acids join to form triglycerides.

32. *Define the term beta-oxidation, and explain how it results in ATP production.*
33. *What are ketone bodies, how are they produced, and for what are they used?*

The Danger of Excessive Amounts of Ketones

Small amounts of ketone bodies in the blood is normal and beneficial. Excessive production of ketone bodies is called **ketosis** (kē-tō′sis). If the increased number of acidic ketone bodies exceeds the capacity of the body's buffering systems, acidosis, a decrease in blood pH, can occur (see chapter 27). Conditions that increase fat metabolism can increase the rate of ketone body formation. Examples are starvation (see Clinical Focus: "Starvation" on p. 954), diets consisting of proteins and fats with few carbohydrates, and untreated diabetes mellitus (see Systems Pathology: "Insulin-Dependent Diabetes Mellitus" in chapter 18). Ketone bodies are excreted by the kidneys and diffuse into the alveoli of the lungs. Ketone bodies in the urine and "acetone breath" are characteristic of untreated diabetes mellitus.

Protein Metabolism

Once absorbed into the body, amino acids are quickly taken up by cells, especially in the liver. Amino acids are used to synthesize needed proteins (see chapter 3) or as a source of energy (figure 25.11). Unlike glycogen and triglycerides, amino acids are not stored in the body.

The synthesis of nonessential amino acids usually begins with keto acids (figure 25.12). A keto acid can be converted into an amino acid by replacing its oxygen with an amine group. Usually this conversion is accomplished by transferring an amine group from an amino acid to the keto acid, a reaction called **transamination** (trans-am′i-nā′shŭn). For example, α-ketoglutaric acid (a keto acid) reacts with an amino acid to form glutamic acid (an amino acid; figure 25.13a). Most amino acids can undergo transamination to produce glutamic acid. The glutamic acid is used as a source of an amine group to construct most of the nonessential amino acids. A few nonessential amino acids are formed in other ways from the essential amino acids.

Figure 25.11   Amino Acid Metabolism

Various entry points for amino acids (*tan ovals*) into carbohydrate metabolism.

(a) Amino acid

(b) Keto acid

Figure 25.12   General Formulas of an Amino Acid and a Keto Acid

(*a*) Amino acid with a carboxyl group (—COOH), an amine group (NH₂), a hydrogen atom (H), and a group called "R" that represents the rest of the molecule. (*b*) Keto acid with a double-bonded oxygen replacing the amine group and the hydrogen atom of the amino acid.

Amino acids can be used as a source of energy. In **oxidative deamination** (dē-am-i-nā′shŭn; deaminization, dē-am′i-nizā′shŭn), an amine group is removed from an amino acid (usually glutamic acid), leaving ammonia and a keto acid (figure 25.13*b*). In the process, NAD^+ is reduced to NADH, which can enter the electron-transport chain to produce ATP. Ammonia is toxic to cells. An accumulation of ammonia to toxic levels is prevented because the liver converts it into urea, which is carried by the blood to the kidneys, where the urea is eliminated (figure 25.13*c*; see chapter 26).

Amino acids are also used as a source of energy by converting them into the intermediate molecules of carbohydrate metabolism (see figure 25.11). These molecules are then metabolized to yield ATP. The conversion of an amino acid often begins with a transamination or oxidative deamination reaction, in which the amino acid is converted into a keto acid (see figure 25.12). The keto acid enters the citric acid cycle or is converted into pyruvic acid or acetyl-CoA.

● **34.** *What is accomplished by transamination and oxidative deamination?*

35. *How are proteins (amino acids) used to produce energy?*

Figure 25.13 Amino Acid Reactions

(a) Transamination reaction in which an amine group is transferred from an amino acid to a keto acid to form a different amino acid. (b) Oxidative deamination reaction in which an amino acid loses an amine group to become a keto acid and to form ammonia. In the process, NADH, which can be used to generate ATP, is formed. (c) Ammonia is converted to urea in the liver. The actual conversion of ammonia to urea is more complex, involving a number of intermediate reactions that constitute the urea cycle.

Interconversion of Nutrient Molecules

Blood glucose enters most cells by facilitated diffusion and is immediately converted to glucose-6-phosphate, which cannot recross the plasma membrane (figure 25.14a). Glucose-6-phosphate then continues through glycolysis to produce ATP. If, however, excess glucose is present (e.g., after a meal), it's used to form glycogen through a process called **glycogenesis** (glī-kō-jen′ĕ-sis). Most of the body's glycogen is contained in skeletal muscle and the liver.

Once glycogen stores, which are quite limited, are filled, glucose and amino acids are used to synthesize lipids, a process called **lipogenesis** (lip-ō-jen′ĕ-sis; see figure 25.10). Glucose molecules can be used to form glyceraldehyde-3-phosphate and acetyl-CoA. Amino acids can also be converted to acetyl-CoA. Glyceraldehyde-3-phosphate is converted to glycerol, and the two-carbon acetyl-CoA molecules are joined together to form fatty acid chains. Glycerol and three fatty acids then combine to form triglycerides.

(a) When blood glucose levels are high, glucose enters the cell and is phosphorylated to form glucose-6-phosphate, which can enter glycolysis or glycogenesis.

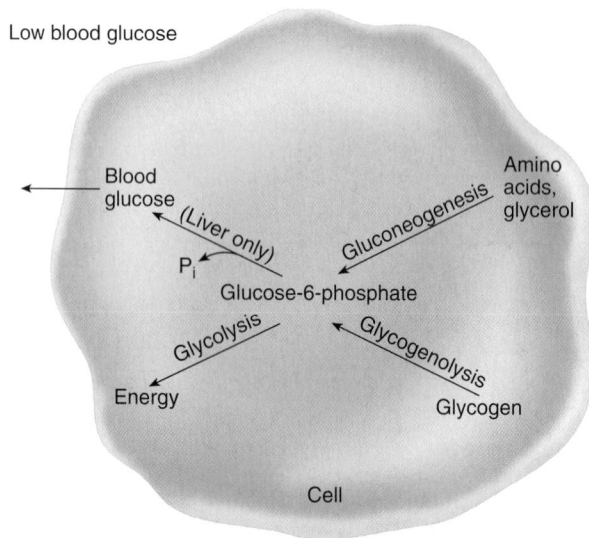

(b) When blood glucose levels drop, glucose-6-phosphate can be produced through glycogenolysis or gluconeogenesis. Glucose-6-phosphate can enter glycolysis, or the phosphate group can be removed in liver tissue and glucose released into the blood.

Figure 25.14 Interconversion of Nutrient Molecules

Alcoholism and Cirrhosis of the Liver

Enzymes in the liver convert ethanol (beverage alcohol) into acetyl-CoA, and in the process two NADH molecules are produced. The NADH molecules enter the electron-transport chain and are used to produce ATP molecules. Each gram of ethanol provides 7 kcal of energy. Because of the high level of NADH in the cell that results from the metabolism of ethanol, the production of NADH by glycolysis and by the citric acid cycle is inhibited. Consequently, sugars and amino acids are not broken down but are converted into fats that accumulate in the liver. Chronic alcohol abuse can, therefore, result in **cirrhosis** (sir-rō′sis) **of the liver,** which involves fat deposition, cell death, inflammation, and scar tissue formation. Death can occur because the liver is unable to carry out its normal functions.

When glucose is needed, glycogen can be broken down into glucose-6-phosphate through a set of reactions called **glycogenolysis** (glī′kō-jĕ-nol′i-sis; figure 25.14*b*). In skeletal muscle, glucose-6-phosphate continues through glycolysis to produce ATP. The liver can use glucose-6-phosphate for energy or can convert it to glucose, which diffuses into the blood. The liver can release glucose, but skeletal muscle cannot because it lacks the necessary enzymes to convert glucose-6-phosphate into glucose.

Release of glucose from the liver is necessary to maintain blood glucose levels between meals. Maintaining these levels is especially important to the brain, which normally uses only glucose for an energy source and consumes about two-thirds of the total glucose used each day. When liver glycogen levels are inadequate to supply glucose, it is synthesized from molecules other than carbohydrates, such as amino acids from proteins and glycerol from triglycerides, in a process called **gluconeogenesis** (gloo′kō-nē-ō-jen′ĕ-sis). Most amino acids can be converted into citric acid cycle molecules, acetyl-CoA, or pyruvic acid (see figure 25.11). Through a series of chemical reactions, these molecules are converted into glucose. Glycerol enters glycolysis by becoming glyceraldehyde-3-phosphate.

● **36.** *Define the terms glycogenesis, lipogenesis, glycogenolysis, and gluconeogenesis.*

Metabolic States

Two major metabolic states have been described in the body. The first is the **absorptive state,** the period immediately after a meal when nutrients are being absorbed through the intestinal wall into the circulatory and lymphatic systems (figure 25.15). The absorptive state usually lasts about 4 hours after each meal, and most of the glucose that enters the circulation is used by cells to provide the energy they require. The remainder of the glucose is converted into glycogen or fats. Most of the absorbed fats are deposited in adipose tissue. Many of the absorbed amino acids are used by cells in protein synthesis, some are used for energy, and others enter the liver and are converted into fats or carbohydrates.

The second state, the **postabsorptive state,** occurs late in the morning, late in the afternoon, or during the night after each absorptive state is concluded (figure 25.16). Normal blood

Figure 25.15   Events of the Absorptive State

Absorbed molecules, especially glucose, are used as sources of energy. Molecules not immediately needed for energy are stored: glucose is converted to glycogen or triglycerides, triglycerides are deposited in adipose tissue, and amino acids are converted to triglycerides or carbohydrates.

Clinical Focus   Starvation

Starvation results from the inadequate intake of nutrients or the inability to metabolize or absorb nutrients. It can have a number of causes, such as prolonged fasting, anorexia, deprivation, or disease. No matter what the cause, starvation follows the same course and consists of three phases. The events of the first two phases occur even during relatively short periods of fasting or dieting, but the third phase occurs only in prolonged starvation and can end in death.

During the first phase of starvation, blood glucose levels are maintained through the production of glucose from glycogen, proteins, and fats. At first, glycogen is broken down into glucose; however, only enough glycogen is stored in the liver to last a few hours. Thereafter, blood glucose levels are maintained by the breakdown of proteins and fats. Fats are decomposed into fatty acids and glycerol. Fatty acids can be used as a source of energy, especially by skeletal muscle, thus decreasing the use of glucose by tissues other

than the brain. Glycerol can be used to make a small amount of glucose, but most of the glucose is formed from the amino acids of proteins. In addition, some amino acids can be used directly for energy.

In the second stage, which can last for several weeks, fats are the primary energy source. The liver metabolizes fatty acids into ketone bodies that can be used as a source of energy. After about a week of fasting, the brain begins to use ketone bodies, as well as glucose, for energy. This usage decreases the demand for glucose, and the rate of protein breakdown diminishes but doesn't stop. In addition, the proteins not essential for survival are used first.

The third stage of starvation begins when the fat reserves are depleted and a switch to proteins as the major energy source takes place. Muscles, the largest source of protein in the body, are rapidly depleted. At the end of this stage, proteins essential for cellular functions are broken down, and cell function degenerates.

In addition to weight loss, symptoms of starvation include apathy, listlessness, withdrawal, and increased susceptibility to infectious disease. Few people die directly from starvation because they usually die of some infectious disease first. Other signs of starvation include changes in hair color, flaky skin, and massive edema in the abdomen and lower limbs, causing the abdomen to appear bloated.

During the process of starvation, the ability of the body to consume normal volumes of food also decreases. Foods high in bulk but low in protein content often cannot reverse the process of starvation. Intervention involves feeding the starving person low-bulk food that provides ample proteins and kilocalories and is fortified with vitamins and minerals. The process of starvation also results in dehydration, and rehydration is an important part of intervention. Even with intervention, a victim may be so affected by disease or weakness that he or she cannot recover.

Stored Nutrients Used
Stored energy molecules are used as sources of energy: glycogen is converted to glucose, and triglycerides are converted to fatty acids. Molecules released from tissues are carried by the blood to the liver.

Nutrients Processed
The liver processes molecules to produce additional energy sources: glycogen and amino acids are converted to glucose and fatty acids to ketones. Glucose and ketones are released into the blood and are transported to tissues.

Figure 25.16   Events of the Postabsorptive State

Stored energy molecules are used as sources of energy: glycogen is converted to glucose; triglycerides are broken down to fatty acids, some of which are converted to ketones; and proteins are converted to glucose.

glucose levels range between 70 and 110 mg/100 mL, and it's vital to the body's homeostasis that this range be maintained, especially for normal functioning of the brain. During the postabsorptive state, blood glucose levels are maintained by the conversion of other molecules to glucose. The first source of blood glucose during the postabsorptive state is the glycogen stored in the liver. This glycogen supply, however, can provide glucose for only about 4 hours. The glycogen stored in skeletal muscles can also be used during times of vigorous exercise. As glycogen stores are depleted, fats are used as an energy source. The glycerol from triglycerides can be converted to glucose. The fatty acids from fat can be converted to acetyl-CoA, moved into the citric acid cycle, and used as a source of energy to produce ATP. In the liver, acetyl-CoA is used to produce ketone bodies that other tissues use for energy. The use of fatty acids as an energy source partly eliminates the need to use glucose for energy, resulting in reduced glucose removal from the blood and maintenance of blood glucose levels at homeostatic levels. Proteins can also be used as a source of glucose or can be used for energy production, again sparing the use of blood glucose.

37. **What happens to glucose, fats, and amino acids during the absorptive state?**

38. **Why is it important to maintain blood glucose levels during the postabsorptive state? Name three sources for this glucose.**

Metabolic Rate

Metabolic rate is the total amount of energy produced and used by the body per unit of time. A molecule of ATP exists for less than 1 minute before it's degraded back to ADP and inorganic phosphate. For this reason, ATP is produced in cells at about the same rate as it's used. Thus, in examining metabolic rate, ATP production and use can be roughly equated. Metabolic rate is usually estimated by measuring the amount of oxygen used per minute because most ATP production involves the use of oxygen. One liter of oxygen consumed by the body is assumed to produce 4.825 kcal of energy.

The daily input of energy should equal the metabolic expenditure of energy; otherwise, a person will gain or lose weight. For a typical 23-year-old, 70 kg (154-pound) male to maintain his weight, the daily input should be 2700 kcal/day; for a typical 58 kg (128-pound) female of the same age 2000 kcal/day is necessary. A pound of body fat provides about 3500 kcal. Reducing kilocaloric intake by 500 kcal/day can result in the loss of 1 pound of fat per week. Clearly, adjusting kilocaloric input is an important way to control body weight.

The Proportion of Fat in the Diet and Body Weight

Not only the number of kilocalories ingested but also the proportion of fat in the diet has an effect on body weight. To convert dietary fat into body fat, 3% of the energy in the dietary fat is used, leaving 97% for storage as fat deposits. On the other hand, the conversion of dietary carbohydrate to fat requires 23% of the energy in the carbohydrate, leaving just 77% as body fat. If two people have the same kilocaloric intake, the one with the higher proportion of fat in his or her diet is more likely to gain weight because fewer kilocalories are used to convert the dietary fat into body fat.

Metabolic energy is used in three ways: for basal metabolism, for the thermic effect of food, and for muscular activity.

Basal Metabolic Rate

The **basal metabolic rate (BMR)** is the energy needed to keep the resting body functional. It is the metabolic rate calculated in expended kilocalories per square meter of body surface area per hour. BMR is determined by measuring the oxygen consumption of a person who is awake but restful and has not eaten for 12 hours. The liters of oxygen consumed are then multiplied by 4.825 because each liter of oxygen used results in the production of 4.825 kcal of energy. A typical BMR for a 70 kg (154-pound) male is 38 kcal/m^2/h.

In the average person, basal metabolism accounts for about 60% of energy expenditure. Basal metabolism supports active-transport mechanisms, muscle tone, maintenance of body temperature, beating of the heart, and other activities. A number of factors can affect the BMR. Muscle tissue is metabolically more active than adipose tissue, even at rest. Younger people have a higher BMR than older people because of increased cell activity, especially during growth. Fever can increase BMR 7% for each degree Fahrenheit increase in body temperature. During dieting or fasting, greatly reduced kilocaloric input can depress BMR, which apparently is a protective mechanism to prevent weight loss. Thyroid hormones can increase BMR on a long-term basis, and epinephrine can increase BMR on a short-term basis (see chapter 18). Males have a greater BMR than females because men have proportionately more muscle tissue and less adipose tissue than women do. During pregnancy, a woman's BMR can increase 20% because of the metabolic activity of the fetus.

Thermic Effect of Food

The second component of metabolic energy concerns the assimilation of food. When food is ingested, the accessory digestive organs and the intestinal lining produce secretions, the motility of the digestive tract increases, active transport increases, and the liver is involved in the synthesis of new molecules. The energy cost of these events is called the **thermic effect of food** and accounts for about 10% of the body's energy expenditure.

Muscular Activity

Muscular activity consumes about 30% of the body's energy. Physical activity resulting from skeletal muscle movement requires the expenditure of energy. In addition, energy must be provided for increased contraction of the heart and of the muscles of respiration. The number of kilocalories used in an activity depends almost entirely on the amount of muscular work performed and on the duration of the activity. Despite the fact that studying can make a person feel tired, intense mental concentration produces little change in the BMR.

Energy loss through muscular activity is the only component of energy expenditure that a person can reasonably control. A comparison of the number of kilocalories gained from food versus the number of kilocalories lost in exercise reveals why losing weight can be difficult. For example, walking (3 mph) for 20 minutes burns the kilocalories supplied by one slice of bread, whereas jogging (5 mph) for the same time eliminates the kilocalories obtained from a soft drink or a beer (see table 25.1). Nonetheless, weight loss through exercise and dieting is possible.

39. *Define the term metabolic rate.*
40. *What is BMR? What factors can alter BMR?*
41. *What is the thermic effect of food?*
42. *BMR, thermic effect of food, and muscular activity each account for what percent of total energy expenditure?*
43. *How are kilocalorie input and output adjusted to maintain body weight?*

Body Temperature Regulation

Humans can maintain a relatively constant internal body temperature despite changes in the temperature of the surrounding environment. Maintenance of a constant body temperature is very important to homeostasis. Most enzymes are very temperature sensitive and function only in narrow temperature ranges. Environmental temperatures are too low for normal enzyme function, and the heat produced by metabolism helps maintain the body temperature at a steady, elevated level that is high enough for normal enzyme function.

Free energy is the total amount of energy liberated by the complete catabolism of food. It's usually expressed in terms of kilocalories (kcal) per mole of food consumed. For example, the complete catabolism of 1 mole of glucose (168 g; see chapter 2) releases 686 kcal of free energy. About 43% of the total energy released by catabolism is used to produce ATP and to accomplish biologic work, such as anabolism, muscular contraction, and other cellular activities. The remaining energy is lost as **heat.**

P R E D I C T   5

Explain why we become warm during exercise and why we shiver when it's cold.

The average normal body temperature usually is considered 37°C (98.6°F) when measured orally and 37.6°C (99.7°F) when measured rectally. Rectal temperature comes closer to the true core body temperature, but an oral temperature is more easily obtained in older children and adults and, therefore, is the preferred measure.

Heat can be exchanged with the environment in a number of ways (figure 25.17). **Radiation** is the loss of heat as infrared radiation, a type of electromagnetic radiation. For example, the coals in a fire give off radiant heat that can be felt some distance away from the fire. **Conduction** is the exchange of heat between objects in direct contact with each other, such as the bottom of the feet and the floor. **Convection** is a transfer of heat between the body and the air or water. A cool breeze results in the movement of air over the body and loss of heat from the body. **Evaporation** is the conversion of water from a liquid to a gaseous form, a process that requires heat. The evaporation of 1 g of water from the body's surface results in the loss of 580 cal of heat.

Body temperature is maintained by balancing heat gain with heat loss. When heat gain equals heat loss, body temperature is maintained. If heat gain exceeds heat loss, body temperature increases, and if heat loss exceeds heat gain, body temperature

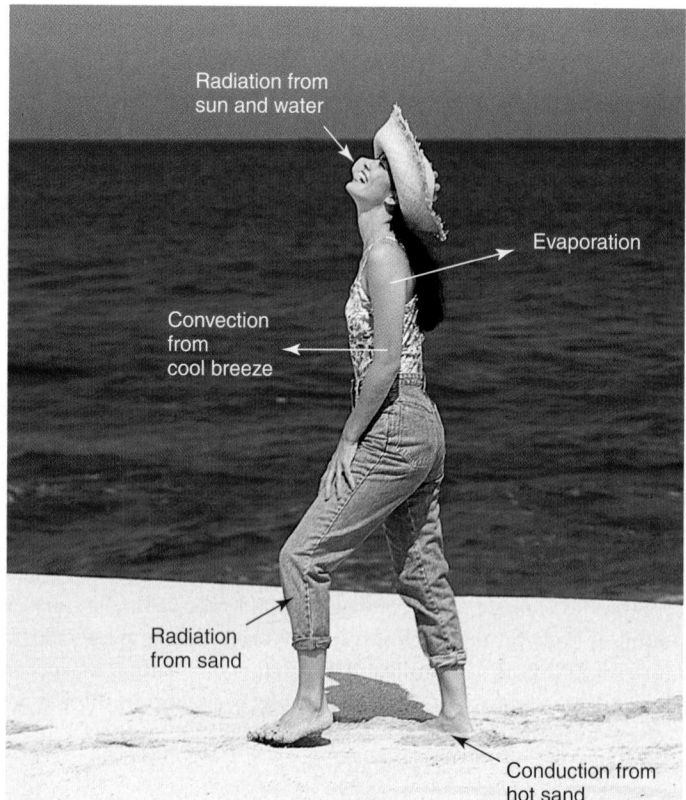

Figure 25.17   Heat Exchange

Heat exchange between a person and the environment occurs by convection, radiation, evaporation, and conduction. *Arrows* show the direction of net heat gain or loss in this environment.

Clinical Focus   Obesity

Obesity is the presence of excess fat, and it results from the ingestion of more food than is necessary for the body's energy needs. Obesity can be defined on the basis of body weight, body mass index, or percent body fat. "Desirable body weight" is listed in the Metropolitan Life Insurance Table and indicates, for any height, the weight that is associated with a maximum life span. Overweight is defined as weighing 10% more than the "desirable weight," and obesity is weighing 20% more than the "desirable weight." Body mass index (BMI) can be calculated by dividing a person's weight (Wt) in kilograms by the square of his or her height (Ht) in meters: $BMI = Wt/Ht^2$. A BMI greater than 25–27 is overweight, and a value greater than 30 is defined as obese. About 10% of people in the United States have a BMI of 30 or greater. In terms of the percent of the total body weight contributed by fat, 15% body fat or less in men and 25% body fat or less in women is associated with reduced health risks. Obesity is defined to be more than 25% body fat in men and 30%–35% in women.

Obesity is classified according to the number and size of fat cells. The greater the amount of lipids stored in the fat cells, the larger their size. In **hyperplastic obesity,** a greater-than-normal number of fat cells occur that are also larger than normal. This type of obesity is associated with massive obesity and begins at an early age. In nonobese children, the number of fat cells triples or quadruples between birth and 2 years of age and then remains relatively stable until puberty, when a further increase in the number occurs. In obese children, however, between 2 years of age and puberty, an increase also occurs in the number of fat cells. **Hypertrophic obesity** results from a normal number of fat cells that have increased in size. This type of obesity is more common, is associated with moderate obesity or being "overweight," and typically develops in adults. People who were thin or of average weight and quite active when they were young become less active as they become older. They begin to gain weight be-

tween age 20 and 40, and, although they no longer use as many kilocalories, they still take in the same amount of food as when they were younger. The unused kilocalories are turned into fat, causing fat cells to increase in size. At one time, it was believed that the number of fat cells did not increase after adulthood. It's now known that the number of fat cells can increase in adults. Apparently, if all the existing fat cells are filled to capacity with lipids, new fat cells are formed to store the excess lipids. Once fat cells are formed, however, dieting and weight loss don't result in a decrease in the number of fat cells—instead, they become smaller in size as their lipid content decreases.

The distribution of fat in obese individuals varies. Fat can be found mainly in the upper body, such as in the abdominal region, or it can be associated with the hips and buttocks. These distribution differences are clinically significant because upper body obesity is associated with an increased likelihood of diabetes mellitus, cardiovascular disease, stroke, and death.

In some cases, a specific cause of obesity can be identified. For example, a tumor in the hypothalamus can stimulate overeating. In most cases, however, no specific cause is apparent. In fact, obesity occurs for many reasons, and obesity in an individual can have more than one cause. A genetic component to obesity seems to exist, and, if one or both parents are obese, their children are more likely to also be obese. Environmental factors, such as eating habits, however, can also play an important role. For example, adopted children can exhibit similarities in obesity to their adoptive parents. In addition, psychologic factors, such as overeating as a means for dealing with stress, can contribute to obesity.

Regulation of body weight is actually a matter of regulating body fat because most changes in body weight reflect changes in the amount of fat in the body. According to the "set point" theory of weight control, the body maintains a certain amount of body

fat. If the amount decreases below or increases above this level, mechanisms are activated to return the amount of body fat to its normal value.

The two factors that affect the amount of adipose tissue in the body are energy intake and energy expenditure. The regulation of energy intake is poorly understood. Apparently, neurons originating in or passing through the hypothalamus continually and spontaneously stimulate appetite and food-seeking behaviors. After food is consumed, several mechanisms are responsible for decreasing further food intake. Neural mechanisms, such as distension of the stomach, are known to inhibit feeding, and a number of hormones released from the gastrointestinal tract or pancreas also decrease appetite. For example, somatostatin, cholecystokinin, glucagon, insulin, and other hormones have been shown to reduce food intake. The level of fatty acids, glucose, or amino acids in the blood also provides the brain with information necessary to adjust appetite. Low levels of fatty acids, glucose, and amino acids stimulate appetite, whereas high levels of these substances inhibit appetite.

Some scientists believe that the number of fat cells in the body also affects appetite. According to this line of reasoning, fat cells maintain their size, and, once a "fat plateau" is attained, the body stays at that plateau. Fat cells accomplish this by effectively taking up triglycerides and converting them to fat. Consequently, less energy is available for muscle and body organs, and, to compensate, appetite increases to provide needed energy. In support of this hypothesis, it's known that obese individuals have an increased amount of the enzyme lipoprotein lipase, which is responsible for the uptake and storage of triglycerides in fat cells. Furthermore, in obese individuals who have lost weight, the levels of lipoprotein lipase increase even more.

It's a common belief that the main cause of obesity is overeating. Certainly for

Continued

Continued

obesity to occur, at some time, energy intake must have exceeded energy expenditure. A comparison of the kilocaloric intake of obese and lean individuals at their usual weights, however, reveals that on a per kilogram basis, obese people consume fewer kilocalories than lean people.

When people lose a large amount of weight, their feeding behavior changes. They become hyperresponsive to external food cues, think of food often, and cannot get enough to eat without gaining weight. It's now understood that this behavior is typical of both lean and obese individuals who are below their relative set point for weight. Other changes, such as a decrease in basal metabolic rate, take place in a person who has lost a large amount of weight. Most of this decrease in BMR probably results from a decrease in muscle mass associated with weight loss. In addition, some evidence exists that energy lost through exercise and the thermic effect of food are also reduced.

Thus, a person who has lost a large amount of weight is a person with an increased appetite and a decreased ability to expend energy. It's no surprise that only a small percentage of obese people maintain weight loss over the long term. Instead, the typical pattern is one of repeated cycles of weight loss followed by a rapid regain of the lost weight.

Current research is attempting to find ways to help manage obesity. Unfortunately, most appetite suppressants can only be used for a short time. Dexfenfluramine (deks-fen-flū′ră-měn), which had been approved by the FDA for long-term use, was recalled because of harmful side effects.

Humans have an *obese (ob)* gene that codes for a protein called **leptin** (lep′tin), which is mainly produced by adipose cells. Leptin is released into the blood and affects appetite and body weight regulation. When energy stores in adipose tissue decrease, leptin levels decrease, resulting in an increased appetite and a decreased metabolism. Decreased leptin levels may be a signal that helps the body to adjust to fasting or starvation by increasing food intake and reducing energy expenditure. Two populations of obese individuals provide supporting evidence for this hypothesis. In some obese people, leptin is inappropriately low as a function of the amount of body fat present. Thus, there is a mismatch between the leptin signal and the amount of energy stored in adipose tissue. Most obesity, however, occurs in the presence of elevated levels of leptin. This observation seems contradictory, because high leptin levels should cause decreased appetite, increased metabolism, and weight loss. It turns out, however, that these obese individuals are leptin-resistant. They may have defective receptors for leptin or in some other way don't respond appropriately to leptin. This is analogous to people with noninsulin-dependent diabetes mellitus (see chapter 18), who have increased levels of insulin but don't respond to it. Future research may determine the mechanism of leptin resistance and the role of leptin in obesity.

The message emerging from current research is that body weight results from many complicated genetic and metabolic factors that go awry in many different ways. Obesity is being regarded as a chronic condition that may respond to medication in much the same way that diabetes does. Nonetheless, medication will only be part of the story. Drugs can help, but eating less and exercising more will still be necessary for optimal health.

---

decreases. Heat gain occurs through metabolism and the muscular contractions of shivering, whereas heat loss occurs through evaporation. Heat gain or loss can occur by radiation, conduction, or convection depending on the skin temperature and the environmental temperature. If the skin temperature is lower than the environmental temperature, heat is gained, but if the skin temperature is higher than the environmental temperature, heat is lost.

The difference in temperature between the body and the environment determines the amount of heat exchanged between the environment and the body. The greater the temperature difference, the greater the rate of heat exchange. Control of the temperature difference is used to regulate body temperature. For example, if environmental temperature is very cold, like on a cold winter day, a large temperature difference exists between the body and the environment, and a large loss of heat occurs. The loss of heat can be decreased by behaviorally selecting a warmer environment, for example, by going inside a heated house. Heat loss can also be decreased by insulating the exchange surface, such as by putting on extra clothes. Physiologically, temperature difference can be controlled through dilation and constriction of blood vessels in the skin. When these blood vessels dilate, they bring warm blood to the surface of the body, raising skin temperature; conversely, vasoconstriction decreases blood flow and lowers skin temperature.

P R E D I C T

Explain why vasoconstriction of the skin's blood vessels on a cool day is beneficial.

When environmental temperature is greater than body temperature, vasodilation brings warm blood to the skin, causing an increase in skin temperature that decreases heat gain from the environment. At the same time, evaporation carries away excess heat to prevent heat gain and overheating.

Body temperature regulation is an example of a negative-feedback system that is controlled by a "set point." A small area in the anterior part of the hypothalamus detects slight increases in body temperature through changes in blood temperature (figure 25.18). As a result, mechanisms are activated that cause heat loss, such as vasodilation and sweating, and body temperature decreases. A small area in the posterior hypothalamus can detect slight decreases in body temperature and can initiate heat gain by increasing muscular activity (shivering) and vasoconstriction.

Under some conditions, the set point of the hypothalamus is actually changed. For example, during a fever, the set point is raised, heat-conserving and heat-producing mechanisms are stimulated, and body temperature increases. In recovery from a fever, the set point is lowered to normal, heat loss mechanisms are initiated, and body temperature decreases.

44. *Define the terms homeotherm and free energy. How much of the free energy is lost as heat from the body?*
45. *What are four ways that heat is exchanged between the body and the environment?*
46. *How is body temperature behaviorally and physiologically maintained in a cold and in a hot environment?*
47. *How does the hypothalamus regulate body temperature?*

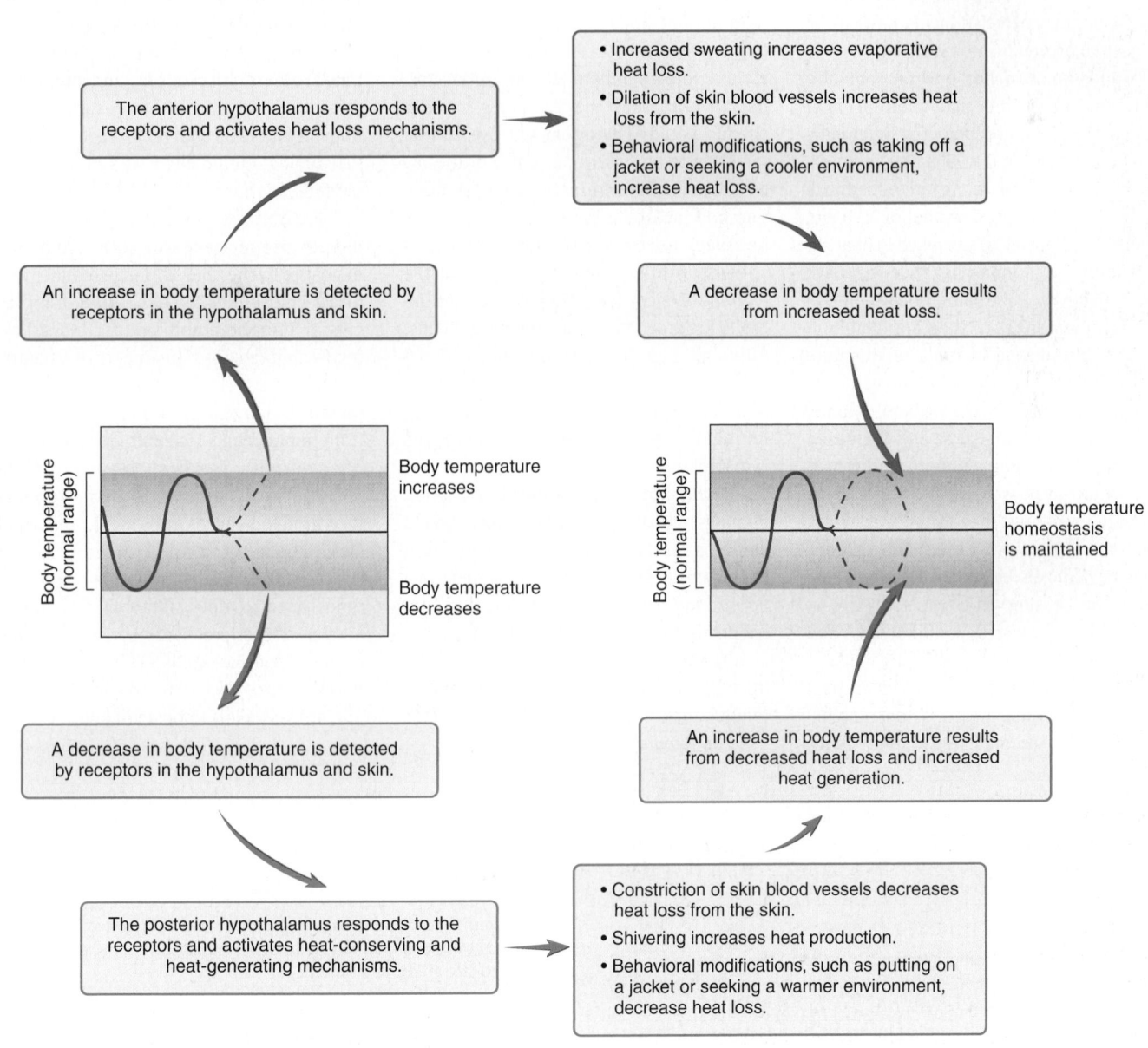

Homeostasis Figure 25.18 **Temperature Regulation**

Clinical Focus   Hyperthermia and Hypothermia

Hyperthermia

If heat gain exceeds the ability of the body to lose heat, body temperature increases above normal levels, a condition called **hyperthermia.** Hyperthermia can result from exercise, exposure to hot environments, fever, and anesthesia.

Exercise increases body temperature because of the heat produced as a by-product of muscle activity (see chapter 9). Normally, vasodilation and increased sweating prevent body temperature increases that are harmful. In a hot, humid environment, the evaporation of sweat is decreased, and exercise levels have to be reduced to prevent overheating.

Exposure to a hot environment normally results in the activation of heat loss mechanisms, and body temperature is maintained at normal levels. This is an excellent example of a negative-feedback mechanism. Prolonged exposure to a hot environment, however, can result in **heat exhaustion.** The normal negative-feedback mechanisms for controlling body temperature are operating, but they are unable to prevent an increase in body temperature above normal levels. Heavy sweating results in dehydration, decreased blood volume, decreased blood pressure, and increased heart rate. Individuals suffering from heat exhaustion have a wet, cool skin because of the heavy sweating. They usually feel weak, dizzy, and nauseated. Treatment includes reducing heat gain by moving to a cooler environment, ceasing activity to reduce heat produced by muscle metabolism, and restoring blood volume by drinking fluids.

Heat stroke is more severe than heat exhaustion because it results from a breakdown in the normal negative-feedback mechanisms of temperature regulation. If the temperature of the hypothalamus becomes too high, it no longer functions appropriately. Sweating stops, and the skin becomes dry and flushed. The person becomes confused, irritable, or even comatose. In addition to the treatment for heat exhaustion, heat loss from the skin should be increased. This can be accomplished by increasing evaporation from the skin by applying wet cloths or by increasing conductive heat loss by immersing the person in a cool bath.

Fever is the development of a higher-than-normal body temperature following the invasion of the body by microorganisms or foreign substances. Lymphocytes, neutrophils, and macrophages release chemicals called **pyrogens** (pī′rō-jenz) that raise the temperature set point of the hypothalamus. Consequently, body temperature and metabolic rate increase. Fever is believed to be beneficial because it speeds up the chemical reactions of the immune system (see chapter 22) and inhibits the growth of some microorganisms. Although beneficial, body temperatures greater than 41°C (106°F) can be harmful. Aspirin lowers body temperature by inhibiting the synthesis of pyrogens (prostaglandins).

Malignant hyperthermia is an inherited muscle disorder. Certain drugs used to induce general anesthesia for surgery cause sustained, uncoordinated muscle contractions in individuals with this disorder. Consequently, body temperature increases.

Therapeutic hyperthermia is an induced local or general body increase in temperature. It's a treatment sometimes used on tumors and infections.

Hypothermia

If heat loss exceeds the ability of the body to produce heat, body temperature decreases below normal levels. **Hypothermia** is a decrease in body temperature to 35°C (95°F) or below. Hypothermia usually results from prolonged exposure to cold environments. At first, normal negative-feedback mechanisms maintain body temperature. Heat loss is decreased by constricting blood vessels in the skin, and heat production is increased by shivering. If body temperature decreases despite these mechanisms, hypothermia develops. The individual's thinking becomes sluggish, and movements are uncoordinated. Heart, respiratory, and metabolic rates decline, and death results unless body temperature is restored to normal. Rewarming should occur at a rate of a few degrees per hour.

Frostbite is damage to the skin and deeper tissues resulting from prolonged exposure to the cold. Damage results from direct cold injury to cells, injury from ice crystal formation, and reduced blood flow to affected tissues. The fingers, toes, ears, nose, and cheeks are most commonly affected. Damage from frostbite can range from redness and discomfort to loss of the affected part. The best treatment is immersion in a warm water bath. Rubbing the affected area and local, dry heat should be avoided.

Therapeutic hypothermia is sometimes used to slow metabolic rate during surgical procedures like heart surgery. Because metabolic rate is decreased, tissues don't require as much oxygen as normal and are less likely to be damaged.

S U M M A R Y

Nutrition   (p. 932)

Nutrition is the taking in and use of food.

Nutrients

1. Nutrients are the chemicals used by the body and consist of carbohydrates, lipids, proteins, vitamins, minerals, and water.

2. Essential nutrients are nutrients that must be ingested because the body cannot manufacture them or is unable to manufacture adequate amounts of them.

Kilocalories

1. A calorie (cal) is the heat (energy) necessary to raise the temperature of 1 g of water 1°C. A kilocalorie (kcal) or Calorie (Cal) is 1000 calories.

2. A gram of carbohydrate or protein yields 4 kcal, and a gram of fat yields 9 kcal.

Carbohydrates

1. Carbohydrates are ingested as monosaccharides (glucose, fructose), disaccharides (sucrose, maltose, lactose), and polysaccharides (starch, glycogen, cellulose).
2. Polysaccharides and disaccharides are converted to glucose. Glucose can be used for energy or stored as glycogen or fats.
3. About 125–175 g of carbohydrates should be ingested each day.

Lipids

1. Lipids are ingested as triglycerides (95%) or cholesterol and phospholipids (5%).
2. Triglycerides are used for energy or are stored in adipose tissue. Cholesterol forms other molecules, such as steroid hormones. Cholesterol and phospholipids are part of the plasma membrane.
3. The recommended daily diet should derive no more than 30% of its kilocalories from lipids, and no more than 300 mg should be in the form of cholesterol.

Proteins

1. Proteins are ingested and broken down into amino acids.
2. Proteins perform many functions: protection (antibodies), regulation (enzymes, hormones), structure (collagen), muscle contraction (actin and myosin), and transportation (hemoglobin, carrier molecules, ion channels).
3. An adult should consume 0.8 g of protein per kilogram of body weight each day.

Vitamins

1. Many vitamins function as coenzymes or as parts of coenzymes.
2. Most vitamins are not produced by the body and must be obtained in the diet. Some vitamins can be formed from provitamins.
3. Vitamins are classified as either fat-soluble or water-soluble.
4. Recommended Dietary Allowances (RDAs) are a guide for estimating the nutritional needs of groups of people based on their age, sex, and other factors.

Minerals

Minerals are necessary for normal metabolism, add mechanical strength to bones and teeth, function as buffers, and are involved in osmotic balance.

Daily Values

1. Daily Values are dietary references that can be used to help plan a healthy diet.
2. Daily Values for vitamins and minerals are based on Reference Daily Intakes, which are generally the highest 1968 RDA value of an age category.
3. Daily Values are based on Daily Reference Values.
   • The Daily Reference Values for energy-producing nutrients (carbohydrates, total fat, saturated fat, and proteins) and dietary fiber are recommended percentages of the total kilocalories ingested daily for each nutrient.
   • The Daily Reference Values for total fats, saturated fats, cholesterol, and sodium are the uppermost limit considered desirable because of their link to diseases.
4. The % Daily Value is the percent of the recommended Daily Value of a nutrient found in one serving of a particular food.

Metabolism   (p. 940)

1. Metabolism consists of anabolism and catabolism. Anabolism is the building up of molecules and requires energy. Catabolism is the breaking down of molecules and gives off energy.
2. The energy in carbohydrates, lipids, and proteins is used to produce ATP through oxidation–reduction reactions.

Carbohydrate Metabolism   (p. 942)

Glycolysis

Glycolysis is the breakdown of glucose into two pyruvic acid molecules. Also produced are two NADH molecules and two ATP molecules.

Anaerobic Respiration

1. Anaerobic respiration is the breakdown of glucose in the absence of oxygen into two lactic acid and two ATP molecules.
2. Lactic acid can be converted to glucose (Cori cycle) using aerobically produced ATP (oxygen debt).

Aerobic Respiration

1. Aerobic respiration is the breakdown of glucose in the presence of oxygen to produce carbon dioxide, water, and 38 (or 36) ATP molecules.
2. The first phase is glycolysis, which produces two ATP, two NADH, and two pyruvic acid molecules.
3. The second phase is the conversion of the two pyruvic acid molecules into two molecules of acetyl-CoA. These reactions also produce two NADH and two carbon dioxide molecules.
4. The third phase is the citric acid cycle, which produces two ATP, six NADH, two FADH2, and four carbon dioxide molecules.
5. The fourth phase is the electron-transport chain. The high-energy electrons in NADH and FADH2 enter the electron-transport chain and are used in the synthesis of ATP and water.

Lipid Metabolism   (p. 949)

1. Adipose triglycerides are broken down and released as free fatty acids.
2. Free fatty acids are taken up by cells and broken down by beta-oxidation into acetyl-CoA.
   • Acetyl-CoA can enter the citric acid cycle.
   • Acetyl-CoA can be converted into ketone bodies.

Protein Metabolism   (p. 950)

1. New amino acids are formed by transamination, the transfer of an amine group to a keto acid.
2. Amino acids are used to synthesize proteins. If used for energy, ammonia is produced as a by-product of oxidative deamination. Ammonia is converted to urea and is excreted.

Interconversion of Nutrient Molecules   (p. 952)

1. Glycogenesis is the formation of glycogen from glucose.
2. Lipogenesis is the formation of lipids from glucose and amino acids.
3. Glycogenolysis is the breakdown of glycogen to glucose.
4. Gluconeogenesis is the formation of glucose from amino acids and glycerol.

Metabolic States   (p. 953)

1. In the absorptive state, nutrients are used as energy or stored.
2. In the postabsorptive state, stored nutrients are used for energy.

Metabolic Rate   (p. 955)

Metabolic rate is the total energy expenditure per unit of time, and it has three components.

Basal Metabolic Rate

Basal metabolic rate is the energy used at rest. It is 60% of the metabolic rate.

Thermic Effect of Food

The thermic effect of food is the energy used to digest and absorb food. It is 10% of the metabolic rate.

Muscular Activity

Muscular energy is used for muscle contraction. It is 30% of the metabolic rate.

Body Temperature Regulation   (p. 956)

1. Body temperature is a balance between heat gain and heat loss.
   - Heat is produced through metabolism.
   - Heat is exchanged through radiation, conduction, convection, and evaporation.

2. The greater the temperature difference between the body and the environment, the greater the rate of heat exchange.
3. Body temperature is regulated by a "set point" in the hypothalamus.

R E V I E W   A N D   C O M P R E H E N S I O N

1. Which of these statements concerning kilocalories is true?
   a. A kilocalorie is the amount of energy required to raise the temperature of 1 g of water 1°C.
   b. There are 9 kcal in a gram of protein.
   c. There are 4 kcal in a gram of fat.
   d. A pound of body fat contains 3500 kcal.

2. Complex carbohydrates include
   a. sucrose.
   b. milk sugar (lactose).
   c. starch, an energy storage molecule in plants.
   d. all of the above.

3. What type of nutrient is recommended as the primary energy source in the diet?
   a. carbohydrates
   b. fats
   c. proteins
   d. cellulose

4. A good source of monounsaturated fats is
   a. fat associated with meat.
   b. egg yolks.
   c. whole milk.
   d. fish oil.
   e. olive oil.

5. A complete protein food
   a. provides the daily amount (grams) of protein recommended in a healthy diet.
   b. can be used to synthesize the nonessential amino acids.
   c. contains all 20 amino acids.
   d. includes beans, peas, and leafy green vegetables.

6. Concerning vitamins,
   a. most can be synthesized by the body.
   b. they are normally broken down before they can be used by the body.
   c. A, D, E, and K are water-soluble vitamins.
   d. many function as coenzymes.

7. Minerals
   a. are inorganic nutrients.
   b. compose about 4%–5% of total body weight.
   c. act as buffers and osmotic regulators.
   d. are components of enzymes.
   e. all of the above.

8. Glycolysis
   a. is the breakdown of glucose to two pyruvic acid molecules.
   b. requires the input of two ATP molecules.
   c. produces two NADH molecules.
   d. does not require oxygen.
   e. all of the above.

9. Anaerobic respiration occurs in the _____ of oxygen and produces _____ energy (ATP) for the cell than aerobic respiration.
   a. absence, more
   b. absence, less
   c. presence, more
   d. presence, less

10. Which of these reactions take place in both anaerobic and aerobic respiration?
    a. glycolysis
    b. citric acid cycle
    c. electron-transport chain
    d. acetyl-CoA formation
    e. all of the above

11. A molecule that moves electrons from the citric acid cycle to the electron-transport chain is
    a. tRNA.
    b. mRNA.
    c. ADP.
    d. NADH.
    e. pyruvic acid.

12. The production of ATP molecules by the electron-transport chain is accompanied by the synthesis of
    a. alcohol.
    b. water.
    c. oxygen.
    d. lactic acid.
    e. glucose.

13. The carbon dioxide you breathe out comes from
    a. glycolysis.
    b. the electron-transport chain.
    c. anaerobic respiration.
    d. the food you eat.

14. Lipids are
    a. stored primarily as triglycerides.
    b. synthesized by beta-oxidation.
    c. broken down by oxidative deamination.
    d. all of the above.

15. Amino acids
    a. are classified as essential or nonessential.
    b. can be synthesized in a transamination reaction.
    c. can be used as a source of energy.
    d. can be converted to keto acids.
    e. all of the above.

16. Ammonia is
    a. a by-product of lipid metabolism.
    b. formed during ketogenesis.
    c. converted into urea in the liver.
    d. produced during lipogenesis.
    e. converted to keto acids.

17. The conversion of amino acids and glycerol into glucose is called
    a. gluconeogenesis.
    b. glycogenolysis.
    c. glycogenesis.
    d. ketogenesis.

18. Which of these events takes place during the absorptive state?
    a. Glycogen is converted into glucose.
    b. Glucose is converted into fats.
    c. Ketones are produced.
    d. Proteins are converted into glucose.

19. The major use of energy by the body is in
    a. basal metabolism.
    b. physical activity.
    c. the thermic effect of food.

20. The loss of heat resulting from the loss of water from the body's surface is
    a. radiation.
    b. evaporation.
    c. conduction.
    d. convection.

Answers in Appendix F

C R I T I C A L   T H I N K I N G

1. One serving of a food has 2 g of saturated fat. What % Daily Value for saturated fat would appear on a food label for this food? (See bottom of figure 25.2 for information needed to answer this question.)

2. An active teenage boy has a daily intake of 3000 kcal/day. What is the maximum amount (weight) of total fats he should consume according to the Daily Values?

3. If the teenager in question 2 eats a serving of food that has a total fat content of 10 g/serving, what is his % Daily Value for total fat?

4. Suppose the food in question 3 is in a package that lists a serving size of 1/2 cup with 4 servings in the package. If the teenager eats half of the contents of the package (1 cup), how much of his % Daily Value does he consume?

5. Why does a vegetarian usually have to be more careful about his or her diet than a person who includes meat in the diet?

6. Explain why a person suffering from copper deficiency feels tired all the time.

7. Some people claim that occasionally fasting for short periods can be beneficial. How can fasts be damaging?

8. Why can some people lose weight on a 1200 kcal/day diet and others cannot?

9. Lotta Bulk, a muscle builder, wanted to increase her muscle mass. Knowing that proteins are the main components of muscle, she began a high-protein diet in which most of her daily kilocalories were supplied by proteins. She also exercised regularly with heavy weights. After 3 months of this diet and exercise program, Lotta increased her muscle mass, but not any more than her friend, who did the same exercises but did not have a high-protein diet. Explain what happened. Was Lotta in positive or negative nitrogen balance?

10. On learning that sweat evaporation results in the loss of calories, an anatomy and physiology student enters a sauna in an attempt to lose weight. He reasons that a liter (about a quart) of water weighs 1000 g, which is equivalent to 580,000 cal or 580 kcal of heat when lost as sweat. Instead of reducing his diet by 580 kcal/day, if he loses a liter of sweat every day in the sauna, he believes he will lose about a pound of fat a week. Will this approach work? Explain.

Answers in Appendix G

A N S W E R S   T O   P R E D I C T   Q U E S T I O N S

1. If vitamins were broken down during the process of digestion, their structures would be destroyed, and, as a result, their ability to function would be lost.

2. The Daily Value for carbohydrate is 300 g/day. One serving of food with 30 g of carbohydrate has a % Daily Value of 10% (30/300 = .10, or 10%).

3. On a 1800 kcal/day diet, the total percentage of Daily Values for energy-producing nutrients should add up to no more than 90%, because 1800/2000 = 0.9, or 90%.

4. If the electron of the electron-transport chain cannot be donated to oxygen, the entire electron-transport chain stops, no ATP can be produced aerobically, and the patient dies because too little energy is available for the body to perform vital functions. Anaerobic respiration is not adequate to provide all the energy needed to maintain human life, except for a short time.

5. When muscles contract, they use ATP. As a result of the chemical reactions necessary to synthesize ATP, heat is also produced. During exercise the large amounts of heat can raise body temperature, and we feel warm. Shivering consists of small, rapid muscle contractions that produce heat in an effort to prevent a decrease in body temperature in the cold.

6. Vasoconstriction reduces blood flow to the skin, which reduces skin temperature because less warm blood from the deeper parts of the body reaches the skin. As the difference in temperature between the skin and the environment decreases, less loss of heat occurs.

Visit the Online Learning Center at www.mhhe.com/seeley7 for chapter quizzes, interactive learning exercises, and other study tools.

Urinary System

CHAPTER 26

The kidneys make up the body's main purification system. They control the composition of blood by removing waste products, many of which are toxic, and conserving useful substances. The kidneys help control blood volume, and consequently play a role in regulating blood pressure. The kidneys also play an essential role in regulation of blood pH. Approximately one-third of one kidney is all that's needed to maintain homeostasis. Even after extensive damage, the kidneys can still perform their life-sustaining function. If the kidneys are damaged further, however, death results unless specialized medical treatment is administered.

The **urinary system** consists of two kidneys; a single, midline urinary bladder; two ureters, which carry urine from the kidneys to the urinary bladder; and a single urethra, which carries urine from the bladder to the outside of the body (figure 26.1).

This chapter explains the *functions of the urinary system* (p. 966), *kidney anatomy and histology* (p. 966), *anatomy and histology of the ureters and urinary bladder* (p. 973), *urine production* (p. 974), *regulation of urine concentration and volume* (p. 990), *clearance and tubular maximum* (p. 993), *and urine movement* (p. 994). We conclude the chapter with a look at the *effects of aging on the kidneys* (p. 996).

Functions of the Urinary System

The kidneys are the major excretory organs of the body. The skin, liver, lungs, and intestines eliminate some waste products, but if the kidneys fail to function, these other excretory organs cannot adequately compensate. The following specific functions are performed by the kidneys:

1. *Excretion.* The kidneys filter blood, and a large volume of filtrate is produced. Large molecules, such as proteins and blood cells, are retained in the blood, while smaller molecules and ions enter the filtrate. Most of the filtrate volume is reabsorbed back into the blood along with useful molecules and ions. Metabolic wastes, toxic molecules, and excess ions remain in the small volume of filtrate. Additional waste products are secreted into the filtrate and the result is urine formation.

2. *Regulation of blood volume and pressure.* The kidneys play a major role in controlling the extracellular fluid volume in the body by producing either a large volume of dilute urine or a small volume of concentrated urine. Consequently, blood volume and blood pressure are regulated by the kidneys.

3. *Regulation of the concentration of solutes in the blood.* The kidneys help regulate the concentration of the major ions such as Na^+, Cl^-, K^+, Ca^{2+}, HCO_3^-, and HPO_4^{-2}.

4. *Regulation of the pH of the extracellular fluid.* The kidneys secrete variable amounts of H^+ to help regulate the extracellular fluid pH.

5. *Regulation of red blood cell synthesis.* The kidneys secrete a hormone, erythropoietin, which regulates the synthesis of red blood cells in bone marrow (see chapter 19).

6. *Vitamin D synthesis.* The kidneys play an important role in controlling blood levels of Ca^{2+} by regulating the synthesis of vitamin D (see chapter 6).

> 1. **List the functions performed by the kidneys, and briefly describe each.**

Kidney Anatomy and Histology

Location and External Anatomy of the Kidneys

The **kidneys** are bean-shaped and about the size of a tightly clenched fist. They lie behind the peritoneum on the posterior abdominal wall on either side of the vertebral column near the lateral borders of the psoas major muscles (figure 26.2). The kidneys extend from the level of the last thoracic (T_{12}) to the third lumbar (L3) vertebrae and the rib cage partially protects them. The liver is superior to the right kidney, causing the right kidney to be slightly lower than the left. Each kidney measures about 11 cm long, 5 cm wide, and 3 cm thick and weighs about 130 g. The **renal capsule,** a layer of fibrous connective tissue, surrounds each kidney. **Perirenal fat,** a dense layer of adipose tissue, in turn, engulfs the renal capsule. This perirenal fat acts as a shock absorber cushioning the kidneys against mechanical shock. A thin layer of connective tissue, the **renal fascia,** surrounds the perirenal fat and helps to anchor the kidneys and surrounding adipose tissue to the abdominal wall. The fat around the renal fascia is between the renal fascia and the parietal peritoneum.

The **hilum** (hī′lŭm) is a small area where the renal artery and nerves enter, and the renal vein and ureter exit, the kidney. It is located on the concave, medial side of the kidney. The hilum opens into the **renal sinus,** a cavity filled with fat and connective tissue. Structures that enter and leave the kidney pass through the renal sinus (figure 26.3).

> 2. **Describe the location, size, and shape of the kidneys.**
> 3. **Describe the renal capsule and the structures that surround the kidney.**
> 4. **List the structures found at the hilum and in the renal sinus of a kidney.**

Internal Anatomy and Histology of the Kidneys

A frontal section of the kidney reveals that the kidney is divided into an outer **cortex** and an inner **medulla,** which surrounds the renal sinus (see figure 26.3). **Renal pyramids** are cone-shaped structures that make up the medulla. **Medullary rays** extend from the renal pyramids into the cortex. **Renal columns** consist of the same tissue as the cortex that projects between the renal pyramids. The bases of the pyramids form the boundary between the cortex and the medulla. The tips of the pyramids, the **renal papillae,** point toward the renal sinus. **Minor calyces** (kal′i-sēz; cup of flower) are funnel-shaped chambers into which the renal papillae extend.

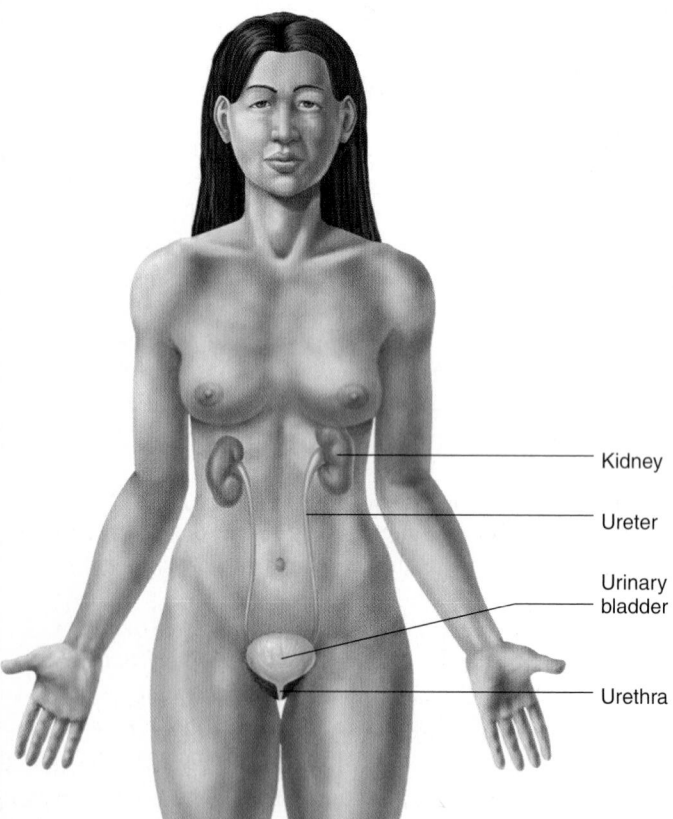

Figure 26.1   **The Urinary System**

The urinary system consists of two kidneys, two ureters, a urinary bladder, and a urethra.

Kidney

Ureter

Urinary bladder

Urethra

Liver

Spleen

Adrenal glands

Tenth rib

Inferior vena cava

Abdominal aorta

Common iliac vein

Common iliac artery

Renal artery

Renal vein

Left kidney

Right kidney

Ureters

Urinary bladder

Urethra

(a) The kidneys are located in the abdominal cavity, with the right kidney just below the liver and the left kidney below the spleen. A ureter extends from each kidney to the urinary bladder within the pelvic cavity. An adrenal gland is located at the superior pole of each kidney.

Liver

Peritoneal cavity

Body wall

Parietal peritoneum

Inferior vena cava

Abdominal aorta

Psoas major muscle

Vertebra

Back muscle

Renal vein

Renal artery

Renal fascia

Perirenal fat

Renal capsule

Kidney

Cross section

(b) The kidneys are located behind the parietal peritoneum. Surrounding each kidney is the perirenal fat. A connective tissue layer, the renal fascia, anchors the kidney to the abdominal wall. Additional fat is found between the renal fascia and the parietal peritoneum. The renal arteries extend from the abdominal aorta to each kidney, and the renal veins extend from the kidneys to the inferior vena cava.

Figure 26.2 Anatomy of the Urinary System

The minor calyces of several pyramids merge to form larger funnels, the **major calyces.** Each kidney contains 8–20 minor calyces and 2 or 3 major calyces. The major calyces converge to form an enlarged chamber called the **renal pelvis,** which is surrounded by the renal sinus. The renal pelvis narrows into a small-diameter tube, the **ureter,** which exits the kidney at the hilum and connects to the urinary bladder. Urine formed within the kidney flows from the renal papillae into the minor calyces. From the

minor calyces, urine flows into the major calyces, collects in the renal pelvis, and then leaves the kidney through the ureter.

The **nephron** (nef′ron) is the histological and functional unit of the kidney (figure 26.4). Each nephron is a tubelike structure with an enlarged terminal end called Bowman's capsule, a proximal tubule, a loop of Henle (nephronic loop), and a distal tubule. The distal tubule empties into a collecting duct, which carries urine from the cortex of the kidney toward the renal papilla. Near the tip

Renal capsule

Cortex

Medulla

Segmental artery

Hilum (indentation)

Renal artery

Renal vein

Renal pelvis

Ureter

Artery and vein
in the renal sinus

Renal sinus
(space)

Renal pyramid

Major calyx

Minor calyx

Renal papilla

Renal column

Medullary rays

(a)

Renal capsule

Cortex

Medulla

Renal pyramid

Renal papilla

Renal column

Hilum
(indentation)

Renal sinus

Renal artery

Renal vein

Renal pelvis

Major calyx

Minor calyx

Ureter

(b)

Figure 26.3   Longitudinal Section of the Kidney and Ureter

(*a*) The cortex forms the outer part of the kidney, and the medulla forms the inner part. A central cavity called the renal sinus contains the renal pelvis. The renal columns of the kidney project from the cortex into the medulla and separate the pyramids. (*b*) Photograph of a longitudinal section of a human kidney and ureter.

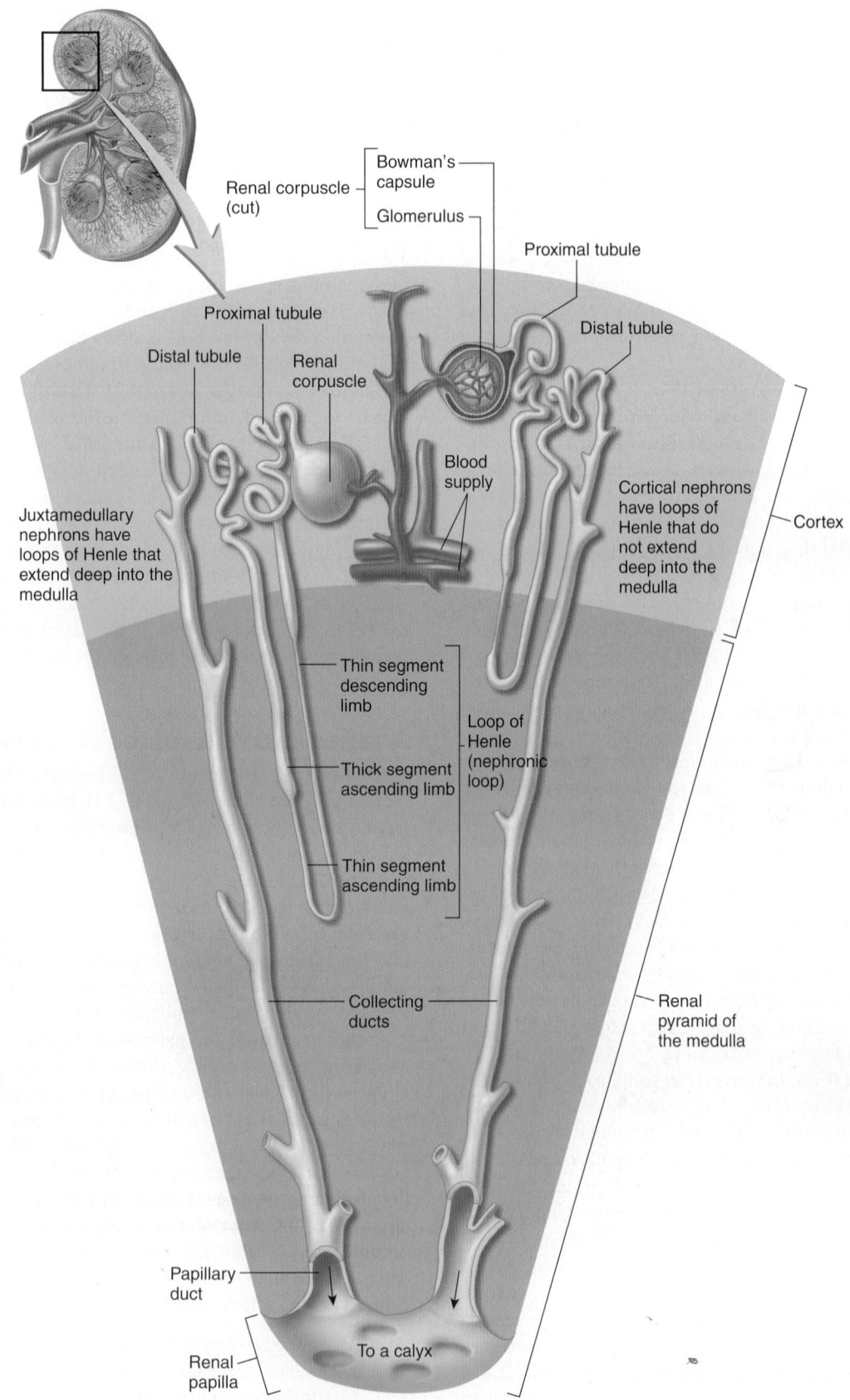

Renal corpuscle (cut) — Bowman's capsule / Glomerulus

Proximal tubule

Distal tubule

Proximal tubule

Distal tubule

Renal corpuscle

Blood supply

Cortex

Juxtamedullary nephrons have loops of Henle that extend deep into the medulla

Cortical nephrons have loops of Henle that do not extend deep into the medulla

Thin segment descending limb

Loop of Henle (nephronic loop)

Thick segment ascending limb

Thin segment ascending limb

Collecting ducts

Renal pyramid of the medulla

Papillary duct

To a calyx

Renal papilla

Figure 26.4   Functional Unit of the Kidney—the Nephron

A nephron consists of a renal corpuscle, proximal tubule, loop of Henle, and distal tubule. The distal tubule empties into a collecting duct. Juxtamedullary nephrons (those near the medulla of the kidney) have loops of Henle that extend deep into the medulla of the kidney, whereas other nephrons do not. Collecting ducts undergo a transition to larger-diameter papillary ducts near the tip of the renal papilla. The papillary ducts empty into a minor calyx.

of the renal papilla, several collecting ducts merge into a larger-diameter tubule called a **papillary duct** which empties into a minor calyx. The renal corpuscles, proximal tubules, and distal tubules are located in the renal cortex, but the collecting tubules, parts of the loops of Henle, and the papillary ducts are in the renal medulla.

There are approximately 1.3 million nephrons in each kidney. Most nephrons measure 50–55 mm in length. Nephrons whose renal corpuscles lie near the medulla are called **juxtamedullary** (juks′ta-med′ŭ-lār-ē; *juxta* is Latin, meaning next to) **nephrons.** They have long loops of Henle, which extend deep into the medulla. Only about 15% of the nephrons are juxtamedullary nephrons. The remainder of the nephrons are called **cortical nephrons,** and their loops of Henle do not extend deep into the medulla (see figure 26.4).

Each **renal corpuscle** consists of the enlarged end of a nephron, called **Bowman's capsule** (glomerular capsule), and a network of capillaries, called the **glomerulus** (glō-mĕr′ū-lŭs) (figure 26.5a and b). The wall of Bowman's capsule is indented to form a double-walled chamber. The glomerulus, which looks like a wad of yarn, fills the indentation. Fluid flows from the glomerulus into Bowman's capsule, and then into the proximal tubule, which carries fluid away from Bowman's capsule.

Bowman's capsule has an outer layer, called the **parietal layer,** and an inner layer, called the **visceral layer** (see figure 26.5b). The parietal layer is constructed of simple squamous epithelium that becomes cube-shaped at the beginning of the proximal tubule. The **visceral layer** is constructed of specialized **podocyte cells,** which wrap around the glomerular capillaries.

Numerous windowlike openings, called **fenestrae** (fe-nes′trē), are in the endothelial cells of the glomerular capillaries. Gaps, called **filtration slits,** are between cell processes of the podocytes that make up the visceral layer of Bowman's capsule. A basement membrane lies sandwiched between the endothelial cells of the glomerular capillaries and the podocytes of Bowman's capsule. Together the capillary endothelium, the basement membrane, and the podocytes of Bowman's capsule form the kidney's **filtration membrane** (figure 26.5c and d), which performs the first major step in urine formation. Urine formation begins when fluid from the glomerular capillaries is filtered across the filtration membrane into the lumen, or space, inside Bowman's capsule.

An **afferent** (af′er-ent) **arteriole** supplies blood to the glomerulus and an **efferent** (ef′er-ent) **arteriole** drains it. A layer of smooth muscle lines both the afferent and efferent arterioles. At the point where the afferent arteriole enters the renal corpuscle, the smooth muscle cells form a cufflike arrangement around the arteriole. These cells are called **juxtaglomerular cells.** Lying between the afferent and efferent arterioles adjacent to the renal corpuscle is part of the distal tubule of the nephron. Specialized tubule cells in this section are collectively called the **macula** (mak′ū-lă) **densa.** The juxtaglomerular cells of the afferent arteriole and the macula densa cells intimately contact each other. Coupled together, they are called the **juxtaglomerular apparatus** (see figure 26.5b). The juxtaglomerular apparatus secretes renin and plays an important role in the autoregulation of filtrate formation (see p. 992).

The **proximal tubule,** also called the **proximal convoluted tubule,** measures approximately 14 mm long and 60 μm in diameter. Simple cuboidal epithelium makes up its wall. The cells rest on a basement membrane, which forms the outer surface of the tubule. Many microvilli project from the luminal surface of the cells (figure 26.6a and b).

The **loops of Henle** (nephronic loops) are continuations of the proximal tubules. Each loop has two limbs, one **descending** and one **ascending.** The first part of the descending limb is similar in structure to the proximal tubules. The loops of Henle that extend into the medulla become very thin near the end of the loop (see figure 26.6a and c). The lumen in the thin part narrows, and an abrupt transition occurs from simple cuboidal epithelium to simple squamous epithelium. Like the descending limb, the first part of the ascending limb is thin and is made of simple squamous epithelium. Soon, however, it becomes thicker, and simple cuboidal epithelium replaces the simple squamous epithelium. The thick part of the loop returns toward the renal corpuscle and ends by giving rise to the distal tubule near the macula densa.

The **distal tubules,** also called the **distal convoluted tubules,** are not as long as the proximal tubules. The epithelium is simple cuboidal, but the cells are smaller than the epithelial cells in the proximal tubules and do not possess a large number of microvilli (figure 26.6a and d). The distal tubules of many nephrons connect to the **collecting ducts,** which are composed of simple cuboidal epithelium. The collecting ducts, which are larger in diameter than other segments of the nephron, form much of the medullary rays and extend through the medulla toward the tips of the renal pyramids.

Arteries and Veins of the Kidneys

A **renal artery** branches off the abdominal aorta and enters the renal sinus of each kidney (figure 26.7a). **Segmental arteries** diverge from the renal artery to form **interlobar arteries,** which ascend within the renal columns toward the renal cortex. Branches from the interlobar arteries diverge near the base of each pyramid and arch over the bases of the pyramids to form the **arcuate** (ar′kū-āt) **arteries. Interlobular arteries** project from the arcuate arteries into the cortex, and afferent arterioles are derived from the interlobular arteries or their branches. The afferent arterioles supply blood to the glomerular capillaries of the renal corpuscles. Efferent arterioles arise from the glomerular capillaries and carry blood away from the glomeruli. After each efferent arteriole exits the glomerulus, it gives rise to a plexus of capillaries called the **peritubular capillaries** around the proximal and distal tubules. Specialized parts of the peritubular capillaries, called **vasa recta** (vā′să rek′tă), course into the medulla along with the loops of Henle (figure 26.7b) and then back toward the cortex. The peritubular capillaries drain into **interlobular veins,** which in turn drain into the **arcuate veins.** The arcuate veins empty into the **interlobar veins,** which drain into the **renal vein.** The renal vein exits the kidney and connects to the inferior vena cava.

5. *What is the functional unit of the kidney? Name its parts.*
6. *Distinguish between cortical and juxtamedullary nephrons.*
7. *List the components of a renal corpuscle.*
8. *Describe the structure of Bowman's capsule, the glomerulus, and the filtration membrane.*
9. *Describe the structure of the afferent and efferent arterioles and the juxtaglomerular apparati.*

Renal corpuscle
- Bowman's capsule
  - Parietal layer
  - Visceral layer (podocyte)
- Glomerular capillary (covered by visceral layer)

Proximal tubule

Capillary

Afferent arteriole

Juxtaglomerular apparatus
- Juxtaglomerular cells
- Macula densa

Distal tubule

Efferent arteriole

(b) Blood flows past the juxtaglomerular apparatus into the glomerulus through the afferent arteriole and leaves the glomerulus through the efferent arteriole. The proximal tubule exits Bowman's capsule.

Renal corpuscle
- Bowman's capsule
- Glomerulus

Proximal tubule

Afferent arteriole

Distal tubule

Efferent arteriole

(a) Bowman's capsule encloses the glomerulus.

Podocyte (visceral layer of Bowman's capsule)
- Cell processes
- Cell body

Filtration slits

Glomerular capillary (cut)

Fenestrae

(c) Podocytes of Bowman's capsule surround the capillaries. Filtration slits between the podocytes allow fluid to pass into Bowman's capsule. The glomerulus is composed of capillary endothelium that is fenestrated. Surrounding the endothelial cells is a basement membrane.

Podocyte

Bowman's capsule

Filtration membrane
- Podocyte cell processes
- Basement membrane
- Capillary endothelium

Capillary

Fenestrae in capillary endothelium

Filtration slits

(d) Capillary endothelial cells, the basement membrane, and podocytes make up the filtration membrane of the kidney.

Figure 26.5 Renal Corpuscle

(a) **Juxtamedullary nephron.**

(b) **Proximal tubule.** The luminal surface of the epithelial cells is lined with numerous microvilli. The basal surface of each cell rests on a basement membrane, and each cell is bound to the adjacent cells by tight junctions. The basal margin of each epithelial cell has deep invaginations, and numerous mitochondria are adjacent to the basal cell membrane. Active reabsorption and secretion are major functions.

(c) **Descending limb of the loop of Henle.** The thin segment of the descending limb is composed of simple squamous epithelial cells that have microvilli and contain a relatively small number of mitochondria. Water easily diffuses from the thin segment into the interstitial fluid.

(d) **Distal tubule.** The cells have sparse microvilli and numerous mitochondria, and they actively reabsorb Na$^+$, K$^+$, and Cl$^-$.

(e) **Collecting duct.** The cells have some microvilli and numerous mitochondria, and they actively reabsorb Na$^+$, K$^+$, and Cl$^-$.

Figure 26.6   Histology of the Nephron

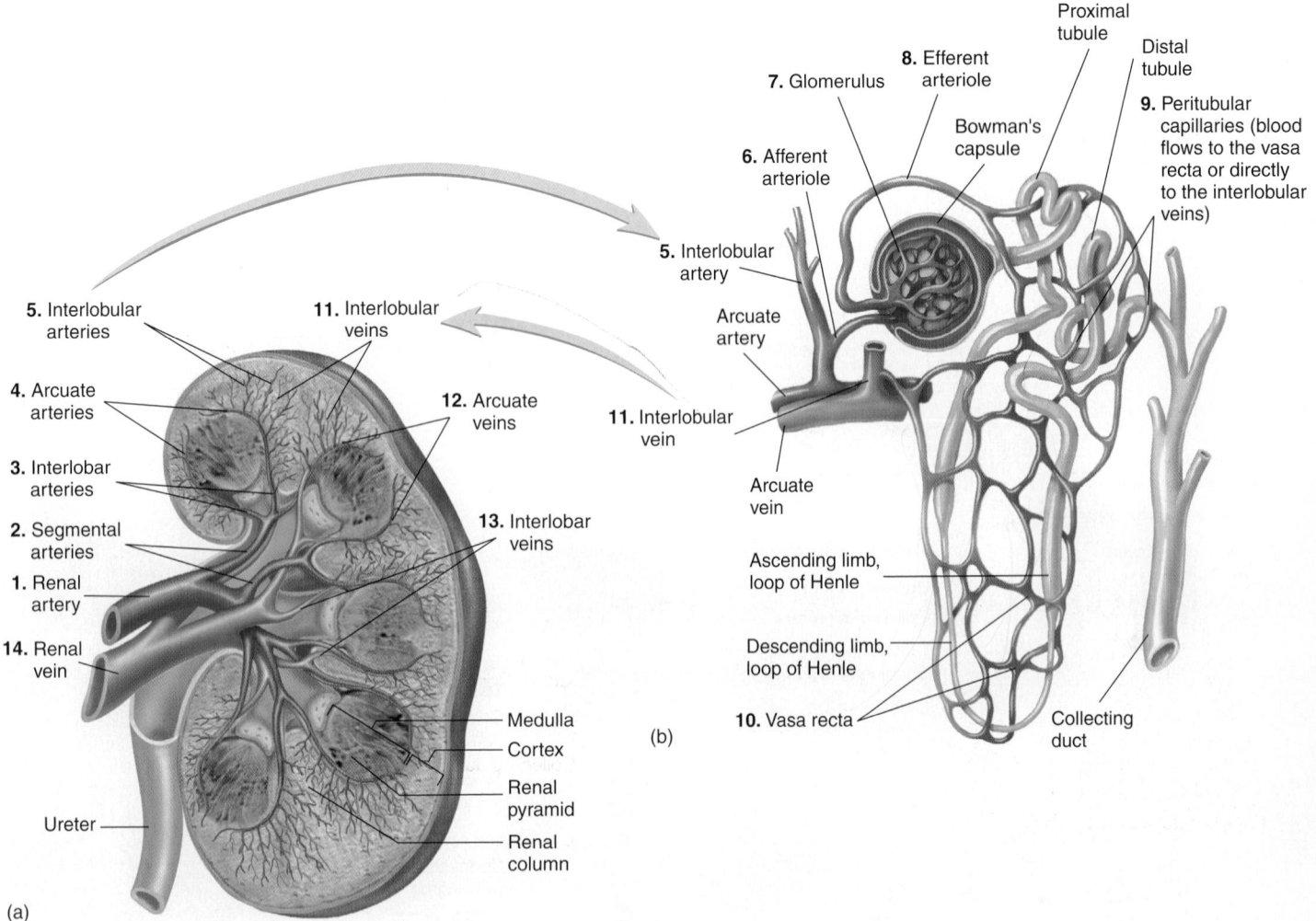

Figure 26.7 Blood Flow Through the Kidney

(*a*) Blood flow through the larger arteries and veins of the kidney. (*b*) Blood flow through the arteries, capillaries, and veins that provide circulation to the nephrons.

10. *Describe the structure and location of the proximal tubules, loops of Henle, distal tubules, collecting ducts, and papillary ducts.*
11. *Describe the blood supply for the kidney.*

Anatomy and Histology of the Ureters and Urinary Bladder

The **ureters** are tubes through which urine flows from the kidneys to the urinary bladder. The ureters extend inferiorly and medially from the renal pelvis at the renal hilum of each kidney to reach the urinary bladder (see figures 26.1, 26.2, and 26.8), which stores urine. The **urinary bladder** is a hollow muscular container that lies in the pelvic cavity just posterior to the symphysis pubis. The ureters enter on its posterolateral surface. In the male, the urinary bladder is just anterior to the rectum, and in the female, it's just anterior to the vagina and inferior and anterior to the uterus. Its volume increases and decreases depending on how much or how little urine is stored in it at any given time.

The urethra, which transports urine to the outside of the body, exits the urinary bladder inferiorly and anteriorly (figure 26.8*a*). The triangular area of its wall between the two ureters posteriorly and the urethra anteriorly is called the **trigone** (trī′gōn). This region differs histologically from the rest of the urinary bladder wall and expands minimally during filling.

Transitional epithelium lines both the ureters and the urinary bladder. The rest of the walls of these structures consists of a lamina propria, a muscular coat, and a fibrous adventitia (figure 26.8*b* and *c*). The wall of the urinary bladder is much thicker than the wall of a ureter. This thickness is caused by layers, composed primarily of smooth muscle, sometimes called the **detrusor** (dē-troo′ser) **muscle,** that are external to the epithelium. Contraction of the urinary bladder smooth muscle forces urine out of the urinary bladder. The epithelium itself ranges from four or five cells thick when the urinary bladder is empty to two or three cells thick when it is distended. Transitional epithelium is specialized so that the cells slide past one another, and the number of cell layers decreases as the volume of the urinary bladder increases. The epithelium of the urethra is stratified or pseudostratified columnar epithelium.

Transitional
epithelium

(a)

Kidney

Ureter

Connective tissue
(lamina propria)

Smooth muscle layer

Connective tissue
(adventitia)

(b)

Parietal peritoneum

Urinary bladder

Opening of ureter

Trigone

Opening of urethra

Location of the
external urinary
sphincter

Transitional epithelium

Connective tissue
(lamina propria)

Smooth muscle layer
(Detrusor muscle)

Connective tissue
(adventitia)

(c)

Figure 26.8   Ureters and Urinary Bladder

(*a*) Ureters extend from the pelvis of the kidney to the urinary bladder. (*b*) The walls of the ureters and the urinary bladder are lined with transitional epithelium, which is surrounded by a connective tissue layer (lamina propria), smooth muscle layers, and a fibrous adventitia. (*c*) Section through the wall of the urinary bladder.

Where the urethra exits the urinary bladder, elastic connective tissue and smooth muscle keep urine from flowing out of the urinary bladder until pressure in the urinary bladder is great enough to force urine to flow from it. The elastic tissue and smooth muscle forms an **internal urinary sphincter** in males (but not in females), which contracts to keep semen from entering the urinary bladder during sexual intercourse (see chapter 28). The **external urinary sphincter** is skeletal muscle that surrounds the urethra as the urethra extends through the pelvic floor. It acts like a valve that controls the flow of urine through the urethra.

In the male, the urethra extends to the end of the penis, where it opens to the outside (see chapter 28). The urethra is much shorter in females than in males, and it opens into the vestibule anterior to the vaginal opening.

> 12. **What are the functions of the ureters, urethra, and urinary bladder? Describe their structure, including the epithelial lining of their inner surfaces. What is the trigone?**

P R E D I C T

Cystitis (sis-tī′tis) is inflammation of the urinary bladder. It typically results from infections that often occur when bacteria from outside the body enter the bladder. Are males or females more prone to cystitis? Explain.

Urine Production

Nephrons are called the **functional units of the kidney** because they are the smallest structural components capable of producing urine. Filtration, reabsorption, and secretion are the three major processes critical to the formation of urine (figure 26.9). **Filtration** is movement of fluid across the filtration membrane as a result of a pressure difference. The fluid entering the nephron is called the **filtrate. Reabsorption** is the movement of substances from the filtrate back into the blood. In general, most of the water and useful solutes are reabsorbed, while waste products, excess solutes, and a small amount of water are not (table 26.1).

Urine formation results from the following three processes:

1. **Filtration.** Filtration (*blue arrow*) results in the movement of fluid from glomerular capillaries across the filtration membrane into Bowman's capsule to form filtrate.

2. **Reabsorption.** Solutes are reabsorbed (*purple arrow*) across the wall of the nephron into the interstitial fluid by transport processes, such as active transport and cotransport. The solutes then enter the peritubular capillaries.
    Water is reabsorbed (*green arrow*) across the wall of the nephron into the interstitial fluid by osmosis. The water then enters the peritubular capillaries.

3. **Secretion.** Solutes are secreted (*orange arrow*) across the wall of the nephron into the filtrate.

Process Figure 26.9   Urine Formation

Table 26.1	Concentrations of Major Solutes				
Substance	**Plasma**	**Filtrate**	**Net Movement of Solute***	**Urine**	**Urine Concentration Plasma Concentration**
Water (L)	180	180	178.6	1.4	—
Organic molecules (mg/100mL)					
Protein	3900–5000	6–11	−100.0	0[†]	0
Glucose	100	100	−100.0	0	0
Urea	26	26	−11.4	1820	70
Uric acid	3	3	−2.7	42	14
Creatinine	1.1	1.1	0.5	196	140
Ions (mEq/L)					
Na^+	142	142	−141.0	128	0.9
K^+	5	5	−4.5	60	12.0
Cl^-	103	103	−101.9	134	1.3
HCO_3^-	28	28	−27.9	14	0.5

*In many cases, solute moves into and out of the nephron. Numbers indicate net movement. Negative numbers are net movement out of the filtrate, and positive numbers are net movement into the filtrate.
[†] Trace amounts of protein can be found in the urine. A value of zero is assumed here.

Secretion is the active transport of solutes into the nephron. Urine produced by the nephrons consists of solutes and water filtered and solutes secreted into the nephron minus the solutes and water that are reabsorbed.

Filtration

Filtration is the movement of fluid, derived from blood flowing through the glomerulus, across the porous filtration membrane. The filtrate includes water and small molecules and ions that easily pass through the filtration membrane. The filtration membrane prevents larger molecules, such as proteins and blood cells, from passing through it. A pressure difference forces the filtrate across the filtration membrane.

    The part of the total cardiac output that passes through the kidneys is called the **renal fraction.** Table 26.2 presents the calculation of renal blood flow rate and other kidney flow rates. The renal fraction varies from 12%–30% of the cardiac output in healthy resting adults, but it averages 21%. This means the average **renal blood flow rate** is 1176 mL/min. The plasma that flows through the kidneys each minute, called the **renal plasma flow rate,** is equal to the renal blood flow rate multiplied by the portion of the blood that is made up of plasma, which is approximately 55% (1176 mL/min × 0.55 = 646.8 mL plasma/min, or approximately 650 mL plasma/min).

    The part of the plasma flowing through the kidney that is filtered through the filtration membranes into the lumen of Bowman's capsules to become filtrate is called the **filtration fraction.** The filtration fraction averages 19% of the plasma flowing through the kidney (650 mL plasma/min × 0.19 = 123.5 mL plasma/min). Thus, approximately 125 mL of filtrate is produced each minute. The amount of filtrate produced each minute is called the **glomerular filtration rate (GFR),** which is equivalent to approximately 180 L of filtrate produced daily. Because only 1–2 L of urine is produced each day by a healthy person, it's obvious that not all of

Table 26.2　Calculation of Renal Flow Rates

Substance	Amount per Minute (mL)	Calculation
Renal blood flow	1176	Amount of blood flowing through the kidneys per minute; equals cardiac output (5600 mL blood/min) times the percent (21%; renal fraction) of cardiac output that enters the kidneys. 5600 mL blood/min × 0.21 = 1176 mL blood/min
Renal plasma flow	650	Amount of plasma flowing through the kidneys per minute; equals renal blood flow times percent of the blood that is plasma. Because the hematocrit is the percent of the blood that consists of formed elements, the percent of the blood that is plasma is 100 minus the hematocrit. Assuming a hematocrit of 45, the percent of the blood that is plasma is 55% (100 − 45). Renal plasma flow is therefore 55% of renal blood flow. 1176 mL blood/min × 0.55 ≈ 650 mL plasma/min
Glomerular filtration rate (GFR)	125	Amount of plasma (filtrate) that enters Bowman's capsule per minute; equals renal plasma flow times the percent (19%; filtration fraction) of the plasma that enters the renal capsule. 650 mL plasma/min × 0.19 ≈ 125 mL filtrate/min
Urine	1	Nonreabsorbed filtrate that leaves the kidneys per minute; equals glomerular filtration rate times the percent (0.8%) of the filtrate that is not reabsorbed into the blood. 125 mL filtrate/min × 0.008 = 1 mL urine/min Milliliters of urine per minute can be converted to liters of urine per day by multiplying by 1.44. 1 mL urine/min × 1.44 = 1.4 L/day

the filtrate becomes urine. Approximately 99% of the filtrate volume is reabsorbed as it passes through the nephron, and less than 1% becomes urine.

P R E D I C T **2**

If the filtration fraction increases from 19% to 22% and if 99.2% of the filtrate is reabsorbed, how much urine is produced in a normal person with a cardiac output of 5600 mL/min? (*Hint:* See table 26.2.)

Filtration Barrier

The filtration membrane is a **filtration barrier,** which prevents the entry of blood cells and proteins into the lumen of Bowman's capsule but allows other blood components to enter. The filtration membrane is many times more permeable than a typical capillary. Water and solutes of a small molecular diameter readily pass from the glomerular capillaries through the filtration membrane into Bowman's capsule, but larger molecules do not. The fenestrae of the glomerular capillary, the basement membrane, and the podocyte cells (see figure 26.5d) prevent molecules larger than 7 nm in diameter or with a molecular mass of 40,000 daltons from passing through. Most plasma proteins are slightly larger than 7 nm in diameter and are retained in the glomerular capillaries. Albumin, which has a diameter just slightly less than 7 nm, enters the filtrate in small amounts so that the filtrate contains about 0.03% protein. Protein hormones are also small enough to pass through the filtration barrier. Proteins that do pass through the filtration membrane are actively reabsorbed by endocytosis and metabolized by the cells in the proximal tubule. Consequently, little protein is found in the urine of healthy people.

P R E D I C T **3**

Hemoglobin has a smaller diameter than albumin, but very little hemoglobin passes from the blood into the filtrate. Explain why. Under what circumstances would large amounts of hemoglobin enter the filtrate?

Hematuria and Glomerulonephritis

Hematuria (hē-mă-too′rē-ă, hem-ă-too′rē-ă) occurs when red blood cells are found in the urine. The source of red blood cells in the urine can be from outside the kidney or from the kidney. Conditions outside the kidney that result in hematuria include kidney stones or tumors in the renal pelvis, ureter, urinary bladder, prostate, or urethra. Infections of the urinary tract, resulting in inflammation of the urinary bladder (cystitis), of the prostate gland (prostatitis, pros-tă-tī′tis), and of the urethra (urethritis, ū-rē-thrī′tis) can also cause hematuria. Conditions inside the kidney include those that affect the filtration membrane or other areas of the kidney. Inflammation of the glomeruli, called **glomerulonephritis** (glō-mār′ū-lō-nef-rī′tis), can increase the permeability of the filtration membrane and allow blood cells to cross. Other areas of the kidney can be a source of blood in response to inflammation of the nephrons due to infections, tumors in the kidney tissue, and infarcted areas of the kidney, where an artery is blocked, resulting in necrosis of part of the kidney tissue.

Filtration Pressure

The formation of filtrate depends on a pressure gradient, called the **filtration pressure,** which forces fluid from the glomerular capillary across the filtration membrane into the lumen of Bowman's capsule. The filtration pressure results from the sum of the forces that move fluid out of the glomerular capillary into the lumen of Bowman's capsule and those that move fluid out of the lumen of Bowman's capsule into the glomerular capillary (figure 26.10). The **glomerular capillary pressure (GCP),** the blood pressure inside the capillary, moves fluid out of the capillary into Bowman's capsule. The glomerular capillary pressure averages approximately 50 mm Hg, which is much higher than that in most capillaries.

Opposing the movement of fluid into the lumen of Bowman's capsule is the **capsule pressure (CP),** which is approximately 10 mm Hg, caused by the pressure of filtrate already inside Bowman's capsule. The **blood colloid osmotic pressure (BCOP)** within the glomerular capillary exists because plasma proteins do not pass through the filtration membrane. Instead, they remain within the glomerular capillary and produce an osmotic force of about 30 mm Hg that favors fluid movement into the glomerular capillary from Bowman's capsule. The BCOP is greater at the end of the glomerular capillary than at its beginning, because as fluid leaves the capillary and enters Bowman's capsule, there is a higher protein concentration in the glomerular capillary. The average BCOP is approximately 30 mm Hg. The filtration pressure is therefore approximately 10 mm Hg.

Filtration capillary pressure (10 mm Hg)	=	Glomerular pressure (50 mm Hg)	−	Capsule pressure (10 mm Hg)	−	Blood colloid osmotic pressure (30 mm Hg)

The colloid osmotic pressure in Bowman's capsule is normally very close to zero because few proteins cross the filtration membrane. However, if the permeability of the filtration membrane increases, as it does in **glomerular nephritis,** the osmotic pressure in Bowman's capsule increases. This results in an elevated filtration pressure and an increase in the filtrate volume.

The high glomerular capillary pressure results from a low resistance to blood flow in the afferent arterioles and glomerular capillaries and from a higher resistance to blood flow in the efferent arterioles. As the diameter of a vessel decreases, resistance to flow through the vessel increases (see chapter 21), and pressure upstream from the point of decreased vessel diameter is higher than the pressure downstream from the point of decreased diameter. For example, in the extreme case of a completely closed vessel, pressure is higher upstream from the constriction and it falls to zero downstream from the constriction. The efferent arterioles have a small diameter, and the blood pressure is higher within the glomerular capillaries because of the low resistance to blood flow in the afferent arterioles and glomerular capillaries and because of the higher resistance to blood flow in the efferent arterioles. Also, the pressure is lower in the peritubular capillaries which are downstream from the efferent arterioles. Consequently, filtrate is forced across the filtration membrane into the lumen of Bowman's capsule. The low pressure in the peritubular capillaries allows fluid to move into them from the interstitial fluid.

The smooth muscle cells in the walls of the afferent and efferent arterioles can alter the vessel diameter and the glomerular

Glomerular capillary pressure: Blood pressure in the glomerular capillary moves fluid from the blood into Bowman's capsule

Capsule pressure: Pressure in Bowman's capsule moves fluid from Bowman's capsule into the glomerular capillary

Blood colloid osmotic pressure: The blood protein concentration moves fluid from Bowman's capsule into the glomerular capillary by osmosis

Glomerular capillary pressure
− Capsule pressure
− Colloid osmotic pressure
─────────────────────
Filtration pressure

Filtration pressure = GCP − BCOP − CP

50 mm Hg GCP (glomerular capillary pressure)
−30 mm Hg BCOP (blood colloid osmotic pressure)
−10 mm Hg CP (capsule pressure)
─────────────────────
10 mm Hg filtration pressure

Figure 26.10 Filtration Pressure

Filtration pressure across the filtration membrane is equal to the glomerular capillary pressure (GCP) minus the blood colloid osmotic pressure (BCOP) in the glomerular capillary minus the pressure in Bowman's capsule (CP).

filtration pressure. For example, dilation of the afferent arterioles or constriction of the efferent arterioles increases glomerular capillary pressure, increasing filtration pressure and glomerular filtration.

13. Name the three general processes involved in the production of urine.

14. Define the terms renal blood flow, renal plasma flow, and glomerular filtration rate. How do they affect urine production?

15. Describe the filtration barrier. What substances do not pass through it?

16. What is filtration pressure? How does glomerular capillary pressure affect filtration pressure and the amount of urine produced?

17. How do systemic blood pressure and afferent arteriole diameter affect glomerular capillary pressure?

P R E D I C T

Karl was pouring gasoline from a small can into his lawnmower when it ignited. He experienced third-degree burns on his chest, neck, arms, and hands. In the hospital emergency room, his blood pressure was found to be in the low normal range, his heart rate was very high, and his pulse was weak. His skin appeared very pale. There was a delay of a couple of hours from the time of the accident until an I.V. was started. For several hours after the accident he produced almost no urine. Explain that response.

Tubular Reabsorption

The filtrate leaves the lumen of Bowman's capsule and flows through the proximal tubule, the loop of Henle, and the distal tubule and then into the collecting ducts. As it passes through these structures, many of the substances in the filtrate undergo **tubular reabsorption.** Tubular reabsorption results from processes, such as diffusion, facilitated diffusion, active transport, cotransport, and osmosis. Inorganic salts, organic molecules, and about 99% of the filtrate volume leave the nephron and enter the interstitial fluid. Because the pressure is low in the peritubular capillaries, these substances enter the peritubular capillaries and flow through the renal veins to enter the general circulation (see figure 26.9).

Solutes reabsorbed from the lumen of the nephron to the interstitial fluid include amino acids, glucose, and fructose, as well as Na^+, K^+, Ca^{2+}, HCO_3^-, and Cl^-. A more complete list is provided in table 26.3 for each part of the nephron.

Water follows the solutes that are reabsorbed across the wall of the nephron. As the solutes in the nephron are reabsorbed, water follows the solutes by the process of osmosis. The transport processes and the permeability characteristics of each portion of the nephron are responsible for reabsorption of the filtrate. The small volume of the filtrate (approximately 1% of the filtrate volume) that forms urine contains a relatively high concentration of urea, uric acid, creatinine, K^+, and other substances that are toxic in high concentrations. Regulation of solute reabsorption and the permeability characteristics of portions of the nephron allow for the production of a small volume of very concentrated urine or a large volume of very dilute urine. The mechanisms in the wall of

Table 26.3	Reabsorption of Major Solutes from the Nephron
Apical Membrane	**Basal Membrane**
Proximal Nephron	
Substances cotransported with Na^+	Active transport Na^+ (exchanged for K^+)
	Facilitated diffusion
K^+	K^+
Cl^-	Cl^-
Ca^{2+}	Ca^{2+}
Mg^{2+}	HCO_3^-
HCO_3^-	PO_4^{3-}
PO_4^{3-}	Amino acids
Amino acids	Glucose
Glucose	Fructose
Fructose	Galactose
Galactose	Lactate
Lactate	Succinate
Succinate	Citrate
Citrate	
Diffusion between nephron cells	
K^+	
Ca^{2+}	
Mg^{2+}	
Thick Ascending Limb of the Loop of Henle	
Substances cotransported with Na^+	Active transport Na^+ (exchanged for K^+)
	Facilitated diffusion
K^+	K^+
Cl^-	Cl^-
Diffusion between nephron cells	
K^+	
Ca^{2+}	
Mg^{2+}	
Distal Nephron and Collecting Duct	
Substances cotransported with Na^+	Active transport Na^+ (exchanged for K^+)
	Facilitated diffusion
Cl^-	K^+
K^+	Cl^-

the nephron that are responsible for reabsorption are described in the following section. Mechanisms that regulate urine concentration are described in a later section of this chapter ("Urine Concentration Mechanism," p. 983).

Reabsorption in the Proximal Tubule

Most reabsorbed substances must pass through the cells that make up the wall of the nephrons. These cells have an **apical surface,** which makes up the inside surface of the nephron, a **basal surface,** which forms the outer wall of the nephron, and **lateral surfaces,** which are bound to the lateral surfaces of other cells of the nephron. Reabsorption of most solute molecules from the proximal tubule is linked to the primary active transport of Na^+ across the **basal membrane** of the nephron epithelial cells from the cytoplasm into the interstitial fluid, thereby creating a low concentration of Na^+ inside the cells (figure 26.11). At the basal

Figure 26.11   Reabsorption of Solutes in the Proximal Nephron

Cotransport of molecules and ions across the epithelial lining of the nephron depends on the active transport of sodium ions, in exchange for potassium ions, across the basal membrane. Cotransport is the process by which carrier proteins move molecules or ions with Na^+ across the apical membrane. The Na^+ concentration gradient provides the energy for cotransport. Amino acids, glucose, K^+, Cl^-, and most other solutes are transported into the cells of the nephron with Na^+. Water enters and leaves the cell by osmosis. Glucose, amino acids, Na^+, Cl^-, and many other solutes leave the cells across the apical membrane by facilitated diffusion.

cell membrane, ATP provides the energy required to move Na^+ out of the cell in exchange for K^+ by countertransport. Because the concentration of Na^+ in the lumen of the tubule is high, a large concentration gradient is present from the lumen of the nephron to the intracellular fluid of the cells lining the nephron. This concentration gradient for Na^+ is responsible for the cotransport of many solute molecules from the lumen of the nephron into the nephron cells.

Carrier molecules that transport amino acids, glucose, and other solutes are located within the apical membrane, which separates the lumen of the nephron from the cytoplasm of epithelial cells. Each of these carrier molecules binds specifically to one of the substances to be transported and to Na^+. The concentration gradient for Na^+ provides the source of energy that moves both the Na^+ and the other molecules or ions bound to the carrier molecule from the lumen into the cell of the nephron. Once the cotransported molecules are inside the cell, they cross the basal membrane of the cell by facilitated diffusion. The number of transport molecules limits the rate at which a substance can be transported. For example, the concentration of glucose in the filtrate exceeds the rate at which glucose can be transported in people suffering from untreated diabetes mellitus. The excess glucose remains in the filtrate and becomes part of the urine (see "Clearance and Tubular Maximum" on p. 993).

Some solutes also diffuse between the cells from the lumen of the nephron into the interstitial fluid. The concentration of these solutes increases as other solutes are cotransported and water moves by osmosis from the lumen into interstitial fluid. As the concentration gradient for these solutes increases above the concentration in the interstitial fluid, they diffuse between the epithelial cells. Some K^+, Ca^{2+}, and Mg^{2+} diffuse between the cells of the proximal tubule wall from the lumen of the tubule to the interstitial fluid. Reabsorption of these solutes by diffusion occurs even though these same ions are also reabsorbed by cotransport processes.

Reabsorption of solutes in the proximal tubule is extensive, and the tubule is permeable to water. As solute molecules are transported from the nephron to the interstitial fluid, water moves by osmosis in the same direction. By the time the filtrate has reached the end of the proximal tubule, its volume has been reduced by approximately 65%. The concentration of the filtrate in the proximal tubule remains about the same as that of the interstitial fluid (300 mOsm/kg) because the wall of the nephron is permeable to water.

Reabsorption in the Loop of Henle

The loop of Henle descends into the medulla of the kidney, where the concentration of solutes in the interstitial fluid is very high. The thin segment of the loop of Henle (figure 26.12) is highly permeable to water and moderately permeable to urea, sodium, and most other ions. As the filtrate passes through the thin segment of the loop of Henle, water moves out of the nephron by osmosis, and some solutes move into the nephron. By the time the filtrate has reached the end of the thin segment of the loop of Henle, the volume of the filtrate has been reduced by another 15%, and the concentration of the filtrate is equal to the high concentration of the interstitial fluid (1200 mOsm/L).

The thin portion of the ascending limb of the loop of Henle is permeable to solutes but is impermeable to water. Therefore, no additional water diffuses from the thin portion of the ascending

limb. The ascending limb of the loop of Henle is surrounded by interstitial fluid that becomes less concentrated toward the cortex. As the filtrate flows through the thin segment of the limb, solutes diffuse into the interstitial fluid, making the filtrate less concentrated.

The thick segment of the ascending limb is not permeable to either water or solutes. Solutes such as Na^+, K^+, and Cl^-, however, are transported from the thick segment of the ascending limb of the loop of Henle into the interstitial fluid. Cotransport is responsible for the movement of K^+ and Cl^- with Na^+ across the apical membrane of the ascending limb of the loop of Henle (figure 26.13). Once inside the cells of the ascending limb, Cl^- and K^+ cross the basal cell membrane into the interstitial fluid from a higher concentration inside the cells to a lower concentration outside the cells by facilitated diffusion.

The concentration of Na^+ in the lumen of the nephron is high and the concentration inside of the nephron cells is low. This concentration gradient is created by the active transport of the Na^+ out of the cell in exchange for K^+ across the basal membrane (see figure 26.13).

Because the thin and thick segments of the ascending limb of the loop of Henle are impermeable to water, and because ions are transported out of the thick portion of the ascending limb, the concentration of solutes in the nephron is reduced to about 100 mOsm/kg by the time the fluid reaches the distal tubule. In contrast, the concentration of the interstitial fluid in the cortex is about 300 mOsm/kg. Thus the filtrate entering the distal tubule is much more dilute than the interstitial fluid surrounding it.

Osmoles, Osmolality, and Osmosis

An **osmole** is a measure of the number of particles in a solution. One osmole is the molecular mass, in grams, of a solute times the number of ions or particles into which it dissociates in solution. A milliosmole (mOsm) is 1/1000 of an osmole. The osmolality of a solution is the number of osmoles in a kilogram of solution. Water moves by osmosis from a solution with a lower osmolality to a solution with a higher osmolality. Thus water moves by osmosis from a solution of 100 mOsm/kg toward a solution of 300 mOsm/kg (see appendix C).

Reabsorption in the Distal Tubule and Collecting Duct

Reabsorption of some solutes occurs in the distal tubule and collecting duct. Chloride ions are cotransported across the apical membrane of the distal tubules and collecting ducts with Na^+. The concentration gradient for Na^+ depends on the active transport of Na^+ across the basal cell membrane, which causes the intracellular Na^+ concentration to be low in comparision to the lumen of the nephron. In addition, the collecting ducts extend from the cortex of the kidney, where the concentration of the interstitial fluid is approximately 300 mOsm/kg, through the medulla of the kidney, where the concentration of the interstitial fluid is very high. Water moves by osmosis from the distal tubule and collecting duct into the more concentrated interstitial fluid. The movement of water is maximal when the distal tubule and collecting duct are permeable to water. Consequently, a small volume of very concentrated urine is produced. Water does not move by osmosis into the interstitial fluid when the distal tubule and collecting duct are not permeable to water. Consequently, a large volume of dilute urine is produced. The production of dilute or

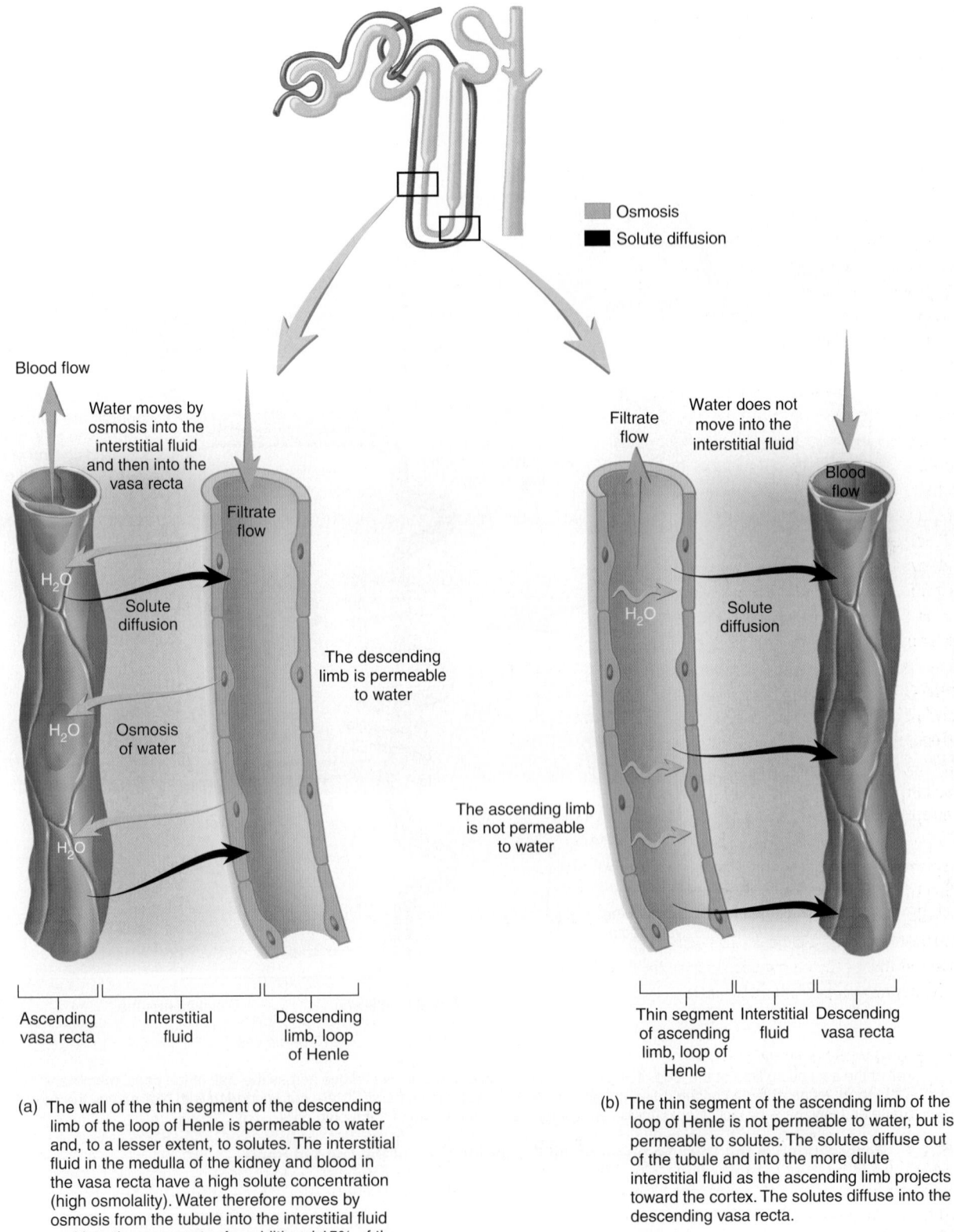

Osmosis
Solute diffusion

Blood flow

Water moves by osmosis into the interstitial fluid and then into the vasa recta

Filtrate flow

H_2O

Solute diffusion

The descending limb is permeable to water

H_2O

Osmosis of water

H_2O

Ascending vasa recta | Interstitial fluid | Descending limb, loop of Henle

Filtrate flow

Water does not move into the interstitial fluid

Blood flow

H_2O

Solute diffusion

The ascending limb is not permeable to water

Thin segment of ascending limb, loop of Henle | Interstitial fluid | Descending vasa recta

(a) The wall of the thin segment of the descending limb of the loop of Henle is permeable to water and, to a lesser extent, to solutes. The interstitial fluid in the medulla of the kidney and blood in the vasa recta have a high solute concentration (high osmolality). Water therefore moves by osmosis from the tubule into the interstitial fluid and into the vasa recta. An additional 15% of the filtrate volume is reabsorbed.

To a lesser extent, solutes diffuse from the vasa recta and interstitial fluid into the tubule.

(b) The thin segment of the ascending limb of the loop of Henle is not permeable to water, but is permeable to solutes. The solutes diffuse out of the tubule and into the more dilute interstitial fluid as the ascending limb projects toward the cortex. The solutes diffuse into the descending vasa recta.

Figure 26.12   Reabsorption in the Loop of Henle: The Descending Limb and the Thin Segment of the Ascending Limb

Filtrate flow

Blood flow

Filtrate flow

Solutes are transported into the tubule cells, but water remains in the ascending limb of the loop of Henle

Solutes are transported out of the cells of the ascending limb of the loop of Henle and enter the vasa recta

Facilitated diffusion
Active transport
Cotransport

H_2O

H_2O

Ascending limb of the loop of Henle is not permeable to water

Transport of solutes

H_2O K^+ Na^+ Cl^-

K^+ Na^+ Cl^-

K^+ Na^+ ATP ADP Cl^-

K^+ Cl^-

Ascending limb, loop of Henle

Interstitial fluid

Descending vasa recta

Apical membrane   Basal membrane

Nephron cell

The wall of the ascending limb of the loop of Henle is not permeable to water. Na^+ move across the wall of the basal membrane by active transport, establishing a concentration gradient for Na^+. Potassium ions and Cl^- are cotransported with Na^+ across the apical membrane and ions pass by facilitated diffusion across the basal membrane of the tubule cells.

Figure 26.13   Reabsorption in the Thick Segment of the Ascending Limb of the Loop of Henle

concentrated urine is under hormonal control, which is described in the upcoming section "Urine Concentration Mechanism," p. 983.

Changes in the Concentration of Solutes in the Nephron

Urea enters the glomerular filtrate and is in the same concentration there as in the plasma. As the volume of filtrate decreases in the nephron, the concentration of urea increases because renal tubules are not as permeable to urea as they are to water. Only 40%–60% of the urea is passively reabsorbed in the nephron, although about 99% of the water is reabsorbed. In addition to urea, urate ions, creatinine, sulfates, phosphates, and nitrates are reabsorbed but not to the same extent as water. They therefore become more concentrated in the filtrate as the volume of the filtrate becomes smaller. These substances are toxic if they accumulate in the body, so their accumulation in the filtrate and elimination in urine help maintain homeostasis.

Conjugation and Excretion

Some drugs, environmental pollutants, and other foreign substances that gain access to the circulatory system are reabsorbed from the nephron. These substances are usually lipid-soluble, nonpolar compounds. They enter the glomerular filtrate and are reabsorbed passively by a process similar to that by which urea is reabsorbed. Because these substances are passively reabsorbed within the nephron, they are not rapidly excreted. The liver cells attach other molecules to them by a process called **conjugation** (kon-jŭ-gā'shŭn), which converts them to more water-soluble molecules. These more water-soluble substances enter the filtrate but do not pass as readily through the wall of the nephron, are not reabsorbed from the renal tubules, and consequently are more rapidly excreted in the urine. One of the important functions of the liver is to convert nonpolar toxic substances to more water-soluble forms, thus increasing the rate at which they are excreted in the urine.

- 18. *Describe what happens to most of the filtrate that enters the nephron?*
- 19. *On what side of the nephron tubule cell does active transport take place during reabsorption of materials?*
- 20. *Describe how cotransport works in the nephron.*
- 21. *Name the substances that are moved by active and passive transport. In what part of the nephron does this movement take place?*

Tubular Secretion

Tubular secretion involves the movement of some substances, such as by-products of metabolism that become toxic in high concentrations and drugs or molecules not normally produced by the body, into the nephron (table 26.4). As with tubular reabsorption, tubular secretion can be either active or passive. Ammonia is produced in the epithelial cells of the nephron as amino groups are removed from amino acids and diffuse into the lumen of the nephron. Hydrogen ions, K^+, penicillin, and ***para*-aminohippuric acid** (par'ă-a-mē'nō-hi-pūr'ik; *p*-aminohippuric acid; **PAH**), among others, are actively secreted by either active transport or countertransport processes into the nephron.

For example, a countertransport process moves H^+ from cells of the nephron into the nephron's lumen. Hydrogen ions bind to carrier molecules on the inside of the plasma membrane, and Na^+ bind to the carrier molecules on the outside of the plasma membrane. As Na^+ move into the cell, hidrogen ions move out of the cell (figure 26.14). The secreted H^+ are produced as a result of carbon dioxide and water reacting to form H^+ and HCO_3^-. The countertransport molecules secrete H^+ into the nephron's lumen, and Na^+ enter the nephron cell. Sodium ions and HCO_3^- are cotransported across the basal membrane of the cell and enter the peritubular capillaries. Hydrogen ions are secreted into the proximal and distal tubules, and K^+ are actively secreted in the distal tubule. Secretion of H^+ by the nephron plays a major role in regulating body fluid pH (see chapter 27). Penicillin and *p*-aminohippuric acid are examples of substances not normally produced by the body that are actively secreted into the proximal tubules.

- 22. *Where does tubular secretion take place? What substances are secreted? Are these substances secreted by active or passive transport?*

Table 26.4	Secretion of Substances into the Nephron
Transport Process	**Substance Transported**
Proximal Convoluted Tubule	
Active transport	H^+
	Hydroxybenzoates
	para-aminohippuric acid
	Neurotransmitters 　Dopamine 　Acetylcholine 　Epinephrine
	Bile pigments
	Uric acid
	Drugs and toxins 　Penicillin 　Atropine 　Morphine 　Saccharin
Passive transport	Ammonia
Distal Convoluted Tubule	
Active transport	K^+
Passive transport	K^+ H^+

Urine Concentration Mechanism

When a large volume of water is consumed, it's necessary to eliminate the excess without losing large amounts of electrolytes or other substances essential for the maintenance of homeostasis. The response of the kidneys is to produce a large volume of dilute urine. On the other hand, when drinking water is not available, producing a large volume of dilute urine would lead to rapid dehydration. When water intake is restricted, the kidneys produce a small volume of concentrated urine that conserves water and contains sufficient waste products to prevent their accumulation in the circulatory system. The kidneys are able to produce urine with concentrations that range from a minimum of 65 to a maximum of 1200 mOsm/kg while maintaining the extracellular fluid concentration very close to 300 mOsm/kg. The maintenance of a high concentration of solutes in the medulla, the countercurrent functions of the loops of Henle, and the mechanism that controls the permeability of the distal nephron to water are essential for the kidneys to control the volume and concentration of the urine they produce.

Medullary Concentration Gradient

The ability of the kidney to concentrate urine depends on maintaining a high concentration of solutes in the medulla. The interstitial fluid concentration is about 300 mOsm/kg in the cortical region of the kidney. Moving from the cortex toward the medulla, the interstitial fluid becomes progressively more concentrated until it achieves a maximum concentration of 1200 mOsm/kg at the tip

(a) Hydrogen ions are secreted into the filtrate by a countertransport mechanism in the proximal tubule, in which H^+ are exchanged for Na^+. The H^+ are derived from two sources. They diffuse from the peritubular capillaries into the interstitial fluid and then into epithelial cells of the tubule, or they are derived from the reaction between carbon dioxide and water in the cells of the tubule. Sodium ions and HCO_3^- are cotransported across the basal membrane into the interstitial fluid and then diffuse into the peritubular capillaries.

(b) Hydrogen ions and K^+ are secreted into the filtrate by countertransport mechanisms in the distal tubule. Sodium ions and K^+ are moved by active transport across the basal membrane of the tubule cell. Sodium ions and HCO_3^- are cotransported across the basal membrane into the interstitial fluid and then diffuse into the peritubular capillaries.

Figure 26.14   Secretion of H^+ and K^+ into the Nephron

of the renal pyramid (figure 26.15). Maintenance of the high solute concentration in the kidney medulla depends on the functions of the loops of Henle, the vasa recta, and on the distribution of urea. The major mechanisms that create and maintain the high solute concentration in the renal medulla are:

1. The active transport of Na^+ and the cotransport of ions such as K^+ and Cl^- and other ions out of the thick portion

of the ascending limb of the loop of Henle into the interstitial fluid of the medulla.

2. The impermeability of the thin and thick parts of the ascending limb of the loop of Henle to water.

3. The vasa recta supply of blood to the medulla of the kidney and removal of excess water and solutes that enter the medulla without destroying the high concentration of solutes in the interstitial fluid of the medulla.

(a) The movement of water and solutes across the wall of the loop of Henle.

(b) The movement of water and solutes across the wall of the vasa recta.

(c) The loop of Henle and the vasa recta function together. Water moves out of the descending limb of the loop of Henle and enters the vasa recta. Solutes diffuse out of the ascending thin segment of the loop of Henle and enter the vasa recta, but water does not. Solutes transported out of the thick segment of the ascending limb of the loop of Henle enter the vasa recta. Excess water and solutes are carried away from the medulla without reducing the high concentration of solutes. The concentration of the filtrate is reduced to 100 mOsm/kg by the time it reaches the distal tubule.

(d) Water and solutes move out of the collecting duct into the vasa recta.

Figure 26.15   Filtrate Concentration and the Medullary Concentration Gradient

The loop of Henle and the vasa recta function together to maintain a high concentration of solutes in the medulla of the kidney. See the text for a description of water and solute movements for each part of this figure.

Clinical Focus   Kidney Dialysis

The artificial kidney (renal dialysis machine) is a machine used to treat patients suffering from renal failure. The use of this machine often allows people with severe acute renal failure to recover without developing the side effects of renal failure, and the machine can substitute for the kidneys for long periods in people suffering from chronic renal failure.

Renal dialysis allows blood to flow though tubes made of a selectively permeable membrane. On the outside of the dialysis tubes is a fluid that contains the same concentration of solutes as the plasma, except for the metabolic waste products. As a consequence, a diffusion gradient exists for the metabolic waste products from the blood to the dialysis fluid. The dialysis membrane has pores that are too small to allow plasma proteins to pass through them. For smaller solutes, the dialysis fluid contains the same beneficial solutes as the plasma, so the net movement of these substances is zero. In contrast, the dialysis fluid contains no metabolic waste products, so metabolic waste products diffuse rapidly from the blood into the dialysis fluid.

Blood usually is taken from an artery, passed through tubes of the dialysis machine, and then returned to a vein. The rate of blood flow is normally several hundred milliliters per minute, and the total surface area of exchange in the machine is close to 10,000–20,000 cm² (figure A). Kidney dialysis is not convenient for those suffering from kidney failure, and it can be emotionally difficult. Clearly, kidney dialysis is not a good substitute for healthy kidneys.

Figure A   Kidney Dialysis

During kidney dialysis blood flows through a system of tubes composed of a selectively permeable membrane. Dialysis fluid, the composition of which is similar to that of blood, except that the concentration of waste products is very low, flows in the opposite direction on the outside of the dialysis tubes. Consequently, waste products such as urea diffuse from the blood into the dialysis fluid. Other substances such as sodium, potassium, and glucose do not rapidly diffuse from the blood into the dialysis fluid because there is no concentration gradient for these substances between the blood and the dialysis fluid.

4. Active transport of ions from the collecting ducts into the interstitial fluid of the medulla.
5. The passive diffusion of urea from the collecting ducts into the interstitial fluid of the medulla.

The roles of each of these mechanisms in the maintenance of the high concentration of solutes in the medulla of the kidney are described in the following sections:

1. *Loops of Henle.* The long loops of Henle of the juxtamedullary nephrons are essential for the maintenance of a high medullary solute concentration. The walls of the descending limbs of the loops of Henle are permeable to water. As filtrate flows into the medulla of the kidney through the descending limbs, water diffuses out of the nephrons into the more concentrated interstitial fluid. The walls of the thick and thin segments of the ascending limbs of the loops of Henle are impermeable to water. Consequently, solutes diffuse out of the thin segment of the ascending limb as it passes through progressively less concentrated interstitial fluid on its way back to the cortex of the kidney. Also, Na⁺, K⁺, and Cl⁻ are actively transported out of the thick segment of the ascending limb into the interstitial fluid. Thus, water enters the interstitial fluid from the descending limbs and solutes enter the interstitial fluid from the ascending limbs of the loops of Henle (figure 26.15a). The solutes that diffuse from the thin

segments, and that are actively transported from the thick segments, add solutes to the medulla.

2. *The vasa recta.* The vasa recta are countercurrent systems that remove excess water and solutes from the medulla of the kidney without changing the high concentration of solutes in the medullary interstitial fluid. The vasa recta also supply blood to the medulla of the kidney. **Countercurrent** (kown′ter-ker′ent) **systems** consist of parallel tubes in which fluid flows, but in opposite directions, and heat or substances such as water or solutes diffuse from tubes carrying fluid in one direction to tubes carrying fluid in the opposite direction so that the fluid in both sets of tubes has nearly the same composition. The vasa recta are a countercurrent system because blood flows through them to the kidney medulla and, after the vessels turn near the tip of the renal pyramid, the blood is carried in the opposite direction. Blood flows slowly through the vasa recta and the walls of the vasa recta are permeable to water and to solutes. As blood flows toward the medulla, water moves out of the vasa recta, and some solutes diffuse into them. As blood flows back toward the cortex, water moves into the vasa recta, and some solutes diffuse out of them (figure 26.15*b*). The slow flow of blood allows time for diffusion of water and solutes to occur. The directions of diffusion are such that slightly more water and slightly more solute are carried from the medulla by the vasa recta than enter it. Thus, the composition of the blood at both ends of the vasa recta is nearly the same, with the volume and osmolality being slightly greater as the blood once again reaches the cortex.

The loops of Henle and the vasa recta are in parallel with one another, and their functions are closely related. The water and solutes that leave the loops of Henle enter the vasa recta. The vasa recta carry away the water and solutes without diminishing the high concentration of solutes in the medulla of the kidney (figure 26.15*c*). Water passes by osmosis from the collecting ducts, and solutes, such as Na$^+$ and Cl$^-$, are actively transported out of the collecting ducts and pass into the interstitial fluid of the medulla of the kidney (figure 26.15*d*). The water and solutes that leave the collecting ducts also enter the vasa recta and are carried away from the medulla.

The loops of Henle and vasa recta function together to maintain a high concentration of solutes in the interstitial fluids of the medulla and to carry away the water and solutes that enter the medulla from the loops of Henle and collecting ducts.

3. *Urea.* **Urea** (ū-rē′ă) molecules are responsible for a substantial part of the high osmolality in the medulla of the kidney (figure 26.16). The walls of the descending limbs of the loops of Henle are permeable to urea, and urea diffuses into the descending limbs from the interstitial fluid. The ascending limbs of the loops of Henle and the distal tubules are impermeable to urea. The collecting ducts are permeable to urea, however, and some urea diffuses out of the collecting ducts into the interstitial fluid of the medulla. Thus, urea flows in a cycle from the interstitial fluid into the descending limbs of the loops of Henle, through the

Figure 26.16   The Medullary Concentration Gradient and Urea Cycling

The concentration of urea in the medulla of the kidney is high and contributes to the overall high concentration of solutes there. The wall of the collecting duct is permeable to urea. Urea diffuses out of the collecting duct into the interstitial fluid of the medulla. The wall of the descending limb of the loop of Henle is also permeable to urea. Urea diffuses from the interstitial fluid into the descending limb. Thus a cycle is produced in which urea flows into the descending limb, through the ascending limb, through the distal nephron, through the collecting duct, out of the collecting duct, and back into the descending limb.

ascending limbs, through the distal tubules, and into the collecting ducts. Urea then diffuses from the collecting ducts back into the interstitial fluid of the medulla. Consequently, a high urea concentration is maintained in the medulla of the kidney.

- 23. List the major mechanisms that create and maintain the high solute concentration in the renal medulla.
- 24. Describe the role of the loops of Henle, vasa recta, and urea in maintaining a high interstitial solute concentration in the kidney medulla.

Summary of Changes in Filtrate Volume and Concentration

In the average person, about 180 L of filtrate enter the proximal tubules daily. Glucose, amino acids, Na$^+$, Ca^{2+}, K$^+$, Cl$^-$, and other substances (see table 26.3) are actively transported, and water moves by osmosis from the lumens of the proximal tubules into the interstitial fluid. The excess solutes and water then enter the peritubular capillaries. Consequently, approximately 65% of the filtrate is reabsorbed as water, and solutes move from the proximal tubules into the interstitial fluid. The osmolality of both the interstitial fluid and the filtrate is maintained at about 300 mOsm/kg.

The filtrate then passes into the descending limbs of the loops of Henle, which are highly permeable to water and solutes. As the descending limbs penetrate deep into the medulla of the kidney, the surrounding interstitial fluid has a progressively greater osmolality. Water diffuses out of the nephrons as solutes slowly diffuse into them. By the time the filtrate reaches the deepest part of the loops of Henle, its volume has been reduced by an additional 15% of the original volume and its osmolality has increased to about 1200 mOsm/kg (figure 26.17). By the time the filtrate has reached the tips of the loops of Henle, at least 80% of the filtrate volume has been reabsorbed.

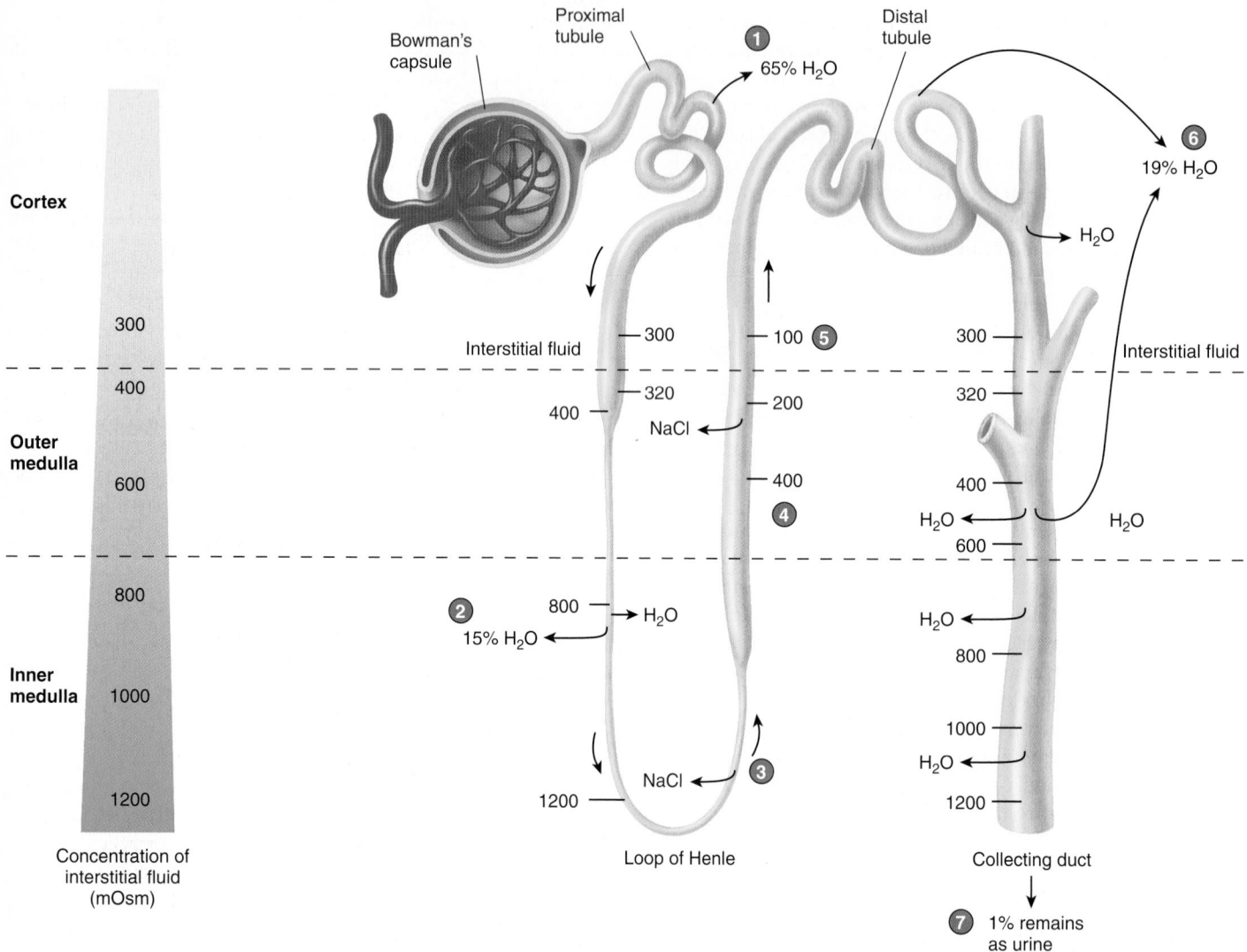

1. Approximately 180 L of filtrate enters the nephrons each day; of that volume 65% is reabsorbed in the proximal tubule. In the proximal tubule, solute molecules move by active transport and cotransport from the lumen of the tubule into the interstitial fluid. Water moves by osmosis because the cells of the tubule wall are permeable to water (see figure 26.11).

2. Approximately 15% of the filtrate volume is reabsorbed in this segment of the descending limb of the loop of Henle. The descending limb passes through the concentrated interstitial fluid of the medulla. Because the wall of the descending limb is permeable to water, water moves by osmosis from the tubule into the more concentrated interstitial fluid (see figure 26.12a). By the time the filtrate reaches the tip of the renal pyramid, the concentration of the filtrate is equal to the concentration of the interstitial fluid.

3. The ascending limb of the loop of Henle is not permeable to water. Solutes diffuse out of the thin segment (see figure 26.12b).

4. Na^+, K^+, and Cl^- are cotransported across the apical membrane of the thick segment. Na^+ are actively transported and K^+ and Cl^- diffuse across the basal membrane of epithelial cells of the thick segment into the interstitial fluid (see figure 26.13).

5. The volume of the filtrate doesn't change as it passes through the ascending limb, but the concentration is greatly reduced (see figure 26.13). By the time the filtrate reaches the cortex of the kidney, the concentration is approximately 100 mOsm/L, which is less concentrated than the interstitial fluid of the cortex (300 mOsm/L).

6. The distal tubule and collecting duct are permeable to water if ADH is present. If ADH is present, water moves by osmosis from the less concentrated filtrate into the more concentrated interstitial fluid. By the time the filtrate reaches the tip of the renal pyramid an additional 19% of the filtrate is reabsorbed.

7. One percent or less remains as urine.

Process Figure 26.17 Urine Concentrating Mechanism

Summary of the mechanisms responsible for the concentration of urine.

After passing through the descending limbs of the loops of Henle, the filtrate enters the ascending limbs. Both the thin and thick segments are impermeable to water, but Na^+, Cl^-, and K^+ are transported from the filtrate into the interstitial fluid in the thick segments (see figure 26.17). The movement of ions, but not water, across the wall of the ascending limbs causes the osmolality of the filtrate to decrease from 1200 to about 100 mOsm/kg by the time the filtrate again reaches the cortex of the kidney. The volume of the filtrate does not change as it passes through the ascending limbs. As a result, the filtrate entering the distal tubules is dilute compared to the concentration of the surrounding interstitial fluid, which has an osmolality of about 300 mOsm/kg.

The changes just described are obligatory; that is, they occur regardless of the concentration and the volume of the urine that is finally produced by the kidney. The mechanisms by which concentrated and dilute urine are formed by the kidney are described in the following sections.

25. Describe how the filtrate volume and concentration change as they flows through the nephrons and collecting ducts.

Formation of Concentrated Urine

Filtrate enters the distal tubules after passing through the loops of Henle and then passes through the collecting ducts. Near the ends of the distal tubules and collecting ducts, the wall of the tubules become very permeable to water, provided antidiuretic hormone (ADH) is present. Water then diffuses from the lumens of the nephrons into the more concentrated interstitial fluid.

ADH increases the permeability of the plasma membranes of the distal nephrons and collecting ducts to water. ADH binds to a membrane-bound receptor, activating a G protein mechanism that increases cAMP synthesis in the cells of the distal tubules and collecting ducts. Cyclic AMP increases the permeability of the plasma membranes of the distal tubules and collecting ducts to water by increasing the number of water channels in the plasma membrane (figure 26.18). When ADH is present, water moves by osmosis out of the distal tubules and collecting ducts, whereas in the absence of ADH, water remains within the nephrons and becomes urine. In the proximal tubules, 65% of the filtrate is reabsorbed, and 15% of the filtrate volume is reabsorbed in the thin segments of the descending limbs of the loops of Henle. About 80% of the volume of the filtrate is therefore reabsorbed in these structures.

The filtrate then flows into the distal tubules and collecting ducts that pass through the medulla of the kidney with its high concentration of solutes. If ADH is present, water diffuses from the collecting ducts into the interstitial fluid. By the time the filtrate passes through the collecting ducts, another 19% of the filtrate is reabsorbed. Thus, 1% of the filtrate remains as urine, and 99% of the

1. ADH binds to ADH receptors in the plasma membranes of the distal tubule cells and the collecting duct cells.

2. When ADH is bound to its receptor, a G protein mechanism is activated, which, in turn, activates adenylate cyclase.

3. The increase in the rate of cyclic adenosine monophosphate (cAMP) synthesis increases the permeability of the epithelial cells to water by increasing the number of water channels in their plasma membranes.

4. Water then moves out of the distal tubule and collecting duct into the interstitial fluid by osmosis, both decreasing the volume of the filtrate and increasing its concentration (osmolality).

Process Figure 26.18 **The Effect of Antidiuretic Hormone (ADH) on the Nephron**

filtrate is reabsorbed. The osmolality of the filtrate at the ends of the collecting ducts is approximately 1200 mOsm/kg (see figure 26.17).

In addition to the dramatic decrease in filtrate volume and the increase in filtrate osmolality, a marked alteration occurs in the filtrate composition. Waste products, such as creatinine and urea as well as K^+, H^+, phosphate, and sulfate ions, occur at a much higher concentration in urine than in the original filtrate because water has been removed from the filtrate. Many substances are selectively reabsorbed from the nephron, and others are secreted into the nephron so that beneficial substances are retained in the body and toxic substances are eliminated.

 26. *Describe how the presence of ADH causes the formation of a small volume of concentrated urine.*

Formation of Dilute Urine

If ADH is not present or if its concentration is reduced, the distal tubules and collecting ducts have a low permeability to water. The amount of water moving by osmosis from the distal tubule and collecting duct, therefore, is low. The concentration of the urine produced is less than 1200 mOsm/kg, and the volume is increased. The volume of this more dilute urine can be much larger than 1% of the filtrate formed each day. If no ADH is secreted, the osmolality of the urine may be close to the osmolality of the filtrate in the distal tubule, and the volume of urine may approach 20–30 L/day.

In a healthy person, even when the kidney produces dilute urine, the concentration of waste products in the urine is large enough to maintain homeostasis in the body. Consequently, beneficial substances are retained, and both toxic substances and excess water are eliminated.

27. *Describe how the absence of ADH causes the formation of a large volume of dilute urine.*

Urine Concentration and Juxtamedullary Nephrons

Only the juxtamedullary nephrons have loops of Henle that descend deep into the medulla, but enough of them exist to maintain a high interstitial concentration of solutes in the interstitial fluid of the medulla. Not all of the nephrons need to have loops of Henle that descend into the medulla to effectively concentrate urine. The cortical nephrons function like the juxtamedullary nephrons, with the exception that their loops of Henle are not as efficient at concentrating urine. Because the filtrate from the cortical nephrons passes through the collecting ducts, however, water can diffuse out of the collecting ducts into the interstitial fluid. Thus the filtrate becomes concentrated. Animals that concentrate urine more effectively than humans have a greater percentage of nephrons that descend into the medulla of the kidney. For example, mammals that live in deserts have many nephrons that descend into the medulla of the kidney, and the renal pyramids are longer than in humans and most other mammals.

Regulation of Urine Concentration and Volume

Urine can be diluted or very concentrated, and it can be produced in large or small amounts. Urine concentration and volume are regulated by mechanisms that maintain the extracellular fluid osmolality and volume within narrow limits.

Filtrate reabsorption in the proximal tubules and the descending limbs of the loops of Henle is obligatory and therefore remains relatively constant. Filtrate reabsorption in the distal tubules and collecting ducts is regulated, however, and can change dramatically, depending on the conditions to which the body is exposed. If homeostasis requires the elimination of a large volume of dilute urine, the dilute filtrate in the distal tubules and collecting ducts can pass through them with little change in concentration. On the other hand, if conservation of water is required to maintain homeostasis, water is reabsorbed from the filtrate as it passes through the distal tubules and collecting ducts. This results in the production of a small volume of very concentrated urine. Regulation of urine volume and concentration involves hormonal mechanisms, autoregulation, and the nervous system.

Hormonal Mechanisms
Antidiuretic Hormone

The distal tubules and the collecting ducts remain relatively impermeable to water in the absence of ADH (see figure 26.18). When little ADH is secreted, a large part of the 19% of the filtrate that's normally reabsorbed in the distal tubules and the collecting ducts becomes part of the urine. People who do not secrete sufficient ADH often produce 10–20 L of urine per day and develop major problems such as dehydration and ion imbalances. Insufficient ADH secretion results in a condition called **diabetes insipidus** (dī-ă-bē′tēz in-sip′i-dŭs); the word *diabetes* implies the production of a large volume of urine, and the word *insipidus* implies the production of a clear, tasteless, dilute urine. This condition is in contrast to **diabetes mellitus** (me-lī′tŭs), which implies the production of a large volume of urine that contains a high concentration of glucose. The word *mellitus* means "honeyed," or "sweet."

The posterior pituitary secretes ADH. Neurons with cell bodies primarily in the supraoptic nuclei of the hypothalamus have axons that course to the posterior pituitary gland. ADH is released into the circulatory system from these neuron terminals. Cells called **osmoreceptor cells** in the supraoptic nuclei are very sensitive to even slight changes in the osmolality of the interstitial fluid. If the osmolality of the blood and interstitial fluid increases, these cells stimulate the ADH-secreting neurons. Action potentials are then propagated along the axons of the ADH-secreting neurons to the posterior pituitary gland, where the axons release ADH from their ends. Reduced osmolality of the interstitial fluid within the supraoptic nuclei inhibits ADH secretion from the posterior pituitary gland (see figure 18.5).

Baroreceptors that monitor blood pressure in the atria of the heart, large veins, carotid sinuses, and aortic arch also influence ADH secretion when the blood pressure increases or decreases in excess of 5%–10%. Decreases in blood pressure are detected by the baroreceptors, which consequently decrease the frequency of action potentials sent along the afferent pathways that ultimately extend to the supraoptic region of the hypothalamus. The result is an increase in ADH secretion.

When blood osmolality increases or when blood pressure declines significantly, ADH secretion increases and acts on the kidneys to increase the reabsorption of water. The reabsorption of water decreases blood osmolality. It also increases blood volume, which increases blood pressure. Conversely, when blood osmolality decreases or when blood pressure increases, ADH secretion declines. The reduced ADH levels cause the kidneys to reabsorb less water and to produce a larger volume of dilute urine. The increased loss of water in the urine increases blood osmolality and decreases blood pressure. ADH secretion occurs in response to small changes in osmolality while a substantial change in blood pressure is required to alter ADH secretion. Thus, ADH is more important in the regulation of blood osmolality than in the regulation of blood pressure.

P R E D I C T 5

Ethyl alcohol inhibits ADH secretion. Given this information, describe the mechanism by which alcoholic beverages affect urine production.

Renin-Angiotensin-Aldosterone

Renin is an enzyme secreted by cells of the juxtaglomerular apparatus. The rate of renin secretion increases if blood pressure in the afferent arteriole decreases, or if the Na^+ concentration of the filtrate decreases as it passes by the macula densa cells of the juxtaglomerular apparatuses. Renin enters the general circulation, acting on **angiotensinogen,** a plasma protein produced by the liver, and converting it to **angiotensin I.** Subsequently, a proteolytic enzyme called **angiotensin-converting enzyme (ACE),** found in capillary beds in organs such as the lungs, converts angiotensin I to **angiotensin II** (figure 26.19). Angiotensin II is a potent vasoconstricting substance that increases peripheral resistance, causing blood pressure to increase. However, angiotensin II is rapidly

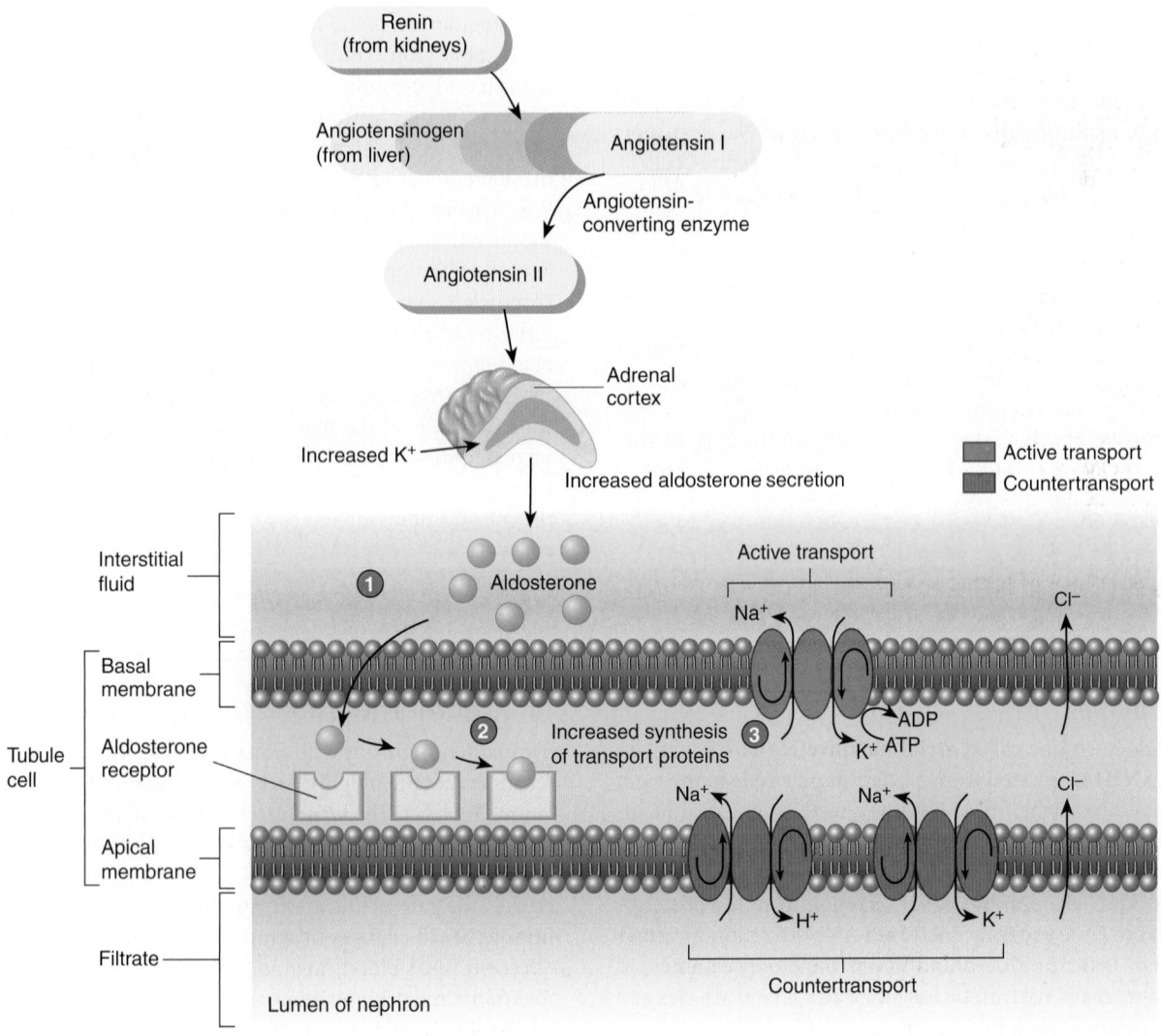

1. Aldosterone secreted from the adrenal cortex enters cells of the distal tubule.

2. Aldosterone binds to intracellular receptors and increases the synthesis of transport proteins of the apical and basal membranes.

3. Newly synthesized transport proteins increase the rate at which Na^+ are absorbed and K^+ and H^+ are secreted. Chloride ions move with the Na^+ because they are attracted to the positive charge of Na^+.

Process Figure 26.19   **Effect of Aldosterone on the Distal Tubule**

broken down so its effect lasts for only a short time. Angiotensin II also increases the rate of aldosterone secretion, the sensation of thirst, salt appetite, and ADH secretion.

The rate of renin secretion decreases if blood pressure in the afferent arteriole increases, or if the Na^+ concentration of the filtrate increases as it passes by the macula densa of the juxtaglomerular apparatuses.

A large decrease in the concentration of Na^+ in the interstitial fluids acts directly on the aldosterone-secreting cells of the adrenal cortex to increase the rate of aldosterone secretion. Angiotensin II is much more important than the blood level of Na^+, however, in regulating aldosterone secretion.

Aldosterone, a steroid hormone secreted by the cortical cells of the adrenal glands (see chapter 18), passes through the circulatory system from the adrenal glands to the cells in the distal tubules and the collecting ducts. Aldosterone molecules diffuse through the plasma membranes and bind to receptor molecules within the cells. The combination of aldosterone molecules with their receptor molecules increases synthesis of the transport protein molecules that increase the transport of Na^+ across the basal and apical membranes of nephron cells. As a result, the rate of Na^+ transport out of the filtrate back into the blood increases (see figure 26.19).

Reduced secretion of aldosterone decreases the rate of Na^+ transport. As a consequence, the concentration of Na^+ in the distal tubules and the collecting ducts remains high. Because the concentration of filtrate passing through the distal tubules and the collecting ducts has a greater-than-normal concentration of solutes, the capacity for water to move by osmosis from the distal tubules and the collecting ducts is diminished, urine volume increases, and the urine has a greater concentration of Na^+.

Increases in blood K^+ levels act directly on the adrenal cortex to stimulate aldosterone secretion, whereas decreases in blood K^+ levels decrease aldosterone secretion (see chapter 27).

P R E D I C T   **6**

Drugs that increase urine volume are called diuretics. Some diuretics inhibit the active transport of Na^+ in the nephron. Explain how these diuretic drugs could cause increased urine volume.

Other Hormones

A polypeptide hormone called **atrial natriuretic** (nā′trē-ū-ret′ik) **hormone (ANH)** is secreted from cardiac muscle cells in the right atrium of the heart when blood volume in the right atrium increases and stretches the cardiac muscle cells (see chapter 21). Atrial natriuretic hormone inhibits Na^+ reabsorption in the kidney tubules. ANH also inhibits ADH secretion from the posterior pituitary gland. Consequently, increased ANH secretion increases the volume of urine produced and lowers the blood volume and blood pressure. Atrial natriuretic hormone also dilates arteries and veins, which reduces peripheral resistance and lowers blood pressure. Thus, a decrease occurs in venous return and blood volume in the right atrium.

Two other substances, prostaglandins and kinins, are formed in the kidneys and affect kidney function. Their roles are unclear, but both substances influence the rate of filtrate formation and Na^+ reabsorption. Prostaglandins probably moderate the sensitivity of the renal blood vessels to neural stimuli and to angiotensin II.

28. *Where is ADH produced? What factors stimulate an increase in ADH secretion?*
29. *What effect does ADH have on urine volume and concentration?*
30. *What are the effects of aldosterone on Na^+ and Cl^- transport? How does aldosterone affect urine concentration, urine volume, and blood pressure?*
31. *Where is aldosterone produced? What factors stimulate aldosterone secretion?*
32. *How is angiotensin II activated? What effects does it produce?*
33. *What factors cause an increase in renin production?*
34. *Where is atrial natriuretic hormone produced, and what effect does it have on urine production?*

Autoregulation

Autoregulation is the maintenance, within the kidneys, of a relatively stable glomerular filtrate rate (GFR) over a wide range of systemic blood pressures. For example, GFR is relatively constant as the systemic blood pressure changes between 90 and 180 mm Hg.

Autoregulation involves changes in the degree of constriction in the afferent arterioles. The precise mechanism by which autoregulation is achieved is unclear, but, as systemic blood pressure increases, the afferent arterioles constrict and prevent an increase in renal blood flow and filtration pressure across the filtration membrane of the renal corpuscle. Conversely, a decrease in systemic blood pressure results in dilation of the afferent arterioles, thus preventing a decrease in renal blood flow and filtration pressure across the filtration membrane.

Autoregulation is also influenced by the rate of flow of filtrate past cells of the macula densa. The macula densa detects an increased flow rate and sends a signal to the juxtaglomerular apparatus to constrict the afferent arteriole. The result is a decrease in the filtration pressure across the filtration membrane of the renal corpuscle.

35. *Describe autoregulation.*

Effect of Sympathetic Stimulation on Kidney Function

Sympathetic neurons with norepinephrine as their neurotransmitter innervate the blood vessels of the kidneys. Sympathetic stimulation constricts the small arteries and afferent arterioles, thereby decreasing renal blood flow and filtrate formation. Intense sympathetic stimulation, such as during shock or intense exercise, decreases the rate of filtrate formation to only a few milliliters per minute. Small changes in sympathetic stimulation have a minimal effect on renal blood flow and filtrate formation. Autoregulation maintains renal blood flow and filtrate formation at a relatively constant rate unless sympathetic stimulation is intense.

In response to severe stress or circulatory shock, renal blood flow can decrease to such low levels that the blood supply to the kidney is inadequate to maintain normal kidney metabolism. As a consequence, kidney tissues can be damaged and thus be unable to perform their normal functions if blood flow is not reestablished. This is one of the reasons why shock should be treated quickly. Reduced blood flow to the kidneys during stress or shock is consistent with homeostasis. Intense vasoconstriction, which maintains blood pressure at levels adequate to sustain blood flow to organs such as the heart and brain, is essential to homeostasis. A short-term reduction in blood flow to organs such as the kidneys is only harmful if the lack of blood flow is prolonged.

 36. *Explain the effect that sympathetic stimulation has on the kidney during rest, exercise, and shock.*

Clearance and Tubular Maximum

Plasma clearance is a calculated value representing the volume of plasma that is cleared of a specific substance each minute. For example, if the clearance value is 100 mL/min for a substance, the substance is completely removed from 100 mL of plasma each minute.

Clearance

The plasma clearance can be calculated for any substance that enters the circulatory system according to the following formula:

$$\text{Plasma clearance (mL/min)} = \text{Quantity of urine (mL/min)} \times \frac{\text{Concentration of substance in urine}}{\text{Concentration of substance in plasma}}$$

Plasma clearance can be used to estimate GFR if the appropriate substance is monitored (see table 26.2). Such a substance must have the following characteristics: (1) it must pass through the filtration membrane of the renal corpuscle as freely as water or other small molecules, (2) it must not be reabsorbed, (3) it must not be secreted into the nephron, and (4) it must not be either metabolized or produced in the kidney. **Inulin** (in′ū-lin) is a polysaccharide that has these characteristics. As filtrate is formed, it has the same concentration of inulin as plasma; but, as the filtrate flows through the nephron, all the inulin remains in the nephron to enter the urine. As a consequence, all the volume of plasma that becomes filtrate is cleared of inulin, and the plasma clearance for inulin is equal to the rate of glomerular filtrate formation.

GFR is reduced when the kidney fails. Measurement of the GFR indicates the degree to which kidney damage has occurred. The clearance value for urea and creatinine can be used clinically. Because these substances are naturally occuring metabolites, foreign substances do not have to be injected. A high plasma concentration and a lower-than-normal clearance value for them indicates a reduced GFR and kidney failure. Creatinine clearance can be used to monitor the progress of changes in GFR in people who suffer from renal failure.

PREDICT 7

During surgery, a patient's blood pressure drops to very low levels, and ischemia of the kidney develops. Within 1 day following the surgery, the GFR and the urine volume decrease to very low levels. Given that the structure of the glomeruli did not dramatically change, but that the epithelium of nephrons suffered from ischemia and sloughed into the nephron to form casts of epithelial cells in the nephron, explain why the GFR was reduced.

Plasma clearance can also be used to calculate renal plasma flow (see table 26.2). Substances with the following characteris-

tics, however, must be used: (1) the substance must pass through the filtration membrane of the renal corpuscle, and (2) it must be secreted into the nephron at a sufficient rate so that very little of it remains in the blood as the blood leaves the kidney. A substance that meets these requirements is *para*-aminohippuric acid (PAH). As blood flows through the kidney, essentially all the PAH is either filtered or secreted into the nephron. The clearance calculation for PAH is therefore a good estimate of the volume of plasma flowing through the kidney each minute. Also, if the hematocrit is known, the total volume of blood flowing through the kidney each minute can be easily calculated.

The concept of plasma clearance can be used to make the measurements described previously, or it can be used to determine the means by which drugs or other substances are excreted by the kidney. A plasma clearance value greater than the inulin clearance value suggests that the substance is secreted by the nephron into the filtrate.

PREDICT 8

A person is suspected of suffering from chronic renal failure. To assess kidney function, urea clearance is measured and found to be very low. Explain what a very low urea clearance indicates in a person suffering from chronic renal failure.

The **tubular load** of a substance is the total amount of the substance that passes through the filtration membrane into the nephrons each minute. Normally, glucose is almost completely reabsorbed from the nephron by the process of active transport. The capacity of the nephron to actively transport glucose across the epithelium of the nephron is limited, however. If the tubular load is greater than the capacity of the nephron to reabsorb it, the excess glucose remains in the urine.

The maximum rate at which a substance can be actively reabsorbed is called the **tubular maximum** (figure 26.20). Each substance

Figure 26.20 Tubular Maximum for Glucose

As the concentration of glucose increases in the filtrate, it reaches a point that exceeds the ability of the nephron to actively reabsorb it. That concentration is called the tubular maximum. Beyond that concentration, the excess glucose enters the urine.

Clinical Focus   Diuretics

Diuretics (dī-ū-ret′iks) are agents that increase the rate of urine formation. Although the definition is simple, a number of different physiologic mechanisms are involved.

Diuretics are used to treat disorders like hypertension and several types of edema that are caused by congestive heart failure, cirrhosis of the liver, and other anomalies. Use of diuretics can lead to complications, however, including dehydration and electrolyte imbalances.

The action of **carbonic anhydrase** (kar-bon′ik an-hī′drās) **inhibitors** reduces the rate of H^+ secretion and the reabsorption of HCO_3^-. As H^+ are secreted into the nephron, they combine with HCO_3^- to form carbonic acid. Carbonic acid forms water and carbon dioxide, which can diffuse across the wall of the nephron. Reduced H^+ secretion causes HCO_3^- to remain in the nephron. The HCO_3^- increase tubular osmotic pressure, causing osmotic diuresis. With long-term use, the diuretic effect of

carbonic anhydrase inhibitors tends to be lost. The diuretic effect of these inhibitors is useful in treating conditions such as glaucoma and altitude sickness.

Sodium ion reabsorption inhibitors include thiazide-type diuretics. They promote the loss of Na^+, Cl^-, and water in the urine. These diuretics are given to some people who have hypertension. Inhibitors of Na^+ reabsorption, such as bumetanide, furosemide, and ethacrynic acid, specifically inhibit transport in the ascending limb of the loop of Henle. These diuretics are frequently used to treat congestive heart failure, cirrhosis of the liver, and renal disease.

Potassium-sparing diuretics are antagonists to aldosterone or directly prevent Na^+ reabsorption in the distal tubules and collecting ducts. Thus they promote Na^+ and water loss in the urine. These diuretics are used to reduce the loss of K^+ in the urine and therefore preserve, or "spare," these ions. They are often used in combination with Na^+ reabsorption inhibitors and

are effective in preventing excess K^+ loss in the urine.

Osmotic diuretics freely pass by filtration into the filtrate, and they undergo limited reabsorption by the nephron. These diuretics increase urine volume by elevating the osmotic concentration of the filtrate, thus reducing the amount of water moving by osmosis out of the nephron. Urea, mannitol, and glycerine have been used as osmotic diuretics. Although they are not commonly used, they are effective in treating people who are suffering from cerebral edema and edema in acute renal failure.

Xanthines (zan′thēnz), including caffeine and related substances, act as diuretics, partly because they increase renal blood flow and the rate of glomerular filtrate formation. They also influence the nephron by decreasing Na^+ and Cl^- reabsorption.

Alcohol acts as a diuretic, although it's not used clinically for that purpose. It inhibits ADH secretion from the posterior pituitary and results in increased urine volume.

---

that's reabsorbed has its own tubular maximum, determined by the number of active transport carrier molecules and the rate at which they are able to transport molecules of the substance. For example, in people suffering from diabetes mellitus the tubular load for glucose can exceed the tubular maximum by a substantial amount, thus allowing glucose to appear in the urine. Urine volume is also greater than normal because the glucose molecules in the filtrate increase the osmolality of the filtrate in the nephron and reduce the effectiveness of water reabsorption by osmosis.

● 37. *Define and explain the significance of the terms plasma clearance, tubular load, and tubular maximum.*

Urine Movement

Urine Flow Through the Nephron and the Ureters

Hydrostatic pressure averages 10 mm Hg in Bowman's capsule and nearly 0 mm Hg in the renal pelvis. This pressure gradient forces the filtrate to flow from Bowman's capsule through the nephron

into the renal pelvis. Because the pressure is 0 mm Hg in the renal pelvis, no pressure gradient exists to force urine to flow to the urinary bladder through the ureters. The circular smooth muscle in the walls of the ureters exhibits peristaltic contractions that force urine to flow through the ureters. The peristaltic waves progress from the region of the renal pelvis to the urinary bladder. They occur from once every few seconds to once every 2–3 minutes. Parasympathetic stimulation increases their frequency, and sympathetic stimulation decreases it.

The peristaltic contractions of each ureter proceed at a velocity of approximately 3 cm/s and can generate pressures in excess of 50 mm Hg. At the point where the ureters penetrate the urinary bladder, they course obliquely through the trigone. Pressure inside the urinary bladder compresses that part of the ureter to prevent the backflow of urine.

When no urine is present in the urinary bladder, internal pressure is about 0 mm Hg. When the volume is 100 mL of urine, pressure is elevated to only 10 mm Hg. Pressure increases slowly as its volume increases to approximately 300 mL, but above urinary bladder volumes of 400 mL the pressure rises rapidly.

Kidney Stones

Kidney stones are hard objects usually found in the pelvis of the kidney. They are normally 2–3 mm in diameter with either a smooth or jagged surface, but occasionally a large branching kidney stone called a **staghorn stone** forms in the renal pelvis. About 1% of all autopsies reveal kidney stones, and many of the stones occur without causing symptoms. The symptoms associated with kidney stones occur when a stone passes into the ureter, resulting in referred pain down the back, side, and groin area. The ureter contracts around the stone, causing the stone to irritate the epithelium and produce bleeding, which appears as blood in the urine, a condition called **hematuria.** In addition to causing intense pain, kidney stones can block the ureter, cause ulceration in the ureter, and increase the probability of bacterial infections.

About 65% of all kidney stones are composed of calcium oxalate mixed with calcium phosphate, 15% are magnesium ammonium phosphate, and 10% are uric acid or cystine; approximately 2.5% of each kidney stone is composed of mucoprotein.

The cause of kidney stones is usually obscure. Predisposing conditions include a concentrated urine and an abnormally high calcium concentration in the urine, although the cause of the high calcium concentration is usually unknown. Magnesium ammonium phosphate stones are often found in people with recurrent kidney infections, and uric acid stones often occur in people suffering from gout. Severe kidney stones must be removed surgically. Instruments that pulverize kidney stones with ultrasound or lasers, however, have replaced most traditional surgical procedures.

Micturition Reflex

The **micturition** (mik-choo-rish′un) **reflex** is activated when the urinary bladder wall is stretched and it results in **micturition,** which is the elimination of urine from the urinary bladder. Integration of the micturition reflex occurs in the sacral region of the spinal cord and it is modified by centers in the pons and cerebrum.

Urine filling the urinary bladder stimulates stretch receptors, which produce action potentials. The action potentials are carried by sensory neurons to the sacral segments of the spinal cord through the pelvic nerves. In response, action potentials are carried to the urinary bladder through parasympathetic fibers in the pelvic nerves (figure 26.21). The parasympathetic action potentials cause smooth muscle of the urinary bladder (the detrusor muscle) to contract. In addition, decreased somatic motor action potentials cause the external urinary sphincter, which consists of skeletal muscle, to relax. Urine flows from the urinary bladder when the pressure there is great enough to force urine to flow through the urethra while the external urinary sphincter is relaxed. The micturition reflex normally produces a series of contractions of the urinary bladder.

Action potentials carried by sensory neurons from stretch receptors in the urinary bladder wall also ascend the spinal cord to a micturition center in the pons and to the cerebrum. Descending action potentials are sent from these areas of the brain to the sacral region of the spinal cord, where they modify the activity of the micturition reflex in the spinal cord. The micturition reflex, integrated in the spinal cord, predominates in infants. The ability to voluntarily inhibit micturition develops at the age of 2–3 years, and subsequently, the influence of the pons and cerebrum on the spinal micturition reflex predominates. The micturition reflex integrated in the spinal cord is automatic, but it is either stimulated or inhibited by descending action potentials. Higher brain centers prevent micturition by sending action potentials from the cerebrum and pons through spinal pathways to inhibit the spinal micturition reflex. Consequently, parasympathetic stimulation of the urinary bladder is inhibited and somatic motor neurons that keep the external urinary sphincter contracted are stimulated.

The slow increase in pressure helps explain why there is little desire to urinate when the bladder contains less than 300 mL. The pressure in the urinary bladder increases rapidly once its volume exceeds approximately 400 mL, and there is an increase in the frequency of action potentials carried by sensory neurons. The increased frequency of action potentials conducted by the ascending spinal pathways to the pons and cerebrum results in an increased desire to urinate.

Voluntary initiation of micturition involves an increase in action potentials sent from the cerebrum to facilitate the micturition reflex and to voluntarily relax the external urinary sphincter. In addition to facilitating the micturition reflex, there is an increased voluntary contraction of abdominal muscles, which causes an increase in abdominal pressure. This enhances the micturition reflex by increasing the pressure applied to the urinary bladder wall.

The desire to urinate normally results from stretch of the urinary bladder wall, but irritation of the urinary bladder or the urethra by bacterial infections or other conditions can also initiate the desire to urinate, even though the urinary bladder may be nearly empty.

- **38.** *What is responsible for the movement of urine through the nephron and ureters?*
- **39.** *Describe the micturition reflex. How is voluntary control of micturition accomplished?*

Automatic, Noncontracting, and Hyperexcitable Urinary Bladder

If the spinal cord is damaged above the sacral region, no micturition reflex exists for a time; but if the urinary bladder is emptied frequently, the micturition reflex eventually becomes adequate to cause it to empty. Some time is generally required for the micturition reflex integrated within the spinal cord to begin to operate. A typical micturition reflex can exist, but no conscious control exists over the onset or duration of it. This condition is called the **automatic bladder.**

Damage to the sacral region of the spinal cord or to the nerves that carry action potentials between the spinal cord and the urinary bladder can result in failure of the urinary bladder to contract although the external urinary sphincter is relaxed. As a result, the micturition reflex cannot occur. The bladder fills to capacity, and urine is forced in a slow dribble through the external urinary sphincter. In elderly people or in patients with damage to the brainstem or spinal cord, a loss of inhibitory action potentials to the sacral region of the spinal cord can occur. Without inhibition, the sacral centers are hyperexcitable, and even a small amount of urine in the bladder can elicit an uncontrollable micturition reflex.

Control of the micturition reflex by higher brain centers

A. Ascending pathways carry an increased frequency of action potentials up the spinal cord to the pons and cerebrum when the urinary bladder becomes stretched. This increases the conscious desire to urinate.

B. Descending pathways carry action potentials to the sacral region of the spinal cord to tonically inhibit the micturition reflex, preventing automatic urination when the bladder is full. Descending pathways carry action potentials from the cerebrum to the sacral region of the spinal cord to stimulate the reflex when stretch of the urinary bladder produces the conscious urge to urinate and when one voluntarily chooses to urinate. This reinforces the micturition reflex.

Micturition reflex

1. Urine in the urinary bladder stretches the bladder wall.

2. Action potentials produced by stretch receptors are carried along pelvic nerves (*green line*) to the sacral region of the spinal cord.

3. Action potentials are carried by parasympathetic nerves (*red line*) to contract the smooth muscles of the urinary bladder. Decreased action potentials carried by somatic motor nerves (*purple line*) cause the external urinary sphincter to relax.

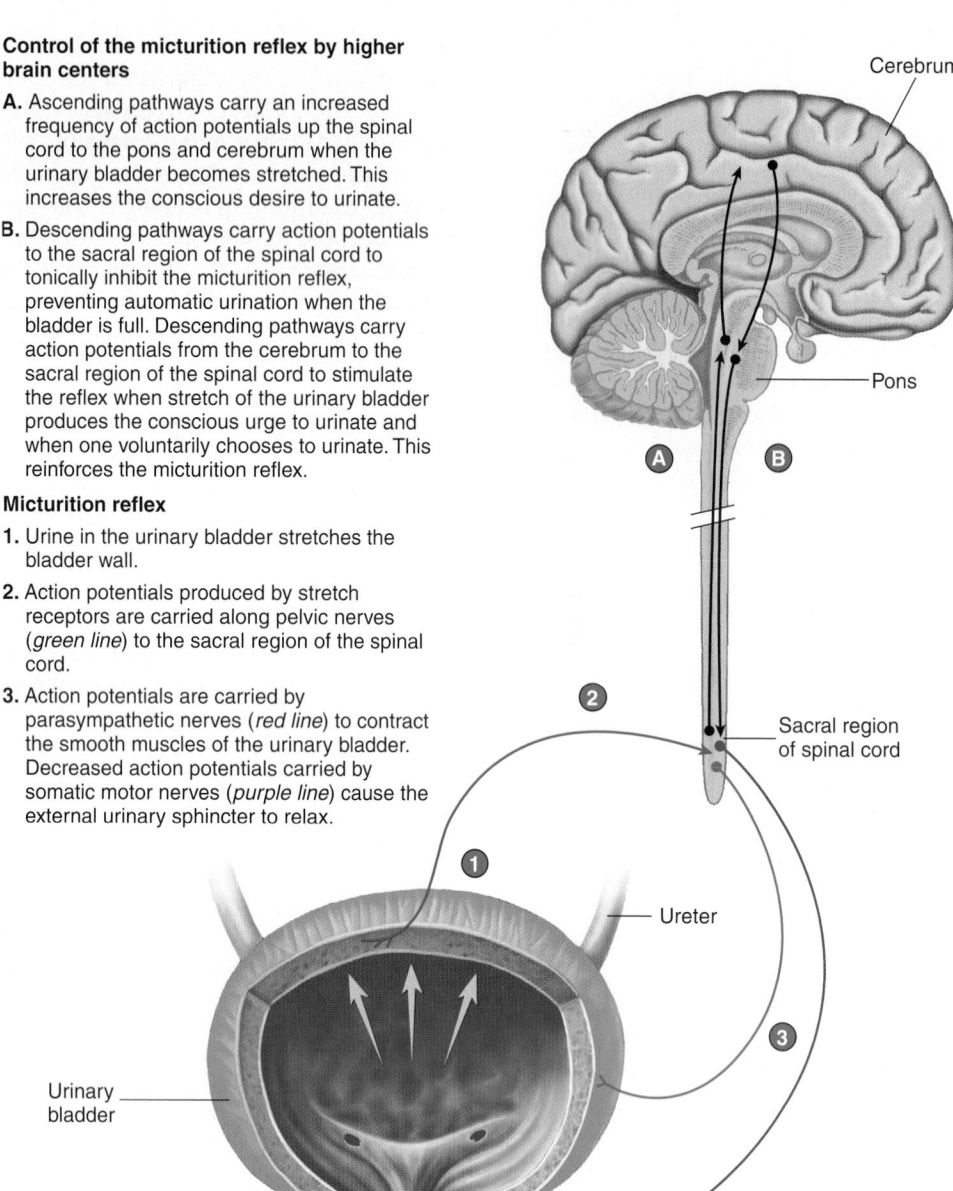

Process Figure 26.21   **Micturition Reflex**

Effects of Aging on the Kidneys

Aging causes a gradual decrease in the size of the kidneys that begins as early as age 20, becomes obvious by age 50, and continues until death. The loss of kidney size appears to be related to changes in the blood vessels of the kidney. The amount of blood flowing through the kidneys gradually decreases. Starting at age 20 there appears to be an approximately 10% decrease every 10 years. Small arteries including the afferent and efferent arterioles become irregular and twisted. Functional glomeruli are lost. By age 80, 40% of the glomeruli are not functioning. About 30% of the glomeruli that stop functioning no longer have a lumen through which blood flows. Other glomeruli thicken and assume a structure similar to arterioles. Some nephrons and collecting ducts become thicker,

Clinical Focus   Renal Pathologies

Glomerular nephritis (glō-măr'ū-lăr ne-frī'tis) results from inflammation of the filtration membrane within the renal corpuscle. It's characterized by an increased permeability of the filtration membrane and the accumulation of numerous white blood cells in the area. As a consequence, a high concentration of plasma proteins enters the filtrate along with numerous white blood cells. A greater-than-normal urine volume accompanies the increase in plasma proteins in the urine.

Acute glomerular nephritis often occurs 1–3 weeks after a severe bacterial infection, such as streptococcal sore throat or scarlet fever. Antigen–antibody complexes associated with the disease become deposited in the filtration membrane and cause its inflammation. This acute inflammation normally subsides after several days.

Chronic glomerular nephritis is long term and usually progressive. The filtration membrane thickens and eventually is replaced by connective tissue. Although in the early stages chronic glomerular nephritis resembles the acute form, in the advanced stages many of the renal corpuscles are replaced by fibrous connective tissue, and the kidney eventually ceases to function.

Pyelonephritis (pī'ě-lō-ne-frī'tis) is inflammation of the renal pelvis, medulla, and cortex. It often begins as a bacterial infection of the renal pelvis and then extends into the kidney itself. It can result from several types of bacteria, including *Escherichia coli*. Pyelonephritis can destroy nephrons and renal corpuscles, but, because the infection starts in the pelvis of the kidney, it affects the medulla more than the cortex. As a consequence, the ability of the kidney to concentrate urine is dramatically affected.

Renal failure can result from any condition that interferes with kidney function. **Acute renal failure** occurs when kidney damage is extensive and leads to the accumulation of urea in the blood and to acidosis (see chapter 27). In complete renal failure, death can occur in 1–2 weeks. Acute renal failure can result from acute glomerular nephritis, or it can be caused by damage to or blockage of the renal tubules. Some poisons, such as mercuric ions or carbon tetrachloride, that are common to certain industrial processes cause necrosis of the nephron epithelium. If the damage doesn't interrupt the basement membrane surrounding the nephrons, extensive regeneration can occur within 2–3 weeks. Severe ischemia associated with circulatory shock resulting from

sympathetic vasoconstriction of the renal blood vessels can cause necrosis of the epithelial cells of the nephron.

Chronic renal failure results when so many nephrons are permanently damaged that those nephrons remaining functional cannot adequately compensate. Chronic renal failure can result from chronic glomerular nephritis, trauma to the kidneys, absence of kidney tissue caused by congenital abnormalities, or tumors. Urinary tract obstruction by kidney stones, damage resulting from pyelonephritis, and severe arteriosclerosis of the renal arteries also cause degeneration of the kidney.

In chronic renal failure, the GFR is dramatically reduced, and the kidney is unable to excrete excess excretory products, including electrolytes and metabolic waste products. The accumulation of solutes in the body fluids causes water retention and edema. Potassium levels in the extracellular fluid are elevated, and acidosis occurs because the distal convoluted tubules and collecting ducts cannot excrete sufficient quantities of potassium and hydrogen ions. Acidosis, elevated potassium levels in the body fluids, and the toxic effects of metabolic waste products cause mental confusion, coma, and finally death when chronic renal failure is severe.

---

shorter, and more irregular in structure. The capacity to secrete and absorb declines, and whole nephrons stop functioning. The ability of the kidney to concentrate urine gradually declines. Eventually changes in the kidney increase the risk of dehydration because of the reduced ability of the kidney to produce a concentrated urine. There's also a decreased ability to eliminate uric acid, urea, creatine, and toxins from the blood.

There is an age-related loss of responsiveness to ADH and to aldosterone. The kidney decreases renin secretion. A reduced ability to participate in vitamin D synthesis occurs, which contributes to Ca^{2+} deficiency, osteoporosis, and bone fractures.

Recall that one-third of one kidney is required to maintain homeostasis. The additional kidney tissue beyond this constitutes a reserve capacity. The age-related changes in the kidney cause a reduction in the reserve capacity of the kidney. As the functional kidney mass is reduced substantially in older people, high blood pressure, atherosclerosis, and diabetes have greater adverse effects.

40. *Describe the effect of aging on the kidneys. Why do the kidneys gradually decrease in size?*

Systems Pathology
Acute Renal Failure

A large piece of machinery overturned at the construction site where Mr. H worked, trapping him beneath it. His legs were severely crushed, although they healed after several months. Mr. H nearly lost his life, however, because of the acute renal failure that developed as a result of his injury. Mr. H was trapped for several hours in a very difficult place to reach. During that time, his blood pressure decreased to very low levels because of the blood loss, the edema in the inflamed tissues, and emotional shock. After he was rescued, he received fluid replacement in the form of both intravenous saline solutions and blood transfusions, and his blood pressure returned to its normal range. Twenty-four hours after the accident, however, his urine volume began to decrease. His urinary Na^+ concentration increased, his urine osmolality decreased, his urine specific gravity decreased, and casts and cellular debris were evident in his urine.

For approximately 7 days, Mr. H exhibited oliguria (reduced urine production). During this period, renal dialysis was required to maintain Mr. H's blood volume and electrolyte concentrations within normal ranges (figure B). After approximately 7 days, his kidneys gradually began to produce large quantities of urine (a diuretic phase). Careful observation was required to keep his blood pressure and electrolyte concentrations within normal ranges. Substantial water, Na^+, and K^+ had to be administered to him. After about 3 weeks, the function of his kidneys slowly began to improve, although many months passed before his kidney function returned to normal.

Figure B   **A Patient Undergoing Dialysis**

Background Information

The events after 24 hours are consistent with acute renal failure caused by prolonged hypotension and ischemia of the kidney. While Mr. H was suffering from hypotension, blood flow to his kidneys was very low. The reduced blood flow was severe enough to result in damage to the epithelial lining of the kidney tubules (tubular necrosis). The period of reduced urine volume resulted from tubular damage. Renal ischemia results in necrosis of tubular cells, which then slough off into the tubules and block them so that filtrate cannot flow through the tubules. In addition, the filtrate leaks from the blocked or partially blocked tubules back into the interstitial spaces and, therefore, back into the circulatory system. As a result, the amount of filtrate that becomes urine is markedly reduced. Blood levels of urea and of creatine increase because of the reduction in filtrate formation and reduced function of the tubular epithelium. The kidneys' ability to eliminate metabolic waste products is therefore reduced. The small amount of urine produced has a high Na^+ concentration but an osmolality that's close to the concentration of the body fluids because the kidneys are not able to reabsorb Na^+ and because the urine-concentrating ability of the kidneys is severely damaged.

System Interactions — The Effect of Acute Renal Failure on Other Systems

System	Interactions
Integumentary	Pallor results from anemia, and bruising results from reduced clotting proteins in the blood because they are lost in the urine. A waxy yellow cast develops to the skin of light-skinned people, an ashen gray cast in black-skinned people, or a yellowish brown cast in brown-skinned people due to accumulation of urinary pigments. When the urea concentration in the blood is very high, white crystals of urea, called uremic frost, may appear on areas of the skin where heavy perspiration occurs.
Skeletal	Changes in the skeletal system are not marked unless kidney damage results in chronic kidney failure. Bone resorption may result during a prolonged diuretic phase because of excessive loss of Ca^{2+} in the urine. Also, vitamin D levels may be reduced during both the oliguric and diuretic phases.
Muscular	Neuromuscular irritability results from the toxic effect of metabolic wastes on the central nervous system and ionic imbalances such as hyperkalemia. Involuntary jerking and twitching may occur as neuromuscular irritability develops. Tremors of the hands are an indication of the toxic effects of metabolic wastes on the cerebrum.
Nervous	Elevated blood K^+ levels and the toxic effects of metabolic wastes result in depolarization of neurons. Slowing of nerve conduction, burning sensations, pain, numbness, or tingling results. Also, decreased mental acuity, reduced ability to concentrate, apathy, and lethargy result. Periods of lethargy may alternate with restlessness and insomnia. In severe cases, the patient may become confused and comatose.
Endocrine	Major predictable hormone deficiencies include vitamin D deficiency. In addition, a decrease occurs in the secretion of reproductive hormones due to the effects of metabolic wastes and ionic imbalances on the hypothalamus.
Cardiovascular	Water and Na^+ retention may result in edema in peripheral tissues and in the lung. Also, hypertension and congestive heart failure may result. Hyperkalemia results in dysrhythmias and may cause cardiac arrest. Anemia due to decreased erythropoietin production by the damaged kidney and decreased half-life of red blood cells may result. Anemia is more likely because of the blood lost as a result of the crushing injury. Nosebleeds and bruising occur due to a reduced concentration of clottting factors because they are lost in the urine.
Lymphatic	No major direct effects occur to the lymphatic system with the exception that increased lymph flow happens as a result of edema.
Respiratory	Early during acute renal failure the depth of breathing increases, and it becomes labored as acidosis develops because the kidney is not able to secrete H^+. Pulmonary edema often develops because of water and Na^+ retention. The likelihood of pulmonary infection increases secondary to pulmonary edema.
Digestive	Anorexia, nausea, and vomiting result from altered gastrointestinal functions due to the effects of ionic imbalances on the nervous sytem. An odor of ammonia may occur on the breath and a metallic taste in the mouth. These effects are the result of the accumulation of metabolic waste products in the gastrointestinal tract, and the action of the normal gastrointestinal microorganisms on the waste products, which convert urea to ammonia. The ammonia and other metabolic waste products predispose the mouth to inflammation and infection.

During the diuretic phase, large quantities of urine were produced because the nephrons were partially healed and could produce urine, but the ability of the nephrons to concentrate urine was not yet normal. Large volumes of urine that contained significant amounts of Na^+ and K^+ were therefore produced. The kidneys were able to produce urine that was more concentrated than the body fluids, but the concentrating ability of the kidneys was still below normal. As time passed, the concentrating ability of the kidneys improved and eventually became normal once again.

P R E D I C T

Nine days after the accident, Mr. H began to appear pale, and he became dizzy and lethargic. His hematocrit was elevated and his heart was arrhythmic. He was very weak. Explain these manifestations.

S U M M A R Y

The excretory system consists of the kidneys, the ureters, the urinary bladder, and the urethra.

Functions of the Urinary System   (p. 966)

The urinary system eliminates wastes; regulates blood volume, ion concentration, and pH; and it is also involved with red blood cell and vitamin D production.

Kidney Anatomy and Histology   (p. 966)

The kidney is surrounded by a renal capsule and perirenal fat and is held in place by the renal fascia.

Location and External Anatomy of the Kidneys

1. The kidneys lie behind the peritoneum on the posterior abdominal wall on either side of the vertebral column.
2. The renal capsule surrounds each kidney, and the perirenal fat and the renal fascia surround each kidney and anchor it to the abdominal wall.
3. The hilum, on the medial side of each kidney, where blood vessels and nerves enter and exit the kidney, opens into the renal sinus containing fat and connective tissue.

Internal Anatomy and Histology of the Kidneys

1. The two layers of the kidney are the cortex and the medulla.
   - The renal columns extend toward the medulla between the renal pyramids.
   - The renal pyramids of the medulla project to the minor calyces.
2. The minor calyces open into the major calyces, which open into the renal pelvis. The renal pelvis leads to the ureter.
3. The functional unit of the kidney is the nephron. The parts of a nephron are the renal corpuscle, the proximal tubule, the loop of Henle, and the distal tubule.
   - The renal corpuscle is Bowman's capsule and the glomerulus. Materials leave the blood in the glomerulus and enter Bowman's capsule through the filtration membrane.
   - The nephron empties through the distal tubule into a collecting duct.
4. The juxtaglomerular apparatus consists of the macula densa (part of the distal tubule) and the juxtaglomerular cells of the afferent arteriole.

Arteries and Veins of the Kidneys

1. Arteries branch as follows: renal artery to segmental artery to interlobar artery to arcuate artery to interlobular artery to afferent arteriole.
2. Afferent arterioles supply the glomeruli.
3. Efferent arteries from the glomeruli supply the peritubular capillaries and vasa recta.
4. Veins form from the peritubular capillaries as follows: interlobular vein to arcuate vein to interlobar vein to renal vein.

Anatomy and Histology of the Ureters and Urinary Bladder   (p. 973)

1. Structure
   - The walls of the ureter and urinary bladder consist of the epithelium, the lamina propria, a muscular coat, and a fibrous adventitia.
   - The transitional epithelium permits changes in size.
2. Function
   - The ureters transport urine from the kidney to the urinary bladder.
   - The urinary bladder stores urine.

Urine Production   (p. 974)

Urine is produced by the processes of filtration, reabsorption, and secretion.

Filtration

1. The renal filtrate is plasma minus blood cells and blood proteins. Most (99%) of the filtrate is reabsorbed.
2. The filtration membrane is fenestrated endothelium, basement membrane, and the slitlike pores formed by podocytes.
3. Filtration pressure is responsible for filtrate formation.
   - Filtration pressure is glomerular capillary pressure minus capsule pressure minus blood colloid osmotic pressure.
   - Filtration pressure changes are primarily caused by changes in glomerular capillary pressure.

Tubular Reabsorption

1. Filtrate is reabsorbed by passive transport, including simple diffusion, facilitated diffusion, active transport, and cotransport from the nephron into the peritubular capillaries.
2. Specialization of tubule segments
   - The thin segment of the loop of Henle is specialized for passive transport.
   - The rest of the nephron and collecting tubules perform active transport, cotransport, and passive transport.
3. Substances transported
   - Active transport moves mainly Na^+ across the wall of the nephron. Other ions and molecules are moved primarily by cotransport.
   - Passive transport moves water, urea, and lipid-soluble, nonpolar compounds.

Tubular Secretion

1. Substances enter the proximal or distal tubules and the collecting ducts.
2. H^+, K^+, and some substances not produced in the body are secreted by countertransport mechanisms.

Urine Concentration Mechanism

1. Countercurrent systems (e.g., vasa recta and loop of Henle) and the distribution of urea are responsible for the concentration gradient in the medulla. The concentration gradient is necessary for the production of concentrated urine.
2. Production of urine
   - In the proximal tubule Na^+ and other substances are removed by active transport. Water follows passively, filtrate volume is reduced 65%, and the filtrate concentration is 300 mOsm/L.
   - In the descending limb of the loop of Henle water exits passively, and solute enters. The filtrate volume is reduced 15%, and the osmolality of the filtrate concentration is 1200 mOsm/kg.
   - In the ascending limb of the loop of Henle, Na^+, Cl^-, and K^+ are transported out of the filtrate, but water remains because this segment of the nephron is impermeable to water. The osmolality of the filtrate concentration is 100 mOsm/kg.
   - In the distal tubules and collecting ducts, water movement out of them is regulated by ADH. If ADH is absent, water is not reabsorbed, and a dilute urine is produced. If ADH is present, water moves out, and a concentrated urine is produced.

Regulation of Urine Concentration and Volume   (p. 990)
Hormonal Mechanisms

1. ADH is secreted by the posterior pituitary and increases water permeability in the distal tubules and the collecting ducts.
   - ADH decreases urine volume, increases blood volume, and thus increases blood pressure.
   - ADH release is stimulated by increased blood osmolality or a decrease in blood pressure.
2. Aldosterone is produced in the adrenal cortex and affects Na^+ and Cl^- transport in the nephron and the collecting ducts.

- A decrease in aldosterone results in less Na^+ reabsorption and an increase in urine concentration and volume. An increase in aldosterone results in greater Na^+ reabsorption and a decrease in urine concentration and volume.
- Aldosterone production is stimulated by angiotensin II, increased blood K^+ concentration, and decreased blood Na^+ concentration.

3. Renin, produced by the kidneys, causes the production of angiotensin II.
   - Angiotensin II acts as a vasoconstrictor and stimulates aldosterone secretion, causing a decrease in urine production and an increase in blood volume.
   - Decreased blood pressure or decreased Na^+ concentration stimulates renin production.

4. Atrial natriuretic hormone, produced by the heart when blood pressure increases, inhibits ADH production and reduces the ability of the kidney to concentrate urine.

Autoregulation

Autoregulation dampens systemic blood pressure changes by altering afferent arteriole diameter.

Effect of Sympathetic Stimulation on Kidney Function

Sympathetic stimulation decreases afferent arteriole diameter.

Clearance and Tubular Maximum   (p. 993)

1. Plasma clearance is the volume of plasma that is cleared of a specific substance each minute.

2. The tubular load is the total amount of substance that enters the nephron each minute.
3. Tubular maximum is the fastest rate at which a substance is reabsorbed from the nephron.

Urine Movement   (p. 994)
Urine Flow Through the Nephron and the Ureters

1. Hydrostatic pressure forces urine through the nephron.
2. Peristalsis moves urine through the ureters.

Micturition Reflex

1. Stretch of the urinary bladder stimulates a reflex that causes the bladder to contract and inhibits the urinary sphincters.
2. Higher brain centers can stimulate or inhibit the micturition reflex.

Effects of Aging on the Kidneys   (p. 996)

1. There is a gradual decrease in the size of the kidney.
2. The decrease in kidney size is associated with a decrease in renal blood flow.
3. The number of functional nephrons decreases.
4. Renin secretion and vitamin D synthesis decrease.
5. The ability of the nephron to secrete and absorb declines.

R E V I E W   A N D   C O M P R E H E N S I O N

---

1. Which of these is *not* a general function of the kidneys?
   a. regulation of blood volume
   b. regulation of solute concentration in the blood
   c. regulation of the pH of the extracellular fluid
   d. regulation of vitamin A synthesis
   e. regulation of red blood cell synthesis

2. The cortex of the kidney contains the
   a. hilus.
   b. glomeruli.
   c. perirenal fat.
   d. renal pyramids.
   e. renal pelvis.

3. A layer of fibrous connective tissue that surrounds each kidney is the
   a. hilum.
   b. renal pelvis.
   c. renal sinus.
   d. renal capsule.
   e. perirenal fat.

4. Given these structures:
   1. major calyx
   2. minor calyx
   3. renal papilla
   4. renal pelvis

   Choose the arrangement that lists the structures in order as urine leaves the collecting duct and travels to the ureter.
   a. 1,4,2,3
   b. 2,3,1,4
   c. 3,2,1,4
   d. 4,1,3,2
   e. 4,3,2,1

5. Which of these structures contains blood?
   a. glomerulus
   b. vasa recta
   c. distal tubule
   d. Bowman's capsule
   e. both a and b

6. Given these vessels:
   1. arcuate artery
   2. interlobar artery
   3. segmental artery

   A red blood cell has just passed through the renal artery. Choose the path the red blood cell must take to reach the interlobular artery.
   a. 1,2,3
   b. 2,1,3
   c. 2,3,1
   d. 3,1,2
   e. 3,2,1

7. The juxtaglomerular cells of the _____ and the macula densa cells of the _____ form the juxtaglomerular apparatus.
   a. afferent arteriole, proximal tubule
   b. afferent arteriole, distal tubule
   c. efferent arteriole, proximal tubule
   d. efferent arteriole, distal tubule

8. Given these blood vessels:
   1. afferent arteriole
   2. efferent arteriole
   3. glomerulus
   4. peritubular capillaries

Choose the correct order as blood passes from an interlobular artery to an interlobular vein.
a. 1,2,3,4
b. 1,3,2,4
c. 2,1,4,3
d. 3,2,4,1
e. 4,3,1,2

9. The urinary bladder
   a. is made up of skeletal muscle.
   b. is lined by simple columnar epithelium.
   c. is connected to the outside of the body by the ureter.
   d. is located in the pelvic cavity.
   e. has two urethras and one ureter attached to it.

10. Kidney function is accomplished by which of these means?
    a. filtration
    b. secretion
    c. reabsorption
    d. both a and b
    e. all of the above

11. The amount of plasma that enters Bowman's capsule per minute is the
    a. glomerular filtration rate.
    b. renal plasma flow.
    c. renal fraction.
    d. renal blood flow.

12. Given these structures:
    1. basement membrane
    2. fenestra
    3. filtration slit

    Choose the arrangement that lists the structures in the order a molecule of glucose encounters them as the glucose passes through the filtration membrane to enter Bowman's capsule.
    a. 1,2,3
    b. 2,1,3
    c. 2,3,1
    d. 3,1,2
    e. 3,2,1

13. If the glomerular capillary pressure is 40 mm Hg, the capsule pressure is 10 mm Hg, and the blood colloid osmotic pressure within the glomerulus is 30 mm Hg, the filtration pressure is
    a. −20 mm Hg.
    b. 0 mm Hg.
    c. 20 mm Hg.
    d. 60 mm Hg.
    e. 80 mm Hg.

14. Which of these conditions reduces filtration pressure in the glomerulus?
    a. elevated blood pressure
    b. constriction of the afferent arterioles
    c. decreased plasma protein in the glomerulus
    d. dilation of the afferent arterioles
    e. decreased capsule pressure

15. Glucose usually is completely reabsorbed from the filtrate by the time the filtrate has reached
    a. the end of the proximal tubule.
    b. the tip of the loop of Henle.
    c. the end of the distal tubule.
    d. the end of the collecting duct.
    e. Bowman's capsule.

16. The greatest volume of water is reabsorbed from the nephron by the
    a. proximal tubule.
    b. loop of Henle.
    c. distal tubule.
    d. collecting duct.

17. Water leaves the nephron by
    a. active transport.
    b. filtration into the capillary network.
    c. osmosis.
    d. facilitated diffusion.
    e. cotransport.

18. Potassium ions enter the _____ by _____.
    a. proximal tubule, diffusion
    b. proximal tubule, active transport
    c. distal tubule, diffusion
    d. distal tubule, counter transport

19. Reabsorption of most solute molecules from the proximal tubule is linked to the primary active transport of Na^+
    a. across the apical membrane and out of the cell.
    b. across the apical membrane and into the cell.
    c. across the basal membrane and out of the cell.
    d. across the basal membrane and into the cell.

20. Which of these ions is used to cotransport amino acids, glucose, and other solutes through the apical membrane of nephron epithelial cells?
    a. K^+
    b. Na^+
    c. Cl^-
    d. Ca^{2+}
    e. Mg^{2+}

21. Which of the following contribute to the formation of a hyperosmotic environment in the medulla of the kidney?
    a. effects of ADH on water permeability of the ascending limb of the loop of Henle
    b. impermeability of the ascending limb of the loop of Henle to water
    c. cotransport of Na^+, K^+, and Cl^- out of the ascending limb of the loop of Henle
    d. both a and c
    e. both b and c

22. At which of these sites is the osmolality lowest (lowest concentration)?
    a. glomerular capillary
    b. proximal tubule
    c. bottom of the loop of Henle
    d. initial section of the distal tubule
    e. collecting duct

23. Increased aldosterone causes
    a. increased reabsorption of Na^+.
    b. decreased blood volume.
    c. decreased reabsorption of Cl^-.
    d. increased permeability of the distal tubule to water.
    e. increased volume of urine.

24. Juxtaglomerular cells are involved in the secretion of
    a. ADH.
    b. angiotensin.
    c. aldosterone.
    d. renin.

25. ADH governs the
    a. Na^+ pump of the proximal tubules.
    b. water permeability of the loop of Henle.
    c. Na^+ pump of the vasa recta.
    d. water permeability of the distal tubules and collecting ducts.
    e. Na^+ reabsorption in the proximal tubule.

26. A decrease in blood osmolality results in which of these?
    a. increased ADH secretion
    b. increased permeability of the collecting ducts to water
    c. decreased urine osmolality
    d. decreased urine output
    e. all of the above

27. Angiotensin II
    a. causes vasoconstriction.
    b. stimulates aldosterone secretion.
    c. stimulates ADH secretion.
    d. increases the sensation of thirst.
    e. all of the above.

28. If blood pressure increases by 50 mm Hg,
    a. the afferent arterioles constrict.
    b. glomerular capillary pressure increases by 50 mm Hg.
    c. glomerular filtration rate increases dramatically.
    d. efferent arterioles constrict.
    e. all of the above.

29. The amount of a substance that passes through the filtration membrane into the nephrons per minute is the
    a. renal plasma flow.
    b. tubular load.
    c. plasma clearance.
    d. tubular maximum.

30. Given these events:
    1. loss of voluntary control of urination
    2. loss of the sensation or desire to urinate
    3. loss of reflex emptying of the urinary bladder

    Which of these events occurs following transection of the spinal cord at level L5?
    a. 1
    b. 2
    c. 1,2
    d. 2,3
    e. 1,2,3

Answers in Appendix F

C R I T I C A L   T H I N K I N G

1. To relax after an anatomy and physiology examination, Mucho Gusto goes to a local bistro and drinks 2 quarts of low-sodium beer. What effect does this beer have on urine concentration and volume? Explain the mechanisms involved.

2. A male eats a full bag of salty potato chips. What effect does this have on urine concentration and volume? Explain the mechanisms involved.

3. During severe exertion in a hot environment, a person can lose up to 4 L of hypoosmotic (less concentrated than plasma) sweat per hour. What effect does this loss have on urine concentration and volume? Explain the mechanisms involved.

4. Harry Macho is doing yard work one hot summer day and refuses to drink anything until he is finished. He then drinks glass after glass of plain water. Assume that he drinks enough water to replace all the water he lost as sweat. How does this much water affect urine concentration and volume? Explain the mechanisms involved.

5. A patient has the following symptoms: slight increase in extracellular fluid volume, large decrease in plasma sodium concentration, very concentrated urine, and cardiac fibrillation. An imbalance of what hormone is responsible for these symptoms? Are the symptoms caused by oversecretion or undersecretion of the hormone?

6. Propose as many ways as you can to decrease the GFR.

7. Design a kidney that can produce hypoosmotic urine, which is less concentrated than plasma, or hyperosmotic urine, which is more concentrated than plasma, by the active transport of water instead of Na^+. Assume that the anatomic structure of the kidney is the same as that in humans. Feel free to change anything else you choose.

8. If only a very small amount of urea were present in the interstitial fluid of the kidney instead of its normal concentration, how would it affect the kidney's ability to concentrate urine?

9. It is well known that some patients with hypertension are kept on a low-salt (low-sodium) diet. Propose an explanation for this therapy.

10. In studies of mammalian kidney function, it is known that animals with kidneys having relatively thicker medullas have a greater ability to conserve water. Explain why this is so.

11. Prednisone is a steroid that reduces inflammation and is prescribed to treat conditions such as autoimmune diseases. It has some aldosterone-like effect on the kidney. Predict the effect of this treatment on blood pressure, blood volume, and edema. Explain.

12. Addison's disease is a consequence of degeneration of the adrenal cortex. Predict the effect of this condition on blood pressure, blood volume, and urine volume. Explain.

Answers in Appendix G

A N S W E R S   T O   P R E D I C T   Q U E S T I O N S

1. The urethra of females is much shorter than the urethra of males. In addition, the opening of the urethra in females is closer to the anus, which is a potential source of bacteria. The female urinary bladder is therefore more accessible to bacteria from the exterior. This accessibility is a major reason that urinary bladder infections are more common in females than in males.

2. If the cardiac output is 5600 mL of blood per minute, and the hematocrit is 45, renal plasma flow is 650 mL of plasma per minute (see table 26.2). If the filtration fraction increased from 19% to 22%, the GFR would be 143 mL of filtrate per minute (650 mL of plasma × 0.22). If 99.2% of the filtrate is reabsorbed, 0.8% becomes urine. Thus the urine produced is 1.14 mL of urine per minute (143 mL of filtrate × 0.008). Compared to the rate of urine production when the filtrate fraction was 19% (i.e., 1 mL per minute), the 3% increase in filtration fraction has caused a 14% increase in urine production. Converting

1.14 mL of urine per minute to liters of urine produced per day yields 1.64 L/day (1.14 mL/min × 1 L/1000 mL × 1440 min/day).

3. Even though hemoglobin is a smaller molecule than albumin, it doesn't normally enter the filtrate because hemoglobin is contained within red blood cells and these cells cannot pass through the filtration membrane. If red blood cells rupture, however, a process called hemolysis, the hemoglobin is released into the plasma, and large amounts of hemoglobin enter the filtrate. Conditions that cause red blood cells to rupture in the circulatory system result in large amounts of hemoglobin entering the urine.

4. Karl was suffering from plasma loss due to the extensive burns, which results in intense vasoconstriction, including vasoconstriction of arteries and arterioles that supply the kidneys. Constriction of the renal arteries and the renal afferent arterioles decreases the blood

pressure in the glomeruli. As a consequence, the filtration pressure decreases. After an I.V. is administered and blood volume increases, the blood pressure returns to normal and renal arteries and arterioles once again dilate, thus blood pressure in the glomeruli increases to normal levels.

5. Inhibition of ADH secretion is one of the numerous effects alcohol has on the body. The lack of ADH secretion causes the distal tubules and the collecting ducts to be relatively impermeable to water. The water cannot therefore move by osmosis from the distal nephrons and collecting ducts and remains in the nephrons to become urine. In addition, because other fluids are normally consumed with the alcohol, the increased water intake also results in an increase in dilute urine production.

6. Without the normal active transport of Na^+, the concentration of Na^+ and ions cotransported with them remains elevated in the nephron. Movement of water by osmosis out of the nephron into the interstitial spaces is decreased, resulting in an increased volume of urine.

7. Anything that reduces the formation of filtrate reduces the GFR. If the epithelium of the nephrons sloughs off and forms casts in the nephrons, normal flow of filtrate through them is blocked. Consequently, the blocked flow of filtrate in the nephron causes the pressure in Bowman's capsule to increase enough so that the pressure inside of Bowman's capsule is close to the pressure in the glomerulus.

Unless the pressure in the glomerulus is higher than the pressure in Bowman's capsule, no filtrate forms, and the GFR is very low. If very little filtrate forms, the volume of urine produced is reduced. Also, filtrate that enters a nephron that is blocked, with the epithelial lining of the nephron disrupted, can flow directly into the peritubular capillaries.

8. Low urea clearance indicates that the amount of blood cleared of urea, a metabolic waste product, per minute is lower than normal. It's consistent with the reduction of the number of functional nephrons that occurs in advanced cases of renal failure. In addition, a low urea clearance is an indication that the GFR is reduced and that the blood levels of urea are increasing.

9. After 7 days, Mr. H's kidneys began to produce a large volume of urine with larger-than-normal Na^+ and K^+ concentrations. The observations are consistent with Mr. H becoming dehydrated by day 10. Dehydration results in reduced blood volume. The pale skin was the result of vasoconstriction, which was triggered by the reduced blood pressure. Dizziness resulted from reduced blood flow to the brain when Mr. H tried to stand and walk. He was lethargic in part because of reduced blood volume but also because of low blood levels of K^+ and Na^+. The arrythmia of his heart was due to low blood levels of K^+ and increased sympathetic stimulation, which also was triggered by low blood pressure.

Visit the Online Learning Center at www.mhhe.com/seeley7 for chapter quizzes, interactive learning exercises, and other study tools.

Color enhanced scanning electron micrograph of a cross section of renal corpuscles in the renal cortex.

Water, Electrolytes, and Acid–Base Balance

Life depends on many complex and highly regulated chemical reactions, all of which occur in water. Many of these reactions are catalyzed by enzymes that can function only within a narrow range of conditions. Changes in the total amount of water, the pH, or the concentration of specific electrolytes can alter chemical reactions on which life depends. Homeostasis requires the maintenance of these parameters within a narrow range of values, and failure to maintain homeostasis can result in illness or death.

The kidneys, along with the respiratory, integumentary, and gastrointestinal systems, regulate water volume, electrolyte concentrations, and pH. The nervous and endocrine systems coordinate the activities of these systems.

This chapter covers *body fluids* (p. 1006), the *regulation of body fluid concentration and volume* (p. 1007), the *regulation of intracellular fluid composition* (p. 1012), the *regulation of specific electrolytes in the extracellular fluid* (p. 1013), and the *regulation of acid–base balance* (p. 1022).

Body Fluids

The proportion of body weight composed of water decreases from birth to old age, with the greatest decrease occurring during the first 10 years of life (table 27.1). Because the water content of adipose tissue is relatively low, the fraction of the body's weight composed of water decreases as the amount of adipose tissue increases. The relatively lower water content of adult females when compared to adult males reflects the greater development of subcutaneous adipose tissue characteristic of women.

For people of all ages and body compositions, the two major fluid compartments are the intracellular and extracellular fluid compartments. The **intracellular** (in-tră-sel′ū-lăr) **fluid compartment** includes all the fluid in the several trillion cells of the body. The intracellular fluid from all cells has a similar composition, and it accounts for approximately 40% of total body weight.

The **extracellular** (eks-tră-sel′ū-lăr) **fluid compartment** includes all of the fluid outside the cells and constitutes nearly 20% of total body weight. The extracellular fluid compartment can be divided into several subcompartments. The major ones are interstitial fluid and plasma; others include lymph, cerebrospinal fluid, and synovial fluid. **Interstitial** (in-ter-stish′ăl) **fluid** occupies the extracellular spaces outside the blood vessels, and **plasma** (plaz′mă) occupies the extracellular space within blood vessels. All the other subcompartments of the extracellular compartment constitute relatively small volumes.

Although the fluid contained in each subcompartment differs somewhat in composition from that in the others, continuous and extensive exchange occurs between the subcompartments. Water diffuses from one subcompartment to another, and small molecules and ions are either transported or diffuse freely between them. Large molecules like proteins are much more restricted in their movement because of the permeability characteristics of the membranes that separate the fluid subcompartments (table 27.2).

The osmotic pressure of most fluid subcompartments is approximately equal. For example, the osmotic pressure of the hyaluronic acid in synovial joints is roughly equal to the osmotic pressure of the proteins in intraocular fluid.

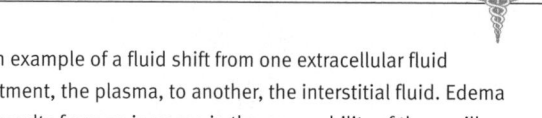

Edema

Edema is an example of a fluid shift from one extracellular fluid subcompartment, the plasma, to another, the interstitial fluid. Edema commonly results from an increase in the permeability of the capillary walls due to inflammation, which allows proteins to diffuse from the plasma into the interstitial fluid. Water moves in the same direction by osmosis. Edema can also result from a change in the hydrostatic pressure across capillary walls. An increased hydrostatic pressure in capillaries due to blockage of veins or heart failure forces fluid from plasma into the interstitial spaces (see chapter 21).

Table 27.1	Approximate Volumes of Body Fluid Compartments*				
Age of Person	Total Body Water	Intracellular Fluid	Extracellular Fluid		
			Plasma	Interstitial	Total
Infants	75	45	4	26	30
Adult males	60	40	5	15	20
Adult females	50	35	5	10	15

*Expressed as percentage of body weight.

Table 27.2	Approximate Concentration of Major Solutes in Body Fluid Subcompartments*		
Solute	Plasma	Interstitial Fluid	Intracellular Fluid†
Cations			
Sodium (Na^+)	153.2	145.1	12.0
Potassium (K^+)	4.3	4.1	150.0
Calcium (Ca^{2+})	3.8	3.4	4.0
Magnesium (Mg^{2+})	1.4	1.3	34.0
TOTAL	162.7	153.9	200.0
Anions			
Chloride (Cl^-)	111.5	118.0	4.0
Bicarbonate (HCO_3^-)	25.7	27.0	12.0
Phosphate (HPO_4^{2-} plus HPO_4^-)	2.2	2.3	40.0
Protein	17.0	0.0	54.0
Other	6.3	6.6	90.0
TOTAL	162.7	153.9	200.0

*Expressed as milliequivalents per liter (mEq/L).
† Data are from skeletal muscle.

Regulation of Body Fluid Concentration and Volume

Regulation of Water Content

The body's water content is regulated so that the total volume of water in the body remains constant. Thus the volume of water taken into the body is equal to the volume lost each day. Changes in the water volume in body fluids alter the osmolality of body fluids, blood pressure, and interstitial fluid pressure. The total volume of water entering the body each day is 1500–3000 mL. Most of that volume (90%) comes from ingested fluids, some comes from food, and a smaller amount, approximately 10%, is derived from the water produced during cellular metabolism (table 27.3 and see figure 24.33).

The movement of water across the wall of the gastrointestinal tract depends on osmosis, and the volume of water entering the body depends, to a large degree, on the volume of water consumed. If a large volume of dilute liquid is consumed, the rate at which water enters the body fluids increases. If a small volume of concentrated liquid is consumed, the rate decreases.

Although fluid consumption is heavily influenced by habit and by social settings, water ingestion does depend, at least in part, on regulatory mechanisms. The sensation of thirst results primarily from an increase in the osmolality of the extracellular fluids and from a reduction in plasma volume. Cells of the supraoptic nucleus within the hypothalamus can detect an increased extracellular fluid osmolality and initiate activity in neural circuits that results in a conscious sensation of thirst (figure 27.1).

Baroreceptors can also influence the sensation of thirst. When they detect a substantial decrease in blood pressure, action potentials are conducted to the brain along sensory neurons to influence the sensation of thirst. Low blood pressure associated with hemorrhagic shock, for example, is correlated with an intense sensation of thirst.

Renin is released from the juxtaglomerular apparatuses of the kidneys in response to reduced blood flow to the kidney. Renin increases the formation of angiotensin II in the circulatory system (see chapters 21 and 26). Angiotensin II opposes a decrease in

Increased osmolality or large decrease in blood pressure

Increased thirst

Hypothalamus

Figure 27.1   Effect of Blood Osmolality and Pressure on Water Intake

Increased blood osmolality affects hypothalamic neurons, and large decreases in blood pressure affect baroreceptors in the aortic arch, carotid sinuses, and atrium. As a result of these stimuli, an increase in thirst results, which increases water intake. Increased water intake reduces blood osmolality.

blood pressure by acting on the brain to stimulate the sensation of thirst, by acting on the adrenal cortex to increase aldosterone secretion, and by acting on blood vessel smooth muscle cells to increase vasoconstriction.

When people who are dehydrated are allowed to drink water, they eventually consume a quantity sufficient to reduce the osmolality of the extracellular fluid to its normal value. They don't normally consume the water all at once. Instead, they drink intermittently until the proper osmolality of the extracellular fluid is established. The thirst sensation is temporarily reduced after the ingestion of small amounts of liquid. At least two factors are responsible for this temporary interruption of the thirst sensation. First, when the oral mucosa becomes wet after it has been dry, sensory neurons conduct action potentials to the thirst center of the hypothalamus and temporarily decrease the sensation of thirst. Second, consumed fluid increases the gastrointestinal tract volume, and stretch of the gastrointestinal wall initiates sensory action potentials in stretch receptors. The sensory neurons conduct action potentials to the thirst center of the hypothalamus, where they temporarily suppress the sensation of thirst. Because absorption of water from the gastrointestinal tract requires time, mechanisms that temporarily suppress the sensation of thirst prevent the consumption of extreme volumes of fluid that would exceed the amount required to reduce blood osmolality. A longer-term suppression of the thirst sensation results when the extracellular fluid osmolality and blood pressure are within their normal ranges.

Learned behavior can be very important in avoiding periodic dehydration through the consumption of fluids either with or without food, even though blood osmolality is not reduced. The volume of fluid ingested by a healthy person usually exceeds the minimum volume required to maintain homeostasis, and the kidneys eliminate the excess water in urine.

Water loss from the body occurs through three major routes (see table 27.3). The greatest amount of water, approximately 61%, is lost through the urine. Approximately 35% of water loss occurs through evaporation from respiratory passages, evaporation of water that diffuses through the skin, and by perspiration. Approximately 4% is lost in the feces.

Table 27.3	Summary of Water Intake and Loss	
Sources of Water	**Routes by Which Water Is Lost**
Ingestion (90%)	Urine (61%)
Cellular metabolism (10%)	Evaporation (35%)
	Perspiration
	Insensible
	Sensible
	Respiratory passages
	Feces (4%)

The volume of water lost through the respiratory system depends on the temperature and humidity of the air, the body temperature, and the volume of air expired. Water lost through the diffusion and evaporation of water from the skin is called **insensible perspiration** (see chapter 25) and it plays a role in heat loss. For each degree that the body temperature rises above normal, an increased volume of 100–150 mL of water is lost each day in the form of insensible perspiration.

Sweat, or **sensible perspiration,** is secreted by the sweat glands (see chapters 5 and 25), and, in contrast to insensible perspiration, it contains solutes. Sweat resembles extracellular fluid in its composition, with sodium chloride as the major component, but it also contains some potassium, ammonia, and urea (table 27.4). The volume of fluid lost as sweat is negligible for a person at rest in a cool environment. The volume of sweat produced is determined primarily by neural mechanisms that regulate body temperature, although some sweat is produced as a result of sympathetic stimulation in response to stress. Under conditions of exercise, elevated environmental temperature, or fever, the volume increases substantially and it plays an important role in heat loss. Sweat losses of 8–10 L/day have been measured in outdoor workers in the summertime.

Adequate fluid replacement during conditions of extensive sweating is important. Sweat is hyposmotic to plasma. The loss of a large volume of hyposmotic sweat causes a decrease in body fluid volume and an increase in body fluid concentration. Fluid volume is lost primarily from the extracellular space, which leads to an increase in extracellular fluid osmolality, a reduction in plasma volume, and an increase in hematocrit. During conditions of severe dehydration, the change can be great enough to cause blood viscosity to increase substantially. The increased workload created for the heart by that increase in viscosity can result in heart failure.

Relatively little water is lost by way of the digestive tract. Although the total volume of fluid secreted into the gastrointestinal tract is large, nearly all the fluid is reabsorbed under normal conditions (see chapter 24). Severe vomiting or diarrhea, however, are exceptions and can result in a large volume of fluid loss.

The kidneys are the primary organs that regulate the composition and volume of body fluids by controlling the volume and concentration of water excreted in the form of urine (see chapter 26). Urine production varies greatly and can range from a small volume of concentrated urine to a large volume of dilute urine in response to mechanisms that regulate the body's water content. The mechanisms that respond to changes in extracellular fluid osmolality and extracellular fluid volume keep the total body water levels within a narrow range of values.

4. *List three factors that increase thirst. Name two things that inhibit the sense of thirst.*
5. *Describe three routes for the loss of water from the body. Contrast insensible and sensible perspiration.*
6. *What are the primary organs that regulate the composition and volume of body fluids?*

Regulation of Extracellular Fluid Osmolality

Adding water to a solution, or removing water from it, changes the osmolality, or concentration, of the solution. Consider a solution contained in a pan on a stove. Adding water to the solution decreases its osmolality, or dilutes it. Boiling the solution in the pan removes water by evaporation. The removal of water from the solution increases its osmolality and makes it more concentrated. Adding or removing water from the body fluids maintains their osmolality between 285 and 300 mOsm/kg.

An increase in the osmolality of the extracellular fluid triggers thirst and antidiuretic hormone (ADH) secretion. Water that is consumed is absorbed from the intestine and enters the extracellular fluid. ADH acts on the distal tubules and collecting ducts of the kidneys to increase reabsorption of water from the filtrate. The increase in the amount of water entering the extracellular fluid causes a decrease in osmolality (figures 27.2 and 27.3). The ADH and thirst mechanisms are sensitive to even small changes in extracellular fluid osmolality and the response is fast (from minutes to a few hours). Larger increases in extracellular fluid osmolality such as during dehydration results in an even greater increase in thirst and in ADH secretion.

A decrease in extracellular fluid osmolality inhibits thirst and ADH secretion. Less water is consumed, and less water is reabsorbed from the filtrate in the kidneys. Consequently, more water is lost as a large volume of dilute urine. The result is an

Increased osmolality or large decrease in BP

Increased ADH release

Kidney

Increased water reabsorption results in decreased osmolality and increased BP

Figure 27.2   **Effect of Blood Osmolality and Blood Pressure on Water Reabsorption in the Kidneys**

Increased blood osmolality affects hypothalamic neurons, and decreased blood pressure affects baroreceptors in the aortic arch, carotid sinuses, and atrium. As a result of these stimuli, an increased rate of antidiuretic hormone (ADH) secretion from the posterior pituitary results, which increases water reabsorption by the kidney.

Table 27.4	Composition of Sweat
Solute	**Concentration (mM)**
Sodium	9.8–77.2
Potassium	3.9–9.2
Chloride	5.5–65.1
Ammonia	1.7–5.6
Urea	6.5–12.1

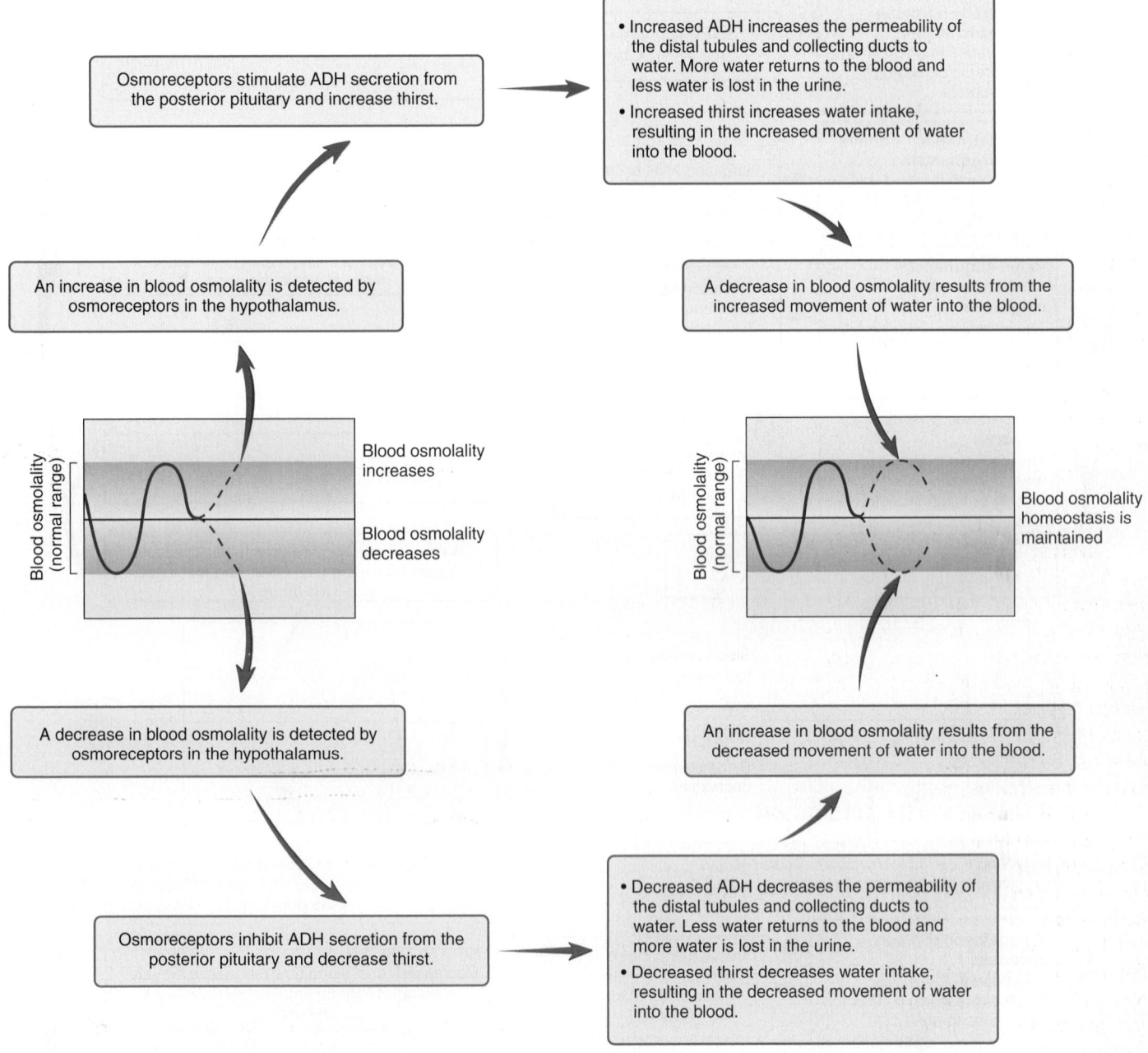

Homeostasis Figure 27.3 Hormonal Regulation of Blood Osmolality

increase in the osmolality of the extracellular fluid (see figure 27.3). For example, consumption of a large volume of water in a beverage results in reduced extracellular fluid osmolality. This results in reduced ADH secretion, less reabsorption of water from the filtrate in the kidneys, and the production of a large volume of dilute urine. This response occurs quickly enough so that the osmolality of the extracellular fluid is maintained within a normal range of values.

7. *What two mechanisms are triggered by an increase in the osmolality of the extracellular fluid?*

8. *Describe how the osmolality of the extracellular fluid is affected when these mechanisms are activated.*

Regulation of Extracellular Fluid Volume

The volume of extracellular fluid can increase, or decrease, even if the osmolality of the extracellular fluid is maintained within a narrow range of values. Sensory receptors that detect changes in blood pressure are important in the regulation of extracellular fluid volume. Carotid sinus and aortic arch baroreceptors monitor blood pressure in large arteries, receptors in the juxtaglomerular apparatuses monitor pressure changes in the afferent arterioles of the kidneys, and receptors in the walls of the atria of the heart and large veins are sensitive to the smaller changes in blood pressure that occur within them. These receptors activate neural mechanisms and three major hormonal mechanisms that regulate extracellular fluid volume (figure 27.4).

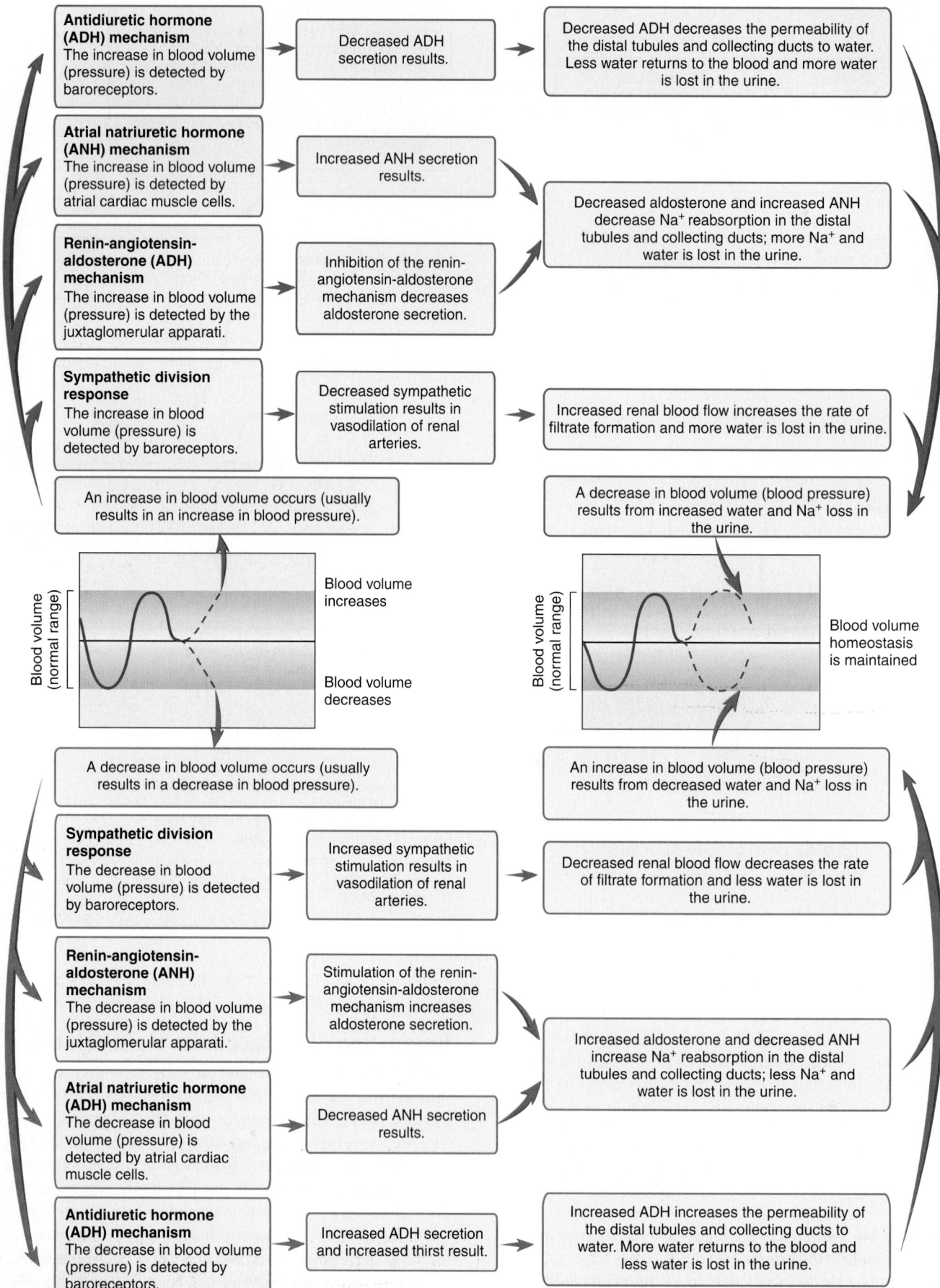

Homeostasis Figure 27.4   Regulation of Blood Volume

1. *Neural mechanisms.* Neural mechanisms change the frequency of action potentials carried by sympathetic neurons to the afferent arterioles of the kidney in response to changes in blood pressure. When baroreceptors detect an increase in arterial and venous blood pressure, the frequency of action potentials carried by sympathetic neurons to the afferent arterioles decreases. Consequently, the afferent arterioles dilate. This increases glomerular capillary pressure, resulting in an increase in the glomerular filtration rate (GFR), an increase in filtrate volume, and an increase in urine volume.

   When baroreceptors detect a decrease in arterial and venous blood pressure, there's an increase in the frequency of action potentials carried by sympathetic neurons to the afferent arterioles. Consequently, the afferent arterioles constrict. This decreases GFR, filtrate volume, and urine volume.

2. *Renin-angiotensin-aldosterone mechanism.* The renin-angiotensin-aldosterone mechanism responds to small changes in blood volume. Increased blood pressure results from increased blood volume. Juxtaglomerular cells detect increases in blood pressure in the afferent arterioles and decrease the rate of renin secretion. The decrease in renin secretion results in a decreased conversion of angiotensinogen to angiotensin II. Reduced angiotensin II causes a decrease in the rate of aldosterone secretion from the adrenal cortex. Decreased aldosterone levels reduce the rate of Na^+ reabsorption, primarily from the distal tubules and collecting ducts. Consequently, more Na^+ remains in the filtrate and fewer Na^+ are reabsorbed. The effect is to increase the osmolality of the filtrate, which reduces the ability of the kidney to reabsorb water. The water remains, with the excess Na^+, in the filtrate. Thus, the volume of urine produced by the kidney increases and the extracellular fluid volume decreases (see figure 27.4).

   A decrease in blood volume causes a decrease in blood pressure in the afferent arterioles, which results in an increased rate of renin secretion by the juxtaglomerular cells. The increase in renin secretion results in an increased conversion of angiotensinogen to angiotensin II. The increased angiotensin II causes an increase in the rate of aldosterone secretion from the adrenal cortex. Increased aldosterone increases the rate of Na^+ reabsorption, primarily from the distal tubules and collecting ducts. Consequently, less Na^+ remains in the filtrate and more Na^+ is reabsorbed. The effect is to decrease the osmolality of the filtrate. This increases the ability of the kidney to reabsorb water and to increase extracellular fluid volume. Thus the volume of urine produced by the kidney decreases and the extracellular fluid volume and blood pressure increase (see figures 27.4 and 27.5).

3. *Atrial natriuretic hormone (ANH) mechanism.* The ANH mechanism responds to increases in extracellular fluid volume. An increase in pressure in the atria of the heart, which usually results from an increase in blood volume, stimulates the secretion of ANH, which decreases Na^+ reabsorption in the distal tubules and collecting ducts. This increases the rate of Na^+ and water loss in the urine. Thus, increased ANH secretion decreases extracellular fluid volume (see figures 27.4 and 27.6).

Figure 27.5 Effect of Blood Pressure on Na^+ and Water Reabsorption in the Kidneys

Low blood pressure (BP) stimulates renin secretion from the kidney. Renin stimulates the production of angiotensin I, which is converted to angiotensin II, which in turn stimulates aldosterone secretion from the adrenal cortex. Aldosterone increases Na^+ and water reabsorption in the kidney.

Figure 27.6 Effect of Blood Pressure in the Right Atrium on Na^+ and Water Excretion

Increased blood pressure in the right atrium of the heart causes increased secretion of atrial natriuretic hormone (ANH), which increases Na^+ excretion and water loss in the form of urine.

ANH doesn't appear to respond strongly to decreases in blood volume. However, a decrease in pressure in the atria of the heart inhibits the secretion of ANH. The decreased ANH decreases the inhibition of Na^+ reabsorption in the distal tubules and collecting ducts. Therefore, the rate of Na^+ reabsorption increases and water reabsorption also increases. Thus, decreased ANH secretion is consistent with a decreased urine volume and an increase in extracellular fluid volume (see figure 27.4).

4. *Antidiuretic hormone (ADH) mechanism.* The ADH mechanism plays an important role in regulating extracellular fluid volume in response to large changes in blood pressure (of 5%–10%). An increase in blood pressure results in a decrease in ADH secretion. As a result, the reabsorption of water from the lumen of the distal tubules and collecting ducts decreases, resulting in a larger volume of dilute urine. This response helps decrease extracellular fluid volume and blood pressure (see figures 27.2 and 27.4).

A decrease in blood pressure results in an increase in ADH secretion. Consequently, the reabsorption of water from the lumen of the distal tubules and collecting ducts increases, resulting in a smaller volume of concentrated urine. This response helps increase extracellular fluid volume and blood pressure (see figures 27.2 and 27.4).

The mechanisms that maintain extracellular fluid concentration and volume function together. However, when mechanisms that maintain fluid volume don't function normally, it's possible to have an increased extracellular fluid volume even though the extracellular concentration of fluids is maintained within a normal range of values. For example, increased aldosterone secretion from an enlarged adrenal cortex increases Na^+ reabsorption by the kidney and the total volume of extracellular fluid increases. Mechanisms, such as the regulation of ADH secretion, keep the concentration of the body fluids constant. The blood pressure can be elevated and edema can result, but the osmolality of the extracellular fluid is maintained between 285 and 300 mOsm/kg. Similarly, in people suffering from heart failure, the resulting reduced blood pressure activates mechanisms that increase blood pressure to its normal range of values. Those mechanisms include the release of renin from the kidneys. Consequently, the renin-angiotensin-aldosterone mechanism is activated as in the case of elevated aldosterone secretion, the result is an increase in the extracellular fluid volume and edema in the periphery, including edema in the lungs (congestive heart failure).

9. *What sensory receptors are responsible for activating neural and hormonal mechanisms that regulate extracellular fluid volume?*

10. *What is the effect on sympathetic stimulation, afferent arterioles, GFR, filtrate volume, urine volume, and extracellular fluid volume when baroreceptors detect an increase in arterial and venous blood pressure?*

11. *Describe the response of the renin-angiotensin-aldosterone mechanism to a decrease in blood pressure. How is extracellular fluid volume and urine volume affected?*

12. *What effect does ANH have on extracellular fluid volume?*

13. *How does an increase in blood pressure affect the secretion of ADH? How does ADH affect extracellular fluid volume?*

Regulation of Intracellular Fluid Composition

The composition of intracellular fluid is substantially different from that of extracellular fluid. Plasma membranes, which separate the two compartments, are selectively permeable—they are relatively impermeable to proteins and other large molecules and have limited permeability to smaller molecules and ions. Consequently, most large molecules synthesized within cells, such as proteins, remain within the intracellular fluid. Some substances, such as electrolytes, are actively transported across the plasma membrane, and their concentrations in the intracellular fluid are determined by the transport processes and by the electric charge difference across the plasma membrane (figure 27.7).

1. Large organic molecules such as proteins, which cannot cross the plasma membrane, are synthesized inside cells and influence the concentration of solutes inside the cells.

2. The transport of ions across the plasma membrane, such as Na^+, K^+, and Ca^{2+}, influences the concentration of ions inside and outside the cell.

3. An electric charge difference across the plasma membrane influences the distribution of ions inside and outside the cell.

4. The distribution of water inside and outside the cell is determined by osmosis.

Process Figure 27.7   Regulation of Intracellular and Extracellular Distribution of Water and Solutes

Water movement across the plasma membrane is controlled by osmosis. Thus, the net movement of water is affected by changes in the concentration of solutes in the extracellular and intracellular fluids. For example, as dehydration develops, the concentration of solutes in the extracellular fluid increases, resulting in the movement of water by osmosis from the intracellular fluid into the extracellular fluid. If dehydration is severe, enough water moves from the intracellular fluid to cause the cells to function abnormally. If water intake increases after a period of dehydration, the concentration of solutes in the extracellular fluids decreases, which results in the movement of water back into the cells.

> **14.** *What factors determine the composition of intracellular fluid? What characteristic of plasma membranes is responsible for maintaining the differences between intracellular and extracellular fluid?*

Regulation of Specific Electrolytes in the Extracellular Fluid

Electrolytes (ē-lek′trō-lītz) are molecules or ions with an electric charge. Major extracellular ions are Na^+, Cl^-, K^+, Ca^{2+}, Mg^{2+}, and phosphate ions. Ingestion of electrolytes in food and water adds them to the body, whereas organs like the kidneys and, to a lesser degree, the liver, skin, and lungs remove them from the body. The concentrations of electrolytes in the extracellular fluid are regulated so that they don't change unless the individual is growing, gaining weight, or losing weight. Regulation of each electrolyte involves the coordinated participation of several organ systems.

Regulation of Sodium Ions

Sodium ions (Na^+) are the dominant extracellular cations. Because of their abundance in the extracellular fluids, they exert substantial osmotic pressure. Approximately 90%–95% of the osmotic pressure of the extracellular fluid is caused by Na^+ and the negative ions associated with them.

Diet and Na^+ Homeostasis

In the United States, the quantity of Na^+ ingested each day is 20–30 times the amount needed. Less than 0.5 g is required to maintain homeostasis, but the average individual ingests approximately 10–15 g of sodium chloride daily. Regulation of the Na^+ content in the body, therefore, depends primarily on the excretion of excess quantities of Na^+. The mechanisms for conserving Na^+ in the body are effective, however, when the Na^+ intake is very low.

The kidneys are the major route by which Na^+ is excreted. Sodium ions readily pass from the glomerulus into the lumen of Bowman's capsule and is present in the same concentration in the filtrate as in the plasma. The concentration of Na^+ excreted in the urine is determined by the amount of Na^+ and water reabsorbed from filtrate in the nephron. If Na^+ reabsorption from the nephron decreases, large quantities are lost in the urine. If Na^+ reabsorption from the nephron increases, only small quantities are lost in the urine.

The rate of Na^+ transport in the proximal tubule is relatively constant, but the Na^+ transport mechanisms of the distal tubule and the collecting duct are under hormonal control. **Aldosterone** increases Na^+ reabsorption from the distal tubule and collecting duct. As little as 0.1 g of sodium is excreted in the urine each day in the presence of high blood levels of aldosterone. When aldosterone is absent, Na^+ reabsorption in the nephron is greatly reduced, and as much as 30–40 g of sodium can be lost in the urine daily.

Sodium ions are also excreted from the body in **sweat.** Normally only a small quantity of Na^+ is lost each day in the form of sweat, but the amount increases during conditions of heavy exercise in a warm environment. The mechanisms that regulate sweating control the quantity of Na^+ excreted through the skin. As the body temperature increases, thermoreceptor neurons within the hypothalamus respond by increasing the rate of sweat production. As the rate of sweat production increases, the quantity of Na^+ lost in the urine decreases to keep the extracellular concentration of Na^+ constant. The loss of Na^+ in sweat is rarely physiologically significant.

The primary mechanisms that regulate Na^+ concentrations in the extracellular fluid don't directly monitor Na^+ levels but are sensitive to changes in extracellular fluid osmolality or changes in blood pressure (see figure 27.4 and table 27.5). The quantity of Na^+ in the body has a dramatic effect on extracellular osmotic pressure and extracellular fluid volume. For example, if the quantity of Na^+ increases, the osmolality of the extracellular fluid increases. An increase in the osmolality of the extracellular fluids stimulates ADH secretion, which increases the reabsorption of water by the kidney and causes a small volume of concentrated urine to be produced. It also increases the sensation of thirst. Consequently there is an increase in extracellular fluid volume. A decrease in the quantity of Na^+ in the body decreases the osmolality of the extracellular fluid. This inhibits ADH secretion, which stimulates a large volume of dilute urine to be produced, and it decreases the sensation of thirst. Thus, extracellular osmolality increases. By regulating extracellular fluid osmolality and extracellular fluid volume, the concentration of Na^+ in the body fluids is maintained within a narrow range of values.

Elevated blood pressure under resting conditions increases Na^+ and water excretion (see figure 27.4 and table 27.5). If blood pressure is low, the total Na^+ content of the body is usually also low. In response to low blood pressure, mechanisms such as the renin-angiotensin-aldosterone mechanism are activated that increase Na^+ concentration and water volume in the extracellular fluid (see figure 27.4 and table 27.5).

P R E D I C T   1

In response to hemorrhagic shock, the kidneys produce a small volume of very concentrated urine. Explain how the rate of filtrate formation changes and how Na^+ transport changes in the distal part of the nephron in response to hemorrhagic shock.

Table 27.5 Homeostasis: Mechanisms Regulating Blood Sodium

Mechanism	Stimulus	Response to Stimulus	Effect of Response	Result
Response to Changes in Blood Osmolality				
Antidiuretic hormone (ADH); the most important regulator of blood osmolality	Increased blood osmolality (e.g., increased Na^+ concentration)	Increased ADH secretion from the posterior pituitary; mediated through cells in the hypothalamus	Increased water reabsorption in the kidney; production of a small volume of concentrated urine	Decreases blood osmolality as reabsorbed water dilutes the blood
	Decreased blood osmolality (e.g., decreased Na^+ concentration)	Decreased ADH secretion from the posterior pituitary; mediated through cells in the hypothalamus	Decreased water reabsorption in the kidney; production of a large volume of dilute urine	Increased blood osmolality as water is lost from the blood into the urine
Response to Changes in Blood Pressure				
Renin-angiotensin-aldosterone	Decreased blood pressure in the kidney's afferent arterioles	Increased renin release from the juxtaglomerular apparatuses; renin initiates the conversion of angiotensinogen to angiotensin; angiotensin I is converted to angiotensin II, which increases aldosterone secretion from the adrenal cortex	Increased Na^+ reabsorption in the kidney (because of increased aldosterone); increased water reabsorption as water follows the Na^+; decreased urine volume	Increased blood pressure as blood volume increases because of increased water reabsorption; blood osmolality is maintained because both Na^+ and water are reabsorbed*
	Increased blood pressure in the kidney's afferent arterioles	Decreased renin release from the juxtaglomerular apparatuses, resulting in reduced formation of angiotensin I; reduced angiotensin I leads to reduced angiotensin II, which causes a decrease in aldosterone secretion from the adrenal cortex	Decreased Na^+ reabsorption in the kidney (because of decreased aldosterone); decreased water reabsorption as fewer Na^+ are reabsorbed; increased urine volume	Decreased blood pressure as blood volume decreases because water is lost in the urine; blood osmolality is maintained because both Na^+ and water are lost in the urine*
Atrial natriuretic hormone (ANH)	Decreased blood pressure in the atria of the heart	Decreased ANH released from the atria	Increased Na^+ reabsorption in the kidney; increased water reabsorption as water follows the Na^+; decreased urinary volume	Increased blood pressure as blood volume increases because of increased water reabsorption; blood osmolality is maintained because both Na^+ and water are reabsorbed*
	Increased blood pressure in the atria of the heart	Increased ANH released from the atria	Decreased Na^+ reabsorption in the kidney; decreased water reabsorption as water is lost with Na^+ in the urine; increased urinary volume	Decreased blood osmolality as blood volume decreases because water is lost in the urine; blood osmolality is maintained because both Na^+ and water are lost in the urine*
ADH—activated by significant decreases in blood pressure; normally regulates blood osmolality (see above)	Decreased arterial blood pressure	Increased ADH secretion from the posterior pituitary; mediated through baroreceptors	Increased water reabsorption in the kidney; production of a small volume of concentrated urine	Increased blood pressure resulting from increased blood volume; decreased blood osmolality
	Increased arterial blood pressure	Decreased ADH secretion from the posterior pituitary; mediated through baroreceptors	Decreased water reabsorption in the kidney; production of a large volume of dilute urine	Decreased blood pressure resulting from decreased blood volume; increased blood osmolality

Abbreviations: ADH = antidiuretic hormone.
*Assumes normal levels of ADH.

Cells in the walls of the atria synthesize ANH, which is secreted in response to an elevation in blood pressure within the right atrium. ANH acts on the kidneys to increase urine production by inhibiting the reabsorption of Na^+ (see figure 27.6 and table 27.5). It also inhibits the effect of ADH on the distal tubules and collecting ducts, and inhibits ADH secretion (see chapter 26, p. 992).

Deviations from the normal concentration range for Na^+ in body fluids result in significant symptoms. Some major causes of **hypernatremia** (hī′per-nă-trē′mē-ă), or an elevated plasma Na^+ concentration, and **hyponatremia** (hī′pō-nă-trē′mē-ă), or a reduced plasma Na^+ concentration, and the major symptoms of each, are listed in table 27.6.

 15. *Name the substance responsible for most of the osmotic pressure of the extracellular fluid.*

16. *How does aldosterone affect the concentration of Na^+ in the urine?*

17. *What role does sweating play in Na^+ balance?*

18. *How does increased blood pressure result in a loss of water and salt? What happens when blood pressure decreases?*

19. *What effect does ANH have on Na^+ and water loss in urine?*

P R E D I C T

If a person consumes an excess amount of Na^+ and water, predict the effect on (a) blood pressure, (b) urine volume, and (c) urine concentration.

Regulation of Chloride Ions

The predominant anions in the extracellular fluid are **chloride ions (Cl^-)**. The electrical attraction of anions and cations makes it difficult to separate these charged particles. Consequently, the regulatory mechanisms that influence the concentration of cations

in the extracellular fluid also influence the concentration of anions. The mechanisms that regulate Na^+, K^+, and Ca^{2+} levels in the body are important in influencing Cl^- levels. Because Na^+ predominate, the mechanisms that regulate extracellular Na^+ concentration are those that are most important in regulating the extracellular Cl^- concentration.

20. *What mechanisms regulate Cl^- concentrations?*

Regulation of Potassium Ions

The extracellular concentration of **potassium ions (K^+)** must be maintained within a narrow range. The concentration gradient of K^+ across the plasma membrane has a major influence on the resting membrane potential, and cells that are electrically excitable are highly sensitive to slight changes in that concentration gradient. An increase in extracellular K^+ concentration leads to depolarization, and a decrease in extracellular K^+ concentration leads to hyperpolarization of the resting membrane potential. **Hyperkalemia** (hī′per-kă-lē′mē-ă) is an abnormally high level of K^+ in the extracellular fluid, and **hypokalemia** (hī′pō-ka-lē′mē-ă) is an abnormally low level of K^+ in the extracellular fluid. Major causes of hyperkalemia and hypokalemia and their symptoms are listed in table 27.7.

K^+ pass freely through the filtration membrane of the renal corpuscle. They are actively reabsorbed in the proximal tubules and actively secreted in the distal tubules and collecting ducts. K^+ secretion into the distal tubule and collecting duct is highly regulated and primarily responsible for controlling the extracellular concentration of K^+.

Aldosterone plays a major role in regulating the concentration of K^+ in the extracellular fluid by increasing the rate of K^+ secretion in the distal tubule and collecting duct. Aldosterone secretion from the adrenal cortex is stimulated by elevated K^+

Table 27.6	Consequences of Abnormal Plasma Levels of Sodium Ions	
Major Causes	**Symptoms**
Hypernatremia	
High dietary sodium rarely causes symptoms	Thirst, fever, dry mucous membranes, restlessness; most serious symptoms—convulsions and pulmonary edema
Administration of hypertonic saline solutions (e.g., sodium bicarbonate treatment for acidosis)	
Oversecretion of aldosterone (i.e., aldosteronism)	When occurring with an increased water volume—weight gain, edema, elevated blood pressure, and bounding pulse
Water loss (e.g., because of fever, respiratory infections, diabetes insipidus, diabetes mellitus, diarrhea)	
Hyponatremia	
Inadequate dietary intake of sodium rarely causes symptoms—can occur in those on low-sodium diets and those taking diuretics	Lethargy, confusion, apprehension, seizures, and coma
Extrarenal losses—vomiting, prolonged diarrhea, gastrointestinal suctioning, burns	When accompanied by reduced blood volume—reduced blood pressure, tachycardia, and decreased urine output
Dilution—intake of large water volume after excessive sweating	When accompanied by increased blood volume—weight gain, edema, and distension of veins
Hyperglycemia, which attracts water into the circulatory system but reduces the concentration of Na^+	

Table 27.7	Consequences of Abnormal Concentrations of Potassium Ions
Major Causes	**Symptoms**
Hyperkalemia	
Movement of K^+ from intracellular to extracellular fluid resulting from cell trauma (e.g., burns or crushing injuries) and alterations in plasma membrane permeability (e.g., acidosis, insulin deficiency, and cell hypoxia)	Mild hyperkalemia (caused mainly by partial depolarization of plasma membranes):
	Increased neuromuscular irritability, restlessness
	Intestinal cramping and diarrhea
Decreased renal excretion of K^+ (e.g., from decreased secretion of aldosterone in persons with Addison's disease)	Electrocardiogram—alterations, including rapid repolarization with narrower and taller T waves and shortened QT intervals
	Severe hyperkalemia (caused mainly by partial depolarization of plasma membranes severe enough to hamper action potential conduction):
	Muscle weakness, loss of muscle tone, and paralysis
	Electrocardiogram—alterations, including changes caused by reduced rate of action potential conduction (e.g., depressed ST segment, prolonged PR interval, wide QRS complex, arrhythmias, and cardiac arrest)
Hypokalemia	
Alkalosis (K^+ shift into cell in exchange for H^+)	Symptoms are mainly due to hyperpolarization of membranes
Insulin administration (promotes cellular uptake of K^+)	Decreased neuromuscular excitability—skeletal muscle weakness
Reduced K^+ intake (especially with anorexia nervosa and alcoholism)	Decreased tone in smooth muscle
Increased renal loss (excessive aldosterone secretion, improper use of diuretics, kidney diseases that result in reduced ability to reabsorb Na^+)	Cardiac muscle—delayed ventricular repolarization, bradycardia, and atrioventricular block

blood levels (figure 27.8; see chapter 26). Aldosterone secretion is stimulated in response to increased angiotensin II. The elevated aldosterone concentrations in the circulatory system increase K^+ secretion into the nephron, thereby lowering blood levels of K^+.

Circulatory system shock can result from plasma loss, dehydration, and tissue damage, such as occurs in burn patients. This shock causes the extracellular K^+ to be more concentrated than normal, which stimulates aldosterone secretion from the adrenal cortex. Aldosterone secretion also occurs in response to decreased blood pressure, which stimulates the renin-angiotensin-aldosterone mechanism. Homeostasis is reestablished as K^+ excretion increases. Also, increased Na^+ and water reabsorption stimulated by aldosterone results in an increase in extracellular fluid volume that dilutes the K^+ in the body fluids. Blood pressure increases toward normal as water reabsorption increases and when vasoconstriction is stimulated by angiotensin II.

21. What effect does an increase or decrease in extracellular K^+ concentration have on resting membrane potential?
22. Where are K^+ secreted in the nephron?
23. How is the secretion of K^+ regulated?

Regulation of Calcium Ions

The extracellular concentration of **calcium ions (Ca^{2+})**, like that of K^+, is regulated within a narrow range. The normal concentration of Ca^{2+} in plasma is 9.4mg/100mL. **Hypocalcemia** (hī′pō-kal-sē′mē-ă) is a below-normal level of Ca^{2+} in the extracellular fluid, and **hypercalcemia** (hī′per-kal-sē′mē-ă) is an

above-normal level of Ca^{2+} in the extracellular fluid. Major symptoms develop when the extracellular concentration of Ca^{2+} declines below 6 mg/100 mL or increases above 12 mg/100 mL. Decreases and increases in the extracellular concentration of Ca^{2+} markedly affect the electrical properties of excitable tissues. Hypocalcemia increases the permeability of plasma membranes to Na^+. As a result, nerve and muscle tissues undergo spontaneous action potential generation. Hypercalcemia decreases the permeability of the plasma membrane to Na^+, thus preventing normal depolarization of nerve and muscle cells. High extracellular Ca^{2+} levels cause the deposition of calcium carbonate salts in soft tissues, resulting in irritation and inflammation of those tissues. Table 27.8 lists the major causes and symptoms of hypocalcemia and hypercalcemia.

The kidneys, intestinal tract, and bones are important in maintaining extracellular Ca^{2+} levels (figure 27.9). Almost 99% of total body calcium is contained in bone. Part of the extracellular Ca^{2+} regulation involves the regulation of Ca^{2+} deposition into and resorption from bone (see chapter 6). Long-term regulation of Ca^{2+} levels, however, depends on maintaining a balance between Ca^{2+} absorption across the wall of the intestinal tract and Ca^{2+} excretion by the kidneys.

Parathyroid (par-ă-thī′royd) **hormone,** secreted by the parathyroid glands, increases extracellular Ca^{2+} levels and reduces extracellular phosphate levels (see figure 27.9). The rate of parathyroid hormone secretion is regulated by extracellular Ca^{2+} levels. Elevated Ca^{2+} levels inhibit and reduced levels stimulate its secretion. Parathyroid hormone causes increased osteoclast activity,

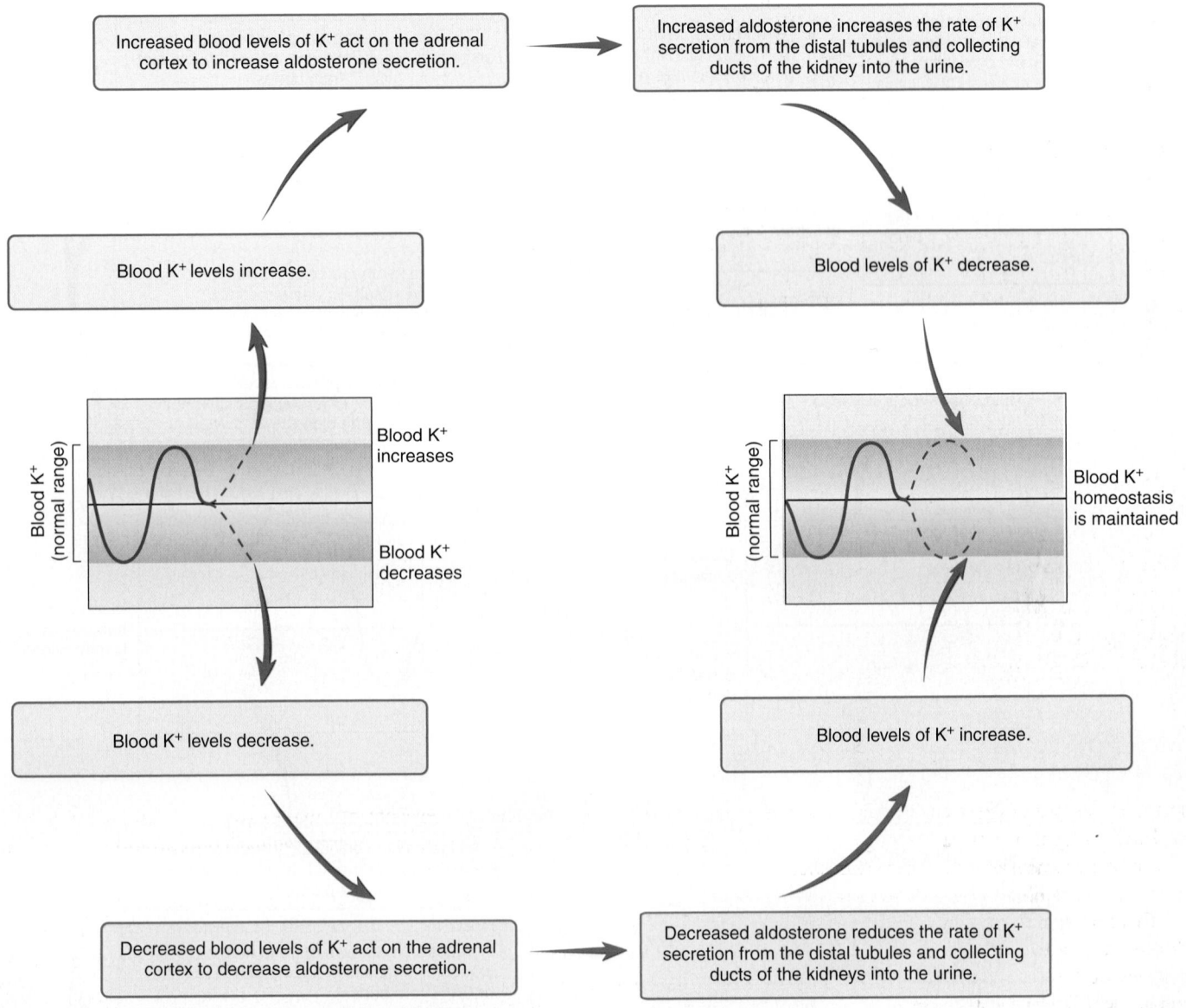

Homeostasis Figure 27.8   **Regulation of Potassium Ions in the Extracellular Fluid**

Aldosterone acts on the kidney to regulate the extracellular concentration of K^+.

Table 27.8	Consequences of Abnormal Concentrations of Calcium
Major Causes	**Symptoms**
Hypocalcemia	
Nutritional deficiencies	Symptoms are mainly due to increased permeability of plasma membranes to Na^+
Vitamin D deficiency	Increase in neuromuscular excitability—confusion, muscle spasms, hyperreflexia, and intestinal cramping
Decreased parathyroid hormone secretion	
Malabsorption of fats (reduced vitamin D absorption)	Severe neuromuscular excitability—convulsions, tetany, inadequate respiratory movements
Bone tumors that increase Ca^{2+} deposition	Electrocardiogram—prolonged QT interval (prolonged ventricular depolarization)
	Reduced absorption of phosphate from the intestine
Hypercalcemia	
Excessive parathyroid hormone secretion	Symptoms are mainly due to decreased permeability of plasma membranes to Na^+
Excess vitamin D	Loss of membrane excitability—fatigue, weakness, lethargy, anorexia, nausea, and constipation
	Electrocardiogram—shortened QT segment and depressed T waves
	Kidney stones

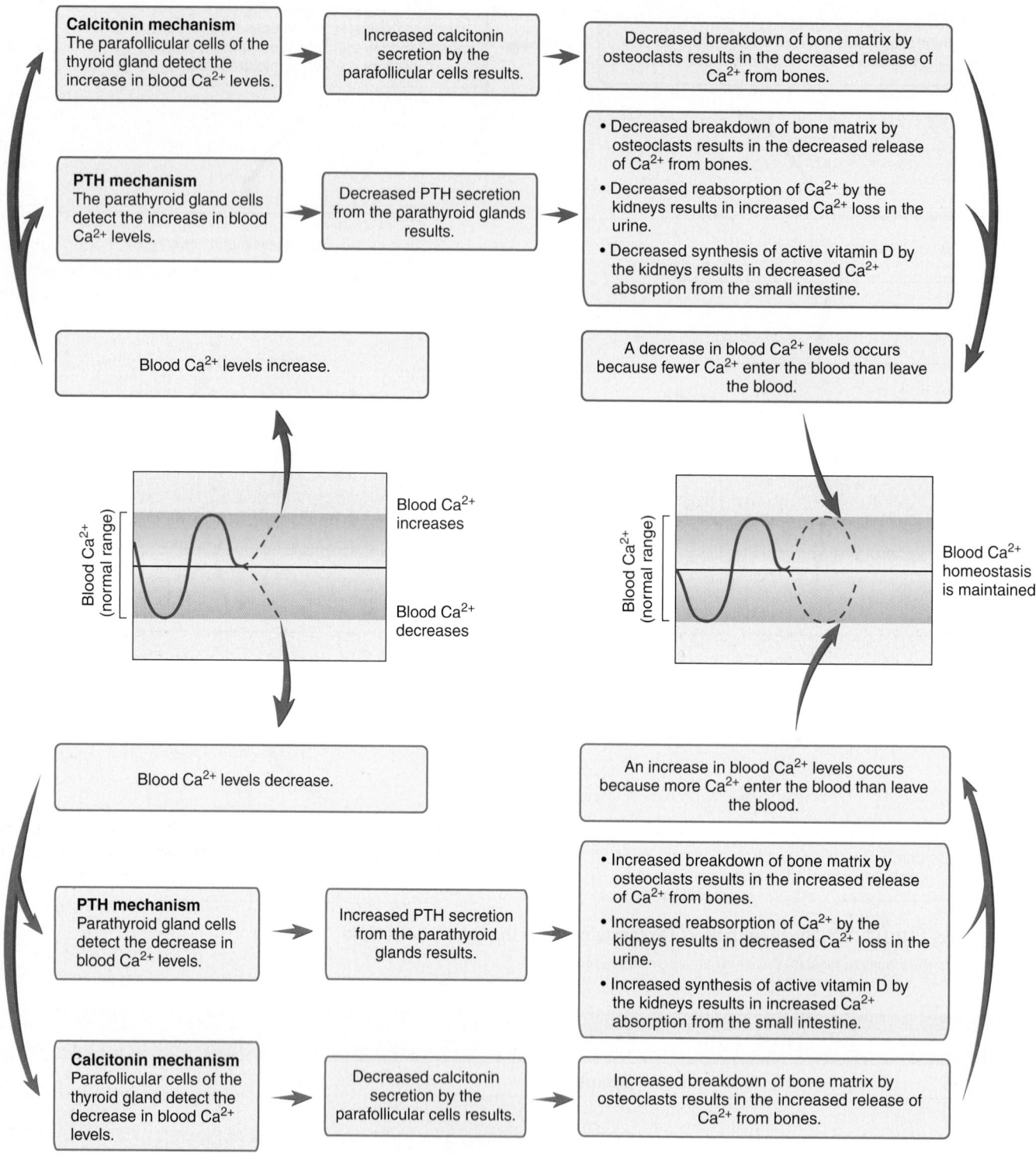

Homeostasis Figure 27.9   Regulation of Ca²⁺ in the Extracellular Fluid

Parathyroid hormone and calcitonin play major roles in regulating the extracellular concentration of Ca^{2+}.

which results in the degradation of bone and the release of Ca^{2+} and phosphate ions into body fluids. Parathyroid hormone increases the rate of Ca^{2+} reabsorption from nephrons in the kidneys and increases the concentration of phosphate ions in the urine. It also increases the rate at which vitamin D is converted to 1,25-dihydroxycholecalciferol, or **active vitamin D.** Active vitamin D

acts on the intestinal tract to increase Ca^{2+} absorption across the intestinal mucosa.

A lack of parathyroid hormone secretion results in a rapid decline in extracellular Ca^{2+} concentration. A reduction in the rate of absorption of Ca^{2+} from the intestinal tract, increased Ca^{2+} excretion by the kidneys, and reduced bone resorption cause this decline. A lack of parathyroid hormone secretion can result in death because of tetany of the respiratory muscles caused by hypocalcemia.

Vitamin D can be obtained from food or from vitamin D biosynthesis. Normally, vitamin D biosynthesis is adequate, but prolonged lack of exposure to sunlight reduces the biosynthesis because ultraviolet light is required for one step in the process (see chapter 5). The consumption of dietary vitamin D can involve the ingestion of active vitamin D or one of its precursors.

Without vitamin D, the transport of Ca^{2+} across the wall of the intestinal tract is negligible. This leads to inadequate Ca^{2+} absorption, even though large amounts of these ions may be present in the diet. Thus, Ca^{2+} absorption depends on both the consumption of an adequate amount of calcium in food and the presence of an adequate amount of vitamin D.

Calcitonin (kal-si-tō′nin), which is secreted by the parafollicular cells of the thyroid gland, reduces extracellular Ca^{2+} levels. It's most effective when Ca^{2+} levels are elevated, although greater-than-normal calcitonin levels in the blood are not consistently effective in causing blood levels of Ca^{2+} to decline below normal values. The major effect of calcitonin is on bone. It inhibits osteoclast activity and prolongs the activity of osteoblasts. Thus, it decreases bone demineralization and increases bone mineralization (see chapter 6).

Elevated Ca^{2+} levels stimulate calcitonin secretion, whereas reduced Ca^{2+} levels inhibit it. Increased calcitonin secretion reduces blood levels of Ca^{2+}, but large doses of calcitonin don't consistently reduce blood levels of Ca^{2+} below normal levels. Although calcitonin reduces the blood levels of Ca^{2+} when they are elevated, it's not as important as parathyroid hormone in the regulation of blood Ca^{2+} levels (see figure 27.9).

> 24. *Describe the effect of hypocalcemia and hypercalcemia on membrane potentials.*
> 25. *Explain the role of parathyroid hormone in regulating the extracellular Ca^{2+} concentration.*
> 26. *Describe the role of vitamin D in regulating the extracellular Ca^{2+} concentration.*
> 27. *Describe the role of calcitonin in the regulation of extracellular Ca^{2+} concentration.*

Regulation of Magnesium Ions

Most of the magnesium in the body is stored in bones or in the intracellular fluid. Less than 1% of the total are ions found in the extracellular fluid. Approximately one-half of those ions are bound to plasma proteins and one-half are free. The free magnesium ion (Mg^{2+}) concentration is 1.8–2.4 mEq/L. Magnesium ions are cofactors for intracellular enzymes, such as the Na^+–K^+ ATPase involved in actively transporting Na^+ out and K^+ into cells. Low and high levels of plasma magnesium produce symptoms (table 27.9) that are associated with the effect of magnesium on Na^+–K^+ active transport.

Free Mg^{2+} passes through the filtration membranes of the kidney into the filtrate. About 85%–90% of those ions are reabsorbed from the filtrate, and only about 10%–15% enter the urine. Of the Mg^{2+} reabsorbed, most is reabsorbed by the loop of Henle. The remainder are reabsorbed by the proximal tubule, distal tubule, and collecting duct.

The capacity of the kidney to reabsorb Mg^{2+} is limited. If the level of free Mg^{2+} increases in the extracellular fluid, the excess Mg^{2+} remains in the filtrate, and there's an increase in the rate of Mg^{2+} loss in the urine. If the level of free Mg^{2+} decreases in the extracellular fluid, nearly all of the Mg^{2+} are reabsorbed, and there's a decrease in the rate of Mg^{2+} loss in the urine. The control of Mg^{2+} reabsorption isn't clear, but a decreased extracellular concentration of Mg^{2+} causes an increase in the rate of reabsorption in the nephron (figure 27.10).

> 28. *In what part of the nephron are most Mg^{2+} reabsorbed?*
> 29. *What effect does a decreased extracellular concentration of Mg^{2+} have on reabsorption in the nephron?*

Regulation of Phosphate Ions

About 85% of phosphate is in the form of calcium phosphate salts found in bone (hydroxyapatite) and teeth. Most of the remaining phosphate is found inside cells. Many of the phosphate ions are covalently bound to other organic molecules. Phosphate ions are bound to lipids (to form phospholipids), proteins, and carbohydrates, and they are important components of DNA, RNA, and ATP. Phosphates also play important roles in the regulation of enzyme activity, and phosphate ions dissolved in the intracellular fluid act as buffers (see phosphate buffers under "Regulation of Acid–Base Balance," p. 1022). The extracellular concentration of phosphate ions is between 1.7 and 2.6

Table 27.9	Consequences of Abnormal Plasma Levels of Magnesium Ions
Major Causes	**Symptoms**
Hypomagnesemia (rare)	
Malnutrition	Symptoms result from increased neuromuscular excitability, and include irritability, increased reflexes, muscle weakness, tetany, and convulsions
Alcoholism	
Reduced absorption of magnesium in the intestine	
Renal tubular dysfunction	
Some diuretics	
Hypermagnesemia (rare)	
Renal failure	Symptoms result from depressed skeletal muscle contractions and nerve functions, and include nausea, vomiting, muscle weakness, hypotension, bradycardia, and reduced respiration
Magnesium-containing antacids	

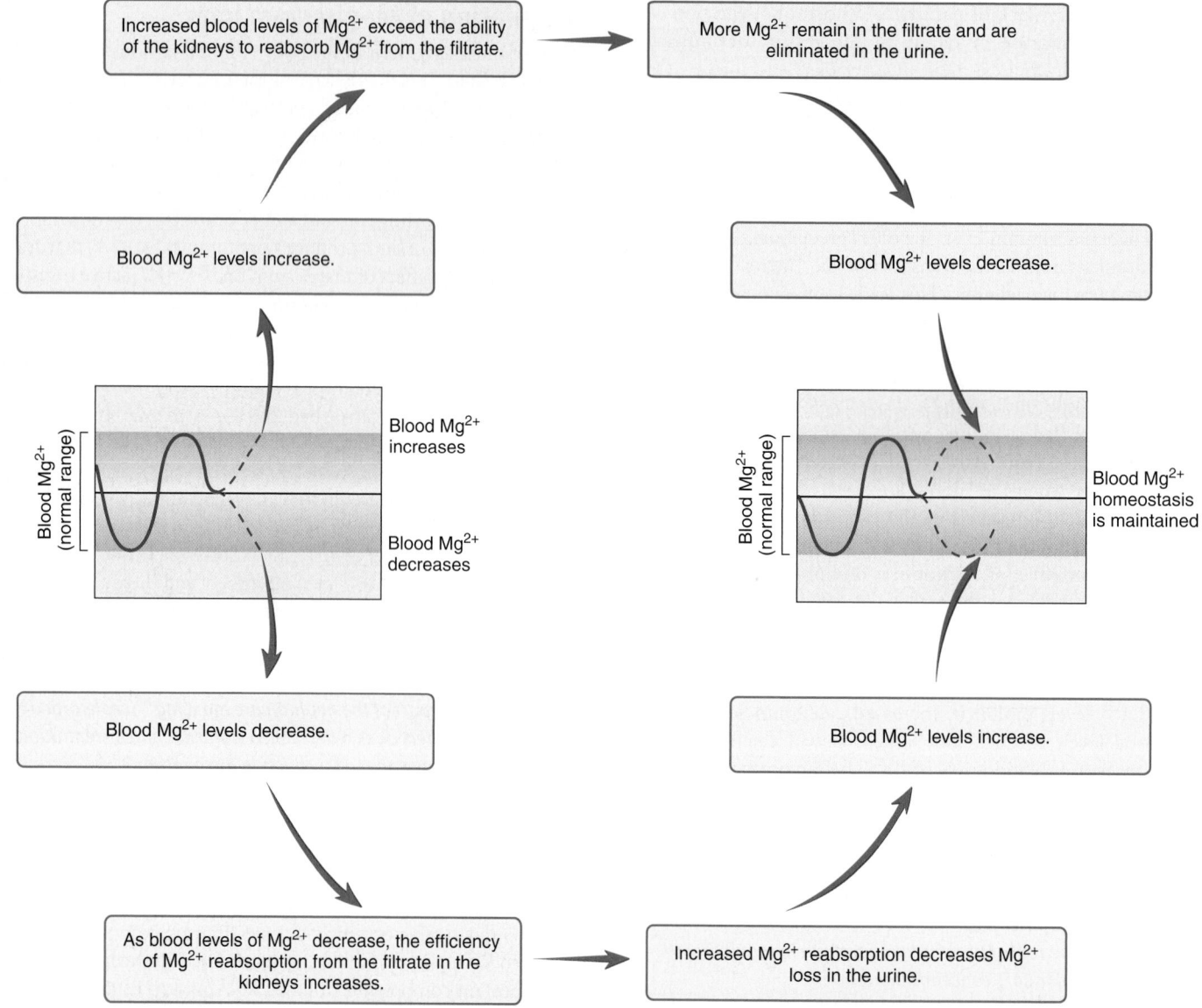

Homeostasis Figure 27.10   Regulation of Blood Mg²⁺ Concentration in the Extracellular Fluid

mEq/L. Phosphate ions are in the form of $H_2PO_4^-$, HPO_4^{2-}, and PO_4^{3-}. The most common phosphate ion is HPO_4^{2-}.

The capacity of the kidneys to reabsorb phosphate ions is limited. If the level of phosphate ions increases in the extracellular fluid, the excess phosphate remains in the filtrate, and there is an increase in the rate of phosphate loss in the urine. If the level of phosphate ions decreases in the extracellular fluid, nearly all of the phosphate ions are reabsorbed, and there is a decrease in the rate of phosphate ion loss in the urine (figure 27.11).

A diet low in phosphate can, over time, increase the rate of phosphate reabsorption. Consequently, most of the phosphate that enters the filtrate is reabsorbed to maintain the extracellular phosphate concentration. Parathyroid hormone (PTH) can play a significant role in regulating extracellular phosphate levels. PTH promotes bone resorption. Thus, large amounts of both Ca^{2+} and phosphate ions are released into the extracellular space. PTH decreases the reabsorption of phosphate ions from renal tubules so that a greater pro-

portion of the tubular phosphate is lost in the urine. Thus, whenever plasma PTH is increased, tubular phosphate reabsorption is decreased, and more phosphate enters the urine. If phosphate levels in the extracellular fluid increase above normal levels, Ca^{2+} and phosphate ions precipitate as calcium phosphate salts in soft tissues.

Elevated blood levels of phosphate may occur with acute or chronic renal failure as a result of a very reduced rate of filtrate formation by the kidney. The rate of phosphate excretion is consequently reduced. Also, the chronic use of laxatives containing phosphates may result in elevated blood levels of phosphate. Symptoms of elevated phosphate levels are related to reduced blood Ca^{2+} levels because phosphate ions and Ca^{2+} precipitate out of solution and are deposited in soft tissues of the body. Prolonged elevation of blood levels of phosphate can result in calcium phosphate deposits in the lungs, kidneys, joints, and other soft tissues. The consequences of increased or reduced plasma levels of phosphates are presented in table 27.10.

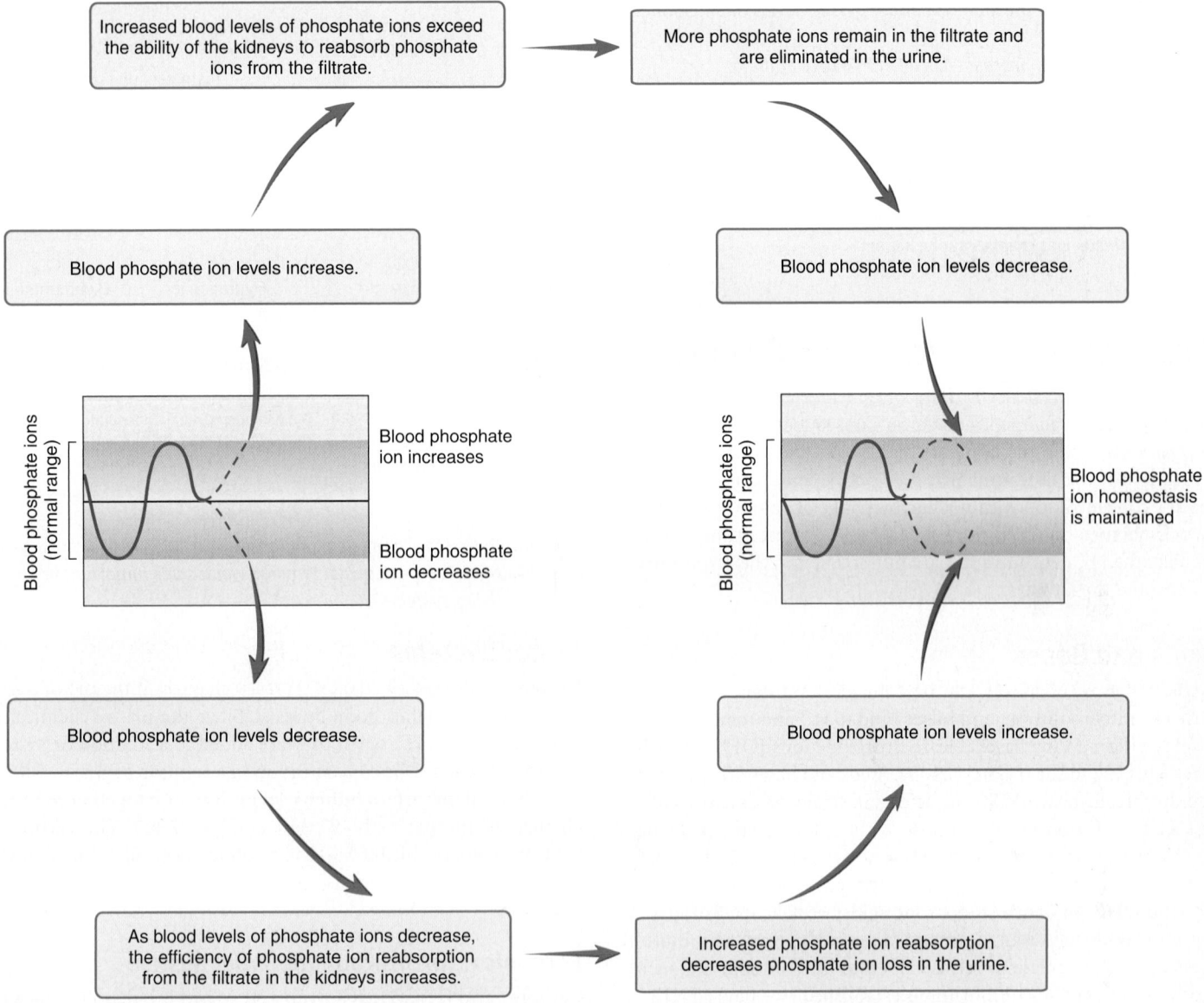

Homeostasis Figure 27.11   Regulation of Blood Phosphate Concentration in the Extracellular Fluid

Table 27.10	Consequences of Abnormal Plasma Levels of Phosphate Ions
Major Causes	**Symptoms**
Hypophosphatemia	
Reduced absorption from the intestine associated with vitamin D deficiency and alcohol abuse	Reduced rate of metabolism, reduced transport of oxygen, reduced white blood cell functions, and blood clotting
Increased renal excretion with hyperparathyroidism	
Hyperphosphatemia	
Renal failure	
Tissue destruction with chemotherapy used to treat metastatic tumors	Symptoms are related to reduced plasma Ca^{2+} concentrations due to calcium phosphate deposited in tissues such as the lungs, kidneys, and joints.
Hyperparathyroidism initially leading to elevated plasma Ca^{2+} combined with reduced excretion of phosphate by the kidney	

30. *Explain how the kidneys control plasma levels of phosphate ions.*

31. *How does an increased level of PTH affect tubular phosphate reabsorption?*

32. *What are the consequences of prolonged elevation of blood levels of phosphate ions?*

P R E D I C T **3**

Mary Thon runs several miles each day. List the mechanisms through which water loss changes during her run.

Regulation of Acid–Base Balance

Hydrogen ions affect the activity of enzymes and interact with many electrically charged molecules. Consequently, most chemical reactions within the body are highly sensitive to the H^+ concentration of the fluid in which they occur. The maintenance of the H^+ concentration within a narrow range of values is essential for normal metabolic reactions. The three major mechanisms that regulate the H^+ concentration are buffer systems, the respiratory system, and the kidneys.

Acids and Bases

Acids, for most purposes, can be defined as substances that release H^+ into a solution, and **bases** bind to H^+ and remove them from solution. Many bases release hydroxide ions (OH^-), which react with H^+ to form water (H_2O). Acids and bases are grouped as either strong or weak. Strong acids and strong bases completely dissociate to form ions in solution. Hydrochloric acid is a strong acid, which dissociates to form H^+ and Cl^- (figure 27.12), and sodium hydroxide is a strong base, which dissociates to form Na^+ and OH^-. In contrast to strong acids, weak acids dissociate, but most molecules remain intact. Many of the weak acid molecules do not dissociate to release H^+ into the solution. For each type of weak acid an equilibrium is established (see figure 27.12). The proportion of weak acid molecules that release H^+ into solution is very predictable, and is influenced by the pH of the solution into which the weak acid is placed. Weak acids are common in living systems and they play important roles in preventing large changes in pH in body fluid pH.

Figure 27.12 **Comparison of Strong and Weak Acids**

Strong acids and bases completely dissociate when dissolved in water. Weak acids do not completely dissociate. Weak acids partially dissociate so that an equilibrium is established between the acid and the ions that are formed when the dissociation occurs.

33. *Define the terms acid and base.*

34. *Describe weak acids. Why are weak acids important in living systems?*

Buffer Systems

Buffers (bŭf′erz) (see chapter 2) resist changes in the pH of a solution. Buffers within body fluids stabilize the pH by chemically binding to excess H^+ when they are added to a solution or by releasing H^+ when their concentration in a solution begins to fall.

Several important buffer systems function together to resist changes in the pH of body fluids (table 27.11). The carbonic acid/bicarbonate buffer system, protein molecules like hemoglobin and plasma proteins, and phosphate compounds all act as buffers.

Carbonic Acid/Bicarbonate Buffer System

Carbonic acid (H_2CO_3) is a weak acid. When it is dissolved in water, the following equilibrium is established:

$$H_2CO_3 \rightleftarrows HCO_3^- + H^+$$

The carbonic acid/bicarbonate buffer system depends on the equilibrium that is quickly established between H_2CO_3 and the

Table 27.11	Buffer Systems
Protein Buffer System	Intracellular proteins and plasma proteins form a large pool of protein molecules that can act as buffer molecules. Because of their high concentration, they provide approximately three-fourths of the buffer capacity of the body. Hemoglobin in red blood cells is an important intracellular protein. Other intracellular molecules like histone proteins and nucleic acids also act as buffers.
Bicarbonate Buffer System	Components of the bicarbonate buffer system are not present in high enough concentrations in the extracellular fluid to constitute a powerful buffer system. Because the concentrations of the components of the buffer system are regulated, however, it plays an exceptionally important role in controlling the pH of extracellular fluid.
Phosphate Buffer System	Concentration of the phosphate buffer components is low in the extracellular fluids compared to the other buffer systems, but it's an important intracellular buffer system.

H^+ and bicarbonate (HCO_3^-). When an amount of H^+ is added to this solution, by adding a small amount of a strong acid, a large proportion of the H^+ binds to HCO_3^- to form H_2CO_3, and only a small percentage remain as free H^+. Thus, a large decrease in pH is resisted by the carbonic acid/bicarbonate buffer system when acidic substances are added to a solution containing H_2CO_3.

When an amount of H^+ is removed from a solution containing H_2CO_3, by adding a small amount of a strong base, many of the H_2CO_3 form HCO_3^- and H^+. Thus, a large increase in pH is resisted when basic substances are added to a solution containing H_2CO_3.

The carbonic acid/bicarbonate buffer system plays an important role in regulating the extracellular pH. It quickly responds to the addition of substances, such as CO_2 or lactic acid produced by increased metabolism during exercise (see chapter 23) and increased fatty acid and ketone body production during periods of elevated fat metabolism (see chapter 25). It also responds to the addition of basic substances, such as the consumption of large amounts of $NaHCO_3$ as an antacid. The carbonic acid/bicarbonate buffer system has a limited capacity to resist changes in pH, but it remains very important because it plays an essential role in the control of pH by both the respiratory system and the kidneys (see Mechanisms of Acid-Base Balance Regulation following).

Protein Buffer System

Intracellular proteins and plasma proteins form a large pool of protein molecules, which act as buffer molecules. They provide approximately three-fourths of the buffer capacity of the body because of their high concentration. Hemoglobin in red blood cells is one of the most important intracellular proteins. Other intracellular molecules like histone proteins associated with nucleic acids also act as buffers. The capacity of proteins to function as buffers is due to the functional groups of amino acids, such as carboxyl ($-COOH$) or amino ($-NH_2$) groups. These functional groups can act like weak acids and bases. Consequently, as the H^+ concentration increases, more H^+ bind to the functional groups, and when the H^+ concentration decreases, H^+ are released from the functional groups (see table 27.11).

Phosphate Buffer System

The concentration of phosphate and phosphate-containing molecules is low in the extracellular fluid compared with the other buffer systems, but it is an important intracellular buffer system. Phosphate-containing molecules such as DNA, RNA, ATP, as well as phosphate ions, such as HPO_4^{2-}, in solution act as buffers. Phosphate ions act as weak acids and, therefore, can bind to H^+, to form $H_2PO_4^-$, when the pH decreases and ions, such as $H_2PO_4^-$, release H^+ into the solution when the pH begins to increase (see table 27.11).

- **35. Define the term buffer. Describe how a buffer works when H^+ ions are added to a solution or when they are removed from a solution.**
- **36. Name the three buffer systems of the body. Which of these systems provides the largest proportion of buffer capacity in the body?**

Mechanisms of Acid–Base Balance Regulation

Buffers and the mechanisms of acid–base balance regulation work together and play essential roles in the regulation of acid–base balance (figure 27.13). Buffers almost instantaneously resist changes in the pH of body fluids. The mechanisms of acid–base regulation depend on the regulation of respiration and on kidney function. The respiratory system responds within a few minutes to changes in pH to bring the pH of body fluids back toward its normal range. Its capacity to regulate pH, however, is not as great as that of the kidneys, nor does the respiratory system have the same ability to return the pH to its precise range of normal values. In contrast, the kidneys respond more slowly, within hours to days, to alterations of body fluid pH, and their capacity to respond is substantial.

- **37. Compare the capacity and rate at which the respiratory system and kidneys function to control body fluid pH.**

Respiratory Regulation of Acid–Base Balance

The respiratory system regulates acid–base balance by influencing the carbonic acid/bicarbonate buffer system. Carbon dioxide (CO_2) reacts with water (H_2O) to form carbonic acid (H_2CO_3), which dissociates to form H^+ and HCO_3^- as follows:

$$H_2O + CO_2 \leftrightarrows H_2CO_3 \leftrightarrows H^+ + HCO_3^-$$

This reaction is in equilibrium. As CO_2 increases, CO_2 combines with H_2O. The higher the concentration of CO_2, the greater the amount of H_2CO_3 formed. Many H_2CO_3 molecules then dissociate to form H^+ and HCO_3^-. If CO_2 levels decline, however, the equilibrium shifts in the opposite direction so that many H^+ and HCO_3^- combine to form H_2CO_3, which then forms CO_2 and H_2O. Thus, H^+ and HCO_3^- decrease in the solution.

The reaction between CO_2 and H_2O is catalyzed by an enzyme, **carbonic anhydrase**, which is found in a relatively high concentration in red blood cells and on the surface of capillary epithelial cells (figure 27.14). This enzyme does not influence equilibrium but accelerates the rate at which the reaction proceeds in either direction so that equilibrium is achieved quickly.

Decreases in body fluid pH, regardless of the cause, stimulate neurons in the respiratory center in the brainstem and cause the rate and depth of ventilation to increase. The increased rate and depth of ventilation cause CO_2 to be eliminated from the body through the lungs at a greater rate, and the concentration of CO_2 in the body fluids decreases. As CO_2 levels decline, the carbonic/bicarbonate buffer system reacts. Hydrogen ions combine with HCO_3^- to form H_2CO_3 which then form CO_2 and H_2O. Consequently, the concentration of H^+ decreases toward its normal range as CO_2 exits through the respiratory system (see figure 27.14).

Increases in body fluid pH, regardless of the cause, inhibit neurons in the respiratory center in the brainstem and cause the rate and depth of ventilation to decrease. The decreased rate and depth of ventilation, causes less CO_2 to be eliminated from the body through the lungs. The concentration of CO_2 in the body

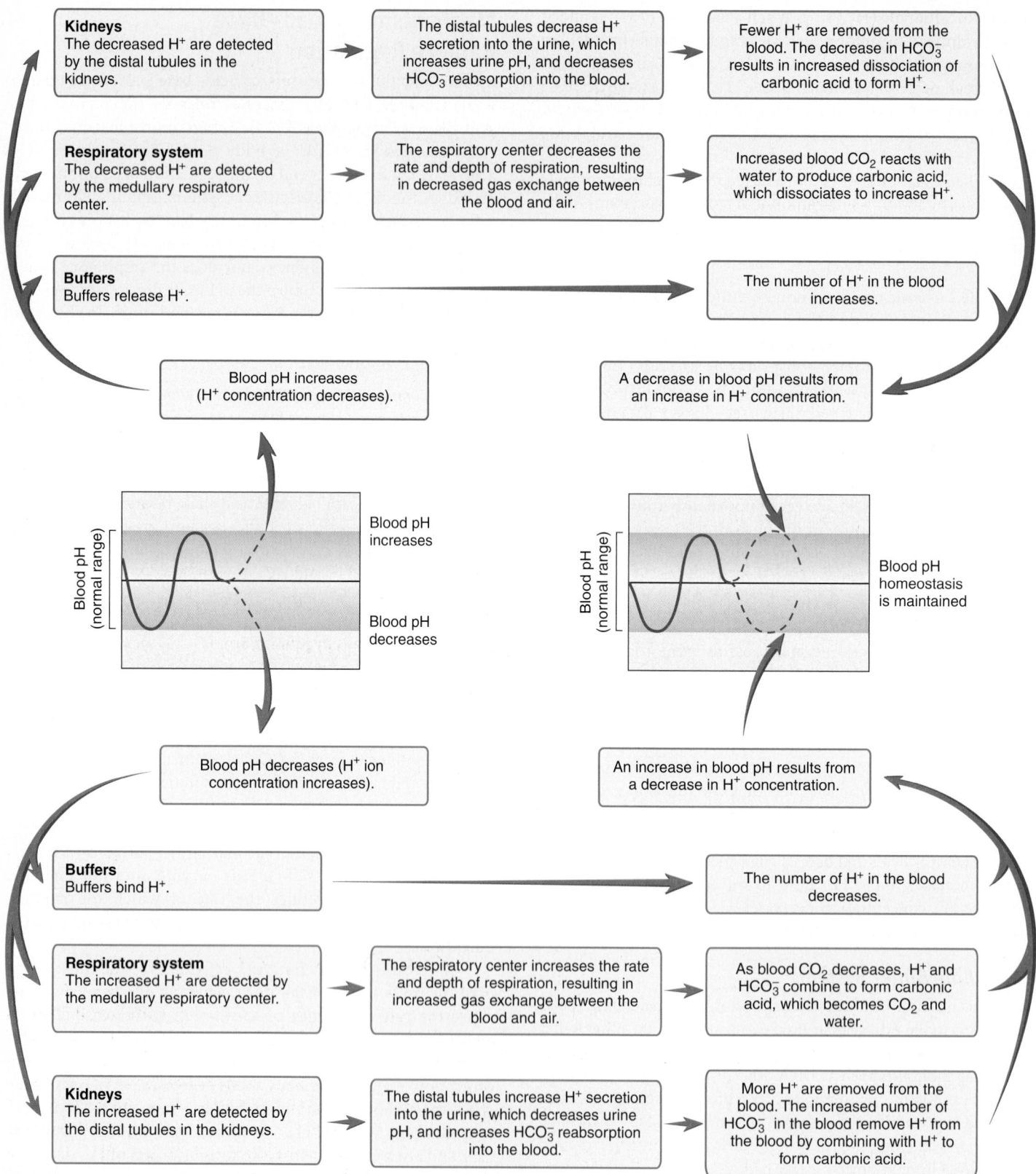

Homeostasis Figure 27.13   Regulation of Acid–Base Balance

Important mechanism through which the pH of the body fluids is regulated by the lungs and the kidneys.

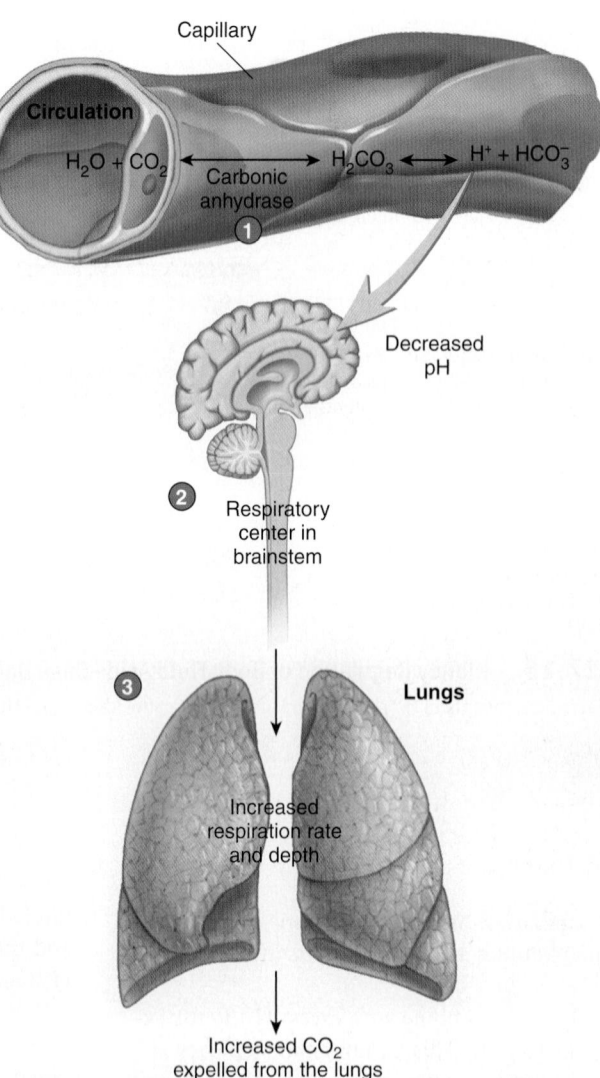

1. CO_2 reacts with H_2O to form H_2CO_3. An enzyme, carbonic anhydrase, found in red blood cells and on the surface of blood vessel epithelium, catalyzed the reaction. H_2CO_3 dissociates to form H^+ and HCO_3^-. An equilibrium is quickly established.

2. A decreased pH of the extracellular fluid stimulates the respiratory center and causes an increased rate and depth of respiration.

3. An increased rate and depth of respiration causes CO_2 to be expelled from the lungs, thus reducing the extracellular CO_2 levels. As CO_2 levels decrease, the extracellular concentration of H^+ decreases, and the extracellular fluid pH increases.

Process Figure 27.14 Respiratory Regulation of Body Fluid Acid–Base Balance

fluids increases because CO_2 is continually produced as a by-product of metabolism in all tissues. Thus, the body fluid concentration of H_2CO_3 increases also. As H_2CO_3 increases, many H_2CO_3 molecules dissociate to form H^+ and HCO_3^-. This results in an increase in the H^+ concentration, and the pH decreases toward its normal range.

P R E D I C T

Under stressful conditions some people hyperventilate. What effect does the rapid rate of ventilation have on the blood pH? Explain why a person who is hyperventilating may benefit from breathing into a paper bag.

Renal Regulation of Acid–Base Balance

Cells of the kidney tubules directly regulate acid–base balance by increasing or decreasing the rate of H^+ secretion into the filtrate (figure 27.15) and the rate of HCO_3^- reabsorption. Carbonic anhydrase is present within nephron cells and catalyzes the formation of H_2CO_3 from CO_2 and H_2O. The carbonic acid molecules dissociate to form H^+ and HCO_3^-. A countertransport system then exchanges H^+ for Na^+ across the apical membrane of the cells. Thus, cells of the kidney tubules secrete H^+ into the filtrate and reabsorbs Na^+. The Na^+ and HCO_3^- are cotransported across the basal membrane. After the Na^+ and HCO_3^- are cotransported from the kidney tubule cells, they diffuse

1. When the filtrate or blood pH decreases, H⁺ combine with HCO_3^- to form carbonic acid that is converted into CO_2 and H_2O. The CO_2 diffuses into tubule cells.

2. In the tubule cells, CO_2 combines with H_2O to form H_2CO_3 that dissociates to form H⁺ and HCO_3^-.

3. A countertransport mechanism secretes H⁺ into the filtrate in exchange for Na⁺ from the filtrate. As a result, filtrate pH decreases.

4. Bicarbonate ions are cotransported with Na⁺ into the interstitial fluid. They then diffuse into capillaries.

5. In capillaries, HCO_3^- combine with H⁺. This decreases the H⁺ concentration and increases blood pH.

□ Countertransport
□ Cotransport

Process Figure 27.15   Kidney Regulation of Body Fluid Acid–Base Balance

As the extracellular pH decreases, the rate of H⁺ secretion by tubule cells of the kidney and HCO_3^- reabsorption increase.

into the peritubular capillaries. As a result, H^+ are secreted into the lumen of the kidney tubules, and HCO_3^- pass into the extracellular fluid.

The reabsorbed HCO_3^- combine with excess H^+ in the extracellular fluid to form H_2CO_3. This combination removes H^+ from the extracellular fluid and increases extracellular pH. The rate of H^+ secretion and HCO_3^- reabsorption increases when the pH of the body fluids decreases, and this process slows when the pH of the body fluids increases (see figure 27.15).

Some of the H^+ secreted by cells of the kidney tubules into the filtrate combine with HCO_3^-, which enters the filtrate through the filtration membrane, in the form of sodium bicarbonate ($NaHCO_3$). Within the kidney tubules H^+ combines with HCO_3^- to form H_2CO_3, which then dissociate to form CO_2 and H_2O. The CO_2 diffuses from the filtrate into the cells of the kidney tubules, where it can react with H_2O to form H_2CO_3, which subsequently dissociates to form H^+ ions and HCO_3^- (see figure 27.15). Once again, H^+ are transported into the lumen of the kidney tubules in exchange for Na^+, whereas HCO_3^- enter the extracellular fluid. As a result, many of the HCO_3^- entering the filtrate reenter the extracellular fluid.

H^+ secreted into the nephron normally exceed the amount of HCO_3^- that enter the kidney tubules through the filtration membrane. Because the H^+ combine with HCO_3^-, almost all the HCO_3^- are reabsorbed from the kidney tubules (see figure 27.15). Few HCO_3^- are lost in the urine unless the pH of the body fluids becomes elevated.

The rate of H^+ secretion into the filtrate and the rate of HCO_3^- reabsorption into the extracellular fluid decrease if the pH of the body fluids increases. As a result, the amount of bicarbonate

filtered into the kidney tubules exceeds the amount of secreted H^+, and the excess of HCO_3^- pass into the urine. Excretion of excess HCO_3^- in the urine diminishes the amount of HCO_3^- in the extracellular fluid. This allows extracellular H^+ to increase and, as a consequence, the pH of the body fluids decreases toward its normal range.

Cushing's Syndrome and Alkalosis

Aldosterone increases the rate of Na^+ reabsorption and K^+ secretion by the kidneys, but in high concentrations aldosterone also stimulates H^+ secretion. Elevated aldosterone levels, such as occur in patients with Cushing's syndrome, can, therefore, elevate body fluid pH above normal (alkalosis). The major factor that influences the rate of H^+ secretion, however, is the pH of the body fluids.

P R E D I C T   5
Predict the effect of aldosterone hyposecretion on body fluid pH.

The secretion of H^+ into the urine can decrease the filtrate pH to approximately 4.5. A filtrate pH below 4.5 inhibits the secretion of additional H^+. The H^+ that pass into the filtrate are greater than the quantity required to decrease the pH of an unbuffered solution below 4.5. Buffers in the filtrate combine with many of the secreted H^+. Bicarbonate ions, phosphate ions (HPO_4^{2-}), and **ammonia** (NH_3) in the filtrate act as buffers. Both HCO_3^- and HPO_4^{2-} enter the kidney tubules through the filtration membrane along with the rest of the filtrate, and NH_3 diffuses

Clinical Focus   Acidosis and Alkalosis

The normal pH value of the body fluids is between 7.35 and 7.45. When the pH value of body fluids is below 7.35, the condition is called **acidosis** (as-i-dō′sis), and when the pH is above 7.45, it is called **alkalosis** (al′kă-lō′sis).

Metabolism produces acidic products that lower the pH of the body fluids. For example, CO_2 is a by-product of metabolism, and CO_2 combines with water to form H_2CO_3. Also, lactic acid is a product of anaerobic metabolism, protein metabolism produces phosphoric and sulfuric acids, and lipid metabolism produces fatty acids. These acidic substances must continuously be eliminated from the body to maintain pH homeostasis. The failure to eliminate acidic products of metabolism results in acidosis. Excess elimination of acidic products of metabolism results in alkalosis.

The major effect of acidosis is depression of the central nervous system. When the pH of the blood falls below 7.35, the central nervous system malfunctions, and the individual becomes disoriented and possibly comatose as the condition worsens.

A major effect of alkalosis is hyperexcitability of the nervous system. Peripheral nerves are affected first, resulting in spontaneous nervous stimulation of muscles. Spasms and tetanic contractions and possibly extreme nervousness or convulsions result. Severe alkalosis can cause death as a result of tetany of the respiratory muscles.

Although buffers help resist changes in the pH of body fluids, the respiratory system and the kidneys regulate the pH of the body fluids. Malfunctions in either the respiratory system or the kidneys can result in acidosis or alkalosis.

Acidosis and alkalosis are categorized by the cause of the condition. **Respiratory acidosis** or **respiratory alkalosis** results from abnormalities in the respiratory system. **Metabolic acidosis** or **metabolic alkalosis** results from all causes other than abnormal respiratory functions.

Inadequate ventilation of the lungs causes respiratory acidosis (table A). The rate at which CO_2 is eliminated from the body fluids through the lungs falls. This increases the concentration of CO_2 in the body fluids. As CO_2 levels increase, CO_2 reacts with water to form H_2CO_3. Carbonic acid forms H^+ and HCO_3^-. The increase in H^+ concentration causes the pH of the body fluids to decrease. If the pH of the body fluids falls below 7.35, symptoms of respiratory acidosis become apparent.

Buffers help resist a decrease in pH, and the kidneys help compensate for failure of the lungs to prevent respiratory acidosis by increasing the rate at which they secrete H^+ into the filtrate and reabsorb HCO_3^-. The capacity of buffers to resist changes in pH can be exceeded, however, and a period of 1–2 days is required for the kidney to become maximally functional. Thus, the kidneys are not effective if respiratory acidosis develops quickly. The kidneys are very effective if respiratory acidosis develops slowly or if it lasts long enough for the kidneys to respond. For example, the kidneys cannot compensate for respiratory acidosis occurring in response to a severe asthma

Continued

Table A   Acidosis and Alkalosis

Acidosis

Respiratory Acidosis

Reduced elimination of CO_2 from the body fluids

Asphyxia

Hypoventilation (e.g., impaired respiratory center function due to trauma, tumor, shock, or renal failure)

Advanced asthma

Severe emphysema

Metabolic Acidosis

Elimination of large amounts of HCO_3^- resulting from mucous secretion (e.g., severe diarrhea and vomiting of lower intestinal contents)

Direct reduction of the body fluid pH as acid is absorbed (e.g., ingestion of acidic drugs like aspirin)

Production of large amounts of fatty acids and other acidic metabolites, such as ketone bodies (e.g., untreated diabetes mellitus)

Inadequate oxygen delivery to tissue resulting in anaerobic respiration and lactic acid buildup (e.g., exercise, heart failure, or shock)

Alkalosis

Respiratory Alkalosis

Reduced CO_2 levels in the extracellular fluid (e.g., hyperventilation due to emotions)

Decreased atmospheric pressure reduces oxygen levels, which stimulates the chemoreceptor reflex, causing hyperventilation (e.g., high altitudes)

Metabolic Alkalosis

Elimination of H^+ and reabsorption of HCO_3^- in the stomach or kidney (e.g., severe vomiting or formation of acidic urine in response to excess aldosterone)

Ingestion of alkaline substances (e.g., large amounts of sodium bicarbonate)

Continued

attack that begins quickly and subsides within hours. If, however, respiratory acidosis results from emphysema, which develops over a long time, the kidneys play a significant role in helping to compensate.

Respiratory alkalosis results from hyperventilation of the lungs (see table A). This increases the rate at which CO_2 is eliminated from the body fluids and results in a decrease in the concentration of CO_2 in the body fluids. As CO_2 levels decrease, H^+ react with HCO_3^- to form H_2CO_3. The H_2CO_3 forms H_2O and CO_2. The resulting decrease in the concentration of H^+ causes the pH of the body fluids to increase. If the pH of body fluids increases above 7.45, symptoms of respiratory alkalosis become apparent.

The kidneys help to compensate for respiratory alkalosis by decreasing the rate of H^+ ion secretion into the filtrate and the rate of HCO_3^- reabsorption. If an increase in pH occurs, the kidneys need 1–2 days to compensate. Thus, the kidneys are not effective if respiratory alkalosis develops quickly. They are very effective, however, if respiratory alkalosis develops slowly. For example, the kidneys are not effective in compensating for respiratory alkalosis that occurs in response to hyperventilation triggered by emotions, which usually begins quickly and subsides within minutes or hours. If alkalosis results, however, from staying at a high altitude over a 2- or 3-day period, the kidneys play a significant role in helping to compensate.

Metabolic acidosis results from all conditions that decrease the pH of the body fluids below 7.35, with the exception of conditions resulting from altered function of the respiratory system (see table A). As H^+ accumulate in the body fluids, buffers first resist a decline in pH. If the buffers cannot compensate for the increase in H^+, the respiratory center helps regulate body fluid pH. The reduced pH stimulates the respiratory center, which causes hyperventilation. During hyperventilation, CO_2 is eliminated at a greater rate. The elimination of CO_2 also eliminates excess H^+ and helps maintain the pH of the body fluids within a normal range.

If metabolic acidosis persists for many hours and if the kidneys are functional, the kidneys can also help compensate for metabolic acidosis by secreting H^+ ions at a greater rate and increasing the rate of HCO_3^- reabsorption. Symptoms of metabolic acidosis appear if the respiratory and renal systems are not able to maintain the pH of the body fluids within its normal range.

Metabolic alkalosis results from all conditions that increase the pH of the body fluids above 7.45, with the exception of conditions resulting from altered function of the respiratory system. As H^+ decrease in the body fluids, buffers first resist an increase in pH. If the buffers cannot compensate for the decrease in H^+, the respiratory center helps regulate body fluid pH. The increased pH inhibits respiration. Reduced respiration allows CO_2 to accumulate in the body fluids. Carbon dioxide reacts with water to produce H_2CO_3. If metabolic alkalosis persists for several hours, and if the kidneys are functional, the kidneys reduce the rate of H^+ secretion to help reverse alkalosis (see table A).

---

across the wall of nephron cells to enter the filtrate. These ions combine with H^+ secreted by the nephron (figure 27.16).

NH_3 is produced in the cells of the nephron when amino acids like glycine are deaminated. Subsequently, NH_3 diffuses from the nephron cells into the filtrate and combines with H^+ in the filtrate to form **ammonium ions** (NH_4^+) (see figure 27.14). The rate of NH_3 production increases when the pH of the body fluids has been depressed for 2–3 days, such as during prolonged respiratory or metabolic acidosis. The elevated ammonia production increases the buffering capacity of the filtrate, allowing secretion of additional H^+ into the urine.

HCO_3^-, HPO_4^{2-}, and NH_3 constitute major buffers within the filtrate, but other weak acids, such as lactic acid in the filtrate, also combine with H^+ and increase the amount of H^+ that can be secreted into the filtrate.

P R E D I C T   **6**

Mr. Puffer suffers from severe emphysema. Gas exchange in his lungs is not adequate, and he must have a supply of oxygen. His blood carbon dioxide level is elevated. Nevertheless his blood pH is close to normal. Explain. Also explain how his blood bicarbonate ion (HCO_3^-) level is likely to differ from normal values.

38. *Name the three mechanisms that play essential roles in the regulation of acid–base balance.*

39. *What happens to blood pH when blood CO_2 levels go up or down? What causes this change?*

40. *What effect do increased CO_2 levels or decreased pH have on respiration? How does this change in respiration affect blood pH?*

41. *Describe the process by which nephron cells move H^+ ions into the kidney tubule lumen and HCO_3^- into the extracellular fluid.*

42. *Describe the process by which HCO_3^- are reabsorbed from the kidney tubule lumen.*

43. *Name the factors that cause an increase and a decrease in H^+ ion secretion.*

44. *What is the purpose of buffers in the urine? Describe how the ammonia buffer system operates.*

1. Hydrogen ions secreted into the filtrate are buffered.

2. Hydrogen ions can react with HCO_3^- that enters the filtrate to form H_2CO_3 which is in equilibrium with H_2O and CO_2.

3. Hydrogen ions can react with HPO_4^{2-} that enters the filtrate to form $H_2PO_4^-$.

4. Hydrogen ions can react with NH_3 formed by amino acid deamination and secreted into the nephron to form NH_4^+.

Process Figure 27.16   Hydrogen Ion Buffering in the Filtrate

The secretion of H^+ into the filtrate decreases filtrate pH. As the concentration of H^+ increases in the filtrate, the ability of tubule cells to secrete additional H^+ becomes limited. Buffering the H^+ in the filtrate decreases their concentration and enables tubule cells to secrete additional H^+.

S U M M A R Y

Water, acid, base, and electrolyte levels are maintained within a narrow range of concentrations. The urinary, respiratory, gastrointestinal, integumentary, nervous, and endocrine systems play a role in maintaining fluid, electrolyte, and pH balance.

Body Fluids   (p. 1006)

1. Intracellular fluid is inside cells.
2. Extracellular fluid is outside cells and includes interstitial fluid and plasma.

Regulation of Body Fluid Concentration and Volume   (p. 1007)

Regulation of Water Content

1. Water crosses the gastrointestinal tract through osmosis.
2. An increase in extracellular osmolality or a decrease in blood pressure stimulates the sense of thirst.
3. Wetting the oral mucosa or stretch of the gastrointestinal tract inhibits thirst.
4. Learned behavior plays a role in the amount of fluid ingested.
5. Routes of water loss
   • Water is lost through evaporation from the respiratory system and the skin (insensible perspiration and sweat).
   • Water loss into the gastrointestinal tract normally is small. Vomiting or diarrhea can significantly increase this loss.
   • The kidneys are the primary regulator of water excretion. Urine output can vary from a small amount of concentrated urine to a large amount of dilute urine.

Regulation of Extracellular Fluid Osmolality

1. Increased water consumption and ADH secretion occur in response to increases in extracellular fluid osmolality. Decreased water consumption and ADH secretion occur in response to decreases in extracellular fluid osmolality.
2. Increased water consumption and ADH decrease extracellular fluid osmolality by increasing water absorption from the intestine and water reabsorption from the nephrons. Decreased water consumption and ADH increase extracellular fluid osmolality by decreasing absorption from the intestine and water reabsorption from the nephrons.

Regulation of Extracellular Fluid Volume

1. Increased extracellular fluid volume results in decreased aldosterone secretion, increased ANH secretion, decreased ADH secretion, and decreased sympathetic stimulation of afferent arterioles. The effect of these changes is to decrease Na^+ reabsorption and to increase urine volume to decrease extracellular fluid volume.
2. Decreased extracellular fluid volume results in increased aldosterone secretion, decreased ANH secretion, increased ADH secretion, and increased sympathetic stimulation of the afferent arterioles. The effect of these changes is to increase Na^+ reabsorption and to decrease urine volume so as to increase extracellular fluid volume.

Regulation of Intracellular Fluid Composition   (p. 1012)

1. Substances used or produced inside the cell and substances exchanged with the extracellular fluid determine the composition of intracellular fluid.
2. Intracellular fluid is different from extracellular fluid because the plasma membrane regulates the movement of materials.
3. The difference between intracellular and extracellular fluid concentrations determines water movement.

Regulation of Specific Electrolytes in the Extracellular Fluid   (p. 1013)

The intake and elimination of substances from the body and the exchange of substances between the extracellular and intracellular fluids determine extracellular fluid composition.

Regulation of Sodium Ions

1. Sodium is responsible for 90%–95% of extracellular osmotic pressure.
2. The amount of Na^+ excreted in the kidneys is the difference between the amount of Na^+ that enters the nephron and the amount that is reabsorbed from the nephron.
   • Glomerular filtration rate determines the amount of Na^+ entering the nephron.
   • Aldosterone determines the amount of Na^+ reabsorbed.
3. Small quantities of Na^+ are lost in sweat.

4. Increased blood osmolality leads to the production of a small volume of concentrated urine and to thirst. Decreased blood osmolality leads to the production of a large volume of dilute urine and to decreased thirst.
5. Increased blood pressure increases water and salt loss.
   • Baroreceptor reflexes reduce ADH secretion.
   • Renin secretion is inhibited, leading to reduced aldosterone production.

Regulation of Chloride Ions

Chloride ions are the dominant negatively charged ions in extracellular fluid.

Regulation of Potassium Ions

1. The extracellular concentration of K^+ affects resting membrane potentials.
2. The amount of K^+ excreted depends on the amount that enters with the glomerular filtrate, the amount actively reabsorbed by the nephron, and the amount secreted into the distal convoluted tubule.
3. Aldosterone increases the amount of K^+ secreted.

Regulation of Calcium Ions

1. Elevated extracellular calcium levels prevent membrane depolarization. Decreased levels lead to spontaneous action potential generation.
2. Parathyroid hormone increases extracellular Ca^{2+} levels and decreases extracellular phosphate levels. It stimulates osteoclast activity, increases calcium reabsorption from the kidneys, and stimulates active vitamin D production.
3. Vitamin D stimulates Ca^{2+} uptake in the intestines.
4. Calcitonin decreases extracellular Ca^{2+} levels.

Regulation of Magnesium Ions

The capacity of the kidney to reabsorb magnesium is limited so that excess magnesium is lost in the urine and decreased extracellular magnesium results in a greater degree of magnesium reabsorption.

Regulation of Phosphate Ions

1. Under normal conditions, reabsorption of phosphate occurs at a maximum rate in the nephron.
2. An increase in plasma phosphate increases the amount of phosphate in the nephron beyond that which can be reabsorbed, and the excess is lost in the urine.

Regulation of Acid–Base Balance   (p. 1022)
Acids and Bases

Acids release H^+ into solution, and bases remove them.

Buffer Systems

1. A buffer resists changes in pH.
   • When H^+ are added to a solution, the buffer removes them.
   • When H^+ are removed from a solution, the buffer replaces them.
2. Carbonic acid/bicarbonate, proteins, and phosphate compounds are important buffers.

Mechanisms of Acid–Base Balance Regulation

Buffers, the respiratory system, and the kidneys regulate acid–base balance.

Respiratory Regulation of Acid–Base Balance

1. Respiratory regulation of pH is achieved through the carbonic acid/bicarbonate buffer system.
   • As carbon dioxide levels increase, pH decreases.
   • As carbon dioxide levels decrease, pH increases.
   • Carbon dioxide levels and pH affect the respiratory centers. Hypoventilation increases blood carbon dioxide levels, and hyperventilation decreases blood carbon dioxide levels.

Renal Regulation of Acid–Base Balance

1. The secretion of H^+ into the filtrate and the reabsorption of HCO_3^- into extracellular fluid cause extracellular pH to increase.
   • Carbonic acid dissociates to form H^+ and HCO_3^- in nephron cells.
   • Active transport pumps H^+ into the nephron lumen and Na^+ into the nephron cell.
   • Na^+ and HCO_3^- diffuse into the extracellular fluid.
2. HCO_3^- in the filtrate are reabsorbed.
   • HCO_3^- combine with H^+ to form carbonic acid, which dissociates to form carbon dioxide and water.
   • Carbon dioxide diffuses into nephron cells and forms carbonic acid, which dissociates to form HCO_3^- and H^+.
   • HCO_3^- diffuse into the extracellular fluid, and H^+ are secreted into the filtrate.
3. The rate of H^+ secretion increases as body fluid pH decreases or as aldosterone levels increase.
4. Secretion of H^+ is inhibited when urine pH falls below 4.5.
   • Carbonic acid/bicarbonate, ammonia, and phosphate buffers in the urine resist a drop in pH.
   • As the buffers absorb H^+, more H^+ are pumped into the urine.

R E V I E W   A N D   C O M P R E H E N S I O N

1. Extracellular fluid
   a. is much like intracellular fluid in composition.
   b. includes interstitial fluid.
   c. osmotic concentration tends to vary greatly in the different fluid compartments of the body.
   d. all of the above.

2. The sensation of thirst increases when
   a. the levels of angiotensin II increase.
   b. the osmolality of the blood decreases.
   c. blood pressure increases.
   d. renin secretion decreases.

3. Insensible perspiration
   a. is lost through sweat glands.
   b. results in heat loss from the body.
   c. increases when ADH secretion increases.
   d. results in the loss of solutes such as Na^+ and Cl^-.

4. The composition and volume of body fluid are regulated primarily by the
   a. skin.
   b. lungs.
   c. kidneys.
   d. heart.
   e. spleen.

5. Which of these conditions *decreases* extracellular fluid volume?
   a. constriction of afferent arterioles
   b. increased ADH secretion
   c. decreased ANH secretion
   d. decreased aldosterone secretion
   e. stimulation of sympathetic nerves to the kidneys

6. A decrease in blood pressure
   a. results in increased aldosterone secretion.
   b. causes decreased ADH secretion.
   c. inhibits sympathetic stimulation.
   d. results in vasodilation.
   e. all of the above.

7. Which of these results in an increased blood Na⁺ concentration?
   a. decrease in ADH secretion
   b. decrease in aldosterone secretion
   c. increase in ANH
   d. decrease in renin secretion

8. Which of these mechanisms is the most important for regulating blood osmolality?
   a. ADH
   b. renin-angiotensin-aldosterone
   c. ANH
   d. parathyroid hormone

9. A decrease in extracellular K^+
   a. produces depolarization of the plasma membrane.
   b. results when aldosterone levels increase.
   c. occurs when tissues are damaged (e.g., in burn patients).
   d. increases ANH secretion.
   e. increases PTH secretion.

10. Ca^{2+} concentration in the blood decreases when
    a. vitamin D levels are lower than normal.
    b. calcitonin secretion decreases.
    c. parathyroid hormone secretion increases.
    d. all of the above.

11. An acid
    a. solution has a pH greater than 7.
    b. is a substance that releases H^+ into a solution.
    c. is considered weak if it completely dissociates in water.
    d. all of the above.

12. Buffers
    a. release H^+ when pH increases.
    b. resist changes in the pH of a solution.
    c. include the proteins of the blood.
    d. all of the above.

13. Which of these is *not* a buffer system in the body?
    a. sodium chloride buffer system
    b. carbonic acid/bicarbonate buffer system
    c. phosphate buffer system
    d. protein buffer system

14. Which of these systems regulating blood pH is the fastest acting?
    a. respiratory
    b. kidney

15. An increase in blood carbon dioxide levels is followed by a(n) _____ in H^+ and a(n) _____ in blood pH.
    a. increase, increase
    b. increase, decrease
    c. decrease, increase
    d. decrease, decrease

16. High levels of bicarbonate ions in the urine indicate
    a. a low level of H^+ secretion into the urine.
    b. that the kidneys are causing blood pH to increase.
    c. that urine pH is decreasing.
    d. all of the above.

17. High levels of ammonium ions in the urine indicate
    a. a high level of H^+ secretion into the urine.
    b. that the kidneys are causing blood pH to decrease.
    c. that urine pH is too alkaline.
    d. all of the above.

18. Blood plasma pH is normally
    a. slightly acidic.
    b. strongly acidic.
    c. slightly alkaline.
    d. strongly alkaline.
    e. neutral.

19. Acidosis
    a. increases neuron excitability.
    b. can produce tetany by affecting the peripheral nervous system.
    c. may lead to coma.
    d. may produce convulsions through the central nervous system.

20. Respiratory alkalosis is caused by _____ and can be compensated for by the production of a more _____ urine.
    a. hypoventilation, alkaline
    b. hypoventilation, acidic
    c. hyperventilation, acidic
    d. hyperventilation, alkaline

Answers in Appendix F

CRITICAL THINKING

1. In patients with diabetes mellitus, not enough insulin is produced; as a consequence, blood glucose levels increase. If blood glucose levels rise high enough, the kidneys are unable to absorb the glucose from the glomerular filtrate, and glucose "spills over" into the urine. What effect does this glucose have on urine concentration and volume? How does the body adjust to the excess glucose in the urine?

2. A patient suffering from a tumor in the hypothalamus produces excessive amounts of ADH, a condition called syndrome of inappropriate ADH (SIADH) production. For this patient, the excessive ADH production is chronic and has persisted for many months. A student nurse keeps a fluid intake–output record on the patient. She is surprised to find that fluid intake and urinary output are normal. What effect was she expecting? Can you explain why urinary output is normal?

3. A patient exhibits the following symptoms: elevated urine ammonia and increased rate of respiration. Does the patient have metabolic acidosis or metabolic alkalosis?

4. Swifty Trotts has an enteropathogenic *Escherichia coli* infection that produces severe diarrhea. What does this diarrhea do to his blood pH, urine pH, and respiratory rate?

5. Acetazolamide is a diuretic that blocks the activity of the enzyme carbonic anhydrase inside kidney tubule cells. This blockage prevents the formation of carbonic acid from carbon dioxide and water. Normally, carbonic acid dissociates to form H^+ and HCO_3^-, and the H^+ are exchanged for Na^+ from the urine. Blocking the formation of H^+ in the cells of the nephron tubule blocks sodium reabsorption, thus inhibiting water reabsorption and producing the diuretic effect. With this information in mind, what effect does acetazolamide have on blood pH, urine pH, and respiratory rate?

6. As part of a physiology experiment, Hardy Breath, an anatomy and physiology student, is asked to breathe through a 3-foot long glass tube. What effect does this action have on his blood pH, urine pH, and respiratory rate?

7. A young girl is suspected of having epilepsy and, therefore, is prone to having convulsions. On the basis of your knowledge of acid–base balance and respiration, propose a hypothetical experiment that might suggest that the girl is susceptible to convulsions.

8. Hardy Explorer climbed to the top of a very high mountain. To celebrate, he drank a glass of whiskey. Alcohol stimulates hydrochloric acid secretion in the stomach. What do you expect to happen to Hardy's respiratory rate and the pH of his urine?

Answers in Appendix G

A N S W E R S   T O   P R E D I C T   Q U E S T I O N S

1. During hemorrhagic shock, blood pressure decreases, and visceral blood vessels constrict (see chapter 21). As a consequence, blood flow to the kidneys and the blood pressure in the glomeruli decrease dramatically. The total filtration pressure decreases, and the amount of filtrate formed each minute decreases. The rate at which Na^+ enter the nephron, therefore, decreases. In addition, renin is secreted from the kidneys in large amounts. Renin causes the formation of angiotensin I from angiotensinogen. Angiotensin I converts to angiotensin II, which stimulates aldosterone secretion. Aldosterone increases the rate at which Na^+ are reabsorbed from the filtrate in the distal tubule and collecting ducts.

2. a. If the amount of Na^+ and water ingested in food exceeds that needed to maintain a constant extracellular fluid composition, it increases total blood volume and also increases blood pressure.

   b. Excessive Na^+ and water intake causes an increase in total blood volume and blood pressure. The elevated blood pressure causes a reflex response that results in decreased ADH secretion. The elevated pressure also causes reduced renin secretion from the kidneys, resulting in a reduction in the rate at which angiotensin II is formed. The reduction in angiotensin II reduces the rate of aldosterone secretion. Together these changes cause increased loss of Na^+ in the urine and an increase in the volume of urine produced. Increased Na^+ and increased blood pressure also cause the secretion of ANH, which inhibits ADH secretion and Na^+ reabsorption in the nephron.

   c. If the amount of water ingested is large, urine concentration is reduced, urine volume is increased, and the concentration of Na^+ in the urine is low. If the amount of salt ingested is great, the concentration of salt in the urine can be high, and the urine volume is larger and contains a substantial concentration of salt.

3. During conditions of exercise, the amount of water lost is increased because of increased evaporation from the respiratory system, increased insensible perspiration, and increased sweat. The amount of water lost in the form of sweat can increase substantially. The amount of urine formed decreases during conditions of exercise.

4. Hyperventilation results in a greater-than-normal rate of carbon dioxide loss from the circulatory system. Because carbon dioxide is lost from the circulatory system, H^+ concentration decreases, and the pH of body fluids increases. Breathing into a paper bag corrects for the effects of hyperventilation because the person rebreathes air that has a higher concentration of carbon dioxide. The result is an increase in carbon dioxide in the body. Consequently, the H^+ concentration increases, and pH decreases toward normal levels.

5. Aldosterone hyposecretion results in acidosis. Aldosterone increases the rate at which Na^+ are reabsorbed from nephrons, but it also increases the rate at which K^+ and H^+ are secreted. Hyposecretion of aldosterone decreases the rate at which H^+ are secreted by the nephrons and, therefore, can result in acidosis.

6. Elevated blood carbon dioxide levels cause an increase in H^+ and a decrease in blood pH due to the following reaction:

$$CO_2 + H_2O \rightleftarrows H_2CO_3 \rightleftarrows H^+ + HCO_3^-$$

However, the kidney plays an important role in the regulation of blood pH. The kidney's rate of H^+ secretion into the urine and reabsorption of HCO_3^- increase. This helps prevent high blood H^+ levels and low blood pH in Mr. Puffer. Because of increased resorption of HCO_3^-, the blood levels of HCO_3^- would be greater than normal.

Visit the Online Learning Center at www.mhhe.com/seeley7 for chapter quizzes, interactive learning exercises, and other study tools.

Reproductive System

Color enhanced scanning electron micrograph of the ciliated epithelial surface of a uterine tubule.

Without the reproductive system, the human species could not survive. However, this system, unlike other organ systems, is not necessary for the survival of individual humans. The reproductive system controls the development of the structural and functional differences between males and females, and it influences human behavior. Most organ systems of the body show little difference between males and females. This isn't the case with the reproductive systems. The male reproductive system produces sperm cells and can transfer them to the female. The female reproductive system produces oocytes and can receive sperm cells, one of which may unite with an oocyte. The female reproductive system is then intimately involved with nurturing the development of a new individual until birth and usually for some considerable time after birth.

Although the male and female reproductive systems show such striking differences, they also share a number of similarities. Many reproductive organs of males and females are derived from the same embryologic structures (see chapter 29). In addition, some hormones are the same in males and females, even though they act in very different ways (table 28.1).

This chapter discusses the *anatomy of the male reproductive system* (p. 1035), the *physiology of male reproduction* (p. 1046), the *anatomy of the female reproductive system* (p. 1050), the *physiology of female reproduction* (p. 1059), and the *effects of aging on the reproductive system* (p. 1071).

Table 28.1   Major Reproductive Hormones

Hormone	Source	Target Tissue	Response
Males			
Gonadotropin-releasing hormone (GnRH)	Hypothalamus	Anterior pituitary	Stimulates secretion of LH and FSH
Luteinizing hormone (LH) (also called interstitial cell-stimulating hormone [ICSH] in males)	Anterior pituitary	Interstitial cells in the testes	Stimulates synthesis and secretion of testosterone
Follicle-stimulating hormone (FSH)	Anterior pituitary	Seminiferous tubules (sustentacular cells)	Supports spermatogenesis
Testosterone	Leydig cells in the testes	Testes and body tissues	Supports spermatogenesis, development and maintenance of reproductive organs and causes development of secondary sexual characteristics
		Anterior pituitary and hypothalamus	Inhibits GnRH, LH, and FSH secretion through negative feedback
Females			
Gonadotropin-releasing hormone (GnRH)	Hypothalamus	Anterior pituitary	Stimulates production of LH and FSH
Luteinizing hormone (LH)	Anterior pituitary	Ovaries	Causes follicles to complete maturation and undergo ovulation; causes the ovulated follicle to become the corpus luteum
Follicle-stimulating hormone (FSH)	Anterior pituitary	Ovaries	Causes follicles to begin development
Prolactin	Anterior pituitary	Mammary glands	Stimulates milk secretion following parturition
Estrogen	Follicles of ovaries	Uterus	Proliferation of endometrial cells
		Mammary glands	Development of the mammary glands (especially duct systems)
		Anterior pituitary and hypothalamus	Positive feedback before ovulation, resulting in increased LH and FSH secretion; negative feedback with progesterone on the hypothalamus and anterior pituitary after ovulation, resulting in decreased LH and FSH secretion
		Other tissues	Causes development of secondary sexual characteristics
Progesterone	Corpus luteum of ovaries	Uterus	Hypertrophy of endometrial cells and secretion of fluid from uterine glands
		Mammary glands	Development of the mammary glands (especially alveoli)
		Anterior pituitary	Negative feedback, with estrogen, on the hypothalamus and anterior pituitary after ovulation, resulting in decreased LH and FSH secretion
		Other tissues	Causes development of secondary sexual characteristics
Oxytocin*	Posterior pituitary	Uterus and mammary glands	Contraction of uterine smooth muscle during intercourse and childbirth; contraction of myoepithelial cells in the breast resulting in milk letdown in lactating women
Human chorionic gonadotropin (HCG)	Placenta	Corpus luteum of ovaries	Maintains the corpus luteum and increases its rate of progesterone secretion during the first one-third (first trimester) of pregnancy

*Covered in chapter 29.

Anatomy of the Male Reproductive System

The male reproductive system consists of the testes (sing., testis), a series of ducts, accessory glands, and supporting structures. The ducts include the epididymides (sing., epididymis), ductus deferentia (sing., deferens; also vas deferens), and urethra. Accessory glands include the seminal vesicles, prostate gland, and bulbourethral glands. Supporting structures include the scrotum and penis (figure 28.1).

Sperm cells are very temperature-sensitive and don't develop normally at usual body temperatures. The testes and epididymides, in which the sperm cells develop, are located outside the body cavity in the scrotum, where the temperature is lower. The ductus deferentia lead from the testes into the pelvis, where they join the ducts of the seminal vesicles to form the ampullae. Extensions of the ampullae, called the ejaculatory ducts, pass through the prostate and empty into the urethra within the prostate. The urethra, in turn, exits from the pelvis and passes through the penis to the outside of the body.

● 1. *List the structures that make up the male reproductive system.*

Scrotum

The **scrotum** (skrō′tŭm) contains the testes and is divided into two internal compartments by an incomplete connective tissue septum. Externally, the scrotum is marked in the midline by an irregular ridge, the **raphe** (rā′fē; a seam), which continues posteriorly to the anus and anteriorly onto the inferior surface of the penis. The outer layer of the scrotum includes the skin, a layer of superficial fascia consisting of loose connective tissue, and a layer of smooth muscle called the **dartos** (dar′tōs; to skin) **muscle.**

When the scrotum is exposed to cool temperatures, the dartos muscle contracts, causing the skin of the scrotum to become

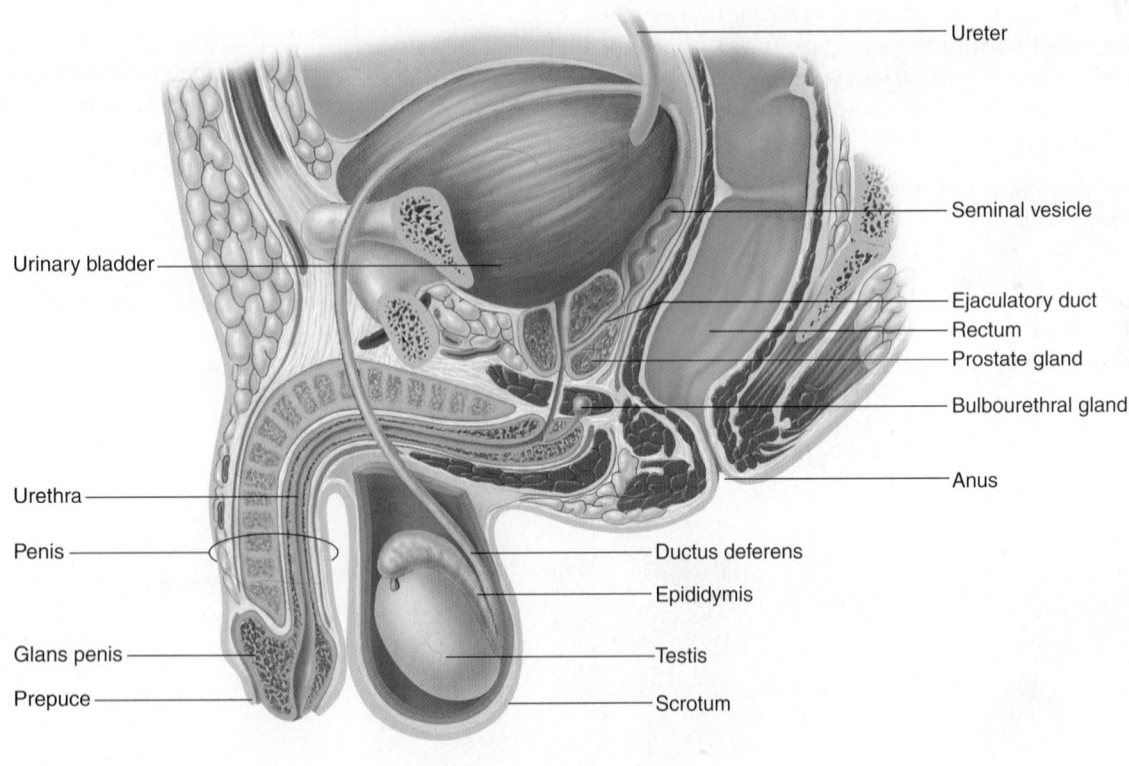

Sagittal view

Figure 28.1 Male Reproductive Structures

Sagittal section of the male pelvis showing the male reproductive structures.

firm and wrinkled and reducing its overall size. At the same time, the **cremaster** (krē-mas′ter) **muscles** (see figure 28.6), which are extensions of abdominal muscles into the scrotum, contract and help pull the testes nearer the body, which helps keep the testes warm. When the scrotum is exposed to warm temperatures or becomes warm because of exercise or fever, the dartos and cremaster muscles relax, and the skin of the scrotum becomes loose and thin, allowing the testes to descend away from the body, which helps keep the testes cool. The response of the dartos and cremaster muscles is important in the regulation of temperature in the testes. If the testes become too warm or too cold, normal sperm cell formation does not occur.

> 2. Describe the structure of the scrotum.
> 3. Describe the role of the dartos and cremaster muscles in temperature regulation of the testes.

Perineum

The area between the thighs, which is bounded by the symphysis pubis anteriorly, the coccyx posteriorly, and the ischial tuberosities laterally, is called the **perineum** (per′i-nē′ŭm). The perineum is divided into two triangles by a set of muscles, the superficial transverse and deep transverse perineal muscles, that runs transversely between the two ischial tuberosities (see figure 10.21). The anterior, or **urogenital** (ū′rō-jen′i-tăl), **triangle,** contains the base of the penis and the scrotum. The smaller posterior, or **anal, triangle,** contains the anal opening (figure 28.2).

> 4. Define the perineum. Describe the two triangles that make up the perineum.

Testes

Testicular Histology

The **testes** (tes′tēz) are small ovoid organs, each about 4–5 cm long, within the scrotum (see figure 28.1). They are both exocrine

and endocrine glands. Sperm cells form a major part of the exocrine secretions of the testes, and testosterone is the major endocrine secretion of the testes.

The outer part of each testis is a thick, white capsule consisting of mostly fibrous connective tissue called the **tunica albuginea** (al-bū-jin′ē-ă; white). Connective tissue of the tunica albuginea enters the testis and forms incomplete **septa** (sep′tă; figure 28.3). The septa divide each testis into about 300–400 cone-shaped **lobules.** The substance of the testis between the septa includes **seminiferous** (sem′i-nif′er-ŭs; seed carriers) **tubules,** in which sperm cells develop, and a loose connective tissue stroma that surrounds the tubules and contains clusters of endocrine cells called **interstitial cells,** or **Leydig cells,** which secrete testosterone.

The combined length of the seminiferous tubules in both testes is nearly half a mile. The seminiferous tubules empty into a set of short, straight tubules, the **tubuli recti,** which in turn empty into a tubular network called the **rete** (rē′tē; net) **testis.** The rete testis empties into 15–20 tubules called **efferent ductules** (dŭk′tools). They have a ciliated pseudostratified columnar epithelium that helps move sperm cells out of the testis. The efferent ductules pierce the tunica albuginea to exit the testis.

> 5. Describe the covering and connective tissue of the testis.
> 6. Where are the seminiferous tubules and interstitial cells located? What are their functions?
> 7. Describe the structures of the tubuli recti, rete testis, and efferent ductules.

Descent of the Testes

The testes develop as retroperitoneal organs in the abdominopelvic cavity by approximately 5 weeks following fertilization. By approximately 14 weeks each testis is connected to a labioscrotal swelling, which becomes the scrotum, by a **gubernaculum**

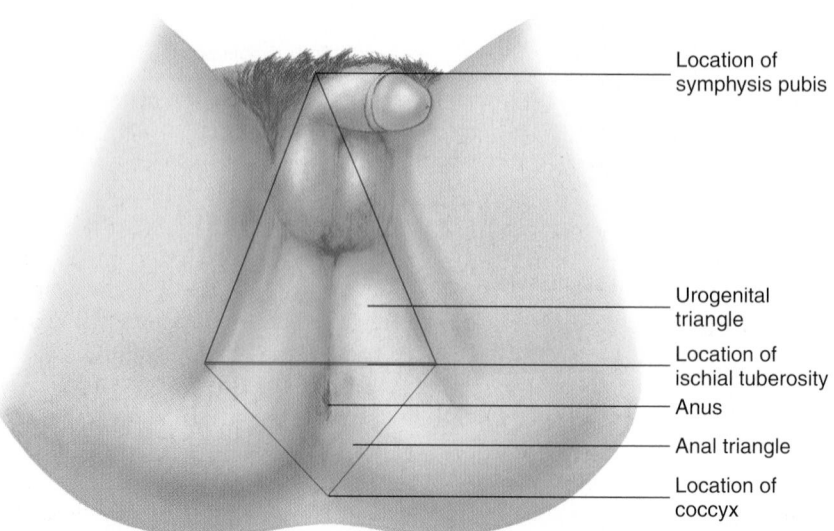

Location of symphysis pubis

Urogenital triangle

Location of ischial tuberosity

Anus

Anal triangle

Location of coccyx

Figure 28.2 Male Perineum
Inferior view of the male perineum.

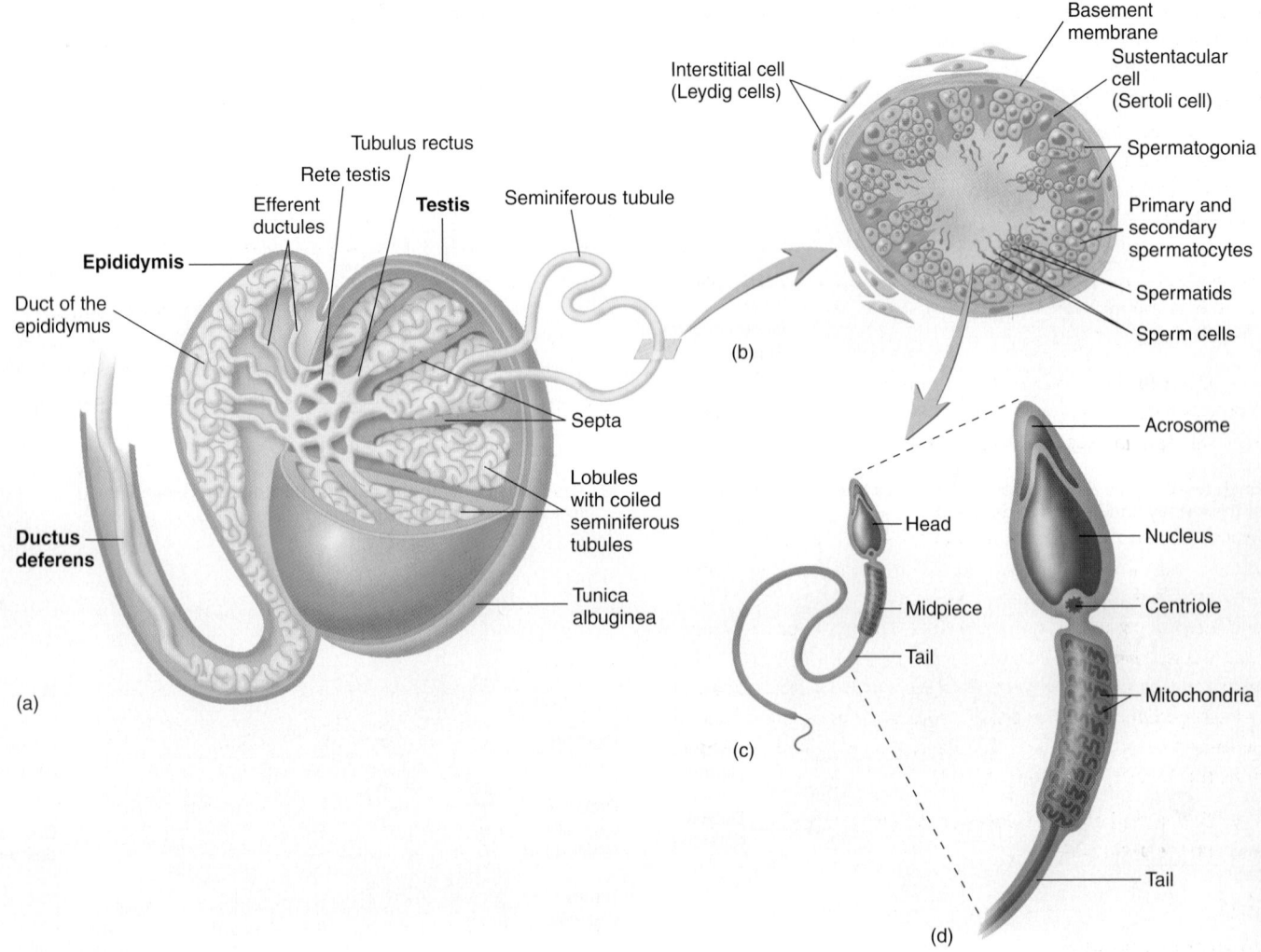

Figure 28.3 **Histology of the Testis**

(*a*) Gross anatomy of the testis with a section cut away to reveal internal structures. (*b*) Cross section of a seminiferous tubule. Spermatogonia are near the periphery, and mature sperm cells are near the lumen of the seminiferous tubule. (*c*) Mature sperm cell. (*d*) Head of a mature sperm cell.

(goo′ber-nak′ū-lŭm), a fibromuscular cord (figure 28.4 *1*; see chapter 29). Between 14 and 28 weeks, the **inguinal** (ing′gwi-năl) **canals** (figure 28.4 *2*) form and the testes move through them into the scrotum (figure 28.4 *3*). As they move into the scrotum, each testis is preceded by an outpocketing of the peritoneum called the **process vaginalis** (vaj′i-nă-lis). The superior part of each process vaginalis usually becomes obliterated, and the inferior part remains as a small, closed sac, the **tunica** (too′ni-kă) **vaginalis** (figure 28.4 *4*). The tunica vaginalis surrounds most of the testis in much the same way that the pericardium surrounds the heart, and the small amount of fluid in it allows the testes to move with little friction. The visceral layer of the tunica vaginalis covers the anterior surface of the testis, and the parietal layer lines the scrotum. The tunica vaginalis is a serous membrane consisting of a layer of simple squamous epithelium that rests on a basement membrane. The testes have descended into the scrotum of approximately 79% of fetuses delivered prior to 28 weeks of development. By 28 weeks

of development, the testes have descended into the scrotum of greater than 97% of fetuses. At 9 months of age the testes have descended into the scrotum of 98.2%, of male infants.

The inguinal canals are bilateral oblique passageways in the anterior abdominal wall. They originate at the **deep inguinal rings,** which open through the aponeuroses of the transversus abdominis muscles. The canals extend inferiorly and obliquely and end at the **superficial inguinal rings,** openings in the aponeuroses of the external abdominal oblique muscles. In females the inguinal canals do develop, but they are much smaller than in males and the ovaries do not normally descend through them.

Cryptorchidism (krip-tōr′ki-dizm) is failure of one or both of the testes to descend into the scrotum. The higher temperature of the abdominal cavity prevents normal sperm cell development and, if it involves both testes, results in no sperm cell production (see chapter 29, page 1099).

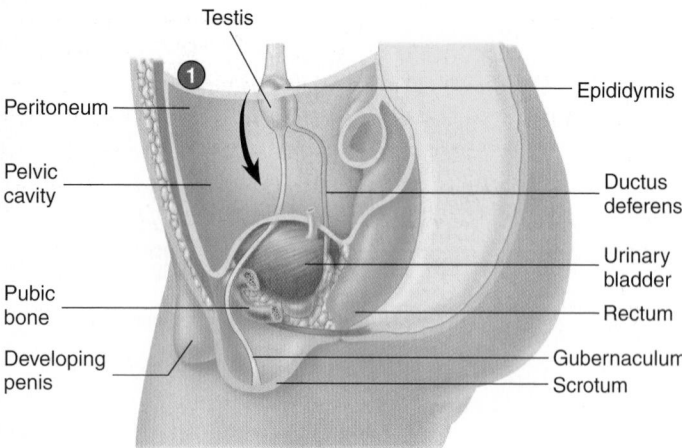

Approximately 8 weeks

(1) Each testis forms as a retroperitoneal structure near the level of the kidney and begins to descend.

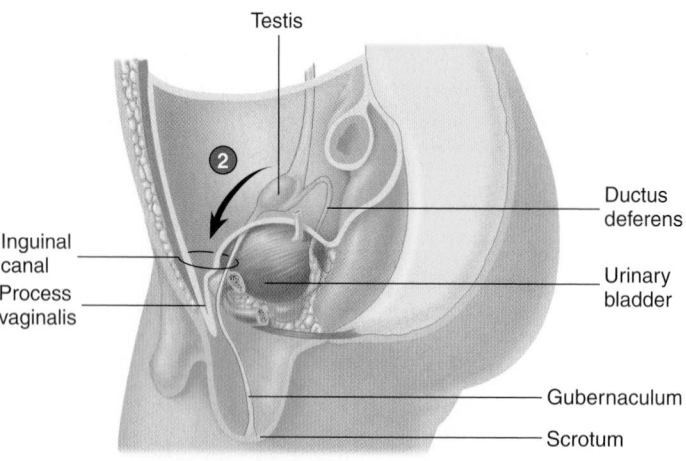

Approximately 12 weeks

(2) The testis beneath the parietal peritoneum descends toward the inguinal canal.

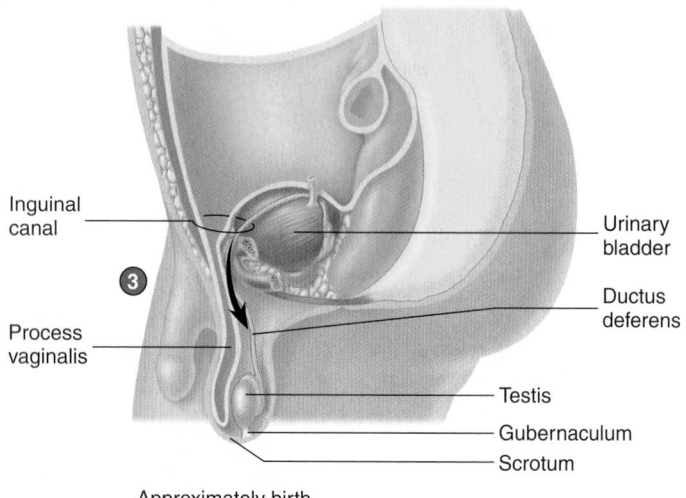

Approximately birth

(3) The testis descends through the inguinal canal into the scrotum.

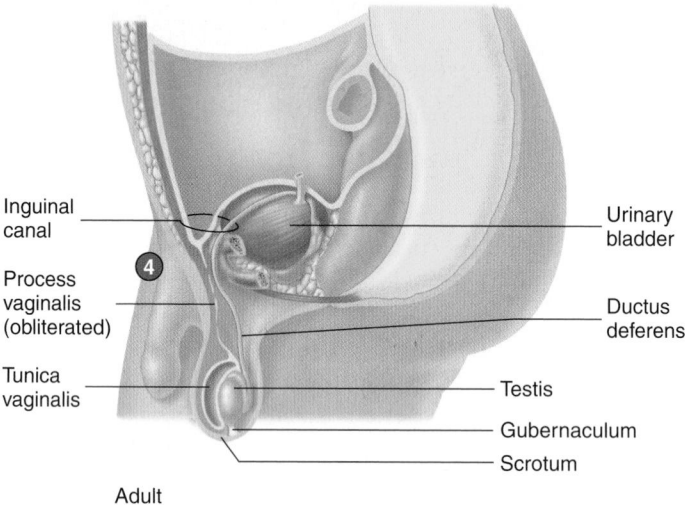

Adult

(4) Between birth and adulthood the process vaginalis is obliterated and its inferior portion becomes the tunica vaginalis.

Process Figure 28.4   Descent of the Testes

Inguinal Hernia

Normally, the inguinal canals are closed, but they do represent weak spots in the abdominal wall. **Inguinal hernias** (her'nē-ăz) are abnormal openings in the abdominal wall in the inguinal region through which structures such as a portion of the small intestine can protrude. If the deep inguinal ring remains open, or if it is weak and enlarges later in life, a loop of intestine can protrude into or even pass through the inguinal canal resulting in indirect inguinal hernia. A direct inguinal hernia results from a tear, or rupture, in a weakened area of the anterior abdominal wall near the inguinal canal, but not through the inguinal canal. These hernias can be quite painful and even very dangerous, especially if a portion of the small intestine is compressed so its blood supply is cut off. Fortunately, inguinal hernias can be repaired surgically. Males are much more prone to inguinal hernias than are females because a male's inguinal canals are larger and weakened because the testes pass through them on their way into the scrotum.

8. When and how do the testes descend into the scrotum?
9. Describe the tunica vaginalis.
10. Define cryptorchidism and its effect on sperm cell production.

Sperm Cell Development

Before puberty, the testes remain relatively simple and unchanged from the time of their initial development. The interstitial cells are not particularly prominent during this period, and the seminiferous tubules lack a lumen and are not yet functional. At 12–14 years of age, the interstitial cells increase in number and size, a lumen develops in each seminiferous tubule, and sperm cell production begins. It takes approximately 74 days for sperm cells to be produced. About 50 of those days are spent in the seminiferous tubules.

A cross section of a mature seminiferous tubule reveals the various stages of sperm cell development, a process called **spermatogenesis** (sper′mă-tō-jen′ĕ-sis; see figures 28.3*b* and 28.5*a*). The seminiferous tubules contain two types of cells, **germ cells** and **sustentacular** (sŭs-ten-tak′ū-lăr), or **Sertoli** (sēr-tō′lē; named for an Italian histologist) **cells.** The sustentacular cells are also sometimes referred to as **nurse cells.**

Sustentacular cells are large cells that extend from the periphery to the lumen of the seminiferous tubule (see figures 28.3*b* and 28.5*b*). They nourish the germ cells and probably produce, together with the interstitial cells, a number of hormones, such as androgens, estrogens, and inhibins. In addition, tight junctions between the sustentacular cells form a **blood–testis barrier** between spermatogonia and sperm cells. It isolates the sperm cells

1. Spermatogonia are the cells from which sperm cells arise. The spermatogonia divide by mitosis. One daughter cell remains a spermatogonium that can divide again by mitosis. One daughter cell becomes a primary spermatocyte.

2. The primary spermatocyte divides by meiosis to form secondary spermatocytes.

3. The secondary spermatocytes divide by meiosis to form spermatids.

4. The spermatids differentiate to form sperm cells.

(a)   (b)

Process Figure 28.5   Spermatogenesis

(*a*) Meiosis during spermatogenesis. A section of the seminiferous tubule illustrating the process of meiosis and sperm cell formation. (*b*) The tight junctions that form between adjacent sustentacular cells form the blood–testis barrier. The sustentacular cells extend from the periphery to the lumen of the seminiferous tubules. Spermatogonia are peripheral to the blood–testis barrier, and spermatocytes are central to it.

Clinical Focus   Meiosis

Sperm cell development and oocyte development involve meiosis (mī-ō′sis). This kind of cell division occurs only in the testes and ovaries. It consists of two consecutive nuclear divisions without a second replication of the genetic material between the divisions. Four daughter cells are produced, and each has half as many chromosomes as the parent cell (figure A).

The normal chromosome number in human cells is 46. This number is called a **diploid** (dip′loyd), or a **2n,** number of chromosomes. The chromosomes consist of 23 pairs, each of which is called a **homologous** (hŏ-mol′ō-gŭs) **pair.** The homologous pairs consist of 22 autosomal pairs and one pair of sex chromosomes. The sex chromosomes are an X and a Y chromosome in males and two X chromosomes in females. One chromosome of each homologous pair is from the male parent, and the other is from the female parent. The chromosomes of each homologous pair look alike, and they contain genes for the same traits.

In sperm cells and oocytes, the number of chromosomes is 23. This number is called a **haploid** (hap′loyd), or **n,** number of chromosomes. Each gamete contains one chromosome from each of the homologous pairs. Reduction in the number of chromosomes in sperm cells or oocytes to an *n* number is important. When a sperm cell and an oocyte fuse to form a fertilized egg, each provides an *n* number of chromosomes, which reestablishes a 2n number of chromosomes. If meiosis did not take place, each time fertilization occurred the number of chromosomes in the fertilized oocyte would double. The extra chromosomal material would be lethal to the developing offspring.

Sperm cells have 22 autosomal chromosomes and either an X or Y chromosome.

Oocytes contain one autosomal chromosome from each of the 22 homologous pairs and an X chromosome. During fertilization, when a sperm cell fuses with an oocyte, the normal number of 46 chromosomes, consisting of 23 pairs of chromosomes, is reestablished. The sex of the baby is determined by the sperm cell that fertilizes the oocyte. The sex is male if the sperm cell that fertilizes the oocyte carries a Y chromosome, and female if the sperm cell carries an X chromosome.

The two divisions of meiosis are called **meiosis I** and **meiosis II.** The stages of meiosis have the same names as these stages in mitosis, that is, prophase, metaphase, anaphase, and telophase; but distinct differences exist between mitosis and meiosis.

Before meiosis begins, all the DNA in the chromosomes is duplicated. At the beginning of meiosis, each of the 46 chromosomes consists of two sister **chromatids** (krō′mă-tid) connected by a **centromere** (sen′trō-mēr; figure A). In prophase of meiosis I, the chromosomes become visible, and the homologous pairs of chromosomes come together. This process is called **synapsis** (si-nap′sis). Because each chromosome consists of two chromatids, the pairing of the homologous chromosomes brings two chromatids of each chromosome close together, an arrangement called a **tetrad.** Occasionally, part of a chromatid of one homologous chromosome breaks off and is exchanged with part of another chromatid from the other homologous chromosome of the tetrad. This exchange of genetic material is called **crossing-over.** Crossing-over allows the exchange of genetic material between maternal and paternal chromosomes.

During metaphase, homologous pairs of chromosomes line up near the center of the cell. For each pair of homologous chromosomes, however, the side of the cell on which the maternal or paternal chromosome is located is random. The way the chromosomes align during synapsis results in the random assortment of maternal and paternal chromosomes in the daughter cells during meiosis. Crossing-over and the random assortment of maternal and paternal chromosomes are responsible for the large degree of diversity in the genetic composition of sperm cells and oocytes produced by each individual.

During anaphase I, the homologous pairs are separated to each side of the cell. During telophase I, new nuclei form and the cell completes division of form two cells. As a consequence, when meiosis I is complete, each daughter cell has one chromosome from each of the homologous pairs. Each of the 23 chromosomes in each daughter cell consists of two chromatids joined by a centromere.

It's during the first meiotic division that the chromosome number is reduced from a 2n number (46 chromosomes, or 23 pairs) to an *n* number (23 chromosomes, or one from each homologous pair). The first meiotic division is, therefore, called a **reduction division.**

The second meiotic division is similar to mitosis. The chromosomes, each consisting of two chromatids, line up near the middle of the cell (figure A). Then, the chromatids separate at the centromere, and each daughter cell receives one of the chromatids from each chromosome. When the centromere separates, each of the chromatids is called a chromosome. Consequently, each of the four daughter cells produced by meiosis contains 23 chromosomes.

---

from the immune system (figure 28.5*b*). This barrier is necessary because, as the sperm cells develop, they form surface antigens that could stimulate an immune response, resulting in their destruction (see figure 28.5*b*).

Testosterone, produced by the interstitial cells, passes into the sustentacular cells and binds to receptors. The combination of testosterone with the receptors is required for the sustentacular cells to function normally. In addition, testosterone is converted to two other steroids in the sustentacular cells: **dihydrotestosterone** (dī-hī′drō-tes-tos′ter-ōn) and **estrogen** (es′trō-jen). The sustentacular cells also secrete a protein called **androgen-binding** (an′drō-jen) **protein** into the seminiferous tubules. Testosterone and dihydrotestosterone bind to androgen-binding protein and are carried along with other secretions of the seminiferous tubules to

First Meiotic Division (Meiosis I)

Second Meiotic Division (Meiosis II)
(continued from the bottom of previous column)

1. Early prophase I
The duplicated chromosomes become visible chromatids (shown separated for emphasis; they actually are so close together that they appear as a single strand).

Centromere

Chromosomes

Nucleus

Chromatids

Centrioles

6. Prophase II
Each chromosome consists of two chromatids.

2. Middle prophase I
Homologous chromosomes synapse to form tetrads. Crossing over may occur at this stage.

Pair of chromosomes

Spindle fibers

3. Metaphase I
Tetrads align at the center of the cell. Random assortment of homologous chromosomes occurs.

Equatorial plane

7. Metaphase II
Chromosomes align along the center of the cell.

4. Anaphase I
Homologous chromosomes move apart to opposite sides of the cell.

8. Anaphase II
Chromatids separate and each is now called a chromosome.

5. Telophase I
New nuclei form, and the cell divides.

Cleavage furrow

9. Telophase II
New nuclei form around the chromosomes. The cells divide to form four daughter cells with a haploid (n) number of chromosomes.

Prophase II (top of next column)

Process Figure A Meiosis

the epididymis. Estradiol and dihydrotestosterone may be the active hormones that promote sperm cell formation.

Scattered between the sustentacular cells are smaller germ cells from which sperm cells are derived. The germ cells are arranged so that the most immature cells are at the periphery and the most mature cells are near the lumen of the seminiferous tubules. The most peripheral cells, those adjacent to the basement membrane of the seminiferous tubules, are **spermatogonia** (sper′mă-tō-gō′nē-ă), which divide by mitosis (see figure 28.5a). Some of the daughter cells produced from these mitotic divisions remain spermatogonia and continue to produce additional spermatogonia. The others divide through mitosis and differentiate to form **primary spermatocytes** (sper′mă-tō-sītz).

Meiosis (see Clinical Focus "Meiosis" on page 1040) begins when the primary spermatocytes divide. Each primary spermatocyte passes through the first meiotic division to become two **secondary spermatocytes.** Each secondary spermatocyte undergoes a second meiotic division to produce two even smaller cells called **spermatids** (sper′mă-tidz). Each spermatid undergoes the last phase of spermatogenesis called **spermiogenesis** (sper′mē-ō-jen′ĕ-sis) to form a mature **sperm cell,** or **spermatozoon** (sper′mă-tō-zō′on; pl., spermatozoa, sper′mă-tō-zō′ă; see figures 28.3c and d and 28.5). Each spermatid develops a head, midpiece, and a tail, or flagellum. The head contains chromosomes, and at the leading end it has a cap, the **acrosome** (ak′rō-sōm), which contains enzymes necessary for the sperm cell to penetrate the female sex cell. The flagellum is similar to a cilium (see chapter 3), and movement of microtubules past one another within the tail causes the tail to move and propel the sperm cell forward. The midpiece has large numbers of mitochondria, which produce the ATP necessary for microtubule movement.

At the end of spermatogenesis, the developing sperm cells gather around the lumen of the seminiferous tubules with their heads directed toward the surrounding sustentacular cells and their tails directed toward the center of the lumen (see figures 28.3b and 28.5). Finally, sperm cells are released into the lumen of the seminiferous tubules.

11. Describe the role of germ cells, sustentacular cells, and the blood–testis barrier in the structure of the seminiferous tubules.
12. Describe the conversion of testosterone to other hormones in the sustentacular cells.
13. Where, specifically, are sperm cells produced in the testis?
14. Describe the role of meiosis in sperm cell formation.
15. Describe and give the function of the parts of a mature sperm cell and explain the events that change a spermatid to a sperm cell.

Ducts

After their release into the seminiferous tubules, the sperm cells pass through the tubuli recti to the rete testis. From the rete testis, they pass through the efferent ductules, which leave the testis and enter the epididymis to join the duct of the epididymis. The sperm cells then leave the epididymis, passing through the ductus epididymis, ductus deferens, ejaculatory duct, and urethra to reach the exterior of the body.

Epididymis

The **efferent ductules** from each testis become extremely convoluted and form a comma-shaped structure on the posterior side of the testis called the **epididymis** (ep-i-did′i-mis; pl., epididymides, ep-i-di-dim′i-dēz; on the twin; "twin" refers to the paired, or twin, testes; figure 28.6).

The final maturation of the sperm cells occurs within the epididymis. It takes 12–16 days to travel through the epididymis and appear in the ejaculate. Changes that occur in sperm cells as they pass through the epididymis include a further reduction in cytoplasm, maturation of the acrosome, and the ability to bind to the zona pellucida of the secondary oocyte (see Oocyte Development and Fertilization on p. 1051). Sperm cells in the head of the epididymis are unable to fertilize secondary oocytes, and they are unable to become motile. Sperm cells from the tail of the epididymis are able to fertilize secondary oocytes and are capable of becoming motile.

Each epididymis consists of a head, a body, and a long tail (see figures 28.3a and 28.6). The head contains the convoluted efferent ductules, which empty into a single convoluted tube, the **duct of the epididymis,** located primarily within the body of the epididymis (see figure 28.3a). This duct alone, if unraveled, would extend for several meters. The duct of the epididymis has a pseudostratified columnar epithelium with elongated microvilli called **stereocilia** (ster′ē-ō-sil′ē-ă). The stereocilia function to increase the surface area of epithelial cells that absorb fluid from the lumen of the duct of the epididymis. The duct of the epididymis ends at the tail of the epididymis, which is located at the inferior border of the testis.

16. Describe how sperm cells move from the epididymis.
17. List the changes that occur in sperm cells while in the epididymis and the functions of the epididymis.

Ductus Deferens and Ejaculatory Duct

The **ductus deferens,** or **vas deferens,** emerges (see figures 28.1, 28.3a, and 28.6) from the tail of the epididymis and ascends along the posterior side of the testis medial to the epididymis and becomes associated with the blood vessels and nerves that supply the testis. These structures constitute the **spermatic cord.** The spermatic cord consists of (1) the ductus deferens, (2) the testicular artery and venous plexus, (3) lymphatic vessels, (4) nerves, and (5) fibrous remnants of the process vaginalis. The coverings of the spermatic cord include the **external spermatic fascia** (fash′ē-ă); the **cremaster muscle,** an extension of the muscle fibers of the internal abdominal oblique muscle of the abdomen; and the **internal spermatic fascia** (see figure 28.6).

The ductus deferens and the rest of the spermatic cord structures ascend and pass through the inguinal canal to enter the pelvic cavity (see figures 28.1 and 28.6). The ductus deferens crosses the lateral and posterior walls of the pelvic cavity, travels over the ureter, and loops over the posterior surface of the urinary bladder to approach the prostate gland. The end of the ductus deferens enlarges to form an **ampulla** (am-pul′lă). The lumen of the ductus deferens is lined by pseudostratified columnar epithelium, which is surrounded by smooth muscle. Peristaltic contractions of this smooth muscle tissue help propel sperm cells through the ductus deferens.

Figure 28.6 **Male Reproductive Structures**

Anterior view of the testes, epididymis, ductus deferens, and glands of the male reproductive system. The urethra is cut open along its dorsal side.

Adjacent to the ampulla of each ductus deferens is a sac-shaped gland called the seminal vesicle. A short duct from the seminal vesicle joins the ampulla of the ductus deferens to form the **ejaculatory** (ē-jak′ū-lă-tōr-ē) **duct.** The ejaculatory ducts are approximately 2.5 cm long. These ducts project into the prostate gland and end by opening into the urethra (see figures 28.1 and 28.6).

- 18. *Describe the structure and functions of the ductus deferens.*
- 19. *List the components and coverings of the spermatic cord.*
- 20. *Describe the route by which the ductus deferens extends from the testis to the prostate gland.*

Urethra

The **male urethra** (ū-rē′thră) is about 20 cm long and extends from the urinary bladder to the distal end of the penis (see figures 28.1, 28.6, and figure 28.7). The urethra is a passageway for both urine and male reproductive fluids. The urethra is divided into three parts: the prostatic part, the membranous part, and the spongy part. The **prostatic** (pros-tat′ik) **urethra** is connected to the bladder and passes through the prostate gland. Fifteen to 30 small ducts from the prostate gland and the two ejaculatory ducts empty into the prostatic urethra. The **membranous urethra** is the shortest part of the urethra and extends

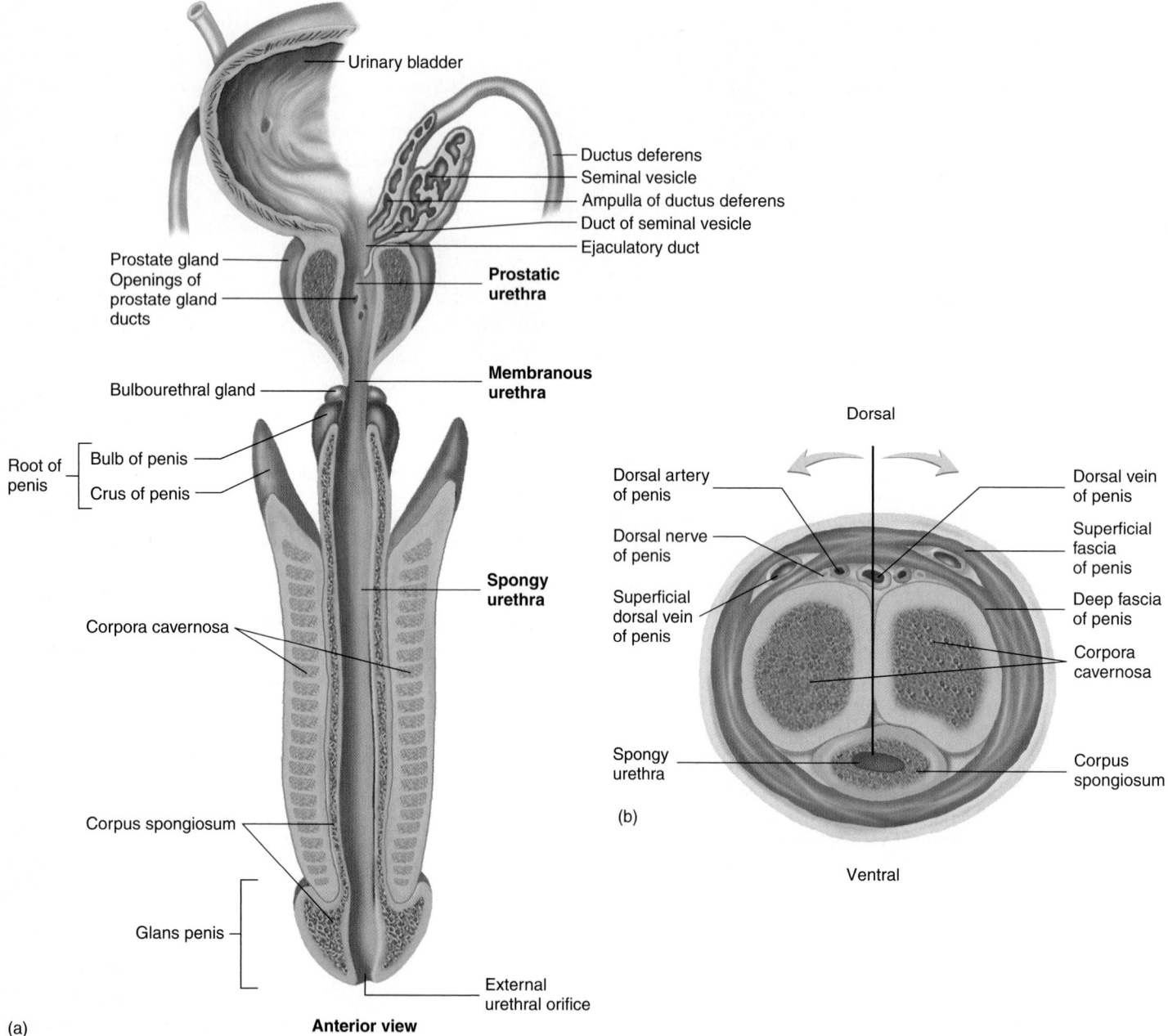

Figure 28.7   Penis

(*a*) Section through the spongy or penile urethra laid open and viewed from above. The prostate is also cut open to show the prostatic urethra. (*b*) Cross section of the penis showing principal nerves, arteries, and veins along the dorsum of the penis. The *line* and *arrows* depict the manner in which (*a*) is cut and laid open.

from the prostate gland through the perineum, which is part of the muscular floor of the pelvis. The **spongy urethra,** also called the **penile** (pē′nīl) **urethra,** is by far the longest part of the urethra and extends from the membranous urethra through the length of the penis. In rare cases, the penis does not develop completely. In these cases the urethra may open to the exterior along the inferior surface of the penis (see discussion of hypospadias, chapter 29, page 1100). Stratified columnar epithe-lium lines most of the urethra, but transitional epithelium is in the prostatic urethra near the bladder, and stratified squamous epithelium is near the opening of the spongy urethra to the outside. Several minute mucus-secreting **urethral glands** empty into the urethra.

● 21. **Distinguish between the ejaculatory duct and the prostatic,**
       membranous, and spongy parts of the male urethra.

Penis

The **penis** contains three columns of erectile tissue (see figure 28.7), and engorgement of this erectile tissue with blood causes the penis to enlarge and become firm, a process called **erection.** The penis is the male organ of copulation through which sperm cells are transferred from the male to the female. Two of the erectile columns form the dorsum and sides of the penis and are called the **corpora cavernosa** (kōr′pōr-ă kav-er-nos′ă). The third column, the **corpus spongiosum** (kōr′pŭs spŭn′jē-ō′sŭm), forms the ventral portion of the penis and it expands to form a cap, the **glans penis,** over the distal end of the penis. The spongy urethra passes through the corpus spongiosum, penetrates the glans penis, and opens as the **external urethral orifice.** At the base of the penis, the corpus spongiosum expands to form the **bulb of the penis,** and each corpus cavernosum expands to form a **crus** (kroos; pl. crura, kroo′ră) **of the penis.** Together, these structures constitute the **root of the penis** and the crura attach the penis to the coxae.

Skin is loosely attached to the connective tissue that surrounds the erectile columns in the shaft of the penis. The skin is firmly attached at the base of the glans penis, and a thinner layer of skin tightly covers the glans penis. The skin of the penis, especially the glans penis, is well supplied with sensory receptors. A loose fold of skin called the **prepuce** (prē′poos), or **foreskin,** covers the glans penis (see figure 28.1).

Circumcision

Circumcision (ser-kŭm-sizh′ŭn) is accomplished by surgically removing the prepuce, usually near the time of birth. Compelling medical reasons for circumcision do not appear to exist. Uncircumcised males have a higher incidence of penile cancer, but the underlying cause appears to be related to chronic infections and poor hygiene. In those few cases in which the prepuce is "too tight" to be moved over the glans penis, circumcision can be necessary to avoid chronic infections and maintain normal circulation.

The primary nerves, arteries, and veins of the penis pass along its dorsal surface (see figure 28.7*b*). Dorsal arteries, with dorsal nerves lateral to them, are found on either side of a single, midline dorsal vein. Additional deep arteries lie within the corpora cavernosa.

- 22. *Describe the erectile tissue of the penis.*
- 23. *Define the terms glans penis, crus, bulb, and prepuce, and describe their structure.*

Accessory Glands

Seminal Vesicles

The **seminal vesicles** (sem′i-năl ves′i-klz) are sac-shaped glands located next to the ampullae of the ductus deferentia (see figure 28.6). Each gland is about 5 cm long and tapers into a short excretory duct that joins the ampulla of the ductus deferens to form the ejaculatory duct. The seminal vesicles have a capsule containing fibrous connective tissue and smooth muscle cells.

Prostate Gland

The **prostate** (pros′tāt; one standing before) **gland** consists of both glandular and muscular tissue and is about the size and shape of a walnut; that is, about 4 cm long and 2 cm wide. It is dorsal to the symphysis pubis at the base of the urinary bladder, where it surrounds the prostatic urethra and the two ejaculatory ducts (see figure 28.1). The gland is composed of a fibrous connective tissue capsule containing distinct smooth muscle cells and numerous fibrous partitions, also containing smooth muscle, that radiate inward toward the urethra. Covering these muscular partitions is a layer of columnar epithelial cells that form saccular dilations into which the columnar cells secrete prostatic fluid. Fifteen to 30 small prostatic ducts carry these secretions into the prostatic urethra.

Cancer of the Prostate

Cancer of the prostate is the second-most common cause of cancer deaths in men in the United States, fewer than from lung cancer and more than from colon cancer. A prostate-specific antigen (PSA) increases in the circulatory system of most men who have prostatic cancer. A blood sample can be taken and an assay performed to test for the presence of the antigen. If the concentration of the antigen in the blood has increased, an examination for prostatic cancer is highly recommended. Because of its prevalence in men older than 50, an annual or biannual analysis for prostatic cancer should be done.

Some debate accompanies the treatment for prostatic cancer. Cancer of the prostate in relatively young men and large tumors in all men generally require treatment; however, treatment in elderly men is controversial. Some evidence suggests that elderly men with small tumors in the prostate are likely to die of causes unrelated to prostatic cancer, even if they receive no treatment. Treatment for prostatic cancer includes radiation, chemotherapy, and surgery. Surgery can result in incontinence and it generally results in the inability to sustain an erection, although new, more successful surgical techniques have been developed.

P R E D I C T

The prostate gland can enlarge for several reasons, including infections, tumor, and old age. The detection of enlargement or changes in the prostate is an important way to detect prostatic cancer. Suggest a way other than surgery that the prostate gland can be examined by palpation for any abnormal changes.

Bulbourethral Glands

The **bulbourethral** (bŭl′bō-ū-rē′thrăl) **glands** are a pair of small glands located near the membranous part of the urethra (see figures 28.1 and 28.6). In young males, each is about the size of a pea, but they decrease in size with age and are almost impossible to see in old men. Each bulbourethral gland is a compound mucous gland (see chapter 4). The small ducts of each gland unite to form a single duct. The single duct from each bulbourethral gland then enters the spongy urethra at the base of the penis.

Semen

Semen (sē′men) is a composite of sperm cells and secretions from the male reproductive glands. The seminal vesicles produce about

60% of the fluid, the prostate gland contributes about 30%, the testes contribute 5%, and the bulbourethral glands contribute 5%. **Emission** is the discharge of semen into the prostatic urethra. **Ejaculation** is the forceful expulsion of semen from the urethra caused by the contraction of the urethra, the skeletal muscles in the floor of the pelvis, and the muscles at the base of the penis.

The bulbourethral glands and urethral mucous glands produce a mucous secretion just before ejaculation. This mucus lubricates the urethra, neutralizes the contents of the normally acidic spongy urethra, provides a small amount of lubrication during intercourse, and helps reduce vaginal acidity.

Testicular secretions include sperm cells, a small amount of fluid, and metabolic by-products. The thick, mucuslike secretions of the seminal vesicles contain large amounts of fructose, citric acid, and other nutrients that nourish the sperm cells. The seminal vesicle secretions also contain fibrinogen, which is involved in a weak coagulation reaction of the semen immediately after ejaculation, and prostaglandins, which can cause uterine contractions.

The thin, milky secretions of the prostate have a rather high pH and with secretions of the seminal vesicles, bulborurethral glands, and urethral mucous glands help to neutralize the acidic urethra. The secretions of the prostate and seminal vesicles also help neutralize the acidic secretions of the testes, and those of the vagina. The prostatic secretions are also important in the transient coagulation of semen because they contain clotting factors that convert fibrinogen from the seminal vesicles to fibrin, resulting in coagulation. The coagulated material keeps the semen as a single, sticky mass for a few minutes after ejaculation, and then fibrinolysin from the prostate causes the coagulum to dissolve, thereby releasing the sperm cells to make their way up the female reproductive tract.

PREDICT 2
Explain a possible reason for having the coagulation reaction.

Before ejaculation, the ductus deferens begins to contract rhythmically to propel sperm cells and testicular and epididymal secretions from the tail of the epididymis to the ampulla of the ductus deferens. Contractions of the ampullae, seminal vesicles, and ejaculatory ducts cause the sperm cells along with testicular and epididymal secretions, to move into the prostatic urethra along with the prostatic secretions. Secretions of the seminal vesicles then enter the prostatic urethra, where they mix with the other secretions.

Normal sperm cell counts in the semen range from 75–400 million sperm cells per milliliter of semen, and a normal ejaculation usually consists of about 2–5 mL of semen. The semen with the highest sperm count is expelled from the penis first because it contains the greater percentage of sperm-containing fluid from the epididymis. Sperm cells become motile as a result of being mixed with secretions of the male accessory glands and with secretions of the female reproductive tract. The alkaline pH (an average of 7.5), nutrients, and removal of inhibitory substances from the surface of sperm cells appear to increase sperm cell motility. Most of the sperm cells (millions) are expended in moving the general group of sperm cells through the female reproductive system. Enzymes carried in the acrosomal cap of each sperm cell help to digest a path through the mucoid fluids of the female reproductive tract and through materials surrounding the oocyte. Once the acrosomal

fluid is depleted from a sperm cell, the sperm cell is no longer capable of fertilization.

24. State where the seminal vesicles, prostate gland, and bulbourethral glands empty into the male reproductive duct system.
25. Define the terms emission and ejaculation.
26. Define the term semen. Describe the contribution to semen of the accessory sex glands.
27. What is the function of the secretions of each of the accessory glands?

Physiology of Male Reproduction

Normal function of the male reproductive system depends on both hormonal and neural mechanisms. Hormones are primarily responsible for the development of reproductive structures and maintenance of their functional capacities, development of secondary sexual characteristics, control of sperm cell formation, and influencing sexual behavior. Neural mechanisms are primarily involved in sexual behavior and controlling the sexual act.

Regulation of Sex Hormone Secretion

Hormonal mechanisms that influence the male reproductive system involve the hypothalamus, the pituitary gland, and the testes (figure 28.8). A small peptide hormone called **gonadotropin-releasing hormone (GnRH)** or **luteinizing hormone-releasing hormone (LHRH)** is released from neurons in the hypothalamus. GnRH passes through the hypothalamohypophyseal portal system to the anterior pituitary gland (see chapter 18). In response to GnRH, cells within the anterior pituitary gland secrete two hormones referred to as **gonadotropins** (gō'nad-ō-trō'pinz, gon'ă-dō-trō'pinz) because they influence the function of the **gonads** (gō'nadz; the testes or ovaries).

The two gonadotropins are **luteinizing hormone (LH)** and **follicle-stimulating hormone (FSH)**. They are named for their functions in females, but they also perform important functions in males. LH in males is sometimes called **interstitial cell-stimulating hormone (ICSH)**. LH binds to the interstitial cells in the testes and causes them to increase their rate of testosterone synthesis and secretion. FSH binds primarily to sustentacular cells in the seminiferous tubules and promotes sperm cell development. Both gonadotropins bind to specific receptor molecules on the membranes of the cells that they influence, and cyclic adenosine monophosphate (cAMP) is an important intracellular mediator in those cells.

For GnRH to stimulate the secretion of large quantities of LH and FSH, the anterior pituitary must be exposed to a series of brief increases and decreases in GnRH. Chronically elevated GnRH levels in the blood cause the anterior pituitary cells to become insensitive to stimulation by GnRH molecules, and little LH or FSH is secreted.

GnRH Pulses and Infertility
GnRH can be produced synthetically, and if administered in small amounts in frequent pulses or surges, it can be useful in treating males who are infertile. GnRH can also inhibit reproduction because chronic administration of it can sufficiently reduce LH and FSH levels to prevent sperm cell production in males or ovulation in females.

Clinical Focus Male Infertility

Infertility (in-fer-til'i-tē) is reduced or diminished fertility. The most common cause of infertility in males is a low sperm cell count. If the sperm cell count drops to below 20 million sperm cells per milliliter, the male is usually infertile.

A decreased sperm cell count can occur because of damage to the testes as a result of trauma, radiation, cryptorchidism, or infections like mumps. Reduced sperm cell counts can result from inadequate secretion of luteinizing hormone and follicle-stimulating hormone, which can be caused by hypothyroidism, trauma to the hypothalamus, infarctions of the hypothalamus or anterior pituitary gland, and tumors. Decreased testosterone secretion also reduces the sperm cell count. Some reports suggest that the average sperm count has decreased substantially since the end of World War II (1945) although there is some controversy about the accuracy of these reports. It's speculated that certain synthetic chemicals are responsible.

Fertility is reduced if the sperm cell count is normal but sperm cell structure is abnormal. Abnormal sperm cell structure can be due to chromosomal abnormalities or genetic factors. Reduced sperm cell motility also results in infertility. A major cause of reduced sperm cell motility is anti-sperm antibodies produced by the immune system, which bind to sperm cells.

Fertility can sometimes be achieved by collecting several ejaculations and concentrating the sperm cells. The more concentrated sperm cells can then be introduced into the female's reproductive tract, a process called **artificial insemination** (in-sem-i-nā′shŭn).

1. GnRH from the hypothalamus stimulates the secretion of LH and FSH from the anterior pituitary.

2. LH stimulates testosterone secretion from the interstitial cells.

3. FSH stimulates sustentacular cells of the seminiferous tubules to increase spermatogenesis and to secrete inhibin.

4. Testosterone has a stimulatory effect on the sustentacular cells of the seminiferous tubules, and it has a stimulatory effect on the development of sex organs and secondary sex characteristics.

5. Testosterone has a negative-feedback effect on the hypothalamus and pituitary to reduce LH and FSH secretion.

6. Inhibin has a negative-feedback effect on the anterior pituitary to reduce FSH secretion.

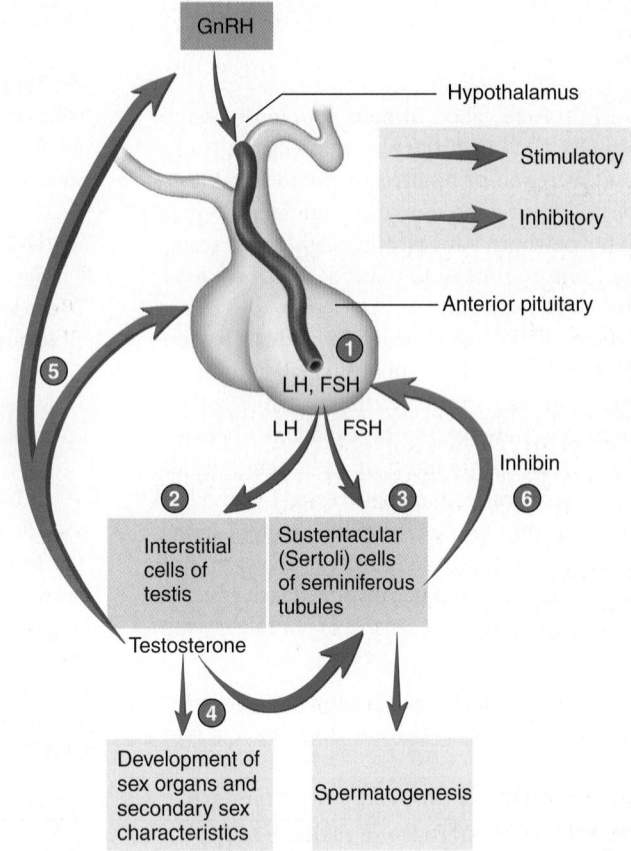

Process Figure 28.8   **Regulation of Reproductive Hormone Secretion in Males**

Testosterone is the major male hormone secreted by the testes. It's classified as an **androgen** (*andros* is Greek for male human being) because it stimulates the development of male reproductive structures (see chapter 29) and male secondary sexual characteristics. The testes secrete other androgens, but they are produced in smaller concentrations and are less potent than testosterone. In addition, the testes secrete small amounts of estrogen and progesterone.

Testosterone has a major influence on many tissues. It plays an essential role in the embryonic development of reproductive structures, their further development during puberty, the development of secondary sexual characteristics during puberty, the maintenance of sperm cell production, and the regulation of gonadotropin secretion. It also influences behavior.

Inside of some target tissue cells, an enzyme converts testosterone to dihydrotestosterone. In these cells, dihydrotestosterone is the active hormone. The scrotum and penis are examples. If the enzyme is not active, these structures don't fully develop normally. In other target tissue cells, an enzyme converts testosterone to estrogen, and estrogen is the active hormone. Some brain cells convert testosterone to estrogen and, in these cells, estrogen may be the active hormone responsible for some aspects of male sexual behavior.

The sustentacular cells of the testes secrete a polypeptide hormone called **inhibin** (in-hib′in). Inhibin inhibits FSH secretion from the anterior pituitary.

- 28. *Where are GnRH, FSH, LH, and inhibin produced? What effects do they produce?*
- 29. *Where is testosterone produced?*
- 30. *Explain the regulation of testosterone secretion.*

Puberty

A gonadotropin-like hormone called **human chorionic** (kō-rē-on′ik) **gonadotropin (HCG),** which the placenta secretes, stimulates the synthesis and secretion of testosterone by the fetal testes before birth. After birth, however, no source of stimulation is present, and the testes of the newborn baby atrophy slightly and secrete only small amounts of testosterone until puberty, which normally begins when a boy is 12–14 years old.

Puberty (pū′ber-tē) is the age at which individuals become capable of sexual reproduction. Before puberty, small amounts of testosterone and other androgens in males inhibit GnRH release from the hypothalamus. At puberty, the hypothalamus becomes much less sensitive to the inhibitory effect of androgens, and the rate of GnRH secretion increases, leading to increased LH and FSH release. Elevated FSH levels promote sperm cell formation, and elevated LH levels cause the interstitial cells to secrete larger amounts of testosterone. Testosterone still has a negative-feedback effect on GnRH secretion after puberty but is not capable of completely suppressing it.

- 31. *What changes in hormone production occur at puberty?*

Effects of Testosterone

Testosterone is by far the major androgen in males. Nearly all of the androgens, including testosterone, are produced by the interstitial cells, with small amounts produced by the adrenal cortex and possibly by the sustentacular cells. Testosterone causes the enlargement and differentiation of the male genitals and reproductive duct system, is necessary for sperm cell formation, and is required for the descent of the testes near the end of fetal development. Testosterone stimulates hair growth in the following regions: (1) the pubic area and extending up the linea alba, (2) the legs, (3) the chest, (4) the axillary region, (5) the face, and (6) occasionally, the back. It causes vellus hair to be converted to terminal hair, which are more pigmented and coarser.

Male Pattern Baldness

Some men have a genetic tendency called **male pattern baldness,** which develops in response to testosterone and other androgens. When testosterone levels increase at puberty, the density of hair on the top of the head begins to decrease. Baldness usually reaches its maximum rate of development when the individual is in the third or fourth decade of life.

Testosterone also causes the texture of the skin to become rougher or coarser. The quantity of melanin in the skin also increases, making the skin darker. Testosterone increases the rate of secretion from the sebaceous glands, especially in the region of the face, frequently resulting near the time of puberty in the development of acne. Beginning near the time of puberty, testosterone also causes hypertrophy of the larynx and reduced tension on the vocal folds. The structural changes can first result in a voice that's difficult to control, but ultimately the voice reaches its normal masculine quality.

Testosterone has a general stimulatory effect on metabolism so that males have a slightly higher metabolic rate than females. The red blood cell count is increased by nearly 20% as a result of the effects of testosterone on erythropoietin production. Testosterone also has a minor mineralocorticoid-like effect, causing the retention of sodium in the body and, consequently, an increase in the volume of body fluids. Testosterone promotes protein synthesis in most tissues of the body; as a result, skeletal muscle mass increases at puberty. The average percentage of the body weight composed of skeletal muscle is greater for males than for females because of the effect of androgens.

Synthetic Androgens and Muscle Mass

Some athletes, especially weightlifters, ingest synthetic androgens in an attempt to increase muscle mass. The side effects of the large doses of androgens are often substantial and include testicular atrophy, kidney damage, liver damage, heart attacks, and strokes. Administration of synthetic androgens is highly discouraged by the medical profession and is a violation of the rules of most athletic organizations.

Testosterone causes rapid bone growth and increases the deposition of calcium in bone, resulting in an increase in height. The growth in height is limited, however, because testosterone also causes early closure of the epiphyseal plates of long bones (see chapter 6). Males who mature sexually at an earlier age grow rapidly but reach their maximum height earlier. Males who mature sexually at a later age do not exhibit a rapid period of growth, but they grow for a longer period and can become taller than those who mature sexually at an earlier age.

- 32. *Describe the effects of testosterone on male reproductive structures and secondary sexual characteristics.*

Male Sexual Behavior and the Male Sex Act

Testosterone is required to initiate and maintain male sexual behavior. Testosterone enters cells within the hypothalamus and the surrounding areas of the brain and influences their function, resulting in sexual behavior. Male sexual behavior may depend, in part, however, on the conversion of testosterone to other steroids, such as estrogen, in the cells of the brain.

The blood levels of testosterone remain relatively constant throughout the lifetime of a male from puberty until about 40 years of age. Thereafter, the levels slowly decline to about 20% of this value by 80 years of age, causing a slow decrease in sex drive and fertility.

The male sex act is a complex series of reflexes that result in erection of the penis, secretion of mucus into the urethra, emission, and ejaculation. Sensations that are normally interpreted as pleasurable occur during the male sexual act and result in a climactic sensation, **orgasm** (ōr′gazm), associated with ejaculation. After ejaculation, a phase called **resolution** occurs, in which the penis becomes flaccid, an overall feeling of satisfaction exists, and the male is unable to achieve erection and a second ejaculation for many minutes to many hours or longer.

Sensory Action Potentials and Integration

Action potentials are conducted by sensory neurons from the genitals through the pudendal nerve to the sacral region of the spinal cord, where reflexes that result in the male sexual act are integrated. Action potentials travel from the spinal cord to the cerebrum to produce conscious sexual sensations.

Rhythmic massage of the penis, especially the glans penis, provides an extremely important source of sensory action potentials that initiate erection and ejaculation. Sensory action potentials, produced in surrounding tissues, such as the scrotum and the anal, perineal, and pubic regions, reinforce sexual sensations. Engorgement of the prostate and seminal vesicles with their secretions also cause sexual sensations. In some cases, mild irritation of the urethra, such as from an infection, can cause sexual sensations.

Psychic stimuli, such as sight, sound, odor, or thoughts, have a major effect on sexual reflexes. Thinking sexual thoughts or dreaming about erotic events tends to reinforce stimuli that trigger sexual reflexes like erection and ejaculation. Ejaculation while sleeping is a relatively common event in young males and is thought to be triggered by psychic stimuli associated with dreaming. Psychic stimuli can also inhibit the sexual act, and thoughts that are not sexual in nature tend to decrease the effectiveness of the male sexual act. The inability to concentrate on sexual sensations is one of the causes of **impotence** (im′pŏ-tens), the inability to achieve or maintain an erection and to accomplish the male sexual act. Impotence can also be caused by nerve damage, such as can result from prostate surgery, or restricted circulation.

Action potentials from the cerebrum that reinforce the sacral reflexes are not absolutely required for the culmination of the male sexual act, and the male sexual act can occasionally be accomplished by males who have suffered spinal cord injuries superior to the sacral region.

Erection, Emission, and Ejaculation

When **erection** (ē-rek′shŭn) occurs, the penis becomes enlarged and rigid. Erection is the first major component of the male sexual act. Action potentials travel from the spinal cord through the pudendal nerve to the arteries that supply blood to the erectile tissues. The nerve fibers release acetylcholine as well as **nitric oxide** (NO) as neurotransmitter substances. Acetylcholine binds to muscarinic receptors and activates a G-protein mechanism that causes smooth muscle relaxation. Nitric oxide diffuses into the smooth muscle cells of blood vessels and combines with the enzyme guanylate cyclase, which converts guanosine triphosphate (GTP) to cyclic guanosine monophosphate (cGMP). The cGMP causes smooth muscle cells to relax and

blood vessels to dilate (figure 28.9). At the same time, other arteries of the penis constrict to shunt blood to the erectile tissues. As a consequence, blood fills the sinusoids of the erectile tissue and compresses the veins. Because venous outflow is partially occluded, the blood pressure in the sinusoids causes inflation and rigidity of the erectile tissue. Nerve action potentials that result in erection come from parasympathetic centers (S2–S4) or sympathetic centers (T2–L1) in the spinal cord. Normally, the parasympathetic centers are more important, but in cases of damage to the sacral region of the spinal cord, it's possible for erection to occur through the sympathetic system.

Parasympathetic action potentials also cause the mucous glands within the penile urethra and the bulbourethral glands at the base of the penis to secrete mucus.

1. Action potentials in parasympathetic neurons cause voltage-gated Ca^{2+} channels to open and Ca^{2+} diffuse into the presynaptic terminals.
2. Calcium ions initiate the release of acetylcholine (ACh) from presynaptic vesicles.
3. Calcium ions also activate nitric oxide synthase, which promotes the synthesis of nitric oxide (NO) from arginine.
4. ACh binds to ACh receptors on the smooth muscle cells and activates a G protein mechanism. The activated G protein causes the relaxation of smooth muscle cells and erection of the penis.
5. NO binds to guanylate cyclase enzymes and activates them. The activated enzymes convert GTP to cGMP, which causes relaxation of the smooth muscle cells and erection of the penis.

Process Figure 28.9 Neural Control of Erection

Erectile Dysfunction

Failure to achieve erections, **erectile dysfunction (ED),** can be a major source of frustration for some men and can contribute to disharmony in relationships. ED can be due to reduced testosterone secretion, which can result from hypothalamic, pituitary, or testicular complications. In other cases, ED can be due to defective stimulation of the erectile tissue by nerve fibers or reduced response of the blood vessels to neural stimulation. Erection can be achieved in some people by oral medication, such as sildenafil (Viagra), or by the injection of specific drugs into the base of the penis, which function to increase blood flow into the sinusoids of the erectile tissue of the penis, resulting in erection for many minutes. Viagra blocks the activity of the enzyme that converts cGMP to GMP. Consequently, it allows cGMP to accumulate in smooth muscle cells in the arteries of erectile tissues and causes them to relax. Although Viagra is effective in causing or enhancing erection in males, its action is not specific to erectile tissue of the penis. It causes vasodilation in other tissues and can increase the workload of the heart.

Emission (ē-mish′ŭn) is the accumulation of sperm cells and secretions of the prostate gland and seminal vesicles in the urethra. Sympathetic centers (T12–L1) in the spinal cord, which are stimulated as the level of sexual tension increases, control emission. Sympathetic action potentials cause peristaltic contractions of the reproductive ducts and stimulate the seminal vesicles and the prostate gland to release their secretions. Consequently, semen accumulates in the prostatic urethra and produces sensory action potentials that pass through the pudendal nerves to the spinal cord.

Integration of these action potentials results in both sympathetic and somatic motor output. Sympathetic action potentials cause constriction of the internal sphincter of the urinary bladder so that semen and urine are not mixed. Somatic motor action potentials are sent to the skeletal muscles of the urogenital diaphragm and the base of the penis, causing several rhythmic contractions that force the semen out of the urethra. The movement of semen out of the urethra is called **ejaculation** (ē-jak-ū-lā′shŭn). In addition, an increase in muscle tension occurs throughout the body.

- 33. *What effects do psychic, tactile, parasympathetic, and sympathetic stimulation have on the male sex act?*
- 34. *Describe the processes of erection, emission, and ejaculation.*

Anatomy of the Female Reproductive System

The female reproductive organs consist of the ovaries, uterine tubes, uterus, vagina, external genital organs, and mammary glands. The internal reproductive organs of the female are within the pelvis between the urinary bladder and the rectum (figure 28.10). The uterus and the vagina are in the midline, with the ovaries to each side of the uterus. A group of ligaments holds the internal reproductive organs in place. The most conspicuous is the **broad ligament,** an extension of the peritoneum that spreads out on both sides of the uterus and to which the ovaries and uterine tubes are attached (figure 28.11).

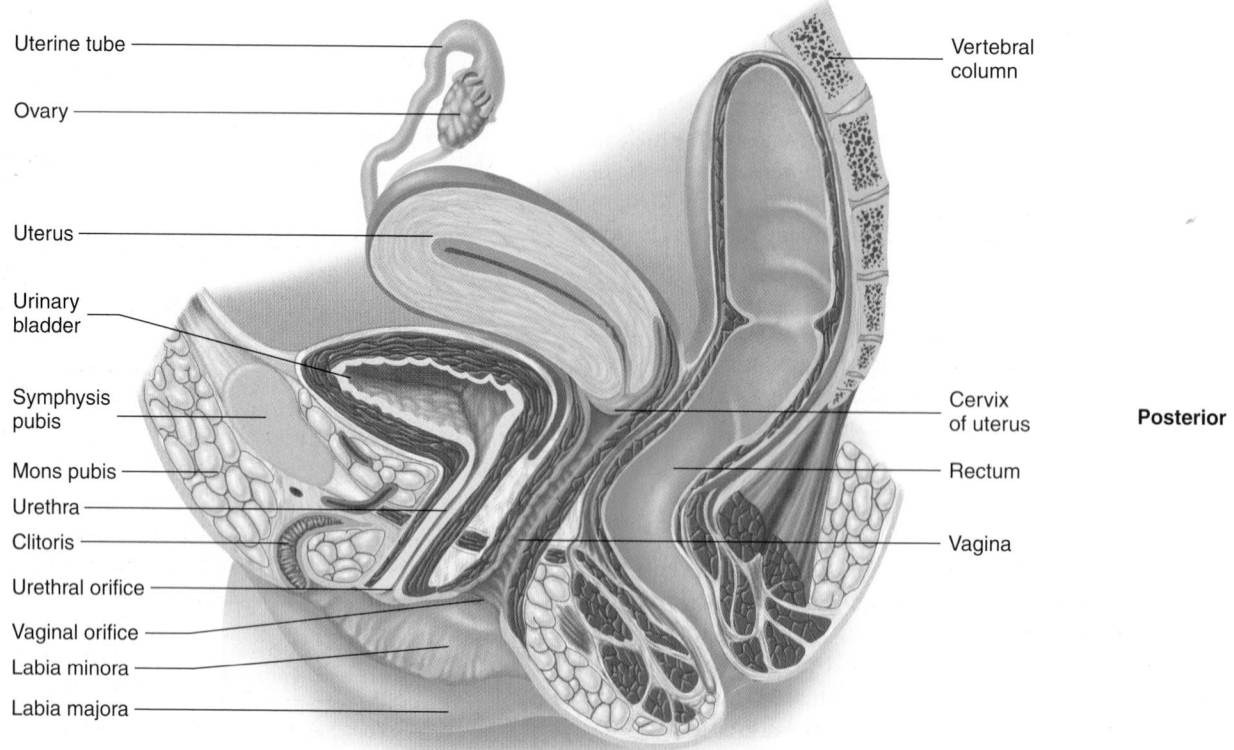

Uterine tube

Ovary

Uterus

Urinary bladder

Symphysis pubis

Mons pubis

Urethra

Clitoris

Urethral orifice

Vaginal orifice

Labia minora

Labia majora

Vertebral column

Cervix of uterus

Rectum

Vagina

Anterior

Posterior

Medial view

Figure 28.10   Sagittal Section of the Female Pelvis

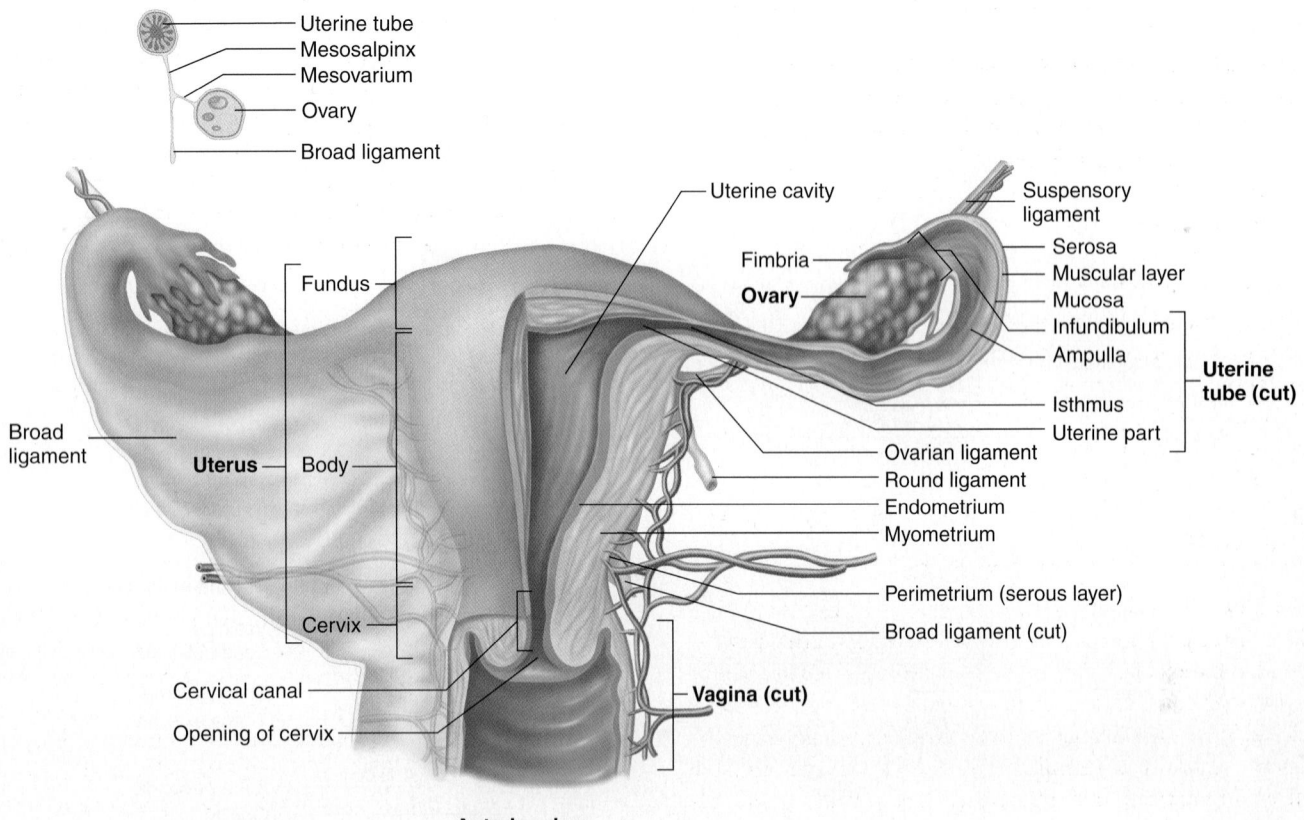

Figure 28.11   Uterus, Vagina, Uterine Tubes, Ovaries, and Supporting Ligaments

Anterior view of the uterus, uterine tubes, and associated ligaments. The uterus and uterine tubes are cut in section (on the left side), and the vagina is cut to show the internal anatomy. The inset shows the relationships between the ovary, uterine tube, and the ligaments that suspend them in the pelvic cavity.

Ovaries

The two **ovaries** (ō'var-ēz) are small organs about 2–3.5 cm long and 1–1.5 cm wide (see figure 28.11). A peritoneal fold called the **mesovarium** (mez'ō-vā'rē-ŭm; mesentery of the ovary) attaches each ovary to the posterior surface of the broad ligament. Two other ligaments are associated with the ovary: the **suspensory ligament,** which extends from the mesovarium to the body wall, and the **ovarian ligament,** which attaches the ovary to the superior margin of the uterus. The ovarian arteries, veins, and nerves traverse the suspensory ligament and enter the ovary through the mesovarium.

Ovarian Histology

The visceral peritoneum, which is made up of simple cuboidal epithelium, where it covers the surface of the ovary, is called the **ovarian,** or **germinal, epithelium.** The term germinal epithelium exists because the cuboidal epithelial cells were once thought to produce oocytes. Immediately below the epithelium is a capsule of dense fibrous connective tissue, the **tunica albuginea** (al-bū-jin'ē-ă). The more dense outer part of the ovary is called the **cortex** and a looser inner part is called the **medulla** (figure 28.12). The connective tissue of the ovary is called the **stroma** of the ovary. Blood vessels, lymphatic vessels, and nerves from the mesovarium enter

the medulla. Numerous small vesicles called ovarian follicles, each of which contains an **oocyte** (ō'ō-sīt), are distributed throughout the stroma of the cortex.

- 35. *Name and describe the ligaments that hold the ovaries in place.*
- 36. *Describe the coverings and structure of the ovary.*

Oocyte Development and Fertilization

The formation of female sex cells begins in the fetus. By the fourth month of development, the ovaries contain 5 million **oogonia** (ō-ō-gō'nē-ă, *oon,* egg + *gone,* generation), the cells from which oocytes develop (figure 28.13). By the time of birth, many of the oogonia have degenerated, and the remaining have begun meiosis. Also, some data indicate that oogonia can form after birth from stem cells, but the extent to which this occurs, and how long it occurs, is not clear. Meiosis stops, however, during the first meiotic division at a stage called prophase I. The cell at this stage is called a **primary oocyte,** and at birth there are about 2 million of them. From birth to puberty, the number of primary oocytes decreases to around 300,000 to 400,000. Only about 400 primary oocytes will complete development and give rise to secondary oocytes that are released from the ovaries.

Ovulation (ov'ū-lā'shŭn, ō'vū-lā'shŭn) is the release of an oocyte from an ovary (see figure 28.13). Just before ovulation, the

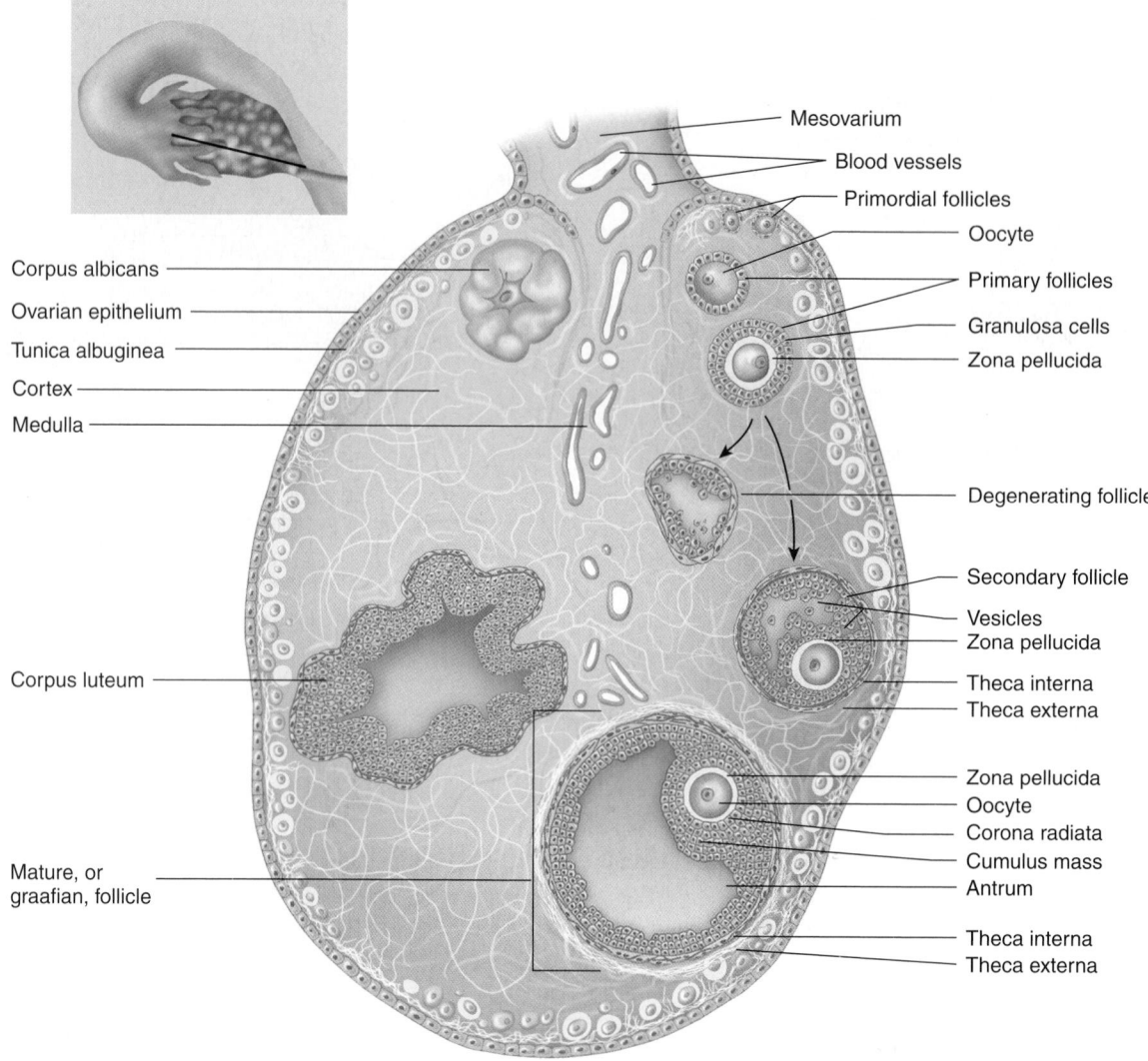

Figure 28.12 Histology of the Ovary

The ovary is sectioned to illustrate its internal structure (inset shows plane of section). Ovarian follicles from each major stage of development are present.

primary oocyte completes the first meiotic division to produce a **secondary oocyte** and a **polar body.** Unlike meiosis in males, cytoplasm is not split evenly between the two cells. Most of the cytoplasm of the primary oocyte remains with the secondary oocyte. The cytoplasm contains organelles, such as mitochondria, and nutrients that increase the viability of the secondary oocyte. The polar body either degenerates or divides to form two polar bodies. Eventually the polar bodies degenerate. The secondary oocyte begins the second meiotic division, but stops in metaphase II.

After ovulation, the secondary oocyte may be fertilized by a sperm cell (see figure 28.13). **Fertilization** (fer′til-i-zā-shŭn) begins when a sperm cell binds to the plasma membrane and penetrates into the cytoplasm of a secondary oocyte. Subsequently, the secondary oocyte completes the second meiotic division to form two cells, each containing 23 chromosomes. One of these cells has very little

cytoplasm and is another polar body that degenerates. In the other, larger cell, the 23 chromosomes from the male join with the 23 from the female to form a **zygote** (zī′gōt, *zygotos,* yolked) and complete fertilization. The zygote has 23 pairs or 46 chromosomes. All cells of the human body contain 23 pairs of chromosomes, except for the male and female sex cells. The zygote divides by mitosis to form two cells, which divide to form four cells, and so on. The mass of cells formed may eventually implant, or attach, to the wall of the uterus and develop into a new individual (see chapter 29).

37. *Starting with oogonia, describe the formation of secondary oocytes by meiosis. What are polar bodies?*
38. *Define ovulation and fertilization.*
39. *Describe the formation of a zygote. How many pairs of chromosomes are in a zygote and where do they come from? What can the zygote become?*

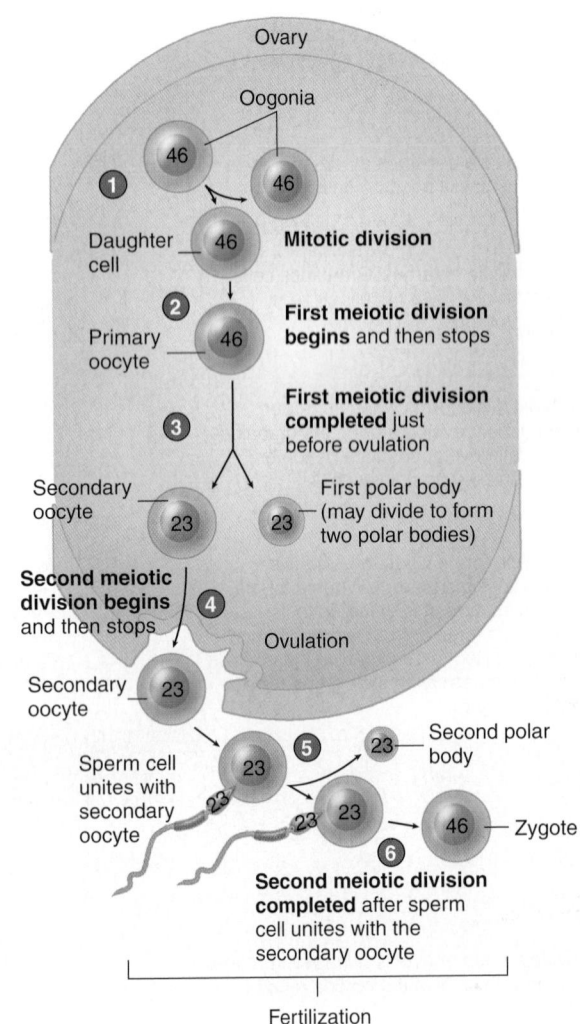

1. Oogonia are the cells from which oocytes arise. The oogonia divide by mitosis to produce other oogonia and primary oocytes.

2. Five million oocytes may be produced by the fourth month of prenatal life. Primary oocytes begin the first meiotic division but stop at prophase I. All of the primary oocytes remain in this state until puberty.

3. The first meiotic division is completed in a single mature follicle just before ovulation during each menstrual cycle. A secondary oocyte and the first polar body result from the unequal division of the cytoplasm.

4. The secondary oocyte begins the second meiotic division but stops at metaphase II.

5. The second meiotic division is completed after ovulation and after a sperm cell unites with the secondary oocyte. A secondary oocyte and a second polar body are formed.

6. Fertilization is completed after the nuclei of the secondary oocyte and the sperm cell unite. The resulting cell is called a zygote.

Process Figure 28.13 Maturation and Fertilization of the Oocyte

The primary oocyte undergoes meiosis and gives off the first polar body to become a secondary oocyte just before ovulation. Sperm cell penetration initiates the completion of the second meiotic division and the expulsion of a second polar body. The nuclei of the oocyte and the sperm cell unite. Fertilization results in the formation of a zygote.

Follicle Development

A **primordial follicle** is a primary oocyte surrounded by a single layer of flat cells, called **granulosa cells** (figure 28.14). Beginning with puberty, some of the primordial follicles are converted to **primary follicles** when the oocyte enlarges and the single layer of granulosa cells becomes enlarged and cuboidal. Subsequently, several layers of granulosa cells form and a layer of clear material is deposited around the primary oocyte called the **zona pellucida** (zō′nă pellū′sid-dă, *zone*, girdle + *pellucidus*, passage of light).

Approximately every 28 days, hormonal changes stimulate some of the primary follicles to continue to develop (see figure 28.14). The primary follicle becomes a **secondary follicle** as fluid-filled spaces called **vesicles** form among the granulosa cells, and a capsule called the **theca** (thē′kă, a box) forms around the follicle. Cells of the **theca interna** surround the granulosa cells and participate with the granulosa cells in the synthesis of ovarian hormones (see page 1062). The **theca externa** is primarily connective tissue that merges with the stroma of the ovary.

The secondary follicle continues to enlarge, and, when the fluid-filled vesicles fuse to form a single fluid-filled chamber called the **antrum** (an′trŭm), the follicle is called the **mature** or **graafian** (graf′ē-ăn, named for the Dutch histologist, Reijnier de Graaf [1641]), **follicle.** The oocyte is pushed off to one side and lies in a mass of granulosa cells called the **cumulus cells,** or **cumulus oophorus** (kū′mū-lŭs ō-of′ōr-ŭs). The innermost cells of this mass resemble a crown radiating from the oocyte and are thus called the **corona radiata.**

The mature follicle forms a lump on the surface of the ovary. During ovulation, the mature follicle ruptures, forcing a small amount of blood, follicular fluid, and the oocyte, surrounded by the cumulus cells, into the peritoneal cavity.

Usually, only one graafian follicle reaches the most advanced stages of development and is ovulated. The other follicles degenerate, a process called **atresia.**

40. List the characteristics that distinguish between primordial, primary, secondary, and mature follicles.

1. The primordial follicle consists of an oocyte surrounded by a single layer of squamous granulosa cells.

2. A primordial follicle becomes a primary follicle as the granulosa cells become enlarged and cuboidal.

3. The primary follicle enlarges. Granulosa cells form more than one layer of cells. A zona pellucida forms around the oocyte.

4. A secondary follicle forms when fluid-filled vesicles (spaces) develop among the granulosa cells and a well-developed theca becomes apparent around the granulosa cells.

5. A mature follicle forms when the fluid-filled vesicles form a single antrum. When a follicle becomes fully mature, it is enlarged to its maximum size, a large antrum is present, and the oocyte is located in the cumulus mass. Cells of the cumulus mass called the corona radiata surround the oocyte.

6. During ovulation the oocyte is released from the follicle, along with cells of the corona radiata.

7. Following ovulation, the granulosa cells divide rapidly and enlarge to form the corpus luteum.

8. When the corpus luteum degenerates, it forms the corpus albicans.

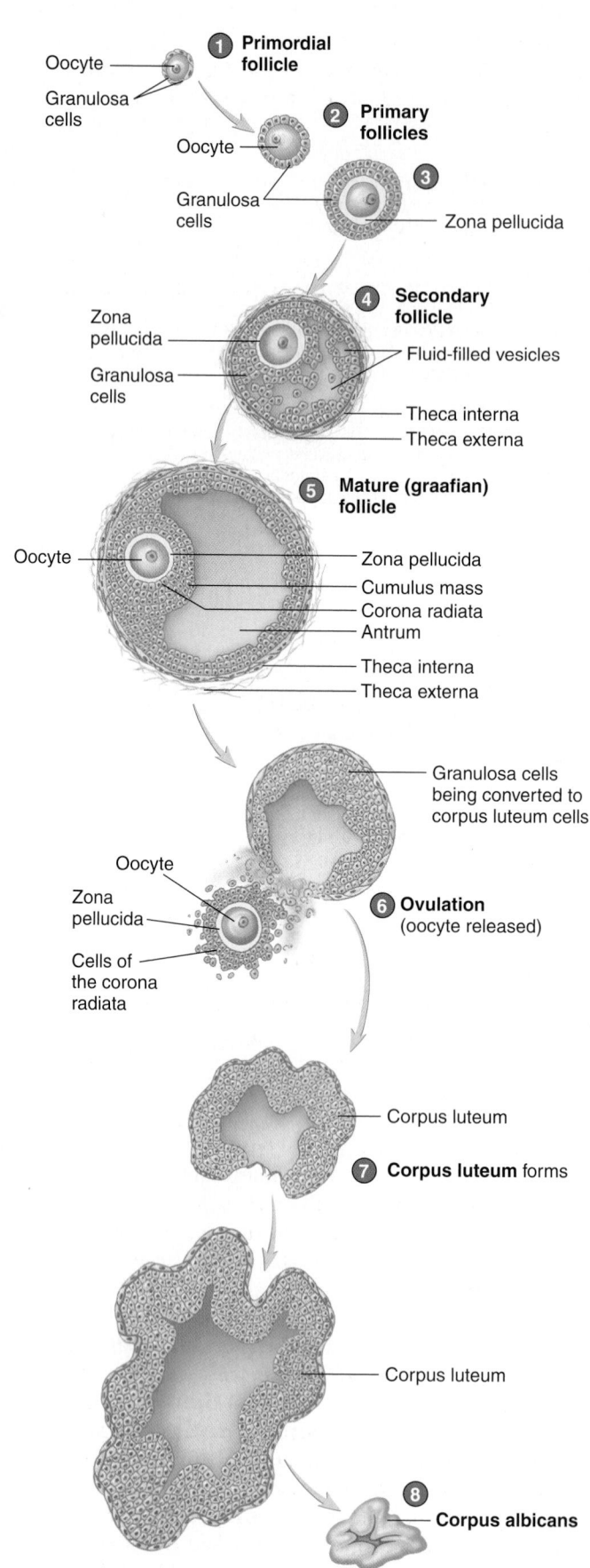

Process Figure 28.14   Maturation of the Follicle and Oocyte

41. *Describe the process of ovulation.*
42. *Starting with the oogonia, describe the development and production of a mature follicle that contains a secondary oocyte.*

Fate of the Follicle

After ovulation, the follicle still has an important function. It becomes transformed into a glandular structure called the **corpus luteum** (kōr′pŭs loo′tē-ŭm; yellow body), which has a convoluted appearance as a result of its collapse after ovulation (see figure 28.14). The granulosa cells and the theca interna, now called **luteal cells,** enlarge and begin to secrete hormones—progesterone and smaller amounts of estrogen.

If pregnancy occurs, the corpus luteum enlarges and remains throughout pregnancy as the **corpus luteum of pregnancy.** If pregnancy does not occur, the corpus luteum remains functional for about 10–12 days and then begins to degenerate. Progesterone and estrogen secretion decreases, and connective tissue cells become enlarged and clear, giving the whole structure a whitish color; it is, therefore, called the **corpus albicans** (al′bĭ-kanz; white body). The corpus albicans continues to shrink and eventually disappears after several months or even years.

43. *What is the corpus luteum? What happens to the corpus luteum if fertilization occurs? If fertilization does not occur?*

Uterine Tubes

Two **uterine tubes,** also called **fallopian** (fa-lō′pē-an) **tubes,** or **oviducts** (ō′vi-dŭkts), are present. There is a uterine tube on each side of the uterus associated with an ovary (see figure 28.11). Each tube is located along the superior margin of the broad ligament. The part of the broad ligament most directly associated with the uterine tube is called the **mesosalpinx** (mez′ō-sal′pinks; mesothelium of the trumpet-shaped uterine tube).

The uterine tube opens directly into the peritoneal cavity to receive the oocyte from the ovary. It expands to form the **infundibulum** (in-fŭn-dib′ū-lŭm; funnel), and long, thin processes called **fimbriae** (fim′brē-ē; fringe) surround the opening of the infundibulum. The inner surfaces of the fimbriae consist of a ciliated mucous membrane.

The part of the uterine tube that is nearest the infundibulum is called the **ampulla.** It's the widest and longest part of the tube and accounts for about 7.5–8 cm of the total 10 cm length of the tube. The part of the uterine tube nearest the uterus, the **isthmus,** is much narrower and has thicker walls than does the ampulla. The **uterine,** or **intramural, part** of the tube passes through the uterine wall and ends in a very small uterine opening.

The wall of each uterine tube consists of three layers (see figure 28.11). The outer **serosa** is formed by the peritoneum, the middle **muscular layer** consists of longitudinal and circular smooth muscle cells, and the inner **mucosa** consists of a mucous membrane of simple ciliated columnar epithelium. The mucosa is arranged into numerous longitudinal folds.

The mucosa of the uterine tubes provides nutrients for the oocyte, or, if fertilization has occurred, to the developing embryonic mass (see chapter 29) as it passes through the uterine tube. The ciliated epithelium helps move the small amount of fluid and the oocyte, or the developing embryonic mass, through the uterine tubes.

44. *Describe the structures of the uterine tube.*
45. *How are the uterine tubes involved in moving the oocyte or the zygote?*

Uterus

The **uterus** (ū′ter-ŭs) is the size and shape of a medium-sized pear and is about 7.5 cm long and 5 cm wide (see figures 28.10 and 28.11). It's slightly flattened anteroposteriorly and is oriented in the pelvic cavity with the larger, rounded part, the **fundus** (fŭn′dŭs; bottom of a rounded flask), directed superiorly and the narrower part, the **cervix** (ser′viks; neck), directed inferiorly. The main part of the uterus, the **body,** is between the fundus and the cervix. A slight constriction called the **isthmus** marks the junction of the cervix and the body. Internally, the uterine cavity continues as the **cervical canal,** which opens through the **ostium** into the vagina.

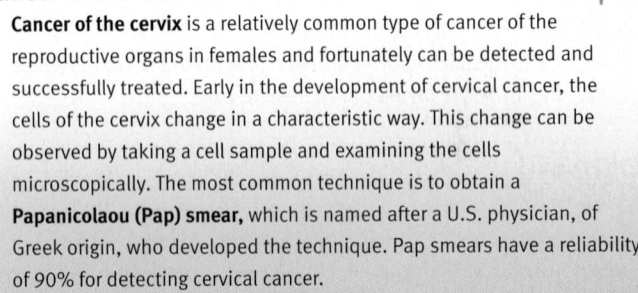

Cancer of the Cervix

Cancer of the cervix is a relatively common type of cancer of the reproductive organs in females and fortunately can be detected and successfully treated. Early in the development of cervical cancer, the cells of the cervix change in a characteristic way. This change can be observed by taking a cell sample and examining the cells microscopically. The most common technique is to obtain a **Papanicolaou (Pap) smear,** which is named after a U.S. physician, of Greek origin, who developed the technique. Pap smears have a reliability of 90% for detecting cervical cancer.

The major ligaments holding the uterus in place are the **broad ligament,** the **round ligaments,** (see figure 28.11) and the **uterosacral ligaments.** The broad ligament is a peritoneal fold extending from the lateral margin of the uterus to the wall of the pelvis on either side. It also ensheaths the ovaries and the uterine tubes. The round ligaments extend from the uterus through the inguinal canals to the labia majora of the external genitalia, and the uterosacral ligaments attach the lateral wall of the uterus to the sacrum. Normally, the uterus is anteverted, meaning that the body of the uterus is tipped slightly anteriorly. In some women, the uterus is retroverted, or tipped posteriorly. In addition to the ligaments, skeletal muscles of the pelvic floor provide much support inferiorly to the uterus. If these muscles are weakened (e.g., in childbirth), the uterus can extend inferiorly into the vagina, a condition called a **prolapsed uterus.**

The uterine wall is composed of three layers: perimetrium (serous), myometrium, and endometrium (see figure 28.11). The **perimetrium** (per-i-mē′trē-ŭm), or **serous layer,** of the uterus is the peritoneum that covers the uterus. The next layer, just deep to the perimetrium, is the **myometrium** (mī′ō-mē′trē-ŭm), or **muscular layer,** which consists of a thick layer of smooth muscle. The myometrium accounts for the bulk of the uterine wall and is the thickest layer of smooth muscle in the body. In the cervix, the muscular layer contains less muscle and more dense connective tissue. The cervix is, therefore, more rigid and less contractile than the rest of the uterus. The innermost layer of the uterus is the **endometrium** (en′dō-mē′trē-ŭm), or **mucous membrane.** The endometrium consists of a simple columnar epithelial lining and a connective tissue, the lamina propria. Simple tubular glands are scattered about the lamina propria and open through the epithelium into the uterine cavity. The endometrium consists of two layers: a thin, deep **basal layer,** which is the deepest part of the lamina propria and is continuous with the myometrium; and a thicker, superficial **functional layer,** which consists of most of the lamina propria and the endothelium, and lines the cavity itself. The functional layer is so named because it undergoes changes and sloughing during the female menstrual cycle. Small **spiral arteries** of the lamina propria provide the blood supply to the functional layer of the endometrium. These blood vessels play an important role in cyclic changes of the endometrium.

Columnar epithelial cells line the cervical canal, which contains **cervical mucous glands.** The mucus fills the cervical canal and acts as a barrier to substances that could pass from the vagina into the uterus. Near the time of ovulation the consistency of the mucus changes, making the passage of sperm cells from the vagina into the uterus easier.

> 46. *Name the parts of the uterus.*
> 47. *Describe the major ligaments holding the uterus in place.*
> 48. *Describe the layers of the uterine wall.*

Vagina

The **vagina** (vă-jī′nă) is a tube about 10 cm long that extends from the uterus to the outside of the body (see figure 28.11). The vagina is the female organ of copulation, functioning to receive the penis during intercourse, and it allows menstrual flow and childbirth. Longitudinal ridges called **columns** extend the length of the anterior and posterior vaginal walls, and several transverse ridges called **rugae** (roo′gē) extend between the anterior and posterior columns. The superior, domed part of the vagina, the **fornix** (fōr′niks; domed), is attached to the sides of the cervix so that a part of the cervix extends into the vagina.

The wall of the vagina consists of an outer muscular layer and an inner mucous membrane. The muscular layer is smooth muscle that allows the vagina to increase in size to accommodate the penis during intercourse and to stretch greatly during childbirth. The mucous membrane is moist stratified squamous epithelium that forms a protective surface layer. The vaginal mucous membrane releases most of the lubricating secretions produced by the female during intercourse.

A thin mucous membrane called the **hymen** (hī′men) covers the **vaginal opening,** or **orifice.** Sometimes, the hymen completely closes the vaginal opening (a condition called imperforate hymen), and it must be removed to allow menstrual flow. More commonly, the hymen is perforated by one or several holes. The openings in the hymen are usually greatly enlarged during the first sexual intercourse. In addition, the hymen can be perforated or torn at some earlier time in a young woman's life, such as during strenuous physical exercise. Thus, the absence of an intact hymen doesn't necessarily indicate that a woman has had sexual intercourse, as was once thought.

> 49. *Where is the vagina located?*
> 50. *Describe the layers of the vaginal wall. What are rugae and columns?*
> 51. *What is the hymen?*

External Genitalia

The external female genitalia, also referred to as the **vulva** (vŭl′vă) or **pudendum** (pū-den′dŭm), consist of the vestibule and its surrounding structures (figure 28.15). The **vestibule** (ves′ti-bool) is the space into which the vagina opens posteriorly and the urethra opens anteriorly. A pair of thin, longitudinal skin folds called the **labia** (lā′bē-ă; lips) **minora** (sing., labium minus) form borders on each side of the vestibule. A small erectile structure called the **clitoris** (klit′ō-ris) is located in the anterior margin of the vestibule. Anteriorly, the two labia minora unite over the clitoris to form a fold of skin called the **prepuce.**

The clitoris is usually less than 2 cm in length and consists of a shaft and a distal glans. It's well supplied with sensory receptors and functions to initiate and intensify levels of sexual tension. The clitoris contains two erectile structures, the **corpora cavernosa,** each of which expands at the base end of the clitoris to form the **crus of the clitoris** and attaches the clitoris to the coxae. The corpora cavernosa of the clitoris are comparable to the corpora cavernosa of the penis, and they become engorged with blood as a result of sexual excitement. In most women, this engorgement results in an increase in the diameter, but not the length, of the clitoris. With increased diameter, the clitoris makes better contact with the prepuce and surrounding tissues and is more easily stimulated.

Erectile tissue that corresponds to the corpus spongiosum of the male lies deep to and on the lateral margins of the vestibular floor on either side of the vaginal orifice. Each erectile body is called a **bulb of the vestibule.** Like other erectile tissue, it becomes engorged with blood and is more sensitive during sexual arousal. Expansion of the bulbs causes narrowing of the vaginal orifice and produces better contact of the vagina with the penis during intercourse.

On each side of the vestibule, between the vaginal opening and the labia minora, is an opening of the duct of the **greater**

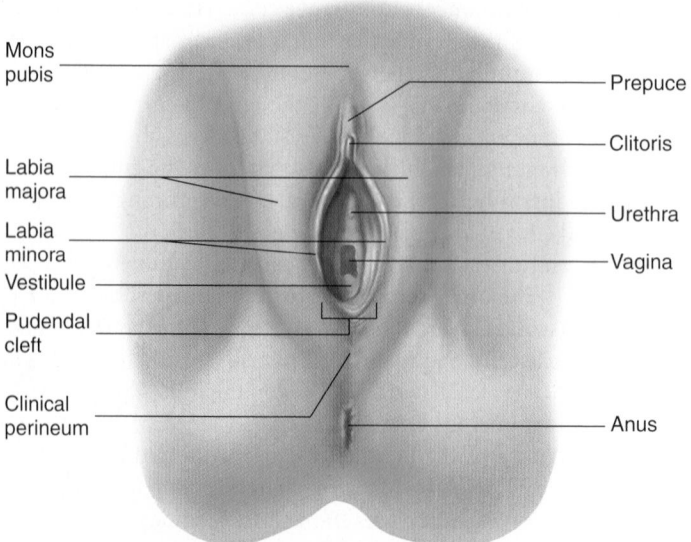

Mons pubis — Prepuce — Clitoris — Labia majora — Urethra — Labia minora — Vagina — Vestibule — Pudendal cleft — Clinical perineum — Anus

Figure 28.15 Female External Genitalia

vestibular gland. Additional small mucous glands, the **lesser vestibular glands,** or **paraurethral glands,** are located near the clitoris and urethral opening. They produce a lubricating fluid that helps to maintain the moistness of the vestibule.

Lateral to the labia minora are two prominent, rounded folds of skin called the **labia majora.** Subcutaneous fat is primarily responsible for the prominence of the labia majora. The two labia majora unite anteriorly in an elevation over the symphysis pubis called the **mons pubis** (monz; mound; pū′bis). The lateral surfaces of the labia majora and the surface of the mons pubis are covered with coarse hair. The medial surfaces are covered with numerous sebaceous and sweat glands. The space between the labia majora is called the **pudendal** (pū-den′dăl) **cleft.** Most of the time, the labia majora are in contact with each other across the midline, closing the pudendal cleft and concealing the deeper structures within the vestibule.

52. What are the vulva, pudendum, and vestibule?
53. What erectile tissue is in the clitoris and bulb of the vestibule?
54. What is the function of the clitoris and bulb of the vestibule?
55. Describe the labia minora, the prepuce, the labia majora, the pudendal cleft, and the mons pubis.
56. Where are the greater and lesser vestibular glands located? What is their function?

Perineum

The **perineum,** as in the male, is divided into two triangles by the superficial and deep transverse perineal muscles (figure 28.16). The anterior, urogenital triangle contains the external genitalia, and the posterior, anal triangle contains the anal opening. The region between the vagina and the anus is the **clinical perineum.**

The skin and muscle of this region can tear during childbirth. To prevent such tearing, an incision called an **episiotomy** (e-piz-ē-ot′ō-mē, e-pis-ē-ot′ō-mē) is sometimes made in the clinical perineum. This clean, straight incision is easier to repair than a tear would be. Alternatively, allowing the perineum to stretch slowly during the delivery may prevent tearing, thereby making an episiotomy unnecessary.

57. Define the term perineum.
58. What is the clinical perineum?
59. Define and give the purpose of an episiotomy.

Mammary Glands

The **mammary glands** are the organs of milk production and are located within the **mammae** (mam′ē), or **breasts** (figure 28.17). The mammary glands are modified sweat glands. Externally, the breasts of both males and females have a raised **nipple** surrounded by a circular, pigmented **areola** (ă-rē′ō-lă). The areolae normally have a slightly bumpy surface caused by the presence of rudimentary mammary glands, called **areolar glands,** just below the surface. Secretions from these glands protect the nipple and the areola from chafing during nursing.

In prepubescent children, the general structure of the breasts is similar, and both males and females possess a rudimentary glandular system, which consists mainly of ducts with sparse alveoli. The female breasts begin to enlarge during puberty, primarily under the influence of estrogen and progesterone. Increased sensitivity or pain in the breasts often accompanies this enlargement. Males often experience these same sensations during early puberty, and their breasts can even develop slight swellings; however, these symptoms usually disappear fairly quickly. On rare occasions, the breasts of a male become enlarged, a condition called **gynecomastia** (gī′nĕ-kō-mas′tē-ă).

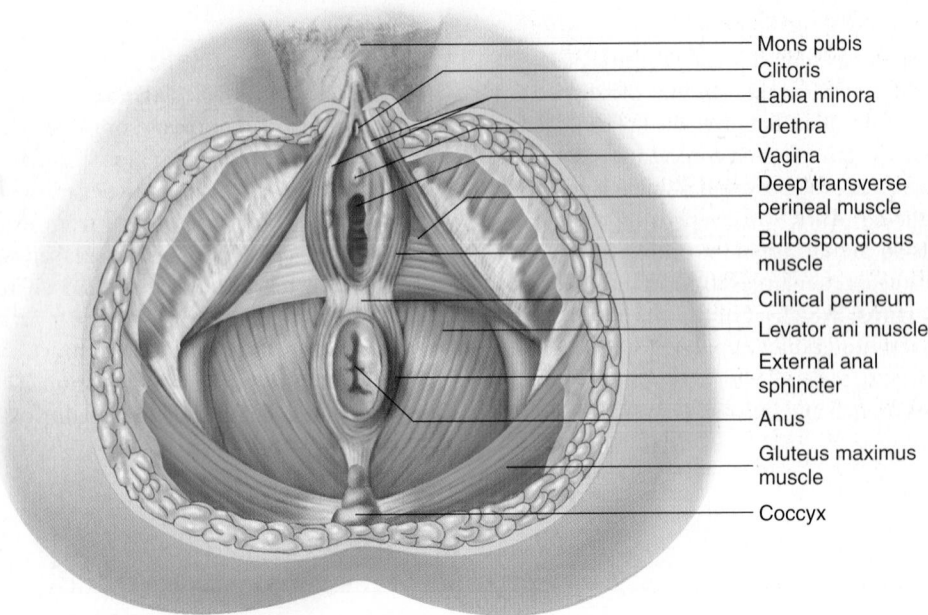

Mons pubis
Clitoris
Labia minora
Urethra
Vagina
Deep transverse perineal muscle
Bulbospongiosus muscle
Clinical perineum
Levator ani muscle
External anal sphincter
Anus
Gluteus maximus muscle
Coccyx

Figure 28.16 Inferior View of the Female Perineum

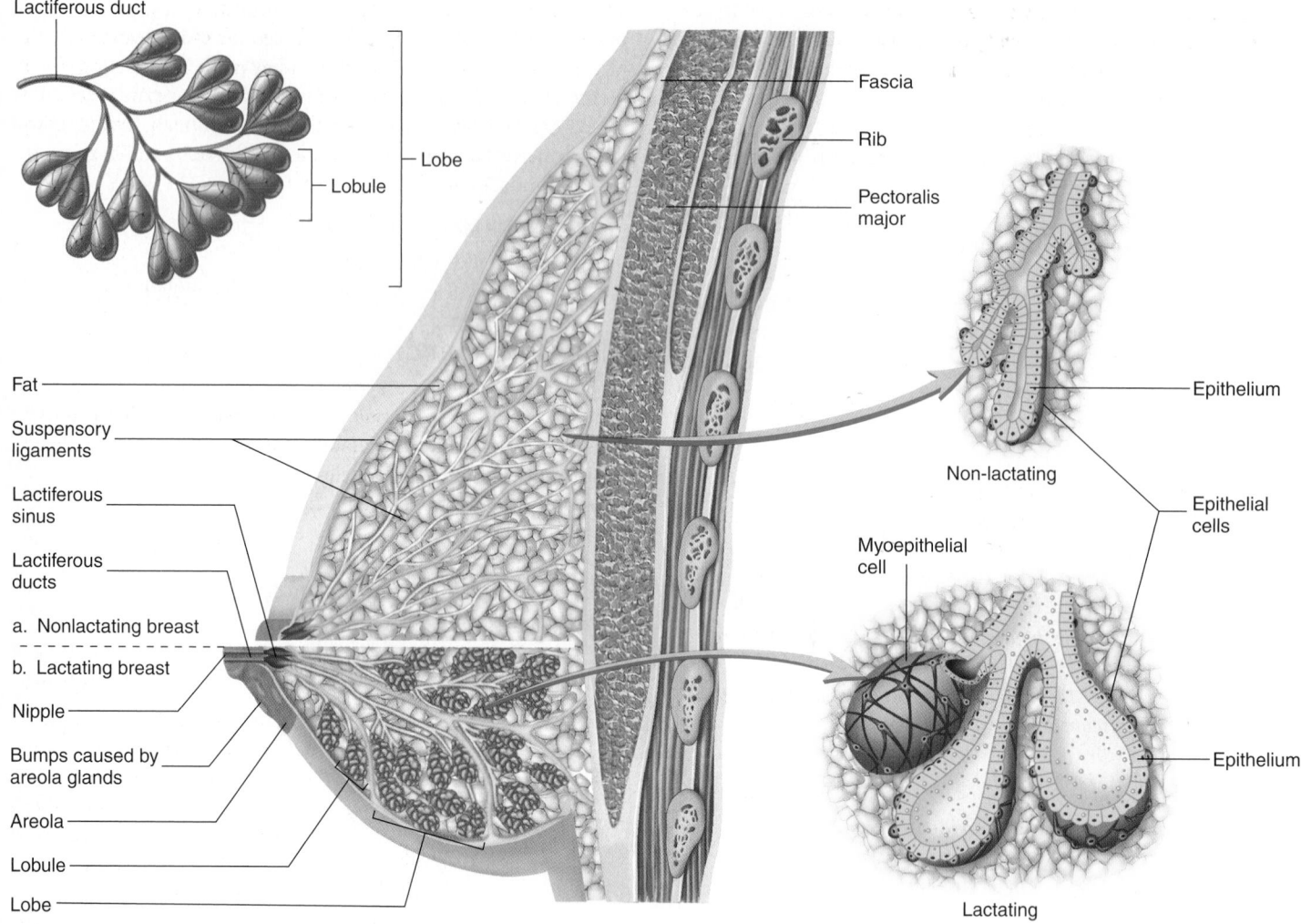

Figure 28.17   Anatomy of the Breast

The section illustrates the blood supply, the mammary glands, and the duct system.

Each adult female mammary gland usually consists of 15–20 glandular **lobes** covered by a considerable amount of adipose tissue. It is primarily this superficial fat that gives the breast its form. The lobes of each mammary gland form a conical mass, with the nipple located at the apex. Each lobe possesses a single **lactiferous** (lak-tif′er-ŭs; milk-producing) **duct,** which opens independently of other lactiferous ducts on the surface of the nipple. Just deep to the surface, each lactiferous duct enlarges to form a small, spindle-shaped **lactiferous sinus,** which accumulates milk during milk production. The lactiferous duct supplying a lobe subdivides to form smaller ducts, each of which supplies a **lobule.** Within a lobule, the ducts branch and become even smaller. In the milk-producing breast, the ends of these small ducts expand to form secretory sacs called **alveoli.**

A group of **mammary,** or **Cooper's, ligaments** support and hold the breasts in place. These ligaments extend from the fascia over the pectoralis major muscles to the skin over the mammary glands and prevent the breasts from excessive sagging. In older adults, however, these ligaments weaken and elongate, allowing the breasts to sag to a greater extent than when the person was younger.

The nipples are very sensitive to tactile stimulation and contain smooth muscle cells that contract, causing the nipple to become erect in response to stimulation. These smooth muscle cells respond, like other erectile tissues, during sexual arousal.

60. *Describe the anatomy of the mammary glands.*
61. *Describe the route taken by a drop of milk from its site of production to the outside of the body.*
62. *What are Cooper's ligaments?*

Fibrocystic Changes and Cancer of the Breast

Fibrocystic changes in the breast are benign changes. They include the formation of fluid filled cysts, hyperplasia of the duct system of the breast, and deposition of fibrous connective tissue. These changes occur in approximately 10% of women who are less than 21 years of age, 25% of women in their reproductive years, and 50% of women who are postmenopausal. The cause of the condition is not known. Major manifestations are breast pain, especially during the luteal phase of the menstrual cycle and continuing until menstruation. Some evidence suggest that some women with certain types of duct hyperplasia, when associated with a family history of breast cancer, have an increased likelihood of developing breast cancer.

Cancer of the breast is a serious, often fatal disease in women. Risk factors include a family history of breast cancer, early menarche, late menopause, obesity, radiation of the chest, abnormal cells in fibrocystic disease, and hormone replacement therapy. The use of mammography and regular self-examination of the breast can help in early detection and effective treatment of breast cancer. Women who have inherited specific gene mutations have increased risk of breast cancer. Prophylactic removal of the breasts in women who are identified, based on genetic analysis, as likely to develop breast cancer may result in a longer life expectancy for these women.

Physiology of Female Reproduction

As in the male, female reproduction is under the control of hormonal and nervous regulation. Development of the female reproductive organs and normal function depend on a number of hormones in the body.

Puberty

During puberty, females experience their first episode of menstrual bleeding, which is called **menarche** (me-nar′ke), which generally begins between 11 and 13 years of age and is generally completed by 16 years of age. The vagina, uterus, uterine tubes, and external genitalia begin to enlarge. Fat is deposited in the breasts and around the hips, causing them to enlarge and assume an adult form. The ducts of the breasts develop, pubic and axillary hair grows, and the voice changes, although this is a more subtle change than in males. Development of sexual drive is also associated with puberty.

Elevated rates of estrogen and progesterone secretion by the ovaries are primarily responsible for the changes associated with puberty. Before puberty, estrogen and progesterone are secreted in very small amounts. LH and FSH levels also remain very low. The low secretory rates are due to a lack of GnRH released from the hypothalamus. At puberty, not only are GnRH, LH, and FSH secreted in greater quantities than before puberty, but the adult pattern is established in which a cyclic pattern of FSH and LH secretion occurs. The cyclic secretion of LH and FSH, ovulation, the monthly changes in secretion of estrogen and progesterone, and the resultant changes in the uterus characterize the menstrual cycle.

63. *Define puberty and describe the changes that occur during puberty.*
64. *Define menarche. What changes occur in LH, FSH, estrogen, and progesterone secretion during puberty?*

Menstrual Cycle

The term **menstrual** (men′stroo-ăl) **cycle** technically refers to the cyclic changes that occur in sexually mature, nonpregnant females and culminate in menses. Typically, the menstrual cycle is about 28 days long, although it can be as short as 18 days in some women and as long as 40 days in others (figure 28.18 and table 28.2). The term **menses** (men′sēz) is derived from a Latin word meaning month. It is a period of mild hemorrhage, which occurs approximately once each month, during which the uterine epithelium is sloughed and expelled from the uterus. **Menstruation** (men-stroo-ā′shŭn) is the discharge of the blood and elements of the uterine mucous membrane. Although the term *menstrual cycle* refers specifically to changes that occur in the uterus, several other cyclic changes are associated with it, and the term is often used to refer to all of the cyclic events that occur in the female reproductive system. These changes include the cyclic changes in hormone secretion, in the ovary, and in the uterus.

The first day of menses is day 1 of the menstrual cycle, and menses typically lasts 4–5 days. Ovulation occurs on about day 14 of a 28-day menstrual cycle, although the timing of ovulation varies from individual to individual and varies within a single individual from one menstrual cycle to the next. The time between ovulation, on day 14, and the next menses is typically 14 days. The time between the first day of menses and the day of ovulation is more variable than the time between ovulation and the next menses. The time between the ending of menses and ovulation is called the **follicular phase,** because of the rapid development of ovarian follicles, or the **proliferative phase,** because of the rapid proliferation of the uterine mucosa. The period after ovulation and before the next menses is called the **luteal phase,** because of the existence of the corpus luteum, or the **secretory phase,** because of maturation of and secretion by uterine glands.

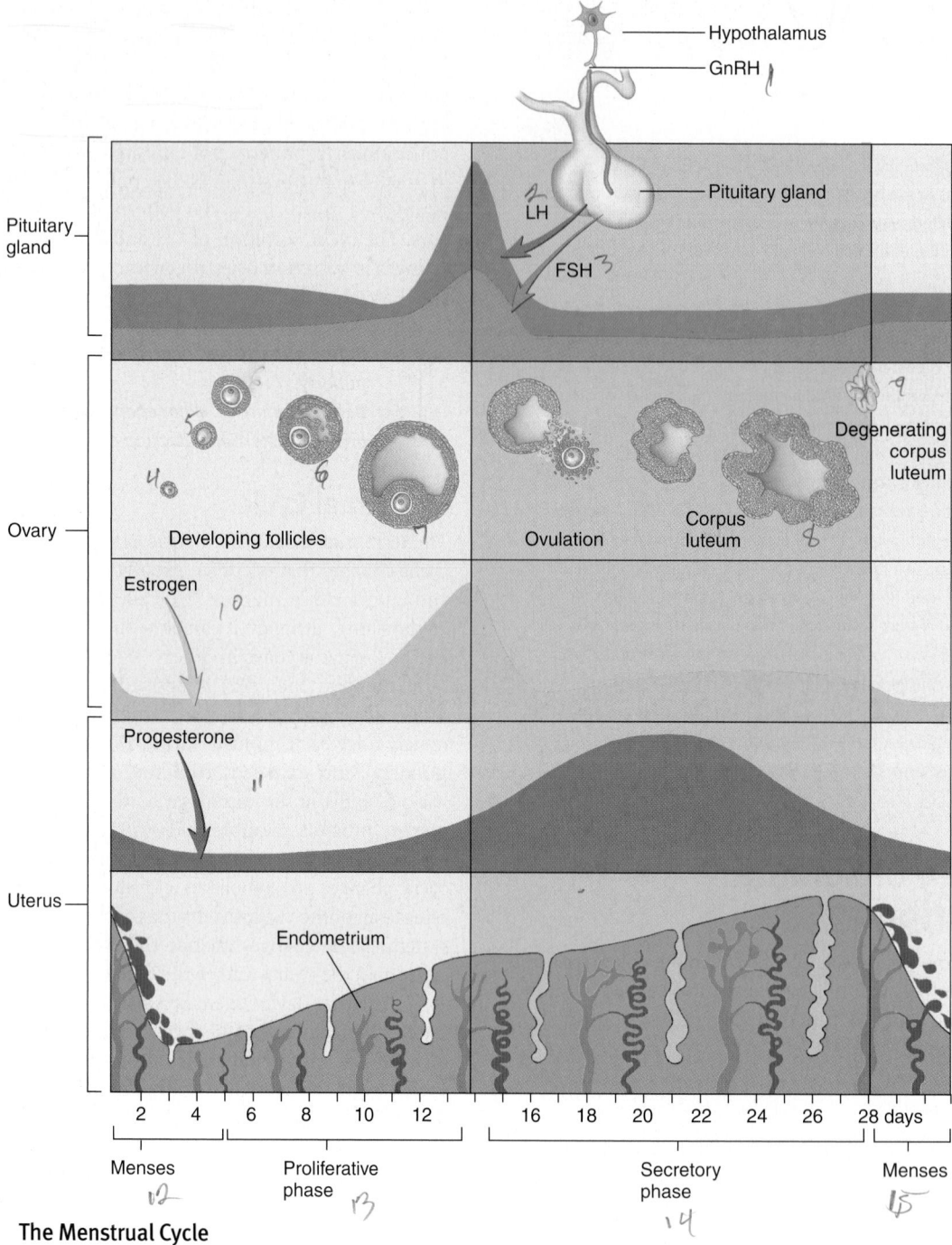

Figure 28.18   **The Menstrual Cycle**

The various lines depict the changes in blood hormone levels, the development of the follicles, and the changes in the endometrium during the cycle.

65. *Define the menstrual cycle.*

66. *What is the length of a typical menstrual cycle and list the major events that occur.*

67. *Which day is ovulation, usually?*

68. *Define the follicular and proliferative phases and the secretory and luteal phases.*

Ovarian Cycle

The **ovarian cycle** specifically refers to the events that occur in a regular fashion in the ovaries of sexually mature, nonpregnant women during the menstrual cycle. The hypothalamus and anterior pituitary release hormones that control these events. FSH from the anterior pituitary is primarily responsible for initiating

Table 28.2 The Menstrual Cycle

Menses	Proliferative Phase	Ovulation	Secretory Phase
Pituitary Hormones			
LH levels are low and remain low; FSH increases somewhat.	LH and FSH levels begin to increase rapidly in response to increases in estrogen near the end of the proliferative phase.	Increasing levels of LH trigger ovulation. Ovulation generally occurs after LH levels have reached their peak. FSH reaches a peak about the time of ovulation and initiates development of follicles that may complete maturation during a later cycle.	LH and FSH levels decline to low levels following ovulation and remain at low levels during the secretory phase in response to increases in estrogen and progesterone.
Developing Follicles			
FSH secreted during menses causes several follicles to begin to enlarge.	Several follicles continue to enlarge. As they enlarge, they begin to secrete estrogen. In addition, many of them degenerate. Only one of the follicles becomes a mature follicle that is capable of ovulating by the end of the proliferative phase.	Normally, a single follicle reaches maturity and ovulates in response to LH. The oocyte and some cumulus cells are released during ovulation.	Following ovulation, the granulosa cells of the ovulated follicle change to luteal cells and begin secreting large amounts of progesterone and some estrogen.
Estrogen			
The ovarian follicles secrete very little estrogen.	Near the end of the proliferative phase, the enlarging follicles begin to secrete increasing amounts of estrogen. The estrogen causes the pituitary gland to secrete increasing quantities of LH and smaller quantities of FSH. The positive-feedback relationship between estrogen and LH results in rapidly increasing LH and estrogen levels several days prior to ovulation. The rapid increase in LH triggers ovulation.	Estrogen, secreted by developing follicles, reaches a peak at ovulation.	Following ovulation, estrogen levels decline. After the luteal cells have been established, smaller amounts of estrogen are secreted by the corpus luteum.
Progesterone			
The ovarian follicles secrete very little progesterone.	Progesterone levels are low during the proliferative phase.	Progesterone levels are low.	Following ovulation, progesterone levels increase due to the secretion of progesterone by the corpus luteum. Progesterone levels remain high throughout the secretory phase and fall rapidly just before menses unless pregnancy occurs.
Uterine Endometrium			
The endometrium of the uterus undergoes necrosis and is eliminated in the menstrual fluid during menses. The necrosis is a result of decreasing progesterone concentrations near the end of the proliferative phase.	In response to estrogen, endometrial cells of the uterus undergo rapid cell division and proliferate rapidly. In addition, the number of progesterone receptors in the endometrial cells increases in response to estrogen.	Ovulation occurs over a short time, and it signals the end of the proliferative phase, as estrogen levels decline, and the onset of the secretory phase, as the progesterone levels begin to increase.	Progesterone causes the endometrial cells to enlarge and secrete a small amount of fluid. The endometrium continues to thicken throughout the secretory phase. Near the end of the secretory phase, declining progesterone levels allow the spiral arteries of the endometrium to constrict, causing ischemia, and the endometrium becomes necrotic unless pregnancy occurs.

the development of primary follicles, and as many as 25 begin to mature during each menstrual cycle. The follicles that start to develop in response to FSH may not ovulate during the same menstrual cycle in which they begin to mature, but they may ovulate one or two cycles later.

Although several follicles begin to mature during each cycle, normally only one is ovulated. The remaining follicles degenerate. Larger and more mature follicles appear to secrete estrogen and other substances that have an inhibitory effect on other less mature follicles.

Early in the menstrual cycle, the release of GnRH from the hypothalamus increases, and sensitivity of the anterior pituitary to GnRH increases. These changes stimulate the production and release of a small amount of FSH and LH by the anterior pituitary.

FSH and LH stimulate follicular growth and maturation. They also cause an increase in estradiol secretion by the developing follicles. FSH exerts its main effect on the granulosa cells. LH exerts its initial effect on the cells of the theca interna and later on the granulosa cells.

LH stimulates the theca interna cells to produce androgens, which diffuse from these cells to the granulosa cells. FSH stimulates the granulosa cells to convert androgens to estrogen. In addition, FSH gradually increases LH receptors in the granulosa cells, and estrogen produced by the granulosa cells increases LH receptors in the theca interna cells. Consequently, theca interna cells and granulosa cells cooperate to produce estrogen. Estrogen, in turn, increases receptors for LH in both theca interna cells and granulosa cells.

After LH receptors in the granulosa cells have increased, LH stimulates the granulosa cells to produce some progesterone, which diffuses from the granulosa cells to the theca interna cells, where it is converted to androgens. These androgens are also converted to estrogen by the granulosa cells. Thus, the production of androgens by the theca interna cells increases, and the conversion of androgens to estrogen by the granulosa cells is responsible for a gradual increase in estrogen secretion by these cells throughout the follicular phase, even though only a small increase in LH secretion occurs. FSH levels actually decrease during the follicular phase because developing follicles produce inhibin, and inhibin has a negative-feedback effect on FSH secretion.

As estrogen levels increase during the follicular phase, they have a negative-feedback effect on the secretion of LH and FSH by the anterior pituitary. However, the gradual increase in estrogen levels, especially late in the follicular phase, begins to have a positive-feedback effect on LH and FSH release from the anterior pituitary. Consequently, as the blood level of estrogen increases, it stimulates greater LH and FSH secretion. The sustained increase in estrogen is necessary for the development of this positive-feedback effect. In response to this positive-feedback effect, LH and FSH secretion increase rapidly and in large amounts just before ovulation (figure 28.19). The increase in blood levels of both LH and FSH is called the **LH surge,** and the increase in FSH is called the **FSH surge.** The LH surge occurs several hours earlier and to a greater degree than the FSH surge, and the LH surge can last up to 24 hours.

The LH surge initiates ovulation and causes the ovulated follicle to become the corpus luteum. FSH can make the follicle more sensitive to the influence of LH by stimulating the synthesis of additional LH receptors in the follicles and by stimulating the development of follicles that may ovulate during later ovarian cycles.

The LH surge causes the primary oocyte to complete the first meiotic division just before or during the process of ovulation. Also, the LH surge triggers several events that are very much like inflammation in a mature follicle. These events result in ovulation. The follicle becomes edematous, proteolytic enzymes cause the degeneration of the ovarian tissue around the follicle, the follicle ruptures, and the oocyte and some surrounding follicle cells are slowly extruded from the ovary.

Shortly after ovulation, production of estrogen by the follicle decreases. The remaining granulosa cells of the ovulated follicle are converted to corpus luteum cells and begin to secrete progesterone. After the corpus luteum forms, progesterone levels become much higher than before ovulation, and some estrogen is produced also. The increased progesterone and estrogen have a negative-feedback effect on GnRH release from the hypothalamus. As a result, LH and FSH release from the anterior pituitary decreases. Estrogen and progesterone also cause downregulation of GnRH receptors in the anterior pituitary, and the anterior pituitary cells become less sensitive to GnRH. Because of the decreased secretion of GnRH and decreased sensitivity of the anterior pituitary to GnRH, the rate of LH and FSH secretion declines to very low levels after ovulation (see figures 28.18 and 28.19).

If fertilization of the ovulated oocyte does take place, the outer layer of the developing embryo begins to secrete the LH-like substance **human chorionic gonadotropin (HCG),** which keeps the corpus luteum from degenerating. As a result, blood levels of estrogen and progesterone do not decrease, and menses does not occur. If fertilization does not occur, HCG is not produced. The cells of the corpus luteum begin to atrophy after day 25 or 26, and the blood levels of estrogen and progesterone decrease rapidly, which results in menses.

- 69. *Describe the events of the ovarian cycle. What role do FSH and LH play in the ovarian cycle?*
- 70. *Describe how the cyclic increase and decrease in FSH and LH is produced.*
- 71. *Where is HCG produced, and what effect does it have on the ovary?*

P R E D I C T

Predict the effect on the ovarian cycle of administering a relatively large amount of estrogen and progesterone just before the preovulatory LH surge. Also predict the consequences of continually administering high concentrations of GnRH.

Uterine Cycle

The term **uterine cycle** refers to changes that occur primarily in the endometrium of the uterus during the menstrual cycle

1. Before ovulation there is an increase in FSH secretion which stimulates follicles to develop and secrete estrogen.

2. Estrogen causes the endometrium to proliferate and the hypothalamus to increase LH secretion, which results in the LH and FSH surges prior to ovulation.

3. The LH surge causes a follicle to mature and ovulate. The corpus luteum develops and secretes progesterone and some estrogen.

4. The progesterone causes hypertrophy of the endometrium and has a negative-feedback effect on LH and FSH secretion. The corpus luteum continues to secrete progesterone for approximately 12 days after ovulation.

Stimulatory

Inhibitory

Hypothalamus

Positive feedback (leads to the LH and FSH surges)

Negative feedback (inhibits LH and FSH secretion)

FSH   LH

① Follicles develop

③ Maturation of follicles and increased estrogen secretion

Causes ovulation

Estrogen secretion increases

Causes corpus luteum formation

Estrogen

Estrogen and progesterone

② Uterine epithelium proliferates

④ Uterine epithelium hypertrophies, forming spiral tubular glands, and secretes a small amount of uterine fluid

Follicular phase (before ovulation)

Luteal phase (after ovulation)

Process Figure 28.19 Regulation of Hormone Secretion During the Menstrual Cycle

The regulation of hormone secretion from the anterior pituitary and the ovary during the menstrual cycle before and after ovulation is depicted.

(see figures 28.18 and 28.20). Other, more subtle changes also occur in the vagina and other structures during the menstrual cycle. Cyclic secretions of estrogen and progesterone are the primary cause of these changes.

The endometrium of the uterus begins to proliferate after menses. The remaining epithelial cells rapidly divide and replace the cells of the functional layer that were sloughed during the last menses. A relatively uniform layer of low cuboidal endometrial cells is produced. It later becomes columnar, and is folded to form tubular **spiral glands.** Blood vessels called **spiral arteries** project through the delicate connective tissue that separates the individual

spiral glands to supply nutrients to the endometrial cells. After ovulation, the endometrium becomes thicker, and the spiral glands develop to a greater extent and begin to secrete small amounts of a fluid rich in glycogen. Approximately 7 days after ovulation, or about day 21 of the menstrual cycle, the endometrium is prepared to receive the developing embryonic mass, if fertilization has occurred. If the developing embryonic mass arrives in the uterus too early or too late, the endometrium does not provide a suitable environment for it.

Estrogen causes the endometrial cells and, to a lesser degree, the myometrial cells to proliferate. It also makes the uterine tissue

1. Menses

The functional layer of the endometrium sluffs off as the spiral arteries remain in a constricted state in response to low levels of progesterone, depriving the functional layer of an adequate blood supply. Functional layer tissue and some blood make up the majority of the menstrual fluid. The basal layer remains.

2. Proliferative phase

The more basal portion of the endometrium is called the basal layer. Epithelial cells of the basal layer of the endometrium proliferate in response to estrogen. As a result, the epithelial cells and loose connective tissue on which they rest form the tubular, spiral-shaped uterine glands. The apical portion of the thickening endometrium is the functional layer. The spiral arteries found in the loose connective tissue between the uterine glands nourish the functional layer.

3. Secretory phase

Epithelial cells of the basal and functional layers undergo hypertrophy in response to progesterone. As a result, the uterine glands become more elongated and more spiral. Consequently, the endometrial layer reaches its greatest thickness. The spiral arteries can be seen in the loose connective tissue between the uterine glands. The spiral arteries remain dilated due to the presence of progesterone.

Process Figure 28.20   **The Uterine Cycle**

more sensitive to progesterone by stimulating the synthesis of progesterone receptor molecules within the uterine cells. After ovulation, progesterone from the corpus luteum binds to the progesterone receptors, resulting in cellular hypertrophy in the endometrium and myometrium and causing the endometrial cells to become secretory. Estrogen increases the tendency of the smooth muscle cells of the uterus to contract in response to stimuli, but progesterone inhibits smooth muscle contractions. When progesterone levels increase while estrogen levels are low, contractions of the uterine smooth muscle are reduced.

PREDICT **4**

Predict the effect on the endometrium of maintaining high progesterone levels in the circulatory system including the period of time during which estrogen normally increases following menstruation.

If pregnancy does not occur by day 24 or 25, progesterone and estrogen levels decline to low levels as the corpus luteum degenerates. As a consequence, the uterine lining also begins to degenerate. The spiral arteries constrict in a rhythmic pattern for longer and longer periods as progesterone levels fall. As a result, all but the basal parts of the spiral glands become ischemic and then necrotic. As the cells become necrotic, they slough into the uterine lumen. The necrotic endometrium, mucous secretions, and a small amount of blood released from the spiral arteries make up the menstrual fluid. Decreases in progesterone levels and increases in inflammatory substances that stimulate myometrial smooth muscle cells cause uterine contractions that expel the menstrual fluid from the uterus through the cervix and into the vagina.

72. *Name the stages of the uterine cycle, and describe the events that take place in each stage.*
73. *What are the effects of estrogen and progesterone on the uterus?*

Female Sexual Behavior and the Female Sex Act

Sexual drive in females, like sexual drive in males, depends on hormones. The adrenal gland and other tissues, such as the liver, convert steroids like progesterone to androgens. Androgens and possibly estrogens affect cells in the brain, especially in the hypothalamus, and influence sexual behavior. Androgens and estrogen alone don't control sexual drive, however. For example, sexual drive cannot be predictably increased simply by injecting these hormones into healthy women or men. Psychologic factors also affect sexual behavior. For example, after removal of the ovaries or after menopause, many women report having an increased sex drive because they no longer fear pregnancy.

The neural pathways, both sensory and motor, involved in controlling sexual responses are the same for males and females. Sensory action potentials are conducted from the genitals to the sacral region of the spinal cord, where reflexes that govern sexual responses are integrated. Ascending pathways, primarily the spinothalamic tracts (see chapter 14), conduct sensory information through the spinal cord to the brain, and descending pathways conduct action potentials back to the sacrum. As a result, cerebral

influences modulate the sacral reflexes. Motor action potentials are conducted from the spinal cord to the reproductive organs by both parasympathetic and sympathetic nerve fibers and to skeletal muscles by the somatic motor nerve fibers.

During sexual excitement, erectile tissue within the clitoris and around the vaginal opening becomes engorged with blood as a result of parasympathetic stimulation. The nipples of the breast often become erect as well. The mucous glands within the vestibule, especially the vestibular glands, secrete small amounts of mucus. Large amounts of mucuslike fluid are also extruded into the vagina through its wall, although no well-developed mucous glands are within the vaginal wall. These secretions provide lubrication that allows for easy entry of the penis into the vagina and easy movement of the penis during intercourse. Tactile stimulation of the female's genitals that occurs during sexual intercourse along with psychologic stimuli, normally triggers an orgasm. The vaginal, uterine, and perineal muscles contract rhythmically, and muscle tension increases throughout much of the body. After the sexual act a period of resolution characterized by an overall sense of satisfaction and relaxation occurs. In contrast to males, females can be receptive to further stimulation and can experience successive orgasms. Although orgasm is a pleasurable component of sexual intercourse, it's not necessary for females to experience an orgasm for fertilization to occur.

Menstrual Cramps, PMS, and Amenorrhea

Menstrual cramps are the result of strong myometrial contractions that occur before and during menstruation. The cramps can result from excessive prostaglandin secretion. Sloughing of the endometrium of the uterus results in an inflammation in the endometrial layer of the uterus, and prostaglandins are produced as part of the inflammatory process. Progesterone inhibits sloughing of the endometrium and contractions of uterine smooth muscle. Estrogen stimulates sloughing of the endometrium because it increases uterine smooth muscle contractions. In some women, menstrual cramps are extremely uncomfortable. Many women can alleviate painful menstruation by taking nonsterioidal anti-inflammatory drugs (NSAIDs), such as aspirin or ibuprofen, which inhibit prostaglandin biosynthesis, just before the onset of menstruation. These treatments, however, are not effective in treating all painful menstruation, especially when the causes of pain are more complicated than inflammation associated with normal menstruation, such as from tumors of the myometrium or obstruction of the cervical canal.

A topic of continuing research concerns **premenstrual syndrome (PMS)**. Some women suffer from severe changes in mood that often result in aggression and other socially unacceptable behaviors prior to menses. It has been hypothesized that hormonal changes associated with the menstrual cycle trigger these mood changes. Some women appear to have been successfully treated with steroid hormones. This treatment however doesn't appear to be effective for everyone with the condition. Similarly, reducing caffeine, alcohol, sugar, and animal fat consumption helps some women. It's unclear how many women are affected by PMS. It's a controversial condition, because a precise definition of the premenstrual period is not well established, the symptoms of the condition vary among individuals and are not easily monitored, and its precise cause and the physiologic mechanisms for the condition are unknown. Whether all women diagnosed as having PMS are suffering from the same condition is unclear. Improvements in defining symptoms associated with PMS and specifically identifying individuals exhibiting characteristics of depression have resulted in improved treatment.

The absence of a menstrual cycle is called **amenorrhea** (ă-men-ō-rē′ă). If the pituitary gland doesn't function properly because of abnormal development, the woman will not begin to menstruate at puberty. This condition is called **primary amenorrhea.** In contrast, if a woman has had normal menstrual cycles and later stops menstruating, the condition is called **secondary amenorrhea.** One cause of secondary amenorrhea is anorexia, a condition in which lack of food causes the hypothalamus to decrease GnRH secretion to levels so low that the menstrual cycle cannot occur. Female athletes or ballet dancers who have rigorous training schedules have a high frequency of secondary amenorrhea. The physical stress that can be coupled with an inadequate food intake also results in very low GnRH secretion. Increased food intake, for anorexic women, and reduced training, for women who exercise intensely, generally restores normal hormone secretion and normal menstrual cycles.

Secondary amenorrhea can result from pituitary tumors that decrease FSH and LH secretion, or from a lack of GnRH secretion from the hypothalamus. Head trauma and tumors that affect the hypothalamus can result in lack of GnRH secretion.

Secondary amenorrhea can result from a lack of normal hormone secretion from the ovaries, which can be caused by autoimmune diseases that attack the ovary or by polycystic ovarian disease, in which cysts in the ovary produce large amounts of androgen that are converted to estrogens by other tissues in the body. The increased estrogen prevents the normal cycle of FSH and LH secretion required for ovulation to occur. Other hormone-secreting tumors of the ovary can also disrupt the normal menstrual cycle and result in amenorrhea.

● **74. Compare the female sexual act with the male sexual act.**
75. Is an orgasm required for fertilization to occur?

Female Fertility and Pregnancy

After the sperm cells are ejaculated into the vagina during sexual intercourse, they are transported through the cervix, the body of the uterus, and the uterine tubes to the ampulla (figure 28.21). The forces responsible for the movement of sperm cells through the female reproductive tract involve the swimming ability of the sperm cells and, possibly, muscular contractions of the uterus and the uterine tubes. During sexual intercourse, oxytocin is released from the posterior pituitary of the female, and the semen introduced into the vagina contains prostaglandins. Both of these hormones stimulate smooth muscle contractions in the uterus and uterine tubes.

While passing through the vagina, uterus, and uterine tubes, the sperm cells undergo **capacitation** (kă-pas′i-tā′shŭn), a process that enables them to penetrate the zona pellucida. The process includes the release of acrosomal enzymes, which allow penetration of the cervical mucus, cumulus mass cells, and the oocyte plasma membrane.

The oocyte can be fertilized for up to 24 hours after ovulation, and some sperm cells remain viable in the female reproductive tract for up to 6 days although most of them have degenerated after 24 hours. For fertilization to occur successfully, sexual intercourse must, therefore, occur between 5 days before and 1 day after ovulation.

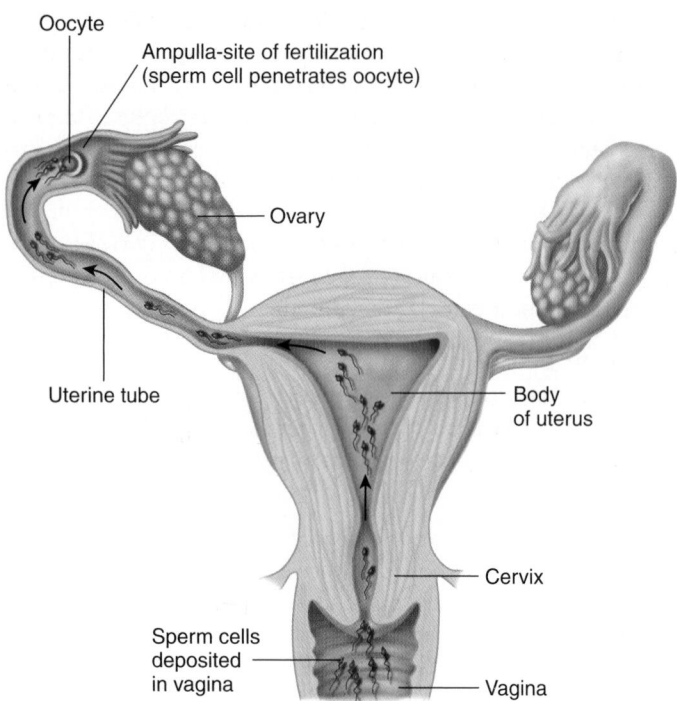

Figure 28.21　**Sperm Cell Movement**

Sperm cells are deposited in the vagina as part of the semen when the male ejaculates. Sperm cells pass through the cervix, the body of the uterus, and the uterine tube. Fertilization normally occurs when the oocyte is in the upper one-third of the uterine tube (the ampulla).

One sperm cell enters the secondary oocyte, and fertilization occurs (see chapter 29). For the next several days, a sequence of cell divisions occurs while the developing cells pass through the uterine tube to the uterus. By 7 or 8 days after ovulation, which is day 21 or 22 of the average menstrual cycle, the endometrium of the uterus is prepared for implantation. Estrogen and progesterone have caused it to reach its maximum thickness and secretory activity, and the developing embryo begins to implant. The outer layer of the developing embryo, the **trophoblast** (trof′ō-blast, trō′fō-blast), secretes proteolytic enzymes that digest the cells of the thickened endometrium (see chapter 29), and the mass digests its way into the endometrium.

Ectopic Pregnancy

An ectopic pregnancy results if implantation occurs anywhere other than in the uterine cavity. The most common site of ectopic pregnancy is the uterine tube. Implantation in the uterine tube eventually is fatal to the fetus and can cause the tube to rupture. The possibility of hemorrhage makes ectopic pregnancy dangerous to the mother. In rare cases, implantation can occur in the mesenteries of the abdominal cavity, and the fetus can develop normally but must be delivered by caesarean section.

The trophoblast secretes HCG, which is transported in the blood to the ovary and causes the corpus luteum to remain functional. As a consequence, both estrogen and progesterone levels continue to increase rather than decrease. The secretion of HCG increases rapidly and reaches a peak about 8–9 weeks after fertilization. Subsequently, HCG levels in the circulatory system decline to a lower level by 16 weeks and remain at a relatively constant level throughout the remainder of pregnancy. Detection of HCG excreted in the urine is the basis for some pregnancy tests.

The estrogen and progesterone secreted by the corpus luteum are essential for the maintenance of pregnancy. After the **placenta** (plă-sen′tă) forms from the trophoblast and uterine tissue, however, it also begins to secrete estrogen and progesterone. By the time the first 3 months of pregnancy are complete, the corpus luteum is no longer needed to maintain pregnancy, the placenta has become an endocrine gland that secretes sufficient quantities of estrogen and progesterone to maintain pregnancy. Estrogen and progesterone levels increase in the woman's blood throughout pregnancy (figure 28.22).

76. *Describe the transport of sperm cells through the female reproductive system.*
77. *What is capacitation of sperm cells?*
78. *Describe the events that occur following fertilization.*
79. *Describe implantation of the embryo and formation of the placenta.*

Menopause

When a female is 40–50 years old, menstrual cycles become less regular, and ovulation often does not occur. Eventually menstrual cycles stop completely. The cessation of menstrual cycles is called **menopause** (men′ō-pawz). The time from the onset of irregular cycles to their complete cessation, which is often 3 to 5 years, is

1. Human chorionic gonadotropin (HCG) increases until it reaches a maximum concentration near the end of the first 3 months of pregnancy and then decreases to a low level thereafter.

2. Progesterone continues to increase until it levels off near the end of pregnancy. Early in pregnancy, progesterone is produced by the corpus luteum in the ovary, later production shifts to the placenta.

3. Estrogen levels increase slowly throughout pregnancy, but they increase more rapidly as the end of pregnancy approaches. Early in pregnancy, estrogen is produced only in the ovary; later production shifts to the placenta.

Process Figure 28.22 Changes in Hormone Concentration During Pregnancy

HCG, progesterone, and estrogen are secreted from the placenta during pregnancy. Early in pregnancy, estrogen and progesterone are secreted by the ovary. During midpregnancy, there is a shift toward estrogen and progesterone secretion by the placenta. Late in pregnancy, these two hormones are secreted by the placenta.

Clinical Focus   Control of Pregnancy

Many methods are used to prevent or terminate pregnancy (figure B), including methods that prevent fertilization (contraception), prevent implantation of the developing embryo (IUDs), or remove the implanted embryo or fetus (abortion). Many of these techniques are quite effective when done properly and used consistently (table A).

Behavorial Methods

Abstinence, or refraining from sexual intercourse, is a sure way to prevent pregnancy when practiced consistently. It's not an effective method when used only occasionally.

Coitus (kō′i-tŭs) **interruptus** is removal of the penis from the vagina just before ejaculation. This is a very unreliable method of preventing pregnancy, because it requires perfect awareness and willingness to withdraw the penis at the correct time. It also ignores the fact that some sperm cells are found in preejaculatory emissions.

Periodic abstinence, the **natural family planning method,** or the **rhythm method** requires abstaining from sexual intercourse near the time of ovulation. A major factor in the success of this method is the ability to predict accurately the time of ovulation. Although the rhythm method provides some protection against becoming pregnant, it has a relatively high rate of failure, resulting from both the inability to predict the time of ovulation and the failure to abstain around the time of ovulation.

Barrier Methods

A **condom** (kon′dom) is a sheath of animal membrane, rubber, or latex. Placed over the erect penis, it is a barrier device, because the semen is collected within the condom instead of within the vagina. Condoms also provide protection against sexually transmitted diseases. A **vaginal condom** also acts as a barrier device. The vaginal condom can be placed into the vagina by the woman before sexual intercourse.

Methods to prevent sperm cells from reaching the oocyte once they are in the vagina include use of a diaphragm, a cervical cap, and spermicidal agents. The diaphragm and cervical cap are flexible plastic or rubber domes that are placed over the cervix within the vagina, where they prevent passage of sperm cells from the vagina through the cervical canal of the uterus. The diaphragm is larger than the cervical cap. The most commonly used **spermicidal agents** are foams or creams that kill the sperm cells. They are inserted into the vagina before sexual intercourse. When used in combination, a condom and foam or cream are much more effective than when they are used alone. A **spermicidal douche** (doosh) is a stream of fluid containing a chemical toxic to sperm cells that is injected into the vagina. The stream of fluid removes and kills sperm cells. Spermicidal douches used alone are not very effective.

Lactation

Lactation (lak-tā′shŭn) prevents the menstrual cycle for a few months after childbirth. Action potentials sent to the hypothalamus in response to suckling that cause the release of oxytocin and prolactin also inhibit FSH and LH release from the anterior pituitary. Lactation, therefore, prevents the development of ovarian follicles and ovulation. Despite continual lactation, the ovarian and uterine cycles eventually resume. Because ovulation occurs before menstruation, relying on lactation to prevent pregnancy is not consistently effective.

Chemical Methods

Synthetic estrogen and progesterone in **oral contraceptives** (birth-control pills) effectively suppress fertility in females. These substances may have more than one action, but they reduce LH and FSH release from the anterior pituitary. Estrogen and progesterone are present in high enough concentrations to have a negative-feedback effect on the pituitary, which prevents the large increase in LH and FSH secretion that triggers ovulation. Over the years, the dose of

Table A	**Effectiveness of Various Methods for Preventing Pregnancy**	
Technique	**Effectiveness When Used Properly (%)**	**Actual Effectiveness (%)**
Abortion	100	Unknown
Sterilization	100	99.9
Combination (estrogens and progesterones) pill	99.9	98
Intrauterine device	98	98
Minipill (low dose of estrogens and progesterones)	99	97
Condom plus spermicide	99	96
Condom alone	97	90
Diaphragm plus spermicide	97	85
Foam	97	80
Rhythm	97	70

estrogen and progesterone in birth-control pills has been reduced. The current lower-dose birth-control pills have fewer side effects than earlier dosages. An increased risk

of heart attack or stroke exists in women using oral contraceptives who smoke or have a history of hypertension or coagulation disorders. For most women, the pill is effective

and has a minimum frequency of complications, until at least age 35.

Continued

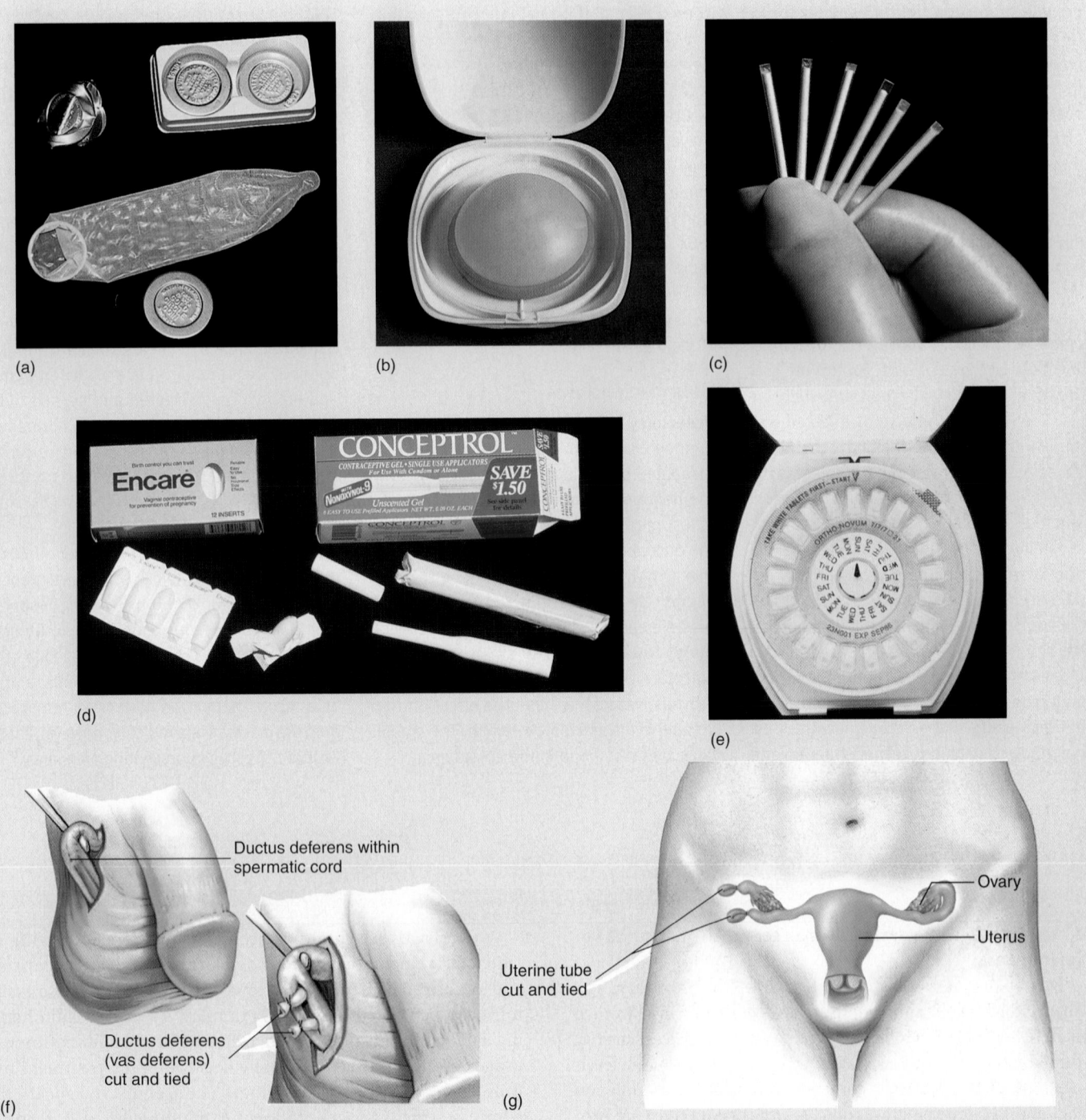

Figure B **Contraceptive Devices and Techniques**

(*a*) Condom. (*b*) Diaphragm used with spermicidal jelly. (*c*) Norplant system. (*d*) Spermicidal foam. (*e*) Oral contraceptives. (*f*) Vasectomy. (*g*) Tubal ligation.

Continued

Progesterone-like chemicals, such as medroxy progesterone (Depo-Provera), which are injected intramuscularly and slowly released into the circulatory system, can act as effective contraceptives. Injected progesterone-like chemicals can provide protection from pregnancy for up to 3 months, depending on the amount injected. A thin silastic tube containing these chemicals, such as the Norplant system, can be implanted beneath the skin, usually in the upper arm, from which they are slowly released into the circulatory system. The implants can be effective for up to 5 years. Menstruation does not normally occur in women using these techniques while the progesterone levels are elevated.

Advantages of injected and implanted progesterone-like contraceptives over other chemical methods of birth control are that they don't require taking pills on a daily basis. The long-term effects of injected and implanted progesterone-like chemicals have not been as thoroughly studied as the long-term effects of birth-control pills have, and they are still being evaluated.

Mifepristone (RU486) blocks the action of progesterone. It causes the endometrium of the uterus to slough off and to be expelled from the uterus as it does at the time of menstruation. Because it blocks progesterone receptors, the endometrium undergoes changes similar to those caused by decreasing progesterone levels. It is, therefore, used to induce menstruation and

reduce the possibility of implantation when sexual intercourse has occurred near the time of ovulation. It can also be used to terminate pregnancies. **Morning-after pills,** similar in composition to birth-control pills, are available. Doubling the number of birth-control pills after sexual intercourse within 3 days and again after 12 more hours is sometimes recommended. This or similar techniques can be used after intercourse but they are only about 75% effective. The elevated estrogen and progesterone levels may inhibit the preovulatory LH surge in some cases, it may alter the rate of transport of the fertilized ovum from the uterine tube to the uterus, or it may inhibit implantation. The precise effect of the elevated estrogen and progesterone-like substances depends on the stage of menstrual cycle when they are taken.

Surgical Methods

Vasectomy (va-sek′tō-mē) is a common method used to render males permanently incapable of fertilization without affecting the performance of the sex act. Vasectomy is a surgical procedure used to cut and tie the ductus deferens from each testis within the scrotal sac. This procedure prevents sperm cells from passing through the ductus deferens and becoming part of the ejaculate. Because such a small volume of ejaculate comes from the testis and epididymis, vasectomy has little effect on the volume of the ejaculated semen. The sperm cells are reabsorbed in the epididymis.

A common method of permanent birth control in females is **tubal ligation** (lī-gā′shŭn), a procedure in which the uterine tubes are tied and cut or clamped through an incision made through the wall of the abdomen. This procedure closes off the pathway between the sperm cells and the oocyte. **Laparoscopy** (lap-ă-ros′kŏ-pē), a procedure in which a special instrument is inserted into the abdomen through a small incision, is commonly used so that only small openings are required to perform the operation.

In some cases, pregnancies are terminated by surgical procedures called **abortions.** The most common method for performing abortions is the insertion of an instrument through the cervix into the uterus. The instrument scrapes the endometrial surface while a strong suction is applied. The endometrium and the embedded embryo are disrupted and sucked out of the uterus. This technique is normally used only in pregnancies that have progressed less than 3 months.

Prevention of Implantation

Intrauterine devices (IUDs) are inserted into the uterus through the cervix to prevent normal implantation of the developing embryonic mass within the endometrium. Some early IUD designs produced serious side effects, such as perforation of the uterus, and, as a result, many IUDs have been removed from the market. Data indicate, however, that IUDs are effective in preventing pregnancy.

---

called the **female climacteric** (klī-mak′ter-ik, klī-mak-ter′ik) or **perimenopause.**

Menopause is associated with changes in the ovary. The number of follicles remaining in the ovaries of menopausal women is small. In addition, the follicles that remain become less sensitive to stimulation by LH and FSH, even though these hormone levels are elevated. As the ovaries become less responsive to stimulation by FSH and LH, fewer mature follicles and corpora lutea are produced. Gradual morphologic changes occur in the female in response to the reduced amount of estrogen and progesterone produced by the ovaries (table 28.3).

A variety of symptoms occur in some females during the climacteric, including "hot flashes," irritability, fatigue, anxiety, and occasionally severe emotional disturbances. These symptoms can be treated successfully in many women by administering small amounts of estrogen, or estrogen in combination with progesterone, and then gradually decreasing the treatment over time or by providing psychologic counseling. It appears that administering estrogen following menopause also helps to prevent osteoporosis and may reduce colorectal cancer. Although estrogen therapy has been successful, in many women it prolongs symptoms associated with menopause. Some potential

Clinical Focus   Causes of Female Infertility

Causes of infertility in females include malfunctions of the uterine tubes, reduced hormone secretion from the pituitary or ovary, and interruption of implantation.

Adhesions from pelvic inflammatory conditions caused by a variety of infections can cause blockage of one or both uterine tubes and is a relatively common cause of infertility in women.

Reduced ovulation can result from inadequate secretion of LH and FSH, which can be caused by hypothyroidism, trauma to the hypothalamus, infarctions of the hypothalamus or anterior pituitary gland, and tumors.

Interruption of implantation may result from uterine tumors or conditions causing abnormal ovarian hormone secretion. For example, premature degeneration of the corpus luteum causes progesterone levels to decline and menses to occur. If the corpus luteum degenerates before the placenta begins to secrete progesterone, the endometrium and the developing embryonic mass degenerate and are eliminated from the uterus. The conditions that result in secondary amenorrhea also reduce fertility (see menstrual cramps, PMS, and amenorrhea, page 1065).

Endometriosis (en′dō-mē-trē-ō′sis), a condition in which endometrial tissue is found in abnormal locations, reduces fertility. Generally, endometriosis is thought to result from some endometrial cells passing from the uterus through the uterine tubes into the pelvic cavity. The endometrial cells invade the peritoneum of the pelvic cavity. Because the endometrium is sensitive to estrogen and progesterone, periodic inflammation of the areas where the endometrial cells have invaded occurs. Endometriosis is a cause of abdominal pain associated with menstruation and it can reduce fertility.

Table 28.3	Possible Changes Caused by Decreased Ovarian Hormone Secretion in Postmenopausal Women
Affected Structures and Functions	**Changes**
Menstrual cycle	Five to 7 years before menopause, the cycle becomes more irregular; finally, the number of cycles in which ovulation does not occur increases, and corpora lutea do not develop
Uterine tubes	Little change
Uterus	Irregular menstruation is gradually followed by no menstruation; chance of cystic glandular hypertrophy of the endometrium increases; the endometrium finally atrophies, and the uterus becomes smaller
Vagina and external genitalia	Dermis and epithelial lining become thinner; vulva becomes thinner and less elastic; labia majora become smaller; pubic hair decreases; vaginal epithelium produces less glycogen; vaginal pH increases; reduced secretion leads to dryness; the vagina is more easily inflamed and infected
Skin	Epidermis becomes thinner; melanin synthesis increases
Cardiovascular system	Hypertension and atherosclerosis occur more frequently
Vasomotor instability	Hot flashes and increased sweating are correlated with vasodilation of cutaneous blood vessels; hot flashes are not caused by abnormal FSH and LH secretion but are related to decreased estrogen levels
Sex drive	Temporary changes, such as either decreases or increases in sex drive, are often associated with the onset of menopause
Fertility	Fertility begins to decline approximately 10 years before the onset of menopause; by age 50, almost all oocytes and follicles are lost; loss is gradual, and no increased follicular degeneration is associated with the onset of menopause

side effects of hormone therapy are of concern, such as an increase in the possibility for the development of breast, ovarian, and uterine cancer. In addition, the increased risk of heart disease for the first few years after the beginning of menopause is not reduced by estrogen or estrogen plus progesterone. Some data indicate that the risk of heart attacks, strokes, and blood clots, is increased.

80. *Define the terms menopause and female climacteric.*
81. *What causes these changes, and what symptoms commonly occur?*

Effects of Aging on the Reproductive System
Age-Related Changes in Males

Several age-related changes occur in the male reproductive system. In some, but not all men, there is an age-related decrease in the size and the weight of the testes. There's an associated decrease in the number of interstitial cells and a thinning of the wall of the seminiferous tubules. These changes may be secondary to a decrease in blood flow to the testes or due to a gradual decrease in sex

Clinical Focus   Infectious Diseases

Sexually Transmitted Diseases

Sexually transmitted diseases (STDs) are a class of infectious diseases spread by intimate sexual contact between individuals. These diseases include the major venereal diseases nongonococcal urethritis, trichomoniasis, gonorrhea, genital herpes, genital warts, syphilis, and acquired immunodeficiency syndrome.

Nongonococcal urethritis refers to any inflammation of the urethra that's not caused by gonorrhea. Factors like trauma or passage of a nonsterile catheter through the urethra can cause this condition, but many cases are acquired through sexual contact. In most cases, the bacterium *Chlamydia trachomatis* is responsible, but other bacteria may be involved. *Chlamydia trachomatis* infection is one of the most common sexually transmitted diseases. It often is unrecognized in people who have it and is responsible for many cases of pelvic inflammatory disease. Left untreated, it can also result in sterility, but antibiotics are usually effective in curing the condition.

Trichomonas vaginalis is a protozoan commonly found in the vagina of females and the urethra of males. If the normal acid-ity of the vagina is disturbed, *Trichomonas* can grow rapidly. *Trichomonas* infection is more common in females than in males because the vagina provides a suitable environment in which these organisms can survive. The rapid growth of these organisms results in inflammation and a greenish yellow discharge characterized by a foul odor.

Gonorrhea (gon-ō-rē′ă) is caused by the bacterium *Neisseria gonorrhoeae*. The organisms attach to the epithelial cells of the vagina or to the male urethra. The invasion of bacteria establishes an inflammatory response in which pus is formed. Males become aware of a gonorrheal infection by painful urination and the discharge of pus-containing material from the urethra. Symptoms appear within a few days to a week. Recovery may eventually occur without complication, but, when complications do occur, they can be serious. The urethra can become partially blocked, or sterility can result from blockage of reproductive ducts with scar tissue. In some cases, other organ systems, such as the heart, meninges of the brain, or joints, may become infected. In females, the early stages of infection may pass unnoticed, but the infection can lead to pelvic inflammatory disease. Gonorrheal eye infections may occur in newborn children of females with gonorrheal infections. Antibiotics are usually effective in treating gonorrheal infections, and the immune system often successfully combats gonorrheal infections in untreated individuals.

Genital herpes (her′pēz) is a viral infection usually caused by herpes simplex type 2. Lesions appear after an incubation period of about 1 week and cause a burning sensation. After this, blisterlike areas of inflammation appear. In males and females, urination can be painful, and walking or sitting can be unpleasant, depending on the location of the lesions. The blisterlike areas heal in about 2 weeks. The lesions may reoccur. The viruses exist in a latent condition in nerve cells and may produce inflamed lesions on the genitals in response to factors like menstruation, emotional stress, or illness. If active lesions are present in the mother's vagina or external genitalia, a caesarean delivery is performed to prevent newborns from becoming infected with the herpesvirus. Because genital herpes is caused by a virus, no effective antibiotic cure for it is available.

---

hormone production. There is a decrease in the rate of sperm cell production and an increase in the number of abnormal sperm cells. However, sperm cell production doesn't stop, and it remains adequate for fertility for most men.

Age-related changes become obvious in the prostate gland by 40 years of age. By 60 years of age there is a clear decrease in blood flow, an increased thickness in the epithelial cell lining of the prostate gland, and a decrease in functional smooth muscle cells in the wall of the prostate. The changes in the prostate gland do not decrease fertility. There's a substantial increase in the incidence of benign prostatic hypertrophy that can create difficulty in urination because it compresses the prostatic urethra. A significant number of men older than 60 years of age (approximately 15%) require medical treatment for benign prostatic hypertrophy. By 80 years this number is 50%. Before 50 years of age prostatic cancer is rare. After 55 years of age it's the third leading cause of death from can-cer in men. Enlargement of the prostate because of prostatic cancer can also create difficulty in urination.

Impotence increases in men with age. By 60 years of age approximately 15% of men have difficulty with impotence and by 80 years of age it is 50%. There's an increase in fibrous connective tissue in the erectile tissue of the penis, which, by the time males are 60 years of age, generally decreases the speed of erection.

Although there's great variation among males, many exhibit a decrease in the frequency of sexual activity and a decrease in sexual performance. Psychological changes, age-related changes in the nervous system, and decreased blood flow explain some of the decrease. Many exhibit decreased sexual activity because of the side effects of medications taken for other conditions.

82. *List major age-related changes that occur in males.*
83. *What are the age-related changes that occur in the prostate gland?*

Genital warts also result from a viral infection (human papillomavirus) and are quite contagious. This disease is common, and its frequency is increasing. Genital warts can also be transmitted from infected mothers to their infants. Genital warts vary from separate, small, warty growths to large cauliflower-like clusters. The lesions are usually not painful, but they can cause painful intercourse, and they bleed easily. For women who have genital warts, an increased risk exists of developing cervical cancer. Treatments for genital warts include topical agents, cryosurgery, or other surgical methods.

Syphilis (sif′i-lis) is caused by the bacterium *Treponema pallidum,* which can be spread by sexual contact of all kinds. Syphilis exhibits an incubation period from 2 weeks to several months. The disease progresses through several recognized stages. In the primary stage, the initial symptom is a small, hard-based **chancre** (shan′ker), or sore, that usually appears at the site of infection. Several weeks after the primary stage, the disease enters the secondary stage, characterized mainly by skin rashes and mild fever. The symptoms of secondary syphilis usually subside after a few weeks, and the disease enters a latent period in which no symptoms are present. In less than half the cases, a tertiary stage develops after many years. In the tertiary stage, many lesions develop that can cause extensive tissue damage and can lead to paralysis, insanity, and even death. Syphilis can be passed on to newborns by an infected mother. Damage to mental development and other neurologic symptoms are among the more serious consequences. Females who have syphilis in the latent phase are most likely to have babies who are infected. Antibiotics are used to treat syphilis, although some strains are very resistant to certain antibiotics.

Acquired immunodeficiency syndrome (AIDS) is caused by infection with the human immunodeficiency virus (HIV), which appears to ultimately result in destruction of the immune system (see chapter 22). The most common mechanisms of transmission of the virus are through sexual contact with a person infected with HIV and through sharing needles with an infected person during the administration of illicit drugs. Although transmission did occur during the early 1980s through tainted blood products, screening techniques now make the transmission of HIV through blood transfusions very rare. Some rare cases of transmission of HIV through accidental needle-sticks in hospitals and other health care facilities have been documented. No evidence exists that casual contact with a person who has AIDS or who is infected with HIV results in transmission of the disease. Transmission appears to require exposure to body fluids of an infected person in a way that allows HIV into the interior of another person. Normal casual contact, including touching an HIV-infected person, doesn't increase the risk of infection.

Other Infectious Diseases

Pelvic inflammatory disease (PID) is a bacterial infection of the female pelvic organs. It usually involves the uterus, uterine tubes, or ovaries. A vaginal or uterine infection may spread throughout the pelvis. Gonorrhea and chlamydia are the most common causes of PID; however, other bacteria can be involved. Early symptoms of PID include increased vaginal discharge and pelvic pain. Early treatment with antibiotics can stop the spread of PID, but lack of treatment results in a life-threatening infection. PID can also lead to sterility.

Age-Related Changes in Females

The most significant age-related change in females is menopause. By age 50 few viable follicles remain in the ovaries. As a result, there's a decrease in the estrogen and progesterone produced by the ovaries.

The uterus decreases in size and the endometrium decreases in thickness. The time between menstruations becomes irregular and longer. Finally menstruations stop. As the uterus decreases in size, it tips posteriorly and assumes a lower position in the pelvic cavity. Occasionally, uterine prolapse, in which the ligaments of the uterus allow it to descend and protrude into the vagina, occurs. Within 15 years after menopause, the uterus is 50% of its original size.

The vaginal wall becomes thinner and less elastic. There is less lubrication of the vagina and the epithelial lining is more fragile. An increased tendency occurs for vaginal infections. Vaginal contractions during intercourse decrease and the vagina narrows with age. In healthy females, sexual excitement requires greater time to develop, the peak levels of sexual activity are lower, and return to the resting state occurs more quickly.

Approximately 10% of all women will have breast cancer. The increase is most rapid between 45 and 65 years of age, and the incidence is greater for those who have a history of breast cancer in their families than for those who do not. Cancer of the endometrium and cancer of the uterine cervix increases between 50 and 65 years of age. Ovarian cancer increases in frequency in older women, and it's the second most common cancer of the reproductive system in older women.

84. *List the major age-related changes that occur in females.*
85. *What are some of the long-term consequences of menopause?*

Systems Pathology
Benign Uterine Tumors

Mrs. M had four children and was 43 years old. She noticed that menstruation was becoming gradually more severe and lasting up to several days longer each time it started. After she menstruated almost continuously for 2 months, she made an appointment with her physician, who performed a pelvic examination, including tests for conditions like cervical cancer and uterine cancer. Palpation of the uterus indicated the presence of enlarged masses in Mrs. M's uterus. The results of a dilation and curettage (D&C—dilation of the cervix and scraping [curettage] of the endometrium to remove growths or other abnormal tissues) indicated that Mrs. M suffered from leiomyomas, or fibroid tumors of the uterus.

Background Information

Uterine **leiomyomas** (lī'ō-mī-ō'măs; figure C), also called uterine fibroids, are enlarged masses of smooth muscle in the myometrium, and they are one of the most common disorders of the uterus. They are the most frequent tumor in women, affecting one of every four. Three-fourths of the women with this condition, however, experience no symptoms. The enlarged mass compresses the uterine lining (endometrium), resulting in ischemia and inflammation and in frequent and severe menstruations. Abdominal cramping because of strong uterine contractions can be present. Constant menstruation is a

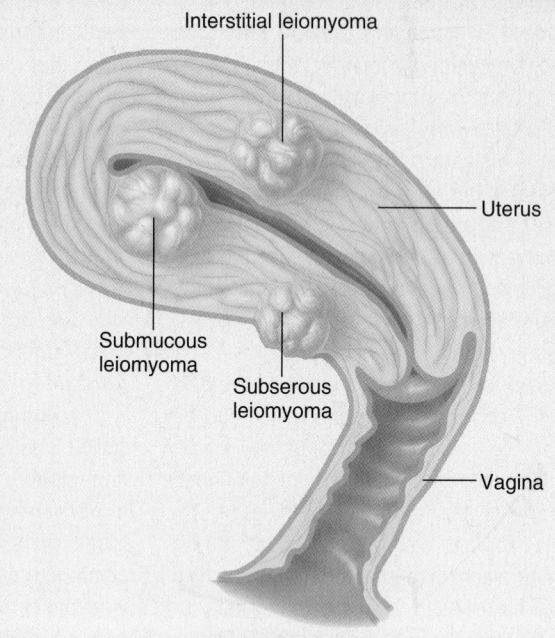

Medial view

Figure C Leiomyomas or Fibroid Tumors

Leiomyomas, or fibroid tumors, are enlarged masses of smooth muscle. They are located near the mucosa (submucous), within the myometrium (interstitial), or near the serosa (subserous).

S U M M A R Y

The male reproductive system produces sperm cells and transfers them to the female. The female reproductive system produces the oocyte and nurtures the developing child.

Anatomy of the Male Reproductive System (p. 1035)
Scrotum

1. The scrotum is a two-chambered sac that contains the testes.
2. The dartos and cremaster muscles help to regulate testicular temperature.

Perineum

The perineum is the diamond-shaped area between the thighs and consists of a urogenital triangle and an anal triangle.

Testes

1. The tunica albuginea is the outer connective tissue capsule of the testes.
2. The testes are divided by septa into compartments that contain the seminiferous tubules and the interstitial cells.
3. The seminiferous tubules become straight to form the tubuli recti which lead to the rete testis. The rete testis opens into the efferent ductules of the epididymis.

4. During development, the testes pass from the abdominal cavity through the inguinal canal to the scrotum.

Sperm Cell Development

1. Sperm cells (spermatozoa) are produced in the seminiferous tubules.
2. Spermatogonia divide (mitosis) to form primary spermatocytes.
3. Primary spermatocytes divide (first division of meiosis) to form secondary spermatocytes, which divide (second division of meiosis) to form spermatids.
4. Spermatids develop an acrosome and a flagellum to become sperm cells.
5. Sertoli cells nourish the sperm cells, form a blood–testes barrier, and produce hormones.

Ducts

1. Efferent ductules extend from the testes to the head of the epididymis.
2. The epididymis is a coiled tube system located on the testis that is the site of sperm cell maturation. It consists of a head, body, and tail.

System Interactions   Effect of Leiomyomas on Other Systems

System	Interaction
Integumentary	If anemia does not develop, skin appearance is normal, but if anemia does develop, the skin can appear pale because of the reduced hemoglobin in red blood cells. The continual loss of blood often results in iron-deficiency anemia. The hemoglobin concentration of blood and the hematocrit are, therefore, reduced.
Muscular	If anemia develops and is severe, muscle weakness may result because of the reduced ability of the cardiovascular system to deliver adequate oxygen to muscles.
Skeletal	The rate of red blood cell synthesis in the red bone marrow increases.
Digestive	An enlarged tumor can put pressure on the rectum or sigmoid colon, resulting in constipation.
Cardiovascular	A chronic loss of blood as in prolonged menstruation over many months to years frequently results in iron-deficiency anemia. Manifestations of anemia include reduced hematocrit, reduced hemoglobin concentration, smaller-than-normal red blood cells (microcytic anemia), and increased heart rate.
Respiratory	Because of anemia, the oxygen-carrying capacity of the blood is reduced. Increased respiration during physical exertion and rapid fatigue are likely to occur if anemia develops.
Urinary	The kidneys increase erythropoietin secretion in response to the loss of red blood cells. The erythropoietin increases red blood cell synthesis in red bone marrow. An enlarged tumor can put pressure on the urinary bladder, resulting in increased frequency of and painful urination.

frequent manifestation of these tumors, and it is one of the most common reasons why women elect to have the uterus removed, a procedure called a **hysterectomy** (his-ter-ek′tō-mē).

P R E D I C T   5

When discussing her condition with her mother, Mrs. M discovered that her mother recalled frequent menstruations that were irregular and prolonged when she was in her late forties. Her mother did not have a hysterectomy, and in a few years the frequency of menstruation began to gradually subside. Explain.

---

3. The ductus deferens passes from the epididymis into the abdominal cavity.
4. The end of the ductus deferens, called the ampulla, and the seminal vesicle join to form the ejaculatory duct.
5. The prostatic urethra extends from the urinary bladder to join with the ejaculatory ducts to form the membranous urethra.
6. The membranous urethra extends through the urogenital diaphragm and becomes the spongy urethra, which continues through the penis.
7. The spermatic cord consists of the ductus deferens, blood and lymphatic vessels, nerves, and remnants of the process vaginalis. Coverings of the spermatic cord consist of the external spermatic fascia, cremaster muscle, and internal spermatic fascia.
8. The spermatic cord passes through the inguinal canal into the abdominal cavity.

Penis

1. The penis consists of erectile tissue.
   • The two corpora cavernosa form the dorsum and the sides of the penis.
   • The corpus spongiosum forms the ventral part and the glans penis.
2. The bulb of the penis and the crura form the root of the penis and the crura attaches the penis to the coxae.
3. The prepuce covers the glans penis.

Accessory Glands

1. The seminal vesicles empty into the ejaculatory ducts.
2. The prostate gland consists of glandular and muscular tissue and empties into the prostatic urethra.
3. The bulbourethral glands are compound mucous glands that empty into the spongy urethra.
4. Semen
   • Semen is a mixture of gland secretions and sperm cells.
   • The bulbourethral glands and the urethral mucous glands produce mucus, which neutralizes the acidic pH of the urethra.
   • The testicular secretions contain sperm cells.
   • The seminal vesicle fluid contains fructose and fibrinogen.
   • The prostate secretions make the seminal fluid more pH-neutral. Clotting factors activate fibrinogen, and fibrinolysin breaks down fibrin.

Physiology of Male Reproduction   (p. 1046)
Regulation of Sex Hormone Secretion

1. GnRH is produced in the hypothalamus and released in surges.
2. GnRH stimulates LH and FSH release from the anterior pituitary.
   • LH stimulates the interstitial cells to produce testosterone.
   • FSH stimulates sperm cell formation.
3. Inhibin, produced by sustacular cells, inhibits FSH secretion.

Puberty

1. Before puberty, small amounts of testosterone inhibit GnRH release.
2. During puberty, testosterone does not completely suppress GnRH release, resulting in increased production of FSH, LH, and testosterone.

Effects of Testosterone

1. Interstitial cells, the adrenal cortex, and possibly the sustentacular cells produce testosterone.
2. Testosterone causes the development of male sex organs in the embryo and stimulates the descent of the testes.
3. Testosterone causes enlargement of the genitals and is necessary for sperm cell formation.
4. Other effects of testosterone
   - Hair growth stimulation (pubic area, axilla, and beard) and inhibition (male pattern baldness)
   - Enlargement of the larynx and deepening of the voice
   - Increased skin thickness and melanin and sebum production
   - Increased protein synthesis (muscle), bone growth, blood cell synthesis, and blood volume
   - Increased metabolic rate

Male Sexual Behavior and the Male Sex Act

1. Testosterone is required for normal sex drive.
2. Stimulation of the sexual act can be tactile or psychologic.
3. Afferent action potentials pass through the pudendal nerve to the sacral region of the spinal cord.
4. Parasympathetic stimulation
   - Erection is due to vasodilation of the blood vessels that supply the erectile tissue.
   - The glands of the urethra and the bulbourethral glands produce mucus.
5. Sympathetic stimulation causes erection, emission, and ejaculation.

Anatomy of the Female Reproductive System (p. 1050)
Ovaries

1. The broad ligament, the mesovarium, the suspensory ligaments, and the ovarian ligaments hold the ovaries in place.
2. The peritoneum (ovarian epithelium) covers the surface of the ovaries.
3. The ovary has an outer capsule (tunica albuginea) and is divided internally into a cortex (contains follicles) and a medulla (receives blood and lymph vessels and nerves).
4. Oocyte development and fertilization
   - Oogonia proliferate and become primary oocytes that are in prophase I of meiosis.
   - Ovulation is the release of an oocyte from an ovary.
   - Prior to ovulation, a primary oocyte continues meiosis and produces a secondary oocyte, which in metaphase II of meiosis, and a polar body, which degenerates or divides to form two polar bodies.
   - Fertilization is the joining of a sperm cell and a secondary oocyte to form a zygote. A sperm cell enters a secondary oocyte, which then completes the second meiotic division and produces a polar body. A zygote is formed when the nuclei of the sperm cell and oocyte fuse to form a diploid nucleus.
   - The haploid nuclei then fuse to form a diploid nucleus.
5. Follicle development
   - Primordial follicles are surrounded by a single layer of flat granulosa cells.
   - Primary follicles are primary oocytes surrounded by cuboidal granulosa cells.
   - The primary follicles become secondary follicles as granulosa cells increase in number and fluid begins to accumulate in the vesicles. The granulosa cells increase in number, and theca cells form around the secondary follicles.
   - Mature follicles are enlarged secondary follicles at the surface of the ovary.

6. Ovulation
   - Ovulation occurs when the follicle swells and ruptures and the secondary oocyte is released from the ovary.
7. Fate of the follicle
   - The mature follicle becomes the corpus luteum.
   - If pregnancy occurs, the corpus luteum persists. If no pregnancy occurs, it becomes the corpus albicans.

Uterine Tubes

1. The mesosalpinx holds the uterine tubes.
2. The uterine tubes transport the oocyte or zygote from the ovary to the uterus.
3. Structures
   - The ovarian end of the uterine tube is expanded as the infundibulum. The opening of the infundibulum is the ostium, which is surrounded by fimbriae.
   - The infundibulum connects to the ampulla, which narrows to become the isthmus. The isthmus becomes the uterine part of the uterine tube and passes through the uterus.
4. The uterine tube consists of an outer serosa, a middle muscular layer, and an inner mucosa with simple ciliated columnar epithelium.
5. Movement of the oocyte
   - Cilia move the oocyte over the fimbriae surface into the infundibulum.
   - Peristaltic contractions and cilia move the oocyte within the uterine tube.
   - Fertilization occurs in the ampulla, where the zygote remains for several days.

Uterus

1. The uterus consists of the body, the isthmus, and the cervix. The uterine cavity and the cervical canal are the spaces formed by the uterus.
2. The uterus is held in place by the broad, round, and uterosacral ligaments.
3. The wall of the uterus consists of the perimetrium (serous membrane), the myometrium (smooth muscle), and the endometrium (mucous membrane).

Vagina

1. The vagina connects the uterus (cervix) to the vestibule.
2. The vagina consists of a layer of smooth muscle and an inner lining of moist stratified squamous epithelium.
3. The vagina is folded into rugae and longitudinal folds.
4. The hymen covers the vestibular opening of the vagina.

External Genitalia

1. The vulva, or pudendum, comprises the external genitalia.
2. The vestibule is the space into which the vagina and the urethra open.
3. Erectile tissue
   - The two corpora cavernosa form the clitoris.
   - The corpora spongiosa form the bulbs of the vestibule.
4. The labia minora are folds that cover the vestibule and form the prepuce.
5. The greater and lesser vestibular glands produce a mucous fluid.
6. When closed, the labia majora cover the labia minora.
   - The pudendal cleft is a space between the labia majora.
   - The mons pubis is an elevated fat deposit superior to the labia majora.

Perineum

The clinical perineum is the region between the vagina and the anus.

Mammary Glands

1. The mammary glands are modified sweat glands located in the breasts.
   - The mammary glands consist of glandular lobes and adipose tissue.
   - The lobes consist of lobules that are divided into alveoli.
   - The lobes connect to the nipple through the lactiferous ducts.
   - The areola surround the nipple.
2. Cooper's ligaments support the breasts.

Physiology of Female Reproduction (p. 1059)

Puberty

1. Puberty begins with the first menstrual bleeding (menarche).
2. Puberty begins when GnRH levels increase.

Menstrual Cycle

1. Ovarian cycle
   - FSH initiates development of the primary follicles.
   - The follicles secrete a substance that inhibits the development of other follicles.
   - LH stimulates ovulation and completion of the first meiotic division by the primary oocyte.
   - The LH surge stimulates the formation of the corpus luteum. If fertilization occurs, HCG stimulates the corpus luteum to persist. If fertilization does not occur, the corpus luteum becomes the corpus albicans.
2. A positive-feedback mechanism causes FSH and LH levels to increase near the time of ovulation.
   - Estrogen produced by the theca cells of the follicle stimulates GnRH secretion.
   - GnRH stimulates FSH and LH, which stimulate more estrogen secretion, and so on.
   - Inhibition of GnRH levels causes FSH and LH levels to decrease after ovulation. Inhibition is due to the high levels of estrogen and progesterone produced by the corpus luteum.
3. Uterine cycle
   - Menses (day 1 to days 4 or 5). The spiral arteries constrict, and endometrial cells die. The menstrual fluid is composed of sloughed cells, secretions, and blood.
   - Proliferation phase (day 5 to day 14). Epithelial cells multiply and form glands, and the spiral arteries supply the glands.
   - Secretory phase (day 15 to day 28). The endometrium becomes thicker, and the endometrial glands secrete.
   - Estrogen stimulates proliferation of the endometrium and synthesis of progesterone receptors.
   - Increased progesterone levels cause hypertrophy of the endometrium, stimulate gland secretion, and inhibit uterine

contractions. Decreased progesterone levels cause the spiral arteries to constrict and start menses.

Female Sexual Behavior and the Female Sex Act

1. Female sex drive is partially influenced by androgens (produced by the adrenal gland) and steroids (produced by the ovaries).
2. Parasympathetic effects
   - The erectile tissue of the clitoris and the bulbs of the vestibule become filled with blood.
   - The vestibular glands secrete mucus, and the vagina extrudes a mucuslike substance.

Female Fertility and Pregnancy

1. Intercourse must take place 5 days before to 1 day after ovulation if fertilization is to occur.
2. Sperm cell transport to the ampulla depends on the ability of the sperm cells to swim and possibly on contractions of the uterus and the uterine tubes.
3. Implantation of the developing embryo into the uterine wall occurs when the uterus is most receptive.
4. Estrogen and progesterone secreted first by the corpus luteum and later by the placenta are essential for the maintenance of pregnancy.

Menopause

The female climacteric begins with irregular menstrual cycles and ends with menopause, the cessation of the menstrual cycle.

Effects of Aging on the Reproductive System (p. 1071)

Age-Related Changes in Males

1. There is an age-related decrease in the size and weight of the testes, decrease in the number of interstitial cells, thinning of the seminiferous tubule wall, and a decrease in sperm production. Sperm cell production is still adequate for fertilization.
2. The prostate gland enlarges, and there is an age-related increase in prostatic cancer.
3. Impotence is age-related and there is a gradual decrease in sexual activity.

Age-Related Changes in Females

1. The most significant age-related change in females is menopause.
2. The uterus decreases in size and the vaginal wall thins.
3. There is an age-related increase in breast, uterine, and ovarian cancer.

R E V I E W   A N D   C O M P R E H E N S I O N

1. If an adult male walked into a swimming pool of cold water, which of these muscles would be expected to contract?
   - a. cremaster muscle
   - b. dartos muscle
   - c. gubernaculum
   - d. prepuce muscle
   - e. both a and b

2. Testosterone is produced in the
   - a. interstitial cells.
   - b. seminiferous tubules of the testes.
   - c. anterior lobe of the pituitary.
   - d. sperm cells.

3. Early in development (4 months after fertilization), the testes
   - a. are found in the peritoneal cavity.
   - b. move through the inguinal canal.
   - c. produce a membrane that becomes the scrotum.
   - d. produce sperm cells.
   - e. all of the above.

4. The site of spermatogenesis in the male is the
   - a. ductus deferens.
   - b. seminiferous tubules.
   - c. epididymis.
   - d. rete testis.
   - e. efferent ductule.

5. The location of final maturation and storage of sperm cells before their ejaculation is the
   - a. seminal vesicles.
   - b. seminiferous tubules.
   - c. glans penis.
   - d. epididymis.
   - e. sperm bank.

6. Given these structures:
   1. ductus deferens
   2. efferent ductule
   3. epididymis
   4. ejaculatory duct
   5. rete testis

   Choose the arrangement that lists the structures in the order a sperm cell passes through them from the seminiferous tubules to the urethra.
   a. 2,3,5,4,1
   b. 2,5,3,4,1
   c. 3,2,4,1,5
   d. 3,4,2,1,5
   e. 5,2,3,1,4

7. Concerning the penis,
   a. the membranous urethra passes through the corpora cavernosa.
   b. the glans penis is formed by the corpus spongiosum.
   c. the penis contains four columns of erectile tissue.
   d. the crus of the penis is part of the corpus spongiosum.
   e. the bulb of the penis is covered by the prepuce.

8. Given these glands:
   1. prostate gland
   2. bulbourethral gland
   3. seminal vesicle

   Choose the arrangement that is in the order the glands contribute their secretions to the formation of semen.
   a. 1,2,3
   b. 2,1,3
   c. 2,3,1
   d. 3,1,2
   e. 3,2,1

9. Which of these glands is correctly matched with the function of its secretions?
   a. bulbourethral gland—neutralizes acidic contents of the urethra
   b. seminal vesicles—contain large amounts of fructose, which nourishes the sperm cells
   c. prostate gland—contains clotting factors that cause coagulation of the semen
   d. all of the above

10. LH in the male stimulates
    a. development of the seminiferous tubules.
    b. spermatogenesis.
    c. testosterone production.
    d. both a and b.
    e. all of the above.

11. Which of these factors causes a decrease in GnRH release?
    a. decreased inhibin
    b. increased testosterone
    c. decreased FSH
    d. decreased LH

12. In the male, before puberty
    a. FSH levels are higher than after puberty.
    b. LH levels are higher than after puberty.
    c. GnRH release is inhibited by testosterone.
    d. all of the above.

13. Testosterone
    a. stimulates the development of terminal hairs.
    b. decreases red blood cell count.
    c. prevents closure of the epiphyseal plate.
    d. decreases blood volume.
    e. all of the above.

14. Which of these is consistent with erection of the penis?
    a. parasympathetic stimulation
    b. dilation of arterioles
    c. engorgement of sinusoids with blood
    d. occlusion of veins
    e. all of the above

15. The first polar body
    a. is normally formed before fertilization.
    b. is normally formed after fertilization.
    c. normally receives most of the cytoplasm.
    d. is larger than an oocyte.
    e. both a and b.

16. After ovulation the mature follicle collapses, taking on a yellowish appearance to become the
    a. degenerating follicle.
    b. corpus luteum.
    c. corpus albicans.
    d. tunica albuginea.
    e. cumulus mass.

17. The ampulla of the uterine tube
    a. is the opening of the uterine tube into the uterus.
    b. has long, thin projections called the ostium.
    c. is connected to the isthmus of the uterine tube.
    d. is lined with ciliated columnar epithelium.

18. The layer of the uterus that undergoes the greatest change during the menstrual cycle is the
    a. perimetrium.
    b. hymen.
    c. endometrium.
    d. myometrium.
    e. broad ligament.

19. The vagina
    a. consists of skeletal muscle.
    b. has ridges called rugae.
    c. is lined with simple squamous epithelium.
    d. all of the above.

20. During sexual excitement, which of these structures fills with blood and causes the vaginal opening to narrow?
    a. bulb of the vestibule
    b. clitoris
    c. mons pubis
    d. labia majora
    e. prepuce

21. Given these vestibular-perineal structures:
    1. vaginal opening
    2. clitoris
    3. urethral opening
    4. anus

    Choose the arrangement that lists the structures in their proper order from the anterior to the posterior aspect.
    a. 2,3,1,4
    b. 2,4,3,1
    c. 3,1,2,4
    d. 3,1,4,2
    e. 4,2,3,1

22. Concerning the breasts,
    a. lactiferous ducts open on the areola.
    b. each lactiferous duct supplies an alveolus.
    c. they are attached to the pectoralis major muscles by mammary ligaments.
    d. even before puberty, the female breast is quite different from the male breast.

23. The major secretory product of the mature follicle is
    a. estrogen.
    b. progesterone.
    c. LH.
    d. FSH.
    e. relaxin.

24. In the average adult female, ovulation occurs at day
    _____ of the menstrual cycle.
    a. 1
    b. 7
    c. 14
    d. 21
    e. 28

25. Which of these processes or phases in the monthly reproductive cycle of the human female occur at the same time?
    a. maximal LH secretion and menstruation
    b. early follicular development and the secretory phase of the uterus
    c. regression of the corpus luteum and an increase in ovarian progesterone secretion
    d. ovulation and menstruation
    e. proliferation stage of the uterus and increased estrogen production

26. During the proliferative phase of the menstrual cycle, one would normally expect
    a. the highest levels of estrogen that occur during the menstrual cycle.
    b. the mature follicle to be present in the ovary.
    c. an increase in the thickness of the endometrium.
    d. both a and b.
    e. all of the above.

27. The cause of menses in the menstrual cycle appears to be
    a. increased progesterone secretion from the ovary, which produces blood clotting.
    b. increased estrogen secretion from the ovary, which stimulates the muscles of the uterus to contract.
    c. decreased progesterone secretion by the ovary.
    d. decreased production of oxytocin, causing the muscles of the uterus to relax.

28. After fertilization the successful development of a mature, full-term fetus depends upon the
    a. release of human chorionic gonadotropin (HCG) by the developing placenta.
    b. production of estrogen and progesterone by the placental tissues.
    c. maintenance of the corpus luteum for all 9 months.
    d. both a and b.
    e. all of the above.

29. A woman with a 28-day menstrual cycle is most likely to become pregnant as a result of coitus on days
    a. 1–3.
    b. 5–8.
    c. 9–14.
    d. 15–20.
    e. 21–28.

30. Menopause
    a. develops when follicles become less responsive to FSH and LH.
    b. results from elevated estrogen levels in 40- to 50-year-old women.
    c. occurs because too many follicles develop during each cycle.
    d. results when follicles develop but contain no oocytes.
    e. occurs because FSH and LH levels decline.

Answers in Appendix F

C R I T I C A L   T H I N K I N G

1. If an adult male were castrated, what would happen to the levels of GnRH, FSH, LH, and testosterone in his blood? What effect would these hormonal changes have on sexual characteristics and behavior?

2. If a 9-year-old boy were castrated, what would happen to the levels of GnRH, FSH, LH, and testosterone in his blood? What effect would these hormonal changes have on sexual characteristics and behavior?

3. Suppose you want to produce a birth-control pill for men. On the basis of what you know about the male hormone system, what do you want the pill to do? Discuss any possible side effects that could be produced by your pill.

4. If the ovaries are removed from a postmenopausal woman, what happens to the levels of GnRH, FSH, LH, estrogen, and progesterone in her blood? What symptoms do you expect to observe?

5. During the secretory phase of the menstrual cycle, you normally expect
   a. the highest levels of progesterone that occur during the menstrual cycle.
   b. a follicle present in the ovary that is ready to undergo ovulation.
   c. that the endometrium reaches its greatest degree of development.
   d. both a and b.
   e. both a and c.

6. If the ovaries are removed from a 20-year-old female, what happens to the levels of GnRH, FSH, LH, estrogen, and progesterone in her blood? What side effects do these hormonal changes have on her sexual characteristics and behavior?

7. A study divides healthy adult females into two groups (A and B). Both groups are composed of females who have been sexually active for at least 2 years and are not pregnant at the beginning of the experiment. The subjects weigh about the same amount, and none smokes cigarettes, although some do drink alcohol occasionally. Group A women receive a placebo in the form of a sugar pill each morning during their menstrual cycles. Group B women receive a pill containing estrogen and progesterone each morning of their menstrual cycles. Then plasma LH levels are measured before, during, and after ovulation. The results are as follows:

Group	4 Days Before Ovulation	The Day of Ovulation	4 Days After Ovulation
A	18 mg/100 mL	300 mg/100 mL	17 mg/100 mL
B	21 mg/100 mL	157 mg/100 mL	15 mg/100 mL

The number of pregnancies in group A was 37/100 females/year. The number of pregnancies in group B was 1.5/100 females/year. What conclusion can you reach on the basis of these data? Explain the mechanism involved.

8. A woman who is taking birth-control pills that consist of only progesterone experiences the hot flash symptoms of menopause. Explain why.

9. GnRH can be used to treat some women who want to have children but have not been able to get pregnant. Explain why it's critical to administer the correct concentration of GnRH at the right time during the menstrual cycle.

Answers in Appendix G

A N S W E R S   T O   P R E D I C T   Q U E S T I O N S

1. The prostate gland is anterior to the wall of the rectum. A finger inserted into the rectum can palpate the prostate gland through the rectal wall.

2. Coagulation may help keep the sperm cells within the female reproductive tract, thereby increasing the likelihood of fertilization.

3. If administered before the preovulatory LH surge, estrogen stimulates the hypothalamus to secrete GnRH. Estrogen and progesterone, in large amounts, inhibit GnRH and LH releases. A large amount of estrogen and progesterone administered at this time should, therefore, reduce the surge of LH. Continual administration of high levels of GnRH causes anterior pituitary cells to become insensitive to GnRH. Thus, LH and FSH levels remain low, and the ovarian cycle stops.

4. High progesterone levels after menses inhibit GnRH secretion from the hypothalamus and, therefore, FSH and LH secretion from the anterior pituitary. Without FSH and LH, the events of the ovarian cycle, including estrogen production, are inhibited. Because estrogen causes proliferation of the endometrium, thickening of the endometrium is not expected. Also, estrogen increases the synthesis of uterine progesterone receptors, and without estrogen the secretory response of the endometrium to the elevated progesterone is inhibited.

5. Mrs. M's mother could have had leiomyomas also, although, without direct data from medical examinations, one cannot be certain. If that were the cause of her irregular menstruations, they may have become less frequent as Mrs. M's mother experienced menopause. During menopause, the uterus gradually becomes smaller, and eventually the cyclic changes in the endometrial lining cease. If the leiomyomas were relatively mild, the onset of menopause could explain the gradual disappearance of the irregular and prolonged menstruations. (*Note:* If the tumors are large, constant and severe menstruations are likely even if regular menstrual cycles stop due to menopause.)

Visit the Online Learning Center at **www.mhhe.com/seeley7** for chapter quizzes, interactive learning exercises, and other study tools.

Development, Growth, Aging, and Genetics

Colorized scanning electron micrograph of an oocyte with sperm cells on the surface.

The stages of life and associated activities are issues of great interest in today's society. We tend to view life stages very differently today than we did just a few years ago. For example, in 1960, 20% of males and 12% of females graduating from high school attended college. Today, over half of all people 25 and older have attended some college. In addition, there are many more nontraditional college students than there were just a few years ago.

In 1900, only 5% of the U.S. population was over age 65. Today, about 16% of the population is over age 65, and by 2030 more than 20% will be older than 65. The average life expectancy in 1900 was about 47 years, in 1940 it was about 63 years, and today it is about 78 years. In 1900, nearly 70% of all males over age 65 were still working; today only about 20% are still working past age 65. Older people are healthier and more active than they have ever been.

The life span is usually considered the period between birth and death; however, the 9 months before birth are a critical part of a person's existence. What happens in these 9 months profoundly affects the rest of a person's life. Although most people develop normally and are born without defects, approximately three out of every 100 people are born with a birth defect so severe that it requires medical attention during the first year of life. Later in life, many more people discover previously unknown problems, such as the tendency to develop asthma, certain brain disorders, or cancer. This chapter discusses *prenatal development* (p. 1082), *parturition* (p. 1103), *the newborn* (p. 1106), *lactation* (p. 1109), the *first year after birth* (p. 1110), *life stages* (p. 1110), *aging* (p. 1111), *death* (p. 1112), and *genetics* (p. 1112).

Prenatal Development

The **prenatal period** is the period from conception until birth, and it is divided into three parts: (1) approximately the first 2 weeks of development, during which the primitive germ layers are formed; (2) from about the second to the end of the eighth week of development, during which the major organ systems come into existence; and (3) the last 30 weeks of the prenatal period, during which the organ systems grow and become more mature.

The developing human between the time of fertilization and 8 weeks of development is called an **embryo** (em′brē-ō). From 8 weeks to birth the developing human is called a **fetus** (fē′tus).

The medical community in general uses the mother's **last menstrual period (LMP)** to calculate the **clinical age** of the unborn child. Most embryologists, on the other hand, use **postovulatory age** to describe the timing of developmental events. Postovulatory age is used in this book. Because ovulation occurs about 14 days after LMP and fertilization occurs near the time of ovulation, it's assumed that postovulatory age is 14 days less than clinical age.

> 1. Describe the three parts of the prenatal period. Give the length of time for each part.
> 2. Define clinical age and postovulatory age, and distinguish between the two.

Fertilization

Fertilization is the process by which a sperm cell attaches to a secondary oocyte, the sperm head enters the oocyte cytoplasm, and joins the oocyte pronucleus to form a new nucleus (figure 29.1). Only a few dozen sperm cells, of the several hundred million deposited in the vagina during sexual intercourse, reach the vicinity of the secondary oocyte in the ampulla of the uterine tube. The **corona radiata,** which are cells of the cumulus mass expelled from the follicle with the oocyte (see figure 28.12), is a barrier to the sperm cells reaching the oocyte. The sperm cells are propelled through the loose matrix between the follicular cells of the corona radiata by the action of their flagella. The **zona pellucida** is an extracellular membrane, comprised mostly of glycoproteins, between the corona radiata and the oocyte. One particular zona

1. Many sperm cells attach to the corona radiata of a secondary oocyte.

Labels: Sperm cell; Corona radiata; Oocyte nucleus; First polar body; Zona pellucida

2. Sperm cells attach to ZP3 receptors on the zona pellucida, and enzymes in the acrosome digest through the zona pellucida.

Label: Sperm cell contacting zona pellucida

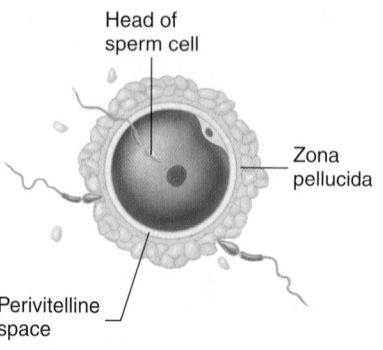

3. The head of one sperm cell penetrates the zona pellucida, attaches to α6β1 integrins on the oocyte plasma membrane, and penetrates the oocyte plasma membrane to enter the oocyte cytoplasm. Changes in the zona pellucida (moving away from the oocyte) form a perivitelline space and prevent additional sperm cells from entering the oocyte.

Labels: Head of sperm cell; Zona pellucida; Perivitelline space

4. In response to the entry of the sperm head into the oocyte, the oocyte nucleus moves to one side of the oocyte where it completes the second meiotic division and gives off a second polar body. When second meiosis is complete, the oocyte nucleus, now called the female pronucleus, moves back toward the center of the oocyte.

Labels: Second polar body; Head of sperm cell; Female pronucleus

5. The sperm head enlarges and becomes the male pronucleus.

Labels: Male pronucleus; Female pronucleus

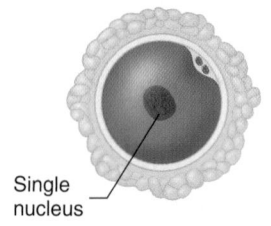

6. The two pronuclei fuse to form a single nucleus. Fertilization is complete and a zygote results.

Label: Single nucleus

Process Figure 29.1   Fertilization

pellucida glycoprotein, called **ZP3,** is a species-specific sperm cell receptor, to which molecules on the acrosomal cap of the sperm cell bind. This binding initiates the **acrosomal reaction,** which results in the activation of digestive enzymes, primarily hyaluronidase, in the acrosome.

The first sperm cell through the zona pellucida attaches to the integrin $\alpha6\beta1$ on the surface of the oocyte plasma membrane. This attachment causes depolarization of the oocyte plasma membrane within 2–3 seconds after sperm cell attachment. This depolarization, called the **fast block to polyspermy,** prevents additional sperm from attaching to the oocyte plasma membrane. Depolarization causes the intracellular release of Ca^{2+}, which, in turn, causes the exocytosis of water and other molecules from secretory vesicles, referred to as cortical granules, on the inner surface of the oocyte plasma membrane. The released fluid causes the oocyte to shrink and the zona pellucida to denature and expand away from the oocyte. As the result of denaturation of the zona pellucida, ZP3 is inactivated, and no additional sperm cells can attach. This reaction is referred to as the **slow block to polyspermy.** The fluid-filled space between the oocyte plasma membrane and the zona pellucida is called the **perivitelline space.**

Entrance of a sperm cell into the oocyte stimulates the female nucleus to undergo the second meiotic division, and the second polar body is formed. The nucleus that remains after the second meiotic division, called the **female pronucleus,** moves to the center of the oocyte, where it meets the enlarged head of the sperm cell, the **male pronucleus.** Both the male and female pronuclei are haploid, each having one-half of each chromosome pair (see chapter 28). Fusion of the pronuclei completes the process of fertilization and restores the diploid number of chromosomes. The product of fertilization is a single cell, the **zygote** (zī′gōt; figure 29.2*a*).

(b)

(c)

(a)

(d)

Figure 29.2 Early Stages of Human Development

(*a*) Zygote (120 μm in diameter). (*b*)–(*d*) During the early cell divisions, the embryo divides into more and more cells, but the total size of the embryo remains relatively constant. (*b*) Two cells. (*c*) Four cells. (*d*) Eight cells.

3. Describe the events of fertilization. Where does fertilization occur?

Cloning

Fusion of the male and female pronuclei forms a new nucleus in the zygote. Alternatively, the nucleus can be removed from an oocyte and the nucleus from some other cell can be introduced. This complete, diploid nucleus forms a zygote. This process of introducing a nucleus from some cell into an oocyte, whose own nucleus has been removed, is called **cloning**. The first cloned mammal was a sheep named Dolly in Scotland in 1996.

Twins

In rare cases, following early cell divisions, the cells may separate and develop to form two individuals, called "identical," or **monozygotic, twins.** Identical twins have identical genetic information in their cells. Other mechanisms that occur a little later in development can also cause identical twins. Occasionally a woman may ovulate two or more secondary oocytes at the same time. Fertilization of two oocytes by different sperm cells results in "fraternal," or **dizygotic, twins.** Multiple ovulations can occur naturally or be stimulated by injection of drugs that stimulate gonadotropin release. These drugs are sometimes used to treat certain forms of infertility. This drug treatment may result in multiple births in women undergoing the treatment.

Early Cell Division

About 18–39 hours after fertilization, the zygote divides to form two cells. Those two cells divide to form four cells, which divide to form eight cells, and so on (figure 29.2*b–d*). The cells of this dividing embryo are referred to as **pluripotent** (ploo-rip′ō-tent; multiple-powered), which means that any cell of the mass has the ability to develop into a wide range of tissues. As a result, the total number of embryonic cells can decrease, increase, or reorganize without affecting the normal development of the embryo.

Morula and Blastocyst

Once the dividing embryo is a solid ball of 12 or more cells, it is a sphere composed of numerous smaller spheres and is, therefore, called a **morula** (mōr′oo-lǎ, mōr′ū-lǎ; mulberry; figure 29.3). Four or 5 days after ovulation, the morula consists of about 32 cells. Near this time, a fluid-filled cavity called the **blastocele** (blas′tō-sēl) begins to appear approximately in the center of the cellular mass. The hollow sphere that results is called a **blastocyst** (blas′tō-sist; see figure 29.3). A single layer of cells, the **trophoblast** (trof′ō-blast, trō′fō-blast; feeding layer), surrounds most of the blastocele,

Figure 29.3   Blastocyst

Green cells are trophoblastic, and *orange* cells will form the embryo proper.

Clinical Focus In Vitro Fertilization and Embryo Transfer

In a small number of women, normal pregnancy is not possible because of some anatomic or physiologic condition. In 87% of these cases, the uterine tubes are incapable of transporting the zygote to the uterus or of allowing sperm cells to reach the oocyte. In vitro fertilization and embryo transfer have made pregnancy possible in hundreds of such women since 1978. **In vitro fertilization (IVF)** involves removal of secondary oocytes from a woman, placing the oocytes into a petri dish, and adding sperm cells to the dish to allow fertilization and early development to occur in vitro, which means "in glass." **Embryo transfer** involves the removal of the developing embryo from the petri dish and its introduction into the uterus of a recipient female.

For IVF and embryo transfer to be accomplished, a woman is first injected with an LH-like substance, which causes more than one follicle to mature at one time. Just before the follicles rupture, the secondary oocytes are surgically removed from the ovary. The oocytes are then incubated in a dish and maintained at body temperature for 6 hours. Then sperm cells are added to the dish.

After 24–48 hours, when the zygotes have divided to form two- to eight-cell masses, several of the embryos are transferred to the uterus. Several embryos are transferred, because only a small percentage of them survives. Implantation and subsequent development then proceed in the uterus as they would for natural implantation.

Typically, three embryos are transferred at a time. The rate of complications, such as multiple pregnancies, miscarriage, and prematurity, however, also increases with the greater numbers of embryos per transfer.

About one-third of transfers of three embryos end in multiple pregnancies. Of triplets born as a result of IVF, 64% require intensive care after birth, and 75% of quadruplets require intensive care, often for several weeks.

Prematurity from IVF pregnancies in the United Kingdom results in newborn mortality in 2.7% of cases, a rate three times that of natural pregnancies. As a result of the possible complications, no more than two to three embryos are now transferred per IVF in the United Kingdom.

The success rate for IVF has dramatically increased through time. The success rate at the best U.S. clinics was 20% in 1982, 30% in 1995, about 50% in 1999, and is now over 70%. This success rate has now exceeded natural limits, because only 50% or less of natural fertilizations result in a successful delivery. The increased success rate probably results from the fact that embryos that appear abnormal are not transferred and that more than one embryo is transferred during IVF.

---

but at one end of the blastocyst the cells are several layers thick. The thickened area is the **inner cell mass** and is the tissue from which the embryo proper develops. The trophoblast forms the placenta and the membranes (chorion and amnion) surrounding the embryo.

Stem Cell Research

Cells of the inner cell mass are referred to as **embryonic stem cells** because they theoretically have the potential to become any type of tissue in the body. Researchers are taking advantage of this pluripotent potential of these stem cells to develop normal tissue lines to supplement defective tissues in people with a wide variety of diseases. Current research focuses on the right combination of growth factors that can drive embryonic stem cell lines toward a specific tissue type. For example, pancreatic islet cells, derived from embryonic stem cells, may be implanted into the pancreas of a person with diabetes to provide the needed supply of insulin.

The issue of stem cell research is very controversial, with opponents arguing that the blastocyst is a living human being and that the blastocyst has to be killed to obtain the stem cells. Their argument is that the sanctity of human life extends well into the prenatal period to include the earliest stages of human development. Proponents of stem cell research argue that the sanctity and quality of human life for those suffering from debilitating, often fatal, and currently incurable diseases is worth the price. Other proponents argue that the definition of a human being does not extend to the blastocyst. The debate will probably not end soon.

Implantation of the Blastocyst and Development of the Placenta

All of the events of the early germinal phase, including the first cell division through formation of the blastocele and the inner cell mass, occur as the embryo moves from the site of fertilization in the ampulla of the uterine tube to the site of implantation in the uterus. About 7 days after fertilization, the blastocyst attaches itself to the uterine wall, usually in the area of the uterine fundus, and begins the process of **implantation,** which is the burrowing of the blastocyst into the uterine wall.

As the blastocyst invades the uterine wall, two populations of trophoblast cells develop and form the embryonic portion of the **placenta** (figure 29.4), the organ of nutrient and waste product exchange between the embryo and the mother. The first is a proliferating population of individual trophoblast cells called the **cytotrophoblast** (sī-tō-trof′ō-blast). The other is a nondividing syncytium, or multinucleated cell, called the **syncytiotrophoblast** (sin-sish′ē-ō-trō′fō-blast). The cytotrophoblast remains nearer the other embryonic tissues, and the syncytiotrophoblast invades the endometrium of the uterus. The syncytiotrophoblast is nonantigenic, which means that as it invades the maternal tissue, no immune reaction is triggered.

As the syncytiotrophoblast encounters maternal blood vessels, it surrounds them and digests the vessel wall, forming pools of maternal blood within cavities called **lacunae** (lă-koo′nē; figure 29.4c). The lacunae are still connected to intact maternal vessels so that blood circulates from the maternal vessels through the lacunae.

1. Frontal section of the uterus and uterine tube showing development 7 days after fertilization.

Ovary

Uterus — Endometrium
— Myometrium

Implantation of blastocyst

2. Implantation of the blastocyst with syncytiotrophoblast beginning to invade the uterine wall (at about 8–12 days).

Uterine endometrium

Uterine gland

Maternal arteriole

Syncytiotrophoblast invading uterine wall

Cytotrophoblast

Cells that will become the embryo proper

Uterine epithelium

Implanted blastocyst

3. Intermediate stage of placental formation (at about 14–20 days). As maternal blood vessels are encountered by the syncytiotrophoblast, lacunae are formed and filled with maternal blood.

Maternal arteriole

Syncytiotrophoblast

Cytotrophoblast

Maternal arteriole wall

Arteriole wall digested by syncytiotrophoblast

Lacuna filled with maternal blood

Connecting stalk to developing embryo

4. Cytotrophoblast cords surround the syncytiotrophoblast and lacunae, and embryonic blood vessels enter the cord (at about 1 month).

Maternal arteriole

Syncytiotrophoblast

Cytotrophoblast

Lacuna filled with maternal blood

Cytotrophoblast cord

Connecting stalk to developing embryo

Mesoderm

Embryonic vessels forming (contain embryonic blood)

Process Figure 29.4 Formation of the Placenta

Implantation of the blastocyst and invasion of the trophoblast to form the placenta.

Cords of cytotrophoblast surround the syncytiotrophoblast and la-cunae (figure 29.4d). Branches, called **chorionic** (kō-rē-on′ik) **villi,** sprout from these cords and protrude into the lacunae, like fingers. The entire embryonic structure facing the maternal tissues is called the **chorion** (kō′rē-on). Embryonic mesoderm and blood vessels grow into the cords and villi as they protrude into the lacunae. In the mature placenta (figure 29.5), the cytotrophoblast disappears, so that the embryonic blood supply is separated from the maternal blood supply by only the embryonic capillary wall, a basement membrane, and a thin layer of syncytiotrophoblast.

Placental Problems

If the blastocyst implants near the cervix, a condition called **placenta previa** (prĕ′vē-ă) occurs. In this condition, as the placenta grows, it may extend partially or completely across the internal cervical opening. As the fetus and placenta continue to grow and the uterus stretches, the region of the placenta over the cervical opening may tear, and hemorrhaging may occur. **Abruptio** (ab-rŭp′shē-o) **placentae** is a tearing away of a normally positioned placenta from the uterine wall accompanied by hemorrhaging. Both of these conditions can result in miscarriage and can also be life-threatening to the mother.

4. What are the events during the first week after fertilization? Define the terms zygote, morula, blastocyst, and blastocele. What is meant by the term pluripotent?

5. Describe the trophoblast and inner cell mass, and explain what develops from each.

6. Describe the implantation of the blastocyst and development of the placenta.

Formation of the Germ Layers

After implantation, a new cavity called the **amniotic** (am-nē-ot′ik) **cavity** forms inside the inner cell mass and is surrounded by a layer of cells called the **amnion** (am′nē-on), or **amniotic sac.** Formation of the amniotic cavity causes part of the inner cell mass near-est the blastocele to separate as a flat disk of tissue called the **embryonic disk** (figure 29.6). This embryonic disk is composed of two layers of cells: an **ectoderm** (ek′tō-derm; outside layer) adja-cent to the amniotic cavity and an **endoderm** (en′dō-derm; inside layer) on the side of the disk opposite the amnion. A third cavity, the **yolk sac,** forms inside the blastocele from the endoderm. The amniotic sac, yolk sac, and intervening double-layered embryonic disk can be thought of as resembling two balloons pushed together.

Figure 29.5 Mature Placenta and Fetus

Fetal blood vessels and maternal blood vessels are in close contact, and nutrients are exchanged between fetal and maternal blood, but fetal and maternal blood don't mix.

Figure 29.6 Embryonic Disk

Embryonic disk consisting of ectoderm (*blue*) and endoderm (*yellow*), with the amniotic cavity and yolk sac.

One balloon represents the amniotic sac, and the other represents the yolk sac. The circular double layer of balloon wall where the two balloons are pressed together represents the embryonic disk. The amniotic sac eventually enlarges to surround the developing embryo, providing it with a protective fluid environment, the "bag of waters," where the embryo forms.

About 13 or 14 days after fertilization, the embryonic disk becomes a slightly elongated oval structure. Proliferating cells of the ectoderm migrate toward the center and the caudal end of the disk, forming a thickened line called the **primitive streak.**

Some ectoderm cells leave the ectoderm, migrate through the primitive streak, and emerge between the ectoderm and endoderm as a new germ layer, the **mesoderm** (mez′o-derm; middle layer; figure 29.7). These three germ layers, the ectoderm, mesoderm, and endoderm, are the beginning of the embryo proper. All tissues of the adult can be traced to them (table 29.1). A cord-like structure called the **notochord** extends from the cephalic end of the primitive streak.

The development of the germ layers and the subsequent development of the organ systems is heavily dependent on cell

1. Cells in the surface ectoderm move toward the primitive streak and migrate through the streak (*blue arrow tails*).

2. Cells of the ectoderm that migrate through the primitive streak become mesodermal cells (*red arrows*).

3. The mesoderm (*red*) lies between the ectoderm (*blue*) and endoderm (*yellow*).

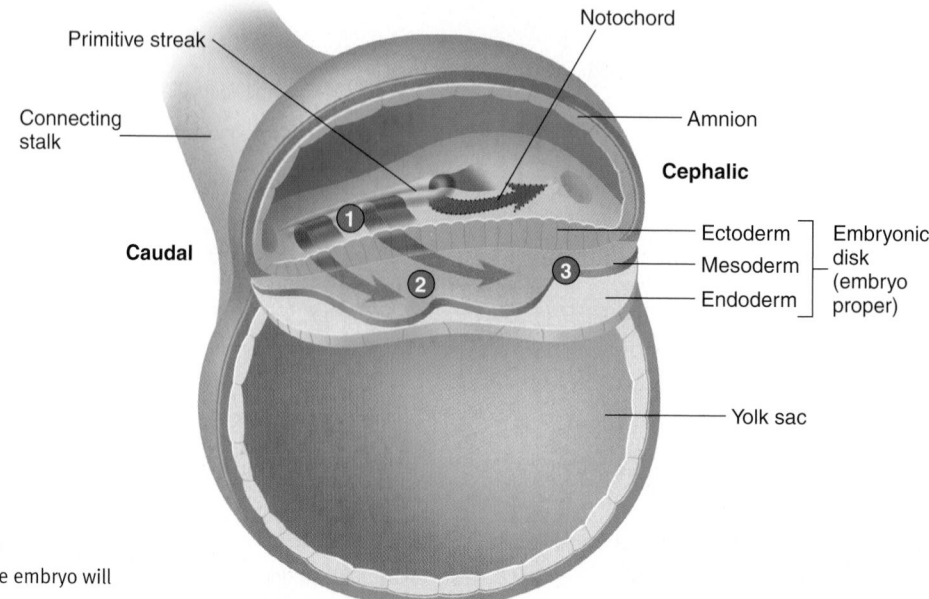

Process Figure 29.7 Primitive Streak

Embryonic disk with a primitive streak. The head of the embryo will develop over the notochord.

Table 29.1	Germ Layer Derivatives

Ectoderm	**Mesoderm**
Epidermis of skin	Dermis of skin
Tooth enamel	Circulatory system
Lens and cornea of eye	Parenchyma of glands
Outer ear	Muscle
Nasal cavity	Bones (except facial)
Anterior pituitary	Microglia
Neuroectoderm	
Brain and spinal cord	**Endoderm**
Somatic motor neurons	Lining of gastrointestinal tract
Preganglionic autonomic neurons	Lining of lungs
Neuroglia cells (except microglia)	Lining of hepatic, pancreatic, and other exocrine ducts
Neural crest cells	Urinary bladder
Melanocytes	Thymus
Sensory neurons	Thyroid
Postganglionic autonomic neurons	Parathyroid
Adrenal medulla	Tonsils
Facial bones	
Teeth (dentin, pulp, and cementum) and gingiva	
A few skeletal muscles in head	

communication. Some of the cell communication depends on direct cell-cell contact, whereas other communication depends on diffusable molecules, such as **growth factors.** Two important families of growth factors are epidermal growth factors (EGF) and fibroblast growth factors (FGF).

 7. Describe the formation of the germ layers and the role of the primitive streak.

P R E D I C T **1**

Predict the results of two primitive streaks forming in one embryonic disk. What if the two primitive streaks are touching each other?

Neural Tube and Neural Crest Formation

The ectoderm near the cephalic end of the primitive streak is stimulated about 18 days after fertilization to form a thickened **neural plate.** The lateral edges of the plate begin to rise like two ocean waves coming together. These edges are called the **neural folds,** and a **neural groove** lies between them (figure 29.8). The underlying notochord stimulates the folding of the neural plate at the neural groove. The crests of the neural folds begin to meet in the midline and fuse into a **neural tube,** which is completely closed by 26 days. The neural tube becomes the brain and the spinal cord, and the cells of the neural tube are called **neuroectoderm** (see table 29.1).

As the neural folds come together and fuse, a population of cells breaks away from the neuroectoderm all along the crests of the folds. These **neural crest cells** migrate down along the side of the developing neural tube to become the sensory and post ganglionic autonomic portions of the peripheral nervous system

and the adrenal medulla and migrate laterally to just below the ectoderm, where they become melanocytes of the skin. In the head, neural crest cells perform additional functions; they contribute to the skull, the dentin of teeth, a few small skeletal muscles, and general connective tissue. Because neural crest cells in the head give rise to many of the same tissues as the mesoderm in the head and trunk, the general term **mesenchyme** (mez′en-kīm) is sometimes applied to cells of either neural crest or mesoderm origin.

Malformations

During the first 2 weeks of development, the embryo is quite resistant to outside influences that may cause malformations. Factors that adversely affect the embryo at this age are more likely to kill it. Between 2 weeks and the next 4–7 weeks (depending on the structure considered), the embryo is more sensitive to outside influences that cause malformations than at any other time.

8. Describe the formation of the neural tube and neural crest.

Somite Formation

As the neural tube develops, the mesoderm immediately adjacent to the tube forms distinct segments called **somites** (sō′mītz; see figure 29.8). In the head, the first few somites never become clearly divided but develop into indistinct segmented structures called **somitomeres.** The somites and somitomeres eventually give rise to a part of the skull, the vertebral column, and skeletal muscle. Most of the head muscles are derived from the somitomeres.

 9. What is a somite?

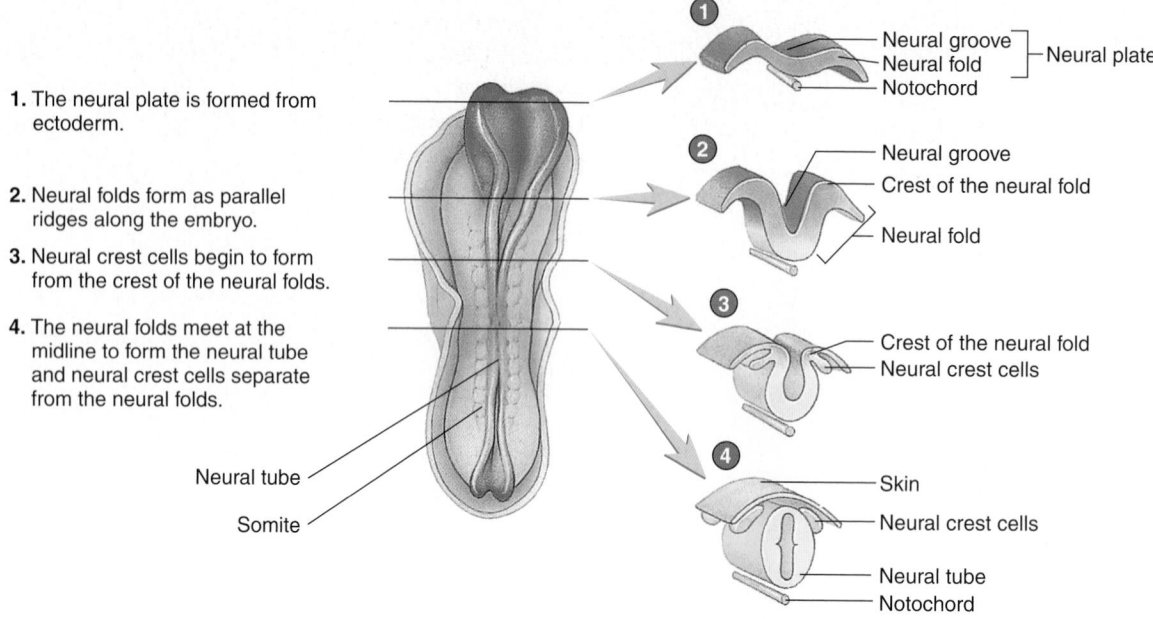

1. The neural plate is formed from ectoderm.

2. Neural folds form as parallel ridges along the embryo.

3. Neural crest cells begin to form from the crest of the neural folds.

4. The neural folds meet at the midline to form the neural tube and neural crest cells separate from the neural folds.

Neural tube

Somite

① Neural groove / Neural fold — Neural plate
Notochord

② Neural groove
Crest of the neural fold
Neural fold

③ Crest of the neural fold
Neural crest cells

④ Skin
Neural crest cells
Neural tube
Notochord

Process Figure 29.8   Formation of the Neural Tube

The neural folds come together in the midline and fuse to form a neural tube. This fusion begins in the center and moves both cranially and caudally. The embryo shown is about 21 days after fertilization. The insets to the *right* show progressive closure of the neural tube.

Formation of the Gut and Body Cavities

At the same time its neural tube is forming, the embryo itself is becoming a tube along the upper part of the yolk sac. The **foregut** and **hindgut** develop as the cephalic and caudal ends of the yolk sac are separated from the main yolk sac. This is the beginning of the digestive tract (figure 29.9*a*). The developing digestive tract pinches off from the yolk sac as a tube but remains attached in the center to the yolk sac by a yolk stalk.

The ends of the foregut and hindgut (figure 29.9*b*) are in close relationship to the overlying ectoderm and form membranes called the oropharyngeal membrane and the cloacal membrane, respectively. The **oropharyngeal membrane** opens to form the mouth, and the **cloacal membrane** opens to form the urethra and anus. Thus, the digestive tract becomes a tube that opens to the outside at both ends.

A considerable number of **evaginations** (ē-vaj-i-nā′shŭnz; outpocketings) occur along the early digestive tract (figure 29.9*c*). The first to form is the allantois (see figure 29.9*b*), part of which will form the urinary bladder. Other evaginations develop into structures like the anterior pituitary, the thyroid gland, the lungs, the liver, the pancreas, and the urinary bladder. At the same time, solid bars of tissue known as **branchial arches** (figures 29.9*c* and 29.10) form along the lateral sides of the head, and the sides of the foregut expand as pockets between the branchial arches. The central expanded foregut is called the **pharynx,** and the pockets along both sides of the pharynx are called **pharyngeal pouches.** Adult derivatives of the pharyngeal pouches include the auditory tubes, tonsils, thymus, and parathyroids.

At about the same time the gut is developing, a series of isolated cavities starts to form within the embryo, thus beginning development of the **celom** (sē′lom; see figure 29.9), or body cavities. The most cranial group of cavities enlarges and fuses to form the **pericardial cavity.** Shortly thereafter, the celomic cavity extends toward the caudal end of the embryo as the **pleural** and **peritoneal cavities.** Initially, all three of these cavities are continuous, but they eventually separate into three distinct adult cavities (see chapter 1).

* **10. Describe the formation of the gut and body cavities.**

Limb Bud Development

Arms and legs first appear at about 28 days as limb buds (see figure 29.10). The **apical ectodermal ridge,** a specialized thickening of the ectoderm, develops on the lateral margin of each limb bud and stimulates its outgrowth. As the buds elongate, limb tissues are laid down in a proximal-to-distal sequence. For example, in the upper limb, the arm is formed before the forearm, which is formed before the hand.

* **11. Describe limb formation. What does a proximal-to-distal growth sequence mean?**

Development of the Face

The face develops by fusion of five embryonic structures: the **frontonasal process,** which forms the forehead, nose, and midportion of the upper jaw and lip; two **maxillary processes,** which form the lateral parts of the upper jaw and lip; and two **mandibular**

Figure 29.9   Formation of the Digestive Tract

Blue arrows show the folding of the digestive tract into a tube. *Dotted lines* show the plane of the section from which cross sections were taken. (*a*) Twenty days after fertilization. (*b*) Twenty-five days after fertilization. (*c*) Thirty days after fertilization. Evaginations are identified along the pharynx and digestive tract in part (*c*).

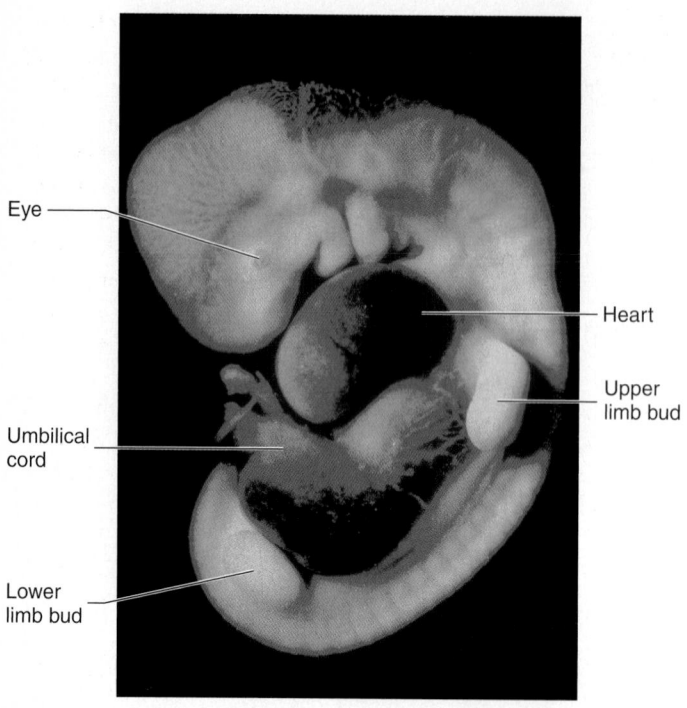

Figure 29.10 Human Embryo 35 Days After Fertilization

processes, which form the lower jaw and lip (figure 29.11 step 1). **Nasal placodes** (plak′ōdz), which develop at the lateral margins of the frontonasal process, develop into the nose and the center of the upper jaw and lip (figure 29.11 step 2).

As the brain enlarges and the face matures, the nasal placodes approach each other in the midline. The medial edges of the placodes fuse to form the midportion of the upper jaw and lip (figure 29.11 steps 3 and 4). This part of the frontonasal process is between the two maxillary processes, which are expanding toward the midline, and fuses with them to form the upper jaw and lip, known as the **primary palate** (see figure 29.11 step 4).

● *12.* *Describe the processes involved in the formation of the face. What clefts may result if these processes fail to fuse?*

Cleft Lip

A **cleft lip** results from failure of the frontonasal and one or both maxillary processes to fuse (see figure 29.11). Because the frontonasal process is a midline structure that normally fuses with the two lateral maxillary processes during formation of the primary palate, cleft lips usually do not occur in the midline but to one side (or both sides). The cleft can vary in severity from a slight indentation in the lip to a fissure that extends from the mouth to the naris (nostril).

At about the same time that the primary palate is forming, the lateral edges of the nasal placodes fuse with the maxillary processes to close off the groove extending from the mouth to the eye (figure 29.11 steps 4 and 5). On rare occasions, these structures fail to meet, resulting in a facial cleft extending from the mouth to the eye. The inferior margins of the maxillary processes

1. **28 days after fertilization**
   The face develops from five processes: frontonasal (*blue*), two maxillary (*yellow*), and two mandibular (*orange*; already fused).

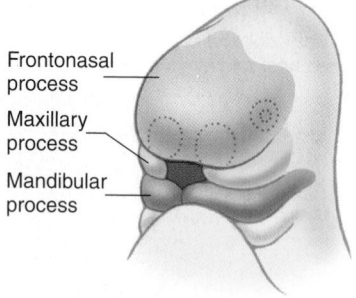

2. **33 days after fertilization**
   Nasal placodes appear in the frontonasal process.

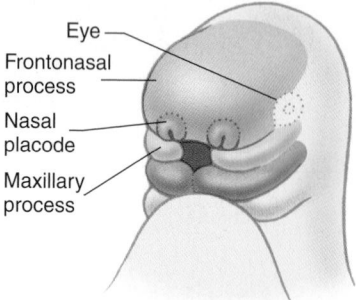

3. **40 days after fertilization**
   Maxillary processes extend toward the midline. The nasal placodes also move toward the midline and fuse with the maxillary processes to form the jaw and lip.

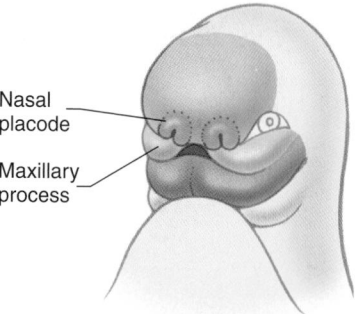

4. **48 days after fertilization**
   Continued growth brings structures more toward the midline.

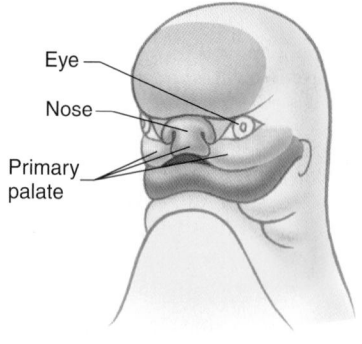

5. **14 weeks after fertilization**
   Colors show the contributions of each process to the adult face.

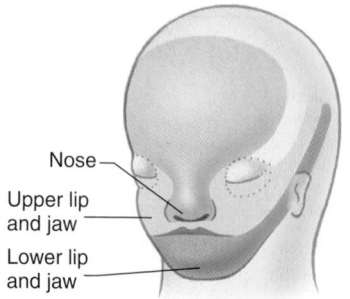

Process Figure 29.11 Development of the Face

fuse with the superior margins of the mandibular processes to decrease the size of the mouth. All of the previously described fusions and the growth of the brain give the face a recognizably "human" appearance by about 50 days.

The roof of the mouth, known as the **secondary palate,** begins to form as vertical shelves, which swing to a horizontal position and begin to fuse with each other at about 56 days of development. Fusion of the entire palate is not completed until about 90 days. If the secondary palate does not fuse, a midline fissure in the roof of the mouth, called a **cleft palate,** results. A cleft palate can range in severity from a slight cleft of the uvula (see figure 24.6) to a fissure extending the entire length of the palate. A cleft lip and cleft palate can occur together, forming a continuous fissure.

Development of the Organ Systems

The major organ systems appear and begin to develop during the embryonic period. The period between 14 and 60 days is, therefore, called the period of organogenesis (table 29.2).

Skin

The **epidermis** of the skin is derived from ectoderm, and the **dermis** is derived from mesoderm, or from neural crest cells in the case of the face. Nails, hair, and glands develop from the epidermis (see chapter 5). Melanocytes and sensory receptors in the skin are derived from neural crest cells.

Skeleton

The skeleton develops from either mesoderm or the neural crest cells through intramembranous or endochondral bone formation (see chapter 6). The bones of the face develop from neural crest cells, whereas the rest of the skull, the vertebral column, and ribs develop from somite- or somitomere-derived mesoderm. The appendicular skeleton develops from limb bud mesoderm.

Muscle

Myoblasts (mī′ō-blastz) are the early, embryonic cells that give rise to skeletal muscle fibers. Myoblasts migrate from somites or somitomeres to sites of future muscle development, where they continue to divide and begin to fuse and form multinuclear cells called **myotubes,** which enlarge to become the muscle fibers of the skeletal muscles. Shortly after myotubes form, nerves grow into the area and innervate the developing muscle fibers. After the basic form of each muscle is established, an increase in the number of muscle fibers causes continued muscle growth. The total number of muscle fibers is established before birth and remains relatively constant thereafter. Muscle enlargement after birth results from an increase in the size of individual fibers.

Nervous System

The nervous system is derived from the neural tube and neural crest cells. Neural tube closure begins at about 21 days of development in the upper cervical region and proceeds into the head and down the spinal cord. Soon after the neural tube has closed at about 25 days of development, the part of the neural tube that becomes the brain begins to expand and develops a series of pouches

(see figure 13.14). The central cavity of the neural tube becomes the ventricles of the brain and the central canal of the spinal cord.

Neural Tube Defects

Anencephaly (an′en-sef′ă-lē; no brain) is a birth defect in which much of the brain fails to form. It results when the neural tube fails to close in the region of the head. A baby born with anencephaly cannot survive. **Spina bifida** (spī′nă bi′fi-dă; split spine) is a general term describing defects of the spinal cord or vertebral column (or both) (figure A). Spina bifida can range from a simple defect with no clinical manifestations and with one or more vertebral spinous processes split or missing to a more severe defect that results in paralysis of the limbs or the bowels and bladder, depending on where the defect occurs.

It has now been well documented that the inclusion of **folic acid** in the diet of a woman during the early stages of her pregnancy significantly reduces the risk of neural tube defects in her developing embryo.

Figure A Spina Bifida

The neuron cell bodies of somatic motor neurons and preganglionic neurons of the autonomic nervous system, which provide axons to the peripheral nervous system, are located within the neural tube. Sensory neurons and postganglionic neurons of the autonomic nervous system are derived from neural crest cells.

Alcohol and Cigarette Smoke

A number of drugs and other chemicals are known to affect the embryo and fetus during development. The two most common are alcohol and cigarette smoke. Alcoholism or binge drinking can result in **fetal alcohol syndrome,** which includes decreased mental function. Exposure of the fetus to **cigarette smoke** throughout pregnancy can stunt the physical growth and mental development of the fetus.

Special Senses

The **olfactory bulb** and **nerve** develop as an evagination from the telencephalon (see figure 13.14). The eyes develop as evaginations from the diencephalon. Each evagination elongates to form an **optic stalk,** and a bulb called the **optic vesicle** develops at its terminal end. The optic vesicle reaches the side of the head and stimulates the overlying ectoderm to thicken into a **lens.**

Table 29.2 Development of the Organ Systems

	Age (Days Since Fertilization)					
	1–5	6–10	11–15	16–20	21–25	26–30
General Features	Fertilization Morula Blastocyst	Blastocyst implants	Primitive streak Three germ layers	Neural plate	Neural tube closed	Limb buds and other "buds" appear
Integumentary System			Ectoderm Mesoderm			Melanocytes from neural crest
Skeletal System			Mesoderm		Neural crest (will form facial bones)	Limb buds
Muscular System			Mesoderm	Somites begin to form		Somites all present
Nervous System			Ectoderm	Neural plate	Neural tube complete Neural crest Eyes and ears begin	Lens begins to form
Endocrine System			Ectoderm Mesoderm Endoderm	Thyroid begins to develop		Parathyroids appear
Cardiovascular System			Mesoderm	Blood islands form Two heart tubes	Single-tubed heart begins to beat	Interatrial septum begins to form
Lymphatic System			Mesoderm			Thymus appears
Respiratory System			Mesoderm Endoderm		Diaphragm begins to form	Trachea forms as single bud Lung buds (primary bronchi)
Digestive System			Mesoderm Endoderm		Neural crest (will form tooth dentin) Foregut and hindgut form	Liver and pancreas appear as buds Tongue bud appears
Urinary System			Mesoderm Endoderm		Pronephros develops Allantois appears	Mesonephros appears
Reproductive System			Mesoderm Endoderm		Primordial germ cells on yolk sac	Mesonephros appears Genital tubercle forms

The sensory part of the ear appears as an ectodermal thickening or placode that invaginates and pinches off from the overlying ectoderm.

Endocrine System

An evagination from the floor of the diencephalon forms the **posterior pituitary gland.** The **anterior pituitary gland** develops from an evagination of ectoderm in the roof of the embryonic oral cavity and grows toward the floor of the brain. It eventually loses its connection with the oral cavity and becomes attached to the posterior pituitary gland (see chapter 18).

The **thyroid gland** originates as an evagination from the floor of the pharynx in the region of the developing tongue and moves into the lower neck, eventually losing its connection with

Age (Days Since Fertilization)					
31–35	**36–40**	**41–45**	**46–50**	**51–55**	**56–60**
Hand and foot plates on limbs	Fingers and toes appear Lips formed Embryo 15 mm	External ear forming Embryo 20 mm	Embryo 25 mm	Limbs elongate to a more adult relationship Embryo 35 mm	Face is distinctly human in appearance
Sensory receptors appear in skin		Collagen fibers clearly present in skin		Extensive sensory endings in skin	
Mesoderm condensation in areas of future bone	Cartilage in site of future humerus	Cartilage in site of future ulna and radius	Cartilage in site of hand and fingers		Ossification begins in clavicle and then in other bones
Muscle precursor cells enter limb buds			Functional muscle		Nearly all muscles appear in adult form
Nerve processes enter limb buds		External ear forming Olfactory nerve begins to form		Semicircular canals in inner ear complete	Eyelids form Cochlea in inner ear complete
Pituitary appears as evaginations from brain and mouth	Gonadal ridges form Adrenal glands forming		Pineal body appears	Thyroid gland in adult position and attachment to tongue lost	Anterior pituitary loses its connection to the mouth
Interventricular septum begins to form		Interventricular septum complete	Interatrial septum complete but still has opening until birth		
Large lymphatic vessels form in neck	Spleen appears			Adult lymph pattern forms	
Secondary bronchi to lobes form	Tertiary bronchi to bronchopulmonary segments form		Tracheal cartilage begins to form		
Oropharyngeal membrane ruptures		Secondary palate begins to form Tooth buds begin to form			Secondary palate begins to fuse (fusion complete by 90 days)
Metanephros begins to develop				Mesonephros degenerates	Anal portion of cloacal membrane ruptures
	Gonadal ridges form	Primordial germ cells enter gonadal ridges	Paramesonephric ducts appear		Uterus forming Beginning of differentiation of external genitalia in male and female

the pharynx. The **parathyroid glands,** which are derived from the third and fourth pharyngeal pouches, migrate inferiorly and become associated with the thyroid gland.

The **adrenal medulla** arises from neural crest cells and consists of specialized postganglionic neurons of the sympathetic division of the autonomic nervous system (see chapter 16). The **adrenal cortex** is derived from mesoderm.

The **pancreas** originates as two evaginations from the duodenum, which come together to form a single gland (see figure 29.9c).

Circulatory System

The heart develops from two endothelial tubes (figure 29.12 step 1), which fuse into a single, midline heart tube (figure 29.12 step 2). Blood vessels and blood cells form from blood islands on the surface

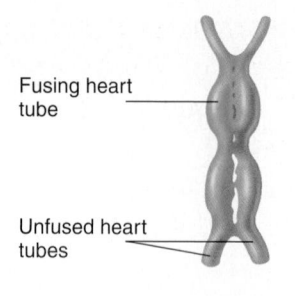

1. **20 days after fertilization**
   At this age, the heart consists of two parallel tubes.

Fusing heart tube

Unfused heart tubes

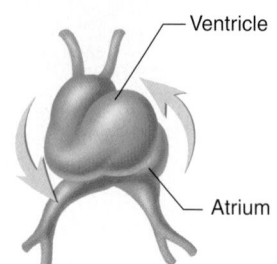

Ventricle

Atrium

2. **22 days after fertilization**
   The two parallel tubes have fused to form one tube. This tube bends as it elongates (*blue arrows* suggest the direction of bending) within the confined space of the pericardium.

Septum primum

Left atrium

Right atrium

Left ventricle

Atrioventricular canals

Interventricular septum

Right ventricle

3. **31 days after fertilization**
   The interatrial septum (septum primum, *green*) and the interventricular septum grow toward the center of the heart.

Septum secundum

Septum primum

Foramen

Interventricular septum

4. **35 days after fertilization**
   The interventricular septum is nearly complete. A foramen opens in the septum primum (*green*) as the septum secundum begins to form (*blue*).

Septum secundum ⎤
                 ⎥ Interatrial
Septum primum ⎦ septum

Right atrium

Left atrium

Foramen ovale

Right ventricle

Left ventricle

5. **The final embryonic condition of the interatrial septum**
   Blood from the right atrium can flow through the foramen ovale into the left atrium. After birth, as blood begins to flow in the other direction, the left side of the interatrial septum is forced against the right side, closing the foramen ovale.

Process Figure 29.12   Development of the Heart

of the yolk sac and inside the embryo. **Blood islands** are small masses of mesoderm that become blood vessels on the outside and blood cells on the inside. These islands expand and fuse to form the circulatory system. A series of dilations appears along the length of the primitive heart tube, and four major regions can be identified: the **sinus venosus,** the site where blood enters the heart; a single **atrium;** a single **ventricle;** and the **bulbus cordis,** where blood exits the heart (see figure 29.12 step 2).

The elongating heart, confined within the pericardium, becomes bent into a loop, the apex of which is the ventricle (see figure 29.12 step 2). The major chambers of the heart, the atrium and the ventricle, expand rapidly. The right part of the sinus venosus becomes absorbed into the atrium, and the bulbus cordis is absorbed into the ventricle. The embryonic sinus venosus initiates contraction at one end of the tubular heart. Later in development, part of the sinus venosus becomes the sinoatrial node, which is the adult pacemaker.

PREDICT ❷
What would happen if the sinus venosus didn't contract before other areas of the primitive heart?

An **interventricular septum** (figure 29.12 steps 3–5) develops, which divides the single ventricle into two chambers. The **interatrial septum,** which separates the two atria in the adult heart, is formed from two parts: the **septum primum** (primary septum) and the **septum secundum** (secondary septum). An opening in the interatrial septum called the **foramen ovale** (ō-val′ē) connects the two atria and allows blood to flow from the right to the left atrium in the embryo and fetus.

Heart Defects

If the interventricular septum does not grow enough to completely separate the ventricles, a **ventricular septal defect (VSD)** results. If the septum secundum fails to grow far enough or if the foramen secundum becomes too large, an **atrial septal defect (ASD)** occurs, allowing blood to flow from the left atrium to the right atrium in the newborn. Both ASDs and VSDs result in abnormal heart sounds called **heart murmurs.** Blood passes through the ASD and VSD from the left to right side of the heart. The right side of the heart usually hypertrophies. In severe cases of VSD (Eisenmenger syndrome), the increased pressure in the pulmonary blood vessels and a decreased blood flow through the systemic blood vessels results in pulmonary edema, cyanosis (a bluish color due to deficient blood oxygenation), or heart failure.

Respiratory System

The lungs begin to develop as a single midline evagination from the foregut in the region of the future esophagus. This evagination branches to form two **lung buds** (figure 29.13 step 1). The lung buds elongate and branch, first forming the bronchi that project to the lobes of the lungs (figure 29.13 step 2) and then the bronchi that project to the bronchopulmonary segments of the lungs (figure 29.13 step 3). This branching continues (figure 29.13 step 4) until, by the end of the sixth month, about 17 generations of branching have occurred. Even after birth, some branching continues as the lungs grow larger, and in the adult about 24 generations of branches have been established.

1. 28 days after fertilization
A single bud forms and divides into two buds, which will become the lungs and primary bronchi.

2. 32 days after fertilization
Primary bronchi branch to form secondary bronchi, which supply the lung lobes.

3. 35 days after fertilization
Secondary bronchi branch to form tertiary bronchi, which supply the bronchopulmonary segments.

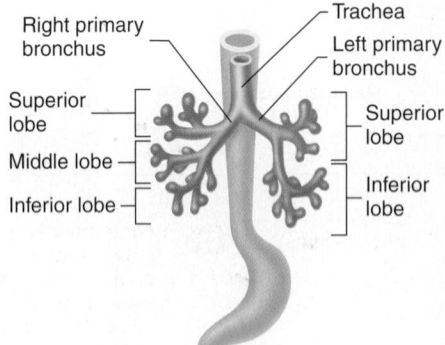
4. 50 days after fertilization
The branching will continue to eventually form the extensive respiratory passages in the lungs.

Process Figure 29.13 Development of the Lung

Urinary System

The kidneys develop from mesoderm located between the somites and the lateral part of the embryo. About 21 days after fertilization, the mesoderm in the cervical region differentiates into a structure called the **pronephros** (the most forward or earliest kidney) (figure 29.14a), which consists of a duct and simple tubules connecting the duct to the open celomic cavity. This type of kidney is the functional adult kidney in some lower chordates, but it's probably not functional in the human embryo and soon disappears.

The **mesonephros** (middle kidney; see figure 29.14a) is a functional organ in the embryo. It consists of a duct, which is a caudal extension of the pronephric duct, and a number of minute tubules, which are smaller and more complex than those of the pronephros. One end of each tubule opens into the mesonephric duct, and the other end forms a glomerulus (see chapter 26).

As the mesonephros is developing, the caudal end of the hindgut begins to enlarge to form the **cloaca** (klō-ā'kă; sewer), the common junction of the digestive, urinary, and genital systems (figure 29.14b). A **urorectal septum** divides the cloaca into two parts: a digestive part called the **rectum** and a urogenital part called the **urethra** (figure 29.14c). The cloaca has two tubes associated with it: the hindgut and the **allantois** (ă-lan'tō-is; sausage), which is a blind tube extending into the umbilical cord (see figures 29.9 and 29.14). The part of the allantois nearest the cloaca enlarges to form the urinary bladder, and the remainder, which is from the bladder to the umbilicus, degenerates (figure 29.14d).

The mesonephric duct extends caudally as it develops and eventually joins the cloaca. At the point of junction, another tube, the **ureter,** begins to form. Its distal end enlarges and branches to form the duct system of the adult kidney, called the **metanephros** (last kidney), which takes over the function of the degenerating mesonephros.

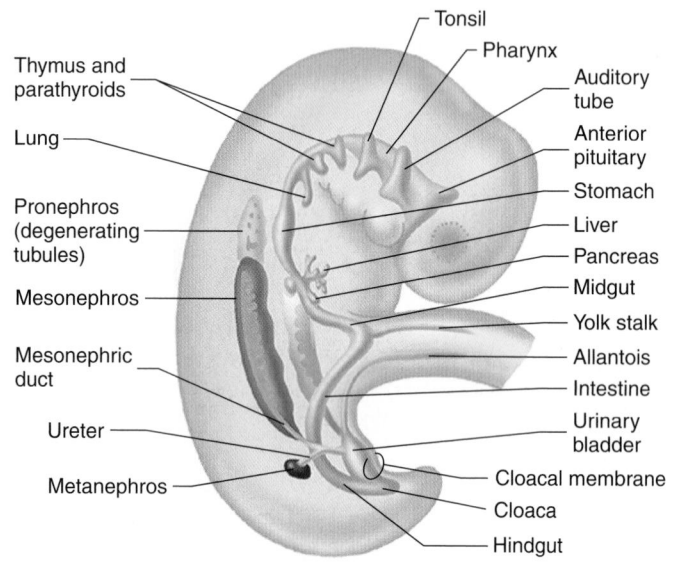

(a) The three parts of the developing kidney: pronephros, mesonephros, metanephros.

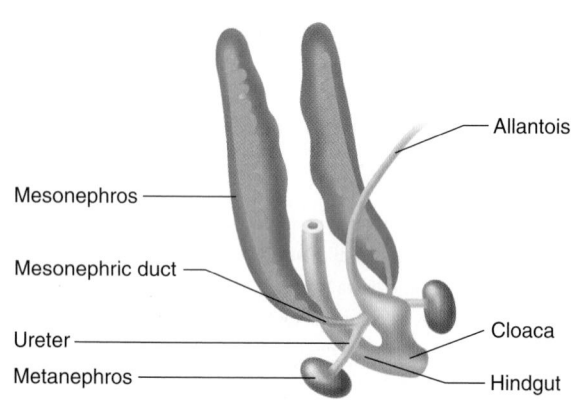

(b) The metanephros (adult kidney) enlarges as the mesonephros degenerates.

(c) The kidney continues to grow and develop.

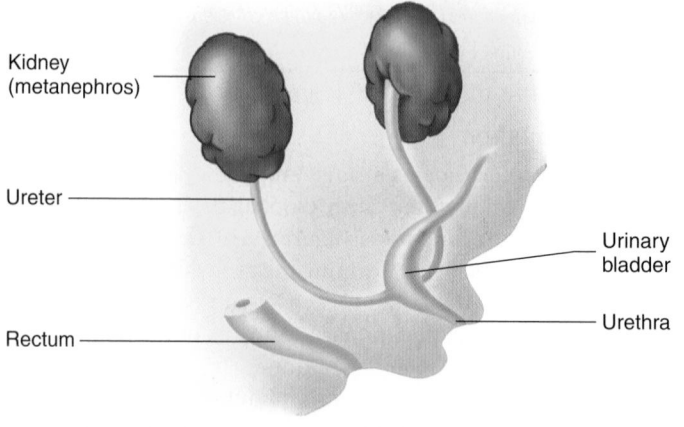

(d) Final duct relationships.

Figure 29.14   Development of the Kidney and Urinary Bladder

Reproductive System

The male and female gonads appear as **gonadal ridges** along the ventral border of each mesonephros (figure 29.15a). **Primordial germ cells,** destined to become oocytes or sperm cells, form on the surface of the yolk sac, migrate into the embryo, and enter the gonadal ridge.

In the female, the ovaries descend from their original position high in the abdomen to a position within the pelvis. In the male, the testes descend even farther. As the testes reach the anteroinferior abdominal wall, a pair of tunnels called the **inguinal canals** form through the abdominal musculature. The testes pass through these canals, leaving the abdominal cavity and coming to lie within the **scrotum** (see figure 28.4). Descent of the testes through the canals begins about 7 months after conception, and the testes enter the scrotum about 1 month before the infant is born.

Cryptorchidism

In approximately 0.5% of male children, one or both testes fail to enter the scrotum. This condition is called undescended testes, or **cryptorchidism** (krip-tōr′ki-dizm). Because testosterone is required for the testes to descend into the scrotum, cryptorchidism is often the result of inadequate testosterone secreted by the fetal testes. If neither testis descends and the defect is not corrected, the male will be infertile because the slightly higher temperature of the body cavity, compared to that of the scrotal sac, causes the spermatogonia to degenerate. Cryptorchidism is treated with hormone therapy, or it may have to be surgically corrected.

Paramesonephric ducts begin to develop just lateral to the **mesonephric ducts** and grow inferiorly to meet one another, where they enter the cloaca as a single, midline tube.

Testosterone, secreted by the testes, causes the mesonephric duct system to enlarge and differentiate to form the epididymis,

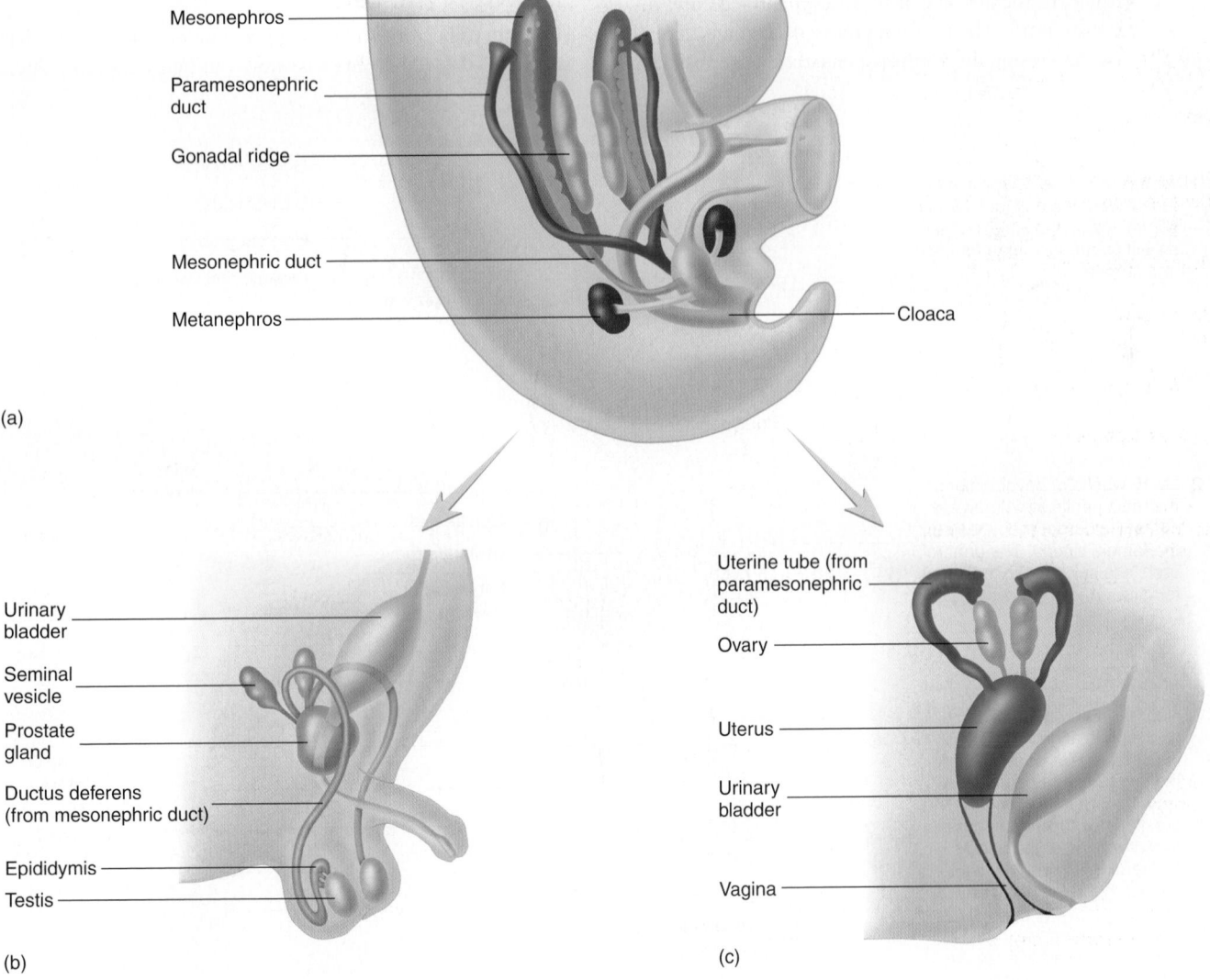

Figure 29.15 Development of the Reproductive System

(a) Indifferent stage. (b) The male, under the influence of male hormones, develops a ductus deferens from the mesonephric duct, and the paramesonephric duct degenerates. (c) The female, without male hormones, develops a uterus and uterine tubes from the paramesonephric duct, and the mesonephros disappears.

ductus deferens, the seminal vesicles, and the prostate gland (figure 29.15*b*). Müllerian-inhibiting hormone, which the testes also secrete, causes the paramesonephric ducts to degenerate. The paramesonephric ducts are also called the müllerian ducts, and they give rise to the uterine tubes, the uterus, and part of the vagina in females (figure 29.15*c*). If neither testosterone nor müllerian-inhibiting hormone is secreted, the mesonephric duct system atrophies, and the paramesonephric duct system develops to form the internal female reproductive structures.

Like the other sexual organs, the external genitalia begin as the same structures in the male and female and then diverge (figure 29.16). An enlargement called the **genital tubercle** develops in the groin of the embryo. **Genital folds** develop on each side of a **urethral groove,** and **labioscrotal swellings** develop lateral to the folds.

In the male, under the influence of dihydrotestosterone, which is derived from testosterone, the genital tubercle and the genital folds close over the urethral groove to form the penis. If this closure does not proceed all the way to the end of the penis, a defect known as **hypospadias** (hī′pō-spā′dē-ăs) results. The testes move into the labioscrotal swellings, which become the scrotum of the male.

In the female, in the absence of testosterone, the genital tubercle becomes the clitoris. The urethral groove disappears; genital folds do not fuse. As a result, the urethra opens somewhat posterior

to the clitoris but anterior to the vaginal opening. The unfused genital folds become the labia minora, and the labioscrotal folds become the labia majora.

13. *Describe the formation of these major organs: skin, bones, skeletal muscles, nervous system, eyes, and respiratory system.*

14. *Explain the formation of the following endocrine glands: anterior pituitary, posterior pituitary, thyroid, parathyroid, adrenal medulla, adrenal cortex, and pancreas.*

15. *Explain the process whereby a one-chambered heart becomes a four-chambered heart.*

16. *Describe the development of the pronephros, mesonephros, and metanephros in the development of the kidneys.*

17. *Describe the effect of hormones on the development of the male and female reproductive systems.*

18. *Compare the male and female structures formed from each of the following: genital tubercle, genital folds, and labioscrotal swellings.*

Growth of the Fetus

The embryo becomes a **fetus** approximately 60 days after fertilization (a 50-day-old embryo is shown in figure 29.17*a*). The major

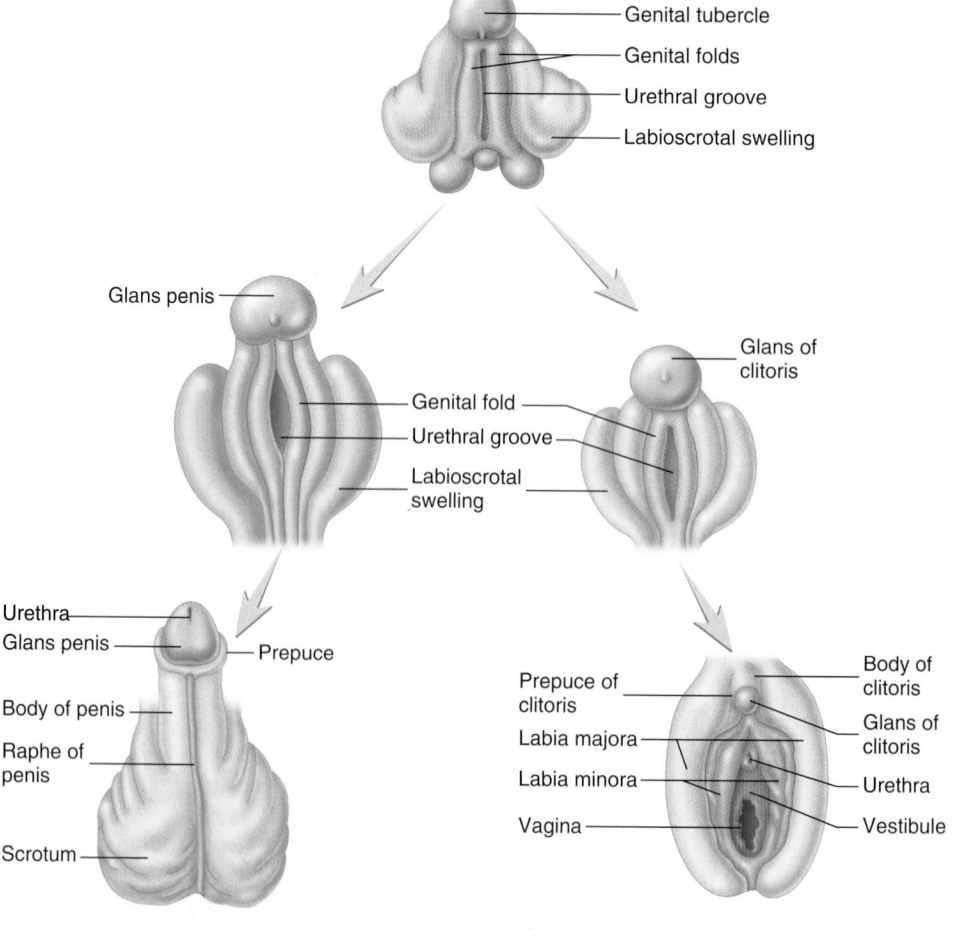

1. At approximately 5 weeks after fertilization, the genital tubercle, genital folds, and labioscrotal swellings are the same for male and female.

Genital tubercle
Genital folds
Urethral groove
Labioscrotal swelling

2. By 10 weeks of development, the male penile structures are somewhat larger than those of the female clitoris. The urethral groove is still open in male and female.

Glans penis
Glans of clitoris
Genital fold
Urethral groove
Labioscrotal swelling

3. Near term, the general adult condition is achieved. The labioscrotal swellings become the scrotum in the male, and labia majora in the female. The genital folds fuse in the male to form the body of the penis, whereas they form the labia minora in the female. The urethral groove in the male is closed by the raphe of the penis. It remains open in the female as the vestibule.

Urethra
Glans penis
Prepuce
Body of penis
Raphe of penis
Scrotum

Prepuce of clitoris
Labia majora
Labia minora
Vagina
Body of clitoris
Glans of clitoris
Urethra
Vestibule

Male   **Female**

Process Figure 29.16   **Development of the External Genitalia**

difference between the embryo and the fetus is that in the embryo most of the organ systems are developing, whereas in the fetus the organs are present. Most morphologic changes occur in the embryonic phase of development, whereas the fetal period is primarily a "growing phase."

The fetus grows within the uterus from about 3 cm and 2.5 g at 60 days to 50 cm and 3300 g at term—more than a 15-fold increase in length and a 1300-fold increase in weight. These changes cause a concomitant increase in the size of the uterus (figure 29.18). Although growth is certainly a major feature of the fetal period, it's not the only feature. The major organ systems still continue to develop during the fetal period.

Fine, soft hair called **lanugo** (la-noo′gō) covers the fetus, and a waxy coat of sloughed epithelial cells called **vernix caseosa** (ver′niks kā-se-ō′să) protects the fetus from the somewhat toxic nature of the amniotic fluid formed by the accumulation of waste products from the fetus.

(a)

(b)

(c)

Figure 29.17 **Embryos and Fetuses at Different Ages**

(*a*) Fifty days after fertilization. (*b*) Three months after fertilization. (*c*) Four months after fertilization.

1 month

4 months

7 months

9 months

Umbilical cord

Mucus plug in cervical canal

Rectum

Vagina

Placenta

Urinary bladder

Symphysis pubis

Urethra

Figure 29.18 **Enlargement of the Uterus During Fetal Development**

Clinical Focus Fetal Monitoring

Amniocentesis (am'nē-ō-sen-tē'sis) is the removal of amniotic fluid from the amniotic cavity (figure B). As the fetus develops, it expels molecules of various types, as well as living cells, into the amniotic fluid. These molecules and cells can be collected and analyzed. A number of normal conditions can be evaluated, and a number of metabolic disorders can be detected by analysis of the types of molecules that the fetus expels. The cells collected by amniocentesis can be grown in culture, and additional metabolic disorders can be evaluated. Chromosome analysis, called a **karyotype** (see page 1112), can also be performed on the cultured cells. Amniocentesis has been done as early as 10 weeks after fertilization, but the success rate at that time is quite low. It is most commonly performed at 13–14 weeks after fertilization.

Fetal tissue samples may also be obtained by **chorionic villus sampling,** in which a probe is introduced into the uterine cavity through the cervix and a small piece of chorion is removed. This technique has an advantage over amniocentesis in that it can be used earlier in development, as early as the seventh to ninth week after fertilization. Furthermore, cells can be used directly for analyses such as karyotyping, rather than having to culture cells as in amniocentesis.

One of the molecules normally produced by the fetus and released into the amniotic fluid is **α-fetoprotein.** If the fetus has tissues exposed to the amniotic fluid that are normally covered by skin, such as nervous tissue, resulting from failure of the neural tube to close, or abdominal tissues, resulting from failure of the abdominal wall to fully form, an excessive amount of α-fetoprotein is lost into the amniotic fluid.

Some of the metabolic by-products from the fetus, such as α-fetoprotein and

Figure B   **Removal of Amniotic Fluid for Amniocentesis**

estriol, a weak form of estrogen produced in the placenta after 20 weeks of gestation, can enter the maternal blood. In some cases, the by-products are processed and passed to the maternal urine. The levels of these fetal products can then be measured in the mother's blood or urine.

The fetus can be seen within the uterus by **ultrasound,** which uses sound waves that are bounced off the fetus like sonar and then analyzed and enhanced by computer; or by **fetoscopy,** in which a fiberoptic probe is introduced into the amniotic cavity. Because of the constantly increasing resolution in ultrasound and because it is noninvasive compared to fetoscopy, it is usually the preferred technique. However, some fetal defects cannot be adequately assessed by ultrasound and require fetoscopy. Ultrasound is used to guide the fetoscope. Ultrasound has not been found to pose any risk to the fetus or mother. It's accomplished by placing a

transducer on the abdominal wall (transabdominal) or by inserting the transducer into the woman's vagina (transvaginal). The latter technique produces much higher resolution because fewer layers of tissue exist between the transducer and the uterine cavity. Transvaginal ultrasound can be used to identify the yolk sac of a developing embryo as early as 17 days after fertilization, and the embryo can be visualized by 25 days. Transabdominal ultrasound allows for fetal monitoring by 6–8 weeks after fertilization.

Fetal heart rate can be detected with an ultrasound stethoscope by the tenth week after fertilization and with a conventional stethoscope by 20 weeks. Fetal heart rate is now more commonly monitored electronically, either indirectly by transducers on the mother's abdomen or by a probe attached to the skin of the fetus. The normal fetal heart rate is 140 bpm (normal range is 110–160).

Subcutaneous fat that accumulates in the older fetus and newborn provides a nutrient reserve, helps insulate the baby, and aids the baby in sucking by strengthening and supporting the cheeks so that negative pressure can be developed in the oral cavity.

Peak body growth occurs late in gestation, but, as placental size and blood supply limits are approached, the growth rate slows. Growth of the placenta essentially stops at about 35 weeks, thus restricting further intrauterine growth.

Fetal Surgery

Fetal surgery performed while the fetus is still in the uterus was first done in the United States in 1979 to drain excess fluid associated with hydrocephalus. These surgeries didn't usually solve the underlying neurologic problems and have been discontinued. Since 1981, in utero surgeries have successfully removed excess fluid from enlarged urinary bladders of male fetuses. The fluid buildup occurs in 1 in every 2000 male fetuses when a flap of tissue grows over the internal opening of the urethra. Without treatment, the amount of amniotic fluid is greatly reduced, and most of those babies die shortly after birth. Since 1989, more advanced surgeries have repaired diaphragmatic hernia, in which part of the abdominal organs push up through a hole in the left side of the diaphragm into the left pleural space, so that the left lung fails to develop fully. The defect occurs in 1 of every 2000 babies, and without surgery, babies with this defect run a 75% risk of dying before or soon after birth. During surgery, the uterus is cut open, and the fetus is pulled far enough out of the opening so that a small incision can be made in its side. The abdominal organs are moved back into the abdomen, the hole in the diaphragm is covered with a surgical patching material called Gore-Tex, the incision in the fetus is closed, and the fetus is tucked back into the uterus. The amniotic fluid, which was removed and saved earlier in the surgery, is replaced, and the incision in the uterus and mother's skin is repaired.

At about 38 weeks of development, the fetus has progressed to the point at which it can survive outside the mother. The average weight at this point is 3250 g for a female fetus and 3300 g for a male fetus.

Exercise During Pregnancy

Healthy women with normal pregnancies are encouraged to engage in moderate exercise during pregnancy. Women who are active before becoming pregnant are able to continue most of their activities during early pregnancy. Later in pregnancy, changes in weight, balance, and joint stability can affect exercise. Exercise during pregnancy is beneficial to the woman and helps her recover more quickly after delivery.

19. What major events distinguish embryonic and fetal development?

Parturition

Parturition (par-toor-ish′ŭn) refers to the process by which a baby is born. Physicians usually calculate the gestation period, or length of the pregnancy, as 280 days (40 weeks) from the last menstrual period (LMP) to the date of delivery of the infant.

P R E D I C T 3
Compare and contrast clinical age and developmental age for fertilization, implantation, beginning of the fetal period, and parturition.

Prematurity

Occasionally, the fetus is delivered before it has sufficiently matured. It is then considered to be **premature.** Prematurity is one of the most significant problems in pediatrics, the branch of medical science dealing with children, because of all the associated complications. The most significant of these complications is **respiratory distress syndrome,** which occurs because very young premature infants cannot produce **surfactant,** a mixture of phospholipids and protein that lines the inner surface of the lungs and allows the lungs to expand as we breathe. Each year, 65,000 premature infants suffer from respiratory distress syndrome in the United States. Until recently, 10% of those infants died. Now, surfactant substitutes are being developed, and glucocorticoid administration can stimulate surfactant production. These therapies have cut the death rate in half, and more effective replacements are being investigated.

Near the end of pregnancy, the uterus becomes progressively more irritable and usually exhibits occasional contractions that become stronger and more frequent until parturition is initiated. The cervix gradually dilates, and strong uterine contractions help expel the fetus from the uterus through the vagina (figure 29.19). Before expulsion of the fetus from the uterus, the amniotic sac ruptures, and amniotic fluid flows through the vagina to the exterior of the woman's body.

Labor is the period during which the contractions occur that result in expulsion of the fetus from the uterus. It occurs as three stages.

1. *First stage.* The first stage begins with the onset of regular uterine contractions and extends until the cervix dilates to a diameter about the size of the fetus's head. This stage of labor commonly lasts from 8–24 hours, but it may be as short as a few minutes in some women who have had more than one child. Normally (95% of the time), the head of the fetus is in an inferior position within the woman's pelvis during labor. The head acts as a wedge, forcing the cervix and vagina to open as the uterine contractions push against the fetus.

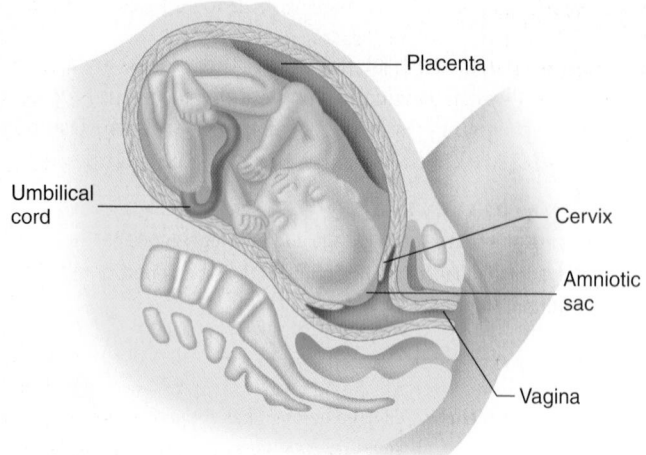

1. **First stage:** the cervix begins to dilate.

2. **Second stage:** further dilation of the cervix and rupture of the amniotic sac occur.

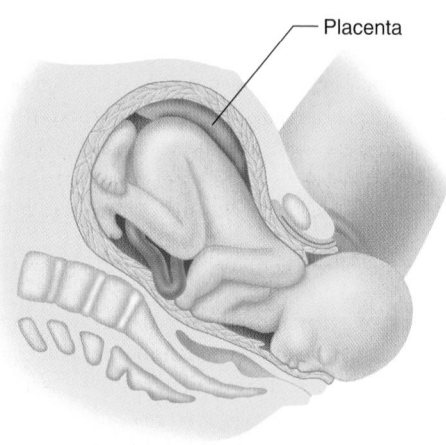

3. **Third stage:** the fetus is expelled from the uterus.

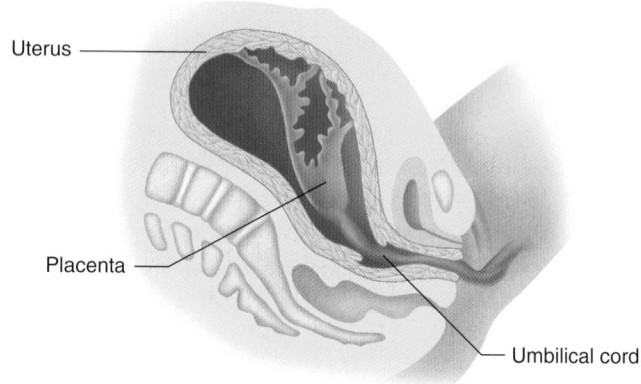

4. **Fourth stage:** the placenta is expelled.

Process Figure 29.19   Parturition

The Perineum in Pregnancy

The **central tendon of the perineum** (see figure 10.21) is very important in supporting the uterus and vagina. Tearing or stretching of the tendon during childbirth may weaken the inferior support of these organs, and prolapse of the uterus may occur. **Prolapse** is a "sinking" of the uterus so that the uterine cervix moves down into the vagina (first degree), moves down near the vaginal orifice (second degree), or may protrude through the vaginal orifice (third degree). During childbirth, an **episiotomy,** a cut through the perineal central tendon, prevents tearing of the perineum. The cut relieves the pressure and heals better than would a tear; however, there appears to be evidence that many episiotomies may be unnecessary.

2. *Second stage.* The second stage of labor lasts from the time of maximum cervical dilation until the baby exits the vagina. This stage may last from a minute to up to an hour. During this stage, contractions of the abdominal muscles assist the uterine contractions. The contractions generate enough pressure to compress blood vessels in the placenta so that blood flow to the fetus is stopped. During periods of relaxation, blood flow to the placenta resumes.

Oxytocin

Occasionally, drugs like oxytocin are administered to women during labor to increase the force of the uterine contractions. Caution must be exercised in the use of this drug, however, so that tetaniclike contractions, which would drastically reduce the blood flow through the placenta, do not occur.

3. *Third stage.* The third stage of labor involves the expulsion of the placenta from the uterus. Contractions of the uterus cause the placenta to tear away from the wall of the uterus. Some bleeding occurs because of the intimate contact between the placenta and the uterus; however, bleeding normally is restricted because uterine smooth muscle contractions compress the blood vessels to the placenta.

Blood levels of estrogen and progesterone fall dramatically after parturition. Once the placenta has been dislodged from the uterus, the source of these hormones is gone. In addition, during the 4 or 5 weeks after parturition, the uterus becomes much smaller, but it remains somewhat larger than it was before pregnancy. The cells of the uterus become smaller, and many of them degenerate. A vaginal discharge composed of small amounts of blood and degenerating endometrium persists for 1 week or more after parturition.

The precise signal that triggers parturition is unknown, but many of the factors that support it have been identified (figure 29.20). Before parturition, the progesterone concentration in the maternal circulation is at its highest level. Progesterone has an inhibitory effect on uterine smooth muscle cells. Near the end of pregnancy, however, estrogen levels rapidly increase in the maternal circulation, and the excitatory influence of estrogens on uterine smooth muscle cells overcomes the inhibitory influence of progesterone.

The adrenal glands of the fetus are greatly enlarged before parturition. The stress of the confined space of the uterus and the limited oxygen supply resulting from a more rapid increase in the size of the fetus than in the size of the placenta increase the rate of adrenocorticotropic hormone (ACTH) secretion by the fetus's anterior pituitary gland. ACTH causes the fetal adrenal cortex to produce glucocorticoids, which travel to the placenta, where they decrease the rate of progesterone secretion and increase the rate of estrogen synthesis. In addition, prostaglandin synthesis is initiated. Prostaglandins strongly stimulate uterine contractions.

During parturition, stretch of the uterine cervix initiates nervous reflexes that cause oxytocin to be released from the woman's posterior pituitary gland. Oxytocin stimulates uterine contractions, which move the fetus farther into the cervix, causing further stretch.

Thus, a positive-feedback mechanism is established in which stretch stimulates oxytocin release and oxytocin causes further stretch. This positive-feedback system stops after delivery, when the cervix is no longer stretched.

Progesterone inhibits oxytocin release and decreases the number of oxytocin receptors. Decreased progesterone levels in the maternal circulation results in increased oxytocin secretion and an increase in the number of oxytocin receptors in the uterus. In addition, estrogens make the uterus more sensitive to oxytocin stimulation by increasing the synthesis of receptor sites for oxytocin. Some evidence suggests that oxytocin also stimulates prostaglandin synthesis in the uterus. All these events support the development of strong uterine contractions.

● **20.** *List the stages of labor, and indicate when each stage begins and its approximate length of time.*

21. *Describe the hormonal changes that take place before and during delivery. How is stretch of the cervix involved in delivery?*

1. The fetal hypothalamus secretes a releasing hormone, corticotropin-releasing hormone (CRH), which stimulates adrenocorticotropic hormone (ACTH) secretion from the pituitary. The fetal pituitary secretes ACTH in greater amounts near parturition.

2. ACTH causes the fetal adrenal gland to secrete greater quantities of adrenal cortical steroids.

3. Adrenal cortical steroids travel in the umbilical blood to the placenta.

4. In the placenta the adrenal cortical steroids cause progesterone synthesis to level off and estrogen and prostaglandin synthesis to increase, making the uterus more irritable.

5. The stretching of the uterus produces action potentials that are transmitted to the brain through ascending pathways.

6. Action potentials stimulate the secretion of oxytocin from the posterior pituitary.

7. Oxytocin causes the uterine smooth muscle to contract.

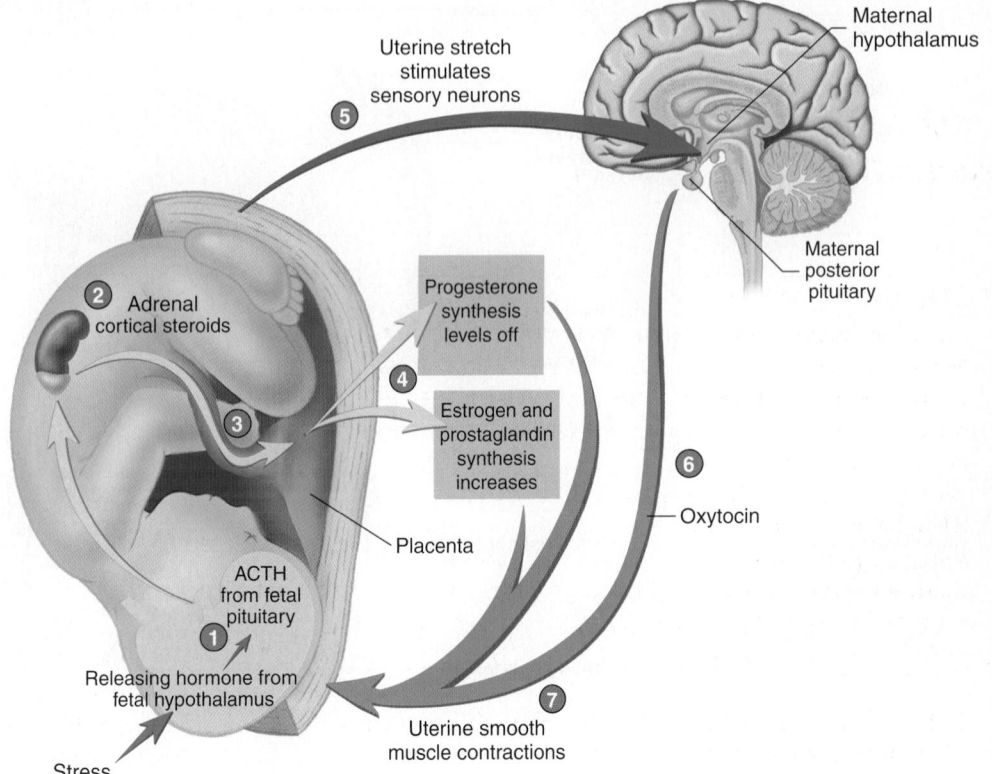

Process Figure 29.20 Factors That Influence the Process of Parturition

Although the precise control of parturition in humans is unknown, these changes appear to play a role.

PREDICT

A woman is having an extremely prolonged labor. From her anatomy and physiology course, she remembers the role of calcium in muscle contraction and asks the doctor to give her a calcium injection to speed the delivery. Explain why the doctor would or would not do as she requests.

The Newborn

The newborn baby, or **neonate,** experiences several dramatic changes at the time of birth. The major and earliest changes in the infant are separation from the maternal circulation and transfer from a fluid to a gaseous environment. The large, forced gasps of air that occur when the infant cries at the time of delivery help inflate the lungs.

Respiratory and Circulatory Changes

The initial inflation of the lungs causes important changes in the circulatory system (figure 29.21). Before birth very little blood flows through the pulmonary arteries to the lungs and back to the heart through the pulmonary veins. As a result, the left atrium has little blood and very low blood pressure. Therefore, blood flows from the right atrium, through the foramen ovale, and into the left atrium.

1. Blood bypasses the lungs by flowing from the pulmonary trunk through the ductus arteriosus to the aorta.

2. Blood also bypasses the lungs by flowing from the right to the left atrium through the foramen ovale.

3. Blood bypasses the liver sinusoids by flowing through the ductus venosus.

4. Oxygen-rich blood is returned to the fetus from the placenta by the umbilical vein.

5. Oxygen-poor blood is carried from the fetus to the placenta through the umbilical arteries.

Superior vena cava

Aortic arch

Ductus arteriosus

Pulmonary trunk

Foramen ovale

Inferior vena cava

Liver

Ductus venosus

Hepatic portal vein

Umbilical vein

Fetal umbilicus

Umbilical cord

Umbilical arteries

Abdominal aorta

Kidney

Common iliac artery

Internal iliac arteries

(a)

Process Figure 29.21   Circulatory Changes at Birth

(a) Circulatory conditions in the fetus.

1. When air enters the lungs, blood is forced through the pulmonary arteries to the lungs. The ductus arteriosus closes and becomes the ligamentum arteriosum (*gray*).

2. The foramen ovale closes and becomes the fossa ovalis. Blood can no longer flow from the right to the left atrium.

3. The ductus venosus degenerates and becomes the ligamentum venosum (*gray*).

4. The umbilical arteries and vein are cut. The umbilical vein becomes the round ligament of the liver (*gray*).

5. The umbilical arteries also degenerate and become the cords of the umbilical arteries (*gray*).

Superior vena cava

Aortic arch

Ligamentum arteriosum (closed ductus arteriosus)

Pulmonary trunk

Fossa ovalis (foramen ovale closed)

Inferior vena cava

Liver

Ligamentum venosum (degenerated ductus venosus)

Hepatic portal vein

Round ligament of liver (degenerated umbilical vein)

Umbilicus

Cords of the umbilical arteries (degenerated umbilical arteries)

Abdominal aorta

Kidney

Common iliac artery

Internal iliac arteries

(b)

Process Figure 29.21 *(continued)*

(*b*) Circulatory changes that occur at birth.

Expansion of the lungs reduces the resistance to blood flow through the lungs, resulting in increased blood flow through the pulmonary arteries. Consequently, more blood flows from the right atrium to the right ventricle and into the pulmonary arteries, and less blood flows from the right atrium through the foramen ovale to the left atrium. In addition, an increased volume of blood returns from the lungs through the pulmonary veins to the left atrium, which increases the pressure in the left atrium. The increased left atrial pressure and decreased right atrial pressure, resulting from decreased pulmonary resistance, forces blood against the septum primum, causing the foramen ovale to close. This action functionally completes the separation of the heart into two pumps: the right side of the heart and the left side of the heart. The closed foramen ovale becomes the **fossa ovalis.**

The **ductus arteriosus,** which connects the pulmonary trunk to the aorta and allows blood to flow from the pulmonary trunk to the systemic circulation, closes off within 1 or 2 days after birth. This closure occurs because of the sphincterlike constriction of the artery and is probably stimulated by local changes in blood pressure and blood oxygen content. Once closed, the ductus arteriosus is replaced by connective tissue and is known as the **ligamentum arteriosum.**

Patent Ductus Arteriosus

If the ductus arteriosus does not close completely, it's said to be **patent.** This is a serious birth defect, resulting in marked elevation in pulmonary blood pressure because blood flows from the left ventricle to the aorta, through the ductus arteriosus to the pulmonary arteries. If not corrected, it can lead to irreversible degenerative changes in the heart and lungs.

The fetal blood supply passes to the placenta through umbilical arteries from the internal iliac arteries and returns through an umbilical vein. The blood passes through the liver via the ductus venosus, which joins the inferior vena cava. When the umbilical cord is tied and cut, no more blood flows through the umbilical vein and arteries, and they degenerate. The remnant of the umbilical vein becomes the **ligamentum teres,** or **round ligament,** of the liver, and the ductus venosus becomes the **ligamentum venosum.**

Digestive Changes

When a baby is born, it's suddenly separated from its source of nutrients provided by the maternal circulation. Because of this separation and the stress of birth and new life, the neonate usually loses 5%–10% of its total body weight during the first few days of life. Although the digestive system of the fetus becomes somewhat functional late in development, it is still very immature compared to that of the adult and can digest only a limited number of food types.

Late in gestation, the fetus swallows amniotic fluid from time to time. Shortly after birth, this swallowed fluid plus cells sloughed from the mucosal lining, mucus produced by intestinal mucous glands, and bile from the liver pass from the digestive tract as a greenish anal discharge called **meconium** (mē-kō′nē-ŭm).

The pH of the stomach at birth is nearly neutral because of the presence of swallowed alkaline amniotic fluid. Within the first 8 hours of life, a striking increase in gastric acid secretion occurs, causing the stomach pH to decrease. Maximum acidity is reached at 4–10 days, and the pH gradually increases for the next 10–30 days.

The neonatal liver is also functionally immature. It lacks adequate amounts of the enzyme required in the production of bilirubin. This enzyme system usually develops within 2 weeks after birth in a healthy neonate, but because this enzyme system is not fully developed at birth, some full-term babies may temporarily develop jaundice, with elevated blood levels of bilirubin. Jaundice often occurs in premature babies.

The newborn digestive system is capable of digesting lactose (milk sugar) from the time of birth. The pancreatic secretions are sufficiently mature for a milk diet, but the digestive system only gradually develops the ability to digest more solid foods over the first year or two; therefore, new foods should be introduced gradually during the first 2 years. It's also advised that only one new food be introduced at a time into the infant's diet so that, if an allergic reaction occurs, the cause is more easily determined.

Amylase secretion by the salivary glands and the pancreas remains low until after the first year. Lactase activity in the small intestine is high at birth but declines during infancy, although the levels still exceed those in adults. Lactase activity is lost in many adults (see chapter 24).

Apgar Scores

The newborn baby may be evaluated soon after birth to assess its physiologic condition. This assessment is referred to as the **Apgar score.** Apgar, which was named for Virginia Apgar, the physician who developed it, also stands for **a**ppearance, **p**ulse, **g**rimace, **a**ctivity, and **r**espiratory effort. Each of these characteristics is rated on a scale of 0–2: 2 denotes normal function, 1 denotes reduced function, and 0 denotes seriously impaired function. The total Apgar score is the sum of the scores from the five characteristics, ranging, therefore, from 0–10 (table 29.3). A total Apgar score of 8–10 at 1–5 minutes after birth is considered normal. Other scoring systems to estimate normal growth and development, including general external appearance and neurologic development, may also be applied to the neonate.

22. What changes take place in the newborn's circulatory system shortly after birth? What do each of the following become: foramen ovale, ductus arteriosus, umbilical vein, and ductus venosus?
23. What changes take place in the newborn's digestive system shortly after birth?
24. Define the term meconium. Why does jaundice often develop after birth?
25. What is the Apgar score?

Table 29.3	Examples of Apgar Rating Scales		
	0	**1**	**2**
Appearance (skin color)	White or blue	Limbs blue, body pink	Pink
Pulse (rate)	No pulse	100 bpm	>100 bpm
Grimace (reflexive grimace initiated by stimulating the plantar surface of the foot)	No response	Facial grimaces, slight body movement	Facial grimaces, extensive body movement
Activity (muscle tone)	No movement, muscles flaccid	Limbs partially flexed, little movement, poor muscle tone	Active movement, good muscle tone
Respiratory effort (amount of respiratory activity)	No respiration	Slow, irregular respiration	Good, regular respiration, strong cry

Lactation

Lactation is the production of milk by the mother's breasts (mammary glands; figure 29.22). It normally occurs in females after parturition and may continue for 2 or 3 years, or even longer, provided suckling occurs often and regularly.

During pregnancy, the high concentration and continuous presence of estrogens and progesterone cause expansion of the duct system and secretory units of the breasts. The ducts grow and branch repeatedly to form an extensive network. Additional adipose tissue is deposited also; thus, the size of the breasts increases substantially throughout pregnancy. Estrogen is primarily responsible for breast growth during pregnancy, but normal development of the breast doesn't occur without the presence of several other hormones. Progesterone causes development of the breasts' secretory alveoli, which enlarge but don't usually secrete milk during pregnancy. The other hormones include growth hormone, prolactin, thyroid hormones, glucocorticoids, and insulin. The placenta secretes a growth hormonelike substance (human somatotropin) and a prolactinlike substance (human placental lactogen), and these substances also help support the development of the breasts.

Prolactin, which is produced by the anterior pituitary gland, is the hormone responsible for milk production. Before parturition, high levels of estrogen stimulate an increase in prolactin production. Milk production is inhibited during pregnancy, however, because high levels of estrogen and progesterone inhibit the effect of prolactin on the mammary gland. After parturition, estrogen, progesterone, and prolactin levels decrease, and, with lower estrogen and progesterone levels,

prolactin stimulates milk production. Despite a decrease in the basal levels of prolactin, a reflex response produces surges of prolactin release. During suckling, mechanical stimulation of the breasts initiates nerve impulses that reach the hypothalamus, causing the secretion of **prolactin-releasing factor (PRF)** and inhibiting the release of **prolactin-inhibiting factor (PIF).** Consequently, prolactin levels temporarily increase and stimulate milk production.

For the first few days after birth, the mammary glands secrete **colostrum** (kō-los′trŭm), which contains little fat and less lactose than milk. The breasts may secrete some colostrum during pregnancy. Eventually, more nutritious milk is produced. Colostrum and milk not only provide nutrition, but they also contain antibodies (see chapter 22) that help protect the nursing baby from infections.

HIV and the Newborn

The human immunodeficiency virus (HIV) can be transmitted from a mother to her child in utero, during parturition, or during breast-feeding. HIV has been isolated from human breast milk and colostrum. In developed countries, HIV patients are advised not to breast-feed but are advised, rather to use formula and bottle-feeding. In underdeveloped countries, bottle-feeding is not yet a viable option. Newborn babies given a single dose of an AIDS drug at birth have half the risk of HIV infection. Most of those gains are lost, however, when babies are breast-fed. The current proposal for underdeveloped countries is that daily AIDS drug treatment for newborns continue as long as the child is breast-feeding. In one such study, treated infants had a risk of 1% of HIV infection, compared to a 15% rate in untreated infants.

1. Stimulation of the nipple by the baby's suckling initiates action potentials in sensory neurons that connect with the hypothalamus.

2. In response, the hypothalamus stimulates the posterior pituitary to release oxytocin and the anterior pituitary to release prolactin.

3. Oxytocin stimulates milk release from the breast. Prolactin stimulates additional milk production.

Process Figure 29.22 Hormonal Control of Lactation

Repeated stimulation of prolactin release, by suckling, makes nursing (breast-feeding) possible for several years. If nursing stops, however, within a few days the ability to produce prolactin ceases, and milk production stops.

Because it takes time to produce milk, an increase in prolactin results in the production of milk to be used in the next nursing period. At the time of nursing, stored milk is released as a result of a reflex response. Mechanical stimulation of the breasts produces nerve impulses that cause the release of oxytocin from the posterior pituitary, which stimulates cells surrounding the alveoli to contract. Milk is then released from the breasts, a process that is called **milk letdown**, or milk ejection. In addition, higher brain centers can stimulate oxytocin release, and such things as hearing an infant cry can result in milk letdown.

> ● 26. *What hormones are involved in preparing the breast for lactation? Describe the events involved in milk production and milk letdown. What is colostrum?*

P R E D I C T ⑤
While nursing her baby, a woman notices that she develops "uterine cramps." Explain what is happening.

First Year After Birth

Many changes occur in the life of the newborn from birth until 1 year of age. The time of these changes may vary considerably from child to child, and the dates given are only rough estimates. The brain is still developing, and much of what the neonate can do depends on how much brain development has occurred. It's estimated that the total adult number of neurons is present in the CNS at birth, but subsequent growth and maturation of the brain involve the addition of new neuroglial cells, some of which form new myelin sheaths, and the addition of new connections between neurons, which may continue throughout life.

By 6 weeks, the infant usually can hold up its head when placed in a prone position and begins to smile in response to people or objects. At 3 months of age, the infant's limbs move apparently aimlessly. The infant has enough control of the arms and hands, however, that voluntary thumb sucking can occur. The infant can follow a moving person with its eyes. At 4 months, the infant begins to raise itself by its arms. It can begin to grasp objects placed in its hand, coo and gurgle, roll from its back to its side, listen quietly when hearing a person's voice or music, hold its head erect, and play with its hands. At 5 months, the infant can usually laugh, reach for objects, turn its head to follow an object, lift its head and shoulders, sit with support, and roll over. At 8 months, the infant recognizes familiar people, sits up without support, and reaches for specific objects. At 12 months, the infant may pull itself to a standing position and may be able to walk without support. It can pick up objects in its hands and examine them carefully. It can understand much of what is said to it and may say several words of its own.

> ● 27. *List the major changes that occur during the first year of life. What do most of these activities depend upon?*

Life Stages

The prenatal and neonatal periods of life previously described are only a small part of the total life span. The life stages from fertilization to death are as follows: (1) the germinal period: fertilization to 14 days; (2) embryo: 14–56 days after fertilization; (3) fetus: 56 days after fertilization to birth; (4) neonate: birth to 1 month after birth; (5) infant: 1 month to 1 or 2 years (the end of infancy is sometimes set at the time that the child begins to walk); (6) child: 1 or 2 years to puberty; (7) adolescent: puberty (age 11–14) to 20 years; and (8) adult: age 20 to death. Adulthood is sometimes divided into three periods: young adult, age 20–40; middle age, age 40–65; and older adult, age 65 to death (figure 29.23).

During childhood, the individual develops considerably. Many of the emotional characteristics that a person possesses throughout life are formed during early childhood.

Major physical and physiologic changes occur during adolescence that also affect the emotions and behavior of the individual. Other emotional changes occur as the adolescent attempts to fit into an adult world. Puberty usually occurs somewhat earlier in females (at about 11–13 years) than in males (at about 12–14 years). A period of rapid growth usually accompanies the onset of puberty. This rapid growth period is followed by a period of slower growth. Full adult stature is usually achieved by age 17 or 18 in females and 19 or 20 years in males.

> ● 28. *Define the different life stages, starting with the germinal stage and ending with the adult.*

Figure 29.23   Active Older Adults
A conspicuous feature of the population of older adults is the range of variability. In some adults over 70 years, many systems are beginning to fail. Others can look forward to at least 10 more years of healthy living.

Aging

Development of a new human being begins at fertilization, as does the process of aging. Cells proliferate at an extremely rapid rate during early development and then the process begins to slow as various cells become committed to specific functions within the body.

Many cells of the body, such as liver and skin cells, continue to proliferate throughout life, replacing dead or damaged tissue. Many other cells, however, such as the neurons in the central nervous system, cease to proliferate once they have reached a certain number, and dead cells are not replaced. After the number of neurons reaches a peak, at about the time of birth, their numbers begin to decline. Neuronal loss is most rapid early in life and later decreases to a slower, steadier rate.

A natural, but as yet unexplained, decline occurs in mitochondrial DNA function with age. If this decline reaches a threshold, which apparently differs from tissue to tissue, the normal function of the mitochondrion is lost, and the tissue or organ may exhibit disease symptoms. In a small number of people, this mitochondrial degeneration occurs very early in life, resulting in premature aging.

The physical plasticity (i.e., the state of being soft and pliable) of young embryonic tissues results largely from the presence of large amounts of hyaluronate and relatively small amounts of collagen; furthermore, the collagen and other related proteins that are present are not highly cross-linked; thus, the tissues are very flexible and elastic. Many of these proteins produced during development are permanent components of the individual, however, and, as the individual ages, more and more cross-links form between these protein molecules, thereby rendering the tissues more rigid and less elastic.

The tissues with the highest content of collagen and other related proteins are those most severely affected by the collagen cross-linking and tissue rigidity associated with aging. The lens of the eye is one of the first structures to exhibit pathologic changes as a result of this increased rigidity. Vision of close objects becomes more difficult with advancing age until most middle-aged people require reading glasses (see chapter 15). Loss of elasticity also affects other tissues, including the joints, blood vessels, kidneys, lungs, and heart, and greatly reduces the functional ability of these organs.

Like nervous tissue, mature muscle cells don't normally proliferate after terminal differentiation occurs before birth. As a result, the total number of skeletal and cardiac muscle fibers declines with age. The strength of skeletal muscle reaches a peak between 20 and 30 years of life and declines steadily thereafter. Furthermore, like the collagen of connective tissue, the macromolecules of muscle undergo biochemical changes during aging and render the muscle tissue less functional. A good exercise program, however, can slow or even reverse this process.

The decline in muscular function also contributes to a decline in cardiac function with advancing age. The heart loses elastic recoil ability and muscular contractility. As a result, total cardiac output declines, and less oxygen and fewer nutrients reach cells, such as neurons in the brain and cartilage cells in the joints, thereby contributing to the decline in these tissues. Reduced cardiac function also may result in decreased blood flow to the kidneys, thus contributing to decreases in their filtration ability. Degeneration of the connective tissues as a result of collagen cross-linking and other factors also decreases the filtration efficiency of the glomerular basement membrane.

Atherosclerosis (ath′er-ō-skler-ō′sis) is the deposit and subsequent hardening of a soft, gruel-like material in lesions of the intima of large- and medium-sized arteries. These deposits then become fibrotic and calcified, resulting in **arteriosclerosis** (ar-tēr′ē-ō-skler-ō′sis; hardening of the arteries). Arteriosclerosis interferes with normal blood flow and can result in a **thrombus,** which is a clot or plaque formed inside a vessel. A piece of the plaque, called an **embolus** (em′bō-lŭs) can break loose, float through the circulation, and lodge in smaller arteries to cause myocardial infarctions or strokes. Atherosclerosis occurs to some extent in all middle-aged and older people and even may occur in certain young people. People with high blood cholesterol, however, appear to be at increased risk for developing atherosclerosis. In addition to dietary influences, this condition seems to have a heritable component, and blood tests are available to screen people for high blood cholesterol levels.

Many other organs, such as the liver, pancreas, stomach, and colon, undergo degenerative changes with age. The ingestion of harmful agents may accelerate such changes. Examples include the degenerative changes induced in the lungs, aside from lung cancer, by cigarette smoke and sclerotic changes in the liver as a result of alcohol consumption.

In addition to the previously described changes associated with aging, cellular wear and tear, or cytologic aging, is another factor that contributes to aging. Progressive damage from many sources like radiation and toxic substances may result in irreversible cellular insults and may be one of the major factors leading to aging. It has been speculated that ingestion of the antioxidant vitamins C and E in combination may help slow this part of aging by stimulating cell repair. Vitamin C also stimulates collagen production and may slow the loss of tissue plasticity associated with aging collagen.

According to the **free radical theory of aging,** free radicals, which are atoms or molecules with an unpaired electron, can react with and alter the structure of molecules that are critical for normal cell function. Alteration of these molecules can result in cell dysfunction, cancer, or other types of cellular damage. Free radicals are produced as a normal part of metabolism and are introduced into the body from the environment through the air we breathe and the food we eat. The damage caused by free radicals may accumulate with age. Antioxidants, such as beta carotene (provitamin A), vitamin C, and vitamin E, can donate electrons to free radicals, without themselves becoming harmful. Thus antioxidants may prevent the damage caused by the free radicals and may ward off age-related disorders, ranging from wrinkles to cancer. Experiments designed to test this hypothesis, however, have not consistently produced positive results.

As a result of poor diet, many people over age 50 do not get the minimum daily allotment of several vitamins and minerals. Feeling "bad" is not necessarily a part of aging but is mostly a result of poor nutrition and lack of exercise. Moderate exercise and avoiding overeating can prolong life. Moderate exercise can reduce the risk of heart attack by as much as 20%. It can also reduce the risk of stroke, high blood pressure, and some forms of cancer. Exercise can also increase a person's ability to reason and remember. Walking 30 minutes a day is recommended.

One characteristic of aging is an overall decrease in ATP production. This decrease is associated with a decline in oxidative phosphorylation, which has been shown in many cases to be associated with **mitochondrial DNA mutations.** Mitochondria lack the DNA repair mechanisms that exist in the nucleus for nuclear DNA. Such mutations are also often associated with Alzheimer's disease. There are also genes in nuclear DNA associated with Alzheimer's.

Immune system changes may also be a major factor contributing to aging. The aging immune system loses its ability to respond to outside antigens but becomes more sensitive to the body's own antigens. These autoimmune changes add to the degeneration of the tissues already described and may be responsible for such things as arthritic joint disorders, chronic glomerular nephritis, and hyperthyroidism. In addition, T lymphocytes tend to lose their functional capacity with aging and cannot destroy abnormal cells as efficiently. This change may be one reason that certain types of cancer occur more readily in older people.

Genetic traits may also cause many of the changes associated with aging. As a general rule, animals with a very high metabolic rate have a shorter life span than those with a lower metabolic rate. In humans, a very small number of exceptional people have a slightly reduced normal body temperature, suggesting a lower metabolic rate. These same people often have an unusually long life span. This tendency appears to run in families and probably has some genetic basis. Studies of the general population suggest that if your parents and grandparents have lived long, so will you.

Another piece of evidence suggesting that a strong genetic component to aging exists comes from a disorder called **progeria** (prō-jēr′ē-ă; premature aging). This apparent genetic trait causes the degenerative changes of aging to occur shortly after the first year of life, and the child may look like a very old person by age 7.

One of the greatest disadvantages of aging is the increasing lack of ability to adjust to stress. Older people have a far more precarious homeostatic balance than younger people, and eventually some stress is encountered that is so great that the body's ability to recover is surpassed, and death results.

29. How does the loss of cells that are not replaced affect the aging process? Give examples.
30. How does loss of tissue plasticity affect the aging process? Give examples.
31. Explain the free radical theory of aging.
32. How does aging affect the immune system?
33. What role does genetics play in aging?

Death

Death is usually not attributed to old age. Some other problem, such as heart failure, renal failure, or stroke, is usually listed as the cause of death.

Death was once defined as the loss of heartbeat and respiration. In recent years, however, more precise definitions of death have been developed because both the heart and the lungs can be kept working artificially, and the heart can even be replaced by an artificial device. Modern definitions of death are based on the permanent cessation of life functions and the cessation of integrated tissue and organ function. The most widely accepted indication of death in humans is **brain death,** which is defined as irreparable brain damage manifested clinically by the absence of response to stimulation, the absence of spontaneous respiration and heart beat, and an isoelectric ("flat") electroencephalogram for at least 30 minutes, in the absence of known CNS poisoning or hypothermia.

34. Define death.

Genetics

Genetics is the study of heredity, that is, those characteristics inherited by children from their parents. Although the environment can influence gene expression, people's physical characteristics and abilities are largely determined by their genetic makeup. Many of a person's abilities, susceptibility to disease, and even life span are influenced by heredity. Because many of the diseases caused by microorganisms now are preventable or treatable, diseases that have a genetic basis are receiving more attention.

Chromosomes

Deoxyribonucleic (dē-oks′ē-rī′bō-noo-klē′ic) **acid (DNA)** is the hereditary material of cells and is responsible for controlling cell activities. DNA molecules and their associated proteins become visible as densely stained bodies, called **chromosomes** (krō′mō-sōmz; colored bodies), during cell division (see figure 3.28). Somatic cells contain 23 pairs of chromosomes, or 46 total chromosomes, and gametes contain 23 chromosomes. **Somatic** (sō-mat′ik) **cells** are all the cells of the body except for the **gametes** (gam′ētz), or sex cells. Examples of somatic cells are epithelial cells, muscle cells, neurons, fibroblasts, lymphocytes, and macrophages. In the male, the gametes are sperm cells, and in the female, the gametes are oocytes (see chapter 28).

A **karyotype** (kar′ē-ō-tīp), or display of the chromosomes in a somatic cell, can be produced by photographing the chromosomes through a microscope, cutting the pictures of the chromosomes out of the photograph, and arranging the chromosomes in pairs (figure 29.24). The 23 pairs of chromosomes are divided into two groups: 22 pairs of **autosomal** (aw-tō-sō′măl) **chromosomes,** all the chromosomes but the sex chromosomes, and one pair of **sex chromosomes,** which determines the sex of the individual. For convenience, the autosomes are numbered in pairs from 1 through 22, and sex chromosomes are denoted as **X** or **Y chromosomes.** A normal female has two X chromosomes (XX) in each somatic cell, whereas a normal male has one X and one Y chromosome (XY) in each somatic cell.

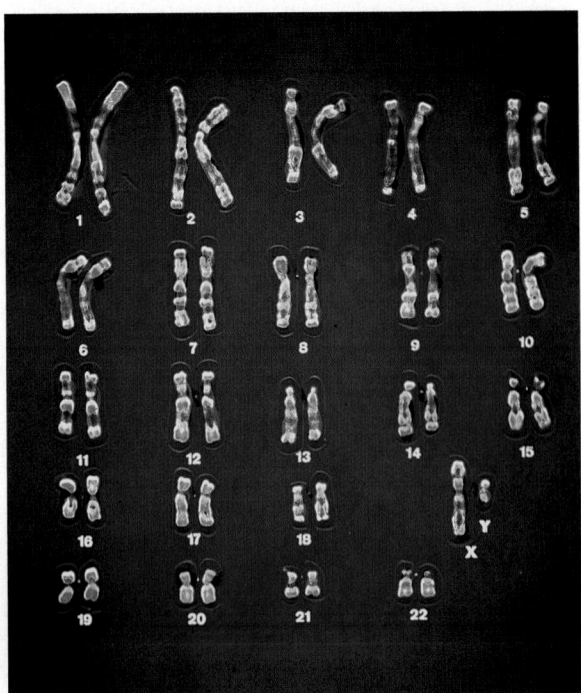

Figure 29.24   **Human Karyotype**

The 23 pairs of chromosomes in humans consist of 22 pairs of autosomal chromosomes (numbered 1–22) and 1 pair of sex chromosomes. This karyotype is of a male and has an X and a Y sex chromosome. A female karyotype would have two X chromosomes.

Sex Chromosome Abnormalities

A wide range of sex chromosome abnormalities exist. The presence of a Y chromosome makes a person genetically a male, and the absence of a Y chromosome makes a person genetically a female, regardless of the number of X chromosomes. As a result of hormonal imbalances, the morphological features (phenotype) may be reversed from the genetic constitution (genotype). Individuals with XO (Turner's syndrome), XX, XXX, or XXXX karyotypes are, therefore, females, and individuals with XY, XXY, XXXY, or XYY karyotypes are males. A YO condition is lethal, because the genes on the X chromosome are necessary for survival. Secondary sexual characteristics are usually underdeveloped in both the XXX female and the XXY male (called Klinefelter's syndrome), and additional X chromosomes (XXXX or XXXY) are often associated with some degree of mental retardation.

Gametes are derived from somatic cells by **meiosis** (mī-ō′sis). In this process, the somatic cells divide twice, and the chromosomes from the somatic cells are distributed to the gametes. Meiosis is called a reduction division because the number of chromosomes in the gametes is half the number in the somatic cells. When a sperm cell and an oocyte fuse during fertilization, each contributes one-half of the chromosomes necessary to produce new somatic cells; therefore, half of an individual's genetic makeup comes from the father and half from the mother.

During meiosis, the chromosomes are distributed in such a way that each gamete receives only one chromosome from each

homologous (hŏ-mol′ō-gŭs) pair of chromosomes. Homologous chromosomes contain the same complement of genetic information. The inheritance of sex illustrates, in part, how chromosomes are distributed during gamete formation and fertilization. During meiosis and gamete formation, the pair of sex chromosomes separates so that each oocyte receives one of a homologous pair of X chromosomes, whereas each sperm cell receives either an X chromosome or a Y chromosome (figure 29.25). When a sperm cell fertilizes an oocyte to form a single cell, the sex of the individual is determined randomly. If a sperm cell with a Y chromosome fertilizes the oocyte, a male results, but if a sperm cell with an X chromosome fertilizes the oocyte, a female results. Estimating the probability of any given zygote being male or female is much like flipping a coin. When all the possible combinations of sperm cells with oocytes are considered, about half the individuals should be female and the other half should be male.

Genes

The functional unit of heredity is the gene. Each **gene** is a portion of DNA coding for a specific polypeptide. A gene is not necessarily a continuous stretch of DNA (see figure 3.42). Each chromosome contains thousands of genes, and each gene occupies a specific **locus** on a chromosome. Both chromosomes of a given pair contain

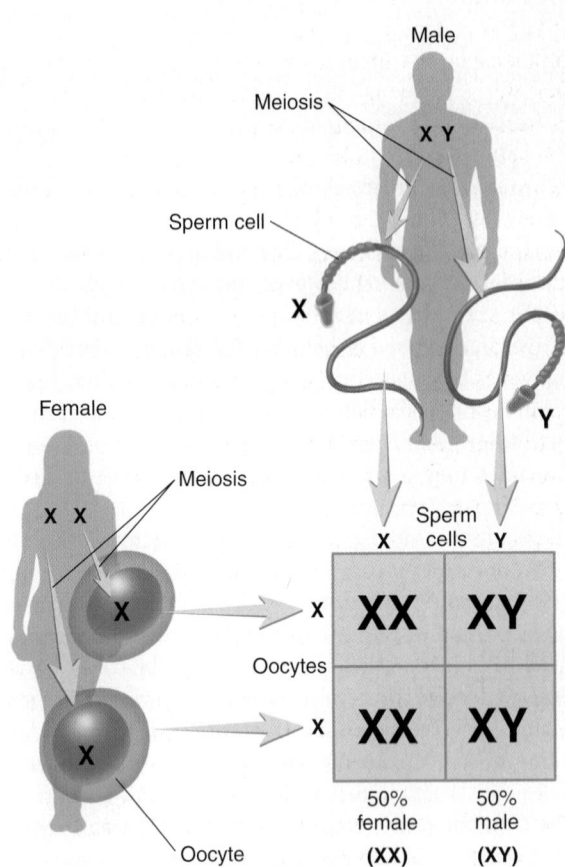

Figure 29.25   **Inheritance of Sex**

The female produces oocytes containing one X chromosome, whereas the male produces sperm cells with either an X or a Y chromosome. There are four possible combinations of an oocyte with a sperm cell, half of which produce females and half of which produce males.

similar but not necessarily identical genes. The genes occupying the same locus on homologous chromosomes are called **alleles** (ă-lēlz′). If the two allelic genes are identical, the person is **homozygous** (hŏ-mŏ-zī′gŭs) for the trait specified by that gene locus. If the two alleles are slightly different, the person is **heterozygous** (het′er-ō-zī′gŭs) for the trait.

Two major types of genes exist: structural and regulatory. **Structural genes** are those DNA sequences that serve as a template for mRNA and code for specific amino acid sequences in proteins like enzymes, hormones, or structural proteins like collagen. **Regulatory genes** are segments of DNA involved in controlling which structural genes are transcribed in a given tissue. By determining the structure of proteins and which proteins are produced by which cells, genes are responsible for the characteristics of cells and, therefore, the characteristics of the entire organism. All the genes in one homologous set of 23 chromosomes in one individual, taken together, is called the **genome** (jē′nōm, jē′nom). The two combined genomes of a person are responsible for all of that person's genetic traits.

Phenylketonuria

Situations in which the alteration of a single gene results in a genetic disorder dramatically illustrate the importance of genes. For example, in **phenylketonuria** (fen′il-kē′tō-noo′rē-ā; **PKU**), the gene responsible for producing an enzyme that converts the amino acid phenylalanine to the amino acid tyrosine is defective. Phenylalanine, therefore, accumulates in the blood and is eventually converted to harmful substances that can cause mental retardation.

Through the processes of meiosis, gamete formation, and fertilization, essentially a random distribution of genes is received from each parent, a process called **independent assortment** of the genes. Several factors, however, influence this random distribution. For example, all of the genes on a given chromosome are **linked,** that is, they tend to be inherited as a set rather than as individual genes, because chromosomes, not individual genes, segregate during meiosis. Sets of linked genes can be broken up, however. When tetrads are formed during meiosis, homologous chromosomes may exchange genetic information by **crossing-over** (see chapter 28).

Furthermore, segregation errors may occur during meiosis. As the chromosomes separate during meiosis, the two members of a homologous pair may become "sticky" and not segregate as they normally do. As a result, one of the daughter cells receives both chromosome pairs and the other daughter cell receives none. This event is called **nondisjunction.** When the gametes are fertilized, the resulting zygote has either 47 chromosomes or 45 chromosomes rather than the normal 46, a condition called **aneuploidy** (an′ū-ploy-dē). Aneuploidies include monosomies, where a chromosome is missing, or trisomies, where an extra chromosome is present. Aneuploidies are usually lethal and is one reason for a high rate of early embryo loss. Some types of aneuploidy are not lethal, however. The sex chromosome abnormalities described in the section dealing with chromosomes are examples. Another example is **Down's syndrome,** or **trisomy 21,** a type of aneuploidy in which three chromosomes 21 are present.

Dominant and Recessive Genes

Most human genetic traits that we are aware of are recognized because defective alleles for those traits exist in the population. For example, on chromosome 11, a gene is present that produces an enzyme necessary for the synthesis of melanin, the pigment responsible for skin, hair, and eye color (see chapter 5). The normal allele for the melanin gene produces a normal, functional enzyme. Another, abnormal allele, however, produces a defective enzyme not capable of catalyzing one of the normal steps in melanin synthesis. If a given person inherits two defective alleles at that melanin-producing enzyme locus, a homozygous condition, the person is unable to produce melanin and, therefore, lacks normal pigment in the skin, hair, and eyes. This condition is referred to as **albinism.** Instead of the normal coloring, the color of a person with albinism consists of shades of pink, blue, and yellow. The pink and blue colors result from blood seen through the skin (see chapter 5), and the yellow color is from the natural accumulation of ingested yellow plant pigments in the skin.

For many genetic traits, the effects of one allele for that trait can mask the effect of another allele for that same trait. For example, a person who is heterozygous for the melanin-producing enzyme gene on chromosome 11 has a normal gene for melanin production on one chromosome 11 and the defective gene for melanin production at the same locus on the other chromosome 11. In the case of this melanin-producing enzyme, one copy of the gene and its resulting enzymes are enough to make normal melanin. As a result, the person who is heterozygous produces melanin and appears normal. In this case, the allele that produces the normal enzyme and is responsible for normal appearance is said to be **dominant,** whereas the allele producing the abnormal enzyme is **recessive.** The lost function of the defective enzyme is masked by the dominant, normal allele. Thus, normal pigmentation is a dominant trait and albinism is a recessive trait. By convention, dominant traits are indicated by uppercase letters, and recessive traits are indicated by lowercase letters. In this example, the letter *A* designates the dominant normal, pigmented condition and the letter *a* the recessive albino condition. It's important to note that not all dominant traits are the normal condition and that not all recessive traits are abnormal. Many examples exist in which the dominant trait is abnormal.

The possible combinations of dominant and recessive alleles for normal melanin production versus albinism are *AA* (homozygous dominant), *Aa* (heterozygous), and *aa* (homozygous recessive). The actual set of alleles that a person has for a given trait is called the **genotype** (jēn′ō-tīp). The person's appearance is called the **phenotype** (fē′nō-tīp). A person with the genotype *AA* or *Aa* has the phenotype of normal pigmentation, whereas a person with the genotype *aa* has the phenotype of albinism. Note that the recessive trait is expressed when it is not masked by the dominant trait.

Genes and the Environment

Genes do not produce a phenotype in isolation. Rather, a phenotype is produced by the combination of genes and the environment. For example, genes establish the basis of skin color, but exposure to the sun stimulates melanin production to establish the phenotype at any given time. Furthermore, height is largely determined by genotype, but poor nutrition can reduce a person's height below its genetic potential.

P R E D I C T 6

Polydactyly (pol-ē-dak'ti-lē) is a condition in which a person has extra fingers or toes. Given that polydactyly is a dominant trait, list all the possible genotypes and phenotypes for polydactyly. Use the letters *D* and *d* for the genotypes.

The inheritance of dominant and recessive traits can be determined if the genotypes of the parents are known. For example, if an albino person (*aa*) mates with a heterozygous normal person (*Aa*), the probability is that half of the children will be albino (*aa*), and half will be normal heterozygous carriers (*Aa*). If two carriers (*Aa*) mate, the probability is that one in four will be homozygous dominant (*AA*), one in four will be homozygous recessive (*aa*), and one in two will be heterozygous (*Aa*). Such a probability can be easily determined by the use of a table called a **Punnett square** (figure 29.26). **Carriers** are heterozygous persons with an abnormal recessive gene but with a normal phenotype because they also have a normal dominant allele for that gene.

35. *What is the number and type of chromosomes in the karyotype of a human somatic cell? How do the chromosomes of a male and female differ from each other?*

36. *How do the chromosomes in somatic cells and gametes differ from each other?*

37. *What is a gene, and how are genes responsible for the structure and function of cells? Define structural genes and regulatory genes.*

38. *What is the cause of the genetic disorder called Down's syndrome?*

39. *Define the terms homozygous dominant, heterozygous, and homozygous recessive.*

40. *What is the difference between genotype and phenotype?*

P R E D I C T 7

If a carrier for albinism mates with a homozygous normal person, what is the likelihood that any of their children will be albinos? Explain.

Sex-Linked Traits

Traits affected by genes on the sex chromosomes are called **sex-linked traits.** Most sex-linked traits are **X-linked,** that is, they are on the X chromosome, whereas, only a few **Y-linked** traits exist, largely because the Y chromosome is very small. An example of an X-linked trait is **hemophilia A** (classic hemophilia) in which the ability to produce one of the clotting factors is not present (see chapter 19). Consequently, clotting is impaired and persistent bleeding occurs either spontaneously or as a result of an injury. Hemophilia A is a recessive trait located on the X chromosome. The possible genotypes and phenotypes are, therefore, $X^H X^H$ (normal homozygous female), $X^H X^h$ (normal heterozygous female), $X^h X^h$ (hemophiliac homozygous female), $X^H Y$ (normal male), and $X^h Y$ (hemophiliac male). Note that a female must have both recessive genes to exhibit hemophilia, whereas a male, because he has only one X chromosome, has hemophilia if he has only one of the recessive genes. An example of the inheritance of hemophilia is illustrated in figure 29.27. If a woman who is a carrier for hemophilia mates with a man who does not have hemophilia, none of their daughters but half of their sons will have hemophilia.

Other Types of Gene Expression

The expression of a dominant over a recessive gene is the simplest manner in which genes determine a person's phenotype. Many other ways exist in which genes influence the expression of a trait. In some cases, the dominant gene doesn't completely mask the effects of the recessive gene, a phenomenon called **incomplete dominance.** An example of incomplete dominance is **sickle-cell disease,** in which a gene responsible for producing hemoglobin in red blood cells is abnormal. Consequently, the hemoglobin produced by the gene is abnormal. The result is red blood cells that are stretched into an elongated sickle shape. These red blood cells tend

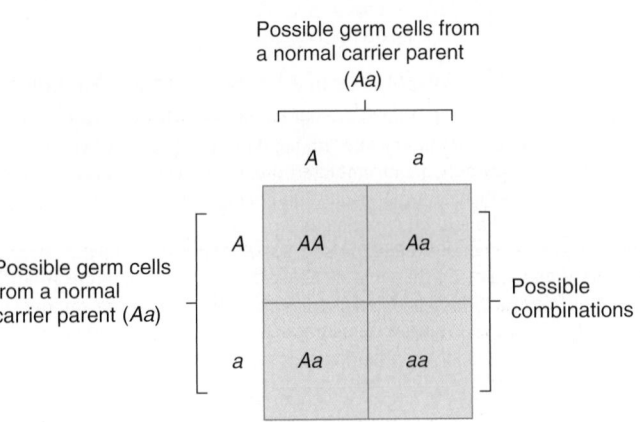

Figure 29.26 Inheritance of a Recessive Trait: Albinism

A represents the normal pigmented condition, and *a* represents the recessive unpigmented condition. The figure shows a Punnett square of a mating between two normal carriers.

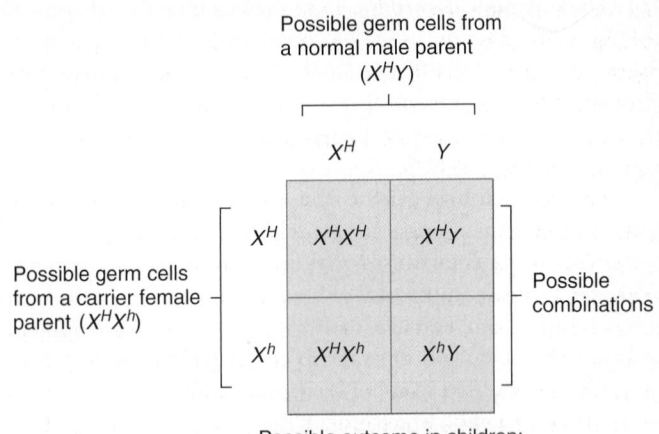

Figure 29.27 Inheritance of an X-Linked Trait: Hemophilia

X^H represents the normal X chromosome condition with all clotting factors, and X^h represents the X chromosome lacking a gene for one clotting factor. The figure shows a Punnett square of a mating between a normal male and a normal carrier female.

to stick in capillaries, thereby blocking blood flow to tissues. In addition, the sickle-shaped cells tend to rupture more easily than normal red blood cells. The normal allele (*S*) for producing normal hemoglobin is dominant over the sickle cell allele (*s*) responsible for producing the abnormal hemoglobin. A person with genotype *SS* has normal hemoglobin. A person with sickle-cell disease has genotype *ss* and has abnormal hemoglobin. A person who is heterozygous has the genotype *Ss,* has half normal hemoglobin and half abnormal hemoglobin, and usually has only a few sickle-shaped red blood cells. This condition is called **sickle-cell trait.** Usually, a person with sickle-cell trait exhibits no adverse symptoms. For this set of alleles, expressing incomplete dominance, however, each genotype presents a unique, recognizable phenotype.

In another type of gene expression, called **codominance,** two alleles can combine to produce an effect without either of them being dominant or recessive. For example, a person with type AB blood has A antigens and B antigens on the surface of the red blood cells (see chapter 19). The antigens result from a gene that causes the production of the A antigen and a different gene that causes the production of the B antigen. In this case, A and B are neither dominant nor recessive in relation to each other.

Polygenic traits are determined by the expression of multiple genes on different chromosomes. Examples are a person's height, intelligence, eye color, and skin color. Polygenic traits typically are characterized by having a great amount of variability. For example, many different shades of eye color and skin color exist (figure 29.28). Because of the many genes involved, it's difficult to predict how a polygenic trait will be passed from one generation to the next. Notice, however, that even though a complex combination of genes determines skin color, one single defective gene can eliminate skin color completely, resulting in albinism, which is inherited as a simple recessive trait.

Genetic Disorders

Abnormalities in a person's genetic makeup, that is, in his or her DNA, cause **genetic disorders.** They may involve a single gene or an entire chromosome, as in the case of aneuploidy (table 29.4). Genetic disorders are often confused with **congenital disorders.** Congenital means "present at birth" and a congenital disorder is commonly referred to as a birth defect, but all congenital disorders are not necessarily genetic. Approximately 15% of all congenital disorders have a known genetic cause, and approximately 70% of all birth defects are of unknown cause. The remaining 15% are the result of environmental causes or a combination of environmental and genetic causes. In the case of environmental causes, the birth defect results from damage to the fetus during development. Agents that cause birth defects are called **teratogens** (ter′ă-tō-jenz). For example, fetal alcohol syndrome results when a pregnant woman drinks alcohol, which crosses the placenta and damages the fetus. The baby is born with a smaller-than-normal head and mental retardation and may exhibit other birth defects.

One cause of genetic disorders is a **mutation,** a change in a gene that usually involves a change in the number or kinds of nucleotides composing the DNA (see chapter 2). Mutations are known to occur by chance (randomly without known cause) or may be caused by chemicals, radiation, or viruses. Agents that cause mutations are called **mutagens** (mū′tă-jenz). In most cases, a

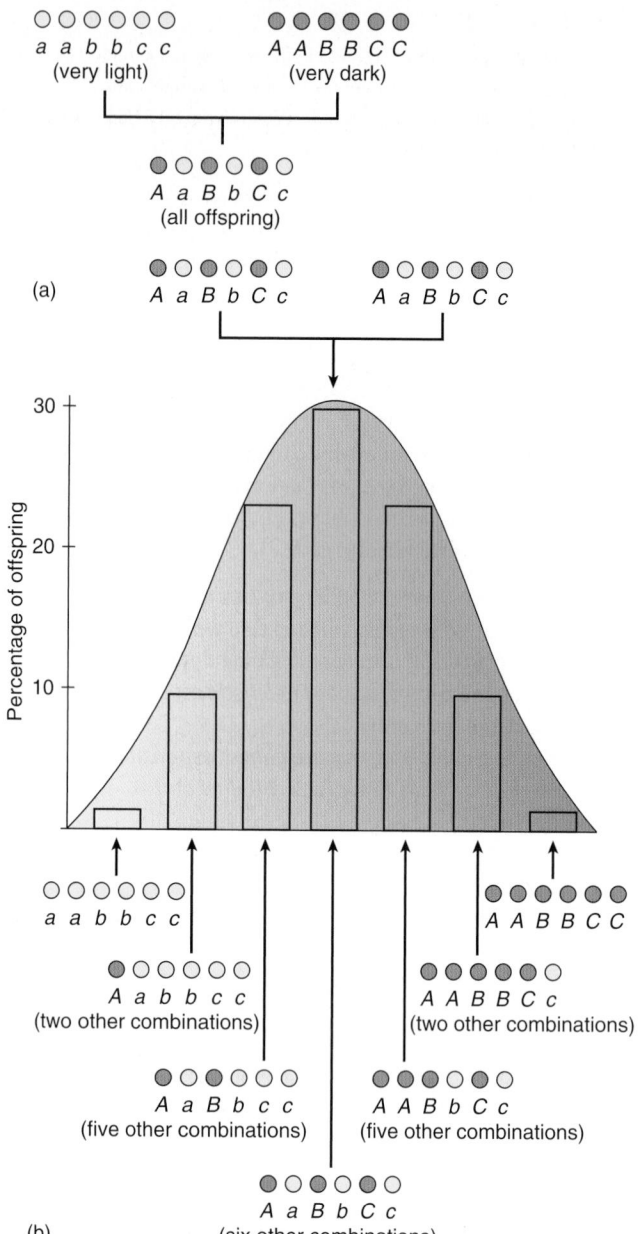

Figure 29.28 Inheritance of a Polygenic Trait: Skin Color

In this example, three genes for skin color are shown. The dominant alleles (*A, B, C*), each of which contributes one "unit of dark color" to the offspring (indicated by a *dark dot*), are incompletely dominant over the recessive alleles (*a, b, c*), each of which contributes one "unit of light color" to the offspring (indicated by a *light dot*). (*a*) A mating between a very light-skinned person (*aabbcc*) and a very dark-skinned person (*AABBCC*) is shown. All the offspring are of intermediate color. (*b*) A mating between two people of intermediate skin color (*AaBbCc*). The possible offspring skin color falls within a normal distribution in which a very low percentage (less than 2%) are either very light or very dark, and most of the offspring are of intermediate color.

specific cause of a mutation cannot be determined. Once a mutation has occurred, however, the abnormal trait can be passed from one generation to the next.

A mutation involving a single nucleotide change is called a point mutation. Mutations that change the sequence or number of nucleotides involve changes in chromosome structure, and therefore

Table 29.4   Genetic Disorders

Disorder	Description
Dominant Traits	
Achondroplasia	Dwarfism characterized by shortening of the upper and lower limbs
Huntington's chorea	Severe degeneration of the basal nuclei and frontal cerebral cortex; characterized by purposeless movements and mental deterioration; onset is usually between 40 and 50 years of age
Hypercholesterolemia	Elevated blood cholesterol levels that contribute to atherosclerosis and cardiovascular disease
Marfan's syndrome	Abnormal connective tissue results in increased height, elongated digits, and weakness in the aortic wall
Neurofibromatosis	Small pigmented lesions (café-au-lait spots) in the skin and disfiguring tumors (noncancerous) caused by proliferation of Schwann cells along nerves
Osteogenesis imperfecta	Abnormal phosphate metabolism results in brittle bones that repeatedly break
Recessive Traits	
Albinism	Lack of the enzyme necessary to produce the pigment melanin; characterized by lack of skin, hair, and eye coloration
Cystic fibrosis	Impaired transport of chloride ions across plasma membranes; results in excessive production of thick mucus that blocks the respiratory and gastrointestinal tract; the most common fatal genetic disorder
Phenylketonuria	Lack of the enzyme necessary to convert the amino acid phenylalanine to the amino acid tyrosine; an accumulation of phenylalanine leads to mental retardation
Severe combined immune deficiency	Inability to form the white blood cells (B cells, T cells, and phagocytes) necessary for an immune system response
Sickle-cell disease	Inability to produce normal hemoglobin; results in abnormally shaped red blood cells that clog capillaries or rupture
Tay-Sachs disease	Lack of the enzyme necessary to break down certain fatty substances; an accumulation of fatty substances impairs action potential propagation, resulting in deterioration of mental and physical functions and death by 3–4 years of age
Thalassemia	Decreased rate of hemoglobin synthesis; results in anemia, enlargement of the spleen, increased cell numbers in red bone marrow, and congestive heart failure
Sex-Linked Traits	
Hemophilia	Most commonly, a recessive gene causes a failure to produce blood clotting factors; resulting in prolonged bleeding
Red-green color blindness	Most commonly, a recessive gene causes a deficiency in functional green-sensitive cones; inability to distinguish between red and green colors
Chromosomal Disorders	
Down's syndrome	Caused by having three chromosomes 21; results in mental retardation, short stature, and poor muscle tone
Duchenne's muscular dystrophy	Caused by deletion or alteration of part of the X chromosome; results in progressive weakness and wasting of muscles
Klinefelter's syndrome	Caused by two or more X chromosomes in a male (XXY); results in small testes, sterility, and development of femalelike breasts
Turner's syndrome	Caused by having only one X chromosome; results in immature uterus, lack of ovaries, and short stature

are called structural mutations. Structural mutations can involve deletions, the loss of part of a chromosome; duplications, the addition of a copied section of a chromosome; translocations, in which part of one chromosome becomes attached to another chromosome; and inversions, where the DNA sequence of a section of a chromosome is reversed.

Cancer is a result of uncontrolled cell divisions. **Oncogenes** (ong'kō-jēnz) are genes associated with cancer. Many oncogenes are actually control genes involved in regulating cell proliferation and differentiation in the embryo and fetus. A change in an oncogene or in the regulation of an oncogene can result in uncontrolled cell proliferation and the development of cancerous tumors. The normal control of oncogenes involves other genes, called **tumor suppressor genes.** Cancer may occur when a mutation activates an oncogene or inactivates a tumor supressor gene. An accumulation of several mutations is necessary for cancer to occur. It's believed that certain chemicals called **carcinogens** (kar-sin'ō-jenz, kar'si-nō-jenz) can induce such mutations and, thereby, initiate the development of cancer. For example, chemicals in cigarette smoke are known to cause lung cancer. Thus, even though everyone has oncogenes, an outside agent may be necessary for a cancer to begin.

A change in somatic cells that results in cancer is not usually inheritable; nonetheless, a genetic basis may exist that allows cancer development, especially under the right environmental conditions. In this sense, the inheritance of cancer and other abnormalities has been described as **genetic susceptibility,** or **genetic predisposition.** For example, if a woman's close relatives, such as her mother or sister, have breast cancer, she has a greater-than-average risk of developing breast cancer. Similar genetic susceptibilities have been found for diabetes mellitus, schizophrenia, and other disorders.

Treatment of genetic disorders usually involves treatment of symptoms but doesn't result in a cure because the treatment doesn't

Clinical Focus   The Human Genome Project

The human genome is all of the genes found in one homologous set of human chromosomes. It's estimated that humans have 20,500–30,000 genes. A **genomic map** is a description of the DNA nucleotide sequences of the genes and their locations on the chromosomes (figure C). Celera, a private corporation working on the **Human Genome Project,** announced on February 12, 2001, that it had completed sequencing the entire genome.

Armed with a knowledge of the human genome and what effects the genome has on a person's physical, mental, and behavioral abilities, medicine and society will be transformed in many ways. Medicine, for example, will shift emphasis from the curative to the preventative. The potential disorders or diseases a person is likely to develop can be prevented or their severity lessened. When prevention is not possible, knowledge of the enzymes or other molecules involved in a disorder may result in new drugs and techniques that can compensate for the genetic disorder. Knowledge of the genes involved in a disorder may result in **gene therapy,** or **genetic engineering,** that repairs or replaces defective genes, resulting in cures of or treatments for genetic disorders.

Despite the great promise of benefits from the Human Genome Project, the knowledge that will be produced has raised a number of ethical and legal questions for society. Should a person's genomic information be public knowledge? Should persons with a genome that predisposes them to cancer or behavioral disorders be barred from certain types of employment or be refused medical insurance because they are a high risk? Can a person demand to know a prospective mate's genome? Should parents know the genome of their fetus and be allowed to make decisions regarding abortion based on this knowledge? Should the same genetic-engineering techniques that provide alteration of the genome to cure genetic disorders be used to create genomes that are deemed to be superior? Such questions raise the specter of genetic discrimination.

---

change the basic genetic material. For example, treating the problems associated with mucus buildup in cystic fibrosis does not cure the disorder. Currently, research is underway to actually alter a person's genetic makeup. Ultimately, it may be possible to insert a normal gene into cells to replace an abnormal or missing gene. Then, if the normal gene functions, the disorder will be cured.

Genetic Counseling

Genetic counseling includes predicting the possible results of matings involving carriers of harmful genes and talking to parents or prospective parents about the possible outcomes and treatments of a genetic disorder. With this knowledge, prospective parents can make informed decisions about having children.

A first step in genetic counseling is to attempt to determine the genotype of the individuals involved. For example, a couple may suspect that they are carriers for a genetic disorder. A family tree, or **pedigree,** provides historical information about family members (figure 29.29). Sometimes, by knowing the phenotypes of relatives, it's possible to determine a person's genotype. Direct means of obtaining genetic information are also available. A karyotype can be taken from white blood cells or the epithelial cells lining the inside of the cheek. Alternatively, the amount of a given substance, such as an enzyme, produced by a carrier can be tested. Sometimes a carrier produces slightly more or less of a given substance because they are heterozygous and have only one dominant gene for the normal or abnormal trait. For example, carriers for cystic fibrosis produce more salt in their sweat than is normal.

Sometimes, it's suspected that a fetus may have a genetic abnormality. Fetal cells can be tested by amniocentesis, which takes cells floating in the amniotic fluid (see figure B on p. 1102), or chorionic villus sampling, which takes cells from the fetal side of the placenta.

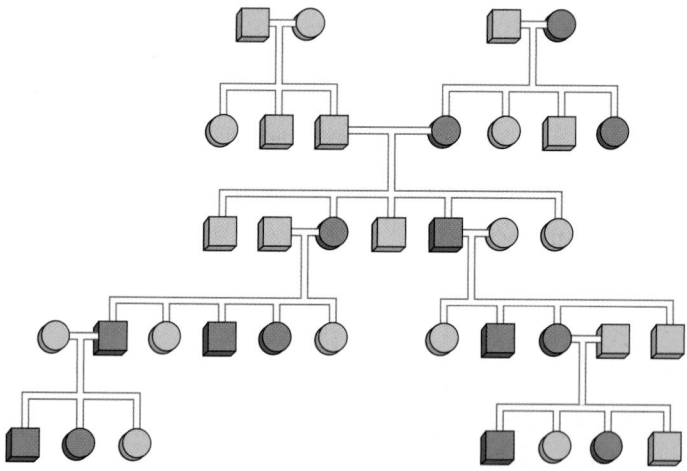

Figure 29.29   Pedigree of a Simple Dominant Trait

Males are indicated by *squares*, females by *circles*. Affected people are indicated in *purple*. The *horizontal line* between symbols represents a mating. The symbols connected to the mating line by vertical and horizontal lines represent the children resulting from the mating in order of birth from *left* to *right*. Matings not related to the pedigree are not shown.

41. *Explain how sex-linked traits are inherited. Give an example.*
42. *How does sickle-cell disease, type AB blood, and a person's height result from the expression of genes?*
43. *What is a mutation?*
44. *Define and give examples of teratogens, mutagens, and carcinogens. What is an oncogene?*
45. *What is genetic susceptibility?*
46. *How are pedigrees, karyotypes, chemical tests, amniocentesis, and chorionic villus sampling used in genetic counseling?*

Chromosome Pairs

1. a. Gaucher's disease
   b. Prostate cancer
   c. Glaucoma
   d. Alzheimer's disease*
2. a. Familial colon cancer*
   b. Waardenburg syndrome
3. a. Lung cancer
   b. Retinitis pigmentosa*
4. a. Huntington's chorea
   b. Parkinson's disease
5. a. Cockayne's syndrome
   b. Familial polyposis of the
      colon
   c. Asthma
6. a. Spinocerebellar ataxia
   b. Diabetes*
   c. Epilepsy*

7. a. Diabetes*
   b. Cystic fibrosis
   c. Obesity*
8. a. Werner's syndrome
   b. Burkitt's lymphoma
9. a. Malignant melanoma
   b. Friedreich ataxia
   c. Tuberous sclerosis
10. a. Multiple endocrine neoplasia,
       type 2
    b. Gyrate atrophy
11. a. Sickle-cell disease
    b. Multiple endocrine neoplasia
12. a. Zellweger syndrome
    b. Phenylketonuria (PKU)

13. a. Breast cancer*
    b. Retinoblastoma
    c. Wilson's disease
14. a. Alzheimer's disease*
15. a. Marfan's syndrome
    b. Tay-Sachs disease
16. a. Polycystic kidney disease
    b. Crohn's disease*
17. a. Peroneal muscular atrophy
    b. Breast cancer*
18. a. Amyloidosis
    b. Pancreatic cancer*
19. a. Familial hypercholesterolemia
    b. Myotonic dystrophy

20. a. Severe combined
       immunodeficiency
21. a. Amyotrophic lateral sclerosis*
22. a. DiGeorge's syndrome
    b. Neurofibromatosis, type 2
X a. Duchenne's muscular
       dystrophy
    b. Menkes's syndrome
    c. X-linked severe combined
       immunodeficiency
    d. Factor VIII deficiency
       (hemophilia A)

*Gene responsible for only some cases.

Figure C   The Human Genomic Map

Representative genetic defects mapped to date. The *green circles* indicate the location of the genes listed for each chromosome.

S U M M A R Y

Prenatal development is an important part of an individual's life.

Prenatal Development   (p. 1082)

1. Prenatal development is divided into three parts: formation of the germ layers, formation of the organ systems, and growth and maturation.
2. Postovulatory age is 14 days less than clinical age.

Fertilization

Fertilization, the union of the oocyte and sperm, results in a zygote.

Early Cell Division

The cells of the early embryo are pluripotent (capable of making any cell of the body).

Morula and Blastocyst

The product of fertilization undergoes divisions until it becomes a mass, called a morula, and then a hollow ball of cells, called a blastocyst.

Implantation of the Blastocyst and Development of the Placenta

The blastocyst implants into the uterus about 7 days after fertilization. The placenta is derived from the trophoblast of the blastocyst.

Formation of the Germ Layers

All tissues of the body are derived from three primary germ layers: ectoderm, mesoderm, and endoderm.

Neural Tube and Neural Crest Formation

The nervous system develops from a neural tube that forms in the ectodermal surface of the embryo and from neural crest cells derived from the developing neural tube.

Somite Formation

Segments called somites that develop along the neural tube give rise to the musculature, vertebral column, and ribs.

Formation of the Gut and Body Cavities

1. The gastrointestinal tract develops as the developing embryo closes off part of the yolk sac.
2. The celom develops from small cavities that fuse within the embryo.

Limb Bud Development

The limbs develop from proximal to distal as outgrowths called limb buds.

Development of the Face

The face develops from the fusion of five major tissue processes.

Development of the Organ Systems

1. The skin develops from ectoderm (epithelium), mesoderm and neural crest (dermis), and the neural crest (melanocytes).
2. The skeletal system develops from mesoderm or neural crest cells.
3. Muscle develops from myoblasts, which migrate from somites.
4. The brain and spinal cord develop from the neural tube, and the peripheral nervous system develops from the neural tube and the neural crest cells.
5. The special senses develop mainly as neural tube or neural crest cell derivatives.
6. Many endocrine organs develop mainly as outpocketings of the brain or digestive tract.
7. The heart develops as two tubes fuse into a single tube that bends and develops septa to form four chambers.
8. The peripheral circulation develops from mesoderm as blood islands become hollow and fuse to form a network.
9. The lungs form as evaginations of the digestive tract. These evaginations undergo repeated branching.
10. The urinary system develops in three stages—pronephros, mesonephros, and metanephros—from the head to the tail of the embryo. The ducts join the allantois, part of which becomes the urinary bladder.
11. The reproductive system develops in conjunction with the urinary system. The presence or absence of certain hormones are very important to sexual development.

Growth of the Fetus

1. The embryo becomes a fetus at 60 days.
2. The fetal period is from day 60 to birth. It is a time of rapid growth.

Parturition   (p. 1103)

1. The total length of gestation is 280 days (clinical age).
2. Uterine contractions force the baby out of the uterus during labor.
3. Increased estrogen levels and decreased progesterone levels help initiate parturition.
4. Fetal glucocorticoids act on the placenta to decrease progesterone synthesis and to increase estrogen and prostaglandin synthesis.
5. Stretching of the uterus and decreased progesterone levels stimulate oxytocin secretion, which stimulates uterine contraction.

The Newborn   (p. 1106)
Respiratory and Circulatory Changes

1. The foramen ovale closes, separating the two atria.
2. The ductus arteriosus closes, and blood no longer flows between the pulmonary trunk and the aorta.
3. The umbilical vein and arteries degenerate.

Digestive Changes

1. Meconium is a mixture of cells from the digestive tract, amniotic fluid, bile, and mucus excreted by the newborn.
2. The stomach begins to secrete acid.
3. The liver does not form adult bilirubin for the first 2 weeks.
4. Lactose can be digested, but other foods must be gradually introduced.

Apgar Scores

1. Apgar represents appearance, pulse, grimace, activity, and respiratory effort.
2. Apgar and other methods are used to assess the physiologic condition of the newborn.

Lactation   (p. 1109)

1. Estrogen, progesterone, and other hormones stimulate the growth of the breasts during pregnancy.
2. Suckling stimulates prolactin and oxytocin synthesis. Prolactin stimulates milk production, and oxytocin stimulates milk letdown.

First Year After Birth   (p. 1110)

1. The number of neuron connections and glial cells increases.
2. Motor skills gradually develop, especially head, eye, and hand movements.

Life Stages   (p. 1110)

The life stages include the following: germinal, embryo, fetus, neonate, infant, child, adolescent, and adult.

Aging   (p. 1111)

1. Loss of cells that are not replaced contributes to aging.
   • A loss of neurons occurs.
   • Loss of muscle cells can affect skeletal and cardiac muscle function.

2. Loss of tissue plasticity results from cross-link formation between collagen molecules. The lens of the eye loses the ability to accommodate. Other organs, such as the joints, kidneys, lungs, and heart, also have reduced efficiency with advancing age.
3. The immune system loses the ability to act against foreign antigens and may attack self-antigens.
4. Many aging changes are probably genetic.

Death (p. 1112)

Death is the loss of brain functions.

Genetics (p. 1112)

Chromosomes

1. Humans have 46 chromosomes in 23 pairs; 22 pairs of autosomes and 1 pair of sex chromosomes.
2. Males have the sex chromosomes XY and females XX.
3. During gamete formation, the chromosomes of each pair of chromosomes separate; therefore, half of a person's genetic makeup comes from the father and half from the mother.

Genes

1. A gene is a portion of a DNA molecule. Genes determine the proteins in a cell.

2. Genes are paired (located on the paired chromosomes).
3. Dominant genes mask the effects of recessive genes.
4. Sex-linked traits result from genes on the sex chromosomes.
5. In incomplete dominance, the heterozygote expresses a trait that is intermediate between the two homozygous traits.
6. In codominance, neither gene is dominant or recessive, but both are fully expressed.
7. Polygenic traits result from the expression of multiple genes.

Genetic Disorders

1. A mutation is a change in the number or kinds of nucleotides in DNA.
2. Some genetic disorders result from an abnormal distribution of chromosomes during gamete formation.
3. Oncogenes are genes associated with cancer.
4. Genetic predisposition makes it more likely a person will develop a disorder.

Genetic Counseling

1. A pedigree (family history) can be used to determine the risk of having children with a genetic disorder.
2. Specific chemical testing or examination of a person's karyotype can be used to determine a person's genotype.

R E V I E W   A N D   C O M P R E H E N S I O N

1. The major development of organ systems takes place in the
   a. first 2 weeks of development
   b. third to eighth week of development
   c. eighth to twentieth week of development
   d. last 30 weeks of development

2. Given these structures:
   1. blastocyst
   2. morula
   3. zygote

   Choose the arrangement that lists the structures in the order in which they are formed during development.
   a. 1,2,3
   b. 1,3,2
   c. 2,3,1
   d. 3,1,2
   e. 3,2,1

3. The embryo proper develops from the
   a. inner cell mass.
   b. trophoblast.
   c. blastocele.
   d. yolk sac.

4. The placenta
   a. develops from the trophoblast.
   b. allows maternal blood to mix with embryonic blood.
   c. invades the lacunae of the embryo.
   d. all of the above.

5. The embryonic disk
   a. forms between the amniotic cavity and the yolk sac.
   b. contains the primitive streak.
   c. becomes a three-layered structure.
   d. all of the above.

6. The brain develops from
   a. ectoderm.
   b. endoderm.
   c. mesoderm.

7. Most of the skeletal system develops from
   a. ectoderm.
   b. endoderm.
   c. mesoderm.

8. Given these structures:
   1. neural crest
   2. neural plate
   3. neural tube

   Choose the arrangement that lists the structure in the order in which they form during development.
   a. 1,2,3
   b. 1,3,2
   c. 2,1,3
   d. 2,3,1
   e. 3,2,1

9. The somites give rise to the
   a. circulatory system.
   b. skeletal muscle.
   c. lungs.
   d. kidneys.
   e. brain.

10. The pericardial cavity forms from
    a. evagination of the early gastrointestinal tract.
    b. the neural tube.
    c. the celom.
    d. the branchial arches.
    e. pharyngeal pouches.

11. The parts of the limbs develop
    a. in a proximal-to-distal sequence.
    b. in a distal-to-proximal sequence.
    c. at approximately the same time.
    d. before the primitive streak is formed.

12. Concerning development of the face,
    a. the face develops by the fusion of five embryonic structures.
    b. the maxillary processes normally meet at the midline to form the lip.
    c. the primary palate forms the roof of the mouth.
    d. clefts of the secondary palate normally occur to one side of the midline.

13. Concerning the development of the heart,
    a. the heart develops from a single tube, which results from fusion of two tubes.
    b. the SA node develops in the wall of the sinus venosus.
    c. the foramen ovale lets blood flow from the right atrium to the left atrium.
    d. the bulbis cordis is absorbed into the ventricle.
    e. all of the above.

14. Given these structures:
    1. mesonephros
    2. metanephros
    3. pronephros

    Choose the arrangement that lists the structures in the order in which they form during development.
    a. 1,2,3
    b. 1,3,2
    c. 2,3,1
    d. 3,1,2
    e. 3,2,1

15. A study of the early embryo indicates that the glans penis of the male develops from the same embryonic structure as which of these female structures?
    a. labia majora
    b. uterus
    c. clitoris
    d. vagina
    e. urinary bladder

16. Which hormone causes differentiation of sex organs in the developing male fetus?
    a. FSH and LH
    b. LH and testosterone
    c. testosterone and dihydrotestosterone
    d. estrogen and progesterone
    e. GnRH and FSH

17. Onset of labor may be a result of
    a. increased estrogen secretion by the placenta.
    b. increased glucocorticoid secretion by the fetus.
    c. increased secretion of oxytocin.
    d. stretch of the uterus.
    e. all of the above.

18. Following birth,
    a. the ductus arteriosus closes.
    b. the pH of the stomach increases.
    c. the fossa ovalis becomes the foramen ovale.
    d. blood flow through the pulmonary arteries decreases.
    e. all of the above.

19. The hormone involved in milk production is
    a. oxytocin.
    b. prolactin.
    c. estrogen.
    d. progesterone.
    e. ACTH.

20. Which of these most appropriately predicts the consequences of removing the sensory neurons from the areola of a lactating rat (or human).
    a. Blood levels of oxytocin decrease.
    b. Blood levels of prolactin decrease.
    c. Milk production and letdown decreases.
    d. All of the above.

21. Which of these life stages is correctly matched with the time that the stage occurs?
    a. neonate—birth to 1 month after birth
    b. infant—1 month to 6 months
    c. child—6 months to 5 years
    d. puberty—10–12 years
    e. middle age—20–40 years

22. Which of these occurs as we get older?
    a. Neurons replicate to replace lost neurons.
    b. Skeletal muscle cells replicate to replace lost muscle cells.
    c. Cross-links between collagen molecules increases.
    d. The immune system become less sensitive to the body's own antigens.
    e. Free radicals help to prevent cancer.

23. A gene is
    a. the functional unit of heredity.
    b. a certain portion of a DNA molecule.
    c. a part of a chromosome.
    d. all of the above.

24. Which of these does *not* contribute to genetic differences between gametes?
    a. crossing-over
    b. independent assortment
    c. linkage
    d. nondisjunction

25. Which of these terms is correctly matched with its definition?
    a. autosome—an X or Y chromosome
    b. phenotype—the genetic makeup of an individual
    c. allele—genes occupying the same locus on homologous chromosomes
    d. heterozygous—having two identical genes for a trait
    e. recessive—a trait expressed when the genes are heterozygous

26. Which of these genotypes is heterozygous?
    a. *DD*
    b. *Dd*
    c. *dd*
    d. both a and c

27. The ABO blood group is an example of
    a. dominant versus recessive alleles.
    b. incomplete dominance.
    c. codominance.
    d. a polygenic trait.
    e. sex-linked inheritance.

28. Assume that a trait is determined by an X-linked dominant gene. If the mother exhibits the trait, but the father does not, then their
    a. sons are more likely than their daughters to exhibit the trait.
    b. daughters are more likely than their sons to exhibit the trait.
    c. sons and daughters are equally likely to exhibit the trait.

29. Which of these could result in a congenital disorder?
    a. a parent has the same disorder
    b. a teratogen
    c. a mutagen
    d. all of the above

Answers in Appendix F

CRITICAL THINKING

1. A woman is told by her physician that she is pregnant and that she is 44 days past her LMP. Approximately how many days has the embryo been developing, and what developmental events are occurring?

2. A high fever can prevent neural tube closure. If a woman has a high fever approximately 35–45 days after her LMP, what kinds of birth defects may be seen in the developing embryo?

3. If the apical ectodermal ridge is damaged during embryonic development when the limb bud is about one-half grown, what kinds of birth defects might be expected? Describe the anatomy of the affected structure.

4. What are the results of exposing a female embryo to high levels of testosterone while she is developing?

5. Three minutes after birth, a newborn has an Apgar score of 5 as follows: A, 0; P, 1; G, 1; A, 1; and R, 2. What are some of the possible causes for this low score? What might be done for this neonate?

6. When a woman nurses, it's possible for milk letdown to occur in the breast that is not being suckled. Explain how this response happens.

7. Dimpled cheeks are inherited as a dominant trait. Is it possible for two parents, each of whom has dimpled cheeks, to have a child without dimpled cheeks? Explain.

8. The ability to roll the tongue to form a "tube" results from a dominant gene. Suppose that a woman and her son can both roll their tongues, but her husband cannot. Is it possible to determine if the husband is the father of her son?

9. A woman who does not have hemophilia mates with a man who has the disorder. Determine the genotype of both parents if half of their children have hemophilia.

Answers in Appendix G

ANSWERS TO PREDICT QUESTIONS

1. Two primitive streaks on one embryonic disk result in the development of two embryos. If the two primitive streaks are touching or are very close to each other, the embryos may be joined. This condition is called conjoined (Siamese) twins.

2. Because the early embryonic heart is a simple tube, blood must be forced through the heart in almost a peristaltic fashion, and the contraction begins in the sinus venosus. If the sinus venosus didn't contract first, blood could flow in the opposite direction.

3.

	Clinical Age	Developmental Age
Fertilization	14 days	0 days
Implantation	21 days	7 days
Fetal Period	70 days	56 days
Parturition	280 days	266 days

4. Elevation of calcium levels might cause the uterine muscles to contract tetanically. This tetanic contraction could compress blood vessels and cut the blood supply to the fetus. Hypercalcemia can also result in arrhythmias and muscle weakness (see chapter 27). The doctor would therefore not administer calcium to the woman in labor but may give oxytocin, which strengthens contractions but is less likely to produce tetany.

5. Nursing stimulates the release of oxytocin from the mother's posterior pituitary gland, which is responsible for milk letdown. Oxytocin can also cause uterine contractions and cramps.

6. Genotype *DD* (homozygous dominant) would have the polydactyly phenotype, genotype *Dd* (heterozygous) would have the polydactyly phenotype, and genotype *dd* (homozygous recessive) would have the normal phenotype.

7. None. One in two will be homozygous normal and one in two will be normal heterozygous carriers.

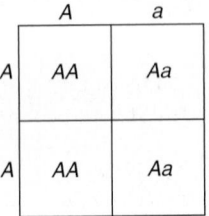

Visit the Online Learning Center at www.mhhe.com/seeley7 for chapter quizzes, interactive learning exercises, and other study tools.

Appendix A

Table of Measurements

Table A.1 — Table of Measurements

Unit	Metric Equivalent	Symbol	U.S. Equivalent
Measures of Length			
1 kilometer	= 1000 meters	km	0.62137 mile
1 meter	= 10 decimeters or 100 centimeters	m	39.37 inches
1 decimeter	= 10 centimeters	dm	3.937 inches
1 centimeter	= 10 millimeters	cm	0.3937 inch
1 millimeter	= 1000 micrometers	mm	
1 micrometer	= 1/1000 millimeter or 1000 nanometers	μm	
1 nanometer	= 10 angstroms or 1000 picometers	nm	
1 angstrom	= 1/10,000,000 millimeter	Å	
1 picometer	= 1/1,000,000,000 millimeter	pm	
Measures of Volume			
1 cubic meter	= 1000 cubic decimeters	m^3	1.308 cubic yards
1 cubic decimeter	= 1000 cubic centimeters	dm^3	0.03531 cubic foot
1 cubic centimeter	= 1000 cubic millimeters or 1 milliliter	cm^3 (cc)	0.06102 cubic inch
Measures of Capacity			
1 kiloliter	= 1000 liters	kL	264.18 gallons
1 liter	= 10 deciliters	L	1.0567 quarts
1 deciliter	= 100 milliliters	dL	0.4227 cup
1 milliliter	= volume of 1 gram of water at standard temperature and pressure	mL	0.3381 ounce
Measures of Mass			
1 kilogram	= 1000 grams	kg	2.2046 pounds
1 gram	= 100 centigrams or 1000 milligrams	g	0.0353 ounce
1 centigram	= 10 milligrams	cg	0.1543 grain
1 milligram	= 1/1000 gram	mg	

Note that a micrometer was formerly called a micron (μ), and a nanometer was formerly called a millimicron (mμ).

Appendix B

Scientific Notation

Very large numbers with many zeros such as 1,000,000,000,000,000 or very small numbers such as 0.0000000000000001 are very cumbersome to work with. Consequently, the numbers are expressed in a kind of mathematical shorthand known as scientific notation. Scientific notation has the following form:

$$M \times 10^n$$

where n specifies how many times the number M is raised to the power of 10. The exponent n has two meanings, depending on its sign. If n is positive, M is multiplied by 10 n times. For example, if $n = 2$ and $M = 1.2$, then

$$1.2 \times 10^2 = 1.2 \times 10 \times 10 = 120$$

In other words, if n is positive, the decimal point of M is moved to the right n times. In this case the decimal point of 1.2 is moved two places to the right.

$$1.\underset{\frown}{2}0.$$

If n is negative, M is divided by 10 n times.

$$1.2 \times 10^{-2} = \frac{1.2}{(10 \times 10)} = \frac{1.2}{100} = 0.012$$

In other words, if n is negative, the decimal point of M is moved to the left n times. In this case the decimal point of 1.2 is moved two places to the left.

$$0.01.2$$

If M is the number 1.0, it often is not expressed in scientific notation. For example, 1.0×10^2 is the same thing as 10^2, and 1.0×10^{-2} is the same thing as 10^{-2}.

Two common examples of the use of scientific notation in chemistry are Avogadro's number and pH. Avogadro's number, 6.023×10^{23}, is the number of atoms in 1 molar mass of an element. Thus

$$6.023 \times 10^{23} = 602,300,000,000,000,000,000,000$$

which is a very large number of atoms.

The pH scale is a measure of the concentration of hydrogen ions in a solution. A neutral solution has 10^{-7} moles of hydrogen ions per liter. In other words

$$10^{-7} = 0.0000001$$

which is a very small amount (1 ten-millionth of a gram) of hydrogen ions.

Appendix C

Solution Concentrations

Physiologists often express solution concentration in terms of percent, molarity, molality, and equivalents.

Percent

The weight-volume method of expressing percent concentrations states the weight of a solute in a given volume of solvent. For example, to prepare a 10% solution of sodium chloride, 10 g of sodium chloride is dissolved in a small amount of water (solvent) to form a salt solution. Then additional water is added to the salt solution to form 100 mL of salt solution. Note that the sodium chloride was dissolved in water and then diluted to the required volume. The sodium chloride was not dissolved directly in 100 mL of water.

Molarity

Molarity determines the number of moles of solute dissolved in a given volume of solvent. A 1 molar (1 M) solution is made by dissolving 1 mole (mol) of a substance in enough water to make 1 L of solution. For example, 1 mol of sodium chloride solution is made by dissolving 58.44 g of sodium chloride in enough water to make 1 L of solution. One mol of glucose solution is made by dissolving 180.2 g of glucose in enough water to make 1 L of solution. Both solutions have the same number (Avogadro's number) of formula units (NaCl) and molecules (glucose) in solution.

Molality

Although 1 M solutions have the same number of solute molecules, they don't have the same number of solvent (water) molecules. Because 58.5 g of sodium chloride occupies less volume than 180 g of glucose, the sodium chloride solution has more water molecules. **Molality** is a method of calculating concentrations that takes into account the number of solute and solvent molecules. A 1 molal solution (1 m) is 1 mol of a substance dissolved in 1 kg of water. Thus all 1-molal solutions have the same number of solvent molecules.

When sodium chloride, which is an ionic compound, is dissolved in water it dissociates to form two ions, a sodium cation (Na^+) and a chloride anion (Cl^-). Glucose does not dissociate when dissolved in water, however, because it's a molecule. Thus, the sodium chloride solution contains twice as many particles as the glucose solution (one Na^+ and one Cl^- for each glucose molecule). To report the concentration of these substances in a way that reflects the number of particles in a given mass of solvent the concept of **osmolality** is used. The osmolality of a solution is the molality of the solution times the number of particles into which the solute dissociates in 1 kg of solvent. Thus 1 mol of sodium chloride in 1 kg of water is a 2 osmolal (osm) solution because sodium chloride dissociates to form two ions.

The osmolality of a solution is a reflection of the number, not the type, of particles in a solution. Thus a 1 osm solution contains 1 osm of particles per kilogram of solvent, but the particles may be all one type or a complex mixture of different types.

The concentration of particles in body fluids is so low that the measurement milliosmole (mOsm), 1/1000 of an osmole, is used. Most body fluids have an osmotic concentration of approximately 300 mOsm and consist of many different ions and molecules. The osmotic concentration of body fluids is important because it influences the movement of water into or out of cells (see chapter 3).

Equivalents

Equivalents are a measure of the concentrations of ionized substances. One equivalent (Eq) is 1 mol of an ionized substance multiplied by the absolute value of its charge. For example, 1 mol of NaCl dissociates into 1 mol of Na^+ and 1 mol of Cl^-. Thus there is 1 Eq of Na^+ (1 mol $\times$ 1) and 1 Eq of Cl^- (1 mol $\times$ 1). One mole of $CaCl_2$ dissociates into 1 mol of Ca^{2+} and 2 mol of Cl^-. Thus there are 2 Eq of Ca^{2+} (1 mol $\times$ 2) and 2 Eq of Cl^- (2 mol $\times$ 1). In an electrically neutral solution the equivalent concentration of positively charged ions is equal to the equivalent concentration of the negatively charged ions. One milliequivalent (mEq) is 1/1000 of an equivalent.

Appendix D

pH

Pure water weakly dissociates to form small numbers of hydrogen and hydroxide ions:

$$H_2O \leftrightarrow H^+ + OH^-$$

At 25°C the concentration of both hydrogen ions and hydroxide ions is 10^{-7} mol/L. Any solution that has equal concentrations of hydrogen and hydroxide ions is considered **neutral**. A solution is an **acid** if it has a higher concentration of hydrogen ions than hydroxide ions, and a solution is a **base** if it has a lower concentration of hydrogen ions than hydroxide ions. In any aqueous solution (at 25°C) the hydrogen ion concentration $[H^+]$ times the hydroxide ion concentration $[OH^-]$ is a constant that is equal to 10^{-14}.

$$[H^+] \times [OH^-] = 10^{-14}$$

Consequently, as the hydrogen ion concentration

decreases, the hydroxide ion concentration increases, and vice versa. For example:

	$[\text{H}^+]$	$[\text{OH}^-]$
Acidic solution	10^{-3}	10^{-11}
Neutral solution	10^{-7}	10^{-7}
Basic solution	10^{-12}	10^{-2}

Although the acidity or basicity of a solution could be expressed in terms of either hydrogen or hydroxide ion concentration, it's customary to use hydrogen ion concentration. The pH of a solution is defined as

$$pH = -\log_{10}(\text{H}^+)$$

Thus a neutral solution with 10^{-7} mol of hydrogen ions per liter has a pH of 7

$$\begin{aligned} pH &= -\log_{10}(\text{H}^+) \\ &= -\log_{10}(10^{-7}) \\ &= -(-7) \\ &= 7 \end{aligned}$$

In simple terms, to convert the hydrogen ion concentration to the pH scale, the exponent of the concentration (e.g., -7) is used, and it's changed from a negative to a positive number. Thus an acidic solution with 10^{-3} mol of hydrogen ions/L has a pH of 3, whereas a basic solution with 10^{-12} hydrogen ions/L has a pH of 12.

Appendix E

Reference Laboratory Values

Table E.1	Blood, Plasma, or Serum Values	
Test	**Normal Values**	**Clinical Significance**
Acetoacetate plus acetone	0.32–2 mg/100 mL	Values increase in diabetic acidosis, fasting, high-fat diet, and toxemia of pregnancy
Ammonia	80–110 µg/100 mL	Values decrease with proteinuria and as a result of severe burns and increase in multiple myeloma
Amylase	4–25 U/mL*	Values increase in acute pancreatitis, intestinal obstruction, and mumps; values decrease in cirrhosis of the liver, toxemia of pregnancy, and chronic pancreatitis
Barbiturate	0	Coma level: phenobarbital, approximately 10 mg/100 mL; most other drugs, 1–3 mg/100 mL
Bilirubin	0.4 mg/100 mL	Values increase in conditions causing red blood cell destruction of biliary obstruction or liver inflammation
Blood volume	8.5%–9% of body weight in kilograms	
Calcium	8.5–10.5 mg/dL	Values increase in hyperparathyroidism, vitamin D hypervitaminosis; values decrease in hypoparathyroidism, malnutrition, and severe diarrhea
Carbon dioxide content	24–30 mEq/L 20–26 mEq/L in infants (as HCO_3^-)	Values increase in respiratory diseases, vomiting, and intestinal obstruction; they decrease in acidosis, nephritis, and diarrhea
Carbon monoxide	0	Symptoms with over 20% saturation
Chloride	100–106 mEq/L	Values increase in Cushing's syndrome, nephritis, and hyperventilation; they decrease in diabetic acidosis, Addison's disease, and diarrhea and after severe burns
Creatine phosphokinase (CPK)	Female 5–35 mU/mL Male 5–55 mU/mL	Values increase in myocardial infarction and skeletal muscle diseases such as muscular dystrophy
Creatinine	0.6–1.5 mg/100 mL	Values increase in certain kidney diseases
Ethanol	0	0.3%–0.4%, marked intoxication 0.4%–0.5%, alcoholic stupor 0.5% or over, alcoholic coma
Glucose	Fasting 70–110 mg/100 mL	Values increase in diabetes mellitus, liver diseases, nephritis, hyperthyroidism, and pregnancy; they decrease in hyperinsulinism, hypothyroidism, and Addison's disease
Iron	50–150 µg/100 mL	Values increase in various anemias and liver disease; they decrease in iron deficiency anemia

*A unit (U) is the quantity of a substance that has a physiologic effect.

continued

Table E.1 *continued*

Test	Normal Values	Clinical Significance
Lactic acid	0.6–1.8 mEq/L	Values increase with muscular activity and in congestive heart failure, severe hemorrhage, shock, and anaerobic exercise
Lactic dehydrogenase	60–120 U/mL	Values increase in pernicious anemia, myocardial infarction, liver diseases, acute leukemia, and widespread carcinoma
Lipids	Cholesterol 120–220 mg/100 mL Cholesterol esters 60%–75% of cholesterol Phospholipids 9–16 mg/100 mL as lipid phosphorus Total fatty acids 190–420 mg/100 mL Total lipids 450–1000 mg/100 mL Triglycerides 40–150 mg/100 mL	Increased values for cholesterol and triglycerides are connected with increased risk of cardiovascular disease, such as heart attack and stroke
Lithium	Toxic levels 2 mEq/L	
Osmolality	285–295 mOsm/kg water	
Oxygen saturation (arterial) see Po_2	96%–100%	
Pco_2	35–43 mm Hg	Values decrease in acidosis, nephritis, and diarrhea; they increase in respiratory diseases, intestinal obstruction, and vomiting
pH	7.35–7.45	Values decrease as a result of hypoventilation, severe diarrhea, Addison's disease, and diabetic acidosis; values increase due to hyperventilation, Cushing's syndrome, and vomiting
Po_2	75–100 mm Hg (breathing room air)	Values increase in polycythemia and decrease in anemia and obstructive pulmonary diseases
Phosphatase (acid)	Male: total 0.13–0.63 U/mL Female: total 0.01–0.56 U/mL	Values increase in cancer of the prostate gland, hyperparathyroidism, some liver diseases, myocardial infarction, and pulmonary embolism
Phosphatase (alkaline)	13–39 IU/L* (infants and adolescents up to 104 IU/L)	Values increase in hyperparathyroidism, some liver diseases, and pregnancy
Phosphorus (inorganic)	3–4.5 mg/100 mL (infants in first year up to 6 mg/100 mL)	Values increase in hypoparathyroidism, acromegaly, vitamin D hypervitaminosis, and kidney diseases; they decrease in hyperparathyroidism
Potassium	3.5–5 mEq/100 mL	
Protein	Total 6–8.4 g/100 mL Albumin 3.5–5 g/100 mL Globulin 2.3–3.5 g/100 mL	Total protein values increase in severe dehydration and shock; they decrease in severe malnutrition and hemorrhage
Salicylate Therapeutic Toxic	0	 20–25 mg/100 mL Over 30 mg/100 mL Over 20 mg/100 mL after age 60
Sodium	135–145 mEq/L	Values increase in nephritis and severe dehydration; they decrease in Addison's disease, myxedema, kidney disease, and diarrhea
Sulfonamide Therapeutic	0	 5–15 mg/100 mL
Urea nitrogen	8–25 mg/100 mL	Values increase in response to increased dietary protein intake; values decrease in impaired renal function
Uric acid	3–7 mg/100 mL	Values increase in gout and toxemia of pregnancy and as a result of tissue damage

Table E.2 Blood Count Values

Test	Normal Values	Clinical Significance
Clotting (coagulation) time	5–10 minutes	Values increase in afibrinogenemia and hyperheparinemia, severe liver damage
Fetal hemoglobin	Newborns: 60%–90% Before age 2: 0%–4% Adults: 0%–2%	Values increase in thalassemia, sickle-cell disease, and leakage of fetal blood into maternal bloodstream during pregnancy
Hemoglobin	Male: 14–16.5 g/100 mL Female: 12–15 g/100 mL Newborn: 14–20 g/100 mL	Values decrease in anemia, hyperthyroidism, cirrhosis of the liver, and severe hemorrhage; values increase in polycythemia, congestive heart failure, obstructive pulmonary disease, high altitudes
Hematocrit	Male: 40%–54% Female: 38%–47%	Values increase in polycythemia, severe dehydration, and shock; values decrease in anemia, leukemia, cirrhosis, and hyperthyroidism
Ketone bodies	0.3–2 mg/100 mL Toxic level: 20 mg/100 mL	Values increase in ketoacidosis, fever, anorexia, fasting, starvation, high-fat diet
Platelet count	250,000–400,000/mm^3	Values decrease in anemias and allergic conditions and during cancer chemotherapy; values increase in cancer, trauma, heart disease, and cirrhosis
Prothrombin time	11–15 seconds	Values increase in prothrombin and vitamin deficiency, liver disease, and hypervitaminosis A
Red blood cell count	Males: 4.6–6.2 million/mm^3 Females: 4.2–5.4 million/mm^3	Values decrease in systemic lupus erythematosus, anemias, and Addison's disease; values increase in polycythemia and dehydration and following hemorrhage
Reticulocyte count	1%–3%	Values decrease in iron-deficiency and pernicious anemia and radiation therapy; values increase in hemolytic anemia, leukemia, and metastatic carcinoma
White blood cell count, differential	Neutrophils 60%–70% Eosinophils 2%–4% Basophils 0.5%–1% Lymphocytes 20%–25% Monocytes 3%–8%	Neutrophils increase in acute infections; eosinophils and basophils increase in allergic reactions; monocytes increase in chronic infections; lymphocytes increase during antigen–antibody reactions
White blood cell count, total	5000–9000/mm^3	Values decrease in diabetes mellitus, anemias, and following cancer chemotherapy; values increase in acute infections, trauma, some malignant diseases, and some cardiovascular diseases

Table E.3 Urine Values

Test	Normal Values	Clinical Significance
Acetone and acetoacetate	0	Values increase in diabetic acidosis and during fasting
Albumin	0 to trace	Values increase in glomerular nephritis and hypertension
Ammonia	20–70 mEq/L	Values increase in diabetes mellitus and liver disease
Bacterial count	Under 10,000/mL	Values increase in urinary tract infection
Bile and bilirubin	0	Values increase in biliary tract obstruction
Calcium	Under 250 mg/24 h	Values increase in hyperparathyroidism and decrease in hypoparathyroidism
Chloride	110–254 mEq/24 h	Values decrease in pyloric obstruction and diarrhea; values increase in Addison's disease and dehydration
Potassium	25–100 mEq/L	Values decrease in diarrhea, malabsorption syndrome, and adrenal cortical insufficiency; values increase in chronic renal failure, dehydration, and Cushing's syndrome
Sodium	75–200 mg/24 h	Values decrease in diarrhea, acute renal failure, and Cushing's syndrome; values increase in dehydration, starvation, and diabetic acidosis
Creatinine clearance	100–140 mL/min	Values increase in renal diseases
Creatinine	1–2 g/24 h	Values increase in infections and decrease in muscular atrophy, anemia, and certain kidney diseases
Glucose	0	Values increase in diabetes mellitus and certain pituitary gland disorders
Urea clearance	Over 40 mL of blood cleared of urea per minute	Values increase in certain kidney diseases
Urea	25–35 g/24 h	Values decrease in complete biliary obstruction and severe diarrhea; values increase in liver diseases and hemolytic anemia
Uric acid	0.6–1 g/24 h	Values increase in gout and decrease in certain kidney diseases
Casts		
Epithelial	Occasional	Increase in nephrosis and heavy-metal poisoning
Granular	Occasional	Increase in nephritis and pyelonephritis
Hyaline	Occasional	Increase in glomerular membrane damage and fever
Red blood cell	Occasional	Values increase in pyelonephritis; blood cells appear in urine in response to kidney stones and cystitis
White blood cell	Occasional	Values increase in kidney infections
Color	Amber, straw, transparent yellow	Varies with hydration, diet, and disease states
Odor	Aromatic	Becomes acetonelike in diabetic ketosis
Osmolality	500–800 mOsm/kg water	Values decrease in aldosteronism and diabetes insipidus; values increase in high-protein diets, heart failure, and dehydration
pH	4.6–8	Values decrease in acidosis, emphysema, starvation, and dehydration; values increase in urinary tract infections and severe alkalosis

Table E.4 Hormone Levels

Test	Normal Values
Steroid hormones	
Aldosterone	Excretion: 5–19 mg/24 h[*]
Fasting at rest, 210 mEq sodium diet	Supine: 48 ± 29 pg/mL[†]
	Upright: 65 ± 23 pg/mL
Fasting at rest, 10 mEq sodium diet	Supine: 175 ± 75 pg/m
	Upright: 532 ± 228 pg/mL
Cortisol	
Fasting	8 A.M.: 5–25 µg/100 mL
At rest	8 P.M.: Below 10 µg/100 mL
Testosterone	Adult male: 300–1100 ng/100 mL[‡]
	Adolescent male: over 100 ng/100 mL
	Female: 25–90 ng/100 mL
Peptide hormones	
Adrenocorticotropin (ACTH)	15–170 pg/mL
Calcitonin	Undetectable in normals
Growth hormone (GH)	
Fasting, at rest	Below 5 ng/mL
After exercise	Children: over 10 ng/mL
	Male: below 5 ng/mL
	Female: up to 30 ng/mL
Insulin	
Fasting	6–26 µU/mL
During hypoglycemia	Below 20 µU/mL
After glucose	Up to 150 µU/mL
Luteinizing hormone (LH)	Male: 6–18 mU/mL
	Preovulatory or postovulatory female: 5–22 mU/mL
	Midcycle peak 30–250 mU/mL
Parathyroid hormone	Less than 10 microl equiv/L
Prolactin	2–15 ng/mL
Renin activity	
Normal diet	
Supine	1.1 ± 0.8 ng/mL/h
Upright	1.9 ± 1.7 ng/mL/h
Low-sodium diet	
Supine	2.7 ± 1.8 ng/mL/h
Upright	6.6 ± 2.5 ng/mL/h
Thyroid-stimulating hormone (TSH)	0.5–3.5 µU/mL
Thyroxine-binding globulin	15.25 µg T_4/100 mL
Total thyroxine	4–12 µg/100 mL

[*]1 microgram (1 µg) is equal to 10^{-6} g.
[†]1 picogram (1 pg) is equal to 10^{-12} g.
[‡]1 nanogram (1 ng) is equal to 10^{-9} g.

Appendix F

Answers To Review and Comprehension Questions

Chapter One

1. a; 2. b; 3. a; 4. c; 5. d; 6. e; 7. a; 8. b; 9. c; 10. d; 11. d; 12. c; 13. d; 14. d; 15. a; 16. b; 17. b; 18. a; 19. e; 20. c; 21. a; 22. b; 23. a; 24. b; 25. e

Chapter Two

1. e; 2. a; 3. b; 4. a; 5. a; 6. d; 7. b; 8. e; 9. c; 10. e; 11. c; 12. d; 13. a; 14. b; 15. c; 16. c; 17. d; 18. c; 19. d; 20. e; 21. b; 22. a; 23. b; 24. d; 25. b; 26. a; 27. c; 28. d; 29. a; 30. e

Chapter Three

1. a; 2. e; 3. c; 4. e; 5. e; 6. c; 7. e; 8. d; 9. b; 10. b; 11. a; 12. b; 13. b; 14. a; 15. d; 16. d; 17. b; 18. b; 19. c; 20. e; 21. c; 22. c; 23. b; 24. c; 25. d; 26. c; 27. a; 28. d; 29. c;

Chapter Four

1. e; 2. c; 3. a; 4. a; 5. b; 6. d; 7. c; 8. d; 9. d; 10. b; 11. a; 12. b; 13. b; 14. e; 15. a; 16. d; 17. b; 18. b; 19. d; 20. d; 21. a; 22. b; 23. e; 24. b; 25. c; 26. c; 27. b; 28. c; 29. b; 30. d

Chapter Five

1. e; 2. a; 3. b; 4. a; 5. b; 6. b; 7. d; 8. e; 9. b; 10. a; 11. c; 12. d; 13. e; 14. c; 15. d; 16. b; 17. c; 18. c; 19. d; 20. b; 21. c; 22. b; 23. a; 24. b.; 25. a; 26. d; 27. c; 28. d; 29. c; 30. c

Chapter Six

1. e; 2. e; 3. d; 4. a; 5, c; 6. e; 7. d; 8. b; 9. d; 10. c; 11. e; 12. a; 13. a; 14. b; 15. e; 16. c; 17. b; 18. c; 19. e; 20. d; 21. a; 22. b; 23. c; 24. e; 25. a; 26. e; 27. c; 28. e; 29. d; 30. a

Chapter Seven

1. c; 2. c; 3. e; 4. d; 5. c; 6. d; 7. c; 8. a; 9. d; 10. a; 11. b; 12. a; 13. a; 14. c; 15. a; 16. b; 17. c; 18. a; 19. d; 20. e; 21. c; 22. b; 23. c; 24. a; 25. b; 26. c; 27. b; 28. c; 29. a; 30. a

Chapter Eight

1. e; 2. b; 3. d; 4. b; 5. a; 6. e; 7. d; 8. e; 9. c; 10. d; 11. b; 12. a; 13. b; 14. e; 15. a; 16. d; 17. c; 18. d; 19. e; 20. a; 21. c; 22. b; 23. c; 24. d; 25. b; 26. b; 27. a; 28. c; 29. c

Chapter Nine

1. c; 2. e; 3. a; 4. d; 5. c; 6. e; 7. b; 8. a; 9. b; 10. b; 11. d; 12. b; 13. d; 14. c; 15. e; 16. e; 17. d; 18. c; 19. c; 20. a; 21. d; 22. b; 23. d; 24. a; 25. d; 26. c; 27. c; 28. c; 29. b; 30. a

Chapter Ten

1. d; 2. b; 3. c; 4. c; 5. c; 6. c; 7. d; 8. d; 9. b; 10. c; 11. a; 12. a; 13. c; 14. d; 15. a; 16. b; 17. d; 18. b; 19. a; 20. d; 21. a; 22. c; 23. a; 24. a; 25. e; 26. c; 27. b; 28. b; 29. d

Chapter Eleven

1. a; 2. b; 3. b; 4. a; 5. c; 6. c; 7. b; 8. a; 9. c; 10. a; 11. b; 12. d; 13. a; 14. c; 15. b; 16. a; 17. a; 18. b; 19. d; 20. c; 21. e; 22. b; 23. d; 24. d; 25. b; 26. b; 27. e; 28. e; 29. e; 30. a

Chapter Twelve

1. d; 2. c; 3. c; 4. d; 5. b; 6. b; 7. d; 8. c; 9. c; 10. b; 11. e; 12. c; 13. d; 14. d; 15. d; 16. e; 17. c; 18. d; 19. c; 20. d

Chapter Thirteen

1. c; 2. b; 3. e; 4. c; 5. d; 6. b; 7. b; 8. b; 9. c; 10. a; 11. e; 12. d; 13. d; 14. a; 15. c; 16. d; 17. b; 18. b; 19. a; 20. a; 21. d; 22. a; 23. b; 24. b; 25. e; 26. b; 27. c; 28. e; 29. c; 30. b; 31. b

Chapter Fourteen

1. c; 2. d; 3. d; 4. c; 5. a; 6. d; 7. b; 8. e; 9. a; 10. b; 11. b; 12. e; 13. c; 14. c; 15. b; 16. d; 17. a; 18. b; 19. b; 20. d; 21. b; 22. b; 23. d; 24. e; 25. d; 26. b; 27. b; 28. d; 29. e; 30. c

Chapter Fifteen

1. e; 2. e; 3. c; 4. a; 5. b; 6. e; 7. b; 8. a; 9. c; 10. d; 11. d; 12. b; 13. c; 14. d; 15. c; 16. b; 17. a; 18. d; 19. c; 20. d; 21. e; 22. a; 23. d; 24. c; 25. a; 26. a; 27. b; 28. d; 29. a; 30. c

Chapter Sixteen

1. e; 2. d; 3. d; 4. a; 5. e; 6. b; 7. d; 8. e; 9. d; 10. d; 11. d; 12. a; 13. a; 14. c; 15. e; 16. a; 17. d; 18. c; 19. e; 20. c

Chapter Seventeen

1. c; 2. c; 3. b; 4. e; 5. d; 6. b; 7. a; 8. e; 9. b; 10. a; 11. c; 12. e; 13. c; 14. e; 15. a; 16. d; 17. c; 18. d; 19. a; 20. c

Chapter Eighteen

1. e; 2. d; 3. e; 4. b; 5. a; 6. c; 7. b; 8. b; 9. e; 10. c; 11. d; 12. e; 13. a; 14. d; 15. b; 16. c; 17. a; 18. a; 19. d; 20. c; 21. d; 22. e; 23. e; 24. d; 25. b; 26. d; 27. a; 28. d; 29. a; 30. b; 31. e; 32. c; 33. c; 34. e; 35. d

Chapter Nineteen

1. e; 2. c; 3. a; 4. e; 5. d; 6. a; 7. e; 8. d; 9. b; 10. c; 11. d; 12. b; 13. b; 14. a; 15. c; 16. b; 17. b; 18. e; 19. e; 20. b; 21. a; 22. a; 23. d; 24. a; 25. d; 26. c; 27. c; 28. c; 29. d; 30. b

Chapter Twenty

1. d; 2. e; 3. c; 4. a; 5. b; 6. d; 7. c; 8. c; 9. e; 10. a; 11. c; 12. d; 13. a; 14. b; 15. e; 16. a; 17. a; 18. b; 19. e; 20. b; 21. c; 22. d; 23. a; 24. a; 25. d; 26. c; 27. c; 28. c; 29. e; 30. c

Chapter Twenty-One

1. b; 2. a; 3. a; 4. d; 5. d; 6. b; 7. c; 8. d; 9. a; 10. a; 11. d; 12. e; 13. c; 14. a; 15. c; 16. b; 17. b; 18. d; 19. a; 20. e; 21. d; 22. b; 23. b; 24. b; 25. d; 26. a; 27. d; 28. c; 29. b; 30. e

Chapter Twenty-Two

1. d; 2. c; 3. a; 4. b; 5. a; 6. e; 7. d; 8. d; 9. e; 10. b; 11. a; 12. d; 13. d; 14. e; 15. d; 16. b; 17. e; 18. e; 19. a; 20. d; 21. c; 22. a; 23. b; 24. c; 25. a; 26. d; 27. d; 28. d; 29. d; 30. c

Chapter Twenty-Three

1. a; 2. b; 3. e; 4. c; 5. e; 6. c; 7. d; 8. b; 9. c; 10. b; 11. d; 12. a; 13. d; 14. c; 15. c; 16. d; 17. c; 18. d; 19. b; 20. d; 21. b; 22. c; 23. b; 24. b; 25. c; 26. a; 27. c; 28. a; 29. d; 30. d

Chapter Twenty-Four

1. a; 2. d; 3. e; 4. d; 5. a; 6. b; 7. a; 8. b; 9. a; 10. c; 11. b; 12. d; 13. c; 14. d; 15. e; 16. b; 17. d; 18. d; 19. e; 20. e; 21. b; 22. b; 23. b; 24. a; 25. e; 26. e; 27. e; 28. a; 29. c; 30. a

Chapter Twenty-Five

1. d; 2. c; 3. a; 4. e; 5. b; 6. d; 7. e; 8. e; 9. b; 10. a; 11. d; 12. b; 13. d; 14. a; 15. e; 16. c; 17. a; 18. b; 19. a; 20. b

Chapter Twenty-Six

1. d; 2. b; 3. d; 4. c; 5. e; 6. e; 7. b; 8. b; 9. d; 10. e; 11. a; 12. b; 13. b; 14. b; 15. a; 16. a; 17. c; 18. d; 19. c; 20. b; 21. e; 22. d; 23. a; 24. d; 25. d; 26. c; 27. e; 28. a; 29. b; 30. c

Chapter Twenty-Seven

1. b; 2. a; 3. b; 4. c; 5. d; 6. a; 7. a; 8. a; 9. b; 10. a; 11. b; 12. d; 13. a; 14. a; 15. b; 16. a; 17. a; 18. c; 19. c; 20. d

Chapter Twenty-Eight

1. e; 2. a; 3. a; 4. b; 5. d; 6. e; 7. b; 8. b; 9. d; 10. c; 11. b; 12. c; 13. a; 14. e; 15. a; 16. b; 17. e; 18. c; 19. b; 20. a; 21. a; 22. c; 23. a; 24. c; 25. e; 26. e; 27. c; 28. d; 29. a; 30. a

Chapter Twenty-Nine

1. b; 2. e; 3. a; 4. a; 5. d; 6. a; 7. c; 8. c; 9. b; 10. c; 11. a; 12. a; 13. e; 14. d; 15. c; 16. c; 17. e; 18. a; 19. b; 20. d; 21. a; 22. c; 23. d; 24. c; 25. c; 26. b; 27. c; 28. c; 29. d

Appendix G

Answers to Critical Thinking Questions

Chapter 1

1. Student B is correct. Body temperature begins to rise as a result of exposure to the hot environment. Sweating eliminates heat from the body and lowers body temperature. Body temperature returning to its ideal normal value is an example of negative feedback. Student A probably thought that it was positive feedback because sweating continued to increase. Sweating, however, is the response. The variable being regulated by sweating is body temperature.

2. Answer *e* is correct. Positive-feedback mechanisms result in movement away from homeostasis and are usually harmful. The continually decreasing blood pressure is an example. Negative-feedback mechanisms result in a return to homeostasis. The elevated heart rate is a negative-feedback mechanism that attempts to return blood pressure back to a normal value. In this case, the negative-feedback mechanism was inadequate to restore homeostasis, and medical intervention (a transfusion) was necessary.

3. When a boy is standing on his head, his nose is superior to his mouth. Remember that directional terms refer to a person in the anatomic position, not to the body's current position.

4. a. Inferior or caudal
   b. Anterior, ventral, or superficial
   c. Proximal, superior, or cephalic
   d. Medial

5. The pancreas is located in the right-upper and left-upper quadrants; it is located in the epigastric and left hypochondriac regions. The top of the urinary bladder is located in the right-lower and left-lower quadrants; it is located in the hypogastric region. The rest of the urinary bladder is located in the pelvic cavity.

6. Answer *a* is correct. The best way to reach the anterior surface of the heart begins with the patient lying on his or her back so that the anterior surfaces of the thorax and heart are facing the surgeon. The heart is located in the anterior portion of the thoracic cavity within the mediastinum and is surrounded by the pericardial cavity. The pericardial cavity is lined with the pericardial serous membranes, through which a cut can be made to reach the heart.

7. The uterus is located in the pelvic cavity. The pelvic cavity, however, is surrounded by the bones of the pelvis and doesn't increase in size during pregnancy. Instead, as the fetus grows the expanding uterus must move into the abdominal cavity, thereby crowding abdominal organs and dramatically increasing the size of the abdominal cavity.

8. After passing through the left thoracic wall, the first membrane encountered is the parietal pleura. Continuing through the pleural cavity the visceral pleura of the left lung and then the left lung are pierced. Leaving the lung the bullet penetrates the visceral pleura, the pleural cavity, and the parietal pleura (remember that the lung is surrounded by a double-membrane sac). Next the parietal pericardium, the pericardial cavity, the visceral pericardium, and the heart are penetrated.

9. After passing through the abdominal wall, the parietal peritoneum is pierced. In passing through the stomach, the visceral peritoneum, the stomach itself, and the visceral peritoneum on the other side of the stomach are penetrated. Because the diaphragm is lined inferiorly by parietal peritoneum and superiorly by parietal pleura, these are the next two membranes pierced. The pole then passes through the pleural space and visceral pleura to enter the lung.

Chapter 2

1. An atom of iron has 26 protons (the atomic number), 30 neutrons (the mass number minus the atomic number), and 26 electrons (because the number of electrons is equal to the number of protons). If an atom of iron loses three electrons, it has three more protons (positive charges) than electrons (negative charges). Therefore the iron ion has an overall charge of $+3$, which is represented symbolically as Fe^{3+}.

2. The formation of free fatty acids and glycerol from a triglyceride is a decomposition reaction because a larger molecule breaks down into smaller molecules. All of the decomposition reactions in the body are collectively referred to as catabolism. This reaction can also be classified as a hydrolysis reaction because as part of the reaction a water molecule is split into hydrogen, which becomes part of the glycerol molecule, and hydroxide, which becomes part of a fatty acid molecule.

3. The slight amount of heat functions as activation energy and starts a chemical reaction. The reaction releases a large amount of heat, thus causing the solution to become hot.

4. Heating (boiling) has destroyed the ability of the molecules in one or both of the solutions to function in the chemical reaction. This is called denaturation. There are two possibilities as to what is denatured. It could be the reactants themselves or an enzyme that catalyzes the reaction.

5. Muscle contains proteins. To increase muscle mass, proteins must be synthesized from amino acids. The synthesis of molecules in living organisms requires the input of energy. That energy comes from the potential energy stored in the chemical bonds of food molecules, which is released during the decomposition of food molecules.

6. Remember that pH is a measure of H^+ concentration. If equal amounts of solutions A and B are mixed, the resulting hydrogen ion concentration is the average value of the two solutions, that is, the pH is $(8 + 2)/2 = 5$. A pH of 5 is acidic. This question illustrates an important point: The pH of a solution can be changed by adding a more acidic or basic solution to it.

7. The $NaHCO_3$ dissociates, thereby increasing the amount of HCO_3^- in the solution. The HCO_3^- combines with H^+ to form H_2CO_3, which becomes CO_2 and H_2O. The decrease in H^+ causes the pH of the solution to increase (become more alkaline).

8. As A and B are added to the solution, the enzyme E catalyzes the formation of C. However, when C binds to the active site of E, the ability of E to catalyze the formation of C is blocked. As more and more C is produced, the rate of formation of C is slowed. Because the reaction of C with E is reversible, there will always be some E that has a functional (not blocked) active site, and some A will therefore always combine with B.

9. One might try heating the substances because proteins can be denatured and can coagulate (like frying an egg). Another possibility is to try dissolving the substances in water. Most lipids are insoluble in water, while many proteins are either soluble in water or form colloids with water.

10. Most proteins (i.e., a typical protein) contain sulfur, which is not found in phospholipids. Typical phospholipids and proteins contain carbon, hydrogen, oxygen, nitrogen, and phosphate.

Chapter 3

1. The cells within the wound swell with water and lyse by the introduction of a hypotonic solution. This kills potentially metastatic cells that may still be present in the wound.

2. Water moves by osmosis from solution B into solution A. Because solution A is hyperosmotic to solution B, solution A has more solutes and less water than does solution B. Water therefore moves from solution B (with more water) to solution A (with less water).

3. Answer *b* is correct. Because the solution is isotonic, there is no exchange of water. Because the solution contains the same concentration of all substances except that it has no urea, only a net movement of urea occurs across the membrane.

4. Answer *b* is correct. At point A on the graph, the extracellular concentration is equal to the intracellular concentration. If movement were by simple diffusion or by facilitated diffusion, at this point the rate of movement would be zero. Because it is not zero, it's reasonable to conclude that the mechanism involved is active transport.

5. A reduced intracellular K^+ concentration reduces the concentration difference for K^+

across the plasma membrane. Thus, the rate at which K^+ diffuses out of the cell is reduced, and a smaller charge difference is required across the plasma membrane to oppose the diffusion of the K^+ out of the cell. The potential difference across the plasma membrane is therefore reduced.

6. Because the drug inhibits mRNA synthesis, protein synthesis is stopped. If the cell releases proteins as they were synthesized, the rate of protein secretion dramatically decreases following the administration of the drug. On the other hand, if the cell releases stored proteins, the rate of secretion at first is normal and then gradually declines.

7. The well-developed rough endoplasmic reticulum is indicative of protein synthesis, and a well-developed Golgi apparatus is indicative of secretion. It's likely that this cell synthesizes and secretes proteins.

Chapter 4

1. The tissue is epithelial tissue, as it is lining a free surface, and the epithelium is stratified because it consists of more than one layer. The types of stratified epithelium are stratified squamous, stratified cuboidal, stratified columnar, or transitional epithelium. The structure of the cells in the surface layers enables the determination of a specific tissue type. Flat cells in the surface layer indicate stratified squamous epithelium. Cuboidal cells in the surface layer indicate stratified cuboidal epithelium, and columnar cells in the surface layer point to stratified columnar epithelium. The surface cells of transitional epithelium are roughly cuboidal with cubelike or columnar cells beneath them. When transitional epithelium is stretched, the surface cells are still roughly cuboidal, but underlying layers can be somewhat flattened.

2. In general, epithelial cells undergo cell division (mitosis) in response to injury, and the newly produced cells replace the damaged cells. If the basement membrane is destroyed, however, nothing is present to provide scaffolding for the newly formed epithelial cells. Without the basement membrane, there's not an effective way for the newly formed epithelial cells to repair a structure such as a kidney tubule. Since the basement membranes appear to be mostly present, the person is likely to survive and the kidney will regain most of its ability to function.

3. Epithelium that functions to resist abrasion is stratified squamous epithelium. The nonkeratinized stratified squamous epithelium lining the mouth and the keratinized stratified squamous epithelium of the skin are examples. The cells at the surface are flattened, and when scraped away due to abrasion they are replaced by the cells beneath them. In contrast, epithelial cells that carry out absorption are either simple cuboidal or simple columnar. Because they are one layer thick, they are more susceptible to damage and are not resistant to abrasion. In addition, these cells are large in volume, which allows them to contain the organelles involved in transport, such as mitochondria to produce ATP in the case of active transport. The surface of the cells that absorb are likely to contain microvilli, which increases the surface area for absorption. The flat cells that resist abrasion have no microvilli.

4. Glands producing merocrine secretions do so with no loss of actual cellular material, whereas glands producing holocrine secretions shed entire cells. The cells rupture and die, and the entire cell becomes part of the secretion. You could chemically analyze the secretions for the types of molecules found in cellular organelles. For example, if phospholipids and proteins normally found in membranes are in the secretion, then the secretion is a holocrine secretion. If the secretion is watery or contains products that are not found in membranes or organelles, it's a merocrine secretion.

5. The tissue described is dense regularly arranged collagenic connective tissue. Injury to this type of tissue affects structures made up of this type of connective tissue, which includes tendons. It does not include ligaments that connect neck vertebrae. These ligaments contain abundant dense regularly arranged elastic connective tissue. The majority of the intervertebral disk consists of dense irregularly arranged collagenic connective tissue.

6. The statement is not appropriate. A tissue capable of contracting is muscle. Both cardiac muscle and smooth muscle cells are mononucleated, although some cardiac muscle cells can have two nuclei, and they are both under involuntary control. Cardiac muscle is striated, and smooth muscle is not, however.

7. Histamine is one of the mediators of inflammation released in response to tissue damage. Several other mediators of inflammation, however, are released during inflammation in addition to histamine. Antihistamines might reduce the inflammatory response somewhat, but it's not likely to have a major effect because of the other mediators of inflammation released at the same time. In certain types of inflammatory responses, such as allergic responses, histamines are released in large amounts. Under these conditions, antihistamines do reduce the inflammatory response.

Chapter 5

1. Yes, the skin (dermis) can be overstretched due to obesity or rapid growth.

2. The stratum corneum, the outermost layer of the skin, consists of many rows of flat, dead epithelial cells. The many rows of cells, which are continuously shed and replaced, are responsible for the protective function of the integument. In infants, there are fewer rows of cells, resulting in skin that's more easily damaged than that in adults.

3. Melanocytes produce melanin, which protects underlying tissue from ultraviolet radiation. Therefore, we expect melanocytes to be as superficial as possible. Also recall that melanin production varies depending on exposure to the sun. Response to stimulation is a characteristic of living cells. Thus, melanocytes should be found in the most superficial living layer of the epidermis, the stratum basale.

4. Alcohol is a solvent that dissolves lipids (see chapter 2). It removes the lipids from the skin, especially in the stratum corneum. The rate of water loss increases after soaking the hand in alcohol because of the removal of the lipids that normally prevent water loss.

5. Carotene, a yellow pigment from ingested plants, accumulates in lipids. The stratum corneum of a callus has more layers of cells than other noncallused parts of the skin, and the cells in each layer are surrounded by lipids. The carotene in the lipids makes the callus appear yellow.

6. When first exposed to the cold temperature just before starting the run, the blood vessels in the skin constrict to conserve heat. This produces a pale skin color. Dilation of the skin blood vessels doesn't occur at this time because the skin has not been exposed to the cold long enough to cause the skin temperature to fall below 15°C. After running for awhile, as a result of the excess heat generated by the exercise, the blood vessels in the skin dilate. This results in heat loss and helps to prevent overheating. Increased blood flow through the skin causes it to turn red. After the run, the body still has excess heat to eliminate, so the skin remains red for some time.

7. Eyelashes have a short growth stage (30 days) and are therefore short. Fingernails grow continuously but are short because they are cut, broken off, or worn down.

8. The hair follicle, but not the hair, is surrounded with nerve endings that can detect movement or pulling of the hair. The hair is dead, keratinized epithelium, so cutting the hair is not painful.

9. Several methods have some degree of success in treating acne: (a) Kill the bacteria. One effective agent is benzoyl peroxide, found in some acne medications. (b) Prevent blockage of the hair follicle. A vitamin A derivative (tretinoin; Retin-A) has proven effective in keeping the follicular epithelial cells and sebum from building up and closing off the hair follicle. (c) Unplug the follicle. Some sulfur compounds (Acnederm) speed up peeling of the skin and thus unplug the follicle.

10. Probably not, because following removal of the nail from the nail fold, it may grow back into the nail fold and the ingrown toenail would reoccur. One solution is to remove the small part of the nail responsible for the ingrown toenail. Prior to this drastic approach, sterile gauze can be placed between the nail and the nail fold to force the nail away from the nail fold. After the nail fold is healed, the gauze can be removed.

11. Rickets is a disease of children resulting from inadequate vitamin D. With inadequate vitamin D there is insufficient absorption of calcium from the intestine, resulting in soft bones. If adequate vitamin D is ingested, rickets is prevented, whether one is dark or fair skinned. However, if dietary vitamin D is inadequate, when the skin is exposed to ultraviolet light, 7-dehydrocholesterol is converted into cholecalciferol, which can be converted to vitamin D. Dark-skinned children are more susceptible to rickets because the additional melanin in their skin screens out the ultraviolet light and they produce less vitamin D.

Chapter 6

1. Normally bone matrix and bone trabeculae are organized to be strongest along lines of stress. Random organization of the collagen fibers of bone matrix results in weaker bones. In addition, the reduced amount of trabecular bone makes the bone weaker. Fractures of the bone can occur when the weakened bone is subjected to stress.

2. Replacement of cartilage of the epiphyseal plate by bone normally occurs on the diaphyseal side of the plate. As growth ceases, the cartilage cells stop dividing and producing new cartilage. Replacement of cartilage with bone continues from the diaphyseal side, and eventually all of the cartilage of the plate is bone.

3. Mechanical stress applied to bone stimulates osteoblast activity, so the patient with a walking cast should heal faster.

4. Osteoporosis is depletion of bone matrix that results when more bone is destroyed than is formed. Because mechanical stress stimulates bone formation (osteoblast activity), running helps to prevent osteoporosis in the bones being stressed. This includes the bones of the lower limbs and the spine.

5. The loss of bone density results because the bones are not bearing weight in the weightless environment. Therefore osteoblasts are not sufficiently stimulated and bone resorption exceeds bone building. Bone loss can be slowed by stressing the bones using exercises against resistance such as cycling.

6. The kidneys are the site of production of active vitamin D (see chapter 5), which is needed for calcium absorption in the small intestine. Kidney failure can result in inadequate vitamin D production, too little uptake of calcium, and therefore osteomalacia.

7. Testosterone normally causes a spurt of growth at puberty followed by slower growth and closure of the epiphyseal plate. Without testosterone, growth is slower, but proceeds longer, resulting in a taller-than-normal person.

8. Blood vessels in central canals run parallel to the long axis of the bone, and perforating canals run at approximately a right angle to the central canals. Thus, perforating canals connect to central canals, which allows blood vessels in the perforating canals to connect with blood vessels in the central canals. After a fracture, blood flow through the central canals stops back to the point where the blood vessels in the central canals connect to the blood vessels in the perforating canals. The regions of bone on either side of the fracture associated with this lack of blood delivery die.

9. Hyperthyroidism stimulates increased bone breakdown and could cause osteitis fibrosa cystica, a condition in which the bone is eaten away as Ca^{2+} are released from the bone. The result can be a deformed bone that is likely to fracture. Vitamin D therapy might help because vitamin D promotes an increase in blood Ca^{2+} levels (see chapter 5) and therefore increased deposition of Ca^{2+} in bone.

Chapter 7

1. An infection in the nasal cavity could spread to adjacent cavities and fossae, including the paranasal sinuses: (1) frontal, (2) maxillary, (3) ethmoidal, and (4) sphenoidal; (5) the orbit (through the nasolacrimal duct); (6) the cranial cavity (through the cribriform plate); and (7) the throat (through the posterior opening of the nasal cavity).

2. Falling on the top of the head could drive the occipital condyles into the superior articulating processes of the atlas, causing a fracture. An uppercut to the jaw would slightly lift the occipital condyles away from the superior articulating processes of the atlas and usually doesn't result in a fractured atlas. Such a blow to the jaw can, however, fracture the temporal bone where it articulates with the mandible.

3. Forceful rotation of the vertebral column is most likely to damage the articular processes, especially in the lumbar region, where the articular processes tend to prevent excessive rotation (the superior articular processes face medially and the inferior articular processes face laterally).

4. Weaker back muscles on one side could cause the vertebral column to bend laterally (scoliosis) toward the opposite side. Lordosis can result from pregnancy. As the fetus causes the abdomen to move anteriorly, the thorax and head tend to pull posteriorly, to restore the center of gravity. This posture increases the lumbar curvature. The same effect can be seen in people who are "pot-bellied."

5. If the ulna and radius become fused, the radius can no longer rotate relative to the ulna, and, as a result, most of the rotation of the forearm and hand is lost.

6. Measure from the anterior superior iliac spine (a "stationary" point relative to the limb, which can be easily found as a surface landmark) to the lateral malleolus. The inferior side of the foot could also be used, if the person is standing on a flat surface. A defect of the foot or ankle may occur, however, in which the ankle on one side is elevated. If the length of the thigh is the only part to be measured, measure to the lateral epicondyle.

7. The ischial tuberosity is the bony protuberance.

8. Women's hips are wider than men's. As the knees are positioned toward the midline the slope of the femur from its proximal end toward its distal end is greater in women, and as a result, women tend to be knock-kneed more often than men.

9. The lateral malleolus extends further distally than does the medial malleolus, thus making it more difficult to turn the foot laterally than medially. The styloid process of the radius extends further distally than the styloid process of the ulna, thus making it more difficult to cock the wrist toward the thumb (laterally) than toward the little finger (medially).

10. Landing on the heels could fracture the calcaneus. Heavy objects, such as Hefty Stomper, landing on the dorsal surface of the foot could fracture the metatarsals or even the tarsals.

Chapter 8

1. If the sternocostal synchondrosis were to ossify, becoming a synostosis, there would no longer be any stretch through the costal cartilage, the thorax could not expand, and, as a result, respiration would be severely hampered.

2. a. Suture, little or no movement.
   b. Syndesmosis, some movement.
   c. Complex synovial joints: the humeroulnoradial joint is a hinge joint, the radioulnar joint is a pivot joint. All have considerable movement.

3. a. Flexion and supination.
   b. Flexion of the hip and extension of the knee.
   c. Abduction of the arm at the shoulder.
   d. Flexion of the knee and plantar flexion of the foot.

4. The anterior drawer test determines the integrity of the anterior cruciate ligament, and the posterior drawer test determines the integrity of the posterior cruciate ligament. Unusual movement during the posterior drawer test indicates damage to the posterior cruciate ligament.

Chapter 9

1. Botulism poisoning results from the consumption of botulism toxin produced by the bacterium *Clostridium botulinum.* The toxin binds to presynaptic nerve terminals and prevents the release of acetylcholine. Thus, action potentials in nerves cannot produce action potentials in skeletal muscles, and the result is paralysis of skeletal muscles, which explains the difficulty in breathing and swallowing. Other reasonable explanations are that the toxin binds to and blocks the receptors for acetylcholine, that the toxin blocks the entry of Ca^{2+} into the presynaptic terminal and thus prevents acetylcholine release, or that the toxin specifically prevents entry of ions through Na^+ channels of skeletal muscle cells.

2. Muscular dystrophy results from gradual atrophy of skeletal muscle fibers and their replacement with connective tissue. Myasthenia gravis results from the degeneration of the receptors for acetylcholine on the postsynaptic membranes of skeletal muscle cells. If an inhibitor of acetylcholinesterase is administered, the result should be an increase in the concentration of acetylcholine in the nerve muscle synapse. Thus, more acetylcholine is available to bind to acetylcholine receptors. In people suffering from myasthenia gravis, the increased concentration of acetylcholine in the synapse allows acetylcholine to bind a greater percentage of the acetylcholine receptors present and causes the muscle contractions to increase in strength. In people who have muscular dystrophy, the muscle contractions don't increase in strength because muscle atrophy is the cause of the weakness. The additional acetylcholine in the neuromuscular synapse has no effect on the weakened muscle fibers.

3. Placing sarcoplasmic reticulum from skeletal muscle cells into the beaker would remove calcium from the solution because sarcoplasmic reticulum transports Ca^{2+} from the solution into the sarcoplasmic reticulum. In addition, ATP would have to be added for two reasons: (1) the sarcoplasmic reticulum actively transports calcium and, therefore, requires ATP; and (2) ATP must bind to the heads of the myosin

molecules before the myosin heads can release from the active sites on the actin molecules.

4. A lower-than-normal temperature decreases the rate of all of the processes that occur in the lag phase of muscular contraction because a lower temperature decreases the rate of all chemical reactions and the rate of ion diffusion. As a consequence, the lag phase requires a longer time.

5. Start with a subthreshold stimulus and increase the stimulus strength by very small increments. Apply the stimulus to the nerve of muscle A and muscle B. If the number of motor units is the same for both preparations, each time the stimulus strength is increased the degree of tension produced by the muscles would also increase to the same degree in each muscle. If one muscle has more motor units than the other, the muscle with the greater number of motor units would exhibit a greater number of separate increases in tension, and the magnitude of the increases in tension would be smaller than those seen in the muscle with fewer motor units.

6. When a muscle slowly lifts an object, the contraction starts with a small number of motor units being stimulated. Each motor unit is stimulated tetanically. As the contraction continues, more and more motor units are recruited to lift the object slowly. To lower the object, the number of motor units stimulated tetanically is reduced slowly and the tension produced by the muscle decreases. In a muscle twitch, a stimulus causes a single action potential in all of the muscle fibers responding to the stimulus. The stimulated muscle fibers contract in an all-or-none fashion and then relax. Both contraction and relaxation occur quickly.

7. The shape of an active tension curve for skeletal muscle can be seen in figure 9.21. In contrast, an active tension curve is much flatter for smooth muscle. That is, for each increase in the length of a muscle fiber there is little change in the active tension produced by the smooth muscle fiber. Smooth muscle has, as one of its major characteristics, the ability to increase in length without much increase in the tension produced by the smooth muscle cells.

8. Both the 100 m run and weight lifting involve rapid and intense contractions of skeletal muscles that are completed quickly. These contractions depend on anaerobic metabolism for a significant amount of the ATP produced. In contrast, the 10,000 m run involves sustained muscular contractions that are not as rapid, but the slower contractions are repeated many times during the run. Aerobic metabolism produces the majority of the ATP for the 10,000 m run. Anaerobic metabolism is associated with a decrease in creatine phosphate, an increase in creatine, an increase in lactic acid, and a decrease in glycogen, and the enzymes responsible for anaerobic metabolism function more rapidly. Aerobic metabolism is associated with increased enzyme activity in the mitochondria and an increase in carbon dioxide production. Oxygen is used more rapidly during aerobic metabolism.

9. During intense exercise it's possible to experience physiologic contracture. Being unable to either contract or relax the muscles for a short time while exercising suggests the existence of physiologic contracture.

10. Smooth muscle contractions in the stomach result from the autorhythmic contraction of visceral smooth muscle tissue. Because stretching this type of muscle doesn't dramatically affect the force of the rhythmic contractions, there is little difference between the contractions 5 minutes and 1 hour after a meal.

11. During the 100 m race Shorty depended on ATP produced by anaerobic metabolism. That produced an oxygen debt at the end of the run, which resulted in an elevated rate of respiration for a time. During the longer and slower run, most of the ATP for muscle contractions was produced by aerobic respiration, and very little oxygen debt developed. Prolonged aerobic respiration is required to "pay back" the oxygen debt. Shorty's rate of respiration was, therefore, prolonged after the 100 m race but not after the longer but slower run.

12. High blood K^+ concentration also results in depolarization of smooth muscle plasma membranes. Depolarization of the smooth muscle plasma membrane results in increased muscle contractions and increased permeability of the plasma membrane to both Na^+ and Ca^{2+}, which results in further depolarization and an increase in the intracellular concentration of Ca^{2+}. These changes result in the production of action potentials and muscle contractions.

13. The muscles would contract. ATP would be available to bind to the myosin heads, thus allowing myosin molecules to be released from actin molecules. The cross-bridges would immediately re-form, and complete cross-bridge cycling would result in contraction of the muscle fibers. As long as Ca^{2+} were present at high concentrations in the sarcoplasm, contraction of the muscles would occur. If the sarcoplasmic reticulum were intact, ATP would be available to drive the active transport of Ca^{2+} into the sarcoplasmic reticulum. As the Ca^{2+} decreased in the sarcoplasm, relaxation would result. If the sarcoplasmic reticulum were not intact, however, and could not transport Ca^{2+} into the sarcoplasmic reticulum as fast as they leak out, the muscle would remain contracted until it fatigued.

14. Hormones can bind to ligand-gated Ca^{2+} channels, and the channels, in response, open. Ca^{2+} diffuse into the cell and cause contraction to occur. Only a small amount of depolarization results as Ca^{2+} diffuse into the cell, and since Na^+ channels don't open, a large change in the resting membrane potential doesn't occur.

15. In experiment A, the students used anaerobic respiration as they started to run in place, but aerobic respiration also increased to meet most of their energy needs. When they stopped, their respiration rate was increased over resting levels because of repayment of the oxygen debt due to anaerobic respiration. In experiment B almost all of the student's respiration came from anaerobic respiration because the students held their breath while in running place. Consequently, the students had a much larger oxygen debt. The student's respiratory rate and depth would be greater than in experiment A, or their respiration rates would be elevated for a longer period of time than in experiment A.

Chapter 10

1.

Muscle	Action	Synergist	Antagonist
Longus capitis	Flexes neck	Rectus capitis anterior Longis coli	Most of the posterior neck muscles
Erector spinae	Extends vertebral column	Interspinales Multifidus Semispinalis thoracis	Most anterior abdominal muscles
Coraco-brachialis	Adducts arm	Latissimus dorsi Pectoralis major Teres major Teres minor	Deltoid Supra-spinatus
	Flexes arm	Deltoid (anterior) Pectoralis major Biceps brachii	Deltoid (posterior) Latissimus dorsi Teres major Teres minor Infra-spinatus Sub-scapularis Triceps brachii

2. Biceps brachii: Pull-ups with hands supinated
Triceps brachii: Push-ups
Deltoid: Abduction of the arms to shoulder height, with weights in the hands (abduction past shoulder height involves mostly scapular rotation by the trapezius)
Rectus abdominis: Sit-ups to 45 degrees (sit-ups past 45 degrees involve mostly the psoas major)
Quadriceps femoris: Extending the legs against a force
Gastrocnemius: Plantar flexion of the feet against a force, such as toe raises with a weight on the shoulders

3. The brachioradialis originates on the humerus and inserts onto the distal end of the radius. The fulcrum of this lever system is the elbow joint. With a weight held in the hand, the force, applied between the weight and the fulcrum, is a class III lever system. With the weight on the forearm, the weight is between the force and the fulcrum and is a class II lever system. A greater weight can be lifted if placed on the forearm rather than in the hand, but weights placed on the forearm cannot be lifted as far.

4. The muscles that flex the head also oppose extension of the neck. In an accident causing hyperextension of the neck, these muscles could be stretched and torn. The muscles involved could include the sternoclei-domastoid, longus capitis, rectus capitis anterior, and longus coli. Automobile headrests are designed so that, if adjusted correctly, the back of the head hits the headrest during a rear-end accident, thereby preventing hyperextension of the neck.

5. The only muscle that elevates the lower eyelid is the orbicularis oculi, which "closes the eye." With this muscle not functioning, the lower eyelid would droop. The levator anguli oris, which elevates the angle of the mouth, was also apparently affected allowing the corner of the mouth to droop. The zygomaticus major may also have been affected, as it inserts onto the corner of the mouth (see figure 10.9).

6. The genioglossus muscle protrudes the tongue. If it becomes relaxed, or paralyzed, the tongue may fall back and obstruct the airway. This can be prevented or reversed by pulling forward and down on the mandible, thus opening the mouth. The genioglossus originates on the genu of the mandible. As the mandible is pulled down and forward, the genioglossus is pulled forward with the mandible, thus pulling the tongue forward also.

7. The rotator cuff muscles are the primary muscles holding the head of the humerus in the glenoid fossa, especially the supraspinatus. In fact, a torn rotator cuff, which usually involves a tear of the supraspinatus muscle, often results in dislocation of the shoulder.

8. With the quadriceps femoris paralyzed, the leg could not be extended, and the lower limb could not bear weight unless the knee were passively extended, such as by pushing back on the distal end of the thigh with the hand. Walking would be almost impossible, except by taking very small steps and by pushing back on the knee with each step, or by bracing the knee in an extended position.

9. Speedy has ruptured the calcaneal tendon, and the gastrocnemius and soleus muscles have retracted, thereby causing the abnormal bulging of the calf muscles. Because the major plantar flexors are no longer connected to the calcaneus, the runner cannot plantar flex the foot, and the foot is abnormally dorsiflexed because the antagonists have been disconnected.

Chapter 11

1. A reduced intracellular concentration of K^+ causes depolarization of the resting membrane potential. Because the intracellular concentration of K^+ is reduced, the concentration gradient for K^+ from the inside to the outside of the plasma membrane is also reduced. Thus the rate at which K^+ diffuse out of the cell is reduced, and a smaller charge difference across the plasma membrane is required to oppose the diffusion of the K^+ out of the cell. Therefore, the potential difference across the plasma membrane is reduced, and the cell is depolarized.

2. Because the plasma membrane is much less permeable to Na^+ than to K^+, changes in the extracellular concentration of Na^+ affect the resting membrane potential less than do changes in the extracellular concentration of K^+. Therefore, increases in extracellular Na^+ have a minimal effect on the resting membrane potential. Because the membrane is much more permeable to Na^+ during the action potential, the elevated concentration of Na^+ in the extracellular fluid results in Na^+ diffusing into the cell at a more rapid rate during the action

potential, resulting in a greater degree of depolarization during the depolarization phase of the action potential.

3. Because lithium ions reduce the permeability of plasma membranes to Na^+, the Na^+ channels in the plasma membrane tend to remain closed. A normal stimulus causes Na^+ channels to open, allowing Na^+ to diffuse into the cell, thereby resulting in depolarization. The cell is less sensitive to stimuli because the membrane is less permeable to Na^+.

4. Smooth muscle cells contract spontaneously in response to spontaneous depolarizations that produce action potentials. One way action potentials can be produced spontaneously is if membrane permeability to Na^+ spontaneously increases. As a result, a few Na^+ enter the smooth muscle cells and cause a slight depolarization of the plasma membrane. The small depolarization can cause voltage-gated Na^+ channels to open, which results in further depolarization, thereby stimulating additional Na^+ voltage-gated ion channels to open. This positive-feedback cycle can continue until the plasma membrane is depolarized to its threshold level and an action potential is produced.

5. Action potential conduction along a myelinated nerve fiber is more energy efficient because the action potential is propagated by saltatory conduction, which produces action potentials at the nodes of Ranvier. Compared to an unmyelinated nerve fiber, only a small portion of the myelinated neuron's membrane has action potentials. Thus there is less flow of Na^+ into the neuron (depolarization) and less flow of K^+ out of the neuron (repolarization). Consequently, the Na^+–K^+ pump has to move fewer ions in order to restore ion concentrations. Because the Na^+–K^+ pump requires ATP, myelinated axons use less ATP than unmyelinated axons.

6. The inhibitory neuromodulator causes the postsynaptic neuron to become less sensitive to excitatory stimuli, probably by causing hyperpolarization of the postsynaptic neuron. As a result, the excitatory neurotransmitter released from the excitatory neuron is less likely to produce postsynaptic action potentials.

7. With aging, there's a decrease in the amount of myelin surrounding axons, which decreases the speed of action potential propagation. At synapses there's also an increase in the time it takes for action potentials in the presynaptic terminal to cause the production of action potentials in the postsynaptic membrane. It's believed this results from a reduced release of neurotransmitter by the presynaptic terminal and a reduced number of receptors in the postsynaptic membrane.

8. Organophosphates inhibit acetylcholinesterase, thereby causing an increase in acetylcholine in the synaptic cleft leading to overproduction of action potentials, tetanus of muscles, and possible death resulting from respiratory failure (see chapter 9). Curare is the best antidote because it blocks the effect of acetylcholine and acts to counteract the organophosphate. Too much curare, however, could cause flaccid paralysis of the respiratory muscles. Injecting

acetylcholine would make the effect of the organophosphate worse. Potassium chloride causes depolarization of muscle cell membranes, thereby making them more sensitive to acetylcholine.

9. If the motor neurons supplying skeletal muscle are innervated by both excitatory and inhibitory neurons, then blocking the activity of the inhibitory neurons with strychnine results in overstimulation of the motor neurons by the excitatory neurons.

Chapter 12

1. If the neuron with its cell body in the cerebrum is an inhibitory neuron and if it also synapses with the motor neuron of a reflex arc, then stimulation of the cerebral neuron could inhibit the reflex.

2. The phrenic nerve is cut in the thorax, and the surgery is performed while the lung is being removed.

3. The ulnar nerve supplies the medial third of the hand, little finger, and medial half of the ring finger. The median nerve supplies the lateral two-thirds of the palm and thumb, and the surface of the index, middle, and lateral half of the ring finger. The radial nerve supplies the lateral two-thirds of the dorsum of the hand.

4. Pulling on the upper limb when it is raised over the head can damage the lower brachial plexus, in this case, the origin of the ulnar nerve. The ulnar nerve innervates muscles that abduct/adduct the fingers and flex the wrist.

5. The sciatic nerve has rootlets from L4–S3. Depending on the rootlet compressed, pain can be felt in different locations.

6. a. Obturator nerve
   b. Femoral nerve
   c. Sciatic (tibial) nerve
   d. Obturator nerve
   e. Obturator nerve, some from femoral nerve

7. The plaster cast is pressing against the common fibular (peroneal) nerve at the neck of the fibula. Tingling is expected along the lateral and anterior leg and the dorsum of the foot.

Chapter 13

1. A condition in which a patient loses all sense of feeling in the left side of the back, below the upper limb, and extending in a band around to the chest, also below the upper limb, but where all sensation on the right is normal, suggests that the patient's dorsal roots have been damaged on the left side adjacent to the part of the spinal cord supplying that part of the body. (The basis of this condition is explained more fully in chapter 14.)

2. If CSF does not drain properly, the fluid accumulates and exerts pressure on the brain (hydrocephalus). In the developing fetus, the ventricles enlarge because of the excess fluid pressure. The head also enlarges because the skull bones have not fused. The expansion of the head is not sufficient, however, to relieve all the pressure exerted on the developing brain by the expanding ventricles. As a result, the cerebral cortex becomes proportionately thinner as it's

compressed between the ventricles and skull. In many cases, less gyri form in the cerebral cortex. Brain damage may or may not result, depending on the amount of excess CSF, the ventricular pressure generated, and the areas of the brain damaged by the pressure.

3. Enlargement of the lateral and third ventricles, without enlargement of the fourth ventricle, suggests a blockage between the third and fourth ventricles in the cerebral aqueduct. This defect is called aqueductal stenosis and is a common congenital problem.

4. Blood in the CSF taken through a spinal tap indicates the presence of blood in the subarachnoid space and suggests that the patient has a damaged blood vessel in the subarachnoid space.

Chapter 14

1. The first sensations that occur when a woman picks up an apple and bites into it are visual (special), tactile (general), and proprioceptive (general). The woman holds the apple in her hand and looks at it. The tactile sensations from mechanoreceptors in the hand tell her that the apple is firm and smooth. The proprioceptive sensations originating in the joints of the hand tell the woman the size and shape of the apple. Visual input also tells her the size of the apple, and that it has a smooth surface, as well as its color. As the woman bites into the apple and begins to chew, proprioceptive sensations from the teeth and jaw provide information as to how widely the jaws must be opened to accommodate the bite and how hard to bite down. Tactile sensations originating in the tongue and cheeks tell the person the location of the bite of apple and its texture as it is moved about in the mouth. Taste sensations (special, chemoreceptor) from the tongue provide information that the apple has characteristics of being both sweet and sour. Olfactory sensations (special) provide more specific information that the "fruity taste" is that of an apple.

2. a. The most likely explanation is that the olfactory neurons accommodate and no longer respond to odor stimulus.

   b. The fact that one can hear the sound when one tries indicates that the hair cells in the spiral organ have not accommodated and are still able to detect the sound stimulus. Many action potentials arriving in the brain are prevented from causing conscious perception, until we consciously "pay attention" to the stimulus. For example, you may not be paying attention to general conversations in a crowded room or hall, until someone says your name. The sound of your name leaps out of the surrounding babble, and you are suddenly interested in what was being said by the person who spoke your name.

3. It is possible that the dorsal-column/medial-lemniscal system within the right side of the spinal cord is damaged. However, it is also possible that this system could be damaged within the medulla oblongata, where neurons synapse and cross over to the left side of the brain, or within the tracts on the left side that ascend from the medulla oblongata to the thalamus. Another possibility is damage to the cerebral cortex on the left side. Additional information is needed to decide exactly where the injury is located.

4. The fibers of the dorsal-column/medial-lemniscal system carry two-point discrimination and proprioceptive information. Primary neurons from the right side of the body ascend the spinal cord in the dorsal column and synapse with secondary neurons in the medulla oblongata. The secondary neurons cross over in the upper medulla and ascend through the left side of the pons to the thalamus. A patient suffering from a loss of two-point discrimination and proprioception on the right side of the body as a result of a lesion in the medial-lemniscal system in the pons has a lesion in the left side of the pons.

5. The fibers of the lateral spinothalamic tract carry impulses for pain and temperature. A lesion in the area where these fibers decussate results in the bilateral loss of pain and temperature sensations only at the level of the lesion, and there is no loss of sensation below the lesion. This occurs because fibers decussating above or below the lesion, as well as tracts that pass lateral to the lesion, are unaffected.

6. The damaged tracts are the lateral corticospinal tract, controlling motor functions on the right side of the body, and the lateral spinothalamic tract for pain and temperature sensations from the left side of the body. Damage to these tracts in the right side of the spinal cord produces the observed symptoms, because, in the cord, the lateral spinothalamic tract crosses over at the level of entry, and is, therefore, located on the opposite side of the cord from its peripheral nerve endings, whereas the corticospinal tract lies on the same side of the cord as its target muscles.

7. Complete unilateral transection of the right side of the spinal cord results in loss of motor function (lateral corticospinal tract), proprioception, and two-point discrimination (dorsal-column/medial-lemniscal system) on the same side of the body as the lesion, below the level of the lesion. Pain and temperature sensations (lateral spinothalamic tract) are lost on the opposite side of the body from below the level of the lesion. These symptoms describe the Brown-Séquard syndrome. Light touch is not greatly affected on either side because of the large number of collateral branches in the anterior spinothalamic tract.

8. The right cerebral cortex controls the left side of the body. The motor cortex has a topographic representation of the opposite side of the body, with the hand, forearm, arm, and shoulder located approximately in the center of the precentral gyrus. The lesion is therefore in the center of the right precentral gyrus of the cerebrum. Some grosser control of the left-upper limb may still exist because of the indirect pathways, but there would be spastic paralysis.

9. Damage to the cerebellum can result in decreased muscle tone, balance impairment, a tendency to overshoot when reaching for or touching something, and an intention tremor. These symptoms are opposite to those seen with basal ganglia dysfunction. Cerebellar dysfunction exhibits very similar symptoms to those seen in an inebriated person, and the same tests could be applied, such as having the person touch their nose or walk a straight line.

10. Memory storage for the 10 minutes prior to the accident was in short-term memory and was disrupted before it could be transferred to long-term memory.

    Additional information: Anytime a person suffers a concussion there's a possibility that he or she may later develop postconcussion syndrome. Symptoms include muscle tension or migraine headaches, reduced alcohol tolerance, difficulty learning new things, reduction in creativity and motivation, fatigue, and personality changes, and the syndrome may last a month to a year. Postconcussion syndrome may be the result of a slowly occurring subdural hematoma, which may be missed by an early examination.

Chapter 15

1. The first sensations that occur when a person picks up an apple and bites into it are visual. The person holds the apple in his hand and looks at it. Visual input (which stimulates light receptors) tells him the size of the apple, and that it has a smooth surface, as well as its color. As the person bites into the apple and begins to chew, taste sensations (chemoreceptors) from the tongue provide information that the apple has characteristics of being both sweet and sour. Olfactory sensations (hemoreceptors) provide more specific information that the "fruity taste" is that of an apple.

2. The lens of the eye is biconvex and causes light rays to converge. If the lens is removed, then the replacement lens should also cause light rays to converge. A biconvex lens or a lens with a single convex surface would work. Bifocals or trifocals could also be recommended because of the loss of accommodation.

3. The light reflected by the tapetum lucidum could stimulate photoreceptors and increase the sensitivity of the eye to light, which could be an advantage when light levels are low. Because the same light image could stimulate different photoreceptors, however, there is a loss of visual acuity and a blurring of vision.

4. Carrots contain vitamin A (retinoic acid), which can be used to form retinal. Retinal and opsin combine to form rhodopsin, which is found in rods. Rhodopsin is necessary for rods to respond to low levels of light. Lack of vitamin A can result in lack of rhodopsin and night blindness.

5. By looking a few inches to the side, the image of the needle and thread is projected to the periphery of the retina, where there is the highest concentration of rods. The rods function better than cones in low-light intensities. If Jean looks directly at the needle and thread, their image falls on the macula, which has few rods and mostly cones, which don't function well in dim light. By looking to the side, however, she is using a part of the retina where the

photoreceptor cells are not as densely packed as in the macula, and the image is fuzzy rather than sharp.

6. This phenomenon is called a negative afterimage. While staring at the clock, the darkest portion of the image (the black clock) causes dark adaptation in part of the retina. That is, part of the retina becomes more sensitive to light. At the same time the lightest part of the image (the white wall) causes light adaptation in the rest of the retina, and that part of the retina becomes less sensitive to light. When the man looks at a black wall, the dark adapted portion of the retina, which is more sensitive to light, produces more action potentials than does the light adapted part of the retina. Consequently, he perceives a light clock against a darker background.

7. A lesion of the optic chiasm results in visual loss in both the right and left temporal fields, a condition called bitemporal hemianopsia, or tunnel vision. Tunnel vision can cause problems for normal functions, such as when driving a car, because the peripheral vision is severely limited. The occurrence of this condition can also suggest a much more serious problem, such as a pituitary tumor, which sits just posterior to the optic chiasm.

8. The most likely area damaged is the spiral organ, where waves result in the production of action potentials. The action is much like ocean waves breaking on the shore during a violent storm as compared to those breaking in from a calm ocean. Specifically, damage likely occurs in the part of the spiral organ near the oval window, because it is this part of the basilar membrane that vibrates the most in response to high-frequency sounds.

9. Normally, as pressure changes, the auditory tubes open to allow an equalization of pressure between the middle ear and the external environment. If this doesn't occur, then the buildup of pressure in the middle ear can rupture the tympanic membrane, or the pressure can be transmitted to the inner ear and cause sensoneural damage.

10. Normally, airborne sounds cause the tympanic membrane to vibrate, resulting in movement of the middle ear ossicles and the production of waves in the perilymph of the scala vestibuli. Vibration of the skull bones can also cause vibration of the perilymph in the scala vestibuli.

Chapter 16

1. The sympathetic division of the ANS is responsible for dilation of the pupil. Preganglionic fibers from the upper thoracic region of the spinal cord pass through spinal nerves (T1 and T2), into the white rami communicantes, and into the sympathetic chain ganglia. The preganglionic fibers ascend the sympathetic chain and synapse with postganglionic neurons in the superior cervical sympathetic chain ganglia. The axons of the postganglionic neurons leave the sympathetic chain ganglia as small nerves that project to the pupil of the eye.

2. Reduced salivary and lacrimal gland secretions could indicate damage to the facial nerves, which innervate the submandibular, sublingual, and lacrimal glands. The glossopharyngeal nerves innervate the parotid glands but not the lacrimal glands.

3. Cutting the preganglionic fibers in the white rami of T2–T3 is the best way to eliminate innervation of the blood vessels in the skin. Cutting the gray rami at levels T2–T3 is inappropriate because the postganglionic fibers that innervate the hand blood vessels exit from the first thoracic and inferior cervical sympathetic chain ganglia. Cutting the spinal nerves is inappropriate because it eliminates all sensory and motor functions to the area supplied.

4. a. Pelvic nerves
   b. Gray rami
   c. Vagus nerves
   d. Cranial nerves
   e. Pelvic nerves

5. The parasympathetic division innervates the heart through the vagus nerves. The postganglionic nerve fibers of the vagus nerves release acetylcholine, which reduces heart rate. Methacholine can bind to the same receptors as acetylcholine and reduce heart rate. Side effects result from stimulating other parasympathetic effector organs. For example, stimulating the salivary glands results in increased salivation. Dilation of the pupils and sweating are effects expected from sympathetic stimulation. The muscles of respiration are not regulated by the ANS, but they do respond to acetylcholine through somatic neurons. Methacholine would be expected to make contractions of respiratory muscles more likely.

6. One would expect mostly parasympathetic effects, because the effects of acetylcholine are enhanced: blurring of vision as a result of contraction of ciliary muscles, excess tear formation because of overstimulation of the lacrimal glands, frequent or involuntary urination because of overstimulation of the urinary bladder. Pallor resulting from vasoconstriction in the skin is a sympathetic effect that would not be expected because skin blood vessels respond to norepinephrine. Muscle twitching or cramps of skeletal muscles might occur because they normally respond to acetylcholine.

7. Epinephrine causes vasoconstriction and confines the drug to the site of administration. This increases the duration of action of the drug locally and decreases systemic effects. Vasoconstriction also reduces bleeding if a dry field (an area clear of blood on its surface) is required.

8. Because normal action potentials are produced, the drug doesn't act at the synapse between the preganglionic and postganglionic neurons. Because injected norepinephrine works, sympathetic receptors in the heart are functioning and are not affected by the drug. Therefore, the drug must somehow affect the postganglionic neurons. Possibly it inhibits neurotransmitter production or release from the postganglionic neurons.

9. Because cutting the white rami of T1–T4 doesn't affect the action of the drug, sympathetic preganglionic neurons in the spinal cord and sympathetic centers in the brain can be ruled out as a site of action. Because cutting the vagus nerves eliminates the effect of the drug, the drug cannot be acting at the synapse between the preganglionic neurons and the postganglionic neurons, or between the synapse of the postganglionic neuron and the effector organ of either division of the ANS. The drug must, therefore, excite parasympathetic centers in the brainstem, resulting in a decrease in heart rate.

10. a. Responses in a person who is extremely angry are primarily controlled by the sympathetic division of the ANS. These responses include increased heart rate and blood pressure, decreased blood flow to the internal organs, increased blood flow to skeletal muscles, decreased contractions of the intestinal smooth muscle, flushed skin in the face and neck region, and dilation of the pupils of the eyes.
    b. For a person who has just finished eating and is now relaxing, the parasympathetic reflexes are more important than sympathetic reflexes. The blood pressure and heart rate are at normal resting levels, the blood flow to the internal organs is greater, contractions of smooth muscle in the intestines are greater, and secretions that achieve digestion are more active. If the urinary bladder or the colon becomes distended, autonomic reflexes that result in urination or defecation can result. Blood flow to the skeletal muscles is reduced.

Chapter 17

1. Liver and kidney disease increases the concentration of this hormone in the blood, and the concentration would remain high for a longer time. The liver modifies the hormone to cause it to be excreted by the kidney more rapidly. In the case of liver disease, the hormone is not modified and excreted rapidly. Therefore, the concentration becomes higher than normal and the concentration of the hormone remains high for longer than normal. A similar result is seen if the kidney is diseased and the hormone cannot be excreted rapidly.

2. Secretion of hormones is usually controlled by a negative-feedback mechanism. If a hormone controls the concentration of a substance in the circulatory system, the hormone is secreted in smaller amounts if the substance increases in the circulatory system. If a tumor begins to secrete the substance in large amounts, the presence of the substance has a negative-feedback effect on the secretion of the hormone and the concentration of the hormone in the circulatory system is very low.

3. Usually intracellular mediator mechanisms respond quickly, and the effect of the hormone is brief. Intracellular receptor mechanisms usually take a long time (several hours) to respond, and their effects last much longer. If the hormone is large and water-soluble, it's probably functioning through an intracellular mediator mechanism, or if the hormone is lipid-soluble, it's probably an

intracellular receptor mechanism. If you have the ability to monitor the concentration of a suspected intracellular mediator and it increases in response to the hormone, or if you can inhibit the synthesis of an intracellular mediator and it prevents the target cells' response to the hormone, it's an intracellular mediator mechanism. If you can inhibit the synthesis of mRNA and this inhibits the action of the hormone, or if you can measure an increase in mRNA synthesis in response to the hormone, then the mechanism is an intracellular receptor mechanism.

4. When the hormone binds to its receptor, the α subunit of the G protein is released. GTP must bind to the α subunit, however, before it can have its normal effect. If the α subunit cannot bind GTP, the hormone, therefore, has no effect on the target tissue.

5. Inhibitors of prostaglandin synthesis reduce prostaglandin synthesis in all tissues, not just in those tissues in which prostaglandins produce undesirable effects. Symptoms such as inflammation, vomiting, and fever are reduced. Because prostaglandins also play a role in producing beneficial effects in some tissues, however, these benefits would not occur normally. Inhibitors of prostaglandin synthesis may cause labor to be delayed or produce other undesirable responses due to their inhibitory effects on the synthesis of prostaglandins.

6. Phosphodiesterase causes the conversion of cAMP to AMP, thus reducing the concentration of cAMP. A drug that inhibits phosphodiesterase, therefore, increases the amount of cAMP in cells where cAMP is produced. Therefore, an inhibitor of phosphodiesterase increases the response of a tissue to a hormone that has cAMP as an intracellular mediator.

7. A short half-life for epinephrine allows epinephrine to produce a short-lived response. The response to a potentially harmful or dangerous situation is terminated shortly after the harmful or dangerous situation passes. If epinephrine had a long half-life, the heart rate and blood glucose would be elevated for a long time, even if the harmful or dangerous situation was very brief.

8. Because thyroid hormones are important in regulating the basal metabolic rate, a long half-life is an advantage. Thyroid hormones are secreted and have a prolonged effect without large fluctuations in the basal metabolic rate. If thyroid hormones had a short half-life, the basal metabolic rate might fluctuate with changes in the rate of secretion of thyroid hormones. Certainly the rate of secretion of thyroid hormones would have to be controlled within narrow limits if it did have a short half-life.

9. If liver disease results in a decrease of plasma proteins to which thyroid hormones bind, higher-than-normal concentrations of free (unbound) thyroid hormones occur in the circulatory system. Because of the higher-than-normal concentration of thyroid hormones that are unbound, the responses to thyroid hormones increase. In addition, the half-life of the thyroid hormones is shortened. Thus, as thyroid hormone secretion increases, the concentration

of thyroid hormone also increases. As the thyroid hormone secretion decreases, the concentration of thyroid hormone also decreases. Thyroid hormones fluctuate in concentration in the circulatory system more than normal.

10. Elevated GnRH levels in the blood as a result of the GnRH-secreting tumor causes down-regulation of GnRH receptors in the anterior pituitary. This decreases the ability of GnRH to stimulate the anterior pituitary, and the rate of luteinizing hormone and follicle-stimulating hormone secretion by the anterior pituitary decreases and remains decreased as long as the GnRH levels are chronically elevated. Therefore, the functions of the reproductive system controlled by luteinizing hormone and follicle-stimulating hormone decrease.

11. Insulin levels normally change in order to maintain normal blood sugar levels, despite periodic fluctuations in sugar intake. A constant supply of insulin from a skin patch might result in insulin levels that are too low when blood sugar levels are high (after a meal) and might be too high when blood sugar levels are low (between meals). In addition, insulin is a protein hormone that would not readily diffuse through the lipid barrier of the skin (see chapter 5). Estrogen is a lipid soluble steroid hormone.

Chapter 18

1. The hypothalamohypophyseal portal system allows neurohormones that function as releasing and inhibiting hormones, which are secreted by neurons in the hypothalamus, to be carried directly from the hypothalamus to the anterior pituitary gland. Consequently, the releasing and inhibiting hormones are not diluted nor are they destroyed by the enzymes, which are abundant in the kidneys, liver, lungs, and general circulation, before they reach the anterior pituitary. Also the time it takes for releasing and inhibiting hormones to reach the anterior pituitary is less than if they were secreted into the general circulation.

2. A hot environment increases ADH secretion. Because the amount of water lost in the form of sweat can be quite large, and because sweat is more dilute than the body fluids, sweating gradually increases the osmolality of the body fluids. The increasing osmolality of body fluids stimulates an increase in ADH secretion. Thus, a hot environment can result in increased ADH secretion because of an increasing osmolality of the body fluids. Increased ADH secretion in a hot environment reduces the amount of water lost in the form of urine. Therefore, water is conserved.

3. Polydipsia and polyuria are consistent with either diabetes mellitus or diabetes insipidus. Diabetes mellitus, however, is consistent with an increased urine osmolality because of the large amount of glucose lost in the urine. Diabetes insipidus is consistent with urine with a low specific gravity because little water is reabsorbed by the kidney. Thus urine has an osmolality close to that of the body fluids, and the rapid loss of dilute urine results in a decrease in blood pressure. Thus polyuria with a low specific gravity is not consistent with diabetes mellitus but is consistent with diabetes insipidus. Administration of ADH

reverses the symptoms of diabetes insipidus. Neither polydipsia nor polyuria results from a lack of glucagon or aldosterone.

4. The symptoms are consistent with acromegaly, which is a consequence of elevated GH secretion after the epiphyses have closed. Increased GH causes enlarged finger bones, growth of bony ridges over the eyes, and increased growth of the jaw. The anterior pituitary tumor increases pressure at the base of the brain near the optic nerves as it enlarges. The pituitary rests in the sella turcica of the sphenoid bone; as it enlarges pressure increases because the pituitary is nearly surrounded by rigid bone and the brain is located just superior to the pituitary. As the anterior pituitary enlarges because of a tumor, it pushes superiorly and pressure is applied to the ventral portion of the brain. The GH also causes bone deposition on the inner surface of skull bones, which also increases the pressure inside the skull.

5. If hyperthyroidism results from a pituitary abnormality, laboratory tests should show elevated TSH levels in the circulatory system in addition to elevated T_3 and T_4 levels. If hyperthyroidism results from the production of a nonpituitary thyroid-stimulating substance, laboratory tests should also show elevated T_3 and T_4 levels, but TSH levels would be low because of the negative-feedback effects of T_3 and T_4 on the hypothalamus and pituitary gland.

6. The second student is correct. Low levels of vitamin D reduce calcium uptake in the gastrointestinal tract, which results in a decreased blood level of calcium ions. As blood calcium levels decrease, the rate of PTH secretion would increase. Parathyroid hormone increases bone breakdown, which maintains blood calcium levels, even if vitamin D deficiency exists for a prolonged time. Osteomalacia results because of the increased bone reabsorption necessary to maintain normal blood calcium levels.

7. A glucose tolerance test can distinguish between these conditions. The person would consume glucose after a period of fasting. Over the next few hours the blood glucose levels in a healthy person increase and then return to fasting levels. The blood glucose levels always remain within the normal range, however. In a person with diabetes, the blood glucose levels increase to above-normal levels and remain elevated for several hours. In a person who secretes large amounts of insulin, blood glucose levels would increase, and then they would decrease to below-normal levels within a relatively short time.

8. Because the person is a diabetic and probably is taking insulin, the condition is more likely to be insulin shock than a diabetic coma. To confirm the condition, however, a blood sample should be taken. If the condition is due to a diabetic coma, then the blood glucose levels will be elevated. If the condition is due to insulin shock, the blood glucose levels will be below normal. In the case of insulin shock, glucose can be administered intravenously. In the case of diabetic coma, insulin should be administered. An isotonic solution containing insulin can be administered to reduce the osmolality of the extracellular fluid.

9. Adrenal diabetes results from elevated and uncontrolled secretion of glucocorticoid hormones, such as cortisol, from the adrenal gland. Because glucocorticoid hormones increase blood glucose levels, elevated secretion of these hormones results in elevated blood glucose levels and symptoms similar to diabetes mellitus. Pituitary diabetes results from elevated secretion of GH from the anterior pituitary. Elevated GH causes an increase in blood glucose levels and, therefore, produces symptoms similar to diabetes mellitus. Prolonged elevation of both glucocorticoids and growth hormone secretion can lead to the development of diabetes mellitus if the insulin-secreting cells of the pancreatic islets degenerate because of the prolonged need to secrete insulin in response to the elevated blood glucose levels.

10. Elevated epinephrine from the adrenal medulla promotes elevated blood pressure and increases the work load on the heart, increases the rate of metabolism, and results in increased sweating and nervousness. The risk of heart attack and stroke are increased. Elevated cortisol causes hyperglycemia and can lead to diabetes mellitus, a depressed immune system with increased susceptibility to infections, and destruction of proteins leading to tissue wasting.

Chapter 19

1. Because of the rapid destruction of the red blood cells we would expect erythropoiesis to increase in an attempt to replace the lost red blood cells. The reticulocyte count would therefore be above normal. Jaundice is a symptom of hereditary hemolytic anemia because the destroyed red blood cells release hemoglobin, which is converted into bilirubin. Removal of the spleen cures the disease because the spleen is the major site of red blood cell destruction.

2. Blood doping increases the number of red blood cells in the blood, thereby increasing its oxygen-carrying capacity. The increased number of red blood cells also makes it more difficult for the blood to flow through the blood vessels, increasing the heart's workload.

3. Symptoms resulting from decreased red blood cells are associated with a decreased ability of the blood to carry oxygen: shortness of breath, weakness, fatigue, and pallor. Symptoms resulting from decreased platelets are associated with a decreased ability to form platelet plugs and clots: small areas of hemorrhage in the skin (petechiae), bruises, and decreased ability to stop bleeding. Symptoms resulting from decreased white blood cells could include an increased susceptibility to infections.

4. Hypoventilation results in decreased blood oxygen levels, which stimulates erythropoiesis. Therefore, the number of red blood cells increases and produces secondary polycythemia.

5. Removal of the stomach removes intrinsic factor, which is necessary for vitamin B_{12} absorption. Therefore, the patient develops pernicious anemia. Lack of stomach acid can decrease iron absorption in the small intestine and results in iron-deficiency anemia.

6. Vitamin B_{12} and folic acid are necessary for blood cell division. Lack of these vitamins results in pernicious anemia. Iron is necessary for the production of hemoglobin. Lack of iron results in iron-deficiency anemia. Vitamin K is necessary for the production of many blood clotting factors. Lack of vitamin K can greatly increase blood clotting time, resulting in excessive bleeding.

Chapter 20

1. The walls of the ventricles are thicker than the walls of the atria because the ventricles must produce a greater pressure to pump blood into the arteries. Only a small pressure is required to pump blood from the atria into the ventricles during diastole. The wall of the left ventricle is thicker than the wall of the right ventricle because the left ventricle produces a much greater pressure to force blood through the aorta than the right ventricle produces to move blood through the pulmonary trunk and pulmonary arteries.

2. During systole, the cardiac muscle in the right and left ventricles contracts, which compresses the coronary arteries. During diastole, the cardiac muscle of the ventricles relaxes and blood flow through the coronary arteries increases. The diastolic pressure is sufficient to cause blood to flow through coronary arteries during diastole.

3. A drug that prolongs the plateau of cardiac muscle cell action potentials prolongs the time each action potential exists and increases the refractory period. Therefore, the drug would slow the heart. A drug that shortens the plateau shortens the length of time each action potential exists and shortens the refractory period. Therefore, the drug could allow the heart rate to increase further.

4. Endurance-trained athletes have decreased heart rates because their cardiac muscle undergoes hypertrophy in response to exercise. The hypertrophied cardiac muscle causes the stroke volume to increase substantially. The increased stroke volume is sufficient to maintain an adequate cardiac output and blood pressure even though the heart rate is slower.

5. The two heartbeats occurring close together can be heard through the stethoscope, because the heart valves open and close normally during each of the heartbeats even if they are close together. The second heartbeat, however, produces a greatly reduced stroke volume because there's not enough time for the ventricles to fill with blood between the first and second contraction of the heart. Thus, the preload is reduced. Because the preload is reduced, the second heartbeat has a greatly reduced stroke volume. The reduced stroke volume fails to produce a normal pulse. The pulse deficit, therefore, results from the reduced stroke volume of the second of the two beats that are very close together.

6. Aerobic training causes hypertrophy of the cardiac muscle in the heart and causes the heart to produce a greater stroke volume as a consequence. The heart rate can decrease while the cardiac output remains the same because cardiac output is equal to the stroke volume times the heart rate. If the stroke volume increases, the heart rate can decrease and the cardiac output can remain the same.

7. Atrial contractions complete ventricular filling, but atrial contractions are not primarily responsible for ventricular filling. Therefore, even if the atria are fibrillating, blood can still flow into the ventricles and ventricular contractions can occur. As long as the ventricles contract rhythmically the heart can pump an adequate amount of blood, even though the atria are not effective pumps. If the ventricles fail to contract forcefully and rhythmically, however, they cannot function as pumps. Thus the stroke volume will become too low to maintain adequate blood flow to tissues. Therefore, atrial transplants are not essential but ventricle transplants are.

8. The results depend on Cee Saw's response to the conditions of the laboratory. First, as Cee Saw's head is lowered, gravity causes blood pressure in the carotid sinuses and aortic arch to increase. The increased blood pressure stimulates baroreceptors, which detect the increased blood pressure and send action potentials indicating that blood pressure increased to the cardioregulatory center in the medulla oblongata along sensory nerve fibers. The cardioregulatory center increases parasympathetic stimulation and reduces sympathetic stimulation of the heart. Thus the heart rate decreases. Second, if, as her head is lowered, Cee Saw becomes excited, the sympathetic division of the ANS becomes more active. The resulting increase in sympathetic stimulation of the heart causes the heart rate to increase.

9. After Cee Saw is tilted so that her head is higher than her feet for a few minutes, the regulatory mechanisms that control blood pressure adjust so that the heart pumps sufficient blood to supply the needs of her tissues. If she is then tilted so that her head is higher than her feet, gravity would cause blood to flow toward her feet, and the blood pressure in the carotid sinus and aortic arch would decrease. The decrease in blood pressure would be detected by the baroreceptors in these vessels and would activate baroreceptor reflexes. The result is increased sympathetic and decreased parasympathetic stimulation of the heart and an increase in the heart rate. The increased heart rate would function to increase the blood pressure to its normal value.

10. An ECG measures the electrical activity of the heart and does not indicate a slight heart murmur. Heart murmurs are detected by listening to the heart sounds. The boy may have a heart murmur, but the mother does not understand the basis for making such a diagnosis.

11. An incompetent aortic semilunar valve does not close completely at the end of systole. Consequently, blood flows from the aorta back into the left ventricle at the end of systole and during diastole. This causes the diastolic blood pressure to be lower than normal, and, at the end of diastole, the volume of blood in the left ventricle is greater than normal. Because of the increased volume of blood in the left ventricle at the end of diastole, the stroke volume during systole is greater than normal. As a result, the systolic blood pressure is greater than normal. The heart sounds are affected by an incompetent aortic semilunar valve. Turbulent blood flow and, consequently, an abnormal heart sound is heard

at the end of systole as blood flows backward through the incompetent aortic semilunar valve. This heart sound follows the second heart sound, which results from closure of the semilunar valves.

12. When both common carotid arteries are clamped, the blood presure within the internal carotid arteries drops dramatically. The decreased blood pressure is detected, and the baroreceptor reflex increases heart rate and stroke volume. The resulting increase in cardiac output causes the increase in blood pressure.

13. Venous return declines markedly in hemorrhagic shock because of the loss of blood volume. With decreased venous return, stroke volume decreases (Starling's law of the heart). The decreased stroke volume results in a decreased cardiac output, which produces a decreased blood pressure. In response to the decreased blood pressure, the baroreceptor reflex causes an increase in heart rate in an attempt to restore normal blood pressure. However, with inadequate venous return the increased heart rate is not able to restore normal blood pressure.

Chapter 21

1. a. Aorta, left coronary artery, circumflex artery, posterior interventricular artery; or aorta, right coronary artery, posterior interventricular artery
   b. Aorta, brachiocephalic artery, right common carotid artery, right internal carotid artery; or aorta, left common carotid artery, left internal carotid artery
   c. Aorta, brachiocephalic artery, right subclavian artery, right vertebral artery, basilar artery; or aorta, left subclavian artery, left vertebral artery, basilar artery
   d. Aorta, left or right common carotid artery, left or right external carotid artery
   e. Aorta, left subclavian artery, axillary artery, brachial artery, radial or ulnar artery, deep or superficial palmar arch, digital artery (on the right: the brachiocephalic artery would be included)
   f. Aorta, common iliac artery, external iliac artery, femoral artery, popliteal artery, anterior tibial artery
   g. Aorta, celiac artery, common hepatic artery
   h. Aorta, superior mesenteric artery, intestinal branches
   i. Aorta, left or right internal iliac artery
2. a. Great cardiac vein, coronary sinus; or anterior cardiac vein
   b. Transverse sinus, sigmoid sinus, internal jugular vein, brachiocephalic vein, superior vena cava
   c. Retromandibular vein, external jugular vein, subclavian vein, brachiocephalic vein, superior vena cava
   d. Deep: vein of hand, radial or ulnar vein, brachial vein, axillary vein, subclavian vein, brachiocephalic vein, superior vena cava
      Superficial: vein of hand, radial or ulnar vein, cephalic or basilic vein, axillary vein, subclavian vein, brachiocephalic vein, superior vena cava

   e. Deep: vein of foot, dorsalis veins of foot, anterior tibial vein, popliteal vein, femoral vein, external iliac vein, common iliac vein, inferior vena cava
      Superficial: vein of foot, great saphenous vein, external iliac vein, common iliac vein, inferior vena cava; or vein of foot, small saphenous vein, popliteal vein, femoral vein, external iliac vein, common iliac vein, inferior vena cava
   f. Gastric vein or gastroepiploic vein, hepatic portal vein, hepatic sinusoids, hepatic vein, inferior vena cava
   g. Renal vein, inferior vena cava
   h. Hemiazygous vein or accessory hemiazygous vein, azygous vein, superior vena cava

3. A superficial vessel would be easiest, such as the right cephalic or basilic vein. The catheter is passed through the cephalic (or brachial) vein and the superior vena cava to the right atrium. Because the pulmonary veins are not readily accessible, dye would not normally be placed directly into them. Instead, the dye would be placed in the right atrium using the procedure just described. The dye passes from the right atrium into the right ventricle, the pulmonary arteries, the lungs, the pulmonary veins, and into the left atrium. If the catheter has to be placed in the left atrium, it could be inserted through an artery, such as the femoral artery, and passed via the aorta to the left ventricle and then into the left atrium.

4. The viscosity of the blood is affected primarily by the hematocrit. As hematocrit increases, the viscosity of the blood increases logarithmically, so that even a small increase in hematocrit results in a large increase in viscosity. Greater force is therefore not needed to cause blood to flow through the blood vessels. With the increased blood volume, blood flow through vessels is adequate without an increase in viscosity.

5. The resistance to blood flow is less in the vena cavae for two reasons: first, the diameter of one vena cava is greater than the diameter of the aorta and second, an increased diameter of a blood vessel reduces resistance to flow (see Poiseuille's law). In addition, there are two venae cavae, the superior vena cava and the inferior vena cava, but only one aorta. The blood flow through the aorta and the venae cavae is about equal, but the velocity of blood flow is much higher in the aorta than it is in the venae cavae.

6. According to Laplace's law, as the diameter of a blood vessel increases, the force applied to the vessel wall increases, even if the pressure remains constant. The increased connective tissue found in the walls of the large blood vessels therefore makes the wall of those vessels stronger and more capable of resisting the force applied to the wall.

7. Veins and lymphatic vessels have one-way valves in them. Massage creates a cycle of increasing and decreasing pressure to the veins, which rhythmically compresses them. The compression of the veins forces fluid to move out of the limb through both veins and lymphatic vessels. The movement of fluid through the veins lowers the pressure within the venous end of the capillary.

Thus the forces that move fluid into the capillaries at their venous ends are greater and they move more interstitial fluid into the capillaries. Compression of the lymphatic capillaries also causes more lymphatic fluid to move into the lymphatic vessels. Because there is less fluid in the limb, the edema decreases.

8. The nursing student's diagnosis was incorrect. Blood pressure measurements are normally made in either the right or left arm, both of which are close to the level of the heart. Blood pressure taken in the leg is influenced by pressure created by the pumping action of the heart, but the effect of gravity on the blood, as it flows into the leg, also influences the blood pressure in a substantial way. In this case gravity increases blood pressure from about 120 mm Hg for the systolic pressure to 200 mm Hg.

9. Decreased liver function includes a decrease in the synthesis of plasma proteins. Consequently, the concentration of plasma proteins decreases, and the colloid osmotic pressure of the blood decreases. Therefore, less water moves by osmosis into the capillaries at the venous ends and the result is edema.

10. Chemoreceptors in the medulla oblongata detect carbon dioxide and the pH of the blood. The normal blood levels of CO_2 and pH stimulate these chemoreceptors, which in turn stimulate the vasomotor center. The vasomotor center keeps blood vessels partially constricted under resting conditions. This basal level of activity is called the vasomotor tone. Blowing off CO_2 reduces the blood levels of carbon dioxide and increases the pH of the body fluids. These changes reduce vasomotor tone and result in vasodilation. If a person hyperventilates and blows off CO_2, the stimulus to the vasomotor center decreases, which results in a decrease in vasomotor tone. The decrease in vasomotor tone results in a decrease in systemic blood pressure. If the blood pressure decreases enough, the blood flow to the brain decreases and can cause a sensation of dizziness or can even cause a person to lose consciousness.

11. Epinephrine is secreted from the adrenal medulla in response to stressful stimuli and the epinephrine stimulates responses that are consistent with increased physical activity. Vasoconstriction of the blood vessels in the skin shunts blood away from the skin to skeletal muscles. Vasodilation occurs in blood vessels of exercising skeletal muscles. Blood flow through the exercising skeletal muscles therefore increases. Because epinephrine causes vasodilation of the blood vessels of cardiac muscle, blood flow through the cardiac muscle increases. This response is consistent with the increased work performed by the heart under conditions of increased physical activity.

12. The hot Jacuzzi increases Skinny's skin and body temperature. As a result, the blood vessels of the skin dilate. Because the blood vessels dilate, peripheral resistance decreases, causing the blood pressure to decrease. The baroreceptors of the carotid sinus and aortic arch detect the decrease in blood pressure and send action potentials to the cardioregulatory center in the

medulla oblongata. As a result, the sympathetic stimulation to the heart increases and the heart rate, in response, increases also. The increased heart rate elevates the blood pressure back to within its normal range of values.

Chapter 22

1. Elevation of the limb reduces blood pressure in the limb, resulting in less fluid movement from the blood into the tissues (see chapter 21). Thus, the edema is reduced as the lymphatic system removes fluid from the tissues faster than it enters them. Massage moves lymph through the lymphatic vessels in the same fashion as contraction of skeletal muscle. The application of pressure periodically to lymphatic vessels forces lymph to flow toward the trunk of the body, but valves prevent the flow of lymph in the reverse direction. The removal of lymph from the tissue helps to relieve edema.

2. Normally T cells are processed in the thymus and then migrate to other lymphatic tissues. Without the thymus this processing is prevented. Because there are normally five T cells for every one B cell, the number of lymphocytes is greatly reduced. The loss of T cells results in an increased susceptibility to infection and an inability to reject grafts because of the loss of cell-mediated immunity. In addition, since helper T cells are involved with activation of B cells, antibody-mediated immunity is also depressed.

3. That there is no immediate effect indicates there is a reservoir of T cells in the lymphatic tissue. As the reservoir is depleted through time, the number of lymphocytes decreases and cell-mediated immunity is depressed, the animal is more susceptible to infections, and the ability to reject grafts decreases. The ability to produce antibodies decreases because of the loss of helper T cells that are normally involved with the activation of B cells.

4. Injection B results in the greatest amount of antibody production. At first, the antigen causes a primary response. A few weeks later, the slowly released antigen causes a secondary response, resulting in a greatly increased production of antibodies. Injection A doesn't cause a secondary response because all of the antigen is eliminated by the primary response.

5. If the patient has already been vaccinated, the booster shot stimulates a memory (secondary) response and rapid production of antibodies against the toxin. If the patient has never been vaccinated, vaccinating now is not effective because there's not enough time for the patient to develop his or her own primary response. Therefore, antiserum is given to provide immediate, but temporary, protection. Sometimes both are given: The antiserum provides short-term protection and the tetanus vaccine stimulates the patient's immune system to provide long-term protection. If the shots are given at the same location in the body, the antiserum (antibodies against the tetanus toxin) could cancel the effects of the tetanus vaccine (tetanus toxin altered to be nonharmful).

6. The infant's antibody-mediated immunity is not functioning properly, whereas his cell-mediated immunity is working properly. This explains the susceptibility to extracellular bacterial infections and the resistance to intracellular viral infections. It took so long to become apparent because IgG from the mother crossed the placenta and provided the infant with protection. The infant began to get sick after these antibodies degraded.

7. Bone marrow is the source of the lymphocytes responsible for adaptive immunity. If successful, the transplanted bone marrow starts producing lymphocytes and the baby has a functioning immune response. In this case, there's a graft versus host rejection in which the lymphocytes in the transplanted red marrow mount an immune attack against the baby's tissues, resulting in death.

8. At the first location an antibody-mediated response results in an immediate hypersensitivity reaction, which produces inflammation. Most likely the response resulted from IgE antibodies. At the second location a cell-mediated response results in a delayed hypersensitivity reaction, which produces inflammation. This probably involves the release of cytokines and the lysis of cells. At the other locations there is neither an antibody-mediated nor a cell-mediated response.

9. The ointment is a good idea for the poison ivy, which causes a delayed hypersensitivity reaction, for example, too much inflammation. For the scrape it's a bad idea, because a normal amount of inflammation is beneficial and helps to fight infection in the scrape.

10. Because antibodies and cytokines both produce inflammation, the fact that the metal in the jewelry results in inflammation is not enough information to answer the question. However, the fact that it took most of the day (many hours) to develop the reaction indicates a delayed hypersensitivity reaction and therefore cytokines.

Chapter 23

1. Minute respiratory volume is equal to the respiration rate times the tidal volume. With a respiration rate of 12 breaths per minute and a tidal volume of 500 mL per breath, normal minute ventilation is 6000 mL/min (12×500). Rapid (24 breaths per minute), shallow (250 mL per breath) breathing results in the same minute ventilation, that is, 6000 mL/min (24×250). Alveolar ventilation rate (V_A) is the respiratory rate (frequency; f) times the difference between the tidal volume (V_T) and dead space (V_D).

$$V_A = f(V_T - V_D)$$

Normal resting $V_A = 12 \times (500 - 150) = 4200 \text{mL/min}$

In this case of rapid shallow breathing,

$$V_A = 24 \times (250 - 150) = 2400 \text{mL/min}$$

Thus, even though the minute ventilation is the same in both cases, the alveolar ventilation rate is less during rapid, shallow breathing because there's less effective exchange of gases between the atmosphere and the dead space. Because there's less exchange of gases, the partial pressures of alveolar gases become closer to the partial pressure of blood gases. Consequently, the alveolar partial pressure of O_2 decreases and the alveolar partial pressure of CO_2 increases. This decreases the concentration gradients for gases, resulting in less gas exchange between alveolar air and blood.

2. We expect vital capacity to be greatest when standing because the abdominal organs move inferiorly, thereby allowing greater depression of the diaphragm and a greater inspiratory reserve volume.

3. The hose increases dead space and therefore decreases alveolar ventilation. Ima Diver has to compensate by increasing respiratory rate or tidal volume. If the hose is too long, she won't be able to compensate. Furthermore, with a long hose, air is simply moved back and forth in the hose with little exchange of air between the atmosphere and the lungs taking place. Another consideration is the effect of water pressure on the thorax, which decreases compliance and increases the work of ventilation. In fact, a few feet underwater there's enough pressure on the thorax to prevent the intake of air even through a short hose connected to the atmosphere.

4. The increase in atmospheric pressure increases the partial pressure of oxygen. According to Henry's law, as the partial pressure of oxygen increases, the amount of oxygen dissolved in the body fluids increases. The increase in dissolved oxygen is detrimental to the gangrene bacteria. Because hemoglobin is already saturated with oxygen, the HBO treatment doesn't increase the ability of hemoglobin to pick up oxygen in the lungs.

5. Compression causes a decrease in thoracic volume and therefore lung volume. Consequently, pressure in the lungs increases over atmospheric pressure and air moves out of the lungs. Raising the arms expands the thorax and lungs. This results in a lower-than-atmospheric pressure in the lungs, and air moves into the lungs.

6. The victim's lungs expand because of the pressure generated by the rescuer's muscles of expiration. This fills the lungs with air that has a greater pressure than atmospheric pressure. Air flows out of the victim's lungs as a result of this pressure difference and because of the recoil of the thorax and lungs. Although the partial pressure of oxygen of the rescuer's expired air is less than atmospheric, enough oxygen can be provided to sustain the victim. The lower partial pressure of oxygen could also activate the chemoreceptor reflex and stimulate the victim to breathe. In addition, the rescuer's partial pressure of carbon dioxide is higher than atmospheric and this could activate the chemosensitive area in the medulla.

7. The left side of the diaphragm moves superiorly. During inspiration, thoracic volume increases as the right side of the diaphragm moves inferiorly and the intercostal muscles move the ribs outward. Increased thoracic volume causes a decrease in pressure in the thoracic cavity (general gas law). As a result, the pressure on the superior surface of the diaphragm is less than on the inferior surface. The paralyzed left side of the diaphragm moves superiorly because of this pressure difference.

8. At the end of expiration, the thoracic wall is not moving. Therefore the forces causing the thoracic wall to move inward and outward must be equal. At the end of expiration, the lungs are adhering to the thoracic wall through the pleurae, and recoil of the lungs is pulling the thoracic wall inward. As a result of the pneumothorax, air enters the pleural cavity, and the visceral and parietal pleurae separate from each other. The recoil of the lungs causes them to collapse. Without the inward force produced by lung recoil, the thoracic wall expands outward.

9. All else being equal (i.e., the thickness of the respiratory membrane, the diffusion coefficient of the gas, and the surface area of the respiratory membrane), diffusion is a function of the partial pressure difference of the gas across the respiratory membrane. The greater the difference in partial pressure, the greater the rate of diffusion. The greatest rate of oxygen diffusion should therefore occur at the end of inspiration when the partial pressure of oxygen in the alveoli is at its highest. The greatest rate of carbon dioxide diffusion should occur at the end of inspiration when the partial pressure of carbon dioxide in the alveoli is at its lowest.

10. Because the partial pressure of oxygen at high altitudes decreases, a shift to the left is advantageous. Such a shift enables hemoglobin to pick up more oxygen at a lower partial pressure of oxygen.

11. Cutting the vagus nerves eliminates the Hering-Breuer reflex and results in a greater-than-normal inspiration. This increases tidal volume. Cutting the phrenic nerves eliminates contraction of the diaphragm. Tidal volume decreases drastically, and death probably results. Cutting the intercostal nerves eliminates raising of the ribs and sternum and decreases tidal volume, unless the diaphragm compensates.

12. While hyperventilating and making ready to leave your instructor behind, you might make the following arguments:
   • Hyperventilation increases the oxygen content of the air in the lungs; therefore, you would have more oxygen to use when holding your breath.
   • It's hemoglobin that is saturated. Hyperventilation increases the amount of oxygen dissolved in the blood plasma.
   • Hyperventilation decreases the amount of carbon dioxide in the blood. This makes it possible to hold one's breath for a longer time because of a decreased urge to take a breath.
   • Hyperventilation activates alveoli not in use because increasing alveolar oxygen and decreasing alveolar carbon dioxide causes lung arterioles to relax, thereby increasing blood flow through the lungs.

Chapter 24

1. With the loss of the swallowing reflex, the vocal folds no longer occlude the glottis. Consequently, vomit can enter the larynx and block the respiratory tract.

2. Without adequate amounts of hydrochloric acid, the pH in the stomach is not low enough for the activation of pepsin. This loss of pepsin function results in inadequate protein digestion. If the food is well chewed, however, proteolytic enzymes in the small intestine (e.g., trypsin, chymotrypsin) can still digest the protein. If the stomach secretion of intrinsic factor decreases, the absorption of vitamin B_{12} is hindered. Inadequate amounts of vitamin B_{12} can result in decreased red blood cell production (pernicious anemia).

3. Even though ulcers are apparently ultimately caused by bacteria, overproduction of hydrochloric acid due to stress is a possible contributing factor. Reducing hydrochloric acid production is recommended. In addition to antibiotic therapy, commonly recommended solutions include relaxation, drugs that reduce stomach acid secretion, and antacids to neutralize the hydrochloric acid. Smaller meals are also advised because distension of the stomach stimulates acid production. Proper diet is also important. The patient is also advised to avoid alcohol, caffeine, and large amounts of protein because they stimulate acid production. Ingestion of fatty acids is recommended because they inhibit acid production by causing release of gastric inhibitory polypeptide and cholecystokinin. Stress also stimulates the sympathetic nervous system, which inhibits duodenal gland secretion. As a result, the duodenum has less of a mucous coating and is more susceptible to gastric acid and enzymes. Relaxing after a meal helps decrease sympathetic activities and increase parasympathetic activities.

4. Lack of bile due to blockage of the common bile duct can result in jaundice (due to an accumulation of bile pigments in the blood) and clay-colored stools (due to lack of bile pigments in the feces). Blockage of the bile duct causes abdominal pain, nausea, and vomiting. Fat absorption is impaired because of the absence of bile salts in the duodenum and a loose, bulky stool would result. Lack of fat absorption reduces the absorption of fat-soluble vitamins such a vitamin K, resulting in lack of normal clotting function.

5. The patient would still be able to defecate. Following a meal, the gastrocolic and duodenocolic reflexes could initiate mass movement of the feces into the rectum. In the rectum, local reflexes and the defecation reflex (integrated in the sacral level of the cord and not requiring connections to high brain centers) would cause defecation. Awareness of the need to defecate would be lost (due to loss of sensory input to the brain) and the ability to voluntarily prevent defecation via the external anal sphincter would also be lost.

6. The accumulation of materials above the site of impaction and the action of bacteria on the material would result in an increase in osmotic pressure in the area. Water would move by osmosis into the colon above the site of impaction. Bowel impaction is very dangerous and must be treated quickly. The increased volume and distention of the digestive tract above the site of impaction causes compression of the mucosa. This compression can occlude blood vessels in the mucosa and lead to necrosis. Necrosis of the mucosa results in increased permeability of the mucosa, thus allowing toxic organisms and substances in the digestive tract to enter the circulation, resulting in septic shock.

Chapter 25

1. In figure 25.2, the Daily Value for saturated fat is listed as less than 20 g for a 2000 kcal/day diet. The % Daily Values appearing on food labels are based on a 2000 kcal/day diet. Therefore, the % Daily Value for saturated fat for one serving of this food is 10% (2/20 = .10, or 10%).

2. According to the Daily Value guidelines, total fats should be no more than 30% of total kilocaloric intake. For someone consuming 3000 kcal/day this is 900 kcal (3000 kcal × 0.30). There are 9 kcal in a gram of fat. Therefore, the maximum amount (weight) of fats the active teenage boy should consume is 100 g (900/9).

3. The % Daily Value is the amount of the nutrient in one serving divided by its Daily Value. Therefore the % Daily Value is 10% (10/100 = .10, or 10%).

4. The % Daily Value for one serving of the food is 10% (see answer to question 3). Since there are four servings in the package, if the teenager eats half of the food in the package, he consumes two servings. Thus, he eats 20% (10% × 2) of the recommended maximum total fat.

5. The protein in meat contains all of the essential amino acids and is a complete protein food. Although plants contain proteins, a variety of different plants must be consumed to ensure that all the essential amino acids are included in adequate amounts. Also, plants contain less protein per unit weight than meat, so a larger quantity of plants must be consumed to get the same amount of protein.

6. Copper is necessary for proper functioning of the electron-transport chain. Inadequate copper in the diet results in reduced ATP production, that is, not enough energy.

7. Fasting can be damaging because proteins are used to produce glucose. The glucose enters the blood and provides an energy source for the brain. This breakdown of proteins can damage tissues such as muscle and disrupt chemical reactions regulated by enzyme systems. A single day without food, however, is unlikely to cause permanent harm.

8. Weight is lost when kilocalories used per day exceeds kilocalories ingested per day. About 60% of the kilocalories used per day is due to basal metabolic rate. A person with a high basal metabolic rate loses weight faster than a person with a low basal metabolic rate, all else being equal. Another factor to consider is the amount of physical activity, which accounts for about 30% of kilocalories used per day. An active person loses more weight than a sedentary person does.

9. Amino acids, derived from ingested proteins, are necessary to build muscles. As Lotta and her friend discovered, excess proteins don't accelerate this process. Excess proteins can be used as an energy source in oxidative deamination, for the formation of the intermediate molecules of carbohydrate metabolism, or in gluconeogenesis. Excess proteins are also converted into storage

molecules through glycogenesis or lipogenesis. Lotta is in positive nitrogen balance because the amount of nitrogen she gains from her diet is greater than the amount she loses by excretion. Some of the nitrogen in the amino acids she ingests is incorporated into the proteins of her muscles as they enlarge.

10. No, this approach doesn't work because he is not losing stored energy from adipose tissue. In the sauna, he gains heat, primarily by convection from the hot air and by radiation from the hot walls. The evaporating sweat is removing heat gained from the sauna. The loss of water will make him thirsty, and he will regain the lost weight from fluids he drinks and food he eats.

Chapter 26

1. The large volume of hypoosmotic fluid ingested increases blood volume and causes blood osmolality to decrease. The increased blood volume is detected by baroreceptors, and the decreased blood osmolality is detected by osmoreceptors in the hypothalamus. The response to these stimuli is inhibition of ADH secretion. The alcohol in the beer also inhibits ADH secretion. The increased volume inhibits the renin-angiotensin-aldosterone mechanism, which, in turn, inhibits aldosterone secretion. The changes in aldosterone, however, take much longer to influence kidney function than changes in ADH. As a result of these changes, a large volume of dilute urine is produced until the blood osmolality and blood volume return to normal.

2. Once the salt is absorbed, the osmolality of the blood increases. The increased osmolality of blood is detected by osmoreceptor neurons in the hypothalamus, thereby stimulating ADH secretion and inhibiting aldosterone secretion. A small volume of concentrate urine is produced as a result, until the excess salt is eliminated and the blood osmolality returns to its normal value.

3. The hypoosmotic sweat loss results in more loss of water than electrolytes. This simultaneously decreases plasma volume and increases blood osmolality, thereby stimulating increased ADH secretion. In addition, the decreased plasma volume stimulates the renin-angiotensin-aldosterone mechanism, resulting in a decreased glomerular filtration rate and increased aldosterone secretion. The effect of the changes is to produce a small amount of concentrated urine.

4. The loss of sweat results in a loss of water and electrolytes. Replacing just the water restores blood volume and also decreases blood osmolality. At first, the decreased osmolality inhibits ADH secretion, and dilute urine is produced. As blood volume decreases as a result of urine production, however, ADH secretion and the renin-angiotensin-aldosterone mechanisms are stimulated. Consequently, urine concentration increases, and only a small amount of urine is produced.

5. As aldosterone levels decrease, sodium reabsorption in the nephron decreases and, consequently, plasma sodium levels decrease. The sodium is lost in the urine, and water follows the sodium by osmosis. Thus, a large amount of urine that has a high concentration of sodium is produced. The loss of water reduces blood volume, which causes the low blood pressure. As aldosterone levels decrease, potassium secretion into the nephron decreases, resulting in an increase in plasma potassium levels. The increased extracellular potassium causes depolarization of nerve and muscle membranes, leading to tremors of skeletal muscles and cardiac arrhythmias including fibrillation.

6. There are several ways to decrease glomerular filtration rate:
   a. Decrease hydrostatic pressure in the glomerulus.
      1. Decrease systemic arterial blood pressure.
         a. Decrease extracellular fluid volume.
         b. Decrease peripheral resistance.
         c. Decrease cardiac output.
      2. Constrict or occlude the afferent arteriole.
      3. Relax the efferent arteriole.
   b. Increase glomerular capsule pressure.
   c. Increase the colloid osmotic pressure of the plasma.
   d. Decrease the permeability of the filtration barrier.
   e. Decrease the total area of the glomeruli available for filtration.

7. Assume that the ascending limb of the loop of Henle and the distal tubules are impermeable to sodium and other ions but actively pump out water. Other characteristics of the kidney are assumed to be unchanged. As the urine moves up the ascending limb, it becomes hyperosmotic, because sodium remains behind as water is pumped out. Assuming that the collecting ducts are impermeable to sodium, upon reaching the collecting ducts the presence or absence of ADH determines the final concentration of the urine. If ADH is absent, there's little or no exchange of water as the urine passes down the collecting ducts and a hyperosmotic urine will be produced. On the other hand, if ADH is present, water moves from the interstitial fluid into the collecting ducts, thus diluting the urine and producing a hypoosmotic urine.

8. Urea is partially responsible for the high osmolality of the interstitial fluid in the medulla of the kidney. Since a high osmolality of the interstitial fluid must exist for the kidney to produce a concentrated urine, a small amount of urea in the kidney results in the production of dilute urine by the kidney.

9. A low-salt diet tends to reduce the osmolality of the blood. Consequently, ADH secretion is inhibited, producing dilute urine and thus eliminating water. This in turn reduces blood volume and blood pressure.

10. As the loops of Henle become longer, the mechanisms that increase concentration of the interstitial fluid of the medulla become more efficient, thus raising the concentration of the interstitial fluid. The maximum concentration for urine is determined by the concentration of the interstitial fluid deep in the medulla of the kidneys. The higher the concentration of interstitial fluid in the medulla of the kidney, the greater the concentration of the urine the kidney is able to produce.

11. Prednisone treatment, because of its aldosterone-like effects, increases the blood volume and it can increase blood pressure. It can also lead to the development of edema as the blood volume increases. The effect of prednisone is to increase Na^+ reabsorption. This increases the Na^+ concentration of the blood and interstitial fluid, and water is retained.

12. Hormones of the adrenal cortex are required to maintain the normal volume of blood and blood pressure. Degeneration of the adrenal cortex results in reduced blood volume and blood pressure. Reduced aldosterone causes less Na^+ to be reabsorbed and leads to an increased urine volume.

Chapter 27

1. When excess glucose is not reabsorbed it osmotically obligates water to remain in the nephron. This results in a large production of urine, called polyuria, with a consequent loss of water, salts, and glucose. The loss of water can be compensated for by increasing fluid intake. The intense thirst that stimulates increased fluid intake is called polydipsia. The loss of salts can be compensated for by increasing the salt intake. The high glucose levels in the blood would increase the blood osmolality, thus stimulating the secretion of ADH. This increases the permeability of the distal convoluted tubule and collecting duct to water. Normally, this would allow reabsorption of water from the collecting ducts and thus conserve water. If glucose levels in the urine are high enough, however, water loss increases even with high levels of ADH being present.

2. When ADH levels first increase, the reabsorption of water increases and urinary output is reduced. This also causes an increase in blood volume and, therefore, an increase in blood pressure. The increased blood pressure increases glomerular filtration rate, which increases urinary output to normal levels. In addition, the increased blood volume inhibits the renin-angiotensin-aldosterone mechanism, inhibits aldosterone secretion, and stimulates natriuretic hormone secretion. These responses also increase urinary output.

3. Elevated ammonia ions in the urine results from an increased secretion of H^+. Increased secretion of H^+ occurs in response to either metabolic or respiratory acidosis. Because an elevated respiratory rate increases blood pH, the most logical conclusion is that the condition is metabolic acidosis, and the observed increase in respiration rate compensates for the metabolic acidosis by lowering H^+ levels.

4. Diarrhea is one of the most common causes of metabolic acidosis, resulting from the loss of bicarbonate ions. Increasing the respiration rate and producing an acidic urine both help to increase the blood pH.

5. Blocking H^+ secretion produces acidosis. Because H^+ are exchanged for Na^+, the Na^+ remain in the urine as sodium bicarbonate. This effectively prevents the reabsorption of HCO_3^- and produces an alkaline urine. The blood pH is reduced because H^+ are not being secreted as rapidly by the nephron. The respiration rate

increases because of the stimulatory effect of decreased blood pH on the respiratory center.

6. Breathing through the glass tube increases the dead air space and decreases the efficiency of gas exchange. Consequently, blood carbon dioxide levels increase and produce a decrease in blood pH. Compensatory responses include an increased respiration rate and the production of acidic urine.

7. A major effect of alkalosis is hyperexcitability of the nervous system. If the girl is prone to having convulsions, then inducing alkalosis might result in a seizure. This could be accomplished by having the girl hyperventilate. The resulting loss of carbon dioxide from the blood causes an increase in blood pH.

8. At high altitudes, we expect stimulation of the chemoreceptor reflex and an increase in respiration rate. This could result in hyperventilation, a decrease in blood carbon dioxide, and respiratory alkalosis. The increased secretion of hydrochloric acid into the stomach could also increase blood pH and contribute to the problem. The kidney produces a more alkaline urine.

Chapter 28

1. Removing the testes would eliminate the major source of testosterone. Blood levels of testosterone would therefore decrease. Because testosterone has a negative-feedback effect on the hypothalamus and pituitary gland, GnRH, FSH, and LH secretion would increase and the blood levels of these hormones would increase.

2. Prior to puberty, the levels of GnRH are very low because the hypothalamus is very sensitive to the inhibitory effects of testosterone. Since GnRH levels are low, so are FSH and LH levels. Loss of the testes and testosterone production would result in an increase in GnRH, FSH, and LH levels. Because little testosterone is produced the boy would not develop sexually and would have no sex drive. Small amounts of androgens would be produced because the adrenal cortex produces some androgens. He would be taller than normal as an adult, with thin bones and weak musculature. His voice would not deepen and the normal masculine distribution of hair would not develop.

3. Ideally the pill would inhibit spermatogenesis. Using the same approach as in females, inhibition of FSH and LH secretion should work. It's known that chronic administration of GnRH suppresses FSH and LH levels enough to cause infertility, through down-regulation. Lack of LH can also result in reduced testosterone levels and a loss of sex drive, however. Some evidence indicates that administration of testosterone in the proper amounts would reduce FSH and LH secretion, thus leading to a reduced sperm cell production. The testosterone, however, maintains normal sex drive. The technique appears to work for a large percentage of males, resulting in a sperm concentration in the semen that's too low to result in fertilization. The technique is not sufficiently precise, however, to be used as a standard birth-control technique.

4. In a postmenopausal woman the ovaries have stopped producing estrogen and progesterone. Without the negative-feedback effect of these

hormones the levels of GnRH, FSH, and LH increase. Removal of the nonfunctioning ovaries in a postmenopausal woman doesn't change the level of any of these hormones or produce any symptoms not already occurring due to the lack of ovarian function.

5. Answer e is correct. The secretory phase of the menstrual cycle occurs after ovulation. It is following ovulation that the corpus luteum forms and produces progesterone. In addition, the progesterone acts on the endometrium of the uterus to cause its maximum development. Progesterone secretion therefore reaches its maximum levels and the endometrium reaches its greatest degree of development during the secretory phase of the menstrual cycle.

6. The removal of the ovaries from a 20-year-old woman eliminates the major site of estrogen and progesterone production, thereby causing an increase in GnRH, FSH, and LH levels due to lack of negative feedback. One expects to see the symptoms of menopause such as cessation of menstruation and reduction in the size of the uterus, vagina, and breasts. There may also be a temporary reduction in sex drive.

7. It's clear that estrogen and progesterone administration resulted in a large decrease in the amount of LH in the plasma the day of ovulation. The differences in plasma LH levels between the groups at other times are very small. The incidence of pregnancies suggests that the reduced plasma LH levels may result in no ovulation.

8. The progesterone inhibits GnRH in the hypothalamus. Consequently, the anterior pituitary is not stimulated to produce LH and FSH. Lack of LH prevents ovulation and lack of FSH prevents development of the follicles. LH also is required for maturation of follicles prior to ovulation. Without follicle development, there's inadequate estrogen production, which causes the hot flash symptoms.

9. GnRH administered either before or after the normal time of ovulation doesn't result in ovulation, because the anterior pituitary is less sensitive to the effect of GnRH during those times. Also, follicles in the ovary are not adequately developed. The concentration of GnRH must be controlled carefully because too little results in inadequate FSH and LH being released from the anterior pituitary. Too little FSH and LH fails to cause ovulation. Too much GnRH given at the proper time results in the maturation of more than one follicle and the release of an oocyte from more than one of the follicles. If the oocytes are fertilized, multiple pregnancies can result.

Chapter 29

1. Postovulatory age, the approximate length of time the embryo has been developing, is 14 days less than the time since the last menstrual period (LMP). In this case, the postovulatory age is 30 days ($44 - 14$). By this time the neural tube has closed, the somites have formed, the digestive tract is developing, the limb buds have appeared, a tubular beating heart is present, and the lungs are developing. Based on reproductive

structures, which are just forming, male and female embryos are indistinguishable at this age.

2. The fever would have occurred on day 21–31 of development, which is during part of the time of neural tube closure (days 18–25). If the fever prevented neural tube closure, the child could be born with anencephalus or spina bifida.

3. The limb buds develop in a proximal-to-distal sequence. If the apical ectodermal ridge is damaged during embryonic development when the limb bud is about one-half grown, the proximal structures, the arm and forearm, develop normally, but the distal structures, the wrist and hand, do not form normally. Depending on the degree of damage, the wrist and hand could be completely absent or underdeveloped.

4. The mesonephric duct system develops, because of testosterone, to form portions of the male reproductive duct system. Without the production of Müllerian-inhibiting hormone, the paramesonephric duct system also develops to form the uterus and uterine tubes. Although ovaries are present, the clitoris may be enlarged because of testosterone to produce somewhat the appearance of male external genitalia. The amount of masculinization would depend on the levels of testosterone and how long it was administered. High levels of testosterone over an extended period would completely masculinize the external genitalia.

5. This total Apgar score of 5 indicates: appearance (A, 0) white or blue; pulse (P, 1) low; grimace (G, 1) slight; activity (A, 1) little movement and poor muscle tone; and respiration (R, 2) normal. The white or blue appearance (A, 0) is consistent with a poor circulation indicated by a reduced pulse (P, 1). The reduced heart rate, resulting in the low pulse, may indicate a circulatory system problem. The reduced reflexes and motor activity (G, 1; A, 1) could result from the lack of oxygen in the muscles resulting from poor circulation. Because the infant has poor circulation despite a normal respiration, clearing the airway (if obstructed) and administering oxygen are in order. This Apgar score could have several causes, and additional information is necessary to determine the actual cause.

6. Suckling the breast stimulates the release of oxytocin from the neurohypophysis (posterior pituitary). Once the oxytocin is in the blood, it travels to both breasts and causes milk letdown.

7. If both parents are heterozygous for dimpled cheeks, then the child could receive a recessive gene for no dimples from each parent, resulting in the homozygous recessive condition with no dimples in the cheeks.

8. It's not possible at present to determine by phenotype if a child is homozygous or heterozygous for tongue rolling. Even if it were possible to determine that the child was heterozygous, that's not very strong evidence that the recessive allele came from the proposed father.

9. Hemophilia is a sex-linked trait. Since the father has hemophilia he must be $X^H Y$. If the mother were $X^H X^H$, all their children would be normal. For half of the children to have hemophilia, she must be $X^H X^h$.

Glossary

Many of the words in this glossary and text are followed by a simplified phonetic spelling showing pronunciation. The pronunciation key reflects standard clinical usage as presented in *Stedman's Medical Dictionary* (27th edition), a leading reference volume in the health sciences. Page numbers indicate where entries are in the text.

The phonetic system used is a basic one and has only a few conventions:

- Two diacritical marks are used; the macron (‾) for long vowels; and the breve (˘) for short vowels.
- Principal stressed syllables are followed by a prime (′); monosyllables do not have a stress mark.
- Other syllables are separated by hyphens.

The following pronunciation key provides examples and consonant sounds encountered in the phonetic system. No attempt has been made to accommodate the slurred sounds common in speech or regional variations in speech sounds. Note that a vowel with a breve (˘) is used for the indefinite vowel sound of the schwa (ə). Native pronunciation of foreign words is approximated as closely as possible.

Pronunciation Key

Vowels

ā	day, care, gauge	f	fit
a	mat, damage	g	got
ă	about, para	h	hit
ah	father	j	jade
aw	fall, cause, raw	k	kept
ē	be, equal, ear	ks	tax
ĕ	taken, genesis	kw	quit
e	term, learn	l	law
ī	pie	m	me
ĭ	pit, sieve, build	n	no
ō	note, for so,	ng	ring
o	not, oncology, ought	p	pan
oo	food	r	rot
ow	cow, out	s	so, miss
oy	troy, void	sh	should
ū	unit, curable	t	ten
ŭ	cut	th	thin, with

Consonants

b	bad	v	very
ch	child	w	we
d	dog	y	yes
dh	this, smooth	z	zero
		zh	azure, measure

In some words the initial sound is not that of the initial letter(s), or the initial letter(s) is not sounded or has a different sound, as in the following examples:

aerobe (ar′ob)	phthalein (thal′e-in)
eimuria (ime′re-a)	pneumonia (nu-mo′ne-a)
gnathic (nath′ik)	psychology (si-kol′o-je)
knuckle (nuk-l)	ptosis (to′sis)
oedipism (ed′i-pizm)	xanthoma (zan-tho′ma)

A

A band   Length of the myosin myofilament in a sarcomere.

abdomen (ab-dō′men, ab′dō-men)   Belly, between the thorax and the pelvis.

abduction (ab-dŭk′shŭn) [L., *abductio,* take away] Movement away from the midline.

absolute refractory period (ab′sŏ-loot rē-frak′tōr-ē) Portion of the action potential during which the membrane is insensitive to all stimuli, regardless of their strength.

absorptive cell (ab-sōrp′tiv)   Cell on the surface of villi of the small intestines and the luminal surface of the large intestine that is characterized by having microvilli; secretes digestive enzymes and absorbs digested materials on its free surface.

absorptive state   Immediately after a meal when nutrients are being absorbed from the intestine into the circulatory system.

accommodation (ă-kom′ŏ-dā′shŭn) [L., *ac + commodo,* to adapt] Ability of electrically excitable tissues, such as nerve or muscle cells, to adjust to a constant stimulus so that the magnitude of the local potential decreases through time.

acetabulum (as-ĕ-tab′ū-lŭm) [L., shallow vinegar vessel or cup] Cup-shaped depression on the external surface of the coxa.

acetylcholine (as-e-til-kō′lēn)   Neurotransmitter substance released from motor neurons, all preganglionic neurons of the parasympathetic and sympathetic divisions, all postganglionic neurons of the parasympathetic division, some postganglionic neurons of the sympathetic division, and some central nervous system neurons.

acetylcholinesterase (as′ē-til-kō-lin-es′ter-ās) Enzyme found in the synaptic cleft that causes the breakdown of acetylcholine to acetic acid and choline, thus limiting the stimulatory effect of acetylcholine.

Achilles tendon   *See* calcaneal tendon.

acid (as′id)   Molecule that is a proton donor; any substance that releases hydrogen ions (H⁺).

acidic   Solution containing more than 10^{-27} mol of hydrogen ions per liter; has a pH less than 7.

acinus; pl., acini (as′i-nŭs, as′ĭ-nī) [L., berry, grape] Grape-shaped secretory portion of a gland. The terms *acinus* and *alveolus* are sometimes used interchangeably. Some authorities differentiate the terms: acini have a constricted opening into the excretory duct, whereas alveoli have an enlarged opening.

acromion (ă-krō′mē-on) [Gr., *akron,* extremity + *omos,* shoulder] Bone comprising the tip of the shoulder.

acrosome (ak′rō-sōm) [Gr., *akron,* extremity + *soma,* body] Cap on the head of the spermatozoon, with hydrolytic enzymes that help the spermatozoon to penetrate the ovum.

actin myofilament (ak′tin)   Thin myofilament within the sarcomere; composed of two F actin molecules, tropomyosin, and troponin molecules.

action potential [L., *potentia,* power, potency] Change in membrane potential in an excitable tissue that acts as an electric signal and is propagated in an all-or-none fashion.

activation energy (ak-ti-vā′shŭn)   Energy that must be added to molecules to initiate a reaction.

active site   Portion of an enzyme in which reactants are brought into close proximity and that plays a role in reducing activation energy of the reaction.

active tension   Tension produced by the contraction of a muscle.

active transport   Carrier-mediated process that requires ATP and can move substances against a concentration gradient.

adaptive immunity   Immune status in which there is an ability to recognize, remember, and destroy a specific antigen.

adenohypophysis (ad′ĕ-nō-hī-pof′ĭ-sis)   Portion of the hypophysis derived from the oral ectoderm; commonly called the anterior pituitary.

adenosine diphosphate (ADP) (ă-den′ō-sēn) Adenosine, an organic base, with two phosphate groups attached to it. Adenosine diphosphate combines with a phosphate group to form adenosine triphosphate.

adenosine triphosphate (ATP)   Adenosine, an organic base, with three phosphate groups attached to it. Energy stored in ATP is used in nearly all of the endergonic reactions in cells.

adenylate cyclase (ad′e-nil sīklās)   An enzyme acting on ATP to form 3′,5′-cyclic AMP plus pyrophosphate (two phosphate groups). A crucial step in the regulation and formation of the intracellular chemical signal 3′,5′-cyclic AMP.

adipocyte (ad′i-pō-sīt)   Fat cell.

adipose (ad′i-pōs) [L., *adeps,* fat]   Fat.

ADP   *See* adenosine diphosphate.

adrenal gland (ă-drē′năl) [L., *ad,* to + *ren,* kidney] Also called the suprarenal gland. Located near the superior pole of each kidney, it is composed of a cortex and a medulla. The adrenal medulla is a highly modified sympathetic ganglion that secretes the hormones epinephrine and norepinephrine; the cortex secretes aldosterone and cortisol as its major secretory products.

adrenaline (ă-dren′ă-lin)   *See* epinephrine.

adrenergic receptor (ad-rĕ-ner′jik)   Receptor molecule that binds to adrenergic agents such as epinephrine and norepinephrine.

adrenocorticotropic hormone (ACTH) (ă-drē′nō-kōr′ti-kō-trō′pik) Hormone of the adenohypophysis that governs the nutrition and growth of the adrenal cortex, stimulates it to functional activity, and causes it to secrete cortisol.

adventitia (ad-ven-tish′ă) [L., *adventicius,* coming from abroad, foreign] Outermost covering of any organ or structure that is properly derived from outside the organ and does not form an integral part of the organ.

aerobic respiration (ār-ō′bik) Breakdown of glucose in the presence of oxygen to produce carbon dioxide, water, and approximately 38 ATPs; includes glycolysis, the citric acid cycle, and the electron-transport chain.

afferent arteriole (af′er-ent) Branch of an interlobular artery of the kidney that conveys blood to the glomerulus.

afferent division Nerve fibers that send impulses from the periphery to the central nervous system.

agglutination (ă-gloo-ti-nā-shŭn) [L., *ad,* to + *gluten,* glue] Process by which blood cells, bacteria, or other particles are caused to adhere to one another and form clumps.

agglutinin (ă-gloo′ti-nin) Antibody that binds to an antigen and causes agglutination.

agglutinogen (ă-gloo-tin′ō-jen) Antigen on surface of red blood cells that can stimulate the production of antibodies (agglutinins) that combine with the antigen and cause agglutination.

agranulocyte (ă-gran′ū-lō-sīt) Nongranular leukocyte (monocyte or lymphocyte).

ala; pl., alae (ā′lă, ā′lē) [L., a wing] Wing-shaped structure.

aldosterone (al-dos′ter-ōn) Steroid hormone produced by the zona glomerulosa of the adrenal cortex that facilitates potassium exchange for sodium in the distal renal tubule, causing sodium reabsorption and potassium and hydrogen secretion.

alkaline (al′kă-līn) Solution containing less than 10^{-7} mol of hydrogen ions per liter; has a pH greater than 7.0.

alkalosis (al-kă-lō′sis) Condition characterized by blood pH of 7.45 or above.

allantois (ă-lan′tō-is) Tube extending from the embryonic hindgut into the umbilical cord; forms the urinary bladder.

allele (ă-lēl′) [Gr., *allelon,* reciprocally] Any one of a series of two or more different genes that may occupy the same position or locus on a specific chromosome.

all-or-none When a stimulus is applied to a cell, an action potential is either produced or not. In muscle cells the cell either contracts to the maximum extent possible (for a given condition) or does not contract.

alternative pathway Part of the nonspecific immune system for activation of complement.

alveolar duct (al-vē′ō-lăr) Part of the respiratory passages beyond a respiratory bronchiole; from it arise alveolar sacs and alveoli.

alveolar gland One in which the secretory unit has a saclike form and an obvious lumen.

alveolar sac Two or more alveoli that share a common opening.

alveolus; pl., alveoli (al-vē′ō-lŭs, al-vē′ō-lī) Cavity. Examples include the sockets into which teeth fit, the endings of the respiratory system, and the terminal endings of secretory glands.

amino acid (ă-mē′nō) Class of organic acids that constitute the building blocks for proteins.

amplitude-modulated signal (am′pli-tood) Signal that varies in magnitude or intensity such as with large versus small concentrations of hormones.

ampulla (am-pul′lă, am-pul′lē) [L., two-handled bottle] Saclike dilatation of a semicircular canal; contains the crista ampullaris. Wide portion of the uterine tube between the infundibulum and the isthmus.

amygdala (a-mig′da-la, -lē). [L. fr. Gr., *amygydale,* almond] A nucleus in the temporal lobe of the brain, amygdaloid nucleus.

amylase (am′il-ās) One of a group of starch-splitting enzymes that cleave starch, glycogen, and related polysaccharides.

anabolism (ă-nab′ō-lizm) [Gr., *anabole,* a raising up] All of the synthesis reactions that occur within the body; requires energy.

anaerobic respiration (an-ār-ō′bik) Breakdown of glucose in the absence of oxygen to produce lactic acid and two ATPs; consists of glycolysis and the reduction of pyruvic acid to lactic acid.

anal canal (ā′năl) Terminal portion of the digestive tract.

anal triangle Posterior portion of the perineal region through which the anal canal opens.

analgesic (an-al-jē′zik) A compound capable of producing analgesia, i.e. one that relieves pain without producing anesthesia or loss of consciousness, characterized by reduced response to painful stimuli.

anaphase (an′ă-fāz) Time during cell division when chromatids divide (or in the case of first meiosis, when the chromosome pairs divide).

anastomoses (ă-nas′tō-mō′sez) A natural communication, direct or indirect, between two blood vessels or other tubular structures. An opening created by surgery, trauma, or disease between two or more normally separate spaces or organs.

anatomic dead air space Volume of the conducting airways from the external environment down to the terminal bronchioles.

androstenedione (an-drō-stēn-dī′ōn) Androgenic steroid of weaker potency than testosterone; secreted by the testis, ovary, and adrenal cortex.

anencephaly (an′en-sef′ă-lē) [Gr., *an* + *enkephalos,* no brain] Defective development of the brain and absence of the bones of the cranium. Only a rudimentary brainstem and some trace of basal ganglia are present.

aneurysm (an′ū-rizm) [Gr., *eurys,* wide] Dilated portion of an artery.

angiotensin I (an-jē-ō-ten′sin) Peptide derived when renin acts on angiotensinogen.

angiotensin II Peptide derived from angiotensin I; stimulates vasoconstriction and aldosterone secretion.

anion (an′ī-on) Ion carrying a negative charge.

antagonist (an-tag′ŏ-nist) Muscle that works in opposition to another muscle.

anterior chamber of eye Chamber of the eye between the cornea and the iris.

anterior interventricular sulcus Groove on the anterior surface of the heart, marking the location of the septum between the two ventricles.

anterior pituitary *See* adenohypophysis.

antibody (an′tē-bod-ē) Protein found in the plasma that is responsible for humoral immunity; binds specifically to antigen.

antibody-mediated immunity Immunity due to B cells and the production of antibodies.

anticoagulant (an′tē-kō-ag′ū-lant) Agent that prevents coagulation.

antidiuretic hormone (ADH) (an′tē-dī-ū-ret′ik) Hormone secreted from the neurohypophysis that acts on the kidney to reduce the output of urine; also called vasopressin because it causes vasoconstriction.

antigen (an′ti-jen) [anti(body) + Gr., *-gen,* producing] Any substance that induces a state of sensitivity or resistance to infection or toxic substances after a latent period; substance that stimulates the specific immune system.

antigenic determinant (an-ti-jen′ik) The specific part of an antigen that stimulates an immune system response by binding to receptors on the surface of lymphocytes.

antithrombin (an-tē-throm′bin) Any substance that inhibits or prevents the effects of thrombin so that blood does not coagulate.

antrum (an′trŭm) [Gr., *antron,* a cave] Cavity of an ovarian follicle filled with fluid containing estrogen.

anulus fibrosus (an′ū-lŭs fī-brō′sus) [L., fibrous ring] Fibrous material forming the outer portion of an intervertebral disk.

anus (ā′nŭs) Lower opening of the digestive tract through which fecal matter is extruded.

aorta (ā-ōr′tă) [Gr., *aorte* from *aeiro,* to lift up] Large elastic artery that is the main trunk of the systemic arterial system; carries blood from the left ventricle of the heart and passes through the thorax and abdomen.

aortic arch (ā-ōr′tik) [L., bow] Curve between the ascending and descending portions of the aorta.

aortic body One of the smallest bilateral structures, similar to the carotid bodies, attached to a small branch of the aorta near its arch; contains chemoreceptors that respond primarily to decreases in blood oxygen; less sensitive to decreases in blood pH or increases in carbon dioxide.

apex (ā′peks) [L., summit or tip] Extremity of a conical or pyramidal structure. The apex of the heart is the rounded tip directed anteriorly and slightly inferiorly.

Apgar score Named for the U.S. anesthesiologist Virginia Apgar (1909–1974). Evaluation of a newborn infant's physical status by assigning numerical values to each of five criteria; appearance (skin color), pulse (heart rate), grimace (response to stimulation), activity (muscle tone), and respiratory effort; a score of 10 indicates the best possible condition.

apical ectodermal ridge Layer of surface ectodermal cells at the lateral margin of the embryonic limb bud; they stimulate growth of the limb.

apical foramen (tooth) [L., aperture] Opening at the apex of the root of a tooth that gives passage to the nerve and blood vessels.

apocrine gland (ap′ō-krin) [Gr., *apo*, away from + *krino*, to separate] Gland whose cells contribute cytoplasm to its secretion (e.g., mammary glands). Sweat glands that produce organic secretions traditionally are called apocrine. These sweat glands, however, are actually merocrine glands. *See also* holocrine and merocrine glands.

appendicular skeleton (ap′en-dik′ū-lăr) The portion of the skeleton consisting of the upper limbs and the lower limbs and their girdles.

appositional growth (ap-ō-zish′ŭn-al) [L., *ap* + *pono*, to put or place] To place one layer of bone, cartilage, or other connective tissue against an existing layer.

aqueous humor (ak′wē-ŭs, ā′kwē-ŭs) Watery, clear solution that fills the anterior and posterior chambers of the eye.

arachnoid (ă-rak′noyd) [Gr., *arachne*, spider, cobweb] Thin, cobweb-appearing meningeal layer surrounding the brain; the middle of the three layers.

arcuate artery (ar′kū-āt) Originates from the interlobar arteries of the kidney and forms an arch between the cortex and medulla of the kidney.

areola (ă-rē′ō-lă, -lē) [L., area] Circular pigmented area surrounding the nipple; its surface is dotted with little projections caused by the presence of the areolar glands beneath.

areolar gland (ă-rē′ō-lăr) Gland forming small, rounded projections from the surface of the areola of the mamma.

arrectores pilorum; pl., arrector pili (ă-rek-tō′rez pī-lōr′um, ă-rek′tōr pī′lī) [L., that which raises; hair] Smooth muscle attached to the hair follicle and dermis that raises the hair when it contracts.

arterial capillary (ar-tē′rē-ăl) Capillary opening from an arteriole or metarteriole.

arteriole (ar-tēr′ē-ōl) Minute artery with all three tunics that transports blood to a capillary.

arteriosclerosis (ar-tēr′ē-ō-skler-ō′sis) [L., *arterio* + Gr., *sklerosis*, hardness] Hardening of the arteries.

arteriovenous anastomosis (ar-tēr′ē-ō-vē′nŭs ă-nas′tō-mō′sis) Vessel through which blood is shunted from an arteriole to a venule without passing through the capillaries.

artery (ar′ter-ē) Blood vessel that carries blood away from the heart.

articular cartilage (ar-tik′ū-lăr kar′ti-lij) Hyaline cartilage covering the ends of bones within a synovial joint.

articulation (ar-tik-ū-lā′shŭn) Place where two bones come together—a joint.

arytenoid cartilages (ar-i-tē′noyd) Small pyramidal laryngeal cartilages that articulate with the cricoid cartilage.

ascending aorta Part of the aorta from which the coronary arteries arise.

ascending colon (kō′lon) Portion of the colon between the small intestine and the right colic flexure.

asthma (az′mă) Condition of the lungs in which widespread narrowing of airways occurs caused by contraction of smooth muscle, edema of the mucosa, and mucus in the lumen of the bronchi and bronchioles.

astrocyte (as′trō-sīt) [Gr., *astron*, star + *kytos*, a hollow, a cell] Star-shaped neuroglia cell involved with forming the blood–brain barrier.

atherosclerosis (ath′er-ō-skler-ō′sis) Arteriosclerosis characterized by irregularly distributed lipid deposits in the intima of large and medium-sized arteries.

atomic number (ă-tom′ik) Number of protons in each type of atom.

ATP *See* adenosine triphosphate.

atrial diastole (ā′trē-ăl dī-as′tō-lē) Dilation of the heart's atria.

atrial natriuretic hormone (ā′trē-ăl nā′trē-ū-ret′ik) Peptide released from the atria when atrial blood pressure is increased; acts to lower blood pressure by increasing the rate of urinary production, thus reducing blood volume.

atrial systole (ā′trē-ăl sis′tō-lē) Contraction of the atria.

atrioventricular bundle (ā′trē-ō-ven-trik′ū-lar) Bundle of modified cardiac muscle fibers that projects from the AV node through the interventricular septum.

atrioventricular (AV) node Small node of specialized cardiac muscle fibers that gives rise to the atrioventricular bundle of the conduction system of the heart.

atrioventricular valve One of two valves closing the openings between the atria and ventricles.

atrium; pl., atria (ā′trē-ŭm, ā′trē-ă) [L., entrance hall] One of two chambers of the heart into which veins carry blood.

auditory cortex (aw′di-tōr-ē kōr′teks) Portion of the cerebral cortex that is responsible for the conscious sensation of sound; in the dorsal portion of the temporal lobe within the lateral fissure and on the superolateral surface of the temporal lobe.

auditory ossicle (os′i-kl) Bone of the middle ear: includes the malleus, incus, and stapes.

auricle (aw′ri-kl) [L., *auris*, ear] Part of the external ear that protrudes from the side of the head. Small pouch projecting from the superior, anterior portion of each atrium of the heart.

auscultatory (aws-kŭltă-tō-rē) Relating to auscultation, listening to the sounds made by the various body structures as a diagnostic method.

autoimmune disease (aw-tō-i-mūn′ di-zēz′) Disease resulting from a specific immune system reaction against self-antigens.

autonomic ganglia (aw-tō-nom′ik gang′glē-ă) Ganglia containing the nerve cell bodies of the autoimmune division of the nervous system.

autonomic nervous system (ANS) Composed of nerve fibers that send impulses from the central nervous system to smooth muscle, cardiac muscle, and glands.

autophagia (aw-tō-fā′jē-ă) [Gr., *auto*, self + *phagein*, to eat] Segregation and disposal of organelles within a cell.

autoregulation (aw′tō-reg-ŭ-lā′shŭn) Maintenance of a relatively constant blood flow through a tissue despite relatively large changes in blood pressure; maintenance of a relatively constant glomerular filtration rate despite relatively large changes in blood pressure.

autorhythmic Spontaneous and periodic; for example, in smooth muscle it implies spontaneous (without nervous or hormonal stimulation) and periodic contractions.

autosome (aw′tō-sōm) [Gr., *auto*, self + *soma*, body] Any chromosome other than a sex chromosome; normally occur in pairs in somatic cells and singly in gametes.

axial skeleton (ak′sē-ăl) Skull, vertebral column, and rib cage.

axillary (ak′sil-ār-ē) Relating to the axilla. The space below the shoulder joint, bounded by the pectoralis major anteriorly, the latissimus dorsi posteriorly, the serratus anterior medially, and the humerus laterally.

axolemma (ak′sō-lem′ă) [Gr., *axo* + *lemma*, husk] Plasma membrane of the axon.

axon (ak′son) [Gr., axis] Main central process of a neuron that normally conducts action potentials away from the neuron cell body.

axon hillock Area of origin of the axon from the nerve cell body.

axoplasm (ak′sō-plazm) Neuroplasm or cytoplasm of the axon.

B

baroreceptor (bar′ō-rē-sep′ter, bar′ō-rē-sep′tōr) (pressoreceptor) Sensory nerve ending in the walls of the atria of the heart, venae cavae, aortic arch, and carotid sinuses; sensitive to stretching of the wall caused by increased blood pressure.

baroreceptor reflex Detects changes in blood pressure and produces changes in heart rate, heart force of contraction, and blood vessel diameter that return blood pressure to homeostatic levels.

basal ganglia (bā′săl gang′glē-ă) Nuclei at the base of the cerebrum involved in controlling motor functions.

base (bās) Molecule that is a proton acceptor; any substance that binds to hydrogen ions.

base [L. and Gr., *basis*] Lower part or bottom of a structure. The base of the heart is the flat portion directed posteriorly and superiorly. Veins and arteries project into and out of the base, respectively.

basement membrane (bās′ment mem′brān) Specialized extracellular material located at the base of epithelial cells and separating them from the underlying connective tissues.

basic (bā′sik) *See* alkaline.

basilar membrane (bas′i-lăr mem′brān) Wall of the membranous labyrinth bordering the scala tympani; supports the organ of Corti.

basophil (bā′sō-fil) [Gr., *basis*, baso + *phileo*, to love] White blood cell with granules that stain specifically with basic dyes; promotes inflammation.

B cell Type of lymphocyte responsible for antibody-mediated immunity.

belly (bel′ē) Largest portion of muscle between the origin and insertion.

beta-oxidation (bā′tă ok-si-dā′shŭn) Metabolism of fatty acids by removing a series of two-carbon units to form acetyl-CoA.

bicarbonate ion (bī-kar′bon-āt) Anion (HCO_3^-) remaining after the dissociation of carbonic acid.

bicuspid [mitral] valve (bī-kŭs′pid) Valve closing the orifice between left atrium and left ventricle of the heart.

bile (bīl)   Fluid secreted from the liver into the duodenum; consists of bile salts, bile pigments, bicarbonate ions, cholesterol, fats, fat-soluble hormones, and lecithin.

bile canaliculus (bīl kan′ă-lik′ū-lŭs)   One of the intercellular channels approximately 1 μm or less in diameter that occurs between liver cells into which bile is secreted; empties into the hepatic ducts.

bile salt   Organic salt secreted by the liver that functions as an emulsifying agent.

bilirubin (bil-i-roo′bin) [L., *bili* + *ruber*, red]   Bile pigment derived from hemoglobin during destruction of red blood cells.

biliverdin (bil-i-ver′din)   Green bile pigment formed from the oxidation of bilirubin.

binocular vision (bin-ok′ū-lăr) [L., *bini*, paired + *oculus*, eye]   Vision using two eyes at the same time; responsible for depth perception when the visual field of each eye overlaps.

bipolar neuron (bī-pō′ler)   One of the three categories of neurons consisting of a neuron with two processes—one dendrite and one axon—arising from opposite poles of the cell body.

blastocele (blas′tō-sēl) [Gr., *blastos*, germ + *koilos*, hollow]   Cavity in the blastocyst.

blastocyst (blas′tō-sist) [Gr., *blastos*, germ + *kystis*, bladder]   Stage of mammalian embryos that consists of the inner cell mass and a thin trophoblast layer enclosing the blastocele.

bleaching   In response to light, retinal separates from opsin.

blind spot (blīnd)   Point in the retina where the optic nerve penetrates the fibrous tunic; contains no rods or cones and therefore does not respond to light.

blood–brain barrier   Permeability barrier controlling the passage of most large-molecular compounds from the blood to the cerebrospinal fluid and brain tissue; consists of capillary endothelium and may include the astrocytes.

blood clot   Coagulated phase of blood.

blood colloid osmotic pressure   Osmotic pressure due to the concentration difference of proteins across a membrane that does not allow passage of the proteins.

blood groups   Classification of blood based on the type of antigen found on the surface of red blood cells.

blood island   Aggregation of mesodermal cells in the embryonic yolk sac that forms vascular endothelium and primitive blood cells.

blood pressure [L., *pressus*, to press]   Tension of the blood within the blood vessels; commonly expressed in units of millimeters of mercury (mm Hg).

blood–thymic barrier   Layer of reticular cells that separates capillaries from thymic tissue in the cortex of the thymus gland; prevents large molecules from leaving the blood and entering the cortex.

Bohr effect   Named for the Danish physiologist Christian Bohr (1855–1911). Shift of the oxygen–hemoglobin dissociation curve to the right or left because of changes in blood pH. The definition sometimes is extended to include shifts caused by changes in blood carbon dioxide levels.

bony labyrinth (lab′i-rinth)   Part of the inner ear; contains the membranous labyrinth that forms the cochlea, vestibule, and semicircular canals.

brachial (brā′kē-āl) [L., *brachium*, arm]   Relating to the arm.

branchial arch   Typically, six arches in vertebrates; in the lower vertebrates they bear gills, but they appear transiently in the higher vertebrates and give rise to structures in the head and neck.

broad ligament   Peritoneal fold passing from the lateral margin of the uterus to the wall of the pelvis on either side.

bronchiole (brong′kē-ōl)   One of the finer subdivisions of the bronchial tubes, less than 1 mm in diameter; has no cartilage in its wall but does have relatively more smooth muscle and elastic fibers.

brush border   Epithelial surface consisting of microvilli.

buffer (bŭf′er)   Mixture of an acid and base that reduces any changes in pH that would otherwise occur in a solution when acid or base is added to the solution.

bulbar conjunctiva (bŭl′bar kon-junk-tī′vă)   Conjunctiva that covers the surface of the eyeball.

bulbourethral gland (bŭl′bō-ū-rē′thrăl)   One of two small compound glands that produce a mucoid secretion; it discharges through a small duct into the spongy urethra.

bulb of the penis   Expanded posterior part of the corpus spongiosum of the penis.

bulb of the vestibule   Mass of erectile tissue on either side of the vagina.

bulbus cordis (bŭl′bŭs) [L., plant bulb]   End of the embryonic cardiac tube where blood leaves the heart; becomes part of the ventricle.

bursa; pl., bursae (ber′să, ber′sē) [L., purse]   Closed sac or pocket containing synovial fluid, usually found in areas where friction occurs.

bursitis (ber-sī′tis) [L., *purse* + Gr., *ites*, inflammation]   Inflammation of a bursa.

C

calcaneal tendon (kal-kā′nē-al)   Common tendon of the gastrocnemius, soleus, and plantaris muscle that attaches to the calcaneus.

calcitonin (kal-si-tō′nin)   Hormone released from parafollicular cells that acts on tissues to cause a decrease in blood levels of calcium ions.

calmodulin (kal-mod′ū-lin) [calcium + modulate]   Protein receptor for Ca^{2+} that plays a role in many Ca^{2+}-regulated processes such as smooth muscle contraction.

calorie (kal′ō-rē) [L., *calor*, heat]   Unit of heat content or energy. The quantity of energy required to raise the temperature of 1 g of water 1°C.

calpain (kal′pān)   Enzyme involved in changing the shape of dendrites; involved with long-term memory.

calyx; pl., calyces (kā′liks, kal′i-sēz) [Gr., cup of a flower]   Flower-shaped or funnel-shaped structure; specifically, one of the branches or recesses of a renal pelvis into which the tips of the renal pyramids project.

canal of Schlemm   Named for the German anatomist Friedrich Schlemm (1795–1858). Series of veins at the base of the cornea that drain excess aqueous humor from the eye.

cancellous bone (kan′sē-lŭs) [L., grating or lattice]   Bone with a latticelike appearance; spongy bone.

cancer (kan′ser)   General term frequently used to indicate any of various types of malignant neoplasms, most of which invade surrounding tissues, may metastasize to several sites, and are likely to recur after attempted removal and to cause death of the patient unless adequately treated.

canine (kā′nīn)   Referring to the cuspid tooth.

cannula (kan′ū-lă) [L., *canna*, reed]   Tube; often inserted into an artery or vein.

capacitation (kă-pas′i-tā′shŭn) [L., *capax*, capable of]   Process whereby spermatozoa acquire the ability to fertilize ova. This process occurs in the female genital tract.

capitulum (kă-pit′ū-lŭm) [L., *caput*, head]   Head-shaped structure.

carbaminohemoglobin (kar-bam′i-nō-hē-mō-glō′bin)   Carbon dioxide bound to hemoglobin by means of a reactive amino group on the hemoglobin.

carbohydrate (kar-bō-hī′drāt)   Monosaccharide (simple sugar) or the organic molecules composed of monosaccharides bound together by chemical bonds, for example, glycogen. For each carbon atom in the molecule there are typically one oxygen molecule and two hydrogen molecules.

carbonic acid/bicarbonate buffer system (kar-bon′ik)   One of the major buffer systems in the body; major components are carbonic acid and bicarbonate ions.

carbonic anhydrase   Enzyme that catalyzes the reaction between carbon dioxide and water to form carbonic acid.

carcinoma (kar-si-nō′mă)   A malignant neoplasm derived from epithelial tissue.

cardiac (kar′dē-ak) [Gr., *kardia*, heart]   Related to the heart.

cardiac cycle [Gr., *kyklos*, circle]   Complete round of cardiac systole and diastole.

cardiac nerve   Nerve that extends from the sympathetic chain ganglia to the heart.

cardiac output (minute volume)   Volume of blood pumped by the heart per minute.

cardiac region   Region of the stomach near the opening of the esophagus.

cardiac reserve [L., *re* + *servo*, to keep back, reserve]   Work that the heart is able to perform beyond that required during ordinary circumstances of daily life.

carotid body (ka-rot′id)   One of the small organs near the carotid sinuses; contains chemoreceptors that respond primarily to decreases in blood oxygen; less sensitive to decreases in blood pH or increases in carbon dioxide.

carotid sinus   Enlargement of the internal carotid artery near the point where the internal carotid artery branches from the common carotid artery; contains baroreceptors.

carpal (kar′păl) [Gr., *karpos*, wrist]   Bone of the wrist.

carrier Person in apparent health whose chromosomes contain a pathologic mutant gene that may be transmitted to his or her children.

cartilage (kar′ti-lij) [L., *cartilage*, gristle] Firm, smooth, resilient, nonvascular connective tissue.

cartilaginous joint (kar-ti-laj′i-nŭs) Bones connected by cartilage; includes synchondroses and symphyses.

catabolism (kă-tab′ō-lizm) [Gr., *katabole*, a casting down] All of the decomposition reactions that occur in the body; releases energy.

catalyst (kat′ă-list) Substance that increases the rate at which a chemical reaction proceeds without being changed permanently.

cataract (kat′a-rakt) Complete or partial opacity of the lens of the eye.

cations (kat′ī-on) [Gr., *kation*, going down] Ions carrying a positive charge.

caveola; pl., caveolae (kav-ē-ō′lă, kav-ē-ō′lē) [L., a small pocket] Shallow invagination in the membranes of smooth muscle cells that may perform a function similar to both the T tubules and sarcoplasmic reticulum of skeletal muscle.

cecum (sē′kŭm, sē′kă) [L., *caecus*, blind] Cul-de-sac forming the first part of the large intestine.

cell-mediated immunity Immunity due to the actions of T cells and null cells.

celom (sē′lom, sē-lō′mă) [Gr., *koilo* + *amma*, a hollow] Principal cavities of the trunk, for example, the pericardial, pleural, and peritoneal cavities. Separate in the adult, they are continuous in the embryo.

cementum (se-men′tŭm) [L., *caementum*, rough quarry stone] Layer of modified bone covering the dentin of the root and neck of a tooth; blends with the fibers of the periodontal membrane.

central nervous system (CNS) Major subdivision of the nervous system consisting of the brain and spinal cord.

central vein Terminal branches of the hepatic veins that lie centrally in the hepatic lobules and receive blood from the liver sinusoids.

centrosome (sen′trō-sōm) Specialized zone of cytoplasm close to the nucleus and containing two centrioles.

cerebellum (ser-e-bel′ŭm) [L., little brain] Separate portion of the brain attached to the brainstem at the pons; important in maintaining muscle tone, balance, and coordination of movement.

cerebrospinal fluid (ser′ĕ-brō-spī-năl) Fluid filling the ventricles and surrounding the brain and spinal cord.

ceruminous glands (sĕ-roo′mi-nŭs) Modified sebaceous glands in the external acoustic meatus that produce cerumen (earwax).

cervical canal (ser′vĭ-kal) Canal extending from the isthmus of the uterus to the opening of the uterus into the vagina.

cervix; pl., cervices (ser′viks, ser-vī′sēz) [L., neck] Lower part of the uterus extending from the isthmus of the uterus into the vagina.

chalazion (ka-lā′zē-on) A chronic inflammation of a meibomian gland.

cheek (chēk) Side of the face forming the lateral wall of the mouth.

chemical signal A molecule that binds to a macromolecule such as receptors or enzymes and alters their function.

chemoreceptor (kē′mō-rē-sep′tor) Sensory cell that is stimulated by a change in the concentration of chemicals to produce action potentials. Examples include taste receptors, olfactory receptors, and carotid bodies.

chemoreceptor reflex Chemoreceptors detect decrease in blood oxygen, increase in carbon dioxide, or decrease in pH and produce an increased rate and depth of respiration, and by means of the vasomotor center, vasoconstriction.

chemosensitive area (kem-ō-sen′si-tiv, kē-mō-sen′si-tiv) Chemosensitive neurons in the medulla oblongata detect changes in blood, carbon dioxide, and pH.

chemotactic factor (kē-mō-tak′tik) Part of a microorganism or chemical released by tissues and cells that act as chemical signals to attract leukocytes.

chemotaxis (kē-mo-tak′sis) [Gr., *chemo* + *taxis*, orderly arrangement] Attraction of living protoplasm (cells) to chemical stimuli.

chief cell Cell of the parathyroid gland that secretes parathyroid hormone. Cell of a gastric gland that secretes pepsinogen.

chloride (klōr′īd) Compound containing chlorine, for example, salts of hydrochloric acid.

chloride shift Diffusion of chloride ions into red blood cells as bicarbonate ions diffuse out; maintains electrical neutrality inside and outside the red blood cells.

choana; pl., choanae (kō′an-ă, kō-ā′nē) *See* internal naris.

cholecystokinin (kō′lē-sis-tō-kī′nin) Hormone liberated by the upper intestinal mucosa on contact with gastric contents; stimulates the contraction of the gallbladder and the secretion of pancreatic juice high in digestive enzymes.

cholinergic neuron (kol-in-er′jik) Refers to nerve fibers that secrete acetylcholine as a neurotransmitter substance.

chondroblast (kon′drō-blast) [Gr., *chondros*, gristle, cartilage + *blastos*, germ] Cartilage-producing cell.

chondrocyte (kon′drō-sīt) [Gr., *chondros*, gristle, cartilage + *kytos*, a cell] Mature cartilage cell.

chorda tympani; pl., chordae (kōr′dă tim′pan-ē, kōr′dē) Branch of the facial nerve that conveys taste sensation from the front two-thirds of the tongue.

chordae tendineae (kōr′dă ten′di-nē-ē) [L., cord] Tendinous strands running from the papillary muscles to the atrioventricular valves.

choroid (ko′royd) Portion of the vascular tunic associated with the sclera of the eye.

choroid plexus [Gr., *chorioeides*, membranelike] Specialized plexus located within the ventricles of the brain that secretes cerebrospinal fluid.

chromatid (krō′mă-tid) One-half of a chromosome; separates from its partner during cell division.

chromatin (krō′ma-tin) Colored material; the genetic material in the nucleus.

chromosome (krō′mō-sōm) Colored body in the nucleus, composed of DNA and proteins and containing the primary genetic information of the cell; 23 pairs in humans.

chronic pain Prolonged or long term pain.

chylomicron (kī-lō-mī′kron) [Gr., *chylos*, juice + *micros*, small] Microscopic particle of lipid surrounded by protein, occurring in chyle and in blood.

chymotrypsin (kī-mō-trip′sin) Proteolytic enzyme formed in the small intestine from the pancreatic precursor chymotrypsinogen.

ciliary body (sil′ē-ar-ē) Structure continuous with the choroid layer at its anterior margin that contains smooth muscle cells and functions in accommodation.

ciliary gland Modified sweat gland that opens into the follicle of an eyelash, keeping it lubricated.

ciliary muscle Smooth muscle in the ciliary body of the eye.

ciliary process Portion of the ciliary body of the eye that attaches by suspensory ligaments to the lens.

ciliary ring Portion of the ciliary body of the eye that contains smooth muscle cells.

circumduction (ser-kŭm-dŭk′shŭn) [L., around + *ductus*, to draw] Movement in a circular motion.

circumferential lamellae (ser-kŭm-fer-en′shē-al lă-mel′ē) Lamellae covering the surface of and extending around compact bone inside the periosteum.

circumvallate papilla (ser-kŭm-val′ăt pă-pil′ă) Type of papilla on the surface of the tongue surrounded by a groove.

cisterna; pl., cisternae (sis-ter′nă, sis-ter′nē) Interior space of the endoplasmic reticulum.

cisterna chyli (kīl′ē) [L., tank + Gr., *chylos*, juice] Enlarged inferior end of the thoracic duct that receives chyle from the intestine.

citric acid cycle (sit′rik) Series of chemical reactions in which citric acid is converted into oxaloacetic acid, carbon dioxide is formed, and energy is released. The oxaloacetic acid can combine with acetyl-CoA to form citric acid and restart the cycle. The energy released is used to form NADH, FADH, and ATP.

classical pathway Part of the specific immune system for activation of complement.

clavicle (klav′i-kl) The collarbone, between the sternum and scapula.

cleavage furrow (klēv′ij) Inward pinching of the plasma membrane that divides a cell into two halves, which separate from each other to form two new cells.

cleft palate (kleft) Failure of the embryonic palate to fuse along the midline, resulting in an opening through the roof of the mouth.

clinical age (klin′i-kl) Age of the developing fetus from the time of the mother's last menstrual period before pregnancy.

clinical perineum (klin′i-kl per′i-nē′ŭm) Portion of the perineum between the vaginal and anal openings.

clitoris (klit′ō-ris) Small cylindrical, erectile body, rarely exceeding 2 cm in length, situated at the most anterior portion of the vulva and projecting beneath the prepuce.

cloaca (klō-ā′kă) [L., sewer] In early embryos the endodermally lined chamber into which the hindgut and allantois empty.

cloning (klōn′ing) Growing a colony of genetically identical cells or organisms.

clot retraction Condensation of the clot into a denser, compact structure; caused by the elastic nature of fibrin.

coagulation (kō-ag-ū-lā′shŭn) Process of changing from liquid to solid, especially of blood; formation of a blood clot.

cochlear duct (kok′lē-ăr) Interior of the membranous labyrinth of the cochlea; cochlear canal.

cochlear nerve Nerve that carries sensory impulses from the organ of Corti to the vestibulocochlear nerve.

cochlear nucleus Neurons from the cochlear nerve synapse within the dorsal or ventral cochlear nucleus in the superior medulla oblongata.

codon (kō′don) Sequence of three nucleotides in mRNA or DNA that codes for a specific amino acid in a protein.

cofactor (kō′fak′ter, kō-fak′ tōr) Nonprotein component of an enzyme such as coenzymes and inorganic ions essential for enzyme action.

collagen (kol′lă-jen) [Gr., *koila*, glue + *gen*, producing] Ropelike protein of the extracellular matrix.

collateral ganglia (ko-lat′er-ăl gang′glē-ă) Sympathetic ganglia that are found at the origin of large abdominal arteries; include the celiac, superior, and inferior mesenteric arteries.

collecting duct Straight tubule that extends from the cortex of the kidney to the tip of the renal pyramid. Filtrate from the distal convoluted tubes enters the collecting duct and is carried to the calyces.

colloid (kol′oyd) [Gr., *kolla*, glue + *eidos*, appearance] Atoms or molecules dispersed in a gaseous, liquid, or solid medium that resist separation from the liquid or gas.

colloidal solution (ko-loyd′ăl) Fine particles suspended in a liquid; particles are resistant to sedimentation or filtration.

colon (kō′lon) Division of the large intestine that extends from the cecum to the rectum.

colostrum (kō-los′trŭm) Thin, white fluid; the first milk secreted by the breast at the termination of pregnancy; contains less fat and lactose than the milk secreted later.

columnar Shaped like a column.

commissure (kom′i-shŭr) [L., *commissura*, a joining together] Connection of nerve fibers between the cerebral hemispheres or from one side of the spinal cord to the other.

common bile duct Duct formed by the union of the common hepatic and cystic ducts; it empties into the small intestine.

common hepatic duct Part of the biliary duct system that is formed by the joining of the right and left hepatic ducts.

compact bone Bone that is more dense and has fewer spaces than cancellous bone.

competition Similar molecules binding to the same carrier molecule or receptor site.

complement (kom′plĕ-ment) Group of serum proteins that stimulates phagocytosis and inflammation.

complement cascade Series of reactions in which each component activates the next component, resulting in activation of complement proteins.

compliance (kom-plī′ans) Change in volume (e.g., in lungs or blood vessels) caused by a given change in pressure.

compound (kom′pownd) A substance composed of two or more different types of atoms that are chemically combined.

concha; pl., conchae (kon′kă, kon′kē) [L., shell] Structure comparable to a shell in shape; the three bony ridges on the lateral wall of the nasal cavity.

conduction (kon-dŭk′shŭn) [L., *con* + *ductus*, to lead, conduct] Transfer of energy such as heat from one point to another without evident movement in the conducting body.

cone (kōn) Photoreceptor in the retina of the eye; responsible for color vision.

congenital (kon-jen′i-tăl) [L., *congenitus*, born with] Occurring at birth; may be genetic or due to some influence (e.g., drugs) during development.

conjunctiva (kon-jŭnk-tī′vă) [L., *conjungo*, to bind together] Mucous membrane covering the anterior surface of the eyeball and lining the lids.

conjunctival fornix (kon-jŭnk-tī′văl fōr′niks) Area in which the palpebral and bulbar conjunctiva meet.

constant region Portion of the antibody that does not combine with the antigen and is the same in different antibodies.

continuous capillary [L., *capillaris*, relating to hair] Capillary in which pores are absent; less permeable to large molecules than other types of capillaries.

contraction phase (kon-trak′shŭn) One of the three phases of muscle contraction; the time during which tension is produced by the contraction of muscle.

convection (kon-vek′shŭn) [L., *con* + *vectus*, to carry or bring together] Transfer of heat in liquids or gases by movement of the heated particles.

coracoid (kōr′ă-koyd) [Gr., *korakodes*, crow's beak] Resembling a crow's beak, for example, a process on the scapula.

Cori cycle Named for the Czech-U.S. biochemist and Nobel laureate, Carl F. Cori (1896–1984). Lactic acid, produced by skeletal muscle, is carried in the blood to the liver, where it is aerobically converted into glucose. The glucose may return through the blood to skeletal muscle or may be stored as glycogen in the liver.

cornea (kōr′nē-ă) Transparent portion of the fibrous tunic that makes up the outer wall of the anterior portion of the eye.

corniculate cartilage (kōr-nik′ū-lāt) Conical nodule of elastic cartilage surmounting the apex of each arytenoid cartilage.

corona radiata A single layer of columnar cells derived from the cumulus mass, which anchor on the zona pellucida of the oocyte in a secondary follicle.

coronary (kōr′o-nār-ē) [L., *coronarius*, a crown] Resembling a crown; encircling.

coronary artery One of two arteries that arise from the base of the aorta and carry blood to the muscle of the heart.

coronary ligament Peritoneal reflection from the liver to the diaphragm at the margins of the bare area of the liver.

coronary sinus Short trunk that receives most of the veins of the heart and empties into the right atrium.

coronoid (kōr′ŏ-noyd) [Gr., *korone*, a crow] Shaped like a crow's beak, for example, a process on the mandible.

corpus; pl., corpora (kōr′pŭs, -pōr-ă) [L., *body*] Any body or mass; the main part of an organ.

corpus albicans (al′bi-kanz) Atrophied corpus luteum leaving a connective tissue scar in the ovary.

corpus callosum (kăl-lō′sŭm) [L., *body* + callous] Largest commissure of the brain, connecting the cerebral hemispheres.

corpus cavernosum; pl., corpora cavernosa One of two parallel columns of erectile tissue forming the dorsal part of the body of the penis or the body of the clitoris.

corpus luteum (lū′tē-ŭm) Yellow endocrine body formed in the ovary in the site of a ruptured vesicular follicle immediately after ovulation; secretes progesterone and estrogen.

corpus luteum of pregnancy Large corpus luteum in the ovary of a pregnant female; secretes large amounts of progesterone and estrogen.

corpus spongiosum (spŭn′jē-ō′sŭm) Median column of erectile tissue located between and ventral to the two corpora cavernosa in the penis; posteriorly it forms the bulb of the penis, and anteriorly it terminates as the glans penis; it is traversed by the urethra. In the female it forms the bulb of the vestibule.

corpus striatum (strī-ā′tŭm) [L., *corpus*, body + *striatus*, striated or furrowed] Collective term for the caudate nucleus, putamen, and globus pallidus; so named because of the striations caused by intermixing of gray and white matter that results from the number of tracts crossing the anterior portion of the corpus striatum.

cortex; pl., cortices (kōr′teks, kōr′ti-sēz) [L., bark] Outer portion of an organ (e.g., adrenal cortex or cortex of the kidney).

corticotropin-releasing hormone (kōr′ti-kō-trō′pin) Hormone from the hypothalamus that stimulates the anterior pituitary gland to release adrenocorticotropic hormone.

cortisol (kōr′ti-sol) Steroid hormone released by the zona fasciculata of the adrenal cortex; increases blood glucose and inhibits inflammation.

cotransport Carrier-mediated simultaneous movement of two substances across a membrane in the same direction.

covalent bond (kō-văl′ent) Chemical bond characterized by the sharing of electrons.

coxa (kok′să) Hipbone.

cranial nerve (krā′nē-ăl) Nerve that originates from a nucleus within the brain; there are 12 pairs of cranial nerves.

cranial vault Eight skull bones that surround and protect the brain; braincase.

craniosacral division (krā′nē-ō-sā′krăl) Synonym for the parasympathetic division of the autonomic nervous system.

cranium (krā′nē-ŭm) [Gr., *kranion*, skull] Skull; in a more limited sense, the braincase.

cremaster muscle (krē-mas′ter) Extension of abdominal muscles originating from the internal oblique muscles; in the male, raises the testicles; in the female, envelops the round ligament of the uterus.

crenation (krē-nā′shŭn) [L., *crena*, notched] Denoting the outline of a shrunken cell.

cricoid cartilage (krī′koyd) Most inferior laryngeal cartilage.

cricothyrotomy (krī′kō-thī-rot′ō-mē) Incision through the skin and cricothyroid membrane for relief of respiratory obstruction.

crista ampullaris (kris′tă am-pul′ăr-ĭs) [L., crest] Elevation on the inner surface of the ampulla of each semicircular duct for dynamic or kinetic equilibrium.

cristae (kris′tă, kris′tē) [L., crest] Shelflike infoldings of the inner membrane of a mitochondrion.

critical closing pressure Pressure in a blood vessel below which the vessel collapses, occluding the lumen and preventing blood flow.

crown (tooth) That part of a tooth that is covered with enamel.

cruciate (kroo′shē-āt) [L., *cruciatus*, cross] Resembling or shaped like a cross.

crus of the penis (krŭs) Posterior portion of the corpus cavernosum of the penis attached to the ischiopubic ramus.

crypt (kript) A pitlike depression or tubular recess.

cryptorchidism (krip-tōr′ki-dizm) Failure of the testis to descend.

crystalline (kris′tă-lēn) Protein that fills the epithelial cells of the lens in the eye.

cuboidal Something that resembles a cube.

cumulus mass (kū′mū-lŭs) *See* cumulus oophorus.

cumulus oophorus (ō-of′ōr-ŭs) [L., a heap] Mass of epithelial cells surrounding the oocyte; also called the cumulus mass.

cuneiform cartilage (kū′nē-i-fōrm) Small rod of elastic cartilage above each corniculate cartilage in the larynx.

cupula; pl., cupulae (koo′poo-lă, kū′pū-lă, koo′poo-lē) [L., *cupa*, a tub] Gelatinous mass that overlies the hair cells of the cristae ampullares of the semicircular ducts.

cuticle (kū′ti-kl) [L., *cutis*, skin] Outer thin layer, usually horny, for example, the outer covering of hair or the growth of the stratum corneum onto the nail.

cystic duct (sis′tik) Duct leading from the gallbladder; joins the common hepatic duct to form the common bile duct.

cytokine (sī′tō-kīn) A protein or peptide secreted by a cell that functions to regulate the activity of neighboring cells.

cytokinesis (sī′tō-ki-nē′sis) [Gr., *cyto*, cell + *kinsis*, movement] Division of the cytoplasm during cell division.

cytology (sī-tol′ō-jē) [Gr., *kytos*, a hollow (cell) + *logos*, study] Study of anatomy, physiology, pathology, and chemistry of the cell.

cytoplasm (sī′tō-plazm) Protoplasm of the cell surrounding the nucleus.

cytoplasmic inclusion (sī-tō-plaz′mik) Any foreign or other substance contained in the cytoplasm of a cell.

cytotoxic reaction (sī′tō-tok′sik) [Gr., *cyto*, cell + L., *toxic*, poison] Antibodies (IgG or IgM) combine with cells and activate complement, and cell lysis occurs.

cytotrophoblast (sī′tō-trof′ō-blast) Inner layer of the trophoblast composed of individual cells.

D

Daily Value Dietary reference values useful for planing a healthy diet. The Daily Values are taken from the Reference Daily Intakes (RDIs) and the Daily Reference Values.

Daily Reference Values Recommended amounts in the diet for total fat, saturated fat, cholesterol, total carbohydrate, dietary fiber, sodium, potassium, and protein. The values for total fat, saturated fat, cholesterol, and sodium are the uppermost limit considered desirable because of their link to certain diseases.

Dalton's law Named for the English chemist John Dalton (1766–1844). In a mixture of gases the portion of the total pressure resulting from each type of gas is determined by the percentage of the total volume represented by each gas type.

dartos muscle (dar′tōs) Layer of smooth muscle in the skin of the scrotum; contracts in response to lower temperature and relaxes in response to higher temperature; raises and lowers testes in the scrotum.

deciduous tooth (dē-sid′ū-ŭs) Tooth of the first set of teeth; primary tooth.

decussate (dē′kŭ-sāt, dē-kŭs′āt) [L., *decusso*, X-shaped, from *decussis*, ten(X)] To cross.

deep inguinal ring (ing′gwi-năl) Opening in the transverse fascia through which the spermatic cord (or round ligament in the female) enters the inguinal canal.

defecation (def-ě-kā′shŭn) [L., *defaeco*, to remove the dregs, purify] Discharge of feces from the rectum.

defecation reflex Combination of local and central nervous system reflexes initiated by distention of the rectum and resulting in movement of feces out of the lower colon.

deglutition (dē-gloo-tish′ŭn) [L., *de* + *glutio*, to swallow] Act of swallowing.

dendrite (den′drīt) [Gr., *dendrites*, tree] Branching processes of a neuron that receives stimuli and conducts potentials toward the cell body.

dendritic cell (den-drit′ik) Large cells with long cytoplasmic extensions that are capable of taking up and concentrating antigens leading to activation of B or T lymphocytes.

dendritic spine Extension of nerve cell dendrites where axons form synapses with the dendrites; also called gemmule.

dental arch (den′tăl) [L., *arcus*, bow] Curved maxillary or mandibular arch in which the teeth are located.

dentin (den′tin) Bony material forming the mass of the tooth.

deoxyhemoglobin (dē-oks′ē-hē-mō-glō′bin) Hemoglobin without oxygen bound to it.

deoxyribonuclease (de-oks′ē-rī-bō-noo′klē-ās) Enzyme that splits DNA into its component nucleotides.

deoxyribonucleic acid (DNA) (dē-oks′ē-rī′bō-noo-klē′ic) Type of nucleic acid containing deoxyribose as the sugar component, found principally in the nuclei of cells; constitutes the genetic material of cells.

depolarization (dē-pō′lăr-i-zā′shŭn) Change in the electric charge difference across the plasma membrane that causes the difference to be smaller or closer to 0 mV; phase of the action potential in which the membrane potential moves toward zero, or becomes positive.

depression (dē-presh′ŭn) Movement of a structure in an inferior direction.

depth perception (per-sep′shun) Ability to distinguish between near and far objects and to judge their distance.

dermatome (der′mă-tōm) Area of skin supplied by a spinal nerve.

dermis (der′mis) [Gr., *derma*, skin] Dense irregular connective tissue that forms the deep layer of the skin.

descending aorta Part of the aorta, further divided into the thoracic aorta and abdominal aorta.

descending colon Part of the colon extending from the left colonic flexure to the sigmoid colon.

desmosome (dez′mō-sōm) [Gr., *desmos*, a band + *soma*, body] Point of adhesion between cells. Each contains a dense plate at the point of adhesion and a cementing extracellular material between the cells.

desquamate (des′kwă-māt) [L., *desquamo*, to scale off] Peeling or scaling off of the superficial cells of the stratum corneum.

diabetes insipidus (dī-ă-bē′tēz in-sip′i-dŭs) Chronic excretion of large amounts of urine of low specific gravity accompanied by extreme thirst; results from inadequate output of antidiuretic hormone.

diabetes mellitus (me-lī′tŭs) Metabolic disease in which carbohydrate use is reduced and that of lipid and protein enhanced; caused by deficiency of insulin or an inability to respond to insulin and is characterized, in more severe cases, by hyperglycemia, glycosuria, water and electrolyte loss, ketoacidosis, and coma.

diapedesis (dī′ă-pě-dē′sis) [Gr., *dia*, through + *pedesis*, a leaping] Passage of blood or any of its formed elements through the intact walls of blood vessels.

diaphragm (dī′ă-fram) Musculomembranous partition between the abdominal and thoracic cavities.

diaphysis (dī-af′i-sis) [Gr., growing between] Shaft of a long bone.

diastole (dī-as′tō-lē) [Gr., *diastole*, dilation] Relaxation of the heart chambers during which they fill with blood; usually refers to ventricular relaxation.

diencephalon (dī-en-sef′ă-lon) [Gr., *dia*, through + *enkephalos*, brain] Second portion of the embryonic brain; in the inferior core of the adult cerebrum.

diffuse lymphatic tissue Dispersed lymphocytes and other cells with no clear boundary; found beneath mucous membranes, around lymph nodules, and within lymph nodes and spleen.

diffusion (di-fū′zhŭn) [L., *diffundo*, to pour in different directions] Tendency for solute molecules to move from an area of high concentration to an area of low concentration in solution; the product of the constant random motion of all atoms, molecules, or ions in a solution.

diffusion coefficient Measure of how easily a gas diffuses through a liquid or tissue.

digestive tract (di-jes′tiv, dī-jes′tiv) Mouth, oropharynx, esophagus, stomach, small intestine, and large intestine.

digit (dij′it) Finger, thumb, or toe.

dilator pupillae (dī′lă-tĕr pū-pil′ē) Radial smooth muscle cells of the iris diaphragm that cause the pupil of the eye to dilate.

diploid (dip′loyd) Normal number of chromosomes (in humans, 46 chromosomes) in somatic cells.

disaccharide (dī-sak′ă-rīd) Condensation product of two monosaccharides by elimination of water.

dissociate (di-sō′sē-āt) [L., *dis* + *socio*, to disjoin, separate] Ionization in which ions are dissolved in water and the cations and anions are surrounded by water molecules.

distal tubule Convoluted tubule of the nephron that extends from the ascending limb of the loop of Henle and ends in a collecting duct.

distributing artery Medium-sized artery with a tunica media composed principally of smooth muscle; regulates blood flow to different regions of the body.

DNA *See* deoxyribonucleic acid.

dominant (dom′i-nant) [L., *dominus*, a master] In genetics a gene that is expressed phenotypically to the exclusion of a contrasting recessive gene.

dorsal root (dōr′săl) Sensory (afferent) root of a spinal nerve.

dorsal root ganglion (gang′glē-on) Collection of sensory neuron cell bodies within the dorsal root of a spinal nerve.

ductus arteriosus (dŭk′tŭs ar-tēr′ē-ō-sŭs) Fetal vessel connecting the left pulmonary artery with the descending aorta.

ductus deferens (def′er-enz) Duct of the testicle, running from the epididymis to the ejaculatory duct; also called the vas deferens.

duodenal gland (doo′ō-dē′năl, doo-od′ĕ-năl) Small gland that opens into the base of intestinal glands; secretes a mucoid alkaline substance.

duodenocolic reflex (doo-ō-dē′nō-kō-lik) Local reflex resulting in a mass movement of the contents of the colon; produced by stimuli in the duodenum.

duodenum (doo-ō-dē′nŭm, doo-od′ĕ-nŭm) [L., *duodeni*] First division of the small intestine; connects to the stomach.

dura mater (doo′ră mā′ter) [L., hard mother] Tough, fibrous membrane forming the outer covering of the brain and spinal cord.

E

eardrum (ēr′drŭm) Tympanic membrane; cellular membrane that separates the external from the middle ear; vibrates in response to sound waves.

ectoderm (ek′tō-derm) Outermost of the three germ layers of an embryo.

ectopic focus; pl., foci (ek-top′ik fōkŭs, fō′sī) Any pacemaker other than the sinus node of the heart; abnormal pacemaker; an ectopic pacemaker.

edema (e-dē′mă) [Gr., *oidema*, a swelling] Excessive accumulation of fluid, usually causing swelling.

effector T cell (ē-fek′tŏr, ē-fek′tōr) Subset of T lymphocytes that is responsible for cell-mediated immunity.

efferent arteriole (ef′er-ent ar-tēr′ē-ōl) Vessel that carries blood from the glomerulus to the peritubular capillaries.

efferent division Nerve fibers that send impulses from the central nervous system to the periphery.

efferent ductule (ef′er-ent dŭk′tool) [L., *ductus*, duct] One of a number of small ducts leading from the testis to the head of the epididymis.

ejaculation (ē-jak-ū-lā′shŭn) Reflexive expulsion of semen from the penis.

ejaculatory duct (ē-jak′ū-lă-tōr-ē) Duct formed by the union of the ductus deferens and the excretory duct of the seminal vesicle; opens into the prostatic urethra.

ejection period (ē-jek′shŭn) Time in the cardiac cycle when the semilunar valves are open and blood is being ejected from the ventricles into the arterial system.

elastin (ĕ-las′tin) A yellow elastic fibrous mucoprotein that is the major connective tissue protein of elastic structures (e.g., large blood vessels and elastic ligaments).

electrocardiogram (ECG) (ē-lek-trō-kar′dē-ō-gram) [Gr., *elektron*, amber + *kardia*, heart + *gramma*, a drawing] Graphic record of the heart's electric currents obtained with the electrocardiograph.

electrolyte (ē-lek′trō-līt) [Gr., *electro* + *lytos*, soluble] Cation or anion in solution that conducts an electric current.

electron (ē-lek′tron) Negatively charged subatomic particle in an atom.

electron-transport chain Series of electron carriers in the inner mitochondrial membrane; they receive electrons from NADH and $FADH_2$, using the electrons in the formation of ATP and water.

element (el′ĕ-ment) [L., *elementum*, a rudiment, beginning] Substance composed of atoms of only one kind.

elevation (el-ĕ-vā′shŭn) Movement of a structure in a superior direction.

embolism (em′bō-lizm) [Gr., *embolisma*, a piece of patch, literally something thrust in] Obstruction or occlusion of a vessel by a transported clot, a mass of bacteria, or other foreign material.

embolus; pl., emboli (em′bō-lŭs, em′bō-lī) [Gr. *embolos*, plug, wedge, or stopper] Plug, composed of a detached clot, mass of bacteria, or other foreign body, occluding a blood vessel.

embryo (em′brē-ō) Developing human from the first to the eighth week of development.

embryonic disk (em-brē-on′ik) Point in the inner cell mass at which the embryo begins to be formed.

embryonic period From approximately the second to the eighth week of development, during which the major organ systems are organized.

emission (ē-mish′ŭn)[L., *emissio*, to send out] Discharge; accumulation of semen in the urethra prior to ejaculation. A nocturnal emission refers to a discharge of semen while asleep.

emmetropia (em-ĕ-trō′pē-ă) [Gr., *emmetros*, according to measure + *ops*, eye] In the eye the state of refraction in which parallel rays are focused exactly on the retina; no accommodation is necessary.

emulsify (ē-mŭl′si-fī) To form an emulsion.

enamel (ē-nam′ĕl) Hard substance covering the exposed portion of the tooth.

endocardium; pl., endocardia (en-dō-kar′dē-ŭm, en-dō-kar′dē-ă) Innermost layer of the heart, including endothelium and connective tissue.

endocrine gland (en′dō-krin) [Gr., *endon*, inside + *krino*, to separate] Ductless gland that secretes a hormone internally, usually into the circulation.

endocytosis (en′dō-sī-tō′sis) Bulk uptake of material through the cell membrane.

endoderm (en′dō-derm) Innermost of the three germ layers of an embryo.

endolymph (en′dō-limf) [Gr., *endo* + L., *lympha*, clear fluid] Fluid found within the membranous labyrinth of the inner ear.

endometrium; pl., endometria (en′dō-mē′trē-ŭm, en′dō-mē′trē-ă) Mucous membrane composing the inner layer of the uterine wall; consists of a simple columnar epithelium and a lamina propria that contains simple tubular uterine glands.

endomysium (en′dō-miz′ē-ŭm, en′dō-mis′ē-ŭm) [Gr., *endo*, within + *mys*, muscle] Fine connective tissue sheath surrounding a muscle fiber.

endoneurium (en-dō-noo′rē-ŭm) [Gr., *endo*, within + *neuron*, nerve] Delicate connective tissue surrounding individual nerve fibers within a peripheral nerve.

endoplasmic reticulum; pl., reticula (en′dō-plas′mik re-tik′ū-lŭm, re-tik′ū-lă) Double-walled membranous network inside the cytoplasm; rough has ribosomes attached to the surface; smooth does not have ribosomes attached.

endorphin (en-dōr′fin) Opiate-like polypeptide found in the brain and other parts of the body; binds in the brain to the same receptors that bind exogenous opiates.

endosteum (en-dos′tē-ŭm) [Gr., *endo*, within + *osteon*, bone] Membranous lining of the medullary cavity and the cavities of spongy bone.

endothelium; pl., endothelia (en-dō-thē′lē-ŭm, en-dō-thē′lē-ă) [Gr., *endo*, within + *thele*, nipple] Layer of flat cells lining blood and lymphatic vessels and the chambers of the heart.

enkephalin (en-kef′ă-lin) Pentapeptide found in the brain; binds to specific receptor sites, some of which may be pain-related opiate receptors.

enterokinase (en′tēr-ō-kī′nās) Intestinal proteolytic enzyme that converts trypsinogen into trypsin.

enzyme (en′zīm) [Gr., *en*, in + *zyme*, leaven] Protein that acts as a catalyst.

eosinophil (ē-ō-sin′ō-fil) [Gr., *eos*, dawn + *philos*, fond] White blood cell that stains with acidic dyes; inhibits inflammation.

epicardium (ep-i-kar′dē-ŭm) [Gr., *epi*, on + *kardia*, heart] Serous membrane covering the surface of the heart. Also called the visceral pericardium.

epidermis (ep-i-derm′is) [Gr., *epi*, on + *derma*, skin] Outer portion of the skin formed of epithelial tissue that rests on or covers the dermis.

epididymis; pl., **epididymides** (ep-i-did′i-mis, -di-dim′i-dēz) [Gr., *epi*, on + *didymos*, twin] Elongated structure connected to the posterior surface of the testis, which consists of the head, body, and tail; site of storage and maturation of the spermatozoa.

epiglottis (ep-i-glot′is) [Gr., *epi*, on + *glottis*, the mouth of the windpipe] Plate of elastic cartilage covered with mucous membrane; serves as a valve over the glottis of the larynx during swallowing.

epimysium (ep-i-mis′ē-ŭm) [Gr., *epi*, on + *mys*, muscle] Fibrous envelope surrounding a skeletal muscle.

epinephrine (ep′i-nef′rin) Hormone (amino acid derivative) similar in structure to the neurotransmitter norepinephrine; major hormone released from the adrenal medulla; increases cardiac output and blood glucose levels.

epineurium (ep-i-noo′rē-ŭm) [Gr., *epi*, upon + *neuron*, nerve] Connective tissue sheath surrounding a nerve.

epiphyseal line (ep-i-fiz′ē-ăl) Dense plate of bone in a bone that is no longer growing, indicating the former site of the epiphyseal plate.

epiphyseal plate Site at which bone growth in length occurs; located between the epiphysis and diaphysis of a long bone; area of hyaline cartilage where cartilage growth is followed by endochondral ossification; also called the metaphysis or growth plate.

epiphysis; pl., **epiphyses** (e-pif′i-sis, e-pif′i-sēz) [Gr., *epi*, on + *physis*, growth] Portion of a bone developed from a secondary ossification center and separated from the remainder of the bone by the epiphyseal plate.

epiploic appendage (ep′i-plō′ik) One of a number of little processes of peritoneum projecting from the serous coat of the large intestine except the rectum; they are generally distended with fat.

epithelium; pl., **epithelia** (ep-i-thē′lē-ŭm, ep-i-thē′lē-ă) [Gr., *epi*, on + *thele*, nipple] One of the four primary tissue types. "Nipples" refers to the tiny capillary-containing connective tissue in the lips, which is where the term was first used. The use of the term was later expanded to include all covering and lining surfaces of the body.

epitope (ep′i-tōp) [Gr., *epi*, upon + *top*, place] *See* antigenic determinant.

eponychium (ep-ō-nik′ē-ŭm) [Gr., *epi*, on + *onyx*, nail] Outgrowth of the skin that covers the proximal and lateral borders of the nail. Cuticle.

erection (ē-rek′shŭn) [L., *erectio*, to set up] Condition of erectile tissue when filled with blood; becomes hard and unyielding; especially refers to this state of the penis.

erythroblastosis fetalis (ē-rith′rō-blas-tō′sis fē-tă′lis) [erythroblast + *osis*, condition] Destruction of erythrocytes in the fetus or newborn caused by antibodies produced in the Rh-negative mother acting on Rh-positive blood of the fetus or newborn.

erythrocyte (ĕ-rith′rō-sīt) [Gr., *erythros*, red + *kytos*, cell] Red blood cell; biconcave disk containing hemoglobin.

erythropoiesis (ĕ-rith′rō-poy-ē′sis) [erythrocyte + Gr., *poiesis*, a making] Production of erythrocytes.

erythropoietin (ĕ-rith′rō-poy′ĕ-tin) Protein that enhances erythropoiesis by stimulating formation of proerythroblasts and release of reticulocytes from bone marrow.

esophagus; pl., **esophagi** (ē-sof′ă-gŭs, ē-sof′ă-gī, ē-sof′ă-jī) [Gr., *oisophagos*, gullet] Portion of the digestive tract between the pharynx and stomach.

essential amino acid Amino acid required by animals that must be supplied in the diet.

estrogen (es′trō-jen) A substance that exerts biologic effects characteristic of estrogen hormone such as stimulating female secondary sexual characteristics, growth, and maturation of long bones, and help controlling the menstrual cycle.

eustachian tube (ū-stā′shŭn, ū-stā′kē-ăn) Named for the Italian anatomist Bartolommeo Eustachio (1524–1574). Auditory canal; extends from the middle ear to the nasopharynx.

evagination (ē-vaj-i-nā′shŭn) [L. *e*, out + *vagina*, sheath] Protrusion of some part or organ from its normal position.

evaporation (ē-vap-ŏ-ra′shŭn) [L., *e*, out + *vaporare*, to emit vapor] Change from liquid to vapor form.

eversion (ē-ver′zhŭn) [L., *everto*, to overturn] Turning outward.

excitation–contraction coupling (ek-sī-tā′shŭn kon-trak′shŭn kŭp′ling) Stimulation of a muscle fiber produces an action potential that results in contraction of the muscle fiber.

excitatory postsynaptic potential (EPSP) (ek-sī′tă-tō-rē pōst-si-nap′tik pō-ten′shăl) Depolarization in the postsynaptic membrane that brings the membrane potential close to threshold.

exocrine gland (ek′sō-krin) [Gr., *exo*, outside + *krino*, to separate] Gland that secretes to a surface or outward through a duct.

exocytosis (ek′sō-sī-to′sis) Elimination of material from a cell through the formation of vacuoles.

expiratory reserve volume Maximum volume of air that can be expelled from the lungs after a normal expiration.

extension (eks-ten′shŭn) [L., *extensio*, to stretch out] To stretch out.

external anal sphincter Ring of striated muscular fibers surrounding the anus.

external acoustic meatus (mē-ā′tŭs) Short canal that opens to the exterior environment and terminates at the eardrum; part of the external ear.

external ear Portion of the ear that includes the auricle and external acoustic meatus; terminates at the eardrum.

external naris; pl., **nares** (nā′ris, nā′res) Nostril; anterior or external opening of the nasal cavity.

external spermatic fascia Outer fascial covering of the spermatic cord.

external urethral orifice Slitlike opening of the urethra in the glans penis.

external urinary sphincter Sphincter skeletal muscle around the base of the urethra external to the internal urinary sphincter.

exteroceptor (eks′ter-ō-sep′ter, eks′ter-ō-sep′tōr) [L., *exterus*, external + *receptor*, receiver] Sensory receptor in the skin or mucous membranes, which responds to stimulation by external agents or forces.

extracellular (eks-tră-sel′ū-lăr) Outside the cell.

extracellular matrix; pl., **matrices** (eks-tră-sel′ū-lăr mā′triks, mā′tri-sēz) Nonliving chemical substances located between connective tissue cells.

extrinsic clotting pathway (eks-trin′sik) Series of chemical reactions resulting in clot formation; begins with chemicals (e.g., tissue thromboplastin) found outside the blood.

extrinsic muscle Muscle located outside the structure being moved.

eyebrow Short hairs on the bony ridge above the eyes.

eyelash Hair at the margins of the eyelids.

eyelid Palpebra; Movable fold of skin in front of the eyeball.

F

facilitated diffusion Carrier-mediated process that does not require ATP and moves substances into or out of cells from a high to a low concentration.

F actin (ak′tin) Fibrous actin molecule that is composed of a series of globular actin molecules (G actin).

falciform ligament (fal′si-fōrm lig′ă-ment) Fold of peritoneum extending to the surface of the liver from the diaphragm and anterior abdominal wall.

fallopian tube (fa-lō′pē-an) *See* uterine tube.

false pelvis Portion of the pelvis superior to the pelvic brim; composed of the bone on the posterior and lateral sides and by muscle on the anterior side.

falx cerebelli (falks ser-ĕ-bel′ī) Dural fold between the two cerebellar hemispheres.

falx cerebri (falx ser′ĕ-brī) Dural fold between the two cerebral hemispheres.

far point of vision Distance from the eye where accommodation is not needed to have the image focused on the retina.

fascia; pl., **fasciae** (fash′ē-ă, fash′ē-ē) [L., band or fillet] Loose areolar connective tissue found beneath the skin (hypodermis) or dense connective tissue that encloses and separates muscles.

fasciculus (fă-sik′ū-lŭs) [L., *fascis*, bundle] Band or bundle of nerve or muscle fibers bound together by connective tissue.

fat [A.S., *faet*] Greasy, soft-solid material found in animal tissues and many plants; composed of two types of molecules: glycerol and fatty acids.

fatigue (fă-tēg′) [L., *fatigo*, to tire] Period characterized by a reduced capacity to do work.

fat-soluble vitamin Vitamin such as A, D, E, and K that is soluble in lipids and absorbed from the intestine along with lipids.

fauces (faw′sēz) [L., throat]   Space between the cavity of the mouth and the pharynx.

female climacteric   Period of life occurring in women, encompassing termination of the reproductive period and characterized by endocrine, somatic, and transitory psychologic changes and ultimately menopause.

female pronucleus   Nuclear material of the ovum after the ovum has been penetrated by the spermatozoon. Each pronucleus carries the haploid number of chromosomes.

fertilization (fer′til-i-zā′shŭn)   Process that begins with the penetration of the secondary oocyte by the spermatozoon and is completed with the fusion of the male and female pronuclei.

fetus (fē′tus)   Developing human following the embryonic period (after 8 weeks of development).

fibrin (fī′brin)   An elastic filamentous protein derived from fibrinogen by the action of thrombin, which releases peptides from fibrinogen in coagulation of the blood.

fibroblast (fī′brō-blast) [L., *fibra*, fiber + Gr., *blastos*, germ]   Spindle-shaped or stellate cells that form connective tissue.

fibrocyte (fī′brō-sīt)   Mature cell of fibrous connective tissue.

fibrous joint (fī′brŭs)   Bones connected by fibrous tissue with no joint cavity; includes sutures, syndesmoses, and gomphoses.

fibrous tunic   Outer layer of the eye; composed of the sclera and the cornea.

filiform papilla (fil′i-fōrm)   Filament-shaped papilla on the surface of the tongue.

filtrate (fil′trāt)   Liquid that has passed through a filter; for example, fluid that enters the nephron through the filtration membrane of the glomerulus.

filtration (fil-trā′shŭn)   Movement, due to a pressure difference, of a liquid through a filter that prevents some or all of the substances in the liquid from passing through.

filtration fraction   Fraction of the plasma entering the kidney that filters into Bowman's capsule. Normally it is around 19%.

filtration membrane   Membrane formed by the glomerular capillary endothelium, the basement membrane, and the podocytes of Bowman's capsule.

filtration pressure   Pressure gradient that forces fluid from the glomerular capillary through the filtration membrane into Bowman's capsule; glomerular capillary pressure minus glomerular capsule pressure minus colloid osmotic pressure.

fimbria; pl., fimbriae (fim′brē-ă, fim′brē-ē) [L., fringe]   Fringelike structure located at the ostium of the uterine tube.

first messenger   *See* intercellular chemical signal.

fixator (fik-sā′ter)   Muscle that stabilizes the origin of a prime mover.

flagellum; pl., flagella (flă-jel′ŭm, flă-jel′ă) [L., whip]   Whiplike locomotory organelle of constant structural arrangement consisting of double peripheral microtubules and two single central microtubules.

flatus (flā′tŭs) [L., a blowing]   Gas or air in the gastrointestinal tract that may be expelled through the anus.

flexion (flek′shŭn) [L., *flectus*]   To bend.

focal point   Point at which light rays cross after passing through a concave lens such as the lens of the eye.

foliate papilla (fō′lē-āt)   Leaf-shaped papilla on the lateral surface of the tongue.

follicle-stimulating hormone (FSH) (fol′i-kl)   Hormone of the adenohypophysis that, in females, stimulates the graafian follicles of the ovary and assists in follicular maturation and the secretion of estrogen; in males, FSH stimulates the epithelium of the seminiferous tubules and is partially responsible for inducing spermatogenesis.

follicular phase (fō-lik′ū-lăr)   Time between the end of menses and ovulation characterized by rapid division of endometrial cells and development of follicles in the ovary; the proliferative phase.

foramen; pl., foramina (fō-rā′men, fō-ram′i-nă)   A hole.

foramen ovale (o-val′ē)   In the fetal heart the oval opening in the septum secundum; the persistent part of septum primum acts as a valve for this interatrial communication during fetal life; postnatally the septum primum becomes fused to the septum secundum to close the foramen ovale, forming the fossa ovale.

force   That which produces a motion in the body; pull.

foregut   Cephalic portion of the primitive digestive tube in the embryo.

foreskin   *See* prepuce.

formed elements   Cells (i.e., red and white blood cells) and cell fragments (i.e., platelets) of blood.

formula unit   A description of the relative number of cations and ions in an ionic compound.

fornix (fōr′niks) [L., arch, vault]   Recess at the cervical end of the vagina; also the recess deep to each eyelid where the palpebral and bulbar conjunctivae meet.

fovea centralis (fō′vē-ă)   Depression in the middle of the macula where there are only cones and no blood vessels.

free energy   Total amount of energy that can be liberated by the complete catabolism of food.

frenulum (fren′ū-lŭm) [L., *frenum*, bridle]   Fold extending from the floor of the mouth to the midline of the undersurface of the tongue.

frequency-modulated signals   Signals, all of which are identical in amplitude, that differ in their frequency; for example, strong stimuli may initiate a high frequency of action potentials and weak stimuli may initiate a low frequency of action potentials.

FSH   *See* follicle-stimulating hormone.

FSH surge   Increase in plasma follicle-stimulating hormone (FSH) levels before ovulation.

fulcrum (ful′krŭm)   Pivot point.

fundus (fŭn′dŭs) [L., bottom]   "Bottom," or rounded end, of a hollow organ, for example, the fundus of the stomach or uterus.

fungiform papilla (fŭn′ji-fōrm)   Mushroom-shaped papilla on the surface of the tongue.

G

G actin (jē ak′tin)   Globular protein molecules that, when bound together, form fibrous actin (F actin).

gallbladder (gawl′blad-er)   Pear-shaped receptacle on the inferior surface of the liver; serves as a storage reservoir for bile.

gamete (gam′ēt)   Ovum or spermatozoon.

gamma globulin (gam′ă glob′ū-lin) [L., *globulus*, globule]   Plasma proteins that include the antibodies.

ganglion; pl., ganglia (gang′glē-on, gang′glē-ă) [Gr., swelling, or knot]   Any group of nerve cell bodies in the peripheral nervous system.

gap junction   Small channel between cells that allows the passage of ions and small molecules between cells; provides means of intercellular communication.

gastric gland (gas′trik)   Gland located in the mucosa of the fundus and body of the stomach.

gastric inhibitory polypeptide   Hormone secreted by the duodenum that inhibits gastric acid secretion.

gastric pit   Small pit in the mucous membrane of the stomach at the bottom of which are the mouths of the gastric glands that secrete mucus, hydrochloric acid, intrinsic factor, pepsinogen, and hormones.

gastrin (gas′trin)   Hormone secreted in the mucosa of the stomach and duodenum that stimulates secretion of hydrochloric acid by the parietal cells of the gastric glands.

gastrocolic reflex (gas′trō-kol′ik)   Local reflex resulting in mass movement of the contents of the colon that occurs after the entrance of food into the stomach.

gastroesophageal (cardiac) opening (gas′trō-ē-sof′ă-jē′ăl)   Opening of the esophagus into the stomach.

gene (jēn) [Gr., *genos*, birth]   Functional unit of heredity. Each gene occupies a specific place, or locus, on a chromosome; is capable of reproducing itself exactly at each cell division; and often is capable of directing the formation of an enzyme or other protein.

general gas law   The pressure of a gas is equal to the number of gram moles of the gas times the gas constant times the absolute temperature divided by the volume of the gas. Assuming a constant temperature, the pressure of a given amount of gas is inversely proportional to its volume. This relationship also is called Boyle's law.

genetics (jĕ-net′iks) [Gr., *genesis*, origin or production]   Branch of science that deals with heredity.

genital fold (jen′i-tăl)   Paired longitudinal ridges developing in the embryo on either side of the urogenital orifice. In the male they form part of the penis; in the female they form the labia minora.

genital tubercle (jen′i-tăl)   Median elevation just cephalic to the urogenital orifice of an embryo; gives rise to the penis of the male or the clitoris of the female.

genotype (jen′ō-tīp) [Gr., *genos*, birth, descent + *typos*, type]   Genetic makeup of an individual.

germ cell (jerm)   Spermatozoon or ovum.

germ layer One of three layers in the embryo (ectoderm, endoderm, or mesoderm) from which the four primary tissue types arise.

germinal center Lighter-staining center of a lymphatic nodule; area of rapid lymphocyte division.

gingiva (jin′ji-vă) Dense fibrous tissue, covered by mucous membrane, that covers the alveolar processes of the upper and lower jaws and surrounds the necks of the teeth.

girdle (ger′dl) Belt or zone; the bony region where the limbs attach to the body.

gland [L., *glans*, acorn] Secretory organ from which secretions may be released into the blood, a cavity, or onto a surface.

glans penis [L., acorn] Conical expansion of the corpus spongiosum that forms the head of the penis.

globin (glō′bin) Protein portion of hemoglobin.

glomerular capillary pressure (glō-mār′ū-lăr) Blood pressure within the glomerulus.

glomerular filtration rate (GFR) Amount of plasma (filtrate) that filters into Bowman's capsules per minute.

glomerulus (glō-mār′ū-lŭs) [L., *glomus*, ball of yarn] Mass of capillary loops at the beginning of each nephron, nearly surrounded by Bowman's capsule.

glottis (glot′is) [Gr., aperture of the larynx] Vocal apparatus; includes vocal folds and the cleft between them.

glucocorticoid (gloo-kō-kōr′ti-koyd) Steroid hormone (e.g., cortisol) released by zonula fasciculata of the adrenal cortex; increases blood glucose and inhibits inflammation.

gluconeogenesis (gloo′kŏ-nē-ō-jen′ĕ-sis) [Gr., *glykys*, sweet + *neos*, new + *genesis*, production] Formation of glucose from noncarbohydrates such as proteins (amino acids) or lipids (glycerol).

glycogenesis (glī′kō-jĕ-nō′-sis) Formation of glycogen from glucose molecules.

glycolysis (glī-kol′i-sis) [Gr., *glykys*, sweet + *lysis*, a loosening] Anaerobic process during which glucose is converted to pyruvic acid; net of two ATP molecules is produced during glycolysis.

goblet cell Mucous-producing epithelial cell that has its apical end distended with mucin.

Golgi apparatus (gol′jē) Named for Camillo Golgi, Italian histologist and Nobel laureate, 1843–1926. Specialized endoplasmic reticulum that concentrates and packages materials for secretion from the cell.

Golgi tendon organ Proprioceptive nerve ending in a tendon.

gomphosis (gom-fō′sis) [Gr., *gomphos*, bolt, nail + *osis*, condition] Fibrous joint in which a peglike process fits into a hole.

gonad (gō′nad) [Gr., *gone*, seed] Organ that produces sex cells; testis of a male or ovary of a female.

gonadal ridge (gō-nad′ăl) Elevation on the embryonic mesonephros; primordial germ cells become embedded in it, establishing it as the testis or ovary.

gonadotropin (gō′nad-ō-trō′pin) Hormone capable of promoting gonadal growth and function. Two major gonadotropins are luteinizing hormone (LH) and follicle-stimulating hormone (FSH).

gonadotropin-releasing hormone (GnRH) Hypothalamic-releasing hormone that stimulates the secretion of gonadotropins (LH and FSH) from the adenohypophysis.

graaffian follicle *See* mature follicle.

granulocyte (gran′ū-lō-sīt) Mature granular white blood cell (neutrophil, basophil, or eosinophil).

granulosa cell (gran-ū-lō′să) Cell in the layer surrounding the primary follicle.

gray matter Collections of nerve cell bodies, their dendritic processes, and associated neuroglial cells within the central nervous system.

greater omentum Peritoneal fold passing from the greater curvature of the stomach to the transverse colon, hanging like an apron in front of the intestines.

greater vestibular gland One of two mucus-secreting glands on either side of the lower part of the vagina. The equivalent of the bulbourethral glands in the male.

growth hormone Somatotropin; stimulates general growth of the individual; stimulates cellular amino acid uptake and protein synthesis.

gubernaculum (goo′ber-nak′ū-lŭm) [L., helm] Column of tissue that connects the fetal testis to the developing scrotum; involved in testicular descent.

gustatory (gŭs′tă-tōr-ē) Associated with the sense of taste.

gustatory hair Microvillus of gustatory cell in a taste bud.

gynecomastia (gī′nĕ-kō-mas′tē-ă) [Gr., *gyne*, woman + *mastos*, breast] Excessive development of the male mammary glands, which sometimes secrete milk.

H

hair [A.S., hear] Columns of dead keratinized epithelial cells.

hair follicle Invagination of the epidermis into the dermis; contains the root of the hair and receives the ducts of sebaceous and apocrine glands.

Haldane effect Named for the Scottish physiologist John S. Haldane (1860–1936). Hemoglobin that is not bound to carbon dioxide binds more readily to oxygen than hemoglobin that is bound to carbon dioxide.

half-life The time it takes for one-half of an administered substance to be lost through biologic processes.

haploid (hap′loyd) Having only one set of chromosomes, in contrast to diploid; characteristic of gametes.

hapten (hap′ten) [Gr., *hapto*, to fasten] Small molecule that binds to a large molecule; together they stimulate the specific immune system.

hard palate Floor of the nasal cavity that separates the nasal cavity from the oral cavity; composed of the palatine processes of the maxillary bones and the horizontal plates of the palatine bones.

haustra (haw′stră) [L., machine for drawing water] Sacs of the colon, caused by contraction of the taeniae coli, which are slightly shorter than the gut, so that the latter is thrown into pouches.

haversian canal (ha-ver′shan) Named for seventeenth-century English anatomist Clopton Havers (1650–1702). Canal containing blood vessels, nerves, and loose connective tissue and running parallel to the long axis of the bone.

haversian system *See* osteon.

heart skeleton Fibrous connective tissue that provides a point of attachment for cardiac muscle cells, electrically insulates the atria from the ventricles, and forms the fibrous rings around the valves.

heat energy Energy that results from the random movement of atoms, ions, or molecules; the greater the amount of heat energy in an object, the higher is the object's temperature.

helicotrema (hel′i-kō-trē′mă) [Gr., *helix*, spiral + *traema*, hole] Opening at the apex of the cochlea through which the scala vestibuli and the scala tympani of the cochlea connect.

helper T cell Subset of T lymphocytes that increases the activity of B cells and T cells.

hematocrit (hē′mă-tō-krit) [Gr., *hemato*, blood + *krin*, to separate] Percentage of blood volume occupied by erythrocytes.

hematoma (he-ma-to′ma) A localized mass of blood released from blood vessels but confined within an organ or space; the blood is usually clotted.

hematopoiesis (hē′mă-tō-poy-ē′sis) [Gr., *haima*, blood + *poiesis*, a making] Production of blood cells.

heme (hēm) Oxygen-carrying, color-furnishing part of hemoglobin.

hemidesmosome (hem-ē-des′mō-sōm) Similar to half a desmosome, attaching epithelial cells to the basement membrane.

hemocytoblast (he′mo-si′to-blast) A blood stem cell derived from mesenchyme that can give rise to red and white blood cells and platelets.

hemoglobin (hē-mō-glō′bin) Red, respiratory protein of red blood cells; consists of 6% heme and 94% globin; transports oxygen and carbon dioxide.

hemolysis (hē-mol′i-sis) [Gr., *haima* + *lysis*, destruction] Destruction of red blood cells in such a manner that hemoglobin is released.

hemopoiesis (hē′mō-poy-ē′sis) [Gr., *haima*, blood + *poiesis*, a making] Formation of the formed elements of blood, that is, red blood cells, white blood cells, and thrombocytes.

hemopoietic tissue (hē′mō-poy-et′ik) [Gr., *haima*, blood + *poiesis*, to make] Blood-forming tissue.

hemostasis (hē′mō-stā-sis) Arrest of bleeding.

Henry's law Named for the English chemist William Henry (1775–1837). The concentration of a gas dissolved in a liquid is equal to the partial pressure of the gas over the liquid times the solubility coefficient of the gas.

heparin (hep′ă-rin) Anticoagulant that prevents platelet agglutination and thus prevents thrombus formation.

hepatic artery (he-pa′tik) Branch of the aorta that delivers blood to the liver.

hepatic cord Plate of liver cells that radiates away from the central vein of a liver lobule.

hepatic portal system System of portal veins that carry blood from the intestines, stomach, spleen, and pancreas to the liver.

hepatic portal vein Portal vein formed by the superior mesenteric and splenic veins and entering the liver.

hepatic sinusoid (si′nŭ-soyd) Terminal blood vessel having an irregular and larger caliber than an ordinary capillary within the liver lobule.

hepatic vein Vein that drains the liver into the inferior vena cava.

hepatocyte (hep′ă-tō-sīt) Liver cell.

hepatopancreatic ampulla Dilation within the major duodenal papilla that normally receives both the common bile duct and the main pancreatic duct.

hepatopancreatic ampullar sphincter Smooth muscle sphincter of the hepatopancreatic ampulla; sphincter of Oddi.

Hering-Breuer reflex (her′ing broy′er) Named for the German physiologist Heinrich Ewald Hering (1866–1948) and the Austrian internist Josef Breuer (1842–1925). Sensory impulses from stretch receptors in the lungs arrest inspiration; expiration then occurs.

heterozygous (het′er-ō-zī′gŭs) [Gr., heteros, other + zygon, yoke] State of having different allelic genes at one or more paired loci in homologous chromosomes.

hiatus (hī-ā′tŭs) [L., aperture, to yawn] Opening.

hilum (hī′lŭm) [L., small bit or trifle] Indented surface on many organs, serving as a point where nerves and vessels enter or leave.

hindgut Caudal or terminal part of the embryonic gut.

histamine (his′tă-mēn) Amine released by mast cells and basophils that promotes inflammation.

histology (his-tol′ō-jē) [Gr., histo, web (tissue) + logos, study] The science that deals with the microscopic structure of cells, tissues, and organs in relation to their function.

holocrine gland (hol′ō-krin) [Gr., holos, complete + krino, to separate] Gland whose secretion is formed by the disintegration of entire cells (e.g., sebaceous gland; see also apocrine and merocrine glands).

homeostasis (hō′mē-ō-stā′sis) [Gr., homoio, like + stasis, a standing] State of equilibrium in the body with respect to functions, composition of fluids and tissues.

homologous (hŏ-mol′ō-gŭs) [Gr., ratio or relation] Alike in structure or origin.

homozygous (hō-mō-zī′gŭs) [Gr., homos, the same + zygon, yoke] State of having identical allelic genes at one or more paired loci in homologous chromosomes.

hormone (hōr′mōn) [Gr., hormon, to set into motion] Substance secreted by endocrine tissues into the blood that acts on a target tissue to produce a specific response.

hormone receptor Protein or glycoprotein molecule of cells that specifically binds to hormones and produces a response.

horn Subdivision of gray matter in the spinal cord. The axons of sensory neurons synapse with neurons in the posterior horn, the cell bodies of motor neurons are in the anterior horn, and the cell bodies of autonomic neurons are in the lateral horn.

human chorionic gonadotropin (HCG) Hormone produced by the placenta; stimulates secretion of testosterone by the fetus; during the first trimester stimulates ovarian secretion from the corpus luteum of the estrogen and progesterone required for the maintenance of the placenta. In a male fetus, stimulates secretion of testosterone by the fetal testis.

humoral immunity (hū′mōr-ăl) [L., humor, a fluid] Immunity due to antibodies in serum.

hyaline cartilage (hī′ă-lin) [Gr., hyalos, glass] Gelatinous, glossy cartilage tissue consisting of cartilage cells and their matrix; contains collagen, proteoglycans, and water.

hyaluronic acid (hī′ă-loo-ron′ik; glassy appearance) A mucopolysaccharide made up of alternating β-(1,4)-linked residues of hyalobiuronic acid, forming a gelatinous material in the tissue spaces and acting as a lubricant and shock absorbant generally throughout the body.

hydrochloric acid (HCl) (hī-drō-klōr′ik) Acid of gastric juice.

hydrogen bond (hī′drō-jen) Hydrogen atoms bound covalently to either N or O atoms have a small positive charge that is weakly attracted to the small negative charge of other atoms such as O or N; can occur within a molecule or between different molecules.

hydrophilic (hi-dro-fil′ik) [Gr., hydro-, water + philos, love] Denoting the property of attracting or associating with water molecules, possessed by polar molecules and ions; the opposite of hydrophobic.

hydrophobic (hi-dro-fob′ik) [Gr., hydro-, water + phobos, fear] Lacking an attraction to water, possessed by nonpolar molecules; the opposite of hydrophilic.

hydroxyapatite (hī-drok′sē-ap′ă-tīt) Mineral with the empiric formula $3\ Ca_3(PO_4)_2 \cdot Ca(OH)_2$; the main mineral of bone and teeth.

hymen (hī′men) [Gr., membrane] Thin, membranous fold partly occluding the vaginal external orifice; normally disrupted by sexual intercourse or other mechanical phenomena.

hyoid (hī′oyd) (Gr., hyoeides, shaped like the Greek letter epsilon [ε]) U-shaped bone between the mandible and larynx.

hypercalcemia (hī′per-kal-sē′mē-ă) Abnormally high levels of calcium in the blood.

hypercapnia (hī′per-kap′nē-ă) Higher-than-normal levels of carbon dioxide in the blood or tissues.

hyperkalemia (hī′per-kă-lē′mē-ă) A greater than normal concentration of potassium ions in the circulating blood.

hypernatremia (hī′per-nă-trē′mē-ă) An abnormally high plasma concentration of sodium ions.

hyperosmotic (hī′per-oz-mot′ĭk) [Gr., hyper, above + osmos, an impulsion] Having a greater osmotic concentration or pressure than a reference solution.

hyperpolarization (hī′per-pō′lăr-i-zā′shŭn) Increase in the charge difference across the plasma membrane; causes the charge difference to move away from 0 mV.

hypertonic (hī-per-ton′ik) [Gr., hyper, above + tonos, tension] Solution that causes cells to shrink.

hypertrophy (hī-per′trō-fē) [Gr., hyper, above + trophe, nourishment] Increase in bulk or size; not due to an increase in number of individual elements.

hypocalcemia (hī-pō-kal-sē′mē-ă) Abnormally low levels of calcium in the blood.

hypocapnia (hī′pō-kap′nē-ă) Lower-than-normal levels of carbon dioxide in the blood or tissues.

hypodermis (hī′pō-der′mis) [Gr., hypo, under + dermis, skin] Loose areolar connective tissue found deep to the dermis that connects the skin to muscle or bone.

hypokalemia (hī′pō-ka-lē′mē-ă) Abnormally small concentration of potassium ions in the blood.

hyponatremia (hī′pō-nă-trē′mē-ă) An abnormally low plasma concentration of sodium ions.

hyponychium (hī-pō-nik′ē-ŭm) [Gr., hypo, under + onyx, nail] Thickened portion of the stratum corneum under the free edge of the nail.

hypophysis (hī-pof′i-sis) [Gr., an undergrowth] Endocrine gland attached to the hypothalamus by the infundibulum. Also called the pituitary gland.

hypopolarization Change in the electric charge difference across the plasma membrane that causes the charge difference to be smaller or move closer to 0 mV.

hyposmotic (hī′pos-mot′ik) [Gr., hypo, under + osmos, an impulsion] Having a lower osmotic concentration or pressure than a reference solution.

hypospadias (hī′pō-spā′dē-ăs) [Gr., one having the orifice of the penis too low; hypospao, to draw away from under] Developmental anomaly in the wall of the urethra so that the canal is open for a greater or lesser distance on the undersurface of the penis; also a similar defect in the female in which the urethra opens into the vagina.

hypothalamohypophysial portal system (hī′pō-thal′ă-mō-hī′pō-fiz′ē-ăl) Series of blood vessels that carry blood from the area of the hypothalamus to the anterior pituitary gland; originate from capillary beds in the hypothalamus and terminate as a capillary bed in the anterior pituitary gland.

hypothalamohypophyseal tract Nerve tract, consisting of the axons of neurosecretory cells, extending from the hypothalamus into the posterior pituitary gland. Hormones produced in the neurosecretory cell bodies in the hypothalamus are transported through the hypothalamohypophyseal tract to the posterior pituitary gland where they are stored for later release.

hypothalamus (hī′pō-thal′ă-mŭs) [Gr., hypo, under + thalamus, bedroom] Important autonomic and neuroendocrine control center beneath the thalamus.

hypothenar (hī-pō-thē′nar) [Gr., hypo, under + thenar, palm of the hand] Fleshy mass of tissue on the medial side of the palm; contains muscles responsible for moving the little finger.

hypotonic (hī-pō-ton′ik) [Gr., *hypo,* under + *tonos,* tension]   Solution that causes cells to swell.

H zone   Area in the center of the A band in which there are no actin myofilaments; contains only myosin.

I

I band   Area between the ends of two adjacent myosin myofilaments within a myofibril; Z disk divides the I band into two equal parts.

ileocecal sphincter (il′ē-ō-sē′kăl)   Thickening of circular smooth muscle between the ileum and the cecum forming the ileocecal valve.

ileocecal valve   Valve formed by the ileocecal sphincter between the ileum and the cecum.

ileum (il′ē-ŭm) [Gr., *eileo,* to roll up, twist]   Third portion of the small intestine, extending from the jejunum to the ileocecal opening into the large intestine; the posterior inferior bone of the coxa.

immunity (i-mū′ni-tē) [L., *immunis,* free from service]   Resistance to infectious disease and harmful substances.

immunization (im-mū′ni-zā′shun)   Process by which a subject is rendered immune by deliberately introducing an antigen or antibody into the subject.

immunoglobulin (im′ū-nō-glob′ū-lin)   Antibody found in the gamma globulin portion of plasma.

implantation (im-plan-tā′shŭn)   Attachment of the blastocyst to the endometrium of the uterus; occurring 6 or 7 days after fertilization of the ovum.

impotence (im′pŏ-tens)   Inability to accomplish the male sexual act; caused by psychologic or physical factors.

incisor (in-sī′zŏr) [L., *incido,* to cut into]   One of the anterior, cutting teeth.

incisura (in′sī-soo′ră) [L., a cutting into]   Notch or indentation at the edge of any structure.

incus (ing′kus) [L., anvil]   Middle of the three ossicles in the middle ear.

inferior colliculus (ko-lik′ū-lŭs) [L., *collis,* hill]   One of two rounded eminences of the midbrain; involved with hearing.

inferior vena cava   Vein that returns blood from the lower limbs and the greater part of the pelvic and abdominal organs to the right atrium.

inflammatory response (in-flam′ă-tōr-ē)   Complex sequence of events involving chemicals and immune cells that results in the isolation and destruction of antigens and tissues near the antigens. *See also* local and systemic inflammation.

infundibulum (in-fŭn-dib′ū-lŭm) [L., funnel]   Funnel-shaped structure or passage, for example, the infundibulum that attaches the hypophysis to the hypothalamus or the funnel-like expansion of the uterine tube near the ovary.

inguinal canal (ing′gwi-năl)   Passage through the lower abdominal wall that transmits the spermatic cord in the male and the round ligament in the female.

inhibin (in-hib′in)   Polypeptide secreted from the testes that inhibits FSH secretion.

inhibitory neuron (in-hib′i-tōr-ē)   Neuron that produces IPSPs and has an inhibitory influence.

inhibitory postsynaptic potential (IPSP)   Hyperpolarization in the postsynaptic membrane that causes the membrane potential to move away from threshold.

innate immunity (i′nāt, i-nāt′)   Immune system response that is the same with each exposure to an antigen; there is no ability for the system to remember a previous exposure to the antigen.

inner cell mass   Group of cells at one end of the blastocyst, part of which forms the body of the embryo.

inner ear   Contains the sensory organs for hearing and balance; contains the bony and membranous labyrinth.

insensible perspiration [L., *per,* through + *spiro,* to breathe everywhere]   Perspiration that evaporates before it is perceived as moisture on the skin; the term sometimes includes evaporation from the lungs.

insertion (in-ser′shŭn)   More movable attachment point of a muscle; usually the lateral or distal end of a muscle associated with the limbs.

inspiratory capacity (in-spī′ră-tō-rē)   Volume of air that can be inspired after a normal expiration; the sum of the tidal volume and the inspiratory reserve volume.

inspiratory reserve volume   Maximum volume of air that can be inspired after a normal inspiration.

insulin (in′sŭ-lin)   Protein hormone secreted from the pancreas that increases the uptake of glucose and amino acids by most tissues.

interatrial septum (in-ter-ā′trē-ăl) [L., *saeptum,* a partition]   Wall between the atria of the heart.

intercalated disk (in-ter′kă-lā-ted)   Cell-to-cell attachment with gap junctions between cardiac muscle cells.

intercalated duct   Minute duct of glands such as the salivary gland and the pancreas; leads from the acini to the interlobular ducts.

intercellular (in-ter-sel′ū-lăr)   Between cells.

intercellular chemical signal   Chemical that is released from cells and passes to other cells; acts as signal that allows cells to communicate with each other.

interferon (in-ter-fēr′on)   Protein that prevents viral replication.

interlobar artery (in-ter-lō′bar)   Branch of the segmental arteries of the kidney; runs between the renal pyramids and gives rise to the arcuate arteries.

interlobular artery (in-ter-lob′ū-lăr)   Artery that passes between lobules of an organ; branches of the interlobar arteries of the kidney pass outward through the cortex from the arcuate arteries and supply the afferent arterioles.

interlobular duct   Any duct leading from a lobule of a gland and formed by the junction of the intercalated ducts draining the acini.

interlobular vein   Parallels the interlobular arteries; in the kidney drains the peritubular capillary plexus, emptying into arcuate veins.

intermediate olfactory area   Part of the olfactory cortex responsible for modulation of olfactory sensations.

internal anal sphincter [Gr., *sphinkter,* band or lace]   Smooth muscle ring at the upper end of the anal canal.

internal naris; pl., nares (nā′ris, nā′res)   Opening from the nasal cavity into the nasopharynx.

internal spermatic fascia   Inner connective tissue covering of the spermatic cord.

internal urinary sphincter   Traditionally recognized as a sphincter composed of a thickening of the middle smooth muscle layer of the bladder around the urethral opening.

interphase (in′ter-fāz)   Period between active cell divisions when DNA replication occurs.

interstitial (in-ter-stish′ăl) [L., *inter,* between + *sisto,* to stand]   Space within tissue. Interstitial growth means growth from within.

interstitial cell (Leydig cell)   Cell between the seminiferous tubules of the testes; secretes testosterone.

interventricular septum (in-ter-ven-trik′ū-lăr)   Wall between the ventricles of the heart.

intestinal gland (in-tes′ti-năl)   Tubular glands in the mucous membrane of the small and large intestines.

intracellular (in-tră-sel′ū-lăr)   Inside a cell.

intracellular mediators   A molecule produced within a cell that binds to a macromolecule such as receptors or enzymes inside of that cell that regulates their activities.

intramural plexus (in′tră-mū′răl plek′sus)   Combined submucosal and myenteric plexuses.

intrinsic clotting pathway (in-trin′sik)   Series of chemical reactions resulting in clot formation that begins with chemicals (e.g., plasma factor XII) found within the blood.

intrinsic factor   Factor secreted by the parietal cells of gastric glands and required for adequate absorption of vitamin B_{12}.

intrinsic muscles   Muscles located within the structure being moved.

intubation (in-too-ba′shun)   Insertion of a tube into an opening, canal, or hollow organ.

inversion (in-ver′zhŭn) [L., *inverto,* to turn about]   Turning inward.

ion (ī′on) [Gr., *ion,* going]   Atom or group of atoms carrying a charge of electricity by virtue of having gained or lost one or more electrons.

ion channel   Pore in the plasma membrane through which ions, such as sodium and potassium, move.

ionic bond (ī-on′ik)   Chemical bond that is formed when one atom loses an electron and another accepts that electron.

iris (ī′ris)   Specialized portion of the vascular tunic; the "colored" portion of the eye that can be seen through the cornea.

ischemia (is-kē′mē-ă) [Gr., *ischo,* to keep back + *haima,* blood]   Reduced blood supply to some area of the body.

ischium (is′kē-ŭm)   Superior bone of the coxa.

isomer (ī′sō-mer) [Gr., *isos,* equal + *meros,* part]   Molecules having the same number and types of atoms but differing in their three-dimensional arrangement.

isometric contraction (ī-sō-met′rik) [Gr., *isos,* equal + *metron,* measure]   Muscle contraction in which the length of the muscle does not change but the tension produced increases.

isosmotic (ī′sō-os-mot′ik) [Gr., *isos*, equal + *osmos*, an impulsion]   Having the same osmotic concentration or pressure as a reference solution.

isotonic solution (ī′sō-ton′ik) [Gr., *isos*, equal + *tonos*, tension]   Solution that causes cells to neither shrink nor swell.

isotope (ī′sō-tōp) [Gr., *isos*, equal + *topos*, part, place]   Either of two or more atoms that have the same atomic number but a different number of neutrons.

isthmus (is′mŭs)   Constriction connecting two larger parts of an organ, such as the constriction between the body and the cervix of the uterus, or the portion of the uterine tube between the ampulla and the uterus.

J

jaundice (jawn′dis) [Fr., *jaune*, yellow]   Yellowish staining of the integument, sclerae, and the other tissues with bile pigments.

jejunum (jĕ-joo′nŭm) [L., *jejunus*, empty]   Second portion of the small intestine; located between the duodenum and the ileum.

juxtaglomerular apparatus (jŭks′tă-glŏ-mer′ū-lăr)   Complex consisting of juxtaglomerular cells of the afferent arteriole and macular densa cells of the distal convoluted tubule near the renal corpuscle; secretes renin.

juxtaglomerular cell   Modified smooth muscle cell of the afferent arteriole located at the renal corpuscle; a component of the juxtaglomerular apparatus.

juxtamedullary nephron (jŭks′tă-med′ū-lăr-ē)   Nephron located near the junction of the renal cortex and medulla.

K

karyotype (kar′ē-ō-tīp)   A display of chromosomes arranged by pairs.

keratinization (ker′ă-tin-i-zā′shŭn)   Production of keratin and changes in the chemical and structural character of epithelial cells as they move to the skin surface.

keratinized (ker′ă-ti-nīzd) [Gr., *keras*, horn]   Word means turned into a horn. In modern usage the term means to become a structure that contains keratin, a protein found in skin, hair, nails, and horns.

keratinocyte (ke-rat′i-nō-sīt) [Gr., *keras*, horn + *kytos*, cell]   Epidermal cell that produces keratin.

keratohyalin (ker′ă-tō-hī′ă-lin)   Nonmembrane-bound protein granules in the cytoplasm of stratum granulosum cells of the epidermis.

ketogenesis (kē-tō-jen′ĕ-sis)   Production of ketone bodies, such as from acetyl-CoA.

ketone body (kē′tōn)   One of a group of ketones, including acetoacetic acid, β-hydrobutyric acid, and acetone.

ketosis (kē-tō′sis) [keone + -osis, condition]   A condition characterized by the enhanced production of ketone bodies, as in diabetes mellitus or starvation.

kidney (kid′nē) [A.S., *cwith*, womb, belly + *neere*, kidney]   One of the two organs that excrete urine. The kidneys are bean-shaped organs approximately 11 cm long, 5 cm wide, and 3 cm thick lying on either side of the spinal column, posterior to the peritoneum, approximately opposite the twelfth thoracic and first three lumbar vertebrae.

kilocalorie (kil′ō-kal-ō-rē)   Quantity of energy required to raise the temperature of 1 kg of water 1°C; 1000 calories. Equal to one dietary calorie.

kinetic energy (ki-net′ik)   Motion energy or energy that can do work.

kinetic labyrinth (lab′i-rinth)   Part of the membranous labyrinth composed of the semicircular canals; detects dynamic or kinetic equilibrium, such as movement of the head.

kinetochore (ki-nĕ′tō-kōr, ki-net′o-) [Gr., *kinēto*, moving + Gr., *chōra*, space]   The structural portion of the chromosome to which microtubules attach.

Korotkoff sounds (kō-rot′kof)   Named for Russian physician Nikolai S. Korotkoff (1874–1920). Sounds heard over an artery when blood pressure is determined by the auscultatory method; caused by turbulent flow of blood.

L

labium majus; pl., labia majora (lā′bē-ŭm, lā′bē-ă)   One of two rounded folds of skin surrounding the labia minora and vestibule; homolog of the scrotum in males.

labium minus; pl., labia minora   One of two narrow longitudinal folds of mucous membrane enclosed by the labia majora and bounding the vestibule; anteriorly they unite to form the prepuce.

lacrimal apparatus (lak′ri-măl)   Lacrimal, or tear, gland in the superolateral corner of the orbit of the eye and a duct system that extends from the eye to the nasal cavity.

lacrimal canaliculus   Canal that carries excess tears away from the eye; located in the medial canthus and opening on a small lump called the lacrimal papilla.

lacrimal gland   Tear gland located in the superolateral corner of the orbit.

lacrimal papilla   Small lump of tissue in the medial canthus or corner of the eye; the lacrimal canal opens within the lacrimal papilla.

lacrimal sac   Enlargement in the lacrimal canal that leads into the nasolacrimal duct.

lactation (lak-tā′shŭn) [L., *lactatio*, suckle]   Period after childbirth during which milk is formed in the breasts.

lacteal (lak′tē-ăl)   Lymphatic vessel in the wall of the small intestine that carries chyle from the intestine and absorbs fat.

lactiferous duct (lak-tif′er-ŭs)   One of 15–20 ducts that drain the lobes of the mammary gland and open onto the surface of the nipple.

lactiferous sinus   Dilation of the lactiferous duct just before it enters the nipple.

lacuna; pl., lacunae (lă-koo′nă, -koo′nē) [L., *lacus*, a hollow, a lake]   Small space or cavity; potential space within the matrix of bone or cartilage normally occupied by a cell that can only be visualized when the cell shrinks away from the matrix during fixation; space containing maternal blood within the placenta.

lag phase   One of the three phases of muscle contraction; time between the application of the stimulus and the beginning of muscular contraction. Also called the latent phase.

lamella; pl., lamellae (lă-mel′ă, lă-mel′ē)   Thin sheet or layer of bone.

lamellated corpuscle (lam′ĕ-lāt-ed)   Pacinian corpuscle. Oval receptor found in the deep dermis or hypodermis (responsible for deep cutaneous pressure and vibration) and in tendons (responsible for proprioception).

lamina; pl., laminae (lam′i-nă, lam′i-nē) [L., *lamina*, plate, leaf]   Thin plate, for example, the thinner portion of the vertebral arch.

lamina propria (prō′prē-ă)   Layer of connective tissue underlying the epithelium of a mucous membrane.

laminar flow (lam′i-nar)   Relative motion of layers of a fluid along smooth concentric parallel paths.

Langerhans cell   Dendritic cell named after the German anatomist Paul Langerhans (1847–1888); found in the skin.

lanugo (lă-noo′gō) [L., *lana*, wool]   Fine, soft, unpigmented fetal hair.

Laplace's law   Named for the French mathematician Pierre S. de Laplace (1749–1827). Force that stretches the wall of a blood vessel is proportional to the radius of the vessel times the blood pressure.

large intestine   Portion of the digestive tract extending from the small intestine to the anus.

laryngitis (lar-in-jī′tis)   Inflammation of the mucous membrane of the larynx.

laryngopharynx (lă-ring′gō-far-ingks)   Part of the pharynx lying posterior to the larynx.

larynx; pl., larynges (lar′ingks, lă-rin′jēz)   Organ of voice production located between the pharynx and the trachea; it consists of a framework of cartilages and elastic membranes housing the vocal folds and the muscles that control the position and tension of these elements.

last menstrual period (LMP)   Beginning of the last menstruation before pregnancy; used clinically to time events during pregnancy.

latent phase   *See* lag phase.

lateral geniculate nucleus (je-nik′ū-lāt)   Nucleus of the thalamus where fibers from the optic tract terminate.

lateral olfactory area (ol-fak′tŏ-rē)   Part of the olfactory cortex involved in the conscious perception of olfactory stimuli.

lens   Transparent biconvex structure lying between the iris and the vitreous humor.

lens fiber   Epithelial cell that makes up the lens of the eye.

lesser omentum (ō-men′tŭm) [L., membrane that encloses the bowels]   Peritoneal fold passing from the liver to the lesser curvature of the stomach and to the upper border of the duodenum for a distance of approximately 2 cm beyond the pylorus.

lesser vestibular gland (ves-tib′ū-lăr)   Paraurethral gland. Number of minute mucous glands opening on the surface of the vestibule between the openings of the vagina and urethra.

leukocyte (loo′kō-sīt)   White blood cell.

leukocytosis (loo′kō-sī-tō′sis)   Abnormally large number of white blood cells in the blood.

leukopenia (loo-kō-pē′nē-ă)   Lower-than-normal number of white blood cells in the blood.

leukotriene (loo-kō-trī′ēn)   Specific class of physiologically active fatty acid derivatives present in many tissues.

lever   Rigid shaft capable of turning about a fulcrum or pivot point.

LH surge Increase in plasma luteinizing hormone (LH) levels before ovulation and responsible for initiating it.

ligamentum arteriosum (lig′ă-men′tŭm) Remains of the ductus arteriosus.

ligamentum venosum Remnant of the ductus venosus.

ligand A molecule that binds to a macromolecule such as receptors or enzymes.

ligand-gated ion channel An ion channel in a plasma membrane caused to either open or close by a ligand or chemical signal binding to a receptor.

limbic system (lim′bik) [L., *limbus*, border] Parts of the brain involved with emotions and olfaction; includes the cingulate gyrus, hippocampus, habenular nuclei, parts of the basal ganglia, the hypothalamus (especially the mammillary bodies, the olfactory cortex, and various nerve tracts (e.g., fornix).

lingual tonsil (ling′gwăl) Collection of lymphoid tissue on the posterior portion of the dorsum of the tongue.

lipase (lip′ās) In general, any fat-splitting enzyme.

lipid (li′pid) [Gr., *lipos*, fat] Substance composed principally of carbon, oxygen, and hydrogen; contains a lower ratio of oxygen to carbon and is less polar than carbohydrates; generally soluble in nonpolar solvents.

lipid bilayer Double layer of lipid molecules forming the plasma membrane and other cellular membranes.

lipochrome (lip′ō-krōm) Lipid-containing pigment that is metabolically inert.

lipotropin (li-pō-trō′pin) One of the peptide hormones released from the adenohypophysis; increases lipolysis in fat cells.

liver (liv′er) Largest gland of the body, lying in the upper-right quadrant of the abdomen just inferior to the diaphragm; secretes bile and is of great importance in carbohydrate and protein metabolism and in detoxifying chemicals.

lobar (secondary) bronchus (brong′kus) Branch from a primary bronchus that conducts air to each lobe of the lungs. There are two branches in the left lung and three branches from the primary bronchus in the right lung.

lobe (lōb) Rounded projecting part, such as the lobe of a lung, the liver, or a gland.

lobule (lob′ūl) Small lobe or a subdivision of a lobe, such as a lobule of the lung or a gland.

local inflammation Inflammation confined to a specific area of the body. Symptoms include redness, heat, swelling, pain, and loss of function.

local potential Depolarization that is not propagated and that is graded or proportional to the strength of the stimulus.

local reflex Reflex of the intramural plexus of the digestive tract that does not involve the brain or spinal cord.

locus; pl., loci (lō′kŭs, lō′sī) Place; usually a specific site.

loop of Henle Named for the German anatomist Friedrich G. J. Henle (1809–1885). U-shaped part of the nephron extending from the proximal to the distal convoluted tubule and consisting of descending and ascending limbs. Some of the loops of Henle extend into the renal pyramids.

lower respiratory tract The larynx, trachea, and lungs.

lung recoil A decrease in the size of an expanded lung as a result of a decrease in the size (volume) of its alveoli; due to elastic recoil of elastic fibers surrounding alveoli and water surface tension of a thin film of water within alveoli.

lunula; pl., lunulae (loo′noo-lă, loo′noo-lē) [L., *luna*, moon] White, crescent-shaped portion of the nail matrix visible through the proximal end of the nail.

luteal phase (loo′tē-ăl) That portion of the menstrual cycle extending from the time of formation of the corpus luteum after ovulation to the time when menstrual flow begins; usually 14 days in length; the secretory phase.

luteinizing hormone (LH) (loo′tē-ĭ-nīz-ing) In females, hormone stimulating the final maturation of the follicles and the secretion of progesterone by them, with their rupture releasing the ovum, and the conversion of the ruptured follicle into the corpus luteum; in males, stimulates the secretion of testosterone in the testes.

luteinizing hormone-releasing hormone (LHRH) *See* gonadotropin-releasing hormone.

lymph (limf) [L., *lympha*, clear spring water] Clear or yellowish fluid derived from interstitial fluid and found in lymph vessels.

lymph capillary Beginning of the lymphatic system of vessels; lined with flattened endothelium lacking a basement membrane.

lymph node Encapsulated mass of lymph tissue found among lymph vessels.

lymph nodule Small accumulation of lymph tissue lacking a distinct boundary.

lymph sinus Channels in a lymph node crossed by a reticulum of cells and fibers.

lymph vessel One of the system of vessels carrying lymph from the lymph capillaries to the veins.

lymphoblast (lim′fō-blast) Cell that matures into a lymphocyte.

lymphocyte (lim′fō-sīt) Nongranulocytic white blood cell formed in lymphoid tissue.

lymphokine (lim′fō-kīn) Chemical produced by lymphocytes that activates macrophages, attracts neutrophils, and promotes inflammation.

lysis (lī′sis) [Gr., *lysis*, a loosening] Process by which a cell swells and ruptures.

lysosome (lī′sō-sōm) [Gr., *lysis*, loosening + *soma*, body] Membrane-bounded vesicle containing hydrolytic enzymes that function as intracellular digestive enzymes.

lysozyme (lī′sō-zīm) Enzyme that is destructive to the cell walls of certain bacteria; present in tears and some other fluids of the body.

M

macrophage (mak′rō-fāj) [Gr., *makros*, large + *phagein*, to eat] Any large mononuclear phagocytic cell.

macula; pl., maculae (mak′ū-lă, mak′ū-lē) [L., a spot] Sensory structures in the utricle and saccule, consisting of hair cells and a gelatinous mass embedded with otoliths.

macula densa Cells of the distal convoluted tubule located at the renal corpuscle and forming part of the juxtaglomerular apparatus.

macula lutea (mak′ū-lă loo′tē-ă) [L., *macula*, a spot + *luteus*, yellow] Small spot different in color from surrounding tissue; spot in the retina directly behind the lens in which densely packed cones are located.

main bronchus; pl., bronchi (brong′kŭs, brong′kī) One of two tubes arising at the inferior end of the trachea; each primary bronchus extends into one of the lungs.

major duodenal papilla Point of opening of the common bile duct and pancreatic duct into the duodenum.

major histocompatibility complex (MHC) Group of genes that control the production of major histocompatibility complex proteins, which are glycoproteins found on the surfaces of cells. The major histocompatibility proteins serve as self-markers for the immune system and are used by antigen-presenting cells to present antigens to lymphocytes.

male pronucleus Nuclear material of the sperm cell after the ovum has been penetrated by the sperm cell.

malignant (mă-lig′nănt) Resistant to treatment; occurring in severe form, and frequently fatal; having the property of locally invasive and destructive growth and metastasis.

malleus; pl., mallei (mal′ē-ŭs, mal′ē-ī) [L., hammer] Largest of the three auditory ossicles; attached to the tympanic membrane.

mamillary bodies (mam′i-lār-ē) [L., breast- or nipple-shaped] Nipple-shaped structures at the base of the hypothalamus.

mamma; pl., mammae (mam′ă, mam′ē) Breast. The organ of milk secretion; one of two hemispheric projections of variable size situated in the subcutaneous layer over the pectoralis major muscle on either side of the chest; it is rudimentary in the male.

mammary ligaments (mam′ă-rē) Cooper's ligaments. Well-developed ligaments that extend from the overlying skin to the fibrous stroma of mammary gland.

manubrium; pl., manubria (mă-noo′brē-ŭm, mă-noo′bre-ă) [L., handle] Part of a bone representing the handle, such as the manubrium of the sternum representing the handle of a sword.

mass movement Forcible peristaltic movement of short duration, occurring only three or four times a day, which moves the contents of the large intestine.

mass number Equal to the number of protons plus the number of neutrons in each atom.

mastication (mas-ti-kā′shŭn) [L., *mastico*, to chew] Process of chewing.

mastication reflex Repetitive cycle of relaxation and contraction of the muscles of mastication that results in chewing of food.

mastoid (mas′toyd) [Gr., *mastos*, breast] Resembling a breast.

mastoid air cells Spaces within the mastoid process of the temporal bone connected to the middle ear by ducts.

mature follicle An ovarian follicle in which the oocyte attains its full size. The follicle contains a fluid-filled antrum and is surrounded by the theca interna and externa.

maximal stimulus Stimulus resulting in a local potential just large enough to produce the maximum frequency of action potentials.

meatus (mē-ā′tŭs) [L., to go, pass] Passageway or tunnel.

mechanoreceptor (mek′ă-nō-rē-sep′tŏr) A sensory receptor that has the role of responding to mechanical pressures. Examples are pressure receptors in the carotid sinus or touch receptors in the skin.

meconium (mē-kō′nē-ŭm) [Gr., *mekon*, poppy] First intestinal discharges of the newborn infant, greenish in color and consisting of epithelial cells, mucus, and bile.

medial olfactory area Part of the olfactory cortex responsible for the visceral and emotional reactions to odors.

medulla oblongata (me-dool′ă ob-long-gah′tă) Inferior portion of the brainstem that connects the spinal cord to the brain and contains autonomic centers controlling such functions as heart rate, respiration, and swallowing.

medullary cavity (med′ul-er-ē, med′oo-lār-ē) Large, marrow-filled cavity in the diaphysis of a long bone.

medullary ray Extension of the kidney medulla into the cortex, consisting of collecting ducts and loops of Henle.

megakaryoblast (meg-ă-kar′ē-ō-blast) [Gr., *mega* + *karyon*, nut (nucleus) + *blastos*, germ)] Cell that gives rise to platelets or thrombocytes.

meibomian cyst (mī-bō′mē-an) Named for German anatomist Hendrik Meibom (1638–1700). A chronic inflammation of a meibomian gland; *see also* chalazion.

meibomian gland Sebaceous gland near the inner margins of the eyelid; secretes sebum that lubricates the eyelid and retains tears.

meiosis (mī-ō′sis) [Gr., a lessening] Process of cell division that results in the formation of gametes. Consists of two divisions that result in one (female) or four (male) gametes, each of which contains one-half the number of chromosomes in the parent cell.

Meissner's corpuscle (mīs′nerz kōr′pŭs-l) Named for Georg Meissner, German histologist (1829–1905). *See* tactile corpuscle.

melanin (mel′ă-nin) [Gr., *melas*, black] A group of related molecules responsible for skin, hair, and eye color. Most melanins are brown to black pigments, some are yellowish or reddish.

melanocyte (mel′ă-nō-sīt) [Gr., *melas*, black + *kytos*, cell] Cell found mainly in the stratum basale that produces the brown or black pigment melanin.

melanocyte-stimulating hormone (MSH) Peptide hormone secreted by the anterior pituitary; increases melanin production by melanocytes, making the skin darker in color.

melanosome (mel′ă-nō-sōm) [Gr., *melas*, black + *soma*, body] Membranous organelle containing the pigment melanin.

melatonin (mel-ă-tōn′in) Hormone (amino acid derivative) secreted by the pineal body; inhibits secretion of gonadotropin-releasing hormone from the hypothalamus.

membrane attack complex (MAC) A channel through a plasma membrane produced by activated complement proteins, primarily complement protein C9; in nucleated cells, water enters the channel causing lysis of cells.

membrane-bound receptor Receptor molecule such as a hormone receptor that is bound to the plasma membrane of the target cell.

membranous labyrinth (mem′bră-nŭs lab′i-rinth) Membranous structure within the inner ear consisting of the cochlea, vestibule, and semicircular canals.

membranous urethra (ū-rē′thră) Portion of the male urethra, approximately 1 cm in length, extending from the prostate gland to the beginning of the penile urethra.

memory cell Small lymphocytes that are derived from B cells or T cells and that rapidly respond to a subsequent exposure to the same antigen.

menarche (me-nar′kē) [Gr., *mensis*, month + *arche*, beginning] Establishment of menstrual function; the time of the first menstrual period or flow.

meninx; pl., meninges (mē′ninks, mě-nin′jes) [Gr., membrane] Connective tissue membranes surrounding the brain.

menisci (sing. meniscus) (me-nis′kus, me-nis′sī) A crescent-shaped intraarticular fibrocartilage found in certain joints such as the crescent-shaped fibrocartilaginous structure of the knee.

menopause (men′ō-pawz) [Gr., *mensis*, month + *pausis*, cessation] Permanent cessation of the menstrual cycle.

menses (men′sēz) [L., *mensis*, month] Periodic hemorrhage from the uterine mucous membrane, occurring at approximately 28-day intervals.

menstrual cycle (men′stroo-ăl) Series of changes that occur in sexually mature, nonpregnant women and result in menses. Specifically refers to the uterine cycle but is often used to include both the uterine and ovarian cycles.

Merkel's disk (mer′kelz) Named for Friedrich Merkel, German anatomist (1845–1919). *See* tactile disk.

merocrine gland (mer′ō-krin) [Gr., *meros*, part + *krino*, to separate] Gland that secretes products with no loss of cellular material; an example is water-producing sweat glands; *see also* apocrine and holocrine glands.

mesencephalon (mez-en-sef′ă-lon) [Gr., *mesos*, middle + *enkephalos*, brain] Midbrain in both the embryo and adult; consists of the cerebral peduncle and the corpora quadrigemini.

mesentery (mes′en-ter-ē) [Gr., *mesos*, middle + *enteron*, intestine] Double layer of peritoneum extending from the abdominal wall to the abdominal viscera, conveying to it its vessels and nerves.

mesoderm (mez′ō-derm) Middle of the three germ layers of an embryo.

mesonephros (mez′ō-nef′ros) One of three excretory organs appearing during embryonic development; forms caudal to the pronephros as the pronephros disappears. It is well developed and is functional for a time before the establishment of the metanephros, which gives rise to the kidney; undergoes regression as an excretory organ, but its duct system is retained in the male as the efferent ductule and epididymis.

mesosalpinx (mez′ō-sal′pinks) [Gr., *mesos*, middle + *salpinx*, trumpet] Part of the broad ligament supporting the uterine tube.

mesothelium (mez-ō-thē′lē-ŭm) A single layer of flattened cells forming an epithelium that lines serous cavities, such as peritoneum, pleura, pericardium.

mesovarium (mez′ō-vā′rē-ŭm) Short peritoneal fold connecting the ovary with the broad ligament of the uterus.

messenger RNA (mRNA) Type of RNA that moves out of the nucleus and into the cytoplasm, where it is used as a template to determine the structure of proteins.

metabolism (mě-tab′ō-lizm) [Gr., *metabole*, change] Sum of all the chemical reactions that take place in the body, consisting of anabolism and catabolism. Cellular metabolism refers specifically to the chemical reactions within cells.

metacarpal (met′ă-kar′păl) Relating to the fine bones of the hand between the carpus (wrist) and the phalanges.

metanephros (met-ă-nef′ros) Most caudally located of the three excretory organs appearing during embryonic development; becomes the permanent kidney of mammals. In mammalian embryos it is formed caudal to the mesonephros and develops later as the mesonephros undergoes regression.

metaphase (met′ă-fās) Time during cell division when the chromosomes line up along the equator of the cell.

metarteriole (met′ar-tēr′ē-ōl) One of the small peripheral blood vessels that contain scattered groups of smooth muscle fibers in their walls; located between the arterioles and the true capillaries.

metastasis (mě-tas′tă-sis) The shifting of a disease or its local manifestations, or the spread of a disease from one part of the body to another as in a malignant neoplasm.

metatarsal (met′ă-tar′sal) [Gr., *meta*, after + *tarsos*, sole of the foot] Distal bone of the foot.

metencephalon (met′en-sef′ă-lon) [Gr., *meta*, after + *enkephalos*, brain] Second-most posterior division of the embryonic brain; becomes the pons and cerebellum in the adult.

micelle (mi-sel′, mī-sel′) [L., *micella*, small morsel] Droplets of lipid surrounded by bile salts in the small intestine.

microfilament (mī-krō-fil′ă-ment) Small fibril forming bundles, sheets, or networks in the cytoplasm of cells; provides structure to the cytoplasm and mechanical support for microvilli and stereocilia.

microglia (mī-krog′lē-ă) [Gr., *micro* + *glia*, glue] Small neuroglial cells that become phagocytic and mobile in response to inflammation; considered to be macrophages within the central nervous system.

microtubule (mī-krō-too′būl) Hollow tube composed of tubulin, measuring approximately 25 nm in diameter and usually several micrometers long. Helps provide support to the cytoplasm of the cell and is a component of certain cell organelles such as centrioles, spindle fibers, cilia, and flagella.

microvillus; pl., microvilli (mī′krō-vil′ŭs, mī′krō-vil′ī)   Minute projection of the cell membrane that greatly increases the surface area.

micturition reflex (mik-choo-rish′ŭn)   Contraction of the urinary bladder stimulated by stretching of the bladder wall; results in emptying of the bladder.

middle ear   Air-filled space within the temporal bone; contains auditory ossicles; between the external and internal ear.

milk letdown   Expulsion of milk from the alveoli of the mammary glands; stimulated by oxytocin.

mineral   Inorganic nutrient necessary for normal metabolic functions.

mineralocorticoid (min′er-al-ō-kōr′ti-koyd)   Steroid hormone (e.g., aldosterone) produced by the zona glomerulosa of the adrenal cortex; facilitates exchange of potassium for sodium in the distal renal tubule, causing sodium reabsorption and potassium and hydrogen ion secretion.

minor duodenal papilla   Site of the opening of the accessory pancreatic duct into the duodenum.

minute ventilation   Product of tidal volume times the respiratory rate.

minute volume   Amount of blood pumped by either the left or right ventricle each minute.

mitochondrion; pl., mitochondria (mī-tō-kon′drē-on, mī-tō-kon′drē-ă) [Gr., *mitos*, thread + *chandros*, granule]   Small, spherical, rod-shaped or thin filamentous structure in the cytoplasm of cells that is a site of ATP production.

mitosis (mī-tō′sis) [Gr., thread]   Cell division resulting in two daughter cells with exactly the same number and type of chromosomes as the mother cell.

M line   Line in the center of the H zone made of delicate filaments that holds the myosin myofilaments in place in the sarcomere of muscle fibers.

modiolus (mō-dī′ō′lŭs) [L., nave of a wheel]   Central core of spongy bone about which turns the spiral canal of the cochlea.

molar (mō′lăr)   Tricuspid tooth; the three posterior teeth of each dental arch.

molecule (mol′ĕ-kūl)   A substance composed of two or more atoms chemically combined to form a structure that behaves as an independent unit.

monoblast (mon′ō-blast)   Cell that matures into a monocyte.

mononuclear phagocytic system (mon-ō-noo′klē-ăr fag-ō-sit′ik)   Phagocytic cells, each with a single nucleus; derived from monocytes.

monosaccharide (mon-ō-sak′ă-rīd)   Simple sugar carbohydrate that cannot form any simpler sugar by hydrolysis.

mons pubis (monz pū′bis) [L., mountain; the genitals]   Prominence caused by a pad of fatty tissue over the symphysis pubis in the female.

morula (mōr′oo-lă, mōr′ū-lă) [L., *morus*, mulberry]   Mass of 12 or more cells resulting from the early cleavage divisions of the zygote.

motor neuron   Neuron that innervates skeletal, smooth, or cardiac muscle fibers.

motor unit   Single neuron and the muscle fibers it innervates.

mucosa (mū-kō′să) [L., *mucosus*, mucous]   Mucous membrane consisting of epithelium and lamina propria. In the digestive tract there is also a layer of smooth muscle.

mucous membrane (mū′kŭs)   Thin sheet consisting of epithelium and connective tissue (lamina propria) that lines cavities that open to the outside of the body; many contain mucous glands that secrete mucus.

mucous neck cell   One of the mucous-secreting cells in the neck of a gastric gland.

mucus (mū′kŭs)   Viscous secretion produced by and covering mucous membranes; lubricates mucous membranes and traps foreign substances.

multiple motor unit summation   Increased force of contraction of a muscle due to recruitment of motor units.

multiple wave summation   Increased force of contraction of a muscle due to increased frequency of stimulation.

multipolar neuron   One of three categories of neurons consisting of a neuron cell body, an axon, and two or more dendrites.

muscarinic receptor (mŭs′kă-rin′ik)   Class of cholinergic receptor that is specifically activated by muscarine in addition to acetylcholine.

muscle fiber   Muscle cell.

muscle spindle   Three to 10 specialized muscle fibers supplied by gamma motor neurons and wrapped in sensory nerve endings; detects stretch of the muscle and is involved in maintaining muscle tone.

muscle tone   Relatively constant tension produced by a muscle for long periods as a result of asynchronous contraction of motor units.

muscle twitch   Contraction of a whole muscle in response to a stimulus that causes an action potential in one or more muscle fibers.

muscular fatigue   Fatigue due to a depletion of ATP within the muscle fibers.

muscularis (mŭs-kū-lā′ris) [Modern L., muscular]   Muscular coat of a hollow organ or tubular structure.

muscularis mucosa   Thin layer of smooth muscle found in most parts of the digestive tube; located outside the lamina propria and adjacent to the submucosa.

musculi pectinati (mŭs-kū-lī pek′tī-nă′tē)   Prominent ridges of atrial myocardium located on the inner surface of much of the right atrium and both auricles.

mutation (mū-tā′shŭn)   A change in the number or kinds of nucleotides in the DNA of a gene.

myelencephalon (mī′el-en-sef′ă-lon) [Gr., *myelos*, medulla, marrow + *enkephalos*, brain]   Most caudal portion of the embryonic brain; medulla oblongata.

myelin sheath (mī′ĕ-lin)   Envelope surrounding most axons; formed by Schwann cell membranes being wrapped around the axon.

myelinated axon (mī′ĕ-li-nāt-ed ak′son)   Nerve fiber having a myelin sheath.

myeloblast (mī′ĕ-lō-blast)   Immature cell from which the different granulocytes develop.

myenteric plexus (mī′en-ter′ik)   Plexus of unmelinated fibers and postganglionic autonomic cell bodies lying in the muscular coat of the esophagus, stomach, and intestines; communicates with the submucosal plexuses.

myoblast (mī′ō-blast) [Gr., *mys*, muscle + *blastos*, germ]   Primitive multinucleated cell with the potential of developing into a muscle fiber.

myofilament (mī-ō-fil′ă-ment)   Extremely fine molecular thread helping to form the myofibrils of muscle; thick myofilaments are formed of myosin, and thin myofilaments are formed of actin.

myometrium (mī′ō-mē′trē-ŭm)   Muscular wall of the uterus; composed of smooth muscle.

myosin myofilament (mī′ō-sin mī-ō-fil′ă-ment)   Thick myofilament of muscle fibrils; composed of myosin molecules.

N

nail (nāl) [A.S., naegel]   Several layers of dead epithelial cells containing hard keratin on the ends of the digits.

nail bed   Epithelial tissue resting on dermis under the nail between the nail matrix and hyponychium; contributes to the formation of the nail. *See* nail matrix.

nail matrix   Epithelial tissue resting on dermis under the proximal end of a nail; produces most of the nail. *See* nail bed.

nasal cavity (nā′zăl)   Cavity between the external nares and the pharynx. It is divided into two chambers by the nasal septum and is bounded inferiorly by the hard and soft palates.

nasal septum   Bony partition that separates the nasal cavity into left and right parts; composed of the vomer, the perpendicular plate of the ethmoid, and hyaline cartilage.

nasolacrimal duct (nā-zō-lak′ri-măl)   Duct that leads from the lacrimal sac to the nasal cavity.

nasopharynx (nā-zō-far′ingks)   Part of the pharynx that lies above the soft palate; anteriorly it opens into the nasal cavity.

near point of vision   Closest point from the eye at which an object can be held without appearing blurred.

neck (tooth)   Slightly constricted part of a tooth, between the crown and the root.

neoplasm (nē′ō-plazm)   An abnormal tissue that grows by cellular proliferation more rapidly than normal and continues to grow after the stimuli that initiated the new growth ceases.

nephron (nef′ron) [Gr., *nephros*, kidney]   Functional unit of the kidney, consisting of the renal corpuscle, the proximal convoluted tubule, the loop of Henle, and the distal convoluted tubule.

nerve tract   Bundles of parallel axons with their associated sheaths in the central nervous system.

neural crest (noor′ăl)   Edge of the neural plate as it rises to meet at the midline to form the neural tube.

neural crest cells   Cells derived from the crests of the forming neural tube in the embryo; together with the mesoderm, form the mesenchyme of the embryo; give rise to part of the skull, the teeth, melanocytes, sensory neurons, and autonomic neurons.

neural plate   Region of the dorsal surface of the embryo that is transformed into the neural tube and neural crest.

neural tube   Tube formed from the neuroectoderm by the closure of the neural groove. The neural tube develops into the spinal cord and brain.

neuroectoderm (noor-ō-ek′tō-derm)   That part of the ectoderm of an embryo giving rise to the brain and spinal cord.

neuroglia (noo-rog′lē-ă) [Gr., *neuro,* nerve + *glia,* glue]   Cells in the nervous system other than the neurons; includes astrocytes, ependymal cells, microglia, oligodendrocytes, satellite cells, and Schwann cells.

neurohormone (noor′ō-hōr′mōn)   Hormone secreted by a neuron.

neurohypophysis (noor′ō-hi-pof′i-sis)   Portion of the hypophysis derived from the brain; commonly called the posterior pituitary. Major secretions include antidiuretic hormone and oxytocin.

neuromodulator   Substance that influences the sensitivity of neurons to neurotransmitters but neither strongly stimulates nor strongly inhibits neurons by itself.

neuromuscular junction   (noor-ō-mŭs′kū-lăr)   Specialized synapse between a motor neuron and a muscle fiber.

neuron (noor′on) [Gr., nerve]   Morphologic and functional unit of the nervous system, consisting of the nerve cell body, the dendrites, and the axon.

neuron cell body   Enlarged portion of the neuron containing the nucleus and other organelles; also called nerve cell body.

neurotransmitter (noor′ō-trans-mit′er) [Gr., *neuro,* nerve + L., *transmitto,* to send across]   Any specific chemical agent released by a presynaptic cell on excitation that crosses the synaptic cleft and stimulates or inhibits the postsynaptic cell.

neutral solution (noo′trăl)   Solution such as pure water that has 10^{-7} mol of hydrogen ions per liter and an equal concentration of hydroxide ions; has a pH of 7.

neutron (noo′tron) [L., *neuter,* neither]   Electrically neutral particle in the nuclei of atoms (except hydrogen).

neutrophil (noo′trō-fil) [L., *neuter,* neither + Gr., *philos,* fond]   Type of white blood cell; small phagocytic white blood cell with a lobed nucleus and small granules in the cytoplasm.

nicotinic receptor (nik-ō-tin′ik)   Class of cholinergic receptor molecule that is specifically activated by nicotine and by acetylcholine.

nipple (nip′l)   Projection at the apex of the mamma, on the surface of which the lactiferous ducts open; surrounded by a circular pigmented area, the areola.

Nissl bodies (nis′l)   Named after the German neurologist Franz Nissl (1860–1919). Areas in the neuron cell body containing rough endoplasmic reticulum.

nociceptor (nō-si-sep′ter) [L., *noceo,* to injure + *capio,* to take]   A sensory receptor that detects painful or injurious stimuli.

nonelectrolyte (non-ē-lek′trō-līt) [Gr., *electro* + *lytos,* soluble]   Molecules that do not dissociate and do not conduct electricity.

norepinephrine (nōr′ep-i-nef′rin)   Neurotransmitter substance released from most of the postganglionic neurons of the sympathetic division; hormone released from the adrenal cortex that increases cardiac output and blood glucose levels.

nose, or **nasus** (nōz, or nā′sŭs)   Visible structure that forms a prominent feature of the face; can also refer to the nasal cavities.

notochord (nō′tō-kōrd) [Gr., *notor,* back + *chords,* cord]   Small rod of tissue lying ventral to the neural tube. A characteristic of all vertebrates, in humans it becomes the nucleus pulposus of the intervertebral disks.

nuchal (noo′kăl)   The back of the neck.

nuclear envelope (noo′klē-er)   Double membrane structure surrounding and enclosing the nucleus.

nuclear pores   Porelike openings in the nuclear envelope where the inner and outer membranes fuse.

nucleic acid (noo-klē′ik, noo-klā′ik)   Polymer of nucleotides, consisting of DNA and RNA, forms a family of substances that comprise the genetic material of cells and control protein synthesis.

nucleolus; pl., **nucleoli** (noo-klē′ō-lŭs, noo-klē′ō-lī)   Somewhat rounded, dense, well-defined nuclear body with no surrounding membrane; contains ribosomal RNA and protein.

nucleotide (noo′klē-ō-tīd)   Basic building block of nucleic acids consisting of a sugar (either ribose or deoxyribose) and one of several types of organic bases.

nucleus; pl., **nuclei** (noo′klē-ŭs, noo′klē-ī) [L., inside of a thing]   Cell organelle containing most of the genetic material of the cell; collection of nerve cell bodies within the central nervous system; center of an atom consisting of protons and neutrons.

nucleus pulposus (pŭl-pō′sŭs) [L., central pulp]   Soft central portion of the intervertebral disk.

nutrient (noo′trē-ent) [L., *nutriens,* to nourish]   Chemicals taken into the body that are used to produce energy, provide building blocks for new molecules, or function in other chemical reactions.

O

olecranon (ō-lek′ră-non, ō′lē-krā′non)   Process on the distal end of the ulna, forming the point of the elbow.

olfaction (ol-fak′shŭn) [L., *olfactus,* smell]   Sense of smell.

olfactory area   Extreme superior region of the nasal cavity.

olfactory bulb (ol-fak′tŏ-rē)   Ganglion-like enlargement at the rostral end of the olfactory tract that lies over the cribriform plate; receives the olfactory nerves from the nasal cavity.

olfactory cortex   Termination of the olfactory tract in the cerebral cortex within the lateral fissure of the cerebrum.

olfactory epithelium   Epithelium of the olfactory recess containing olfactory receptors.

olfactory tract   Nerve tract that projects from the olfactory bulb to the olfactory cortex.

oligodendrocyte (ol′i-gō-den′drō-sīt)   Neuroglial cell that has cytoplasmic extensions that form myelin sheaths around axons in the central nervous system.

oncogene (ong′kō-jĕn)   A gene that can change or be activated to cause cancer.

oncology (ong-kol′ō-jē)   The study of neoplasms.

oocyte (ō′ō-sīt) [Gr., *oon,* egg + *kytos,* a hollow (cell)]   Immature ovum.

oogenesis (ō-ō-jen′ĕ-sis)   Formation and development of a secondary oocyte or ovum.

oogonium (ō-ō-gō′nē-ŭm) [Gr., *oon,* egg + *gone,* generation]   Primitive cell from which oocytes are derived by meiosis.

opposition   Movement of the thumb and little finger toward each other; movement of the thumb toward any of the fingers.

opsin (op′sin)   Protein portion of the rhodopsin molecule. A class of proteins that bind to retinal to form the visual pigments of the rods and cones of the eye.

opsonin (op′sŏ-nin) [Gr., *opsonein,* to prepare food]   Substance such as antibody or complement that enhances phagocytosis.

optic chiasma (op′tik kī′az′mă) [Gr., two crossing lines; *chi,* the letter χ]   Point of crossing of the optic tracts.

optic disc   Point at which axons of ganglion cells of the retina converge to form the optic nerve, which then penetrates through the fibrous tunic of the eye.

optic nerve   Nerve carrying visual signals from the eye to the optic chiasm.

optic stalk   Constricted proximal portion of the optic vesicle in the embryo; develops into the optic nerve.

optic tract   Tract that extends from the optic chiasma to the lateral geniculate nucleus of the thalamus.

optic vesicle   One of the paired evaginations from the walls of the embryonic forebrain from which the retina develops.

oral cavity (ōr′ăl)   The mouth; consists of the space surrounded by the lips, cheeks, teeth, and palate; limited posteriorly by the fauces.

orbit (ōr′bit)   Eye socket; formed by seven skull bones that surround and protect the eye.

organ of Corti (ōr′găn)   Named for the Italian anatomist Marquis Alfonso Corti (1822–1888). Spiral organ; rests on the basilar membrane and supports the hair cells that detect sounds.

organelle (or′gă-nel) [Gr., *organon,* tool]   Specialized part of a cell serving one or more specific individual functions.

orgasm (ōr′gazm) [Gr., *orgao,* to swell, be excited]   Climax of the sexual act, associated with a pleasurable sensation.

origin (ōr′i-jin)   Less movable attachment point of a muscle; usually the medial or proximal end of a muscle associated with the limbs.

oropharynx (ōr′ō-far′ingks)   Portion of the pharynx that lies posterior to the oral cavity; it is continuous above with the nasopharynx and below with the laryngopharynx.

oscillating circuit   Neuronal circuit arranged in a circular fashion that allows action potentials produced in the circuit to keep stimulating the neurons of the circuit.

osmolality (os-mō-lal′i-tē)   Osmotic concentration of a solution; the number of moles of solute in 1 kg of water times the number of particles into which the solute dissociates.

osmoreceptor cell (os-mō-rē-sep′ter, os′mō-rē-sep′tōr) [Gr., *osmos,* impulse]   Receptor in the central nervous system that responds to changes in the osmotic pressure of the blood.

osmosis (os-mō′sis) [Gr., *osmos,* thrusting or an impulsion]   Diffusion of solvent (water) through a membrane from a less concentrated solution to a more concentrated solution.

osmotic pressure (os-mot′ik)   Force required to prevent the movement of water across a selectively permeable membrane.

ossification (os′i-fi-kā′shŭn) [L., *os*, bone + *facio*, to make]   Bone formation.

osteoblast (os′tē-ō-blast) [Gr., *osteon*, bone + *blastos*, germ]   Bone-forming cell.

osteoclast (os′tē-ō-klast) [Gr., *osteon*, bone + *klastos*, broken]   Large multinucleated cell that absorbs bone.

osteocyte (os′tē-ō-sīt) [Gr., *osteon*, bone + *kytos*, cell]   Mature bone cell surrounded by bone matrix.

osteomalacia (os′tē-ō-mă-lā′shē-ă)   Softening of bones due to calcium depletion. Adult rickets.

osteon (os′tē-on)   A central canal containing blood capillaries and the concentric lamellae around it; occurs in compact bone.

osteoporosis (os′tē-ō-pō-rō′sis) [Gr., *osteon*, bone + *poros*, pore + *osis*, condition]   Reduction in quantity of bone, resulting in porous bone.

ostium (os′tē-ŭm) [L., door, entrance, mouth]   Small opening, for example, the opening of the uterine tube near the ovary or the opening of the uterus into the vagina.

otolith (ō′tō-lith)   Crystalline particles of calcium carbonate and protein embedded in the maculae.

oval window (ō′văl)   Membranous structure to which the stapes attaches; transmits vibrations to the inner ear.

ovarian cycle (ō-var′ē-an)   Series of events that occur in a regular fashion in the ovaries of sexually mature, nonpregnant females; results in ovulation and the production of the hormones estrogen and progesterone.

ovarian epithelium   (germinal epithelium)   Peritoneal covering of the ovary.

ovarian ligament   Bundle of fibers passing to the uterus from the ovary.

ovary (ō′vă-rē)   One of two female reproductive glands located in the pelvic cavity; produces the secondary oocyte, estrogen, and progesterone.

oviduct (ō′vi-dŭkt)   *See* uterine tube.

ovulation (ov′ū-lā′shun)   Release of an ovum, or secondary oocyte, from the vesicular follicle.

oxidation (ok-si-dā′shŭn)   Loss of one or more electrons from a molecule.

oxidation–reduction reaction   Reaction in which one molecule is oxidized and another is reduced.

oxidative deamination (ok-si-dā′tiv)   Removal of the amine group of an amino acid to form a keto acid, ammonia, and NADH.

oxygen debt (ok′sē-jen)   Oxygen necessary for the synthesis of the ATP required to remove lactic acid produced by anaerobic respiration.

oxygen–hemoglobin dissociation curve   Graph describing the relationship between the percentage of hemoglobin saturated with oxygen and a range of oxygen partial pressures.

oxyhemoglobin (ox′sē-hē-mō-glō′bin)   Oxygenated hemoglobin.

P

pacinian corpuscle (pa-sin′ē-an)   Named for Filippo Pacini, Italian anatomist (1812–1883). *See* lamellated corpuscle.

palate (pal′ăt) [L., *palatum*, palate]   Roof of the mouth.

palatine tonsil (pal′ă-tīn)   One of two large oval masses of lymphoid tissue embedded in the lateral wall of the oral pharynx.

palpebra; pl., palpebrae (pal-pē′bră, pal-pē′brē) [L., eyelid]   An eyelid.

palpebral conjunctiva (pal-pē′brăl kon-jŭnk-tī′vă)   Conjunctiva that covers the inner surface of the eyelids.

palpebral fissure   Space between the upper and lower eyelids.

pancreas (pan′krē-as) [Gr., *pankreas*, the sweetbread]   Abdominal gland that secretes pancreatic juice into the intestine and insulin and glucagon from the pancreatic islets into the bloodstream.

pancreatic duct (pan-krē-at′ik)   Excretory duct of the pancreas that extends through the gland from tail to head, where it empties into the duodenum at the greater duodenal papilla.

pancreatic islet   Islets of Langerhans; cellular mass varying from a few to hundreds of cells lying in the interstitial tissue of the pancreas; composed of different cell types that make up the endocrine portion of the pancreas and are the source of insulin and glucagon.

pancreatic juice [L., *jus*, broth]   External secretion of the pancreas; clear, alkaline fluid containing several enzymes.

papilla (pă-pil′ă) [L., nipple]   A small nipplelike process. Projection of the dermis, containing blood vessels and nerves, into the epidermis. Projections on the surface of the tongue.

papillary muscle (pap′i-lār′ē)   Nipplelike conical projection of myocardium within the ventricle; the chordae tendineae are attached to the apex of the papillary muscle.

parafollicular cell (par-ă-fo-lik′ū-lăr)   Endocrine cell scattered throughout the thyroid gland; secretes the hormone calcitonin.

paramesonephric duct (par-ă-mes-ō-nef′rik)   One of two embryonic tubes extending along the mesonephros and emptying into the cloaca; in the female the duct forms the uterine tube, the uterus, and part of the vagina; in the male it degenerates.

paranasal sinus (par-ă-nā′săl)   Air-filled cavities within certain skull bones that connect to the nasal cavity; located in the frontal, maxillary, sphenoid, and ethmoid bones.

parasympathetic (par-ă-sim-pa-thet′ik)   Subdivision of the autonomic nervous system; characterized by having the cell bodies of its preganglionic neurons located in the brainstem and the sacral region of the spinal cord (craniosacral division); usually involved in activating vegetative functions such as digestion, defecation, and urination.

parathyroid gland (par-ă-thī′royd)   One of four glandular masses imbedded in the posterior surface of the thyroid gland; secretes parathyroid hormone.

parathyroid hormone   Peptide hormone produced by the parathyroid gland; increases bone breakdown and blood calcium levels.

parietal (pă-rī′ĕ-tăl) [L., *paries*, wall]   Relating to the wall of any cavity.

parietal cell   Gastric gland cell that secretes hydrochloric acid.

parietal pericardium   Serous membrane lining the fibrous portion of the pericardial sac.

parietal peritoneum   Layer of peritoneum lining the abdominal walls.

parietal pleura   Serous membrane that lines the different parts of the wall of the pleural cavity.

parotid gland (pă-rot′id)   Largest of the salivary glands; situated anterior to each ear.

partial pressure   Pressure exerted by a single gas in a mixture of gases.

passive tension   Tension applied to a load by a muscle without contracting; produced when an external force stretches the muscle.

patella (pa-tel′ă) [L., *patina*, shallow disk]   Kneecap.

pectoral girdle (pek′tŏ-răl)   Site of attachment of the upper limb to the trunk; consists of the scapula and the clavicle.

pedicle (ped′ĭ-kl) [L., *pes*, feet]   Stalk or base of a structure, such as the pedicle of the vertebral arch.

pelvic brim (pel′vik)   Imaginary plane passing from the sacral promontory to the pubic crest.

pelvic girdle   Site of attachment of the lower limb to the trunk; ring of bone formed by the sacrum and the coxae.

pelvic inlet   Superior opening of the true pelvis.

pelvic outlet   Inferior opening of the true pelvis.

pelvis; pl., pelves (pel′vis, pel′vēz) [L., basin]   Any basin-shaped structure; cup-shaped ring of bone at the lower end of the trunk, formed from the ossa coxae, sacrum, and coccyx.

pennate (pen′āt) [L., *penna*, feather]   Muscles with fasciculi arranged like the barbs of a feather along a common tendon.

pepsin (pep′sin) [Gr., *pepsis*, digestion]   Principal digestive enzyme of the gastric juice, formed from pepsinogen; digests proteins into smaller peptide chains.

pepsinogen (pep-sin′ō-jen) [pepsin + Gr. *gen*, producing]   Proenzyme formed and secreted by the chief cells of the gastric mucosa; the acidity of the gastric juice and pepsin itself converts pepsinogen into pepsin.

peptidase (pep′ti-dās)   An enzyme capable of hydrolyzing one of the peptide links of a peptide.

peptide bond (pep′tīd)   Chemical bond between amino acids.

Percent Daily Value   The percent of the recommended daily value of a nutrient found in one serving of a particular food.

perforating canal   Canal containing blood vessels and nerves and running through bone perpendicular to the haversian canals.

periarterial sheath (per′ē-ar-tē′rē-ăl)   Dense accumulations of lymphocytes (white pulp) surrounding arteries within the spleen.

pericapillary cell   One of the slender connective tissue cells in close relationship to the outside of the capillary wall; relatively undifferentiated and may become a fibroblast, macrophage, or smooth muscle cell.

pericardial cavity (per-i-kar′dē-ăl)   Space within the mediastinum in which the heart is located.

pericardial fluid   Viscous fluid contained within the pericardial cavity between the visceral and parietal pericardium; functions as a lubricant.

pericardium (per-i-kar′dē-ŭm) [Gr., *pericardion*, the membrane around the heart] Membrane covering the heart.

perichondrium (per-i-kon′drē-ŭm) [Gr., *peri*, around + *chondros*, cartilage] Double-layered connective tissue sheath surrounding cartilage.

perilymph (per′i-limf) [Gr., *peri*, around + L., *lympha*, a clear fluid (lymph)] Fluid contained within the bony labyrinth of the inner ear.

perimetrium (per-i-mē′trē-ŭm) Outer serous coat of the uterus.

perimysium (per-i-mis′ē-ŭm, per-i-miz′ē-ŭm) [Gr., *peri*, around + *mys*, muscle] Fibrous sheath enveloping a bundle of skeletal muscle fibers (muscle fascicle).

perineum (per′i-nē′ŭm) Area inferior to the pelvic diaphragm between the thighs; extends from the coccyx to the pubis.

perineurium (per-i-noo′rē-ŭm) [L., *peri*, around + Gr., *neuron*, nerve] Connective tissue sheath surrounding a nerve fascicle.

periodontal ligament (per′ē-ō-don′tăl) Connective tissue that surrounds the tooth root and attaches it to its bony socket.

periosteum (per-ē-os′tē-ŭm) [Gr., *peri*, around + *osteon*, bone] Thick, double-layered connective tissue sheath covering the entire surface of a bone except the articular surface, which is covered with cartilage.

peripheral nervous system (PNS) (pĕ-rif′ĕ-răl) Major subdivision of the nervous system consisting of nerves and ganglia.

peripheral resistance Resistance to blood flow in all the blood vessels.

peristaltic wave (per-i-stal′tik) Contraction in a tube such as the intestine characterized by a wave of contraction in smooth muscle preceded by a wave of relaxation that moves along the tube.

peritubular capillary The capillary network located in the cortex of the kidney; associated with the distal and proximal convoluted tubules.

permanent tooth One of the 32 teeth belonging to the second, or permanent, dentition.

peroneal (per-ō-nē′ăl) [Gr., *perone*, fibula] Associated with the fibula.

peroxisome (per-ok′si-sōm) Membrane-bounded body similar to a lysosome in appearance but often smaller and irregular in shape; contains enzymes that either decompose or synthesize hydrogen peroxide.

Peyer's patch Named for the Swiss anatomist Johann K. Peyer (1653–1712). Lymph nodule found in the lower half of the small intestine and the appendix.

phagocyte (fag′ō-sīt) Cell possessing the property of ingesting bacteria, foreign particles, and other cells.

phagocytosis (fag′ō-sī-tō′sis) [Gr., *phagein*, to eat + *kytos*, cell + *osis*, condition] Process of ingestion by cells of solid substances, such as other cells, bacteria, bits of necrosed tissue, and foreign particles.

phalange; pl., phalanges (fă-lanj′, fă-lan′jēz) [Gr., *phalanx*, line of soldiers] Bone of the fingers or toes.

pharyngeal pouch (fă-rin′jē-ăl) Paired evagination of embryonic pharyngeal endoderm between the brachial arches that gives rise to the thymus, thyroid gland, tonsils, and parathyroid glands.

pharyngeal tonsil (fă-rin′jē-ăl) One of two collections of aggregated lymphoid nodules on the posterior wall of the nasopharynx.

pharynx (far′ingks) [Gr., *pharynx*, throat, the joint opening of the gullet and windpipe] Upper expanded portion of the digestive tube between the esophagus below and the oral and nasal cavities above and in front.

phenotype (fē′nō-tīp) [Gr., *phaino*, to display, show forth + *typos*, model] Characteristic observed in an individual due to expression of his genotype.

phosphodiesterase (fos′fō-dī-es′ter-ās) Enzymes that split phosphodiester bonds, that is, that break down cyclic AMP to AMP.

phospholipid (fos-fō-lip′id) Lipid with phosphorus, resulting in a molecule with a polar end and a nonpolar end; main component of the lipid bilayer.

phosphorylation (fos′fōr-i-lā′shŭn) Addition of phosphate to an organic compound.

photoreceptor (fō′tō-rē-sep′ter, fō′tō-rē-sep′tōr) [L., *photo*, light + *ceptus*, to receive] A sensory receptor that is sensitive to light. Examples are rods and cones of the retina.

phrenic nerve (fren′ik) Nerve derived from spinal nerves C3–C5; supplies the diaphragm.

physiologic contracture (fiz-ē-ō-loj′ik kon-trak′chūr) Temporary inability of a muscle to either contract or relax because of a depletion of ATP so that active transport of calcium ions into the sarcoplasmic reticulum cannot occur.

physiologic dead space Sum of anatomic dead air space plus the volume of any nonfunctional alveoli.

physiologic shunt Deoxygenated blood from the alveoli plus deoxygenated blood from the bronchi and bronchioles.

pia mater (pī′ă mā′ter, pē′a mah′ter) [L., tender mother] Delicate membrane forming the inner covering of the brain and spinal cord.

pigmented retina Pigmented portion of the retina.

pineal body (pin′ē-ăl) [L., *pineus*, relating to pine trees] A small pine cone–shaped structure that projects from the epiphysis of the diencephalon; produces melatonin.

pinna (pin′ă) [L., *pinna* or *penna*, feather (in plural, wing)] *See* auricle.

pinocytosis (pin′ō-sī-tō′sis, pī′no-sī-tō′sis) [Gr., *pineo*, to drink + *kytos*, cell + *osis*, condition] Cell drinking; uptake of liquid by a cell.

pituitary gland (pi-too′i-tār-ē) *See* hypophysis.

plane (plān) [L., *planus*, flat] A flat surface. An imaginary surface formed by extension through any axis or two points. Examples include a midsagittal plane, a coronal plane, and a transverse plane.

plasma (plaz′mă) [Gr., something formed] Fluid portion of blood.

plasma cell Cell derived from B cells; produces antibodies.

plasma clearance Volume of plasma per minute from which a substance can be completely removed by the kidneys.

plasmin (plaz′min) Enzyme derived from plasminogen; dissolves clots by converting fibrin into soluble products.

plateau phase of action potential Prolongation of the depolarization phase of a cardiac muscle cell membrane; results in a prolonged refractory period.

platelet (plāt′let) Irregularly shaped disk found in blood; contains granules in the central part and clear protoplasm peripherally but has no definite nucleus.

platelet plug Accumulation of platelets that stick to each other and to connective tissue; functions to prevent blood loss from damaged blood vessels.

pleural cavity (ploor′ăl) Potential space between the parietal and visceral layers of the pleura.

plexus; pl., plexuses (plek′sŭs, plek′sŭs-ez) [L., a braid] Intertwining of nerves or blood vessels.

plicae circulares (plī′kă, plī′sē) (circular folds) Numerous folds of the mucous membrane of the small intestine.

pluripotent (ploo-rip′ō-tent) [L., *pluris*, more + *potentia*, power] In development the term refers to a cell or group of cells that have not yet become fixed or determined as to what specific tissues they are going to become.

podocyte (pod′ō-sīt) [Gr., *pous, podos*, foot + *kytos*, a hollow (cell)] Epithelial cell of Bowman's capsule attached to the outer surface of the glomerular capillary basement membrane by cytoplasmic foot processes.

Poiseuille's law (pwah-zuh′yez) Named for the French physiologist and physicist Jean Léonard Marie Poiseuille (1797–1869). The volume of a fluid passing per unit of time through a tube is directly proportional to the pressure difference between its ends and to the fourth power of the internal radius of the tube and inversely proportional to the tube's length and the viscosity of the fluid.

polar body (pō′lăr) One of the two small cells formed during oogenesis because of unequal division of the cytoplasm.

polar covalent bond Covalent bond in which atoms do not share their electrons equally.

polarization Development of differences in potential between two points in living tissues, as between the inside and outside of the plasma membrane.

polycythemia (pol′ē-sī-thē′mē-ă) Increase in red blood cell number above the normal.

polygenic (pol-ē-jen′ik) Relating to a hereditary disease or normal characteristic controlled by interaction of genes at more than one locus.

polysaccharide (pol-ē-sak′ă-rīd) Carbohydrate containing a large number of monosaccharide molecules.

polyunsaturated Fatty acid that contains two or more double covalent bonds between its carbon atoms.

pons (ponz) [L., bridge] That portion of the brainstem between the medulla and midbrain.

popliteal (pop-lit′ē-ăl, pop-li-tē′ăl) [L., ham] Posterior region of the knee.

porta (pōr'tă) [L., gate]   Fissure on the inferior surface of the liver where the portal vein, hepatic artery, hepatic nerve plexus, hepatic ducts, and lymphatic vessels enter or exit the liver.

portal system (pōr'tal)   System of vessels in which blood, after passing through one capillary bed, is conveyed through a second capillary network.

portal triad   Branches of the portal vein, hepatic artery, and hepatic duct bound together in the connective tissue that divides the liver into lobules.

postabsorptive state   Following the absorptive state; blood glucose levels are maintained because of conversion of other molecules to glucose.

posterior chamber of the eye (pos-tēr'ē-ŏr)   Chamber of the eye between the iris and the lens.

posterior interventricular sulcus   Groove on the diaphragmatic surface of the heart, marking the location of the septum between the two ventricles.

posterior pituitary   See neurohypophysis.

postganglionic neuron (pōst'gang-glē-on'ik)   Autonomic neuron that has its cell body located within an autonomic ganglion and sends its axon to an effector organ.

postovulatory age   Age of the developing fetus based on the assumption that fertilization occurs 14 days after the last menstrual period before the pregnancy.

postsynaptic (pōst-si-nap'tik)   Refers to the membrane of a nerve, muscle, or gland that is in close association with a presynaptic terminal. The postsynaptic membrane has receptor molecules within it that bind to neurotransmitter molecules.

potential difference (pō-ten'shăl)   Difference in electrical potential, measured as the charge difference across the plasma membrane.

potential energy [Gr., en, in + ergon, work]   Energy in a chemical bond that is not being exerted or used to do work.

PQ interval   Time elapsing between the beginning of the P wave and the beginning of the QRS complex in the electrocardiogram; also called PR interval.

precapillary sphincter (prē-kap'i-lăr-ē sfingk'ter)   Smooth muscle sphincter that regulates blood flow through a capillary.

preganglionic neuron (prē'gang-glē-on'ik)   Autonomic neuron that has its cell body located within the central nervous system and sends its axon through a nerve to an autonomic ganglion, where it synapses with postganglionic neurons.

premolar (prē-mō'lăr)   Bicuspid tooth.

prenatal (prē-nā'tal) [pre- + L., natus, born]   Preceding birth.

prepuce (prē'poos)   In males, the free fold of skin that more or less completely covers the glans penis; the foreskin. In females, the external fold of the labia minora that covers the clitoris.

pressoreceptor (pres'ō-rē-sep'ter, pres'ō-rē-sep'tōr)   See baroreceptor.

presynaptic terminal (prē'si-nap'tik)   Enlarged axon terminal or terminal bouton.

primary palate   In the early embryo, gives rise to the upper jaw and lips.

primary response   Immune response that occurs as a result of the first exposure to an antigen.

primary spermatocyte (sper'mă-tō-sīt)   Spermatocyte arising by a growth phase from a spermatogonium; gives rise to secondary spermatocytes after the first meiotic division.

prime mover   Muscle that plays a major role in accomplishing a movement.

primitive streak (prim'i-tiv)   Ectodermal ridge in the midline of the embryonic disk from which arises the mesoderm by inward and then lateral migration of cells.

primordial germ cell (prī-mōr'dē-ăl)   Most primitive undifferentiated sex cell, found initially outside the gonad on the surface of the yolk sac.

PR interval   See PQ interval.

process (pros'es, prō'ses)   Projection on a bone.

processus vaginalis (prō-ses'ŭs vaj'i-năl-ŭs)   Peritoneal outpocketing in the embryonic lower anterior abdominal wall that traverses the inguinal canal; in the male it forms the tunica vaginalis testis and normally loses its connection with the peritoneal cavity.

progeria (prō-jēr'ē-ă) [Gr., pro, before + ge + amras, old age]   Severe retardation of growth after the first year accompanied by a senile appearance and death at an early age.

progesterone (prō-jĕs'tē-rōn)   A steroid hormone secreted by the corpus luteum and one of the hormones secreted by the placenta.

prolactin (prō-lak'tin)   Hormone of the adenohypophysis that stimulates the production of milk.

prolactin-inhibiting hormone (PIH)   Neurohormone released from the hypothalamus that inhibits prolactin release from the adenohypophysis.

prolactin-releasing hormone (PRH)   Neurohormone released from the hypothalamus that stimulates prolactin release from the adenohypophysis.

proliferative phase (prō-lif'er-ă-tiv)   See follicular phase.

pronation (prō-nā'shŭn) [L., pronare, to bend forward]   Rotation of the forearm so that the anterior surface is down (prone).

pronephros (prō-nef'ros)   In the embryos of higher vertebrates, a series of tubules emptying into the celomic cavity. It is a temporary structure in the human embryo, followed by the mesonephros and still later by the metanephros, which gives rise to the kidney.

prophase (prō'fāz)   First stage in cell division when chromatin strands condense to form chromosomes.

proprioception (prō-prē-ō-sep'shun) [L., proprius, one's own + capio, to take]   Information about the position of the body and its various parts.

proprioceptor (prō'prē-ō-sep'ter)   Sensory receptor associated with joints and tendons.

prostaglandin (pros'tă-glan'din)   Class of physiologically active substances present in many tissues; among effects are those of vasodilation, stimulation and contraction of uterine smooth muscle, and promotion of inflammation and pain.

prostate gland (pros'tāt) [Gr., prostates, one standing before]   Gland that surrounds the beginning of the urethra in the male. The secretion of the gland is a milky fluid that is discharged by 20–30 excretory ducts into the prostatic urethra as part of the semen.

prostatic urethra (pros-tat'ik)   Part of the male urethra, approximately 2.5 cm in length, that passes through the prostate gland.

protease (prō'tē-ās)   Enzyme that breaks down proteins.

protein (prō'tēn, prō'tē-ĭn) [Gr., proteios, primary]   Macromolecule consisting of long sequences of amino acids linked together by peptide bonds.

protein kinase (kin-āz)   A class of enzymes that phosphorylates other proteins. Many of these kinases are responsive to other chemical signals (e.g., cAMP, cGMP, insulin, epidermal growth factor, calcium and calmodulin).

proteoglycan (prō'tē-ō-glī'kan)   Macromolecule consisting of numerous polysaccharides attached to a common protein core.

prothrombin (prō-throm'bin)   Glycoprotein present in blood that, in the presence of prothrombin activator, is converted to thrombin.

proton (prō'ton) [Gr., protos, first]   Positively charged particle in the nuclei of atoms.

protraction (prō-trak'shŭn) [L., protractus, to draw forth]   Movement forward or in the anterior direction.

provitamin (prō-vī'tă-min)   Substance that may be converted into a vitamin.

proximal tubule (prok'si-măl)   Part of the nephron that extends from the glomerulus to the descending limb of the loop of Henle.

pseudostratified epithelium   Epithelium consisting of a single layer of cells but having the appearance of multiple layers.

psychologic fatigue (sī-kō-loj'-ik)   Fatigue caused by the central nervous system.

ptosis (tō'sis) [G., ptosis, a falling]   Falling down of an organ, for example, drooping of the upper eyelid.

puberty (pū'ber-tē) [L., pubertas, grown up]   Series of events that transform a child into a sexually mature adult; involves an increase in the secretion of GnRH.

pubis (pū'bis)   Anterior inferior bone of the coxa.

pudendal cleft (pū-den'dăl)   Cleft between the labia majora.

pudendum (pū-den'dŭm)   See vulva.

pulmonary artery (pŭl'mō-nār-ē)   One of the arteries that extend from the pulmonary trunk to the right or left lungs.

pulmonary capacity   Sum of two or more pulmonary volumes.

pulmonary trunk   Large elastic artery that carries blood from the right ventricle of the heart to the right and left pulmonary arteries.

pulmonary vein   One of the veins that carry blood from the lungs to the left atrium of the heart.

pulp (tooth) (pŭlp) [L., pulpa, flesh]   The soft tissue within the pulp cavity, consisting of connective tissue containing blood vessels, nerves, and lymphatics.

pulse pressure (pŭls)   Difference between systolic and diastolic pressure.

pupil (pū'pĭl)   Circular opening in the iris through which light enters the eye.

Purkinje fiber (pŭr-kĭn′jē) Named for the Bohemian anatomist Johannes E. von Purkinje (1787–1869). Modified cardiac muscle cells found beneath the endocardium of the ventricles. Specialized to conduct action potentials.

pus (pŭs) Fluid product of inflammation; contains white blood cells, the debris of dead cells, and tissue elements liquefied by enzymes.

P wave First complex of the electrocardiogram representing depolarization of the atria.

pyloric opening (pī-lōr′ik) Opening between the stomach and the superior part of the duodenum.

pyloric sphincter Thickening of the circular layer of the gastric musculature encircling the junction between the stomach and duodenum.

pylorus; pl., pylori (pī-lōr′us, pī-lōr′ī) [L. pylōros, a gatekeeper] A muscular sphincter surrounding and controlling the outlet of the stomach.

pyrogen (pī′rō-jen) Chemical released by microorganisms, neutrophils, monocytes, and other cells that stimulates fever production by acting on the hypothalamus.

Q

QRS complex Principal deflection in the electrocardiogram, representing ventricular depolarization.

QT interval Time elapsing from the beginning of the QRS complex to the end of the T wave, representing the total duration of electrical activity of the ventricles.

R

radial pulse (rā′dē-ăl) Pulse detected in the radial artery.

radiation (rā′dē-ā′shŭn) [L., radius, ray, beam] The sending forth of light, short radiowaves, ultraviolet or x-rays, or any other rays for treatment or diagnosis or for other reasons; radiant heat.

radioactive isotope (rā′dē-ō-ak′tiv) Isotope with a nuclear composition that is unstable from which subatomic particles and electromagnetic waves are emitted.

ramus; pl., rami (rā′mŭs, rā′mī) [L., branch] One of the primary subdivisions of a nerve or blood vessel. The part of a bone that forms an angle with the main body of the bone.

raphe (rā′fē) [Gr., rhaphe, suture, seam] Central line running over the scrotum from the anus to the root of the penis.

receptor A structural protein or glycoprotein molecule on the cell surface or within the cytoplasm that binds to a specific factor (ligand or chemical signal).

recessive (rē-ses′iv) In genetics, a gene that may not be expressed because of suppression by a contrasting dominant gene.

Recommended Dietary Allowances (RDAs) A guide for estimating the nutritional needs of groups of people based on their age, sex, and other factors. First established in 1941.

Recommended Daily Intake (RDIs) Generally, the highest RDA value in each of four categories: infants, toddlers, people over 4 years of age, and pregnant or lactating women. The RDIs used to determine the Daily Values are based on the 1968 RDAs.

rectum (rek′tŭm) [L., rectus, straight] Portion of the digestive tract that extends from the sigmoid colon to the anal canal.

red bone marrow (mar′o) A soft, pulpy connective tissue filling the cavities of bones; consists of reticular fibers and the development stages of blood cells and platelets; gradually replaced by yellow marrow in long bones and the skull.

red pulp [L., pulpa, flesh] Reddish brown substance of the spleen consisting of venous sinuses and the tissues intervening between them called pulp cords.

reduction (rē-dŭk′shŭn) Gain of one or more electrons by a molecule.

refraction (rē-frak′shŭn) Bending of a light ray when it passes from one medium into another of different density.

refractory period (rē-frak′tōr-ē) [Gr., periodos, a way around, a cycle] Period following effective stimulation during which excitable tissue such as heart muscle fails to respond to a stimulus of threshold intensity.

regeneration (rē′jen-er-ā′shŭn) Reproduction or reconstruction of a lost or injured part.

regulatory gene Gene involved with controlling the activity of structural genes.

relative refractory period Portion of the action potential following the absolute refractory period during which another action potential can be produced with a greater-than-threshold stimulus strength.

relaxation phase (rē-lak-sā′shŭn) Phase of muscle contraction following the contraction phase; the time from maximal tension production until tension decreases to its resting level.

relaxin A polypeptide hormone secreted by the corpus luteum and placenta during pregnancy. Facilitates the birth process by causing a softening and lengthening of the pubic symphysis and cervix.

renal artery (rē′năl) Originates from the aorta and delivers blood to the kidney.

renal blood flow rate Volume at which blood flows through the kidneys per minute; an average of approximately 1200 mL/min.

renal column Cortical substance separating the renal pyramids.

renal corpuscle Glomerulus and Bowman's capsule that encloses it.

renal fascia (făsh′i-ā) Connective tissue surrounding the kidney that forms a sheath or capsule for the organ.

renal fat pad Fat layer that surrounds the kidney and functions as a shock-absorbing material.

renal fraction Portion of the cardiac output that flows through the kidneys; averages 21%.

renal pelvis Funnel-shaped expansion of the upper end of the ureter that receives the calyces.

renal pyramid One of a number of pyramidal masses seen on longitudinal section of the kidney; they contain part of the loops of Henle and the collecting tubules.

renin (rē′nin) Enzyme secreted by the juxtaglomerular apparatus that converts angiotensinogen to angiotensin I.

renin-angiotensin-aldosterone mechanism Renin, released from the kidneys in response to low blood pressure, converts angiotensinogen to angiotensin I. Angiotensin I is converted by angiotensin-converting enzyme to angiotensin II, which causes vasoconstriction, resulting in increased blood pressure. Angiotensin II also increases aldosterone secretion, which increases blood pressure by increasing blood volume.

repolarization (rē′pō-lăr-i-zā′shŭn) Phase of the action potential in which the membrane potential moves from its maximum degree of depolarization toward the value of the resting membrane potential.

reposition Return of a structure to its original position.

residual volume (rē-zid′ū-ăl) Volume of air remaining in the lungs after a maximum expiratory effort.

resolution (rez-ō-loo′shŭn) [L., resolutio, a slackening] Phase of the male sexual act after ejaculation during which the penis becomes flaccid; feeling of satisfaction; inability to achieve erection and second ejaculation. Last phase of the female sexual act, characterized by an overall sense of satisfaction and relaxation.

respiration (res-pi-ră′shŭn) [L., respiratio, to exhale, breathe] Process of life in which oxygen is used to oxidize organic fuel molecules, providing a source of energy, carbon dioxide, and water. Movement of air into and out of the lungs, the exchange of gases with blood, the transportation of gases in the blood, and gas exchange between the blood and the tissues.

respiratory bronchiole (res′pi-ră-tōr-ē, rĕ-spīr′ă-tōr-ē) Smallest bronchiole (0.5 mm in diameter) that connects the terminal bronchiole to the alveolar duct.

respiratory membrane Membrane in the lungs across which gas exchange occurs with blood.

resting membrane potential Electric charge difference inside a plasma membrane, measured relative to just outside the plasma membrane.

reticular (re-tik′ū-lăr) [L., rete, net] Relating to a fine network of cells or collagen fibers.

reticular cell Cell with processes making contact with those of other similar cells to form a cellular network; along with the network of reticular fibers, the reticular cells form the framework of bone marrow and lymphatic tissues.

reticulocyte (re-tik′ū-lō-sīt) Young red blood cell with a network of basophilic endoplasmic reticulum occurring in larger numbers during the process of active red blood cell synthesis.

reticuloendothelial system (re-tik′ū-lō-en-dō-thē′lē-ăl) See mononuclear phagocytic system.

retina (ret′i-nă) Nervous tunic of the eyeball.

retinaculum (ret-i-nak′ū-lŭm) [L., band, halter, to hold back] Dense regular connective tissue sheath holding down the tendons at the wrist, ankle, or other sites.

retraction (rē-trak′shŭn) [L., retractio, a drawing back] Movement in the posterior direction.

retroperitoneal (re′trō-per′i-tō-nē′ăl) Behind the peritoneum.

rhodopsin (rō-dop′sin) Light-sensitive substance found in the rods of the retina; composed of opsin loosely bound to retinal.

ribonuclease (rī-bō-nū′klē-ās) Enzyme that splits RNA into its component nucleotides.

ribonucleic acid (RNA) (rī′bō-noo-klē′ik) Nucleic acid containing ribose as the sugar component; found in all cells in both nuclei and cytoplasm; helps direct protein synthesis.

ribosomal RNA (rRNA) (rī′bō-sōm-ăl) RNA that is associated with certain proteins to form ribosomes.

ribosome (rī′bō-sōm) Small, spherical, cytoplasmic organelle where protein synthesis occurs.

right lymphatic duct Lymphatic duct that empties into the right subclavian vein; drains the right side of the head and neck, the right-upper thorax, and the right-upper limb.

rigor mortis (rig′er mōr′tĭs) Increased rigidity of muscle after death due to cross-bridge formation between actin and myosin as calcium ions leak from the sarcoplasmic reticulum.

rod Photoreceptor in the retina of the eye; responsible for noncolor vision in low-intensity light.

root of the penis Proximal attached part of the penis, including the two crura and the bulb.

root of the tooth That part below the neck of a tooth covered by cementum rather than enamel and attached by the periodontal ligament to the alveolar bone.

rotation (rō-tā′shun) Movement of a structure about its axis.

rotator cuff muscle (rō-tā′ter, rō-tā′tor) One of four deep muscles that attach the humerus to the scapula.

round ligament Fibromuscular band that is attached to the uterus on either side in front of and below the opening of the uterine tube; it passes through the inguinal canal to the labium majus.

round ligament of the liver Remains of the umbilical vein.

round window Membranous structure separating the scala tympani of the inner ear from the middle ear.

Ruffini's end-organ (roo-fē′nēz) Named for Angelo Ruffini, Italian histologist (1864–1929); receptor located deep in the dermis and responding to continuous touch or pressure.

ruga; pl., rugae (roo′gă, roo′gē) [L., a wrinkle] Fold or ridge; fold of the mucous membrane of the stomach when the organ is contracted; transverse ridge in the mucous membrane of the vagina.

S

saccule (sak′yūl) Part of the membranous labyrinth; contains sensory structure, the macula, that detects static equilibrium.

salivary amylase (sal′i-văr-ē am′il-ās) Enzyme secreted in the saliva that breaks down starch to maltose and isomaltose.

salivary gland Gland that produces and secretes saliva into the oral cavity. The three major pairs of salivary glands are the parotid, submandibular, and sublingual glands.

salt Molecule consisting of a cation other than hydrogen and an anion other than hydroxide.

sarcolemma (sar′kō-lem′ă) [Gr., *sarco*, muscle + *lemma*, husk] Plasma membrane of a muscle fiber.

sarcoma (sar-kō′mă) A malignant neoplasm derived from connective tissue.

sarcomere (sar′kō-mēr) [Gr., *sarco*, muscle + *meros*, part] Part of a myofibril between adjacent Z disks.

sarcoplasm (sar′kō-plazm) [Gr., *sarco*, muscle + *plasma*, a thing formed] Cytoplasm of a muscle fiber, excluding the myofilaments.

sarcoplasmic reticulum (sar′kō-plaz′mik) [Gr., *sarco*, muscle + *plasma*, a thing formed + *reticulum*, net] Endoplasmic reticulum of muscle.

satellite cell (sat′ĕ-līt) Specialized cell that surrounds the cell bodies of neurons within ganglia.

saturated (satch′ŭ-rāt-ĕd) Fatty acid in which the carbon chain contains only single bonds between carbon atoms.

saturation (satch-ŭ-rā′shun) Point when all carrier molecules or enzymes are attached to substrate molecules and no more molecules can be transported or reacted.

scala tympani (skā′lă tim′pă-nī) [L., stairway] Division of the spiral canal of the cochlea lying below the spiral lamina and basilar membrane.

scala vestibuli (skā′lă ves-tib′ū-lī) Division of the cochlea lying above the spiral lamina and vestibular membrane.

scapula (skap′ū-lă) Bone forming the shoulder blade.

scar (skar) [Gr., *eschara*, scab] Fibrous tissue replacing normal tissue; cicatrix.

sciatic nerve (sī-at′ik) Tibial and common peroneal nerves bound together.

sclera (sklēr′ă) White of the eye; white, opaque portion of the fibrous tunic of the eye.

scrotum; pl., scrota, scrotums (skrō′tŭm, skrō′tă, skrō′tŭmz) Musculocutaneous sac containing the testes.

sebaceous gland (sē-bā′shŭs) [L., *sebum*, tallow] Gland of the skin, usually associated with a hair follicle, that produces sebum.

second messenger See intracellular mediator.

secondary follicle Follicle in which the secondary oocyte is surrounded by granulosa cells at the periphery; contains fluid-filled antral spaces.

secondary (memory) response Immune response that occurs when the immune system is exposed to an antigen against which it has already produced a primary response.

secondary oocyte (ō′ō-sīt) Oocyte in which the second meiotic division stops at metaphase II unless fertilization occurs.

secondary palate Roof of the mouth in the early embryo that gives rise to the hard and the soft palates.

secondary spermatocyte (sper′mă-tō-sīt) Spermatocyte derived from a primary spermatocyte by the first meiotic division; each secondary spermatocyte gives rise by the second meiotic division to two spermatids.

secretin (se-krē′tin) Hormone formed by the epithelial cells of the duodenum; stimulates secretion of pancreatic juice high in bicarbonate ions.

secretion (se-krē′shŭn) General term for a substance produced inside a cell and released from the cell.

secretory phase (se-krēt′ĕ-rē, sē′krĕ-tōr-ē) See luteal phase.

segmental artery One of five branches of the renal artery, each supplying a segment of the kidney.

segmental (tertiary) bronchus (brong′kŭs) Extends from the secondary bronchus and conducts air to each lobule of the lungs.

self-antigen Antigen produced by the body that is capable of initiating an immune response against the body.

semen (sē′men) [L., seed (of plants, men, animals)] Penile ejaculate; thick, yellowish white, viscous fluid containing spermatozoa and secretions of the testes, seminal vesicles, prostate, and bulbourethral glands.

semicircular canal (sem′ē-sir′kū-lăr) Canal in the petrous portion of the temporal bone that contains sensory organs that detect kinetic or dynamic equilibrium. Three semicircular canals are within each inner ear.

seminal vesicle (sem′i-năl) One of two glandular structures that empty into the ejaculatory ducts; its secretion is one of the components of semen.

seminiferous tubule (sem′i-nif′er-ŭs) Tubule in the testis in which spermatozoa develop.

sensible perspiration (sen′si-bl pers-pi-rā′shŭn) Perspiration excreted by the sweat glands that appears as moisture on the skin; produced in large quantity when there is much humidity in the atmosphere.

sensory retina (sen′sŏ-rē) Portion of the retina containing rods and cones.

septum primum (sep′tŭm prī′mŭm) First septum in the embryonic heart that arises on the wall of the originally single atrium of the heart and separates it into right and left chambers.

septum secundum (sek′ŭn-dŭm) Second of two major septal structures involved in the partitioning of the atrium, arising later than the septum primum and located to the right of it; it remains an incomplete partition until after birth, with its unclosed area constituting the foramen ovale.

serosa (se-rō′să) [L., *serosus*, serous] Outermost covering of an organ or structure that lies in a body cavity; *see also* adventitia.

serous fluid (ser′ŭs) Fluid similar to lymph that is produced by and covers serous membrane; lubricates the serous membrane.

serous membrane Thin sheet composed of epithelial and connective tissues; lines cavities that do not open to the outside of the body or contain glands but do secrete serous fluid.

serous pericardium Lining of the pericardial sac composed of a serous membrane.

Sertoli cell (ser-tō′lē) Named for the Italian histologist Enrico Sertoli (1842–1910). Elongated cell in the wall of the seminiferous tubules to which spermatids are attached during spermatogenesis.

serum (sēr′ŭm) [L., whey] Fluid portion of blood after the removal of fibrin and blood cells.

sesamoid bone (ses′ă-moyd) [Gr., *sesamoceies*, like a sesame seed] Bone found within a tendon; such as the patella.

sex chromosomes Pair of chromosomes responsible for sex determination; XX in female and XY in male.

sex-linked trait   Characteristic resulting from the expression of a gene on a sex chromosome.

sigmoid colon (sig′moyd)   Part of the colon between the descending colon and the rectum.

sigmoid mesocolon   Fold of peritoneum attaching the sigmoid colon to the posterior abdominal wall.

simple epithelium   Epithelium consisting of a single layer of cells.

sinoatrial (SA) node (si′nō-ā′trē-ǎl)   Mass of specialized cardiac muscle fibers; acts as the "pacemaker" of the cardiac conduction system.

sinus (sī′nŭs) [L., cavity]   Hollow in a bone or other tissue; enlarged channel for blood or lymph.

sinus venosus   End of the embryonic cardiac tube where blood enters the heart; becomes a portion of the right atrium, including the SA node.

sinusoid (si′nŭ-soyd) [L., sinus + Gr., *eidos*, resemblance]   Terminal blood vessel having a larger diameter than an ordinary capillary.

sinusoidal capillary (si-nŭ-soy′dǎl)   Capillary with caliber of from 10–20 μm or more; lined with a fenestrated type of endothelium.

small intestine [L., *intestinus*, the entrails]   Portion of the digestive tube between the stomach and the cecum; consists of the duodenum, jejunum, and ileum.

sodium–potassium (Na⁺–K⁺) pump   Biochemical mechanism that uses energy derived from ATP to achieve the active transport of potassium ions opposite to that of sodium ions. Also called **sodium potassium ATP-ase.**

soft palate   Posterior muscular portion of the palate, forming an incomplete septum between the mouth and the oropharynx and between the oropharynx and the nasopharynx.

solute (sol′ūt, sō′loot) [L., *solutus*, dissolved]   Dissolved substance in a solution.

solution (sō-loo′shŭn) [L., *solutio*]   Homogenous mixture formed when a solute is dissolved in a solvent.

solvent (sol′vent) [L., *solvens*, to dissolve]   Liquid that holds another substance in solution.

soma (sō′mǎ) [Gr., body]   Neuron cell body or the enlarged portion of the neuron containing the nucleus and other organelles.

somatic (sō-mat′ik) [Gr., *somatikos*, bodily]   Relating to the body; the cells of the body except the reproductive cells.

somatic nervous system   Composed of nerve fibers that send impulses from the central nervous system to skeletal muscle.

somatomedin (sō′mǎ-tō-mē′din)   Peptide synthesized in the liver capable of stimulating certain anabolic processes in bone and cartilage such as synthesis of DNA, RNA, and protein.

somatotropin (sō′mǎ-tō-trō′pin)   Protein hormone of the anterior pituitary gland; it promotes body growth, fat mobilization, and inhibition of glucose utilization.

somite (sō′mīt) [Gr., *soma*, body + *ite*]   One of the paired segments consisting of cell masses formed in the early embryonic mesoderm on either side of the neural tube.

somitomere (sō′mīt-ō-mēr)   An indistinct somite in the head region of the embryo.

spatial summation   Summation of the local potentials in which two or more action potentials arrive simultaneously at two or more presynaptic terminals that synapse with a single neuron.

specific heat   Heat required to raise the temperature of any substance 1°C compared to the heat required to raise the same volume of water 1°C.

speech   Use of the voice in conveying ideas.

spermatic cord (sper-mat′ik)   Cord formed by the ductus deferens and its associated structures; extends through the inguinal canal into the scrotum.

spermatid (sper′mǎ-tid) [Gr., *sperma*, seed + *id*]   Cell derived from the secondary spermatocyte; gives rise to a spermatozoon.

spermatogenesis (sper′mǎ-tō-jen′ě-sis)   Formation and development of the spermatozoon.

spermatogonium (sper′mǎ-tō-gō′nē-ŭm) [Gr., *sperma*, seed + *gone*, generation]   Cell that divides by mitosis to form primary spermatocytes.

spermatozoon; pl., spermatozoa (sper′mǎ-tō-zō′on, sper′mǎ-to-zō′ǎ) [Gr., *sperma*, seed + *zoon*, animal]   Sperm cell. Male gamete or sex cell, composed of a head and a tail. The spermatozoon contains the genetic information transmitted by the male.

sphenoid (sfē′noyd) [Gr., *shen*, wedge]   Wedge-shaped.

sphincter pupillae (sfingk′ter pū-pil′ē)   Circular smooth muscle fibers of the iris's diaphragm that constrict the pupil of the eye.

sphygmomanometer (sfig′mō-mǎ-nom′ě-ter) [Gr., *sphygmos*, pulse + *manos*, thin, scanty + *metron*, measure]   Instrument for measuring blood pressure.

spinal nerve (spī′nǎl)   One of 31 pairs of nerves formed by the joining of the dorsal and ventral roots that arise from the spinal cord.

spindle fiber (spin′dl)   Specialized microtubule that develops from each centrosome and extends toward the chromosomes during cell division.

spiral artery (spī′rǎl)   One of the corkscrewlike arteries in premenstrual endometrium; most obvious during the secretory phase of the uterine cycle.

spiral ganglion   Cell bodies of sensory neurons that innervate hair cells of the organ of Corti are located in the spiral ganglion.

spiral lamina   Attached to the modiolus and supports the basilar and vestibular membranes.

spiral ligament   Attachment of the basilar membrane to the lateral wall of the bony labyrinth.

spiral organ   Organ of Corti; rests on the basilar membrane and consists of the hair cells that detect sound.

spiral tubular gland   Well-developed simple or compound tubular glands that are spiral in shape within the endometrium of the uterus; prevalent in the secretory phase of the uterine cycle.

spirometer (spī-rom′ě-ter) [L., *spiro*, to breathe + Gr., *metron*, measure]   Gasometer used for measuring the volume of respiratory gases; usually understood to consist of a counterbalanced cylindrical bell sealed by dipping into a circular trough of water.

spirometry (spī-rom′ě-trē)   Making pulmonary measurements with a spirometer.

spleen (splēn)   Large lymphatic organ in the upper part of the abdominal cavity on the left side between the stomach and diaphragm, composed of white and red pulp. It responds to foreign substances in the blood, destroys worn out red blood cells, and is a storage site for blood cells.

spongy urethra   Portion of the male urethra, approximately 15 cm in length, that traverses the corpus spongiosum of the penis.

squamous (skwā′mŭs) [L., *squama*, a scale]   Scalelike, flat.

stapedius (stā-pē′dē-ŭs)   Small skeletal muscles attached to the stapes.

stapes (stā′pēz) [L., stirrup]   Smallest of the three auditory ossicles; attached to the oval window.

Starling's law of the heart   Named for the English physiologist Ernest H. Starling (1866–1927). Force of contraction of cardiac muscle is a function of the length of its muscle fibers at the end of diastole; the greater the ventricular filling, the greater the stroke volume produced by the heart.

sternum (ster′nŭm) [L., *sternon*, chest]   Breastbone.

steroid (stēr′oyd, ster′oyd)   Large family of lipids, including some reproductive hormones, vitamins, and cholesterol.

stomach (stŭm′ŭk)   Large sac between the esophagus and the small intestine, lying just beneath the diaphragm.

stratified epithelium (strat′i-fīd ep-i-thē′lē-ŭm)   Epithelium consisting of more than one layer of cells.

stratum basale (strat-ŭm bǎh-sǎl′ē) [L., layer; basal]   Basal or deepest layer of the epidermis.

stratum corneum (kōr′nē-ŭm) [L., layer; *corneus*, horny]   Most superficial layer of the epidermis consisting of flat, keratinized, dead cells.

stratum germinativum   *See* stratum basale.

stratum granulosum (gran′ū-lō′sŭm) [L., layer; *granulum*, a small grain]   Layer of cells in the epidermis filled with granules of keratohyalin.

stratum lucidum (lū′sid-ŭm) [L., layer; *lucidus*, clear]   Clear layer of the epidermis found in thick skin between the stratum granulosum and the stratum corneum.

stratum spinosum (spī′nōs-ŭm) [L., layer; *spina*, spine]   Layer of many-sided cells in the epidermis with intercellular connections (desmosomes) that give the cells a spiny appearance.

stria; pl., striae (strī′ǎ, strī′ē) [L., channel]   Line or streak in the skin that is a different texture or color from the surrounding skin. Stretch mark.

striated (strī′āt-ēd) [L., *striatus*, furrowed]   Striped; marked by stripes or bands.

stroke volume [L., *volumen*, something rolled up, scroll, from *volvo*, to roll]   Volume of blood pumped out of one ventricle of the heart in a single beat.

structural gene   Gene with the function of determining the structure of a specific protein or peptide.

sty (stī)   Inflamed ciliary gland of the eye.

subcutaneous (sŭb′-koo-tā′nē-ŭs) [L., *sub*, under + *cutis*, skin]   Under the skin; same tissue as the hypodermis.

sublingual gland (sŭb-ling′gwǎl)   One of two salivary glands in the floor of the mouth beneath the tongue.

submandibular gland (sŭb-man-dib′ū-lǎr)   One of two salivary glands in the neck, located in the space bounded by the two bellies of the digastric muscle and the angle of the mandible.

submucosa (sŭb-moo-kō′sǎ)   Layer of tissue beneath a mucous membrane.

submucosal plexus (sŭb-mū-kō′săl) [L., a braid] Gangliated plexus of unmyelinated nerve fibers in the intestinal submucosa.

substantia nigra (sŭb-stan′shē-ă nī′gră) [L., substance; black] Black nuclear mass in the midbrain; involved in coordinating movement and maintaining muscle tone.

subthreshold stimulus Stimulus resulting in a local potential so small that it does not reach threshold and produce an action potential.

sucrose (soo′krōs) Disaccharide composed of glucose and fructose; table sugar.

sulcus; pl., sulci (sool′kŭs, sŭl′sī) [L., furrow or ditch] Furrow or groove on the surface of the brain between the gyri; may also refer to a fissure.

superficial inguinal ring (ing′gwi-năl) Slitlike opening in the aponeurosis of the external oblique muscle of the abdominal wall through which the spermatic cord (round ligament in the female) emerges from the inguinal canal.

superior colliculus (ko-lik′ū-lŭs) [L., *collis*, hill] One of two rounded eminences of the midbrain; aids in coordination of eye movements.

superior vena cava (vē′nă cā′vă) Vein that returns blood from the head and neck, upper limbs, and thorax to the right atrium.

supination (soo′pi-nā′shŭn) [L., *supino*, to bend backward, place on back] Rotation of the forearm (when the forearm is parallel to the ground) so that the anterior surface is up (supine).

supramaximal stimulus Stimulus of greater magnitude than a maximal stimulus; however, the frequency of action potentials is not increased above that produced by a maximal stimulus.

suppressor T cell Subset of T lymphocytes that decreases the activity of B cells and T cells.

surfactant (ser-fak′tănt) Lipoproteins forming a monomolecular layer over pulmonary alveolar surfaces; stabilizes alveolar volume by reducing surface tension and the tendency for the alveoli to collapse.

suspension (sŭs-pen′shŭn) Liquid through which a solid is dispersed and from which the solid separates unless the liquid is kept in motion.

suspensory ligament (sŭs-pen′sŏ-rē) Band of peritoneum that extends from the ovary to the body wall; contains the ovarian vessels and nerves. Small ligament attached to the margin of the lens in the eye and the ciliary body to hold the lens in place.

suture (soo′choor) [L., *sutura*, a seam] Junction between flat bones of the skull.

sweat (swet) [A.S., *swat*] Perspiration; secretions produced by the sweat glands of the skin. *See also* sensible and insensible perspiration.

sweat gland Usually means structures that produce a watery secretion called sweat. Some sweat glands, however, produce viscous organic secretions.

sympathetic chain ganglion (sim-pă-thet′ik) Collection of sympathetic postganglionic neurons that are connected to each other to form a chain along both sides of the spinal cord.

sympathetic division Subdivision of the autonomic division of the nervous system characterized by having the cell bodies of its preganglionic neurons located in the thoracic and upper lumbar regions of the spinal cord (thoracolumbar division); usually involved in preparing the body for physical activity.

symphysis; pl., symphyses (sim′fi-sis, sim′fă-sēz) [Gr., a growing together] Fibrocartilage joint between two bones.

synapse; pl., synapses (sin′aps, sĭ-naps′, sĭ-nap′sēz) [Gr., *syn*, together + *haptein*, to clasp] Functional membrane-to-membrane contact of a nerve cell with another nerve cell, muscle cell, gland cell, or sensory receptor; functions in the transmission of action potentials from one cell to another.

synaptic cleft (si-nap′tik) Space between the presynaptic and the postsynaptic membranes.

synaptic fatigue Fatigue due to depletion of neurotransmitter vesicles in the presynaptic terminals.

synaptic vesicle Secretory vesicle in the presynaptic terminal containing neurotransmitter substances.

synchondrosis; pl., synchondroses (sin′kon-drō′sis, -sēz) [Gr., *syn*, together + *chondros*, cartilage + *osis*, condition] Union between two bones formed by hyaline cartilage.

syncytiotrophoblast (sin-sish′ē-ō-trō′fō-blast) Outer layer of the trophoblast composed of multinucleated cells.

syndesmosis; pl., syndesmoses (sin′dez-mō′sis, sin′dez-mō′sēz) [Gr., *syndeo*, to bind + *osis*, condition] Form of fibrous joint in which opposing surfaces that are some distance apart are united by ligaments.

synergist (sin′er-jist) Muscle that works with other muscles to cause a movement.

synovial (si-nō′vē-ăl) [Gr., *syn*, together + *oon*, egg] Relating to or containing synovia (a substance that serves as a lubricant in a joint, tendon sheath, or bursa).

synovial fluid Slippery fluid found inside synovial joints and bursae; produced by the synovial membranes.

systemic inflammation (sis-tem′ik) Inflammation that occurs in many areas of the body. In addition to symptoms of local inflammation, increased neutrophil numbers in the blood, fever, and shock can occur.

systole (sis′tō-lē) [Gr., *systole*, a contracting] Contraction of the heart chambers during which blood leaves the chambers; usually refers to ventricular contraction.

T

tactile corpuscle (tak′til) Oval receptor found in the papillae of the dermis; responsible for fine, discriminative touch; Meissner's corpuscle.

tactile disk Cuplike receptor found in the epidermis; responsible for light touch and superficial pressure; Merkel's disk.

talus (tā′lŭs) [L., ankle bone, heel] Tarsal bone contributing to the ankle.

target tissue Tissue on which a hormone acts.

tarsal (tar′săl) [Gr., *tarsos*, sole of foot] One of seven ankle bones.

tarsal plate (tar′săl) Crescent-shaped layer of connective tissue that helps maintain the shape of the eyelid.

taste (tāst) Sensations created when a chemical stimulus is applied to the taste receptors in the tongue.

taste bud Sensory structure, mostly on the tongue, that functions as a taste receptor.

T cell Thymus-derived lymphocyte of immunologic importance; it is of long life and is responsible for cell-mediated immunity.

tectum (tek′tŭm) Roof of the midbrain.

tegmentum (teg-men′tŭm) Floor of the midbrain.

telencephalon (tel-en-sef′ă-lon) [Gr., *telos*, end + *enkephalos*, brain] Anterior division of the embryonic brain from which the cerebral hemispheres develop.

telophase (tel′ō-fāz) Time during cell division when the chromosomes are pulled by spindle fibers away from the cell equator and into the two halves of the dividing cell.

temporal summation (tem′pŏ-răl) Summation of the local potential that results when two or more action potentials arrive at a single synapse in rapid succession.

tendon (ten′dŏn) Band or cord of dense connective tissue that connects a muscle to a bone or other structure.

tensor tympani (ten′sōr tim′pa-nī) Small skeletal muscle attached to the malleus.

tentorium cerebelli (ten-tō′rē-ŭm ser′ĕ-bel′ī) Dural folds between the cerebrum and the cerebellum.

terminal bouton (bū-ton′) [Fr., button] Enlarged axon terminal or presynaptic terminal.

terminal cisterna (sis-ter′nă) [L., *terminus*, limit + *cista*, box] Enlarged end of the sarcoplasmic reticulum in the area of the T tubules.

terminal hair [L., *terminus*, a boundary, limit] Long, coarse, usually pigmented hair found in the scalp, eyebrows, and eyelids and replacing vellus hair.

terminal sulcus (sŭl′kŭs) [L., furrow or ditch] V-shaped groove on the surface of the tongue at the posterior margin.

testis; pl., testes (tes′tis, tes′tēz) One of two male reproductive glands located in the scrotum; produces spermatozoa, testosterone, and inhibin.

testosterone (tes-tos′tĕ-rōn) Steroid hormone secreted primarily by the testes; aids in spermatogenesis, maintenance and development of male reproductive organs, secondary sexual characteristics, and sexual behavior.

tetraiodothyronine (T₄) (tet′ră-ī-ō′dō-thī′rō-nēn) One of the iodine-containing thyroid hormones; also called thyroxine.

thalamus (thal′ă-mŭs) [Gr., *thalamos*, a bed, a bedroom] Large mass of gray matter that forms the larger dorsal subdivision of the diencephalon.

theca (thē′kă) [Gr., *theke*, a box] Sheath or capsule.

theca externa External fibrous layer of the theca or capsule of a vesicular follicle.

theca interna Inner vascular layer of the theca or capsule of the secondary and mature follicle; produces estrogen and contributes to the formation of the corpus luteum after ovulation.

thenar (thē′nar) [Gr., palm of the hand] Fleshy mass of tissue at the base of the thumb; contains muscles responsible for thumb movements.

thick skin Found in the palms, soles, and tips of the digits and has all five epidermal strata.

thin skin Found over most of the body, usually without a stratum lucidum, and has fewer layers of cells than thick skin.

thoracic cavity (thō-ras′ik) Space within the thoracic walls, bounded below by the diaphragm and above by the neck.

thoracic duct Largest lymph vessel in the body, beginning at the cisterna chyli and emptying into the left subclavian vein; drains the left side of the head and neck, the left upper thorax, the left upper limb, and the inferior half of the body.

thoracolumbar division (thōr′ă-kō-lŭm′bar) Synonym for the sympathetic division of the autonomic nervous system.

thoroughfare channel Channel for blood through a capillary bed from an arteriole to a venule.

threshold potential (thresh′ōld) Value of the membrane potential at which an action potential is produced as a result of depolarization in response to a stimulus.

threshold stimulus Stimulus resulting in a local potential just large enough to reach threshold and produce an action potential.

thrombocyte (throm′bō-sīt) Platelet.

thrombocytopenia (throm′bō-sī′tō-pē′nē-ă) [*thrombocyte* + Gr., *penia*, poverty] Condition in which there is an abnormally small number of platelets in the blood.

thromboxane (throm-bok′sān) Specific class of physiologically active fatty acid derivatives present in many tissues.

thrombus; pl., thrombi (throm′bŭs, throm′bī) [Gr., *thrombos*, a clot] Clot in the cardiovascular system formed from constituents of blood; may be occlusive or attached to the vessel or heart wall without obstructing the lumen.

thymus (thī′mŭs) [Gr., *thymos*, sweetbread] Bilobed lymph organ located in the inferior neck and superior mediastinum; secretes the hormone thymosin.

thyroid cartilage (thī′royd) Largest laryngeal cartilage. It forms the laryngeal prominence, or Adam's apple.

thyroid gland [Gr., *thyreoeides*, shield] Endocrine gland located inferior to the larynx and consisting of two lobes connected by the isthmus; secretes the thyroid hormones triiodothyronine (T$_3$) and tetraiodothyronine (T$_4$).

thyroid-stimulating hormone (TSH) Glycoprotein hormone released from the hypothalamus; stimulates thyroid hormone secretion from the thyroid gland.

thyrotropin (thī-rot′rō-pin, thī-rō-trō′pin) *See* thyroid-stimulating hormone.

thyroxine (thi-rok′sēn, thi-rok′sin) *See* tetraiodothyronine.

tidal volume (tī′dăl) Volume of air that is inspired or expired in a single breath during regular, quiet breathing.

tissue repair (tish′ū) Substitution of viable cells for damaged or dead cells by regeneration or replacement.

tolerance (tol′er-ăns) Failure of the specific immune system to respond to an antigen.

tongue (tŭng) Muscular organ occupying most of the oral cavity when the mouth is closed; major attachment is through its posterior portion.

tonicity (tō-nis′ĭ-tē) [Gr., *tonos*, tone] The osmotic pressure or tension of a solution, usually relative to that of blood. A state of continuous activity or tension caused by muscle contraction beyond the tension related to physical properties of muscle.

tonsil; pl., tonsils (ton′sil, ton′silz) [L., *tonsilla*, stake] Any collection of lymphoid tissue; usually refers to large collections of lymphatic tissue beneath the mucous membrane of the oral cavity and pharynx; lingual, pharyngeal, and palatine tonsils.

total lung capacity Volume of air contained in the lungs at the end of a maximum inspiration; equals vital capacity plus residual volume.

total tension Sum of active and passive tension.

trabecula; pl., trabeculae (tră-bek′ū-lă, tră-bek′ū-lē) [L., *trabs*, beam] One of the supporting bundles of fibers traversing the substance of a structure, usually derived from the capsule or one of the fibrous septa, such as trabeculae of lymph nodes, testes; a beam or plate of cancellous bone.

trachea (tră′kē-ă) [Gr., *tracheia arteria*, rough artery] Air tube extending from the larynx into the thorax, where it divides to form the bronchi; composed of 16–20 rings of hyaline cartilage.

tracheostomy (tra′ke-os′to-me) [tracheo-, trachea + Gr., *stoma*, mouth] An operation to make an opening into the trachea; usually the opening is intended to be permanent, and a tube is inserted into the trachea to allow airflow. See tracheotomy.

tracheotomy (tra′ke-ot′o-me) [tracheo-, trachea + Gr., *tome*, incision] The act of cutting into the trachea. See tracheostomy.

triacylglycerol (trī-as′il-glis′er-ol) *See* triglyceride.

transcription (tran-skrip′shŭn) Process of forming RNA from a DNA template.

transfer RNA (tRNA) RNA that attaches to individual amino acids and transports them to the ribosomes, where they are connected to form a protein polypeptide chain.

transfusion (trans-fū′zhŭn) [L., *trans*, across + *fundo*, to pour from one vessel to another] Transfer of blood from one person to another.

transitional epithelium (tran-sish′ŭn-ăl) Stratified epithelium that may be either cuboidal or squamouslike, depending on the presence or absence of fluid in the organ (as in the urinary bladder).

translation (trans-lā′shŭn) Synthesis of polypeptide chains at the ribosome in response to information contained in mRNA molecules.

transverse colon (trans-vers′ kō′lon) Part of the colon between the right and left colic flexures.

transverse mesocolon (mez′ō-kō′lon) Fold of peritoneum attaching the transverse colon to the posterior abdominal wall.

transverse tubule [L., *tubus*, tube] Tubule that extends from the sarcolemma to a myofibril of striated muscles.

treppe (trep′eh) [Ger., staircase] Series of successively stronger contractions that occur when a rested muscle fiber receives closely spaced stimuli of the same strength but with a sufficient stimulus interval to allow complete relaxation of the fiber between stimuli.

triad (trī′ad) Two terminal cisternae and a T tubule between them.

tricuspid valve (trī-kŭs′pid) Valve closing the orifice between the right atrium and the right ventricle of the heart.

trigone (trī′gōn) [Gr., *trigonon*, triangle] Triangular smooth area at the base of the bladder between the openings of the two ureters and that of the urethra.

triglyceride (tri-glis′er-id) A three carbon glycerol molecule with a fatty acid attached to each carbon; constitute approximately 95% of the fats in the human body. Also called a triacylglycerol.

triiodothyronine (trī-ī′ō-dō-thī′rō-nēn) One of the iodine-containing thyroid hormones.

trochlea (trok′lē-ă) [L., pulley] Structure shaped like or serving as a pulley or spool.

trochlear nerve (trok′lē-ar) [L., *trochlea*, pulley] Cranial nerve IV, to the muscle (superior oblique) turning around a pulley.

trophoblast (trof′ō-blast) [Gr., *trophe*, nourishment + *blastos*, germ] Cell layer forming the outer layer of the blastocyst, which erodes the uterine mucosa during implantation; the trophoblast does not become part of the embryo but contributes to the formation of the placenta.

tropomyosin (trō-pō-mī′ō-sin) Fibrous protein found as a component of the actin myofilament.

troponin (trō′pō-nin) Globular protein component of the actin myofilament.

true pelvis Portion of the pelvis inferior to the pelvic brim.

true, or vertebrosternal, rib (ver-tĕ′brō-ster′năl) Rib that attaches by an independent costal cartilage directly to the sternum.

trypsin (trip′sin) Proteolytic enzyme formed in the small intestine from the inactive pancreatic precursor trypsinogen.

T tubule (tū′bul) Tubelike invagination of the sarcolemma that conducts action potentials toward the center of the cylindrical muscle fibers.

tubercle (too′ber-kl) Lump on a bone.

tubular load (too′bū-lăr) Amount of a substance per minute that crosses the filtration membrane into Bowman's capsule.

tubular maximum Maximum rate of secretion or reabsorption of a substance by the renal tubules.

tubular reabsorption Movement of materials, by means of diffusion, active transport, or cotransport, from the filtrate within a nephron to the blood.

tubular secretion Movement of materials, by means of active transport, from the blood into the filtrate of a nephron.

tumor (too′mŏr) Any swelling or growth; a neoplasm.

tunic (too′nik) [L., coat] One of the enveloping layers of a part; one of the coats of a blood vessel; one of the coats of the eye; one of the coats of the digestive tract.

tunica adventitia (too′ni-kă ad-ven-tish′ă) Outermost fibrous coat of a vessel or an organ that is derived from the surrounding connective tissue.

tunica albuginea (al-bū-jin′ē-ă) Dense, white, collagenous tunic surrounding a structure; such as the capsule around the testis.

tunica intima (in′ti-mă) Innermost coat of a blood vessel; consists of endothelium, a lamina propria, and an inner elastic membrane.

tunica media Middle, usually muscular, coat of an artery or other tubular structure.

turbulent flow Flow characterized by eddy currents exhibiting nonparallel blood flow.

T wave Deflection in the electrocardiogram following the QRS complex, representing ventricular repolarization.

tympanic membrane (tim-pan′ik)   Eardrum; cellular membrane that separates the external from the middle ear; vibrates in response to sound waves.

U

unipolar neuron (oo-ni-pō′lar)   One of the three categories of neurons consisting of a nerve cell body with a single axon projecting from it; also called a pseudounipolar neuron.

unmyelinated axon (ŭn-mī′ĕ-li-nā-ted)   Nerve fibers lacking a myelin sheath.

unsaturated (ŭn-sach′ŭr-āt-ed)   Carbon chain of a fatty acid that possesses one or more double or triple bonds.

upper respiratory tract   The nasal cavity, pharynx, and associated structures.

up-regulation   An increase in the concentration of receptors in response to a signal.

ureter (ū-rē′ter, ū′rē-ter) [Gr., *oureter,* urinary canal]   Tube conducting urine from the kidney to the urinary bladder.

urethral gland (ū-rē′thrăl)   One of numerous mucous glands in the wall of the spongy urethra in the male.

urogenital triangle   Anterior portion of the perineal region containing the openings of the urethra and vagina in the female and the urethra and root structures of the penis in the male.

uterine cycle (ū′ter-in, ū′ter-īn)   Series of events that occur in a regular fashion in the uterus of sexually mature, nonpregnant females; prepares the uterine lining for implantation of the embryo.

uterine part   Portion of the uterine tube that passes through the wall of the uterus.

uterine tube   One of the tubes leading on either side from the uterus to the ovary; consists of the infundibulum, ampulla, isthmus, and uterine parts; also called the fallopian tube or oviduct.

uterus (ū′ter-ŭs)   Hollow muscular organ in which the fertilized ovum develops into a fetus.

utricle (oo′tri-kl)   Part of the membranous labyrinth; contains sensory structure, the macula, that detects static equilibrium.

uvula (ū′vŭ-lă) [L., *uva,* grape]   Small grapelike appendage at posterior margin of soft palate.

V

vaccination (vak′si-nā′shŭn)   Deliberate introduction of an antigen into a subject to stimulate the immune system and produce immunity to the antigen.

vaccine (vak′sēn, vak-sēn′) [L., *vaccinus,* relating to a cow]   Preparation of killed microbes, altered microbes, or derivatives of microbes or microbial products intended to produce immunity. The method of administration is usually inoculation, but ingestion is preferred in some instances, and nasal spray is used occasionally.

vagina (vă-jī′nă) [L., sheath]   Genital canal in the female, extending from the uterus to the vulva.

vapor pressure   Partial pressure exerted by water vapor.

variable region   Part of the antibody that combines with the antigen.

vas deferens (vas def′er-enz)   *See* ductus deferens.

vasa recta (vā′să rek′tă)   Specialized capillary that extends from the cortex of the kidney into the medulla and then back to the cortex.

vasa vasorum (vā′sor-ŭm) [L., vessel, dish]   Small vessels distributed to the outer and middle coats of larger blood vessels.

vascular tunic (vas′kū-lăr)   Middle layer of the eye; contains many blood vessels.

vasoconstriction (vā′sō-kon-strik′shŭn, vas′ō-kon-strik′shŭn)   Decreased diameter of blood vessels.

vasodilation (vā′sō-dī-lā′shŭn)   Increased diameter of blood vessels.

vasomotion (vā-sō-mō′shŭn)   Periodic contraction and relaxation of the precapillary sphincter, resulting in cyclic blood flow through capillaries.

vasomotor center (vā-sō-mō′ter, vas-ō-mō′ter)   Area within the medulla oblongata that regulates the diameter of blood vessels by way of the sympathetic nervous system.

vasomotor tone   Relatively constant frequency of sympathetic impulses that keep blood vessels partially constricted in the periphery.

vasopressin (vā-sō-pres′in, vas-ō-pres′in)   Hormone secreted from the neurohypophysis that causes vasoconstriction and acts on the kidney to reduce urinary volume; also called antidiuretic hormone.

vellus (vel′ŭs) [L., fleece]   Short, fine, usually unpigmented hair that covers the body except for the scalp, eyebrows, and eyelids. Much of the vellus is replaced at puberty by terminal hairs.

venous capillary (vē′nŭs)   Capillary opening into a venule.

venous return   Volume of blood returning to the heart.

venous sinus   Endothelium-lined venous channel in the dura mater that receives cerebrospinal fluid from the arachnoid granulations.

ventilation (ven-ti-lā′shŭn) [L., *ventus,* the wind]   Movement of gases into and out of the lungs.

ventral root (ven′trăl)   Motor (efferent) root of a spinal nerve.

ventricle (ven′tri-kl) [L., *venter,* belly]   Chamber of the heart that pumps blood into arteries (i.e., the left and right ventricles). In the brain, a fluid-filled cavity.

ventricular diastole (ven-trik′ū-lăr)   Dilation of the heart ventricles.

ventricular systole   Contraction of the ventricles.

venule (ven′ool, vē′nool)   Minute vein, consisting of endothelium and a few scattered smooth muscles, that carries blood away from capillaries.

vermiform appendix (ver′mi-fōrm) [L., *vermis,* worm + *forma,* form; appendage]   Wormlike sac extending from the blind end of the cecum.

vesicle (ves′i-kl) [L., *vesica,* bladder]   Small sac containing a liquid or gas, such as a blister in the skin or an intracellular, membrane-bounded sac.

vestibular fold (ves-tib′ū-lăr) (false vocal cord)   One of two folds of mucous membrane stretching across the laryngeal cavity from the angle of the thyroid cartilage to the arytenoid cartilage superior to the vocal cords; helps close the glottis; false vocal cord.

vestibular membrane   Membrane separating the cochlear duct and the scala vestibuli.

vestibule (ves′ti-bool) [L., antechamber, entrance court]   Anterior part of the nasal cavity just inside the external nares that is enclosed by cartilage; space between the lips and the alveolar processes and teeth; middle region of the inner ear containing the utricle and saccule; space behind the labia minora containing the openings of the vagina, urethra, and vestibular glands.

vestibulocochlear nerve (ves-tib′ū-lō-kok′lē-ăr)   Formed by the cochlear and vestibular nerves and extends to the brain.

villus; pl., villi (vil′ŭs, vil′ī) [L., shaggy hair (of beasts)]   Projections of the mucous membrane of the intestine; they are leaf-shaped in the duodenum and become shorter, more finger-shaped, and sparser in the ileum.

visceral (vis′er-ăl)   Relating to the internal organs.

visceral pericardium (per′i-kar′dē-ŭm)   Serous membrane covering the surface of the heart. Also called the epicardium.

visceral peritoneum (per′i-tō-nē′ŭm) [Gr., *periteino,* to stretch over]   Layer of peritoneum covering the abdominal organs.

visceral pleura (vis′er-ăl plūr′ă)   Serous membrane investing the lungs and dipping into the fissures between the several lobes.

visceroreceptor (vis′er-ō-rē-sep′tŏr)   Sensory receptor associated with the organs.

viscosity (vis-kos′i-tē) [L., *viscosus,* viscous]   In general, the resistance to flow or alteration of shape by any substance as a result of molecular cohesion.

visual cortex (vizh′oo-ăl)   Area in the occipital lobe of the cerebral cortex that integrates visual information and produces the sensation of vision.

visual field   Area of vision for each eye.

vital capacity (vīt-ăl)   Greatest volume of air that can be exhaled from the lungs after a maximum inspiration.

vitamin (vīt′ă-min) [L., *vita,* life + amine]   One of a group of organic substances present in minute amounts in natural foodstuffs that are essential to normal metabolism; insufficient amounts in the diet may cause deficiency diseases.

vitamin D   Fat-soluble vitamin produced from precursor molecules in skin exposed to ultraviolet light; increases calcium and phosphate uptake from the intestines.

vitreous humor (vit′rē-ŭs)   Transparent jellylike material that fills the space between the lens and the retina.

Volkmann's canal   Named for the German surgeon Richard Volkmann (1830–1889). Canal in bone containing blood vessels; not surrounded by lamellae; runs perpendicular to the long axis of the bone and the haversian canals, interconnecting the latter with each other and the exterior circulation.

vulva (vŭl′vă) [L., wrapper or covering, seed covering, womb]   External genitalia of the female composed of the mons pubis, the labia majora and minora, the clitoris, the vestibule of the vagina and its glands, and the opening of the urethra and of the vagina; the pudendum.

W

water-soluble vitamin Vitamin such as B complex and C that is absorbed with water from the intestinal tract.

white matter Bundles of parallel axons with their associated sheath in the central nervous system.

white pulp That part of the spleen consisting of lymphatic nodules and diffuse lymphatic tissue; associated with arteries.

white ramus communicans; pl., rami communicantes (rā′mŭs kŏ-mū′nĭ-kans, rā′mī kŏ-mū-nĭ-kan′tēz) Connection between spinal ganglia through which myelinated preganglionic axons project.

wisdom tooth Third molar tooth on each side in each jaw.

X

xiphoid (zi′foyd) [Gr., *xiphos*, sword] Sword-shaped, with special reference to the sword tip; the inferior part of the sternum.

X-linked Gene located on an X chromosome.

Y

yellow bone marrow (mar′o) Connective tissue filling the cavities of bones; consists primarily of reticular fibers and fat cells; replaces red marrow in long bones and the skull.

Y-linked Gene located on a Y chromosome.

yolk sac (yōk, yōlk) Highly vascular layer surrounding the yolk of an embryo.

Z

Z disk Delicate membranelike structure found at either end of a sarcomere to which the actin myofilaments attach.

zona fasciculata (zō′nă fa-sik′ū-lă′tă) [L., *zone*, a girdle, one of the zones of the sphere] Middle layer of the adrenal cortex that secretes cortisol.

zona glomerulosa (glō-măr-ū-lōs-ă) Outer layer of the adrenal cortex that secretes aldosterone.

zona pellucida (pe-lū′sĭ-dă) Layer of viscous fluid surrounding the oocyte.

zona reticularis (rē-tik′ū-lar′is) Inner layer of the adrenal cortex that secretes androgens and estrogens.

zonula adherens (zō′nū-lă ad-her′enz) [L., a small zone; adhering] Small zone holding or adhering cells together.

zonula occludens (ō-klūd′enz) [L., occluding] Junction between cells in which the plasma membranes may be fused; occludes or blocks off the space between the cells.

zygomatic (zī-gō-mat′ik) [Gr., *zygon*, yoke] To yoke or join. Bony arch created by the junction of the zygomatic and temporal bones.

zygote (zī′gōt) [Gr., *zygotos*, yoked] Diploid cell resulting from the union of a sperm cell and an oocyte.

Credits

Photo Credits

Chapter 1:
Opener: © Quest/Science Photo Library/Photo Researchers, Inc.; **Figure A:** © Omikron/Photo Researchers, Inc.; **Figure B:** © Bernard Benoit/Science Photo Library/Photo Researchers, Inc.; **Figure C:** © Scott Camazine & Sue Trainor; **Figure D:** © D. P. M. Ribotsky/Custom Medical Stock Photo; **Figure E:** © ISM/Phototake NYC; **Figure F:** © Photodisc; **Figure G:** © McGill University/CNRI/Phototake; **1.1:** © Bart Harris/CORBIS; **1.10 left & right, 1.11a & b, 11.13a:** © McGraw-Hill Higher Education, Inc./Eric Wise, photographer; **1.13b–d:** © R. T Hutchings.

Chapter 2:
Opener: © Prof. P. Motta/Dept. of Anatomy/University "La Sapienza", Rome/Science Photo Library/Custom Medical Stock Photo; **2.4c:** © Trent Stephens; **2.15c:** © Barry King/Tom Stack & Assocs.

Chapter 3:
Opener: © Dennis Kunkel Microscopy, Inc.; **3.2b:** © J. David Robertson, from Charles Flickinger, Medical Cell Biology, Philadelphia; **3.22b:** © Manfred Kage/Peter Arnold, Inc.; **3.25b:** © Courtesy of Dr. Birgit H. Satir; **3.26b & 3.27b:** © Don Fawcett/Photo Researchers, Inc.; **3.27c:** © Bernard Gilula/Photo Researchers, Inc.; **3.29b:** © Biology Media/Photo Researchers, Inc.; **3.30a:** © E. de Harven/Photo Researchers, Inc.; **3.30b:** © Dr. Gopal Murti/Science Photo Library/Photo Researchers, Inc.; **3.32b:** © Courtesy Susumi Ito, Ph.D.; **3.34b:** © J. David Robertson, from Charles Flickinger, Medical Cell Biology, Philadelphia; **3.35b:** © Robert Bollender/Don Fawcett/Visuals Unlimited; **3.38b:** © Don Fawcett/Visuals Unlimited; **3.44:** © Ed Reschke.

Chapter 4:
Opener: © Prof. P. Motta & E. Vizza/Science Photo Library/Photo Researchers, Inc.; **4.1b:** © Victor Eroschenko; **4.1c:** © Ed Reschke; **4.2a2:** © McGraw-Hill Higher Education, Inc./Dennis Strete, photographer; **4.2b2, 4.2c2, 4.3a2 & 4.3b2:** © Victor Eroschenko; **4.3c2:** © R. Kessel/Visuals Unlimited; **4.4a2, 4.4b2, 4.4b3, 4.7a2 & 4.7b2:** © Victor Eroschenko; **4.8a:** © Ed Reschke; **4.9a2, 4.9b2 & 4.9c2:** © Victor Eroschenko; **4.9d2:** © Ed Reschke; **4.10a2:** © Carolina Biological Supply/Phototake; **4.10b2:** © Victor Eroschenko; **4.11a2:** © Carolina Biological Supply/Phototake; **4.11b2, 4.11c2 & 4.12a2:** © Victor Eroschenko; **4.12b2:** © Trent Stephens; **4.13a2, 4.13b2, 4.15a2 & 4.15b2:** © Ed Reschke; **4.15c2:** © Victor Eroschenko; **4.16a2, 4.16b2 & Figure 4.5:** © Trent Stephens.

Chapter 5:
Opener: © Dennis Kunkel Microscopy, Inc.; **5.2a:** © Alan Stevens, University of Nottingham, Med School, UK; **5.2b:** © The Bergman Collection; **Figure Ca, b & c:** © Thomas B. Habif; **Figure Da:** © James Stevenson/Science Photo Library/Photo Researchers, Inc.; **Figure Db:** © Stan Levy/Photo Researchers, Inc.; **5.10a & b:** Courtesy of A. M. Kligman, Professor of Dermatology, University of Pennsylvania School of Medicine, Philadelphia, PA.

Chapter 6:
Opener: © Prof. P. Motta/Dept. of Anatomy/University "La Sapienza", Rome/Science Photo Library/Photo Researchers, Inc.; **6.1:** © Ed Reschke; **6.6a–c:** © Trent Stephens; **6.7c:** © Bio-Photo Assocs/Photo Researchers, Inc.; **6.9:** © Robert Caladine/Visuals Unlimited; **6.10b:** © Trent Stephens; **6.11 (1):** © Victor Eroschenko; **6.11 (2):** R. Kessel/Visuals Unlimited; **6.11 (3):** © Victor Eroschenko; **6.12:** © Visuals Unlimited; **6.14a:** © Science Photo Library/Photo Researchers, Inc.; **6.14b:** © Bio-Photo Assocs/Photo Researchers, Inc.; **6.16:** © J. M. Booher; **Figure Aa:** © Ewing Galloway, Inc.; **Figure Ab:** © Hilt & Cogburn, Manual of Orthopedics; **Figure Ac:** © Brashear & Beverly, Handbook of Orthopedic Surgery; **Figure Ad:** © Hilt & Cogburn, Manual of Orthopedics; **Figure Ca:** © Princess Margaret Rose Orthopaedic Hospital/Science Photo Library/Photo Researchers, Inc.; **Figure Cb:** © Dr. Michael Klein/Peter Arnold, Inc.

Chapter 7:
Opener: © Fred Hossler/Visuals Unlimited; **7.5 & 7.7:** © McGraw-Hill Higher Education, Inc./Eric Wise, photographer; **7.10c:** © CNRI/Science Photo Library/Photo Researchers, Inc.; **Figure A:** © Brashear & Beverly, Handbook of Orthopedic Surgery; **7.15a&b:** © Trent Stephens; **7.16:** © McGraw-Hill Higher Education, Inc./Eric Wise, photographer; **7.17e, 7.18a, 7.19c & 7.21c:** © Trent Stephens; **7.22:** © McGraw-Hill Higher Education, Inc./Eric Wise, photographer; **7.24d:** © Trent Stephens; **7.28, 7.32 & 7.37:** © McGraw-Hill Higher Education, Inc./Eric Wise, photographer; **7.39:** © Trent Stephens.

Chapter 8:
Opener: © CNRI/Phototake; **8.14, 8.15a–c, 8.16, 8.17, 8.18a & b, 8.19a & b, 8.20, 8.21, 8.22a & b, 8.23a & b, 8.24a & b, 8.25a & b & 8.26:** © McGraw-Hill Higher Education, Inc./Eric Wise, photographer; **8.31e:** © R. T. Hutchings; **Figure Ca:** © James Stevenson/Science Photo Library/Photo Researchers, Inc.; **Figure Cb:** © CNRI/Science Photo Library/Photo Researchers, Inc.

Chapter 9:
Opener: © Professors P. M. Motta, P. M. Andrews, K. R. Porter & J.Vial/Science Photo Library/Photo Researchers, Inc.; **9.1:** © Ed Reschke; **9.5a:** © Richard Rodewald; **9.11c:** © Fred Hossler/Visuals Unlimited; **9.17b:** © Don Fawcett/Photo Researchers, Inc.; **9.22:** © Victor Eroschenko; **Figure A:** © Phillippe Plailly/Science Photo Library/Photo Researchers, Inc.

Chapter 10:
Opener © Eye of Science/Science Photo Library/Photo Researchers, Inc.; **10.6b, 10.7b, 10.8a–d & 10.19b:** © McGraw-Hill Higher Education, Inc./Eric Wise, photographer; **10.23b:** © McGraw-Hill Higher Education, Inc./Rebecca Gray, photographer/Don Kincaid, dissections; **10.23c:** © McGraw-Hill Higher Education, Inc./Eric Wise, photographer; **10.24b:** © McGraw-Hill Higher Education, Inc./Rebecca Gray, photographer/Don Kincaid, dissections; **10.24c:** © McGraw-Hill Higher Education, Inc./Eric Wise, photographer; **10.26b:** © McGraw-Hill Higher Education, Inc./Rebecca Gray, photographer/Don Kincaid, dissections; **10.26c & 10.28d:** © McGraw-Hill Higher Education, Inc./Eric Wise, photographer; **10.29c:** © McGraw-Hill Higher Education, Inc./Rebecca Gray, photographer/Don Kincaid, dissections; **10.29d:** © McGraw-Hill Higher Education, Inc./Eric Wise, photographer; **10.30c & 10.33:** © Trent Stephens; **10.34:** © McGraw-Hill Higher Education, Inc./Rebecca Gray, photographer/Don Kincaid, dissections; **10.35:** © Trent Stephens; **10.38:** © McGraw-Hill Higher Education, Inc./Eric Wise, photographer; **10.40c:** © McGraw-Hill Higher Education, Inc./Rebecca Gray, photographer/Don Kincaid, dissections; **10.40d & 10.41d:** © McGraw-Hill Higher Education, Inc./Eric Wise, photographer; **Figure A:** © Duomo/Paul J. Sutton.

Chapter 11:
Opener: © Dennis Kunkel Microscopy, Inc..

Chapter 12:
Opener: © Loyola University Medical Education Network (LUMEN). (http://www.meddean.luc.edu/lumen/); **12.3b:** © Ed Reschke/Peter Arnold, Inc.; **12.15b:** © McGraw-Hill Higher Education, Inc./Rebecca Gray, photographer/Don Kincaid, dissections.

Chapter 13:
Opener: © Quest/Science Photo Library/Photo Researchers, Inc.: **13.1:** © Courtesy of Branislav Vidic; **13.4a–c:** © McGraw-Hill Higher Education, Inc./Rebecca Gray, photographer/Don Kincaid, dissections; **13.6a:** © R. T. Hutchings; **13.6b & 13.7b:** © McGraw-Hill Higher Education, Inc./Rebecca Gray, photographer/Don Kincaid, dissections; **13.8b:** © Trent Stephens.

Chapter 14:
Opener: © Omikron/Photo Researchers, Inc.; **14.2a–c:** © McGraw-Hill Higher Education, Inc./Eric Wise, photographer; **14.17:** © Marcus Raichle, MD, Washington University School of Medicine; **Figure Ba:** © Howard J. Radzyner/ Phototake; **Figure Bb:** © CNRI/Science Photo Library/Photo Researchers, Inc.

Chapter 15:
Opener: © Mireille Lavigne-Rebillard, INSERM Unit 254, Montpellier ("from the site Promenade around the cochlea (http://www.iurc.montp.inserm.fr/cric/audition/fran%E7ais/cochlea/fcochlea.htm)" by R. Pujol, S Blatrix and T. Pujol, CRIC, University of Montpellier/INSERM; **15.4g:** © Omikran/Photo Researchers, Inc.; **15.7:** © McGraw-Hill Higher Education, Inc./Eric Wise, photographer; **15.11:** © McGraw-Hill Higher Education, Inc./Rebecca Gray, photographer/Don Kincaid, dissections; **15.14a:** © A. L. Blum/Visuals Unlimited; **15.22d:** © McGraw-Hill Higher Education, Inc./Rebecca Gray, photographer/Don Kincaid, dissections; **Figure C:** Reproduced from Isihara's *Tests for Color Blindness* published by Kanehara & Co., Ltd., Tokyo, Japan, but tests for color blindness cannot be conducted with this material. For accurate testing, the original plates should be used.; **Figure Ca–d:** © National Eye Institute, Bethedsa, MD; **15.24:** © McGraw-Hill Higher Education, Inc./Eric Wise, photographer; **15.27a:** Courtesy of R. A. Jacobs and A. J. Hudspeth; **15.27b:** © Fred Hossler/Visuals Unlimited; **15.28a:** Courtesy of A. J. Hudspeth; **15.28c:** Courtesy of David P. Corey and John A. Assad, Harvard Medical School; **15.36a & b:** © Trent Stephens; **15.38b:** © Jerry Wachter/Photo Researchers, Inc.

Chapter 16:
Opener: © Innserspace Imaging/Science Photo Library/Photo Researchers, Inc.

Chapter 17:
Opener: © Quest/Science Photo Library/Photo Researchers, Inc.

Chapter 18:
Opener: © Dr. L. Orci, University of geneva/Science Photo Library/Photo Researchers, Inc.; **18.8c, 18.11b & 18.13b:** © Victor Eroschenko; **Figure A & 18.16:** © Bio-Photo Assocs/Photo Researchers, Inc.; **Figure B (Boy):** © Mark Clark/Science Photo Library/Photo Researchers, Inc.; **Figure B (Kit):** © Visuals Unlimited.

Chapter 19:
Opener: © Dennis Kunkel Microscopy, Inc.; **19.1a:** © McGraw-Hill Higher Education, Inc./Eric Wise, photographer; **19.3a:** © National Cancer Institute/Science Photo Library/Photo Researchers, Inc.; **19.7:** © Ed Reschke; **19.8a–e:** © Victor Eroschenko; **19.10:** © Oliver Meckes/Photo Researchers, Inc.; **Figure A:** © Omikron/Science Source/Photo Researchers, Inc.

Chapter 20:
Opener: © David M. Phillips/Photo Researchers, Inc.; **20.5b:** © R. T. Hutchings; **20.8a & b:** © McMinn & Hutchings, *Color Atlas of Human Anatomy*/Mosby; **20.12b:** © Ed Reschke; **20.20:** © Terry Cockerham/Cynthia Alexander/Synapse Media Productions; **Figure A:** © Hank Morgan/Science Source/Photo Researchers, Inc.

Chapter 21:
Opener: © Prof. P. Motta/Dept. of Anatomy/University "La Sapienza", Rome/Science Photo Library/Photo Researchers, Inc. **21.5:** © Visuals Unlimited.

Chapter 22:
Opener: © Dennis Kunkel Microscopy, Inc.; **22.4:** © Victor Eroschenko; **22.6b:** © Trent Stephens; **22.7d:** © Victor Eroschenko; **22.8c:** © Trent Stephens; **Figure A:** © Kenneth Greer/Visuals Unlimited.

Chapter 23:
CO 23: © David Phillips/Photo Researchers, Inc.; **23.2b:** © R. T. Hutchings; **23.4b:** © CNRI/Phototake; **23.5a:** © John Cunningham/Visuals Unlimited; **23.5b:** © Manfred Kage/Peter Arnold, Inc.; **23.5c:** © Ed Reschke/Peter Arnold, Inc.; **23.6b:** © Walter Reiter/Phototake; **23.7b:** © Carolina Biological Supply/Phototake; **23.9b1 & 2:** © Branislav Vidic; **23.15a:** © Custom Medical Stock Photo; **Figure A:** © Deni McIntyre/Photo Researchers, Inc.

Chapter 24:
Opener: © Dennis Kunkel Microscopy, Inc.; **24.5b & c:** © McGraw-Hill Higher Education, Inc./Rebecca Gray, photographer/Don Kincaid, dissections; **24.9c:** © Ed Reschke; **24.11c:** © Victor Eroschenko; **24.16d:** © David M. Phillips/Visuals Unlimited; **Figure A:** © SIU Biomedical/Photo Researchers, Inc.; **24.25b:** © CNRI/Science Photo Library/Photo Researchers, Inc. **Figure B:** © Maxine Cass, 2003.

Chapter 25:
Opener: © Professors P. Motta & T. Naguro/Science Photo Library/Photo Researchers, Inc; **25.17:** © Kyle Rothenborg/Pacific Stock.

Chapter 26:
Opener: © Dennis Kunkel Microscopy, Inc.; **26.3b:** © McGraw-Hill Higher Education, Inc./Rebecca Gray, photographer/Don Kincaid, dissections; **Figure B:** © Hank Morgan/Photo Researchers, Inc.

Chapter 27:
Opener: © Custom Medical Stock Photo

Chapter 28:
Opener: © Prof. P. Motta & S. Makabe/Science Photo Library/Photo Rsearchers, Inc.; **28.20a:** © Educational Images/Custom Medical Stock Photo **28.20b & c:** © Biophoto Associates/Photo Researchers, Inc.; **FigureBa:** © M. Long/Visuals Unlimited; **Figure Bb:** © Ray Ellis/ Photo Researchers, Inc.; **Figure Bc:** © Hank Morgan/Photo Researchers, Inc.; **Figure Bd:** © McGraw-Hill Education; **Figure Be:** © McGraw-Hill Higher Education Group, Inc./Bob Coyle photographer.

Chapter 29:
Opener & 29.2a-d: © Dr. Yorgus Nikas/Jason Burns/Phototake; **29.10:** © John Giannicchi/Photo Researchers, Inc.; **Figure A:** © NMSB/Custom Medical Stock Photo **29.17a–c:** © Petit Format/Nestle/Photo Researchers, Inc.; **29.23:** © Will & Deni McIntyre/Photo Researchers, Inc.; **29.24:** © CNRI/Science Photo Library/Photo Researchers, Inc.

Index